The Elements

Name	Symbol	Atomic Number	Relative Atomic Mass*	Name	Symbol	Atomic Number	Relative Atomic Mass*
Actinium	Ac	89	227.028	Meitnerium	Mt	109	(266)
Aluminum	Al	13	26.9815	Mendelevium	Md	101	(258)
Americium	Am	95	(243)	Mercury	Hg	80	200.59
Antimony	Sb	51	121.757	Molybdenum	Mo	42	95.94
Argon	Ar	18	39.948	Neodymium	Nd	60	144.24
Arsenic	As	33	74.9216	Neon	Ne	10	20.1797
Astatine	At	85	(210)	Neptunium	Np	93	237.048
Barium	Ba	56	137.327	Nickel	Ni	28	58.693
Berkelium	Bk	97	(247)	Niobium	Nb	41	92.9064
Beryllium	Be	4	9.01218	Nitrogen	N	7	14.0067
Bismuth	Bi	83	208.980	Nobelium	No	102	(259)
Bohrium	Bh	107	(262)	Osmium	Os	76	190.23
Boron	B	5	10.811	Oxygen	O	8	15.9994
Bromine	Br	35	79.904	Palladium	Pd	46	106.42
Cadmium	Cd	48	112.411	Phosphorus	P	15	30.9738
Calcium	Ca	20	40.078	Platinum	Pt	78	195.08
Californium	Cf	98	(251)	Plutonium	Pu	94	(244)
Carbon	C	6	12.011	Polonium	Po	84	(209)
Cerium	Ce	58	140.115	Potassium	K	19	39.0983
Cesium	Cs	55	132.905	Praseodymium	Pr	59	140.908
Chlorine	Cl	17	35.4527	Promethium	Pm	61	(145)
Chromium	Cr	24	51.9961	Protactinium	Pa	91	231.036
Cobalt	Co	27	58.9332	Radium	Ra	88	226.025
Copper	Cu	29	63.546	Radon	Rn	86	(222)
Curium	Cm	96	(247)	Rhenium	Re	75	186.207
Darmstadtium	Ds	110	(281)	Rhodium	Rh	45	102.906
Dubnium	Db	105	(262)	Rubidium	Rb	37	85.4678
Dysprosium	Dy	66	162.50	Ruthenium	Ru	44	101.07
Einsteinium	Es	99	(252)	Rutherfordium	Rf	104	(261)
Erbium	Er	68	167.26	Samarium	Sm	62	150.36
Europium	Eu	63	151.965	Scandium	Sc	21	44.9559
Fermium	Fm	100	(257)	Seaborgium	Sg	106	(263)
Fluorine	F	9	18.9984	Selenium	Se	34	78.96
Francium	Fr	87	(223)	Silicon	Si	14	28.0855
Gadolinium	Gd	64	157.25	Silver	Ag	47	107.868
Gallium	Ga	31	69.723	Sodium	Na	11	22.9898
Germanium	Ge	32	72.61	Strontium	Sr	38	87.62
Gold	Au	79	196.967	Sulfur	S	16	32.066
Hafnium	Hf	72	178.49	Tantalum	Ta	73	180.948
Hassium	Hs	108	(265)	Technetium	Tc	43	(98)
Helium	He	2	4.00260	Tellurium	Te	52	127.60
Holmium	Ho	67	164.930	Terbium	Tb	65	158.925
Hydrogen	H	1	1.00794	Thallium	Tl	81	204.383
Indium	In	49	114.818	Thorium	Th	90	232.038
Iodine	I	53	126.904	Thulium	Tm	69	168.934
Iridium	Ir	77	192.22	Tin	Sn	50	118.710
Iron	Fe	26	55.847	Titanium	Ti	22	47.88
Krypton	Kr	36	83.80	Tungsten	W	74	183.84
Lanthanum	La	57	138.906	Uranium	U	92	238.029
Lawrencium	Lr	103	(260)	Vanadium	V	23	50.9415
Lead	Pb	82	207.2	Xenon	Xe	54	131.29
Lithium	Li	3	6.941	Ytterbium	Yb	70	173.04
Lutetium	Lu	71	174.967	Yttrium	Y	39	88.9059
Magnesium	Mg	12	24.3050	Zinc	Zn	30	65.39
Manganese	Mn	25	54.9381	Zirconium	Zr	40	91.224

* These atomic masses are relative to a value of exactly 12 assigned to the carbon isotope $^{12}_{6}C$. Values in parentheses are the mass numbers of the most common or most stable isotopes of radioactive elements.

General Chemistry

FOURTH EDITION

John W. Hill

University of Wisconsin-River Falls

Ralph H. Petrucci

California State University, San Bernardino

Terry W. McCreary

Murray State University

Scott S. Perry

University of Houston

PEARSON

Prentice
Hall

Upper Saddle River, New Jersey 07458

Library of Congress Cataloging-in-Publication Data

General chemistry -- 4th ed. / John W. Hill ... [et al].
    p. cm.
  Rev. ed of: General chemistry / John W. Hill, Ralph H. Petrucci 3rd ed. c2002.
  Includes index
  ISBN 0-13-140283-8
    1. Chemistry. I. Hill, John William - II. Hill, John William - General chemistry.

QD33.2.H55 2005
540--dc22   2003068981

Senior Editor: *Kent Porter Hamann*
Editor in Chief, Science: *John Challice*
Editor in Chief, Development: *Ray Mullaney*
Development Editor: *Irene M. Nunes*
Executive Managing Editor: *Kathleen Schiaparelli*
Media Editor, Physical Sciences: *Michael J. Richards*
Media Development Editor and Project Manager: *Edward Dodd, III*
Project Manager : *Kristen Kaiser*
Assistant Managing Editor, Science Media: *Nicole M. Bush*
Media Supervisor: *Craig Hough*
Media Production Editors: *Elizabeth Klug, Elizabeth Wright*
Media Illustrators: *Chris Kayes, David Lynch, Brandon Gilbert, Jonathan Derk*
Contributing Media Studio: *OpenTech*
Senior Marketing Manager: *Steve Sartori*
Manufacturing Buyer: *Alan Fischer*
Manufacturing Manager: *Trudy Pisciotti*
Vice President of Production and Manufacturing: *David W. Riccardi*
Director of Creative Services: *Paul Belfanti*
Creative Director: *Carole Anson*
Art Editor: *Connie Long*
Managing Editor, Audio/Visual Assets and Production: *Patricia Burns*
Art Studio: *Artworks*
    Manager, Art Production Technologies: *Matt Haas*
    Production Manager: *Ronda Whitson*
    Illustration Supervisor: *Kathryn Anderson*
    Illustrators: *Royce Copenheaver, R. Jay McElroy, Daniel Knopsnyder, Mark Landis,*
        Audrey Simonetti, Scott Weiber, Nathan Storck, Ryan Currier, Stacy Smith
    Art Quality Assurance: *Pamela Taylor, Timothy Nguyen, Cathy Shelley, Kenneth Mooney,*
        Anna Whalen
Art Director: *Jonathan Boylan*
Cover and Interior Designers: *Tom Nery & Jonathan Boylan*
Cover Credit: *Image of MgO(100) surface obtained by atomic force microscopy,* Hyun I. Kim
    and Scott S. Perry, University of Houston; Art Studio: *Artworks*; Illustrator: *Royce Copenheaver*
Photo Researcher: *Yvonne Gerin*
Photo Editor: *Charles Morris*
Development Editor: *Irene Nunes*
Production Assistant: *Nancy Bauer*
Editorial Assistant: *Jackie Howard*
Production Supervision/Composition: *Preparé, Inc.*

© 2005, 2002, 1999, 1996 Pearson Education, Inc.
Pearson Prentice Hall
Pearson Education, Inc.
Upper Saddle River, New Jersey 07458

Printed in the United States of America

10 9 8 7 6 5 4 3

ISBN 0-13-140283-8 (Student edition)
ISBN 0-13-192015-4 (School edition)

Pearson Education LTD., *London*
Pearson Education Australia PTY, Limited, *Sydney*
Pearson Education Singapore, Pte. Ltd
Pearson Education North Asia Ltd, *Hong Kong*
Pearson Education Canada, Ltd., *Toronto*
Pearson Educación de Mexico, S.A. de C.V.
Pearson Education—Japan, *Tokyo*
Pearson Education Malaysia, Pte. Ltd

Brief Contents

iii

Contents

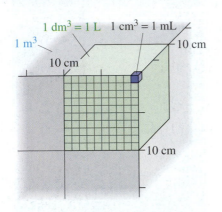
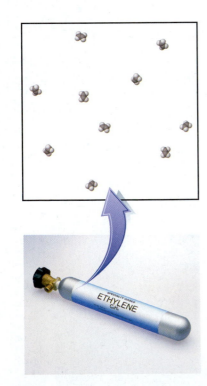

4   Chemical Reactions in Aqueous Solutions   125

5   Gases   169

6   Thermochemistry   215

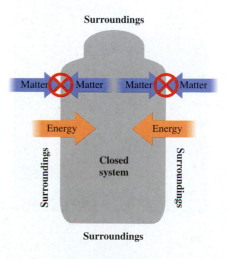

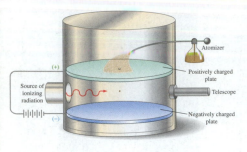

Liquid oxygen is paramagnetic and is held between the poles of the magnet.

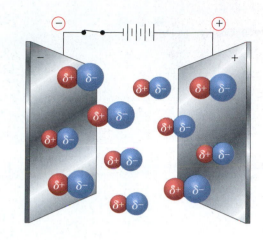

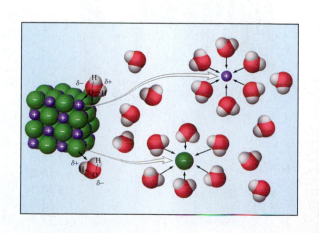

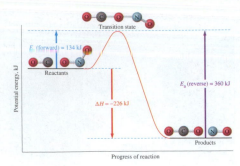

13   Chemical Kinetics: Rates and Mechanisms of Chemical Reactions   527

14   Chemical Equilibrium   575

15   Acids, Bases, and Acid–Base Equilibria   615

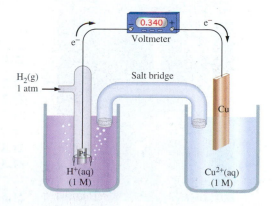

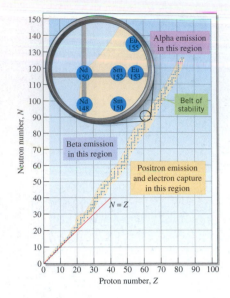

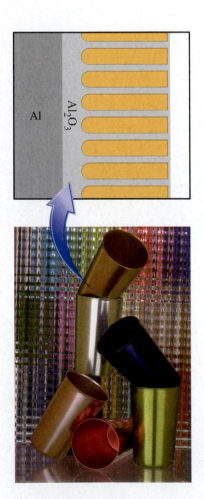

22 The *d*-Block Elements and Coordination Chemistry 903

23 Chemistry and Life: More on Organic, Biological, and Medicinal Chemistry 941

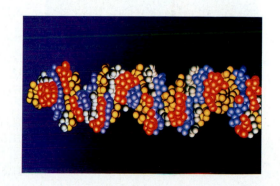

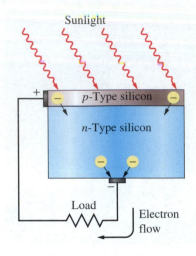

Sunlight

p-Type silicon

n-Type silicon

Load   Electron flow

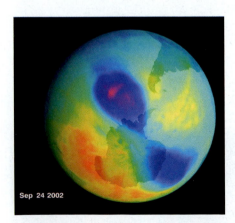

Sep 24 2002

Appendices

Photo Credits   C1

Index   I1

About the Authors

John W. Hill

John W. Hill received his B.S. in Chemistry and Mathematics from Middle Tennessee State University in 1957 and a Ph.D. in Organic Chemistry from the University of Arkansas in 1961. Following three years at Northeast Louisiana State College, he joined the faculty of the University of Wisconsin-River Falls in 1963. He served 13 years as Chair of the Department of Chemistry. In 1985, he was chosen Outstanding Faculty Member, Sciences and Mathematics. In 1989, he received the Robert C. Brasted Award for Outstanding College Teaching. A longtime member of the College Chemistry Consultants Service, he has served as a consultant at more than 40 colleges and universities. He is the author of several books, including *Chemistry for Changing Times* (with Doris K. Kolb) and *Chemistry and Life: An Introduction to General, Organic, and Biological Chemistry* (with Stuart Baum and Rhonda Scott-Ennis). He has also written a children's mystery book, *The Crimecracker Kids and the Bake Shop Break-in* (with Marilyn D. Duerst), and a children's picture book, *Grandpa: A Semi-autobiographical Story* (illustrated by Nora L. Koch).

Ralph H. Petrucci

Ralph Petrucci received his B.S. in Chemistry from Union College and his Ph.D. from the University of Wisconsin-Madison. Following several years of teaching, research, consulting, and directing the NSF Institutes for Secondary School Science Teachers at Case Western Reserve University, Dr. Petrucci joined the planning staff of the new California State University campus at San Bernardino in 1964. There, in addition to his faculty appointment, he served as Chairman of the Natural Sciences Division and Dean of Academic Planning. Professor Petrucci, now retired from teaching, is the author of several books, including *General Chemistry: Principles and Modern Applications* with William S. Harwood and Geoffrey Herring.

Terry W. McCreary

Terry McCreary received his B.S. in Chemistry from St. Francis University and his M.S. in Analytical Chemistry from the University of Georgia. After teaching at Cumberland College for six years, he traveled to Virginia Tech for a Ph.D. in Analytical Chemistry. At Murray State University since 1988, he is a member of the American Chemical Society, the Kentucky Academy of Science, the American Association of Aeronautics and Astronautics, and has served as Technical Editor for the *Journal of Pyrotechnics.* In his spare time, he indulges in pyrotechnics activities with Pyrotechnic Guild International and builds and flies large rockets with the Tripoli Rocketry Association. He is the author of several laboratory manuals, as well as *Experimental Composite Propellant*, a manual on the properties and preparation of rocket propellants.

Scott S. Perry

Scott Perry received his B.S. in Chemistry from Furman University and his Ph.D. from the University of Texas at Austin. Following postdoctoral research at the University of California, Berkeley, he joined the faculty at the University of Houston and now holds the position of Professor of Chemistry and Professor of Chemical Engineering. In 2001 and 2003, he served as Visiting Professor within the Department of Materials at the Swiss Federal Institute of Technology (ETH) in Zurich. At the University of Houston, he was the recipient of the 1997 Cooper Award for Teaching Excellence, the 2000 Teaching Excellence Award for the College of Natural Science and Mathematics, and the 2002 Research Excellence Award. He is the author of more than 80 publications dealing with surface chemistry and related interfacial phenomena.

Preface

STUDENTS COME TO a general chemistry course with a variety of backgrounds and interests. Most plan to become scientists, engineers, or professionals in medicine or other areas of the life sciences. Part of the challenge to the chemistry instructor is to convince students that knowledge of chemistry is essential to the understanding of other disciplines, ranging from cell biology to medicine to materials science. Indeed, the concepts that students learn in this course will come up again in many aspects of their personal and professional lives.

In this text, our goal is to provide students with core principles; to help them formulate strategies for solving problems—the basis on which their knowledge will so often be tested—and to teach them problem-solving skills. We hope that the text will also satisfy several other needs of the typical student: the need for background material and a second voice for students to hear; help in visualizing chemical phenomena, both what students can see with their eyes and what they must learn to see with their minds' eyes; and interesting applications of chemistry that will help them in their professions and also enrich their lives. In crafting the fourth edition of this text, we have strived to strike a necessary balance in meeting these basic needs.

Approach to Problem Solving

Problem-solving skills and the ability to think critically are essential for success in today's world. We provide ample opportunities for practicing these skills. For every type of problem, we provide *Examples* that are carefully worked out, many by an approach that first outlines a strategy for solving the problem, followed by a step-by-step solution, and often culminating in an assessment. The assessments may focus on the theoretical or practical significance of the Example, assumptions that are implicitly or explicitly made in solving a problem, means of checking the accuracy of results, and alternate ways of solving a problem.

Some *Problem-Solving Notes* in the text provide additional background information for specific in-text Examples or Exercises. Others have a more general applicability, and their presence in the margin makes them available for ready reference. Also, in several instances in the early chapters, particularly Chapter 3 (Stoichiometry), the various terms in a series of calculations are annotated. These annotations outline the rationale for each step in a multistep calculation. They are meant to help students see where they are and what they are doing in the calculation at all times.

The in-text Examples are followed by *Exercises* that students can use to test their understanding of the methods illustrated. Two Exercises, labeled *A* and *B*, are given for each Example. The goal in an *A* Exercise is to apply to a similar situation the method outlined in the Example. In a *B* Exercise, students often must combine that method with other ideas previously learned. Many of the *B* Exercises provide a context closer to that in which chemical knowledge is applied, and they thus serve as a bridge between the worked Examples and the more challenging problems at the end of the chapter. The *A* and *B* Exercises provide a simple way for the instructor to assign homework that is closely related to the Examples.

The ability to plug numbers into an equation and get an answer is, in itself, seldom enough to attain mastery of a concept. For example, students should generally be able to judge whether an answer is reasonable and, in some cases, to obtain a reasonable estimate of an answer without doing a detailed calculation. To assist in the acquisition of these skills, we offer worked-out *Estimation Examples* followed by *Estimation Exercises*. Examples and Exercises of this type are found throughout the text.

Students also need to develop insights into chemical concepts that are often best demonstrated by solving problems of a qualitative or conceptual nature. To emphasize this aspect of problem solving, we provide guided *Conceptual Examples* followed by *Conceptual Exercises.*

Through the different types of Examples and Exercises described, students using this text should gain a balanced set of skills in chemical problem solving. As additional reinforcement, the text offers four kinds of end-of-chapter exercises:

- *Self-Assessment Questions* are intended to provide a qualitative measure of student understanding of the main ideas introduced in the chapter, ideas that are most likely to appear in quizzes or examinations. The Self-Assessment Questions include discussion, calculation, and multiple-choice selections, and answers to many of these questions are given in Appendix F (Answers to Selected Problems).

- *Problems* are arranged by topic; they test mastery of the problem-solving techniques discussed in the chapter. The Problems are arranged in matched pairs, with answers to odd-numbered problems given in Appendix F.

- *Additional Problems* are not grouped by type. Some are more challenging than the Problems, often requiring a synthesis of ideas from more than one chapter. Others pursue an idea further than is done in the text, or introduce new ideas. Answers to more than one-third of the Additional Problems are given in Appendix F.

- The problems in the *Apply Your Knowledge* section offer a selection of applied, multiconceptual, and/or challenging problems carrying labels such as Laboratory, Environmental, Biochemical, and Historical. Some are designed for the collaborative effort of small groups of students. Answers to selected Apply Your Knowledge problems are given in Appendix F.

- *e-Media Problems* specifically address and require reference to interactive activities that can be viewed through the Interactive Student Tutorial module, available in OneKey, Blackboard, WebCT, and the Companion Website.

Some of the end-of-chapter Problems and Additional Problems are of an estimation or conceptual type, mirroring similar types of exercises within each chapter. We do not specifically label questions of these types as we do in the body of the chapter, however, because we want to give students experience in recognizing different types of problems as well as solving them.

As a whole, the vast problem-solving resources allow students to build their skills at a pace with which they are most comfortable and to check their understanding at every step in the learning process.

Visualization and Media Presentation

Difficulty in somehow "seeing" what cannot be seen and in visualizing objects in three dimensions is cited among the top three barriers confronting students in a general chemistry course. (The other two are poor study habits and poor math skills, both of which are addressed by specific print and media supplements to this text that we will describe shortly.) In this book, we use drawings, computer graphics, and photographs to help students visualize chemical phenomena at both the microscopic (molecular) and macroscopic (visible) levels. Instructors using this text may wish to take advantage of the Interactive Student Tutorial, an electronic text that includes hundreds of animations, simulations, exercises, and molecular models that students can interactively explore on their computers.

Coverage of the Major Areas of General Chemistry

Our goal is to provide a truly general course that integrates all the major areas of chemistry. Physical principles, inorganic compounds, and analytical techniques are addressed throughout. As in the previous editions, organic chemistry is incorporated where appropriate, as we think that students are not adequately served if they are sheltered from organic compounds until late in their study of general chemistry. We therefore introduce some simple organic chemistry in Chapter 2 and use it thereafter to describe physical properties of substances, aspects of chemical bonding, acid–base chemistry, and oxidation–reduction reactions. We introduce biochemistry in Chapter 6 in a discussion of carbohydrates and fats as fuels for our bodies and use biochemistry frequently in the following chapters where appropriate. We have continued to amplify our treatment of analytical chemistry in this fourth edition, with a new section in Chapter 4 that ties together different types of titrations, as well as descriptions of analytical tools and a large number of problems that are analytical-specific.

As in previous editions, Chapter 23 (Chemistry and Life: More About Organic, Biological, and Medicinal Chemistry) brings together the core organic chemistry concepts from earlier chapters, expands on those whose earlier introduction was necessarily brief, and discusses the chemistry of selected biomolecules and medicinal compounds. Thus, we have tried to provide useful core material to those who will never take an organic chemistry course, while also offering a broader-than-usual preparation for students who will enroll in such courses.

Organization

The first 18 chapters of the text emphasize chemical principles, but the principles are illustrated throughout with significant applications. Chapter 20 (The s-Block Elements), Chapter 21 (The p-Block Elements), and Chapter 22 (The d-Block Elements and Coordination Chemistry) provide a systematic treatment of descriptive chemistry, but emphasize how the properties of substances relate to the principles learned earlier in the text. Chapter 19 (Nuclear Chemistry), Chapter 23 (Chemistry and Life), Chapter 24 (Chemistry of Materials), and Chapter 25 (Environmental Chemistry) are fairly independent, freestanding chapters. They can serve as capstones to a general chemistry course because each revisits some basic principles of earlier chapters, applying them to topics in which students generally have a strong interest. These four chapters can be studied, in whole or in part, in just about any order, following coverage of background material in earlier chapters.

Changes in This New Edition

We have revised the previous edition in specific ways, partly in response to reviewer suggestions.

- Material has been updated throughout the text to reflect new developments, while retaining and enhancing the relaxed, easy-to-read style of the third edition. We have retained learning aids such as voice balloons and added new problem-solving strategies. We have continued the use of Conceptual Examples and Exercises that require the student to reflect on what has been learned. We have also retained the Estimation Exercises, which minimize the "plug-and-chug" approach to problem solving and encourage student analysis of numerical answers.

- A number of new and more challenging end-of-chapter problems has been added to each chapter. These include conceptual problems, visualization problems with accompanying art, data-based problems that require the student to analyze and select the appropriate data to be used, real-world problems that include raw data from actual laboratory experiments, and problems requiring students to synthesize concepts and skills acquired in earlier chapters.

- New and revised art has been carefully selected, much of which visually relates macroscopic, observable phenomena to microscopic (molecular/atomic) behavior. Phenomena so represented include the limiting-reactant concept, solution dilution, electrolyte behavior, gas behavior, the critical point, equilibrium, the common ion effect, and ion exchange. All of the macro-micro art from the third edition has been carefully reviewed for effectiveness and modified accordingly.

- The format of the end-of-chapter review material has been changed substantially. It begins with a *Cumulative Example* that integrates new concepts from the chapter with important ideas from previous chapters. This is followed by a *Concept Review with Key Terms* that provides section-by-section summaries of the chapter, in which all the key terms in the chapter are used in context and some of the key equations and figures are recalled for students. Next comes a list of a dozen or so *Assessment Goals* informing students what they should understand and be able to do as a result of studying the chapter. Finally, students are encouraged to test their attainment of the Assessment Goals through a set of *Self-Assessment Questions*.

- Chapter 24 has been reorganized and revised to include a chemical description of some modern materials being used in advanced technologies. The goal of this discussion is to highlight the fundamental chemical aspects of different materials that are enabling the development of new technologies.

- The accompanying media elements have been extensively revised, placing a significant emphasis on directed study and follow-up assessment questions in order to maximize the impact of these activities. In addition, we are excited by the incorporation of many of the media activities into quiz questions in OneKey, Blackboard, and WebCT, as well as PH GradeAssist, where some are algorithmically generated.

In summary, we have integrated the collective experience of the original authors (John Hill and Ralph Petrucci) in teaching and writing for various audiences with the classroom and laboratory expertise of the new authors (Terry McCreary and Scott Perry) to produce a textbook that strikes a balance between the principles that give meaning to chemistry and the applications that make it come alive.

At every step of the learning process, this textbook supports students' efforts and gives them ample opportunity to check and build their understanding.

Supplements

For the Instructor

Annotated Instructor's Edition (ISBN 0-13-140313-3), with annotations by Terry W. McCreary, Murray State University. This special edition of the text includes the entire student text plus marginal icons and annotations to aid instructors in preparing their lectures. Included are suggestions for lecture demonstrations, teaching tips, common student misconceptions, and indications of which graphics in the textbook are available as overhead transparencies. Some of the notes relate to Apply Your Knowledge problems that can be taken up in advance of the chapter where they appear. The AIE also includes cross-references to all figures, demonstrations, and animations available in electronic form on the Instructor's Resource Center CD.

Instructor's Resource Manual (ISBN 0-13-140316-8), prepared by Marie Hankins, University of Southern Indiana. This book provides chapter-by-chapter lecture outlines, teaching tips, common student misconceptions, background references, and suggested lecture demonstrations for in-class use.

Solutions Manual (ISBN 0-13-140349-4), by C. Alton Hassell, Baylor University. Contains worked-out solutions to all in-chapter, end-of-chapter, review, conceptual, and estimation exercises and problems.

Transparencies (ISBN 0-13-140315-X). Over 250 full-color transparencies chosen from the text put principles into visual perspective and save you time when preparing lectures.

Test Item File (ISBN 0-13-140317-6), by Michael Mosher, University of Nebraska, Kearney. This printed test bank includes over 2000 questions written exclusively for this text, with all answers section-referenced to the text.

Instructor's Resource Center on CD/DVD (ISBN 0-13-140318-4), by Jennifer R. Chase, Northwest Nazarene University. This lecture resource provides a fully searchable and integrated collection of resources to help you make efficient and effective use of your lecture preparation time, as well as to enhance your classroom presentations and assessment efforts. This resource features almost all the art from the text, including tables; two pre-built PowerPoint™ presentations; PDF files of the art; movies; animations; interactive activities; and the Instructor's Resource Manual Word™ files. This CD also features a search-engine tool that lets you find relevant resources via a number of different parameters, such as key terms, learning objectives, figure numbers, and resource type (e.g., Media Activities). This CD/DVD set also contains the TestGen, a computerized version of the Test Item File that allows professors to create and tailor exams to their needs.

PH GradeAssist (www.prenhall.com/phga)

PH GradeAssist is a timesaving homework and assessment system that allows students unlimited practice with algorithmically generated problems in an on-line environment. Instructors can administer quizzes and assignments, control the content and assignment parameters, and receive assignments and view performance statistics with the built-in grade book. This customizable system has a volume built specifically for this text.

OneKey

OneKey offers the best teaching and learning resources all in one place. OneKey for General Chemistry, Fourth Edition is all your students need for anytime anywhere access to your course materials. OneKey is all you need to plan and administer your course. Conveniently organized by textbook chapter, these compiled resources help you save time and help your students reinforce and apply what they have learned in class. Available resources include Concept Review with Key Terms, Web Destinations, Tools, Research Navigator, Math Review, Self-Quiz, and Media Gallery, plus a full electronic version of the text, with imbedded animations, movies, and molecules as the Interactive Student Tutorial. Assessment content includes Quiz, Master Quiz, MCAT Study Guide, and the Test Item File. Resources from the Instructor's Resource Center on CD/DVD (above) are also included.

Course Management

All of the content in OneKey is also available in WebCT and Blackboard. Prentice Hall offers content cartridges for these text-specific Course Management Systems. Visit *www.prenhall.com/demo* for details. These courses also offer just the Test Item File separately.

Web Assign (www.webassign.net)

WebAssign, an on-line homework system, offers selected end-of-chapter questions from the fourth edition. You can order access codes directly through your Prentice Hall sales representative.

For the Student

Companion Website for General Chemistry, Fourth Edition (*http://chem.prenhall.com/hillpetrucci*) For any students who want access to a free, open-access website, the Companion Website contains the Concept Review with Key Terms, Web Destinations, Tools, Research Navigator, Math Review, and Self-Quiz modules included in the Course Management Systems above. It also contains the animations, molecules, and movies from the Interactive Student Tutorial in the e-Media Activities module, and the Interactive Student Tutorial as a password-protected module.

Student Accelerator CD-ROM, with contributions from Scott S. Perry, University of Houston, and Mark E. Ott, Jackson Community College. This book-specific CD contains animations, molecules, and movies from the Companion Website, which enhances the performance of the Website when students download high-bandwidth media. The *General Chemistry, Fourth Edition,* Accelerator CD comes packaged with every new student textbook. By using the CD and the Course Management Systems or Companion Website simultaneously, many of the media objects on the site will load much more quickly, so that students are not restricted if they have slow connections. The CD can also be used as a library of media objects, apart from the Companion Website, if a student does not have a live connection.

Study Guide (ISBN 0-13-140347-8), by Dixie J. Goss, Hunter College. This book is keyed to the main text and provides further learning material for students: chapter-by-chapter overviews, learning goals, numerous examples and exercises, parallel text material, worked-out solutions, and practice tests with answers. This book serves as an excellent diagnostic tool and also helps sharpen students' skills in test taking.

Selected Solutions Manual (ISBN 0-13-140346-X), by C. Alton Hassell, Baylor University. Contains worked-out solutions to over half of the text's problems. The answers to these problems also appear in the text as Appendix F, Answers to Selected Problems.

Math Review Toolkit (ISBN 0-13-150824-5), by Gary Long, Virginia Tech. This brief paperback is engineered for students who find math a significant challenge in this course. The book provides a chapter-by-chapter review of the mathematics used throughout the text; a guide to preparing for a career in chemistry; and a review of some of the special writing requirements often needed in the general chemistry course, focusing particularly on the laboratory notebook. This supplement is free to qualified adopters; please speak with your local Prentice Hall representative.

Student Lecture Notebook (ISBN 0-13-146996-7). This lecture notebook contains the art from the transparency set with note-taking sections to obviate the need for students to spend time redrawing figures in the lecture, allowing them to concentrate on taking notes.

Acknowledgments

John W. Hill would like to thank his colleagues at the University of Wisconsin-River Falls for so many ideas that made their way into this text. He is especially indebted to Ina Hill and Cynthia Hill for library research, proofreading, trips to the express mail drop box, and for their unfailing support throughout this project.

Ralph H. Petrucci would like to thank his colleagues at California State University, San Bernardino, for the interest they have shown and the helpful suggestions they have contributed to his textbooks over the years. His greatest debt is to his wife, Ruth, for her love, encouragement, and continuing support of a husband who has been preoccupied and neglectful for so long.

Terry W. McCreary would like to thank his colleagues and his students at Murray State University for their aid and support. In particular, he would like to thank Dr. David Owen, who worked many of the end-of-chapter Problems and provided helpful commentary on them. Most of all, he wishes to thank his wife, Geniece, and their children, Corinne and Yvette, for their unflagging support, understanding, and love.

Scott S. Perry acknowledges the support of the faculty and administration of the University of Houston in his exploration of new teaching methodologies, many of which have been the basis for developments in this project. He wishes to express his deepest gratitude to his wife, Julie, and children, Austin and Lauren, for their love and patience throughout this project. He also wishes to thank Professor Nic Spencer for his support and wisdom in seeking a balance between research and education endeavors.

We are especially indebted to our students, who have challenged us to be better teachers, and to the reviewers of this and our other books, who have challenged us to write more clearly and accurately. We also owe a debt of gratitude to the many creative people at Prentice Hall who have contributed their talents to this edition. Kent Porter Hamann, our chemistry editor, has provided imaginative guidance throughout the project. As editorial assistant, Jackie Howard provided invaluable organizational skills and kept the entire team up to date on developments. Irene Nunes, our development editor, patiently reviewed and improved our occasionally opaque language, and she challenged us throughout to improve the pedagogy. Her every thought was to increase student understanding of the concepts. We are grateful to Ray Mullaney, editor in chief of Engineering, Science, and Mathematics Development, who played the critical role of keeping the entire project on schedule; to Kristen Kaiser, project manager, who provided careful management of the supplements for the text; to our photo researcher, Yvonne Gerin, who thoroughly searched and reviewed many of the photographic illustrations; to our copyeditor Fay Ahuja, whose attention to detail improved the consistency of the text; to Kathleen Schiaparelli in Production and Jonathan Boylan and Carole Anson in Design for their diligence and patience in bringing all the parts together to yield a finished work; and to Steve Sartori, senior marketing manager, for his market perspective and support. The creative resourcefulness and tenacity of our media editor, Michael J. Richards, and media project manager, Ed Dodd, have brought a new dimension to the media suite accompanying our text. Mark E. Ott, Louis Kirschenbaum, and Stacey Buchanan have been an integral part of the development of the media package to accompany the text.

This edition has benefited from the thoughtful and careful input of users of the previous edition and reviewers of the new manuscript. To these colleagues we convey our sincere appreciation; we are pleased to acknowledge them here:

Tom Berke
Brookdale Community College

Ralph Berni
University of New Orleans

Nancy E. Byrnes
Emory University

Charles R. Cornett
University of Wisconsin-Platteville

Sandra Etheridge
Gulf Coast Community College

Robert Evans
Hanover College

David A. Franz
Lycoming College

Jack T. Gill
Texas Woman's University

M. Dale Hawley
Kansas State University

Kirk Kawagoe
Fresno City College

Louis Kirschenbaum
University of Rhode Island

Richard H. Langley
Stephen F. Austin State University

Roderick M. Macrae
Marian College

Mary W. Mumper
Frostburg State University

Chip Nataro
Lafayette College

Steve Rowley
Middlesex County College

Kenneth D. Schlecht
SUNY-Brockport

James Smith
University of Rhode Island

David Spurgeon
The University of Arizona

Karen Stephens
Baton Rouge Community College

Kathy Thrush
Villanova University

James Tyrrell
Southern Illinois University

Philip R. Watson
Oregon State University

Sarah West
University of Notre Dame

Special thanks to Professors Richard H. Langley, Philip R. Watson, Charles R. Cornett, and James Tyrell, who also served as accuracy checkers for the fourth edition, as well as to Dr. Kathy Thrush, who checked the accuracy of our end-of-chapter problems.

John W. Hill
jwhill602@comcast.net

Ralph H. Petrucci
rhpetrucci@earthlink.net

Terry W. McCreary
terry.mccreary@murraystate.edu

Scott S. Perry
perry@uh.edu

PROBLEM SOLVING AND ACTIVE SELF-ASSESSMENT

For you to become a better problem solver, you need to practice your problem-solving skills often and assess your understanding at every stage in the learning process. *General Chemistry, Fourth Edition,* introduces a problem-solving strategy that will build your understanding of each step in the problem-solving process. You will have ample opportunity to check your understanding of the problem and thereby improve your skills with every problem you solve.

Example 3.6

Calculate, to four significant figures, the mass percent of each element in ammonium nitrate.

STRATEGY

First, we will determine the molar mass of ammonium nitrate, based on the formula unit NH_4NO_3. Then, for one mole of compound, we can determine mass ratios and percentages.

SOLUTION

$$\text{Formula mass} = (2 \times \text{atomic mass N}) + (4 \times \text{atomic mass H})$$
$$+ (3 \times \text{atomic mass O})$$
$$= (2 \times 14.01)u + (4 \times 1.008)u + (3 \times 16.00)u$$
$$= 28.02\ u + 4.032\ u + 48.00\ u = 80.05\ u$$

$$\text{Molar mass} = 80.05\ \text{g/mol}\ NH_4NO_3$$

Now we establish the mass ratios and convert them to percentages:

$$\% \text{N} = \frac{28.02\ \text{g N}}{80.05\ \text{g}\ NH_4NO_3} \times 100\% = 35.00\%\ \text{N}$$

$$\% \text{H} = \frac{4.032\ \text{g H}}{80.05\ \text{g}\ NH_4NO_3} \times 100\% = 5.037\%\ \text{H}$$

$$\% \text{O} = \frac{48.00\ \text{g O}}{80.05\ \text{g}\ NH_4NO_3} \times 100\% = 59.96\%\ \text{O}$$

ASSESSMENT

To check, we add the percentages to ensure that they add up to 100.00%. (Sometimes the total may differ from 100.00% by ±0.01% due to rounding.)

EXERCISE 3.6A

Calculate the mass percent of each element in (a) ammonium sulfate and (b) urea, $CO(NH_2)_2$. Which compound has the greatest mass percent nitrogen: ammonium nitrate (see Example 3.6), ammonium sulfate, or urea?

EXERCISE 3.6B

Calculate the mass percent of

(a) N in triethanolamine, $N(CH_2CH_2OH)_3$ (used in dry-cleaning agents and household detergents);

NEW Strategy/Solution/ Assessment Sections Are Added to Many Examples

By outlining a smart strategy for solving the problem, you learn the strategic thought process necessary to successfully break down and solve problems. The assessment step allows you to confirm the accuracy of your calculations.

A and B Exercises Accompany Every Example

A Exercises pose questions that are similar to those in the Example, allowing you to create a problem-solving strategy based on the Example. *B* Exercises encourage conceptual understanding by asking you to create a strategy that includes some variation of the technique in the Example. *A* and *B* Exercises follow all examples—so you can immediately practice the techniques learned.

NEW Two-Column Examples

Two-column examples place the explanation beside the mechanics of a problem for additional clarity. One-third of the worked-examples are two-column.

SOLUTION

The dissociation of the ionic solutes in the solution is represented by the equations describing the dissolution of the two salts.

$$Na_2SO_4(s) \xrightarrow{H_2O} 2\ Na^+(aq) + SO_4^{2-}(aq)$$
$$\text{Salt} \qquad\qquad \text{Cation} \qquad \text{Anion}$$

$$NaCl(s) \xrightarrow{H_2O} Na^+(aq) + Cl^-(aq)$$
$$\text{Salt} \qquad\qquad \text{Cation} \qquad \text{Anion}$$

Because *one* mole of $Na_2SO_4(s)$ produces *one* mole of $SO_4^{2-}(aq)$, the molarity of the SO_4^{2-} is the same as that given for $Na_2SO_4(aq)$.

$$[SO_4^{2-}] = \frac{0.00384\ \text{mol}\ Na_2SO_4}{1\ \text{L}} \times \frac{1\ \text{mol}\ SO_4^{2-}}{1\ \text{mol}\ Na_2SO_4} = 0.00384\ \text{M}$$

The situation is similar for $Cl^-(aq)$. The molarity of $Cl^-(aq)$ is the same as that of the $NaCl(aq)$.

$$[Cl^-] = \frac{0.00202\ \text{mol}\ NaCl}{1\ \text{L}} \times \frac{1\ \text{mol}\ Cl^-}{1\ \text{mol}\ NaCl} = 0.00202\ \text{M}$$

NEW Cumulative Worked Example at the end of Every Chapter

These examples show you how to analyze and solve problems that integrate concepts from the present chapter with those of previous chapters—showing you how to solve more difficult problems, helping you increase your retention of earlier material, and giving you an understanding of how concepts are connected.

NEW End-of-Chapter Problems

In addition to the categories of problems retained from previous editions, interdisciplinary problems are now incorporated in end-of-chapter problems to further reveal chemistry's relevance to other disciplines.

Challenging problems are noted with an asterisk for easy reference.

Flow Charts

The flow charts are placed directly beside the accompanying Example to give you a graphic picture of the required problem-solving steps.

Cumulative Example

The combustion in oxygen of 1.5250 g of an alkane-derived compound composed of carbon, hydrogen, and oxygen yields 3.047 g CO_2 and 1.247 g H_2O. The molecular mass of this compound is 88.1 u. Draw a plausible structural formula for this compound. Is there more than one possibility? Explain.

STRATEGY

Example 3.12 provides much of the initial guidance for this problem. We can use the masses of CO_2 and H_2O from the combustion analysis to determine first the number of moles of carbon and hydrogen in the sample and then the masses of carbon and hydrogen. Next, we can find the mass of oxygen by subtraction and convert that mass to moles of oxygen. Now, we can use the numbers of moles of the elements to establish the empirical formula of the sample. Once we have the empirical formula, we can use the molecular mass to lead us to the molecular formula. Finally, using our knowledge of alkanes and functional groups (Section 2.9), we can draw structural formulas and determine whether more than one structure is possible.

SOLUTION

We begin by finding the moles of carbon in 3.047 g of CO_2 and the moles of hydrogen in 1.247 g of H_2O.

$$? \text{ mol C} = 3.047 \text{ g } CO_2 \times \frac{1 \text{ mol } CO_2}{44.010 \text{ g } CO_2} \times \frac{1 \text{ mol C}}{1 \text{ mol } CO_2} = 0.06923 \text{ mol C}$$

$$? \text{ mol H} = 1.247 \text{ g } H_2O \times \frac{1 \text{ mol } H_2O}{18.015 \text{ g } H_2O} \times \frac{2 \text{ mol H}}{1 \text{ mol } H_2O} = 0.1384 \text{ mol H}$$

(d) It is desired to make a 20-20-20 fertilizer using KNO_3, NH_4NO_3, NaH_2PO_4, and an inert ingredient (sand). If this can be done, give the relative mass of each of the ingredients present. If this cannot be done, show why this is the case.

* **129. [Environmental]** Hydrogen produced by the decomposition of water has considerable potential as a fuel. Key to its development is finding an appropriate series of chemical reactions that has as its overall reaction: $2 H_2O \longrightarrow 2 H_2 + O_2$. Demonstrate that this requirement is met by the Fe/Cl cycle described as follows: (1) A chloride of iron with 44.19% Fe by mass reacts with steam [$H_2O(g)$] at 500 °C, producing an oxide of iron with 72.36% Fe by mass, together with hydrogen and hydrogen chloride gases. (2) The oxide of iron reacts with hydrogen chloride and chlorine gases at 200 °C to produce another chloride of iron with 34.43% Fe, together with steam, and oxygen gases. (3) At 420 °C, This second chloride of iron produces the first chloride of iron, together with chlorine gas. (*Hint:* The equation for the overall reaction is the sum of the equations for the three reactions described, after their coefficients have been properly adjusted.)

130. **[Environmental]** Calculate the atom % economy (page 113) in the production of carbon tetrachloride carried out in the follow-

* **134. [Laboratory]** In a student experiment, the empirical formula of a copper halide was found by adding aluminum metal to an aqueous solution of the halide, displacing copper metal. The copper metal was filtered, washed with distilled water, dried, and weighed; three separate determinations were performed. If the copper halide solution contained 42.62 g of copper chloride per liter, find the empirical formula that should be reported from the data below. Comment on whether the formula seems reasonable.

Volume of copper chloride solution	Mass of filter paper	Mass of filter paper plus copper
49.6 mL	0.908 g	1.694 g
48.3 mL	0.922 g	1.693 g
42.2 mL	0.919 g	1.588 g

* **135. [Biochemical]** The human body is a complex mixture of many substances. In a typical human, 62.8% of the atoms are H,

---

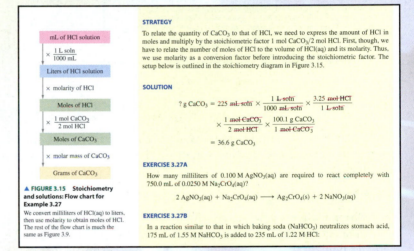

▲ **FIGURE 3.15 Stoichiometry and solutions: Flow chart for Example 3.27**
We convert milliliters of HCl(aq) to liters, then use molarity to obtain moles of HCl. The rest of the flow chart is much the same as Figure 3.9.

STRATEGY

To relate the quantity of $CaCO_3$ to that of HCl, we need to express the amount of HCl in moles and multiply by the stoichiometric factor 1 mol $CaCO_3$/2 mol HCl. First, though, we have to relate the number of moles of HCl to the volume of HCl(aq) and its molarity. Thus, we use molarity as a conversion factor before introducing the stoichiometric factor. The setup below is outlined in the stoichiometry diagram in Figure 3.15.

SOLUTION

$$? \text{ g } CaCO_3 = 225 \text{ mL soln} \times \frac{1 \text{ L soln}}{1000 \text{ mL soln}} \times \frac{3.25 \text{ mol HCl}}{1 \text{ L soln}}$$
$$\times \frac{1 \text{ mol } CaCO_3}{2 \text{ mol HCl}} \times \frac{100.1 \text{ g } CaCO_3}{1 \text{ mol } CaCO_3}$$
$$= 36.6 \text{ g } CaCO_3$$

EXERCISE 3.27A

How many milliliters of 0.100 M $AgNO_3$(aq) are required to react completely with 750.0 mL of 0.0250 M Na_2CrO_4(aq)?

$$2 AgNO_3(aq) + Na_2CrO_4(aq) \longrightarrow Ag_2CrO_4(s) + 2 NaNO_3(aq)$$

EXERCISE 3.27B

In a reaction similar to that in which baking soda ($NaHCO_3$) neutralizes stomach acid, 175 mL of 1.55 M $NaHCO_3$ is added to 235 mL of 1.22 M HCl:

CONCEPTUALIZING CHEMISTRY

Too often a student simply memorizes material without gaining a conceptual understanding of chemistry. *General Chemistry, Fourth Edition,* encourages you to think critically about the concepts being presented, to make connections to knowledge already acquired, and to draw conclusions about possible outcomes. As a result, you build both a quantitative and a qualitative understanding of chemistry.

Conceptual Examples

You may not be used to solving non-quantitative problems. These examples show you how to do so. By emphasizing concepts over calculations, these examples assist in building a thorough understanding of the most important concepts in the course.

Example 3.16 A Conceptual Example

Write a plausible chemical equation for the reaction between water and a liquid molecular chloride of phosphorus to form an aqueous solution of hydrochloric acid and phosphorus acid. The phosphorus-chlorine compound is 77.45% Cl by mass.

ANALYSIS AND CONCLUSIONS

In this problem, balancing a chemical equation is the last step. First, we must apply some earlier ideas:

Establishing the formula of the chloride of phosphorus.
We can apply the method of Examples 3.9 and 3.10 to a compound that is 22.55% P and 77.45% Cl. A 100.00-g sample of the compound consists of 22.55 g P and 77.45 g Cl, corresponding to 0.728 mol P and 2.185 mol Cl. The formula $P_{0.728}Cl_{2.185}$ reduces to PCl_3.

Establishing the formulas of hydrochloric and phosphorus acids.
We described the relationship between names and formulas on pages 57–58. Hydrochloric

NEW Concept Review with Key Terms

Appearing at the end of each chapter, the Concept Review with Key Terms section gives a unique section-by-section illustrated summary of the main concepts of the chapter. Key Terms are used in context and are highlighted for emphasis.

Concept Review with Key Terms

4.1 Some Electrical Properties of Aqueous Solutions—Soluble ionic compounds are completely dissociated into ions in aqueous solution and are therefore **strong electrolytes**. A few water-soluble molecular compounds are completely ionized in aqueous solution and are also **strong electrolytes**. Most molecular compounds exist in solution either as molecules (**nonelectrolytes**) or as a mixture of molecules and ions (**weak electrolytes**).

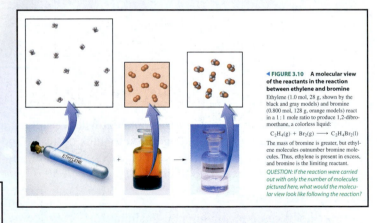

4.2 Reactions of Acids and Bases—A few acids are **strong acids**; these acids are strong electrolytes. However, most acids are **weak acids** (weak electrolytes). The common **strong bases** are water-soluble ionic hydroxides. **Weak bases**, like the weak acids, are molecular compounds that exist as a mixture of molecules and ions in aqueous solution. Many common weak bases are related to ammonia.

Neutralization reactions between acids and bases are conveniently represented by ionic equations and **net ionic equations**. Net ionic equations include only those ions that undergo a chemical reaction in solution; **spectator ions** are eliminated. The neutralization

the **reducing agent** is oxidized. In a **disproportionation reaction**, the same substance acts as both oxidizing agent and reducing agent. Strong oxidizing agents include a few nonmetals and some species having atoms with high oxidation numbers. Strong reducing agents include the active metals and some species having atoms with low oxidation numbers. The **activity series of the metals** ranks metals in order of their strength as reducing agents. It can be used to predict reactions between a metal and other metal ions in solution.

4.5 Applications of Oxidation and Reduction—In everyday life, peroxides and hypochlorites are encountered as oxidizing agents. In industry, oxygen gas, chlorine, and chlorine-containing compounds are used as oxidizing agents. Antioxidants such as vitamin C are reducing agents that can scavenge reactive free radicals. Photosynthesis is an important redox reaction. Redox reactions are used in organic chemistry to oxidize alcohols to **aldehydes** and **ketones**.

Estimation Examples

These exercises cement your conceptual understanding by having you estimate your answer to a problem and determine the reasonableness of that answer rather than relying solely on your calculator.

Example 3.8   An Estimation Example

Without doing detailed calculations, determine which of these compounds contains the greatest mass of sulfur per gram of compound: barium sulfate, lithium sulfate, sodium sulfate, or lead sulfate.

ANALYSIS AND CONCLUSIONS

To make this comparison, we need formulas of the compounds, which we can get from their names:

$$BaSO_4 \quad Li_2SO_4 \quad Na_2SO_4 \quad PbSO_4$$

The compound with the greatest mass of sulfur per gram of compound also has the greatest mass of sulfur per 100 g of compound—in other words, the greatest % S by mass. From the formulas, we see that in one mole of each compound there is one mole of sulfur, which means 32.066 g S. Thus, the compound with the greatest % S is the one with the *smallest* formula mass. Because each formula unit has one SO_4^{2-} ion, all we have to do is compare some atomic masses: that of barium to twice that of lithium, and so on. With just a glance at an atomic mass table, we see the answer must be lithium sulfate, Li_2SO_4.

VISUALIZING CHEMISTRY

Through an enhanced art program, *General Chemistry, Fourth Edition*, enables you to view the world from the unique perspective of a chemist. Allowing you to see the world as a chemist builds your conceptual understanding.

Enhanced Art Program

This edition includes new Macro/Micro art that helps you make connections between macroscopic observations and the underlying molecular behavior of matter.

◄ **FIGURE 3.10   A molecular view of the reactants in the reaction between ethylene and bromine**
Ethylene (1.0 mol, 28 g, shown by the black and gray models) and bromine (0.800 mol, 128 g, orange models) react in a 1 : 1 mole ratio to produce 1,2-dibromoethane, a colorless liquid:

$$C_2H_4(g) + Br_2(g) \longrightarrow C_2H_4Br_2(l)$$

The mass of bromine is greater, but ethylene molecules outnumber bromine molecules. Thus, ethylene is present in excess, and bromine is the limiting reactant.

QUESTION: If the reaction were carried out with only the number of molecules pictured here, what would the molecular view look like following the reaction?

3.10   Yields of Chemical Reactions

In many industrial and commercial chemical processes, the quantity of product actually obtained—the *yield*—is one of the important measures of a successful chemical reaction. The calculated quantity of product in a reaction is called the **theoretical yield** of the reaction. In Example 3.20, the calculated quantity of product is 48.45 g Mg_3N_2, and this is the theoretical yield of magnesium nitride. The quantity of magnesium nitride actually formed in the reaction described in Example 3.20—the **actual yield**—might well be less than this theoretical yield. Actual yields of chemical reactions are often less than theoretical yields for a variety of reasons (Figure 3.11).

◄ **FIGURE 3.11   A reaction that has less than 100% yield:**

$$8 \, Zn(s) + S_8(s) \longrightarrow 8 \, ZnS(s)$$

The actual yield of ZnS(s) obtained, shown in part (c), is less than that calculated for the starting mixture shown in part (a) for several reasons:

- Neither the powdered zinc nor the powdered sulfur is pure.
- The Zn(s) can combine with $O_2(g)$ in air to produce ZnO(s), and some of the sulfur burns in air to produce $SO_2(g)$.
- As suggested in (b), some of the product escapes from the reaction mixture as small lumps and as a fine dust.

(a)   (b)   (c)

Voice Balloons

These features are used to highlight and identify the most important aspects of a photograph or illustration.

APPLICATIONS OF CHEMISTRY

Discussions of how chemistry is applied to a variety of real-world situations appear throughout
General Chemistry, Fourth Edition, to generate your interest in the sciences.

Application Notes

Application Notes highlight the intriguing ways in which we can put our knowledge of chemistry to work, touching on fields as diverse as medicine, engineering, and agriculture.

$$\frac{1 \text{ mol CO}}{2 \text{ mol H}_2} \quad \text{or} \quad \frac{2 \text{ mol H}_2}{1 \text{ mol CO}}$$

It is necessary to keep in mind that the chemical equivalence between H_2 and CO depends on the particular reaction. For another reaction between these same two substances,

$$CO + 3 H_2 \longrightarrow CH_4 + H_2O$$

the corresponding fixed ratio is

$$\frac{1 \text{ mol CO}}{3 \text{ mol H}_2} \quad \text{or} \quad \frac{3 \text{ mol H}_2}{1 \text{ mol CO}}$$

Figure 3.7 illustrates a commonplace example of an equivalence encountered when parking automobiles. The equivalence between automobile and curb length is different depending on whether the parking arrangement is parallel or perpendicular.

Application Note

Alkanes containing more carbon atoms than CH_4, such as the alkanes in gasoline, can be synthesized by reactions that incorporate different proportions of CO and H_2. These mixtures of CO and H_2, called *synthesis gas,* can be produced by a reaction between coal and steam.

Application Boxes

These vignettes appear throughout the text to provide deeper insight into real-world applications of chemistry. These boxes capture your attention—by showing how chemistry is relevant to your life.

Oxidation–Reduction and the Color of Glass

Glassmaking and glassblowing are among the oldest of art forms. Modern museums display glass bottles and vases that are thousands of years old. Stained-glass windows constructed in the Middle Ages still decorate many churches throughout Europe. In these windows, artisans have made use of intricate combination of colored glass to generate their mosaics. The color of many forms of glass is directly related to the oxidation states of impurities found throughout the glass matrix.

Ordinary (soda-lime) glass is made from sand (SiO_2), sodium carbonate ("soda"), and calcium oxide ("lime"). Additives confer other properties, including color. Often the sand

◀ This composite image shows a glass electrical insulator before irradiation with gamma rays (left) and after (right), illustrating the oxidation of Mn(II) to Mn(VII).

contains a small proportion of iron, usually in the +3 oxidation state, which imparts a pale tan color. During processing the iron attains the +2 oxidation state. Look at a piece of window glass end-on, or at a "clear" glass bottle, and you can see the green color that is characteristic of iron(II) ion.

Colorless glass is more difficult to make than "bottle-green" glass. Small batches of colorless optical glass are made using sand that is essentially free of impurities. For large batches, manganese(IV) oxide can be added. The MnO_2 oxidizes the iron back to Fe(III) ion. The manganese is converted to the +2 oxidation state, which is pale pink. The right combination of Mn(II) and Fe(III) ions makes the glass nearly colorless.

Another redox reaction occurs when manganese-containing glass is exposed to intense sunlight. Ultraviolet light is energetic enough to convert manganese gradually from the +2 to the +7 oxidation state, which is purple. The +7 manganese is locked into the structure of the glass, so it does not come into contact with other species (as it could in an aqueous solution) and therefore cannot be reduced back to a lower oxidation state. After years of exposure to sunlight, the colorless manganese-containing glass appears purple. Brief exposure to energetic radiation such as gamma rays (Chapter 19) can cause the same transition.

MEDIA RESOURCES
Hill/Petrucci/McCreary/Perry
General Chemistry, Fourth Edition

Classroom Management Systems: Student Practice, Tutorial, and Assessment

 One Key is Pearson Education's chapter-specific online resource, featuring:

Available Resources:
- Concept Review with Key Terms from the text.
- Online resources such as Web Destinations and online Tools, and the new Research Navigator search tool providing students useful web-related resources.
- Self-Quiz of multiple choice questions with hints and feedback.
- Useful Math Review.
- The new Interactive Student Tutorial (IST) module featuring a full electronic version of the text, with imbedded animations, movies, and molecules.

Assessment Content:
- New Media Gallery including animations and movies from the IST, with follow-up questions.
- Quiz and Master Quiz of multiple choice questions with hints and feedback.
- New Kaplan MCAT Study Guide providing 200 multiple choice questions.
- Test Item File questions.
- Resources from the Instructor's Resource Center.

WEBCT/BLACKBOARD

All of the content in OneKey is also available in WebCT and Blackboard. Prentice Hall offers content cartridges for these text-specific Classroom Management Systems. Visit **www.prenhall.com/demo** for details.

COMPANION WEBSITE

For any students who want access to a free, open-access website, the Companion Website contains the **Concept Review with Key Terms, Web Destinations, Tools, Research Navigator, Self-Quiz,** and **Math Review** modules included in the Classroom Management Systems above. It also contains the animations, movies, and molecules from the IST module in a separate **e-Media Activities** module, and the Interactive Student Tutorial (IST) in a password-protected module.

All of the resources on this page can be accessed at http://chem.prenhall.com/hill

STUDENT ACCELERATOR CD-ROM

The student CD – which contains animations, movies, and molecules from the Classroom Management Systems and the Companion Website – accelerates the performance of these tools when students download high-bandwidth media, so that students are not restricted by slow connections. The *General Chemistry, Fourth Edition,* Accelerator CD comes packaged for no additional charge with every new student textbook.

On-line Homework

PH GRADEASSIST *(www.prenhall.com/phga)*

This time-saving practice and assessment system makes assigning, distributing, and grading homework easier. PH GradeAssist for *General Chemistry, Fourth Edition,* features static and algorithmic versions of many end-of-chapter problems, as well as other questions written for the Fourth Edition. Many questions include links to relevant portions of the text, to provide perspective on such questions. Questions are in a variety of formats, including free-response and multiple choice. Contact your Prentice Hall sales representative for more information.

WEBASSIGN *(www.webassign.net)*

WebAssign, an online homework system, offers selected end-of-chapter questions from the Fourth Edition. You can order access codes directly through your Prentice Hall sales representative.

Classroom Presentation Tools

INSTRUCTOR'S RESOURCE CENTER on CD/DVD (0-13-140318-4)

This flexible, easy-to-use tool contains a wealth of instructor resources specific to *General Chemistry, Fourth Edition.* Organized by chapter, the IRC features almost all the art from the text, two pre-built PowerPoint presentations for each chapter, PDF files of all the art, animations, movies, molecules, and the lecture outlines in Word format from the Instructor's Resource Manual—all in one convenient resource.

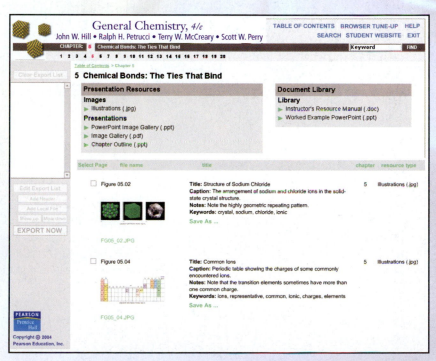

Chapter 1

Chemistry: Matter and Measurement

TODAY, MOST PEOPLE in the industrialized nations have a higher standard of living than the human race has ever known: more nutritious food, better health, and greater wealth. Much of this prosperity is due to chemistry. Chemistry enables us to design all sorts of materials. We make drugs to fight disease; computer chips to enhance communication; pesticides to protect our health and crops; fertilizers to grow abundant food; fuels for transportation; fibers to provide comfort and variety in clothes; building materials for affordable housing; plastics to package food, replace worn-out body parts, and stop bullets; sports equipment to enrich our leisure time; and much more.

Chemistry also helps us to comprehend the nature of our environment, our universe, and ourselves. It provides essential information about issues such as atmospheric ozone depletion, acid rain, and global warming. Chemistry plays a vital role in our understanding and treatment of diseases such as cancer and AIDS, and it helps unravel the mysteries of the human mind. In fact, the theories of chemistry illuminate our understanding of the material world from tiny atoms to giant galaxies. It is a journey we have barely begun, but we can be sure that a knowledge of chemistry will light the path toward a better understanding of our natural world.

1.1 Chemistry: Principles and Applications

The principles of chemistry that we consider in this book have many useful applications. Exploring the practical side of chemistry has stimulated the discovery of many new principles. In chemistry, theory and applications are interwoven like the threads of a fine fabric. To illustrate, let's look briefly at chlorine, a familiar element.

Chlorine is a pale yellow-green gas at room temperature, but most people never see it as a gas. In nature, chlorine is found only combined with other elements, such as with sodium in sodium chloride—common table salt. Seawater is 3% sodium chloride, and blood serum is about 0.8%. Sodium chloride, and hence the chlorine it contains, are essential to life. Elemental chlorine was discovered in 1774, but it did not become

◄ False-color image, obtained by a technique called scanning tunneling microscopy (STM), confirms the atomic nature of matter. The image represents a 1,000,000× magnification of a crystal of palladium and reflects the arrangement of palladium atoms on the surface. Palladium is an element that is often used as an industrial catalyst. The idea that all matter is made of atoms is over 2000 years old, but images of atoms have been available for only a few decades. *Source:* © 2004 RGB Research Ltd. www.element-collection.com.

▲ **Chlorine gas**
Chlorine is one of a relatively small number of fundamental kinds of matter called chemical *elements*. A chemical *compound* is composed of two or more elements in fixed proportions. Elements and compounds are discussed further in Section 1.2.

Physical Properties of Halogens movie

an important commercial chemical until late in the nineteenth century when chemists developed an inexpensive method for obtaining it from sodium chloride. Currently, the chemical industry produces some 10,000 chlorine-containing substances that are used in such varied materials as bleaches, flame retardants, pesticides, drugs, solvents, and plastics.

Perhaps the most familiar use of chlorine is in disinfecting water. Treatment of water supplies with chlorine all but eliminates waterborne diseases such as typhoid fever (responsible for 35,000 deaths in the United States in 1900). However, this use of chlorine has a disadvantage in that it converts some dissolved substances into tiny quantities of chlorinated compounds that are suspected of causing cancer. These trace quantities can be detected only by highly sophisticated methods of analysis, and potential problems with the use of chlorine have become apparent only in recent decades.

Trace quantities of toxic substances called dioxins are formed when chlorine-containing materials are burned and when chlorine is used as a bleach in the pulp and paper industry. Dioxins harm fish and other wildlife and perhaps even humans. Other chlorine compounds that cause environmental concerns are chlorofluorocarbons (CFCs), once widely used in refrigerators and air conditioners and in making foamed plastics. In the stratosphere, CFCs release chlorine atoms that react with and thus destroy some of the ozone found there. Ozone, a form of oxygen, protects life on Earth by absorbing harmful ultraviolet light from sunlight. Interestingly, by using chemical reactions in the laboratory, scientists were able to *predict* the ozone problem several years before ozone depletion in the stratosphere was actually detected and confirmed.

Scientists and others have carefully thought further about many of chlorine's applications, and certain chlorine-containing substances are no longer used. The once-common insecticide DDT has been phased out in developed countries, and substances known as PCBs (polychlorinated biphenyls) are no longer used in printing inks and electrical transformers. CFCs are being replaced by other refrigerants that are less destructive of stratospheric ozone. The pulp and paper industry has replaced chlorine bleach with other types of bleach that do not form objectionable chlorinated compounds. Ozone, which has fewer undesirable side effects, is a potential replacement for chlorine in water treatment.

Scientists know that it is neither possible nor desirable to ban chlorine-containing materials totally from our lives. For example, even though ozone is as effective as

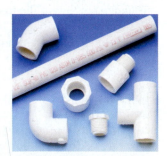

▶ An important current use of chlorine is in the manufacture of polyvinyl chloride (PVC). PVC plastics are highly versatile and are used in many consumer and industrial goods, including clothing, coatings, pipe, flexible tubing, flooring, and toys.

chlorine in killing microorganisms, water treated with ozone can become contaminated again as it enters the distribution system. Chlorine, in contrast, has a residual disinfectant action that extends beyond the treatment plant. And for some uses, such as in the cancer-fighting drug cisplatin, there are no known replacements for chlorine and its compounds.

Chemical knowledge was used to make the chlorine products that have provided so many benefits. Chemical knowledge played a large role in revealing the negative side of these products, and even more chemical knowledge will be required in the search for substitutes. The applications of chemistry, much like the science itself, undergo constant change.

1.2 Getting Started: Some Key Terms

To get started, let's define a few key terms, some of which may already be part of your vocabulary. We will introduce new terms only as we need them in the text. The few we define in this section are those we need to begin our study.

Chemistry is a study of the composition, structure, and properties of matter and of changes that occur in matter. What is **matter?** Matter is anything that has mass and occupies space. It is the stuff of which things are made. Mass, the quantity of matter in an object, is discussed in Section 1.3.

Sometimes we study matter at the *macroscopic* level, that is, we deal with quantities large enough to be seen by the unaided eye. At other times, we deal with matter at the *microscopic* level, where particles are so small that they can be viewed only with special instruments, such as the scanning tunneling microscope that was used to create the image at the beginning of this chapter. It is easy to see that macroscopic objects (made of matter) occupy space and that no two macroscopic objects can occupy the same space at the same time. At the microscopic level, we can also see that no two particles (also made of matter), regardless of how tiny, can occupy the same space at the same time.

Wood, sand, people, water, and air are all examples of matter. Heat and light are not matter; they are forms of energy. The amount of iron in the burner on a kitchen range does not change as the burner transfers heat to the water in a teakettle. Nor does the amount of matter in a pane of window glass change as sunlight passes through. We will examine the nature of energy in Chapter 6.

The tiny, microscopic building blocks of matter known as atoms and molecules are central concerns of chemists. **Atoms** are the smallest distinctive units in a sample of matter; **molecules** are larger units in which two or more atoms are joined together. What a sample of matter is and how it behaves depend ultimately on the particular atoms that are present and on the ways in which they are joined together. Walls built of concrete blocks do not look like those built of bricks because the building blocks are different. A brick fireplace does not look like a brick wall. The building blocks (bricks) are the same, but they are joined together in different ways. **Composition** refers to the types of atoms and the relative proportions of the different atoms in a sample of matter.

The ability to predict natural phenomena is a special characteristic of all the natural sciences.

▲ This model shows two oxygen atoms joined into an oxygen molecule.

Properties

Suppose we need a beaker for an experiment and find just the right one, but it contains a clear, colorless liquid. We can safely pour the liquid down the drain if it is water, but perhaps not if it is something else. How can we determine if the liquid is water? If the liquid has no odor, it could be water—water is odorless, but so are some other liquids. However, if the liquid has a distinctive odor, it cannot be water. Thus, we could easily distinguish between water, ethyl alcohol, and acetic acid (vinegar) by odor alone.

Odor is an example of a **physical property,** a characteristic displayed by a sample of matter without undergoing any change in its composition. When ethyl alcohol is identified by its odor, there is no change in its composition. No changes occur in copper when we observe its color or its ability to conduct electric current, or in diamond when we observe its brilliance and hardness.

▲ A model of a water molecule.

▲ **FIGURE 1.1   Properties of matter**

Copper (left) and ethyl alcohol (right) are easily distinguished by their properties. Copper is a solid; ethyl alcohol is a liquid. Copper is opaque and red-brown; ethyl alcohol is transparent and colorless. Copper does not burn; ethyl alcohol does.

  Water 3D model

  Physical vs. Chemical Change animation

Another way to distinguish between ethyl alcohol and water is that ethyl alcohol burns and water does not. But when ethyl alcohol burns, it is converted to carbon dioxide gas and water. A **chemical property** is a characteristic displayed by a sample of matter as it undergoes a change in composition. Consider the chemical change that takes place as ethyl alcohol burns, a change that results in the formation of carbon dioxide and water. The composition of the molecules formed in the process are quite different from each other and from the composition of ethyl alcohol molecules. Carbon dioxide molecules are composed of two oxygen atoms and one carbon atom; water molecules are composed of two hydrogen atoms and one oxygen atom; and ethyl alcohol molecules are composed of six hydrogen atoms, two carbon atoms, and one oxygen atom. Thus, we conclude that flammability, the ability to burn, is a chemical property. We will encounter many other examples of chemical properties in the text, such as the ability to react with acids and the ability to react with chlorine.

Table 1.1 gives some examples of physical and chemical properties. The physical properties are classified as *qualitative* (describing an essential feature or characteristic) or *quantitative* (numerical). Figure 1.1 illustrates some physical and chemical properties of copper and ethyl alcohol.

When ice melts, solid water is changed to liquid water. The water undergoes a rather profound change at the macroscopic level—what we see—but the change at the microscopic level is more subtle. Water molecules are composed of two hydrogen atoms and one oxygen atom in both solid and liquid water. In a **physical change,** a sample of matter usually undergoes a noticeable change at the macroscopic level but no change in composition. When a chunk of ice changes from a solid to a liquid, the arrangement of the water molecules is changed but their composition is not. By contrast, in a **chemical change,** also called a **chemical reaction,** a sample of matter undergoes a change in composition and/or a change in the structure of its molecules. In many cases, chemical changes also produce effects observable at the macroscopic level, such as the flame observed when ethyl alcohol burns. The changes that take place when food is cooked and when food spoils are common

Table 1.1 Some Examples of Physical and Chemical Properties	
Physical Properties	
Property	**Example**
Qualitative	
Color	Sulfur is *yellow.*
Odor	Hydrogen sulfide *stinks.*
Solubility	Table salt *dissolves in water.*
Hardness	Diamond is *exceptionally hard.*
Electrical conductivity	Copper *conducts electricity.*
Quantitative	
Mass	A nickel has a mass of *5 grams.*
Temperature	Water for the bath is at *40 °C.*
Melting point	Lead melts at *327.5 °C.*
Density	At 20 °C, water has a density of *0.998 grams per milliliter.*
Chemical Properties	
Substance	**Typical Chemical Property**
Iron	*Rusts* (combines with oxygen to form iron oxide)
Carbon	*Undergoes combustion* (combines with oxygen to form carbon dioxide)
Silver	*Tarnishes* (combines with sulfur to form silver sulfide)
Sodium	*Reacts* violently *with water* to form hydrogen gas and a solution of sodium hydroxide.
Nitroglycerin	*Explodes* (decomposes, when detonated, to a mixture of gases)

examples of chemical changes. Figure 1.2 illustrates a situation involving both physical and chemical changes.

Classifying Matter

A **substance** is a type of matter that has a definite, or fixed, composition that does not vary from one sample of the substance to another. All substances are either elements or compounds. An **element** is a substance that cannot be broken down into other simpler substances by chemical reactions. Viewed at the microscopic level, an element is made up of atoms of only a single type. (We will specify just what we mean by a "type" of atom in Chapter 2.) At the present time, more than 110 elements are known. You are most likely already familiar with such common elements as oxygen, nitrogen, carbon, iron, aluminum, copper, silver, and gold, but many other elements are less well known.

A **compound** is a substance made up of atoms of two or more elements, with the different kinds of atoms combined in fixed proportions. In the compound water, the fundamental units are molecules having two hydrogen atoms joined to an oxygen atom. Carbon dioxide, sodium chloride (table salt), sucrose (cane sugar), and iron oxide (rust) are also compounds. Compounds can be broken down into simpler substances—elements—by chemical reactions. The possible number of compounds is essentially limitless. At present, scientists have recorded more than 30 million compounds, and more than a million new ones are added each year. If this seems like a huge number of compounds from such a limited number of elements, think of the 26 letters of the English alphabet and the vast number of words formed from them.

Because elements and compounds are so fundamental to our study, it is convenient to use symbols to represent them. A **chemical symbol** is a one- or two-letter designation derived from the name of an element. Most symbols are based on English names; a few are based on the Latin name of the element or one of its compounds (Table 1.2). Although the name of an element is written in lower case, the first letter of a chemical symbol is *always* capitalized and the second is *never* capitalized. The names and symbols of all the elements are listed inside the front cover of this book. Chemists designate compounds by combinations of chemical symbols called *chemical formulas*. Writing chemical formulas is a little more complex than writing symbols. For example, ethyl alcohol has the formula CH_3CH_2OH. We will discuss how to write chemical formulas in Chapter 2.

In the scheme for classifying matter shown in Figure 1.3, we see two broad categories of matter: substances and mixtures. A **mixture** has no fixed composition; its composition may vary over a broad range. Ordinary table salt and liquid water form a mixture, and we can vary the proportions of salt and water from sample to sample.

A **homogeneous mixture**—which is also called a **solution**—is a mixture that has the same composition and properties throughout. Any particular solution of sodium chloride in water, sometimes called a saline solution, has the same "saltiness" throughout the solution. In contrast, a **heterogeneous mixture** varies in composition and/or

▲ **FIGURE 1.2   Physical and chemical changes**

White gas, a type of gasoline used in portable stoves, is stored as a liquid in the red fuel tank. When the tank valve is opened, the liquid vaporizes to form a gas—a physical change. The gas mixes with oxygen in air and burns—a chemical change. The products of the combustion are carbon dioxide and water.

QUESTION: Name another physical change that appears to be occurring here.

Capitalization can be very important in chemical symbols. For example, Co is the symbol for the element cobalt, but CO represents the poisonous compound carbon monoxide.

 Mixtures and Compounds movie

Table 1.2 Some Elements with Symbols Derived from Latin Names		
Usual Name	**Latin Name**	**Symbol**
Copper	Cuprum	Cu
Gold	Aurum	Au
Iron	Ferrum	Fe
Lead	Plumbum	Pb
Mercury	Hydrargyrum	Hg
Potassium	Kalium*	K
Silver	Argentum	Ag
Sodium	Natrium*	Na
Tin	Stannum	Sn

* The elements potassium and sodium were unknown in ancient times. The names kalium and natrium shown here are the names that were used for the compounds potassium carbonate and sodium carbonate.

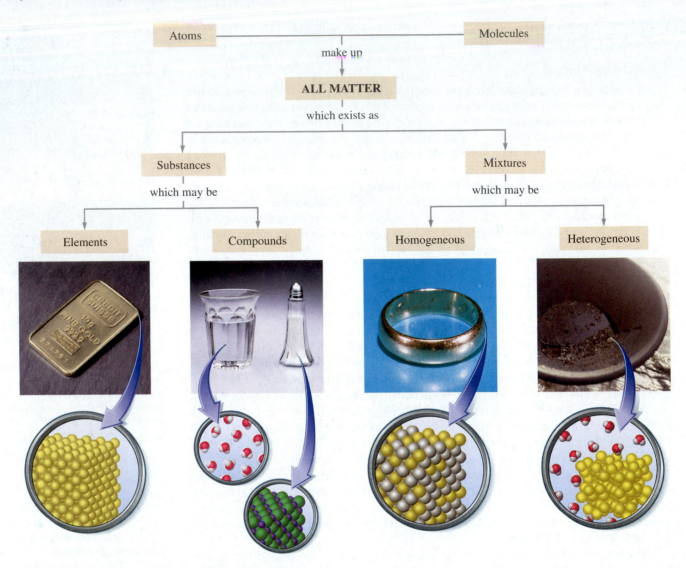

▲ **FIGURE 1.3**   **A scheme for classifying matter**

From left to right: At the molecular level, gold—an *element*—is made up of only one kind of atom, represented here as spheres. In each of the *compounds*, water and sodium chloride (table salt), there are two types of atoms in fixed proportions (2 : 1 for water, 1 : 1 for sodium chloride). Twelve-karat gold—a *homogeneous mixture* of gold and silver—has gold atoms and silver atoms distributed at random, but the composition is the same throughout the mixture. A gold miner's pan contains a *heterogeneous mixture* of gold and water; the composition of this mixture differs vastly from one place to another in the mixture.

Classification of Matter activity

properties from one part of the mixture to another, as in a glass of ice water. Although ice and liquid water have the same composition (both are made of water molecules), the physical properties of the ice and the liquid on which it floats are different. In a mixture of coarse sand and water, both the properties and the composition vary within the mixture. The sand is opaque, tan, and solid, while the water is liquid and transparent. Both homogeneous and heterogeneous mixtures can be separated into their individual components by *physical* changes—chemical reactions are not required. The dissolved sodium chloride can be separated from a saline solution by allowing the water to evaporate away. Sand can be recovered from a sand–water mixture by passing the mixture through filter paper similar to that used in coffeemakers.

Scientific Methods

Chemists and other scientists use certain terms to describe the way in which they conduct their studies. We will briefly consider some of these terms, but the key idea to keep in mind is that scientific knowledge is *testable, reproducible, explanatory, predictive*, and *tentative*.

◀ **FIGURE 1.4   A scientific model of a gas**
The chlorine gas in this container is made up of chlorine molecules. Each molecule is a combination of two chlorine atoms. Chlorine molecules, like all other gas molecules, are in constant random, chaotic motion and undergo frequent collisions with each other and with the container walls. This model is used in Chapter 5 to explain several properties of gases.

Chlorine Gas animation

Scientists often begin by making observations and then formulating a hypothesis. A **hypothesis** is a *tentative* explanation or prediction concerning some phenomenon. It may be just an educated guess, but it must be a guess that can be tested. Scientists *test* a hypothesis through a carefully controlled procedure called an **experiment**. The facts obtained through careful observation and measurements made during experiments are called scientific **data.** Examples of scientific data are the melting point of iron (1535 °C) and the speed of light (2.99792458×10^8 meters per second). Further experiments may refine these data to some degree, but other scientists can verify the basic facts in similar experiments; the data are *reproducible*.

Scientists try to identify patterns in large collections of data. They summarize these patterns in brief statements called **scientific laws.** Many of these laws are stated mathematically. Scientists also use scientific *models*—tangible items or pictures—to represent invisible processes and explain complicated phenomena. For example, the invisible particles (atoms and molecules) of solids, liquids, and gases can be visualized as billiard balls, marbles, or dots or circles on paper (Figure 1.4). The ultimate goal of scientists is to formulate theories. A scientific **theory** provides *explanations* of observed natural phenomena and *predictions* that can be tested by further experiments. Theories often serve as a framework for organizing scientific knowledge.

Contrary to some popular notions, scientific knowledge is not absolute. No hypothesis or theory can ever be proved completely true; it can only be *dis*proved. The most promising hypothesis can be destroyed by one stubborn fact. Thus, the body of scientific knowledge is growing, changing, and never final. Scientists discard old concepts when new tools and new techniques reveal new data and generate new concepts. However, well-established theories, such as the atomic nature of matter, may be modified, but it is highly unlikely that they will be discarded.

There is no single scientific method that serves as a guideline for all to follow. Sometimes scientists make great leaps of the imagination in coming up with new concepts. Usually, though, they proceed methodically, checking their ideas with carefully designed experiments.

Although conceivably any theory can be disproved, extraordinary claims require extraordinary evidence. The evidence that would require scientists to discard the theory that matter is composed of atoms would have to be compelling indeed.

1.3   Scientific Measurements

To make it easier to gather and check data—activities essential to the methods of science—scientists worldwide use a common system of measurement, called *Système Internationale d'Unités* (International System of Units) and abbreviated SI. This system, adopted in 1960, is a modernized version of the metric system established in France in 1791.

In the original metric system, the standard unit of length was the meter, taken to be 1/10,000,000th of the distance from the equator to the North Pole as measured along the meridian passing through Paris. Other standards of length, volume, and mass were ultimately related to the meter.

Table 1.3 The Seven SI Base Units

Physical Quantity	Name of Unit	Symbol of Unit
Length	Meter*	m
Mass	Kilogram	kg
Time	Second	s
Temperature	Kelvin	K
Amount of substance	Mole	mol
Electric current	Ampere	A
Luminous intensity	Candela	cd

* Spelled *metre* in most countries other than the United States.

SI Prefix activity

Table 1.4 Some Common SI Prefixes

Multiple	Prefix
10^{12}	*tera* (T)
10^9	*giga* (G)
10^6	*mega* (M)
10^3	*kilo* (k)
10^2	*hecto* (h)
10^1	*deca* (da)
10^{-1}	*deci* (d)
10^{-2}	*centi* (c)
10^{-3}	*milli* (m)
10^{-6}	*micro* (μ)*
10^{-9}	*nano* (n)
10^{-12}	*pico* (p)

* The Greek letter μ (spelled "mu" and pronounced "mew").

In modern scientific work, all measured quantities can be expressed in terms of the seven base units listed in Table 1.3. We will use the first six of the seven base units in this text, starting with the first four base units in this chapter. An essential aspect of SI is the use of exponential ("powers of ten") notation for numbers. If you wish to review this notation, you will find a discussion of it in Appendix A.

Length

The SI base unit of length is the **meter (m)**, a unit about 10% longer than the yard. Units larger and smaller than the base unit are expressed through prefixes (Table 1.4). For example, to measure lengths much larger than the meter, such as distances along a highway, we often use the kilometer (km), with *kilo*- meaning 1000:

$$1 \text{ km} = 10^3 \text{ m} = 1000 \text{ m}$$

In the laboratory, it is often more convenient to use lengths smaller than the meter. For example, consider the centimeter (cm), with *centi*- meaning 1/100, and the millimeter (mm), with *milli*- meaning 1/1000. A Popsicle stick is roughly 1 cm wide and 1 mm thick:

$$1 \text{ cm} = 10^{-2} \text{ m} = 0.01 \text{ m}$$
$$1 \text{ mm} = 10^{-3} \text{ m} = 0.001 \text{ m}$$

For measurements at the atomic and molecular level, we use the micrometer (μm), the nanometer (nm), and the picometer (pm):

$$1 \text{ μm} = 10^{-6} \text{ m} \qquad 1 \text{ nm} = 10^{-9} \text{ m} \qquad 1 \text{ pm} = 10^{-12} \text{ m}$$

For example, a chlorophyll molecule is about 0.1 μm long (a length that can also be expressed as 100 nm), and the diameter of a sodium atom is about 372 pm.

Area and Volume

The units for area and volume are related to the base unit of length. The SI unit of area is the square meter (m^2), although for laboratory work we often find it more convenient to work with square centimeters (cm^2), or square millimeters (mm^2):

$$1 \text{ cm}^2 = (10^{-2} \text{ m})^2 = 10^{-4} \text{ m}^2 \qquad 1 \text{ mm}^2 = (10^{-3} \text{ m})^2 = 10^{-6} \text{ m}^2$$

A square centimeter is easy to picture; it is about the area of a button on many touchtone telephones. A square millimeter is about the size of a typesetting "bullet" (•).

The SI unit of volume is the cubic meter (m^3), but Figure 1.5 pictures two units that are more commonly used in the laboratory: the cubic centimeter (cm^3)—about the volume of a sugar cube—and the cubic decimeter (dm^3)—slightly larger than 1 quart.

$$1 \text{ cm}^3 = (10^{-2} \text{ m})^3 = 10^{-6} \text{ m}^3 \qquad 1 \text{ dm}^3 = (10^{-1} \text{ m})^3 = 10^{-3} \text{ m}^3$$

Although it is not an SI unit, the old metric unit *liter* is also commonly used. A **liter (L)** is the same volume as 1 cubic decimeter, or 1000 cubic centimeters:

$$1 \text{ L} = 1000 \text{ mL} = 1 \text{ dm}^3 = 1000 \text{ cm}^3$$

The milliliter (mL) is one-thousandth of a liter and the same as a cubic centimeter:

$$1 \text{ mL} = 0.001 \text{ L} = 1 \text{ cm}^3$$

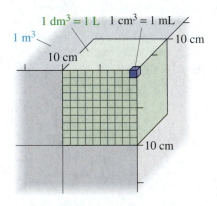

$1 \text{ dm}^3 = 1 \text{ L}$ $1 \text{ cm}^3 = 1 \text{ mL}$
1 m^3
10 cm
10 cm
10 cm

▲ **FIGURE 1.5 Some volume units compared**

The largest volume, shown only in part, is the SI standard of 1 cubic meter (m^3). A cube 10 cm (1 dm) on an edge (the green cube in this drawing) has a volume of 1000 cm^3 (1 dm^3), which is equal to a volume of 1 liter (1 L). The dark blue cube is 1 cm on edge and has a volume of 1 $cm^3 = 1$ mL.

QUESTION: What portion of this art represents a volume of one deciliter?

Where Smaller Is Better

A modern ink-jet printer operates by ejecting tiny droplets of ink onto the paper. Manufacturers of such printers strive for the smallest possible ink droplets because the smaller the droplet, the better the appearance of text and photographs. When the ink-jet printer was first introduced, typical droplet volume was about 200 picoliters—200 pL, two *ten-billionths* of a liter. Advances in technology have reduced this volume to a mere 4 pL! A cartridge holding just 50 mL of ink ejects more than ten billion droplets in its lifetime.

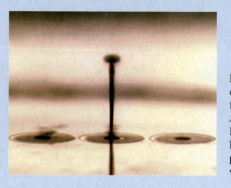

Photomicrograph of a droplet being ejected from an ink-jet orifice. *Source:* © 1998 Hewlett-Packard Development Company, L.P. Reproduced with permission.

Mass

Mass is the quantity of matter in an object. Mass can be measured in many ways, but the most common is through weighing (Figure 1.6). The *weight* of an object is the force of Earth's gravity on the object, and this force is directly proportional to the mass of the object. Two objects of the same mass will weigh the same at any given location on Earth. If the same two objects are at different locations, they may have slightly different weights because of slight variations in Earth's gravitational pull. Therefore, we use mass and not weight as the fundamental measure of quantity of matter.

The SI base quantity of mass is the **kilogram (kg)**, which is about the mass of 1 liter (slightly more than 1 quart) of water. This base quantity is unique in that it already has a prefix (the *kilo-* part). A more convenient mass unit for most laboratory work is the gram (g):

$$1 \text{ kg} = 10^3 \text{ g} = 1000 \text{ g}$$

The milligram (mg) is a suitable unit for small quantities of materials, such as some drug dosages:

$$1 \text{ mg} = 10^{-3} \text{ g} = 0.001 \text{ g}$$

Chemists can now detect masses in the microgram (μg), the nanogram (ng), and even the picogram (pg) range.

Time

The SI base unit for measuring intervals of time is the **second (s)**. Extremely short periods of time are expressed through the usual SI prefixes: *milli*seconds, *micro*seconds, *nano*seconds, and *pico*seconds. Long time intervals, in contrast, are usually expressed in traditional, non-SI units: minute (min), hour (h), day (d), and year (y).

▲ **FIGURE 1.6   Measuring mass by weighing**
Weight is the force of gravity on an object. This force is proportional to the object's mass. In this photograph, the force of gravity on the two objects on the balance pan—the large black cylinder and the black wedge-shaped portion that was cut from it—is counterbalanced by a magnetic force. The magnitude of this magnetic force is registered as a mass 1000.0 in the digital readout. The mass being weighed here is 1.0000 kg.

Example 1.1

Convert the unit of each of the following measurements to a unit that replaces the power of ten by a prefix.

(a) 9.56×10^{-3} m   **(b)** 1.07×10^3 g

SOLUTION

Our goal is to replace each power of ten with the appropriate prefix from Table 1.4. For example, the table tells us that 10^{-3} means we should use the prefix *milli*.

(a) 10^{-3} corresponds to the prefix *milli*; 9.56 mm
(b) 10^3 corresponds to the prefix *kilo*; 1.07 kg

◀ A 1-L bottle of soft drink has a mass of about 1 kg.

EXERCISE 1.1A*

Restate each of the following measurements by attaching an appropriate prefix to the unit to eliminate the power of ten.

(a) 2.05×10^{-6} m   (b) 4.03×10^3 g   (c) 7.06×10^{-9} s   (d) 5.15×10^{-2} m

EXERCISE 1.1B

Restate each of the following measurements by changing the numerical value given to a quantity with a coefficient greater than 1 and less than 10, followed by the appropriate power of ten.

(a) 6217 g   (b) 0.0016 s   (c) 0.0717 g   (d) 387 m

Example 1.2

Use exponential notation to express each of the following measurements in terms of an SI base unit.

(a) 1.42 cm   (b) 645 μs

SOLUTION

(a) Our goal is to find the power of ten that relates the given unit to the SI base unit. Here, the letter c (for centi), used as a prefix with the base unit meter (m), means the same thing as multiplying the base unit by 10^{-2}.

$$1.42 \text{ cm} = 1.42 \times 10^{-2} \text{ m}$$

(b) To change microsecond to the base unit second, we need to replace the prefix *micro* by 10^{-6}. To get our answer in the conventional form of exponential notation, that is, with the coefficient of the power of ten having a value greater than 1 and less than 10, we also need to replace the coefficient 645 by 6.45×10^2. The result of these two changes is

$$645 \text{ }\mu\text{s} = 645 \times 10^{-6} \text{ s} = 6.45 \times 10^2 \times 10^{-6} \text{ s} = 6.45 \times 10^{-4} \text{ s}$$

EXERCISE 1.2A

Use exponential notation to express each of the following measurements in terms of its SI base unit.

(a) 355 μs   (b) 1885 km   (c) 1350 cm   (d) 425 nm

EXERCISE 1.2B

Use exponential notation to express each of the following measurements in terms of its SI base unit.

(a) 2.28×10^5 g   (b) 0.083 cm   (c) 4.05×10^2 μm   (d) 20.25 min

Temperature

Temperature Conversion activity

Temperature is difficult to define. We can say that it is a measure of "hotness." A hot object is at a higher temperature than a cold one, but hot and cold are relative terms. We can refine our first attempt at a definition by considering what happens when two objects at different temperatures are brought together: Heat flows from the warmer to the colder object. The temperature of the warmer object drops and the temperature of the colder object increases until the two objects reach the same temperature. Temperature is therefore a property that tells us in what direction heat will flow. For example, if you touch a hot test tube, heat will flow from the tube to your hand. If the tube is hot enough, your hand will be burned. We will present some additional ideas about temperature in Chapter 5.

*You will see Exercises A and B following each example. The goal in an A exercise is to apply the method outlined in the example to a similar situation. In a B exercise, you often need to combine the method just illustrated in the example with other ideas learned previously. The B exercises provide the kind of context in which we usually apply chemical knowledge.

The SI base unit of temperature is the **kelvin (K)**. For routine laboratory work, we often use the more familiar Celsius temperature scale. On this scale, the freezing point of water is 0 degrees Celsius (0 °C) and the boiling point is 100 °C. Another temperature scale widely used in the United States, but probably unfamiliar to most people in the rest of the world, is the Fahrenheit scale. As pointed out in Figure 1.7, the Celsius and Fahrenheit scales differ in

- their assignment of the numerical value of an important physical property—the freezing point of water: 32 °F and 0 °C;
- the temperature interval called a *degree* (°) The 180-degree interval between freezing and boiling water on the Fahrenheit scale equals the 100-degree interval between those points on the Celsius scale, a ratio of 1.8 : 1.

We can use these two facts to develop two versions of the equation that relates temperatures on the two scales. In one relationship, we must multiply the degrees of Celsius temperature by the factor 1.8 to obtain degrees of Fahrenheit temperature and then add 32 to account for the fact that 0 °C is the same as 32 °F. In the other equation, we first subtract 32 from the Fahrenheit temperature to get the number of degrees Fahrenheit above the freezing point of water. Then we divide this quantity by 1.8.

$$T_F = 1.8\,T_C + 32 \qquad T_C = \frac{T_F - 32}{1.8}$$

You will not often need to convert between Celsius and Fahrenheit temperature, but Example 1.3 illustrates a practical situation where this conversion would be necessary.

Example 1.3

At home you like to keep the thermostat at 72 °F. While traveling in Canada, you find the room thermostat calibrated in degrees Celsius. To what Celsius temperature would you need to set the thermostat to get the same temperature you enjoy at home?

STRATEGY

The general approach to conversions between the Fahrenheit and Celsius temperature scales is to use the appropriate form of the equations written above. The equation for converting a Fahrenheit temperature to the Celsius scale is

$$T_C = \frac{T_F - 32}{1.8}$$

SOLUTION

You merely need to substitute the given Fahrenheit temperature and solve for T_C:

$$T_C = \frac{72 - 32}{1.8} = \frac{40}{1.8} = 22\ °C$$

ASSESSMENT

One simple check of this calculation is to see that the numerical value of the temperature in degrees Celsius is less than the original temperature in degrees Fahrenheit, as suggested in Figure 1.7. (Note that this statement is true only for temperatures greater than −40 °F.)

EXERCISE 1.3A

Carry out the following temperature conversions.

(a) 85.0 °C to °F
(b) −12.2 °C to °F
(c) 335 °F to °C
(d) −20.8 °F to °C

EXERCISE 1.3B

The lowest possible temperature on the Kelvin scale—0 K—is at −273.15 °C. What is this lowest possible temperature on the Fahrenheit scale?

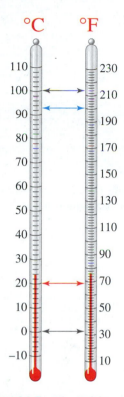

▲ **FIGURE 1.7 The Celsius and Fahrenheit temperature scales compared**

The thermometer on the left is marked in degrees Celsius, and the one on the right in degrees Fahrenheit. The freezing point of water is at 0 °C and 32 °F; the boiling point is at 100 °C and 212 °F. An interval of 18 °F corresponds to an interval of 10 °C, which gives rise to the factor 18/10 = 1.8 in the equations that relate the two scales.

QUESTION: What are the Celsius and Fahrenheit temperatures at the points indicated by the red and blue arrows?

We will define the Kelvin temperature scale in Chapter 5, where it will be used extensively.

To help organize your thoughts when solving a numerical problem, consider following the three-step approach **Strategy ⟶ Solution ⟶ Assessment**, which we introduce here.

▲ **FIGURE 1.8** **Poster board used to gather data given in Table 1.5**

1.4   Precision and Accuracy in Measurements

Counting can be exact: We can count exactly 18 students in a room. All measurements, on the other hand, are subject to error and uncertainty. One source of error is the measuring instruments themselves. A bathroom scale, for example, may always read five pounds low if it is set to read that way. Even when set correctly, most bathroom scales can give weight only to the nearest pound. Other errors may result from the experimenter's lack of skill or care in using measuring instruments.

Suppose you were one of five students asked to measure the dimensions of the poster board in Figure 1.8, using a meter stick. Table 1.5 presents the five sets of measurements, all converted to meters. The **precision** of a set of measurements refers to how closely individual measurements agree with one another. The precision is good (or high) if each of the measurements in a set is close to the average of the set. The precision is poor (or low) if some (or all) of the measurements deviate widely from the average value. How would you describe the precision of the data in Table 1.5? Examine the individual data for the length and width, note the average values, and determine how much the individual data differ from the averages. You will find that the maximum deviations from average values are 0.003 m for the length and 0.001 m for the width. Thus, we describe the precision as good for both sets of measurements.

Table 1.5 Five Measurements of the Dimensions of a Poster Board

Student	Length, m	Width, m
1	1.827	0.761
2	1.824	0.762
3	1.826	0.763
4	1.828	0.762
5	1.829	0.762
Average:	1.827	0.762

The **accuracy** of a set of measurements refers to how close the average of the set comes to the true, or most probable, value. Measurements of high precision are more likely to be accurate than those of poor precision, but even highly precise measurements are sometimes inaccurate. For example, what if the meter sticks used to obtain the data in Table 1.5 were actually 1005 mm long but still marked as if they were one meter? The accuracy of the measurements would be rather poor even though the precision would remain high. Figure 1.9 provides another comparison of precision and accuracy.

(a) Low accuracy
Low precision

(b) Low accuracy
High precision

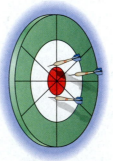
(c) High accuracy
Low precision

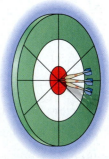
(d) High accuracy
High precision

▲ **FIGURE 1.9**   **Comparing precision and accuracy: a dartboard analogy**
(a) The darts are both scattered far from one another (low precision) and off center (low accuracy). **(b)** The darts are in a tight cluster (high precision) but still off center (low accuracy). **(c)** The darts are somewhat scattered from one another (low precision) but evenly distributed about the center (high accuracy). **(d)** The darts are in a tight cluster (high precision) and well centered (high accuracy).

QUESTION: If two of the darts from part **(d)** *were removed, would* **(d)** *still represent high accuracy? High precision? Explain.*

Sampling Errors

No matter how accurate it is, a measurement will not mean much unless it was performed on valid, representative samples. Consider determining the level of dissolved oxygen in the water of a lake. Dissolved oxygen is vital to fish. Measurements would yield varying results, depending on several factors, such as *where* the sample was taken—at the surface or near the bottom, near the mouth of a swiftly flowing entering stream or in a stagnant bay. Results would also depend on *when* the sample was taken—on a warm, still day or when the wind was whipping up whitecap waves. Dissolved oxygen levels also depend on other factors, such as the temperature of the water. There is no *single* value for the dissolved oxygen level that is valid for the whole lake or valid for any length of time. An average level could be obtained by extensive random sampling. Even then, scientists would usually record the conditions under which measurements were made.

Significant Figures

Look again at Table 1.5. Notice that the five measurements of length agree in the first three digits (1.82); they differ only in the fourth digit. We say the fourth digit is uncertain. All digits known with certainty (three in this case) *plus* the first uncertain one are called significant digits, or **significant figures.** We use significant figures to reflect the precision of a measurement—the more significant figures, the more precise the measurement. The length measurements in Table 1.5 have *four* significant figures. In other words, we are quite sure that the length of the board is between 1.82 m and 1.83 m. Our best estimate of the average value, including the uncertain digit, is 1.827 m.

It is easy to establish that 1.827 m has four significant figures: We simply count the number of digits. In any measurement that is properly reported, all *nonzero* digits are significant. *Zeros* present problems because they can be used in two ways: as a part of the measured value or to position a decimal point.

As scientists, we do not want to claim in our work more precision than we actually obtained. That is why we use the significant figure convention.

- Zeros between two other significant digits (enclosed zeros) are significant. Examples: 1107 km (four significant figures); 50.002 g (five significant figures).
- A lone zero preceding a decimal point is written for cosmetic purposes. It is not significant. Example: 0.762 g (three significant figures).
- Zeros to the right of the decimal point that precede the first nonzero digit (leading zeros) and are used only to position the decimal point are also *not* significant. Example: 0.000163 m (three significant figures).
- Zeros at the end of a number are significant if they are to the *right* of the decimal point. Examples: 0.2000 g (four significant figures); 0.050120 kg (five significant figures).

These four situations all conform to the general rule that when we read a number from left to right, all the digits starting with the first nonzero digit are significant. Numbers written without a decimal point and ending in zeros are a special case, however.

- Zeros at the end of a number may or may not be significant if the number is written *without* a decimal point. Example: 400

The measurement 0.2000 g means that the mass is known to be between 0.1999 and 0.2001 g. A measurement of 0.20 g means that the mass is anywhere between 0.19 and 0.21 g.

We do not know whether the number 400 was measured to the nearest one, ten, or hundred. To avoid this confusion, we use exponential notation (see Appendix A). In exponential notation, 400 is expressed as 4×10^2 or 4.0×10^2 or 4.00×10^2 to indicate one, two, or three significant figures, respectively. The only significant digits are those in the coefficient, not in the power of ten.

The concept of significant figures applies only to *measurements*—which are quantities subject to error. It does not apply to any quantity that is (a) inherently an integer, such as 4 sides to a square or 12 items in a dozen, (b) inherently a fraction, such as the radius of a circle $= \frac{1}{2}$ the diameter of the circle, or (c) obtained by an exact count, such as 127 students in a class. It also does not apply to *defined* quantities, such as 1 km = 1000 m. In these contexts, the numbers 4, 12, $\frac{1}{2}$, 127, and 1000 are not limited in their numbers of significant figures. In effect, each has an unlimited number of significant figures (4.000..., 12.000..., 0.5000...) or, more properly, each is an *exact* value.

Significant Figures in Calculations: Multiplication and Division

In an item on its Web site, the U.S. Census Bureau, extrapolating from data it had collected, states that as of July 7, 2000, the world population was 6,081,590,240. Is it possible that all the digits in this number are truly significant and that the extrapolation could account for every single individual? Clearly this number is only an estimate based on assumptions regarding such factors as rates of births and deaths.

Just as a chain is only as strong as its weakest link, a calculation is only as precise as the least precise measurement included in the calculation. A strict application of this principle involves a sophisticated method of analyzing the relevant data that is beyond the scope of this text, but we can do a pretty good job through a practical rule involving significant figures.

In multiplication and division, the reported result should have no more significant figures than the factor with the fewest significant figures.

To write a numerical answer with the proper number of significant figures, we often have to round off numbers. In rounding, we drop all digits that are not significant and, if necessary, adjust the last reported digit. In this textbook, these are the rules that we will follow in rounding.*

- If the leftmost digit to be dropped is 0, 1, 2, 3, or 4, leave the final remaining digit *unchanged*. Example: 369.443 rounds to 369.44 if we need five significant figures, and to 369.4 if we need four.

- If the leftmost digit to be dropped is 5, 6, 7, 8, or 9, *increase* the final digit by *one*. Example: 538.768 rounds to 538.77 if we need five significant figures and to 538.8 if we need four. Similarly, 74.395 rounds to 74.40 if we need four significant figures, and to 74.4 if we need three.

Example 1.4

Calculate the area, in square meters, of the poster board whose dimensions are given in Table 1.5. Report the correct number of significant figures in your answer.

STRATEGY

The area of the rectangular poster board is the product of its length and width. In expressing the result of this multiplication, we can show only as many significant figures as are found in the least precisely stated dimension: the 0.762-m width (three significant figures).

SOLUTION

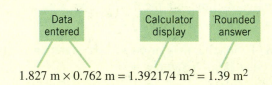

$$1.827 \text{ m} \times 0.762 \text{ m} = 1.392174 \text{ m}^2 = 1.39 \text{ m}^2$$

ASSESSMENT

When using a calculator, the display often has more digits than are significant (Figure 1.10). We use the rules for rounding off numbers as the basis for dropping the digits "2174."

EXERCISE 1.4A

Calculate the volume, in cubic meters, of the poster board described in Table 1.5, given that its thickness is 6.4 mm. Report your answer to the correct number of significant figures.

EXERCISE 1.4B

Calculate the volume of the poster board in cubic centimeters, expressing your answer to the appropriate number of significant figures.

▲ **FIGURE 1.10** **Significant figures**
The calculator shows 1.392174, suggesting seven significant figures, but the rule for multiplication tells us there are only three significant figures: 1.39.

*C. J. Guare, *J. Chem. Educ.*, 68, 818 (1991).

Example 1.5

For a laboratory experiment, a teacher wants to divide all of a 453.6-g sample of sulfur equally among the 21 members of her class. How many grams of sulfur should each student receive?

STRATEGY

Here we need to recognize the number "21" as a counted number. It is therefore an exact number and not subject to significant figure rules. The answer should carry *four* significant figures, the same as in 453.6 g.

SOLUTION

Divide the total mass by the number of students.
$$? g = \frac{453.6 \text{ g}}{21} = 21.60 \text{ g}$$

ASSESSMENT

In this calculation, a calculator displays the result "21.6." We need to add the digit "0" to indicate that the result is precise to four significant figures.

EXERCISE 1.5A

The experiment described in Example 1.5 also requires that each student have available 2.04 times as much zinc as sulfur (by mass). What total mass of zinc will the teacher need, expressed with the appropriate number of significant figures?

EXERCISE 1.5B

Calculate the total surface area, in square centimeters, of a sugar cube that is 9.2 mm on edge, and express this area with the appropriate number of significant figures.

Significant Figures in Calculations: Addition and Subtraction

When we add or subtract, our concern is not with the number of significant figures but with digits to the right of the decimal point. If the quantities being added or subtracted differ in the number of significant digits they have to the right of their decimal point, the quantity with the *fewest* such digits determines the number of significant figures in the answer. The number of significant digits to the right of the decimal point in the answer should be the same as the number of such digits in the added/subtracted quantity with the fewest. The idea is that if you are adding several lengths and one of them is measured only to the nearest *centi*meter, the total length cannot be stated to the nearest *milli*meter, no matter how precise the other measurements are.

We apply this idea in Example 1.6 and illustrate another point as well: In a calculation involving several steps, slightly different answers are possible, depending on the way in which rounding off is done.

Example 1.6

Perform the following calculation, and round off the answer to the correct number of significant figures.
$$49.146 \text{ m} + 72.13 \text{ m} - 9.1434 \text{ m} = ?$$

STRATEGY

We must add two numbers and then subtract a third from their sum. We do this in two ways below. In both cases, we express the answer to two decimal places, the number of decimal places in "72.13 m."

SOLUTION

Intermediate rounding

```
    49.146 m                      49.146 m
  + 72.13  m                    + 72.13  m
  ─────────                     ─────────
   121.276 m = 121.28 m          121.276 m
            −  9.1434 m          − 9.1434 m
            ──────────           ──────────
            112.1366 m = 112.14 m  112.1326 m = 112.13 m
```

Calculator display   Rounded answer   Calculator display   Rounded answer

(a)   (b)

Problem-Solving Note

Remember that all three dimensions must be expressed in the same unit, as is the case in this Example.

Problem-Solving Note

In solving problems, whenever possible, you should perform calculations in a single setup, storing intermediate results in a calculator and rounding off only the final answer. In this text, to make solutions to worked examples easier to follow, we will often use a stepwise approach. In these cases we will write down and round off intermediate answers, as well as the final answer. At times this may yield a slightly different result than the preferred single-step approach.

Problem-Solving Note

In addition and subtraction, all terms must be expressed in the same unit.

Table 1.6 Some Conversions Between Common (U.S.) and Metric Units

Metric	Common
Mass	
1 kg	= 2.205 lb
453.6 g	= 1 lb
28.35 g	= 1 ounce (oz)
Length	
1 m	= 39.37 in.
1 km	= 0.6214 mi
2.54 cm	= 1 in.*
Volume	
1 L	= 1.057 qt
3.785 L	= 1 gal
29.57 mL	= 1 fluid ounce (fl oz)

* U.S. inch is defined as exactly 2.54 cm. The other equivalencies are rounded off.

ASSESSMENT

The slight difference in the answers in (a) and (b) stems from a difference in rounding off. In method (a) we round off an intermediate result—121.276. Then we round off the final answer: 112.1366 m to 112.14 m. In method (b) we do *not* round off the intermediate result and round off only the final answer: 112.1326 m to 112.13 m. When we use a calculator, we generally do not need to write down or otherwise take note of an intermediate result.

EXERCISE 1.6A

Perform the indicated operations and give answers with the proper number of significant figures.

(a) 48.2 m + 3.82 m + 48.4394 m

(b) 148 g + 2.39 g + 0.0124 g

(c) 451 g − 15.46 g − 20.3 g

(d) 15.436 L + 5.3 L − 6.24 L − 8.177 L

EXERCISE 1.6B

Perform the indicated operations, and give answers with the proper number of significant figures. Do these calculations in two different ways: First, round off intermediate results involving addition/subtraction, followed by rounding off the final answer. Then, repeat the calculation and round off only the final answer. Are there any differences in the results in these particular cases?

(a) $(51.5 \text{ m} + 2.67 \text{ m}) \times (33.42 \text{ m} - 0.124 \text{ m})$

(b) $\dfrac{125.1 \text{ g} - 1.22 \text{ g}}{52.5 \text{ mL} + 0.63 \text{ mL}}$

(c) $\dfrac{47.5 \text{ kg} - 1.44 \text{ kg}}{10.5 \text{ m} \times 0.35 \text{ m} \times 0.175 \text{ m}}$

(d) $\dfrac{0.307 \text{ g} - 14.2 \text{ mg} - 3.52 \text{ mg}}{(1.22 \text{ cm} - 0.28 \text{ mm}) \times 0.752 \text{ cm} \times 0.51 \text{ cm}}$

1.5 A Problem-Solving Method

Much of the study of chemistry involves solving problems. To solve them, you generally need to apply some basic principles to describe a new situation based on the information given. Many chemistry problems require calculations and yield quantitative results. In this section, we will describe a useful method for doing such calculations, a method called unit conversion. Additional ideas on problem solving are presented in Section 1.6.

The Unit-Conversion Method

Imagine that an American orders a leather belt from a foreign country. The customer knows her waist size in common units but must give it in metric units. Suppose her waist measurement is 24 inches, so that she will need a belt that is 26 inches long. What is this length in centimeters? The length 26 inches is a fixed length, whether we express it in inches, feet, centimeters, or millimeters. Thus, when we measure something in one unit and then convert that unit to another one, we must not change the measured quantity in any fundamental way.

In mathematics, multiplying a quantity by 1 does not change the value of the quantity, and we can therefore use a factor that is equivalent to 1 to convert between inches and centimeters. We find our factor in the definition of the inch in Table 1.6:

$$1 \text{ in.} = 2.54 \text{ cm}$$

If we divide both sides of this equation by 1 in. we obtain a ratio of quantities that is equal to 1:

$$\frac{1 \text{ in.}}{1 \text{ in.}} = \frac{2.54 \text{ cm}}{1 \text{ in.}} = 1$$

If we divide both sides of the equation by 2.54 cm, we also obtain a ratio of quantities that is equal to 1:

$$\frac{1 \text{ in.}}{2.54 \text{ cm}} = \frac{2.54 \text{ cm}}{2.54 \text{ cm}} = 1$$

The two ratios, one printed in red and the other in blue, are conversion factors. A **conversion factor** is a ratio of terms—equivalent to the number 1—used to change the unit in which we express a quantity. A conversion factor can be generated from any equality, such as the one above relating inches to centimeters.

What happens when we multiply a known quantity by a conversion factor? The original unit used to express the quantity cancels out and is replaced by the desired unit. To put the matter another way, our general approach to using conversion factors is

Desired quantity and unit = given quantity and unit × conversion factor(s).

Now let's return to the question about the length of the belt. Its length—the measured quantity—is 26 in. The appropriate conversion factor (in red) must have the desired unit (cm) in the *numerator* and the unit to be replaced (in.) in the *denominator*. In this arrangement, we can cancel the unit *in.* so that only the unit *cm* remains:

$$? \text{ cm} = 26 \text{ in.} \times \frac{2.54 \text{ cm}}{1 \text{ in.}} = 66 \text{ cm}$$

It is not hard to see why the blue conversion factor will not work. It gives a nonsensical unit, $\text{in.}^2/\text{cm}$:

$$26 \text{ in.} \times \frac{1 \text{ in.}}{2.54 \text{ cm}} = 10 \frac{\text{in.}^2}{\text{cm}}$$

Example 1.7 is a more typical illustration of the *unit-conversion method* of problem solving because we need to use more than one conversion factor. We also use this example to point out some important practices that we will follow in writing conversion factors and dealing with significant figures.

Example 1.7

What is the length in millimeters of a 1.25-ft rod?

STRATEGY

We could make this conversion with a single conversion factor if we had a relationship between feet and millimeters. However, Table 1.6 does not provide such a relationship. We therefore need to do three successive conversions:

1. From feet to inches using the fact that 1 ft = 12 in.
2. From inches to meters with data from Table 1.6.
3. From meters to millimeters based on the prefixes in Table 1.4.

SOLUTION

We can write three answers, one for each conversion, with the third answer being our final result. It is just as easy, however, to combine the three conversions into a single setup without writing any intermediate answers. This is the procedure sketched below.

| We want the unit mm and the number that goes with it (?). | We start here. | This converts ft to in. | This converts in. to m. | We get the desired unit here. | Our answer: the number   the unit |

$$? \text{ mm} = 1.25 \text{ ft} \times \frac{12 \text{ in.}}{1 \text{ ft}} \times \frac{1 \text{ m}}{39.37 \text{ in.}} \times \frac{1000 \text{ mm}}{1 \text{ m}} = 381 \text{ mm}$$

 **Conversion Example activity**

ASSESSMENT

Now let's look at the use of significant figures in this problem. The measured quantity, 1.25 ft, is given with *three* significant figures. The relationship 12 in. = 1 ft is *exact*, and

this will not affect the precision of our calculation. It is understood that the relationship
1 m = 39.37 in. is stated to *four* significant figures; there is no need to write 1 m as
1.000 m. Finally, the relationship 1000 mm = 1 m is *exact*; this defines the way we relate
millimeters and meters. Our answer should therefore have *three* significant figures.

EXERCISE 1.7A

Carry out the following conversions.

(a) 76.3 mm to meters   (b) 0.0856 kg to milligrams
(c) 0.556 km to feet

EXERCISE 1.7B

Carry out the following conversions.

(a) 1.95×10^{-3} oz to µg   (b) 3.50 gal to fluid ounces
(c) 7.75×10^{4} ft to kilometers

Because a conversion factor has an intrinsic value of 1 (the numerator and denominator are equivalent), we can raise the factor to a power and it still has an intrinsic value
of 1. For example,

$$\frac{2.54 \text{ cm}}{1 \text{ in.}} = 1 \quad \text{and} \quad \left(\frac{2.54 \text{ cm}}{1 \text{ in.}}\right)^2 = 1^2 = 1 \quad \text{and} \quad \left(\frac{2.54 \text{ cm}}{1 \text{ in.}}\right)^3 = 1^3 = 1$$

Example 1.8

What is the volume, in cubic centimeters, of the block of wood pictured here?

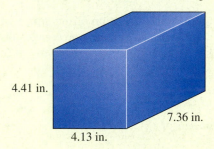

4.41 in.

7.36 in.

4.13 in.

STRATEGY

The volume of the block is the product of the three dimensions given in the drawing. We can
proceed in one of two ways: (1) change each dimension from inches to centimeters and
obtain the product in cubic centimeters or (2) obtain the volume in cubic inches and then
convert to cubic centimeters.

SOLUTION

In method (1), the factor 2.54 cm/1 in. appears three times in the setup:

$$? \text{ cm}^3 = 4.13 \text{ in.} \times \left(\frac{2.54 \text{ cm}}{1 \text{ in.}}\right) \times 4.41 \text{ in.} \times \left(\frac{2.54 \text{ cm}}{1 \text{ in.}}\right) \times 7.36 \text{ in.} \times \left(\frac{2.54 \text{ cm}}{1 \text{ in.}}\right)$$

$$= 2.20 \times 10^3 \text{ cm}^3$$

In method (2), the three conversion factors are combined into a single factor $(2.54 \text{ cm}/1 \text{ in.})^3$:

$$? \text{ cm}^3 = (4.13 \text{ in.} \times 4.41 \text{ in.} \times 7.36 \text{ in.})^3 \times \frac{(2.54)^3 \text{ cm}^3}{(1)^3 \text{ in.}^3} = 2.20 \times 10^3 \text{ cm}^3$$

ASSESSMENT

Whenever two or more approaches to a problem are possible, one way of assessing your
answer is to verify that the same answer is obtained, regardless of the approach. Another
way of assessing the correctness of your answer is to check that the unit cancellation you
used yields the final unit you need.

EXERCISE 1.8A

Carry out the following conversions.

(a) 476 cm^2 to square inches   (b) 1.56×10^4 in.^{3} to cubic meters

Example 1.9

The commonly accepted measurement now used by dietary specialists in assessing whether a person is overweight is the *body mass index* (BMI), which is based on a person's mass and height. It is the mass, in kilograms, divided by the *square* of the height in meters. Thus, the units for BMI are kg/m^2. Generally speaking, if the BMI exceeds 25, a person is considered overweight. What is the BMI of a person who is 69.0 inches tall and weighs 158 lb?

STRATEGY

We must convert *two* units in the original measured quantities: from pounds to kilograms and from inches to meters (or, more specifically, from inches squared to meters squared). We will solve the problem in two ways. In the first approach, we begin by expressing the BMI in $lb/in.^2$ Then we can convert from lb to kg in the numerator and from $in.^2$ to m^2 in the denominator. In an alternate approach, we do the conversions in the numerator and denominator separately and then divide the numerator by the denominator.

SOLUTION

The conversion factors required in the first approach are

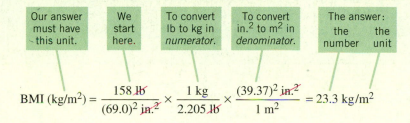

$$\text{BMI } (kg/m^2) = \frac{158 \text{ lb}}{(69.0)^2 \text{ in.}^2} \times \frac{1 \text{ kg}}{2.205 \text{ lb}} \times \frac{(39.37)^2 \text{ in.}^2}{1 \text{ m}^2} = 23.3 \text{ kg/m}^2$$

The three steps in the alternate approach are

$$\text{Numerator: } 158 \text{ lb} \times \frac{1 \text{ kg}}{2.205 \text{ lb}} = 71.7 \text{ kg}$$

$$\text{Denominator: } (69.0)^2 \text{ in.}^2 \times \frac{1 \text{ m}}{39.37 \text{ in.}} \times \frac{1 \text{ m}}{39.37 \text{ in.}} = 3.07 \text{ m}^2$$

$$\text{Division: BMI } (kg/m^2) = \frac{71.7 \text{ kg}}{3.07 \text{ m}^2} = 23.4 \text{ kg/m}^2$$

Note:
in. × in. = in.²

ASSESSMENT

Both methods are correct, even though the answers differ slightly. The source of the difference is in rounding off the intermediate results in the second method (71.655329 to 71.7 in the numerator and 3.071619 to 3.07 in the denominator). If we had stored these intermediate results in the memory of a calculator and divided the stored quantities, we would have obtained 23.328196, which rounds off to 23.3, just as in the first method.

Density: A Physical Property and Conversion Factor

People often say that iron is "heavy" and aluminum is "light." As you know, they do not mean to suggest that an iron nail weighs more than an aluminum ladder. What they mean is that iron is more *dense* than aluminum. If we compare the masses of *equal volumes* of the two metals, we find that the iron has a greater mass. **Density (*d*)** is the mass per unit volume of a substance. This important physical property is defined by the equation

$$d = \frac{m}{V}$$

The SI unit of density is kilograms per cubic meter (kg m^{-3}), but chemists more often measure densities in grams per cubic centimeter g/cm^3 or grams per milliliter g/mL. The density of aluminum is 2.70 g/cm^3, and that of iron is 7.87 g/cm^3. Densities of several other familiar substances are given in Table 1.7.

Although the mass of an object remains constant as the temperature is raised, the volume generally increases—the object expands—and therefore its density decreases. As a consequence, it is usually necessary to state the temperature at which a density is measured. Ethylene glycol (the principal ingredient in many antifreeze solutions), water, and ethanol are all colorless liquids. We can distinguish these liquids by their densities. At 20 °C, the density of ethylene glycol is 1.114 g/mL, that of ethanol is 0.789 g/mL, and that of water is 0.998 g/mL.

Density activity

Table 1.7 Densities of Some Common Substances at Specified Temperatures		
Substance[a]	**Density**	**Temperature (°C)**
Solids		
Aluminum (Al)	2.698 g/cm^3	20
Copper (Cu)	8.94 g/cm^3	25
Gold (Au)	19.3 g/cm^3	25
Iron (Fe)	7.784 g/cm^3	20
Magnesium (Mg)	1.738 g/cm^3	20
Water (ice) (H$_2$O)	0.917 g/cm^3	0
Liquids		
Blood plasma (a mixture)	1.027 g/mL	25
Bromine (Br$_2$)	3.123 g/mL	20
Chloroform (CHCl$_3$)	1.483 g/mL	20
Ethyl alcohol (C$_2$H$_5$OH)	0.789 g/mL	20
Hexane (C$_6$H$_{14}$)	0.660 g/mL	20
Mercury (Hg)	13.534 g/mL	25
Urine (a mixture)	1.003–1.030 g/mL	25
Water (H$_2$O)	0.99984 g/mL	0
	0.999973 g/mL	3.98
	0.99820 g/mL	20
	0.98804 g/mL	50
	0.95836 g/mL	100
Gases[b]		
Air (a mixture)	1.185 g/L	25
Argon (Ar)	1.7824 g/L	0
Carbon dioxide (CO$_2$)	1.976 g/L	0
Chlorine (Cl$_2$)	3.214 g/L	0
Helium (He)	0.194 g/L	25
Hydrogen (H$_2$)	0.08988 g/L	0
Oxygen (O$_2$)	1.429 g/L	0

[a] Formulas are provided for possible future reference.

[b] Densities at 1 atmosphere pressure.

For many purposes, we can assume the density of water to be 1.00 g/mL near room temperature. That assumption can sometimes be useful in solving problems.

If we combine two liquids that are *immiscible* (which means that they do not mix together to form a solution), the liquid with the lower density will float on top of the liquid of higher density. Density also determines how combinations of solids and liquids interact. A solid that does not dissolve in a liquid will float on the liquid if the solid is *less* dense than the liquid; and the floating solid displaces a *mass* of liquid equal to the mass of the solid. If the solid is *more* dense than the liquid, the solid *sinks* in the liquid and displaces a *volume* of liquid equal to the volume of the solid. We illustrate some of these ideas with Figure 1.11 and topics presented later in the chapter.

Example 1.10

A beaker has a mass of 85.2 g when empty and 342.4 g when it contains 325 mL of liquid methanol. What is the density of the methanol?

STRATEGY

We must obtain the mass of methanol as the difference in mass between the filled and empty beaker. Then we can divide the calculated mass of the methanol by its volume to obtain its density.

SOLUTION

The mass of the methanol is

$$? \text{ g methanol} = 342.4 \text{ g} - 85.2 \text{ g} = 257.2 \text{ g}$$

The number of milliliters of methanol is given. The density is the ratio of mass to volume:

$$d = \frac{m}{V} = \frac{257.2 \text{ g}}{325 \text{ mL}} = 0.791 \text{ g/mL}$$

ASSESSMENT

We have correctly accounted for the mass of methanol; that is, it is neither of the masses stated in the problem but rather the difference between the two. We have obtained the proper unit for density, g/mL, and the numerical value is comparable to those for other liquids in Table 1.7

EXERCISE 1.10A

What is the density of a 4.085-g steel ball having a 5.00-mm radius?

EXERCISE 1.10B

A 25.0-lb cast iron boat anchor displaces 1.62 qt of water when used. What is the density of the cast iron, in g/cm^3?

In Example 1.10, we found that the density of methanol is 0.791 g/mL. We can also use the following equivalence to describe this density:

$$1 \text{ mL methanol} = 0.791 \text{ g methanol}$$

This equivalence then allows us to formulate two conversion factors containing the mass and volume of methanol:

(a)	(b)
$\dfrac{0.791 \text{ g methanol}}{1 \text{ mL methanol}}$	$\dfrac{1 \text{ mL methanol}}{0.791 \text{ g methanol}}$

We can use density, which is conversion factor (a), to convert a volume of methanol to a mass of methanol. To convert from mass to volume, we can use the inverse of density, which is conversion factor (b).

Application Note

The volume of a floating aluminum canoe (mostly air) is much greater than the volume of the metal from which the canoe is made. The *effective* density of the canoe (determined by both $d_{air} = 0.0.0012 \text{ g/cm}^3$ and $d_{metal} = 2.7 \text{ g/cm}^3$) is *less than* that of water ($d_{water} = 1.0 \text{ g/cm}^3$). If the side of the canoe slips below the surface of the water, however, water displaces the air in the canoe and the canoe sinks.

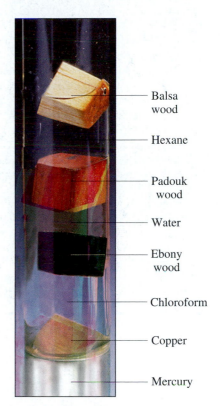

▲ **FIGURE 1.11 A comparison of some densities**

The liquids are as follows: Chloroform ($d = 1.48 \text{ g/cm}^3$) floats on the mercury ($d = 13.6 \text{ g/cm}^3$). Water ($d = 1.00 \text{ g/cm}^3$), which does not mix with chloroform, floats on the chloroform. Hexane ($d = 0.66 \text{ g/cm}^3$), which does not mix with water, floats on the water. Liquid mercury is so dense that copper ($d = 8.94 \text{ g/cm}^3$) floats on it. Wood, a material of variable composition, has a range of densities. The densities listed here are representative for three types of wood. Balsa wood ($d = 0.11 \text{ g/cm}^3$) has such a low density that it floats on hexane; padouk wood ($d = 0.86 \text{ g/cm}^3$) floats on water but not on hexane; and ebony wood ($d = 1.2 \text{ g/cm}^3$) sinks in water but floats on chloroform.

Fat Floats: Body Density and Fitness

Are you physically fit? If you are an athlete, or if you simply want to get in better physical condition through exercise, your first step might well be a fitness evaluation. One part of the evaluation is the determination of your percent body fat, a procedure that is best done by measuring your body density. The density of human fat is 0.903 g/mL, whereas that of water is 1.00 g/mL. Will fat float or sink if placed in water?

Because fat is less dense than water, it floats on water. The higher your proportion of body fat, the more buoyant you are in water. Body density is determined from body mass (weight) and volume. Body volume is determined by weighing a person when submerged in water (see Figure 1.12). Once body density is determined, the percent body fat is estimated either from a graph or from a table of compiled data. The average percent body fat is about 16% for an adult male and 25% for an adult female. Male athletes in superb condition have less than 7% body fat, and female athletes less than 12% body fat.

Project: You can measure your own body density. Fill a bathtub to a level high enough that you will be able to completely submerse yourself when you get in. Mark the level of the water on the tub. Then get in and submerge yourself. Mark the new water level. (A friend would help a lot here.) The difference in water level represents your volume. You can measure that by using a measuring cup to fill the tub with water between the low- and high-water marks. Keep track of the amount! Now weigh yourself out of water. Convert your weight in pounds to grams and then divide the weight in grams by the volume you found from the tub. That is your density! You should get a number that is close to 1 gram per cubic centimeter.

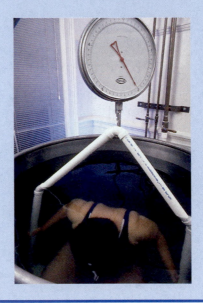

◀ **FIGURE 1.12**
Determining body volume

It is easy to measure a person's mass (weight), but what about a person's volume? When submerged in water, a body displaces its own volume of water. The difference in the person's weight in air and when submerged in water equals the mass of displaced water. This mass, divided by the density of water, yields the volume of displaced water and—with an appropriate correction for the volume of air in the lungs and gases in the intestine—the person's volume (see Problem 103).

Example 1.11

How many kilograms of methanol does it take to fill the 15.5-gal fuel tank of an automobile modified to run on methanol?

STRATEGY

Our goal is to determine the mass of a certain volume of methanol. For this, we need to use conversion factor (a) previously given. Before using the factor, however, we must convert a volume in gallons to one in milliliters. Finally, we need to convert from grams to kilograms of methanol.

Density can be thought of as simply another conversion factor, one that relates mass (grams) and volume (mL or cm³).

SOLUTION

We can write all the required factors in a single setup:

$$? \text{ kg} = 15.5 \text{ gal} \times \frac{3.785 \text{ L}}{1 \text{ gal}} \times \frac{1000 \text{ mL}}{1 \text{ L}} \times \frac{0.791 \text{ g}}{1 \text{ mL}} \times \frac{1 \text{ kg}}{1000 \text{ g}} = 46.4 \text{ kg}$$

ASSESSMENT

When using a single setup, ensure that the all unwanted units cancel.

EXERCISE 1.11A

What is the volume in gallons of 10.0 kg of methanol ($d = 0.791$ g/mL)?

EXERCISE 1.11B

What mass of ethyl alcohol occupies the same volume as 15.0 kg of gasoline ($d = 0.690$ g/mL)?

1.6   Further Remarks on Problem Solving

In Section 1.5, we considered a method of solving *quantitative* chemistry problems—those that can be answered with numbers. At times, though, we need only a *qualitative* answer: a statement, a picture, symbols, or a diagram to represent what is expected. Even when the answer is a simple "yes" or "no," substantial thought may be required to arrive at that answer.

Regardless of the type of problem, it often helps to visualize how to solve it, perhaps by drawing a diagram or sketch. The next action is vital: Outline *a* strategy, a step-by-step approach to solving the problem. We emphasize the "*a*" because often more than one strategy will work. In general, any logical scheme you choose will be valid as long as it leads to an appropriate conclusion. You demonstrate your mastery of the underlying principles by devising a problem-solving strategy that works. On occasion, you may devise a strategy that was not anticipated by the person who posed the problem. In doing so, you may demonstrate a special insight into the principles involved.

To assist you in developing a broad range of problem-solving skills, from time to time throughout the remaining chapters we will present three types of examples that differ somewhat from those introduced earlier—estimation examples, conceptual examples, and cumulative examples. We close this chapter by describing these types.

Estimation Examples

A calculator will always give you an answer, but the answer may not be correct. You may have entered a wrong number or executed the wrong function. If you learn to *estimate* answers, you at least will know whether your answer is a reasonable one. And at times, an approximate answer is quite acceptable because there are not enough data to make a precise calculation. Moreover, the ability to estimate answers is important in everyday life as well as in chemistry.

As you consider the following illustrations, keep in mind that the idea behind an estimation is to perform only a rough calculation, *not* to provide a complete worked solution. We will designate as "An Estimation Example" those in-chapter examples in which estimations form a major part of the solution. However, we will not use such notation for the end-of-chapter problems. When working those problems, try to identify from the context of the problem situations in which an estimated answer is appropriate.

Example 1.12—An Estimation Example

A small storage tank for liquefied petroleum gas (LPG) appears to be spherical and to have a diameter of about 1 ft. Suppose that some common volumes for LPG tanks are 1 gal, 2 gal, 5 gal, and 10 gal. Which is the most probable volume of this particular tank?

ANALYSIS AND CONCLUSIONS

Draw a sketch similar to Figure 1.13, which shows how the tank will just fit into a cubical box about 12 in. on each side. Each edge of the box is a little over 30 cm (12 in. × 2.54 cm/1 in.). Therefore, the volume of the box is slightly larger than $V = 30 \text{ cm} \times 30 \text{ cm} \times 30 \text{ cm} = 27 \times 10^3 \text{ cm}^3 = 27$ L. Because one gallon is slightly less than 4 L, the volume of the box is about 7 gal. The volume of the tank is clearly less than that of the box, perhaps about one-half the volume, say 3.5–4 gal. Of the volume choices given, 5 gal is the best answer.

EXERCISE 1.12A

Which of the following is a reasonable estimate of the mass of a 20-qt pail full of water?

5 kg   10 kg   15 kg   20 kg

EXERCISE 1.12B

With data from Example 1.12 and Figure 1.11 and a minimum of calculation, determine which of the following should have the greatest mass: (a) a 1-ft cube of balsa wood; (b) 2.00 L of water at 25 °C; (c) a 1850-g sample of liquid mercury; (d) a mixture of 1.50 L of hexane and 1.00 L of water at 25 °C.

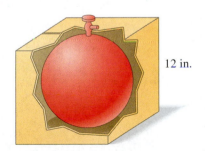
12 in.

▲ FIGURE 1.13   Example 1.12 visualized

Conceptual Examples

A common pattern in this book is to follow the introduction of an idea, concept, or technique with a basic in-text example, often presented in quantitative terms. From time to time, we also try to show, through conceptual examples, how fundamental knowledge of a concept may be applied to novel situations.

Conceptual examples may be either qualitative or quantitative; their solutions may require calculations, simple estimates, or verbal explanations. Sometimes a conceptual example will require you to assess what information is relevant to a problem and what information is immaterial, as illustrated in Example 1.13.

Example 1.13—A Conceptual Example

A sulfuric acid solution at 25 °C has a density of 1.27 g/mL. A 20.0-mL sample of this acid is measured out at 25 °C, introduced into a 50-mL flask, and allowed to cool to 21 °C. The mass of the flask plus solution is then measured at 21 °C. Use the stated data, as necessary, to calculate the mass of the acid sample.

ANALYSIS AND CONCLUSIONS

The density concept tells us how to relate the mass and volume of a sample of matter. Thus, the mass of a 20.0-mL sample of the solution that is measured out at 25 °C is

$$m = d \times V = \frac{1.27 \text{ g}}{\text{mL}} \times 20.0 \text{ mL} = 25.4 \text{ g}$$

The fact that the sample is placed in a 50-mL flask is immaterial (the flask could have any volume as long as it is greater than 20 mL). Likewise, it makes no difference that the mass was measured at 21 °C because mass is not a function of temperature. That is, the sample will have a mass of 25.4 g regardless of the temperature. Finally, the fact that the solution is one of sulfuric acid is also immaterial. A 20.0-mL sample of any liquid with a density of 1.27 g/mL would also have a mass of 25.4 g under the conditions stated here.

EXERCISE 1.13A

In Example 1.13, why wasn't the 20.00-mL volume of solution measured at 21 °C instead of 25 °C?

EXERCISE 1.13B

Saturn orbits the Sun in 29.46 Earth years and rotates on its axis in 0.436 Earth days. Saturn has an average surface temperature of −139.2 °C, an average radius of 58232 km, and a mass 95.2 times that of Earth, which is 6.0×10^{24} kg. Use these data, as necessary, to calculate the average density of Saturn.

Example 1.14—A Conceptual Example

The sketches in Figure 1.14 show observations made on a small block of plastic material in four situations. What does each observation tell you about the density of the plastic?

ANALYSIS AND CONCLUSIONS

Here is what each sketch tells us:

(a) The block has a mass of 50.0 g, but this does not tell us anything about the density of the plastic material of which the block is made. For that, we also have to know the block's volume.

(b) Because the block sinks to the bottom of the ethyl alcohol, the plastic must be denser than the ethyl alcohol. That is, $d_{\text{plastic}} > 0.789$ g/mL.

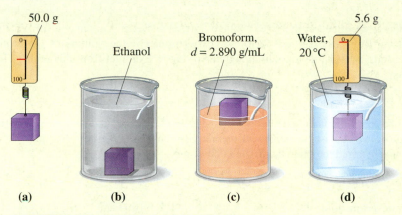

▲ FIGURE 1.14   Example 1.14 visualized

(c) Because the block floats on bromoform, the density of the plastic must be less than that of bromoform. That is, $d_{plastic} < 2.890$ g/mL. Moreover, because the block is about 40% submerged, the volume of bromoform having the same 50.0-g mass as the block is only about 40% of the volume of the block. Thus, using the expression $V = m/d$, we can write

$$\text{Volume of displaced bromoform} \approx 0.40 \times \text{volume of block}$$

$$\frac{\text{Mass of bromoform}}{\text{Density of bromoform}} = \frac{0.40 \times \text{mass of block}}{d_{plastic}}$$

$$\frac{50.0 \text{ g}}{2.890 \text{ g/mL}} = 0.40 \times \frac{50.0 \text{ g plastic}}{d_{plastic}}$$

$$d_{plastic} \approx 0.40 \times 2.890 \text{ g/mL} \approx 1.2 \text{ g/mL}$$

(d) When submerged in water, the block displaces its volume of water. The mass of water displaced is the difference between the block's mass measured in air (from step a) and the block's mass measured in water: $(50.0 - 5.6)$ g = 44.4 g. The volume of this mass of water = mass/density = 44.4 g/1.00 g/mL = 44.4 mL. Thus the density of the plastic is 50.0 g/44.4 mL = 1.13 g/mL.

EXERCISE 1.14A

Modify Figure 1.11 by removing the appropriate object and replacing it with the block of plastic described in Example 1.14, floating at the appropriate depth.

EXERCISE 1.14B

Suppose the experiment in Example 1.14 were repeated with a 75.0-g block of plastic that has a density of 1.22 g/mL and with the bromoform replaced by chloroform $(d = 1.483$ g/mL). Describe the observations you would make corresponding to (a), (b), (c), and (d) in Example 1.14.

Cumulative Examples

The typical in-text example illustrates a specific technique or a limited aspect of a concept. By contrast, many problems of special significance require the simultaneous application of several basic concepts; that is, the problems are based on cumulative knowledge. The concluding example in each chapter of the text—a Cumulative Example—is of this type. One of the challenges in solving a cumulative problem is to identify the several concepts bearing on the problem and devise a strategy that brings them together in an appropriate way to obtain an acceptable answer.

Cumulative Example

The volume of droplets generated by ink-jet printers is described in the essay "Where Smaller Is Better" on page 9. **(a)** What is the diameter, in micrometers, of a spherical ink droplet from an early version of an ink-jet printer if the volume of the droplet is 200 ± 10 pL? **(b)** If the ink has a density of 1.1 g/mL, what is the mass in milligrams of ink in a droplet from this printer? Is milligrams an appropriate unit for describing this mass?

STRATEGY

In part (a), we can use the formula for the volume of a sphere, along with appropriate conversion factors, to find the dimensions (radius and then diameter) of the spherical droplet. In part (b), we can use the density of the ink as a conversion factor to find mass from the volume of the ink droplet. An additional conversion is required to obtain the desired unit of mg.

SOLUTION

(a) To obtain the volume in units of (length)^3, convert picoliters to cubic centimeters, using the equivalence of one liter to 1000 cm^3.

$$200 \text{ pL} \times \frac{10^{-12} \text{ L}}{1 \text{ pL}} \times \frac{1000 \text{ cm}^3}{1 \text{ L}} = 2.0 \times 10^{-7} \text{ cm}^3$$

To calculate the radius of the spherical drop in centimeters, use the formula for the volume of a sphere.

$$V = \frac{4\pi r^3}{3}$$

$$2.0 \times 10^{-7} \text{ cm}^3 = \frac{4(3.14)r^3}{3}$$

$$r^3 = 3(2.0 \times 10^{-7} \text{ cm}^3)/12.56 = 4.8 \times 10^{-8} \text{ cm}^3$$

$$r = 3.6 \times 10^{-3} \text{ cm}$$

Next convert the radius from centimeters to micrometers.

$$r = 3.6 \times 10^{-3} \text{ cm} \times \frac{10^{-2} \text{ m}}{1 \text{ cm}} \times \frac{1 \text{ μm}}{10^{-6} \text{ m}} = 36 \text{ μm}$$

Finally, find the diameter by multiplying the radius by 2.

$$d = 2r = 72 \text{ μm}$$

(b) To calculate the mass of a droplet, use the volume obtained in part (a) and the density of the ink (1.1 g/mL), expressed as 1.1 g/cm^3.

$$2.0 \times 10^{-7} \text{ cm}^3 \times \frac{1.1 \text{ g}}{1 \text{ cm}^3} = 2.2 \times 10^{-7} \text{ g}$$

Finally, a simple conversion gives mass in mg.

$$2.2 \times 10^{-7} \text{ g} \times \frac{1 \text{ mg}}{10^{-3} \text{ g}} = 2.2 \times 10^{-4} \text{ mg}$$

Although milligrams is an acceptable unit for mass, a smaller unit might be more appropriate for this tiny mass—for example, either 0.22 μg or 220 ng rather than 2.2×10^{-4} mg. Note that whichever unit we choose, the mass is known only to two significant figures.

ASSESSMENT

Because the ink has a density of about 1 g/cm^3 and because 1 cm^3 = 1000 mm^3, the mass of 1 cm^3 of ink is about 1 g, and the mass of 1 mm^3 of ink is about 1/1000 g = 1 mg. A droplet of ink is much smaller than 1 mm^3 and thus should have a mass considerably smaller than 1 mg—which it does in our answer.

Concept Review with Key Terms

1.1 Chemistry: Principles and Applications—The fundamental principles of chemistry described in this book have widespread applications in many of the products, practices, and issues already familiar to you. Look for linkages between principles and applications throughout the text.
1.2 Getting Started: Some Key Terms—**Matter** is made up of **atoms** or **molecules** and can be subdivided into two broad categories: substances and mixtures. **Substances** have fixed **compositions**; they are either **elements** or compounds. The compositions of elements are commonly designated through one- or two-letter **chemical symbols**. **Compounds** can be broken down into their constituent elements through chemical reactions, but elements cannot be subdivided into simpler substances. **Mixtures** of substances are either homogeneous or heterogeneous. Some substances can be mixed in varying proportions to produce **homogeneous mixtures** (also called **solutions**). The composition and properties are uniform throughout a solution. The composition and/or properties of a **heterogeneous mixture** vary from one part of the mixture to another.

Substances exhibit characteristics called **physical properties** without undergoing a change in composition. In displaying a **chemical property**, a substance undergoes a change in composition—new substances are formed during a **chemical reaction**. A **physical change** produces a change in the appearance of a sample of matter but no change in its microscopic structure and composition. In a **chemical change**, the composition and/or microscopic structure of matter changes.

Scientific knowledge is testable, reproducible, explanatory, predictive, and tentative. The scientific method, involves making observations, forming **hypotheses**, gathering **data** through **experiments**, and formulating **scientific laws** and **theories**.

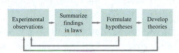

1.3 Scientific Measurements—Four basic physical quantities of measurement are introduced in this section: **mass**, length, time, and temperature. Their respective SI base units are the **kilogram (kg)**, the **meter (m)**, the **second (s)**, and the **kelvin (K)**. Volume measurements are derived from length and have the SI base unit of cubic meter (m^3); however the **liter (L)**, equal to 1 cubic decimeter (dm^3), is more commonly used. Two temperature scales other than the Kelvin scale—the Celsius and Fahrenheit scales—are considered and compared in this section.

In the SI system, measured quantities may be reported in the base unit or as multiples or submultiples of the base unit. Multiples and submultiples are based on powers of ten and reflected through prefixes in names and abbreviations.

*nano*meter (nm)	10^{-9} m
*micro*meter (μm)	10^{-6} m
*milli*meter (mm)	10^{-3} m
meter (m)	m
*kilo*meter (km)	10^3 m

High accuracy
Low precision

1.4 Precision and Accuracy in Measurements—**Precision** describes the agreement among a set of individual measurements of a quantity. **Accuracy** describes the agreement between the average of a set of measurements and the "correct," or "true," value of the measured quantity. To indicate its precision, a measured quantity must be expressed with the proper number of **significant figures**. Furthermore, special attention must be paid to the concept of significant figures in reporting calculated quantities. Several rules for dealing with significant figures are described in this section.

1.5 A Problem-Solving Method—Calculations frequently can be done by the **unit-conversion** method. In this method, a given quantity is multiplied by a **conversion factor**, which changes the given unit of the quantity to a different, desired unit. The conversion factor is a ratio having its numerator and denominator equal to one another but expressed in different units.

The **density** (*d*) of a material is its mass per unit volume:

$$d = m/V$$

Balsa wood

Hexane

Padouk wood

Water

Density is an important conversion factor. When the volume of a substance or homogeneous mixture (cm^3) is multiplied by its density (g/cm^3), volume is converted to mass. When the mass of a substance or homogeneous mixture (g) is multiplied by the *inverse* of density (cm^3/g), mass is converted to volume.

1.6 Further Remarks on Problem Solving—Three special types of problems are introduced in this section and used throughout the text. Estimation examples provide methods for obtaining approximate answers without detailed calculations or for checking the reasonableness of calculated answers. Conceptual examples are used in problem-solving situations of a nonroutine type, that is, in situations where a clear understanding of underlying concepts is generally more important than computational skills. Cumulative examples summarize and apply a number of chemical principles, generally from two or more chapters, within a given problem.

Assessment Goals

When you have mastered the material in this chapter, you will be able to:

- Describe the usual method used to study science.

- Classify a sample of matter as being substance or mixture, element or compound, homogeneous or heterogeneous.

- Classify properties and changes as chemical or physical, and understand the differences between them.

- Recognize and use the common SI units and prefixes.

- Express and convert measurements using the common SI prefixes and exponential notation.

- Convert, when relevant data are given, measurements between the SI and other systems of measurement.

- Use the appropriate number of significant figures in writing numbers and expressing calculation results.

- Explain the difference between accuracy and precision.

- Solve problems, including those involving density, using the unit-conversion method.

Self-Assessment Questions

1. Which of the following is/are *not* an example of matter?
   (a) steel   (c) clouds   (e) water
   (b) nitrogen   (d) heat   (f) music

2. What are some of the requirements of scientific activity, as compared, for example, with artistic activity?

3. What is a hypothesis? How are hypotheses tested?

4. What are scientific data, and what characterizes good data?

5. What is a theory? What must one be able to do with a theory? Can a theory be proved?

6. Frequently, people use the phrase "I have a theory about that" when describing an observation they have made. Comment on the appropriateness of the word "theory" in this context.

7. How do physical and chemical properties differ?

8. Which of the following is *not* a physical property?
   (a) Metals reflect light.
   (b) Water has a density of 1.0 g/mL.
   (c) Gasoline is burned in a combustion engine.
   (d) Chlorine is a gas at room temperature.

9. An example of a chemical change is
   (a) the melting of an ice cube
   (b) the boiling of gasoline
   (c) the frying of an egg
   (d) all of these

10. Refer to the listing of the elements inside the front cover, and determine which of the following represent elements and which do not. Explain.
    (a) C   (c) Cl   (e) Na
    (b) CO   (d) $CaCl_2$ (f) KI

11. An example of a homogeneous mixture or solution is
    (a) maple syrup   (c) distilled water
    (b) liquid oxygen   (d) chicken soup

12. Describe briefly the distinction between the following pairs of terms.
    (a) element and compound
    (b) homogeneous and heterogeneous mixture
    (c) mass and density

13. What are the names and symbols of the SI base units for length, mass, and time?

14. Express the SI units for area, volume, and density in terms of SI *base* units.

15. What is the difference between the precision and the accuracy of a set of measurements?

16. Of the following numbers, the one with three significant figures is
    (a) 22.03   (c) 1.070
    (b) 0.0260   (d) 2000

17. How does the number of significant figures in a measured quantity relate to the precision of the measurement? to the accuracy?

18. If the numerical value of a measured quantity includes zeros, are these zeros significant? Explain.

19. What is a conversion factor? How must the numerator and denominator of a conversion factor be related?

20. A fertilizer contains 20.0% nitrogen, by mass. To provide a fruit tree with an equivalent of 1.0 lb of nitrogen, the quantity of *fertilizer* required is
    (a) 20.0 lb   (c) 0.050 lb
    (b) 0.20 lb   (d) 5.0 lb

21. Why are the terms "light" and "heavy" imprecise when used to refer to low-density and high-density materials?

22. Of the following substances, the greatest density is that of
    (a) 1000 g of water at 4 °C
    (b) 100.0 cm^3 of chloroform weighing 148.9 g
    (c) a 10.0-cm^3 piece of wood that weighs 7.72 g
    (d) an ethyl alcohol–water mixture with a density of 0.83 cm^3/g

Problems

A word of advice: You cannot learn to work problems by reading them or watching your instructor work them, any more than you can become a piano player solely by reading about piano-playing skills or attending a performance. As you work through problems, you will find many opportunities to improve your understanding of the ideas presented in the chapter and also to practice your estimation skills and your ability to synthesize concepts. Plan to work through the great majority of these problems.

Some Basic Ideas

23. Which of the following are examples of matter?
    (a) iron   (d) automobile exhaust
    (b) the human body   (e) philosophy
    (c) blue light

24. Which of the following are not examples of matter?
    (a) air   (d) a meteor
    (b) an infrared lamp   (e) the balance in a checking account
    (c) an idea

25. Identify the following as physical or chemical properties.
    (a) Solid iron melts at a temperature of 1535 °C.
    (b) Solid sulfur is yellow.
    (c) Natural gas burns.
    (d) Diamond is extremely hard.

26. Which of the following describe a chemical change, and which a physical change?
    (a) Sheep are sheared, and the wool is spun into yarn.
    (b) A cake is baked from a mixture of flour, baking powder, sugar, eggs, shortening, and milk.
    (c) Milk turns sour when left out of the refrigerator for many hours.

    (d) Silkworms feed on mulberry leaves and produce silk.
    (e) An overgrown lawn is manicured by mowing it with a lawn mower.

27. Which of the following are substances and which are mixtures? Explain.
    (a) helium gas used to fill a balloon
    (b) the juice squeezed from a lemon
    (c) a premium red wine
    (d) salt used to deice roads

28. Which of the following mixtures are homogeneous, and which are heterogeneous?
    (a) gasoline   (c) Italian salad dressing
    (b) raisin pudding   (d) an intravenous glucose solution

29. A formerly chubby but now trim person completes a successful diet. Has the person's weight changed? Has the person's mass changed? Explain.

30. Sample A, which is on the Moon, has exactly the same mass as sample B on Earth. Do the two samples weigh the same? Explain your answer.

31. Can a set of measurements be precise without being accurate? Can the average of a set of measurements be accurate even if the individual measurements are imprecise? Explain.

32. Describe several ways in which a scientific law differs from a legislative law.

33. In Europe, most recipes specify the mass of flour used to make baked goods. In the United States, flour is usually specified by volume. Comment on whether one method may yield more consistent results than the other.

34. Explain why the common saying, "The exception proves the rule," is incompatible with the scientific method.

Measurement and Unit Conversion

35. Change the unit used to report each of the following measurements by replacing the power of ten by an appropriate SI prefix.
  (a) 8.01×10^{-6} g   (c) 1.05×10^3 m
  (b) 7.9×10^{-3} L

36. Use exponential notation to express each of the following measurements in terms of an SI base unit.
  (a) 45 mg   (b) 125 ns   (c) 10.7 μL

37. Carry out the following temperature conversions.
  (a) 98.6 °C to °F
  (b) 1069 °F to °C   (c) −52 °C to °F

38. The melting point of a solid is the temperature at which the solid undergoes a physical change to become a liquid. The following melting points are expressed on one temperature scale. Convert them to temperatures on the scale indicated.
  (a) Convert to °F: cesium, 28.5 °C; germanium, 937 °C; silver, 961.93 °C; mercury, −38.36 °C
  (b) Convert to °C: gold, 1948.57 °F; lead, 621.37 °F; ethyl alcohol, −174.1°F

39. The metal indium melts at 156 °C. The boiling point of the organic liquid cyclohexanol is 321 °F. Will indium melt in boiling cyclohexanol?

40. Ray Bradbury chose *Fahrenheit 451* as the title of his novel about book burning in a future society. A firefighter told him that is the temperature at which paper ignites. What would his title have been had he been given the temperature in degrees Celsius?

41. Carry out the following conversions.
  (a) 37.4 mL to liters   (d) 1.19 m^2 to cm^2
  (b) 1.55×10^2 km to m   (e) 78 μs to ms
  (c) 198 mg to grams   (f) 88 km/h to m/s

42. Carry out the following conversions.
  (a) 546 mm to meters   (d) 46.3 m^3 to liters
  (b) 1.00 h to microseconds   (e) 181 pm to micrometers
  (c) 87.6 mg to kilograms   (f) 55 mi/h to km/min

43. (a) A block of wood weighs 15 lb, 2 oz. Determine its mass in kilograms.
  (b) A standard size for lumber used in construction is 1.5 in. × 3.5 in. × 92 5/8 in. Convert each dimension to cm.
  (c) A diesel truck has a rectangular tank 16 in. × 24 in. × 42 in. Determine the volume of the tank in cubic meters.

44. (a) A book weighs 3.7 lb. Find its mass in grams.
  (b) A CD-ROM disk is 0.120 m in diameter. Find its diameter in millimeters and in inches.
  (c) What is the volume, in milliliters and cubic meters, of a 12-fl oz soft drink?

45. *Without doing detailed calculations*, arrange the following in order of increasing length (shortest first): (1) a 1.21-m chain, (2) a 75-in. rope, (3) a 3-ft-5-in. rattlesnake, (4) a yardstick.

46. *Without doing detailed calculations*, arrange the following in order of increasing mass (lightest first): (1) a 5-lb bag of potatoes, (2) a 1.65-kg cabbage, (3) 2500 g of sugar.

47. The *furlong* is a unit used in horse racing, and the units *chain* and *link* are used in surveying land. There are 8 furlongs in 1 mi, 10 chains in 1 furlong, and 100 links in 1 chain. Calculate the length of 1 link, in inches (1 mi = 5280 ft).

48. A horse is found to run 2.00 furlongs in 39.4 seconds. Use the information given in Problem 47 to find the horse's speed in miles per hour and in km/min.

Significant Figures

49. How many significant figures are there in each of the following measured quantities?
  (a) 104 m   (d) 4.200×10^5 s
  (b) 2.10×10^8 mi^3 (e) 0.0890 g
  (c) 0.00003975 ms   (f) 1.050 °C

50. How many significant figures are there in each of the following measured quantities?
  (a) 4051 m   (d) 5.00×10^9 m
  (b) 0.0169 s   (e) 1.60×10^{-9} s
  (c) 0.0430 g   (f) 0.0150 °C

51. Express each of the following measured quantities in exponential notation.
  (a) 2804 m   (c) 0.00090 cm
  (b) 901 s   (d) 221.0 s

52. Express each of the following measured quantities in exponential notation.
  (a) 8352 m   (c) 0.0885 cm
  (b) 300.0 s   (d) 122.2 s

53. Express each of the following measured quantities in common decimal form (for example, $3.11 \times 10^2 = 311$). Comment on the significance of any terminal zeros that you write.
  (a) 5.055×10^2 m   (c) 6.10×10^{-3} g
  (b) 2.12×10^3 s   (d) 4.00×10^4 mL

54. Express each of the following measured quantities in common decimal form (for example, $6.375 \times 10^3 = 6375$). Comment on the significance of any terminal zeros that you write.
  (a) 3.18×10^5 m   (c) 4.1×10^{-4} s
  (b) 7.50×10^2 mL   (d) 9.200×10^4 m

55. Perform the indicated operations, and give answers with the proper number of significant figures.
  (a) 36.5 m − 0.132 m + 9.44 m
  (b) 154.44 cm − 93 cm + 105.1 cm
  (c) 4.61 g + 39.9 g − 0.0220 g
  (d) 12.52 L + 1.5 L − 3.18 L − 0.72 L

56. Perform the indicated operations, and give answers in the indicated unit and with the proper number of significant figures.
  (a) 13.25 cm + 26 mm − 7.8 cm − 0.186 m (in centimeters)
  (b) 48.834 g + 717 mg − 0.166 g + 1.0251 kg (in kilograms)

57. Perform the indicated operations, and give answers with the proper number of significant figures.

    (a) 73.0 mm × 1.340 mm × (25.31 mm − 1.6 mm)

    (b) $\dfrac{33.58 \text{ cm} \times 1.007 \text{ cm}}{0.00705 \text{ g}}$

    (c) $\dfrac{418.7 \text{ mm} \times 31.8 \text{ mm}}{(19.27 \text{ mg} - 18.98 \text{ mg})}$

    (d) $\dfrac{2.023 \text{ g} - (1.8 \times 10^{-3} \text{ g})}{1.05 \times 10^{4} \text{ mL}}$

58. Perform the indicated operations, and give answers with the proper number of significant figures. Assume whole numbers are exact counts.

    (a) 265.02 mm × 0.000581 mm × 12.18 mm

    (b) $\dfrac{22.61 \text{ cm} \times 0.0587 \text{ cm}}{135 \times 28}$

    (c) $\dfrac{(33.62°C + 12.2°C - 48.36°C)}{26 \times 12.13 \text{ s}}$

    (d) $\dfrac{(4.6 \times 10^{3} \text{ mg} + 2.2 \times 10^{2} \text{ mg})}{3.11 \times 10^{4} \text{ cm} \times 7.12 \times 10^{2} \text{ cm}}$

Density

For some of the following problems, you may need to refer to Table 1.7 for data. Express all your answers with the proper number of significant figures.

59. What is the density of a salt solution if 50.0 mL has a mass of 57.0 g?

60. Blood plasma can be donated by a typical adult human about every 48 hours. What is the mass in grams of 477 mL of blood plasma?

61. Some metal chips having a volume of 3.29 cm³ are placed on a piece of paper and weighed. The combined mass is found to be 18.432 g. The paper itself weighs 1.214 g. Calculate the density of the metal.

62. A glass container weighs 48.462 g. A sample of 4.00 mL of antifreeze solution is added, and the container plus the antifreeze weigh 54.513 g. Calculate the density of the antifreeze solution.

63. A rectangular block of balsa wood is 7.6 cm × 7.6 cm × 94 cm. Use data from Figure 1.11 to find the mass of the block in grams.

64. A rectangular block of gold-colored material measures 3.00 cm × 1.25 cm × 1.50 cm and has a mass of 28.12 g. Can the material be gold? Explain.

65. A 5.79-mg piece of gold is hammered into gold leaf of uniform thickness with an area of 44.6 cm². What is the thickness of the gold leaf?

66. A roll of steel "shim stock" used in machine shop work is 6.0 in. wide and 103 in. long and has a mass of 14.2 oz. If the density of the steel is 7.97 g/cm³, how thick is the shim stock?

67. A box with a square base measuring 0.80 m on each side and having a height of 1.20 m is filled with 3.2 kg of expanded polystyrene packing material. What is the bulk density, in grams per cubic centimeter, of the packing material? (The bulk density includes the air between the pieces of polystyrene foam.)

68. Hylon VII, a starch-based substitute for polystyrene packing material, has a bulk density of 12.8 kg/m³. How many grams of the material are needed to fill a volume of 2.00 ft³?

69. Which of the following items would be most difficult to lift onto the back of a pickup truck: (1) a 100-lb bag of potatoes, (2) a 15-gal plastic bottle filled with water, (3) a 3.0-L flask filled with mercury?

70. Liquid mercury is commonly shipped in iron flasks that contain 76 lb of mercury. Will one of these flasks fit inside a wooden box that has inside dimensions of 4.0 in. × 4.0 in. × 8.0 in.?

71. Several irregularly shaped pieces of zinc weighing 30.0 g are dropped into the graduated cylinder on the left. The water level rises as shown in the right cylinder. Calculate the density of zinc, and express your result with the maximum number of significant figures permitted in this experiment.

72. The container pictured below on the left is filled with water at 20 °C, just to the overflow spout. A cube of wood with edges of 1.0 in. is floated on the water, and 10.8 mL of water is collected, as shown on the right. Calculate the density of the wood, and express your result with the maximum number of significant figures permitted in this experiment.

Additional Problems

Problems marked with an * may be more challenging than others.

73. Sample B, which is on the Moon, has the same mass as does sample A on Earth. Do samples A and B weigh the same? If not, which weighs more? Explain. Sample C, also on the Moon, weighs the same as does sample A on Earth. Do samples A and C have the same mass? If not, which has the greater mass? Explain.

74. Can there be a temperature(s) at which the Fahrenheit and Celsius scales have the same numerical value(s)? If so, what is (are) the temperature value(s)? If this is not possible, why not?

75. Must it always be true that the result of a calculation can carry no more significant figures than the term in the calculation having the smallest number of significant figures? Explain.

76. If the meter stick used in the measurements in Table 1.5 on page 12 were actually 1005 mm long, how much error would this produce in calculating the area of the poster board in Example 1.4?

77. Which device do you think would be more precise for measuring the length of the poster board in Table 1.5 on page 12: a meter stick or a 2-m steel tape measure, each marked in millimeter increments? Which do you think would be more accurate? Explain your choices. Would the situation be the same for measuring the width? Explain.

78. The speed of light is 3.00×10^8 m s^{-1}. Calculate the speed of light in furlongs per fortnight (a fortnight is 14 days; see Problem 47 for the furlong).

79. In its nonstop, round-the-world trip in 1986, the aircraft *Voyager* traveled 25,102 mi in 9 days, 3 min, and 44 s. Using the maximum number of significant figures permitted, calculate the average speed of *Voyager* in mi/h.

80. In the United States, land area is commonly measured in acres: 640 acre = 1 mi^2. In most of the rest of the world, land area is measured in hectares: 1 hectare = 1 km^2. Which is the larger area, the acre or the hectare? Write a conversion factor that relates the acre and the hectare. (1 mi = 5280 ft; 1 m = 39.37 in.)

81. In scientific work, densities usually are expressed in grams per cubic centimeter. Engineers often use the unit pounds per cubic foot (lb/ft^3) What is the density of water at 20 °C in pounds per cubic foot?

82. An empty 3.00-L bottle weighs 1.70 kg. Filled with a homemade wine, the bottle weighs 4.72 kg. The wine contains 11.0% ethyl alcohol by mass. How many ounces of ethyl alcohol are present in 275 mL of this wine?

*83. A roll of aluminum wire is used to make finishing nails (nails that do not have heads). If the nails are 5.00 cm long and 0.300 cm in diameter and the roll of wire has a mass of 225 kg, how many nails can be made from one roll?

84. Aerogels consist of a solid framework with most of their volume occupied by air. Silica (silicon dioxide) powder has a density of 2.2 g/cm^3. Silica can be expanded into an aerogel with a bulk density of 0.015 g/cm^3. Calculate the volume of silica aerogel that can be made from 125 cm^3 of silica powder.

*85. Calculate the total surface area of the aerogel described in Problem 84, when there is 470 m^2 of surface per gram of silica in the aerogel.

86. Dust from air that is not significantly polluted is deposited ("dustfall") at a typical rate of 10 tons/mi^2 per month. The density of dust is about 1.7 g/cm^3. Use the preceding data, as needed, to express this dustfall in milligrams per square meter per hour.

87. Refer to Figure 1.11. When a block of padouk wood floats on water, what fraction of the block is above the waterline and what fraction is below the waterline? Explain your reasoning.

*88. A 500-mL beaker and 1.0-L beaker are both completely filled with water. A brass cube, 2.0 cm on edge, is gently placed on the water in the 1.0-L beaker and a rectangular block of cork, 5.0 cm × 4.0 cm × 2.0 cm, on the water in the 500-mL beaker. The density of brass is 8.40 g/cm^3 and that of cork is 0.22 g/cm^3. From which vessel will the greater volume of water overflow? Explain.

*89. The square metal nut pictured is 14.0 mm on edge, 6.0 mm thick, and has a 7.0-mm diameter hole. The density of the metal is 7.87 g/cm^3. Approximately how many of these nuts are present in a 1.00-lb package?

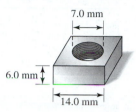

*90. The significant figure rules outlined in the chapter give a first approximation of the precision of a calculation based on measurements. A more refined approach looks at the absolute and percentage errors in measured quantities. The data in Table 1.5 on page 12, for example, include a length of 1.827 ± 0.003 m and a width of 0.762 ± 0.001 m. The absolute errors are ±0.003 m and ±0.001 m, respectively, and the percent errors are

$$\frac{0.003 \text{ m}}{1.827 \text{ m}} \times 100\% = 0.2\% \quad \text{and} \quad \frac{0.001 \text{ m}}{0.762 \text{ m}} \times 100\% = 0.1\%$$

In addition and subtraction, the absolute error in the result is the sum of the absolute errors in the measurements. In multiplication and division, the accumulated percent error is the sum of the percent errors.

(a) Use data in Table 1.5 to determine the perimeter of the poster board and the absolute error in the perimeter.

(b) Use data in Table 1.5 to determine the area of the poster board, the percent error, and the absolute error in the area.

(c) Show that by the method outlined here, you are permitted to carry one more significant figure in the area of the poster board than you are by the significant figure rules introduced in the chapter. Do you think that this would be the case in all calculations? Explain.

91. The density of a liquid is to be found by weighing a dry 100-mL volumetric flask, then filling it with the liquid and weighing it again. The volumetric flask has a mass of about 70 g and an uncertainty of ±0.1 mL. The liquid has a density that is very nearly 1 g/mL. Four balances are available, reproducible to 0.1 g, 0.01 g, 0.001 g, and 0.0001 g, respectively. Which balance(s) will give the maximum precision possible for the density measurement? Is there an advantage to using the 0.0001-g balance for this work? Explain.

*92. Glass microballoons are hollow spheres, averaging about 0.070 mm in diameter, used as filler for certain types of lightweight composite material. A 1.00-L package of microballoons has a mass of only about 62 g, even though the microballoons occupy 68% of the total volume available when packed tightly.

(a) If the glass itself has a density of 2.2 g/cm^3 about how thick is the glass in the average microballoon?

(b) A composite material is made by filling the space between microballoons with epoxy glue (density = 1.67 g/mL). Calculate the density of this composite mixture.

*93. We want to determine whether the object pictured, which has a mass of 5.15 kg, can be used as a float on pure water.

(a) Without doing detailed calculations, (that is, by estimation), show that the object will not float.

(b) Suppose that we completely fill the cylindrical hole with a plug of balsa wood ($d = 0.11$ g/cm^3). Show that the object will not float.

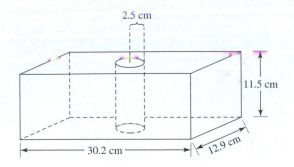

(c) What minimum number of cylindrical holes like that shown would have to be drilled and filled with balsa plugs to make the object float?

94. Above each square inch of Earth's surface is 14.7 pounds of air. The density of air at sea level is about 1.27 g/L. The Earth is 7930 miles in diameter and has a mass of 5.98×10^{24} kg.

(a) Use the above data, as needed, to determine the mass in kilograms of Earth's atmosphere.

(b) Determine the average density of Earth in grams per milliliter.

95. On July 23, 1983, Air Canada Flight 143 required 22,300 kg of jet fuel to fly from Montreal to Edmonton. The density of jet fuel is 0.803 g/mL, or 1.77 lb/L. The plane had 7682 L of fuel on board in Montreal. The ground crew there multiplied the 7682 L by the factor 1.77 (without units) and concluded that they needed an additional 8703 kg or 4916 L of fuel for the trip to Edmonton. They added 5000 L. On its flight, the plane ran out of fuel and safely crash-landed near Winnipeg, hundreds of kilometers short of its destination. What mistake did the ground crew make? How much fuel should they have added before takeoff?

* 96. Refer to Example 1.14. You have available quantities of chloroform ($d = 1.483$ g/mL) and bromoform ($d = 2.890$ g/mL) and a piece of plastic ($d = 1.950$ g/mL). Chloroform and bromoform form a homogeneous mixture (solution) when they are mixed. Assume that the volume of the mixture is the sum of the volumes of the two pure liquids. How many milliliters of each liquid are required to produce 100.0 mL of a solution on which the piece of plastic will just barely float?

* 97. Refer to the essay on ink-jet printers on page 9. **(a)** The photograph shows that the ink is actually ejected in a roughly cylindrical shape. If the length of a cylinder of ink is about ten times its diameter, what is the diameter of the cylinder, in micrometers, from a modern printer? **(b)** Resolution of a printer is often reported as "dots per inch." Assuming a resolution of 600 circular dots per inch, what would be the approximate thickness in micrometers of ink deposited on paper from a modern printer? **(c)** If a droplet 2.0 μm in diameter could be produced, approximately what resolution in dots per inch could be obtained, assuming the same thickness of ink as found in (b)?

Apply Your Knowledge

98. **[Historical]** "Galileo's thermometer" is shown below. It is filled with water. The temperature is that engraved on the metal disc of the lowest floating glass bulb. Describe the operation of this thermometer. Assuming that all of the bulbs have the same volume, which one is heaviest? Is the bulb at the very top the "highest temperature" or the "lowest temperature" bulb? The thermometer works properly only near room temperature. What problems might be encountered in constructing such a thermometer to work over the range from 0–100 °C? Explain, and verify your explanation using appropriate data from a handbook.

99. **[Biochemical]** The food section of a newspaper presented a hamburger recipe requiring the following ingredients for six burgers: $1\frac{1}{2}$ pounds ground beef, 2 to 3 ounces of bleu cheese, vegetable oil, six slices of French-style bread, 4 tablespoons (or so) of butter, salt, black pepper, 1/4 cup minced shallots, 1 cup red wine. The following nutritional information was given for each burger: 1765 calories, 58 g total fat (23 g saturated fat), 62 g protein, 238 g carbohydrates, 128 mg cholesterol, 3117 mg sodium. Criticize this reporting of nutritional information in relation to the information given about the ingredients. Rewrite the nutritional information in a way that better conforms to the reporting of measured and calculated quantities, as described in Section 1.4.

100. **[Environmental]** Satellite pictures show that coverage of Arctic sea ice in winter has decreased by 6% since 1978, equivalent to an area the size of Texas, and submarine sonar measurements have shown a decline in average ice thickness in late summer of 42% since the 1950s. The Antarctic ice cap also is melting, with large icebergs forming by splitting off (calving) from glaciers where the glaciers meet the ocean. Which, if either, of these events can raise mean sea level? Explain.

101. **[Environmental]** In an article on contamination near an abandoned lead smelter, a major metropolitan newspaper stated that "the federal government has reduced the amount of lead that is considered safe by a factor of four. Now, a blood level of as little as 10 milligrams per 10 liters is enough to call for prompt intervention." The actual maximum safe level is 10 micrograms per deciliter (10 μg/dL). Use unit conversions to show whether or not the two quantities are the same. If not, how might the reporter have come up with the "10 milligrams per 10 liters" value?

* 102. **[Laboratory]** The device shown, called a pycnometer, can be used to determine the densities of liquids very precisely. The pycnome-

Empty:
32.105 g

Filled with water
at 20 °C: 42.062 g
Density of water:
0.9982 g/mL

Filled with benzene
at 20 °C: 40.873 g
Density of benzene:
?

ter is weighed empty and again when filled with a liquid of known density, usually water. The volume of the pycnometer can be obtained from these two measurements. Finally, the pycnometer is weighed when filled with a liquid of unknown density. Now the unknown density can be calculated.

(a) From the data given, determine the density of benzene at 20 °C.

(b) In experiments performed at 20 °C, the mass of a dry, empty pycnometer is found to be 36.2142 g. When the pycnometer is filled with water, its mass is 46.1894 g. When a small sample of zinc is added to the dry, empty pycnometer, the mass is 38.1055 g. Finally, when the pycnometer with the sample of zinc is again filled with water, the measured mass is 47.8161 g. Use these data, together with the density of water at 20 °C—0.99823 g/mL—to determine the density of the zinc.

* **103. [Biochemical]** Following is a table of data used to determine an individual's percent body fat. Make a graph of the data (see Appendix A), and then estimate the percent body fat of the following.

Body Density g/cm^3	% Body Fat
1.010	38.3
1.030	29.5
1.050	21.0
1.070	12.9
1.090	5.07

(a) A person with a body density of 1.037 g/cm^3.

(b) A person who weighs 165 lb in air and 14 lb when submerged in water.

(c) Which person is more likely to be a well-trained athlete?

(d) A 20-year-old male weighs 89.34 kg on land and 5.30 kg underwater. The water has a density of 0.9951 kg/L. His residual lung volume is 1.57 L, and his maximum lung volume is 6.51 L. Calculate his body density. What mass of lead ($d = 11.34$ g/cm^3) could the young man have on his person and just barely float in the water with his lungs fully inflated? What volume would this lead occupy? Why does a person float in spite of the fact that his or her body density is greater than that of water?

104. [Collaborative] Steel, an iron-based alloy is used in the construction of ocean-going ships, as in the ocean liner *Titanic*. Though declared unsinkable by its owners, the *Titanic* sank on its maiden voyage in 1912.

(a) How can a material denser than water be made to float on water?

(b) How might an ocean liner be constructed to be "unsinkable"?

(c) Why do you suppose the *Titanic* sank despite the claims of its unsinkability?

105. [Collaborative] Volumes of liquids are not always additive; that is, the sum of the volumes is not necessarily the volume of the mixture. When 53.8 mL of ethyl alcohol is added to 52.6 mL of water, the volume of that which is produced is 100.2 mL. Calculate the density of the product. Discuss whether you think this represents a physical change or a chemical change, and suggest means of testing your hypothesis.

106. [Environmental] Most scientists think that global warming is under way and will result in a rise in sea level. Which of the following will cause the sea level to rise?

(a) melting of mountain glaciers

(b) a decrease in size of the Antarctic ice sheet

(c) a breakup of the Antarctic Larsen Ice Shelf

(d) a decrease in size of the Arctic ice sheet

(e) melting of the Arctic permafrost

107. [Collaborative] Disney cartoon character Scrooge McDuck was supposed to have a money bin, nearly full, with a volume of 3.0 cubic acres. **(a)** What do you think "3.0 cubic acres" means? That is, what is the most likely interpretation of the term? **(b)** Using this interpretation, what would the money bin's volume be in cubic feet and in cubic meters?

 # e-Media Problems

The activities described in these problems can be found in the e-Media Activities and Interactive Student Tutorial (IST) modules of the Companion Website, *http://chem.prenhall.com/hillpetrucci*.

108. Observe the **Water 3D** model *(Section 1–2)*. By using the pop-up menu (right-click for Windows, click and hold for Macintosh), you can change the representation of the model from ball-and-stick to space-filling. Describe the general shape and character of the molecule in the two representations. How do they differ in the two schemes?

109. Using the **Classification of Matter** activity *(Section 1–2)*, identify each of the following as an element, compound, heterogeneous mixture, or homogeneous mixture. Propose a method of separating mixtures into their component substances.

(a) carbonated water

(b) sodium chloride (table salt)

(c) krypton

(d) pyrite (FeS$_2$, or fool's gold)

110. View the **Chlorine Gas** animation *(Section 1–2)*. What new observations about the molecular nature of this gas arise that are not evident in the static figure in the text?

111. Using the **Temperature Conversion** activity *(Section 1–3)*, convert the temperatures 37.2 °C and 47.2 °C into °F. By comparing the two sets of temperatures, determine which temperature scale is more precise when citing only three significant figures.

112. Use the **Density** activity *(Section 1–5)* to determine an algebraic relationship between density and volume for samples of the same mass.

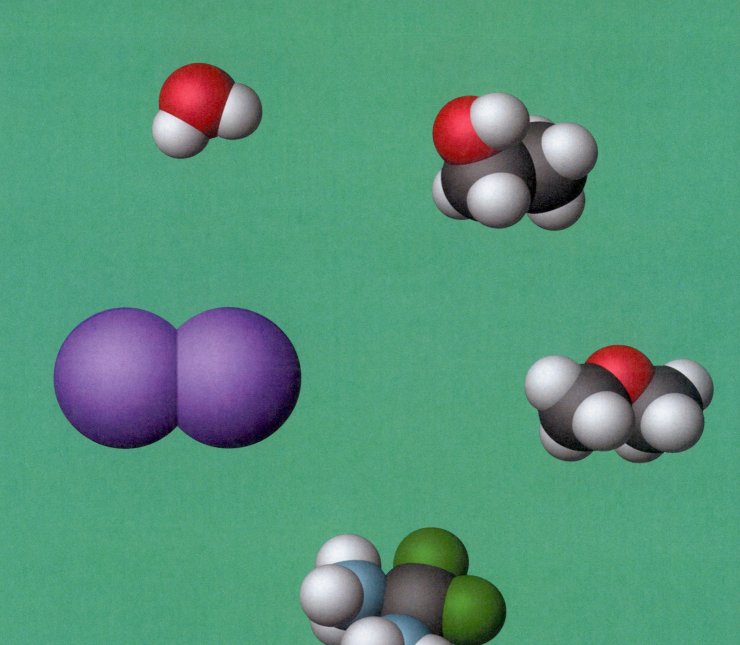

Chapter 2

Atoms, Molecules, and Ions

THE VOCABULARY OF chemistry is specialized. It includes such terms as atom, ion, molecule, isotope, acid, base, salt, and saturated hydrocarbon. Chemistry also involves symbolic notation. Atoms of elements are represented by *symbols*, such as H, C, N, O, Na, and Cl. These symbols are the alphabet of chemistry. A list of the elements and their symbols appears inside the front cover of this book. To represent compounds, symbols are combined into *chemical formulas*, such as NaCl, H_2O, CO_2, and C_5H_{12}. Formulas are the words of chemistry. In this chapter, we will explore both the specialized vocabulary of chemistry and the alphabet and words of the chemist's symbolic language. In the next chapter, we will extend the symbolic language to include sentences—chemical equations.

In the first part of this chapter, we will focus on some basic ideas about atoms and the laws that govern how they combine to form compounds. In the second part, we will introduce two broad categories of compounds: molecular and ionic. Our emphasis in that discussion will be on *chemical nomenclature*, the relationship between the names of chemical compounds and their formulas. Specifically, we will learn to write the formulas of compounds if we know their names and the names of compounds if we know their formulas. A working knowledge of nomenclature is useful because the name of a compound often provides clues about its structure, chemical behavior, or both.

In the context of this nomenclature overview, we will introduce *organic* compounds, which are those based on the element carbon. Organic compounds vastly outnumber all other compounds, which collectively are called *inorganic* compounds. More than 95% of the tens of millions of known compounds are organic. Although our introduction will be limited in scope, it should still provide a glimpse of the enormous diversity of organic compounds.

CONTENTS

◄ Much of our study of chemistry is focused on molecules, which are often represented by space-filling models. Clockwise from left center are iodine, I_2; water, H_2O; ethyl alcohol, CH_3CH_2OH; dimethyl ether, CH_3OCH_3; and cisplatin, $PtCl_2(NH_3)_2$.

▲ Antoine Lavoisier (1743–1794) was a chemist, biologist, and economist. He used the return on investments in a much hated private tax-collecting agency of King Louis XVI to equip a private laboratory and finance his research. Lavoisier's experience as an accountant led him to apply principles of the balance sheet to everything, including chemical reactions. As a royal tax collector, though, he earned the enmity of leaders of the French Revolution and was beheaded on a guillotine. Little did his executioners realize that their victim would later become known as the Father of Modern Chemistry.

Laws and Theories: A Brief Historical Introduction

Early practitioners of alchemy, medicine, and related fields accumulated a lot of useful information about chemicals and their behavior. However, it wasn't until the laws of chemical combination were firmly established about two centuries ago that chemistry became a scientific study. In this part of the chapter, we look at the historical development of some of these laws. We also look at several other basic ideas that we will use throughout the text.

2.1   Laws of Chemical Combination

We noted in Chapter 1 that data obtained by experiment can often be summarized into laws. Scientific theories are then formulated to explain these laws. We present John Dalton's theory of the atomic structure of matter in Section 2.2, but first let's consider two laws that he attempted to explain with his theory.

Lavoisier: The Law of Conservation of Mass

Modern chemistry dates from the eighteenth century, when scientists began to make *quantitative* observations. Antoine Lavoisier made mass measurements precise to about 0.0001 g and established that total mass does not change during a chemical reaction. For example, Lavoisier heated the red oxide of mercury, causing it to decompose (break down) to two new products: mercury metal and oxygen gas. By measuring carefully, he found that the total mass of the products was exactly the same as the mass of the mercury oxide he started with. Lavoisier summarized his findings through the **law of conservation of mass.**

> *The total mass remains constant during a chemical reaction.**

Example 2.1   A Conceptual Example

Jan Baptista van Helmont (1579–1644) first measured the mass of a young willow tree and, separately, the mass of a bucket of soil and then planted the tree in the bucket. After five years, he found that the tree had gained 75 kg in mass even though the soil had lost only 0.057 kg. He had added only water to the bucket, and so he concluded that all the mass gained by the tree had come from the water. Explain and criticize his conclusion.

ANALYSIS AND CONCLUSIONS

Van Helmont's explanation anticipated the law of conservation of mass by dismissing the possibility that the increased mass of the tree could have been created from nothing. His main failure, however, was in not identifying all the substances involved. He did not know about the role of carbon dioxide gas in the growth of plants. In applying the law of conservation of mass, we must focus on *all* the substances involved in a chemical reaction.

EXERCISE 2.1A

A sealed photographic flashbulb containing magnesium and oxygen has a mass of 45.07 g. On firing, a brilliant flash of white light is emitted and a white powder is formed inside the bulb. What should the mass of the bulb be after firing? Explain.

EXERCISE 2.1B

Is the mass of the burned match pictured in the photograph the same as, less than, or more than the mass of the burning match? Explain your answer.

---

*A twentieth-century modification of this law, based on the work of Albert Einstein, allows for the possibility of converting mass to energy (and vice versa). This need not concern us, however, because in chemical reactions the amounts of mass converted to energy are too small to detect.

Proust: The Law of Definite Proportions

By the end of the eighteenth century, Lavoisier and other scientists had succeeded in decomposing many compounds into the elements that form them. One of these scientists, Joseph Proust (1754–1826), did careful quantitative studies by which he established the **law of constant composition** (also called the *law of definite proportions*).

> *All samples of a compound have the same composition; that is, all samples have the same proportions, by mass, of the elements present in the compound.*

Proust based this law in part on his studies of a substance that we now call basic copper carbonate (Figure 2.1); all samples of this compound have the same composition: 57.48% copper, 5.43% carbon, 0.91% hydrogen, and 36.18% oxygen by mass.

A compound not only has a constant, or fixed, composition; it also has fixed properties. For example, under normal atmospheric pressure, pure water always freezes at 0 °C and boils at 100 °C. Furthermore, at 20 °C, 100.0 g of pure water always dissolves a maximum of 35.9 g of sodium chloride (table salt). The physical and chemical properties of chemical substances—both elements and compounds—depend on their composition; they are not a matter of chance. We will learn more about how and why this is true in later chapters.

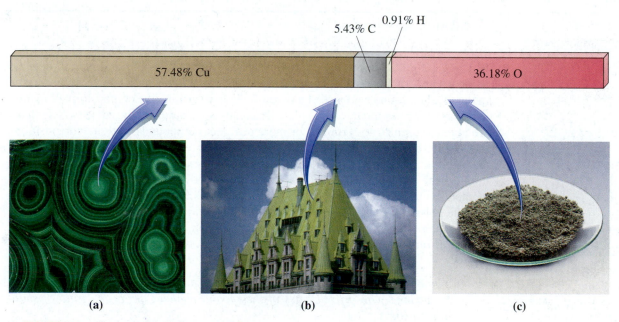

▲ FIGURE 2.1   The law of definite proportions
A substance known as basic copper carbonate (**a**) occurs in nature as the mineral *malachite*, (**b**) forms as a patina on copper roofs and bronze statues, and (**c**) can be synthesized in the laboratory. Regardless of its source, basic copper carbonate always has the same composition.

Example 2.2

The mass ratio of oxygen to magnesium in the compound magnesium oxide is 0.6583 : 1. What mass of magnesium oxide will form when 2.000 g of magnesium is completely converted to magnesium oxide by burning in pure oxygen gas?

STRATEGY

We usually write mass ratios in a form such as "the ratio of O to Mg is 0.6583 : 1." The first number represents the mass of the first element named—in this case, a mass of *oxygen*, say 0.6583 g oxygen—and the second number represents the mass of the second element named—here a mass of *magnesium*. Although written only as "1," we assume that the second number is as precisely known as the first, that is, 1.0000 g magnesium. According to the law of constant composition, the mass of oxygen that combines with the 2.000 g of magnesium must be just the right amount that makes the mass ratio of oxygen to magnesium in the product 0.6583 : 1. According to the law of conservation of mass, the mass of magnesium oxide must equal the sum of the masses of the magnesium and oxygen that react.

SOLUTION

We can state the mass ratio in the form of a conversion factor and then determine the required mass of oxygen.

$$?\text{g oxygen} = 2.000 \text{ g magnesium} \times \frac{0.6583 \text{ g oxygen}}{1 \text{ g magnesium}} = 1.317 \text{ g oxygen}$$

The mass of the sole product, magnesium oxide, equals the total of the masses of the substances entering into the reaction.

$$?\text{g magnesium oxide} = 2.000 \text{ g magnesium} + 1.317 \text{ g oxygen}$$
$$= 3.317 \text{ g magnesium oxide}$$

ASSESSMENT

As a simple check, compare the calculated mass with the starting mass of magnesium. From the law of conservation of mass, the product of the reaction of magnesium with oxygen must have a mass greater than the starting mass of magnesium; and 3.317 g magnesium oxide is indeed greater than 2.000 g magnesium.

EXERCISE 2.2A

What mass of magnesium oxide is formed when 1.500 g of oxygen combines with magnesium?

EXERCISE 2.2B

When a strip of magnesium metal was burned in pure oxygen gas, 1.554 g of oxygen was consumed and the only product formed was magnesium oxide. What must have been the masses of magnesium metal burned and magnesium oxide formed?

▲ John Dalton (1766–1844) was not principally an experimenter (perhaps because he was color-blind), but he skillfully used the results of other scientists in formulating his atomic theory. Dalton also kept records of the weather throughout his life. His studies of humidity as measured by dew-point temperatures led to his discovery of an important law concerning mixtures of gases (Section 5.10).

 Multiple Proportions animation

2.2   John Dalton and the Atomic Theory of Matter

In 1803, John Dalton proposed a theory to explain the laws of conservation of mass and constant composition. As he developed what would become known as his atomic theory, Dalton found evidence of a scientific law describing the composition of matter, a law his theory would have to explain: In some cases, atoms of the same two elements are able to combine to form two or more different compounds, and the compositions of the different compounds are related. Let's illustrate Dalton's reasoning with a simple example: two compounds composed of the elements carbon and oxygen. In carbon dioxide, the familiar gas produced in respiration and in the burning of fuels, the two elements combine in the ratio of 8.0 g of oxygen to 3.0 g of carbon. In carbon monoxide, the poisonous gas formed when a fuel is burned in limited air, the elements combine in the ratio of 4.0 g of oxygen to 3.0 g of carbon. Based on evidence such as this, Dalton formulated the **law of multiple proportions,** which we can state in the following way:

In two or more compounds of the same two elements, the masses of one element that combine with a fixed mass of the second element are in the ratio of small whole numbers.

Thus for carbon dioxide and carbon monoxide, the masses of oxygen (8.0 g and 4.0 g) that combine with a fixed mass of carbon (3.0 g) are in a small whole-number ratio, 8.0 : 4.0 = 2 : 1. Figure 2.2 shows a scheme for establishing this 2 : 1 ratio.

Dalton's Atomic Theory

The atomic model Dalton developed to explain the laws of chemical combination is based on four ideas:

- *All matter is composed of extremely small, indivisible particles called atoms.*
- *All atoms of a given element are alike in mass and other properties, but the atoms of one element differ from the atoms of every other element.*
- *Compounds are formed when atoms of different elements unite in fixed proportions.* That is, the relative numbers of each kind of atom in a compound form a simple ratio, such as one atom of A to one of B in AB, two atoms of A to one of B in A_2B, and so on.
- *A chemical reaction involves a rearrangement of atoms to produce new compounds. No atoms are created, destroyed, or broken apart in a chemical reaction.*

	Carbon monoxide (CO)	Carbon dioxide (CO$_2$)
The elements	3.0 g carbon (C) + 4.0 g oxygen (O)	3.0 g carbon (C) + 8.0 g oxygen (O)
The compound	7.0 g carbon monoxide (CO)	11.0 g carbon dioxide (CO$_2$)
Oxygen-to-carbon mass ratio	$\dfrac{4.0 \text{ g oxygen}}{3.0 \text{ g carbon}}$	$\dfrac{8.0 \text{ g oxygen}}{3.0 \text{ g carbon}}$

Comparing two mass ratios:

$$\text{Mass ratio for CO}_2 \searrow \quad \frac{\dfrac{8.0 \text{ g oxygen}}{3.0 \text{ g carbon}}}{\dfrac{4.0 \text{ g oxygen}}{3.0 \text{ g carbon}}} = \frac{8.0 \text{ g oxygen}}{4.0 \text{ g oxygen}} = 2:1$$

Mass ratio for CO

◀ **FIGURE 2.2   The law of multiple proportions**

The oxygen-to-carbon mass ratio in carbon dioxide is twice that in carbon monoxide.

Explanations Using Dalton's Theory

Let's first explain the law of constant composition and the law of conservation of mass as Dalton might have. For simplicity, we will consider the compound hydrogen fluoride (HF), although it was not known in Dalton's time, and use the following reasoning suggested by Figure 2.3.

All HF molecules are made up of one hydrogen atom and one fluorine atom. Because a fluorine atom has 19 times the mass of a hydrogen atom, 1/20th, or 5.0%, of the mass of hydrogen fluoride is hydrogen; 19/20th, or 95%, is fluorine. This is true whether we consider one HF molecule, the four HF molecules shown in Figure 2.3, or many, many HF molecules. This is consistent with the law of constant composition.

The law of conservation of mass is also illustrated in Figure 2.3. The figure shows that six fluorine atoms (in the form of three F$_2$ molecules) and four hydrogen atoms (as two H$_2$ molecules) are available for reaction. Four of the F atoms combine with the four H atoms, producing four HF molecules. Two fluorine atoms do not react and are left unchanged as one F$_2$ molecule. Thus, after the reaction there are four hydrogen atoms and six fluorine atoms, just as there were before. The total mass remains unchanged. With Dalton's atomic theory, we can restate the law of conservation of mass in this way:

Atoms can neither be created nor destroyed in a chemical reaction; thus, the total mass remains unchanged.

Dalton's theory also nicely explains the law of multiple proportions. Recall that in carbon dioxide, two oxygen atoms are combined with each carbon atom (CO$_2$), in carbon monoxide, there is one oxygen atom per carbon atom (CO). The ratio of the numbers of oxygen atoms per carbon atom in the two compounds is 2:1. Because the atoms of oxygen have the same mass, this 2:1 ratio based on numbers of atoms is the same as the 2:1 ratio based on masses that we saw in Figure 2.2.

Application Note

The law of conservation of mass tells us that we cannot create materials from nothing, but that we can make new materials only by changing the way atoms are combined in existing materials. It also tells us that we cannot destroy materials in the sense of making them disappear—the mass of any new materials we create in a chemical process must equal the mass of the original materials. However, chemistry offers the opportunity to transform one material into another, for example, to change potentially hazardous wastes to less harmful materials. Such transformations of matter are what chemistry is all about.

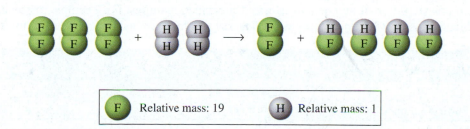

◀ **FIGURE 2.3   Dalton's atomic theory and the laws of constant composition and conservation of mass**

Dalton's theory explained these two basic laws of chemical combination as described in the text.

QUESTION: According to these laws, how many H$_2$ and HF molecules would be present after the reaction of three F$_2$ and four H$_2$ molecules?

2.3   The Divisible Atom

Dalton's concept of indivisible atoms fueled scientific work for nearly all of the nineteenth century. However, like most other scientific theories, Dalton's model eventually had to be modified in light of later discoveries. Toward the end of the nineteenth century, certain experiments that we will describe in subsequent chapters began to reveal that atoms are made up of smaller parts. Of the dozens of subatomic (smaller than atomic) particles now known, three are of special importance in the study of chemistry. They are the proton, neutron, and electron.

Subatomic Particles

The mass and charge of subatomic particles are so small that they are conveniently expressed in *relative* units. The **proton** has a relative mass of 1. The proton also carries one fundamental unit of positive electric charge, denoted 1+. The **neutron,** as its name implies, is an electrically neutral particle; it has no charge. Although its mass is slightly greater than that of the proton, for many purposes we can consider the neutron also to have a relative mass of 1. The third particle, the **electron,** has a mass that is 1/1836 (or 0.0005447) of the mass of a proton. An electron has the same quantity of charge as a proton, but it is a negative charge, denoted as 1−. Protons, neutrons, and electrons are *fundamental* particles. This means that all protons are alike, all neutrons are alike, and all electrons are alike in whatever element they are found. These three subatomic particles and their properties are listed in Table 2.1. (Appendix B contains a review of some essential ideas about electric charge and electricity.)

Protons and neutrons are densely packed into a tiny, positively charged core of the atom known as the **nucleus.** The extremely lightweight electrons are widely dispersed around the nucleus. An atom as a whole is electrically neutral: It has no net charge because the negative and positive charges balance each other.

Although the numbers differ from one element to another, the atoms of every element have equal numbers of electrons and protons.

An atom is mostly empty space. Picture it as something like this: If an entire atom were represented by a room 5 m × 5 m × 5 m, the nucleus would be only about as big as the period at the end of this sentence. The electrons would be tiny specks, too.

Dalton believed that the identity of an element is determined by the *mass* of one of its atoms. We now know that it is not the mass of an atom but rather the *number of protons* in the nucleus that determines the kind of atom and therefore the identity of an element. The **atomic number (Z)** is the number of protons in the nucleus of an atom of a given element, and it is this number of protons that *defines* the element. Every atom having two protons in its nucleus has $Z = 2$ and is an atom of helium. Every atom with 92 protons has $Z = 92$ and is an atom of uranium. A list of all the known elements with their symbols and atomic numbers is located inside the front cover of this book.

 **Atomic Notation activity**

Isotopes

Dalton's belief that all atoms of a given element have the same mass isn't quite true. Any two atoms of a given element do have the same number of protons and electrons, but they may have different numbers of neutrons. Atoms that have the same number of protons but different numbers of neutrons are called **isotopes.** For example, there are three isotopes of hydrogen. The most abundant isotope, occasionally called protium,

Table 2.1   Subatomic Particles

Particle	Symbol	Approximate Relative Mass	Relative Charge	Location in Atom
Proton	p^+	1	1+	Inside nucleus
Neutron	n	1	0	Inside nucleus
Electron	e^-	0.000545	1−	Outside nucleus

has a single proton and no neutrons in its nucleus. About one in every 6700 hydrogen atoms, however, has a neutron as well as a proton. Because the mass of a neutron is essentially the same as the mass of a proton, the mass of this hydrogen isotope, called deuterium, is about twice that of protium. A third, very rare isotope of hydrogen, called tritium, has two neutrons and one proton in the nucleus. A tritium atom has about three times the mass of a protium atom.

Let's introduce a term closely related to the relative mass of an atom: The **mass number (A)*** is the sum of the numbers of protons and neutrons in an atom. The mass number is always an integral (whole) number. By contrast, the actual relative mass of an atom, though close in value to the mass number, is a nonintegral number. We will learn why this is so in Chapter 19, but it is not a matter that need concern us until we reach that point.

The isotopes of a given element have the same atomic number but different mass numbers. As a result, isotopes are often identified by the name of the element followed by the mass number, such as carbon-14, cobalt-60, and uranium-235. Only the hydrogen isotopes have special names.

Some elements have only one naturally occurring isotope; these include fluorine-19, sodium-23, and phosphorus-31. Most elements, however, have two or more isotopic forms. Tin has the greatest number of naturally occurring isotopes: ten. For all but a few elements, the naturally occurring isotopes are always found in certain precise proportions. In natural sources of chlorine, for example, 75.77% of the atoms are chlorine-35 and 24.23% are chlorine-37.

Chemical symbols for isotopes are commonly written in the form $^{A}_{Z}E$, with A being the mass number and Z the atomic number of the element E. For the two naturally occurring isotopes of chlorine, we can write $^{35}_{17}Cl$ and $^{37}_{17}Cl$, indicating mass numbers of 35 and 37 for chlorine-35 and chlorine-37, respectively. Because the atomic number of an element is implied by its chemical symbol, we sometimes choose to omit the atomic number, as in the simplified forms ^{35}Cl and ^{37}Cl.

The number of neutrons in an atom is easily calculated from the values of A and Z:

$$\text{Number of neutrons} = A - Z \qquad (2.1)$$

Isotopes activity

Example 2.3

How many protons, neutrons, and electrons are present in a ^{81}Br atom?

STRATEGY

We can use the identity of the element and several simple relationships between the subatomic particles described above to determine the numbers of these particles.

SOLUTION

The atomic number of bromine is not given here, but we can obtain it from the list of elements on the inside front cover.	$Z = 35 = $ number of protons
In the bromine atom, the number of positively charged protons equals the number of negatively charged electrons.	Number of protons = number of electrons = 35
From mass number 81 and atomic number 35, we calculate the number of neutrons, using Equation (2.1).	Number of neutrons = $A - Z = 81 - 35 = 46$

EXERCISE 2.3A

Use the notation $^{A}_{Z}E$ to represent the isotope of tin having 66 neutrons.

EXERCISE 2.3B

Isobars are atoms with the same mass number but different atomic numbers. Indicate the numbers of protons, neutrons, and electrons and the $^{A}_{Z}E$ notation for a cadmium atom that is an isobar of tin-116.

---

*The mass number is sometimes called the nucleon number, where *nucleon* is a collective name for protons and neutrons.

2.4 Atomic Masses

We are familiar with measuring the mass of a macroscopic sample of matter. We can weigh ourselves on a bathroom scale, a bag of apples at the checkout stand in a supermarket, or a letter on a postal scale. But how can we weigh an individual atom, a microscopic sample of matter that we can't even see?

We can do what John Dalton did: *arbitrarily* assign a mass to one atom and determine the masses of other atoms relative to it. Dalton called these relative masses *atomic weights*, and let's do the same for the time being. Suppose we assign a mass of 1 to the hydrogen atom, just as Dalton did. Now, to establish the mass of an oxygen atom, we can measure the mass ratio of hydrogen to oxygen in water. The best value available in Dalton's time was 1 g H : 7 g O. (The modern value is 1 g H : 8 g O.) At this point, Dalton *assumed* that H and O atoms in water were combined in the numerical ratio 1 : 1. Then, if a given number of oxygen atoms weighed seven times as much as the same number of hydrogen atoms, one oxygen atom must also weigh seven times as much as a hydrogen atom. Thus, Dalton assigned oxygen an atomic weight of 7. However, as outlined in Figure 2.4, if we apply our current understanding that the mass ratio of H to O in water is 1 : 8 and that there are *two* H atoms for every O atom in water, we get a value of 16 for the atomic weight of oxygen. Once chemists figured out how to determine the combining ratios of atoms in compounds, they were able to measure atomic weights rather easily.

By international agreement, the current atomic mass standard is the *pure* isotope carbon-12, which is assigned a mass of *exactly* 12 atomic mass units (12 u). Based on this standard, we can define an **atomic mass unit** (abbreviated amu and having the unit u) as exactly one-twelfth the mass of a carbon-12 atom. In more familiar units of mass, $1\text{ u} = 1.66054 \times 10^{-24}$ g. Because we are interested in the *masses* of atoms and not their weights, we will now shift to the term *atomic mass* in place of the older *atomic weight*,* except in a few cases of historical interest.

The **atomic mass** of an element is defined as the *weighted average* of the masses of the naturally occurring isotopes of that element, and here is why. Consider carbon, which consists of a *mixture* of two naturally occurring isotopes. The much more abundant isotope is carbon-12. The other is carbon-13, with a mass of 13.00335 u. Both isotopes are present in substances containing carbon atoms, and in proportions that generally do not vary from one carbon-containing substance to another. To describe the atomic mass of carbon, then, we need to use an average value, but not the simple average $(12 + 13)/2 = 12.5$. Because carbon-12 is much more abundant than carbon-13, the average we seek lies much closer to the mass of carbon-12 than to that of carbon-13. We say that it is "weighted" toward the mass of carbon-12.

To calculate the atomic mass of an element, we need two quantities—(1) the atomic masses of the isotopes of the element and (2) the naturally occurring fractional abundances of the isotopes. We explain how these quantities are obtained experimen-

▶ **FIGURE 2.4 Dalton's atomic weight problem**

Dalton assumed a combining ratio of hydrogen to oxygen atoms of 1 : 1. Data at the time led him to think that the hydrogen-to-oxygen mass ratio in water was 1 : 7. In this model, assigning a value of 1 for the atomic weight of hydrogen meant that the atomic weight of oxygen had to be 7. Modern data indicate that the combining ratio of hydrogen to oxygen atoms is 2 : 1 and that the mass ratio is 1 : 8 (or 2 : 16). If the atomic weight of hydrogen is taken to be 1, that of oxygen must be 16.

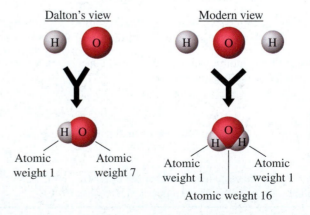

*The term *atomic weight* is still widely used, however, as by the *Commission on Atomic Weights* of the International Union of Pure and Applied Chemistry (IUPAC).

tally in Chapter 7, but for now let's just describe how they are used. To illustrate, let's return to the atomic mass of carbon.

As noted earlier, the masses of the two* carbon isotopes are 12.0000 u for carbon-12 and 13.00335 u for carbon-13. The term *fractional abundances* refers to the fraction of all carbon atoms that are carbon-12 and the fraction of all carbon atoms that are carbon-13. The two fractions must add up to 1. However, relative abundances of isotopes are often expressed as percentages rather than fractions. If this is the case, first we need to convert these percentages to fractions. Percentages are actually fractions expressed on a per-hundred basis, that is,

$$\text{Percent} = \text{fraction} \times 100\%$$

In turn, fractions are percentages divided by 100%:

$$\text{Fraction} = \frac{\text{percent}}{100\%}$$

The percentage abundances and fractional abundances of the carbon isotopes are as follows:

Isotope	Percent Abundance	Fractional Abundance
Carbon-12	98.892%	0.98892
Carbon-13	1.108%	0.01108

To obtain a weighted average atomic mass, we calculate the contribution of each isotope to the weighted average from the relationship

$$\text{Contribution of isotope} = \text{fractional abundance} \times \text{mass of isotope} \qquad \textbf{(2.2)}$$

Example 2.4

Use the data cited above to determine the weighted average atomic mass of carbon.

STRATEGY

The contribution each isotope makes to the weighted average atomic mass is given by Equation (2.2). The weighted average atomic mass is the sum of the two contributions.

SOLUTION

The contributions are

$$\text{Contribution of carbon-12} = 0.98892 \times 12.00000\ u = 11.867\ u$$
$$\text{Contribution of carbon-13} = 0.01108 \times 13.00335\ u = 0.1441\ u$$

The weighted average mass is

$$\text{Atomic mass of carbon} = 11.867\ u + 0.1441\ u = 12.011\ u$$

ASSESSMENT

This is the value listed in a table of atomic masses. As expected, the atomic mass of carbon is much closer to 12 u than to 13 u.

EXERCISE 2.4A

There are three naturally occurring isotopes of neon. Their percent abundances and atomic masses are neon-20, 90.51%, 19.99244 u; neon-21, 0.27%, 20.99395 u; neon-22, 9.22%, 21.99138 u. Calculate the weighted average atomic mass of neon.

EXERCISE 2.4B

The two naturally occurring isotopes of copper are copper-63, mass 62.9298 u, and copper-65, mass 64.9278 u. What must be the percent natural abundances of the two isotopes if the atomic mass of copper listed in a table of atomic masses is 63.546 u?

Problem-Solving Note

Your *grade-point average* is a weighted average in which letter grades are converted to points (A = 4.0, B = 3.0, and so on). These grade points are equivalent to the masses of isotopes, and the fraction of the total credit units of enrollment assigned to each course is equivalent to the fractional abundance of an isotope. Thus, an A in a five-hour course counts more toward a GPA than does an A in a one-hour course, just as, for a given element, the mass of an isotope having a 50% abundance counts more toward the element's weighted average atomic mass than does the mass of an isotope having only a 10% abundance.

Problem-Solving Note

The fractional abundances of the two isotopes are unknowns. Can you see that if one of them is x, the other must be $1.000 - x$?

*A tiny amount of carbon-14 also occurs in nature, but in such vanishingly small amounts that it does not affect the weighted average atomic mass of carbon.

Example 2.5   An Estimation Example

Indium has *two* naturally occurring isotopes and a weighted average atomic mass of 114.82 u. One of the isotopes has a mass of 112.9043 u. Which is likely to be the second isotope: ^{111}In, ^{112}In, ^{114}In, or ^{115}In?

ANALYSIS AND CONCLUSIONS

The masses of isotopes differ only slightly from whole numbers, which tells us that the isotope with mass 112.9043 u is ^{113}In. To account for the observed weighted average atomic mass of 114.82 u, the second isotope must have a mass number greater than 114. It can be only ^{115}In.

EXERCISE 2.5A

The masses of the three naturally occurring isotopes of magnesium are ^{24}Mg, 23.98504 u; ^{25}Mg, 24.98584 u; ^{26}Mg, 25.98259 u. Use the atomic mass given inside the front cover to determine which of the three is the most abundant. Can you determine which is the second most abundant? Explain.

EXERCISE 2.5B

For the three magnesium isotopes described in Exercise 2.5A, **(a)** could the percent natural abundance of ^{24}Mg be 60.00%? **(b)** What is the smallest possible value for the percent natural abundance of ^{24}Mg?

2.5   The Periodic Table: Elements Organized

In the nineteenth century, chemists discovered dozens of new elements. By 1830, 55 elements were recognized, but there was no apparent pattern in their properties. Chemists badly needed a way to organize the growing collection of chemical data. One way to do this was to arrange the elements in a manner that would establish categories of elements having similar physical and chemical properties. Dmitri Mendeleev published the first successful arrangement, called a *periodic table,* in 1869. In its modern form, the periodic table organizes a vast array of chemical knowledge. The periodic table is so important that we will devote most of Chapter 8 to it, and it is a recurring theme in Chapters 20, 21, and 22. For now, though, we will deduce a few ideas from it to help us name and write formulas of chemical compounds.

Mendeleev's Periodic Table

Mendeleev arranged the elements in order of increasing atomic weight, from left to right in rows and from top to bottom in columns (or groups). In this arrangement, elements that most closely resemble one another in physical and chemical properties tend to fall in the same vertical group. This group similarity recurs *periodically*, hence the name **periodic table.**

So that there would be no exceptions to the principle that all the elements in a group display similar properties, Mendeleev placed some elements out of order, that is, not in the strict order of increasing atomic weight. For example, he correctly placed tellurium (atomic weight 127.6) ahead of iodine (atomic weight 126.9) so that tellurium would be in the same column as the similar elements sulfur and selenium.

When Mendeleev placed elements with similar properties in the same vertical group, a few gaps were created in his table. Instead of seeing these gaps as defects, he boldly predicted the existence of undiscovered elements to fill the gaps. Furthermore, because the table was based on patterns of *properties,* he was able to predict some properties of the missing elements. For example, he left a blank space for an undiscovered element that he called "eka-silicon" and used its location between silicon and tin to predict an atomic weight of 72 and other properties. Table 2.2 shows just how accurate his predictions were, when compared to the properties of the actual element germanium discovered 15 years later. The *predictive* nature of Mendeleev's periodic table led to its wide acceptance as a tremendous scientific accomplishment.

▲ Dmitri Ivanovich Mendeleev (1834–1907) developed a periodic table while trying to systematize the properties of the elements for presentation in a chemistry textbook. His highly influential text lasted for 13 editions, including five published after his death.

Table 2.2 Properties of Germanium: Predicted and Observed

Property	Predicted: Eka-silicon[a] (1871)	Observed: Germanium (1886)
Atomic weight	72	72.6
Density, g/cm^3	5.5	5.47
Color	Dirty gray	Grayish white
Density of oxide, g/cm^3	$EsO_2 : 4.7$	$GeO_2 : 4.703$
Boiling point of chloride	$EsCl_4$: below 100 °C	$GeCl_4$: 86 °C
Density of chloride, g/cm^3	$EsCl_4 : 1.9$	$GeCl_4 : 1.887$

[a] The term "eka" is derived from Sanskrit and means "first." Literally, eka-silicon means "first comes silicon" (and then comes the unknown element).

The Modern Periodic Table

There are more than 110 elements listed in the modern periodic table shown in Figure 2.5. Each box, or entry, in the table gives the chemical symbol, atomic number (Z), and atomic mass of an element (illustrated for iron). The elements are placed in order of increasing *atomic number,* a property more responsible for the behavior of an element than is its atomic mass.

The periodic table is divided into groups and periods. *Groups,* or families, are the vertical columns of elements that have similar properties. We place a group number (for example, 1A, 2A, 3B, and so on) at the top of each column. *Periods* are the horizontal rows of elements. The periods vary in length from two elements (the first period) to 32 elements (the sixth and seventh periods). A number at the left of the row identifies each period. If all the elements were included in the periodic table, it would have to be quite wide to accommodate the periods containing 32 elements, and there would be a large blank space above these long periods. By extracting a 14-element series from the sixth and seventh periods and listing these two series separately at the bottom of the table, we can use a more compact form that limits the width of the table to 18 elements. The two 14-element series are called the *lanthanide* series (atomic numbers 58–71) and the *actinide* series (90–103).

Interactive Periodic Table activity

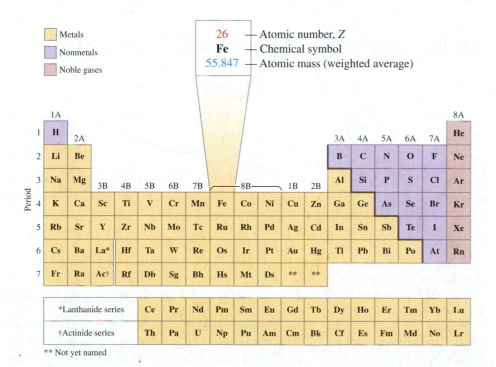

▼ **FIGURE 2.5   The modern periodic table**

▶ Copper, a metal shown in various forms on the left in this photograph, can be obtained as pellets that can be hammered into a thin foil or drawn into a wire. Sulfur, the yellow nonmetal shown on the right, can be obtained as lumps that crumble into a powder when hammered.

Elements are divided into two main classes by the heavy, stepped gray line shown in Figure 2.5. Except for hydrogen, those elements to the left of this line are **metals.** Metals have a characteristic luster and are generally good conductors of heat and electricity. Most metals are *malleable,* which means they can be hammered into thin sheets or foil, and *ductile,* meaning they can be drawn into wires. Except for mercury (a liquid at room temperature), all the metallic elements are solids at room temperature.

Elements to the right of the stepped line are **nonmetals,** elements that lack metallic properties. For example, nonmetals generally are poor conductors of heat and electricity. At room temperature, several nonmetals, including oxygen, nitrogen, fluorine, and chlorine, are gases. Others, such as carbon, sulfur, phosphorus, and iodine, are brittle solids. Bromine is the only nonmetal that is a liquid at room temperature. Some of the elements bordering the stepped line resemble metals in some of their properties and nonmetals in others. They are sometimes considered in a special category called *semimetals* or **metalloids.**

We will consider the theoretical basis of the periodic table in Chapter 8. For example, we will discover that the group numbers and letters (A and B) are related to the way electrons are arranged about the nuclei of atoms. We will also discuss why periods are not all the same length, why the nonmetal hydrogen appears in group 1A with a group of metals, and other fundamental ideas about the periodic table. In the following sections and chapters, we will use the arrangement of elements in the periodic table as a guide in understanding the formation of chemical compounds. In addition, we will use the atomic masses recorded in the periodic table in many calculations.

Introduction to Chemical Compounds

We have already mentioned chemical compounds several times, but we have just scratched the surface of this broad subject. The remainder of this chapter serves as an introduction to chemical compounds, and we will be dealing with this topic throughout the text. In later chapters, we will explore the compositions, properties, and reactions of compounds and how macroscopic properties are related to the structure of matter at the microscopic level. For example, it is the atomic structure of an element—a microscopic property—that determines whether the element displays the macroscopic properties of a metal or those of a nonmetal. The sections that follow include some useful information about classifying, naming, and writing formulas of compounds. You may want to return to this information from time to time; it may be helpful each time you encounter an unfamiliar chemical compound.

Chemists classify compounds in different ways. As noted at the beginning of this chapter, the two broadest categories are *organic compounds* and *inorganic compounds.* Although there are borderline cases, most chemical compounds can be placed in one category or the other.

Chemists also use other terms to describe and classify chemical compounds. Some of these additional terms that are introduced in this chapter are molecular compounds, ionic compounds, acids, bases, and salts. At times we may use more than one term to classify a compound. For example, when we say that sulfuric acid is an *inorganic acid,* we indicate that it is both an inorganic compound and an acid.

We can represent any compound symbolically. A **chemical formula** is a symbolic representation of the composition of a compound expressed in terms of its constituent elements. A chemical formula uses symbols to indicate which elements are present, and it uses subscripts to indicate the relative numbers of atoms of the different elements present. For example, the formula B_2O_3 signifies that the compound boron oxide contains two boron atoms for every three oxygen atoms.

Ammonia has nitrogen and hydrogen atoms in the ratio of 1 to 3, but we write the formula of ammonia NH_3, *not* N_1H_3. If there is *no* subscript number following a symbol in a formula, the subscript "1" is implied.

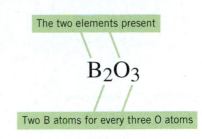

The two elements present

$$B_2O_3$$

Two B atoms for every three O atoms

2.6 Molecules and Molecular Compounds

A **molecule** is a group of two or more atoms held together in a definite spatial arrangement by forces called *covalent* bonds.* Molecules are the smallest characteristic entities of a **molecular compound,** and these molecules determine the properties of the substance. Considered separately, the component atoms of molecules do not determine the properties of a substance. An analogy: Carrot cake does not taste like flour, nor like carrots, butter, or eggs. The flavor of carrot cake is determined by all of those ingredients collectively.

Depending on how much information we want to convey, we have several choices in the type of chemical formula we use to represent a compound. An **empirical formula** is the simplest formula we can write for a compound. It lists the elements present and indicates the smallest integral (whole-number) ratio in which atoms of these elements are combined. The empirical formula CH_2O indicates that the elements C, H, and O are present in the atom ratios 1:2:1, respectively. However, several different compounds have this same empirical formula, for example, acetic acid (found in vinegar) and glucose (a sugar).

A **molecular formula** shows the differences between compounds with identical empirical formulas by giving the symbol and the *actual* number of each kind of atom in a molecule. The molecular formula of acetic acid is $C_2H_4O_2$ and that of glucose is $C_6H_{12}O_6$. The molecular formula $C_2H_4O_2$ has a total number of atoms *twice* that of the empirical formula, that is, $2 \times (1:2:1) = 2:4:2$. The formula $C_6H_{12}O_6$ has a total 6 times that of the empirical formula: $6 \times (1:2:1) = 6:12:6$. Water ($H_2O$), ammonia ($NH_3$), and carbon dioxide ($CO_2$) are familiar substances whose molecular formulas are the same as their empirical formulas.

Most of the molecules we encounter are those of compounds, but some elements also exist in molecular form, as shown in Figure 2.6. Notice that chlorine occurs not as individual Cl atoms but as pairs of atoms joined into Cl_2 molecules. Molecules containing only two atoms are called *diatomic* molecules. Other diatomic elements include hydrogen (H_2), nitrogen (N_2), oxygen (O_2), and the group 7A elements, the *halogens* (F_2, Cl_2, Br_2, I_2). A few elements exist as *polyatomic* (many-atom) molecules. These include sulfur (S_8, as shown in Figure 2.6) and phosphorus, P_4. The *noble gases* of group 8A exist in *monatomic* (one-atom) form. Thus, we represent helium simply as He and neon as Ne.

The atoms in a molecule are not arranged at random. They are attached in a definite order. In water molecules, the order is always H—O—H, never H—H—O. A **structural formula** is a chemical formula that shows how atoms are attached to one another. The structural formulas of ammonia, methane, and acetic acid are:

Cl_2, a diatomic molecule

S_8, a polyatomic molecule

▲ **FIGURE 2.6   Molecular forms of two common elements**

$$
\begin{array}{ccc}
\begin{array}{c} H \\ | \\ H-N-H \end{array} &
\begin{array}{c} H \\ | \\ H-C-H \\ | \\ H \end{array} &
\begin{array}{c} \quad H \quad O \\ | \quad || \\ H-C-C-O-H \\ | \\ H \end{array} \\
\text{Ammonia } (NH_3) & \text{Methane } (CH_4) & \text{Acetic acid } (CH_3COOH)
\end{array}
$$

The lines in structural formulas represent the bonds between atoms. A single line represents a single bond. All the bonds in ammonia, NH_3, and in methane, CH_4, are

---

*Covalent bonds are discussed in Chapters 9 and 10. In this section, we concentrate on the atoms present in molecules and not on the forces between atoms.

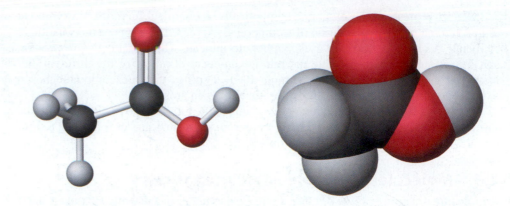

▶ **FIGURE 2.7**   **Two types of molecular models**
The ball-and-stick model of acetic acid (left) and the space-filling model (right) correspond to the structural formula of the acetic acid molecule shown on page 47. The models better show the three-dimensional shape of the molecule.

Acetic Acid 3D model

Computer software (CHIME and RASMOL) is widely available to help you visualize molecular shape in three dimensions.

single bonds. All but one of the bonds in acetic acid, CH_3COOH, are also single bonds. The exception is the bond between a C atom and the lone O atom. That bond, represented by a two closely spaced parallel lines, is a double bond. Still another bond type found in some molecules is a triple bond, represented by three closely spaced parallel lines. We will discuss these different bond types in detail in Chapters 9 and 10. For now, simply look upon a double bond as being stronger than a single bond and a triple bond as being stronger still.

The geometric shapes of molecules are difficult to represent through structural formulas. These shapes are best represented by three-dimensional molecular models. Figure 2.7 shows two types of models of the acetic acid molecule. The *ball-and-stick* model shows the spatial arrangements of the bonds, and the *space-filling* model shows that atoms in a molecule occupy space and are in direct contact with one another. Figure 2.8 presents the color scheme used for atoms in molecular models in this book.

Writing Formulas and Names of Binary Molecular Compounds

The molecules of a *binary* molecular compound are made up of *two* elements. Most often, both elements are nonmetals.

In writing formulas for binary molecular compounds, we must decide which element symbol to write first. As a general rule, we write first the symbol of the element that lies farthest to the left in its period on the periodic table and/or lowest in its group. Hydrogen, oxygen, and a few other cases are exceptions. The scheme based on the

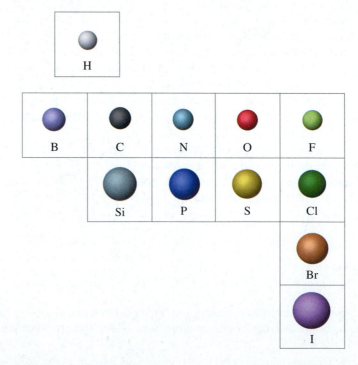

▶ **FIGURE 2.8**   **The color scheme for atoms in molecular models used throughout this book**

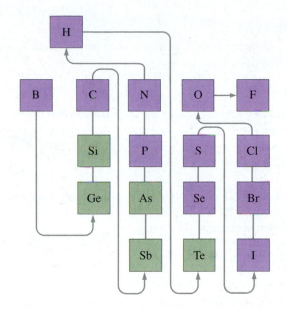

◀ **FIGURE 2.9** **A scheme based on the periodic table to assist in writing formulas and names of binary molecular compounds**
The lines trace a continuous path from boron (B) to fluorine (F). The element closer to the beginning of this path is generally written first in the formula for a binary molecular compound.

QUESTION: In what order would the elemental symbols be written for binary compounds formed from nitrogen and (a) carbon, (b) silicon, and (c) oxygen?

portion of the periodic table shown in Figure 2.9 will meet most of our current needs. For example, it shows that we put the nitrogen atom first in NH_3 but the hydrogen atom first in H_2Te, and that we write Cl_2O but OF_2.

Once we have written the formula for a binary molecular compound, we apply the following ideas in establishing the name of the compound:

- The name consists of two words, one for each element in the compound.
- The first word is the name of the element that appears first in the formula.
- The second word is an altered version of the name of the second element. This word retains the stem of the element name and replaces the ending by *-ide*. Thus, the element name chlor*ine* becomes chlor*ide*.
- The names are further modified by adding prefixes such as *mono-*, *di-*, *tri-*, and so on to denote the numbers of atoms of each element in the molecule (see Table 2.3). Thus, P_4S_3 is called *tetra*phosphorus *tri*sulfide. The prefix *mono-* is treated in a special way, however. We do not use it for the first-named element, but we do for the second, as we see in the name for CO, which is carbon monoxide and *not* monocarbon monoxide.

Application Note

Is P_4S_3 just a curiosity chemical with a tongue-twister name? No, indeed. It is the chemical on the head of a strike-anywhere match that is ignited through frictional heat. P_4S_3 does not spontaneously ignite at temperatures below 100 °C.

Binary compound nomenclature activity

Table 2.3 Numeric Prefixes in Names of Binary Molecular Compounds

Number of Atoms	Prefix	Examples[a]
1	mono	NO nitrogen monoxide
2	di	NO_2 nitrogen dioxide
3	tri	N_2O_3 dinitrogen trioxide
4	tetra	N_2O_4 dinitrogen tetroxide
5	penta	N_2O_5 dinitrogen pentoxide
6	hexa	SF_6 sulfur hexafluoride
7	hepta	IF_7 iodine heptafluoride
8	octa	P_4O_8 tetraphosphorus octoxide
9	nona	P_4S_9 tetraphosphorus nonasulfide
10	deca	As_4O_{10} tetraarsenic decoxide

[a] When the prefix ends in "a" or "o" and the element name begins with "a" or "o," the final vowel of the prefix is usually dropped for ease of pronunciation. For example, nitrogen *mon*oxide and not nitrogen *mono*oxide, and dinitrogen *tetr*oxide, not dinitrogen *tetra*oxide.

Example 2.6

Write the molecular formula and name of a compound for which each molecule contains six oxygen atoms and four phosphorus atoms.

STRATEGY

We need to write the chemical symbols of the two elements and use the stated numbers of atoms as subscripts following the symbols. We then determine which element to place first in the molecular formula.

SOLUTION

We represent the six atoms of oxygen as O_6 and the four atoms of phosphorus as P_4. According to the scheme in Figure 2.9, the element O is followed only by F. Phosphorus therefore comes first in the formula; we write P_4O_6.

The name of a compound with four (tetra-) P atoms and six (hexa-) oxygen atoms in its molecules is *tetra*phosphorus *hex*oxide.

EXERCISE 2.6A

Write the molecular formula and name of a compound for which each molecule contains four fluorine atoms and two nitrogen atoms.

EXERCISE 2.6B

Write the molecular formula and name of a compound the molecules of which each contain one oxygen atom and eight sulfur atoms.

Example 2.7

Write **(a)** the molecular formula of phosphorus pentachloride and **(b)** the name of S_2F_{10}.

SOLUTION

(a) *Choosing which element symbol goes first.* The order of elements shown in the molecular formula must be the same as the order in the name.
Writing subscripts. The lack of a prefix on *phosphorus* signifies one P atom per molecule. The prefix *penta-* indicates five chlorine atoms. The molecular formula is PCl_5.

(b) The subscripts indicate two (*di-*) sulfur atoms and ten (*deca-*) fluorine atoms. The compound is disulfur decafluoride.

EXERCISE 2.7A

Write **(a)** the molecular formula of tetraphosphorus decoxide and **(b)** the name of S_7O_2.

EXERCISE 2.7B

Write a plausible molecular formula for a compound that has one sulfur atom, two oxygen atoms, and two fluorine atoms in each of its molecules. Comment on any ambiguity that exists in this case.

2.7   Ions and Ionic Compounds

In an isolated atom, the number of protons equals the number of electrons, and the atom is therefore electrically neutral. In some chemical reactions, however, an individual atom or a group of bonded atoms may lose or gain one or more electrons, thereby acquiring a net electric charge and becoming an **ion.** Ions are formed *only* through the loss or gain of electrons; there is *no* change in the number of protons in the nucleus of the atom(s). If electrons are *lost,* there are more protons than electrons in the resulting ion and so it has a *positive* charge. If electrons are *gained,* there are more electrons than protons in the resulting ion and so it has a *negative* charge.

Ions differ greatly from the atoms from which they are formed. Sodium *atoms* make up an element that, although quite reactive, can exist individually. Sodium *ions* are generally unreactive, and they can't exist alone. They can occur only in the presence of negatively charged ions.

Monatomic ions (also called simple ions) are formed when a single atom loses or gains one or more electrons. If a sodium atom loses one electron, it acquires a charge of 1+ and is represented as Na^+. If a chlorine atom gains one electron, it acquires a charge of 1− and is written as Cl^-. Positively charged ions are called **cations,** and negatively charged ions are called **anions.** In general, metal atoms form cations and nonmetal atoms form anions. Some groupings of bonded atoms can also lose or gain electrons to form *polyatomic ions.* For example, when the group of bonded atoms consisting of one sulfur atom and four oxygen atoms gains two electrons, the resulting ion is $SO_4{}^{2-}$.

Oppositely charged ions (cations and anions) attract one another and are held together in huge clusters by electrostatic attractions. Such clusters are called **ionic compounds.** In ionic compounds, there are no identifiable smallest entities comparable to the molecules of a molecular compound.

Monatomic Ions

We can use the periodic table to predict the charges on some **monatomic** ions (Figure 2.10). For metal atoms in the A groups, the number of electrons given up is usually equal to the periodic table group number. Thus, the highly reactive metal atoms of group 1A—the *alkali metals*—give up one electron to form cations with charge 1+. Group 2A atoms—the *alkaline earth metals*—give up two electrons to form cations with 2+ charge. Aluminum, the most common and commercially important metal in group 3A, forms cations with 3+ charge.

To name monatomic cations, we add the word *ion* to the name of the parent element: sodium ion, magnesium ion, and so on. In writing chemical symbols of cations, we write the ionic charge as an Arabic numeral followed by a plus sign. For a 1+ charge, however, we usually omit the 1 and write just +. Thus, Mg^{2+} and Al^{3+}, but Na^+ rather than Na^{1+}.

The periodic table is less useful in predicting the most likely charge on cations formed by B-group elements. In a few cases, the magnitude of the charge is equal to the group number, but in most cases it is not. Moreover, you will notice from Figure 2.10 that some B-group elements form ions with different charges, such as Fe^{2+} and Fe^{3+}. We must give these two ions different names, and we do so by using Roman numerals to indicate charge, such as *iron(II) ion* for Fe^{2+} and *iron(III) ion* for Fe^{3+}. In an older system of nomenclature, still encountered occasionally, the names of certain ions combine Latin stems and the endings *-ous* and *-ic*. The *-ous* indicates the lower of two possible charges on the ion, and the *-ic* indicates the higher charge. In this

1A	2A											3A	4A	5A	6A	7A	8A
Li^+														N^{3-}	O^{2-}	F^-	
Na^+	Mg^{2+}						8B					Al^{3+}		P^{3-}	S^{2-}	Cl^-	
		3B	4B	5B	6B	7B			1B	2B							
K^+	Ca^{2+}	Sc^{3+}	Ti^{2+} Ti^{4+}	V^{2+} V^{3+}	Cr^{2+} Cr^{3+}	Mn^{2+} Mn^{4+}	Fe^{2+} Fe^{3+}	Co^{2+} Co^{3+}	Ni^{2+}	Cu^+ Cu^{2+}	Zn^{2+}				Se^{2-}	Br^-	
Rb^+	Sr^{2+}									Ag^+	Cd^{2+}		Sn^{2+}			I^-	
Cs^+	Ba^{2+}									Au^+ Au^{3+}			Pb^{2+}				

▲ **FIGURE 2.10   Symbols and periodic table locations of some monatomic ions**

Three general observations can be made of these data: (1) Aluminum and the metals of groups 1A and 2A form just one cation, which carries a positive charge equal in magnitude to the A-group number. (2) Most of the metals of the B groups form two or more cations of different charges, though in some cases only one of these cations is commonly encountered. (3) The nonmetals of groups 7A and 6A, along with nitrogen and phosphorus of group 5A, form anions that have a charge equal to the group number minus 8.

system, Fe^{2+} is called *ferrous ion* and Fe^{3+} is *ferric ion*. For example, ferrous sulfate is listed as the active ingredient in a mineral supplement pill.

When they combine with metal atoms, nonmetal atoms generally gain electrons to form anions with a charge equal to "the periodic table group number minus 8." The nonmetal atoms of group 7A—the *halogens*—gain one electron to form anions with charge -1: $(7 - 8 = -1)$. Two examples are F^- and Cl^-. Group 6A atoms gain two electrons to form anions such as O^{2-} and S^{2-} $(6 - 8 = -2)$. Nitrogen and phosphorus, in group 5A, form the anions N^{3-} and P^{3-}, respectively. To name monatomic anions, we use an *-ide* ending on the name of the parent element and add the word *ion*. When a chlor*ine* atom gains one electron, it becomes a chlor*ide* ion. An ox*ygen* atom, by gaining two electrons, becomes an ox*ide* ion.

Formulas and Names for Binary Ionic Compounds

Binary (two-element) ionic compounds are assemblages of enormous numbers of monatomic cations and anions. These assemblages of ions must be electrically neutral—that is, the ionic compounds must have no net charge, neither positive nor negative. This requirement dictates how we write chemical formulas of ionic compounds.

Because ionic compounds do not consist of discrete molecules, we base the chemical formula on the simplest collection of cations and anions that represents an electrically neutral unit. We call this *hypothetical* collection of ions a **formula unit** of the compound. The formula unit is hypothetical because the assemblage it describes does not exist as a separate entity. Figure 2.11 shows the arrangement of a few dozen representative ions in the binary ionic compound sodium chloride, and from that drawing it is not hard to imagine that the formula unit consists of just one Na^+ ion and one Cl^- ion. The formula of sodium chloride is therefore NaCl. The name sodium chloride highlights a general rule: *The name of an ionic compound is a combination of two words—the name of the cation followed by that of the anion.*

Now consider the formula for aluminum oxide. We cannot simply combine one Al^{3+} and one O^{2-}, because this would produce a formula unit with a net charge of $1+$. The combination of *two* Al^{3+} ions and *three* O^{2-} ions, though, is an electrically neutral formula unit:

$$2(3+) + 3(2-) = +6 - 6 = 0$$

Therefore, the formula for aluminum oxide is Al_2O_3.

Why do we call Al_2O_3 aluminum oxide instead of *di*aluminum *tri*oxide? It is simply because we don't use prefixes if we don't need them. If we know the charges on the cations and anions, we can always figure out how many cations and anions are needed to produce an electrically neutral formula unit. Then we can name the formula unit in terms of the cations and anions present, regardless of their relative numbers. You should generally find it easy to write the names and formulas of compounds formed by combinations of the ions listed in Figure 2.10.

NaCl 3D model

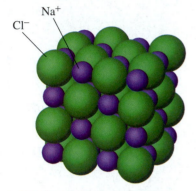

Na^+
Cl^-

▲ **FIGURE 2.11** **Sodium Chloride** The ionic compound sodium chloride consists of Na^+ and Cl^- ions held together by electrostatic attractions in an extensive, ordered network called a *crystal*. The hypothetical combination of the one Na^+ and one Cl^- ion indicated is a formula unit. It is the smallest collection of ions from which we can deduce the formula NaCl.

QUESTION: Could this model be used to represent a possible structure of potassium bromide? Magnesium iodide?

Because of the way we define a formula unit, the formula of an ionic compound is always the simplest formula possible, that is, an empirical formula.

Example 2.8

Determine the formula for **(a)** calcium chloride and **(b)** magnesium oxide.

SOLUTION

(a) First we write the symbols for the ions, with the cation first: Ca^{2+} and Cl^-. The simplest combination of these ions that gives an electrically neutral formula unit is *one* Ca^{2+} ion for every *two* Cl^- ions. The formula is $CaCl_2$.

$$Ca^{2+} + 2\,Cl^- = CaCl_2$$

(b) Figure 2.10 tells us that the ions are Mg^{2+} and O^{2-}. The simplest ratio for an electrically neutral formula unit is $1:1$. The formula of this binary ionic compound is MgO.

$$Mg^{2+} + O^{2-} = MgO$$

EXERCISE 2.8A

Give the formula for each of the following ionic compounds:

(a) potassium sulfide **(b)** lithium oxide **(c)** aluminum fluoride

EXERCISE 2.8B

Give the formula for each of the following ionic compounds:

(a) chromium(III) oxide **(b)** iron(II) sulfide **(c)** lithium nitride

Example 2.9

What are the names of **(a)** MgS and **(b)** $CrCl_3$?

SOLUTION

(a) MgS is made up of Mg^{2+} and S^{2-} ions. Its name is magnesium sulfide.

(b) From Figure 2.10 we see that there are two simple ions of chromium, Cr^{3+} and Cr^{2+}. Because there are three Cl^- ions in the formula unit $CrCl_3$, the cation must have a charge of 3+, that is, it must be Cr^{3+}. Because there are two chromium cations, there are two chlorides, $CrCl_2$ and $CrCl_3$, and we must assign a different name to each. Therefore, the name of our compound cannot be simply chromium chloride; instead, to indicate the 3+ cation, we say the name is chromium(III) chloride.

EXERCISE 2.9A

Name the following compounds:

(a) $CaBr_2$ **(b)** Li_2S **(c)** $FeBr_2$ **(d)** CuI

EXERCISE 2.9B

Write the name and formula for each of the following compounds:

(a) the sulfide of copper(I) **(b)** the oxide of cobalt(III) **(c)** the nitride of magnesium

Polyatomic Ions

As noted earlier, a **polyatomic ion** is a charged group of bonded atoms—for example, NH_4^+ or SO_4^{2-}. Table 2.4 lists some of the more common polyatomic ions. Notice that polyatomic anions are more common than cations, that the suffixes *-ite* and *-ate* are frequently used, and that the prefixes *hypo-* and *per-* are occasionally used. For any given element, an anion with a name ending in *-ite* generally has one fewer oxygen atom and the same charge as the anion with a name ending in *-ate*. If an element has more than two polyatomic anions, the prefix *hypo-* represents one fewer oxygen atom than the *-ite* anion and the prefix *per-* represents one more oxygen atom than the *-ate* anion. Here we illustrate the scheme, using the oxo (oxygen-containing) anions of chlorine and sulfur:

		Example	Name	Example	Name
Increasing ↓	hypo__ite	ClO^-	hypochlorite ion	—	—
number of ↓	__ite	ClO_2^-	chlorite ion	SO_3^{2-}	sulfite ion
oxygen ↓	__ate	ClO_3^-	chlorate ion	SO_4^{2-}	sulfate ion
atoms ↓	per__ate	ClO_4^-	perchlorate ion	—	—

If a polyatomic anion has hydrogen as a third element, we indicate its presence in the name: HPO_4^{2-} is the *hydrogen* phosphate ion, and $H_2PO_4^-$ is the *dihydrogen* phosphate ion.

We can write formulas and names for compounds containing polyatomic ions by combining information from Figure 2.10 and Table 2.4. Note that we have to use parentheses in some of these formulas. Consider a formula unit of magnesium nitrate,

Table 2.4 Some Common Polyatomic Ions

Name	Formula	Typical Compound
Cation		
Ammonium ion	NH_4^+	NH_4Cl
Anions		
Acetate ion	$^aC_2H_3O_2^-$	$NaC_2H_3O_2$
Carbonate ion	CO_3^{2-}	Li_2CO_3
Hydrogen carbonate ion (or bicarbonate ion)[b]	HCO_3^-	$NaHCO_3$
Hypochlorite ion	ClO^-	$Ca(ClO)_2$
Chlorite ion	ClO_2^-	$NaClO_2$
Chlorate ion	ClO_3^-	$NaClO_3$
Perchlorate ion	ClO_4^-	$KClO_4$
Chromate ion	CrO_4^{2-}	K_2CrO_4
Dichromate ion	$Cr_2O_7^{2-}$	$(NH_4)_2Cr_2O_7$
Cyanate ion	OCN^-	$KOCN$
Thiocyanate ion[c]	SCN^-	$KSCN$
Cyanide ion	CN^-	KCN
Hydroxide ion	OH^-	$NaOH$
Nitrite ion	NO_2^-	$NaNO_2$
Nitrate ion	NO_3^-	$NaNO_3$
Oxalate ion	$C_2O_4^{2-}$	CaC_2O_4
Permanganate ion	MnO_4^-	$KMnO_4$
Phosphate ion	PO_4^{3-}	Na_3PO_4
Hydrogen phosphate ion	HPO_4^{2-}	Na_2HPO_4
Dihydrogen phosphate ion	$H_2PO_4^-$	NaH_2PO_4
Sulfite ion	SO_3^{2-}	Na_2SO_3
Hydrogen sulfite ion (or bisulfite ion)[b]	HSO_3^-	$NaHSO_3$
Sulfate ion	SO_4^{2-}	Na_2SO_4
Hydrogen sulfate ion (or bisulfate ion)[b]	HSO_4^-	$NaHSO_4$
Thiosulfate ion[c]	$S_2O_3^{2-}$	$Na_2S_2O_3$

[a] The acetate ion is also represented as CH_3COO^-.

[b] The prefix "bi-" means that the ion contains a replaceable H atom. This should not be confused with the prefix "di-," which means two (usually used to represent a doubling of a simpler unit).

[c] The prefix "thio-" means that a sulfur atom has replaced an oxygen atom.

Charge Balancing in Binary Compounds activity

which consists of one Mg^{2+} ion and two NO_3^- ions. We can't simply write a subscript "2" following NO_3 because the result would be $MgNO_{32}$. If we write MgN_2O_6, we would fail to make clear the presence of nitrate ions. Rather, we enclose NO_3 in parentheses, followed by the subscript 2: $(NO_3)_2$. The formula for magnesium nitrate is therefore $Mg(NO_3)_2$.

Example 2.10

Write the formula for (**a**) sodium sulfite and (**b**) ammonium sulfate.

SOLUTION

(**a**) It is possible to identify the sulfite ion without memorizing all the ions in Table 2.4. If you remember the name and formula of one of the sulfur–oxygen polyatomic anions, you should be able to deduce the names of others. Suppose you remember that sulf*ate* is SO_4^{2-}. The -*ite* anion has one fewer oxygen atom, 3 instead of 4, and so it is SO_3^{2-}. The charges of the two ions in a formula unit must balance, which means the formula unit of sodium sulfite must have Na^+ and SO_3^{2-} in the ratio 2 : 1. The formula is therefore Na_2SO_3.

(**b**) The ammonium ion is NH_4^+, and the sulfate ion is SO_4^{2-}. A formula unit of ammonium sulfate has two NH_4^+ ions and one SO_4^{2-}. ion. To represent the two NH_4^+ ions, we place parentheses around the NH_4, followed by a subscript 2, $(NH_4)_2$, and thus arrive at the formula $(NH_4)_2SO_4$.

EXERCISE 2.10A

What is the formula for **(a)** ammonium carbonate, **(b)** calcium hypochlorite, and **(c)** chromium(III) sulfate?

EXERCISE 2.10B

Write a plausible formula for

(a) potassium aluminum sulfate   **(b)** magnesium ammonium phosphate

Example 2.11

What is the name of **(a)** $NaCN$ and **(b)** $Mg(ClO_4)_2$?

SOLUTION

(a) The ions in this compound are Na^+, sodium ion, and CN^-, cyanide ion (see Table 2.4). The name of the compound is sodium cyanide.

(b) The ions present are Mg^{2+}, magnesium ion, and ClO_4^-, perchlorate ion. The name of the compound is magnesium perchlorate.

EXERCISE 2.11A

Name each of the following compounds:

(a) $KHCO_3$ **(b)** $FePO_4$ **(c)** $Mg(H_2PO_4)_2$

EXERCISE 2.11B

Give a plausible name for the following:

(a) Na_2SeO_4 **(b)** $FeAs$ **(c)** Na_2HPO_3

Hydrates

If you scan the labels in a chemical storeroom, you are likely to find some formulas that differ from the ones we have written. For example, a bottle of calcium chloride may carry the label $CaCl_2 \cdot 6\ H_2O$ instead of $CaCl_2$. This label indicates that the substance is a *hydrate*. A **hydrate** is an ionic compound in which the formula unit includes a fixed number of water molecules associated with the cations and anions. The hydrate $CaCl_2 \cdot 6\ H_2O$ is called calcium chloride *hexa*hydrate. *Calcium chloride* pertains to the Ca^{2+} ion and two Cl^- ions in the formula unit, *hydrate* signifies the inclusion of H_2O molecules in the formula unit, and *hexa-* indicates that there are *six* of them (recall Table 2.3). Other examples of hydrates are barium chloride *di*hydrate, $BaCl_2 \cdot 2\ H_2O$; lithium perchlorate *tri*hydrate, $LiClO_4 \cdot 3\ H_2O$; and magnesium carbonate *penta*hydrate, $MgCO_3 \cdot 5\ H_2O$:

A center dot in a formula denotes a compound made up of simpler compounds. That is, $CaCl_2 \cdot 6\ H_2O$ signifies that $CaCl_2$ and H_2O are both capable of independent existence. Basic copper carbonate (Figure 2.1) is made up of two simpler compounds. Its formula is $CuCO_3 \cdot Cu(OH)_2$.

Adding water to white anhydrous copper(II) sulfate produces brilliant blue copper(II) sulfate pentahydrate.

Copper(II) sulfate pentahydrate ($CuSO_4 \cdot 5\ H_2O$) is a brilliant blue, whereas anhydrous copper(II) sulfate ($CuSO_4$) is white.

Cations, Anions, and the Human Body

Dietary minerals are mainly inorganic ions. They are essential nutrients that must be obtained from the diet. Following is a list of some of the more important ions in the cells and fluids of our bodies:

- Sodium ions (Na^+) are the principal cations found outside cells in the body. They help regulate and control the level of body fluids. Too little Na^+ leads to diarrhea, anxiety, a decrease in body fluids, and circulatory failure. However, most people have the opposite problem—too much sodium ion—ingested mainly as table salt and salty snack foods. Too much Na^+ increases water retention, leading to high blood pressure (hypertension). About 50 million people in the United States suffer from hypertension. Uncontrolled hypertension can lead to stroke, heart attack, kidney failure, or heart failure. Antihypertensives are among the most prescribed drugs in the United States.

- Potassium ions (K^+) are the principal cations found inside cells in the body. Bananas, orange juice, and potatoes are good sources of K^+. Potassium ions help regulate cellular functions, including nerve impulses and heartbeats, and the level of body fluids.

- Chloride ions (Cl^-) are the principal anions found outside cells in the body. They serve as counterions (ions necessary to balance electrical charge) for Na^+ in the extracellular fluid and for H^+ in gastric juice. Like Na^+, chloride ions are ingested mainly as table salt. Like sodium and potassium ions, chloride ions are involved in maintaining acid–base and fluid balances. It is difficult to separate the effect of too much Cl^- from that of too much Na^+; both seem to be involved in hypertension. Too little dietary Cl^- is rare, but it can result from heavy sweating, chronic diarrhea, and vomiting.

- Calcium ions (Ca^{2+}) occur mainly in the skeleton and account for 1.5–2% of body mass. Ca^{2+} is therefore essential for building and maintaining bones and teeth. Also, Ca^{2+} plays a crucial role in blood clotting, muscle contraction, and the transmission of nerve signals to cells. An adequate supply of Ca^{2+} is especially important during pregnancy and in growing children. It helps to prevent osteoporosis in older people. Good sources of calcium are milk and other dairy products, nuts, and legumes.

- Magnesium ions (Mg^{2+}), like Ca^{2+}, are found mainly in the bones, but they are also vital components of many enzymes, which are substances our bodies need in order to release energy from the food we eat. Good sources of Mg^{2+} are green vegetables (Mg^{2+} is a component of the chlorophyll in all green plants), milk, bread, cereals, and potatoes.

- Phosphate ions exist mainly as $H_2PO_4^-$ and HPO_4^{2-} in body fluids. About 85% of the phosphorus-containing ions in the body are in the bones, where they act as the counterions for Ca^{2+}. Also, they play an important role in energy production from food. Good sources of phosphate are milk and other dairy products, cereals, and meat.

- In addition to the above ions, the body needs smaller amounts of ions found in trace minerals. These include the ions iron(II), chromium(III), copper(II), zinc, fluoride, iodide, and bicarbonate, as well as the hydrogen ion. Also required are compounds of manganese, molybdenum, and selenium, although these are not necessarily in the form of simple ions. These trace minerals play a variety of roles, several of which are discussed in other chapters.

Knowledge of ions is important not only to your success in a chemistry course but also to an understanding of many critical life processes.

Application Note

Anhydrous calcium chloride, $CaCl_2$, is a good *desiccant*—a drying agent. In a tightly closed container, it removes much of the water vapor from the air and is successively converted to the hydrates $CaCl_2 \cdot H_2O$, $CaCl_2 \cdot 2\,H_2O$, and $CaCl_2 \cdot 6\,H_2O$. This taking up of water provides an environment in which wet substances can be dried without heating.

Introduction to Acids movie

Introduction to Bases movie

Some *anhydrous* (nonhydrated) compounds form hydrates when exposed to atmospheric water vapor. More commonly, hydrates are obtained when certain ionic compounds are derived from their solutions in water. Although hydrates are common, by no means do all ionic compounds form them, and there is no need to try to learn which ones do. You need only to recognize a hydrate formula when you see one.

2.8   Acids, Bases, and Salts

When we digest food, shed tears, bake bread, or take medicine for an upset stomach, we participate in complicated chemical processes. Central to many of these processes are two special kinds of compounds called acids and bases. We eat them and drink them, and our bodies produce them. Some common acids and bases used around the home are shown in Figure 2.12.

Historically, acids and bases have been classified according to some distinctive properties. **Acids** are substances that have the following characteristics when dissolved in water:

- Acids taste sour if diluted with enough water to be tasted safely.
- Acids produce a pricking or stinging sensation on the skin.

- Acids turn the color of litmus, an indicator dye, from blue to red.
- Acids react with many metals to produce ionic compounds and hydrogen gas.
- Acids also react with bases, thereby losing their acidic properties.

Bases are substances that have the following characteristics when dissolved in water:

- Bases taste bitter if diluted with enough water to be tasted safely.*
- Bases feel slippery or soapy on the skin.
- Bases turn the color of the indicator dye litmus from red to blue.
- Bases react with acids, thereby losing their basic properties.

Acids and Bases: The Arrhenius Concept

In 1887, the Swedish chemist Svante Arrhenius proposed that an **acid** is a molecular compound (such as HBr) that ionizes, or breaks up, in water to form a solution containing H^+ cations plus anions (such as Br^-). He viewed a **base** as a compound that ionizes in water to form a solution containing OH^- anions and related cations. Some bases, such as NaOH and KOH, are ionic compounds. When they dissolve in water, the hydroxide ions and cations *dissociate* (separate) from one another. Most bases, however, are not ionic compounds, nor do they contain hydroxide ions. The hydroxide ions and cations are formed in a reaction between the base and water, as we will see in later chapters.

Arrhenius proposed that the essential reaction between an acid and a base is the combination of H^+ ions from the acid and OH^- ions from the base to form water (that is, HOH, or H_2O). This reaction is called *neutralization*. The cation from the base and the anion from the acid make up an ionic compound called a **salt.** For example, the reaction between HCl (an acid) and NaOH (a base) to form NaCl (a salt) and H_2O (water) is a neutralization reaction.

Like many other scientific theories, the Arrhenius theory has been supplanted by newer ones that better explain all the available data. We will study more modern acid–base theories in Chapter 15, but for now, this older Arrhenius theory will help us to identify acids and bases and to write their names and formulas.

Formulas and Names of Acids, Bases, and Salts

Let's begin with the simplest case, salts. Because salts are ionic compounds, we write their formulas and names just as we do for ionic compounds in general. The only new idea is that salts are ionic compounds formed in the reaction between an acid and a base.

Next, consider *ionic* bases. We also write their formulas and names like those of other ionic compounds. In Arrhenius bases, however, the anions are always hydroxide ions, OH^-. The principal ionic bases are those of the group 1A and 2A cations. The three most common are

NaOH	sodium hydroxide
KOH	potassium hydroxide
$Ca(OH)_2$	calcium hydroxide

Most bases are *molecular* compounds and do not contain hydroxide ion. Rather, hydroxide ions are produced when these molecular bases react with water. The principal molecular bases are ammonia (NH_3) and compounds related to ammonia.

The naming of acids is a bit more complex but still not difficult if we take a systematic approach. Let's first consider *binary* acids. These are certain molecular compounds in which hydrogen is combined with a second nonmetallic element. We noted earlier that HCl is hydrogen chloride, and we will continue to use this name for the pure *gaseous compound*. When hydrogen chloride dissolves in and reacts with water, however, H^+ and Cl^- ions are formed. We call this *solution* an acid, and we name it by

*According to the biblical account, the Israelites, in their journey from Egypt to Canaan, came upon the bitter waters at Marah (Exodus 15:23). Although the writers may not have meant to, they thus recorded for posterity the existence of a base.

(a)

(b)

▲ **FIGURE 2.12   Some common acids and bases**

(a) Some common acids: toilet bowl cleaner, vinegar, aspirin, tomato juice, other vegetable and fruit juices. (b) Some common bases: ammonia; drain cleaner, antacid tablets, baking soda, washing soda, oven cleaners.

▲ In addition to his work on acids and bases, Svante Arrhenius (1859–1927) derived a mathematical expression that relates reaction rates to several factors (Chapter 13). He was also the first to relate carbon dioxide in the atmosphere to the greenhouse effect (Chapter 25).

changing *hydrogen* to *hydro* and *chloride* to *chloric*. The name is hydrochloric acid. A few other examples are as follows:

Pure Gaseous Compound	Formula	Dissolved in Water
Hydrogen bromide	HBr	hydrobromic acid
Hydrogen iodide	HI	hydroiodic acid
Hydrogen sulfide	H_2S	hydrosulfuric acid

Not all binary compounds of hydrogen and a nonmetal are acids by any means—only those that ionize in water to produce H^+ cations and anions. Methane, CH_4, is a binary compound of hydrogen, and it is not an acid. We will learn to recognize acids later on. For now, our emphasis is on how to write formulas and names for substances identified as acids.

As Table 2.5 suggests, most acids are *ternary* acids, which means their molecules are made up of atoms of *three* elements: hydrogen and two other nonmetals. Many of the ternary acids are *oxoacids*, which have atoms of hydrogen, *oxygen*, and a third nonmetal in their molecules. Some nonmetals form a set of ternary oxoacids. We base the names of the members of the set on the number of oxygen atoms per molecule, as illustrated here for ternary oxoacids of chlorine and sulfur:

		Example	Name		Example	Name
Increasing	hypo__ous	HClO	hypochlorous acid		—	—
number of	__ous	$HClO_2$	chlorous acid		H_2SO_3	sulfurous acid
oxygen atoms	__ic	$HClO_3$	chloric acid		H_2SO_4	sulfuric acid
	per__ic	$HClO_4$	perchloric acid		—	—

To name the salts of these acids, we change the acid name endings as follows:

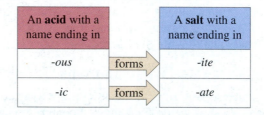

An **acid** with a name ending in		A **salt** with a name ending in
-ous	forms	-ite
-ic	forms	-ate

Table 2.5 Formulas and Names of Some Common Acids and Their Salts

		Sodium Salt	
Formula of Acid	Name of Acid	Formula	Name
HCl	*Hydro*chlor*ic* acid	NaCl	Sodium chlor*ide*
HClO	*Hypo*chlor*ous* acid	NaClO	Sodium *hypo*chlor*ite*
$HClO_2$	Chlor*ous* acid	$NaClO_2$	Sodium chlor*ite*
$HClO_3$	Chlor*ic* acid	$NaClO_3$	Sodium chlor*ate*
$HClO_4$	*Per*chlor*ic* acid	$NaClO_4$	Sodium *per*chlor*ate*
H_2S	*Hydro*sulfur*ic* acid	Na_2S	Sodium sulf*ide*
H_2SO_3[a]	Sulfur*ous* acid	Na_2SO_3	Sodium sulf*ite*
H_2SO_4[a]	Sulfur*ic* acid	Na_2SO_4	Sodium sulf*ate*
HNO_2	Nitr*ous* acid	$NaNO_2$	Sodium nitr*ite*
HNO_3	Nitr*ic* acid	$NaNO_3$	Sodium nitr*ate*
H_3PO_4[a]	Phosphor*ic* acid	Na_3PO_4	Sodium phosph*ate*
H_2CO_3[a]	Carbon*ic* acid	Na_2CO_3	Sodium carbon*ate*

[a] Table 2.4 lists anions found in some salts of these acids in which not all of the available H atoms are replaced. If one or more H atoms remains unreplaced, formulas and names must be written accordingly; for example, $NaHSO_4$ is sodium hydrogen sulfate and NaH_2PO_4 is sodium dihydrogen phosphate.

As a result, we obtain the same names as on page 54: NaClO is sodium *hypochlorite*, $NaClO_2$ is sodium *chlorite*, $NaClO_3$ is sodium *chlorate*, and $NaClO_4$ is sodium *perchlorate*.

Finally, with a few exceptions—such as carbonic acid, H_2CO_3—the naming of acids containing carbon differs somewhat from that for other acids. We will categorize organic acids in a different way in the next section.

2.9 Organic Compounds

We now turn our attention to a category of compounds so vast and important that its study constitutes an entire branch of chemistry—*organic chemistry*. Organic compounds, as noted at the beginning of this chapter, are based on the element carbon. We briefly introduce organic chemistry now for several reasons:

- The vast majority of the tens of millions of known compounds are organic compounds.

- Many familiar substances are organic compounds, such as the fuels propane and butane; the beverage ingredient ethyl alcohol; and indigo, the dye in blue jeans. Additional familiar organic materials are the gasoline that powers our cars, the fuels that heat our homes, synthetic substances such as nylon, vinyl plastics, acetylsalicylic acid (aspirin), and a host of other medicines. Carbon-based compounds, such as carbohydrates, fats, and proteins, are vital components of our bodies—indeed of all living creatures.

- Although the labels "inorganic chemistry" and "organic chemistry" suggest two different disciplines, chemistry is a seamless whole. The same physical laws apply to organic compounds as to inorganic compounds.

- All told, the diversity of organic compounds is astonishing. In this section, we present a few of the details that will help you recognize some simple organic compounds and their structures. Later in this text, as more organic compounds are introduced and discussed, we will use a combination of names, molecular formulas, and models to describe them. Three common classes of organic compounds are described in this way in this section: alkanes, alcohols, and carboxylic acids.

Wider coverage of the structures and names of these and other organic compounds is provided in Chapter 23 and Appendix D.

The rich red, yellow, and orange colors of watermelons, tomatoes, carrots, and other vegetables and fruits are due to naturally occurring organic compounds. These same compounds are present whether the crops are grown "organically" or by conventional farming.

Alkanes

The simplest organic compounds, the **hydrocarbons,** contain only carbon and hydrogen atoms. There are several types of hydrocarbons, and we will start with a discussion of the variety known as alkanes. **Alkanes** are also called *saturated hydrocarbons* because their molecules contain the maximum number of hydrogen atoms possible for the number of carbon atoms; in other words, the molecules are *saturated* with H atoms. *Unsaturated hydrocarbons,* which have molecules with fewer than the maximum number of hydrogen atoms possible for the number of carbon atoms, are discussed in Chapter 9.

The three simplest alkanes are shown in Figure 2.13. *Methane* is the principal component of natural gas. In methane, four hydrogen atoms are attached to a central carbon atom. The next member of the alkane series is *ethane*, a minor component of natural gas. In this molecule, two carbon atoms are joined together and three hydrogen atoms are attached to each carbon atom. *Propane,* the familiar bottled gas used as a fuel in portable torches, gas grills, and stoves, is the third member of the alkane series.

Perhaps you see a pattern emerging: Each member of the alkane series differs from the preceding member by a CH_2 unit, that is, one carbon atom and two hydrogen atoms. All alkanes conform to the general formula C_nH_{2n+2}, where *n* is the number of carbon atoms in the molecule. For propane, for instance, with its three C atoms, the molecular formula is $C_3H_{(2\times3)+2} = C_3H_8$.

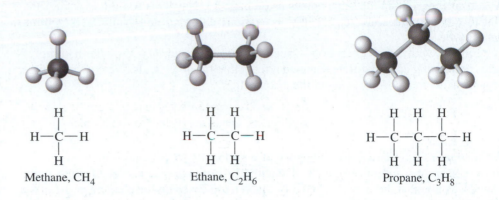

▶ **FIGURE 2.13** **Alkanes**
Methane, ethane, and propane are the first three members of the alkane series.

Methane, CH$_4$ Ethane, C$_2$H$_6$ Propane, C$_3$H$_8$

Table 2.6 Word Stems Indicating the Number of Carbon Atoms in Simple Organic Molecules

Stem	Number of C Atoms
meth-	1
eth-	2
prop-	3
but-	4
pent-	5
hex-	6
hept-	7
oct-	8
non-	9
dec-	10

The names of simple alkanes are composed of two parts. A word stem (Table 2.6) indicates the number of carbon atoms, and the ending *-ane* indicates that the hydrocarbon belongs to the alk*ane* family. Thus, C$_5$H$_{12}$ is *pent*ane, and C$_6$H$_{14}$ is *hex*ane. Note that for C$_5$ and beyond, the stem has the same root as the prefixes in Table 2.3.

In nearly all carbon compounds, each carbon atom forms *four* bonds and each hydrogen atom forms *one* bond. We can often decide whether a given structural formula is plausible simply by counting the number of bonds to each C and H atom. Carbon compounds can exist as continuous chains (sometimes called *straight chains*), as branched chains, and as ring structures.

When we consider a structural formula for C$_4$H$_{10}$, the fourth member of the alkane series, we find that there are *two* possibilities. One has four carbon atoms bonded together in a continuous chain; in the other, a —CH$_3$ group branches off a three-carbon chain.

Compounds that have the same molecular formula but different structural formulas are called **isomers.** When there are only a few isomers for a given molecular formula, we can use prefixes to give them different names. Here we use the names *butane* and *isobutane*. However, when large numbers of isomers are possible for a given molecular formula, we need to use a more systematic approach. (Imagine trying to learn 18 such individual names for the 18 isomeric alkanes with the formula C$_8$H$_{18}$.) Such a system is outlined in Appendix D.

Isomers are distinctly different compounds. Butane and isobutane are both gases at room temperature, but butane boils at about 0 °C and isobutane at −12 °C. Differences in the structures of isomers are illustrated most clearly by ball-and-stick models (Figure 2.14).

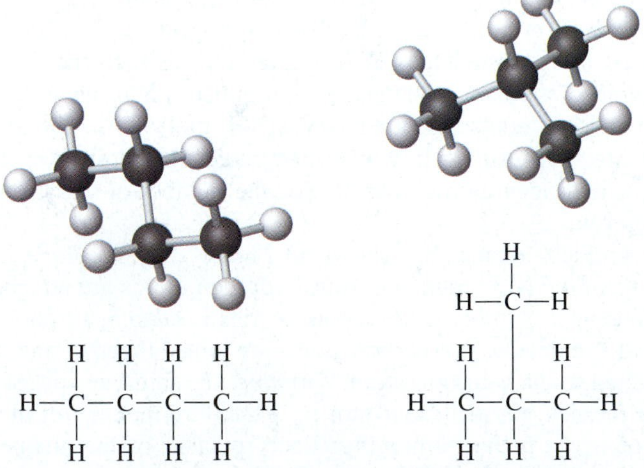

▶ **FIGURE 2.14** **Butane and isobutane**
Ball-and-stick and structural models illustrate the difference between the two isomers of butane.

Butane, C$_4$H$_{10}$ Isobutane, C$_4$H$_{10}$

On paper, we can use the structural formulas of the type shown in Figure 2.14 for butane and isobutane, or, to save space, we often use *condensed* structural formulas:

$$CH_3$$
$$|$$
$$CH_3CH_2CH_2CH_3 \qquad CH_3CHCH_3$$
$$\text{Butane} \qquad\qquad \text{Isobutane}$$

These can be further condensed and written on a single line by using parentheses to set off groups that branch from a longer chain, as in this representations of isobutane:

$$CH_3CH(CH_3)CH_3 \quad \text{or} \quad CH_3CH(CH_3)_2$$
$$\text{Isobutane}$$

The number of isomers increases rapidly with the number of carbon atoms: three C_5H_{12} alkanes, five C_6H_{14} alkanes, nine C_7H_{16} alkanes, eighteen C_8H_{18} alkanes, and so on. Only one compound in each set of isomers has straight-chain molecules; all the others have branched chains.

A hydrocarbon branch off either a carbon chain or a carbon ring is often called an *alkyl group*. An **alkyl group** is derived from an alkane by removing a hydrogen atom. The methyl group (CH_3—), for example, is derived from methane (CH_4), and the ethyl group (CH_3CH_2—) from ethane (CH_3CH_3). Propane yields two different alkyl groups, depending on whether the missing hydrogen atom is from the end or from the middle carbon atom:

$$CH_3CH_2CH_2— \qquad\qquad CH_3CHCH_3$$
$$|$$

$$\text{Propyl group} \qquad\qquad \text{Isopropyl group}$$

Alkyl groups are also found in many other simple organic molecules.

You can visualize a cyclic alkane molecule by imagining the two ends of a straight-chain alkane molecule coming together, the end C atoms each shedding one H atom, and the ends joining to each other. The result is a ringlike structure with the generic formula C_nH_{2n}. Alkane molecules having a ring structure are named with the prefix *cyclo-*. The smallest number of carbon atoms in a ring structure is three, as in the compound cyclopropane.

The cyclic alkane with a six-carbon-atom ring is cyclohexane. The leftmost representation in Figure 2.15 is a normal structural formula. The middle representation is a *line-angle formula* (also called a *stick figure*), in which the the lines denote bonds between atoms. By comparing this line-angle formula with the structural formula to its left, you can see that any C—H bonds are understood to be present and are not shown. Any intersection of two lines, as well as the end or beginning of a line, stands for a carbon atom. The ring C atoms are not labeled, and the H atoms are not shown. The ring of C atoms in these representations of cyclohexane appears to be flat, but the ball-and-stick model on the right in Figure 2.15 shows the three-dimensional shape.

It isn't critical that you be able to determine the number of different isomers for a formula, especially because that number increases rapidly with the number of carbon atoms. You *should*, however, be able to determine from their structures whether two compounds are isomers.

 **Organic Nomenclature activity**

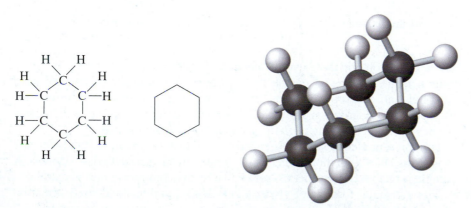

◀ **FIGURE 2.15 Cylcohexane**
The molecular and line-angle formulas for cyclohexane indicate the bonding of atoms within the molecule. The ball-and-stick model further indicates that all of the carbon atoms of the cyclohexane molecule do not lie in the same plane. Rather, it can assume several different arrangements or conformations. The most stable arrangement, shown here, is called the *chair conformation* because the six carbon atoms outline a structure that somewhat resembles a reclining chair.

QUESTION: What characteristic of the carbon atom, apparent in the ball-and-stick model, prevents the molecule from existing in a planar structure?

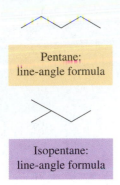

Pentane:
line-angle formula

Isopentane:
line-angle formula

Line-angle formulas can also be used for open-chain compounds. We assume in all cases that each carbon atom has enough H atoms to give it a total of four bonds. Thus, we can represent pentane, $CH_3CH_2CH_2CH_2CH_3$, and isopentane, $(CH_3)_2CHCH_2CH_3$, as shown in the margin.

Cyclic structures can have other atoms or groups of atoms attached to the ring. Methylcyclopropane has a three-carbon ring with a $-CH_3$ (methyl) group attached. It can be represented with either a partial line-angle formula—on the left here, where letters for the methyl group are combined with the line-angle formula for the ring—or a complete line-angle formula, as on the right:

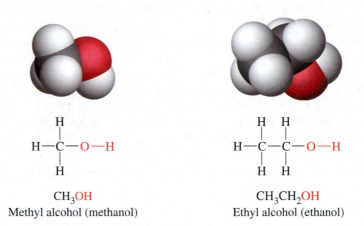

Methylcyclopropane

Alcohols

An important organizing principle in organic chemistry is the concept of the functional group. A **functional group** is an atom or group of atoms, attached to or inserted in a hydrocarbon chain or ring, that confers characteristic properties to the molecule as a whole. Many simple organic molecules are composed of two parts: a functional group, where most of the reactions of the molecule occur, and a hydrocarbon chain that usually is unreactive. Molecules with the same functional group generally have similar properties. We will briefly consider two common functional groups in this chapter, and we will introduce others in subsequent chapters. A list of some common functional groups is provided in Table D.1 of Appendix D.

The functional group common to **alcohols** is the *hydroxyl* group, $-OH$. The two simplest alcohols are shown in Figure 2.16. Alcohols are named in two ways. The common name is that of the alkyl group followed by the family name *alcohol*. Thus CH_3OH, consisting of a methyl group and a hydroxyl group, is methyl alcohol. In the systematic name, methanol, the stem *meth-* indicates a compound based on methane and the ending *-ol* signifies that this compound is an alcohol. The next higher alcohol, ethanol, is based on the two-carbon alkane, ethane.

Application Note

Methanol is widely used as a solvent, and because it burns more cleanly than gasoline, methanol is used in some fleet vehicles in smoggy areas. Methanol is fairly toxic. Ingestion of as little as 30 mL can cause blindness and even death. Ethanol is the alcohol in alcoholic beverages. It is considerably less toxic than methanol. (The lethal dose of ethanol, when rapidly ingested, is about 500 mL.) Like methanol, ethanol is widely used as a solvent. It is also used as a gasoline additive and, in some instances, even as a gasoline substitute.

Organic Functionality Nomenclature activity

▶ **FIGURE 2.16** **Alcohols**
Methanol and ethanol are the two simplest alcohols.

H
|
H—C—O—H
|
H

CH_3OH
Methyl alcohol (methanol)

H H
| |
H—C—C—O—H
| |
H H

CH_3CH_2OH
Ethyl alcohol (ethanol)

Application Note

Rubbing alcohol is a solution containing 70% isopropyl alcohol in water.

There are *two* isomeric three-carbon alcohols (Figure 2.17).

Common names are based on the three-carbon alkyl groups, so that we have propyl alcohol and isopropyl alcohol. In the systematic names, the prefix "1-" indicates that the OH group is on the first, or end, carbon atom, and the prefix "2-" indicates that the OH group is on the second carbon from the end.

We can represent alcohols by the general formula ROH, where R stands for any alkyl group. Molecules with more than one OH group are also classified as alcohols. A good example is one used in cosmetics and known by the common names glycerol and glycerin. The molecular formula of glycerol is $C_3H_8O_3$, and its condensed structural formula is $CH_2OHCHOHCH_2OH$; its space-filling model is shown in Figure 2.18.

The systematic name of glycerol is *1,2,3-propanetriol*. In this name, *propane* signifies a chain of three carbon atoms. The ending *-ol* indicates the presence of OH

H—C—C—C—O—H

(H atoms above and below each C)

CH₃CH₂CH₂OH
Propyl alcohol (1-propanol)

H—C—C—C—H

(with OH on middle carbon)

CH₃CHOHCH₃
Isopropyl alcohol (2-propanol)

▲ **FIGURE 2.17** **Propanol and isopropanol**
The alcohol containing three carbon atoms exists as two isomers.

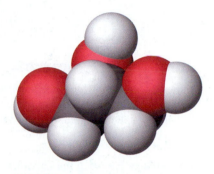

▲ **FIGURE 2.18** **The glycerol molecule**
Note that two of the hydrogen atoms are hidden in this model.

groups. *Tri*ol means that there are three of these groups, and the numbers "1,2,3" signify that they are attached to the first, second, and third carbon atoms of propane.

Carboxylic Acids

The functional group that most commonly confers acidic properties to an organic compound is the *carboxyl group:*

$$-\overset{\displaystyle O}{\overset{\|}{C}}-O-H \quad \text{or} \quad -COOH$$

The carbon atom in a carboxyl group forms a *double* bond to one of the oxygen atoms and a single bond to the other. In the condensed form (—COOH), the double bond to one of the oxygen atoms is understood. The presence of a carboxyl group in its molecules identifies a compound as a **carboxylic acid.** Carboxylic acids are often represented by the generic formula RCOOH. The two simplest carboxylic acids are shown in Figure 2.19.

The simplest carboxylic acid has one carbon atom per molecule. The common name, formic acid, is derived from the Latin *formica,* meaning "ant." An ant bite hurts in part because the ant injects formic acid when it bites. In the systematic name, *methan-* indicates one carbon atom and *-oic acid* tells us that the compound is a carboxylic acid. Just as with the alkanes and alcohols, there is a series of carboxylic acids. The two-carbon carboxylic acid has the common name acetic acid. Acetic acid is a component of vinegar. It is commonly used in chemistry laboratories, mainly to neutralize bases but also to make other organic compounds.

Application Note

Acetic acid can be made by the fermentation of cider or wine, producing vinegar, which is 4–10% acetic acid.

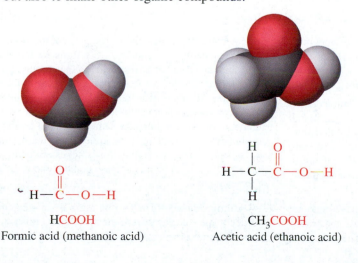

HCOOH
Formic acid (methanoic acid)

CH₃COOH
Acetic acid (ethanoic acid)

◄ **FIGURE 2.19** **Carboxylic acids**
Formic acid and acetic acid are the common names of the two simplest organic acids.

When a carboxylic acid dissolves in water, some of the hydrogen atoms of the carboxyl groups become H^+ ions. Hydrogen atoms attached to the *carbon* atoms of carboxylic acids do *not* split off as H^+. When an organic acid is neutralized with a base, a salt is produced, just as with an inorganic acid. Thus, acetic acid and sodium hydroxide react to form the salt sodium acetate, CH_3COONa, an ionic compound consisting of Na^+ and CH_3COO^- ions.

We will encounter carboxylic acids several times in later chapters, especially in Chapters 4 and 15.

Cumulative Example

Show that the following experiment is consistent with the law of conservation of mass (within the limits of experimental error): A 10.00-g sample of calcium carbonate was dissolved in 100.0 mL of hydrochloric acid solution ($d = 1.148$ g/mL). The products were 120.40 g of solution (a mixture of hydrochloric acid and calcium chloride) and 2.22 L of carbon dioxide gas ($d = 0.0019769$ g/mL).

STRATEGY

This problem may give the initial impression of being quite formidable because several data are given. Upon examination, it is much less challenging than you might think. We must show that mass is conserved in the experiment. That means that we must compare the mass of the starting materials to the mass of the end products of the chemical change. If mass is conserved, the two masses should be identical, within the limits of experimental error. Our job then is to find the masses of the starting materials and of the end products.

SOLUTION

Let's begin by identifying the starting materials and end products. The context of the problem makes it clear that calcium carbonate reacts with a hydrochloric acid solution, and so calcium carbonate and the HCl solution are the starting materials. The end products are another solution and carbon dioxide gas.

Starting mass: The mass of calcium carbonate is given. We can use the density and the volume of the HCl solution to find its mass.

$$100.0 \text{ mL HCl solution} \times \frac{1.148 \text{ g}}{1 \text{ mL}} = 114.8 \text{ g HCl solution}$$

Then we can add the masses of the two starting materials.

$$114.8 \text{ g HCl solution} + 10.00 \text{ g CaCO}_3 = 124.8 \text{ g (start)}$$

Mass of products: This time, the mass of the solution is given. We must use volume and density of carbon dioxide gas to find its mass. However, we must first convert the volume in liters to milliliters because the density is given in grams per milliliter. Then we can add the masses of the two products.

$$2.22 \text{ L CO}_2 \times \frac{1000 \text{ mL}}{1 \text{ L}} \times \frac{0.0019769 \text{ g}}{1 \text{ mL}} = 4.39 \text{ g CO}_2$$

$$120.40 \text{ g solution} + 4.39 \text{ g CO}_2 = 124.79 \text{ g products}$$

ASSESSMENT

We note that the masses of reactants and products are not exactly the same. However, the 124.8 g of reactants is reported to four significant figures, and so it is precise only to 0.1 g. The difference between the masses of the starting materials and end products is less than 0.1 g. Clearly, the difference in masses is smaller than the uncertainty in the mass of the starting materials. We can therefore conclude that this experiment is consistent with the law of conservation of mass.

Concept Review with Key Terms

2.1 Laws of Chemical Combination—The basic laws of chemical combination are the **law of conservation of mass**, the **law of constant composition**, and the **law of multiple proportions**. Each played an important role in Dalton's development of the atomic theory.

2.2 John Dalton and the Atomic Theory of Matter—Dalton developed his atomic theory to account for the basic laws of chemical combination. The theory centered around the existence of indivisible small particles of matter called atoms and addressed the unique nature of chemical elements, the formation of chemical compounds from atoms of different elements, and the atomic nature of chemical reactions.

2.3 The Divisible Atom—Of the fundamental particles found in atoms, the three of most concern to chemists are **protons**, **neutrons**, and **electrons**. Protons and neutrons make up the **nucleus**, and their combined number is the **mass number**, A, of the atom. The number of protons is the **atomic number**, Z. Electrons are found outside the nucleus, and their number is also equal to the atomic number. The negative charge on an electron is equal in magnitude to the positive charge on a proton. All atoms of an element have the same atomic number, but they may have different numbers of neutrons and hence different mass numbers. Atoms containing the same number of protons (atomic number) but different numbers of neutrons (mass number) are **isotopes** of an element. Chemical symbols for isotopes are commonly written in the form $^A_Z E$, with A being the mass number and Z the atomic number of the element E.

2.4 Atomic Masses—The **atomic mass** of an element is a weighted average value calculated from the masses and relative abundances of its naturally occurring isotopes. The **atomic mass unit** represents the standard unit of measure of atomic masses; it is exactly 1/12 of the mass of a carbon-12 atom.

2.5 The Periodic Table: Elements Organized—The **periodic table** is an arrangement of the elements by atomic number into rows and columns. This arrangement places elements having similar properties in the same vertical groups (families). This arrangement also allows for the classification of elements as **metal**, **nonmetal**, or **metalloid**.

2.6 Molecules and Molecular Compounds—A **chemical formula**, the generic term for the various notations used to represent compounds, indicates the relative numbers of atoms of each type in a compound. An **empirical formula** expresses the simplest atom ratio, and a **molecular formula** reflects the actual composition of a **molecule**. **Structural formulas** describe the arrangement of atoms within molecules. Molecular models are also used to represent the structure and shape of molecules. For example, for acetic acid:

CH_2O	$C_2H_4O_2$	$H-\overset{\displaystyle H}{\underset{\displaystyle H}{C}}-\overset{\displaystyle O}{C}-O-H$	CH_3COOH
Empirical formula	Molecular formula	Structural formula	Condensed structural formula

A **molecular compound** consists of molecules. In a binary molecular compound, the molecules are made up of atoms of two elements. In naming these compounds, the numbers of atoms in the molecules are denoted by prefixes; the names also feature *-ide* endings.

Examples:

    NI_3 = nitrogen *tri*iodide S_2F_4 = *di*sulfur *tetra*fluoride

2.7 Ions and Ionic Compounds—**Ions** are formed by the loss or gain of electrons by single atoms or groups of atoms. Positive ions are **cations**, and negative ions are **anions**. An **ionic compound** is made up of cations and anions held together by electrostatic attractions. Chemical formulas of ionic compounds are based on an electrically neutral combination of cations and anions called a **formula unit**, such as NaCl.

The names of some monatomic cations include roman numerals to designate the charge on the ion. The names of **monatomic** anions are those of the nonmetallic elements, modified to an *-ide* ending. **Polyatomic ions** contain more than one atom. For **polyatomic anions**,

the prefixes *hypo-* and *per-* and the endings *-ite* and *-ate* are commonly used. A **hydrate** is an ionic compound that includes a fixed number of water molecules associated with the formula unit.

Examples:

    MgF_2 = magnesium fluor*ide*

    Li_2S = lithium sulf*ide*

    Cu_2O = copper(I) ox*ide*

    CuO = copper(II) ox*ide*

    $Ca(ClO)_2$ = calcium *hypo*chlor*ite*

    KIO_4 = potassium *per*iod*ate*

    $CuSO_4 \cdot 5\,H_2O$ = copper(II) sulf*ate* penta*hydrate*

2.8 Acids, Bases, and Salts—According to the Arrhenius theory, an **acid** produces H^+ in water and a **base** produces OH^-. A neutralization reaction between an acid and a base forms water and an ionic compound called a **salt**. Binary acids have hydrogen and a nonmetal as their constituent elements. Their names feature the prefix *hydro-* and the ending *-ic* attached to the stem of the name of the nonmetal. Ternary oxoacids have oxygen as an additional constituent element, and their names use prefixes (*hypo-* and *per-*) and endings (*-ous* and *-ic*) to indicate the number of O atoms per molecule.

Examples: HI = *hydro*iod*ic* acid HIO_3 = iod*ic* acid

    $HClO_2$ = chlor*ous* acid $HClO_4$ = *per*chlor*ic* acid

2.9 Organic Compounds—Organic compounds are based on the element carbon. **Hydrocarbons** contain only hydrogen and carbon. **Alkanes** have carbon atoms joined together by single bonds into chains or rings, with hydrogen atoms attached to the carbon atoms. Alkanes with four or more carbon atoms can exist as **isomers**, which are molecules that have the same molecular formula but different structures and properties.

An **alkyl group** is derived from an alkane by removing a hydrogen atom. **Functional groups** confer distinctive physical and chemical properties to an organic molecule. **Alcohols** feature the hydroxyl group, $-OH$. **Carboxylic acids** have a carboxyl group, $-COOH$.

Assessment Goals

When you have mastered the material in this chapter, you will be able to

- State and apply the laws of constant composition (definite proportions), multiple proportions, and conservation of mass.
- Relate Dalton's atomic theory to these fundamental laws of chemical combination.
- Describe the important properties of protons, neutrons, and electrons.
- Define *isotope*, and determine the number of neutrons, protons, and electrons from atomic and mass numbers.
- Define *atomic mass unit* and *atomic mass*.
- Calculate the weighted average atomic mass of an element from isotopic masses and natural abundances.
- Distinguish metals from nonmetals, using the periodic table.
- Write empirical and molecular formulas.

- Relate the names and formulas of binary molecular compounds.
- Define *ion* and *ionic substance*, and be able to write the names and formulas of monatomic ions and their compounds.
- Relate the names and formulas for polyatomic ions and name compounds containing them.
- Describe the characteristics of acids, bases, and salts according to the Arrhenius concept.
- Relate names and formulas of simple acids and bases.
- Recognize alkane hydrocarbons, relate the names and formulas of simple alkanes, and represent them with molecular, structural, condensed structural, and line-angle formulas.
- Recognize the alcohol and carboxylic acid functional groups, and relate the names and formulas of simple molecules containing them.
- Identify organic molecules that are isomers.

Self-Assessment Questions

1. When 10.0 g zinc and 8.0 g sulfur are allowed to react, all the zinc is used up, 14.9 g of zinc sulfide is formed, and some unreacted sulfur remains. The mass of *unreacted* sulfur is **(a)** 2.0 g, **(b)** 3.1 g, **(c)** 4.9 g, **(d)** impossible to determine from this information alone.

2. One oxide of rubidium has 0.187 g O per gram Rb (relative atomic masses: O = 16.0; Rb = 85.5). A possible O-to-Rb mass ratio for a second oxide of rubidium is **(a)** 16:1, **(b)** 16:85.5, **(c)** 0.374:1, **(d)** all of these.

3. Polychlorinated biphenyls (PCBs) are environmental contaminants composed of carbon, hydrogen, and chlorine atoms. A news report states that a new method of disposal of PCBs converts them completely to carbon dioxide and water. The necessary oxygen atoms come from oxygen in the air. Do you think this method will work? Explain.

4. Are the following findings, expressed to the nearest atomic mass unit, in agreement with Dalton's atomic theory? Explain your answers. **(a)** An atom of calcium has a mass of 40 u and one of vanadium, 50 u. **(b)** An atom of calcium has a mass of 40 u and one of potassium, 40 u. **(c)** One atom of calcium has a mass of 40 u, and another calcium atom has a mass of 44 u.

5. What is an atomic nucleus? What subatomic particles are found in the nucleus?

6. What are isotopes, and what is meant by the mass number of an isotope?

7. Which of the following pairs of symbols represent isotopes? Which are isobars? (Recall Exercise 2.3B.)
   (a) $^{70}_{33}E$ and $^{70}_{34}E$ **(c)** $^{186}_{74}E$ and $^{186}_{74}E$ **(e)** $^{22}_{11}E$ and $^{44}_{22}E$
   (b) $^{57}_{28}E$ and $^{66}_{28}E$ **(d)** $^{7}_{3}E$ and $^{8}_{4}E$

8. Use the symbolism $^{A}_{Z}E$ to represent each of the following atoms. You may refer to the periodic table.
   (a) boron-8   **(c)** uranium-235
   (b) carbon-14   **(d)** cobalt-60

9. What do the listed atomic mass values inside the front cover represent?

10. Explain why a chemist calls the compound $MgCl_2$ magnesium chloride and not magnesium dichloride. Would a chemist similarly call SCl_2 sulfur chloride? Explain.

11. In what two ways might we use assigned names to distinguish between $FeCl_2$ and $FeCl_3$? Which is the systematic and which is the common name?

12. Which of the following names refer to actual substances that you might find in containers on a storeroom shelf? What do the other names represent?
    (a) magnesium   **(d)** ammonia
    (b) methyl   **(e)** ammonium
    (c) chloride   **(f)** ethane

13. What type of information is conveyed by each of the following representations of a molecule?
    (a) empirical formula   **(d)** ball-and-stick model
    (b) molecular formula   **(e)** space-filling model
    (c) structural formula

14. A substance has the molecular formula $C_4H_8O_2$. **(a)** What is the empirical formula of this substance? **(b)** Can you write a structural formula from the molecular formula given? Explain.

15. According to the Arrhenius theory, all acids have one element in common. What is that element? Are all compounds containing that element acids? Explain.

16. Suggest some ways in which you might determine whether a particular solution contains an acid or a base.

17. Can a substance with the molecular formula C_3H_4 be an *alkane?* Explain.

18. Are hexane and cyclohexane isomers? Explain.

19. For which of the following is the *molecular* formula alone enough to identify the type of compound? For which must you have the *structural* formula?
    (a) an organic compound   **(d)** an alkane
    (b) a hydrocarbon   **(e)** a carboxylic acid
    (c) an alcohol

20. Explain the difference in meaning between each pair of terms:
    (a) a group and a period of the periodic table
    (b) an ion and an ionic substance
    (c) an acid and a salt
    (d) an isomer and an isotope

Problems

Laws of Chemical Combination

21. A student heats 1.000 g of zinc and 0.600 g of sulfur in a closed container. She obtains 1.490 g of zinc sulfide and recovers 0.110 g of unreacted sulfur. Are these results consistent with the law of conservation of mass? Explain.

22. When 0.2250 g of magnesium is heated with 0.5331 g of nitrogen in a closed container, the magnesium is completely converted to 0.3114 g of magnesium nitride. What mass of unreacted nitrogen must remain?

23. When 24.3 g of magnesium is burned in 16.0 g of oxygen, 40.3 g of magnesium oxide is formed. When 24.3 g of magnesium is burned in 80.0 g of oxygen, **(a)** what is the total mass of substances present after the reaction? **(b)** What mass of magnesium oxide is formed? **(c)** What law(s) is (are) illustrated by this reaction? **(d)** If 48.6 g of magnesium is burned in 80.0 g of oxygen, what mass of magnesium oxide is formed? Explain.

24. When 3.06 g of hydrogen was allowed to react with an excess of oxygen, 27.35 g of water was obtained. In a second experiment, electric current was used to break down a sample of water into 1.45 g of hydrogen and 11.51 g of oxygen. Are these results consistent with the law of constant composition? Demonstrate why or why not.

25. A colorless liquid is thought to be a pure compound of carbon, hydrogen, and oxygen. Analyses of three samples give the following results.

	Mass of Sample	Mass of Carbon	Mass of Hydrogen
Sample 1	1.000 g	0.625 g	0.0419 g
Sample 2	1.549 g	0.968 g	0.0649 g
Sample 3	2.774 g	1.734 g	0.116 g

Could the material be a pure compound?

26. Azulene, a blue solid, is thought to be a pure compound. Analyses of three samples of the material give the following results.

	Mass of Sample	Mass of Carbon	Mass of Hydrogen
Sample 1	1.000 g	0.937 g	0.0629 g
Sample 2	0.244 g	0.229 g	0.0153 g
Sample 3	0.100 g	0.094 g	0.0063 g

Could the material be a pure compound?

27. Two experiments were performed in which sulfur was burned completely in pure oxygen gas, producing sulfur dioxide and leaving some unreacted oxygen. In the first experiment, 0.312 g of sulfur produced 0.623 g of sulfur dioxide. What mass of sulfur dioxide would have been produced in the second experiment in which 1.305 g of sulfur was burned?

28. Refer to Example 2.2 to determine the mass ratios: mass oxygen/ mass magnesium oxide and mass magnesium/mass magnesium oxide. Then determine the masses of magnesium and oxygen that must combine to form 1.000 g of magnesium oxide.

29. One oxide of nitrogen is found to have an oxygen-to-nitrogen mass ratio of 1.142 : 1 (that is, 1.142 g of oxygen for every 1.000 g of nitrogen). Which of the following oxygen-to-nitrogen mass ratios are possible for different oxides of nitrogen? (*Hint:* Recall the scheme outlined in Figure 2.2.)

(**a**) 0.571 : 1   (**c**) 2.285 : 1

(**b**) 1.000 : 1   (**d**) 2.500 : 1

30. A sample of one oxide of tin, SnO, is found to consist of 0.742 g of tin and 0.100 g of oxygen. A sample of a second oxide of tin consists of 0.555 g of tin and 0.150 g of oxygen. What must be the formula of this second oxide?

Atoms and Atomic Masses

31. Indicate the numbers of protons, neutrons, and electrons in the following atoms.

(**a**) $^{64}_{30}Zn$ (**b**) $^{109}_{47}Ag$ (**c**) $^{14}_{6}C$ (**d**) $^{243}_{95}Am$

32. Indicate the numbers of protons, neutrons, and electrons in the following atoms.

(**a**) ^{232}Th (**b**) ^{55}Fe (**c**) ^{39}K (**d**) ^{197}Au

Problems 33 and 34 are based on the following table of atomic species.

	#1	#2	#3	#4	#5	#6	#7
No. protons	16	20	19	18	20	22	20
No. neutrons	18	20	21	22	24	26	28
No. electrons	18	20	18	17	18	22	20

33. For each of the numbered species that is a neutral atom, represent its composition in the form $^{A}_{Z}E$. Which of these species are isotopes?

34. Use the $^{A}_{Z}E$ symbolism to represent each of the ions in the table. Include the net ionic charge as a superscript.

35. The weighted average atomic mass of bromine, 79.904 u, suggests that the principal isotope of bromine might be ^{80}Br. However, this isotope does not occur naturally. How do you account for the observed weighted average atomic mass?

36. There are two naturally occurring isotopes of silver that exist in approximately equal proportions. One of these is ^{107}Ag. Which of the following must be the other: ^{106}Ag, ^{108}Ag, ^{109}Ag? Explain.

37. In nature, gallium consists of two isotopes, gallium-69 (mass = 68.926 u and fractional abundance = 0.601), and gallium-71 (mass = 70.925 u and fractional abundance = 0.399). Calculate the weighted average atomic mass of gallium.

38. In nature, europium consists of two isotopes, europium-151 (mass = 150.92 u and fractional abundance = 0.478) and europium-153 (mass = 152.92 u and fractional abundance = 0.522). Calculate the weighted average atomic mass of europium.

39. Silicon in nature consists of the following isotopes.

Isotope	Atomic Mass, u	Percent Abundance
Silicon-28	27.97693	92.21
Silicon-29	28.97649	4.70
Silicon-30	29.97376	3.09

Calculate the weighted average atomic mass of silicon.

40. Naturally occurring strontium consists of the following isotopes.

Isotope	Atomic Mass, u	Percent Abundance
Strontium-84	83.913	0.56
Strontium-86	85.909	9.86
Strontium-87	86.909	7.00
Strontium-88	87.906	82.58

Calculate the weighted average atomic mass of strontium.

41. The two naturally occurring isotopes of rubidium are rubidium-85, with an atomic mass of 84.91179 u; and rubidium-87, with an atomic mass of 86.90919 u. What are the percent natural abundances of these isotopes?

42. The two naturally occurring isotopes of nitrogen are nitrogen-14, with an atomic mass of 14.003074 u; and nitrogen-15, with an atomic mass of 15.000108 u. What are the percent natural abundances of these isotopes?

The Periodic Table

43. Indicate the group and period numbers for each of the following elements, and classify the element as a metal or nonmetal.

(**a**) Ga   (**c**) I   (**e**) Li   (**g**) Xe

(**b**) P   (**d**) Ra   (**f**) La

44. Identify the elements represented by the following group and period numbers. State whether each element is a metal or a nonmetal.

(**a**) group 3A, period 4   (**d**) group 1A, period 2

(**b**) group 1B, period 4   (**e**) group 4A, period 2

(**c**) group 7A, period 5   (**f**) group 4B, period 4

Chemical Formulas: Elements

45. Give molecular formulas for the following elements.
(a) nitrogen   (c) iodine
(b) hydrogen   (d) phosphorus

46. Give either an atomic symbol or a molecular formula for the following, whichever best represents how the element exists in the natural state.
(a) manganese   (c) potassium   (e) argon
(b) fluorine   (d) sulfur

Chemical Formulas: Binary Molecular Compounds

47. Which of the following are binary molecular compounds? Explain.
(a) HCl   (c) $BrCl_3$ (e) HIO
(b) MgF_2 (d) NH_3

48. Which of the following are binary molecular compounds? Explain.
(a) hydrogen sulfide   (d) silver chloride
(b) silicon tetrabromide (e) cycloalkanes
(c) strontium carbonate

49. Supply either a formula to match the name or a name to match the formula for each of the following binary molecular compounds.

(a) silicon tetrachloride   (e) N_2O
(b) sulfur hexafluoride   (f) I_2O_5
(c) carbon disulfide   (g) PCl_5
(d) dichlorine trioxide   (h) P_2O_3

50. Supply either a formula to match the name or a name to match the formula for each of the following binary molecular compounds.

(a) NO_2 (e) phosphorus trichloride
(b) XeO_3 (f) sulfur dioxide
(c) IF_5 (g) dichlorine heptoxide
(d) SiS_2 (h) tetraphosphorus decoxide

Chemical Symbols and Formulas: Ions

51. Supply either a formula to match the name or a name to match the formula for each of the following monatomic ions.
(a) magnesium ion   (d) I^-
(b) bromide ion   (e) S^{2-}
(c) titanium(IV) ion   (f) Cr^{2+}

52. Supply either a formula to match the name or a name to match the formula for each of the following monatomic ions.
(a) Li^+ (d) iron(II) ion
(b) Cu^{2+} (e) oxide ion
(c) N^{3-} (f) selenide ion

53. Supply either a formula to match the name or a name to match the formula for each of the following polyatomic ions.
(a) NH_4^+ (e) hydroxide ion
(b) ClO^- (f) sulfite ion
(c) PO_4^{3-} (g) acetate ion
(d) HSO_4^- (h) carbonate ion

54. Supply either a formula to match the name or a name to match the formula for each of the following polyatomic ions.
(a) HCO_3^- (e) nitrate ion
(b) NO_2^- (f) sulfate ion
(c) $Cr_2O_7^{2-}$ (g) thiosulfate ion
(d) HPO_4^{2-} (h) perchlorate ion

Chemical Formulas: Ionic Compounds

55. Name the following ionic compounds.
(a) KI   (f) Al_2O_3
(b) MgF_2 (g) $BaCl_2 \cdot 2\,H_2O$
(c) $NiSO_4$ (h) $Li_2C_2O_4$
(d) $TiBr_4$ (i) $Na_2SO_4 \cdot 10\,H_2O$
(e) NH_4HSO_4

56. Name the following ionic compounds.
(a) Li_2S (g) $Mg(HCO_3)_2$
(b) $FeCl_3$ (h) $Na_2S_2O_3 \cdot 5\,H_2O$
(c) Cr_2O_3 (i) $K_2Cr_2O_7$
(d) $BaSO_3$ (j) $Ca(ClO_2)_2$
(e) NH_4CN (k) CuI
(f) $Cr(NO_3)_2 \cdot 9\,H_2O$ (l) $Mg(H_2PO_4)_2$

57. Write formulas for the following ionic compounds.
(a) potassium chloride (a salt substitute)
(b) calcium carbonate (chalk)

(c) chromium(III) oxide (a green pigment)
(d) potassium perchlorate (used in fireworks)
(e) sodium chlorate (a weedkiller)
(f) iron(II) sulfate heptahydrate (iron supplement for cattle)

58. Write formulas for the following ionic compounds.
(a) titanium(IV) oxide (a white pigment)
(b) calcium oxalate monohydrate (found in rhubarb leaves)
(c) ammonium bicarbonate (used in baking)
(d) ammonium hydrogen phosphate (a fertilizer)
(e) magnesium sulfate heptahydrate (a cathartic)
(f) copper(II) sulfate pentahydrate (an algicide)

59. Give the name or formula, as is appropriate.
(a) sodium oxide   (e) $NaBrO_3$
(b) copper(I) hydroxide   (f) $LiHSO_4$
(c) barium iodide   (g) $(NH_4)_2Cr_2O_7$
(d) aluminum sulfate   (h) NiC_2O_4

60. Give the name or formula, as is appropriate.
   (a) K_2SO_3
   (b) $ZnSO_4 \cdot 7\,H_2O$
   (c) $Mn(CN)_2$
   (d) $AgNO_3$
   (e) magnesium dihydrogen phosphate
   (f) cobalt(III) oxide
   (g) calcium permanganate
   (h) iron(III) chloride nonahydrate

61. The formula given for each name is *incorrect*. Indicate the nature of the error(s) in the formula, and give the correct formula for the compound.
   (a) ammonium chlorate, NH_4Cl
   (b) potassium nitrate, KNO_2
   (c) sodium sulfate, $NaSO_4$
   (d) barium hydroxide, $BaOH$
   (e) zinc oxalate, ZnO
   (f) manganese(IV) oxide, MnO_4

   (g) strontium chromate, $Sr(CrO_4)_2$
   (h) copper(II) phosphate, Cu_2PO_4

62. The name given for each of the following formulas is either incorrect or not the simplest name possible. Give the correct and most acceptable name for each compound.
   (a) $CaCl_2$, calcium dichloride
   (b) $MgBr_2$, magnesium(II) bromide
   (c) $KClO$, potassium chloroxide
   (d) $MgSO_3$, magnesium sulfur trioxide
   (e) $Ti(CO_3)_2$, titanium(IV) bicarbonate
   (f) FeO, iron oxide
   (g) $Ca(ClO)_2$, calcium chlorite
   (h) $Zn(H_2PO_4)_2$, zinc hydrogen phosphate

Acids and Bases

63. Supply either a formula to match the name or a name to match the formula for each of the following acids and bases.
   (a) HBr
   (b) HNO_2
   (c) H_3PO_3
   (d) chlorous acid
   (e) potassium hydroxide
   (f) sulfurous acid
   (g) $Ba(OH)_2$
   (h) hydroiodic acid

64. Supply either a formula to match the name or a name to match the formula for each of the following acids and bases.

   (a) $HClO_2$
   (b) $Ba(OH)_2$
   (c) NH_3
   (d) H_2CO_3
   (e) H_3PO_4
   (f) $H_2C_2O_4$
   (g) hydrosulfuric acid
   (h) periodic acid
   (i) bromic acid
   (j) strontium hydroxide

65. Give the formula for the iodate ion and the formula and name of its corresponding acid.

66. An ion with the formula $H_2PO_2^-$ is derived from an acid. Give the formula and name of the corresponding acid.

Organic Compounds: Formulas, Structures, and Functional Groups

67. Write structural and condensed structural formulas for each of the following organic compounds.
   (a) pentane
   (b) butanoic acid
   (c) 1-propanol
   (d) cyclobutane
   (e) isobutane
   (f) propionic acid

68. Write structural and condensed structural formulas for each of the following organic compounds.
   (a) hexane
   (b) pentanoic acid
   (c) isopropyl alcohol
   (d) cyclopentane
   (e) isopentane
   (f) 1-octanol

69. There are two carboxylic acids with the molecular formula $C_4H_8O_2$. For each, (a) write a structural formula, showing all bonds to hydrogen atoms, (b) write a condensed structural formula, and (c) draw a line-angle formula. (d) What is the relationship between the two compounds?

70. Which of the following structures represent isomers of $CH_3CH_2C(CH_3)_2CH_2CH_2OH$?
   (a) $CH_3(CH_2)_5CH_2OH$
   (b) $CH_3CH(OH)CH_2CH_2CH(CH_3)_2$
   (c) $CH_3CHCH_2CH_2CHCH_3$
   $\qquad\quad |\qquad\qquad\;\; |$
   $\qquad\;\; CH_3 \qquad\;\; OH$
   (d) $CH_3CHCH_2CHCH_2CH_2OH$
   $\qquad\quad |\qquad\;\; |$
   $\qquad\;\; CH_3\quad CH_3$

71. Each of the following formulas belongs to one of the following classes of substances: straight-chain alkane, branched-chain alkane, cyclic alkane, hydrocarbon, alcohol, carboxylic acid, or inorganic compound. Match each formula with the term that identifies it *most specifically*.
   (a) $CH_3(CH_2)_2CH_3$
   (b) $CH_3CH_2CHOHCH_3$
   (c) C_5H_{10}
   (d) C_2H_2
   (e) $CH_3(CH_2)_2COOH$
   (f) $Cu(NH_3)_6Cl_2$
   (g) $HOOC(CH_2)_4COOH$
   (h) Na_2CO_3

72. Each of the following formulas belongs to one of the following classes of substances: straight-chain alkane, branched-chain alkane, cyclic alkane, hydrocarbon, alcohol, carboxylic acid, or inorganic compound. Match each formula with the term that identifies it *most specifically*.
   (a) $CH_3(CH_2)_3CHCH_2CH_3$
   $\qquad\qquad\qquad\quad |$
   $\qquad\qquad\qquad\; CH_3$
   (b) $(CH_3)_2COHCH_2CH_3$
   (c) $KHCO_3$
   (d) $CH_3(CH_2)_3CHCH_2CH_3$
   $\qquad\qquad\qquad\quad |$
   $\qquad\qquad\qquad\; OH$
   (e) C_6H_6
   (f) HCOOH
   (g) $CH_3(CH_2)_4CH_3$
   (h) C_8H_{16}
   (i) $(CH_3)_2CHCH_2CH(CH_3)_2$
   (j) $CrCl_2(NH_3)_4NO_3$
   (k) HOOCCOOH

73. Which of the following formulas applies to the molecular model below?

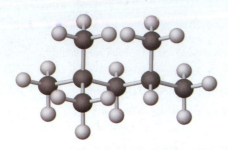

(a) $CH_3(CH_2)_6CH_3$

(b) $CH_3\overset{\underset{|}{CH_3}}{C}CH_2CH\overset{\underset{|}{}}{CH_2}$
$\qquad\qquad\quad\underset{CH_3}{}\ \underset{CH_3}{}$

(c) $CH_3\overset{\underset{|}{CH_3}}{C}\!\!-\!\!CHCH_3$
$\qquad\quad\underset{CH_3}{}\ \underset{CH_3}{}$

(d) $CH_3\overset{\underset{|}{CH_3}}{C}\!-\!CH_2\!-\!\overset{\underset{|}{CH_3}}{C}CH_3$
$\qquad\quad\underset{CH_3}{}\qquad\underset{CH_3}{}$

74. Write a structural formula for the molecular model of this alcohol:

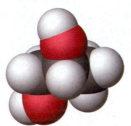

Problems 75–78 deal with functional groups that will be encountered in later chapters. They are offered here as a preview for interested students.

75. An *ether* (ROR′) may be considered a derivative of an alcohol (ROH) in which the hydrogen atom of the OH group has been replaced by a second hydrocarbon group (R′). Ethers are named by naming the groups attached to the oxygen, listed alphabetically, and then adding the generic name *ether*. If the two groups are the same, the group name should be preceded by the prefix *di-*. Name each of the following ethers, and indicate the alcohol(s) from which it is formed.

(a) CH_3-O-CH_3 (c) $CH_3CH_2-O-CH_2CH_3$

(b) $CH_3-O-CH_2CH_3$

76. An *amine* is a derivative of ammonia in which one or more H atoms are replaced by hydrocarbon groups. Which of the structures written below represent isomers of the amine $(CH_3)_2CHCH(NH_2)CH_2CH_3$?

(a) $CH_3(CH_2)_3CH_2NHCH_3$

(b) $CH_3CH(NHCH_3)CH_2CH_2CH_2CH_3$

(c) $CH_3\overset{\underset{|}{}}{C}HCH_2\overset{\underset{|}{}}{C}HCH_3$
$\qquad\ \underset{CH_3}{}\quad\ \underset{NH_2}{}$

(d) $CH_3CH_2CH(NH_2)\overset{\underset{|}{}}{C}HCH_3$
$\qquad\qquad\qquad\qquad\underset{CH_3}{}$

77. An *ether* (ROR′) has two alkyl groups attached to an oxygen atom. Write structural formulas for three ether isomers having the molecular formula $C_4H_{10}O$.

78. A *ketone* [R(C=O)R′] has two alkyl groups attached to the carbon atom of a carbonyl group (C=O). Write structural formulas for three ketone isomers having the molecular formula $C_5H_{10}O$.

Additional Problems

Problems marked with an * may be more challenging than others.

79. Convert each of the following condensed structural formulas to a line-angle formula:

(a) $(CH_3)_2CHCH(CH_3)_2$

(b) $(CH_3)_2CHCH(CH_3)CH(CH_3)_2$

(c) $CH_2-CH-CH_3$
$\ \ |\qquad\ |$
$\ \ CH_2-C(CH_3)_2$

80. Draw a line-formula structure for each of the following:

(a) $CH_3CH_2CH(CH_2CH_3)_2$

(b) $CH_3CH_2C(CH_3)_2CH_2CH_2CH(CH_3)CH_2CH_3$

(c)
$\qquad\ CH_2$
$\quad\ \diagup\quad\diagdown$
$CH_2\qquad CH-CH(CH_3)_2$
$\ |\qquad\quad |$
$CH_2\quad\ CH_2$
$\quad\diagdown\quad\diagup$
$\qquad CH_2$

81. What total number of atoms is represented by the formula $Ge[S(CH_2)_4CH_3]_4$?

82. Lavoisier listed magnesia as a chemical element in his table of 33 known elements. Which of the following observations best shows that magnesia cannot be an element? (a) Magnesia reacts with excess water, forming a milky suspension. (b) Magnesia is produced when magnesite is roasted. (c) When a certain bright metal strip is burned in oxygen, magnesia is produced (with no other products). (d) Magnesia melts at a temperature of 2800 °C.

83. Certain elements whose isotopes are all radioactive are indicated by a value in parentheses on the periodic table; that value is the mass number for the most stable isotope of the element. Find the number of protons and neutrons in the nucleus of the most stable isotope of each element after lawrencium, $Z = 103$.

84. A student in your study group reasons as follows: Selenium has an atomic number of 34 and an atomic mass of 78.96 u, which rounds to 79 as the mass number. Therefore, an atom of selenium must have 34 protons and $79 - 34 = 45$ neutrons. Do you agree with this line of reasoning? Explain.

85. A biology student needs phosphate ion to make a buffer solution (Chapter 15). Can she go to the stockroom and obtain a bottle of phosphate ion? If so, explain. If not, suggest what she might ask for instead.

* **86.** The relative abundances of hydrogen-1 (protium) and hydrogen-2 (deuterium) are ~99.99% and ~0.01%, respectively.

Write formulas for the different types of hydrogen molecules found in nature. Which type would be most abundant, and which would be least abundant? If a container held one billion molecules of hydrogen, about how many of them would be of the least abundant type?

87. How many isotopes would you expect to have masses that are whole-number values? For which elements? Explain.

88. Students in a study group come up with the following different formulas for dinitrogen pentoxide. Which, if any, is correct?

(a) $2 NO_5$ (b) N_2O_5 (c) $N_2(O_2)_5$ (d) N_5O_2

89. Compounds called amino acids have an amino (NH_2) group as well as a carboxyl (COOH) group attached to a hydrocarbon stem. Analyses of a newly synthesized amino acid yielded the following results. Can these results be used to verify the law of constant composition? Explain.

	Mass of Sample	Mass of Carbon	Mass of Hydrogen
Sample 1	0.2450 g	0.1141 g	0.0216 g
Sample 2	0.3005 g	0.1400 g	0.0264 g
Sample 3	0.1371 g	0.0639 g	0.0121 g

90. In addition to CO and CO_2, another known but uncommon oxide of carbon is the colorless, pungent gas C_3O_2, often called carbon suboxide. Expand Figure 2.2 to include carbon suboxide, and show that it conforms to the law of multiple proportions, as do the other two oxides.

91. One hundred grams of cobalt may combine with either 27.2 g or 40.7 g or 36.2 g of oxygen, forming three different oxides. Show that these data conform to the law of multiple proportions, and name two of the three oxides.

92. Three oxides of nitrogen, N_2O, NO, and NO_2, were known to Dalton and figured prominently in his proposal of the law of multiple proportions. Explain how he would have been able to use this law to verify the existence of N_2O_5 but not N_2O_4, two oxides of nitrogen that are now well known.

93. One oxide of iron is found to consist of 22.36% oxygen by mass. Which of the following represents a possible mass percent oxygen of a second iron oxide? Explain.

(a) 27.64%   (c) 50.00%

(b) 44.72%   (d) 67.08%

*****94.** The principal naturally occurring isotope of magnesium makes up 78.99% of all Mg atoms and has an atomic mass of 23.98504 u. The other two naturally occurring isotopes have atomic masses of 24.98584 u and 25.98259 u, respectively. What are the percent natural abundances of these other two isotopes of magnesium?

95. A local newspaper article describing a groundwater pollution problem states, "Wells in Redlands, Loma Linda, and Riverside may contain perchlorate, a chemical that can affect the thyroid gland." Elsewhere, the article states that, "Perchlorate . . . is used in manufacturing explosives." Can you identify precisely what substance is being referred to? Do you think a chemist would be able to? Can you write a formula for the substance? Explain.

96. The following name and formula suggest compounds that do not exist. Explain why they do not.

(a) isoethanol   (b) $CH_3CH_2C(CH_3)_3CH_2CH_3$

97. Are the following pairs of molecules isomers or not? Explain.

(a) $CH_3\underset{\overset{|}{OH}}{CH}CH_2CH_3$ and $CH_3CH_2\underset{\overset{|}{OH}}{CH}CH_3$

(b) $CH_3(CH_2)_5CH(CH_3)_2$ and $CH_3(CH_2)_4CH(CH_3)CH_2CH_3$

(c) $CH_3(CH_2)_4CH(CH_3)_2$ and $(CH_3)_2CHCH_2CH_2CH_3$

(d) $CH_3CH_2C(CH_3)_2(CH_2)_3CH_3$ and $C_2H_5C(CH_3)_2C_4H_9$

98. Write the *general* formula for (a) a straight-chain or branched-chain alcohol having one OH group per molecule, (b) a straight-chain or branched-chain carboxylic acid having one COOH group per molecule, and (c) a cyclic alkane alcohol having two OH groups per molecule.

The molecular models shown here refer to Problems 99–104. The colors used in the models are given in Figure 2.8.

(a)

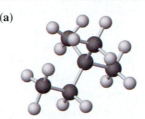

(b)

(c)

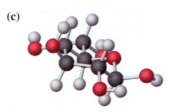

(d)

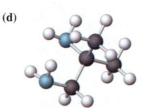

(e)

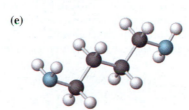

(f)

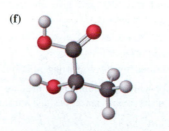

99. Which model represents a cyclic alkane? What is the name of the compound?

100. Which pair of models represents isomers? Draw their structural formulas.

101. Which model represents a compound with a carboxylic acid functional group? Write its condensed structural formula.

102. Write structural formulas for two isomers of **(a)**.

103. What are the empirical and molecular formulas of the compound represented by **(e)**?

104. Which model represents a compound that has the same empirical formula as **(c)** but is not an isomer of **(c)**?

* **105.** When carboxylic acids react with bases, they can form *salts,* which are ionic compounds. One such compound is sodium acetate, formed from acetic acid and sodium hydroxide. The formula for sodium acetate can be written as CH_3COONa. The following salts are better known as *soaps.* Write formulas for them.

   (a) ammonium laurate (lauric acid is a straight-chain, saturated carboxylic acid with 12 carbon atoms)

   (b) calcium stearate (stearic acid is a straight-chain, saturated carboxylic acid with 18 carbon atoms)

   (c) potassium oleate (oleic acid is a straight-chain carboxylic acid with 18 carbon atoms and one carbon-to-carbon double bond in the middle of the chain)

106. An *amide* is derived from a carboxylic acid by replacing the OH of the carboxylic acid with an NH_2 group. Draw the structural formula for **(a)** acetamide and **(b)** formamide.

107. An *ester* (RCOOR′) may be considered a derivative of a carboxylic acid (RCOOH) and an alcohol (R′OH) in which the OH group of the acid has been replaced by the OR′ group of the alcohol. Esters are named by naming the alkyl group from the alcohol first, followed by the name of the acid portion named as if it were part of a salt. That is, the *-ic* ending of the parent acid is replaced by the suffix *-ate.* Name each of the following esters, and indicate the alcohol and the carboxylic acid from which it is formed:

   (a) $CH_3 - \overset{\displaystyle O}{\overset{\|}{C}} - O - CH_3$

   (b) $CH_3 - \overset{\displaystyle O}{\overset{\|}{C}} - O - CH_2CH_2CH_3$

   (c) $H - \overset{\displaystyle O}{\overset{\|}{C}} - O - CH(CH_3)_2$

108. An *amine* may be considered a derivative of ammonia in which one (RNH_2), two (RR′NH), or all three (RR′R″N) H atoms are replaced by hydrocarbon groups. The common names for simple aliphatic amines merely specify the alkyl groups attached to nitrogen and add the suffix *-amine.* Name each of the following amines:

   (a) $(CH_3)_2CHNH_2$ **(c)** $CH_3NHCH_2CH_2CH_3$

   (b) $CH_3CH_2NHCH_2CH_3$

109. Write condensed structural formulas and give an acceptable name for the three isomeric pentanes (C_5H_{12}).

110. Write condensed structural formulas and give an acceptable name for the five isomeric hexanes (C_6H_{14}).

* **111.** Deduce the isotopes described by the following relationships.

   (a) Identify the isotope for which the numbers of protons, neutrons, and electrons in its atoms are equal and their *total* is 60.

   (b) Identify the isotope for which the mass number of its atoms is 234 and its atoms have 60.0% more neutrons than protons.

   (c) An isotope $_Z^A E$ has 7.50 times the mass of ^{16}O. The isotope also has 3 times the number of neutrons as a second isotope having one-half the atomic number of $_Z^A E$ and 4 times the mass number of carbon-12. What is the isotope $_Z^A E$?

   (d) The species $_Z^A E^{n+}$ is a commonly encountered ion of the most abundant naturally occurring isotope of the element E. (The isotope has an abundance of nearly 50%.) What is the most likely ion that this might be, given the following conditions that must also apply: (i) The mass number of the ion is twice its atomic number plus twice the ionic charge. (ii) The atomic number of the ion is twice the atomic number of a particular nonmetal atom. (iii) The number of electrons in the ion is the sum of the atomic numbers of two different noble gas atoms.

* **112.** In mass spectrometry (Chapter 7), molecules are ionized, and some of the ionized molecules are fragmented. Masses of the molecular ions and ionized fragments are found using magnetic and/or electric fields of known strength, and the relative numbers of fragments are plotted as a function of their mass-to-charge ratios, m/z. Because the ions typically carry a charge of $1+$, the m/z ratios also represent mass numbers for most fragments.

   (a) The mass spectrum of elemental bromine (Br_2) is shown below. Refer to Problem 35, and explain what species the peaks at 79, 81, 158, 160, and 162 represent. Why is the peak at 160 about twice the height of the ones at 158 and 162?

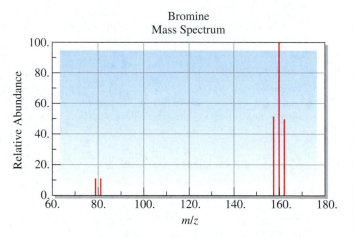

Bromine
Mass Spectrum

   (b) In the mass spectrum of 2-methylpentane, peaks are noted at the $m/z = 43, 57, 71,$ and 86, with the most intense peak coming at $m/z = 43$. The same four peaks are seen in the mass spectrum of 3-methylpentane, but here the highest peak is at $m/z = 57$. Draw structural formulas of the $1+$ ions responsible for these four peaks, and explain why the principal peak is not the same in the two mass spectra.

   (c) There are two naturally occurring isotopes of chlorine (34.9689 u, 75.77%; 36.9659 u, 24.23%) and two of bromine (78.9183 u, 50.69%; 80.9163 u, 49.31%). Bromine and chlorine combine to form bromine monochloride. How many different molecular masses are possible for bromine monochloride molecules? Calculate the weighted average molecular mass of bromine monochloride, using just the data given here. How does this result compare with the weighted average molecular mass obtained from a table of atomic masses?

113. There are four butyl groups: Two—butyl and *secondary*-butyl (abbreviated *sec*-butyl)—are derived from butane, and two—isobutyl and *tertiary*-butyl (*tert*-butyl)—are derived from isobutane. (a) Write condensed structural formulas for the four butyl groups. Then write the (b) condensed structural formula, (c) draw a line-angle formula, and (d) give the common name of each of the four butyl alcohols.

114. Replacement of one hydrogen atom of an alkane with a halogen atom gives an *alkyl halide*. The common names of these compounds consist of two parts, the name of the alkyl group and the stem of the name of the halogen, with the ending -*ide*. Name the following compounds. (*Hint:* See Problem 113.)

(a) $CH_3CH_2CH_2Br$ (c) CH_3CH_2I

(b) $CH_3CH_2CHCH_3$ (d) $(CH_3)_3CF$
$\qquad\quad\ |$
$\qquad\quad Cl$

Apply Your Knowledge

115. [Collaborative] People often refer to organic food, fertilizer, medicines, cosmetics, and other products. To you and some of your classmates, what does *organic* mean in each case? Which of the following is organic? In what sense? (a) cane sugar (sucrose, $C_{12}H_{22}O_{11}$), (b) *E. coli* O 157 : H 7 (a bacterial strain that causes food poisoning), (c) cabbage grown without synthetic fertilizers or pesticides, (d) cow manure, (e) shampoo made with synthetic detergents. Explain your answer in each case. You may use an ordinary dictionary and a dictionary of scientific terms.

116. [Historical] Before 1961, physicists used an atomic mass scale based on oxygen-16, to which they assigned a mass of exactly 16 u, whereas chemists used a mass of exactly 16 u for the naturally occurring mixture of ^{16}O, ^{17}O, ^{18}O.

(a) Before 1961, which atomic masses were greater, those on the physicists' scale or the chemists' scale? Explain.

(b) To establish a new international scale of atomic masses in 1961, physicists required that it be based on a pure isotope with mass number divisible by 4 and chemists required that their existing atomic masses would change by no more than 1 part in 10,000. Show that the carbon-12 standard meets these requirements.

* **117.** [Collaborative] Together with a group of your classmates:

(a) Draw structural formulas of all the possible C_7H_{16} alkane isomers.

(b) Draw structural formulas of all 18 possible C_8H_{18} alkane isomers.

(c) Draw structural formulas of all the possible isomers of the saturated hydrocarbon, C_8H_{16}, that have one or more methyl groups attached to the ring, but limited to one methyl group per C atom in the ring.

(d) Use line-angle formulas to represent 16 isomers of the saturated hydrocarbon, C_7H_{14}, that have no substituent larger than an ethyl group.

118. [Collaborative] Write the following positive ions in a horizontal row across the top of a page and the negative ions in a vertical column down the side of the page. Then, (a) construct a table containing the formulas of all possible compounds that can be formed by combining a positive ion with a negative ion. (b) Name the compounds. (c) This exercise in formula writing will generate some formulas that correspond to hypothetical or nonexistent compounds. What types of data and observations constitute evidence for the existence of a chemical compound?

Positive ions: K^+, Cr^{3+}, Sr^{2+}, Cu^+, Co^{3+}, Pb^{2+}, Ag^+, NH_4^+

Negative ions: Br^-, P^{3-}, S^{2-}, F^-, CO_3^{2-}, PO_4^{3-}, NO_3^-, HSO_4^-, $Cr_2O_7^{2-}$, OCN^-, OH^-, HCO_3^-

119. [Collaborative] The following named substances (actually, some may be mixtures) are found in products from the home, kitchen, garden, grocery store, and so forth. Find a product that is or contains each named material. Which of the names appear to correspond to systematic names? Divide the list among classmates, and compare your results.

Polysorbate 80, sodium silicoaluminate, potassium iodide, sodium hydroxide, guar gum, calcium carbonate, aluminum hydroxide, sodium bicarbonate, potassium aluminum sulfate, ammonium sulfate, calcium hypochlorite, chlorinol, potassium chloride, sodium lauryl sulfate, SD alcohol 40, ferrous sulfate, dextrose, silicon carbide.

120. [Historical] William Prout (1815) hypothesized that all other atoms are built up of hydrogen atoms. If so, all elements should have integral (whole-number) atomic masses relative to an atomic mass of 1 for hydrogen. However, the hypothesis appeared inconsistent with atomic masses such as 24.3 for magnesium and 35.5 for chlorine. Based on modern knowledge, Prout's hypothesis seems more reasonable. Explain why.

🌐 e-Media Problems

The activities described in these problems can be found in the e-Media Activities and Interactive Student Tutorial (IST) modules of the Companion Website, *http://chem.prenhall.com/hillpetrucci*.

121. From the **Multiple Proportions** animation (*Section 2–2*), determine the mass of carbon present in samples of carbon monoxide and carbon dioxide if 10.0 g of oxygen are contained in each sample. (Remember that the mass ratio of oxygen to carbon will remain fixed for all samples of a given compound.)

122. Explore the **Interactive Periodic Table** (*Section 2–5*) to answer the following questions:

(a) Can more elements be categorized as metals or nonmetals?

(b) In general, which elements have the greater number of isotopes, metals or nonmetals?

(c) Which element has the lowest boiling point? Which element has the highest boiling point?

123. View the **Introduction to Acids** and **Introduction to Bases** animations (*Section 2–8*). Describe the species produced in solution in addition to H^+ and OH^- ions for each example of an acid and base.

124. A compound commonly used to attract mosquitoes has the name 1-octen-3-ol. Use the **Organic Nomenclature** and **Organic Functionality Nomenclature** activities (*Section 2–9*) to deduce a structure for this compound. (The "1-" and "en" indicate a carbon–carbon double bond present on the first carbon.)

Chapter 3

Stoichiometry: Chemical Calculations

WE CAN DESCRIBE many important physical and chemical properties with words alone: Ammonium nitrate is a white solid used both as a fertilizer and as an explosive. Chlorine is a poisonous, green-yellow gas used to disinfect water. Potassium is a soft, silvery metal that reacts violently with water to produce hydrogen gas and potassium hydroxide. These are *qualitative* descriptions.

However, chemists often face questions that require *quantitative* answers:

- How much ammonium nitrate should be used to provide an avocado tree with the desired amount of the element nitrogen?
- How much chlorine gas is required to establish a level of two parts per million of chlorine in a swimming pool?
- The reaction of the chlorine with the pool water makes the water acidic. How much soda ash (Na_2CO_3) is required to neutralize the acidity?
- How much hydrogen gas is produced when 1.00 kg of potassium metal reacts with water?

Quantitative answers require mathematics, and indeed some chemical questions require quite sophisticated mathematics. However, the questions just posed require only arithmetic or simple algebra. In this chapter, we will consider (1) calculations based on chemical formulas, (2) the use of *chemical equations* to represent chemical reactions symbolically, and (3) calculations based on chemical equations. Collectively, these topics constitute a broad subject known as **stoichiometry.**

Stoichiometry of Chemical Compounds

In Chapter 2, we emphasized writing the names and formulas of chemical compounds. We also noted some of the information that is embodied in a formula: Empirical and molecular formulas give the ratios in which atoms or ions combine, and structural

 Sodium and Potassium in Water movie

◀ Climbers at Everest Base Camp in Nepal, May 2003, marking the 50th anniversary of the first conquest of the world's highest mountain by Sir Edmund Hillary and Tenzing Norgay. Even in this bitter cold, ventilation is required for both camp stoves and the human machine—but how much oxygen must be provided for the necessary chemical and biological processes? Stoichiometry explores the mass relationships in chemical reactions.

formulas tell us the order in which atoms are joined together in molecules. As we will see in the opening sections of this chapter, chemical formulas provide a lot of additional useful information.

3.1   Molecular Masses and Formula Masses

We noted in Chapter 2 that each element has a characteristic atomic mass. Because chemical compounds are made up of two or more elements, the masses we associate with molecules or formula units represent the sum of atomic masses. Atomic masses, molecular masses, and formula masses all enter into the calculations we do in this chapter.

Molecular Masses

Recall that the masses assigned to the atoms of the different elements are relative to the mass of a carbon-12 atom. We call these *atomic masses* or, in an older but still widely used terminology, *atomic weights*. For a molecular substance, we can similarly assign a **molecular mass** (or molecular weight), which is the mass of one molecule of the substance relative to the mass of a carbon-12 atom. Moreover, we can calculate this molecular mass from atomic masses:

> *Molecular mass is the sum of the masses of the atoms represented in a molecular formula.*

For example, because the formula O_2 specifies two O atoms per molecule of oxygen, the molecular mass of O_2 is twice the atomic mass of a single oxygen atom, O:

$$\text{Molecular mass of } O_2 = 2 \times \text{atomic mass of O}$$
$$= 2 \times 15.9994 \text{ u} = 31.9988 \text{ u}$$

The molecular mass of carbon dioxide, CO_2, is the sum of the atomic mass of one carbon atom and two times the atomic mass of one oxygen atom:

$$1 \times \text{atomic mass of C} = 1 \times 12.011 \text{ u} = 12.011 \text{ u}$$
$$2 \times \text{atomic mass of O} = 2 \times 15.9994 \text{ u} = 31.9988 \text{ u}$$
$$\text{Molecular mass of } CO_2 = 44.010 \text{ u}$$

The atomic masses we use in computing a molecular mass are those listed on the inside front cover. The isotopes of an element occur in the same proportions in a compound as they do in the free element, and as a result, the calculated molecular mass is usually an average value.

Example 3.1

Calculate the molecular mass of glycerol (1,2,3-propanetriol).

STRATEGY

To determine a molecular mass, we must start with the molecular formula, which is not stated, but we can deduce the formula either from the name or from the molecular model (see Figure 2.18).

SOLUTION

One —OH group replaces one H atom on each of the three C atoms in propane, leading to the condensed structural formula $CH_2OHCHOHCH_2OH$, which translates to the molecular formula $C_3H_8O_3$.

To obtain the molecular mass, we must add together *three* times the atomic mass of carbon, *eight* times the atomic mass of hydrogen, and *three* times the atomic mass of oxygen:

$$3 \times \text{atomic mass of C} = 3 \times 12.011 \text{ u} = 36.033 \text{ u}$$
$$8 \times \text{atomic mass of H} = 8 \times 1.00794 \text{ u} = 8.06352 \text{ u}$$
$$3 \times \text{atomic mass of O} = 3 \times 15.9994 \text{ u} = 47.9982 \text{ u}$$
$$\text{Molecular mass of } C_3H_8O_3 = 92.095 \text{ u}$$

EXERCISE 3.1A

Calculate, to *three* significant figures, the molecular mass of **(a)** P_4, **(b)** N_2O_4, **(c)** H_2SO_4, and **(d)** CH_3CH_2COOH.

Problem-Solving Note

You do not need to write down intermediate results as we have done here. Simply store 36.033, 8.06352, and 47.9982 in your calculator and record their sum, rounded to the third place following the decimal point.

EXERCISE 3.1B

Calculate, to *five* significant figures, the molecular mass of **(a)** phosphorus penta-chloride, **(b)** dinitrogen pentoxide, **(c)** butanoic acid, and **(d)** methyl butyl ketone ($CH_3COCH_2CH_2CH_2CH_3$).

Formula Masses

As we noted in Chapter 2, there are no molecules in sodium chloride, just large clusters of Na^+ and Cl^- ions. For this reason, the term "molecular mass" is not appropriate for sodium chloride. We base the formula of an ionic compound on a hypothetical entity called a *formula unit*. For an ionic compound or any other compound in which there are no discrete molecules, we use the term **formula mass** (or formula weight); this is the mass of a formula unit relative to the mass of a carbon-12 atom. Just as with molecular mass, we can calculate a formula mass from atomic masses:

Formula mass is the sum of the masses of the atoms or ions present in a formula unit.

Thus, for the ionic compound $BaCl_2$,

$$\text{Formula mass of } BaCl_2 = \text{atomic mass of Ba} + (2 \times \text{atomic mass of Cl})$$
$$= 137.327 \text{ u} \qquad + (2 \times 35.4527 \text{ u})$$
$$= 208.232 \text{ u}$$

Note that we used the atomic masses of the *atoms* Ba and Cl in this calculation, but should we have used the masses of the *ions* Ba^{2+} and Cl^- instead? It is true that the mass of an atom increases ever so slightly when it gains an electron, but that increase is exactly offset by the tiny decrease in mass of an atom as it loses an electron. Thus, the formula mass based on atomic masses is the same as if we used masses of the ions.

Example 3.2

Calculate the formula mass of ammonium sulfate, a fertilizer commonly used by home gardeners.

STRATEGY

Before we can determine a formula mass, we must first write the correct chemical formula. Then we can sum the masses of the atoms represented in the formula.

SOLUTION

Ammonium sulfate is an ionic compound consisting of ammonium ions (NH_4^+) and sulfate ions (SO_4^{2-}); its formula is therefore $(NH_4)_2SO_4$. To derive a formula mass from a complex formula like this one, we must make certain that all the atoms in the formula unit are accounted for, which means paying particular attention to all the subscripts and parentheses in the formula. Let's first note the relevant atomic masses and the way in which they must be combined:

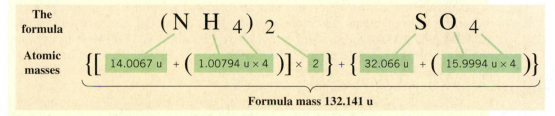

Formula Mass Calculation activity

Summing the atomic masses

$$(NH_4)_2: [14.0067 \text{ u} + (4 \times 1.00794 \text{ u})] \times 2 = 36.0769 \text{ u}$$
$$SO_4: \qquad 32.066 \text{ u} + (4 \times 15.9994 \text{ u}) = 96.064 \text{ u}$$
$$\text{Formula mass of } (NH_4)_2SO_4 = 132.141 \text{ u}$$

ASSESSMENT

As long as every atom in the formula unit is accounted for, we can check our answer by using a different summation, for example, by considering each element separately:

$$\text{Formula mass} = (2 \times \text{atomic mass N}) + (8 \times \text{atomic mass H})$$
$$+ (1 \times \text{atomic mass of S}) + (4 \times \text{atomic mass O})$$
$$= (2 \times 14.0067\text{ u}) + (8 \times 1.00794\text{ u})$$
$$+ (1 \times 32.066\text{ u}) + (4 \times 15.9994\text{ u})$$
$$= 132.141\text{ u}$$

EXERCISE 3.2A

Calculate, to *three* significant figures, the formula mass of **(a)** Li_2O, **(b)** $Mg(NO_3)_2$ **(c)** $Ca(H_2PO_4)_2$ and **(d)** K_2SbF_5.

EXERCISE 3.2B

Calculate, to *five* significant figures, the formula mass of **(a)** sodium hydrogen sulfite, **(b)** ammonium perchlorate, **(c)** chromium(III) sulfate, and **(d)** copper(II) sulfate pentahydrate.

3.2 The Mole and Avogadro's Number

In Chapter 2, we saw the importance of relative numbers of atoms in the formation of compounds. We also learned how relative masses of atoms can be based on the arbitrary choice of the carbon-12 atom as a standard (Section 2.4). Now, we introduce a concept that enables us to deal with actual rather than relative numbers of atoms and masses of substances. This will pave the way to many quantitative applications in chemistry.

We noted in Section 2.2 that with a plentiful supply of oxygen, carbon burns to form carbon dioxide. With a limited supply of oxygen, carbon monoxide is formed. How can we find out what minimum quantity of oxygen is needed to ensure that when carbon burns, it is completely converted to carbon dioxide? The carbon dioxide molecule, CO_2, consists of one C atom and two O atoms. The oxygen molecule, O_2, consists of two O atoms. Thus, we need at least as many O_2 molecules as C atoms to ensure that we have two O atoms for every C atom. The problem, though, is that atoms and molecules are so small that we cannot actually count them.

Our dilemma can be resolved by relating the mass of a substance, which we can measure, to the countless number of atoms or molecules present. To do this, we introduce the SI base unit for an amount of substance.

*A **mole (mol)** is an amount of substance that contains as many elementary entities as there are atoms in exactly 12 g of carbon-12.*

The "elementary entities" in solid carbon are atoms of C; in oxygen gas, they are molecules of O_2; and in carbon dioxide gas, they are molecules of CO_2.

Although the mass of 1 mol of carbon-12 is exactly 12 g, the mass of 1 mol of carbon obtained from natural sources is 12.011 g. This is because the carbon contains a small amount (1.108%) of the heavier carbon-13 isotope. Because naturally occurring oxygen atoms are more massive than carbon-12 atoms by the factor 15.9994 u/12.0000 u, the mass of 1 mol of oxygen atoms is 15.9994 g. In turn, the mass of 1 mol of O_2 molecules is 2×15.9994 g $= 31.9988$ g. The way in which all these values relate to the reaction of carbon atoms and oxygen molecules to form carbon dioxide molecules is depicted in Figure 3.1.

We can do a great deal with the mole concept without knowing the actual number of elementary entities in a mole of substance. However, at times we will need to work with this number, which is called **Avogadro's number, N_A**, named after Amedeo Avogadro, the first person to sense the significance of the mole. Later in the text, we will describe ways in which Avogadro's number can be established, but for now we will just state it.

$$N_A = 6.02214199 \times 10^{23}\text{ mol}^{-1}$$

Application Note

Even a sample of carbon as small as a pencil-mark period at the end of a sentence contains about 10^{18} C atoms—that is, about 1,000,000,000,000,000,000 C atoms.

The unit mole is abbreviated mol. The symbols M and m are reserved for other quantities.

Avogadro's number is almost beyond imagination. If you had 6.02×10^{23} dollars, you could spend a billion dollars a second for your entire lifetime and still have used less than 0.001% of your money. If carbon atoms were the size of peas, 6.02×10^{23} of them would cover the entire surface of Earth to a depth of more than 100 m.

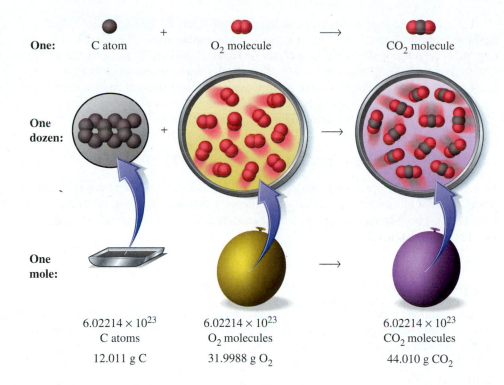

One: C atom   + O_2 molecule   ⟶ CO_2 molecule

One dozen:

One mole:

6.02214×10^{23} C atoms	6.02214×10^{23} O_2 molecules	6.02214×10^{23} CO_2 molecules
12.011 g C	31.9988 g O_2	44.010 g CO_2

◀ **FIGURE 3.1   Microscopic and macroscopic views of the combination of carbon and oxygen to form carbon dioxide**

At the microscopic (molecular) level, chemical reactions occur between atoms and molecules (top), but we can show only a few atoms and molecules (middle) to represent the enormous numbers that actually make up the samples. In reality, we usually observe substances at the macroscopic level, and the artist's sketch of the substances at that level (bottom) suggests one mole each (Avogadro's number) of C atoms, O_2 molecules, and CO_2 molecules.

If we don't need all those significant figures in a calculation, we can round off Avogadro's number to 6.022×10^{23} mol^{-1}, or even 6.02×10^{23} mol^{-1}. The unit "mol^{-1}," which we read as "per mole," signifies that a collection of N_A elementary entities (atoms, molecules, or formula units) is equivalent to one mole at the macroscopic level. For example, a mole of carbon contains 6.02×10^{23} atoms of C; a mole of oxygen gas contains 6.02×10^{23} molecules of O_2; and a mole of sodium chloride contains 6.02×10^{23} formula units of NaCl.

A dozen is the same *number* whether we have a dozen oranges or a dozen watermelons. However, a dozen oranges and a dozen watermelons do not have the same *mass*. Similarly, a mole of magnesium and a mole of iron contain the same number of atoms—6.022×10^{23}—but have *different* masses. Figure 3.2 is a photograph of one mole each of several elements.

Views of a Chemical Reaction activity

◀ **FIGURE 3.2   One mole each of four elements**

The watch glass on the left contains one mole of sulfur atoms, and the watch glass in the middle contains one mole of copper atoms. The weighing bottle on the right contains one mole of liquid mercury, and the balloon contains one mole of helium gas.

QUESTION: Do one-mole samples of all four substances have the same mass?

The **molar mass** of a substance is the mass of 1 mol of that substance. Because 1 mol of carbon-12 has a mass of exactly 12 g, the molar mass of carbon-12 (exactly 12 g/mol) is *numerically* equal to the atomic mass of carbon-12 (exactly 12 u) even though the two have different units. For other elements, the molar mass is also numerically equal to the weighted average atomic mass, but the molar mass is expressed in the unit *grams per mole (g/mol)*. For example, sodium has an atomic mass of 22.99 u and a molar mass of 22.99 g/mol. Thus, we can use the definitions of mole, Avogadro's number, and molar mass to write relationships of this type:

$$1 \text{ mol Na} = 6.022 \times 10^{23} \text{ Na atoms} = 22.99 \text{ g Na} \tag{3.1}$$

We can use relationships such as these to derive factors for converting between mass in grams, amount in moles, and number of elementary entities (atoms, molecules, or formula units). These conversion factors are then used in the unit-conversion method first presented in Section 1.5.

> Molar mass can be thought of as another conversion factor, one that relates mass (grams) and moles. It can be used to convert mass to moles or to convert moles to mass.

Example 3.3

Determine **(a)** the mass of a 0.0750-mol sample of Na, **(b)** the number of moles of Na in a 62.5-g sample, **(c)** the mass of a sample of Na containing 1.00×10^{25} Na atoms, and **(d)** the mass of a single Na atom.

STRATEGY

To perform these calculations, we will use a conversion factor derived from molar mass to relate moles and grams, and a conversion factor derived from Avogadro's number to relate moles and number of atoms. These are the relationships embodied in Equation (3.1).

SOLUTION

(a) To convert from moles to grams, we need to use the first and third terms in Equation (3.1), written as a conversion factor: 22.99 g Na/1 mol Na:

$$? \text{ g Na} = 0.0750 \text{ mol Na} \times \frac{22.99 \text{ g Na}}{1 \text{ mol Na}} = 1.72 \text{ g Na}$$

(b) Again, we need the first and third terms in Equation (3.1), but this time we write the conversion factor as the inverse—1 mol Na/22.99 g Na—because we are to convert from grams to moles:

$$? \text{ mol Na} = 62.5 \text{ g Na} \times \frac{1 \text{ mol Na}}{22.99 \text{ g Na}} = 2.72 \text{ mol Na}$$

(c) Here we can use all three terms in Equation (3.1), written as two conversion factors, to convert first from number of atoms to number of moles, and then from moles to the mass in grams:

$$? \text{ g Na} = 1.00 \times 10^{25} \text{ Na atoms} \times \frac{1 \text{ mol Na}}{6.022 \times 10^{23} \text{ Na atoms}} \times \frac{22.99 \text{ g Na}}{1 \text{ mol Na}} = 382 \text{ g Na}$$

Alternatively, we can use the second and third terms in Equation (3.1), written as the conversion factor 22.99 g Na/6.022 × 10²³ Na atoms, to convert directly from number of atoms to mass in grams:

$$? \text{ g Na} = 1.00 \times 10^{25} \text{ Na atoms} \times \frac{22.99 \text{ g Na}}{6.022 \times 10^{23} \text{ Na atoms}} = 382 \text{ g Na}$$

(d) The answer must have the unit grams per sodium atom (g/Na atom). Thus, if we know the mass of a certain number of Na atoms, our answer is simply that mass divided by the corresponding number of atoms. And we know these quantities from the molar mass

(22.99 g Na/mol Na) and Avogadro's number (1 mol Na/6.022 × 10²³ Na atoms). Our answer is the product of these two factors:

| Molar mass | 1/Avogadro's number |

$$? \text{ g/Na atom} = \frac{22.99 \text{ g Na}}{1 \text{ mol Na}} \times \frac{1 \text{ mol Na}}{6.022 \times 10^{23} \text{ Na atoms}}$$

$$= 3.818 \times 10^{-23} \text{ g/Na atom}$$

ASSESSMENT

In these examples, note the common practice of expressing molar mass and Avogadro's number with at least one significant figure more than the number of significant figures in the least precisely known quantity. Doing this ensures that the precision of the calculated results is limited only by the least precisely known quantity.

In part (d), this simple fact should deter you from mistakenly multiplying instead of dividing by Avogadro's number: Individual atoms are exceedingly small and possess masses that are many orders of magnitude less than one gram. It is also worth noting that the calculated mass is the true mass of a sodium atom—^{23}Na has no isotopes. For elements with two or more isotopes, the mass calculated for an atom is a *weighted average*.

EXERCISE 3.3A

Calculate **(a)** the mass in milligrams of 1.34×10^{-4} mol Ag and **(b)** the number of oxygen *atoms* in 20.5 mol O_2.

EXERCISE 3.3B

Calculate **(a)** the number of moles of Al in a cube of aluminum metal 5.5 cm on an edge $(d = 2.70 \text{ g/cm}^3)$ and **(b)** the volume occupied by 4.06×10^{24} Br atoms present as Br_2 molecules in liquid bromine $(d = 3.12 \text{ g/mL})$.

Problem-Solving Note

You will need to use the relationships between mass, volume, and density described in Section 1.5.

3.3 The Mole and Molar Mass

In the preceding section, we emphasized either individual atoms or diatomic molecules of the elements as elementary entities (for example, C, Na, and O_2). Now let's consider more complex entities—from Example 3.1, molecules of glycerol, $(C_3H_8O_3)$, or from Example 3.2, formula units of ammonium sulfate, $(NH_4)_2SO_4$. You can visualize one mole of each of these compounds this way:

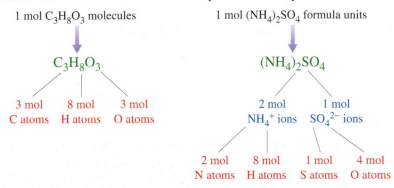

When we apply the concept of molar mass to a compound, we mean the mass of one mole of molecules or formula units of the compound. Furthermore, the molar mass in grams per mole is numerically equal to the molecular mass or the formula mass in atomic mass units. Thus, in Example 3.1 we saw that the molecular mass of glycerol, $C_3H_8O_3$, is 92.095 u, and so its molar mass is 92.095 g/mol. In Example 3.2, we found that the formula mass of $(NH_4)_2SO_4$ is 132.141 u; its molar mass is therefore 132.141 g/mol. (Think of Avogadro's number and the mole as a "scaling up" factor from the molecular level to the macroscopic level.)

Stoichiometry of Molecules and Compounds activity

For quantitative problem solving of the type illustrated in Example 3.4, we need relationships similar to those given in expression (3.1)—relationships such as

$$1 \text{ mol } C_3H_8O_3 = 6.022 \times 10^{23} C_3H_8O_3 \text{ molecules} = 92.095 \text{ g } C_3H_8O_3 \qquad (3.2)$$

$$1 \text{ mol } (NH_4)_2SO_4 = 6.022 \times 10^{23} (NH_4)_2SO_4 \text{ formula units}$$
$$= 132.141 \text{ g } (NH_4)_2SO_4 \qquad (3.3)$$

Example 3.4

Determine **(a)** the number of NH_4^+ ions in a 145-g sample of $(NH_4)_2SO_4$ and **(b)** the volume of 1,2,3-propanetriol (glycerol, $d = 1.261$ g/mL) that contains 1.00 mol O atoms.

STRATEGY

As before, we will use a conversion factor derived from molar mass to relate moles and grams, and a conversion factor derived from Avogadro's number to relate moles and number of ions or atoms.

SOLUTION

(a) In this calculation, we need the relationships given in Equation (3.3). Also, we must use the relationship between the NH_4^+ ion and the $(NH_4)_2SO_4$ formula unit shown on page 81; that is, 2 mol NH_4^+/1 mol $(NH_4)_2SO_4$. The path to the desired answer is

$$\text{g } (NH_4)_2SO_4 \longrightarrow \text{mol } (NH_4)_2SO_4 \longrightarrow \text{mol } NH_4^+ \longrightarrow \text{number of } NH_4^+ \text{ ions}$$

$$? \, NH_4^+ \text{ ions} = 145 \text{ g } (NH_4)_2SO_4 \times \frac{1 \text{ mol } (NH_4)_2SO_4}{132.14 \text{ g } (NH_4)_2SO_4} \times \frac{2 \text{ mol } NH_4^+}{1 \text{ mol } (NH_4)_2SO_4}$$

$$\times \frac{6.022 \times 10^{23} \, NH_4^+ \text{ ions}}{1 \text{ mol } NH_4^+} = 1.32 \times 10^{24} \, NH_4^+ \text{ ions}$$

(b) The conversion factors needed for this calculation are (1) the relationship between moles of O atoms and $C_3H_8O_3$ molecules (diagram on page 81), (2) the molar mass of $C_3H_8O_3$ (Equation 3.2), and (3) the inverse of the density of $C_3H_8O_3$:

$$? \text{ mL } C_3H_8O_3 = 1.00 \text{ mol O atoms} \times \frac{1 \text{ mol } C_3H_8O_3}{3 \text{ mol O atoms}} \times \frac{92.095 \text{ g } C_3H_8O_3}{1 \text{ mol } C_3H_8O_3}$$

Mol atoms   Mol molecules / Mol atoms   Molar mass (g/mol)

$$\times \frac{1 \text{ mL } C_3H_8O_3}{1.261 \text{ g } C_3H_8O_3} = 24.3 \text{ mL } C_3H_8O_3$$

1/density (mL/g)   Answer: mL

ASSESSMENT

In Example 3.3d, we cited the need to divide by Avogadro's number because the mass of an individual atom (or ion or molecule) is exceedingly small. By contrast, in part **(a)** of this example we must multiply (not divide) by Avogadro's number. The number of atoms, ions, or molecules in a macroscopic sample of matter is exceedingly large.

EXERCISE 3.4A

Determine **(a)** the total number of Cl atoms in 125 mL of liquid CCl_4 ($d = 1.589$ g/mL) and **(b)** the number of grams of carbon in 215 g of sucrose, $C_{12}H_{22}O_{11}$.

EXERCISE 3.4B

You need to obtain 1.00 mol C atoms, and your source for these C atoms is the sucrose ($C_{12}H_{22}O_{11}$) contained in an aqueous solution having a density of 1.0181 g/mL and 5.05% of its mass present as sucrose. How many milliliters of the solution are required?

Example 3.5   An Estimation Example

Which of the following is a reasonable value for the number of atoms in 1.00 g of helium?

(a) 4.1×10^{-23}

(b) 4.0

(c) 1.5×10^{23}

(d) 1.5×10^{24}

ANALYSIS AND CONCLUSIONS

Response (a) is what we get if, in error, we *divide* the number of moles of He, 0.25, by Avogadro's number. Clearly, we can't have anything less than one atom. Response (b) expressed as 4.0 u is the atomic mass of He; expressed as 4.0 g/mol He, it is the molar mass. In either case, it is far too small to be the number of atoms for any macroscopic sample. Response (c) is the correct order of magnitude, as is response (d). However, (d) is *larger* than Avogadro's number, while 1.00 g is *less than* one mole of helium. We would expect a fraction of a mole of helium to have fewer than Avogadro's number of atoms. Response (c) is the correct answer, obtained from the calculation: $0.25 \times N_A$.

EXERCISE 3.5A

Which of the following is a reasonable value for the mass of 1.0×10^{23} magnesium atoms? (Try to reason through the problem, and avoid using your calculator if possible. Your goal is to find a reasonable answer rather than to calculate a specific number.)

(a) 2.4×10^{-22} g

(b) 0.17 g

(c) 2.4 g

(d) 4.0 g

EXERCISE 3.5B

Without doing detailed calculations, determine which of the following has the greatest number of carbon atoms per gram of compound:

(a) CO_2

(b) $CH_3CH_2CH_3$

(c) CH_3CH_2COOH

(d) $CH_3CH_2OCH_2CH_3$

3.4 Mass Percent Composition from Chemical Formulas

Most scientists think that carbon dioxide released into the atmosphere when fossil fuels are burned contributes to global warming (Chapter 25). One way to limit the production of carbon dioxide gas is to use fuels that produce less CO_2 upon combustion. Methane and butane are both hydrocarbon fuels, but which one produces less carbon dioxide when equal masses are burned? There are several ways to answer this question, but we would have an immediate answer if we knew the *percent* by mass of carbon in each of these compounds. The one with the smaller percent carbon produces the lesser amount of CO_2 (and this proves to be CH_4). The **mass percent composition** of a compound refers to the proportion of each constituent element expressed as the number of grams of that element per 100 g of the compound.

We can determine the molar mass of a compound from its chemical formula, as we did in Section 3.3. If we determine the contribution of each element to that molar mass, we can establish a ratio of the mass of each element to the mass of the compound as a whole. This gives us the fractional composition of the compound, by mass. Multiplying each fraction by 100% gives the percentage of each element in the compound. For the element carbon in butane, C_4H_{10}, as we show in Figure 3.3, these quantities are

$$Mass\ fraction = \frac{mass\ carbon}{mass\ butane} = \frac{48.044\ g\ C}{58.123\ g\ C_4H_{10}} = 0.8266 \qquad (3.4)$$

$$Mass\ percent = mass\ fraction \times 100\% = 82.66\% \qquad (3.5)$$

Application Note

Carbon dioxide is one of several greenhouse gases found in the atmosphere. These gases trap heat radiated by Earth's surface and may contribute to global warming (Chapter 25).

Molecular formula of butane	C_4H_{10}
Mass of C in 1 mol C_4H_{10}	4 mol C × 12.011 g C/mol C = 48.044 g C
Mass of H in 1 mol C_4H_{10}	10 mol H × 1.0079 g H/mol H = 10.079 g H
Molar mass of C_4H_{10}	48.044 g C + 10.079 g H = 58.123 g/mol C_4H_{10}
Mass percent C in C_4H_{10}	$\dfrac{48.044 \text{ g C}}{58.123 \text{ g } C_4H_{10}} \times 100\% = 82.66\% \text{ C}$
Mass percent H in C_4H_{10}	$\dfrac{10.079 \text{ g C}}{58.123 \text{ g } C_4H_{10}} \times 100\% = 17.34\% \text{ H}$

▶ **FIGURE 3.3   Determining the mass percent composition of butane**

Example 3.6

Calculate, to four significant figures, the mass percent of each element in ammonium nitrate.

STRATEGY

First, we will determine the molar mass of ammonium nitrate, based on the formula unit NH_4NO_3. Then, for one mole of compound, we can determine mass ratios and percentages.

SOLUTION

$$\text{Formula mass} = (2 \times \text{atomic mass N}) + (4 \times \text{atomic mass H})$$
$$+ (3 \times \text{atomic mass O})$$
$$= (2 \times 14.01)u + (4 \times 1.008)u + (3 \times 16.00)u$$
$$= 28.02 \text{ u} + 4.032 \text{ u} + 48.00 \text{ u} = 80.05 \text{ u}$$
$$\text{Molar mass} = 80.05 \text{ g/mol } NH_4NO_3$$

Now we establish the mass ratios and convert them to percentages:

$$\% \text{ N} = \frac{28.02 \text{ g N}}{80.05 \text{ g } NH_4NO_3} \times 100\% = 35.00\% \text{ N}$$

$$\% \text{ H} = \frac{4.032 \text{ g H}}{80.05 \text{ g } NH_4NO_3} \times 100\% = 5.037\% \text{ H}$$

$$\% \text{ O} = \frac{48.00 \text{ g O}}{80.05 \text{ g } NH_4NO_3} \times 100\% = 59.96\% \text{ O}$$

ASSESSMENT

To check, we add the percentages to ensure that they add up to 100.00%. (Sometimes the total may differ from 100.00% by ±0.01% due to rounding.)

EXERCISE 3.6A

Calculate the mass percent of each element in **(a)** ammonium sulfate and **(b)** urea, $CO(NH_2)_2$. Which compound has the greatest mass percent nitrogen: ammonium nitrate (see Example 3.6), ammonium sulfate, or urea?

EXERCISE 3.6B

Calculate the mass percent of

(a) N in triethanolamine, $N(CH_2CH_2OH)_3$ (used in dry-cleaning agents and household detergents);

Problem-Solving Note

We can also determine the mass percent of one element from the mass percentages of all the others. For example, in NH_4NO_3 we get the percent oxygen by subtracting the percentages of the other two elements from 100.00%:

%O = 100.00% − 35.00% N

−5.037% H = 59.96% O

However, when we do this, we lose the opportunity to check our result.

(b) O in glyceryl tristearate (a saturated fat):

$$CH_2OCO(CH_2)_{16}CH_3$$
$$|$$
$$CHOCO(CH_2)_{16}CH_3$$
$$|$$
$$CH_2OCO(CH_2)_{16}CH_3$$

(c) H_2O in copper(II) sulfate pentahydrate (a fungicide and algacide).

We can use mass percent (parts per hundred) as a conversion factor to determine the mass of an element in any quantity of a compound. For example, we can use the fact that ammonium nitrate is 35.00% N by mass (Example 3.6) to formulate the conversion factor (red) and find the mass of nitrogen in 46.34 g NH_4NO_3.

$$? \text{ g N} = 46.34 \text{ g } NH_4NO_3 \times \frac{35.00 \text{ g N}}{100.00 \text{ g } NH_4NO_3} = 16.22 \text{ g N}$$

However, there is a simpler approach to determining the mass of a given element in a compound. This approach, illustrated in Example 3.7, uses the chemical formula as the source of factors for converting from the mass of a compound (NH_4NO_3) to the mass of a constituent element (N). We don't have to evaluate the mass percent N first. Similar conversion factors are compared in Example 3.8, where no detailed calculations are required.

Example 3.7

How many grams of nitrogen are present in 46.34 g ammonium nitrate?

STRATEGY

We first convert the mass of ammonium nitrate to moles, then use the formula NH_4NO_3 to obtain the ratio of moles of N to moles of NH_4NO_3, and finally we use the molar mass of nitrogen to calculate the required mass.

SOLUTION

The central factor (shown in red) in the conversion is based on the chemical formula NH_4NO_3. The other factors in the following setup are based on molar masses:

| We want (?) and the unit g N. | We start here. | This converts g NH_4NO_3 to mol NH_4NO_3. | This converts mol NH_4NO_3 to mol N. | This converts mol N to g N. |

$$? \text{ g N} = 46.34 \text{ g } NH_4NO_3 \times \frac{1 \text{ mol } NH_4NO_3}{80.05 \text{ g } NH_4NO_3} \times \frac{2 \text{ mol N}}{1 \text{ mol } NH_4NO_3} \times \frac{14.01 \text{ g N}}{1 \text{ mol N}}$$

$$= 16.22 \text{ g N}$$

EXERCISE 3.7A

People with hypertension (high blood pressure) are advised to limit the amount of sodium (actually sodium ion, Na^+) in their diet. Sodium hydrogen carbonate, $NaHCO_3$, packaged as baking soda, is one of many familiar products containing sodium ions. Calculate the number of milligrams of Na^+ in 5.00 g of $NaHCO_3$.

EXERCISE 3.7B

A fertilizer mixture contains 12.5% ammonium nitrate, 20.3% ammonium sulfate, and 11.4% urea [$CO(NH_2)_2$] by mass. How many grams of nitrogen are present in a 1.00-kg bag of this fertilizer?

▲ The "5-10-5" designation on this box of fertilizer is indicative of its nitrogen, phosphorus, and potassium content, but in a special way. See Problem 128 at the end of the chapter.

Example 3.8   An Estimation Example

Without doing detailed calculations, determine which of these compounds contains the greatest mass of sulfur per gram of compound: barium sulfate, lithium sulfate, sodium sulfate, or lead sulfate.

ANALYSIS AND CONCLUSIONS

To make this comparison, we need formulas of the compounds, which we can get from their names:

$$BaSO_4 \quad Li_2SO_4 \quad Na_2SO_4 \quad PbSO_4$$

The compound with the greatest mass of sulfur per gram of compound also has the greatest mass of sulfur per 100 g of compound—in other words, the greatest % S by mass. From the formulas, we see that in one mole of each compound there is one mole of sulfur, which means 32.066 g S. Thus, the compound with the greatest % S is the one with the *smallest* formula mass. Because each formula unit has one SO_4^{2-} ion, all we have to do is compare some atomic masses: that of barium to twice that of lithium, and so on. With just a glance at an atomic mass table, we see the answer must be lithium sulfate, Li_2SO_4.

EXERCISE 3.8A

Without doing detailed calculations, determine which of these compounds has the greatest percent phosphorus by mass: lithium dihydrogen phosphate, calcium dihydrogen phosphate, or ammonium hydrogen phosphate.

EXERCISE 3.8B

Without doing detailed calculations, determine which of these compounds has the greatest mass percent of carbon: methanol, acetic acid, butane, or octane.

3.5   Chemical Formulas from Mass Percent Composition

As we saw in the preceding section, there are practical reasons we may need to determine the mass percent composition of a compound from its formula. The reverse situation—deducing the formula of a compound from its mass percent composition—is of even more fundamental importance. As we will see, mass composition data yield only an empirical formula, and we usually want a molecular formula. However, determining the empirical formula is often an important step toward obtaining a molecular formula.

Determining Empirical Formulas

When we determine the mass percent composition of a compound by experiment, we deal with masses of the constituent elements. In an empirical formula, we must represent the atoms of the constituent elements based on their relative numbers, that is, on a mole basis. The first step in finding an empirical formula is to convert the mass of each element in a sample of a compound to an amount in moles.

We could base our calculation on a sample of any mass, but the task is simpler if we choose 100.00 g. This makes the masses of the elements numerically equal to their mass percentages. We can use a five-step procedure for converting mass percent composition data to an empirical formula. We will use butane as a simple example because we already know its molecular formula and percent composition from Figure 3.3.

Step 1: *Convert the percent of each element to a mass:*
Butane is 82.66% C and 17.34% H. A 100.00-g sample therefore consists of 82.66 g C and 17.34 g H.

Step 2: *Convert the mass of each element to an amount in moles:*

$$? \text{ mol C} = 82.66 \text{ g C} \times \frac{1 \text{ mol C}}{12.011 \text{ g C}} = 6.882 \text{ mol C}$$

$$? \text{ mol H} = 17.34 \text{ g H} \times \frac{1 \text{ mol H}}{1.0079 \text{ g H}} = 17.20 \text{ mol H}$$

Determining Empirical Formulas activity

Problem-Solving Note

If we are given masses of the elements rather than percentages, we can usually begin the calculation at Step 2.

Step 3: *Use the number of moles of each element as that element's subscript in a tentative formula:*

$$C_{6.882}H_{17.20}$$

Step 4: *Attempt to get <u>integers</u> as subscripts by dividing each subscript in Step 3 by the smallest subscript:*
We divide 6.882 and 17.20 each by 6.882, the smaller of the two subscripts:

$$C_{6.822/6.822}H_{17.20/6.882} \longrightarrow C_1 H_{2.499}$$

Step 5: *If any subscripts obtained in Step 4 are fractional quantities, multiply all the subscripts by the smallest integer that will convert all the subscripts to integers. The result is an <u>empirical</u> formula.*
The subscript of H in $CH_{2.499}$ is very close to a fractional quantity: $5/2 = 2.500$. The smallest multiple that will convert it to an integer is 2:

$$\text{Empirical formula: } C_{(2\times1)}H_{(2\times2.500)} = C_2H_5$$

You might think we have determined an incorrect formula for butane, which you know is C_4H_{10}. Remember, however, that C_4H_{10} is a *molecular* formula; the empirical or simplest formula on which it is based is indeed $C_2H_5(C_{2\times2}H_{2\times5} = C_4H_{10})$.

Problem-Solving Note

A common error here is to divide each number of moles *by itself*. That, of course, gives a formula where the subscripts are all ones.

Example 3.9

Phenol, a general disinfectant, has the composition 76.57% C, 6.43% H, and 17.00% O by mass. Determine its empirical formula.

STRATEGY

We can follow the steps outlined above to determine first the number of moles of each element and then the empirical formula.

SOLUTION

Step 1: A 100.00-g sample of phenol contains 76.57 g C, 6.43 g H, and 17.00 g O.

Step 2: We convert the masses of C, H, and O to amounts in moles:

$$? \text{ mol C} = 76.57 \text{ g C} \times \frac{1 \text{ mol C}}{12.011 \text{ g C}} = 6.375 \text{ mol C}$$

$$? \text{ mol H} = 6.43 \text{ g H} \times \frac{1 \text{ mol H}}{1.0079 \text{ g H}} = 6.38 \text{ mol H}$$

$$? \text{ mol O} = 17.00 \text{ g O} \times \frac{1 \text{ mol O}}{15.999 \text{ g O}} = 1.063 \text{ mol O}$$

Step 3: Now we use the numbers of moles in Step 2 as subscripts in a tentative formula:

$$C_{6.375}H_{6.38}O_{1.063}$$

Step 4: Next we divide each subscript by the smallest (1.063) to try to get integral subscripts:

$$C_{6.375/1.063}H_{6.38/1.063}O_{1.063/1.063} \longrightarrow C_{5.997}H_{6.00}O_{1.000} \longrightarrow C_6H_6O$$

Step 5: The subscripts in Step 4 are all integers. We need do nothing further. The empirical formula of phenol is C_6H_6O.

Problem-Solving Note

In rounding off 5.997 to 6, we are following the practice of rounding off to an integer any subscript that is within one or two hundredths of an integral value.

EXERCISE 3.9A

Cyclohexanol, used in the manufacture of plastics, has the composition 71.95% C, 12.08% H, and 15.97% O by mass. Determine its empirical formula.

EXERCISE 3.9B

Mebutamate, a diuretic (water pill) used to treat high blood pressure, has the composition 51.70% C, 8.68% H, 12.06% N, and 27.55% O by mass. Determine its empirical formula.

Example 3.10

Diethylene glycol, used in antifreeze, as a softening agent for textile fibers and some leathers, and as a moistening agent for glues and paper, has the composition 45.27% C, 9.50% H, and 45.23% O by mass. Determine its empirical formula.

STRATEGY

Following our five-step procedure, we determine first the number of moles of each element and then the empirical formula.

SOLUTION

Step 1: A 100.00-g sample of diethylene glycol contains 45.27 g C, 9.50 g H, and 45.23 g O.

Step 2: We convert the masses of C, H, and O to amounts in moles:

$$? \text{ mol C} = 45.27 \text{ g C} \times \frac{1 \text{ mol C}}{12.011 \text{ g C}} = 3.769 \text{ mol C}$$

$$? \text{ mol H} = 9.50 \text{ g H} \times \frac{1 \text{ mol H}}{1.008 \text{ g H}} = 9.42 \text{ mol H}$$

$$? \text{ mol O} = 45.23 \text{ g O} \times \frac{1 \text{ mol O}}{15.999 \text{ g O}} = 2.827 \text{ mol O}$$

Step 3: Now we use the numbers of moles in Step 2 as subscripts in a tentative formula:

$$C_{3.769}H_{9.42}O_{2.827}$$

Step 4: Next we must divide each subscript by the smallest (2.827) in an attempt to get integral subscripts:

$$C_{3.769/2.827}H_{9.42/2.827}O_{2.827/2.827} \longrightarrow C_{1.333}H_{3.33}O_{1.000}$$

Step 5: Finally we multiply all the subscripts from Step 4 by a common factor to convert them all to integers. By recognizing that $1.333 = 4/3$ and $3.33 = 10/3$, we can see that the common factor we need is 3:

$$C_{(1.333 \times 3)}H_{(3.33 \times 3)}O_{(1.000 \times 3)} \longrightarrow C_4H_{10}O_3$$

EXERCISE 3.10A

Anthracene, used in the manufacture of dyes, has the composition 94.34% C and 5.66% H by mass. Determine its empirical formula.

EXERCISE 3.10B

Determine the empirical formula of **(a)** a compound composed of 72.4% iron and 27.6% oxygen by mass; **(b)** a compound composed of 9.93% carbon, 58.6% chlorine, and 31.4% fluorine by mass; and **(c)** trinitrotoluene (TNT), an explosive that has the composition 37.01% C, 2.22% H, 18.50% N, and 42.27% O by mass.

Problem-Solving Note

Common decimals and their fractional equivalents are

$0.500 = 1/2$	$0.333 = 1/3$
$0.250 = 1/4$	$0.200 = 1/5$
$0.167 = 1/6$	$0.143 = 1/7$
$0.125 = 1/8$ and so on.	

The numbers shown in red are the factors we must multiply by in order to convert subscripts having these decimal endings to integers.

Relating Molecular Formulas to Empirical Formulas

The formulas for ionic compounds are empirical formulas. In a molecular formula, the subscript integers are either the same as those in the empirical formula or simple multiples of them. For example, in glucose, $C_6H_{12}O_6$, we multiply the subscripts of the empirical formula CH_2O by 6 to convert them to those of the molecular formula.

To establish a molecular formula, we need to know both the molecular mass and the empirical formula mass. The relationship between them can be stated as

$$\text{Empirical formula mass} \times \text{integral factor} = \text{molecular mass} \tag{3.6}$$

$$\text{Integral factor} = \frac{\text{molecular mass}}{\text{empirical formula mass}} \tag{3.7}$$

The integral factor (a simple whole number) is the multiplier that converts the subscripts of the empirical formula to those of the molecular formula. We can obtain the empirical formula mass from the empirical formula, and—as we have just seen—we can establish the empirical formula from mass percent composition data. In later chapters, we will consider several experimental methods of determining molecular masses.

Example 3.11

The empirical formula of hydroquinone, a chemical used in photography, is C_3H_3O, and its molecular mass is 110 u. What is its molecular formula?

STRATEGY

We can determine the molecular formula by multiplying the subscripts of the empirical formula by the integral factor obtained by using Equation (3.7).

SOLUTION

The empirical formula mass is $(3 \times 12.0\,u) + (3 \times 1.0\,u) + 16.0\,u = 55.0\,u$. The multiplier we need to convert the subscripts in the empirical formula to those in the molecular formula is the integral factor from Equation (3.7):

$$\text{Integral factor} = \frac{\text{molecular mass}}{\text{empirical formula mass}} = \frac{110\,u}{55.0\,u} = 2$$

The molecular formula is $(C_3H_3O)_2$, or $C_6H_6O_2$.

EXERCISE 3.11A

Ethylene (molecular mass 28.0 u), cyclohexane (84.0 u), and 1-pentene (70.0 u) all have the empirical formula CH_2. Give the molecular formula of each compound.

EXERCISE 3.11B

Give the empirical formula for **(a)** P_4O_6, **(b)** C_6H_9, **(c)** $C_6H_8O_6$, **(d)** $HOCH_2COOH$, **(e)** CH_2OHCH_2OH, and **(f)** $Cu_2C_2O_4$.

Problem-Solving Note

Because the integral factor is a whole number, high precision in these masses is not required.

3.6 Elemental Analysis: Experimental Determination of Mass Percent Composition

When a chemist in a research laboratory synthesizes a new compound, he or she most likely has some idea of its composition and structure. However, to publish this discovery in the chemical literature or to obtain a patent covering this work, the chemist must first present evidence of the identity of the compound. An important initial piece of evidence is an elemental analysis establishing the mass percent composition of the compound. The methods of analysis vary, depending on the elements present. We will limit our discussion to the analysis of organic compounds that contain only carbon, hydrogen, and oxygen. However, because more than 95% of all compounds are carbon-based, and because nearly all of them contain hydrogen and many contain oxygen as well, this covers a vast number of compounds.

In the apparatus pictured in Figure 3.4, a weighed sample of a compound is placed in a combustion "boat" and burned in a stream of oxygen gas in a high-temperature furnace. The carbon dioxide gas and water vapor produced in the combustion pass into sections of the apparatus containing absorbents. A substance such as magnesium perchlorate absorbs water vapor, and sodium hydroxide absorbs carbon dioxide. The masses of carbon dioxide and water are determined as the differences in masses of the absorbents after and before the combustion.

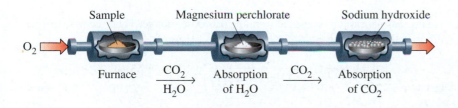

◄ FIGURE 3.4 Apparatus for combustion analysis

Oxygen gas enters the combustion apparatus and streams over the sample under analysis in a high-temperature furnace. The absorbers collect the water vapor and carbon dioxide produced in the combustion.

▶ **FIGURE 3.5**   **The law of conservation of mass in the combustion of methanol, CH₃OH**

In this molecular view, the two CH₃OH molecules are replaced by two CO₂ molecules and four H₂O molecules. Four of the original seven O₂ molecules remain after the combustion, which is carried out in an excess of oxygen.

QUESTION: How many O₂ molecules would remain if four molecules of methanol reacted with the seven O₂ molecules?

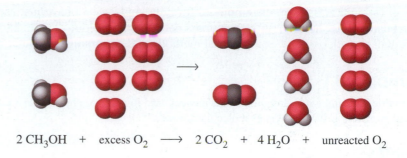

$$2\ CH_3OH\ +\ \text{excess}\ O_2\ \longrightarrow\ 2\ CO_2\ +\ 4\ H_2O\ +\ \text{unreacted}\ O_2$$

Consider the combustion of methanol (CH_3OH) in excess oxygen, which yields only carbon dioxide and water. The law of conservation of mass tells us that all the carbon atoms in the compound end up in the carbon dioxide molecules and all the hydrogen atoms end up in the water molecules. In the molecular view, depicted in Figure 3.5, we see that for every molecule of CH_3OH burned, one molecule of CO_2 and two molecules of H_2O are formed. By determining the mass of carbon in the CO_2 and the mass of hydrogen in the H_2O, we effectively determine the masses of carbon and hydrogen present in the original sample of methanol. From these masses and the mass of methanol burned, we can determine the mass percent of C and H in the methanol. Because the oxygen atoms in the combustion products come partly from the methanol and partly from the oxygen used in the combustion, we can determine the mass percent of O only indirectly: % O = 100.00% − % C − % H.

The conversion steps required to establish the mass percent composition of a compound from combustion analysis data are

$$g\ CO_2\ \longrightarrow\ mol\ CO_2\ \longrightarrow\ mol\ C\ \longrightarrow\ g\ C\ \longrightarrow\ \%\ C\ \text{in original sample}$$

$$g\ H_2O\ \longrightarrow\ mol\ H_2O\ \longrightarrow\ mol\ H\ \longrightarrow\ g\ H\ \longrightarrow\ \%\ H\ \text{in original sample}$$

Example 3.12

Burning a 0.1000-g sample of a carbon–hydrogen–oxygen compound in oxygen yields 0.1953 g CO_2 and 0.1000 g H_2O. A separate experiment shows that the molecular mass of the compound is 90 u. Determine **(a)** the mass percent composition, **(b)** the empirical formula, and **(c)** the molecular formula of the compound.

STRATEGY

First, we use conversions like those described above to calculate the mass percents in the compound. From this information, we can determine the empirical formula, integral factor, and molecular formula. Note that there is a specific order in which the calculations must be performed in order to arrive at the molecular formula.

SOLUTION

(a) First, we do the conversions outlined to calculate the mass of carbon in the CO_2 produced:

We want (?) and the unit g C.	Mass of CO₂ produced in the combustion	1/molar mass as factor to convert g CO₂ to mol O₂	Factor relating mol C to mol CO₂	Molar mass as factor to convert mol C to g C	Our answer: the number the unit

$$?\ g\ C\ =\ 0.1953\ g\ CO_2\ \times\ \frac{1\ mol\ CO_2}{44.010\ g\ CO_2}\ \times\ \frac{1\ mol\ C}{1\ mol\ CO_2}\ \times\ \frac{12.011\ g\ C}{1\ mol\ C}\ =\ 0.05330\ g\ C$$

This mass of carbon originated from the 0.1000-g sample. Thus, the mass percent carbon in the compound is

$$\text{Mass \% C}\ =\ \frac{0.05330\ g\ C}{0.1000\ g\ sample}\ \times\ 100\%\ =\ 53.30\%\ C$$

We can use similar calculations to determine first the mass of hydrogen and then the mass percent of hydrogen in the compound:

$$? \text{ g H} = 0.1000 \text{ g H}_2\text{O} \times \frac{1 \text{ mol H}_2\text{O}}{18.015 \text{ g H}_2\text{O}} \times \frac{2 \text{ mol H}}{1 \text{ mol H}_2\text{O}} \times \frac{1.0079 \text{ g H}}{1 \text{ mol H}} = 0.01119 \text{ g H}$$

$$\text{Mass \% H} = \frac{0.01119 \text{ g H}}{0.1000 \text{ g sample}} \times 100\% = 11.19\% \text{ H}$$

Finally, we can calculate the mass percent oxygen by subtracting the mass percents of C and H from 100.00%:

$$\% \text{ O} = 100.00\% - 53.30\% - 11.19\% = 35.51\%$$

(b) Here we apply the method of Examples 3.9 and 3.10, but we need only the first four steps.

Step 1: A 100.00-g sample of the compound contains 53.30 g C, 11.19 g H, and 35.51 g O.

Step 2: We convert the masses of C, H, and O to numbers of moles:

$$? \text{ mol C} = 53.30 \text{ g C} \times \frac{1 \text{ mol C}}{12.011 \text{ g C}} = 4.438 \text{ mol C}$$

$$? \text{ mol H} = 11.19 \text{ g H} \times \frac{1 \text{ mol H}}{1.0079 \text{ g H}} = 11.10 \text{ mol H}$$

$$? \text{ mol O} = 35.51 \text{ g O} \times \frac{1 \text{ mol O}}{15.999 \text{ g O}} = 2.220 \text{ mol O}$$

Step 3: Next, we use the numbers of moles in Step 2 as subscripts in a tentative formula:

$$C_{4.438}H_{11.10}O_{2.220}$$

Step 4: We divide all subscripts by the smallest subscript (2.220) to get integral subscripts:

$$C_{4.438/2.220}H_{11.10/2.220}O_{2.220/2.220} \longrightarrow C_{1.999}H_{5.000}O_{1.000} \longrightarrow C_2H_5O$$

(c) The empirical formula mass is $(2 \times 12.0 \text{ u}) + (5 \times 1.0 \text{ u}) + 16.0 \text{ u} = 45.0 \text{ u}$. The multiplier we need to convert the subscripts in the empirical formula to those in the molecular formula is the integral factor in Equation (3.7):

$$\text{Integral factor} = \frac{\text{molecular mass}}{\text{empirical formula mass}} = \frac{90 \text{ u}}{45.0 \text{ u}} = 2$$

The molecular formula is $(C_2H_5O)_2$, or $C_4H_{10}O_2$.

ASSESSMENT

As in previous examples, we check to see that the sum of the mass percentages is equal to 100% and that the calculation of the integral factor truly yields an integer.

EXERCISE 3.12A

Complete combustion of 0.255 g of an alcohol produces 0.561 g of CO_2 and 0.306 g of H_2O. Calculate **(a)** the mass percent composition and **(b)** the empirical formula for this alcohol.

EXERCISE 3.12B

A 0.3629-g sample of tetrahydrocannabinol (THC), the principal active component of marijuana, is burned to yield 1.0666 g of carbon dioxide and 0.3120 g of water. Calculate **(a)** the mass percent composition and **(b)** the empirical formula of THC.

Stoichiometry of Chemical Reactions

So far in this chapter, we have investigated stoichiometric calculations based on chemical formulas. We have shown how we can calculate mass percent compositions from formulas and how we can determine empirical formulas from mass percent compositions. In the following sections, we consider chemical reactions and some of the various stoichiometric calculations that aid our understanding of these processes. We start by considering chemical equations, which are symbolic representations of chemical reactions that provide considerable useful information about all the substances involved in the reactions.

3.7   Writing and Balancing Chemical Equations

A **chemical equation** is a shorthand description of a chemical reaction, using symbols and formulas to represent the elements and compounds involved. This shorthand description, based on *experiments*, shows that a reaction occurred, identifies all the substances involved in the reaction, and establishes the formulas of these substances.

For the reaction of carbon with a plentiful source of oxygen, as we have noted, the sole product is carbon dioxide. We can write an equation for this reaction:

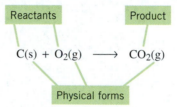

$$C(s) + O_2(g) \longrightarrow CO_2(g)$$

The plus sign indicates that carbon and oxygen react, and the arrow—usually read as "yields"—points to the result of their reaction: carbon dioxide. We generally call the starting substances in a reaction the **reactants** and the substances formed in the reaction, the **products.** The reactants appear to the *left* of the arrow in an equation, and the products appear to the *right* of the arrow.

Occasionally, we may need to indicate the physical form of the reactants and products, and we use these symbols to do this:

(g) = gas; (l) = liquid; (s) = solid; (aq) = aqueous (water) solution

These parenthetical symbols are attached to the formulas of the reactants and products:

$$C(s) + O_2(g) \longrightarrow CO_2(g)$$

If it is necessary to heat a mixture of reactants to bring about a chemical reaction, we sometimes denote this by placing a capital Greek letter delta, Δ, above the yield arrow. Sometimes the temperature or other conditions under which the reaction is carried out are also noted above the yield arrow:

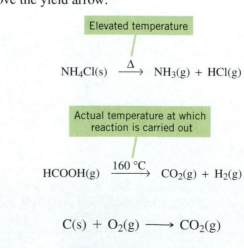

$$NH_4Cl(s) \xrightarrow{\Delta} NH_3(g) + HCl(g)$$

$$HCOOH(g) \xrightarrow{160\ °C} CO_2(g) + H_2(g)$$

The equation

$$C(s) + O_2(g) \longrightarrow CO_2(g)$$

can be interpreted in several ways. We can use it as a qualitative description of the reaction, as in "solid carbon and gaseous oxygen react to form gaseous carbon dioxide." We can give a microscopic interpretation, as in "one carbon atom reacts with one oxygen molecule to form one molecule of carbon dioxide gas." However, because we must usually work at the macroscopic level, the most useful interpretation of the equation is based on enormously large numbers of atoms and molecules, specifically the numbers found in a mole of substance: 1 mol (12.01 g) of carbon reacts with 1 mol (32.00 g) of oxygen gas to produce 1 mol (44.01 g) of carbon dioxide. This molar interpretation is at the heart of the quantitative calculations based on chemical equations, as we will see in Section 3.8.

The equation for the reaction of carbon and oxygen to form carbon dioxide is deceptively easy to write. If we try something similar for the reaction of hydrogen and oxygen to form water, however, we run into a bit of a problem. The following equation does not conform to the law of conservation of mass, and therefore it is not *balanced*:

$$H_2(g) + O_2(g) \longrightarrow H_2O(l) \quad \textit{(not balanced)}$$

Two O atoms in the form of an O_2 molecule are shown on the left side of the equation, but there is only one O atom, in the H_2O molecule, on the right side. More O atoms are present on the reactant side than on the product side, but we know that atoms cannot be created or destroyed in a chemical reaction. We should not assume that H_2 and O_2 molecules react in a 1 : 1 ratio. They don't, as we show in the molecular interpretation in Figure 3.6, where we illustrate how to balance the equation so that it agrees with the law of conservation of mass. However, we don't need to draw molecular pictures to balance equations; we can work directly with the equation and use stoichiometric coefficients to adjust the ratios of the reactants and products.

Stoichiometric coefficients are numbers placed in front of formulas in a chemical equation to balance the equation; they indicate the combining ratios of the reactants and the ratios in which products are formed. A stoichiometric coefficient multiplies everything in the formula that follows it. If there is no coefficient before a formula, the coefficient 1 is understood.

To balance the equation

$$H_2(g) + O_2(g) \longrightarrow H_2O(l) \quad \textit{(not balanced)}$$

we can begin by placing the stoichiometric coefficient 2 in front of the formula H_2O in order to balance the oxygen:

$$H_2(g) + O_2(g) \longrightarrow 2\,H_2O(l) \quad \textit{(O balanced, H not balanced)}$$

Balancing an Equation activity

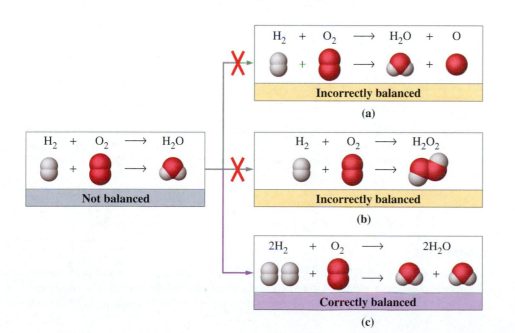

◀ **FIGURE 3.6** **Balancing the chemical equation for the reaction of hydrogen with oxygen to form water.**

(a) *Incorrect:* There is no evidence for the presence of atomic oxygen as a product. *A reactant or product having a chemical formula different from the formula of any substance in the original equation cannot be introduced for the purpose of balancing an equation.* (b) *Incorrect:* The product of the reaction is water, H_2O, not hydrogen peroxide, H_2O_2. *A formula cannot be changed in order to balance a chemical equation.* (c) *Correct:* An equation can be balanced only through the use of *correct formulas and coefficients.*

Now we have two oxygen atoms on each side of the equation, but our added 2 increases the number of H atoms on the right to *four* at the same time that it increases the number of O atoms to two. This is a problem because there are only *two* H atoms on the left. We can correct this imbalance by placing another stoichiometric coefficient 2 in front of the H_2 on the left. The equation is now balanced:

$$2\,H_2(g) + O_2(g) \longrightarrow 2\,H_2O(l) \quad (balanced)$$

with four H atoms and two O atoms on each side of the arrow.

The point made in Figure 3.6 is extremely important: We can balance an equation *only* by adjusting coefficients. We *cannot* do so by changing any subscripts in the formulas or by adding or removing a reactant or product from the equation. This might balance the equation, but *it would no longer describe the desired reaction.*

The method we just described is called *balancing by inspection.* The following simple strategies reduce the trial-and-error aspect of the method.

- If an element is present in just one compound on each side of the equation, try balancing that element first.
- In some reactions, certain groupings of atoms (such as in polyatomic ions) remain unchanged. In such cases, balance these groupings as a unit.
- Ordinarily, balance reactants or products existing as free elements last.
- At times, an equation can be balanced most readily by first using a fractional coefficient(s) and then clearing the fractions by multiplying all coefficients by a common multiplier.

In any case, the most important step in any strategy is to check an equation to ensure that it is indeed balanced: *For each element, the same number of atoms must appear on each side of the arrow.* To phrase this most important point another way, atoms are conserved in chemical reactions.

Example 3.13

Balance the equation

$$Fe + O_2 \longrightarrow Fe_2O_3 \quad (not\ balanced)$$

SOLUTION

It seems that the easiest place to begin is with the iron atoms. There is one on the left (Fe) and two on the right (Fe_2O_3), and we can balance them by placing the coefficient 2 on the left:

$$2\,Fe + O_2 \longrightarrow Fe_2O_3 \quad (Fe\ balanced,\ O\ not\ balanced)$$

To balance the oxygen atoms, we begin by noting that there are two of them on the left and three on the right. An easy way to get three O atoms on the left, thereby balancing the equation, is to use the *fractional* coefficient 3/2 before O_2 on the left $\left\{\frac{3}{2} \times 2 = 3\right\}$:

$$2\,Fe + \tfrac{3}{2}O_2 \longrightarrow Fe_2O_3 \quad (balanced,\ fractional\ coefficient)$$

Problem-Solving Note

Ordinarily, when a balanced equation is requested, it is balanced according to the smallest whole-number coefficients.

Fractional coefficients are not only acceptable in equations, sometimes they are desirable. If we don't want them, however, we can easily clear an equation of them by multiplying *every* coefficient by the smallest integer required to clear the fractions, in this case, 2:

$$2 \times \left\{2\,Fe + \tfrac{3}{2}O_2 \longrightarrow Fe_2O_3\right\}$$

becomes

$$4\,Fe + 3\,O_2 \longrightarrow 2\,Fe_2O_3 \quad (balanced,\ coefficients\ integral)$$

ASSESSMENT

To verify that an equation is balanced, we multiply each subscript in a formula by the stoichiometric coefficient for that formula. This provides a count of the number of atoms of each type on both sides of the equation. For each element, the number of atoms must be the same on the two sides of a *balanced* equation. In the equation above, 4 Fe and 6 O atoms are seen on each side of the equation; it is balanced.

EXERCISE 3.13A

Balance the following equations using integral coefficients:

(a) $SiCl_4 + H_2O \longrightarrow SiO_2 + HCl$

(b) $PCl_5 + H_2O \longrightarrow H_3PO_4 + HCl$

(c) $CaO + P_4O_{10} \longrightarrow Ca_3(PO_4)_2$

EXERCISE 3.13B

Use integral coefficients to write a balanced equation for each of the following reactions. Use the (g), (l), (s), (aq) symbols to indicate the form of each reactant and product:

(a) The formation of solid lead iodide and aqueous potassium nitrate from aqueous solutions of lead nitrate and potassium iodide.

(b) The reaction of gaseous hydrogen chloride and oxygen to form gaseous water and chlorine.

Example 3.14

Balance the equation

$$C_2H_6 + O_2 \longrightarrow CO_2 + H_2O$$

SOLUTION

Oxygen appears as the free element on the left, so let's leave it for last and balance the other two elements first. To balance carbon, we place the coefficient 2 in front of CO_2:

$$C_2H_6 + O_2 \longrightarrow 2\,CO_2 + H_2O \quad \textit{(C balanced, H and O not balanced)}$$

To balance hydrogen, we need the coefficient 3 before H_2O:

$$C_2H_6 + O_2 \longrightarrow 2\,CO_2 + 3\,H_2O \quad \textit{(C and H balanced, O not balanced)}$$

Now, if we count oxygen atoms, we find two on the left and seven on the right. We can get seven on each side by using the *fractional* coefficient $\frac{7}{2}$ on the left $\left(\frac{7}{2} \times 2 = 7\right)$:

$$C_2H_6 + \tfrac{7}{2}O_2 \longrightarrow 2\,CO_2 + 3\,H_2O \quad \textit{(balanced)}$$

To obtain integral coefficients, we multiply each coefficient by 2:

$$2 \times \{C_2H_6 + \tfrac{7}{2}O_2 \longrightarrow 2\,CO_2 + 3\,H_2O\}$$

That leads to

$$2\,C_2H_6 + 7\,O_2 \longrightarrow 4\,CO_2 + 6\,H_2O \quad \textit{(balanced)}$$

ASSESSMENT

We ensure that the equation is balanced by multiplying the subscript for each element by the appropriate integral coefficient to get the number of atoms on the two sides of the equation. Note that oxygen atoms are present in two product molecules. On each side of the equation are four C atoms, twelve H atoms, and fourteen O atoms; the equation is balanced.

EXERCISE 3.14A

Balance the following equations:

(a) $C_4H_{10} + O_2 \longrightarrow CO_2 + H_2O$

(b) $CH_3CH_2CH_2CH(OH)CH_2OH + O_2 \longrightarrow CO_2 + H_2O$

EXERCISE 3.14B

Write a balanced equation to represent the complete combustion of the compound represented by the molecular model in the margin. (Recall Figures 2.7 and 2.8.)

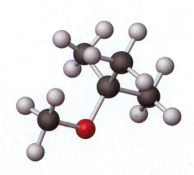

Example 3.15

Balance the equation

$$H_3PO_4 + NaCN \longrightarrow HCN + Na_3PO_4$$

SOLUTION

Notice that the PO_4 and CN *groups* remain unchanged in the reaction. For purposes of balancing equations, we can often treat such groups as a whole rather than breaking them down into their constituent atoms. To balance hydrogen atoms, we place a 3 before HCN:

$$H_3PO_4 + NaCN \longrightarrow 3\,HCN + Na_3PO_4 \quad \textit{(not balanced)}$$

To balance cyanide ions, we put a 3 in front of the NaCN. Doing this also balances the sodium ions:

$$H_3PO_4 + 3\,NaCN \longrightarrow 3\,HCN + Na_3PO_4 \quad \textit{(balanced)}$$

Because the PO_4 was balanced to begin with and we did nothing to upset that balance, we are finished; the equation is balanced.

ASSESSMENT

Note that in the balanced equation, there are three Na atoms, three H atoms, one PO_4 group, and three CN groups on each side of the equation.

EXERCISE 3.15A

Balance the following equations.

(a) $FeCl_3 + NaOH \longrightarrow Fe(OH)_3 + NaCl$

(b) $Ba(NO_3)_2 + Al_2(SO_4)_3 \longrightarrow BaSO_4 + Al(NO_3)_3$

EXERCISE 3.15B

Write a balanced equation for the reaction between solid calcium hydroxide and aqueous phosphoric acid to form solid calcium phosphate and water. Include symbols indicating the physical form of each reactant and product.

Example 3.16   A Conceptual Example

Write a plausible chemical equation for the reaction between water and a liquid molecular chloride of phosphorus to form an aqueous solution of hydrochloric acid and phosphorus acid. The phosphorus-chlorine compound is 77.45% Cl by mass.

ANALYSIS AND CONCLUSIONS

In this problem, balancing a chemical equation is the last step. First, we must apply some earlier ideas:

Establishing the formula of the chloride of phosphorus.
We can apply the method of Examples 3.9 and 3.10 to a compound that is 22.55% P and 77.45% Cl. A 100.00-g sample of the compound consists of 22.55 g P and 77.45 g Cl, corresponding to 0.728 mol P and 2.185 mol Cl. The formula $P_{0.728}Cl_{2.185}$ reduces to PCl_3.

Establishing the formulas of hydrochloric and phosphorus acids.
We described the relationship between names and formulas on pages 57–58. Hydrochloric acid, a binary acid, has the formula HCl. Table 2.5 gives H_3PO_4 as the formula of phosphoric acid. Phosphorus acid should have one O atom fewer per molecule than the "-*ic*" acid, giving the formula H_3PO_3.

Writing and balancing the equation.
The unbalanced equation, including an indication of the physical form of each substance, is

$$PCl_3(l) + H_2O(l) \longrightarrow HCl(aq) + H_3PO_3(aq)$$

The balanced equation requires the coefficient 3 for $H_2O(l)$ and for $HCl(aq)$:

$$PCl_3(l) + 3\,H_2O(l) \longrightarrow 3\,HCl(aq) + H_3PO_3(aq)$$

EXERCISE 3.16A

Write a plausible chemical equation for the combustion of liquid triethylene glycol in an abundant supply of oxygen gas. Triethylene glycol is 47.99% C, 9.40% H, and 42.62% O by mass and has a molecular mass of 150.2 u.

EXERCISE 3.16B

Upon heating, solid lead(II) nitrate yields solid lead(II) oxide and two gaseous products, one of which is nitrogen dioxide. Write a balanced equation for this reaction.

Our main interest in this section has been simply balancing equations. It is much more important, however, to be able to *predict* whether a chemical reaction will occur and then to write an equation to represent it. We will introduce a number of new ideas in later chapters to show how to make such predictions.

3.8 Reaction Stoichiometry

Whether making medicines, obtaining metals from their ores, studying the combustion of a rocket fuel, synthesizing new compounds, or simply testing a hypothesis, chemists need to consider the relative amounts of materials involved. Material quantities are often expressed in terms of their masses; however, it is the number of *moles* that relates quantities of atoms or molecules to one another. The necessary mole/mass relationships can be derived from chemical formulas.

Consider the reaction of carbon monoxide and hydrogen to form methanol:

$$CO + 2\,H_2 \longrightarrow CH_3OH$$

At the microscopic level, the stoichiometric coefficients mean that for every *one* molecule of CO that reacts, *two* molecules of H_2 react and *one* molecule of CH_3OH is formed. In the reaction of *10* molecules of CO, *20* molecules of H_2 also react and *10* molecules of CH_3OH are formed. The reactants and product retain this 1 : 2 : 1 ratio no matter how many molecules we choose. If we work in the range of Avogadro's number ($N_A = 6.022 \times 10^{23}$), we are at the macroscopic level and can switch to a molar basis: *One* mole of CO reacts with *two* moles of H_2 to produce *one* mole of CH_3OH. These amounts of CO, H_2, and CH_3OH are said to be *stoichiometrically equivalent*, which we can represent in the following way:

$$1\ mol\ CO \,\approx\, 2\ mol\ H_2$$

$$1\ mol\ CO \,\approx\, 1\ mol\ CH_3OH$$

$$2\ mol\ H_2 \,\approx\, 1\ mol\ CH_3OH$$

The symbol $\approx$ means "is stoichiometrically equivalent to." To say that 1 mol CO is stoichiometrically equivalent to 2 mol H_2 in this reaction is not the same as saying that these two substances are equal. There is no way CO and H_2 can be thought of as equal or identical—they are two completely different substances. Nevertheless, the two combine in a fixed ratio. We can express this ratio as either of two fractions:

$$\frac{1\ mol\ CO}{2\ mol\ H_2} \quad or \quad \frac{2\ mol\ H_2}{1\ mol\ CO}$$

It is necessary to keep in mind that the chemical equivalence between H_2 and CO depends on the particular reaction. For another reaction between these same two substances,

$$CO + 3\,H_2 \longrightarrow CH_4 + H_2O$$

the corresponding fixed ratio is

$$\frac{1\ mol\ CO}{3\ mol\ H_2} \quad or \quad \frac{3\ mol\ H_2}{1\ mol\ CO}$$

Figure 3.7 illustrates a commonplace example of an equivalence encountered when parking automobiles. The equivalence between automobile and curb length is different depending on whether the parking arrangement is parallel or perpendicular.

Problem-Solving Note

We can read a balanced equation in terms of molecules and atoms *or* in terms of moles. For much of what we do in chemistry, moles will be more useful.

Stoichiometric Equivalence activity

Application Note

Alkanes containing more carbon atoms than CH_4, such as the alkanes in gasoline, can be synthesized by reactions that incorporate different proportions of CO and H_2. These mixtures of CO and H_2, called *synthesis gas*, can be produced by a reaction between coal and steam.

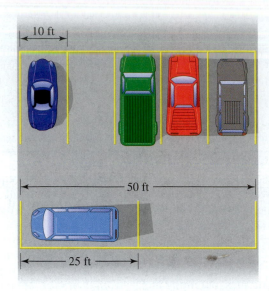

▶ **FIGURE 3.7   The concept of equivalence**

With parallel parking (bottom), each vehicle *is equivalent* to 25 ft of curb space; that is, 1 vehicle/25 ft. The number of vehicles that can be parked is

$$50 \text{ ft} \times \frac{1 \text{ vehicle}}{25 \text{ ft}} = 2 \text{ vehicles}$$

QUESTION: With perpendicular parking (top), what is the conversion factor to be used, and how many vehicles can be parked along the 50-ft section of curb?

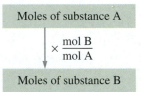

▲ **FIGURE 3.8   The mole ratio in stoichiometry**

Substances A and B may be either reactants or products. Substance A is the one about which information is given, and substance B is the one about which we seek information (the answer). We use the mole ratio of B to A as a conversion factor.

Conversion factors formed from the stoichiometric coefficients in a chemical equation are called **stoichiometric factors** or **mole ratios.** These mole ratios are key conversion factors in solving stoichiometry problems, entering into such calculations as shown in Figure 3.8 and illustrated in Example 3.17.

Example 3.17

When 0.105 mol propane is burned in an excess of oxygen, how many moles of oxygen are consumed? The reaction is

$$C_3H_8 + 5\,O_2 \longrightarrow 3\,CO_2 + 4\,H_2O$$

STRATEGY

We will relate the moles of consumed oxygen to moles of propane burned, using as our conversion factor the ratio of the number of moles of each species present in the balanced chemical equation.

SOLUTION

The stoichiometric coefficients in the equation allow us to write the equivalence

$$1 \text{ mol } C_3H_8 \,\Leftrightarrow\, 5 \text{ mol } O_2$$

From this equivalence, we can derive two conversion factors:

$$\frac{5 \text{ mol } O_2}{1 \text{ mol } C_3H_8} \quad \text{and} \quad \frac{1 \text{ mol } C_3H_8}{5 \text{ mol } O_2}$$

Because we are given the number of moles of propane (substance A in Figure 3.8) and are seeking the number of moles of O_2 consumed (substance B in Figure 3.8), we need the factor that has the unit "mol O_2" in the numerator and the unit "mol C_3H_8" in the denominator. This is the conversion factor shown in red above:

$$? \text{ mol } O_2 = 0.105 \text{ mol } C_3H_8 \times \frac{5 \text{ mol } O_2}{1 \text{ mol } C_3H_8} = 0.525 \text{ mol } O_2$$

EXERCISE 3.17A

For the combustion of propane in Example 3.17:

(a) How many moles of carbon dioxide are formed when 0.529 mol C_3H_8 is burned?

(b) How many moles of water are produced when 76.2 mol C_3H_8 is burned?

(c) How many moles of carbon dioxide are produced when 1.010 mol O_2 is consumed?

EXERCISE 3.17B

For the combustion of 55.6 g of butane in an excess of oxygen, **(a)** how many moles of carbon dioxide are formed and **(b)** how many moles of oxygen are consumed?

Although the mole is essential in calculations based on chemical equations, we cannot measure out molar amounts directly. We have to relate them to quantities that we can measure: mass in grams or kilograms, volume in milliliters or liters, and so on. For calculations involving mass, we can use the molar mass of the substance in question as a conversion factor. Figure 3.9 provides a general flow chart for mass calculations. Notice that we've created this flow chart simply by adding steps to Figure 3.8. As we encounter more complex problems, we will further modify the basic flow chart.

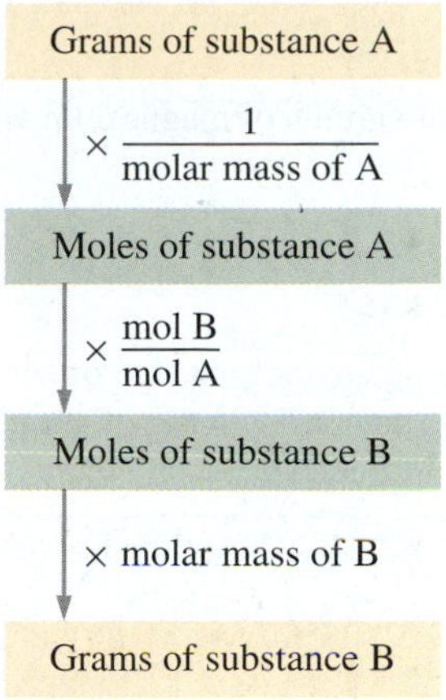

Grams of substance A

$\times \dfrac{1}{\text{molar mass of A}}$

Moles of substance A

$\times \dfrac{\text{mol B}}{\text{mol A}}$

Moles of substance B

$\times$ molar mass of B

Grams of substance B

▶ **FIGURE 3.9   Stoichiometry involving mass**
Preliminary and follow-up calculations are required in addition to the mole-ratio conversion factor shown in Figure 3.8. We begin by using the inverse of the molar mass of substance A to convert mass of A to moles of A and finish by using the molar mass of B to convert moles of B to mass of B.

Example 3.18

The final step in the production of nitric acid involves the reaction of nitrogen dioxide with water; nitrogen monoxide is also produced. How many grams of nitric acid are produced for every 100.0 g of nitrogen dioxide that reacts?

STRATEGY

Working from a balanced chemical equation and using the appropriate stoichiometric factor, we determine the moles of product produced from the given moles of reactant. The conversions from grams to moles and moles to grams are based on conversion factors associated with the respective molar masses.

SOLUTION

Because no equation is given, we must first write a chemical equation from the description of the reaction. Recall that we related the names and formulas of these reactants and products in Chapter 2.

$$NO_2 + H_2O \longrightarrow HNO_3 + NO \quad \textit{(not balanced)}$$

To balance the equation, let's first balance H atoms because H appears in one reactant and one product. Then we can balance N atoms because they appear on the reactant side only in NO_2. At this point, O atoms will also be balanced.

$$3\,NO_2 + H_2O \longrightarrow 2\,HNO_3 + NO \quad \textit{(balanced)}$$

Now, we convert the mass of the given substance, NO_2, to an amount in moles, using the molar mass in the manner shown in Figure 3.9.

$$?\text{ mol } NO_2 = 100.0 \text{ g } NO_2 \times \frac{1 \text{ mol } NO_2}{46.006 \text{ g } NO_2} = 2.174 \text{ mol } NO_2$$

Next, we use coefficients from the balanced equation to establish the stoichiometric equivalence of NO_2 and HNO_3.

$$3 \text{ mol } NO_2 \backsimeq 2 \text{ mol } HNO_3$$

Because we need to convert from moles of NO_2 to moles of HNO_3, we should use this equivalence to write the stoichiometric factor needed for the conversion.

$$\frac{2 \text{ mol } HNO_3}{3 \text{ mol } NO_2}$$

$$?\text{ mol } HNO_3 = 2.174 \text{ mol } NO_2 \times \frac{2 \text{ mol } HNO_3}{3 \text{ mol } NO_2} = 1.449 \text{ mol } HNO_3$$

Finally, we can convert from moles of HNO_3 to grams of HNO_3, using its molar mass.

$$?\text{ g } HNO_3 = 1.449 \text{ mol } HNO_3 \times \frac{63.013 \text{ g } HNO_3}{1 \text{ mol } HNO_3} = 91.31 \text{ g } HNO_3$$

ASSESSMENT

As you gain experience with reaction stoichiometry calculations, you will likely switch to doing all that we have done here in a combined setup that obviates the need to write intermediate answers.

$$?\text{ g } HNO_3 = 100.0 \text{ g } NO_2 \times \frac{1 \text{ mol } NO_2}{46.006 \text{ g } NO_2} \times \frac{2 \text{ mol } HNO_3}{3 \text{ mol } NO_2} \times \frac{63.013 \text{ g } HNO_3}{1 \text{ mol } HNO_3}$$

$$= 91.31 \text{ g } HNO_3$$

EXERCISE 3.18A

How many grams of magnesium are required to convert 83.6 g $TiCl_4$ to titanium metal? The balanced equation representing this reaction is

$$TiCl_4 + 2\,Mg \xrightarrow{\Delta} Ti + 2\,MgCl_2$$

EXERCISE 3.18B

Upon being strongly heated or subjected to severe mechanical shock, ammonium nitrate decomposes into nitrogen gas, oxygen gas, and water vapor. If 75.5 g of ammonium nitrate decomposes in this way, how many grams of nitrogen and how many grams of oxygen are produced?

Example 3.19

Ammonium sulfate, a common fertilizer used by gardeners, is produced commercially by passing gaseous ammonia into an aqueous solution that is 65% H_2SO_4 by mass and has a density of 1.55 g/mL. How many milliliters of this sulfuric acid solution are required to convert 1.00 kg NH_3 to $(NH_4)_2SO_4$?

STRATEGY

Here we want to determine the amount of one reactant (sulfuric acid) needed to react completely with a second reactant (ammonia). As before, we will use a stoichiometric factor based on a balanced chemical equation to determine the stoichiometric equivalence between the two reactants. We will use the respective molar masses in calculating the number of moles of NH_3 initially present and the mass of H_2SO_4 consumed. Finally, we will use the solution density and mass percentage to calculate the volume of $H_2SO_4(aq)$ needed.

SOLUTION

First, we must write the balanced equation for the reaction:

$$2\,NH_3(g) + H_2SO_4(aq) \longrightarrow (NH_4)_2SO_4(aq)$$

The required setup has 1.00 kg NH_3—the given substance—as its starting point. Because we are seeking a *volume* of $H_2SO_4(aq)$, the setup has the general form

$$? \text{ mL } H_2SO_4(aq) = 1.00 \text{ kg } NH_3 \times \text{conversion factors}$$

We begin by converting kilograms of NH_3 to grams of NH_3. Then we can apply the flow chart of Figure 3.9 to find grams of H_2SO_4. Finally, we use the percent composition and then the density of the $H_2SO_4(aq)$ to convert from grams of H_2SO_4 to milliliters of $H_2SO_4(aq)$:

Problem-Solving Note

Careful here! The notation "g H_2SO_4" means the mass of sulfuric acid, and "g $H_2SO_4(aq)$" means the mass of the *aqueous solution.* Don't mix them up.

$$kg\ NH_3 \longrightarrow g\ NH_3 \longrightarrow mol\ NH_3 \longrightarrow mol\ H_2SO_4$$

> 1000 g/kg

$$\longrightarrow g\ H_2SO_4 \longrightarrow g\ H_2SO_4(aq) \longrightarrow ml\ H_2SO_4(aq)$$

> 100.0 g/65 g   1 mL/1.55 g

In the combined setup, the conversions are done in the order in which they are described above:

$$? \text{ mL } H_2SO_4(aq) = 1.00 \text{ kg } NH_3 \times \frac{1000 \text{ g } NH_3}{1 \text{ kg } NH_3} \times \frac{1 \text{ mol } NH_3}{17.03 \text{ g } NH_3} \times \frac{1 \text{ mol } H_2SO_4}{2 \text{ mol } NH_3}$$

$$\times \frac{98.08 \text{ g } H_2SO_4}{1 \text{ mol } H_2SO_4} \times \frac{100.0 \text{ g } H_2SO_4(aq)}{65 \text{ g } H_2SO_4} \times \frac{1 \text{ mL } H_2SO_4(aq)}{1.55 \text{ g } H_2SO_4(aq)}$$

$$= 2.9 \times 10^3 \text{ mL } H_2SO_4(aq)$$

ASSESSMENT

Note that the percent composition (65% H_2SO_4) and density (1.55 g/mL) are written in the *inverted* form. This provides for the proper cancellation of units, and it makes sense. The mass of $H_2SO_4(aq)$ should be greater than that of the pure H_2SO_4 from which the solution is made, and the volume of $H_2SO_4(aq)$ should be a smaller number than its mass because the density of the solution is greater than 1.

EXERCISE 3.19A

How many milliliters of liquid water are produced by the combustion in abundant oxygen of 775 mL of octane, $C_8H_{18}(l)$? Assume that the volumes of both liquids are measured at 20.0 °C, where the densities are 0.7025 g/mL for octane and 0.9982 g/mL for water.

$$2\ C_8H_{18}(l) + 25\ O_2(g) \longrightarrow 16\ CO_2(g) + 18\ H_2O\ (l)$$

EXERCISE 3.19B

How many milliliters of water at 20.0 °C ($d = 0.9982$ g/mL) are formed in the combustion of 25.0 g of a methanol–propanol mixture that is 62.5% methanol by mass? Assume there is an excess of oxygen available.

3.9   Limiting Reactants

Suppose, in running a reaction in the laboratory, we measure out our reactants such that the number of moles of each reactant is the same as the reactant's stoichiometric coefficient in the balanced equation. In this case, we say that the reactants are in *stoichiometric proportions*, and we find that if the reaction goes to completion, all the reactants are totally consumed. In practice, however, we often carry out reactions with a limited amount of one reactant and plentiful amounts of the others.

The reactant that is completely consumed in a chemical reaction limits the amounts of products formed and is called the **limiting reactant** (or, sometimes, the *limiting reagent*, where "reagent" is a general term for a chemical used in a reaction). In the combustion of octane described in Exercise 3.19A,

$$2\ C_8H_{18}(l) + 25\ O_2(g) \longrightarrow 16\ CO_2(g) + 18\ H_2O\ (l)$$

if we allow 2 mol C_8H_{18} to react with 25 mol O_2, the reactants are in stoichiometric proportions. On the other hand, if we allow the 2 mol C_8H_{18} to burn in a plentiful supply of $O_2(g)$—more than 25 mol—then the C_8H_{18} is the limiting reactant. It is completely consumed, and some unreacted O_2 remains; the O_2 is a reactant present in excess. Figure 3.10 is a microscopic view of a chemical reaction in which one of the reactants is the limiting reactant and the other is in excess.

 **Limiting Reagent animation**

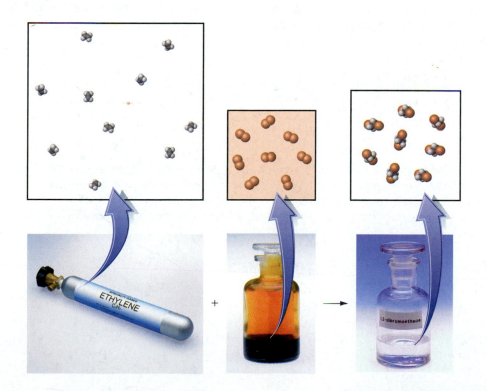

◀ **FIGURE 3.10   A molecular view of the reactants in the reaction between ethylene and bromine** Ethylene (1.0 mol, 28 g, shown by the black and gray models) and bromine (0.800 mol, 128 g, orange models) react in a 1 : 1 mole ratio to produce 1,2-dibromoethane, a colorless liquid:

$$C_2H_4(g) + Br_2(g) \longrightarrow C_2H_4Br_2(l)$$

The mass of bromine is greater, but ethylene molecules outnumber bromine molecules. Thus, ethylene is present in excess, and bromine is the limiting reactant.

QUESTION: If the reaction were carried out with only the number of molecules pictured here, what would the molecular view look like following the reaction?

A limiting reactant problem is much like the task of packaging meals for a school outing. Let's say each package consists of a sandwich, two cookies, and an orange:

1 sandwich : 2 cookies : 1 orange   yields   1 packaged meal

Limiting Reagent Analogy activity

The "stoichiometric proportions" of the "reactants" are 1 : 2 : 1. If we have 100 sandwiches, 200 cookies, and 100 oranges, we can prepare 100 packages and have nothing left over—the components needed for the packages are in stoichiometric proportions. Suppose we have available 105 sandwiches, 202 cookies, and 107 oranges. How many packages can we prepare? One way to answer the question is to consider each component separately and determine how many packages can be made from the available quantity of that component, assuming for the moment that there is enough of the other components available:

$$? \text{ Packages} = 105 \; \cancel{\text{sandwiches}} \times \frac{1 \text{ package}}{1 \; \cancel{\text{sandwich}}} = 105 \text{ packages}$$

$$? \text{ Packages} = 202 \; \cancel{\text{cookies}} \times \frac{1 \text{ package}}{2 \; \cancel{\text{cookies}}} = 101 \text{ packages}$$

$$? \text{ Packages} = \cancel{107 \text{ oranges}} \times \frac{1 \text{ package}}{1 \; \cancel{107 \text{ orange}}} = 107 \text{ packages}$$

Problem-Solving Note

General approach for limiting reactant problems: Find the moles of product produced by each reactant. The limiting reactant will form the smallest number of moles of product; base the calculation on that number.

Problem-Solving Note

In Example 3.20, comparing the starting *masses*—35.00 g Mg and 15.00 g N_2—can actually be misleading. We need to work on a molar basis and with stoichiometric factors as well. The substance present in greater mass, Mg, is actually the limiting reactant in this instance.

Only one of these answers can be correct: the *smallest*. We have enough sandwiches to make 105 packages and enough oranges for 107 packages, but enough cookies for only 101 packages. When we have prepared 101 packages, we will have used up all the cookies and have some sandwiches and oranges left over. The cookies are the limiting reactant, and the sandwiches and oranges are in excess.

Be sure to keep in mind that the limiting reactant is not necessarily the one present in smallest quantity. In the snack analogy, the cookies were definitely not the component present in smallest quantity. The situation required twice as many cookies as sandwiches and oranges, however; and so even with 202 cookies, we didn't have twice as many cookies as sandwiches or twice as many cookies as oranges. A similar approach is used in the following examples to determine the limiting reactant and the amount of product produced from specific amounts of reactants.

Example 3.20

Magnesium nitride can be formed by the reaction of magnesium metal with nitrogen gas. **(a)** How many grams of magnesium nitride can be made in the reaction of 35.00 g of magnesium and 15.00 g of nitrogen? **(b)** How many grams of the excess reactant remain after the reaction?

STRATEGY

In this problem, we need to determine which of the reactants is completely consumed and is therefore the limiting reactant. The quantity of this reactant, in turn, will determine the quantity of magnesium nitride produced. We will need a grams-to-moles conversion factor to convert from the given reactant masses and a moles-to-grams factor to convert to the desired product mass. The quantity of excess reactant can be calculated as the difference between the given mass of this reactant and the mass consumed in the reaction.

SOLUTION

(a) As usual, we must first write a balanced equation, and we can do this by using ideas from this chapter and Chapter 2.

$$3 \text{ Mg (s)} + N_2(g) \longrightarrow Mg_3N_2(s)$$

We can identify the limiting reactant by finding the number of moles of $Mg_3N_2(s)$ produced by assuming first one reactant, and then the other is the limiting reactant.

Assuming Mg is the limiting reactant and N_2 is in excess:

$$? \text{ mol Mg}_3\text{N}_2 = 35.00 \text{ g Mg} \times \frac{1 \text{ mol Mg}}{24.305 \text{ g Mg}} \times \frac{1 \text{ mol Mg}_3\text{N}_2}{3 \text{ mol Mg}}$$

$$= 0.4800 \text{ mol Mg}_3\text{N}_2$$

Assuming N_2 is the limiting reactant and Mg is in excess:

$$? \text{ mol Mg}_3\text{N}_2 = 15.00 \text{ g N}_2 \times \frac{1 \text{ mol N}_2}{28.013 \text{ g N}_2} \times \frac{1 \text{ mol Mg}_3\text{N}_2}{1 \text{ mol N}_2}$$

$$= 0.5355 \text{ mol Mg}_3\text{N}_2$$

Because the amount of product in the first calculation (0.4800 mol Mg_3N_2) is smaller than that in the second (0.5355 mol Mg_3N_2), we know that magnesium is the limiting reactant. When 0.4800 mol Mg_3N_2 has been formed, the Mg is completely consumed and the reaction stops, producing a specific mass of Mg_3N_2.

$$? \text{ g Mg}_3\text{N}_2 = 0.4800 \text{ mol Mg}_3\text{N}_2 \times \frac{100.93 \text{ g Mg}_3\text{N}_2}{1 \text{ mol Mg}_3\text{N}_2} = 48.45 \text{ g Mg}_3\text{N}_2$$

(b) Having found that the amount of product is 0.4800 mol Mg_3N_2, we can now calculate how much N_2 must have been consumed.

$$? \text{ g N}_2 = 0.4800 \text{ mol Mg}_3\text{N}_2 \times \frac{1 \text{ mol N}_2}{1 \text{ mol Mg}_3\text{N}_2} \times \frac{28.013 \text{ g N}_2}{1 \text{ mol N}_2}$$

$$= 13.45 \text{ g N}_2$$

From this, we calculate the mass of excess N_2.

$$15.00 \text{ g N}_{2(\text{initially})} - 13.45 \text{ g N}_{2(\text{consumed})} = 1.55 \text{ g N}_{2(\text{excess})}$$

EXERCISE 3.20A

One way to produce hydrogen sulfide gas is by the reaction of iron(II) sulfide with hydrochloric acid:

$$\text{FeS(s)} + 2 \text{ HCl(aq)} \longrightarrow \text{FeCl}_2\text{(aq)} + \text{H}_2\text{S(g)}$$

If 10.2 g HCl is added to 13.2 g FeS, how many grams of H_2S can be formed? What is the mass of the excess reactant remaining?

EXERCISE 3.20B

A convenient laboratory source of hydrogen gas is the reaction of an aqueous hydrochloric acid solution with aluminum metal. An aqueous solution of aluminum chloride is the other product of the reaction. How many grams of hydrogen are produced in the reaction between 12.5 g of aluminum and 250.0 mL of an aqueous hydrochloric acid solution that is 25.6% HCl by mass and has a density of 1.13 g/mL?

3.10 Yields of Chemical Reactions

In many industrial and commercial chemical processes, the quantity of product actually obtained—the *yield*—is one of the important measures of a successful chemical reaction. The calculated quantity of product in a reaction is called the **theoretical yield** of the reaction. In Example 3.20, the calculated quantity of product is 48.45 g Mg_3N_2, and this is the theoretical yield of magnesium nitride. The quantity of magnesium nitride actually formed in the reaction described in Example 3.20—the **actual yield**—might well be less than this theoretical yield. Actual yields of chemical reactions are often less than theoretical yields for a variety of reasons (Figure 3.11).

Mixture of Zn(s) and S_8(s)

(a)

(b)

Mixture of ZnS(s), ZnO(s), unreacted Zn, and impurities from Zn and S

(c)

◀ **FIGURE 3.11** **A reaction that has less than 100% yield:**

$$8 \text{ Zn(s)} + \text{S}_8\text{(s)} \longrightarrow 8 \text{ ZnS(s)}$$

The actual yield of ZnS(s) obtained, shown in part **(c)**, is less than that calculated for the starting mixture shown in part **(a)** for several reasons:

- Neither the powdered zinc nor the powdered sulfur is pure.
- The Zn(s) can combine with O_2(g) in air to produce ZnO(s), and some of the sulfur burns in air to produce SO_2(g).
- As suggested in **(b)**, some of the product escapes from the reaction mixture as small lumps and as a fine dust.

The starting materials may not be pure, meaning that the actual quantities of reactants are less than what we weighed out. Some of the product may be left behind in the process of separating it from excess reactants. Side reactions may occur in addition to the main reaction, converting some of the original reactants into different, undesired products. (In the reaction in Example 3.20, for instance, if there is oxygen present, some of the Mg could be converted to the by-product MgO, reducing the yield of Mg_3N_2.)

Yields are usually expressed as percentages. The **percent yield** is the ratio of the actual yield to the theoretical yield multiplied by 100%:

$$\text{Percent yield} = \frac{\text{actual yield}}{\text{theoretical yield}} \times 100\% \qquad \textbf{(3.8)}$$

 **Percent Yield activity**

If the actual yield of Mg_3N_2 in Example 3.20 had been 47.87 g, the percent yield would have been

$$\text{Percent yield} = \frac{47.87\ \text{g}}{48.45\ \text{g}} \times 100\% = 98.80\%$$

Yields in an Organic Reaction

In some areas of chemistry—for example, quantitative analytical chemistry—only those reactions with virtually 100% yield are useful. In other areas, 100% yield is rare, and improving the percent yield of a reaction can be an important consideration. Yield calculations are almost always important in synthesis reactions, especially those in organic chemistry. Consider, for example, the reaction by which ethanol is converted to diethyl ether. The reaction is carried out in the presence of sulfuric acid, a fact that we indicate by writing the formula H_2SO_4 above the yield arrow:

$$2\ CH_3CH_2OH \xrightarrow{\ H_2SO_4\ } CH_3CH_2OCH_2CH_3 + H_2O$$
$$\quad\text{Ethanol} \qquad\qquad\qquad \text{Diethyl ether}$$

On paper, the reaction seems to be straightforward, but in the laboratory there are several complications. In an important *side reaction*, some of the ethanol is converted to ethylene, a hydrocarbon with a double bond (described in Chapter 9):

$$CH_3CH_2OH \longrightarrow CH_2{=}CH_2 + H_2O$$
$$\quad\text{Ethanol} \qquad\qquad \text{Ethylene}$$

Any molecules of ethanol converted to ethylene obviously cannot also form diethyl ether, and the yield of diethyl ether is accordingly reduced.

There are also practical problems. For example, the diethyl ether is purified by distilling it from the reaction mixture, and some will always remain in the distillation glassware. Also, some of the ethanol may distill with the ether, effectively removing that ethanol as a reactant.

For reactions in which the product is a solid, filtration is often necessary. Some product will remain behind on the filter paper, and some will usually remain dissolved in the solvent. Again, the yield is reduced below 100%. Even under the best of conditions, actual yields above 80–85% are difficult to achieve. Chemists often have to settle for 50%—and sometimes even less than that.

◀ Organic reactions often have less than 100 percent yield. Here, reaction of isopropyl alcohol, $CH_3CHOHCH_3$, with potassium dichromate produces acetone, CH_3COCH_3, with chromium(III) compounds (the gray-green solid) as by-products. To obtain pure acetone, the crude product would have to be worked up (purified), and some would be left behind at each stage of the process.

Example 3.21

Ethyl acetate is a solvent used as fingernail polish remover. What mass of acetic acid is needed to prepare 252 g ethyl acetate if the expected percent yield is 85.0%? Assume that the other reactant, ethanol, is present in excess. The equation for the reaction, carried out in the presence of H_2SO_4, is

$$CH_3COOH + HOCH_2CH_3 \xrightarrow{H_2SO_4} CH_3COOCH_2CH_3 + H_2O$$
Acetic acid   Ethanol   Ethyl acetate

STRATEGY

The mass we are given in this problem—252 g ethyl acetate—is an *actual* yield, but the stoichiometric calculation requires that we first calculate the *theoretical* yield of the reaction. For this we use Equation (3.8).

SOLUTION

First, we can solve Equation (3.8) for the theoretical yield and substitute known quantities for the actual and percent yields.

$$\text{Theoretical yield} = \frac{\text{actual yield}}{\text{percent yield}} \times 100\%$$

$$= \frac{252 \text{ g ethyl acetate}}{85.0\%} \times 100\%$$

$$= 296 \text{ g ethyl acetate.}$$

Then, we can determine the mass of acetic acid required to produce 296 g ethyl acetate.

$$? \text{ g } CH_3COOH = 296 \text{ g } CH_3COOCH_2CH_3 \times \frac{1 \text{ mol } CH_3COOCH_2CH_3}{88.11 \text{ g } CH_3COOCH_2CH_3}$$

$$\times \frac{1 \text{ mol } CH_3COOH}{1 \text{ mol } CH_3COOCH_2CH_3} \times \frac{60.05 \text{g } CH_3COOH}{1 \text{ mol } CH_3COOH}$$

$$= 202 \text{ g } CH_3COOH$$

ASSESSMENT

A common source of error in a problem of this type is to misuse Equation (3.8). If we had multiplied 252 g ethyl acetate by 85% rather than dividing by it, we would have obtained 214 g ethyl acetate as the theoretical yield. Clearly, this cannot be so—the actual yield can never be greater than the theoretical yield, that is, never greater than 100%.

EXERCISE 3.21A

Isopentyl acetate is the main component of banana flavoring. Calculate the theoretical yield of isopentyl acetate that can be made from 20.0 g isopentyl alcohol and 25.0 g acetic acid.

$$CH_3COOH + HOCH_2CH_2CH(CH_3)_2 \longrightarrow CH_3COOCH_2CH_2CH(CH_3)_2 + H_2O$$
Acetic acid   Isopentyl alcohol   Isopentyl acetate

If the percent yield of the reaction is 90.0%, what is the actual yield of isopentyl acetate?

EXERCISE 3.21B

How many grams of isopentyl alcohol are needed to make 433 g isopentyl acetate in the reaction described in Exercise 3.21A if the expected yield is 78.5%? Assume the acetic acid is in excess.

Example 3.22   A Conceptual Example

What is the maximum yield of $CO(g)$ obtainable from 725 g of $C_6H_{14}(l)$ regardless of the reaction(s) used, assuming no other carbon-containing reactant or product?

ANALYSIS AND CONCLUSIONS

If the maximum yield is independent of the reaction(s) used, then we ought to be able to determine this quantity without writing any chemical equations and without using stoichiometric factors based on chemical equations. While at first this may not seem feasible, recall that in Section 3.6 we dealt with a similar idea. There we needed to relate the mass of CO_2

produced in combustion analysis to the mass of a carbon-containing compound from which it formed. According to the law of conservation of mass, all the C atoms in the C_6H_{14} have to be accounted for, and the maximum yield results if all the C atoms end up in CO.

Let's rephrase the original question. What mass of CO contains the same number of C atoms as 725 g C_6H_{14}? We can start the calculation by converting grams of C_6H_{14} to moles of C_6H_{14}, and we can end it by converting moles of CO to grams of CO. The critical link between these two "ends" of the calculation are the factors that relate moles of C to moles of C_6H_{14} on the one hand and moles of CO to moles of C on the other. These are the factors shown in red in the following equation.

$$? \text{ g CO} = 725 \text{ g } C_6H_{14} \times \frac{1 \text{ mol } C_6H_{14}}{86.18 \text{ g } C_6H_{14}} \times \frac{6 \text{ mol C}}{1 \text{ mol } C_6H_{14}} \times \frac{1 \text{ mol CO}}{1 \text{ mol C}} \times \frac{28.01 \text{ g CO}}{1 \text{ mol CO}}$$

$$= 1.41 \times 10^3 \text{ g CO}$$

EXERCISE 3.22A

What is the maximum yield of the important commercial fertilizer ammonium hydrogen phosphate that can be obtained per kilogram of phosphoric acid?

EXERCISE 3.22B

Without doing detailed calculations, determine whether calcium, magnesium, or aluminum yields the most hydrogen per gram of metal when the metal reacts with an excess of hydrochloric acid. An aqueous solution of the metal chloride is the other product of the reaction.

3.11   Solutions and Solution Stoichiometry

Many chemical reactions are rapid and reproducible when carried out in the liquid state. As a result, chemists often dissolve solids in a liquid and carry out reactions in a liquid medium. Most of the reactions in our bodies occur in aqueous (water) solutions.

In Chapter 1, we noted that a solution is a homogeneous mixture of two or more substances. A solution of sugar in water does not consist of tiny particles of solid sugar dispersed among droplets of liquid water. Rather, individual sugar molecules are randomly distributed among water molecules in a uniform liquid medium. A solution is *homogeneous* right down to the molecular level. The composition and the physical and chemical properties are identical in all portions of a solution.

The components of a solution are the **solute**(s)—the substance(s) being dissolved—and the **solvent**—the substance doing the dissolving. The solutes are usually the components present in lesser amounts, and the solvent is usually present in the greater amount. There are many common solvents. *Hexane* dissolves grease. *Ethanol* is the solvent for many medicines. *Isopentyl acetate*, a component of banana oil, is a solvent for the glue used in making model airplanes. *Water* is the most familiar solvent, dissolving many common substances, such as sugar, salt, and ethanol. Solutions in which water is the solvent are called *aqueous* solutions.

 **Components of a Solution activity**

The *concentration* of a solution refers to the quantity of solute in a given quantity of either solvent or solution. A *dilute* solution is one that contains relatively little solute in a large quantity of solvent. A *concentrated* solution contains a relatively large amount of solute in a given quantity of solvent. These terms are imprecise, however. For example, a dilute sugar solution tastes faintly sweet, and a concentrated solution has a sickeningly sweet taste, but the terms "dilute" and "concentrated" do not in themselves tell us the precise proportions of sugar and water.

For commercially available acids and bases, the term *concentrated* generally signifies the highest concentration, usually expressed as a mass percent, that is commonly available. Commercial concentrated hydrochloric acid is about 38% HCl by mass; the rest is water. Commercial concentrated sulfuric acid is about 98% H_2SO_4 by mass, with the remainder water.

Molar Concentration

Mass percent composition, as in 38% HCl by mass, is one way to describe the concentration of a solution, but there are several other ways as well. One widely used concentration unit is **molarity (M),** or **molar concentration,** which is the amount of solute, in moles, per liter of solution:*

$$\text{Molarity (M)} = \frac{\text{moles of solute}}{\text{liters of solution}} \qquad (3.9)$$

The molarity of a solution made by dissolving 3.50 mol NaCl in enough water to produce 2.00 L of solution is

$$\text{Molarity} = \frac{\text{moles of solute}}{\text{liters of solution}} = \frac{3.50 \text{ mol NaCl}}{2.00 \text{ L solution}} = 1.75 \text{ M NaCl}$$

We read the notation "1.75 M NaCl" as "1.75 molar NaCl." Chemists generally choose to work with molarity because

- Substances enter into chemical reactions according to certain *molar* ratios.
- Volumes of solutions are more convenient to measure than masses of solutions.

Keep in mind that molarity signifies moles of solute per liter of *solution,* not per liter of solvent. Thus, we must prepare the solution in a vessel that holds the precise volume we need, such as a *volumetric flask*. Figure 3.12 illustrates the preparation of 0.01000 M $KMnO_4$ using such a flask. The solute (0.01000 mol, or 1.580 g) is weighed and added to a 1.000-L volumetric flask partially filled with water. After the solid has completely dissolved, just enough water is added to bring the volume up to the 1.000-L mark, followed by thorough mixing.

Solution Formation from a Solid animation

Volumetric flasks are available from 1 mL to 2 L or larger. They are labeled "TC" to indicate that they are calibrated *to contain* the stated volume to within about 0.1% or better.

(a)   (b)   (c)

◀ **FIGURE 3.12   Preparation of a 0.01000 M $KMnO_4$ solution**

In a step not shown, the balance is set to zero (tared) with just the weighing paper present. **(a)** The sample of $KMnO_4$ has a mass of 1.580 g, which is equivalent to 0.01000 mol. **(b)** The $KMnO_4$ is dissolved in water in the partially filled 1.000-L volumetric flask. In another step not shown, more water is added, and the solution is thoroughly mixed. **(c)** Finally, the flask is filled to the mark by adding a small quantity of water one drop at a time.

QUESTION: Note that we do not add exactly 1.000 L of water to prepare the solution. Why not?

Example 3.23

What is the molarity of a solution in which 333 g potassium hydrogen carbonate is dissolved in enough water to make 10.0 L of solution?

STRATEGY

According to Equation (3.9), the molarity of a solution is calculated from the number of moles of solute and the volume of solution in liters. Because we are given the volume of solution (10.0 L), the only requirement prior to substitution into Equation (3.9) is to convert the quantity of solute from mass in grams to number of moles.

---

*Recall from Chapter 1 that a liter is the same as a cubic decimeter: $1 \text{ L} = 1 \text{ dm}^3$. The derived SI unit for molarity is moles per cubic decimeter (mol/dm^3). The liter isn't a basic SI unit, but the unit mol/L is still widely used.

SOLUTION

First, let's prepare the setup that converts mass to number of moles of $KHCO_3$.

$$333 \text{ g KHCO}_3 \times \frac{1 \text{ mol KHCO}_3}{100.1 \text{ g KHCO}_3}$$

Now, without solving this expression, let's use it as the *numerator* in the defining equation for molarity. The solution volume, 10.0 L, is the *denominator*.

$$\text{Molarity} = \frac{333 \text{ g KHCO}_3 \times \dfrac{1 \text{ mol KHCO}_3}{100.1 \text{ g KHCO}_3}}{10.0 \text{ L solution}}$$

$$= 0.333 \text{ M KHCO}_3$$

EXERCISE 3.23A

Calculate the molarity of the solute in each of the following solutions.

(a) 3.00 mol KI in 2.39 L of solution   **(b)** 0.522 g HCl in 0.592 L of solution   **(c)** 2.69 g $C_{12}H_{22}O_{11}$ in 225 mL of solution

EXERCISE 3.23B

Calculate the molarity of **(a)** glucose, $C_6H_{12}O_6$, in 100.0 mL of solution containing 126 mg of glucose; **(b)** ethanol, CH_3CH_2OH, in a solution containing 10.5 mL of ethanol ($d = 0.789$ g/mL) in 25.0 mL of solution; and **(c)** urea, $CO(NH_2)_2$ in a solution of urea whose concentration is expressed as 9.5 mg N/mL solution.

Problem-Solving Note

In **(c)**, you need to determine the mass of urea having a nitrogen content of 9.5 mg. You will then have the concentration in mg urea/mL solution, which you must convert to mol urea/L solution.

At times we need to determine the number of *moles of solute* required to prepare a given volume of solution of a specified molarity. Other times we want to determine the *volume of solution* of a specified molarity containing a given number of moles of solute. We illustrate both types of calculations in Example 3.24, where the central conversion factors are those derived from this equivalency for 6.68 M NaOH:

$$6.68 \text{ mol NaOH} \backsimeq 1 \text{ L soln}$$

When we recast this equivalency into the usual two conversion factors, we find that one is simply the definition of molarity and the other is its inverse:

$$\frac{6.68 \text{ mol NaOH}}{1 \text{ L soln}} \quad \text{and} \quad \frac{1 \text{ L soln}}{6.68 \text{ mol NaOH}}$$

Example 3.24

We want to prepare a 6.68 molar solution of NaOH (6.68 M NaOH).

(a) How many moles of NaOH are required to prepare 0.500 L of 6.68 M NaOH?

(b) How many liters of 6.68 M NaOH can we prepare with 2.35 kg NaOH?

SOLUTION

(a) This calculation requires only a one-step conversion from liters of solution to moles of NaOH, with the molarity of the solution as the conversion factor:

$$? \text{ mol NaOH} = 0.500 \text{ L soln} \times \frac{6.68 \text{ mol NaOH}}{1 \text{ L soln}} = 3.34 \text{ mol NaOH}$$

(b) The central conversion factors in this calculation are the *inverse* of the molar mass of NaOH to convert from grams of NaOH to moles of NaOH and the *inverse* of the molarity—1 L soln/6.68 mol NaOH—to convert from moles of NaOH to liters of solution. We must also convert from kilograms of NaOH to grams of NaOH. In all, the required conversions are

$$\text{kg NaOH} \longrightarrow \text{g NaOH} \longrightarrow \text{mol NaOH} \longrightarrow \text{L soln}$$

which are set up as follows:

$$? \text{ L soln} = 2.35 \text{ kg NaOH} \times \frac{1000 \text{ g NaOH}}{1 \text{ kg NaOH}} \times \frac{1 \text{ mol NaOH}}{40.00 \text{ g NaOH}} \times \frac{1 \text{ L soln}}{6.68 \text{ mol NaOH}}$$

$$= 8.79 \text{ L soln}$$

EXERCISE 3.24A

How many grams of potassium hydroxide are required to prepare each of the following solutions?

(a) 2.00 L of 6.00 M KOH

(b) 10.0 mL of 0.100 M KOH

(c) 35.0 mL of 2.50 M KOH

EXERCISE 3.24B

How many milliliters of 1-butanol, $CH_3CH_2CH_2CH_2OH$, ($d = 0.810$ g/mL) are required to prepare 725 mL of a 0.350 M aqueous solution of this solute?

Labels on bottles of stock solutions of acids and bases often indicate concentrations only in percent solute by mass. If we want to know the molarity of such a solution, we must either know or measure the density of the solution. Density provides the conversion factor from mass of solution in grams to volume of solution in milliliters. We need to convert from milliliters to liters of solution, and we also need conversion factors based on mass percent and molar mass. We illustrate how all these factors enter into the calculation in Example 3.25.

Example 3.25

The label of a stock bottle of aqueous ammonia indicates that the solution is 28.0% NH_3 by mass and has a density of 0.898 g/mL. Calculate the molarity of the solution.

STRATEGY

We will find it most convenient to base the calculation on a 1.00-L volume of solution. When we have found the number of moles of NH_3 in this 1.00 L of solution, we will have found the molarity. The way in which these factors enter into the calculation is outlined below.

SOLUTION

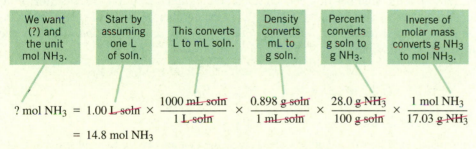

| We want (?) and the unit mol NH_3. | Start by assuming one L of soln. | This converts L to mL soln. | Density converts mL to g soln. | Percent converts g soln to g NH_3. | Inverse of molar mass converts g NH_3 to mol NH_3. |

$$? \text{ mol } NH_3 = 1.00 \text{ L soln} \times \frac{1000 \text{ mL soln}}{1 \text{ L soln}} \times \frac{0.898 \text{ g soln}}{1 \text{ mL soln}} \times \frac{28.0 \text{ g } NH_3}{100 \text{ g soln}} \times \frac{1 \text{ mol } NH_3}{17.03 \text{ g } NH_3}$$

$$= 14.8 \text{ mol } NH_3$$

Because 14.8 mol NH_3 are present in 1.00 L, the solution is 14.8 M. That is,

$$\text{Molarity} = \frac{14.8 \text{ mol } NH_3}{1.00 \text{ L soln}} = 14.8 \text{ M } NH_3$$

EXERCISE 3.25A

A stock bottle of aqueous formic acid indicates that the solution is 90.0% HCOOH by mass and has a density of 1.20 g/mL. Calculate the molarity of the solution.

EXERCISE 3.25B

A concentrated solution of perchloric acid, $HClO_4$, is 11.7 M and has a density of 1.67 g/mL. What is the mass percent perchloric acid in this solution?

Problem-Solving Note

What is the mass of 1.00 L of the perchloric acid solution? What is the mass of perchloric acid in the 1.00 L of solution?

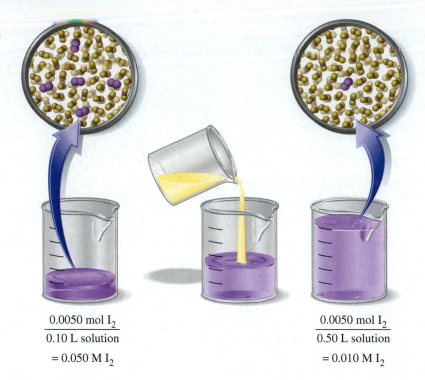

► **FIGURE 3.13** **Visualizing the dilution of a solution of I₂(s) in CS₂(l)**

The solution to be diluted (left beaker) initially contains 0.0050 mol of solute in 0.10 L of solution, making the concentration 0.0050 mol/0.10 L = 0.050 mol/L = 0.050 M. Then 0.40 L of additional solvent is added to this solution (middle beaker). Because the number of solute molecules in the beaker does not change as more solvent is added, the number of solute molecules *per unit volume* of solution decreases. As the right beaker shows, now we have the same 0.005 mol of solute but in 0.10 L + 0.40 L = 0.50 L of solution. The concentration of the diluted solution is 0.0050 mol/0.50 L = 0.01 mol/L = 0.010 M.

QUESTION: What happens to the concentration of a solution when solvent molecules evaporate?

$$\frac{0.0050 \text{ mol I}_2}{0.10 \text{ L solution}}$$

$$= 0.050 \text{ M I}_2$$

$$\frac{0.0050 \text{ mol I}_2}{0.50 \text{ L solution}}$$

$$= 0.010 \text{ M I}_2$$

Solution Formation by Dilution animation

Dilution of Solutions

We can generally find concentrated solutions, often ones that are commercially available, in a chemical storeroom, and we can use them to prepare solutions of lower concentrations. The process of preparing a dilute solution by adding solvent to a concentrated solution is called **dilution.** A basic principle of dilution suggested by Figure 3.13 is that

> *Addition of solvent does not change the amount of solute in a solution but does change the solution concentration.*

Suppose we let M_{conc} and M_{dil} represent the molar concentrations and V_{conc} and V_{dil} represent the volumes of a concentrated and a diluted solution. Because the product of a molarity (mol/L) and a volume (L) is the number of moles of solute in a solution, and *because the amount of solute does not change during dilution*, we can write these simple equations:

$$M_{conc} \times V_{conc} = \text{moles of solute} = M_{dil} \times V_{dil}$$

and $\quad M_{conc} \times V_{conc} = M_{dil} \times V_{dil}$ **(3.10)**

Example 3.26

How many milliliters of a 2.00 M $CuSO_4$ stock solution are needed to prepare 0.250 L of 0.400 M $CuSO_4$?

STRATEGY

We have considered two ways of viewing situations in which a dilute solution is prepared from a more concentrated one. One way is based on the principle of dilution illustrated in Figure 3.13, and the other employs Equation (3.10). If we work the same problem both ways, we expect to obtain the same result, thus providing us with an answer and a check of the answer at the same time.

SOLUTION

Applying the principle of dilution:
The key is to note that all the solute in the unknown volume of the stock solution appears in the 0.250 L of 0.400 M $CuSO_4$. First, let's calculate that amount of solute:

$$? \text{ mol CuSO}_4 = 0.250 \text{ L} \times \frac{0.400 \text{ mol CuSO}_4}{1 \text{ L}} = 0.100 \text{ mol CuSO}_4$$

Now we need to answer the question, "What volume of 2.00 M $CuSO_4$ contains 0.100 mol $CuSO_4$?" In doing so, we will have answered the original question.

$$? \text{ mL} = 0.100 \text{ mol } CuSO_4 \times \frac{1 \text{ L}}{2.00 \text{ mol } CuSO_4} \times \frac{1000 \text{ mL}}{1 \text{ L}} = 50.0 \text{ mL}$$

Of course, we could have done all of this in a single setup:

$$? \text{ mL} = 0.250 \text{ L} \times \frac{0.400 \text{ mol } CuSO_4}{1 \text{ L}} \times \frac{1 \text{ L}}{2.00 \text{ mol } CuSO_4} \times \frac{1000 \text{ mL}}{1 \text{ L}} = 50.0 \text{ mL}$$

Using the dilution equation:
First, we can identify the terms we need for Equation (3.10):

$$M_{conc} = 2.00 \text{ M}; \; V_{conc} = ?; \; M_{dil} = 0.400 \text{ M}; \; V_{dil} = 0.250 \text{ L}$$

Then, we can substitute these terms into the equation:

$$M_{conc} \times V_{conc} = M_{dil} \times V_{dil}$$

$$2.00 \text{ M} \times V_{conc} = 0.400 \text{ M} \times 0.250 \text{ L}$$

$$V_{conc} = \frac{0.400 \text{ M}}{2.00 \text{ M}} \times 0.250 \text{ L} = 0.0500 \text{ L}$$

$$V_{conc} = 0.0500 \text{ L} \times \frac{1000 \text{ mL}}{1 \text{ L}} = 50.0 \text{ mL}$$

ASSESSMENT

As expected, the two methods yield the same result: 50.0 mL 2.00 M $CuSO_4(aq)$. To prepare the dilute solution, we should measure out 50.0 mL of 2.00 M $CuSO_4$ and add it to enough water to make 0.250 L of solution, as illustrated in Figure 3.14.

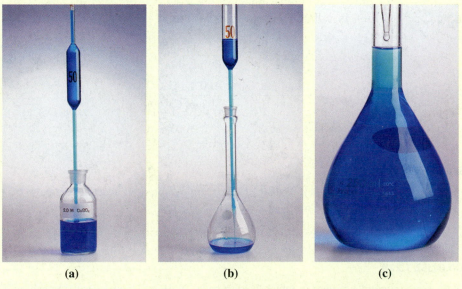

(a)   (b)   (c)

▲ **FIGURE 3.14   Dilution of a copper(II) sulfate solution: Example 3.26 illustrated**
(a) The pipet is being filled with 50.0 mL of 2.00 M $CuSO_4$. The amount of $CuSO_4$ in the filled pipet will be 0.100 mol. **(b)** The 50.0 mL of 2.00 M $CuSO_4$ solution is transferred to a 250.0-mL volumetric flask, water is added, and the solution is thoroughly mixed. **(c)** Finally, the flask is filled to the mark as the remaining water is added dropwise.

EXERCISE 3.26A

How many milliliters of a 10.15 M NaOH stock solution are needed to prepare 15.0 L of 0.315 M NaOH?

EXERCISE 3.26B

How many milliliters of a 5.15 M CH_3OH stock solution are needed to prepare 375 mL of a solution having 7.50 mg of methanol per mL of solution?

Solutions in Chemical Reactions

Molarity provides an important additional tool for reaction stoichiometry calculations. Specifically, it gives us two conversion factors: one to convert from liters of solution to moles of solute and another to convert from moles of solute to liters of solution. We use these conversion factors in the early and/or late stages of the setup of a stoichiometric calculation. The heart of the calculation, however, is still a stoichiometric factor (mole ratio) derived from the chemical equation. We illustrate these points in Example 3.27 and Exercises 3.27A and B, and we will explore reaction stoichiometry in solutions in greater detail in Chapter 4.

Example 3.27

A chemical reaction familiar to geologists is that used to identify limestone. The reaction of hydrochloric acid with limestone, which is largely calcium carbonate, is seen through an effervescence—a bubbling due to the liberation of gaseous carbon dioxide:

$$CaCO_3(s) + 2\ HCl(aq) \longrightarrow CaCl_2(aq) + H_2O(l) + CO_2(g)$$

How many grams of $CaCO_3(s)$ are consumed in a reaction with 225 mL of 3.25 M HCl?

STRATEGY

To relate the quantity of $CaCO_3$ to that of HCl, we need to express the amount of HCl in moles and multiply by the stoichiometric factor 1 mol $CaCO_3$/2 mol HCl. First, though, we have to relate the number of moles of HCl to the volume of HCl(aq) and its molarity. Thus, we use molarity as a conversion factor before introducing the stoichiometric factor. The setup below is outlined in the stoichiometry diagram in Figure 3.15.

SOLUTION

$$? \text{ g } CaCO_3 = 225 \text{ mL soln} \times \frac{1 \text{ L soln}}{1000 \text{ mL soln}} \times \frac{3.25 \text{ mol HCl}}{1 \text{ L soln}}$$

$$\times \frac{1 \text{ mol } CaCO_3}{2 \text{ mol HCl}} \times \frac{100.1 \text{ g } CaCO_3}{1 \text{ mol } CaCO_3}$$

$$= 36.6 \text{ g } CaCO_3$$

EXERCISE 3.27A

How many milliliters of 0.100 M $AgNO_3(aq)$ are required to react completely with 750.0 mL of 0.0250 M $Na_2CrO_4(aq)$?

$$2\ AgNO_3(aq) + Na_2CrO_4(aq) \longrightarrow Ag_2CrO_4(s) + 2\ NaNO_3(aq)$$

EXERCISE 3.27B

In a reaction similar to that in which baking soda ($NaHCO_3$) neutralizes stomach acid, 175 mL of 1.55 M $NaHCO_3$ is added to 235 mL of 1.22 M HCl:

$$NaHCO_3(aq) + HCl(aq) \longrightarrow NaCl(aq) + H_2O(l) + CO_2(g)$$

(a) How many grams of CO_2 are liberated?

(b) What is the molarity of the NaCl(aq) produced? Assume that the solution volume is 175 mL + 235 mL = 410 mL.

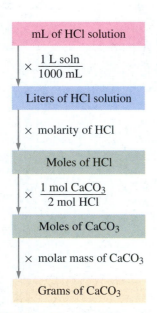

▲ FIGURE 3.15 Stoichiometry and solutions: Flow chart for Example 3.27

We convert milliliters of HCl(aq) to liters, then use molarity to obtain moles of HCl. The rest of the flow chart is much the same as Figure 3.9.

Green Chemistry: Atom Economy

Like most other people, chemists are concerned with environmental pollution. This concern has given rise to a field called *green chemistry*, which involves the design of chemical products and processes that minimize environmental problems. P. T. Anastas and J. C. Warner describe 12 principles of green chemistry in their book *Green Chemistry: Theory and Practice* (Oxford University Press: New York, 1998). One of the principles, called *atom economy*, involves—among other things—designing reactions in such a way that the quantity of reactants that end up in the desired final product is the maximum possible. (We will consider other principles in later chapters.)

We can define the percent atom economy (% AE) of a reaction by the formula

$$\% \ AE = \frac{\text{formula mass of desired final product}}{\text{sum of formula masses of all the reactants}} \times 100\%$$

Atom economy also must consider substances such as solvents, separation agents, and drying agents that are used in the process but are not directly part of the chemical reaction. Using a green chemistry approach, chemists attempt to either reduce the amount of or eliminate completely as many of these substances as possible. Those that cannot be eliminated are reused or recycled when possible.

As an example, consider two ways to make ethylene oxide, which is used to sterilize medical devices and as an intermediate in the synthesis of ethylene glycol and other chemicals. The old way, called the chlorhydrin route, starts with the hydrocarbon ethylene (Section 9.11), chlorine, and calcium hydroxide, and involves several steps. (The process is named for one of the intermediates, $HOCH_2CH_2Cl$, called chlorhydrin.) The overall reaction produces the desired ethylene oxide but has calcium chloride and water as by-products:

$$C_2H_4 + Cl_2 + Ca(OH)_2 \longrightarrow C_2H_4O + CaCl_2 + H_2O$$

Assuming 100% yield for all the reactions involved, the atom economy for the overall reaction is

$$\% \ AE = \frac{\text{formula mass of } C_2H_4O}{\substack{\text{sum of formula masses} \\ \text{of } C_2H_4 + Cl_2 + Ca(OH)_2}} \times 100\%$$

$$\% \ AE = \frac{44.05 \ u}{28.05 \ u + 70.91 \ u + 74.09 \ u} \times 100\% = 25.46\%$$

The newer process uses a catalyst (a substance that speeds up a reaction without itself being used up). Ethylene and oxygen react to give ethylene oxide as the sole product:

$$C_2H_4 + \tfrac{1}{2}O_2 \longrightarrow C_2H_4O$$

Assuming a 100% yield in this reaction, the atom economy is also 100%:

$$\% \ AE = \frac{\text{formula mass of } C_2H_4O}{\text{sum of formula masses of } C_2H_4 + \tfrac{1}{2}O_2} \times 100\%$$

$$\% \ AE = \frac{44.05 \ u}{28.05 \ u + 16.00 \ u} \times 100\% = 100\%$$

Manufacturing processes can be highly complex. At times, by-products can either be used directly or serve as starting materials for other processes. Such factors should be considered in the overall atom economy. In the case of calcium chloride, however, the current need is not great, and so it is not a valuable by-product. Clearly, the catalytic route to ethylene oxide, with no by-products to worry about, is "greener" than the chlorohydrin path.

Cumulative Example

The combustion in oxygen of 1.5250 g of an alkane-derived compound composed of carbon, hydrogen, and oxygen yields 3.047 g CO_2 and 1.247 g H_2O. The molecular mass of this compound is 88.1 u. Draw a plausible structural formula for this compound. Is there more than one possibility? Explain.

STRATEGY

Example 3.12 provides much of the initial guidance for this problem. We can use the masses of CO_2 and H_2O from the combustion analysis to determine first the number of moles of carbon and hydrogen in the sample and then the masses of carbon and hydrogen. Next, we can find the mass of oxygen by subtraction and convert that mass to moles of oxygen. Now, we can use the numbers of moles of the elements to establish the empirical formula of the sample. Once we have the empirical formula, we can use the molecular mass to lead us to the molecular formula. Finally, using our knowledge of alkanes and functional groups (Section 2.9), we can draw structural formulas and determine whether more than one structure is possible.

SOLUTION

We begin by finding the moles of carbon in 3.047 g of CO_2 and the moles of hydrogen in 1.247 g of H_2O.

$$? \ \text{mol C} = 3.047 \ \text{g CO}_2 \times \frac{1 \ \text{mol CO}_2}{44.010 \ \text{g CO}_2} \times \frac{1 \ \text{mol C}}{1 \ \text{mol CO}_2} = 0.06923 \ \text{mol C}$$

$$? \ \text{mol H} = 1.247 \ \text{g H}_2O \times \frac{1 \ \text{mol H}_2O}{18.015 \ \text{g H}_2O} \times \frac{2 \ \text{mol H}}{1 \ \text{mol H}_2O} = 0.1384 \ \text{mol H}$$

We will need these numbers of moles for the empirical formula determination later.

Next, we convert moles of each element to mass.

$$0.06923 \ \text{mol C} \times \frac{12.011 \ \text{g C}}{1 \ \text{mol C}} = 0.8315 \ \text{g C}$$

$$0.1384 \ \text{mol H} \times \frac{1.00794 \ \text{g H}}{1 \ \text{mol H}} = 0.1395 \ \text{g H}$$

We find the mass of oxygen by subtracting the masses of carbon and hydrogen from the sample mass.

$$? \ \text{g O} = 1.5250 \ \text{g} - 0.8315 \ \text{g} - 0.1395 \ \text{g} = 0.5540 \ \text{g O}$$

Then we convert the mass of oxygen to moles.

$$? \ \text{mol O} = 0.5540 \ \text{g O} \times \frac{1 \ \text{mol O}}{15.999 \ \text{g O}} = 0.03463 \ \text{mol O}$$

Now we use the moles of carbon, hydrogen, and oxygen to construct a tentative formula.

$$C_{0.06923}H_{0.1384}O_{0.03463}$$

Then we divide each subscript by the subscript for oxygen (because this is the smallest of the three subscripts).

$$C_{0.06923/0.03463}H_{0.1384/0.03463}O_{0.03463/0.03463} \longrightarrow C_{2.00}H_{4.00}O_{1.000} \longrightarrow C_2H_4O$$

The empirical formula mass is 44.053 u.

$$2(12.011 \ \text{u}) + 4(1.0079 \ \text{u}) + 15.999 \ \text{u} = 44.053 \ \text{u}.$$

Next, we use Equation (3.7) to calculate the ratio of the molecular mass to the empirical formula mass.

$$\text{Integral factor} = \frac{\text{molecular mass}}{\text{empirical formula mass}} = \frac{88.1 \ \text{u}}{44.053 \ \text{u}} = 2$$

We multiply each subscript by the factor 2 to obtain the molecular formula.

$$\text{Molecular formula} = (C_2H_4O)_2 = C_4H_8O_2$$

Recall from Section 2.9 that a carboxylic acid has two oxygen atoms. Butanoic acid fits the molecular formula (I). However, there are many other possibilities. The three-carbon chain of butanoic acid could be replaced by a branched group, as shown in structure (II).

$$\text{CH}_3\text{CH}_2\text{CH}_2 - \overset{\overset{\displaystyle O}{\|}}{\text{C}} - \text{OH} \qquad \text{or} \qquad (\text{CH}_3)_2\text{CH} - \overset{\overset{\displaystyle O}{\|}}{\text{C}} - \text{OH}$$

(I)   (II)

The compounds (III) and (IV) also fit the molecular formula.

$$\text{CH}_3\text{CH}_2 - \overset{\overset{\displaystyle O}{\|}}{\text{C}}\text{OCH}_3 \qquad\qquad \text{CH}_3 - \overset{\overset{\displaystyle O}{\|}}{\text{C}}\text{OCH}_2\text{CH}_3$$

(III)   (IV)

These are but a few of the possible structures.

ASSESSMENT

Because the masses of CO_2 and H_2O formed in the combustion are considerably less than their molar masses, we expect the moles of carbon and of hydrogen to be less than 1, and they are. Also, although it is not compelling evidence, the fact that we obtained small integral values for the subscripts in the empirical formula suggests that it is a reasonable formula, as is the molecular formula. Another interesting observation is that alkane-based alcohols (ROH) and ethers [(ROR′), where R and R′ represent alkyl groups] are not possible structures for the formula $C_4H_8O_2$. There are not enough hydrogen atoms in the molecule for this to be the case.

Concept Review with Key Terms

The subject of **stoichiometry** involves quantitative calculations based on chemical formulas and chemical equations.

3.1 Molecular Masses and Formula Masses—**Molecular masses** and **formula masses** are the masses, expressed in atomic mass units (u), of individual molecules and formula units. They are calculated from the masses of the atoms represented in the molecular or empirical formulas, respectively. Molecular mass applies only to molecular compounds; formula mass is appropriate for ionic compounds.

3.2 The Mole and Avogadro's Number—A **mole (mol)** is an amount of substance containing a number of elementary entities

(atoms, molecules, formula units, etc.) equal to the number of atoms in exactly 12 g of carbon-12. This number is **Avogadro's number**, $N_A = 6.022 \times 10^{23} \ \text{mol}^{-1}$. The mole is the SI unit for the amount of a substance and is used extensively in chemical equations and calculations.

3.3 The Mole and Molar Mass—The mass, in grams, of one mole of substance is called the **molar mass**; it is numerically equal to an atomic, molecular, or formula mass but carries the unit g/mol. Conversions between the number of moles and the number of grams of a substance require molar mass as a conversion factor. Calculations involving volume, density, and numbers of atoms or molecules can also be used to determine molar quantities.

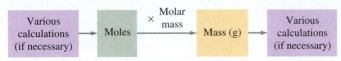

3.4 Mass Percent Composition from Chemical Formulas—The

mass percentages of individual elements in a compound can be determined from the chemical formula and molar mass with the following equation:

$$\text{Mass percent} = \frac{\begin{array}{c}\text{atomic mass} \\ \text{of element}\end{array} \times \begin{array}{c}\text{number of atoms of} \\ \text{that element in formula}\end{array}}{\text{molar mass}} \times 100\%$$

The collection of these mass percentages represents the **mass percent composition** of the compound.

3.5 Chemical Formulas from Mass Percent Composition—An em-

pirical formula can be established from the mass percent composition of a compound by calculating molar ratios of the different elements present in the compound. Empirical formulas calculated in this way may or may not be equivalent to the molecular formula. To establish a molecular formula, we must also know the molecular mass.

3.6 Elemental Analysis: Experimental Determination of Mass Percent Composition—The mass percents of carbon, hydrogen, and

oxygen in organic compounds can be determined by combustion analysis. Other methods are required to determine the mass percent composition of inorganic compounds.

3.7 Writing and Balancing Chemical Equations—A **chemical**

equation uses symbols and formulas for the elements and/or compounds involved in a reaction. A chemical equation portrays the progress of a reaction, indicated by an arrow, from **reactants** to **products**. **Stoichiometric coefficients** are placed before the symbols or formulas in the equation to balance the equation. As required by the law of conservation of mass, for each element in a balanced chemical equation the number of atoms on the product side of the equation will be the same as on the reactant side.

3.8 Reaction Stoichiometry—Stoichiometry involves quantitative re-

lationships in a chemical reaction. **Stoichiometric factors**—also called **mole ratios**—are based on the coefficients in the balanced equation and are used to relate moles of one reactant or product to another.

Molar masses and stoichiometric factors, together with other factors, are used to determine information about one reactant or product in a chemical reaction from known information about another. The strategy for reaction stoichiometry calculations can be outlined diagrammatically, as suggested below.

3.9 Limiting Reactants—The **limiting reactant** is the reactant that

is completely consumed in a reaction. Its quantity determines the theoretical quantity of the products formed. Other reactants are said to be present in excess. In some stoichiometry problems the limiting reactant must first be identified through a preliminary calculation.

3.10 Yields of Chemical Reactions—The calculated quantity of a

product is the **theoretical yield** of a reaction. The quantity physically obtained from a chemical reaction, called the **actual yield**, is often less and is commonly expressed as a percentage of the theoretical yield.

$$\textbf{Percent yield} = \frac{\text{actual yield}}{\text{theoretical yield}} \times 100\%$$

3.11 Solutions and Solution Stoichiometry—Solutions are formed

by dissolving one substance, the **solute**, into another substance, the **solvent**. The solute is usually the component present in a lesser amount. The **molarity (M)**, or **molar concentration**, of a solution is the number of moles of solute per liter of solution.

$$\text{Molarity (M)} = \frac{\text{moles of solute}}{\text{liters of solution}}$$

Common calculations involving solutions include relating an amount of solute to solution volume and molarity. Solutions of a desired concentration are often prepared from more concentrated solutions by dilution. **Dilution** increases the volume of a solution, but the amount of solute is unchanged. As a consequence, the concentration decreases. Stoichiometric calculations for reactions in solution often use molarity or its inverse as conversion factors, in addition to the stoichiometric factor and possibly other conversion factors.

Assessment Goals

When you have mastered the material in this chapter, you will be able to

- Define and determine molecular mass and formula mass.
- Identify the number of elementary entities in a mole of any substance and calculate the molar masses of elements and compounds.
- Perform calculations involving moles, masses, and numbers of atoms, molecules, or formula units.
- Determine the mass percentages of individual elements in compounds from chemical formulas.
- Determine empirical formulas from mass percent composition.
- Relate molecular formulas to empirical formulas, using molar masses.

- Describe the process of combustion analysis and perform calculations based on combustion data.
- Balance chemical equations, and construct stoichiometric factors from them.
- Determine, through stoichiometric calculations, quantities such as the number of moles or the mass of any reactant or product in a chemical reaction.
- Determine the limiting reactant in a reaction.
- Calculate theoretical and percentage yields in reactions.
- Calculate the molar concentration of a solution, or quantities of solutes and volumes of solution that are related through molarity.
- Perform dilution calculations.
- Use molarity as a conversion factor in stoichiometric calculations.

Self-Assessment Questions

1. Explain the difference between the *atomic mass* of oxygen and the *molecular mass* of oxygen. Explain how each is determined from data in the periodic table.

2. How many oxygen molecules and how many oxygen atoms are in 1.00 mol O_2?

3. Complete the phrase: One *mole* of fluorine gas, F_2
   (a) weighs 19.0 g
   (b) contains 6.02×10^{23} F atoms
   (c) contains 1.20×10^{24} F atoms
   (d) weighs 6.02×10^{23} g

4. For calcium nitrate, how many (a) calcium ions and how many nitrate ions are there in 1.00 mol $Ca(NO_3)_2$ (b) How many nitrogen atoms and how many oxygen atoms are there in 1.00 mol $Ca(NO_3)_2$

5. What are the empirical formulas of the compounds with the following molecular formulas?
   (a) H_2O_2 (b) C_8H_{16} (c) $C_{10}H_8$ (d) $C_6H_{12}O_2$

6. Consider the following equation. (a) Explain its meaning at the molecular level. (b) Interpret it in terms of moles. (c) State the mass relationships conveyed by the equation.

   $$CH_4 + 2 O_2 \longrightarrow CO_2 + 2 H_2O$$

7. Translate the following chemical equations into words:
   (a) $2 H_2(g) + O_2(g) \longrightarrow 2 H_2O(l)$
   (b) $4 NaClO_3(s) \longrightarrow 3 NaClO_4(s) + NaCl(s)$
   (c) $2 Al(s) + 6 HCl(aq) \longrightarrow 2 AlCl_3(aq) + 3 H_2(g)$

8. The decomposition of potassium chlorate to produce potassium chloride and oxygen gas is expressed symbolically as which of the following?
   (a) $KClO_3(s) \longrightarrow KCl(s) + O_2(g) + O(g)$
   (b) $2 KClO_3(s) \longrightarrow 2 KCl(s) + 3 O_2(g)$
   (c) $KClO_3(s) \longrightarrow KClO(s) + O_2(g)$
   (d) either (b) or (c), but not (a)

9. Write a balanced chemical equation to represent (a) the decomposition, by heating, of solid mercury(II) nitrate to produce pure liquid mercury, nitrogen dioxide gas, and oxygen gas, (b) the reaction of aqueous sodium carbonate with aqueous hydrochloric acid (hydrogen chloride) to produce water, carbon dioxide gas, and aqueous sodium chloride, (c) the complete combustion of

malonic acid, a compound with 34.62% C, 3.88% H, and 61.50% O, by mass.

10. In a reaction of 2.0 mol CCl_4 with an excess of HF, 1.70 mol CCl_2F_2 is obtained. Which statement is correct?

    $$CCl_4 + 2 HF \longrightarrow CCl_2F_2 + 2 HCl$$

    (a) The theoretical yield is 1.70 mol CCl_2F_2.
    (b) The theoretical yield is 1.0 mol CCl_2F_2.
    (c) The percent yield of the reaction is 85%.
    (d) The theoretical yield of the reaction depends on how large an excess of HF is used.

11. If the reaction of 1.00 mol $NH_3(g)$ and 1.00 mol $O_2(g)$ is carried out to completion:

    $$4 NH_3(g) + 5 O_2(g) \longrightarrow 4 NO(g) + 6 H_2O(l)$$

    (a) all of the $O_2(g)$ is consumed.
    (b) 4.0 mol $NO(g)$ is produced.
    (c) 1.5 mol $H_2O(l)$ is produced.
    (d) None of these.

12. Is the volume of a solution changed by dilution? Is the concentration? Is the number of moles of solute? Explain.

13. Some handbooks list the concentration of solute in an aqueous solution in units of g solute/100 mL H_2O, and others use g solute/100 mL solution. Are these concentration units the same? Explain.

14. For some applications, solution concentrations are expressed as grams of solute per liter of solution, and for some, as milligrams of solute per milliliter of solution. How are these two concentrations units related? Explain.

15. Explain why a stoichiometric calculation gives the same result, regardless of the coefficients used in the equation for a chemical reaction, as long as the equation is balanced.

16. What is green chemistry? What is meant by the atom economy of a reaction?

Problems

Molecular and Formula Masses

17. Calculate the molecular mass or formula mass of each of the following. Which ones are molecular masses?

    (a) C_3H_7Br (e) $Ti_2(SO_4)_3$
    (b) $Mg(HCO_3)_2$ (f) disulfur dichloride
    (c) aluminum perchlorate   (g) $(CH_3CH_2CH_2)_2O$
    (d) iron(III) nitrate nonahydrate (h) isooctane

18. Calculate the molecular or formula mass of each of the following. Which ones are molecular masses?

    (a) $KHSO_3$ (e) tetrasulfur dinitride
    (b) $Fe_2(Cr_2O_7)_3$ (f) dinitrogen pentoxide
    (c) titanium(III) oxalate   (g) silver hydrogen phosphate
        decahydrate
    (d) $KAl(SO_4)_2 \cdot 12 H_2O$ (h) 1-pentanol

19. Calculate the molecular mass of (a) the substance fensulfothion, an insecticide, with the condensed structural formula $(CH_3CH_2O)_2PSOC_6H_4SOOCH_3$, and (b) methyl salicylate represented above right. (*Hint:* Refer to the color code in Figure 2.8.)

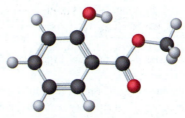

Methyl salicylate

20. Calculate the molecular mass of (a) trimethobenzamide, a substance used to suppress nausea and vomiting, having the condensed structural formula $(CH_3)_2NCH_2CH_2OC_6H_4CH_2NHCOC_6H_2(OCH_3)_3$, and (b) the substance represented by the space-filling model below. (*Hint:* Refer to the color code in Figure 2.8.)

Avogadro's Number and Molar Masses

21. Calculate the mass, in grams, of (a) 0.773 mol of KOH, (b) 0.250 mol of silicon tetrachloride, (c) 0.158 mol of strontium hydrogen sulfate, and (d) 3.91×10^{-4} mol of $(CH_3)_4Si$.

22. Calculate the mass, in grams, of (a) 1.37 mol of $AgNO_3$, (b) 0.314 mol of ethylene glycol, $HOCH_2CH_2OH$, (c) 0.615 mol of potassium permanganate, and (d) 8.92×10^5 mol of phosphoric acid.

23. Calculate the amount, in moles, of (a) 647 g of $Ba(OH)_2$, (b) 16.3 g of sulfur trioxide, (c) 35.6 g of nickel(II) chloride heptahydrate, (d) 218 mg ethyl mercaptan, CH_3CH_2SH, and (e) 3.32×10^4 kg of iron(II) sulfide.

24. Calculate the amount, in moles, of (a) 37.0 g of $H_4P_4O_7$, (b) 18.92 g tetraphosphorus hexoxide, (c) 42.5 g manganese(III) fluoride, (d) 3.39 mg octene, C_8H_{16}, (e) 7.71×10^4 kg of NO_2, and (f) 0.453 lb of lithium nitrate trihydrate.

25. How many oxide ions and iron(III) ions are there in 1.00 mol Fe_2O_3?

26. How many of each type of atom are present in 0.25 mol of $[CrCl_2(NH_3)_4]Cl$?

27. Calculate (a) the number of molecules in 4.68 mol H_2S, (b) the number of sulfate ions in 86.2 g of barium sulfate, (c) the average mass of an atom of tellurium, and (d) the average mass of a permanganate ion.

28. Calculate (a) the number of chloride ions in 1.75 mol of iron(III) chloride; (b) the number of molecules in 42.4 mL of ethanol, CH_3CH_2OH, $(d = 0.789 \text{ g/mL})$; (c) the average mass of an isopentyl acetate molecule, $CH_3COOC_5H_{11}$; and (d) the number of chloride ions in 45.3 mg of $BaCl_2 \cdot 2 H_2O$.

29. *Without doing detailed calculations*, place the following in order from *fewest* number of atoms to *greatest* number of atoms, and justify your answer. (a) 0.86 mol Al, (b) 6.1×10^{23} molecules of O_2, (c) 250.0 g of uranium, (d) 0.76 mol of HCl

30. *Without doing detailed calculations*, determine which of the following samples has the greatest mass, and explain your choice. (a) 0.80 mol Fe, (b) 1.1×10^{24} S atoms, (c) 50.0 mL H_2O, or (d) 41.0 g $Al_2(SO_4)_3$.

Percent Composition and Empirical Formulas

31. Calculate the mass percent of each element in (a) KNO_2 (b) C_4H_9OH (c) $Al(PO_3)_3$ and (d) $HOOCCOOH \cdot 2 H_2O$

32. Calculate the mass percent of each element in (a) $Ba(ClO_3)_2$, (b) NH_4HSO_4, (c) $(C_5H_5)_2Fe$, and (d) C_6H_5COONa.

33. What is the mass percent of (a) oxygen in the compound having the condensed structural formula $HOOCCH_2CH(CH_3)COOH$, (b) nitrogen in the compound having the condensed structural formula $CH_3CH_2CH(CH_3)CONH_2$, and (c) beryllium in the mineral beryl, $Be_3Al_2Si_6O_{18}$?

34. (a) What is the mass percent of uranium in the mineral carnotite, $K_2(UO_2)_2(VO_4)_2 \cdot 3 H_2O$? (b) What is the maximum mass of uranium obtainable from 1.00 kg of carnotite? (c) What is the smallest mass of carnotite from which 1.00 kg of uranium can be obtained?

35. What are the empirical formulas of the compounds with the following molecular formulas?

    (a) N_2O_5 (b) $C_{10}H_{22}$

36. What are the empirical formulas of the compounds with the following molecular formulas?

    (a) $C_4H_8O_2$ (b) $C_8H_{18}O_2$

37. What is the molecular formula of each of the following?

    (a) *para*-dichlorobenzene, used as a moth repellent, empirical formula C_3H_2Cl, molecular mass 147 u

    (b) a compound that is 40.00% C, 6.71% H, and 53.29% O, by mass; molecular mass about 180 u

38. What is the molecular formula of each of the following?

    (a) benzene, empirical formula CH, molecular mass 78.0 u

    (b) vitamin C, which is 40.92% C, 4.58% H, and 54.50% O, by mass; molecular mass 176 u

39. Determine the empirical formula of each of the following.

    (a) the painkiller codeine, mass percent composition 72.22% C, 7.07% H, 4.68% N, and 16.03% O

    (b) a compound whose mass percent composition is 21.9% Mg, 27.8% P, and 50.3% O

40. Determine the empirical formula of each of the following.

    (a) urea, used as a fertilizer and in the manufacture of plastics, mass percent composition 20.00% C, 6.71% H, 46.65% N, and 26.64% O

    (b) a compound of which a 9.2-g sample gives 2.8 g N and 6.4 g O

41. Resorcinol, used in manufacturing resins, drugs, and other products, is 65.44% C, 5.49% H, and 29.06% O by mass. Its molecular mass is 110 u. What is its molecular formula?

42. Sodium tetrathionate, an ionic compound formed when sodium thiosulfate reacts with iodine, is 17.01% Na, 47.46% S, and 35.52% O by mass, and has a formula mass of 270 u. What is its formula?

43. A hydrate is found to have the mass percent composition: 4.33% Li, 22.10% Cl, 39.89% O, and 33.69% H_2O by mass. What is its formula?

44. A hydrate of magnesium bromide is found to contain 37% H_2O by mass. What is its formula?

45. *Without doing detailed calculations*, determine which of these compounds has the greatest mass percent nitrogen: $(NH_4)_2SO_4$, NH_4NO_2, NH_4NO_3, NH_4Cl.

46. *Without doing detailed calculations*, determine which of the following hydrates has the greatest mass percent water: $LiNO_3 \cdot 3 H_2O$, $MgCl_2 \cdot 6 H_2O$, $ZnSO_4 \cdot 7 H_2O$, $Al(NO_3)_3 \cdot 9 H_2O$.

47. Elemental analysis of 5.000 g of a compound showed that it contained 1.278 g of carbon, 0.318 g of hydrogen, and 3.404 g of sulfur. The molecular mass of the compound was determined in a separate experiment and found to be 94.19 u. Determine the empirical and molecular formulas of the compound.

48. When 1.019 g of potassium reacts with oxygen in air, it forms 1.860 g of a rather strange compound. Find the empirical formula of this compound, and explain what is strange about its formula.

49. Thiophene is a carbon–hydrogen–sulfur compound used in the manufacture of pharmaceuticals. When burned completely in excess oxygen, its combustion products are CO_2, H_2O, and SO_2. Combustion of a 0.535-g sample yields 1.119 g CO_2, 0.229 g H_2O, and 0.407 g SO_2. What is the empirical formula of thiophene?

50. Dimethylhydrazine is a carbon–hydrogen–nitrogen compound used in rocket fuels. When burned completely in excess oxygen, a 0.312-g sample produces 0.458 g CO_2 and 0.374 g H_2O. The nitrogen content of a separate 0.525-g sample is converted to 0.244 g N_2. What is the empirical formula of dimethylhydrazine?

51. To decrease carbon monoxide emissions from automobiles in some geographic areas, gasoline is required to include oxygen-containing additives (oxygenates) such as methanol (CH_3OH), ethanol (CH_3CH_2OH), and (formerly) methyl *tertiary*-butyl ether (MTBE), $CH_3OC(CH_3)_3$. **(a)** *Without doing detailed calculations*, arrange these three compounds in the order of *increasing* mass percent oxygen. **(b)** The 1990 U.S. Clean Air Amendment requires fuels to contain 2.7% O. Does a fuel that contains 10.5% methanol by mass as its only oxygen-containing component meet this standard? **(c)** What mass percent of MTBE should be present as the sole oxygenate in gasoline if the gasoline is to contain 2.7% O?

52. Morton Lite Salt® has 290 mg of sodium (as Na^+) and 340 mg of potassium (as K^+) per 0.25 teaspoon. Assume that 1.0 teaspoon of the Lite Salt has a mass of 6.0 g and that the Na^+ comes from NaCl and the K^+ from KCl. Calculate the mass percent of NaCl and of KCl in Lite Salt.

Chemical Equations

53. Balance the following equations.

(a) $Cl_2O_5 + H_2O \longrightarrow HClO_3$

(b) $V_2O_5 + H_2 \longrightarrow V_2O_3 + H_2O$

(c) $Al + O_2 \longrightarrow Al_2O_3$

(d) $TiCl_4 + H_2O \longrightarrow TiO_2 + HCl$

(e) $Sn + NaOH \longrightarrow Na_2SnO_2 + H_2$

(f) $PCl_5 + H_2O \longrightarrow H_3PO_4 + HCl$

(g) $CH_3SH + O_2 \longrightarrow CO_2 + SO_2 + H_2O$

(h) $Zn(OH)_2 + H_3PO_4 \longrightarrow Zn_3(PO_4)_2 + H_2O$

(i) $CH_3CH_2OH + PCl_3 \longrightarrow CH_3CH_2Cl + H_3PO_3$

54. Balance the following equations.

(a) $TiCl_4 + H_2O \longrightarrow TiO_2 + HCl$

(b) $WO_3 + H_2 \longrightarrow W + H_2O$

(c) $C_5H_{12} + O_2 \longrightarrow CO_2 + H_2O$

(d) $Al_4C_3 + H_2O \longrightarrow Al(OH)_3 + CH_4$

(e) $Al_2(SO_4)_3 + NaOH \longrightarrow Al(OH)_3 + Na_2SO_4$

(f) $Ca_3P_2 + H_2O \longrightarrow Ca(OH)_2 + PH_3$

(g) $Cl_2O_7 + H_2O \longrightarrow HClO_4$

(h) $MnO_2 + HCl \longrightarrow MnCl_2 + Cl_2 + H_2O$

(i) $CH_4 + O_2 \longrightarrow C_2H_2 + H_2O$

55. Write a balanced chemical equation to represent the reaction of solid magnesium and gaseous oxygen to form solid magnesium oxide. In the equation, one mole of oxygen gas is chemically equivalent to how many moles of magnesium? To how many moles of magnesium oxide?

56. Write a balanced chemical equation to represent **(a)** the decomposition of solid ammonium nitrate into dinitrogen monoxide gas and liquid water, and **(b)** the combustion of liquid heptane, C_7H_{16}, in oxygen gas to produce carbon dioxide gas and liquid water as the sole products. What is the chemical equivalence between ammonium nitrate and dinitrogen monoxide in equation (a)? Between heptane and carbon dioxide in equation (b)?

57. Write a balanced equation to represent **(a)** the reaction of hydrochloric acid with zinc metal to form hydrogen gas and an aqueous solution of zinc chloride; **(b)** the reaction of the gases ethane, C_2H_6, and water to form the gases carbon monoxide and hydrogen; **(c)** the reaction of solid tetraphosphorus decoxide and liquid water to form phosphoric acid; and **(d)** the reaction that produces electricity in a lead-acid storage battery. (The solids lead and lead(IV) oxide react with an aqueous solution of sulfuric acid to produce solid lead(II) sulfate and liquid water.)

58. Write a balanced equation to represent **(a)** the decomposition of solid mercury(II) oxide upon heating to form liquid mercury and oxygen gas; **(b)** the combustion of liquid isopropanol, $(CH_3)_2CHOH$, in oxygen to produce gaseous carbon dioxide and liquid water; **(c)** the reaction of the gases ammonia and oxygen to produce the gases nitrogen monoxide and water; and **(d)** the reaction of the gases chlorine and ammonia with an aqueous solution of sodium hydroxide to form water, an aqueous solution containing sodium chloride, and hydrazine, N_2H_4 (a chemical used in the synthesis of pesticides).

59. At 400 °C, hydrogen gas is passed over iron(III) oxide. Water vapor is formed, together with a black residue—a compound that is 72.3% Fe and 27.7% O by mass. Write a balanced equation for this reaction.

60. An aluminum–carbon compound that is 74.97% Al and 25.03% C by mass, reacts with water to produce aluminum hydroxide and a hydrocarbon that is 74.87% C and 25.13% H by mass. Write a balanced equation for this reaction.

61. Write a balanced chemical equation for the reaction represented by the molecular models shown here:

62. Write a balanced chemical equation for the reaction represented by the molecular models shown here:

Stoichiometry of Chemical Reactions

63. Consider the combustion in excess oxygen of octane, a major component of gasoline.

$$2\,C_8H_{18} + 25\,O_2 \longrightarrow 16\,CO_2 + 18\,H_2O$$

(a) How many moles of CO_2 are produced when 451 mol C_8H_{18} is burned?

(b) How many moles of oxygen are consumed in the combustion of 585 mol C_8H_{18}?

(c) How many moles of H_2O are produced when 188 mol C_8H_{18} is burned?

(d) How many moles of C_8H_{18} are consumed when 2.2×10^4 mol O_2 is consumed?

64. Lead(II) oxide reacts with ammonia as follows:

$$PbO(s) + NH_3(g) \longrightarrow Pb(s) + N_2(g) + H_2O(l)$$
$$\textit{(not balanced)}$$

(a) How many grams of NH_3 are consumed in the reaction of 8.16 g PbO?

(b) If 928 g Pb(s) are produced in this reaction, how many grams of nitrogen are also formed?

(c) How many moles of $N_2(g)$ are produced when 907 kg PbO(s) are consumed?

(d) How many grams of $H_2O(l)$ are made when 14.4 g $N_2(g)$ are formed?

65. Two solids, calcium cyanamide and magnesium nitride, both react with water to produce ammonia gas. *Without doing detailed calculations*, determine which of the two produces the greater amount of ammonia per kilogram of solid when it reacts with an excess of water.

$$CaCN_2(s) + H_2O(l) \longrightarrow CaCO_3(s) + NH_3(g)$$
$$\textit{(not balanced)}$$

$$Mg_3N_2(s) + H_2O(l) \longrightarrow Mg(OH)_2(s) + NH_3(g)$$
$$\textit{(not balanced)}$$

66. The two solids ammonium nitrate and potassium chlorate both produce oxygen gas when decomposed by heating. *Without doing detailed calculations*, determine which of the two yields the greater (a) number of moles of O_2 per mole of solid and (b) number of grams of O_2 per gram of solid.

$$NH_4NO_3(s) \longrightarrow N_2(g) + O_2(g) + H_2O(g) \quad \textit{(not balanced)}$$

$$KClO_3(s) \longrightarrow KCl(s) + O_2(g) \qquad \textit{(not balanced)}$$

67. Kerosene, a mixture of hydrocarbons, is used in domestic heating and as a jet fuel. Assume that kerosene can be represented as $C_{14}H_{30}$ and that it has a density of 0.763 g/mL.

$$C_{14}H_{30}(l) + O_2(g) \longrightarrow CO_2(g) + H_2O(l)$$
$$\textit{(not balanced)}$$

How many grams of CO_2 are produced by the combustion of 7.53 L of kerosene in an indoor heater?

68. Acetaldehyde, CH_3CHO ($d = 0.788$ g/mL), a liquid used in the manufacture of perfumes, flavors, dyes, and plastics, can be produced by the reaction of ethanol with oxygen.

$$CH_3CH_2OH + O_2 \longrightarrow CH_3CHO + H_2O$$
$$\textit{(not balanced)}$$

How many liters of liquid ethanol ($d = 0.789$ g/mL) must be consumed to produce 25.0 L of acetaldehyde?

69. Ordinary chalkboard chalk is a solid mixture, with limestone ($CaCO_3$) and gypsum ($CaSO_4$) as its principal ingredients. Limestone dissolves in dilute HCl(aq) but gypsum does not.

$$CaCO_3(s) + HCl(aq) \longrightarrow CaCl_2(aq) + CO_2(g) + H_2O(l)$$
$$\textit{(not balanced)}$$

(a) If a 12.3-g piece of chalk that is 69.7% $CaCO_3$ is dissolved in excess HCl(aq), what mass of $CO_2(g)$ will be produced?

(b) Determine the mass percent of $CaCO_3$ in a 4.38-g piece of chalk that yields 1.31 g CO_2 when it reacts with excess HCl(aq).

70. Use the following equation to determine

(a) how many milliliters of dilute HCl(aq) ($d = 1.045$ g/mL) that is 9.50% HCl by mass are required to react completely with 4.97 g Al, and

(b) how many grams of hydrogen can be produced by the reaction of 3.23 L of the HCl(aq) with an excess of aluminum.

$$Al(s) + HCl(aq) \longrightarrow AlCl_3(aq) + H_2(g) \quad \textit{(not balanced)}$$

Limiting Reactant and Yield Calculations

71. What is meant by the limiting reactant in a chemical reaction? Under what circumstances might we say that a reaction has two limiting reactants? Explain.

72. Is the limiting reactant always the reactant present in smaller mass? Explain. Is the limiting reactant always the reactant with fewer moles present? Explain.

73. Lithium hydroxide absorbs carbon dioxide to form lithium carbonate and water. If a reaction vessel contains 4.40 mol LiOH and 3.20 mol CO_2, which compound is the limiting reactant? How many moles of Li_2CO_3 can be produced?

74. Boron trifluoride reacts with water to produce boric acid (H_3BO_3) and fluoroboric acid (HBF_4). If a reaction vessel contains 0.496 mol BF_3 and 0.313 mol H_2O,

(a) which compound is the limiting reactant?

(b) How many moles of HBF_4 can be produced?

75. Liquid mercury and oxygen gas react to form mercury(II) oxide. *Without doing detailed calculations*, decide which of the following should result from the reaction of 0.200 mol Hg(l) and 4.00 g $O_2(g)$.

(a) 4.00 g HgO(s) and 0.200 mol Hg(l);

(b) 0.100 mol HgO(s), 0.100 mol Hg(l), and 2.40 g $O_2(g)$;

(c) 0.200 mol HgO(s) and no $O_2(g)$; or

(d) 0.200 mol HgO(s) and 0.80 g $O_2(g)$? Explain.

76. The purification of titanium(IV) oxide, an important step in the commercial production of titanium metal, involves its reaction with carbon and chlorine gas to form titanium tetrachloride and carbon monoxide gas. *Without doing detailed calculations,* decide which of the following initial conditions will result in the production of the maximum amount of $TiCl_4(g)$.

(a) 1.5 mol TiO_2, 2.1 mol C, and 4.4 mol Cl_2;

(b) 1.6 mol TiO_2, 2.5 mol C, and 3.6 mol Cl_2;

(c) 2.0 mol each of TiO_2, C, and Cl_2; or

(d) 3.0 mol each of TiO_2, C, and Cl_2? Explain.

77. Potassium iodide, a dietary supplement used to prevent goiter, an iodine-deficiency disease, is prepared by the reaction of hydroiodic acid and potassium hydrogen carbonate. Water and carbon dioxide are also produced. In the reaction of 398 g HI and 318 g $KHCO_3$,

(a) how many grams of KI are produced;

(b) which reactant is in excess, and how many grams of it remain after the reaction?

78. Sodium nitrite, used as a preservative in meat (to prevent botulism), is prepared by passing nitrogen monoxide and oxygen gases into an aqueous solution of sodium carbonate. Carbon dioxide gas is another product of the reaction. In the reaction of 154 g Na_2CO_3, 105 g NO, and 75.0 g $O_2(g)$,

(a) how many grams of $NaNO_2$ are produced;

(b) which reactants are in excess, and how many grams of each remain after the reaction?

79. Electronic circuit boards are sometimes made by allowing some of the copper metal that coats a special plastic sheet to react with iron(III) chloride. The products are copper(II) chloride and iron

metal. If 3.72 g of copper metal must be dissolved, will 8.48 g of iron(III) chloride be enough to do the job?

80. You find that 1.13 g of acetylene, C_2H_2, burns in an unspecified amount of oxygen gas to form 3.61 g of carbon dioxide, plus water. Was acetylene or oxygen the limiting reactant?

81. Calculate the theoretical yield of ZnS, in grams, from the reaction of 0.488 g Zn and 0.503 g S_8.

$$8 Zn + S_8 \longrightarrow 8 ZnS$$

If the actual yield is 0.606 g ZnS, what is the percent yield?

82. Calculate the theoretical yield of CH_3CH_2Cl, in grams, from the reaction of 11.3 g of ethanol and 13.48 g PCl_3.

$$3 CH_3CH_2OH + PCl_3 \longrightarrow 3 CH_3CH_2Cl + H_3PO_3$$

If the actual yield is 12.4 g CH_3CH_2Cl, what is the percent yield?

83. A student prepares ammonium bicarbonate by the reaction

$$NH_3 + CO_2 + H_2O \longrightarrow NH_4HCO_3$$

She uses 14.8 g NH_3 and 41.3 g CO_2. Water is present in excess. What is her actual yield of ammonium bicarbonate if she obtains a 74.7% yield in the reaction?

84. A student who needs 625 g of zinc sulfide, a white pigment, for an art project can synthesize it using the reaction

$$Na_2S(aq) + Zn(NO_3)_2(aq) \longrightarrow ZnS(s) + 2 NaNO_3(aq)$$

How many grams of zinc nitrate will he need if he can make the zinc sulfide in 85.0% yield? Assume that he has plenty of sodium sulfide.

Solution Stoichiometry

85. Calculate the molarity of each of the following aqueous solutions.

(a) 2.60 mol $CaCl_2$ in 1.15 L of solution

(b) 0.000700 mol Li_2CO_3 in 10.0 mL of solution

(c) 6.631 g $NaNO_3$ in 100.0 mL of solution

(d) 412 g sucrose, $C_{12}H_{22}O_{11}$, in 1.25 L of solution

(e) 15.50 mL glycerol, $C_3H_8O_3$ ($d = 1.265$ g/mL) in 225.0 mL of solution

(f) 35.0 mL 2-propanol, $CH_3CHOHCH_3$, ($d = 0.786$ g/mL) in 250 mL of solution

86. Calculate the molarity of each of the following aqueous solutions.

(a) 2.50 mol H_2SO_4 in 5.00 L of solution

(b) 0.200 mol C_2H_5OH in 35.0 mL of solution

(c) 44.35 g KOH in 125.0 mL of solution

(d) 2.46 g $H_2C_2O_4$ in 750.0 mL of solution

(e) 22.00 mL triethylene glycol, $(CH_2OCH_2CH_2OH)_2$, ($d = 1.127$ g/mL) in 2.125 L of solution

(f) 15.0 mL isopropylamine, $CH_3CH(NH_2)CH_3$, ($d = 0.694$ g/mL) in 225 mL of solution.

87. How much solute is required to prepare each of the following solutions?

(a) moles of NaOH for 1.25 L of 0.0235 M NaOH

(b) grams of $C_6H_{12}O_6$ for 10.0 mL of 4.25 M $C_6H_{12}O_6$

(c) grams of $CuSO_4 \cdot 5 H_2O$ for 3.00 L of a 0.275 molar solution

(d) milliliters of 2-butanol, $CH_3CHOHCH_2CH_3$, ($d = 0.808$ g/mL) for 715 mL of a 1.34 molar solution

88. How much solute is required to prepare each of the following solutions?

(a) moles of $K_2Cr_2O_7$ for 315 mL of 2.50 M $K_2Cr_2O_7$

(b) grams of $KMnO_4$ for 20.0 mL of 0.0100 M $KMnO_4$

(c) grams of KBr and of NaCl for 1.50 L of a solution that is 0.250 M KBr and 0.350 M NaCl

(d) milliliters of ethyl acetate, $CH_3COOCH_2CH_3$, ($d = 0.902$ g/mL) for 315 mL of a 0.0150 molar solution

89. How many milliliters of 0.215 molar solution are required to contain

(a) 0.0867 mol NaBr,

(b) 32.1 g $CO(NH_2)_2$, and

(c) 715 mg methanol, CH_3OH?

90. How many milliliters of 0.0886 molar solution are required to contain

(a) 3.52×10^{-2} mol $Al_2(SO_4)_3$,

(b) 15.6 g $C_{12}H_{22}O_{11}$, and

(c) 35.4 mg diethyl ether, $CH_3CH_2OCH_2CH_3$?

91. A stock bottle of phosphoric acid indicates that the solution is 85.0% H_3PO_4 by mass and has a density of 1.689 g/mL. Calculate the molarity of the solution.

92. More dilute sodium hydroxide solutions are often prepared from a saturated NaOH solution that is 52.0% NaOH by mass and has a density of 1.48 g/mL. Calculate the volume of this solution needed to prepare 35.0 L of 0.125 M NaOH.

93. If 14.00 mL of 1.04 M Na_2CO_3 is diluted to 0.500 L, what is the molarity of Na_2CO_3 in the diluted solution?

94. If 211 mL of 19.1 M NaOH is diluted to 2.00 L, what is the molarity of NaOH in the diluted solution?

95. How many milliliters of 6.052 M HCl are required to make
(a) 2.000 L of 0.5000 M HCl, and
(b) 500.0 mL of a solution containing 7.150 mg HCl per milliliter?

96. How many milliliters of 3.124 M KOH are required to make
(a) 250.0 mL of 1.200 M KOH, and
(b) 15.0 L of a solution containing 0.245 g KOH per liter?

97. *Without doing detailed calculations,* decide which of the following is the most likely concentration of an aqueous solution obtained by mixing 0.100 L of 0.100 M NH_3 and 0.200 L of 0.200 M NH_3:
(a) 0.13 M NH_3 (c) 0.17 M NH_3
(b) 0.15 M NH_3 (d) 0.30 M NH_3.
Explain your answer.

98. The Acculute™ solution in the vial in the photograph, when diluted with water to 1.000 L, produces 0.1000 M HCl. *Without doing detailed calculations,* decide which of the following is a plausible approximate concentration of the solution in the vial:
(a) 0.5 M HCl   (c) 10.0 M HCl
(b) 2.0 M HCl   (d) 12.0 M HCl.
Explain your answer.

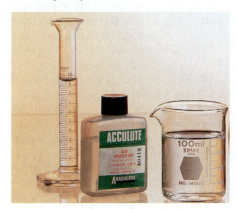

◄ The contents of the vial, when diluted to 1.000 L with water, produce a solution that is 0.1000 M HCl.

99. Suppose you need about 80 mL of 0.100 M $AgNO_3$. You have available about 150 mL of 0.04000 M $AgNO_3$ and also about 1.0 g of solid $AgNO_3$. Assume that you have available standard laboratory equipment such as an analytical balance, 10.00-mL and 25.00-mL pipets, 100.0-mL and 250.0-mL volumetric flasks, and so on. Describe how you would prepare the desired $AgNO_3$ solution, including actual masses or volumes required.

100. Two sucrose solutions, 125 mL of 1.50 M $C_{12}H_{22}O_{11}$ and 275 mL of 1.25 M $C_{12}H_{22}O_{11}$, are mixed. Assuming a final solution volume of 400 mL, what is the molarity of $C_{12}H_{22}O_{11}$ in the final solution?

101. How many grams of $BaSO_4(s)$ are formed when an excess of $BaCl_2(aq)$ is added to 635 mL of 0.314 M $Na_2SO_4(aq)$?

$$BaCl_2(aq) + Na_2SO_4(aq) \longrightarrow BaSO_4(s) + 2\,NaCl(aq)$$

102. How many milliliters of 3.84 M HCl are required to consume 4.12 grams of zinc in the following reaction?

$$Zn(s) + 2\,HCl(aq) \longrightarrow ZnCl_2(aq) + H_2(g)$$

103. After 2.02 g aluminum has reacted completely with 0.400 L of 2.75 M HCl (the excess reactant), what is the molarity of the remaining HCl(aq)?

$$2\,Al(s) + 6\,HCl(aq) \longrightarrow 2\,AlCl_3(aq) + 3\,H_2(g)$$

104. How many grams of aluminum should be added to 0.400 L of 2.75 M HCl to reduce the concentration of the acid to 2.50 M HCl as a result of the reaction in Problem 103?

105. The reaction of calcium carbonate and hydrochloric acid produces calcium chloride, carbon dioxide, and water. How many grams of carbon dioxide are produced when 4.35 g of calcium carbonate is added to 75.0 mL of 1.50 M hydrochloric acid?

106. In aqueous solution, the reaction of silver nitrate and potassium chromate yields solid silver chromate and aqueous potassium nitrate. How many grams of silver chromate are produced when 37.5 mL of a 0.625 M potassium chromate solution are added to 145 mL of 0.0525 M silver nitrate solution?

107. A drop (0.05 mL) of 12.0 M HCl is spread over a thin sheet of aluminum ($d = 2.70\ g/cm^3$) foil (see photograph).

(a) What will be the maximum area, in cm^2, of the hole produced if the thickness of the foil is 0.10 mm?

$$Al(s) + HCl(aq) \longrightarrow AlCl_3(aq) + H_2(g) \quad \text{(not balanced)}$$

(b) Suppose the foil is 0.065 mm thick and is made of copper ($d = 8.96\ g/cm^3$) rather than aluminum. What is the minimum number of drops of 6.0 M HNO_3 needed to dissolve a 1.50-cm^2 hole in the metal?

$$3\,Cu(s) + 8\,HNO_3(aq)$$
$$\longrightarrow 3\,Cu(NO_3)_2(aq) + 2\,NO(g) + 4\,H_2O(l)$$

108. A sheet of iron with a surface area of 525 cm^2 is covered with a coating of rust that has an average thickness of 0.0021 cm. What minimum volume of an HCl solution having a density of 1.07 g/mL and consisting of 14% HCl by mass is required to clean the surface of the metal by reacting with the rust? Assume that the rust is $Fe_2O_3(s)$, that it has a density of 5.2 g/cm^3, and that the reaction is

$$Fe_2O_3(s) + 6\,HCl(aq) \longrightarrow 2\,FeCl_3(aq) + 3\,H_2O(l)$$

Additional Problems

Problems marked with an * may be more challenging than others.

109. Calcium tablets for use as dietary supplements are available in the form of several different compounds. Calculate the mass of each required to furnish 875 mg Ca^{2+}.
(a) calcium carbonate, $CaCO_3$
(b) calcium lactate, $Ca(C_3H_5O_3)_2$
(c) calcium gluconate, $Ca(C_6H_{11}O_7)_2$
(d) calcium citrate, $Ca_3(C_6H_5O_7)_2$

110. Iron, as Fe^{2+}, is an essential nutrient. Pregnant women often take 325-mg ferrous sulfate ($FeSO_4$) tablets as a dietary supplement. Yet iron tablets are the leading cause of poisoning deaths in children. As little as 550 mg Fe^{2+} can be fatal to a 22-lb child. How many 325-mg ferrous sulfate tablets would it take to constitute a lethal dose to a 22-lb child?

111. Chlorophyll, found in plant cells and essential to the process of photosynthesis, contains 2.72% Mg by mass. Assuming one magnesium atom per chlorophyll molecule, calculate the molecular mass of chlorophyll.

112. A 0.507-g sample of a compound containing only carbon, hydrogen, and oxygen is burned in oxygen gas to produce 0.698 g of CO_2 and 0.571 g of H_2O. A member of your study group solves this problem by finding that the CO_2 and H_2O together contain 0.190 g C, 0.0639 g H, and 1.015 g O; this gives a mole ratio of 1:4:4 and a formula of CH_4O_4. What conceptual error did he make? Determine the correct empirical formula for this compound.

113. A sample of a compound of Br and Cl is allowed to react with excess hydrogen gas to give 0.210 g of HCl and 0.155 g of HBr. Determine the empirical formula of the compound.

114. In a common chemistry experiment, copper metal is converted to copper(II) nitrate by dissolving it in nitric acid; the other products are water and NO gas. The copper(II) nitrate is converted to copper(II) hydroxide by addition of sodium hydroxide; the other product is an aqueous solution of sodium nitrate. The filtered copper(II) hydroxide is converted to copper(II) oxide by intense heating; the other product is water vapor. The copper(II) oxide is converted back to copper metal by heating the oxide in a stream of hydrogen gas; the other product is water. Calculate the masses of copper(II) nitrate, copper(II) hydroxide, and copper(II) oxide that should be obtained from an original 0.412 g of copper metal. What mass of copper should be obtained in the final reaction? Explain.

* 115. A 0.8150-g sample of the compound MCl_2 reacts with an excess of $AgNO_3(aq)$ to produce 1.8431 g AgCl(s) and $M(NO_3)_2(aq)$. What is the atomic mass of the element M, and what is the element M?

116. When burned in oxygen in combustion analysis, a 0.1888-g sample of a hydrocarbon produced 0.6260 g CO_2 and 0.1602 g H_2O. The molecular mass of the compound is 106 u. Calculate

   (a) the mass percent composition,

   (b) the empirical formula, and

   (c) the molecular formula of the hydrocarbon.

* 117. Explain whether in the combustion analysis of an alkane,

   (a) the mass of CO_2 obtained can be greater than the mass of H_2O,

   (b) the mass of CO_2 obtained can be less than the mass of H_2O, and

   (c) the masses of CO_2 and H_2O obtained can be equal.

118. A laboratory manual calls for 13.0 g of 1-butanol, 21.6 g of sodium bromide, and 33.8 g H_2SO_4 as reactants in this reaction.

   $$C_4H_9OH + NaBr + H_2SO_4 \longrightarrow C_4H_9Br + NaHSO_4 + H_2O$$

   A student carrying out this reaction obtains 16.8 g of butyl bromide (C_4H_9Br). What are the theoretical yield and the percent yield of this reaction?

119. The urea [$CO(NH_2)_2$] solutions pictured are mixed, and the resulting solution is evaporated to a final volume of 825 mL. What is the molarity of $CO(NH_2)_2$ in this final solution?

655 mL 0.852 M $CO(NH_2)_2$

432 mL 0.487 M $CO(NH_2)_2$

825 mL ? M $CO(NH_2)_2$

Mixing   **Evaporation**

120. What is the simplest formula of a compound containing 37.51% C, 3.15% H, and 59.34% F by mass? The molecular mass of the compound is 96.052 u. What is the molecular formula? When the compound is burned in excess oxygen, the products are CF_4, CO_2, and H_2O. Write a balanced chemical equation for the reaction.

* 121. Two binary compounds of phosphorus and an unknown element X have X/P mass ratios of 1.84 and 3.06. If both compounds are made up of molecules that have only one atom of phosphorus, what is the identity of element X and what are the formulas of the two compounds?

* 122. Perchloric acid dihydrate, $HClO_4 \cdot 2 H_2O$, is a stable liquid with a density of 1.65 g/mL. It may be considered as a compound or as a solution of $HClO_4$ in water. If the latter, what would be the molarity of $HClO_4$ in the solution?

* 123. In Example 3.20 (page 102), we described the reaction of magnesium and nitrogen to form magnesium nitride. Magnesium reacts even more readily with oxygen to form magnesium oxide. The noble gas argon, on the other hand, is inert; it does not react with magnesium. Suppose that for the conditions described in the example—35.00 g Mg and 15.00 g N_2—the nitrogen gas was not pure. Calculate the number of grams of product you would expect if the nitrogen contained **(a)** 5% of argon by mass, **(b)** 15% argon by mass, and **(c)** 25% oxygen by mass.

* 124. A 10.000-g sample of a compound with a molar mass of 60 g/mol and the mass percent composition 40.00% C, 6.71% H, and 53.29% O reacts with 7.621 g of a compound with the mass percent composition 22.56% P and 77.44% Cl. Two products are formed. The mass of one product is 4.552 g and its composition is 3.69% H, 37.77% P, and 58.53% O. From the information given, write a balanced equation for the reaction.

* 125. A particular natural gas has the following mass percent composition: 81.29% methane, 8.18% ethane, 4.18% propane, 2.12% butanes, 1.06% pentanes, 0.61% hexanes, 0.61% heptanes, and 1.95% of noncombustible N_2. How many kilograms of CO_2 are produced by the complete combustion of 1.00 kg of this natural gas?

Apply Your Knowledge

126. **[Collaborative]** Use the information from Problem 63 to determine the mass of carbon dioxide produced by the typical driving done by members of your chemistry class in one week. Use C_8H_{18} as a representative formula of gasoline ($d = 0.703$ g/mL) and assume complete combustion of the gasoline. Poll your class to obtain the additional data necessary.

* 127. **[Laboratory]** A mixture of $BaCl_2 \cdot 2 H_2O$ and NaCl with a mass of 1.6992 g is placed in a vial weighing 3.3531 g. The mixture is

heated at 105 °C to drive off water. The residue of the mixture now has a mass of 1.4804 g. Use the preceding data, as needed, to calculate the % NaCl in the mixture.

* 128. **[Historical]** The photograph on page 85 refers to the labeling of fertilizers. The three numbers on the label are the *NPK designation.* They indicate the nitrogen content expressed as % N, the phosphorus content expressed as % P_2O_5, and the potassium content expressed as % K_2O. The use of mass percentages based

on oxides is a holdover from the way compositions were reported in the early days of analytical chemistry. That is, the 5-10-5 fertilizer shown on page 85 is 5% N, contains phosphorus equivalent to 10% P_2O_5, and contains potassium equivalent to 5% K_2O.

(a) What are the percentages of phosphorus and potassium in 5-10-5 fertilizer?

(b) What is the NPK designation for pure ammonium nitrate?

(c) A mixture of two compounds in a mole ratio of 1:1 carries an NPK rating of about 10-53-18. What combination of two of the following compounds might this be: KH_2PO_4, K_2HPO_4, K_3PO_4, $NH_4H_2PO_4$, $(NH_4)_2HPO_4$, $(NH_4)_3PO_4$?

(d) It is desired to make a 20-20-20 fertilizer using KNO_3, NH_4NO_3, NaH_2PO_4, and an inert ingredient (sand). If this can be done, give the relative mass of each of the ingredients present. If this cannot be done, show why this is the case.

* **129. [Environmental]** Hydrogen produced by the decomposition of water has considerable potential as a fuel. Key to its development is finding an appropriate series of chemical reactions that has as its overall reaction: $2 H_2O \longrightarrow 2 H_2 + O_2$. Demonstrate that this requirement is met by the Fe/Cl cycle described as follows: (1) A chloride of iron with 44.19% Fe by mass reacts with steam $[H_2O(g)]$ at 500 °C, producing an oxide of iron with 72.36% Fe by mass, together with hydrogen and hydrogen chloride gases. (2) The oxide of iron reacts with hydrogen chloride and chlorine gases at 200 °C to produce another chloride of iron with 34.43% Fe, together with steam, and oxygen gases. (3) At 420 °C, This second chloride of iron produces the first chloride of iron, together with chlorine gas. (*Hint:* The equation for the overall reaction is the sum of the equations for the three reactions described, after their coefficients have been properly adjusted.)

130. [Environmental] Calculate the atom % economy (page 113) in the production of carbon tetrachloride carried out in the following two-step process. First, carbon disulfide reacts with chlorine gas to form carbon tetrachloride and disulfur dichloride. Then disulfur dichloride reacts with more carbon disulfide, producing carbon tetrachloride and elemental sulfur.

131. [Environmental] An environmental newsletter states that 11 billion gallons of a sulfuric acid solution that is 70% H_2SO_4 by mass is scheduled for shipment to the White Pine Mine over a one-year period. The current annual production of pure H_2SO_4 in the United States is just under 100 billion pounds. Do you think the statement in the newsletter is accurate? Explain. Assume a density of 1.61 g/mL for the sulfuric acid.

132. [Laboratory] An old book for preparation of laboratory solutions describes a method of making pure sodium hydroxide, using sodium metal. How many grams of sodium metal must react with 250.0 mL of water to produce a solution that is 0.315 M NaOH? (Assume the final solution volume is 250.0 mL.)

$$Na(s) + H_2O(l) \longrightarrow NaOH(aq) + H_2(g) \quad (not\ balanced)$$

* **133. [Laboratory]** A 0.6118-g sample containing only $MgCl_2$ and NaCl was analyzed by adding 145 mL of 0.1006 M $AgNO_3$. The precipitate of AgCl(s) formed had a mass of 1.7272 g. Calculate the mass of each component in the original sample.

* **134. [Laboratory]** In a student experiment, the empirical formula of a copper halide was found by adding aluminum metal to an aqueous solution of the halide, displacing copper metal. The copper metal was filtered, washed with distilled water, dried, and weighed; three separate determinations were performed. If the copper halide solution contained 42.62 g of copper chloride per liter, find the empirical formula that should be reported from the data below. Comment on whether the formula seems reasonable.

Volume of copper chloride solution	Mass of filter paper	Mass of filter paper plus copper
49.6 mL	0.908 g	1.694 g
48.3 mL	0.922 g	1.693 g
42.2 mL	0.919 g	1.588 g

* **135. [Biochemical]** The human body is a complex mixture of many substances. In a typical human, 62.8% of the atoms are H, 25.4% O, 9.4% C, and 1.4% N. The remaining 1% are almost all from these seven elements: sodium, potassium, calcium, magnesium, phosphorus, sulfur, and chlorine. About 70% of the typical body mass is water. Suppose, simply as a mental exercise, we try to write a hypothetical "chemical formula" for the typical human body. (a) What would be its empirical formula, in the form $C_VH_WO_XN_Y(E)_Z$, where V, W, X, Y, and Z are subscripts and (E) comprises the seven minor elements; and (b) what would be its formula with the 70% mass percent water expressed as "hydrate" water? (c) Discuss why it is not appropriate to speak of the chemical formula of a human body.

e-Media Problems

The activities described in these problems can be found in the e-Media Activities and Interactive Student Tutorial (IST) modules of the Companion Website, *http://chem.prenhall.com/hillpetrucci*.

136. View the **Sodium and Potassium in Water** movie (*Section 3-Intro*). In what ways can the reactions be described *qualitatively?* What aspects of the way in which the reactions are carried out in the movie would make it difficult to describe the reaction *quantitatively?*

137. In the **Views of Chemical Reactions** activity (*Section 3-2*), which "view" or quantity is most representative of a laboratory setting? Which view is most helpful in describing the molecular pathway of a chemical reaction?

138. In the last experiment of the **Limiting Reagent** animation (*Section 3-9*), how much zinc will remain on the bottom of the flask at the completion of the reaction?

139. View the **Solution Formation from a Solid** animation (*Section 3-11*). How many grams of copper sulfate pentahydrate should be added to the 250-mL flask to prepare 0.25 L of a 2.5 M solution? How can this answer be deduced from the information in the movie? What steps are necessary to directly calculate the answer?

140. In the **Solution Formation by Dilution** animation (*Section 3-11*), what volume of 1.0 M $CuSO_4$ should be transferred to the 250-mL vessel to prepare this volume of a 0.50 M solution?

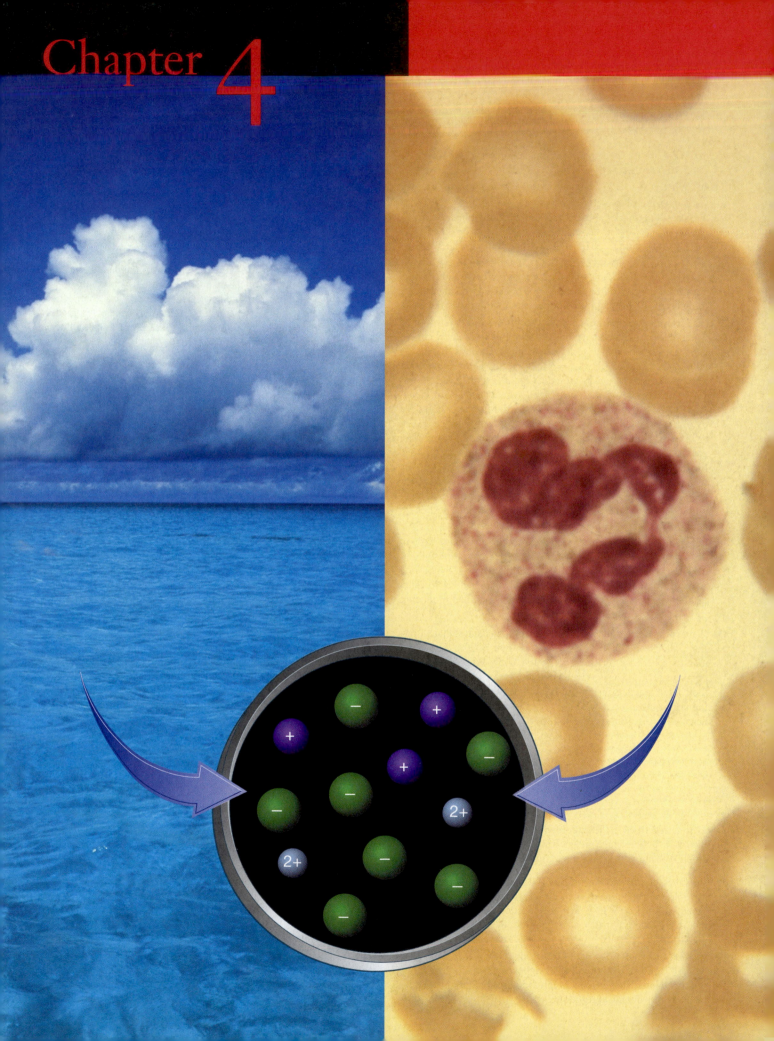

Chapter 4

Chemical Reactions in Aqueous Solutions

THREE-QUARTERS OF Earth's surface is covered with water, but this natural water is not chemically pure. It contains dissolved substances. Solutions are formed as liquid water comes into contact with gases in the atmosphere, materials in Earth's solid crust, and materials released into the environment by human activities. The water in living organisms is also in the form of aqueous solutions. In fact, water is such a good solvent for so many ionic and molecular substances that it has been called the *universal solvent*.

Reactions that take place in aqueous solutions can be grouped into a few basic categories, all of which have important applications, as we will see in this chapter. In exploring these reactions, we will continue to stress ideas from previous chapters. Later in the text, we will examine each reaction type in more detail.

CONTENTS

4.1   Some Electrical Properties of Aqueous Solutions

The outward appearance of an aqueous solution does not tell us what is present at the microscopic level, that is, whether the solute particles are ions, molecules, or a mixture of the two. Nevertheless, some of the earliest insights into the microscopic nature of aqueous solutions came through macroscopic observations on the ability of solutions to conduct electricity. To understand why, let's first look at some significant early discoveries about electricity.

Static electricity, such as that produced by running a comb through your hair, has been recognized since ancient times. By the end of the eighteenth century, two types of electric charge—*positive* and *negative*—had been identified, and the interactions between positively and negatively charged objects were well understood (Figure 4.1).

At the beginning of the nineteenth century, it was discovered that electricity could flow through metal wires. We now know that an electric current is a *flow of charged particles*. In solid and liquid metals, those charged particles are *electrons*. Metals are good electrical conductors, that is, they transmit electric current with relatively little

◀ Many of the chemical reactions that we study take place in aqueous solution. Seawater is a solution of ions, principally Na^+ and Cl^-, but it also contains dissolved CO_2 and other gases from the atmosphere. The fluids of living cells, such as our blood cells, are also solutions of ions and other dissolved substances; their composition is quite close to that of seawater. The three types of reactions that we consider in this chapter all occur in the sea and in our bodies.

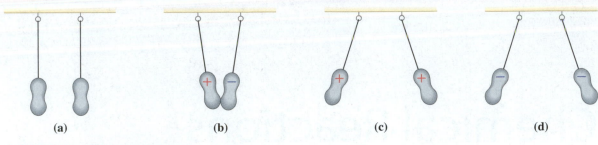

▲ **FIGURE 4.1** **Electrostatic forces**

The objects pictured here are plastic-foam peanuts, the type used as packing material. (a) The peanuts carry no electric charge, and therefore no electrostatic force acts between them. (b) The peanuts are oppositely charged and therefore attract each other. One has a positive charge; the other has a negative charge. In both (c) and (d), the peanuts carry like charges and therefore repel each other.

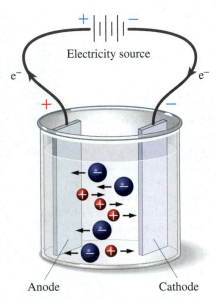

▲ **FIGURE 4.2** **Conduction of electric current through a solution**

The electricity source directs electrons from the electrodes through the wires from the anode to the cathode. In the solution, cations (+) are attracted to the cathode (−), and anions (−) are attracted to the anode (+). This migration of ions represents the flow of electricity through the solution.

Dissolution of NaCl in Water animation

resistance. The electrons in a metal wire can be set in motion in different ways—by an electric generator in a power plant, for example, or by a chemical reaction in an electric battery.

Molten (liquid) ionic compounds and aqueous solutions of ionic compounds are also good electrical conductors, but in these cases *ions* are the charged particles that flow. Michael Faraday (1791–1867) did much of the early work on the conduction of electricity through solutions, the topic of interest in this section. Faraday coined a number of terms that are still used by chemists today:

- *Electrodes* are electrical conductors (such as wires or metal plates) partially immersed in a solution and connected to a source of electricity. The *anode* is the electrode connected to the *positive* pole of the source of electricity, and the *cathode* is the electrode connected to the *negative* pole.

- An *ion* is a carrier of electricity through a solution. (*Ion* is derived from Greek and means "wanderer.") *Anions* have a negative charge (−) and are attracted to the anode (+); *cations* carry a positive charge (+) and are attracted to the cathode (−).

Figure 4.2 suggests how electricity passes through a solution. The external source of electricity—a battery—withdraws electrons from the positive anode and forces them onto the negative cathode. The ions in solution then migrate to the electrodes—*cat*ions to the negatively charged *cat*hode, *an*ions to the positively charged *an*ode—carrying electricity through the solution and completing the electric circuit.

Arrhenius's Theory of Electrolytic Dissociation

Faraday did not speculate about how the ions that conduct electricity are formed in a solution. Other scientists at the time thought that as it entered a solution via the electrodes, the electric current caused solute molecules to break apart, or *dissociate*, into cations and anions. In his doctoral dissertation in 1884, however, Svante Arrhenius presented the hypothesis that certain substances, such as NaCl and HCl, dissociate into cations and anions *when they dissolve in water*. In other words, electricity does not produce ions in an aqueous solution; rather, the ions that already exist in solution allow electricity to flow. And, of course, if there are no ions, there is no electric current. We call a solute that produces enough ions to make a solution an electrical conductor an *electrolyte*. Arrhenius's ideas are now called the *theory of electrolytic dissociation*.

Figure 4.3 illustrates a way to demonstrate the relative abilities of solutions to conduct electric current. We place two electrodes made of graphite (the material in pencil "lead") in the solution to be tested and connect them by wires to a source of electric current. We also place an electric lightbulb in the circuit. Electrons can pass freely through the wires and the bulb, but they cannot pass through the solution. Unless the solution contains enough ions to carry electric charge, no electric current passes. Suppose that the solutions in the beakers are all 1 M in a solute. No matter what the solute is, we will always observe one of the following three possibilities.

(a)
1 M CH$_3$OH
Nonelectrolyte
Solute consists
of molecules;
no ions

(b)
1 M NaCl(aq)
Strong electrolyte
Solute consists of ions:

$+$ Na$^+$ $-$ Cl$^-$

(c)
1 M CH$_3$COOH(aq)
Weak electrolyte
Solute consists
mostly of molecules;
some ions:

$-$ CH$_3$COO$^-$ $+$ H$_3$O$^+$

▲ **FIGURE 4.3   Electrolytic properties of aqueous solutions**
In order for there to be an electric current through each solution, cations and anions must be present and must be free to migrate between the two graphite electrodes. In the microscopic view of each solution, solvent molecules are not shown because we want to focus on the solutes; the blue background suggests the presence of water as the solvent. In (a), there are no ions. In (b), all the solute particles are ions: Na$^+$ and Cl$^-$. In (c), about one in every 200 molecules ionizes, producing an acetate ion, CH$_3$COO$^-$, and a hydrogen ion, which attaches itself to a water molecule, forming H$_3$O$^+$.

Figure 4.3a: The bulb does not light. This observation signifies that there are essentially no ions in the solution, certainly not nearly enough to carry a significant amount of charge through the solution. This is what we see when the liquid in the beaker is either *pure* water* or an aqueous solution of a molecular substance that does not ionize, such as ethanol, CH$_3$CH$_2$OH, or sucrose, C$_{12}$H$_{22}$O$_{11}$.

> A **nonelectrolyte** *is a solute that is present in solution almost exclusively as molecules. A solution of a nonelectrolyte is a nonconductor of electricity.*

Figure 4.3b: The bulb lights brightly. This indicates that a large number of ions are present in the solution. As in NaCl(s), there are no NaCl molecules in NaCl(aq), but only separate Na$^+$ and Cl$^-$ ions. We say that NaCl(aq) is a strong electrolyte.

> A **strong electrolyte** *is a solute that is present in solution almost exclusively as ions. A solution of a strong electrolyte is a good electrical conductor.*

Electrolytes and Nonelectrolytes animation

*Ordinary tap water is not pure. It contains enough dissolved ionic compounds to conduct electricity to a limited extent. This is why we must exercise great care when handling electrical equipment near pools of water.

Hydrogen chloride is also a strong electrolyte in an aqueous solution. In this case, however, HCl is a molecular substance in the pure state. It forms H^+ and Cl^- ions when dissolved in water.

Figure 4.3c: The bulb lights but only dimly. This is what we expect if a solute exists partly in molecular form and partly in ionic form in solution. Acetic acid, CH_3COOH, behaves this way in aqueous solution.

*A **weak electrolyte** is a solute that is only partially ionized in solution. A solution of a weak electrolyte is a poor conductor of electricity.*

Strong and Weak Electrolytes movie

Note that the terms "strong" and "weak" refer to the extent to which an electrolyte produces ions in solution. If an electrolyte exists almost exclusively as ions in solution, it is a *strong electrolyte;* if much of it remains in molecular form, it is a *weak electrolyte.* Thus, even though an extremely dilute solution of sodium chloride might not conduct electricity as well as a somewhat more concentrated solution of acetic acid, NaCl is still a strong electrolyte and CH_3COOH is still a weak electrolyte. The following generalizations classify water-soluble solutes according to their electrolytic properties:

- Essentially all soluble ionic compounds are *strong electrolytes.*
- Only a few molecular compounds (mainly a few acids, such as HCl) are *strong electrolytes.*
- Most molecular compounds are either *nonelectrolytes* or *weak electrolytes.*
- Most organic compounds are molecular and *nonelectrolytes;* carboxylic acids and a class of organic compounds called amines (page 132) are *weak electrolytes.*

We can use these generalizations to decide how best to represent the solute in an aqueous solution. Because a solute that is a *nonelectrolyte* exists only in molecular form, we use its molecular formula. For example, we represent an aqueous solution of ethanol as $CH_3CH_2OH(aq)$. When we dissolve a water-soluble ionic compound—a *strong electrolyte*—the cations and anions dissociate completely from one another and appear in solution as independent solute particles*. We can represent the dissociation of NaCl as

$$NaCl(s) \xrightarrow{H_2O} Na^+(aq) + Cl^-(aq)$$

The H_2O above the arrow signifies that the dissociation occurs when NaCl(s) dissolves in water, but H_2O is not a reactant in the usual sense. In many cases, we can continue to represent aqueous solutions of ionic compounds as we did in Chapter 3, that is, without indicating their dissociation, as in NaCl(aq). However, in other situations the ionic form—$Na^+(aq) + Cl^-(aq)$—gives a better representation of the matter at hand, as we will see next.

Calculating Ion Concentrations in Solution

Later in the text, we will find these bracket symbols to be indispensable for representing molar concentrations in algebraic equations.

At times, we may need to determine the concentrations of the individual ions in an aqueous solution of some ionic compound. Thus, because one mole each of Na^+ ions and Cl^- ions appear for every mole of NaCl(s) dissolved, the ion concentrations in a 0.010 M NaCl(aq) solution are 0.010 M $Na^+(aq)$ and 0.010 M $Cl^-(aq)$. A common convention for representing molar concentrations uses a set of *brackets* enclosing the solute symbol: $[Na^+]$ = 0.010 M $Na^+(aq)$ and $[Cl^-]$ = 0.010 M $Cl^-(aq)$.

Problem-Solving Note

From this point on in the text, we will not routinely show the cancellation of units in calculations, although you should continue to use them as an aid in problem-solving. However, we will retain cancel marks for emphasis in certain equations and derivations.

Because the dissociation of Na_2SO_4 produces *two* Na^+ ions and *one* SO_4^{2-} ion per formula unit, the ion concentrations in a 0.010 M Na_2SO_4 solution are 2×0.010 mol Na^+ per liter and 0.010 mol SO_4^{2-} per liter. That is, $[Na^+]$ = 0.020 M and $[SO_4^{2-}]$ = 0.010 M. Example 4.1 illustrates an additional idea about ion concentrations in solution: There is only one concentration for any given ion in a solution, even if the ion has more than one source. Additionally, the total ion concentration in a solution is the sum of the molarities of the individual ions.

*Dissociation is complete in very dilute solutions of ionic substances. In more concentrated solutions this will likely not be the case. For our purposes in this text, we will generally assume that dissociation is complete.

Example 4.1

Calculate the molarity of each ion in an aqueous solution that is 0.00384 M Na_2SO_4 and 0.00202 M NaCl. In addition, calculate the total ion concentration of the solution.

STRATEGY

First, note that there are three types of ions in the solution—Na^+, SO_4^{2-}, and Cl^-. Of these three types, SO_4^{2-} comes only from the dissolved Na_2SO_4, Cl^- comes only from the NaCl, and Na^+ comes from *both* solutes. Thus, we can establish $[SO_4^{2-}]$ just from the molarity of $Na_2SO_4(aq)$ and $[Cl^-]$ just from the molarity of NaCl(aq). For $[Na^+]$, we will have to work with both molarities. Finally, we sum the molarities of the individual ions to obtain the total ion concentration.

SOLUTION

The dissociation of the ionic solutes in the solution is represented by the equations describing the dissolution of the two salts.

$$Na_2SO_4(s) \xrightarrow{H_2O} \underset{\text{Cation}}{2\,Na^+(aq)} + \underset{\text{Anion}}{SO_4^{2-}(aq)}$$
$$\underset{\text{Salt}}{}$$

$$NaCl(s) \xrightarrow{H_2O} \underset{\text{Cation}}{Na^+(aq)} + \underset{\text{Anion}}{Cl^-(aq)}$$
$$\underset{\text{Salt}}{}$$

Because *one* mole of $Na_2SO_4(s)$ produces *one* mole of $SO_4^{2-}(aq)$, the molarity of the SO_4^{2-} is the same as that given for $Na_2SO_4(aq)$.

$$[SO_4^{2-}] = \frac{0.00384\ \text{mol Na}_2\text{SO}_4}{1\ \text{L}} \times \frac{1\ \text{mol SO}_4^{2-}}{1\ \text{mol Na}_2\text{SO}_4} = 0.00384\ \text{M}$$

The situation is similar for $Cl^-(aq)$. The molarity of $Cl^-(aq)$ is the same as that of the NaCl(aq).

$$[Cl^-] = \frac{0.00202\ \text{mol NaCl}}{1\ \text{L}} \times \frac{1\ \text{mol Cl}^-}{1\ \text{mol NaCl}} = 0.00202\ \text{M}$$

The molarity of the Na^+ is the sum of the two terms in the right column, one for each source of Na^+.

$$[Na^+] = \left[\frac{0.00384\ \text{mol Na}_2\text{SO}_4}{1\ \text{L}} \times \frac{2\ \text{mol Na}^+}{1\ \text{mol Na}_2\text{SO}_4} \right]$$
$$+ \left[\frac{0.00202\ \text{mol NaCl}}{1\ \text{L}} \times \frac{1\ \text{mol Na}^+}{1\ \text{mol NaCl}} \right]$$
$$= 0.00768\ \text{M} + 0.00202\ \text{M} = 0.00970\ \text{M}$$

The sum of the individual ion molarities gives the total ion concentration.

$$\text{Total ion concentration} = [Na^+] + [Cl^-] + [SO_4^{2-}]$$
$$= 0.00970\ \text{M} + 0.00202\ \text{M} + 0.00384\ \text{M} = 0.01556\ \text{M}$$

ASSESSMENT

As you become more familiar with calculating ion concentrations, you should be able to work a problem of this sort just by inspecting the formulas of the solutes, thereby arriving directly at the conclusions; $[SO_4^{2-}] = 0.00384$ M, $[Cl^-] = 0.00202$ M, and $[Na^+] = (2 \times 0.00384) + 0.00202 = 0.00970$ M.

EXERCISE 4.1A

Seawater is essentially 0.438 M NaCl and 0.0512 M $MgCl_2$, together with several other minor solutes. What are the molarities of Na^+, Mg^{2+}, and Cl^- in seawater?

EXERCISE 4.1B

Each year, oral rehydration therapy (ORT)—feeding a person an electrolyte solution—saves the lives of a million children worldwide who become dehydrated from diarrhea. Each liter of the electrolyte solution contains 3.5 g sodium chloride, 1.5 g potassium chloride, 2.9 g sodium citrate ($Na_3C_6H_5O_7$), and 20.0 g glucose ($C_6H_{12}O_6$). Calculate the molarity of each species present in the solution. (*Hint:* Sodium citrate is a strong electrolyte, and glucose is a nonelectrolyte.)

4.2 Reactions of Acids and Bases

When we first encountered Arrhenius's theory of acids and bases in Chapter 2, we focused on names and formulas. In Chapter 3, we used acids and/or bases as reactants in a few of the reactions. In this section, we will look at acid–base reactions in more detail.

Strong and Weak Acids

We will adopt a more general acid–base theory in Chapter 15. However, for now we will continue to use the Arrhenius view that an acid is a substance that produces hydrogen ions (H^+)* in aqueous solution, and we will expand the definition of an acid in a way that explains the ability of aqueous acid solutions to conduct electricity.

Introduction to Acids animation

Acids that are completely ionized in water produce aqueous solutions that are good electrical conductors. These acids are *strong electrolytes* and are called **strong acids.** As just one example, the equation representing the ionization of hydrogen chloride in water to produce hydrochloric acid, a strong acid, is

$$HCl(g) \xrightarrow{\text{H}_2\text{O}} H^+(aq) + Cl^-(aq)$$

We can therefore represent an aqueous solution of hydrochloric acid either as HCl(aq) or as $H^+(aq) + Cl^-(aq)$, and we can calculate the concentrations of cations and anions as in Example 4.1. Specifically, the ionization equation tells us that 1 mol HCl(g) dissociates to yield 1 mol $H^+(aq)$ and 1 mol $Cl^-(aq)$. Thus, in 0.0010 M HCl,

$$[H^+] = 0.0010 \text{ M} \quad \text{and} \quad [Cl^-] = 0.0010 \text{ M}$$

Because essentially all the HCl molecules have ionized, we take the concentration of HCl *molecules* to be zero:

$$[HCl] = 0$$

▲ A ball-and-stick model of acetic acid, CH_3COOH, a weak acid.

Acetic Acid 3D model

The vast majority of acids are *weak electrolytes*; that is, only some of their molecules ionize in aqueous solution. The rest remain as intact molecules. These acids are called **weak acids.** The carboxylic acids (Section 2.9) containing one to four carbon atoms are soluble in water and are weak acids. In 1.0 M CH_3COOH, for example, only about 0.5% of the molecules ionize. Because most weak acids remain largely in molecular form, we use a molecular formula, such as $CH_3COOH(aq)$, to represent them. We can write the following equation to represent the limited ionization of acetic acid:

$$\textit{Ionization:} \quad \underset{\text{Acetic acid}}{CH_3COOH(aq)} \longrightarrow H^+(aq) + \underset{\text{Acetate ion}}{CH_3COO^-(aq)}$$

Taken by itself, this equation is misleading because it falsely implies that the ionization goes to completion. Ionization does not go to completion, however. The ions are able to recombine to form neutral molecules in a reverse reaction:

$$\textit{Recombination of ions:} \quad H^+(aq) + CH_3COO^-(aq) \longrightarrow CH_3COOH(aq)$$

The best way to represent the limited ionization of a weak acid is to write a single equation with a *double arrow* ($\rightleftharpoons$):

$$CH_3COOH(aq) \rightleftharpoons H^+(aq) + CH_3COO^-(aq)$$

A double arrow in a chemical equation signifies that the reaction is *reversible:* A forward reaction and a reverse reaction occur simultaneously. Instead of going to completion, a reversible reaction reaches a state of *equilibrium* in which the concentrations of reactants and products remain constant over time. We will discuss reversible reactions and the nature of equilibrium in more detail in Chapter 14, but for now, we have simply noted the significance of the double arrow.

Some acids can produce two or more hydrogen ions per molecule of the acid. Sulfuric acid, H_2SO_4, and phosphoric acid, H_3PO_4, are two common examples. Sulfuric acid is interesting in that it is a *strong* acid in its first ionization and a *weak* acid in its second:

$$\textit{First ionization:} \quad H_2SO_4(aq) \longrightarrow H^+(aq) + HSO_4^-(aq)$$

$$\textit{Second ionization:} \quad HSO_4^-(aq) \rightleftharpoons H^+(aq) + SO_4^{2-}(aq)$$

▲ A space-filling model of sulfuric acid, a strong acid in its first ionization and a weak acid in its second.

Sulfuric Acid 3D model

---

*The simple hydrogen ion, H^+, does not exist in aqueous solutions. Instead, it is associated with several H_2O molecules; that is, it is present as $H(H_2O)_n^+$, where n is an integer. The most important of these hydrated hydrogen ions has $n = 1$. Its formula is H_3O^+, and it is called the *hydronium ion.* We will use the abbreviation H^+ here and switch to H_3O^+ in Chapter 15.

Phosphoric acid, on the other hand, is a *weak* acid in each of its three ionization steps. We will discuss sulfuric acid and phosphoric acid further in Chapter 15.

Can we look at a chemical formula and tell whether a substance is an acid? If it is, can we tell whether it is a strong or weak acid? We will tackle questions such as these in our in-depth treatment of acids in Chapter 15, but even at this point we can get some partial answers by applying the following ideas.

1. *The molecular formula of an acid is generally written with ionizable H atoms appearing first.* Thus, HNO_3, H_2SO_4, and H_3PO_4 are acids containing one, two, and three ionizable H atoms, respectively. Methane, CH_4, has four H atoms, but they are *not* ionizable, and therefore CH_4 is *not* an acid. From its name, we know that acetic acid is an acid; and if we see its formula written as $HC_2H_3O_2$, we know that it has one H atom that is ionizable and three H atoms that are not.

2. *Structural and condensed structural formulas show where H atoms are found in a molecule; their locations usually help identify ionizable H atoms.* The condensed structural formula for acetic acid, CH_3COOH, indicates that three of the H atoms are bonded to the C atom. Just like the four H atoms in CH_4, the three in the CH_3 group are *not* ionizable. Only the H atom bonded to one of the O atoms in the carboxylic acid group, —COOH, is ionizable.

The simplest way to tell whether an acid is strong or weak is to note that there aren't many strong acids; the most common ones are those in Table 4.1. Unless you have information to the contrary, you can assume that any other acid is a weak acid.

Strong and Weak Bases

By the Arrhenius definition, a base is a substance that produces hydroxide ions, OH^-, in aqueous solution. Moreover, because ionic compounds are *strong* electrolytes, ionic hydroxides, such as NaOH, are **strong bases.**

$$NaOH(s) \xrightarrow{H_2O} Na^+(aq) + OH^-(aq)$$

 **Introduction to Bases movie**

Some compounds produce OH^- ions by *reacting* with water, not just by dissolving in it. These substances are also bases. Gaseous ammonia dissolves in water, and the $NH_3(aq)$ formed reacts with water to produce some ions:

$$NH_3(aq) + H_2O(l) \rightleftharpoons NH_4^+(aq) + OH^-(aq)$$

As in the case of acetic acid, most of the NH_3 molecules remain nonionized at equilibrium, and ammonia is therefore a **weak base.*** Most molecular substances that act as bases are *weak* bases.

Table 4.1 Common Strong Acids and Strong Bases

Acids		Bases	
Binary Hydrogen Compounds	**Oxoacids**	**Group 1A hydroxides**	**Group 2A hydroxides**
HCl	HNO_3	LiOH	$Mg(OH)_2$
HBr	H_2SO_4[a]	NaOH	$Ca(OH)_2$
HI	$HClO_4$	KOH	$Sr(OH)_2$
		RbOH	$Ba(OH)_2$
		CsOH	

[a] H_2SO_4 is a strong acid in its first ionization step but weak in its second ionization step.

*In Arrhenius's time, chemists generally believed that a substance must contain OH groups in order to be a base. Thus, $NH_3(aq)$ was thought to be NH_4OH (ammonium hydroxide). Even though there is no compelling evidence for the existence of NH_4OH molecules, this formula is still often seen as a representation of $NH_3(aq)$.

The organic substances known as *amines* are molecular compounds in which one or more of the H atoms in NH_3 are replaced by a hydrocarbon group. In these two amines, one of the H atoms has been replaced:

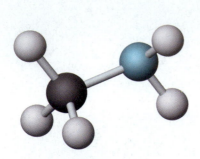

H—C—N—H   or CH_3NH_2 H—C—C—N—H   or $CH_3CH_2NH_2$

Methylamine   Ethylamine

The replacement of two and three H atoms, respectively, is seen in dimethylamine $[(CH_3)_2NH]$ and trimethylamine $[(CH_3)_3N]$.

Amines containing one to four carbon atoms are water soluble, and like NH_3 they are *weak* bases. For example,

$$CH_3NH_2(aq) + H_2O(l) \rightleftharpoons CH_3NH_3^+(aq) + OH^-(aq)$$

Methylamine   Methylammonium ion

As with acids, we can often determine whether a substance is a base from its formula. For example, if the formula indicates an *ionic* compound containing metal ions and OH^- ions, we expect it to be a strong base. NaOH and KOH are strong bases, and Table 4.1 lists several others. In contrast, methanol, CH_3OH, is *not* an Arrhenius base. There are only nonmetals in the compound—no metals—and therefore CH_3OH is a *molecular* compound. The OH group is covalently bonded to the C atom, *not* present as OH^-. Similarly, acetic acid (CH_3COOH) is not a base because the OH group in it is bonded to the C atom. Instead, acetic acid produces H^+ and is therefore an acid. To identify a weak base, you usually need to see a chemical equation for the ionization reaction. However, you can identify many weak bases by using these facts:

There are only a few common strong bases (Table 4.1), and the most common weak bases are ammonia and the amines.

Acid–Base Reactions: Neutralization

In any reaction between an aqueous acid and base, the identifying characteristics of the acid and base (Section 2.8) cancel out, or neutralize, each other. As noted in Chapter 2, this process is called **neutralization,** and one of the products of the process is a **salt.** Generally, we don't see anything happen in a neutralization reaction at the macroscopic level because both the acid and base are usually colorless, as is the salt solution formed from them. How then do we know when a solution is neutral, that is, neither acidic nor basic? As we mentioned in Section 2.8, one characteristic of acids and bases is their ability to affect the color of litmus, a natural dye. A substance whose color is affected by the relative amounts of H^+ and OH^- ions in solution is called an acid–base **indicator.** For example, as shown in Figure 4.4, the indicator phenol red is yellow in acidic solutions, orange in neutral solutions, and red in basic solutions.

If we use conventional formulas for the acid and base, we can write what we might call a *complete-formula equation** for a neutralization reaction:

$$HCl(aq) + NaOH(aq) \longrightarrow NaCl(aq) + H_2O(l)$$

Acid   Base   Salt   Water

However, this complete-formula equation is not always the best way to show what happens in the neutralization. To show that the HCl(aq), a strong acid; NaOH(aq), a strong base; and NaCl(aq), a salt, exist in solution as ions, we can write the equation in *ionic* form:

$$\underbrace{H^+(aq) + Cl^-(aq)}_{Acid} + \underbrace{Na^+(aq) + OH^-(aq)}_{Base} \longrightarrow \underbrace{Na^+(aq) + Cl^-(aq)}_{Salt} + \underbrace{H_2O(l)}_{Water}$$

A ball-and-stick model of methylamine, CH_3NH_2, a weak base.

Methylamine 3D model

Application Note

Phenol red is familiar to people who maintain a home swimming pool. It is the indicator commonly used to ensure that the pool water is kept about neutral, that is, neither acidic nor basic.

Neutralization Reactions activity

---

*The complete-formula equation is often called a *molecular equation*, but this term is misleading because many of the chemical formulas written in such an equation, for example, NaCl(aq), represent not molecules but formula units of ionic compounds.

(a)

(b)

(c)

◄ **FIGURE 4.4**   **Phenol red—an acid–base indicator**

Phenol red is (a) *yellow* in an acidic solution (0.10 M HCl), (b) *orange* in a neutral solution (0.10 M NaCl), and (c) *red* in a basic solution (0.10 M NaOH).

QUESTION: What would happen to the color if ammonia were added to the beaker on the right?

Moreover, we can eliminate **spectator ions**—those that just "look on" and appear in the same form on the two sides of the ionic equation. Eliminating these ions reduces the equation to a simple form called a **net ionic equation.** In the equation we just wrote, Na^+ and Cl^- are spectator ions, and the net ionic equation is

$$H^+(aq) + OH^-(aq) \longrightarrow H_2O(l)$$

The net ionic equation conveys the essence of this neutralization: H^+ and OH^- ions combine to form water. In general, net ionic equations represent the gist of a reaction. If the spectator ions in a neutralization reaction are those of a soluble salt, they remain in solution. Evaporation of the water leaves the pure salt, such as NaCl (table salt).

Example 4.2

Barium nitrate, used to produce a green color in fireworks, can be made by the reaction of nitric acid with barium hydroxide. Write **(a)** a complete-formula equation, **(b)** an ionic equation, and **(c)** a net ionic equation for this neutralization reaction.

SOLUTION

(a) Write chemical formulas for the substances involved in the reaction, and then balance the equation:

    Nitric acid   Barium hydroxide   Barium nitrate   Water

$$HNO_3(aq) + Ba(OH)_2(aq) \longrightarrow Ba(NO_3)_2(aq) + H_2O(l) \quad \textit{(not balanced)}$$

$$2\,HNO_3(aq) + Ba(OH)_2(aq) \longrightarrow Ba(NO_3)_2(aq) + 2\,H_2O(l) \quad \textit{(balanced)}$$

(b) Now, represent the strong electrolytes by the formulas of their ions and the nonelectrolyte water by its molecular formula:

    A strong acid   A strong base   A salt   Water

$$2\,H^+(aq) + 2\,NO_3^-(aq) + Ba^{2+}(aq) + 2\,OH^-(aq) \longrightarrow Ba^{2+}(aq) + 2\,NO_3^-(aq) + 2\,H_2O(l)$$

(c) Cancel the spectator ions (Ba^{2+} and NO_3^-) in the ionic equation:

$$2\,H^+(aq) + 2\,\cancel{NO_3^-(aq)} + \cancel{Ba^{2+}(aq)} + 2\,OH^-(aq)$$
$$\longrightarrow \cancel{Ba^{2+}(aq)} + 2\,\cancel{NO_3^-(aq)} + 2\,H_2O(l)$$

This gives the equation

$$2\,H^+(aq) + 2\,OH^-(aq) \longrightarrow 2\,H_2O(l)$$

or, more simply, the net ionic equation.

$$H^+(aq) + OH^-(aq) \longrightarrow H_2O(l)$$

EXERCISE 4.2A

Calcium hydroxide is used to neutralize a waste stream of hydrochloric acid. Write **(a)** a complete-formula equation, **(b)** an ionic equation, and **(c)** a net ionic equation for this neutralization reaction.

EXERCISE 4.2B

Write the **(a)** complete-formula equation, **(b)** ionic equation, and **(c)** net ionic equation for the following neutralization reaction.

$$KHSO_4(aq) + NaOH(aq) \longrightarrow ?$$

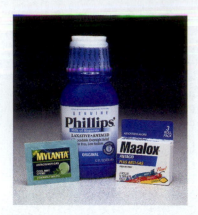

▲ Water-insoluble hydroxides such as $Mg(OH)_2(s)$ (milk of magnesia) and $Al(OH)_3(s)$ are used as antacids. Using ionic hydroxides that are soluble in water would be quite dangerous. In high concentrations, OH^-(aq) is highly basic and causes severe tissue burning and scarring.

Acid–Base Reactions: Additional Examples

While some acid–base reactions can be described by the simple net ionic equation involving H^+(aq), OH^-(aq), and H_2O(l), other acid–base reactions have different net ionic equations. For example, magnesium hydroxide, though only slightly soluble in water, is a strong base because what little does dissolve is completely dissociated into ions. In the net ionic equation that shows how a slurry of $Mg(OH)_2(s)$ in water neutralizes excess stomach acid, we should use the formula of solid magnesium hydroxide on the left. On the right side, hydroxide ion from $Mg(OH)_2$ has combined with H^+ to form H_2O(l), leaving magnesium ion as Mg^{2+}(aq):

$$Mg(OH)_2(s) + 2\,H^+(aq) \longrightarrow Mg^{2+}(aq) + 2\,H_2O(l)$$

This equation helps us understand why magnesium hydroxide, which is only very slightly soluble in pure water, dissolves readily in an acidic solution. Initially, H^+(aq) combines with the relatively few OH^- ions present in the $Mg(OH)_2$ slurry, forming H_2O(l). However, the removed OH^- ions are immediately replaced by new OH^- ions, released as more $Mg(OH)_2(s)$ dissolves. The replacement OH^- ions suffer the same fate as the original—combination with H^+ and removal as H_2O(l). Rather quickly, all the $Mg(OH)_2(s)$ in the acid solution dissolves.

The net ionic equation for the neutralization of HCl(aq) with NH_3(aq),

$$H^+(aq) + NH_3(aq) \longrightarrow NH_4^+(aq)$$

which shows that H^+ from the acid combines directly with NH_3 molecules, looks rather different from that of a strong acid and strong base,

$$H^+(aq) + OH^-(aq) \longrightarrow H_2O(l)$$

and for two reasons:

- NH_3 is a weak base, and we should therefore write its complete formula.
- The NH_3 molecule contains no OH^-, which means it cannot be satisfactorily described as an Arrhenius base.

In the same way that H^+ can combine with OH^- to form H_2O and with NH_3 to form NH_4^+, it can combine with certain other anions to produce either weak electrolytes or nonelectrolytes. In a broad sense, these reactions are also acid–base reactions.

In baking, carbon dioxide gas causes dough to rise. The process involves a reaction between baking soda, $NaHCO_3$, and some acidic ingredient in the dough (Figure 4.5). Specifically, H^+ from a weak acid (let's call it HA) reacts with the hydrogen carbonate ions from the baking soda to form the weak acid carbonic acid, H_2CO_3. Carbonic acid is unstable and decomposes to H_2O(l) and CO_2(g). The equations for these two reactions and the net ionic equation are

▲ **FIGURE 4.5 The leavening action of baking soda**

When acidified, here with citric acid ($H_3C_6H_5O_7$) from a lemon, baking soda ($NaHCO_3$) reacts to produce carbonic acid (H_2CO_3), which decomposes to carbon dioxide and water. The carbon dioxide gas produces a lift in the dough being baked.

$$HA(aq) + HCO_3^-(aq) \longrightarrow \cancel{H_2CO_3(aq)} + A^-(aq)$$
$$\cancel{H_2CO_3(aq)} \longrightarrow H_2O(l) + CO_2(g)$$

Net: $HA(aq) + HCO_3^-(aq) \longrightarrow A^-(aq) + H_2O(l) + CO_2(g)$

The reaction of carbonate ion (CO_3^{2-}) with an acid also produces carbonic acid that decomposes to H_2O(l) and CO_2(g). In similar reactions, sulfite ion (SO_3^{2-}) and hydrogen sulfite ion (HSO_3^-) react with an acid to produce unstable sulfurous acid (H_2SO_3), which decomposes to H_2O(l) and SO_2(g). Net ionic equations for a few gas-forming reactions are summarized in Table 4.2.

Table 4.2 Some Common Gas-Forming Acid–Base Reactions

Anion	Reaction with H^+
HCO_3^-	$HCO_3^- + H^+ \longrightarrow CO_2(g) + H_2O(l)$
CO_3^{2-}	$CO_3^{2-} + 2 H^+ \longrightarrow CO_2(g) + H_2O(l)$
HSO_3^-	$HSO_3^- + H^+ \longrightarrow SO_2(g) + H_2O(l)$
SO_3^{2-}	$SO_3^{2-} + 2 H^+ \longrightarrow SO_2(g) + H_2O(l)$
HS^-	$HS^- + H^+ \longrightarrow H_2S(g)$
S^{2-}	$S^{2-} + 2 H^+ \longrightarrow H_2S(g)$

Example 4.3   A Conceptual Example

Explain the observations illustrated in Figure 4.6.

ANALYSIS AND CONCLUSIONS

The bulb in Figure 4.6(a) is only dimly lit because acetic acid is a weak acid and therefore a weak electrolyte [recall Figure 4.3(c)]. The situation in (b) is similar because ammonia is a weak base and therefore also ionizes only slightly. When the two solutions are mixed, which is what has been done in (c), the H^+ ions from the CH_3COOH readily combine with NH_3 molecules to form NH_4^+ ions:

$$H^+(aq) + NH_3(aq) \longrightarrow NH_4^+(aq)$$

By removing H^+ ions from the solution, this reaction causes more CH_3COOH molecules to ionize, producing more H^+ to react with more NH_3, and so on. Soon all the CH_3COOH molecules ionize, and the neutralization goes to completion:

$$CH_3COOH(aq) + NH_3(aq) \longrightarrow \underbrace{NH_4^+(aq) + CH_3COO^-(aq)}$$

Weak acid   Weak base   Salt

The original weak acid and weak base are replaced by an aqueous solution of a salt—an ionic compound and strong electrolyte. The solution is now a good electrical conductor, as seen in Figure 4.6(c).

EXERCISE 4.3A

In a situation similar to that in Figure 4.6, describe the observations you would expect to make if the original solutions were $CH_3NH_2(aq)$ and $HNO_3(aq)$.

EXERCISE 4.3B

Describe the observations you would expect in a situation similar to that in Figure 4.6 if you start with a slurry of $Mg(OH)_2(s)$ in water (milk of magnesia) and add vinegar [essentially 1 M $CH_3COOH(aq)$] until all the $Mg(OH)_2(s)$ dissolves and some vinegar remains in excess.

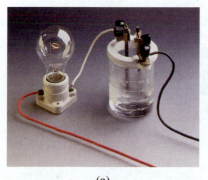
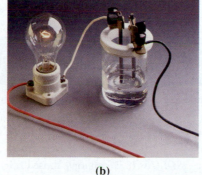
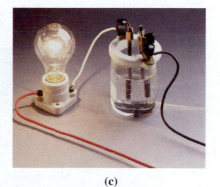

(a)   (b)   (c)

▲ FIGURE 4.6   **Change in electrical conductivity as a result of a chemical reaction**
(a) When the beaker contains a 1 M solution of acetic acid, CH_3COOH, the bulb in the electric circuit glows only very dimly.
(b) When the beaker contains a 1 M solution of ammonia, NH_3, the bulb again glows only dimly. (c) When the two solutions are in the same beaker, the bulb glows brightly. What happens when the two solutions are mixed is described in Example 4.3.

This solubility guideline is only qualitative and somewhat arbitrary. We discuss solubility more quantitatively in Chapter 16.

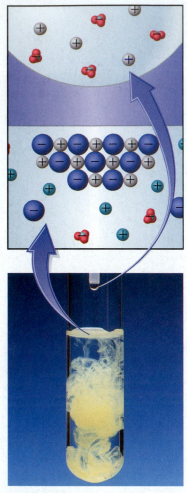

▲ **FIGURE 4.7   The precipitation of silver iodide, AgI(s)**

When clear, colorless silver nitrate solution is added to clear, colorless potassium iodide solution, the product is a yellow precipitate of silver iodide. The microscopic view shows a droplet of $AgNO_3$ solution about to strike the surface of the KI solution. Notice that the only substance that forms is AgI (represented by the cluster of gray and blue ions); the K^+ and NO_3^- (and excess I^-) remain as ions in solution. This means that the net reaction is between silver ions and iodide ions: $Ag^+(aq) + I^-(aq) \longrightarrow AgI(s)$

 Precipitation Reactions movie

4.3   Reactions that Form Precipitates

Earlier in the chapter, we noted that many ionic compounds dissolve in water, but there is a limit to how much will dissolve in a given quantity of water. For NaCl(aq) at 25 °C, the limit is a concentration of about 5.47 M NaCl. Because the maximum solute concentration for NaCl is relatively high, we say that NaCl(s) is readily soluble in water. If the maximum concentration of solute is less than about 0.01 M, we generally refer to this sparingly soluble solute as *insoluble* in water.

Combinations of certain cations and anions, then, yield ionic compounds that are insoluble in water. If these ions are taken from separate sources and brought together in an aqueous solution, the insoluble ionic compound, a **precipitate,** comes out of the solution and settles to the bottom of the container as a solid. We say that the insoluble compound *precipitates* from the solution, and a chemical reaction between ions that produces a precipitate is called a *precipitation reaction*. Figure 4.7 shows that a precipitate of insoluble silver iodide forms when solutions of the soluble compounds silver nitrate and potassium iodide are mixed.

Predicting Precipitation Reactions

Let's now consider how we might have *predicted* that a precipitate of silver iodide should form in the reaction illustrated in Figure 4.7. When asked to *predict* a chemical reaction, you will have information about the reactants—the left side of an equation—and will need to provide information about the products—the right side of the equation. You will also need to recognize when no reaction occurs, noted by writing "no reaction" on the right side of the equation.

In predicting precipitation reactions, it generally helps to write the equation in its ionic form, as in this equation for the addition of $AgNO_3$(aq) to KI(aq):

$$Ag^+(aq) + NO_3^-(aq) + K^+(aq) + I^-(aq) \longrightarrow ?$$

The only simple compounds that could form by combinations of these ions that are different from the reactants are KNO_3 and AgI. A precipitation reaction will occur *only* if a potential product is insoluble.

In short, to make predictions, we need to know which ionic compounds are soluble in water and which are not. We can look up solubility data in handbooks, but that may not be necessary. Memorizing solubilities is easier than it might first seem because a lot of data for common ionic compounds can be summarized in a few *solubility guidelines,* which are listed in Table 4.3.

The guidelines indicate that all nitrates are soluble in water, and so we conclude that KNO_3 is soluble. Similarly, Table 4.3 tells us that all chlorides, bromides, and iodides are soluble except those of Pb^{2+}, Ag^+, and Hg_2^{2+}, and thus we conclude that AgI is insoluble. We can complete the equation by showing the AgI precipitate as a solid and the KNO_3 as dissociated into ions.

$$Ag^+(aq) + \cancel{NO_3^-(aq)} + \cancel{K^+(aq)} + I^-(aq) \longrightarrow AgI(s) + \cancel{K^+(aq)} + \cancel{NO_3^-(aq)}$$

Table 4.3   General Guidelines for the Water Solubilities of Common Ionic Compounds

Almost all nitrates, acetates, perchlorates, group 1A metal salts, and ammonium salts are *SOLUBLE*.

Most chlorides, bromides, and iodides are *SOLUBLE*. Exceptions: those of Pb^{2+}, Ag^+, and Hg_2^{2+}.

Most sulfates are *SOLUBLE*. Exceptions: those of Sr^{2+}, Ba^{2+}, Pb^{2+}, and Hg_2^{2+} ($CaSO_4$ is slightly soluble).

Most carbonates, hydroxides, phosphates, and sulfides are *INSOLUBLE*. Exceptions: ammonium and group 1A metal salts of any of those anions are soluble; hydroxides and sulfides of Ca^{2+}, Sr^{2+}, and Ba^{2+} are slightly to moderately soluble.

Because our interest is usually in the net ionic equation, we can eliminate the spectator ions and write

$$Ag^+(aq) + I^-(aq) \longrightarrow AgI(s)$$

This equation tells us that, regardless of their sources, if Ag^+ and I^- are placed in the same solution, they will form a precipitate of $AgI(s)$.

Example 4.4

Predict whether a precipitation reaction will occur in each of the following cases. If so, write a net ionic equation for the reaction.

(a) $Na_2SO_4(aq) + MgCl_2(aq) \longrightarrow$?   (c) $K_2CO_3(aq) + ZnCl_2(aq) \longrightarrow$?

(b) $(NH_4)_2S(aq) + Cu(NO_3)_2(aq) \longrightarrow$?

STRATEGY

We must consider the simple compounds that could be formed from the combinations of ions present. Then we can use the solubility guidelines to determine whether any potential products are insoluble compounds. If one or more of these potential products are insoluble, a reaction will occur. If all potential products are soluble, there will be no reaction.

SOLUTION

(a) The initial reactants are in aqueous solution. We can begin by writing the equation in ionic form:

$$2\,Na^+(aq) + SO_4^{2-}(aq) + Mg^{2+}(aq) + 2\,Cl^-(aq) \longrightarrow ?$$

Then we can identify the possible products: $NaCl$ and $MgSO_4$. From Table 4.3, we see that sodium compounds (group 1A) are soluble, as are most sulfates, including that of magnesium. Thus, we conclude

$$Na_2SO_4(aq) + MgCl_2(aq) \longrightarrow \textit{no precipitate}\quad \textit{(no reaction)}$$

(b) We can represent the reactants in solution by an ionic equation:

$$2\,NH_4^+(aq) + S^{2-}(aq) + Cu^{2+}(aq) + 2\,NO_3^-(aq) \longrightarrow ?$$

Here the possible products are NH_4NO_3 and CuS. According to the solubility guidelines, all nitrates are soluble, but most sulfides, including CuS, are *not*. Thus, the reaction is

$$2\,\cancel{NH_4^+(aq)} + S^{2-}(aq) + Cu^{2+}(aq) + 2\,\cancel{NO_3^-(aq)}$$
$$\longrightarrow CuS(s) + 2\,\cancel{NH_4^+(aq)} + 2\,\cancel{NO_3^-(aq)}$$

The NH_4^+ and NO_3^- are spectator ions and can be canceled, yielding the net ionic equation

$$S^{2-}(aq) + Cu^{2+}(aq) \longrightarrow CuS(s)$$

(c) The ionic equation is

$$2\,K^+(aq) + CO_3^{2-}(aq) + Zn^{2+}(aq) + 2\,Cl^-(aq) \longrightarrow ?$$

The potential products are KCl and $ZnCO_3$. Potassium compounds (group 1A) are soluble, and among carbonates, only those of the group 1A metals are soluble. Because zinc is not in group 1A, we expect a precipitate of $ZnCO_3(s)$ and an aqueous solution containing K^+ and Cl^- ions. The net ionic equation is

$$CO_3^{2-}(aq) + Zn^{2+}(aq) \longrightarrow ZnCO_3(s)$$

EXERCISE 4.4A

Predict whether a reaction will occur in each of the following cases. If so, write a net ionic equation for the reaction.

(a) $MgSO_4(aq) + KOH(aq) \longrightarrow$?   (c) $Sr(NO_3)_2(aq) + Na_2SO_4(aq) \longrightarrow$?

(b) $FeCl_3(aq) + Na_2S(aq) \longrightarrow$?

EXERCISE 4.4B

In each of the following cases, predict whether a reaction will occur, and, if so, write the net ionic equation for the reaction.

(a) $ZnSO_4(aq) + BaS(aq) \longrightarrow$?   (c) $NaHCO_3(aq) + Ca(OH)_2(aq) \longrightarrow$?

(b) $Mg(OH)_2(s) + NaOH(aq) \longrightarrow$?

▲ **FIGURE 4.8 Predicting the product of a precipitation reaction**

Addition of $NH_3(aq)$ to $FeCl_3(aq)$ produces a precipitate.

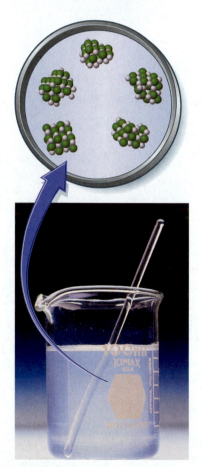

▲ **FIGURE 4.9 Chloride ion in tap water**

When a drop or two of $AgNO_3(aq)$ is added to tap water, the solution turns cloudy. Eventually, a trace of white precipitate settles to the bottom of the beaker. Most likely the precipitate is $AgCl(s)$, indicating the presence of Cl^- in the water.

QUESTION: What other ions in tap water might be responsible for this precipitate?

Example 4.5 A Conceptual Example

Figure 4.8 shows that the dropwise addition of $NH_3(aq)$ to $FeCl_3(aq)$ produces a precipitate. What is the precipitate?

ANALYSIS AND CONCLUSIONS

The reactants, NH_3 and $FeCl_3$, are both soluble in water. That all ammonium compounds are soluble means the precipitate is not likely to contain NH_4^+. It must therefore contain Fe^{3+}. But what is the anion? Recall that NH_3 is a weak base and that it produces OH^- ions in aqueous solution:

$$\text{(a) } NH_3(aq) + H_2O(l) \rightleftharpoons NH_4^+(aq) + OH^-(aq)$$

From the solubility guidelines, we expect $OH^-(aq)$ to combine with $Fe^{3+}(aq)$ to form *insoluble* $Fe(OH)_3(s)$:

$$\text{(b) } Fe^{3+}(aq) + 3\ OH^-(aq) \longrightarrow Fe(OH)_3(s)$$

To get the net ionic equation for the precipitation reaction, we need to multiply Equation **(a)** by 3 to get three OH^- ions and add the resulting equation to Equation **(b)**:

$$3\ NH_3(aq) + 3\ H_2O(l) \rightleftharpoons 3\ NH_4^+(aq) + \cancel{3\ OH^-(aq)}$$
$$Fe^{3+}(aq) + \cancel{3\ OH^-(aq)} \longrightarrow Fe(OH)_3(s)$$

$$\textit{Net: } \quad Fe^{3+}(aq) + 3\ NH_3(aq) + 3\ H_2O(l) \longrightarrow 3\ NH_4^+(aq) + Fe(OH)_3(s)$$

The cancellation slashes show that the hydroxide ions formed in the ionization of NH_3 are consumed in the precipitation of $Fe(OH)_3(s)$ and that no free $OH^-(aq)$ appears in the net ionic equation.

EXERCISE 4.5A

Suppose that after all the $Fe(OH)_3(s)$ is precipitated, a large quantity of $HCl(aq)$ is added to the beaker in Figure 4.8. Describe what you would expect to see, and write a net ionic equation for this change.

EXERCISE 4.5B

Potassium palmitate is a typical water-soluble soap. It is formed by the neutralization of palmitic acid, $CH_3(CH_2)_{14}COOH$, with potassium hydroxide. When calcium chloride is added to an aqueous solution of potassium palmitate, a gray precipitate is observed. What is the likely precipitate? Write ionic and net ionic equations for its formation.

Some Applications of Precipitation Reactions

In many cases, precipitation reactions are an attractive method of preparing chemical substances because they generally can be done with simple equipment and give a high percent yield of product. Table 4.4 lists a few industrially important precipitation reactions.

In other cases, precipitation reactions are used in chemical analysis. Analysis of a sample of matter to find out what it contains, but with no concern for how much, is called a *qualitative* analysis. For example, as shown in Figure 4.9, a sample of municipal water typically becomes cloudy when $AgNO_3(aq)$ is added to it, signaling that chloride ion is probably present in the water.

$$Cl^-(\text{from sample under analysis}) + Ag^+(aq) \longrightarrow AgCl(s)$$

Within the precision of the measurements used, the actual yield of a precipitation reaction is often equal to the theoretical yield. In this case, the reaction can be used for a *quantitative* analysis. For example, if the source of the chloride ion in a water-soluble material is NaCl, we can determine the actual quantity of NaCl present, as shown in Example 4.6.

Table 4.4 Some Precipitation Reactions of Practical Importance

Reaction in Aqueous Solution	Application
$Al^{3+}(aq) + 3\ OH^-(aq) \longrightarrow Al(OH)_3(s)$	Water purification. (The gelatinous precipitate carries down suspended matter.)
$Al^{3+}(aq) + PO_4^{3-}(aq) \longrightarrow AlPO_4(s)$	Removal of phosphates from wastewater in sewage treatment.
$Mg^{2+}(aq) + 2\ OH^-(aq) \longrightarrow Mg(OH)_2(s)$	Precipitation of magnesium ion from seawater. (First step in the Dow process for extracting magnesium from seawater.)
$Ag^+(aq) + Br^-(aq) \longrightarrow AgBr(s)$	Preparation of AgBr for use in photographic film.
$Zn^{2+}(aq) + SO_4^{2-}(aq) + Ba^{2+}(aq) + S^{2-}(aq) \longrightarrow ZnS(s) + BaSO_4(s)$	Production of *lithopone*, a mixture used as a white pigment in both water paints and oil paints.
$H_3PO_4(aq) + Ca(OH)_2(aq) \longrightarrow CaHPO_4 \cdot 2\ H_2O(s)$	Preparation of calcium hydrogen phosphate dihydrate, used as a polishing agent in toothpastes.

Example 4.6

One cup (about 240 g) of a certain clear chicken broth yields 4.302 g AgCl when excess $AgNO_3(aq)$ is added to it. Assuming that all the Cl^- is derived from NaCl, what is the mass of NaCl in the sample of broth?

STRATEGY

In this problem, chloride ions derived from NaCl are precipitated as AgCl. From the given mass of AgCl, we can determine the number of moles of chloride ions initially present. From the moles of Cl^-, we can calculate first the moles of NaCl and then the mass of NaCl.

SOLUTION

The equation for the precipitation reaction is

$$Ag^+(aq) + Cl^-(aq) \longrightarrow AgCl(s)$$

The stoichiometric equivalence from which we derive a stoichiometric factor is

$$1\ mol\ Cl^-(aq) \rightleftharpoons 1\ mol\ AgCl(s)$$

and the series of conversions required in the stoichiometric calculation is

$$g\ AgCl \longrightarrow mol\ AgCl \longrightarrow mol\ Cl^- \longrightarrow mol\ NaCl \longrightarrow g\ NaCl$$

In the usual manner, we can combine these conversions into a single setup:

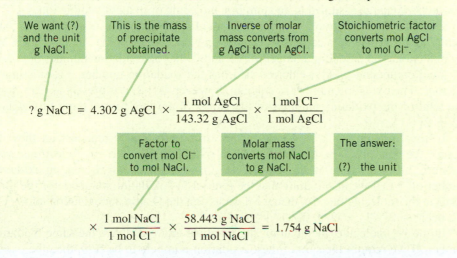

4.4 Reactions Involving Oxidation and Reduction

The third category of chemical reactions that we consider in this chapter, *oxidation–reduction* reactions, is perhaps the largest of all. It includes all combustion processes, most metabolic reactions in living organisms, extraction of metals from their ores, manufacture of countless chemicals, and many of the reactions occurring in our natural environment. Chemists often refer to oxidation–reduction reactions by reversing the words and shortening them to *redox* reactions.

The term *oxidation* was originally used to describe reactions in which a substance combines with oxygen. The opposite process, the removal of oxygen, was called reduction. To encompass a wide range of reactions, however, we need much broader definitions of oxidation and reduction. To assist us in formulating these more comprehensive definitions, let's first consider the concept of oxidation numbers.

Oxidation Numbers

As an aid to understanding, we have tried, whenever possible, to relate observations at the macroscopic level to behavior at the molecular level. Oxidation numbers, however, don't precisely reflect anything at the microscopic or molecular level. They are an arbitrary construction that chemists find useful when dealing with oxidation and reduction.

The related term *oxidation state* refers to the state, or condition, corresponding to a given oxidation number. For example, Cl⁻ ion is an oxidation state of chlorine that has an oxidation number of −1. The terms *oxidation number* and *oxidation state* are often used interchangeably.

Oxidation numbers are easier to illustrate than to define, but after we have considered some practical examples, you should be able to relate them to this definition: An **oxidation number** represents either the actual charge on a monatomic ion or a *hypothetical* charge assigned to an atom in a molecule or in a polyatomic ion.

Consider the formation of sodium chloride. Each Na atom loses one electron, and each Cl atom gains one electron. The compound is made up of Na^+ and Cl^- ions. We say that Na has an oxidation number of +1 and Cl has an oxidation number of −1. In the ionic compound $CaCl_2$, chlorine also has the oxidation number −1, existing as Cl^- ions. The oxidation number of calcium, however, is +2; it is present as Ca^{2+} ions. The total of the oxidation numbers of the atoms (ions) in a formula unit of $CaCl_2$ is $+2 + 2(-1) = 0$.

In the formation of a molecule, no electrons are transferred; instead they are shared. We can, however, *arbitrarily* assign oxidation numbers *as if* electrons were transferred. For example, in the molecule H_2O, we assign each H atom an oxidation number of +1. If we also require that the total of the oxidation numbers for the three atoms in the molecule be *zero*, then we must assign the O atom an oxidation number of −2, because $2(+1) - 2 = 0$.

In the H_2 molecule, the H atoms are identical and must have the same oxidation number. If we require the sum of these oxidation numbers to be *zero*, then the oxidation number of each H atom must also be zero.

From these examples, you can see that we must assign oxidation numbers systematically. We can deal with the great majority of compounds with the following rules. Important exceptions are listed in notes in the margin. The rules are listed *by priority;* if two rules contradict each other, use the one with the higher priority. This generally takes care of exceptions. For each rule, we have provided some examples. Additional examples are illustrated in Example 4.7.

1. *For the atoms in a neutral species—an isolated atom, a molecule, or a formula unit—the sum of all the oxidation numbers is 0.*
   Examples: The oxidation number of an uncombined Fe atom is 0. The sum of the oxidation numbers of all the atoms in Cl_2, S_8, and $C_6H_{12}O_6$ is 0. The sum of the oxidation numbers of the ions in $MgBr_2$ is 0.

   > Because all the atoms in a molecule of an element are alike, each Cl atom in Cl_2 and each S atom in S_8 has an oxidation number of 0.

2. *For the atoms in an ion, the sum of the oxidation numbers is equal to the charge on the ion.*
   Examples: The oxidation number of Cr in the Cr^{3+} ion is +3. The sum of the oxidation numbers in PO_4^{3-} is −3, and the sum in NH_4^+ is +1.

3. *In compounds, the group 1A metals all have an oxidation number of +1 and the group 2A metals all have an oxidation number of +2.*
   Examples: The oxidation number of Na in Na_2SO_4 is +1 and that of Ca in $Ca_3(PO_4)_2$ is +2.

4. *In compounds, the oxidation number of fluorine is −1.*
   Examples: The oxidation number of F is −1 in HF, ClF_3, and SO_2F_2.

   > The principal exception to Rule 5 occurs when H is bonded to a metal, as it is in compounds called metal hydrides. In these cases, H has an oxidation number of −1. Examples are NaH and CaH_2.

5. *In compounds, hydrogen has an oxidation number of +1.*
   Examples: The oxidation number of H is +1 in HCl, H_2O, NH_3, and CH_4.

6. *In most compounds, oxygen has an oxidation number of −2.*
   Examples: The oxidation number of O is −2 in CO, CH_3OH, $C_6H_{12}O_6$, and ClO_4^-.

7. *In binary compounds with metals, group 7A elements have an oxidation number of −1, group 6A elements have an oxidation number of −2, and group 5A elements have an oxidation number of −3.*
   Examples: The oxidation number of Br is −1 in $CaBr_2$, that of S is −2 in Na_2S, and that of N is −3 in Mg_3N_2.

   > The principal exceptions to Rule 6 occur when O atoms are bonded to one another, as in *peroxides* (for example, in H_2O_2 the oxidation number of O is −1) and in *superoxides* (for example, in KO_2 the oxidation number of O is −1/2).

Example 4.7

What are the oxidation numbers assigned to the atoms of each element in

  (a) $KClO_4$ **(b)** $Cr_2O_7^{2-}$ **(c)** CaH_2 **(d)** Na_2O_2 **(e)** Fe_3O_4

STRATEGY

In applying the rules for determining oxidation numbers, we must adhere to the priority order in which they are listed above. Note, however, that except in the case of uncombined elements, we cannot apply Rule 1 first.

SOLUTION

(a) The oxidation number of K is +1 (Rule 3). The oxidation number of O is −2 (Rule 6), and the total for *four* O atoms is −8. For these two elements, the total is +1 − 8 = −7. The oxidation number of the Cl atom in this ternary compound must be +7, to give a total of zero (+1 − 8 + 7 = 0) for all atoms in the formula unit (Rule 1).

(b) The oxidation number of O is −2 (Rule 6), and the total for *seven* O atoms is −14. The total of the oxidation numbers in this *ion* must be −2 (Rule 2). Therefore the total of the oxidation numbers of *two* Cr atoms is +12, and that of *one* Cr atom is +6.

(c) Keeping in mind that the total for the formula unit must be 0 (Rule 1) and that the oxidation number of Ca is +2 (Rule 3), the oxidation number of H must be −1 rather than its usual +1 (Rule 5). Thus, for CaH_2 the sum of the oxidation numbers is +2 + (2 × −1) = 0.

(d) The oxidation number of Na is +1 (Rule 3), and for the *two* Na atoms, +2. The total for the formula unit must be 0 (Rule 1). Even though the oxidation number of O is usually −2 (Rule 6), here it must be −1 so that the total for the *two* O atoms is −2. Rule 3 takes priority over Rule 6.

(e) The oxidation number of O is −2 (Rule 6). For *four* O atoms, the total is −8. The total for the formula unit must be 0 (Rule 1). The total for *three* Fe atoms must be +8, and for each Fe atom, +8/3.

ASSESSMENT

Remember that the sum of the oxidation numbers of all the atoms present must add up to equal the total charge on the molecular or ionic formula. Notice how Rules 1 and 2 were used at one point or another in each part of this example.

Usually, fractional oxidation numbers, as seen in part (e) signify an *average*. The compound Fe_3O_4 is actually $Fe_2O_3 \cdot FeO$. Two of the Fe atoms have oxidation numbers of +3, and one has an oxidation number of +2. The average is $(3 + 3 + 2)/3 = +8/3$.

EXERCISE 4.7A

Assign an oxidation number to each atom in

(a) Al_2O_3, (b) P_4, (c) CH_3F, (d) $HAsO_4^{2-}$, (e) $NaMnO_4$, (f) ClO_2^-, (g) CsO_2

EXERCISE 4.7B

Assign an oxidation number to each underlined atom:

(a) $H\underline{Sb}F_6$, (b) $\underline{C}HCl_3$, (c) $\underline{P}_3O_{10}^{5-}$, (d) $\underline{S}_4O_6^{2-}$, (e) $\underline{C}_3O_2$, (f) $\underline{N}O_2^+$, (g) $\underline{C}_2O_4^{2-}$

Problem-Solving Note

The compounds CH_3F, $CHCl_3$, C_3O_2, and the polyatomic ion $C_2O_4^{2-}$ demonstrate how the oxidation number of carbon can vary in organic compounds.

▲ **FIGURE 4.10   The thermite reaction**

Thermite Reaction movie

Identifying Oxidation–Reduction Reactions

The spectacular reaction pictured in Figure 4.10, called the *thermite* reaction, is used to produce liquid iron for welding large iron objects:

$$2\,Al(s) + Fe_2O_3(s) \longrightarrow 2\,Fe(l) + Al_2O_3(s)$$

Even by the limited definitions we gave at the start of this section, we can call this an oxidation–reduction reaction. The reactant Al is *oxidized* to Al_2O_3; aluminum atoms *gain* oxygen atoms. The reactant Fe_2O_3 is *reduced* to Fe; iron(III) oxide *loses* oxygen atoms.

Now consider how we can use oxidation numbers to identify an oxidation–reduction reaction. In the equation for the thermite reaction, the oxidation numbers of the Al, Fe, and O atoms are assigned according to the conventions just established and written as small numbers above the chemical symbols:

$$\overset{0}{2\,Al(s)} + \overset{+3\ -2}{Fe_2O_3(s)} \longrightarrow \overset{0}{2\,Fe(l)} + \overset{+3\ -2}{Al_2O_3(s)}$$

In the thermite reaction, the oxidation number of Al atoms (red) *increases* from 0 to +3, and the oxidation number of Fe atoms (blue) *decreases* from +3 to 0, illustrating this oxidation-number view of oxidation–reduction:

In an oxidation–reduction reaction, the oxidation number of one or more elements increases—an **oxidation** *process—and the oxidation number of one or more elements decreases—a* **reduction.**

The reaction pictured in Figure 4.11 differs strikingly from the thermite reaction, but the expanded definition identifies this as an oxidation–reduction reaction, as we can see from the oxidation numbers noted in the following equation:

$$\overset{0}{Mg(s)} + \overset{+2}{Cu^{2+}(aq)} \longrightarrow \overset{+2}{Mg^{2+}(aq)} + \overset{0}{Cu(s)}$$

Mg(s) is *oxidized* to $Mg^{2+}(aq)$, and $Cu^{2+}(aq)$ is *reduced* to Cu(s).

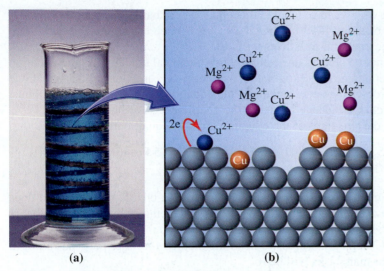
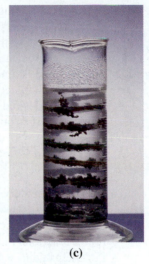

(a)   (b)   (c)

▲ **FIGURE 4.11   An oxidation–reduction reaction**

(a) A coil of magnesium ribbon immersed in a solution of $CuSO_4$(aq). (b) A microscopic view shows electrons being transferred from metallic magnesium to a Cu^{2+} ion, which is thereby converted to an atom of Cu metal. The transfer of electrons also produces a Mg^{2+} ion in solution. (For clarity, sulfate ions have been left out of this microscopic view.) (c) After a few hours, all of the Cu^{2+} has been displaced from the solution, leaving a deposit of red-brown copper metal, some unreacted magnesium, and clear, colorless $MgSO_4$(aq).

QUESTION: Does the mass of the coil increase, decrease, or remain the same during the course of this oxidation–reduction reaction? Why?

   The simple reaction shown in Figure 4.11 suggests the most fundamental definition of oxidation–reduction reactions. When you look closely at part (b) of the figure, you see that two things happen simultaneously:

Oxidation-Reduction Reactions—Part 1 animation

Oxidation: $Mg(s) \longrightarrow Mg^{2+}(aq) + 2e^-$

Reduction: $Cu^{2+}(aq) + 2e^- \longrightarrow Cu(s)$

The oxidation number of Mg increases from 0 to +2—an oxidation—as the Mg atom loses two electrons. The Cu^{2+} ion gains the two electrons, and its oxidation number decreases from +2 to 0—a reduction. Because electrons, particles of matter, can be neither created nor destroyed, oxidation and reduction must always occur together. This fundamental definition of oxidation and reduction reflects these ideas:

   An oxidation–reduction reaction consists of two processes that occur simultaneously. In one process, called oxidation, electrons are lost, and in the other, called reduction, they are gained.

Oxidation–Reduction Equations

Permanganate ion, MnO_4^-, in an acidic aqueous solution, oxidizes Fe^{2+} ion to Fe^{3+} and is itself reduced to Mn^{2+} ion. Let's see if we can write an equation describing this reaction. From the description of the reaction, we see that MnO_4^- and Fe^{2+} should both appear on the left side and Mn^{2+} and Fe^{3+} on the right:

   $MnO_4^-(aq) + Fe^{2+}(aq) \longrightarrow Mn^{2+}(aq) + Fe^{3+}(aq)$ *(incomplete)*

   Can you see why we have labeled this equation incomplete? There are four O atoms on the left that are not accounted for on the right. Recall, however, that the description of the reaction also indicates the presence of both H^+ and H_2O ("an acidic aqueous solution"). If we include these species in the equation, specifically, with 4 H_2O on the right to balance the four O atoms in MnO_4^- and 8 H^+ on the left to balance eight H atoms in the 4 H_2O, we obtain

   $MnO_4^-(aq) + Fe^{2+}(aq) + 8 H^+(aq) \longrightarrow Mn^{2+}(aq) + Fe^{3+}(aq) + 4 H_2O(l)$
                                                                              (not balanced)

Problem-Solving Note

For an algebraic derivation, let the unknown coefficient of Fe^{2+} and Fe^{3+} be x. The net charge on the left is $2x + 8 - 1$, and that on the right is $3x + 2$:

$$2x + 8 - 1 = 3x + 2$$
$$2x - 3x = 2 - 8 + 1$$
$$-x = -5$$
$$x = 5$$

Are you wondering why we have labeled this equation not balanced? Clearly it is balanced for numbers of atoms—1 Mn, 4 O, 1 Fe, and 8 H on each side. The important requirement that the equation fails to meet is that *electric charge must be balanced.* Electric charge cannot be created or destroyed in a chemical reaction. As the equation stands, there is a net charge of 9+ on the left and only 5+ on the right. As usual, we can, by trial and error, change coefficients in the equation to achieve a balance. Thus, we get the desired charge balance by using the coefficient *5* for both Fe^{2+} and Fe^{3+}. The net charges on the two sides of the equation are now equal; both are 17+:

$$MnO_4^-(aq) + 5\,Fe^{2+}(aq) + 8\,H^+ \longrightarrow Mn^{2+}(aq) + 5\,Fe^{3+}(aq) + 4\,H_2O(l)$$

(balanced)

Unfortunately, the trial-and-error method of achieving charge balance often does not work. What we need is a systematic method of balancing redox equations that is more likely to succeed. We present such a method based on the transfer of electrons in redox reactions in Chapter 18, where we need the method for other purposes as well. For the present, your main tasks will be to

- Identify oxidation–reduction reactions.
- Balance certain simple redox equations by inspection.
- Recognize, in all cases, whether a redox equation is properly balanced.

Oxidizing and Reducing Agents

 Oxidation-Reduction Reactions—Part 2 animation

An inorganic chemistry treatise lists dinitrogen tetroxide, N_2O_4, as a "fairly strong oxidizing agent" and hydrazine, N_2H_4, as a "powerful reducing agent." Such terms describe the way substances participate in redox reactions, and we will find it helpful to understand their meanings.

In an oxidation–reduction reaction, the substance that is *oxidized*—because it causes some other substance to be reduced—is called a **reducing agent.** Similarly, the substance that is *reduced* is called an **oxidizing agent** because it causes another substance to be oxidized. The financial practice of lending or borrowing money provides an analogy. In such a transaction, there must be both a lending agent and a borrower in order for the transaction to occur. Furthermore, the amount of money gained by the borrower must equal the amount of money given up by the lender. Likewise, both a reducing agent and an oxidizing agent are needed for a redox reaction, and the number of electrons obtained by the oxidizing agent must equal the number given up by the reducing agent.

Based on our description of oxidizing and reducing agents, we might well predict that nitrogen tetroxide and hydrazine should react with each other. They do indeed. The reaction between these two substances, which is accompanied by the release of large quantities of heat, is the basis of a rocket propulsion system:

$$N_2O_4(l) + 2\,N_2H_4(l) \longrightarrow 3\,N_2(g) + 4\,H_2O(g)$$

In the reaction, N_2O_4 is reduced to N_2 (oxidation number of N decreases from +4 to 0), making N_2O_4 the oxidizing agent, and N_2H_4 is oxidized to N_2 (oxidation number of N increases from −2 to 0), making N_2H_4 the reducing agent. Although the changes in oxidation numbers occur in N atoms, we do not call the atoms themselves the oxidizing or reducing agents. The *compounds* in which these atoms are found (N_2O_4 and N_2H_4) are the oxidizing and reducing agents, respectively.

Oxidation Numbers of Nonmetals

Some compounds and ions that contain the nonmetallic elements nitrogen, sulfur, or chlorine are listed in Figure 4.12. They are arranged, in order of decreasing oxidation number of the nonmetal atoms, in columns that correspond to the periodic table. We can use this figure to illustrate some additional ideas about oxidizing and reducing agents:

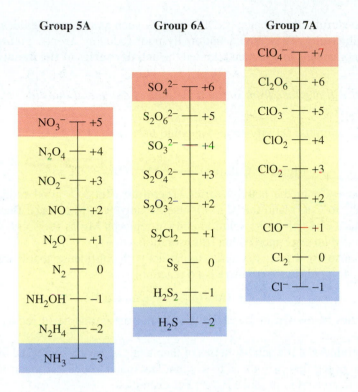

Group 5A

NO_3^- — +5
N_2O_4 — +4
NO_2^- — +3
NO — +2
N_2O — +1
N_2 — 0
NH_2OH — −1
N_2H_4 — −2
NH_3 — −3

Group 6A

SO_4^{2-} — +6
$S_2O_6^{2-}$ — +5
SO_3^{2-} — +4
$S_2O_4^{2-}$ — +3
$S_2O_3^{2-}$ — +2
S_2Cl_2 — +1
S_8 — 0
H_2S_2 — −1
H_2S — −2

Group 7A

ClO_4^- — +7
Cl_2O_6 — +6
ClO_3^- — +5
ClO_2 — +4
ClO_2^- — +3
— +2
ClO^- — +1
Cl_2 — 0
Cl^- — −1

◀ **FIGURE 4.12 Oxidation numbers in some nitrogen-, sulfur-, and chlorine-containing species**
The species in red can act only as oxidizing agents; those in blue can act only as reducing agents. Those in between can act as either, depending on the reaction.

- The *maximum* oxidation number for a nonmetal atom is equal to the atom's group number in the periodic table: +5 for group 5A atoms, +6 for group 6A, and +7 for group 7A. Oxygen and fluorine are exceptions (recall rules 4 and 6 for assigning oxidation numbers).

- The *minimum* oxidation number for a nonmetal atom is equal to the group number minus eight: −3 for group 5A atoms, −2 for group 6A, and −1 for group 7A.

- Species in which a nonmetal atom has its maximum oxidation number are invariably oxidizing agents because the oxidation number of the nonmetal atom can only *decrease* in a redox reaction. Thus, in a redox reaction, NO_3^- can only be an oxidizing agent.

- Species in which a nonmetal atom has its minimum oxidation number are reducing agents. Thus, in a redox reaction, H_2S can only be a reducing agent.

- In principle, species in which a nonmetal atom has an intermediate oxidation number can be either an oxidizing or a reducing agent, depending on the particular reaction. Generally, one role or the other is more common. For example, N_2O_4, with the oxidation number +4 for N, is almost always an oxidizing agent, and N_2H_4, with the oxidation number −2 for N, is almost always a reducing agent. Even though it is high on the oxidation-number scale for sulfur, SO_3^{2-} usually acts as a reducing agent. Conversely, Cl_2, though low on the oxidation number scale for chlorine, is generally an oxidizing agent.

Redox Chemistry of Tin and Zinc movie

- A redox reaction in which the same substance is both an oxidizing and a reducing agent is called a **disproportionation reaction.** Thus, when H_2O_2 disproportionates, half the H_2O_2 is oxidized to $O_2(g)$ and half is reduced to $H_2O(l)$:

$$2\ H_2O_2(aq) \longrightarrow 2\ H_2O(l) + O_2(g)$$

Metals As Reducing Agents

In all common metallic compounds, the metal atom has a positive oxidation number. Elemental metals, of course, have atoms with oxidation number 0, their lowest common oxidation state. Metals are *reducing* agents, but their strengths as reducing agents vary widely. The atoms of some metals, such as those of groups 1A and 2A, lose electrons easily, which means the atoms are readily oxidized to metal cations and are

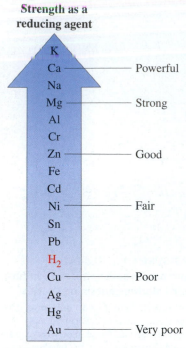

Strength as a reducing agent

Powerful

Strong

Good

Fair

Poor

Very poor

▲ FIGURE 4.13   **Activity series of some metals**

Formation of Silver Crystals movie

therefore powerful reducing agents. Other metals, such as silver and gold, are oxidized with great difficulty. They are exceptionally poor reducing agents. Figure 4.13 lists some common metals in a sequence called the **activity series of the metals.** As a general rule,

A metal will displace from solution the ions of any metal that lies below it in the activity series.

For example, with the activity series we could have predicted the reaction pictured in Figure 4.11:

$$Mg(s) + Cu^{2+}(aq) \longrightarrow Mg^{2+}(aq) + Cu(s)$$

Mg lies above Cu in the activity series. Therefore Mg(s), a good reducing agent, reduces Cu^{2+} to Cu(s), and the Mg(s) is itself oxidized to $Mg^{2+}(aq)$. Because Mg^{2+} takes the place of Cu^{2+} in the solution and Cu replaces Mg as the solid, we say that magnesium *displaces* copper(II) ion from solution.

We can also use the activity series to predict with confidence that if we add Ag(s) rather than Mg(s) to $Cu^{2+}(aq)$, there is no reaction:

$$Ag(s) + Cu^{2+}(aq) \longrightarrow \text{no reaction}$$

Because it lies below Cu in the activity series, silver is unable to reduce $Cu^{2+}(aq)$, to Cu(s).

The usefulness of the activity series of the metals is greatly increased by including H_2, because doing this permits us to say that any metal *above* hydrogen in the series can react with an acid to produce $H_2(g)$. For example,

$$2\,Al(s) + 6\,H^+(aq) \longrightarrow 2\,Al^{3+}(aq) + 3\,H_2(g)$$

Any metal *below* hydrogen *cannot* react with an acid to produce $H_2(g)$. For example,

$$Ag(s) + H^+(aq) \longrightarrow \text{no reaction}$$

Finally, there are a few special circumstances in which a metal that lies below hydrogen in the activity series may still undergo reaction in an acidic solution. One such case is described in Example 4.8.

Example 4.8   A Conceptual Example

Explain the difference in what happens when a copper-clad penny is immersed in **(a)** hydrochloric acid and **(b)** nitric acid, as shown in Figure 4.14.

▲ FIGURE 4.14   **The action of HCl(aq) and HNO₃(aq) on copper**
A copper-clad penny does not react with hydrochloric acid (left), but it does react with concentrated nitric acid, producing red-brown fumes and a green-blue solution (right).

ANALYSIS AND CONCLUSIONS

(a) Because copper lies below hydrogen in the activity series of the metals, Cu(s) cannot reduce $H^+(aq)$ to $H_2(g)$ and be oxidized to $Cu^{2+}(aq)$. Looking at it the other way, H^+ is not a strong enough oxidizing agent to oxidize Cu(s) to $Cu^{2+}(aq)$. Chloride ion in

HCl(aq) can only be a *reducing* agent. As neither H^+ nor Cl^- can oxidize Cu, we expect no reaction between Cu(s) and HCl(aq).

(b) In contrast, we clearly observe a reaction between copper and nitric acid. The Cu(s) is oxidized to $Cu^{2+}(aq)$ (blue color). From the HCl analysis, we know that $H^+(aq)$ is not strong enough as an oxidizing agent to oxidize Cu(s) to $Cu^{2+}(aq)$, and we know that $NO_3^-(aq)$ is an oxidizing agent here because the N atom has its highest possible oxidation number, +5. Therefore, copper is being oxidized to $Cu^{2+}(aq)$ by $NO_3^-(aq)$. Figure 4.12 suggests that $NO_3^-(aq)$ might be reduced to any one of several products. We have no way at this point of predicting what that product might be, but the red-brown gas proves to be nitrogen dioxide, NO_2. The oxidation number of nitrogen in NO_2 is the same as in N_2O_4, that is, +4. Using this information, we can write a net ionic equation:

$$Cu(s) + H^+(aq) + NO_3^-(aq) \longrightarrow Cu^{2+}(aq) + NO_2(g) + H_2O(l) \quad \text{(not balanced)}$$

Finally, the equation is balanced by inspection.

$$Cu(s) + 4\,H^+(aq) + 2\,NO_3^-(aq) \longrightarrow Cu^{2+}(aq) + 2\,NO_2(g) + 2\,H_2O(l)$$

EXERCISE 4.8A

The oxidizing agent potassium dichromate, $K_2Cr_2O_7(s)$ reacts one way when heated with hydrochloric acid and another way when heated with nitric acid. With one of the acids, a gas is evolved and the solution color changes from red-orange to green. With the other acid, the original red-orange color remains unchanged; that is, no reaction occurs. Explain this difference in behavior.

EXERCISE 4.8B

Since 1982, U.S. pennies have been made of zinc with a thin copper coating. If the edge of a new cent is notched with a knife and dropped in hydrochloric acid overnight, only a hollow shell of copper remains the next morning. How would the resulting solution differ from the two solutions in Figure 4.14?

Oxidation–Reduction and the Color of Glass

Glassmaking and glassblowing are among the oldest of art forms. Modern museums display glass bottles and vases that are thousands of years old. Stained-glass windows constructed in the Middle Ages still decorate many churches throughout Europe. In these windows, artisans have made use of intricate combination of colored glass to generate their mosaics. The color of many forms of glass is directly related to the oxidation states of impurities found throughout the glass matrix.

Ordinary (soda-lime) glass is made from sand (SiO_2), sodium carbonate ("soda"), and calcium oxide ("lime"). Additives confer other properties, including color. Often the sand contains a small proportion of iron, usually in the +3 oxidation state, which imparts a pale tan color. During processing the iron attains the +2 oxidation state. Look at a piece of window glass end-on, or at a "clear" glass bottle, and you can see the green color that is characteristic of iron(II) ion.

Colorless glass is more difficult to make than "bottle-green" glass. Small batches of colorless optical glass are made using sand that is essentially free of impurities. For large batches, manganese(IV) oxide can be added. The MnO_2 oxidizes the iron back to Fe(III) ion. The manganese is converted to the +2 oxidation state, which is pale pink. The right combination of Mn(II) and Fe(III) ions makes the glass nearly colorless.

Another redox reaction occurs when manganese-containing glass is exposed to intense sunlight. Ultraviolet light is energetic enough to convert manganese gradually from the +2 to the +7 oxidation state, which is purple. The +7 manganese is locked into the structure of the glass, so it does not come into contact with other species (as it could in an aqueous solution) and therefore cannot be reduced back to a lower oxidation state. After years of exposure to sunlight, the colorless manganese-containing glass appears purple. Brief exposure to energetic radiation such as gamma rays (Chapter 19) can cause the same transition.

◀ This composite image shows a glass electrical insulator before irradiation with gamma rays (left) and after (right), illustrating the oxidation of Mn(II) to Mn(VII).

4.5 Applications of Oxidation and Reduction

As we noted at the beginning of the previous section, redox reactions have many important applications. Let's look at a few of these.

In Everyday Life

Oxygen is undoubtedly the most important oxidizing agent in all aspects of life. We use it to oxidize fuels for heating our homes and propelling our automobiles. Oxygen corrodes metals by oxidizing them to cations, as in the rusting of iron. Oxygen even "burns" the foods we eat, to release the energy we need for life.

A common household oxidizing agent is hydrogen peroxide, H_2O_2, usually found in the form of an aqueous solution containing 3% H_2O_2. An advantage of hydrogen peroxide over other oxidizing agents is that in most reactions it is converted to water, an innocuous product. The 3% hydrogen peroxide solution is used in medicine as an *antiseptic* to treat minor cuts and abrasions. An enzyme in blood catalyzes the decomposition of hydrogen peroxide which, as noted on page 145, is a disproportionation reaction:

$$2\ H_2O_2(aq) \longrightarrow 2\ H_2O(l) + O_2(g)$$

The escaping bubbles of oxygen gas help to carry dirt and germs out of a wound.

Benzoyl peroxide $(C_6H_5COO)_2$ is a powerful oxidizing agent that has long been used at 5% and 10% concentrations for treating acne. In addition to its antibacterial action, benzoyl peroxide acts as a skin irritant, causing old skin to flake off and be replaced by newer, fresher-looking skin. However, when used on areas exposed to sunlight, benzoyl peroxide is thought to promote skin cancer.

Chlorine and its compounds are commonly encountered as oxidizing agents in daily life. Elemental Cl_2 is used to kill microorganisms, both in drinking water and in wastewater treatment. Swimming pools are usually disinfected by "chlorination." In large pools, the chlorine is often introduced from steel cylinders in which it is stored as a liquid. The chlorine disproportionates in water to form HCl and hypochlorous acid, HOCl, which is the actual sanitizing agent:

$$Cl_2(g) + H_2O(l) \longrightarrow H^+(aq) + Cl^-(aq) + HOCl(aq)$$

The HCl(aq) soon makes the pool water too acidic, and it must be neutralized by adding a base such as sodium carbonate.

Small swimming pools are sometimes chlorinated with sodium hypochlorite, NaOCl, formed by dissolving Cl_2 in NaOH(aq). The added NaOH causes these pools to become too basic, and hydrochloric acid is added to neutralize the excess basicity. Calcium hypochlorite, $Ca(OCl)_2$, is used to disinfect clothing and bedding in hospitals and nursing homes.

Nearly any oxidizing agent can be used as a bleach to remove unwanted color from fabrics, hair, or other materials. However, some are too expensive, some harm fabrics, some produce undesirable products, and some are simply unsafe.

In Foods and Nutrition

In food chemistry, the substances known as *antioxidants* are reducing agents. Ascorbic acid (vitamin C), which is soluble in water, is thought to retard potentially damaging oxidation of living cells. Tocopherol (vitamin E) is a fat-soluble antioxidant. In the body, it is thought to act by scavenging harmful by-products of metabolism, such as the highly reactive molecular fragments called *free radicals*. In foods, vitamin E acts to prevent fats from being oxidized and thus becoming rancid.

When it acts as an antioxidant, vitamin C $(C_6H_8O_6)$ is oxidized to dehydroascorbic acid, $C_6H_6O_6$. For example, when nitrite ions from foods get into the bloodstream, they oxidize iron in hemoglobin, destroying its ability to carry oxygen. In the stomach, however, ascorbic acid reduces nitrite ion to NO(g):

$$C_6H_8O_6(aq) + 2\ H^+(aq) + 2\ NO_2^-(aq) \longrightarrow C_6H_6O_6(aq) + 2\ H_2O(l) + 2\ NO(g)$$

Ascorbic acid
(vitamin C)

Dehydroascorbic
acid

▲ Hydrogen peroxide "bubbles" when poured over a cut because of the production of gaseous oxygen.

▲ Swimming pool "chlorine" used to treat small home pools is an alkaline solution of sodium hypochlorite, formed by the reaction of chlorine gas with aqueous sodium hydroxide:

$Cl_2(g) + 2\ NaOH(aq)$
$\longrightarrow NaOCl(aq) + NaCl(aq) + H_2O(l)$

Oxidation–Reduction in Bleaching and in Stain Removal

Many colored organic materials are made up of large molecules that contain alternate double and single carbon-to-carbon bonds. An example is lycopene, the compound that gives tomatoes their bright red color:

$$\left[\begin{array}{c} CH_3 \\ | \\ CH_3-C=CH-CH_2-CH_2-C=CH-CH=CH-C=CH-CH=CH-C=CH-CH= \end{array} \right]_2$$

The brackets and subscript 2 mean that the formula within the brackets is doubled. Think of the structure shown here as the complete formula folded over from right to left; the right-hand bracket is like a hinge. When you open up the hinge, you get the complete formula. Chemists often use space-saving devices like this to represent complex formulas. The interesting feature of the lycopene molecule is the series of alternating double and single bonds that are shown in color. This feature is called a *conjugated system,* and such systems often confer the property of color to organic compounds.

Bleaching agents oxidize colored molecules such as lycopene by destroying the conjugated system. Thus, lycopene is oxidized by aqueous hypochlorite ions to colorless organic compounds, and the hypochlorite ion is reduced to chloride ion:

$$C_{40}H_{56} + OCl^-(aq) \longrightarrow Cl^-(aq) + \text{colorless organic products}$$

Hypochlorite bleaches are safe and effective for cotton and linen fabrics because they do not oxidize the cellulose that makes up these fabrics. However, hypochlorites do oxidize the protein and protein-like molecules that make up wool, silk, and nylon and therefore should not be used on those fabrics. Chlorine dioxide (ClO_2) and sodium chlorite ($NaClO_2$) are equally good oxidizing agents that do less damage to fabrics than do chlorine and hypochlorites.

Some other bleaching agents are hydrogen peroxide, "oxygen" bleaches containing sodium percarbonate (often represented as $2\ Na_2CO_3 \cdot 3\ H_2O_2$ to indicate an association between Na_2CO_3 and H_2O_2), sodium perborate ($NaBO_2 \cdot H_2O_2$), and a variety of chlorine-containing organic compounds that release Cl_2 in water.

Stain removal is not nearly so simple a process as bleaching. A few stain removers are oxidizing agents or reducing agents; others have quite different chemical natures. Nearly all stains require rather specific stain removers.

Hydrogen peroxide in cold water removes bloodstains from cotton and linen fabrics. Potassium permanganate can be used to remove most stains from white fabrics (except rayon). The purple permanganate stain then can be removed in a redox reaction with oxalic acid:

$$5\ H_2C_2O_4(aq) + 2\ MnO_4^-(aq) + 6\ H^+(aq) \longrightarrow 2\ Mn^{2+}(aq) + 8\ H_2O + 10\ CO_2(g)$$

Iodine, used as a disinfectant, often stains clothing it contacts. The iodine stain is readily removed in a redox reaction with sodium thiosulfate:

$$I_2 + 2\ S_2O_3^{2-}(aq) \longrightarrow 2\ I^-(aq) + S_4O_6^{2-}(aq)$$

▲ Pure water (left) has little ability to remove a dried tomato sauce stain. Sodium hypochlorite, $NaOCl(aq)$ (right), easily bleaches the stain away by oxidizing the colored pigments of the sauce to colorless products.

Green plants carry out the redox reaction that makes possible almost all life on Earth. They do this through a process called photosynthesis, in which carbon dioxide and water are converted to glucose, a simple sugar. The synthesis of glucose requires a variety of proteins called enzymes and a green pigment called chlorophyll that converts sunlight into chemical energy. The overall change that occurs is

$$6\ CO_2 + 6\ H_2O \longrightarrow C_6H_{12}O_6 + 6\ O_2$$

In this reaction, CO_2 is reduced to glucose and H_2O is oxidized to oxygen gas. Other reactions convert the simple sugar to more complex carbohydrates and to plant proteins and oils.

In Industry

The most widely used oxidizing agent in industrial processes, and certainly the cheapest and least objectionable environmentally, is oxygen itself. In the first step of the conversion of iron to steel, high-pressure oxygen gas is blown over molten impure iron

▲ An oxyacetylene torch being used to weld air-conditioning equipment in Malaysia.

Application Note

Ozone, O_3, is a strong oxidizing agent that finds significant use in water purification. Gaseous chlorine dioxide, ClO_2, was used in December 2001 to destroy possible anthrax contamination in the Hart Senate Office Building in Washington, DC.

Application Note

Tungsten has a high electrical conductivity and the highest melting point of all the metals (3410 °C). These properties make it ideal for use as the light-emitting filaments in electric lightbulbs.

Application Note

When a person drinks an alcoholic beverage, enzymes in the liver cause the ethanol to be oxidized to acetaldehyde. If the person drinks moderately, the acetaldehyde is further oxidized to acetic acid and then to carbon dioxide and water. The liver can handle about 1 oz of ethanol an hour, the quantity in one average drink. If a person drinks more than that, the acetaldehyde concentration builds up and the individual gets intoxicated. Acetaldehyde is thought to be responsible for many of the harmful effects of ethanol, such as hangovers and fetal alcohol syndrome in babies born to women who drink heavily while pregnant.

(pig iron). This process burns off carbon and sulfur as gaseous oxides. The oxygen also converts the elements Si, P, and Mn to their solid oxides.

Oxygen is used to oxidize hydrogen, H_2, or acetylene, C_2H_2, in torches for welding and cutting metals. The heat released in such reactions provides the high temperatures needed to melt metals. In the oxyacetylene torch, for example, the reaction is

$$2\,C_2H_2(g) + 5\,O_2(g) \longrightarrow 4\,CO_2(g) + 2\,H_2O(g)$$

Another group of important industrial oxidizing agents includes chlorine gas and chlorine compounds in which the chlorine atoms have positive oxidation numbers. As noted above, chlorine gas and solutions containing hypochlorite ion, OCl^-, are used in water-treatment plants to kill pathogenic (disease-causing) microorganisms, and these agents are also used in the paper and textile industries for bleaching.

The principal industrial reducing agents are carbon and hydrogen. Carbon is often used in a solid form known as *coke*, which is produced by driving off the volatile matter from coal. In the blast furnace method for manufacturing iron from iron ore, the actual reducing agent is carbon monoxide, which is produced from coke:

$$C(s) + O_2(g) \longrightarrow CO_2(g)$$
$$C(s) + CO_2(g) \longrightarrow 2\,CO(g)$$

The $CO(g)$ then reduces the iron ions in the ore to elemental iron:

$$Fe_2O_3(s) + 3\,CO(g) \longrightarrow 2\,Fe(l) + 3\,CO_2(g)$$

Hydrogen is used as a reducing agent in smaller-scale processes and in cases where a metal might react with carbon to form an objectionable metal carbide. As an example, passing a stream of $H_2(g)$ over WO_3 at 1200 °C produces tungsten metal:

$$WO_3(s) + 3\,H_2(g) \longrightarrow W(s) + 3\,H_2O(g)$$

In Organic Chemistry

Dichromate ion, $Cr_2O_7^{2-}$, is a widely used oxidizing agent in many areas of chemistry. In organic chemistry, it is used to oxidize alcohols (ROH). The product of the reaction depends on the location of the OH group in the alcohol molecule, on the relative proportions of alcohol and dichromate ion, and on reaction conditions. If the OH group is on a terminal carbon atom and the product is distilled off as formed, the product is an **aldehyde.** The characteristic functional group of an aldehyde is

$$\overset{\displaystyle O}{\underset{\displaystyle }{\overset{\displaystyle \|}{-C-H}}}$$

It is often written on one line as CHO, in which case the carbon–oxygen double bond is understood. As an example, consider the reaction

$$3\,CH_3CH_2OH + Cr_2O_7^{2-} + 8\,H^+ \longrightarrow 3\,CH_3CHO + 2\,Cr^{3+} + 7\,H_2O$$

Ethanol   Acetaldehyde
Boiling point 78 °C   Boiling point 21 °C

Because it has a lower boiling point, the acetaldehyde can be boiled off from the reaction mixture, leaving the ethanol behind. If the acetaldehyde is not removed as it forms, it is further oxidized to acetic acid. In this case, the overall reaction is

$$3\,CH_3CH_2OH + 2\,Cr_2O_7^{2-} + 16\,H^+ \longrightarrow 3\,CH_3COOH + 4\,Cr^{3+} + 11\,H_2O$$

From the oxidation number of Cr in $Cr_2O_7^{2-}$ (+6) and in Cr^{3+} (+3), we see that $Cr_2O_7^{2-}$ is reduced in this reaction. This indicates that ethanol is oxidized to acetaldehyde in the first reaction and to acetic acid in the second. The average oxidation numbers of the C atoms are −2 in ethanol, −1 in acetaldehyde, and 0 in acetic acid. The oxidations described here are pictured in Figure 4.15.

When the OH group of an alcohol is bonded to an interior carbon atom, the product of oxidation of the alcohol is called a **ketone,** featuring the functional group known as the *carbonyl* group with two alkyl groups attached to the carbonyl carbon atom:

$$\overset{\displaystyle O}{\underset{\displaystyle }{\overset{\displaystyle \|}{-C-}}}$$

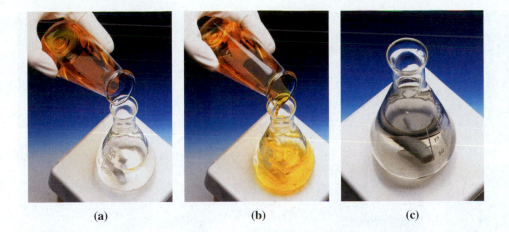

(a)　　　　　(b)　　　　　(c)

◀ **FIGURE 4.15** **The oxidation of ethanol by dichromate ion in acidic solution**

(a) Orange $K_2Cr_2O_7$(aq) about to be added to colorless CH_3CH_2OH(aq) that has been acidified with H_2SO_4. (b) The ethanol solution becomes colored because of the added $Cr_2O_7^{2-}$(aq). (c) After the reaction, the solution turns pale violet, signifying that the $Cr_2O_7^{2-}$ is gone and that Cr^{3+}(aq) is now present.

The simplest ketone, derived from 2-propanol, is the common solvent *acetone,* $(CH_3)_2CO$, used in applications as varied as varnishes and lacquers, rubber cement, and nail polish remover:

$$3\ CH_3\overset{OH}{\underset{|}{CH}}CH_3 + Cr_2O_7^{2-} + 8\ H^+ \longrightarrow 3\ CH_3\overset{O}{\overset{\|}{C}}CH_3 + 2\ Cr^{3+} + 7\ H_2O$$

Isopropyl alcohol / 2-Propanol　　Acetone / 2-Propanone

As we have just seen, aldehydes and ketones can be formed by the oxidation of alcohols. Conversely, aldehydes and ketones can be reduced to alcohols. Reduction of the carbonyl group is important in living organisms. For example, in *anaerobic* exercise (exercise in which the supply of oxygen is limited), pyruvic acid is reduced to lactic acid in the muscles:

$$CH_3{-}\overset{O}{\overset{\|}{C}}{-}\overset{O}{\overset{\|}{C}}{-}OH \xrightarrow{\text{reduction}} CH_3{-}\overset{OH}{\underset{|}{CH}}{-}\overset{O}{\overset{\|}{C}}{-}OH$$

Pyruvic acid　　　　Lactic acid

(Pyruvic acid is both a carboxylic acid and a ketone; only the ketone group is reduced.) The buildup of lactic acid during vigorous exercise is responsible in large part for the muscle pain that we experience.

4.6 Titrations

In Section 4.2, we described the neutralization of HCl(aq) with NaOH(aq). By what process is neutralization achieved? In a procedure called **titration,** one reactant is carefully added to another reactant until the two have combined in their exact stoichiometric proportions. Although titrations may differ in the nature of the chemical reaction that occurs, all titrations have this in common: Both reactants are fully consumed at the end of the titration.

The objective of a titration is usually to find the number of moles or grams, the concentration, or the percentage of the **analyte** (the substance we are looking for) in a sample. This is usually done by measuring the precise volume and concentration of a **titrant** (the solution being added) needed to react completely with the analyte. Stoichiometric relationships are then used to determine the quantity of analyte present from the number of moles (volume × molarity) of titrant added. Let's look at some common types of titrations.

Acid–Base Titrations

The most frequently encountered titration is one in which one reactant is an acid and the other is a base. An example of an *acid–base titration* is pictured in Figure 4.16. The central piece of equipment in a titration is a *buret,* a long

 Acid-Base Titration animation

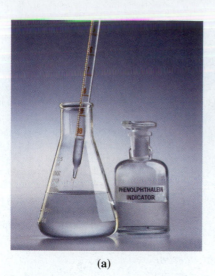
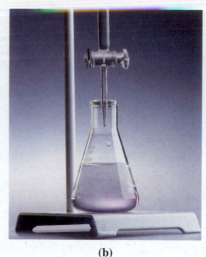

(a)   (b)   (c)

▲ **FIGURE 4.16**   **The technique of titration**

The steps in the technique corresponding to the photographs (a), (b), and (c) are described in the text.

QUESTION: Adding water to increase the volume of the analyte solution before beginning the titration changes the concentration of the analyte solution in the flask but does not affect the volume of titrant needed to reach the equivalence point. Why not?

graduated tube with a stopcock (valve) at one end. The buret is filled with the titrant, and the solution containing the analyte is placed in a small flask. A few drops of an indicator are usually added to the flask. The indicator is chosen to signal, through its change in color, when the **equivalence point** in the titration has been reached. The equivalence point in acid–base reactions occurs when the titrant completely neutralizes the analyte. Consider the titration pictured in Figure 4.16:

(a) A precisely measured volume of HCl(aq) of unknown concentration is delivered into the flask (with a device called a *pipet*). This is the analyte. A few drops of the acid–base indicator *phenolphthalein* are added. Phenolphthalein is colorless in acidic solution and pink in basic solution. Thus, the acidic solution in the flask is initially colorless.

(b) NaOH(aq) from the buret is slowly added to the flask. At first, the HCl is in excess and the NaOH is the limiting reactant. After it is stirred, the solution remains colorless, indicating that it is still acidic.

(c) When just enough NaOH has been added to react with all the HCl, the amounts of NaOH and HCl have been brought to their stoichiometric proportions, and the equivalence point has been reached. A tiny amount (less than one drop) of excess NaOH(aq) makes the solution basic, and we see a color change to pink; this color change marks the **endpoint.** At the endpoint, the titration is stopped and the volume of NaOH(aq) solution delivered from the buret is recorded. From this information, the concentration of the HCl(aq) solution can be determined.

From this example, it should be apparent that a successful titration requires an indicator that changes color at (or very closely after) the equivalence point. We will discuss how to choose such an indicator in Chapter 15.

Acid–base titrations see widespread use. Swimming pool water, wastewater, coal and coal ash, protein in foods, some pollutants, and raw industrial ingredients can all be analyzed by acid–base titrations. Happily, calculations involved in these titrations are not really new calculations. They are simply different applications of the solution stoichiometry calculations we already know.

Example 4.9

What volume (mL) of 0.2010 M NaOH is required to neutralize 20.00 mL of 0.1030 M HCl in an acid–base titration?

STRATEGY

We need to do four things to solve this problem. In general, other titration problems can be solved with a similar approach.

Step 1: Write an equation describing the neutralization and obtain a stoichiometric factor relating moles of NaOH and moles of HCl.

Step 2: Determine how many moles of HCl are to be neutralized.

Step 3: Find the number of moles of NaOH required in the neutralization.

Step 4: Determine the volume of the solution containing this number of moles of NaOH.

SOLUTION

Step 1: First write an equation for the neutralization reaction.

$$NaOH(aq) + HCl(aq) \longrightarrow NaCl(aq) + H_2O(l)$$

From this equation, we relate the amount of the titrant, NaOH(aq), to a given amount of the analyte, HCl(aq), and write the appropriate stoichiometric factor as a ratio.

$$\frac{1 \text{ mol NaOH}}{1 \text{ mol HCl}}$$

Step 2: The number of moles of HCl to be titrated is the product of the volume, in liters, and the molarity of the HCl(aq).

$$? \text{ mol HCl} = 0.02000 \text{ L HCl(aq)} \times \frac{0.1030 \text{ mol HCl}}{1 \text{ L HCl(aq)}}$$
$$= 0.002060 \text{ mol HCl}$$

Step 3: Now, we use the stoichiometric factor to convert moles of HCl to moles of NaOH required for the titration.

$$? \text{ mol NaOH} = 0.002060 \text{ mol HCl} \times \frac{1 \text{ mol NaOH}}{1 \text{ mol HCl}}$$
$$= 0.002060 \text{ mol NaOH}$$

Step 4: Finally, we use the inverse of the molarity of NaOH as a conversion factor to find the volume of the NaOH(aq) and convert from liters to milliliters.

$$? \text{ mL NaOH(aq)} = 0.002060 \text{ mol NaOH} \times \frac{1 \text{ L NaOH(aq)}}{0.2010 \text{ mol NaOH}} \times \frac{1000 \text{ mL}}{1 \text{ L}}$$
$$= 10.25 \text{ ml NaOH(aq)}$$

As in most stoichiometric calculations, we can combine these steps into a single setup:

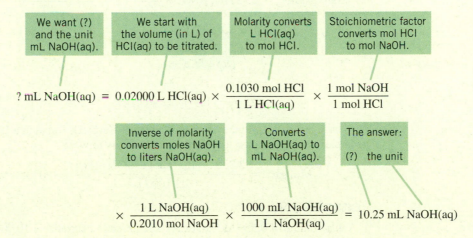

ASSESSMENT

The molarity of the NaOH(aq) is just about twice that of the HCl(aq), and the combining mole ratio of the acid and base is 1 : 1. This suggests that the volume of NaOH(aq) required should be just about one-half that of the HCl(aq), and it is: 10.25 mL of the NaOH(aq) neutralizes 20.00 mL of the HCl(aq).

EXERCISE 4.9A

What volume of 0.01060 M HBr(aq) is required to neutralize 25.00 mL of 0.01380 M Ba(OH)$_2$

$$2 \text{ HBr(aq)} + \text{Ba(OH)}_2\text{(aq)} \longrightarrow \text{BaBr}_2\text{(aq)} + 2 \text{ H}_2\text{O(l)}$$

EXERCISE 4.9B

A 2.000-g sample of a sulfuric acid solution that is 96.5% H_2SO_4 by mass is dissolved in a quantity of water and titrated. What volume of 0.3580 M KOH(aq) is required for the titration? Assume that at the equivalence point the solution in the flask is K_2SO_4(aq).

Example 4.10

A 10.00-mL sample of an aqueous solution of calcium hydroxide is neutralized by 23.30 mL of 0.02000 M HNO_3(aq). What is the molarity of the calcium hydroxide solution?

STRATEGY

To determine the molarity of the $Ca(OH)_2$(aq), we need to determine how many moles of $Ca(OH)_2$ are consumed in the titration and divide this number by the volume of the sample (0.01000 L, or 10.00 mL). To determine the number of moles of $Ca(OH)_2$, we can proceed through much of the problem as we did in Example 4.9.

SOLUTION

Let's start with a balanced equation for this reaction:

$$2 \text{ HNO}_3\text{(aq)} + \text{Ca(OH)}_2\text{(aq)} \longrightarrow \text{Ca(NO}_3)_2\text{(aq)} + 2 \text{ H}_2\text{O(l)}$$

| Nitric acid | Calcium hydroxide | Calcium nitrate | Water |

We can determine the number of moles of $Ca(OH)_2$ through a series of conversions:

$$\text{mL HNO}_3\text{(aq)} \longrightarrow \text{L HNO}_3\text{(aq)} \longrightarrow \text{mol HNO}_3 \longrightarrow \text{mol Ca(OH)}_2$$

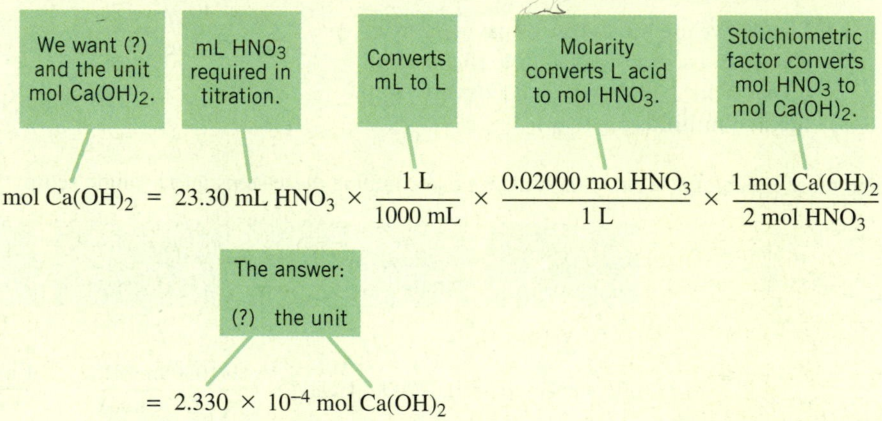

We want (?) and the unit mol Ca(OH)$_2$. | mL HNO$_3$ required in titration. | Converts mL to L | Molarity converts L acid to mol HNO$_3$. | Stoichiometric factor converts mol HNO$_3$ to mol Ca(OH)$_2$.

$$? \text{ mol Ca(OH)}_2 = 23.30 \text{ mL HNO}_3 \times \frac{1 \text{ L}}{1000 \text{ mL}} \times \frac{0.02000 \text{ mol HNO}_3}{1 \text{ L}} \times \frac{1 \text{ mol Ca(OH)}_2}{2 \text{ mol HNO}_3}$$

The answer:

(?) the unit

$$= 2.330 \times 10^{-4} \text{ mol Ca(OH)}_2$$

We have just calculated the number of moles of $Ca(OH)_2$ found in a 10.00-mL (0.01000 L) sample. Now we can use the definition of molarity to write

$$\text{Molarity} = \frac{2.330 \times 10^{-4} \text{ mol Ca(OH)}_2\text{(aq)}}{0.01000 \text{ L}} = 0.02330 \text{ M Ca(OH)}_2$$

EXERCISE 4.10A

What volume of 0.550 M NaOH(aq) is required to neutralize a 10.00-mL sample of vinegar that is 4.12% by mass acetic acid, CH_3COOH? Assume that the vinegar has a density of 1.01 g/mL.

EXERCISE 4.10B

A sample of battery acid is to be analyzed for its sulfuric acid content. A 1.00-mL sample has a mass of 1.239 g. This 1.00-mL sample is diluted to 250.0 mL with water, and 10.00 mL of the diluted acid is neutralized by 32.44 mL of 0.00986 M NaOH. What is the mass percent H_2SO_4 in the battery acid?

Application Note

Vinegar can contain from 4 to 10% acetic acid. The exact amount of acetic acid in a particular vinegar can be determined by titration with a base of known concentration.

Precipitation Titrations

A reaction that forms a precipitate may also be the basis for a titration. The most common precipitation titrations involve an aqueous solution of silver ion, either as the titrant or as the analyte. For example, in one type of precipitation titration the amount of chloride ion in an unknown solution is found by titrating the sample with silver nitrate solution. Insoluble silver chloride is formed during the titration. Potassium chromate can be used as the indicator; red-brown silver chromate forms only after all of the chloride ions have precipitated from solution. The endpoint of this titration is indicated by the first appearance of this colored precipitate, as shown in Figure 4.17.

(a)

(b)

(c)

◀ **FIGURE 4.17**

A precipitation titration

(Left) Silver nitrate solution in the buret reacts with chloride ion in the flask, producing a precipitate of AgCl (Center). The indicator is yellow CrO_4^{2-}. (Right) At the endpoint, when the Cl^- is consumed, we see the formation of a brick-red precipitate of Ag_2CrO_4.

Example 4.11

Suppose a 0.4096-g sample from a box of commercial table salt is dissolved in water and requires 49.57 mL of 0.1387 M $AgNO_3(aq)$ to completely precipitate the chloride ion. If the chloride ion present in solution comes only from the sodium chloride, find the mass of NaCl in the sample. Is commercial table salt pure sodium chloride?

STRATEGY

As with other solution stoichiometry problems, we can determine first the number of moles of NaCl and then the NaCl mass from the volume and molarity of the $AgNO_3(aq)$, the appropriate stoichiometric factor from the balanced equation for the precipitation reaction, and the molar mass of NaCl.

SOLUTION

We start with the balanced equation for the precipitation reaction.

$$AgNO_3(aq) + NaCl(aq) \longrightarrow NaNO_3(aq) + AgCl(s)$$

An outline of the problem illustrates our strategy for calculating the mass of NaCl in the sample of commercial table salt.

mL $AgNO_3(aq) \longrightarrow$ L $AgNO_3(aq) \longrightarrow$ mol $AgNO_3$
$\longrightarrow$ mol NaCl $\longrightarrow$ g NaCl

In a first step, we calculate the moles of $AgNO_3$ from the given volume and molarity.

$$? \text{ mol } AgNO_3 = 49.57 \text{ mL } AgNO_3(aq) \times \frac{1 \text{ L}}{1000 \text{ mL}} \times \frac{0.1387 \text{ mol } AgNO_3}{1 \text{ L } AgNO_3(aq)}$$

$$= 6.875 \times 10^{-3} \text{ mol } AgNO_3$$

In the second step, the mass of the analyte (NaCl) is calculated from the stoichiometric ratio of NaCl to $AgNO_3$ and the molar mass of NaCl.

$$? \text{ g NaCl} = 6.875 \times 10^{-3} \text{ mol } AgNO_3 \times \frac{1 \text{ mol NaCl}}{1 \text{ mol } AgNO_3} \times \frac{58.44 \text{ g NaCl}}{1 \text{ mol NaCl}}$$

$$= 0.4018 \text{ g NaCl}$$

ASSESSMENT

The calculated mass of the NaCl is *less* than the mass of the sample of table salt, as it should be because the sample is not pure NaCl. Also, notice that we did not need to use the mass of table salt in our calculation, although we would need that information if we wanted to find the mass percent of NaCl in the table salt.

Note that we could have combined the two steps into a single setup in which we would not have had to record the intermediate result: 6.875×10^{-3} mol $AgNO_3$. The final answer proves to be the same by either method: 0.4018 g NaCl.

EXERCISE 4.11A

A solution containing a water-soluble salt of the radioactive element thorium can be titrated with oxalic acid solution to form insoluble thorium(IV) oxalate. If 25.00 mL of a solution of thorium(IV) ion requires 19.63 mL of 0.02500 M $H_2C_2O_4$ for complete precipitation, find the molar concentration of thorium(IV) ion in the unknown solution.

EXERCISE 4.11B

A mixture of 0.1015 g of NaCl and 0.1324 g KCl is dissolved in water and titrated with 0.1500 M $AgNO_3(aq)$. What volume of the $AgNO_3(aq)$ will be needed to reach the endpoint?

Redox Titrations

In a redox titration, one reactant—often the titrant—is an oxidizing agent and the other reactant is a reducing agent. Permanganate ion, usually from $KMnO_4$, is one of the most commonly used oxidizing agents in the chemical laboratory and makes an excellent titrant. When $MnO_4^-(aq)$ is used as a titrant, a separate indicator often is not required because even a slight excess of $MnO_4^-(aq)$ will produce a visible pink color in the analyte solution when the endpoint of the titration is reached.

In a titration for determining the percent iron in an iron ore sample, $MnO_4^-(aq)$ can be used to oxidize Fe^{2+} to Fe^{3+} in an acidic solution:

$$5\,Fe^{2+}(aq) + MnO_4^-(aq) + 8\,H^+(aq) \longrightarrow 5\,Fe^{3+}(aq) + Mn^{2+}(aq) + 4\,H_2O(l)$$

The stoichiometric equivalence between the reactants is

$$5\text{ mol }Fe^{2+} \backsimeq 1\text{ mol }MnO_4^-$$

The titration is illustrated in Figure 4.18, and sample data are presented in Example 4.12.

(a)   (b)   (c)

▶ **FIGURE 4.18   A redox titration using permanganate ion as the oxidizing agent**

(a) The acidic solution in the flask (clear) contains an unknown amount of Fe^{2+}, and the buret contains $KMnO_4(aq)$ of a known concentration (dark purple). (b) As the titrant is added to the Fe^{2+} solution, the intense color of each drop quickly disappears as MnO_4^- reacts with Fe^{2+}. (c) When all the Fe^{2+} has been oxidized to Fe^{3+}, the next drop of $KMnO_4(aq)$ produces a lasting pink (very light purple) color in the solution.

Example 4.12

A 0.2865-g sample of an iron ore is dissolved in acid, and the iron is converted entirely to $Fe^{2+}(aq)$. To titrate the resulting solution, 0.02645 L of 0.02250 M $KMnO_4(aq)$ is required. What is the mass percent of iron in the ore?

STRATEGY

We can work this problem in two parts. The first requires finding the number of grams of Fe in the unknown solution from the titration data. This calculation is similar to ones we have

Problem-Solving Note

In Examples 4.11 and 4.12, the mass of a sample is given. A common error in titration calculations is to begin with this mass rather than a volume and molarity. Additionally, note that in Example 4.11 this mass is not needed, and in Example 4.12 the sample is not a pure substance.

done for acid–base and precipitation titrations. The second part is a straightforward calculation of a percentage, using the sample mass of 0.2865 g.

SOLUTION

The purpose of each factor in the calculation involving titration data is indicated in the following outline:

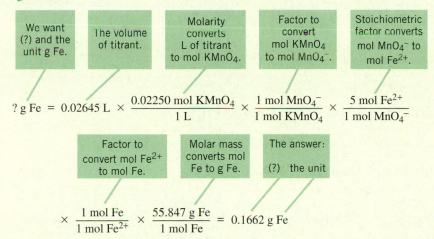

We want (?) and the unit g Fe.

The volume of titrant.

Molarity converts L of titrant to mol $KMnO_4$.

Factor to convert mol $KMnO_4$ to mol MnO_4^-.

Stoichiometric factor converts mol MnO_4^- to mol Fe^{2+}.

$$? \text{ g Fe} = 0.02645 \text{ L} \times \frac{0.02250 \text{ mol } KMnO_4}{1 \text{ L}} \times \frac{1 \text{ mol } MnO_4^-}{1 \text{ mol } KMnO_4} \times \frac{5 \text{ mol } Fe^{2+}}{1 \text{ mol } MnO_4^-}$$

Factor to convert mol Fe^{2+} to mol Fe.

Molar mass converts mol Fe to g Fe.

The answer: (?) the unit

$$\times \frac{1 \text{ mol Fe}}{1 \text{ mol } Fe^{2+}} \times \frac{55.847 \text{ g Fe}}{1 \text{ mol Fe}} = 0.1662 \text{ g Fe}$$

Finally, the mass percent of iron is

$$\% \text{ Fe} = \frac{0.1662 \text{ g Fe}}{0.2865 \text{ g iron ore}} \times 100\% = 58.01\% \text{ Fe}$$

ASSESSMENT

Because iron is only one component of iron ore, the mass of iron in the ore sample must be less than 0.2865 g and the percent iron must be less than 100%. These facts can be used to assess the plausibility of the results obtained in the calculation.

EXERCISE 4.12A

Suppose the titration in Example 4.12 was carried out with 0.02250 M $K_2Cr_2O_7(aq)$ rather than $KMnO_4(aq)$. What volume of the $K_2Cr_2O_7(aq)$ would be required? The net ionic equation is

$$6 \text{ Fe}^{2+}(aq) + Cr_2O_7^{2-}(aq) + 14 \text{ H}^+(aq) \longrightarrow 6 \text{ Fe}^{3+}(aq) + 2 \text{ Cr}^{3+}(aq) + 7 \text{ H}_2O(l)$$

EXERCISE 4.12B

A 20.00-mL sample of $KMnO_4(aq)$ is required to titrate 0.2378 g sodium oxalate in an acidic solution. How many milliliters of this same $KMnO_4(aq)$ are required to titrate a 25.00-mL sample of 0.1010 M $FeSO_4$ in an acidic solution?

$$2 \text{ MnO}_4^-(aq) + 16 \text{ H}^+(aq) + 5 \text{ C}_2O_4^{2-}(aq)$$
$$\longrightarrow 2 \text{ Mn}^{2+}(aq) + 8 \text{ H}_2O(l) + 10 \text{ CO}_2(g)$$

Cumulative Example

Sodium nitrite is used in the production of fabric dyes, as a meat preservative, as a bleach for fibers, and in photography. It is prepared by passing nitrogen monoxide gas and oxygen gas into an aqueous solution of sodium carbonate. Carbon dioxide gas is the other product of the reaction. (a) Write a balanced equation for the reaction. (b) What mass of sodium nitrite should be produced in the reaction of 748 g of Na_2CO_3, with the other reactants in excess? (c) In another preparation, the reactants are 225 mL of 1.50 M $Na_2CO_3(aq)$, 22.1 g nitrogen monoxide, and excess O_2. What mass of sodium nitrite should be produced if the reaction has a yield of 95.1%?

STRATEGY

For part (a), we can write the formulas for reactants and products (Sections 2.6 and 2.7), and then construct and balance the equation (Section 3.7). Using the balanced equation, we can solve part (b), starting with grams of Na_2CO_3 and ending with grams of $NaNO_2$ (Section 3.8) Part (c) can be broken down into three steps: Convert volume of Na_2CO_3 to moles using molarity (Section 3.11); solve as a limiting reactant problem (Section 3.9) for the mass of $NaNO_2$ produced from NO and Na_2CO_3; use that theoretical yield of product along with percentage yield (Section 3.10) to find the actual yield.

SOLUTION

(a) First, we write formulas for the reactants and products, then place the formulas in their proper places in the equation.

Reactants   Products

$$NO + O_2 + Na_2CO_3 \longrightarrow NaNO_2 + CO_2$$

To balance, begin with an element that appears in only one compound on each side (Na, for example).

$$NO + O_2 + Na_2CO_3 \longrightarrow 2\,NaNO_2 + CO_2$$

Then balance the N atoms.

$$2\,NO + O_2 + Na_2CO_3 \longrightarrow 2\,NaNO_2 + CO_2$$

Next balance the O atoms.

$$2\,NO + \tfrac{1}{2}O_2 + Na_2CO_3 \longrightarrow 2\,NaNO_2 + CO_2$$

Finally, multiply through by 2 to remove the fraction.

$$4\,NO + O_2 + 2\,Na_2CO_3 \longrightarrow 4\,NaNO_2 + 2\,CO_2$$

(b) A series of conversions is required for this stoichiometric calculation.

$$g\ Na_2CO_3 \longrightarrow mol\ Na_2CO_3 \longrightarrow mol\ NaNO_2 \longrightarrow g\ NaNO_2$$

We can perform the three steps either individually or in a single setup as shown here.

$$?\ g\ NaNO_2 = 748\ g\ Na_2CO_3 \times \frac{1\ mol\ Na_2CO_3}{106.0\ g\ Na_2CO_3} \times \frac{4\ mol\ NaNO_2}{2\ mol\ Na_2CO_3} \times \frac{69.00\ g\ NaNO_2}{1\ mol\ NaNO_2}$$

$$= 974\ g\ NaNO_2$$

(c) Convert volume to moles of Na_2CO_3.

$$225\ mL \times \frac{1\ L}{1000\ mL} \times \frac{1.50\ mol\ Na_2CO_3}{1\ L} = 0.338\ mol\ Na_2CO_3$$

Determine which is the limiting reactant by comparing the yield from the Na_2CO_3 with that from the NO.

$$0.338\ mol\ Na_2CO_3 \times \frac{4\ mol\ NaNO_2}{2\ mol\ Na_2CO_3} = 0.676\ mol\ NaNO_2$$

$$22.1\ g\ NO \times \frac{1\ mol\ NO}{30.01\ g\ NO} \times \frac{4\ mol\ NaNO_2}{4\ mol\ NO} = 0.736\ mol\ NaNO_2$$

Calculate the theoretical yield based on the limiting reactant, Na_2CO_3.

$$0.676\ mol\ NaNO_2 \times \frac{69.00\ g\ NaNO_2}{1\ mol\ NaNO_2} = 46.6\ g\ NaNO_2$$

Use the percentage yield (95.1%) to calculate the actual or expected yield.

$$95.1\% = \frac{actual\ yield\ (g)}{46.6\ g\ NaNO_2} \times 100\%$$

Actual yield (g) $= 95.1\% \times 46.6\ g\ NaNO_2/100\% = 44.3\ g\ NaNO_2$

ASSESSMENT

The balanced equation has four N atoms, twelve O atoms, four Na atoms, and two C atoms on each side. Because the molar mass of $NaNO_2$ is roughly half that of Na_2CO_3 and because the mole ratio between $NaNO_2$ and Na_2CO_3 is 2 : 1, the mass of $NaNO_2$ we get in part (b) should be comparable to the mass of Na_2CO_3 used. In part (c), because the yields of product from the two reactants are so similar, it is difficult to determine whether the limiting reactant is correct except by close inspection of the setup. However, we can note that the actual yield of $NaNO_2$ is indeed less than the theoretical yield, as it should be.

Concept Review with Key Terms

4.1 Some Electrical Properties of Aqueous Solutions—Soluble ionic compounds are completely dissociated into ions in aqueous solution and are therefore **strong electrolytes**. A few water-soluble molecular compounds are completely ionized in aqueous solution and are also strong electrolytes. Most molecular compounds exist in solution either as molecules (**nonelectrolytes**) or as a mixture of molecules and ions (**weak electrolytes**).

4.2 Reactions of Acids and Bases—A few acids are **strong acids**; these acids are strong electrolytes. However, most acids are **weak acids** (weak electrolytes). The common **strong bases** are water-soluble ionic hydroxides. **Weak bases**, like the weak acids, are molecular compounds that exist as a mixture of molecules and ions in aqueous solution. Many common weak bases are related to ammonia.

    Neutralization reactions between acids and bases are conveniently represented by ionic equations and **net ionic equations**. Net ionic equations include only those ions that undergo a chemical reaction in solution; **spectator ions** are eliminated. The neutralization reaction of an acid and a base produces water and an ionic compound called a **salt**. The color of an **indicator** can be used to determine whether a solution is acidic, basic, or neutral.

4.3 Reactions that Form Precipitates—Another important type of reaction in solution is one in which ions combine to form an insoluble solid—a **precipitate**. Solubility guidelines can often be used to predict precipitation reactions (review Table 4.3). Qualitative and quantitative chemical analyses and industrial processes frequently make use of precipitation reactions.

4.4 Reactions Involving Oxidation and Reduction—The **oxidation number** is the charge on a monatomic ion; for species other than monatomic ions, it is a hypothetical charge on an atom, assigned by a set of rules. **Oxidation** is an increase in oxidation number accompanied by loss of electrons. **Reduction** is a decrease in oxidation number accompanied by gain of electrons. The two processes occur simultaneously in a redox reaction. In a redox reaction, the **oxidizing agent** is reduced and the **reducing agent** is oxidized. In a **disproportionation reaction**, the same substance acts as both oxidizing agent and reducing agent. Strong oxidizing agents include a few nonmetals and some species having atoms with high oxidation numbers. Strong reducing agents include the active metals and some species having atoms with low oxidation numbers. The **activity series of the metals** ranks metals in order of their strength as reducing agents. It can be used to predict reactions between a metal and other metal ions in solution.

4.5 Applications of Oxidation and Reduction—In everyday life, peroxides and hypochlorites are encountered as oxidizing agents. In industry, oxygen gas, chlorine, and chlorine-containing compounds are used as oxidizing agents. Antioxidants such as vitamin C are reducing agents that can scavenge reactive free radicals. Photosynthesis is an important redox reaction. Redox reactions are used in organic chemistry to oxidize alcohols to **aldehydes** and **ketones**.

4.6 Titrations—A **titration** is a type of chemical analysis in which two reactants are combined in exact stoichiometric proportions. In a titration, the volume of a solution of known concentration (the **titrant**) is added to a solution containing the **analyte** or sought-for substance. The **equivalence point** occurs when the reactants are present in stoichiometric proportions. Often an indicator is used to render the equivalence point visible through a color change. That color change indicates the **endpoint** of the titration. Titrations can involve acid–base reactions, precipitation reactions, or redox reactions.

Assessment Goals

When you have mastered the material in this chapter, you will be able to

- Identify strong electrolytes, weak electrolytes, and nonelectrolytes.

- Calculate ion concentrations in solutions of strong electrolytes.

- Classify a substance as an acid or base.

- Describe and identify strong and weak acids and strong and weak bases.

- Describe neutralization reactions.

- Identify spectator ions in a solution. Write a net ionic equation for a reaction in solution.

- Cite and use the solubility guidelines in Table 4.3.

- Predict whether precipitation will occur when certain ionic compounds are present in the same solution.

- Solve stoichiometry problems based on precipitation reactions.

- Assign oxidation numbers to elements in compounds or ions.

- Recognize a redox reaction, and determine whether the equation for the reaction is balanced.

- Define and recognize oxidizing and reducing agents.

- Use the activity series to predict the products of redox reactions involving metals and metal ions.

- Describe some oxidation–reduction reactions of practical significance.

- Describe how a titration is performed, and carry out calculations related to titrations.

Self-Assessment Questions

1. The *best* electrical conductor of the following aqueous solutions is (a) 0.10 M NaCl; (b) 0.10 M CH_3CH_2OH; (c) 0.15 M CH_3COOH; (d) 0.10 M $CH_2OHCHOHCH_2OH$.

2. Identify each of the following substances as either a strong acid, a weak acid, a strong base, a weak base, or a salt.

   (a) H_3PO_4 (d) CH_3NH_2

   (b) HCl   (e) NH_4NO_3

   (c) LiOH   (f) CH_3CH_2COOH

3. Which of the following solutions has the *highest* and which has the *lowest* concentration of NO_3^-?

   (a) 0.10 M KNO_3 (c) 0.040 M $Al(NO_3)_3$

   (b) 0.040 M $Ca(NO_3)_2$ (d) 0.050 M $Mg(NO_3)_2$

4. Which of the following solutions has the highest *total* concentration of ions?

   (a) 0.012 M $Al_2(SO_4)_3$ (c) 0.040 M $Al(NO_3)_3$

   (b) 0.030 M KCl   (d) 0.025 M K_2SO_4

5. Which of the following aqueous solutions is the *best* electrical conductor? Explain.

   (a) 0.08 M NaCl   (c) 0.10 M CH_3COOH

   (b) 1.0 M CH_3CH_2OH (d) 2.0 M $C_6H_{12}O_6$

6. The greatest $[H^+]$ will be found in which of the following aqueous solutions? (a) 0.10 M HNO_3, (b) 0.10 M H_2CO_3, (c) 0.10 M NaOH, (d) 0.10 M CH_3NH_2.

7. According to the Arrhenius theory, (a) are all hydrogen-containing compounds acids? (b) Are all compounds containing OH groups bases? Explain.

8. In each of the following pairs of mixtures, reaction occurs in one mixture but not the other. Explain why this is so, and write a net ionic equation for the reaction that occurs.

   (a) $HCl(aq) + CH_3CH_2NH_2(aq)$

      or $NaOH(aq) + CH_3NH_2(aq)$

   (b) $ZnCl_2(aq) + MgSO_4(aq)$

      or $ZnCl_2(aq) + KOH(aq)$

9. When treated with dilute HCl(aq), which compound produces a gaseous product? (a) ZnO, (b) $NaNO_3$, (c) $BaSO_3$, (d) Na_2SO_4.

10. When ammonium carbonate is added to a zinc chloride solution, zinc ions will precipitate from solution. Write the (a) complete-formula equation, (b) ionic equation, and (c) net ionic equation for this reaction.

11. Only one of the following compounds is *insoluble* in water. Which one must it be? Explain.

    (a) $Ba(NO_3)_2$ (c) $CuSO_4$

    (b) $ZnCl_2$ (d) $PbSO_4$

12. Which of the following compounds reacts to precipitate Mg^{2+} from an aqueous solution of $MgCl_2$? Write an equation for the reaction.

    (a) Na_2S (c) Na_2CO_3

    (b) NaI   (d) $NaNO_3$

13. What simple chemical test can you perform to determine whether a particular barium compound is $BaSO_4(s)$ or $BaCO_3(s)$?

14. What is the usual oxidation number of hydrogen atoms in compounds? What is that of oxygen atoms in compounds? What are some exceptions?

15. What happens to the oxidation number of one of its elements when a compound is oxidized, and when it is reduced?

16. Of the following metals, all will react with HCl(aq) except (a) Ca, (b) Cu, (c) Fe, (d) Zn.

17. In the reaction, $Cu(s) + 4 H^+(aq) + SO_4^{2-}(aq) \longrightarrow Cu^{2+}(aq) + 2 H_2O(l) + SO_2(g)$, which statement is true? (a) Cu is the oxidizing agent; (b) SO_2 is the oxidizing agent; (c) H^+ is the oxidizing agent; (d) SO_4^{2-} is reduced.

18. Indicate the oxidation number of the underlined atom in each of the following.

    (a) $\underline{Cr}$ (f) $Ca\underline{Ru}O_3$

    (b) $\underline{Cl}O_2^-$ (g) $Sr\underline{Ti}O_3$

    (c) $K_2\underline{Se}$ (h) $\underline{P}_2O_7^{4-}$

    (d) $Te\underline{F}_6$ (i) $\underline{S}_4O_6^{2-}$

    (e) $\underline{P}H_4^+$ (j) $\underline{N}H_2OH$

19. Use the conventions on page 141 to determine the oxidation numbers of the carbon atoms in the following organic compounds.

    (a) C_2H_6 (d) C_2H_6O

    (b) CH_2O_2 (e) $C_2H_2O_4$

    (c) $C_2H_2O_2$

20. In the reaction

    $$Cu(s) + 2 H_2SO_4(aq) \longrightarrow CuSO_4(aq) + 2 H_2O(l) + SO_2(g)$$

    is the $H_2SO_4(aq)$ oxidized or reduced or neither? Explain.

21. A reaction occurs in one of these mixtures but not the other. Explain why this is so, and write a net ionic equation for the reaction that occurs.

    $$Zn(s) + CH_3COOH(aq) \text{ or } Au(s) + HCl(aq)$$

22. Both magnesium and aluminum react with an acidic solution to produce hydrogen. Why is it that only one of the following equations correctly describes the reaction?

    $$Mg(s) + 2H^+(aq) \longrightarrow Mg^{2+}(aq) + H_2(g)$$
    $$\text{and } Al(s) + 2H^+(aq) \longrightarrow Al^{3+}(aq) + H_2(g)$$

23. What is the equivalence point in an acid–base titration? What is the function of the indicator in the titration? What is the relationship between an *equivalence* point and an *end*point?

24. The complete neutralization of 10.00 mL of 0.100 M $H_2SO_4(aq)$ requires (a) 10.00 mL of 0.100 M NaOH; (b) 100.0 mL of 0.0200 M NaOH; (c) 5.00 mL of 0.100 M KOH; or (d) 20.00 mL of 0.100 M $Ba(OH)_2$.

Problems

Types of Electrolytes

25. Identify each substance as a strong electrolyte, a weak electrolyte, or a nonelectrolyte.

   (a) HBr   **(d)** HCOOH

   (b) KCl   **(e)** NaOH

   (c) HI   **(f)** $Ca(NO_3)_2$

26. Identify each substance as either a strong acid, a weak acid, a strong base, a weak base, or a salt.

   (a) Na_2SO_4 **(d)** CH_3CH_2COOH

   (b) KOH   **(e)** NH_4I

   (c) $BaCl_2$ **(f)** $CH_3CH_2NH_2$

27. Which, if any, of the following potassium compounds when dissolved in water could cause the bulb in the apparatus shown in Figure 4.3 to glow brightly: the sulfate, the hydroxide, the carbonate, the chloride, the acetate? Explain.

28. A solution of a substance causes the bulb in the apparatus shown in Figure 4.3 to glow weakly. Is the substance necessarily a weak electrolyte? Explain.

Concentrations in Aqueous Solutions

29. Determine the molarity of each of the following.

   (a) Li^+ and NO_3^- in 0.647 M $LiNO_3$

   (b) Ca^{2+} and I^- in 0.035 M CaI_2

   (c) Al^{3+} and SO_4^{2-} in 1.07 M $Al_2(SO_4)_3$

30. Determine the molarity of each of the following.

   (a) K^+ and HCO_3^- in 0.231 M $KHCO_3$

   (b) Mg^{2+} and CH_3COO^- in 0.850 M $Mg(CH_3COO)_2$

   (c) Fe^{2+} and SO_4^{2-} in 0.2000 M $Fe(NH_4)_2(SO_4)_2$

31. A solution is 0.0554 M NaCl and 0.0145 M Na_2SO_4. What are $[Na^+]$, $[Cl^-]$, and $[SO_4^{2-}]$ in this solution?

32. A solution is 0.015 M each in LiCl, MgI_2, Li_2SO_4, and $AlCl_3$. What is the molarity of each ion in this solution?

33. In what volume of solution must 16.11 g of $MgCl_2$ be dissolved to make a solution that is 0.1000 M in chloride ion?

34. In what volume of solution must 31.7 g of oxalic acid dihydrate, $H_2C_2O_4 \cdot 2 H_2O$, be dissolved to make a solution that is 0.0859 M in oxalic acid?

35. *Without doing detailed calculations,* place the following solutions in order from highest to lowest $[NO_3^-]$: 0.10 M KNO_3, 0.040 M $Al(NO_3)_3$, 0.047 M $Ca(NO_3)_2$.

36. *Without doing detailed calculations,* place the following solutions in order from highest to lowest *total* concentration of ions: 0.030 M KCl, 0.025 M K_2SO_4, 0.040 M $Fe(NO_3)_3$.

37. An aqueous solution is prepared by dissolving 18.3 g $MgSO_4 \cdot 7 H_2O$ in water and diluting to 285 mL of solution. What is $[SO_4^{2-}]$ in this solution?

38. What is the total concentration of ions in the solution in Problem 37?

39. The components of seawater are sometimes expressed in milligrams per liter (mg/L). Use the description of seawater given in Exercise 4.1A to determine the chloride ion content of seawater in mg Cl^-/L seawater.

40. A unit commonly used to describe low concentrations of a solute is *ppm* (parts per million, meaning, for example, grams solute per million grams of solution). If the concentration of chloride ion in a municipal water supply is given as 30.6 ppm Cl^-, what is the molarity of Cl^- in the water? (Assume the density of the water is 1.00g/mL.)

41. What volume of 0.0250 M $MgCl_2$ should be diluted to 250.0 mL to obtain a solution with $[Cl^-] = 0.0135$ M?

42. A solution is prepared by mixing 100.0 mL 0.438 M NaCl, 100.0 mL 0.0512 M $MgCl_2$, and 250.0 mL of water. What are $[Na^+]$, $[Mg^{2+}]$, and $[Cl^-]$ in the resulting solution?

43. *Without doing detailed calculations,* place these solutions in order from lowest to highest $[Cl^-]$.

   (a) 0.21 M RbCl

   (b) 0.45 M $FeCl_3 \cdot 9 H_2O$

   (c) 1.20 moles of $MgCl_2$ dissolved in 2.00 L solution

   (d) a solution that is 0.15 M KCl and 0.35 M NaCl

44. *Without doing detailed calculations,* determine which of the following contains the greatest mass of the element nitrogen.

   (a) 1.00 L of 0.0020 M aluminum nitrate

   (b) 500 mL of an ammonium nitrate solution containing 80 mg N/L

   (c) 100 mL of a 0.10% by mass magnesium nitrate solution ($d = 1.00$ g/mL).

Acid–Base Reactions

45. Write equations to show the ionization of the following aqueous acids and bases.

   (a) HBr   **(d)** HIO_3

   (b) LiOH   **(e)** $(CH_3)_2NH$

   (c) HF   **(f)** HCOOH

46. Write equations to show the ionization of the following aqueous acids and bases.

   (a) HNO_2 **(d)** $CH_3CH_2NH_2$

   (b) $CH_3(CH_2)_2COOH(aq)$ **(e)** HSO_4^-

   (c) $Ba(OH)_2$ **(f)** $HClO_4$

47. *Without performing detailed calculations*, place the following aqueous solutions in order from *lowest* to *highest* concentration of H^+ ion.

(a) 0.10 M HCl   (c) 0.10 M CH_3COOH

(b) 0.10 M H_2SO_4 (d) 0.15 M NH_3

48. Consider the four solutions in Problem 47. Might the order be affected if (a) the HCl was changed to 0.11 M; (b) the NH_3 was changed to 0.050 M; (c) the CH_3COOH was changed to 0.090 M? Explain.

49. Which of the following have the same net ionic reaction as HCl(aq) and NaOH(aq), (a) CH_3COOH(aq) and HNO_3(aq). (b) $Ba(OH)_2$(aq) and $HClO_4$(aq). (c) NH_3(aq) and HBr(aq)? Write a net ionic equation for the reaction that occurs in each case.

50. Strontium iodide can be made by the reaction of solid strontium carbonate with hydroiodic acid. Write the (a) complete-formula equation, (b) ionic equation, and (c) net ionic equation for this reaction.

51. Lime deposits on brass faucets are mostly $CaCO_3$. Though usually white, the deposits may be slightly green from copper in the brass [recall Figure 2.1(b)]. The deposits may be removed by soaking the faucet in hydrochloric acid. (The commercial product is often called muriatic acid.) Write a net ionic equation for the reaction that occurs.

52. A paste of sodium hydrogen carbonate (sodium bicarbonate) and water can be used to relieve the pain of an ant bite. The irritant in the ant bite is formic acid (HCOOH). Write a net ionic equation for the reaction that occurs.

Ionic Equations

53. Each equation represents the mixing of two aqueous solutions. Complete each as a *net ionic equation*. If no reaction occurs, write NR.

(a) K^+(aq) + I^-(aq) + Pb^{2+}(aq) + 2 NO_3^-(aq) $\longrightarrow$

(b) Mg^{2+}(aq) + 2 Br^-(aq) + Zn^{2+}(aq) + SO_4^{2-}(aq) $\longrightarrow$

(c) Cr^{3+}(aq) + 3 Cl^-(aq) + Li^+(aq) + OH^-(aq) $\longrightarrow$

(d) H^+(aq) + Cl^-(aq) + CH_3COOH(aq) $\longrightarrow$

(e) Ba^{2+}(aq) + 2 OH^-(aq) + H^+(aq) + I^-(aq) $\longrightarrow$

(f) K^+(aq) + HSO_4^-(aq) + Na^+(aq) + OH^-(aq) $\longrightarrow$

54. Each equation represents the mixing of two aqueous solutions. Complete each as a *net ionic equation*. If no reaction occurs, write NR.

(a) Ba^{2+}(aq) + 2 Cl^-(aq) + 2 Na^+(aq) + CO_3^{2-}(aq) $\longrightarrow$

(b) Pb^{2+}(aq) + 2 NO_3^-(aq) + Mg^{2+}(aq) + SO_4^{2-}(aq) $\longrightarrow$

(c) 2 Na^+(aq) + SO_4^{2-}(aq) + Cu^{2+}(aq) + 2 Cl^-(aq) $\longrightarrow$

(d) K^+(aq) + OH^-(aq) + Na^+(aq) + HSO_4^-(aq) $\longrightarrow$

(e) Na^+(aq) + OH^-(aq) + Mg^{2+}(aq) + 2 Cl^-(aq) $\longrightarrow$

(f) CH_3CH_2COOH(aq) + Ba^{2+}(aq) + 2 OH^-(aq) $\longrightarrow$

55. Predict whether a reaction is likely to occur in each of the following cases. If so, write a net ionic equation for the reaction.

(a) $Mg(OH)_2$(s) + HI(aq) $\longrightarrow$

(b) HCOOH(aq) + NH_3(aq) $\longrightarrow$

(c) CH_3COOH(aq) + H_2SO_4(aq) $\longrightarrow$

(d) $CuSO_4$(aq) + Na_2CO_3(aq) $\longrightarrow$

(e) KBr(aq) + $Zn(NO_3)_2$(aq) $\longrightarrow$

56. Predict whether a reaction is likely to occur in each of the following cases. If so, write a net ionic equation for the reaction.

(a) BaS(aq) + $CuSO_4$(aq) $\longrightarrow$

(b) $Cr(OH)_3$(s) + HBr(aq) $\longrightarrow$

(c) NH_3(aq) + H_2SO_4(aq) $\longrightarrow$

(d) $MgBr_2$(aq) + $ZnSO_4$(aq) $\longrightarrow$

(e) NaOH(aq) + $Mg(NO_3)_2$(aq) $\longrightarrow$

Solubility Guidelines and Precipitation Reactions

57. Classify the following as being soluble or insoluble in water:

(a) $Sr(NO_3)_2$ (c) $CuSO_4$

(b) $CuCl_2$ (d) PbS

58. Which of the following compounds reacts to precipitate Fe^{2+} from an aqueous solution of $FeCl_2$? Write an equation for the reaction.

(a) Na_2SO_4 (c) Na_2CO_3

(b) KBr   (d) $NaNO_3$

59. You suspect that a certain white powder is either $MgSO_4$(s) or $Mg(OH)_2$(s). You add dilute HCl(aq) and obtain a clear solution as shown on the right. Does the test indicate what the powder is? If not, what test would you perform instead? Explain.

60. You suspect that a particular solution is either $CuCl_2(aq)$ or $Cu(NO_3)_2(aq)$. You add dilute $KOH(aq)$ and obtain the result shown. Does the test indicate what the solution is? If not, what test would you perform instead? Explain.

61. When aqueous solutions of copper(II) nitrate and potassium carbonate are mixed, a precipitate forms. Write the net ionic equation for this reaction.

62. When aqueous solutions of iron(III) chloride and sodium sulfide are mixed, a precipitate forms. Write the net ionic equation for this reaction.

63. You suspect that a particular unlabeled aqueous solution is one of the following: $Na_2SO_4(aq)$, $NH_3(aq)$ or $Ba(NO_3)_2(aq)$. Explain how you can use precipitation reactions on small test samples of the solution to determine its identity. You have available the variety of aqueous solutions usually found in a general chemistry laboratory.

64. The addition of $MgCl_2(g)$ to an aqueous solution containing a single unknown ionic compound as a solute produces a white precipitate. List three compounds that the unknown might be and three that it cannot be. Explain your choices.

Oxidation–Reduction Reactions

65. Indicate whether the first-named substance in each change undergoes an oxidation, a reduction, or neither. Explain your reasoning.

    (a) Blue $CrCl_2(aq)$ changes to green $CrCl_3(aq)$ when exposed to air.

    (b) Yellow $K_2CrO_4(aq)$ changes to orange $K_2Cr_2O_7(aq)$ when acidified.

    (c) Dinitrogen pentoxide produces nitric acid when it reacts with water.

66. Indicate whether the first-named substance in each change undergoes an oxidation, a reduction, or neither. Explain your reasoning.

    (a) Sulfur trioxide gas produces sulfuric acid when passed into water.

    (b) Nitrogen dioxide converts to dinitrogen tetroxide when cooled.

    (c) Carbon monoxide is converted to methane in the presence of hydrogen.

67. Balance the following redox equations by inspection.

    (a) $HCl + O_2 \longrightarrow Cl_2 + H_2O$

    (b) $NO + H_2 \longrightarrow NH_3 + H_2O$

    (c) $CH_4 + NO \longrightarrow N_2 + CO_2 + H_2O$

    (d) $Ag + H^+ + NO_3^- \longrightarrow Ag^+ + H_2O + NO$

    (e) $IO_4^- + I^- + H^+ \longrightarrow I_2 + H_2O$

68. Balance the following redox equations by inspection, except in any case where the reaction is not possible.

    (a) $CH_4 + NO_2 \longrightarrow N_2 + CO_2 + H_2O$

    (b) $Ca(ClO)_2 + HCl \longrightarrow CaCl_2 + H_2O + Cl_2$

    (c) $SeO_3^{2-} + I^- + H^+ \longrightarrow Se + I_2 + H_2O$

    (d) $Fe^{2+} + NO_3^- + H^+ \longrightarrow Fe^{3+} + H_2O + NO$

    (e) $Zn + Cr_2O_7^{2-} + H^+ \longrightarrow Zn^{2+} + Cr^{3+} + H_2O$

69. Identify the oxidizing and reducing agents in Problem 67.

70. Identify the oxidizing and reducing agents in Problem 68.

71. The reactions represented by the following equations cannot occur as written. Why?

    (a) $PbO + V^{3+} + H_2O \longrightarrow PbO_2 + VO^{2+} + H^+$

    (b) $Fe_2S_3(s) + H_2O(l) \longrightarrow Fe(OH)_3(s) + S(s)$

72. The reactions represented by the following equations cannot occur as written. Why?

    (a) $O_3(g) + ClO_3^-(aq) + OH^-(aq) \longrightarrow H_2O(l) + Cl^-(aq)$

    (b) $NH_3(g) + H_2O(g) \longrightarrow N_2H_4(g) + NO_2(g)$

73. Use the activity series of the metals to predict chemical reactions in the following cases. Write a plausible balanced equation for each reaction that does occur and NR for those that do not.

    (a) $Zn(s) + H^+(aq) \longrightarrow$

    (b) $Cu(s) + Zn^{2+}(aq) \longrightarrow$

    (c) $Fe(s) + Ag^+(aq) \longrightarrow$

    (d) $Au(s) + H^+(aq) \longrightarrow$

74. An unknown metal M gives the following results in some laboratory tests. Use these data to find the approximate location of M in the activity series on page 146.

$$M(s) + 2\,H^+(aq) \longrightarrow M^{2+}(aq) + H_2(g)$$
$$M(s) + Cu^{2+}(aq) \longrightarrow M^{2+}(aq) + Cu(s)$$
$$M(s) + Fe^{2+}(aq) \longrightarrow M^{2+}(aq) + Fe(s)$$
$$2\,Al(s) + 3\,M^{2+}(aq) \longrightarrow 2\,Al^{3+}(aq) + 3\,M(s)$$
$$M(s) + Zn^{2+}(aq) \longrightarrow \text{no reaction}$$

Titrations

75. How many milliliters of 0.0195 M HCl are required to titrate **(a)** 25.00 mL of 0.0365 M KOH(aq), **(b)** 10.00 mL of 0.0116 M $Ca(OH)_2(aq)$, **(c)** 20.00 mL of 0.0225 M $NH_3(aq)$?

76. How many milliliters of 0.0108 M $Ba(OH)_2(aq)$ are required to titrate **(a)** 20.00 mL of 0.0265 M $H_2SO_4(aq)$, **(b)** 25.00 mL of 0.0213 M HCl(aq), **(c)** 10.00 mL of 0.0868 M $CH_3COOH(aq)$?

77. Vinegar is an aqueous solution of acetic acid, CH_3COOH. A 10.00-mL sample of a particular vinegar requires 31.45 mL of 0.2560 M KOH for its titration. What is the molarity of acetic acid in the vinegar?

78. Most window cleaners are aqueous solutions of ammonia. A 10.00-mL sample of a particular window cleaner requires 39.95 mL of 0.1008 M HCl for its titration. What is the molarity of ammonia in the window cleaner?

79. A tablet of a dietary supplement containing calcium carbonate is found to neutralize 38.8 mL of 0.251 M HCl, forming calcium chloride, water, and CO_2. Calculate the number of milligrams of **(a)** $CaCO_3$ and **(b)** Ca^{2+} in the tablet.

80. A 5.00% NaOH solution by mass has a density of 1.054 g/mL. What is the minimum molarity of an HCl(aq) solution that can be used to titrate a 5.00-mL sample of the NaOH(aq) if the titration is to be accomplished without having to refill a 50.00-mL buret used in the titration?

81. *Without doing detailed calculations,* determine which of the following popular antacids is able to neutralize more stomach acid [dilute HCl(aq)] if equal masses are compared: Alka-Seltzer® (sodium hydrogen carbonate) or Tums® (calcium carbonate).

82. *Without doing detailed calculations,* determine which of the following is more effective in reducing the acidity of a home swimming pool if equal masses are compared: caustic soda (sodium hydroxide) or soda ash (sodium carbonate).

83. Which of the following points in the titration of 10.00 mL of 1.00 M CH_3COOH with 0.500 M NaOH would produce a solution with the "molecular view" like the figure? After the volume of 0.5000 M NaOH added is **(a)** 0.00 mL, **(b)** 5.00 mL, **(c)** 20.00 mL, **(d)** 22.00 mL, **(e)** 30.00 mL. Explain.

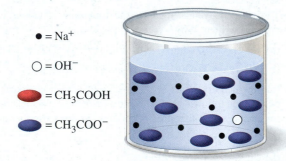

- $\bullet$ = Na^+
- $\bigcirc$ = OH^-
- = CH_3COOH
- = CH_3COO^-

84. Refer to Problem 83. In a similar manner to the figure shown, draw a sketch to represent the titration mixture after 10.00 mL of 0.5000 M NaOH has been added.

85. How many milliliters of 0.02091 M $AgNO_3$ would be needed to titrate **(a)** 25.00 mL of 0.1235 M KI, **(b)** 40.00 mL of 0.01944 M $FeCl_3$, **(c)** 0.0323 g of Na_2CO_3?

86. How many milliliters of 0.1467 M $Pb(NO_3)_2$ would be needed to titrate each of the samples given in Problem 85?

87. Titration of 0.2558 g of Na_2SO_4 that has been dissolved in water requires 41.60 mL of a solution of $Ba(ClO_4)_2$. Write the balanced net ionic equation for the reaction, and calculate the molar concentration of the $Ba(ClO_4)_2$ solution.

88. A sample of NaCl weighing 1.4477 g is dissolved in water and diluted to 250.0 mL. What volume of this solution will be needed to titrate 25.00 mL of 0.1000 M $AgNO_3$(aq)?

89. How many milliliters of 0.1050 M $KMnO_4$(aq) are required for the titration of **(a)** 20.00 mL of 0.3252 M Fe^{2+}(aq), and **(b)** a 1.065-g sample of KNO_2? The balanced equations are:

$$5\,Fe^{2+} + MnO_4^- + 8\,H^+ \longrightarrow 5\,Fe^{3+} + Mn^{2+} + 4\,H_2O$$

$$5\,NO_2^- + 2\,MnO_4^- + 6\,H^+ \longrightarrow 2\,Mn^{2+} + 5\,NO_3^- + 3\,H_2O$$

90. For the reactions described in Problem 89, what is the molarity of $KMnO_4$(aq) if **(a)** 22.55 mL $KMnO_4$(aq) is required to titrate 10.00 mL of 0.2434 M $FeSO_4$(aq), and **(b)** if a 567.4 mg sample of KNO_2 requires 31.61 mL $KMnO_4$(aq) for its titration?

91. The concentration of Mn^{2+}(aq) can be determined by titration with MnO_4^-(aq) in basic solution. The balanced equation is

$$3\,Mn^{2+} + 2\,MnO_4^- + 4\,OH^- \longrightarrow 5\,MnO_2(s) + 2\,H_2O$$

A 25.00-mL sample of Mn^{2+}(aq) requires 34.77 mL of 0.05876 M $KMnO_4$(aq) for its titration. What is the molarity of the Mn^{2+}(aq)?

92. The concentration of $KMnO_4$(aq) is to be determined by titration against As_2O_3(s). A 0.1156-g sample of As_2O_3(s) requires 27.08 mL of the $KMnO_4$(aq) for its titration. The balanced equation is given below. What is the molarity of the $KMnO_4$(aq)?

$$5\,As_2O_3 + 4\,MnO_4^- + 9\,H_2O + 12\,H^+$$
$$\longrightarrow 10\,H_3AsO_4 + 4\,Mn^{2+}$$

Additional Problems

Problems marked with an * may be more challenging than others.

93. A sample of human blood serum has a density of 1.022 g/mL and contains 18.9 mg of K^+ and 365 mg of Cl^- per 100 mL. Calculate the molar concentrations of K^+ and Cl^- in the sample.

94. Would nitric acid and acetic acid be equally effective and useful for removing lime deposits on a brass faucet? (See Problem 51.) Explain.

95. A sample of ordinary table salt is 98.8% NaCl and 1.2% $MgCl_2$ by mass. What is $[Cl^-]$ if 6.85 g of this mixture is dissolved in 500.0 mL of an aqueous solution?

96. A solution is 0.0240 M KI and 0.0146 M MgI_2. What volume of water should be added to 100.0 mL of this solution to produce a solution with $[I^-] = 0.0500$ M?

97. A white solid is known to be either $MgCl_2$, $MgSO_4$, or $Mg(NO_3)_2$. An aqueous solution prepared from the solid yields a white precipitate when treated with $Ba(NO_3)_2$. What must the solid be?

98. What reagent solution (including pure water) would you use to separate the cations in the following pairs, that is, with one cation appearing in solution and the other in a precipitate?

(a) $BaCl_2$(s) and NaCl(s)

(b) $MgCO_3$(s) and Na_2CO_3(s)

(c) $AgNO_3$(s) and KNO_3(s)

(d) $PbSO_4$(s) and $CuCO_3$(s)

(e) $Mg(OH)_2$(s) and $BaSO_4$(s)

99. A railroad tank car carrying 1.5×10^3 L of concentrated sulfuric acid derails and spills its load. The acid is 93.2% H_2SO_4 and has a density of 1.84 g/mL. How many kilograms of sodium carbonate (soda ash) are needed to neutralize the acid? (*Hint:* What is the neutralization reaction?)

100. To 125 mL of 1.05 M $Na_2CO_3(aq)$ is added 75 mL of 4.5 M $HCl(aq)$. Then the solution is evaporated to dryness. What mass of $NaCl(s)$ is obtained?

101. A 15,000-gallon home swimming pool is disinfected by the daily addition of 0.50 gal of a "chlorine" solution—NaOCl in NaOH(aq). To maintain the proper acidity in the pool, the basic components in the chlorine solution must be neutralized. By experiment, it is found that about 220 mL of an HCl(aq) solution that is 31.4% HCl by mass (d = 1.16 g/mL) is required to neutralize 0.50 gal of the chlorine solution. What is the $[OH^-]$ of the chlorine solution? (1 gal = 3.785 L)

102. What is the molarity of OH^- in the solution formed by mixing 25.10 mL of 0.2455 M NaOH and 35.05 mL of 0.1524 M HNO_3?

103. A 0.235-g sample of a solid that is 92.5% NaOH and 7.5% $Ca(OH)_2$ requires 45.6 mL of an HCl(aq) solution for its titration. What is the molarity of the HCl(aq)?

* 104. To titrate a 5.00-mL sample of a saturated aqueous solution of sodium oxalate, $Na_2C_2O_4$, requires 25.82 mL of 0.02140 M $KMnO_4(aq)$. How many grams of $Na_2C_2O_4$ would be present in 250.0 mL of the saturated solution? The balanced equation is

$$5\,C_2O_4{}^{2-}(aq) + 2\,MnO_4{}^-(aq) + 16\,H^+(aq) \longrightarrow$$
$$2\,Mn^{2+}(aq) + 8\,H_2O(l) + 10\,CO_2(g)$$

105. The compound $NdCaMn_2O_6$ has half its Mn atoms with oxidation number +3 and half with oxidation number +4. What is the oxidation number of the neodymium?

106. Biochemists sometimes describe reduction as the gain of H atoms. Use examples from the text to show that this definition conforms to that based on oxidation numbers.

107. A 25.00-mL sample of 0.1996 M V^{2+} solution requires 48.97 mL of 0.3000 M Ce^{4+} for titration. The Ce^{4+} is converted to Ce^{3+} in the titration reaction. What is the oxidation number for vanadium ion after the titration is complete?

* 108. A stock buffer solution is prepared by dissolving 31.5 g of NH_4Cl and 282 mL of concentrated ammonia in water and diluting to 1.00 liter. The concentrated ammonia solution (d = 0.899 g/mL) is 17.0 M NH_3. Ten milliliters of the buffer solution is added to 40.0 mL of $MgCl_2(aq)$. This mixture is titrated with 0.01000 M ethylenediamine tetraacetate ion, symbolized here as Y^{4-}.

$$Mg^{2+} + Y^{4-} \longrightarrow MgY^{2-}$$

It takes 38.26 mL of the Y^{4-} solution to reach the equivalence point. Use the data given, as needed, to calculate **(a)** the mass of $MgCl_2$ in the $MgCl_2(aq)$, **(b)** the concentration of chloride ion in the final solution after titration. Assume that the volumes of solution may be added to obtain the total volume.

109. Methanol (CH_3OH) can reduce chlorate ion to chlorine dioxide in an acidic solution. The methanol is oxidized to carbon dioxide. What volume of methanol (d = 0.791 g/mL), in milliliters, is needed to produce 125 kg $ClO_2(g)$? The balanced equation for the reaction is

$$CH_3OH + 6\,H^+ + 6\,ClO_3{}^- \longrightarrow 6\,ClO_2 + CO_2 + 5\,H_2O$$

110. What volume of 0.185 M $MgCl_2$ must be added to 235 mL of 0.206 M KCl to produce a solution with a concentration of 0.250 M Cl^-?

* 111. The exact concentration of an aqueous solution of oxalic acid (HOOCCOOH, that is, $H_2C_2O_4$) is determined by an acid–base titration. Then the oxalic acid solution is used to determine the concentration of $KMnO_4(aq)$ by a redox titration in acidic solution. The titration of 25.00-mL samples of the oxalic acid solution requires 32.15 mL of 0.1050 M NaOH and 28.12 mL of the $KMnO_4(aq)$. What is the molarity of the $KMnO_4(aq)$? (*Hint:* The balanced equation can be derived from Problem 104)

* 112. A piece of marble (assume it to be pure $CaCO_3$) reacts with 2.00 L of 2.52 M HCl. After dissolution of the marble, a 10.00-mL sample of the remaining HCl(aq) is withdrawn, added to some water, and titrated with 24.87 mL of 0.9987 M NaOH. What must have been the mass of the piece of marble? Comment on the precision of this method; that is, how many significant figures are justified in the result?

* 113. A sample of impure potassium hydroxide consists of 92.25% KOH, 2.25% K_2CO_3, and 5.50% H_2O by mass. A 1.250-g sample of this solid is allowed to react with 25.00 mL of 1.840 M HCl. The excess HCl is neutralized with 1.050 M KOH, and the solution is then evaporated to dryness. What solid residue is obtained, and what is its mass?

* 114. Feldspars are a group of rock-forming silicate minerals that make up over half of Earth's crust. Two elements commonly found in feldspars are sodium and potassium. In the analysis of feldspars, the percentages of sodium and potassium are frequently given as % Na_2O and % K_2O (see Chapter 3, Problem 128). The sodium and potassium content of a 0.7500-g sample of a feldspar mineral is obtained as a 0.2250 g mixture of NaCl(s) and KCl(s). The chloride mixture is dissolved in water and requires 22.61 mL of 0.1525 M $AgNO_3(aq)$ for complete precipitation of the chloride ion as AgCl(s). Determine the mass percent of Na_2O and K_2O in the mineral.

* 115. For use as a fuel for generating electrical power, natural gas must be freed of sulfur impurities to comply with environmental standards. The sulfur content of a 4.476-g sample of natural gas was determined by burning the sample in excess oxygen. The gases produced were bubbled through a 3% solution of H_2O_2, which oxidized the SO_2 to sulfuric acid. Then 25.00 mL of 0.00923 M NaOH, an amount in excess of that needed to neutralize the sulfuric acid, was added to the solution. The excess NaOH was titrated, requiring 13.33 mL of 0.01007 M HCl for the titration. Calculate the percentage of sulfur in the sample.

Apply Your Knowledge

116. [Biochemical] The Kjeldahl analysis is used to determine the protein content of foods. The food analyzed is heated with sulfuric acid in the presence of a catalyst. The carbon in the protein is converted to $CO_2(g)$, the hydrogen to water, and the nitrogen to $(NH_4)_2SO_4$. The mixture is then treated with concentrated NaOH(aq) and heated to drive off $NH_3(g)$. The $NH_3(g)$ is neutralized by an excess of a standard acid, and the acid present after the neutralization is titrated with a standard base. From the titration data and the known mass of the sample, the % N can be calculated. Based on the fact that the average percentage of nitrogen in most plant and animal protein is about 16%, the percent protein is obtained by multiplying the % N by the factor 6.25. A 2.500-g sample of meat is subjected to Kjeldahl analysis. The liberated $NH_3(g)$ is absorbed in 50.00 mL of H_2SO_4(aq). The excess acid requires 19.90 mL of 0.5510 M NaOH for its complete neutralization. A separate 25.00-mL sample of the H_2SO_4(aq) requires 22.65 mL of the 0.5510 M NaOH for its titration. What is the percent protein in the meat?

* 117. [Laboratory] *Back titration* may be performed when a reaction is too slow for direct titration. An antacid tablet weighing 0.7023 grams and containing calcium carbonate plus starch and sweeteners was added to 50.00 mL of 0.3000 M HCl ($d = 1.03$ g/mL) in a flask. After the reaction was complete, the excess HCl in the flask required 28.06 mL of 0.1200 M NaOH ($d = 1.01$ g/mL) for titration to the equivalence point. Use the data given, as needed, to calculate the mass of $CaCO_3$ in the tablet.

118. [Laboratory] In a *Volhard* precipitation titration, a piece of sterling silver alloy weighing 0.5039 grams was dissolved in 10.0 mL of concentrated (15 M) nitric acid. The solution was diluted to 60.0 mL with water, and 5.00 mL of 0.100 M Fe $(NO_3)_3$ was added as an indicator. The contents of the flask were titrated with 43.56 mL of 0.1005 M KSCN to reach the red endpoint. Write the net ionic equation for the titration reaction (the product of the titration reaction was AgSCN precipitate), and use the data given as needed to calculate the percentage of silver in the alloy.

119. [Environmental] Incineration of a chlorine-containing toxic waste such as a polychlorinated biphenyl (PCB) produces CO_2 and HCl.

$$C_{12}H_4Cl_6 + O_2 \longrightarrow CO_2 + HCl + H_2O$$

Balance the equation for this combustion reaction. Comment on the advantages and disadvantages of incineration as a method of disposal of such wastes.

* 120. [Environmental] Algal protoplasm is a complex mixture of many substances, but its composition can be represented by a "formula," $C_{106}H_{263}O_{110}N_{16}P$. The protoplasm is produced during photosynthesis by a process that we can represent by an "equation."

$$CO_2 + H_2O + H^+ + NO_3^- + HPO_4^{2-} + \text{trace elements}$$
$$+ \text{ energy} \longrightarrow C_{106}H_{263}O_{110}N_{16}P + O_2$$

(a) Balance the equation. **(b)** Which is the limiting nutrient, NO_3^- or HPO_4^{2-}, in a lake that has the following concentrations: NO_3^-, 434 mg/L; and HPO_4^{2-}, 65.0 mg/L? **(c)** What mass of algal protoplasm would be produced in a pond wih an area of 2.55 hectares (1 hectare = 1 square hectometer;

1 hm = 100 m) and an average depth of 4.45 m? **(d)** What mass of oxygen would be consumed by the decay of the amount of algal protoplasm in **(c)**, assuming the decay is the reverse of the above reaction? **(e)** If the algae die when the dissolved oxygen concentration reaches 15.0 mg O_2/L, would the algal decay deplete the dissolved oxygen in the pond? **(f)** If phosphates from sewage entering the lake raise the HPO_4^{2-} concentration to 110 mg/L, what mass of algal protoplasm would be formed, assuming that the HPO_4^{2-} is the limiting reactant?

* 121. [Laboratory] Although Figure 4.6 gives a dramatic picture, chemists use precise measurements to determine how well a solution conducts electric current and how the conductance changes as the result of a chemical reaction. For our purposes, we do not need the details of the method used; we just need to note the ideas that follow.

The conductivity of a solution depends on the total concentration of ions.

Ions differ in their individual abilities to carry electric current.

H^+ and OH^- are much better electrical conductors than other ions.

Shown here are four idealized graphs of the electrical conductance of a solution during the course of a titration. For example, (a) represents the titration of HCl(aq) with NaOH(aq)

$$H^+ + Cl^- + Na^+ + OH^- \longrightarrow Na^+ + Cl^- + H_2O$$

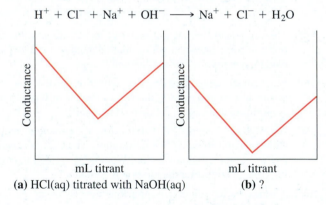

(a) HCl(aq) titrated with NaOH(aq)   **(b)** ?

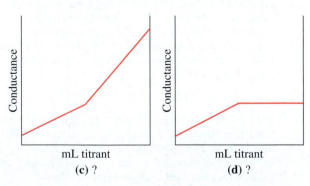

(c) ?   **(d)** ?

The conductance starts high because $[H^+]$ is greatest at the start. The conductance decreases during the titration because H^+ reacts with OH^- to form the nonelectrolyte H_2O, and H^+ is replaced by Na^+, which does not conduct current nearly as well. The conductance falls to a minimum at the equivalence point, where the solution is simply NaCl(aq). Beyond the equivalence point, the conductance again rises because of the accumulation of excess OH^-, which is a very good electrical conductor.

Match each of the following titrations to the appropriate remaining graph: (b), (c), or (d). In the manner used for graph (a), explain the general shape of each graph.

$NH_3(aq)$ with $HCl\ (aq)$ as the titrant

$CH_3COOH(aq)$ with $NH_3\ (aq)$ as the titrant

$Ba(OH)_2$ with $(NH_4)_2SO_4$ as the titrant

* **122. [Laboratory]** To be most effective as an antiseptic solution, $H_2O_2(aq)$ should contain 3.0% H_2O_2 by mass. The following experiment is done to determine if a particular hydrogen peroxide solution is at full strength. A 10.00-mL sample of the aqueous solution is treated with an excess of $KI(aq)$. The liber-

ated I_2 forms the triiodide ion, I_3^-, and the triiodide-ion solution requires 28.91 mL of 0.1522 M $Na_2S_2O_3$ for its titration. Is the $H_2O_2(aq)$ at full strength? (Assume that the density of the $H_2O_2(aq)$ is 1.00 g/mL.)

$$H_2O_2(aq) + H^+(aq) + I^-(aq) \longrightarrow H_2O(l) + I_3^-(aq)$$
(not balanced)

$$I_3^-(aq) + S_2O_3^{2-}(aq) \longrightarrow S_4O_6^{2-}(aq) + I^-(aq)$$
(not balanced)

123. [Laboratory] In the discussion of redox titrations, it was stated that the titrant is usually the oxidizing agent. In such a titration, the analyte must first be converted entirely to a single low oxidation state. Once the analyte is so converted, it is usually titrated immediately and as quickly as possible. Suggest practical reasons for this procedure.

 # e-Media Problems

The activities described in these problems can be found in the e-Media Activities and Interactive Student Tutorial (IST) modules of the Companion Website, *http://chem.prenhall.com/hillpetrucci*.

124. View the **Electrolytes and Nonelectrolytes** and **Dissolution of NaCl in Water** animations (*Section 4–1*). Describe on an atomic scale the difference between solid sodium chloride and an aqueous solution of sodium chloride. What property of the solution is responsible for its being categorized as an electrolyte?

125. From the **Strong and Weak Electrolytes** movie (*Section 4–1*), what can you deduce from the experimental observation (brightness of the lightbulb) about the relative fraction of ions in solution for the hydrogen chloride and acetic acid solutions?

126. In the **Precipitation Reactions** movie (*Section 4–3*), what factors will influence the amount of precipitate formed in the test

tube for each of the two reactions shown? Write a balanced chemical equation for each case, and discuss the properties of the solutions involved in the two reactions.

127. Consider the reaction that is depicted in the **Formation of Silver Crystals** movie (*Section 4–5*). Predict the effect of changing the wire placed in the solution from copper to zinc. Would changing the wire from copper to gold produce a similar effect? Describe what would be observed if a solution of lead nitrate were instead added to the beaker containing the copper wire.

128. For the reaction shown in the **Acid–Base Titrations** animation (*Section 4–6*), calculate the concentration of each species found in solution after 20.0 mL of the 0.100 M sodium hydroxide solution has been added to the acidic solution being titrated. How does this differ from the solution concentrations after 60.0 mL of the sodium hydroxide solution has been added?

Gases

EARTH'S LIFE-SUPPORT system relies on a thin blanket of gases that we call air. Air is so insubstantial that it is difficult to think of it as matter. But air is matter. Air has mass and occupies space; it is matter in the gaseous state.

In the macroscopic world, we probably have a more intuitive feeling for liquids and solids than for gases. In the microscopic realm, however, gases are much simpler conceptually than are liquids and solids. This is because the behavior of liquids and solids is critically dependent on forces between molecules, which in turn are closely related to molecular structure. In contrast, the simple laws governing the behavior of gases are based on the *absence* of such forces. Thus, we can consider gases now and return to liquids and solids in Chapter 11.

In this chapter, we will introduce one of the most successful scientific theories—the kinetic-molecular theory. In Chapter 13, we will find that this theory not only explains the behavior of gases but also provides important insights into the rates of chemical reactions.

5.1   Gases: What Are They Like?

Gases are composed of widely separated particles (molecules in most cases, atoms in some) that are in constant random motion. Gases flow readily and occupy the entire volume of their container, regardless of its shape. If someone opens a bottle of household ammonia in one part of a room, we can soon smell the ammonia throughout the room.

Unlike a liquid or solid, in which molecules or atoms are already quite close to one another, a gas is easily compressed because of its widely separated particles. As a result, it is possible to store enough air in a small portable tank for an hour or more of breathing during underwater diving. When air is compressed, the particles are forced closer together, but they are still much farther apart than those in a liquid or solid.

At room temperature, nearly all ionic substances, even those of low molar mass, are solids (for example, NaCl: 58.44 g/mol). In contrast, most molecular substances

◀ Balloons filled with helium gas "float" in air and are often used for decoration. The air or atmosphere in which they float is composed of a mixture of gases. Why do helium balloons float in the atmosphere? Why do they appear to deflate after a period of time? What properties govern a mixture of gases? In this chapter we will examine the behavior of both pure gases and gas mixtures.

Table 5.1 Some Common Gases[a]

Substance	Formula	Typical Use(s)
Acetylene	C_2H_2	Fuel for welding metals
Ammonia	NH_3	Fertilizer, manufacture of plastics
Argon	Ar	Filling gas for specialized lightbulbs
Butane	C_4H_{10}	Fuel for heating (LPG)
Carbon dioxide	CO_2	Beverage carbonation
Carbon monoxide	CO	Reducing agent in metallurgy
Chlorine	Cl_2	Disinfectant, bleach
Ethylene	C_2H_4	Manufacture of plastics
Helium	He	Lifting gas for balloons
Hydrogen	H_2	Chemical reagent, fuel for fuel cells
Hydrogen sulfide	H_2S	Chemical reagent
Methane	CH_4	Fuel, manufacture of hydrogen
Nitrogen	N_2	Manufacture of ammonia
Nitrous oxide	N_2O	Anesthetic
Oxygen	O_2	Support of combustion, respiration
Propane	C_3H_8	Fuel for heating (LPG)
Sulfur dioxide	SO_2	Preservative, disinfectant, bleach

[a] All of these substances are gases at room temperature (about 25 °C) and at pressures comparable to atmospheric pressure, but they can be converted to liquids and solids by cooling or an increase in pressure.

▲ Several gases are among the top industrial chemicals produced in the United States and are often shipped as liquids in tanker trucks. Among all the industrial chemicals, typical annual rankings in the United States, by mass, have nitrogen in the no. 2 position; oxygen, no. 3; ethylene, no. 4; ammonia, no. 6; and chlorine, no. 10. (Sulfuric acid is the top U.S. chemical overall.) The tanker truck here is being filled with liquid oxygen.

of low molar mass are either gases or else liquids that are easily vaporized. For example, nitrogen is a gas (N_2: 28.01 g/mol) and ethanol is a liquid (CH_3CH_2OH: 46.07 g/mol). We use the term **vapor** to denote the gaseous state of a substance that is more commonly encountered as a liquid (such as water or ethanol) or as a solid (such as *para*-dichlorobenzene, used as a moth repellent and room deodorizer).

Table 5.1 lists some common gases. These gases and many others are commercially available in tanks, either as highly compressed gases or as liquids that vaporize when the pressure on them is released. Perhaps the most familiar and important gases are those found mixed together in ordinary air—N_2, O_2, Ar, and CO_2. The gases that make up the air we breathe are so significant that we will devote considerable attention to a study of the atmosphere in Chapter 25.

5.2   An Introduction to the Kinetic-Molecular Theory

The **kinetic-molecular theory** was developed in the mid-nineteenth century. It provides a model for gases at the microscopic level that explains the physical properties we observe at the macroscopic level. The theory treats gases as collections of particles in rapid random motion. The "kinetic" in the name conveys the idea of motion, and "molecular" indicates that the motion is that of molecules. The theory refers to particles of a gas as molecules even though the particles are monatomic in some cases. The particles are molecules (N_2) in nitrogen gas, for instance, and atoms (Ar) in argon gas.

The molecules of a gas are in such rapid motion that they seem to defy the force of gravity. They do not fall and collect at the bottom of a container, as do the molecules in a liquid. Gas molecules are distributed uniformly throughout the entire volume of a container. Yet, because the distances between gas molecules are generally much greater than the dimensions of the molecules themselves, the typical gas is mostly empty space. For example, a molecule of N_2 in room-temperature air travels an average distance of about 200 times its molecular diameter before it encounters another molecule. The movement of gaseous molecules through three-dimensional space is called *translational* motion, and it is *random*. That is, it is not possible to predict, for any given molecule, its speed, direction, or energy of motion (called kinetic energy). The speeds and kinetic energies can vary widely, and all directions of motion are equally probable. A gas molecule moves in a straight line until it strikes either another molecule of the gas or the container wall. Then it bounces off and travels along a new

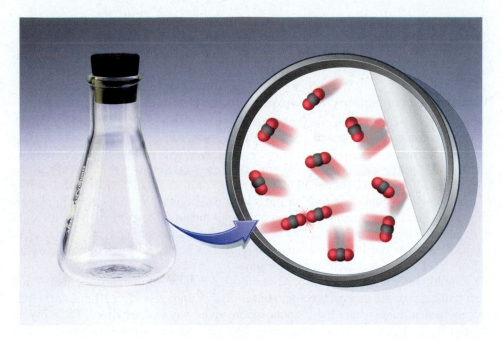

◀ FIGURE 5.1   Visualizing molecular motion in a gas
Molecules of a gas (here, carbon dioxide) are in constant random, straight-line motion. They undergo elastic collisions with one another and with the walls of the container.

 Motion of a Gas animation

straight-line path until its next collision (Figure 5.1). Some molecules lose energy and slow down as a result of collisions, but others gain energy and speed up. There is no *net* loss or gain of energy in a collision, and therefore the total translational kinetic energy of the molecules of a gas is not changed by collisions. Because kinetic energy is conserved, we say that the collisions of gas molecules are *elastic*.

The kinetic-molecular theory explains what we measure when we measure temperature. According to the theory, temperature is a measure of the average translational kinetic energy of the molecules of a sample of matter, whether solid, liquid, or gas. The higher the average translational kinetic energy is, the higher the temperature. On average, molecules in a cold sample of a gas move more slowly than those in a hot sample of the same gas. The temperature is related to the *average* translational energy because the individual molecules of a gas move at different speeds and have different kinetic energies.

The kinetic-molecular theory also explains the origin of gas pressure, an important property of gases that we explore in the next section. Consider a gas-filled balloon: When a molecule of the gas strikes the wall of the balloon, it gives the wall a little push. When we measure gas pressure, we assess the net result of these molecular pushes.

We will return to a more quantitative discussion of the kinetic-molecular theory in Section 5.11 after we have studied some of the natural laws that describe gas behavior on a macroscopic level.

5.3   Gas Pressure

The push exerted by an individual gas molecule as it bounces off our skin is too tiny for us to feel. However, we can note the collective effect of very large numbers of gas molecules by measuring the pressure they exert. **Pressure** is defined as force per unit area, that is, a force divided by the area over which the force is exerted:

$$\text{Pressure} = \frac{\text{force}}{\text{area}} = \frac{F}{A}$$

In SI units, force is expressed in *newtons* (N) and area in square meters (m^2). Therefore, the derived SI unit for pressure is newton per square meter, also called a **pascal (Pa).***

$$1 \text{ Pa} = 1 \text{ N/m}^2$$

A newton has the unit $kg \, m \, s^{-2}$. For a review of fundamental physical quantities, such as the newton, see Appendix B.

---

*Units that are named after individuals are not capitalized, but the symbols for such units are. Thus, the pressure unit named for Blaise Pascal (1623–1662) is called a pascal, but it is represented by the symbol Pa.

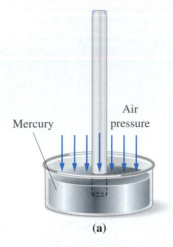

▲ FIGURE 5.2   Measuring air pressure with a mercury barometer
(a) With an *open-end* tube, the mercury levels are equal inside and outside the tube because the tube is open to the atmosphere and filled with air. (b) In a mercury barometer, a column of mercury 760 mm high is maintained in a *closed-end* tube by standard atmospheric pressure. The small space above the mercury is devoid of air (a vacuum), containing only a tiny trace of mercury vapor.

Table 5.2 The Standard Atmosphere of Pressure in Different Units		
1 atm =	760 mmHg	
	760 Torr	
	1.01325 bar	
	1013.25 mb	
	14.696 lb/in.^{2}	
	101,325 N/m^2	
	101,325 Pa	
	101.325 kPa	

The pascal is such a small unit that the **kilopascal (kPa)** is usually used instead. As we shall see next, the easiest way to measure a gas pressure is to compare it with the pressure exerted by a column of liquid. Units of pressure stated in terms of liquid heights are still widely used.

Measuring Atmospheric Pressure: Barometers

We use a device called a **barometer** to measure the pressure of the atmosphere. In the mercury barometer, invented in 1643 by Evangelista Torricelli (1608–1647), a glass tube, about a meter long and closed at one end, is filled with mercury and inverted into a shallow dish that also contains mercury (Figure 5.2). Some of the mercury in the tube drains into the dish, but not all of it. The level of mercury in the tube falls because of gravity, while at the same time air pressure pushes down on the surface of the mercury in the dish. The mercury drains out only until the pressure exerted on the mercury in the dish by the mercury remaining in the tube exactly balances the pressure exerted on the mercury in the dish by the atmosphere. At this point, these two opposing forces balance each other and reach an equilibrium.

Liquid mercury is quite dense. At sea level, a column of mercury about 760 mm high will balance the push of a column of air many kilometers high, from Earth's surface up to an altitude where Earth's atmosphere gives way to outer space. The pressure exerted by a column of mercury *exactly* 760 mm high is a pressure unit called **1 atmosphere (atm).*** Another pressure unit, the **millimeter of mercury (mmHg)** is also known by the name *torr* and the symbol **Torr** (after Torricelli). That is,

$$1 \text{ atm} = 760 \text{ mmHg} = 760 \text{ Torr}$$

Rather than the SI unit pascal, we will use the pressure units torr and atm for the most part in this text. The relationship between the pascal and the unit atmosphere is

$$1 \text{ atm} = 101,325 \text{ Pa} = 101.325 \text{ kPa}$$

For approximate work, remember that 1 atm is about 100 kPa.

Several other pressure units are widely used (Table 5.2). Weather reports in the United States, for instance, often include atmospheric pressure in *millibars* (*mb*) or *inches of mercury* (*in. Hg*):

$$1 \text{ atm} = 29.921 \text{ in. Hg} = 1.01325 \text{ bar} = 1013.25 \text{ mb}$$

Engineers often use *pounds per square inch* (lb/in.^{2}) and the symbol *psi* for practical applications like steam pressure in boilers and turbines:

$$1 \text{ atm} = 14.696 \text{ lb/in.}^2 = 14.696 \text{ psi}$$

Example 5.1

A Canadian weather report gives the atmospheric pressure as 100.2 kPa. What is the pressure expressed in the unit torr?

STRATEGY

To convert a pressure from one unit to another, we need a relationship between the two units that we can use as a conversion factor. Generally we can find such a relationship in Table 5.2.

SOLUTION

To convert from kPa to Torr, we need this relationship from Table 5.2: 1 atm = 760 Torr = 101.325 kPa. That is,

$$\frac{760 \text{ Torr}}{101.325 \text{ kPa}} = 1$$

---

*For the pressure to be exactly 1 atm, the 760-mm column of mercury must be at 0 °C ($d = 13.59508$ g/cm^3) and at a location where the acceleration due to gravity (g) is 9.80665 m/s^2.

With this conversion factor, we find that

$$? \text{ Torr} = 100.2 \text{ kPa} \times \frac{760 \text{ Torr}}{101.325 \text{ kPa}} = 751.6 \text{ Torr}$$

ASSESSMENT

Note that the units kPa cancel and that the pressure expressed in the unit torr is a larger number (by a factor of about 7.5) than the pressure in kilopascals, as expected from the form of the conversion factor.

EXERCISE 5.1A

Carry out the following conversions of pressure units.

(a) 0.947 atm to mmHg

(b) 98.2 kPa to Torr

(c) 29.95 in. Hg to Torr

(d) 768 Torr to atm

EXERCISE 5.1B

What is the pressure, in kilopascals and in atmospheres, if a force of 1.00×10^2 N is exerted on an area of 5.00 cm^2?

Although gas molecules do not settle under the force of gravity in laboratory-size containers, they do settle in that huge repository we call the atmosphere. As a result, the density of air—the mass per unit volume—decreases rapidly with increased height above Earth's surface. Because the atmospheric pressure falls off rapidly with altitude, we must define the standard atmosphere in terms of a mercury column at sea level.

When driving up a mountain or riding an express elevator to the top of a tall building, our ears may pop. This popping sensation is caused by an unequal air pressure on the two sides of our eardrums. The popping stops as soon as the higher-pressure, low-altitude air behind the eardrum escapes, reducing the pressure inside the ear to that of the outside lower-pressure, high-altitude air.

Measuring Other Gas Pressures: Manometers

A mercury barometer is fine for measuring the pressure of the atmosphere, but we often need to know the pressure of a gas in a small closed container. One common device for measuring gas pressures is called a **manometer.**

The manometer in Figure 5.3 is a *closed-end* manometer. It is like a barometer that has been bent into a U-shape. When gases are evacuated from the bulb connected to the manometer, the mercury columns in the two arms are at the same level and the gas pressure in the bulb is essentially zero. When the bulb is filled with a gas, the gas pressure causes the mercury column in the closed-end arm to rise. The gas pressure is the *difference* in the levels of the mercury columns in the two arms.

With an *open-end* manometer (Figure 5.4), the gas pressure, P_{gas}, is not zero when the mercury columns are at the same level. Instead, P_{gas} is equal to the prevailing atmospheric pressure, as measured with a barometer and called *barometric pressure*, P_{bar}. When there is a difference in the mercury levels, P_{gas} differs from P_{bar}. In Figure 5.4a, the mercury level is higher in the arm open to the atmosphere, indicating that P_{gas} is *greater* than P_{bar}. In Figure 5.4b, the mercury level is higher in the closed

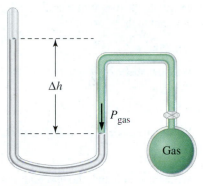

▲ **FIGURE 5.3   Measuring gas pressure with a closed-end manometer**

The gas pressure is equal to the *difference* in height (Δh) of the mercury columns in the two arms of the manometer.

Application Note

For gas pressures that are close to atmospheric pressure, liquids much less dense than mercury are used in manometers. Respiration therapists, for example, measure pressure differences in the unit *centimeters of water.*

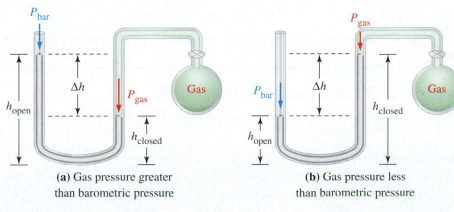

(a) Gas pressure greater than barometric pressure

(b) Gas pressure less than barometric pressure

◀ **FIGURE 5.4   Measuring gas pressure with an open-end manometer**

The measured gas pressure, P_{gas}, is the barometric pressure, P_{bar}, plus the difference in mercury levels between the open and closed arms of the of the manometer (Δh). That is, $P_{gas} = P_{bar} + \Delta h$; where Δh is either positive (a) or negative (b).

QUESTION: For part (b), what is P_{gas} if $\Delta h = 100$ mm and $P_{bar} = 760$ mmHg?

arm, indicating that P_{gas} is *smaller* than P_{bar}. Suppose we define the difference in the mercury levels in the two columns (Δh) as follows:

$$\Delta h = h_{open} - h_{closed}$$

Now, by noting that Δh may be either positive or negative, we see that the gas pressure, in mmHg, is always given by the equation

$$P_{gas} = P_{bar} + \Delta h$$

Let's extend the idea of the pressures exerted by liquid columns. First, we note that force is the product of mass and acceleration. If the acceleration is that due to gravity ($g = 9.807 \text{ m s}^{-2}$), we can write

$$P = \frac{F}{A} = \frac{g\,m}{A}$$

The mass (m) of a liquid column is the product of its density (d) and volume (V). In turn, the volume is equal to the product of the area at the base of the liquid column (A) and the height of the column (h). With these ideas, we can obtain the general expression

$$P = \frac{g\,m}{A} = \frac{g\,d\,V}{A} = \frac{g\,d\,h\,\cancel{A}}{\cancel{A}} = g\,d\,h$$

Example 5.2

Calculate the height of a column of water ($d = 1.00 \text{ g/cm}^3$) that exerts the same pressure as a column of mercury ($d = 13.6 \text{ g/cm}^3$) 760 mm high.

STRATEGY

With the above equation, we can establish the pressures exerted by the columns of mercury and water:

Mercury	Water
$P_{Hg} = g\,d_{Hg}h_{Hg}$	$P_{H_2O} = g\,d_{H_2O}\,h_{H_2O}$

When we equate these two pressures, we will find that there is only one unknown, h_{H_2O}, for which we can solve.

SOLUTION

We start by equating the two liquid pressures.

$$g\,d_{Hg}\,h_{Hg} = g\,d_{H_2O}\,h_{H_2O}$$

Next we cancel the factor g and substitute the known numerical values into the resulting expression.

$$13.6 \text{ g/cm}^3 \times 760 \text{ mm} = 1.00 \text{ g/cm}^3 \times h_{H_2O}$$

Then we solve the expression for h_{H_2O}.

$$h_{H_2O} = \frac{13.6 \text{ g/cm}^3 \times 760 \text{ mm}}{1.00 \text{ g/cm}^3} = 10{,}300 \text{ mm} = 10.3 \text{ m}$$

ASSESSMENT

Because water has a much lower density than mercury (about 1/14), it should take a much taller water column to produce the same pressure as a mercury column, and that is what we found. To measure normal atmospheric pressure with a water-filled barometer, we would have to use one that is more than 10 m tall—as tall as a three-story building!

EXERCISE 5.2A

Calculate the height of a column of carbon tetrachloride, CCl_4 ($d = 1.59 \text{ g/cm}^3$), that exerts the same pressure as a column of mercury ($d = 13.6 \text{ g/cm}^3$) 760 mm high.

EXERCISE 5.2B

A diver reaches a depth of 30.0 m. What is the pressure in atmospheres that is exerted by this depth of water? What is the *total* pressure the diver experiences at this depth? Explain.

Example 5.3   A Conceptual Example

Without doing calculations, arrange the drawings in Figure 5.5 so that the pressures denoted in red are in increasing order.

ANALYSIS AND CONCLUSIONS

The pressure in (a) is expressed as the depth of the liquid mercury, that is, 745 mmHg. The pressure of helium in the open-end manometer (b) is slightly *greater* than P_{bar}, which is

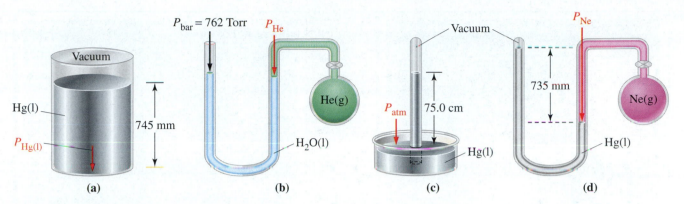

▲ FIGURE 5.5   **Example 5.3 illustrated**

762 Torr = 762 mmHg and, therefore, greater than the pressure in (a). Although there is much less mercury in (c) than in (a), the *pressure* of the mercury column in (c) is greater than in (a) because of its greater height: 75.0 cm = 750 mm. Finally, in the closed-end manometer in (d), the difference in the two mercury levels, 735 mm, is the actual gas pressure.

(d)	<	(a)	<	(c)	<	(b)
P_{Ne}	<	$P_{Hg}(l)$	<	P_{atm}	<	P_{He}
735 mmHg	<	745 mmHg	<	750 mmHg	<	above 762 mmHg

EXERCISE 5.3A

Place a barometric pressure of (e) 101 kPa into the order of increasing pressures established in Example 5.3. If possible, also place a barometric pressure of (f) 103 kPa in the order. If not possible, explain why.

EXERCISE 5.3B

Use principles from this section to explain **(a)** how a beverage is transferred from glass to mouth when one uses a soda straw and **(b)** why an old-fashioned hand-operated pump cannot raise water from a well if the water level is more than about 30 feet below the pump.

5.4   Boyle's Law: The Pressure–Volume Relationship

We use four variables to specify a sample of gas in calculations: its amount in moles (n), its volume (V), its temperature (T), and its pressure (P). These variables are related through some simple gas laws that show how one of the variables (for example, V) changes as a second variable (for example, P) changes and the other two (for example, n and T) remain constant.

The first of these simple gas laws, discovered by Robert Boyle in 1662, concerns the relationship between the pressure and volume of a gas. **Boyle's law** states that

For a fixed amount of gas at a constant temperature, the volume of the gas varies inversely with its pressure.

That is, when the pressure increases, the volume decreases; when the pressure decreases, the volume increases. Conversely, when the volume increases, the pressure decreases; when the volume decreases, the pressure increases.

Let's think of a gas in terms of the kinetic-molecular theory, as suggested in Figure 5.6. Gas molecules bounce off the container walls below the piston with a certain frequency, creating a particular pressure. As the piston rises, the volume of gas and the area of the container walls both increase, while the amount of gas inside the container remains unchanged. This means that molecules strike the container walls less often, and the gas pressure decreases.

Mathematically, for a fixed amount of gas at a constant temperature, we can express Boyle's law as

$$V \propto \frac{1}{P}$$

▲ Experiments on air by Robert Boyle (1627–1691), along with his textbook *The Sceptical Chemist*, helped to establish modern chemistry.

 Boyle's Law simulation

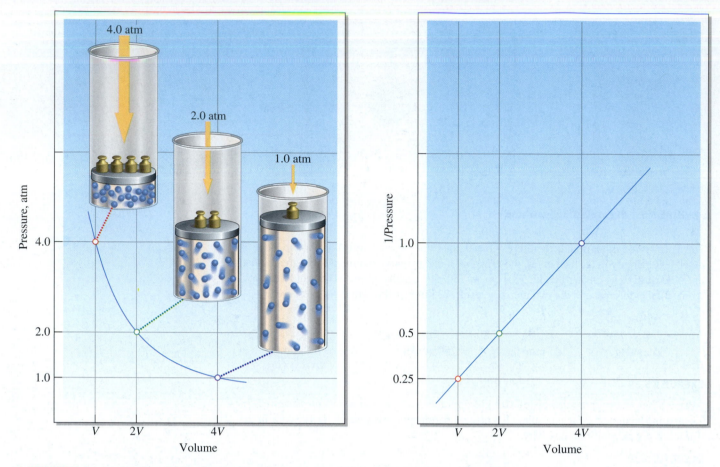

▲ **FIGURE 5.6** **Boyle's law: A kinetic-theory view and a graphical representation**
As the pressure is reduced from 4.00 atm to 2.00 atm and then to 1.00 atm, the volume doubles and then doubles again. (Left) The volume is *inversely* proportional to the pressure, and a plot of P versus V is a curved line (a hyperbola). (Right) By plotting 1/P versus V, we obtain a straight line instead of a curved one.

QUESTION: Estimate the pressure at 3V and at 5V. Which plot is easier to use for making these estimations?

By analogy, the time required for a trip on a freeway is inversely proportional to the speed. The faster you drive, the shorter your driving time is. This doesn't allow you to calculate an actual time, however. In order to make that calculation, you need to change the proportionality to an equation by introducing a proportionality constant—the distance to be driven.

Application Note

Gases are usually stored under high pressure even though they will be used at atmospheric pressure. This allows a large amount of gas to be stored in a small volume.

Pressure-Volume Relationships animation

where the symbol $\propto$ means "is proportional to." However, we cannot use a proportionality expression for quantitative calculations. For such calculations, we need a *proportionality constant* and an equals symbol in the expression. Let's denote the proportionality constant in Boyle's law by the symbol a and then write the equation

$$V = \frac{a}{P}$$

Multiplying both sides of the equation by P, we get

$$P \times V = a \qquad\qquad (5.1)$$

Another way to state Boyle's law, then, is that

For a fixed amount of gas at a constant temperature, the product of the pressure and volume is a constant.

This is an elegant and precise way of summarizing a lot of experimental data. If the product $P \times V$ is to be constant, P must decrease as V increases and vice versa. This relationship is demonstrated in the pressure–volume graphs in Figure 5.6.

Boyle's law has a number of practical applications perhaps best illustrated by some examples. In these applications, both the amount of gas and the temperature must be held constant, and any units can be used for pressure and volume as long as the same units are used throughout a calculation. Also note that in the typical problem, we deal with the same gas under two different conditions, which we call the initial and

Boyle's Law and Breathing

The pressure–volume relationship for gases helps explain the mechanics of breathing. When we breathe in (inspire), the diaphragm is lowered and the chest wall is expanded, increasing the volume of the chest cavity (Figure 5.7a). Boyle's law tells us that the pressure inside the cavity must decrease momentarily. Outside air then enters the lungs because that air is at a higher pressure than the air in the chest cavity. When we breathe out (expire), the diaphragm rises and the chest wall contracts, decreasing the volume of the chest cavity (Figure 5.7b). The pressure is increased, and some air is forced out.

During normal inspiration, the pressure inside the lungs drops about 3 Torr below atmospheric pressure. During expiration, the internal pressure is about 3 Torr above atmospheric pressure. About 0.5 L of air is moved in and out of the lungs in this process, and this normal breathing volume is referred to as the *tidal volume*. The *vital capacity* is the maximum volume of air that can be forced from the lungs and ranges from 3 to 7 L, depending on the individual. A pressure inside the lungs 100 Torr greater than the external pressure is not unusual during such a maximum expiration.

The lungs are never emptied completely, however. The space around the lungs is maintained at a slightly lower pressure than are the lungs themselves, causing the lungs to be kept partially inflated by the higher pressure within them. If a lung, the diaphragm, or the chest wall is punctured, allowing the two pres-

sures to equalize, the lung will collapse. Sometimes a medical doctor will collapse a patient's damaged lung intentionally to give it time to heal. Closing the opening to the lung reinflates it.

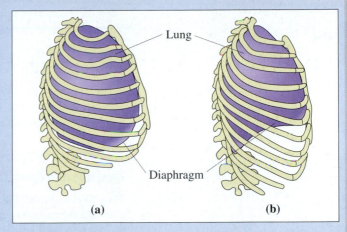

▲ **FIGURE 5.7** **The mechanics of breathing**
(a) Inspiration. The diaphragm is pulled down, and the rib cage is lifted up and out, increasing the volume of the chest cavity. (b) Expiration. The diaphragm is relaxed, the rib cage is down, and the volume of the chest cavity decreases.

final conditions. The pressure–volume product of the gas is a constant under these two conditions, leading to the expression

$$P_{initial} \times V_{initial} = a = P_{final} \times V_{final}$$

or

$$P_{initial} \times V_{initial} = P_{final} \times V_{final} \qquad (5.2)$$

Example 5.4

A helium-filled party balloon has a volume of 4.50 L at sea level, where the atmospheric pressure is 748 Torr. Assuming that the temperature remains constant, what will be the volume of the balloon when it is taken to a mountain resort at an altitude of 2500 m, where the atmospheric pressure is 557 Torr?

STRATEGY

We can solve the mathematical expression of Boyle's law for V_{final} and calculate its value from the information given.

SOLUTION

Using Equation (5.2), we derive the equation for V_{final}:

$$V_{final} = V_{initial} \times \frac{P_{initial}}{P_{final}}$$

Then we can substitute known values for the three variables on the right side of the equation:

$$V_{final} = 4.50 \text{ L} \times \frac{748 \text{ Torr}}{557 \text{ Torr}} = 6.04 \text{ L}$$

ASSESSMENT

Because the final pressure is less than the initial pressure, we expect the final volume to be greater than the initial volume, and it is.

EXERCISE 5.4A

A sample of helium occupies 535 mL at 988 Torr and 25 °C. If the sample is transferred to a 1.05-L flask at the same temperature, what will be the gas pressure in the flask?

EXERCISE 5.4B

A sample of air occupies 73.3 mL at 98.7 kPa and 0 °C. What volume will the air occupy at 4.02 atm and the same temperature?

Example 5.5   An Estimation Example

A gas is enclosed in a cylinder fitted with a piston. The volume of the gas is 2.00 L at 398 Torr. The piston is moved to increase the gas pressure to 5.15 atm. Which of the following is a reasonable value for the volume of the gas at the greater pressure?

0.20 L   0.40 L   1.00 L   16.0 L

ANALYSIS AND CONCLUSIONS

The initial pressure (398 Torr) is about 0.5 atm. An increase in pressure to 5.15 atm is about a tenfold increase. As a result, the volume should drop to about one-tenth of its initial value of 2.00 L. A final volume of 0.20 L is the most reasonable estimate. (The calculated value is 0.203 L.)

EXERCISE 5.5A

A gas is enclosed in a 10.2-L tank at 1208 Torr. Which of the following is a reasonable value for the pressure when the gas is transferred to a 30.0-L tank?

0.40 atm   25 lb/in^2 400 mmHg   3600 Torr

EXERCISE 5.5B

Which of the following is a reasonable estimate of the pressure-volume product of gas in the 30.0-L tank of Exercise 5.5A? (a) 3.6×10^4 Torr L, (b) 4×10^3 mmHg L, (c) 1.6×10^4 kg/m^2, (d) 1.6 kPa m^3

5.5   Charles's Law: The Temperature–Volume Relationship

 **Charles's Law simulation**

In 1787, Jacques Charles (1746–1823) studied the relationship between the volume and the temperature of gases. He found that when a fixed mass of gas is cooled at constant pressure, its volume decreases. When the fixed mass of gas is heated at constant pressure, its volume increases. Temperature and volume are directly proportional; that is, they rise or fall together. However, the relationship between volume and temperature is not as neat as it may seem on first impression. For example, consider the following: If a quantity of gas that occupies 1.00 L is heated from 100 °C to 200 °C at constant pressure, the volume does not double; instead, it increases to only about 1.27 L.

We can learn more about this relationship by considering a plot of volume versus temperature for a gas. As Charles noted, for each Celsius degree rise in temperature, the volume of a gas increases by 1/273 of its volume at 0 °C. For each Celsius degree drop in temperature, the volume of a gas decreases by 1/273 of its value at 0 °C. Figure 5.8 shows that, for a fixed amount of gas at constant pressure, a plot of the volume as a function of temperature gives a straight line. We can *extrapolate* the line beyond the range of measured temperatures to the temperature at which the volume of the gas *appears* to become zero. Before a gas ever reaches this temperature,

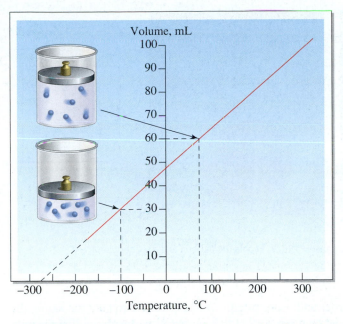
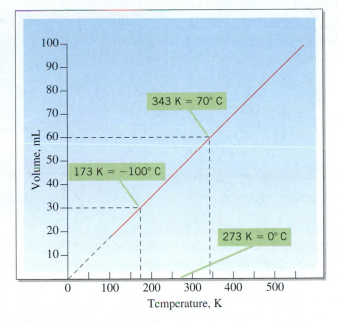

▲ FIGURE 5.8   Charles's law: Gas volume as a function of temperature

The 60-mL volume of gas at about 70 °C drops to 30 mL at about −100 °C. The gas volume continues to decrease linearly as the temperature is lowered. The extrapolated line intersects the temperature axis at −273.15 °C. When the same volume data are plotted versus the Kelvin temperature, the point of zero volume corresponds to 0 K.

QUESTION: How would a plot of volume versus temperature differ from the one shown here if the pressure were increased by a factor of 2?

however, it liquefies, and then the liquid freezes. Therefore this is an exercise for the imagination.

The temperature obtained by extrapolation to zero volume is −273.15 °C. In 1848, William Thomson (Lord Kelvin) made this temperature the **absolute zero** on a temperature scale now called the **Kelvin scale**. On the Kelvin scale, negative temperatures are not possible. The unit of temperature on this scale is the **kelvin (K)**, which is equal to a degree of Celsius temperature. Note that the degree sign is *not* used for the kelvin. Temperatures on the Kelvin scale are 273.15 degrees higher than on the Celsius scale, a fact that we can express through the equation

$$T(\text{K}) = T(^{\circ}\text{C}) + 273.15 \qquad (5.3)$$

A modern statement of **Charles's law** reflects the significance of the Kelvin temperature scale and the absolute zero of temperature on which it is based:

The volume of a fixed amount of a gas at a constant pressure is directly proportional to its Kelvin temperature.

Mathematical statements of Charles's law are of the form

$$V \propto T \quad \text{and} \quad V = bT \quad \text{(where } b \text{ is a constant)} \qquad (5.4)$$

As we did with Boyle's law, we can express Charles's law for initial and final conditions of a sample of gas at constant pressure, leading to the expressions

$$\frac{V_{\text{initial}}}{T_{\text{initial}}} = b = \frac{V_{\text{final}}}{T_{\text{final}}}$$

and

$$\frac{V_{\text{initial}}}{T_{\text{initial}}} = \frac{V_{\text{final}}}{T_{\text{final}}} \qquad (5.5)$$

The kinetic-molecular model readily accounts for the relationship between gas volume and temperature. When we heat a gas, we supply the gas molecules with increased energy, and they begin to move faster. These speedier molecules strike the

Zero pressure or zero volume really means *zero*—no pressure or volume to be measured. The same is true for temperatures *only* if expressed on an absolute (Kelvin) scale. Zero degrees Celsius (0 °C) signifies only an arbitrary value for the freezing point of water.

Problem-Solving Note

In using Equation (5.3), according to the significant figure rule on addition and subtraction (page 15), the two-significant-digit 27 °C, when added to 273, becomes the three-significant-digit 300 K. Similarly, 27.0 °C becomes 300.2 K (that is, 27.0 + 273.15 = 300.15, which rounds to 300.2).

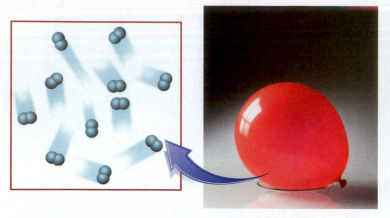

▲ **FIGURE 5.9** **Charles's law: A dramatic illustration**
(a) At room temperature the balloon is fully inflated. (b) Liquid nitrogen (boiling point, −196 °C) cools the balloon and its contents to a temperature far below room temperature, and the balloon collapses.

walls of their container harder and more often. For the pressure to stay the same, the volume of the container must increase so that the increased molecular collisions with the container walls will be distributed over a greater surface area. In this way, the pressure exerted by the faster molecules in the larger volume (high temperature) is the same as that of the slower-moving molecules in the smaller volume (low temperature). Figure 5.9 illustrates the dramatic change in gas volume that occurs with a large drop in temperature.

Example 5.6

A balloon indoors, where the temperature is 27 °C, has a volume of 2.00 L. What will its volume be outdoors, where the temperature is −23 °C? (Assume no change in the gas pressure.)

STRATEGY

We can use Charles's law to determine how the change in temperature affects the volume, but we must consider the temperature change on the absolute (Kelvin) temperature scale.

SOLUTION

First, we convert both temperatures to the Kelvin scale, using Equation (5.3):

$$T_{initial} = 27 + 273 = 300 \text{ K}$$

$$T_{final} = -23 + 273 = 250 \text{ K}$$

Now we apply Charles's law in the form of Equation (5.5) and solve for V_{final}:

$$V_{final} = V_{initial} \times \frac{T_{final}}{T_{initial}} = 2.00 \text{ L} \times \frac{250 \text{ K}}{300 \text{ K}} = 1.67 \text{ L}$$

ASSESSMENT

Because the final temperature is lower than the initial temperature, we expect the final volume to be smaller than the initial volume, and it is.

EXERCISE 5.6A

A sample of hydrogen gas occupies 692 L at 602 °C. If the pressure is held constant, what volume will the gas occupy after cooling to 23 °C?

EXERCISE 5.6B

The balloon described in Example 5.6 must be at a particular temperature in order to have a volume of 2.25 L. Assuming that the pressure remains constant, find this temperature in kelvins and in degrees Celsius.

Example 5.7   An Estimation Example

A sample of nitrogen gas occupies a volume of 2.50 L at $-120\ °C$ and 1.00 atm pressure. To which of the following approximate temperatures should the gas be heated in order to double its volume while maintaining a constant pressure?

$$-240\ °C \qquad -60\ °C \qquad -12\ °C \qquad 30\ °C$$

ANALYSIS AND CONCLUSIONS

The initial temperature is about 150 K $(-120 + 273.15 \approx 150\ K)$. The final temperature must be twice the initial temperature—about 300 K—if the volume is to double. The Celsius temperature corresponding to 300 K is about $30\ °C(30 + 273.15 \approx 300\ K)$. Notice that the four temperatures from which to choose are all multiples or fractions of $-12\ °C$, but this has *nothing* to do with our answer because the relationship between volume and temperature must be based on the Kelvin scale and not the Celsius scale.

EXERCISE 5.7A

If the gas in Example 5.7—initially 2.50 L at $-120\ °C$ and 1.00 atm—is brought to a temperature of $180\ °C$ while a constant pressure is maintained, which of the following is the approximate final volume?

(a) 3.75 L   (b) 5.0 L   (c) 7.5 L   (d) 10.0 L

EXERCISE 5.7B

A sample of gas is initially 1.50 L at $50\ °C$ and 0.60 atm pressure. If this gas is brought to a final temperature of $150\ °C$ while the volume is held constant, which of the following is a close estimate of the final pressure?

(a) 1.8 atm   (b) 0.8 atm   (c) 0.5 atm   (d) 0.2 atm

5.6   Avogadro's Law: The Mole–Volume Relationship

As we will describe in Section 5.9, in 1811 Amedeo Avogadro proposed an important hypothesis to explain some observations about the ratios in which volumes of gases combine during chemical reactions. **Avogadro's hypothesis** states that equal numbers of molecules of different gases at the same temperature and pressure occupy equal volumes. Thus, Avogadro's hypothesis relates an amount of gas (numbers of molecules) and gas volume when temperature and pressure remain constant. We call the simple gas law implied by this relationship **Avogadro's law:**

> At a fixed temperature and pressure, the volume of a gas is directly proportional to the amount of gas (that is, to the number of moles of gas, n, or to the number of molecules of gas).

If we double the number of moles of gas at a fixed temperature and pressure, the volume of the gas doubles. Because the mass of a gas is proportional to the number of moles, doubling the *mass* of a gas also doubles its volume. Mathematically, we can state Avogadro's law as

$$V \propto n \quad \text{or} \quad V = cn \quad \text{(where } c \text{ is a constant)} \tag{5.6}$$

When we use Avogadro's hypothesis to compare different gases, the gases must be at the same temperature and pressure. As we will illustrate next, a convenient temperature/pressure combination for such comparisons is $0\ °C$ (273.15 K) and 1 atm (760 Torr), known as the **standard temperature and pressure (STP).**

Molar Volume of a Gas

Suppose that in comparing different gases at STP, we use Avogadro's number as the number of molecules present. Avogadro's hypothesis states that under these conditions, 1 mol $(6.022 \times 10^{23}$ molecules) of *any* gas occupies the same volume as 1 mol of any other gas. The **molar volume of a gas** is the volume occupied by one mole of the gas. By experiment, the molar volumes at STP of some common gases are 22.428 L of H_2, 22.404 L of N_2, 22.394 L of O_2, and 22.360 L CH_4.

▲ Amedeo Avogadro (1776–1856) did not live to see his ideas accepted by the scientific community. This acceptance finally came in 1860 at an international conference where Stanislao Cannizzaro (1826–1910) effectively put forth Avogadro's ideas from five decades before. See Problem 124.

Avogadro's Law simulation

▶ **FIGURE 5.10** **Molar volume of a gas visualized**
The wooden cube has a volume of 22.4 L, the same volume as 1 mol of gas at STP. The volumes of some familiar objects offer a contrast to the molar volume of a gas.

In Section 5.12, we will discover why these values are not identical; but for the present, to *three* significant figures, we can state that for any gas,

$$1 \text{ mol gas} = 22.4 \text{ L gas (at STP)}$$

Figure 5.10 pictures a volume of 22.4 L and relates it to some familiar objects. The 22.4-L container would hold 2.02 g H_2, 28.0 g N_2, 32.0 g O_2, or 44.0 g CO_2.

Example 5.8

Calculate the volume occupied by 4.11 kg of methane gas, $CH_4(g)$, at STP.

STRATEGY

We must first convert the mass of gas to an amount in moles and then use the molar volume as a conversion factor to get the volume of the gas at STP.

SOLUTION

We can do all this in a single setup, where the conversion factor based on molar volume is shown in red.

$$? \text{ L } CH_4 = 4.11 \text{ kg } CH_4 \times \frac{1000 \text{ g } CH_4}{1 \text{ kg } CH_4} \times \frac{1 \text{ mol } CH_4}{16.04 \text{ g } CH_4} \times \frac{22.4 \text{ L } CH_4}{1 \text{ mol } CH_4}$$

$$= 5.74 \times 10^3 \text{ L } CH_4$$

EXERCISE 5.8A

What is the mass of propane, C_3H_8, in a 50.0-L container of the gas at STP?

EXERCISE 5.8B

Solid carbon dioxide, called dry ice, is useful in maintaining frozen foods because it vaporizes to $CO_2(g)$ rather than melting to a liquid. How many liters of $CO_2(g)$, measured at STP, will be produced by the vaporization of a block of dry ice ($d = 1.56 \text{ g/cm}^3$) that measures 12.0 in. × 12.0 in. × 2.00 in.?

5.7 The Combined Gas Law

From the three simple gas laws,

$$V = \frac{a}{P}, \quad V = bT, \quad \text{and} \quad V = cn$$

it seems reasonable that the volume of a gas (V) should be *directly* proportional to the Kelvin temperature (T) and to the amount of gas (n), and *inversely* proportional to the pressure (P). That is indeed the case:

$$V \propto \frac{nT}{P}$$

Or, expressed as an equation rather than a proportionality,

$$\frac{PV}{nT} = \text{constant}$$

This so-called **combined gas law** is most useful when we want to describe the final conditions for a gas from a knowledge of the initial conditions and the changes to which the gas is subjected. In these cases, we write

Initial *Final*

$$\frac{P_1V_1}{n_1T_1} = \text{constant} = \frac{P_2V_2}{n_2T_2} \quad \text{or simply} \quad \frac{P_1V_1}{n_1T_1} = \frac{P_2V_2}{n_2T_2} \qquad \textbf{(5.7)}$$

The subscript 1 represents the initial condition of the gas with respect to pressure (P_1), volume (V_1), amount (n_1), and temperature (T_1); the subscript 2 represents the final condition. If one or more of the gas properties remain constant during a change from initial to final conditions, we can simplify this expression by canceling these constant terms from the two sides of the equation. To derive Boyle's law from the combined gas law, for instance, we would fix n and T (that is, assume that the final amount and temperature of the gas are identical to the initial amount and temperature). To derive Charles's law, we would fix n and P.

Now, let's look at one more law by considering the case of a fixed amount of gas $(n_1 = n_2)$ in a fixed volume $(V_1 = V_2)$. Under these conditions, Equation (5.7)

$$\frac{P_1 \cancel{V_1}}{\cancel{n_1} T_1} = \frac{P_2 \cancel{V_2}}{\cancel{n_2} T_2}$$

simplifies to

$$\frac{P_1}{T_1} = \frac{P_2}{T_2} \qquad \textbf{(5.8)}$$

This equation shows that the pressure of a fixed amount of gas in a constant volume is proportional to its Kelvin temperature. This relationship, a simple gas law sometimes called *Amontons's law,* certainly seems reasonable from the standpoint of the kinetic-molecular theory, as is illustrated in Figure 5.11.

 **Pressure-Temperature Relationship simulation**

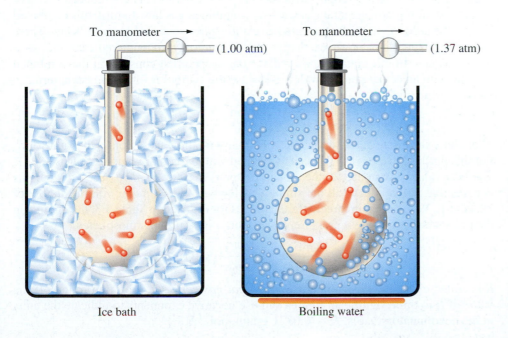

To manometer ⟶ (1.00 atm)

To manometer ⟶ (1.37 atm)

Ice bath

Boiling water

◀ **FIGURE 5.11   The effect of temperature on the pressure of a fixed amount of gas in a constant volume: A kinetic-theory interpretation**

The amount of gas and its volume are the same in both flasks, but if the gas in the ice bath (0 °C) exerts a pressure of 1.00 atm, the gas in the boiling-water bath (100 °C) exerts a pressure of 1.37 atm. Both the frequency and the force of the molecular collisions with the container walls are greater at the higher temperature.

QUESTION: What pressure would be exerted if the vessel were transferred to an oil bath at 200 °C?

Example 5.9

The flasks pictured in Figure 5.11 contain $O_2(g)$, the one on the left at STP and the one on the right at 100 °C. What is the pressure at 100 °C?

STRATEGY

We start with Equation (5.8),

$$\frac{P_1}{T_1} = \frac{P_2}{T_2}$$

where the subscript 1 represents the initial condition (STP) and the subscript 2 represents the final condition. Then we solve for the final pressure, P_2.

SOLUTION

$$P_2 = P_1 \times \frac{T_2}{T_1} = 1.00 \text{ atm} \times \frac{(100 + 273) \text{ K}}{273 \text{ K}} = 1.37 \text{ atm}$$

ASSESSMENT

Because the final temperature is higher than the initial temperature, we expect the final pressure to be greater than the initial pressure; and it is. Note that, although we need to know that the amount and volume of $O_2(g)$ remain constant, we do not need to know the value of either. Also, we do not need to know that the gas is O_2; it could be some other gas.

EXERCISE 5.9A

Aerosol containers often carry the warning that they should not be heated. Suppose an aerosol container was filled with a gas at 2.5 atm and 22 °C. If there is the possibility that the container may rupture if the pressure exceeds 8.0 atm, at what temperature is rupture likely to occur?

EXERCISE 5.9B

In the manner we used to derive Amontons's law relating gas pressure and temperature, derive an expression relating the pressure and amount of gas when both the temperature and volume remain constant. Rationalize your result in terms of the kinetic-molecular theory and a sketch in the manner of Figure 5.11.

5.8   The Ideal Gas Law and Its Applications

We have seen how three simple gas laws—Boyle's, Charles's, and Avogadro's—serve as the basis for a more general gas law—the combined gas law. In turn, this combined law can be used to establish other simple relationships, such as Amontons's law. There is another expression, however, that is more commonly encountered than the combined gas law. In this equation, we replace the unspecified constant in the combined gas law with an actual numeric value. Let's use the symbol R for the constant term:

$$\frac{PV}{nT} = \text{constant} = R$$

Then, let's substitute data into the equation that will allow us to obtain the value of R. For this purpose, we can use the data associated with the molar volume of a gas at STP. The best value we can use for the molar volume is that for an *ideal gas*. An **ideal gas** is a gas that *strictly* obeys all the simple gas laws and has a molar volume at STP of 22.4141 L. We can obtain three different values for R, listed in Table 5.3, depending on the units we use to express pressure. If we choose pressure in atmospheres,

$$R = \frac{PV}{nT} = \frac{1 \text{ atm} \times 22.4141 \text{ L}}{1 \text{ mol} \times 273.15 \text{ K}} = 0.082058 \frac{\text{L atm}}{\text{mol K}}$$

The gas constant, R, has widespread use in chemistry. We will use it often in later chapters, usually in the form 8.3145 J mol^{-1} K^{-1}, where J represents joules, an energy unit.

R is called either the **gas constant** or the *ideal gas constant*. We often round off the value of R to three significant figures and use negative exponents to indicate the units in the denominator, that is, $R = 0.0821$ L atm mol^{-1} K^{-1}.

Table 5.3 Units for the Gas Constant, *R*	
R has the value	When
0.082058 L atm mol^{-1} K^{-1}	*P* is in atm
62.364 L Torr mol^{-1} K^{-1}	*P* is in torr
8.3145 J mol^{-1} K^{-1}	*P* is in Pa; *V* is in m^3

The equation, called either the **ideal gas law** or the **ideal gas equation,** is generally written in the form

$$PV = nRT \qquad (5.9)$$

An ideal gas is a *hypothetical* gas. *Real* gases (Table 5.1) approximate ideal gas behavior under many conditions, so that we can describe real gases fairly well using the ideal gas law. In Section 5.12, we will examine the conditions (generally, high pressures and low temperatures) under which real gases depart most seriously from ideal gas behavior. For now, we will simply avoid such conditions.

In a typical calculation using the ideal gas equation, we will determine any one of the four quantities *P*, *V*, *n*, or *T* from known values of the other three. We begin each calculation in the following examples by solving the ideal gas equation for the unknown variable. We also make certain that we express quantities in the units needed to match the units of *R*, for example: L (volume), atm (pressure), mol (amount of gas), and K (temperature) to match $R = 0.0821$ L atm mol^{-1} K^{-1}.

 **Ideal Gas Behavior simulation**

$$V = \frac{nRT}{P}$$

$$n = \frac{PV}{RT}$$

$$T = \frac{PV}{nR}$$

Example 5.10

What is the pressure exerted by 0.508 mol O_2 in a 15.0-L container at 303 K?

STRATEGY

Knowing *n*, *V*, and *T* of a gas, we can use the ideal gas law to solve for its pressure.

SOLUTION

First, we solve the ideal gas law, Equation (5.9), for pressure, *P*, by dividing both sides of the equation by the volume, *V*, and then canceling *V* on the left side of the equation.

$$\frac{P\cancel{V}}{\cancel{V}} = \frac{nRT}{V}$$

$$P = \frac{nRT}{V}$$

Because data are given in the units mol, L, and K, we can use $R = 0.0821$ L atm mol^{-1} K^{-1} and substitute the data directly into the equation.

$$P = \frac{nRT}{V} = \frac{0.508 \text{ mol} \times 0.0821 \text{ L atm} \times 303 \text{ K}}{15.0 \text{ L mol K}} = 0.842 \text{ atm}$$

Our result will be a pressure in atmospheres.

ASSESSMENT

An important check is for the proper cancellation of units, leading to an answer in atm. Also, compare the O_2(g) with one mole of gas at STP. The 0.508 mol O_2(g), if confined to 22.4 L at 273 K, would exert a pressure of about 0.5 atm. Because the gas is confined to 15.0 L, its pressure should exceed 0.5 atm, but still be less than 1 atm. The pressure increase on raising the temperature to 303 K is small (increasing by 303/273). The answer (0.842 atm) is reasonable.

EXERCISE 5.10A

How many moles of nitrogen gas, N_2, are there in a sample that occupies 35.0 L at a pressure of 3.15 atm and a temperature of 852 K?

EXERCISE 5.10B

If volume and temperature are held constant, how many moles of N_2(g) should be added to the gas described in Exercise 5.10A to increase the pressure to 5.00 atm?

Example 5.11

What is the volume occupied by 16.0 g ethane gas (C_2H_6) at 720 Torr and 18 °C?

STRATEGY

We will need to solve the ideal gas equation for volume, V, and use an appropriate unit for each variable in the equation.

SOLUTION

We can use the units that are consistent with the value $R = 0.0821$ L atm mol^{-1} K^{-1}, as shown here.

$$n = 16.0 \text{ g } C_2H_6 \times \frac{1 \text{ mol } C_2H_6}{30.07 \text{ g } C_2H_6} = 0.532 \text{ mol } C_2H_6$$

$$T = (273 + 18) \text{ K} = 291 \text{ K}$$

$$P = 720 \text{ Torr} \times \frac{1 \text{ atm}}{760 \text{ Torr}} = 0.947 \text{ atm}$$

Finally, we can substitute the converted data for the variables n, T, and P into the ideal gas equation, which we have solved for V.

$$V = \frac{nRT}{P} = \frac{0.532 \text{ mol} \times 0.0821 \text{ L atm} \times 291 \text{ K}}{0.947 \text{ atm mol K}} = 13.4 \text{ L}$$

ASSESSMENT

For 0.5 mol C_2H_6 at STP, the volume would be 11.2 L. The actual amount of C_2H_6 is slightly more than 0.5 mol; the temperature is somewhat higher than 273 K; and the pressure (720 Torr) is somewhat less than 1 atm. All of these factors would make the answer somewhat larger than 11.2 L, and we find that it is (13.4 L).

EXERCISE 5.11A

What is the temperature, in degrees Celsius, at which 15.0 g O_2 exerts a pressure of 785 Torr in a volume of 5.00 L?

EXERCISE 5.11B

How many grams of $N_2(g)$, at 25.0 °C and 734 Torr, will occupy the same volume as 25.0 g $O_2(g)$ at 30.0 °C and 755 Torr?

Molecular Mass Determination

Suppose we have a *fixed quantity* of gas in a known volume at a known temperature and pressure. By the methods used in Examples 5.10 and 5.11, we can readily calculate the number of moles of the gas, n:

$$n = \frac{PV}{RT}$$

Suppose we also measure the mass, m, of this *fixed quantity* of gas. We now know the quantity of a sample of substance both in grams and in moles. Recall that in Chapter 3 we often had to convert from moles to grams of a substance, using molar mass, M, as a conversion factor. That is,

Mass in grams = number of moles × molar mass in g/mol

or

$$m = n \times M$$

When we know both the mass and the number of moles in a specific sample of a substance, we can calculate the molar mass of the substance:

$$M = \frac{m}{n}$$

We can use this relationship, together with ideal gas law measurements, to obtain unknown molecular masses, as shown in Example 5.12. The method works for any gas under conditions where it behaves as an ideal gas. It usually also works well for the vapor of an easily vaporized liquid.

Example 5.12

If 0.550 g of a gas occupies 0.200 L at 0.968 atm and 289 K, what is the molecular mass of the gas?

STRATEGY

Here, the mass (m) of gas is given, and in an approach similar to that used in Example 5.11, we will solve the ideal gas equation for the number of moles (n). The molar mass is $M = m/n$, and the molecular mass (in atomic mass units) is numerically equal to the molar mass (in g/mol).

SOLUTION

First, let's calculate the amount of gas, in moles, from the ideal gas equation.

$$n = \frac{PV}{RT} = \frac{0.968 \text{ atm} \times 0.200 \text{ L}}{0.0821 \text{ L atm mol}^{-1} \text{ K}^{-1} \times 289 \text{ K}} = 0.00816 \text{ mol}$$

Now we can use the known mass and the calculated number of moles to determine the molar mass.

$$\text{Molar mass} = \frac{0.550 \text{ g}}{0.00816 \text{ mol}} = 67.4 \text{ g/mol}$$

The molar mass of the gas is 67.4 g/mol, and the molecular mass is therefore 67.4 u.

ASSESSMENT

Our answers have the correct units for molar mass and molecular mass. Moreover, the values seem to be in the correct range. The molecular mass of a gaseous substance at about STP cannot be less than 1 u (that of H atoms), nor is it likely to be very large (say, greater than 100–200 u). At about STP, substances that have high molecular masses are most likely to be liquids or solids (as discussed in Chapter 11).

EXERCISE 5.12A

Calculate the molar mass of a gas if 0.440 g occupies 179 mL at 741 mmHg and 86 °C.

EXERCISE 5.12B

If 8.0 g $CH_4(g)$ occupies the same volume at 0 °C as 8.0 g $O_2(g)$ at STP, what is the pressure of the $CH_4(g)$?

An alternative approach to the one used in Example 5.12 is to combine the two equations

$$n = \frac{PV}{RT} \quad \text{and} \quad M = \frac{m}{n}$$

into a variation of the ideal gas equation:

$$n = \frac{m}{M} = \frac{PV}{RT} \qquad mRT = MPV$$

$$M = \frac{mRT}{PV} \qquad\qquad\qquad \textbf{(5.10)}$$

We illustrate this alternative approach in Example 5.13.

In Chapter 3, we learned how to establish the empirical formula of a compound from its experimentally determined mass percent composition. And if we also knew the molecular mass of the compound, we were able to determine its true molecular formula. We can combine these earlier ideas with what we have learned here to determine the molecular formula of a gaseous compound.

Example 5.13

Calculate the molecular mass of a liquid that, when vaporized at 100 °C and 755 Torr, yields 185 mL of vapor that has a mass of 0.523 g.

STRATEGY

We can use Equation (5.10) to calculate molar mass from our knowledge of m, T, P, and V. Then we can establish molecular mass from molar mass.

SOLUTION

We must convert from mL to L and from °C to K, but we can leave the pressure in Torr and use the corresponding value of R from Table 5.3, that is, 62.364 L Torr mol^{-1} K^{-1}.

$$V = 185 \text{ mL} \times \frac{1 \text{ L}}{1000 \text{ mL}} = 0.185 \text{ L}$$

$$T = (273 + 100)\text{K} = 373 \text{ K}$$

Now we can substitute these data into Equation (5.10).

$$M = \frac{mRT}{PV} = \frac{0.523 \text{ g} \times 62.364 \text{ L Torr mol}^{-1} \text{ K}^{-1} \times 373 \text{ K}}{755 \text{ Torr} \times 0.185 \text{ L}}$$

$$= 87.1 \text{ g mol}^{-1}$$

The molar mass is 87.1 g/mol, and the molecular mass is 87.1 u.

EXERCISE 5.13A

Calculate the molecular mass of a liquid that, when vaporized at 98 °C and 715 mmHg, yields 121 mL of vapor having a mass of 0.471 g.

EXERCISE 5.13B

Diacetyl, a substance that contributes to the characteristic flavor and aroma of butter, consists of 55.80% C, 7.03% H, and 37.17% O by mass. In the gaseous state at 100 °C and 747 mmHg, a 0.3060-g sample of diacetyl occupies a volume of 111 mL. Establish the molecular formula of diacetyl.

Gas Densities

Gases are much less dense than liquids and solids, and we generally express their densities in grams per liter rather than grams per milliliter. To establish the density of $O_2(g)$ at STP, we simply divide the molar mass of O_2 by its molar volume at STP:

$$d = \frac{m}{V} = \frac{32.00 \text{ g O}_2}{22.4 \text{ L}} = 1.43 \text{ g O}_2/\text{L}$$

However, the density of $O_2(g)$ is different from this value under conditions other than STP. Much more so than the densities of liquids and solids, the densities of gases are critically dependent on both temperature and pressure because the volume of a fixed mass of gas depends so strongly on temperature and pressure. We can best see how the different variables affect gas density by another rearrangement of the ideal gas equation. First, we substitute $n = m/M$ to obtain

$$PV = \frac{mRT}{M}$$

which rearranges to

$$MPV = mRT$$

We solve this equation for m,

$$m = \frac{MPV}{RT}$$

and divide both sides by V to get

$$\frac{m}{V} = \frac{MP}{RT}$$

Problem-Solving Note

It is not really necessary to use five figures for the value of R in Example 5.13, though we do want to include more significant figures than our data have, in order to minimize round-off error.

Connections: Balloons, the Gas Laws, and Chemistry

Early knowledge of the behavior of gases was stimulated by the achievement of human flight using balloons. In June 1782, two brothers, Joseph Michel and Jacques Étienne Montgolfier, launched the first such balloon by lighting a fire under an opening in a large bag. The less-dense hot air in the bag allowed the balloon to rise slowly through the denser, cooler atmosphere. By August of that year, Jacques Charles filled a balloon with hydrogen, a gas discovered 16 years earlier by Henry Cavendish. Charles made hydrogen on a scale never before attempted, using the reaction of about 500 kg of iron with acid:

$$Fe(s) + 2 HCl(aq) \longrightarrow FeCl_2(aq) + H_2(g)$$

When they are compared at the same temperature and pressure, hydrogen is only one-fourteenth as dense as air. Therefore each kilogram of hydrogen can carry aloft a payload of 13 kg. In December 1782, Charles and a companion flew for 25 km in a balloon filled with hydrogen. They landed in a small village where they were attacked by frightened farmers who tore the balloon with pitchforks.

By 1804, Joseph Gay-Lussac had ascended to an altitude of 7 km and brought back samples of the rarefied air at that altitude. Within just a few years, scientists had used balloons to learn a great deal about the nature of gases, about Earth's atmosphere, and about weather.

▲ A contemporary drawing depicting the first ascent of the Montgolfier brothers in a hot-air balloon in 1782.

Then, substituting this value of m/V into the equation for gas density, $d = m/V$, we get

$$d = \frac{m}{V} = \frac{MP}{RT} \qquad (5.11)$$

From this equation, we see that the density of a gas is *directly* proportional to its molar mass (M) and pressure (P) and *inversely* proportional to its Kelvin temperature (T). In this case, $1/R$ is the proportionality constant needed for the equation to be true.

Example 5.14

Calculate the density of methane gas, CH_4, in grams per liter at 25 °C and 0.978 atm.

STRATEGY

We could do this as a two-step problem in which we (1) determine the number of moles of CH_4 in 1.00 L of the gas at the stated temperature and pressure and then (2) determine the mass of this gas. To do so, however, is really the same as solving Equation (5.11). So let's do it the (simpler) latter way.

SOLUTION

$$d = \frac{MP}{RT} = \frac{16.04 \text{ g } CH_4 \text{ mol}^{-1} \times 0.978 \text{ atm}}{0.0821 \text{ L atm mol}^{-1} \text{ K}^{-1} \times 298 \text{ K}}$$

$$= 0.641 \text{ g } CH_4/L$$

EXERCISE 5.14A

Calculate the density of ethane gas (C_2H_6) in grams per liter at 15 °C and 748 Torr.

EXERCISE 5.14B

What is the molar mass of a gaseous alkane hydrocarbon having a density of 2.42 g/L at 20.0 °C and 762 Torr? What is the molecular formula of the gas?

Example 5.15

Under what pressure must $O_2(g)$ be maintained at 25 °C to have a density of 1.50 g/L?

STRATEGY

We can use Equation (5.11), $d = MP/RT$, to solve for any one of the five quantities (here, P) if the other four are known.

SOLUTION

First, we solve Equation (5.11) for the unknown pressure:

$$P = \frac{dRT}{M}$$

Then, we substitute the known quantities into the right side:

$$P = \frac{1.50 \text{ g L}^{-1} \times 0.0821 \text{ L atm mol}^{-1} \text{ K}^{-1} \times 298 \text{ K}}{32.00 \text{ g mol}^{-1}}$$

$$= 1.15 \text{ atm}$$

ASSESSMENT

To ensure that we have used a correct variation of the ideal gas equation, we can use unit cancellation as a check. Here, all units cancel except for the desired atm.

EXERCISE 5.15A

To what temperature must propane gas (C_3H_8) at 785 Torr be heated to have a density of 1.51 g/L?

EXERCISE 5.15B

At what temperature will the density of a sample of $O_2(g)$ at 725 Torr be the same as the density of $NH_3(g)$ at 22.5 °C and 1.45 atm?

5.9   Gases in Reaction Stoichiometry

With what we have learned about gases, we can greatly expand our scope of stoichiometry calculations.

The Law of Combining Volumes

In 1809, Joseph Gay-Lussac published an experimental result known as the **law of combining volumes:**

> *In a reaction involving gases, the volumes of gaseous reactants and products are in small whole-number ratios as long as they are measured at the same temperature and pressure.*

For example, two volumes of hydrogen unite with one volume of oxygen to produce two volumes of steam (water vapor) when the volumes of these gases are measured at the same temperature and pressure. Thus, the combining ratio for the volumes is 2 : 1 : 2.

One hypothesis that was successful in explaining a few of Gay-Lussac's results was that equal volumes of different gases at the same temperature and pressure contained equal numbers of atoms. John Dalton rejected this idea, however. His view was that the volume ratios should have been 1 : 1 : 1, corresponding to the reaction:

$$H(g) + O(g) \longrightarrow OH(g) \quad \textit{(incorrect)}$$

Even if he had accepted the formula H_2O for water, the two volumes of hydrogen would have reacted with one volume of oxygen to produce only *one* volume of steam.

$$2 H(g) + O(g) \longrightarrow H_2O(g) \quad \textit{(incorrect)}$$

Avogadro gave the correct explanation of Gay-Lussac's law in 1811. He accepted the equal volumes-equal numbers hypothesis but suggested that the equal numbers were not of atoms but of *molecules*—in other words, Avogadro proposed that gases

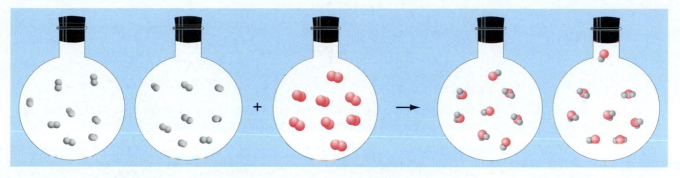

▲ **FIGURE 5.12 Avogadro's explanation of Gay-Lussac's law of combining volumes**
When all the gas volumes are measured at the same temperature and pressure, each of the identical flasks contains the same number of molecules. Notice how the combining ratio: 2 volumes H_2 to 1 volume O_2 to 2 volumes H_2O leads to a result in which all the *atoms* present initially are accounted for in the product.

may exist in *molecular* form. By postulating the existence of the molecules H_2 and O_2, he explained the reaction in the same way we do today. *Two molecules* of H_2 react with *one molecule* of O_2 to form *two molecules* of water. The combining ratios determined from the number of molecules, the number of moles, and the volumes are all $2:1:2$.

$$2 H_2(g) + O_2(g) \longrightarrow 2 H_2O(g)$$

Figure 5.12 suggests Avogadro's line of reasoning.

In applying the law of combining volumes in Example 5.16, we will use the following ideas:

- The law of combining volumes relates the volumes of any two *gaseous* species in a reaction, even if some reactants or products are liquid or solid.

- A reaction need not be carried out at the particular temperature and/or pressure at which the gaseous species are compared.

- We don't even need to know the actual temperature and pressure at which the volumes of the individual gases are measured, as long as all are at the *same* temperature and the *same* pressure.

Example 5.16

How many liters of $O_2(g)$ are consumed for every 10.0 L of $CO_2(g)$ produced in the combustion of liquid pentane, C_5H_{12}, if all volumes are measured at STP?

STRATEGY

The fact that pentane and water are liquids does not affect our calculation because the substances of interest, CO_2 and O_2, are both gases. The actual temperatures and pressures before, during, and after the combustion reaction are not relevant, as long as each gas volume is measured at the same temperature and pressure. The condition of STP is relevant only in establishing the particular temperature and pressure at which the two gas volumes are compared.

SOLUTION

We begin by writing a balanced equation for the combustion reaction:

$$C_5H_{12}(l) + 8 O_2(g) \longrightarrow 5 CO_2(g) + 6 H_2O(l)$$

The fundamental equivalence in the reaction is

$$8 \text{ mol } O_2 \backsim 5 \text{ mol } CO_2$$

Based on the equal volume–equal numbers hypothesis, we can substitute a volume unit for the unit mol:

$$8 \text{ L } O_2 \backsim 5 \text{ L } CO_2$$

This equivalence provides the stoichiometric factor we need. Thus,

$$? \text{ L O}_2(g) = 10.0 \text{ L CO}_2(g) \times \frac{8 \text{ L O}_2(g)}{5 \text{ L CO}_2(g)} = 16.0 \text{ L O}_2(g)$$

EXERCISE 5.16A

What volume of oxygen gas is required to burn 0.556 L of propane gas, C_3H_8, if both gas volumes are measured at the same temperature and the same pressure?

EXERCISE 5.16B

What volume of $O_2(g)$ is consumed in the combustion of 125 g of gaseous dimethyl ether, $(CH_3)_2O$, if both gas volumes are measured at the temperature and pressure at which the density of dimethyl ether is 1.81 g/L?

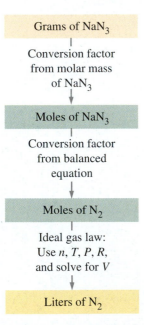
Ideal Gas Law and Chemical Reactions activity

The Ideal Gas Equation in Reaction Stoichiometry

We can use the law of combining volumes *only* for gases and *only* if the gases are at the same temperature and pressure. Often the amount of a gaseous reactant or product needs to be related to that of a solid or liquid, however. In these cases, we must first work with mole ratios to relate moles of solid or liquid to moles of the gaseous species and then turn to the ideal gas equation to relate the moles of gas to other gas properties, such as volume and pressure.

Example 5.17

In the chemical reaction used in automotive air-bag safety systems, $N_2(g)$ is produced by the decomposition of sodium azide, NaN_3, at a somewhat elevated temperature:

$$2 \text{ NaN}_3(s) \longrightarrow 2 \text{ Na}(l) + 3 \text{ N}_2(g)$$

What volume of $N_2(g)$, measured at 25 °C and 0.980 atm, is produced by the decomposition of 62.5 g NaN_3?

STRATEGY

For this reaction, we must relate the number of moles of $N_2(g)$ to the number of moles of $NaN_3(s)$, using their molar ratio. Then, using the ideal gas equation, we can determine the volume of nitrogen from the number of moles of nitrogen.

SOLUTION

First, we can determine the amount of $N_2(g)$ produced when 62.5 g NaN_3 decomposes. Here we use the molar mass (1 mol NaN_3/65.01 g NaN_3) and a stoichiometric factor from the balanced equation (3 mol N_2/2 mol NaN_3) as conversion factors:

$$? \text{ mol N}_2 = 62.5 \text{ g NaN}_3 \times \frac{1 \text{ mol NaN}_3}{65.01 \text{ g NaN}_3} \times \frac{3 \text{ mol N}_2}{2 \text{ mol NaN}_3} = 1.44 \text{ mol N}_2$$

Then, we can use the ideal gas equation to determine the volume of 1.44 mol $N_2(g)$ under the stated conditions:

$$V = \frac{nRT}{P} = \frac{1.44 \text{ mol} \times 0.0821 \text{ L atm mol}^{-1} \text{ K}^{-1} \times (273 + 25) \text{ K}}{0.980 \text{ atm}}$$

$$= 35.9 \text{ L N}_2(g)$$

The solution is also outlined in the stoichiometry diagram of Figure 5.13.

ASSESSMENT

We can establish an approximate answer rather easily. Because the quantity of NaN_3 is about 1 mol, the number of moles of N_2 should be about equal to the stoichiometric factor, that is, $3/2 = 1.5$. Because the gas conditions are not far removed from STP, the 1.5 mol of gas should occupy a volume of about $1.5 \times 22.4 \text{ L} = 33.6 \text{ L}$. This is close to our calculated value of 35.9 L.

▲ **FIGURE 5.13   Stoichiometry diagram for Example 5.17**

Note the similarity to other stoichiometry diagrams such as Figure 3.9.

Problem-Solving Note

Although we have solved this problem in two steps, there is no need to reenter the intermediate result (1.44 mol N_2) into your calculator. The number displayed at the end of the first step is the first entry of the next step, resulting in 36.0 L $N_2(g)$.

EXERCISE 5.17A

Quicklime (CaO), used in the construction industry, is manufactured by heating limestone ($CaCO_3$) to decompose it:

$$CaCO_3(s) \xrightarrow{\Delta} CaO(s) + CO_2(g)$$

How many liters of $CO_2(g)$ at 825 °C and 754 Torr are produced in the decomposition of 45.8 kg $CaCO_3(s)$?

EXERCISE 5.17B

How many liters of cyclopentane, C_5H_{10} ($d = 0.7445$ g/mL), must be burned in excess $O_2(g)$ to produce 1.00×10^6 L of $CO_2(g)$ measured at 25.0 °C and 736 Torr?

5.10   Mixtures of Gases: Dalton's Law of Partial Pressures

Early experimenters did not make a strong distinction between air and other gases. Carbon dioxide was first called "fixed air"; oxygen, "dephlogisticated air"; and hydrogen, "flammable air." We now know that air is a *mixture* of gases. Early experimenters succeeded in much of the research leading to the ideal gas equation only because the equation applies to mixtures of gases just as it does to individual gases. Dalton was among the first to consider the properties of gaseous mixtures.

Partial Pressures and Mole Fractions

Although best known for his atomic theory, Dalton had wide-ranging interests. His interest in meteorology led him to perform experiments on atmospheric gases in an attempt to understand weather. In fact, he formulated his atomic theory, in part, to explain the results of those meteorological experiments. In one experiment, he found that if he added water vapor at a certain pressure to dry air, the pressure exerted by the air increased by an amount equal to the pressure of the water vapor. Based on this and other experiments, Dalton concluded that each gas in a mixture behaves independently of the other gases. **Dalton's law of partial pressures** states that in a mixture of gases, each gas expands to fill the container and exerts its own pressure; this individual gas pressure is called a **partial pressure.** Moreover,

> *The total pressure exerted by a mixture of gases is equal to the sum of the partial pressures exerted by the separate gases.*

The meaning of this statement at both the macroscopic and microscopic levels is suggested pictorially in Figure 5.14. The 5.0-L gas samples shown separately in Figures 5.14a and 5.14b are forced into a single 5.0-L volume, forming the mixture

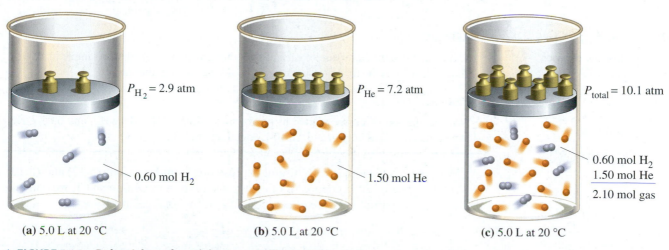

(a) 5.0 L at 20 °C   (b) 5.0 L at 20 °C   (c) 5.0 L at 20 °C

▲ **FIGURE 5.14   Dalton's law of partial pressures illustrated**
Each gas expands to fill the container and exerts a pressure that is readily calculated using the ideal gas equation. The total pressure of the mixture of gases is equal to the sum of the partial pressures of the individual gases.
QUESTION: What would be the total pressure if 0.90 mol of N_2 were added to the gas mixture in (c)?

of gases shown in Figure 5.14c. The number of moles of gas in the mixture is the sum of the number of moles of each gas, and the total pressure in the mixture (10.1 atm) is the sum of the individual gas pressures. Thus, we can state Dalton's law mathematically as

$$P_{\text{total}} = P_1 + P_2 + P_3 + \cdots \tag{5.12}$$

The terms on the right side of this equation refer to the partial pressures of gases 1, 2, 3, ... (The three dots signify that there may be more gases in the mixture than the three noted.) Applying Equation 5.12 to the mixture of gases in Figure 5.14, we find

$$P_{\text{total}} = P_{\text{H}_2} + P_{\text{He}} = 2.9 \text{ atm} + 7.2 \text{ atm} = 10.1 \text{ atm}$$

We can obtain the individual partial pressures by assuming that each gas can be described by the ideal gas law. For gas 1, of which there are n_1 moles, gas 2, of which there are n_2 moles, and so on, we have

$$P_1 = \frac{n_1 RT}{V}; \quad P_2 = \frac{n_2 RT}{V}; \text{ and so on.} \tag{5.13}$$

Example 5.18 illustrates the idea that a component (N_2) of a mixture of gases (air) expands to fill its container and exerts a distinctive partial pressure.

Example 5.18

A 1.00-L sample of dry air at 25 °C contains 0.0319 mol N_2, 0.00856 mol O_2, 0.000381 mol Ar, and 0.00002 mol CO_2. Calculate the partial pressure of $N_2(g)$ in the mixture.

STRATEGY

We are given a rather complete description of the composition of air, but since each component gas expands to fill the 1.00-L volume, the basic question becomes, "What is the pressure exerted by 0.0319 mol N_2 in a 1.00-L container at 25 °C?"

SOLUTION

We apply Equation 5.13 for $N_2(g)$ in the dry air mixture,

$$P_{N_2} = \frac{n_{N_2} RT}{V}$$

$$= \frac{0.0319 \text{ mol} \times 0.0821 \text{ L atm mol}^{-1} \text{ K}^{-1} \times 298 \text{ K}}{1.00 \text{ L}} = 0.780 \text{ atm}$$

Problem-Solving Note

Of course, for a mixture of gases, $P_{\text{total}} \times V = n_{\text{total}} \times RT$.

EXERCISE 5.18A

Calculate the partial pressure of each of the other components of the air sample in Example 5.18. What is the total pressure exerted by all the gases in the sample?

EXERCISE 5.18B

What is the total pressure exerted by a mixture of 4.05 g N_2, 3.15 g H_2, and 6.05 g He when the mixture is confined to a 6.10-L container at 25 °C?

We can obtain another set of useful expressions from Equations (5.12) and (5.13). Designating one of the components of a gaseous mixture as component 1, we can take the *ratio* of the partial pressure of component 1 to the total gas pressure:

$$\frac{P_1}{P_{\text{total}}} = \frac{P_1}{P_1 + P_2 + \cdots}$$

We can then write corresponding ratios for the other components: 2, 3, and so on. When we apply the ideal gas equation to each partial pressure, we find that the ratio of the partial pressure of any component in a gas mixture to the total gas pressure is equal to the ratio of the number of moles of that gas to the total number of moles of gas in the mixture:

$$\frac{P_1}{P_{total}} = \frac{\dfrac{n_1 RT}{V}}{\dfrac{n_1 RT}{V} + \dfrac{n_2 RT}{V} + \cdots} = \frac{n_1}{n_1 + n_2 + \cdots} = \frac{n_1}{n_{total}}$$

The ratio n_1/n_{total} is the *mole fraction* of component 1 in the mixture and is represented by x_1. The **mole fraction** (x_i) of any component (i) of a mixture is the fraction of all the molecules in the mixture that are of type (i):

$$x_i = \frac{n_i}{n_{total}} \tag{5.14}$$

The sum of the mole fractions of all the components in a mixture is 1.

$$x_1 + x_2 + x_3 + \cdots = 1$$

As with other fractional parts of the whole, we can multiply mole fractions by 100% to obtain *mole percents*.

Expressions of the type

$$\frac{P_1}{P_{total}} = \frac{n_1}{n_{total}} = x_1 \tag{5.15}$$

have a special significance: They allow us to relate the partial pressures of the components of a gaseous mixture to the total gas pressure:

$$P_1 = x_1 P_{total} \tag{5.16}$$

The compositions of gaseous mixtures are often given in percent by volume, but these compositions are also mole percents because in a gas at a fixed temperature and pressure, the volume and moles of a gas are directly proportional to one another. In Example 5.19, we calculate the partial pressures of the components of air from its volume percent composition.

Example 5.19

The main components of dry air, by volume, are N_2, 78.08%; O_2, 20.95%; Ar, 0.93%; and CO_2, 0.04%. What is the partial pressure of each gas in a sample of air at 1.000 atm?

STRATEGY

Volume percent is the same as mole percent, and from the mole percents we can write the mole fractions. Thus, 78.08 volume percent N_2 is the same as 78.08 mole percent N_2, which, in turn, is the same as a 0.7808 mole fraction of N_2. We can apply this reasoning to the other gases in the mixture, too: The mole fraction for O_2 is 0.2095; for Ar, 0.0093; and for CO_2, 0.0004.

SOLUTION

Each partial pressure can be obtained using Equation (5.16):

$$P_{N_2} = 0.7808 \times 1.000 \text{ atm} = 0.7808 \text{ atm}$$
$$P_{O_2} = 0.2095 \times 1.000 \text{ atm} = 0.2095 \text{ atm}$$
$$P_{Ar} = 0.0093 \times 1.000 \text{ atm} = 0.0093 \text{ atm}$$
$$P_{CO_2} = 0.0004 \times 1.000 \text{ atm} = 0.0004 \text{ atm}$$

EXERCISE 5.19A

A sample of expired (exhaled) air is composed, by volume, of the following main gaseous components: N_2, 74.1%; O_2, 15.0%; H_2O, 6.0%; Ar, 0.9%; and CO_2, 4.0%. What is the partial pressure of each gas in the expired air at 37 °C and 1.000 atm?

EXERCISE 5.19B

A 75.0-L sample of natural gas at 23.5 °C contains methane (CH_4) at a partial pressure of 505 Torr, ethane (C_2H_6) at 201 Torr, propane (C_3H_8) at 43 Torr, and 10.5 g of butane (C_4H_{10}). Calculate the mole fraction of each component.

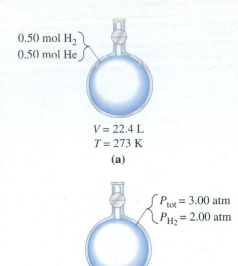

0.50 mol H_2
0.50 mol He

$V = 22.4$ L
$T = 273$ K
(a)

$P_{tot} = 3.00$ atm
$P_{H_2} = 2.00$ atm

$V = 22.4$ L
$T = 273$ K
(b)

▲ **FIGURE 5.15**   **Example 5.20 illustrated**

Example 5.20   A Conceptual Example

Describe what must be done to change the gaseous mixture of hydrogen and helium shown in Figure 5.15a to the conditions illustrated in Figure 5.15b.

ANALYSIS AND CONCLUSIONS

First, we assess the situation in (a): Because the chosen volume is 22.4 L, the temperature is 273 K (0 °C), and the total amount of gas is 1.00 mol, the total pressure must be 1.00 atm. This condition is the familiar molar volume of an ideal gas at STP (22.4 L).

Next, we appraise the situation in (b): The volume and temperature remain as they were in (a), but the pressure increases to 3.00 atm. This tripling of the pressure while the volume and temperature are fixed requires a tripling of the amount of gas. Therefore, there must be 3.00 mol gas in (b) instead of the 1.00 mol in (a). Because the partial pressure P_{H_2} is 2.00 atm, the partial pressure of the other gas or gases must be 1.00 atm. The corresponding mole fractions are 2/3 for H_2 and 1/3 for the other gas or gases. The 3.00-mol mixture of gases in (b) must consist of 2.00 mol H_2 and 1.00 mol of another gas or gases.

In order to get from (a), where we have 0.50 mol H_2 and 0.50 mol He, to (b), where we have 2.00 mol H_2 and 1.00 mol of some other gas or gases, we must add 1.50 mol H_2 and 0.50 mol of the other gas or gases to the flask. Note that this 0.50 mol could be 0.50 mol He, but it could also be any single gas or combination of gases other than hydrogen.

EXERCISE 5.20A

Why is it not possible to achieve the change between (a) and (b) in Figure 5.15 by adding hydrogen gas alone? Why can it not be achieved by adding helium gas alone? Why is it necessary that some hydrogen be added but not necessary that any helium be added?

EXERCISE 5.20B

Which of the following actions would you take to change the pressure of 2.2 L of N_2 at STP to 2.0 atm at 400 °C while maintaining a constant volume: (a) add 0.10 mol N_2; (b) add 0.020 mol N_2; (c) release 0.10 mol N_2; (d) release 0.46 L of N_2, measured at STP?

Table 5.4 Vapor Pressure of Water as a Function of Temperature

Temp. (°C)	Pressure (mmHg)
15	12.8
16	13.6
17	14.5
18	15.5
19	16.5
20	17.5
21	18.7
22	19.8
23	21.1
24	22.4
25	23.8
30	31.8
40	55.3

Collection of Gases over Water

Many gases—including oxygen, nitrogen, and hydrogen—are only slightly soluble in water. This limited solubility allows us to collect such gases over water, as pictured in Figure 5.16. As the essentially insoluble gas is passed into the water in the inverted collection bottle, the gas rises in the bottle because the gas density is much less than that of water. To make room for the incoming gas, water must be displaced or pushed out of the bottle.

When a gas is collected by displacement, water vapor produced by evaporation gathers in the collection bottle as well. The total pressure in the bottle is that of the gas (P_{gas}) plus that of water vapor ($P_{H_2O(g)}$).

$$P_{total} = P_{gas} + P_{H_2O(g)}$$

Assuming the gas is saturated with water vapor, the partial pressure of the water vapor is a quantity called the *vapor pressure* of the water. The vapor pressure depends only on the temperature: The warmer the water is, the higher its vapor pressure (Table 5.4). To find the partial pressure of the collected gas in Figure 5.16, we need only make the total pressure in the bottle equal to atmospheric pressure P_{bar} as

▶ **FIGURE 5.16**   **Collection of a gas over water**

If we raise or lower the bottle slightly so that the water levels inside and outside the bottle are the same, the pressure inside the bottle is the same as the outside pressure, P_{bar}. The partial pressure of the collected gas is given by

$$P_{gas} = P_{bar} - P_{H_2O}$$

The vapor pressure of water is obtained from tabulated data such as in Table 5.4.

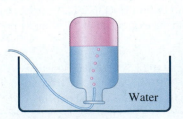
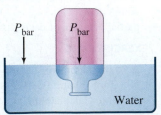

P_{bar}

P_{bar}

Water

Water

described in the figure and then subtract the vapor pressure of the water from the total pressure in the bottle:

$$P_{gas} = P_{total} - P_{H_2O(g)}$$
$$= P_{bar} - P_{H_2O(g)}$$

Thus, if the barometric pressure is 735 Torr and the water temperature is 22 °C, the water vapor pressure is 19.8 Torr (from Table 5.4) and the gas pressure is 735 Torr − 19.8 Torr = 715 Torr.

Example 5.21

Hydrogen produced in the following reaction is collected over water at 23 °C when the barometric pressure is 742 Torr:

$$2\, Al(s) + 6\, HCl(aq) \longrightarrow 2\, AlCl_3(aq) + 3\, H_2(g)$$

What volume of the "wet" gas will be collected in the reaction of 1.50 g Al(s) with excess HCl(aq)?

STRATEGY

The "wet" gas is the mixture of hydrogen gas and water vapor. However, because the two gases are found in the same collection vessel, they occupy the same volume, which means we need to calculate the volume of only one of them. Although we can easily get the pressure of the water vapor from Table 5.4, that information alone is not enough to calculate the volume of the water vapor using the ideal gas law because, in order to work the problem that way, we also need to know the number of moles of water vapor, and we do not. We can, however, gather the data necessary to determine the volume of *hydrogen*, by using Dalton's law of partial pressures and the ideal gas law.

SOLUTION

To begin, we know the temperature, 23 °C, and we can calculate the partial pressure of hydrogen using Dalton's law of partial pressures.

$$P_{total} = P_{H_2} + P_{H_2O}$$
$$742\ \text{Torr} = P_{H_2} + 21.1\ \text{Torr}$$
$$P_{H_2} = 742\ \text{Torr} - 21.1\ \text{Torr} = 721\ \text{Torr}$$

We can determine the number of moles of hydrogen from the stoichiometry of the reaction.

$$?\ \text{mol}\ H_2 = 1.50\ \text{g Al} \times \frac{1\ \text{mol Al}}{26.98\ \text{g Al}} \times \frac{3\ \text{mol}\ H_2}{2\ \text{mol Al}} = 0.0834\ \text{mol}\ H_2$$

With this information, we can solve the ideal gas equation for V and then calculate the volume of hydrogen, which is equal to the volume of the "wet" gas.

$$V = \frac{nRT}{P} = \frac{0.0834\ \text{mol} \times 0.0821\ \text{L atm mol}^{-1}\ \text{K}^{-1} \times (273 + 23)\ \text{K}}{(721/760)\ \text{atm}}$$
$$= 2.14\ \text{L}$$

EXERCISE 5.21A

Hydrogen gas is collected over water at 18 °C. The total pressure inside the collection jar is set at the barometric pressure of 738 Torr. If the volume of the gas is 246 mL, what mass of hydrogen is collected? What is the mass of the wet gas? (*Hint:* What is the mass of the water vapor?)

EXERCISE 5.21B

A sample of $KClO_3$ was heated to decompose it to potassium chloride and oxygen. The O_2 was collected over water at 21 °C and a barometric pressure of 746 mmHg. A 155-mL volume of the gaseous mixture was obtained. What mass of $KClO_3$ was decomposed?

5.11 The Kinetic-Molecular Theory: Some Quantitative Aspects

We introduced the kinetic-molecular theory in Section 5.2. Now let's examine the theory in a more quantitative fashion. Recall that a theory is developed to explain data but that successful theories also allow for predictions. We begin by restating the principal assumptions of the kinetic-molecular theory.

• A gas is made up of molecules (atoms in monatomic gases) that are in constant, random, straight-line motion.

- Molecules of a gas are far apart, which means that a gas is mostly empty space.
- There are no forces between molecules except during the instant of collision. Each molecule acts independently of all the others and is unaffected by their presence, except during collisions.
- Individual molecules may gain or lose energy as a result of collisions, but in a collection of molecules at constant temperature, *the total energy remains constant*.

With these assumptions, it is possible to calculate the pressure exerted by a collection of gas molecules confined to a given volume at a constant temperature. The derivation is too complex for us to consider in detail, but the final result is

$$P = \frac{1}{3} \times \frac{N}{V} \times m \times \overline{u^2} \tag{5.17}$$

Here, V is the volume containing N molecules, each having mass m; $\overline{u^2}$ is the *average* of the *squares* of their speeds (the bar above a term indicates an average value); and P is the pressure exerted by the gas. The important conclusions of the kinetic-molecular theory stem from this equation.

The Kinetic-Molecular Theory and Temperature

The kinetic-molecular theory gives us insight into the meaning of the temperature of a gas. Suppose we modify Equation (5.17) by multiplying both sides by V and replacing the fraction $\frac{1}{3}$ by the product $\frac{2}{3} \times \frac{1}{2}$:

$$PV = \frac{1}{3} \times N \times \overline{mu^2} = \frac{2}{3} \times N \times \left(\frac{1}{2}\overline{mu^2}\right) \tag{5.18}$$

We do this to isolate the term $\frac{1}{2}\overline{mu^2}$, which represents the average translational kinetic energy, e_k, of the gas molecules. Recall from page 171 that the translational kinetic energy of a molecule is that associated with the movement of the molecule in three-dimensional space.

Now let's assume we have 1 mol of gas, which means N will be Avogadro's number, N_A. We can then use the ideal gas equation for one mole of gas:

$$PV = nRT \quad \text{(where } n = 1 \text{ mol)}$$

Because $n = 1$ here, we see that $PV = RT$ and substitute RT for PV and e_k for $\frac{1}{2}\overline{mu^2}$ in Equation (5.18), to obtain

$$RT = \frac{2}{3} \times N_A \times e_k$$

Finally, we can solve for e_k:

$$e_k = \frac{3}{2} \times \frac{R}{N_A} \times T \tag{5.19}$$

Because $3/2$, R, and N_A are all fixed quantities, this equation is equivalent to the expression

$$e_k = \text{constant} \times T \tag{5.20}$$

We have arrived at this important conclusion:

> *The average translational kinetic energy of the molecules of a gas, e_k, is directly proportional to the Kelvin temperature.*

Thus, the molecules of *any* gas at a given temperature have the same average translational kinetic energy as the molecules of *any other* gas at the same temperature.

When we heat a gas, we increase the average translational kinetic energy of the gas molecules and the temperature increases. If we were to cool a gas to 0 K, the average translational kinetic energy of the molecules would drop to zero. This, then, is the kinetic-molecular interpretation of absolute zero: At 0 K, translational molecular motion ceases; molecules stop moving. Absolute zero temperature is unattainable, but recent attempts have produced temperatures as low as a few nanokelvins.

Molecular Speeds: Faster than a Speeding Bullet

The molecules of a gas have a wide distribution of speeds. We used an *average* molecular speed in the equations on page 198 to calculate the *average* translational kinetic energy. The average speed $\bar{u}$ is defined as

$$\bar{u} = \frac{\text{sum of the speeds of all the molecules}}{\text{total number of molecules}}$$

$$= \frac{u_1 + u_2 + u_3 + \cdots + u_n}{N} \quad (5.21)$$

For most applications, however, the root-mean-square speed is a more useful expression of molecular speed than is the average speed. The **root-mean-square speed,** u_{rms}, is the *square root* of the *average* of the *squares* of the speeds of a large number (N) of molecules:

$$u_{rms} = \sqrt{\overline{u^2}} = \sqrt{\frac{u_1^2 + u_2^2 + \cdots + u_n^2}{N}} \quad (5.22)$$

Because we cannot measure the speeds of individual molecules, we cannot use Equations (5.21) and (5.22) to calculate either the average molecular speed or the root-mean-square speed. One of the triumphs of the kinetic-molecular theory is that it permits us to derive equations for calculating average speeds. We will not attempt any of these derivations, although that of the root-mean-square speed is not difficult (see Problem 116). The result of that derivation is

$$u_{rms} = \sqrt{\overline{u^2}} = \sqrt{\frac{3RT}{M}} \quad (5.23)$$

Typical u_{rms} speeds of molecules are quite high. For hydrogen at 25 °C, for example, $u_{rms} = 1.92 \times 10^3$ m/s. This is about 4300 mi/h, or twice as fast as the speed of a bullet fired from an M-16 rifle.

To determine u_{rms} in meters per second with this equation requires that we express the gas constant, R, as 8.3145 joules (J)* per mole per kelvin ($J\ mol^{-1}\ K^{-1}$) and the molar mass, M, in kilograms per mole. For now, however, let's just apply this equation in a qualitative way.

First, we see that the root-mean-square speed is *inversely* proportional to the *square root* of the molar mass. Thus, the *lower* the molar mass, the faster a molecule moves. In Figure 5.17a the relative numbers of different molecules having given speeds are plotted against the speed. The peaks of the curves represent the *most-probable speeds* for molecules of the gas (average speeds are proportional to most-probable speeds). Hydrogen, having the lightest molecules (lowest molar mass), has the highest molecular speeds, while oxygen has the lowest. Second, we see in Figure 5.17b that u_{rms} is *directly* proportional to the *square root* of the Kelvin temperature: Doubling the temperature from 1000 K to 2000 K does not double the speed but increases it only by a factor of about 1.4($\sqrt{2} = 1.414$).

Although it is easy to calculate u_{rms} values using Equation (5.23), we can make some significant comparisons without having to calculate molecular speeds. We consider such a case in Example 5.22.

Example 5.22   A Conceptual Example

Without doing detailed calculations, determine which of the following is a likely value for u_{rms} of O_2 molecules at 0 °C, if u_{rms} of H_2 at 0 °C is 1838 m/s.

   (a) 115 m/s   **(b)** 460 m/s   **(c)** 1838 m/s   **(d)** 7352 m/s   **(e)** 29,400 m/s

---

*The *joule* is the SI unit of work and energy. It is discussed in Appendix B and in Section 6.1.

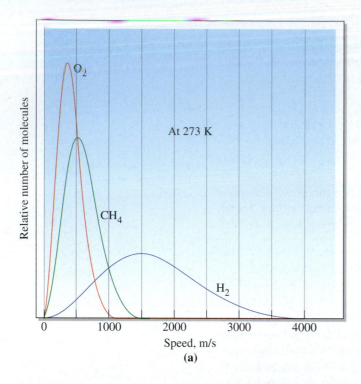

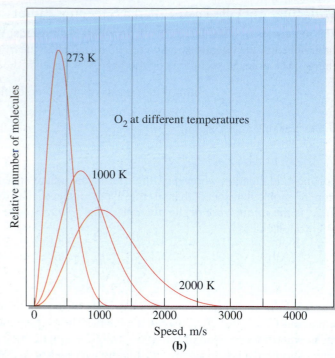

▲ **FIGURE 5.17** **Distribution of molecular speeds: Effects of molar mass and temperature**
(a) The most probable speed increases as molar mass decreases. At 273 K, the most probable speeds for O_2, CH_4, and H_2 molecules are 377, 532, and 1500 m/s, respectively. (b) The most probable speed increases with temperature. Notice that the temperature must be roughly 2000 K before O_2 molecules have the same most probable speed as do H_2 molecules at 273 K.

QUESTION: What would the curve for the distribution of molecular speeds of water vapor molecules look like on the graph of part (a)?

Distribution of Molecular Speeds simulation

ANALYSIS AND CONCLUSIONS

We could calculate u_{rms} for O_2 at 0 °C from Equation (5.23), but let's see how we can come up with a reasonable answer with a minimum of calculation.

We must not be misled by response (c)—that the H_2 and O_2 molecules travel at the same speed. Because they are at the same temperature, they *do* have equal average translational kinetic energies $1/2(mu)$. However, because O_2 molecules have a greater mass (m) than H_2 molecules, the O_2 molecules must have a lesser velocity (u). Thus O_2 molecules, on average, should move more slowly. This fact eliminates responses (d) and (e), as well as (c).

The molar mass ratio of O_2 to H_2 is $32/2 = 16$. Response (a), 115 m/s, is one-sixteenth of 1838 m/s, but this answer is incorrect because the molecular speeds are inversely related not to the molar mass ratio but to the *square root* of this ratio: $\sqrt{1/16} = 1/4$. Thus the u_{rms} speed of O_2 molecules is one-quarter that for hydrogen, or about 460 m/s. The correct response is (b).

EXERCISE 5.22A

Without doing detailed calculations, determine which of the following gases has the greatest u_{rms}:

(a) $CH_4(g)$ at 0 °C,

(b) $O_2(g)$ at 250 °C,

(c) $SO_2(g)$ at 750 °C,

(d) $NH_3(g)$ at 35 °C.

EXERCISE 5.22B

At which temperature(s) listed here will u_{rms} of O_2 be *greater* than u_{rms} of H_2 at 0 °C, that is, greater than 1838 m/s?

(a) 1000 K

(b) 2000 K

(c) 3000 K

(d) 4000 K

(e) 5000 K

Diffusion and Effusion

If a worker in one part of a laboratory accidentally releases some hydrogen sulfide, a gas with the distinctive odor of rotten eggs, other people in the laboratory soon detect the odor. The migration of the H_2S molecules through the air eventually leads to an even distribution of the gas molecules throughout the room. **Diffusion** is the process by which one substance mixes with one or more other substances as a result of the translational motion of molecules. Gaseous diffusion is relatively rapid because the molecules in a gas are far apart. Diffusion of liquids is much slower than that of gases, and that of solids is extremely slow.

Even though much faster than in solids and liquids, the diffusion of gases is slower than we might expect from molecular speeds. If molecules move at several hundred meters per second, we could expect to smell substances almost instantly when they are released in a far corner of a large room, but this is not what is commonly observed. When a bottle of ammonia is opened in the kitchen, it may be some time before we smell it in the bathroom. Molecules do move rapidly, true, but they collide with one another frequently and therefore are constantly changing direction. Those molecules of ammonia travel tortuous, zigzag paths on the way from the mouth of the bottle to our noses. The diffusion of $NH_3(g)$ and of $HCl(g)$ to form solid NH_4Cl (white "smoke") is depicted in Figure 5.18.

The average distance a molecule travels between collisions is called its *mean free path*. At sea level, the mean free path of an N_2 or O_2 molecule is only about 6×10^{-8} m. Although the rate of diffusion of a gas does depend on the average molecular speed of the gas molecules, frequent collisions make precise calculations quite complicated. Let's look at a somewhat simpler phenomenon for some insight into the process.

Effusion is the process in which a gas escapes from its container through a tiny hole, or orifice. Because lighter molecules have greater average speeds, we expect them to escape more quickly through an orifice than heavier molecules do. We can relate effusion rates of gases to their root-mean-square speeds. For two gases, 1 and 2, whose molar masses are M_1 and M_2, the effusion rates are related by the expression

$$\frac{\text{rate}_1}{\text{rate}_2} = \frac{\sqrt{\dfrac{3RT}{M_1}}}{\sqrt{\dfrac{3RT}{M_2}}} = \sqrt{\frac{M_2}{M_1}} \qquad (5.24)$$

Application Note

A small amount of the compound methyl mercaptan, CH_3SH, is added to commercial natural gas. Natural gas is essentially odorless, but the distinctive odor of the mercaptan makes a gas leak detectable even when the mercaptan is present only in the parts-per-billion range. Its diffusion from the site of a natural gas leak warns of a fire or explosion hazard.

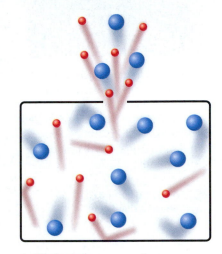

▲ *Effusion* is the escape of gas molecules from an orifice. As suggested here, lighter molecules effuse more rapidly than heavier molecules, due to the difference in molecular velocity. Because molecular collisions need not be taken into account in effusion calculations, effusion is much easier to describe mathematically (Equation 5.24) than diffusion.

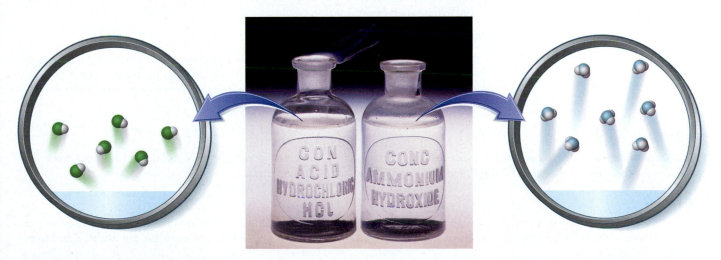

▲ **FIGURE 5.18   Diffusion of gases**
Molecules of HCl diffuse from the bottle of hydrochloric acid on the left, while NH_3 molecules diffuse from the ammonia solution on the right. Where the two gases meet, they react to form white ammonium chloride, which appears as a very faint "smoke." The microscopic view shows that the lighter ammonia molecules (mass = 17 u) move—and diffuse—faster than the heavier HCl molecules (mass = 36.5 u).

QUESTION: Why is the "smoke" a little closer to the HCl bottle than to the NH_3 bottle?

This equation is a kinetic-molecular description of a nineteenth-century law proposed by Thomas Graham (1805–1869) and known as *Graham's law of effusion:**

> *At a given temperature, the rates of effusion of gas molecules are inversely proportional to the square roots of their molar masses.*

This dependence of rate on molar mass is suggested in the microscopic views of Figure 5.18 and in the margin figure illustrating effusion of blue and red gas molecules.

In Example 5.23, we compare the effusion rates of two gases. In Example 5.24, instead of working with effusion rates, we work with effusion *times*. Effusion times and rates are inversely related:

$$\text{Effusion time} \propto \frac{1}{\text{effusion rate}} \qquad (5.25)$$

This inverse relationship should seem reasonable. For example, if a car traveling at 100 km/h makes a particular trip in 1.00 h, a car traveling 50 km/h will require 2.00 h for the same trip.

▲ In the Manhattan Project to develop nuclear weapons during World War II, uranium-235 and uranium-238 isotopes were separated by diffusion of $UF_6(g)$ through porous barriers in a long series of cylinders such as these.

Gas Diffuson and Gas Effusion simulation

Diffusion of Bromine Vapor movie

Example 5.23

If compared under the same conditions, how much faster than helium does hydrogen effuse through a tiny hole?

STRATEGY

To answer this question, we can set up a ratio of the two effusion rates in accordance with Graham's law (Equation 5.24).

SOLUTION

If we place the rate of effusion of hydrogen, r_{H_2}, in the numerator on the left, its molar mass, M_{H_2}, must go into the denominator under the square-root sign on the right. The situation is reversed for helium:

$$\frac{r_{H_2}}{r_{He}} = \sqrt{\frac{M_{He}}{M_{H_2}}} = \sqrt{\frac{4.00}{2.02}} = 1.41$$

The ratio of the two rates is 1.41, which means $r_{H_2} = 1.41 \times r_{He}$. Hydrogen effuses 1.41 times faster than helium.

EXERCISE 5.23A

Which effuses faster, N_2 or Ar, when the two gases are compared under the same conditions? How much faster?

EXERCISE 5.23B

Which effuses faster, nitrogen monoxide or dimethyl ether, $(CH_3)_2O$, when the two gases are compared under the same conditions? How much faster, expressed as a percentage?

Example 5.24

One percent of a measured amount of Ar(g) escapes through a tiny hole in 77.3 s. One percent of the same amount of an unknown gas escapes under the same conditions in 97.6 s. Calculate the molar mass of the unknown gas.

STRATEGY

Here we are given information regarding effusion times. From Equation (5.25), we can relate effusion times to effusion rates and, in turn, establish a relationship between effusion time and molar mass. Thus, the molar mass of the unknown gas can be calculated from the ratio of the effusion times.

*Graham's law adequately describes effusion only if the gas pressure is very low, so that molecules escape because of their random motion and not as a jet of gas. Also, the orifice must be tiny, so that no molecular collisions occur in the orifice as the gas effuses.

SOLUTION

First, let's get the appropriate form for an expression relating effusion times and molar masses. To do this, we can combine Equation (5.25), relating effusion time to effusion rate, and Equation (5.24), relating effusion rate to molar mass.

$$\frac{(\text{effusion time})_{\text{unk}}}{(\text{effusion time})_{\text{Ar}}} = \frac{1/(\text{effusion rate})_{\text{unk}}}{1/(\text{effusion rate})_{\text{Ar}}} = \sqrt{\frac{M_{\text{unk}}}{M_{\text{Ar}}}}$$

This result tells us that effusion time is directly proportional to the square root of molar mass.

$$\frac{(\text{effusion time})_{\text{unk}}}{(\text{effusion time})_{\text{Ar}}} = \sqrt{\frac{M_{\text{unk}}}{M_{Ar}}}$$

Now the only variable that is not known is M_{unk}.

$$\frac{97.6\ \text{s}}{77.3\ \text{s}} = 1.26 = \sqrt{\frac{M_{\text{unk}}}{39.95\ \text{g/mol}}}$$

We solve for M_{unk} by squaring both sides of the final equation.

$$\left(\sqrt{\frac{M_{\text{unk}}}{39.95\ \text{g/mol}}}\right)^2 = (1.26)^2$$

$$M_{\text{unk}} = 39.95\ \text{g/mol} \times (1.26)^2$$

$$= 63.4\ \text{g/mol}$$

ASSESSMENT

We can see that this is a reasonable answer. Because the unknown gas effuses more slowly than does Ar, the unknown molar mass must be greater than that of Ar.

EXERCISE 5.24A

Two percent of a sample of $N_2(g)$ effuses from a tiny opening in 57 s. Two percent of the same amount of an unknown gas escapes under the same conditions in 83 s. Calculate the molar mass of the unknown gas.

EXERCISE 5.24B

Five percent of a sample of $O_2(g)$ effuses from an orifice in 123 s. How long should it take five percent of the same amount of butane, C_4H_{10}, to effuse under the same conditions?

5.12 Real Gases

Under many conditions, real gases do not follow the ideal gas law. We can account for their deviations from ideal behavior by reexamining the assumptions of the kinetic-molecular theory: There are no intermolecular forces between molecules and gas molecules are *point masses*, that is, having mass but *no volume*. When gas molecules are far apart from one another, we can safely assume minimal forces of attraction between them. However, at high pressures, molecules are closer together, and at low temperatures, they are both closer together and moving by one another more slowly. Under these conditions, we must reckon with two factors:

- Intermolecular forces of attraction cause the measured pressure of a real gas to be less than we would expect, as suggested by Figure 5.19.
- When molecules are close together, the volume of the molecules themselves becomes a significant fraction of the total volume of a gas.

 Real Gas Behavior simulation

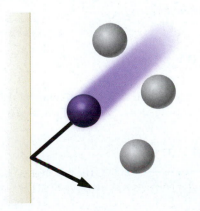 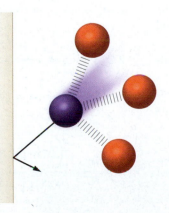

◄ **FIGURE 5.19 Pressure and intermolecular forces of attractions**
(Left) With no intermolecular attractions, the purple molecule strikes the wall with considerable force. (Right) Attractive forces between the orange molecules and the purple molecule cause the purple molecule to exert less force in its collision with the wall.

Table 5.5 van der Waals Constants for Selected Gases

Substance	a (L^2 atm mol^{-2})	b (L mol^{-1})
He	0.0341	0.02370
Ar	1.34	0.0322
H_2	0.244	0.0266
O_2	1.36	0.0318
CO_2	3.59	0.0427
CCl_4	20.4	0.1383

The ideal gas law makes no allowance for intermolecular forces or for the volumes of the molecules themselves. In 1873, Johannes van der Waals proposed an equation that does. In the *van der Waals equation*, the term n^2a/V^2 adjusts for intermolecular forces of attraction and the term nb adjusts for the volume of the gas molecules:

$$\left(P + \frac{n^2a}{V^2}\right) \times (V - nb) = nRT \qquad (5.26)$$

Whereas the ideal gas equation is perfectly general and can be applied without regard to the particular gas, the van der Waals equation includes two parameters, a and b, that are different for different gases and must be determined by experiment (Table 5.5).

For a real gas to approach ideal gas behavior, the term n^2a/V^2 must be much smaller than P and nb must be much smaller than V. These conditions are most likely to be met when a small amount of gas (n) is found in a large volume (V) or, generally speaking, when a gas is at a *high temperature* and a *low pressure*. At room temperature or above and at pressures less than a few atmospheres, most gases obey the ideal gas equation reasonably well.

Cumulative Example

Two cylinders of gas are used in welding. One cylinder is 1.2 m high and 18 cm in diameter, containing oxygen gas at 2550 psi and 19 °C. The other is 0.76 m high and 28 cm in diameter, containing acetylene gas (C_2H_2) at 320 psi and 19 °C. Assuming complete combustion, which tank will be emptied, leaving unreacted gas in the other?

STRATEGY

Our basic task is to determine the limiting reactant in the combustion, that is, C_2H_2 or O_2. For this, we must compare the numbers of moles of reactants in the two cylinders to the requirements dictated by the stoichiometric relationship between them (that is, the stoichiometric factor). We obtain the stoichiometric factor from the balanced equation for the combustion reaction. To determine the number of moles of O_2 and of C_2H_2 available for reaction, we must use the ideal gas law; but to obtain the necessary gas volumes we need first to calculate the volumes of the cylinders from the cylinder dimensions. When we have both the stoichiometric factor and the numbers of moles of the two gases, we can determine the limiting reactant. The cylinder containing the limiting reactant will be emptied in the reaction.

SOLUTION

We begin with the unbalanced equation for the combustion, which forms carbon dioxide and water.

$$C_2H_2 + O_2 \longrightarrow CO_2 + H_2O \quad \textit{(unbalanced)}$$

Stoichiometric coefficients are added to balance the equation.

$$2\,C_2H_2 + 5\,O_2 \longrightarrow 4\,CO_2 + 2\,H_2O$$

Beginning with the oxygen cylinder, the volume in cubic centimeters is calculated from the data given.

$$V = \pi r^2 h = 3.14 \times (9.0\text{ cm})^2 \times 1.2 \times 10^2\text{ cm} = 3.1 \times 10^4\text{ cm}^3 = 31\text{ L}$$

We now solve for the number of moles of oxygen that occupies this volume. Table 5.2 gives the conversion factor we need between psi (pounds per square inch) and atmospheres.

$$n = \frac{PV}{RT} = \frac{\left(2550\text{ psi} \times \dfrac{1\text{ atm}}{14.696\text{ psi}}\right) \times 31\text{ L}}{0.0821\text{ L atm mol}^{-1}\text{ K}^{-1} \times 292\text{ K}} = 2.2 \times 10^2\text{ mol O}_2$$

We repeat the calculations of volume and number of moles for acetylene.

$$V = \pi r^2 h = 3.14 \times (14 \text{ cm})^2 \times (76 \text{ cm}) = 4.7 \times 10^4 \text{ cm}^3 = 47 \text{ L}$$

$$n = \frac{PV}{RT} = \frac{\left(320 \text{ psi} \times \dfrac{1 \text{ atm}}{14.696 \text{ psi}}\right) \times 47 \text{ L}}{0.0821 \text{ L atm mol}^{-1} \text{ K}^{-1} \times 292 \text{ K}} = 43 \text{ mol C}_2\text{H}_2$$

Because the question merely asks which gas is completely consumed, we do not need to determine the amount of product. We just need to find which reactant is limiting.

A simple calculation of the number of moles of C_2H_2 needed to react with all the available O_2 will suffice.

$$\text{mol C}_2\text{H}_2 = 2.2 \times 10^2 \text{ mol O}_2 \times \frac{2 \text{ mol C}_2\text{H}_2}{5 \text{ mol O}_2} = 88 \text{ mol C}_2\text{H}_2$$

We have available only 43 mol of C_2H_2, meaning all the C_2H_2 will be used up.

ASSESSMENT

We can check several parts of our work. The volume of the oxygen cylinder (30.5 L) is somewhat more than the molar volume at STP (22.4 L), and the pressure is much higher than 150 atm, telling us there should be much more than 150 mol of oxygen gas in the cylinder. Likewise, the acetylene cylinder holds about two molar volumes at roughly 20 atm, or approximately 40 mol of gas. The amount of C_2H_2 that could be consumed by the available O_2 is more than 60 mol ($\frac{2}{5} \times 150$), but with only about 40 mol C_2H_2 available, the acetylene tank will be emptied.

Concept Review with Key Terms

5.1 Gases: What Are They Like?—Gases consist of widely separated molecules that are moving constantly and randomly throughout their containers. The term **vapor** denotes the gaseous state of a substance that is more commonly encountered as a liquid.

5.2 An Introduction to the Kinetic-Molecular Theory—This microscopic theory describes the random translational motion of gas molecules. The theory states that gas molecules are distributed uniformly throughout the entire volume of a container and that elastic molecular collisions with the container walls create a gas pressure.

5.3 Gas Pressure—Atmospheric **pressure** can be measured with a **barometer**, and other gas pressures, with a **manometer**. The SI unit of pressure is the **pascal** (Pa), but more commonly used are the standard **atmosphere** (atm), the **millimeter of mercury** (mmHg), and the **torr**: 1 atm = 760 mmHg = 760 Torr. The unit **kilopascal** (kPa) is a commonly used form of the SI unit.

5.4 Boyle's Law: The Pressure–Volume Relationship—For a given amount of gas at a constant temperature, the volume and pressure are *inversely* proportional:

$$V \propto \frac{1}{P}$$

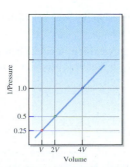

5.5 Charles's Law: The Temperature–Volume Relationship—For a given amount of gas at a constant pressure, the volume and absolute temperature are *directly* proportional:

$$V \propto T$$

The temperature obtained by extrapolation to zero volume of a plot of volume versus temperature is −273.15 °C. This temperature is defined as **absolute zero** on the **Kelvin scale**. Temperatures on this scale are expressed in the unit **kelvin (K)**. Temperatures can be converted from degrees Celsius to kelvins by using the equation

$$T(\text{K}) = T(^\circ\text{C}) + 273.15$$

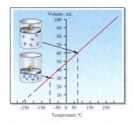

5.6 Avogadro's Law: The Mole–Volume Relationship—Avogadro's **hypothesis** implied the law also known by his name: At a fixed temperature and pressure, the volume and amount of a gas are *directly* proportional:

$$V \propto n$$

At 0 °C and 1 atm pressure, defined as the **standard temperature and pressure** (STP), the **molar volume of a gas** (volume of one mole) is about 22.4 L.

5.7 The Combined Gas Law—The three simple gas laws can be consolidated into the **combined gas law**:

$$\frac{P_1 V_1}{n_1 T_1} = \text{constant} = \frac{P_2 V_2}{n_2 T_2}$$

5.8 The Ideal Gas Law and Its Applications—Gases that have properties that conform to the simple gas laws are said to be **ideal gases**. The **ideal gas law** is usually written as

$$PV = nRT$$

where R is defined as the **gas constant**. The ideal gas law can be used in place of the simple or combined gas laws, especially for calculating molecular mass (M) or gas density (d):

$$M = \frac{mRT}{PV} \qquad d = \frac{m}{V} = \frac{MP}{RT}$$

5.9 Gases in Reaction Stoichiometry—The **law of combining volumes** states that when gases measured at the same temperature and pressure react, the volumes of gaseous reactants and products are in small whole-number ratios. The ideal gas law is often used in stoichiometric calculations for reactions involving gases.

5.10 Mixtures of Gases: Dalton's Law of Partial Pressures—Each gas in a mixture expands to fill the container and exerts its own **partial pressure**. **Dalton's law of partial pressures** states that the total pressure in a gaseous mixture is the sum of the partial pressures:

$$P_{total} = P_1 + P_2 + P_3 + \cdots$$

$$\text{where} \quad P_1 = \frac{n_1 RT}{V}; P_2 = \frac{n_2 RT}{V}; \text{and so on.}$$

The partial pressure of a gas can also be calculated from the product of the **mole fraction** of the gas and the total pressure:

$$P_1 = x_1 \times P_{total}$$

A gas collected over water is a mixture. It also contains water vapor at a partial pressure equal to the vapor pressure of water at the collection temperature. Stoichiometric calculations involving gases collected over water use stoichiometric factors from the balanced equation, Dalton's law of partial pressures, and the ideal gas equation.

Water

5.11 The Kinetic-Molecular Theory: Some Quantitative Aspects—The postulates of the kinetic-molecular theory lead to an equation for gas pressure and a relationship between Kelvin temperature and the average translational kinetic energy of gas molecules. A collection of gas molecules will possess a range of molecular speeds. The *square root* of the *average* of the *squares* of the molecular speeds is defined as the **root-mean-square speed** (u_{rms}):

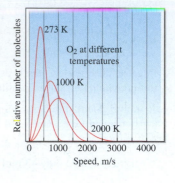

$$u_{rms} = \sqrt{\overline{u^2}} = \sqrt{\frac{3\,RT}{M}}$$

The kinetic-molecular theory also provides a basis for describing **diffusion**, the mixing of one gas with another, and **effusion**, the process in which a gas escapes from its container through a tiny hole, or orifice.

5.12 Real Gases—Real gases are most likely to exhibit ideal behavior at high temperatures and low pressures. When the ideal gas equation fails, it generally can be replaced by another equation, such as that of van der Waals:

$$\left(P + \frac{n^2 a}{V^2}\right) \times (V - nb) = nRT$$

Assessment Goals

When you have mastered the material in this chapter, you will be able to:

- Describe the physical properties of a gas.
- Define pressure and convert between various pressure units.
- Calculate the pressure exerted by a column of liquid, and describe how atmospheric pressure is measured in a barometer and other gas pressures in manometers.
- Use Boyle's law to relate pressure and volume of a fixed amount of gas at a fixed temperature.
- Use Charles's law to relate temperature and volume of a fixed amount of gas at a fixed pressure.
- State Avogadro's law, define standard temperature and pressure (STP), and use the molar volume of a gas at STP.
- Write the combined gas law and use it to calculate initial and final values of a gas variable (n, P, V, T) for a gas undergoing a change.
- Relate the ideal gas law to the combined gas law and use it to calculate unknown properties of a gas (n, P, V, T).

- Use the mass of a gas and the ideal gas equation to determine the molecular mass of a gas.
- Calculate the density of a gas using appropriate gas data and the ideal gas equation.
- State and apply Gay-Lussac's law of combining volumes.
- Perform stoichiometric calculations involving gases as reactants or products.
- Apply Dalton's law of partial pressures to determine the total pressure or partial pressures of individual gases in a mixture of gases.
- Calculate the mole fractions of components of a gas mixture.
- List the principles of the kinetic-molecular theory of gases and describe their relationship to the molecular properties of gases and to the simple gas laws.
- State Graham's law and apply it to calculating the relative rates of effusion of different gases.
- Explain why gases may exhibit nonideal behavior, and describe the conditions under which nonideal behavior is likely to be seen.

Self-Assessment Questions

1. What properties define a gas at the molecular level?
2. How does the operation of a barometer differ from that of a manometer?
3. Which of the following laws is incorrectly described by the mathematical proportionality?
   (a) Charles's law $V \propto T$ (c) Avogadro's law $V \propto n$
   (b) Boyle's law $V \propto P$

4. Why must an absolute temperature scale (Kelvin) be used for Charles's law calculations rather than the Celsius scale?
5. Convert each of the following pressures to units of atmospheres:
   (a) 21.92 in. Hg   (c) 525 Torr
   (b) 77.2 kPa   (d) 892 mmHg

6. What effect will the following changes have on the volume of a fixed amount of gas?

   (a) an increase in pressure at constant temperature

   (b) a decrease in temperature at constant pressure

   (c) an increase in pressure coupled with an increase in temperature

   (d) a decrease in pressure coupled with an increase in temperature

7. What effect will the following changes have on the pressure of a fixed amount of a gas?

   (a) an increase in temperature at constant volume

   (b) a decrease in volume at constant temperature

   (c) an increase in temperature coupled with a decrease in volume

   (d) a decrease in temperature coupled with a decrease in volume

8. According to kinetic-molecular theory, what happens to average molecular speeds when temperature drops? What happens to the observed gas pressure when the walls of the container are struck less often by molecules?

9. Which of the following would not result in an increase in the volume of a fixed amount of gas?

   (a) an increase in temperature at constant pressure

   (b) an increase in pressure at constant temperature

   (c) a decrease in temperature at constant pressure

   (d) a threefold increase in pressure, together with a twofold increase in Kelvin temperature.

10. Of the following gases, the one with the greatest density at STP is

    (a) Cl_2 **(c)** N_2O

    (b) SO_3 **(d)** PF_3

11. For a fixed amount of gas at a fixed pressure, changing the temperature from 100 °C to 200 K causes the gas volume **(a)** to decrease; **(b)** to double; **(c)** to increase, but not to twice its original value; **(d)** not to change.

12. Complete the statement: At 0 °C and 0.500 atm, 4.48 L $NH_3(g)$ **(a)** contains 0.20 mol NH_3; **(b)** weighs 3.40 g; **(c)** contains 6.02×10^{22} molecules; **(d)** contains 0.40 mol NH_3.

13. For each of the following, indicate whether a given gas would have the same or different densities in the two containers. If the densities are different, in which container is the density greater?

    (a) Containers A and B have the same volume and are at the same temperature, but the gas in A is at a higher pressure.

    (b) Containers A and B are at the same pressure and temperature, but the volume of A is greater than that of B.

    (c) Containers A and B are at the same pressure and volume, but the gas in A is at a higher temperature.

14. What is the pressure within a 10.0-L vessel containing 0.250 mol each of oxygen, hydrogen, and carbon dioxide at 298 K?

    (a) 1.83 atm   **(c)** 0.038 atm

    (b) 0.611 atm   **(d)** 1.64 atm

15. What is the mole fraction of oxygen in the mixture containing 0.25 mol each of oxygen, hydrogen, and carbon dioxide?

    (a) 0.25   **(c)** 0.75

    (b) 0.33   **(d)** 3.0

16. To establish a pressure of 2.00 atm in a 2.24-L cylinder containing 1.60 g $O_2(g)$ at 0 °C, we must **(a)** add 1.60 g O_2, **(b)** release 0.80 g O_2, **(c)** add 2.00 g He, **(d)** add 0.60 g He.

17. A sample of $O_2(g)$ is collected over water at 23 °C at a barometric pressure of 751 Torr (vapor pressure of water at 23 °C = 21 Torr). What is the partial pressure of $O_2(g)$ in the sample? **(a)** 21 mmHg, **(b)** 751 Torr, **(c)** 0.96 atm, **(d)** 1.02 atm.

18. A comparison is made at STP of 0.50 mol $H_2(g)$ and 1.0 mol He(g). Complete the statement: the two gases will **(a)** have equal average molecular kinetic energies, **(b)** have equal average molecular speeds, **(c)** occupy equal volumes, **(d)** have equal effusion rates.

19. A mixture of 5.0×10^{-3} mol $H_2(g)$ and 5.0×10^{-3} mol $SO_2(g)$ at 25 °C is introduced into a 10.0-L container having a "pinhole" leak. After a period of time, the mole fraction of $H_2(g)$ in the remaining gas mixture **(a)** is the same as in the original mixture, **(b)** is less than in the original mixture, **(c)** is greater than in the original mixture, **(d)** can not be determined relative to the original mixture.

20. The temperature and pressure at which Cl_2 is most likely to behave like an ideal gas are **(a)** 100 °C and 10.0 atm, **(b)** 200 °C and 0.50 atm, **(c)** 0 °C and 0.50 atm, **(d)** −100 °C and 25 atm.

Problems

(Assume that the simple and ideal gas laws apply in all cases, unless the problem states otherwise.)

Pressure

21. Carry out the following conversions between pressure units.

    (a) 0.934 atm to Torr   **(c)** 698 mmHg to kPa

    (b) 767 Torr to atm   **(d)** 33.5 lb/in.^{2} to mmHg

22. Carry out the following conversions between pressure units.

    (a) 1.87 atm to Torr   **(c)** 996 mb to kPa

    (b) 748 Torr to mmHg   **(d)** 26.88 in. Hg to atm

23. Elephant seals dive to depths as great as 1250 m. What is the pressure, in atmospheres, exerted by water at that depth? The densities of water and mercury are 1.00 g/mL and 13.6 g/mL, respectively.

24. The maximum working pressure (MWP) of a scuba tank is a rating that represents the highest pressure the tank can safely hold at a given temperature. If the person filling the tank exceeds this pressure, he or she is legally liable to the diver who uses it if an accident occurs. The MWP for a tank with certain valves and fittings is 207.0 bar at 20.0 °C. What is that pressure in pounds per square inch? In millimeters of mercury?

25. Calculate the *height*, in meters, of a column of $CCl_4(l)$ ($d = 1.59$ g/mL) that exerts the same pressure as a 11.2-cm column of Hg(l) ($d = 13.6$ g/mL).

26. Calculate the *pressure*, in atmospheres, exerted by a 2.50-m column of carbon disulfide having a volume of 1.25 L and a mass of 1.58 kg. Note that for Hg(l), $d = 13.6$ g/mL.

27. The open-end manometer pictured in Figure 5.4a is filled with an oil that has a density of 0.901 g/mL. Calculate the pressure, in Torr, of the gas in the flask if barometric pressure is 751 Torr and the oil level on the right side, h_{closed}, is 83 mm below that on the left side.

28. The open-end manometer pictured in Figure 5.4b is filled with glycerol ($d = 1.261$ g/mL). Calculate the pressure, in Torr, of the gas in the flask if barometric pressure is 722 Torr and the glycerol level on the left side, h_{open}, is 22.7 cm below that on the right side.

29. We have noted that to measure atmospheric pressure, a mercury barometer must be nearly 1 m in height. What is the height requirement for the closed-end mercury manometer pictured in Figure 5.3? Explain.

30. How would you explain to a young child the "trick" suggested by the accompanying photograph of a filled water glass being held in the air?

Boyle's Law

31. A sample of xenon is held at a constant temperature and occupies 882 mL at 752 Torr. Determine the volume of the xenon at **(a)** 719 Torr, **(b)** 1.38 atm, **(c)** 125 kPa.

32. A sample of fluorine gas occupies 7.08 L at 732 mmHg. Assume that the temperature is held constant, and determine the pressure of the fluorine if the volume is changed to **(a)** 12.5 L, **(b)** 435 mL, **(c)** 0.150 m^3.

33. A decompression chamber used by deep-sea divers has a volume of 10.3 m^3 and operates at an internal pressure of 4.50 atm. What volume, in cubic meters, would the air in the chamber occupy if it were at normal atmospheric pressure, assuming no temperature change?

34. A novel energy storage system involves storing air under high pressure. (Energy is released when the air is allowed to expand.) How many cubic feet of air, measured at standard atmospheric pressure of 14.7 lb/in.^{2}, can be compressed into a 19-million-ft^3 underground cavern at a pressure of 1070 lb/in.^{2}?

35. Oxygen used in respiration therapy is stored at room temperature under a pressure of 1.50×10^2 atm in a gas cylinder with a volume of 43.0 L.

(a) What volume would the gas occupy at a pressure of 750.0 Torr? Assume no temperature change.

(b) If the oxygen flow to the patient is adjusted to 8.00 L per minute at room temperature and 750.0 Torr, how long will the tank of gas last?

36. The figure shows a cylinder with 1.20 L of a gas at the standard temperature and pressure. What will the gas pressure be if the piston is depressed 5.25 cm further into the cylinder?

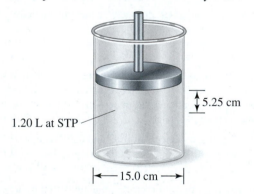

1.20 L at STP

5.25 cm

15.0 cm

Charles's Law

37. A gas at a temperature of 99.8 °C occupies a volume of 641 mL. What will the volume be at a temperature of 5.0 °C, assuming no change in pressure?

38. A 32.3-L sample of a gas at 305 °C and 1.20 atm is to be cooled at constant pressure until its volume becomes 28.4 L. What will be the new gas temperature?

39. If a 15.0 °C temperature increase causes a 10.0% increase in the volume of a 832-mL sample of He(g) while the gas pressure is held constant, what was the original temperature?

40. What increase in the Celsius temperature will produce a 5.0% increase in the volume of a sample of gas originally at 25.0 °C if the gas pressure is held constant?

Avogadro's Law and Molar Volume at STP

41. What is the mass, in kilograms, of 4.55×10^3 L of methane gas (CH_4) at STP?

42. What is the volume, in milliliters, of 88.3 mg of carbon dioxide gas at STP?

43. If 125 mg of Ar(g) is added to a 505-mL sample of Ar(g) at STP, what volume will the sample occupy when the conditions of STP are restored?

44. How many grams of gas must be released from a 45.2-L sample of N_2(g) at STP to reduce the volume to 45.0 L at STP?

45. *Without doing detailed calculations,* determine which of the following samples contains the greatest number of molecules.
(a) 5.0 g H_2 **(b)** 50 L SF_6(g) at STP
(c) 1.0×10^{24} molecules of CO_2
(d) 67 L of a gaseous hydrocarbon at STP

46. *Without doing detailed calculations,* determine which of the following samples occupies the greatest volume when measured at STP.
(a) 30.0 g O_2(g)   **(b)** 1.10 mol SO_2(g)
(c) 24.0 L CO(g) at 22 °C and 745 mmHg
(d) 7.2×10^{23} molecules of Cl_2(g)

The Combined Gas Law

47. A sealed can with an internal pressure of 721 Torr at 25 °C is thrown into an incinerator operating at 755 °C. What will be the pressure inside the heated can, assuming the container remains intact during incineration?

48. A fixed amount of He exerts a pressure of 775 mmHg in a 1.05-L container at 26 °C. At what new temperature would the gas pressure equal 725 mmHg? Assume the volume of gas remains constant.

49. What volume container would hold a 1.00-mol sample of gas at a pressure of 1.00 atm and a temperature of 25 °C?

50. At 25 °C, the pressure in a gas cylinder containing 8.00 mol O_2 is 5.05 atm. To maintain a constant pressure of 5.05 atm, how many moles of $O_2(g)$ should be released when the temperature is raised to 235 °C?

51. If a fixed amount of gas occupies 2.53 m^3 at a temperature of -15 °C and 191 Torr, what volume will it occupy at 25 °C and 1142 Torr?

52. What is the molar volume of a gas at **(a)** 25.0 °C and a pressure of 55.0 psi, and **(b)** at -78 °C and a pressure of 100.0 kPa?

The Ideal Gas Law

53. Calculate **(a)** the volume, in liters, of 1.88 mol of an ideal gas at 55 °C and 1.08 atm; **(b)** the pressure, in atmospheres, of 137 g $CO(g)$ in a 4.49-L tank at 28 °C; **(c)** the mass, in milligrams, of 97.4 mL $H_2(g)$ at 744 Torr and -2 °C; and **(d)** the pressure, in kilopascals, of 19.6 g $N_2(g)$ in a 6.41-L flask at 0 °C.

54. Calculate **(a)** the pressure, in Torr, of 15.7 mol of an ideal gas in a volume of 268 L at 17 °C; **(b)** the volume, in milliliters, of 42.4 mg $H_2(g)$ at 0.768 atm and -45 °C; **(c)** the mass, in kilograms, of 84.4 m^3 of $Cl_2(g)$ at 726 Torr and 6.5 °C; and **(d)** the pressure, in pounds per square inch (psi), of 306 mg of $CCl_2F_2(g)$ in a 1073-mL flask at 21 °C.

55. If the $O_2(g)$ gas present in 7.30 L at STP is changed to a temperature of 17 °C and a pressure of 788 Torr, what will be the new volume?

56. If 4.42 g $SO_2(g)$ is added, at constant temperature and volume, to 25.5 L of $SO_2(g)$ at 733 mmHg and 21.3 °C, what will be the new gas pressure?

57. A hyperbaric chamber is an enclosure containing oxygen at higher-than-normal pressures used in the treatment of certain heart and circulatory conditions. What volume of $O_2(g)$ from a cylinder at 25 °C and 151 atm is required to fill a 4.20×10^3-L hyperbaric chamber to a pressure of 2.50 atm at 17 °C?

58. The best laboratory vacuum systems can pump down to as few as 1.0×10^9 molecules per cubic meter of gas. Calculate the corresponding pressure, in atmospheres, assuming a temperature of 25 °C.

59. Calculate the molecular mass of a liquid that, when vaporized at 98 °C and 756 Torr, gives 139 mL of vapor with a mass of 0.808 g.

60. Calculate the molecular mass of a liquid that, when vaporized at 99 °C and 716 Torr, gives 285 mL of vapor with a mass of 1.102 g.

61. A liquid hydrocarbon is found to be 8.75% H by mass. A 1.261-g vaporized sample of the hydrocarbon has a 435-mL volume at 115 °C and 761 Torr. What is the molecular formula of this hydrocarbon?

62. A liquid hydrocarbon is found to be 16.37% H by mass. A 1.158-g vaporized sample of the hydrocarbon has a 385-mL volume at 71.0 °C and 749 mmHg. What is the molecular formula of this hydrocarbon?

63. Calculate the density, in grams per liter, of **(a)** $Ne(g)$ at STP and **(b)** $C_2H_4(g)$ at 1.19 atm and 147 °C.

64. Calculate the density, in grams per liter, of **(a)** $AsH_3(g)$ at STP and **(b)** $N_2(g)$ at 715 Torr and 98 °C.

65. What pressure in atm must be applied to $O_2(g)$ at 37 °C to give a density of 1.01 g/L?

66. At what temperature will $N_2(g)$ have a density of 0.985 g/L at 753 Torr?

67. The density of sulfur vapor at 445 °C and 755 mmHg is 4.33 g/L. What is the molecular formula of sulfur vapor?

68. A gaseous hydrocarbon contains 14.37% H by mass and has a density of 1.69 g/L at 24 °C and 743 mmHg. What is the molecular formula of this hydrocarbon?

69. *Without doing detailed calculations*, determine which gas has the greatest density, **(a)** $H_2(g)$ at -15 °C and 745 Torr, **(b)** $He(g)$ at STP, **(c)** $CH_4(g)$ at -10 °C and 1.15 atm, or **(d)** $C_2H_6(g)$ at 50 °C and 435 Torr.

70. *Without doing detailed calculations*, determine which gas is (are) *more* dense than $O_2(g)$ at STP, **(a)** N_2 at STP, **(b)** $CO(g)$ at 0 °C and 1100 Torr, **(c)** $SO_2(g)$ at 300 °C and 750 Torr, and/or **(d)** $H_2(g)$ at 25 °C and 10 atm.

Gases and Stoichiometry

71. What volume of nitrogen gas can be produced from the decomposition of 37.6 L of ammonia, with both gases measured at 725 °C and 5.05 atm pressure?

$$2\,NH_3(g) \longrightarrow N_2(g) + 3\,H_2(g)$$

72. In the reaction in Problem 71, what total volume of gas at 667 °C and 11.3 atm can be produced by the decomposition of 1.19×10^4 L of ammonia at that same temperature and pressure?

73. How many liters of $SO_3(g)$ can be produced by the reaction of 6.06 L $SO_2(g)$ and 2.25 L $O_2(g)$ if all three gases are measured at the same temperature and pressure?

$$2\,SO_2(g) + O_2(g) \longrightarrow 2\,SO_3(g)$$

74. How many liters of $CO_2(g)$ can be produced in the reaction of 5.24 L $CO(g)$ and 2.65 L $O_2(g)$ if all three gases are measured at the same temperature and pressure?

75. How many milligrams of magnesium metal must react with excess $HCl(aq)$ to produce 28.50 mL of $H_2(g)$, measured at 26 °C and 758 Torr?

$$Mg(s) + 2\,HCl(aq) \longrightarrow MgCl_2(aq) + H_2(g)$$

76. A 100.0-g sample of aqueous hydrogen peroxide solution decomposes over time, producing 3.31 L $O_2(g)$ at 21 °C and 715 Torr.

$$2\,H_2O_2(aq) \longrightarrow 2\,H_2O(l) + O_2(g)$$

What must have been the mass percent H_2O_2 in the solution?

77. How many liters of $CO_2(g)$ measured at 26 °C and 767 Torr are produced in the complete combustion of 125 mL of 1-propanol ($d = 0.804$ g/mL)?

$$CH_3CH_2CH_2OH(l) + O_2(g) \longrightarrow CO_2(g) + H_2O(l)$$
(not balanced)

78. How many liters of $O_2(g)$ measured at 22 °C and 763 Torr are consumed in the complete combustion of 2.55 L of dimethyl ether measured at 25 °C and 748 Torr?

$$CH_3OCH_3(g) + O_2(g) \longrightarrow CO_2(g) + H_2O(l)$$
(not balanced)

79. A laboratory experiment requires you to prepare at least 565 mL of carbon dioxide gas at 748 Torr and 21 °C. If the reactants are 0.375 M HCl and solid calcium carbonate, what minimum volume of HCl will be required?

80. For Problem 79, if the reactants are 0.375 M H_2SO_4 and solid sodium hydrogen carbonate, what volume of acid and what mass of sodium hydrogen carbonate will be needed? Assume the salt produced is sodium sulfate.

Mixtures of Gases

81. A gas sample has 76.8 mol percent N_2, 20.1 mol percent O_2, and 3.1 mol percent CO_2. If the total pressure is 762 mmHg, what are the partial pressures of the three gases?

82. A sample of intestinal gas was collected and found to consist of 44% CO_2, 38% H_2, 17% N_2, 1.3% O_2, and 0.003% CH_4, by volume. (The percentages do not total 100% because of rounding.) What is the partial pressure of each gas if the total pressure in the intestine is 818 Torr? (*Hint:* Recall that volume percent is the same as mole percent for ideal gas mixtures.)

83. Mixtures of helium and oxygen are used in scuba diving. What are **(a)** the mole fractions of the two gases, **(b)** their partial pressures, and **(c)** the total pressure in a mixture of 1.96 g He and 60.8 g O_2 confined in a 5.00-L tank at 25.0 °C?

84. A 267-mL sample of a mixture of noble gases at 25.0 °C contains 0.354 g Ar, 0.0521 g Ne, and 0.0049 g Kr. What are **(a)** the mole fractions of the three gases, **(b)** their partial pressures, and **(c)** the total gas pressure?

85. Oxygen is collected over water at 22 °C and a barometric pressure of 756 Torr. What is the partial pressure and mole fraction of $O_2(g)$ in the container?

86. A gas sample containing methane and carbon monoxide is collected over water at 18 °C. If the total pressure of the mixture is 1.014 atm, calculate the mole fraction of water vapor in the sample.

87. *Elodea* is a green plant that carries out photosynthesis under water.

$$6\,CO_2(g) + 6\,H_2O(l) \longrightarrow C_6H_{12}O_6(aq) + 6\,O_2(g)$$

In an experiment, some *Elodea* produce 122 mL of $O_2(g)$, collected over water at 743 Torr and 21 °C. What mass of oxygen is produced? What mass of glucose ($C_6H_{12}O_6$) is produced concurrently?

88. A 2.02-g sample of aluminum reacts with an excess of HCl(aq), and the liberated hydrogen is collected over water at a temperature of 24 °C. What is the *total* volume of the gas collected at a barometric pressure of 760.0 Torr?

$$2\,Al(s) + 6\,H^+(aq) \longrightarrow 2\,Al^{3+}(aq) + 3\,H_2(g)$$

Kinetic-Molecular Theory

89. A gaseous mixture with equal numbers of molecules of H_2 and He is allowed to effuse through an orifice for a certain period of time. Which of the conditions pictured here, (a), (b), (c), or (d) is most likely to result?

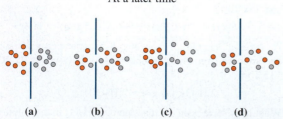

90. Concerning the molecular-level sketches in Problem 89, propose two gases that could be substituted for the He and H_2 that would yield the result pictured in sketch (d). Explain your reasoning.

91. Explain why two different gases in the same container *may* exert different pressures but *may not* have different temperatures or occupy different volumes.

92. Explain how Avogadro's equal volumes-equal numbers of molecules hypothesis for gases can be rationalized by using the basic equation of the kinetic-molecular theory (Equation 5.18).

93. It takes 44 s for a sample of $N_2(g)$ to effuse through a tiny orifice. Determine the molecular masses of gases whose effusion time under exactly the same conditions are **(a)** 75 s and **(b)** 42 s.

94. At a certain temperature, the root-mean-square speed of CH_4 molecules is 1610 km/h. What is the root-mean-square speed of CO_2 molecules at the same temperature?

Additional Problems

Problems marked with an * may be more challenging than others.

95. Automobile tire pressure is *gauge* pressure, the difference between the prevailing atmospheric pressure and the absolute or total pressure. (When a tire gauge reads zero, the absolute pressure is equal to atmospheric pressure.) A 21.1-L tire is inflated to 32 lb/in^2 gauge pressure. What is the mass of air in the tire at 21 °C? (Air may be considered to have an average molar mass of 28.96 g/mol.) After driving for half an hour, the temperature increases to 44 °C. What is the new gauge pressure of the air in the tire, assuming no change in volume?

96. Consider the following two simple gas laws not specifically mentioned in the chapter. Diver's law—so called because of its importance to scuba divers—relates pressure to amount of gas with temperature and volume held constant. Another (unnamed) law relates amount of gas and temperature with pressure and volume held constant. Write each law in the form of mathematical equations **(a)** with a proportionality constant and **(b)** using subscripts to indicate initial and final conditions. **(c)** Explain each of these laws using the kinetic-molecular theory.

97. A refrigerator has been developed that uses compressed helium as a refrigerant gas. A typical system uses 5.00 in^3 of He compressed to 195 psi at 20 °C. What mass of helium, in grams, is needed for one refrigerator?

98. In terms of pressure (*P*), volume (*V*), Kelvin temperature (*T*), and amount of gas (*n*), and in the manner of Figures 5.6 and 5.8, sketch a graph of **(a)** *V* as a function of *P*, with *T* and *n* held constant; **(b)** *n* as a function of *P*, with *T* and *V* held constant; **(c)** *T* as a function of *P*, with *V* and *n* held constant; and **(d)** *n* as a function of *T*, with *P* and *V* held constant.

99. In an attempt to verify Avogadro's hypothesis, small quantities of several different gases were weighed in 100.0-mL syringes. The following masses were obtained using an analytical balance: 0.0080 g H_2, 0.1112 g N_2, 0.1281 g O_2, 0.1770 g CO_2, 0.2320 g C_4H_{10}, and 0.4824 g CCl_2F_2. Within 1%, are these results consistent with Avogadro's hypothesis? Explain.

* 100. The $O_2(g)$ produced in the decomposition of a 3.275-g mixture of potassium chlorate and potassium chloride with 65.82% $KClO_3$ is collected over water at 21 °C and 753.5 mmHg atmospheric pressure.

$$2 \ KClO_3(s) \longrightarrow 2 \ KCl(s) + 3 \ O_2(g)$$

How many milliliters of the gas are collected for the conditions shown in the figure?

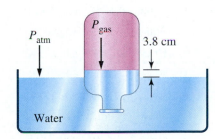

* 101. A 2.135-g sample of a gaseous chlorofluorocarbon (a type of gas implicated in the depletion of stratospheric ozone) occupies a volume of 315.5 mL at 739.2 mmHg and 26.1 °C. Analysis of the compound shows it to be 14.05% C, 41.48% Cl, and 44.46% F, by mass. What is the molecular formula of this compound?

102. The gaseous hydrocarbon 1,3-butadiene is used to make synthetic rubber. The following measurements were made to determine its molecular mass: A glass container weighed 45.0143 g when evacuated; 192.8273 g when filled with Freon-113™, a liquid with a density of 1.576 g/mL; and 45.2217 g when filled with butadiene at 751.2 mmHg and 21.48 °C. What is the molecular mass of 1,3-butadiene?

103. Project *Echo*, a prototype communications satellite from the early 1960s, was made of aluminized polyester. This spherical balloon, which was 30.5 m in diameter, reflected broadcast radio waves back to Earth. The pressure needed to inflate the balloon was 1.0×10^{-5} atm at a temperature of about −70 °C. Calculate the volume in liters of argon at 35 atm and 25 °C needed to inflate *Echo*.

104. A 1.000-kg cylinder has a diameter of 4.10 cm and a height of 10.18 cm. *Without doing detailed calculations*, determine which of the following most likely represents the pressure (exclusive of barometric pressure) exerted by the cylinder on the surface beneath it: (a) 102 mmHg; (b) 56 Torr; (c) 74 kPa.

* 105. Gaseous mixtures of the anesthetic cyclopropane and air with between 2.4 and 10.3% cyclopropane by volume are explosive. A 1.50-L cylinder of gaseous cyclopropane at 25 °C and 25.0 atm is placed in a fume hood of volume 72 ft^3. Would an explosive mixture form if the valve on the cylinder were to break and the cyclopropane mix thoroughly with the air in the fume hood, which is at 755 mmHg and 25 °C?

* 106. A 98.5-L tank of argon gas at 2940 psi is allowed to vent into an airtight room that is 3.0 m × 6.0 m × 6.0 m and in which the temperature is 20.0 °C and the air pressure is 756 mmHg. Assuming ideal behavior, what is the volume percent argon in the air in the room after the argon has thoroughly mixed with the air?

107. Calculate the mass of Earth's atmosphere, in tons, given that the surface area of Earth is 1.95×10^8 mi^2.

108. Rescue squads, to extricate people trapped under heavy objects such as automobiles and trains, use an inflatable air bag made of steel-reinforced rubber. A square bag 3.0-ft on a side can lift a 73-ton object. What must be the air pressure in the bag, in pounds per square inch, to make this possible?

* 109. Calculate the volume of $H_2(g)$ required to react with 15.0 L CO(g) in the reaction

$$3 \ CO(g) + 7 \ H_2(g) \longrightarrow C_3H_8(g) + 3 \ H_2O(l)$$

(a) if both gases are measured at STP; **(b)** if the CO(g) is measured at STP, and the $H_2(g)$ at 22 °C and 745 mmHg; **(c)** if both gases are measured at 22 °C and 745 mmHg; and **(d)** if the CO(g) is measured at 25 °C and 757 mmHg, and the $H_2(g)$ at 22 °C and 745 mmHg.

* 110. Magnesium–lithium alloy is used where its extremely low density is a desirable property. A piece of this alloy weighing 0.0297 g is analyzed by adding HCl(aq). The liberated hydrogen gas was collected over water at 19 °C and at atmospheric pressure of 746 Torr and was found to occupy 40.71 mL. Calculate the percentages of magnesium and lithium in the alloy.

111. Most recipes for peanut brittle use baking soda, $NaHCO_3$ to foam the hot candy. The baking soda decomposes into $Na_2CO_3(s)$, $H_2O(g)$, and $CO_2(g)$. How many liters of gas will be produced at 310 °F and 753 Torr by the complete decomposition of 3.2 g of $NaHCO_3$?

* **112.** Use C_8H_{18} as the formula of gasoline and 0.71 g/mL for its density. If a car gets 31.2 mi/gal (1 gal = 3.785 L), what volume of $CO_2(g)$ measured at 28 °C and 732 mmHg is produced in a trip of 265 mi? Assume complete combustion of the gasoline.

* **113.** What volume of *air*, measured at 23.0 °C and 741 mmHg, is required for the complete combustion of 1.00×10^3 L of a particular natural gas, measured at STP? The composition of the natural gas is 77.3% CH_4, 11.2% C_2H_6, 5.8% C_3H_8, 2.3% C_4H_{10}, and 3.4% noncombustible gases, by volume. What volume of $CO_2(g)$, measured at 35.0 °C and 985 mb, is produced? Use the composition of air given in Example 5.19.

114. Use the definitions of u and u_{rms} on page 199 to calculate $\bar{u}$ and u_{rms} for a group of six particles with the speeds: 9.83×10^3, 9.05×10^3, 8.33×10^3, 6.48×10^3, 3.67×10^3, and 1.75×10^3 m/s, respectively.

115. Calculate the root-mean-square speed of SO_2 molecules at 27 °C.

116. Use other appropriate equations from the text to derive the kinetic-molecular theory equation for the root-mean-square speed of a gas given on page 199.

117. Calculate the pressure exerted by 1.00 mol $CO_2(g)$ when it is confined to a volume of 2.50 L at 298 K by using **(a)** the ideal gas equation; **(b)** the van der Waals equation (page 204). Use data from Table 5.5. **(c)** Compare the two results, and comment on the reason(s) for the difference between them.

118. A vacuum pump sold by a scientific company lists "13.3 MPa $(1.0 \times 10^{-4}$ Torr)" as the ultimate vacuum that can be attained. What is wrong with the specification? Suppose that you wish to measure the pressure in a vacuum chamber attached to this pump. Will it be practical to use the closed-end manometer pictured in Figure 5.3, if dibutyl phthalate ($d = 1.058$ g/mL) is substituted for mercury? Explain.

* **119.** A 1.405-g sample of an alkane yields 4.305 g CO_2 and 2.056 g H_2O on combustion. A 0.403-g sample of the gaseous hydrocarbon occupies a volume of 145 mL at 99.8 °C and 749 Torr. Another type of analysis reveals one methyl group as a side group on the main hydrocarbon chain. Draw structural formulas for all the possible isomers that fit this description.

120. Pictured are molecular models of four gases. (Recall Figure 2.8, the color scheme for molecular models.)

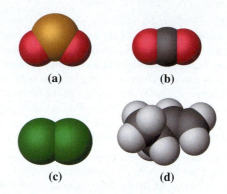

(a)   (b)

(c)   (d)

(a) Determine the gas having the maximum and the gas having the minimum root-mean-square speed at 0 °C and 1 atm.

(b) Which of the four gases have a u_{rms} greater than the speed of sound, which at 0 °C and 1 atm is 1086 ft/s?

Apply Your Knowledge

121. [**Laboratory**] The *Dumas* method for determining molar mass of a volatile liquid uses a Dumas bulb, a round glass vessel with a very narrow bent neck, as the container for a vapor. A small sample of an unknown volatile liquid was placed in a bulb weighing 103.868 g, and the bulb was immersed in boiling water at 99.2 °C and 739 Torr until the liquid had vaporized completely. The neck of the bulb was sealed in a flame, and the bulb was allowed to cool to 23.2 °C. The bulb then weighed 104.772 g. The sealed tip was placed into a vessel of distilled water and the tip broken, so that the bulb nearly filled with water. The filled bulb, plus the bits of glass that had fallen off, weighed 300.623 g. Calculate the molar mass of the liquid to the maximum number of significant figures permitted.

* **122.** [**Biochemical**] As we have seen, the various gas laws can be used to describe air, a mixture of gases. In some cases, these gas laws have direct application to the air that we breathe.

(a) How long does it take a person at rest to breathe "one mole of air" if the person breathes 80 mL per second of air that is measured at 25 °C and 755 mmHg?

(b) Typically, when a person coughs, he or she first inhales about 2.0 L of air at 1.0 atm and 25 °C. The epiglottis and the vocal cords then shut, trapping the air in the lungs, where it is warmed to 37 °C and compressed to a volume of about 1.7 L by the action of the diaphragm and chest muscles. The sudden opening of the epiglottis and vocal cords releases this air explosively. Just prior to this release, what is the approximate pressure of the gas inside the lungs?

(c) Helium–oxygen mixtures are used by divers to avoid the bends (page 501), and used in medicine to treat some respiratory ailments. What mass percent of He must be present in a helium–oxygen mixture having a density of 0.518 g/L at 25 °C and 721 mmHg?

* **123.** [**Environmental**] A sounding balloon for atmospheric studies is a bag filled with $H_2(g)$ that carries a set of instruments (the payload). Because this combination of bag, gas, and payload has a smaller mass than a corresponding volume of air, the balloon rises and expands as it does so. From the following data, estimate the maximum height to which the balloon can rise: mass of empty balloon, 1200 g; mass of payload, 1700 g; quantity of $H_2(g)$ in balloon, 120 ft^3 at STP; diameter of spherical balloon at maximum height, 25 ft. Air pressure and temperature as a function of altitude are:

0 km, 1.0×10^3 mb, 290 K

5 km, 5.4×10^2 mb, 266 K

10 km, 2.7×10^2 mb, 235 K

20 km, 5.5×10^1 mb, 217 K

30 km, 1.2×10^1 mb, 239 K

40 km, 2.9×10^0 mb, 267 K

50 km, 8.1×10^{-1} mb, 280 K

60 km, 2.3×10^{-1} mb, 260 K

Treat air as if it were a single gas with a molar mass of 28.96 g/mol air (referred to as the *apparent* molar mass of air).

* **124.** [**Historical**] In 1860, Stanislao Cannizzaro showed how Avogadro's hypothesis could be used to establish the atomic masses

of elements in *gaseous* compounds. Cannizzaro took the atomic mass of hydrogen to be exactly 1 and assumed that hydrogen exists as H_2 molecules (molecular mass, 2). Next, he determined the volume of $H_2(g)$ at STP that has a mass of exactly 2 g. This volume is 22.4 L. Then, he assumed that 22.4 L of any other gas would have the same number of molecules as in 22.4 L of $H_2(g)$. (Here is where Avogadro's hypothesis entered.) Finally, he reasoned that the ratio of the mass of 22.4 L of any other gas to the mass of 22.4 L of $H_2(g)$ should be the same as the ratio of their molecular masses. The gases in the table all contain the element X. Their molecular masses were determined by Cannizzaro's method. Use the percent composition data to deduce the atomic mass of X, the number of atoms of X in each of the gas molecules, and the identity of X.

Compound	Molecular mass, u	Mass percent X, %
Nitryl fluoride	65.01	49.4
Nitrosyl fluoride	49.01	32.7
Thionyl fluoride	86.07	18.6
Sulfuryl fluoride	102.07	31.4

* **125. [Collaborative]** The following experiment was once commonly done in introductory science classes to show that air is one-fifth oxygen: A candle was placed in a pan half filled with water. The candle was lighted and a glass cylinder was inverted over the flame, trapping a certain quantity of air. The candle burned and water rose in the cylinder. By the time the candle went out, water had replaced about 20% of the enclosed gas. Together with other members of your study group, answer the following questions. **(a)** Does this experiment, in fact, prove that air is 20% oxygen, by volume? Why or why not? (Assume the "formula" $C_{25}H_{52}$ for candle wax.) **(b)** Propose variations in the experiment that might support or disprove your conjectures. **(c)** Propose an alternative combustion reaction that might be used to determine the volume percent of oxygen in air. What are the advantages or disadvantages of the alternative reaction?

* **126. [Historical]** Based on the essay on page 189, calculate the approximate diameter of the balloon used by Jacques Charles in his flight. Note any assumptions you have made in your calculation.

 # e-Media Problems

The activities described in these problems can be found in the e-Media Activities and Interactive Student Tutorial (IST) modules of the Companion Website, *http://chem.prenhall.com/hillpetrucci*.

127. View the **Motion of a Gas** animation *(Section 5-2)*. **(a)** Follow an individual gas particle as it travels throughout the vessel and describe the types of interactions that occur. **(b)** How would the animation differ if fewer gas particles were present in the vessel?

128. Observe the reaction of sodium azide in the **Air Bags** animation *(Section 5-9)*. **(a)** What mass of sodium azide must react to produce 25.6 L of gas at a pressure of 1.10 atm and a temperature of 30 °C? **(b)** If the heat of reaction raised the temperature of the 25.6 L of gas to 200 °C, what would be the resulting pressure within the air bag?

129. Using the **Ideal Gas Behavior** simulation *(Section 5-8)*, **(a)** list the pairs of gas properties that exhibit a direct proportionality between the two (an *x–y* plot of the two properties produces a straight line). **(b)** For each pair, write an expression for the slope of the line in terms of other quantities present in the ideal gas law.

130. In the **Diffusion of Bromine Vapor** movie *(Section 5-11)*, the gaseous molecules experience a resistance to "flow" up the tube. **(a)** Predict the effect of lowering the atmospheric pressure in the tube. **(b)** If the same experiment were carried out with water, would water vapor reach the clamp faster or more slowly than bromine vapor? (Ignore the influence of attractive intermolecular forces.)

131. In the **Real Gas Behavior** simulation *(Section 5-12)*, "tune" each variable of the simulation, one variable at a time, over the full range of available values. In each case, hold all but one of the other variables constant and observe the behavior of the dependent variable. Determine the range of the tuned variable that causes the dependent variable to deviate from ideal gas behavior by more than one percent.

Chapter 6

Thermochemistry

WE CARRY OUT many chemical reactions just for the associated energy changes, not to obtain material products. Consider the combustion of methane, the principal component of natural gas:

$$CH_4(g) + 2 O_2(g) \longrightarrow CO_2(g) + 2 H_2O(l)$$

We burn methane for cooking and to warm our homes, using the heat that is released when methane burns.

In fact, we sometimes use energy-producing reactions *in spite of* the products. Modern industrialized nations run mainly on fossil fuels: natural gas, petroleum, and coal. Unfortunately, burning these fuels produces more than heat. The formation of such air pollutants as carbon monoxide, oxides of nitrogen, and particulate matter accompanies the combustion of fuels. The product carbon dioxide is a greenhouse gas; it and certain other gases trap heat in Earth's atmosphere, and most atmospheric scientists think that a buildup of these gases will lead to global warming, an important environmental issue.

In this chapter, after first examining a few fundamental concepts, we will look at some quantitative relationships involving chemical reactions and energy changes. We will conclude the chapter with a discussion about the foods that sustain life and the fuels that enrich it.

CONTENTS

6.1 Energy

Because we will use the concepts of energy and energy changes in nearly everything we do in this chapter, let's briefly consider some basic ideas about energy. Also, you should read Appendix B for more detail on some of these matters.

The term *energy* is derived from Greek and means, literally, "work within." However, no object of matter *contains* work. Work is done only when matter is moved. We might say, then, that energy is the capacity to do work, the capacity to move something. Objects that can do work either because of their composition or because of their location possess **potential energy.** Moving objects also have the capacity to do work, such as moving other objects they may strike; moving objects possess **kinetic energy.** In Chapter 5, we discussed the kinetic energy associated with the motion of gas molecules.

◄ In one part of the steelmaking process, oxygen gas is blown into the molten metal. Although the oxygen is cold compared to the liquid steel, the steel remains molten, in part because the chemical reactions that occur with oxygen evolve energy as a byproduct. Thermochemistry is a study of the heat and work associated with chemical reactions.

▲ **FIGURE 6.1** **Energy conversion**
The potential energy of the water stored behind this dam is converted to kinetic energy as the water falls through conduits called penstocks. These conduits deliver the flowing water to turbines, where the flowing water turns rotors in the turbines, thereby producing electricity.

Let's use the practical example shown in Figure 6.1 to compare these two types of energy. Water behind a dam has *potential* energy due to its position and gravitational attraction to Earth's center. The water therefore has the capacity to do work, but as long as it remains behind the dam, it does none. When the water is allowed to flow through a pipe to a lower level, some of its potential energy is converted to *kinetic* energy. Water rushing through the pipe can be made to turn the blades of a turbine (a waterwheel), which in turn can rotate a coil of wire in an electrical generator, producing electricity. The net result is that some of the potential energy originally stored in the water is converted to electrical work.

To get an idea of the magnitudes and units of quantities of energy, let's use the mathematical definition of kinetic energy (E_k) that we introduced in Section 5.11:

$$E_k = \tfrac{1}{2}mv^2 *$$ **(6.1)**

If we express mass (m) in kilograms and speed (v) in meters per second, the units of kinetic energy are

$$(kg) \times (m/s)^2 = kg\ m^2\ s^{-2}$$

For example, a 2-kg object moving at a speed of 1 m/s has kinetic energy

$$E_k = \tfrac{1}{2} \times 2\ kg \times (1\ m/s)^2 = 1\ kg\ m^2\ s^{-2} = 1\ J$$

The **joule (J),** which is defined as $1\ kg\ m^2\ s^{-2}$, is the SI unit of energy.

From a macroscopic point of view, the joule is a tiny energy unit. A bowling ball rolling slowly down a bowling lane has a kinetic energy of a few joules. A 60-watt lightbulb uses 60 J of energy per second. To describe an energy change during a chemical reaction, the kilojoule (kJ) is commonly used: 1 kJ = 1000 J.

Before we can establish the relationship between energy and work, we need to consider a couple of other quantities. The **work** required to move an object is the product of a force acting in the direction of motion and the distance the object is moved:

$$Work = force \times distance$$ **(6.2)**

As described in Appendix B, a force is required to change the speed of an object, that is, to *accelerate* it. In SI base units of kilogram, meter, and second, one *newton* of force imparts an acceleration of one meter per second per second to a 1-kg object:

$$Force = 1\ kg \times 1\ m\ s^{-2} = 1\ newton\ (N)$$ **(6.3)**

The work done by a force of 1 N acting over a distance of 1 m is 1 newton-meter (N m), which leads to the result

$$Work = 1\ N \times 1\ m = 1\ kg\ m^2\ s^{-2} = 1\ J$$

Thus, work has the same unit as energy, the joule.

To further illustrate the relationship between energy and work, let's consider the bouncing tennis ball in Figure 6.2. First, we have to do work to raise the ball to its starting position. That is, we have to apply an upward force on the ball to overcome the force of gravity that tends to pull the ball straight down. This work of lifting is stored in the ball as potential energy.

When the ball is released, it falls as it is pulled toward Earth's center by the force of gravity. As the ball falls, some of its stored potential energy is converted to kinetic energy of the moving ball. The kinetic energy reaches a maximum just as the ball strikes the surface. As the ball rebounds, its kinetic energy decreases (the ball slows down) as its potential energy increases (the ball rises).

If its collisions with the immovable surface were perfectly *elastic* (like those we assume for gas molecules in the kinetic-molecular theory), the ball would always rebound with the same energy it had before the collision. At every point on its path, the *sum* of the potential and kinetic energies would be constant: The ball would return to its original highest position on each bounce, and it would go on bouncing forever. We know, however, that a real ball behaves otherwise. In reality, the maximum height of

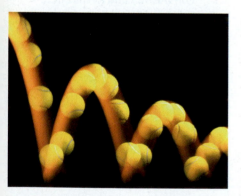

▲ **FIGURE 6.2** **Potential energy and kinetic energy**
A bouncing ball illustrates the interconversion of potential energy and kinetic energy.

QUESTION: At what point in each bounce is the potential energy of the ball at a maximum?

Potential and Kinetic Energy animation

---

*In Section 5.11, we used the symbol u to describe a molecule's speed and the symbol e_k to describe its kinetic energy. For macroscopic quantities of matter, we will use the symbols v and E_k.

each bounce is less than the height of the previous one, and the ball eventually comes to rest on the surface. The potential energy of the ball at its original height is lost, and it no longer appears as kinetic energy of the ball itself, which is at rest. Instead, it appears as additional kinetic energy in the atoms and molecules that make up the ball, the surface, and the surrounding air, and results in a slight increase in the temperature of the ball and its surroundings. Thus, we see that some energy originally on the macroscopic level has changed to energy on the microscopic level and that the situation is a bit more complicated than it might first appear. We need to explore a few basic ideas in more detail.

6.2   Thermochemistry: Some Basic Terms

Thermochemistry is the study of energy changes that occur during physical processes and chemical reactions. It is a branch of the broader study of the relationship between heat and work called *thermodynamics* (see Chapter 17). Usually we assess energy changes in terms of something losing energy and something else gaining it. When we warm our cold hands over a campfire, the burning wood gives off energy (as heat) and our hands gain energy, raising their temperature. For scientific purposes, however, we need to describe the two "somethings" that exchange energy a little more precisely.

System and Surroundings

In thermochemistry and thermodynamics, we define the **system** as that part of the universe we are studying. The system may be as simple as a solution in a beaker, a gas in a cylinder, or a block of frozen spinach in a dish, or it may be as complicated as a polluted lake or even Earth's atmosphere. The **surroundings** are the rest of the universe, although we can typically limit our concern only to those parts of the surroundings that interact with the system. *Interactions* refer to the exchange of energy or matter or both between a system and its surroundings. The surroundings of a solution in a beaker, for example, might be the beaker itself, the surface on which the beaker is placed, and the surrounding air with which the solution effectively exchanges heat, matter, or both.

We commonly encounter three types of systems. An *open system* can exchange either matter or energy (or both) with its surroundings. A steaming hot cup of coffee is an open system; the coffee loses heat and water vapor to its surroundings, as seen by its gradual cooling and the appearance of condensed water vapor (steam) in the cooler air above the coffee. A *closed system* may exchange energy but not matter with the surroundings. An example is tea bags in a capped bottle of water placed in sunlight to make sun tea, represented in Figure 6.3. Only energy (as heat) is transferred from the surroundings to the system. An *isolated system* exchanges neither matter nor energy

(a)

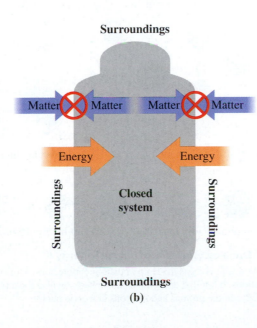

(b)

◀ **FIGURE 6.3   A real-world closed system as compared to the chemist's conception**

(a) Preparation of sun tea. (b) The sketch suggests that the system is closed: Energy enters the system from the surroundings, but no matter leaves or enters.

QUESTION: After the lid of the jar is unscrewed, which kind of system is it?

with its surroundings. A tightly stoppered thermos flask with hot coffee is, at least for a short time, a good approximation of an isolated system. In our study of thermochemistry, we will focus on simple systems, often considering them as being isolated from their surroundings. We know from experience, however, that we cannot isolate a system completely. Thus, the stoppered thermos flask is not a truly isolated system; the contents of the flask slowly cool, indicating that heat is being lost to the air surrounding the flask.

Internal Energy

The **internal energy** (U) of a system is the total energy contained within the system, partly as kinetic energy and partly as potential energy. As Figure 6.4 suggests, the kinetic energy component comes from various types of motion at the molecular level. Such movements include straight-line motion, or *translational motion,* of molecules (Figure 6.4a); the spinning, or *rotational motion,* of molecules (Figure 6.4b); and the displacements of atoms within molecules, called *vibrational motion* (Figure 6.4c). Collectively, the kinetic energy contributions to internal energy are sometimes called *thermal energy.* The potential energy component of internal energy comes from interactions between particles of matter. Some of these interactions, such as those that hold protons and neutrons together in atomic nuclei, remain unchanged in chemical reactions, and we do not concern ourselves with them. The most important interactions that chemists consider are the electrostatic attractions that produce chemical bonds between atoms. These are known as *intramolecular forces* (Figure 6.4d). Also important are electrostatic attractions between molecules, called *intermolecular forces* (Figure 6.4e). Collectively, these potential energy contributions to internal energy are sometimes referred to as *chemical energy.*

Heat

Heat (q) is a quantity of energy *transferred* between a system and its surroundings as a result of a temperature difference between them. Heat passes spontaneously from the region of higher temperature to the region of lower temperature. Heat transfer stops when the system and surroundings reach the same temperature, at which point the system and surroundings are said to be at *thermal equilibrium.* Although we often use expressions such as "heat flows," "heat is lost," and "heat is gained," *a system does not contain heat.* Rather, it contains energy—internal energy.

To consider heat transfer at the molecular level, recall that temperature is a measure of the average translational kinetic energy of the molecules in a sample. As suggested in Figure 6.5, kinetic energy is transferred during molecular collisions at the

Components of Internal Energy animation

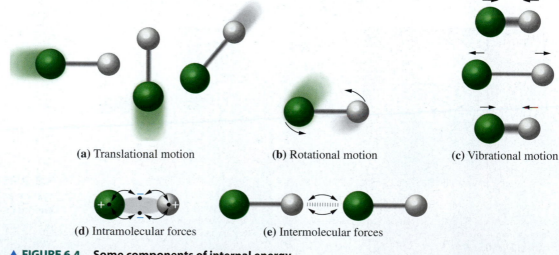

(a) Translational motion   **(b)** Rotational motion   **(c)** Vibrational motion

(d) Intramolecular forces   **(e)** Intermolecular forces

▲ **FIGURE 6.4** **Some components of internal energy**
The system suggested here is a gas made up of diatomic molecules. Most real-world systems are more complex. Other important contributions to internal energy, not shown here, involve the attractions between the electrons and protons in atoms and those between the protons and neutrons in atomic nuclei.

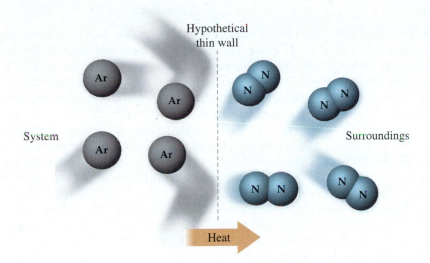

System

Hypothetical
thin wall

Surroundings

Heat

▲ FIGURE 6.5   **Heat transfer between a system and surroundings**

In their collisions with the hypothetical thin wall, the more energetic gaseous argon atoms of the system transfer some kinetic energy to the less energetic nitrogen molecules in the cooler surroundings. The average translational kinetic energy of the argon atoms in the system decreases and that of the nitrogen molecules in the surroundings increases until the average translational kinetic energies of the two gases become equal. At this point of thermal equilibrium, the temperatures of the system and surroundings have become equal and heat no longer flows.

QUESTION: How do the root-mean-square speeds of the argon atoms and nitrogen molecules compare at the point of thermal equilibrium?

 Heat Transfer animation

interface between a system and its surroundings. The more energetic molecules (the ones at the higher temperature) lose energy to the less energetic molecules (the ones at the lower temperature). In effect, heat is simply a transfer of molecular kinetic energy that brings a system and its surroundings to the same temperature.

Now, let's consider a system consisting of methane gas burning in oxygen. This is the principal reaction that takes place in a laboratory Bunsen burner (Figure 6.6). We will define the methane and oxygen gases and the reaction products as the system and the burner and everything else around it as the surroundings. When the methane burns, some of the energy associated with chemical bonds in the methane and oxygen molecules—chemical energy—is converted to thermal energy, and the temperature of the system rises. Because the temperature of the system is now above that of the surroundings, heat is transferred from the system to the surroundings as the gaseous molecules of the system collide with the molecules in the surrounding air.

Next consider a system composed of water initially in the form of a block of ice. When the ice is placed in surroundings above 0 °C, heat passes from the surroundings into the ice (Figure 6.7). The heat absorbed by the system is used to melt the ice.

▲ FIGURE 6.6   **A reaction evolving heat**
In the combustion of methane (natural gas), heat is given off by the system—the burning gas—to the surroundings—the air around the flame, the burner, the bench top, and so on. The thermometer closer to the system is at a higher temperature (130.1 °C) than is the one farther away (24.4 °C). These temperature readings indicate that heat flows from the system to the surroundings.

▲ FIGURE 6.7   **Another reaction involving heat**
In the melting of ice, heat is given off by the surroundings (the air around the ice) to the system (the block of ice). The temperature in the system (0 °C) is lower than the temperature in the surroundings (22.9 °C and 19.4 °C). These temperature readings trace the heat flow from the surroundings into the system. As we know from common experience, the heat absorbed by the system causes ice to melt.

Work

Work, (w), like heat, is an energy transfer between a system and its surroundings; *a system does not contain work.* There are several types of work, but for now we will consider only *pressure–volume work*—work done when gases either expand or are compressed. Later in the text we will discuss another type—electrical work.

In Figure 6.8, a gas is confined in a cylinder by a freely moving, massless piston. At first, the gas is maintained in a fixed volume by the two weights atop the piston. When one of these weights is removed, the gas expands and the remaining weight is lifted through the distance h. At this point, the volume has increased and the piston no longer moves. To raise the weight, the gas had to exert the force necessary to offset the force of gravity on the weight. This force acted through the distance h, and the product of a force and a distance is an amount of work.

In Chapter 5, we described forces exerted by gas molecules through the concept of pressure. Pressure is a force divided by the area over which the force is distributed: $P = F/A$. The force exerted by the gas, then, is given by $F = P \times A$. We can combine these ideas into the expression

$$\text{Work } (w) = \text{force } (F) \times \text{distance } (h) = P \times A \times h \qquad \textbf{(6.4)}$$

The product of the cross-sectional area A of the cylinder and the height h represents the *change* in volume of the gas, which is designated by the symbol ΔV. (The Greek letter *delta,* Δ, is commonly used to represent a change in some quantity.) Mathematically, we can represent a change as some final value minus an initial value: $\Delta V = V_{\text{final}} - V_{\text{initial}}$. By replacing $A \times h$ in Equation (6.4) with its equivalent, ΔV, we see that the work associated with a gas expanding against a constant external pressure is

$$\text{Work } (w) = \text{pressure } (P) \times \text{change in volume } (\Delta V) = P\Delta V$$

This product of pressure and a change in volume is usually called pressure–volume work. In Figure 6.8, the constant external pressure is that exerted by the single weight in the cylinder on the right, and the change in volume (ΔV) is indicated.

As a final adjustment to the equation for pressure–volume work, we need to add a negative sign:

$$\text{Work } (w) = -P\Delta V \qquad \textbf{(6.5)}$$

Thus, when a gas expands, ΔV is positive and the work is negative. A negative quantity of work signifies that the system loses energy, that is, that energy passes from the system to the surroundings. When a gas is compressed by its surroundings, ΔV is negative, the quantity of work is positive, and energy is gained by the system. We will say more about this and other sign conventions in the next section.

Work of Gas Expansion animation

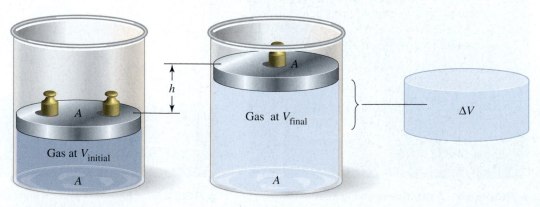

▲ **FIGURE 6.8** **Pressure–volume work**
The gas expands when one of the weights is removed. The remaining weight is then lifted through the distance h by the expanding gas. The gas volume increases by an amount ΔV, and the gas has done work on the surroundings (the weight).

QUESTION: How would the magnitude of ΔV compare to the original gas volume if the two weights were identical?

6.3 Internal Energy (*U*), State Functions, and the First Law of Thermodynamics

Earlier (page 218), we described some of the kinds of energy that collectively make up the internal energy of a system. However, we did not say how one goes about determining an actual value of this quantity, and for good reason: We cannot measure the absolute value of the internal energy of a system because we cannot account exactly for all the energy quantities.

Fortunately, internal energy has one crucial property that enables us to use it successfully despite our inability to measure it. *Internal energy is a function of state*. The **state** of a system refers to its exact condition, determined by the kinds and amounts of matter present, the structure of this matter at the molecular level, and the prevailing temperature and pressure. A **state function** is a property that has a unique value that depends *only* on the present state of the system and *not* on how the state was reached.

Interestingly, heat, *q*, and work, *w*, are *not* state functions. Because they are not contained in a system, they are quantities of energy that we observe *only when a system changes from one state to another*. Consequently, the values of *q* and *w* depend on the way in which a change is brought about—they depend on the *path* chosen.

Perhaps we can better understand state functions by considering the analogy to climbing a mountain, presented in Figure 6.9. Think of the base of the mountain as the initial state and the summit as the final state. Climbing the mountain is like a thermodynamic process that proceeds from an initial to a final state. The *elevation gain*, the difference in elevation between the base and summit of the mountain, is analogous to the *change* in internal energy, ΔU. We can measure elevation gain without knowing the actual elevation of either the base or the summit of the mountain. The gain in elevation has a unique value regardless of the path we climb. Similarly, because the internal energies of both the initial and final states of a system have unique values, so does the ΔU of the system. That is,

$$\text{Initial state } (U_{\text{initial}}) \longrightarrow \text{final state } (U_{\text{final}})$$

$$\Delta U = U_{\text{final}} - U_{\text{initial}} \tag{6.6}$$

In the mountain-climbing analogy, when we return to the base of the mountain from the summit, our net elevation gain Δh is *zero* for the round-trip. The positive elevation gain we make in climbing to the summit is wiped out by the negative elevation gain when we return to the base. This analogy illustrates an important property of state functions:

A state function returns to its initial value if a process is reversed and a system is returned to its initial state.

As we have described, a change in internal energy—the difference in internal energy between two states of a system—has a unique value. But how can we measure this difference? Here, we are aided by another fundamental scientific law—the **law of conservation of energy:**

In a physical or chemical change, energy can be exchanged between a system and its surroundings, but no energy can be created or destroyed.

This law implies that if we can account for all the energy exchanges between a system and its surroundings, we should be able to determine how the internal energy of the system changed in a physical or chemical process. That is, the *change* in internal energy of a system, ΔU, must be related to the energy exchanges that occur as heat, *q*, and work, *w*. When we recast the law of conservation of energy in this way, it is called the **first law of thermodynamics,** expressed mathematically as

$$\Delta U = q + w \tag{6.7}$$

We can see from this equation that in an *isolated* system the internal energy remains constant. That is, if a system can exchange neither heat nor work with its surroundings, *q* and *w* are both *zero* and ΔU for the system is also *zero*. It certainly seems that ΔU for a *closed* system should also be zero if the system *gains* energy as heat but *loses* an equal quantity as work. However, the defining equation gives this result only

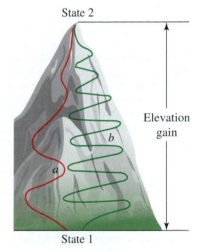

▲ **FIGURE 6.9 Mountain-climbing analogy to state functions**

Path *a* (red) is shorter than path *b* (green), and so a shorter distance is walked along path *a*. The gain in elevation, however, is independent of which path is chosen. The internal energy of a system is also independent of the path by which a change occurs. Therefore internal energy, *U*, is a *state function*.

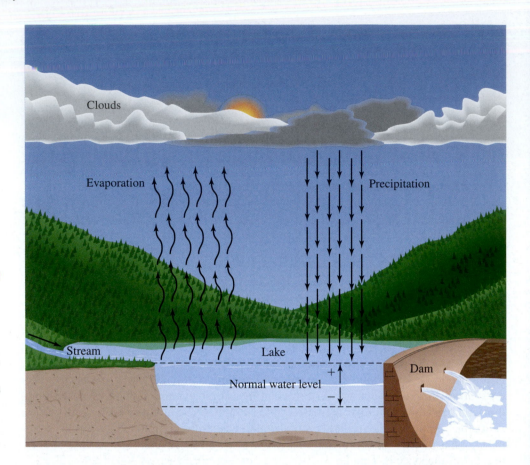

▶ **FIGURE 6.10   Analogy to internal energy changes**
Throughout the year, the balance between processes adding water (through stream runoff and precipitation) and processes removing water (through evaporation and dam release) determines the change in the water level of a mountain reservoir. Similarly, the summation of all the interactions that involve the transfer of energy (in the form of heat and work) between a system and its surroundings determines the change in internal energy for the system.

Sign Conventions for Work and Heat animation

if one of the energy exchanges has a positive sign and the other a negative sign. This suggests that we need to establish some sign conventions to be used with the first law of thermodynamics. The conventions are

- Energy *entering* a system carries a *positive* sign: If heat is *absorbed* by the system, $q > 0$. If work is done *on* the system, $w > 0$.
- Energy *leaving* a system carries a *negative* sign: If heat is *given off* by the system, $q < 0$. If work is done *by* the system, $w < 0$.

Accounting for energy exchanges between a system and its surroundings is much like monitoring the water level in a mountain reservoir, as suggested by Figure 6.10. The reservoir level, which changes during the year, is determined both by how much water is introduced into the reservoir and by how much is removed over the course of the year. The balance (summation) of these processes determines the water level much in the same way that both heat and work contribute to the change in internal energy of a system.

Example 6.1

A gas does 135 J of work while expanding, and at the same time it absorbs 156 J of heat. What is the change in internal energy?

STRATEGY

The key to getting the correct value of ΔU in Equation (6.7) is to use the correct signs for the given quantities, w and q. Note that heat is *absorbed by* the system; this means q is a *positive* quantity: $q = +156$ J. Work is *done by* the system; this means w is a *negative* quantity: $w = -135$ J.

SOLUTION

We add the values for q and w, each with its correct sign. Note that because more heat is absorbed than work done, the internal energy increases.

$$\Delta U = q + w = +156 \text{ J} + (-135 \text{ J}) = +21 \text{ J}$$

EXERCISE 6.1A

In a process in which 89 J of work is done on a system, 567 J of heat is given off. What is ΔU of the system?

EXERCISE 6.1B

In a particular process, the internal energy of a system increases by 41.4 J and the quantity of work the system does on its surroundings is 81.2 J. Is heat absorbed or given off by the system? What is the value of q?

Example 6.2   A Conceptual Example

The internal energy of a fixed quantity of an ideal gas depends only on its temperature. If a sample of an ideal gas is allowed to expand against a *constant* pressure at a *constant* temperature, **(a)** what is ΔU for the gas? **(b)** Does the gas do work? **(c)** Is any heat exchanged with the surroundings?

ANALYSIS AND CONCLUSIONS

(a) Because the expansion occurs at a constant temperature, the expanded gas (state 2) is at a lower pressure than the compressed gas (state 1) but the temperature is *unchanged*. Because the internal energy of the ideal gas depends only on the temperature, $U_2 = U_1$ and $\Delta U = U_2 - U_1 = 0$.

(b) The gas does work in expanding against the confining pressure, P. The pressure–volume work is $w = -P\Delta V$, as was illustrated in Figure 6.8. The work is negative because it is done *by* the system.

(c) The work done by the gas represents energy leaving the system. If this were the only energy exchange between the system and its surroundings, the internal energy of the system would decrease, and so would the temperature. However, because the temperature remains constant, the internal energy does not change. This means that the gas must absorb enough heat from the surroundings to compensate for the work that it does in expanding: $q = -w$. And, according to the first law of thermodynamics, $\Delta U = q + w = -w + w = 0$.

EXERCISE 6.2A

In an *adiabatic* process, a system is thermally insulated from its surroundings so that there is no exchange of heat ($q = 0$). If an ideal gas undergoes an adiabatic expansion against a constant pressure, **(a)** does the gas do work? **(b)** Does the internal energy of the gas increase, decrease, or remain unchanged? **(c)** What happens to the temperature?

EXERCISE 6.2B

A sample of an ideal gas does 152 kJ of work as it very slowly expands at constant temperature. To compress the gas very quickly back to its original condition, 209 J of work must be done. How is it possible for this expansion/compression process to occur? What will be ΔU for the overall process? Explain.

We have now examined in some detail the three quantities that are related by the first law of thermodynamics: $\Delta U = q + w$. Of these three, the easiest to measure experimentally is q. When the thermodynamic process is a chemical reaction, the heat involved, q, is called the *heat of reaction*. We will consider heats of reaction in the next section and also introduce another important thermodynamic function called *enthalpy*.

6.4   Heats of Reaction and Enthalpy Change, ΔH

We can think of a chemical reaction as a process in which a thermodynamic system changes from an *initial* state (the reactants) to a *final state* (the products). A quantity of heat is almost always associated with the change, and sometimes work as well. We classify thermochemical processes by the descriptive terms *exothermic* and *endothermic*. First, we will focus on how exothermic and endothermic reactions relate to the first law of thermodynamics.

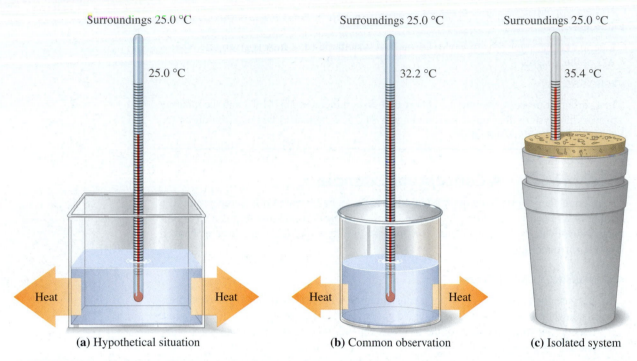

(a) Hypothetical situation **(b)** Common observation **(c)** Isolated system

▲ **FIGURE 6.11** **Conceptualizing an exothermic reaction**
In an exothermic reaction, chemical energy in a system is converted to thermal energy. (a) In a hypothetical situation, the thermal energy is instantly released as heat to the surroundings. There is no temperature change in the system. (b) A typical system is not isolated, nor is the flow of heat instantaneous. Some heat flows to the surroundings, and the remainder increases the temperature of the system. (c) A perfectly isolated system, suggested by the plastic-foam cups, retains all the heat in the system. A maximum temperature increase occurs.

QUESTION: How does the system temperature change as a function of time following the exothermic reaction carried out in the system shown in (b)? Why does it change in this way?

During an **exothermic reaction**, chemical energy in a system is converted to thermal energy. Changes characteristic of exothermic reactions are

- An exothermic reaction in an *isolated* system produces a temperature increase in the system.
- An exothermic reaction in a nonisolated (*open* or *closed*) system causes heat to be given off to the surroundings. Thus, $q < 0$—a *negative* quantity.

These situations are pictured in Figure 6.11. In a nonisolated system shown in Figure 6.11a, the temperature of the system, initially 25 °C, would not change at all if the heat evolved during the exothermic reaction could escape instantaneously. Part (a) depicts a hypothetical situation, however, and what we commonly observe instead is the situation in Figure 6.11b, where the temperature of the nonisolated system does rise to some extent initially because the exothermic heat of reaction cannot escape fast enough. In the isolated system in Figure 6.11c, no significant amount of heat escapes and the temperature of the system rises to the greatest extent possible.

In an **endothermic reaction**, thermal energy in a system is converted to chemical energy. Changes characteristic of endothermic reactions are

- An endothermic reaction in an *isolated* system produces a temperature decrease in the system.
- An endothermic reaction in a nonisolated (*open* or *closed*) system causes heat to be *absorbed* from the surroundings. Thus, $q > 0$—a *positive* quantity.

From arguments parallel to those for exothermic reactions, we can infer that an endothermic reaction in a perfectly isolated system causes the greatest possible temperature *decrease*. In a nonisolated system, the temperature drop will be *less* than in the isolated case, with heat passing from the surroundings into the system.

Application Note

Solid NaOH, sold in markets as *lye,* is a common drain cleaner. Dissolving lye in water is an *exothermic* process, and the liberated heat may melt congealed grease, allowing it to be flushed from a clogged drainpipe.

Application Note

Instant cold packs, used to treat sports injuries, have separate compartments of water and solid ammonium nitrate in a plastic bag. When the barrier between the two is broken, the NH_4NO_3 dissolves in the water endothermically. Heat is absorbed from the surroundings—including the injured area of the athlete's body.

The **heat of reaction**, sometimes denoted by the symbol q_{rxn}, is the quantity of heat exchanged between a reaction system and its surroundings when the reaction occurs *at a fixed temperature*. This definition implies measuring the quantity of heat represented in Figure 6.11a for an exothermic reaction. We will see how to do this when we describe the experimental measurement of heats of reactions. In addition to energy exchanges as heat, a reaction can produce work in several different forms, but most commonly as the pressure–volume work associated with the expansion or compression of gaseous reactants and products.

Before proceeding further with heats of reaction, let's pause to summarize a few ideas: We can treat a chemical reaction as a thermodynamic system in which the reactants represent the initial state (state 1) and the products represent the final state (state 2):

$$\text{Reactants} \longrightarrow \text{Products}$$
$$\text{State 1: } U_1 \qquad\qquad \text{State 2: } U_2$$

The internal energies of these two states have unique values, as does their difference:

$$\Delta U = U_2 - U_1$$

The change in internal energy of the system is related to the exchanges of heat and work between the system and its surroundings, as given by the first law of thermodynamics:

$$\Delta U = q + w$$

Even though q and w depend on the path chosen, ΔU has a unique value for a given reaction; that is, the value of ΔU does not depend on the method used to carry out the reaction. Recall also that pressure–volume work is given by Equation (6.5):

$$w = -P\Delta V$$

Now, let us return to our discussion of heats of reaction. Most reactions are carried out either at constant volume or at constant pressure. If we carry out a reaction in a system of constant volume, the volume cannot change during the reaction, and therefore $\Delta V = 0$. If $\Delta V = 0$, however, then, so too must $P\Delta V = 0$. This means that no pressure–volume work is done. And, if no other type of work takes place during the reaction—as is often the case—then $w = 0$. Now notice the implication of this fact on the heat of reaction, q:

$$\Delta U = q + w$$
$$\Delta U = q + 0$$

The heat of reaction q is equal to ΔU for this reaction. A heat of reaction *at constant volume* is represented by the symbol q_V (the subscript V indicates that the volume remains constant), and so we can write

$$\Delta U = q_V \tag{6.8}$$

Now consider a reaction carried out at *constant pressure* and with the only work done being pressure–volume work. In this case, we can use the subscript P to represent a heat of reaction at constant pressure: q_P. Then, when we substitute $-P\Delta V$ for w, the first law of thermodynamics becomes

$$\Delta U = q + w = q_P - P\Delta V$$

Finally, let's rearrange the equation to isolate the term q_P:

$$q_P = \Delta U + P\Delta V \tag{6.9}$$

Now we can use Figure 6.12 to help us visualize two ways of carrying out an exothermic reaction. When the reaction is carried out at constant volume, as in the tightly sealed reaction vessel of Figure 6.12a, where the piston cannot move, all the thermal energy produced by conversion from chemical energy is released as heat. That is, $q_V = \Delta U$. And, because the reaction is exothermic, both q_V and ΔU are *negative* quantities.

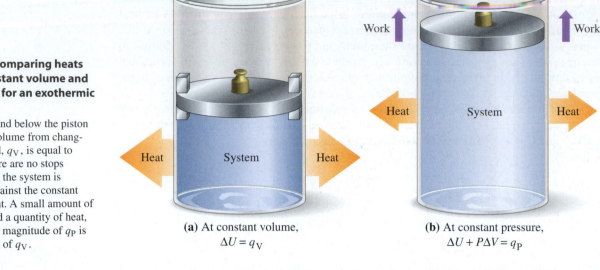

(a) At constant volume,
$\Delta U = q_V$

(b) At constant pressure,
$\Delta U + P\Delta V = q_P$

▶ **FIGURE 6.12 Comparing heats of reaction at constant volume and constant pressure for an exothermic reaction**

(a) The stops above and below the piston prevent the system volume from changing. The heat evolved, q_V, is equal to ΔU. (b) Because there are no stops restricting the piston, the system is allowed to expand against the constant pressure of the weight. A small amount of work, w, is done, and a quantity of heat, q_P, is generated. The magnitude of q_P is slightly less than that of q_V.

 Heats of Reaction at Constant Volume and Constant Pressure animation

When the reaction is carried out at constant pressure, as in Figure 6.12b where the piston is free to move, most of the thermal energy is also released as heat, but a small amount may appear as the work required to expand the system against the surroundings. The quantity of heat liberated is somewhat less than in the constant-volume case. Notice how the equation $q_P = \Delta U + P\Delta V$ leads us to the same conclusion: ΔU is negative, the $P\Delta V$ term is positive, and their sum, q_P, is not quite as negative as is ΔU.

Some systems, of course, may contract in volume when an exothermic reaction is carried out at constant pressure. In this case, the surroundings do work on the system. Now the $P\Delta V$ term is negative and q_P is more negative than ΔU.

Enthalpy

For a reaction carried out at constant volume, we can determine ΔU simply by measuring q_V. However, most chemical reactions are carried out at constant pressure, not at constant volume. They take place in vessels open to the atmosphere, and the pressure in the reaction system is the prevailing atmospheric pressure. Figure 6.13 shows one such atmospheric-pressure reaction, that of magnesium metal with hydrochloric acid, as it is carried out in a chemistry laboratory.

$$Mg(s) + 2\,HCl(aq) \longrightarrow MgCl_2(aq) + H_2(g)$$

Because the vast majority of heats of reaction that chemists measure are q_P values, it is useful to have a thermodynamic function whose change in a chemical reaction is exactly equal to q_P. Chemists have defined a function called *enthalpy* for just this purpose. **Enthalpy (H)** is defined as the sum of the internal energy and the pressure–volume product of a system:

$$H = U + PV \tag{6.10}$$

From this definition and Equation (6.9), $q_P = \Delta U + P\Delta V$, we see that the heat of reaction q_P is just the **enthalpy change (ΔH)** for a process carried out at constant temperature and pressure and with work limited to pressure–volume work:

$$q_P = \Delta H = \Delta U + P\Delta V \tag{6.11}$$

Some Properties of Enthalpy

Enthalpy (H) is especially useful because of certain properties it shares with internal energy (U) and certain other thermodynamic functions.

1. **Enthalpy is an *extensive* property.** An *extensive* property, such as mass or volume, depends on the quantity of matter observed. An *intensive* property, such as density, is independent of the quantity observed. The enthalpy of a system depends on the quantities of substances present, making enthalpy an extensive

▲ **FIGURE 6.13 Visualizing a reaction carried out at atmospheric pressure**

In the reaction

$Mg(s) + 2\,HCl(aq) \longrightarrow$
$\qquad MgCl_2(aq) + H_2(g)$

the evolved $H_2(g)$ pushes air out of the flask and into the surroundings. The expanding H_2 gas does work, represented by the blue arrows.

property. Even though we cannot measure the absolute enthalpy of a substance any more than we can measure its internal energy, we can say, for example, that the enthalpy of 2.00 mol CO_2 is exactly twice the enthalpy of 1.00 mol CO_2.

2. **Enthalpy is a state function.** The enthalpy of a system depends only on its present state and not on the route by which it got there. Because U, P, and V are all state functions and because H is a function of only those variables (that is, $H = U + PV$), H is also a state function.

3. **Enthalpy changes have unique values.** Because the enthalpy in each of two states of a system has a unique value, the difference in enthalpy between the two states—the enthalpy *change*, ΔH—also has a unique value. This change in enthalpy is equal to the heat of reaction at constant pressure: $\Delta H = q_P$.

Now, let's examine some practical consequences of these statements about enthalpy and enthalpy change.

Representing ΔH for a Chemical Reaction

First, we can expand our symbolic description of a chemical reaction, as shown here for the combustion of methane at 25 °C:

$$CH_4(g) + 2\ O_2(g) \longrightarrow CO_2(g) + 2\ H_2O(l) \qquad \Delta H = -890.3\ kJ$$

The enthalpy change in the combustion of 1 mol $CH_4(g)$ at 25 °C is -890.3 kJ. The quantity ΔH is equal to q_P, and its negative value shows that the combustion of methane at constant pressure is an exothermic reaction.

The equation representing the decomposition of mercury(II) oxide at 25 °C has a positive value of ΔH, indicating an endothermic reaction:

$$2\ HgO(s) \longrightarrow 2\ Hg(l) + O_2(g) \qquad \Delta H = +181.66\ kJ$$

Enthalpies and enthalpy changes are extensive properties. Therefore, the decomposition of 1 mol of $HgO(s)$ is accompanied by one-half the enthalpy change noted above:

$$HgO(s) \longrightarrow Hg(l) + \tfrac{1}{2} O_2(g) \qquad \Delta H = \tfrac{1}{2} \times 181.66\ kJ = +90.83\ kJ$$

We can use a chemical equation to represent enthalpy change, or we can use a graphical representation known as an **enthalpy diagram,** as illustrated in Figure 6.14.

If a process is carried out first in one direction and then brought back to its initial state, the total enthalpy change is zero because enthalpies are state functions. In any such *cyclic process,* the initial and final states are the same. This means that ΔH changes sign when a process is reversed (Figure 6.15). Thus, when we reverse the mercury(II) oxide decomposition reaction and form $HgO(s)$ from its elements at 25 °C, the enthalpy change for the reaction is

$$Hg(l) + \tfrac{1}{2} O_2(g) \longrightarrow HgO(s) \qquad \Delta H = -90.83\ kJ$$

Application Note

Lavoisier used this reaction in the discovery of the law of conservation of mass (page 36). To supply the necessary heat, he used a magnifying glass to focus sunlight on HgO(s) in sealed glass containers.

Problem-Solving Note

According to the text on page 14, after dividing the five-digit 181.66 kJ by 2, we should report the result as the four-digit 90.83 kJ. Otherwise, the 0.0011 percent uncertainty in 90.830 [that is, (0.001/90.830) × 100%] would be much less than the 0.0055% uncertainty in 181.66 [that is, (0.01/181.66) × 100%].

Reversing a Chemical Reaction animation

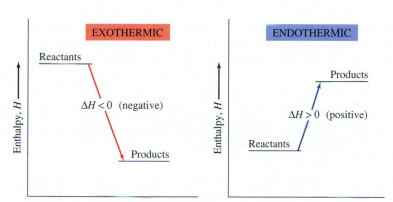

▲ **FIGURE 6.14   Enthalpy diagrams**
No numbers are shown on the enthalpy axis because absolute values of enthalpy cannot be measured. For an exothermic reaction (red), the products have a lower enthalpy than the reactants and ΔH is negative ($\Delta H < 0$). For an endothermic reaction (blue), the products have a higher enthalpy than the reactants and ΔH is positive ($\Delta H > 0$).

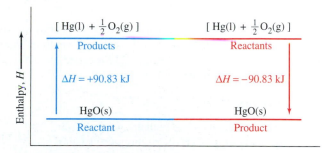

▲ **FIGURE 6.15   Reversing a chemical reaction**
The *forward* reaction (shown in blue by the upward-pointing arrow) is the dissociation at 25 °C of 1.00 mol HgO(s) into its elements. It is accompanied by an increase in enthalpy of 90.83 kJ; $\Delta H = +90.83$ kJ. In the *reverse* reaction (shown in red by the downward-pointing arrow), 1.00 mol HgO(s) is formed from its elements. This reaction is accompanied by a *decrease* in enthalpy of 90.83 kJ; $\Delta H = -90.83$ kJ. When a reaction is reversed, the magnitude of ΔH remains the same, but its sign changes.

Example 6.3

Given Equation (a), calculate ΔH for Equation (b).

(a) $H_2(g) + I_2(s) \longrightarrow 2\ HI(g)$ $\Delta H = 152.96\ kJ$

(b) $HI(g) \longrightarrow \frac{1}{2} H_2(g) + \frac{1}{2} I_2(s)$.

STRATEGY

Our first requirement is to determine how Equation (b) is related to Equation (a). Then we must adjust the given enthalpy change to reflect the ways in which Equation (a) was modified to get Equation (b).

SOLUTION

$HI(g)$ is on the right in Equation (a) but on the left in Equation (b). We therefore first reverse Equation (a) and change the sign of the ΔH value.

(c) $2\ HI(g) \longrightarrow H_2(g) + I_2(s)$ $\Delta H = -52.96\ kJ$

Because the coefficient of $HI(g)$ is 2 in Equation (c) but only 1 in Equation (b), we must multiply the enthalpy change for the reaction by $\frac{1}{2}$.

$HI(g) \longrightarrow \frac{1}{2} H_2(g) + \frac{1}{2} I_2(s)$ $\Delta H = \frac{1}{2} \times (-52.96) = -26.48\ kJ$

EXERCISE 6.3A

Given the equation

$$3\ O_2(g) \longrightarrow 2\ O_3(g) \qquad \Delta H = +285.4\ kJ$$

calculate ΔH for the reaction

$$\tfrac{3}{2} O_2(g) \longrightarrow O_3(g)$$

EXERCISE 6.3B

Given the equation

$$2\ Ag_2S(s) + 2\ H_2O(l) \longrightarrow 4\ Ag(s) + 2\ H_2S(g) + O_2(g) \qquad \Delta H = +595.5\ kJ$$

calculate ΔH for the reaction

$$Ag(s) + \tfrac{1}{2} H_2S(g) + \tfrac{1}{4} O_2(g) \longrightarrow \tfrac{1}{2} Ag_2S(s) + \tfrac{1}{2} H_2O(l)$$

Example 6.4

The complete combustion of liquid octane, C_8H_{18}, to produce gaseous carbon dioxide and liquid water at 25 °C and at a constant pressure gives off 47.9 kJ of heat per gram of octane. Write a chemical equation to represent this information.

STRATEGY

Because the reaction is carried out at constant pressure, the heat of reaction is a q_P value and therefore an enthalpy change ΔH. Moreover, the fact that heat is given off signifies that the reaction is exothermic and that ΔH is a negative quantity. Finally, the enthalpy change is given for 1 g of reactant; we must convert this to a molar basis because the coefficients in the equation represent moles of substances.

SOLUTION

First, let's convert the enthalpy change from the unit kJ/g C_8H_{18} to the unit kJ/mol C_8H_{18}.

$$?\ kJ/mol\ C_8H_{18} = \frac{-47.9\ kJ}{1\ g\ C_8H_{18}} \times \frac{114.2\ g\ C_8H_{18}}{1\ mol\ C_8H_{18}} = -5.47 \times 10^3\ kJ/mol\ C_8H_{18}$$

Now, we can write a balanced equation for the reaction and enter the ΔH value.

$$C_8H_{18}(l) + \tfrac{25}{2} O_2(g) \longrightarrow 8\ CO_2(g) + 9\ H_2O(l) \qquad \Delta H = -5.47 \times 10^3\ kJ$$

Note that if we remove the fractional coefficient by multiplying all the coefficients by 2, as we so often do in balancing an equation, we must also multiply the ΔH value by 2.

$$2\ C_8H_{18}(l) + 25\ O_2(g) \longrightarrow 16\ CO_2(g) + 18\ H_2O(l) \qquad \Delta H = -1.09 \times 10^4\ kJ$$

EXERCISE 6.4A

Write a chemical equation to express the fact that, at 0 °C, ice melts by absorbing 334 J of heat per gram of ice.

EXERCISE 6.4B

Express the following information as a chemical equation. At 25 °C and constant pressure, dinitrogen trioxide gas decomposes to nitrogen monoxide and nitrogen dioxide gases with the absorption of 0.533 kJ of heat for every gram of dinitrogen trioxide that decomposes.

ΔH in Stoichiometric Calculations

We can calculate the quantity of heat involved in a chemical reaction in much the same way that we calculate masses of reactants and products. That is, we can think of heat absorbed in an endothermic reaction as being a reactant and heat evolved from an exothermic reaction as being a product. The ΔH value then becomes the basis for a conversion factor. Consider, for example, the exothermic reaction

$$H_2(g) + Cl_2(g) \longrightarrow 2\,HCl(g) \qquad \Delta H = -184.6\text{ kJ}$$

This equation indicates that 184.6 kJ of heat is released by the reaction system to the surroundings for every 1 mol H_2 consumed. We can use such reasoning to write the conversion factors

$$\frac{-184.6\text{ kJ}}{1\text{ mol }H_2} \qquad \frac{-184.6\text{ kJ}}{1\text{ mol }Cl_2} \qquad \frac{-184.6\text{ kJ}}{2\text{ mol HCl}}$$

We can use such conversion factors to determine the enthalpy change for a reaction, given a specific quantity of a product or reactant, as in Example 6.5 and Exercise 6.5A.

Another typical situation is presented in Exercise 6.5B. There, we need a conversion factor in order to relate a reactant amount to a quantity of heat through a ΔH value. Then we use the ideal gas equation to convert the amount of the gaseous reactant to a gas volume at a given temperature and pressure.

Example 6.5

What is the enthalpy change associated with the formation of 5.67 mol HCl(g) in this reaction?

$$H_2(g) + Cl_2(g) \longrightarrow 2\,HCl(g) \qquad \Delta H = -184.6\text{ kJ}$$

STRATEGY

The quantity we are seeking, ΔH, has the unit kilojoules. The quantity on which the calculation is based is the amount of HCl(g) produced, and the only conversion factor required relates kilojoules of heat to moles of HCl.

SOLUTION

The conversion factor from the chemical equation is shown in red here:

$$?\text{ kJ} = 5.67\text{ mol HCl} \times \frac{-184.6\text{ kJ}}{2\text{ mol HCl}} = -523\text{ kJ}$$

The formation of 5.67 mol HCl in this exothermic reaction results in the release of 523 kJ of heat to the surroundings.

EXERCISE 6.5A

What is the enthalpy change when 12.8 g $H_2(g)$ reacts with excess $Cl_2(g)$ to form HCl(g)?

$$H_2(g) + Cl_2(g) \longrightarrow 2\,HCl(g) \qquad \Delta H = -184.6\text{ kJ}$$

EXERCISE 6.5B

What volume of $CH_4(g)$, measured at 25 °C and 745 Torr, must be burned in excess oxygen to release 1.00×10^6 kJ of heat to the surroundings in the reaction

$$CH_4(g) + 2\,O_2(g) \longrightarrow CO_2(g) + 2\,H_2O(l) \qquad \Delta H = -890.3\text{ kJ}$$

6.5   Calorimetry: Measuring Quantities of Heat

We have seen how ΔU and ΔH can be related to heats of reaction, how ΔH can be incorporated into chemical equations, and how we can include quantities of heat in stoichiometric calculations. Now, we need to consider another question: How do we *measure* a heat of reaction? We do so by measuring quantities of heat.

The process of measuring quantities of heat is called calorimetry, and the device in which the measurements are made is called a **calorimeter.** Calorimetry is based on the law of conservation of energy: Whatever heat is *lost* by a system must be *gained* by its surroundings (or vice versa). Successful calorimetry depends on our ability to measure quantities of heat accurately and to use the concepts of heat capacity and specific heat. We begin by introducing and illustrating these terms.

Heat Capacity

Some systems absorb a given quantity of heat from the surroundings and experience a large increase in temperature; they are said to have a low heat capacity. Other systems may absorb the same quantity of heat but experience a smaller temperature increase; they are said to have a higher heat capacity. The **heat capacity** (C) of a system is the quantity of heat required to change the temperature of the system by 1 °C (or by 1 K). We can determine a heat capacity by dividing the quantity of heat (q) by the temperature change it produces (ΔT):

$$C = \frac{q}{\Delta T} \tag{6.12}$$

The units of heat capacity are joules per degree Celsius (J/°C) or joules per kelvin (J/K). Because a change of 1 K is equal to a change of 1 °C, heat capacity has the same numerical value in either case.

Example 6.6

Calculate the heat capacity of an aluminum block that must absorb 629 J of heat from its surroundings in order for its temperature to rise from 22 °C to 145 °C.

SOLUTION

We need the ratio of q to ΔT. Because $q = 629$ J and ΔT is final temperature minus initial temperature, we have

$$C = \frac{q}{\Delta T} = \frac{629 \text{ J}}{(145 - 22)\ °C} = \frac{629 \text{ J}}{123\ °C} = 5.11 \text{ J/°C}$$

EXERCISE 6.6A

Calculate the heat capacity of a sample of brake fluid if the sample must absorb 911 J of heat in order for its temperature to rise from 15 °C to 100 °C.

EXERCISE 6.6B

A burner on an electric range has a heat capacity of 345 J/K. What is the value of q, in kilojoules, as the burner cools from 467 °C to 23 °C?

Molar Heat Capacity and Specific Heat

The heat capacity of a system depends on the quantity and type(s) of matter in the system. A large block of aluminum has a higher heat capacity than does a small piece of the metal. To avoid working with this extensive property, we can use either of two alternative intensive properties. The first, **molar heat capacity,** is the heat capacity of one mole of a substance. The second, **specific heat,** is the heat capacity of a 1-g sample; it is the quantity of heat required to change the temperature of 1 g of a substance by 1 K (or by 1 °C). We get the specific heat for a substance when we divide its heat capacity by its mass:

$$\text{Specific heat} = \frac{\text{heat capacity}}{\text{mass}} = \frac{C}{m} \tag{6.13}$$

Because the heat capacity is $C = q/\Delta T$, we have

$$\text{Specific heat} = \frac{q}{m \times \Delta T} \qquad (6.14)$$

If the aluminum block described in Example 6.6 has a mass of 5.7 g, then the specific heat of aluminum is

$$\text{Specific heat} = \frac{629 \text{ J}}{5.7 \text{ g} \times 123 \text{ °C}} = 0.90 \text{ J g}^{-1} \text{ °C}^{-1}$$

and the molar heat capacity of aluminum is

$$\text{Specific heat} \times \text{molar mass} = 0.90 \text{ J g}^{-1} \text{ °C}^{-1} \times 27.0 \text{ g mol}^{-1} = 24 \text{ J mol}^{-1} \text{ °C}^{-1}$$

The specific heats of some familiar substances are given in Table 6.1.

In a typical calorimetric calculation, we relate a quantity of heat, a temperature change, and the mass and specific heat of a substance. For this purpose, we can rearrange the specific-heat Equation (6.14) into the form

$$q = \text{mass} \times \text{specific heat} \times \Delta T \qquad (6.15)$$

When we raise the temperature of a system, the final temperature, T_f, is higher than the initial temperature, T_i. The temperature change, which is $\Delta T = (T_f - T_i)$, is positive, and therefore so is q, which means heat is gained by the system. Lowering the temperature of a system means that T_f is smaller than T_i, ΔT is negative, q is also negative, and heat is lost by the system.

The specific heat of a substance varies with temperature, and some of our calculations will yield only approximate answers. For example, the specific heat of silver is 0.235 J g^{-1} K^{-1} at 298 K (25 °C), but it increases steadily to 0.278 J g^{-1} K^{-1} at 1000 K. The most important specific heat value we will use in calculations is that of water. At about room temperature, the specific heat of water is 4.18 J g^{-1} °C^{-1}, and over the temperature range from 0 to 100 °C, it remains within 1% of this value.

The specific heat of water at 15 °C is 4.184 J g^{-1} °C. At one time, the quantity of heat required to raise the temperature of 1 g of water from 14.5 to 15.5 °C was defined as one *calorie* (cal). The calorie is not an SI unit, but it was widely used in the past and is still used to some extent. Now we simply *define* the calorie through the expression 1 cal = 4.184 J. Like the joule, the calorie is a relatively small energy unit, and the *kilo*calorie is often used.

Table 6.1 Specific Heats of Some Substances at 25 °C

Substance	Specific heat, J g^{-1}°C^{-1}
Aluminum (Al)	0.902
Copper (Cu)	0.385
Ethanol (CH_3CH_2OH)	2.46
Iron (Fe)	0.449
Lead (Pb)	0.128
Mercury (Hg)	0.139
Silver (Ag)	0.235
Sulfur (S)	0.706
Water (H_2O)	4.180

Application Note

The fact that the specific heat of water is so much higher than the specific heats of other substances (see Table 6.1) helps to account for the observation that the climate near large bodies of water is generally more moderate than in interior locations. The daily temperature rises and falls more slowly near a lake, for example, than in a desert region.

Example 6.7

How much heat, in joules and in kilojoules, does it take to raise the temperature of 225 g of water from 25.0 to 100.0 °C?

STRATEGY

This calculation is a direct application of Equation (6.15) once we have replaced ΔT by the difference in two temperatures. The conversion from joules to kilojoules requires dividing the first answer by 1000.

SOLUTION

The specific heat of water is 4.18 J g^{-1} °C^{-1}. The temperature change is $(100.0 - 25.0)$ °C = 75.0 °C, and the quantity of water to be heated is 225 g. We substitute these data in Equation (6.15):

$$q = \text{mass} \times \text{specific heat} \times \Delta T$$

$$q = 225 \text{ g H}_2\text{O} \times \frac{4.18 \text{ J}}{\text{g H}_2\text{O °C}} \times (100.0 - 25.0) \text{ °C}$$

$$= 7.05 \times 10^4 \text{ J} = 70.5 \text{ kJ}$$

EXERCISE 6.7A

How much heat, in calories and in kilocalories, does it take to raise the temperature of 814 g of water from 18.0 to 100.0 °C?

EXERCISE 6.7B

What mass of water, in kilograms, can be heated from 5.5 to 55.0 °C by 9.09×10^{10} J of heat?

Example 6.8

What will be the final temperature if a 5.00-g silver ring at 37.0 °C gives off 25.0 J of heat to its surroundings? Use the specific heat of silver listed in Table 6.1.

STRATEGY

The loss of heat from the system—the ring—means that q is negative: $q = -25.0$ J. This loss is not compensated for by any energy input into the system, and so the temperature must fall, meaning that $\Delta T < 0$. We must solve the specific-heat Equation (6.15), first for the temperature change $(T_f - T_i)$, and then for T_f.

SOLUTION

We can rearrange Equation (6.15) to the form shown here, isolating the term ΔT.

$$\Delta T = \frac{q}{\text{mass} \times \text{specific heat}}$$

Next, we substitute the relevant data: $q = -25.0$ J; mass = 5.00 g; specific heat (from Table 6.1) = 0.235 J g^{-1} °C^{-1}, and solve for ΔT.

$$\Delta T = \frac{-25.0 \text{ J}}{5.00 \text{ g} \times 0.235 \text{ J g}^{-1} \text{ °C}^{-1}} = -21.3°C$$

Finally, we replace ΔT by $T_f - T_1$; substitute $T_i = 37.0$ °C; and solve for T_f.

$$T_f - T_i = -21.3 \text{ °C}$$
$$T_f = T_i - 21.3 \text{ °C} = 37.0 \text{ °C} - 21.3 \text{ °C} = 15.7 \text{ °C}$$

ASSESSMENT

When using Equation (6.15), always check for the correct signs on quantities of heat and temperature changes. Here, the temperature change must be *negative* because heat is lost to the surroundings. The final temperature of the ring must be lower than the initial temperature, and it is $(15.7 \text{ °C} < 37.0 \text{ °C})$.

EXERCISE 6.8A

A 454-g block of lead is at an initial temperature of 22.5 °C. What will be the temperature of the lead after it absorbs 4.22 kJ of heat from its surroundings?

EXERCISE 6.8B

How many grams of copper can be heated from 22.5 to 35.0 °C by the same quantity of heat that can raise the temperature of 145 g H_2O from 22.5 to 35.0 °C?

Measuring Specific Heats

The simple calorimeter pictured in Figure 6.16 is made from plastic-foam cups. It is quite suitable for determining the specific heats of solid substances that are insoluble in water. One approach is to consider the solid as the system and the water as the surroundings.

A measured mass of the solid is heated to a given temperature and then dropped into the calorimeter, which contains a fixed mass of water at a known temperature. Heat is transferred from the hot solid to the cooler water. The temperature of the solid falls and that of the water rises until the solid and water reach the same final temperature. The basic principle of the method is that

All the heat lost by the hot solid is gained by the water in the cup.

Plastic-foam cups are excellent for the simple calorimeter in Figure 6.16 for two reasons. (1) The very low heat capacity of the cup allows us to ignore the small amount of heat it gains as the solid cools, and (2) the foam is a good thermal insulator that permits very little heat to escape from the calorimeter.

From the mass and temperature change of the water, we can determine the quantity of heat transferred. We then use this quantity of heat and the mass and temperature change of the solid to determine its specific heat. The method is illustrated in Example 6.9.

Application Note

An old-time application of this technique is to heat a poker over an open fire and then plunge it into a mug of cider. The result is instant hot cider.

Example 6.9

A 15.5-g sample of a metal alloy is heated to 98.9 °C and then dropped into 25.0 g of water in a calorimeter. The temperature of the water rises from 22.5 to 25.7 °C. Calculate the specific heat of the alloy.

STRATEGY

We need to use Equation (6.15) twice. First, we use it as written to find the heat absorbed by the water, and then we rearrange it to solve for the unknown specific heat.

SOLUTION

We begin by calculating the quantity of heat absorbed by the water (the surroundings):

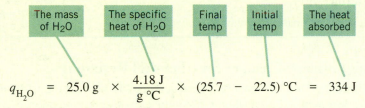

$$q_{H_2O} = 25.0 \text{ g} \times \frac{4.18 \text{ J}}{\text{g °C}} \times (25.7 - 22.5) \text{ °C} = 334 \text{ J}$$

As suggested in Figure 6.17, we assume that only the water (and not the container) absorbs heat and that the only source of this heat is the alloy (the system) as it cools. Because the alloy *loses* heat, we write

$$q_{alloy} = -334 \text{ J}$$

Now, we can calculate the specific heat of the alloy:

$$\text{Specific heat}_{alloy} = \frac{q_{alloy}}{\text{mass} \times \Delta T} = \frac{-334 \text{ J}}{15.5 \text{ g} \times (25.7 - 98.9 \text{ °C})}$$
$$= 0.29 \text{ J g}^{-1} \text{ °C}^{-1}$$

ASSESSMENT

We really are not justified in writing the intermediate result (334 J) to three significant figures, because the ΔT term is precise only to two significant figures. However, we ordinarily would not record this intermediate result and would simply round off to the appropriate two significant figures in the final answer. We must be careful not to make errors in signs when using Equation (6.15), but in the calculation here we can catch these by noting that specific heats must always be positive. Also, the data in Table 6.1 give us a good indication of the range of possible values for specific heats.

EXERCISE 6.9A

A 23.9-g sample of iridium is heated to 89.7 °C and then dropped into 20.0 g of water in a calorimeter. The temperature of the water rises from 20.1 °C to 22.6 °C. Calculate the specific heat of iridium.

EXERCISE 6.9B

A 135-g piece of iron is heated to 225 °C in an oven. It is then dropped into a calorimeter containing 250.0 mL of glycerol ($d = 1.261$ g/mL) at 23.5 °C. The temperature of the glycerol rises to a maximum value of 44.7 °C. Use these data to determine the specific heat of glycerol.

▲ FIGURE 6.16   A simple calorimeter made from plastic-foam cups

The inner cup is closed off with a cork stopper through which a thermometer and stirrer are immersed into the calorimeter. The outer cup provides additional thermal insulation from the surroundings.

▲ FIGURE 6.17   Example 6.9 illustrated

Example 6.10   An Estimation Example

Without doing detailed calculations, determine which of the following is a likely approximate final temperature when 100 g of iron at 100 °C is added to 100 g of 20 °C water in a calorimeter: **(a)** 20 °C **(b)** 30 °C **(c)** 60 °C **(d)** 70 °C.

ANALYSIS AND CONCLUSIONS

Because we have the same mass of each substance, *if* the water and iron had the same specific heat, the rise in temperature of the water would be the same as the drop in temperature of the iron. The final temperature would be the average of 20 °C and 100 °C, or 60 °C. However, from Table 6.1 we see that the specific heat of water ($4.18 \, J \, g^{-1} \, °C^{-1}$) is much greater than that of iron ($0.45 \, J \, g^{-1} \, °C^{-1}$). It takes nearly ten times more heat to change the temperature of a given mass of water than it takes to change the temperature of the same mass of iron. The final water temperature must be below 60 °C, but it also has to be above 20 °C, the initial temperature. The only possible approximate temperature of the four values we are given is 30 °C.

EXERCISE 6.10A

Without doing detailed calculations, determine the final temperature if 200.0 mL of water at 80 °C is added to 100.0 mL of water at 20 °C.

EXERCISE 6.10B

Calculate the exact final temperature in Example 6.10.

Measuring Enthalpy Changes for Chemical Reactions

By our definition of heat of reaction, the reactants and products must be at the same temperature. We pictured a constant-temperature exothermic reaction in a schematic fashion in Figure 6.11a. In addition, recall that we also usually observe the result pictured in Figure 6.11b: The temperature does not remain constant as heat leaves or enters the system in which a chemical reaction occurs. However, if we measure a heat of reaction in an isolated system, as pictured in Figure 6.11c, the change in chemical energy will appear solely as a change in thermal energy of the system. We will observe a corresponding temperature increase if the reaction is exothermic or a temperature decrease if the reaction is endothermic. Then we can use measured masses of substances, specific heats, and a temperature change to calculate the heat of reaction.

Consider the exothermic neutralization reaction in which the acid HCl(aq) neutralizes the base NaOH(aq) to produce the salt NaCl(aq) and $H_2O(l)$:

$$HCl(aq) + NaOH(aq) \longrightarrow NaCl(aq) + H_2O(l)$$

Suppose we allow stoichiometric proportions of the acid and base at temperature T_i to react in an isolated system, such as a plastic-foam cup calorimeter. After the reaction, an aqueous solution of NaCl is all that is present. The heat of the reaction remains in the system and raises the temperature to T_f. The heat of reaction (q_{rxn}) is the quantity of heat that would be given off to the surroundings by the NaCl(aq) *if* we allowed it to cool down to the initial temperature (T_i). But q_{rxn} is also the *negative* of the quantity of heat that produces the rise in temperature in the calorimeter from T_i to T_f in the first place, and this is the quantity of heat that we do measure. It is the product of the mass of the NaCl(aq) formed, the specific heat of the solution, and the temperature change ($\Delta T = T_f - T_i$). Let's call this quantity of heat $q_{calorim}$:

$$q_{calorim} = \text{mass of NaCl(aq)} \times \text{specific heat of NaCl(aq)} \times \Delta T$$

The heat of the reaction, then, is the *negative* of $q_{calorim}$, that is,

$$q_{rxn} = -q_{calorim} \tag{6.16}$$

For reactions carried out in systems under the constant pressure of the atmosphere, the quantity we measure is a heat of reaction at constant pressure, q_P. Thus, we can write

$$q_{rxn} = q_P = \Delta H \tag{6.17}$$

In Example 6.11, we carry out the calculation just outlined for the reaction between HCl(aq) and NaOH(aq).

Example 6.11

A 50.0-mL sample of 0.250 M HCl at 19.50 °C is added to 50.0 mL of 0.250 M NaOH, also at 19.50 °C, in a calorimeter. After mixing, the solution temperature rises to 21.21 °C. Calculate the heat of this reaction.

STRATEGY

To identify the quantities that enter into the calculation and to keep it simple, we can make four assumptions:

- The solution volumes are additive. The volume of $NaCl(aq)$ that forms is equal to 50.0 mL $+ 50.0$ mL $= 100.0$ mL.
- The $NaCl(aq)$ is sufficiently dilute that its density and specific heat are about the same as those values for pure water: 1.00 g/mL and 4.18 J g^{-1} $°C^{-1}$, respectively.
- The system is completely isolated: No heat escapes from the calorimeter.
- The heat required to warm any part of the calorimeter other than the $NaCl(aq)$ is negligible.

SOLUTION

The heat produced by the reaction and retained in the calorimeter is

$$q_{calorim} = mass \times specific\ heat \times \Delta T$$

$$= 100.0\ mL \times \frac{1.00\ g}{mL} \times \frac{4.18\ J}{g\ °C} \times (21.21 - 19.50)\ °C = 715\ J$$

$$q_{rxn} = q_P = -q_{calorim} = -715\ J$$

EXERCISE 6.11A

A 100.0-mL portion of 0.500 M HBr at 20.29 °C is added to 100.0 mL of 0.500 M KOH, also at 20.29 °C, in a calorimeter. After mixing, the temperature rises to 23.65 °C. Calculate the heat of this reaction. You may make the same assumptions we made in Example 6.11.

EXERCISE 6.11B

Suppose the acid used in Example 6.11 was HI(aq) instead of HCl(aq). Based on the same assumptions used in Example 6.11, how would you expect the result of the hypothetical experiment using HI(aq) to compare with that found in Example 6.11? Explain.

Problem-Solving Note

The first assumption allows us to work with the volume of $NaCl(aq)$ and its density rather than having to measure its mass.

Problem-Solving Note

The second assumption allows us to avoid having to measure the density and specific heat of the $NaCl(aq)$ or having to find these data somewhere in the chemical literature.

Example 6.12

Express the result of Example 6.11 for molar amounts of the reactants and products. That is, determine the value of ΔH that should be written in the equation for the neutralization reaction:

$$HCl(aq) + NaOH(aq) \longrightarrow NaCl(aq) + H_2O(l) \qquad \Delta H = ?$$

STRATEGY

With data from Example 6.11, we can first determine the number of moles of each reactant. Next, we can determine the number of moles formed of one of the products (for example, moles of H_2O). As implied by the above equation, the value of ΔH we are seeking is that for the formation of 1 mol H_2O. This we can obtain by dividing the heat of reaction in Example 6.11 by the number of moles of H_2O found here.

SOLUTION

We first calculate the number of moles of reactants involved in Example 6.11.

$$?\ mol\ HCl = 0.0500\ L\ HCl(aq) \times \frac{0.250\ mol\ NaOH}{1\ L\ HCl(aq)} = 0.0125\ mol\ HCl$$

$$?\ mol\ NaOH = 0.0500\ L\ NaOH(aq) \times \frac{0.250\ mol\ NaOH}{1\ L\ NaOH(aq)} = 0.0125\ mol\ NaOH$$

The number of moles of products (NaCl and H_2O) is the same as the number of moles of reactants (HCl and NaOH). For example, consider the number of moles of H_2O.

$$?\ mol\ H_2O = 0.0125\ mol\ HCl \times \frac{1\ mol\ H_2O}{1\ mol\ HCl} = 0.0125\ mol\ H_2O$$

Now, we calculate the enthalpy change per mole of H_2O formed, based on the fact (from Example 6.11) that $\Delta H = -715$ J for the formation of 0.0125 mol H_2O.

$$\Delta H = 1.00\ mol\ H_2O \times \frac{-715\ J}{0.0125\ mol\ H_2O} = -5.72 \times 10^4\ J\ (-57.2\ kJ)$$

Finally, we write the reaction equation based on the formation of 1 mol H_2O with the corresponding value of ΔH.

$$HCl(aq) + NaOH(aq) \longrightarrow NaCl(aq) + H_2O(l) \qquad \Delta H = -57.2 \text{ kJ}$$

EXERCISE 6.12A

Express the result of Exercise 6.11A for molar amounts of reactants and products.

$$HBr(aq) + KOH(aq) \longrightarrow KBr(aq) + H_2O(l) \qquad \Delta H = \text{?}$$

EXERCISE 6.12B

A 125-mL sample of 1.33 M HCl and 225 mL of 0.625 M NaOH, both initially at 24.4 °C, are allowed to react in a calorimeter. What is the final temperature in the calorimeter? (Use the type of assumptions stated in Example 6.11 and the value of ΔH obtained in Example 6.12 to determine the final temperature.)

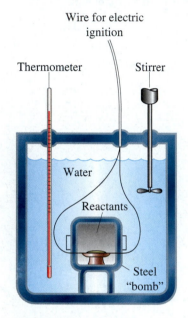

▲ **FIGURE 6.18   A bomb calorimeter**

The sample to be burned is placed in a small cup in the steel bomb, which is then filled with oxygen. An electric current is applied to ignite the sample. The heat of the reaction is determined from the temperature rise in the water that surrounds the bomb. The steel bomb confines the reactants and products to a constant volume.

QUESTION: What role does the stirrer play in bomb calorimetry measurements?

    Bomb Calorimetry activity

Bomb Calorimetry: Reactions at Constant Volume

Reactions carried out in plastic-foam calorimeters and in most other laboratory vessels are open to the atmosphere and therefore under constant pressure. (The lid on the calorimeter simply minimizes heat loss; it is not airtight.) We cannot carry out reactions involving gases in these calorimeters because the gases would escape; the system would no longer be isolated.

For combustion reactions and others involving gases, we ordinarily use a device called a *bomb calorimeter.* One principal part of this type of calorimeter (Figure 6.18) is a steel-walled container in which an exothermic reaction is carried out at a near-explosive rate. The container—a "bomb" that does not explode—confines the reactants and products to a constant volume. To perform a bomb calorimetry experiment, we place a small sample of known mass in a metal cup in the bomb. We then fill the bomb with oxygen at a pressure of about 30 atm and place the bomb in a thermally insulated container filled with a known quantity of water. We initiate the reaction with an electric ignition wire and record the highest water temperature reached.

From the temperature rise and the heat capacity of the calorimeter (established in a separate experiment), we can determine the increase in thermal energy of the calorimeter contents, q_{calorim}:

$$q_{\text{calorim}} = \text{heat capacity of calorimeter} \times \Delta T \qquad \textbf{(6.18)}$$

Notice that, unlike with foam cup calorimeters, where we worked with the product (mass $\times$ specific heat), we use the heat capacity of the calorimeter as a whole in bomb calorimetry. (We do it this way because a bomb calorimeter has too many components for us to determine their separate masses and specific heats.) We treat the contents of the calorimeter as an isolated system. As in Example 6.11, we also apply Equation (6.16):

$$q_{\text{rxn}} = -q_{\text{calorim}}$$

The pressure in the bomb does not necessarily remain constant during the reaction, but the *volume* does. This means that $\Delta V = 0$ and $w = -P\Delta V = 0$. The heat of reaction we measure is q_V. Thus, for a reaction carried out in a bomb calorimeter,

$$q_{\text{rxn}} = q_V = \Delta U \qquad \textbf{(6.19)}$$

We measure heats of combustion in bomb calorimeters, but we usually carry out combustion reactions in the open atmosphere, such as in a natural-gas heater or a propane torch. The more useful property of combustion reactions is therefore ΔH (the heat of reaction at constant pressure), not ΔU (the heat of reaction at constant volume). It is not difficult to calculate a ΔH value from a ΔU value, but we need not do so here. Actually, in some cases, ΔH and ΔU are equal because $P\Delta V$ is zero; in most others, ΔH and ΔU are nearly equal because the $P\Delta V$ term is quite small. The principal exceptions are reactions that involve a large change in the number of moles of gas.

In examples and exercises like those that follow, we will treat the results of bomb calorimetry experiments as if they were ΔH values. To do a bomb calorimetry calculation, we must know the heat capacity of the calorimeter assembly, and Exercise 6.13B suggests how this quantity can be determined by experiment.

Example 6.13

In a preliminary experiment, the heat capacity of a bomb calorimeter assembly is found to be 5.15 kJ/°C. In a second experiment, a 0.480-g sample of graphite (carbon) is placed in the bomb with an excess of oxygen. The water, bomb, and other contents of the calorimeter are in thermal equilibrium at 25.00 °C. The graphite is ignited and burned, and the water temperature rises to 28.05 °C. Calculate ΔH for the reaction

$$C(graphite) + O_2(g) \longrightarrow CO_2(g) \qquad \Delta H = ?$$

STRATEGY

First, we must use the heat capacity of the calorimeter and the temperature change to calculate the quantity of heat absorbed by the calorimeter ($q_{calorim}$). The heat of the reaction is the negative of $q_{calorim}$. Finally, we can convert q_{rxn} for a 0.480-g sample of graphite to a per-mole basis (similar to what we did in Example 6.12).

SOLUTION

The heat absorbed by the calorimeter is $q_{calorim} = C \times \Delta T$, where C is the heat capacity of the calorimeter.

$q_{calorim} = 5.15$ kJ °$C^{-1} \times (28.05 - 25.00)$ °C $= 15.7$ kJ

The heat of reaction is related to $q_{calorim}$ through Equation (6.16).

$q_{rxn} = -q_{calorim} = -15.7$ kJ

To get ΔH, we need to determine the heat released by the combustion of 1 mol (12.01 g) of graphite.

$$\Delta H = \frac{12.01 \text{ g C}}{1 \text{ mol C}} \times \frac{-15.7 \text{ kJ}}{0.480 \text{ g C}} = -393 \text{ kJ/mol C}$$

Finally, we write the reaction equation with the appropriate value of ΔH.

$$C(graphite) + O_2(g) \longrightarrow CO_2(g) \qquad \Delta H = -393 \text{ kJ}$$

EXERCISE 6.13A

A 0.250-g sample of diamond (another form of carbon) is burned with an excess of $O_2(g)$ in a bomb calorimeter having a heat capacity of 6.52 kJ/°C. The calorimeter temperature rises from 20.00 °C to 21.26 °C. Calculate ΔH for the reaction

$$C(diamond) + O_2 \longrightarrow CO_2(g) \qquad \Delta H = ?$$

EXERCISE 6.13B

A 0.8082-g sample of glucose ($C_6H_{12}O_6$) is burned in a bomb calorimeter assembly, and the temperature is noted to rise from 25.11 °C to 27.21 °C. Determine the heat capacity of the assembly.

$$C_6H_{12}O_6(s) + 6\ O_2 \longrightarrow 6\ CO_2(g) + 6\ H_2O(l) \qquad \Delta H = -2803 \text{ kJ}$$

6.6   Hess's Law of Constant Heat Summation

What if we are interested in the enthalpy change for the combustion of carbon to carbon monoxide (CO)? Here we *cannot* use bomb calorimetry in the same way we used that procedure for the combustion of carbon to carbon dioxide in Example 6.13, because CO(g), once formed, will undergo combustion to $CO_2(g)$. Carbon monoxide is an intermediate in the combustion of carbon to $CO_2(g)$, but we cannot stop the reaction at the CO stage. However, if we start with pure CO(g), we can carry out its combustion to $CO_2(g)$ in a bomb calorimeter. Let's summarize what we can and cannot measure directly, and the values we would obtain, expressed to four significant figures:

(a) $\qquad C(graphite) + \frac{1}{2} O_2(g) \longrightarrow CO(g) \qquad \Delta H_{(a)} = ?$

(b) $\qquad CO(g) + \frac{1}{2} O_2(g) \longrightarrow CO_2(g) \qquad \Delta H_{(b)} = -283.0 \text{ kJ}$

(c) $\qquad C(graphite) + O_2(g) \longrightarrow CO_2(g) \qquad \Delta H_{(c)} = -393.5 \text{ kJ}$

Suppose we look upon these three situations as different states that appear on the enthalpy diagram of Figure 6.19. The three enthalpy changes are represented by arrows, and we see that the red arrow representing $\Delta H_{(a)}$ is the difference between the red arrows representing $\Delta H_{(c)}$ and $\Delta H_{(b)}$:

$$\Delta H_{(a)} = \Delta H_{(c)} - \Delta H_{(b)}$$

$$\Delta H_{(a)} = -393.5 \text{ kJ} - (-283.0 \text{ kJ}) = -110.5 \text{ kJ}$$

Problem-Solving Note

You will find that the result in Exercise 6.13A is slightly different from that in Example 6.13 because the initial states of the systems differ. [C(diamond) is different from C(graphite).]

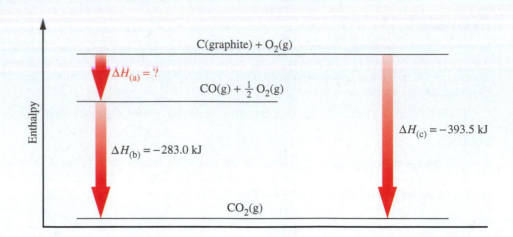

▶ **FIGURE 6.19** **Determining an unknown enthalpy change through an enthalpy diagram**

This diagram illustrates the relationship between the unknown $\Delta H_{(a)}$ and the known values of $\Delta H_{(b)}$ and $\Delta H_{(c)}$:
$\Delta H_{(a)} = \Delta H_{(c)} - \Delta H_{(b)}$.

QUESTION: What is the difference between ΔH for the combustion of C(graphite) to CO(g) (step a) and ΔH for the combustion of C(graphite) to CO_2(g) (step c)?

We could use this graphical method to evaluate the enthalpy change for almost any reaction that we cannot measure directly by relating the unknown change to enthalpy changes we can measure. After doing this a few times, we would probably discover that we really do not have to draw graphs. In this way, we would repeat the discovery of a principle first stated by Germain Hess (1802–1850). According to **Hess's law** of constant heat summation,

The heat of a reaction is constant, whether the reaction is carried out directly in one step or indirectly through a number of steps.

The basic principle in using Hess's law is to express the equation for the reaction of interest as the sum of two or more other equations. At the same time that we add the equations, we add their ΔH values to obtain the unknown ΔH. Thus, referring to Equations (a), (b), and (c) on page 237, to get Equation (a) and its ΔH value, we must reverse Equation (b) and change the sign of its ΔH value. Then we can add it to Equation (c):

$$-(b) \qquad \cancel{CO_2(g)} \longrightarrow CO(g) + \tfrac{1}{2} O_2(g) \qquad \Delta H_{(b)} = -(-283.0 \text{ kJ})$$

$$\underline{(c) \quad C(graphite) + O_2(g) \longrightarrow \cancel{CO_2(g)} \qquad\qquad \Delta H_{(c)} = -393.5 \text{ kJ}}$$

$$(a) \quad C(graphite) + \tfrac{1}{2} O_2(g) \longrightarrow CO(g) \qquad\qquad \Delta H_{(a)} = -110.5 \text{ kJ}$$

In adding the equations, CO_2(g) cancels out. Also, the sum of $\tfrac{1}{2} O_2$(g) on the right of equation $-$(b) and 1 O_2(g) on the left of Equation (c) leaves a net of $\tfrac{1}{2} O_2$(g) on the left of Equation (a).

A strategy for combining two or more chemical equations and their ΔH values to obtain the equation of interest and its ΔH value is outlined in Example 6.14. As part of this strategy, you can generally expect to do the following:

Hess's Law animation

• Reverse certain equations and change the signs of their ΔH values.
• Multiply certain equations and their ΔH values by appropriate factors. The factors may be either whole numbers or fractions $\left(\tfrac{1}{4}, \tfrac{1}{2} \dots \right)$.

Example 6.14

Calculate the enthalpy change for reaction (a), given the data in equations (b), (c), and (d).

(a)	$2\,C(graphite) + 2\,H_2(g) \longrightarrow C_2H_4(g)$	$\Delta H = ?$
(b)	$C(graphite) + O_2(g) \longrightarrow CO_2(g)$	$\Delta H = -393.5 \text{ kJ}$
(c)	$C_2H_4(g) + 3\,O_2(g) \longrightarrow 2\,CO_2(g) + 2\,H_2O(l)$	$\Delta H = -1410.9 \text{ kJ}$
(d)	$H_2(g) + \tfrac{1}{2} O_2(g) \longrightarrow H_2O(l)$	$\Delta H = -285.8 \text{ kJ}$

STRATEGY AND SOLUTION

First, we look for an equation that will place the term 2 C(graphite) on the left, to match this term in Equation (a). Only Equation (b) has C(graphite) on the left, and to get 2 C(graphite), we need to double Equation (b) and its ΔH value.

2(b) $2\,C(graphite) + 2\,O_2(g) \longrightarrow 2\,CO_2(g)$

$$\Delta H = 2 \times (-393.5)\ kJ = -787.0\ kJ$$

Next, we identify an equation that will place the term $C_2H_4(g)$ on the right. For this, we need Equation (c), and we need to reverse it and change the sign of its ΔH value.

$-$(c) $2\,CO_2(g) + 2\,H_2O(l) \longrightarrow C_2H_4(g) + 3\,O_2(g)$

$$\Delta H = -(-1410.9)\ kJ = 1410.9\ kJ$$

Finally, we see that Equation (a) has the term 2 H$_2$(g) on the left. Therefore to get 2 H$_2$(g) into our final equation, we need to double Equation (d) and its ΔH value.

2(d) $2\,H_2(g) + O_2(g) \longrightarrow 2\,H_2O(l)$

$$\Delta H = 2 \times (-285.8\ kJ) = -571.6\ kJ$$

We have used all the information given to us in order to correctly account for the 2 C(graphite) and 2 H$_2$(g) on the left of Equation (a) and the C$_2$H$_4$ on the right. If we now add together Equations 2(b), $-$(c), and 2(d), we expect O$_2$(g), CO$_2$(g), and H$_2$O(l) to cancel out in the summation. They do.

2(b)	$2\,C(graphite) + 2\,O_2(g) \longrightarrow 2\,CO_2(g)$	$\Delta H = -787.0\ kJ$
$-$(c)	$2\,CO_2(g) + 2\,H_2O(l) \longrightarrow C_2H_4(g) + 3\,O_2(g)$	$\Delta H = 1410.9\ kJ$
2(d)	$2\,H_2(g) + O_2(g) \longrightarrow 2\,H_2O(l)$	$\Delta H = -571.6\ kJ$
(a)	$2\,C(graphite) + 2\,H_2(g) \longrightarrow C_2H_4(g)$	

$$\Delta H = -787.0\ kJ + 1410.9\ kJ - 571.6\ kJ = 52.3\ kJ$$

ASSESSMENT

It does not matter in what order we adjust the equations before combining them into the required final equation, as long as each adjustment is a step toward our goal of getting the correct terms on each side of the equation. In addition to achieving this correctly balanced final equation, we must also carefully check all the adjustments made to ΔH values in the individual equations to ensure that we get the correct ΔH for the reaction of interest.

EXERCISE 6.14A

Calculate the enthalpy change for the reaction

$$2\,CH_4(g) + 3\,O_2(g) \longrightarrow 2\,CO(g) + 4\,H_2O(l) \qquad \Delta H = ?$$

from the following data:

$$CO(g) + \tfrac{1}{2}\,O_2(g) \longrightarrow CO_2(g) \qquad \Delta H = -283.0\ kJ$$

$$CH_4(g) + 2\,O_2(g) \longrightarrow CO_2(g) + 2\,H_2O(l) \qquad \Delta H = -890.3\ kJ$$

EXERCISE 6.14B

Calculate the enthalpy change for the reaction

$$C_2H_4(g) + H_2(g) \longrightarrow C_2H_6(g) \qquad \Delta H = ?$$

given the following data:

$$C_2H_4(g) + 3\,O_2(g) \longrightarrow 2\,CO_2(g) + 2\,H_2O(l) \qquad \Delta H = -1410.9\ kJ$$

$$2\,C_2H_6(g) + 7\,O_2(g) \longrightarrow 4\,CO_2(g) + 6\,H_2O(l) \qquad \Delta H = -3119.4\ kJ$$

$$2\,H_2(g) + O_2(g) \longrightarrow 2\,H_2O(l) \qquad \Delta H = -571.6\ kJ$$

6.7 Standard Enthalpies of Formation

We can determine the enthalpy changes for some reactions by calorimetry. Other enthalpy changes we can calculate with Hess's law, using ΔH values that can be measured. Of course, it would be best if we could simply calculate ΔH values directly, using the relationship

$$\Delta H = H_{products} - H_{reactants} \qquad \qquad (6.20)$$

We cannot do this, however, because there is no way to obtain *absolute* values of enthalpies. However, this situation is similar to that of determining the gain in elevation in climbing a mountain. As noted in Section 6.3, we can express the elevation gain as

$$\text{Elevation gain } (\Delta h) = \text{elevation}_{top} - \text{elevation}_{base}$$

We do not have absolute values for the elevations, but we can use elevations relative to sea level.

Chemists have devised a similar scale of relative enthalpies called *enthalpies of formation*. To explain how this scale works, we need first to describe the standard states for elements and compounds. The **standard state** of a solid or liquid substance is defined as being the pure element or compound at 1 atm pressure* and the temperature of interest. For a gaseous substance, the standard state is the (hypothetical) pure gas behaving as an ideal gas at 1 atm pressure and at the temperature of interest. The **standard enthalpy of reaction** ($\Delta H°$) is the enthalpy *change* for a reaction in which the reactants in their standard states yield products in their standard states. We use a superscript degree symbol (°) to denote a standard enthalpy change. Most of the enthalpy changes that we have used to this point are in fact standard enthalpies of reaction, and henceforth we will use the symbol $\Delta H°$ where appropriate.

The **standard enthalpy of formation** ($\Delta H_f°$) of a substance is the enthalpy change that occurs as 1 mol of the substance forms from its elements when both products and reactants are in their standard states. Moreover, the elements must be in their *reference* forms. With few exceptions,[†] the *reference form* of an element is the most stable form of the element at the given temperature and 1 atm pressure. The degree sign labels the enthalpy change as a *standard* enthalpy change, and the subscript f refers to the reaction in which a compound is *formed* from its elements. The standard enthalpy of formation is often called the standard heat of formation, or more simply, the *heat of formation*.

What should we use for the standard enthalpy of formation for the reference form of an element? Because formation of the reference form of an element from itself is not an actual change, its $\Delta H_f°$ value is *zero*. That is,

The standard enthalpy of formation of a pure element in its reference form is 0.

Scientists have compiled extensive tables of standard enthalpies of formation that include both compounds and elements; these values are usually given for 298.15 K (25.00 °C) and expressed in kJ/mol. You will find a short list in Table 6.2 and a longer tabulation in Appendix C.

Two forms of carbon—graphite and diamond—are readily attainable at 25 °C and 1 atm. The two forms are different states of carbon and must have different enthalpies of formation. As we demonstrated in Example 6.13 and Exercise 6.13A, they have different enthalpies of combustion, given here to four significant figures:

$$C(\text{graphite}) + O_2(g) \longrightarrow CO_2(g) \qquad \Delta H° = -393.5 \text{ kJ}$$
$$C(\text{diamond}) + O_2(g) \longrightarrow CO_2(g) \qquad \Delta H° = -395.4 \text{ kJ}$$

If we reverse the second equation and add it to the first, then by Hess's law we get

$$C(\text{graphite}) \longrightarrow C(\text{diamond}) \qquad \Delta H° = -393.5 \text{ kJ} - (-395.4 \text{ kJ}) = 1.9 \text{ kJ}$$

Graphite has been chosen as the reference form of carbon at 298.15 K, and so $\Delta H_f°$ [C(graphite)] = 0. When graphite is converted to diamond, there is an increase in enthalpy of 1.9 kJ/mol, indicating that $\Delta H_f°$[C(diamond)] = 1.9 kJ/mol.

Table 6.2 Some Standard Enthalpies of Formation at 25 °C	
Substance	**$\Delta H_f°$, kJ/mol**
$CO(g)$	−110.5
$CO_2(g)$	−393.5
$CH_4(g)$	−74.81
$C_2H_2(g)$	226.7
$C_2H_4(g)$	52.26
$C_2H_6(g)$	−84.68
$C_3H_8(g)$	−103.8
$C_4H_{10}(g)$	−125.7
$C_6H_6(l)$	48.99
$CH_3OH(l)$	−238.7
$CH_3CH_2OH(l)$	−277.7
$HBr(g)$	−36.40
$HCl(g)$	−92.31
$HF(g)$	−271.1
$HI(g)$	26.48
$H_2O(g)$	−241.8
$H_2O(l)$	−285.8
$NH_3(g)$	−46.11
$NO(g)$	90.25
$N_2O(g)$	82.05
$NO_2(g)$	33.18
$N_2O_4(g)$	9.16
$SO_2(g)$	−296.8
$SO_3(g)$	−395.7

---

*The International Union of Pure and Applied Chemistry (IUPAC) has recommended that the standard state pressure be changed from 1 atm (101,325 Pa) to 1 bar (1×10^5 Pa). The effects of this change are quite small, and we will continue to use 1 atm.

[†]A notable exception is phosphorus, for which the reference form at 25 °C and 1 atm is solid white phosphorus, even though solid red phosphorus is the more stable of the two forms. Over a long period of time, a sample of white phosphorus slowly converts to the more stable red phosphorus.

Calculations Based on Standard Enthalpies of Formation

Now let's explore how we can use standard enthalpies of formation to calculate standard enthalpy changes of chemical reactions. Suppose we need to know the standard enthalpy change for the conversion of nitrogen dioxide to dinitrogen tetroxide at 25 °C and 1 atm.

Formation of Aluminium Bromide movie

$$2 \, NO_2(g) \longrightarrow N_2O_4(g) \qquad \Delta H° = \, ?$$

The standard enthalpies of formation of these two gases can be found in Table 6.2:

(a) $\frac{1}{2} N_2(g) + O_2(g) \longrightarrow NO_2(g) \qquad \Delta H° = \Delta H_f°[NO_2(g)]$

(b) $N_2(g) + 2 \, O_2(g) \longrightarrow N_2O_4(g) \qquad \Delta H° = \Delta H_f°[N_2O_4(g)]$

With Hess's law, we can combine these two equations and arrive at the equation we want. To do this, we first reverse and then double equation (a):

$-2(a) \qquad 2 \, NO_2(g) \longrightarrow \cancel{N_2(g)} + \cancel{2 \, O_2(g)} \quad \Delta H° = -2 \, \Delta H_f°[NO_2(g)]$

$\underline{(b) \quad \cancel{N_2(g)} + \cancel{2 \, O_2(g)} \longrightarrow N_2O_4(g) \qquad\qquad \Delta H° = \Delta H_f°[N_2O_4(g)]}$

$\qquad\qquad 2 \, NO_2(g) \longrightarrow N_2O_4(g)$

$$\Delta H° = \Delta H_f°[N_2O_4(g)] - 2 \times \Delta H_f°[NO_2(g)]$$

We have just derived a specific example of a more general expression that allows us to relate a standard enthalpy of reaction and standard enthalpies of formation *without* formally going through Hess's law:

$$\Delta H° = \sum \nu_p \times \Delta H_f°(\text{products}) - \sum \nu_r \times \Delta H_f°(\text{reactants}) \qquad \textbf{(6.21)}$$

The symbol Σ signifies that we take a sum of several terms. The symbol ν (Greek letter "nu") is the stoichiometric coefficient used in front of a symbol or formula in a chemical equation. Equation (6.21) includes one term for each substance in the chemical reaction, and we obtain this term by multiplying the stoichiometric coefficient by the standard enthalpy of formation of the substance. First, we take the sum of terms for the products of a reaction, and then we subtract the sum of terms for the reactants. The result is the standard enthalpy change, $\Delta H°$. Example 6.15 shows how we can use this equation to calculate a standard enthalpy of reaction, $\Delta H°$, from tabulated standard enthalpies of formation, $\Delta H_f°$. Example 6.16 illustrates the reverse: how we can establish an unknown standard enthalpy of formation from a measured standard enthalpy of reaction. With organic compounds, the measured $\Delta H°$ is often the standard enthalpy of combustion, $\Delta H_{comb}°$.

Application Note

A chemical equation simply notes the initial and final conditions, but often there are rather specific requirements of temperature, pressure, and the use of catalysts for bringing about a chemical reaction. The reaction in Example 6.15 certainly will not occur just by passing methane and carbon dioxide gases into water.

Example 6.15

Synthesis gas is a mixture of carbon monoxide and hydrogen that is used to synthesize a variety of organic compounds. One reaction for producing synthesis gas is

$$3 \, CH_4(g) + 2 \, H_2O(l) + CO_2(g) \longrightarrow 4 \, CO(g) + 8 \, H_2(g) \qquad \Delta H° = \, ?$$

Use standard enthalpies of formation from Table 6.2 to calculate the standard enthalpy change for this reaction.

STRATEGY

A good way to begin a calculation based on Equation (6.21) is to list, under the formula of each substance in the balanced chemical equation, the value of its standard molar enthalpy of formation. Next, we can substitute these $\Delta H_f°$ values into Equation (6.21), being careful in each case to multiply the $\Delta H_f°$ value by the appropriate stoichiometric coefficient (ν) from the balanced equation. Throughout the calculation, we must pay particular attention to the proper use of algebraic signs.

SOLUTION

First, we list the $\Delta H_f°$ value, in kJ/mol, under each substance in the balanced equation.

$$3 \, CH_4(g) + 2 \, H_2O(l) + CO_2(g) \longrightarrow 4 \, CO(g) + 8 \, H_2(g)$$
$$-74.81 \qquad -285.8 \qquad -393.5 \qquad\quad -110.5 \qquad\quad 0$$

We can then calculate the two parts of Equation (6.21) separately.

$$\sum \nu_p \times \Delta H_f^\circ(\text{products}) = 4 \text{ mol CO} \times \frac{-110.5 \text{ kJ}}{1 \text{ mol CO}} + 8 \text{ mol H}_2 \times \frac{0 \text{ kJ}}{1 \text{ mol H}_2}$$

$$= -442.0 \text{ kJ}$$

$$\sum \nu_r \times \Delta H_f^\circ(\text{reactants}) = 3 \text{ mol CH}_4 \times \frac{-74.81 \text{ kJ}}{1 \text{ mol CH}_4}$$

$$+ 2 \text{ mol H}_2\text{O} \times \frac{-285.8 \text{ kJ}}{1 \text{ mol H}_2\text{O}} + 1 \text{ mol CO}_2 \times \frac{-393.5 \text{ kJ}}{1 \text{ mol CO}_2}$$

$$= -224.4 \text{ kJ} - 571.6 \text{ kJ} - 393.5 \text{ kJ} = -1189.5 \text{ kJ}$$

Now, we substitute the two summation terms into Equation (6.21) to calculate the standard enthalpy change.

$$\Delta H^\circ = \sum \nu_p \times \Delta H_f^\circ(\text{products}) - \sum \nu_r \times \Delta H_f^\circ(\text{reactants})$$

$$= -442.0 \text{ kJ} - [-1189.5 \text{ kJ}]$$

$$= -442.0 \text{ kJ} + 1189.5 \text{ kJ} = +747.5 \text{ kJ}$$

EXERCISE 6.15A

Ethylene, derived from petroleum, is used to make ethanol for use as a fuel or solvent. The reaction is

$$C_2H_4(g) + H_2O(l) \longrightarrow CH_3CH_2OH(l)$$

Use data from Table 6.2 to calculate ΔH° for this reaction.

EXERCISE 6.15B

Use data from Table 6.2 to calculate ΔH° for the combustion of butane gas, C_4H_{10}, to produce gaseous carbon dioxide and liquid water.

Example 6.16

The combustion of isopropyl alcohol, common rubbing alcohol, is represented by the equation

$$2 (CH_3)_2CHOH(l) + 9 O_2(g) \longrightarrow 6 CO_2(g) + 8 H_2O(l) \qquad \Delta H^\circ = -4011 \text{ kJ}$$

Use this equation and data from Table 6.2 to establish the standard enthalpy of formation for isopropyl alcohol.

STRATEGY

We need essentially the same strategy as in Example 6.15, except that we enter the standard enthalpy change for the reaction on the left side of Equation (6.21) and the unknown standard enthalpy of formation, ΔH_f°, with the appropriate other data, on the right side. Then we solve for the desired ΔH_f°.

SOLUTION

$$2 (CH_3)_2CHOH(l) + 9 O_2(g) \longrightarrow 6 CO_2(g) + 8 H_2O(l)$$

ΔH_f°, kJ/mol: $\qquad \Delta H_f^\circ \qquad\qquad 0 \qquad\qquad -393.5 \qquad -285.8$

$$\Delta H^\circ = \sum \nu_p \times \Delta H_f^\circ(\text{products}) - \sum \nu_r \times \Delta H_f^\circ(\text{reactants})$$

$$-4011 \text{ kJ} = \left[6 \text{ mol CO}_2 \times \frac{-393.5 \text{ kJ}}{1 \text{ mol CO}_2} + 8 \text{ mol H}_2\text{O} \times \frac{-285.8 \text{ kJ}}{1 \text{ mol H}_2\text{O}} \right]$$

$$- \left[2 \text{ mol (CH}_3)_2\text{CHOH} \times \Delta H_f^\circ + 9 \text{ mol O}_2 \times \frac{0 \text{ kJ}}{1 \text{ mol O}_2} \right]$$

$$= [-2361 \text{ kJ} - 2286 \text{ kJ}] - 2 \text{ mol (CH}_3)_2\text{CHOH} \times \Delta H_f^\circ$$

$$2 \text{ mol(CH}_3)_2\text{CHOH} \times \Delta H_f^\circ = 4011 \text{ kJ} - 2361 \text{ kJ} - 2286 \text{ kJ} = -636 \text{ kJ}$$

$$\Delta H_f^\circ = \frac{-636 \text{ kJ}}{2 \text{ mol(CH}_3)_2\text{CHOH}(l)}$$

$$= -318 \text{ kJ/mol (CH}_3)_2\text{CHOH}(l)$$

Example 6.17 A Conceptual Example

Without performing a calculation, determine which of these two substances should yield the greater quantity of heat *per mole* upon complete combustion: ethane, $C_2H_6(g)$, or ethanol, $CH_3CH_2OH(l)$.

ANALYSIS AND CONCLUSIONS

Each combustion has precisely the same final state: 2 mol $CO_2(g)$ + 3 mol $H_2O(l)$.

$$C_2H_6(g) + \tfrac{7}{2}O_2(g) \longrightarrow 2\,CO_2(g) + 3\,H_2O(l)$$

$$CH_3CH_2OH(l) + 3\,O_2(g) \longrightarrow 2\,CO_2(g) + 3\,H_2O(l)$$

The initial state is 1 mol of either ethane or ethanol and enough oxygen to complete the combustion. Because $\Delta H_f°[O_2(g)] = 0$, the enthalpy of the initial state is simply the standard enthalpy of formation of 1 mol of ethane in one case and 1 mol of ethanol in the other. The combustion reaction that releases more heat is the one with the more negative $\Delta H_{comb}°$, and this will be the reaction that has in its initial state the compound with the higher standard enthalpy of formation. The standard enthalpies of formation are -84.68 kJ/mol C_2H_6 and -277.7 kJ/mol CH_3CH_2OH. Ethane has the higher (that is, less negative) standard enthalpy of formation, and its combustion will liberate more heat than the combustion of ethanol. The sketch in Figure 6.20 should help you to visualize this conclusion.

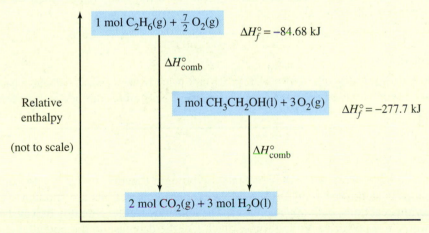

▲ **FIGURE 6.20** **Example 6.17 illustrated**

Table 6.3 Some Standard Enthalpies of Formation of Ions in Aqueous Solution at 25 °C

Ion	ΔH_f°, kJ/mol
H^+	0
Na^+	-240.1
K^+	-252.4
NH_4^+	-132.5
Ag^+	$+105.6$
Mg^{2+}	-466.9
Ca^{2+}	-542.8
Ba^{2+}	-537.6
OH^-	-230.0
Cl^-	-167.2
NO_3^-	-205.0
CO_3^{2-}	-677.1
SO_4^{2-}	-909.3

 Enthalpy of Solution activity

Ionic Reactions in Solutions

In Chapter 4, we saw the importance of reactions involving ions, especially in aqueous solutions. How can we apply thermochemical concepts to such reactions? Consider the reaction between HCl(aq) and NaOH(aq) in Example 6.11. It is an ionic reaction, expressed here in the form of a net ionic equation and a more accurate $\Delta H°$ value:

$$H^+(aq) + OH^-(aq) \longrightarrow H_2O(l) \qquad \Delta H° = -55.8 \text{ kJ}$$

It is somewhat difficult to extend the concept of enthalpy of formation to ions because there is no reaction in aqueous solution by which we can form cations alone or anions alone; we always obtain both. We resolve the difficulty by arbitrarily assigning $H^+(aq)$ an enthalpy of formation of zero. Example 6.18 illustrates how we can use this assignment to establish ΔH_f° for $OH^-(aq)$. By similar calculations, we can tabulate data such as those in Table 6.3.

Example 6.18

Use the net ionic equation just given, together with $\Delta H_f^\circ = 0$ for $H^+(aq)$, to obtain ΔH_f° for $OH^-(aq)$.

SOLUTION

By proceeding exactly as we did in Example 6.16, we can write

$$H^+(aq) + OH^-(aq) \longrightarrow H_2O(l) \qquad \Delta H° = -55.8 \text{ kJ}$$

ΔH_f°, kJ/mol 0 ΔH_f° -285.8

$$-55.8 \text{ kJ} = 1 \text{ mol } H_2O \times (-285.8 \text{ kJ/mol } H_2O)$$
$$-[1 \text{ mol } H^+ \times 0 \text{ kJ/mol } H^+ + 1 \text{ mol } OH^- \times \Delta H_f^\circ]$$
$$-55.8 \text{ kJ} = -285.8 \text{ kJ} - 1 \text{ mol } OH^- \times \Delta H_f^\circ$$
$$\Delta H_f^\circ = \frac{-285.8 \text{ kJ} + 55.8 \text{ kJ}}{1 \text{ mol } OH^-} = -230.0 \text{ kJ/mol } OH^-$$

EXERCISE 6.18A

Use data from Table 6.3 and the following data about the precipitation of $BaSO_4(s)$ to determine $\Delta H_f^\circ[BaSO_4(s)]$.

$$Ba^{2+}(aq) + SO_4^{2-}(aq) \longrightarrow BaSO_4(s) \qquad \Delta H° = 26 \text{ kJ}$$

EXERCISE 6.18B

Given that $\Delta H_f^\circ[Mg(OH)_2(s)] = -924.5$ kJ/mol, what is the standard enthalpy change, $\Delta H°$, for the reaction between aqueous solutions of magnesium chloride and potassium hydroxide?

Looking Ahead

A reaction that occurs when the reactants are brought together under the appropriate conditions is said to be *spontaneous*. Most exothermic reactions are spontaneous. In exothermic reactions, enthalpy decreases. The tendency for the enthalpy of a system to be lowered is a powerful driving force, much like the decrease in the potential energy of water as it flows downhill. However, some endothermic processes, such as the dissolving of many ionic solids in water, also occur spontaneously, even though they involve an enthalpy increase. We need to consider a second factor, called *entropy*, before we can fully answer the question of why some processes occur spontaneously and others do not. We will do this in Chapter 17.

In describing spontaneous change, we must also distinguish between the *tendency* of a reaction to occur and the *rate* at which it occurs. Some reactions that we expect to occur from thermodynamic considerations take place too slowly to be observed. We will discuss rates of chemical reactions in Chapter 13.

6.8 Combustion and Respiration: Fuels and Foods

Two types of exothermic reactions are of utmost importance: (1) The combustion of fuels provides the energy that sustains our modern way of life. (2) The foods we eat are "burned" in the respiratory process to give us the energy to stay alive and perform our many activities.

Fossil Fuels: Coal, Natural Gas, and Petroleum

A *fuel* is a substance that burns with the release of heat. In much of the world today, wood is the principal fuel for cooking food and keeping warm, just as it was in ancient times. However, the modern industrial world is powered mainly by the fossil fuels: coal, petroleum, and natural gas. These materials were formed over a period of millions of years from organic matter that became buried and compressed under mud and water. It is inevitable that Earth's fossil fuel reserves will be exhausted someday because we are using them 50,000 times faster than they were formed. Renewable energy sources, such as solar energy and wind power, are likely to become increasingly important in the future.

Coal is a complex organic material with carbon as the primary element. Coal also contains significant quantities of H, O, N, and S. Several varieties of coal are listed in Table 6.4. The fuel value of coal lies mainly in its carbon content. Complete combustion of carbon produces carbon dioxide.

$$C(s) + O_2(g) \longrightarrow CO_2(g)$$

High-grade coal yields about 30 kJ of heat per gram of coal. Burning coal also produces carbon monoxide and soot (unburned carbon) from incomplete combustion. Noncombustible matter is left unburned as ashes and enters the atmosphere as fly ash. Sulfur oxides are formed from sulfur compounds in the coal, and at the high temperatures at which coal burns, nitrogen oxides are produced in the reaction between the nitrogen and oxygen present in the air.

Natural gas is mainly methane (CH_4), with some ethane (C_2H_6), propane (C_3H_8), and butane (C_4H_{10}). Complete combustion of methane gives carbon dioxide and water:

$$CH_4(g) + 2\,O_2(g) \longrightarrow CO_2(g) + 2\,H_2O(l)$$

The enthalpy change in the combustion of methane (the heat of combustion) is -890.3 kJ/mol, or -55.49 kJ/g. On a mass basis, the heat of combustion of natural gas is almost twice that of coal. With far fewer impurities than coal, natural gas also burns more cleanly. Some carbon monoxide, soot, and nitrogen oxides are formed, but there are only traces of sulfur oxides and no ash.

Petroleum is an exceedingly complex mixture of hydrocarbons. Petroleum products such as gasoline burn more cleanly than coal, although not as cleanly as natural gas. Like petroleum itself, gasoline is a mixture of hydrocarbons with molecules having five to twelve carbon atoms. Octane, C_8H_{18}, is often used as a representative gasoline molecule, with complete combustion of octane yielding carbon dioxide and water:

$$C_8H_{18}(l) + \tfrac{25}{2}\,O_2(g) \longrightarrow 8\,CO_2(g) + 9\,H_2O(l) \qquad \Delta H° = -5450 \text{ kJ}$$

Foods: Fuels for the Body

Our bodies require food for the growth and repair of tissue and as a source of energy. The three principal classes of foods are carbohydrates, fats, and proteins. The body's preferred fuel is carbohydrates (starches and sugars). During digestion, starches and sugars are converted to the simple sugar glucose, $C_6H_{12}O_6$.

If glucose is burned in a calorimeter, its heat of combustion at 25 °C is found to be -2803 kJ/mol, or -15.56 kJ/g:

$$C_6H_{12}O_6(s) + 6\,O_2(g) \longrightarrow 6\,CO_2(g) + 6\,H_2O(l) \qquad \Delta H° = -2803 \text{ kJ}$$

Table 6.4 Mass Percent Carbon in Some Typical Coal Samples

Grade of Coal	Mass Percent C
Lignite	41.2
Subbituminous	53.9
Bituminous	76.7
Semianthracite	78.3
Anthracite	79.8

Regular or Premium?

Most gasoline stations have three pumps for gasoline: regular, mid-grade, and premium. What is the difference? All three fuels have about the same enthalpy of combustion per gram. The differences lie in the specific types of hydrocarbons and additives present. The hydrocarbons that make up gasoline vary widely in how smoothly they burn. Isooctane, for example, burns very smoothly, while heptane tends to burn explosively, causing a rattling sound called *knocking* and damaging the engine. The *octane number* of gasoline tells how smoothly it will burn, on a scale of 100 for isooctane and 0 for heptane.

$$CH_3-C-CH_2-C-CH_3$$

Isooctane
(2,2,4-trimethylpentane)
Octane number: 100

$$CH_3CH_2CH_2CH_2CH_2CH_2CH_3$$

Heptane
Octane number: 0

For many years, about 0.1% tetraethyllead, $(CH_3CH_2)_4Pb$, was added to gasoline to increase the octane number. Since the 1970s, tetraethyllead has been phased out because of environmental concerns, and today in the United States, "leaded" gasoline is no longer used. Methanol, ethanol, and other *oxygenates* (oxygen-containing compounds) are now used to raise the octane number of gasoline.

Some consumers think that premium gasoline (91-octane) will provide significantly better gas mileage than regular gasoline (87-octane). In fact, premium gasoline is generally recommended only for those vehicles that require its improved antiknock properties. Most modern automobiles are designed to run on regular gasoline, and premium gasoline does not provide significant additional benefits.

When it fuels our bodies instead of burning in a calorimeter, glucose produces the same products and yields about the same quantity of energy. The energy released is not exactly the same as in a calorimeter because the reactions in our bodies occur at body temperature, about 37 °C, rather than 25 °C. Also, rather than being burned rapidly, the glucose in our bodies is metabolized through a sequence of steps. Still, Hess's law applies. The overall enthalpy change is essentially the same.

Fats belong to a family of organic compounds called esters. An **ester** is derived from a carboxylic acid (RCOOH) and an alcohol (R'OH). An OH group of the carboxylic acid is replaced by an OR' group of the alcohol. We can represent an ester in general by

$$R-\overset{\overset{\displaystyle O}{\|}}{C}-O-R' \quad \text{or} \quad RCOOR'$$

Fats are esters formed from the three-carbon alcohol glycerol and long-chain carboxylic acids (*fatty acids*). A typical fat, glyceryl trilaurate, has the molecular formula $C_{39}H_{74}O_6$. Its structure is shown in Figure 6.21. Complete combustion of glyceryl trilaurate gives carbon dioxide and water:

$$C_{39}H_{74}O_6(s) + 54.5\ O_2(g) \longrightarrow 39\ CO_2(g) + 37\ H_2O(l)$$

$$\Delta H° = -2.39 \times 10^4\ kJ$$

The heat of combustion is -2.39×10^4 kJ/mol $C_{39}H_{74}O_6$, or -37.4 kJ/g $C_{39}H_{74}O_6$. Note that per gram, this heat of combustion is more than twice that of glucose. In general, fats yield more than twice as much energy per gram as carbohydrates. Fats are the body's preferred material for energy storage because more energy can be packed into a given mass of fat than of carbohydrate. The body can store at most about 500 g of carbohydrate as glycogen, a form of starch. The ability to store fat appears to be almost without limit.

Proteins are found mainly in such structural materials as muscle, connective tissue, skin, and hair. Enzymes that regulate the many chemical reactions that occur in living cells are also proteins. Proteins eaten in excess of the body's need for growth,

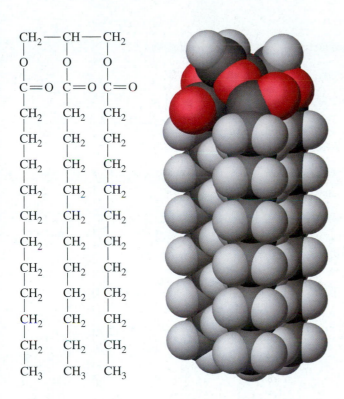

◀ **FIGURE 6.21   Glyceryl trilaurate, a typical fat**

Glycerol makes up the three-carbon backbone (at the top), and reaction with lauric acid molecules produces the three tails.

▲ **FIGURE 6.22   A nutrition label from a box of pasta**

QUESTION: Can you verify, per serving, the total calories and the calories from fat shown on the label? Use other information in the label and in the text.

repair, and replacement can be used for energy. Proteins have about the same energy value as carbohydrates.

In nutrition, the energy value of foods (Figure 6.22) is measured in kilocalories. The energy value of glucose is 3.72 kcal/g, and that of glyceryl trilaurate is 8.94 kcal/g. In everyday life, the well-known "food calorie" (Cal) is one kilocalorie:

$$1 \text{ kcal} = 1 \text{ Cal} = 1000 \text{ cal} = 4.18 \text{ kJ}$$

The capital C is sometimes used to distinguish the food Calorie (a kilocalorie) from the calorie used in science. The term *large calorie* has also been used to designate the food Calorie. Because sometimes people are not careful to make a clear distinction between the two meanings, we must rely on the context in which the word is used.

In general, the energy value of carbohydrates and proteins is about 4.0 kcal/g, and that of fats is more than twice as high, about 9.0 kcal/g.

Cumulative Example

The human body is about 67% water by mass. What mass of sucrose, $C_{12}H_{22}O_{11}(s)$, for which $\Delta H_f^\circ = -2225$ kJ/mol, must be metabolized by a 55-kg person with hypothermia (low body temperature) to raise the temperature of the body water from 33.5 °C to the normal body temperature of 37.0 °C? Assume the products of metabolism are in their most stable states at 25 °C. What volume of air at 37 °C having a partial pressure of O_2 of 151 Torr is required for the metabolism?

STRATEGY

We can treat the first question in two parts and then combine the parts. In the first part, we write an equation for the metabolism of sucrose similar to the one for glucose in Section 6.8, and we use standard enthalpies of formation to find the enthalpy change for the reaction. In the second part, we determine the mass of water in the body, and we use the specific heat of water and the temperature change to find the heat required to raise the body temperature (Equation 6.15). At this point, we can combine the results of the first two parts to find the mass of sucrose needed. To answer the second question, the volume of air needed, we use the mass of sucrose in a stoichiometric calculation, together with the ideal gas law and the concept of partial pressure.

SOLUTION

Because the metabolism of sucrose is essentially a combustion reaction, we can write and balance the equation for its combustion.

$$C_{12}H_{22}O_{11}(s) + 12 \text{ O}_2(g) \longrightarrow 11 \text{ H}_2O(l) + 12 \text{ CO}_2(g)$$

We apply Equation (6.21), using ΔH_f° values from Table 6.2, and solve for the enthalpy change.

$$\Delta H^\circ = \sum \nu_p \times \Delta H_f^\circ(\text{products}) - \sum \nu_r \times \Delta H_f^\circ(\text{reactants})$$

$$\Delta H^\circ = \left[(11 \times \Delta H_f^\circ[\text{H}_2\text{O(l)}]) + (12 \times \Delta H_f^\circ[\text{CO}_2(\text{g})]) \right]$$
$$- \left[(12 \times \Delta H_f^\circ[\text{O}_2(\text{g})]) + (\Delta H_f^\circ[\text{C}_{12}\text{H}_{22}\text{O}_{11}(\text{s})]) \right]$$
$$= \left[(11 \times [-285.8])\text{kJ} + (12 \times [-393.5])\text{kJ} \right] - \left[(-2225)\text{kJ} + (12 \times 0)\text{kJ} \right]$$
$$= -5641 \text{ kJ}$$

Next, we need to find the mass of water in a 55-kg $(5.5 \times 10^4 \text{ g})$ body that is 67% water, and then apply Equation (6.15).

Mass of water $= 5.5 \times 10^4 \text{ g} \times 0.67 = 3.7 \times 10^4 \text{ g water}$

$$q = \text{mass} \times \text{specific heat} \times \Delta T$$

$$q = 3.7 \times 10^4 \text{ g H}_2\text{O} \times \frac{4.18 \text{ J}}{\text{g H}_2\text{O} \cdot {}^\circ\text{C}} \times (37.0 - 33.5) \,{}^\circ\text{C}$$

$$= 5.4 \times 10^5 \text{ J} = 5.4 \times 10^2 \text{ kJ}$$

We have found the heat *absorbed* (positive sign) by the water, which is the same quantity as the heat *evolved* (negative sign) by the desired "combustion" reaction. We next use the enthalpy of combustion and the molar mass of sucrose to determine the number of moles and the mass of sucrose needed:

$$? \text{ mol sucrose} = -5.4 \times 10^2 \text{ kJ} \times \frac{1 \text{ mol C}_{12}\text{H}_{22}\text{O}_{11}}{-5641 \text{ kJ}} = 0.096 \text{ mol C}_{12}\text{H}_{22}\text{O}_{11}$$

$$? \text{ g sucrose} = 0.096 \text{ mol C}_{12}\text{H}_{22}\text{O}_{11} \times \frac{342.3 \text{ g}}{1 \text{ mol C}_{12}\text{H}_{22}\text{O}_{11}} = 33 \text{ g C}_{12}\text{H}_{22}\text{O}_{11}$$

Now, we convert mol sucrose to mol $O_2(\text{g})$, using a stoichiometric factor from the balanced equation.

$$? \text{ mol O}_2 = 0.096 \text{ mol C}_{12}\text{H}_{22}\text{O}_{11} \times \frac{12 \text{ mol O}_2}{1 \text{ mol C}_{12}\text{H}_{22}\text{O}_{11}} = 1.2 \text{ mol O}_2$$

Finally, we use the temperature, moles of oxygen, and its partial pressure to find the volume occupied by the gas. Note that the volume of air is the same as the volume of oxygen; it is the partial pressures of the gases that differ.

$$V = \frac{nRT}{P} = \frac{1.2 \text{ mol O}_2 \times 0.0821 \text{ L atm mol}^{-1} \text{ K}^{-1} \times (273 + 37) \text{ K}}{[151 \text{ Torr} \times (1 \text{ atm}/760 \text{ Torr})]}$$

$$= 154 \text{ L} = 1.5 \times 10^2 \text{ L}$$

ASSESSMENT

The mass of sucrose calculated—33 g, somewhat over an ounce—is not an unreasonable amount of sucrose for a person to consume. The volume of gas may appear rather large at first glance, but the balanced equation requires 12 mol O_2 per mole of sucrose, corresponding to about 1.2 mol O_2, which is about 30 L O_2 at 1.0 atm pressure and 37 °C. The partial pressure of oxygen is about 0.2 atm, and the volume of air would be about five times greater than calculated for pure oxygen, that is, about 150 L.

Concept Review with Key Terms

6.1 Energy—Common forms of energy include **kinetic energy** and **potential energy**. The SI unit of energy is the **joule (J)**. When a force causes an object to move, **work** is done:

Work = force × distance

6.2 Thermochemistry: Some Basic Terms—**Thermochemistry** is the study of energy changes in physical processes or chemical reactions. Basic thermochemical ideas include the notion of a **system** and its **surroundings** and of closed, open, and isolated systems; the concepts of kinetic energy, potential energy, and **internal energy**; and the distinction between two types of energy exchanges, **heat** (q) and **work**

(w). Work involves an energy transfer resulting from the movement of an object, and heat involves an energy transfer resulting from a difference in temperature between a system and its surroundings.

6.3 Internal Energy (U), State Functions, and the First Law of Thermodynamics—Internal energy (U) is a **state function**, which means it has a unique value once the **state**, or condition, of a system is defined. The change in the internal energy of a system is equal to the difference between the internal energy of the final and initial states. According to the **law of conservation of energy**, energy can be neither created nor destroyed, but energy can be transferred to or from the surroundings. **The first law of thermodynamics**—$\Delta U = q + w$— relates the heat and work exchanged between a system and its surroundings to changes in the internal energy of a system. This law is the basis for many of the calculations of thermochemistry.

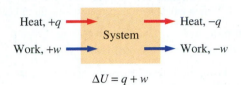

Heat, $+q$ Heat, $-q$

System

Work, $+w$ Work, $-w$

$$\Delta U = q + w$$

6.4 Heats of Reaction and Enthalpy Change, ΔH—For any reaction occurring at constant temperature, the **heat of reaction** is the quantity of heat exchanged between the reaction system and its surroundings. If a reaction is carried out at constant volume, the heat of reaction at constant volume (q_V) is equal to the change in internal energy (ΔU). For constant-pressure processes, **enthalpy** (H) is a more useful property than internal energy, to which it is related. At a constant pressure and with work limited to pressure–volume work, the heat of reaction (q_P) is equal to the **enthalpy change** (ΔH):

$$q_P = \Delta H = \Delta U + P\Delta V$$

Like internal energy, enthalpy is a function of state. Enthalpy changes can be written into chemical equations and incorporated into conversion factors that relate amounts of substances to quantities of heat released or absorbed in chemical reactions. In an **exothermic reaction**, enthalpy decreases and heat is given off to the surroundings. In an **endothermic reaction**, enthalpy increases and heat is absorbed from the surroundings. These changes can be represented graphically in the form of an **enthalpy diagram**.

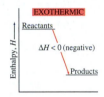

6.5 Calorimetry: Measuring Quantities of Heat—Calorimetry is the process of measuring quantities of heat, and a **calorimeter** is the device used for the procedure. The actual measurements are of masses and temperature changes. Other data required are heat capacities and/or specific heats. The **heat capacity** (C) of a system is the quantity of heat needed to change the temperature of the system by 1 °C (or by 1 K):

$$C = \frac{q}{\Delta T}$$

The **molar heat capacity** is the heat capacity of one mole of substance. The **specific heat** is the heat capacity of a 1-g sample of the substance:

$$\text{Specific heat} = \frac{\text{heat capacity}}{\text{mass}} = \frac{C}{m} = \frac{q}{m \times \Delta T}$$

Many reactions can be carried out under the constant pressure of the atmosphere in a plastic-foam calorimeter. The results obtained are q_P or ΔH values. Combustion reactions are generally carried out under constant-volume conditions in a bomb calorimeter. The results are q_V or ΔU values; ΔH values can be calculated from ΔU values when necessary.

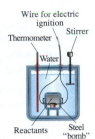

6.6 Hess's Law of Constant Heat Summation—Because enthalpy is a state function, the value of ΔH is the same whether a reaction is carried out in one step or several. We can write the equation for a reaction as several steps, and use **Hess's law** to evaluate ΔH by summing the enthalpy changes of the steps.

6.7 Standard Enthalpies of Formation—At the temperature of interest, the **standard state** of a pure solid or liquid is the substance at 1 atm; for a gas, it is the pure gas behaving ideally at 1 atm. The **standard enthalpy of reaction, $\Delta H°$**, is the enthalpy change for a reaction in which both the reactants and the products are in their standard states. The **standard enthalpy of formation**, $\Delta H_f°$, is equal to zero for an element in its reference form, which is usually the most stable form at 1 atm and the temperature of interest. The standard enthalpy of formation of a compound is the standard enthalpy change of a reaction in which the compound is formed from the reference forms of its elements in their standard states. Standard enthalpies of formation (usually listed at 25.00 °C) can be used to calculate standard enthalpies of reaction through the following equation (where the symbols ν_r and ν_p are stoichiometric coefficients):

$$\Delta H° = \sum \nu_p \times \Delta H_f°(\text{products}) - \sum \nu_r \times \Delta H_f°(\text{reactants})$$

6.8 Combustion and Respiration: Fuels and Foods—In this section, two aspects of life in which thermochemical ideas play key roles are explored: the combustion of fossil fuels—coal, natural gas, and petroleum—and the metabolism of foods—carbohydrates, fats, and proteins.

Assessment Goals

When you have mastered the material in this chapter, you should be able to

- Define and distinguish between kinetic and potential energy.

- Identify systems and surroundings, describe the three types of thermodynamic systems, and give examples of each.

- Describe the transfer of heat between a system and its surroundings and what is meant by thermal equilibrium.

- List and use the sign conventions for energy changes associated with work and heat.

- State and apply the law of conservation of energy and the first law of thermodynamics.

- Describe what is meant by a state function.

- Define the terms exothermic reaction and endothermic reaction. Give examples of each and indicate the sign of q_{rxn} for each.

- Describe the relationship between internal energy (U) and enthalpy (H) and how the quantities ΔU and ΔH are related to heats of reaction.

- Perform calculations involving enthalpy changes in chemical reactions.

• Explain the difference between heat capacity, molar heat capacity, and specific heat.

• Describe how plastic-foam cup and bomb calorimeters work and the type of reactions carried out in them.

• Perform calorimetric calculations.

• State Hess's law of constant heat summation, explain its significance, and apply it.

• Understand the meanings of standard state, standard enthalpy of reaction, and standard enthalpy of formation, ΔH_f°.

• Use ΔH_f° values to determine standard enthalpy changes for chemical reactions, including ionic reactions.

Self-Assessment Questions

1. An aqueous solution of copper(II) sulfate is a thermodynamic system under study. What type of system is this if the solution is contained in a beaker on a laboratory table?

2. How is heat transferred from a system to its surroundings?

3. Which of the following sign conventions is incorrectly matched with the associated energy change?
   (a) $q > 0$, heat is absorbed by the system
   (b) $q < 0$, heat is absorbed by the surroundings
   (c) $w > 0$, work is done by the surroundings
   (d) $w < 0$, work is done by a system

4. Why must we indicate the physical state of all reactants and products when designating the enthalpy change for a reaction?

5. How is the enthalpy of a system related to its internal energy? Under what conditions is the heat of reaction equal to the change in internal energy in a reaction? Under what conditions is the heat of reaction equal to the change in enthalpy in a reaction?

6. Which of the following represents an incorrect statement regarding enthalpy?
   (a) $\Delta H < 0$ for an endothermic reaction.
   (b) Enthalpy is a state function.
   (c) ΔH_f° of an element in its reference form is zero.
   (d) ΔH is an extensive property.

7. What quantity is physically measured in bomb calorimetry? How is this related to a property of the chemical system being studied?

8. A cold pack works by the dissolving of ammonium nitrate in water. Cold packs are carried by athletic trainers when transporting ice is not possible. Which of the following is true of this process?
   (a) $\Delta H < 0$, process is exothermic
   (b) $\Delta H > 0$, process is exothermic
   (c) $\Delta H < 0$, process is endothermic
   (d) $\Delta H > 0$, process is endothermic

9. What is the temperature increase (°C) if 236 J of heat is added to a 14.5-g block of nickel given that the specific heat of nickel is 0.444 J/g °C?
   (a) 36.7 °C   (c) 7.23 °C
   (b) 0.027 °C   (d) 41.6 °C

10. Pieces of copper, aluminum, silver, and lead of equal mass and at room temperature are dropped into a beaker of water at 100 °C. Which will absorb the greatest amount of heat?
    (a) copper   (c) silver
    (b) aluminum   (d) lead

11. What do we mean when we say a substance is in its standard state? When we say a substance is in its reference form?

12. How much heat (kJ) is given off in the consumption of 4.301 grams of $H_2(g)$ in the presence of excess oxygen according to the following reaction?

    $$2\,H_2(g) + O_2(g) \longrightarrow 2\,H_2O(g) \qquad \Delta H = -483.6\ \text{kJ}$$

    (a) 1934.4 kJ   (c) 259.9 kJ
    (b) 483.6 kJ   (d) 515.9 kJ

13. Which one of the following does not correspond to the standard enthalpy of formation for the product?
    (a) $H_2(g) + O(g) \longrightarrow H_2O(l)$
    (b) $\frac{1}{2} H_2(g) + \frac{1}{2} Cl_2(g) \longrightarrow HCl(g)$
    (c) $6\,C(\text{graphite}) + 3\,H_2(g) \longrightarrow C_6H_6(l)$
    (d) $6\,C(\text{graphite}) + 6\,H_2(g) + 3\,O_2(g) \longrightarrow C_6H_{12}O_6(s)$

14. Write the equation for a reaction that has as its heat of reaction the standard enthalpy of formation of $Fe_2O_3(s)$.

15. Explain why it is necessary to assign standard enthalpies of formation of *zero* both to $H_2(g)$ and to $H^+(aq)$. Is the same true for $Cl_2(g)$ and $Cl^-(aq)$? Explain.

16. What is a fuel? List the three principal fossil fuels and some advantages of gaseous and liquid fuels over solid fuels.

17. How is the food Calorie related to the calorie used in science? List the three major classes of foods and indicate which type has the highest energy value per gram.

18. Use data from page 247 to estimate the total food calories in a slice of bread that has 9.0 g of carbohydrates, 2.0 g of protein, and 1.0 g of fat. What percentage of total calories is from fat?

Problems

First Law of Thermodynamics

19. What is the change in internal energy of a system that gives off 89 J of heat and has 531 J of work done on it?

20. What is the change in internal energy of a system that does 7.02 kJ of work and absorbs 888 J of heat?

21. How much heat, in joules, must be involved in a process in which a system does 123 cal of work if the internal energy of the system is to remain unchanged?

22. If ΔU for a system is -68.3 kJ in a process in which the system absorbs 68.3 kcal of heat, how much work, in kJ, must have been involved?

23. Can a system do work and absorb heat at the same time? Can it do so while maintaining a constant internal energy?

24. Can a system do work and give off heat at the same time? Can it do so while maintaining a constant internal energy?

Enthalpy and Enthalpy Changes

25. At high temperatures, water is decomposed to hydrogen and oxygen.

$$H_2O(g) \longrightarrow H_2(g) + \tfrac{1}{2} O_2(g)$$

Decomposition of 67.3 g H_2O at constant pressure requires that 902 kJ of heat be absorbed by the system. Is the reaction endothermic or exothermic? What is the value of q for the reaction, per mole of water? Is the value of q equal to ΔU or ΔH? Explain.

26. Combustion of 0.144 g of sucrose (table sugar, $C_{12}H_{22}O_{11}$) in the open air results in the release of 2.38 kJ of heat.

$$C_{12}H_{22}O_{11}(s) + 12\,O_2(g) \longrightarrow 12\,CO_2(g) + 11\,H_2O(l)$$

Is the reaction endothermic or exothermic? What is the value of q for the reaction, per mole of sucrose? Is this a value of ΔU or ΔH? How would you expect the values of ΔU and ΔH to compare for this reaction? Explain.

27. When 0.1500 mol of solid calcium oxide is mixed with 0.1500 mol of liquid water, solid calcium hydroxide is formed and 9.78 kJ of heat is released. Write a chemical equation for the reaction producing 1 mol of calcium hydroxide, including the physical states of all the substances and the value of ΔH.

28. When 0.125 mol of ethane gas, C_2H_6, is burned in excess oxygen, 195 kJ of heat is evolved. Write a chemical equation for the combustion of 1 mol of ethane, including the physical states of all the substances and the value of ΔH. (*Hint:* What are the products of the reaction, and in what forms do they appear at 25 °C and 1 atm pressure?)

29. Given the reaction

$$3\,Fe_2O_3(s) + CO(g) \longrightarrow 2\,Fe_3O_4(s) + CO_2(g)$$

$$\Delta H = -46 \text{ kJ}$$

determine ΔH for the following reactions.

(a) $2\,Fe_2O_3(s) + \tfrac{2}{3} CO(g) \longrightarrow \tfrac{4}{3} Fe_3O_4(s) + \tfrac{2}{3} CO_2(g)$

(b) $6\,Fe_3O_4(s) + 3\,CO_2(g) \longrightarrow 9\,Fe_2O_3(s) + 3\,CO(g)$

30. Given the reaction

$$2\,Na_2O_2(s) + 2\,H_2O(l) \longrightarrow 4\,NaOH(s) + O_2(g)$$

$$\Delta H = -109 \text{ kJ}$$

determine ΔH for the following reactions.

(a) $16\,Na_2O_2(s) + 16\,H_2O(l) \longrightarrow 32\,NaOH(s) + 8\,O_2(g)$

(b) $2\,NaOH(s) + \tfrac{1}{2} O_2(g) \longrightarrow Na_2O_2(s) + H_2O(l)$

31. The reaction of solid white phosphorus (P_4) with chlorine gas produces phosphorus trichloride as the sole product. The reaction of 1.00 g P_4 with an excess of $Cl_2(g)$ gives off 9.27 kJ of heat. Write the equation representing the formation of 1 mol of phosphorus trichloride by this reaction and the accompanying ΔH value.

32. A handbook gives the heat of combustion of liquid carbon disulfide as -3.24 kcal/g of the liquid. Write the equation representing the combustion of 1 mol of carbon disulfide and the accompanying ΔH value. (*Hint:* The sulfur-containing combustion product is sulfur dioxide.)

Heats of Reaction and Reaction Stoichiometry

33. Calcium oxide (lime) reacts with carbon dioxide to form calcium carbonate (chalk).

$$CaO(s) + CO_2(g) \longrightarrow CaCO_3(s) \qquad \Delta H = -178.4 \text{ kJ}$$

How many kilojoules of heat are evolved in the reaction of 0.500 kg $CaO(s)$ with an excess of carbon dioxide?

34. Calcium carbide (CaC_2) can be made by heating calcium oxide (lime) with carbon (charcoal).

$$CaO(s) + 3\,C(s) \xrightarrow{\Delta} CaC_2(s) + CO(g)$$

$$\Delta H = +464.8 \text{ kJ}$$

How many kilojoules of heat are absorbed in a reaction in which 323 kg $C(s)$ is consumed?

35. The reaction of sodium peroxide and water is a source of $O_2(g)$.

$$2\,Na_2O_2(s) + 2\,H_2O(l) \longrightarrow 4\,NaOH(aq) + O_2(g)$$

$$\Delta H = -287 \text{ kJ}$$

How many kilojoules of heat are evolved in the reaction of 226 g Na_2O_2 with 98.5 g H_2O?

36. Calcium carbide (CaC_2) reacts with water to form acetylene (C_2H_2), a gas used as a fuel in welding.

$$CaC_2(s) + 2\,H_2O(l) \longrightarrow C_2H_2(g) + Ca(OH)_2(s)$$

$$\Delta H = -128.0 \text{ kJ}$$

How many kilojoules of heat are evolved in the reaction of 3.50 kg CaC_2 with 1.25 L H_2O?

37. How many liters of ethane, measured at 17 °C and 714 Torr, must be burned to give off 2.75×10^4 kJ of heat?

$$2\,C_2H_6(g) + 7\,O_2(g) \longrightarrow 4\,CO_2(g) + 6\,H_2O(l)$$

$$\Delta H = -3.12 \times 10^3 \text{ kJ}$$

38. How many liters of $CO_2(g)$, measured at 23 °C and 779 Torr, are produced when 4.45×10^7 kJ of heat is evolved in the burning of butane?

$$2\,C_4H_{10}(g) + 13\,O_2(g) \longrightarrow 8\,CO_2(g) + 10\,H_2O(l)$$

$$\Delta H = -5.76 \times 10^3 \text{ kJ}$$

39. How many grams of $CaO(s)$ must react with an excess of water to liberate the same quantity of heat as does the combustion of 24.4 L of $CH_4(g)$, measured at 24.7 °C and 753 Torr?

$$CaO(s) + H_2O(l) \longrightarrow Ca(OH)_2(s) \qquad \Delta H = -65.2 \text{ kJ}$$

$$CH_4(g) + 2\,O_2(g) \longrightarrow CO_2(g) + 2\,H_2O(l)$$

$$\Delta H = -890.3 \text{ kJ}$$

40. A reaction mixture of $N_2O_4(g)$ and $NO_2(g)$ absorbs the heat given off in the combustion of 6.35 L $CH_4(g)$ measured at 24.7 °C and 812 Torr. How many moles of N_2O_4 can be converted to NO_2 as a result?

$$CH_4(g) + 2\,O_2(g) \longrightarrow CO_2(g) + 2\,H_2O(l)$$

$$\Delta H = -890.3 \text{ kJ}$$

$$N_2O_4(g) \longrightarrow 2\,NO_2(g) \qquad \Delta H = +57.20 \text{ kJ}$$

41. *Without doing detailed calculations,* determine whether ethane or butane gives off the greater quantity of heat on combustion, when the comparison is made on an equal-mass basis. The combustion reactions and their enthalpy changes are given in Problems 37 and 38.

42. *Without doing detailed calculations,* determine whether ethane or butane produces the *lesser* mass of $CO_2(g)$ on combustion when the comparison is made on the basis of an equal quantity of heat liberated. The combustion reactions and their enthalpy changes are given in Problems 37 and 38.

Calorimetry: Heat Capacity and Specific Heat

(Use data from Table 6.1, as necessary.)

43. Calculate the heat capacity of a piece of iron if a temperature rise from 18 to 69 °C requires 672 J of heat.

44. Calculate the heat capacity of a sample of radiator coolant if a temperature rise from 5 to 107 °C requires 932 J of heat.

45. How much heat, in kilojoules, is required to raise the temperature of **(a)** 320.0 g water from 15.0 to 96.0 °C, and **(b)** 74.3 g ethanol from −36.4 to 44.5 °C?

46. How much heat, in kilojoules, is released when the temperature of **(a)** 47.0 g water drops from 45.4 to 10.0 °C, and **(b)** 209 g iron drops from 400.0 to 22.6 °C?

47. A 638-g block of lead initially at 27.0 °C absorbs 2044 J of heat. What is the final temperature of the lead?

48. To bring 36.2 g of sulfur from 25.0 °C to its melting point, the sulfur must absorb 2402 J of heat. What is the melting point of sulfur?

49. A 9.13-g sample of vanadium is heated to 99.10 °C and is then dropped into 20.0 g water in a calorimeter. The water temperature rises from 20.51 to 24.46 °C. Calculate the specific heat of vanadium.

50. A 2.05-g sample of a metal alloy is heated to 98.88 °C. It is then dropped into 28.0 g water in a calorimeter. The water temperature rises from 19.73 to 21.23 °C. Calculate the specific heat of the alloy.

51. A 2.29-kg iron skillet (specific heat = $0.449 \text{ J g}^{-1} \text{ °C}^{-1}$) is heated to 375 °F, and 453 g of water at 11.5 °C is poured into the skillet. Assuming no loss of heat, will the water reach its boiling point?

52. A piece of stainless steel (specific heat = $0.50 \text{ J g}^{-1} \text{ °C}^{-1}$) is taken from an oven at 178 °C and immersed in 225 mL of water at 25.9 °C. The water temperature rises to 42.4 °C. What is the mass of the piece of steel? How precise is this method of mass determination? Explain.

53. *Without doing detailed calculations,* determine which of the following metal samples must absorb the most heat if it is to be brought from room temperature to the boiling point of water: 2.25 g copper, 4.50 g lead, or 1.87 g silver.

54. *Without doing detailed calculations,* determine which of the following metal samples will be raised to the highest temperature when a 100.0-g sample at 22 °C absorbs 1.00 kJ of heat: aluminum, iron, or silver.

Calorimetry: Measuring Heats of Reaction

55. A 500.0-mL sample of 0.500 M NaOH at 20.00 °C is mixed with an equal volume of 0.500 M HCl at the same temperature in a plastic-foam cup calorimeter. The reaction

$$\text{HCl(aq)} + \text{NaOH(aq)} \longrightarrow \text{NaCl(aq)} + \text{H}_2\text{O(l)} \qquad \Delta H = \text{?}$$

takes place, and the temperature rises to 23.21 °C. Calculate ΔH for the reaction. (You may make the same assumptions as in Example 6.11.)

56. A 65.0-mL sample of 0.600 M HI at 18.46 °C is mixed with 84.0 mL of a solution containing excess potassium hydroxide, at 18.46 °C in a plastic-foam cup calorimeter. The reaction

$$\text{HI(aq)} + \text{KOH(aq)} \longrightarrow \text{KI(aq)} + \text{H}_2\text{O(l)} \qquad \Delta H = \text{?}$$

takes place, and the temperature rises to 21.96 °C. Calculate ΔH for the reaction. (You may make the same assumptions as in Example 6.11.)

57. A 1.50-g sample of $NH_4NO_3(s)$ is added to 35.0 g of water in a foam cup and stirred until it dissolves. The temperature of the solution drops from 22.7 to 19.4 °C. What is the heat of solution of NH_4NO_3, expressed in kJ/mol NH_4NO_3; that is, what is ΔH for the following process?

$$\text{NH}_4\text{NO}_3\text{(s)} \xrightarrow{\text{H}_2\text{O}} \text{NH}_4\text{NO}_3\text{(aq)} \qquad \Delta H = \text{?}$$

58. What is the maximum temperature that can be reached in a plastic-foam cup containing 105 mL of water at 20.12 °C fol-

lowing the addition and dissolving of a small pellet of KOH(s) weighing 0.215 g?

$$\text{KOH(s)} \xrightarrow{\text{H}_2\text{O}} \text{KOH(aq)} \qquad \Delta H = -57.6 \text{ kJ}$$

59. A 0.7971-g sample of coal is burned in a bomb calorimeter with a heat capacity of 4.62 kJ/°C. The temperature in the calorimeter rises from 20.11 to 25.05 °C. Calculate the heat of combustion of the coal, in kilojoules per gram.

60. A 0.828-g sample of gasoline is burned in a bomb calorimeter with a heat capacity of 9.89 kJ/°C. The temperature in the calorimeter rises from 22.75 to 26.37 °C. Calculate the heat of combustion of the gasoline, in kilojoules per gram of gasoline.

61. A pure substance of known heat of combustion can be used to determine the heat capacity of a bomb calorimeter. When 1.83 g sucrose is burned in a particular calorimeter, the temperature rises from 18.64 to 21.88 °C. What is the heat capacity of the calorimeter? The heat of combustion of sucrose is −16.5 kJ/g.

62. Benzoic acid (C_6H_5COOH) is sometimes used as a standard to determine the heat capacity of a bomb calorimeter. When 1.22 g C_6H_5COOH is burned in a calorimeter that is being calibrated, the temperature rises from 21.13 °C to 22.93 °C. What is the heat capacity of the calorimeter? The heat of combustion of benzoic acid is −26.42 kJ/g.

63. A 1.108-g sample of naphthalene, $C_{10}H_8(s)$, is burned in a bomb calorimeter and a temperature increase of 5.92 °C is noted. When a 1.351-g sample of thymol, $C_{10}H_{14}O(s)$ (a preservative and mold and mildew inhibitor), is burned in the same calorimeter, the temperature increase is 6.74 °C. If the heat of combustion of naphthalene is -5153.5 kJ/mol $C_{10}H_8$, what is the heat of combustion of thymol, in kJ/mol $C_{10}H_{14}O$?

64. A 1.148-g sample of benzoic acid is burned in an excess of oxygen in a bomb calorimeter. The temperature of the calorimeter rises from 24.96 to 30.25 °C. The heat of combustion of benzoic acid is -26.42 kJ/g. In a second experiment, a 0.895-g powdered coal sample is burned in the same calorimeter assembly. The temperature rises from 24.98 to 29.73 °C. How many kilograms of this coal would have to be burned to liberate 1.00×10^9 kJ of heat?

Hess's Law of Constant Heat Summation

65. Use the following equations

$$C(\text{graphite}) + O_2(g) \longrightarrow CO_2(g) \qquad \Delta H = -393.5 \text{ kJ}$$
$$2\,CO(g) + O_2(g) \longrightarrow 2\,CO_2(g) \qquad \Delta H = -566.0 \text{ kJ}$$

to calculate the enthalpy change for the reaction

$$2\,C(\text{graphite}) + O_2(g) \longrightarrow 2\,CO(g) \qquad \Delta H = ?$$

66. Use the following equations

$$N_2(g) + 2\,O_2(g) \longrightarrow N_2O_4(g) \qquad \Delta H = +9.2 \text{ kJ}$$
$$N_2(g) + 2\,O_2(g) \longrightarrow 2\,NO_2(g) \qquad \Delta H = +33.2 \text{ kJ}$$

to calculate the enthalpy change for the reaction

$$2\,NO_2(g) \longrightarrow N_2O_4(g) \qquad \Delta H = ?$$

67. Use the following equations

$$N_2H_4(l) + O_2(g) \longrightarrow N_2(g) + 2\,H_2O(l)$$
$$\Delta H = -622.2 \text{ kJ}$$
$$H_2(g) + \tfrac{1}{2}O_2(g) \longrightarrow H_2O(l) \qquad \Delta H = -285.8 \text{ kJ}$$
$$H_2(g) + O_2(g) \longrightarrow H_2O_2(l) \qquad \Delta H = -187.8 \text{ kJ}$$

to calculate the enthalpy change for the reaction

$$N_2H_4(l) + 2\,H_2O_2(l) \longrightarrow N_2(g) + 4\,H_2O(l) \qquad \Delta H = ?$$

68. Use the following equations

$$C_3H_8(g) + 5\,O_2(g) \longrightarrow 3\,CO_2(g) + 4\,H_2O(l)$$
$$\Delta H = -2219.9 \text{ kJ}$$
$$CO(g) + \tfrac{1}{2}O_2(g) \longrightarrow CO_2(g) \qquad \Delta H = -283.0 \text{ kJ}$$

to calculate the enthalpy change for the reaction

$$C_3H_8(g) + \tfrac{7}{2}O_2(g) \longrightarrow 3\,CO(g) + 4\,H_2O(l) \qquad \Delta H = ?$$

69. Determine the enthalpy change for the oxidation of ammonia

$$4\,NH_3(g) + 5\,O_2(g) \longrightarrow 4\,NO(g) + 6\,H_2O(l) \qquad \Delta H = ?$$

from the following data:

$$N_2(g) + 3\,H_2(g) \longrightarrow 2\,NH_3(g) \qquad \Delta H = -99.22 \text{ kJ}$$
$$N_2(g) + O_2(g) \longrightarrow 2\,NO(g) \qquad \Delta H = +180.5 \text{ kJ}$$
$$2\,H_2(g) + O_2(g) \longrightarrow 2\,H_2O(l) \qquad \Delta H = -571.6 \text{ kJ}$$

70. Calculate the enthalpy change for the reaction

$$BrCl(g) \longrightarrow Br(g) + Cl(g) \qquad \Delta H = ?$$

by using the following data:

$$Br_2(l) \longrightarrow Br_2(g) \qquad \Delta H = +30.91 \text{ kJ}$$
$$Br_2(g) \longrightarrow 2\,Br(g) \qquad \Delta H = +192.9 \text{ kJ}$$
$$Cl_2(g) \longrightarrow 2\,Cl(g) \qquad \Delta H = +243.4 \text{ kJ}$$
$$Br_2(l) + Cl_2(g) \longrightarrow 2\,BrCl(g) \qquad \Delta H = +29.2 \text{ kJ}$$

Standard Enthalpies of Formation and Standard Enthalpies of Reaction

71. Use standard enthalpies of formation from Appendix C to calculate the standard enthalpy change for each of the following reactions.
(a) $Al_2O_3(s) + 3\,H_2O(l) \longrightarrow 2\,Al(OH)_3(s)$
(b) $C_2N_2(g) + 2\,O_2(g) \longrightarrow 2\,CO_2(g) + N_2(g)$
(c) $NH_4HCO_3(s) \longrightarrow NH_3(g) + H_2O(g) + CO_2(g)$
(d) $3\,Si(s) + 2\,Fe_2O_3(s) \longrightarrow 3\,SiO_2(s) + 4\,Fe(s)$

72. Use standard enthalpies of formation from Appendix C to calculate the standard enthalpy change for each of the following reactions.
(a) $NH_3(g) + HCl(g) \longrightarrow NH_4Cl(s)$
(b) $NH_3(g) + HNO_3(l) \longrightarrow NH_4NO_3(s)$
(c) $Zn(s) + 2\,HCl(g) \longrightarrow ZnCl_2(s) + H_2(g)$
(d) $FeO(s) + CO(g) \longrightarrow Fe(s) + CO_2(g)$

73. Use the following equation and data from Appendix C to calculate the enthalpy of formation, per mole, of $ZnS(s)$.

$$2\,ZnS(s) + 3\,O_2(g) \longrightarrow 2\,ZnO(s) + 2\,SO_2(g)$$
$$\Delta H° = -878.2 \text{ kJ}$$

74. Use the following equation and data from Appendix C to calculate the enthalpy of formation, per mole, of sucrose, $C_{12}H_{22}O_{11}(s)$.

$$C_{12}H_{22}O_{11}(s) + 12\,O_2(g) \longrightarrow 12\,CO_2(g) + 11\,H_2O(l)$$
$$\Delta H° = -5.65 \times 10^3 \text{ kJ}$$

75. When it undergoes complete combustion in oxygen, a 1.050-g sample of the industrial solvent, diethylene glycol, $C_4H_{10}O_3$, gives off 23.50 kJ of heat to the surroundings. Calculate the standard enthalpy of formation of liquid diethylene glycol. Assume that the initial reactants and the products of the combustion are at 25 °C and 1 atm pressure.

76. When it undergoes complete combustion in oxygen, a 658.0-mg sample of adipic acid, $HOOC(CH_2)_4COOH$, a substance used in the manufacture of nylon, gives off 12.63 kJ of heat to the surroundings. Calculate the standard enthalpy of formation of solid adipic acid. Assume that the reactants and the products of the combustion are at 25 °C and 1 atm pressure.

77. Use data from Appendix C to determine the standard enthalpy change at 25 °C for the reaction

$$NH_4^+(aq) + OH^-(aq) \longrightarrow NH_3(g) + H_2O(l)$$

78. Use data from Appendix C to determine how many kilojoules of heat are given off per gram of aluminum consumed in the following reaction.

$$2\,Al(s) + 6\,H^+(aq) \longrightarrow 2\,Al^{3+}(aq) + 3\,H_2(g)$$

Fuels and Foods

79. The heats of combustion, in kilojoules per mole, of the first eight straight-chain alkanes are: methane (-890), ethane (-1560), propane (-2220), butane (-2879), pentane (-3536), hexane (-4163), heptane (-4811), octane (-5450).

(a) Use these data to make a graph of the heat of combustion, *per mol C*, versus the number of C atoms per molecule for these eight alkanes.

(b) Use this graph to estimate the heats of combustion of decane ($C_{10}H_{22}$) and dodecane ($C_{12}H_{26}$).

80. *Without doing detailed calculations*, determine which of the alkanes listed in Problem 79 will evolve the greatest quantity of heat upon combustion (a) per mole of alkane, (b) per gram of alkane, (c) per gram of carbon content, and (d) per gram of $CO_2(g)$ produced.

81. A person on a 2500-Cal low-fat diet attempts to maintain a fat intake of no more than 20% of food Calories from fat. What percent of the daily fat allowance is represented by a tablespoon of peanut butter (15.0 g and 50.1% fat)? Recall that fat has a food value of about 9.0 Cal/g.

82. Verify the claim by sugar manufacturers that a teaspoon (about 4.8 g) of sugar (sucrose, $C_{12}H_{22}O_{11}$) contains only 19 Cal. Use the heat of combustion from Problem 74, and recall the definition of a food Calorie.

Additional Problems

Problems marked with an * may be more challenging than others.

83. Which of the following quantities of energy is the largest: (a) the kinetic energy of a hydrogen molecule at STP, (b) the kinetic energy of a 1.0-g BB shot traveling at a speed of 100 m/s, or (c) the heat required to raise the temperature of 10 mL of water from 20 to 21 °C?

84. For the melting of 1 mol of ice at 0 °C, we can write the expression

$$H_2O(s, d = 0.917 \text{ g/cm}^3) \longrightarrow H_2O(l, d = 0.9998 \text{ g/mL})$$

$$\Delta H = +6.01 \text{ kJ}$$

Would you expect ΔU for the melting of ice at 0 °C to be greater, less than, or equal to +6.01 kJ/mol? Explain.

85. Calculate the final temperature for the situation described in Example 6.10 on page 233.

86. While on his honeymoon in Switzerland, James Joule (whose discoveries led to the first law of thermodynamics) measured the temperature of the water at the top of a waterfall and again at the bottom. What observation do you think he made? Explain.

87. The sample of water in the smaller plastic-foam cup is added to that in the larger cup. What is the water temperature when thermal equilibrium is reached?

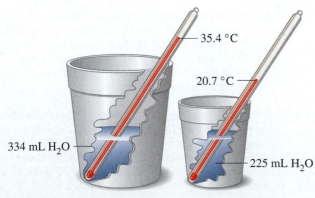

* **88.** The accompanying sketch shows a sample of hot water in a plastic-foam cup. Also shown are three objects, all at the same room temperature of 25.0 °C: (a) a steel ball, 2.20 cm in diameter ($d = 7.83$ g/cm³, specific heat = 0.45 J g⁻¹ °C⁻¹), (b) a tightly wound roll of aluminum foil, 2.00 m long, 5.0 cm wide, and 0.10 mm thick ($d = 2.70$ g/cm³, specific heat = 0.902 J g⁻¹ °C⁻¹), and (c) a graduated cylinder with 5.0 mL of water. Which would you expect to produce the greatest lowering of the temperature of the water in the cup: submerging the steel ball, submerging the roll of aluminum foil, or adding the 5.0 mL of water? Explain. (*Hint:* Do you have to calculate a final temperature?)

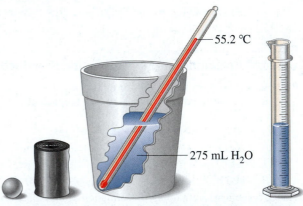

89. A British thermal unit (Btu) is defined as the quantity of heat required to change the temperature of 1 lb of water by 1 °F. Assuming the specific heat of water to be independent of temperature, how much heat is required to raise the temperature of the water in a 30-gal water heater from 66 to 145 °F: (a) in Btu; (b) in kcal; (c) in kJ?

* **90.** The combustion of hydrogen–oxygen mixtures is used to produce very high temperatures (about 2500 °C) needed for certain types of welding. Consider the combustion reaction to be

$$H_2(g) + \tfrac{1}{2}O_2(g) \longrightarrow H_2O(g) \qquad \Delta H° = -241.8 \text{ kJ}$$

How much heat is evolved when a 100.0-g mixture containing 15.5% $H_2(g)$, by mass, is burned?

91. The thermite reaction is highly exothermic.

$$Fe_2O_3(s) + 2\,Al(s) \longrightarrow Al_2O_3(s) + 2\,Fe(s)$$
$$\Delta H° = -852\ kJ$$

The reaction is started in a room-temperature (25 °C) mixture of 1.00 mol $Fe_2O_3(s)$ and 2.00 mol $Al(s)$. The liberated heat is retained within the products, whose combined specific heat over a broad temperature range is about 0.8 J g^{-1} $°C^{-1}$. Show that the quantity of heat liberated is sufficient to raise the temperature of the products to the melting point of iron (1530 °C).

* 92. The composition of a particular natural gas, expressed on a mole fraction basis, is CH_4, 0.830; C_2H_6, 0.112; C_3H_8, 0.058. A 215-L sample of this natural gas, measured at 24.5 °C and 744 mmHg, is burned in an excess of oxygen. How much heat is evolved in the combustion? (*Hint:* What are the heats of combustion of the individual gases?)

93. Care must be taken in preparing solutions of solutes that liberate heat on dissolving. The heat of solution of NaOH is -42 kJ/mol NaOH. What should be the approximate maximum temperature reached in the preparation of 500.0 mL of 6.0 M NaOH from NaOH(s) and water at 20 °C?

* 94. A particular gasohol fuel is 85% gasoline and 15% ethanol by mass. If we approximate gasoline by the formula for octane C_8H_{18}, determine which gives off more heat on combustion, straight gasoline or the same mass of gasohol mixture. (*Hint:* Use data from Table 6.2 and Problem 79.)

* 95. The heat of neutralization of HCl(aq) by NaOH(aq) is -55.90 kJ/mol H_2O produced. If 50.00 mL of 1.16 M NaOH at 25.15 °C is added to 25.00 mL of 1.79 M HCl at 26.34 °C in a plastic-foam cup calorimeter, what will the solution temperature be immediately after the neutralization reaction has occurred? You can make the same assumptions as in Example 6.11.

96. Write two equations to represent the combustion of dodecane, $C_{12}H_{26}(l)$, in an excess of oxygen gas. In each case, the products are carbon dioxide gas and water, but in one reaction, the water is obtained as a liquid and in the other, as a gas. *Without doing detailed calculations*, determine which of these two combustion reactions gives off the greater quantity of heat. Also determine the *difference* in the two heats of combustion, per mole of dodecane burned.

* 97. Substitute natural gas (SNG) is a gaseous mixture containing $CH_4(g)$ that can be used as a fuel. One reaction for the production of SNG is

$$4\,CO(g) + 8\,H_2(g) \longrightarrow 3\,CH_4(g) + CO_2(g) + 2\,H_2O(l)$$
$$\Delta H = ?$$

Show how the following data, as necessary, can be used to determine $\Delta H°$ for this SNG reaction.

$$C(graphite) + \tfrac{1}{2}O_2(g) \longrightarrow CO(g) \qquad \Delta H = -110.5\ kJ$$
$$CO(g) + \tfrac{1}{2}O_2(g) \longrightarrow CO_2(g) \qquad \Delta H = -283.0\ kJ$$
$$H_2(g) + \tfrac{1}{2}O_2(g) \longrightarrow H_2O(l) \qquad \Delta H = -285.8\ kJ$$
$$C(graphite) + 2\,H_2(g) \longrightarrow CH_4(g) \qquad \Delta H = -74.81\ kJ$$

Compare this result with that obtained by using enthalpy of formation data directly on the SNG reaction.

98. Under the entry H_2SO_4, a handbook lists several values for the enthalpy of formation, $\Delta H_f°$. For example, for pure $H_2SO_4(l)$, $\Delta H_f° = -814$ kJ/mol; for 1.0 M $H_2SO_4(aq)$, $\Delta H_f° = -888$ kJ/mol; for 0.25 M $H_2SO_4(aq)$, $\Delta H_f° = -890$ kJ/mol; for 0.020 M $H_2SO_4(aq)$, $\Delta H_f° = -897$ kJ/mol.

(a) Explain why these values are not all the same.

(b) When concentrated aqueous solutions are diluted, does the solution temperature increase or decrease? Explain.

* 99. A 1.103-g sample of a gaseous compound that occupies a volume of 582 mL at 765.5 Torr and 25.00 °C is burned in an excess of $O_2(g)$ in a bomb calorimeter. The products of the combustion are 2.108 g $CO_2(g)$, 1.294 g $H_2O(l)$, and enough heat to raise the temperature of the calorimeter assembly from 25.00 to 31.94 °C. The heat capacity of the calorimeter is 5.015 kJ/°C. Write an equation for the combustion reaction, and indicate $\Delta H°$ for this reaction at 25.00 °C.

* 100. Use the fact that $\Delta H = \Delta U + P\Delta V$ to determine whether ΔH is equal to, greater than, or less than ΔU for the following reactions. Recall that "greater than" means more positive or less negative, and "less than" means less positive or more negative. Assume that the only significant change in volume is that associated with gases, and assume further that all gases obey the ideal gas law.

(a) the complete combustion of one mol of butanol

(b) the complete combustion of one mol of glucose, $C_6H_{12}O_6$

(c) the decomposition of one mol of solid ammonium nitrate into liquid water and gaseous dinitrogen monoxide.

* 101. The following sketch is an enthalpy diagram based on these reactions.

$$3\,H_2(g) + \tfrac{3}{2}O_2(g) \longrightarrow 3\,H_2O(l)$$
$$N_2(g) + O_2(g) \longrightarrow 2\,NO(g)$$
$$N_2(g) + 3\,H_2(g) \longrightarrow 2\,NH_3(g)$$
$$2\,NH_3(g) + \tfrac{5}{2}O_2(g) \longrightarrow 2\,NO(g) + 3\,H_2O(l)$$

Determine the enthalpy changes for these reactions, using data from Table 6.2, and enter these values in place of the red question marks on the sketch. Also, write on the blue lines the substances present, starting with the information provided on one of those lines.

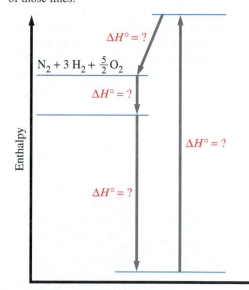

* 102. A 0.7782-g sample of methylcyclohexane is burned in an excess of oxygen in a bomb calorimeter with a heat capacity of 9.96 kJ/°C. The temperature of the calorimeter rises from 19.44 to 23.10 °C. Assuming that the products of combustion are in their standard states, calculate the standard enthalpy of formation of methylcyclohexane in kJ/mol.

Apply Your Knowledge

* **103.** [Environmental] Several books and movies have dealt with the scenario of an asteroid or comet striking the Earth. Assume that an iron asteroid 4.00 km in diameter strikes the Earth at a speed of 5.0×10^4 km/h, in the Pacific Ocean. The Pacific has an average depth of 4.2×10^3 m and an average temperature of 17 °C. Determine the approximate diameter in km of the Pacific that will be raised to the boiling point by the impact of the asteroid. State the nature of any assumptions you make.

* **104.** [Laboratory] A 1.075-g sample of a component of the light petroleum distillate called naphtha is found to yield 3.294 g CO_2 and 1.573 g H_2O on complete combustion. This particular component is also found to be an alkane with one methyl group attached to a longer carbon chain, to have a molecular formula twice its empirical formula, and to have the following properties: melting point, −154 °C; boiling point, 60.3 °C; density, 0.6532 g/mL at 20 °C; specific heat, 2.25 J g^{-1} °C^{-1}; and $\Delta H_f^\circ = -204.6$ kJ/mol.

   (a) Write an equation for the combustion reaction, including a value of $\Delta H°$ for the reaction.

   (b) Write a plausible condensed structural formula for the compound(s) that this component might be.

   (c) How many gallons of this substance, measured at 20 °C, must be burned to provide enough heat to warm 25.0 m^3 of water from 19.2 to 33.0 °C, assuming that all the heat of combustion is transferred to the water?

105. [Historical] James Joule published his definitive work related to the first law of thermodynamics in 1850, stating that "the quantity of heat capable of increasing the temperature of one pound of water by 1 °F requires for its evolution the expenditure of a mechanical force represented by the fall of 772 lb through the space of one foot." Validate this statement by relating it to information given in this text.

106. [Biochemical] The metabolism of glucose, $C_6H_{12}O_6$, yields $CO_2(g)$ and $H_2O(l)$ as products. Heat released in the process is converted to useful work with about 70% efficiency. Calculate the mass of glucose metabolized by a 58.0-kg person in climbing a mountain with an elevation gain of 1450 m. Assume that the work performed in the climb is four times that required to simply lift 58.0 kg by 1450 m. ΔH_f° of $C_6H_{12}O_6$ is −1273.3 kJ/mol.

107. [Historical] In 1818, Dulong and Petit observed that the molar heat capacities of the elements in their solid states were approximately the same. A modern statement of their law is

$$\text{molar mass} \times \text{specific heat} \ (J\,g^{-1}\,°C^{-1}) = \text{constant}$$

   (a) Use data from Table 6.1 to establish a value of the constant in the given equation.

   (b) What form of a graph of molar masses and specific heats would you plot to obtain a straight line? Explain.

   (c) The element cobalt (specific heat = 0.421 J g^{-1} °C^{-1}) was known at the time of Dulong and Petit. What approximate value would they have obtained for the atomic weight of cobalt with their equation?

   (d) When a 25.5-g sample of titanium at 99.7 °C was added to a quantity of water in a foam cup calorimeter, the water temperature was found to increase from 24.6 to 27.4 °C. What must have been the approximate volume of the water?

108. [Collaborative] A common thermochemical lecture demonstration is the determination of the enthalpy of solution of an ionic substance. Discuss within your study group criteria for a useful substance for such an experiment, and select two substances that would be good candidates according to your criteria.

* **109.** [Laboratory] To determine the enthalpy of hydration of sodium sulfate, 106.6 g of water at 23.28 °C was weighed into a plastic-foam cup and 8.56 g of sodium sulfate was added. The final temperature of the resulting solution was 24.97 °C. The experiment was repeated with 105.4 g of water at 24.40 °C and 19.3 g of sodium sulfate decahydrate, and the solution achieved a final temperature of 17.34 °C. Based on these data, and assuming that the solution in each case has a specific heat the same as that of water, calculate the molar enthalpy of hydration of sodium sulfate:

$$Na_2SO_4(s) + 10\,H_2O(l) \longrightarrow Na_2SO_4 \cdot 10\,H_2O(s)$$

Also determine the theoretical value for this process, and compare it to the laboratory determination. Suggest a reason for any error seen.

e-Media Problems

The activities described in these problems can be found in the e-Media Activities and Interactive Student Tutorial (IST) modules of the Companion Website, *http://chem.prenhall.com/hillpetrucci.*

110. From the **Potential and Kinetic Energy** animation (*Section 6-1*), describe the difference in motion of objects possessing high kinetic, thermal, and potential energy. Animated "objects" in this animation include both the ball and the atoms of the surface.

111. View the **Components of Internal Energy** animation (*Section 6-2*). Use your general knowledge about the physical properties of solids, liquids, and gases to predict the dominant components of internal energy for each state.

112. Consider the process occurring in the **Heat Transfer** animation (*Section 6-2*). **(a)** Describe, with a graph, the influence of increasing the thickness of the wall on the rate of heat transfer. **(b)** A thermos bottle is a double-walled vessel with an evacuated space between the walls. Using the principles illustrated in this animation, describe why a thermos bottle works as an insulator on a molecular scale.

113. From the **Bomb Calorimetry** activity (*Section 6-5*), estimate which of the compounds, naphthalene or trinitrotoluene, has a higher heat of combustion. (*Hint*: Consider the reaction of equal numbers of moles of material.)

114. For the reaction described in the **Formation of Aluminum Bromide** movie (*Section 6-7*), **(a)** calculate the amount of Al used if its complete reaction led to the production 435 kJ of heat. [The heat of formation of $AlBr_3(s)$ is -527 kJ/mol.] **(b)** What aspects of the reaction shown in the movie make it difficult to quantify the reaction yield in terms of both products (heat and aluminum bromide)?

115. The dissolution of several different ionic compounds is depicted in the **Enthalpy of Solution** activity (*Section 6-7*). Although the simulated process involves the formation of both aqueous anions and aqueous cations, comparative studies can be used to deduce the energy required to form a specific species in solution. Compare the dissolution of NaCl and NaOH, and determine the relative enthalpies of formation of the Cl^- and OH^- in solution.

Chapter 7

Atomic Structure

SOME DISCOVERIES IN chemistry have had a great impact on other disciplines. For example, chemists determine the structures of proteins and other biochemical substances and thus enable biologists to study life processes at the molecular level. As a further example, progress in genetics was once made solely by studying patterns of inheritance in populations. Today, a more fundamental approach involves determining the structure of DNA molecules in minute detail. By the same token, discoveries in other sciences have had dramatic effects on chemistry, changing forever the way chemists look at the world. In this chapter, we will consider developments in physics leading to our current knowledge of the structure of atoms.

We began the story of atomic structure in Chapter 2 with a brief introduction. There we focused on the atomic nucleus with its protons and neutrons, including the concept of atomic masses. So far, we have limited our interest in electrons mainly to their loss or gain in ion formation and in redox reactions. In this chapter, we focus mostly on electrons because their behavior is at the heart of our modern view of atomic structure.

This chapter consists of three parts. The first part emphasizes the discovery of the electron and determination of its properties, providing experimental evidence for the picture of the atom we adopted in Chapter 2. The second part is mainly about light. At first glance, this might seem unrelated to atomic structure. However, a complete understanding of the nature of light requires some totally new ideas. We use these ideas in the third part of the chapter to explain the behavior of electrons in atoms, and that leads to the modern view of atomic structure.

◄ These representations of the Crab Nebula are (clockwise from the upper left) an X-ray image (here in the false color blue), an optical image, a radio-wave image (violet), and an infrared image (orange and yellow). The Crab Nebula is the remnant of a supernova explosion that was seen on Earth in 1054 AD. At its center is a rapidly spinning neutron star. All that we know about objects in outer space comes from energy given off or absorbed by atoms. This energy also reveals the structure of atoms, the subject of this chapter.

Classical View of Atomic Structure

A body of knowledge, often called *classical physics*, was accumulated over several centuries up until the early part of the twentieth century. This classical view of nature included some ideas about atomic structure. We explore the classical view of atomic structure in the next section.

7.1 The Electron: Experiments by Thomson and Millikan

Our story of the discovery of the electron begins about the middle of the nineteenth century. By that time, static electricity, current electricity, and magnetism were rather well understood (recall Section 4.1, and see Appendix B).

Cathode Rays

Beginning with Michael Faraday in 1838 and for many years thereafter, physicists experimented with the passage of electricity between metal electrodes fused into glass tubes from which most of the air had been removed. They found that high-voltage electricity passed readily through these highly evacuated tubes. The mysterious, invisible carriers of the electric current seemed to emanate from the cathode and flow to the anode, and so became known as **cathode rays**.

The electric discharge tube shown in Figure 7.1 is of a type devised by William Crookes in 1879. The Crookes tube uses a metal slit to concentrate cathode rays into a narrow beam. The beam of cathode rays then passes along a screen that gives off a green fluorescence, thereby marking the path of the otherwise invisible rays. (*Fluorescence* is light emission by a phosphor when it is struck by energetic radiation, much as light is emitted when electricity passes through a fluorescent lighting tube.) The photograph also shows that cathode rays are deflected from a straight-line path as they pass near a magnet. The magnet exerts a force on the cathode rays much like a strong crosswind deflects a tennis ball from its straight path.

By the 1890s, scientists had compiled many observations about cathode rays, including the following.

1. The rays are emitted from the cathode when electricity is passed through an evacuated tube.
2. The rays are emitted in a straight line, perpendicular to the cathode surface.
3. The rays cause glass and other materials to fluoresce.
4. The rays are deflected by a magnet in the direction expected for negatively charged particles.*

Application Note

Television picture tubes and computer monitor tubes are cathode ray tubes (CRT). Cathode rays are deflected by magnetic fields to appropriate spots on a screen, where the fluorescence they cause creates an image.

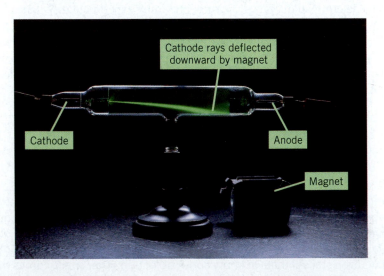

▶ **FIGURE 7.1 Cathode rays and their deflection in a magnetic field**
The beam of cathode rays, made visible by the green fluorescence of a screen coated with zinc sulfide, originates at the cathode and is deflected by the magnet located slightly behind the anode.

*The phenomena in which electricity and magnetism are interrelated, called electromagnetism, are further described in Appendix B.

5. The properties of cathode rays *do not* depend on the composition of the cathode. For example, the cathode rays from an aluminum cathode are the same as those from a silver cathode.

The German and British scientists who conducted most of the cathode ray research agreed on the facts, but they developed strikingly different hypotheses to explain them. In general, German scientists thought that cathode rays were a form of electromagnetic radiation, much like light. In contrast, most British scientists thought the rays were particles of matter—probably residual gas molecules that had acquired a negative charge from the cathode. When J. J. Thomson assessed all the available cathode ray data in 1897, he leaned firmly toward the particle hypothesis. He then settled the question unequivocally in a landmark series of experiments.

The Experiments of J. J. Thomson

In his first experiments, Thomson measured the deflections of cathode rays in magnetic *and* electric fields and confirmed that the rays are negatively charged particles. In another crucial set of experiments, he showed that the magnetic deflections were the same no matter what the residual gas in the cathode ray tube was—whether it was hydrogen, air, carbon dioxide, or any other gas. This observation strongly suggested that cathode rays are not ions formed from gaseous atoms or molecules; instead, they are negatively charged particles found in *all* matter. To strengthen his argument, Thomson designed an experiment to

▲ J. J. Thomson (1856–1940) entered Cambridge University as an undergraduate in 1875 and a scant eight years later was named Cavendish Professor there. In 1893, Thomson's scientific reputation was greatly enhanced with the publication of the book *Recent Researches in Electricity and Magnetism,* and he soon attracted outstanding students from throughout the world. Thomson won the Nobel Prize in 1906 in Physics for characterizing the electron.

Two Accidental Discoveries with Far-Reaching Consequences

A number of cathode ray researchers, apparently including J. J. Thomson himself, observed that sometimes objects *outside* a cathode ray tube glowed when the tube was in use. It remained for Wilhelm Roentgen (1845–1923) to show that a highly penetrating radiation is emitted when cathode rays strike a surface. In turn, this emitted radiation causes objects at some distance from the cathode ray tube to fluoresce. Because the nature of this radiation was not understood at first, Roentgen called the radiation *X rays,* a name still used today.

Within a few weeks, Roentgen was able to characterize X rays quite thoroughly. They were undeflected by electric and magnetic fields and had all the qualities of electromagnetic radiation (light), including an ability to penetrate through matter. Roentgen announced his discovery of X rays on December 28, 1895, and this announcement was quickly followed by practical applications. On January 20, 1896, X rays were used in a medical procedure for the first time—in Dartmouth, New Hampshire—to assist in setting a broken arm bone. X rays later played a vital role in crystal structure determinations (Chapter 11).

Antoine Henri Becquerel (1852-1908) thought that X rays and the phenomenon of fluorescence were related and sought to find out whether naturally fluorescent materials would produce X rays. In an experiment performed in 1896, Becquerel wrapped a photographic plate with black paper, placed a coin on the paper, covered the coin with a fluorescent mineral, and exposed this assembly to sunlight. Sure enough, the developed plate revealed a clear image of the coin. The exposure of the photographic plate was apparently caused by X rays from the fluorescent mineral.

In repeating the experiment, Becquerel had to set aside the assembly in a desk drawer because the day was overcast and the mineral did not fluoresce. On resuming the experiment several days later, Becquerel decided to replace the original photographic plate, fearing that it might have become slightly exposed. He did develop the original plate, however, and much to his surprise observed as sharp an image of the coin as in the original experiment. He soon concluded that the fluorescent mineral had emitted radiation continuously, without the need of sunlight, and discovered that the source of this radiation was uranium, one of the elements in the mineral. Becquerel had discovered *radioactivity*—a phenomenon in which the nuclei of certain unstable heavy atoms emit radiation as they disintegrate. We devote all of Chapter 19 to a further exploration of radioactivity.

▲ The discovery of X rays by Wilhelm Roentgen (above) was accidental but fortunate.

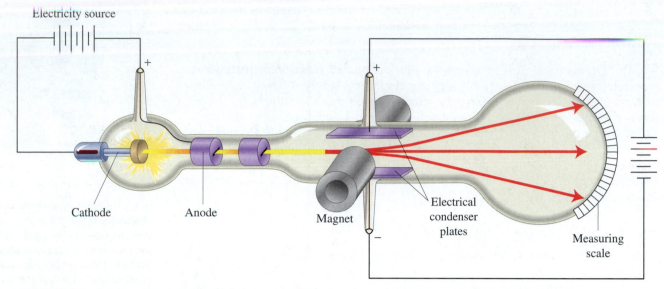

▲ FIGURE 7.2 Thomson's apparatus for determining the mass-to-charge ratio, m_e/e, of cathode rays

obtain an easily measured property of cathode rays: the ratio of their mass (m_e) to charge (e):

$$\frac{\text{Mass of cathode ray particle}}{\text{Charge on cathode ray particle}} = \frac{m_e}{e} \qquad \textbf{(7.1)}$$

In the apparatus outlined in Figure 7.2, a beam of cathode rays passes first through a slit in the anode and then between the oppositely charged plates of an electrical condenser. Superimposed on the electric field between the condenser plates is a magnetic field; one pole of the magnet is in front of the cathode ray tube, and the other is behind it. The cathode rays are bent upward toward the positive plate by the electric field and downward by the magnetic field. The electric and magnetic fields can be adjusted so that the cathode ray beam strikes the fluorescent screen undeflected. The strengths of the two fields producing this undeflected condition are used to calculate the mass-to-charge ratio of the cathode rays. The best available measurements give the value

$$m_e/e = -5.686 \times 10^{-12} \text{ kilogram per coulomb (kg/C)}$$

[The coulomb (C) is the SI unit of electric charge (see Appendix B).]

This value of m_e/e is about 2000 times smaller than the smallest previously known mass-to-charge ratio—that of hydrogen ions, ($+1.045 \times 10^{-8}$ kg/C). This observation suggests the following possibilities.

1. If the charge on a cathode ray particle is comparable to that on a H^+ ion, the mass of a cathode ray particle (m_e) is much smaller than the mass of H^+; or

2. If the mass of a cathode ray particle is comparable to that of a H^+ ion, the charge of a cathode ray particle (e) is much larger than the charge on H^+; or

3. The situation is somewhere between the extremes described in the first two statements.

Thomson thought the first statement would prove to be true, but he had no precise way to determine either m_e or e. Once either could be measured, the other could be calculated through the value of m_e/e.

Based on Faraday's early work on electrolysis, the Irish physicist George Stoney proposed, in 1874, the existence of a fundamental unit of electric charge that he later called an *electron*. After Thomson's work was published, *electron* became the commonly accepted term for cathode ray particles. Next, we will consider how e, the charge on the electron, was established.

"Thus we have in the cathode rays matter in . . . a state in which the sub-division of matter is carried very much further than in the ordinary gaseous state: a state in which all matter . . . is of one and the same kind."

J. J. Thomson

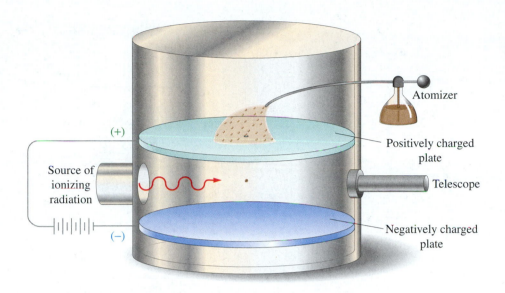

Millikan Oil Drop Experiment animation

◀ **FIGURE 7.3   Millikan's oil-drop experiment**
Oil droplets from an atomizer enter the apparatus through a tiny hole. Friction causes some droplets to acquire an electric charge as they escape from the atomizer. A source of ionizing radiation, such as X rays, also produces ions. The absorption of these ions sporadically changes the electric charges on the droplets. The droplets are observed through a telescope with a measuring scale in the eyepiece.

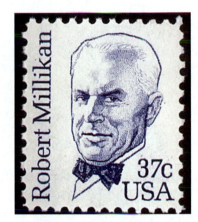

▲ Robert A. Millikan (1868–1953) was a professor at the University of Chicago at the time of his famous experiments on the electronic charge (1909–1913). He won the 1923 Nobel Prize in Physics for these experiments and other experiments in which he confirmed Einstein's explanation of the photoelectric effect (page 273). In 1921, Millikan moved to the California Institute of Technology. There, he developed an interest in a newly discovered type of radiation that enters Earth's atmosphere from outer space, for which he coined the term "cosmic radiation."

"Here, then, is direct, unimpeachable proof that . . . the electrical charges found on ions all have either exactly the same value or else small exact multiples of that value."

Robert Millikan

Electron Charge: Millikan's Oil-Drop Experiment

In 1909, Robert Millikan devised an elegant experiment to determine the charge on an electron. A version of his apparatus is shown in Figure 7.3. The idea is to produce tiny oil droplets, have them acquire an electric charge, and measure the velocity of a falling droplet both in the absence of an electric field and in the presence of an electric field. The velocity is measured by observing the time required for the droplet to fall between two fine lines in the eyepiece of an observing telescope.

Without the electric field, each droplet falls through air under the force of gravity and quickly reaches a constant terminal velocity, v_g. This is analogous to the terminal velocity reached by a parachutist falling through air. When an electric field is imposed, the velocity of an electrically charged droplet changes to a new value, v_e. Consider the situation in Figure 7.3. A droplet carrying a negative charge would be attracted to the positively charged plate and be slowed down ($v_e < v_g$), just as the descent of a parachutist is slowed if the parachute is caught in an updraft.

By analyzing the data from hundreds of experiments, Millikan found that droplets carried an electric charge equal either to a fundamental unit of charge, e, or to some multiple of e. The fundamental unit of *negative* charge is that carried by an electron. The charge on an electron is

$$e = -1.602 \times 10^{-19} \text{ coulomb (C)} \tag{7.2}$$

With this value and Thomson's value of m_e/e, we can calculate the mass of an electron, m_e:

$$m_e = \frac{m_e}{e} \times e = -5.686 \times 10^{-12} \text{ kg/C} \times \frac{-1.602 \times 10^{-19} \text{ C}}{1 \text{ electron}}$$

$$= 9.109 \times 10^{-31} \text{ kg/electron} \tag{7.3}$$

7.2   Atomic Models: Thomson and Rutherford

Once he had established electrons as fundamental particles of all matter, Thomson became keenly aware that an atom could not consist of electrons alone. The negative charges of the electrons had to be neutralized by an equivalent amount of positive charge, and the electrons had to be organized into some kind of stable configuration, or else they would fly apart.

Thomson's "Raisin Pudding" Atomic Model

Because Thomson did not know exactly how positive electric charge is distributed in an atom, he considered the case that was easiest to describe mathematically. He developed a model in which the positive charge is uniformly distributed in a spherically shaped "cloud" of charge and the electrons are embedded in the cloud in such a way

Uniformly distributed positive charge

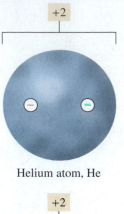

+2

Helium atom, He

+2

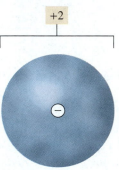

Helium ion, He⁺

▶ **FIGURE 7.4** **Thomson's raisin pudding model of the atom**
For a helium atom, the model proposes a large spherical cloud in which two units of positive charge are distributed uniformly. The two electrons lie on a line through the center of the cloud. The loss of one electron produces the He⁺ ion, with the remaining electron having moved to the center of the cloud.

▲ Ernest Rutherford (1871–1937) is said to have been digging potatoes on his father's farm in New Zealand when notified of a scholarship to work with J. J. Thomson at Cambridge. He threw down his spade, proclaiming, "That's the last potato I'll dig." Although a physicist, Rutherford was awarded the Nobel Prize in Chemistry in 1908 for his pioneering work in radioactivity.

"It is about as incredible as if you had fired a 15-inch shell at a piece of tissue paper and it came back and hit you."

Ernest Rutherford

that their attraction for the positive charge just offsets the repulsions among the electrons. This arrangement somewhat resembles raisin pudding.

For a hydrogen atom, Thomson proposed one electron at the exact center of the spherical cloud. For an atom with two electrons (helium), he proposed two electrons along a straight line passing through the center of the cloud, with each electron halfway between the center and the outer surface of the cloud (Figure 7.4). Thomson applied this kind of analysis to atoms containing up to 100 electrons.

Rutherford's Nuclear Model of the Atom

Ernest Rutherford was a pioneer in the study of radioactivity. He characterized one type of radiation, called *alpha* (α) particles, as identical to doubly ionized helium atoms, He^{2+}. Based on Thomson's model of the atom, Rutherford expected that most α particles would pass through atoms undeflected. However, he also expected that any of the positively charged α particles that came close to an electron should be deflected to some extent. By measuring such deflections, he hoped to gain information about the distribution of electrons in an atom.

Rutherford assigned the experiment illustrated in Figure 7.5 to his assistant, Hans Geiger, and an undergraduate student, Ernest Marsden. When they bombarded very thin foils of metals such as gold, silver, and platinum with α particles, Geiger and Marsden found that most of the particles went right through the foil either undeflected or deflected only slightly. This is just what Rutherford had expected. However, much to Rutherford's surprise, a few particles were deflected sharply, and once in a while, an α particle would bounce right back toward the source.

These α-particle scattering experiments could not be explained by the Thomson model, in which positive charge is spread throughout an atom. Thus, Rutherford concluded that all the positive charge of an atom is concentrated at the center of an atom in a tiny core called the *nucleus*. When a positively charged α particle approaches a positively charged nucleus, the particle is strongly repelled and therefore sharply deflected. Because only a few α particles were deflected, Rutherford concluded that the nucleus occupies only a tiny fraction of the volume of an atom. Most of the α

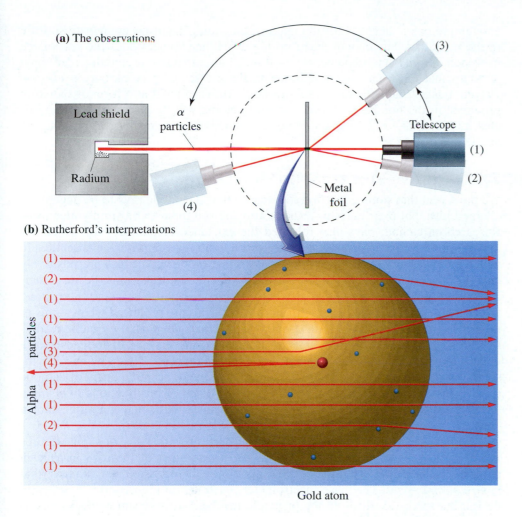

(a) The observations

Lead shield

Radium

α particles

(4)

(3)

Telescope

(1)

(2)

Metal foil

(b) Rutherford's interpretations

Alpha particles

(1)
(2)
(1)
(1)
(1)
(3)
(4)
(1)
(1)
(2)
(1)
(1)

Gold atom

◀ **FIGURE 7.5   The scattering of alpha particles by a thin metal foil**
(a) Alpha particles produce light flashes whenever they strike the zinc sulfide-coated screen at the end of the telescope. Geiger and Marsden observed the following: (1) Most of the α particles pass through the foil undeflected. (2) Some α particles are deflected slightly as they penetrate the foil. (3) A few (about 1 in 20,000) are greatly deflected. (4) A similar small number do not penetrate the foil at all but are reflected back toward the source. (b) Rutherford interpreted the trajectories observed at the different positions: (1) An α particle passes through the atom undeflected (a fate shared by most of the particles); (2) an α particle is deflected slightly as it passes near an electron; (3) an α particle is deflected strongly by passing close to the nucleus; and (4) an α particle bounces back as it approaches the nucleus head-on. Rutherford concluded that each metal atom in the foil consists of a massive, positively charged nucleus plus lightweight, negatively charged electrons outside the nucleus. Note that both the nucleus and the electrons are much smaller than depicted here.

particles passed right through because, except for the tiny nucleus and a small number of electrons outside, an atom is empty space.

To picture Rutherford's model of a nuclear atom, visualize the atom as the size of a gigantic indoor football stadium. A pea at the middle of the structure represents the nucleus. A few bees buzzing here and there throughout the stadium represent the electrons. The roof of the stadium prevents the bees from leaving, but the atom has no comparable covering. Electrons remain in the atom because they are strongly attracted to the positively charged nucleus.

If all portions of Figure 7.5 were drawn to relative scale, the gold atom would be of the size pictured, but the nucleus, electrons, and α particles would be invisible to the naked eye.

 **Rutherford Experiment animation**

7.3   Protons and Neutrons

The experiments that led to the idea of a nuclear atom also provided data that could be used to determine the amount of positive nuclear charge. Rutherford believed that this positive charge was carried by particles called *protons*, that the proton charge was the fundamental unit of positive charge, and that the nucleus of a hydrogen atom consisted of a single proton. These ideas were proved correct some years later through experiments in which protons were ejected from atomic nuclei and found to be identical to the nuclei of hydrogen atoms.

In the early 1900s, scientists commonly used the notion of an *atomic number*, but it was just a numerical ranking of the elements in order of increasing atomic mass; it seemed to have no underlying fundamental significance. By 1914, however, experiments we will describe in the next chapter showed that the atomic number is equal to the number of units of positive charge on the nucleus. And, when protons were shown to be the particles responsible for the positive charge on the nucleus, the atomic number was seen to be equal to the number of protons in the nucleus.

If we assume that all protons have the same mass (which proves to be a completely valid assumption), then, except for hydrogen, an atom does not have enough protons to account for the mass of the atom. Electrons contribute so little to the mass that they cannot make up the difference. Where does the rest of the mass come from? One hypothesis was that the atomic nucleus contains additional particles that have masses comparable to the masses of protons but *no electric charge*. James Chadwick discovered these particles in 1932 as a form of radiation produced in the bombardment of beryllium atoms with α particles. These particles, called *neutrons*, did indeed prove to have about the same mass as protons and no electric charge.

7.4   Positive Ions and Mass Spectrometry

We have seen that some early investigators believed cathode rays to be negatively charged ions, but their mass-to-charge ratio showed them to be simply electrons. Research on cathode rays also revealed the existence of particles for which the deflection direction in magnetic and electric fields is opposite the direction in which cathode rays are deflected, indicating that the new particles are positively charged. These positively charged particles were shown to be ions, because their mass-to-charge ratios are (1) significantly larger than those of cathode rays and (2) different from one type of gas to another. The ions are produced when atoms or molecules of the residual gas in a cathode ray tube are struck by cathode rays (electrons). In such collisions, one or more electrons may be dislodged from a neutral atom or molecule, leaving a positive ion. In Figure 7.2, we saw how Thomson cut a slit through the anode in a cathode ray tube to permit a beam of cathode rays to pass through. Similarly, we can obtain a beam of positive ions by using a slitted cathode, and we can study these beams in an experimental technique known as *mass spectrometry*.

A **mass spectrometer** is a device that separates positive gaseous ions according to their mass-to-charge ratio. Actually, because most of the ions in a beam carry a 1+ charge, we can think of the separation as being based just on mass (that is $m/e = m/1 = m$). The mass spectrometer illustrated in Figure 7.6 is of an early design but serves to illustrate the principle involved. If a stream of positive ions having equal velocities is brought into a magnetic field,

- all the ions experience a force that deflects them from straight-line paths into circular paths, and
- the lightest ions are deflected the most (into the tightest circle) and the heaviest ones are deflected the least.

Think of this analogous situation: A Ping-Pong ball, a tennis ball, and a baseball are all thrown with the same speed in the same direction but into a strong crosswind. In moving forward a certain distance, the Ping-Pong ball, being the lightest of the three, is

Application Note

A technique closely related to mass spectrometry, called *ion mobility spectrometry*, is used at many airports to screen luggage for explosives.

Mass Spectrometer simulation

▶ **FIGURE 7.6   A mass spectrometer**

A gaseous sample is ionized by bombarding it with electrons in the lower part of the apparatus (not shown), producing positive ions. The ions are accelerated to a particular velocity and pass through a narrow slit into the curved chamber. A magnetic or electric field is applied to the beam of ions and causes their deflection from a linear trajectory. All the ions with the same mass-to-charge ratio are deflected into the same circular path. (In most cases, the ionic charge is 1+ and thus the mass-to-charge ratio has the same numeric value as the mass.) Modern spectrometers use electronic detection schemes rather than photographic plates or film to establish mass-to-charge ratios and relative numbers of ions.

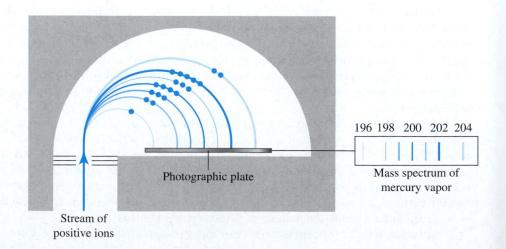

Photographic plate

Stream of positive ions

196 198 200 202 204

Mass spectrum of mercury vapor

most strongly deflected. Next comes the tennis ball. The heaviest object, the baseball, makes the same forward progress with the least deflection.

Each circular path in the mass spectrometer of Figure 7.6 represents ions having a specific mass-to-charge ratio, and the point where the ions strike the photographic plate is related to the mass-to-charge ratio. The density (or darkness) of the image on the plate is related to the relative number of ions with this ratio. A record of the separation of ions in a mass spectrometer is called a *mass spectrum*. Figure 7.7 shows the form in which a mass spectrum is usually recorded by modern instruments. Mass spectral data such as these are used to calculate the weighted average atomic masses we discussed in Chapter 2.

Light and the Quantum Theory

Space is vast and dark. Here and there, a star lights a small volume of the void. Our Sun is one such star, lighting Earth and the other parts of the solar system. Our eyes perceive sunlight to be white, but under proper conditions, we can separate white light into a rainbow of colors. Later in the chapter, we will see that to describe the arrangement of electrons outside the nuclei of atoms, we need to analyze the light emitted by energetic atoms. To do this analysis, we need to know something about the physical nature of light.

7.5   The Wave Nature of Light

Many types of waves exist in nature. Waves of water break on the shores of oceans and lakes. A concentric pattern of waves is formed when a pebble is tossed into a pond. Earthquakes send waves through Earth's crust. Sound waves carry music to our ears. With microwaves, we can thaw a TV dinner or heat a cup of soup. But just what is a wave?

A *wave* is a progressive, repeating disturbance that spreads through a medium from a point of origin to more distant points. The material that makes up the medium moves little in the direction of the disturbance. It is rather like a whispered message ("disturbance") passed along by seated people ("little movement") from one end of a row ("point of origin") to the other ("distant point"). A person holding a long rope fastened to a wall can start a wave motion in the rope by jerking the end of the rope up and down (Figure 7.8).

Electromagnetic waves originate from the movement of electric charges. This movement produces oscillations (fluctuations) in electric and magnetic fields that are propagated over distances. Unlike water waves and sound waves, electromagnetic waves require no medium for their propagation. It is the ability of electromagnetic radiation to travel through empty space that makes it possible for the Sun to transmit some of its energy to Earth as sunlight or for scientists to send a command from Earth to a computerized robot on the surface of Mars.

We characterize electromagnetic radiation by its wavelength, frequency, and amplitude (Figure 7.9). The **wavelength** is the distance between any two identical points in consecutive cycles; we usually choose peaks (also called crests) for convenience. Wavelength is denoted by the Greek letter λ (lambda). A common SI unit of wavelength, especially for visible light, is the nanometer (1 nm = 10^{-9} m). A non-SI unit that is still used to some extent in scientific work is the angstrom, Å (Å = 10^{-10} m).

The **frequency** of a wave is the number of cycles of the wave that pass through a point in some unit of time. For example, for a wave with a 60-cycle frequency, 60 crests (or 60 troughs) pass a fixed point in 1 s. That is, 60 cycles of the wave pass the point each second. Frequency is denoted by the Greek letter ν (nu, pronounced the same as "new"). The SI unit of frequency is the *hertz* (*Hz*). A hertz is 1 cycle/s, but generally the word cycle is understood. That is, 1 Hz = 1 s^{-1}.

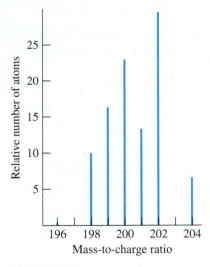

▲ **FIGURE 7.7   Mass spectrum for mercury**

The photographic record of Figure 7.6 has been converted to a scale of relative numbers of atoms. The percent natural abundances for the mercury isotopes are ^{196}Hg, 0.146%, ^{198}Hg, 10.02%, ^{199}Hg, 16.84%, ^{200}Hg, 23.13%, ^{201}Hg, 13.22%, ^{202}Hg, 29.80%, ^{204}Hg, 6.85%.

QUESTION: Where would ^{200}Hg^{2+} appear in the mass spectrum of mercury?

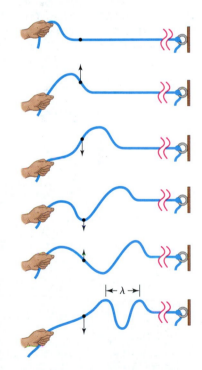

▲ **FIGURE 7.8   The simplest wave motion: A traveling wave in a rope**

Imagine an infinitely long rope. Up-and-down hand motion at the left end of the rope causes waves to pass along the rope from left to right. The up-and-down motion of a typical point (dot) on the rope as the wave passes is shown. This one-directional moving wave is called a *traveling wave.*

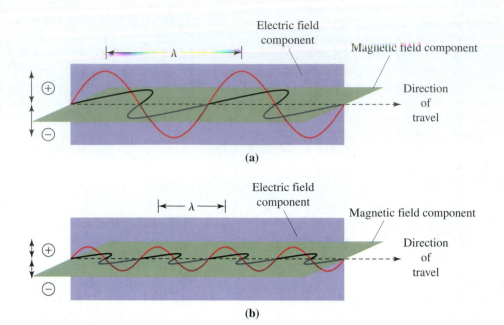

▶ **FIGURE 7.9 An electromagnetic wave**
This artist's rendering visualizes an electromagnetic wave as a superposition of two wave motions, that of an alternating electric field and, perpendicular to it, that of an alternating magnetic field. (These alternating fields are generated by the relative motion of electric charges.) Note that the wave in (b) requires four cycles to travel the same distance from left to right across the drawing as covered by only two cycles of the wave in (a). This means that the wave in (b) has a *higher* frequency and a *shorter* wavelength than the wave in (a). The wave in (a) is shown with a greater amplitude than in (b), but this need not be the case. The amplitude of a wave is independent of frequency and wavelength.

Application Note

A commercial radio station broadcasts at a fixed frequency. An FM station that is "101 on the dial" is broadcasting radio waves that have a frequency of 101 MHz (megahertz). An AM station at "710" is broadcasting waves at 710 kHz (kilohertz).

Properties of a Wave animation

The *amplitude* of a wave is its height: the distance from a line of no disturbance that runs horizontally through the center of the wave to a peak. Imagine yourself on a raft at sea. The number of times you bob up and down per unit time is the frequency of the ocean waves. How far up or down you go relative to a calm sea is the amplitude of the waves. An electromagnetic wave differs from an ocean wave in that the electromagnetic wave does not bob up and down as it moves. Rather, the strengths of its magnetic and electric fields oscillate, and the amplitude corresponds to the strengths of those fields.

Consider again the wave in the rope shown in Figure 7.8. If ν cycles of the wave pass through a point on the rope each second—say, for instance, the point marked with a black dot in the drawing—and if the length of each cycle is the wavelength λ, the total distance the wave travels in 1 s is

$$\nu\lambda = c \qquad (7.4)$$

If we express the frequency ν in seconds^{-1} and the wavelength λ in meters, we see that the unit for c is s$^{-1} \times$ m = m s^{-1}, a distance per unit time. Thus c is a *speed*—the speed of the wave.

In a vacuum, light travels at a constant speed of 2.99792458×10^8 m/s (often rounded to 3.00×10^8 m/s). In air, the speed of light is slightly less, but in calculations, we will generally assume that the light waves travel in a vacuum. The speed of light is the fastest speed attainable. A beam of light travels from London to San Francisco in just 0.03 s and from Earth to the Moon in 1.28 s.

Because the speed of light in any given medium is constant, wavelength and frequency are inversely related to each other. The longer the wavelength, the lower the frequency, or conversely, the shorter the wavelength, the higher the frequency. The amplitude is unrelated to wavelength and frequency.

Example 7.1

Calculate the frequency of an X ray that has a wavelength of 8.21 nm.

STRATEGY

When using Equation (7.4), the value of c will always be 3.00×10^8 m s^{-1}. We must solve Equation (7.4) for the unknown variable, either ν or λ, and substitute a known value for the other variable, expressed in the appropriate unit. Here, ν is the unknown; λ is the known value, expressed in meters.

SOLUTION

First, we solve Equation (7.4) for ν.	$\nu = \dfrac{c}{\lambda}$
Next, we convert the wavelength from nanometers to meters.	$\lambda = 8.21 \text{ nm} \times \dfrac{1 \times 10^{-9} \text{ m}}{1 \text{ nm}} = 8.21 \times 10^{-9} \text{ m}$
Then, we can substitute the known values of c and λ.	$\nu = \dfrac{3.00 \times 10^8 \text{ m s}^{-1}}{8.21 \times 10^{-9} \text{ m}} = 3.65 \times 10^{16} \text{ s}^{-1}$

ASSESSMENT

Besides checking for correct units when applying Equation (7.4), expect ν generally to be a large number, often a large positive power of ten, and expect λ (in meters) generally to be a small number, often a large negative power of ten.

EXERCISE 7.1A

Calculate the frequency, in hertz, of a microwave that has a wavelength of 1.07 mm.

EXERCISE 7.1B

Calculate the wavelength, in nanometers, of infrared radiation that has a frequency of 9.76×10^{13} Hz.

The Electromagnetic Spectrum

The types of electromagnetic radiation range from the extremely short-wavelength, high-frequency *gamma* (γ) rays, which are produced in certain radioactive decay processes, to the very-long-wavelength, very-low-frequency waves emitted by electric power transmission lines. This complete range of wavelengths and frequencies is known as the **electromagnetic spectrum** and is illustrated in Figure 7.10.

The electromagnetic spectrum is largely *invisible*; we see only a tiny portion near the middle of the range. The visible spectrum ranges from about 760 nm, where red light blends into the infrared, to about 390 nm, where violet light fades into the ultraviolet. Our senses can detect some radiation other than visible. We feel warmth from the energy content of infrared radiation when this radiation comes through a car window on a cold winter day. Sunburned skin is a sign that we have been

Regions of the Electromagnetic Spectrum animation

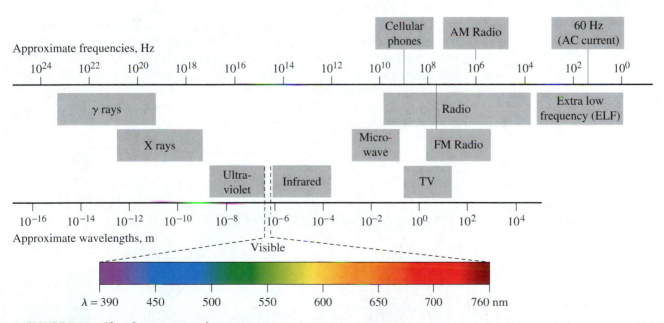

▲ **FIGURE 7.10   The electromagnetic spectrum**

The visible region, which extends from red at the longest wavelength to violet at the shortest wavelength, is only a small portion of the spectrum. The approximate wavelength, frequency ranges, and uses of some other forms of electromagnetic radiation are also indicated.

QUESTION: What are the approximate frequencies of blue, green, and red visible radiation?

(a)

(b)

▲ The eyes of bees and other insects are sensitive to ultraviolet light. In ordinary (white) light, a flower is yellow (a). Under ultraviolet light, the petals lose their color and the nectar guides become visible to the human eye (b). Bees see these nectar guides as prominent dark patches on the petals leading to the center of the flower, helping the bees to find the flower nectaries.

Visible Spectrum activity

exposed to ultraviolet radiation too long. We can also detect the different forms of electromagnetic radiation with appropriate devices: radios to detect radio waves, heat sensors to respond to infrared radiation, ordinary photographic film to record images of objects in visible light, and so on.

Materials vary in their ability either to absorb or let pass (transmit) different parts of the electromagnetic spectrum. Our bodies absorb visible light but transmit most X rays. That is, most parts of our bodies, except for bones, are *transparent to* X rays; the X rays pass right through. Ordinary window glass is transparent to visible light and some infrared radiation, but it effectively blocks most ultraviolet radiation by absorbing it. (You cannot get a suntan from sunlight coming through window glass.) Thus, we use some materials to shield us from harmful rays and others to transmit beneficial rays.

Example 7.2   A Conceptual Example

Which light has the higher frequency: the bright red brake light of an automobile or the faint green light of a distant traffic signal?

ANALYSIS AND CONCLUSIONS

The difference in brightness is immaterial because, as we have noted, the amplitude of a light wave has no bearing on the frequency. From Figure 7.10, we see that there is a relationship between the color of light and its frequency and wavelength. The order of increasing frequency in the visible spectrum proceeds from red to violet:

Red < orange < yellow < green < blue < violet

The green light has a higher frequency than the red light.

EXERCISE 7.2A

Which source produces electromagnetic radiation of the longer wavelength: a microwave oven or the fluorescent screen of a color television set?

EXERCISE 7.2B

Which of the following has the highest frequency: (a) green light, (b) radiation with wavelength 4610 Å, (c) 91.9 MHz radiation, (d) 622-nm radiation?

Continuous and Line Spectra

When white light from an incandescent lamp passes through a narrow slit and then through a glass prism, the light separates into a *spectrum*; that is, the various components of the light are spread out into a rainbow of colors (Figure 7.11). The spectrum is a **continuous spectrum**, so called because each color merges into the next without a break. Sunlight interacting with raindrops forms a rainbow in the sky because each raindrop acts as a prism. The different colors of light correspond to different wavelengths and frequencies, as we saw in Figure 7.10.

▶ **FIGURE 7.11   The spectrum of white light**

Red light is bent the least and violet light the most when white light is passed through a glass prism. The other colors of the visible spectrum are found between red and violet.

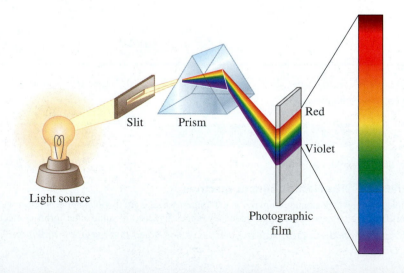

Slit Prism Red Violet

Light source

Photographic film

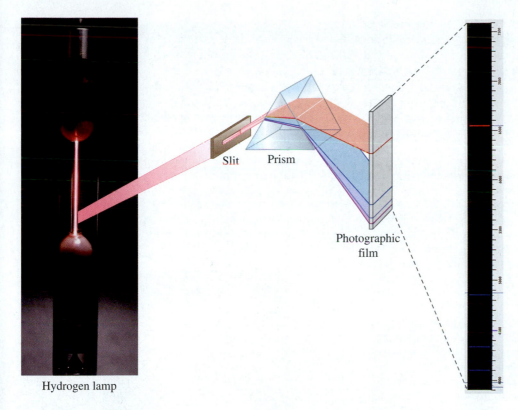

Hydrogen lamp

◄ **FIGURE 7.12   The emission spectrum of hydrogen**

Passage of an electric current through a lamp containing hydrogen gas produces a pink glow of light. When that light is passed through a slit and a prism, it generates four images of the slit—four lines in this *line spectrum* of hydrogen.

Now consider light from a lamp containing hydrogen gas at low pressure. When an electric current passes through the lamp, some of the hydrogen atoms are energized as they are struck by electrons. These energized atoms—we say they have become *excited*—give off energy in the form of light. If we pass this light from a hydrogen lamp through a slit and prism, we observe not a continuous spectrum but only a few images of the slit. These images appear as narrow colored lines separated by dark regions (Figure 7.12); this is a *discontinuous* spectrum. Each line corresponds to electromagnetic radiation of a specific frequency and wavelength. The pattern of lines produced by the light emitted by excited atoms of an element is called a **line spectrum**.

The analysis of the light emitted when an element is strongly heated or energized by an electric spark or in an electric discharge lamp is called *emission spectroscopy*. The emitted light is dispersed into individual wavelength components as in Figure 7.12. A photograph or other record of the emitted light, called the **emission spectrum** of the element, is prepared. Every element has a unique emission spectrum, as illustrated by Figure 7.13. The frequencies or wavelengths of its spectral lines identify the element. Thus, the emission spectrum is like an "atomic fingerprint."

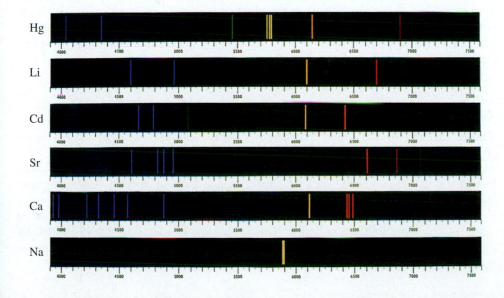

◄ **FIGURE 7.13   Line emission spectra of selected elements**

Atomic spectra of six elements. Wavelengths are given in angstroms, where $1\ \text{Å} = 10^{-10}$ m. Each element has its own distinctive spectrum that can be used to identify the element. In addition to their practical use in analyzing matter, atomic spectra have led to many of the ideas concerning atomic structure.

QUESTION: A mixture, when vaporized and heated, emits light at about 6100Å and 6700Å. What element might be present in the mixture?

Fireworks and Spectra

In the early days of spectroscopy, there was need for a steady, clean-burning gas flame. In 1855, Robert Bunsen, a pioneer chemist and spectroscopist, developed the well-known *Bunsen burner* to produce such a flame. With this new burner, line spectra such as those shown in Figure 7.13 could be easily obtained. Using his new burner and a spectroscope, Bunsen and his co-worker Gustav Kirchhoff discovered a new element in 1860. They knew it was a new element because it produced a line spectrum different from that of any known element. Several of the lines were in the blue region of the visible portion of the electromagnetic spectrum, and so the new element was named *cesium* (Latin *caesius*, sky blue). In 1861, they discovered another new element, *rubidium* (Latin *rubidius*, deepest red).

The brilliant colors of certain emission spectra have been put to use for hundreds of years—in fireworks. Strontium produces a brilliant red, barium produces green, sodium produces yellow, and copper may produce either blue or purple.

Of course, there is more to coloring fireworks than just adding the desired element to the mixture. For example, although strontium atoms emit some red light, it is the transitory species SrCl that produces the brilliant red associated with strontium. Ordinary strontium chloride would seem a logical ingredient for fireworks, but it is not, because $SrCl_2$ is difficult to maintain as a dry powder. Instead, oxidizing and reducing agents, strontium carbonate, and a chlorine-containing compound are incorporated into the mixture. A redox reaction produces a flame; the chlorine-containing compound decomposes and reacts with the strontium carbon-

ate to form SrCl, which is then excited by the flame and emits its brilliant red light.

The flame temperature must also be controlled. At low temperatures, not enough of the SrCl may be formed or excited and the color may be pale or nonexistent. At very high temperatures, the SrCl may decompose excessively. Temperature is controlled by using different oxidizing agents and by including aluminum or magnesium metal (auxiliary fuel) in the firework. By combining science and art, the pyrotechnician uses emission spectra to paint the sky with color.

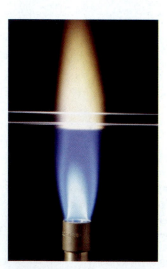

Atoms of certain elements—Li, Na, K, Rb, Cs, Ca, Sr, and Ba, for example—strongly emit visible light when the elements or their compounds are heated in a gas flame. The flame takes on a distinctive color determined by the particular element, and the procedure is referred to as a *flame test* for the element. For example, two closely spaced, exceptionally bright lines in the yellow region dominate the emission spectrum of sodium, as seen in the bottom spectrum of Figure 7.13. These spectral lines are produced in the flame test for sodium and sodium compounds. Figure 7.14 shows this yellow light being emitted by a piece of glass tubing heated in a gas flame. (The raw materials used to make ordinary soda-lime glass are sodium carbonate, calcium carbonate, and silicon dioxide.)

7.6   Photons: Energy by the Quantum

Toward the end of the nineteenth century, some scientists thought that classical physics was capable of dealing with all scientific questions. Others, however, were disturbed by the seeming inability of classical physics to explain some rather fundamental natural phenomena, such as the existence of line spectra. Let's consider another phenomenon that classical physics could not explain.

All solids emit some electromagnetic radiation at all temperatures, but in most cases it is invisible infrared radiation. This is the type of radiation that is emitted by Earth's surface and trapped by CO_2 and certain atmospheric gases, causing the greenhouse effect (Chapter 25).

At high temperatures, solids emit radiation that we can see because the wavelengths of such radiation are in the visible range. At about 750 °C, for example, a solid emits considerable red-light (think of a red-hot fireplace poker). As the temperature rises further, more light in the yellow and blue portions of the spectrum blends with the red, until at about 1200 °C, the solid glows white (hence, the term "white-hot"). This radiation depends only on the temperature of a solid and not on the particular elements present. It is called *black-body radiation* and differs in a fundamental way from the light emitted by energetically excited *gaseous* atoms, which *does* depend on the particular elements.

The difference between the light emitted by solids and that emitted by gases is based on the fact that atoms in a gas are largely independent of one another, whereas those in a solid are close together and interact strongly with one another.

Planck's Quantum Hypothesis

According to classical physics, the atoms in a solid vibrate about fixed points. As the temperature increases, the atoms vibrate more vigorously. Black-body radiation by a solid is simply the release of some of the energy of a system of vibrating atoms as electromagnetic radiation. One hypothesis of classical physics successfully explained the relationship between the quantity of energy emitted and the radiation frequency for high frequencies, and a different hypothesis worked reasonably well for low frequencies.

In 1900, Max Planck derived a single relationship between the energy and frequency of the emitted radiation that worked for *all* frequencies. To do so, however, he had to introduce a new fundamental constant into his formulation. He could find no justification for this constant in classical physics, and he therefore felt compelled to make a radical hypothesis. Planck proposed that the vibrating atoms in a heated solid could absorb or emit electromagnetic energy only in certain discrete amounts. The smallest amount of energy, called a **quantum**, is defined by the relationship

$$E = h\nu \qquad (7.5)$$

where h is a constant called **Planck's constant** and has the value of 6.626×10^{-34} J s.

Planck's quantum hypothesis states that energy can be absorbed or emitted only as a quantum or as some whole multiple of a quantum. That is, an energy change can be $h\nu$, $2h\nu$, $3h\nu$, and so on, but it cannot be $1.5h\nu$ or $3.06h\nu$. According to classical physics, energy varies *continuously*—there are no limitations on the amount of energy that a system can gain or lose. Quantum theory, in contrast, allows energy changes to occur only by discrete amounts—variations in energy are *discontinuous*. In a way, the quantization of energy is like an "atomization" of energy.

Planck's constant, h, is an extremely small quantity, and energy quanta are therefore also very small. As a consequence, the quantization of energy is of special importance in the microscopic world, where we deal with only tiny quantities of energy.

Scientists in Planck's time, including Planck himself, found the quantum theory quite strange and had difficulty accepting it. Albert Einstein and Niels Bohr soon changed all that by successfully applying Planck's theory to other areas in which classical physics had failed. Planck was awarded the 1918 Nobel Prize in Physics for his part in forever changing the way scientists view the world.

Application Note

Some situations in daily life involve "quantized" quantities. Consider a vending machine that accepts only nickels, dimes, and quarters. Items in the machine can have prices only at five-cent intervals. Thus, a price of $0.55 is possible, but a price of $0.57 is not.

The Photoelectric Effect: Einstein and Photons

In 1905, Albert Einstein extended Planck's quantum theory and used it to explain the phenomenon known as the *photoelectric effect*, illustrated in Figure 7.15. A vacuum tube called a *phototube* contains a metal cathode of specific composition. When light ("photo") shines on the cathode, electrons may be ejected, creating an electric current ("electric"). However, in order for the photoelectric effect to occur, the frequency of the light must be above a certain minimum value, called the *threshold frequency*. Light below the threshold frequency does not cause the photoelectric effect, no matter how intense the light. Classical physics cannot explain the existence of a threshold frequency.

Einstein viewed Planck's quantization of electromagnetic radiation in a special way. He considered the electromagnetic energy to be bundled into little packets called **photons.** The energy of a photon is Planck's quantum of energy. Thus, for light of frequency ν,

$$\text{Energy of photon} = E = h\nu$$

Einstein explained the photoelectric effect as follows: Photons of light strike atoms on the cathode surface and give up their energy to certain electrons in these atoms.

Application Note

In some applications of the photoelectric effect, a beam of light shines across an opening, striking a metal in a photoelectric cell. The metal gives off electrons, and the moving electrons then constitute an electric current. When a person interrupts the beam of light, the current is cut off and a switching device operates a door opener or rings an alarm. In some smoke detectors, smoke serves to interrupt the light beam.

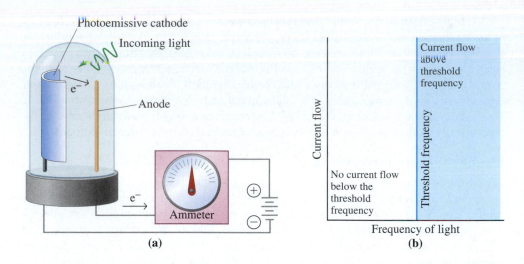

Photoemissive cathode

Incoming light

Anode

e^-

e^-

Ammeter

(a)

Current flow

No current flow below the threshold frequency

Threshold frequency

Current flow above threshold frequency

Frequency of light

(b)

▶ **FIGURE 7.15   The photoelectric effect and the frequency of light**

(a) A voltage is applied between the cathode and anode of a phototube. When light strikes the cathode, electrons are ejected, producing an electric current that is registered by the ammeter, but only if the frequency of the light is high enough. That is, there is a *threshold frequency* for the photoemissive material. (b) With light below the threshold frequency, electrons are not ejected, no matter how intense the light.

QUESTION: With light above the threshold frequency, what role does the intensity of the light play in the results?

Photoelectric Effect animation

"It not only prohibits the killing of two birds with one stone, but also the killing of one bird with two stones."

James Jeans, commenting on Einstein's explanation of the photoelectric effect

The energized electrons overcome their attraction to the atomic nuclei and escape from the surface. To be effective, the photons must have more than a certain minimum of energy, accounting for a threshold frequency (recall Equation 7.5). Photons that have energies above the threshold frequency will also eject electrons, with the excess energy appearing as kinetic energy of the electrons. But repeated hits by photons with too little energy will be ineffective. Ejection of an electron must be secured through the action of a *single* photon acting alone; the photons cannot join forces.

To understand the photoelectric effect, consider the analogy of a truck stuck in the mud. A small garden tractor fails to dislodge it. A farm tractor, on the other hand, is able to get the truck out. Larger tractors exert even more push than necessary, and give the freed truck some kinetic energy.

To emphasize that the expression $E = h\nu$ gives the energy of a single photon of light, we can express Planck's constant as $h = 6.626 \times 10^{-34}$ J s photon^{-1}. In Example 7.3, we see that the energy of a single photon is extremely small. The quantity of energy associated with 1 mol of photons, on the other hand, is comparable to that of enthalpy changes in reactions. The energy of 1 mol of photons, as we see in Example 7.4, is Planck's equation multiplied by Avogadro's number:

$$E = N_A h\nu \qquad (7.6)$$

To summarize, we can best describe some phenomena involving light with the classical view of electromagnetic waves. However, to describe other phenomena, we need the quantum view and its photons. If we think of photons as "particles" of light, then we might say that light has a wave-particle duality. The idea of a wave-particle duality is an important one for topics that come later in the chapter.

Example 7.3

Calculate the energy, in joules, of a photon of violet light that has a frequency of 6.15×10^{14} s^{-1}.

STRATEGY

This problem involves a direct application of Planck's Equation (7.5), where both Planck's constant, h, and the frequency, ν, are known.

SOLUTION

$$E = h\nu$$
$$= 6.626 \times 10^{-34} \text{ J s photon}^{-1} \times 6.15 \times 10^{14} \text{ s}^{-1}$$
$$= 4.07 \times 10^{-19} \text{ J photon}^{-1}$$

EXERCISE 7.3A

Calculate the energy, in joules per photon, of microwave radiation that has a frequency of 2.89×10^{10} s^{-1}.

EXERCISE 7.3B

Calculate the energy, in joules per photon, of ultraviolet light with a wavelength of 235 nm.

Problem-Solving Note

Keep in mind that

High energy (E) means *high* frequency (*large* value of ν) and *short* wavelength (*small* value of λ).

Low energy (E) means *low* frequency (*small* value of ν) and *long* wavelength (*large* value of λ).

Example 7.4

A laser produces red light of wavelength 632.8 nm. Calculate the energy, in kilojoules, of 1 mol of photons of this red light.

STRATEGY

We know that Equation (7.5) yields the energy of one photon, but for 1 mol photons we must use Equation (7.6). However, we first need to find the frequency of the red light; and to obtain this, we need Equation (7.4), remembering also that wavelength must be expressed in meters. At this point, we will have all the necessary data to apply Equation (7.6).

SOLUTION

First, we convert a 632.8-nm wavelength to meters.

$$\lambda = 632.8 \text{ nm} \times \frac{1 \times 10^{-9} \text{ m}}{1 \text{ nm}} = 6.328 \times 10^{-7} \text{ m}$$

Next, we use this wavelength in Equation (7.4) to obtain the frequency of the red light.

$$\nu = \frac{c}{\lambda} = \frac{2.998 \times 10^8 \text{ m s}^{-1}}{6.328 \times 10^{-7} \text{ m}} = 4.738 \times 10^{14} \text{ s}^{-1}$$

Finally, we substitute Avogadro's number, Planck's constant, and the light frequency into Equation (7.6).

$$E = N_A h\nu$$
$$= 6.022 \times 10^{23} \text{ photon mol}^{-1} \times 6.626 \times 10^{-34} \text{ J s photon}^{-1} \times 4.738 \times 10^{14} \text{ s}^{-1}$$
$$= 1.891 \times 10^5 \text{ J/mol} = 189.1 \text{ kJ/mol}$$

ASSESSMENT

In this problem it is very important to check the signs of the powers of ten to be certain that they conform to the quantities they represent. That is, the wavelength of red light in meters is a very small number; the frequency is a large number; Avogadro's number is very large; and Planck's constant is very small. Using powers of ten with the wrong sign can cause errors of many orders of magnitude.

EXERCISE 7.4A

The lower wavelength limit of visible light is about 400 nm. What is the energy of this radiation, expressed in kilojoules per mole of photons?

EXERCISE 7.4B

Use Figure 7.10 to help you determine in which region of the electromagnetic spectrum you would find radiation having an energy of 100 kJ/mol.

Quantum View of Atomic Structure

According to the laws of classical physics, the negatively charged electrons in Thomson's raisin pudding model could be stationary because Thomson presumed electrons to be surrounded by a cloud of positive charge. In Rutherford's nuclear model, however, the electrons could *not* be stationary; they would have to be in motion to overcome their attraction to a positive charge confined to a small nucleus. Classical physics predicts that such electron motion would have to be accompanied by the *continuous* emission of light. As they lost energy as light, the electrons would be drawn ever closer to the nucleus and eventually spiral into it (Figure 7.16). Such atoms would be unstable. Obviously, classical physics could not explain atomic structure, just as it failed to explain emission spectra, black-body radiation, and the photoelectric effect. A satisfactory explanation required the quantum theory.

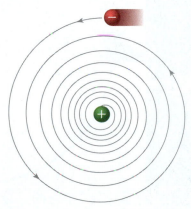

▶ **FIGURE 7.16   An unsatisfactory model for the hydrogen atom**
According to classical physics, light should be emitted as the electron circles the nucleus. A loss of energy would cause the electron to be drawn ever closer to the nucleus and eventually spiral into it.

▲ Niels Bohr (1885–1962) proposed his theory of the hydrogen atom early in his career, and he received a Nobel Prize in Physics for this work in 1922. Later, he directed the Institute of Theoretical Physics in Copenhagen, a center of attraction for theoretical physicists in the 1920s and 1930s. He worked on the atomic bomb project in World War II, but after the war he became one of the strongest proponents for peaceful uses of atomic energy.

7.7   Bohr's Hydrogen Atom: A Planetary Model

In 1913, Niels Bohr combined ideas from classical physics and the new quantum theory to explain the structure of the hydrogen atom. In doing so, he was also able to explain the spectrum of light emitted by hydrogen atoms.

Basing his work on that of Planck and Einstein, Bohr assumed that the angular momentum of the electron in a hydrogen atom is quantized, that is, that it can have only certain specified values. (Momentum is the product of the mass and velocity of a particle. A particle traveling a curved path has angular momentum.) With this basic assumption, Bohr was able to use classical physics to calculate other properties. In particular, he found that the energy (E_n) was also quantized. Each specified energy value (E_1, E_2, E_3, ...) is called an **energy level** of the atom, and the only allowable values are given by the relationship

$$E_n = \frac{-B}{n^2} \qquad (7.7)$$

In this equation, n is an integer (that is, $n = 1, 2, 3, \ldots$) and B is a constant based on quantities such as Planck's constant and the mass and charge of an electron: $B = 2.179 \times 10^{-18}$ J. The energy is arbitrarily assigned a value of *zero* when the electron is located infinitely far from the nucleus, and as a result, energies associated with forces of attraction are taken to be negative, accounting for the minus sign in the equation.

An especially important part of Bohr's theory was the postulate that as long as an electron remains in a given energy level, it cannot emit energy as electromagnetic radiation. This prevents the electron in a hydrogen atom from spiraling into the nucleus. Bohr imagined the electron to orbit about the nucleus much as planets orbit the Sun. According to this model, different energy levels for the electron correspond to different orbits, and only a discrete set of energy levels, or orbits, is possible (Figure 7.17).

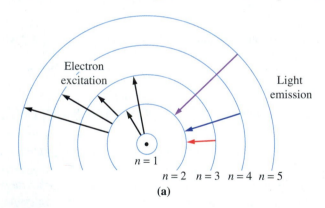

Electron excitation

Light emission

$n = 1$

$n = 2$ $n = 3$ $n = 4$ $n = 5$

(a)

▲ **FIGURE 7.17   The Bohr model of the hydrogen atom**

(a) A portion of the hydrogen atom model, with the nucleus at the center of the atom and the set of discrete orbits $n = 1, 2, 3, 4, \ldots$. When a hydrogen atom is excited, the electron moves to some higher orbit (black arrows). When the electron falls to a lower level (violet, blue, and red arrows), light is emitted. (b) Microscopic views of a few of the electronic transitions associated with the red-violet glow of a hydrogen lamp. Clockwise, beginning at the top, these are excitation of an electron from $n = 1$ level to $n = 4$ level; dropping of an electron from $n = 5$ level to $n = 2$ level; dropping of an electron from $n = 4$ level to $n = 2$ level; and dropping of an electron from $n = 3$ level to $n = 2$ level.

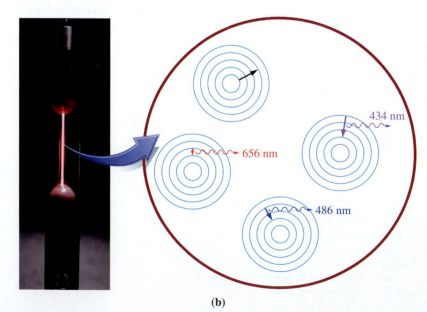

434 nm

656 nm

486 nm

(b)

The lowest (most negative) energy level has $n = 1$ and represents the orbit closest to the nucleus; the next higher energy level has $n = 2$; and so on.

Example 7.5

Calculate the energy of an electron in the second energy level of a hydrogen atom.

SOLUTION

We can use Equation (7.7) for the hydrogen atom, with $B = 2.179 \times 10^{-18}$ J and $n = 2$:

$$E_2 = \frac{-B}{n^2} = \frac{-2.179 \times 10^{-18} \text{ J}}{2^2}$$

$$= \frac{-2.179 \times 10^{-18} \text{ J}}{4} = -5.448 \times 10^{-19} \text{ J}$$

EXERCISE 7.5A

Calculate the energy of an electron in the energy level $n = 6$ of a hydrogen atom.

EXERCISE 7.5B

Is there a hydrogen-atom energy level at -2.179×10^{-19} J? Explain.

Bohr's Explanation of Line Spectra

When we use Equation (7.7), we are generally interested in the energy change (ΔE_{level}) that accompanies the leap of an electron from one energy level to another in the hydrogen atom. We will define ΔE_{level} as the energy in the final level (E_f) minus the energy in the initial level (E_i):

$$\Delta E_{level} = E_f - E_i$$

For the final and initial levels, we can write

$$E_f = \frac{-B}{n_f{}^2} \quad \text{and} \quad E_i = \frac{-B}{n_i{}^2}$$

The energy difference between n_f and n_i is therefore

$$\Delta E_{level} = \frac{-B}{n_f{}^2} - \frac{-B}{n_i{}^2} = B\left(\frac{1}{n_i{}^2} - \frac{1}{n_f{}^2}\right) \quad \text{(7.8)}$$

If $n_f > n_i$, the electron *absorbs* a quantum of energy and moves farther away from the nucleus—that is, from the energy level n_i to the higher level n_f—and ΔE_{level} is positive. This energy absorption can occur, for example, when electric current flows through a tube filled with a gas at low pressure. If $n_f < n_i$, the electron drops from a higher energy level n_i to a lower energy level n_f. The electron moves closer to the nucleus, and a quantum of energy is *emitted* as a photon of light. In this case, ΔE_{level} is negative. In all changes in energy level—called *transitions*—the electron makes a leap from one allowable level to another; it cannot stop between levels. Every hydrogen atom in which an electron makes the same transition emits a photon of the same energy. Together, all the photons of this energy produce one spectral line. The collection of lines for all the possible transitions is the observed emission spectrum.

As an analogy to discrete energy levels, consider running up a flight of steps. One can land on every step, every other step, perhaps every third step, or some combination of these possibilities, but one cannot land between steps.

Example 7.6

Calculate the energy change, in joules, that occurs when an electron falls from the $n_i = 5$ to the $n_f = 3$ energy level in a hydrogen atom.

STRATEGY

We find the initial and final energy levels of the electron in the statement of the problem. To determine the energy change during an electron transition, we then substitute the numbers of the levels ($n_i = 5$ and $n_f = 3$) and the value of the constant B (2.179×10^{-18} J) into Equation (7.8).

SOLUTION

$$\Delta E_{\text{level}} = 2.179 \times 10^{-18}\left(\frac{1}{5^2} - \frac{1}{3^2}\right) = 2.179 \times 10^{-18}\left(\frac{1}{25} - \frac{1}{9}\right)$$

$$= 2.179 \times 10^{-18}(0.04000 - 0.1111) = -1.550 \times 10^{-19}\,\text{J}$$

ASSESSMENT

The minus sign indicates that the atom has given up energy in the form of a photon of light, corresponding to the drop from the higher to lower energy level. When using Equation (7.8), always compare your result with your intuitive expectation.

EXERCISE 7.6A

Calculate the energy change that occurs when an electron is raised from the $n_i = 2$ energy level to the $n_f = 4$ level of a hydrogen atom.

EXERCISE 7.6B

Can there be an energy difference of 4.269×10^{-20} J between energy levels for the hydrogen atom? Explain.

We can easily calculate the frequency and wavelength of the photons released when an electron drops from one energy level to a lower one. We first calculate the energy change, as in Example 7.6. The magnitude of this energy change, ΔE_{level}, becomes the E value that we use in Planck's equation:

$$\Delta E_{\text{level}} = h\nu \qquad (7.9)$$

Now, we can determine the frequency of the light corresponding to this energy. Finally, if necessary, we can use the relationship $c = \nu\lambda$ to calculate the wavelength of the light. This is the strategy employed in Example 7.7 and Exercises 7.7A and 7.7B.

Example 7.7

Calculate the frequency of the radiation released by the transition of an electron in a hydrogen atom from the $n = 5$ level to the $n = 3$ level, the transition we looked at in Example 7.6.

SOLUTION

From Example 7.6, we know that $\Delta E_{\text{level}} = 1.550 \times 10^{-19}$ J. Using this quantity as the value of E_{level} in Equation (7.9), we can solve the equation for ν:

$$\nu = \frac{E_{\text{level}}}{h} = \frac{1.550 \times 10^{-19}\,\text{J}}{6.626 \times 10^{-34}\,\text{J s}} = 2.339 \times 10^{14}\,\text{s}^{-1}$$

ASSESSMENT

In Example 7.6, we obtained $\Delta E_{\text{level}} = -1.550 \times 10^{-19}$ J. The negative sign indicated that energy was emitted. Here in Example 7.7, we dropped the negative sign and used only the magnitude: 1.550×10^{-19} J. We did this so that ν would be a positive quantity. We can accomplish the same objective by considering that the excited atom loses energy (negative) and the emitted photon carries it to the surroundings (positive).

EXERCISE 7.7A

Calculate the frequency of the radiation emitted during the transition of an electron from the $n = 4$ energy level in a hydrogen atom to the $n = 1$ level.

EXERCISE 7.7B

Calculate the wavelength, in nanometers, of the radiation emitted as the electron in a hydrogen atom moves from the $n = 5$ level to the $n = 2$ level. In what region of the electromagnetic spectrum is the spectral line produced by this radiation?

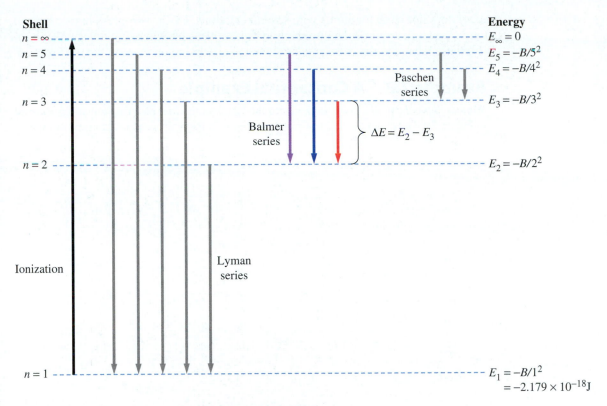

▲ FIGURE 7.18   Energy levels and spectral lines for hydrogen
The distance between energy levels is not to scale. Three of the four visible lines in the Balmer series are shown. Each series is named for the scientist who either discovered or characterized it.

QUESTION: What is the longest-wavelength line in the Balmer series?

The Line Spectrum of Hydrogen

The emission spectrum of hydrogen consists of several series of lines. The most common series are in the ultraviolet, visible, and near infrared portions of the electromagnetic spectrum. The electron transitions that produce these spectral series are shown in Figure 7.18. The spectral series in which each electron transition terminates at $n = 1$, called the Lyman series, is in the ultraviolet range. In the Balmer series, all transitions terminate at $n = 2$. Three of the four lines in the visible region of the Balmer series (recall Figure 7.12) are shown in Figure 7.18. Other lines in the series are in the ultraviolet region. In the Paschen series, found in the infrared range, electron transitions terminate at $n = 3$.

We can also use Figure 7.18 to explain ionization of the hydrogen atom, which requires complete removal of the electron. In ionization, the electron is boosted from the $n = 1$ level to a level having an infinitely large value of n, that is, to the $n = \infty$ level. Thus, the zero value for energy (E_∞) corresponds to the ionized atom ($n = \infty$).

Spectral Lines of Hydrogen activity

Ground States and Excited States

The electron in a hydrogen atom is usually in the lowest energy level (the level closest to the nucleus). When an atom has its electrons in their lowest possible energy levels, the atom is said to be in its **ground state.** When an electric current, a flame, or some other source has supplied the energy to promote an electron from the lowest possible level to a higher level, the atom is in an **excited state.** An atom in an excited state eventually emits energy in the form of photons as the electron drops back to one of the lower energy levels and ultimately reaches the ground state.

Bohr's theory introduced the important idea of energy levels for electrons in atoms and was a great success in explaining the line spectra of the hydrogen atom. It also gives good results for other one-electron species, such as He^+ and Li^{2+}, but the theory does not work for atoms with several electrons. Further progress in the study of atomic structure required the introduction of new ideas into quantum theory. We introduce these ideas and the "new" quantum mechanics based on them in the next section.

Example 7.8   A Conceptual Example

Without doing detailed calculations, determine which of the four electron transitions shown in Figure 7.19 produces the shortest-wavelength line in the hydrogen emission spectrum.

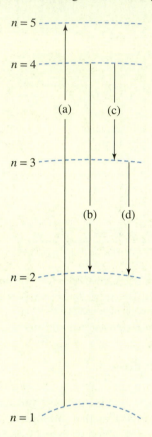

▲ **FIGURE 7.19**   **Four electronic transitions in a hydrogen atom: Example 7.8 illustrated**

ANALYSIS AND CONCLUSIONS

First, we should recognize that transition (**a**), even though it involves the greatest span of energy levels, requires energy *absorption*, not emission. We can therefore eliminate (**a**). The other three transitions do involve light emission. Let's compare the energy changes in these transitions, recognizing that the larger the energy change, the *greater* the frequency and the *shorter* the wavelength of the spectral line produced. Transitions (**b**) and (**d**) both terminate at the same level, $n = 2$, but the drop in energy in (**b**) is greater than in (**d**). Transition (**c**) begins at the same level as transition (**b**), but terminates at a higher energy level, $n = 3$. The energy change for transition (**c**) is therefore less than that for transition (**b**). Thus, the shortest-wavelength spectral line of the three is that produced by transition (**b**).

EXERCISE 7.8A

Without doing detailed calculations, determine which of the following electron transitions in a hydrogen atom requires absorption of the greatest amount of energy: (**a**) from $n = 1$ to $n = 2$, (**b**) from $n = 3$ to $n = 8$, (**c**) from $n = 4$ to $n = 1$, (**d**) from $n = 2$ to $n = 3$.

EXERCISE 7.8B

Can the electron in a hydrogen atom drop from the level $n = 4$ with the emission of 4.90×10^{-20} J of energy? Explain.

Atomic Absorption Spectroscopy

We have described emission spectroscopy, in which light emitted by excited atoms is dispersed into line spectra. Knowing that the atoms become excited by *absorbing* quanta that have just the right energy to promote ground-state atoms to an excited state, we can now discuss *absorption* spectroscopy.

An excited state can be produced by striking the atoms in a high-temperature gaseous sample with a beam of light whose frequency matches the frequency of the light the excited gas atoms themselves would emit. Ground-state atoms absorb photons from the light beam, and the beam leaves the high-temperature region with a reduced intensity. The more atoms present that can absorb the photons, the lower the intensity of the transmitted light. In *atomic absorption (AA) spectroscopy*, we analyze light that is absorbed rather than emitted.

Suppose we gradually increase the wavelength of a light beam passing through gaseous sodium. When we get to the wavelength 589.00 nm, the light is absorbed strongly for the first time, and at 589.59 nm, we see another strong absorption. These absorptions occur at the same wavelengths as the two bright yellow lines in the sodium emission spectrum. The transition producing a line in an emission spectrum emits the same amount of energy as is absorbed in the transition producing the corresponding line in the absorption spectrum.

More than 70 elements can be detected by AA spectroscopy, even at concentrations as low as a few parts per billion in some cases. AA spectroscopy is routinely used in the food industry to test for metals such as calcium, copper, and iron; in environmental analyses for metals such as cadmium, lead, and mercury; and in astronomy to study the chemical makeup of the stars and the atmospheres of the planets.

Sunlight is a white light because the black body radiation from the extremely hot surface of the Sun contains all the wavelength components of the visible electromagnetic spectrum. However, some gaseous atoms in the cooler outer atmosphere of the Sun absorb portions of this radiation, creating dark lines in the solar spectrum called Fraunhofer lines. Figure 7.20 represents the absorption spectrum of sunlight; the elements responsible for the dark lines are identified.

During the solar eclipse of 1868, a line was observed at 587.6 nm. Although quite close to the two sodium lines, this new line was not in the sodium spectrum. It corresponded to no known element on Earth. When a new element was discovered in 1895, its emission spectrum produced the same line as the unknown line seen on the Sun. This element was therefore named helium, from the Greek *helios*, "the sun."

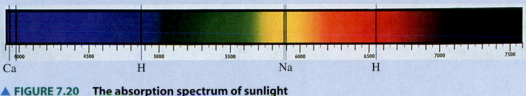

▲ FIGURE 7.20   **The absorption spectrum of sunlight**

7.8   Wave Mechanics: Matter as Waves

As we have noted, light exhibits *wave-particle duality;* it can act as either waves or particles. The wave nature is evident when light is dispersed into a spectrum by a prism. The particle nature is displayed when photons displace electrons from a metal in the photoelectric effect. We usually regard matter as consisting of particles. Is it possible that under the proper circumstances, matter can behave as waves? Such speculation led Louis de Broglie to propose a startling new theory as part of his doctoral thesis in 1923. De Broglie's theory, in turn, led to a new mathematical description of atoms that has broad application in modern chemistry.

De Broglie's Equation

De Broglie proposed that a particle of mass m moving at speed v will have a wave nature consistent with a wavelength given by the following equation, in which h, once again, is Planck's constant:

$$\lambda = \frac{h}{mv} \qquad\qquad (7.10)$$

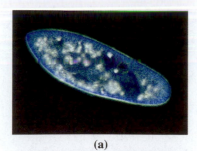

(a)

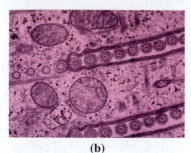

(b)

▲ **FIGURE 7.21   Images from optical and electron microscopes**

In an electron microscope, electric and magnetic fields direct and focus electron beams, much as lenses and prisms focus light in optical microscopes. Because of the short de Broglie wavelengths of electrons, however, the resolving power of an electron microscope is thousands of times greater than that of an optical microscope. (a) A paramecium as seen with an optical microscope. (b) An image of a paramecium made by an electron microscope shows structural details that cannot be seen through the optical microscope.

▲ Erwin Schrödinger (1887–1961) was at the University of Zurich when he promulgated his wave equation (1926). He moved to the University of Berlin in 1928 to succeed Max Planck. Although he was not Jewish, Schrödinger left Nazi Germany in 1933, the same year he received the Nobel Prize in Physics. In his later years, Schrödinger's interest turned to biology, with this new interest expressed in his famous 1944 book, *What Is Life?*

Even large objects presumably have wave properties, but their associated wavelengths are so short that we cannot observe the waves. For example, a 1000-kg car moving at 100 km/h has an associated wavelength of 2.39×10^{-38} m. This wavelength is far shorter than any part of the electromagnetic spectrum and is therefore undetectable (see again Figure 7.10). In contrast, the wave properties of the far less massive subatomic particles are much longer and readily observable.

De Broglie's prediction of matter waves was verified six years after he proposed it, and his work soon led to the development of the electron microscope. This instrument makes use of the wave nature of electrons, and it is now a standard piece of equipment in many scientific laboratories. As can be seen in Figure 7.21, an electron microscope can provide images of objects as tiny as a few hundred picometers in diameter.

Example 7.9

Calculate the wavelength, in meters and nanometers, of an electron moving at a speed of 2.74×10^6 m/s. The mass of an electron is 9.11×10^{-31} kg, and 1 J = 1 kg m^2 s^{-2}.

STRATEGY

To use de Broglie's equation (7.10), we must first substitute the units kg m^2 s^{-2} for the energy unit J. This changes the units of Planck's constant, h, from J s to kg m^2 s^{-2} s. Then we must make sure that the particle mass is in kilograms and the particle velocity is in meters per second. These required units are used in the statement of this particular problem.

SOLUTION

We can substitute the given data directly into the de Broglie equation:

$$\lambda = \frac{h}{mv} = \frac{6.626 \times 10^{-34} \text{ kg m}^2 \text{ s}^{-2} \text{ s}}{9.11 \times 10^{-31} \text{ kg} \times 2.74 \times 10^6 \text{ m s}^{-1}}$$

$$= 2.65 \times 10^{-10} \text{ m}$$

The wavelength of the electron is 2.65×10^{-10} m or 0.265 nm.

ASSESSMENT

As we have implied, the cancellation of units leading to a wavelength in meters will not occur unless we have converted the other data in Equation (7.10) to the proper units. A further check on the reasonableness of the answer is that the wavelength must be short, for example, in the nanometer range.

EXERCISE 7.9A

Calculate the wavelength, in nanometers, of a proton moving at a speed of 3.79×10^3 m/s. The mass of a proton is 1.67×10^{-27} kg.

EXERCISE 7.9B

Use data from Example 7.9 and Exercise 7.9A to determine the speed a proton must have if its wavelength is to be equal to that of an electron moving at 2.74×10^6 m/s.

Wave Functions

Although the Bohr model of the hydrogen atom invoked quantum theory, it was still primarily based on classical physics, and we could picture the atom in fairly concrete terms involving particles: An electron orbiting around a nucleus is not unlike Earth orbiting the Sun. In contrast, the treatment of atomic structure using the wave properties of the electron is called either **quantum mechanics** or *wave mechanics*. In 1926, Erwin Schrödinger developed an equation, called a wave equation, to describe the hydrogen atom. An acceptable solution to Schrödinger's wave equation is called a

Electron Probabilities and a Close-up Look at Atoms

In the Bohr atom, we considered how much energy is required to move an electron from one orbit to a higher orbit. In wave mechanics, we replaced definite orbits with the probabilities of finding an electron in different regions outside the nucleus. The wave mechanical interpretation even includes the extremely small but nonzero possibility that an atom may transfer an electron to an adjacent atom without first ionizing. This transfer can occur when an electron has a significant probability of being closer to the nucleus of a neighboring atom than to the nucleus of its parent atom. This transfer is called *tunneling* and requires much less energy than ionization.

The tunneling effect was put to good use in the early 1980s when Gerd Binnig and Heinrich Rohrer developed the technique of *scanning tunneling microscopy* to obtain atomic-scale images of a solid surface. A scanning tunneling microscope (STM) uses a metallic probe with an extraordinarily sharp tip that is carefully placed within 0.5 nm of the surface being studied. A minute electric potential is applied across the gap separating the tip from the surface, increasing the probability of electrons tunneling between the probe tip and the surface. The flow of these tunneling electrons constitutes a small electric current. The magnitude of this current is proportional to the distance between the tip and the surface—as the tip comes closer to the surface, the tunneling current increases, and as the tip moves away from the surface, the tunneling current decreases. The instrument can be set up to keep this current constant, which in turn requires that the probe tip move up and down as it follows the contours of the surface. The surface is scanned repeatedly, and a computer is used to process the results to give a three-dimensional map of the surface. As seen in the accompanying photographs, on a coarse scale (top), the contours represent the topography of the surface. On a fine scale (bottom), the contours represent regions of high or low electron density, depending on whether electrons are tunneling into or out of the surface. The regions of electron density, similar to the atomic orbitals discussed in this chapter, are related to the positions of individual atoms within the surface plane, but note that the density regions are *not* necessarily equivalent to the positions of the nuclei.

STM images tell us nothing about the internal structure of atoms. For that, we still use indirect evidence such as atomic spectra.

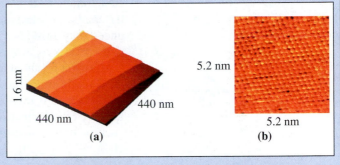

▲ These false-color images of a palladium surface obtained with a scanning tunneling microscope reveal (a) steps between individual layers of palladium atoms in a 440 nm × 440 nm area of the surface and (b) points of high tunneling current associated with single palladium atoms in a 5.2 nm × 5.2 nm area.

wave function. A wave function, denoted by the Greek letter ψ (psi, pronounced "sigh"), represents an energy state of the atom.

In contrast to the precise planetary orbits of the Bohr atom, wave mechanics provides a less certain picture of the hydrogen atom. Instead of talking about the exact location of the electron, we can speak only of the *probability* of the electron's being located in certain regions of the atom. A useful interpretation of a wave function is the one first proposed by Max Born (1882–1970):

The square of a wave function (ψ^2) gives the probability of finding an electron in a particular infinitesimally small volume of space in an atom.

The de Broglie equation treats the electron as a wave, not a discrete particle. This means that the concept of finding the electron wave at a given location is a bit misleading because we cannot pinpoint the location of a wave. If we adopt the view that the electron is a cloud of negative electric charge rather than a particle, we can speak only of the *charge densities* in various parts of the atom. This fuzzier atomic model is the only type acceptable according to an important principle of science established by Werner Heisenberg in 1927.

Heisenberg's **uncertainty principle** essentially states that we cannot *simultaneously* know exactly where a tiny particle like an electron is and exactly how it is moving. One way to visualize the principle is to recognize that when we try to make measurements on an extremely small particle, the act of measuring interferes with the particle.

▲ Werner Heisenberg (1901–1976) devised an approach to atomic structure while on a North Sea vacation in 1927. His approach, called matrix mechanics, was purely mathematical but proved to be equivalent to Schrödinger's wave equation. Later that year, he deduced his famous uncertainty principle. Unlike Schrödinger, Heisenberg remained in Nazi Germany during World War II, directing an unsuccessful atomic bomb project.

Application Note

An everyday analogy to uncertainty of position and momentum involves two photographs of a thrown baseball. A photograph taken at very high shutter speed freezes the motion; the location of the baseball is well known, but its direction of motion and its speed are uncertain. With a longer exposure, we see a blur. Because of this blurring, the location of the ball is uncertain, but the blur shows its direction of motion and an indication of its speed.

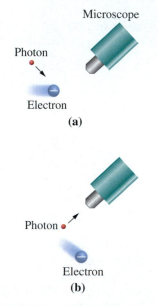

▲ FIGURE 7.22   The uncertainty principle
(a) A free electron moves into the focus of a hypothetical microscope and is struck by a photon of light; the photon transfers momentum to the electron, deflecting the electron from its original path. (b) The reflected photon is seen in the microscope, but by this time the electron has moved. In other words, the electron is not where the reflected photon would indicate.

That is, we produce changes in one measured quantity in the act of measuring some other quantity (Figure 7.22).

As we noted on page 276, the momentum (p) of a particle is the product of its mass (m) and its velocity (v): $p = mv$. If we focus on measuring a particle's momentum, we lose a sense of its exact position (x); and if we focus on measuring the particle's position, we lose a sense of its exact momentum (p). Heisenberg related the product of the uncertainty in position (Δx) and the uncertainty in momentum (Δp) to Planck's constant, h,

$$\Delta x \, \Delta p \geq \frac{h}{4\pi} \tag{7.11}$$

where the term $h/4\pi$ has the magnitude 5.3×10^{-35} J s, an extremely small number.

This uncertainty relationship need not concern us in the macroscopic world because we do not have the means to measure the position and momentum of a large object precisely enough to come anywhere the uncertainty limit of 5.3×10^{-35}. This means, for example, that we can readily and accurately describe the orbit of Earth about the Sun or of the smallest artificial satellite about Earth without dealing with Heisenberg's expression. The situation is entirely different when we deal with a particle as small as an electron, however, whose mass is only 9.11×10^{-31} kg. Now even the smallest uncertainties in position and momentum that meet the requirements of the uncertainty principle are *relatively* large. We cannot measure both these properties with great precision. As a result, the uncertainty in the position of an electron in an atom might be about as large as the atom itself.

In light of the uncertainty principle, Bohr's model of the hydrogen atom fails, in part, because it tells us *more* than we can know with certainty. It gives us a precise location for the electron—the particular orbit—and it enables us to make a precise calculation of the electron speed (and thus its momentum, mv) in this orbit.

7.9   Quantum Numbers and Atomic Orbitals

The wave functions for the hydrogen atom contain three parameters that must have specific integral values called **quantum numbers.** A wave function with a given set of these three quantum numbers is called an **atomic orbital.** Atomic orbitals, although they are only mathematical expressions, allow us to visualize, in any atom, a three-dimensional region in which there is a significant probability of finding an electron. Consequently, we tend to think of orbitals as geometric regions as well as mathematical expressions.

Because orbitals are characterized by them, we first say a bit more about quantum numbers.

Quantum Numbers

Let's consider the three quantum numbers and their possible values. When we specify a set of values for the three quantum numbers, we define a specific atomic orbital. We might think of the three quantum numbers as similar to a residential address. One part of the address locates the state; a second, the city; and a third, the street address. Table 7.1 summarizes the discussion that follows, and you may find it helpful to refer to the table as we proceed.

Table 7.1 Electronic Shells, Orbitals, and Quantum Numbers													
Principal Shell	**First**	**Second**			**Third**								
n	1	2	2	2	2	3	3	3	3	3	3	3	3	3
l	0	0	1	1	1	0	1	1	1	2	2	2	2	2
m_l	0	0	−1	0	+1	0	−1	0	+1	−2	−1	0	+1	+2
Subshell and orbital designation	1s	2s	2p	2p	2p	3s	3p	3p	3p	3d	3d	3d	3d	3d
Number of orbitals in the subshell	1	1	3			1	3			5				

1. The **principal quantum number** (*n*) is assigned first because the allowable values of the other two quantum numbers depend on the value assigned to *n*. The value of *n* can only be a *positive integer* (whole number):

$$n = 1, 2, 3, 4, 5, 6, 7, \ldots$$

The quantum number *n* is analogous to *n* in Bohr's planetary model. The size of an atomic orbital and its electron energy depend mainly on this principal quantum number. As *n* increases, the electron is at a higher energy level and spends more time farther from the nucleus. Orbitals having the same value of *n* are said to be in the same **principal shell.**

2. The **orbital angular momentum quantum number** (*l*) determines the *shape* of an orbital. For any given orbital, this quantum number can have positive integral values from zero up to one less than the value of *n* for the orbital:

$$l = 0, 1, 2, \ldots, (n - 1)$$

All orbitals having the same value of *n* and the same value of *l* are said to be in the same principal shell and in the same **subshell.** Orbitals and subshells are also designated by letter (*s*, *p*, *d*, *f*) according to the following scheme:

Value of *l*	0	1	2	3
Orbital and subshell designations	*s*	*p*	*d*	*f*

Each of these four orbital types describes a region of space that has a unique shape. Also note that the number of different kinds of orbitals and subshells in a principal shell is equal to the principal quantum number *n*. For example, the third principal shell (*n* = 3) has three subshells and three kinds of orbitals—*s*, *p*, and *d*, corresponding to *l* values of 0, 1, and 2, respectively.

3. The **magnetic quantum number** (m_l) determines the orientation in space of the orbitals of a given type in a subshell. This quantum number can have any integral value ranging from −*l* to +*l*, including 0:

$$m_l = 0, \pm 1, \pm 2, \ldots \pm l$$

Thus, if *l* = 0, the only permitted value of m_l is also 0; if *l* = 1, m_l may be −1, 0, or +1; and so on. In all, the number of possible values for $m_l = 2l + 1$, and this total number determines the number of orbitals allowed in a subshell.

Let's summarize the relationship between the *l* and m_l quantum numbers:

Quantum numbers activity

s orbitals	*p* orbitals	*d* orbitals	*f* orbitals
l = 0	*l* = 1	*l* = 2	*l* = 3
$m_l = 0$	$m_l = 0, \pm 1$	$m_l = 0, \pm 1, \pm 2$	$m_l = 0, \pm 1, \pm 2, \pm 3$
one s orbital	*three p* orbitals	*five d* orbitals	*seven f* orbitals
in an *s* subshell	in a *p* subshell	in a *d* subshell	in an *f* subshell

To indicate the principal shell in which an orbital or subshell is found, we combine the principal quantum number *n* of that shell with the orbital or subshell designation. Thus, the symbol 2*p* represents either the *p* subshell of the second principal shell (*n* = 2) or any one of the three *p* orbitals that make up the subshell.

Example 7.10 tests your understanding of relationships among the quantum numbers. Example 7.11 applies this understanding to questions about principal shells, subshells, and atomic orbitals.

Example 7.10

Considering the limitations on values for the various quantum numbers, state whether an electron can be described by each of the following sets. If a set is not possible, state why not.

(a) $n = 2, l = 1, m_l = -1$ **(c)** $n = 7, l = 3, m_l = +3$

(b) $n = 1, l = 1, m_l = +1$ **(d)** $n = 3, l = 1, m_l = -3$

SOLUTION

(a) All the quantum numbers are allowed values.

(b) Not possible. The value of l must be less than the value of n.

(c) All the quantum numbers are allowed values.

(d) Not possible. The value of m_l must be in the range $-l$ to $+l$ (in this case, -1, 0, or $+1$).

EXERCISE 7.10A

Considering the limitations on values for the various quantum numbers, state whether an electron can be described by each of the following sets. If a set is not possible, state why not.

(a) $n = 2, l = 1, m_l = -2$
(b) $n = 3, l = 2, m_l = +2$

(c) $n = 4, l = 3, m_l = +3$
(d) $n = 5, l = 2, m_l = +3$

EXERCISE 7.10B

Replace the question marks by suitable responses in the following quantum number assignments.

(a) $n = 3, l = 1, m_l = ?$ (b) $n = 4, l = ?, m_l = -2$ (c) $n = ?, l = 3, m_l = ?$

Example 7.11

Consider the relationship among quantum numbers and orbitals, subshells, and principal shells to answer the following. (a) How many orbitals are there in the $4d$ subshell? (b) What is the first principal shell in which f orbitals can be found? (c) Can an atom have a $2d$ subshell? (d) Can a hydrogen atom have a $3p$ subshell?

SOLUTION

(a) The d subshell corresponds to a value of $l = 2$. With $l = 2$, there are five possible values of m_l: -2, -1, 0, $+1$, and $+2$. Thus, there are five d orbitals in the $4d$ subshell.

(b) The f orbital corresponds to a value of $l = 3$. Because the maximum value of l is $n - 1$, the allowable values of n are 4, 5, 6, and so on. The first shell to contain f orbitals is therefore the fourth principal shell ($n = 4$).

(c) No. For a d subshell, $l = 2$. The maximum value of l is $n - 1$, and so if $n = 2$, l cannot be 2. There cannot be a $2d$ subshell.

(d) Yes. Although the electron in a hydrogen atom is usually in the $1s$ orbital, it can be excited to one of the higher energy states, such as one of the $3p$ orbitals.

EXERCISE 7.11A

Consider the relationship among quantum numbers and orbitals, subshells, and principal shells to answer the following. (a) How many orbitals are there in the $5p$ subshell? (b) What subshells of the hydrogen atom consist of a total of seven orbitals?

EXERCISE 7.11B

What is the total number of orbitals in the principal shell for which $n = 4$? Derive an equation that relates the total number of orbitals in a principal shell to the principal quantum number n.

Electron Probabilities and the Shapes of Orbitals

What do the regions of high electron probabilities described by atomic orbitals look like? We cannot see these regions, of course, but we can translate the mathematical forms of the orbitals into geometric pictures. Strictly speaking, an atomic orbital must have an infinite size to guarantee a 100% certainty of finding an electron in it, and we cannot picture that. For practical purposes, though, we can outline a region in which there is a certain probability (say, a 90% probability) that the electron will be found within the region. Alternatively, if we think of the

Radial Distribution of Electron Density activity

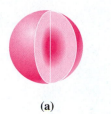

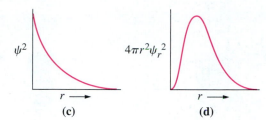

(a) (b) (c) (d)

orbital as a charge cloud, we can say that this region contains a certain percentage of the electron's charge (say, 90%).

The wave mechanical picture of an electron in an *s* orbital looks like a fuzzy ball; that is, the orbital has *spherical* symmetry. The value of *n* determines the size of the orbital and the electron probability distribution. Figure 7.23 shows us four ways to look at a 1*s* orbital. Figure 7.23a is a cutaway view of a spherical shell within which there is a high probability (say, 90%) of finding the 1*s* electron. Figure 7.23b is a two-dimensional cross-section through the center of the sphere. The dots within the circular area represent the probability of the 1*s* electron's being at a particular point—the heavier the concentration of dots, the greater the probability of finding an electron. Because the heaviest concentration of dots is quite near the center of the sphere, the greatest probability of finding an electron in a 1*s* orbital is very near the nucleus. Figure 7.23c represents these probabilities as a function of distance *r* from the nucleus through the function ψ^2 (recall the relationship between ψ^2 and probability described on page 283).

Of equal importance is the total probability $4\pi r^2 \psi_r^2$ of an electron's being at a given distance from the nucleus, and this is different from ψ^2 alone. Finding the probability $4\pi r^2 \psi_r^2$ is comparable to adding up all the dots that are a given distance from the nucleus in Figure 7.23b. These dots would lie on the circumference of a circle that has the nucleus at its center. In the three-dimensional picture, this means summing the probabilities at all the points that lie in a very thin spherical shell that has the nucleus at its center. Of all the possible shells we might choose, the one with the greatest probability of containing the 1*s* electron has a radius of 52.9 pm, a distance identical to the first Bohr orbit of hydrogen. Figure 7.23d shows a two-dimensional cross-section of this probability distribution. The dartboard in Figure 7.24 is a two-dimensional analogy to this description of a three-dimensional 1*s* orbital.

The 2*s* orbital has two regions of high electron probability, both spherical. The region nearer the nucleus is separated from the outer region by a spherical *node*—a spherical shell in which the electron probability is zero. The 2*s* electron is more likely to be found in the outer region. Figure 7.25 represents a 2*s* orbital in much the same way Figure 7.23 represents a 1*s* orbital.

The second principal shell ($n = 2$) consists of two subshells—2*s* and 2*p*. The 2*s* subshell has only one orbital, the 2*s* orbital just described. The 2*p* subshell has three 2*p* orbitals, corresponding to $l = 1$ combined with the three possibilities for m_l.

▶ **FIGURE 7.23** **The 1s orbital**
(a) The 1*s* orbital represented as a sphere, within which there is a 90% probability that the 1*s* electron will be found. (b) A two-dimensional cross section (a circle) of the 1*s* orbital in (a). The dots represent the probability of the electron's being found; that is, where the density of dots is greatest, so is the probability that the electron will be found in that region. (c) The probability that the electron will be found in some infinitesimally small volume of space as a function of the distance *r* from the nucleus. (d) The probability of the electron's being somewhere in a thin spherical shell of radius *r* as a function of the distance *r* from the nucleus.

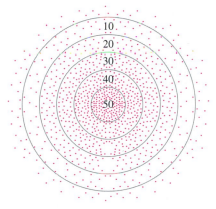

▲ **FIGURE 7.24** **An analogy to a 1s orbital**
Imagine that the hole left by each dart throw represents the probability that an electron is located at that point. If 900 of 1000 dart throws hit the board, the dartboard is analogous to the 90% outline of a 1*s* orbital. The greatest number of holes per unit area (say, per 1 cm²) is in the 50 region—at which the dart thrower aims. But what is the most likely score the average thrower will make on a given throw? Can you see that the answer is 30? Although the number of holes per square centimeter in the 30 region is smaller than in the 50 region, the total area of the 30 region is much greater. The 30 ring of the dartboard is analogous to the 52.9-pm spherical shell of the 1*s* orbital in the hydrogen atom.

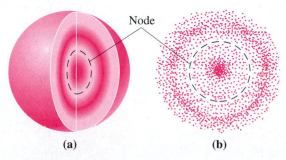
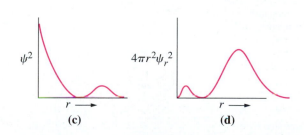

Node

(a) (b) (c) (d)

▲ **FIGURE 7.25** **The 2s orbital**
The 2*s* orbital differs from the 1*s* (Figure 7.23) in having a larger volume that encompasses two regions of high electron probability separated by a spherical node.

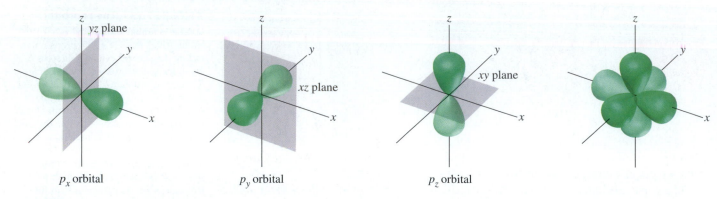

p_x orbital p_y orbital p_z orbital

▲ **FIGURE 7.26** **The three 2p orbitals**
The regions of high electron probabilities are dumbbell-shaped and oriented along the x-, y-, and z- axes. Each p orbital has a *nodal plane*, which is a planar region of zero electron probability that passes through the nucleus of the atom. The rightmost drawing is a composite showing all three p orbitals in a single atom.

The 2p orbitals describe dumbbell-shaped regions (Figure 7.26); two lobes lie along a line with the atomic nucleus centered between the lobes. In a 2p orbital, there is a planar node between the two lobes of the orbital. To distinguish among the different 2p orbitals, we sometimes refer to them as the $2p_x$, $2p_y$, and $2p_z$ orbitals because they are perpendicular to one another and can be drawn along the x, y, and z coordinate axes, as is done in Figure 7.26.

Higher p orbitals—that is, 3p, 4p, and so on—have similar overall shapes, but they have spherical and planar nodes that separate regions of higher electron density. Their sizes increase with increasing values of n.

The third principal energy shell ($n = 3$) is divided into three subshells: the 3s subshell has one 3s orbital, the 3p subshell has three 3p orbitals, and the 3d subshell has five 3d orbitals (corresponding to five possible values of m_l). The d orbitals are pictured in Figure 7.27.

The fourth principal energy shell ($n = 4$) is divided into four subshells: a 4s subshell with one orbital, a 4p subshell with three orbitals, a 4d subshell with five orbitals, and a 4f subshell with seven orbitals. The seven f orbitals have complex shapes that we will not consider here.

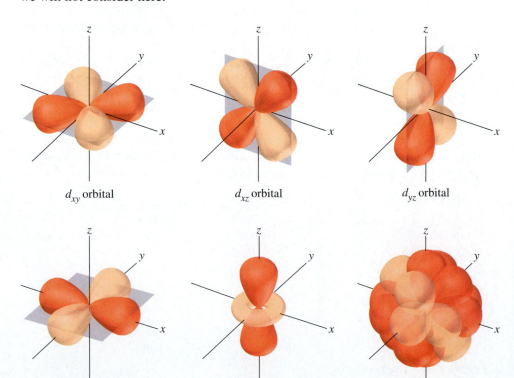

d_{xy} orbital d_{xz} orbital d_{yz} orbital

$d_{x^2-y^2}$ orbital d_{z^2} orbital

▶ **FIGURE 7.27** **The five d orbitals**
The designations xy, xz, yz, $x^2 - y^2$, and z^2 refer to the directional characteristics of certain combinations of the orbitals having the allowed values of m_l when $l = 2$. The rightmost drawing is a composite showing all five d orbitals in a single atom.

QUESTION: Which two d orbitals together occupy a region of space that is similar to the region occupied by the three p orbitals shown in the composite drawing in the rightmost part of Figure 7.26?

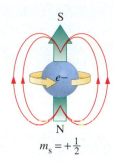 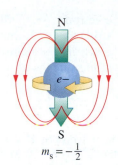

▶ FIGURE 7.28   Electron spin visualized

Two possibilities for electron spin, with the coexisting magnetic fields indicated by the red lines. (A magnetic field is visualized in Appendix B.) The magnetic fields of two electrons having opposite spins cancel each other, with the result that there is no net magnetic field for the pair.

$m_s = +\frac{1}{2}$ $m_s = -\frac{1}{2}$

Electron Spin: A Fourth Quantum Number

The quantum numbers n, l, and m_l arise from Schrödinger's treatment of hydrogen's electron as a matter wave. This concept can be extended to electrons in other atoms as well. Although these three quantum numbers fully characterize the orbitals in an atom, we need an additional quantum number to describe the electrons that occupy these orbitals. This fourth quantum number, the **electron spin quantum number** (m_s), was proposed in 1925 by Samuel Goudsmit and George Uhlenbeck to explain some of the finer features of atomic emission spectra. The two possible values of the spin quantum number are $+\frac{1}{2}$ (usually represented by an up arrow ↑) and $-\frac{1}{2}(\downarrow)$.

The name *electron spin quantum number* suggests that electrons have a spinning motion, as suggested in Figure 7.28. However, we cannot attach a precise physical reality to electron spin. All we know is that certain observations made in experiments can best be explained in terms of electron spin.

In 1921, Otto Stern and Walter Gerlach performed the experiment illustrated in Figure 7.29. When they shot a beam of gaseous silver atoms into a powerful magnetic field, the beam was split in two by the field, indicating that the atoms themselves act like small magnets.

It has long been known that a moving electric charge induces a magnetic field about itself, and therefore a spinning electron should have an associated magnetic field. The direction of the magnetic field produced by an electron with $m_s = +\frac{1}{2}$ is opposite the direction of the field produced by an electron with $m_s = -\frac{1}{2}$. As a result, the magnetic effect is canceled for a pair of electrons with opposing spins.

We can explain the Stern–Gerlach experiment as follows: Of the 47 electrons in a silver atom, 23 have a spin in one direction and 24 have the opposite spin. If 46 of the electrons have their spins paired, their magnetic fields cancel. The direction of deflection of a given silver atom depends on the type of spin on the 47th electron. In a collection of silver atoms, there is an equal chance that the "odd" electron will have a spin of $+\frac{1}{2}$ or $-\frac{1}{2}$. The beam of silver atoms is thus split into two beams. A beam of hydrogen atoms would be similarly split in a magnetic field, indicating that half the electrons have a spin of $+\frac{1}{2}$ and half have a spin of $-\frac{1}{2}$.

Looking Ahead

We have now described the four quantum numbers used to characterize electrons in atoms. In the next chapter, we will see how these quantum numbers, together with a few simple rules, allow us to establish which orbitals in an atom are occupied by

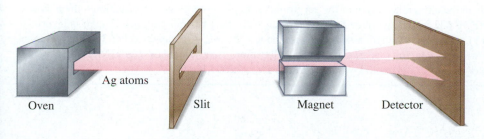

◀ FIGURE 7.29   The Stern–Gerlach experiment: Demonstration of electron spin

Silver atoms vaporized in the oven are shaped into a beam by the slit, and the beam is passed through a nonuniform magnetic field. The beam splits in two. (The field has to be stronger in some directions than in others because the beam of atoms would not experience a force if the field were uniform.)

Oven   Ag atoms   Slit   Magnet   Detector

electrons. We will also discover a close relationship between the electronic structures of atoms and the periodic table. In Chapters 9 and 10, we will see how electronic structure, in turn, determines the nature of chemical bonds between atoms and the structures of molecules.

Cumulative Example

Which will produce more energy per gram of hydrogen: H atoms undergoing an electronic transition from the $n = 4$ level to the $n = 1$ level or hydrogen gas burned in the reaction $2\,H_2(g) + O_2(g) \rightarrow 2\,H_2O(l)$?

STRATEGY

We can use Equation (7.8) to find the photon energy emitted by one hydrogen atom as its electron undergoes a transition from $n = 4$ to $n = 1$. We can multiply the energy of one photon by Avogadro's number to determine the energy emitted by 1 mol of hydrogen atoms. Then we can use the molar mass to find the energy produced per gram of hydrogen. In a separate calculation, we can determine the enthalpy change first for the combustion reaction as written and then per gram of $H_2(g)$ burned. Our final step is to compare the two results.

SOLUTION

Energy emitted as electromagnetic radiation:

We first apply Equation (7.8) to the transition from $n = 4$ to $n = 1$.

$$\Delta E_{\text{level}} = 2.179 \times 10^{-18}\left(\frac{1}{4^2} - \frac{1}{1^2}\right) = 2.179 \times 10^{-18}\left(\frac{1}{16} - 1\right)$$

$$= 2.179 \times 10^{-18}(-0.9375) = -2.043 \times 10^{-18}\,\text{J}$$

The calculated energy is for one atom. We multiply by Avogadro's number to obtain the energy emitted per mole of H, and then convert from moles of H to grams of H.

$$?\,\text{kJ/g H} = \frac{-2.043 \times 10^{-18}\,\text{J}}{1\,\text{atom H}} \times \frac{6.022 \times 10^{23}\,\text{atoms}}{1\,\text{mol H}} \times \frac{1\,\text{mol H}}{1.00794\,\text{g H}} \times \frac{1\,\text{kJ}}{1 \times 10^3\,\text{J}}$$

$$= -1.221 \times 10^3\,\text{kJ/g H}$$

Enthalpy of combustion of $H_2(g)$:

To obtain this enthalpy of combustion, we can use Equation (6.21), but because the enthalpies of formation of $H_2(g)$ and $O_2(g)$ are zero, we note that the enthalpy of combustion is simply twice the enthalpy of formation of $H_2O(l)$.

$$\Delta H° = 2\,\text{mol H}_2\text{O} \times \Delta H_f°[H_2O(l)] - 2\,\text{mol H}_2 \times \Delta H_f°[H_2(g)] - 1\,\text{mol O}_2 \times \Delta H_f°[O_2(g)]$$

$$\Delta H° = 2\,\text{mol H}_2\text{O} \times \Delta H_f°[H_2O(l)]$$

$$= 2\,\text{mol H}_2\text{O} \times (-285.8\,\text{kJ/mol H}_2\text{O(l)}) = -571.6\,\text{kJ}$$

Now we convert this enthalpy of combustion to a per-gram $H_2(g)$ basis.

$$?\,\text{kJ/g H}_2 = \frac{-571.6\,\text{kJ}}{2\,\text{mol H}_2} \times \frac{1\,\text{mol H}_2}{2.01588\,\text{g H}_2} = -141.8\,\text{kJ/g H}_2$$

Comparison of results:

The energy emitted per gram of hydrogen as electromagnetic radiation by H atoms (1221 kJ) is over eight times as great as that released in the combustion of $H_2(g)$ (141.8 kJ).

ASSESSMENT

Figure 7.18 shows that the emitted light falls in the ultraviolet region of the spectrum. Example 7.4 provides information to help assess our results. In that example, light of wavelength 632.8 nm was found to have an energy of 189.1 kJ/mol photons. Energy is inversely proportional to wavelength ($E = h\nu = hc/\lambda$), and the wavelength associated with the $n = 4 \rightarrow n = 1$ transition, 97.3 nm, is less than one-sixth of 632.8 nm. Thus, the energy we calculate for this transition should be more than six times 189 kJ/mol, or roughly 1200 kJ/mol. Hydrogen has a molar mass very nearly 1 g H/mol H, which corresponds to 1200 kJ/g H.

Concept Review with Key Terms

7.1 The Electron: Experiments of Thomson and Millikan— **Cathode rays** are formed when electricity passes through gases at very low pressures. Thomson's extensive studies of cathode rays proved that they are negatively charged fundamental particles—now called electrons—found in all matter. He also established the mass-to-charge ratio of electrons by observing their deflections in electric and magnetic fields. The charge on an electron was established by Millikan in a study of the behavior of electrically charged oil droplets in an electric field. An electron bears one fundamental unit of negative electric charge.

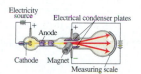

7.2 Atomic Models: Thomson and Rutherford—Thomson's raisin pudding model viewed an atom as electrons ("raisins") embedded in a positively charged "pudding." In contrast, Rutherford's interpretation of the scattering of alpha particles by metal foils suggested that the atom had a tiny, massive, positively charged nucleus that is surrounded by electrons.

7.3 Protons and Neutrons—Later experiments by Rutherford and others confirmed his atomic model and showed that the nucleus consists of protons and neutrons, which account for practically all the mass of an atom.

7.4 Positive Ions and Mass Spectrometry—Weighted atomic masses of the elements and relative abundances of the isotopes of an element can be established by **mass spectrometry**, a technique that separates charged particles according to their mass-to-charge ratios.

7.5 The Wave Nature of Light—Electromagnetic radiation is an energy transmission in the form of oscillating electric and magnetic fields. The oscillations produce **electromagnetic waves** that are characterized by their **frequencies** (ν), **wavelengths** (λ), and velocity (c).

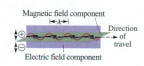

These characteristics are related through the equation

$$c = \nu\lambda = 3.00 \times 10^8 \text{ m s}^{-1} \text{ in vacuum}$$

The complete span of possibilities for frequency and wavelength is described as the **electromagnetic spectrum**, of which visible light is only a very small part. Light consisting of essentially all wavelength components in a region of the electromagnetic spectrum is a **continuous spectrum** (for example, the rainbow of colors obtained when a beam of sunlight is dispersed by a glass prism or raindrops in the sky). Energetically excited gaseous atoms typically emit an **emission spectrum** containing only a discrete set of wavelength components in the form of a **line spectrum**.

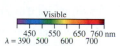

7.6 Photons: Energy by the Quantum—Planck explained the spectral characteristics of black-body radiation by proposing that the energy of electromagnetic radiation is limited to integral multiples of a fundamental quantity that is called a **quantum** and has the value

$$E = h\nu$$

where h is defined as **Planck's constant**. The photoelectric effect—the emission of electrons by illuminated surfaces—is explained by thinking of quanta of energy as concentrated into particles of light called **photons**.

7.7 Bohr's Hydrogen Atom: A Planetary Model—Bohr's theory requires the electron in a hydrogen atom to be in one of a discrete set of **energy levels**. The fall of an electron from an **excited state** to a **ground state** releases a discrete amount of energy as a photon of light with a characteristic frequency. Bohr's theory accounts for the observed atomic spectrum of hydrogen and is also applicable to the spectra of other one-electron species, such as He^+ and Li^{2+}.

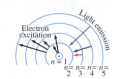

7.8 Wave Mechanics: Matter as Waves—A more satisfactory quantum mechanical treatment of atomic structure is based on two additional fundamental concepts—the wavelike properties of small particles and the Heisenberg **uncertainty principle**. These concepts transform the precisely described Bohr atom into an atomic model that replaces certainty with probability. Specifically, De Broglie proposed that a particle of mass m moving at speed v will have a wavelength λ, as described by the equation

$$\lambda = \frac{h}{mv}$$

Thus, the electron in a hydrogen atom can be viewed as a matter wave enveloping the nucleus. The representation of the wavelike properties of the electron through a wave equation is called **quantum mechanics**, and solutions of the wave equation are wave functions.

7.9 Quantum Numbers and Atomic Orbitals—Wave functions require the assignment of three **quantum numbers**: the **principal quantum number, n**, the **orbital angular momentum quantum number, l**, and the **magnetic quantum number, m_l**. Wave functions with acceptable values of the three quantum numbers are called **atomic orbitals**. Orbitals describe regions in an atom that have a high probability of containing an electron or a high electronic charge density. Orbitals having the same value of n are in the same principal energy level, or **principal shell**. Those with the same value of n and of l are in the same **subshell**. The shapes associated with orbitals depend on the value of l. An s orbital ($l = 0$) is spherical, and a p orbital ($l = 1$) is dumbbell-shaped.

$1s$ $2p_x$

The n, l, and m_l quantum numbers define an orbital, but a fourth quantum number is also required to characterize an electron in an orbital—the **electron spin quantum number, m_s**. This quantum number may have either of two values: $+\frac{1}{2}$ or $-\frac{1}{2}$.

Assessment Goals

When you have mastered the material in this chapter, you will be able to:

- Describe how the cathode ray tube works and some of the results of Thomson's experiments.

- Describe how Millikan's experiments were used to establish the charge on an electron.

- Explain the difference between Thomson's raisin pudding model of the atom and the nuclear model proposed by Rutherford.

- Explain how α particle scattering experiments led to Rutherford's nuclear model of the atom.
- Describe the properties of protons and neutrons and how they were discovered.
- Cite the functions and uses of a mass spectrometer.
- List the characteristic properties of electromagnetic waves.
- Describe the electromagnetic spectrum in terms of the wavelength and frequency range of various types of electromagnetic radiations.
- Calculate the wavelength, frequency, and/or energy of electromagnetic radiation.
- Describe the difference between continuous and line spectra.

- Use the concept of energy quantization to explain the photoelectric effect, line spectra, and the emission and absorption of photons by atoms.
- Describe the Bohr model of the hydrogen atom.
- Calculate the energy of an electron in a given energy level of a hydrogen atom and the energy differences in the levels.
- Describe the wave properties of matter and the nature of wave mechanics and wave functions.
- List the designations of the four quantum numbers, their possible values, and their relationship to atomic orbitals.
- Describe the meaning of atomic orbitals and the shapes of s, p, and d orbitals.

Self-Assessment Questions

1. Which is *not* a property of cathode rays?
   - (a) They are negatively charged.
   - (b) They differ in characteristics from element to element.
   - (c) They can be deflected by electric fields.
   - (d) They cause other substances to fluoresce.

2. Millikan's experiments hinged on the absorption of ions by oil droplets. How did he translate these observations into a determination of the fundamental unit of electric charge, and specifically the charge on an electron?

3. Answer the following questions in describing the atomic models of Thomson, Rutherford, Bohr, and quantum mechanics. (a) How did Rutherford's interpretation of the Geiger–Marsden metal-foil experiments contradict Thomson's raisin pudding model of the atom? (b) Why did Rutherford's model of the atom require that electrons be in motion, whereas Thomson's model did not? (c) How does the quantum mechanical orbital differ from a Bohr orbit for an electron in a hydrogen atom?

4. Both negative and positive particles were discovered in the passage of electric discharges through gases at low pressure. Explain why the negative particles proved to be fundamental particles of matter but the positive particles did not.

5. Which property of waves is *not* mathematically related to the other three?
   - (a) wavelength
   - (b) frequency
   - (c) speed
   - (d) amplitude

6. Which of the following wavelengths of electromagnetic radiation corresponds to the highest frequency?
   - (a) 735 nm
   - (b) 6.3×10^{-5} cm
   - (c) 1.05 mm
   - (d) 3.5×10^{-6} m

7. A particular electromagnetic radiation with wavelength 200 nm
   - (a) has a higher frequency than radiation with wavelength 400 nm.
   - (b) is in the visible region of the electromagnetic spectrum.
   - (c) has a greater speed in vacuum than does radiation of wavelength 400 nm.
   - (d) has a greater energy content per photon than does radiation with wavelength 100 nm.

8. What is black-body radiation? Can we detect it with our senses?

9. Which of the following statements is never true regarding the photoelectric effect?

   - (a) Electrons are produced from a metal surface irradiated with white light.
   - (b) Brighter light produces electrons of greater kinetic energy.
   - (c) Below a threshold frequency, no electrons are ejected.
   - (d) Higher frequency light produces electrons of greater kinetic energy.

10. Why is the quantum theory of such importance in describing behavior at the microscopic level? Why is it of little importance in describing events at the macroscopic level, for example, the speed and trajectory of a bullet fired from a rifle?

11. What is meant by the wave-particle duality of light? Name two ways in which light can be said to act as a wave and one way in which it acts as a particle.

12. Explain why it is reasonable to expect the energy levels of the Bohr hydrogen atom to be given by an equation of the form $E_n = -B/n^2$, rather than $E_n = -B \times n^2$. What is the significance of the negative sign?

13. Which hydrogen atom has absorbed the most energy? One in which the electron has moved from
   - (a) the first to third energy level.
   - (b) the second to the third energy level.
   - (c) the second to the fourth energy level.
   - (d) the third to the fourth energy level.

14. How does Heisenberg's uncertainty principle place a limitation on our ability to learn about the structure of an atom?

15. The line at 434 nm in the Balmer series of the hydrogen spectrum corresponds to a transition of an electron from the nth to the second Bohr orbit. What is the value of n?

16. What is a matter wave? Is there any physical evidence to support the existence of matter waves? What is the significance of matter waves in describing atomic structure?

17. Give the subshell notation for each of the following sets of quantum numbers.
   - (a) $n = 3$ and $l = 2$
   - (b) $n = 2$ and $l = 0$
   - (c) $n = 4$ and $l = 1$
   - (d) $n = 4$ and $l = 3$

18. The set of quantum numbers, $n = 2$, $l = 2$, $m_l = 0$
   - (a) describes an electron in a 2d orbital.
   - (b) describes an electron in a 2p orbital.
   - (c) describes one of five orbitals of a similar type.
   - (d) is not allowed.

19. How many subshells are there in a principal shell with (**a**) $n = 3$, (**b**) $n = 2$, and (**c**) $n = 4$?

20. What is the shape of the geometric region described by an s orbital? How does a $2s$ orbital differ from a $1s$ orbital?

21. What is the shape of the geometric region described by a p orbital?
    (**a**) circular   (**b**) elliptical   (**c**) spherical   (**d**) dumbbell

22. Describe the property called electron spin and an experiment that can be explained in terms of electron spin.

Problems

Subatomic Particles

23. The mass-to-charge ratio of the proton is 1.044×10^{-8} kg/C. What is the charge on a proton? What is its mass?

24. The mass-to-charge ratio of the positron (a product of the radioactive decay of certain isotopes) is 5.686×10^{-12} kg/C. The charge on a positron is the same as that on a proton. Calculate the mass of a positron. Does any other fundamental particle have this same mass? Explain.

25. An oil-drop experiment yielded the following charges on an oil droplet at various times: 6.4×10^{-19} C, 3.2×10^{-19} C, 8.0×10^{-19} C, 4.8×10^{-19} C. What value for the electronic charge can you deduce from these data?

26. An oil-drop experiment yielded the following charges on an oil droplet at various times: 9.6×10^{-19} C, 3.2×10^{-19} C, 16.0×10^{-19} C, 6.4×10^{-19} C. What value of the electronic charge can you deduce from these data?

27. Determine approximate mass-to-charge ratios for the following ions.
    (**a**) $^{80}Br^-$ (**b**) $^{18}O^{2-}$ (**c**) $^{40}Ar^+$
    Why are these values only approximate? (*Hint:* What are the masses of these ions?)

28. Which of the following ions should exhibit the same mass-to-charge ratios: $^{40}Ca^{2+}$, $^{10}B^+$, $^{60}Co^{2+}$, $^{20}Ne^+$, $^{120}Sn^{2+}$, $^{80}Br^-$? Would you expect the ratios to be exactly the same or only approximately so? Explain.

Mass Spectrometry

29. The atomic masses of the five naturally occurring isotopes of germanium are germanium-70, 69.9243 u; germanium-72, 71.9217 u; germanium-73, 72.9234 u; germanium-74, 73.9219 u; germanium-76, 75.9214 u. Use these values and data from the accompanying bar graph mass spectrum of germanium to determine a weighted average atomic mass of germanium. (*Hint:* Recall Example 2.4.)

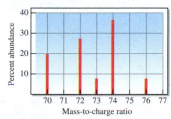

30. The atomic masses and percent natural abundances of the four naturally occurring isotopes of lead are lead-204, 203.9731 u, 1.4%; lead-206, 205.9745 u, 24.1%; lead-207, 206.9759 u, 22.1%; lead-208, 207.9766 u, 52.3%. In the manner of Figure 7.7, sketch a bar graph mass spectrum of lead.

Electromagnetic Radiation

31. What is the wavelength and in what part of the electromagnetic spectrum are the following found?
    (**a**) a broadcast frequency of 725 kHz
    (**b**) a nuclear magnetic resonance (NMR) spectrometer frequency of 555 MHz
    (**c**) radiation that has a frequency of 1.05×10^{19} s^{-1}

32. What is the frequency and in what part of the electromagnetic spectrum are the following found?
    (**a**) radiation that has a wavelength of 11.5 nm
    (**b**) radiation that has a wavelength of 9.32 cm
    (**c**) radiation that has a wavelength of 1.93×10^6 m

33. The accompanying sketch suggests a particular light wave. What are the wavelength and frequency of this light? What color of light is represented by this wave?

34. On the same scale as the light wave in Problem 33, sketch a reasonable representation of (**a**) a wave of yellow light and (**b**) an infrared light wave with a frequency of 1.0×10^{14} s^{-1}.

35. When the *Voyager* space probe radioed information back from Neptune to Earth, the electromagnetic signals had to travel about 4.4×10^9 km. How long did it take for these signals to reach Earth?

36. It took 11 min for a command from the controllers on Earth to reach the *Pathfinder* space vehicle on the surface of Mars on July 4, 1997. How many miles did the message have to travel?

Photons and the Photoelectric Effect

37. Calculate the energy, in joules, of a photon of violet light that has a frequency of 7.42×10^{14} s^{-1}. Is this energy greater or less than that of light of 655-nm wavelength? Explain.

38. Calculate the energy, in joules, of a photon of red light that has a frequency of 3.73×10^{14} s^{-1}. Is this energy greater or less than that of light of 530-nm wavelength? Explain.

39. The laser in a compact disc player produces light with a wavelength of 780 nm. Calculate the energy of this radiation, in joules per photon and in kilojoules per mole of photons.

40. A microwave oven operates with radiation having a wavelength of 12.0 cm. Calculate the energy of this radiation, in joules per photon and in kilojoules per mole of photons.

41. The minimum energy necessary to remove an electron from a lithium atom is 4.65×10^{-19} J. What is the maximum wave-

length of light, in nanometers, that will show a photoelectric effect with lithium?

42. The minimum energy necessary to remove an electron from an iron atom is 7.21×10^{-19} J. What is the maximum wavelength of light, in nanometers, that will show a photoelectric effect with iron?

43. What is the wavelength, in nanometers, of light that has an energy content of 487 kJ/mol photons? In what portion of the electromagnetic spectrum will this light be found?

44. What is the maximum wavelength, in nanometers, of light capable of dissociating gaseous oxygen molecules into oxygen atoms: $O_2(g) \rightarrow 2\,O(g)$? (*Hint:* Use appropriate data from Appendix C.)

The Bohr Model of the Hydrogen Atom

45. What is the energy of the photon emitted when the electron in a hydrogen atom drops from the energy level $n = 6$ to the energy level **(a)** $n = 1$, and **(b)** $n = 4$?

46. How much energy must a hydrogen atom absorb to raise its electron from the energy level $n = 1$ to the energy level **(a)** $n = 2$, and **(b)** $n = 7$?

47. Calculate the frequency, in s^{-1}, of the electromagnetic radiation emitted by a hydrogen atom when its electron drops from **(a)** the $n = 5$ to the $n = 2$ energy level, and **(b)** from the $n = 3$ to the $n = 1$ energy level.

48. Calculate the wavelength, in nanometers, of the electromagnetic radiation emitted by a hydrogen atom when its electron drops from **(a)** the $n = 6$ to the $n = 2$ energy level, and **(b)** from the $n = 5$ to the $n = 1$ energy level.

49. *Without doing detailed calculations*, determine whether there can be an energy level in the Bohr hydrogen atom with $E_n = -1.00 \times 10^{-17}$ J. Explain your reasoning.

50. *Without doing detailed calculations*, determine whether it is likely that one of the energy levels in the Bohr hydrogen atom has $E_n = -2.179 \times 10^{-21}$ J. Explain your reasoning.

51. The energy required to raise the energy of a hydrogen atom from its ground state to the state E_n in which $n = \infty$ is called the *ionization energy*. What is the ionization energy, in kilojoules per mole, of hydrogen atoms?

52. Would you expect to find energy states of the hydrogen atom in which the energy is a *positive* quantity? Explain.

Matter as Waves

53. Calculate the wavelength, in nanometers, associated with a proton traveling at a speed of 1.96×10^5 m s^{-1}. Use 1.67×10^{-27} kg as the proton mass.

54. Calculate the wavelength, in picometers, associated with an electron traveling 60.0% of the speed of light.

55. Electrons from an electron microscope have an associated wavelength of 0.00510 nm. To what speed, in meters per second, are the electrons accelerated?

56. To what speed, in meters per second, must electrons be accelerated to have an associated wavelength of 223 pm?

Quantum Numbers and Atomic Orbitals

57. What is the lowest numbered principal shell in which **(a)** p orbitals are found, and **(b)** the f subshell can be found?

58. Is there a 3d subshell in a hydrogen atom? Are there 3f orbitals? Explain.

59. If $n = 4$, what are the possible values of l? If $l = 1$, what are the possible values of m_l?

60. If $n = 5$, what are the possible values of l? If $l = 3$, what are the possible values of m_l?

61. Indicate whether each of the following is a permissible set of quantum numbers. If the set is not permissible, state why it is not.

    (a) $n = 2, l = 0, m_l = +1$ **(c)** $n = 4, l = 3, m_l = -2$

    (b) $n = 5, l = 3, m_l = -3$ **(d)** $n = 0, l = 0, m_l = 0$

62. Indicate whether each of the following is a permissible set of quantum numbers. If the set is not permissible, state why it is not.

    (a) $n = 3, l = 2, m_l = +2$ **(c)** $n = 3, l = 1, m_l = +2$

    (b) $n = 4, l = 4, m_l = 0$ **(d)** $n = 7, l = 0, m_l = 0$

63. Consider the electronic structure of an atom: **(a)** What are the n, l, and m_l quantum numbers corresponding to the 4p orbital? **(b)** List all the possible quantum number values for an orbital in the 4f subshell. **(c)** In what specific subshell will an electron having the quantum numbers $n = 3, l = 2$, and $m_l = -1$ be found?

64. Consider the electronic structure of an atom: **(a)** What are the n, l, and m_l quantum numbers corresponding to the 4s subshell? **(b)** List all the possible quantum number values for an orbital in the 3d subshell. **(c)** In what specific subshell will an electron having the quantum numbers $n = 4, l = 3$, and $m_l = 0$ be found?

65. In the following assignments of the four quantum numbers, write acceptable values for the missing ones. What are the orbitals corresponding to these quantum numbers?

(a) $n = ?, l = 1, m_l = 0, m_s = ?$

(b) $n = 3, l = ?, m_l = 1, m_s = -\frac{1}{2}$

(c) $n = 5, l = ?, m_l = 2, m_s = ?$

(d) $n = ?, l = 0, m_l = ?, m_s = \frac{1}{2}$

66. Which of the following statements is (are) correct? An electron having the quantum numbers $n = 5$ and $m_l = -3$; (a) may be in a d subshell; (b) must have $m_s = +\frac{1}{2}$; (c) must have $l = 3$;

(d) may have $l = 0, 1, 2,$ or 3; (e) may have $m_s = +\frac{1}{2}$ or $m_s = -\frac{1}{2}$; (f) cannot exist.

67. A certain atomic orbital contains an electron with quantum numbers $n = 3, l = 2, m_l = 0, m_s = +\frac{1}{2}$. If a second electron were to be found in the same orbital, what would be its quantum numbers? If a second electron were to be found in another orbital in the same subshell, what would be an acceptable set of quantum numbers for that electron?

68. An atomic orbital has a combination of its $n, l,$ and m_l quantum numbers totaling 6. Indicate the possible combinations of the three quantum numbers and the corresponding orbital designations, with the orbitals limited to the $s, p, d,$ and f types.

Additional Problems

Problems marked with an * may be more challenging than others.

69. Use the data presented in Figure 7.7 to establish an approximate weighted average atomic mass of mercury. Why is the result only approximately correct?

70. In Problems 25 and 26, data from oil-drop experiments are used to deduce the value of the electronic charge. The values we get in those two problems are different. Can they both be correct? If not, is one of them correct? Explain.

71. The *period* of a wave is the time required for one cycle. Household electricity has a frequency of 60 s^{-1}. What is the period of this electricity? How are period and frequency related?

72. One photon of ultraviolet light can dislodge an electron from the surface of a metal. Two photons of red light with the same total energy as the one photon of ultraviolet do not produce any photoelectrons. Explain why this is so.

73. The imaginary star Fronkensteen consists primarily of hot helium gas. A planet that revolves around Fronkensteen has an atmosphere that is mostly helium gas. Astronauts who land on this planet can see other stars, but Fronkensteen is almost invisible to them. Explain.

*74. The energy from electromagnetic radiation can be used to break chemical bonds. A minimum energy of 946 kJ/mol is required to break the nitrogen-to-nitrogen bond in N_2. What is the lowest frequency of radiation having the necessary energy to break this bond? Can a photon emitted from a hydrogen atom break the N_2 bond? Explain.

75. In everyday usage, the term "quantum jump" describes a change of significant magnitude as compared to more gradual, incremental changes. Does the term have the same meaning when used by a scientist to describe events in the microscopic world? Explain.

76. Show that there should be only *four* lines in the visible spectrum of hydrogen and at the wavelengths indicated in Figure 7.12.

77. What are the maximum and minimum wavelengths of lines in the Lyman series of the hydrogen spectrum (recall Figure 7.18)?

78. From what energy level must an electron fall to produce a line at 1094 nm in the Paschen series of the hydrogen spectrum (recall Figure 7.18)?

79. Between which two energy levels of a hydrogen atom must an electron fall to account for a spectral line at 486.1 nm?

80. An electron in a hydrogen atom drops from the energy level $n = 3$ to $n = 2$, followed by a drop from $n = 2$ to $n = 1$. Show that the total energy emitted is the same as if the electron had fallen directly from the energy level $n = 3$ to $n = 1$.

81. The energy difference between the first and second energy levels of the Bohr hydrogen atom is 3/4 B. Which of the following represents the energy difference between the first and third energy levels? Explain your reasoning.

(a) $3/2 \times (3/4 B)$ (c) $32/27 \times (3/4 B)$

(b) $2/3 \times (3/4 B)$ (d) $(3/4 B) + (4/9 B)$

*82. The Pfund series of the hydrogen spectrum has as its *longest* wavelength component a line at 7400 nm. Describe the electron transitions that produce this series. Specifically, give a Bohr quantum number that is common to this series.

83. The Bohr theory can be extended to one-electron species such as He^+ and Li^{2+}, in which case the energy levels become $E_n = -Z^2B/n^2$, where Z is the atomic number. How much energy, in joules, is required to ionize the electron from a ground-state He^+ ion?

84. Calculate the wavelength, in meters, associated with a bullet with a mass of 25.0 g traveling with a speed of 110 m/s.

*85. Einstein's equation for the equivalence of matter and energy is $E = mc^2$. Use this equation, together with Planck's equation, to derive a relationship similar to de Broglie's equation.

86. Which must have the greater velocity to produce matter waves of the same wavelength (for example, 1 nm), protons or electrons? Explain your reasoning.

*87. What must be the speed of electrons if the matter wave associated with them is to have the same wavelength as the spectral line in the Lyman series of the hydrogen spectrum in which the electron transition is from $n = 5$?

88. On page 284, we make the statement that the Heisenberg uncertainty principle has no practical importance when dealing with large objects. Show that this should be the case for a 1000-kg automobile traveling at 100 km/h. That is, show that we could measure the mass and speed of this automobile with very high precision and still be able to locate its position precisely.

89. Radio signals from *Voyager 1* spacecraft on its trip to Jupiter in the late 1970s were broadcast at a frequency of $8.4 \times 10^9 \text{ s}^{-1}$. These signals were intercepted on Earth by an antenna capable of detecting 4×10^{-21} watt (1 watt = 1 J/s). At a minimum, approximately how many photons per second of electromagnetic radiation was the antenna capable of intercepting?

90. Before the discovery of neutrons, some scientists thought that the nucleus contained electrons to balance some or all the protons. How could such a model explain the following observations?
(a) Fluorine has an atomic number of 9 and a mass number of 19.
(b) Calcium has an atomic number of 20 and a mass number of 40.

91. The greatest probability of finding the electron in a small volume element of the 1s orbital of the hydrogen atom is at the nucleus. Yet, the most probable distance of the electron from the nucleus is 52.9 pm. How can you reconcile these two statements?

92. Suppose that the 1000-dart hit represented in Figure 7.24 was distributed as shown in the following summary.

Summary of scoring (1000 darts)

50 darts score	"50"
150	"40"
350	"30"
200	"20"
150	"10"
100	off the board

Plot the number of hits versus the score, and describe which illustration in the text the plot most resembles. Explain the similarities and differences between the two.

93. Anton Zeilinger (University of Vienna) and his colleagues suggest that fullerenes, hollow, spherical molecules made of carbon atoms, provide an ideal system for further exploration of the boundary between quantum and classical physics. Calculate the de Broglie wavelength for a C_{60} molecule traveling at 220 m/s. Compare the calculated value with the observed value of about 0.025 nm as determined by Zeilinger's group.

94. An atom in which just one of the outer-shell electrons is excited to a very high quantum level (n) is called a "high Rydberg-state" atom. In some ways, all these atoms resemble a Bohr hydrogen atom with its electron in a high-numbered orbit. Explain why you might expect this to be the case.

* 95. Is there a spectral line in the hydrogen spectrum in the range **(a)** 750 ± 25 nm; **(b)** 1850 ± 50 nm?

* 96. There are two naturally occurring isotopes of chlorine (34.9689 u, 75.77%; 36.9659 u, 24.23%) and two of bromine (78.9183 u, 50.69%; 80.9163 u, 49.31%). Sketch the expected mass spectrum of bromine monochloride in a plot similar to Figure 7.7. Include only peaks attributable to $BrCl^+$ ions.

* 97. The *work function* (ϕ) of a photoelectric material is the energy that a photon of light must possess to just expel an electron from the surface of the material. The corresponding frequency of the light is the *threshold frequency* (ν_{th}), and $\phi = h\nu_{th}$. The higher the energy of the incident radiation, the more kinetic energy that ejected electrons have in moving away from the surface. The work function for cesium is 3.42×10^{-19} J.

(a) What is the threshold frequency for the photoelectric effect for cesium?

(b) Will 1000-nm infrared radiation produce the photoelectric effect on cesium? Explain.

(c) A 425-nm blue light shines on a piece of cesium. What should be the speed of the electrons emitted from the metal? (*Hint:* Think of the energy of the photons as used partly to overcome the work function and partly to give the photoelectrons kinetic energy.)

* 98. Microwave ovens have become popular in kitchens everywhere. They are also used in the laboratory, particularly in drying samples for chemical analysis. A typical microwave oven uses radiation with a 12.2-cm wavelength.

(a) How many moles of photons of this radiation are required to raise the temperature of 345 g water from 26.5 °C to 99.8 °C?

(b) The *watt* is a unit of power—the rate at which energy is delivered or consumed: $1 \text{ W} = 1 \text{ J s}^{-1}$. Assume that all the energy of a 700-W microwave oven is delivered to the heating of water described in (a). How long will it take to heat the water?

(c) Do you think that hydrogen atoms are able to emit microwave radiation of 12.2-cm wavelength? If so, between which two successive energy levels, ($n + 1$) and n, would the electron in the atom have to fall?

Apply Your Knowledge

99. [Historical] In 1885, Johann Balmer derived the following equation (in which n is an integer greater than 2) for the lines in the Balmer series of the hydrogen spectrum (Figure 7.18).

$$\nu = 3.2881 \times 10^{15} \text{ s}^{-1} \times \left(\frac{1}{2^2} - \frac{1}{n^2} \right)$$

Balmer apparently derived his equation by trial and error, but a more common scientific procedure is to deduce an equation from experimental data. Show that Balmer's equation is that of a straight line. What variables should be plotted in this graph, and what are the numerical values of the slope and intercept of this straight line?

100. [Historical/Collaborative] Although Albert Einstein made some early contributions to quantum theory, he was never able to accept the Heisenberg uncertainty principle. He stated, "God does not play dice with the Universe." What do you think Einstein meant by this remark? In reply to Einstein's remark, Niels Bohr is supposed to have said, "Albert, stop telling God what to do." What do you suppose Bohr meant by this remark?

* 101. [Historical] The angular momentum of an electron in the Bohr hydrogen atom is mvr, where m is the mass of the electron, v is its velocity, and r is the radius of the Bohr orbit. The angular momentum can have values equal only to $nh/2\pi$ (where n is an integer), and the radii of the allowable orbits in the hydrogen atom predicted by Bohr's theory are

$$r_n = n^2 a_0, \text{ where } n = 1,2,3,\ldots \text{ and}$$
$$a_0 = 0.53 \text{ Å } (53 \text{ pm})$$

Combine these ideas with other data from the text to determine for an electron in the third orbit ($n = 3$) of the Bohr hydrogen atom **(a)** its velocity, **(b)** the number of revolutions per second that the electron makes about the nucleus.

102. [Biochemical] What is the approximate range (minimum to maximum) of energy, in kJ/mol photons, to which our eyes can respond? Suggest a reason why it would be difficult to genetically engineer eyes to respond to a range a hundred times wider than your calculated range. The human eye can respond to as few as seven photons. How much more sensitive than Voyager's antenna (Problem 89) is the human eye?

* **103.** [Laboratory] Most of the electrons in a sample of matter at room temperature are in the ground state. However, just as there is a distribution of molecular velocities in a gas (Section 5.11), there is also a distribution of energy states in a sample of matter, and a certain fraction of the atoms in matter have energy states high enough to be excited states. The fraction of atoms in an excited state is given by a form of the *Boltzmann* equation:

$$N*/N° = (g_m/g_n)e^{-\Delta E/kT}$$

where $N*/N°$ is the ratio of excited-to-ground-state particles, g_m/g_n is a quantum mechanical factor for the particular ground and excited states, e is the base of the natural logarithms, ΔE is the energy difference between the states, k is the Boltzmann constant $(1.38 \times 10^{-23}\,\text{J K}^{-1})$, and T is the temperature in kelvins. Consider the sodium line at 589.0 nm, for which $g_m/g_n = 2$ and the energy difference between ground and excited states is 2.10 electronvolts. (See inside back cover.)

   (a) What percentage of the sodium atoms at 25 °C will be in the excited state? What percentage will be in the ground state?

   (b) What percentage of the sodium atoms at 2700 K (the approximate temperature of an acetylene flame) will be in the excited state? In the ground state?

   (c) What percentage of the sodium atoms at 2200 K (the approximate temperature of a methane flame) will be in the excited state? In the ground state?

   (d) In atomic emission spectrometry, we record emission of photons from excited atoms. Use your answers to **(b)** and **(c)** to explain why atomic emission spectrometry is rather sensitive to changing flame temperatures.

   (e) In atomic absorption spectrometry, we measure the absorbance of photons by ground-state atoms. Using your answers to **(b)** and **(c)**, explain why atomic absorption spectrometry is rather tolerant to changing flame temperatures.

* **104.** [Laboratory] *Inductively coupled plasma/mass spectrometry* (ICP-MS) is a relatively recent instrumental method for determining extremely low concentrations of elements in aqueous samples. In ICP-MS, a sample is heated to temperatures approaching 6000 K in an electrical discharge in argon gas. The sample breaks down and is ionized, and the ions are detected by a mass spectrometer. Concentrations as low as a few ng/L can be detected. However, there are several interferences that can occur. For example, the argon gas interferes with calcium, and the species ArO can interfere with Na determinations. Suggest a reason for these two interferences.

e-Media Problems

The activities described in these problems can be found in the e-Media Activities and Interactive Student Tutorial (IST) modules of the Companion Website, *http://chem.prenhall.com/hillpetrucci*.

105. In the **Millikan Oil-Drop Experiment** animation *(Section 7-1)*, oil drops are observed to stop falling and "float" at a constant height in the chamber. **(a)** Sketch the forces on a particle in this condition. **(b)** Notice the sign (+ or −) of the charged plates on top and bottom of the chamber. How would the behavior of the drops differ if the signs of the charges on the plates were switched? What is learned about the sign of the charge on electrons from this observation?

106. From the **Regions of the Electromagnetic Spectrum** activity *(Section 7-5)*, list the range of wavelengths corresponding to red and blue visible light. Which of these two colors consists of photons of higher energy?

107. Estimate the frequency and energy of light emitted for each of the three metals tested in the **Flame Tests for Metals** movie

(Section 7-5). What additional observations or experimental procedures would be necessary to fully measure the emission spectra of these elements?

108. View the **Photoelectric Effect** animation *(Section 7-6)*. **(a)** Describe the influence of increasing the *frequency* of incident light from low to intermediate to high as shown in the animation. **(b)** For each frequency, describe the effect of increasing the *intensity* of the incident light. **(c)** Why is it important to perform the measurements in an *evacuated* tube?

109. For each of the **Atomic Orbital** 3D models *(Section 7-9)*, rotate the orbital in space by moving the mouse while holding down the (*left* on Windows computers) mouse button. **(a)** What do the patterns illustrated for these orbitals represent? **(b)** For which of the orbitals does rotation produce no overall change in orientation? (Which orbitals are completely symmetric?) **(c)** In what ways does the radial distribution of electron density differ between the types of orbitals?

1		2		
H		2A		
0794				
3		4		
Li		**B**		
941		9.01		
11	3	12		3
Na		**Mg**		3B
22.9898		24.3050		
19	Period 4	20	21	
K		**Ca**	**Sc**	
39.0983		40.078	4 9559	47
37		3		
Rb		**S**		
4678		87.		
55		5		
Cs		**B**		
2.905		137. 27	58.90	17
87		88	89	

Electron Configurations, Atomic Properties, and the Periodic Table

IN CHAPTER 2, we described how Dmitri Mendeleev constructed the periodic table in 1869 by arranging elements with similar properties into groups. Recall that to make his classification scheme work, Mendeleev corrected some atomic mass values and boldly left gaps in his table for undiscovered elements. Mendeleev's approach to the periodic table was *empirical;* that is, he based his classification scheme on the observed facts. In this chapter, we will see that Mendeleev's scheme also makes sense from a *theoretical* standpoint. We will first note how electrons are distributed among the orbitals and subshells in the energy-level diagram of an atom, a description called the *electron configuration* of the atom. Then, we will explore electron configurations as a theoretical underpinning for the periodic table. We will also introduce some properties of individual atoms—atomic properties—and see how the periodic table helps us to summarize trends in these properties among the elements.

8.1 Multielectron Atoms

In Chapter 7, we described Schrödinger's wave mechanical treatment of the atomic structure of the hydrogen atom and of other species consisting of a positively charged nucleus and a single electron, that is, species such as H, He^+, and Li^{2+}. Wave mechanics works fairly well for these simple species, but things become complicated when we consider *multielectron* species, because now several electrons are attracted to the nucleus while simultaneously repelling one another. It is possible to write a wave equation that accounts for all the interactions in a multielectron atom—repulsions as well as attractions—but the equation cannot be solved exactly. The usual approach is to assume that, with appropriate adjustments, all atoms can be described through *hydrogen-like* orbitals. An important difference between the orbitals of the hydrogen atom and those of multielectron atoms is the orbital energies.

CONTENTS

◄ Sodium (top left) and potassium (bottom left) both react readily and dramatically with water, liberating hydrogen. Calcium (bottom right) also reacts with water, but much less readily. Magnesium (top right) does not react with cold water, but only with hot water or steam. We can see from these photographs that reactivity may vary significantly from group to group. We can also see that reactivity may vary within a group . Reactivity is just one of the many periodic properties of the elements.

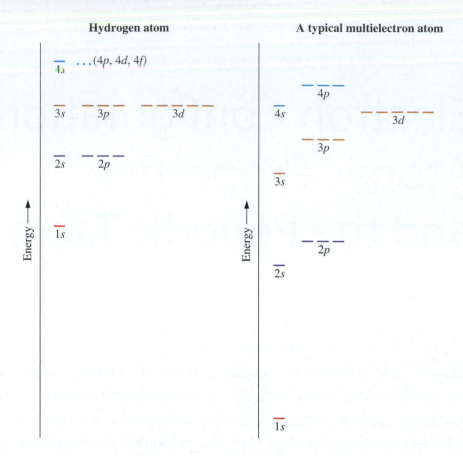

▶ **FIGURE 8.1   Orbital energy diagrams**

Some orbital energies of the first four principal shells are compared for a hydrogen atom and a typical multielectron atom. The energies are not plotted to scale.

QUESTION: What is the order of increasing energy of the subshells in a typical multielectron atom?

Orbital Energy Diagrams activity

Figure 8.1 presents one orbital energy-level diagram for hydrogen and another for a typical but unspecified multielectron atom. Orbital energy diagrams can be obtained experimentally. Figure 8.1 illustrates four points that are relevant to the discussion in several sections of this chapter:

1. *In the hydrogen atom, all subshells of a principal shell are at the same energy level.* The orbital energies in the hydrogen atom depend only on the principal quantum number, *n*. This is what we predict with the Bohr model: $E_n = -B/n^2$.

2. *Orbital energies are lower in multielectron atoms than in the hydrogen atom.* All energy states associated with forces of attraction are negative. Because the attractive force between the nucleus and an electron in any orbital increases with increasing nuclear charge, the orbital energy becomes more negative (lower) as the atomic number increases. Thus, the energies of the 1*s*, 2*s*, 3*s*, and 4*s* orbitals in Figure 8.1 are lower in all multielectron atoms than in the hydrogen atom.

3. *In a multielectron atom, the various subshells of a principal shell are at <u>different</u> energy levels, but all orbitals within a subshell are at the <u>same</u> energy level.*

4. *In higher-numbered principal shells of a multielectron atom, some subshells of different principal shells have nearly identical energies.* In Figure 8.1, we see that the energy-level difference between 3*d* and 4*s* is quite small. This small energy difference has important consequences that we explore in Section 8.4.

Regarding the third point in this list, one way to think about the repulsive forces between electrons in a multielectron atom is that all the electrons located between a given electron and the nucleus *screen,* or *shield,* that electron from the full attractive force of the nucleus. The effectiveness of the shielding depends on the orbitals in which the shielding electrons and the shielded electron are found.

An *s* electron from any principal shell has some probability of being found near the nucleus (Figure 7.23). We say that an *s* electron *penetrates* through inner-shell electrons to approach the nucleus. In contrast, the two lobes of a *p* orbital are separated by a nodal plane through the nucleus, at which point the electron probability falls to zero (Figure 7.26). A *p* electron is less penetrating than an *s* electron, and electrons in

d and f orbitals are less penetrating still. Highly penetrating electrons are less effectively shielded by inner-shell electrons, are more strongly attracted to the nucleus, and are at lower energies than less penetrating electrons.

The increasing order of subshell energies is

$$s < p < d < f$$

and in Figure 8.1, we see this pattern for the multielectron atom. Of the subshells shown, the order of increasing energy is

$$2s < 2p; 3s < 3p < 3d; 4s < 4p$$

In an isolated multielectron atom, energy levels within a subshell are not split further. For example, the three $3p$ orbitals are **degenerate** (at the same energy level). Likewise, the five $3d$ orbitals are degenerate; they all have the same energy, but the energy of the $3d$ orbitals is higher than that of the $3p$ orbitals.

In relation to quantum numbers, orbital energy is determined primarily by n: The lower the value of n, the lower the orbital energy. Orbital energy is further affected by l: In a given shell, the lower the value of l, the lower the orbital energy. The orbital energy does not depend on the value of m_l.

8.2   An Introduction to Electron Configurations

The scientific view of where electrons are found in atoms has evolved since the discovery of the electron in 1897. J. J. Thomson saw them as embedded in a cloud of positive charge. Rutherford hypothesized electrons in motion outside a positively charged atomic nucleus. Bohr proposed discrete circular orbits for electrons around the nucleus. Schrödinger, Heisenberg, and others spoke of regions of high electron probability or high electron charge density that are defined by mathematical functions called atomic orbitals.

Although we often say that an electron is "in a $1s$ orbital" or "in a $2p$ orbital" and so on, remember from Chapter 7 that the orbitals are not actual regions in an atom. They are mathematical expressions used to determine the probabilities of finding an electron in various regions of an atom. That is the view of orbitals we have in mind when we say the **electron configuration** of an atom describes the distribution of electrons among the various orbitals in the atom. We can represent electron configurations in two quite similar ways.

The *spdf* **notation** uses a number to designate a principal shell and a letter—s, p, d, or f—to identify a subshell. A superscript number following the letter indicates the number of electrons in the subshell. Empty subshells are *not* included in the notation; that is, we do not include terms such as $3d^0$. Thus, the notation $1s^2 2s^2 2p^3$ indicates an atom with two electrons in the $1s$ subshell, two in the $2s$ subshell, and three in the $2p$ subshell. An atom having this electron configuration has an atomic number of seven; it is an atom of the element nitrogen.

This *spdf* notation leaves an unanswered question: How are the three $2p$ electrons of a nitrogen atom distributed among the three orbitals in the $2p$ subshell? We can show this distribution through an *expanded spdf notation*:

$$1s^2 2s^2 2p_x^{\;1} 2p_y^{\;1} 2p_z^{\;1}$$

This representation shows that each of the three $2p$ orbitals contains a single electron.

Another way to represent an electron configuration is with an **orbital diagram,** which uses boxes to represent orbitals within subshells and arrows to represent electrons. As described on page 289, the directions of the arrows represent the sign of the spin quantum number. The orbital diagram for nitrogen is

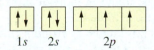

It indicates that the nitrogen atom has two electrons of opposing spins in the $1s$ subshell and two more of opposing spins in the $2s$ subshell. Electrons with opposing spins

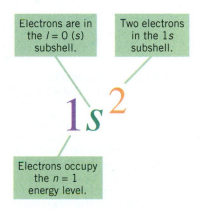

Electrons are in the $l = 0$ (s) subshell.

Two electrons in the $1s$ subshell.

Electrons occupy the $n = 1$ energy level.

Electron Configuration Symbolism activity

are *paired*. Each orbital of the 2*p* subshell has one electron, and all three electrons have spins in the same direction; they are said to have *parallel* spins.

Several questions arise when we look carefully at the electron configurations we have written:

- Why are there never more than two electrons in an atomic orbital?
- When there are two electrons in an orbital, why do they have opposing spins?
- Why are all orbitals in a subshell occupied singly before pairing of electrons occurs?
- Why do the electrons in singly occupied orbitals have parallel spins?

8.3   Rules for Electron Configurations

To answer these questions and to lay the groundwork for predicting probable electron configurations of the elements, we consider three basic principles that govern the distribution of electrons among atomic orbitals:

1. Electrons occupy orbitals of the lowest energy available.
2. No two electrons in the same atom may have all four quantum numbers alike.
3. In a group of orbitals of identical energy, electrons enter empty orbitals whenever possible. Electrons in half-filled orbitals have parallel spins.

Let's examine these rules individually. Regarding statement 1, the order in which orbitals of the first three principal shells fill in a multielectron atom is suggested by Figure 8.1. Note that the 1*s* orbital is at a lower energy level than the 2*s*; electrons therefore enter the *1s* before the 2*s*. Electrons enter the three orbitals in the 2*p* subshell before the 3*s* orbital, and so on. With some exceptions, electrons occupy the subshells of atoms in the order

$$1s, 2s, 2p, 3s, 3p, 4s, 3d, 4p, 5s, 4d, 5p, 6s, 4f, 5d, 6p, 7s, 5f, 6d, 7p$$

This order need not be memorized. If need be, it can be constructed readily, using the scheme shown in Figure 8.2. Later in the chapter, we will learn to relate electron configurations to the periodic table, which is probably the best guide to the order of subshell filling.

Overlapping of energy levels for certain orbitals in the higher-numbered principal shells causes a few exceptions to the above order of filling. Moreover, electrons do not occupy orbitals merely according to orbital energies; rather, the orbitals are occupied in a way that leads to the lowest possible energy state for the *atom* when all electrons are in place. We can use quantum mechanics to determine the lowest energy state for an atom among possible alternatives. In the final analysis, however, the order of subshell filling and other details of electron configurations are established by experiments, such as spectroscopic and magnetic studies.

Statement 2 is known as the **Pauli exclusion principle**, developed by Wolfgang Pauli in 1926 to explain the complex features of emission spectra in a magnetic field. This principle has an important consequence for electron configurations: Because electrons in a given orbital must have the same values of n, l, and m_l (for example, $n = 3$, $l = 0$, $m_l = 0$ in the 3*s* orbital), they must have different values of m_s. Only two values of m_s are possible: $+\frac{1}{2}$ and $-\frac{1}{2}$. As a result, we conclude that

An atomic orbital can accommodate only two electrons, and these electrons must have opposing spins.

Each principal shell consists of a given number of subshells, and each subshell contains a given number of orbitals. The Pauli exclusion principle therefore limits the number of electrons that can be found in individual orbitals, subshells, and principal shells. These limits are set forth in Table 8.1.

The first part of statement 3 is known as **Hund's rule** (after the quantum physicist H. Friedrich Hund). We can rationalize Hund's rule as follows: Two electrons carry identical charge and therefore repel each other; they tend not to occupy the same region of space. As a result, electrons go into separate orbitals as long as empty orbitals of the same energy are available. This is rather like the inclination of bus pas-

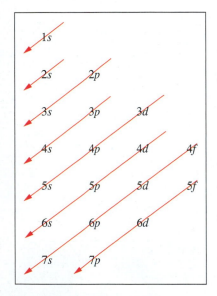

▲ **FIGURE 8.2   The order in which subshells are filled with electrons**

Start at the upper left of the grid, and follow the arrows. The order in which the arrows slice through the subshell designations is the order in which the subshells fill. The blank space at the upper right corresponds to nonexistent subshells (1*p*, 1*d*, 2*d*, and so on). The blank space at the bottom corresponds to subshells that are not filled in the known elements (7*d*, 6*f*, and so on).

Table 8.1 Maximum Capacities of Subshells and Principal Shells

n	1	2	3	4	$...n$
l	0	0 1	0 1 2	0 1 2 3	
Subshell designation	s	s p	s p d	s p d f	
Orbitals in subshell	1	1 3	1 3 5	1 3 5 7	
Subshell capacity	2	2 6	2 6 10	2 6 10 14	
Principal shell capacity	2	8	18	32	$...2n^2$

sengers to sit singly, which gives each person maximum seat space. Only when all seats have one occupant do passengers begin to sit two to a seat.

Hund's rule leads us to write the electron configuration of nitrogen just as we showed it in Section 8.2:

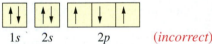

$$1s \quad 2s \quad 2p$$

The following electron configuration for nitrogen is in conflict both with experimental evidence and with Hund's rule:

$$1s \quad 2s \quad 2p \qquad (incorrect)$$

It is not so easy to explain why electrons in singly occupied orbitals have parallel spins. However, both experiment and theory indicate that an electron configuration in which all unpaired electrons have parallel spins represents a lower energy state of an atom than any other electron configuration that we can write. This fact tells us, for example, that we should *not* write the electron configuration of nitrogen as

$$1s \quad 2s \quad 2p \qquad (incorrect)$$

Electron Configurations animation

8.4   Electron Configurations: The Aufbau Principle

The electron configurations we write in this section are for atoms in their lowest-energy state; that is, in the *ground state*. To write these ground-state electron configurations, we will use the rules of Section 8.3 and an idea known as the *aufbau principle*. (Aufbau is a German word that means "building up.") The **aufbau principle** describes a hypothetical process (we cannot actually do this) in which we think about building up each atom from the one that precedes it in atomic number. To do so, we imagine adding a proton and some neutrons to the nucleus and one electron to an atomic orbital. That is, we think about building a helium atom from a hydrogen atom; a lithium atom from a helium atom; and so on. In this process, we focus on the particular atomic orbital into which the added electron goes in order to form a ground-state atom.

Let's illustrate the aufbau principle by building a few atoms. A neutral hydrogen atom has one proton ($Z = 1$) and thus only one electron. That electron goes into the orbital that has the lowest possible energy, the $1s$ orbital:

$$(Z = 1) \text{ H } 1s^1$$

Based on the rule for the order in which subshells fill, the electron we add in building a helium atom from a hydrogen atom goes into the lowest-energy orbital available, the $1s$. The electron configuration of helium is therefore

$$(Z = 2) \text{ He } 1s^2$$

In the lithium atom, the first two electrons are in the $1s$ orbital, just as in the helium atom. The Pauli exclusion principle tells us that the third electron cannot enter the fully occupied $1s$ orbital. The added electron must therefore go into the $2s$ orbital, the vacant orbital next lowest in energy. The first principal shell is filled, and the second begins to fill:

$$(Z = 3) \text{ Li } 1s^2 2s^1$$

In the beryllium atom, the added electron also goes into the $2s$ orbital:

$$(Z = 4) \text{ Be } 1s^2 2s^2$$

With the boron atom, the Pauli exclusion principle dictates that the added electron enter a $2p$ orbital:

$$(Z = 5) \text{ B } 1s^2 2s^2 2p^1$$

The $2p$ subshell fills as the atomic number increases from $Z = 6$ to $Z = 10$ in the series carbon, nitrogen, oxygen, fluorine, neon. In these atoms, Hund's rule requires that electrons enter $2p$ orbitals singly and with parallel spins before electrons pair up. This fact is not explicitly shown in the $spdf$ notation, but it is clearly shown in orbital diagrams:

$(Z = 6)$ C $1s^2 2s^2 2p^2$

$(Z = 7)$ N $1s^2 2s^2 2p^3$

$(Z = 8)$ O $1s^2 2s^2 2p^4$

$(Z = 9)$ F $1s^2 2s^2 2p^5$

$(Z = 10)$ Ne $1s^2 2s^2 2p^6$

Problem-Solving Note

A common student error in writing electron configurations is to skip a subshell entirely. Noble-gas-core abbreviations generally make it easier to avoid this error.

For any element after helium in the periodic table, we can abbreviate the electron configuration by showing a bracketed chemical symbol in place of the portion of the $spdf$ notation that corresponds to the electron configuration of the noble gas preceding the element in the table. This representation is called a *noble-gas-core abbreviated* electron configuration. We can therefore represent the electron configuration of lithium $(1s^2 2s^1)$ as $[\text{He}]2s^1$ by replacing $1s^2$ with $[\text{He}]$. Likewise, we can use $[\text{He}]2s^2 2p^3$ in place of $1s^2 2s^2 2p^3$ to represent the electron configuration of nitrogen.

With two electrons in the first shell and eight in the second shell, neon has both the first and second principal shells filled (recall Table 8.1). In extending the aufbau process from neon to sodium, the added electron goes into the available orbital of lowest energy, the $3s$:

$$(Z = 11) \text{ Na } 1s^2 2s^2 2p^6 3s^1 \quad or \quad (Z = 11) \text{ Na } [\text{Ne}]3s^1$$

Other atoms with electrons entering the third principal shell are featured in Example 8.1 and Exercise 8.1, where we illustrate the following general strategy for writing electron configurations.

Step 1: *Determine the number of electrons to appear in the electron configuration.* For a neutral atom, this is simply the atomic number of the element.

Step 2: *Add electrons to subshells in order of increasing subshell energy.* This will be the order of filling of subshells outlined on page 302.

Step 3: *Observe the Pauli exclusion principle.* There can be no more than two electrons in an orbital, and they must have opposing spins.

Step 4: *Observe Hund's rule.* Orbitals in a subshell are singly occupied whenever possible. Also, all unpaired electrons have parallel spins.

Example 8.1

Write electron configurations for sulfur, using both the *spdf* notation and an orbital diagram.

SOLUTION

We can follow the four-step strategy just outlined.

Step 1: *Determine the number of electrons to appear in the electron configuration.* The atomic number of sulfur is 16. The S atom therefore has 16 electrons, all of which must appear in the electron configuration.

Step 2: *Add electrons to subshells in order of increasing subshell energy.* We must place the electrons into the lowest-energy subshells available, using the order of subshell filling illustrated in Figure 8.2. Two electrons go into the $1s$ subshell, two into the $2s$ subshell, six into the $2p$ subshell, and two into the $3s$ subshell. That leaves four electrons to be placed into the $3p$ subshell.

Step 3: *Observe the Pauli exclusion principle.* Each orbital in the subshells $1s$, $2s$, $2p$, and $3s$ is filled to its limit of two electrons with opposing spins, accounting for a total of 12 electrons. The remaining four electrons go into the $3p$ subshell.

Step 4: *Observe Hund's rule.* In adding four electrons to the three orbitals in the $3p$ subshell, two of the orbitals must have a single electron each, and one must have a pair of electrons. The unpaired electrons have parallel spins.

$$(Z = 16)\ \text{S}\ \ 1s^2 2s^2 2p^6 3s^2 3p^4$$

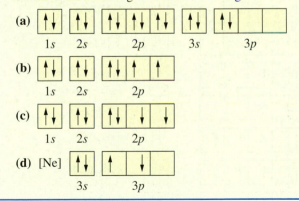

In the noble-gas-core abbreviated electron configuration, we substitute the symbol [Ne] for $1s^2 2s^2 2p^6$, the portion of the electron configuration that corresponds to that of neon:

$$(Z = 16)\ \text{S [Ne]}\ \ 3s^2 3p^4$$

$(Z = 16)$ S [Ne]

EXERCISE 8.1A

Write electron configurations for chlorine using **(a)** *spdf* notation, **(b)** a noble-gas-core abbreviated electron configuration, and **(c)** an orbital diagram.

EXERCISE 8.1B

Which of the following are valid orbital diagrams?

(a) $1s$ $2s$ $2p$ $3s$ $3p$

(b) $1s$ $2s$ $2p$

(c) $1s$ $2s$ $2p$

(d) [Ne] $3s$ $3p$

Problem-Solving Note

It is immaterial whether we represent a pair of electrons that have opposing spins as ↑↓ or ↓↑. It is also immaterial whether we designate the parallel spins of unpaired electrons, such as the three in the nitrogen atom, as ↑↑↑ or ↓↓↓.

The electron configurations predicted by the aufbau principle for the rest of the third-period elements follow a regular pattern through argon, in which the $3p$ subshell fills:

$$(Z = 18)\ \text{Ar}\ 1s^2 2s^2 2p^6 3s^2 3p^6 \quad or \quad (Z = 18)\ \text{Ar [Ne]}3s^2 3p^6$$

Now we need to be careful. In proceeding to potassium $(Z = 19)$, we must be guided by the filling order shown in Figure 8.2 rather than by Table 8.1, which states that the maximum capacity of the third principal shell is 18. According to Figure 8.2, the nineteenth electron goes into the 4s subshell, not the 3d:

$$(Z = 19) \text{ K } 1s^2 2s^2 2p^6 3s^2 3p^6 4s^1 \quad or \quad (Z = 19) \text{ K } [Ar]4s^1$$

In calcium, the twentieth electron pairs up with the nineteenth electron in the 4s orbital.

$$(Z = 20) \text{ Ca } 1s^2 2s^2 2p^6 3s^2 3p^6 4s^2 \quad or \quad (Z = 20) \text{ Ca } [Ar]4s^2$$

Main-Group and Transition Elements

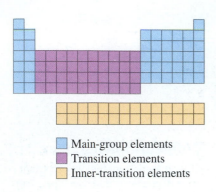

Main-group elements
Transition elements
Inner-transition elements

Elements in which the orbitals being filled in the aufbau process are either *s* or *p* orbitals of the outermost occupied shell (the electron-containing shell with the highest principal quantum number) are called either **main-group elements** or *representative elements*. The main-group elements are the A-group elements in the periodic table. The first 20 elements are all main-group elements. Scandium ($Z = 21$) is the first of the *transition elements*. In **transition elements**—the B-group elements—the subshell being filled in the aufbau process is an inner principal shell, a shell with principal quantum number smaller than the outermost occupied shell. For example, in the fourth-period transition elements, the *d* subshell of the third principal shell (3d) fills, after the outermost fourth principal shell has acquired two 4s electrons.

The electron configuration of the transition element scandium is shown here in two commonly used representations:

(a) $\text{Sc } 1s^2 2s^2 2p^6 3s^2 3p^6 3d^1 4s^2$ (b) $\text{Sc } 1s^2 2s^2 2p^6 3s^2 3p^6 4s^2 3d^1$

In method (a), all subshells in the same principal shell are grouped together, and the principal shells are arranged according to increasing principal quantum number. In method (b), the subshells are arranged in the order in which they appear to acquire electrons. We will generally use method (a).

Electron configurations of the elements from scandium through zinc, known as the first transition series, are summarized in Table 8.2. Following zinc, we return to a series of main-group elements, beginning with gallium (Ga). Because the third principal shell, which can hold a maximum of 18 electrons, is filled in zinc, the added elec-

Electron Configuration activity

Table 8.2 Electron Configurations of the First Transition Series Elements

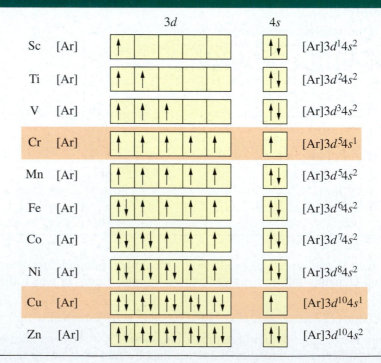

tron enters the lowest-energy subshell available, the $4p$ subshell. The six atoms in which this subshell fills range from gallium to krypton:

$$(Z = 31) \text{ Ga } [\text{Ar}]3d^{10}4s^24p^1 \quad \text{to} \quad (Z = 36) \text{ Kr } [\text{Ar}]3d^{10}4s^24p^6$$

Exceptions to the Aufbau Principle

Consider the electron configurations of chromium (Cr) and copper (Cu). The experimentally determined electron configurations listed in Table 8.2. are not the configurations that we might expect based on the order of filling of subshells in Figure 8.2 or other aspects of the aufbau process. Measurements of the emission spectra and magnetic properties of Cr and Cu indicate the following differences between what is expected and what is actually observed.

	Expected	Observed
Cr ($Z = 24$)	$[\text{Ar}]3d^44s^2$	$[\text{Ar}]3d^54s^1$
Cu ($Z = 29$)	$[\text{Ar}]3d^94s^2$	$[\text{Ar}]3d^{10}4s^1$

An electron configuration with a half-filled $3d$ subshell ($3d^5$) and a half-filled $4s$ subshell ($4s^1$)—resulting in six unpaired electrons instead of the expected four— represents a lower energy state for chromium than does the configuration based on a strict application of the aufbau principle. In copper, the lower energy state corresponds to a filled $3d$ subshell ($3d^{10}$) and half-filled $4s$ subshell ($4s^1$). Actually, because there is little difference between the $4s$ and $3d$ orbital energies, the expected and observed electron configurations are quite close to each other in energy.

At higher principal quantum numbers, the energy difference between certain subshells ($4f$ and $5d$, $5f$ and $6d$) is even smaller than that between the $3d$ and $4s$ subshells. As a result, there are additional exceptions to the aufbau principle among the heavier transition elements.

8.5 Electron Configurations: Periodic Relationships

As we have just seen, we can use the aufbau principle to assign electron configurations to the elements in order of increasing atomic number. In the modern periodic table, the elements are classified by similarities in physical and chemical properties in an arrangement based on atomic number. In Figure 8.3, where we show the electron configurations of the elements arranged in the periodic table format, it is easy to spot some distinctive patterns.

Let's focus on the portion of the electron configuration describing the occupied principal shell having the highest principal quantum number, n. This outermost occupied shell is called the **valence shell.** For example, the valence shell of the chlorine atom has the electron configuration $3s^23p^5$. From Figure 8.3, we find that, with few exceptions, all the elements in a vertical group of the periodic table have the same number of electrons in the valence shells of their atoms. Moreover, we can make the following generalization:

For main-group elements (A-group elements), the number of valence-shell electrons is the same as the periodic table group number.

Thus, except for helium, which has only two electrons, all the noble gases (group 8A) have *eight* valence-shell electrons in the configuration ns^2np^6; that is, $2s^22p^6$ for neon, $3s^23p^6$ for argon, and so on. As we will see later in the chapter, this valence-shell electron configuration seems to impart to the noble gases a stability, and hence a lack of reactivity, that is not found in any other group of elements.

The atoms in group 1A (the alkali metals) have one electron in the valence-shell electron configuration ns^1: $2s^1$ for Li, $3s^1$ for Na, and so on. Similarly, the group 2A atoms have two electrons in the valence-shell electron configuration ns^2. As a further example, the seven valence-shell electrons of the group 7A atoms are in the configuration ns^2np^5. These valence-shell electron configurations relate to the kinds of ions formed by the group 1A, 2A, and 7A elements, as we will see shortly.

FIGURE 8.3 Electron configurations and the periodic table

1A	2A	3B	4B	5B	6B	7B	8B	8B	8B	1B	2B	3A	4A	5A	6A	7A	8A
1 **H** $1s^1$																	2 **He** $1s^2$
3 **Li** $2s^1$	4 **Be** $2s^2$											5 **B** $2s^2 2p^1$	6 **C** $2s^2 2p^2$	7 **N** $2s^2 2p^3$	8 **O** $2s^2 2p^4$	9 **F** $2s^2 2p^5$	10 **Ne** $2s^2 2p^6$
11 **Na** $3s^1$	12 **Mg** $3s^2$											13 **Al** $3s^2 3p^1$	14 **Si** $3s^2 3p^2$	15 **P** $3s^2 3p^3$	16 **S** $3s^2 3p^4$	17 **Cl** $3s^2 3p^5$	18 **Ar** $3s^2 3p^6$
19 **K** $4s^1$	20 **Ca** $4s^2$	21 **Sc** $3d^1 4s^2$	22 **Ti** $3d^2 4s^2$	23 **V** $3d^3 4s^2$	24 **Cr** $3d^5 4s^1$	25 **Mn** $3d^5 4s^2$	26 **Fe** $3d^6 4s^2$	27 **Co** $3d^7 4s^2$	28 **Ni** $3d^8 4s^2$	29 **Cu** $3d^{10} 4s^1$	30 **Zn** $3d^{10} 4s^2$	31 **Ga** $4s^2 4p^1$	32 **Ge** $4s^2 4p^2$	33 **As** $4s^2 4p^3$	34 **Se** $4s^2 4p^4$	35 **Br** $4s^2 4p^5$	36 **Kr** $4s^2 4p^6$
37 **Rb** $5s^1$	38 **Sr** $5s^2$	39 **Y** $4d^1 5s^2$	40 **Zr** $4d^2 5s^2$	41 **Nb** $4d^4 5s^1$	42 **Mo** $4d^5 5s^1$	43 **Tc** $4d^5 5s^2$	44 **Ru** $4d^7 5s^1$	45 **Rh** $4d^8 5s^1$	46 **Pd** $4d^{10}$	47 **Ag** $4d^{10} 5s^1$	48 **Cd** $4d^{10} 5s^2$	49 **In** $5s^2 5p^1$	50 **Sn** $5s^2 5p^2$	51 **Sb** $5s^2 5p^3$	52 **Te** $5s^2 5p^4$	53 **I** $5s^2 5p^5$	54 **Xe** $5s^2 5p^6$
55 **Cs** $6s^1$	56 **Ba** $6s^2$	57 ***La** $5d^1 6s^2$	72 **Hf** $5d^2 6s^2$	73 **Ta** $5d^3 6s^2$	74 **W** $5d^4 6s^2$	75 **Re** $5d^5 6s^2$	76 **Os** $5d^6 6s^2$	77 **Ir** $5d^7 6s^2$	78 **Pt** $5d^9 6s^1$	79 **Au** $5d^{10} 6s^1$	80 **Hg** $5d^{10} 6s^2$	81 **Tl** $6s^2 6p^1$	82 **Pb** $6s^2 6p^2$	83 **Bi** $6s^2 6p^3$	84 **Po** $6s^2 6p^4$	85 **At** $6s^2 6p^5$	86 **Rn** $6s^2 6p^6$
87 **Fr** $7s^1$	88 **Ra** $7s^2$	89 †**Ac** $6d^1 7s^2$	104 **Rf** $6d^2 7s^2$	105 **Db** $6d^3 7s^2$	106 **Sg** $6d^4 7s^2$	107 **Bh**	108 **Hs**	109 **Mt**	110 **Ds**	111	112	Unknown	114	Unknown	116		

* Lanthanides

58 **Ce** $4f^2 6s^2$	59 **Pr** $4f^3 6s^2$	60 **Nd** $4f^4 6s^2$	61 **Pm** $4f^5 6s^2$	62 **Sm** $4f^6 6s^2$	63 **Eu** $4f^7 6s^2$	64 **Gd** $4f^7 5d^1 6s^2$	65 **Tb** $4f^9 6s^2$	66 **Dy** $4f^{10} 6s^2$	67 **Ho** $4f^{11} 6s^2$	68 **Er** $4f^{12} 6s^2$	69 **Tm** $4f^{13} 6s^2$	70 **Yb** $4f^{14} 6s^2$	71 **Lu** $4f^{14} 5d^1 6s^2$

† Actinides

90 **Th** $6d^2 7s^2$	91 **Pa** $5f^2 6d^1 7s^2$	92 **U** $5f^3 6d^1 7s^2$	93 **Np** $5f^4 6d^1 7s^2$	94 **Pu** $5f^6 7s^2$	95 **Am** $5f^7 7s^2$	96 **Cm** $5f^7 6d^1 7s^2$	97 **Bk** $5f^9 7s^2$	98 **Cf** $5f^{10} 7s^2$	99 **Es** $5f^{11} 7s^2$	100 **Fm** $5f^{12} 7s^2$	101 **Md** $5f^{13} 7s^2$	102 **No** $5f^{14} 7s^2$	103 **Lr** $5f^{14} 6d^1 7s^2$

▲ FIGURE 8.3 **Electron configurations and the periodic table**

Subshells that lie at lower energies than those listed are filled with electrons. For example, the electron configuration of arsenic is $[Ar]3d^{10}4s^2 4p^3$, that of iodine is $[Kr]4d^{10}5s^2 5p^5$, that of lead is $[Xe]4f^{14}5d^{10}6s^2 6p^2$, and that of uranium is $[Rn]5f^3 6d^1 7s^2$.

Atomic Number: The Work of Henry G. J. Moseley

Henry G. J. Moseley (1887–1915) was one of several brilliant students who studied with Ernest Rutherford just before World War I. Moseley's most important work was finding a relationship between the wavelengths of X rays and the elements giving rise to them. X rays are a form of electromagnetic radiation produced when cathode rays strike a metal anode, called a *target* (Figure 8.4).

Moseley reasoned that X rays are emitted when electrons in excited atoms drop down to lower energy levels—the same way as other emission processes. Knowing that the energy levels of the electrons depend primarily on the magnitude of the nuclear charge on an atom, as do differences between energy levels, ΔE_{level}, Moseley hypothesized that the energy and frequency of the emitted X rays $(\Delta E_{\text{level}} = h\nu)$, and thus the wavelength, should also depend on the magnitude of the nuclear charge. He used different metals as the targets in an X-ray tube and determined the wavelengths of the emitted X rays.

Moseley's startling results are seen in the regular pattern in the photographic images produced by the X rays, shown in Figure 8.5. The two lines shown for each metal represent two different X-ray emissions from the target in the X-ray tube. Starting with calcium (Ca), the pairs of lines are displaced progressively to the left, toward shorter wavelengths and higher fre-

quencies. Moseley derived an equation that linked each X-ray frequency to the metal producing it, by assigning to the element a number equal to the positive charge on the atomic nucleus—the *atomic number*. In this way, Moseley accounted for all the atomic numbers from 13 (aluminum) to 79 (gold). This included predicting three new elements at $Z = 43(\text{Tc})$, $Z = 61(\text{Pm})$, and $Z = 75(\text{Re})$. He also demonstrated that there could be no other elements in this portion of the periodic table.

Moseley was only in his mid-twenties when he performed this work. Within a year of its conclusion, he was killed in battle in World War I at Gallipoli, Turkey.

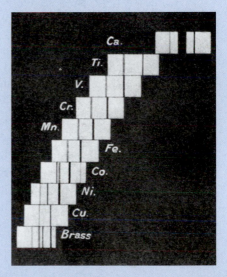

▲ **FIGURE 8.5   Moseley's X-ray spectra of several metals**
Each pure metal displays a pair of lines. The line on the left is the shorter-wavelength line, and the one on the right is the longer-wavelength line. The spectrum of cobalt has four lines, but only two of them are produced by cobalt; one of the other two corresponds to iron as an impurity, and the other to nickel. Brass is an alloy of copper and zinc. Two of the four lines in its spectrum are identical to those in the copper spectrum shown above it. The other two lines are produced by zinc.

Electricity source

Cathode rays

Target

Anode (+)

Cathode (−)

X rays

▲ **FIGURE 8.4   The production of X rays**

The correlation between group number and number of valence-shell electrons does not extend to the B-group elements, the transition elements. Atoms of the transition elements typically have only two valence-shell electrons (ns^2), some have only one (ns^1), and palladium has none.*

Another fundamental relationship we can infer from Figure 8.3 and the periodic table inside the front cover is that

The period number is the same as the principal quantum number, n, of the electrons in the valence shell of an element.

Interactive Periodic Table activity

---

*Palladium $(Z = 46)$ has no electrons in its $5s$ subshell. Thus, even though it is in the fifth period, only four of its principal shells are occupied.

All elements in the fourth period, for example, have one or more electrons with $n = 4$ and no electrons with a higher value of n. This period begins with potassium (K), which has the electron configuration $[Ar]4s^1$, and ends with krypton: Kr ($[Ar]3d^{10}4s^24p^6$). The next subshell to fill after $4p$ is $5s$, and so the element rubidium (Rb), with the electron configuration $[Kr]5s^1$, begins the fifth period.

Each period of the periodic table begins with a group 1A element and ends with a group 8A element. The periods differ in length because of the different numbers of subshells that must fill to get from the valence-shell electron configuration ns^1 to the valence-shell configuration ns^2np^6. In the two-member first period, we go directly from $1s^1$ to $1s^2$ (there is no p subshell in the first principal shell). In the eight-member second period, the $2s$ and $2p$ subshells fill; in the eight-member third period, the $3s$ and $3p$ subshells fill. The fourth and fifth periods have 18 members each. As we progress to higher atomic numbers, electrons fill the $4s$, $3d$, and $4p$ subshells in the fourth period, and the $5s$, $4d$, and $5p$ subshells in the fifth period. Finally, the 32-member sixth and seventh periods require the filling of the $6s$, $4f$, $5d$, and $6p$, and the $7s$, $5f$, $6d$, and $7p$ subshells, respectively.

Using the Periodic Table to Write Electron Configurations

We do not have to memorize an order of orbital filling (Figure 8.2) or use an orbital energy diagram (Figure 8.1) to write probable electron configurations. We can deduce the general form of the electron configurations given in Figure 8.3 (but without some of the details) directly from the periodic table. All we need to know is which subshells fill in different regions of the periodic table. To help make such deductions, we refer to the four blocks of elements shown in Figure 8.6.

s-block: The ns subshell (the s subshell of the valence shell) fills by the aufbau process. These are main-group elements.

p-block: The np subshell (the p subshell of the valence shell) fills. These are also main-group elements.

d-block: The $(n - 1)d$ subshell (the d subshell of the shell below the n shell) fills. These are transition elements found in the main body of the periodic table.

f-block: The $(n - 2)f$ subshell (the f subshell of the second shell below the n shell) fills. To keep the periodic table at a maximum width of 18 members, these elements are placed below the main body of the table. The $4f$ subshell fills in the **lanthanide** series, and the $5f$ subshell fills in the **actinide** series. Because they fall within series of d-block elements, the f-block elements are sometimes called the *inner-transition* elements. The lanthanide series follows lanthanum ($Z = 57$) in the periodic table, and the actinide series follows actinium ($Z = 89$).

Order of Filling activity

▶ **FIGURE 8.6** **The periodic table and the order of filling of subshells**
Read through this periodic table, starting at the upper left, and you will discover the same order of filling of subshells as shown in Figure 8.2. Note that helium ($Z = 2$) is an s-block element, but it is grouped with the p-block elements because we place it in group 8A with the other noble-gas elements that it so strongly resembles.

QUESTION: What subshell fills after the 5d, and what subshell should fill after the 6d?

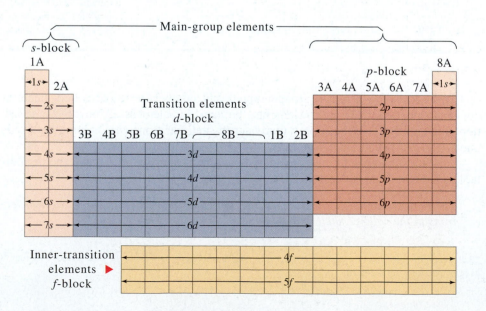

Example 8.2

Give the complete ground-state electron configuration of a strontium atom **(a)** in the *spdf* notation and **(b)** in the noble-gas-core abbreviated notation.

STRATEGY

The four-step general strategy for the aufbau process outlined on page 304 is applicable here, but instead we can use Figure 8.6 to quickly get through the electron configurations for all the elements up to the noble gas that precedes strontium in the periodic table. Then our emphasis shifts to the portion of the electron configuration specific to strontium's period and group.

SOLUTION

(a) We proceed through Figure 8.6 until we reach Sr ($Z = 38$). The $1s$ subshell fills as we go through the first period. As we continue through the second period (Li through Ne), the $2s$ and $2p$ subshells fill. Moving through the third period, the $3s$ and $3p$ subshells fill. In the fourth period, the $4s$, $3d$, and $4p$ subshells fill in succession. Strontium's location in group 2A of the fifth period indicates two electrons in the $5s$ orbital. We can write the electron configuration either (i) in the order of filling or (ii) by grouping the subshells into principal shells:

(i) $1s^2 2s^2 2p^6 3s^2 3p^6 4s^2 3d^{10} 4p^6 5s^2$ (ii) $1s^2 2s^2 2p^6 3s^2 3p^6 3d^{10} 4s^2 4p^6 5s^2$

(b) In the noble-gas-core abbreviated notation of the electron configuration, we replace $1s^2 2s^2 2p^6 3s^2 3p^6 3d^{10} 4s^2 4p^6$ with [Kr], which gives us [Kr]$5s^2$.

EXERCISE 8.2A

Refer to Figure 8.6, and give the ground-state electron configuration, in the *spdf* notation and in the noble-gas-core abbreviated *spdf* notation, for **(a)** Mo and **(b)** Bi.

EXERCISE 8.2B

Refer only to the periodic table on the inside front cover, and give the ground-state electron configuration **(a)** for tin, using the noble-gas-core abbreviated *spdf* notation, and **(b)** for zirconium, using an orbital diagram.

Valence Electrons and Core Electrons

Valence electrons are those with the highest principal quantum number, n, that is, the ones in the outermost occupied principal shell of an atom.* Electrons in inner shells are called **core electrons;** their principal quantum numbers are less than n. Thus, in a calcium atom, which has the electron configuration [Ar]$4s^2$, the $4s$ electrons are valence electrons and those in the [Ar] configuration are core electrons. Bromine, with the electron configuration [Ar]$3d^{10}4s^2 4p^5$, has seven valence electrons; the core electrons are in the configuration [Ar]$3d^{10}$.

Electron Configurations of Ions

To obtain the electron configuration of an anion by the aufbau process, we introduce additional electrons to the valence shell of the neutral nonmetal atom, *without* adding protons or neutrons to the nucleus. The number of electrons added is usually the number needed to fill the valence shell. Thus, a nonmetal atom usually gains one or two electrons (sometimes three) and thereby attains the electron configuration of a noble gas atom. Some examples are

$$Br\ ([Ar]3d^{10}4s^2 4p^5) + e^- \longrightarrow Br^-\ ([Ar]3d^{10}4s^2 4p^6\ \ or\ \ [Kr])$$

$$S\ ([Ne]3s^2 3p^4) + 2\ e^- \longrightarrow S^{2-}\ ([Ne]3s^2 3p^6\ \ or\ \ [Ar])$$

$$N\ ([He]2s^2 2p^3) + 3\ e^- \longrightarrow N^{3-}\ ([He]2s^2 2p^6\ \ or\ \ [Ne])$$

---

*Some chemists view valence electrons as those involved in chemical reactions, and core electrons as those that are not involved. From this perspective, some inner-shell electrons of the transition elements might be considered valence electrons. However, we will limit our use of the term valence electrons to those in the outermost occupied principal shell of an atom.

Disputed Issues Concerning the Periodic Table

The periodic table has been around for well over a century, and many variations of the table have been proposed. Yet, in all that time, no single version of the table has gained universal acceptance.

For many years, a particular state of confusion has existed because of the different ways the letters A and B are used. In the United States, A has been used to designate the main-group elements, and B the transition elements. In Europe, the groups to the left of Fe/Ru/Os were typically labeled as A groups, and those to the right of Ni/Pd/Pt as B groups.

In order to eliminate confusion over the use of A and B, the International Union of Pure and Applied Chemistry (IUPAC) has recommended that the groups simply be numbered from 1 to 18. Still, the A/B system that we use in this text is helpful in that the main-group numbers are equal to the numbers of valence-shell electrons in atoms of the main-group elements.

Placing hydrogen properly is difficult. Although hydrogen is a member of group 1A, it is a gaseous nonmetal and not at all like an alkali metal. Most periodic tables put hydrogen in group 1A because it has the electron configuration $1s^1$. We could just as legitimately place it in group 7A, however, because, like the halogen atoms, it is one electron short of a noble-gas electron configuration. A few periodic tables place it in both groups, and some place it alone at the top of the table near the center.

Although chemists may disagree over just what form of the periodic table is most correct, all agree that the table is tremendously useful.

To obtain the electron configuration of a main-group metal cation, we start with the electron configuration of the element and remove electrons from the valence shell, starting with electrons with the greatest l value, until the appropriate cationic charge is reached. Thus, for the cations of the main-group metals of the third period, we obtain the following electron configurations, each corresponding to the noble gas neon:

$$\text{Na ([Ne]}3s^1) \longrightarrow \text{Na}^+(\text{[Ne]}) + e^-$$

$$\text{Mg ([Ne]}3s^2) \longrightarrow \text{Mg}^{2+}(\text{[Ne]}) + 2\,e^-$$

$$\text{Al ([Ne]}3s^2 3p^1) \longrightarrow \text{Al}^{3+}(\text{[Ne]}) + 3\,e^-$$

The electron configurations of cations of the p-block metals of the fourth period and beyond do not correspond to that of a noble gas. For example, we have

$$\text{Ga ([Ar]}3d^{10}4s^2 4p^1) \longrightarrow \text{Ga}^{3+}(\text{[Ar]}3d^{10}) + 3\,e^-$$

$$\text{Sn ([Kr]}4d^{10}5s^2 5p^2) \longrightarrow \text{Sn}^{2+}(\text{[Kr]}4d^{10}5s^2) + 2\,e^-$$

For cations formed from transition metal atoms, the removal of electrons may be limited to the electronic shell of highest principal quantum number, n, as in Fe^{2+}, with the loss of the $4s^2$ electrons,

$$\text{Fe([Ar]}3d^6 4s^2) \longrightarrow \text{Fe}^{2+}(\text{[Ar]}3d^6) + 2\,e^-$$

or it may continue into the $(n-1)$ shell with the loss of one or more electrons having the greatest l value. Thus, in the formation of Fe^{3+} from Fe, one $3d$ electron is lost following the $4s^2$:

$$\text{Fe([Ar]}3d^6 4s^2) \longrightarrow \text{Fe}^{3+}(\text{[Ar]}3d^5) + 3\,e^-$$

Notice that neither Fe^{2+} nor Fe^{3+} has a noble-gas electron configuration.

Some possibilities for cation electron configurations are summarized in Table 8.3.

Electron Configuration of Metal Ions activity

Table 8.3 Electron Configurations of Some Metal Ions							
Noble Gas			**Pseudo–Noble Gas**[a]		**18 + 2**[b]	**Various**	
Li^+	Be^{2+}	Al^{3+}	Cu^+	Zn^{2+}	In^+	Cr^{2+}:	$[Ar]3d^4$
Na^+	Mg^{2+}		Ag^+	Cd^{2+}	Tl^+	Cr^{3+}:	$[Ar]3d^3$
K^+	Ca^{2+}		Au^+	Hg^{2+}	Sn^{2+}	Mn^{2+}:	$[Ar]3d^5$
Rb^+	Sr^{2+}				Pb^{2+}	Mn^{3+}:	$[Ar]3d^4$
Cs^+	Ba^{2+}				Sb^{3+}	Fe^{2+}:	$[Ar]3d^6$
					Bi^{3+}	Fe^{3+}:	$[Ar]3d^5$
						Co^{2+}:	$[Ar]3d^7$
						Co^{3+}:	$[Ar]3d^6$
						Ni^{2+}:	$[Ar]3d^8$

[a] In the pseudo–noble gas configuration, all valence electrons are lost and the remaining $(n - 1)$ shell has 18 electrons in the configuration $(n - 1)s^2(n - 1)p^6(n - 1)d^{10}$.

[b] In the 18 + 2 configuration, $(n - 1)s^2(n - 1)p^6(n - 1)d^{10}ns^2$, two valence electrons remain.

Example 8.3

Write the electron configuration of the ion Co^{3+} in a noble-gas-core abbreviated *spdf* notation.

STRATEGY

First, we can obtain the electron configuration of cobalt in the manner described in Example 8.2. Then, we can remove three electrons in the manner just described for Fe^{3+}. This will leave us with the electron configuration of Co^{3+}.

SOLUTION

Using Figure 8.6 as an aid, we see that in the cobalt atom $(Z = 27)$ the $1s$, $2s$, $2p$, $3s$, $3p$, and $4s$ subshells fill progressively in the aufbau process. This accounts for 20 electrons (18 in the argon core and an additional two $4s$ electrons). The remaining seven electrons partially fill the $3d$ subshell, giving

$$Co: [Ar]3d^7 4s^2$$

To form Co^{3+} from Co, we must remove three electrons, the $4s^2$ pair and one $3d$ electron. The noble-gas-core abbreviated *spdf* notation of Co^{3+} is $[Ar]3d^6$.

EXERCISE 8.3A

Write electron configurations in the noble-gas-core abbreviated *spdf* notation for the ions Se^{2-} and Pb^{2+}.

EXERCISE 8.3B

Write noble-gas-core abbreviated orbital diagrams for the ions I^- and Cr^{3+}.

8.6 Magnetic Properties: Paired and Unpaired Electrons

When considering electron spin in Chapter 7, we noted that each electron in an atom creates a magnetic field—each electron acts like a tiny magnet. The magnetic fields of a *pair* of electrons with opposite spins cancel each other, and the pair shows no resultant magnetism.

Most substances do not exhibit magnetic properties until they are placed in a magnetic field. Then we find two possible types of behavior:

- If all the electrons in the atoms, ions, or molecules of a substance are paired, the substance is weakly repelled by a magnetic field. This weak repulsion associated with paired electrons is called **diamagnetism.**

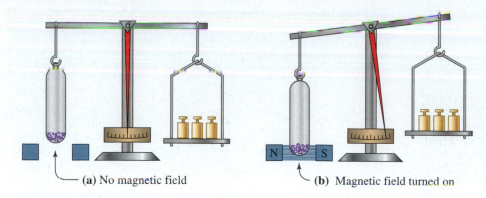

▶ **FIGURE 8.7   Paramagnetism illustrated**

(a) A sample (left side of balance) is weighed in the absence of a magnetic field. (b) When the field is turned on, the balanced condition is upset. The sample appears to gain weight because it is now subjected to two attractive forces—the force of gravity and the force exerted by the magnetic field.

(a) No magnetic field   **(b)** Magnetic field turned on

- Atoms, ions, or molecules having at least one unpaired electron are attracted by a magnetic field. This attraction associated with unpaired electrons is called **paramagnetism.** The paramagnetism of unpaired electrons is a much stronger effect than the weak diamagnetism of paired electrons.

The exceptionally strong attractions of a magnetic field for iron and a few other substances is a third type of magnetic behavior, called *ferromagnetism,* which we will consider in Chapter 22.

The magnetic properties of a substance can be determined by weighing the substance in the absence and in the presence of a magnetic field, as indicated in Figure 8.7. Such measurements would show, for example, a stronger paramagnetic effect for a sample containing Fe^{3+} than one containing Fe^{2+}. The measurements indicate that the number of unpaired electrons in Fe^{3+} is greater than in Fe^{2+}, thereby providing experimental verification for the orbital diagrams of these two ions:

$$Fe^{2+} \; [Ar] \quad \boxed{\uparrow\downarrow \;|\; \uparrow \;|\; \uparrow \;|\; \uparrow \;|\; \uparrow} \quad \text{and} \quad Fe^{3+} \; [Ar] \quad \boxed{\uparrow \;|\; \uparrow \;|\; \uparrow \;|\; \uparrow \;|\; \uparrow}$$
$$\qquad\qquad\qquad 3d \qquad\qquad\qquad\qquad\qquad\qquad 3d$$

Example 8.4

A sample of chlorine gas is found to be diamagnetic. Can this gaseous sample be composed of individual Cl atoms?

STRATEGY

We can deduce the electron configuration of Cl from its position in the periodic table and represent it by an orbital diagram.

SOLUTION

$$(Z = 17) \; Cl \quad \boxed{\uparrow\downarrow} \;\; \boxed{\uparrow\downarrow} \;\; \boxed{\uparrow\downarrow \;|\; \uparrow\downarrow \;|\; \uparrow\downarrow} \;\; \boxed{\uparrow\downarrow} \;\; \boxed{\uparrow\downarrow \;|\; \uparrow\downarrow \;|\; \uparrow}$$
$$\qquad\qquad\quad 1s \quad\; 2s \qquad\quad 2p \qquad\quad 3s \qquad\quad 3p$$

The diagram shows one unpaired electron per atom. We would therefore expect atomic chlorine (Cl) to be paramagnetic. Because the gas is diamagnetic, it cannot be composed of individual Cl atoms. (In fact, chlorine gas consists of diatomic molecules, Cl_2, in which all electrons are paired.)

EXERCISE 8.4A

How many unpaired electrons would you expect to find in each of the following?

(a) a nickel atom   **(c)** a sulfide ion   **(e)** a cadmium ion

(b) a tungsten atom   **(d)** a calcium ion   **(f)** a neodymium (II) ion

EXERCISE 8.4B

Which of the following would you expect to exhibit paramagnetism? Explain.

(a) a K atom   **(d)** a N atom   **(g)** a Cu^{2+} ion

(b) a Hg atom   **(e)** a F^- ion

(c) a Ba^{2+} ion   **(f)** a Ti^{2+} ion

8.7 Periodic Atomic Properties of the Elements

In developing his periodic table, Mendeleev emphasized the similarities in chemical properties for groups of elements, such as those elements whose oxides and hydrides had similar formulas. Lothar Meyer, a contemporary of Mendeleev's, developed a classification scheme based primarily on similarities in physical properties, such as density. The underlying principle in both cases is the **periodic law,** which in its modern form states that certain sets of physical and chemical properties recur at regular intervals (that is, periodically) when the elements are arranged according to increasing atomic number.

Some physical properties, such as thermal and electrical conductivity, hardness, and melting point, are associated only with *bulk* matter, that is, aggregations of atoms large enough to see and measure at the macroscopic level. Other properties, called *atomic properties,* are associated with individual atoms. In this section, we examine four atomic properties: atomic radii, ionic radii, ionization energies, and electron affinities.

Atomic Radii

We cannot measure the exact size of an isolated atom, because its outermost electrons have a remote chance of being found quite far from the nucleus. However, we can measure the distance between the nuclei of two atoms, and we can derive a property called the **atomic radius** from this distance. Internuclear distances depend on the particular environment in which the atoms are found, and therefore there can be more than one value for the atomic radius of an element. Let's consider two specific cases.

- The **covalent radius** of an atom is one-half the distance between the nuclei of two identical atoms joined into a molecule. Consider, for example, the I_2 molecule (Figure 8.8). The distance between the two iodine nuclei is 266 pm, and the covalent radius of iodine is one-half this distance, or 133 pm.*

- The **metallic radius** of an atom is half the distance between the nuclei of adjacent atoms in a solid metal.

In this text, whenever we use the simple expression "atomic radius," you can assume we mean *covalent* radius for nonmetals and *metallic* radius for metals.

Figure 8.9 shows graphically that atomic radius is an excellent example of a periodic atomic property of the elements. Each red line plots the atomic radii for elements within a single period. Notice that the first member of each period, a group 1A metal,

Atomic radii are sometimes expressed in the non-SI unit *angstrom* (Å); $1 \text{ Å} = 10^{-10} \text{ m} = 100 \text{ pm}$. The covalent radius of iodine shown in Figure 8.8 is 1.33 Å.

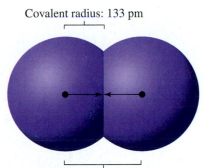

Covalent radius: 133 pm

Internuclear distance: 266 pm

▲ **FIGURE 8.8 Atomic radius illustrated by the covalent radius of iodine**

The covalent radius is half the distance between the nuclei of the two I atoms in the I_2 molecule.

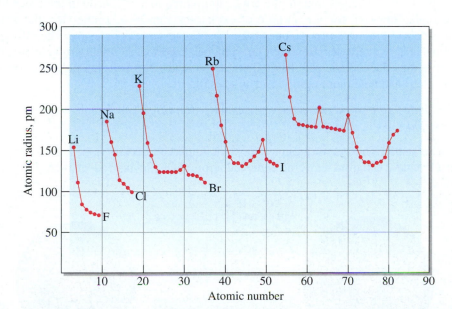

◄ **FIGURE 8.9 Atomic radii of the elements**

The values shown here, in picometers, are metallic radii for metals and covalent radii for nonmetals. Data are not included for the noble gases because it is difficult to assess their covalent radii (few noble-gas compounds are known). Explanations have been offered for the small peaks in the middle of some periods and for other irregularities, but they are beyond the scope of this book.

QUESTION: What accounts for the jump in atomic radius in moving from fluorine to sodium?

---

*The value of a covalent radius depends on whether the bond between the two atoms is single, double, or triple. The covalent radii used in this chapter are single-bond covalent radii.

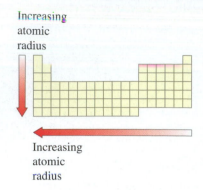

Increasing
atomic
radius

Increasing
atomic
radius

**Periodic Trends: Atomic Radii
animation**

has the largest atomic radius for the period (Li for period 2, Na for period 3, and so on). The atomic radii then generally decline through the period, reaching a low value with the group 7A nonmetal at the end of the period. To explain these trends, let's think of an atomic radius as roughly equal to the distance from the nucleus to the electrons in the valence shell. Any factor that causes this distance to increase also increases the atomic radius.

Within a given group of the periodic table, each succeeding member has one more occupied principal shell than its immediate predecessor. Thus, the sodium atom has electrons in the principal shells $n = 1, 2,$ and 3, whereas the lithium atom has electrons only in $n = 1$ and 2. The electrons in the valence shell are farther from the nucleus as n increases, and therefore

Atomic radii increase from top to bottom within a group of the periodic table.

The trend in atomic radii within a period may seem a bit unexpected, as you might assume that size increases with increasing numbers of electrons. To explain the left-to-right trend, we find a new concept helpful: The **effective nuclear charge** (Z_{eff}) acting on an electron in an atom is the actual nuclear charge minus the *screening effect* of other electrons in the atom.

Figure 8.10 illustrates the concept of effective nuclear charge. In the sodium atom, if the $3s$ valence electron were at all times outside the region in which the 10 core electrons ($1s^2 2s^2 2p^6$) are found, the core electrons would perfectly screen the $3s$ electron from the positively charged nucleus. Thus, the $3s$ electron would experience an attraction to a net positive charge of only $+11 - 10 = +1$. The magnesium atom also has 10 core electrons but has 12 protons. There are two $3s$ electrons outside the neon core and a net positive charge of $+12 - 10 = +2$ acting on each $3s$ electron. Proceeding in this fashion, we find that the net positive charge acting on the valence electrons progressively increases across the third period.

Evaluation of an effective nuclear charge also requires the following ideas:

1. Valence electrons can penetrate inner shells and approach the nucleus. On page 300, we described how differences in the penetrating abilities of *s, p, d,* and *f* electrons cause a splitting between subshell energies.

2. Core electrons are not all equally effective in shielding valence electrons.

3. To some extent, valence electrons shield one another from the nuclear charge.

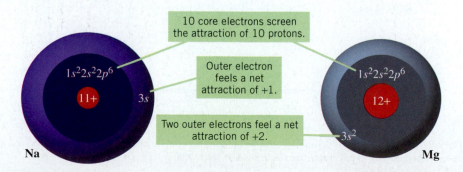

By taking these three factors into account, we can get much better estimates of effective nuclear charges than in the crude approximations we made in the preceding paragraph. However, for the purpose of explaining trends in atomic size, the crude estimates and the more precise ones both lead to the same conclusion: The effective nuclear charge increases from left to right in a period of main-group elements. Because the effective nuclear charge increases, valence electrons are pulled in toward the nucleus and held more tightly, and as a result,

Atomic radii of the main-group elements decrease from left to right across a period.

The restriction to main-group elements is important because in a series of transition elements, electrons enter an *inner* electron shell. So instead of increasing, the

Effect of Shielding and Nuclear Charge activity

▶ **FIGURE 8.10** **A simplified view of shielding and effective nuclear charge**

In the sodium atom ($Z = 11$), the ten core electrons shield the $3s$ valence electron from the charge of ten of the protons. The net charge experienced by the valence electron is therefore $11 - 10 = +1$. The magnesium atom ($Z = 12$) also has ten core electrons but has *twelve* protons. The net attraction felt by its valence electrons is therefore $12 - 10 = +2$. The valence electrons in magnesium experience a greater effective nuclear charge than in sodium.

QUESTION: In this simplified representation, what should be the effective nuclear charge on an aluminum atom?

10 core electrons screen the attraction of 10 protons.

Outer electron feels a net attraction of +1.

Two outer electrons feel a net attraction of +2.

$1s^2 2s^2 2p^6$

$11+$ $3s$

Na

$1s^2 2s^2 2p^6$

$12+$

$3s^2$

Mg

effective nuclear charge therefore remains essentially constant as electrons are added across a period. For example, consider the effective nuclear charges of iron, cobalt, and nickel, by the crude method just outlined (core electrons are shown in green):

$$\text{Fe: } [Ar]3d^64s^2 \qquad \text{Co: } [Ar]3d^74s^2 \qquad \text{Ni: } [Ar]3d^84s^2$$

$$Z_{eff} \approx 26 - 24 \approx 2 \qquad Z_{eff} \approx 27 - 25 \approx 2 \qquad Z_{eff} \approx 28 - 26 \approx 2$$

Because the effective nuclear charges are about the same, we conclude that the radii should also be about the same. The actual values for Fe, Co, and Ni are 124, 124, and 125 pm, respectively. The difference in trends in atomic radii in the main-group and transition element regions of a period of elements is clearly illustrated in Figure 8.9, where the large differences in main-group element radii (indicated by the steep lines following K, Rb, and Cs) contrast the smaller differences between the transition element radii.

Effective Nuclear Charge animation

Example 8.5

With reference only to a periodic table, arrange each set of elements in order of increasing atomic radius:

   (a) Mg, S, Si   **(b)** As, N, P   **(c)** As, Sb, Se

SOLUTION

(a) All three are main-group elements in the same period (third). Atomic radii within a period decrease from left to right. The order of increasing radius is therefore

$$\text{S (smallest)} < \text{Si} < \text{Mg (largest)}$$

(b) All three elements are in the same group (5A), and atomic radii increase from top to bottom. The order of increasing radius is therefore

$$\text{N (smallest)} < \text{P} < \text{As (largest)}$$

(c) Of this set, As and Se are in the same period (fourth); As is to the left and therefore larger than Se. Because Sb is below As in the same group (5A), it is larger than As. The overall order is

$$\text{Se (smallest)} < \text{As} < \text{Sb (largest)}$$

EXERCISE 8.5A

With reference only to a periodic table, arrange each set of elements in order of increasing atomic radius.

   (a) Be, F, N   **(b)** Ba, Be, Ca   **(c)** Cl, F, S   **(d)** Ca, K, Mg

EXERCISE 8.5B

With reference only to a periodic table, arrange the following elements in the expected order of increasing atomic radius. Comment on any ambiguous rankings.

$$\text{Sn, Ge, Ca, P, Pb, Cs.}$$

Ionic Radii

Like atomic radii, ionic radii are based on an internuclear distance—in this case, the distance between the nuclei of two ions. Figure 8.11 shows a Mg^{2+} and an O^{2-} ion in contact. The **ionic radius** of each ion is the portion of the distance between the nuclei occupied by that ion. Ionic radii are determined by crystal structure studies of the type we describe in Chapter 11. When atoms of metals react, they often lose all their valence-shell electrons. With one fewer electronic shell, a metal ion is smaller than the atom from which it comes. Also, because the nuclear charge is now greater than the number of electrons, the nucleus attracts the remaining electrons more strongly and holds them more closely in a cation than in the corresponding atom. We conclude therefore that

Cations are <u>smaller</u> than the atoms from which they are formed.

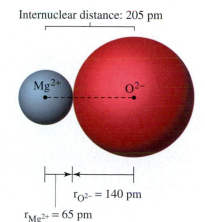

▲ **FIGURE 8.11   The ionic radii of Mg^{2+} and O^{2-}**

The distance between the centers of the two ions (205 pm) is apportioned between the Mg^{2+} (65 pm) and O^{2-} (140 pm) ions.

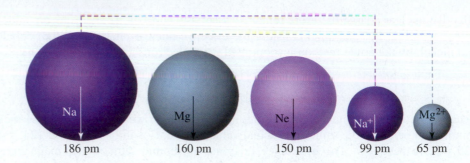

Metallic radii are shown for Na and Mg, and ionic radii for Na⁺ and Mg²⁺. The radius for Ne is for a nonbonded atom.

QUESTION: Where should an aluminum ion fall in this comparison?

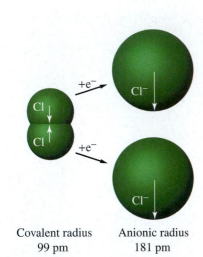

Covalent radius
99 pm

Anionic radius
181 pm

▲ **FIGURE 8.13**   **A comparison of covalent atomic radii and anionic radii**

The two chlorine atoms in a Cl₂ molecule gain one electron each to form two chloride ions (2 Cl⁻), which are larger than the atoms that formed them. The atomic and ionic radii are marked by the white arrows.

Gain and Loss of Electrons animation

Figure 8.12 compares the radii of five species: a sodium atom (Na), a magnesium atom (Mg), a neon atom (Ne), a sodium ion (Na^+), and a magnesium ion (Mg^{2+}). The Mg atom is smaller than the Na atom, and the ions are smaller than the corresponding atoms. The species Ne, Na^+, and Mg^{2+} all have the same number (10) of electrons: They are **isoelectronic.** They also have identical electron configurations ($1s^2 2s^2 2p^6$). Neon has a nuclear charge of +10. Because the sodium ion has a nuclear charge of +11, Na^+ is *smaller* than Ne. Because its nuclear charge is +12, Mg^{2+} is smaller still.

When a nonmetal atom gains an electron to form an anion, the positive nuclear charge remains constant, but there are more electrons to repel one another. Increased repulsions cause the electrons to spread out more, and the size increases. The formation of two Cl^- ions from a Cl_2 molecule and the large increase from the covalent atomic radius in Cl_2 to the anionic radius in Cl^- are suggested in Figure 8.13. In general, we can say that

- *Anions are <u>larger</u> than the atoms from which they are formed.*
- *For a series of isoelectronic species (ions or atoms) that have the same electron configuration, the greater the nuclear charge, the smaller the species.*

Figure 8.14 summarizes much of our discussion of trends in atomic and ionic radii:

1. Atomic radii vary with position in a group and period of the periodic table.
2. The radii of cations and anions are related to the radii of the atoms from which they are formed—cations smaller and anions larger.
3. The magnitude of the ionic charge affects the radii of cations and anions.

You may need to refer to this figure from time to time.

Example 8.6

Refer to a periodic table but not to Figure 8.14, and arrange the following species in the expected order of increasing radius: Ca^{2+}, Fe^{3+}, K^+, S^{2-}, Se^{2-}.

STRATEGY AND SOLUTION

Three of these ions (Ca^{2+}, K^+, and S^{2-}) are isoelectronic: They all have the same electron configuration ($1s^2 2s^2 2p^6 3s^2 3p^6$). According to our generalization, their comparative radii should be $Ca^{2+} < K^+ < S^{2-}$. Because Fe is in the same period as Ca and farther to the right, we expect the Fe atom to be smaller than the Ca atom. Also, we expect the loss of three electrons (two 4s and one 3d) by an Fe atom to produce a smaller ion than the loss of two 4s electrons by a Ca atom. The ion Fe^{3+} should therefore be smaller than Ca^{2+}.

To get a feel for the size of the Se^{2-} ion, note that it is in the same group as S^{2-}. Because Se^{2-} has the same charge but more occupied shells than S^{2-}, we expect Se^{2-} to be larger than S^{2-}. We conclude that the overall order is $Fe^{3+} < Ca^{2+} < K^+ < S^{2-} < Se^{2-}$.

EXERCISE 8.6A

Refer to a periodic table but not to Figure 8.14, and arrange the following species in the expected order of increasing radius: Br^-, Rb^+, Se^{2-}, Sr^{2+}, Y^{3+}.

EXERCISE 8.6B

Without reference to Figure 8.14, arrange the following species in the expected order of increasing radius: Ca^{2+}, Cr^{2+}, Cs^+, Cr^{3+}, K^+.

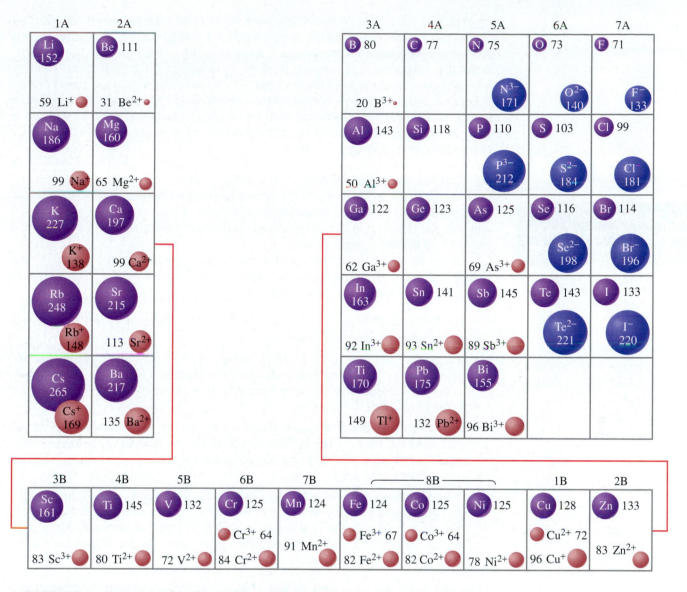

▲ **FIGURE 8.14   Some representative atomic and ionic radii**
The values, in picometers, are metallic radii for metals, single covalent radii for nonmetals, and ionic radii for ions.

Ionization Energy

When they undergo chemical reactions, metal atoms tend to lose valence electrons. However, isolated atoms do not eject electrons spontaneously. Work must be done to remove an electron from an atom, and the amount of work depends on the size of the atom.

Ionization Energy animation

Ionization energy is the energy required to remove an electron from a ground-state atom (or ion) in the gaseous state. The quantity of energy is usually expressed in terms of 1 mol of atoms. Atoms with more than one electron can ionize in successive steps. Consider the boron atom, which has five electrons—two in an inner core $(1s^2)$ and three valence electrons $(2s^2 2p^1)$. The five steps and their successive ionization energies, I_1 through I_5, are

$$B(g) \longrightarrow B^+(g) + e^- \qquad I_1 = 801 \text{ kJ/mol}$$
$$B^+(g) \longrightarrow B^{2+}(g) + e^- \qquad I_2 = 2,427 \text{ kJ/mol}$$
$$B^{2+}(g) \longrightarrow B^{3+}(g) + e^- \qquad I_3 = 3,660 \text{ kJ/mol}$$
$$B^{3+}(g) \longrightarrow B^{4+}(g) + e^- \qquad I_4 = 25,025 \text{ kJ/mol}$$
$$B^{4+}(g) \longrightarrow B^{5+}(g) + e^- \qquad I_5 = 32,822 \text{ kJ/mol}$$

Periodic Trends: Ionization
Energies animation

Table 8.4 Ionization Energies of Group 1A and Group 2A Elements, kJ/mol

	1A	2A
	Li	Be
I_1	520	900
I_2	7298	1757
	Na	Mg
I_1	496	738
I_2	4562	1451
	K	Ca
I_1	419	590
I_2	3051	1145
	Rb	Sr
I_1	403	550
I_2	2633	1064
	Cs	Ba
I_1	376	503
I_2	2230	965

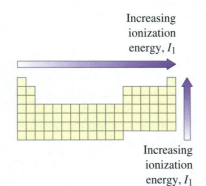

Increasing ionization energy, I_1

Increasing ionization energy, I_1

The first electron to be removed is from the highest-energy subshell ($2p$). It is by far the easiest to remove, and the energy required is called the *first* ionization energy, I_1. The *second* ionization energy, I_2, is more than three times greater than the first. Why? The first electron is removed from the $2p$ orbital of a neutral B atom. The second electron is removed from the $2s$ orbital of a B^+ ion. The second ionization energy is larger than the first, partly because the $2s$ orbital is at a lower energy than the $2p$ but mostly because the second electron must be stripped from a positive ion to which it is strongly attracted. Removal of the third electron must occur from a more highly charged ion, B^{2+}, making I_3 larger still.

Compared to the first three, the fourth and fifth ionization energies, I_4 and I_5, of boron are extremely large. Note that the first three electrons are *valence* electrons and the last two are *core* electrons. For the main-group elements, removing a core electron takes much more work than removing an additional valence electron. This is consistent with the observation that only valence electrons are associated with the chemical reactivity of the main-group elements.

Tables 8.4 and 8.5 list some ionization energies for main-group elements, and Figure 8.15 is a graph of first ionization energy versus atomic number. These sources are the basis of some useful generalizations:

- The first ionization energy of an atom is its lowest. Compare I_1 and I_2 values for the members of either group in Table 8.4.

- A large increase in ionization energy occurs between the removal of the last valence electron and the removal of the first core electron. Compare I_1 and I_2 values in Table 8.4. Notice that the jump from I_1 to I_2 is not nearly as large for group 2A as for group 1A.

- Ionization energies decrease down a group, as you can see by comparing, for instance, all the 1A I_1 values in Table 8.4 or all the 2A I_2 values. The trend is also apparent in Figure 8.15; see how the peaks for the noble gases decrease with increasing atomic number.

- Generally speaking, ionization energies increase from left to right through a period. For example, compare the I_1 values for a period across Table 8.5. In Figure 8.15, this trend is seen in the steady rise in I_1 (with some notable exceptions) from the low value for the group 1A element of any period to the high value for the group 8A element of the same period.

Other factors contribute to these patterns, but we can explain general trends most easily in terms of atomic size. The greater the distance between the atomic nucleus and the electron to be removed, the less tightly that electron is held to the nucleus and the more readily ionization occurs. As we have already seen, the general trends in atomic radii are an increase down a group and a decrease across a period.

We can explain the irregularity in I_1 values as we move from group 2A to group 3A in Table 8.5 as follows. It is easier to remove a $2p$ electron from a boron atom than it is to remove a $2s$ electron from a beryllium atom because the $2p$ electron is at a higher energy than the $2s$ electron. As a result, I_1 is smaller for B (801 kJ/mol) than for Be (900 kJ/mol).

Table 8.5 Ionization Energies of the Second- and Third-Period Elements, kJ/mol

	1A	2A	3A	4A	5A	6A	7A	8A
	Li	Be	B	C	N	O	F	Ne
I_1	520	900	801	1086	1402	1314	1681	2081
I_2	7298	1757	2427	2352	2856	3388	3374	3952
	Na	Mg	Al	Si	P	S	Cl	Ar
I_1	496	738	578	787	1012	999	1251	1521
I_2	4562	1461	1817	1577	1904	2251	2298	2666

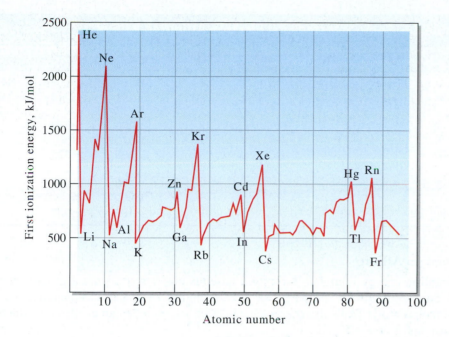

◀ **FIGURE 8.15** **First ionization energy as a function of atomic number**

QUESTION: How would a plot of second ionization energy versus atomic number differ from this figure?

To explain the irregularity between groups 5A and 6A in Table 8.5, we can consider repulsions between electrons. This repulsive tendency makes it easier to remove one of the *paired* electrons in a filled oxygen 2p orbital than to remove an *unpaired* electron from a nitrogen half-filled 2p orbital. The I_1 values are 1314 kJ/mol for O and 1402 kJ/mol for N.

Another way to account for the fact that I_1 for N exceeds I_1 for O is the special stability of a half-filled subshell (that is, $2p^3$). This is similar to the half-filled 3d subshell ($3d^5$) that accounts for the "anomalous" electron configuration of chromium (page 307).

Example 8.7

Without reference to Figure 8.15, arrange each set of elements in the expected order of increasing first ionization energy.

  (a) Mg, S, Si   **(b)** As, N, P   **(c)** As, Ge, P

SOLUTION

  (a) All three elements are in the same period. Within a period, I_1 increases from left to right as the atoms become smaller. The order of increasing I_1 is therefore

$$\text{Mg (lowest)} < \text{Si} < \text{S (highest)}$$

  (b) All three elements are in the same group. Within a group, I_1 decreases from top to bottom as atoms become larger. The order of increasing I_1 is therefore

$$\text{As (lowest)} < \text{P} < \text{N (highest)}$$

  (c) Of this set, As and Ge are in the same period. Because Ge is to the left of As, Ge has the lower I_1 value. Because it is in the same group as P and just below P, As has a lower I_1 than P. The order of increasing I_1 is thus

$$\text{Ge (lowest)} < \text{As} < \text{P (highest)}$$

EXERCISE 8.7A

Without reference to Figure 8.15, arrange each set of elements in the expected order of increasing first ionization energy.

  (a) Be, F, N   **(b)** Ba, Be, Ca   **(c)** F, P, S   **(d)** Ca, K, Mg

EXERCISE 8.7B

Arrange the following ionization energies in their expected order of increasing magnitude.

$$I_1 \text{ for S}; \ I_2 \text{ for Mg}; \ I_2 \text{ for K}; \ I_1 \text{ for Ar}; \ I_1 \text{ for Al}; \ I_1 \text{ for H}$$

Table 8.6 Some Selected First Electron Affinities, kJ/mol

1A	2A	3A	4A	5A	6A	7A	8A
Li −60	Be >0	B −27	C −154	N ≈0	O −141	F −328	Ne >0
Na −53					S −200	Cl −349	
K −48					Se −195	Br −325	
Rb −47					Te −190	I −295	
Cs −46					Po −183	At −270	

Electron Affinity animation

Electron Affinity

Ionization energy relates to forming a gaseous positive ion from a gaseous atom. An analogous atomic property for the formation of a gaseous negative ion is **electron affinity** (*EA*), the energy change that occurs when an electron is added to a gaseous atom.

An electron approaching a neutral atom is attracted to the positively charged nucleus. Repulsion of the incoming electron by electrons in the atom tends to offset this attraction. Still, in many cases, the incoming electron is incorporated into the atom and energy is given off, as in the process

$$F(g) + e^- \longrightarrow F^-(g) \qquad EA = -328 \text{ kJ/mol}$$

When a fluorine atom gains an electron, energy is given off. The process is exothermic, and so the electron affinity is a negative quantity.* Table 8.6 lists a number of electron affinities. Note that there are fewer clear-cut trends and more irregularities than in the ionization energies of Tables 8.4 and 8.5. Some of the data in Table 8.6 suggest a rough correlation between electron affinity and atomic size: Smaller atoms have more negative electron affinities. It seems reasonable that the smaller the atom, the closer an added electron can approach the atomic nucleus and the more strongly it is attracted to the nucleus. This certainly seems to be the case for the group 1A elements, for group 6A from S to Po, and for group 7A from Cl to At. The second-period elements present some problems, though. The electron affinity of O is not as negative as that of S, nor is that of F as negative as that of Cl. Electron repulsions in the small, compact O and F atoms may keep the added electron from being as strongly attracted as we might expect.

In most cases in Table 8.6, the added electron goes into a subshell that is already partly filled. For the group 2A and 8A atoms, however, the added electron would have to enter the next higher energy subshell. For group 2A atoms, the added electron enters the empty *p* subshell of the valence shell; for group 8A atoms, the electron enters the *s* orbital of the next available principal shell. The electron affinities of the atoms in groups 2A and 8A have positive values, and thus the atoms do not form stable anions.

Similar to the stepwise loss of electrons in the formation of positive ions, we can consider a stepwise addition of electrons in anion formation. For example, the first and second electron affinities for the oxygen atom are

$$O(g) + e^- \longrightarrow O^-(g) \qquad EA_1 = -141 \text{ kJ/mol}$$

$$O^-(g) + e^- \longrightarrow O^{2-}(g) \qquad EA_2 = +744 \text{ kJ/mol}$$

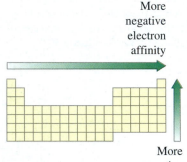

More negative electron affinity

More negative electron affinity

---

*Some scientists define electron affinity as the reverse of the process we described. That is, as

$$F^-(g) \longrightarrow F(g) + e^- \qquad EA = +328 \text{ kJ/mol.}$$

As we learned in Chapter 6, when we reverse a process, we change the sign of thermodynamic functions. The plus sign on this *EA* value tells us that energy must be absorbed to remove an electron from a gaseous fluoride ion. You may see electron affinities written both ways in the chemical literature.

It's not hard to see why the second electron affinity is a positive quantity. Here, an electron approaches an ion that has a net charge of -1. The electron is strongly repelled, and work must be done to force the extra electron onto the $O^-(g)$ ion. In fact, the O^{2-} ion forms only in situations where other energetically favorable processes offset the large energy expense to create it. This happens, for example, in the formation of ionic oxides such as MgO.

Example 8.8   A Conceptual Example

Which of the values given is a reasonable estimate of the second electron affinity (EA_2) for sulfur?

$$S^-(g) + e^- \longrightarrow S^{2-}(g) \qquad EA_2 = ?$$

$\quad -200 \text{ kJ/mol} \qquad +450 \text{ kJ/mol} \qquad +800 \text{ kJ/mol} \qquad +1200 \text{ kJ/mol}$

ANALYSIS AND CONCLUSIONS

Because of the repulsion between the $S^-(g)$ anion and the approaching electron, EA_2 must be *positive* (the process is endothermic). This eliminates -200 kJ/mol as a possibility. To choose among the other three values, consider the value of EA_2 for oxygen: $+744$ kJ/mol. Because of its smaller size, we should expect the $O^-(g)$ anion to exert a stronger repulsive force on the incoming electron than would the $S^-(g)$ anion. Thus the EA_2 value for $S^-(g)$ should be less than $+744$ kJ/mol. The only reasonable estimate is $+450$ kJ/mol.

EXERCISE 8.8A

Based on the result of Example 8.8, which of the values given is a reasonable estimate of the second electron affinity for selenium?

$$Se^-(g) + e^- \longrightarrow Se^{2-}(g) \qquad EA_2 = ?$$

$\quad -390 \text{ kJ/mol} \qquad +400 \text{ kJ/mol} \qquad +460 \text{ kJ/mol} \qquad +500 \text{ kJ/mol}$

EXERCISE 8.8B

Given the following first electron affinities (in kJ/mol), Al -50.2, Si -138.1, and P -75.3, offer a plausible reason, based on electron configurations, (a) why the EA of Si is more negative than that of Al and (b) why the EA of P is less negative than that of Si.

8.8   Metals, Nonmetals, Metalloids, and Noble Gases

In Section 2.5, we distinguished between metals, metalloids, and nonmetals in terms of general appearance and bulk physical properties. Now we can consider them in relation to atomic properties and positions in the periodic table.

Generally speaking, atoms of **metals** have only a few electrons (one, two, or three) in their valence shells and tend to form positive ions. For example, an aluminum atom, the electron configuration of which is $[Ne]3s^2 3p^1$, loses the three valence electrons when forming the ion Al^{3+}. All s-block elements except hydrogen and helium are metals, all d- and f-block elements are metals, and a few p-block elements are metals.

Atoms of **nonmetals** generally have more electrons in their valence shell than do metals, and consequently many nonmetals tend to form negative ions. Except for the special cases of the s-block elements hydrogen and helium, nonmetals are all p-block elements. A heavy, stepped, diagonal line in the periodic table separates the metals from the nonmetals. Several of the elements along this line have the physical appearance of metals but have some nonmetallic properties as well. As we noted in Chapter 2, these borderline elements are sometimes called **metalloids.** The gaseous nonmetal elements in the column at the extreme right of the periodic table (group 8A) are often singled out as a special group, the **noble gases.** Each period of the periodic table ends with a noble gas. This information is summarized in the format of the periodic table in Figure 8.16.

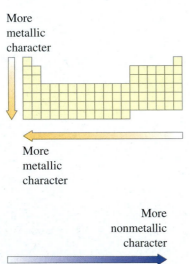

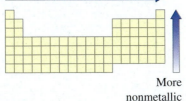

▲ **FIGURE 8.16**   **Metals, nonmetals, metalloids, and noble gases**

There is lack of agreement on just which elements to label as metalloids. However, they are generally considered to lie either immediately adjacent to the stair-step line or close by.

Metallic character is closely related to atomic radius and ionization energy. The easier it is to remove electrons from an atom, the more metallic the element. Easy removal of electrons corresponds to large atomic radius and low ionization energy.

Metallic character increases from top to bottom in a group and decreases from left to right in a period.

The more readily an atom gains an electron, the more nonmetallic the character of the atom. A strong tendency to gain electrons corresponds to a large negative electron affinity, a property found among the smaller nonmetal atoms.

Nonmetallic character decreases from top to bottom in a group and increases from left to right in a period.

From the first generalization, we can identify the alkali metals (group 1A) as highly metallic, and from the second generalization, we can identify the halogens (group 7A) as highly nonmetallic. In the middle of the periodic table, we see both metallic and non-metallic behavior. Group 4A has carbon, a nonmetal, at the top and two metals, tin and lead, at the bottom. In between are two metalloids, silicon and germanium.

Figure 8.17 summarizes trends in atomic properties and metallic/nonmetallic behavior in the periodic table.

Example 8.9

In each set, indicate which is the more metallic element.

(a) Ba, Ca   (b) Sb, Sn   (c) Ge, S

SOLUTION

(a) Ba is below Ca in group 2A. The Ba atom is larger than the Ca atom, and its first and second ionization energies are lower than those of Ca. As a result, Ba is more metallic than Ca.

(b) Sn is to the left of Sb in the fifth period. We therefore expect Sn to be a larger atom with lower ionization energies than those of Sb. Sn is more metallic than Sb.

(c) Ge is to the left of S and in the following period. Because of its larger atoms, we expect Ge to be more metallic than S.

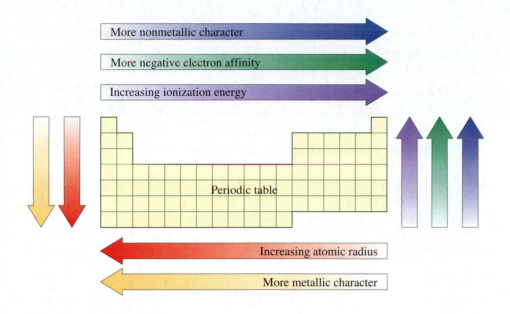

▲ **FIGURE 8.17 Atomic properties—A summary of trends in the periodic table**
A horizontal arrow points in the direction of a trend within a period. The vertical arrow of the same color points in the direction of the same trend within a group.

EXERCISE 8.9A

In each set, indicate which is the more nonmetallic element.

(a) O, P **(b)** As, S **(c)** P, F

Periodic Trends activity

EXERCISE 8.9B

Arrange the following series in the expected order of metallic behavior, with the most metallic first, and indicate any cases in which there is some doubt: Se, Sb, Fe, Co, Na, Ba, S.

Example 8.10 A Conceptual Example

Using only a blank periodic table such as the one in Figure 8.17, state the atomic number of **(a)** the element that has the electron configuration $4s^2 4p^6 4d^5 5s^1$ for its fourth and fifth principal shells and **(b)** the most metallic of the fifth-period *p*-block elements.

ANALYSIS AND CONCLUSIONS

(a) The $4d^5 5s^1$ identifies this as a transition element in the *d*-block (the 4*d* subshell is being filled) and in the fifth period ($n = 5$ is the highest principal quantum number). The underlying noble-gas electron configuration is that of krypton ($Z = 36$): $1s^2 2s^2 2p^6 3s^2 3p^6 3d^{10} 4s^2 4p^6$. Replacing $1s^2 2s^2 2p^6 3s^2 3p^6\ 3d^{10} 4s^2 4p^6$ with [Kr] gives $[Kr]4d^5 5s^1$. This is the element with atomic number $Z = 36 + 5 + 1 = 42$ (molybdenum). It is the sixth element in the fifth period, as shown in Figure 8.18.

(b) In general, the more metallic elements are those with large atomic radii and low ionization energies. These elements tend to be found toward the left side of their respective periods. The elements in group 3A are farthest to the left in the *p*-block, and therefore the most metallic of the fifth-period *p*-block elements has the valence-shell electron configuration $5s^2 5p^1$. Its noble-gas-core abbreviated electron configuration is $[Kr]4d^{10} 5s^2 5p^1$. This is the element with atomic number $Z = 36 + 10 + 2 + 1 = 49$, indium, as shown in Figure 8.18.

EXERCISE 8.10A

Using only the blank periodic table in Figure 8.17, enter the following information into the proper position in the table.

(a) The atomic number of the element having $4s^2 4p^6 4d^{10} 5s^2 5p^4$ as the electron configuration of its fourth and fifth principal shells.

(b) The atomic number of the largest atom in the first transition series.

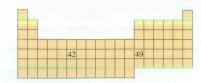

▲ **FIGURE 8.18   Example 8.10 illustrated**

(c) The atomic numbers of the *d*-block elements of the fifth period having the lowest and highest I_1 values.

(d) The atomic number of the most *nonmetallic* element of group 5A.

EXERCISE 8.10B

Using only the blank periodic table in Figure 8.17, enter the following information into the proper position in the table.

(a) The atomic numbers of the fifth-period *p*-block elements having the lowest and highest I_1 values.

(b) The atomic numbers of all the fourth-period elements that have the valence-shell electron configuration ns^1.

(c) Asterisks (*) to indicate all the third-period elements whose gaseous atoms are diamagnetic.

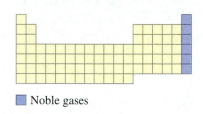

■ Noble gases

Application Note

Most of the world's helium supply is obtained from natural gas. The helium is thought to be the product of the decay of radioactive uranium and thorium (Chapter 19). In some natural gas wells, the concentration of helium is so high (up to 8%) that the presence of this inert component can interfere with the combustion process.

The Noble Gases

In the last decade of the nineteenth century, a group of elements was discovered that made up an entirely new group, one completely unexpected by Mendeleev and his contemporaries. This new group, called the noble gases, was placed between the highly reactive nonmetals of group 7A and the very active alkali metals (group 1A). In modern periodic tables, the noble gases are placed to the far right as group 8A.

The six noble gases are helium, neon, argon, krypton, xenon, and radon. All are found to some extent in the atmosphere. Argon is abundant, making up nearly 1% of the atmosphere by volume. Xenon, on the other hand, is quite rare, accounting for only 91 parts per billion of the atmosphere. (Its mole fraction is 0.000000091.) Radon, a radioactive element produced by the decay of heavier elements such as uranium, makes up only an inconsequential portion of the atmosphere. Radon is a potential health problem, however, when it seeps from the ground into a poorly ventilated building (Chapter 19).

The noble gases rarely enter into chemical reactions. This lack of reactivity is a reflection of their electron configurations, ionization energies, and electron affinities. They have particularly stable electronic configurations, comprising filled *s* and *p* valence orbitals. For example, the helium atom has a filled first principal shell ($1s^2$) and an exceptionally high first ionization energy (2372 kJ/mol). Helium gives up an electron only with great difficulty, and the helium atom has no affinity for an additional electron. The added electron would have to be accommodated in the $2s$ orbital at a much higher energy than the filled $1s$ orbital. Thus, helium does not form an anion. The other noble gas atoms have the valence-shell electron configuration ns^2np^6 and show a similar aversion to both losing and gaining an electron.

Because of their lack of reactivity, noble-gas atoms tend not to combine with other atoms, even of their own kind. Therefore the noble gases occur naturally only in elemental form and only as monatomic species. Since 1962, a few compounds of the heavier noble gases have been prepared. Synthesis in 2000 of HArF, the first stable neutral argon compound, leaves helium and neon as the only two noble gases for which, as yet, no compounds have been made. While this group of elements is no longer called inert, as it once was, its nobility is unquestioned.

8.9   Using Atomic Properties and the Periodic Table to Explain the Behavior of the Elements

We conclude this chapter by looking again at a few matters discussed earlier in the text, but now with the added insights gained through the concepts presented in this chapter.

Flame Colors

As we noted in Chapter 7, atoms absorb energy when electrons move from lower to higher energy states and emit energy when electrons fall from higher to lower energy states. If the energy change in the latter transition is in the range of visible light, the

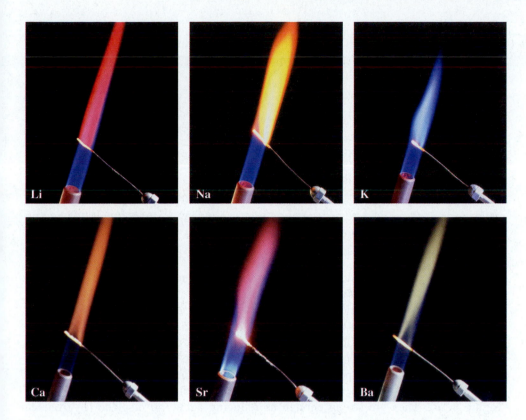

◀ **FIGURE 8.19** Flame colors of some alkali metals (group 1A) and alkaline earth metals (group 2A)

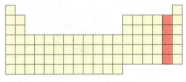

Flame Tests for Metals movie

emitted light is colored. Because the quantity of energy available from a gas flame is not very high, only atoms that have low first ionization energies are excited in a Bunsen burner flame. The group 1A metals and the heavier group 2A metals have the lowest ionization energies, and all these elements display a colored flame. Beryllium and magnesium have higher first ionization energies and do not exhibit a flame color. The flame test for sodium, for example, results when the valence electron, after having been excited to some higher energy level, drops back to the $3s$ subshell. The transition from the $3p$ to the $3s$ subshell, for instance,

$$\text{Na}([\text{Ne}]3p^1) \longrightarrow \text{Na}([\text{Ne}]3s^1)$$

produces a yellow color. The colors exhibited by three group 1A and three group 2A elements are shown in Figure 8.19.

The Halogens as Oxidizing Agents

In general, the halogens (group 7A) are good oxidizing agents, but their strength decreases sharply from F_2 down the group to I_2. The relative oxidizing strengths of the halogens are illustrated in a series of reactions in aqueous solution. For example, when $Cl_2(g)$ is bubbled into a solution containing iodide ions (Figure 8.20), chlorine oxidizes I^- to I_2. Chlorine atoms gain electrons and are reduced to chloride ions. The oxidation number of the chlorine drops from 0 in Cl_2 to -1 in Cl^-.

$$Cl_2(g) + 2\,I^-(aq) \longrightarrow 2\,Cl^-(aq) + I_2(s)$$

This redox reaction should seem reasonable because Cl atoms have a more negative electron affinity (-349 kJ/mol) than I atoms (-295 kJ/mol). In the competition for electrons, the Cl atoms in Cl_2 extract electrons from the I^- ions. Reasoning in this same way, we would predict that $Cl_2(g)$ also reacts with $Br^-(aq)$, and that $Br_2(l)$ reacts with $I^-(aq)$.

$$Br_2(l) + 2\,I^-(aq) \longrightarrow 2\,Br^-(aq) + I_2(s)$$

In contrast, the reverse of this reaction does not occur. When iodine is added to a solution of bromide ions, the atom with the greater affinity for an additional electron— bromine—already has the electron.

$$I_2(s) + 2\,Br^-(aq) \longrightarrow \text{no reaction}$$

■ Group 7A: The halogens

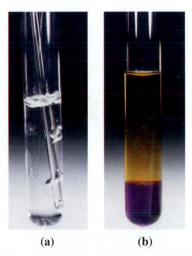

(a)   (b)

▲ **FIGURE 8.20** Displacement of $I^-(aq)$ by $Cl_2(g)$

(a) Chlorine gas is bubbled through a colorless aqueous solution containing iodide ions. (b) $Cl_2(g)$ is reduced to $Cl^-(aq)$, and $I^-(aq)$ is oxidized to I_2. Dense $CCl_4(l)$ is added, and the I_2 concentrates in the CCl_4 (purple layer), in which it is much more soluble than in water.

Astatine is radioactive and exceedingly rare. Little is known about its chemistry, and it is generally excluded from a discussion of the halogens.

We might think that F_2, being a stronger oxidizing agent than Cl_2, would react with Cl^- in aqueous solution to form F^- and Cl_2. However, F_2 is the most powerful of all oxidizing agents, and in aqueous solutions it oxidizes water to oxygen gas (rather than Cl^- to Cl_2):

$$2\ F_2(g) + 2\ H_2O(l) \longrightarrow 4\ HF(aq) + O_2(g)$$

Electron affinities do correlate somewhat with the oxidizing power of the halogens, but other factors are also involved. We will take another look at the halogens as oxidizing agents later in the text.

The s-Block Metals as Reducing Agents

In Chapter 4 (page 74), we related the activity series of the metals to the tendency of metal atoms in the solid state to become cations in aqueous solution, that is, to undergo oxidation, as does magnesium in the reaction

$$Mg(s) + 2\ H^+(aq) \longrightarrow Mg^{2+}(aq) + H_2(g)$$

As $Mg(s)$ is oxidized, it reduces $H^+(aq)$ to $H_2(g)$. By undergoing oxidation, a metal acts as a reducing agent, as we noted in Chapter 4.

Ionization energy is a measure of the tendency of gaseous atoms to lose electrons to form gaseous ions, rather than ions in aqueous solution. Even so, ionization energy is roughly related to the oxidation tendencies expressed by the activity series. Thus, the lower the ionization energy of a metal atom, the more easily it is oxidized and the better reducing agent we expect it to be. All the s-block elements except hydrogen and helium are metals, and all these metals are above hydrogen in the activity series (Figure 4.13). They are strong enough as reducing agents to displace H_2 from an acidic solution.

In fact, the s-block metals are among the most powerful reducing agents known. All the alkali metals (group 1A) and the heavier alkaline earth metals (group 2A) are able to displace H^+ ions as $H_2(g)$, not only from acidic solutions but even from neutral or basic solutions, in which the concentration of $H^+(aq)$ is extremely low. Thus, calcium and potassium both displace even the relatively few H^+ ions in water as $H_2(g)$.

$$2\ K(s) + 2\ H_2O(l) \longrightarrow 2\ K^+(aq) + 2\ OH^-(aq) + H_2(g)$$

$$Ca(s) + 2\ H_2O(l) \longrightarrow Ca^{2+}(aq) + 2\ OH^-(aq) + H_2(g)$$

These two reactions are pictured in Figure 8.21, which also shows that magnesium apparently does not react with cold water. There is a rough correlation between the vigor (or lack of vigor) of these reactions and ionization energies: for K ($I_1 = 419$ kJ/mol), for Ca ($I_1 = 590$, $I_2 = 1145$), for Mg ($I_1 = 738$, $I_2 = 1451$). Other factors are also involved, and we will consider reactions of this sort again in later chapters.

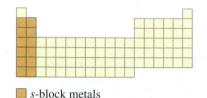

s-block metals

▶ **FIGURE 8.21** **The behavior of potassium, calcium, and magnesium in cold water**

(a) Potassium is less dense than water and floats on the surface as it enters into a vigorous exothermic reaction with the water. The evolved hydrogen gas spontaneously ignites, producing the observed flame. (b) Calcium is denser than water. It sinks to the bottom of the test tube and reacts more slowly with the water. The evolved hydrogen gas slowly bubbles to the surface and does not ignite. A few drops of phenolphthalein indicator have been added to the water to show that the solution is basic. (c) Magnesium metal shows no evidence of reaction in cold water.

(a)

(b)

(c)

Acidic, Basic, and Amphoteric Oxides

The name *oxygen* was coined by Lavoisier because he thought all acids contained this element; oxygen means "acid former" in Greek. As we noted in Chapter 4, hydrogen, not oxygen, is the element common to typical acids. Nevertheless, most acids do contain oxygen, and some acids are formed by reaction of an element oxide with water. Oxides that produce acids in this way are **acidic oxides.** They are molecular substances and are generally the oxides of nonmetals, for example, SO_3 and P_4O_{10}:

$$SO_3(g) + H_2O(l) \longrightarrow H_2SO_4(aq)$$

$$P_4O_{10}(s) + 6\,H_2O(l) \longrightarrow 4\,H_3PO_4(aq)$$

Like acids, acidic oxides can react directly with bases in neutralization reactions such as

$$SO_2(g) + 2\,NaOH(aq) \longrightarrow Na_2SO_3(aq) + H_2O(l)$$

In contrast to the oxides of nonmetals, metal oxides are ionic. Many metal oxides, because they react with water to form bases, are called **basic oxides.** Two examples are

$$Li_2O(s) + H_2O(l) \longrightarrow 2\,LiOH(aq)$$

$$BaO(s) + H_2O(l) \longrightarrow Ba(OH)_2(aq)$$

Basic oxides can also react directly with acids in neutralization reactions, for example, as in the reaction

$$MgO(s) + 2\,HCl(aq) \longrightarrow MgCl_2(aq) + H_2O(l)$$

Now let's assess the character of the element oxides in the third period of the periodic table. Starting at the left, the oxides of the group 1A and 2A third-period elements, Na_2O and MgO, are both basic. Starting at the right, Ar does not form an oxide, but those of Cl, S, P, and Si are all acidic. The oxide of the group 3A element Al is an interesting case. Consider these reactions of Al_2O_3, first with the acid $HCl(aq)$ and then with the base $NaOH(aq)$:

$$\underset{\text{Base}}{Al_2O_3(s)} + \underset{\text{Acid}}{6\,HCl(aq)} \longrightarrow 2\,AlCl_3(aq) + 3\,H_2O(l)$$

$$\underset{\text{Acid}}{Al_2O_3(s)} + \underset{\text{Base}}{2\,NaOH(aq)} + 3\,H_2O(l) \longrightarrow \underset{\text{Sodium aluminate}}{2\,Na[Al(OH)_4](aq)}$$

In the first reaction, Al_2O_3 acts like a base and aluminum appears in solution as a cation (Al^{3+}). In the second reaction, Al_2O_3 acts like an acid and the aluminum appears in solution in the anion ($[Al(OH)_4]^-$).* An oxide like Al_2O_3 that can react with either an acid or a base is said to be **amphoteric.**

Figure 8.22 categorizes the main-group elements according to the acidic, basic, or amphoteric nature of their oxides. Not unexpectedly, we see that amphoterism is found in a belt of elements located along the stairstep diagonal line separating the metals from the nonmetals.

Periodic Trends: Acid-Base Behavior of Oxides animation

Application Note

The amphoterism of $Al_2O_3(s)$ is not just a laboratory curiosity. It plays an important role in the commercial extraction of aluminum metal from bauxite ore (Section 21.3).

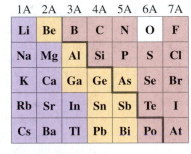

▲ **FIGURE 8.22   Acidic, basic, and amphoteric oxides of the main-group elements**

Cumulative Example

Given that the density of solid sodium is 0.968 g/cm³, estimate the atomic (metallic) radius of a Na atom. Assess the value obtained, indicating why the result is only an estimate and whether the actual radius should be larger or smaller than the estimate.

STRATEGY

The inverse of the density (1 cm³/0.968 g Na) describes the volume occupied by 1 g of Na atoms. If we multiply this quantity by the molar mass of Na, we obtain the volume occupied by 1 mol of Na atoms. Dividing this molar volume of Na by Avogadro's number yields the volume of a single Na atom. At this point, we can calculate the radius of the spherical atom. To assess the validity of our estimate, we need to consider carefully where assumptions may enter into the calculation and whether these assumptions are justifiable.

*The anion $[Al(OH)_4]^-$ is an example of a *complex ion*. We will describe this and other complex ions in Chapters 16 and 22.

SOLUTION

Using the density, molar mass, and Avogadro's number, we determine the volume per Na atom.

$$? \text{ cm}^3/\text{Na atom} = \frac{1 \text{ cm}^3}{0.968 \text{ g Na}} \times \frac{22.99 \text{ g Na}}{1 \text{ mol Na}} \times \frac{1 \text{ mol Na}}{6.022 \times 10^{23} \text{ Na atom}}$$

$$= 3.94 \times 10^{-23} \text{ cm}^3/\text{Na atom}$$

Next, we use the formula for the volume of a sphere to calculate the radius of a Na atom.

$$V = 4(\pi r^3)/3$$

$$r = \sqrt[3]{\frac{3V}{4\pi}} = \sqrt[3]{\frac{3 \times 3.94 \times 10^{-23} \text{ cm}^3}{4 \times 3.14}} = 2.11 \times 10^{-8} \text{ cm}$$

To express this atomic radius in the usual unit, we convert to picometers.

$$r = 2.11 \times 10^{-8} \text{ cm} \times \frac{1 \text{ m}}{100 \text{ cm}} \times \frac{1 \text{ pm}}{1 \times 10^{-12} \text{ m}} = 211 \text{ pm}$$

ASSESSMENT

The estimate of the atomic radius—211 pm—falls within the range of atomic radii seen in this chapter, indicating that the mathematical calculations are probably correct. The result is only an estimate because of an implicit assumption made in the calculation—that a volume comprising 1 mol of Na atoms is completely filled with matter. However, this cannot be the case for spherical atoms because there must be empty spaces (voids) no matter how closely the spheres are packed. The actual amount of matter in 1 mol of atoms is the volume calculated here *less* the volume of the voids. The actual atomic radius of Na must therefore be less than 211 pm; we see in Figure 8.14 that it is 186 pm. (We learn how to account for the volume of the voids in Chapter 11.)

Concept Review with Key Terms

8.1 Multielectron Atoms—The wave mechanical treatment of the hydrogen atom can be extended to multielectron atoms, but with two essential differences: (1) Energy levels are lower in multielectron atoms than in the hydrogen atom, and (2) the energy levels in multielectron atoms are split; that is, the different subshells of a principal shell have different energies. The order of increasing subshell energy in a principal shell is $s < p < d < f$. All the orbitals within a subshell, however, have the same energy—they are **degenerate** orbitals.

8.2 An Introduction to Electron Configurations—**Electron configuration** is the distribution of electrons in orbitals among the subshells and principal shells in an atom. Three types of notation of electron configuration are the *spdf* **notation**, the noble-gas-core abbreviated notation, and the **orbital diagram**.

$$\text{N } 1s^2 2s^2 2p^3 \qquad \text{N[He]} 2s^2 2p^3 \qquad \text{N}$$

| *spdf* notation | Noble-gas-core abbreviated notation | Orbital diagram |

8.3 Rules for Electron Configurations—To write a probable electron configuration, we note that (a) electrons tend to occupy the lowest-energy orbitals available, (b) no two electrons can have all four quantum numbers alike (**Pauli exclusion principle**), and (c) where possible, electrons occupy orbitals singly and with parallel spins, rather than in pairs (**Hund's rule**).

8.4 Electron Configurations: The Aufbau Principle—The **aufbau principle** describes a process of hypothetically building up one atom from the atom having the preceding atomic number. With this principle and the other ideas cited in the chapter, it is possible to predict probable electron configurations for many of the elements. In the aufbau process for the **main-group elements**, electrons are added to either the *s* or the *p* subshell of the principal shell of highest principal quantum number (n). In **transition elements**, electrons

go into the *d* subshell that is one shell in from the outermost occupied principal shell.

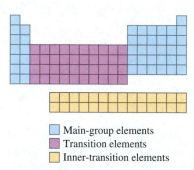

☐ Main-group elements
☐ Transition elements
☐ Inner-transition elements

8.5 Electron Configurations: Periodic Relationships—Elements with similar electron configurations fall in the same group of the periodic table. For main-group elements, the group number corresponds to the number of electrons in the principal shell of highest quantum number. The period number is the same as the principal quantum number of the **valence shell**. The division of the periodic table into *s-*, *p-*, *d-*, and *f*-blocks assists in the assignment of probable electron configurations. Within the *f*-block, the **lanthanide** series consists of elements in which the 4*f* subshell is the one being filled; the **actinide** series consists of elements in which the 5*f* subshell is being filled. **Valence electrons** occupy the outermost occupied principal shell and have the highest principal quantum number of the electrons within the atom or ion. **Core electrons** occupy the inner shells, those with principal quantum numbers less than that of the valence shell.

8.6 Magnetic Properties: Paired and Unpaired Electrons—An atom with all electrons paired is **diamagnetic**, and an atom with one or more unpaired electrons is **paramagnetic**. Experimentally determined magnetic properties help to verify electron configurations.

8.7 Periodic Atomic Properties of the Elements—The **periodic law** describes the regular recurrence of certain physical and chemical properties of atoms when they are arranged by increasing atomic number. Clear trends are seen in plots of **atomic radius** versus atomic number. The specific definition of radius dimensions depends on the bonding environment in which elements are found; in molecules, we speak of the **covalent radius** of the atoms, and in metals, we speak of the **metallic radius** of the atoms. For ionic species, the size is expressed in terms of the **ionic radius**.

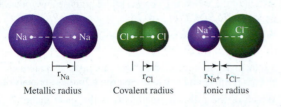

r_{Na}	r_{Cl}	r_{Na^+} r_{Cl^-}
Metallic radius	Covalent radius	Ionic radius

Various species can be **isoelectronic** (contain the same number of electrons) but still have different atomic or ionic radii because the effective nuclear charge is different in each. The **effective nuclear charge** (Z_{eff}) felt by an outer electron in an atom is the actual nuclear charge less the screening effect of all other electrons in the atom. **Ionization energy** is the energy required to remove an electron from a ground-state atom (or ion) in the

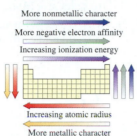

More nonmetallic character

More negative electron affinity

Increasing ionization energy

Increasing atomic radius

More metallic character

gaseous state. **Electron affinity** is the energy change associated with the addition of an electron to a gaseous atom or ion. Periodic trends are observed for both of these properties.

8.8 Metals, Nonmetals, Metalloids, and Noble Gases—The regions of the periodic table designated **metals**, **nonmetals**, **metalloids**, and the **noble gases** are related to the values of atomic properties. In general, metallic properties are associated with the ease with which atoms lose electrons; and nonmetallic properties, with the ease with which they gain electrons. The most metallic elements are located toward the left and bottom of the periodic table, and the most nonmetallic elements are located toward the right and top of the table.

8.9 Using Atomic Properties and the Periodic Table to Explain the Behavior of the Elements—The colors produced from metal flame tests, oxidation and reduction properties, and the acid–base behavior of oxides all exhibit trends consistent with the periodic law. For example, oxides of nonmetals are generally **acidic oxides**; that is, they produce an acid upon reaction with water. **Basic oxides**, oxides that react with water to form bases, are generally oxides of metals. Oxides of elements having both metallic and nonmetallic character are **amphoteric**; they can react with either an acid or a base.

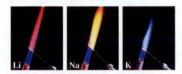

Assessment Goals

When you have mastered the material in this chapter, you will be able to:

- Explain the origin of energy differences between orbitals of a hydrogen atom and those of a multielectron atom and between different subshells within a principal shell of a multielectron atom.
- Write electron configurations using *spdf* notation and orbital diagrams.
- Apply the Pauli exclusion principle, Hund's rule, and the aufbau principle to the writing of electron configurations.
- Relate the electron configurations of the elements to their positions in the periodic table.
- Deduce electron configurations by using the four principal blocks of elements in the periodic table: *s*-block, *p*-block, *d*-block, and *f*-block.

- Identify valence and core electrons.
- Write the electron configurations of ions.
- Predict the magnetic properties of atoms and ions.
- Define atomic radius, ionization energy, and electron affinity.
- State how atomic radii, ionic radii, ionization energy, and electron affinity vary within a group and across the periodic table.
- State how the radius of an ion differs from that of its parent atom.
- Describe the characteristic properties of metals, nonmetals, and noble gases.
- Explain trends in macroscopic properties of element oxides—for example, flame test colors, strengths of oxidizing and reducing agents, and acid–base behavior—in terms of atomic properties and the periodic table.

Self-Assessment Questions

1. Select the subshell(s) that consist(s) of degenerate orbitals.

   (a) $1s$ (b) $2p$ (c) $3d$ (d) $4s$

2. Which set(s) of hydrogen orbitals have identical energies? How would your answer change for an atom other than hydrogen?

   (a) $1s, 2s, 2p$ (b) $3s, 3p, 3d$ (c) $3p_x, 3p_y, 3p_z$

3. According to Hund's rule which of the following represents the correct electron configuration for the carbon atom?

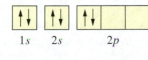

4. What element corresponds to each configuration?

   (a) $1s^2 2s^2 2p^5$

   (b) $1s^2 2s^2 2p^6 3s^2 3p^6 3d^{10} 4s^1$

   (c) $1s^2 2s^2 2p_x{}^1 2p_y{}^1 2p_z{}^1$

   (d) $[Ne]3s^1$

   (e) [Ar]

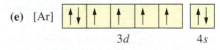

5. What subshell(s) fill in the aufbau process in each of the following regions of the periodic table?

   (a) groups 1A and 2A   (c) the transition elements

   (b) groups 3A through 7A (d) the lanthanides and actinides

6. What similarity in electron configuration is shared by lithium, sodium, and potassium? By beryllium, magnesium and calcium?

7. How many valence and core electrons are in an atom of (a) C, (b) Ne, (c) F, (d) Al, and (e) Mg?

8. Referring to the periodic table, why are there 2 elements in the first period, 8 in the third, 18 in the fifth, and 32 in the seventh?

9. In the periodic table, why is argon (Ar), with an atomic mass of 39.948, placed ahead of potassium, with an atomic mass of 39.098?

10. How many electrons are described by the notation $2p^6$? What is the general shape of the orbitals described in the notation? How many orbitals are included in the notation?

11. Identify the *d*-block elements among the following:
    (a) Ca   (b) Al   (c) Fe   (d) Pb   (e) Pt

12. Must all atoms having an odd atomic number be paramagnetic? Must all atoms having an even atomic number be diamagnetic? Explain.

13. Why don't the sizes of atoms increase uniformly with increasing atomic number?

14. Why are cations smaller than the atoms from which they are formed, whereas anions are larger?

15. Which of the following is (are) isoelectronic with Se^{2-}? (a) S^{2-} (b) I^- (c) Xe (d) Sr^{2+}

16. Which of the following possesses the highest first ionization energy? (a) Cs (b) Cl (c) I (d) Li

17. Which of the following has the most negative electron affinity? (a) Br (b) Sn (c) Ba (d) Li

18. Explain why all ionization energies are positive quantities, whereas both positive and negative electron affinities are possible.

19. Which of the following elements is the most metallic: (a) Mg (b) Li (c) K or (d) Ca?

20. What is the characteristic behavior of an oxide that permits us to call it (a) acidic, (b) basic, and (c) amphoteric?

Problems

The Rules for Electron Configurations

21. Explain why the following orbital diagrams are not possible for a ground-state electron configuration.

    (a)
        1s   2s   2p

    (b)
        1s   2s   2p

    (c)
        1s   2s   2p

22. Explain why the following orbital diagrams are not possible for a ground-state electron configuration.

    (a)
        1s   2s   2p   3s

    (b)
        1s   2s   2p   3s

    (c)
        1s   2s   2p   3s   3p

23. Explain why none of the following electron configurations is reasonable for a ground-state atom.
    (a) $1s^2 2s^2 2p^6 3s^1 3p^1$ (c) $1s^2 2s^2 2p^6 2d^5$
    (b) $1s^2 2s^2 2p^6 3s^2 3p^6 3d^1$

24. Explain why none of the following electron configurations is reasonable for a ground-state atom.
    (a) $1s^2 2s^2 2p^2 3s^1$ (c) $1s^2 2s^2 2p^6 3s^2 3p^6 3d^8 4s^2 4p^1$
    (b) $1s^2 2s^2 3s^2$

25. What principle(s) or rule(s) does each of the following electron configurations violate?
    (a) $1s^2 2s^6 3s^2$ (b) $1s^2 2s^2 2p^7 3s^1$ (c) $1s^2 2s^2 2p^6 2d^3$

26. What principle(s) or rule(s) does each of the following electron configurations violate?
    (a) $[Ar]2d^{10}$ (b) $[Ar]3f^3 4s^2$ (c) $[Kr]4d^{10} 4f^{14} 5s^2$

Electron Configurations: The Aufbau Principle

For Problems 27–36, refer only to the periodic table inside the front cover.

27. Using *spdf* notation, write the ground-state electron configuration of (a) Mg, (b) F, (c) O, (d) Be, (e) P, (f) Si, (g) C, (h) K, and (i) Sc.

28. Using *spdf* notation, write the ground-state electron configuration of (a) Ar, (b) Al, (c) Ne, (d) B, (e) Ca, (f) N, (g) Cl, (h) S, and (i) Zn.

29. Using a noble-gas-core abbreviated *spdf* notation, write the ground-state electron configuration for (a) Cs, (b) Sr, (c) Ti, (d) Sb, (e) Br, and (f) Pb.

30. Using a noble-gas-core abbreviated *spdf* notation, write the ground-state electron configuration for (a) Al, (b) Ga, (c) Zr, (d) I, (e) As, and (f) Ba.

31. Give an orbital diagram for the ground-state electron configuration of (a) N, (b) B, (c) Si, (d) Ca, (e) Cl, and (f) Sc.

32. Give an orbital diagram for the ground-state electron configuration of **(a)** C, **(b)** O, **(c)** K, **(d)** Al, **(e)** S, and **(f)** Mg.

33. Give the orbital diagram for the electrons beyond the xenon core of the rhenium (Re) atom. Relate this electron configuration to the position of rhenium in the periodic table.

34. Give the orbital diagram for the electrons beyond the radon core of the rutherfordium (Rf) atom, element 104. Relate this electron configuration to the position of rutherfordium in the periodic table.

35. Use orbital diagrams to represent the electron configurations of **(a)** Cl^-, **(b)** Zn^{2+}, **(c)** Pb^{2+}, and **(d)** Mg^{2+}. (Part of your diagram may be a noble-gas-core abbreviation.)

36. Use orbital diagrams to represent the electron configurations of **(a)** Ga^{3+}, **(b)** V^{3+}, **(c)** I^-, and **(d)** Sb^{3+}. (Part of your diagram may be a noble-gas-core abbreviation.)

37. Your study group is considering the identity of the species that has the electron configuration $1s^2 2s^2 2p^6 3s^2 3p^6 3d^7$. The students offer the following as possible answers: **(a)** a Mn atom; **(b)** a Co^{3+} ion; **(c)** a Ni^{2+} ion; and **(d)** a Cu^{2+} ion. Which, if any, is correct?

38. Your study group is considering the identity of the species that has the electron configuration $1s^2 2s^2 2p^6 2d^1$. The students offer the following as possible answers: **(a)** an excited state of neon, **(b)** a ground-state sodium atom, **(c)** a Na^+ ion, and **(d)** an excited-state sodium atom. Which, if any, is correct?

39. Arsenic is capable of forming two different cations. What charges would you expect on those cations? Explain.

40. The only common ion of silver is the 1+ ion. Explain.

Electron Configurations and the Periodic Table

41. Give the period number and group number for the element whose atoms have the electron configuration

(a) $1s^2 2s^2 2p^6$ **(d)** $1s^2 2s^2$

(b) $1s^2 2s^2 2p^6 3s^2 3p^2$ **(e)** $1s^2 2s^2 2p^3$

(c) $1s^2 2s^2 2p^6 3s^1$ **(f)** $1s^2 2s^2 2p^6 3s^2 3p^1$

42. Write the noble-gas-core abbreviated electron configuration for the element in **(a)** group 5A and period 4, **(b)** group 3A and period 6, and **(c)** group 3B and period 5.

43. Use the relationship between electron configurations and the periodic table to determine the number of **(a)** valence-shell electrons in an atom of Bi, **(b)** electrons in the fourth principal shell of Au, **(c)** elements whose atoms have five valence-shell electrons, **(d)** unpaired electrons in an atom of Se, and **(e)** transition elements in the fifth period.

44. Based on the periodic table and rules for electron configurations, indicate the number of **(a)** 3p electrons in an atom of P, **(b)** 4s electrons in an atom of Cs, **(c)** 4d electrons in an atom of Se, **(d)** 4f electrons in an atom of Bi, **(e)** unpaired electrons in an atom of Ga, **(f)** elements in group VA of the periodic table, **(g)** elements in the sixth period of the periodic table.

Magnetic Properties

45. Which of the following are diamagnetic and which are paramagnetic? Explain.

(a) a Sr atom   **(c)** a V^{5+} ion   **(e)** a Zn^{2+} ion

(b) a F atom   **(d)** an Al atom

46. Which of the following are diamagnetic and which are paramagnetic? Explain.

(a) a Sn atom   **(c)** a Tl^{3+} ion   **(e)** a Rn atom

(b) a W atom   **(d)** an O^{2-} ion

47. Identify all the fourth-period elements that are diamagnetic.

48. There are only two third-period elements whose atoms are diamagnetic, while there are six ions from that period that are diamagnetic. Explain.

Atomic and Ionic Radii

For Problems 49–52, refer only to the periodic table inside the front cover.

49. Which member of the following pairs has the *larger* radius? Explain.

(a) Al or S   **(c)** Ba or Sn

(b) Cl^- or Ca^{2+} **(d)** Na^+ or K

50. Which member of the following pairs has the *smaller* radius? Explain.

(a) Ca or C   **(c)** Si or F

(b) Mg^{2+} or Fe^{2+} **(d)** Zn^{2+} or Ca^{2+}

51. Arrange each set in order of increasing radius, and explain the basis for this order.

(a) Al, B, K, Mg   **(b)** P, Br, Br^-, Cl

52. Arrange each set in order of increasing radius, and explain the basis for this order.

(a) Ga, In, S, Se   (b) In^{3+}, Rb, Rb^+, Sr^{2+}

53. Where in the periodic table would you expect to find the two or three elements having the largest atoms? Explain.

54. Explain why there is a large difference in atomic radius between the elements $Z = 11$ (Na, 186 pm) and $Z = 12$ (Mg, 160 pm), whereas that between $Z = 24$ (Cr, 125 pm) and $Z = 25$ (Mn, 124 pm) is quite small.

55. Without reference to a table of data, should you be able to predict (a) whether atoms of Ca are larger than atoms of Cl and (b) whether the ionic radius of K^+ is greater than that of F^-? Explain.

56. Without reference to a table of data, should you be able to predict (a) whether atoms of I are larger than those of Li and (b) whether the ionic radius of Cl^- is greater than that of Ca^{2+}? Explain.

Ionization Energy

57. Arrange each set of the following elements in order of increasing first ionization energy, and explain the basis for this order.

(a) Ca, Mg, Ba   (b) P, Cl, Al   (c) F, Na, Fe, Cl, Ne

58. Arrange each set of the following elements in order of increasing first ionization energy, and explain the basis for this order.

(a) Ca, Na, As   (b) S, As, Sn   (c) Kr, Ba, Zn, Sc, Al, Br

59. Describe the trend in successive ionization energies as electrons are removed one at a time from an aluminum atom. Why is there a big jump between I_3 and I_4?

60. Which should have the lower first ionization energy—sulfur or phosphorus—and why?

Electron Affinity

61. Which group of elements has electron affinities with the largest negative values? Explain.

62. Which of the main groups of elements do not form stable negative ions? Explain, using electron configurations.

63. Silicon has an electron affinity of -134 kJ/mol. The electron affinity of phosphorus is -72 kJ/mol. Give a plausible reason for this difference.

64. Lithium has an electron affinity of -60 kJ/mol. That of boron is -27 kJ/mol. Give a plausible reason for this difference.

Atomic Properties and the Periodic Table

65. One of the first periodic properties studied was that of *atomic volume*, the atomic mass of an element divided by its density (as a solid). Draw a graph to show that atomic volume is a periodic property of the following elements. Densities, in g/cm^3, are Na, 0.97; Mg, 1.74; Al, 2.70; Si, 2.33; P, 2.20; S, 2.07; Cl, 2.03; Ar, 1.66; K, 0.86; Ca, 1.55; Sc, 2.99; Cr, 7.19; Co, 8.90; Zn, 7.13; Ga, 5.91; As, 4.70; Br, 4.05; Kr, 2.82; Rb, 1.53; and Sr, 2.54. To what atomic property described in this chapter does the atomic volume seem most closely related? Explain.

66. Draw a graph to show that melting point is a periodic property of these elements (melting points are in °C): Al, 660; Ar, −189; Be, 1278; B, 2300; C, 3350; Cl −101; F, −220; Li, 179; Mg, 651; Ne, −249; N, −210; O, −218; P, 590; Si, 1410; Na, 98; S, 119. For this purpose, does it matter whether melting points are expressed on the Celsius scale or the Kelvin scale? Explain.

67. What properties could you use to assess the degree of metallic character in an element? Arrange the following elements in the expected order of *increasing* metallic character: K, P, Al, Rb, Bi, Ca, Ge. Explain your arrangement.

68. What properties could you use to assess the degree of nonmetallic character in an element? Arrange the following elements in the expected order of *increasing* nonmetallic character: Pb, Sb, N, Br, As, F, O, Si, and explain your arrangement.

69. Following are the periodic table locations of certain elements. Arrange them in the expected order of *increasing* ionization energy, I_1. (a) group 4A, period 4; (b) group 6A, period 3;

(c) group 3A, period 6; (d) group 8A, period 2; (e) group 6A, period 4.

70. Classify each of the following substances as an acidic oxide, a basic oxide, or neither; explain your classification: (a) CO_2, (b) O_2, (c) SrO, (d) HCOOH, (e) P_4O_6, (f) $Ba(OH)_2$.

71. Complete and balance the following equations. If no reaction occurs, so indicate.

(a) $Cl_2(g) + Br^-(aq) \longrightarrow$

(b) $I_2(s) + F^-(aq) \longrightarrow$

(c) $Br_2(l) + I^-(aq) \longrightarrow$

72. Complete and balance the following equations. If no reaction occurs, so indicate.

(a) $K(s) + H_2O(l) \longrightarrow$

(b) $Ca(s) + H^+(aq) \longrightarrow$

(c) $Be(s) + H_2O(l) \longrightarrow$

73. Complete and balance the following equations.

(a) $N_2O_5(s) + H_2O(l) \longrightarrow$

(b) $MgO(s) + CH_3COOH(aq) \longrightarrow$

(c) $Li_2O(s) + H_2O(l) \longrightarrow$

74. Complete and balance the following equations.

(a) $SO_3(g) + KOH(aq) \longrightarrow$

(b) $Al_2O_3(s) + H^+(aq) \longrightarrow$

(c) $CaO(s) + H_2O(l) \longrightarrow$

Additional Problems

Problems marked with an * may be more challenging than others.

75. Without referring to any tables or listing in the text, mark an appropriate location for each of the following in the blank periodic table provided: (a) the fourth-period noble gas, (b) a fifth-period element whose atoms have three unpaired electrons, (c) the d-block element having one 3d electron, (d) a p-block element that is a metalloid, (e) a metal that forms the oxide M_2O_3.

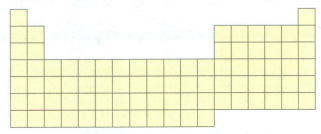

76. Propose a probable electron configuration for the yet-to-be-discovered element with $Z = 117$.

*77. Use ideas presented in this chapter to indicate (a) three metals that you would expect to exhibit the photoelectric effect with visible light, and three that you would not; (b) the approximate first ionization energy of fermium ($Z = 100$); (c) the approximate atomic radius of francium.

78. Arrange the following ionization energies in the most probable order of increasing value, and explain your reasoning: I_1 for B, I_1 for Cs, I_2 for In, I_2 for Sr, I_2 for Xe, and I_3 for Ca.

*79. The designations given below are for the energy changes associated with the variety of ion-formation processes discussed in this chapter. Write a chemical equation for each process. Then arrange the equations in what you expect to be the order of increasing magnitude of the energy change: (a) EA_2 for P; (b) I_1 for O; (c) EA_1 for O; (d) I_1 for Ar; (e) I_2 for Sr; (f) I_3 for Be.

80. The first and second ionization energies, in kilojoules per mole, for gold, mercury, and thallium, are Au ($I_1 = 890.1$, $I_2 = 1980$), Hg ($I_1 = 1007.1$, $I_2 = 1810$), and Tl ($I_1 = 589.4$, $I_2 = 1971$). (a) Which element has the largest difference between I_1 and I_2, and why would you expect this to be the case? (b) Why is I_1 for thallium lower than that for gold? Lower than that for mercury?

*81. If all other rules governing electron configurations were valid, what would be the electron configuration of rubidium if (a) there were *three* values of m_s instead of two and (b) the quantum number l could have the value n, as well as its other values. Would sodium and rubidium fall in the same group of elements in a periodic table based on electron configurations in either case? Explain.

82. In a hydrogen atom, the 1s subshell is at an energy of -1.31×10^3 kJ/mol, whereas in a helium atom it is at

-2.37×10^3 kJ/mol. Explain why the level is not the same in the two cases. For hydrogen-like atoms, Bohr's equation can be written as $E_n = -B\, Z^2/n^2$ (see Chapter 7, Problem 83). Calculate the energy of the 1s subshell for He^+ using this equation, and explain why the value you get is not the same as that given for helium.

83. The fourth ionization energy of aluminum is over four times its value for I_3. However, I_4 for gallium is only about twice its value for I_3. In light of the fact that both aluminum and gallium are group 3A elements, suggest an explanation.

84. Magnesium exhibits *two* large "jumps" in its ionization energies. One is between I_2 and I_3. Where would you expect to find the other?

85. Calculate I_1 for hydrogen (a) using Bohr's equation for the energy levels of the hydrogen atom, $E_n = -B/n^2$, and (b) using the shortest wavelength line in the Balmer series of the hydrogen spectrum and the longest wavelength line in the Lyman series. Explain why they give the same result.

*86. The trend in I_1 values (Table 8.5) and explanations for small irregularities were presented in the text for the second-period elements. In a similar way, discuss the trend and irregularities in the I_2 values for the second-period elements found in Table 8.5.

87. The production of gaseous chloride ions from chlorine molecules can be considered a two-step process in which the first step is

$$Cl_2(g) \longrightarrow 2\,Cl(g) \qquad \Delta H = +242.8 \text{ kJ}$$

What is the second step? Is the overall process endothermic or exothermic?

*88. Use ionization energies and electron affinities to determine whether the following reaction is endothermic or exothermic.

$$Mg(g) + 2\,Cl(g) \longrightarrow Mg^{2+}(g) + 2\,Cl^-(g)$$

*89. With the help of Figure 8.14, determine which of the following combinations can best be represented by the atomic and/or ionic models shown here: (a) Fe, Y, Mn; (b) K, Zr, Be; (c) B, Mg, Ru^{3+}; (d) N, C, In^+.

90. In this text, ionization energies are given in the unit kJ/mol. Another way of expressing these quantities is in terms of a single atom rather than a mole of atoms, using the unit electron volts per atom, eV/atom. Use physical constants and other data from the appendices to show that 1 eV/atom = 96.49 kJ/mol.

Apply Your Knowledge

*91. **[Laboratory]** Gaseous atoms can be ionized when struck by sufficiently energetic photons of electromagnetic radiation. At photon energies just corresponding to I, electrons are ejected with no energy to spare. At higher photon energies, the ejected electrons acquire some kinetic energy. The relationship between photon energy ($h\nu$), ionization energy (I), and the kinetic energies ($\frac{1}{2}mv^2$) of the ejected electrons is

$$h\nu = I + \frac{1}{2}mv^2$$

By knowing the frequency of the radiation used and measuring the kinetic energy of the ejected electrons, we can calculate the ionization energy.

In the equation, we have written I rather than I_1 because the method is not limited to ejecting an electron from the highest-energy subshell. A graph of the number of electrons ejected (or a property related to it) versus the calculated ionization energies consists of a series of peaks corresponding to ionization from different subshells. The technique of producing such graphs is called *photoelectron spectroscopy*. The graph shown is for the element boron.

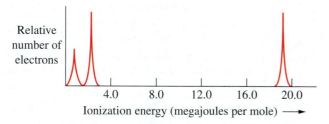

(a) What subshell is represented by each peak in the graph?

(b) Why are two of the peaks similar in size, but one peak smaller than the other two?

(c) Is there a relationship between the ionization energies shown here and the successive ionization energies for boron on page 319? Explain.

(d) What is the maximum wavelength of radiation that can be used to obtain the complete photoelectron spectrum of boron?

(e) Sketch a graph similar to the one provided (but not to scale) for the expected photoelectron spectrum of aluminum.

*92. **[Historical]** In Mendeleev's time, indium and uranium were both known, and atomic weights had been assigned to them. Mendeleev corrected both of these atomic weights to conform to his periodic table. In his work, Mendeleev used an empirical formula established by Dulong and Petit in 1818 for the elements in their solid states.

$$\text{Atomic weight} \times \text{specific heat } (\text{cal g}^{-1}\,{}^{\circ}\text{C}^{-1}) \approx 6.4$$

(a) Indium oxide was known to be 82.5% In and was thought to have the formula InO. Mendeleev thought that this formula was in error. Also, he measured the specific heat of indium and found it to be 0.055 cal/g. Where would indium be placed in the periodic table on the inside front cover if the correct formula of its oxide was InO? What do you suppose Mendeleev thought the formula of the oxide to be? Where do you suppose he placed indium in his table?

(b) In 1885, a chloride of uranium was found to be 37.34% Cl by mass. The specific heat of uranium was found to be 0.0276 cal g^{-1} °C^{-1}. Show that these data are in agreement with Mendeleev's assignment of 240 as the atomic weight of uranium.

*93. **[Collaborative]** In an alternate universe, the s energy subshells hold 3 electrons; p subshells, 9; d subshells, 15; and f subshells, 21. Assume the same order of filling of subshells as on Earth. (a) Sketch a periodic table for the elements in this universe, using the atomic numbers for elements through $Z = 177$. (b) Denote the s-, p-, d-, and f-blocks in the new table. (c) Do any of the atomic numbers in this hypothetical table fall in the same relative positions as in our periodic table? (d) What are the atomic numbers of the most metallic group of elements? Of the noble gases? (e) Write the electron configurations of (i) two main-group metals, (ii) two nonmetals, and (iii) two transition elements that should closely resemble each other in physical and chemical properties. (f) What is the atomic number of the largest atom in this table? The most nonmetallic element?

*94. **[Historical]** Some of the X-ray data that H. G. J. Moseley obtained, as represented in Figure 8.5, are summarized below.

Element	λ (cm $\times$ 10^8)
Ca	3.368
Ti	2.758
V	2.519
Cr	2.301
Mn	2.111
Fe	1.946
Co	1.798
Ni	1.662
Cu	1.549
Zn	1.445

Moseley was able to fit these data to the equation, $\nu = A(Z - b)^2$, where ν is the X-ray frequency, Z is the atomic number, and A and b are constants. (a) Use a graphical method to determine the values of A and b in this equation. (b) Use an algebraic method to determine these same values and compare the results. (c) Moseley predicted a new element with atomic number 43, because the X-ray wavelength corresponding to this element was missing from the series of wavelengths he compiled. What is the value of λ for this element?

e-Media Problems

The activities described in these problems can be found in the e-Media Activities and Interactive Student Tutorial (IST) modules of the Companion Website, *http://chem.prenhall.com/hillpetrucci*.

95. As shown in the **Electron Configurations** animation (*Section 8-3*), **(a)** how do the orbital energy diagrams for one-electron atoms or ions differ from those of atoms or ions containing more than one electron? **(b)** How might this difference be observed qualitatively in the emission spectrum of hydrogen and helium?

96. How do the principles presented in the **Gain and Loss of Electrons** animation (*Section 8-7*) allow you to account for the difference in trends of atomic and ionic radii? For the second-period elements, write out the equation for the most probable ionization process for each element. Use this information to describe the difference in atomic and ionic trends.

97. **(a)** What causes the dramatic difference in the third and fourth ionization energies of aluminum illustrated in the **Ionization Energy** (*Section 8-7*) animation? **(b)** Which family of elements is expected to show a dramatic difference between the first and second ionization energies?

98. **(a)** Describe the periodic trend observed in the color of emitted light in the **Flame Tests for Metals** movie (*Section 8-9*). **(b)** According to this trend, what color would characterize the light emitted by rubidium?

99. **(a)** From the principles outlined in the **Periodic Trends: Acid–Base Behavior of Oxides** animation (*Section 8-9*), predict whether N_2O_3 or N_2O_5 will be more acidic. **(b)** Where on the periodic table are the elements whose oxides exhibit amphoteric behavior, relative to the position of elements whose oxides exhibit acidic or basic behavior?

Chapter 9

Chemical Bonds

FORCES CALLED chemical bonds hold atoms together in molecules and keep ions in place in solid ionic compounds. Ultimately, most properties of a material can be traced to the nature of its chemical bonds.

We will use the concepts of chemical bonding described in this and the next two chapters to answer such important questions as

- Why do carbon atoms seem always to form four bonds in their compounds with other atoms?
- Why is a double bond between two carbon atoms stronger than a single bond, but not twice as strong?
- Why are some compounds solids at room temperature, whereas others are liquids or even gases at room temperature?
- Why is water (H_2O) a liquid at room temperature, whereas a substance having a similar formula—hydrogen sulfide (H_2S)—is a gas?

Later in the text we will find answers to other bonding questions, including

- Why are carbon atoms the key atoms of the molecules in living things—fats, proteins, carbohydrates, and nucleic acids?
- Although both carbon monoxide and oxygen molecules form bonds to hemoglobin, why does oxygen sustain life, whereas carbon monoxide kills?
- How is the energy required to maintain a heartbeat, breathing, and other physical functions stored in chemical bonds?

9.1 Chemical Bonds: A Preview

Chemical bonds are electrical forces; they reflect a balance in the forces of attraction and repulsion between electrically charged particles. In Figure 9.1a, we represent the attractions and repulsions in the molecular ion H_2^+. The two nuclei (protons) repel each other, and the single electron is simultaneously attracted to both nuclei.

◀ In the 1960s, the U.S. National Cancer Institute started a program of screening the extracts taken from a variety of natural sources for biological activity. An extract from the bark of the Pacific yew tree (*Taxus brevifolia*) was found to be effective against a broad range of rodent tumors. Researchers isolated the active compound, paclitaxel (Taxol) and determined its structure (insert). Paclitaxel is now widely used against ovarian and breast cancers. The locations and types of bonds in paclitaxel dictate, in part, its useful properties.

H₂ Bond Formation animation

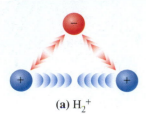

(a) H_2^+

(b) H_2

▲ **FIGURE 9.1**
Electrostatic attractions and repulsions in two simple molecular species

The electrostatic attractions (red) and repulsions (blue) in H_2^+ and H_2 are described in the text.

▲ Gilbert Newton Lewis (1875–1946) was one of the foremost American chemists of the first half of the twentieth century. In addition to his pioneer work in describing chemical bonding, Lewis was a driving force in bringing thermodynamics into mainstream chemistry; he also made important contributions to acid–base theory.

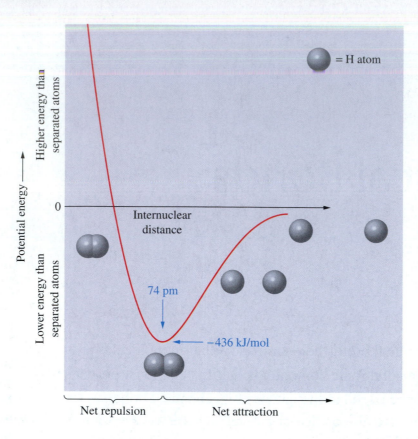

◄ **FIGURE 9.2   The energy of interaction of two hydrogen atoms**

The molecular models and graph together show that the lowest energy for two H atoms comes at the internuclear distance in the H_2 molecule, 74 pm.

QUESTION: How might the potential energy diagram for the interaction of two Cl atoms differ from the one shown here?

Figure 9.1b represents the attractions and repulsions in the H_2 molecule. Here the two nuclei also repel each other, and each of the two electrons is simultaneously attracted to both nuclei. In addition, the two electrons repel each other.

We can assess the relative importance of the attractive and repulsive forces represented in Figure 9.1b by examining the changes in potential energy that occur as two H atoms come together. This process is suggested in Figure 9.2. The point of zero in potential energy occurs when two unbonded H atoms are so far apart that no forces of attraction or repulsion exist between them. The curve representing the potential energy of the system of two H atoms begins at the far right of Figure 9.2, where the internuclear distance is large. At that point, the potential energy is slightly below zero because the attractive forces between the electrons and the atomic nuclei slightly outweigh repulsive forces. Moving to the left along the curve, the excess of attractive over repulsive forces steadily increases as the two H atoms approach each other, with the attractive forces reaching a maximum at an internuclear distance of 74 pm. At this distance, the potential energy of the pair of H atoms reaches a minimum. At shorter internuclear distances, repulsive forces lead to an increase in potential energy. As a result, two H nuclei brought closer together than 74 pm tend to move back to the internuclear distance of 74 pm. Two H atoms at an internuclear distance of 74 pm and a potential energy of −436 kJ/mol correspond to an H_2 molecule in its lowest energy state.

Through appropriate measurements, scientists can determine the internuclear distances that correspond to the lowest energy states of molecules. Then they can use quantum mechanical calculations to develop a theoretical model that fits the experimental measurements. We will explore such descriptions of chemical bonding in the next chapter. For most of this chapter, however, we will use a simpler approach to chemical bonding, an approach that predates quantum mechanics but is often still used by chemists.

9.2   The Lewis Theory of Chemical Bonding: An Overview

In the years during and following World War I, two Americans, G. N. Lewis and Irving Langmuir, and a German, Walther Kossel, independently published similar ideas about chemical bonding. These ideas, presented here in modern form, are generally referred to as the *Lewis theory*:

- Electrons, particularly valence electrons, play a fundamental role in chemical bonding.

- When metals and nonmetals combine, valence electrons usually are transferred from the metal atoms to the nonmetal atoms. Cations and anions are formed, and the electrostatic forces of attraction between the ions give rise to *ionic bonds*. Examples of compounds whose constituent atoms exist as ions joined through ionic bonds include $NaCl$, KBr, and MgO.

- In combinations involving only nonmetal atoms, one or more *pairs* of valence electrons are *shared* between the bonded atoms, producing *covalent bonds*. Examples of molecular compounds in which H atoms are bonded to another nonmetal atom through covalent bonds include H_2O, NH_3, and CH_4.

- In losing, gaining, or sharing electrons to form chemical bonds, atoms tend to acquire the electron configurations of noble gases. We might refer to this as a noble-gas rule. In forming chemical bonds, H, Li, and Be tend to follow a *duet rule*, in which they acquire a helium electron configuration, $1s^2$. The other main-group elements follow an **octet rule,** meaning they tend to acquire an electron configuration like that of the noble gases, with eight electrons in the valence shell: ns^2np^6 (where $n = 2, 3, \dots$). The octet rule is less applicable to the transition elements, primarily because most transition metal ions do not have a noble-gas electron configuration (recall Table 8.3).

In preparing to apply Lewis's theory to specific examples of ionic and covalent bonding, we will next consider a special set of symbols Lewis used to represent his ideas.

Lewis Symbols

In a **Lewis symbol,** the chemical symbol for the element represents the nucleus and core electrons of the atom, and dots around the symbol represent the valence electrons. Because Lewis theory is closely linked to noble-gas electron configurations, we generally write Lewis symbols only for the elements that acquire such configurations when they form bonds. For the most part, these are main-group elements, and their numbers of valence electrons are equal to their group numbers in the periodic table. For the second-period elements, for example,

1A	2A	3A	4A	5A	6A	7A	8A
Li·	Be·	·B·	·C̈·	·N̈:	·Ö:	:F̈:	:N̈e:

In writing Lewis symbols, we will follow Lewis's practice of placing the first four dots one on each of the four sides of the chemical symbol and then pairing dots as we add the next four. Lewis symbols do not specifically reflect electron pairing; the idea of electron spin had not yet been established when he proposed the theory. Thus, although the atoms Be, B, and C have zero, one, and two unpaired electrons in their electron configurations, respectively, the Lewis symbols of Be, B, and C have two, three, and four separate dots. Interestingly, as we will see in Chapter 10, Lewis symbols sometimes predict chemical bonding better than ground-state electron configurations do.

Example 9.1

Give Lewis symbols for magnesium, silicon, and phosphorus.

 **Periodic Trends: Lewis Structures activity**

STRATEGY

The two main ideas we must apply are (1) the Lewis symbol of an atom depends only on the group of the periodic table in which that atom is found and (2) the group number tells us the number of valence-shell electrons that must appear as dots in the Lewis symbol.

SOLUTION

Each of these three elements has the same valence-shell electron configuration and thus the same distribution of dots in its Lewis symbol as the second-period element preceding it in its

Problem-Solving Note

It makes no difference on which side of the symbol we begin adding dots. Here we have placed them around the symbol in a clockwise direction, starting at the three-o'clock position.

Electron Dot Structures II activity

▲ **FIGURE 9.3   The reaction of sodium and chlorine**

Sodium metal and chlorine gas provide striking visual evidence of their reaction to produce the solid ionic substance sodium chloride.

Formation of Sodium Chloride movie

group. Consequently, the Lewis symbols of magnesium (group 2A), silicon (group 4A), and phosphorus (group 5A) resemble those of Be, C, and N, respectively:

$$\text{Mg·} \qquad \text{·Si·} \qquad \text{·P·}$$

EXERCISE 9.1A

Give the Lewis symbol for **(a)** Ar, **(b)** Br, and **(c)** K.

EXERCISE 9.1B

Give the Lewis symbol for **(a)** arsenic, **(b)** rubidium, and **(c)** tellurium.

Ionic Bonding

Figure 9.3 shows the reaction between sodium and chlorine. Sodium is a soft, low-density, silvery metal, and chlorine is a yellow-green, toxic, gaseous nonmetal. They react to produce sodium chloride, a white crystalline solid. One way to demonstrate that sodium chloride is an *ionic* compound is to measure the electrical conductivity of its aqueous solutions. In Chapter 4, we saw that NaCl is completely dissociated into ions in aqueous solutions and that NaCl(aq) is a good electrical conductor—a strong electrolyte. We begin our discussion of chemical bonding with ionic bonding because conceptually this is the easiest type to describe.

9.3   Ionic Bonds and Ionic Crystals

Let's use the electron configurations of Na and Cl atoms to interpret the reaction between sodium and chlorine. By losing an electron, a sodium atom, Na, forms the cation Na$^+$, which has the same electron configuration as the noble gas neon:

$$\text{Na} \longrightarrow \text{Na}^+ + e^-$$

Electron configuration: $\underbrace{1s^22s^22p^63s^1}_{\text{Na}} \qquad \underbrace{1s^22s^22p^6 = [\text{Ne}]}_{\text{Na}^+}$

By gaining an electron, a chlorine atom, Cl, forms the anion Cl$^-$, which has the same electron configuration as the noble gas argon:

$$\text{Cl} + e^- \longrightarrow \text{Cl}^-$$

Electron configuration: $\underbrace{[\text{Ne}]3s^23p^5}_{\text{Cl}} \qquad \underbrace{[\text{Ne}]3s^23p^6 = [\text{Ar}]}_{\text{Cl}^-}$

These two processes occur together in the reaction shown in Figure 9.3. That is, the sodium atoms lose electrons and the chlorine atoms gain them. In writing a chemical equation for the reaction, we can represent the Na atoms in solid sodium as Na(s), but the Cl atoms in gaseous chlorine exist as diatomic molecules, Cl_2(g). Thus, our equation should be based on the transfer of one electron from each of two Na atoms to the two Cl atoms in one Cl_2 molecule:

$$2\,\text{Na(s)} + \text{Cl}_2\text{(g)} \longrightarrow 2\,\text{Na}^+\text{Cl}^-\text{(s)}$$

In giving up an electron, a sodium atom does not become a neon atom. The sodium ion and neon atom are isoelectronic, but the sodium ion has 11 protons in its nucleus and a charge of 1+, whereas the neon atom has 10 protons in its nucleus and is electrically neutral. Similarly, a chlorine atom does not become an argon atom.

Because the two ions formed in the reaction between a sodium atom and a chlorine atom have opposite charges, they are strongly attracted to each other and form an *ion pair* (Na$^+$Cl$^-$). However, a given sodium ion in solid sodium chloride is not attracted to just one chloride ion. Rather, each sodium ion is surrounded by and attracted strongly to six neighboring chloride ions and also attracted, more weakly, to the more distant chloride ions. Likewise each chloride ion strongly attracts (and is attracted by) six neighboring sodium ions and also attracts the more distant sodium ions much less strongly. Although ions of like charge repel one another, the attractive

Sodium Chloride 3D model

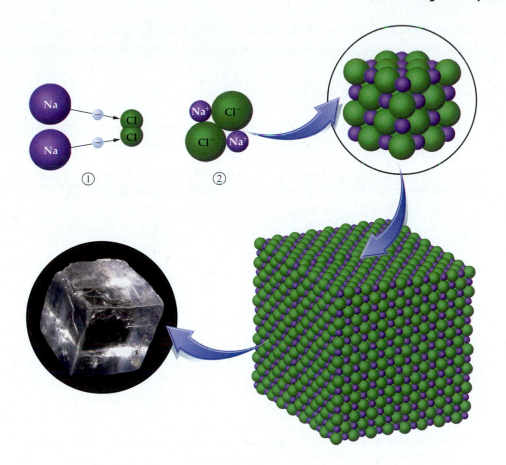

◀ **FIGURE 9.4   Formation of a crystal of sodium chloride**
The formation of two ion pairs from two atoms of sodium and a molecule of chlorine is depicted in steps 1 and 2. In the formation of crystalline NaCl, each Na^+ ion (purple) is surrounded by six Cl^- ions (green). In turn, each Cl^- ion is surrounded by six Na^+ ions. This arrangement repeats itself many times over, ultimately resulting in a crystal of sodium chloride.

QUESTION: Why have we changed the sizes of the purple and green spheres in going from step 1 to step 2?

forces predominate. The net result of all the interactions is an extensive arrangement of ions in a regular pattern of alternating cations and anions, as seen in Figure 9.4.

The net attractive electrostatic forces that hold the cations and anions together are called **ionic bonds,** and the highly ordered solid collection of ions is called an *ionic crystal.* In general, a **crystal** is a distinctive pattern of particles that repeats over and over at the *microscopic* level to produce a solid structure with a regular geometric shape at the *macroscopic* level. Figure 9.4 suggests the progressive stages by which an ionic crystal of sodium chloride is formed from individual ion pairs. We have illustrated ionic bonding and crystal formation with sodium chloride, but these processes are relevant to ionic compounds in general.

9.4   Using Lewis Symbols to Represent Ionic Bonding

Lewis developed his bonding theory primarily to describe covalent bonding, but we can also use it to depict ionic bonding. However, because we limit our use of Lewis symbols to atoms that can acquire noble-gas electron configurations, we will use Lewis symbols to represent ionic bonding only between nonmetals and the *s*-block metals, the *p*-block metal aluminum, and a few *d*-block metals.

Suppose that instead of using complete electron configurations to represent the loss and gain of electrons, we use Lewis symbols:

$$\left[Na\cdot\right] \longrightarrow \left[Na\right]^+ + e^-$$

$$\left[:\overset{\cdot}{\underset{\cdot\cdot}{Cl}}:\right] + e^- \longrightarrow \left[:\overset{\cdot\cdot}{\underset{\cdot\cdot}{Cl}}:\right]^-$$

Because these processes occur together, we can use an equation to represent the net result:

$$\left[Na\cdot\right] + \left[:\overset{\cdot}{\underset{\cdot\cdot}{Cl}}:\right] \longrightarrow \left[Na\right]^+ + \left[:\overset{\cdot\cdot}{\underset{\cdot\cdot}{Cl}}:\right]^-$$

As another example, consider the reaction of magnesium, a group 2A metal, with oxygen, a group 6A nonmetal, to form a stable white crystalline solid, magnesium oxide (MgO):

$$\left[Mg\cdot \right] + \left[:\ddot{O}: \right] \longrightarrow \left[Mg \right]^{2+} + \left[:\ddot{\ddot{O}}: \right]^{2-}$$

In order to acquire the electron configuration of the noble gas neon, a magnesium atom must give up two electrons and an oxygen atom must gain two electrons.

Oxygen atoms, which need two electrons to complete an octet, can react with lithium atoms, which have only one valence electron to give. In this case, two atoms of lithium are needed for each oxygen atom. The product is lithium oxide, Li_2O:

$$\begin{bmatrix} Li\cdot \\ Li\cdot \end{bmatrix} + \left[:\ddot{O}: \right] \longrightarrow \begin{bmatrix} Li \end{bmatrix}^{+} \\ \begin{bmatrix} Li \end{bmatrix}^{+} + \left[:\ddot{\ddot{O}}: \right]^{2-}$$

or

$$2\left[Li\cdot \right] + \left[\cdot\ddot{O}: \right] \longrightarrow 2\left[Li \right]^{+} + \left[:\ddot{\ddot{O}}: \right]^{2-}$$

Lithium atoms have only three electrons. In the reaction with oxygen, each lithium atom loses one electron to become Li^+, and in doing so it acquires the $1s^2$ electron configuration of helium.

Example 9.2

Use Lewis symbols to show the formation of ionic bonds between magnesium and nitrogen. What are the name and formula of the compound that results?

STRATEGY

First, we determine the number of electrons that must be lost by magnesium atoms and gained by nitrogen atoms so that the resulting ions have noble-gas electron configurations. Then we can combine Lewis symbols for the ions in the appropriate proportions to produce a neutral formal unit, from which we can deduce the compound's name and formula.

SOLUTION

Mg atoms (group 2A) lose their two valence electrons, and N atoms (group 5A) gain three additional valence electrons. To produce an electrically neutral formula unit, three Mg atoms must lose a total of six electrons and two N atoms must gain a total of six:

$$\begin{bmatrix} \cdot\ddot{N}\cdot \\ \cdot \end{bmatrix} \begin{bmatrix} \cdot\ddot{N}\cdot \\ \cdot \end{bmatrix} \\ \left[Mg\cdot \right] \left[Mg\cdot \right] \left[Mg\cdot \right] \longrightarrow 3\left[Mg \right]^{2+} + 2\left[:\ddot{N}: \right]^{3-}$$

The compound is magnesium nitride, Mg_3N_2.

EXERCISE 9.2A

Use Lewis symbols to show the formation of ionic bonds between barium and iodine. What are the name and formula of the compound that results?

EXERCISE 9.2B

Use Lewis symbols to show the formation of ionic bonds in aluminum oxide.

9.5 Energy Changes in Ionic Compound Formation

In Figure 9.2, we illustrated the tendency for two H atoms to reach a lower energy state by forming the molecule H_2. Metal and nonmetal atoms show a similar tendency to reach a lower energy state by forming ionic bonds. Let's see how this works with sodium chloride.

Recall from Chapter 8 that the energy needed to remove the valence electron from a gaseous sodium atom is the first ionization energy (I_1). The energy released when an electron is added to a gaseous chlorine atom is the electron affinity (*EA*) of chlorine:

$$Na(g) \longrightarrow Na^+(g) + e^- \qquad I_1 = +496 \text{ kJ/mol}$$

$$Cl(g) + e^- \longrightarrow Cl^-(g) \qquad EA = -349 \text{ kJ/mol}$$

The net energy change for the transfer of an electron from an isolated sodium atom to an isolated chlorine atom is $(496 - 349)$ kJ/mol $= +147$ kJ/mol. Based on this calculation alone, the simultaneous formation of separate sodium cations and chloride anions from gaseous atoms is not energetically favorable. However, there are several other issues to consider.

To begin, we note that the reaction in Figure 9.3 is between *solid* sodium, Na(s), and *gaseous* chlorine, $Cl_2(g)$, to form *solid* sodium chloride, NaCl(s). The enthalpy change in the reaction is the standard enthalpy of formation of NaCl(s):

$$Na(s) + \tfrac{1}{2} Cl_2(g) \longrightarrow NaCl(s) \qquad \Delta H_f^\circ = -411 \text{ kJ}$$

This negative enthalpy change indicates that the reaction is energetically favorable. That is, the system loses energy, and the ionic compound NaCl(s) is in a lower energy state than the elements. We can reach the same conclusion by considering a *hypothetical* multistep process known as a *Born–Haber cycle*.

We need to imagine a way to convert Na(s) and $Cl_2(g)$ to NaCl(s) in a series of steps for which we can obtain ΔH values. When we start with 1 mol of Na(s) and $\tfrac{1}{2}$ mol of $Cl_2(g)$, the combination of the steps results in the overall reaction in which 1 mol of NaCl(s) is obtained as the sole product. The sum of ΔH values for the individual processes is ΔH_f° for NaCl(s), in accordance with Hess's law (Section 6.6).

In the following outline, each process is described in the column of text on the left; an equation for that process is written in the center column; and the energy change for the process, written as an enthalpy change (ΔH), is given in the column on the right. At the end of the outline, we obtain the overall reaction by combining the reactions of the individual processes; the enthalpy of the overall reaction is the sum of the individual ΔH values. The cumulative process is also represented schematically by the enthalpy diagram in Figure 9.5.

A Born–Haber cycle is simply a useful application of Hess's law to ionic solids.

 **Born–Haber Cycle activity**

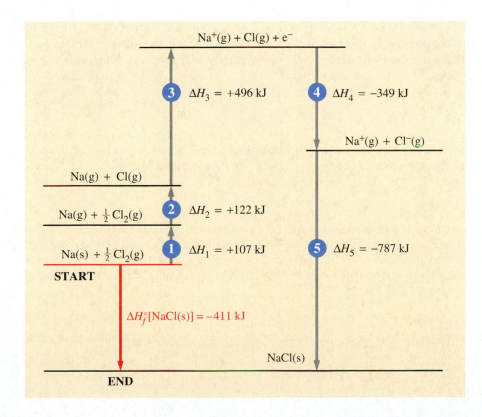

◀ **FIGURE 9.5 Starting point and five steps of Born–Haber cycle for 1 mol of sodium chloride**

The sum of the five enthalpy changes (ΔH_1 to ΔH_5) gives ΔH_f°. (The ΔH values are not to scale.) The equivalent one-step reaction for the formation of NaCl(s) directly from Na(s) and $\tfrac{1}{2} Cl_2(g)$ is shown in red.

QUESTION: Suppose ΔH_1 were +207 kJ (it cannot be, but this is just a hypothetical question). How would the lattice energy ΔH_5 change if all other enthalpies remained the same?

Process	Equation	Enthalpy Change
1 *Conversion of solid Na to gaseous Na atoms.* The energy required to convert 1 mol of solid to gas is the *enthalpy of sublimation.* For 1 mol Na(s), this is 107 kJ.	$Na(s) \longrightarrow Na(g)$	$\Delta H_1 = +107$ kJ
2 *Dissociation of Cl_2 molecules to Cl atoms.* The energy required to break 1 mol of bonds in $Cl_2(g)$ molecules is called the *bond-dissociation energy,* a concept we will look at more closely in Section 9.10. For $Cl_2(g)$, this is 243 kJ/mol. To break the bonds in $\frac{1}{2}$ mol $Cl_2(g)$ requires $\frac{1}{2}$ mol $\times$ 243 kJ/mol = 122 kJ.	$\frac{1}{2} Cl_2(g) \longrightarrow Cl(g)$	$\Delta H_2 = +122$ kJ
3 *Ionization of* Na(g) *atoms to form* $Na^+(g)$ *ions.* 1 mol Na(s) is converted to 1 mol $Na^+(g)$. The energy requirement is the *first ionization energy,* $I_1 = 496$ kJ/mol.	$Na(g) \longrightarrow Na^+(g) + e^-$	$\Delta H_3 = +496$ kJ
4 *Conversion of* Cl(g) *atoms to* $Cl^-(g)$ *ions.* 1 mol Cl(g) is converted to 1 mol $Cl^-(g)$. The energy requirement is the *electron affinity, EA* = -349 kJ/mol.	$Cl(g) + e^- \longrightarrow Cl^-(g)$	$\Delta H_4 = -349$ kJ
5 *Assembly of* Na^+ *and* Cl^- *ions into a crystal.* 1 mol NaCl(s) is formed from 1 mol $Na^+(g)$ and 1 mol $Cl^-(g)$. The energy of formation of 1 mol of an ionic solid from its separated gaseous ions is the **lattice energy.** For NaCl(s), this is -787 kJ.	$Na^+(g) + Cl^-(g) \longrightarrow NaCl(s)$	$\Delta H_5 = -787$ kJ
Overall: Combine the equations into an overall equation.	$Na(s) + \frac{1}{2} Cl_2(g) \longrightarrow NaCl(s)$	$\Delta H_f^\circ = \Delta H_1 + \Delta H_2 + \Delta H_3 + \Delta H_4 + \Delta H_5$ $\Delta H_f^\circ = -411$ kJ

A commonly used convention defines lattice energy in terms of the breakup of a crystal rather than its formation. By that convention, all lattice energies are positive quantities and $\Delta H_5 = -$(lattice energy).

From this analysis, we see that the large negative value of the lattice energy (ΔH_5) is the major factor that makes ionic compound formation an energetically favorable process. In practice, we cannot measure lattice energies directly. In fact, we generally use the Born–Haber cycle to calculate a lattice energy from other measured quantities, as illustrated in Example 9.3.

Example 9.3

Use the following data to determine the lattice energy of $MgF_2(s)$: enthalpy of sublimation of magnesium, +146 kJ/mol; I_1 for Mg, +738 kJ/mol; I_2 for Mg, +1451 kJ/mol; bond-dissociation energy of $F_2(g)$, +159 kJ/mol F_2; electron affinity of F, -328 kJ/mol F; enthalpy of formation of $MgF_2(s)$, -1124 kJ/mol.

STRATEGY

The approach we need here differs in three ways from the one used for NaCl(s). (1) The compound $MgF_2(s)$ has two anions for each cation in the crystal. Therefore, in the step where $F_2(g)$ dissociates, we need the bond-dissociation energy based on 1 mol of $F_2(g)$ in order to get 2 mol of F(g). Similarly, in the electron affinity step, we must produce 2 mol of $F^-(g)$ rather than 1 mol. (2) Because the magnesium cation carries a 2+ charge, we must include two ionization steps and both I_1 and I_2 in our calculation. (3) The enthalpy of formation of $MgF_2(s)$ is given, and our unknown is the lattice energy.

SOLUTION

The setup that follows incorporates all of the steps needed to determine the unknown lattice energy of $MgF_2(s)$ from the given data.

Sublimation:	$Mg(s) \longrightarrow Mg(g)$	$\Delta H_1 = +146 \text{ kJ}$
Bond dissociation:	$F_2(g) \longrightarrow 2F(g)$	$\Delta H_2 = +159 \text{ kJ}$
First ionization:	$Mg(g) \longrightarrow Mg^+(g) + e^-$	$\Delta H_3 = +738 \text{ kJ}$
Second ionization:	$Mg^+(g) \longrightarrow Mg^{2+}(g) + e^-$	$\Delta H_4 = +1451 \text{ kJ}$
Electron gain:	$2F(g) + 2e^- \longrightarrow 2F^-(g)$	$\Delta H_5 = 2(-328 \text{ kJ})$
Unknown:	$Mg^{2+}(g) + 2F^- \longrightarrow MgF_2(s)$	$\Delta H_6 = \text{lattice energy}$
Overall:	$Mg(s) + F_2(g) \longrightarrow MgF_2(s)$	$\Delta H_f^\circ = -1124 \text{ kJ}$

The enthalpy of formation is equal to the sum of the energies of the individual processes.

$$\Delta H_f^\circ = -1124 \text{ kJ} = (146 + 159 + 738 + 1451 - 656) \text{ kJ} + \text{lattice energy}$$

Rearranging the summed expression, we solve for the lattice energy.

Lattice energy $= (-1124 - 146 - 159 - 738 - 1451 + 656) \text{ kJ} = -2962 \text{ kJ}$

The lattice energy is $-2962 \text{ kJ/mol } MgF_2(s)$.

ASSESSMENT

As in other applications of Hess's law, we ensure that each enthalpy change is written with the correct sign and that all the proper species cancel when individual equations are summed to give the overall equation. An additional check we have in this particular calculation is that the lattice energy must have a negative value because we defined lattice energy as the energy released when gaseous ions assemble into a crystal.

EXERCISE 9.3A

For lithium, the enthalpy of sublimation is $+159 \text{ kJ/mol}$ and the first ionization energy is $+520 \text{ kJ/mol}$. The bond-dissociation energy of fluorine is $+159 \text{ kJ/mol } F_2$, and the electron affinity of fluorine is -328 kJ/mol. The lattice energy of LiF is -1047 kJ/mol. Calculate the overall enthalpy change for the reaction

$$Li(s) + \tfrac{1}{2} F_2(g) \longrightarrow LiF(s) \qquad \Delta H_f^\circ = ?$$

EXERCISE 9.3B

Use the data provided in this section and the enthalpy of formation of lithium chloride, $\Delta H_f^\circ = -409 \text{ kJ/mol } LiCl(s)$, to determine the lattice energy of LiCl.

Covalent Bonding

As we noted in Figure 9.2, the energy state of two H atoms joined by a chemical bond in a molecule of H_2 is lower than the energy state of two isolated H atoms. However, the bond cannot be ionic, because all hydrogen atoms have an equal electron affinity, and therefore one hydrogen atom is unlikely to accept an electron from another. They also have a high ionization energy ($I_1 = 1312 \text{ kJ/mol}$), indicating that hydrogen atoms lose electrons only with great difficulty. Lewis proposed that in such cases, the atoms are bonded by a pair of electrons *shared* between them. This arrangement is called a **covalent bond** and is represented by a pair of dots (electrons) between two Lewis symbols:

Covalent bond
(shared pair
of electrons)

$$H\cdot \; + \; \cdot H \; \longrightarrow \; H\!:\!H$$

9.6   Lewis Structures of Some Simple Molecules

A representation such as H : H for the H_2 molecule is called a **Lewis structure,** defined as a combination of Lewis symbols that represents the formation of covalent bonds between atoms. Note that

- A Lewis structure indicates the proportions in which atoms combine.
- In most cases, a Lewis structure shows the bonded atoms as having the electron configuration of a noble gas; that is, the atoms obey the octet rule (except for H atoms, which obey the duet rule).

Notice that if we count both electrons in the Lewis structure H : H as "belonging" first to one of the H and then to the other—a process of double-counting—each H atom appears to have two electrons in its valence shell, the electron configuration of helium.

Consider next the case of chlorine, which also exists in the diatomic form, Cl_2. The chlorine atoms in Cl_2 are joined by a covalent bond:

$$:\overset{..}{\underset{..}{Cl}}\cdot \ + \ \cdot\overset{..}{\underset{..}{Cl}}: \ \longrightarrow \ :\overset{..}{\underset{..}{Cl}}:\overset{..}{\underset{..}{Cl}}:$$

Again, by double-counting the two electrons shared by the two atoms, we find eight electrons around each Cl atom in the Lewis structure of Cl_2; each Cl atom follows the octet rule.

The shared pairs of electrons in a molecule are called **bonding pairs.** The other electron pairs, which stay with one atom and are not shared, are called either *nonbonding pairs* or **lone pairs.** Bonding pairs (:) and lone pairs (:) in the Cl_2 molecule are shown here. We also illustrate the common practice of representing a bonding pair by a dash (—):

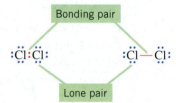

Except for boron, the nonmetals of the second period tend to form a number of covalent bonds equal to *8 minus the group number.* Thus, fluorine (group 7A) forms *one* bond (8 − 7 = 1), oxygen (group 6A), *two;* nitrogen (group 5A), *three;* and carbon (group 4A), *four.* This idea is illustrated for the molecules CH_4, NH_3, H_2O, and HF in Figure 9.6. Although the Lewis structures in Figure 9.6 are written in a way that

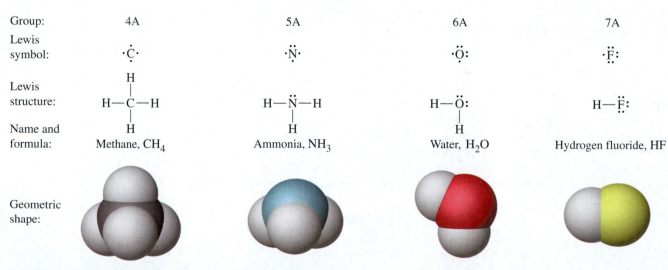

▲ FIGURE 9.6   Four hydrogen compounds of second-period nonmetals

The formulas of the four molecules can be deduced from their Lewis structures, but their geometric shapes cannot. Geometric shapes of molecules can be predicted by methods presented in Chapter 10, and they can be precisely established by experiment.

QUESTION: Why do the numbers of hydrogen atoms bonded to the central atom decrease regularly from left to right for the four molecules shown here?

suggests the shapes of the molecules, Lewis structures in themselves do *not* predict molecular shapes. A straight-line Lewis structure for water is as acceptable as the angular one shown. We will consider how to predict molecular shapes in Chapter 10.

Multiple Covalent Bonds

The covalent bonds we have considered so far have involved one shared pair of electrons, a bond type called a **single bond.** Two bonded atoms can also share more than one pair of electrons between them, resulting in a *multiple bond.* In a **double bond,** the bonded atoms share two pairs of electrons, and a **triple bond** involves the sharing of three pairs of electrons.

Let's see how to write Lewis structures involving multiple bonds. Our first attempt at a Lewis structure for carbon dioxide, CO_2, might look like this:

$$:\!\ddot{O}\!\cdot \; + \; \cdot\dot{C}\cdot \; + \; \cdot\ddot{O}\!: \; \longrightarrow \; :\!\ddot{O}\!:\!\dot{C}\!:\!\ddot{O}\!: \quad \text{or} \quad :\!\dot{O}\!-\!\dot{C}\!-\!\ddot{O}\!: \quad (not\ correct)$$

This structure is unsatisfactory because none of the atoms has acquired a valence-shell octet. However, we can obtain a satisfactory structure by shifting the four unpaired electrons into the regions between the C and O atoms, which, with the appropriate double-counting, gives each atom a valence-shell octet:

$$:\!\ddot{O}\!\!\curvearrowright\!\!\ddot{C}\!\!\curvearrowleft\!\!\ddot{O}\!: \; \longrightarrow \; :\!\ddot{O}\!=\!C\!=\!\ddot{O}\!:$$

Each O atom is joined to the C atom by a double bond. We can represent each electron pair by a single dash, and two parallel dashes therefore represent a double bond.

Our first attempt at a Lewis structure for the nitrogen molecule, N_2, looks equally bad, with neither of the nitrogen atoms having eight electrons in its valence shell:

$$:\!\dot{N}\!\cdot \; + \; \cdot\dot{N}\!: \; \longrightarrow \; :\!\dot{N}\!:\!\dot{N}\!: \quad \text{or} \quad :\!\dot{N}\!-\!\dot{N}\!\cdot \quad (incorrect)$$

Here we can produce an octet for each nitrogen atom by shifting all the unpaired electrons into the region between the two N atoms:

$$:\!N\!\!\curvearrowright\!\!\curvearrowleft\!\!N\!: \; \longrightarrow \; :\!N\!\equiv\!N\!:$$

In doing so, we obtain a triple bond. As in the single and double bond, each electron pair is represented by a dash.

The Importance of Experimental Evidence

Using the same approach we used for CO_2 and N_2, we can write a good Lewis structure for the O_2 molecule. Doing so, we obtain the structure

$$:\!\ddot{O}\!\cdot \; + \; \cdot\ddot{O}\!: \; \longrightarrow \; :\!\ddot{O}\!=\!\ddot{O}\!:$$

This Lewis structure conforms to the octet rule (each atom has eight valence electrons when we double-count the four shared electrons), and it properly designates the oxygen-to-oxygen bond as a double bond. Thus, this Lewis structure is totally consistent with the observed bond-dissociation energy of oxygen (listed in Table 9.1 and discussed in Section 9.10). However, there is one piece of experimental evidence that the Lewis structure fails to account for. As shown in Figure 9.7, oxygen gas is paramagnetic. That is, the O_2 molecule contains unpaired electrons. To explain this property of the oxygen molecule, which the Lewis structure cannot do, we need to turn to a quantum mechanical approach to chemical bonding, which we do in Chapter 10.

Liquid oxygen is paramagnetic and is held between the poles of the magnet.

▲ **FIGURE 9.7 Paramagnetism of oxygen**

Liquid oxygen is attracted into a magnetic field, the region between the poles of this large magnet. This is the expected behavior for paramagnetic substances (recall page 314).

9.7 Polar Covalent Bonds and Electronegativity

Metal atoms can transfer electrons to nonmetal atoms to form ionic bonds. Identical atoms combine by sharing pairs of electrons to form covalent bonds. What kind of bond forms between two atoms that are different from each other but not different enough to form ionic bonds? Let's consider the bond between a hydrogen atom and a chlorine atom as an example.

Electronegativity

In the HCl molecule, a hydrogen atom and a chlorine atom share a pair of electrons:

$$\text{H·} + \text{·}\overset{..}{\underset{..}{\text{Cl}}}: \longrightarrow \text{H}:\overset{..}{\underset{..}{\text{Cl}}}: \quad \text{or} \quad \text{H}—\overset{..}{\underset{..}{\text{Cl}}}:$$

However, the Lewis structure does not reveal this fact: The chlorine atom has a greater attraction for the electrons than does the hydrogen atom, and the sharing of electrons is unequal.

In Chapter 8, we encountered two atomic properties related to the attraction a given atom has for electrons: ionization energy and electron affinity. The greater the ionization energy of an atom, the greater its tendency to retain its electrons. The more negative its electron affinity, the more inclined the atom is to acquire an additional electron. However, these properties apply only to isolated gaseous atoms and not directly to atoms in molecules. **Electronegativity (EN),** which is related to ionization energy and electron affinity, is a measure of the ability of an atom to compete for electrons with other atoms to which it is bonded.

The greater the electronegativity of an atom in a molecule, the more strongly it attracts electrons to itself when bonded to other atoms.

For example, in an HCl molecule, the chlorine atom displays a greater attraction for electrons than does the hydrogen atom, and the electronegativity of chlorine is greater than that of hydrogen.

Atoms of the elements in the upper right of the periodic table—relatively small, nonmetal atoms—attract bonding electrons most strongly: They have the greatest electronegativities. Atoms of the elements toward the lower left of the table—relatively large metal atoms—have a weaker hold on electrons: They have the smallest electronegativities. Several electronegativity scales have been devised, each defined a little differently from the others. As a consequence, the same element might have somewhat different electronegativity values on the different scales. On the scale devised by Linus Pauling and shown in Figure 9.8, the most electronegative element, fluorine, is assigned a value of 4.0. The most metallic elements have electronegativities of about 1.0 or lower. All the electronegativity scales show two general trends among the elements:

Within a period, electronegativity generally increases from left to right.

Within a group, electronegativity generally increases from bottom to top.

Periodic Trends: Electronegativity animation

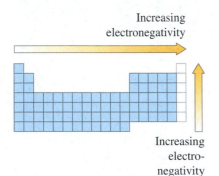

Increasing electronegativity

Increasing electronegativity

In Chapter 2, we used a portion of the periodic table (Figure 2.9) to determine which element symbol to write first when we write the formula of a compound. We can now rationalize the rule: We generally write first the symbol of the element that has the lower electronegativity.

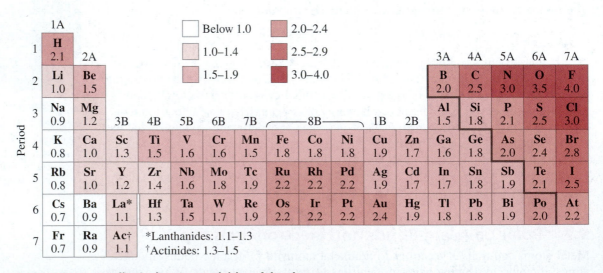

▲ **FIGURE 9.8** **Pauling's electronegativities of the elements**

Values are from L. Pauling, *The Nature of the Chemical Bond,* 3rd edition, Cornell University, Ithaca, NY, 1960, p. 93. (Some values have been modified by later investigators.) Because noble-gas compounds are limited to just a few compounds of Kr and Xe, electronegativities for the group 8A elements are not included.

In Figure 9.8, we see that the first trend is completely regular for the second period. Electronegativity increases by 0.5 per element as we move from lithium to fluorine. In other periods, the differences in electronegativity are less regular. The second trend tells us that chlorine is less electronegative than fluorine, sulfur is less electronegative than oxygen, and so on.

A comparison of electronegativities is not quite so straightforward when we consider two elements that are neither in the same period nor in the same group. As we suggest in Figure 9.9, generally the element above or to the right (or both) is more electronegative than one below or to the left (or both).

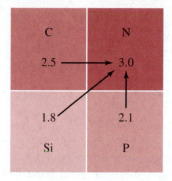

◀ **FIGURE 9.9   Electronegativities in relation to position in the periodic table**

In general, electronegativities increase in the directions of the arrows.

▲ Linus Pauling (1901–1994) used quantum theory to extend Lewis's theory. Pauling's work was summarized in the 1939 text *The Nature of the Chemical Bond,* and he was awarded the Nobel Prize in Chemistry in 1954. He also won the Nobel Peace Prize in 1962 for his fight to control nuclear weapons. His efforts were influential in the establishment of the 1963 Nuclear Test Ban Treaty. Later in life, Pauling became interested in the medical value of megadoses of vitamins, particularly vitamin C. His ideas on vitamins are controversial, but they have stimulated continuing research.

Example 9.4

Referring only to the periodic table inside the front cover, arrange the following sets of atoms in the expected order of increasing electronegativity.

(a) Cl, Mg, Si   **(b)** As, N, Sb   **(c)** As, Se, Sb

STRATEGY

Our essential task is to apply electronegativity trends relative to positions in the periodic table. Part (a) deals with the trend within a period (the third), and part (b), with the trend within a group (5A). In part (c), both period and group trends are involved.

SOLUTION

(a) Because the order of increasing electronegativity within a period is from left to right, we have Mg < Si < Cl.

(b) Because the order of increasing electronegativity within a group is from bottom to top, we have Sb < As < N.

(c) As and Se are in the same period; Se is to the right of As, and so we expect Se to be the more electronegative. As and Sb are in the same group; As is above Sb, and so we expect As to be more electronegative. Thus, As is in the center of the set of three, and the order of increasing electronegativity is Sb < As < Se.

EXERCISE 9.4A

Referring only to the periodic table inside the front cover, arrange the following sets of atoms in the expected order of increasing electronegativity.

(a) Ba, Be, Ca   **(b)** Ga, Ge, Se   **(c)** Cl, S, Te

EXERCISE 9.4B

Referring only to the periodic table inside the front cover, arrange the following elements in the expected order of decreasing electronegativity: scandium, iron, rubidium, sulfur, chlorine, sodium, boron.

Electronegativity Difference and Bond Type

The *difference* in the electronegativities of bonded atoms is relevant to chemical bonding. Electronegativity differences (ΔEN) characterize covalent bonds as being either *nonpolar* or *polar,* and the polarity of bonds can greatly influence the properties of

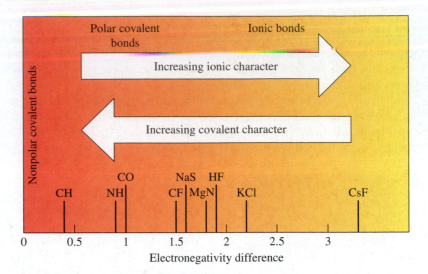

▶ FIGURE 9.10 **Electronegativity difference and bond type**
Several bonds and their ΔEN values are shown.

We classify the bond between two metal atoms in the gaseous state, as in Na₂(g), as nonpolar covalent because the electronegativity difference is zero. However, in *solid* sodium, which consists of many, many Na atoms bonded together, we observe something quite different—a metallic bond (Chapter 24). This type of bond accounts for the ability of the solid metal to conduct electric current.

molecular substances. Two identical atoms have the same electronegativity and share a bonding electron pair equally. That is, the bonding electrons are not drawn any closer to one atom than to the other, and the bond is a **nonpolar covalent** bond. The covalent bonds in H—H and Cl—Cl are nonpolar.

Even if two atoms are not identical, the covalent bond they form may be essentially nonpolar if the electronegativity difference is small. For example, the ΔEN between C (EN = 2.5) and H (EN = 2.1) is only 0.4, and therefore the C—H bonds in hydrocarbons are nearly nonpolar.

In hydrogen chloride, the ΔEN between H (EN = 2.1) and Cl (EN = 3.0) is 0.9 and the electrons in the H—Cl bond are drawn closer to chlorine, the atom of higher electronegativity. Such bonds are said to be **polar covalent.** With still larger differences in electronegativity, electrons may be completely transferred from metal atom to nonmetal atom to form *ionic* bonds.

As Figure 9.10 suggests, there is no distinct boundary between covalent and ionic bonds. Some bonds are clearly nonpolar covalent, and some are nearly 100% ionic, but many bonds have an intermediate character—they are polar covalent bonds.

Example 9.5

Use electronegativity values to arrange the following bonds in order of increasing polarity: Br—Cl, Cl—Cl, Cl—F, H—Cl, I—Cl

STRATEGY

We begin with the knowledge that the polar character of a bond is determined by electronegativity differences. For the individual electronegativities of the atoms, we turn to Figure 9.8.

SOLUTION

The electronegativities (EN) and electronegativity differences (ΔEN) are

EN:	2.8 3.0	3.0 3.0	3.0 4.0	2.1 3.0	2.5 3.0
	Br—Cl	Cl—Cl	Cl—F	H—Cl	I—Cl
ΔEN:	0.2	0.0	1.0	0.9	0.5

The order of increasing polarity is therefore Cl—Cl < Br—Cl < I—Cl < H—Cl < Cl—F.

Bond Polarity activity

Depicting Polar Covalent Bonds

According to modern quantum theory, which we will discuss in Chapter 10, we can picture the electron-pair bond between two atoms as a cloud of negative electric charge that encompasses both atoms. In a nonpolar covalent bond, such as H—H, the contribution of the electron-pair bond to the overall negative charge density in the molecule is greater between the bonded atoms than elsewhere, but otherwise the charge is uniformly distributed. In a polar covalent bond, such as H—Cl, the contribution of the electron-pair bond to the overall negative charge density is strongly displaced toward the more electronegative chlorine atom. Both these cases are shown in Figure 9.11.

Two methods are widely used to indicate the polar nature of a bond. One method uses the lowercase Greek letter delta (δ) to indicate partial charges:

$$\overset{\delta+}{H}\!-\!\overset{\delta-}{Cl}$$

The $\delta+$ and $\delta-$ (read "delta plus" and "delta minus") signify that one end of the molecule (the H end in this case) is partially positive and the other end (Cl) is partially negative. The terms *partial positive charge* and *partial negative charge* indicate something less than the full charges that would result from complete electron transfer, making ions.

The second method uses a cross-based arrow:

$$\overset{+\longrightarrow}{H\!-\!Cl}$$

This arrow indicates the direction of displacement of negative charge, from the less electronegative element to the more electronegative element. Note that the + of the $\delta+$ and the + on the tail of the cross-based arrow always indicate the *positive* end of the bond, that is, the less electronegative element.

9.8 A Strategy for Writing Lewis Structures

Now let's use some of the concepts of covalent bonding we have developed—especially the octet rule, multiple bonds, and electronegativity—together with one or two new concepts to extend the range of Lewis structures we can write.

Whenever possible, we should check a Lewis structure to see if it is consistent with experimental evidence. Lacking such evidence, we can still use the strategy developed over the next several pages to write plausible Lewis structures.

Skeletal Structures

We can use the duet and octet rules alone to conclude that the Lewis structure of H_2O is H—Ö—H (not H—H—Ö:). In the correct structure, each H atom has a valence-shell duet and the O atom has an octet. In the incorrect structure, the left H atom has a duet but the other H atom has four valence-shell electrons and the O atom has only six. In many cases, however, the way to arrange atoms within a structure may not be as evident. We need some sort of strategy to deal with these cases.

Nonpolar covalent bond

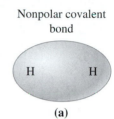

(a)

Polar covalent bond

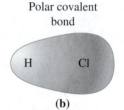

(b)

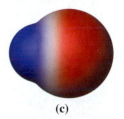

(c)

▲ **FIGURE 9.11 Charge distribution in nonpolar and polar covalent bonds**

(a) In the H—H molecule, there is an even distribution of electron charge density between the atoms. (b) In the H—Cl molecule, electron charge density is displaced toward the Cl atom. (c) This electrostatic potential model is a common way of showing regions of a molecule possessing partial positive charge (blue) and regions possessing partial negative charge (red).

The arrangement of atoms within a molecule or polyatomic ion is its **skeletal structure**; this type of structure shows the order in which the atoms are attached to one another. The skeletal structure consists of one or more central atoms plus terminal atoms. A **central atom** in a skeletal structure is any atom bonded to two or more atoms in the structure, and a **terminal atom** is any atom bonded only to one other atom. In the ammonia molecule, for instance, we have

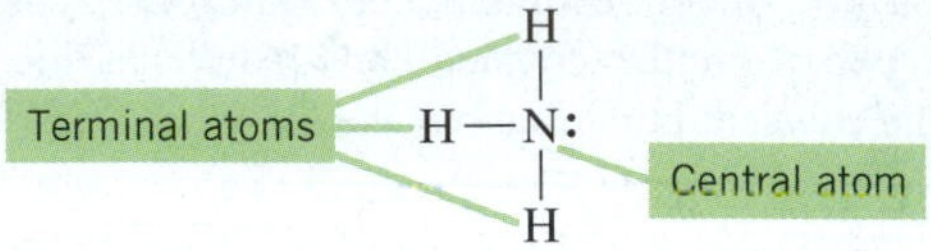

$$
\text{Terminal atoms} \quad
\begin{array}{c}
\text{H} \\
| \\
\text{H} - \text{N} : \\
| \\
\text{H}
\end{array}
\quad \text{Central atom}
$$

In writing a skeletal structure, we must attach every atom to the rest of the structure by at least one bond. We do this by joining the bonded atoms by single dashes and making no attempt at this point to account for all the valence electrons. The skeletal structure is simply the *first step* in deducing a plausible Lewis structure. In the absence of specific information about a skeletal structure, we can use the following observations to devise a likely one:

1. *Hydrogen atoms are terminal atoms.* A bonded H atom can have only two electrons in its valence shell and thus forms only one bond. Exceptions to this rule are rare, and we will not encounter any in this chapter. As a typical example, the ethane molecule, C_2H_6, or CH_3CH_3, has two C atoms as central atoms and six H atoms as terminal atoms:

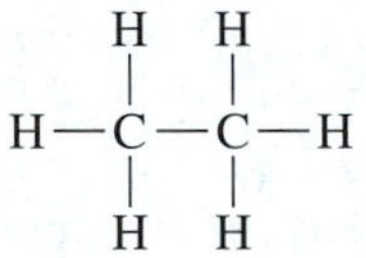

$$
\begin{array}{cc}
\text{H} & \text{H} \\
| & | \\
\text{H} - \text{C} - \text{C} - \text{H} \\
| & | \\
\text{H} & \text{H}
\end{array}
$$

In chemical formulas, the central atom is often, but not always, written first, as in NH_3, CH_4, $COCl_2$, and SO_2F_2.

2. *The central atom(s) of a structure usually has the <u>lowest</u> electronegativity, and the terminal atoms generally have <u>higher</u> electronegativities.* Because hydrogen atoms must always be terminal atoms, they will often be an exception to this rule (as they are in H_2O, NH_3, and C_2H_6). Also, because it has a higher electronegativity than any other element, we expect fluorine always to be a terminal atom. We will consider the basis for this rule later in the section, but for now we can see its application to the skeletal structure of the poisonous gas phosgene, $COCl_2$, used in the manufacture of plastics:

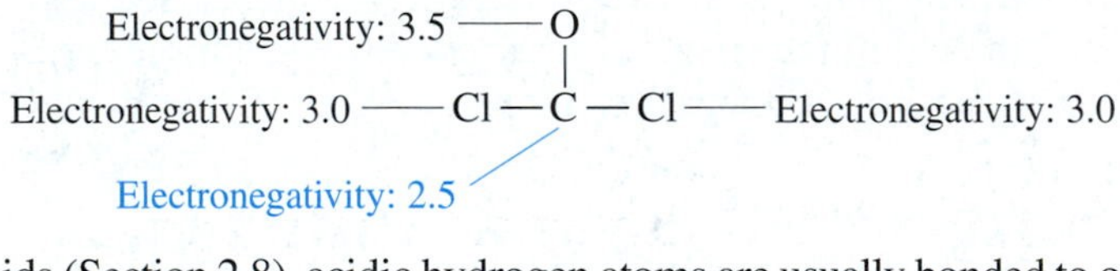

Electronegativity: 3.5 ———— O
Electronegativity: 3.0 ———— Cl — C — Cl ———— Electronegativity: 3.0
Electronegativity: 2.5

Even though Cl is the central atom in chloric acid, this atom does not appear first in the formula. Instead, there are two customary ways of writing the formula for this compound. The formula $HClO_3$ places the H atom first, emphasizing that this is an acid. The formula $HOClO_2$ signifies that the ionizable H atom is bonded to an O atom, not to the Cl atom.

3. In oxoacids (Section 2.8), acidic hydrogen atoms are usually bonded to oxygen atoms:

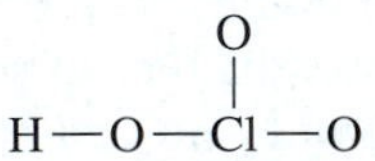

$$
\begin{array}{c}
\text{O} \\
| \\
\text{H} - \text{O} - \text{Cl} - \text{O}
\end{array}
\qquad
\begin{array}{c}
\text{O} \\
| \\
\text{H} - \text{O} - \text{S} - \text{O} - \text{H} \\
| \\
\text{O}
\end{array}
$$

Chloric acid ($HOClO_2$) 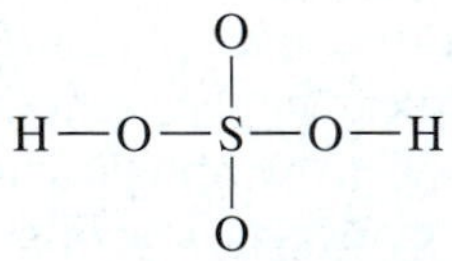   Sulfuric acid [$(HO)_2SO_2$]

4. *Molecules and polyatomic ions usually have compact, symmetrical structures.* A typical application of this idea is to the molecule SO_2F_2:

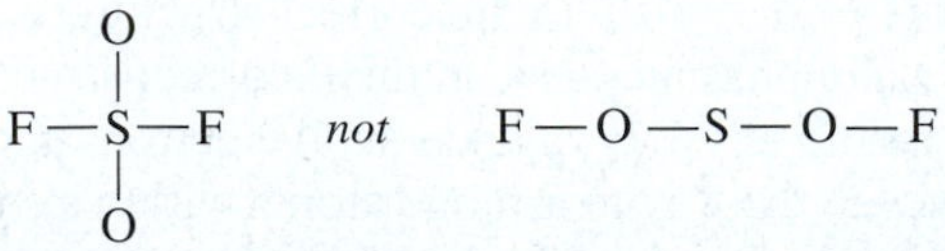

$$
\begin{array}{c}
\text{O} \\
| \\
\text{F} - \text{S} - \text{F} \\
| \\
\text{O}
\end{array}
\qquad not \qquad
\text{F} - \text{O} - \text{S} - \text{O} - \text{F}
$$

Sulfuryl fluoride (SO_2F_2)

Organic molecules, which can be based on long chains of carbon atoms, are a major exception to this idea.

A Method for Writing Lewis Structures

When we write Lewis structures through the remainder of the chapter, we will apply a strategy that minimizes false starts and gets us directly to a plausible structure. The following five-step approach is designed for that purpose.

Step 1: *Determine the total number of valence electrons; these and no other electrons must appear in the Lewis structure.* The total number of valence electrons in the Lewis structure of a molecule is the sum of the valence electrons of all the atoms. For a polyatomic anion, *add* to the sum of the valence electrons one electron for each unit of negative charge. For a polyatomic cation, *subtract* from the sum of the valence electrons one electron for each unit of positive charge. Examples: N_2O_4 has $(2 \times 5) + (4 \times 6) = 34$ valence electrons; NO_3^- has $5 + (3 \times 6) + 1 = 24$ valence electrons; and NH_4^+ has $5 + (4 \times 1) - 1 = 8$ valence electrons.

Step 2: *Use the four ideas listed on page 354 to write a skeletal structure; connect the bonded atoms by dashes representing single covalent bonds.*

Step 3: *Place pairs of electrons as lone pairs around the terminal atoms to give each terminal atom except hydrogen an octet and each hydrogen a duet.*

Step 4: *Assign any remaining electrons as lone pairs around the central atom(s).*

Step 5: *If necessary, move one or more lone pairs from a terminal atom(s) to form a multiple bond to a central atom(s).* If the number of valence electrons is just sufficient that, once step 4 is completed, all atoms in the structure have an octet (duet for H), the structure has only single bonds. If there are not enough electrons to form octets, it is necessary to form one or more multiple bonds. The atoms most commonly involved in double bonds are carbon, nitrogen, oxygen, and sulfur; those most commonly involved in triple bonds are carbon and nitrogen.

Problem-Solving Note

Because electrons in Lewis structures are almost always placed in pairs, you may find it helpful to divide the valence electrons into pairs in step 1. Thus, in N_2O_4 there are 17 pairs of valence electrons, and in NO_3^- and NH_4^+ there are 12 and 4 pairs, respectively.

Example 9.6

Write the Lewis structure of nitrogen trifluoride, NF_3.

STRATEGY

We must apply, in consecutive order, the first four steps of the general strategy outlined above. We will proceed to the fifth step only if necessary.

SOLUTION

Step 1: *Determine the number of valence electrons.*

In one N atom (group 5A) and three F atoms (group 7A), there are $5 + (3 \times 7) = 26$ valence electrons.

Step 2: *Write a skeletal structure.* The electronegativity of N is 3.0; that of F is 4.0. We expect a skeletal structure with N as a central atom and F as terminal atoms. The three nitrogen-to-fluorine bonds in this structure account for six electrons.

$$F—N—F$$
$$|$$
$$F$$

Step 3: *Complete the octets of terminals atoms.* We complete the octets of the F atoms by placing three lone pairs of electrons around each. This accounts for 18 additional electrons.

$$:\ddot{F}—N—\ddot{F}:$$
$$|$$
$$:\ddot{F}:$$

Step 4: *Assign lone pairs to central atom(s).* We have now assigned $6 + 18 = 24$ of the 26 valence electrons. We place the remaining two as a lone pair on the N atom.

$$:\ddot{F}—\ddot{N}—\ddot{F}:$$
$$|$$
$$:\ddot{F}:$$

ASSESSMENT

Because each atom in the structure shown in step 4 has an octet, that is the Lewis structure of NF_3. We have no need to go to step 5.

EXERCISE 9.6A

Write the Lewis structure of hydrazine, N_2H_4.

EXERCISE 9.6B

Write the Lewis structure of chloroethane, C_2H_5Cl.

Example 9.7

Write a plausible Lewis structure for phosgene, $COCl_2$.

STRATEGY

We must apply, in consecutive order, the first four steps of our general strategy, proceeding to the fifth step only if necessary.

SOLUTION

Step 1: *Determine the number of valence electrons.*

In one C atom (group 4A), one O atom (group 6A), and two Cl atoms (group 7A), there are $4 + 6 + (2 \times 7) = 24$ valence electrons.

Step 2: *Write a skeletal structure.* The electronegativities are 2.5 for C, 3.5 for O, and 3.0 for Cl. We expect C, the least electronegative atom, to be the central atom, and O and Cl to be terminal atoms attached to it. This skeletal structure accounts for six of the valence electrons.

$$Cl-\underset{\underset{\displaystyle Cl}{\displaystyle |}}{\overset{\overset{\displaystyle O}{\displaystyle |}}{C}}$$

Step 3: *Complete the octets of terminal atoms.* Place three lone pairs around the O atom and three lone pairs around each of the Cl atoms, for a total of 18 electrons. This completes the octets of the terminal atoms.

$$:\overset{\displaystyle \ddot{O}:}{\underset{|}{}}$$
$$:\ddot{C}l-C-\ddot{C}l:$$
$$\ddot{}\ddot{}$$

Step 4: *Assign lone pairs to central atom(s).* Up to this point, we have assigned $6 + 18 = 24$ of the 24 valence electrons, so there are none available to complete the octet of the central C atom. Therefore we must go to step 5.

Step 5: *Form multiple bonds to complete octets of central atom(s).* Complete the octet on the C atom by shifting a lone pair of electrons from the O atom to form a carbon-to-oxygen double bond. (C and O are two atoms that are able to form a double bond.)

$$:\ddot{C}l-\overset{\overset{\displaystyle :\ddot{O}:}{\displaystyle |}}{C}-\ddot{C}l: \longrightarrow :\ddot{C}l-\overset{\overset{\displaystyle :\ddot{O}}{\displaystyle \|}}{C}-\ddot{C}l:$$

ASSESSMENT

The structure obtained in step 5 has the properly assigned numbers of valence electrons (24) and an octet for each atom; it is therefore a plausible Lewis structure. In Example 9.9 we will see why the double bond is carbon-to-oxygen and not carbon-to-chlorine.

EXERCISE 9.7A

Write a plausible Lewis structure for carbonyl sulfide, COS.

EXERCISE 9.7B

A compound known as nitrosyl chloride, NOCl, is present in *aqua regia,* a mixture of concentrated nitric and hydrochloric acids capable of reacting with gold. Write a plausible Lewis structure for NOCl. (*Hint:* The Cl atom does not form a double bond.)

Example 9.8

Write a plausible Lewis structure for the chlorate ion, ClO_3^-.

STRATEGY

Again, we must apply, in consecutive order, the first four steps of the general strategy introduced earlier and the fifth step if necessary. In determining the total number of valence electrons in step 1, we must not forget the extra electron that conveys the -1 charge to the ion.

SOLUTION

Step 1: *Determine the number of valence electrons.*

For this polyatomic anion, the number of valence electrons that must appear in the Lewis structure are those for one Cl atom (group 7A), three O atoms (group 6A), and one additional electron to convey the -1 ionic charge: $7 + (3 \times 6) + 1 = 26$.

Step 2: *Write a skeletal structure.* The electronegativities are 3.0 for Cl and 3.5 for O. We expect the less electronegative chlorine atom to be the central atom in the skeletal structure.

$$O—Cl—O$$
$$\mid$$
$$O$$

Step 3: *Complete the octets of terminal atoms.* We complete the octets on the O atoms by placing three lone pairs of electrons on each.

$$:\ddot{O}—Cl—\ddot{O}:$$
$$\mid$$
$$:\ddot{O}:$$

Step 4: *Assign lone pairs to central atom(s).* We have now assigned $6 + 18 = 24$ of the 26 valence electrons and must place the remaining two as a lone pair on Cl.

$$\left[:\ddot{O}—\ddot{Cl}—\ddot{O}:\right]^{-}$$
$$\mid$$
$$:\ddot{O}:$$

ASSESSMENT

All 26 valence electrons were assigned in step 4, and all atoms have an octet; our result is therefore a plausible Lewis structure for the chlorate ion.

EXERCISE 9.8A

Write a plausible Lewis structure for the perchlorate ion.

EXERCISE 9.8B

Write a plausible Lewis structure for the nitronium ion, NO_2^+.

Formal Charge

The ideas we have used to this point in our discussion of Lewis structures still leave some unanswered questions. For example, why do we choose the atom(s) of lowest electronegativity for the central atom(s) of a skeletal structure? And, given two different atoms of equal electronegativity as the possible central atom in a Lewis structure, which do we choose? And, in writing the Lewis structure of $COCl_2$ in Example 9.7, why did we make a carbon-to-oxygen double bond rather than a carbon-to-chlorine double bond?

The concept of *formal charge* sheds some light on these issues and also gives us an additional tool to use in writing Lewis structures. **Formal charge** is the difference between the number of valence electrons in a free (uncombined) atom and the number of valence electrons assigned to that atom when bonded to others in a Lewis structure:

$$\text{Formal charge (FC)} = \left(\begin{array}{c}\text{number of valence}\\ \text{electrons in an}\\ \text{uncombined atom}\end{array}\right) - \left(\begin{array}{c}\text{number of valence electrons}\\ \text{assigned to the bound atom}\\ \text{in a Lewis structure}\end{array}\right) \quad \textbf{(9.1)}$$

 **Formal Charges animation**

To evaluate formal charges, we use a form of electron bookkeeping that goes beyond just accounting for the total number of valence electrons in a Lewis structure. The first term on the right in Equation (9.1) is easy to determine because the number of valence electrons for a main-group element is equal to its group number. To get the second term, we use the following simple rules to assign electrons to a given atom in a Lewis structure:

- All lone-pair electrons on an atom are assigned to that atom.
- Electrons in a bond are assigned equally to the two bonded atoms—half to one atom and half to the other.

We can combine these rules into an equation for the formal charge on a given atom in a Lewis structure.

Formal charge (FC) =

$$\left(\begin{array}{c}\text{number of valence}\\ \text{electrons in the}\\ \text{uncombined atom}\end{array}\right) - \left(\begin{array}{c}\text{number of lone-}\\ \text{pair electrons}\\ \text{on the bound atom}\end{array}\right) - \frac{1}{2}\left(\begin{array}{c}\text{number of}\\ \text{electrons in}\\ \text{bonds to the atom}\end{array}\right) \quad \textbf{(9.2)}$$

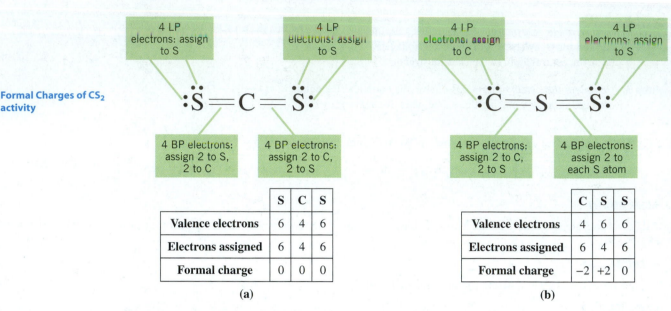

▲ **FIGURE 9.12**   **The concept of formal charge illustrated**

Because it has formal charges of zero, Lewis structure (a) is more plausible than Lewis structure (b).
(LP = lone pair; BP = bonding pair.)

We can use this equation to evaluate formal charges, or we can use the simple method outlined in Figure 9.12.

Where we do find non-zero formal charges, we can designate them as shown here for structure (b) in Figure 9.12:

$$:\overset{\underset{\displaystyle\cdot\cdot}{}}{\overset{\displaystyle\boxed{-2}}{C}}=\overset{\displaystyle\boxed{+2}}{S}=\overset{\underset{\displaystyle\cdot\cdot}{}}{S}:$$

We use small encircled numbers for formal charges to distinguish them from authentic ionic charges. Formal charges are *hypothetical;* the individual atoms in a covalent molecule do not actually carry these charges.

An atom in a Lewis structure with a formal charge of zero actually contributes half the electrons to the bonds it forms, just as we assume when assigning the bonding electrons. If an atom is found to have a positive formal charge, this means that the atom actually contributes more than half the electrons to a bond that it forms, and the receiving atom acquires a negative formal charge. A bond formed as a result of unequal electron contributions from the bonded atoms, although indistinguishable from other covalent bonds, is sometimes called a *coordinate covalent bond.* An example is seen in the carbon-to-sulfur double bond in Figure 9.12b.

Often we can write two or more nonequivalent Lewis structures for a molecule or polyatomic ion. We then look for the Lewis structure with an optimal set of formal charges, which we can find with the help of the following ideas:

- Usually, the most plausible Lewis structure is one with formal charges of zero on all atoms.
- Where non-zero formal charges are required, they should be as small as possible, and negative formal charges should appear on the most electronegative atoms.
- Adjacent atoms in a structure should not carry formal charges of the same sign.
- The total of formal charges on the atoms in a Lewis structure must be zero for a neutral molecule and must equal the net charge for a polyatomic ion.

These ideas point to structure (a) in Figure 9.12 as the acceptable Lewis structure for CS_2. When we apply the ideas to the $COCl_2$ molecule of Example 9.7, as we will do next in Example 9.9, we confirm that the most likely Lewis structure has carbon, the least electronegative atom, as its central atom and that the double bond in the structure is carbon-to-oxygen and not carbon-to-chlorine.

Example 9.9

In Example 9.7, we wrote a Lewis structure for the molecule $COCl_2$, shown here as structure (a). Show that structure (a) is more plausible than (b) or (c).

$$:\ddot{O}=C-\ddot{C}l: \qquad :\ddot{O}-C=\underset{\cdot\cdot}{C}l: \qquad :\ddot{C}=O-\ddot{C}l:$$

$$\quad\;\; | \qquad\qquad\qquad | \qquad\qquad\qquad |$$

$$\quad :\underset{\cdot\cdot}{C}l: \qquad\qquad\; :\underset{\cdot\cdot}{C}l: \qquad\qquad\quad :\underset{\cdot\cdot}{C}l:$$

$$\qquad\textbf{(a)} \qquad\qquad\qquad \textbf{(b)} \qquad\qquad\qquad \textbf{(c)}$$

STRATEGY

We can use Equation (9.2) to assign formal charges to the atoms in each structure, and we then apply the list of ideas cited on page 358 to evaluate the results.

SOLUTION

Formal charge (FC) =

$$\left(\begin{array}{c}\text{number of valence}\\\text{electrons in the}\\\text{uncombined atom}\end{array}\right) - \left(\begin{array}{c}\text{number of lone-}\\\text{pair electrons}\\\text{on the bound atom}\end{array}\right) - \frac{1}{2}\left(\begin{array}{c}\text{number of}\\\text{electrons in}\\\text{bonds to the atom}\end{array}\right)$$

Analyzing Structure (a)

The O atom: FC = 6 valence e^- − 4 lone pair e^- − $\left(\frac{1}{2} \times 4$ bond pair $e^-\right)$ = 0

The C atom: FC = 4 valence e^- − 0 lone pair e^- − $\left(\frac{1}{2} \times 8$ bond pair $e^-\right)$ = 0

Each Cl atom: FC = 7 valence e^- − 6 lone pair e^- − $\left(\frac{1}{2} \times 2$ bond pair $e^-\right)$ = 0

Because all of the atoms have a formal charge of zero, structure (a) is entirely plausible for $COCl_2$.

Analyzing Structure (b)

The O atom: FC = 6 valence e^- − 6 lone pair e^- − $\left(\frac{1}{2} \times 2$ bond pair $e^-\right)$ = −1

The C atom: FC = 4 valence e^- − 0 lone pair e^- − $\left(\frac{1}{2} \times 8$ bond pair $e^-\right)$ = 0

The —Cl atom: FC = 7 valence e^- − 6 lone pair e^- − $\left(\frac{1}{2} \times 2$ bond pair $e^-\right)$ = 0

The =Cl atom: FC = 7 valence e^- − 4 lone pair e^- − $\left(\frac{1}{2} \times 4$ bond pair $e^-\right)$ = +1

The formal charges add up to zero, as they should. However, a Cl atom in structure (b) has a +1 formal charge even though it is more electronegative than a C atom, which has a formal charge of zero:

$$\textbf{(b)} \;\; :\overset{\boxed{-1}}{\underset{\cdot\cdot}{O}}-C=\overset{\boxed{+1}}{\underset{\cdot\cdot}{C}}l:$$

$$\qquad\qquad | $$

$$\qquad\quad :\underset{\cdot\cdot}{C}l:$$

Therefore structure (b) is not as plausible as structure (a).

Analyzing Structure (c)

The C atom: FC = 4 valence e^- − 4 lone pair e^- − $\left(\frac{1}{2} \times 4$ bond pair $e^-\right)$ = −2

The O atom: FC = 6 valence e^- − 0 lone pair e^- − $\left(\frac{1}{2} \times 8$ bond pair $e^-\right)$ = +2

Each Cl atom: FC = 7 valence e^- − 6 lone pair e^- − $\left(\frac{1}{2} \times 2$ bond pair $e^-\right)$ = 0

Structure (c) is the least plausible of the three because it has the largest formal charges. Also, the O atom, with the highest electronegativity, has a positive rather than a negative formal charge:

$$\textbf{(c)} \;\; :\overset{\boxed{-2}}{\ddot{C}}=\overset{\boxed{+2}}{O}-\ddot{C}l:$$

$$\qquad\qquad | $$

$$\qquad\quad :\underset{\cdot\cdot}{C}l:$$

ASSESSMENT

Structure (a), with formal charges of zero, is the most plausible Lewis structure for $COCl_2$.

EXERCISE 9.9A

A student has proposed two condensed structural formulas—H_2NOH and H_2ONH—for a compound with the molecular formula H_3NO. Write a Lewis structure corresponding to each formula, assign formal charges, and select the more plausible Lewis structure.

EXERCISE 9.9B

Methyl acetate, CH_3COOCH_3, is used as a paint remover, a solvent for lacquers, and an artificial flavoring. Write the best Lewis structure that you can for this molecule.

Resonance: Delocalized Bonding

Application Note

Near ground level, ozone, O_3, is an environmental pollutant found in smog. In the stratosphere, it provides a crucial shield against harmful ultraviolet radiation.

When we apply our general strategy for Lewis structures to the ozone molecule, O_3, we get the structure

$$:\ddot{\underset{-1}{O}}—\underset{+1}{\ddot{O}}=\ddot{O}:$$

This structure implies that one of the oxygen-to-oxygen bonds should be a single bond and the other a double bond. However, experimental data show that the two oxygen-to-oxygen bonds in the ozone molecule are identical and intermediate between single and double. We cannot write a single Lewis structure consistent with this fact. The best we can do is to write some plausible Lewis structures and try to average them into a composite, or hybrid.

This description of the O_3 molecule is based on the **resonance** theory, which states that whenever a molecule or ion can be represented by two or more plausible Lewis structures that differ *only in the distribution of electrons,* the true structure is a composite of them. The different plausible structures are called **resonance structures.** All the atoms are located in exactly the same place in each resonance structure; the only difference between them is the distribution of electrons among the atoms. The actual molecule or ion that is a hybrid of the resonance structures is called a **resonance hybrid.** One way to represent the resonance hybrid is to write the resonance structures and join them by double-headed arrows. The resonance structures of ozone are

$$:\underset{-1}{\ddot{O}}—\underset{+1}{\ddot{O}}=\ddot{O}: \longleftrightarrow :\ddot{O}=\underset{+1}{\ddot{O}}—\underset{-1}{\ddot{O}}:$$

Although we might take the double-headed arrow to suggest that the structure of the O_3 molecule shifts back and forth from one resonance structure to the other, this is *not* the case. There is a single, actual structure—the resonance hybrid—that is a blend of the two resonance structures.

Actually, the printing analogy is the reverse of resonance. That is, the colors magenta and cyan are real colors, and purple, their blend (hybrid), is not. With resonance, the hybrid is real and the contributing resonance structures are not.

We find a similar situation in the printing of this page. No purple ink is used in the printing, but we can get *purple* by blending *magenta* and *cyan*. The page is run through the press twice, first in magenta and then in cyan. It's as if magenta and cyan were resonance structures and purple were their resonance hybrid.

In molecules such as H_2O, NH_3, and CH_4, in which resonance is not involved, bonding electron pairs are considered to exist in fairly well-defined regions between two atoms; when this is the case, we say the electrons are *localized*. In O_3, to produce oxygen-to-oxygen bonds that are intermediate between single and double bonds, some of the electrons in the resonance hybrid are *delocalized*. Delocalized electrons are bonding electrons that are spread out over several atoms. We can attempt to picture a resonance hybrid by using a dotted line to represent delocalized electrons, as in this resonance hybrid of O_3:

$$:\ddot{O}\cdots\ddot{O}\cdots\ddot{O}:$$

Resonance hybrid

Four electrons shared among the three oxygen atoms.

The dotted line across the top represents four electrons shared among three O atoms—one electron from each terminal atom and two from the central atom.

Example 9.10

Write three equivalent Lewis structures for the SO_3 molecule that conform to the octet rule, and describe how the resonance hybrid is related to the three structures.

STRATEGY

Whether resonance is involved or not, in writing a plausible Lewis structure, we follow the same five-step strategy first applied in Example 9.6. In fact, we often do not realize that resonance is involved until we have examined one of the resonance structures.

SOLUTION

Step 1: *Determine the number of valence electrons.*

There are $6 + (3 \times 6) = 24$ valence electrons.

Step 2: *Write a skeletal structure.* The skeletal structure has sulfur, the atom of lower electronegativity, as the central atom.

Step 3: *Complete the octets of terminal atoms.* Place three lone pairs of electrons on each O atom to complete its octet.

Step 4: *Assign lone pairs to central atom(s).* We have already assigned all 24 valence electrons; there are none left to place as lone pairs on the central atom. Because the S atom does not yet have an octet, we must go on to step 5.

Step 5: *Form multiple bonds to complete octets of central atom(s).* Move a lone pair of electrons from any terminal O atom to form a double bond to the central S atom. Because the double bond can go to any one of the three terminal O atoms, we get three equivalent structures that differ only in the position of the double bond.

Resonance structures

ASSESSMENT

The SO_3 molecule is the resonance hybrid of the three resonance structures. The three sulfur-to-oxygen bonds are all identical; each is intermediate between a single and a double bond. Note also the formal charges, a common feature in resonance structures.

EXERCISE 9.10A

Write three Lewis structures for the nitrate ion, NO_3^-, and describe how the resonance hybrid is related to the three structures.

EXERCISE 9.10B

Determine the number of Lewis structures required to represent adequately **(a)** hydrogen carbonate ion, **(b)** chlorous acid, **(c)** cyanide ion, **(d)** carbide ion, C_2^{2-}, **(e)** formate ion.

9.9   Molecules that Do Not Follow the Octet Rule

Molecules made of atoms of the main-group elements generally have Lewis structures that follow the octet rule, but there are exceptions. These fall into three categories that are readily identified by some structural characteristic.

Molecules with an Odd Number of Valence Electrons

In all Lewis structures so far, we have dealt with electron pairs, either as bonding pairs or lone pairs. However, in Lewis structures with an odd number of valence electrons, it is not possible to have all the electrons in pairs, nor is it possible to satisfy the octet rule for all the atoms. Here are three examples: nitrogen monoxide, NO, with $5 + 6 = 11$ valence electrons; nitrogen dioxide, NO_2, with $5 + 6 + 6 = 17$ valence electrons; and chlorine dioxide, ClO_2, with $7 + 6 + 6 = 19$ valence electrons.

Application Note

Both NO and NO_2 are major components of smog. Made in multiton quantities, ClO_2 is used for bleaching flour and paper.

$$\cdot \ddot{N}\!=\!\ddot{O}\!:$$

$$:\ddot{O}\!-\!\ddot{N}\!=\!\ddot{O}\!: \longleftrightarrow :\ddot{O}\!=\!\ddot{N}\!-\!\ddot{O}\!:$$

$$\cdot\ddot{O}\!-\!\ddot{Cl}\!-\!\ddot{O}\!: \longleftrightarrow :\ddot{O}\!-\!\ddot{Cl}\!-\!\ddot{O}\!\cdot$$

Application Note

The free radicals H·, O·, HO·, and HO$_2$· are all believed to be present as intermediates in the explosive reaction of hydrogen and oxygen gases:
2 H$_2$ + O$_2$ $\longrightarrow$ 2 H$_2$O

There are relatively few stable molecules with odd numbers of electrons. Most odd-electron species, called **free radicals,** are fragments of molecules, are highly reactive, and have only a fleeting existence as intermediates in chemical reactions. An important free radical in the atmosphere is the *hydroxyl* radical (·OH). We usually use a single dot (·) to represent an unpaired electron.

One example of a reaction involving free radicals is that between a hydroxyl radical and a methane molecule to produce a methyl radical:

$$\cdot OH(g) \; + \; CH_4(g) \longrightarrow \cdot CH_3(g) \; + \; H_2O(g)$$

Molecules with Incomplete Octets

Sometimes when we write a Lewis structure, there are too few electrons to give every atom a valence-shell octet. Such electron-deficient molecules generally have some unusual bonding characteristics and are often quite reactive. You can expect to encounter electron-deficient molecules mainly when the central atom is Be, B, or Al. As an example, let's consider the bonding of boron and fluorine. Boron atoms have three valence electrons, and fluorine atoms have seven. In the molecule boron trifluoride, the central boron atom shares its three valence electrons with three fluorine atoms:

$$\begin{array}{c} :\ddot{F}: \\ | \\ B\!-\!\ddot{F}: \\ | \\ :\ddot{F}: \end{array}$$

Application Note

A catalyst in many organic chemical reactions, BF$_3$ is an important industrial chemical, but its main uses are related to its electron-deficient nature, not to its boron or fluorine content.

This structure accounts for all 24 valence electrons, but the central B atom has only six electrons, two short of an octet. Nevertheless, this structure is quite acceptable according to the formal-charge rules because all four atoms have a formal charge of 0. To write a Lewis structure that conforms to the octet rule, we must use a boron-to-fluorine double bond:

$$\begin{array}{c} :\ddot{F}: \\ | \\ \ominus B\!=\!F\!:\!\oplus \\ | \\ :\ddot{F}: \end{array}$$

In fact, there are three equivalent structures containing a boron-to-fluorine double bond. However, we might be inclined to disfavor all of them because each places a positive formal charge on a fluorine atom, the most electronegative of all atoms. What, then, is the real structure of the BF$_3$ molecule? Experimental evidence indicates that the best representation is probably a hybrid of these four resonance structures:

$$\begin{array}{cccc} :\ddot{F}: & :\ddot{F}: & :\ddot{F}: & :\ddot{F}: \\ | & | & | & \| \\ B\!-\!\ddot{F}: & B\!=\!\ddot{F}: & B\!-\!\ddot{F}: & B\!-\!\ddot{F}: \\ | & | & \| & | \\ :\ddot{F}: & :\ddot{F}: & :\ddot{F}: & :\ddot{F}: \end{array}$$

The large electronegativity difference between B and F (ΔEN = 2.0) suggests appreciable ionic character in the boron-to-fluorine bonds, suggesting the possibility of such *ionic* resonance structures as

$$\begin{array}{c} :\ddot{F}\!-\!B^+ \;\; ^-\!:\ddot{F}: \\ | \\ :\ddot{F}: \end{array}$$

The fact that the central boron atom is deficient in electrons is consistent with the observed high chemical reactivity of BF$_3$. For example, BF$_3$ readily reacts to form a coordinate covalent bond with a lone pair of electrons, such as those provided by a fluoride ion:

$$\begin{array}{ccc} & :\ddot{F}: & :\ddot{F}: \\ & | & | \\ :\ddot{F}:^- \;\; + & B\!-\!\ddot{F}: \longrightarrow & \left[:\ddot{F}\!-\!B\!-\!\ddot{F}: \right]^- \\ & | & | \\ & :\ddot{F}: & :\ddot{F}: \end{array}$$

The measured boron-to-fluorine internuclear distance in BF_3 is shorter than in the single-bonded BF_4^- ion (130 pm versus 145 pm). As we shall see in Section 9.10, this shorter distance is consistent with some double-bond character in the boron-to-fluorine bonds.

In conclusion, we might say that BF_3 exists as the resonance hybrid of the four resonance structures shown on page 362, with perhaps the major contributor being the leftmost structure, the one with an incomplete octet on the boron atom. This conclusion highlights two other important facets of resonance theory:

1. Structures that contribute to a resonance hybrid need not be completely equivalent (as they are in O_3 and SO_3).

2. Experimental evidence may indicate that one or more resonance structures make a greater contribution to the resonance hybrid structure than others do.

Structures with Expanded Valence Shells

The second-period elements carbon, nitrogen, oxygen, and fluorine nearly always obey the octet rule. Electron-deficient and odd-electron molecules are obvious exceptions. Because the valence shell of the second-period elements holds a maximum of eight electrons $(2s^2 2p^6)$, these elements do not exceed an octet. However, for elements in the third and higher-numbered periods, the situation is different. Although the third period ends with argon $(3s^2 3p^6)$, in principle the third principal shell can hold up to 18 electrons $(3s^2 3p^6 3d^{10})$.

For example, we cannot write Lewis structures based on the octet rule for the molecules PCl_5 and SF_6 because the octet rule allows for a maximum of only four bonds between a central and a terminal atom. In PCl_5, we need 10 electrons surrounding the P in five bonds, and in SF_6, we need 12 electrons surrounding the S in six bonds. To accommodate more than eight valence electrons for a central atom in the third or higher-numbered period, we can use **expanded valence shells,** as shown in these structures:

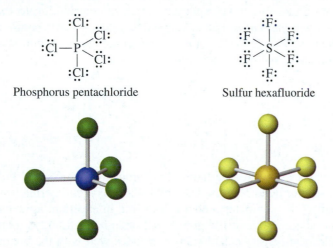

Phosphorus pentachloride   Sulfur hexafluoride

In some cases, even though we can write Lewis structures that conform to the octet rule, structures based on expanded valence shells appear to agree better with experimental data than do those based on the octet rule. Consider, for instance, dichlorine monoxide, Cl_2O, and perchlorate ion, ClO_4^-. A Lewis structure of Cl_2O with O as the central atom has valence-shell octets for each atom and formal charges of zero:

$$:\ddot{C}l-\ddot{O}-\ddot{C}l:$$

Let's assume that this plausible Lewis structure for Cl_2O corresponds to the normal chlorine-to-oxygen internuclear distance in a Cl—O single bond. Experimental evidence indicates this distance is 170 pm. The Lewis structure for the perchlorate ion conforms to the octet rule but has formal charges, with the positive formal charge on the Cl atom being particularly high:

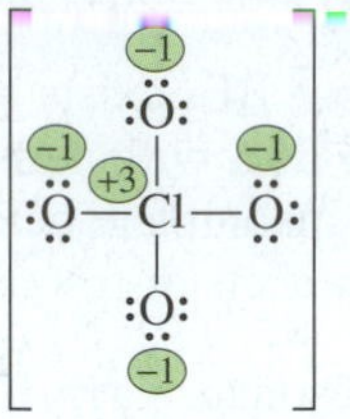

$$
\left[\begin{array}{c}
\overset{-1}{:\ddot{O}:} \\
| \\
\overset{-1}{:\ddot{O}} - \overset{+3}{Cl} - \overset{-1}{\ddot{O}:} \\
| \\
\overset{-1}{:\ddot{O}:}
\end{array} \right]
$$

Experimental evidence indicates that the chlorine-to-oxygen internuclear distance in ClO_4^- is 144 pm. These shorter distances suggest that the Cl—O bonds in ClO_4^- have some multiple-bond character. With the concept of expanded valence shells, we can write a number of Lewis structures having double bonds and reduced formal charges. The true structure is a resonance hybrid of many structures, not all of which are equally important. One example of a possible contributing structure is

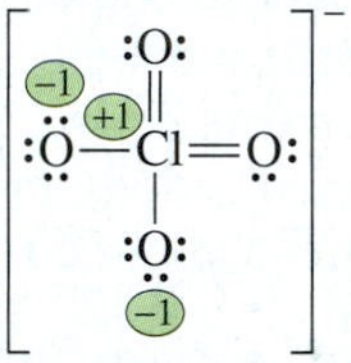

$$
\left[\begin{array}{c}
:O: \\
\| \\
\overset{-1}{:\ddot{O}} - \overset{+1}{Cl} = \ddot{O}: \\
| \\
:\ddot{O}: \\
\overset{-1}{}
\end{array} \right]^-
$$

We have just made the case for using formal charge and expanded valence shells to obtain the "best" Lewis structures when the central atoms are in the third period, but there is also evidence that simpler Lewis structures based on the octet rule may in fact be better. With the aid of high-speed computers, chemists can do quantum mechanical calculations that assess the relative importance of various resonance structures to a resonance hybrid. A 1995 study* suggests that the octet structure, even with its high formal charge on Cl, is the most important structure contributing to the resonance hybrid of ClO_4^-, with structures having expanded valence shells being of negligible importance. The study suggests that most of the remaining contributions to the resonance hybrid are due to such ionic structures as

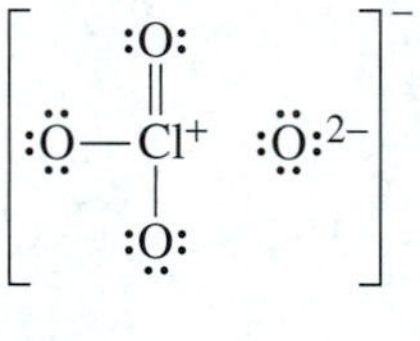

$$
\left[\begin{array}{c}
:O: \\
\| \\
:\ddot{O} - Cl^+ \quad :\ddot{O}:^{2-} \\
| \\
:\ddot{O}:
\end{array} \right]^-
$$

These ionic structures suggest a partial ionic character in the Cl—O bonds that can account for the shorter bond lengths. The chlorine-to-oxygen internuclear distance in ClO_4^- predicted by the computer model is 145 pm, in excellent agreement with the experimentally obtained value of 144 pm. On the other hand, a study reported in 2001[†] and based on electron-density calculations correlates well with the bonding deduced from Lewis structures that use expanded valence shells and minimal formal charges.

How should we deal with these conflicting views? Our approach will be to use expanded valence shells in cases where we are unable to write a Lewis structure without them (as with PCl_5 and SF_6) but otherwise to continue to rely mostly on the octet rule. In Chapter 10, we will find that this approach works quite well when it comes to predicting the shapes of molecules.

In Example 9.11, we find that an expanded valence shell may need to accommodate lone-pair electrons as well as bonding pairs. In Exercise 9.11B, you cannot tell from the name and formula alone whether to use an expanded valence shell. However, you will find that this is necessary.

---

*L. Suidan, et al. *J. Chem. Educ.* **72**, 583 (1995).

[†]G. H. Purser, *J. Chem. Educ.* **78**, 981 (2001).

Example 9.11

Write the Lewis structure for bromine pentafluoride, BrF_5.

STRATEGY

The formula BrF_5 tells us that we must use an expanded valence shell because five Br—F single bonds require a minimum of 10 electrons in the valence shell of Br. With this thought in mind, we can then follow the general strategy for writing Lewis structures to arrive at an acceptable Lewis structure.

SOLUTION

Step 1: *Determine the number of valence electrons.*

Both Br and F are in group 7A. All the atoms in the structure have seven valence electrons, and $7 + (5 \times 7) = 42$ valence electrons must appear in the Lewis structure.

Step 2: *Write a skeletal structure.* The skeletal structure has Br, the atom of lower electronegativity, as the central atom.

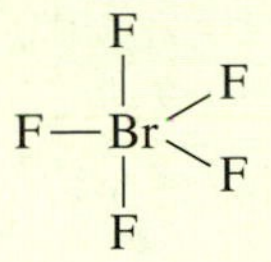

Step 3: *Complete the octets of terminal atoms.* To complete the octets of the F atoms, we place three lone pairs of electrons around each one.

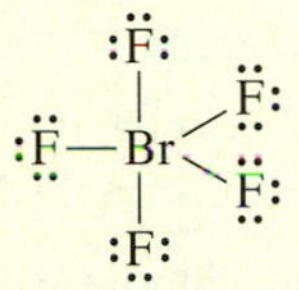

Step 4: *Assign lone pairs to central atom(s).* Through step 3, we have assigned 40 of the 42 valence electrons; two remain to be placed. They can be shown as a lone pair on the bromine atom. In this representation, the Br atom has an expanded valence shell containing 12 electrons.

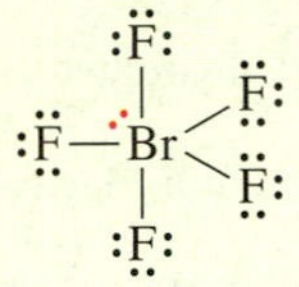

ASSESSMENT

There is no need to proceed beyond step 4, because the Lewis structure we obtained in that step is a plausible structure with no non-zero formal charges.

EXERCISE 9.11A

Write the Lewis structure of phosphorus trichloride.

EXERCISE 9.11B

Write the Lewis structure of **(a)** chlorine trifluoride and **(b)** sulfur tetrafluoride.

Example 9.12   A Conceptual Example

Indicate the error in each of the following Lewis structures. Replace each by a more acceptable structure(s).

a. :C≡N: **b.** H—Ö=N—Ö: with :Ö: above N

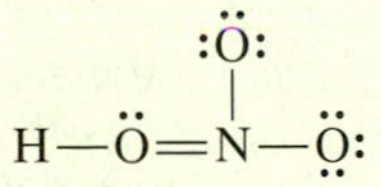

ANALYSIS AND CONCLUSIONS

(a) This structure conforms to the octet rule, but let's back up to the first step in the general strategy for writing Lewis structures: determining the total number of valence electrons. This number is 4 (from C) + 5 (from N) = 9, not the 10 electrons shown. The structure shown here is not for a molecular species but for the cyanide *ion*, CN^-. It should be written $[:C≡N:]^-$.

(b) The number of valence electrons required in the structure is $1 + 5 + (3 \times 6) = 24$. If we count the number of valence electrons in the structure, we do indeed find 24. So far, there is no problem. When we assign formal charges to the atoms in the structure, however, we find non-zero formal charges on all the atoms except H:

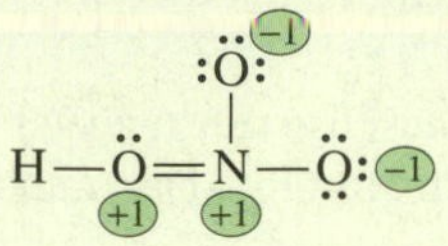

$$H-\overset{..}{\underset{+1}{O}}=\overset{|}{\underset{+1}{N}}-\overset{..}{\underset{}{O}}:_{(-1)} \qquad :\overset{..}{O}:_{(-1)}$$

Further, we have formal charges of the same sign (+1) on adjacent atoms, the N atom and one of the O atoms. We should be able to write a better structure. To do so, we need to have the nitrogen-to-oxygen double bond to a *terminal* O atom rather than another central atom. When we do this, we find that there are two equivalent structures, each of which has only two atoms with non-zero formal charges. The correct structure is a resonance hybrid of these two resonance structures:

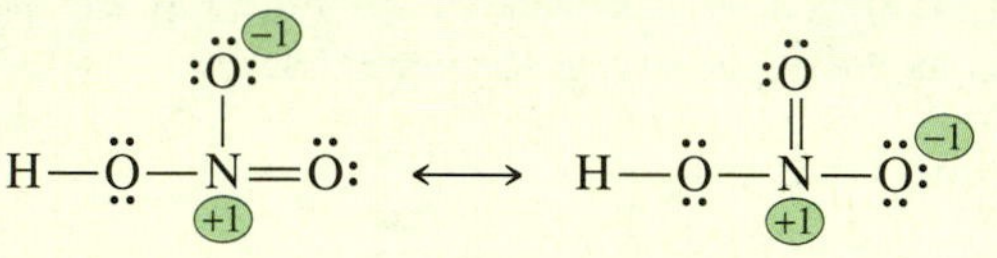

$$H-\overset{..}{\underset{+1}{O}}-\overset{:\overset{..}{O}:_{(-1)}}{\underset{}{N}}=\overset{..}{\underset{}{O}}: \longleftrightarrow H-\overset{..}{\underset{+1}{O}}-\overset{:\overset{..}{O}}{\underset{}{N}}=\overset{..}{\underset{}{O}}:_{(-1)}$$

EXERCISE 9.12A

Only one of the following Lewis structures is correct. Identify that one, and indicate the error(s) in the others.

a. chlorine dioxide, $:\overset{..}{\underset{..}{O}}-\overset{..}{\underset{..}{Cl}}-\overset{..}{\underset{..}{O}}:$

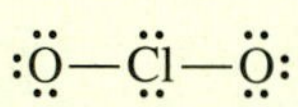

b. hydrogen peroxide, $H-\overset{..}{\underset{..}{O}}-\overset{..}{\underset{..}{O}}-H$

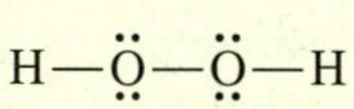

c. dinitrogen difluoride, $:\overset{..}{\underset{..}{F}}-\overset{..}{\underset{}{N}}-\overset{..}{\underset{}{N}}-\overset{..}{\underset{..}{F}}:$

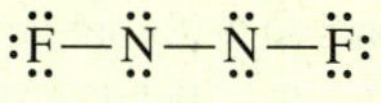

EXERCISE 9.12B

Write Lewis structures of the sulfuric acid molecule using two alternative views of expanded valence shells, as shown for the perchlorate ion on page 364. Describe the differences between the two structures.

9.10   Bond Lengths and Bond Energies

The electron charge density associated with shared electron pairs in covalent bonds is concentrated in the region between the nuclei of the bonded atoms. The greater the attractive forces between the positively charged nuclei and the bonding electrons, the more tightly the atoms are joined. Thus, shared electrons are the "glue" that binds atoms together in molecules and polyatomic ions. Generally, bond strengths vary for different combinations of atoms and for different numbers of shared electrons. The usual pattern for a bond between two particular atoms is that *the more electrons in the bond, the more tightly the atoms are held together.* Think of the electron charge between bonded atoms as partially neutralizing the repulsion between the positively charged nuclei. The greater the electron charge density in the bond, the closer the nuclei can approach each other.

Bond length is the distance between the nuclei of two atoms joined by a covalent bond. Bond length depends on the particular atoms in the bond and on the *bond order* (that is, whether the bond is single, double, or triple). For a given pair of atoms, there is little variation in bond length from one molecule to another. Thus, in alkane hydrocarbons, all the C—C single bonds have a length of 154 pm. Because the atoms in a double bond are more tightly bound than in a single bond, a double bond between two atoms is shorter than a single bond between the same two atoms. For example, the C=C bond length in ethylene, $H_2C=CH_2$, is 134 pm, versus the 154-pm bond length in ethane, H_3C-CH_3. A triple bond is even shorter than a double bond between the same two atoms. Thus the C≡C bond length in ethyne, HC≡CH, is 120 pm. Some single, double, and triple bond lengths are listed in Table 9.1.

Table 9.1 Some Representative Bond Lengths and Bond Energies

Bond	Bond Length, pm	Bond Energy,[a] kJ/mol	Bond	Bond Length, pm	Bond Energy,[a] kJ/mol
H—H	74	436	C—O	143	360
H—C	110	414	C=O	120	736[b]
H—N	100	389	C—Cl	178	339
H—O	97	464	N—N	145	163
H—S	132	368	N=N	123	418
H—F	92	565	N≡N	110	946
H—Cl	127	431	N—O	136	222
H—Br	141	364	N=O	120	590
H—I	161	297	O—O	145	142
C—C	154	347	O=O	121	498
C=C	134	611	F—F	143	159
C≡C	120	837	Cl—Cl	199	243
C—N	147	305	Br—Br	228	193
C=N	128	615	I—I	266	151
C≡N	116	891			

[a] Bond-dissociation energy for the bonds in diatomic molecules (H_2, HF, HCl, HBr, HI, N_2, O_2, F_2, Cl_2, Br_2 and I_2) and average bond energies for the other bonds.

[b] The value for the CO bond in CO_2 is considerably different: 799 kJ/mol.

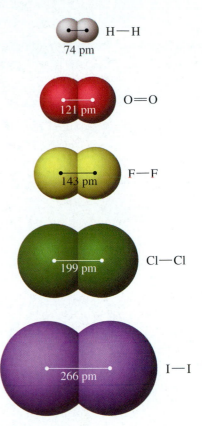

▲ Bond lengths of five common diatomic molecules.

Now we can appreciate the meaning of a *single* covalent radius, introduced in Section 8.7. Because it is one-half the distance between the nuclei of identical atoms when they are joined by a *single* covalent bond, the single covalent radius is one-half the bond length. Thus, the single covalent radius of an iodine atom, 133 pm (Figure 8.8), is one-half the I—I bond length of 266 pm (Table 9.1). Furthermore, even for unlike atoms of comparable electronegativities, we can make the following rough generalization:

The length of the covalent bond joining unlike atoms is the sum of the covalent radii of the two atoms.

Exceptions to this generalization occur when differences between bonded atoms are appreciable. That is, the partial ionic character of polar covalent bonds produces greater bond strengths and shorter bond lengths than we would otherwise expect. Thus, the H—Cl bond length we calculate from the H—H and Cl—Cl bond lengths is $\left[\left(\frac{1}{2} \times 74\right) + \left(\frac{1}{2} \times 199\right)\right]$ pm = 137 pm, but the measured bond length is only 127.4 pm.

At times, experimentally determined bond lengths can help us choose the best Lewis structure for a molecule, but the reverse is also true. As illustrated in Example 9.13, we can estimate a bond length from a plausible Lewis structure.

Example 9.13

Estimate the length of **(a)** the nitrogen-to-nitrogen bond in N_2H_4 and **(b)** the bond in BrCl.

STRATEGY

In part (a), the first thing we must do is draw a plausible Lewis structure for N_2H_4 and determine the bond order. Then we can select the appropriate bond length from Table 9.1. In part (b), we will determine the bond order in BrCl from its Lewis structure and use the approximation that the bond length is the sum of the covalent radii of the bonded atoms.

SOLUTION

(a) The Lewis structure

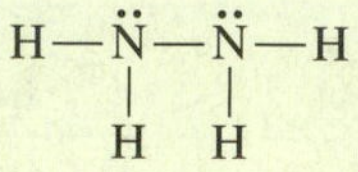

$$H-\overset{..}{N}-\overset{..}{N}-H$$
$$\quad\ \ \ |\quad\ \ |$$
$$\quad\ \ \ H\quad H$$

is a plausible one: valence-shell octets for the N atoms, duets for the H atoms, and formal charges of 0. The nitrogen-to-nitrogen bond is single, and from Table 9.1 we find its bond length to be 145 pm.

(b) To get the Lewis structure of BrCl, we imagine substituting one Br atom for one Cl atom in the Lewis structure of Cl_2, arriving at the structure

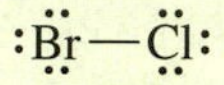

$$:\overset{..}{\underset{..}{Br}}-\overset{..}{\underset{..}{Cl}}:$$

The BrCl molecule contains a Br—Cl single bond with a length approximately one-half the Cl—Cl bond length *plus* one-half the Br—Br bond length. From Table 9.1, we get $\left[\left(\frac{1}{2} \times 199\right) + \left(\frac{1}{2} \times 228\right)\right] = 214$ pm.

ASSESSMENT

In part (b), the estimated bond length is in excellent agreement with the measured value of 213.8 pm, suggesting that the small difference in electronegativity (0.2) between Br (2.8) and Cl (3.0) is not an important consideration.

EXERCISE 9.13A

Estimate the oxygen-to-fluorine bond length in OF_2.

EXERCISE 9.13B

Estimate the nitrogen-to-nitrogen bond length in nitramide, NH_2NO_2.

Bond Energy

Energy must be *absorbed* to break a covalent bond. As noted in our discussion of the Born–Haber cycle, we define **bond-dissociation energy (D)** as the quantity of energy required to break one mole of covalent bonds between two atoms in a molecule in the gas phase. Bond-dissociation energies are measured for specific bonds in molecules and are generally expressed in kilojoules per mole of bonds.

It is quite simple to calculate bond-dissociation energies for diatomic molecules. Because there is only one bond (be it single, double, or triple) per molecule, we can represent the bond-dissociation energy as the enthalpy change ΔH for the dissociation reaction. When we do things this way, the enthalpy change for the reverse reaction, in which a bond is formed, is the negative of the bond-dissociation energy. For example,

Bond breaking: $\quad Cl_2(g) \longrightarrow 2\,Cl(g) \quad \Delta H = D(Cl-Cl) = +243$ kJ/mol

Bond forming: $\quad 2\,Cl(g) \longrightarrow Cl_2(g) \quad \Delta H = -D(Cl-Cl) = -243$ kJ/mol

With a polyatomic molecule, such as H_2O, the situation is different. As shown in Figure 9.13, the energy required to dissociate one mole of H atoms by breaking one

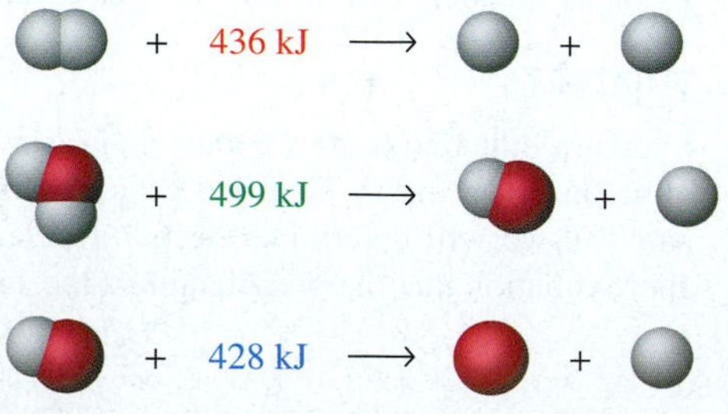

▶ **FIGURE 9.13   Some bond energies compared**

The same quantity of energy (436 kJ/mol) is required to break H—H bonds in all diatomic H_2 molecules. The situation is more complicated in polyatomic molecules. In water, for instance, more energy is required to break the first O—H bond (499 kJ/mol) than to break the second O—H bond (428 kJ/mol).

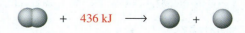

O—H bond per $H_2O(g)$ molecule is different from the energy required to dissociate a second mole of H atoms by breaking the remaining O—H bond in OH(g):

$$H—OH(g) \longrightarrow H(g) + OH(g) \qquad \Delta H = +499 \text{ kJ/mol}$$

$$O—H(g) \longrightarrow H(g) + O(g) \qquad \Delta H = +428 \text{ kJ/mol}$$

As we can readily see from Table 9.1, there is a relationship between bond order and bond-dissociation energy. A double bond between a given pair of atoms has a higher bond-dissociation energy than does a single bond between the same atoms, and a triple bond has a still higher bond-dissociation energy. Generally, the higher the bond order, the higher the bond-dissociation energy for bonds between a given pair of atoms.

Bond-dissociation energies also depend on the *environment* of a given bond, that is, on the other atoms in the molecule that are near the bond. For example, the bond-dissociation energy of the O—H bond in H—O—H is different from that in H—O—O—H or in H_3C—O—H. For this reason, it is customary to use average values. An **average bond energy** is the average of the bond-dissociation energies for a number of different molecules containing the particular bond. In Table 9.1, the energy values shown for bonds in stable diatomic molecules, such as H—H, H—Cl, and Cl—Cl, are precisely known *bond-dissociation energies*. For other bonds, such as H—C, H—N, and H—O, the values are *average bond energies*.

Calculations Involving Bond Energies

Let's imagine carrying out a gas-phase reaction in the following way. First, while keeping track of the energy required, we break all the bonds in the reactant molecules to produce a gas made up of free, uncombined atoms. Then, we recombine these atoms into the product molecules and note how much energy is released.

$$\text{Gaseous reactants} \longrightarrow \text{gaseous atoms} \longrightarrow \text{gaseous products}$$

The sum of the enthalpy changes for breaking the old bonds and forming the new bonds is the enthalpy change ΔH for the reaction:

$$\Delta H = \Delta H_{\text{bonds broken}} + \Delta H_{\text{bonds formed}} \qquad (9.3)$$

Because some of the bond energies (BE)* that we use in calculations are average bond energies rather than the more precise bond-dissociation energies, the calculated enthalpy changes are only approximately correct. That is, the calculations are based on the following relationships:

$$\Delta H_{\text{bonds broken}} \approx \sum BE(\text{reactants})$$

$$\Delta H_{\text{bonds formed}} \approx \sum - BE(\text{products})$$

$$\Delta H \approx \sum BE(\text{reactants}) - \sum BE(\text{products}) \qquad (9.4)$$

In Figure 9.14, we illustrate these ideas from a molecular viewpoint by considering the formation of gaseous hydrazine, N_2H_4, from nitrogen and hydrogen gases. In Example 9.14, we use bond energies to estimate the enthalpy of formation for $N_2H_4(g)$.

Breaking and Forming Bonds in a Chemical Reaction activity

Bond energies are defined as positive quantities and specifically refer to *breaking* bonds, as occurring in the reactants. The *forming* of bonds, as occurring in the products, is described by the negative of bond energies. This explains the form of Equation (9.4), which differs from the more familiar form for enthalpy changes, which employs ΔH of *products* minus ΔH of *reactants*.

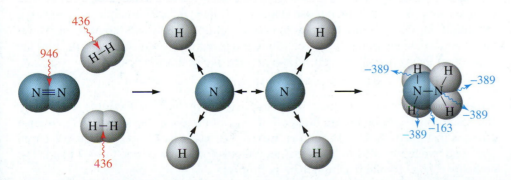

◀ **FIGURE 9.14   Bond breakage and bond formation in a reaction**

The values shown, in kilojoules per mole, are the quantities of energy absorbed in breaking bonds in the reactants N_2 and H_2 and released in forming bonds in the product N_2H_4.

*These energies are more properly called *bond enthalpies*, but we will use the more familiar term *bond energies*.

Example 9.14

Use bond energies from Table 9.1 to estimate the enthalpy of formation of gaseous hydrazine. Compare the result with the value of $\Delta H_f^\circ [N_2H_4(g)]$ from Appendix C.

STRATEGY

To choose the proper bond energies from Table 9.1, we first need to write Lewis structures for the substances involved in the reaction. Then, using data from Table 9.1, we evaluate ΔH for bonds broken and for bonds formed. Finally, we apply Equations (9.3) and (9.4) to obtain the enthalpy of formation we seek.

SOLUTION

We write the Lewis structures in the form of reactants and products in a balanced chemical equation.

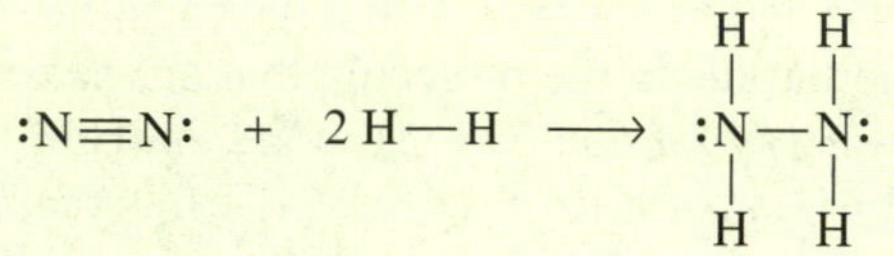

$$:N{\equiv}N: \ + \ 2\,H{-}H \ \longrightarrow \ \begin{matrix} H & H \\ | & | \\ :N{-}N: \\ | & | \\ H & H \end{matrix}$$

Now we evaluate ΔH for bonds broken and bonds formed.

ΔH for bond breakage:

1 mol N≡N bonds = 946 kJ

2 mol H—H bonds = (2 × 436) kJ = 872 kJ

Total for bonds broken = 1818 kJ

ΔH for bond formation:

1 mol N—N bonds = −163 kJ

4 mol N—H bonds = (4 × −389) kJ = −1556 kJ

Total for bonds formed = −1719 kJ

The enthalpy of formation is the sum of the enthalpies for bond breakage and bond formation.

$$\Delta H_f = \Delta H_{\text{bonds broken}} + \Delta H_{\text{bonds formed}}$$

$$\Delta H_f = 1818 \text{ kJ} - 1719 \text{ kJ} = 99 \text{ kJ}$$

ASSESSMENT

For the reaction in which gaseous hydrazine is formed from its elements, the estimate that $\Delta H_f^\circ [N_2H_4(g)] = 99$ kJ/mol $N_2H_4(g)$ agrees reasonably well with the value of 95.40 kJ/mol listed in Appendix C. To avoid errors in an estimation of this type, remember that the energies listed in Table 9.1 are all positive quantities. Some data (for bond breakage) are used as is, but other data (for bond formation) require a change in sign.

EXERCISE 9.14A

Estimate ΔH for the reaction

$$C_2H_6(g) + Cl_2(g) \longrightarrow C_2H_5Cl(g) + HCl(g)$$

EXERCISE 9.14B

The standard enthalpy of formation of nitrosyl fluoride, NOF(g), is −66.5 kJ/mol. Use this fact, together with data from Table 9.1, to estimate the bond energy of the N—F bond. (*Hint:* What is the best Lewis structure for nitrosyl fluoride?)

As the result of Example 9.14 suggests, enthalpy changes calculated from bond energies are less accurate than those calculated from standard enthalpies of formation (Section 6.7). We use bond energies mostly in applications where the necessary thermodynamic data are not available or where rough estimates will suffice. Consider the reaction of the hydroxyl radical (·OH) with methane in the atmosphere:

$$\cdot OH + CH_4 \longrightarrow \cdot CH_3 + H_2O \qquad \Delta H = ?$$

To assess ΔH for this reaction, let's simplify the method of Example 9.14 a bit. Because four C—H bonds are broken in CH_4 and three are formed in ·CH_3, the net effect is that only one C—H bond is broken. Because one O—H bond is broken in ·OH and two are formed in H_2O, the net effect is formation of one O—H bond. The enthalpy change for the reaction, then, is simply

$$\Delta H = \text{BE(C—H)} - \text{BE(O—H)} = +414 \text{ kJ} - 464 \text{ kJ} = -50 \text{ kJ}.$$

Unsaturated Hydrocarbons

In Chapter 2, we learned to write structural formulas for the alkanes. In these compounds, which have the general formula C_nH_{2n+2}, all the bonds are single bonds and there are no lone-pair electrons. The Lewis structures of the alkanes are the same as their structural formulas, as in this representation of ethane:

$$H—\underset{\underset{H}{|}}{\overset{\overset{H}{|}}{C}}—\underset{\underset{H}{|}}{\overset{\overset{H}{|}}{C}}—H$$

Because alkane molecules cannot form bonds to additional atoms, the alkanes are called *saturated* hydrocarbons.

Multiple bonds are common in organic compounds. Collectively, hydrocarbons with double or triple bonds between carbon atoms are called **unsaturated hydrocarbons.** In contrast to the saturated hydrocarbons, unsaturated hydrocarbons can bond to additional atoms under the appropriate conditions. A carbon atom can also form a double bond with a nitrogen, oxygen, or sulfur atom, or a triple bond with a nitrogen atom.

9.11   Alkenes and Alkynes

Some hydrocarbons and many other organic compounds have one or more double bonds between carbon atoms. Compounds with carbon-to-carbon triple bonds are less common but by no means rare.

Alkenes

The **alkenes** (note the *-ene* ending) are hydrocarbons that contain at least one carbon-to-carbon double bond. *Simple alkenes* have just one double bond per molecule and have the general formula C_nH_{2n}. Hydrocarbons with this formula make up a series that parallels the straight-chain and branched-chain alkanes, C_nH_{2n+2}. The simplest alkene is *ethene*, C_2H_4. When we construct its Lewis structure from Lewis symbols, we can see that the ethene molecule must have a carbon-to-carbon double bond:

We can also use a structural formula or condensed structural formula to represent ethene. The structural formula is the same as the Lewis structure. It shows that the double bond is shared by the two carbon atoms and does not involve the hydrogen atoms. The condensed structural formula does not make this point quite as obvious, but it is the easier formula to set in type, requires less space on a printed page, and is consequently more widely used:

$$\underset{\underset{H}{\diagup}\quad\underset{H}{\diagdown}}{C=C}$$

Structural formula

$$CH_2{=}CH_2 \quad \text{or} \quad H_2C{=}CH_2$$

Condensed structural formula

A space-filling molecular model of ethene is shown in Figure 9.15.

Many simple alkenes have common, nonsystematic names. Ethene, the two-carbon alkene, is most often called *ethylene*. The three-carbon alkene, propene, $CH_3CH{=}CH_2$, is frequently called *propylene*. Three different four-carbon alkenes (isomers) have the formula C_4H_8, and the number of isomers increases rapidly with the number of carbon atoms. Whenever there are large numbers of isomers, it is difficult to remember the many possible common names, and chemists resort to the IUPAC system described in Appendix C.

Testing for Unsaturated Hydrocarbons with Bromine movie

▲ **FIGURE 9.15   Molecular model of ethene (ethylene)**

Application Note

Ethylene is produced in greater quantity than any other organic chemical in the United States; propylene is second in production. Both are used in the manufacture of polymers (see page 373).

Fats and Oils: Hydrogenation

A fat (*triglyceride*) is made by the reaction of three molecules of long-chain carboxylic acids (*fatty acids*) with a molecule of glycerol, $CH_2OHCHOHCH_2OH$, to form an ester (see Figure 6.21). A triglyceride is called a *fat* if it is a solid at 25 °C and an *oil* if it is a liquid at that temperature. These differences in melting points are due to the degree of unsaturation—the number of carbon-to-carbon double bonds—of their fatty acids.

Saturated fatty acids such as stearic acid contain no carbon-to-carbon double bonds:

$$CH_3CH_2CH_2CH_2CH_2CH_2CH_2CH_2CH_2CH_2CH_2CH_2CH_2CH_2CH_2CH_2CH_2COOH$$

Stearic acid (saturated)

Monounsaturated fatty acids such as oleic acid contain one carbon-to-carbon double bond per molecule:

$$CH_3CH_2CH_2CH_2CH_2CH_2CH_2CH_2CH\!=\!CHCH_2CH_2CH_2CH_2CH_2CH_2CH_2COOH$$

Oleic acid (monounsaturated)

Polyunsaturated fatty acids have two or more carbon-to-carbon double bonds per molecule. For example, linolenic acid has three such bonds:

$$CH_3CH_2CH\!=\!CHCH_2CH\!=\!CHCH_2CH\!=\!CHCH_2CH_2CH_2CH_2CH_2CH_2CH_2COOH$$

Linolenic acid (polyunsaturated)

Saturated fats contain mostly saturated fatty acids, while *polyunsaturated fats* incorporate mainly unsaturated fatty acids. Triglycerides obtained from animal sources are usually solids, whereas those of plant origin are generally oils.

Margarine and vegetable shortening are made by partially *hydrogenating* inexpensive, abundant vegetable oils such as cotton-seed, corn, or soybean oil. The consistency of the product can be controlled by the degree of hydrogenation. Fats and oils are broken down in the digestive tract by *enzymes*—biological catalysts—into smaller molecules that can be absorbed and used for energy. A typical fat yields about 37 kJ per gram.

Calorie-free fats such as *olestra*, made from fatty acids and sucrose (a sugar), actually yield about the same energy as regular fats when oxidized in a bomb calorimeter (Chapter 6). Unlike ordinary fats, however, olestra molecules are not broken down by enzymes in the human body and are therefore not absorbed into the body. Because it is not digested, olestra contributes no calories when used to fry potato chips, tortilla chips, and other foods. However, in large quantities olestra may cause gastrointestinal discomfort and diarrhea in some people. Also, because it is not absorbed, it carries fat-soluble vitamins A, D, E, and K through the digestive tract, lessening their absorption.

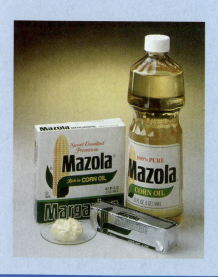

◄ About 85% of the fatty acid residues in corn oil are unsaturated. To convert the oil to a solid (margarine), it is hydrogenated. The process converts some of the unsaturated molecules in the oil to saturated molecules in the product.

▲ The reddish-brown color of bromine vapor (left) is removed when a strip of uncooked bacon is placed in the beaker (right). Bromine molecules react with unsaturated fats in the bacon in a reaction similar to a hydrogenation reaction.

Alkenes are chemically reactive. These unsaturated molecules can use two electrons from one of the electron pairs in a double bond to form bonds to two other atoms. In the resulting product, all the covalent bonds are single bonds and the molecule is saturated. A commercially important reaction in which the H atoms of H_2 are added to the double-bonded carbon atoms in unsaturated molecules is called a *hydrogenation* reaction. The reaction requires a *catalyst,* a substance that greatly increases the speed of a reaction while remaining unchanged by the reaction. The catalysts most often used in hydrogenation reactions are nickel, palladium, and platinum. In the hydrogenation of ethene represented here, the nickel catalyst is indicated above the yield arrow, and the red bond lines show the two new carbon-to-hydrogen bonds in the product:

$$\text{Ethene} + \text{H—H} \xrightarrow{Ni} \text{Ethane}$$

The product of this reaction, ethane, is an alkane having the same carbon skeletal structure as the original alkene, ethene.

Alkynes

In the double bond of an alkene molecule, carbon atoms share two pairs of electrons. Carbon atoms can also share three pairs of electrons, forming triple bonds. Hydrocarbon molecules containing carbon-to-carbon triple bonds are called **alkynes.** The simplest alkyne is ethyne, commonly called acetylene (Figure 9.16):

$$\text{H:C:::C:H} \quad \text{or} \quad \text{H—C} \equiv \text{C—H}$$

The IUPAC nomenclature for alkynes parallels that for alkenes, except that the family ending is *-yne* rather than *-ene.*

Alkyne molecules are unsaturated; they can add atoms and form additional bonds just as alkene molecules can. However, because either one or two pairs of electrons in the triple bond can participate, it is possible to add twice as much of a reagent to an alkyne as to an alkene. For example, in hydrogenation under carefully controlled conditions, acetylene can react with one molecule of hydrogen to form ethene:

$$\text{H—C} \equiv \text{C—H} + \text{H—H} \xrightarrow{Ni} \text{C=C}$$

More likely, however, an acetylene molecule will react with two molecules of hydrogen to form first ethylene and then ethane. The overall reaction is

$$\text{H—C} \equiv \text{C—H} + 2\,\text{H—H} \xrightarrow{Ni} \text{H—C—C—H}$$

9.12 Polymers

If we attempt to write a Lewis structure for ethylene simply by bringing together two Lewis symbols of carbon, $\cdot\ddot{C}\cdot$, and four of hydrogen, H·, this is what we might get as a first attempt:

$$\cdot\text{C—C}\cdot$$

▲ FIGURE 9.16 Molecular model of ethyne (acetylene)

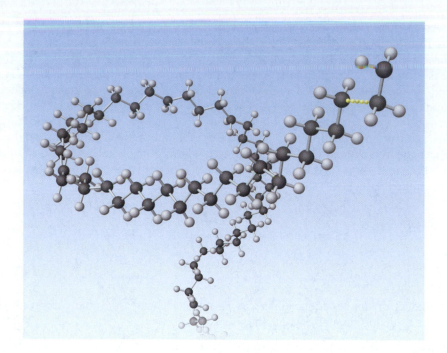

► **FIGURE 9.17   The formation of polyethylene**

In the synthesis of polyethylene, many monomer units join together to form gigantic polymer molecules. In this computer-generated representation, the yellow dots at the upper right indicate a new bond forming as an ethylene molecule is added to the growing polymer chain.

QUESTION: The molecule of poly-ethylene is no longer an alkene. What kind of organic compound is it?

Of course, we would reject this structure because the C atoms have incomplete octets. Suppose, however, that instead of creating a second bond between the carbon atoms, we consider joining two more such structures to the previous one. We would get this, where the red indicates the newly added segments:

$$
\underset{\underset{H}{|}}{\overset{\overset{H}{|}}{\cdot C}} - \underset{\underset{H}{|}}{\overset{\overset{H}{|}}{C}} - \underset{\underset{H}{|}}{\overset{\overset{H}{|}}{C}} - \underset{\underset{H}{|}}{\overset{\overset{H}{|}}{C}} - \underset{\underset{H}{|}}{\overset{\overset{H}{|}}{C}} - \underset{\underset{H}{|}}{\overset{\overset{H}{|}}{C}}\cdot
$$

And if we add two more structures (blue), we get this:

$$
\cdot C - C - C - C - C - C - C - C - C - C\cdot
$$

We could repeat the process many times over and ultimately end up with a very long molecule (Figure 9.17).

We have just described a process in which a large number of simple molecules called **monomers** (Greek *monos*, single; and *meros*, parts) are joined together into a giant, long-chain molecule called a **polymer** (Greek *poly*, many; and *meros*, parts). The process is called *polymerization*. In this case, the monomers are ethylene molecules and the polymer is called *polyethylene*. To represent the polymer accurately, we would have to continue the structure off the page, across the room, and (perhaps) out the door. Rather than attempt that, we will represent it through the repeating unit:

$$
\left[\begin{array}{c} H \quad H \\ | \quad \; | \\ -C - C - \\ | \quad \; | \\ H \quad H \end{array} \right]_n
$$

▲ Polyethylene is quite inert chemically and sees widespread use in the food industry. Milk jugs, plastic wrap, and orange juice bottles as shown here are just a few of the items made from polyethylene.

Here, *n* indicates the number of monomer units joined together in the polymer; typically *n* has a value from several hundred to more than one thousand for polyethylene. Even though it is called polyethylene, the polymer is *not* an alkene. The double bonds of the alkene starting material are converted to single bonds.

Table 9.2 A Selection of Addition Polymers

Monomer	Polymer	Polymer Name	Some Uses
$H_2C{=}CH_2$	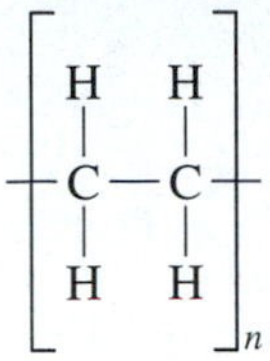 $\left[\begin{array}{c} \text{H} \quad \text{H} \\ -\text{C}-\text{C}- \\ \text{H} \quad \text{H} \end{array}\right]_n$	Polyethylene	Plastic bags, milk bottles, toys, electrical insulation, plastic wrap, paper coatings, food containers and lids, storage containers, laboratory bottles, wastebaskets
$H_2C{=}CH{-}CH_3$	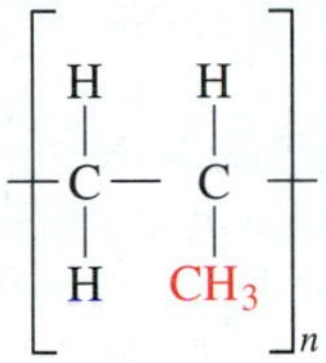 $\left[\begin{array}{c} \text{H} \quad \text{H} \\ -\text{C}-\text{C}- \\ \text{H} \quad \text{CH}_3 \end{array}\right]_n$	Polypropylene	Indoor–outdoor carpeting, bottles, insulating sportswear, tear-resistant envelopes, labels, disposable clothing and protective outerwear, moisture-resistant packaging, chemical-resistant plumbing pipe
$H_2C{=}CH{-}Cl$	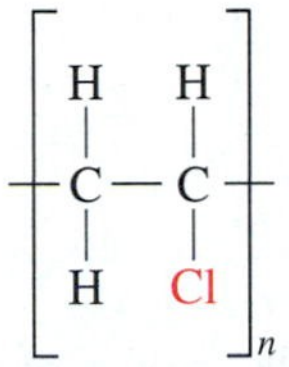 $\left[\begin{array}{c} \text{H} \quad \text{H} \\ -\text{C}-\text{C}- \\ \text{H} \quad \text{Cl} \end{array}\right]_n$	Poly(vinyl chloride), PVC	Plastic wrap, simulated leather (Naugahyde), T-shirt printing, garden hoses, clear flexible tubing, "soft" toys, plumbing fixtures and pipe, vinyl flooring, rainwear
$F_2C{=}CF_2$	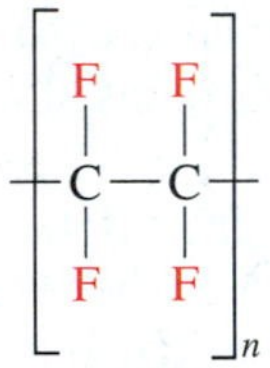 $\left[\begin{array}{c} \text{F} \quad \text{F} \\ -\text{C}-\text{C}- \\ \text{F} \quad \text{F} \end{array}\right]_n$	Polytetrafluoroethylene, Teflon	Nonstick coating for cooking utensils, electrical insulation, chemical-resistant laboratory ware, coatings for clothing and carpet

A few common alkene monomers, and the polymers made from them, are listed in Table 9.2. In Chapter 24, we will provide some additional details about polymerization, such as how the double bond in ethylene opens up to get the polymerization started and how unpaired electrons are eliminated to terminate a polymer chain.

Biological Polymers

The word "polymer" usually brings to mind synthetic plastics like polyethylene and polystyrene, but that connotation is too restrictive because in fact all living systems are based on polymer molecules. Proteins are an example of biological polymers. Proteins have molecular masses ranging from several thousand to several million atomic mass units. Their monomer units are amino acids, molecules that contain both an amino group (NH_2) and a carboxyl group (COOH):

$$RCH(NH_2)COOH$$

An amino acid

Simple polymers, such as polyethylene and polypropylene, have only one type of monomer unit. In contrast, proteins of all living species, from bacteria to humans, are constructed from a basic set of 20 amino acid monomers. A typical protein has 200–300 amino acid units, but some proteins are much smaller than this. Those with only a few amino acid units are often called *peptides*. The largest protein known has about 27,000 amino acid units in a single chain.

Almost all amino acids have their amino group (NH_2) bonded to the alpha (α) carbon atom, which is the carbon atom adjacent to the carbon atom in the carboxyl group.

The individual amino acid groups have different R groups. Two amino acid molecules join by eliminating a molecule of water between them and forming a *peptide* bond:

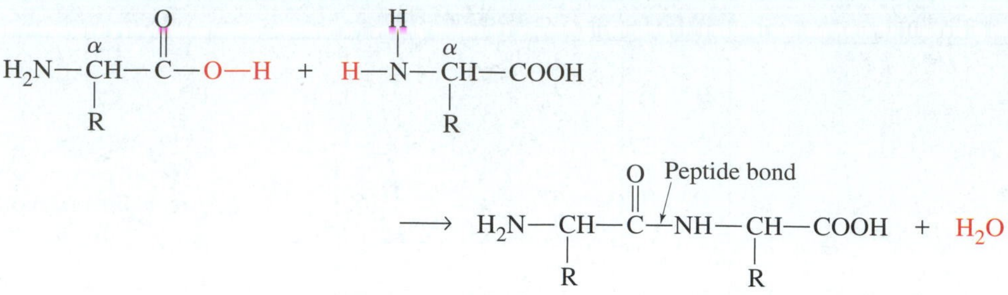

A molecule formed in this manner from two amino acid units is called a dipeptide. The groups at the ends of a dipeptide can each join with another amino acid molecule, forming a chain. Two more amino acid molecules can be added to the ends of that chain, and so on:

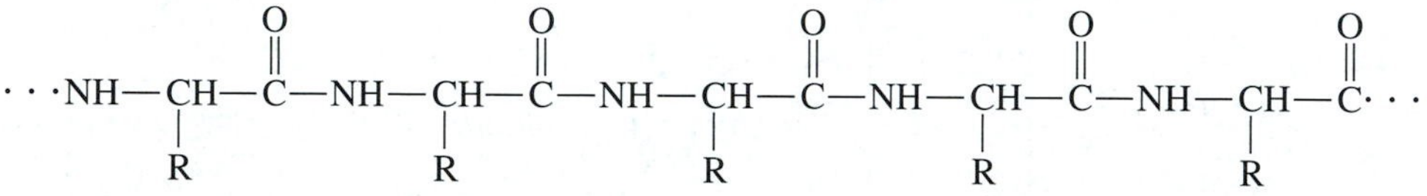

An organism's genetic code specifies the *sequence* of amino acids in each kind of protein (Chapter 23). This sequence is called the *primary structure* of the protein. Proteins differ from one another in the sequence of amino acids along the protein chain, in chain length, and in the number of chains.

The overall structure of a protein molecule is determined by additional factors, some of which we will consider in subsequent chapters. These factors include covalent linkages through a pair of sulfur atoms (disulfide bridges), ionic bonds sometimes called salt bridges, and intermolecular forces. We explore the latter vital topic in Chapter 11.

Cumulative Example

Use data from this chapter and elsewhere in the text to estimate a value for the standard enthalpy of formation at 25 °C of HNCO(g). Assess the most likely source of error in this estimation.

STRATEGY

We can use bond energies to estimate the enthalpy change in a reaction, as long as all the reactants and products are gases. In the formation reaction for HNCO(g), all the reactant species are gaseous *except* for C(graphite). However, to obtain the desired formation reaction and its enthalpy change, we can use bond energies for a hypothetical reaction in which all the species are gaseous, together with the enthalpy of formation of C(g) from Appendix C (716.7 kJ/mol) and Hess's law (Section 6.6). To evaluate the enthalpy change for the hypothetical reaction, we will also have to write an acceptable Lewis structure for HNCO.

SOLUTION

When added together, equations (a) and (b) yield reaction (c), which has ΔH_f° as its standard enthalpy change (recall Section 6.7).

(a) $\quad \frac{1}{2} H_2(g) + \frac{1}{2} N_2(g) + \frac{1}{2} O_2(g) + C(g) \longrightarrow HNCO(g) \qquad \Delta H^\circ = ?$

(b) $\qquad\qquad\qquad\qquad\qquad\qquad C(graphite) \longrightarrow C(g) \qquad \Delta H^\circ = 716.7 \text{ kJ}$

(c) $\quad \frac{1}{2} H_2(g) + \frac{1}{2} N_2(g) + \frac{1}{2} O_2(g) + C(graphite) \longrightarrow HNCO(g) \qquad \Delta H_f^\circ = ?$

The enthalpy change for reaction (b) comes from Appendix C; it is 716.7 kJ. The enthalpy change for reaction (a) can be obtained from Equation (9.4) based on these Lewis structures:

$$H\!-\!H \qquad :\ddot{O}\!=\!\ddot{O}: \qquad :N\!\equiv\!N: \qquad H\!-\!\ddot{N}\!=\!C\!=\!\ddot{O}:$$

To evaluate ΔH for bond breakage in reaction (a), we sum the bond energies for the bonds broken; those energies are positive values.

$\frac{1}{2}$ mol H—H bonds $= \frac{1}{2} \times 436 \text{ kJ} = 218 \text{ kJ}$

$\frac{1}{2}$ mol N≡N bonds $= \frac{1}{2} \times 946 \text{ kJ} = 473 \text{ kJ}$

$\frac{1}{2}$ mol O=O bonds $= \frac{1}{2} \times 498 \text{ kJ} = 249 \text{ kJ}$

no bonds in C(g)

Total $= 940 \text{ kJ}$

To evaluate ΔH for bond formation in reaction (a), we sum the bond energies for the bonds formed; those energies are negative values.

1 mol N—H bonds $= -389 \text{ kJ}$

1 mol N=C bonds $= -615 \text{ kJ}$

1 mol C=O bonds $= -736 \text{ kJ}$

Total $= -1740 \text{ kJ}$

The enthalpy change for reaction (a) is the sum of the energies for bond breakage and bond formation.

$\Delta H° = 940 \text{ kJ} - 1740 \text{ kJ} = -800 \text{ kJ}$

Finally, the enthalpy change for reaction (c), the standard enthalpy of formation of HNCO(g), is found by adding the enthalpy changes for reactions (a) and (b).

$\Delta H_f° = -800 \text{ kJ} + 716.7 \text{ kJ} = -83 \text{ kJ}$

ASSESSMENT

The enthalpy of vaporization of graphite (reaction b) is a precisely known quantity, as are the bond-dissociation energies of the diatomic molecules H_2, N_2, and O_2. In contrast, the three bond energies in HNCO(g) are average bond energies. Their values are less precisely known and cause the resulting $\Delta H_f°$ to be only an approximate value.

Concept Review with Key Terms

9.1 Chemical Bonds: A Preview—**Chemical bonds** form when the attractive forces between negatively charged electrons and positively charged nuclei exceed the repulsions between the nuclei to the maximum extent possible. The nature of the chemical bonds in a material determines many of its properties.

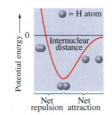

9.2 The Lewis Theory of Chemical Bonding: An Overview—A **Lewis symbol** represents the valence electrons of an element, shown through dots distributed about the chemical symbol of the element. Lewis symbols of main-group elements are related to their locations in the periodic table. In forming compounds, main-group elements generally tend to follow an **octet rule**, which states that bonded atoms tend to acquire electron configurations having eight electrons in the valence shell.

1A	2A	3A	4A	5A	6A	7A	8A
Li·	Be·	·B·	·Ċ·	·N̈:	·Ö:	:F̈:	:N̈e:

9.3 Ionic Bonds and Ionic Crystals—An **ionic bond** forms through the transfer of electrons from metal to nonmetal atoms and the subsequent electrostatic attraction between the resulting ions. An ionic **crystal** consists of an extended regular arrangement of ions.

9.4 Using Lewis Symbols to Represent Ionic Bonding—Ionic bonds between nonmetals and s-block, p-block, and a few d-block metals can be represented through the Lewis symbols of the respective anions and cations. This representation indicates the nature of electron transfer and the basis for electrostatic attraction.

$$\left[\text{Na·}\right] + \left[:\ddot{\text{C}}\text{l}:\right] \longrightarrow \left[\text{Na}\right]^+ + \left[:\ddot{\text{C}}\ddot{\text{l}}:\right]^-$$

9.5 Energy Changes in Ionic Compound Formation—The net energy decrease in the formation of an ionic crystal from its gaseous ions is the **lattice energy**. A Born–Haber cycle relates the lattice energy and enthalpy of formation of an ionic compound, together with other atomic and molecular properties.

9.6 Lewis Structures of Some Simple Molecules—A **covalent bond** forms when two atoms share electrons. A **Lewis structure** of a molecule shows the arrangement of atoms in the molecule and the covalent bonds between them. The Lewis structure depicts electron pairs in the molecule as either **bonding pairs** or **lone pairs**:

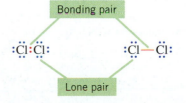

Usually, each atom in a Lewis structure acquires a noble-gas electron configuration, which for most atoms is a valence-shell octet of electrons. A covalent bond involving one electron pair is called a **single bond**. A **double bond** involves two shared electron pairs, and a **triple bond** involves three.

9.7 Polar Covalent Bonds and Electronegativity—Electronegativity (EN) is a measure of the ability of an atom to attract the electrons of a chemical bond to itself and is related to the position of an element in the periodic table. In a covalent bond between atoms of different EN values, electrons are displaced toward the atom with the higher EN. Chemical bonds vary from **nonpolar covalent** (zero or very small ΔEN) to **polar covalent** to ionic (large ΔEN).

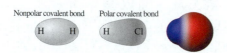

Nonpolar covalent bond   Polar covalent bond

9.8 A Strategy for Writing Lewis Structures—Writing plausible Lewis structures for molecules or polyatomic ions involves two general tasks: (1) writing the **skeletal structure**, which consists of one or more **central atoms** bonded to a number of **terminal atoms**, and (2) distributing the valence electrons among the atoms so as to (usually) follow the octet rule. The **formal charge** on an atom can help determine the best Lewis structure for a molecule:

$$\text{Formal charge} = \begin{pmatrix} \text{number of valence} \\ \text{electrons in the} \\ \text{uncombined atom} \end{pmatrix} - \begin{pmatrix} \text{number of lone-} \\ \text{pair electrons} \\ \text{on the bound atom} \end{pmatrix}$$

$$- \frac{1}{2}\begin{pmatrix} \text{number of} \\ \text{electrons in} \\ \text{bonds to the atom} \end{pmatrix}$$

   Resonance describes a phenomenon in which two or more Lewis structures have the same skeletal structure but different distributions of electrons among the bonded atoms. The best description of the actual structure, the **resonance hybrid**, is a combination of plausible **resonance structures**. In a resonance hybrid, some of the bonding

electrons are **delocalized** over several atoms. The resonance hybrid should conform to experimental data, such as bond lengths and/or bond energies, if these data are available.

9.9 Molecules that Do Not Follow the Octet Rule—Exceptions to the octet rule are found in odd-electron molecules and in molecular fragments called **free radicals**. A few structures appear to have too few electrons to complete the octets of all atoms in the molecule or ion. Some structures appear to have too many. In the latter case, a central atom may have an **expanded valence shell** involving five, six, or even more electron pairs.

Sulfur hexafluoride

9.10 Bond Lengths and Bond Energies—Bond length, which in some cases can be related to atomic radii, is the internuclear distance between two bonded atoms. Bond length is greatly affected by bond order, that is, whether the bond is single, double, or triple. **Bond-dissociation energy** is the energy required to break one mole of bonds in a gaseous species; its value depends on (1) the atoms in the bond, (2) the bond order, and (3) the particular molecule in which the bond is found. Tabulated values are often averaged over a number of molecules containing the same bond. Bond-dissociation energies and **average bond energies** can be used to estimate enthalpy changes of reactions via the relationship

$$\Delta H = \Delta H_{\text{bonds broken}} + \Delta H_{\text{bonds formed}}$$

9.11 Alkenes and Alkynes—Unsaturated hydrocarbons have one or more multiple bonds between carbon atoms. **Alkenes** have double bonds, and **alkynes** have triple bonds. A characteristic reaction of alkenes and alkynes is their ability to add atoms across the multiple bonds, thereby converting the multiple bonds to single bonds.

9.12 Polymers—Some molecules containing multiple bonds undergo polymerization, a reaction in which small molecules (**monomers**) join together in large numbers to produce a gigantic molecule (**polymer**). Polymers typically are made up of long carbon chains, have high molecular masses, and often are represented through the structure of only the repeating unit.

Assessment Goals

When you have mastered the material in this chapter, you will be able to:

- Describe the energetics of chemical bond formation.

- Cite the characteristics of ionic and covalent bonds, and list the types of elements that typically form these bonds.

- Write Lewis symbols for elements, and represent ionic and covalent bonds in Lewis structures of compounds, using the duet and octet rules where appropriate.

- Describe the nature of multiple covalent bonds in terms of shared electron pairs.

- Define electronegativity, know its periodic trends, and describe its influence on chemical bonding.

- Identify nonpolar covalent, polar covalent, and ionic bonds.

- Apply a general strategy for writing Lewis structures for molecules and polyatomic ions.

- Calculate formal charges, use them in assessing the plausibility of Lewis structures, and explain the relationship between formal charges and coordinate covalent bonds.

- Where appropriate, write resonance structures for a given molecule or ion and describe the resonance hybrid.

- Recognize exceptions to the octet rule, and write Lewis structures for these exceptions.

- Define bond length, describe the general relationship between bond order and bond length, and deduce probable bond lengths from Lewis structures.

- Calculate reaction enthalpy changes from bond-dissociation energies, and recognize the limitations of these calculations.

- Describe the general properties of alkanes, alkenes, alkynes and the general nature of organic polymers, including biopolymers.

Self-Assessment Questions

1. What are the essential propositions of the Lewis theory of chemical bonding? To which elements does the *octet* rule apply? To which elements does the *duet* rule apply?

2. What are the structural differences and similarities in each of the following sets?

   (a) a sodium ion and a neon atom

   (b) chlorine atoms, chlorine molecules, and chloride ions

3. Using the periodic table, write the Lewis symbol for an atom of each of the following elements.

   (a) sodium   (d) bromine

   (b) oxygen   (e) calcium

   (c) silicon   (f) arsenic

4. Use *spdf* notation *and* Lewis symbols to represent the electron configuration of each of the following.

   (a) potassium ion   (d) chloride ion

   (b) sulfide ion   (e) magnesium ion

   (c) aluminum ion   (f) nitride ion

5. Use Lewis symbols to represent formation of the ionic compound between

   (a) calcium and bromine atoms

   (b) barium and oxygen atoms

   (c) aluminum and sulfur atoms

6. Use Lewis symbols to show the sharing of electrons between iodine atoms to form an iodine molecule. Label all electron pairs as bonding pairs or lone pairs.

7. Use Lewis symbols to show the sharing of electrons between a hydrogen atom and a fluorine atom. Label the molecule with symbols that indicate polarity.

8. What is the electronegativity of an atom? How does it differ from electron affinity?

9. If atoms of the two elements in each of the following sets are joined by a covalent bond, which atom will more strongly attract the electrons in the bond?

   (a) N and S   (c) As and F

   (b) B and Cl   (d) S and O

10. Using only the periodic table (inside front cover), indicate which element in each set is more electronegative.

    (a) Br or F   (c) Cl or As

    (b) Br or Se   (d) N or H

11. Classify the following bonds as ionic or covalent. For those bonds that are covalent, indicate whether they are polar or nonpolar.

    (a) KF   (d) NO   (g) Br_2

    (b) IBr   (e) CaO   (h) F_2

    (c) MgS   (f) NaBr   (i) HCl

12. How many single covalent bonds do each of the following atoms usually form in molecules that have only single covalent bonds? You may refer to the periodic table.

    (a) H   (c) O   (e) N

    (b) C   (d) F   (f) Br

13. Which of the structural formulas shown below corresponds to the most plausible skeletal structure for a molecule? Why?

    $$
    \begin{array}{c}
    \text{O} \\
    |
    \end{array}
    $$
    (a) F—C—F

          F

    $$
    \begin{array}{c}
    \text{F} \\
    |
    \end{array}
    $$
    (b) C—O—F

    (c) F—O—C—F

14. The formal charge of the O atoms in the ion NO_2^+ is

    (a) -2 (b) -1 (c) 0   (d) $+1$

15. Write the Lewis structures for each of the following that involve an exception to the octet rule.

    (a) PF_5 (b) CS_2 (c) NO   (d) O_3 (e) BrO_2

16. Which of the following species exhibit resonance?

    (a) O_3 (b) CO_3^{2-} (c) CO_2 (d) NO_3^-

17. Of the following species, which one contains a *triple* covalent bond?

    (a) NO_3^- (b) CO_2 (c) CN^- (d) $AlCl_3$

18. Without looking up the values, place the following species in order of increasing bond length.

    (a) O_2 (b) N_2 (c) I_2 (d) ICl

19. The highest bond-dissociation energy among the following diatomic molecules is found in

    (a) O_2 (b) N_2 (c) Cl_2 (d) HF

20. Given the bond energies, N-to-O bond in NO, 628 kJ/mol; H—H, 435 kJ/mol; N—H, 389 kJ/mol; O—H, 464 kJ/mol, calculate ΔH for the reaction

    $$2\,NO(g) + 5\,H_2(g) \longrightarrow 2\,NH_3(g) + 2\,H_2O(g)$$

21. Classify the following compounds as saturated hydrocarbons, alkenes, or alkynes.

    (a) $CHCCH_3$ (c) $CHCCH_2CH_3$

    (b) $CH_3CH_2CHCH_2$ (d) $CH_3(CH_2)_3CH_3$

22. What is the essential structural feature of the alkene monomer molecules that enables them to polymerize?

Problems

Ions and Ionic Bonding

23. Which of the following ions have noble-gas electron configurations? What are the electron configurations of the ions that do not?

(a) Mo^{6+} (c) Ti^{4+} (e) La^{3+}

(b) Te^{2-} (d) Zn^{2+} (f) Hg^{2+}

24. Use noble-gas-core abbreviated *spdf* notation to give the electron configuration for the most likely simple ion formed by each of the following elements.

(a) Br   (c) Sr   (e) Rb

(b) S   (d) Cd   (f) O

25. Write Lewis structures to represent the following ionic compounds.

(a) LiBr   (c) Na_3N

(b) cesium sulfide   (d) yttrium(III) sulfide

26. Write Lewis structures to represent the following ionic compounds.

(a) CaO   (c) $BaCl_2$

(b) potassium bromide   (d) strontium nitride

Energetics of Ionic Bond Formation

For Problems 27–32, use the given values and (as needed) values from tables in Chapter 8, from Section 9.5 (including Example 9.3), and from Appendix C.

27. The lattice energy of potassium chloride is −701 kJ/mol KCl. The enthalpy of sublimation of K(s) is 89.24 kJ/mol, and the first ionization energy of K(g) is 419 kJ/mol. Determine the enthalpy of formation of KCl(s). Compare your result with the value listed in Appendix C.

28. The lattice energy of sodium fluoride is −914 kJ/mol NaF. Determine the enthalpy of formation of NaF(s). Compare your result with the value listed in Appendix C.

29. The enthalpy of formation of sodium iodide is

$$Na(s) + \tfrac{1}{2} I_2(s) \longrightarrow NaI(s) \qquad \Delta H_f^\circ = -288 \text{ kJ/mol}$$

The enthalpy of sublimation of iodine is

$$I_2(s) \longrightarrow I_2(g) \qquad \Delta H^\circ = +62 \text{ kJ/mol}$$

Calculate the lattice energy of NaI(s).

30. The enthalpy of formation of cesium bromide is

$$Cs(s) + \tfrac{1}{2} Br_2(l) \longrightarrow CsBr(s) \qquad \Delta H_f^\circ = -405.9 \text{ kJ/mol}$$

The enthalpy of sublimation of cesium is 76.1 kJ/mol, and the enthalpy of vaporization of liquid bromine is

$$Br_2(l) \longrightarrow Br_2(g) \qquad \Delta H^\circ = 30.9 \text{ kJ/mol}$$

Calculate the lattice energy of CsBr(s).

31. The enthalpy of sublimation of calcium is 178.2 kJ/mol, and the enthalpy of vaporization of liquid bromine is

$$Br_2(l) \longrightarrow Br_2(g) \qquad \Delta H^\circ = +30.9 \text{ kJ/mol}$$

Calculate the lattice energy of calcium bromide.

32. The enthalpy of sublimation of lithium is 159.4 kJ/mol. Calculate the lattice energy of lithium oxide.

Lewis Structures

33. Write Lewis structures for the simplest covalent molecules formed by the following atoms, assuming that the octet rule (duet rule for hydrogen) is followed in each case.

(a) Si and H   (b) N and Cl

34. Write Lewis structures for the simplest covalent molecules formed by the following atoms, assuming that the octet rule (duet rule for hydrogen) is followed in each case.

(a) P and H   (b) C and F

35. Write plausible Lewis structures for the following covalent molecules.

(a) NF_3 (c) C_2H_4 (e) H_2SiO_3

(b) C_2H_2 (d) CH_3NH_2 (f) HCN

36. Write plausible Lewis structures for the following covalent molecules.

(a) CH_3OH (c) NH_2OH (e) COF_2

(b) CH_2O (d) N_2H_4 (f) PCl_3

37. Indicate what is wrong with each of the following Lewis structures. Replace each with a more acceptable structure.

(a) Mg $:\ddot{O}:$

(b) $[:\ddot{\underset{..}{C}l}]^+ [:\ddot{\underset{..}{O}}:]^{2-} [\ddot{\underset{..}{C}l}:]^+$

(c) $[:\ddot{O}-\ddot{N}=\ddot{O}:]^+$

(d) $[:\ddot{S}-C\equiv\ddot{N}:]^-$

38. Only one of the following Lewis structures is correct. Select that one, and indicate the errors in the others.

(a) cyanate ion, $[:\ddot{O}-C=\ddot{N}:]^-$

(b) acetylide ion, $[C\equiv C:]^{2-}$

(c) hypochlorite ion, $[:\ddot{\underset{..}{C}l}-\ddot{O}:]^-$

(d) nitrogen monoxide, $:\ddot{N}=\ddot{O}:$

Electronegativity: Polar Covalent Bonds

39. Without referring to figures or tables in the text (other than the periodic table), arrange each of the following sets of atoms in their expected order of increasing electronegativity.

(a) As, Br, K   (b) Be, Ca, Cs   (c) Cl, Pb, Sb

40. Without referring to figures or tables in the text (other than the periodic table), arrange each of the following sets of atoms in their expected order of increasing electronegativity.

(a) B, Ba, In   (b) Al, Mg, Sr   (c) Ge, Si, Sn

41. Use differences in electronegativity values to arrange each of the following sets of bonds in order of increasing polarity. Use symbols $\delta+$ and $\delta-$ to indicate partial charges, if any, in the bonds.

(a) H—C, H—F, H—H, H—N, H—O

(b) C—Br, C—C, C—Cl, C—F, C—I

42. Use differences in electronegativity values to arrange each of the following sets of bonds in order of increasing polarity. Use the symbols $\delta+$ and $\delta-$ to indicate partial charges, if any, in the bonds.

(a) Cl—F, F—F, Br—F, H—F, I—F

(b) H—Br, H—Cl, H—F, H—H, H—I

43. Each of the following bonds is polar. Identify the bonds for which a cross-based arrow can be drawn merely by referring to a periodic table, and identify those for which actual values of electronegativity are required to determine the direction of the arrow.

(a) As—P   (b) Al—P   (c) I—P   (d) Cl—P

44. Identify the bonds that can be classified as polar bonds by merely referring to a periodic table, and identify those for which actual values of electronegativity are required to classify them as polar or nonpolar.

(a) Ga—S   (b) N—Br   (c) Sn—Br   (d) C—S

Formal Charges

45. Assign formal charges to each atom in the following structures.

(a) :C≡O: (c) [:N̈=N=N̈:]⁻

(b) H—C≡N: (d) [H—Ö—Ö:]⁻

46. Assign formal charges to each atom in the following structures.

(a) :Ö:
      |
   :O̤=S:

(b) :Ö:
      |
   ·N=O̤:

(c) [:O̤—S—O̤:]²⁻ with :Ö: above S

(d) [:F̈—B—F̈:]⁻ with :F̈: above and :F̈: below B

47. Assign formal charges to each atom in the following structures. Based on these formal charges, which is the preferred Lewis structure for this molecule, the dinitrogen monoxide molecule?

(a) :N̈=N=Ö: (b) :N≡N—Ö:

48. Assign formal charges to each atom in the following structures. Based on these formal charges, which is the preferred Lewis structure for this ion, the cyanate ion?

(a) [:C̈=N=Ö:]⁻   (b) [:N≡C—Ö:]⁻

Resonance Structures

49. Write two equivalent resonance structures for the bicarbonate ion, $HOCO_2^-$. Describe the resulting resonance hybrid structure for the bicarbonate ion.

50. Write three equivalent resonance structures for the carbonate ion, CO_3^{2-}. Describe the resulting resonance hybrid structure for the carbonate ion.

51. Write Lewis structures for (a) nitrous acid and (b) nitric acid. In which of these substances is resonance more important? Explain.

52. Write Lewis structures for (a) acetic acid and (b) acetate ion. In which of these substances is resonance more important? Explain.

More Lewis Structures

53. Write the simplest Lewis structure for each of the following. Comment on any unusual features of the structures.

(a) NO   (b) ClF₃   (c) BCl₃   (d) SeF₄

54. Write the simplest Lewis structure for each of the following. Comment on any unusual features of the structures.

(a) ClO₂   (b) IF₅   (c) Be(CH₃)₂   (d) XeF₆

55. Write Lewis structures for the following. Where appropriate, use the concepts of formal charge, expanded valence shells, and resonance to choose the most likely structure(s).

(a) SSF₂   (c) H₂CO₃   (e) SF₅⁻

(b) I₃⁻   (d) CN⁻   (f) BrO₃⁻

56. Write Lewis structures for the following. Where appropriate, use the concepts of formal charge, expanded valence shells, and resonance to choose the most likely structure(s).

(a) HNO₃   (c) XeO₄   (e) CH₂SF₄

(b) IF₄⁻   (d) ICl₂⁻   (f) IO₄⁻

57. Write Lewis structures for the following organic molecules.
   (a) isopropyl alcohol
   (b) methanoic (formic) acid
   (c) dimethyl ether

58. Write Lewis structures for the following organic molecules.
   (a) propyne
   (b) dimethylamine
   (c) propionaldehyde

59. Indicate what is wrong with each of the following Lewis structures. Replace each with a more acceptable structure(s).

   (a)

$$\left[:\ddot{O}-\ddot{N}-\ddot{O}: \atop \quad\quad \overset{:\ddot{O}:}{|} \right]^{-}$$

   (b) $H-\overset{\,}{\underset{..}{N}}=C=N:$

   (c) $H-\overset{..}{\underset{..}{N}}-C\equiv O:$

60. Indicate what is wrong with each of the following Lewis structures. Replace each with a more acceptable structure(s).

   (a) $[:\ddot{S}-C=\ddot{N}:]^{-}$

   (b) $:\ddot{O}=N=\ddot{O}:$

   (c) $:\ddot{C}l-\overset{..}{N}=\ddot{C}l:$
                $\overset{|}{:\ddot{C}l:}$

61. A carbon–hydrogen–oxygen compound is 53.31% C, 11.18% H, and 35.51% O. Write a Lewis structure based on the empirical formula of this compound. Is it a satisfactory structure? If not, substitute one that is.

62. A carbon–hydrogen–nitrogen compound is 39.97% C, 13.42% H, and 46.61% N. Write a Lewis structure based on the empirical formula of this compound. Is it a satisfactory structure? If not, substitute one that is.

Bond Lengths and Bond Energies

In Problems 63–76, use the data provided, together with bond lengths and bond energies from Table 9.1 and enthalpies of formation from Appendix C, as needed.

63. Estimate bond lengths for the following bonds. For which bond is your estimate likely to be too high? Explain.

   (a) I—Cl   **(b)** C—F

64. Estimate bond lengths for the following bonds. Do you think your estimates are likely to be too high? Explain.

   (a) O—Cl   **(b)** N—I

65. On page 367, we compared the experimentally determined bond length of HCl with a value determined from atomic radii. For which of the hydrogen halides would you expect the *smallest* percent difference between the calculated and experimental values of the bond length: HF, HCl, HBr, or HI? Explain.

66. A diatomic interhalogen molecule XZ has a bond between two different halogen elements (F, Cl, Br, I). In a manner similar to that of Problem 65, for which of the interhalogen compounds XZ would you expect to find the *greatest* percent difference between the calculated and experimental values of the bond length? Explain.

67. Dinitrogen difluoride has a nitrogen-to-nitrogen bond length of 123 pm and nitrogen-to-fluorine bond lengths of 141 pm. Write a Lewis structure consistent with these data.

68. Hydrogen peroxide has an oxygen-to-oxygen bond length of 147.5 pm and two oxygen-to-hydrogen bonds with lengths of 95.0 pm. Write a Lewis structure consistent with these data.

69. Estimate the enthalpy change (ΔH) for the reaction

$$H_2(g) + F_2(g) \longrightarrow 2 HF(g)$$

70. Estimate the enthalpy change (ΔH) for the reaction

$$CH_4(g) + Cl_2(g) \longrightarrow CH_3Cl(g) + HCl(g)$$

71. *Without doing detailed calculations,* indicate whether you expect the following reaction to be endothermic or exothermic.

$$C_3H_8(g) + Cl_2(g) \longrightarrow C_3H_7Cl(g) + HCl(g)$$

72. *Without doing detailed calculations,* indicate whether you expect the following reaction to be endothermic or exothermic.

$$N_2H_4(g) + H_2(g) \longrightarrow 2 NH_3(g)$$

73. In the stratosphere, the reaction of ozone and atomic oxygen helps to maintain the heat balance on Earth.

$$O_3(g) + O(g) \longrightarrow 2 O_2(g) \qquad \Delta H = -391.9 \text{ kJ}$$

Estimate a value for the oxygen-to-oxygen bond energy in the O_3 molecule.

74. The oxidation of carbon monoxide to carbon dioxide by oxygen can be represented by the equation

$$2 CO(g) + O_2(g) \longrightarrow 2 CO_2(g) \qquad \Delta H = -566 \text{ kJ}$$

Estimate a value for the bond-dissociation energy of the CO molecule. (*Hint:* See the footnote to Table 9.1.)

75. The enthalpy of vaporization of carbon is

$$C(\text{graphite}) \longrightarrow C(g) \qquad \Delta H = 717 \text{ kJ}$$

Use this value, the enthalpy of formation of methane, and the bond-dissociation energy of hydrogen to calculate a value for the carbon-to-hydrogen single bond energy. Compare your result with the value listed in Table 9.1.

76. The following reaction has been observed in the atmosphere.

$$CO(g) + O_3(g) \longrightarrow CO_2(g) + O_2(g)$$

Use a bond-dissociation energy of 1072 kJ/mol for data from Table 9.1 and its footnote, and data from Appendix C to obtain a value of the oxygen-to-oxygen bond energy in the O_3 molecule. Also, obtain an estimate based on the Lewis structure(s) of O_3 and data from Table 9.1. How well do the two values agree?

Alkenes and Alkynes

77. Give the condensed structural formula for each of the following.

   (a) propene   **(c)** 1-pentene

   (b) 1-butyne   **(d)** 3-hexyne

78. Give the condensed structural formula for each of the following.

   (a) ethene   **(c)** ethyne

   (b) 1-butene   **(d)** 2-pentyne

79. Write equations to represent the reaction of **(a)** hydrogen with ethene and **(b)** two molecules of hydrogen with ethyne.

80. Write equations to represent each of the following reactions.

   (a) hydrogen with 1-butene

   (b) one molecule of hydrogen with ethyne

81. Determine the enthalpy change for the following reaction by using **(a)** average bond energies and **(b)** enthalpy of formation data. Compare the results.

$$C_2H_4(g) + H_2(g) \longrightarrow C_2H_6(g) \qquad \Delta H = ?$$

82. Determine the enthalpy change for the following reaction by using **(a)** average bond energies and **(b)** enthalpy of formation data. Compare the results.

$$C_2H_4(g) + H_2O(g) \longrightarrow CH_3CH_2OH(g) \qquad \Delta H = ?$$

83. With the aid of Table 9.2, draw a section of polymer chain that is at least eight carbon atoms long for each of the following.

   (a) polyethylene

   (b) poly(vinyl chloride)

   (c) polypropylene

84. Draw four-monomer-unit sections of the polymers formed from the following monomers.

   (a) tetrafluoroethylene, $F_2C{=}CF_2$

   (b) vinylidene chloride, $H_2C{=}CCl_2$

   (c) acrylonitrile, $H_2C{=}CH{-}CN$

Additional Problems

Problems marked with an * may be more challenging than others.

85. The Lewis structures and structural formulas of the alkane hydrocarbons are identical. Explain why this is the case but why the same statement is not true for all organic compounds.

86. The concepts of oxidation state and formal charge both deal with the *arbitrary* assignment of electrons between bonded atoms. How do these concepts differ? Are there ways in which they are similar? Explain.

87. Two different molecules have the formula C_2H_6O. Write Lewis structures for the two molecules.

88. Draw four resonance structures for the oxalate ion, $C_2O_4{}^{2-}$.

89. Draw the structural formulas of five hydrocarbons that have the formula C_4H_6. (*Hint:* Use multiple bonds, cyclic structures, and combinations of these.)

90. The aldehyde propynal has the formula HCCCHO. Write a Lewis structure for propynal, and estimate all the bond lengths in the molecule.

91. Hydrogen azide, HN_3, is an acid in aqueous solutions. Its salts are quite reactive [sodium azide, NaN_3, is used in air-bag systems, and lead azide, $Pb(N_3)_2$, is used in detonators]. The structure of hydrogen azide is indicated here. Write a Lewis structure(s) consistent with the data given.

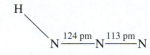

*92. Estimate the nitrogen-to-fluorine and nitrogen-to-oxygen bond lengths in the molecule nitryl fluoride, NO_2F. To do so, draw plausible resonance structures for this molecule and use data from Table 9.1.

93. Use data from the text to determine a value for the enthalpy of formation of the ethyl radical, $\cdot C_2H_5$.

*94. Calculate ΔH_f° for the hypothetical compound MgCl(s) by using the following data (in kilojoules per mole): enthalpy of sublimation of Mg, +150; first ionization energy of Mg(g), +738; enthalpy of dissociation of $Cl_2(g)$, +243; electron affinity of Cl(g), 349; and lattice energy of MgCl(s), 676.

*95. Use the data of Problem 94, I_2 of Mg(g) (1451 kJ/mol), and the lattice energy of $MgCl_2(s)$ (2500 kJ/mol) to calculate ΔH_f° for $MgCl_2(s)$. Explain why you would expect $MgCl_2(s)$ to be more stable than MgCl(s).

*96. Hydrogen forms ionic compounds with certain metals in which it is present as hydride ion, H^-. Determine the electron affinity of H through a Born–Haber calculation, using a lattice energy of -812 kJ/mol NaH and data from page 346, Table 9.1, and Appendix C.

97. The formula for hypochlorous acid is often given as HClO. If you did not know that hydrogen atoms in oxoacids are almost always bonded to an oxygen atom, how might you otherwise come to this conclusion for hypochlorous acid?

*98. Of the species NO_2, $NO_2{}^+$, and $NO_2{}^-$, which two should have bonds of nearly the same length? Estimate the bond length and bond energy in those two species.

*99. The second electron affinity of oxygen cannot be measured directly.

$$O^-(g) + e^- \longrightarrow O^{2-}(g) \qquad EA_2 = ?$$

The O^{2-} ion can exist in the solid state, however, where the high energy requirement for its formation is offset by the large lattice energies of ionic oxides.

(a) Show that EA_2 can be calculated from the enthalpy of formation and lattice energy of MgO(s), enthalpy of sublimation of Mg(s), ionization energies of Mg(g), bond energy of O_2(g), and EA_1 of O(g).

(b) The lattice energy of MgO(s) is 3795 kJ/mol. Combine this value with other data from Chapters 8 and 9 and Appendix C to calculate a value of EA_2 for oxygen.

* **100.** A 1.450-g sample of an organic compound that occupies 766 mL when in the gaseous state at 99.8 °C and 758 Torr yields 3.296 g CO_2 and 1.349 g H_2O as the only products when it is burned in an excess of oxygen.

  (a) Write at least four plausible Lewis structures that might be used to represent this compound.

(b) If the compound is formed by the oxidation of an alcohol and can itself be oxidized to a carboxylic acid, what is its Lewis structure (recall Section 4.5)?

* **101.** A 0.507-g sample of a gaseous oxide of carbon has 47.04% O by mass. It occupies a volume of 184 mL at 25 °C and 752 mmHg. The sample of oxide reacts with water to produce as the sole product an acid having a molar mass of 104 g/mol and requiring 29.52 mL of 0.5050 M NaOH for its complete neutralization. **(a)** Write an equation for the reaction of the oxide with water, and **(b)** write plausible Lewis structures for the reactants and products of this reaction.

102. Calculate the bond-dissociation energy of the pollutant molecule NO(g), using various data found in the text. Then compare your result to a value found in Table 9.1.

Apply Your Knowledge

103. [Collaborative] In Problem 93 in Chapter 8 we proposed an alternate universe in which the s energy subshells hold 3 electrons; p subshells, 9; d subshells, 15; and f subshells, 21. Now consider that when elements form chemical bonds, they tend to acquire a dodecet (12 valence electrons) instead of an octet. When they form covalent bonds, they share a trio (3) of valence electrons.

  (a) Draw the Lewis symbol for elements Dd($Z = 4$), and Ad($Z = 52$).

  (b) Draw the Lewis structure for the compound formed when Mm reacts with Zz.

  (c) Draw the Lewis structure for the simplest compound formed when Zz reacts with Ad.

104. [Environmental] One reason for representing organic molecules through condensed structural formulas rather than complete structural formulas is to minimize the space required to print them. A two-carbon segment of a polyethylene molecule is about 310 pm long. The representation $-CH_2-CH_2-$ in print is about 1.3 cm long. **(a)** What would be the length of a paper representation of a linear polyethylene molecule with a molecular mass of 75,000 u? **(b)** The major column on this page is 12.2 cm wide. Across how many such pages would the print structure extend?

105. [Environmental] The free radical ·ClO is intimately involved in reactions that lead to the destruction of ozone in the stratosphere. Use an enthalpy of formation from Appendix C, a bond energy from Table 9.1, a chlorine-to-oxygen bond energy of 243 kJ/mol, and Hess's law to obtain a value for the enthalpy of formation of ·ClO.

* **106.** [Historical] In developing his scale of electronegativities, Linus Pauling originally thought that AB bond energy should be the geometric mean of the AA and BB bond energies. (The geometric mean is the square root of the product of the two bond energies.) He designated the *difference* between the measured bond energy and this mean of bond energies by the symbol Δ'.

  (a) Use the bond energies in Table 9.1 to calculate a value of Δ' for HCl.

  (b) Using bond energies in *kcal/mol*, Pauling devised the square root function, $x = 0.18\sqrt{\Delta}$ to give "a convenient range of electronegativity values." Calculate the value of x for HCl.

  (c) Next, taking the electronegativity of hydrogen to be zero, Pauling assigned the value of x to the other element (X) in the HX bond. Then, he added 2.05 to all values to give "the first row elements C to F the values 2.5 to 4.0." What is the adjusted value of the electronegativity for Cl? (Pauling held that the values were reliable to only one decimal place.)

* **107.** [Collaborative] Draw Lewis structures for **(a)** all ketones having the formula $C_5H_{10}O$, **(b)** three carboxylic acids having the formula $C_5H_{10}O_2$. **(c)** Is it possible to draw a Lewis structure for a ketone that has the formula $C_5H_{12}O$? Explain. **(d)** Is it possible to draw a Lewis structure for a carboxylic acid that has the formula $C_5H_8O_2$.

e-Media Problems

The activities described in these problems can be found in the e-Media Activities and Interactive Student Tutorial (IST) modules of the Companion Website, *http://chem.prenhall.com/hillpetrucci*.

108. List the types of forces described in the **H₂ Bond Formation** animation (*Section 9-1*). Describe the positions along the H—H distance axis where each type of force dominates.

109. In the **Periodic Trends: Lewis Structures** activity (*Section 9-2*), (a) how do the Lewis symbols of the different elements vary across a period (row)? How do the Lewis symbols vary down a given group (column)? (b) How could these observations be used to predict similarities or differences in types of compounds formed between a specific element and elements of a particular group? (For example, consider the molecules HF, HCl, HBr, and HI.)

110. View the **Sodium Chloride** 3D model (*Section 9-3*) and select the ball-and-stick display. (Right-click [Windows] or click-hold [Macintosh] to see the pop-up menu. Choose Display > Ball and Stick.) (a) Determine from this representation the number of chloride ions surrounding each sodium ion in the crystal. How many sodium ions surround a chloride ion in the crystal? (b) Use this atomic-level view to explain the portion of the Born–Haber cycle in which ions are assembled into a crystal.

111. By identifying all of the nonpolar molecules in the **Bond Polarity** activity (*Section 9-7*), what is one general feature that is common to these nonpolar species?

112. (a) Use the approach presented in **Breaking and Forming Bonds in a Chemical Reaction** activity (*Section 9-10*) to describe the reaction of methane and molecular chlorine to form chloroform ($CHCl_3$) and molecular hydrogen. (b) Use the information of Table 9.1 to calculate the enthalpy of this reaction.

Chapter 10

Sulfanilamide   *para*-aminobenzoic acid

Bonding Theory and Molecular Structure

IN THIS CHAPTER, we will use quantum mechanics to extend our knowledge of chemical bonding and molecular structure beyond what we attained using the Lewis theory (Chapter 9). Recall that in Chapter 7 we used modern quantum mechanics to extend our knowledge of atomic structure beyond the Bohr model.

With added insights into chemical bonding, we will be able to explain some matters that we cannot explain with simple Lewis structures. For example, we will be able to give plausible explanations of the following observations:

- The three atoms in a CO_2 molecule are arranged in a straight line, whereas the three atoms in a H_2O molecule form a bent molecule.
- Chloroform ($CHCl_3$) molecules are polar, but carbon tetrachloride (CCl_4) molecules are nonpolar.
- There is only one butane ($CH_3CH_2CH_2CH_3$) molecule, but there are two isomers of 2-butene, $CH_3CH{=}CHCH_3$.
- The Ne_2 molecule does not exist, but the F_2 molecule is well known, and the Na_2 molecule exists in sodium vapor.
- The O_2 molecule has a double covalent bond yet has unpaired electrons that account for its paramagnetism.

Molecular Geometry

Just what do we mean by the shape of a molecule? Like every other sample of matter, a molecule occupies space and is therefore three-dimensional in shape. A simple verbal description of the overall shape of a molecule can be difficult to formulate and is actually not the most useful one. Instead, the **molecular geometry,** or the shape, of a molecule is described by the geometric figure formed when we imagine using straight lines to join all the atomic nuclei of the molecule.

◄ Navy Corpsman Lavern L. Hamer of Jordan, MN, pours sulfa powder onto the wound of a Marine on a battlefield in the Marshall Islands in 1944; the Marine was later removed to safety and to a hospital. The antibiotic sulfa drugs—sulfonamides—are credited with saving thousands of lives in World War II. Their antibacterial powers were discovered when it was determined that sulfonamides were biochemically similar to naturally occurring para-aminobenzoic acid (PABA). The chemical similarity of the two compounds arises from their similar molecular shape (see inset on photograph), one of the topics explored in this chapter.

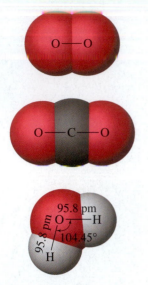

▲ **FIGURE 10.1** **Molecular geometry and space-filling models**
The molecular geometry of the diatomic O_2 molecule is linear, as is the molecular geometry of the triatomic CO_2 molecule. The molecular geometry of H_2O is angular, or bent. Each O — H bond has a length of 95.8 pm, and the angle between the two bonds is 104.45°.

🌐 **Oxygen, Carbon Dioxide, Water 3D models**

Diatomic molecules, such as O_2, have only two nuclei. Because the two points representing the nuclei determine a straight line, we describe the molecular geometry as *linear*. If the three nuclei of a triatomic molecule fall along a straight line, as they do in CO_2, the molecular geometry is also linear. If the three nuclei are not in a straight line, the molecular geometry is *angular*, or *bent*. (We will see how to predict these shapes—linear and angular—in the next section). The water molecule is a good example of an angular molecule. Molecular models of O_2, CO_2, and H_2O are shown in Figure 10.1, where the framework of the H_2O molecule, displayed within the molecular model, emphasizes two key features of the geometric structure of a molecule: bond lengths and bond angles. We have this framework in mind when we say that the H_2O molecule is angular. Most polyatomic molecules have molecular geometries that are more complex than linear or angular, as we shall see shortly.

Precise molecular geometry can be determined only by experiment, but the shapes of many molecules and polyatomic ions can be predicted fairly well, and we now consider a method for making these predictions.

10.1 Valence-Shell Electron-Pair Repulsion (VSEPR) Method

As the name implies, the **valence-shell electron-pair repulsion,** or **VSEPR,** method is based on the idea that pairs of valence electrons in bonded atoms repel one another. These mutual repulsions push electron pairs as far from one another as possible. This maximum separation minimizes the energy of repulsion and represents the lowest energy configuration of the molecule or polyatomic ion. Depending on the number of valence electron pairs and on whether they are bonding pairs or lone pairs, the terminal atoms in a molecule adopt specific orientations about the central atom to which they are bonded. This gives the molecule or polyatomic ion a distinctive shape.

Electron-Group Geometries

Before we attempt to predict molecular shapes by the VSEPR method, we should broaden our view of repulsions to involve *groups* of valence electrons rather than just pairs. An **electron group** is any collection of valence electrons, localized in a region around a central atom, that repels other groups of valence electrons. An electron group can be any of the following:

- a single unpaired electron
- a lone pair of electrons
- one bonding pair of electrons in a single covalent bond
- two bonding pairs of electrons in a double covalent bond
- three bonding pairs of electrons in a triple covalent bond

Most commonly, there are two, three, four, five, or six electron groups about a central atom. The mutual repulsions among electron groups lead to an orientation of the groups that we call the **electron-group geometry.** The electron-group geometries are

- two electron groups: *linear*
- three electron groups: *trigonal planar*
- four electron groups: *tetrahedral*
- five electron groups: *trigonal bipyramidal*
- six electron groups: *octahedral*

Figure 10.2 offers an analogy that may help you visualize electron-group repulsions. When balloons are twisted together, the lobes of the balloons seek positions that cause the least interference from other lobes, just as electron groups tend to remain as far apart as possible.

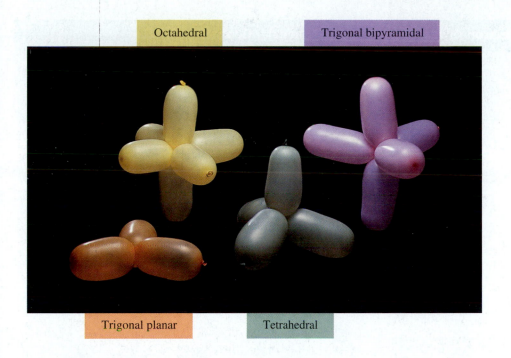

Octahedral   Trigonal bipyramidal

Trigonal planar   Tetrahedral

◄ **FIGURE 10.2**   **Balloon analogy to electron-group geometries for three, four, five, and six electron groups**

Each lobe, which represents an electron group, is positioned to maximize the distance from neighboring lobes in the molecular structures.

VSEPR Notation and Molecular Geometry

In the VSEPR notation used to describe molecular geometries, the central atom under consideration is denoted as A, terminal atoms as X, and lone pairs of electrons as E. In this notation, AX_2E_2 describes a structure with two terminal atoms and two lone pairs of electrons around a central atom. For example, the water molecule is described by the AX_2E_2 notation.

Before proceeding, let's clarify the difference between two related terms: *electron-group geometry* and *molecular geometry*. An electron-group geometry describes how, as a result of their mutual repulsions, groups of valence electrons are arranged about a central atom. A molecular geometry describes how *bonded atoms* are arranged about the same central atom. If all the electron groups are bonding groups, the orientations of bonded atoms and electron groups are the same. Thus, for structures with no lone-pair electrons (AX_n), the molecular geometry and electron-group geometry are identical. If there are lone-pair electrons, there will be one or more positions around the central atom where an electron group *but no atom* is found. In these cases, the molecular geometry is related to and determined by the electron-group geometry, but the two are *not* the same.

Table 10.1 summarizes various possibilities for molecular geometries in relation to electron-group geometries. Over the next few pages, we will describe several of the entries in this table.

Structures with No Lone-Pair Electrons, AX_n

AX_2   The molecules $BeCl_2$ and CO_2 are of the type AX_2. From its Lewis structure, we see that in electron-deficient $BeCl_2$, each of the two electron groups around the Be atom is an electron pair involved in a single covalent bond. In CO_2, each of the two electron groups around the C atom consists of two electron pairs involved in a double covalent bond:

Both the electron-group geometry and the molecular geometry for two electron groups are *linear.*

AX_3   The molecules BF_3 and SO_3 are of the type AX_3. In the electron-deficient BF_3, the three electron groups surrounding the central atom are electron pairs in single covalent bonds. Three electron groups result in a trigonal planar electron-group geometry; the absence of additional lone pairs means the molecular geometry is the

VSEPR animation

Table 10.1 activity

Table 10.1   VSEPR Notation, Electron-Group Geometry, and Molecular Geometry

Number of Electron Groups	Electron-Group Geometry	Number of Lone Pairs	VSEPR Notation	Molecular Geometry	Ideal Bond Angles	Example	Molecular Model
2	Linear	0	AX_2	X—A—X Linear	180°	$BeCl_2$	
3	Trigonal planar	0	AX_3	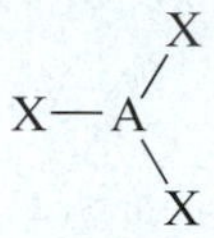 X—A Trigonal planar	120°	BF_3	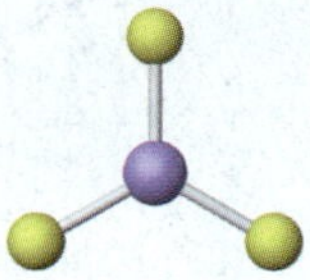
3	Trigonal planar	1	AX_2E	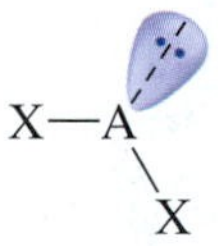 X—A Angular	120°	SO_2	
4	Tetrahedral	0	AX_4	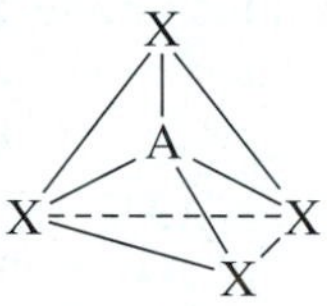 Tetrahedral	109.5°	CH_4	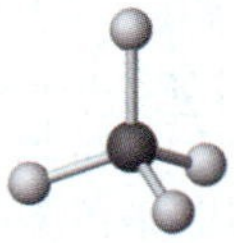
4	Tetrahedral	1	AX_3E	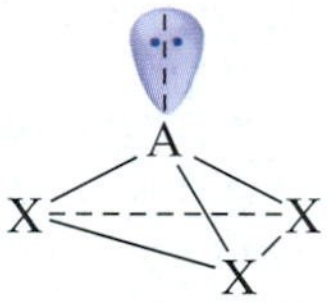 Trigonal pyramidal	109.5°	NH_3	
4	Tetrahedral	2	AX_2E_2	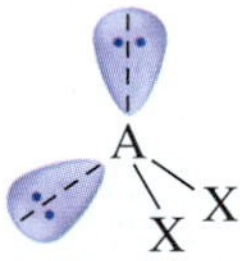 Angular	109.5°	OH_2	
5	Trigonal bipyramidal	0	AX_5	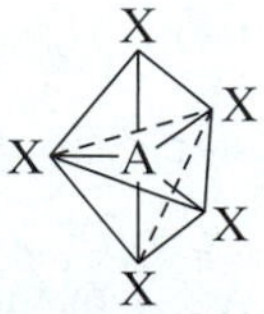 Trigonal bipyramidal	90°, 120°, 180°	PCl_5	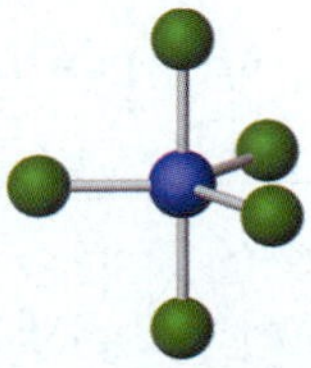

Number of Electron Groups	Electron-Group Geometry	Number of Lone Pairs	VSEPR Notation	Molecular Geometry	Ideal Bond Angles	Example	Molecular Model
5	Trigonal bipyramidal	1	AX_4E	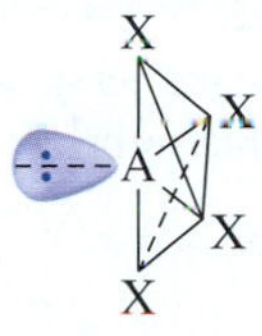 Seesaw	90°, 120°, 180°	SF_4	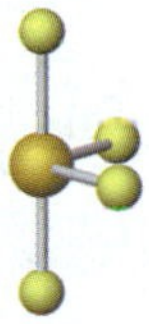
5	Trigonal bipyramidal	2	AX_3E_2	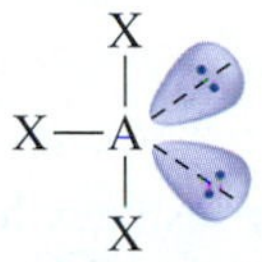 T-shaped	90°, 180°	ClF_3	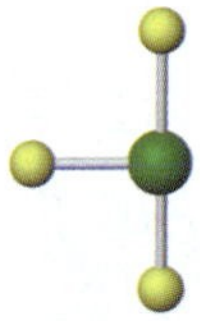
5	Trigonal bipyramidal	3	AX_2E_3	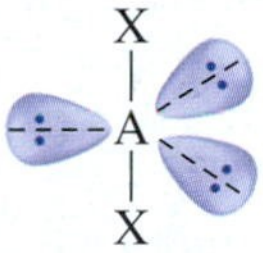 Linear	180°	XeF_2	
6	Octahedral	0	AX_6	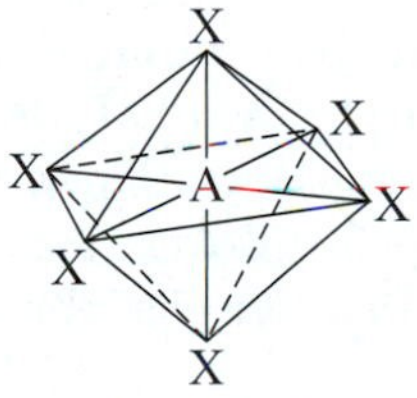 Octahedral	90°, 180°	SF_6	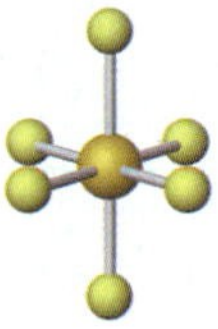
6	Octahedral	1	AX_5E	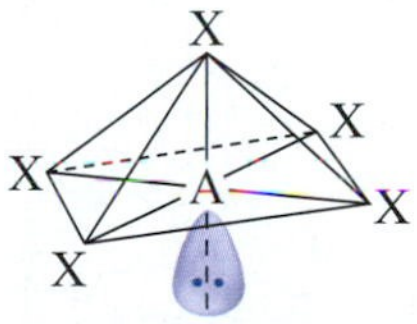 Square pyramidal	90°	BrF_5	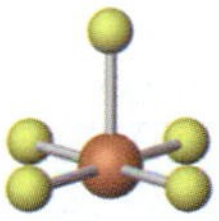
6	Octahedral	2	AX_4E_2	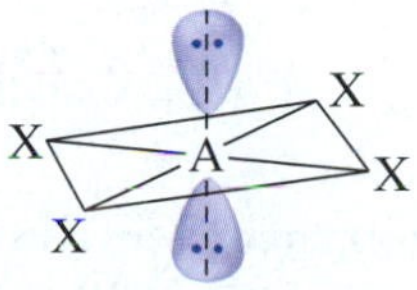 Square planar	90°	XeF_4	

same as the electron-group geometry. Therefore the F atoms are distributed around the B atom in a trigonal planar molecular geometry:

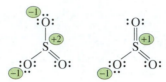

There are several possible Lewis structures for SO_3; we show two of them here. The first is one of three resonance structures conforming to the octet rule. The second is one of four possible structures based on an expanded valence shell (the true structure is a resonance hybrid):

These structures illustrate an important point:

All the plausible Lewis structures that can be written for a molecule or polyatomic ion should give the same electron-group geometry.

Both structures shown for SO_3 indicate a trigonal planar distribution of three electron groups and hence of the three O atoms. In this molecular geometry, the central atom and three terminal atoms are all in the same plane, and the bond angles are all 120°.

AX_4 Methane, CH_4, is a prime example of a molecule of the type AX_4:

The energy of the molecule is at a minimum when the four bonding pairs of electrons are as far from one another as possible. The carbon atom is at the center of a geometric shape called a *regular tetrahedron,* which has four faces, each an equilateral triangle. The electron pairs are directed to the four corners of the tetrahedron, where the hydrogen atoms reside. Both the predicted and experimentally determined H-C-H bond angles in CH_4 are 109.5°.

As Figure 10.3 illustrates, both the electron-group geometry and the molecular geometry in CH_4 are *tetrahedral.* In addition, we portray a few details of the molecular geometry as shown in part (b) of the figure: A solid line represents a bond in the plane of the page, broken lines are used for bonds that extend *behind* the plane of the page, and wedges are used for bonds projecting *out* from the plane of the page:

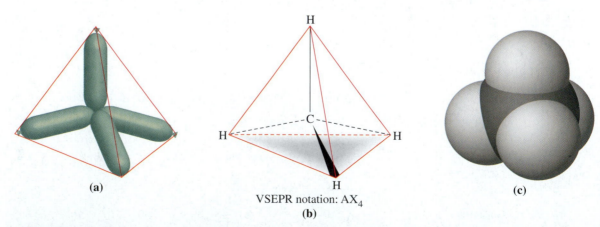

(a)   VSEPR notation: AX_4 (c)
                                    (b)

▲ **FIGURE 10.3   The electron-group geometry and molecular geometry of methane**

(a) The electron-group geometry is represented by two balloons twisted together. The balloons separate into four lobes directed to the corners of a tetrahedron. Each lobe represents an electron pair in a C—H bond. (b) The molecular geometry, which is identical to the electron-group geometry, has the four H atoms in CH_4 situated at the corners of a tetrahedron. The C—H bonds are represented by black lines, and the edges of the tetrahedron by red lines. (c) A space-filling model of methane.

AX₅ and AX₆ In Table 10.1, PCl_5 is shown as an example of an AX_5 molecule and SF_6 as an example of an AX_6 molecule. Both structures require that the central atom be surrounded by more than four electron pairs. This means that the central atom has an expanded valence shell, and therefore we encounter such structures only when the central atom is a third-period element or higher.

The trigonal bipyramidal electron-group geometry of PCl_5 has three electron pairs directed to the vertices of an equilateral triangle. The two remaining pairs are directed along a line perpendicular to the plane defined by the triangle, one pair above and one below that plane. As seen in Table 10.1, the trigonal bipyramidal molecular geometry has a P atom at the center of the equilateral triangle and Cl atoms at the five corners of the trigonal bipyramid. (A trigonal bipyramid is a geometric figure with six triangular faces.)

The octahedral electron-group geometry of SF_6 has four electron pairs directed to the vertices of a square. The remaining pairs are directed along a line that is perpendicular to the plane of the square and runs through the center of the square. One pair is above the plane of the square and the other below. As seen in Table 10.1, the octahedral molecular geometry has a S atom at the center of the square and F atoms at the six corners of the octahedron. (A regular octahedron is a geometric figure with eight faces, each face an equilateral triangle.)

A Strategy for Applying the VSEPR Method

To predict the shape of a molecule or polyatomic ion, we can generally follow the four-step strategy described and illustrated below for sulfur trioxide.

 **Predicting Molecular Geometry Using VSEPR Method activity**

Step 1: *Draw a Lewis structure of the molecule or polyatomic ion.* The structure must be plausible, but it does not need to be the "best" one. That is, it may have non-zero formal charges and may be only one of several structures contributing to a resonance hybrid.

Lewis structure

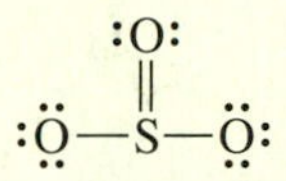

Electron groups

Three;
all bonding groups

Step 2: *Determine the number of electron groups around the central atom, and identify each as either a bonding group or a lone pair.* Keep in mind that a double bond and a triple bond each count as *one* bonding group.

Step 3: *Identify the electron-group geometry.* This may be linear, trigonal planar, tetrahedral, trigonal bipyramidal, or octahedral, corresponding to two, three, four, five, and six electron groups, respectively.

Electron-group geometry

Trigonal planar

Step 4: *Identify the molecular geometry.* This is based on the positions around the central atom that are occupied by other atoms (not by lone-pair electrons). Use information from Table 10.1 if necessary.

Molecular geometry

Trigonal planar: S atom at center; 120° O-S-O bond angles.

Example 10.1

Use the VSEPR method to predict the shape of the nitrate ion.

STRATEGY

The problem can be broken down into two parts. First, we must obtain an acceptable Lewis structure for the nitrate ion, using the general strategy developed in Chapter 9. With this Lewis structure in hand, we focus on the central nitrogen atom to establish first the electron-group geometry and then the molecular geometry.

SOLUTION

The formula for the nitrate ion is NO_3^-.

Step 1: *Draw a plausible Lewis structure.*

The distribution of 24 valence electrons [that is, $5 + (3 \times 6) + 1$] gives each terminal O atom in the skeletal structure an octet, but the central N atom is lacking two electrons for a valence-shell octet. We must shift a lone pair of electrons from one of the O atoms into its bond with the nitrogen atom, forming a nitrogen-to-oxygen double bond.

$$\left[\ddot{\underset{..}{O}} - N - \ddot{\underset{..}{O}}: \atop \overset{\displaystyle :\underset{\|}{\overset{..}{O}}:}{} \right]^-$$

Step 2: *Determine the number of electron groups.*

There are three electron groups around the N atom; all are bonding groups (one N-to-O double bond and two N-to-O single bonds). The VSEPR notation is therefore AX_3.

Step 3: *Identify the electron-group geometry.*

The distribution of the three electron groups is trigonal planar.

Step 4: *Identify the molecular geometry.*

For the structure AX_3, the molecular geometry is the same as the electron-group geometry: trigonal planar. The N and O atoms are all in the same plane, and each O-N-O bond angle is 120° (Figure 10.4).

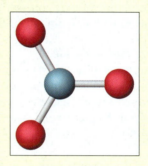

◀ **FIGURE 10.4**
The nitrate ion

ASSESSMENT

The Lewis structure obtained in step 1 is just one of three possible resonance structures, but it is clear that we reach the same conclusion about the molecular geometry no matter which of the three we use.

EXERCISE 10.1A

Predict the molecular geometry of **(a)** silicon tetrachloride and **(b)** antimony pentachloride.

EXERCISE 10.1B

Predict the molecular geometry of **(a)** BF_4^- and **(b)** N_3^-.

Structures with Lone-Pair Electrons, AX_nE_m

As we mentioned earlier, in structures that have lone-pair electrons (AX_nE_m) the molecular geometry differs from the electron-group geometry. Let's examine some of the differences.

AX_2E Consider two resonance structures of the SO_2 molecule:

$$:\ddot{O}::\ddot{S}:\ddot{O}: \longleftrightarrow :\ddot{O}:\ddot{S}::\ddot{O}:$$

There are three electron groups around the central atom, and the electron-group geometry is therefore trigonal planar. Oxygen atoms form bonds with two of the electron groups, but the third remains as a lone pair. However, because molecular geometry is based on the geometric arrangement of *atoms,* not electron groups, the SO_2

molecule is not trigonal planar but angular. As a result, we predict a bond angle of 120°, which is close to the observed bond angle of 119°:

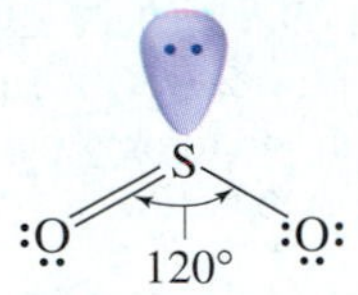

AX₃E In the NH_3 molecule, three electron pairs in the valence shell around the N atom are bonding pairs, and the fourth is a lone pair. Because there are four electron groups, the electron group geometry is tetrahedral, but one of the corners of the tetrahedron is occupied by the lone-pair electrons. The corresponding molecular geometry is pictured in Figure 10.5 and is called *trigonal pyramidal.* The N atom is situated at the top of the pyramid, and the three H atoms are at the vertices of the triangular base. We predict H-N-H bond angles of 109.5° (the same as in CH_4 in Figure 10.3). The actual angle is found to be 107°. Repulsions between the lone-pair electrons and the bonding electrons push the bonding pairs closer together, reducing the bond angle slightly from the regular tetrahedral angle.

AX₂E₂ The VSEPR notation for the H_2O molecule is AX_2E_2, as we see from the Lewis structure:

$$H:\overset{..}{\underset{..}{O}}:H$$

Again, the electron-group geometry for four electron groups is tetrahedral, but only two of the tetrahedral positions around the central atom (O) are occupied by terminal atoms (H). This triatomic molecule has an angular shape, as shown in Figure 10.6. Recall also the representation of the H_2O molecule in Figure 10.1. There, we gave the H-O-H bond angle as 104.45°. This angle deviates from the tetrahedral angle (109.5°) somewhat more than does the N-H angle in NH_3. The two lone pairs of electrons on the O atom of H_2O exert an even stronger repulsion on the bonding electrons than does the one lone pair on the N atom of NH_3.

Other structures By considering some of the remaining structures in Table 10.1 and those in Example 10.2 and Exercises 10.2A and 10.2B, we can expand our view of electron-group repulsions to include the following ideas.

- *The closer together two groups of electrons are, the stronger is the repulsion between them.* The force of repulsion between two electron groups increases dramatically as the groups are forced very close together. Thus, repulsions between two groups of bonding electrons increase significantly when a bond angle is reduced from 180° to 120° to 90°.

- *Lone-pair (LP) electrons spread out more than do bonding-pair (BP) electrons.* Because BP electrons are simultaneously attracted to two nuclei, the charge cloud associated with them is pulled into a compact shape. In contrast, because LP electrons are associated with just one nucleus, their charge cloud is much more spread out. As a result, the repulsion of one lone pair of electrons for another lone pair is greater than the repulsion between two bonding pairs. In general, the strength of repulsive forces, from strongest to weakest, is

    LP–LP repulsions > LP–BP repulsions > BP–BP repulsions

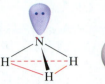

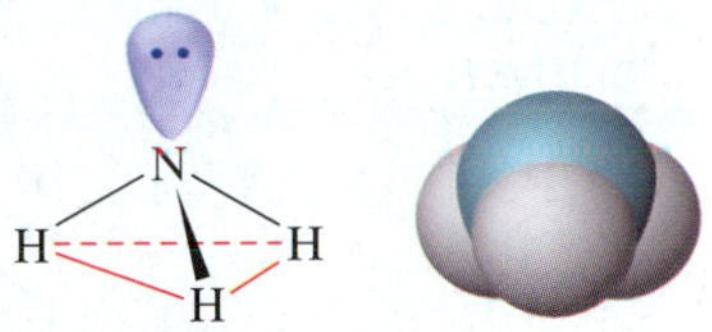

VSEPR notation: AX_3E

▲ **FIGURE 10.5 Electron-group geometry and molecular geometry of ammonia**

The molecular geometry—trigonal pyramidal—is outlined by the solid black bond lines and the red lines. The lone pair of electrons (shown in blue) is directed to a corner of the tetrahedron corresponding to the electron-group geometry. (The tetrahedron is not shown.)

QUESTION: If we removed one H^+ from this structure, what would be the shape of the three-atom structure NH_2^-?

The notation AXE_3 is trivial. It is the notation that applies to $H-\overset{..}{\underset{..}{F}}:$. The F atom has four electron groups—one bonding pair and three lone pairs—but the molecule must be linear because there are only two atoms present.

🌐 **Ammonia and Water 3D models**

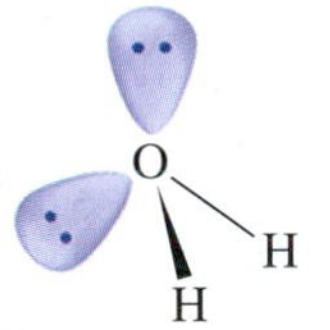

VSEPR notation: AX_2E_2

▲ **FIGURE 10.6 Electron-group geometry and molecular geometry of water**

The bonds in H_2O are represented by solid black lines and outline the molecular geometry—angular. The two lone pairs of electrons (shown in blue) are directed to corners of the tetrahedron corresponding to the electron-group geometry. (The tetrahedron is not shown.)

QUESTION: If we added one H^+ to this structure, what would be the shape of the four-atom structure H_3O^+?

Example 10.2

Use the VSEPR method to predict the molecular geometry of XeF_2.

STRATEGY

We can try the four-step method outlined on page 393 and applied in Example 10.1 to see if it leads us to the molecular geometry. If it does not, we will need to consider the two additional factors cited above as well.

SOLUTION

Step 1: *Draw a plausible Lewis structure.*

There are $8 + (2 \times 7) = 22$ valence electrons in XeF_2. We need only 20 electrons to attach two F atoms to a central Xe atom and to provide each of the three atoms with an octet. We must place the additional pair in an expanded valence shell on the Xe atom. The Lewis structure is therefore

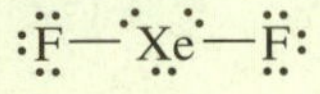

$$:\ddot{F}\!-\!\ddot{X}\!\ddot{e}\!-\!\ddot{F}:$$

Step 2: *Determine the number of electron groups.*

There are five electron groups around the Xe atom, two bonding pairs and three lone pairs. The VSEPR notation is AX_2E_3.

Step 3: *Identify the electron-group geometry.*

The five electron groups signify that the distribution of the electron groups is trigonal bipyramidal.

Step 4: *Identify the molecular geometry.*

We cannot decide on the molecular geometry at this point. There seem to be three possibilities, but only one can be correct:

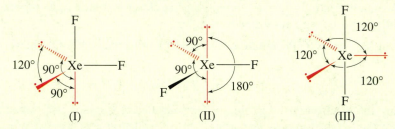

(I)   (II)   (III)

Structure I has one 120° LP–LP repulsion and two 90° LP–LP repulsions. Structure II has one 180° LP–LP repulsion and two 90° LP–LP repulsions. Structure III has no 90° LP–LP repulsions but three 120° LP–LP repulsions. Based on the fact that repulsions are strongest for a 90° bond angle and decrease as the bond angle increases (90° > 120° > 180°), we predict that structure III represents the lowest energy configuration. Thus, we expect the Xe and two F atoms to lie in a straight line; the molecular geometry is linear.

ASSESSMENT

The predicted linear molecular geometry is also what is observed experimentally. This result suggests a general rule. In molecules containing five electron groups, lone pairs are found in equatorial ("side") positions 120° apart, as in structure III, rather than in axial ("top/bottom") positions 180° apart.

EXERCISE 10.2A

Use the VSEPR method to explain why SF_4 has the seesaw molecular geometry shown in Table 10.1 rather than the tetrahedral molecular geometry shown here:

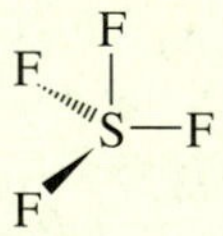

EXERCISE 10.2B

Use the VSEPR method to explain why ClF_3 has the T-shaped molecular geometry shown in Table 10.1 rather than the trigonal planar geometry shown here:

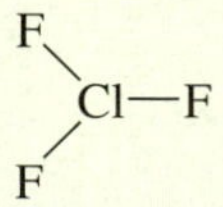

VSEPR Treatment of Structures with More than One Central Atom

So far, we have used the VSEPR method only in structures with one central atom, such as CH_4, NH_3, H_2O, NO_3^-, and XeF_2. However, we already have seen many molecules and polyatomic ions that have more than one central atom—for example, C_2H_6, C_2H_4, CH_3OH, and H_2O_2. For these more complex species, we first describe the molecular geometry around each central atom and then combine the results to give an overall molecular geometry.

Molecular Shape, Drug Action, and Drug Design

Most drugs act on protein molecules in the tissues or organs in the body. The drugs do not act just anywhere on the molecules, however. They must attach to specific locations called *receptor sites*. Receptor sites have three-dimensional shapes designed to accommodate certain molecules produced naturally in the body. An effective drug molecule, by virtue of its size, shape, and bond polarities, also fits a receptor site. The drug acts either by replacing natural molecules on the site or by blocking the site from the natural molecules.

Drug molecules that fit a receptor site and cause their own characteristic response are called *agonists*. Those that bind to the site and prevent the action of the intended molecules are called *antagonists*. If an agonist and an antagonist both are present, they compete for the receptor sites. In general, an antagonist binds more tightly than an agonist, and a small quantity of an antagonist can block the action of a larger quantity of an agonist. For example, the human brain has receptor sites for morphine, the principal narcotic in opium poppies. A 1-mg intravenous dose of naloxone, a morphine antagonist, can block the action of 25 mg of heroin, a morphine-related agonist.

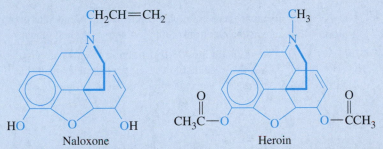

Naloxone   Heroin

▲ Substances produced in the body during strenuous exercise may create the "runner's high" by activating receptor sites on nerve cells.

Why should the human brain have receptors for a plant-derived drug like morphine? The answer seems to lie in several morphinelike substances formed naturally by the human body, substances called *endorphins*. (The word "endorphin" is a contraction of endogenous morphine—Greek, *endon*, within; *gen-*, produced—thus "morphine produced in the body.") Endorphins are peptides. β-Endorphin, the most potent human form, is composed of a chain of 30 amino acid residues. Morphine is a potent painkiller, and so the production of these compounds during, for example, strenuous athletic activity may account for the fact that athletes can sometimes continue to compete after an injury. They do not feel the pain until the event is over. (The phenomenon called "runner's high," experienced by some runners, may also be due to endorphin production.)

For much of human history, drugs were discovered by accident. Beginning in the nineteenth century, though, chemists learned how to isolate active compounds from plant materials. Morphine, for instance, was first isolated in 1805. Later, chemists were able to determine the chemical structures of these drugs. Then they synthesized related compounds and tested them, hoping to find safer and more effective drugs. Naloxone and heroin are among the hundreds of derivatives of morphine that have been tested. Even today, this is the main approach to drug discovery, but things are changing. Scientists are beginning to determine the structures of the receptor molecules and to design prospective drugs that will fit the receptor sites. They use computers to make models that predict molecules that will fit. In this way, only a few compounds have to be sent through the enormously expensive process of evaluation that leads to approval for use in medicine.

Example 10.3

Use the VSEPR method to describe, as best you can, the molecular geometry of the nitric acid molecule, HNO_3.

STRATEGY

We can use our four-step approach, starting with a plausible Lewis structure that will enable us to identify the central atoms. In the first three steps, we can describe the geometry about each central atom. At this point, we should be able to formulate a description of the overall molecular geometry.

SOLUTION

Step 1: *Draw a plausible Lewis structure.*

There are $1 + 5 + (3 \times 6) = 24$ valence electrons in HNO_3. Of the two structures shown here, only the first is plausible. The second, which has only N as a central atom, would not have enough electrons to give each atom an octet, and it has the hydrogen atom bonded to a nitrogen atom rather than to an oxygen atom.

$$H-\ddot{\underset{..}{O}}-N \overset{:O:}{\underset{\ddot{O}:}{\parallel}} \text{ (correct)} \qquad :\ddot{\underset{..}{O}}-N \overset{:\ddot{O}:}{\underset{H}{|}} -\ddot{\underset{..}{O}}: \text{ (incorrect)}$$

There are two central atoms: the N atom and the O atom bonded to the H atom.

Step 2: *Determine the number of electron groups around each central atom.*

Central N atom: This atom has three electron groups involved in two single bonds and one double bond. Its VSEPR notation is AX_3.

Central O atom: This atom has four electron groups. Two of the groups are bonding pairs, and the other two are lone pairs. The VSEPR notation for the O atom is therefore AX_2E_2.

Step 3: *Identify the electron-group geometry for each central atom.*

Central N atom: trigonal planar.

Central O atom: tetrahedral.

Step 4: *Identify the molecular geometry around each central atom.*

Central N atom: Because there are no LP electrons, the molecular geometry around this atom is the same as the electron-group geometry: trigonal planar.

Central O atom: The atoms in an AX_2E_2 structure take on an angular orientation. The H-O-N portion of the molecule has the same general shape as the H-O-H molecule, although not necessarily the same bond angles.

ASSESSMENT

As suggested in Figure 10.7, our description of the overall molecular geometry of HNO_3 has the N and three O atoms in the same plane and O-N-O bond angles of about 120°. We know that the H atom is situated to produce a H-O-N bond angle of about 109°, but we cannot say whether the H atom lies above, below, or in the plane of the other atoms. (Experimental evidence indicates that it lies in about the same plane as the other atoms.)

Problem-Solving Note

We cannot assign a unique name to the geometry of HNO_3. Instead, we describe the geometries around the central N and O atoms to arrive at an overall molecular picture.

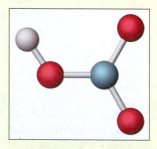

◄ **FIGURE 10.7   Predicted molecular geometry of the HNO_3 molecule by the VSEPR method**

The prediction is that the N and the three O atoms all lie in the same plane. The trigonal planar electron-group geometry around the central N atom leads us to predict bond angles of about 120°. The tetrahedral electron-group geometry around the central O atom leads us to predict a H—O—N bond angle of about 109°.

 Nitric Acid 3D model

EXERCISE 10.3A

Use the VSEPR method to describe the molecular geometry of dimethyl ether, $(CH_3)_2O$, as completely as you can.

EXERCISE 10.3B

Use the VSEPR method to describe the molecular geometry of the propanal (propionaldehyde) molecule, CH_3CH_2CHO.

10.2   Polar Molecules and Dipole Moments

In Section 9.7, we noted that covalent bonds in which the electronegativities of the bonded atoms differ are *polar covalent*. The opposite ends of the bond have *partial charges* that we indicate in Lewis structures by $\delta+$ and $\delta-$. Thus, we can represent the polar covalent bond in the HCl molecule as

$$\overset{\delta+ \ \ \delta-}{H\!:\!\ddot{\underset{\cdot\cdot}{Cl}}\!:}$$

A molecule with separate centers of partial positive and partial negative charge is called a **polar molecule.** A quantity called the dipole moment describes the significance of this charge separation. The **dipole moment (μ)** of a molecule is the product of the magnitude of the partial charge (δ) and the distance (d) that separates the centers of positive and negative partial charge:

$$\mu = \delta d \qquad\qquad (10.1)$$

Thus, a polar molecule is one having a nonzero dipole moment, and a nonpolar molecule is one with dipole moment $\mu = 0$.

Dipole moments are based on units that reflect the electric charge times the distance between the two charge centers: coulomb $\times$ meter (C m). Dipole moments are generally expressed in a unit called the *debye*. One debye (D) is equal to 3.34×10^{-30} C m. For example, the measured dipole moment of HCl is $\mu = 1.07$ D. Figure 10.8 suggests the behavior of polar molecules in an electric field. When the metal plates are electrically charged, the average orientation of polar molecules is aligned with the field. Nonpolar molecules would be largely unaffected by the field.

The apparatus suggested in Figure 10.8 can be used to measure dipole moments. When the current is turned on, the polar molecules increase the charge-storing capacity of the plates to an extent that is related to the dipole moment of the molecules.

Bond Dipoles and Molecular Dipoles

In continuing our discussion of polar molecules and dipole moments, we need to distinguish between a bond dipole and a molecular dipole. Every polar covalent bond has a *bond dipole*—a separation of centers of positive and negative charge in an individual bond. A charge separation in the molecule as a whole, considering *all* the bonds, is a *molecular dipole*. Although molecular dipoles can be measured quantitatively, we will be concerned mainly with qualitative determinations: (a) does a given molecule have a dipole moment, and (b) how do the dipole moments of similar molecules compare?

In a diatomic molecule formed by identical atoms (*homonuclear* molecules), there is no electronegativity difference and thus no separation of charge. There is neither a bond dipole nor a molecular dipole. The bond is nonpolar covalent, and the molecule is nonpolar and has no dipole moment. In a diatomic molecule formed by different atoms (*heteronuclear* molecules), there is generally an electronegativity difference and a corresponding charge separation in the bond. In this type of molecule, the bond dipole and molecular dipole are the same because there is only one bond. The molecule is polar and has a dipole moment, as we saw with HCl.

Simple calculations involving dipole moments can yield some interesting results. (See Problems 90 and 92.)

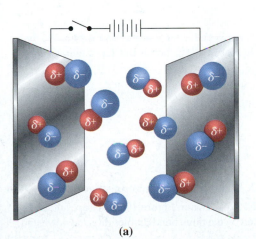

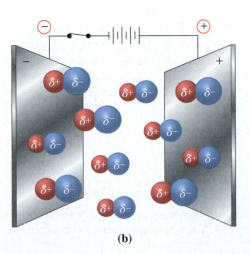

(a)   (b)

◀ **FIGURE 10.8   Behavior of polar molecules in an electric field**

(a) In the absence of an electric field, polar molecules are oriented randomly. (b) When the metal plates are electrically charged, the molecules orient themselves as shown.

QUESTION: How would NH_3 molecules be oriented in (b)?

In polyatomic molecules, the situation is more complex. Consider the bonds in the CO_2 molecule:

$$O = C = O$$
$$\mu = 0$$

There are two bond dipoles in this molecule, one for each C—O bond. The cross-based arrows (recall page 353) point from the positive to the negative end of each bond dipole, from the C to the more electronegative O atoms. Bond dipoles have both a *magnitude* and a *direction*. In CO_2, the bond dipoles are equal in magnitude and point in opposite directions, leading to a cancellation of their effects. The result is that there is no molecular dipole, consequently no dipole moment ($\mu = 0$), and the CO_2 molecule is nonpolar. The situation is analogous to a tug-of-war in which two evenly matched teams (the O atoms) pull on a rope equally hard. A knot in the center of the rope (the C atom), though subjected to strong pulls, does not move at all.

If we start with the experimental fact that water has the dipole moment $\mu = 1.84$ D, we must conclude that H_2O cannot be a linear molecule. The two H—O bond dipoles must combine in such a way as not to cancel each other. A structure consistent with the experimental evidence is

$$\overset{O - H}{\underset{H}{}} \quad 104.5°$$

In this structure, the black cross-based arrows represent the bond dipoles and the red cross-based arrow depicts the molecular dipole that results when the two bond dipoles are combined.

Molecular Shapes and Dipole Moments

We can generally determine whether a molecule is polar or nonpolar by following the strategy employed in Example 10.4.

Example 10.4

Explain whether you expect the following molecules to be polar or nonpolar.

(a) CCl_4 (b) $CHCl_3$

STRATEGY

We can follow this three-step approach:

Step 1: Use electronegativity values (from trends or Figure 9.8) to predict the individual bond dipoles.

Step 2: Use the VSEPR method to predict the molecular geometry.

Step 3: From the molecular geometry, determine whether all the bond dipoles cancel to give a nonpolar molecule or whether they combine to produce a resultant molecular dipole.

SOLUTION

Step 1: There are electronegativity differences between C and H atoms and between C and Cl atoms. As a result, there are bond dipoles in all the bonds in CCl_4 and in $CHCl_3$.

Step 2: The Lewis structures of the two molecules are

$$\begin{array}{ccc} :\overset{..}{C}l: & & H \\ | & & | \\ :\overset{..}{C}l-C-\overset{..}{C}l: & & :\overset{..}{C}l-C-\overset{..}{C}l: \\ | & & | \\ :\overset{..}{C}l: & & :\overset{..}{C}l: \\ \textbf{(a)} & & \textbf{(b)} \end{array}$$

Both molecules (VSEPR notation: AX_4) have tetrahedral molecular geometry.

Step 3: The distribution of bond dipoles in CCl_4 is shown in Figure 10.9a. The resultant dipole in the downward direction produced by the three bond dipoles directed down and

---

Quantities having both a magnitude and a direction are called vectors.

Molecular Polarity activity

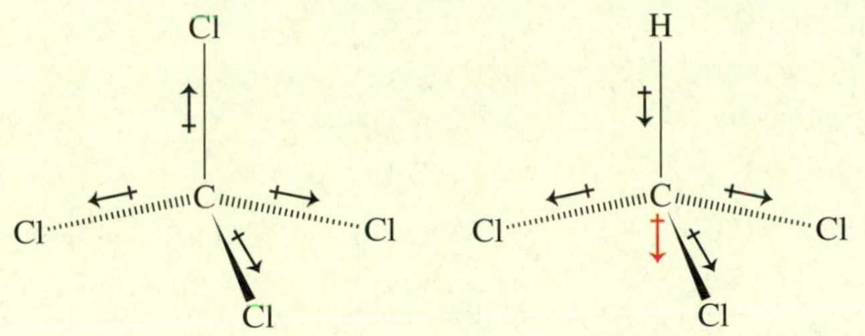

▲ **FIGURE 10.9 Molecular geometry and molecular dipoles**
The bond dipoles are represented by black cross-based arrows ↤→
(a) The individual bond dipoles cancel, and there is no molecular dipole.
(b) All the bond dipoles point downward and combine to form a molecular dipole (red arrow).

away from the center of the structure is just matched by the straight upward bond dipole at the top. Thus all the bond dipoles cancel, so there is no molecular dipole, and CCl_4 is a nonpolar molecule.

Things are dramatically different when we substitute a H atom for the Cl atom at the top of the structure, as shown in Figure 10.9b. The resultant downward dipole of the three Cl atoms at the bottom remains the same as in CCl_4, but there is no counterbalancing upward dipole. Instead, there is an additional slight downward dipole because the C atom is more electronegative than the H atom. Thus, the four bond dipoles combine to form a molecular dipole pointed toward the bottom of the structure, making $CHCl_3$ a polar molecule. (Its dipole moment is 1.01 D.)

ASSESSMENT

A further interesting conclusion is that whatever atom (call it X) we substitute for one of the Cl atoms in CCl_4, the molecule $CXCl_3$ will have a resultant dipole moment, so long as the electronegativity of X is different from that of Cl.

EXERCISE 10.4A

Are BF_3, SO_2, BrCl, and N_2 polar or nonpolar? Explain each of your answers.

EXERCISE 10.4B

Are SO_3, SO_2Cl_2, ClF_3, and BrF_5 polar or nonpolar? Explain each of your answers.

Problem-Solving Note

General rule: An AX_n molecule where all X atoms are the same will be nonpolar.

Example 10.5   A Conceptual Example

Of the two compounds NOF and NO_2F, one has $\mu = 1.81$ D, and the other has $\mu = 0.47$ D. Which dipole moment do you predict for each compound? Explain.

ANALYSIS AND CONCLUSIONS

We need to identify bond dipoles and determine how they combine to produce a molecular dipole and hence a dipole moment. To do this, we first need to use the VSEPR method to determine the geometric shape of the two molecules. Let's begin with Lewis structures. Because it has the lowest electronegativity of the three kinds of atoms, we expect N to be the central atom in both structures.

	NOF	**NO₂F**
Valence electrons	18	24

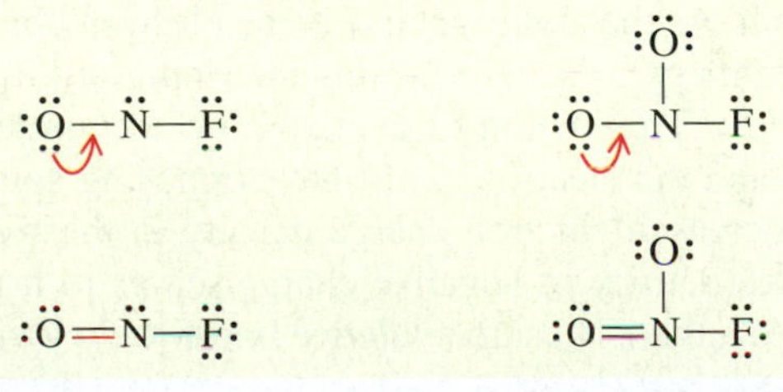

Electron groups around central atom	3	3
Electron-group geometry	Trigonal planar	Trigonal planar
VSEPR notation	AX_2E	AX_3
Molecular geometry	Angular	Trigonal planar
Bond dipoles (electronegativities F > O > N)		

In NOF, there are two bond dipoles and both point downward, leading to a net downward molecular dipole. In NO_2F, the upward-pointing N—O bond dipole opposes the other two, and consequently we expect a smaller molecular dipole. Our prediction is therefore NOF, $\mu = 1.81$ D, and NO_2F, $\mu = 0.47$ D.

EXERCISE 10.5A

Of the two molecules NOF and NO_2F, one has a measured F-N-O bond angle of 110°, and the other has the same atoms arranged in a 118° angle. Which bond angle do you predict for each molecule? Explain.

EXERCISE 10.5B

Which of the following molecules would you expect to have the largest dipole moment and why? CS_2, COF_2, NOF_2, NO, SO_3

We have shown how electronegativity differences lead to bond dipoles and how bond dipoles can combine to produce a resultant molecular dipole. Lone-pair electrons can also make a contribution to dipole moments. Consider the molecule NF_3. Electrons in the N—F bonds are displaced toward the F atoms and away from the N atom. To a considerable extent, however, this displacement is counteracted by the spreading out of the charge cloud of the lone-pair electrons on the N atom. As a result, the dipole moment of NF_3 is rather small: $\mu = 0.24$ D. In contrast, the NH_3 molecule has the same shape as NF_3, but the displacement of bonding-pair electrons is *toward* the N atom. Because this displacement is in the same general direction as that produced by the lone-pair electrons, NH_3 has a much larger dipole moment: $\mu = 1.47$ D.

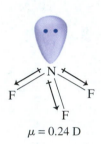

$\mu = 0.24$ D

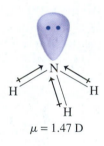

$\mu = 1.47$ D

Valence Bond Theory

Soon after Schrödinger applied quantum mechanics to the hydrogen atom, other scientists began to apply quantum mechanics to molecules, beginning with H_2. We will consider two quantum mechanical approaches to molecular structure. In *valence bond theory*, we continue to think about the atoms in a molecule in terms of their atomic orbitals, and we focus on the particular orbitals involved in covalent bond formation. In the second approach, *molecular orbital theory*, we describe bonding in terms of molecular orbitals that are constructed from atomic orbitals.

10.3 Atomic Orbital Overlap

Imagine two hydrogen atoms approaching each other, each with a single electron in its 1s orbital. As the atoms get close, the electron charge clouds represented by the orbitals begin to merge. We refer to this intermingling as the *overlap* of the 1s orbitals of the two atoms. This region of overlap now accommodates two electrons—one from each atom—and the electrons must have opposing spins. The atomic orbital overlap results in an increased electron charge density in the region between the atomic nuclei. The increased density of negative charge serves to hold the two positively charged atomic nuclei together. Thus, the **valence bond theory** views a covalent bond in this way:

A covalent bond is formed by the pairing of two electrons with opposing spins in the region of overlap of atomic orbitals between two atoms. This overlap region has a high electron charge density.

In general, the more extensive the overlap between two orbitals, the stronger the bond between the two atoms. If two atoms are forced too closely together, however, the repulsive forces the atomic nuclei exert on each other become more important than the attractions between electrons and nuclei, and the bond becomes unstable. For each bond, then, there is a condition of optimal orbital overlap that leads to maximum bond strength (bond energy) at a particular internuclear distance (bond length). The valence bond theory attempts to find the best approximation of this condition for all the bonds in a molecule. Figure 10.10 depicts the bonding of two hydrogen atoms into a hydrogen molecule through the overlap of their 1s orbitals

Next, let's consider H_2S, a more representative example of a molecule that can be described by the valence bond theory. First, let's focus on the isolated atoms in Figure 10.11a. The orbitals that contain an unpaired electron are those that will overlap to form bonds. For the hydrogen atoms, this is the 1s orbital (red). For the sulfur atom, we need consider only the valence shell ($n = 3$). As seen in the orbital diagram, the sulfur orbitals containing a single electron are the $3p_y$ and $3p_z$; the $3p_x$ orbital is filled. The 3s orbital, also filled, is not shown. Thus, the 1s orbitals of the two H atoms overlap with the $3p_y$ and $3p_z$ orbitals of the sulfur atom to form the molecule H_2S. Several important points are brought out by this example.

- Most of the electrons in a molecule remain in the same orbital locations they occupied in the isolated atoms.

- Bonding electrons are localized in the region of atomic orbital overlap; the greatest probability of finding the bonding electron pair is in this region.

- For orbitals that have directional lobes, maximum overlap occurs when atomic orbitals overlap end to end; that is, a hypothetical line joining the nuclei of the two bonded atoms passes through the region of maximum overlap. (Recall that p orbitals are directed along perpendicular axes through the nucleus of an atom, but s orbitals are spherically symmetric—they have no directional lobes.)

- The molecular geometry depends on the geometric relationships among the central-atom orbitals that participate in bonding. In H_2S, for example, because the two bonding 3p orbitals of the S atom are perpendicular to each other, the predicted H-S-H bond angle in H_2S is 90° (Figure 10.11b).

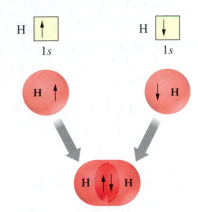

▲ **FIGURE 10.10 Atomic orbital overlap and bonding in H_2**

Each 1s atomic orbital contains one electron. As a result of the overlap of the two orbitals, the electrons become paired and produce a region of high electron charge density (high electron probability)—a covalent bond.

H_2 Bond Formation animation

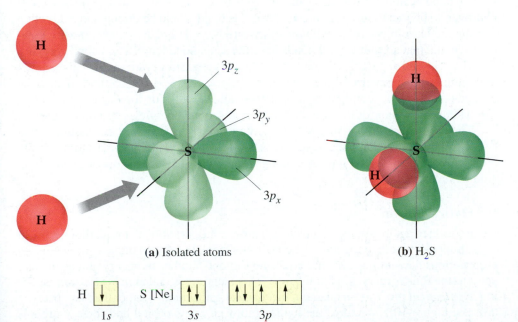

(a) Isolated atoms   (b) H_2S

◀ **FIGURE 10.11 Atomic orbital overlap and bonding in H_2S**

(a) For S, only 3p orbitals are shown. The $3p_x$ orbital is filled with an electron pair, and the $3p_y$ and $3p_z$ orbitals contain one electron each. (b) The 1s orbitals of the two H atoms overlap with the $3p_y$ and $3p_z$ orbitals of the S atom, producing an H_2S molecule with a predicted bond angle of 90°.

QUESTION: Based on this figure, explain why only two hydrogen atoms bond to the sulfur atom.

As a first approximation, the VSEPR method predicts a tetrahedral bond angle (109.5°) in H_2S. However, by taking into account the strong repulsions between the lone pairs of electrons on the S atom and the bonding pairs, we expect the bonds in H_2S to be forced into a somewhat smaller angle. The measured bond angle in the H_2S molecule is 92.1°. This rather large discrepancy between measured bond angle and tetrahedral angle suggests that the valence bond theory describes covalent bonding in H_2S better than does the VSEPR method. Unfortunately, the valence bond theory using unmodified atomic orbitals produces good results for relatively few molecules. We will be most successful in describing molecular geometries by using a combination of the VSEPR method and a modification of the valence bond theory.

10.4   Hybridization of Atomic Orbitals

Let's use the valence bond theory to describe the simplest hydrocarbon molecule possible. We start with the ground-state electron configuration of carbon and consider only the valence-shell orbitals:

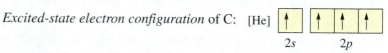

Ground-state electron configuration of C: [He]

Hybridization animation

We see two unpaired electrons in the $2p$ subshell and might predict the simplest hydrocarbon molecule to be CH_2 with a bond angle of 90°, just as we saw for H_2S. However, the CH_2 "molecule" does not follow the octet rule because the central C atom is surrounded by only six electrons. Experiments show that CH_2 is not a stable molecule. The simplest stable hydrocarbon is methane, CH_4. To account for the four covalent bonds in this molecule, we need for carbon an orbital diagram that shows all four valence-shell electrons unpaired—in other words, with each electron in a separate orbital. To get such a diagram, we might imagine that one of the $2s$ electrons is *promoted* to the empty $2p$ orbital. To boost the $2s$ electron to a higher-energy subshell requires that energy be absorbed. The resulting electron configuration is that of an *excited state*:

Energy and Hybridization activity

Excited-state electron configuration of C: [He]

The three mutually perpendicular $2p$ orbitals of this excited-state configuration lead us to predict a molecule with three C—H bonds at angles of 90°. The fourth C—H bond would be based on the overlap of carbon's spherical $2s$ orbital with hydrogen's spherical $1s$ orbital. Presumably, this bond would be oriented in a direction causing the least interference with the other three C—H bonds. *By experiment,* however, we find that all four C—H bonds have the same bond lengths and bond energies. The H-C-H bond angles are all the same, 109.5° (see again Figure 10.3). This tetrahedral molecular geometry is the one predicted by the VSEPR method. Thus, the excited-state electron configuration of carbon allows only for the correct number of C—H bonds, but not for the correct bond lengths, bond energies, and bond angles.

The problem with this analysis of bonding in CH_4 is that we have assumed that bonded atoms have the same kinds of orbitals $(s, p, \ldots)$ as isolated atoms. Quite often, this seems not to be the case, though, and there is a straightforward way to get around the problem.

sp^3 Hybridization

Let's consider again the excited-state electron configuration of the carbon atom in CH_4. Suppose we blend together the one $2s$ and three $2p$ orbitals of the carbon atom to produce four new orbitals that are equivalent to one another in energy and shape and point to the four corners of a tetrahedron. This blending is called *hybridization,* and it is a *hypothetical* process—a quantum mechanical calculation—rather than an observed process. Figure 10.12 pictures the hybridization of one s orbital and three p orbitals into the new set of four **hybrid orbitals.**

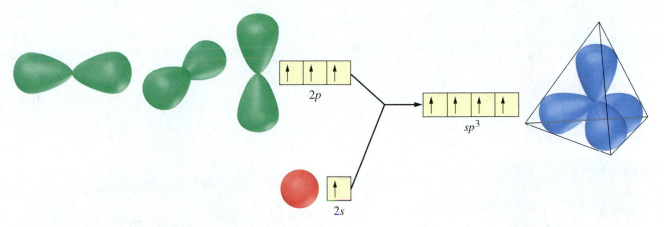

▲ FIGURE 10.12   The sp^3 hybridization scheme for C
The one $2s$ and three $2p$ orbitals shown on the left combine to yield the four sp^3 orbitals on the right. The $2s$ orbital makes about a 25% contribution to the hybridized orbitals, and the $2p$ orbitals contribute about 75%.

The new orbitals are called sp^3 **hybrid orbitals** because they were formed from one s orbital and three p orbitals.

Whenever a hybridization scheme provides the best description of a molecular structure in the valence bond theory, it is important to keep the following points in mind:

* Although hybridization schemes can be used for any atoms in covalent bonds, they are most often employed for central atoms only.
* The number of hybrid orbitals produced in a hybridization scheme is equal to the total number of atomic orbitals combined.
* To form covalent bonds, hybrid orbitals may overlap either with pure atomic orbitals or with other hybrid orbitals.
* Molecular geometry is determined by the shapes and orientations of hybrid orbitals, which are different from the shapes and orientations of atomic orbitals.

▲ An sp^3 hybrid orbital involves one small and one large lobe located on opposite sides of the nucleus. To add clarity to drawings of sets of hybrid orbitals, we usually omit the small lobe and elongate the large lobe, as in Figure 10.12.

Most important of all, we must remember that a molecular structure is determined *by experiment* and that hybridization is just a way to rationalize that structure. The hybridization scheme must assess the energy changes in a series of hypothetical processes: promoting a valence electron in an atom from the ground state to an excited state, hybridizing orbitals in the excited state, and using the hybrid orbitals to form bonds. A successful hybridization scheme minimizes the energy of the molecular structure and satisfactorily accounts for the observed molecular geometry.

To illustrate the role of sp^3 hybrid orbitals in bond formation in methane, we can show hybridization in the central carbon atom with an orbital diagram:

sp^3 *hybridization* in C: [He]   ↑ ↑ ↑ ↑
                                         sp^3

Overlaps between H $1s$ atomic orbitals and C sp^3 hybrid orbitals to form CH_4 are illustrated in Figure 10.13.

We can use sp^3 hybridization not only for structures of the type AX_4 (as in CH_4) but also for AX_3E (as in NH_3) and AX_2E_2 (as in H_2O). For example, suppose we form sp^3 hybrid orbitals from the valence-shell atomic orbitals of the central N atom in NH_3. When we assign nitrogen's five valence electrons to these four hybrid orbitals, we have a lone pair of electrons in one of the orbitals and unpaired electrons in the other three, as seen below and in Figure 10.14:

sp^3 *hybridization* in N: [He]   ↑↓ ↑ ↑ ↑
                                         sp^3

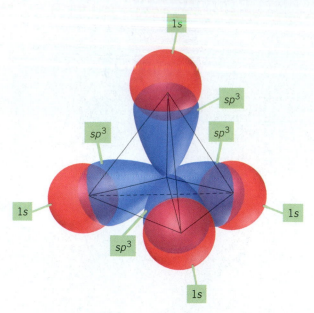

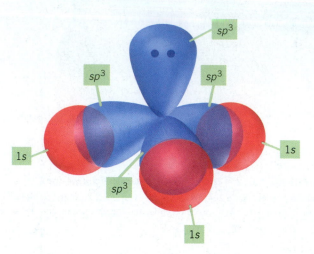

▲ **FIGURE 10.13** *sp^3* **Hybrid orbitals and bonding in CH$_4$**
The four *sp^3* hybrid orbitals of the C atom (blue) have been modified to eliminate the small lobes directed toward the center of the structure; these lobes are not involved in orbital overlaps. The hydrogen atomic orbitals are 1*s* (red). The molecular geometry of the CH$_4$ is tetrahedral; the H-C-H bond angles are 109.5°.

▲ **FIGURE 10.14** *sp^3* **Hybrid orbitals and bonding in NH$_3$**
The three N—H bonds are formed by the overlap of the 1s atomic orbitals of the three H atoms with three of the four *sp^3* hybrid orbitals of the N atom. The molecular geometry of the NH$_3$ is trigonal pyramidal. The lone-pair electrons on the N atom reside in the fourth *sp^3* orbital of the N atom.

🌐 **Methane and Ammonia 3D models**

The three hybrid orbitals containing one unpaired electron each overlap with the 1*s* atomic orbitals of three H atoms to form the three N—H bonds. The predicted H-N-H bond angles of 109.5° are close to the experimentally observed angles of 107°.

A similar hybridization scheme for the O atom in H$_2$O accounts for the formation of two O—H bonds and the presence of two lone pairs of electrons on the O atom:

sp^3 hybridization in O: [He] ↑↓ ↑↓ ↑ ↑
sp^3

The predicted H-O-H bond angle of 109.5° is also reasonably close to the observed 104.5°. As in our discussion of the VSEPR method, we can explain the somewhat smaller-than-tetrahedral bond angles in NH$_3$ and H$_2$O through repulsions involving lone-pair electrons.

sp^2 Hybrid Orbitals

The *sp^2* hybridization scheme, as we will see in Section 10.5, is especially useful in describing double covalent bonds. However, let's first look at a simpler application by describing boron compounds. The first step in the *sp^2* hybridization scheme involves promoting a boron 2*s* electron to an empty 2*p* orbital:

▲ Like *sp^3* hybrid orbitals, *sp^2* hybrid orbitals involve one small and one large lobe located on opposite sides of the nucleus. The small lobe is larger here than in an *sp^3* orbital, indicating more *s* character in an *sp^2* orbital (about 33%) than in an *sp^3* orbital (about 25%).

Ground-state electron configuration of B: [He] ↑↓ | ↑ | | |
2*s*   2*p*

Excited-state electron configuration of B: [He] ↑ | ↑ ↑ |
2*s*   2*p*

As shown in Figure 10.15, boron's one 2*s* orbital and its two 2*p* orbitals, each occupied by an unpaired electron, are hybridized into three *sp^2* **hybrid orbitals**, and the empty 2*p* orbital remains unhybridized:

sp^2 hybridization in B: [He] ↑ ↑ ↑ | |
sp^2   2*p*

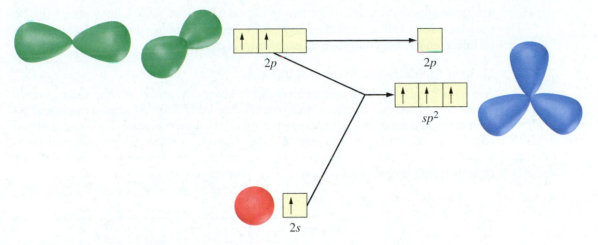

▲ **FIGURE 10.15   The sp^2 hybridization scheme for boron**
The $2s$ orbital makes about a 33% contribution to the hybridized orbitals, and the $2p$ orbitals contribute about 67%.

The geometric distribution of the three sp^2 hybrid orbitals is in a plane, with the lobes directed at 120° angles. In the bonding of such a hybridized B atom with three F atoms, valence bond theory predicts a trigonal planar molecule with 120° F-B-F bond angles, exactly as is observed experimentally.

sp Hybrid Orbitals

The sp hybridization scheme is especially useful in describing triple covalent bonds (Section 10.5). However, let's look first at a simpler compound. Beryllium and chlorine form the triatomic molecule $BeCl_2$, which is a gas at high temperatures. To describe bonding in this molecule, once again the first step is to promote a $2s$ electron to a $2p$ orbital, followed by hybridization of the orbitals in the excited-state atom:

Ground-state electron configuration of Be: [He] ↑↓ ▢▢▢
                                                      $2s$ $2p$

Excited-state electron configuration of Be: [He] ↑ ↑▢▢
                                                     $2s$ $2p$

The $2s$ orbital and the $2p$ orbital occupied by an unpaired electron are hybridized into two **sp hybrid orbitals**, and the two empty $2p$ orbitals remain unhybridized. This scheme is illustrated here in an orbital diagram and also in Figure 10.16:

sp hybridization in Be: [He] ↑ ↑ ▢▢
                                   sp $2p$

▲ Like sp^3 and sp^2 hybrid orbitals, sp hybrid orbitals involve one small and one large lobe located on opposite sides of the nucleus. The small lobe is larger here than in an sp^3 or sp^2 orbital, indicating more s character in an sp orbital (about 50%) than in an sp^2 orbital (about 33%) or in an sp^3 orbital (about 25%).

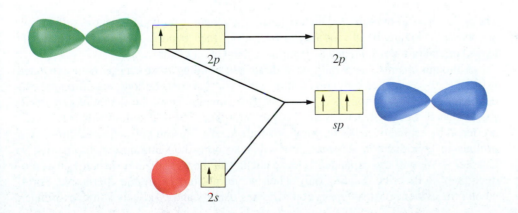
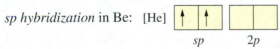

◀ **FIGURE 10.16   The sp hybridization scheme for beryllium**
The $2s$ and $2p$ orbitals make about equal contributions to the sp hybridized orbitals.

The geometric distribution of the two *sp* hybrid orbitals is along a line through the Be nucleus, with a 180° angle between them. We predict that the $BeCl_2$ molecule should be linear, and this is confirmed by experimental evidence.

Hybrid Orbitals Involving *d* Subshells

Any hybridization scheme that involves only *s* and *p* orbitals can accommodate a maximum of eight valence electrons, that is, a valence-shell octet. To apply a hybridization scheme to structures involving expanded valence shells, we need additional orbitals, and we can find these in the *d* subshell. For example, we need five hybrid orbitals to describe bonding in PCl_5. We get these by combining one *s*, three *p*, and one *d* atomic orbitals from the central phosphorus atom, as suggested by these orbital diagrams:

Ground-state electron configuration of P:

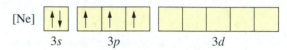

Excited-state electron configuration of P:

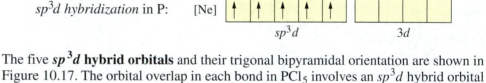

sp^3d *hybridization* in P: [Ne]

▲ FIGURE 10.17   The *sp³d* hybrid orbitals

These sp^3d hybrid orbitals in a trigonal bipyramidal arrangement are deployed by a phosphorus atom in the molecule PCl_5. (See Table 10.1.)

The five **sp^3d hybrid orbitals** and their trigonal bipyramidal orientation are shown in Figure 10.17. The orbital overlap in each bond in PCl_5 involves an sp^3d hybrid orbital of the P atom and a 3*p* atomic orbital of a Cl atom.

The SF_6 molecule also features an expanded valence shell. Here, six hybrid orbitals for the central sulfur atom are required to describe bonding. We obtain them through the hybridization scheme sp^3d^2, represented by these orbital diagrams:

Ground-state electron configuration of S:

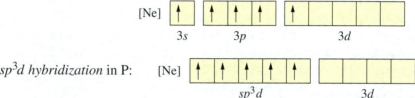

Excited-state electron configuration of S:

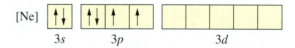

sp^3d^2 *hybridization* in S: [Ne]

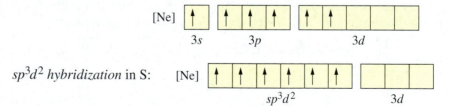

▲ FIGURE 10.18   The *sp³d²* hybrid orbitals

In the SF_6 molecule, the sulfur atom deploys sp^3d^2 hybrid orbitals in an octahedral arrangement. (See Table 10.1.)

QUESTION: In a species in which only four of the six hybrid orbitals in this figure are involved in bond formation, what is the VSEPR notation for the central atom?

The **sp^3d^2 hybrid orbitals** and their octahedral orientation around the central sulfur atom are shown in Figure 10.18. The orbital overlap in each bond in SF_6 involves an sp^3d^2 hybrid orbital of the S atom and a 2*p* atomic orbital of an F atom.

Although chemists generally have described bonding in structures with expanded valence shells as involving hybrid orbitals with *d*-orbital contributions, experimental evidence for these descriptions is rather weak. The situation is similar to that of expanded-valence-shell Lewis structures in Chapter 9 (page 364). The lowering of energy produced by orbital overlaps may not offset the energy cost in promoting electrons to *d* orbitals in hybridization schemes. We will take here the same approach we took in Chapter 9: We will use expanded valence shells and the hybridization schemes that produce them to describe bonding only when we do not have a simple alternative. Fortunately, the evidence for hybridization schemes using *s* and *p* orbitals is much stronger,

Table 10.2 Hybrid Orbitals and Their Geometric Orientation

Hybrid Orbitals	Geometric Orientation	Example
sp	Linear	$BeCl_2$
sp^2	Trigonal planar	BF_3
sp^3	Tetrahedral	CH_4
sp^3d	Trigonal bipyramidal	PCl_5
sp^3d^2	Octahedral	SF_6

and sp, sp^2, and sp^3 hybrid orbitals are the ones that we will encounter most frequently. In Section 10.5, for example, these are the only types of hybrid orbitals we will use.

Predicting Hybridization Schemes

In hybridization schemes, one hybrid orbital is produced for every simple atomic orbital involved. In a molecule, each hybrid orbital of the central atom acquires an electron pair, either a bonding pair or a lone pair. Also, the hybrid orbitals have the same symmetrical orientations as the electron-group geometries predicted by the VSEPR method and summarized in Table 10.2.

If we have experimental evidence relating to a molecular structure, the hybridization scheme we choose to describe bonding in the structure must conform to the evidence. Often, however, we do not have experimental evidence at hand and therefore need to predict a probable hybridization scheme. In these cases, the four-step approach illustrated in Example 10.6 generally works.

Example 10.6

Iodine pentafluoride, IF_5, is used commercially as a fluorinating agent—a substance that, via a chemical reaction, introduces fluorine into other compounds. Describe a hybridization scheme for the central atom, and sketch the molecular geometry of the IF_5 molecule.

STRATEGY

We can follow this four-step approach:

Step 1: Draw a plausible Lewis structure for the molecule or ion.
Step 2: Use the VSEPR method to predict the electron-group geometry of the central atom.
Step 3: Select the hybridization scheme that corresponds to the VSEPR prediction.
Step 4: Describe the orbital overlap and molecular geometry.

SOLUTION

Before we even begin with the four-step approach, we can see that the hybridization scheme cannot be sp, sp^2, or sp^3 because the central iodine atom forms five bonds, requiring an expanded valence shell.

Step 1: Both I and F are in group 7A of the periodic table, and the six atoms therefore contribute $(6 \times 7) = 42$ valence-shell electrons to the Lewis structure. We draw the five I—F bonds and then complete the octets of the F atoms:

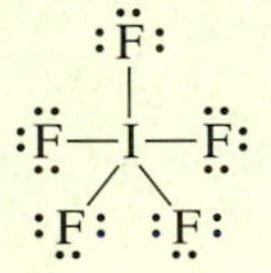

This accounts for 40 of the 42 valence electrons. The remaining electron pair (red) must go on the central iodine atom:

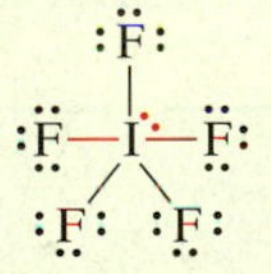

Step 2: The electron-group geometry for six electron groups is octahedral, with the six electron groups directed to the corners of the octahedron.

Step 3: The hybridization scheme that gives an octahedral distribution of hybridized orbitals is sp^3d^2.

Step 4: Each I—F single covalent bond needs one electron from the I atom. The LP electrons belong entirely to the I atom. We therefore assign $(5 + 2) = 7$ valence electrons to the six sp^3d^2 orbitals of iodine (the LP electrons are in red):

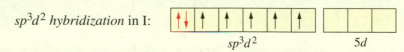

sp^3d^2 *hybridization* in I:

$$sp^3d^2 \qquad 5d$$

As shown in Figure 10.19, five of the hybrid orbitals for iodine overlap with $2p$ orbitals of the fluorine atoms. Because the six hybrid orbitals are equivalent, we can place the LP electrons in any one of them. The molecular geometry that corresponds to this distribution of BP and LP electrons is that of a *square pyramid*. We expect F-I-F bond angles of about 90°.

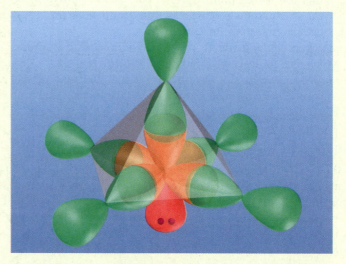

 Iodine pentafluoride 3D model

▲ **FIGURE 10.19** **Bonding scheme for iodine pentafluoride, IF₅**
The orbitals around the central I atom are six sp^3d^2 hybrid orbitals. One of them (red) is occupied by the LP electrons. The other five are the bonding orbitals. Each bond involves the overlap of an I sp^3d^2 hybrid orbital with a $2p$ atomic orbital of a terminal F atom. Because of repulsions between the LP electrons and the bonding pairs, the (imaginary) plane at the base of the molecule is raised slightly above the I atom.

ASSESSMENT

Because its formula is similar to that of PCl_5, we might have excepted IF_5 to have the same molecular geometry as PCl_5—trigonal bipyramidal (Table 10.1). This is not so, however, because PCl_5 is VSEPR type AX_5 and IF_5 is type AX_5E.

EXERCISE 10.6A

Describe a hybridization scheme for the central atom and the molecular geometry of the silicon tetrachloride molecule.

EXERCISE 10.6B

Describe a hybridization scheme for the central atom and the molecular geometry of the triiodide ion, I_3^-.

10.5 Hybrid Orbitals and Multiple Covalent Bonds

We have described ways to predict the geometric structures of molecules and poly-atomic ions containing double and triple covalent bonds. We can gain additional insight into some essential characteristics of multiple covalent bonds, such as their bond energies, by combining this earlier knowledge with the valence bond theory.

In Chapter 9, we found the Lewis structure of ethene (ethylene), C_2H_4, to have a double bond between the two carbon atoms:

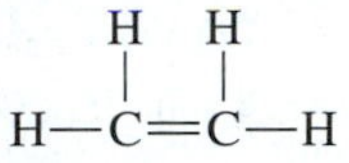

$$\begin{array}{cc} H & H \\ | & | \\ H-C & =C-H \end{array}$$

With the VSEPR method, we predict that the electron-group geometry around each C atom is trigonal planar. Thus, each CH_2 group lies in a plane with an H-C-H bond angle of 120°. Each H-C-C bond angle is also 120°. As we can see from Figure 10.20, however, the VSEPR method does not tell us how the two CH_2 groups are oriented with respect to each other. Are they both in the same plane, in planes perpendicular to each other, or at some other angle? We can show that the valence bond theory accounts for both the 120° bond angles *and* the orientation of the CH_2 groups. Let's start with orbital diagrams for sp^2 hybridization of the two C atoms:

sp^2 *hybridization* in first C: [He] ↑ | ↑ | ↑ | ↑ 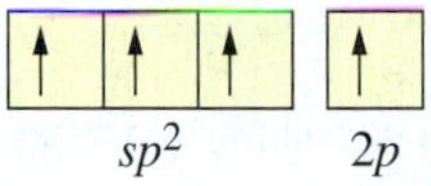
                                        sp^2 $2p$

sp^2 *hybridization* in second C: [He] ↑ | ↑ | ↑ | ↑
                                          sp^2 $2p$

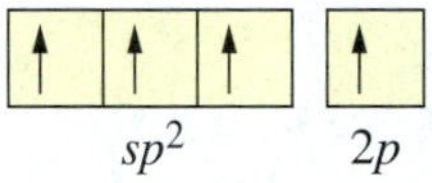

Figure 10.21 shows that all the C—H bonds in C_2H_4 are formed by the overlap of sp^2 hybrid orbitals of the C atoms with $1s$ atomic orbitals of the H atoms. The region of maximum overlap occurs along hypothetical lines drawn between the nuclei of the bonded atoms. We say that the orbitals overlap end to end. Covalent bonds formed by the end-to-end overlap of orbitals, regardless of orbital type, are called **sigma (σ) bonds.** All single covalent bonds are σ bonds.

The double bond between the two C atoms in C_2H_4, in contrast, has two compo-nents: One of the bonds involves sp^2 hybrid orbitals overlapping end to end along the line joining the two carbon nuclei; like the C—H bonds, this bond is a σ bond. The other bond between the two C atoms results from the overlap of the half-filled, unhy-bridized $2p$ orbitals that extend above and below the plane of the C and H atoms. These orbitals overlap in a parallel, or side-by-side, fashion, not along a line joining the carbon nuclei. A bond formed by this parallel orbital overlap is called a **pi (π) bond.** A double covalent bond consists of one σ bond and one π bond. Valence bond

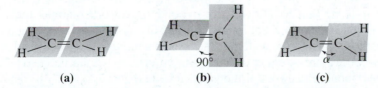

(a)   **(b)**   **(c)**

▲ **FIGURE 10.20   Lewis structure and VSEPR description of C_2H_4**
Lewis theory predicts a double bond between the C atoms in ethylene. The VSEPR method predicts that the three atoms in a CH_2 group lie in the same plane, but it does not indicate how the two planes are oriented with respect to each other. That is, we do not know whether the two CH_2 groups are (a) copla-nar, (b) perpendicular to each other, or (c) at some angle $α$ to each other.

Ethene 3D model

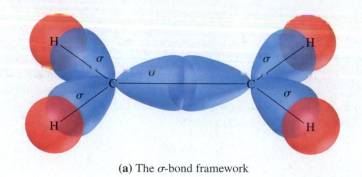

(a) The σ-bond framework

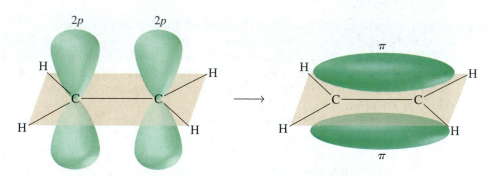

(b) The formation of a π-bond by the overlap of the half-filled 2p orbitals

▶ **FIGURE 10.21** **Bonding in ethylene, C_2H_4, by the valence bond theory**

(a) The overlap of the C sp^2 hybrid orbitals between the carbons and with H $1s$ atomic orbitals generates the σ-bond framework. (b) The two separate atomic $2p$ orbitals shown on the left come together to form the π bond represented by the two bullet-shaped orbitals on the right. Note that even though the π bond has two lobes, it is only one bond.

QUESTION: If one hydrogen atom were replaced by a fluorine atom, what would be the overlap designation for that new bond?

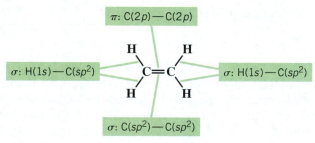

(c) Hybridization and bonding scheme

Multiple Bond Formation activity

theory predicts that all six atoms in C_2H_4 lie in the same plane, as shown by the tan shading in parts (b) and (c) of Figure 10.21. This is the arrangement that allows for the maximum overlap of the $2p$ atomic orbitals and consequently the strongest π bond. If one CH_2 group were twisted out of the plane of the other, the extent of p-orbital overlap would be lessened, the π bond would be weakened, and the molecule would be less stable. The point of minimum overlap would occur when the planes of the two CH_2 groups were mutually perpendicular.

The positions of terminal atoms with respect to a central atom are determined by end-to-end orbital overlaps, the σ bonds. Collectively, the σ bonds in a structure constitute the *σ-bond framework*, and the σ-bond framework outlines the molecular geometry. Electron pairs in π bonds do not affect the positions of bonded atoms. In the VSEPR method, when we treat all the electrons in a multiple covalent bond as a single electron group, we are in effect constructing an electron-group geometry that is the same as the σ-bond framework of the valence bond theory.

We can describe a triple covalent bond in much the same way as a double bond. Consider the ethyne (acetylene) molecule, C_2H_2, with the Lewis structure $H—C≡C—H$. The molecule is linear, with 180° $H—C—C$ bond angles, as predicted by the VSEPR method and confirmed by experiment. To account for these bond

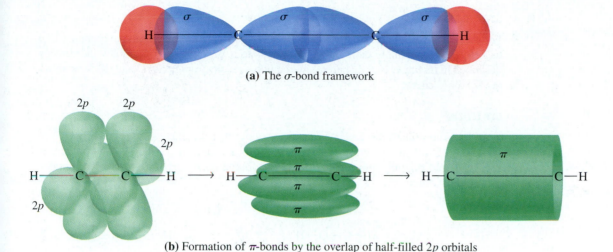

(a) The σ-bond framework

H—C—C—H ⟶ H—C—C—H ⟶ H—C—C—H

(b) Formation of π-bonds by the overlap of half-filled 2*p* orbitals

σ: C(*sp*)—C(*sp*)   π: C(2*p*)—C(2*p*)

σ: C(*sp*)—H(1*s*)

H—C≡C—H

σ: H(1*s*)—C(*sp*)

π: C(2*p*)—C(2*p*)

(c) Hybridization and bonding scheme

▲ **FIGURE 10.22   Bonding in acetylene, C₂H₂, by the valence bond theory**

Ethyne 3D model

(a) The σ-bond framework joins the atoms in a linear structure through the overlap of 1*s* atomic orbitals of the H atoms and *sp* hybrid orbitals of the C atoms. (b) Each π bond can be thought of as two parallel cigar-shaped segments. In fact, however, when two π bonds are present, the segments merge into a hollow and symmetric cylindrical shell with the carbon-to-carbon σ bond as its axis. (c) Summary of hybridization and the bonding scheme.

angles with the valence bond theory, we assume *sp* hybridization of the valence-shell orbitals of the two C atoms:

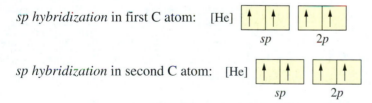

In the C≡C bond in C_2H_2, as in all other triple bonds, one bond is a σ bond and the other two are π bonds. The bonding scheme in acetylene is illustrated in Figure 10.22.

Example 10.7

Formic acid, HCOOH, is the simplest carboxylic acid.

  (a) Predict a plausible molecular geometry for this molecule.

  (b) Propose a hybridization scheme for the central atoms that is consistent with that geometry.

  (c) Sketch a bonding scheme for the molecule.

STRATEGY

For part (a), we follow the general strategy outlined on page 393, as applied in Example 10.3 to a molecule containing more than one central atom. For part (b), we employ the strategy described in Example 10.6. In part (c), we combine results from parts (a) and (b) to establish the bonding scheme.

SOLUTION

(a) To write a plausible Lewis structure, we note that there are $(2 \times 1) + (2 \times 6) + 4 = 18$ valence electrons. All 18 electrons are needed in the skeletal structure, and, additionally, one carbon-to-oxygen double bond is required for an acceptable Lewis structure:

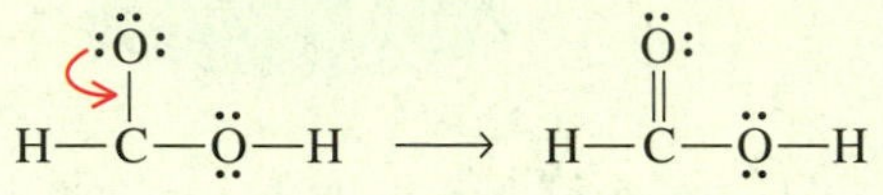

The orientation of the three electron groups about the C atom—two single bonds and a double bond—is trigonal planar. The orientation of the four electron groups about the central O atom—two BP and two LP—is tetrahedral.

The molecular geometry around the C atom is the same as the electron-group geometry (trigonal planar). We expect the H-C-O and the O-C-O bond angles to be about 120°. The molecular geometry around the central O atom is angular, and we therefore expect a C-O-H bond angle of about 109.5° (the tetrahedral bond angle for the VSEPR notation AX_2E_2).

(b) The hybridization schemes for the electron-group geometries found in part (a) are sp^2 for the central C atom and sp^3 for the central O atom.

(c) One of the bonds in the C=O double bond is a π bond. All the other bonds in the molecule are σ bonds. The bonding scheme for these bonds and the orbital overlaps producing them is

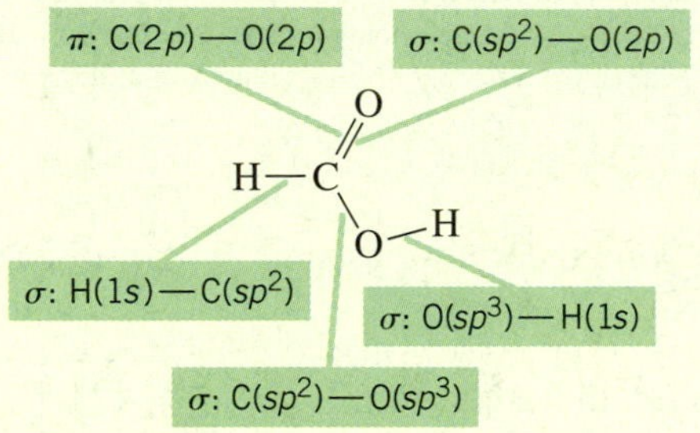

EXERCISE 10.7A

Methanol, CH_3OH, is the simplest alcohol.

(a) Predict a plausible molecular geometry for this molecule.

(b) Propose a hybridization scheme for the central atoms that is consistent with that geometry.

(c) Sketch a bonding scheme for the molecule.

EXERCISE 10.7B

Cyanogen, C_2N_2, is a highly toxic gas used as a fumigant and in synthesizing organic compounds.

(a) Predict a plausible molecular geometry for this molecule.

(b) Propose a hybridization scheme for the central atoms that is consistent with the predicted geometry.

(c) Sketch a bonding scheme for the molecule.

Geometric Isomerism

In discussing Figure 10.20, we concluded that the two CH_2 groups in ethylene, $CH_2=CH_2$, must lie in the same plane to produce maximum overlap of the $2p$ orbitals in the π bond between the carbon atoms. Also, to maintain maximum orbital overlap in the π bond, one CH_2 group is prevented from rotating (spinning) with respect to the other. As we will now see, this restricted rotation about a double bond has an important consequence.

There are four C_4H_8 alkene isomers. One of these isomers is 1-butene (which is related to butane), and a second is isobutylene (related to isobutane):

$$CH_2=CHCH_2CH_3 \qquad\qquad CH_2=\overset{\displaystyle \overset{CH_3}{|}}{C}-CH_3$$

1-Butene   Isobutylene

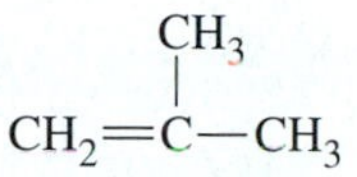

A third isomer readily comes to mind, also related to butane, in which the double bond connects the second and third carbon atoms:

$$CH_3CH=CHCH_3$$

2-Butene

2-Butene 3D model

In Latin, *cis* means on this side and *trans* means across. With these prefixes, it is not difficult to figure out the meanings of the adjectives cisatlantic and transatlantic.

But what is the fourth structure? Although it may not be immediately apparent, there are two isomers of 2-butene (Figure 10.23). If we could hold one end of one of the structures in Figure 10.23 fixed and rotate the other end by 180° about the double bond, we would obtain the other structure shown in Figure 10.23, and there would be no isomerism. However, this rotation would decrease the overlap of $2p$ orbitals on the C atoms and effectively break the π bond; this conversion does not occur. The two isomers of 2-butene are *distinctly different* compounds. To distinguish between them, we call one *cis*-2-butene and the other *trans*-2-butene:

H   H   H   CH₃
  C=C   C=C
CH₃ CH₃   CH₃   H

cis-2-Butene   *trans*-2-Butene

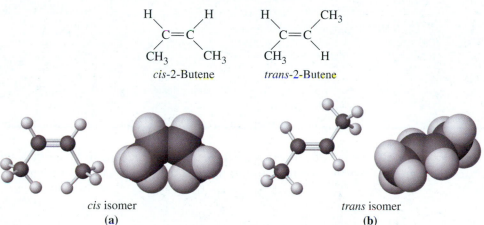

cis isomer   *trans* isomer
(a)   (b)

◄ **FIGURE 10.23   Geometric (cis–trans) isomerism in 2-butene**

The **cis** isomer has both methyl groups (CH_3) on the same side of the molecule. In the **trans** isomer, the methyl groups are on opposite sides, or across the double bond, from each other. (To figure out what represents the two "sides" of the molecule, draw a straight line that passes through the double-bond carbon atoms. If the methyl groups fall on the same side of the line, the compound is *cis*. If they fall on opposite sides, the compound is *trans*.) Because cis and trans isomers differ only in the geometric arrangement of certain substituent groups, they are called **geometric isomers.**

If either carbon atom in a double bond has two identical atoms or groups bonded to it, cis–trans isomerism is not possible. Propene (propylene) has two hydrogen atoms bonded to one of the carbon atoms in the double bond. As a result, there are no cis–trans isomers of propylene. Although we can draw two structural formulas,

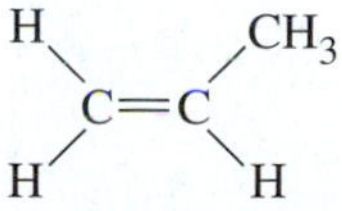

H   CH₃   H   H
 C=C   C=C
H   H   H   CH₃

(I) 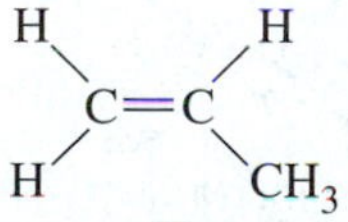   (II)

Geometric Isomerism and Vision

The sense of vision requires that light convert one geometric isomer to another. The molecule that undergoes conversion is a light-sensitive pigment known as rhodopsin, which is found in the receptor cells of the retina of the eye. Rhodopsin is a complex of a compound called 11-*cis*-retinal and a protein called opsin. When light strikes rhodopsin, a reaction called *isomerization* occurs, converting the cis isomer to the trans isomer:

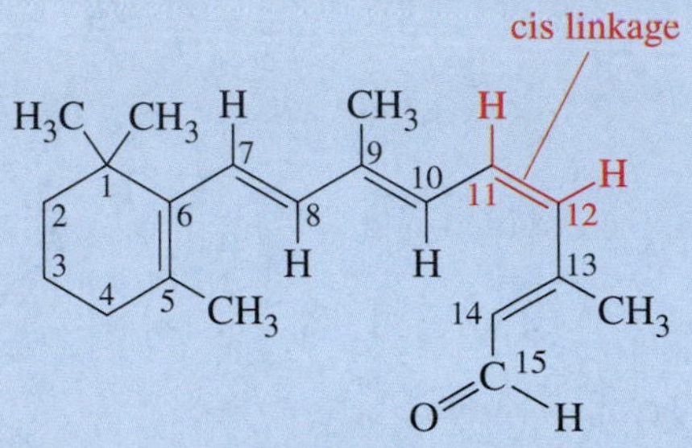

11-*cis*-Retinal

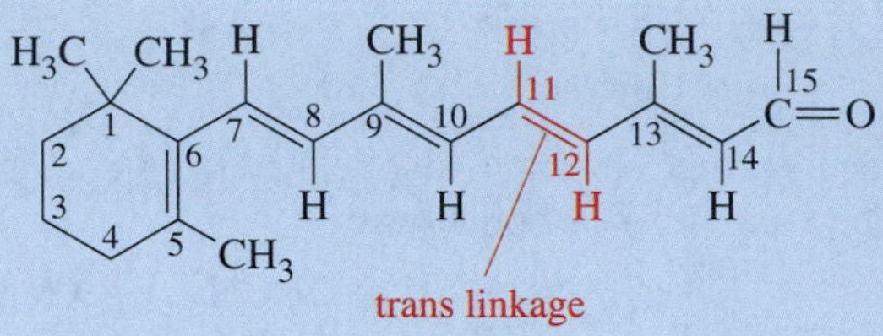

11-*trans*-Retinal

Cis to trans isomerization of 11-*cis*-retinal initiates an electrical impulse that is transmitted to the brain through the optic nerve. The brain translates the nerve impulse into an image, and we see. Following transmission of the impulse, the opsin protein releases the 11-*trans*-retinal, and another enzyme converts 11-*trans*-retinal back to 11-*cis*-retinal. The 11-*cis*-retinal again complexes with opsin and is primed for the next pulse of light.

Some retinal is lost during the regeneration of opsin from rhodopsin and must be replaced by vitamin A from the bloodstream, making vitamin A an essential vitamin for proper vision. Vitamin A differs from 11-*trans*-retinal only in that the vitamin A molecule has a terminal alcohol group, CH_2OH, and *trans*-retinal has an aldehyde group, CHO:

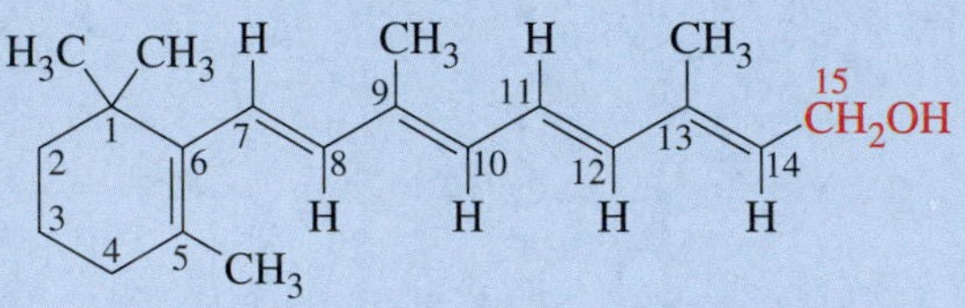

Vitamin A

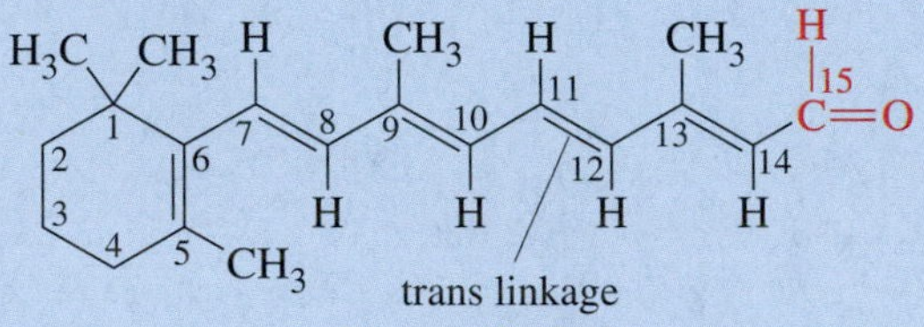

11-*trans*-Retinal

structure II is really the same as structure I. To see this, imagine picking structure II up from the page and flipping it top to bottom to see that the two structures are identical:

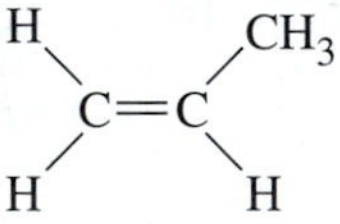

(I)

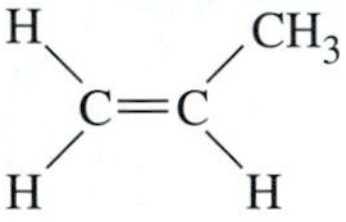

(II: flipped over)

Example 10.8   A Conceptual Example

Is it possible to write a unique structural formula for 1,2-dichloroethene if we are told that the molecule is nonpolar?

ANALYSIS AND CONCLUSIONS

From the name 1,2-dichloroethene alone, we cannot write a unique structural formula because two isomers have this name:

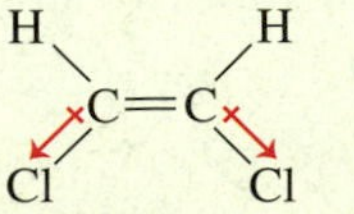

(I)   (II)

cis-1,2-Dichloroethene   *trans*-1,2-Dichloroethene

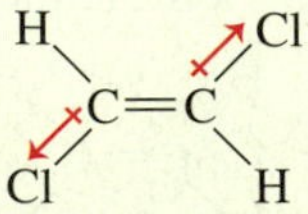

In structure I, a cis isomer, both Cl atoms are on the same side of the double bond, giving rise to a resultant dipole moment and therefore a polar molecule. In structure II, a trans isomer, the bond dipole moments associated with the C—Cl bonds cancel each other, as do those associated with the C—H bonds, which means this isomer is nonpolar. Thus, the nonpolar 1,2-dichloroethene is *trans*-1,2-dichloroethene.

EXERCISE 10.8A

With reference to Example 10.8, there is a third isomeric dichloroethene. Write its structural formula and predict whether it is polar or nonpolar.

EXERCISE 10.8B

Which of the following molecules would you expect to be polar: **(a)** fluoroethene; **(b)** *trans*-2-butene; **(c)** acetylene; **(d)** 2,3-dichloro-*cis*-2-butene? Explain.

Molecular Orbital Theory

In the valence bond theory, some of the features of isolated atoms are retained when the atoms are bonded together. Only the valence orbitals involved in the bonding are modified. In an alternative quantum mechanical approach, we start from scratch. In **molecular orbital theory,** we use an arrangement of appropriately placed atomic nuclei and place electrons in *molecular orbitals* in a way that leads to an energetically favorable, stable molecule.

10.6 Characteristics of Molecular Orbitals

Molecular orbitals (MOs) are mathematical equations that describe the regions in a molecule where there is a high probability of finding electrons. In this regard, molecular orbitals are like the atomic orbitals used to describe the electrons in atoms. In fact, one way to derive molecular orbitals is by an appropriate combination of the atomic orbitals of the atoms being united into a molecule.

Figure 10.24 describes the formation of molecular orbitals by the combination of two $1s$ atomic orbitals. One combination leads to a **bonding molecular orbital,** σ_{1s},

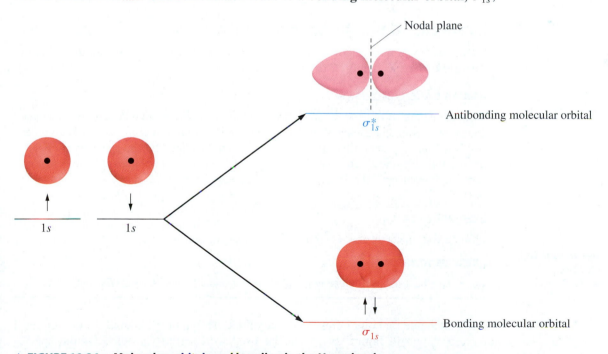

▲ **FIGURE 10.24 Molecular orbitals and bonding in the H$_2$ molecule**

The short horizontal lines represent the relative energy levels of the atomic and molecular orbitals. The small arrows stand for electrons in the hydrogen atoms and in the hydrogen molecule. The red (occupied) and pink (unoccupied) shapes above the energy-level lines denote the electron charge density of the molecular orbitals. Note that, if occupied, the electron charge density in the antibonding molecular orbital would fall to zero in the nodal plane between the atomic nuclei (black dots).

Two atomic orbitals, when combined, form two molecular orbitals. In general, combining n atomic orbitals gives n molecular orbitals.

which is at a lower energy level than the separate atomic orbitals. It corresponds to a high electron probability, or electron charge density, *between* the bonded atoms. The other combination leads to an **antibonding molecular orbital**, σ_{1s}^*, which is at a higher energy level than the separate atomic orbitals. It corresponds to a high electron probability *away from* the region between the bonded atoms. We use an asterisk (*) to designate an antibonding orbital.

In Figure 10.24, electrons are represented by arrows. The electrons in the separated H atoms are unpaired, but when the atoms join to form a molecule, the electrons pair up and occupy the lowest available energy level, the σ_{1s} bonding molecular orbital. The higher energy σ_{1s}^* antibonding orbital is unoccupied. This assignment conforms to an aufbau process, similar to that used in writing electron configurations of atoms and subject to these rules:

- Electrons seek the lowest-energy molecular orbitals available to them.
- A maximum of two electrons can be present in a molecular orbital (Pauli exclusion principle).
- Electrons enter molecular orbitals of identical energies singly with parallel spins before they pair up (Hund's rule).

As discussed in Chapter 9, a single bond has a bond order of 1; a double bond, a bond order of 2; and a triple bond, a bond order of 3.

Electrons in bonding molecular orbitals contribute to the strength of a bond between atoms. Any electrons found in antibonding molecular orbitals detract from the strength of the bond. (In some cases, there is a third type of molecular orbital, a *nonbonding* molecular orbital, which neither contributes to nor detracts from bonding.) The **bond order** in a molecule is one-half the difference between the number of electrons in bonding molecular orbitals and the number in antibonding molecular orbitals.

$$\text{Bond order} = \frac{\left(\begin{array}{c}\text{number of } e^- \text{ in} \\ \text{bonding MOs}\end{array}\right) - \left(\begin{array}{c}\text{number of } e^- \text{ in} \\ \text{antibonding MOs}\end{array}\right)}{2} \quad (10.2)$$

For the H_2 molecule, the bond order is $(2 - 0)/2 = 1$, signifying the same single bond that we represent in the Lewis structure.

Example 10.9 A Conceptual Example

Molecular orbital theory allows for species with a one-electron bond. What does the term "one-electron bond" signify? Cite an example of such a bond.

ANALYSIS AND CONCLUSIONS

From the definition of bond order, we can think of a one-electron bond as a bond that has one more electron in bonding MOs than in antibonding MOs, in which case the bond order is $\frac{1}{2}$. The simplest example has one electron in the σ_{1s} bonding MO and an empty antibonding σ_{1s}^* MO. It is the hydrogen molecule ion, H_2^+, formed by combining a hydrogen atom and a hydrogen ion. The energy-level diagram of Figure 10.25 suggests the formation of H_2^+.

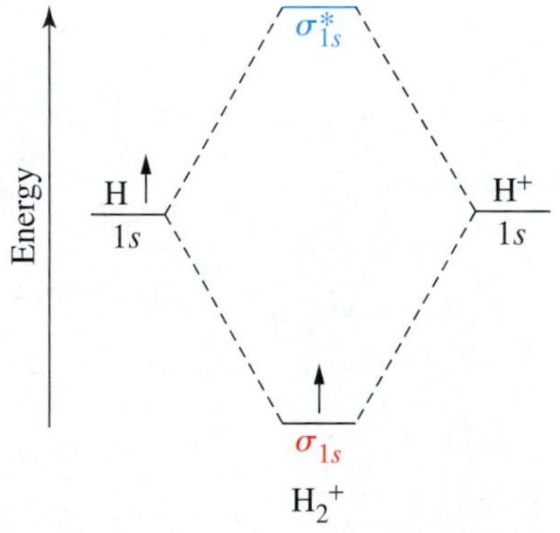

EXERCISE 10.9A

Would you expect the ion H_2^- to exist as a stable species? Explain.

EXERCISE 10.9B

Can there be a stable species involving He atoms and a He—He bond? Explain.

▲ **FIGURE 10.25** **Molecular orbital energy-level diagram for the hydrogen molecule ion, H_2^+**

The scheme for combining $1s$ atomic orbitals in Figure 10.24 and the energy-level diagram shown in Figure 10.25 can be used to describe various diatomic species involving the first-period elements, H and He. We next consider molecular orbitals associated with the second-period elements.

10.7   Homonuclear Diatomic Molecules of the Second-Period Elements

We can treat the diatomic species of the first two members of the second period, Li and Be, much as we did the first-period elements. First, we fill the σ_{1s} and σ_{1s}^* molecular orbitals with electrons, and then we concentrate on the molecular orbitals formed from the valence-shell atomic orbitals, the $2s$. We obtain another pair of molecular orbitals, σ_{2s} and σ_{2s}^*, at a higher energy than the first-shell molecular orbitals.

To continue with the second period, proceeding from boron to neon, we need to consider molecular orbitals formed by combining $2p$ atomic orbitals. These molecular orbitals are pictured in Figure 10.26. They share two features with $1s$ orbitals we considered in Figure 10.24:

- For every two atomic orbitals that are combined, two molecular orbitals result. A total of six molecular orbitals are formed from the six $2p$ atomic orbitals.

- Of each pair of molecular orbitals, one is a bonding molecular orbital at a lower energy than the separate atomic orbitals, and the other is an antibonding orbital at a higher energy.

However, a new feature is that we obtain two types of molecular orbitals from $2p$ atomic orbitals. One type of $2p$ molecular orbitals, designated σ_{2p} and σ_{2p}^*, results from what amounts to end-to-end overlap of the atomic orbitals (top row of Figure 10.26). The other type results from the side-by-side overlap of the atomic orbitals, leading to two degenerate (equal-energy) π_{2p} bonding molecular orbitals and two degenerate π_{2p}^* antibonding molecular orbitals.

We have already noted that the σ_{2s} and σ_{2s}^* molecular orbitals are at higher energies than the underlying σ_{1s} and σ_{1s}^* orbitals. Figure 10.27 suggests the energy levels of all the valence-shell molecular orbitals of the homonuclear diatomic molecules of the second period—Li_2, Be_2, B_2, and so on. For each molecule, the energy levels are

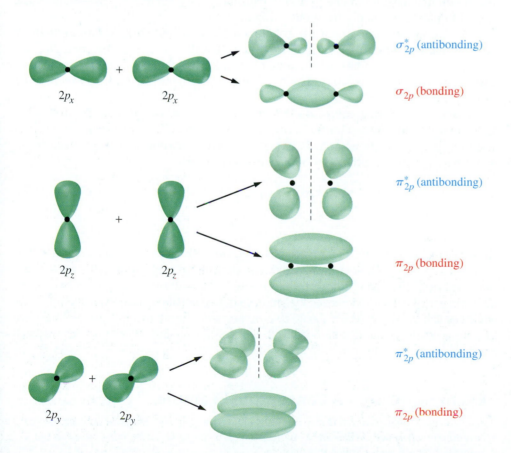

| $2p_x$ | $+$ | $2p_x$ | | σ_{2p}^* (antibonding) |
| | | | | σ_{2p} (bonding) |

| $2p_z$ | $+$ | $2p_z$ | | π_{2p}^* (antibonding) |
| | | | | π_{2p} (bonding) |

| $2p_y$ | $+$ | $2p_y$ | | π_{2p}^* (antibonding) |
| | | | | π_{2p} (bonding) |

◀ **FIGURE 10.26   Molecular orbitals formed by combining 2p atomic orbitals**

These figures suggest the distribution of electron charge density for the six molecular orbitals formed by the $2p$ atomic orbitals. The broken lines represent regions of zero electron density (nodal planes) in the antibonding orbitals.

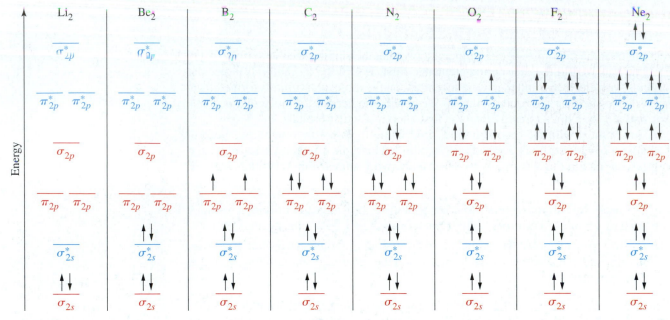

▲ **FIGURE 10.27** **Relative energy levels of molecular orbitals obtained from 2*s* and 2*p* atomic orbitals for some actual and hypothetical diatomic molecules**

The σ_{1s} and σ_{1s}^{*} molecular orbitals are filled, and the second-shell orbitals are occupied as shown here. The crossover in the energy levels of the σ_{2p} and π_{2p} bonding molecular orbitals that occurs between N_2 and O_2 is predicted by molecular orbital theory and has been confirmed by experiment.

QUESTION: Which are the hypothetical molecules and what are their bond orders?

Filling Molecular Orbitals animation

Molecular Orbital animation

arranged according to increasing energy, but the drawing is not to scale. For example, the continuous decline in energy levels with increasing nuclear charge of the atoms (recall Figure 8.1) is not brought out in Figure 10.27.

We can draw some important conclusions from Figure 10.27. One is that we can use the definition of bond order to show that the molecules Be_2 and Ne_2 are unlikely to exist. Each has the same number of electrons in antibonding molecular orbitals as in bonding orbitals. The bond orders are therefore zero, and there is no bond at all.

Another interesting conclusion has to do with the O_2 molecule. The following molecular orbital diagram for the second shell is arranged in a manner similar to that used for electron configurations in atoms. We have to assign 12 valence electrons to this diagram, following the rules of the aufbau principle:

O_2 ⊞ ⊞ ⊞ ⊞ ⊞ ⊞ ☐

       σ_{2s} σ_{2s}^{*} σ_{2p} π_{2p} π_{2p}^{*} σ_{2p}^{*}

To assess the bond order in O_2, we note that there are eight electrons in bonding MOs (two in σ_{2s}, two in σ_{2p}, and four in π_{2p}) and four electrons in antibonding MOs (two in σ_{2s}^{*} and two in π_{2p}^{*}). The bond order is therefore $\frac{1}{2}(8 - 4) = 2$, and so we predict an oxygen-to-oxygen double bond. Moreover, the presence of two unpaired electrons in the π_{2p}^{*} MOs indicates that the O_2 molecule should be *paramagnetic*. Recall that in Chapter 9 we were able to account for the double bond with the Lewis structure: $:\ddot{O}=\ddot{O}:$, but we were unable to account for the paramagnetism.

Example 10.10   A Conceptual Example

When an electron is removed from a N_2 molecule, forming an N_2^{+} ion, the bond between the N atoms is weakened. When an O_2 molecule is ionized to O_2^{+}, the bond between the O atoms is strengthened. Explain this difference.

ANALYSIS AND CONCLUSIONS

To explain the difference, we need to identify the electron lost in each case. We expect this to be an electron from the highest-energy occupied molecular orbital because electrons from such orbitals should be lost most easily. If we look at the molecular orbital diagrams in Figure 10.27, we see that for N_2, the ionized electron is from a bonding molecular orbital (σ_{2p}). The loss of this bonding electron *reduces* the bond order, and the bond becomes weaker. In contrast, with O_2, the ionized electron is from an antibonding orbital (one of the π_{2p}^* degenerate orbitals). The loss of this antibonding electron *increases* the bond order, and the bond becomes stronger.

EXERCISE 10.10A

Figure 10.27 shows that in O_2, F_2, and Ne_2 the σ_{2p} energy level is below the π_{2p} levels. Suppose that was also true for Li_2, Be_2, B_2, C_2, and N_2. How would this change affect the bond orders and magnetic properties of these five molecules?

EXERCISE 10.10B

Predict the bond order in nitrogen monoxide, NO.
[*Hint:* Which energy-level diagram(s) in Figure 10.27 will be useful in making this prediction?]

10.8   Bonding in Benzene

Benzene, C_6H_6, a liquid that smells somewhat like gasoline, was discovered by Michael Faraday in 1825. In 1865, F. A. Kekulé proposed a *cyclic* structure for the benzene molecule that features six carbon atoms in a hexagonal ring. He proposed that a hydrogen atom was attached to each carbon atom and that alternating single and double bonds joined the carbon atoms together:

These Kekulé structures are often abbreviated so that the C and H are not written but are understood to be present:

Kekulé structures of C_6H_6

(If atoms of other elements substitute for H atoms in C_6H_6, we do write their symbols.)

It was more than 100 years after the discovery of benzene that its true structure was recognized as a resonance hybrid to which the Kekulé structures are the most important contributors. The carbon-to-carbon bonds in the benzene molecule are all equivalent and intermediate in length and strength between single and double bonds. However, even with its partial double-bond character, the benzene molecule does not behave like unsaturated hydrocarbon molecules. For one thing, it undergoes a hydrogenation reaction (recall page 373) only under extreme conditions. The six electrons that we ordinarily would place between carbon atoms to make three double bonds are spread out over all six carbon atoms; in other words, they are *delocalized*. The modern symbol for benzene—a hexagon with an inscribed circle—incorporates this idea:

Resonance hybrid of C_6H_6

▲ Friedrich August Kekulé (1829–1896) claimed to have discovered the cyclic structure of benzene while dozing by the fire. (In some versions of the story, he was dozing on an omnibus.) He dreamed of atoms and molecules as snakes. Suddenly one of the twisting snakes seized its own tail, forming a ring. Kekulé's other contributions included the idea that carbon is tetravalent (1858) and an influential textbook in which he defined organic chemistry as the chemistry of carbon compounds.

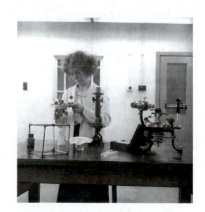

▲ In the 1920s, Kathleen Londsdale (1903–1971) used a technique called X-ray diffraction to prove that the benzene ring is planar and not puckered. She also determined the three-dimensional structure of diamond and graphite (Chapter 11).

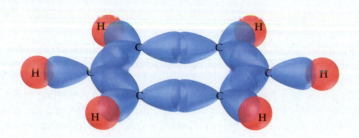

▶ **FIGURE 10.28** The σ-bond framework for benzene, C_6H_6

▶ **FIGURE 10.28** The σ-bond framework for benzene, C_6H_6 The carbon atoms in benzene use sp^2 hybrid orbitals to form σ bonds. Each carbon atom bonds to two other carbon atoms and to a hydrogen atom. All the atoms in the σ-bond framework are in the same plane.

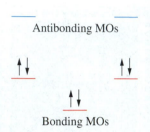

Antibonding MOs

Bonding MOs

▲ **FIGURE 10.29** Molecular orbital diagram for benzene, C_6H_6

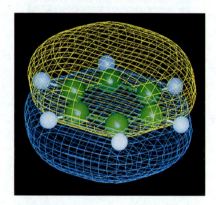

▲ **FIGURE 10.30** Computer-generated representation of bonding in benzene, C_6H_6

Bonding of Benzene activity

Many aromatic compounds, including benzene, are carcinogenic (cancer causing). Modern workplace regulations severely restrict exposure to many aromatic compounds.

Now let's use features of both the valence bond theory and the molecular orbital theory to describe the benzene molecule. To account for the planar shape of the molecule and the 120° bond angles, we presume that all six carbon atoms have sp^2 hybridization:

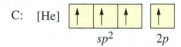

The set of sp^2 hybrid orbitals overlap to produce the σ-bond framework shown in Figure 10.28. The six half-filled $2p$ atomic orbitals combine to form six π molecular orbitals. As shown in Figure 10.29, three of these molecular orbitals are bonding and three are antibonding. All six $2p$ electrons are found in the bonding MOs. These six electrons produce a total of three π bonds, occupied by two electrons each. The three π bonds are delocalized over the six carbon atoms. The molecule also has six carbon-to-carbon σ bonds. Thus, the total number of bonds joining the six carbon atoms is nine, and the average bond order of the carbon-to-carbon bonds is $9/6 = 1.5$. If we average the two Kekulé structures, we see that each carbon-to-carbon bond is halfway between a single and a double bond, also a bond order of 1.5.

Figure 10.30 is a computer-generated model of the benzene molecule that emphasizes the delocalized electrons in π molecular orbitals.

10.9 Aromatic Compounds

Many of the first benzenelike compounds discovered had pleasant odors and hence acquired the name *aromatic*. In modern chemistry, the term **aromatic compound** simply refers to a substance that has a ring structure and bonding characteristics and properties related to those of benzene, which means all aromatic compounds have some delocalized bonding. Alkanes, alkenes, alkynes, and their cyclic analogues are called *aliphatic hydrocarbons*, and all organic compounds that are not aromatic are called **aliphatic compounds**.

Benzene is perhaps the best-known aromatic compound. More than 7 billion kg of benzene are produced in the United States each year, most of it from petroleum. Benzene is used as a solvent and as a starting material for the synthesis of many other organic compounds, including drugs, detergents, dyes, pesticides, and plastics. It is also used as a component in some gasoline.

Table 10.3 lists some important aromatic compounds related to benzene. Methylbenzene, $C_6H_5CH_3$, more commonly called *toluene*, is both aromatic (the ring) and aliphatic (the CH_3 group that substitutes for one of the H atoms of C_6H_6). More than 3 billion kg of toluene are produced in the United States each year.

Halogen atoms may also be substituted for one or more of the hydrogen atoms on a benzene ring. These compounds, known as *halogenated benzenes*, are important intermediates for the synthesis of many other aromatic compounds. We name them by using prefixes to indicate the halogen, as with *bromo*benzene, C_6H_5Br, which is listed in Table 10.3. No number is needed to indicate the position of the substituent in *mono*substituted benzenes because the hexagonal ring is symmetrical; all the positions are equivalent.

Ethylbenzene, $C_6H_5CH_2CH_3$, is another interesting aromatic compound. About 6 billion kg of ethylbenzene are also produced worldwide annually, nearly all of which is converted to styrene, $C_6H_5CH=CH_2$, a monomer used to produce the common

TABLE 10.3   Some Representative Aromatic Compounds

Name	Structure	Typical Use(s)
Aniline	⟨benzene⟩—NH$_2$	Starting material for the synthesis of dyes, drugs, resins, varnishes, perfumes; solvent; vulcanizing rubber
Benzoic acid	⟨benzene⟩—COOH	Food preservative; starting material for the synthesis of dyes and other organic compounds; curing of tobacco
Bromobenzene	⟨benzene⟩—Br	Starting material for the synthesis of many other aromatic compounds; solvent; motor oil additive
Nitrobenzene	⟨benzene⟩—NO$_2$	Starting material for the synthesis of aniline; solvent for cellulose nitrate; in soaps and shoe polish
Phenol	⟨benzene⟩—OH	Disinfectant; starting material for the synthesis of resins, drugs, and other organic compounds
Toluene	⟨benzene⟩—CH$_3$	Solvent; gasoline octane booster; starting material for the synthesis of benzoic acid, benzaldehyde, and many other organic compounds

plastic polystyrene. You can think of this polymerization as proceeding just like that of ethylene to polyethylene (Section 9.12). Simply replace one of the H atoms of ethylene, $CH_2=CH_2$, with the *phenyl* group, C_6H_5.

We saw earlier that no number is needed in the name when there is only one substituent on a benzene ring. When there are two or more substituents, however, positions are no longer equivalent, and we must designate relative positions. There are three possible *di*substituted benzenes, and we can use numbers to distinguish them. For example, the dichlorobenzenes are

1,2-Dichlorobenzene   1,3-Dichlorobenzene   1,4-Dichlorobenzene
(*o*-dichlorobenzene)   (*m*-dichlorobenzene)   (*p*-dichlorobenzene)

Common names are also used: *ortho-* (*o-*) designates a 1,2 substitution, *meta-* (*m-*) indicates a 1,3 substitution, and *para-* (*p-*) indicates a 1,4 arrangement. *p*-Dichlorobenzene is used as an insecticide, especially as a moth repellent, and as a bathroom deodorant.

With three or more substituents, numbers are almost always used to indicate the location of substituents:

1,2,4-Trichlorobenzene

Application Note

Styrene is a component of the polyester resin used to repair automobile bodies and build fiberglass-reinforced boats. It is responsible for the strong odor of that resin.

In the compounds toluene, aniline, and phenol, the groups CH_3, NH_2, and OH, respectively, substitute for a H atom in C_6H_6 (Table 10.3). The C atom to which any of these three substituents is attached is numbered carbon 1 on the benzene ring. Then substances derived from toluene, aniline, and phenol are named as illustrated here:

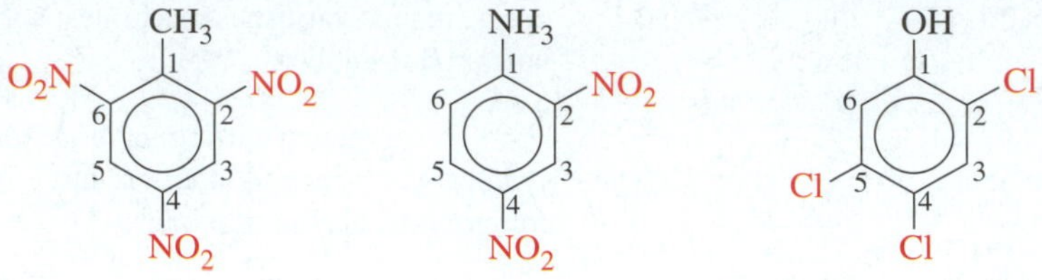

2,4,6-Trinitrotoluene   2,4-Dinitroaniline   2,4,5-Trichlorophenol
(TNT, an explosive)   (used to make dyes)   (a fungicide)

There are many organic compounds that have atoms other than carbon in an aromatic ring. These *heterocyclic* compounds, discussed in Chapter 23, include amines related to purine and pyrimidine that are important constituents of the nucleic acids DNA and RNA.

Cumulative Example

Methyl isocyanate (MIC), used in the manufacture of pesticides and polymers, is a carbon–hydrogen–oxygen–nitrogen compound with a molecular mass of 57.05 u; it is 5.29% H by mass. The nitrogen in a 0.7500-g sample of the compound is converted to $NH_3(g)$, which is neutralized by passing it into 50.00 mL of 0.2800 M $H_2SO_4(aq)$. After neutralization of the $NH_3(g)$, the excess $H_2SO_4(aq)$ requires 36.49 mL of 0.4070 M NaOH(aq) for complete neutralization. Indicate hybridization for the central atoms in the methyl isocyanate molecule, and draw a sketch of the molecule with appropriate bond angles.

STRATEGY

There are four essential tasks in this problem: (1) Determine the molecular formula of methyl isocyanate from the analytical data. (2) Write the most plausible Lewis structure of the molecule. (3) Propose a hybridization scheme that corresponds to the Lewis structure. (4) Make a sketch of the molecule based on the bond angles determined using the VSEPR method.

SOLUTION

We find the number of moles of H in 1 mol of MIC from the percent H and the molar mass of H.

$$\frac{? \text{ mol H}}{1 \text{ mol MIC}} = \frac{5.29 \text{ g H}}{100 \text{ g MIC}} \times \frac{57.05 \text{ g MIC}}{1 \text{ mol MIC}} \times \frac{1 \text{ mol H}}{1.00794 \text{ g H}}$$

$$= \frac{2.994 \text{ mol H}}{1 \text{ mol MIC}} \approx \frac{3 \text{ mol H}}{1 \text{ mol MIC}}$$

We need to find the number of moles of N per mol MIC but can do this only after we have used the titration data to determine the amount of N present. The first piece of data is the total amount of H_2SO_4 in the 50.00 mL used—the product of the molarity and volume of the $H_2SO_4(aq)$.

$$? \text{ mol } H_2SO_4(\text{total}) = 50.00 \text{ mL } H_2SO_4 \times \frac{0.2800 \text{ mol } H_2SO_4}{1000 \text{ mL } H_2SO_4}$$

$$= 0.01400 \text{ mol } H_2SO_4$$

The amount of excess H_2SO_4 remaining after neutralization of the $NH_3(g)$, which is the amount consumed in the titration with NaOH(aq), is calculated next.

$$? \text{ mol } H_2SO_4(\text{rxn with NaOH}) = 36.49 \text{ mL NaOH(aq)} \times \frac{0.4070 \text{ mol NaOH}}{1000 \text{ mL NaOH}} \times \frac{1 \text{ mol } H_2SO_4}{2 \text{ mol NaOH}}$$

$$= 0.007426 \text{ mol } H_2SO_4$$

The amount of H_2SO_4 that reacted with the $NH_3(g)$ is the difference in the total amount of H_2SO_4 in the original 50.00 mL and the amount consumed in the titration with NaOH(aq).

$$0.01400 \text{ mol } H_2SO_4 - 0.007426 \text{ mol } H_2SO_4 = 0.006574 \text{ mol } H_2SO_4 \text{ reacts w/NH}_3$$

Now, with suitable conversion factors, we calculate the number of moles of N in 1 mol of MIC.

$$\frac{0.006574 \text{ mol } H_2SO_4}{0.7500 \text{ g MIC}} \times \frac{2 \text{ mol } NH_3}{1 \text{ mol } H_2SO_4} \times \frac{1 \text{ mol } N}{1 \text{ mol } NH_3} \times \frac{57.05 \text{ g MIC}}{1 \text{ mol MIC}} = \frac{1.000 \text{ mol } N}{1 \text{ mol MIC}}$$

At this point, we have established that 1 mol of MIC contains 3 mol of H and 1 mol of N. The remainder of the molar mass can be found by difference.

$$57.05 \text{ g MIC} - 14.0067 \text{ g N} - 3(1.00794 \text{ g H}) = 40.02 \text{ g remaining}$$

The only combination of carbon and oxygen that gives 40 g/mol is 2 mol of C and 1 mol of O. This gives the molecular formula.

C_2H_3NO

The name "methyl isocyanate" indicates the presence of a CH_3 group. The isocyanate residue must contain one each of N, C, and O atoms. From this information, we write a suitable Lewis structure; assign hybridization schemes to the nitrogen and each carbon atom; and determine the bond angles from the electron-group geometries. These angles allow us to sketch the molecule. (The C-N-C bond angle might actually be slightly less than 120° because of LP–BP repulsion.)

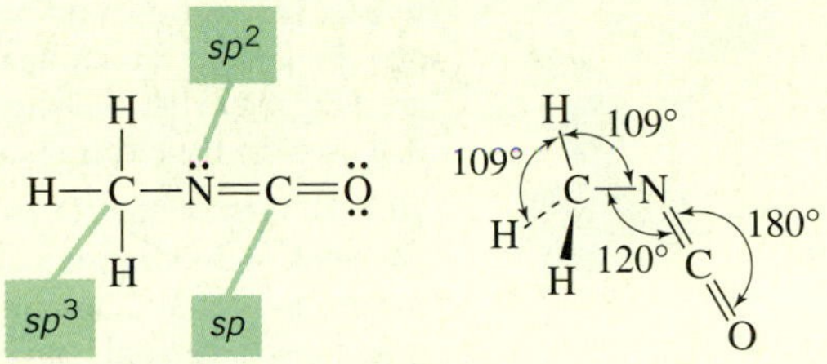

ASSESSMENT

If the analytical data are used carefully, especially the titration data, there is little chance of arriving at the wrong molecular formula. A distinct possibility for error is the selection of a less plausible Lewis structure, such as the ones shown below. Note the presence of non-zero formal charges in these alternatives. The structure used in the Solution has formal charges of 0 (and it corresponds to the experimentally determined structure). Also, the alternate structures have different (incorrect) bond angles.

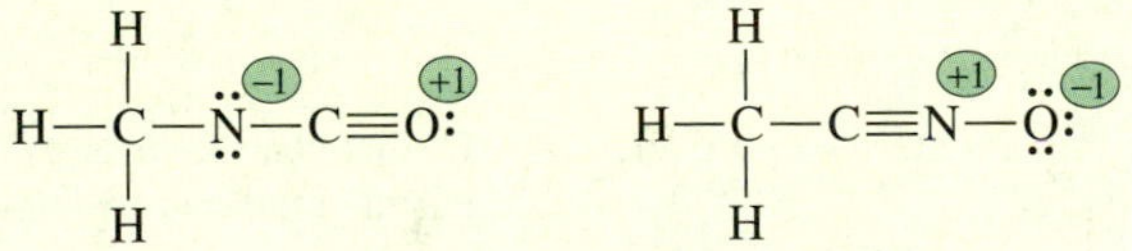

Concept Review with Key Terms

The **molecular geometry**, or shape, of molecules is pictured by imagining the three-dimensional arrangement of atoms in space, with straight lines connecting the nuclei. Both bonding electrons and lone pairs of electrons influence molecular geometry.

10.1 Valence-Shell Electron-Pair Repulsion (VSEPR) Method—The **VSEPR** method is used to predict the shapes of molecules and polyatomic ions based on the mutual repulsions among valence-shell **electron groups**. An electron group is a single unpaired electron, an unshared pair of electrons, or the electrons in a single, double, or triple bond. The number of electron groups is found for each central atom in the molecule or polyatomic ion, and the geometric distribution of these electron groups is assessed. If all electron groups are bonding groups, the molecular geometry is the same as the **electron-group geometry**. If some of the electron groups are lone pairs, the molecular geometry is derived from (but is different from) the electron-group geometry.

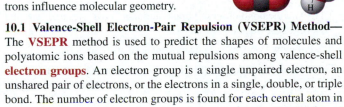

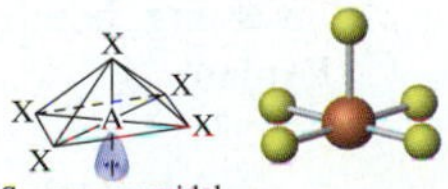

Square pyramidal

10.2 Polar Molecules and Dipole Moments—A polar covalent bond has separate centers of partial positive and partial negative charge, creating a bond dipole. Both bond dipoles and molecular geometry determine whether a molecule as a whole is polar. A **polar molecule** is one having a nonzero **dipole moment** (μ). A polar molecule has bond dipoles that do not cancel. A symmetrical distribution of identical bond dipoles about a central atom leads to cancellation of all bond dipoles, resulting in a nonpolar molecule.

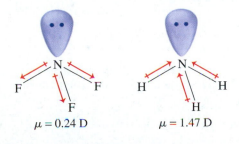

$\mu = 0.24 \text{ D}$ $\mu = 1.47 \text{ D}$

10.3 Atomic Orbital Overlap—In the **valence bond theory**, a covalent bond is formed by the overlap of atomic orbitals of the bonded atoms, producing a region of high electron density between the atomic nuclei. Molecular geometry is determined by the spatial orientations of the atomic orbitals involved in bonding.

10.4 Hybridization of Atomic Orbitals—Valence bond theory often requires that bonding atomic orbitals be hybridized. Hybridization represents the blending of the character (in terms of energy and spatial orientation) of different atomic orbitals. **Hybrid orbitals** include sp, sp^2, sp^3, sp^3d, and sp^3d^2 orbitals. The geometric distribution of the hybrid orbitals in the valence bond theory is the same as the electron-group geometry predicted by the VSEPR method.

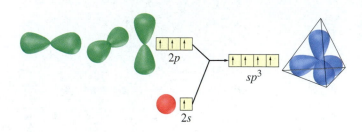

10.5 Hybrid Orbitals and Multiple Covalent Bonds—To apply the valence bond theory to structures containing multiple bonds, hybridization schemes must leave some orbitals unhybridized, as in the set $sp^2 + p$. Hybrid orbitals overlap in the usual way (end to end) to form **sigma (σ) bonds**. Unhybridized p orbitals overlap in a side-by-side manner to form **pi (π) bonds**. A double bond consists of one σ bond and one π bond, and a triple bond consists of one σ bond and two π bonds. The lack of rotation around a double bond leads to the possibility of **geometric isomers**. **Cis** isomers have the substituent groups located on the same side of the molecule; **trans**

isomers have the substituent groups located across the double bond from one another.

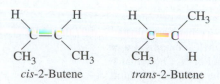

cis-2-Butene   *trans*-2-Butene

10.6 Characteristics of Molecular Orbitals—In **molecular orbital theory**, atomic orbitals of separated atoms are combined into **molecular orbitals (MOs)**. A pair of molecular orbitals, consisting of a **bonding orbital** and an **antibonding orbital**, is formed for every pair of atomic orbitals combined. The **bond order** is one-half the difference between the number of electrons in bonding molecular orbitals and the number in antibonding molecular orbitals. Valence electrons are assigned to the molecular orbitals in an aufbau procedure similar to that used for the electron configurations of atoms.

10.7 Homonuclear Diatomic Molecules of the Second-Period Elements—Bonding and magnetic properties can be predicted through molecular orbital diagrams, as shown in the chapter for the series of molecules (some only hypothetical) Li_2 through Ne_2.

10.8 Bonding in Benzene—the benzene molecule can be represented by two Kekulé structures, which are resonance structures of a resonance hybrid. A symbolic representation of the resonance hybrid is also used. Bonding in benzene can most easily be explained by a combination of valence bond theory for the σ-bond framework and molecular orbital theory for the π bonds.

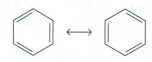

10.9 Aromatic Compounds—Compounds whose molecules have ring structures based on the benzene molecule are called **aromatic compounds**. All open-chain organic compounds and all ring compounds that are not aromatic are called **aliphatic compounds**.

Assessment Goals

When you have mastered the material in this chapter, you will be able to:

- Identify the geometrical shapes of molecules and polyatomic ions according to the principles of VSEPR theory.

- Describe the difference between electron-group geometry and molecular geometry.

- Predict the bond angles within molecules and the influence of lone-pair electrons on bond angles.

- Describe the shapes of molecules having more than one central atom.

- Define a dipole moment and categorize molecules as polar or nonpolar.

- Describe the valence bond theory view of covalent bond formation.

- Describe hybridization of atomic orbitals in terms of orbital energy and spatial orientation and the effect of these spatial orientations in establishing bond angles.

- Predict central-atom hybridization and atomic orbital overlap associated with bond formation within a molecule.

- Describe the hybridization schemes that lead to multiple bond formation, and identify the σ and π bonds in bonding schemes for molecular structures.

- Describe cis–trans isomers and the bonding that leads to them.

- Discuss the formation of energetically stable molecules in terms of bonding and antibonding molecular orbitals.

- Generate molecular orbital diagrams for simple diatomic molecules and ions.

- Calculate the bond order of simple diatomic molecules and ions and describe their magnetic properties.

- Describe the bonding in benzene and aromatic compounds.

- Relate names and formulas of compounds related to benzene.

Self-Assessment Questions

1. Explain why we describe all diatomic molecules as linear. Is it possible for a molecule consisting of three or more atoms to be linear? Explain.

2. In the VSEPR method, which of the electron-group repulsions involving bonding pairs and lone pairs are the strongest? Explain.

3. What is the hybridization scheme of Xe in XeF_2?

   (a) sp (c) sp^3d

   (b) sp^3 (d) sp^3d^2

4. Which of the following involves a hybridization scheme for the central atom that includes a d orbital contribution?

   (a) I_3^- (c) NO_3^-

   (b) PCl_3 (d) CF_4

5. What is the hybridization of the nitrogen atom in each of the following: NO_2, $NOCl$, and NH_2OH?

6. What approximate bond angles would you expect in triatomic molecules that have the following electron-group geometries about the central atom?

   (a) linear   (c) tetrahedral

   (b) trigonal planar

7. Which of the following contains a bond angle of slightly less than 90°?

   (a) PH_3 (c) OCl_2

   (b) ClF_3 (d) $HCOOH$

8. Must every chemical bond have a bond dipole? Must every molecule have a dipole moment? Explain.

9. Explain why SO_2 is a polar molecule, whereas SO_3 is not.

10. Which one of the following molecules is *not* polar?

    (a) BCl_3 (c) NH_3

    (b) CH_2Cl_2 (d) FNO

11. Explain why the H-X-H bond angles decrease from methane to ammonia to water. (X = C, N, and O, respectively, for the three compounds.)

12. Use the VSEPR model to explain the O-N-O bond angles in NO_2^+ (180°) and NO_2^- (115°).

13. Which, if either, of the following statements is valid? "All bonds in an alkane are σ bonds." "All bonds in an alkene are π bonds." Explain.

14. Which one of the following will exhibit geometric isomerism?

    (a) CH_2O

    (b) C_2H_2

    (c) C_2H_2ClF (where one hydrogen is bonded to each carbon)

    (d) C_2H_2ClF (where both hydrogens are bonded to a single carbon)

15. What is the total number of (a) σ bond and (b) π bonds in the molecule $HCONH_2$?

16. How do *antibonding* molecular orbitals differ spatially from *bonding* molecular orbitals? How do they differ energetically?

17. Which of the following species possesses a bond order of 1?

    (a) H_2^+ (c) He_2

    (b) Li_2 (d) H_2^-

18. What are the Kekulé structures of benzene? Why is a single Kekulé structure inadequate for describing the benzene molecule?

19. What are the similarities in the covalent bonding in ethene, C_2H_4, and in benzene, C_6H_6? What are the differences?

20. What is the chemical distinction between the terms *aliphatic* compound and *aromatic* compound?

Problems

VSEPR Method

21. What is the VSEPR notation of the central atom in each of the following?

    (a) $HOCl$ (c) NCl_3

    (b) SiF_4

22. What is the VSEPR notation of the central atom in each of the following?

    (a) $COCl_2$ (c) ClO_2^-

    (b) PH_3

23. Predict the molecular geometry of each species in Problem 21.

24. Predict the molecular geometry of each species in Problem 22.

25. Predict the molecular geometry of each of the following.

    (a) CS_2 (c) NH_4^+

    (b) SCN^- (d) H_2CO

26. Predict the molecular geometry of each of the following.

    (a) CO_3^{2-} (c) N_2O

    (b) PCl_4^+ (d) CCl_2F_2

27. Predict the molecular geometry for each of the following.

    (a) SF_2 (c) XeF_4

    (b) NI_3 (d) BrO_3^-

28. Predict the molecular geometry for each of the following.

    (a) H_3O^+ (c) IF_3

    (b) $BiCl_3$ (d) S_3^{2-}

29. Describe the shape of each of the following.

    (a) CH_3CN

    (b) N_2O_4

    (c) CH_3SH

30. Describe the shape of each of the following.

    (a) H_2CO_3

    (b) $(CH_3)_2CO$

    (c) C_2N_2

31. Explain why the molecule BF_3 is trigonal planar, whereas a molecule with a similar formula, ClF_3, is T-shaped.

32. Explain why the ion ICl_4^- is square planar, whereas an ion with a similar formula, BF_4^-, is tetrahedral.

33. In which of the molecules, CH_4 or $COCl_2$, do you think the actual bond angles are closer to those predicted by the VSEPR method? Explain.

34. Describe the shape of the sulfur dichloride molecule. Does the predicted Cl-S-Cl bond angle agree with the experimental angle of 103°? Explain.

35. The molecules NF_3 and PCl_3 are both trigonal pyramidal, and we can conclude that their bond angles are about the same. The molecules H_2O and SO_2 are both angular, but we cannot draw the same conclusion about them. Why not?

36. Determine each bond angle in 1-octanol, $CH_3(CH_2)_6CH_2OH$. Why is this task not as formidable as it might sound?

Molecular Shape and Dipole Moments

37. Identify the following molecules as polar or nonpolar. Explain.

(a) N_2O (c) SF_2

(b) IF_3 (d) SF_6

38. Indicate which of the following molecules you would expect to be polar. Explain.

(a) XeF_4 (c) $HOCl$

(b) NI_3 (d) $BiCl_3$

39. Which of the molecules, H_2O or OF_2, would you expect to have the larger dipole moment? Explain.

40. Arrange the following molecules in order of increasing dipole moment: bromomethane, chloromethane, fluoromethane, methane, tetrafluoromethane. Explain, noting any molecules with a dipole moment of zero.

41. Draw structural formulas and use cross-based arrows to represent bond dipoles and any resultant molecular dipole in molecules of (a) $GeCl_4$ and (b) SF_4.

42. Draw structural formulas and use cross-based arrows to represent bond dipoles and any resultant molecular dipole in molecules of (a) $ClNO$ and (b) COS.

Valence Bond Theory

43. Describe the bonding in the molecules $Li_2(g)$ and $F_2(g)$ by the valence bond theory. How can the theory account for the difference in bond energies, 106 kJ/mol for the Li—Li bond and 157 kJ/mol for the F—F bond?

44. The molecule phosphine, PH_3, has H-P-H bond angles of 93.6°. Use the valence bond theory to describe the bonding in this molecule.

45. Indicate the hybridization scheme expected for the central atom in each of the following.

(a) SO_3 (c) SF_2

(b) CO_2 (d) XeF_4

46. Indicate the hybridization scheme expected for the central atom in each of the following.

(a) NH_4^+ (c) $COCl_2$

(b) NO_2^- (d) $HOCl$

47. Indicate the hybridization scheme expected for each central atom in the following.

(a) $HNCO$ (c) CH_3CCCH_3

(b) CH_3CN (d) CH_3NH_2

48. Indicate the hybridization scheme expected for each central atom in the following.

(a) C_2N_2 (c) CH_3CHO

(b) NH_2OH (d) $(CH_3)_2O$

49. Propose a simple Lewis structure (or resonance structures), the molecular geometry, and a bonding scheme of the type used in Example 10.7 for each of the following.

(a) HNO_3 (c) CH_3CCH

(b) AsF_6^-

50. Propose a simple Lewis structure (or resonance structures), the molecular geometry, and a bonding scheme of the type used in Example 10.7 for each of the following.

(a) $ClNO_2$ (c) CO_3^{2-}

(b) OF_2

51. In both of the ions, ICl_2^+ and ICl_2^-, an iodine atom is bonded to two Cl atoms. Do you expect the same hybridization scheme for the central I atom in each case? Explain.

52. In both the molecule OSF_4 and the ion SF_5^-, a sulfur atom is bonded to five other atoms. Do you expect the same hybridization scheme for the central S atom in each case? Explain.

53. The structure of the oxalate ion, $C_2O_4^{2-}$, is represented as follows.

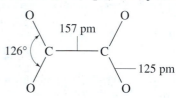

Propose hybridization and bonding schemes consistent with this structure. (*Hint:* Use data from Table 9.1.)

54. The structure of hydrazoic acid, HN_3, is indicated as follows.

124 pm   113 pm
N————N————N
113°
H

Propose hybridization and bonding schemes consistent with this structure. (*Hint:* See Table 9.1.)

55. Draw hybridization and bonding schemes like those in Example 10.7 that are consistent with the molecular models shown (white = H, red = O, blue = N, and black = C).

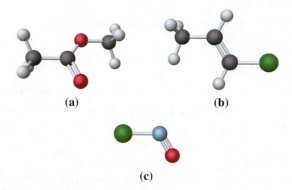

(a)   (b)

(c)

56. Draw hybridization and bonding schemes like those in Example 10.7 that are consistent with the molecular models shown (white = H, red = O, blue = N, black = C, and green = Cl).

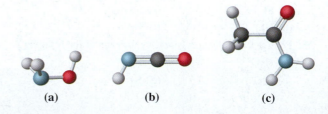

(a)   (b)   (c)

Geometric Isomerism

57. Which of the following compounds can exist as cis–trans isomers? Explain.

(a) CH_3CH_2CH＝$CHCH_2CH_3$

(b) CH_2＝$CHCH_2CH_2CH_3$

(c) CH_3CH_2CH＝$CHCH_3$

58. Which of the following compounds can exist as cis–trans isomers? Explain.

(a) CH_3C＝$CHCH_2CH_3$
  |
  CH_3

(b) CH_3CH＝$CHCH_2CH_2CH_3$

(c) CH_3CHCH＝$CHCH_3$
  |
  CH_3

59. Refer to the structures of 1-butene and isobutylene on page 415. Would you expect either of these substances to exhibit cis–trans isomerism? Explain. Imagine substituting a Cl atom for one H atom on the first carbon atom in each structure. Would the resulting substances exhibit cis–trans isomerism? Explain.

60. Arrange the following molecules in the expected order of increasing dipole moment: (a) chloroethane, (b) *cis*-2-butene, (c) *cis*-1,2-dichloroethene, (d) *trans*-1,2-dibromoethene. Explain your arrangement.

Molecular Orbital Theory

61. Which ion would you expect to have the greater fluorine-to-fluorine bond energy—F_2^+ or F_2^-? Explain.

62. The text mentions that Be_2 is not likely to exist as a stable molecule. Would you expect the ions Be_2^+ or Be_2^- to be any more stable? Explain.

63. Compare the bond order of the carbon-to-carbon bond in the diatomic molecule C_2 that you would establish from a Lewis structure and from a molecular orbital diagram. Why are they not the same?

64. In contrast to the situation in Problem 63, show that both a Lewis structure and a molecular orbital diagram give the same bond order for the carbide ion, C_2^{2-}.

65. Assume that the energy-level diagrams of Figure 10.27 apply to the ions CN^+ and CN^-.

(a) Which ion has the stronger carbon-to-nitrogen bond? Explain.

(b) Is either of these ions paramagnetic? Explain.

66. Assume that the energy-level diagrams of Figure 10.27 apply to the diatomic molecules BN and CN.

(a) Which molecule has the stronger bond between the atoms? Explain.

(b) Is either of these molecules paramagnetic? Explain.

67. These two ions derived from O_2 are commonly encountered in the oxides of the group 1A metals: *peroxide* ion, O_2^{2-}, and *superoxide* ion, O_2^-. Use molecular orbital theory to compare these ions with respect to bond order and magnetic properties.

68. Explain how molecular orbital theory makes it possible to describe a structure of benzene without the use of resonance structures and a resonance hybrid.

Aromatic Compounds

69. Draw structural formulas for (a) o-dinitrobenzene, (b) 2,4-dibromophenol, and (c) 1,2,4-trimethylbenzene.

70. Draw structural formulas for (a) p-bromotoluene, (b) m-diethylbenzene, and (c) 2,4-diiodotoluene.

71. Name the following compounds by the IUPAC system.

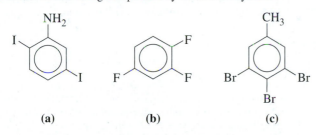

(a)   (b)   (c)

72. Name the following compounds by the IUPAC system.

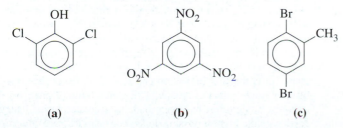

(a)   (b)   (c)

73. The aromatic compound azobenzene, $(C_6H_5N)_2$, is an important starting material in the production of dyes (called azo dyes). (a) Write two Lewis structures for azobenzene that demonstrate a type of cis–trans isomerism in this substance. (b) Propose hybridization schemes for the C and N atoms in these structures. (c) What are the bond angles in these structures?

74. Toluene-2,4-diisocyanate is widely used in the manufacture of polyurethane foam. Its structure is suggested by the following formula.

$$CH_3$$
（benzene ring with NCO substituents）

(a) How many different hybridization schemes must be used for the C atoms in this structure? What are they? Explain.

(b) What is the hybridization scheme for the two N atoms?

(c) What are the N-C-O bond angles? What are the C-N-C bond angles?

Additional Problems

Problems marked with an * may be more challenging than others.

75. Do you think the following is a valid statement? "The greater the electronegativity difference between the atoms in a molecule, the larger the dipole moment of that molecule." Explain.

76. We examined three possible structures for the XeF_2 molecule in Example 10.2 and then selected the correct one. Why is the actual structure not just a resonance hybrid of the three structures?

77. The NO_2F molecule depicted in Example 10.5 is a symmetrical molecule with bond angles of about 120°. Why does it not have a dipole moment of zero?

78. Draw a Lewis structure for the N_2 molecule and then predict a hybridization scheme consistent with this structure. Could bonding in N_2 be described by the use of pure (unhybridized) atomic orbitals? Explain.

79. Apply the VSEPR method to predict the shape of the (a) BrF_4^+ ion and (b) XeF_5^+ ion.

80. In the manner of Example 10.7, construct a bonding scheme for methyl isocyanate (see the Cumulative Example).

81. Phosphorus pentachloride is molecular in the gas phase, but in the solid phase it is an ionic compound consisting of the ions $[PCl_4]^+$ and $[PCl_6]^-$. Propose plausible hybridization and bonding schemes to represent these molecular and ionic forms of phosphorus pentachloride.

82. Arrange the following compounds in the expected order of increasing dipole moment, and explain the basis of your answer.

    ethane   fluoromethane   methanol   methylamine

* **83.** Describe the geometric shapes for each of the following molecules: cyclopropane, cyclobutane, and cyclopentane. Which of these molecules would you expect to be most stable (least reactive)? Explain. How does the geometric shape of cyclohexane differ from that of benzene shown in Figure 10.30? Explain.

* **84.** Carbon suboxide, C_3O_2, is a foul-smelling gas. The C—O bond lengths in the molecule are 116 pm, and the C—C bond lengths are 128 pm. Draw a plausible Lewis structure for this molecule, propose hybridization and bonding schemes, and predict its geometric shape. (*Hint:* Use data from Table 9.1.)

* **85.** Dicyandiamide, $NCNHC(NH)NH_2$, is used in the manufacture of melamine plastics. (a) Write a plausible Lewis structure for this molecule. (b) Indicate a plausible hybridization scheme for the central atoms in this structure. (c) Indicate the approxi-

mate bond angles in the molecule. (d) Make a rough sketch of the molecule.

* **86.** The molecule disulfur tetrafluoride has the skeletal structure F_3SSF. Two of the F-S-F bond angles are 90°, and one is 180°. Write a Lewis structure for the molecule, and propose hybridization schemes for the sulfur atoms that are consistent with this structure. Make a rough sketch of the molecule.

87. The structure of the allene molecule, CH_2CCH_2, is indicated here. Propose hybridization schemes for the C atoms in this molecule.

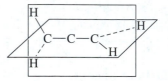

88. From each pair of diatomic molecules/ions, select the one with the greater bond energy: (a) B_2 or B_2^+, (b) O_2^+ or O_2^-, (c) Be_2 or Be_2^+, (d) F_2 or F_2^+, (e) F_2 or F_2^-.

89. For the molecules N_2O_3 and N_2O_5, draw the most plausible Lewis structures and describe the hybridization scheme at each central atom. Is either molecule planar?

* **90.** Use the definitions of dipole moment and the Debye unit given in the text (page 399) to estimate a hypothetical resultant dipole moment of a HCl molecule if the charges $\delta+$ and $\delta-$ were equal to the charge on an electron. The experimentally determined dipole moment for HCl is $\mu = 1.07$ D. What is the percent ionic character in the H—Cl bond? (*Hint:* What is the bond length in HCl?)

* **91.** The following resultant dipole moments have been measured: HF, $\mu = 1.82$ D; cis-N_2F_2, $\mu = 0.16$ D; $trans$-N_2F_2, $\mu = 0$. Explain these comparative values of the dipole moments based on the molecular structures of the three molecules.

* **92.** The text states that for the H_2O molecule, the resultant dipole moment is 1.84 D and the H-O-H bond angle is 104.5°. (a) By a geometric calculation, show that the value of the O—H bond dipoles is 1.50 D. (b) Use the same method as in part (a) to estimate the bond angle in H_2S, given that the H—S bond dipole is 0.67 D and the resultant dipole moment of H_2S is 0.93 D.

* **93.** The synthesis of a salt containing the N_5^+ cation was reported in 1999. What is the likely shape of this ion: linear, bent, zigzag, tetrahedral, seesaw, or square planar? Explain.

Apply Your Knowledge

* **94.** [**Collaborative**] In Problem 93 (Chapter 8) and Problem 103 (Chapter 9) we hypothesize a fictitious alternate universe, with elements based on triads of electrons instead of pairs. Consult those problems, then determine the molecular shapes of (a) the hypothetical compound formed when Zz reacts with Ad, and (b) the hypothetical compound formed when Ad reacts with A (atomic number = 1).

* **95.** [**Laboratory**] Analysis of a substance Q finds it to be 85.7% carbon and 14.3% hydrogen by mass. A gaseous sample of the compound has a density of 1.72 g/L and a molar volume of 24.45 L at 1 atm and 25.0 °C, and a density of 1.38 g/L and a molar volume of 30.6 L at 1 atm and 100.0 °C. (a) Give two possible Lewis structures for molecules of Q. (b) Give names to the

two molecules. (c) Sketch ball-and-stick models for the two molecules. (d) Are there other possible molecules that fit the data? If so, what are they?

* **96.** [**Laboratory**] Concerning the eight gaseous molecular species in Figure 10.27, (a) write equations to represent their first ionization energies. (b) Might Ne_2^+ exist as a stable species? Explain. (c) Explain the observation that F_2 has a lower first ionization energy than does F. (d) Is the first ionization energy of O_2 greater than, less than, or equal to that of atomic O? Explain. (e) Is the first ionization energy of N_2 greater than, less than, or equal to that of atomic N? Explain. (f) In plasma spectrometry, some effort has been made to generate plasma using gases with the highest possible ionization energies. Use your re-

sponses to (d) and (e) to explain why helium and molecular nitrogen should be better candidates for this application than molecular oxygen.

97. **[Biochemical]** When vegetable oils are hydrogenated to make margarine, some of the cis fatty acids are converted to saturated fatty acids, but others are converted to trans unsaturated fatty acids. Margarines have 11 to 49% trans fatty acids. Fried foods and pastries may have 35 to 38% trans fatty acids. On the right are ball-and-stick models of three fatty acids. **(a)** Identify the saturated, cis unsaturated, and trans unsaturated molecules. **(b)** Why might trans fatty acids have the same effect as saturated fatty acids in the development of heart disease? **(c)** What simple chemical test can be used to distinguish between a saturated fatty acid and an unsaturated fatty acid?

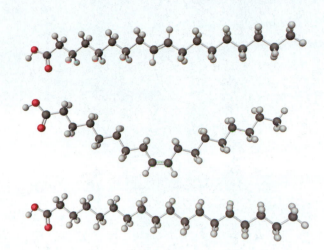

 # e-Media Problems

The activities described in these problems can be found in the e-Media Activities and Interactive Student Tutorial (IST) modules of the Companion Website, *http://chem.prenhall.com/hillpetrucci.*

98. In the **Hybridization** animation (*Section 10-4*), **(a)** what is the relationship between the number of atomic orbitals involved in hybridization and the number of resulting hybridized orbitals? **(b)** What determines the number of atomic orbitals involved in hybridization? Consider the hybridization of the central nitrogen atom of ammonia as shown in the animation.

99. View the **Energy and Hybridization** activity (*Section 10-4*) describing bonding in methane and in ammonia. Using a similar approach for water, sketch the hybridization scheme for the central oxygen atom and draw an orbital model describing the bonding in water. Label all atomic and hybrid orbitals in your drawing.

100. View the **Methane 3-D** and **Ammonia 3-D** models (*Section 10-4*). How do the three-dimensional structures of these molecules compare to the orbital bonding models of Figures 10.13 and 10.14? Account for any differences.

101. View the **Ethene 3-D** and **Ethyne 3-D** models (*Section 10-5*). Rotate each of the molecules so that all carbon atoms of the molecule lie in the plane of the screen. [Turn off the continuous rotation of the molecule by right-clicking (Windows) or click-holding (Macintosh) and deselecting Rotation in the menu. Use the mouse to drag the molecule while holding the mouse button to rotate it by hand.] Next, rotate each molecule so that all atoms are in a plane perpendicular to the screen. **(a)** Can this be done for the **Ethane 3-D** model? Why or why not? **(b)** What does this exercise reveal about the unique structural properties of alkenes and alkynes?

102. View the **2-Butene 3-D** models (*Section 10-5*). Rotate the two geometric isomers in space by holding down the (left on Windows computers) mouse button while moving the mouse over the molecule. **(a)** How do the three-dimensional geometries of the molecules differ? **(b)** What property of the carbon-to-carbon double bond gives rise to this structural difference? **(c)** Describe the types of atomic orbitals involved in the formation of the carbon-to-carbon bond and how these relate to the structures of 2-butene.

Chapter 11

States of Matter and Intermolecular Forces

As we noted in Chapter 5, we can describe many gases with a single general equation—the ideal gas equation—because intermolecular forces between gas molecules are often small enough to ignore. With liquids and solids, however, intermolecular forces are of prime importance. Because these forces differ in kind and strength from one substance to another, we cannot use general equations to describe liquids and solids. Thus, our primary goal in this chapter will be to achieve a *qualitative* understanding of natural phenomena. That is, we seek to answer many questions of the following type:

- Why does ice float on liquid water?
- Why does dry ice (solid CO_2) fail to melt under normal conditions?
- Why does it take longer to boil an egg on a mountaintop than at the seashore?
- Why is diamond, one form of carbon, so hard that it can scratch glass, whereas graphite, another form of carbon, is soft enough to use in pencils?

11.1 Intermolecular Forces and the States of Matter: A Chapter Preview

In the previous two chapters, we focused on the forces that bind atoms to one another *within* molecules. We called these forces chemical bonds, but we could also have called them *intra*molecular forces.* Intramolecular forces determine such molecular properties as molecular geometries and dipole moments. Forces that exist *between* molecules—*inter*molecular forces—determine the macroscopic *physical* properties of liquids and solids. In fact, if there were no intermolecular attractive forces, there would be no liquids or solids—everything would be in a gaseous state. Moreover,

*Latin, *intra*, within; *inter*, between. Notice that these are the meanings conveyed by the terms *intra*mural sports and *inter*collegiate athletics.

CONTENTS

◄ Water exists in three physical states in this view of Paradise Bay in Antarctica. Water vapor is invisible, but when condensed into liquid droplets, water is visible as clouds. The liquid water in the foreground and the solid water in the snow and ice on the mountains and in the glacier represent the two condensed states that are the focus of this chapter.

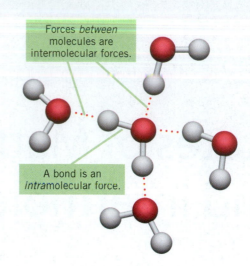

▶ **FIGURE 11.1 Intermolecular and intramolecular forces compared**
In the hypothetical situation described here, the gray "sticks" represent *intra*molecular forces—in other words, chemical bonds. The broken red lines suggest forces between molecules—*inter*molecular forces.

States of Matter activity

Gas

Liquid

Solid

▲ **FIGURE 11.2 Some comparisons of the states of matter**
In the gaseous state, molecules are widely spaced, motion is chaotic, and disorder is at a maximum. In the liquid state, molecules continue to undergo translational motion but are much closer together. In the solid state, there is a high degree of order among the structural particles; the motion of the particles is vibrational only.

intermolecular forces are themselves related to the kinds of chemical bonds found in substances. Figure 11.1 compares intermolecular and intramolecular forces.

We encounter substances in three different physical forms called the **states of matter:** gas, liquid, and solid. Let's briefly compare these three states, first at the macroscopic level:

- A gas expands to fill its container, has neither a fixed volume nor a fixed shape, and is easily compressed.
- A liquid has a fixed volume, flows to cover the bottom and assumes the shape of that portion of the container it fills, and is only slightly compressible.
- A solid has a fixed volume, maintains a definite shape, and is even more difficult to compress than a liquid.

The best explanations for these observations are based on the behavior of particles at the microscopic level:

- In a gas, speedy, energetic, widely spaced atoms or molecules are undergoing frequent collisions but never coming to rest or clumping together.
- In a liquid, atoms or molecules are close together, and intermolecular forces are strong enough to hold them in a fixed volume but not in a definite shape.
- In a solid, the structural particles (atoms, ions, or molecules) are in direct contact with one another, and intermolecular forces hold them into a fixed volume and definite shape.

The microscopic view in Figure 11.2 suggests that in each of the three states of matter, the atoms, ions, or molecules move. In gases and liquids, this motion includes a translational component; the structural particles move from point to point in three-dimensional space. In solids, the motion is mainly vibrational; that is, the structural units move back and forth about fixed points.

In this chapter, we will first describe changes from one state of matter to another. Then, we will explore the types of intermolecular forces that underlie the physical properties of substances. Finally, we will examine the structures of solids.

Phase Changes

When ice melts, water undergoes a change in state from a solid to a liquid. A block of dry ice (solid carbon dioxide) at room temperature changes directly from a solid to a gas. These changes in state belong to a larger category of physical changes called **phase changes,** of which we will consider six main types:

- *Melting*   solid $\longrightarrow$ liquid
- *Freezing*   liquid $\longrightarrow$ solid
- *Vaporization*   liquid $\longrightarrow$ gas
- *Condensation*   gas $\longrightarrow$ liquid
- *Sublimation*   solid $\longrightarrow$ gas
- *Deposition*   gas $\longrightarrow$ solid

We will first describe each of these phase changes, and then we will learn how to represent them collectively in a graph called a *phase diagram*.

11.2 Vaporization and Vapor Pressure

Evaporation, or **vaporization,** is the conversion of a liquid to a gas. As we noted in Chapter 5, the name *vapor* implies the gaseous state of a substance that is more commonly encountered in the liquid or solid state.

Let's try to picture the process of vaporization at the molecular level. As with gases, the molecules of a liquid have a range of speeds and kinetic energies, with the temperature of the liquid determining their average kinetic energy. Surface molecules having enough kinetic energy in excess of this average to overcome the intermolecular attractive forces of neighboring molecules in and below the surface fly off into the gaseous state: they *vaporize.*

As the more energetic molecules leave the surface of a liquid through vaporization, the average kinetic energy of the remaining molecules of liquid decreases. Consequently, the temperature of the liquid decreases, as in the familiar cooling sensation on your skin when a volatile (readily vaporized) liquid, such as rubbing alcohol (isopropyl alcohol), evaporates.

Enthalpy (Heat) of Vaporization

To maintain a constant temperature while it vaporizes, a liquid must absorb heat to replace the kinetic energy carried away by the vaporizing molecules, which means that vaporization is an endothermic process. The **enthalpy (heat) of vaporization,** ΔH_{vapn}, is the quantity of heat that must be absorbed to vaporize a given amount of liquid at a constant temperature. We usually express this quantity in kilojoules per mole, as shown in this equation for the vaporization of 1 mol of water at 298 K:

$$H_2O(l) \longrightarrow H_2O(g) \qquad \Delta H_{vapn} = 44.0 \text{ kJ/mol}$$

Several ΔH_{vapn} values are listed in Table 11.1.

Table 11.1 Some Enthalpies (Heats) of Vaporization at 298 K[a]	
Liquid	ΔH_{vapn}, kJ/mol
Carbon disulfide, CS_2	27.4
Carbon tetrachloride, CCl_4	37.0
Methanol, CH_3OH	38.0
Octane, C_8H_{18}	41.5
Ethanol, CH_3CH_2OH	43.3
Water, H_2O	44.0
Aniline, $C_6H_5NH_2$	52.3

[a] ΔH_{vapn} values are somewhat temperature-dependent.

The conversion of a gas to a liquid—the reverse of vaporization—is called *condensation.* A familiar example is the condensation of water on the mirror in a steamy bathroom. Because enthalpy is a function of state, if we vaporize a quantity of

Changes of State animation

Two or more physically distinct forms called phases may exist in the solid state of pure substances and mixtures and in the liquid state of mixtures. For example, at 25 °C most mixtures of triethylamine and water separate into two liquid phases—one a solution of triethylamine in water and the other, of water in triethylamine. At temperatures below 18.5 °C these two phases merge into a single liquid phase.

At about room temperature, we commonly refer to oxygen, nitrogen, and hydrogen as *gases* and water and methanol as *vapors.*

liquid and then condense the vapor back to liquid at the same temperature, the total enthalpy change must be zero. As a consequence,

$$\Delta H_{vapn} + \Delta H_{condn} = 0$$

$$\Delta H_{condn} = -\Delta H_{vapn} \qquad (11.1)$$

$$H_2O(g) \longrightarrow H_2O(l) \qquad \Delta H_{condn} = -44.0 \text{ kJ/mol (at 298 K)}$$

Vaporization is an endothermic process, and condensation is therefore exothermic.

Most refrigeration and air-conditioning systems are based on a repeated cycle of vaporization and condensation. The refrigerant is allowed to vaporize in a closed system, and the interior of the refrigerator becomes colder by giving up heat to vaporize the refrigerant. The gaseous refrigerant is compressed and condensed back to a liquid, and the heat of condensation of the refrigerant is absorbed by the surroundings, that is, the air outside the refrigerator.

Application Note

Chlorofluorocarbons (CFCs), such as CCl_3F (bp 24 °C), were once commonly used as refrigerants. Some of the CFCs have been implicated in the depletion of stratospheric ozone discussed in Chapter 25.

Example 11.1

How much heat, in kilojoules, is required to vaporize 175 g methanol, CH_3OH, at 25 °C?

STRATEGY

We can find a value of ΔH_{vapn} for 1 mol CH_3OH at 298 K in Table 11.1. We must express the quantity of methanol in moles and multiply by ΔH_{vapn}.

SOLUTION

We use molar mass to convert from grams to moles. When we multiply the number of moles of methanol by its ΔH_{vapn} value, we obtain the required quantity of heat:

$$? \text{ kJ} = 175 \text{ g } CH_3OH \times \frac{1 \text{ mol } CH_3OH}{32.04 \text{ g } CH_3OH} \times \frac{38.0 \text{ kJ}}{1 \text{ mol } CH_3OH} = 208 \text{ kJ}$$

ASSESSMENT

Recognizing that 175 g CH_3OH is somewhat more than five moles, we expect a result somewhat more than 5 times the value of ΔH_{vapn} found in Table 11.1, which is what we got.

EXERCISE 11.1A

To vaporize a 1.50-g sample of liquid benzene, C_6H_6, requires 652 J of heat. What is ΔH_{vapn} of benzene in kilojoules per mole?

EXERCISE 11.1B

How many kilojoules of heat are required to raise the temperature of 0.750 L of ethanol $(d = 0.789 \text{ g/mL})$ from 0.0 to 25.0 °C and then to vaporize 10.0% of the sample? The specific heat of ethanol is $2.46 \text{ J g}^{-1} \text{ °C}^{-1}$.

Example 11.2   An Estimation Example

Without doing detailed calculations, determine which liquid in Table 11.1 requires the greatest quantity of heat for the vaporization of 1 kg of liquid.

ANALYSIS AND CONCLUSIONS

The liquid requiring the greatest quantity of heat *per mole* is the one with the largest ΔH_{vapn}: aniline, $C_6H_5NH_2$, 52.3 kJ/mol. Next greatest on a per-mole basis is water, H_2O, with $\Delta H_{vapn} = 44.0$ kJ/mol. The molar mass of H_2O is about 18 g/mol and that of $C_6H_5NH_2$ is approximately 93 g/mol. In a 93-g sample, there is 1 mol of $C_6H_5NH_2$, but $93/18 \approx 5$ mol H_2O. Thus, much more heat is required to vaporize a given mass (1 kg) of H_2O than the same mass of $C_6H_5NH_2$. Of the remaining liquids in Table 11.1, none has a ΔH_{vapn} as great as that of H_2O or a molar mass as small. Therefore, water requires the greatest quantity of heat for the vaporization of 1 kg.

EXERCISE 11.2A

Without doing detailed calculations, determine which liquid in Table 11.1 requires the smallest quantity of heat for the vaporization of 1 kg of liquid.

EXERCISE 11.2B

Enthalpies of vaporization are a function of temperature. For example, ΔH_{vapn} for water is 44.0 kJ/mol at 25.0 °C but 40.7 kJ/mol at water's normal boiling point of 100.0 °C. Approximately how many grams of liquid water can be heated from 0 °C to 100 °C with the same quantity of heat required to vaporize 1.0 mol H_2O at 100 °C?

Vapor Pressure

When we place a liquid in an open container, eventually all the liquid molecules escape into the gaseous state and are dispersed in the atmosphere. When we place a liquid in a closed container, at first the volume of liquid decreases, even if only slightly, as some of the liquid is converted to gas. Then evaporation seems to stop, the volume of liquid remains constant, and it appears that nothing further is happening. To explain what really happens, we need to consider the system at the molecular level.

In Figure 11.3a, a liquid begins to vaporize. However, as soon as molecules appear in the vapor state, some of them strike the liquid surface, are captured, and return to the liquid state (Figure 11.3b). Condensation and vaporization occur simultaneously. We can represent these two concurrent and opposing processes by arrows pointing in opposite directions.

$$\text{Liquid} \underset{\text{condensation}}{\overset{\text{vaporization}}{\rightleftharpoons}} \text{vapor} \qquad\qquad (11.2)$$

Liquid-Vapor Equilibrium animation

At first, many more molecules pass from the liquid to the vapor than in the reverse direction because the rate of vaporization is greater than the rate of condensation.

 ↑ Molecules undergoing vaporization   ↓ Molecules undergoing condensation

(a) Vaporization

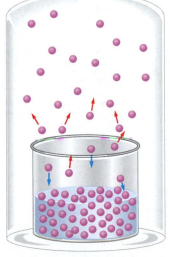

(b) Vaporization rate > condensation rate

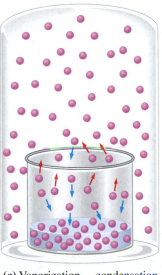

(c) Vaporization rate = condensation rate

▲ **FIGURE 11.3   Liquid–vapor equilibrium and vapor pressure**

(a) A liquid begins to vaporize. (b) Condensation begins as soon as the first vapor molecules appear; here, the rate of condensation is still less than the rate of vaporization. (c) The rates of vaporization and condensation have become equal. The partial pressure exerted by vapor molecules on the container walls—the vapor pressure of the liquid—remains constant.

QUESTION: How would the rates illustrated in (c) change if the temperature of the liquid were increased?

As the number of vapor molecules increases, however, so does the rate of condensation. Eventually the rate of condensation becomes equal to the rate of vaporization (Figure 11.3c). At this point, the maximum number of molecules that can be accommodated in the vapor state has been reached, and the number of molecules per unit volume in the vapor state remains constant with time.

In a *dynamic equilibrium*, no change is seen in the system at the macroscopic level, but there is activity at the molecular level. This is what we mean by *dynamic*. In liquid–vapor equilibrium, molecules of liquid continue to vaporize and molecules of vapor continue to condense, as suggested in Figure 11.3c. The **vapor pressure** of a liquid is the partial pressure exerted by its vapor in dynamic equilibrium with the liquid at a constant temperature.

The time required to reach liquid–vapor equilibrium at a particular temperature depends on several factors. For example, the smaller the vapor volume and the greater the surface area of the liquid, the faster equilibrium is reached. The equilibrium vapor pressure, in contrast, depends only on what the liquid is and on the equilibrium temperature.

Look again at Figure 11.3. An increase in temperature increases the average kinetic energy of the molecules in the liquid. More molecules have enough kinetic energy to escape from the liquid state, and the rate of vaporization increases. At equilibrium, the rates of vaporization and condensation are again equal, but the pressure exerted by the vapor is higher at the higher temperature. *The vapor pressure of a liquid increases with temperature.*

Table 11.2 lists the vapor pressure of water at various temperatures. Recall that we used data from a table like this when we applied Dalton's law of partial pressures to the collection of gases over water (page 196). In Example 11.3, we show how a vapor pressure may enter into an ideal gas law calculation.

Equilibrium Vapor Pressure activity

Table 11.2 Vapor Pressure of Water at Various Temperatures

Temperature, °C	Pressure, mmHg	Temperature, °C	Pressure, mmHg	Temperature, °C	Pressure, mmHg
0.0	4.6	29.0	30.0	93.0	588.6
10.0	9.2	30.0	31.8	94.0	610.9
20.0	17.5	40.0	55.3	95.0	633.9
21.0	18.7	50.0	92.5	96.0	657.6
22.0	19.8	60.0	149.4	97.0	682.1
23.0	21.1	70.0	233.7	98.0	707.3
24.0	22.4	80.0	355.1	99.0	733.2
25.0	23.8	90.0	525.8	100.0	760.0
26.0	25.2	91.0	546.0	110.0	1074.6
27.0	26.7	92.0	567.0	120.0	1489.1
28.0	28.3				

Example 11.3

Suppose the equilibrium illustrated in Figure 11.3 is between liquid hexane, C_6H_{14}, and its vapor at 298.15 K. A sample of the equilibrium vapor is found to have a density of 0.701 g/L. What is the vapor pressure of hexane, in mmHg, at 298.15 K?

STRATEGY

We will assume that the intermolecular forces in the gaseous hexane are negligible and that we can therefore apply the ideal gas equation: $PV = nRT$. Moreover, because we are given a vapor density rather than the amount and volume of gas separately, we will find Equation (5.11) to be a more useful form of the ideal gas equation. We can solve Equation (5.11) for P.

SOLUTION

First, let's rearrange Equation (5.11) to solve for P.

$$d = \frac{MP}{RT} \quad \text{and} \quad P = \frac{dRT}{M}$$

Because we seek the vapor pressure in the unit mmHg, let's use the value $R = 62.364 \text{ L Torr mol}^{-1} \text{ K}^{-1}$ from Table 5.3. The molar mass of C_6H_{14} is 86.18 g/mol.

$$P = \frac{0.701 \text{ g L}^{-1} \times 62.364 \text{ L Torr mol}^{-1} \text{ K}^{-1} \times 298.15 \text{ K}}{86.18 \text{ g mol}^{-1}}$$

$$= 151 \text{ Torr}$$

Recalling that 1 Torr = 1 mmHg, we can express the pressure in the required units.

$$P_{C_6H_{14}} = 151 \text{ mmHg, at 298.15 K}$$

ASSESSMENT

The vapor pressures of liquids exist over a broad range. When we have learned more about intermolecular forces, we may have a good idea of a reasonable value for a calculated vapor pressure—for example, whether it should be greater or less than 1 atm. For now, our principal check on a vapor pressure calculation must be the proper cancellation of units.

EXERCISE 11.3A

A 335-mL sample of Br_2 vapor has a mass of 1.100 g and is in equilibrium with liquid bromine at 39 °C. Calculate the vapor pressure of bromine at 39 °C.

EXERCISE 11.3B

Suppose the equilibrium illustrated in Figure 11.3 is between liquid water and its vapor at 22 °C. What mass of water vapor would be present in a vapor volume of 275 mL at 22 °C? Use data from Table 11.2.

A graph of vapor pressure as a function of temperature is called a **vapor pressure curve** (Figure 11.4). Liquid and vapor coexist in equilibrium at all points on the curve, and each point on the curve represents the vapor pressure of the liquid at a given temperature.

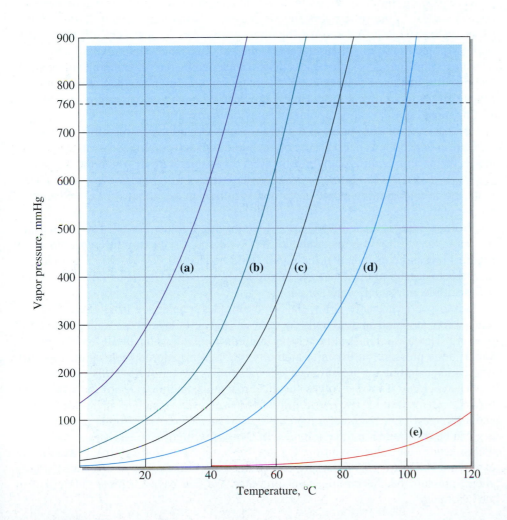

Vapor Pressure vs. Temperature animation

◀ **FIGURE 11.4   Vapor pressure curves of several liquids**

(a) Carbon disulfide, CS_2; (b) methanol, CH_3OH; (c) ethanol, CH_3CH_2OH; (d) water, H_2O; (e) aniline, $C_6H_5NH_2$. The point at which any curve intersects the broken horizontal line representing $P = 760$ mmHg is the normal boiling point for that liquid.

If intermolecular forces within a liquid are weak, many molecules escape from the liquid surface before equilibrium is established. The resulting vapor pressure is high, and we say that the liquid is *volatile*. If the intermolecular forces are strong, the equilibrium vapor pressure is low, and we say that the liquid is *nonvolatile*. Diethyl ether, with a vapor pressure of 534 mmHg at 25 °C, is highly volatile. At 25 °C, water has a vapor pressure of 23.8 mmHg; it is moderately volatile. Relative to diethyl ether and water, mercury would be considered nonvolatile. Nevertheless, mercury vapor in equilibrium with liquid mercury is present at a high enough partial pressure in air to be toxic (see Problem 104). Gasoline is a mixture of hydrocarbons, some of which are quite volatile and others less so. Gasoline for use in cold weather is formulated to have a higher proportion of volatile components so that the vapor produced by the cold gasoline can be ignited more readily.

Vapor Pressure Equations

Vapor pressure curves help us understand the concept of vapor pressure, but you are not likely to find them in chemistry reference books. Nor are you likely to find many tables of vapor pressure. Instead you will find mathematical equations that summarize experimental vapor pressure data. In this way, vapor pressure data for hundreds of liquids can be presented in just a page or two. Moreover, data can be calculated from these equations more precisely than they can be read from a graph.

Boiling Point

When we heat a liquid in an open container, we observe a special vaporization phenomenon. At a particular temperature, vaporization occurs not only at the surface but *throughout the liquid*. Vapor produced in the interior of the liquid forms bubbles that rise to the surface and escape. The liquid *boils*.

The **boiling point** of a liquid is the temperature at which its vapor pressure becomes equal to prevailing atmospheric pressure. No boiling occurs at lower vapor pressures because the pressure of the atmosphere would cause any incipient vapor bubbles in the liquid to collapse. That is, the pressure inside the bubbles—the vapor pressure of the liquid—would be less than that exerted by the surrounding atmosphere. Once boiling begins, the temperature of a liquid cannot rise above the boiling point. As long as some liquid remains, the energy provided by continued heating goes into converting more liquid to vapor, not into raising the temperature.

The **normal boiling point** of a liquid is its boiling point at 1 atm, that is, the temperature at which the vapor pressure of the liquid is exactly 1 atm. On a vapor pressure curve, the normal boiling point of a liquid is the temperature at which the line representing $P = 1$ atm (760 mmHg) intersects the curve. From Figure 11.4, we can estimate the normal boiling point of carbon disulfide to be about 46 °C and that of methanol to be about 65 °C.

Boiling point is a useful property in identifying liquids. For example, if we find that an unknown liquid boils at about 78 °C, we could refer to a reference book (or Figure 11.4) and conclude that the liquid could be ethanol. We would have to do further tests to be sure of this, but from Figure 11.4 we could be sure the liquid is not carbon disulfide, methanol, water, or aniline.

Atmospheric pressure is lower at high altitudes than at sea level. Therefore, the temperature at which a liquid boils decreases as the altitude increases and atmospheric pressure decreases. Thus, although the boiling point of water is 100 °C at normal atmospheric pressure, in the mile-high city of Denver, Colorado (elevation 1609 m), where the prevailing air pressure is about 630 mmHg, water boils at 95 °C (203 °F).

The cooking of food involves chemical reactions, and the rates of these reactions depend on temperature. The reactions go faster with even a small temperature increase and slower with even a small temperature decrease. When water boils at 100 °C, an egg can be soft-boiled in three minutes. In Denver, it might take five to six minutes to boil an egg to the same condition. On top of Mount Everest (8848 m), it would take much longer. Figure 11.5 illustrates an extreme case of the effect of a reduced atmospheric pressure on the boiling point of water. We can easily achieve this effect in the laboratory.

Some liquids undergo decomposition at their normal boiling point, and in such cases the boiling point may be recorded at a pressure lower than 760 mmHg. For example, the boiling point of lauric acid, a fatty-acid component of vegetable oils, may be listed as $225^{100\,mm}$: The liquid boils at 225 °C under a pressure of 100 mmHg.

▲ **FIGURE 11.5   The effect of pressure on the boiling point of water**

Reducing the air pressure in the bell jar can cause water to boil at room temperature.

We exploit the increase in boiling point that is caused by increased external pressure when we use a pressure cooker. In a pressure cooker, water boils at higher-than-normal temperatures. At such temperatures, the chemical reactions involved in cooking potatoes, beets, or a tough piece of meat proceed faster than in an open pot.

The Critical Point

If we heat a liquid in a *closed* container, boiling does not occur. The gas pressure above the liquid increases continuously as vapor accumulates above the liquid, and incipient bubbles of vapor in the liquid cannot overcome this pressure. Instead of boiling, as we raise the temperature (Figure 11.6) we observe that the density of the liquid decreases, the density of the vapor increases, and the boundary (meniscus) between the liquid and vapor becomes blurred. Finally, the densities of liquid and vapor become the same, and the boundary disappears, leaving a single fluid phase. The **critical temperature,** T_c, is the highest temperature at which a liquid and vapor can coexist in equilibrium as physically distinct states of matter. The vapor pressure at this temperature is called the *critical pressure,* P_c. The condition corresponding to a temperature of T_c and a pressure of P_c is called the **critical point.**

▲ In an autoclave, bacteria (even resistant spores) are killed more rapidly than in boiling water, not directly by the increased pressure but by the higher temperatures attained.

Vapor

Liquid

Room temperature

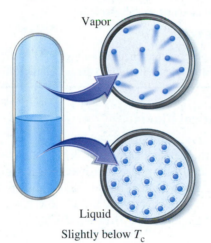

Vapor

Liquid

Slightly below T_c

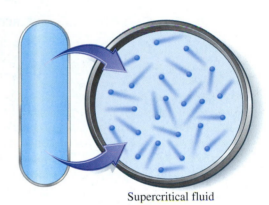

Supercritical fluid

Critical temperature T_c

▲ **FIGURE 11.6 The critical point**
The meniscus separating a liquid from its vapor disappears at the critical point. The liquid state does not exist above the critical temperature, regardless of the pressure that might be applied to the substance.

Critical Point simulation

Alternatively, the critical temperature is the highest temperature at which a gas can be condensed into a liquid solely by increasing the pressure of the gas. If T_c for a gas is above room temperature, we can liquefy the gas at room temperature just by applying enough pressure to it. However, if T_c is below ambient temperature (usually about 300 K), we must both apply pressure and lower the temperature to a value below T_c. Sometimes we use the term *vapor* to refer to the gaseous state of a substance below its T_c, and *gas* to the gaseous state above T_c. In these terms, a vapor can be condensed to a liquid simply by applying pressure; a gas cannot. Table 11.3 lists critical temperatures and pressures for some common substances.

Table 11.3 Critical Temperature and Pressure of Various Substances

	T_c, K	P_c, atm
H_2	33.0	12.8
N_2	126.3	33.5
O_2	154.8	50.1
CH_4	190.6	45.4
CO_2	304.2	72.9
C_2H_6	305.4	48.2
HCl	324.6	81.5
C_3H_8	369.8	41.9
NH_3	405.6	111.3
SO_2	430.6	77.9
H_2O	647.3	218.3

Example 11.4   A Conceptual Example

To keep track of how much gas remains in a cylinder, we can weigh the cylinder when it is empty, when it is full, and after each use. In some cases, though, we can equip the cylinder with a pressure gauge and simply relate the amount of gas to the measured gas pressure. Which method should we use to keep track of the bottled propane, C_3H_8, in a gas barbecue?

ANALYSIS AND CONCLUSIONS

The propane is at a temperature below its T_c (369.8 K) and therefore exists in the cylinder as a mixture of a liquid and vapor. As the fuel is consumed, the volume of liquid in the cylinder decreases and that of the vapor increases. However, the vapor pressure of the liquid propane does not depend on the amounts of liquid and vapor present. The pressure will remain constant (assuming a constant temperature) as long as some liquid remains. Only after the last of the liquid has vaporized will the pressure drop. Measuring the pressure does not tell us how much propane is left in the cylinder until it is almost gone. We would have to monitor the contents of the cylinder by weighing.

EXERCISE 11.4A

Which of the two methods described in Example 11.4 could you use if the fuel in the cylinder were methane, CH_4? Explain.

EXERCISE 11.4B

Are the observations pictured in Figure 11.6 affected in any way by the volume of liquid that is initially introduced into the sealed tube? Explain.

Supercritical Fluids

Because liquids and gases both flow readily—they are *fluids*—and because there is only one fluid phase above the critical temperature and critical pressure, it is customary to call this phase a **supercritical fluid** (SCF). The density of this fluid is close to that of a typical liquid, but its ability to flow and diffuse is more like that of a typical gas. Supercritical fluids now have many practical applications (see the essay on green chemistry on page 443).

11.3   Phase Changes Involving Solids

Now let's consider phase changes that involve solids. In particular, we will emphasize energy changes that accompany transitions from solid to liquid and vapor.

Melting, Melting Point, and Heat of Fusion

The structural units of a solid (atoms, ions, or molecules) move very little except for vibration about fixed points. As the temperature of a solid is raised, these vibrations become more vigorous. Finally, the vibrations are strong enough to cause the structural units to leave their fixed points, and a liquid is formed. The conversion of a solid to a liquid is called either **melting** or **fusion.** The temperature at which a solid melts is its **melting point.** The reverse of melting—the conversion of a liquid to a solid—is called *freezing;* the temperature at which it occurs is the *freezing point.* Whether we think of it as a melting solid or a freezing liquid, solid and liquid coexist at equilibrium, and the freezing point and melting point are identical for pure crystalline solids. The freezing point of water and the melting point of ice are the same temperature, 0 °C. Melting points vary slightly with pressure. The data in tables of melting points are usually *normal* melting points, the temperatures at which melting occurs when the pressure on the solid and liquid states is 1 atm.

The quantity of heat required to melt a given amount of solid is called the **enthalpy (heat) of fusion,** ΔH_{fusion}. As shown here for the melting of 1 mol of ice, melting is an endothermic process:

$$H_2O(s) \longrightarrow H_2O(l) \qquad \Delta H = +6.01 \text{ kJ/mol}$$

In almost all cases, solid mixtures and solutions do not undergo complete melting or freezing at a constant temperature.

Green Chemistry: Supercritical Fluids in the Food-Processing Industry

In Chapter 3, we introduced the concept of green chemistry, which involves the design of chemical products and processes that minimize environmental problems, and discussed one of the principles, called *atom economy*. Another principle of green chemistry involves the use of safer solvents and auxiliary substances, such as separation agents. Supercritical fluids are now increasingly used as solvents, serving as prime examples of this category of green chemistry.

Supercritical fluids are versatile solvents because their properties vary significantly with changes in temperature and pressure. By choosing an appropriate temperature and pressure, a supercritical fluid can be made to dissolve or extract one component of a mixture while leaving others unaffected. Supercritical fluids are now widely used to perform extractions both in laboratories and in industrial processes. Supercritical fluid extraction can be used to measure the fat content of foods, to remove environmental contaminants such as diesel fuel or polychlorinated biphenyls (PCBs) from soil, and to prepare samples (particularly nonvolatile ones such as polymers) for analysis.

Supercritical carbon dioxide is particularly attractive as a solvent in the food-processing industry. In the past, for example, solvents such as dichloromethane, CH_2Cl_2, were used to remove caffeine from coffee. At high concentrations, CH_2Cl_2 can cause health problems. Even though no CH_2Cl_2 has been found in the beverage brewed from decaffeinated coffee beans, dichloromethane can be a workplace hazard. The risk is eliminated by using supercritical CO_2 to dissolve and carry off the caffeine.

Supercritical fluids also are used to extract cholesterol from eggs, butter, lard, and other fatty foods, making them more suitable for low-cholesterol diets. Potato chips, made by cooking paper-thin slices of potato in oil, can be treated with supercritical CO_2 to reduce their normally high fat content. This change improves both the nutritional value and the shelf life of the chips. Supercritical fluids can also be used to extract the chemical compounds that impart flavor and fragrance to such products as lemons, black pepper, almonds, and nutmeg. These extracts can be used to flavor other foods or give a fragrance to household products. After CO_2 has served as a solvent, its pressure is reduced and the CO_2 reverts to the gaseous state. It is then recovered and reused.

In a typical dry-cleaning establishment, such chlorinated hydrocarbons as perchloroethylene, $Cl_2C = CCl_2$, are used as solvents. These substances are potentially carcinogenic and a health risk to workers. In addition, escaping vapors are an important source of urban air pollution. These undesirable consequences are avoided when supercritical CO_2 is used as the dry-cleaning solvent, a technology currently being implemented in more and more dry-cleaning plants.

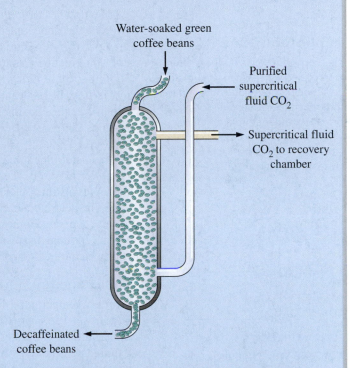

▲ Water-soaked coffee beans enter the vessel from the top and slowly drop through supercritical carbon dioxide at 160 to 220 atm pressure, which enters the vessel from the bottom. The caffeine content of the beans drops from an initial 1 to 3% before processing to 0.02%. The CO_2 passes into an absorption chamber where a spray of water leaches the caffeine from the supercritical fluid. The purified CO_2 is then recycled.

Freezing, the reverse of melting, is an exothermic process, and the enthalpy of freezing is the negative of the enthalpy of fusion:

$$H_2O(l) \longrightarrow H_2O(s) \qquad \Delta H = -6.01 \text{ kJ/mol}$$

Table 11.4 lists some typical enthalpies of fusion.

We can determine the freezing point of a pure liquid by measuring its temperature as it slowly cools. The temperature falls regularly with time, but when freezing begins, the temperature remains constant until all the liquid has frozen. Then the temperature falls regularly as the solid cools. If we plot temperature versus time with data obtained in this way, we get a graph called a *cooling curve*. A typical cooling curve is shown in

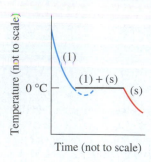

▲ **FIGURE 11.7 Cooling curve for water**

The dashed line portion represents the phenomenon of supercooling described in the text. Also note the symbols: (l) = liquid and (s) = solid.

Application Note

The droplets in high clouds are often supercooled water. When an airplane flies through these clouds, the surface of the aircraft provides sites for crystal formation, and the airplane suffers from a phenomenon called *rime icing*. If allowed to accumulate, the ice can cause the airplane to crash.

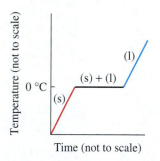

▲ **FIGURE 11.8 Heating curve for water**

This curve traces the phase changes that occur as solid water (ice) is heated from below its melting point to above its melting point.

Application Note

Foods and other materials are often preserved by freeze-drying. The material is frozen under a high vacuum so that the water is rapidly removed by sublimation. Harmful microorganisms cannot grow without water. Freeze-dried foods are especially favored for backpacking and camping trips; they are rendered light in weight but retain their nutritive value.

Sublimation activity

Table 11.4 Some Enthalpies (Heats) of Fusion

Substance	Melting Point, °C	ΔH_{fusion}, kJ/mol
Mercury, Hg	−38.9	2.30
Ethanol, CH_3CH_2OH	−114	5.01
Water, H_2O	0.0	6.01
Benzene, C_6H_6	5.5	9.87
Silver, Ag	960.2	11.95
Iron, Fe	1537	15.19

Figure 11.7. The freezing point of a liquid is the temperature along the horizontal constant-temperature portion of the cooling curve. Actually, if our interest is only in the freezing-point temperature, we don't need to plot an entire cooling curve. We just need to observe the liquid as it cools and record the constant temperature as it freezes.

Sometimes the temperature of a cooling liquid may drop below the freezing point without the appearance of a solid, a condition called **supercooling**. This condition may occur if a liquid is cooled quickly to below its normal freezing point. Formation of a solid requires a site, such as a dust particle, on which the crystal can grow. Supercooling is more likely to occur if a liquid is kept free of such sites. A supercooled liquid is unstable; it cannot persist indefinitely and may begin to freeze at any time. When freezing begins, the temperature of the liquid rises to the freezing point and remains there until freezing is completed, a process suggested by the dashed line in Figure 11.7.

To determine a melting point, we start with the pure solid and add heat. The temperature rises until melting begins, remains constant until melting is completed, and then rises again. A plot of temperature versus time is called a *heating curve*. In appearance, the heating curve for a substance resembles its cooling curve but is flipped right to left (compare Figure 11.8 to Figure 11.7). As expected, the freezing point obtained from a cooling curve and the melting point obtained from a heating curve are identical for the same substance.

Sublimation

At room temperature, few solids are as volatile as liquids, but some solids do have a significant vapor pressure. Mothballs, solid-stick room deodorizers, and dry ice are three familiar examples. The passage of molecules directly from the solid state to the vapor state is called **sublimation**. The reverse process, the condensation of a vapor to a solid, is generally called *deposition*. The sublimation of solid iodine and the deposition of iodine vapor are pictured in Figure 11.9.

Vapor condenses back to crystals of I_2

I_2 vapor

Solid I_2

◀ **FIGURE 11.9 Sublimation of iodine**

Even at 70 °C, well below its melting point (114 °C), solid iodine has a vapor pressure that is high enough to allow the solid to sublime. Once sublimation has begun, deposition starts up, and the purple iodine vapor deposits as $I_2(s)$ on the colder walls of the flask.

A dynamic equilibrium is reached when the rates of sublimation and deposition become equal. Like vapor in equilibrium with a liquid, vapor in equilibrium with a solid exerts a characteristic vapor pressure, often called the *sublimation pressure* A plot of the vapor pressure of a solid versus temperature is called a **sublimation curve.**

Although no liquid phase is involved when a solid sublimes, we can think of sublimation as equivalent to a two-step process: melting followed by vaporization. Using this idea and Hess's law, we see that the **enthalpy (heat) of sublimation** is simply the sum of the enthalpies of fusion and vaporization measured at the same temperature:

$$\Delta H_{\text{subln}} = \Delta H_{\text{fusion}} + \Delta H_{\text{vapn}} \tag{11.3}$$

People living in cold climates are especially familiar with the phenomenon of sublimation. Because the vapor pressure of ice at 0 °C is 4.58 mmHg, snow disappears from the ground and ice from the windshield of an automobile even though the temperature stays below 0 °C. This occurs through sublimation, not melting; no liquid water is present at any point in the process.

The Triple Point

We have now discussed all the phase changes listed on page 435, but we should consider one additional situation, that in which solid, liquid, and vapor exist together at equilibrium. It is the point at which the vapor pressure curve for a substance (extending from higher temperatures) and its sublimation curve (extending from lower temperatures) meet. The solid–liquid equilibrium curve, called the *fusion curve*, also originates here. This point is called the **triple point;** it defines the *only* temperature and pressure at which the three states of matter—solid, liquid, and gas—can coexist at equilibrium.

The triple point for water is at 0.0098 °C and 4.58 mmHg. Why isn't the triple-point temperature for water exactly 0 °C? Can solid water (ice), liquid water, and water vapor not exist together in the open atmosphere at 0 °C? They can, but this system involves two different pressures. The vapor exists at a pressure of 4.58 mmHg, its partial pressure, but the ice and liquid water are under atmospheric pressure (760 mmHg). Moreover, at its normal melting point, liquid water contains some dissolved air, and this affects the equilibrium temperature slightly. To have a true triple point, a system must consist of a *pure* substance existing *only under the pressure of its own vapor.* There can be no extraneous substances present (such as the gases in air).

11.4 Phase Diagrams

A **phase diagram** is a graphical representation of the conditions of temperature and pressure under which a substance exists as a solid, a liquid, a gas, or some combination of these in equilibrium. The phase diagram summarizes information about phase changes. Figure 11.10 is a simple (hypothetical) phase diagram. The various areas denote the range of temperatures and pressures in which a substance exists as a solid, a liquid, or a gas. The supercritical fluid area represents the gas at temperatures above T_c and pressures above P_c.

Phase Diagram activity

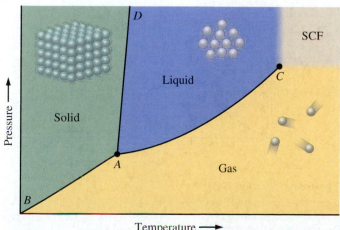

◄ **FIGURE 11.10** **A generalized phase diagram, representing temperatures, pressures, and states of matter of a substance**

Because of the extreme range of pressures and temperatures involved, a phase diagram is usually not drawn to scale.

QUESTION: If this diagram were for a particular substance, where would the normal melting point and normal boiling point of the substance be represented on the diagram?

The lines that separate adjoining areas in a phase diagram represent conditions of temperature and pressure in which two states of matter are in equilibrium. We can summarize the phases present for points and curves in Figure 11.10 as follows:

- Point *A* is the triple point (solid + liquid + gas).
- Curve *AD* is the fusion curve (solid + liquid).
- Curve *AB* is the sublimation curve (solid + gas).
- Curve *AC* is the vapor pressure curve (liquid + gas).
- Point *C* is the critical point (liquid and gas become indistinguishable).

We present the phase diagrams of three substances in this section: mercuric iodide, carbon dioxide, and water. Each adds important new ideas to what we have outlined here.

Mercury(II) Iodide, HgI$_2$

The phase diagram of mercury(II) iodide (Figure 11.11) illustrates *polymorphism,* a phenomenon in which a solid exists in two or more forms. Red HgI$_2$(s) is the form we see in a storeroom bottle of mercury(II) iodide. It persists up to 127 °C, at which temperature red HgI$_2$(s) converts to yellow HgI$_2$(s). Only the yellow form is stable from 127 °C up to 259 °C. Red and yellow HgI$_2$(s) are two distinct *phases* of solid mercury(II) iodide.

There can be only one triple point involving the three *states* of matter in a phase diagram. For mercury(II) iodide, this is the equilibrium between yellow HgI$_2$(s), liquid, and gas at 259 °C. However, there can be additional triple points that involve three *phases* of matter. For mercury(II) iodide the additional triple point, pictured in the photograph in Figure 11.11, has two solid phases and a gas phase: red HgI$_2$(s), yellow HgI$_2$(s), and HgI$_2$(g), at 127 °C. The deposition of a mixture of the red and yellow solids on the colder walls of the flask indicates that HgI$_2$ vapor is present.

Carbon Dioxide, CO$_2$

Now let's examine the phase diagram for carbon dioxide (Figure 11.12). First, we have specifically drawn the diagram to suggest that, in general, if a solid–liquid fusion curve is followed to sufficiently high pressures, the curve extends beyond the critical temperature. (For CO$_2$, a pressure of several thousand atmospheres is required.) If at first this observation seems unlikely, think about how close together molecules are

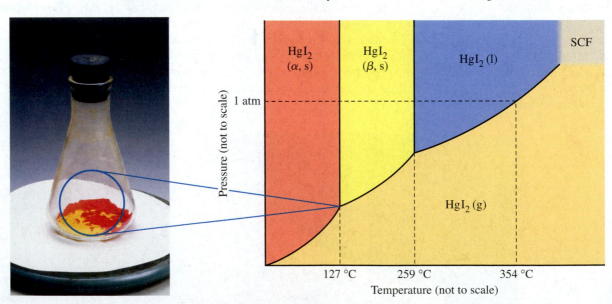

▲ **FIGURE 11.11   Phase diagram of mercury(II) iodide, HgI$_2$**

The photograph shows equilibrium among the phases: red solid HgI$_2$ (alpha form, α), yellow solid HgI$_2$ (beta form, β), and gaseous HgI$_2$ at 127 °C. Also highlighted on the temperature axis of the phase diagram are the HgI$_2$(β, s)/HgI$_2$(l)/HgI$_2$(g) triple point at 259 °C and the normal boiling point of HgI$_2$(l) at 354 °C.

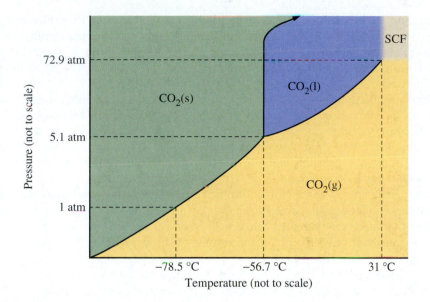

◀ **FIGURE 11.12** **Phase diagram of carbon dioxide, CO_2**

Noted in the diagram are the normal sublimation temperature of $CO_2(s)$, $-78.5\ °C$, 1 atm; the triple point, $-56.7\ °C$, 5.1 atm; and the critical point, $31\ °C$, 72.9 atm. The fusion curve slopes away from the pressure axis and at very high pressures reaches temperatures above the critical temperature.

QUESTION: What change occurs as the pressure is increased from 1 atm to 6 atm while the temperature is held at $-58\ °C$?

 Molecular Details of a Phase Diagram activity

forced at extremely high pressures. The solid state is the one we should expect in all cases. Furthermore, although it may seem like the critical temperature should be high for all substances, T_c for carbon dioxide (304.2 K) is only a little above room temperature (about 293 K).

Another interesting feature of the CO_2 phase diagram is that the triple-point pressure, 5.1 atm, is well above usual atmospheric pressures. Liquid carbon dioxide can be maintained only at pressures greater than 5.1 atm, and certainly not at normal atmospheric pressure. Liquid CO_2 does not have a normal boiling point, and solid CO_2 lacks a normal melting point. When solid CO_2 is heated at 1 atm pressure, it sublimes at $-78.5\ °C$. Solid CO_2, called dry ice, is a useful coolant for two reasons. First, because no melting occurs when solid CO_2 is maintained at normal atmospheric pressure, the chilled materials do not contact any liquid and remain dry. Second, the temperature of the remaining dry ice stays at $-78.5\ °C$, no matter how much dry ice has already sublimed. Thus, the cooling effect of the dry ice can be maintained for a relatively long time.

Water, H_2O

The notable feature of the phase diagram for water (Figure 11.13) is that the solid–liquid equilibrium curve (the fusion curve) has a negative slope; it tilts to the left, toward the pressure axis. The broken black lines represent the normal melting point of ice (1 atm, 0 °C). In contrast, notice the pressure-temperature condition represented by the broken red lines. If solid water (ice) is maintained at a pressure greater than 1 atm,

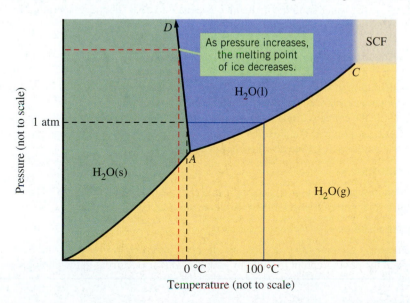

◀ **FIGURE 11.13** **Phase diagram of water, H_2O**

The triple point, A, is at $+0.0098\ °C$ and 0.00603 atm. The critical point, C, is at 374.1 °C and 218.2 atm. The negative slope of the fusion curve AD is greatly exaggerated. Not shown are the several polymorphic forms of ice that exist at pressures in excess of 2045 atm. We describe the significance of the broken black and red lines in the text.

QUESTION: What is the significance of the solid blue lines?

the melting point is below 0 °C. The melting point of solid water (ice) *decreases* as the pressure *increases*. Water is unusual in this respect. In the phase diagrams for almost all other substances, the fusion curve has a positive slope; it tilts to the right, away from the pressure axis. The melting points of most solids *increase* as the pressure *increases*.

Let's use these ideas about pressure and melting point to examine some hypotheses about a common activity: ice-skating. People have claimed that ice-skating is possible because the high pressure exerted by the skate blades causes the ice to melt. Thus, the skater may skim along on a thin film of liquid water from the melted ice. This explanation is probably not correct, however. The last point (*D*) on the fusion curve (*AD*) is at 2045 atm and −22.0 °C (−8 °F). At higher pressures and lower temperatures, ice exists in several polymorphic forms, and liquid water does not exist at all. Skaters cannot produce pressures nearly as high as 2045 atm, and moreover, it is possible to skate at temperatures below −22 °C.

A second hypothesis for why ice-skating is possible is that the frictional resistance of the ice to the skate blades causes an increase in temperature, which melts some of the ice. However, a more recent proposal is backed by experimental measurements. This model suggests that H_2O molecules on the surface of ice retain the vibrational freedom found in liquid water. The vibrational freedom of these molecules gives the ice surface a liquidlike quality.

Example 11.5 A Conceptual Example

In Figure 11.14, 50.0 mol $H_2O(g)$ (steam) at 100.0 °C and 1.00 atm is added to an insulated cylinder that contains 5.00 mol $H_2O(s)$ (ice) at 0 °C. *With a minimum of calculation*, use the data provided to determine which of the following will describe the final equilibrium condition: (a) ice and liquid water at 0 °C, (b) liquid water at 50 °C, (c) steam and liquid water at 100 °C, (d) steam at 100 °C. Data you will need are $\Delta H_{fusion} = 6.01$ kJ/mol, $\Delta H_{vapn} = 40.6$ kJ/mol (at 100 °C), molar heat capacity of $H_2O(l) = 76$ J mol^{-1} °C^{-1}.

ANALYSIS AND CONCLUSIONS

Two changes must occur: Ice must melt, and steam must condense. Let's calculate the quantities of heat involved if each change were to go to completion. Note that the heat of condensation of the steam is simply the negative of the heat of vaporization of $H_2O(l)$.

Melting of ice: $\qquad 5.00 \text{ mol} \times \dfrac{+6.01 \text{ kJ}}{1 \text{ mol}} = +30.1 \text{ kJ}$

Condensation of steam: $\quad 50.0 \text{ mol} \times \dfrac{-40.6 \text{ kJ}}{1 \text{ mol}} = -2030 \text{ kJ}$

Because the heat evolved in the condensation of the steam is far more than is required to melt the ice, no ice can remain. We can eliminate condition (a).

Suppose the water from the melted ice is heated to 50.0 °C, the temperature for condition (b). The quantity of heat required is

$$? \text{ kJ} = 5.00 \text{ mol} \times \frac{76 \text{ J}}{\text{mol °C}} \times (50.0 - 0.0) \text{ °C} \times \frac{1 \text{ kJ}}{1000 \text{ J}} = 19 \text{ kJ}$$

The condensation of steam provides much more heat than is needed to melt the ice *and* raise its temperature to 50.0 °C. We can eliminate condition (b).

Because the condensation of steam produces $H_2O(l)$, liquid water must be present; this eliminates condition (d). The final condition, then, must be one of liquid and gaseous water at 100 °C; condition (c). (See also Problem 84.)

EXERCISE 11.5A

A 1.05-mol sample of $H_2O(g)$ is compressed into a 2.61-L flask at 30.0 °C. Use a point on the phase diagram of Figure 11.13 to indicate the final condition reached.

EXERCISE 11.5B

When chunks of solid CO_2 (dry ice) are dropped into a beaker of water, the result is that shown in the photograph in the margin. Explain what is occurring in the photograph.

Initial condition

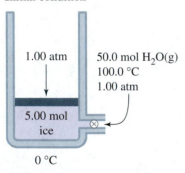

1.00 atm — 5.00 mol ice — 0 °C — 50.0 mol $H_2O(g)$ 100.0 °C 1.00 atm

▲ **FIGURE 11.14** Visualizing Example 11.5

 Heating Curves activity

▲ **Exercise 11.5B illustrated**

Liquid Crystals

The cholesterol molecule ($C_{27}H_{45}OH$) has an alcohol group and forms an ester with benzoic acid (C_6H_5COOH). The ester, cholesteryl benzoate ($C_{27}H_{45}OCOC_6H_5$), has some interesting properties. It melts sharply at 145.5 °C to produce a milky fluid. Heating the fluid to 178.5 °C causes it to change abruptly to a clear liquid. Between 145.5 and 178.5 °C, cholesteryl benzoate has the fluid properties of a liquid, the optical properties of a crystalline solid, and some unique properties of its own. It is in a form commonly called *liquid crystal*. Liquid crystals are familiar to almost everyone. We see them in liquid crystal display (LCD) devices such as digital watches, calculators, thermometers, and computer screens.

Liquid crystals often form from organic compounds that have rodlike molecules and molar masses in the range of a few hundred grams per mole. Figure 11.15 suggests three possibilities for the orientation of these rodlike molecules in liquid crystals. In *nematic* (threadlike) liquid crystals, the molecules are arranged in parallel. The molecules can move in any direction, and they can rotate on their long axes in the same way that a single pencil can be moved in a loosely packed box of pencils. In *smectic* (greaselike) liquid crystals, molecules are arranged in layers, with the long axes of the molecules perpendicular to the planes of the layers. The molecules can rotate on their long axes and move within the layers.

Cholesteric liquid crystals are related to the smectic form, but the orientation of the molecules in each layer is different from that in the layer above and below. A collection of layers forms a repeating sequence of orientations. The distance between two layers with the same orientation of molecules is a distinctive property of a cholesteric liquid crystal. When a beam of white light strikes a film of cholesteric liquid crystal, the color of the reflected light depends on this characteristic distance. Because this distance changes with temperature, so does the color of the reflected light. Some liquid crystal temperature-sensing devices exhibit color changes for temperature changes as small as 0.01 °C.

The orientation of molecules in a thin film of a nematic liquid crystal changes in the presence of an electric field. Liquid crystal displays exploit this change in orientation, which produces changes in optical properties of the film. In a digital watch or calculator display, electrodes coated with films of liquid crystals are arranged in patterns in the shape of numbers. When an electric field is imposed on the electrodes, the patterns of the electrodes (the numbers) become visible.

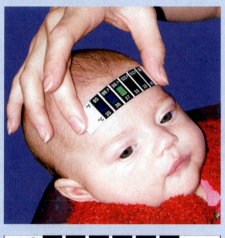

▲ A forehead thermometer made of cholesteric liquid crystals provides a fast, accurate, noninvasive way to measure body temperature.

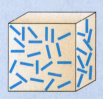

(a) Orientation of molecules in liquid

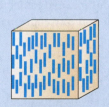

Nematic liquid crystal

Smectic liquid crystal

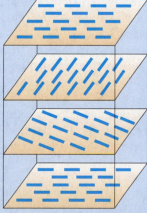

Cholesteric liquid crystal

(b) Orientation of molecules in liquid crystals

◀ **FIGURE 11.15**
The liquid crystalline state

Intermolecular Forces

Figures 9.1 and 9.2 illustrate repulsive forces at very short distances.

Dispersion Forces activity

(a) Unpolarized molecule

(b) Instantaneous dipole

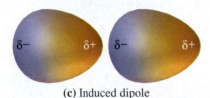

(c) Induced dipole

▲ **FIGURE 11.16 Dispersion forces**
(a) *Unpolarized molecule*. The electron charge distribution is symmetrical.
(b) *Instantaneous dipole*. A displacement of electron charge density (to the left) produces an instantaneous dipole.
(c) *Induced dipole*. The instantaneous dipole on the left induces charge separation in the molecule to its right, creating an induced dipole. The attraction between any two dipoles constitutes an intermolecular force.

As mentioned earlier, an **intermolecular force** exists *between* molecules. Although intermolecular forces are repulsive at extremely small separations, we find that they are predominately attractive at ordinary distances between molecules. When a gas is cooled sufficiently, it condenses to a liquid or solid; this could not occur without forces of attraction to make the gas molecules stick together. Even the noble gas helium can be obtained as a liquid at temperatures below about 5 K.

11.5 Intermolecular Forces of the van der Waals Type

In this section, we will describe two types of intermolecular forces that belong to a class called **van der Waals forces.** They are so called because they are the kind of forces we take into account in modifying the ideal gas equation into the van der Waals equation (page 204).

Dispersion Forces

To visualize an attractive molecular force among helium atoms, we must keep in mind that the electron charge density pictures we have used since Chapter 7 are for average situations only. For example, on average, the electron charge density associated with helium's two 1*s* electrons is evenly distributed in a spherical region about the nucleus. However, at any given instant, the actual location of the two electrons relative to the nucleus can lead to a fleeting separation of charge, creating an *instantaneous dipole*. This transitory dipole, in turn, can influence the distribution of electrons in neighboring helium atoms, producing *induced dipoles* in those atoms. Figure 11.16 illustrates the formation of an instantaneous dipole and an induced dipole. The force of attraction between an instantaneous dipole and an induced dipole is known as a **dispersion force** (also called a *London* force, named for Fritz London, who offered a theoretical explanation of these forces in 1928). Figure 11.17 shows a familiar example of the induction of electric charge from the everyday world.

The **polarizability** of an atom or molecule is a measure of the ease with which electron charge density is distorted by an external electric field; it reflects the facility with which a dipole can be induced. In general:

> *The greater the polarizability of molecules, the stronger the intermolecular forces between them.*

Large atoms have more electrons and larger electron clouds than small atoms. In a large atom, the outer electrons are more loosely bound; they can shift toward another atom more readily than can the more tightly bound electrons in a small atom. Large atoms and molecules are therefore more polarizable than small ones. Atomic and molecular sizes are also closely related to atomic and molecular masses, which means that both polarizability and intermolecular force strength increase with increased atomic and molecular mass.

We can readily note this trend in the physical properties of the group 7A elements, the halogens. All are nonpolar. The first member, fluorine (F_2), is a gas at room temperature (boiling point -188 °C). The second member, chlorine (Cl_2), is also a gas (boiling point -34 °C), but it is more easily liquefied than is fluorine. At room temperature, bromine (Br_2) is a liquid (boiling point 58.8 °C) and iodine (I_2) is a solid (melting point 184 °C). Because large molecules are easily polarizable, intermolecular forces between them are strong enough to form liquids or even solids.

◄ **FIGURE 11.17 The phenomenon of induction**
The balloon develops a static electric charge when it is rubbed with a cloth. Then, when it is brought close to a surface, the charged balloon induces an electric charge on the surface, with the sign of that surface charge being opposite the sign of the charge on the balloon. The balloon is then attracted to the surface and maintained there by an electrostatic force of attraction.

(a) Octane
$CH_3(CH_2)_6CH_3$
melting point −56.8 °C
boiling point 125.7 °C

◄ FIGURE 11.18
Molecular shape and polarizability
Octane's higher melting point and boiling point arise because the elongated octane molecules have stronger dispersion forces than do the more compact isooctane molecules.

QUESTION: Where should the boiling point of 2,4-dimethylhexane fall relative to the boiling points of octane and isooctane?

(b) (Isooctane)
2,2,4-Trimethylpentane
$CH_3C(CH_3)_2CH_2CH(CH_3)_2$
melting point −104.7 °C
boiling point 99.2 °C

Another factor that affects the strength of dispersion forces is molecular shape. Elongated molecules make contact with neighboring molecules over a greater surface area than do more compact molecules. As a result, the dispersion forces among elongated molecules are greater than among more compact molecules. Figure 11.18 describes a pair of isomers found in gasoline: octane and isooctane. They have identical molecular masses but different molecular shapes. Intermolecular forces are stronger among the more elongated octane molecules than among the more compact isooctane molecules. As a result, octane has a higher melting point and a higher boiling point than isooctane.

 Octane and Isooctane 3D models

The Gecko's Toe

The gecko can walk across a glass ceiling and hang from a single toe. This adhesive power of the gecko's toe exceeds that of some commercial adhesives. Attempts to explain this behavior have included suction, sticky glands, even Velcro® like action. In 1999, scientists at Lewis and Clark College and at the University of California, Berkeley, suggested that the secret resides in *seta*—millions of tiny hairlike projections on each foot, like tiny bristles on a soft brush. When the gecko's foot is pressed against a surface, the ends of the seta can closely approach the surface, close enough for van der Waals attractive forces to come into play. Because the forces involved are intermolecular forces and not covalent bonds, the adhesion is not permanent. The gecko can manipulate its feet to release the contact and unstick the foot very quickly.

The concept is now being investigated for commercial use. For instance, scientists have succeeded in creating a surface covered with synthetic microscopic fibers similar to gecko hairs, attaching this surface to an object, and sticking the object to the underside of a glass plate. Ultimately, it might be possible to produce an adhesive tape that would be stronger than almost any common glue, yet easily released from a surface simply by peeling it off at the correct angle.

▲ Scientists have recently developed an adhesive tape (such as the one attaching this toy to the surface overhead) that contains an array of microscopic, flexible polymer hairs that act through dispersion forces, much as the seta of the gecko's toe.

◄ The giant Madagascar day gecko can walk on a wall or ceiling as readily as on the floor.

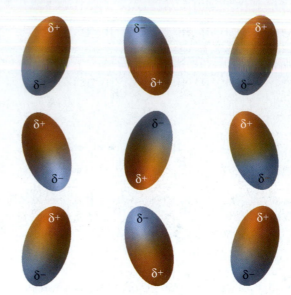

▶ **FIGURE 11.19** **Dipole–dipole interactions**
Thermal motion of the molecules prevents a perfect alignment of these permanent dipoles. Nevertheless, the dipoles do maintain a general arrangement leading to the attractions $\delta+ \cdots \delta-$.

Nitrogen, Nitrogen Monoxide, and Oxygen 3D models

Propane and Acetaldehyde 3D models

Dipole–Dipole Forces

As we have just shown, instantaneous and induced dipoles can form in a nonpolar substance. Recall that a *polar* substance, because of its molecular shape and differences in electronegativities between bonded atoms, has permanent dipoles. As shown in Figure 11.19, permanent dipoles tend to align themselves with the positive end of one dipole directed toward the negative ends of neighboring dipoles, giving rise to *dipole–dipole forces*. This preferred alignment of permanent dipoles is partially offset by the random thermal motion of the molecules, which occurs in liquids more than in solids. Also, when dipolar molecules come close to one another, repulsions occur between like-charged regions. However, a permanent dipole in one molecule can induce a dipole in a neighboring molecule, giving rise to a *dipole-induced dipole force*. This type of attraction exists even when the permanent dipoles are not perfectly aligned.

There are *net* attractive forces in a collection of polar molecules. Moreover, the dipole–dipole and dipole-induced dipole forces are *in addition to* the dispersion forces found between all molecules. As a result, the sum of the intermolecular forces is greater in a polar substance than in a nonpolar substance of about the same molar mass. Let's compare nitrogen, nitrogen monoxide, and oxygen to see how polarity affects intermolecular forces and hence the boiling points of these substances:

	N_2	NO	O_2
Molar mass, g/mol	28.0	30.0	32.0
Dipole moment, μ (D)	0	0.15	0
Boiling point, K	77	121	90

Stronger intermolecular forces are present in NO(l) than in N_2(l) or O_2(l); liquid NO must therefore be warmed to a higher temperature than the other two before it boils.

The more polar a molecule—that is, the greater its dipole moment—the more pronounced the effect of dipole–dipole forces on physical properties. We can see this by comparing two substances of nearly identical molar masses: *propane*, C_3H_8 (44.10 g/mol), and *acetaldehyde*, CH_3CHO (44.05 g/mol). The electronegativity difference between C and H atoms is very small, and therefore propane is nonpolar. In acetaldehyde, the electronegativity difference between C and O is large. This produces a bond dipole that is not offset by other bond dipoles and consequently a large resultant dipole moment ($\mu = 2.69$ D). As we expect, acetaldehyde has a considerably higher boiling point (20.2 °C) than propane (−42.1 °C).

Predicting Physical Properties of Molecular Substances

We can use knowledge of the intermolecular forces that exist in various substances to make predictions about trends in such properties as melting points, boiling points, and enthalpies of vaporization. The following summary can help us to do so:

- Dispersion forces become stronger with increasing molar mass and increasing elongation in molecules. *In comparing nonpolar substances, molar mass and molecular shape are the essential factors.*

- Dipole–dipole and dipole-induced dipole forces are found in polar substances. *In comparing polar substances with nonpolar substances of comparable molar masses, intermolecular forces are usually stronger in the polar substances. The more polar the substance, that is, the larger its dipole moment (μ), the greater we expect the intermolecular force to be.*

- Because they occur in *all* molecular substances, dispersion forces must always be considered. Often they predominate.

Example 11.6

Arrange the following substances in the expected order of increasing boiling point: carbon tetrabromide, CBr_4; butane, $CH_3CH_2CH_2CH_3$; fluorine, F_2; acetaldehyde, CH_3CHO.

STRATEGY

We should expect to be generally successful in making comparisons of the sort required here, by carefully applying the three ideas described in the list above.

SOLUTION

The first three substances are nonpolar: F_2 because the atoms in the molecules are identical, CBr_4 because of the symmetrical tetrahedral molecular structure, and butane because the electronegativities of the C and H atoms are so nearly alike. We expect the boiling points of the three to increase with increasing molar mass:

$$F_2(38.00 \text{ g/mol}) < CH_3CH_2CH_2CH_3(58.12 \text{ g/mol}) < CBr_4(331.6 \text{ g/mol})$$

As we noted on page 452, the carbon-to-oxygen bond dipole in acetaldehyde is not offset by any other bond dipoles. Even though the molar mass of acetaldehyde (44.05 g/mol) is somewhat smaller than that of butane, we expect acetaldehyde to have the higher boiling point because this molecule is quite polar ($\mu = 2.69$ D).

Comparing acetaldehyde and CBr_4 is more difficult. The polar character of acetaldehyde suggests that it should have the higher boiling point. Because the molar mass of CBr_4 is so very much greater than that of acetaldehyde, though, we should expect CBr_4 to have the higher boiling point. We predict the following order of increasing boiling points:

$$F_2 < CH_3CH_2CH_2CH_3 < CH_3CHO < CBr_4$$

ASSESSMENT

The observed boiling points of 20.2 °C for acetaldehyde and 189.5 °C for CBr_4 confirm our expectation that the dispersion forces associated with massive CBr_4 molecules are more significant than the dipole–dipole forces in acetaldehyde.

EXERCISE 11.6A

Consider the two substances, BrCl and IBr. One is a solid at room temperature, and one is a gas. Which is which? Explain your choice.

EXERCISE 11.6B

Which member of each pair would you expect to have the *lower* boiling point? Explain.

(a) toluene, $C_6H_5CH_3$, or aniline, $C_6H_5NH_2$

(b) *cis*-1,2-dichloroethene or *trans*-1,2-dichloroethene (see page 416)

(c) *ortho*-dichlorobenzene or *para*-dichlorobenzene (see page 423)

Problem-Solving Note

We can't say precisely how large a difference in molar mass is required in order for a nonpolar substance to have a higher boiling point than a polar one. As we see here, however, a difference of several hundred g/mol is more than enough, and a difference of 10 or 20 g/mol is not enough.

Homology

A series of compounds whose formulas and structures vary in a regular manner also have properties that vary in a predictable manner, a principle called **homology**. For example, both the densities and the boiling points of the straight-chain alkanes increase in a continuous and regular fashion with increasing numbers of carbon atoms in the chain. Such trends occur because the regular increase in molar mass produces a fairly regular increase in the strength of dispersion forces. Homology aids our study of organic chemistry in much the same way that the periodic table provides an organizing principle for the chemistry of the elements. Instead of studying the properties of a bewildering array of individual organic compounds, we can usually study a few members of a *homologous series* and deduce properties of the others.

Example 11.7   An Estimation Example

The boiling points of the straight-chain alkanes pentane, hexane, heptane, and octane are 36.1, 68.7, 98.4, and 125.7 °C, respectively. Estimate the boiling point of the straight-chain alkane decane.

ANALYSIS AND CONCLUSIONS

We need to figure out the pattern of increasing boiling point with increasing length of the carbon chain. The data are for carbon-chain lengths of five, six, seven, and eight C atoms. We want to estimate the boiling point of the 10-carbon alkane, *dec*ane.

Alkane formula	C_5H_{12}	C_6H_{14}	C_7H_{16}	C_8H_{18}	C_9H_{20}	$C_{10}H_{22}$
Boiling point, °C	36.1	68.7	98.4	125.7	?	?
Increase per CH$_2$ unit, °C		32.6	29.7	27.3	≈25	≈23

Note the trend in boiling-point increase per added CH$_2$ unit: 32.6, 29.7, 27.3. This trend suggests that the boiling point of C_9H_{20} is about 25 °C above that of C_8H_{18}, and that the boiling point of $C_{10}H_{22}$ is about 23 °C greater than that of C_9H_{20}. We therefore estimate the boiling point of decane to be 174 °C. (The observed boiling point is 174.1 °C.)

EXERCISE 11.7A

Use data from Example 11.7 to estimate the boiling point of butane (C_4H_{10}).

EXERCISE 11.7B

The kerosene component of petroleum consists of hydrocarbons with boiling points ranging from about 200 to 260 °C. Based on the data in Example 11.7, what are the formulas of the straight-chain alkanes you would expect to find in kerosene?

11.6   Hydrogen Bonds

Suppose we are asked to predict the boiling points of water and acetaldehyde. If we reason as we did in Example 11.6, we would conclude that acetaldehyde, CH_3CHO ($\mu = 2.69$ D), should have a higher boiling point than water, H_2O ($\mu = 1.84$ D). Both are polar, but acetaldehyde has a larger dipole moment and also a higher molar mass (44.05 g/mol) than water (18.02 g/mol). However, our prediction would be wrong. Water boils at 100 °C, whereas acetaldehyde boils at 20.2 °C. Why is our prediction incorrect? A likely explanation is that there is an additional kind of intermolecular force that is either not found in acetaldehyde or is not nearly so important in acetaldehyde as it is in water. There is indeed such a force—a *hydrogen bond*.

A **hydrogen bond** between molecules is an intermolecular force in which a hydrogen atom covalently bonded to a nonmetal atom in one molecule is *simultaneously* attracted to a nonmetal atom of a neighboring molecule. Although in rare cases the nonmetal atoms may include chlorine and sulfur, the strongest hydrogen bonds are formed if the nonmetal atoms are *small* and *highly electronegative*.

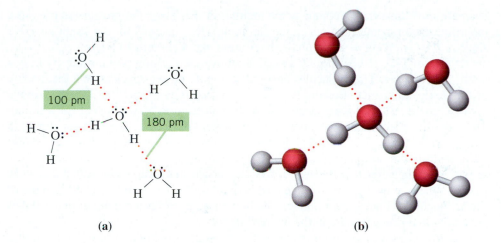

(a)   **(b)**

◀ **FIGURE 11.20   Hydrogen bonds in water**
As suggested through (a) Lewis structures and (b) ball-and-stick models, each water molecule is linked to four others through hydrogen bonds. Each H atom lies along a line that joins two O atoms. The shorter distances (100 pm) correspond to covalent bonds, and the longer distances (180 pm) correspond to the hydrogen bonds.

This limits hydrogen bonding mainly to species containing nitrogen, oxygen, or fluorine atoms.

Think of a hydrogen bond in this way: In a covalent bond between an H atom and, for example, an O atom, an electron cloud joins the hydrogen atom to the oxygen atom. The electron cloud is much denser at the oxygen end of the bond, making the bond polar with $\delta-$ on the O atom and $\delta+$ on the H atom. This leaves the hydrogen nucleus somewhat exposed. As a result, an oxygen atom of a neighboring molecule can approach that hydrogen nucleus rather closely. This shorter internuclear distance makes for a strong force of attraction. Hydrogen bonding in water is suggested in Figure 11.20, where we follow the common convention of using dotted lines to represent hydrogen bonds.

Although "bond" is part of the name, we should think of a hydrogen bond as a special type of intermolecular force.

   Hydrogen Bonding activity

Water: Some Unusual Properties

The model for ice in Figure 11.21 shows how hydrogen bonds hold water molecules in a rigid but open structure. As ice melts, some of the hydrogen bonds are overcome, and water molecules move into the holes that were in the ice structure. As a result, the H_2O molecules are closer together in liquid water than in ice. When ice melts, there is about a 10% decrease in volume and a corresponding increase in density. Water is most unusual in this regard because for most substances the liquid state is *less dense* than the solid state. High pressures facilitate the disruption of hydrogen bonding and the

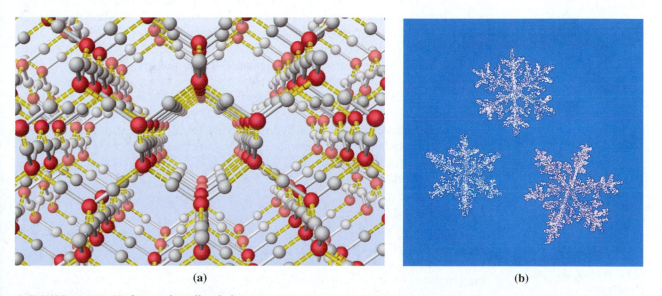

(a)   **(b)**

▲ **FIGURE 11.21   Hydrogen bonding in ice**
(a) Oxygen atoms are arranged in layers of distorted hexagonal rings. Hydrogen atoms lie between pairs of O atoms, closer to one O (covalent bond) than to the other (hydrogen bond). (b) At the macroscopic level, this structural pattern is revealed in the hexagonal shapes of snowflakes.

decrease in volume that accompany the melting of ice. Thus, the greater the pressure, the lower the temperature at which ice can melt. This accounts for the slight negative slope of the fusion curve in the phase diagram of water (Figure 11.13).

If we continue to heat liquid water just above the melting point, more hydrogen bonds are ruptured. The molecules become still more closely packed, and the density of the liquid water increases until a maximum density is reached at 3.98 °C. Above 3.98 °C, the density of water decreases with temperature, as we expect for a liquid. These density phenomena explain why a freshwater lake freezes from the top down. In winter, when the surface water temperature reaches 4 °C, some of this more dense water sinks to the bottom of the lake as the remaining surface water continues to cool to 0 °C and then freezes. Because ice is less dense than water, the water that does freeze remains at the top to cover the lake with a layer of ice. The ice covering reduces further heat loss. Except for relatively shallow lakes in extremely cold climates, lakes generally do not freeze solid in the wintertime.

Hydrogen bonding is also the reason for the unusually high boiling point of water (100 °C). Many substances having considerably higher molar masses than water (for example, CO_2 and SF_6) are gases at room temperature because the only intermolecular forces in these nonpolar gases are dispersion forces. Moreover, even some polar substances, such as SO_2, are gases at room temperature. This clearly illustrates the strength of hydrogen bonding as compared to dispersion and dipole–dipole forces.

Hydrogen Bonding in Organic Substances

Hydrocarbons form no hydrogen bonds because carbon atoms, though small, are not highly electronegative. Many organic compounds contain oxygen or nitrogen, however, and so hydrogen bonding is common in much of organic chemistry.

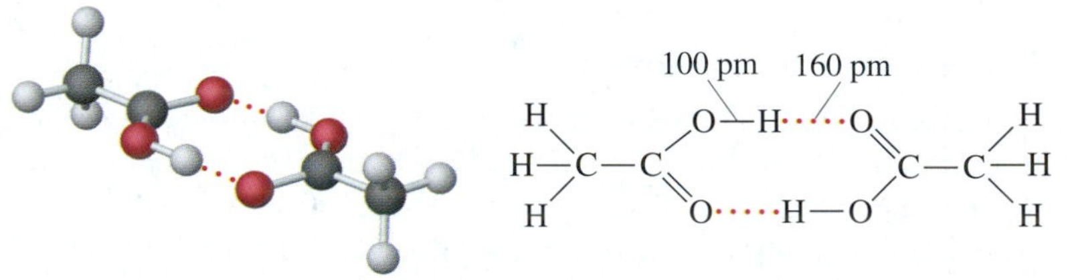

▶ **FIGURE 11.22   Hydrogen bonding in acetic acid**

The two acetic acid molecules are joined through two intermolecular hydrogen bonds (dotted lines) into a "double" molecule, or dimer. Note that the hydrogen bond lengths (H $\cdots$ O) are longer than the H–O covalent bond lengths.

Acetic acid, CH_3COOH, meets the requirements that we have set forth for hydrogen bonding. However, it has a much lower heat of vaporization than we would expect for a substance with strong intermolecular forces. Why should this be? We find that there is indeed hydrogen bonding in acetic acid. In fact, the hydrogen bonds are strong enough to produce double molecules called *dimers* (Figure 11.22). When acetic acid vaporizes, many of the dimers hold together. Because the hydrogen bonds in these dimers do not need to be broken in order for vaporization to occur, it takes less energy to convert a given quantity of liquid to vapor than we might expect, and the heat of vaporization is abnormally low.

In some organic molecules, a hydrogen bond forms between two atoms in the same molecule, and these molecules have an *intra*molecular hydrogen bond. An example is seen in salicylic acid (Figure 11.23), an analgesic (pain reliever) and antipyretic (fever reducer) related to aspirin.

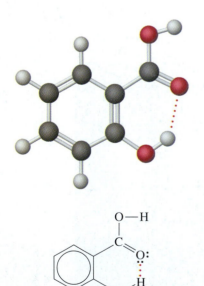

▲ **FIGURE 11.23   Hydrogen bonding in salicylic acid**

An intramolecular hydrogen bond (dotted line) joins the OH group to the doubly bonded oxygen atom of the carboxyl group on the same molecule.

Example 11.8

In which of these substances is hydrogen bonding an important intermolecular force: N_2, HI, HF, CH_3CHO, and CH_3OH? Explain.

STRATEGY

We must examine each molecule to determine whether the essential criteria for hydrogen bonding are met, namely, H atoms bonded to *small* and *highly electronegative* nonmetal atoms—N, O, or F.

SOLUTION

N$_2$: Nitrogen atoms are small and highly electronegative, but we cannot have hydrogen bonds without hydrogen atoms. *Hydrogen bonding does not occur.*

HI: Hydrogen atoms are present, but iodine atoms are large and only moderately electronegative. *Hydrogen bonding does not occur.*

HF: Hydrogen atoms are bonded to small, highly electronegative nonmetal atoms (fluorine). *Hydrogen bonding is a significant intermolecular force.*

CH$_3$CHO: Both hydrogen atoms and highly electronegative nonmetal atoms (oxygen) are present, but as we see from the structural formula, the hydrogen atoms are bonded to *carbon,* not oxygen:

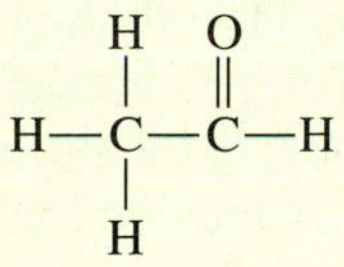

$$\begin{array}{ccc} H & O & \\ | & \| & \\ H - C - C - H \\ | \\ H \end{array}$$

Hydrogen bonding is not a significant intermolecular force.

CH$_3$OH: Again, both hydrogen atoms and a small, highly electronegative nonmetal oxygen atom are present, and one of the hydrogen atoms is bonded to the oxygen atom:

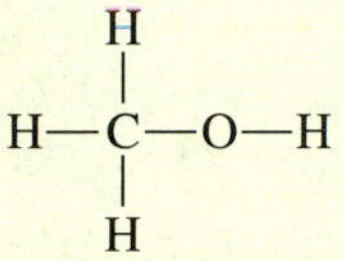

$$\begin{array}{c} H \\ | \\ H - C - O - H \\ | \\ H \end{array}$$

Hydrogen bonding is a significant intermolecular force.

EXERCISE 11.8A

In which of these substances is hydrogen bonding an important intermolecular force: NH$_3$, CH$_4$, C$_6$H$_5$OH, CH$_3$COOH, H$_2$S, H$_2$O$_2$? Explain.

EXERCISE 11.8B

Arrange the liquids represented by the following molecular models in the expected order of increasing normal boiling point.

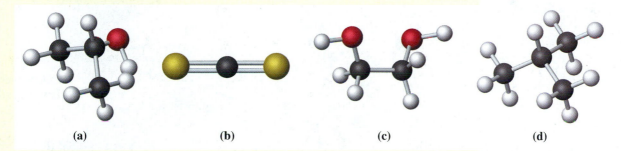

(a)   (b)   (c)   (d)

Hydrogen Bonding and Life Processes

The hydrogen bond may seem to be merely an interesting piece of chemical theory, but we cannot overstate its importance to life and health. The structure of proteins, substances essential to life, is determined partly by hydrogen bonding. The action of enzymes, the protein molecules that catalyze the reactions that sustain life, depends in part on the forming and breaking of hydrogen bonds. The hereditary information passed from one generation to the next is carried in nucleic acid molecules joined by hydrogen bonds into an elegant structure. In DNA and proteins, certain bonds must be easy to break and re-form. Of all the kinds of intermolecular forces, only hydrogen bonds have exactly the right amount of energy for this—from about 15 to 40 kJ/mol. In contrast, covalent chemical bonds have energies that range from about 150 to several hundred kJ/mol, and van der Waals forces have energies in the range of only 2 to 20 kJ/mol.

Proteins and Amino Acids animation

Hydrogen Bonding in Proteins

Proteins are polymers constructed from a basic set of 20 amino acid monomers. As we noted in Chapter 9 (page 375), the amino acid units are joined though peptide bonds. The sequence of amino acids in a protein chain is the *primary structure* of the protein. The chains, in turn, can twist and fold to give a variety of shapes to protein molecules. These shapes are called the *secondary structure* of the protein. As Figure 11.24 suggests, hydrogen bonding is of great importance in determining the two main kinds of secondary structures found in all proteins.

Another level of organization, called the *tertiary structure* of a protein, is its complete three-dimensional shape. The tertiary structure is directly related to the function of the protein, whether as an enzyme (Chapter 13), a structural protein as is found in muscle and tendon, or in other forms. The folding of a protein chain brings otherwise distant parts of its primary structure close together. Interactions between side chains of the amino acids in these regions stabilize the tertiary structure. These interactions are of four types (Figure 11.25), and they include ionic bonds, covalent bonds, and intermolecular forces (both hydrogen bonds and dispersion forces).

DNA Segment 3D model

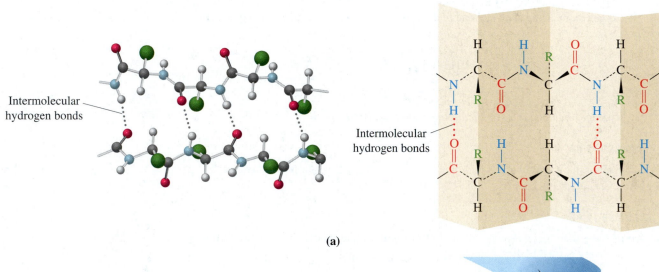

Intermolecular hydrogen bonds

Intermolecular hydrogen bonds

(a)

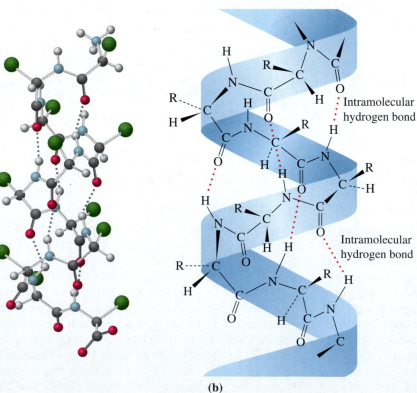

Intramolecular hydrogen bond

Intramolecular hydrogen bond

(b)

▶ **FIGURE 11.24** **The two principal secondary structures of proteins**

(a) In the pleated sheet arrangement, protein chains run parallel and in alternating directions. The chains are held together by hydrogen bonds that join an NH group on one chain with a CO group on a neighboring chain. The beige background emphasizes the pleats (imagine that the formula is written on a pleated sheet of paper). (b) In the α-helical arrangement, the protein chain is coiled into a helix. Each NH group is hydrogen bonded to a CO group one helical turn (3.6 amino acid units) away in the same chain, giving a fairly rigid cylindrical structure with side chains on the outside.

In some proteins, the chains are twisted about one another into larger cables and fibers, which are used for connections, support, and structure. These *fibrous proteins* are found, for example, in hair, skin, and muscle and in insect fibers such as silk. In other proteins, the chains are folded back on themselves to make compact *globular proteins.* Hemoglobin, enzymes, and the gamma globulins that serve as protective antibodies are globular proteins. Enzymes are protein catalysts that make nearly all the reactions in living cells possible.

A protein molecule functions only when it is in the proper configuration. Such factors as heat, ultraviolet light, and some chemical substances *denature* proteins, that is, change them in ways that destroy their function. These factors work by disrupting the bonds and intermolecular forces that hold the molecules in their proper arrangement. Most proteins are denatured when heated above 50 °C. Heat, ultraviolet light, and organic compounds such as alcohols and phenols are used to disinfect things by denaturing the proteins in bacteria and thus killing the bacteria. We cook most of our protein-containing food in part to kill harmful microorganisms and also because denatured proteins are usually easier to chew and more easily broken down by digestive enzymes.

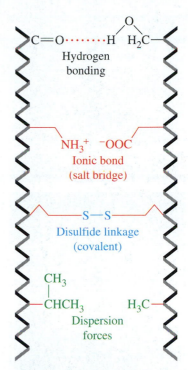

▲ **FIGURE 11.25   The four types of interactions that determine the tertiary structure of proteins.**

11.7   Intermolecular Forces and Two Liquid Properties

So far, we have seen that the strengths of intermolecular forces affect densities, melting points, boiling points, and enthalpies of vaporization. Other physical properties are also affected. In this section, we look at two familiar properties that reflect intermolecular forces quite well: surface tension and viscosity.

Surface Tension

Regardless of the nature of the intermolecular forces in a liquid, molecules within the bulk of a liquid are attracted to more neighboring molecules than are surface molecules (Figure 11.26). Because the molecules within the liquid experience a greater net intermolecular attractive force than do surface molecules, the molecules in the bulk of a liquid are in a lower energy state than are surface molecules. Also, given that molecules tend toward as low an energy state as possible, they crowd into the interior of a liquid to the greatest extent possible, leaving a minimum of surface area. A sphere has a smaller ratio of surface area to volume than any other three-dimensional figure, and free-falling liquids therefore tend to form spherical drops.

It takes energy to increase the surface area of a liquid because to do so requires that some molecules pass from the lower-energy bulk liquid to the higher-energy surface state. **Surface tension** is the amount of work required to extend a liquid surface. Surface tension (γ) is usually expressed in the unit joule per square meter (J/m^2)—that is, the work required to extend the surface area of a liquid by one square meter. Consider these two values of surface tension at 20 °C:

$$\text{Hexane, } CH_3(CH_2)_4CH_3 \qquad \gamma = 0.0184 \ J/m^2$$
$$\text{Water, } H_2O \qquad \gamma = 0.0729 \ J/m^2$$

We can account for the fact that the surface tension of water is significantly greater than that of hexane by comparing intermolecular forces. Strong intermolecular hydrogen bonding makes it more difficult to extend the surface of liquid water than to

◄ **FIGURE 11.26   Intermolecular forces in a liquid**

Molecules at the surface of a liquid are attracted only by other molecules at the surface and by molecules below the surface. Molecules in the interior of a liquid experience forces from neighboring molecules in all directions.

▲ FIGURE 11.27 Surface tension of water

The surface tension of water enables the steel needle, though denser than water, to float on the surface of the water. It also supports the insect known as a water strider.

Intermolecular Forces and Interfaces activity

extend the surface of liquid hexane, where the intermolecular forces are dispersion forces only.

When liquids are heated, the intermolecular forces are more easily overcome by the increased thermal energy of the molecules. Surface tension therefore decreases with an increase in temperature.

The tendency of molecules to be drawn toward the interior of a liquid makes the surface act as if it were covered with a tight skin. This allows the steel needle in Figure 11.27 to float on water, even though the needle is much denser than water and "should" sink. The force of gravity on the needle (its weight) is not enough to break the skin, that is, to spread the surface of the water over the needle.

When a drop of liquid spreads across a surface, we say that the liquid "wets" the surface. The two major factors that determine whether a liquid will spread across a surface are the strengths of adhesive and cohesive forces. **Adhesive forces** are intermolecular forces between unlike molecules, and **cohesive forces** are those between like molecules. If the adhesive forces between a liquid and a surface are stronger than the cohesive forces within the liquid, the liquid will wet the surface. If the adhesive forces are weaker, the liquid will not wet the surface.

In order to clean a surface with water, the water must wet the surface. Glass and some fabrics, for example, are easily wet by water. If glass is coated with a film of grease or oil, however, water won't wet the surface because the forces of attraction between the polar water molecules and nonpolar oil molecules are weak. Instead, water droplets bead on the greasy film (Figure 11.28). A detergent acts to disperse the grease in water. To waterproof a fabric (a canvas tent, for example), we purposely coat the material with an oil that prevents water from wetting the fabric.

If a liquid wets the interior surface of its container, the liquid is drawn up the walls of the container to a slight extent. The interface between the liquid and the air above it, called a **meniscus,** has a *concave,* or saucerlike, shape ($\smile$). In contrast, if a liquid does not wet the walls of its container, the liquid pulls away from the wall and exhibits a *convex* meniscus, like an inverted saucer ($\frown$). Water in a glass container produces a concave meniscus, and mercury forms a convex one (Figure 11.29). Meniscus formation is greatly exaggerated in tubes of very small diameter, called *capillary tubes.* We call this phenomenon *capillary action.* As shown in Figure 11.30, water is drawn up several centimeters into a small glass capillary tube. A sponge soaks up water through capillary action.

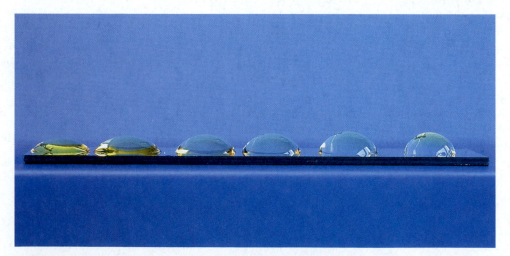

▲ FIGURE 11.28 Adhesive and cohesive forces

When water droplets are placed on a surface coated with an organic film that varies in composition from left to right, the varying strength of interaction between water molecules and the film is reflected in the wetting behavior of the water. Strong adhesive forces between water and the film cause water droplets to spread on the film containing OH groups at the surface (far left). When water droplets are placed on the film containing CH_3 groups at the surface (far right), adhesive forces between the film and water cannot overcome the cohesive forces in water; drops of water "bead up."

**▲ FIGURE 11.29
Meniscus formation**

Because it wets glass, water forms a concave meniscus in a glass container (left). Mercury does not wet glass (right), and for that reason forms a convex meniscus.

QUESTION: Which meniscus results from adhesive forces being weaker than cohesive forces?

Water rises because of capillary action.

◄ **FIGURE 11.30   Capillary action**
The spread of a thin film of water up the walls of a capillary tube produces a slight drop in pressure below the meniscus. Atmospheric pressure then pushes a column of water up the capillary to offset the pressure difference. The force of gravity on the liquid column opposes this push and thereby limits the height to which the water rises.

Viscosity

When a liquid flows, one portion of it moves with respect to neighboring portions, and cohesive forces within the liquid create an internal friction that reduces the rate of flow. The **viscosity** of a liquid is a measure of its resistance to flow. The unit of viscosity is the *poise* (P).* Liquids such as molasses, honey, and heavy motor oil do not flow readily; they have a high viscosity and are said to be *viscous*. On the other hand, hexane, water, and ethanol flow easily and are *mobile* liquids.

The viscosity of a liquid is related in part to the strength of intermolecular forces—the stronger these forces, the greater the viscosity. We can see this rather clearly by comparing three alcohols:

$CH_3CH_2CH_2OH$	$CH_3CHOHCH_2OH$	$CH_2OHCHOHCH_2OH$
Propyl alcohol	Propylene glycol	Glycerol
(1-propanol)	(1,2-propanediol)	(1,2,3-propanetriol)
0.02 P	0.5 P	10 P

For these three liquids, we would expect increasing molar mass to produce only a gradual increase in viscosity from left to right. However, the dramatic increases we see are attributed to hydrogen bonding. As the number of OH groups in an alcohol increases, the possibilities for hydrogen bonding increase and the resistance to flow (the viscosity) also increases.

The viscosity of a liquid can be determined by timing its flow under carefully controlled conditions. The flows of two motor oils of widely different viscosities are compared in Figure 11.31.

Intermolecular forces become less effective with an increase in temperature, and as is the case with surface tension, viscosity decreases with an increase in temperature.

▲ **FIGURE 11.31   Measuring viscosity**
The flow of automotive motor oil is started at the same time from the two identical funnels. The highly viscous oil (SAE 40) on the left flows much more slowly than the less viscous oil (SAE 10) on the right.

QUESTION: Which of the two motor oils would likely perform better in Alaska in the winter? In the Arabian desert in the summer?

The Structures of Solids

The constituent atoms, ions, or molecules of a solid are in close contact with one another. If these structural units are in disorganized clusters with no long-range order, a solid is *amorphous*. In many solids, however, the structural units exist in highly ordered assemblies called *crystals*. Recall that we described the formation of a crystal of NaCl in Section 9.3. Table 11.5 lists some characteristics of the major types of crystalline solids. In the remainder of this chapter, we will discuss several of these types of solids.

---

*The poise (named for J. L. Poiseuille) is a non-SI unit. The SI unit of viscosity is $1\ N\ s\ m^{-2} = 10\ P$.

Table 11.5 Some Characteristics of Crystalline Solids

Type	Structural Particles	Intermolecular Forces	Typical Properties	Examples
Molecular				
Nonpolar	Atoms or nonpolar molecules	Dispersion forces	Extremely low to moderate melting points; soluble in nonpolar solvents	$Ar, H_2, I_2,$ CCl_4, CH_4, CO_2
Polar	Polar molecules	Dispersion forces, dipole–dipole and dipole-induced dipole attractions	Low to moderate melting points; soluble in some polar and some nonpolar solvents	$HCl, H_2S,$ $CHCl_3, (CH_3)_2O,$ $(CH_3)_2CO$
Hydrogen-bonded	Molecules with H bonded to N, O, or F	Hydrogen bonds	Low to moderate melting points; soluble in some hydrogen-bonded and some polar liquids	$H_2O, HF,$ $NH_3, CH_3OH,$ CH_3COOH
Network Covalent	Atoms	Covalent bonds	Most are very hard; sublime or melt at very high temperatures; most are nonconductors of electricity	C(diamond), C(graphite) SiC, SiO_2, BN
Ionic	Cations and anions	Electrostatic attractions	Hard; brittle; moderate to very high melting points; nonconductors as solids, but electrical conductors as liquids; many are soluble in water	$NaCl, CaF_2,$ K_2S, MgO
Metallic	Cations and delocalized electrons	Metallic bonds	Hardness varies from soft to very hard; melting points vary from low to very high; lustrous; ductile; malleable; good to excellent conductors of heat and electricity	Na, Mg, Al, Fe, Cu, Zn, Mo, Ag, Cd, W, Pt, Hg, Pb

11.8 Network Covalent Solids

In general, *intra*molecular forces (bonds between atoms) are quite strong. In contrast, *inter*molecular forces are much weaker, perhaps only a few percent as strong as intramolecular forces. As a result, many molecular substances exist as gases at room temperature, and the rest are mostly liquids or solids with low to moderate melting points. However, in a few covalently bonded substances, called **network covalent solids,** a network of covalent bonds extends throughout a crystalline solid, holding it firmly together. Two prime examples are diamond and graphite, the two principal forms of carbon.

Diamond

Knowing that the carbon-to-carbon bonds in diamond are single covalent bonds, we can attempt to write a Lewis structure:

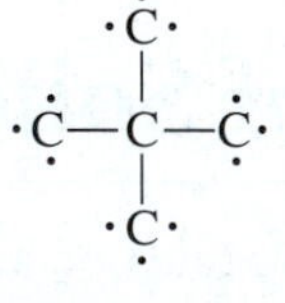

$$\cdot \overset{\displaystyle \cdot \overset{\textstyle \cdot}{C} \cdot}{\underset{\displaystyle \cdot \underset{\textstyle \cdot}{C} \cdot}{C}} \cdot C \cdot$$

An unsatisfactory Lewis structure for diamond

In this structure, only the central C has a valence-shell octet. Even though the Lewis structure will still be unsatisfactory, we can increase the proportion of C atoms with valence-shell octets by adding more carbon atoms. In this way, we can build up a gigantic "molecule" that consists of all the atoms in the diamond crystal. In other words, the C atoms of a diamond form a network covalent solid.

As shown in the tiny portion of a diamond crystal represented in Figure 11.32, each carbon atom is bonded to four other carbon atoms in a tetrahedral arrangement. This bonding arrangement is consistent with sp^3 hybridization of the carbon atoms. To scratch or break a diamond crystal, we have to break many covalent bonds. Because this is quite difficult to do, diamond is the hardest substance known. No other substance can scratch a diamond, but a diamond can scratch other solids, such as glass. Diamond is the best abrasive available. To melt a diamond, we must also break covalent bonds; this accounts for its exceptionally high melting point of more than 3500 °C.

Because silicon is in the same group as carbon (group 4A), we might expect Si atoms to be able to substitute for some of the C atoms in the diamond structure. This actually occurs in the compound silicon carbide, SiC. Silicon carbide is best known as the abrasive Carborundum®, widely used to make grindstones.

Diamond does not conduct electricity. The two requirements for electrical conductivity are (1) charged particles must be present and (2) the particles must be free to move in an electric field. In diamond, the charged particles (electrons) are present, but they are all *localized* in covalent bonds. An electric field does not set them in motion, and consequently diamond is a *nonconductor* of electricity.

Graphite

Another bonding scheme for carbon that uses all the valence electrons has each C atom bonded to three other C atoms in the same plane. According to this model, three of the four valence electrons of each C atom are localized in sp^2 hybrid orbitals, but the fourth is in a $2p$ orbital perpendicular to the plane of the sp^2 orbitals. These $2p$ electrons are delocalized. The bonding is similar to that in benzene, except that the delocalized electrons in graphite are spread throughout planes of C atoms rather than in individual hexagonal rings of C atoms. Overall, bonds within the planes are strong covalent bonds, and attractions between the planes are much weaker van der Waals forces. The crystal structure of graphite is shown in Figure 11.33.

▲ **FIGURE 11.32   The crystal structure of diamond**

Each carbon atom is bonded to four others in a tetrahedral arrangement.

 **Diamond, Graphite, and Buckminsterfullerene 3D models**

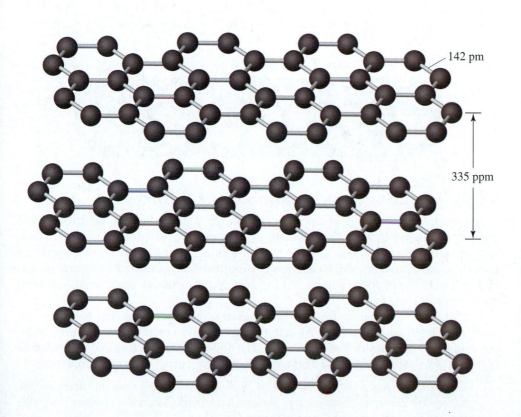

142 pm

335 ppm

◄ **FIGURE 11.33   A ball-and-stick model showing the crystal structure of graphite**

Note the difference between the bond length between any two C atoms in one plane and the distance between two C atoms in separate planes.

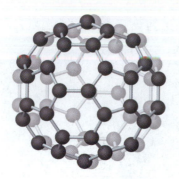

▲ **FIGURE 11.34** **A ball-and-stick model of C₆₀, a buckyball**

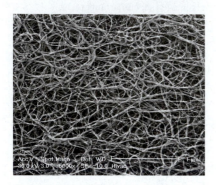

▲ An electron microscope image (magnification 35000 X) of carbon nanotubes confirms the general structure shown in Figure 11.35. Many of the structures found in samples of carbon nanotubes will contain multiple sheets of carbon atoms in the wall of the tube.

▶ **FIGURE 11.35** **Carbon nanotubes**

A ball-and-stick model. The break in the structure indicates that nanotube molecules vary in length from a few nanometers to a micrometer or more. Diameters are generally in the range of a few nanometers.

Graphite has some interesting properties that are consistent with the bonding scheme just described:

1. The carbon-to-carbon bond lengths *within* layers (142 pm) are comparable to those in benzene (139 pm). The carbon-to-carbon distances *between* layers are much greater (335 pm).

2. The large distances and weak interactions between layers allow the layers to glide over one another relatively easily. These features are related to graphite's use as a lubricant and in writing utensils (graphite pencils).

3. Graphite is a good electrical conductor because its delocalized *p* electrons can be set in motion by an external electric field. Graphite is extensively used as electrodes in batteries and in electrolysis reactions.

Diamond and graphite are polymorphic forms of carbon, but they are more than that. Two or more forms of an *element* that differ in their basic *molecular* structure are called **allotropes.** Diamond and graphite are allotropes of carbon.

Other Allotropes of Carbon

In 1985, a number of previously unknown carbon molecules were discovered in the products formed when graphite was vaporized. The predominant species had a molecular mass of 720 u, corresponding to the formula C_{60}. The structure proposed for the molecule is a roughly spherical collection of atoms in the shape of hexagons and pentagons, very much like a soccer ball, as seen in Figure 11.34. The C_{60} molecule was named "buckminsterfullerene" because it resembles the geodesic-domed structures pioneered by architect R. Buckminster "Bucky" Fuller. The general name *fullerenes* is now used both for C_{60} and for similar molecules discovered subsequently, such as C_{70}, C_{74}, and C_{82}. They are often colloquially called "buckyballs."

The fullerenes are allotropic forms of carbon in that they have a molecular structure different from that of diamond or that of graphite. Still other allotropic forms of carbon called *nanotubes* are also network covalent solids. We can visualize a nanotube as a graphitelike sheet that has been rolled into a tube. The tubular array of carbon atoms is capped at each end by half of a C_{60} molecule (Figure 11.35). These nanotubes have unusual mechanical and electrical properties that are of great interest in current research.

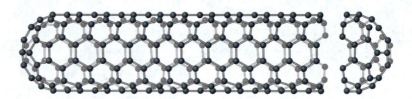

11.9 Ionic Bonds as "Intermolecular" Forces

When we introduced ionic compounds in Chapter 2 and when we described ionic bonding in Chapter 9, we noted that there are no molecules of a solid ionic compound, and therefore there cannot be any intermolecular forces. Instead, we speak of *interionic attractions* in which each ion is simultaneously attracted to several neighboring ions of the opposite charge. These interionic attractions—ionic bonds—extend throughout an ionic crystal.

In Chapter 9, we identified the *lattice energy* as a property that measures the strength of interionic attractions. We can make qualitative comparisons, however, without using actual values of lattice energies. The following generalization, illustrated in Figure 11.36, describes fairly well how we relate lattice energy to atomic properties:

The attractive force between a pair of oppositely charged ions increases as the charges on the ions increase and as the ionic radii decrease. Lattice energies increase as the attractive forces increase.

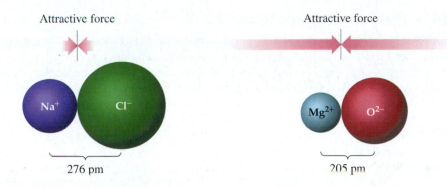

Attractive force   Attractive force

Na⁺   Cl⁻   Mg²⁺   O²⁻

276 pm   205 pm

◄ **FIGURE 11.36   Interionic forces of attraction**

Because of the higher charges on the ions and the reduced interionic distance, the attractive force between Mg^{2+} and O^{2-} is about 7 times as great as that between Na^+ and Cl^-. The interionic distances are the sums of the ionic radii listed in Figure 8.14.

QUESTION: Which is expected to have greater interionic forces, MgF_2 or KF?

Most lattice energies are high enough that ionic solids do not readily sublime. We can melt ionic solids, however, if we supply enough thermal energy to break down the crystalline lattice. In general, the greater the lattice energy, the higher the melting point of an ionic solid.

Solid ionic compounds meet only one of the two requirements for electrical conductivity cited on page 463. They have charged particles (ions). However, because the ions are fixed in a crystalline lattice, they do not meet the requirement of mobility. Consequently, ionic solids do not conduct electricity. However, when we melt an ionic solid or dissolve it in a suitable solvent, such as water, the ions are free to move. Solutions of ionic compounds are good electrical conductors, as we noted in Chapter 4.

Example 11.9

Arrange the ionic solids MgO, NaBr, and NaCl in the expected order of increasing melting point.

STRATEGY

To make the necessary comparisons here, we need to apply two ideas: (1) The greater the lattice energy of an ionic compound, the higher its melting point; and (2) lattice energy is related to ionic charges and sizes.

SOLUTION

Mg^{2+} and O^{2-} ions have higher charges than do Na^+, Cl^-, and Br^-. Also, Mg^{2+} is a smaller ion than Na^+, and O^{2-} is smaller than both Cl^- and Br^-. As a result, we expect the lattice energy of MgO to be much greater than that of either NaCl or NaBr. Because it has the highest lattice energy, MgO should have the highest melting point of the three compounds.

Now we need to determine whether NaCl or NaBr has the greater lattice energy. We see that the compounds have the same cation and that their anions have the same charge. The only difference is in the anion radii. The Cl^- ion is smaller than the Br^- ion, and therefore the interionic attractions in NaCl are stronger than in NaBr. We expect NaCl to have a higher lattice energy and a higher melting point than NaBr. The expected order is therefore NaBr < NaCl < MgO.

ASSESSMENT

We find confirmation of our qualitative reasoning in the observed melting points of 747 °C for NaBr, 801 °C for NaCl, and 2832 °C for MgO.

EXERCISE 11.9A

Arrange the ionic solids CsBr, KCl, KI, and MgF_2 in the expected order of increasing melting point.

EXERCISE 11.9B

One of the four ionic solids in Exercise 11.9A is insoluble in water, and the other three are water-soluble. Which do you think is the insoluble one, and why?

11.10   The Structure of Crystals

In a macroscopic view, a **crystal** is a solid substance with a regular shape, having planar surfaces and sharp edges that intersect at fixed angles. The microscopic view has a small number of atoms, ions, or molecules making up a unit that is repeated over and over. An essential feature of a crystal is that we must be able to figure out its entire structure from this tiny unit.

Crystal Lattices

Most of us have dealt with a repeating pattern at one time or another. Usually, we need to deal with a pattern in only one dimension, as in stringing beads, or in two dimensions, as in laying floor tiles. These two everyday examples are illustrated in Figure 11.37.

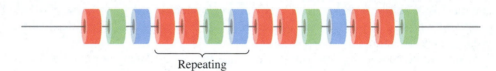

Repeating
unit
(a) A string of beads represents a one-dimensional pattern.

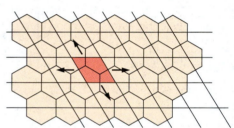

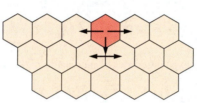

▶ **FIGURE 11.37   One- and two-dimensional patterns**

(b) A hexagonal floor tile layout is a two-dimensional pattern.

(c) Two-dimensional pattern overlain with a grid of parallel lines. The repeating unit is in red.

In stringing beads (Figure 11.37a), we can single out a group of four beads, say red–red–green–blue, as a repeating unit. By repeating this unit many times, we could generate strings of beads of any length, all of which would have the same pattern. The situation with the hexagonal floor tiles (Figure 11.37b) is a bit more complex. If we use the red hexagon as a repeating unit, we can generate its neighbors in the same row by a single displacement to the left and a single displacement to the right. To generate its neighbors in the row below, however, requires *two* displacements: first down and then to the left or to the right. However, we can select a different repeating unit that will generate the entire pattern with just single displacements in all directions and without overlapping any of the pattern previously formed as the unit is moved. This is the parallelogram shown in red in Figure 11.37c.

To describe crystals, we need to work with three-dimensional patterns. The framework on which we outline the pattern is called the *lattice*. There are 14 different lattices that describe all the various known crystalline solids, but we will limit our discussion to the *cubic* types.

The lattice shown in Figure 11.38 consists of three sets of equidistant, mutually perpendicular planes. A geometric figure called a *parallelepiped* is outlined in color. It has six faces formed by the intersection of three pairs of parallel planes. This particular parallelepiped is a *cube* because the intersecting planes form 90° angles, and all edges of the parallelepiped have the same length.

The highlighted cube in Figure 11.38 can be used to generate the entire lattice by simple, straight-line displacements. This repeating unit of the lattice is called the **unit cell.** The simplest type of unit cell has structural particles (atoms, ions, or molecules) centered only at its corners; it is a **simple cubic cell** (Figure 11.39a).

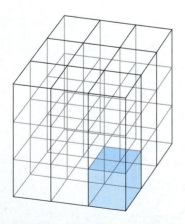

▲ **FIGURE 11.38   The cubic lattice**
The entire lattice can be generated by straight-line displacements (left and right, front and back, up and down) of the unit cell shaded in blue.

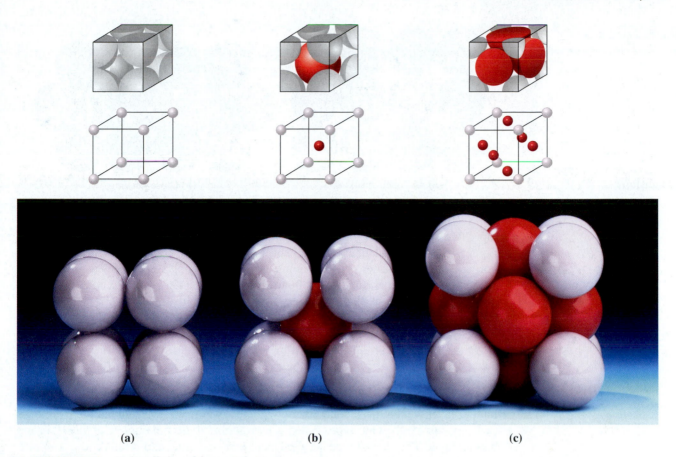

▲ **FIGURE 11.39   Unit cells in cubic crystal structures**

The bottom photo shows that some spheres (atoms) are in direct contact. The upper row of illustrations emphasizes that most of the spheres are shared by neighboring unit cells. In the middle row, only the locations of the centers of the spheres are shown. Although all spheres are identical, some are colored red for easier visibility. (a) A simple cubic cell. (b) A body-centered cubic cell. (c) A face-centered cubic cell.

However, we can usually better describe a crystal structure by using a unit cell that has more structural particles. The **body-centered cubic (bcc)** structure has an additional structural particle at the center of the cube (Figure 11.39b), and the **face-centered cubic (fcc)** structure has an additional particle at the center of each face (Figure 11.39c).

A given structural particle in a crystal is in contact with a certain number of neighboring particles. The number of neighbors it is in contact with is called the *coordination number* of the lattice and is an important concept in describing crystals. For example, in the body-centered unit cell, we can see from Figure 11.39b that the atom in the center of the cell is in contact with each of the eight corner atoms; its coordination number is 8. The coordination number in the simple cubic cell (Figure 11.39a) is just a bit harder to assess. However, if we visualize several unit cells, as in Figure 11.40, we see that the coordination number is 6. The coordination number in a face-centered cubic cell (Figure 11.39c) is 12, as is most readily established in the next section.

Close-Packed Structures

In a crystal made up of metal atoms, we view the atoms as a collection of identical spheres, much like a collection of marbles in a box. We cannot pack marbles in a box in a way that will fill all the space. Rather, there must always be some open spaces, or *voids,* among the marbles. Arrangements of identical spheres known as *close-packed structures* are more efficient than other arrangements in filling an available volume.

In these cubic types of crystal structure, most structural particles are only partly inside a given unit cell.

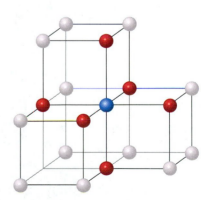

▲ **FIGURE 11.40   Coordination number 6 in a simple cubic cell**

The blue atom is shared by eight unit cells (four are shown) and has six red atoms as nearest neighbors. All atoms are identical and color is for emphasis only.

Close-Packing movie

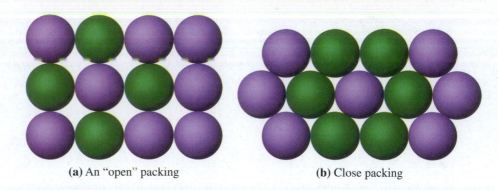

▶ **FIGURE 11.41** **Two arrangements for packing spheres in two dimensions**

(a) An "open" packing   **(b)** Close packing

To illustrate, let's first consider arranging spheres in a single layer on a table-top. Two possible arrangements, both viewed from above, are suggested by Figure 11.41. In the open arrangement, each sphere is in contact with four neighboring spheres (in green) and the amount of space among the spheres is greater than in the close-packed arrangement of part (b), in which each sphere touches six neighboring spheres (in green). Now let's focus on the close-packed arrangement and begin to build multiple layers of spheres. Figure 11.42 portrays an example of a multilayered, close-packing order.

The voids among the spheres in the close-packed array in Figure 11.41b are all alike. This layer is also shown in Figure 11.42 as the bottom layer (layer A, red). It makes no difference where we begin the process of adding spheres in a second layer (layer B, yellow), because each added sphere will rest in the hollow above a void in the bottom layer. When the second layer of spheres is in place, we see that there are now two types of voids. One type, called a *tetrahedral hole,* falls above a *sphere* in the bottom layer. The second type is called an *octahedral hole,* and it falls above a *void* in the bottom layer. The two types of holes lead to two possibilities for adding a third layer of spheres. If we cover all the tetrahedral holes, the third layer (effectively layer A again, red) is identical to the bottom layer and the pattern begins to repeat itself.

Assembly of Close-Packed Structures activity

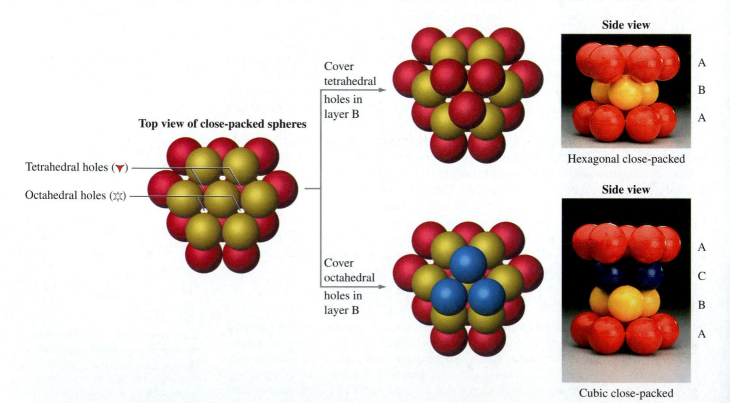

▲ **FIGURE 11.42** **Close-packing of spheres in three dimensions**

This possibility is pictured in the ABA side-view photograph of Figure 11.42 and is called a **hexagonal close-packed (hcp)** arrangement. In contrast, if we cover the octahedral holes, the third layer (layer C, blue) is *not* identical with the bottom layer, and only when we add a fourth layer does the pattern begins to repeat itself. This possibility is shown in the ACBA side-view photograph and is called the **cubic close-packed (ccp)** arrangement. We can also use Figure 11.42 to help us assess the coordination numbers in the close-packed arrangements. Within any single layer of the structure, a given atom is in contact with six others. It is also in contact with three atoms in the layer below and three in the layer above, giving it a coordination number of 12.

If we now return to the three unit cells shown in Figure 11.39, we see that the simple cubic structure is not a close-packed structure; in fact, 47.64% of the volume in this structure is voids. The body-centered cubic structure is a more tightly packed arrangement, with only 31.98% of the total volume being voids. The face-centered cubic structure is the one that results from the cubic close-packed arrangement of Figure 11.42; it has only 25.96% of its volume in voids. Figure 11.43 shows that the ccp arrangement does indeed correspond to a face-centered cubic unit cell.

Many metals crystallize in the close-packed arrangements. For example, Cu, Ag, and Au solidify in the ccp arrangement, and Mg, Zn, and Cd solidify in the hcp arrangement. However, some metals do not use the closest packed arrangements possible. To name a few, Fe and Cr, as well as the alkali metals—Li, Na, K, Rb, and Cs—form body-centered cubic crystals.

Apportioning the Atoms in a Unit Cell

We will shortly introduce some simple calculations that are possible with such data as atomic radii and unit-cell dimensions. In these calculations, we will need to know the number of atoms in a unit cell, but this is *not* the same as the number we use when we draw a picture of a unit cell.

Consider the simple cubic cell in Figure 11.39a. Our picture shows eight atoms, but all eight are shared with neighboring unit cells. Look again at Figure 11.40. The blue atom having six red atoms as nearest neighbors is shared by the four unit cells pictured, as well as by four more that are not pictured. (Can you sketch in these four?) We can therefore allot only *one-eighth* of this atom to any particular unit cell. Because one-eighth of each of the eight corner atoms belongs to a given unit cell, the simple cubic unit cell contains the equivalent of one atom, that is, $1/8 \times 8 = 1$.

As we show in Figure 11.44, only one-eighth of each of the eight corner atoms in the body-centered unit cell belongs to the unit cell. The atom in the center of the cube, however, belongs entirely to the unit cell. The equivalent number of atoms in a bcc unit cell is therefore $(1/8 \times 8) + 1 = 2$. Example 11.10 illustrates how we determine the number of atoms in an fcc unit cell.

▲ **FIGURE 11.43   Cubic close-packing of spheres and the face-centered cubic unit cell**

The cluster of spheres shown in the two photographs represents a portion of a larger array that has a cubic close-packed arrangement. In the upper photograph, the two middle layers have six atoms each, and the top and bottom layers have one atom each, accounting for the 14 spheres that make up one fcc unit cell.

QUESTION: Can you see that rotating the group of 14 spheres from the top view to the bottom view reveals the fcc unit cell?

 **Cubic Crystal Structures 3D models**

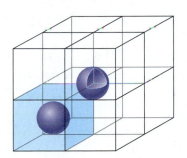

◀ **FIGURE 11.44   Apportioning atoms in a body-centered unit cell**

The eight unit cells outlined are body-centered cubic. Two atoms are shown in color. One is in the exact center of a unit cell. The other, a corner atom, is shared among all eight unit cells. The effective number of atoms in a bcc unit cell is therefore two.

  Apportioning Atoms of a Unit Cell activity

Calculations Based on Atomic Radii and Unit-Cell Dimensions

Now that we have seen the different ways that atoms are arranged in crystal lattices and their unit cells, we can perform a variety of calculations. We will illustrate some calculations here and present additional ones in the end-of-chapter problems.

Example 11.10

Copper crystallizes in the cubic close-packed arrangement. The atomic (metallic) radius of a Cu atom is 127.8 pm. **(a)** What is the length, in picometers, of the unit cell in a sample of copper metal? **(b)** What is the volume of that unit cell, in cubic centimeters? **(c)** How many atoms belong to the unit cell?

STRATEGY

(a) For a ccp arrangement of spheres, the crystal structure is face-centered cubic. Each face of the unit cell is a square; the length of the square (l) is also the length of the unit cell. We need to apply geometry to calculate this length, and we can best see how to do this through a sketch (Figure 11.45).

(b) We can obtain the volume of the unit cell directly from its length (l),

(c) Determining the number of atoms per unit cell requires an analysis similar to that used for the bcc unit cell in Figure 11.44.

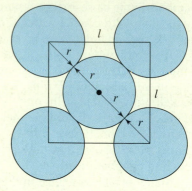

▲ **FIGURE 11.45** **Example 11.10 illustrated**

SOLUTION

(a) Figure 11.45 represents one face of the fcc unit cell. The length of the unit cell is the length of the square, marked l, and the atomic radius of Cu is r. Therefore the diagonal of the face has a length of $4r$. The diagonal divides the square into two right triangles and is the hypotenuse of each one.

At this point, we apply the theorem of Pythagoras.

$$c^2 = a^2 + b^2$$
$$(4r)^2 = l^2 + l^2$$
$$16r^2 = 2l^2$$
$$l^2 = 8r^2$$
$$l = r\sqrt{8} = 127.8 \text{ pm} \times \sqrt{8} = 361.5 \text{ pm}$$

(b) We can determine the length, l, in centimeters, by using conversion factors.

$$l = 361.5 \text{ pm} \times \frac{10^{-12} \text{ m}}{1 \text{ pm}} \times \frac{100 \text{ cm}}{1 \text{ m}} = 3.615 \times 10^{-8} \text{ cm}$$

The volume of the unit cell, V, is given by the formula $V = l^3$.

$$V = (3.615 \times 10^{-8})^3 \text{ cm}^3 = 4.724 \times 10^{-23} \text{ cm}^3$$

(c) The four atoms at the corners of the square in Figure 11.45 are also at the four of the corners of the face-centered cubic unit cell pictured in Figure 11.39(c). The unit cell has eight corner atoms. Each corner atom belongs only in 1/8 part to a given unit cell, so collectively the corner atoms contribute the equivalent of $1/8 \times 8 = 1$ atom. Two unit cells share the atom in the center of a face. A cube has six faces, and the total contribution of the six atoms in the faces is $1/2 \times 6 = 3$ atoms. The total number of atoms in the unit cell is thus $1 + 3 = 4$.

EXERCISE 11.10A

The unit cell of crystalline iron is body-centered cubic, and the atomic radius of iron is 124.1 pm. Determine the length of a unit cell of iron.

EXERCISE 11.10B

Starting with the result of Exercise 11.10A, determine the volume of a unit cell of iron and the number of Fe atoms per unit cell.

Example 11.11

Use the results of Example 11.10, the molar mass of copper, and Avogadro's number to calculate the density of metallic copper.

STRATEGY

The density of copper is the mass of the copper atoms in a unit cell divided by the volume of the cell. The mass of the unit cell is the product of the number of Cu atoms in the cell (from Example 11.10c) and the mass per atom. To determine the mass per Cu atom, we must use the molar mass of copper and Avogadro's number. We determined the volume of the cell in Example 11.10b.

SOLUTION

The mass of a copper atom is the molar mass, 63.546 g/mol Cu, divided by Avogadro's number.

$$? \text{ g/Cu atom} = \frac{63.546 \text{ g}}{1 \text{ mol Cu}} \times \frac{1 \text{ mol Cu}}{6.0221 \times 10^{23} \text{ Cu atoms}}$$
$$= 1.0552 \times 10^{-22} \text{ g/Cu atom}$$

The unit cell contains the equivalent of four copper atoms. The mass of the unit cell is therefore equivalent to the mass of four copper atoms.

$$4 \text{ Cu atoms} \times \frac{1.0552 \times 10^{-22} \text{ g}}{1 \text{ Cu atom}} = 4.2208 \times 10^{-22} \text{ g}$$

Now we can calculate the density of copper from the mass and the volume of the unit cell.

$$d = \frac{m}{V} = \frac{4.2208 \times 10^{-22} \text{ g}}{4.724 \times 10^{-23} \text{ cm}^3} = 8.935 \text{ g/cm}^3$$

ASSESSMENT

This calculation involves both a very large number (Avogadro's number) and very small numbers (the mass and volume of a unit cell), creating the potential for a wide range of computational errors. However, the final result, the density of the metal, must fall into a very small range—about 1 to 20 g/cm^3.

EXERCISE 11.11A

Starting with the results of Exercises 11.10A and 11.10B, together with a density of 7.874 g/cm^3 for iron and other data, estimate a value of Avogadro's number.

EXERCISE 11.11B

On page 469, we stated that the percentage of voids in the simple cubic unit cell (Figure 11.39a) is 47.64%. Verify this fact by calculation. (*Hint:* This is a simple geometry problem. For spheres with radius *r* in a simple cubic arrangement, calculate the volume of the unit cell and the volume of the portions of the spheres contained within the unit cell.)

Ionic Crystal Structures

Ionic crystal structures are somewhat more complicated than those of metals for two reasons: (1) Ionic crystals have two types of structural particles, cations and anions, rather than atoms of a single kind; and (2) the size of the cations in a crystal lattice is usually different from the size of the anions. One approach to ionic crystals is to consider that some voids in a close-packed arrangement of anions are filled by the smaller cations. In this arrangement, cations and anions can be in contact with one another, but the structure would not be stable if the anions were also in direct contact with one another. Therefore the cations must be large enough to fill voids and also to force some separation between the anions. The particular packing arrangement of anions in an ionic crystal therefore depends on the ratio of the cation radius (r_c) to the anion radius (r_a), that is r_c/r_a.

Recall that a close-packed arrangement of spheres has two type of voids: tetrahedral holes and octahedral holes. Tetrahedral holes are the smaller of the two and can accommodate only certain small cations in contact with larger anions. The following range of radius ratios leads to the filling of tetrahedral holes:

$$\textit{Tetrahedral:} \quad 0.225 < r_c/r_a < 0.414$$

The larger octahedral holes can accommodate somewhat larger cations in an array of anions. The range of radius ratios for the filling of octahedral holes is

$$\textit{Octahedral:} \quad 0.414 < r_c/r_a < 0.732$$

If the cations and anions are more nearly of equal size, there isn't enough room for the cations in the voids in a close-packed arrangement of anions. The anions are forced into a more open cubic structure, and the ratio we are interested in must have values

$$\textit{Cubic:} \quad r_c/r_a > 0.732$$

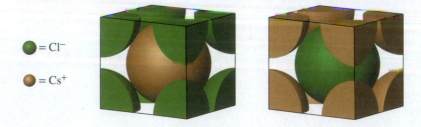

The space-filling model on the left shows a simple cube of Cl^- ions with a Cs^+ ion in the interior. In the model on the right, the simple cube is of Cs^+ ions and a Cl^- ion is in the interior. In this crystal structure, both the Cs^+ and the Cl^- ions have a coordination number of eight.

Cesium Chloride 3D model

In this case, think of a sphere that is resting in one of the hollows of the open packing of spheres in Figure 11.41a and surmounted by a layer of spheres identical to the bottom layer shown.

The radius ratio for cesium chloride is $r_{Cs^+}/r_{Cl^-} = 169$ pm/181 pm = 0.934. This ratio suggests a crystal structure based on a set of interpenetrating simple cubic lattices, one of Cs^+ ions and one of Cl^- ions. In this structure, each Cs^+ ion is in contact with eight Cl^- ions, and each Cl^- ion is in contact with eight Cs^+ ions, as suggested in Figure 11.46.

The radius ratio for sodium chloride is $r_{Na^+}/r_{Cl^-} = 99$ pm/181 pm = 0.55. The Na^+ ions fill the octahedral holes in a close-packed array of Cl^- ions, producing the unit cell shown in Figure 11.47. To show that this unit cell is consistent with the formula NaCl, notice that only the single Na^+ ion at the center belongs entirely to the cell and that each of the 12 Na^+ ions at the centers of the edges is shared among four unit cells. The number of Na^+ ions in the cell is thus $1 + (1/4 \times 12) = 4$. One-eighth of each of the eight Cl^- ions at the corners and one-half of each of the six Cl^- ions in the centers of the faces belong to the unit cell, making the total number of Cl^- ions in the cell $(1/8 \times 8) + (1/2 \times 6) = 4$. The unit cell contains the equivalent of four formula units of NaCl.

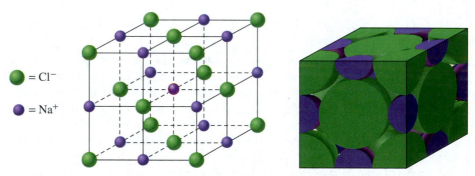

▲ FIGURE 11.47   **A unit cell of sodium chloride**

The relative locations of the centers of the ions are shown on the left; a space-filling model of the sodium chloride structure on the right illustrates how oppositely charged ions are in contact throughout the unit cell. Here, Cl^- ions are at the eight corners of the unit cell and in the centers of the six faces. Twelve Na^+ ions are distributed on the 12 edges of the cell, one Na^+ ion at the center of each edge, and a single Na^+ ion (highlighted) is located at the center of the cell. Note that an alternate unit cell can be drawn with a single Cl^- ion located at the center of the cell.

QUESTION: What is the coordination number of the Na^+ ions in NaCl? Of the Cl^- ions?

Experimental Determination of Crystal Structures

We cannot see the patterns of atoms, ions, or molecules in crystalline solids, even with the help of ordinary light microscopes, because these particles are much too small. To figure out these patterns, we must use radiation that has wavelengths comparable to the dimensions of unit cells. X rays are ideal for this purpose.

Figure 11.48 suggests the interaction of X rays with a crystal and a geometric analysis of the results. Wave a is reflected by one plane of atoms, ions, or molecules in the crystal, and wave b is reflected by the next plane below. Both reflected waves arrive at the detector on the right, but wave b travels farther than does wave a to get

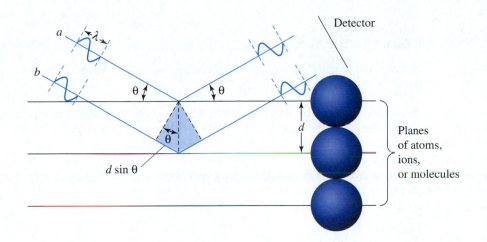

The hypotenuse of each blue-shaded right triangle is equal to the interatomic distance, d. The side opposite the angle θ has length $d \sin \theta$. Wave b travels farther than a by the distance $2d \sin \theta$, and this distance is an integral multiple of the X-ray wavelength λ. Thus, $n\lambda = 2d \sin \theta$, where $n = 1,2,3,\ldots$.

there. To achieve a maximum intensity of the reflected radiation, waves a and b must reinforce each other—that is, their crests and troughs must line up. For this to happen, the additional distance traveled by wave b must be an integral multiple n of the wavelength, λ, of the X rays:

$$n\lambda = 2d \sin \theta \qquad (11.4)$$

By measuring the angle θ at which scattered X rays of known wavelength have their greatest intensity, one can calculate the spacing d between atomic planes. If the process is repeated for different orientations of the crystal, eventually the complete crystal structure can be worked out.

Cumulative Example

Here are some data about an organic compound: Its normal boiling point is a few degrees below that of $H_2O(l)$, and its vapor density at 99.0 °C and 752 Torr is within 5% of 2.0 g/L. Using these data and other information from this and previous chapters, determine which one of the following compounds it is most likely to be: (a) $(CH_3)_2O$ (b) $CH_3CH_2CH_2OH$ (c) $CH_3CH_2OCH_3$ (d) $HOCH_2CH_2OH$.

STRATEGY

We have to make comparisons of a property (boiling point) that depends on intermolecular forces of attraction. In turn, these forces are influenced by molecular masses and molecular structures. We can first determine the molecular masses of the four possible compounds from their formulas. Then, using Equation (5.11), we can determine, within a few percent, the molar mass of the unknown compound. At this point, we will reduce the number of possibilities to the extent possible. Following this, we will see whether we can isolate the answer by using the comparison to the boiling point of $H_2O(l)$ and other information from the text.

SOLUTION

First, we determine the molecular masses of the four compounds.

(a) $(CH_3)_2O$, 46.07 u   (c) $CH_3CH_2OCH_3$, 60.10 u

(b) $CH_3CH_2CH_2OH$, 60.10   (d) $HOCH_2CH_2OH$, 62.07 u

We substitute the given data about the compound we seek into Equation (5.11) and solve for the molar mass, M.

$$M = \frac{dRT}{P}$$

$$= \frac{2.0 \text{ g L}^{-1} \times 0.08206 \text{ L atm mol}^{-1} \text{ K}^{-1} \times (273 + 99) \text{ K}}{(752/760) \text{ atm}} = 62 \text{ g/mol}$$

The range of M values is ±5% of the calculated molar mass.

This allows us to eliminate one choice.

Range = 59–65 g/mol
(a) $(CH_3)_2O$, 46.07 u is eliminated.

Compounds (b) and (c) have equal molar masses, but we see that compound (b) can hydrogen bond; it therefore should have a much higher boiling point than compound (c).

(b) $CH_3CH_2CH_2OH$

(c) $CH_3CH_2OCH_3$

At this point, we turn to Figure 11.4 for key pieces of data:

Methanol, CH_3OH, bp ≈ 65 °C
Ethanol, CH_3CH_2OH, bp ≈ 80 °C

Next, we can apply the principle of homology (page 454).

The addition of one CH_2 group to ethanol suggests 1-propanol, which is compound (b), $CH_3CH_2CH_2OH$, bp ≈ 95 °C

We eliminated compound (a) from further consideration because its molecular mass fell outside the required range. Compound (b) certainly seems to be the correct choice since it meets all the criteria so well. Compound (c) is not a suitable choice because its boiling point should be considerably below that of compound (b), that is below 95 °C. However, there is still one additional choice possible—(d) $HOCH_2CH_2OH$. What can we say about this compound—ethylene glycol? Its molar mass is slightly greater than that of 1-propanol, and on this basis it might also have a boiling point near that of water. However, notice the two OH groups per molecule. This allows for much more extensive hydrogen bonding than in ethanol or 1-propanol, and the boiling point of ethylene glycol should be much greater than that of 1-propanol.

ASSESSMENT

The best assessment of the reasoning that led us to our choice of 1-propanol would be to turn to a handbook of chemical data. Upon doing so, we would find that the experimentally determined boiling points are as follows: methanol, 65.15 °C; ethanol, 78.5 °C; compound (a) −25 °C, (b) 97.4 °C, (c) 10.8 °C, (d) 198.9 °C.

Concept Review with Key Terms

11.1 Intermolecular Forces and the States of Matter: A Chapter Preview—Gases, solids, and liquids comprise the three **states of matter**. When a substance undergoes a change from one state to another, it is said to undergo a **phase change**. **Intermolecular forces** are forces between molecules that determine the physical properties of liquids and solids.

Solid   Liquid   Gas

11.2 Vaporization and Vapor Pressure—**Vaporization** is the conversion of a liquid to a gas (vapor), and the quantity of heat associated with this phase change is known as the **enthalpy (heat) of vaporization**. Condensation is the conversion of a gas (vapor) to a liquid, and dynamic equilibrium is established when vaporization and condensation occur at the same rate in a closed system. The partial pressure of the vapor under these conditions is called the **vapor pressure**, a temperature-dependent property. A plot of vapor pressure versus temperature produces a **vapor pressure curve**. The **boiling point** of a liquid is the temperature at which its vapor pressure is equal to the prevailing atmospheric pressure; a liquid's boiling point at 1 atm it is called the **normal boiling point**. The **critical temperature**, T_c, is the highest temperature at which a liquid and its vapor can coexist, and the pressure-temperature point corresponding to this condition is the **critical point**. Above the critical point, the single phase that exists is called a **supercritical fluid**.

Dynamic Equilibrium

Vaporization = condensation
rate   rate

11.3 Phase Changes Involving Solids—The conversion of a solid to a liquid is called either **fusion** or **melting**; the temperature at which this change occurs is the **melting point**. The quantity of heat required to melt a given amount of a solid is the **enthalpy (heat) of fusion**. A plot of temperature versus time as a solid is slowly heated is known as a heating curve; a similar plot for a liquid that is slowly cooled is known as a cooling curve. In some cases, it is possible to cool a liquid below its freezing point without having a solid form, a process known as **supercooling**. The conversion of a solid directly to a gas (vapor) is called **sublimation**. A plot of the vapor pressure of a solid versus temperature is known as a **sublimation curve**. The quantity of heat required to convert a given amount of solid directly to a gas is the

enthalpy (heat) of sublimation, which is equal to the sum of the enthalpy of fusion and the enthalpy of vaporization:

$$\Delta H_{subln} = \Delta H_{fusion} + \Delta H_{vapn}$$

The **triple point** of a substance is the unique temperature and pressure at which the solid, liquid, and vapor states of the substance coexist.

11.4 Phase Diagrams—A **phase diagram** is a plot showing the phases of a substance at various temperatures and pressures. Curves represent two-phase equilibria in the diagram: the vaporization (vapor pressure) curve, the sublimation curve, and the fusion curve. Polymorphism describes the existence of a solid, such as $HgI_2(s)$, in two or more different solid phases.

Temperature →

11.5 Intermolecular Forces of the van der Waals Type—Fluctuations in electron charge density in a molecule produce an instantaneous dipole, which in turn creates induced dipoles in neighboring molecules. The ease with which this occurs in a substance is known as its **polarizability**. Attractions between instantaneous and induced dipoles, called **dispersion forces**, are found in all substances. Polar molecules also have dipole–dipole and dipole-induced dipole intermolecular forces, arising from permanent dipoles in the molecules. Collectively known as **van der Waals forces**, dispersion forces, dipole-dipole forces, and dipole—induced dipole forces affect such physical properties as melting points and boiling points. A series of compounds with regularly varying structures and formulas also has regularly varying properties; this is the principle of **homology**.

Dipole-dipole or dipole-induced dipole interaction

11.6 Hydrogen Bonds—A strong intermolecular force—the **hydrogen bond**—forms when a H atom covalently bonded to an O, a N, or a F atom of one molecule is simultaneously attracted to an O, a N, or a F atom of another molecule. Hydrogen bonding accounts for some of the unusual properties of water (for example, an unusually high boiling point and the fact that the density of liquid water is greater than the density of ice). Hydrogen bonding also governs aspects of the behavior of biological molecules such as proteins and the nucleic acids.

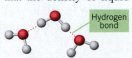
Hydrogen bond

11.7 Intermolecular Forces and Two Liquid Properties—**Surface tension** (γ) is defined as the amount of work required to increase the surface area of a liquid and is related to the **cohesive forces** between liquid molecules. When a liquid drop is placed on a solid surface, the relative strengths of cohesive forces (those between like molecules) and **adhesive forces** (those between unlike molecules) determine the degree to which the drop will spread. At the interface of a liquid and its container, these forces give rise to a **meniscus**, or curvature of the liquid surface. The intermolecular forces in a liquid partially determine its **viscosity**, or resistance to flow.

11.8 Network Covalent Solids—In some solids, covalent bonds extend throughout a crystal. Such solids, known as **network covalent solids**, generally are harder, have much higher melting points, and are less volatile than other molecular solids. Diamond and graphite, two **allotropes** of carbon, are examples of these solids.

11.9 Ionic Bonds as "Intermolecular" Forces—In ionic solids, interionic attractions bind all the ions together in a crystal. The strengths of these attractions depend primarily on the ionic charges and the sizes of the ions and partly determine physical properties, such as melting points.

11.10 The Structure of Crystals—The structure of a **crystal**, which is a solid substance composed of an extended periodic array of atoms, is described by a three-dimensional pattern called a lattice. There are three types of cubic lattices: **simple cubic cell**, **body-centered cubic (bcc)**, and **face-centered cubic (fcc)**.

A **unit cell**, when repeated in three dimensions, generates the entire crystal lattice. Unit cell properties and dimensions, often measured by X-ray crystallography, can be used to find atomic radii and the densities of crystalline substances. The crystal structures of metals can be described as the packing of spheres. The two types of close-packing of spheres, **hexagonal close-packed (hcp)** and **cubic close-packed (ccp)**, both reduce to the smallest possible fraction the volume occupied by holes or voids. A model in which cations occupy the voids present among a close-packed array of anions generally works well for the crystal structures of ionic substances.

Assessment Goals

When you have mastered the material in this chapter, you will be able to:

- Describe the distinguishing characteristics of the three states of matter.
- Identify the six main types of phase changes.
- Describe the processes of vaporization and condensation.
- Calculate the heat associated with the transformation of a liquid to a gas or vice versa, using the enthalpy of vaporization or the enthalpy of condensation.
- Define vapor pressure, identify the factors affecting vapor pressure, and extract data from vapor pressure curves.
- Define boiling point and melting point, and distinguish them from the normal boiling point and normal melting point.
- Calculate the heat required for the transformation of a solid to a liquid or vice versa, using the enthalpy of fusion.
- Describe the process of sublimation, and relate the enthalpy of sublimation to the enthalpies of fusion and vaporization.
- Define the triple point and the critical point of a substance.

- Construct a phase diagram given information regarding the triple point, critical point, sublimation curve, fusion curve, vapor pressure curve, and supercritical fluid region,
- Describe the origin of intermolecular forces, and predict physical properties of substances from the nature of their intermolecular forces.
- For a given pair of molecules, classify intermolecular forces as dispersion forces, dipole–dipole forces, dipole-induced dipole forces, or hydrogen bonding.
- Describe phenomena related to surface tension and viscosity and factors influencing them.
- Compare the properties of network covalent solids and other types of solids.
- Predict trends in the melting points of ionic solids based on the strength of interionic attractions.
- Recognize the three types of cubic crystal structures and the two types of close-packed crystal structures.
- Calculate density, Avogadro's number, and unit cell dimensions from crystal-structure data.

Self-Assessment Questions

1. What is meant by **(a)** a condition of *dynamic equilibrium*, **(b)** the triple point of a substance, **(c)** the critical point of a substance, **(d)** a dispersion force?

2. Distinguish between **(a)** an intramolecular force and an intermolecular force; **(b)** volatile and nonvolatile substances; **(c)** vaporization and boiling; **(d)** the boiling point of a liquid and normal boiling point of a liquid; **(e)** the terms gas and vapor; **(f)** the terms dipole moment and polarizability of a molecule; **(g)** an instantaneous and an induced dipole; **(h)** a formula unit and a unit cell of an ionic compound; **(i)** the terms crystal lattice and repeating unit in a crystal lattice.

3. Which one of the following variables do you expect to affect the vapor pressure of a liquid?

   (a) temperature

   (b) volume of liquid in the liquid–vapor equilibrium

   (c) volume of vapor in the liquid–vapor equilibrium

   (d) area of contact between liquid and vapor

4. In a phase diagram for a pure substance, at a minimum, how many different phases are present? What do the curves in the diagram represent?

5. Of these quantities, the one that we expect to be largest for a substance (expressed in kJ/mol) is which of the following?

   (a) molar heat capacity of a liquid

   (b) enthalpy of fusion

   (c) enthalpy of vaporization

   (d) enthalpy of sublimation

6. How are different polymorphic solids represented in phase diagrams?

7. If the triple-point pressure of a substance is greater than 1 atm, we expect which of the following?

   (a) the solid to sublime without melting

   (b) the boiling-point temperature to be lower than the triple-point temperature

   (c) the melting point of the solid to come at a lower temperature than the triple point

   (d) that the substance cannot exist as a liquid

8. Which of the following possesses the highest normal boiling point?

   (a) $O_2(l)$ (c) $SO_3(l)$

   (b) Ne (l)   (d) $Br_2(l)$

9. State the principal reasons why CH_4 is a gas at room temperature, whereas H_2O is a liquid.

10. Of the compounds HF, CH_4, CH_3OH, and N_2H_4, hydrogen bonding as an important intermolecular force is expected in which of these?

    (a) none of these   (c) all but one of these

    (b) two of these   (d) all of these

11. What determines the shape of a meniscus and whether it is convex or concave?

12. How do ionic charges and sizes affect lattice energy? How does lattice energy affect the melting point of an ionic solid?

13. A metal that crystallizes in the body-centered cubic (bcc) structure has a crystal coordination number of

    (a) 6   (c) 12

    (b) 8   (d) any value between 4 and 12

14. Which of the following is true of a unit cell of an ionic crystal?

    (a) It is the same as the formula unit.

    (b) It is any portion of the ionic crystal that has a cubic shape.

    (c) It shares a portion of its ions with other unit cells.

    (d) It always contains the same number of cations and anions.

15. Argon, copper, sodium chloride, and carbon dioxide all crystallize in the fcc structure. How can this be when their physical properties are so different?

16. The boiling point of $O_2(l)$ is -183 °C. Which of the following is a reasonable boiling point of $N_2(l)$?

    (a) -170 °C   (b) -196 °C   (c) -216 °C

Problems

Heat of Vaporization

For Problems 17–24, you may need data from Table 11.1 and Table 6.1.

17. How much heat, in kilojoules, is required to vaporize 3.530 kg of octane, C_8H_{18}?

18. How much heat, in joules, is required to vaporize 275 μL of $CS_2(l)$ ($d = 1.26$ g/mL)?

19. How much heat, in kilojoules, is required to convert 79.8 g H_2O from liquid at 11.3 °C to vapor at 25.0 °C?

20. How much heat, in kilojoules, is released when 1.25 mol $CH_3OH(g)$ at 25.0 °C is condensed and cooled to 15.5 °C? The specific heat of $CH_3OH(l)$ is 2.53 J g^{-1} $°C^{-1}$.

21. How many grams of propane must be burned to supply the heat required to vaporize 6.75 L of water at 298 K?

    $$C_3H_8(g) + 5\,O_2(g) \longrightarrow 3\,CO_2(g) + 4\,H_2O(l)$$

    $$\Delta H = -2.22 \times 10^3 \text{ kJ}$$

22. The combustion of 1.25 g of pentane produces just enough heat to vaporize 165 g of hexane. What is the molar enthalpy of vaporization of hexane?

    $$C_5H_{12}(l) + 8\,O_2(g) \longrightarrow 5\,CO_2(g) + 6\,H_2O(l)$$

    $$\Delta H = -3.51 \times 10^3 \text{ kJ}$$

    $$C_6H_{14}(l) \longrightarrow C_6H_{14}(g) \qquad \Delta H_{vapn} = ?$$

23. You can hold your hand in an oven at 100 °C for some time without feeling great discomfort. However, you can hold your hand a centimeter or two above rapidly boiling water in a kettle for only a second or less. Explain why the effects are different.

24. A double boiler is a device sometimes used in cooking. Cooking takes place in an inner container in contact with the vapor from boiling water in an outside container. (A related laboratory device is called a steam bath.) (a) How is energy conveyed to the food being cooked? (b) What is the maximum temperature that can be reached in the inside container? Explain.

Vaporization, Vapor Pressure

25. With reference to Figure 11.4, estimate the approximate value of (a) the vapor pressure of aniline at 100 °C; (b) the boiling point of methanol when the barometric pressure is 680 mmHg.

26. By referring to Figure 11.4, estimate the approximate value of (a) the vapor pressure of carbon disulfide at 30 °C; (b) the boiling point of ethanol when the barometric pressure is 720 mmHg.

27. What vapor volume must be in equilibrium with 125 g $CS_2(l)$ at 20 °C if 1.5×10^{22} CS_2 molecules are to be present in the vapor? (*Hint:* Refer to Figure 11.4.)

28. How many silver atoms are present in a vapor volume of 486 mL when Ag(l) and Ag(g) are in equilibrium at 1360 °C? (Vapor pressure of silver at 1360 °C = 1.00 mmHg.)

29. Equilibrium is established between a small quantity of $CCl_4(l)$ and its vapor at 40.0 °C in a flask having a volume of 285 mL. The total mass of vapor present is 0.480 g. What is the vapor pressure of CCl_4, in mmHg, at 40.0 °C?

30. The density of acetone vapor in equilibrium with liquid acetone, $(CH_3)_2CO$, at 32 °C is 0.876 g/L. What is the vapor pressure of acetone, in mmHg, at 32 °C?

31. A 0.195-g sample of H_2O is injected into a 2.55-L flask at 30.0 °C. At equilibrium, what phase(s) of H_2O will be present, that is, solid and/or liquid and/or vapor? Explain.

32. A 0.625-g sample of methanol is injected into a 8.77-L flask at 0 °C. At equilibrium, what phase(s) of methanol will be present, that is, solid and/or liquid and/or vapor? Explain.

33. How would you explain to a young schoolchild the phenomenon depicted in the photograph? Water is boiled in a paper cup over an open flame.

34. How would you explain to a young schoolchild the following phenomenon? A small quantity of water is boiled in a can; the can is capped off and cooled; the can collapses.

The Critical Point

35. Is it always possible to liquefy a gas by applying sufficient pressure? By sufficiently reducing the temperature? By a combination of a pressure and temperature change? Explain.

36. The states of matter in bottled propane at room temperature are liquid and gas. Would the same be true for bottled methane? Explain.

Phase Changes

For Problems 37–42, you may need data from Table 11.4 and other sources as indicated.

37. How much heat, in kilojoules, is required to melt an ice cube that is 4.2 cm × 2.1 cm × 2.9 cm? The density of ice is 0.92 g/cm^3.

38. An ice calorimeter measures quantities of heat by the quantity of ice melted. How many grams of ice would be melted by the heat released in the combustion of 3.12 L of $CH_4(g)$, measured at 25 °C and 753 mmHg?

$$CH_4(g) + 2 O_2(g) \longrightarrow CO_2(g) + 2 H_2O(l)$$
$$\Delta H = -890.3 \text{ kJ}$$

39. An ice cube with a mass of 25.5 g at a temperature of 0.0 °C is added to 125 mL of water at 26.5 °C in an insulated container. What will be the final temperature after the ice has melted?

40. A 0.506-kg chunk of ice at 0.0 °C is added to an insulated container holding 315 mL of water at 20.2 °C. Will any ice remain after thermal equilibrium is established?

41. *Without doing detailed calculations,* use data from this chapter and Chapter 6 to determine which of the following would evolve the greatest quantity of heat: **(a)** freezing 2.00 kg Hg(l) at −38.9 °C, **(b)** condensing 0.50 mol of steam [$H_2O(g)$] at 100 °C, **(c)** forming 10.0 g of frost [$H_2O(s)$] from water vapor at −2.0 °C, **(d)** cooling 100.0 mL of $H_2O(l)$ from 51.0 °C to 1.0 °C.

42. *Without doing detailed calculations,* use data from this chapter and Chapter 6 to determine which of the following would require the greatest input of heat: **(a)** melting 3.0 mol of ice at 0 °C, **(b)** evaporating 10.0 g $H_2O(l)$ at 298 K, **(c)** melting 2.0 mol of ice at 0 °C and heating the resulting $H_2O(l)$ to 10 °C, **(d)** subliming 1.0 mol of ice at 0 °C.

Phase Diagrams

43. The drawing on the right is a portion of the phase diagram for sulfur. The stable form of solid sulfur at room temperature is rhombic sulfur, S_α; at the normal melting point, it is monoclinic sulfur, S_β. **(a)** Indicate the phases present in the portions of the diagram marked (?). **(b)** Identify the *three* triple points, and indicate the phases at equilibrium at each one. **(c)** Describe the phase changes that occur as the pressure of a sample is raised, at constant temperature, from point X to point Y.

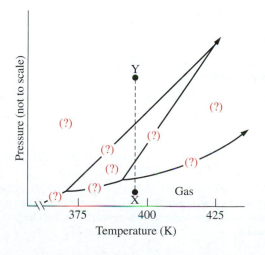

44. The figure below is a phase diagram for iodine. **(a)** Indicate the phases present in the portions of the diagram marked (?). **(b)** Use the letters *A*, *B*, *C*, and *D* to represent the triple point, the normal melting point, the normal boiling point, and the critical point, respectively. **(c)** Describe the phase changes that occur as the temperature of a sample is raised, at constant pressure, from point X to point Y.

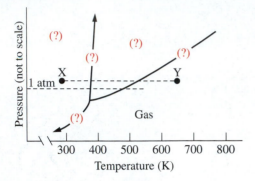

45. Sketch a phase diagram for hydrazine, and locate these points: the triple point (2.0 °C and 3.4 mmHg), the normal boiling point (113.5 °C), and the critical point (380 °C and 145 atm).

46. Sketch a phase diagram for benzene, and locate these points: the triple point (5.5 °C and 35.8 mmHg), the normal boiling point (80.1 °C), and the critical point (288.5 °C and 47.7 atm).

47. Describe the phase changes that occur when a sample of solid carbon dioxide is heated in a device similar to that pictured in Figure 11.14 at a constant pressure of 10.0 atm from −100 °C to 100 °C. Also draw a heating curve to represent these phase changes.

48. Indicate where each of the following points should fall on the phase diagram for water, that is, whether in a one-phase region (solid, liquid, or gas) or on a two-phase curve (vaporization, sublimation, or fusion). Use information supplied with Figure 11.13 and elsewhere in the chapter.

(a) 88.15 °C and 0.954 atm pressure

(b) 25.0 °C and 0.0313 atm pressure

(c) 0 °C and 2.50 atm pressure

(d) −10 °C and 0.100 atm pressure

Intermolecular Forces

49. Which of the following would you expect to have the *lower* boiling point, phosphorus pentabromide or carbon tetrafluoride? Why?

50. Of the compounds sulfur hexafluoride and carbon tetraiodide, one is a solid and one is a gas at room temperature. Which is the solid?

51. Which of the following would you expect to have the *higher* melting point, 1-pentanol, $CH_3CH_2CH_2CH_2CH_2OH$, or 3,3-dimethylpentane, $CH_3CH_2C(CH_3)_2CH_2CH_3$? Why?

52. Which of the following would you expect to have the *higher* melting point, pentane, $CH_3CH_2CH_2CH_2CH_3$, or diethyl ether, $(CH_3CH_2)_2O$? Why?

53. Which one of these substances would you expect to be a gas at STP: NI_3, BF_3, PCl_3, CH_3COOH? Why?

54. Which one of these substances would you expect to be a solid at STP: C_6H_5COOH, $CH_3(CH_2)_8CH_3$, C_6H_{14}, $(CH_3CH_2)_2O$? Why?

55. Arrange the following substances in the expected order of *increasing* boiling point: H_2O, NH_3, CH_4, CH_3CH_3. Give the reasons for your ranking.

56. Arrange the following substances in the expected order of *increasing* boiling point: C_4H_9OH, NO, C_6H_{14}, N_2, $(CH_3)_2O$. Give the reasons for your ranking.

57. Arrange the following substances in the expected order of *increasing* melting point: NaOH, CH_3OH, LiOH, C_6H_5OH. Give the reasons for your ranking.

58. Arrange the following substances in the expected order of *increasing* melting point: Cl_2, CsCl, CCl_4, $MgCl_2$. Give the reasons for your ranking.

Surface Tension

59. Explain which of the following liquids you would expect to have the higher surface tension, octane, C_8H_{18}, or 1-octanol, $C_8H_{17}OH$?

60. Which of the following liquids would you expect to have the higher surface tension, isopropyl alcohol, $(CH_3)_2CH_2OH$, or ethylene glycol, CH_2OHCH_2OH? Explain.

61. What is meant by the statement that water wets glass? Does water "wet" all materials? Can additives to water make it wetter, as is sometimes claimed in advertisements for cleaning agents? Explain.

62. Consider a typical sponge with a rectangular shape of 4 cm × 15 cm and a thickness of 2 cm. It can soak up more than its own mass of water. If the water-filled sponge is held in air with its 2-cm dimension vertical, all the water is retained. With its 4-cm dimension held vertically, the sponge also retains all the water. However, when its 15-cm dimension is held vertically, water drains from the sponge. Explain these observations in terms of the natural phenomena involved.

Crystal Structures

63. In the two-dimensional pattern pictured, **(a)** can both of the outlined portions be repeating units (in the same sense that a unit cell is the repeating unit of a crystal structure)? If not, why not; and which one is a repeating unit? **(b)** What are the numbers of ♥, ♦, and ♣, that should be assigned to the repeating unit?

64. In the two-dimensional pattern pictured, **(a)** can both of the outlined portions be repeating units (in the same sense that a unit cell is the repeating unit of a crystal structure)? If not, why not, and which one is a repeating unit? **(b)** What are the numbers of ♦, ♣, and ♥ that should be assigned to the repeating unit?

65. Barium has an atomic radius of 217 pm and a bcc crystal structure. **(a)** What is the length of the unit cell, in picometers? **(b)** What is the volume of the unit cell, in cubic centimeters? **(c)** Calculate the density of barium.

66. Lead has an atomic radius of 175 pm and an fcc crystal structure. **(a)** What is the length of the unit cell, in picometers?

(b) What is the volume of the unit cell, in cubic centimeters? **(c)** Calculate the density of lead.

67. Use 169 pm for the radius of Cs^+, 181 pm for Cl^-, the unit cell pictured in Figure 11.46, and a density of $CsCl(s)$ of 3.988 g/cm^3 to determine **(a)** the length of the unit cell, and **(b)** an estimate of Avogadro's number, N_A.

68. Potassium chloride has the same type of crystal structure as sodium chloride. The internuclear distance between K^+ and Cl^- is 314.54 pm. The density of $KCl(s)$ is 1.9893 g/cm^3. Determine **(a)** the length of the unit cell, and **(b)** the value of Avogadro's number, N_A.

69. Show that the unit cell of titanium dioxide is consistent with the formula TiO_2. What are the coordination numbers of Ti^{4+} and O^{2-}? Would you expect them to be the same? Explain.

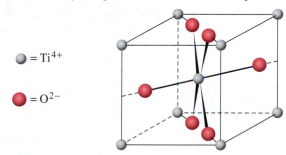

70. Show that the unit cell of calcium fluoride is consistent with the formula CaF_2. What are the coordination numbers of Ca^{2+} and F^-? Would you expect them to be the same? Explain.

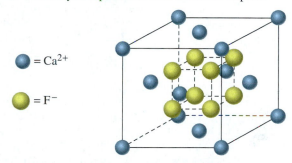

Additional Problems

Problems marked with an * may be more challenging than others.

71. Which of the following is a reasonable estimate of the number of moles of water in the vapor phase when equilibrium is established after 0.10 mL of liquid water is placed in a 1.00-L flask at 25 °C? The vapor pressure of water is given in Table 11.2.

    1.2×10^{-3} 2.3×10^{-3} 5.5×10^{-3} 4.5×10^{-2}

72. A 150.0-mL sample of $N_2(g)$ at 25.0 °C and 750.0 mmHg is passed through benzene, $C_6H_6(l)$, until the gas becomes saturated with $C_6H_6(g)$. The new volume of the gas is 172 mL at a total pressure of 750.0 mmHg. What is the vapor pressure of benzene at 25.0 °C?

73. The following data are for CCl_4: normal melting point, -23 °C; normal boiling point, 77 °C; density of liquid, 1.59 g/mL; heat of fusion, 3.28 kJ/mol; vapor pressure at 25.0 °C, 110 mmHg. **(a)** How much heat must be absorbed to convert 10.0 g of solid CCl_4 to liquid at -23 °C? **(b)** What volume is occupied by 1.00 mol of the saturated vapor at 77 °C? **(c)** What phases—

solid, liquid, and/or vapor—are present if 3.5 g CCl_4 is kept in an 8.21-L volume at 25.0 °C?

74. State *several* reasons why you would expect the boiling point of hexanoic acid, $CH_3(CH_2)_4COOH$, to be *higher* than that of 2-methylbutane, $CH_3CH(CH_3)CH_2CH_3$.

75. A newspaper article describing the unusual properties of water includes the statement, "Water is the only kind of matter that can exist as solid, liquid, and gas all at the same temperature." Is this an accurate statement? Explain.

76. What is the most important intermolecular force involved at the freezing point of each of the following substances? What other intermolecular force(s) is are involved?

  (a) CH_3CH_2OH **(b)** CH_3CH_2Cl **(c)** HCOONa

* 77. Sketch a graph that shows how the pressure changes inside a cylinder of carbon dioxide, initially at a pressure of 70 atm and a temperature of 20 °C, as $CO_2(g)$ leaks from the cylinder at constant temperature. Sketch a similar graph for a leak at a constant temperature of 40 °C.

78. Over a certain small temperature range, liquid water can have the same density at *two* different temperatures. Explain how this is possible. At approximately what temperatures would you expect this to occur?

79. What is the *lowest* temperature at which liquid water can be made to boil? Explain how this can be done.

* **80.** The vapor pressure of liquid ammonia is given by the following equation, where *p* is the vapor pressure in mmHg and *t* is the Celsius temperature.

$$\log p = 7.3605 - \frac{1617.9}{t + 240.17}$$

(a) What is the vapor pressure of $NH_3(l)$ at $-75.0\ ^\circ C$?

(b) What is the normal boiling point of $NH_3(l)$?

(c) The critical temperature of ammonia is $T_c = 405.6\ K$. Calculate the critical pressure, P_c.

81. The graphs show the normal boiling points of hydrides of the elements in groups 4A–7A. Explain the trends represented in these graphs.

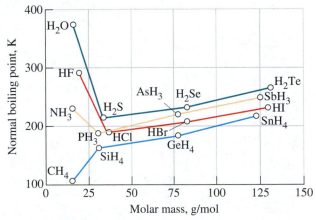

82. Following are several characteristic enthalpies of transition: ΔH_{vapn}, ΔH_{fusion}, ΔH_{condn}, ΔH_{subln}. Arrange them in order of increasing magnitude (that is, increasing in numerical value *without* regard for signs). In general, how do these values compare to the enthalpy changes of chemical reactions? Explain. Would you expect exceptions to this generalization? If so, cite some.

* **83.** A mixture of 1.00 g H_2 and 10.00 g O_2 is ignited in a 3.15-L flask. What is the pressure in the flask when it is cooled to 25 °C?

84. In the final condition reached in Example 11.5, that is, condition (c), what mass of water is present as liquid and what mass as vapor?

* **85.** A 1.00-g sample of $H_2O(g)$ is injected into a 40.0-L flask at a temperature of 35.0 °C and cooled. To the nearest degree Celsius, determine the temperature at which the first liquid water will condense.

86. Intramolecular bonding in salicylic acid was illustrated on page 456. Salicylic acid actually forms both intramolecular and intermolecular hydrogen bonds. Draw the structure of a dimer of salicylic acid that shows four hydrogen bonds altogether.

87. Sketch a phase diagram for tin. Label significant points in the diagram, given that there are three polymorphic forms of the solid: α, gray tin below 19 °C; β, white tin between 19 and 161 °C; and γ, brittle tin from 161 °C to the melting point at 232 °C. The normal boiling point of tin is 2623 °C. Sketch the cooling curve you would expect to obtain if a sample of liquid tin at 250 °C is slowly cooled to 0 °C.

88. Would you expect the crystalline structure of magnesium oxide to be of the CsCl type or NaCl type? Explain. (*Hint:* Use data from Chapter 8, as necessary.)

* **89.** In a manner similar to that used in Exercise 11.11B, verify that the percent voids in the bcc and fcc unit cells are 31.98% and 25.96%, respectively.

* **90.** The unit cell of diamond is shown here. Use the methods of Examples 11.10 and 11.11 and a carbon-to-carbon bond length of 154.45 pm to calculate the density of diamond.

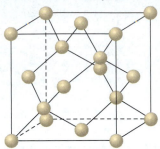

91. Assume that gasoline is approximated as a mixture of isooctane and octane. Should gasoline that is to be sold in Minnesota in November have a higher or lower proportion of isooctane than gasoline to be sold in North Carolina in May? Explain.

* **92.** The boiling points of pure HCl, HBr, and HI (not the aqueous solutions) are $-85\ ^\circ C$, $-67\ ^\circ C$, and $-35\ ^\circ C$, respectively. How do the relative strengths of the different types of intermolecular forces in these three gases compare? Explain.

93. Using the phase diagrams in this chapter as a guide, sketch a phase diagram and label four points, W, X, Y, and Z so that (a) the transition from W to X is from liquid to solid at constant temperature; (b) the transition from X to Y is from solid to gas at constant pressure; (c) the transition from Y to Z involves *no* phase change, although both temperature and pressure change.

* **94.** A 525-cm³ sample of Hg(l) at 20.0 °C is added to a large quantity of $N_2(l)$ kept at its boiling point in a thermally insulated container. What mass of $N_2(l)$ is vaporized as the Hg is brought to the temperature of the $N_2(l)$? For the specific heat of Hg(l) from 20 to $-39\ ^\circ C$, use 0.033 cal $g^{-1}\ ^\circ C^{-1}$, and for Hg(s) from -39 to $-196\ ^\circ C$, 0.030 cal $g^{-1}\ ^\circ C^{-1}$. The density of Hg(l) is 13.6 g/cm³; the melting point of Hg is $-39\ ^\circ C$, and its heat of fusion is 2.30 kJ/mol. The boiling point of $N_2(l)$ is $-196\ ^\circ C$ and its $\Delta H_{vapn} = 5.58$ kJ/mol.

* **95.** In the same manner as Figure 11.26, sketch microscopic views of the surface of the slide in Figure 11.28, corresponding to the left and right sides of the slide. Prepare an explanation for the details of the micro views, and comment on those features that cannot be shown easily in a sketch.

* **96.** Calculate the approximate length and diameter, in nanometers, of the capped nanotube shown in Figure 11.35.

* **97.** Sketch "micro" views of the contents of a sealed 5.00-L vessel at 25 °C, similar to the views in Figure 11.3, to represent the following. (Use data from Figure 11.4 and numbers of "molecules" in roughly the proportions you would expect to find them.)

(a) A 10.00-g sample of $CS_2(l)$ ($d = 1.26$ g/mL) is introduced into the evacuated vessel at 25 °C and $CS_2(g)$ has reached one-half of its equilibrium value.

(b) The same view as (a) but after liquid–vapor equilibrium has been reached.

(c) After an additional 2.00 g $CS_2(l)$ has been added to the vessel in (b) and equilibrium reestablished.

(d) After 3.00 L of $CS_2(g)$, measured at 750 Torr and 25 °C, has been removed from the vessel in (c).

*98. At 0 °C, barometric pressure decreases by 1/30 of its value for every 900 ft increase in altitude. Assuming a constant temperature of 0 °C and a barometric pressure of 1.00 atm at sea level, determine atop 1917-m Mt. Washington in New Hampshire **(a)** the barometric pressure and **(b)** the approximate boiling point of water.

99. The phenomenon illustrated in the two photographs is called the *regelation* of ice. How is the thin wire pulled through the block of ice without cutting the block in two? Would you expect the same phenomenon to occur with dry ice [$CO_2(s)$]? Explain.

*100. The vapor pressure of liquid chlorine is given by the following equation

$$\log p = 6.9379 - \frac{861.34}{t + 246.33}$$

and that of solid chlorine by the equation

$$\log p = 9.7051 - \frac{1444.2}{t + 267.13}$$

where p is the vapor pressure in mmHg and t is the Celsius temperature. Estimate the temperature and pressure at the triple point of chlorine.

Apply Your Knowledge

101. **[Environmental]** A heat pump is a device that is used to provide cooling during hot weather and heating during cold weather. It is based on the same condensation/evaporation cycle as a refrigerator. Describe how the heat pump works. Why do you think heat pumps are of limited value in very cold climates?

*102. **[Laboratory]** One method of determining the surface tension of a liquid is to measure the rise of the liquid in a capillary tube (recall Figure 11.30). The height to which the liquid rises (h) is related to the density (d) and surface tension (γ) of the liquid, the acceleration due to gravity (g), and the radius of the capillary tube (r) through the equation $\gamma = hdgr/2$. Liquid ethanol ($d = 0.789$ g/mL) rises 1.10 cm in a capillary tube having a radius of 0.50 mm. What is the surface tension of the ethanol, in J/m^2?

103. **[Collaborative]** Students in your laboratory group are told that an opaque bottle fitted with a one-hole stopper contains a sample of a single unknown substance. You cannot pick up the bottle or shake it. You can sniff at the hole in the stopper at the beginning of the laboratory period and again an hour later. All of you agree that the substance has a pungent odor that is equally strong at the second sniffing. What physical state(s) is/are likely to be present? Defend your choice(s), and dispute all other choices.

104. **[Environmental]** The National Institute of Occupational Safety and Health (NIOSH) limit for exposure to mercury vapor is 0.050 mg Hg/m^3 of air over a 10-hour period. Use the following data about a mercury spill, as needed, to determine how many times greater than the NIOSH concentration limit the mercury vapor concentration became. The spill of 0.40 g of Hg(l) at 22 °C and 751 Torr occurred in a laboratory 10.0 × 12.0 × 3.0 meters. The equilibrium vapor pressure of mercury at 22 °C is 0.0012 Torr.

e-Media Problems

The activities described in these problems can be found in the e-Media Activities and Interactive Student Tutorial (IST) modules of the Companion Website, *http://chem.prenhall.com/hillpetrucci*.

105. In the **States of Matter** activity (*Section 11-1*), **(a)** estimate the number of atoms per square centimeter for each phase, assuming that each illustration represents an area of 170 Å^2. **(b)** Arrange the phases in order of increasing number density. **(c)** Describe the types of motion of individual atoms in the different phases.

106. In the **Liquid–Vapor Equilibrium** activity (*Section 11-2*), **(a)** what characterizes the point of physical equilibrium? **(b)** The activity illustrates the dynamics of atoms or molecules at the liquid–gas interface as a function of time up to the point of equilibrium. Describe on a molecular scale what would take place if the process were repeated at a higher temperature.

107. View the **Sublimation** activity (*Section 11-3*). **(a)** How can sublimation be used to purify immiscible solid substances? **(b)** Describe what happens at the molecular level as the desired compound is separated from impurities and collected as purified

material. **(c)** How does the procedure depend on the relative vapor pressures of the desired solid compound and of the impurity?

108. In the **Molecular Details of a Phase Diagram** activity (*Section 11-4*), describe the change in molecular motion as the pressure is increased from 1.0 atm to 5.0 atm at a constant temperature of −65 °C. What phase changes, if any, are encountered during this pressure change?

109. View the **Propane and Acetaldehyde** 3-D models (*Section 11-5*). In analogy to Figures 11.16 and 11.19, sketch the types of intermolecular interactions in pure liquid samples of these molecules. Illustrate the influence of *induced* and/or *permanent* dipoles.

110. In the **Hydrogen Bonding** activity (*Section 11-6*), describe how hydrogen bonds between molecules differ from the covalent bonds between oxygen and hydrogen.

111. View the **Close-Packing** movie (*Section 11-10*). **(a)** Sketch the appearance of the bubble raft if it were possible to create bubbles with the shape of an equilateral triangle. **(b)** Why would the resulting 2-D structure be potentially less ordered than the structure of close-packed circles (spherical bubbles)?

Chapter 12

Physical Properties of Solutions

ONE OF THE broad categories of matter we described in Chapter 1 was homogeneous mixtures, or *solutions,* and we encountered solutions often in the chapters that followed. In Chapter 3, we considered some basic terminology, introduced the concentration unit molarity, and used molarity in stoichiometric calculations. We then devoted Chapter 4 to an overview of the main types of chemical reactions that occur in aqueous solutions.

In this chapter, we will examine some *physical* properties of solutions and phenomena related to them. Consider the coolant in an automobile engine. We could use water alone to cool the engine, but we usually prefer a solution of a substance such as ethylene glycol ($HOCH_2CH_2OH$) in water because it has both a higher *boiling point* and a lower *freezing point* than does water. The solution is therefore less likely than water to boil away in a hot engine or to freeze at low temperatures; it acts as both a coolant and an antifreeze.

Consider petroleum, a solution composed of hundreds of compounds, most of them hydrocarbons. To make gasoline from petroleum, we take advantage of differences in the *vapor pressures* of the various hydrocarbons to separate them from one another. The hydrocarbon mixture boils over a range of temperatures, enabling us to separate them by a process called *fractional distillation.* Also consider how much health-care professionals need to know about solutions. They inject fluids intravenously to restore fluids to the body, but they cannot use pure water for this purpose. The water would enter blood cells, causing them to swell and burst. Instead, they need aqueous solutions with the correct value of a property known as *osmotic pressure.*

In this chapter, we will explore interrelated physical properties of solutions, including boiling points, freezing points, vapor pressures, and osmotic pressures. We will also consider some fundamental aspects of solution formation that will help us to understand which substances dissolve in one another and why.

◄ Like so many natural liquids, seawater is not a pure substance, but a solution. Most people are aware of the sodium ions and chloride ions that are dissolved in seawater and give it a characteristic salty taste. However, seawater contains many other less-familiar solutes, including molecular, ionic, solid, liquid, and gaseous substances, which also influence its properties. The physical properties of solutions explored in this chapter are in large part responsible for many life-sustaining processes observed in natural systems.

Application Note

Solutions of perfluorocarbons, such as perfluorodecalin ($C_{10}F_{18}$) saturated with $O_2(g)$, are examples of a solution of a gas in a liquid. Although not yet approved for general use in the United States, animal studies and clinical trials have demonstrated the utility of this solution for liquid ventilation in the lungs. For trials involving use in the bloodstream, a milky emulsion as shown here is employed.

Dissolution of NaCl in Water animation

12.1 Some Types of Solutions

The *solvent* is the solution component that determines the state of matter of the solution; it is usually the component present in the greatest amount. The substance(s) dissolved in the solvent is the *solute(s)*. In a solution of sucrose (table sugar) in water, solid sucrose is the solute and liquid water is the solvent. The proportion of the solute can be small (dilute solution) or large (concentrated solution). In carbonated water, gaseous CO_2 is the solute and, again, water is the solvent. Note, however, that solutions are not limited to the liquid phase. All gaseous mixtures are solutions, and even some mixtures of solids are solutions. The various possibilities for types of solutions and their solutes and solvents are listed in Table 12.1.

To study the physical properties of solutions, we need several concentration units. In the next section, we review some units we have seen before and introduce some new ones.

Table 12.1 Some Common Types of Solutions

Solute	Solvent	Solution Phase	Examples [solute(s) listed prior to solvent]
Gas	Gas	Gas	Air (O_2, Ar, CO_2, ... in N_2) Natural gas (C_2H_6, C_3H_8, ... in CH_4)
Gas	Liquid	Liquid	Club soda (CO_2 in H_2O) Blood substitute (O_2 in perfluorodecalin)
Liquid	Liquid	Liquid	Vodka (CH_3CH_2OH in H_2O) Vinegar (CH_3COOH in H_2O)
Solid	Liquid	Liquid	Saline solution (NaCl in H_2O) Racing fuel (naphthalene in gasoline)
Gas	Solid	Solid	Hydrogen (H_2) in palladium (Pd)
Solid	Solid	Solid	14-karat gold (Ag in Au) Yellow brass (Zn in Cu)

12.2 Solution Concentration

The *molarity* of a solution relates an amount of solute in moles and a solution volume in liters:

$$\text{Molarity (M)} = \frac{\text{amount of solute (mol)}}{\text{volume of solution (L)}} \tag{12.1}$$

Thus, if we measure the volume of a solution and know its molarity, we can readily calculate the number of moles of solute present. This makes molarity the ideal concentration unit to use in titrations, as we saw in Section 4.6. However, molarity is not as useful for some of the applications in this chapter.

Solution Formation from a Solid animation

Percent by Mass, Percent by Volume, and Mass/Volume Percent

For many practical applications, we can use percent composition to express solution concentration. Then, if we require a precise quantity of solute, we just measure out a mass or volume of solution. For use in storage batteries, for example, sulfuric acid is supplied as a solution that is 35.7% H_2SO_4 *by mass*. Every 100.0 g of this solution contains 35.7 g H_2SO_4 as solute and 64.3 g H_2O as solvent. If we want a quantity of solution containing 135.0 g H_2SO_4, we can calculate the mass of the battery acid needed:

$$\text{Mass} = 135.0 \text{ g } H_2SO_4 \times \frac{100.0 \text{ g battery acid}}{35.7 \text{ g } H_2SO_4} = 378 \text{ g battery acid}$$

Example 12.1

How would you prepare 750 g of an aqueous solution that is 2.5% NaOH by mass?

STRATEGY

We need to calculate how many grams of NaOH as solute and grams of H_2O as solvent are required to make 750 g of solution. First, we can determine the required mass of NaOH by using the percent composition as a conversion factor. Then we can obtain the mass of H_2O by difference.

SOLUTION

The percent composition allows us to convert grams of solution to grams of NaOH.

$$? \text{ g NaOH} = 750 \text{ g solution} \times \frac{2.5 \text{ g NaOH}}{100 \text{ g solution}} = 19 \text{ g NaOH}$$

The mass of water, the solvent, is the difference between the solution mass and the mass of NaOH, the solute.

$$? \text{ g } H_2O = 750 \text{ g solution} - 19 \text{ g NaOH} = 731 \text{ g } H_2O$$

To make the 750 g of solution, we measure out 19 g NaOH and dissolve it in 731 g H_2O.

EXERCISE 12.1A

Glucose is a sugar found abundantly in nature—for example, in ripe grapes. What is the percent by mass of glucose in a solution made by dissolving 163 g glucose in 755 g water? Do you need to know the formula of glucose to do that calculation? Explain.

EXERCISE 12.1B

What is the mass percent sucrose in the solution obtained by mixing 225 g of an aqueous solution that is 6.25% sucrose by mass with 135 g of an aqueous solution that is 8.20% sucrose by mass?

Because it is so easy to measure liquid volumes if both the solute and solvent are liquids, the concentration of such a solution is often expressed as *percent by volume.* Ethanol, CH_3CH_2OH, for medicinal purposes is generally known as USP ethanol; it is 95% CH_3CH_2OH by volume.* That is, it consists of 95 mL $CH_3CH_2OH(l)$ per 100 mL of aqueous solution.

▲ In tests for intoxication, blood alcohol levels are expressed as percent by volume. A blood alcohol level of 0.08% by volume means 0.08 mL ethanol per 100 mL of blood, and is considered proof of intoxication in many states. (Left) This test device measures ethanol concentration using the reduction of yellow-orange $K_2Cr_2O_7$ to blue-green chromium(III) compounds. (Right) An electronic Draeger Breathalyzer uses an electrochemical fuel cell (Section 18.7) to determine ethanol concentration in the breath. In both cases, readings are calibrated against actual blood alcohol levels.

*USP is an abbreviation for *United States Pharmacopoeia,* the official publication of standards for pharmaceutical products.

Example 12.2

At 20 °C, pure ethanol has a density of 0.789 g/mL and USP ethanol has a density of 0.813 g/mL. What is the mass percent ethanol in USP ethanol?

STRATEGY

Because percent by volume means parts per hundred by volume, for convenience we can base our calculation on a 100.0-mL sample of the USP ethanol. We can get the mass of the USP alcohol from its volume and density. Similarly, we can get the mass of ethanol solute in the USP solution from the solution volume and the density of pure ethanol. We obtain the mass percent composition from these two masses.

SOLUTION

We use the density of the USP ethanol solution to convert from milliliters to grams of solution.

$$\text{Mass of solution} = 100.0 \text{ mL} \times \frac{0.813 \text{ g}}{1 \text{ mL}} = 81.3 \text{ g}$$

To determine the mass of ethanol in the 100.0 mL of USP ethanol, we use the volume percent composition of the solution and the density of pure ethanol as conversion factors.

$$? \text{ g ethanol} = 100.0 \text{ mL solution} \times \frac{95.0 \text{ mL ethanol}}{100.0 \text{ mL solution}} \times \frac{0.789 \text{ g ethanol}}{1 \text{ mL ethanol}}$$

$$= 75.0 \text{ g ethanol}$$

Now we form the mass ratio of ethanol to solution, and multiply this ratio by 100%.

$$\text{Mass percent ethanol} = \frac{75.0 \text{ g ethanol}}{81.3 \text{ g solution}} \times 100\% = 92.3\%$$

ASSESSMENT

Because ethanol is less dense than water, we should expect the mass contribution of the ethanol to be less than its volume contribution to the USP alcohol, and it is—95.0% on a volume basis but only 92.3% on a mass basis.

EXERCISE 12.2A

Assume that the volumes are additive, and determine the volume percent toluene, $C_6H_5CH_3$, in a solution made by mixing 40.0 mL of toluene with 75.0 mL of benzene, C_6H_6.

EXERCISE 12.2B

At 20 °C, the density of benzene is 0.879 g/mL and that of toluene is 0.866 g/mL. For the solution described in Exercise 12.2A, what are (a) the mass percent toluene and (b) the density?

Another concentration unit widely used in medicine is *mass/volume percent*. For example, a solution of sodium chloride used in intravenous injections has the composition 0.92% (mass/vol) NaCl; that is, it contains 0.92 g NaCl per 100 mL solution. This unit is most commonly used for fairly dilute solutions. A volume of 100 mL is also one-tenth of a liter, or one *deciliter*. If we express the mass of solute in *milligrams*, we then have a mass/volume concentration unit in *milligrams per deciliter* (mg/dL). A blood cholesterol reading of 187, for example, means 187 mg cholesterol/dL blood.

Parts per Million, Parts per Billion, and Parts per Trillion

For solutions that are extremely dilute, concentrations are often expressed in parts per million (ppm), parts per billion (ppb), or even parts per trillion (ppt). Fluoridated drinking water, for instance, has a fluoride ion concentration of about 1 ppm, and a typical level of the contaminant chloroform ($CHCl_3$) in a municipal drinking water sample from the lower Mississippi River is 8 ppb.

We get a sense of the meanings of these terms by considering that one person in the city of San Diego, California, represents about 1 ppm of the city's population, and one person in India represents about 1 ppb of that country's population. Figure 12.1 demonstrates the extreme dilution implied by the units ppm, ppb, and ppt.

The units ppm, ppb, and ppt are somewhat ambiguous. For gaseous solutions, such as air, ppm, ppb, and ppt are generally based on numbers of molecules or on volumes. Thus, 1 ppm of some substance in air is one molecule of the substance per

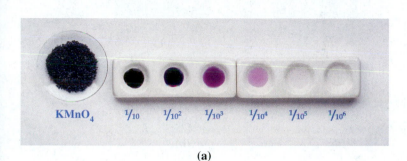

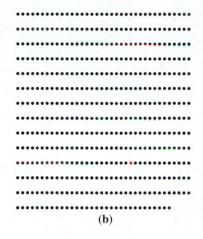

(a)   (b)

▲ FIGURE 12.1   **Visualizing one part per million**
(a) The intensely colored solid is potassium permanganate ($KMnO_4$). In the first solution, a sample of $KMnO_4$ is dissolved in 9 times its mass of water, making its concentration 1 part in 10 by mass. The second solution is a tenfold dilution of the first; it has 1 part of solute per 100. The third solution has 1 part per 1000, and so on. The color of the solution is no longer discernible when the concentration has been reduced to 1 part per 1,000,000 (1 ppm). (b) The display has one red dot in 500 total dots, equivalent to 2 parts per thousand. To show 1 ppm, we would need a display 2000 times as large—about 56 pages of dots with only one red dot. To show 1 ppb, we would need more than 50 books the size of this one—still with only one red dot in all those books of black dots.

1×10^6 molecules of air or 1 L of that substance per 1×10^6 L of air. For liquid solutions, these units are generally based on mass. Thus, 1 ppm of solute in a solution is the same as 1 g solute per 1×10^6 g solution. Because dilute aqueous solutions have a density of about 1.00 g/L, 1 liter has a mass of about 10^6 mg, or 10^9 μg, or 10^{12} ng. Thus, we can derive the following useful equivalents (see Example 12.3):

$$1 \text{ ppm} = 1 \text{ mg/L} \qquad 1 \text{ ppb} = 1 \text{ } \mu\text{g/L} \qquad 1 \text{ ppt} = 1 \text{ ng/L}$$

Example 12.3

The maximum allowable level of nitrates in drinking water in the United States is 45 mg NO_3^-/L. What is this level expressed in parts per million (ppm)?

STRATEGY

The expression ppm signifies a ratio having parts of the target constituent in the numerator and 1,000,000 parts of the other constituent(s) in the denominator. The starting point is already in the form of a ratio, 45 mg NO_3^-/1 L. Thus, what we need to do is convert the numerator and denominator to the same unit, with the denominator being 1,000,000 × unit.

SOLUTION

The density of water, even with traces of dissolved substances, is essentially 1.00 g/mL. One liter of water has a mass of 1000 g. The allowable nitrate level can be expressed as a ratio of masses.

$$\frac{45 \text{ mg NO}_3^-}{1000 \text{ g water}}$$

If we convert the denominator to milligrams, the result is the number of mg NO_3^- per 1 million milligrams of water, the ppm NO_3^-.

$$\frac{45 \text{ mg NO}_3^-}{1000 \text{ g water}} \times \frac{1 \text{ g water}}{1000 \text{ mg water}} = \frac{45 \text{ mg NO}_3^-}{1,000,000 \text{ mg water}} = 45 \text{ ppm NO}_3^-$$

EXERCISE 12.3A

What is the concentration in (a) ppb and (b) ppt corresponding to a maximum allowable level in water of 0.1 μg/L of the pesticide chlordane?

EXERCISE 12.3B

What is the concentration in ppm of Na^+ in 0.00152 M Na_2SO_4?

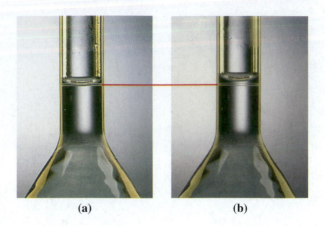

▶ **FIGURE 12.2   The effect of temperature on molarity**
(a) A 0.1000 M HCl solution prepared at 20.0 °C. (b) The solution expands on being warmed to 25.0 °C. At this higher temperature, because the same amount of solute is present in a larger volume of solution than at 20.0 °C, the molarity is slightly less than 0.1000 M HCl.

(a)   **(b)**

🌐 **Solution Concentration activity**

Molality

The *molarity* of a solution varies with temperature, a fact illustrated in Figure 12.2. Because the volume of the solution increases as its temperature increases from 20 °C to 25 °C, the molarity of the solution *decreases*. To be *independent* of temperature, a concentration unit must be based on mass only, not volume. Percent by mass is such a unit; percent by volume and mass/volume percent are not.

Molality is another unit independent of temperature. The **molality (*m*)** of a solution is the number of moles of solute per *kilogram of solvent* (not of solution):

$$\text{Molality } (m) = \frac{\text{amount of solute (mol)}}{\text{mass of solvent (kg)}} \tag{12.2}$$

For example, a solution of 0.22 mol ethanol in 550 g of water is 0.40 *m* ethanol:

$$\frac{0.22 \text{ mol ethanol}}{0.550 \text{ kg water}} = 0.40 \; m \text{ ethanol}$$

Example 12.4

What is the molality of a solution prepared by dissolving 5.05 g naphthalene [$C_{10}H_8(s)$] in 75.0 mL of benzene, C_6H_6 ($d = 0.879$ g/mL)?

STRATEGY

Molality has the units *mol solute/kg solvent*. We need to convert the mass of naphthalene to mol $C_{10}H_8$ and the volume of benzene to kg C_6H_6, making use of density as a conversion factor. The molality of the solution will be the ratio of solute to solvent, when each is expressed in the appropriate units.

SOLUTION

We use the molar mass of naphthalene to convert from grams to moles.

$$? \text{ mol } C_{10}H_8 = 5.05 \text{ g } C_{10}H_8 \times \frac{1 \text{ mol } C_{10}H_8}{128.2 \text{ g } C_{10}H_8} = 0.0394 \text{ mol } C_{10}H_8$$

To convert from mL C_6H_6 to kg C_6H_6, we need two conversion factors.

$$? \text{ kg } C_6H_6 = 75.0 \text{ mL } C_6H_6 \times \frac{0.879 \text{ g } C_6H_6}{1 \text{ mL } C_6H_6} \times \frac{1 \text{ kg } C_6H_6}{1000 \text{ g } C_6H_6} = 0.0659 \text{ kg } C_6H_6$$

The molality of the solution is calculated from the ratio of these two values.

$$\frac{0.0394 \text{ mol } C_{10}H_8}{0.0659 \text{ kg } C_6H_6} = 0.598 \; m \; C_{10}H_8$$

EXERCISE 12.4A

What is the molality of a solution prepared by dissolving 225 mg glucose ($C_6H_{12}O_6$) in 5.00 mL ethanol ($d = 0.789$ g/mL)?

EXERCISE 12.4B

What is the molality of a solution composed of 25.0 mL of toluene, $C_6H_5CH_3$ ($d = 0.866$ g/mL), in 75.0 mL of benzene ($d = 0.879$ g/mL)?

Example 12.5

How many grams of benzoic acid, C_6H_5COOH, must be dissolved in 50.0 mL of benzene, C_6H_6 ($d = 0.879$ g/mL), to produce 0.150 m C_6H_5COOH?

SOLUTION

Here we use the molality as a conversion factor between the mass of solvent (in kilograms) and the number of moles of solute. As shown in the outline that follows, we first need two conversions to get the quantity of solvent in kilograms. Then we apply molality as the pivotal factor (in red). Finally, we convert moles to grams of solute.

> Start with the volume of solvent, in mL.

> Density converts mL to g solvent.

> Mass of solvent is converted from g to kg.

$$? \text{ g } C_6H_5COOH = 50.0 \text{ mL } C_6H_6 \times \frac{0.879 \text{ g } C_6H_6}{1 \text{ mL } C_6H_6} \times \frac{1 \text{ kg } C_6H_6}{1000 \text{ g } C_6H_6}$$

> Molality converts kg of solvent to moles of solute.

> Molar mass converts moles to grams of solute.

> The answer: ? unit

$$\times \frac{0.150 \text{ mol } C_6H_5COOH}{1 \text{ kg } C_6H_6} \times \frac{122.1 \text{ g } C_6H_5COOH}{1 \text{ mol } C_6H_5COOH} = 0.805 \text{ g } C_6H_5COOH$$

EXERCISE 12.5A

How many milliliters of ethanol (CH_3CH_2OH, $d = 0.789$ g/mL) must be mixed with 125 mL of benzene (C_6H_6, $d = 0.879$ g/mL) in order to produce a solution that is 0.0652 m CH_3CH_2OH?

EXERCISE 12.5B

How many milliliters of water ($d = 0.998$ g/mL) are required to dissolve 25.0 g of urea and thereby produce a 1.65 m solution of urea, $CO(NH_2)_2$?

Mole Fraction and Mole Percent

Calculations we will do later in the chapter require that solution concentrations be expressed in a particular unit. For example, vapor pressure calculations (Section 12.6) require concentrations expressed in mole fractions. As with gases (page 195), the **mole fraction** (x_i) of a solution component i is the fraction of all the molecules in the solution that are molecules of i:

$$x_i = \frac{\text{amount of component } i \text{ (mol)}}{\text{total amount of solution components (mol)}} = \frac{n_i}{n_{\text{total}}} \qquad (12.3)$$

Note that mole fraction has no unit; it is dimensionless. Note also that the sum of the mole fractions of all the components in a solution is equal to 1:

$$x_1 + x_2 + x_3 + \cdots = 1 \qquad (12.4)$$

The **mole percent** of a solution component is its mole fraction multiplied by 100%. Thus, a solution made by mixing 4 mol methanol and 6 mol ethanol has the mole fractions $x_{CH_3OH} = 4/(4 + 6) = 0.40$ and $x_{CH_3CH_2OH} = 6/(4 + 6) = 0.60$. The corresponding mole percents are 40 mol% CH_3OH and 60 mol% CH_3CH_2OH.

Later in the chapter, when we are doing calculations on some property of a solution, we may need to do a preliminary calculation to convert from a given concentration unit to another unit specifically required by that calculation. Some typical conversions are illustrated in Example 12.6.

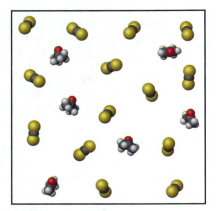

$x_{C_3H_6O} = 0.3$

$x_{CS_2} = 0.7$

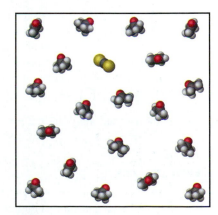

$x_{C_3H_6O} = 0.95$

$x_{CS_2} = 0.05$

▲ Mole fractions represent relative numbers of moles or of molecules.

Example 12.6

An aqueous solution of ethylene glycol used as an automobile engine coolant is 40.0% $HOCH_2CH_2OH$ by mass and has a density of 1.05 g/mL. What are the **(a)** molarity, **(b)** molality, and **(c)** mole fraction of $HOCH_2CH_2OH$ in this solution?

STRATEGY

Let's start by listing what we need for each calculation.

(a) Molarity: number of moles of solute and volume of solution in liters

(b) Molality: number of moles of solute and mass of solvent in kilograms

(c) Mole fraction: number of moles of solute and number of moles of solvent

When we are not given a particular quantity of solution to consider, we can pick a convenient one, usually 1.00 L or 1.00 kg. Let's choose 1.00 L for this problem. We will need conversion factors based on density, mass percent composition, and molar masses.

SOLUTION

(a) The mass of solution is found from its volume and density.

$$? \text{ g solution} = 1.00 \text{ L} \times \frac{1000 \text{ mL}}{1 \text{ L}} \times \frac{1.05 \text{ g}}{1 \text{ mL}} = 1050 \text{ g}$$

We can use the mass percent composition to convert from mass of solution to mass of $HOCH_2CH_2OH$.

$$? \text{ g } HOCH_2CH_2OH = 1050 \text{ g solution} \times \frac{40.0 \text{ g } HOCH_2CH_2OH}{100 \text{ g solution}}$$

$$= 420 \text{ g } HOCH_2CH_2OH$$

Then, we can use the molar mass of $HOCH_2CH_2OH$ to convert from grams to moles of $HOCH_2CH_2OH$.

$$? \text{ mol } HOCH_2CH_2OH = 420 \text{ g } HOCH_2CH_2OH \times \frac{1 \text{ mol } HOCH_2CH_2OH}{62.07 \text{ g } HOCH_2CH_2OH}$$

$$= 6.77 \text{ mol } HOCH_2CH_2OH$$

Finally, we calculate the molarity of the solution, based on our chosen 1.00-L volume.

$$\frac{6.77 \text{ mol } HOCH_2CH_2OH}{1 \text{ L}} = 6.77 \text{ M } HOCH_2CH_2OH$$

(b) From part (a), we have the number of moles of $HOCH_2CH_2OH$ in 1.00 L of solution (6.77 mol) and the mass of solution (1050 g) and solute (420 g). The mass of water, the *solvent,* is the difference in mass between that of the solution and the solute.

$$? \text{ kg } H_2O = (1050 - 420) \text{ g } H_2O \times \frac{1 \text{ kg } H_2O}{1000 \text{ g } H_2O} = 0.630 \text{ kg } H_2O$$

Now, we can express the molality as moles of $HOCH_2CH_2OH$ per kg H_2O.

$$\frac{6.77 \text{ mol } HOCH_2CH_2OH}{0.630 \text{ kg}} = 10.7 \text{ m } HOCH_2CH_2OH$$

(c) To get the mole fractions of the solution components, we need the number of moles of each. We have the number of moles of $HOCH_2CH_2OH$ (6.77 mol) from part (a), and we can get the number of moles of H_2O in the 630 g H_2O calculated in part (b).

$$? \text{ mol } H_2O = 630 \text{ g } H_2O \times \frac{1 \text{ mol } H_2O}{18.02 \text{ g } H_2O} = 35.0 \text{ mol } H_2O$$

Now we can apply the definition of mole fraction.

$$x_{HOCH_2CH_2OH} = \frac{6.77 \text{ mol } HOCH_2CH_2OH}{6.77 \text{ mol } HOCH_2CH_2OH + 35.0 \text{ mol } H_2O} = 0.162$$

EXERCISE 12.6A

Calculate **(a)** the molality of CH_3OH in a solution that is 7.50% by mass CH_3OH in CH_3CH_2OH and **(b)** the mole percent of urea, $CO(NH_2)_2$, in an aqueous solution that is 1.05 m $CO(NH_2)_2$.

EXERCISE 12.6B

A 2.90 m solution of methanol (CH_3OH) in water has a density of 0.984 g/mL. What are the **(a)** mass percent, **(b)** molarity, and **(c)** mole percent of methanol in this solution?

Example 12.7   An Estimation Example

Without doing detailed calculations, determine which aqueous solution has the greatest mole fraction of CH_3OH: (a) 1.0 m CH_3OH, (b) 10.0% CH_3OH by mass, (c) $x_{CH_3OH} = 0.10$.

ANALYSIS AND CONCLUSIONS

The mole fraction of solution (c) is given: 0.10. The ratio of moles of CH_3OH to moles of H_2O in solution (c) must therefore be 1/9. Let's estimate the corresponding ratios for the other solutions and compare them with 1/9. We will need to use the molar masses of H_2O and CH_3OH, but these can be just rough estimates: 18 g H_2O/mol and 32 g CH_3OH/mol.

Solution (a) contains 1.0 mol CH_3OH per kilogram of H_2O. One kilogram of water is 1000 g H_2O, and 1000 g H_2O is about 50 mol H_2O (1000/18). The ratio of mol CH_3OH/mol H_2O is about 1/50.

Solution (b) has 10.0 g CH_3OH for every 90.0 g H_2O. The 90.0 g H_2O is almost exactly 5.0 mol H_2O, and the 10.0 g CH_3OH is about $10/32 \approx 0.3$ mol CH_3OH. The ratio mol CH_3OH/mol H_2O in solution (b) is about 0.3/5.0. If we multiply both factors in this ratio by 3, we get 0.9/15, which we can round to 1/15. Thus, there is about 1 mol CH_3OH for each 15 mol H_2O.

Solution (c) has the highest ratio (1/9) of mol CH_3OH to mol H_2O, and thus the greatest mole fraction CH_3OH.

EXERCISE 12.7A

Without doing detailed calculations, determine which of the following aqueous solutions has the greatest mole percent CH_3CH_2OH: (a) 0.50 M CH_3CH_2OH ($d = 0.994$ g/mL), (b) 5.0% CH_3CH_2OH by mass, (c) 0.50 m CH_3CH_2OH, (d) 5.0% CH_3CH_2OH by volume ($d = 0.991$ g/mL). (The density of pure liquid CH_3CH_2OH is 0.789 g/mL.)

EXERCISE 12.7B

Without doing detailed calculations, determine which of the following aqueous solutions has the greatest concentration of Na^+, expressed in ppm by mass: (a) 0.010 m NaCl(aq), (b) 0.010 M Na_2SO_4, (c) 1.0% $NaNO_3$(aq) by mass, (d) an aqueous solution with mol fractions of Na^+ and Cl^- each equal to 0.010.

12.3   Energetics of Solution Formation

Why do some substances dissolve in a given solvent, but others do not? For example, why does ethanol dissolve in water, but the hydrocarbon ethane does not? Why does sodium carbonate, but not magnesium carbonate, dissolve in water? Why does grease dissolve in kerosene but not in water? To answer questions like these, let us consider two significant factors in solution formation: the enthalpy of solution and intermolecular forces in mixtures.

Enthalpy of Solution

We can gain some insight into solution formation by considering it as taking place in three *hypothetical* steps: (1) Move the molecules of solvent apart to make room for the solute molecules. This requires work to overcome the intermolecular forces, and the enthalpy of the solvent increases: $\Delta H_1 > 0$. (2) Separate the molecules of solute to the distances found between them in the solution. Again, work is required and $\Delta H_2 > 0$. (3) Allow the separated solute and solvent molecules to mix randomly. For this process, we expect energy to be released, $\Delta H_3 < 0$, because now there are forces of attraction between the solute and solvent molecules. These steps and the net change are summarized here:

(1) Pure solvent $\longrightarrow$ separated solvent molecules ΔH_1

(2) Pure solute $\longrightarrow$ separated solute molecules ΔH_2

(3) Separated solvent and solute molecules $\longrightarrow$ solution ΔH_3

Net: Pure solvent + pure solute $\longrightarrow$ solution $\Delta H_{soln} = \Delta H_1 + \Delta H_2 + \Delta H_3$

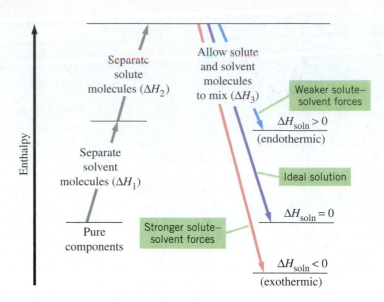

 **Enthalpy of Solution Diagram activity**

▶ **FIGURE 12.3** **Visualizing an enthalpy of solution with an enthalpy diagram**

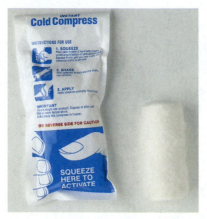

▲ An instant cold pack contains a salt (ammonium chloride or ammonium nitrate) that dissolves endothermically.

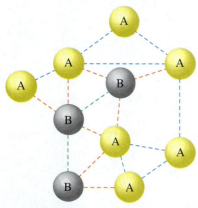

▲ **FIGURE 12.4** **Intermolecular forces in solution formation**
The intermolecular forces between solvent molecules, A, are represented in blue; between solute molecules, B, in green; and between solvent molecules, A, and solute molecules, B, in red.

Whether the formation of a solution is endothermic ($\Delta H_{soln} > 0$) or exothermic ($\Delta H_{soln} < 0$) depends on the relative values of the enthalpy changes of the three hypothetical steps. Figure 12.3 presents a generalized enthalpy diagram for an endothermic solution process, an exothermic process, and a special case in which $\Delta H_{soln} = 0$.

As we shall see next, we can relate the enthalpy changes in the three-step hypothetical solution process to intermolecular forces of attraction.

Intermolecular Forces in Solution Formation

Figure 12.4 suggests three types of intermolecular forces in a solution. The forces A–A (blue) are between solvent molecules; their strength determines the enthalpy change ΔH_1 in the hypothetical three-step solution process. The forces B–B (green) are between solute molecules and determine the value of ΔH_2. The intermolecular forces A–B (red) are between molecules of solvent and molecules of solute; these forces establish ΔH_3.

The comparative strengths of these three types of intermolecular forces are an important factor in determining whether a given solvent and solute form a solution. Let's consider the following four possibilities.

1. *All intermolecular forces are of comparable strength.* If the three different intermolecular forces depicted in Figure 12.4, A–A, B–B, and A–B, are of the same type and comparable in strength, we expect the molecules of solute and solvent to intermingle randomly into a solution. Moreover, this situation might well lead to the case in Figure 12.3 in which $\Delta H_{soln} = 0$. There is no net energy change in the formation of the solution, and we expect the volume of the solution to be the sum of the volumes of the solvent and the solute ($\Delta V_{soln} = 0$). A solution that meets these requirements is called an **ideal solution.**

Mixtures of ideal gases are ideal solutions. It is unlikely that any liquid solution meets these requirements perfectly, but some come close enough that we consider them to be ideal. One example is the mixture of hydrocarbons that make up gasoline. Another is a solution of toluene in benzene. Because of the similarities in molecular structure (Figure 12.5), we expect the magnitude of the intermolecular forces between benzene molecules to be quite similar to the magnitude of those between toluene molecules.

2. *Intermolecular forces between solute and solvent molecules are <u>stronger</u> than other intermolecular forces.* When this condition is met (Figure 12.3, red arrow), the magnitude of ΔH_3 can exceed the sum of ΔH_1 and ΔH_2. We expect a solution to form with the evolution of energy as heat. For example, in forming a solution of ethanol in water, $\Delta H_3 > \Delta H_1 + \Delta H_2$; and therefore solution formation is

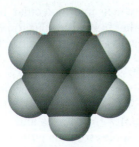

Benzene, C_6H_6

Toluene, $C_6H_5CH_3$

▲ **FIGURE 12.5   Two molecules with similar structures**
Benzene is a six-carbon ring with a H atom attached to each C atom. Toluene differs only in that a CH_3 group substitutes for one of the H atoms.
QUESTION: Why do these two form an ideal solution?

100 mL

▲ **FIGURE 12.6   A non-ideal solution: Ethanol and water**
Identical volumetric flasks are filled to the 50.0-mL mark, one with ethanol and the other with water. When the two liquids are mixed, the volume is less than the expected 100.0 mL; it is only about 95 mL.

QUESTION: What would you expect to see if one of the two 50.0-mL flasks contained water and the other contained methanol? If one 50.0-mL flask contained methanol and the other contained ethanol?

exothermic. Moreover, as shown in Figure 12.6, in ethanol–water solutions, the volume of the solution is *less than* the sum of the volumes of the ethanol and water. We expect solutions of this type to be *nonideal* because $\Delta H_{soln} < 0$ and $\Delta V_{soln} < 0$.

3. *Intermolecular forces between solute and solvent molecules are <u>weaker</u> than other intermolecular forces.* In this situation, the magnitude of ΔH_3 in Figure 12.3 is less than the sum of ΔH_1 and ΔH_2. A solution may form, but it will be nonideal; $\Delta H_{soln} > 0$, and the solution process is endothermic.

How can a solution form in an endothermic process, with the solution in a higher energy (enthalpy) state than the separate components? As you might well suspect, an additional factor is involved. This factor is a change in a thermodynamic property called *entropy*, which we will describe in Chapter 17.

4. *Intermolecular forces between solute and solvent molecules are <u>much weaker</u> than other intermolecular forces.* In this case, the magnitude of ΔH_3 in Figure 12.3 may be so much smaller than the sum of ΔH_1 and ΔH_2 that the solution is at too high an energy state to be attained. The solute does not dissolve in the solvent; the two remain as separate phases in a heterogeneous mixture. For example, the dominant intermolecular forces among the nonpolar hydrocarbon molecules in gasoline are dispersion forces; among water molecules, they are hydrogen bonds. However, no hydrogen bonds form between water and hydrocarbon molecules, and consequently a solution of gasoline and water does not form.

Two rules of thumb help us summarize three of the four possibilities just described. The rule that *like dissolves like* works well for cases **1** and **2**, that is, when all intermolecular forces are of comparable strength and when intermolecular forces between solute and solvent molecules are stronger than other intermolecular forces. In contrast, the old saying "Oil and water do not mix" is another way of explaining case **4**: Substances with structurally dissimilar molecules, such as gasoline and water, tend not to dissolve in each other.

▲ This oil slick produced by the massive Exxon *Valdez* oil spill in Alaska in 1989 reminds us that hydrocarbon oils and water do not mix.

Neither description helps with case **3**, in which intermolecular forces between solute and solvent molecules are weaker than other intermolecular forces. As we have already noted, we will learn about the factors involved here in Chapter 17.

Aqueous Solutions of Ionic Compounds

Now let us consider ionic compounds dissolving in water. Interionic attractions hold ions together in an ionic solid. The forces causing the solid to dissolve are *ion–dipole* forces—the attraction of water dipoles for cations and anions. As suggested by Figure 12.7, the attraction of water dipoles for ions pulls the ions out of a crystalline lattice and into aqueous solution. Thus, ion–dipole forces break down the crystal of a soluble ionic compound. The ion–dipole forces in the solution lessen the tendency for ions to return to the crystalline state. All ions in an aqueous solution have a certain number of water molecules associated with them. Such ions are *hydrated*.

Dissolution of KMnO₄ animation

The extent to which an ionic solid dissolves in water is determined largely by the competition between the interionic attractions that hold the ions in the crystal lattice and the ion–dipole attractions that pull them into solution. If the ion–dipole attractions predominate, the solid is soluble in water. This is the case with the common compounds of the group 1A metals and ammonium compounds (Table 4.3). If interionic attractions are the dominant factor, as is the case with many carbonates, hydroxides, and sulfides, the solid is insoluble.

Enthalpy of Solution simulation

It is often difficult to predict the water solubility of a specific compound. For example, we would expect the strong interionic attractions between the highly charged Al^{3+} and SO_4^{2-} ions to lead to limited solubility in water for $Al_2(SO_4)_3$. Yet this compound is quite soluble in water. It may be that ion–dipole attractions are especially strong in the solution, possibly because the negative charge centers of water dipoles can closely approach the small, highly charged Al^{3+} ions. In general, we can most readily predict the solubilities of ionic compounds by using the solubility guidelines listed in Table 4.3. There, we see that the high solubility of $Al_2(SO_4)_3$ in water conforms to the solubility rule that most sulfates are soluble.

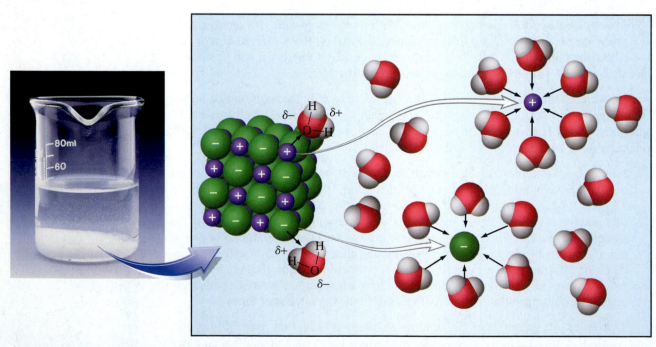

▲ **FIGURE 12.7   Ion–dipole forces in the dissolving of an ionic crystal**
The attraction of water dipoles for the ions pulls the ions out of the crystalline lattice and into aqueous solution. The ion–dipole forces exist in the solution as well, lessening the tendency for ions to return to the crystalline state. The combination of an ion in solution and the neighboring water dipoles to which it is attracted is called a *hydrated ion*. Two hydrated ions are represented here.

Example 12.8

Predict whether each combination is likely to be a solution or a heterogeneous mixture:

(a) methanol, CH_3OH, and water, HOH

(b) pentane, $CH_3(CH_2)_3CH_3$, and octane, $CH_3(CH_2)_6CH_3$

(c) sodium chloride, NaCl, and carbon tetrachloride, CCl_4

(d) 1-decanol, $CH_3(CH_2)_8CH_2OH$, and water, HOH

SOLUTION

(a) Both molecules feature an OH group. This group is attached to a H atom in water and to a CH_3 group in methanol. The molecules are similar in that hydrogen bonding is the major intermolecular force in both pure liquids and in their mixtures. We expect methanol and water to form a solution.

(b) The hydrocarbons pentane and octane are quite similar; they differ by only three CH_2 groups. Intermolecular forces are dispersion forces and are all of similar magnitude. We expect pentane and octane to form a solution (like dissolves like).

(c) Sodium chloride and carbon tetrachloride are both chlorides, but the similarity ends there. NaCl is an ionic compound, but CCl_4 is molecular. Moreover, because the tetrahedral CCl_4 molecule is nonpolar, there are no ion–dipole forces available to separate the ions in NaCl(s). There is no dissolving, and we expect the mixture to be heterogeneous.

(d) This case may seem similar to (a)—a hydrocarbon chain substituted for a H atom in HOH—but here the hydrocarbon chain is 10 carbon atoms long. This long chain is the principal structural feature of 1-decanol, and dispersion forces are its dominant intermolecular forces. We therefore expect 1-decanol and water mixtures to be heterogeneous (no significant dissolving occurs). This is a case of oil and water not mixing, with 1-decanol being the "oil."

Problem-Solving Note

Alcohols become less and less soluble in water as the number of carbon atoms increases beyond three.

EXERCISE 12.8A

In which solvent would you expect nitrobenzene, $C_6H_5NO_2$, to be more soluble: H_2O or C_6H_6? Explain.

EXERCISE 12.8B

Figure 12.8 shows molecular models of four substances. Identify the substances, and arrange them in the expected order of increasing solubility in water.

  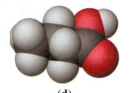

| (a) | (b) | (c) | (d) |

▲ **FIGURE 12.8**   **Exercise 12.8B illustrated**

12.4 Equilibrium in Solution Formation

Some substances can dissolve in each other without limit. Liquids that mix in all proportions are said to be *miscible*. For example, we can prepare an ethanol–water solution with a mass percent ethanol ranging essentially from 0 to 100% ethanol. Ethanol and water are completely miscible. Most substances, though, have limited solubilities. This limit varies with the nature of the solute and of the solvent, and with temperature.

What happens when we place 40 g of NaCl in 100 g water at 20 °C? Initially, many of the Na^+ and Cl^- ions are plucked from the surface of crystals by ion–dipole forces, and the ions then wander about at random through the solution; some of the

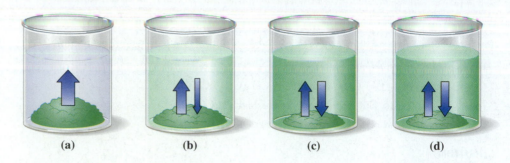

(a)   **(b)**   **(c)**   **(d)**

▶ **FIGURE 12.9   Formation of a saturated solution**
(a) A solid solute is added to a fixed quantity of water. (b) After a few minutes, the solution is colored by the dissolved solute and there is less undissolved solute than in (a). (c) After a longer time, the solution color has deepened and the quantity of undissolved solute is further diminished from that in (b). The solution in (b) must be unsaturated. (d) Still later, the solution color and the quantity of undissolved solute appear to be the same as that in (c). Dynamic equilibrium must have been attained in (c) and persists in (d). In both (c) and (d), the solution is saturated.

QUESTION: What change in the system would occur if 90% of the solid in (c) were removed? If more solid were added to (d)?

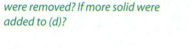

Equilibrium in Solution Formation animation

NaCl *dissolves*. In their wanderings, some of the ions pass near a crystal surface and are attracted back to the surface by ions of the opposite charge; some NaCl *crystallizes* out of the solution. As more and more NaCl dissolves, the number of "wanderers" that return to the crystals increases. Eventually (when 36 g NaCl has dissolved) the number of ions returning to the crystals equals the number of ions leaving the crystals. At this point, the rate of crystallization is equal to the rate of dissolving, and a condition of dynamic equilibrium is established. The net amount of NaCl in solution remains constant despite the fact that there is a lot of activity as ions come and go from the surface of the crystals. The net quantity of undissolved NaCl also remains constant (in this example, 4 g), even though individual crystals may change in size and shape. Some small crystals may disappear as others grow larger. This process is illustrated in Figure 12.9.

When there is a dynamic equilibrium between an undissolved solute and a solution, the solute concentration has attained its maximum value for the given temperature; the solution is **saturated.** The concentration of the solute in a saturated solution is the **solubility** of the solute. The solubility of NaCl in water at 20 °C, for example, is 36 g NaCl/100 g H_2O. Any solution containing less solute than can be held at equilibrium is **unsaturated.** At 20 °C, a solution with 24 g NaCl/100 g H_2O is unsaturated, as is one with 32 g NaCl/100 g H_2O.

The terms unsaturated and saturated are not related to whether a solution is dilute or concentrated. For example, a saturated aqueous solution of $Ca(OH)_2$ at 20 °C is 0.023 M, quite a dilute solution. In contrast, at 20 °C, a 10 M NaOH solution, though quite concentrated, is still unsaturated.

Solubility as a Function of Temperature

Solubility varies with temperature, so when we cite solubility data, we should indicate the temperature. Later in the text, we will comment further on the relationship of solubility to temperature, but for now, let's just use this generalization regarding the solubilities of ionic compounds in water:

About 95% of all ionic compounds have aqueous solubilities that increase significantly with increasing temperature. Most of the remainder have solubilities that change little with temperature. A very few—for example, some sulfates—have solubilities that decrease with increasing temperature.

A graph of solubility as a function of temperature is called a *solubility curve*. Figure 12.10 shows the solubility curves of a number of ionic compounds in water, expressed in a unit often used in reference books: g solute/100 g solvent.

If we cool a saturated solution of lead nitrate that is in contact with excess $Pb(NO_3)_2(s)$, some solute crystallizes from solution until equilibrium is once again established at the lower temperature. For example, referring to Figure 12.10, consider a saturated solution of lead nitrate at 90 °C. The solution contains about 122 g of $Pb(NO_3)_2$ for each 100 g water. If we cool the solution to 20 °C, the solution can contain only about 54 g $Pb(NO_3)_2$ per 100 g water. The excess of 68 g crystallizes out.

Now consider what would happen if we started to cool a saturated solution of lead nitrate with no excess solute present. Would lead nitrate crystallize? It should. Then again, there is no dynamic equilibrium because there are no crystals present to capture wandering ions. If we are able to cool the solution without crystallization occurring,

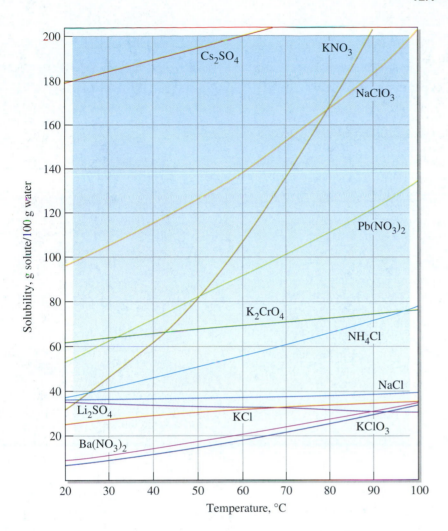

◀ **FIGURE 12.10** **The solubility curves for some salts in water**

QUESTION: What changes, if any, would occur if a solution at 60 °C containing 65.0 g $Pb(NO_3)_2$/100 g water were cooled to 20 °C?

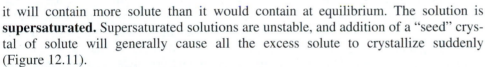

it will contain more solute than it would contain at equilibrium. The solution is **supersaturated.** Supersaturated solutions are unstable, and addition of a "seed" crystal of solute will generally cause all the excess solute to crystallize suddenly (Figure 12.11).

Crystallizing a solute from a concentrated solution by lowering the temperature can be an effective method of purifying a substance. In this technique, the impure solute is dissolved in a minimum quantity of a hot solvent. The concentrated solution is then cooled to a temperature at which the solubility of the solute is much smaller than in the hot solvent. The excess solute crystallizes from solution, but the impurities, being present in much smaller quantities, remain in solution. The crystals are filtered off and washed with small quantities of the cold solvent to remove any adhering solution.

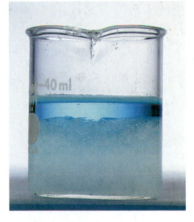

▲ When $KNO_3(s)$ is crystallized from an aqueous solution of KNO_3 containing $CuSO_4$ as an impurity, the $CuSO_4$ (blue) remains in solution.

◀ **FIGURE 12.11** **Seeding a supersaturated solution**

A tiny seed crystal is added to a supersaturated solution of sodium acetate (left). Crystallization begins (center) and continues until all the excess solute has crystallized from the solution (right).

Some Supersaturated Solutions in Nature

Supersaturated solutions are not just laboratory curiosities; they occur naturally. Honey is one example; the principal solute is the sugar glucose, and the solvent is water. If honey is left to stand, the glucose eventually crystallizes. We say, not very scientifically, that the honey has turned to sugar. Supersaturated sucrose solutions include jellies and many candies. Sucrose often crystallizes from jelly that has been standing for a long time. Fudge that is made incorrectly may turn grainy as sucrose crystals separate. Hard candies often include corn syrup as an ingredient to impede crystallization.

Some wines have a high concentration of potassium hydrogen tartrate, $KHC_4H_4O_6$. When chilled, the solution may become supersaturated. After a time, crystals may form and settle out if the wine is stored in a refrigerator. Modern wineries solve this problem—and render the wine less acidic—by a process known as cold stabilization. The wine is chilled to near 0 °C, a temperature below usual refrigerator temperatures. Tiny seed crystals of $KHC_4H_4O_6$ are added to the supersaturated wine. When crystallization is complete, the excess crystals are filtered off. At one time, winemaking was the principal source of potassium hydrogen tartrate, the "cream of tartar" used in baking.

▲ Imagine a child's lunch without the supersaturated solution we call jelly!

12.5   The Solubilities of Gases

Solutions of gases in liquids, especially in water, are quite common. Consider the familiar case of carbonated beverages. All are solutions of $CO_2(g)$ in water, usually with added flavors and sweeteners. Other examples of solutions of gases include blood, which contains dissolved $O_2(g)$ and $CO_2(g)$; formalin, an aqueous solution of formaldehyde gas (HCHO) that is used as a biological preservative; and a variety of household cleaners that are aqueous solutions of $NH_3(g)$. In addition, *all* natural waters contain dissolved $O_2(g)$ and $N_2(g)$ plus traces of other gases.

As with other solutes, the solubilities of gases depend on temperature, but they depend even more significantly on pressure.

The Effect of Temperature

Most gases become less soluble in liquids as the temperature increases, as Figure 12.12 illustrates for the solubility of air in water at standard atmospheric pressure and different temperatures. Nitrogen and oxygen together make up about 99% of air. Note that the solubility curve for air (black) is essentially the sum of the curves for N_2 in air (red) and O_2 in air (blue). Although oxygen makes up only about 23% of air by mass, it constitutes about 35% of the air dissolved in water, because $O_2(g)$ is more soluble in water than is $N_2(g)$.

The solubility of air in water, limited as it is, is essential to aquatic life. Fish depend on dissolved air for $O_2(g)$. Moreover, the fact that the solubility decreases with temperature explains why many fish (trout, for example) cannot survive in warm water—they do not get enough oxygen. At 30 °C, for example, the amount of dissolved $O_2(g)$ in water is only about one-half of that at 0 °C.

Application Note

The world's major ocean fisheries are located in *cold* regions such as the Bering Sea off the coast of Alaska and the Grand Banks off Newfoundland.

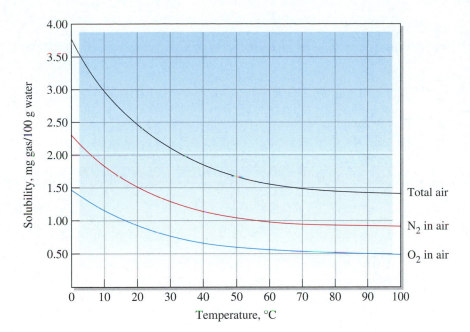

◀ **FIGURE 12.12** **Solubility of air in water as a function of temperature at 1 atm pressure**

The Effect of Pressure

At constant temperature, the solubility (S) of a gas is directly proportional to the pressure of the gas (P_{gas}) in equilibrium with the solution. Thus, doubling the pressure of a gas doubles its solubility; a tenfold increase in pressure produces a tenfold increase in solubility; and so on. This relationship is expressed through this equation for a straight line of slope k:

$$S = k\,P_{gas} \qquad (12.5)$$

The value of k depends on the gas being dissolved and on the solvent.

The effect of pressure on the solubility of a gas was stated by William Henry in 1803 and is known as **Henry's law.** Figure 12.13 suggests a way to rationalize this law: If they are to dissolve in a liquid, gas molecules must first collide with the liquid surface. Increasing the pressure of a gas in contact with a saturated solution increases the number of molecules per unit volume in the gas. This increases the collision rate with the liquid surface. More gas molecules dissolve, and their concentration in solution increases. However, the number of molecules returning to the gaseous state from this more concentrated solution is also increased. In the new dynamic equilibrium at the higher gas pressure, both the concentration of gas molecules in the gas phase and the concentration of gas molecules in the solution are greater than in the original saturated solution. Thus, the solubility increases with increased gas pressure.

If a gaseous solute reacts with the solvent, as do HCl(g) and NH_3(g), the situation is more complicated, and Henry's law may not accurately describe the solubility of the gas.

▲ A moderate pressure of CO_2(g) above the beverage in a soft-drink bottle (left) keeps a significant quantity of the gas dissolved in the water. When the bottle is opened (right), this pressure is released and dissolved CO_2(g) escapes, causing the familiar fizzing.

◀ **FIGURE 12.13** **The effect of pressure on the solubility of a gas in a liquid solvent**

As the gas is compressed at constant temperature into a smaller volume, increasing the number of molecules per unit volume, the number of dissolved molecules per unit volume—in other words, the concentration of the gas in the solution—also increases.

QUESTION: What difference(s), if any, would you expect in this illustration if the temperature were raised significantly?

Henry's Law animation

Henry's Law simulation

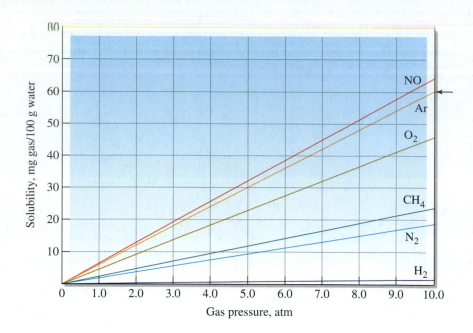

▶ **FIGURE 12.14   The effect of gas pressure on aqueous solubilities of gases at 20 °C**
We refer in Example 12.9 to the small arrow that appears at the right terminus of the yellow line labeled Ar.

The straight-line plots in Figure 12.14 represent the solubilities in water of different gases as a function of gas pressure. We use data from this figure in the application of Henry's law in Example 12.9.

Example 12.9

A 225-g sample of pure water is shaken with air under a pressure of 0.95 atm at 20 °C. How many milligrams of Ar(g) will be present in the water when solubility equilibrium is reached? Use data from Figure 12.14 and the fact that the mole fraction of Ar in air is 0.00934.

STRATEGY

Let us plan an approach in which we use Henry's law, $S = k P_{gas}$. To do so, we need values of P_{gas} and k. We can get P_{gas} by using the relationship expressed in Equation (5.15), and we can get k from data in Figure 12.14.

SOLUTION

Determining the partial pressure of argon in air:

Using Equation (5.15), we can express the mole fraction of argon as the ratio of the partial pressure of argon to the total pressure.

$$x_{Ar} = \frac{P_{Ar}}{P_{total}} = \frac{P_{Ar}}{P_{air}} = \frac{P_{Ar}}{0.95\ \text{atm}} = 0.00934$$

From this expression, we calculate the partial pressure of argon.

$$P_{Ar} = 0.00934 \times 0.95\ \text{atm} = 0.0089\ \text{atm}$$

Determining k:

To derive a value of k, let's choose a point on the argon solubility curve in Figure 12.14 for which we can read the relevant data with some precision—namely, the point identified by the arrow.

$$P_{Ar} = 10.0\ \text{atm} \quad \text{and} \quad S = 60\ \text{mg Ar/100 g water}$$

We can solve Henry's law for k and substitute values for S and P_{Ar}.

$$k = \frac{S}{P_{gas}} = \frac{S}{P_{Ar}} = \frac{60\ \text{mg Ar/100 g water}}{10.0\ \text{atm}} = \frac{0.060\ \text{mg Ar}}{\text{g water atm}}$$

Determining the solubility of argon at 0.0089 atm:

Now we use the values for k and the partial pressure of argon in air to solve for the solubility, S.

$$S = k P_{Ar} = \frac{0.060\ \text{mg Ar}}{\text{g water atm}} \times 0.0089\ \text{atm} = \frac{5.3 \times 10^{-4}\ \text{mg Ar}}{\text{g water}}$$

Determining the mass of argon in 225 g of water:

The value of S just derived is expressed per gram of water. We just need to multiply S by the mass of water, 225 g.

$$\text{Mass of Ar} = 225 \text{ g water} \times \frac{5.3 \times 10^{-4} \text{ mg Ar}}{\text{g water}} = 0.12 \text{ mg Ar}$$

EXERCISE 12.9A

At 25 °C and 1 atm gas pressure, the solubility of $CO_2(g)$ is 149 mg/100 g water. When air at 25 °C and 1 atm is in equilibrium with water, what is the concentration of dissolved CO_2, in mg/100 g water? Air contains 0.037 mol % $CO_2(g)$.

EXERCISE 12.9B

How many milligrams of a mixture containing equal numbers of moles of CH_4 and N_2 at 10.0 atm total pressure will dissolve in 1.00 L of water at 20 °C?

Deep-Sea Diving: Applications of Henry's Law

Scuba divers must carry their own supply of air, about 800 L of air for every hour of a dive, compressed into a small tank. The regulator on the tank provides the diver with air that is at the same pressure as the surrounding water. As we should expect from Henry's law, the air consumed at depths, and thus at higher pressure, is much more soluble in blood and other body fluids than is air at normal pressures. One of the dissolved gases, nitrogen, can cause two kinds of problems: nitrogen narcosis and decompression sickness.

Nitrogen narcosis is experienced by divers breathing air at depths of 100 ft (30 m) or more. Ordinarily, the nitrogen we breathe in air has no physiological effect, but at the high pressures at these depths, nitrogen acts as a narcotic. It often produces pleasurable sensations similar to those of narcotics, accounting for the alternate name "rapture of the deep." Nitrogen narcosis impairs judgment and often causes serious diving accidents. Divers can minimize the risk by using a compressed helium–oxygen mixture as a substitute for compressed air. Because helium is considerably less soluble in blood than is nitrogen, absorption of helium into the bloodstream is quite limited, and it has no narcotic effect.

Decompression sickness, caused by bubbles of nitrogen gas in the blood and other tissues, occurs when a person is exposed to a very rapid drop in air pressure. Tiny bubbles form, just as they do when a bottled carbonated beverage is opened. These bubbles block blood flow in capillaries, and the person suffers joint and muscle pain severe enough to cause him or her to curl up, a reaction that explains the common name of the illness, "the bends." Divers must be careful to return to the surface slowly or to spend considerable time in a decompression chamber where the pressure is gradually lowered. If they do not, the bubbles of excess $N_2(g)$ escaping from solution in blood can cause unconsciousness, deafness, paralysis, even death. A slow return to normal pressure allows the excess gases to leave the blood gradually. During this process, excess O_2 is used in metabolism, and excess N_2 is removed to the lungs and expelled by normal breathing. People who work in deep mines or tunnels, where compressed air is used to keep water from infiltrating, must take similar precautions.

▲ Underwater divers who surface too quickly may experience decompression sickness, a painful and dangerous condition also known as "the bends."

The solution properties that we have discussed so far in this chapter depend on the identity of the solute. Recall from Figure 12.10 the differences in solubility for different ionic solutes. A solution at 20 °C with a total ion concentration of 0.50 M would be unsaturated for some solutes, supersaturated for others, and perhaps just saturated for a few (NaCl has a substantially lower solubility in water than $NaClO_3$ does). In contrast, we will consider in the next four sections a number of physical properties of solutions, called **colligative properties,** that depend on the number of solute particles present but *not* on the identity of the solute. For example, aqueous 0.10 *m* sucrose ($C_{12}H_{22}O_{11}$) and aqueous 0.10 *m* urea [$CO(NH_2)_2$] have the same number of molecules (0.10 mol) in the same mass of solvent (1.00 kg H_2O). The solute molecules differ significantly from each other in composition and structure, but the two solutions have the same colligative properties, such as freezing point and boiling point. All the solutes and solvents in the solutions that we describe in Sections 12.6, 12.7, and 12.8 exist in molecular form—that is, they are solutions of nonelectrolytes. In Section 12.9, we will consider solutions in which the solutes are electrolytes.

12.6 Vapor Pressures of Solutions

In 1887, F. M. Raoult reported the results of his studies on the vapor pressures of solutions. He found that the presence of a solute lowers the vapor pressure of the solvent in a solution. We begin with a modern statement of **Raoult's law:**

> *The vapor pressure of the solvent above a solution (P_{solv}) is the product of the vapor pressure of the pure solvent (P°_{solv}) and the mole fraction of the solvent in the solution (x_{solv}):*

$$P_{solv} = x_{solv} \cdot P^\circ_{solv} \qquad \textbf{(12.6)}$$

Note that this expression conforms to Raoult's observation of the lowering of vapor pressure. The presence of a solute in a solvent means that $x_{solv} < 1$, and consequently P_{solv} is less than P°_{solv}. Thus, the solute has lowered the vapor pressure of the solvent. Raoult's law is strictly followed only in an ideal solution, but it often works for nonideal solutions that are sufficiently dilute. A solution in which $x_{solv} > 0.97$, for example, might meet this requirement. Although there are various "molecular" explanations of Raoult's law, we will present a more satisfactory thermodynamic explanation in Chapter 17.

Example 12.10

The vapor pressure of pure water at 20.0 °C is 17.5 mmHg. What is the vapor pressure at 20.0 °C above a solution that has 0.250 mol $C_{12}H_{22}O_{11}$ (sucrose) and 75.0 g $CO(NH_2)_2$ (urea) dissolved per kilogram of water?

STRATEGY

We are given one of the terms required in Raoult's law, $P^\circ_{solv} = 17.5$ mmHg. The other term we need is $x_{solv} = x_{H_2O}$. To get x_{H_2O} from Equation (12.3), we need to know the number of moles of water in the solution as well as the number of moles of dissolved sucrose and urea.

SOLUTION

The number of moles of sucrose is given.

$$0.250 \text{ mol } C_{12}H_{22}O_{11}$$

We determine the number of moles of urea from its mass and molar mass.

$$? \text{ mol } CO(NH_2)_2 = 75.0 \text{ g } CO(NH_2)_2 \times \frac{1 \text{ mol } CO(NH_2)_2}{60.06 \text{ g } CO(NH_2)_2}$$

$$= 1.25 \text{ mol } CO(NH_2)_2$$

Similarly, we can convert from mass to number of moles of water.

$$? \text{ mol } H_2O = 1000 \text{ g } H_2O \times \frac{1 \text{ mol } H_2O}{18.02 \text{ g } H_2O}$$

$$= 55.5 \text{ mol } H_2O$$

Now, we determine the mole fraction of water.

As a preliminary step, we can establish the total number of moles of solute.

$$0.250 \text{ mol } C_{12}H_{22}O_{11} + 1.25 \text{ mol } CO(NH_2)_2 = 1.50 \text{ mol solute}$$

Then, we substitute into Equation (12.3) to solve for x_{H_2O}.

$$x_{H_2O} = \frac{55.5 \text{ mol } H_2O}{1.50 \text{ mol solute} + 55.5 \text{ mol } H_2O} = 0.974$$

Finally, we use Raoult's law to calculate the vapor pressure of water above the solution.

$$P_{H_2O} = (x_{H_2O})P^{\circ}_{H_2O} = 0.974 \times 17.5 \text{ mmHg} = 17.0 \text{ mmHg}$$

ASSESSMENT

Our result is a reasonable one by two measures. The calculated vapor pressure is less than that of pure water, which it must be for a solution of nonvolatile solutes in a volatile solvent. Moreover, because the solution is relatively dilute in solutes (mole fraction less than 0.03), the partial pressure of $H_2O(g)$ should be only slightly less than the vapor pressure of pure water (that is, 17.0 mmHg versus 17.5 mmHg).

Another noteworthy point is that the greater mass of the 0.250 mol $C_{12}H_{22}O_{11}$ (about 86 g) is much less effective than the smaller mass of the 75.0 g $CO(NH_2)_2$ (1.25 mol) in lowering the vapor pressure of the water. This observation emphasizes that it is the number of solute particles and not their mass or identity that determines a colligative property.

EXERCISE 12.10A

The vapor pressure of benzene (C_6H_6) at 25 °C is 95.1 mmHg. What is the vapor pressure of benzene above a solution in which 5.05 g C_6H_5COOH (benzoic acid) is dissolved in 245 g C_6H_6?

EXERCISE 12.10B

The density of toluene ($C_6H_5CH_3$) at 25.0 °C = 0.862 g/mL, and its vapor pressure at 25.0 °C = 28.44 mmHg. At 25.0 °C, the vapor pressure of toluene above a solution of naphthalene ($C_{10}H_8$) in 500.0 mL toluene is found to be 27.92 mmHg. How many grams of naphthalene are present in the solution?

Raoult's law is not limited to the solvent if a solution also contains *volatile* solutes. We can write a Raoult's law expression for *any* volatile component in a solution, whether it is the solvent or a solute, as long as the solution is nearly ideal. The benzene–toluene solution in Examples 12.11 and 12.12 is an essentially ideal solution in which both components are volatile. In Example 12.11, we calculate the vapor pressures of the two components in the solution; and in Example 12.12, we calculate the composition of the vapor in equilibrium with the solution.

Example 12.11

At 25 °C, the vapor pressures of pure benzene (C_6H_6) and pure toluene (C_7H_8) are 95.1 and 28.4 mmHg, respectively. A solution is prepared that has equal mole fractions of C_6H_6 and C_7H_8. Determine the vapor pressures of C_6H_6 and C_7H_8 and the total vapor pressure above this solution. Consider the solution to be ideal.

STRATEGY

The key data are the vapor pressures of the pure liquids and the fact that their mole fractions in the solution are equal. We apply Raoult's law twice to obtain the partial pressures of the components in the vapor, and we sum these partial pressures to get the total vapor pressure.

SOLUTION

Because $x_{C_6H_6} = x_{C_7H_8}$ and because $x_{C_6H_6} + x_{C_7H_8} = 1$, it follows that $x_{C_6H_6} = x_{C_7H_8} = 0.500$. Now we can apply Raoult's law.

$$P_{benzene} = x_{benzene}P^{\circ}_{benzene} = 0.500 \times 95.1 \text{ mmHg} = 47.6 \text{ mmHg}$$

$$P_{toluene} = x_{toluene}P^{\circ}_{toluene} = 0.500 \times 28.4 \text{ mmHg} = 14.2 \text{ mmHg}$$

The total pressure is the sum of the two partial pressures.

$$P_{total} = P_{benzene} + P_{toluene} = 47.6 \text{ mmHg} + 14.2 \text{ mmHg} = 61.8 \text{ mmHg}$$

ASSESSMENT

As a rough check on our answer, note that the total vapor pressure above this ideal solution must lie between the vapor pressure of the less volatile liquid (toluene) and that of the more volatile liquid (benzene), that is, $28.4 \text{ mmHg} < P_{total} < 95.1 \text{ mmHg}$.

EXERCISE 12.11A

At 25 °C, what are the partial pressures of benzene and toluene and the total pressure above a solution containing equal masses of the two components?

EXERCISE 12.11B

The vapor above a benzene–toluene solution at 25 °C is found to have equal mole fractions of benzene and toluene. What must be the mole fraction composition of the liquid that is in equilibrium with this vapor? (*Hint*: Try a straightforward algebraic solution.)

Problem-Solving Note

It does not matter what particular mass of the components you choose to work with.

Example 12.12

What is the composition, expressed as mole fractions, of the vapor in equilibrium with the benzene–toluene solution of Example 12.11?

STRATEGY

The vapor is a mixture of gaseous benzene and gaseous toluene at the partial pressures calculated in Example 12.11. The relationship between the mole fraction and partial pressure of a component of a gaseous mixture is given by Equation (5.15).

SOLUTION

Equation (5.15) states that

$$x_i = \frac{n_i}{n_{total}} = \frac{P_i}{P_{total}}$$

Substituting the results of Example 12.11 into Equation (5.15), we obtain

$$x_{benzene \, vapor} = \frac{47.6 \text{ mmHg}}{61.8 \text{ mmHg}} = 0.770$$

$$x_{toluene \, vapor} = \frac{14.2 \text{ mmHg}}{61.8 \text{ mmHg}} = 0.230$$

ASSESSMENT

Note that, as they must, the sum of the mole fractions in the mixture is equal to one:

$$x_{benzene \, vapor} + x_{toluene \, vapor} = 0.770 + 0.230 = 1.000$$

EXERCISE 12.12A

Above which of these solutions does the vapor have the greater mole fraction of benzene at 25 °C: a solution with equal masses of benzene and toluene or a solution with equal numbers of moles of benzene and toluene? (*Hint:* You will need information from Examples 12.11 and 12.12, but you do not need to do detailed calculations.)

EXERCISE 12.12B

Suppose the vapor mixture from Example 12.12 is condensed to a liquid. At 25 °C, what will be the vapor pressures of benzene and toluene above this condensed liquid?

Consider the following: Pure benzene, C_6H_6, has a higher vapor pressure (95.1 mmHg at 25 °C) than pure toluene, C_7H_8 (28.4 mmHg at 25 °C). In benzene–toluene solutions, benzene is therefore the more volatile component. In Example 12.11, we started with a benzene–toluene solution that had $x_{C_6H_6} = 0.500$. In Example 12.12, we found that the vapor in equilibrium with that solution had $x_{C_6H_6} = 0.770$. These observations lead to an important idea about liquid–vapor equilibrium in a solution of two volatile components:

> *The vapor in equilibrium with an ideal solution of two volatile components has a higher mole fraction of the more volatile component than is found in the liquid.*

As illustrated by Exercise 12.12B, if vapor having the mole fraction $x_{C_6H_6} = 0.770$ were condensed to a liquid and again vaporized, the new vapor would be richer still in benzene. A series of vaporizations and condensations should ultimately allow us to obtain essentially pure benzene from the vapor. In practice, rather than carry out the vaporizations and condensations at a constant temperature (here, 25 °C), we would use a constant pressure (atmospheric pressure) and boil the benzene–toluene solution in an apparatus like that shown in Figure 12.15. In this method, called **fractional distillation,** essentially pure benzene would be condensed from the vapor at the top of the column and the less volatile toluene would be retained as a liquid in the warmer flask at the bottom of the column.

Equilibrium Vapor Pressure simulation

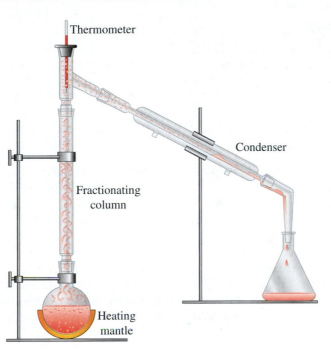

◀ **FIGURE 12.15   Fractional distillation of a two-component mixture**

The upright fractionating column is packed with an inert material such as glass beads. As the solution in the flask boils, vapor rises up the column and condenses on the cooler beads. As more hot vapor rises, it heats the beads and revaporizes the condensed liquid. This vapor then condenses farther up the column. The process is repeated many times as the vapor moves to the top of the column, becoming successively richer in the more volatile, lower-boiling component. If the column is tall enough, a pure sample of the lower-boiling component will leave the top of the column and enter the water-cooled condenser, where it condenses to the liquid state and is collected in the receiving flask.

Example 12.13   A Conceptual Example

Figure 12.16 shows two different aqueous solutions placed in the same enclosure. After a time, the solution level has risen in container A and dropped in container B. Explain how and why this happens.

ANALYSIS AND CONCLUSIONS

Water can move from one container to the other only as the vapor, $H_2O(g)$. A net vaporization from container B and condensation into container A must have taken place, indicating that the vapor pressure of H_2O above solution B was initially greater than that above solution A. According to Raoult's law, the mole fraction of water in solution initially was greater in container B than in container A; in other words, the solution in B was more dilute. Water vapor passed from the more dilute to the more concentrated solution. Notice that this transfer of water was unrelated to the initial volumes of the solutions.

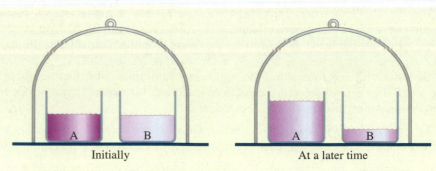

▲ **FIGURE 12.16** **A phenomenon related to the vapor pressures of solutions**

Note that the identity of the solute in solutions A and B is immaterial; the solutions can even have different solutes, as long as they have the same *solvent*. The vapor pressure lowering depends only on the concentration of the solute and not on other characteristics of the solute molecules. Vapor pressure lowering is a prime example of a colligative property.

EXERCISE 12.13A

Will the process depicted in Figure 12.16 continue until the solution in container B evaporates to dryness? Explain.

EXERCISE 12.13B

If container A in Figure 12.16 holds pure water and container B holds some other liquid, will the change observed over time be the same as that shown in Figure 12.16 or will the change be different in some way? Explain.

12.7 Freezing Point Depression and Boiling Point Elevation

In this section, we focus on solutions made up of a solvent that is volatile and a solute that is (a) nonvolatile, (b) a nonelectrolyte, and (c) soluble in the liquid form of the solvent but not in the solid form of the solvent. In solutions with these characteristics, the vapor pressure of the solution is simply that of the solvent in the solution, and at all temperatures this vapor pressure is lower than that of the pure solvent. This statement is illustrated in Figure 12.17, in which a partial phase diagram for a solution (red) is superimposed on the phase diagram for the pure solvent (blue).

Let's examine Figure 12.17. The dotted line at $P = 1$ atm intersects the fusion curve (red) of the solution at a *lower* temperature (fp) than it does the fusion curve (blue) of the pure solvent (fp$_0$). The presence of the solute lowers or *depresses* the

▶ **FIGURE 12.17** **Vapor pressure lowering by a nonvolatile solute**

In a solution, the vapor pressure curve of the solvent is displaced to lower pressures, and the fusion curve is displaced to lower temperatures (red curves). As a consequence, the freezing point of the solvent is lowered by the amount ΔT_f and the boiling point is raised by the quantity ΔT_b. Because the solute is assumed to be insoluble in the solid solvent, the sublimation curve of the solid solvent is unaffected by the presence of a solute in the liquid solution.

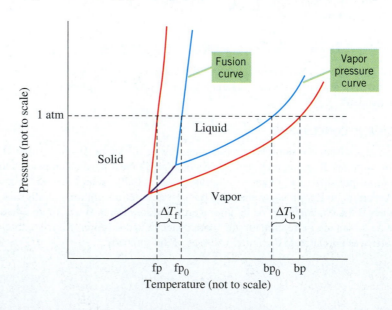

freezing point by the quantity ΔT_f. The dotted line at $P = 1$ atm intersects the vapor pressure curve (red) for the solution at a *higher* temperature (bp) than it does the vapor pressure curve of the pure solvent (bp$_0$). The presence of a solute raises, or *elevates*, the boiling point by the amount ΔT_b.

How much the freezing point is depressed and how much the boiling point is elevated depend on the mole fraction of solute present in a solution. However, in *dilute* solutions, the molality of solute is proportional to its mole fraction, and because of this we can write

$$\Delta T_f = T_f(\text{solution}) - T_f(\text{solvent}) = -K_f \times m \qquad \textbf{(12.7)}$$

$$\Delta T_b = T_b(\text{solution}) - T_b(\text{solvent}) = K_b \times m \qquad \textbf{(12.8)}$$

In these equations, ΔT_f and ΔT_b are the *freezing point depression* and *boiling point elevation*, respectively, and m is the molality of solute. The proportionality constants K_f and K_b represent the freezing point depression and boiling point elevation values when the solute concentration in a solution is 1 m. (For example, $\Delta T_b = K_b \times m = K_b \times 1 = K_b$.) For this reason, K_f and K_b are often referred to as the *molal freezing point depression constant* and the *molal boiling point elevation constant*, respectively (Table 12.2).* (In the freezing point depression equation, K_f is preceded by a minus sign because ΔT_f must be a negative quantity.)

In *dilute aqueous* solutions, molarity and molality are nearly equal and *both* are proportional to the mole fraction of solute.

 Boiling-Point Elevation and Freezing-Point Depression simulation

 Boiling-Point Elevation and Freezing-Point Depression II simulation

Table 12.2 Molal Freezing Point Depression and Molal Boiling Point Elevation Constants

Solvent	Normal Freezing Point, °C	K_f, °C m^{-1}	Normal Boiling Point, °C	K_b, °C m^{-1}
Acetic acid	16.63	3.90	117.90	3.07
Benzene	5.53	5.12	80.10	2.53
Cyclohexane	6.55	20.0	80.74	2.79
Nitrobenzene	5.8	8.1	210.8	5.24
Water	0.00	1.86	100.00	0.512

▲ According to the University of Vermont Extension, maple syrup is made by boiling maple sap until the temperature of the boiling sap is 7 °F above the boiling point of water (see Problem 111).

For the solutions we are describing, only the solvent freezes and only the solvent escapes as vapor during boiling; the solute(s) remains in solution. As a result, the solute concentration increases as freezing or boiling occurs, causing the freezing point to drop still more and the boiling point to rise higher. Thus we see that the freezing and boiling points of solutions are *not* constant, as they are for pure liquids. When we speak of the freezing point of a solution, we mean the temperature at which the first bit of solid freezes from solution. In the case of boiling point, we mean the temperature at which boiling first occurs.

Example 12.14

What is the freezing point of an aqueous sucrose solution that has 25.0 g $C_{12}H_{22}O_{11}$ per 100.0 g H_2O?

STRATEGY

The freezing point of the solution is that of pure water (0.00 °C) plus the freezing point depression (ΔT_f). We can find ΔT_f with Equation (12.7). To do so, we will need to obtain the value of K_f for water from Table 12.2. We also need to determine the value of m, the molality of the solution.

SOLUTION

First, let us determine the number of moles of sucrose in 25.0 g.

$$? \text{ mol } C_{12}H_{22}O_{11} = 25.0 \text{ g } C_{12}H_{22}O_{11} \times \frac{1 \text{ mol } C_{12}H_{22}O_{11}}{342.3 \text{ g } C_{12}H_{22}O_{11}}$$

$$= 0.0730 \text{ mol } C_{12}H_{22}O_{11}$$

*These constants are also called the *cryoscopic* (K_f) and *ebullioscopic* (K_b) constants.

Now, we can calculate the molality of the sucrose from the number of moles of sucrose and the quantity of water.

$$100.0 \text{ g H}_2\text{O} = 0.1000 \text{ kg H}_2\text{O}$$

$$? \, m = \frac{0.0730 \text{ mol C}_{12}\text{H}_{22}\text{O}_{11}}{0.1000 \text{ kg H}_2\text{O}} = 0.730 \, m$$

We use the molality and the constant K_f to determine the depression of the freezing point, ΔT_f.

$$\Delta T_f = -K_f \times m = -1.86 \text{ °C } m^{-1} \times 0.730 \, m = -1.36 \text{ °C}$$

The freezing point of the solution is the sum of ΔT_f and the freezing point of the pure solvent, fp_0.

$$T_f = fp_0 + \Delta T_f = 0.00 \text{ °C} - 1.36 \text{ °C} = -1.36 \text{ °C}$$

ASSESSMENT

Because the value of K_f is the freezing point depression for a 1 m solution and the sucrose solution is less than 1 m, we should expect the freezing point to be below 0 °C but not as low as -1.86 °C, as indeed we found to be the case.

EXERCISE 12.14A

What is the freezing point of a solution in which 10.0 g of naphthalene [$C_{10}H_8(s)$] is dissolved in 50.0 g of benzene [$C_6H_6(l)$]?

EXERCISE 12.14B

What mass of sucrose, $C_{12}H_{22}O_{11}$, should be added to 75.0 g H_2O to raise the boiling point to 100.35 °C ?

Example 12.15

Sorbitol is a sweet substance found in fruits and berries and sometimes used as a sugar substitute. An aqueous solution containing 1.00 g sorbitol in 100.0 g water is found to have a freezing point of -0.102 °C. Elemental analysis indicates that sorbitol consists of 39.56% C, 7.75% H, and 52.70% O by mass. What are the **(a)** molar mass and **(b)** molecular formula of sorbitol?

STRATEGY

(a) We can use the freezing point depression Equation (12.7) to determine the molality of the sorbitol solution. From the molality, we can determine the number of moles of sorbitol, and from the number of moles of sorbitol and its mass (1.00 g), we can get the molar mass.

(b) We could use the methods of Examples 3.9 and 3.11 to determine the molecular formula from the mass percent data and the molar mass. However, an alternative approach may be easier in this case: Start with a 1-mol sample of sorbitol (182 g), and use the mass percent data to determine the mass of each element in the sample. Convert these masses to amounts in moles, and use those amounts as the subscripts in a molecular formula.

SOLUTION

(a) The freezing point depression is the difference between the freezing point of the solution and that of the pure solvent.

$$\Delta T_f = -0.102 \text{ °C} - 0.000 \text{ °C} = -0.102 \text{ °C}$$

We can get the molality of the solution from Equation (12.7).

$$\Delta T_f = -K_f \times m.$$

$$\text{Molality} = \frac{\Delta T_f}{-K_f} = \frac{-0.102 \text{ °C}}{-1.86 \text{ °C } m^{-1}} = 0.0548 \, m$$

Now, we use the molality to determine the number of moles of sorbitol in the solution.

$$? \text{ mol sorbitol} = 0.1000 \text{ kg H}_2\text{O} \times \frac{0.0548 \text{ mol sorbitol}}{\text{kg H}_2\text{O}}$$

$$= 0.00548 \text{ mol sorbitol}$$

The molar mass is the number of grams of sorbitol divided by the number of moles.

$$\text{Molar mass} = \frac{1.00 \text{ g sorbitol}}{0.00548 \text{ mol sorbitol}} = 182 \text{ g/mol sorbitol}$$

(b) We use the percent composition data to determine the numbers of moles of C, H, and O in 1 mol of sorbitol.

$$? \text{ mol C} = 182 \text{ g sorbitol} \times \frac{39.56 \text{ g C}}{100 \text{ g sorbitol}} \times \frac{1 \text{ mol C}}{12.01 \text{ g C}} = 5.99 \text{ mol C}$$

$$? \text{ mol H} = 182 \text{ g sorbitol} \times \frac{7.75 \text{ g H}}{100 \text{ g sorbitol}} \times \frac{1 \text{ mol H}}{1.008 \text{ g H}} = 13.99 \text{ mol H}$$

$$? \text{ mol O} = 182 \text{ g sorbitol} \times \frac{52.70 \text{ g O}}{100 \text{ g sorbitol}} \times \frac{1 \text{ mol O}}{16.00 \text{ g O}} = 5.99 \text{ mol O}$$

The molecular formula of sorbitol is $C_6H_{14}O_6$.

EXERCISE 12.15A

A 1.065-g sample of an unknown substance is dissolved in 30.00 g of benzene; the freezing point of the solution is 4.25 °C. The compound is 50.69% C, 4.23% H, and 45.08% O by mass. Use these data and data from Table 12.2 to determine the molecular formula of the substance.

EXERCISE 12.15B

A crystalline white solid is a mixture of glucose ($C_6H_{12}O_6$) and sucrose ($C_{12}H_{22}O_{11}$). Is it possible that a 10.00-g sample of the solid dissolved in 100.0 g H_2O might have a freezing point of −1.25 °C? Explain.

Freezing point depression and boiling point elevation have practical applications. Both are involved, for example, in the action of an antifreeze such as ethylene glycol ($HOCH_2CH_2OH$) in an automotive cooling system. Water alone would boil over when the engine overheats in hot weather, and the water would freeze in the depths of northern winters. Addition of antifreeze raises the boiling point of the coolant and also lowers its freezing point. With the proper ratio of ethylene glycol and water, it is possible to protect an automotive cooling system to temperatures as low as −48 °C, and from boilover to temperatures of about 113 °C.

Citrus growers are keenly aware of the role that freezing point depression plays in protecting their crop in cold weather. If the temperature drops a degree or two below the freezing point of water for only a few hours, citrus fruit is well protected from freezing. Solutes in the fruit juice, mainly sugars, lower the freezing point of the juice below 0 °C. Also, growers know that lemons are more prone to freezing than oranges because lemon juice is less concentrated in solutes than is orange juice.

Salts such as NaCl and $CaCl_2$ are scattered on icy sidewalks and streets to melt ice. The ice melts as long as the outdoor temperature is above the lowest freezing point of the salt-and-water mixture. Below this temperature, the ice will not melt. Sodium chloride can melt ice at temperatures as low as −21 °C (−6 °F), but no lower. Calcium chloride is effective at temperatures as low as −55 °C (−67 °F). In Section 12.9, we will discover one of the factors accounting for the greater effectiveness of $CaCl_2$ over NaCl in lowering the freezing point of water.

▲ Alcohols having more than one hydroxy group (OH) per molecule are commonly used for deicing aircraft. Because of its toxicity, ethylene glycol is being replaced by safer substances, such as propylene glycol ($CH_3CHOHCH_2OH$).

12.8 Osmotic Pressure

Everyday experience tells us that we can separate coffee grounds from brewed coffee by passing the mixture through filter paper. Filter paper is *permeable* to water and other solvents, but it is *impermeable* to the insoluble coffee grounds. However, filter paper will not separate the caffeine from brewed coffee because the paper is permeable both to water, the solvent, and caffeine, a solute. As we might expect, there are materials that are *semipermeable;* these materials allow solvent molecules to pass but not solute molecules.

Semipermeable membranes are usually sheets or films of a material containing a network of microscopic holes or pores through which small solvent molecules can pass, but larger solute molecules cannot. Some of these membranes are composed of

Semipermeable Membrane animation

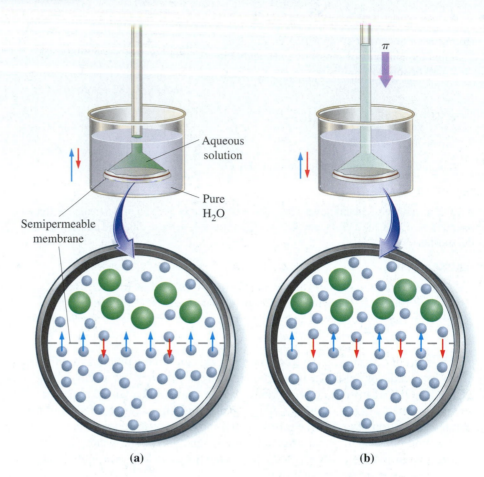

(a)   (b)

▶ **FIGURE 12.18** **Osmosis and osmotic pressure**
(a) An aqueous solution (green) in a flared-end tube is separated from pure water (blue) by a membrane permeable to H_2O molecules but not to solute particles. (b) A net flow of water through the semipermeable membrane from the beaker to the tube dilutes the green solution and causes the liquid level in the tube to rise. The liquid column now exerts a downward pressure (a *hydrostatic* pressure) that helps push H_2O molecules through the membrane and back into the beaker. When the flow of H_2O through the membrane is the same in both directions, there is no further change, and the hydrostatic pressure at that point is called the *osmotic pressure*, π.

QUESTION: *Why might the height of the column in (b) depend on the identity of the solvent?*

▲ In spring, as the days grow longer, starch stored in the leaf buds of trees and shrubs is converted to glucose. Each starch molecule forms dozens or even hundreds of glucose molecules. The molality of the water (sap) in the bud becomes higher than that in the stem. Then, by osmosis, water flows from the stem into the bud, causing it to swell and eventually to burst.

natural materials of animal or vegetable origin, such as pig's bladder or parchment. Others are made of synthetic materials such as cellophane. In the human body, cell membranes, the lining of the digestive tract, and the walls of blood vessels are all semipermeable membranes.

Figure 12.18 illustrates a phenomenon first observed in 1748 and called *osmosis*. The flared end of a tube is covered with a semipermeable membrane. An aqueous solution is placed in the tube, and the tube is placed in a beaker of pure water. After some time, the liquid level in the tube has risen above that of the water in the beaker. Figure 12.18 also gives a molecular-level explanation. The larger solute molecules are unable to pass through the holes in the semipermeable membrane, but the small molecules of water can pass in both directions. The rise in solution level in the tube signifies that there is a *net* flow of water molecules into the tube. This observation suggests a natural tendency for solvent molecules to migrate from a region where they exist in greater numbers (low solute mole fraction) into a region where the solvent molecules are less numerous (higher solute mole fraction).

Osmosis is the net flow of solvent molecules through a semipermeable membrane from a solution of lower concentration to a solution of higher concentration. The more dilute solution, of course, can simply be pure solvent. If the tube in Figure 12.18 contained a 20% aqueous sucrose solution by mass, a net flow of H_2O molecules from pure water could lift a column of the solution to a height of 150 m. We can reduce osmosis by increasing the flow of solvent molecules *out* of the solution. We can do this by applying pressure to the solution. At a sufficiently high pressure, there is no net flow of solvent through the semipermeable membrane—in other words, no osmosis. The pressure required to stop osmosis is called the **osmotic pressure** of the solution. For a 20% aqueous sucrose solution, the osmotic pressure is about 15 atm.

To understand that osmotic pressure is a colligative property, think of the solute only in terms of reducing the mole fraction of solvent in a solution. The lower the mole fraction of solvent, the greater the net flow of solvent molecules into the solution and the greater the osmotic pressure of the solution. The identity of the solute is immaterial.

An expression that works quite well for calculating osmotic pressures of *dilute* solutions of nonelectrolytes bears a striking resemblance to the ideal gas equation:

$$\pi = \frac{n}{V} RT \qquad (12.9)$$

In this equation, π is the osmotic pressure in atm, n is the number of moles of solute in V liters of a solution, R is the universal gas constant, and T is the Kelvin temperature. Because the ratio of number of moles of solute to a solution volume in liters is the *molarity* (M), we can simply write

$$\pi = \text{M } RT \qquad (12.10)$$

This equation suggests that we can use osmotic pressure measurements to determine molar masses of substances, but for most solutions, osmotic pressures are so high that they are difficult to measure. However, for solutions of *macromolecular* substances (polymers, proteins), where solute molar mass is high, molarity is low and osmotic pressures are fairly low and are easily measured. Osmotic pressure and freezing point depression measurements complement each other nicely for molar mass determinations. Generally, in cases where an osmotic pressure measurement works well, freezing point depression does not, and vice versa, as we illustrate in Example 12.16.

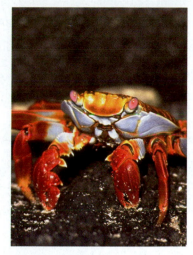

▲ Crabs have blood that contains the protein hemocyanin.

Example 12.16

An aqueous solution is prepared by dissolving 1.50 g of hemocyanin, a protein obtained from crabs, in 0.250 L of water. The solution has an osmotic pressure of 0.00342 atm at 277 K. **(a)** What is the molar mass of hemocyanin? **(b)** What should the freezing point of the solution be?

STRATEGY

(a) We can find the molarity of the solution by using Equation (12.10) for osmotic pressure, and from the molarity and solution volume, we can determine the number of moles of hemocyanin. We establish the molar mass of hemocyanin from the mass and number of moles.

(b) We can assume that the molarity and molality of this dilute solution are equal and then use Equation (12.7) for freezing point depression.

SOLUTION

(a) We rearrange Equation (12.10), substitute the relevant data, and solve for molarity.

$$\pi = \text{M } RT \quad \text{and} \quad \text{M} = \pi/RT$$

$$\text{M} = \frac{0.00342 \text{ atm}}{0.08206 \text{ L atm mol}^{-1} \text{ K}^{-1} \times 277 \text{ K}} = 1.50 \times 10^{-4} \text{ mol L}^{-1}$$

From the molarity and volume of solution, we determine the number of moles of hemocyanin.

$$? \text{ mol hemocyanin} = 0.250 \text{ L} \times \frac{1.50 \times 10^{-4} \text{ mol hemocyanin}}{\text{L}}$$

$$= 3.75 \times 10^{-5} \text{ mol hemocyanin}$$

We calculate the molar mass from the calculated number of moles and the given mass of hemocyanin.

$$\text{Molar mass of hemocyanin} = \frac{1.50 \text{ g}}{3.75 \times 10^{-5} \text{ mol}} = 4.00 \times 10^{4} \text{ g/mol}$$

(b) Equating the molarity from part (a) with the molality of the aqueous solution, we solve for the freezing point using K_f for water.

$$\Delta T_f = -K_f \times m = -1.86 \text{ }^{\circ}\text{C } m^{-1} \times 1.50 \times 10^{-4} m = -2.79 \times 10^{-4} \text{ }^{\circ}\text{C}$$

The freezing point of the solution should be $-0.000279 \text{ }^{\circ}\text{C}$.

ASSESSMENT

The freezing point of the solution would be extremely difficult to measure with precision and accuracy. Common laboratory thermometers cannot be read more precisely than to the nearest 0.05 °C. Freezing point depression is not a good method for determining large molar masses. However, the osmotic pressure corresponds to 2.60 mmHg, which is easily measured.

EXERCISE 12.16A

An aqueous solution is prepared by dissolving 1.08 g of human serum albumin, a protein obtained from blood plasma, in 50.0 mL of water. The solution has an osmotic pressure of 0.00770 atm at 298 K. What is the molar mass of the albumin?

What is the osmotic pressure, expressed in the unit mmH_2O, of a solution containing 125 μg of vitamin B_{12} ($C_{63}H_{88}CoN_{14}O_{14}P$) in 2.50 mL H_2O at 25 °C?

Problem-Solving Note

Low osmotic pressures are often measured and expressed in mmH_2O. Use the method of Example 5.2 to relate a pressure in atmospheres to millimeters of mercury and then to millimeters of H_2O.

Practical Applications of Osmosis

Osmosis occurs in living organisms everywhere: Every cell is surrounded by a semipermeable membrane and filled with solutions of ions, small and large molecules, and still larger cell components. If we place red blood cells in pure water, a net flow of water into the cells causes them to burst. In contrast, if we place the cells in a solution that is 0.92% NaCl (mass/vol), there is no net flow of water through the cell membranes, and the cells are stable. The fluids inside the cells have the same osmotic pressure as the sodium chloride solution. A solution having the same osmotic pressure as body fluids is an **isotonic** solution. If the concentration of NaCl in a saline solution is greater than 0.92%, a net flow of water *out* of the cells causes them to shrink. The saline solution is a **hypertonic** solution; it has a higher osmotic pressure than red blood cells. If the concentration of NaCl in the solution surrounding the cells is less

Medical Applications of Osmosis

The rupture of a cell by a hypotonic solution is called *plasmolysis*. If the cell is a red blood cell, the more specific term is *hemolysis*. The shrinkage of a cell in a hypertonic solution, called *crenation,* can lead to the death of the cell.

When replacing body fluids or providing nourishment intravenously, it is important to use an isotonic fluid. Otherwise, hemolysis or crenation results, and the patient's health is jeopardized. As we have already described, a 0.92% NaCl (mass/vol) solution, called physiological saline, is isotonic with the fluid in red blood cells. The "D5W" so often referred to by television's doctors and paramedics is a 5.5% (mass/vol) solution of glucose (also called dextrose, D) in water (W). It, too, is isotonic with the fluid in red blood cells. The 0.92% NaCl is about 0.16 M, and the 5.5% glucose solution is approximately 0.31 M. (In Section 12.9, we will see why these two solutions have different molarities.)

There are shortcomings to using D5W for intravenous feeding. A patient can accommodate only about 3 L of water in a day.

If an isotonic solution of 5.5% glucose is used, 3.0 L of this solution supplies only about 160 g glucose. The glucose yields about 640 kcal (640 food Calories) per day, a woefully inadequate amount of energy because even a resting patient requires about 1400 kcal/day. And for a person suffering from serious burns, for example, requirements may be as high as 10,000 kcal/day. With carefully formulated solutions containing other vital nutrients as well as glucose, the nourishment a patient receives can be increased to about 1200 kcal/day, but this still falls short of the requirements of many seriously ill people.

One possibility is to use solutions that are about six times as concentrated as isotonic solutions. However, instead of administering them through a vein in an arm or a leg, the solution is infused directly through a tube into the superior vena cava, a large blood vessel leading to the heart. The large volume of blood flowing through this vein quickly dilutes the solution to levels that do not damage the blood. Using this technique, patients have been given 5000 kcal/day and have even gained weight.

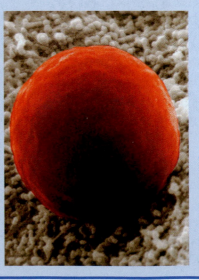

◄ Red blood cells in a hypertonic solution (left), an isotonic solution (center), and a hypotonic solution (right).

than 0.92%, water flows *into* the cells. We call the saline solution a **hypotonic** solution; it has a lower osmotic pressure than red blood cells.

One modern application of osmosis, called *reverse osmosis,* is based on reversing the normal net flow of water molecules through a semipermeable membrane. That is, by applying to a solution a pressure exceeding the osmotic pressure, water can be driven from a solution into pure water. In this way, pure water can be extracted from brackish water, seawater, or industrial wastewater. Reverse osmosis requires membranes that can withstand high pressures. This process is widely used in ships at sea and in arid nations of the Middle East.

12.9 Solutions of Electrolytes

So far, our discussion of colligative properties has focused on solutions of nonelectrolytes. Do electrolyte solutions have colligative properties? We have already implied that they do by noting the osmotic properties of isotonic saline solutions and the use of NaCl and $CaCl_2$ in deicing roads. The major difference in treating solutions of electrolytes lies in how we assess solute concentrations.

The solute particles of nonelectrolyte solutions are molecules. Strong electrolyte solutions have ions as solute particles, and weak electrolyte solutions have *both molecules and ions* as solute particles. When we use the formula of a solute in establishing the molarity or molality of a solution, we generally disregard these facts. However, when assessing colligative properties, we must base the solute concentration on the *total number of particles* in solution (ions and/or molecules). For example, 0.010 M NaCl(aq) is a strong electrolyte solution and thus has $[Na^+] = 0.010$ M, $[Cl^-] = 0.010$ M, and a total ion concentration of 0.020 M. In 0.010 M CH_3COOH(aq), a weak electrolyte solution, most of the CH_3COOH molecules remain intact, but a small fraction of them do ionize to form H^+ and CH_3COO^- ions. This makes the total particle concentration slightly greater than 0.010 M but less than 0.020 M.

We can use a factor called the **van't Hoff factor** (*i*) to modify the equations for colligative properties by accounting for the presence of ions in solution:

Freezing point depression: $\qquad \Delta T_f = -i\, K_f m$ (**12.11**)

Boiling point elevation: $\qquad \Delta T_b = i\, K_b m$ (**12.12**)

Osmotic pressure: $\qquad \pi = i\, M\, RT$ (**12.13**)

For nonelectrolyte solutions, $i = 1$, and the molarity or molality of the solute is also the molarity of the solute particles. For a solution of the strong electrolyte NaCl, we should expect $i = 2$; because we have *two* moles of ions for each mole of solute. For Na_2SO_4 and $CaCl_2$, we should expect $i = 3$. For a solution of a weak electrolyte, *i* is somewhat greater than 1 but less than 2 because a weak electrolyte is only partially ionized. The extent to which a weak electrolyte ionizes depends on the concentration of the solution, which means that *i* also depends on the concentration. For example, $i \approx 1.04$ for 0.010 M CH_3COOH, which signifies that about 4% of the molecules ionize; in 0.10 M CH_3COOH, $i \approx 1.01$.

The value of *i* for strong electrolytes also depends on the concentration. Only if the solution is quite dilute will *i* be an integer (that is, 2, 3, ...); otherwise it will be a noninteger. The values of *i* for various solutions of NaCl, $MgSO_4$, and $Pb(NO_3)_2$ are listed in Table 12.3. In concentrated solutions of strong electrolytes, attractive forces

▲ In 1901, Jacobus Henricus van't Hoff (1852–1911) received the first Nobel Prize in Chemistry for his work on the physical properties of solutions. However, he is also known for proposing (simultaneously with the French chemist Joseph Le Bel) the tetrahedral geometries of carbon compounds (for example, CH_4). Van't Hoff was 22 years old at the time of this proposition.

Table 12.3 Variation of the van't Hoff Factor, *i*, with Solution Molality

| Solute | Molality, *m* | | | | | Infinite Dilution |
	1.0	0.10	0.010	0.0010	⋯	
NaCl	1.81	1.87	1.94	1.97	⋯	2
$MgSO_4$	1.09	1.21	1.53	1.82	⋯	2
$Pb(NO_3)_2$	1.31	2.13	2.63	2.89	⋯	3

The limiting values $i = 2$, 2, and 3 are reached when the solution is infinitely dilute. In general, a solute with singly charged ions (for example, NaCl) approaches its limiting value more quickly than does a solute with ions of higher charge, because interionic attractions are greater in solutes with more highly charged ions.

between cations and anions cause some of them to associate into *ion pairs* and to otherwise behave as if the solute were not completely dissociated. In this text, we will assume solutions are dilute enough that strong electrolytes are completely dissociated into ions.

Example 12.17   A Conceptual Example

Without doing detailed calculations, place the following solutions in order of decreasing osmotic pressure: **(a)** 0.01 M $C_{12}H_{22}O_{11}$(aq) at 25 °C, **(b)** 0.01 M CH_3CH_2COOH(aq) at 37 °C, **(c)** 0.01 *m* KNO_3(aq) at 25 °C, **(d)** a solution of 1.00 g polystyrene (molar mass: 3.5×10^5 g/mol) in 100 mL of benzene at 25 °C.

ANALYSIS AND CONCLUSIONS

Let's begin by writing the most general equation for the osmotic pressure of a solution, $\pi = i\, M\, RT$, and then applying it to each solution.

(a) Suppose we designate the osmotic pressure of 0.01 M $C_{12}H_{22}O_{11}$(aq), a nonelectrolyte solution ($i = 1$), as π_a. Then we can compare the other osmotic pressures with π_a.

(b) This compound is a carboxylic acid. It must be a weak acid because it is not among the strong acids listed in Table 4.1. The osmotic pressure of this solution should be slightly greater than that in (a) because i should be somewhat greater than 1: $\pi_b > \pi_a$. Also, the difference in temperature between 37 and 25 °C would cause π_b to be greater then π_a by a factor of 310/298, but this is inconsequential relative to the comparison made in (c).

(c) The solution is 0.01 *m* KNO_3 rather than 0.01 M KNO_3. However, in dilute aqueous solutions, because $d \approx 1.0$ g/mL, molality and molarity are essentially equal. This is a solution of an electrolyte for which $i \approx 2$. The osmotic pressure is about twice as great as that of solution (a) and also almost twice that of solution (b): $\pi_c > \pi_b > \pi_a$.

(d) Each of the other three solutions has at least 0.001 mol of solute particles in a 100-mL sample. The *mass* of 0.001 mol of the polystyrene is 0.001 mol $\times$ 3.5×10^5 g/mol = 350 g, but the mass of polystyrene in the given solution is only 1.00 g. The solution is therefore much more dilute than 0.01 M. The fact that the solute is a polymer and the solvent is an aromatic hydrocarbon has no bearing on the situation. Solution (d) has the lowest osmotic pressure by far; solution (c) has the highest. Our ranking therefore is $\pi_c > \pi_b > \pi_a > \pi_d$.

EXERCISE 12.17A

Without doing detailed calculations, determine which of the following solutions has the lowest freezing point and which has the highest: 0.010 *m* $C_6H_{12}O_6$(aq), 0.0080 M HCl(aq), 0.0050 *m* $MgCl_2$(aq), 0.0030 M $Al_2(SO_4)_3$(aq).

EXERCISE 12.17B

Hydrogen chloride is soluble both in water and in benzene. The freezing point depression in 0.01 *m* HCl(aq) is about 0.04 °C, and in 0.01 *m* HCl (in benzene) it is about 0.05 °C. Are these the results you would expect for the two solutions? Explain.

12.10   Colloids

Once a solute and solvent are thoroughly mixed, the atoms, ions, or molecules of the solute remain randomly distributed in the solvent; the *solute does not settle out*. The mixture is homogeneous—a *solution*. If we try to dissolve sand (silica, SiO_2) in water, for example, we will not form a solution. The best we can achieve is a temporary dispersion of the sand in water, called a *suspension*. Grains of sand quickly settle to the bottom of the container. The sand–water mixture is *heterogeneous*. The question we want to look at now is, Are there mixtures that lie between true solutions and suspensions? There are such mixtures, and they are called *colloids*.

A mixture is a colloid not because of the kind of matter present, but because of the *size* of the dispersed particles. True solutions have solute and solvent particles less than about 1 nm in diameter. Ordinary suspensions have particle dimensions of about 1000 nm or more. Such particles are generally visible to the eye or through an ordinary microscope. A **colloid** is a dispersion in an appropriate medium of particles ranging in size from about 1 nm to 1000 nm. Colloidal particles generally cannot be seen through an ordinary microscope.

The shapes of the particles in a colloid can vary widely, but in every case, at least one dimension (length, width, thickness) is in the range of about 1 to 1000 nm. The particles in an aqueous colloid of silica (SiO_2) are spheres. Some particles in colloids are rod-shaped, for example, certain viruses. Still other colloidal particles are disk-shaped, such as the blood plasma protein gamma globulin. Some natural and synthetic polymers form long, randomly coiled filaments.

We can classify colloids according to the state of matter of the dispersed particles and the state of matter of the medium in which they are dispersed. In all, eight combinations are possible (Table 12.4), and each is given a different name. For example, a *sol* is a dispersion of solid particles in a liquid medium; an *emulsion* is a dispersion of liquid particles (microscopic droplets) in a liquid medium; and an *aerosol* is a dispersion of either solid or liquid particles in a gas (usually air).

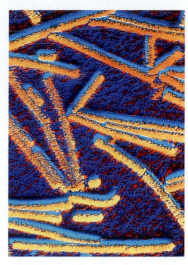

▲ These rod-shaped tobacco mosaic viruses are colloidal-sized particles.

Table 12.4 Some Common Types of Colloids

Dispersed Phase	Dispersion Medium	Type	Examples
Solid	Liquid	Sol	Starch solution, clay sols, jellies
Liquid	Liquid	Emulsion	Milk, mayonnaise
Gas	Liquid	Foam	Whipped cream, meringues
Solid	Gas	Aerosol	Fine dust or soot in air
Liquid	Gas	Aerosol	Fog, hair sprays
Solid	Solid	Solid sol	Ruby glass, black diamond
Liquid	Solid	Solid emulsion	Pearl, opal, butter
Gas	Solid	Solid foam	Floating soaps, pumice, lava

Much of living matter is in the form of sols and emulsions. Substances with high molecular masses, such as starches and proteins, generally form colloidal mixtures rather than true solutions with water. Smog is an aerosol in part, and its suspended particles are both solid (smoke) and liquid (fog): *smo*ke + *fog* = smog. However, other constituents of smog are of molecular size, for example, SO_2, CO, NO, and O_3.

As you might expect, the properties of colloids are different from those of true solutions and suspensions. Colloids often appear milky or cloudy, and even those that appear clear reveal their colloidal nature by scattering a beam of light passing through them. This phenomenon, first studied by John Tyndall in 1869, is known as the **Tyndall effect** (Figure 12.19). The spectacular sunsets often seen in desert areas are caused, at least in part, by the preferential scattering of blue light as sunlight passes through dust-laden air (an aerosol). The transmitted light is deficient in the color blue and thus has a reddish color.

◀ **FIGURE 12.19   The Tyndall effect**
The light beam is not visible as it passes through a true solution (left), but it is readily visible as it passes through colloidal iron (III) oxide in water.

Colloidal Suspensions activity

▶ **FIGURE 12.20** **A heterogeneous mixture and a colloid**

(Left) Fine sand (silica) added to water initially forms a suspension that quickly settles, producing a heterogeneous mixture with water on top and silica on the bottom. (Right) A colloidal dispersion (Ludox®) prepared from a sodium silicate solution, $Na_2SiO_3(aq)$, has the same proportions of silica and water as on the left. The dispersed particles of hydrated silica, $SiO_2 \cdot xH_2O$, are much larger than atoms and typical molecules. During their formation, the silica particles acquire a negative charge by adsorbing OH^- ions. The similarly charged silica particles repel one another and thereby stay suspended indefinitely.

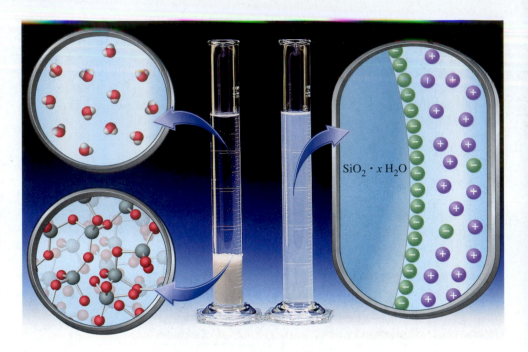

Application Note

In plants, the xylem vessels, which are composed of dead cells, are rigid tubes that transport water and dissolved minerals from roots to leaves. The vessel walls have a gelatinous coating on the inside that impedes flow. An increase in the concentration of trace ions from the soil flowing through the conduits rapidly increases the rate of flow, presumably by breaking down the gel.

We have noted that the properties of surfaces are in some ways different from those of bulk matter (recall our discussion of surface tension on page 459). A special property of surfaces is their ability to *adsorb*, or attach, ions from a solution. This ability has important consequences for colloids. The particles in a colloid generally adsorb certain cations or anions in preference to others. For example, the SiO_2 particles in colloidal silica adsorb OH^- (Figure 12.20). As a result, even though the mixture as a whole is electrically neutral, the attached OH^- ions give the particles a net negative charge. Because all the silica particles are negatively charged, they repel one another. These repulsions are strong enough to overcome the force of gravity, and the particles do not settle. However, a high concentration of an electrolyte can cause a colloid to *coagulate*, or *precipitate* (Figure 12.21), by neutralizing the charges on the colloidal particles.

Removing excess ions can prevent coagulation and make a colloid more stable. Excess ions can be removed through *dialysis*, a process similar to osmosis. In dialysis, ions pass from the colloid through a semipermeable membrane into pure water, but the much larger colloidal particles do not. In *electrodialysis*, the process is facilitated by

▶ **FIGURE 12.21** **Formation and coagulation of a colloid**

The red colloidal hydrous iron(III) oxide, $Fe_2O_3 \cdot xH_2O$ (left), is produced by adding concentrated $FeCl_3(aq)$ to boiling water. When a few drops of $Al_2(SO_4)_3(aq)$ are added, the suspended particles rapidly coagulate into a precipitate of $Fe_2O_3 \cdot xH_2O$ (right).

QUESTION: Would you expect NaCl(aq) to be as effective as Al₂(SO₄)₃(aq) in coagulating the colloidal iron(III) oxide? Explain.

the attractions of ions to an electrode having the opposite charge. A human kidney acts as a dialyzer by removing excess electrolytes from blood, which is a colloid. In some diseases, the kidney loses its dialyzing ability, but fortunately this function can be served to a degree by a dialysis machine.

Cumulative Example

A 375-mL sample of hexane vapor in equilibrium with liquid hexane, C_6H_{14} ($d = 0.6548$ g/mL), at 25.0 °C is dissolved in 50.0 mL of liquid cyclohexane, C_6H_{12} ($d = 0.7739$ g/mL; vp = 97.58 Torr), at 25.0 °C. Use information found elsewhere in the text (such as Example 11.3) to calculate the total vapor pressure above the solution at 25.0 °C. How reliable is this calculation?

STRATEGY

We must sum the partial pressures of the two components to obtain the total vapor pressure of the solution. Assuming ideal solution behavior, we can obtain these partial pressures with Raoult's law (Equation 12.6). In that equation, concentrations must be expressed in mole fractions, which requires that we first determine the number of moles of each solution component and then use Equation (12.3). We can use the ideal gas Equation (5.9) to calculate the number of moles of hexane vapor; this is the amount of hexane dissolved in the hexane–cyclohexane solution. We can then determine the number of moles of cyclohexane from its volume, density, and molar mass.

SOLUTION

First, we rearrange the ideal gas equation to solve for number of moles and substitute the vapor equilibrium data, using the value 151 Torr from Example 11.3 for the vapor pressure of hexane at 25 °C.

$$n = \frac{PV}{RT}$$

$$= \frac{151 \text{ Torr} \times \dfrac{1 \text{ atm}}{760 \text{ Torr}} \times 0.375 \text{ L hexane}}{0.08206 \text{ L atm mol}^{-1} \text{ K}^{-1} \times (273.2 + 25.0) \text{ K}}$$

$$= 0.00304 \text{ mol hexane}$$

Then, we calculate the number of moles of cyclohexane in the solution from the given volume and density.

$$? \text{ mol } C_6H_{12} = 50.0 \text{ mL } C_6H_{12} \times \frac{0.7739 \text{ g}}{1 \text{ mL}} \times \frac{1 \text{ mol } C_6H_{12}}{84.16 \text{ g}}$$

$$= 0.460 \text{ mol } C_6H_{12}$$

We get the mole fraction of hexane and cyclohexane from their molar quantities.

$$x_{C_6H_{14}} = \frac{0.00304 \text{ mol } C_6H_{14}}{0.00304 \text{ mol } C_6H_{14} + 0.460 \text{ mol } C_6H_{12}} = 0.00657$$

$$x_{C_6H_{12}} = \frac{0.460 \text{ mol } C_6H_{12}}{0.00304 \text{ mol } C_6H_{14} + 0.460 \text{ mol } C_6H_{12}} = 0.994$$

Next, we use Equation (12.6) to determine the partial pressures of hexane and cyclohexane.

$$P_{C_6H_{14}} = 0.00657 \times 151 \text{ Torr} = 0.992 \text{ Torr}$$

$$P_{C_6H_{12}} = 0.994 \times 97.58 \text{ Torr} = 97.0 \text{ Torr}$$

Finally, we calculate the total vapor pressure of the solution from Dalton's law of partial pressures.

$$P_{total} = P_{C_6H_{12}} + P_{C_6H_{14}} = 97.0 \text{ Torr} + 0.992 \text{ Torr} = 98.0 \text{ Torr}$$

ASSESSMENT

Our treatment of hexane vapor as an ideal gas should be valid because of the relatively low pressure of this gas—151 Torr (0.20 atm). Our assumption of ideal behavior for the hexane–cyclohexane solution should also be valid. The nonpolar hydrocarbons have compositions and molar masses that are nearly the same, suggesting that the intermolecular forces (dispersion forces) in the two substances should be very similar. This ensures the validity of Raoult's law.

The total vapor pressure of the solution (98.0 Torr) is nearly equal to that of pure cyclohexane (97.58 Torr)—as we might expect for this very dilute solution. Significantly, however, the solution vapor pressure is slightly higher than for pure cyclohexane; and this is consistent with the fact that hexane, the solute (vp = 151 Torr), is noticeably more volatile than cyclohexane, the solvent (vp = 97.58 Torr).

Concept Review with Key Terms

12.1 Some Types of Solutions—Many types of solutions can be formed by dissolving one substance (the solute) in another substance (the solvent). Solutions can be formed from all three states of matter (solid, liquid, and gas) and in various combinations.

12.2 Solution Concentration—Several concentration units are used in describing solutions. For many applications, concentration can be expressed as percent by mass, percent by volume, and mass per volume percent. Other concentration units include molarity, moles of solute per liter of solution, and **molality**, (*m*), moles of solute per kilogram of solvent:

$$\text{Molality } (m) = \frac{\text{amount of solute (mol)}}{\text{mass of solvent (kg)}}$$

The concentration of a component can also be specified as a **mole fraction** (x_i) or given as a **mole percent** (mole fraction multiplied by 100%):

$$x_i = \frac{\text{amount of component } i \text{ (mol)}}{\text{total amount of solution components (mol)}} = \frac{n_i}{n_{\text{total}}}$$

Some related units used for very dilute solutions are parts per million, parts per billion, and parts per trillion.

12.3 Energetics of Solution Formation—In an **ideal solution**, all the intermolecular forces between molecules are of the same type and are essentially equal in magnitude. In a nonideal solution, the types and magni- tudes of solute–solvent intermolecular forces differ appreciably from the types and magnitudes of solute–solute and solvent–solvent intermolecular forces. If these differences are great enough, components remain segregated in a heterogeneous mixture.

12.4 Equilibrium in Solution Formation—The **solubility** of a solute is its concentration in a **saturated solution**, a solution in which the undissolved solute exists in a dynamic equilibrium with the solution. Solutions contain- ing less solute than the saturation concentration are called **unsaturated**, and those containing more (made possible through special handling) are called **supersaturated**. The solubility of most solid solutes in water increases with temperature, but a few become less soluble.

12.5 The Solubilities of Gases—The solubilities of gases in water generally decrease with an increase in tem- perature, whereas an increase in the pressure of the gas above the solution leads to an increase in solubility, as described by **Henry's law** and the equation

$$S = k\, P_{\text{gas}}$$

12.6 Vapor Pressures of Solutions—Vapor pressure lowering, freezing point depression, boiling point elevation, and osmotic pressure are **colligative properties**—properties that depend on the particular sol-

vent and on the number of solute particles present, but not on the identity of the solute. The presence of a solute lowers the vapor pressure of the solvent in a solution. For ideal solutions and dilute nonideal solutions, the lowering of vapor pressure follows **Raoult's law**:

$$P_{\text{solv}} = x_{\text{solv}} \cdot P_{\text{solv}}^{\circ}$$

The vapor in equilibrium with a solution of two volatile components contains a higher mole fraction of the more volatile component than does the solution. This behavior is the basis of the separation method known as **fractional distillation**.

12.7 Freezing Point Depression and Boiling Point Elevation—The introduction of a solute typically lowers the freezing point and raises the boiling point of a solvent. The extent of these changes is related to the molality of the solute and a solvent-specific constant:

$$\Delta T_{\text{f}} = T_{\text{f}}(\text{solution}) - T_{\text{f}}(\text{solvent}) = -K_{\text{f}} \times m$$
$$\Delta T_{\text{b}} = T_{\text{b}}(\text{solution}) - T_{\text{b}}(\text{solvent}) = K_{\text{b}} \times m$$

12.8 Osmotic Pressure—**Osmosis** is the net flow of solvent molecules through a **semipermeable membrane** from a solution of lower solute concentration to a solution of higher solute concentration. **Osmotic pressure** (π) is the pressure that must be applied to a solution to prevent such a flow.

$$\pi = \text{M}\,RT$$

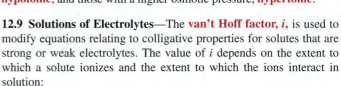

Osmotic pressure measurements can be used to establish molar masses, especially of macromolecular materials such as polymers. Reverse osmosis involves applying to a solution pressures greater than its osmotic pressure, thereby driving solvent out of the solution; this process is used, for example, to purify water. Solutions with an osmotic pressure equal to that of body fluids are said to be **isotonic**, those with a lower osmotic pressure are **hypotonic**, and those with a higher osmotic pressure, **hypertonic**.

12.9 Solutions of Electrolytes—The **van't Hoff factor, *i*,** is used to modify equations relating to colligative properties for solutes that are strong or weak electrolytes. The value of *i* depends on the extent to which a solute ionizes and the extent to which the ions interact in solution:

$$\Delta T_{\text{f}} = -i\, K_{\text{f}}m \qquad \Delta T_{\text{b}} = i\, K_{\text{b}}m \qquad \pi = i\,\text{M}\,RT$$

12.10 Colloids—In **colloids**, the particles of dispersed matter range in size from about 1 to 1000 nm. One way that colloids differ from solutions is that colloids scatter light, a phenomenon known as the **Tyndall effect**, and true solutions do not. Many common materials are colloidal.

Assessment Goals

When you have mastered the material in this chapter, you will be able to:

- Recognize the types of solutions that can be formed from the three states of matter.
- Express solution concentrations in molarity, molality, mole fraction, and mole percent, and perform conversions between different concentration units.

- Express the concentration of solutes in percent by mass, percent by volume, mass/volume percent, ppm, ppb, and ppt.
- Describe the influence of intermolecular forces on the energetics of solution formation.
- Distinguish between ideal and non-ideal solutions.
- Predict if a mixture is likely to be homogeneous or heterogeneous, based on information about intermolecular forces.

- Describe the process by which ionic compounds dissolve in water.
- Define solubility; distinguish between saturated, supersaturated, and unsaturated solutions; and extract solubility data from solubility curves.
- Describe the effects of temperature and pressure on the solubility of gases.
- Calculate the solubility of gases using Henry's law.
- Calculate the vapor pressure and vapor composition of solutions using Raoult's law.
- Describe the process of fractional distillation in relation to the vapor pressures of solution components.

- Calculate the freezing points and boiling points of solutions.
- Describe the movement of solvent in the processes of osmosis and reverse osmosis.
- Calculate osmotic pressure from solution concentrations, and determine molar masses from osmotic pressure values.
- Estimate the colligative properties for solutions of electrolytes, using the van't Hoff factor, i, if necessary.
- Describe the physical characteristics and some unique properties of colloids.

Self-Assessment Questions

1. An *aqueous* solution is 0.01 M CH_3OH. Which of the following can be used to describe the approximate concentration?
   (a) 0.01% CH_3OH (mass/volume)
   (b) 0.01 m CH_3OH
   (c) $x_{CH_3OH} = 0.01$
   (d) 0.99 M H_2O

2. Explain why molarity is temperature-dependent, whereas molality is not. Is mole fraction temperature-dependent? Is volume percent? Is mass percent?

3. Precipitation is induced in a supersaturated solution by adding a seed crystal. When no more solid crystallizes, is the solution saturated, unsaturated, or supersaturated? Explain.

4. Would you expect a solution of ethanol (CH_3CH_2OH) in water to be ideal? Explain.

5. Which of the following are most likely to form solutions? Explain.
   (a) NaCl and benzene, C_6H_6
   (b) motor oil and water
   (c) motor oil and benzene
   (d) 2-decanol and water
   (e) pentane and decane

6. The most water soluble of the following is
   (a) $C_6H_6(l)$ (c) $NH_2OH(s)$
   (b) C(s)   (d) $C_{10}H_8(s)$

7. Of the following compounds, which of these is moderately soluble both in water [$H_2O(l)$] and benzene [$C_6H_6(l)$]?
   (a) phenol, C_6H_5OH
   (b) naphthalene, $C_{10}H_8$
   (c) hexane, C_6H_{14}
   (d) sodium chloride, NaCl

8. When the cap is removed from a bottle of carbonated beverage, carbon dioxide gas escapes from the liquid. Explain why.

9. The *solubility* of a nonreactive gas in water increases with which of the following?
   (a) an increase in temperature
   (b) an increase in gas pressure

   (c) increases both in temperature and gas pressure
   (d) an increase in the volume of gas in equilibrium with the available water.

10. The vapor in equilibrium with a pure liquid has the same composition as the liquid. Is this statement also true for solutions? Explain.

11. What is a colligative property? Name two properties of a solution that are *not* colligative. What are the chief colligative properties discussed in this chapter? Is the density of a solution a colligative property? Explain.

12. Explain why a nonvolatile solute *raises* the boiling point of a solvent, whereas it *lowers* the freezing point of the same solvent.

13. The freezing point will be depressed most by dissolving which of the following in 250.0 grams of water?
    (a) 25.0 g CH_3OH
    (b) 30.0 g CH_3CH_2OH
    (c) 35.0 g $C_6H_{12}O_6$

14. Which laboratory measurement gives a more accurate result for the molar mass of a polymeric substance: freezing point depression or osmotic pressure? Explain.

15. How do the terms *osmosis* and *osmotic pressure* differ in meaning?

16. Do all *isotonic* solutions have the same concentration? Explain.

17. The net direction in which gaseous molecules diffuse is from *higher* to lower pressure. In osmosis, the net direction in which molecules diffuse is from *lower* to higher solution concentration. Do these statements seem incompatible? Explain.

18. Contrast a true solution and a colloid with regard to (a) the size of the solute particles, (b) the nature of the distribution of solute and solvent particles, (c) the color and clarity of the solution, and (d) the Tyndall effect.

19. Compare and contrast osmosis and dialysis with respect to the kinds of particles that pass through a semipermeable membrane.

20. For each pair of solutions at 25 °C, indicate which member has the higher osmotic pressure.
    (a) 0.1 M $NaHCO_3$, 0.05 M $NaHCO_3$
    (b) 1 M NaCl, 1 M glucose
    (c) 1 M NaCl, 1 M $CaCl_2$
    (d) 1 M NaCl, 3 M glucose

Problems

Percent Concentration

21. Describe how you would prepare 2.30 kg of an aqueous solution that is 4.85% $NaNO_3$ by mass.

22. Describe how you would prepare exactly 2.00 L of an aqueous solution that is 9.77% acetic acid by volume.

23. What is the mass percent of solute in each of the following solutions?

(a) 175 mg NaCl/g solution

(b) 0.275 L methanol ($d = 0.791$ g/mL)/kg water

(c) 4.5 L ethylene glycol ($d = 1.114$ g/mL) in 6.5 L propylene glycol ($d = 1.036$ g/mL)

24. What is the mass percent of solute in each of the following solutions?

(a) 4.12 g NaOH in 100.0 g water

(b) 5.00 mL ethanol ($d = 0.789$ g/mL) in 50.0 g water

(c) 1.50 mL glycerol ($d = 1.324$ g/mL) in 22.25 mL H_2O ($d = 0.998$ g/mL)

25. What is the volume percent of the first-named component in each of the following solutions?

(a) 58.0 mL water in 625 mL of an ethanol–water solution

(b) 10.00 g methanol ($d = 0.791$ g/mL) in 75.00 g ethanol ($d = 0.789$ g/mL) (*Hint:* Assume the volumes are additive.)

(c) 24.0% by mass ethanol ($d = 0.789$ g/mL) in an aqueous solution with $d = 0.963$ g/mL

26. What is the volume percent of the first-named component in each of the following solutions?

(a) 35.0 mL water in 725 mL of an ethanol–water solution

(b) 10.00 g acetone ($d = 0.789$ g/mL) in 1.55 L of an acetone–water solution

(c) 1.05% by mass 1-butanol ($d = 0.810$ g/mL) in ethanol ($d = 0.789$ g/mL) (*Hint:* Assume the volumes are additive.)

27. Cholesterol readings are described on page 486. What mass in grams of cholesterol is contained in an adult with a cholesterol number of 234? (The average human adult has about 5.0 L of blood.)

28. Nitrocellulose lacquer is used as a protective coating on wood. What mass of nitrocellulose is contained in one gallon (3.8 L) of lacquer that is 5.1% nitrocellulose by mass? The density of the lacquer is 0.79 g/mL.

Parts per Million, Parts per Billion, Parts per Trillion

29. Express the following aqueous concentrations in the unit indicated.

(a) 5 μg trichloroethylene/L water, as ppb trichloroethylene

(b) 0.0025 g KI/L water, as ppm of KI

(c) 37 ppm SO_4^{2-} as molarity of sulfate ion, $[SO_4^{2-}]$

30. Express the following aqueous concentrations in the unit indicated.

(a) 1 μg benzene/L water, as ppb benzene

(b) 0.0035% NaCl, by mass, as ppm of NaCl

(c) 2.4 ppm F^-, as molarity of fluoride ion, $[F^-]$

31. *Without doing detailed calculations*, arrange *aqueous* solutions with the following concentrations in the order of increasing mass percent of solute: (a) 1% by mass, (b) 1 mg solute/dL solution, (c) 1 ppb, (d) 1 ppm, (e) 1 ppt.

32. *Without doing detailed calculations*, determine which has the greater mass percent of solute: 0.0010 M $MgCl_2(aq)$ or $MgCl_2(aq)$ with 105 ppm Mg^{2+}.

Molarity and Molality

33. Calculate the molality of a solution that has 1.02 kg sucrose, $C_{12}H_{22}O_{11}$, dissolved in 554 g water.

34. Laboratory ethanol is sometimes *denatured* (rendered unfit to drink) by adding methanol. Calculate the molality of a solution prepared by dissolving 225 mL pure methanol, CH_3OH, ($d = 0.791$ g/mL) in 4.00 kg ethanol.

35. An aqueous solution is prepared by diluting 3.30 mL acetone, CH_3COCH_3, ($d = 0.789$ g/mL) with water to a final volume of 75.0 mL. The density of the solution is 0.993 g/mL. What are the molarity and molality of acetone in this solution?

36. A typical commercial-grade phosphoric acid is 75% H_3PO_4, by mass, and has $d = 1.57$ g/mL. What are the molarity and the molality of H_3PO_4 in this acid?

37. How many moles of ethylene glycol are present in 2.30 L of 6.27 m $HOCH_2CH_2OH$ ($d = 1.035$ g/mL)?

38. A dilute sulfuric acid solution that is 3.39 m H_2SO_4 has a density of 1.18 g/mL. How many moles of H_2SO_4 are there in 375 mL of this solution?

Mole Fraction and Mole Percent

39. What is the mole fraction of naphthalene, $C_{10}H_8$, in (a) a solution prepared by dissolving 23.5 g $C_{10}H_8(s)$ in 315 g benzene, $C_6H_6(l)$, and (b) a 0.250 m solution of $C_{10}H_8$ in C_6H_6?

40. What is the mole fraction of (a) urea, $CO(NH_2)_2$, in a solution prepared by dissolving 25.2 g $CO(NH_2)_2$ in 125 mL of water at 20 °C ($d = 0.998$ g/mL), and (b) ethanol in an aqueous solution that is 22.4% ethanol by mass?

41. *Without doing detailed calculations*, determine which of the following *aqueous* solutions has the greatest mole fraction of solute: (a) 1.00 m CH_3OH, (b) 5.0% CH_3CH_2OH, by mass, (c) 10.0% $C_{12}H_{22}O_{11}$, by mass. Explain.

42. *Without doing detailed calculations*, determine which of the following *aqueous* solutions has the greater mass percent urea, $CO(NH_2)_2$: 0.10 m urea or $x_{urea} = 0.010$? Explain your reasoning.

Solubilities

43. Predict which of the following substances is highly soluble in water, which is slightly soluble, and which is insoluble.

    (a) chloroform, $CHCl_3$

    (b) benzoic acid, C_6H_5COOH

    (c) propylene glycol, $CH_3CHOHCH_2OH$

44. One of the following substances is soluble in *both* water and in benzene, C_6H_6. Which is it? Explain.

    (a) chlorobenzene, C_6H_5Cl

    (b) ethylene glycol, $HOCH_2CH_2OH$

    (c) phenol, C_6H_5OH

Saturated, Unsaturated, and Supersaturated Solutions

For Problems 45–48, refer to Figure 12.10 as necessary.

45. Is a mixture of 25 g NH_4Cl and 55 g H_2O at 60 °C saturated, unsaturated, or supersaturated? Explain.

46. Can an aqueous solution be prepared that is 30.0% by mass of KNO_3 at 20 °C? Explain.

47. Determine (a) the additional mass of water that should be added to a mixture of 35 g K_2CrO_4 and 35 g H_2O to dissolve all the solute at 25 °C; and (b) the approximate temperature to which a mixture of 50.0 g KNO_3 and 75.0 g water must be heated to complete the dissolving of the KNO_3.

48. Determine (a) the molality of saturated $Li_2SO_4(aq)$ at 50 °C; and (b) the additional mass of $Pb(NO_3)_2$ that can be dissolved in an aqueous solution containing 325 mg $Pb(NO_3)_2/g$ solution at 30 °C.

49. Can a solute be crystallized from an unsaturated solution without changing the solution temperature? Explain.

50. Are there any exceptions to the general rule that a supersaturated solution can be made to deposit excess solute by cooling? Explain.

Solubilities of Gases

51. The solubility of $O_2(g)$ in water is 4.43 mg $O_2/100$ g H_2O at 20 °C and a gas pressure of 1 atm. (a) What is the molarity of the saturated solution? (b) What pressure of $O_2(g)$ would be required to produce a saturated solution that is 0.010 M O_2?

52. When 250 g of water saturated with air is heated from 20 °C to 80 °C, estimate the quantity of air released in terms of its (a) mass, in milligrams; (b) volume, in milliliters (at STP). Use data from Figure 12.12, and assume 28.96 g/mol as the molar mass of air.

Liquid–Vapor Equilibrium

53. A solution has a 1:4 *mole* ratio of pentane to hexane. The vapor pressures of the pure hydrocarbons at 20 °C are 441 mmHg for pentane and 121 mmHg for hexane. (a) What are the partial pressures of the two hydrocarbons above the solution? (b) What is the mole fraction composition of the vapor?

54. A solution has a 2:3 *mass* ratio of toluene (C_7H_8) to benzene (C_6H_6). The vapor pressures of toluene and benzene at 25 °C are 28.4 mmHg and 95.1 mmHg, respectively. (a) What are the par-

tial pressures of the two hydrocarbons above the solution? (b) What is the mole fraction composition of the vapor?

55. What is the vapor pressure of water above a 0.20 *m* glucose ($C_6H_{12}O_6$) solution at 20 °C? Use vapor pressure data from Table 11.2.

56. What is the vapor pressure of water above 3.49 M glycerol ($C_3H_8O_3$) solution ($d = 1.073$ g/mL) at 20 °C? Use vapor pressure data from Table 11.2.

Freezing Point Depression and Boiling Point Elevation

For Problems 57–64, use data from Table 12.2, as necessary.

57. Determine the freezing points of (a) 0.25 *m* urea in water, and (b) 5.0% *para*-dichlorobenzene, $C_6H_4Cl_2$, by mass, in benzene.

58. Determine the boiling points of (a) 0.44 *m* naphthalene in benzene, and (b) 1.80 M sucrose ($C_{12}H_{22}O_{11}$) in water ($d = 1.23$ g/mL).

59. What is the molality of a naphthalene-in-benzene solution with the same freezing point as 0.55 *m* sucrose in water?

60. How many grams of naphthalene, $C_{10}H_8$, would you add to 50.0 g of benzene, $C_6H_6(l)$, to produce a solution that has the same freezing point as pure water?

61. A 2.11-g sample of naphthalene ($C_{10}H_8$) dissolved in 35.00 g of *para*-xylene has a freezing point of 11.25 °C. The pure solvent has a freezing point of 13.26 °C. What is the K_f of *para*-xylene?

62. A 1.45-g sample of an unknown compound is dissolved in 25.00 mL benzene ($d = 0.879$ g/mL). The solution freezes at 4.25 °C. What is the molar mass of the unknown?

63. An unknown compound consists of 33.81% C, 1.42% H, 45.05% O, and 19.72% N by mass. A 1.505-g sample of the compound, when dissolved in 50.00 mL benzene ($d = 0.879$ g/mL), lowers the freezing point of the benzene to 4.70 °C. What is the molecular formula of the compound?

64. Coniferin, a sugar derivative found in conifers such as fir trees, has a composition of 56.13% C, 6.48% H, and 37.39% O, by mass. A 2.216-g sample is dissolved in 48.68 g of H_2O, and the solution is found to have a boiling point of 100.068 °C. What is the molecular formula of coniferin?

Osmotic Pressure

65. Pickles are made by soaking cucumbers in a salt solution. Which has the higher osmotic pressure, the salt solution or the liquid in the cucumber? Explain.

66. Describe how the phenomenon discussed in Example 12.13 is related to osmosis.

67. What is the osmotic pressure at 37 °C of an aqueous solution that is 1.80% CH_3CH_2OH (mass/vol)? Is this solution isotonic, hypotonic, or hypertonic? [*Hint:* Compare the osmotic pressure to that of a 5.5% (mass/vol) solution of glucose, $C_6H_{12}O_6$.]

68. The osmotic pressure exerted by a 0.325-g sample of the polymer polystyrene in 50.00 mL of benzene at 25 °C is capable of supporting a column of the solution ($d = 0.88$ g/mL) 5.3 mm in height. What is the molar mass of the polystyrene? (*Hint:* What is the height of mercury [$d = 13.6$ g/mL] equivalent to that of the polymer solution?)

69. In the illustration, if aqueous solution A is 0.15 M sucrose ($C_{12}H_{22}O_{11}$) and solution B has 5.15 g urea [$CO(NH_2)_2$] in

75.0 mL of H_2O, will the net flow of water through the semipermeable membrane be to the left or to the right? Explain.

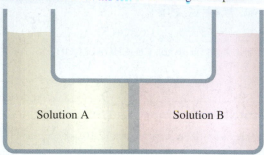

Solution A   Solution B

70. In the illustration above, if aqueous solution A is 0.18 M sucrose ($C_{12}H_{22}O_{11}$) and solution B is 0.22 M glucose ($C_6H_{12}O_6$), what pressure must be applied and to which arm of the apparatus—left or right—so that there is no net flow of water through the semipermeable membrane?

Colligative Properties of Strong Electrolytes, Weak Electrolytes, and Nonelectrolytes

71. Predict approximate freezing points of **(a)** 0.10 m glucose ($C_6H_{12}O_6$), **(b)** 0.10 m $CaCl_2$, **(c)** 0.10 m CH_3COOH, and **(d)** 0.10 m KI. Which of these predictions is probably the most accurate? Explain.

72. The text describes an isotonic solution of NaCl as being about 0.16 M, whereas an isotonic solution of glucose is about 0.31 M. Explain why the concentrations of these isotonic solutions are not the same.

73. What is the maximum van't Hoff factor that would be expected for 0.0010 mole of CaO in 1 kg of water? Explain.

74. What would be the effect on red blood cells if they were placed in **(a)** 5.5% (mass/vol) NaCl(aq), **(b)** in 0.92% (mass/vol) glucose (aq)? Explain.

75. Arrange the following aqueous solutions in order of *decreasing* freezing point, and state your reasons: **(a)** 0.15 m CH_3COOH,

(b) 0.15 m $CO(NH_2)_2$ (urea), **(c)** 0.10 m H_2SO_4, **(d)** 0.10 m $Mg(NO_3)_2$, **(e)** 0.10 m NaBr.

76. Arrange the following five solutions in the order of *increasing* boiling point elevation, ΔT_b, and state your reasons: **(a)** 0.25 m sucrose(aq), **(b)** 0.15 m KNO_3(aq), **(c)** 0.048 m $C_{10}H_8$ (naphthalene) in benzene, **(d)** 0.15 m CH_3COOH(aq), **(e)** 0.15 m H_2SO_4(aq).

77. At a barometric pressure of 744 mmHg, the boiling point of water is 99.4 °C. What approximate mass of NaCl would you add to 3.50 kg of the boiling water to raise the boiling point to 100.0 °C?

78. The text lists −21 °C as the lowest temperature at which NaCl can melt ice. What is the approximate mass percent of NaCl in an aqueous solution having this freezing point? Why is the calculation only approximate?

Colloids

79. Which 0.005 M solution would be most effective in coagulating the colloidal silica represented in Figure 12.20: KCl(aq), $MgCl_2$(aq), or $AlCl_3$(aq)? Explain.

80. Aluminum sulfate is commonly used to coagulate or precipitate colloidal suspensions of clay particles in municipal water-treatment plants. Why do you suppose aluminum sulfate is more effective than sodium chloride for this purpose?

Additional Problems

Problems marked with an * may be more challenging than others.

81. An aqueous solution with density 0.980 g/mL at 20 °C is prepared by dissolving 11.3 mL CH_3OH ($d = 0.793$ g/mL) in enough water to produce 75.0 mL of solution. What is the percent CH_3OH, expressed as **(a)** volume percent; **(b)** mass percent; **(c)** mass/volume percent; and **(d)** mole percent?

82. Methanol and water are miscible. If 50 mL of methanol is mixed with 50 mL of water, the resulting solution has a volume of 94 mL. From this information, what can be deduced about the relative strengths of methanol–methanol intermolecular forces,

water–water forces, and methanol–water forces? Do you expect the dissolving process to be exothermic or endothermic? Explain.

83. Calculate the minimum pressure, in atmospheres, needed to produce pure water from seawater by reverse osmosis at 25 °C. Use the composition of seawater given in Exercise 4.1A.

*84. The liquids water and triethylamine [$(CH_3CH_2)_3N$] are only partially miscible at 20 °C. That is, triethylamine dissolves in water to some extent, and water dissolves in triethylamine to some extent. In a mixture of 50.0 g water and 50.0 g triethylamine, 40.0 g of a phase consisting of 84.5% water and 15.5% triethylamine is obtained—a saturated solution of triethylamine in water. What is the percent by mass of water in the second phase—a saturated solution of water in triethylamine?

* **85.** A solution is prepared by mixing 25.00 mL methanol, CH_3OH, $(d = 0.791$ g/mL) and 25.00 mL water $(d = 0.998$ g/mL). The solution can be described as **(a)** the solute methanol in water or **(b)** the solute water in methanol. Explain why the molality is different depending on which of these two descriptions is used. In which case, **(a)** or **(b)**, is the molality greater? (*Hint:* An actual calculation of molalities is not required.)

* **86.** What mass of sucrose, $C_{12}H_{22}O_{11}$, must be dissolved per liter of water $(d = 0.998$ g/mL) to obtain a solution with 2.50 mole percent $C_{12}H_{22}O_{11}$?

* **87.** An aqueous solution has a freezing point of -0.28 °C. Which of the following could it be: **(a)** 0.15 M sodium chloride, **(b)** 0.15 M sucrose, **(c)** a solution that is 0.10 M NaCl and 0.05 M sucrose, **(d)** a solution that is 0.05 M NaCl and 0.05 M sucrose? Explain. If more than one of these solutions could have the observed freezing point, what additional information could be used to determine the identity of the solution?

* **88.** Nitrobenzene and benzene are completely miscible in each other. With these two liquids, it is possible to prepare two different solutions having a freezing point of 0.0 °C. What are the compositions of these two solutions, expressed as mass percent nitrobenzene? Will these two solutions also have the same boiling point? Explain. (*Hint:* Use data from Table 12.2, as needed.)

89. Example 12.2 shows that the mass percent ethanol in an ethanol–water solution is less than the volume percent. Would you expect this to be the case for all ethanol–water solutions? For all solutes in water solution? Explain.

90. Use data from Figure 12.12 to establish the solubility of air in water at 20 °C in the units mL air(STP)/100 g water. This is the unit commonly used in listing gas solubilities in handbooks. (*Hint:* Use 28.96 g/mol as the apparent molar mass of air.)

91. Consider three different solutions: saturated NaCl(aq) in contact with NaCl(s); 0.10 m NaCl(aq); and 0.20 m $(CH_3)_2CO$ $[(CH_3)_2CO$ is acetone, a liquid with a boiling point of 56.5 °C].
  (a) Which solution would you expect to have the greatest total vapor pressure? Explain.
  (b) Which solution would you expect to have the lowest freezing point? Explain.
  (c) When in an open container, which solution(s) has a total vapor pressure that changes with time, and which solution(s) has a vapor pressure that does not? Explain.

* **92.** Commercial grades of stearic acid $(C_{17}H_{35}COOH)$ usually contain some palmitic acid $(C_{15}H_{31}COOH)$ as well. A 1.115-g sample of a commercial-grade stearic acid is dissolved in 50.00 mL benzene $(d = 0.879$ g/mL), and the freezing point of the solution is found to be 5.072 °C. The freezing point of pure benzene is 5.533 °C What is the approximate mass percent palmitic acid in the stearic acid sample?

93. Given that the vapor pressures of benzene and toluene at 100.0 °C are 1351 mmHg and 556.3 mmHg, respectively, calculate, at 100 °C the mole fraction composition of **(a)** the benzene–toluene solution having a normal boiling point of 100.0 °C, and **(b)** the vapor in equilibrium with this solution.

94. The solubility of $CO_2(g)$ is 149 mg $CO_2/100$ g H_2O at 20 °C when the $CO_2(g)$ pressure is maintained at 1 atm. What is the concentration of CO_2 in water that is saturated with air at 20 °C? Express this concentration as mL CO_2 (STP)/100 g water. Use the fact that the mole percent of CO_2 in air is 0.037%.

95. An aqueous, isotonic solution can be described in terms of freezing point depression as well as osmotic pressure. It has a freezing point of -0.52 °C. Show that this definition describes isotonic NaCl(aq) reasonably well (that is, 0.92% NaCl [mass/vol]).

* **96.** Suppose that initially, solution A in Figure 12.16 is 200.0 g of an aqueous solution with a mole fraction of urea, $CO(NH_2)_2$, of 0.100 and solution B is 100.0 g of an aqueous solution with a mole fraction of urea of 0.0500. What will be the mass and mole fraction concentration of each solution when equilibrium is established, that is, when there is no longer a net transfer of water between the two solutions?

* **97.** A 1.684-g sample of an unknown oxygen derivative of a hydrocarbon yields 3.364 g CO_2 and 1.377 g H_2O upon complete combustion. A 0.605-g sample of the same compound dissolved in 34.89 g water lowers the freezing point of the water to -0.244 °C. What is the molecular formula of the compound?

* **98.** When phenol, C_6H_5OH, is dissolved in bromoform, $CHBr_3$, it is partially associated into dimers (double molecules). The freezing point depression in a solution of 2.58 g phenol in 100.0 g bromoform is 2.374 °C. The freezing point of bromoform is 8.1 °C and $K_f = 14.1$ °C m^{-1}. What fraction of the phenol is present as dimers?

99. An important test for the purity of a substance is to measure its melting point. If the sample tested melts at a lower temperature than the known melting point of the pure substance, the sample is judged to be impure. Why should we expect this to be the case? Must this always be the case? Explain.

* **100.** Refer to Exercise 12.15B. If the 10.00-g sample of a glucose–sucrose mixture lowers the freezing point of 100.0 g H_2O to -0.795 °C, what must be the mass percent composition of the solid?

101. Demonstrate that **(a)** for a *dilute aqueous* solution, the numerical value of the molality is essentially equal to that of the molarity; **(b)** in a *dilute* solution, the solute mole fraction is proportional to the molality; and **(c)** in a *dilute aqueous* solution, the solute mole fraction is proportional to the molarity.

* **102.** The photograph suggests what happens, at times, when $CaCl_2 \cdot 6 H_2O(s)$ is taken from a closed bottle and exposed to the atmosphere: the solid becomes wet with a liquid. What is this liquid? How does it form? Would you expect all solids to behave in this way? If not, what are the conditions that favor formation of a liquid?

103. Use Figure 12.10, as needed, to determine what will precipitate when 1 mol of KCl and 1 mol of $NaClO_3$ are dissolved in 200 g of water at 100 °C and the solution is cooled to 20 °C. How much precipitate will form? Assume that the solubilities in the mixture are the same as the individual solubilities.

104. The illustration suggests the successive subdivision of a cube of material, 1.0 cm on edge. First, the cube is cut in half; then, each of the two sections is cut in half; and the four sections are again cut in half. In this first subdivision, there are eight cubes, each 0.50 cm on edge. Now the subdivision into cubes is repeated many times, as suggested by the series of arrows in the bottom row. **(a)** How many successive subdivisions are required before the material enters the colloidal domain, that is, with a particle dimension less than 1000 nm? **(b)** At the point described in (a), what are the volume and total surface area of the particles of material and how do these compare with the volume and surface area of the starting material? Explain the idea that colloidal particles have a much larger surface area-to-volume ratio than does bulk matter.

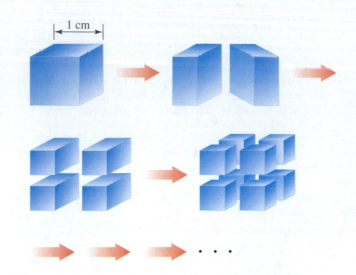

Apply Your Knowledge

105. [Laboratory] A graphite-furnace atomic absorption spectrometer can perform an analysis on a sample of only 10.0 microliters of solution. The instrument is especially sensitive to zinc; about 0.1 ppb Zn by mass can be detected. How many zinc atoms are present in such a sample?

106. [Historical] In 1912, Victor Grignard won the Nobel Prize for his discovery of a widely useful class of *organometallic compound*, today called the *Grignard reagent*. A common undergraduate preparation of a Grignard reagent involves the reaction of magnesium metal and bromobenzene in diethyl ether solvent.

$$Mg + C_6H_5Br \longrightarrow C_6H_5MgBr$$

The phenylmagnesium bromide produced is then allowed to react further to form alcohols, hydrocarbons, or carboxylic acids. In a typical student laboratory preparation, 2.5 g of magnesium and 14.9 g of bromobenzene are mixed with 175 mL of diethyl ether ($d = 0.706$ g/mL). What is the maximum molality of phenylmagnesium bromide that can be prepared?

* **107. [Laboratory]** A *weight titration* can be a convenient alternative to a volumetric titration. An ordinary squeeze bottle is used in place of a buret, and the bottle is weighed on an electronic balance before and after the titration. The unit *weight molarity* is used for such titrations:

$$M_{(wt)} = \text{moles of solute/kg of solution}$$

(a) Does the weight molarity depend on the temperature at which the solution is used? Explain.

(b) A solution is prepared by dissolving 12.654 g of NaOH in 898.3 g of water. It takes 69.26 g of this solution to titrate 3.641 g of battery acid. Calculate the weight molarity of NaOH in the titrant, and the mass percent of H_2SO_4 in the battery acid.

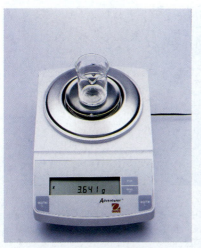

▲ Mass of battery acid

Three drops of phenolphthalein added

End point of titration

108. [Biochemical] In Chapter 4, we described a solution that is used in oral rehydration therapy (ORT) (Exercise 4.1B). Show that this solution is essentially isotonic. Use the definition of an isotonic solution given in Problem 95.

109. [Biochemical] The solubility of nitrogen gas in blood at 37 °C and at a partial pressure of 0.78 atm is 5.5×10^{-4} M. A diver with a blood volume of 5.0 L breathes compressed air with a partial pressure of nitrogen of 4.0 atm. Calculate the volume, in liters, of nitrogen gas released from the blood when the diver returns to the surface, where the partial pressure of nitrogen is 0.78 atm.

110. [Collaborative] Have your study group devise several everyday analogies to the concentration units ppm, ppb, and ppt, similar to the one discussed on page 486 (population of San Diego and of India) and the one represented by Figure 12.1(b).

111. [Biochemical] Refer to the photograph of maple syrup processing shown on page 507. Assuming the boiling point elevation is due entirely to glucose, what is the approximate molal concentration of glucose in maple syrup? It takes about 35 gallons of sap to make one gallon of maple syrup. What is the approximate molal concentration of glucose in the sap?

e-Media Problems

The activities described in these problems can be found in the e-Media Activities and Interactive Student Tutorial (IST) modules of the Companion Website, *http://chem.prenhall.com/hillpetrucci.*

112. Analogous to the process depicted in the **Solution Formation from a Solid** animation (*Section 12-2*), determine the quantity, in grams, of $CuSO_4$ needed to prepare 250 mL of a 1.0 M solution if anhydrous $CuSO_4$ were used instead of copper sulfate pentahydrate.

113. View the **Dissolution of KMnO₄** animation (*Section 12-3*). **(a)** Describe, on a molecular scale, all of the intermolecular forces that must be overcome to form the solution. Which species are involved in the different types of intermolecular interactions? **(b)** Describe, on a molecular scale, the new intermolecular interactions that arise in the dissolution process. Which species are involved?

114. **(a)** Categorize each of the solutes in the **Enthalpy of Solution** simulation (*Section 12-3*) as having an endothermic or exothermic ΔH_{soln} in water. **(b)** Which of the solutes has the most exothermic molar enthalpy of solution? (*Hint:* Consider the number of moles of solute added to similar quantities of water.)

115. Consider the data presented in the **Equilibrium Vapor Pressure** simulation (*Section 12-6*). **(a)** What is the minimum temperature at which the vapor pressure of benzene and ethanol differ by 10%? **(b)** Why would it be difficult to carry out a fractional distillation of these two liquids below this temperature?

116. Consider the addition of 3.0 g of each solute to 100 g of water in the **Boiling Point Elevation and Freezing Point Depression** simulation (*Section 12-7*). **(a)** Which of the solutes has the greatest effect on the boiling point and freezing point of water? **(b)** Is the effect greater for boiling or freezing? Why? **(c)** What solute properties account for the differences in the boiling point elevations and the freezing point depressions produced by the several 3.0-g samples?

117. View the **Semipermeable Membrane** animation (*Section 12-8*). **(a)** Describe, on a molecular scale, the similarities and differences between this process and that of the physical equilibrium (solid–liquid or liquid–gas) of a single substance. **(b)** In analogy to the animation, sketch the interface that would exist for a blood cell in a hypertonic solution. Label the cell wall, the cell interior, the identity of the species crossing the membrane, and the direction of a net flow of that species.

Chapter 13

Chemical Kinetics: Rates and Mechanisms of Chemical Reactions

AS WE HAVE seen, we can write chemical equations and use them to relate quantities of reactants and products in a chemical reaction (Chapter 3). We also have considered three important categories of reactions—acid–base, precipitation, and oxidation–reduction (Chapter 4), and we have studied the heat effects that accompany chemical reactions (Chapter 6). You might wonder what more there is to learn about chemical reactions.

Actually, there are several important questions we have yet to answer, and we consider some of them in this chapter:

- How fast does a reaction go—that is, how much reactant is consumed or how much product is formed in a given time?
- Are there ways we can speed up or slow down a chemical reaction?
- What happens at the molecular level when reactions occur—that is, what is the step-by-step mechanism that leads from reactants to products?

13.1 Chemical Kinetics—A Preview

These questions are the essence of **chemical kinetics,** the study of the rates of chemical reactions, the factors that affect these rates, and the sequences of molecular steps—called *reaction mechanisms*—by which reactions occur.

Reactions proceed at different rates. Some continue for years before the reactants are consumed. For example, the disintegration of an aluminum can by atmospheric oxidation or of a plastic bottle by the action of sunlight can take years, decades, or even centuries. Such reactions are exceedingly slow. In contrast, some reactions occur much faster than the blink of an eye. A neutralization reaction, for example, occurs just as rapidly as we can mix the acid and base. Moreover, a given reaction can take place at widely different rates depending on conditions. Iron rusts rather rapidly in a humid environment, but in a desert it corrodes so slowly that objects discarded half a century

 Rates of Change animation

◄ Luciferin (inset, opposite) is the complex, naturally occurring molecule responsible for the firefly's glow. Production of luciferin involves a series of chemical reactions. The rates at which fireflies flash and crickets chirp (as seen here and in Problem 94) are affected by temperature in much the same way that the rates of ordinary chemical reactions are affected, indicating that the underlying physiological processes are of a chemical nature.

▲ **FIGURE 13.1** **The reaction of hydrogen and oxygen to form water**
The reaction of hydrogen in a balloon with oxygen in the air is extremely rapid once the gases are ignited.

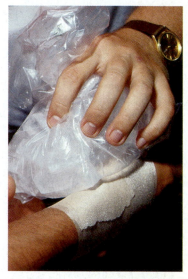

▲ Athletic injuries often are treated by applying an ice pack as soon as possible. The biochemical reactions that cause inflammation are slower at the lower temperature, thus presumably lessening damage to the tissues. After a day or so, heat packs are sometimes applied to the injured area. Perhaps the higher temperature speeds up the reactions of the healing process.

ago may show only slight signs of rust. Hydrogen and fluorine at room temperature form hydrogen fluoride in a highly exothermic and extremely rapid reaction. The reaction of hydrogen and oxygen to form water is also highly exothermic, but the reaction is immeasurably slow at room temperature. When the hydrogen–oxygen mixture is ignited at a high temperature, however, the reaction occurs with explosive speed (Figure 13.1). Reaction rates often increase dramatically as the temperature increases.

Carbon monoxide and nitrogen monoxide are both serious air pollutants that are produced in large quantities by automobile engines. Interestingly, these gases can react to form carbon dioxide and nitrogen, which are generally safer products:

$$2\ CO(g) + 2\ NO(g) \longrightarrow 2\ CO_2(g) + N_2(g)$$

Unfortunately, this reaction is extremely slow, even at high temperatures. However, we can speed up the reaction by using a catalyst, and some automobile catalytic converters do just that. A *catalyst* is a substance that speeds up a reaction but emerges from the reaction unchanged. Obviously, a catalyst is not just a "spectator" (like the spectator ions in an ionic reaction). Later in the chapter, we will consider how catalysts work, including the action of enzymes, perhaps the most important catalysts of all.

Two goals of chemical kinetics that we present in the first several sections of this chapter are to *measure* and to *predict* the rates of chemical reactions. Thus, atmospheric scientists can use experiments in laboratory smog chambers to acquire enough data to predict pollution levels during smog episodes. Scientists also use reaction rates to establish plausible mechanisms for chemical reactions. For example, scientists used data on the rates of a variety of atmospheric reactions to develop a plausible mechanism for the ozone depletion observed in the stratosphere in polar regions. This mechanism indicates that chlorine atoms from chlorofluorocarbons (CFCs) act as catalysts in ozone destruction. Confidence in this proposed mechanism has led the nations of the world to take measures to reduce and eventually eliminate their use of CFCs.

If we can predict reaction rates, often we can also control them. We can speed up or slow down reactions by adjusting certain variables, such as

- *Concentrations of reactants.* Reaction rates generally increase as the concentrations of the reactants are increased.
- *Temperature.* Reaction rates generally increase rapidly as the temperature is increased and decrease as the temperature is lowered.
- *Surface area.* For reactions that occur on a surface rather than in a solution, the rate increases as the surface area is increased (Figure 13.2).
- *Catalysis.* Catalysts speed up reactions, and *inhibitors* slow them down.

◄ **FIGURE 13.2** **The effect of surface area on reaction rate**
Finely divided and dispersed flour dust undergoes rapid combustion because of the large surface area on which the combustion reaction occurs. This swift flaming of flour has been responsible for many flour mill explosions.

13.2   The Meaning of Reaction Rate

Rate, or *speed,* refers to how much something changes in a unit of time. Consider a runner moving at a speed of 16 km/h. The runner's position changes by 16 km in one hour. In a chemical reaction, the change is in the concentration of a reactant or product, which we can express as moles per liter $(mol\ L^{-1})$, which is equal to molarity (M). Thus, the rate of reaction has the units *moles per liter per (unit of) time.* If we choose 1 s as the unit of time, the rate of reaction is expressed either as *mol L^{-1} s^{-1}* or as *M s^{-1}*.

Consider the reaction in which sucrose is broken down into the simpler sugars glucose and fructose. The reaction, called the inversion of sucrose, takes place in an aqueous solution with H^+ as a catalyst:

$$C_{12}H_{22}O_{11} + H_2O \xrightarrow{\ H^+\ } C_6H_{12}O_6 + C_6H_{12}O_6$$

$$\text{Sucrose} \qquad\qquad\qquad \text{Glucose} \quad\ \text{Fructose}$$

Application Note

Honeybees use sucrose inversion to produce honey, which is a mixture of sucrose, glucose, and fructose. The bees use the enzyme *invertase* as a catalyst.

We can express the rate of reaction as the change in concentration of the reactant, sucrose, in a unit of time, or we can use the change in concentration of either of the products. As is customary in chemical kinetics, we will use the bracket notation introduced in Chapter 4 to represent the concentration of a reactant or product. For example, we represent the concentration of sucrose as [sucrose].

Suppose [sucrose]$_1$ is the molarity of sucrose at an initial time t_1 and [sucrose]$_2$ is the molarity at a later time t_2. The change in concentration is Δ[sucrose] = [sucrose]$_2$ − [sucrose]$_1$. The change in time is $\Delta t = t_2 - t_1$. Then the rate at which sucrose is consumed—that is, the rate at which sucrose disappears—can be expressed by the equation

$$\text{Rate of disappearance of sucrose} = \frac{\text{change in concentration of sucrose}}{\text{change in time}}$$

$$= \frac{[\text{sucrose}]_2 - [\text{sucrose}]_1}{t_2 - t_1} = \frac{\Delta[\text{sucrose}]}{\Delta t}$$

Because sucrose is *consumed* in the reaction, [sucrose]$_2$ is smaller than [sucrose]$_1$ and Δ[sucrose] is a negative quantity. By convention, we express rates of reactions as positive quantities, and so the rate of the reaction is the negative of the rate of sucrose disappearance:

$$\text{Rate} = -\text{rate of disappearance of sucrose} = -\frac{\Delta[\text{sucrose}]}{\Delta t}$$

Now let's consider glucose, a product of the reaction. The concentration of glucose increases with time, and therefore [glucose]$_2$ is larger than [glucose]$_1$, making the rate of formation of glucose a positive quantity:

$$\text{Rate} = \text{rate of formation of glucose} = \frac{\Delta[\text{glucose}]}{\Delta t}$$

To summarize, the **rate of a reaction** is the change in concentration of a product per unit of time (rate of formation of product) or the *negative* of the change in concentration of a reactant per unit of time (−rate of disappearance of reactant). If, for example, we start with 0.100 M sucrose and find that two hours (7200 s) later the concentration of sucrose is 0.080 M, we can determine that the average rate of the reaction over this two-hour period was $-(-0.020\ \text{M}/7200\ \text{s}) = 2.8 \times 10^{-6}\ \text{M}\ \text{s}^{-1}$.

The General Rate of Reaction

Now let us consider the rate of the reaction by which the common household antiseptic hydrogen peroxide, $H_2O_2(aq)$, decomposes:

$$2\ H_2O_2(aq) \longrightarrow 2\ H_2O(l) + O_2(g)$$

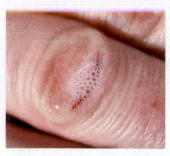

▲ Household hydrogen peroxide decomposes rapidly on a cut because of a catalyst in the blood.

We can base the rate on either the disappearance of H_2O_2 or the formation of O_2.

$$\text{Rate} = -\text{rate of disappearance of } H_2O_2 = -\frac{\Delta[H_2O_2]}{\Delta t}$$

$$\text{Rate} = \text{rate of formation of } O_2 = \frac{\Delta[O_2]}{\Delta t}$$

However, these two rates are not the same. From the balanced equation, we note that 2 mol of H_2O_2 are consumed to produce 1 mol of O_2, which means H_2O_2 disappears twice as fast as O_2 forms. Thus, the reaction rate based on the disappearance of H_2O_2 is twice the rate based on the formation of O_2. This relationship is illustrated diagrammatically in Figure 13.3. It seems that if we want to describe a rate of reaction, we must do so in terms of a specific reactant or product.

A better alternative is to define a *general* rate of reaction that has the same value regardless of which reactant or product we study. In the decomposition of hydrogen peroxide, to say that H_2O_2 disappears twice as fast as O_2 forms is the same as saying that O_2 forms only half as fast as H_2O_2 disappears:

$$\text{Rate} = -\frac{1}{2}\frac{\Delta[H_2O_2]}{\Delta t} = \frac{\Delta[O_2]}{\Delta t}$$

This expression is called the *general rate of reaction* for the reaction

$$2\,H_2O_2(aq) \longrightarrow 2\,H_2O(l) + O_2(g)$$

The general rate of reaction is obtained by dividing the rate of disappearance of a reactant or the rate of formation of a product by the stoichiometric coefficient of that reactant or product in the balanced chemical equation. For example, if the rate of disappearance of H_2O_2 is found to be $-2 \times 10^{-5}\,M\,s^{-1}$, the rate of formation of O_2 will be $1 \times 10^{-5}\,M\,s^{-1}$. The rate of reaction will be either $-\frac{1}{2}(-2 \times 10^{-5})\,M\,s^{-1}$ or $\frac{1}{1}(1 \times 10^{-5})\,M\,s^{-1}$. That is, the general rate of reaction $= 1 \times 10^{-5}\,M\,s^{-1}$.

Applied to the more general reaction

$$a\,A + b\,B \longrightarrow c\,C + d\,D$$

the general rate of reaction is

$$\text{Rate} = -\frac{1}{a}\frac{\Delta[A]}{\Delta t} = -\frac{1}{b}\frac{\Delta[B]}{\Delta t} = \frac{1}{c}\frac{\Delta[C]}{\Delta t} = \frac{1}{d}\frac{\Delta[D]}{\Delta t} \qquad (13.1)$$

Kinetics and Stoichiometry animation

In a reaction rate study, we usually monitor the concentration of a particular reactant or product. In this text, we will generally write chemical equations so that the general rate of reaction is the same as the rate based on the reactant or product being monitored. For example, in the decomposition of hydrogen peroxide, we will monitor

▶ **FIGURE 13.3 Stoichiometry and rate of decomposition of H_2O_2**

The coefficients in a balanced equation determine the relationship between the rates of decomposition of reactants and rates of formation of products. In this example, one H_2O molecule is formed for each H_2O_2 molecule that decomposes, and so the rates of formation of H_2O and of decomposition of H_2O_2 are numerically equal.

QUESTION: When H_2O_2 decomposes at the rate of 1 M per minute, how fast is O_2 being formed?

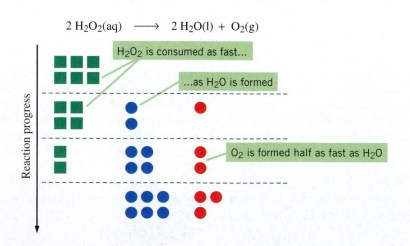

$$2\,H_2O_2(aq) \longrightarrow 2\,H_2O(l) + O_2(g)$$

H_2O_2 is consumed as fast...

...as H_2O is formed

O_2 is formed half as fast as H_2O

Reaction progress

the concentration of H_2O_2 as a function of time, and we will therefore write the chemical equation in a form that shows 1 mol H_2O_2:

$$H_2O_2(aq) \longrightarrow H_2O(l) + \tfrac{1}{2} O_2(g)$$

We can then express the rate of reaction in the form

$$\text{Rate} = -\frac{\Delta[H_2O_2]}{\Delta t} \qquad (13.2)$$

The Average Rate of Reaction

The decomposition of $H_2O_2(aq)$ starts off rapidly, but the rate slows down as more and more of the reactant decomposes. For this reason, when we use the expression rate $= -\Delta[H_2O_2]/\Delta t$, what we calculate is an *average* rate of reaction during a time interval Δt. At the beginning of the interval, the rate is faster than this average rate; and at the end of the interval, it is slower. The situation is rather like taking your foot off the accelerator when driving an automobile at 50 mph and coming to a stop at a red light. Your average speed in this time interval may be 25 mph, but the actual speed in the interval would decrease through the range from 50 mph to 0 mph.

We calculate an average rate of reaction in Example 13.1 and illustrate different ways of expressing reaction rates in Exercises 13.1A and 13.1B. We will determine more exact rates of reaction in the next section.

Example 13.1

Consider the hypothetical reaction

$$A + 2B \longrightarrow 3C + 2D$$

Suppose that at one point in the reaction, $[A] = 0.4658$ M, and 125 s later $[A] = 0.4282$ M. During this time period, what is the average **(a)** rate of reaction expressed in M s^{-1} and **(b)** rate of formation of C, expressed in M $\min^{-1}$.

STRATEGY

(a) The relevant data are $t_1 = 0$ s, $[A]_1 = 0.4658$ M and $t_2 = 125$ s, $[A]_2 = 0.4282$ M. We can use these data to determine the rate of disappearance of A and then relate the general rate of reaction to the rate of disappearance of A.

(b) Here we must use stoichiometric coefficients from the equation for the hypothetical reaction to relate the rate of formation of C to the rate of disappearance of A found in part (a).

SOLUTION

(a) The rate of reaction is equal to the negative of the rate of disappearance of A. The rate of disappearance of A is equal to the difference in concentrations at times 1 and 2, divided by $\Delta t = t_2 - t_1$. We substitute the relevant data into Equation (13.1).

$$\text{Rate} = -\text{rate of disappearance of A}$$

$$= -\left(\frac{[A]_2 - [A]_1}{\Delta t}\right)$$

$$= -\left(\frac{0.4282 \text{ M} - 0.4658 \text{ M}}{125 \text{ s} - 0 \text{ s}}\right) = -\left(\frac{-0.0376 \text{ M}}{125 \text{ s}}\right)$$

$$\text{Rate} = 3.01 \times 10^{-4} \text{ M s}^{-1}$$

(b) Three moles of C are formed for every mole of A that disappears. To express the rate in M $\min^{-1}$, we use the conversion factor 60 s/1 min.

$$\text{Rate of formation of C} = \frac{3 \text{ mol C}}{1 \text{ mol A}} \times 3.01 \times 10^{-4} \frac{\text{mol A}}{\text{L s}}$$

$$= 9.03 \times 10^{-4} \text{ M s}^{-1}$$

$$\text{Rate of formation of C} = 9.03 \times 10^{-4} \text{ M s}^{-1} \times \frac{60 \text{ s}}{1 \text{ min}}$$

$$= 5.42 \times 10^{-2} \text{ M min}^{-1}$$

ASSESSMENT

Note that an alternative approach to obtaining the rate of formation of C in part (b) is to use Equation (13.1) directly. Because the rate of reaction $= \tfrac{1}{3}$ (rate of formation of C), the rate of formation of C $= 3 \times$ (rate of reaction), that is, $3 \times 3.01 \times 10^{-4}$ M s^{-1} $= 9.03 \times 10^{-4}$ M s^{-1}.

EXERCISE 13.1A

Consider the hypothetical reaction

$$2 A + B \longrightarrow 2 C + D$$

Suppose that at some point during the reaction [D] = 0.2885 M and that 2.55 min (that is, 2 min, 33 s) later [D] = 0.3546 M. **(a)** What is the average rate of reaction during this time period, expressed in M min^{-1}? **(b)** What is the rate of formation of C, expressed in M s^{-1}?

EXERCISE 13.1B

In the reaction 2 A + B $\longrightarrow$ 3 C + D, $-\Delta[A]/\Delta t$ is 2.10×10^{-5} M s^{-1}. Using the idea of a general rate of reaction, determine **(a)** the rate of reaction and **(b)** the rate of formation of C.

13.3 Measuring Reaction Rates

Over a period of time, the 3% aqueous solution of hydrogen peroxide that many of us keep in our medicine cabinets loses its effectiveness because the H_2O_2 decomposes to oxygen gas and water:

$$H_2O_2(aq) \longrightarrow H_2O(l) + \tfrac{1}{2} O_2(g)$$

Figure 13.4 suggests a simple method of determining the rate at which $H_2O_2(aq)$ decomposes: Allow $O_2(g)$ to escape and determine the mass of the reaction mixture at various times. For example, the difference in mass for the 60-s time interval between Figure 13.4a and 13.4b is the mass of oxygen produced, 2.960 g. Some typical data are given in Table 13.1. A plot of the molarities of H_2O_2 from Table 13.1 at the corresponding times gives the black curve in Figure 13.5. The red, blue, and dashed green lines convey information about the rate of decomposition of H_2O_2 that we explore next.

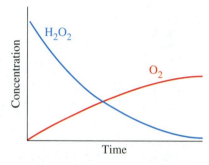

▲ As hydrogen peroxide decomposes and its concentration decreases, the rate of decomposition decreases. Oxygen gas concentration increases as the hydrogen peroxide concentration decreases.

- *In general, the greater the concentration of a reactant, the faster the reaction goes.* We can confirm this statement by comparing data points in Table 13.1 or by inspecting the black curve in Figure 13.5. For example, in the first 60 s, $[H_2O_2]$ drops by 0.185 M (from 0.882 to 0.697 M). In contrast, in the 60-s interval from 540 to 600 s, $[H_2O_2]$ drops by only 0.026 M (from 0.120 to 0.094 M).

- The average rate of reaction during the experiment (from $t = 0$ to $t = 600$ s) is the *negative* of the slope of the dashed green line. That is, the slope of the line is

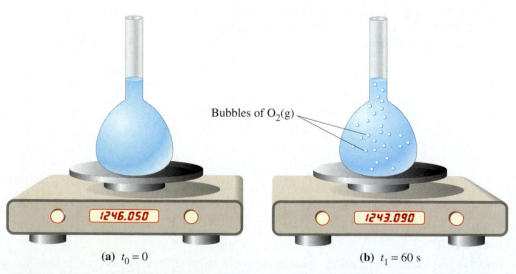

(a) $t_0 = 0$ **(b)** $t_1 = 60$ s

▲ **FIGURE 13.4**   **An experimental setup for determining the rate of decomposition of H_2O_2**

The difference between (a) the initial mass and (b) the mass at a later time t_1 is the mass of $O_2(g)$ that was produced in the time interval. Because the reaction would otherwise occur too slowly, $I^-(aq)$ has been added as a catalyst. The mechanism by which I^- catalyzes the reaction is discussed on page 553.

Table 13.1 Decomposition of H_2O_2[a]

Time, s	Accumulated Mass O_2, g	$[H_2O_2]$, M[b]
0	0	0.882
60	2.960	0.697
120	5.056	0.566
180	6.784	0.458
240	8.160	0.372
300	9.344	0.298
360	10.336	0.236
420	11.104	0.188
480	11.680	0.152
540	12.192	0.120
600	12.608	0.094

[a] The decomposition of H_2O_2 in 1.00 L of 0.882 M H_2O_2, with I^- as the catalyst.
[b] Values of $[H_2O_2]$ are calculated in Problem 91.

the average rate of disappearance of H_2O_2, and we change the sign to get a positive rate of reaction.

$$\text{Average rate} = -\text{average rate of disappearance of } H_2O_2 = -\frac{(0.094 - 0.882)\ M}{(600 - 0)\ s}$$

$$= 1.31 \times 10^{-3}\ M\ s^{-1}$$

- Because the rate constantly decreases during the reaction, to get a precise value at a particular point in the reaction, we must use a very short time interval in the rate calculation. The rate of reaction at a given time is called an **instantaneous rate of reaction:**

$$\text{Instantaneous rate} = -\text{instantaneous rate of disappearance of } H_2O_2 = -\frac{\Delta[H_2O_2]}{\Delta t}$$

(where Δt is very small, that is, $\Delta t \longrightarrow 0$).

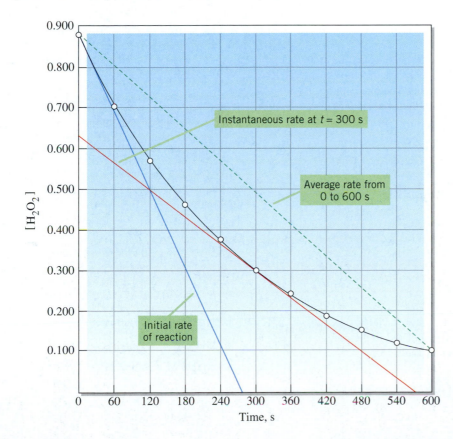

Reaction Rates activity

◀ **FIGURE 13.5** **Kinetic data for the reaction** $H_2O_2(aq) \longrightarrow H_2O(l) + \frac{1}{2} O_2(g)$ Reaction rates are obtained from the slopes of the straight lines: an average rate from the dashed green line, the instantaneous rate at $t = 300$ s from the red line, and the initial rate from the blue line.

QUESTION: Over what time interval are the instantaneous rates greater than the average rate measured for the entire 600-s period?

Alternatively, we can use the red line, a tangent line to the curved black concentration-time graph. The slope of the tangent is the instantaneous rate of disappearance of H_2O_2, and the instantaneous reaction rate is the *negative of the slope of the tangent*.

- The instantaneous rate at the beginning of a reaction is called the **initial rate of reaction**. The initial rate is the negative of the slope of the blue tangent line in Figure 13.5. Note that the blue line and the black curve essentially coincide in the earliest stages of the reaction. At the very start of the reaction, the average rate taken over a short time interval and the instantaneous rate are practically the same.

- When interpreting a plot of the concentration of a *product* versus time, all the preceding points still apply, with one important exception. Average, instantaneous, and initial rates are equal to the slopes of tangent lines, with no change of sign required because the slopes of the tangent lines are positive.

Example 13.2

Use data from Table 13.1 and/or Figure 13.5 to **(a)** determine the initial rate of reaction and **(b)** calculate $[H_2O_2]$ at $t = 30$ s.

STRATEGY

(a) Because the tangent line and concentration-versus-time curve coincide at the start of a reaction, we can determine the initial rate in two ways: (1) Substitute concentration and time data for $t = 0$ s and $t = 60$ s into Equation (13.2), or (2) obtain the rate from the slope of the blue tangent line in Figure 13.5.

(b) We will apply Equation (13.2) to the initial rate of the reaction, solving for the change in concentration, $\Delta[H_2O_2]$. It is the difference between $[H_2O_2]$ at $t = 30$ s and $[H_2O_2]$ at $t = 0$ s.

SOLUTION

(a) *Method one.* Let us use the first two data points in Table 13.1.

$$\text{Initial rate} = -\frac{\Delta[H_2O_2]}{\Delta t}$$

$$= -\frac{(0.697 \text{ M} - 0.882 \text{ M})}{(60 - 0) \text{ s}}$$

$$= 3.08 \times 10^{-3} \text{ M s}^{-1}$$

Method two. To obtain the initial rate from the blue tangent line in Figure 13.5, we need the slope of the line. Let us base the slope on the points where the line intersects the two axes. The intercept on the y-axis is 0.882 M at 0 s. The intercept on the x-axis is 0 M at an estimated time of 275 s.

$$\text{Slope of tangent} = \frac{y_2 - y_1}{x_2 - x_1} = \frac{(0 - 0.882) \text{ M}}{(275 - 0) \text{ s}}$$

$$= -3.21 \times 10^{-3} \text{ M s}^{-1}$$

$$\text{Initial rate} = -\text{slope of tangent line}$$

$$= -(-3.21 \times 10^{-3} \text{ M s}^{-1})$$

$$= 3.21 \times 10^{-3} \text{ M s}^{-1}$$

(b) Now we can substitute the initial rate from (a) and $\Delta t = 30$ s into Equation (13.2) and solve for $\Delta[H_2O_2]$.

$$\text{Initial rate} = -\frac{\Delta[H_2O_2]}{\Delta t}$$

$$3.21 \times 10^{-3} \text{ M s}^{-1} = -\frac{\Delta[H_2O_2]}{30 \text{ s}}$$

$$\Delta[H_2O_2] = -3.21 \times 10^{-3} \text{ M s}^{-1} \times 30 \text{ s}$$

$$= -0.096 \text{ M}$$

From the expression for $\Delta[H_2O_2]$, we can solve for $[H_2O_2]_{(30\text{s})}$. We then substitute the value of $\Delta[H_2O_2]$ calculated in the previous step and the value of $[H_2O_2]_{(0\text{s})}$ given in Table 13.1.

$$\Delta[H_2O_2] = [H_2O_2]_{(30\text{s})} - [H_2O_2]_{(0\text{s})}$$

$$[H_2O_2]_{(30\text{s})} = \Delta[H_2O_2] + [H_2O_2]_{(0\text{s})}$$

$$= -0.096 \text{ M} + 0.882 \text{ M} = 0.786 \text{ M}$$

ASSESSMENT

(a) The initial rates agree rather well between the two methods, but method (2) based on the slope of the tangent line is likely to be the better one. The blue and black lines in Figure 13.5 begin to diverge before $t = 60$ s.

(b) A calculation of the type employed here is most reliable when only a small percentage of the initial reactant has been consumed (say, a few percent). Here the decline in $[H_2O_2]$ is from 0.882 M to 0.786 M, about 11%, somewhat beyond the range of maximum reliability.

EXERCISE 13.2A

From Figure 13.5, (a) determine the instantaneous rate of reaction at $t = 300$ s. (b) Use the result of (a) to calculate a value of $[H_2O_2]$ at $t = 310$ s.

EXERCISE 13.2B

Estimate the time at which the instantaneous rate of reaction is the same as the average rate given by the dashed line in Figure 13.5. (*Hint:* Draw a tangent line that has the same slope as the dashed line. Is more than one tangent possible?)

13.4 The Rate Law of a Chemical Reaction

In Section 13.3, we used experimental data to obtain reaction rates during the decomposition of H_2O_2. We used both calculation and a graphical method (that is, from the slope of a tangent line), but there is a serious limitation: Our results applied only to a solution that was initially 0.882 M H_2O_2. What if we want to know the initial rates for the decomposition of 0.225 M H_2O_2 or 0.500 M H_2O_2 or some other concentration of H_2O_2? Do we have to do an experiment with each of these solutions to gather the necessary data?

On page 528 we noted that, in general, the greater the reactant concentrations, the faster a reaction goes. This suggests that the initial rates for 0.225 M H_2O_2 and 0.500 M H_2O_2 aqueous solutions should both be smaller than the one in Example 13.2 because both of these solutions are less concentrated than 0.882 M H_2O_2. We can deal with this and other matters more definitively by using the *rate law* for the decomposition of H_2O_2.

The Rate Law and Its Meaning

The **rate law** for a chemical reaction relates the reaction rate to the concentrations of reactants. Consider again a general reaction of the type

$$a\,A + b\,B \cdots \longrightarrow c\,C + d\,D \cdots$$

where the dots suggest that there may be additional reactants and products. We can express the general reaction rate in terms of the rate at which the reactants disappear:

$$\text{Rate} = -\frac{1}{a}\frac{\Delta[A]}{\Delta t} = -\frac{1}{b}\frac{\Delta[B]}{\Delta t} = \cdots \tag{13.3}$$

 **Rate Law Expressions activity**

The rate law for this reaction is of the form

$$\text{Rate} = k[A]^m[B]^n \cdots \tag{13.4}$$

where [A], [B], $\cdots$ are molarities of the reactants at a particular time and the exponents $m, n, \cdots$ are generally small positive integers (0, 1, 2), but they may be negative or occasionally nonintegral.

The exponents in a rate law must be determined by <u>experiment</u>. *They are not derived from the stoichiometric coefficients in an overall chemical equation, though in some instances they may be the same as those coefficients.*

It is important to note that the exponents in a rate law can be determined only by experiment.

The values of the exponents in a rate law establish the **order of a reaction.** If $m = 1$, the reaction is *first order* in A. If $n = 2$, the reaction is *second order* in B, and so on. The *overall order* of a reaction is the sum of the exponents in a rate law: $m + n + \cdots$.

The proportionality constant k is the **rate constant.** The numerical value of k depends on (1) the particular reaction, (2) the temperature, and (3) the presence of a catalyst (if any). As we will see shortly, the units of k depend on the values of the exponents $m, n, \cdots$.

Now consider again the decomposition of H_2O_2. The rate law is simple; the reaction is first order in H_2O_2:

$$\text{Rate} = k[H_2O_2]^1 = k[H_2O_2]$$

Shortly we will see how to determine this order from experimental data. For now, we will simply accept that the reaction is first order and that it remains first order, regardless of the set of stoichiometric coefficients we use for the balanced equation.

We usually write the rate law in the form shown above with rate alone on the left side of the equation, but we can also solve it for another quantity. For example, we can use an experimentally measured rate and the corresponding concentration(s) to calculate the rate constant k. In Example 13.2, we found the initial rate to be 3.21×10^{-3} M s^{-1} when $[H_2O_2] = 0.882$ M, and we can calculate the rate constant from those values:

$$k = \frac{\text{rate}}{[H_2O_2]} = \frac{3.21 \times 10^{-3} \text{ M s}^{-1}}{0.882 \text{ M}} = 3.64 \times 10^{-3} \text{ s}^{-1}$$

We expect the unit s^{-1} for k in a first-order reaction. The product of s^{-1} and the molarity unit M is M s^{-1}, the units for a reaction rate.

With the rate law and a value of k, we can calculate instantaneous rates of reaction for any given concentration. For instance, the instantaneous rate of the H_2O_2 decomposition reaction when the reactant concentration is 0.775 M H_2O_2(aq) is

$$\text{Rate} = k[H_2O_2] = 3.64 \times 10^{-3} \text{ s}^{-1} \times 0.775 \text{ M} = 2.82 \times 10^{-3} \text{ M s}^{-1}$$

More About the Rate Constant, k

We just confirmed that s^{-1} is an appropriate unit for k for a first-order reaction. What are the units for reactions of other orders? For the hypothetical zero-order reaction, A $\longrightarrow$ B, the rate law is

$$\text{Rate} = k[A]^0 = k \qquad (13.5)$$

Because any quantity raised to the zero power is equal to one, the rate of a zero-order reaction must be equal to the rate constant k. Thus, for zero-order reactions, the rate constant has the same units as a rate of reaction, M s^{-1}.

The reaction of hydrogen and iodine monochloride to produce iodine and hydrogen chloride is first order in H_2, first order in ICl, and second order overall:

$$H_2(g) + 2 \text{ ICl}(g) \longrightarrow I_2(g) + 2 \text{ HCl}(g)$$

$$\text{Rate} = k[H_2][\text{ICl}]$$

Note that the units of k for a second-order reaction must be M^{-1} s^{-1} so that the product of units in the rate law gives the expected units of M s^{-1} for the reaction rate:

$$\text{Rate} = \underbrace{k}_{\text{M}^{-1}\text{s}^{-1}} \times \underbrace{[H_2]}_{\text{M}} \times \underbrace{[\text{ICl}]}_{\text{M}} = \text{M s}^{-1}$$

From these examples and the table in the margin, we see that the unit of k for a reaction of any overall order is M$^{(1-\text{overall order})}$ s^{-1}.

Sometimes the distinction between the *rate* of a reaction and the *rate constant* of a reaction can be confusing. It helps to remember these points:

- The two terms have quite different meanings. The *rate* of a reaction gives the change in a concentration with time, whereas the *rate constant* is the proportionality constant that relates the rate of a reaction to the concentrations of reactants.

- The rate constant remains *constant* throughout a reaction and has the same value regardless of the initial concentrations of the reactants. Except for zero-order reactions, the rate of a reaction *varies* as concentrations vary.

Note that in this reaction there is not a one-to-one correspondence between the stoichiometric coefficients and the exponents in the rate law. This is often the case.

Overall Reaction Order	Units for k
Zero	M s^{-1}
First	s^{-1}
Second	M^{-1} s^{-1}
Third	M^{-2} s^{-1}

• The rate of reaction and the rate constant of the reaction have the same numerical values and units *only* in zero-order reactions. For reaction orders other than zero, the rate and rate constant can be numerically equal *only* when the concentrations of all reactants are 1 M. Even then, their units *remain* different.

The Method of Initial Rates

Key to determining the rate law for a reaction is to devise experiments that establish the exponents in the rate law. One way to do this is to determine the initial rate for different initial concentrations of the reactants, a procedure called the **method of initial rates.** This involves a series of experiments in which the initial concentrations of some reactants are held constant and others are varied in convenient multiples. The initial rate of the reaction, the instantaneous rate, and the average rate are all equal at the very start of a reaction. Also, we know the concentrations of reactants with certainty at the beginning of the reaction. If there is any tendency for a reverse reaction to occur, that is, for products to react to re-form the reactants, the effect is minimal at the very start of the reaction. We illustrate the method using the reaction

$$2 \, NO(g) + Cl_2(g) \longrightarrow 2 \, NOCl(g)$$

Table 13.2 Initial Rates of the Reaction $2 \, NO(g) + Cl_2(g) \longrightarrow 2 \, NOCl(g)$

Experiment	Initial [NO]	Initial [Cl_2]	Initial Rate, M s^{-1}
1	0.0125 M	0.0255 M	2.27×10^{-5}
2	0.0125 M	0.0510 M	4.55×10^{-5}
3	0.0250 M	0.0255 M	9.08×10^{-5}

Table 13.2 lists data for three experiments. Each experiment differs from the others only in the initial concentration of one reactant. For example, the initial [Cl_2] is the same for Experiments 1 and 3, but the initial [NO] in Experiment 3 is twice that of Experiment 1. In contrast, the initial rate in Experiment 3 is four times that of Experiment 1. Let's put these facts in mathematical form by writing the rate law for each experiment and obtaining a ratio of the two initial rates. In the following expression, the subscript numbers 1 and 3 refer to Experiments 1 and 3, respectively. Quantities that are the same for each experiment cancel.

$$\frac{(\text{Initial rate})_3}{(\text{Initial rate})_1} = \frac{\cancel{k}[NO]_3^m [Cl_2]_3^n}{\cancel{k}[NO]_1^m [Cl_2]_1^n} = \frac{(2 \times 0.0125)^m}{(0.0125)^m} = \frac{2^m \times \cancel{(0.0125)^m}}{\cancel{(0.0125)^m}} = 2^m$$

Thus, the ratio of the initial rates is

$$\frac{(\text{Initial rate})_3}{(\text{Initial rate})_1} = 2^m$$

However, we can also calculate the actual value of this ratio:

$$\frac{(\text{Initial Rate})_3}{(\text{Initial Rate})_1} = \frac{9.08 \times 10^{-5} \, \cancel{M \, s^{-1}}}{2.27 \times 10^{-5} \, \cancel{M \, s^{-1}}} = 4.00$$

The left sides of the two expressions we have just written are the same, and therefore so are the right sides: $2^m = 4.00$. The value of $m = 2$, that is, $2^2 = 4$, and thus the reaction is second order in NO.

We can determine the exponent n by comparing Experiments 1 and 2:

$$\frac{(\text{Initial rate})_2}{(\text{Initial rate})_1} = \frac{k[NO]_2^m [Cl_2]_2^n}{k[NO]_1^m [Cl_2]_1^n} = \frac{(2 \times 0.0255)^n}{(0.0255)^n}$$

$$= \frac{2^n \times (0.0255)^n}{(0.0255)^n} = 2^n = \frac{4.55 \times 10^{-5} \, M \, s^{-1}}{2.27 \times 10^{-5} \, M \, s^{-1}} = 2$$

We see that $2^n = 2$, that is, $2^n = 2^1$. Thus $n = 1$; the reaction is first order in Cl_2.

This is an example of a reaction in which—by coincidence—reaction orders and stoichiometric coefficients coincide.

The rate law for the reaction is third order overall $(m + n = 2 + 1 = 3)$:

$$\text{Rate} = k[NO]^2[Cl_2]$$

We will obtain the value of k for this reaction in Example 13.3.

The following list summarizes the effects on the initial rate caused by doubling the concentration of one reactant while the concentrations of the others are held constant. If the reaction is

- *Zero* order in the reactant—the initial rate is *unchanged*
- *First* order in the reactant—the initial rate *doubles*
- *Second* order in the reactant—the initial rate *quadruples*
- *Third* order in the reactant—the initial rate increases *eightfold*

Example 13.3

For the reaction $2\,NO(g) + Cl_2(g) \longrightarrow 2\,NOCl(g)$ described in the text and in Table 13.2, **(a)** what is the initial rate for a hypothetical Experiment 4, which has [NO] = 0.0500 M and [Cl$_2$] = 0.0255 M? **(b)** What is the value of k for the reaction?

STRATEGY

We established the rate law for the reaction as rate $= k[NO]^2[Cl_2]$. We can most easily answer part (a) by using this rate law and comparing the initial rates in Experiments 3 and 4. To answer part (b), we can use the initial rate and initial concentrations for any of the four experiments to solve the rate law for k.

SOLUTION

(a) Obtaining the Experiment 3 data from Table 13.2, we compare the values of [Cl$_2$] and [NO$_2$] in Experiment 4 to the values in Experiment 3.

Experiment 4	**Experiment 3**
[NO] = 0.0500 M	0.0250 M
[Cl$_2$] = 0.0255 M	0.0255 M

In going from Experiment 3 to Experiment 4, the value of [Cl$_2$] remains the same and that of [NO] doubles. Because the reaction is second order in NO, doubling its initial concentration causes a *fourfold* increase in the initial rate.

$$(\text{Initial rate})_4 = 4 \times (\text{initial rate})_3$$
$$= 4 \times 9.08 \times 10^{-5}\ \text{M s}^{-1}$$
$$= 3.63 \times 10^{-4}\ \text{M s}^{-1}$$

(b) We can solve the rate law for k and then substitute data from Table 13.2, for example, for Experiment 1.

$$k = \frac{(\text{initial rate})_1}{[NO]^2[Cl_2]}$$
$$= \frac{2.27 \times 10^{-5}\ \text{M s}^{-1}}{(0.0125\ \text{M})^2(0.0255\ \text{M})}$$
$$= 5.70\ \text{M}^{-2}\ \text{s}^{-1}$$

ASSESSMENT

(a) Let's compare other data in Table 13.2 to the new Experiment 4 to see if we get the same value of k. For instance, Experiment 4/Experiment 2 leads to a fourfold increase in [NO] and a reduction by one-half in [Cl$_2$]. The increase in [NO] should lead to a 16-fold (4^2) increase in the reaction rate, and the reduction in [Cl$_2$] to a decrease by a factor of one-half. The rate should increase eightfold $\left(16 \times \frac{1}{2}\right)$, to a value of $8 \times 4.55 \times 10^{-5}\ \text{M s}^{-1} = 3.64 \times 10^{-4}\ \text{M s}^{-1}$, in good agreement with the answer we obtained.

(b) We can use other data points from Table 13.2, such as from Experiment 2, for which we find that $k = 4.55 \times 10^{-5}\ \text{M s}^{-1}/ (0.0125\ \text{M})^2(0.0510\ \text{M}) = 5.71\ \text{M}^{-2}\ \text{s}^{-1}$, again in good agreement with the answer we obtained.

EXERCISE 13.3A

Refer to Example 13.3, and calculate the initial rate if the initial concentrations are [NO] = 0.200 M and [Cl$_2$] = 0.400 M. (*Hint:* Recall that we now know a value of k.)

EXERCISE 13.3B

In the thermal decomposition of acetaldehyde,

$$CH_3CHO(g) \longrightarrow CH_4(g) + CO(g)$$

the rate of reaction increases by a factor of about 2.8 when the initial concentration of acetaldehyde is doubled. What is the order of the reaction?

13.5 First-Order Reactions

In this section, we focus our discussion on **first-order reactions** in which a single reactant yields products—that is, reactions of the type A ⟶ products.

The rate law for such a reaction is simply

$$\text{Rate} = k[A]^1 = k[A] \qquad (13.6)$$

Concentration as a Function of Time: The Integrated Rate Law

In the preceding section, we noted that decomposition of $H_2O_2(aq)$ is a first-order reaction obeying the rate law rate $= k[H_2O_2]$, and we used this rate law to evaluate k and to calculate the rate of decomposition at various concentrations. However, such calculations will not tell us how long 3% $H_2O_2(aq)$ can be stored before decomposition progresses to the point at which the solution loses its antiseptic qualities. In this and many other practical situations, what we want to know is

How can we determine what the concentration of a reactant will be at a later time if we know its concentration initially?

We can answer this question by using an equation derived from the rate law. The **integrated rate law** is an equation that describes the concentration of a reactant as a function of time. The form of the equation depends on the overall order of the reaction. For a first-order reaction, the integrated rate law has the form

$$\ln \frac{[A]_t}{[A]_0} = -kt \qquad (13.7)$$

In this equation, k has the same meaning as in the rate law; it is the rate constant. The elapsed time in the reaction is represented by t. The concentrations of A are $[A]_t$ at time t and $[A]_0$ at $t = 0$. The initial concentration is $[A]_0$. We express the concentration terms as a ratio, and we take the natural logarithm of the ratio, denoted by the symbol ln.*

This integrated rate law for a first-order reaction is the equation of a *straight line*. We can get the more familiar form of the equation for a straight line by replacing ln $[A]_t/[A]_0$ by the equivalent term $\ln [A]_t - \ln [A]_0$ and solving for $\ln[A]_t$:

$$\ln [A]_t = -kt + \ln [A]_0 \qquad (13.8)$$

When Equation (13.8) is compared with the general equation for a straight line, $y = mx + b$, we see that y corresponds to $\ln[A]_t$, m to $-k$, x to t, and b to $\ln[A]_0$. If we plot $\ln[A]_t$ against time t and the result is a straight line, the reaction is first order and the slope of the line is the negative of k. If the plot is not a straight line, the reaction is not first order.

Data for the first-order decomposition of H_2O_2 from Table 13.1 are presented again in Table 13.3, this time also listing $\ln [H_2O_2]$. The straight-line plot of $\ln [H_2O_2]$ versus time in Figure 13.6 indicates that the reaction is first order. We can

In the language of calculus, a rate law is called a *differential* equation. The equation is solved by a technique called *integration*, which we apply to a first-order rate law in Appendix A.

Refer to Appendix A for a review of the characteristics of logarithms.

First-Order Process animation

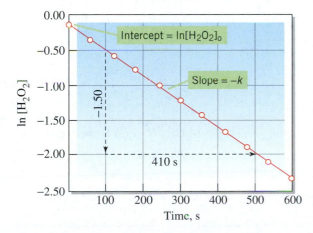

◀ **FIGURE 13.6** **Test for a first-order reaction: Decomposition of $H_2O_2(aq)$**

The data plotted are from Table 13.3. The straight-line plot of $\ln[H_2O_2]$ versus time proves the reaction to be first order.

QUESTION: At approximately what time in this reaction would $[H_2O_2] = 0.600\ M$?

Table 13.3 Decomposition of H_2O_2: Data Required to Test for a First-Order Reaction		
Time, s	**$[H_2O_2]$, M**	**$\ln [H_2O_2]$**
0	0.882	−0.126
60	0.697	−0.361
120	0.566	−0.569
180	0.458	−0.781
240	0.372	−0.989
300	0.298	−1.21
360	0.236	−1.44
420	0.188	−1.67
480	0.152	−1.88
540	0.120	−2.12
600	0.094	−2.36

*The units of molarity cancel in the numerator and denominator of the ratio $[A]_t/[A]_0$, yielding a dimensionless quantity. We can take the logarithms only of pure numbers.

derive the value of the rate constant k from the slope (m) of the straight-line plot as follows:

$$m = (y_2 - y_1)/(x_2 - x_1) = -1.50/410 \text{ s} = -3.66 \times 10^{-3} \text{ s}^{-1}$$

$$k = -(\text{slope}) = -(-3.66 \times 10^{-3} \text{ s}^{-1}) = 3.66 \times 10^{-3} \text{ s}^{-1}$$

Example 13.4 and Exercise 13.4A illustrate two types of calculations that are possible with the integrated rate law for a first-order reaction. Exercise 13.4B applies the integrated rate law in a broader context.

Example 13.4

For the first-order decomposition of $H_2O_2(aq)$, given $k = 3.66 \times 10^{-3} \text{ s}^{-1}$ and $[H_2O_2]_0 = 0.882$ M, determine **(a)** the time at which $[H_2O_2] = 0.600$ M and **(b)** $[H_2O_2]$ after 225 s.

STRATEGY

We must use the *integrated* rate law for this first-order reaction, that is, Equation (13.7). In each part, we know three of the four quantities in the equation and can solve for the fourth. In part (a), two concentrations and the rate constant are known and we solve for the time, t. In part (b), the rate constant and time are known and we solve for the ratio $\ln [A]_t/[A]_0$. We know $[A]_0$, and $[A]_t$ is the final result we seek.

SOLUTION

(a) Let's evaluate the left side of Equation (13.7), using the two known concentrations and then solve for t.

$$\ln \frac{0.600 \text{ M}}{0.882 \text{ M}} = -3.66 \times 10^{-3} \text{ s}^{-1} \times t$$

$$\ln 0.680 = -0.386 = -3.66 \times 10^{-3} \text{ s}^{-1} \times t$$

$$t = \frac{-0.386}{-3.66 \times 10^{-3} \text{ s}^{-1}} = 105 \text{ s}$$

(b) We now know the quantities k and t on the right side of the integrated rate law and so can solve for the left side.

$$\ln \frac{[H_2O_2]_t}{[H_2O_2]_0} = -kt$$

$$= -\{3.66 \times 10^{-3} \text{ s}^{-1} \times 225 \text{ s}\}$$

$$= -0.824$$

Next we seek the number whose natural logarithm is -0.824. This number is equal to the ratio of concentration terms.

$$\frac{[H_2O_2]_t}{[H_2O_2]_0} = e^{-0.824} = 0.439$$

Finally, we can use the known value of $[H_2O_2]_0$ to calculate $[H_2O_2]_t$.

$$[H_2O_2]_t = 0.439 \times [H_2O_2]_0$$
$$= 0.439 \times 0.882 \text{ M}$$
$$= 0.387 \text{ M}$$

ASSESSMENT

Here is a way to check the answer to part (b): In part (a), we found that it took 105 s for $[H_2O_2]$ to fall from 0.885 M to 0.600 M. Suppose we follow the reaction for another 120 s from this point (that is, 105 s + 120 s = 225 s). When we redo the calculation in part (b) to determine $[H_2O_2]_t$ at 120 s with $[H_2O_2]_0 = 0.600$ M, we get the same answer as previously: 0.387 M.

$$\ln[H_2O_2]_t/0.600 \text{ M} = -3.66 \times 10^{-3} \text{ s}^{-1} \times 120 \text{ s} = -0.439$$

$$e^{-0.439} = 0.645 \quad \text{and} \quad [H_2O_2]_t = 0.645 \times 0.600 \text{ M} = 0.387 \text{ M}$$

EXERCISE 13.4A

The decomposition of nitramide, NH_2NO_2, is a first-order reaction:

$$NH_2NO_2(aq) \longrightarrow H_2O(l) + N_2O(g)$$

The rate law is rate $= k[NH_2NO_2]$, with $k = 5.62 \times 10^{-3} \text{ min}^{-1}$ at 15 °C. Starting with 0.105 M NH_2NO_2, **(a)** at what time will $[NH_2NO_2] = 0.0250$ M and **(b)** what is $[NH_2NO_2]$ after 6.00 h?

EXERCISE 13.4B

Refer to the reaction in Exercise 13.4A. Starting with 0.0750 M NH_2NO_2, what is the rate of the reaction after 35.0 min? (*Hint:* You will need to use both the rate law and the integrated rate law.)

Variations of the Integrated Rate Law

At times, it is convenient to replace molarities in an integrated rate law by quantities that *are proportional to concentration*. For example, if we multiply the molarity (M) by the volume of a reaction mixture (V), we get the number of moles (n) of reactant. Further multiplication by the molar mass gives the mass of the reactant. Thus, mass is proportional to molarity.

In gas-phase reactions, it is easy to measure pressures. In a constant-volume mixture at a constant temperature, the partial pressure of a gas is proportional to its molarity. We can see this by rearranging the ideal gas equation, $PV = nRT$, to $P = (n/V) \times RT$. Replace n/V by the molarity, M, and, because R and T are constants, we get $P = \text{constant} \times \text{M}$.

We will make use of these variations of the integrated rate law next.

Half-life of a Reaction

We often describe a first-order reaction in terms of the time required to reduce a reactant concentration (or mass or partial pressure) to a *fraction* of its initial value. The convenient fraction one-half is used to define a *half-life*. The **half-life ($t_{1/2}$)** of a reaction is the time in which one-half of the reactant originally present is consumed. At this time in the first-order reaction A $\longrightarrow$ products, we say that $[A]_t = \frac{1}{2}[A]_0$, or $(P_A)_t = \frac{1}{2}(P_A)_0$ or $(m_A)_t = \frac{1}{2}(m_A)_0$, depending on whether we are measuring molarity, partial pressure, or simply mass of the reactant.

To show the relationship between half-life and rate constant for a first-order reaction, let us begin with the integrated rate law (Equation 13.7) and substitute the values $[A]_{t_{1/2}}$ for $[A]_t$ and $t_{1/2}$ for t.

$$\ln \frac{[A]_{t_{1/2}}}{[A]_0} = \ln \frac{\frac{1}{2}\cancel{[A]_0}}{\cancel{[A]_0}} = -kt_{1/2}$$

We can cancel $[A]_0$ from both numerator and denominator, leading to the simple equation

$$\ln \left(\tfrac{1}{2}\right) = -kt_{1/2}$$

Next, we solve for $t_{1/2}$ and also take $\ln \left(\frac{1}{2}\right)$ (that is, $\ln 0.500$).

$$t_{1/2} = -\frac{\ln \left(\tfrac{1}{2}\right)}{k} = -\frac{-0.693}{k} = \frac{0.693}{k} \tag{13.9}$$

We can summarize the relationship as follows:

For a __first-order__ reaction, the half-life is a constant; it depends only on the rate constant, k, and __not__ on the concentration of reactant. If k is known, $t_{1/2}$ can be calculated, and if $t_{1/2}$ is known, k can be calculated.

This summary statement holds only for a *first-order* reaction; in reactions of other orders, $t_{1/2}$ does depend on concentration.

The idea of a half-life can be extended throughout a reaction. That is, if the concentration of reactant is reduced to $\left(\frac{1}{2}\right)$ its initial value in $t_{1/2}$, it is reduced to $\left(\frac{1}{2}\right) \times \left(\frac{1}{2}\right) = \left(\frac{1}{4}\right)$ in the time $2 \times t_{1/2}$, and to $\left(\frac{1}{2}\right) \times \left(\frac{1}{2}\right) \times \left(\frac{1}{2}\right) = \left(\frac{1}{8}\right)$ in $3 \times t_{1/2}$ and so on. After n half-lives, $\left(\frac{1}{2}\right)^n$ of the initial concentration remains.

The half-life concept for a first-order reaction is illustrated in Figure 13.7 and Example 13.5.

Problem-Solving Note

To obtain ln 0.680 using a simple calculator in part (a) of Example 13.4, enter 0.680, followed by the ln key; the value in the display, −0.385662, is the natural logarithm of 0.680. Finding the number having −0.824 as its natural logarithm in part (b) means evaluating $e^{-0.824}$. That is, a number N and $\ln N$ are related as follows: $N = e^{\ln N}$. Enter −0.824 into your calculator, followed by the e^x key (or the INV key, then the ln key); the value displayed, 0.43867, is equal to $e^{-0.824}$. If you use a graphing calculator, a slightly different procedure may be required.

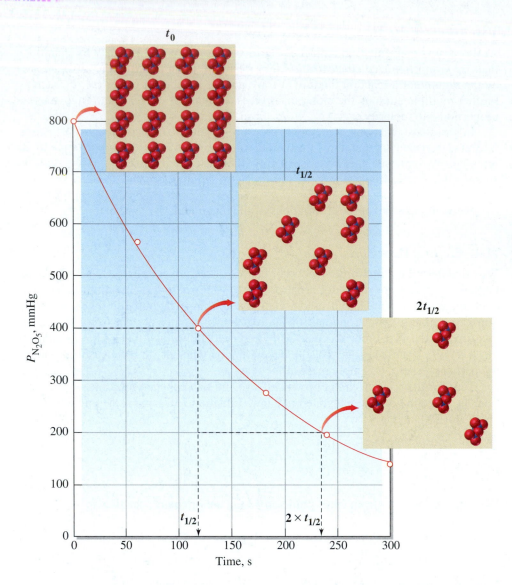

▶ **FIGURE 13.7** **Decomposition of N_2O_5 at 67 °C**

The significance of the time periods $t_{1/2}$ and $2 \times t_{1/2}$ is discussed in the text and illustrated in Example 13.6.

QUESTION: If the plot were extended to the point $4t_{1/2}$, how many molecules of N_2O_5 should be shown at this point to be consistent with the rest of the figure?

Example 13.5   A Conceptual Example

Use data from Figure 13.7 to evaluate the **(a)** half-life and **(b)** rate constant for the first-order decomposition of N_2O_5 at 67 °C:

$$N_2O_5(g) \longrightarrow 2\,NO_2(g) + \tfrac{1}{2}\,O_2(g)$$

Problem-Solving Note

You do not need to start at $t = 0$ in a graph such as Figure 13.7 when applying the half-life concept. The difference in time between *any* two points that differ in molarity or pressure by a factor of 2 is a half-life, $t_{1/2}$. For example, between $P = 600$ mmHg at about 50 s and $P = 300$ mmHg at about 170 s, one half-life has elapsed (about 120 s).

ANALYSIS AND CONCLUSIONS

(a) When the molarity of N_2O_5 drops to one-half its initial value, $P_{N_2O_5}$ also drops to one-half its initial value because partial pressure and molarity are proportional. As the pressure drops from 800 mmHg to 400 mmHg, one-half the N_2O_5 originally present is consumed. This occurs in about 120 s, thus $t_{1/2} \approx 120$ s. The pressure drops to 200 mmHg at about 240 s: $2 \times t_{1/2} \approx 240$ s. Again, $t_{1/2} \approx 120$ s, confirming that the decomposition of N_2O_5 is first order.

(b) We determine the rate constant k using $t = 120$ s in Equation (13.9).

$$k = \frac{0.693}{t_{1/2}} = \frac{0.693}{120\ \text{s}} = 5.8 \times 10^{-3}\ \text{s}^{-1}$$

Example 13.6   An Estimation Example

Which seems like a probable approximate time for 90% of a sample of N_2O_5 to undergo decomposition at 67 °C: (a) 200 s, (b) 300 s, (c) 400 s, or (d) 500 s?

ANALYSIS AND CONCLUSIONS

We could, of course, use the expression $\ln P_t/P_0 = -kt$ and substitute 0.10 for the ratio P_t/P_0. (When 90% of the reactant is consumed, only 10% remains.) However, we can make a reasonable estimate by using the half-life concept. The remaining fraction of N_2O_5 that we are looking for is 0.10. The fraction left after one half-life is 0.50; after two half-lives, 0.25; after three, 0.125; and after four, 0.0625. The time we are seeking is between three and four half-lives. If we use a half-life of about 120 s, which we found in Example 13.5, we see that our answer is

$$3 \times 120 \text{ s} < t < 4 \times 120 \text{ s}$$

$$360 \text{ s} < t < 480 \text{ s}$$

Of the values we have to choose from, only (c) is probable: 400 s.

EXERCISE 13.6A

Without doing detailed calculations or extrapolating the graph, estimate $P_{N_2O_5}$ at 475 s in the graph of Figure 13.7.

EXERCISE 13.6B

The decomposition of di-*t*-butyl peroxide (DTBP) into acetone and ethane is a first-order reaction.

$$C_8H_{18}O_2(g) \longrightarrow 2 \text{ CH}_3\text{COCH}_3(g) + CH_3CH_3(g)$$

$$\quad\text{DTBP}\qquad\qquad\quad\text{Acetone}\qquad\qquad\text{Ethane}$$

In a reaction carried out at 147 °C, the partial pressure of DTBP is found to be 320 mmHg after 80.0 min and 80 mmHg after 240 min. *Without doing detailed calculations*, determine the *initial* partial pressure of DTBP.

13.6   Reactions of Other Orders

The reactions that we will consider in this section are zero order and second order.

Zero-Order Reactions

It may seem strange at first that the rates of some reactions are independent of the concentrations of the reactants. However, sometimes factors other than reactant concentrations control how fast reactants can enter into a reaction. For example, in a reaction that requires the absorption of light, the intensity of light determines the reaction rate. In a surface-catalyzed reaction, the available surface area determines the rate. In the rate law for a reaction that is **zero order** overall, the sum of the exponents is zero: $m + n + \cdots = 0$.

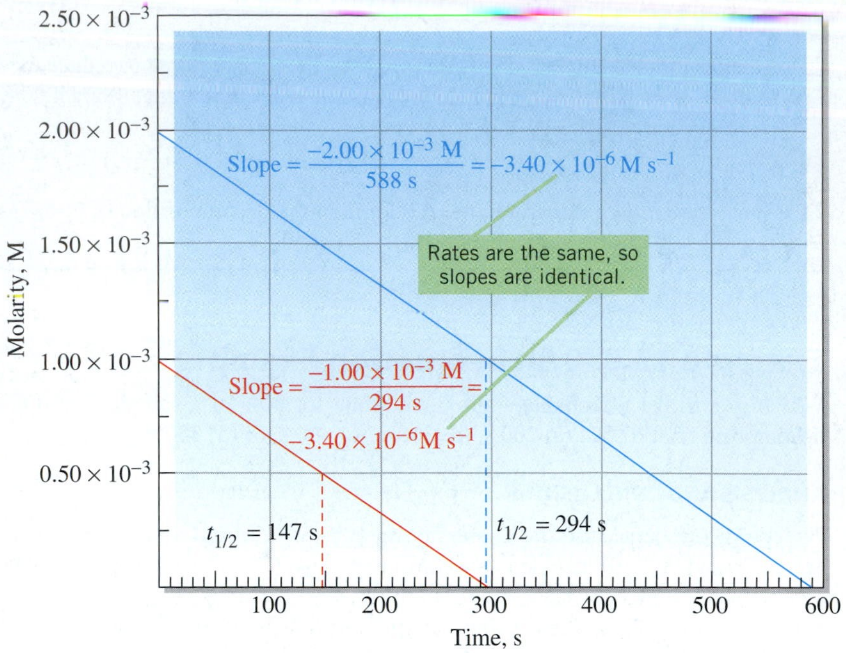

▶ **FIGURE 13.8** **The decomposition of ammonia on a tungsten surface at 1100 °C, a zero-order reaction**

The concentration of $NH_3(g)$ is plotted as a function of time for two initial concentrations, 2.00×10^{-3} M (blue) and 1.00×10^{-3} M (red). The parallel, straight-line graphs show this to be a zero-order reaction. The slopes of the lines are indicated, as are the half-lives for the two experiments.

The reaction by which ammonia, $NH_3(g)$, decomposes on a tungsten (W) surface at 1100 °C is zero order:

$$NH_3(g) \xrightarrow{\text{W}} \tfrac{1}{2} N_2(g) + \tfrac{3}{2} H_2(g)$$

$$\text{Rate} = k[NH_3]^0 = k$$

In Figure 13.8, the concentration of $NH_3(g)$ is plotted as a function of time for two initial concentrations: 2.00×10^{-3} M (blue) and 1.00×10^{-3} M (red). Note the following:

- The graph of concentration versus time for each experiment is a straight line with a negative slope. The blue and red lines are *parallel* to each other and thus have the same slope: -3.40×10^{-6} M s^{-1}.

Plotting Kinetic Data simulation

- The rate of reaction remains constant throughout and is equal to the rate constant k and to the *negative* of the slope:

$$\text{Rate} = k = -(-3.40 \times 10^{-6} \text{ M s}^{-1}) = 3.40 \times 10^{-6} \text{ M s}^{-1}.$$

- Because the reaction rate has the same value at all points in a zero-order reaction, the rate is also independent of the initial concentration of the reactant.

- Expressed in the form, $y = mx + b$, the integrated rate law for a zero-order reaction is $[A]_t = -kt + [A]_0$. When the reaction is half completed, $[A]_t = \tfrac{1}{2}[A]_0$, and $t_{1/2} = [A]_0/2k$. Thus, the half-life is directly proportional to the initial concentration, as we see in Figure 13.8, where the half-life doubles as the initial concentration is doubled.

Second-Order Reactions

A **second-order reaction** has a rate law in which the sum of the exponents is 2: $m + n + \cdots = 2$. An example is the reaction of $NO(g)$ and $O_3(g)$.

$$NO(g) + O_3(g) \longrightarrow NO_2(g) + O_2(g)$$

$$\text{Rate} = k[NO][O_3]$$

This reaction is first order in NO, first order in O_3, and second order overall.

A simpler possibility for a second-order reaction that is sometimes seen is a reaction involving a single reactant: $A \longrightarrow$ products.

$$\text{Rate} = k[A]^2 \tag{13.10}$$

The integrated rate law that expresses [A] as a function of time for this type of second-order reaction has the form

$$\frac{1}{[A]_t} = kt + \frac{1}{[A]_0} \tag{13.11}$$

From this equation, we see that a graph of $1/[A]$ versus time is a straight line. The slope of the straight line is equal to the rate constant, k, and the intercept $(t = 0)$ is $1/[A]_0$. We can obtain the half-life by substituting $[A]_t = \frac{1}{2}[A]_0$ into Equation (13.11) and simplifying:

A brief derivation of this integrated rate law is given in Appendix A.

$$t_{1/2} = \frac{1}{k[A]_0} \tag{13.12}$$

As with a zero-order reaction, *the half-life of a second-order reaction depends on the initial concentration as well as on the rate constant, k.*

Example 13.7

The second-order decomposition of HI(g) at 700 K is represented in Figure 13.9.

$$HI(g) \longrightarrow \tfrac{1}{2} H_2(g) + \tfrac{1}{2} I_2(g) \qquad Rate = k[HI]^2$$

What are the **(a)** rate constant and **(b)** half-life of the decomposition of 1.00 M HI(g) at 700 K?

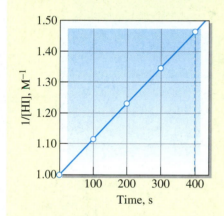

◀ **FIGURE 13.9   The decomposition of hydrogen iodide at 700 K, a second-order reaction**

The *reciprocal* of the concentration of HI(g), that is, $1/[HI]$, is plotted as a function of time. The rate constant for this second-order reaction is equal to the slope of the line: $k = 1.2 \times 10^{-3} \, M^{-1} \, s^{-1}$.

STRATEGY

In this problem, $[HI]_0 = 1.00$ M. This makes $1/[HI]_0 = 1.00 \, M^{-1}$ at $t = 0$ s. For a second-order reaction, we can obtain k from the slope of the straight-line graph of $1/[A]_t$ versus t, and $t_{1/2}$ from the equation $t_{1/2} = 1/k[A]_0$.

SOLUTION

(a) We can base the slope of the line in Figure 13.9 on any two of the points. Here we choose the first and last of the recorded data points.

$$(1.46 \, M^{-1}, 400 \, s) \quad \text{and} \quad (1.00 \, M^{-1}, 0 \, s)$$

In the equation for the slope, m is equal to $\Delta y / \Delta x$, where Δy is the difference in concentration and Δx is the difference in time.

$$\Delta y = 1.46 \, M^{-1} - 1.00 \, M^{-1} = 0.46 \, M^{-1}$$
$$\Delta x = 400 \, s - 0 \, s = 400 \, s$$

The rate constant, k, is *equal* to the slope.

$$k = \text{slope} = \frac{\Delta y}{\Delta x}$$

$$= \frac{0.46 \, M^{-1}}{400 \, s} = 1.2 \times 10^{-3} \, M^{-1} \, s^{-1}$$

(b) We now have values of $[HI]_0$ and k needed to solve for the half-life using Equation (13.12).

$$t_{1/2} = \frac{1}{k[A]_0} = \frac{1}{1.2 \times 10^{-3} \, M^{-1} \, s^{-1} \times 1.00 \, M}$$

$$= 8.3 \times 10^2 \, s$$

ASSESSMENT

We can verify the likely correctness of the answers by examining the units. In part (a), we obtained the same units for k as established on page 536 for a second-order reaction. In part (b), the equation for a half-life yielded a unit of time (s), which it must do regardless of the order of the reaction.

EXERCISE 13.7A

If, in the second-order reaction: A $\longrightarrow$ products, it takes 55 s for the concentration of reactant A to fall to 0.40 M from an initial concentration of 0.80 M, what is the rate constant k for the reaction?

EXERCISE 13.7B

In the reaction described in Exercise 13.7A, how long would it take for [A] to fall to 0.20 M? To 0.10 M? Describe how the half-life varies from one half-life period to the next in the reaction.

Example 13.8   A Conceptual Example

Shown here are graphs of [A] versus time for two different experiments dealing with the reaction A $\longrightarrow$ products. What is the order of this reaction?

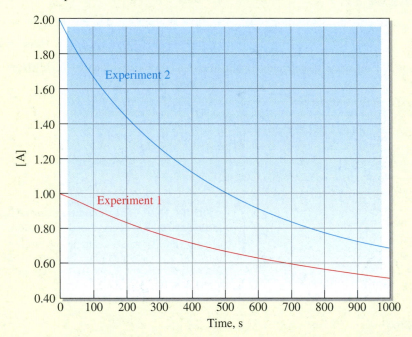

Problem-Solving Note

We cannot deduce the order of a reaction from the form of the chemical equation alone. A reaction of the type A $\longrightarrow$ products might be of any order, including a nonintegral one. We can establish the order only by experiment.

ANALYSIS AND CONCLUSIONS

Our task is to find the value of the exponent m in the rate law

$$\text{Rate} = k[A]^m$$

To assist in our task, let's refer to Table 13.4, a compilation of facts about zero-, first-, and second-order reactions of the type: A $\longrightarrow$ products.

Table 13.4 A Summary of Kinetic Data for Reactions of the Type: A $\longrightarrow$ products

Order	Rate Law	Integrated Rate Law	Straight-Line Plot	$k =$	Units of k	Half-life
0	Rate = k	$[A]_t = -kt + [A]_0$	[A] vs. t	$-$slope	M s^{-1}	$\dfrac{[A]_0}{2k}$
1	Rate = $k[A]$	$\ln \dfrac{[A]_t}{[A]_0} = -kt$	ln [A] vs. t	$-$slope	s^{-1}	$\dfrac{0.693}{k}$
2	Rate = $k[A]^2$	$\dfrac{1}{[A]_t} = kt + \dfrac{1}{[A]_0}$	1/[A] vs. t	slope	M^{-1} s^{-1}	$\dfrac{1}{k[A]_0}$

Test for Zero-Order Reaction   If the reaction were zero order, the rate would be constant (rate $= k$), and the graphs would be straight lines. The reaction is not zero order.

Test for First-Order Reaction   One way to test for first order is to see if the half-life is constant. In Experiment 1, [A] falls from 1.00 M to one-half this value, 0.50 M, in 1000 s. In Experiment 2, [A] falls from 2.00 M to 1.00 M in 500 s. The half-life is not constant; the reaction is not first order.

Test for Second-Order Reaction   There are three methods that we might use: (1) Obtain the initial reaction rates, either from the slopes of tangent lines or by evaluating $-(\Delta[A]/\Delta t)$, and then apply the method of initial rates. (2) Determine if the graph of 1/[A] versus time is a straight line. (3) Obtain the value of k from the half-life in each experiment, using the equation $t_{1/2} = 1/k[A]_0$. If the value of k is the same for the two, the reaction is second order.

Let's use the third method and begin by relating k to $t_{1/2}$, using the half-lives we found when applying the test for a first-order reaction.

$$k = \frac{1}{t_{1/2}[A]_0}$$

Experiment 1

When $[A]_0 = 1.00$ M, $t_{1/2} = 1000$ s: $k = \dfrac{1}{1000 \text{ s} \times 1.00 \text{ M}} = 1.00 \times 10^{-3} \text{ M}^{-1} \text{ s}^{-1}$

Experiment 2

When $[A]_0 = 2.00$ M, $t_{1/2} = 500$ s: $k = \dfrac{1}{500 \text{ s} \times 2.00 \text{ M}} = 1.00 \times 10^{-3} \text{ M}^{-1} \text{ s}^{-1}$

The two values of k are the same, indicating that the reaction is second order.

EXERCISE 13.8A

Use data from the graphs in Example 13.8 to show that the reaction is second order, by using the first two of the three methods outlined.

EXERCISE 13.8B

Consider two hypothetical reactions, A $\longrightarrow$ products: One is a first-order reaction with $k = 1.00 \times 10^{-3}$ s^{-1} and an initial concentration, $[A]_0 = 2.00$ M. The other is a second-order reaction with $k = 1.00 \times 10^{-3}$ M^{-1} s^{-1} and an initial concentration, $[A]_0 = 1.00$ M. Will graphs of [A] as a function of time have any common points for these two reactions? If so, at approximately what time(s)?

13.7   Theories of Chemical Kinetics

So far, we have explored some practical matters related to the rates of chemical reactions, none of which required us to look at what happens at the molecular level. However, important questions remain: Why are some reactions first order, others second order, and so on? Why are some reactions so fast and others so slow? How do catalysts work? To answer such questions, we need to shift our focus to the molecular level. We will do so for much of the rest of the chapter.

Collision Theory

Before atoms, molecules, or ions can react with one another, they must first come together, or *collide*. However, if all collisions between molecules produced a chemical reaction, reaction rates would be much greater than what we generally observe. From this observation, we can conclude that some of the collisions must be ineffective. The reaction rate therefore is proportional to both the frequency of molecular collisions and the fraction of these collisions that are effective ones.

An effective collision between two molecules puts enough energy into certain key bonds to break them. Such a collision may occur between two fast-moving molecules or perhaps when an especially fast molecule collides with a slower one. A collision between two slow-moving molecules will most likely fail to break bonds.

Distribution of Molcular Speeds simulation

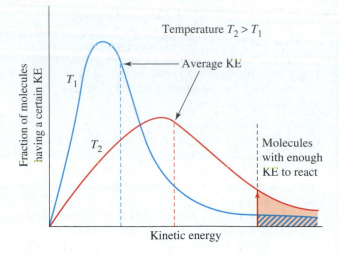

Temperature $T_2 > T_1$

Average KE

T_1

T_2

Molecules with enough KE to react

Fraction of molecules having a certain KE

Kinetic energy

▶ **FIGURE 13.10**
Distribution of kinetic energies of molecules

The fraction of molecules having kinetic energy in excess of the value marked by the red arrow is much smaller than the total number of molecules. However, this fraction increases rapidly with temperature, corresponding to the blue crosshatched area at T_1 and the red-shaded areas at T_2.

The **activation energy** (E_a) is the minimum energy that must be supplied by collisions in order for a reaction to occur.

The kinetic-molecular theory allows us to calculate the fraction of all the molecules in a collection that possess a certain kinetic energy. Figure 13.10 represents the situation at two temperatures. (Figure 5.17, in which the fraction of all the molecules is plotted as a function of molecular speeds, is similar.) As the graph shows, the average kinetic energy at T_1, the lower temperature, is less than the average kinetic energy at T_2. Expressed another way, the fraction of low-kinetic-energy molecules in the sample at T_1 is larger than the fraction of low-kinetic-energy molecules in the sample at T_2. If we assume that in order to react, molecules require kinetic energies in excess of the value marked by the red arrow in Figure 13.10, we are led to two conclusions:

- Only a small fraction of the molecules at either temperature are energetic enough to react.
- This energetic fraction becomes larger as the temperature increases.

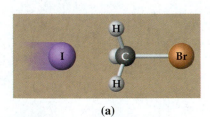

▲ **FIGURE 13.11   A reaction in which the orientation of colliding molecules is unimportant**

No matter from which direction an incoming hydrogen atom (blue) approaches the target hydrogen atom (red), the approaching atom's "view" of the impending collision is the same. We conclude that there is no preferred direction for one hydrogen atom to approach another when they react to form a hydrogen molecule.

In many cases, we cannot account for a reaction rate just by considering the collision frequency and the fraction of activated species because the *orientation* of the colliding species also affects the rate. One case where orientation is not a factor is shown in Figure 13.11. Because of the symmetrical distribution of the electron cloud of a hydrogen atom, every approach of one H atom to another (front, back, top, bottom, and so forth) is the same. As a result, the orientation of the colliding atoms is not a factor in the rate of the reaction

$$H\cdot \ + \ \cdot H \longrightarrow H_2$$

In most cases, however, the orientation of the colliding species is a factor. Consider the reaction of iodide ion with methyl bromide to produce methyl iodide and bromide ion:

$$I^- + CH_3Br \longrightarrow CH_3I + Br^-$$

As shown in Figure 13.12, a collision between an I^- ion and the C atom of CH_3Br can lead to a reaction, but a collision between an I^- ion and the Br atom of CH_3Br cannot.

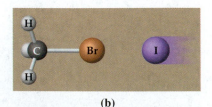

▲ **FIGURE 13.12   The importance of orientation of colliding molecules**

(a) The I^- ion collides with the C atom of CH_3Br, a collision that can lead to reaction. (b) The I^- ion collides with the Br atom of CH_3Br, a collision unlikely to lead to reaction.

Transition State Theory

We can picture a chemical reaction as more than just the result of a collision. We can imagine, in slow motion, the gradual breaking of bonds in the reactants during the collision, followed by the formation of bonds in the products. However, during the collision there is, so to speak, a point of no return. Before this point is reached, the colliding species rebound from each other unchanged. Beyond this point, the colliding species proceed to products. The configuration of the atoms of the colliding species at this crucial point is called the **transition state**, and the transitory species having this configuration is called the **activated complex**.

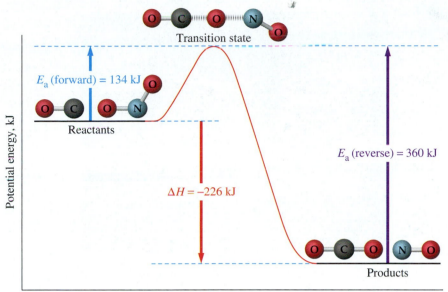

Transition state

E_a (forward) = 134 kJ

Reactants

E_a (reverse) = 360 kJ

$\Delta H = -226$ kJ

Products

Potential energy, kJ

Progress of reaction

◀ **FIGURE 13.13   A reaction profile for the reaction** $CO(g) + NO_2(g) \longrightarrow CO_2(g) + NO(g)$ This reaction profile follows energy changes during the course of the reaction. From left to right, we proceed from reactants through the transition state to products.

 **Bimolecular Reaction animation**

For the reaction in which iodide ion displaces bromide ion from methyl bromide, we can represent the progress of the reaction in this way:

$$I^- + CH_3{-}Br \longrightarrow \overset{\delta^-}{I}\cdots\cdots CH_3 \cdots\cdots \overset{\delta^-}{Br} \longrightarrow I{-}CH_3 + Br^-$$

Reactants   Activated complex   Products

In the reactants, there is no bond between the I^- ion and the C atom and a full bond between the C and Br atoms. In the activated complex, a partial bond ($\cdots\cdots$) is formed between the I and C atoms, the bond between the C and Br atoms is partially broken ($\cdots\cdots$), and a unit of negative charge is divided into partial charges (δ^-) on the I and Br atoms. When the activated complex proceeds to products, there is a full bond between the I and C atoms and no bond between the C atom and the Br^- ion.

Another way of thinking about a reaction is illustrated in Figure 13.13. This representation, called a **reaction profile**, shows potential energy plotted as a function of a parameter called the progress of the reaction. (Think of the progress of the reaction as a representation of how far the reaction has gone. That is, the reaction begins with reactants on the left, passes through a transition state, and ends with products on the right.) This reaction profile is for the reaction

$$CO(g) + NO_2(g) \longrightarrow CO_2(g) + NO(g) \qquad \Delta H = -226 \text{ kJ}$$

at temperatures above 600 K. The difference between the potential energies of the products and the potential energies of the reactants is ΔH for the reaction,* which is exothermic. An energy barrier separates the products from the reactants, and reactant molecules must have enough energy to surmount this barrier if a reaction is to occur.

In transition state theory, the activation energy (E_a) is the difference between the potential energy of the transition state at the top of the barrier and the potential energy of the reactants. The value of E_a (forward) in Figure 13.13 signifies that, collectively, 1 mol of CO molecules and 1 mol of NO_2 molecules must bring 134 kJ of energy into their collisions in order to form one mole of activated complex. The activated complex then dissociates into product molecules.

Figure 13.13 also describes the reverse reaction, in which CO_2 and NO react to form CO and NO_2. The activation energy of the reverse reaction, E_a(reverse) = 360 kJ, is greater than that of the forward reaction.

---

*The difference in potential energy between reactants and products is ΔU of a reaction. For this reaction, $\Delta H = \Delta U$, but as we learned in Section 6.4, even for reactions in which these terms are not equal, the difference between ΔH and ΔU is usually quite small.

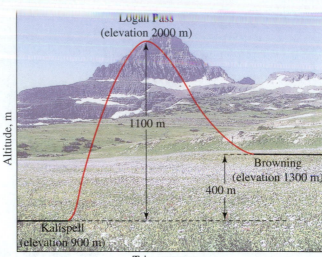

▶ **FIGURE 13.14** **An analogy to a reaction profile and activation energy**

If you were in the Montana town of Kalispell (the reactants) and wished to travel to Browning (the products), you could choose to drive the scenic Going-to-the-Sun Highway (the reaction profile) through Glacier National Park. To do this, you would have to cross the Continental Divide at Logan Pass (the transition state). To get to the pass, you would have to climb 1100 m (the activation energy), but then it would be downhill the rest of the way.

Two important ideas are illustrated in Figure 13.13. First, the enthalpy of a reaction, ΔH, is equal to the difference in activation energies of the forward and reverse reactions.

$$\Delta H = E_a(\text{forward}) - E_a(\text{reverse}) \tag{13.13}$$

Second, for an *endothermic* reaction, the activation energy is always at least equal to, and generally greater than, the enthalpy of reaction, ΔH. Figure 13.14 suggests an analogy to a reaction profile and activation energy.

13.8   Effect of Temperature on Reaction Rates

Charcoal (carbon) reacts so slowly with oxygen at room temperature that there is no visible change. When we raise the temperature of the charcoal with flames from burning lighter fluid, though, the charcoal begins to burn more rapidly. As it does so, the heat of combustion raises the temperature even more, and the charcoal burns faster still. To slow down reactions, we generally lower the temperature. We refrigerate butter to slow down the reactions that cause it to go rancid, for instance, and in the laboratory we usually store $H_2O_2(aq)$ in the refrigerator to slow down its decomposition.

From a theoretical point of view, it seems reasonable that raising the temperature should speed up a reaction because the average kinetic energies of the molecules increase, leading to more frequent collisions. However, increased collision frequency is not the most important factor. As we saw by the increase in the shaded areas under the curves in Figure 13.10, higher temperatures increase the fraction of molecules with enough kinetic energy to produce a reaction. Thus, not only are there more collisions but the percentage of *effective* collisions is also greater, and this causes the increase in reaction rate.

Reaction rate increases as the temperature increases for both exothermic and endothermic reactions. The reaction rate is governed by the height of the potential energy barrier (E_a) and not by the potential energy difference (ΔH) between reactants and products (Figure 13.13).

In 1889, Svante Arrhenius proposed this mathematical expression for the effect of temperature on the rate constant, k:

$$k = Ae^{-E_a/RT} \tag{13.14}$$

In the *Arrhenius equation*, e is the base number for natural logarithms, E_a is the activation energy, and $e^{-E_a/RT}$ represents the fraction of molecular collisions sufficiently energetic to produce a reaction. (R is the gas constant, 8.3145 J mol^{-1} K^{-1}, and T is the Kelvin temperature.) The constant A, called the *frequency factor,* is the product of the collision frequency (Z) and a probability factor (p) that takes into account the orientations required for effective molecular collisions. The product Zp is the number of collisions per unit time capable of leading to a reaction.

Rates of Reaction activity

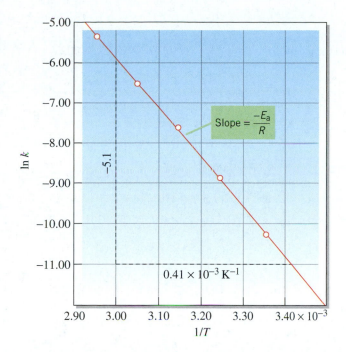

▲ **FIGURE 13.15 A plot of ln k versus $1/T$ for the decomposition of dinitrogen pentoxide:**

$$N_2O_5(g) \longrightarrow 2\,NO_2(g) + \tfrac{1}{2}O_2(g)$$

Evaluation of E_a:

$$\text{Slope} = \frac{-5.1}{0.41 \times 10^{-3}\ \text{K}} = -1.2 \times 10^4\ \text{K}$$

$$E_a = -(\text{slope}) \times R$$

$$E_a = 1.2 \times 10^4\ \text{K} \times 8.3145\ \text{J mol}^{-1}\ \text{K}^{-1}$$

$$= 1.0 \times 10^5\ \text{J/mol}$$

QUESTION: How could you obtain the frequency factor, A, from this graph?

We can represent the Arrhenius equation in a form in which we take the logarithms of both sides:

$$\ln k = \ln A e^{-E_a/RT}$$

$$\ln k = \ln A - \frac{E_a}{RT}$$

$$\ln k = -\frac{E_a}{RT} + \ln A \qquad (13.15)$$

This final equation is that of a straight line and is in the form $y = mx + b$, where $y = \ln k$, $x = 1/T$, the slope of the line is $m = -E_a/R$, and the intercept is $b = \ln A$. A typical straight-line plot of ln k versus $1/T$ is shown in Figure 13.15, and the activation energy, E_a, is evaluated from the slope.

An alternative method for determining E_a, one that requires no graphing, is to measure the rate constant at two temperatures and apply the Arrhenius equation in the form

$$\ln \frac{k_2}{k_1} = \frac{E_a}{R}\left[\frac{1}{T_1} - \frac{1}{T_2}\right] \qquad (13.16)$$

In this equation, k_1 and k_2 are the rate constants at the Kelvin temperatures T_1 and T_2, and E_a is the activation energy in joules per mole. We can use the equation not only to calculate E_a but also to calculate any one of the five quantities from known values of the other four. The required form of the gas constant R is 8.3145 J mol^{-1} K^{-1}.

The derivation of this form of the Arrhenius equation is given in Appendix A. Some prefer the following form of the Arrhenius equation for solving problems:

$$\ln \frac{k_2}{k_1} = \frac{E_a}{R}\left[\frac{T_2 - T_1}{T_1 \times T_2}\right]$$

 **Temperature Influence on Kinetics activity**

Example 13.9

Estimate a value of k at 375 K for the decomposition of dinitrogen pentoxide illustrated in Figure 13.15, given that $k = 2.5 \times 10^{-3}$ s^{-1} at 332 K.

STRATEGY

First, notice that we cannot just read off a value of ln k from the straight-line graph of Figure 13.15 because the point corresponding to 375 K ($1/T = 2.67 \times 10^{-3}$ K^{-1}) is far outside the range shown. Instead, we can use the Arrhenius equation (13.16). It does not matter whether we denote the unknown rate constant as k_1 or k_2, but typically we have used 1 for the starting condition and 2 for the final, or unknown, condition. We can do the same here. That is, let us call the unknown rate constant k_2 and solve Equation (13.16) for its value.

SOLUTION

We know k_1, T_1, and T_2 from the problem statement, and the legend to Figure 13.15 gives us our E_a value. The rate constant at T_2 is unknown.

$$k_1 = 2.5 \times 10^{-3}\,s^{-1} \qquad T_1 = 332\,K \qquad T_2 = 375\,K$$
$$E_a = 1.0 \times 10^5\,J\,mol^{-1} \quad \text{and} \quad R = 8.3145\,J\,mol^{-1}\,K^{-1}$$
$$k_2 = ?$$

We substitute our known information into the Arrhenius equation and solve for the unknown rate constant.

$$\ln\frac{k_2}{k_1} = \frac{E_a}{R}\left[\frac{1}{T_1} - \frac{1}{T_2}\right]$$

First, we solve for the value of $\ln k_2/k_1$ and then take the antilogarithm of both sides of the equation.

$$\ln\frac{k_2}{k_1} = \frac{1.0 \times 10^5\,J\,mol^{-1}}{8.3145\,J\,mol^{-1}\,K^{-1}}\left[\frac{1}{332\,K} - \frac{1}{375\,K}\right]$$
$$= 1.2 \times 10^4\,\cancel{K}(0.00301 - 0.00267)\,\cancel{K^{-1}} = 4.1$$

$$\frac{k_2}{k_1} = e^{4.1} = 60$$

Finally, we solve for the value of k_2.

$$k_2 = 60 \times k_1 = 60 \times 2.5 \times 10^{-3}\,s^{-1} = 0.15\,s^{-1}$$

ASSESSMENT

We expect a rate constant to increase with temperature, as the result here certainly indicates—0.15 s^{-1} at 375 K versus 2.5 × 10^{-3} s^{-1} at 332 K. Moreover, we see just how strongly temperature affects the reaction rate. The temperature increase from 59 to 102 °C causes the reaction to go 60 times faster. As a very rough rule of thumb, in the vicinity of room temperature, a reaction rate doubles for each 10 °C increase in temperature.

Note also that our result would have been the same if we had let the unknown be k_1 instead of k_2: $\ln k_2/k_1$ would have been −4.1; $k_2/k_1 = 0.017$; $k_1 = k_2/0.017 = 2.5 \times 10^{-3}\,s^{-1}/0.017 = 0.15\,s^{-1}$.

EXERCISE 13.9A

Use data from Example 13.9 to determine the temperature at which the rate constant for the decomposition of dinitrogen pentoxide is $k = 1.0 \times 10^{-5}\,s^{-1}$.

EXERCISE 13.9B

Di-*tert*-butyl peroxide (DTBP) is used as a catalyst in the manufacture of polymers. In the gaseous state, DTBP decomposes to acetone and ethane by a first-order reaction.

$$(C_4H_9)_2O_2(g) \longrightarrow 2\,(CH_3)_2CO(g) + C_2H_6(g)$$

The half-life of DTBP is 17.5 h at 125 °C and 1.67 h at 145 °C. What is the activation energy, E_a, of the decomposition reaction?

13.9   Reaction Mechanisms

A few simple reactions proceed in the single step suggested by the balanced equation for the reaction. The reaction between methyl bromide and iodide ion (page 548) is an example. That reaction occurs as the direct result of one effective collision of I^- and CH_3Br:

$$I^- + CH_3Br \longrightarrow CH_3I + Br^-$$

Now consider the reaction between nitrogen monoxide and oxygen, a contributing reaction in the formation of smog:

$$2\,NO(g) + O_2(g) \longrightarrow 2\,NO_2(g)$$

It is highly unlikely that two NO molecules and one O_2 molecule would collide simultaneously, just as it is extremely unlikely that three basketballs might simultaneously collide in midair as basketball players take warm-up shots at the basket. Instead, the overall reaction is accomplished in simpler steps.

A **reaction mechanism** is a series of simple steps that ultimately lead from the initial reactants to the final products of a reaction. An **elementary reaction** represents, at the molecular level, a single stage in the progress of the overall reaction. An elementary reaction involves either an altering of the energy or geometry of starting molecules or the formation of new molecules. Two requirements must be met by any plausible reaction mechanism:

- The mechanism must account for the *experimentally determined* rate law.
- The mechanism must be consistent with the stoichiometry of the net reaction.

In proposing a reaction mechanism, we attempt to describe how countless molecules act to produce the observable changes in the reaction. We do this by tracing the likely path of a few molecules through a series of elementary reactions. We assume—as seems quite likely—that all the reactant molecules follow the same pathway. In contrast, we cannot begin to predict the behavior of an entire human population by observing just a few people. And even at the molecular level, we cannot establish a reaction mechanism with total certainty. We can determine only that a reaction mechanism is *plausible*. Occasionally, two or more equally plausible mechanisms have been proposed for the same reaction.

Elementary Reactions

Before turning to specific reaction mechanisms, let's consider a few ideas about the elementary reactions that make up a reaction mechanism. The **molecularity** of an elementary reaction refers to the number of free atoms, ions, or molecules that enter into the reaction. An elementary reaction in which a single molecule dissociates is a *unimolecular reaction;* one in which two molecules collide effectively is a *bimolecular reaction*. A *termolecular reaction*, requiring the simultaneous collision of three molecules, is much less likely than unimolecular or bimolecular reactions.

The exponents in the rate law for an elementary reaction are *the same as the stoichiometric coefficients in the chemical equation for the elementary reaction.* (Recall that this is usually *not* the case with the rate law for the *overall* reaction.) Elementary reactions are reversible, that is, the forward and reverse reactions occur simultaneously. Some elementary reactions reach a state of equilibrium in which the rates of forward and reverse reactions are equal. Also, in the series of elementary reactions that make up an overall reaction, one elementary reaction may be much slower than all the others. In many cases, this is the **rate-determining step**, the crucial step in establishing the rate of the overall reaction. In order to begin to develop a feel for rate-determining steps, consider a cook who makes burgers at a fast-food restaurant. The cooking of the burger is the slowest step and therefore is the one that determines the rate of burger making. Putting the burgers into buns faster will not speed up the overall process because you cannot put them on buns faster than they are cooked.

A Mechanism with a Slow Step Followed by a Fast Step

As we mentioned in Figure 13.4, a catalyst can be used to speed up a reaction to a more easily measurable rate. For example, the catalyst $I^-(aq)$ speeds the decomposition of hydrogen peroxide:

$$2\ H_2O_2(aq) \xrightarrow{I^-} 2\ H_2O(l) + O_2(g)$$

We have noted that a reaction mechanism must give a description at the molecular level that is consistent with the facts we acquire in studying the rate of a reaction. In the decomposition of hydrogen peroxide, the facts are these: (1) The rate of decomposition of H_2O_2 is first order in both H_2O_2 and I^-, or second order overall. (2) The reactant I^- is unchanged in the reaction and hence does not appear in the equation for the net reaction. A reaction mechanism that is plausible and consistent with these facts is

Slow step:	$H_2O_2 + I^- \longrightarrow H_2O + OI^-$
Fast step:	$H_2O_2 + OI^- \longrightarrow H_2O + O_2 + I^-$
Net equation:	$2\ H_2O_2 \longrightarrow 2\ H_2O + O_2$

Biomolecular Reaction animation

Unimolecular

Bimolecular

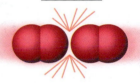

Termolecular

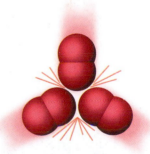

Fire and Fire Suppression

A fire is a self-sustaining chain reaction with most of the important reactions occurring in the gas phase (Figure 13.16). Examine a candle flame and you will see a central pale-blue area topped by a bright yellow region that dims slightly at the tip. In the blue region, the heat radiated from the yellow region melts and vaporizes the wax. As the vapor moves upward into the hotter regions, it decomposes into reactive molecular fragments, free radicals, and other species, including microscopic soot particles and C_2 molecules. When excited, the C_2 molecules produce some of the illumination from the flame. Oxygen diffuses into the region, and the free radicals combine with it rapidly and exothermically. The heat evolved excites C_2 species, heats soot to incandescence, and melts and vaporizes more wax, continuing the process.

Carbon dioxide "smothers" a fire, primarily by absorbing heat that would otherwise go into vaporizing fuel. The CO_2 also dilutes the reaction mixture, reducing the concentration of the free radicals and other reactive species. That means fewer effective collisions and a lower rate of reaction.

It takes considerable CO_2 to cool and dilute a fire. More effective for fire suppression is an *inhibitor*, a species that selectively reacts with and removes free radicals. One of the oldest inhibitors is carbon tetrachloride, CCl_4. This substance is toxic, however, and produces toxic by-products in a flame. It has gradually been replaced by *halons,* such as Halon 1211 ($CBrClF_2$) and Halon 1301 (CF_3Br). Halons partially decompose in a flame, producing bromine atoms that react with the free radicals needed for one of the elementary reactions of the combustion (Figure 13.17).

$$Br\cdot + X\cdot \longrightarrow \text{even-electron species (less reactive than } X\cdot)$$

By removing the species required for one elementary step of the flame reaction, halon interrupts the flame mechanism and extinguishes the flame.

◄ FIGURE 13.16   Acetylene combustion: Mechanism for a combustion reaction

The extremely hot flame of an oxyacetylene torch is used to cut and weld metals. Combustion mechanisms are highly complex. For example, one proposed mechanism for acetylene combustion contains 52 species and 367 elementary reactions. In one part of the mechanism, shown below, (a) a reactive CH_2 species is generated; (b) the CH_2 reacts with O_2 to form free radicals; and (c) free radicals react further to generate other free radicals that then propagate the reaction. Other prominent radicals formed in flames include $H\cdot$ and $O\cdot$.

(a) $C_2H_2 + O_2 \longrightarrow CH_2 + CO_2$

(b) $CH_2 + O_2 \longrightarrow HCO\cdot + \cdot OH$

(c) $HCO\cdot + O_2 \longrightarrow CO + HOO\cdot$

◄ FIGURE 13.17   Mechanism by which a halon suppresses a fire

Halons—carbon compounds that contain bromine and other halogen atoms—have long been used in sprays to extinguish fires, particularly those on aircraft. Although halons, like CFCs, have been implicated in ozone layer depletion, to date no substitutes have been found that have all the desirable properties of the halons: essentially nontoxic, nonconducting, noncorrosive, and leaving no visible residue.

Halons suppress fires by scavenging free radicals. In the reaction scheme outlined below, (a) heat causes the halon CF_3Br to decompose, forming heavy bromine atoms, which are free radicals that are less reactive than most of the other radicals formed in the fire. (b) The bromine atom reacts with a hydrocarbon molecule RH to form HBr and a hydrocarbon radical $R\cdot$. (c) The HBr reacts with hydrogen atoms to regenerate bromine atoms, thus propagating the chain reaction. After steps (b) and (c) have been repeated about 80 times, (d) the chain is terminated by the combination of two radicals. As little as 3% by volume of a halon in the fire-suppressing mixture can effectively suppress a fire.

(a) $CF_3Br \longrightarrow CF_3\cdot + Br\cdot$

(b) $Br\cdot + RH \longrightarrow HBr + R\cdot$

(c) $H\cdot + HBr \longrightarrow H_2 + Br\cdot$

(d) $Br\cdot + HCO\cdot \longrightarrow HCOBr$

The OI^- ion formed in the first step is consumed in the second step and therefore does not appear in the equation for the overall reaction; OI^- is an *intermediate*. In contrast, I^- ion, which also does not appear in the net equation, is consumed in the first step and produced in the second step—in other words, it is recycled through the mechanism; I^- ion is a *catalyst*. The first step, being the slower one, almost exclusively determines the rate of the reaction; it is the *rate-determining step*. In the rate law, we set the reaction rate equal to the rate of the slow step:

$$\text{Rate} = \text{rate of slow step} = k[H_2O_2][I^-]$$

Earlier in the chapter, we treated the catalyzed decomposition of H_2O_2 as a first-order reaction, without going into any detail about why we could do this. Now you can see why: Because $[I^-]$ remains constant throughout the reaction, the product $k \times [I^-]$ is also a constant that we can call $k' = k[I^-]$:

$$\text{Rate} = k'[H_2O_2]$$

Because the value of k' depends on $[I^-]$, the more of the catalyst used, the faster the reaction goes.

As an analogy to a slow first step followed by a fast second step, consider driving to a market 500 m away, a trip you usually complete in a couple of minutes. To cover this very short distance today, however, you must cross a bridge under construction, where traffic is limited to one lane controlled by a road worker. The average waiting time is 15 minutes. The trip to the market, on average, will take just over 15 minutes. The time to complete the trip is determined almost entirely by the rate of the slow step: crossing the bridge.

A Mechanism with a Fast Reversible Step Followed by a Slow Step

Many reactions occur by a mechanism that has a fast reversible step followed by a slow step. The first step is an elementary reaction in which the reactants form an intermediate (rate constant k_1), but the intermediate readily decomposes back into the initial reactants in a reverse reaction (rate constant k_{-1}). We assume that the rates of the forward and reverse reactions quickly become equal, in a condition called a *rapid equilibrium*. However, a small quantity of the intermediate is removed by reacting in a slow second elementary reaction, which is the rate-determining step, to form the final products (rate constant k_2).

Consider the smog-forming reaction of $NO(g)$ and $O_2(g)$ producing $NO_2(g)$:

$$2\ NO(g) + O_2(g) \longrightarrow 2\ NO_2(g)$$

The rate law for the reaction is investigated experimentally and found to be

$$\text{Rate} = k[NO]^2[O_2]$$

A simple mechanism with only one step would account for the observed rate law, but this would be a highly unlikely *termolecular* reaction. A more plausible mechanism is

Fast step: $\qquad 2\ NO \underset{k_{-1}}{\overset{k_1}{\rightleftharpoons}} N_2O_2$

Slow step: $\qquad N_2O_2 + O_2 \xrightarrow{k_2} 2\ NO_2$

$\rule{6cm}{0.4pt}$

Net equation: $\quad 2\ NO + O_2 \longrightarrow 2\ NO_2$

Note that by canceling out the intermediate, N_2O_2, we obtain the net equation for the reaction.

So far, the mechanism looks good, but what does it predict for a rate law? If we base the rate law on the slow, rate-determining step, we obtain a rate that depends on $[N_2O_2]$ and $[O_2]$.

$$\text{Rate} = k_2[N_2O_2][O_2]$$

Application Note

Cholesterol is synthesized in the body from acetyl coenzyme A in a series of dozens of reactions.
The rate-determining step in the sequence is the conversion of hydroxymethylglutaryl coenzyme A (HMG-CoA) to mevalonic acid, a reaction catalyzed by the enzyme HMG-CoA reductase. Drugs such as lovastatin act as inhibitors of HMG-CoA reductase, thereby limiting the synthesis of cholesterol.

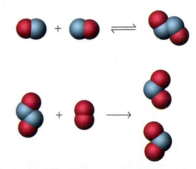

▲ Molecular representation of the mechanism for the reaction $2\ NO + O_2 \longrightarrow 2\ NO_2$.

Remember, though, that we can state the experimentally determined rate law only in terms of substances that appear in the net equation, and therefore we must eliminate $[N_2O_2]$. To do this, we begin with the equation for the first elementary reaction and write the rate laws associated with the forward and reverse elementary reactions.

$$2\,NO \underset{k_{-1}}{\overset{k_1}{\rightleftharpoons}} N_2O_2$$

$$Rate_{forward} = k_1[NO]^2$$
$$Rate_{reverse} = k_{-1}[N_2O_2]$$

Next, we assume that equilibrium is rapidly established in this elementary reaction, meaning that the forward and reverse reactions occur at the same rate.

$$Rate_{forward} = rate_{reverse}$$
$$k_1[NO]^2 = k_{-1}[N_2O_2]$$

Now, we can solve the equation relating the two rates for $[N_2O_2]$.

$$[N_2O_2] = \frac{k_1[NO]^2}{k_{-1}}$$

Finally, we can substitute this expression for $[N_2O_2]$ into the original rate law expression for the rate-determining step.

$$\begin{aligned} Rate &= k_2[N_2O_2][O_2] \\ &= k_2\frac{k_1[NO]^2}{k_{-1}}[O_2] \\ &= \frac{k_2k_1}{k_{-1}}[NO]^2[O_2] \\ &= k[NO]^2[O_2] \end{aligned}$$

    Thus, the proposed mechanism is consistent with the observed rate law. The observed rate constant, k, is a combination of three elementary reaction rate constants.

$$k = k_1k_2/k_{-1}$$

Example 13.10

For the reaction $H_2(g) + I_2(g) \longrightarrow 2\,HI(g)$, a proposed mechanism is

$$Fast\ step: \qquad I_2 \underset{k_{-1}}{\overset{k_1}{\rightleftharpoons}} 2\,I$$

$$Slow\ step: \qquad 2\,I + H_2 \overset{k_2}{\longrightarrow} 2\,HI$$

What is the net equation for the overall reaction, and what is the order of the reaction according to this mechanism?

STRATEGY

We can see that when we add the two steps of the proposed mechanism, the terms "2 I" cancel, leading to the anticipated net equation:

$$H_2 + I_2 \longrightarrow 2\,HI$$

To get the order of the reaction, we must establish the rate law for the overall reaction, and we can do so in a manner similar to that just shown for the reaction of $NO(g)$ and $O_2(g)$ to form $NO_2(g)$.

SOLUTION

To obtain the rate law, we begin with the rate law for the slow, rate-determining step.

$$Rate = k_2[H_2][I]^2$$

We can eliminate the concentration of the intermediate, [I], by assuming rapid equilibrium in the first elementary reaction. That is, we assume that the rates of the forward and reverse reactions are equal.

$$k_1[I_2] = k_{-1}[I]^2$$

Next, we solve for the term $[I]^2$.

$$[I]^2 = \frac{k_1}{k_{-1}}[I_2]$$

Then we make a substitution for $[I]^2$ in the rate law based on the slow step.

$$\text{Rate} = k_2[H_2][I]^2 = k_2\frac{k_1}{k_{-1}}[H_2][I_2]$$

$$= k[H_2][I_2]$$

The reaction is first order in H_2, first order in I_2, and second order overall.

ASSESSMENT

If we were to do a kinetic study, we would find that the reaction is indeed first order in H_2, first order in I_2, and second order overall, as we have just shown. Our conclusion, then, is that the proposed mechanism is *plausible*. Note, however, that until this mechanism was proposed in 1967, the prevailing view was also a plausible mechanism—a single-step mechanism based on the bimolecular reaction between H_2 and I_2.

EXERCISE 13.10A

The decomposition of nitrosyl chloride,

$$2\,NOCl(g) \longrightarrow 2\,NO(g) + Cl_2(g)$$

is a first-order reaction. Propose a mechanism for this reaction consisting of one fast step and one slow step.

EXERCISE 13.10B

An alternative to the mechanism described in the text for the reaction

$$2\,NO(g) + O_2(g) \longrightarrow 2\,NO_2(g)$$

involves NO_3 as an intermediate instead of N_2O_2. **(a)** Write the steps for this alternative mechanism, and **(b)** show that the alternative mechanism is also consistent with the observed rate law.

13.10   Catalysis

Raising the temperature generally speeds up a reaction, sometimes dramatically. Raising the temperature in a living organism can have a disastrous effect, however, as doing so can kill the organism. Living things generally rely on catalysis rather than temperature increases to speed up reactions. Many industrial and laboratory processes require catalysis as well.

The decomposition of potassium chlorate is sometimes used to produce small quantities of oxygen in the laboratory:

$$2\,KClO_3(s) \longrightarrow 2\,KCl(s) + 3\,O_2(g)$$

Without a catalyst, the $KClO_3(s)$ must be heated to above 400 °C in order to produce $O_2(g)$ at a useful rate. However, if we add a small quantity of $MnO_2(s)$, we can get the same rate of oxygen evolution at just 250 °C. Further, after the reaction is complete, we can recover essentially all the manganese dioxide, unchanged. As we have noted in several instances in this chapter, a **catalyst** increases the reaction rate without itself being changed in a chemical reaction. In general, a catalyst works by changing the mechanism of a chemical reaction. The pathway of a catalyzed reaction has a lower activation energy than that of an uncatalyzed reaction. If the activation energy is lowered, then more molecules have sufficient energies to engage in effective collisions. Figure 13.18 extends the analogy of Figure 13.14 to describe the difference between the reaction profile of a catalyzed reaction and the reaction profile of an uncatalyzed reaction.

 **Catalysis movie**

▶ **FIGURE 13.18** **An analogy to the reaction profile and activation energy in a catalyzed reaction**

To return to our analogy of a trip from Kalispell to Browning (Figure 13.14), it is possible to take an alternate route because U.S. Highway 2 crosses the Continental Divide here at Marias Pass (catalyzed reaction). This route involves a climb of only 700 m (lower activation energy) rather than the 1100-m climb needed when we travel via Logan Pass (uncatalyzed reaction). This alternate route is analogous to the different pathway provided by a catalyst in a chemical reaction.

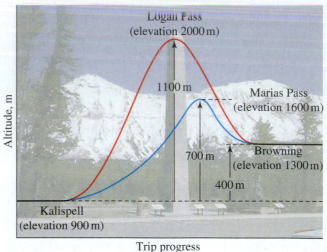

Homogeneous Catalysis

A catalyzed reaction that occurs in a *homogeneous* (one-phase) mixture involves *homogeneous catalysis*. One simple mechanism for homogeneous catalysis in a reaction in which two reactants A and B yield two products C and D involves two steps:

Step 1:	A + ~~catalyst~~ ⟶ ~~intermediate~~ + C
Step 2:	B + ~~intermediate~~ ⟶ D + ~~catalyst~~
Net equation:	A + B ⟶ C + D

As expected, neither the intermediate nor the catalyst appears in the net equation.

A practical example of this mechanism has O_3 and O as reactants and O_2 as the product, with Cl the catalyst and ClO an intermediate.

Step 1:	O_3 + ~~Cl·~~ ⟶ ~~ClO·~~ + O_2	E_a = 2.1 kJ
Step 2:	~~ClO·~~ + O· ⟶ ~~Cl·~~ + O_2	E_a = 0.4 kJ
Net equation:	O_3 + O· ⟶ 2 O_2	

Figure 13.19 presents reaction profiles for both the uncatalyzed and the catalyzed decomposition of ozone. The two activation energies for the catalyzed route

Application Note

The seasonal depletion of ozone observed in the stratosphere over the polar regions is thought to occur by this mechanism involving Cl atoms as catalysts. We will explore this matter further in Chapter 25.

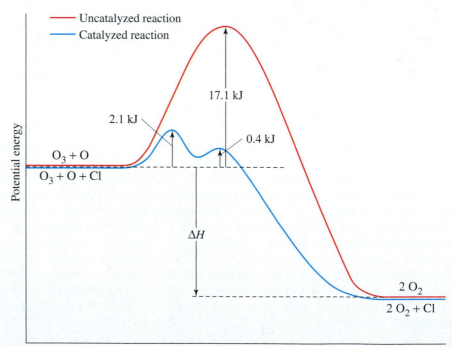

▶ **FIGURE 13.19** **Reaction profile for the uncatalyzed (red) and catalyzed (blue) decomposition of ozone**

The net reaction in both cases is O_3 + O ⟶ 2 O_2. In the catalyzed reaction, Cl is the catalyst.

correspond to the two steps in the mechanism, but even the higher of the two (2.1 kJ) is much lower than that of the uncatalyzed reaction (17.1 kJ). As a consequence, the rate constant for the catalyzed reaction is thousands of times greater than that of the uncatalyzed reaction. Except that the catalyst (Cl) appears in the catalyzed reaction, the initial states for the two reaction pathways are the same, as are the final states, meaning that the enthalpy of reaction, ΔH, is identical in the catalyzed and uncatalyzed reaction.

Heterogeneous Catalysis

Many reactions can be catalyzed by the surfaces of appropriate solids. An essential aspect of this catalytic activity is the ability of surfaces to *adsorb* (bind) reactant molecules from the gaseous or liquid state. Because a surface-catalyzed reaction occurs in a *heterogeneous* mixture, the catalytic action is called *heterogeneous catalysis*. There are five basic steps in heterogeneous catalysis:

1. Reactant molecules are adsorbed.
2. Reactant molecules diffuse along the surface.
3. Adsorbed molecules react with surface atoms to form reaction intermediates.
4. Adsorbed intermediates react to form product molecules.
5. Product molecules are desorbed (that is, released from the surface).

Figure 13.20 shows a hypothetical reaction profile for a surface-catalyzed reaction and compares it with the profile for an uncatalyzed, homogeneous gas-phase reaction.

The zero-order decomposition of ammonia on tungsten described on page 544 is a surface-catalyzed reaction, as is the hydrogenation of food oils to produce solid or semisolid fats described in Section 9.11. A simplified mechanism of the surface-catalyzed conversion of oleic acid to stearic acid,

$$CH_3(CH_2)_7CH{=}CH(CH_2)_7COOH + H_2 \xrightarrow{\text{Ni}} CH_3(CH_2)_7CH_2CH_2(CH_2)_7COOH$$

$$\underset{\text{Oleic acid}}{\qquad\qquad} \qquad\qquad\qquad\qquad \underset{\text{Stearic acid}}{\qquad\qquad\qquad\qquad}$$

is suggested in Figure 13.21.

▲ The decomposition of hydrogen peroxide, H_2O_2(aq), to H_2O(l) and O_2(g) is a highly exothermic reaction that is strongly catalyzed by the surface of the platinum metal in this wire mesh electrode.

Surface Reaction-Hydrogenation animation

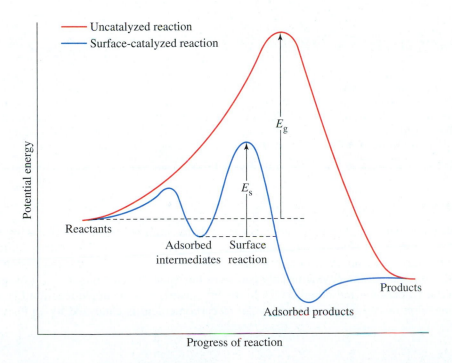

◀ **FIGURE 13.20   Reaction profile for a surface-catalyzed reaction**

In the reaction profile (blue) for the surface-catalyzed reaction, the activation energy for the reaction step, E_s, is considerably less than in the reaction profile (red) for the uncatalyzed gas-phase reaction, E_g.

QUESTION: Where along the reaction profile do reaction intermediates appear? Where are activated complexes present?

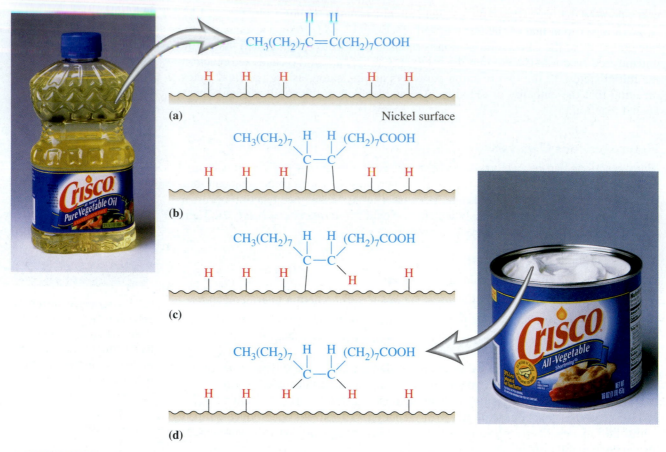

▲ FIGURE 13.21 **Heterogeneous catalysis on a nickel surface**

A molecule of oleic acid (a) adsorbs to a nickel surface through C atoms at the double bond (b); and in the process, the double bond is converted to a single bond. The red H atoms are the result of the dissociative adsorption of H_2 molecules on the nickel surface. First one (c) and then a second adsorbed H atom react with the surface-bonded C atoms. (d) The stearic acid molecule formed in this step desorbs from the surface and escapes. This hydrogenation process converts a liquid oil to a saturated fat, which is a solid at room temperature.

QUESTION: Where would each organic species shown in this figure appear on a reaction profile such as the one shown in Figure 13.20?

Catalysis in Green Chemistry

The classic preparation of aldehydes, ketones, and carboxylic acids involves the uncatalyzed oxidation of alcohols using potassium dichromate (page 150). Reactions involving water-insoluble alcohols are often run in chlorinated hydrocarbon solvents, a method requiring the recovery or disposal of toxic chromium compounds and the chlorinated solvents. A new method, catalyzed by a water-soluble, recyclable palladium(II) compound, uses water as a reaction medium and oxygen from air as the oxidant. This green chemistry is more economical and friendlier to the environment.

13.11   Enzyme Catalysis

Substances such as platinum and nickel are able to catalyze a variety of reactions, but the catalysts for the chemical reactions in living organisms are usually quite specific in their catalytic action. Almost all of these catalysts are high-molecular-mass proteins called **enzymes.** An enzyme usually catalyzes one specific reaction—or a set of quite similar reactions—and no others. In living organisms, even a simple reaction like the conversion of carbon dioxide and water to carbonic acid is catalyzed by an enzyme *(carbonic anhydrase)*. Each enzyme molecule can convert 100,000 CO_2 molecules per second in the reaction

$$CO_2(g) + H_2O(l) \underset{\text{carbonic anhydrase}}{\rightleftharpoons} H_2CO_3(aq)$$

Biochemists often use the model pictured in Figure 13.22, called the *induced-fit* model, to explain enzyme action. The reacting substance, called the **substrate** (S), attaches itself to an area on the enzyme (E) called the **active site,** to form an enzyme–substrate complex (ES). The enzyme–substrate complex decomposes to form products (P), and the enzyme is regenerated. Using the symbolism that biochemists often use, we can represent the process as

$$E + S \longrightarrow ES \longrightarrow E + P$$

Factors Influencing Enzyme Activity

The rates of enzyme-catalyzed reactions are influenced by several factors, including substrate concentration, enzyme concentration, medium acidity, and temperature.

Figure 13.23 shows how the reaction rate changes as the concentration of substrate increases while the enzyme concentration remains constant. The reaction rate increases as the concentration of substrate increases, but only up to a point. Eventually, a concentration is reached that saturates all the active sites on the enzyme molecules. The rate levels off, and a further increase in substrate concentration leaves the rate unchanged.

At low substrate concentrations, the reaction is first order in the substrate because the rate of enzyme–substrate complex formation is proportional to [S]:

$$\text{Rate} = k[S] \qquad (13.17)$$

At high substrate concentrations, adding more substrate cannot accelerate the reaction. The rate of the reaction is independent of substrate concentration, and therefore the reaction is zero order:

$$\text{Rate} = k'[S]^0 = k'' \qquad (13.18)$$

An analogy is 10 taxis (enzyme molecules) waiting at a taxi stand to take people (substrate) on a 10-minute trip to a concert hall, one passenger at a time (corresponding to one active site per enzyme molecule). If only 5 people are present at the stand, the rate of their arrival at the concert hall is 5 people in 10 minutes. If the number of people at the stand is increased to 10, the rate increases to 10 arrivals in 10 minutes. With 20 people at the taxi stand, the rate would still be 10 arrivals in 10 minutes. The taxis have been saturated.

If the concentration of substrate is held constant and *in excess* (if we always have more people than taxis), the reaction rate is proportional to the concentration of enzyme. The more enzyme there is, the faster the reaction goes (the more taxis, the more people transferred). This relationship holds over a wide range of enzyme concentration (Figure 13.24).

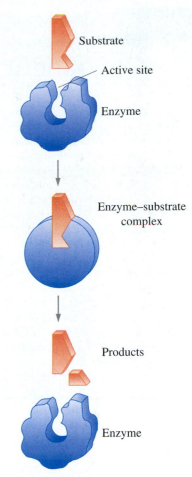

▲ **FIGURE 13.22 Model of enzyme action**

The *induced-fit model* holds that the shapes of the substrate and the active site are not perfectly complementary, but the active site adapts to fit the substrate, much like a glove molds to fit the hand that is inserted into it.

 **Enzyme Catalysis animation**

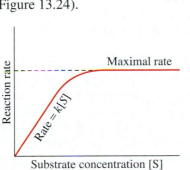

▲ **FIGURE 13.23 The effect of substrate concentration on reaction rate with enzyme concentration held constant**

When (E_a) is low, the reaction rate increases with increasing [S]. At high values of [S], the reaction rate is independent of [S], and a maximum rate is observed.

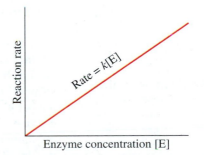

▲ **FIGURE 13.24 The effect of enzyme concentration on reaction rate with substrate concentration held constant and in excess**

Kinetics and Home Pregnancy Tests

Enzymes are so specific in their response to a particular substrate that a class of chemical analysis known as *kinetic methods* is based on the use of enzymes. In one type of kinetic analysis, tiny depressions in a plastic "spot plate" are coated with an enzyme. Each depression is filled with a sample to be analyzed or with a solution of substrate of known concentration. The enzyme interacts with the substrate, and a colored product is the result. After a certain time, the reaction is stopped. The depth of color at that point is directly related to the concentration of substrate.

Of course, sometimes we are simply interested in a qualitative answer: Is the substrate present or not? A well-known example of just such an analysis is the inexpensive and reliable home pregnancy test. A typical test uses a strip of material coated with an enzyme specific for human chorionic gonadotrophin (HCG) hormone. When the strip is wetted with a urine sample, interaction between HCG and the enzyme produces a red-colored product after several minutes. Pregnancy tests of 50 years ago required 48 hours or more, and results were rather uncertain until the second or even third month of pregnancy. In sharp contrast, the modern home pregnancy test takes only a few minutes and is so sensitive that it can determine pregnancy within a week or two of conception.

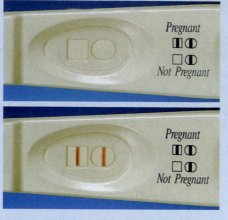

◀ An enzyme-catalyzed process produces the red color that marks a positive result in a home pregnancy test.

Enzymes are proteins. The molecules contain both acidic groups (such as COOH) and basic groups (such as NH_2). Enzyme activity depends on the concentration of H^+ ions present in the medium surrounding the enzyme. Each enzyme has its own *optimum acidity.* Those that operate in the stomach work best in highly acidic media. Those that catalyze reactions in the small intestine, in contrast, show optimum activity in slightly basic media.

As an example, consider *lysozyme,* an enzyme that breaks certain bonds in complex structural molecules found in the cell walls of bacteria. Lysozyme is active only when two conditions are met simultaneously:

1. Aspartic acid, one of the amino acid components of lysozyme, is ionized. That is, the free COOH group of aspartic acid must be converted to COO^-.

2. Glutamic acid, another amino acid component of lysozyme, is *not* ionized. That is, the COOH group of glutamic acid is intact.

When the medium is slightly acidic, both conditions are met and lysozyme is active. However, in either highly acidic or highly basic solutions, the enzyme cannot function. In the former case, both acid groups are present in their COOH form. In the latter case, both groups are present as COO^-.

Enzyme action is sensitive to temperature changes. For living cells, there is often a rather narrow range of optimum temperatures. Both higher and lower temperatures can disable an enzyme or even kill an organism. A change in temperature changes the shape of the protein molecule so that the substrate no longer fits the active site. The temperature range for the activity of enzymes generally is 10 to 50 °C. However, some enzymes found in *thermophilic* (heat-loving) bacteria operate well near 100 °C. The *optimum temperature* for many enzymes in the human body is normal body temperature of 37 °C (Figure 13.25).

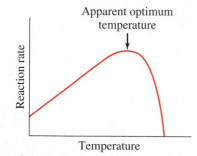

▲ **FIGURE 13.25** **Enzyme activity as a function of temperature**

If the hypothetical reaction whose rate is suggested here occurs in humans, the optimum temperature is likely to be 37 °C—normal body temperature.

Enzyme Inhibition

An *inhibitor* is a substance that makes an enzyme either less active or completely inactive. Some inhibitors bind at the active site of the enzyme and block the substrate. Others bind elsewhere in the enzyme molecule but change the shape of the active site or otherwise hinder a substrate molecule's access to the site. Inhibition is an important, naturally occurring process that is essential in growth and metabolism, but some substances that we ingest—in food or otherwise—also act as inhibitors. Some inhibitors, called poisons, make us sick or even kill us. Others, called drugs, alleviate or even cure diseases.

Enzymes, like many other proteins, contain the amino acid *cysteine,* which has a sulfhydryl group (SH) attached to each unit. Heavy metal ions such as Hg^{2+} and Pb^{2+} can deactivate enzymes by reacting with sulfhydryl groups to form sulfides. This poisoning can occur at a position removed from the active site, distorting or destroying the site (Figure 13.26).

Enzymes play a key role in the chemistry of the nervous system. When an electric signal from the brain reaches the end of a nerve cell, the substance *acetylcholine* is liberated and carries the signal across a tiny gap to a second nerve cell. This second cell acts as a receptor for the signal. Once it has acted on the receptor, acetylcholine is rapidly broken down into acetic acid and choline in a reaction catalyzed by the enzyme *cholinesterase*:

$$CH_3COOCH_2CH_2N^+(CH_3)_3 + H_2O \underset{\text{acetylase}}{\overset{\text{cholinesterase}}{\rightleftharpoons}} CH_3COOH + HOCH_2CH_2N^+(CH_3)_3$$

$\quad\quad$ Acetylcholine $\quad\quad\quad\quad\quad\quad\quad\quad\quad\quad\quad\quad$ Acetic acid $\quad\quad\quad$ Choline

In effect, the breakdown of acetylcholine resets the receptor cell to the "off" position, making it ready to receive the next impulse. Other enzymes, such as acetylase, convert the acetic acid and choline back to acetylcholine, completing the cycle.

Anticholinesterase poisons block the action of cholinesterase. The insecticides *malathion* and *parathion* and the warfare agents *tabun* and *sarin* are well-known examples of such nerve poisons. The molecule of poison binds tightly to the cholinesterase (Figure 13.27), preventing the enzyme from performing its normal function. If the breakdown of acetylcholine is blocked, the concentration of this messenger compound builds up, causing the receptor nerve cell to fire repeatedly, that is, to be continuously "on." This overstimulates the muscles, glands, and organs. The heart beats wildly and irregularly. The victim goes into convulsions and dies quickly.

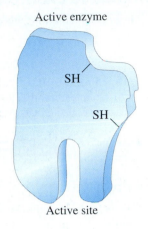

Active enzyme

SH

SH

Active site

Hg^{2+}

S

Hg

S

Inactive enzyme

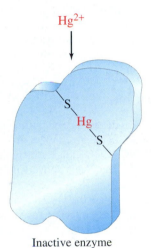

▲ **FIGURE 13.26** **Mercury poisoning as an example of enzyme inhibition**

Mercury(II) ions react with sulfhydryl groups to change the conformation of the enzyme and destroy the active site.

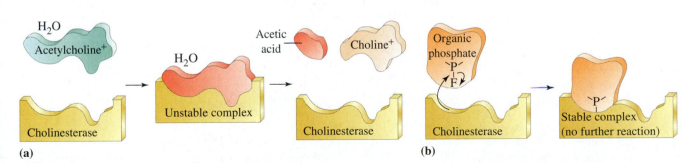

(a) $\quad\quad\quad\quad\quad\quad\quad\quad\quad\quad\quad\quad\quad\quad\quad\quad\quad\quad\quad\quad\quad\quad\quad\quad\quad\quad\quad\quad\quad$ (b)

▲ **FIGURE 13.27** **Action of cholinesterase and its inhibition**

(a) Cholinesterase catalyzes the hydrolysis of acetylcholine to acetic acid and choline. (b) An organic phosphate (as in the nerve gas sarin) binds to cholinesterase, preventing it from breaking down acetylcholine.

Cumulative Example

In acidic aqueous solutions, nitroacetic acid decomposes to nitromethane and carbon dioxide gas:

$$CH_2(NO_2)COOH \longrightarrow CH_3NO_2 + CO_2(g)$$

Nitroacetic acid   Nitromethane

In one experiment, 0.1051 g nitroacetic acid was allowed to decompose, and the evolved $CO_2(g)$ was collected at 25.0 °C over $CaCl_2(aq)$ saturated with $CO_2(g)$ and having a water vapor pressure of 9.7 Torr. The barometric pressure was 758.2 Torr. Use the following data on collected gas volume as a function of time to determine the half-life of this reaction at 25.0 °C.

Time, min	0	1.64	3.64	6.14	9.64	15.19
Volume, mL	0	3.96	7.93	11.92	15.93	19.93

STRATEGY

The time–volume tabulation is similar to what we saw in Table 13.1. Moreover, in a fashion similar to that shown in Table 13.1, we will need to relate the rate at which a reactant (nitroacetic acid) disappears to the rate at which a product (carbon dioxide) appears. We can readily convert the initial mass of the reactant to a mole basis. We can convert the gas volumes to moles of CO_2, using the ideal gas equation (5.9) with the necessary adjustment of pressure for a gas collected over an aqueous solution. From the number of moles of $CO_2(g)$ produced at various times throughout the reaction, we can calculate the number of moles of nitroacetic acid remaining at these times. Following this, either by graphing the kinetic data or by using the appropriate integrated rate law, we can determine the rate constant and half-life.

SOLUTION

First, we determine the initial number of moles in the given mass of $CH_2(NO_2)COOH$, using its molar mass (105.05 g/mol) in a conversion factor.

$$? \text{ mol} = 0.1051 \text{ g} \times \frac{1 \text{ mol}}{105.05 \text{ g}} = 1.000 \times 10^{-3} \text{ mol}$$

Next, we relate the partial pressure of $CO_2(g)$ in the collected gas to the barometric pressure and the partial pressure of the water vapor.

$$P_{CO_2} = P_{bar} - P_{H_2O}$$

$$= (758.2 - 9.7) \text{ Torr} \times \frac{1 \text{ atm}}{760 \text{ Torr}}$$

$$= 0.9849 \text{ atm}$$

Also, we can represent the volume, v, of collected $CO_2(g)$ in liters.

$$v \text{ ml} \times \frac{1 \text{ L}}{1000 \text{ mL}} = \frac{v}{1000} \text{ L}$$

Now we substitute the pressure and volume just derived, together with the temperature, into the ideal gas equation (5.9).

$$n_{CO_2} = \frac{PV}{RT}$$

$$= \frac{0.9849 \text{ atm} \times (v/1000) \text{ L}}{0.08206 \text{ L atm mol}^{-1} \text{ K}^{-1} \times (273.2 + 25.0) \text{ K}}$$

$$= 4.025 \times 10^{-5} \times v \text{ mol}$$

The number of moles of nitroacetic acid at any time is equal to the difference between the initial amount and the amount consumed. One mole of nitroacetic acid is consumed for every mole of $CO_2(g)$ produced.

$$\text{mol } CH_2(NO_2)COOH_{remaining} = \text{mol } CH_2(NO_2)COOH_{initial} - \text{mol } CO_{2\,produced}$$

We can combine the above results into an expression for the number of moles of nitroacetic acid remaining at a time t (min) at which the total volume of $CO_2(g)$ collected is v (mL). To make our use of this expression a little easier to follow, we have changed the initial amount of nitroacetic acid from 1.000×10^{-3} mol to 10.00×10^{-4} mol.

$$\text{mol } CH_2(NO_2)COOH_{remaining} = 10.00 \times 10^{-4} \text{ mol} - (4.025 \times 10^{-5} \times v) \text{ mol}$$

We calculate the number of moles of nitroacetic acid remaining at $t = 1.64$ min by substituting the value $v = 3.96$ mL.

$$\text{mol CH}_2(\text{NO}_2)\text{COOH}_{\text{remaining}} = 10.00 \times 10^{-4}\text{ mol} - (4.025 \times 10^{-5} \times 3.96)\text{ mol}$$
$$= (10.00 \times 10^{-4} - 1.59 \times 10^{-4})\text{ mol}$$
$$= 8.41 \times 10^{-4}\text{ mol}$$

Similar calculations can be done for all the other data points, and the results tabulated.

Time, min	0	1.64	3.64	6.14	9.64	15.19
Volume, mL	0	3.96	7.93	11.92	15.93	19.93
Mol $CO_2 \times 10^4$	0	1.59	3.19	4.80	6.41	8.02
Mol $CH_2(NO_2)COOH \times 10^4$	10.00	8.41	6.81	5.20	3.59	1.98

Let's see if our data conform to the integrated rate equation (13.7) for a first-order reaction. We will use $[A]_0 = 10.00 \times 10^{-4}$ mol. In the first calculation, $[A]_t = 8.41 \times 10^{-4}$ mol, in the second, $[A]_t = 6.81 \times 10^{-4}$ mol, and so on.

$$k = -\frac{1}{1.64\text{ min}} \times \ln(8.41/10.00) = 0.106\text{ min}^{-1}$$

$$= -\frac{1}{3.64\text{ min}} \times \ln(6.81/10.00) = 0.106\text{ min}^{-1}$$

$$= -\frac{1}{6.14\text{ min}} \times \ln(5.20/10.00) = 0.107\text{ min}^{-1}$$

$$= -\frac{1}{9.64\text{ min}} \times \ln(3.59/10.00) = 0.106\text{ min}^{-1}$$

$$= -\frac{1}{15.19\text{ min}} \times \ln(1.98/10.00) = 0.107\text{ min}^{-1}$$

The value of $k = 0.106$ min^{-1}, and the half-life is found using Equation (13.9).

$$t_{1/2} = \frac{0.693}{k} = \frac{0.693}{0.106\text{ min}^{-1}} = 6.54\text{ min}$$

ASSESSMENT

The constant value of k tells us that the reaction is indeed first order and that the half-life is given by Equation (13.9). An alternate method of establishing this as a first-order reaction would be to plot the data as in Figure 13.6 and deduce k from the slope of the straight line. As still another possibility, we might have surmised that the reaction was first order by estimating $t_{1/2}$ from the tabulated data. For example, in the 6.14-min time period, the amount of nitroacetic acid declined by nearly one-half, from 10.00×10^{-4} mol to 5.20×10^{-4} mol, suggesting a $t_{1/2}$ of a little more than 6.14 min. During the 6.00-min time interval from 3.64 to 9.64 min, the amount also declined by nearly one-half, again suggesting a half-life of somewhat more than 6.00 min. Of course, if these methods had failed to confirm this as a first-order reaction, tests for other reaction orders, as in Example 13.8, would have been required.

Two points are worth noting. If the $CO_2(g)$ had been collected over water rather than in $CaCl_2(aq)$ previously saturated with CO_2, some of the first CO_2 produced would have dissolved and not appeared in the collected gas. Also, we were able to use moles rather than concentrations in Equation (13.7) because the volume term needed to convert from moles to molarity would have canceled from the numerator and denominator of the left side of equation.

Concept Review with Key Terms

13.1 Chemical Kinetics—A Preview—Chemical kinetics encompasses the measurement of reaction rates, the study of the factors affecting these rates, and the determination of the mechanisms by which chemical reactions occur.

13.2 The Meaning of Reaction Rate—Rates of reactions are based on the rate of disappearance of a reactant or formation of a product; typically rates are expressed in concentration per time (M s^{-1}). For the reaction $a\text{ A} \longrightarrow c\text{ C}$,

$$\text{Rate} = -\frac{1}{a}\frac{\Delta[A]}{\Delta t} = \frac{1}{c}\frac{\Delta[C]}{\Delta t}$$

13.3 Measuring Reaction Rates—From experimental data, the *average* rate of reaction, the **initial rate of reaction** (rate at $t = 0$),

and the **instantaneous rate of reaction** (rate at any specified time) can be determined. In general, these rates differ from one another because reaction rates depend on the concentrations of reactants, which decrease with time.

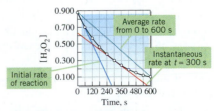

13.4 The Rate Law of a Chemical Reaction—The **rate law** for a reaction is

$$\text{Rate} = k[A]^m[B]^n \cdots$$

where A, B, . . . are reactants. The constant k is the **rate constant**, m is the **order of the reaction** in A, n is the order in B, and $m + n + \cdots$ is the overall order. Values of m, n, $\cdots$ must be obtained by experiment. Once established, the rate law can be used to calculate a value of k from known concentrations and an observed rate of reaction. Orders of reaction and rate constants are often determined through the **method of initial rates**.

13.5 First-Order Reactions—An **integrated rate law** relates concentration and time. The **integrated rate law** for a **first-order reaction** is

$$\ln ([A]_t/[A]_0) = -kt$$

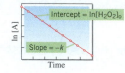

The rate constant, k, is the negative of the slope of the straight-line graph: $\ln [A]_t$ versus t.

The **half-life** $(t_{1/2})$ of a reaction is the time in which one-half of the reactant initially present is consumed. For a first-order reaction, the half-life is constant:

$$t_{1/2} = 0.693/k$$

13.6 Reactions of Other Orders—In **a zero-order reaction**, the rate is independent of the concentration of reactant; a graph of reactant concentration as a function of time is a straight line. For a **second-order reaction** with rate $= k[A]^2$, a plot of $1/[A]$ versus t yields a straight line. For both zero-order and second-order reactions, the half-life depends on concentration as well as on the value of k. The integrated rate law and half-life expression for a second order reaction are

$$\frac{1}{[A]_t} = kt + \frac{1}{[A]_0} \qquad t_{1/2} = \frac{1}{k[A]_0}$$

13.7 Theories of Chemical Kinetics—Chemical reactions occur through collisions between sufficiently energetic molecules in the proper orientation. The point in the collision where the reactants are converted to products is the **transition state**. The reaction species at this transition state is called an **activated complex**. A **reaction profile** charts the progress of a reaction through changes in the potential energy of the reaction mixture. The potential energy rises from that of the reactants to that of the transition state and then falls back to the poten-

tial energy of the products. The excess energy of the transition state over that of the reactants is the **activation energy** (E_a) of the reaction. The reaction profile also shows the difference in potential energy between reactants and products, that is, the enthalpy change (ΔH) of a reaction.

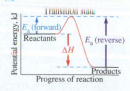

13.8 Effect of Temperature on Reaction Rate—Reactions generally go faster at higher temperatures because of the large increase in the fraction of molecules having energies in excess of the activation energy. The effect of temperature and activation energy on the rate constant of a reaction can be expressed through the Arrhenius equation,

$$k = Ae^{-E_a RT}$$

and a straight-line graph of $\ln k$ versus $1/T$ has a slope $-E_a/R$.

13.9 Reaction Mechanisms—A plausible **reaction mechanism**, describing the steps by which reactants are transformed to products, may be deduced from the experimentally determined rate law and generally consists of a series of **elementary reactions**. These elementary reactions represent the **molecularity** of a reaction—the number of atoms, ions, or molecules directly involved in the reaction step. The slowest elementary reaction is often found to be the **rate-determining step**. A reaction mechanism must yield the observed net equation, and the rate law expressed by the mechanism must be the same as that found experimentally.

13.10 Catalysis—A **catalyst** speeds up a reaction by changing the mechanism to one of lower activation energy; the catalyst is regenerated in the reaction. Some catalysts function in homogeneous (single-phase) mixtures; others function in heterogeneous (multiphase) mixtures, such as in surface-catalyzed reactions.

13.11 Enzyme Catalysis—An **enzyme** is a biological molecule, such as a high-molecular-mass protein, that functions as a catalyst. In enzyme catalysis, the **substrate** (S) (reactant) binds to an **active site** on the enzyme (E) to form a complex (ES) that dissociates into product molecules (P).

$$E + S \longrightarrow ES \longrightarrow E + P$$

Enzyme activity depends on the concentrations of substrate and enzyme, acidity of the medium, and temperature.

Assessment Goals

When you have mastered the material in this chapter, you will be able to:

- Express reaction rates in terms of concentrations of reactants and/or products and use experimental data to determine the rate of a reaction.

- Predict the influence of different experimental factors on reaction rates.

- Determine average rates of reaction, the initial rate of reaction, and instantaneous rates of reaction from experimental data and from plots of concentration versus time.

- Use the method of initial rates to determine the rate law and rate constant of a chemical reaction.

- Use rate laws to calculate rates of a reaction at different concentrations.

- Relate concentrations of reactants and time, using the integrated rate laws for zero-order, first-order, and second-order reactions.

- Explain the meaning of half-life, and derive and use equations for the half-life of zero-order, first-order, and second-order reactions.

- Calculate half-lives and/or rate constants from half-life equations.

- Describe the kinetics of chemical reactions in terms of collision theory and transition state theory.

- Construct reaction profiles and identify activation energies, transition states, and reaction enthalpies on these profiles.

- Calculate the effects of temperature on rates of reaction, using the Arrhenius equation.

- Describe the properties of elementary reactions and rate-determining steps.

- Deduce rate laws from simple reaction mechanisms, and establish the plausibility of reaction mechanisms based on kinetic data.

- Describe how a catalyst increases the rate of reaction.

- Explain how enzymes function and the factors that influence that function.

Self-Assessment Questions

1. State two quantities that must be measured to establish the rate of a chemical reaction and cite several factors that affect the rate of a chemical reaction.

2. Explain why the rate of disappearance of NO and the rate of formation of N_2 are not the same in the reaction $2\,CO(g) + 2\,NO(g) \longrightarrow 2\,CO_2(g) + N_2(g)$.

3. What is the difference among the initial rate, the average rate, and the instantaneous rate of a reaction? Can the average rate ever be the same as the instantaneous rate? Can the initial rate and an instantaneous rate be the same? Explain.

4. How does the rate of a reaction differ from the rate constant of the reaction? Can they ever be the same? Explain.

5. What variables are related through the *rate law* of a reaction, and what variables are related through the *integrated rate law?* Write these two equations for the first-order reaction: $N_2O_5(g) \longrightarrow 2\,NO_2(g) + \frac{1}{2}\,O_2(g)$.

6. For the reaction A $\longrightarrow$ products, a plot of [A] versus time is found to be a straight line. The order of this reaction is

   (a) first

   (b) second

   (c) zero

   (d) impossible to determine from this graph

7. The reaction $2\,A + B \longrightarrow C + 2\,D$ is *first* order in A and *first* order in B. For this reaction

   (a) rate of reaction $= k[A]^2[B]$

   (b) rate of reaction $= k[A]^2$

   (c) rate of disappearance of A = rate of disappearance of B

   (d) rate of formation of C = −(rate of disappearance of B)

8. A *first-order* reaction has a half-life of 13.9 min. The rate at which this reaction proceeds when [A] = 0.40 M is

   (a) $0.020\ \text{M min}^{-1}$ (c) $8.0\ \text{M min}^{-1}$

   (b) $0.050\ \text{M min}^{-1}$ (d) $0.125\ \text{M min}^{-1}$

9. The rate of a chemical reaction generally increases rapidly, even for small temperature increases, because of a rapid increase with temperature in the

   (a) collision frequency

   (b) fraction of molecules with energies in excess of the activation energy

   (c) activation energy

   (d) average kinetic energy of molecules

10. Why is the orientation of the colliding molecules an important factor in determining the rate of a reaction in some reactions but not in all reactions?

11. What is the reaction profile of a chemical reaction? In relation to the reaction profile, what are the transition state and the activated complex?

12. The reaction A + B $\longrightarrow$ C + D has a reaction enthalpy equal to +25 kJ. The activation energy, E_a, of this reaction is equal to which of the following?

    (a) −25 kJ

    (b) less than +25 kJ

    (c) more than +25 kJ

    (d) either less than +25 kJ or more than +25 kJ, which can only be determined by experiment.

13. What plot of experimental data can you use to evaluate the activation energy, E_a, of a reaction? How is E_a related to this plot?

14. What are the chief requirements that must be met by a plausible reaction mechanism? Why do we say "plausible" mechanism rather than "correct" mechanism?

15. For the net reaction A + B $\longrightarrow$ C, which proceeds by a single-step bimolecular mechanism, which one of the following equations holds true?

    (a) $t_{1/2} = 0.693/k$

    (b) Rate of reaction $= k[A][B]$

    (c) Rate of appearance of C = rate of disappearance of A

    (d) $\ln[A] = -kt + \ln[A]_0$

16. In a reaction mechanism, (a) what is the difference between an *activated complex* and an *intermediate?* (b) What is meant by the rate-determining step? Which elementary reaction is often the rate-determining step?

17. State *two* important requirements for a substance to be considered a catalyst in a chemical reaction.

18. Neither a catalyst nor an intermediate appears in the net equation for a chemical reaction. Are they the same? If not, how do they differ?

19. What is the difference between *homogeneous* and *heterogeneous* catalysis?

20. What is the substrate in an enzyme-catalyzed reaction?

21. Describe the induced-fit model of an enzyme-catalyzed reaction.

22. What factors other than substrate concentration affect the rate of an enzyme-catalyzed reaction?

Problems

Rate of Reaction and Reaction Order

Problems 23 and 24 refer to the reaction
$H_2O_2(aq) \longrightarrow H_2O(l) + \frac{1}{2}\,O_2(g)$.

23. The initial concentration of H_2O_2 is 0.2546 M, and the initial rate of reaction is $9.32 \times 10^{-4}\ \text{M s}^{-1}$. What will be $[H_2O_2]$ at $t = 35$ s?

24. The initial concentration of H_2O_2 is 0.1108 M, and 12 s later the concentration is 0.1060 M. What is the initial rate of this reaction expressed in (a) M s^{-1} and (b) M min^{-1}?

25. In the reaction A + 2 B $\longrightarrow$ C + 3 D, the rate of disappearance of B is $-6.2 \times 10^{-4}\ \text{M s}^{-1}$. What is (a) the rate of disap-

pearance of A? (b) The rate of formation of D? (c) The general rate of reaction as described on page 530?

26. In the reaction $2\,A + 2\,B \longrightarrow C + 2\,D$, the rate of disappearance of A is $-2.2 \times 10^{-4}\ \text{M s}^{-1}$. What is (a) the rate of disappearance of B? (b) The rate of formation of C? (c) The general rate of reaction as described on page 530?

27. For the reaction, A $\longrightarrow$ products, a graph of [A] versus time is a curve. What can be concluded about the order of this reaction?

28. What would be the unit of k for a reaction A $\longrightarrow$ products for which the order of the reaction was one-half?

29. For the reaction $2\ A \longrightarrow D + C$, at least one of the following statements is true. Label the statements as true or false, and explain your reasoning.

(a) The half-life of this reaction can be determined with the expression $t_{1/2} = 1/k[A]_0$, where k is the rate constant and $[A]_0$ is the initial concentration of A.

(b) The rate at which B is produced is not the same as the rate at which A is consumed.

30. Following are two statements pertaining to the reaction $2\ A + B \longrightarrow 2\ C$, for which the rate law is $rate = k[A][B]$. Identify which statement is true and which is false, and explain your reasoning.

(a) The value of k is *independent* of the initial concentrations, $[A]_0$ and $[B]_0$.

(b) The unit of the rate constant for this reaction can be expressed either as s^{-1} or min^{-1}.

31. The rate of the following reaction in aqueous solution is monitored by measuring the rate of formation of I_3^-. Data obtained are listed in the table.

$$S_2O_8^{2-} + 3\ I^- \longrightarrow 2\ SO_4^{2-} + I_3^-$$

Experiment	$[S_2O_8^{2-}]$, M	$[I^-]$, M	Initial Rate, M s^{-1}
1	0.038	0.060	1.4×10^{-5}
2	0.076	0.060	2.8×10^{-5}
3	0.076	0.120	5.6×10^{-5}

(a) Determine the order of the reaction with respect to $S_2O_8^{2-}$, with respect to I^-, and overall.

(b) What is the value of the rate constant k?

(c) What would be the initial rate of reaction if $[S_2O_8^{2-}] = 0.083$ M and $[I^-] = 0.115$ M?

32. The rate of the following reaction in aqueous solution is monitored by measuring the number of moles of Hg_2Cl_2 that precipitate per liter per minute. The data obtained are listed in the table.

$$2\ HgCl_2 + C_2O_4^{2-} \longrightarrow 2\ Cl^- + 2\ CO_2(g) + Hg_2Cl_2(s)$$

Experiment	$[HgCl_2]$, M	$[C_2O_4^{2-}]$, M	Initial Rate, mol L^{-1} min^{-1}
1	0.105	0.15	1.8×10^{-5}
2	0.105	0.30	7.1×10^{-5}
3	0.052	0.30	3.5×10^{-5}
4	0.052	0.15	8.9×10^{-6}

(a) Determine the order of the reaction with respect to $HgCl_2$, with respect to $C_2O_4^{2-}$, and overall.

(b) What is the value of the rate constant k?

(c) What would be the initial rate of reaction if $[HgCl_2] = 0.094$ M and $[C_2O_4^{2-}] = 0.19$ M?

(d) Are all four experiments necessary to answer parts (a)–(c)? Explain.

33. In the reaction $A \longrightarrow$ products, we find that when [A] has fallen to half its initial value, the reaction proceeds at the same rate as its initial rate. Is the reaction zero order, first order, or second order? Explain.

34. In the reaction $A \longrightarrow$ products, with the initial concentration $[A]_0 = 1.512$ M, [A] is found to be 1.496 M at $t = 30$ s. With the initial concentration $[A]_0 = 2.584$ M, [A] is 2.552 M at $t = 1.0$ min. What is the order of this reaction?

35. Use a value of $k = 3.66 \times 10^{-3}$ s^{-1} to establish the instantaneous rate of the first-order decomposition of 2.05 M H_2O_2(aq).

36. What should be the instantaneous rate of the first-order decomposition of 3% H_2O_2(aq), by mass? Assume that the solution has a density of 1.00 g/mL and that $k = 3.66 \times 10^{-3}$ s^{-1}.

First-Order Reactions

37. A first-order reaction, $A \longrightarrow$ products, has a rate constant, k, of 0.0462 min^{-1}. (a) What is [A] at the time when the reaction is proceeding at a rate of 0.0150 M min^{-1}? (b) It takes 12 min for the reaction to go to 12% completion and 48 min to go to 56% completion. *Without doing detailed calculations,* determine how long it should take the reaction to go to 78% completion. (*Hint:* What is the percent remaining at each of these times?)

38. A first-order reaction, $A \longrightarrow$ products, has a rate of reaction of 0.00250 M s^{-1} when [A] = 0.484 M. (a) What is the rate constant, k, for this reaction? (b) Does $t_{3/4}$ depend on the initial concentration? Does $t_{4/5}$? Explain.

39. The half-life for the first-order decomposition of sulfuryl chloride at 320 °C is 8.75 h.

$$SO_2Cl_2(g) \longrightarrow SO_2(g) + Cl_2(g)$$

(a) What is the value of the rate constant, k, in s^{-1}? (b) What is the pressure of sulfuryl chloride 3.00 h after the start of the reaction, if its initial pressure is 722 mmHg? (c) How long after the start of the reaction will the pressure of sulfuryl chloride become 125 mmHg?

40. In the first-order decomposition of dinitrogen pentoxide at 335 K,

$$N_2O_5(g) \longrightarrow 2\ NO_2(g) + \tfrac{1}{2} O_2(g)$$

If we start with a 2.50-g sample of N_2O_5 at 335 K and have 1.50 g remaining after 109 s, (a) What is the value of the rate constant k? (b) What is the half-life of the reaction? (c) What mass of N_2O_5 will remain after 5.0 min?

41. The smog constituent peroxyacetyl nitrate (PAN) dissociates into peroxyacetyl radicals and NO_2(g) in a first-order reaction with a half-life of 32 min.

$$\underset{\text{PAN}}{CH_3COONO_2} \longrightarrow \underset{\substack{\text{Peroxyacetyl} \\ \text{radical}}}{CH_3COO\cdot} + NO_2$$

If the initial concentration of PAN in an air sample is 2.7×10^{15} molecules/L, what will be the concentration 2.24 h later?

42. The decomposition of dimethyl ether at 504 °C is a first-order reaction with a half-life of 27 min.

$$(CH_3)_2O(g) \longrightarrow CH_4(g) + H_2(g) + CO(g)$$

(a) What will be the partial pressure of $(CH_3)_2O$(g) after 1.00 h if its initial partial pressure was 749 mmHg?

(b) What will be the total gas pressure after 1.00 h? [*Hint:* $(CH_2)_2O$ and its decomposition products are the only gases present in the reaction vessel.]

Reactions of Various Orders

43. The rate constant for the second-order decomposition of HI at 700 K is $k = 1.2 \times 10^{-3} \, M^{-1} \, s^{-1}$. **(a)** In a reaction in which $[HI]_0 = 0.56 \, M$, what will $[HI]$ be at $t = 2.00 \, h$? **(b)** At what time will $[HI] = 0.28 \, M$ for the reaction?

44. Refer to Figure 13.8, which describes the decomposition of ammonia on a tungsten surface at 1100 °C. **(a)** Determine $[NH_3]$ at $t = 335 \, s$ if initially $[NH_3] = 0.0452 \, M$. **(b)** What is the half-life for the reaction in which $[NH_3] = 0.0452 \, M$?

45. The half-lives of zero-order and second-order reactions both depend on the initial concentration as well as on the rate constant k. In one case, the half-life gets longer as the initial concentration increases, and in the other case, it gets shorter. Which is which, and why is the situation not the same for both?

46. For the reaction $A \longrightarrow 2\,B + C$, the following data are obtained for $[A]$ as a function of time: $t = 0 \, min$, $[A] = 0.80 \, M$; 8 min, $[A] = 0.60 \, M$; 24 min, $[A] = 0.35 \, M$; 40 min, $[A] = 0.20 \, M$.

   (a) Establish the order of the reaction.

   (b) What is the value of the rate constant, k?

   (c) Calculate the rate of formation of B at 22 min.

47. The following data were obtained in two separate experiments in the reaction: $A \longrightarrow$ products. Determine the rate law for this reaction, including the value of k.

Experiment 1		Experiment 2	
[A], M	**Time, s**	**[A], M**	**Time, s**
0.800	0	0.400	0
0.775	40	0.390	64
0.750	83	0.380	132
0.725	129	0.370	203
0.700	179	0.360	278

48. Listed in the following table are initial rates, expressed in terms of the rate of decrease of partial pressure of a reactant in the following reaction at 826 °C. Determine the rate law for this reaction, including the value of k.

$$NO(g) + H_2(g) + \longrightarrow \tfrac{1}{2} N_2(g) + H_2O(g)$$

With Initial $P_{H_2} = 400 \, mmHg$		With Initial $P_{NO} = 400 \, mmHg$	
Initial P_{NO}, mmHg	**Rate, mmHg/s**	**Initial P_{H_2}, mmHg**	**Rate, mmHg/s**
359	0.750	289	0.800
300	0.515	205	0.550
152	0.125	147	0.395

49. In a process called flash photolysis, a flash of strong light decomposes Cl_2 into Cl atoms. The Cl atoms recombine according to a second-order rate law. The first half-life is 12 ms. How long will it take for the $[Cl]$ to drop to one-eighth its initial value?

50. A reaction that is second order in reactant A has $[A]_0 = 0.200 \, M$. The half-life is 45.6 sec. What is $[A]$ after 3.00 min?

Collision Theory; Activation Energy

51. Chemical reactions occur as a result of molecular collisions, and the frequency of molecular collisions can be calculated with the kinetic-molecular theory. Calculations of the rates of chemical reactions, however, are generally not very successful. Explain why.

52. A temperature increase of 10 °C causes an increase in collision frequency of only a few percent, yet it can cause the rate of a chemical reaction to increase by a factor of two or more. How do you explain this apparent discrepancy?

53. The activation energy for an endothermic reaction is never less than the enthalpy change for the reaction. Can we make a similar statement about exothermic reactions? Explain.

54. A mixture of hydrogen and oxygen gases is indefinitely stable at room temperature, but if struck by a spark, the mixture immediately explodes. What explanation can you offer for this observation?

55. Rate constants for the first-order decomposition of acetonedicarboxylic acid are $k = 4.75 \times 10^{-4} \, s^{-1}$ at 293 K and $k = 1.63 \times 10^{-3} \, s^{-1}$ at 303 K.

$$CO(CH_2COOH)_2(aq) \longrightarrow CO(CH_3)_2(aq) + 2\,CO_2(g)$$

Acetonedicarboxylic acid   Acetone

What is the activation energy, E_a, of this reaction?

56. The decomposition of di-*tert*-butyl peroxide (DTBP) is a first-order reaction with a half-life of 320 min at 135 °C and 100 min at 145 °C. Calculate E_a for this reaction.

$$(C_4H_9)_2O_2(g) \longrightarrow 2\,(CH_3)_2CO(g) + C_2H_6(g)$$

DTBP   Acetone   Ethane

57. The decomposition of ethylene oxide at 652 K is a first-order reaction with $k = 0.0120 \, min^{-1}$ and an activation energy of 218 kJ/mol.

$$(CH_2)_2O(g) \longrightarrow CH_4(g) + CO(g)$$

Calculate **(a)** the rate constant of the reaction at 525 K and **(b)** the temperature at which the rate constant $k = 0.0100 \, min^{-1}$.

58. A reaction has first-order rate constants of $4.82 \times 10^{-5} \, s^{-1}$ at 25 °C and $1.41 \times 10^{-2} \, s^{-1}$ at 70 °C. Calculate the rate constant at 90 °C.

Reaction Mechanisms

59. It is easy to see how a bimolecular elementary reaction in a reaction mechanism can occur as a result of a collision between two molecules. How do you suppose a *unimolecular* process is able to occur?

60. Why should we not necessarily expect the rate law of a net reaction to be the same as the rate law of one of the elementary reactions in a plausible reaction mechanism? Cite *two* situations, however, in which this may indeed be the case.

61. The following is proposed as a plausible reaction mechanism:

$$A + B \longrightarrow I \quad \text{(slow)}$$

$$I + B \longrightarrow C + D \quad \text{(fast)}$$

What is **(a)** the net reaction described by this mechanism, and **(b)** a plausible rate law for the reaction?

62. The reaction $A + 2B \longrightarrow C + 2D$ is found to be first order in A and first order in B. A proposed mechanism for the reaction involves the following first step:

$$A + B \longrightarrow I + D \quad \text{(slow)}$$

(a) Write a plausible second step in a two-step mechanism. **(b)** Is the second step slow or fast? Explain.

63. At temperatures below 600 K, the following reaction exhibits the rate law: Rate = $k[NO_2]^2$.

$$NO_2(g) + CO(g) \longrightarrow NO(g) + CO_2(g)$$

Propose a two-step mechanism involving one fast step and one slow step that is consistent with the net equation and the observed rate law.

64. Explain why the following mechanism is *not* plausible for the reaction in Problem 63.

Fast: $\qquad 2\,NO_2 \rightleftharpoons N_2O_4$

Slow: $\qquad N_2O_4 + 2\,CO \longrightarrow 2\,NO + 2\,CO_2$

65. The following reaction exhibits the rate law: Rate = $k[NO]^2[Cl_2]$.

$$2\,NO(g) + Cl_2(g) \longrightarrow 2\,NOCl(g)$$

Explain why the following mechanism is *not* plausible for this reaction.

Fast: $\qquad NO + Cl_2 \rightleftharpoons NOCl + Cl$

Slow: $\qquad NO + Cl \longrightarrow NOCl$

66. Propose a two-step mechanism involving one fast reversible step and one slow step that is consistent with the net equation and the observed rate law for the reaction in Problem 65.

67. Show that the proposed mechanism is consistent with the rate law for the following reaction in aqueous solution,

$$Hg_2^{2+} + Tl^{3+} \longrightarrow 2\,Hg^{2+} + Tl^+$$

for which the observed rate law is

$$\text{Rate} = k \times \frac{[Hg_2^{2+}][Tl^{3+}]}{[Hg^{2+}]}$$

Proposed mechanism:

Fast: $\qquad Hg_2^{2+} \rightleftharpoons Hg^{2+} + Hg$

Slow: $\qquad Hg + Tl^{3+} \longrightarrow Hg^{2+} + Tl^+$

68. Show that the proposed mechanism is consistent with the rate law for the following reaction having the rate law: Rate = $k[H_2][NO]^2$.

$$2\,H_2(g) + 2\,NO(g) \longrightarrow N_2(g) + 2\,H_2O(g)$$

Proposed mechanism:

$$2\,NO \rightleftharpoons N_2O_2$$

$$N_2O_2 + H_2 \longrightarrow N_2O + H_2O$$

$$N_2O + H_2 \longrightarrow N_2 + H_2O$$

Which must be the rate-determining step in this mechanism? Explain.

Catalysis

69. The decomposition of $H_2O_2(aq)$ is usually studied in the presence of $I^-(aq)$. The reaction is first order in both H_2O_2 and in I^-. Why can we treat the reaction as if it were only first order overall rather than second order?

70. Describe in a general way how the reaction profile for the surface-catalyzed reaction of SO_2 and O_2 to form SO_3 differs from the reaction profile for the noncatalyzed, homogeneous gas-phase reaction.

71. Describe ways in which an enzyme inhibitor may function.

72. A bacterial enzyme has an optimum temperature of 35 °C. Will the enzyme be more or less active at the normal human body temperature? Will it be more active or less active if the patient has a fever of 40 °C?

73. The kinetics of some surface-catalyzed reactions are similar to enzyme-catalyzed reactions. Explain this connection.

74. What happens to the rate of an enzyme-catalyzed reaction if the concentration of substrate X is doubled, when **(a)** the concentration of X is low, **(b)** the concentration of X is very high?

Additional Problems

Problems marked with an * may be more challenging than others.

75. In the first-order reaction $A \longrightarrow$ products, the following concentrations were found at the indicated times: $t = 0$ s, $[A] = 0.88$ M; 25 s, 0.74 M; 50 s, 0.62 M; 75 s, 0.52 M; 100 s, 0.44 M; 125 s, 0.37 M; 150 s, 0.31 M. Calculate the instantaneous rate of reaction at $t = 125$ s.

76. The decomposition of $H_2O_2(aq)$ can be followed by removing samples from the reaction mixture and titrating them with $MnO_4^-(aq)$.

$$2\,MnO_4^- + 5\,H_2O_2 + 6\,H^+ \longrightarrow 2\,Mn^{2+} + 8\,H_2O + 5\,O_2(g)$$

Assume that at each of the times listed in Table 13.1, a 5.00-mL sample of the remaining $H_2O_2(aq)$ is removed and titrated with

0.0500 M $KMnO_4$. Add a column to the table listing the volume of $KMnO_4$ required for the titration, initially, then at $t = 60$ s, then at $t = 120$ s, and so on.

77. Refer to Problem 76. Because the volumes of 0.0500 M $KMnO_4$ required for the titrations are proportional to the remaining $[H_2O_2]$, you can plot ln (mL $KMnO_4$) versus time and determine a value of k for the decomposition of $H_2O_2(aq)$. Show that the same value is obtained as in Figure 13.6.

* **78.** Benzenediazonium chloride decomposes in water, yielding $N_2(g)$.

$$C_6H_5N_2Cl \longrightarrow C_6H_5Cl + N_2(g)$$

The data in the following table were obtained for the decomposition of a 0.071 M solution at 50 °C ($t = \infty$ corresponds to the completed reaction). To obtain $[C_6H_5N_2Cl]$ as a function of time, note that during the first 3 min, the volume of $N_2(g)$ produced was 10.8 mL of a total 58.3 mL, corresponding to this fraction of the total reaction: 10.8 mL/58.3 mL = 0.185. An equal fraction of the available $C_6H_5N_2Cl$ was consumed during the same time.

Time, min	$N_2(g)$, mL	Time, min	$N_2(g)$, mL
0	0	18	41.3
3	10.8	21	44.3
6	19.3	24	46.5
9	26.3	27	48.4
12	32.4	30	50.4
15	37.3	∞	58.3

(a) Plot graphs showing the disappearance of $C_6H_5N_2Cl$ and the formation of $N_2(g)$ as a function of time. **(b)** What is the initial rate of formation of $N_2(g)$? **(c)** What is the rate of disappearance of $C_6H_5N_2Cl$ at $t = 20$ min? **(d)** What is the half-life, $t_{1/2}$, of the reaction? **(e)** Write the rate law for this reaction, including a value for k.

79. Refer to Problem 56. At 147 °C, the decomposition of DTBP is a first-order reaction with a half-life of 80.0 min. If a 4.50-g sample of DTBP is introduced into a 1.00-L flask at 147 °C, **(a)** what is the initial gas pressure in the flask? **(b)** What is the *total* gas pressure in the flask after 80.0 min? **(c)** the *total* gas pressure after 125 min?

80. Refer to Problem 41. The half-life for the decomposition of PAN at 25 °C is 32 min, and the activation energy of the reaction is 113 kJ/mol. **(a)** What is the half-life at 35 °C? **(b)** What is the initial rate of decomposition of PAN if 6.0×10^{14} molecules of PAN are injected into an air sample at 35 °C?

* **81.** The rate constant for the first-order dissociation of cyclobutane to ethylene,

$$cyclo\text{-}C_4H_8(g) \longrightarrow 2\ C_2H_4(g)$$

can be represented as $k = 4.0 \times 10^{16}\ s^{-1} \times e^{-E_a/RT}$ where $E_a = 262$ kJ/mol. Determine the temperature at which the half-life of the reaction is 1.00 min.

* **82.** The conversion of *tert*-butyl bromide to *tert*-butyl alcohol is achieved in the first-order reaction

$$(CH_3)_3CBr + H_2O \longrightarrow (CH_3)_3COH + HBr$$

The half-life of this reaction is 14.1 h at 25 °C and 48.8 min at 50 °C. How long will it take for this conversion to go to 90% completion at 65 °C?

83. A rule of thumb in chemical kinetics states that for many reactions, the rate of reaction approximately doubles for a temperature rise of 10 °C. What must be the activation energy of a

reaction if the rate is indeed found to double between 25 °C and 35 °C?

84. For which type of reaction would you expect the rate to increase more rapidly with increasing temperature, one with high or one with low activation energy? Explain.

* **85.** Hydroxide ion is involved in the mechanism but not consumed in this reaction in aqueous solution.

$$OCl^- + I^- \xrightarrow{OH^-} OI^- + Cl^-$$

(a) From the data in the table, determine the order of the reaction with respect to OCl^-, I^-, and OH^-, and the overall order.

$[OCl^-]$, M	$[I^-]$, M	$[OH^-]$, M	Rate of Formation of OI^-, mol $L^{-1}\ s^{-1}$
0.0040	0.0020	1.00	4.8×10^{-4}
0.0020	0.0040	1.00	5.0×10^{-4}
0.0020	0.0020	1.00	2.4×10^{-4}
0.0020	0.0020	0.50	4.6×10^{-4}
0.0020	0.0020	0.25	9.4×10^{-4}

(b) Write the rate law, and determine a value of the rate constant, k.

(c) Show that the following mechanism is consistent with the net equation and with the rate law. Which is the rate-determining step?

$$OCl^- + H_2O \rightleftharpoons HOCl + OH^-$$

$$I^- + HOCl \longrightarrow HOI + Cl^-$$

$$HOI + OH^- \longrightarrow H_2O + OI^-$$

(d) Is it appropriate to refer to OH^- as a catalyst in this reaction? Explain.

86. The observed rate law for the reaction $2\ O_3 \longrightarrow 3\ O_2$ is Rate $= k[O_3]^2/[O_2]$. Propose a two-step mechanism for this reaction.

87. For the first-order reaction $2\ N_2O_5 \longrightarrow 2\ N_2O_4 + O_2$, the activation energy is 106 kJ/mol. How many times faster will the reaction go at 100 °C than at 25 °C?

* **88.** Consider the reaction A $\longrightarrow$ products, and three hypothetical reactions all having the same numerical value for the rate constant, k. One of the reactions is zero order; one is first order; and one is second order. What must be the initial concentration of reactant, $[A]_0$, if **(a)** the zero- and first-order; **(b)** the zero- and second-order; **(c)** the first- and second-order reactions are to have the same half-life?

89. A student prepared a mixture moistened with water and placed it in a desiccator (drying chamber) with silica gel to dry at 22 °C. The silica gel absorbed water vapor as the mixture dried. The mixture was weighed periodically, giving the data in the table below. Determine the order of the drying process and find the rate constant.

Time, hr	Grams of Mixture
0	21.70
3.58	21.14
7.67	20.86
24.33	20.45
44.58	20.27
∞	19.99

90. The process of radioactive decay is first order. The rate of disintegration—called the *activity A*, or the decay rate—is directly proportional to the number of atoms, *N*, of the radioactive isotope present. Thus, *A* is analogous to a rate of reaction and *N*, to a molar concentration. A common means of determining a decay rate is to record ("count") the number of particles emitted by the radioisotope per unit time.

A sample containing radium-224, which decays by emitting α particles, has an activity of 648 counts per minute (cpm)

when first observed ($t = 0$). Its activities are 633 cpm at $t = 3.00$ h; 589 cpm at $t = 12.0$ h; and 562 cpm at $t = 18.0$ h. How long after first being observed will the activity of the sample have dropped to 25% of its initial value?

91. Use the data of accumulated mass of $O_2(g)$ as a function of time in Table 13.1 to derive the values of $[H_2O_2]$ shown. (*Hint*: This calculation bears some similarities to the Cumulative Example.)

Apply Your Knowledge

92. **[Biochemical]** The enzyme acetylcholinesterase has COOH and NH_2 groups on side chains of amino acids at the active site. The enzyme is inactive in acidic solution, but the activity increases as the solution becomes more basic. Explain.

* 93. **[Laboratory]** The iodination of acetone in aqueous solution is catalyzed by hydrogen ion:

$$I_2 + CH_3COCH_3 \xrightarrow{H^+} HI + CH_3COCH_2I$$

The reaction can be followed visually by adding starch. The purple starch–iodine complex forms immediately, then disappears when the iodine has been consumed. The kinetics of the reaction may be determined by recording the time required for the color to disappear, which is the time required for the iodine to be consumed. Given the data in the table below, **(a)** determine the order of the reaction with respect to I_2; **(b)** determine the order with respect to H^+; **(c)** determine the order with respect to acetone; **(d)** find the rate constant at 25.0 °C and at 42.4 °C; **(e)** find the activation energy in kJ/mol.

Solution	mL of 0.0010 M I_2	mL of 0.05 M HCl	mL of 1.00 M Acetone	mL of Water	Temperature, °C	Time for Color to Disappear (s)
A	5.0	10.0	10.0	25.0	25.0	130
B	10.0	10.0	10.0	20.0	25.0	249
C	10.0	20.0	10.0	10.0	25.0	128
D	10.0	10.0	20.0	10.0	25.0	131
E	10.0	10.0	10.0	20.0	42.4	38

94. **[Biochemical]** The chirping of tree crickets (*Oecanthus*) can be described by the Arrhenius rate equation. Use the experimental measurement of 179 chirps/min at 25.0 °C and 142 chirps/min at 21.7 °C to determine **(a)** the activation energy for the chirping process; **(b)** the number of chirps/min at 20.0 °C; and **(c)** how closely the result in (b) conforms to the rule of thumb that the Fahrenheit temperature is "40 plus the number of chirps in 15 s."

* 95. **[Laboratory]** A clock reaction is commonly used to demonstrate principles of chemical kinetics. In a clock reaction, an initially colorless solution changes color at a precise time during the course of the reaction. One such clock reaction is based on reaction (a).

$$\text{(a)} \quad S_2O_8^{2-}(aq) + 3\,I^-(aq) \longrightarrow 2\,SO_4^{2-}(aq) + I_3^-(aq)$$

Also present in the reaction mixture is thiosulfate ion, which reacts with I_3^- (reaction b) just as fast as the I_3^- is formed.

$$\text{(b)} \quad 2\,S_2O_3^{2-}(aq) + I_3^-(aq) \longrightarrow S_4O_6^{2-}(aq) + 3\,I^-(aq)$$

As soon as all the thiosulfate ion originally present is consumed, reaction (c) occurs between I_3^- and starch, which is also present in the original reaction mixture.

$$\text{(c)} \quad I_3^-(aq) + \text{starch} \longrightarrow \text{deep blue complex}$$

The color of the complex is discernible even when $[I_3^-]$ is extremely small. The table gives some representative data.

	Initial Concentrations, M		
Experiment	$(NH_4)_2S_2O_8$	KI	Time, s
1	0.20	0.20	21
2	0.10	0.20	42
3	0.050	0.20	81
4	0.20	0.10	42
5	0.20	0.050	79

Reaction conditions at 24 °C are 25.0 mL of $(NH_4)_2SO_4(aq)$, 25.0 mL of KI(aq), 10.0 mL of 0.010 M $Na_2S_2O_3(aq)$, and 5.0 mL starch solution are mixed. The time is for the first appearance of starch–iodine complex.

(a) Show that the rate of reaction is inversely related to the length of time that it takes for the blue color to appear.

(b) Use the data provided to determine the order of reaction (a) with respect to $S_2O_8^{2-}$, to I^-, and the overall order of the reaction.

(c) Calculate the initial rate of reaction (a) in Experiment 1, expressed in M s^{-1}. (*Hint*: Do not forget to take into account the dilutions that occur when the various solutions are mixed.)

(d) Determine a value for the rate constant, *k*, for reaction (a).

(e) The following mechanism has been proposed for reaction (a).

$$S_2O_8^{2-} + I^- \longrightarrow IS_2O_8^{3-} \qquad \text{(slow)}$$
$$IS_2O_8^{3-} \longrightarrow 2\,SO_4^{2-} + I^+ \qquad \text{(fast)}$$
$$I^+ + I^- \longrightarrow I_2 \qquad \text{(fast)}$$
$$I_2 + I^- \longrightarrow I_3^- \qquad \text{(fast)}$$

Write the balanced equation for the net reaction, and give the overall rate law based on this mechanism. Why should the first step be slow and the third step be fast?

* **96.** [**Environmental**] The hydroxyl radical ($\cdot$OH) is the principal chemical oxidant in the atmosphere. It reacts with organic molecules, carbon monoxide, nitrogen dioxide, and other pollutants, facilitating their removal from the atmosphere. Following are some important reactions in these processes.

1. The ($\cdot$OH) radical extracts a H atom from organic molecules:

$$RH + \cdot OH \longrightarrow R\cdot + HOH$$

2. Certain radicals react with O_2 to form peroxyl radicals:

$$R\cdot + O_2 \longrightarrow RO_2\cdot$$

3. Hydroperoxyl radicals ($H\cdot + O_2 \longrightarrow HO_2\cdot$) act by donating an O atom to molecules (X) capable of accepting them:

$$HO_2\cdot + X \longrightarrow XO + \cdot OH$$

4. The above reactions proceed until either a stable product (such as CO_2 or H_2O) or a rain-soluble substance is formed.

Use these facts and other information from the text in the remainder of this problem.

(a) With bond energies from Figure 9.13, calculate ΔH for the reaction of $\cdot$OH with $H_2(g)$.

(b) With bond energies from Table 9.1, estimate ΔH for the reaction of $\cdot$OH with $CH_4(g)$.

(c) Write equations for plausible reactions by which $\cdot$OH can remove (1) CO, (2) NO_2, and (3) HCHO (formaldehyde) from the atmosphere.

(d) Write equations for plausible reactions by which $\cdot$OH can convert $CH_4(g)$ to HCHO(g).

(e) Write a plausible mechanism for the reaction in which $\cdot$OH catalyzes the decomposition of ozone.

(f) Hydrocarbons react with both O_2 and $\cdot$OH, yielding CO_2 and H_2O in each case. Why is $\cdot$OH more effective than O_2 at removing pollutants from air?

(g) In reacting with a hydrocarbon, for which of the following radicals should the orientation during collisions affect the reaction rate: $\cdot$OH, $\cdot$Cl, $HO_2\cdot$, $CH_3O\cdot$, $CH_3O_2\cdot$?

(h) The reaction of $\cdot$OH with ozone has a frequency factor of $A = 1.6 \times 10^{-12}$ cm^3 molecule^{-1} s^{-1} and an activation energy of $E_a \approx 7.8$ kJ mol^{-1}. Calculate the rate constant, k, for the reaction at 220 K.

e-Media Problems

The activities described in these problems can be found in the e-Media Activities and Interactive Student Tutorial (IST) modules of the Companion Website, *http://chem.prenhall.com/hillpetrucci*.

97. The simplistic process of changing blue balls into red balls is illustrated in the **Rates of Change** animation (*Section 13-1*). (a) Suggest a quantitative way to describe the rate of this process. (b) Is the rate of disappearance of the blue balls the same as the rate of appearance of the red balls? (c) What must happen in this animation for change to occur? (d) Suggest a way to alter the animation to increase the rate of change.

98. For the reaction described in the **First-Order Process** animation (*Section 13-5*), (a) what is the length of the fourth half-life? (b) What is the total time required for the initial concentration to decrease to a value of 0.0625 M?

99. Consider the kinetic data presented in the **Plotting Kinetic Data** activity (*Section 13-6*). For each reaction, describe the shapes of

plots of (a) concentration, (b) the reciprocal of concentration, and (c) the natural log of concentration versus time. What do these shapes indicate regarding the kinetics of the reaction?

100. The reaction of OH^- and CH_3Cl is illustrated in the **Bimolecular Reaction** animation (*Section 13-9*). (a) How does this reaction mechanism qualitatively differ from a unimolecular and a termolecular reaction? (b) Identify the activated complex in this reaction. (c) Why would this species be less stable than either CH_3Cl or CH_3OH?

101. The reaction of hydrogen and ethylene on a metal surface is shown in the **Surface Reaction–Hydrogenation** animation (*Section 13-10*). (a) Write the balanced equation for this reaction. (b) What role does the metal surface play in the reaction? (c) Sketch a reaction profile for the catalyzed chemical reaction labeling the positions of the surface hydrogen (M—H) and absorbed hydrocarbon (M—CH_2CH_3) intermediates.

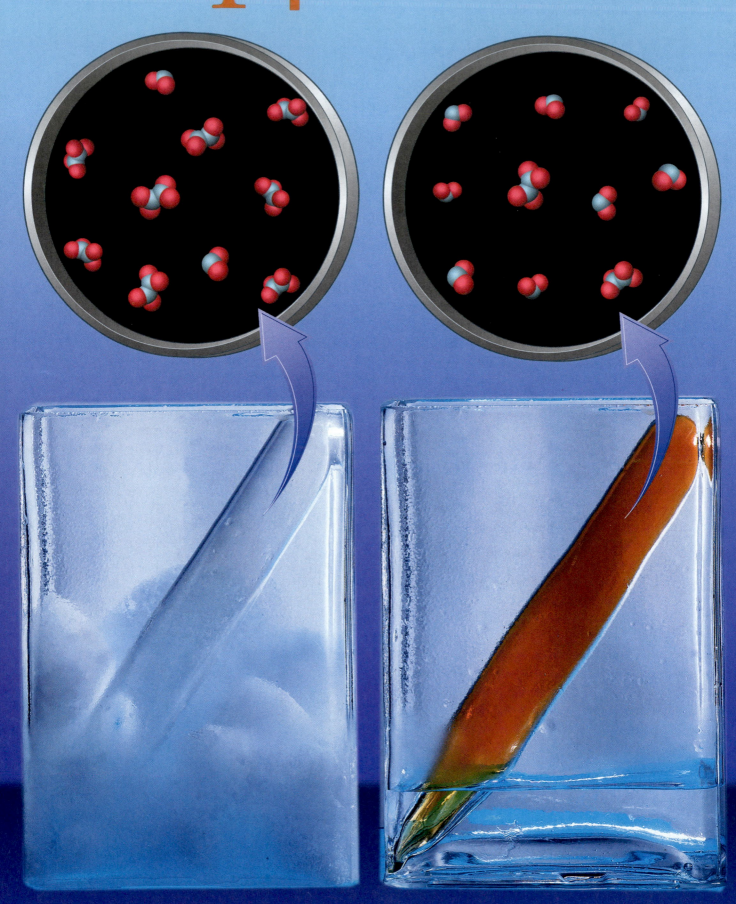

Chapter 14

Chemical Equilibrium

UP TO NOW, we have limited our calculations concerning chemical reactions to reactions that go to completion and that can therefore be described by stoichiometry alone. However, not all reactions go to completion. Reversible reactions reach an equilibrium state in which forward and reverse reactions proceed at the same rate. To calculate the amounts of reactants and products present at equilibrium, we need a new quantity, called the equilibrium constant.

Our study of equilibrium extends to some degree over the next five chapters and will address a number of fundamental questions. How do we determine equilibrium constants by experiment? How can we calculate equilibrium constants from other quantities? How do changes in pressure, volume, temperature, and amounts of reactants affect the equilibrium state? What effect does a catalyst have on equilibrium? As we proceed through these chapters, you will see that the principles of dynamic equilibrium are encountered in the laboratory, in chemical industry, in living organisms, and in other natural phenomena.

CONTENTS

 NO$_2$-N$_2$O$_4$ Equilibrium animation

14.1 The Dynamic Nature of Equilibrium

Equilibrium involves opposing processes occurring at equal rates. In vapor–pressure equilibrium (Chapter 11), the rate of evaporation of a liquid is equal to the rate of condensation of its vapor. In solubility equilibrium (Chapter 12), the rate of dissolution of a solid is equal to its rate of crystallization from solution. Moreover, these equilibria are *dynamic* (not *static* like a teaspoon balanced on the lip of a teacup). We can use radioactivity to demonstrate dynamic equilibrium, as is illustrated in Figure 14.1. If we add a small amount of NaCl(s) containing a trace of radioactive sodium-24 to a saturated NaCl(aq) solution, radioactivity shows up immediately in the saturated solution as well as in the undissolved solid, indicating that some solid dissolves. And because the concentration of a saturated solution remains constant, the amount of NaCl(s) that dissolves must be matched by the amount of NaCl(s) that crystallizes from solution.

◀ Red-brown nitrogen dioxide and colorless dinitrogen tetroxide exist in equilibrium with one another. When cooled, NO$_2$ reacts to form N$_2$O$_4$, and the gas mixture appears almost colorless (left). When warmed, NO$_2$ predominates and the mixture is dark red-brown (right). In this chapter we will examine some of the many chemical reactions that involve this sort of dynamic equilibrium.

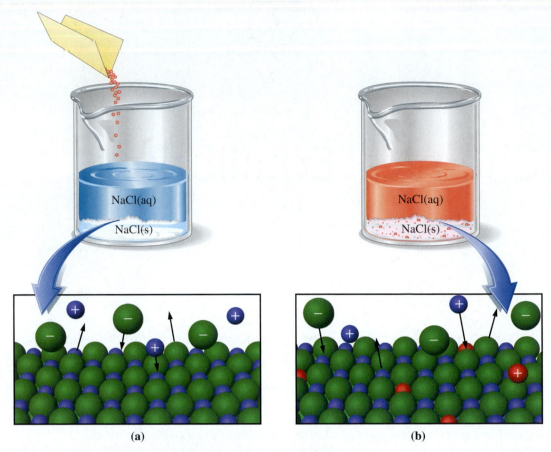

▲ **FIGURE 14.1   Dynamic equilibrium in a saturated solution**

(a) A trace of radioactive NaCl(s) (red) is added to a saturated NaCl(aq) solution. (b) Radioactivity immediately appears in the solution phase (pink). Radioactive sodium chloride dissolves, and at the same time, both radioactive and nonradioactive sodium chloride crystallize. Dissolution and crystallization do not stop when a solution becomes saturated.

Figure 14.2 depicts the rates of the forward and reverse reactions for the gas-phase equilibrium between dinitrogen tetroxide and nitrogen dioxide:

$$N_2O_4(g) \rightleftharpoons 2 NO_2(g)$$

As reactants are consumed, the rate of the forward reaction decreases; as products are formed, the rate of the reverse reaction increases. The curves representing these rates meet, and the rates become identical when the reversible reaction reaches equilibrium, at the time marked t_e. Beyond this point, although both forward and reverse reactions

🌐 **Dynamic Equilibrium animation**

🌐 **Rate vs. Time animation**

▶ **FIGURE 14.2   The concept of dynamic equilibrium**

As the reversible reaction $N_2O_4(g) \rightleftharpoons 2 NO_2(g)$ progresses, the forward and reverse rates approach each other. At equilibrium, the two reaction rates are equal.

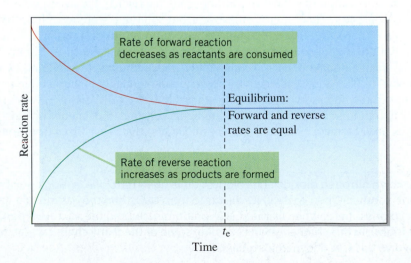

Rate of forward reaction decreases as reactants are consumed

Equilibrium:
Forward and reverse rates are equal

Rate of reverse reaction increases as products are formed

Reaction rate

t_e

Time

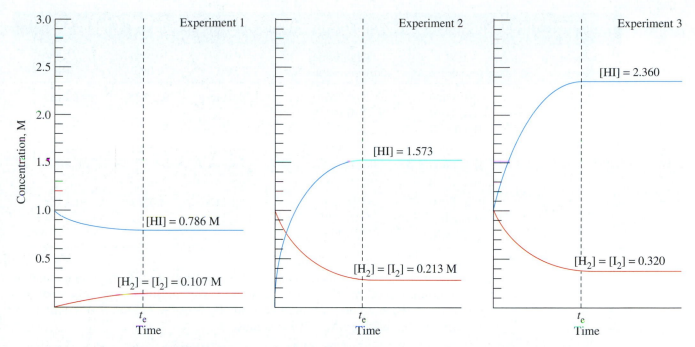

▲ **FIGURE 14.3**   **Concentration-versus-time graph for the reversible reaction 2 HI(g) $\rightleftharpoons$ H₂(g) + I₂(g) at 698 K**

After time t_e, the reaction is at equilibrium and the concentrations of reactant and products undergo no further change. The data for these three experiments are listed in Table 14.1.

QUESTION: Why do the reactant and product curves not meet as they do in Figure 14.2?

continue to occur, there is no net change in the reaction mixture. In chemical kinetics, we focus on the portion of the concentration-versus-time graph *before* t_e. In chemical equilibrium, we focus on what transpires *after* t_e.

Concentration vs. Time activity

Chemical Equilibrium activity

In a condition of **equilibrium,** *a forward reaction and a reverse reaction proceed at equal rates, and the concentrations of reactants and products remain constant.*

As an example of dynamic equilibrium, think of bailing out a leaking rowboat. Water leaking into the boat is analogous to a forward reaction, and our pouring buckets of water overboard is analogous to the reverse reaction. As we bail water out, we reach a point of equilibrium at which the water is bailed out just as fast as it leaks in. The level of water in the boat remains constant, analogous to the constant concentrations of reactants and products at equilibrium.

Figure 14.3 shows how concentrations vary with time in the reaction involving HI(g), H₂(g) and I₂(g) at 698 K:

$$2\ \text{HI(g)} \rightleftharpoons \text{H}_2\text{(g)} + \text{I}_2\text{(g)}$$

The curves are similar to some of the concentration-versus-time graphs in Chapter 13, but note an important difference: After t_e, the curves level off. In any reaction in which reactants and products attain constant, nonzero concentrations, we know that the reaction is reversible and that the forward reaction does not go to completion. That is why we use a double arrow when writing a chemical equation for a reversible reaction.

We could show experimentally that the equilibrium in this reaction is dynamic by introducing into the equilibrium mixture some I₂(g) containing a trace of radioactive iodine-131. The radioactivity would soon show up in the HI(g) as well as in the I₂(g).

14.2   The Equilibrium Constant Expression

The three parts of Figure 14.3 present data for Experiments 1, 2, and 3 in Table 14.1. Each experiment involves a different set of initial concentrations for the same reaction. Consider the data in the third column of Table 14.1—the equilibrium concentrations of HI, H₂ and I₂. Note that these data have nothing in common for the three experiments.

Experiment Number	Initial Concentrations, M	Equilibrium Concentrations, M	$\dfrac{[H_2][I_2]}{[HI]}$	$\dfrac{[H_2][I_2]}{2\,[HI]}$	$\dfrac{[H_2][I_2]}{[HI]^2}$
1	[HI]: 1.000 [H$_2$]: 0.000 [I$_2$]: 0.000	0.786 0.107 0.107	0.0146	0.00728	0.0185
2	[HI]: 0.000 [H$_2$]: 1.000 [I$_2$]: 1.000	1.573 0.213 0.213	0.0288	0.0144	0.0183
3	[HI]: 1.000 [H$_2$]: 1.000 [I$_2$]: 1.000	2.360 0.320 0.320	0.0434	0.0217	0.0184

Table 14.1 Three Experiments Involving the Reaction $2\,HI(g) \rightleftharpoons H_2(g) + I_2(g)$ at 698 K

Let us use trial and error (called a *heuristic approach*) to see if there is some ratio of the equilibrium concentrations that yields the same result for all three of the approaches to the equilibrium state. For example, consider the ratios of equilibrium concentrations in the fourth and fifth columns:

$$\frac{[H_2][I_2]}{[HI]} \quad \text{and} \quad \frac{[H_2][I_2]}{2[HI]}$$

Neither ratio has the same value in the three experiments; the first ratio changes from 0.0146 to 0.0288 to 0.0434, and the second changes from 0.00728 to 0.0144 to 0.0217. In the ratio in the sixth column of Table 14.1, concentrations are raised to powers given by the stoichiometric coefficients. Now, allowing for slight variations due to experimental errors, we see that this ratio does have the same value for all three experiments. This ratio of *equilibrium* concentrations raised to powers equal to the stoichiometric coefficients is called the **equilibrium constant expression.** It has a constant value regardless of the initial concentrations of reactants and products. This constant is denoted by the symbol K_c and is called the **concentration equilibrium constant:**

The *inverse* of the equilibrium constant expression also has a constant value. We will consider its significance in Section 14.3.

$$K_c = \frac{[H_2][I_2]}{[HI]^2} = 1.84 \times 10^{-2} \text{ (at 698 K)}$$

The subscript c in K_c signifies that concentrations (molarities) are used. We note the temperature because equilibrium constants are temperature-dependent. Thus, the value $K_c = 1.84 \times 10^{-2}$ applies only to the reaction $2\,HI(g) \rightleftharpoons H_2(g) + I_2(g)$ and only at 698 K.

Keep in mind that the equilibrium state can be approached from an initial condition in which only reactants are present (Experiment 1 in Table 14.1), or from an initial condition in which only products are present (Experiment 2), or from an initial condition in which both reactants and products are present (Experiment 3). In every case, the value of K_c is the same.

Another example of a reversible reaction is the oxidation of NO(g) to NO$_2$(g), a reaction that contributes to the formation of photochemical smog:

$$2\,NO(g) + O_2(g) \rightleftharpoons 2\,NO_2(g)$$

Equilibrium Constant Expression activity

For this reaction, the ratio of equilibrium concentrations that has a constant value is

$$K_c = \frac{[NO_2]^2}{[NO]^2[O_2]}$$

From these two examples, we can begin to see the general nature of an equilibrium constant expression:

- Concentrations of the products appear in the numerator, and concentrations of the reactants appear in the denominator. (Concentrations are expressed as molarities, but, as noted on page 580, units are generally omitted in the K_c expression.)

- The exponents of the concentrations are identical to the stoichiometric coefficients in the chemical equation.

These general ideas are illustrated by the following hypothetical reaction and its associated equilibrium constant expression:

$$a\,A + b\,B + \cdots \rightleftharpoons g\,G + h\,H + \cdots$$

$$K_c = \frac{[G]^g[H]^h\cdots}{[A]^a[B]^b\cdots} \qquad (14.1)$$

We can get a sense of the significance of an equilibrium constant expression by considering a case in which the decomposition of HI(g) at 698 K produces equilibrium concentrations of H_2 and I_2 of 0.0250 M. To find the equilibrium concentration of HI, we first write the equilibrium constant expression and then substitute these concentrations:

$$K_c = \frac{[H_2][I_2]}{[HI]^2} = \frac{(0.0250)(0.0250)}{[HI]^2} = 1.84 \times 10^{-2}$$

Then we solve for [HI]:

$$[HI]^2 = \frac{(0.0250)(0.0250)}{1.84 \times 10^{-2}}$$

$$[HI] = \sqrt{\frac{(0.0250)(0.0250)}{1.84 \times 10^{-2}}} = 0.184 \text{ M}$$

We will consider more calculations based on equilibrium constant expressions later in the chapter.

One method of determining $[I_2]$ in the equilibrium mixture is to extract a sample from the mixture and titrate the sample with $Na_2S_2O_3$(aq) (see Problem 92):

$$I_2(aq) + 2\,S_2O_3^{2-}(aq) \longrightarrow S_4O_6^{2-}(aq) + 2\,I^-(aq)$$

Example 14.1

If the equilibrium concentrations of Cl_2 and $COCl_2$ are the same at 395 °C, find the equilibrium concentration of CO in the reaction

$$CO(g) + Cl_2(g) \rightleftharpoons COCl_2(g) \qquad K_c = 1.2 \times 10^3 \text{ at } 395\,°C$$

STRATEGY

When asked to determine the concentration of a substance involved in a reversible reaction at equilibrium, we must write the equilibrium constant expression, enter the data given, and solve for the unknown concentration.

SOLUTION

Because $[Cl_2]$ equals $[COCl_2]$ in the equilibrium mixture at 395 °C, these two terms cancel in the equilibrium constant expression.

$$K_c = \frac{\cancel{[COCl_2]}}{[CO]\cancel{[Cl_2]}} = \frac{1}{[CO]} = 1.2 \times 10^3$$

We can then solve this expression for [CO].

$$[CO] = \frac{1}{1.2 \times 10^3} = 8.3 \times 10^{-4} \text{ M}$$

ASSESSMENT

This purely algebraic solution leads to the conclusion that there is only one possible value of [CO] for the condition of equal concentrations of Cl_2 and $COCl_2$, regardless of what those concentrations are.

EXERCISE 14.1A

If [CO] equals $[Cl_2]$ at equilibrium for the reaction in Example 14.1, is there just one possible value of $[COCl_2]$? Explain.

EXERCISE 14.1B

Suppose $[O_2]$ is fixed at a certain constant value when equilibrium is reached in the reversible reaction

$$2\,SO_2(g) + O_2(g) \rightleftharpoons 2\,SO_3(g) \qquad K_c = 1.00 \times 10^2$$

Do $[SO_2]$ and $[SO_3]$ have unique values? Does the ratio $[SO_2]/[SO_3]$ have a unique value? Explain. Describe the equilibrium state when $[O_2] = 1.00$ M.

The Condition of Equilibrium—A Kinetics View

Let us consider again the reaction

$$2\,HI(g) \rightleftharpoons H_2(g) + I_2(g)$$

If we assume that the reaction occurs in a single bimolecular elementary step, we can see that the rate laws for the forward and reverse reactions are

$$\text{Rate of forward reaction} = k_f[HI]^2$$

$$\text{Rate of reverse reaction} = k_r[H_2][I_2]$$

At equilibrium, the rates of the forward and reverse reaction are equal, and so we can set the right sides of the two equations equal:

$$k_f[HI]^2 = k_r[H_2][I_2]$$

We can then gather the two rate constants on the same side of the equation

$$\frac{k_f}{k_r} = \frac{[H_2][I_2]}{[HI]^2} = K_c$$

to see that the ratio k_f/k_r is equal to the equilibrium constant, K_c.

You may object to this derivation on the grounds that we assumed a particular reaction mechanism that might not be the most plausible. However, as you can see in Problem 89 of this chapter, we could derive the same expression for K_c by using a different reaction mechanism (recall Example 13.10). In fact, we can say the same for any plausible multistep mechanism in which each step reaches equilibrium and the sum of the steps yields the correct overall reaction.

Although it is of theoretical interest, because the kinetics approach to chemical equilibrium also requires experimental data, we usually just evaluate equilibrium constants directly by experiments.

> This derivation brings out the relationship between rate constants and equilibrium constants for chemical reactions, but you must be careful not to confuse the two. Lowercase italic k is the symbol for the rate constant of a reaction, and uppercase K_c is the symbol for the equilibrium constant of a reaction.
>
> In practice the kinetics approach to chemical equilibrium is not frequently used. The kinetic data are usually harder to obtain than the direct measurement of equilibrium concentrations.

The Condition of Equilibrium—The Thermodynamic View

In Chapter 17, we will show that the equilibrium constant can be related to other fundamental thermodynamic properties, and we will then call it the *thermodynamic equilibrium constant, K_{eq}*. Moreover, we will show how to use tabulated thermodynamic data to *predict* values of equilibrium constants.

In anticipation of the switch to K_{eq}, we have written equilibrium constants as dimensionless numbers because the thermodynamic equilibrium constant expression uses dimensionless quantities known as *activities*. In this chapter, we will use molarities and partial pressures (in atmospheres) in equilibrium constant expressions. In both cases, we will omit the units. In Chapter 17 we will see why it is permissible to do so.

14.3 Modifying Equilibrium Constant Expressions

Sometimes we need to modify an equilibrium constant expression to make it applicable to a particular situation. We consider a few important modifications in this section.

Modifying the Chemical Equation The following equation is one way to describe the formation of $NO_2(g)$ at 298 K:

$$2\,NO(g) + O_2(g) \rightleftharpoons 2\,NO_2(g)$$

Using appropriate experimental data similar to those in Table 14.1, we could establish the numerical value of K_c as being

$$K_c = \frac{[NO_2]^2}{[NO]^2[O_2]} = 4.67 \times 10^{13} \text{ (at 298 K)}$$

If we were interested in the decomposition of $NO_2(g)$ at 298 K, we would likely write the chemical equation as the reverse of that for its formation:

$$2\,NO_2(g) \rightleftharpoons 2\,NO(g) + O_2(g) \qquad K'_c = ? \text{ (at 298 K)}$$

However, we do not need to do another set of experiments to establish the value of the new equilibrium constant, designated K'_c, because the equilibrium constant for the decomposition of $NO_2(g)$ is the *inverse* of the equilibrium constant for its formation:

$$K'_c = \frac{[NO]^2[O_2]}{[NO_2]^2} = \frac{1}{\dfrac{[NO_2]^2}{[NO]^2[O_2]}} = \frac{1}{K_c} = \frac{1}{4.67 \times 10^{13}} = 2.14 \times 10^{-14}$$

This modification illustrates a general rule:

When we <u>reverse</u> the equation for a chemical reaction for which the equilibrium constant is K_c, we <u>invert</u> the equilibrium constant. That is, the reverse reaction has the equilibrium constant $1/K_c$.

Modifying Chemical Equilibrium Expressions activity

Suppose we decide to describe the decomposition of $NO_2(g)$ based on one mole of reactant instead of two:

$$NO_2(g) \rightleftharpoons NO(g) + \tfrac{1}{2}O_2(g) \qquad K''_c = ? \text{ (at 298 K)}$$

Again, we do not need any additional experimental data because we can use the relationship

$$K''_c = \frac{[NO][O_2]^{1/2}}{[NO_2]} = \left[\frac{[NO]^2[O_2]}{[NO]^2}\right]^{1/2} = (K'_c)^{1/2}$$

$$= \left[\frac{1}{K_c}\right]^{1/2} = \sqrt{2.14 \times 10^{-14}} = 1.46 \times 10^{-7}$$

The preceding illustrates another general rule:

When the coefficients of an equation are <u>multiplied</u> by a common factor <u>n</u> to produce a new equation, we <u>raise</u> the original K_c value to the <u>power n</u> to obtain the new equilibrium constant.

In the preceding example, $n = \tfrac{1}{2}$. If we double the coefficients in an equation, the factor $n = 2$, and so on.

In summary, the form of an equilibrium constant expression and the value of K_c depend on exactly how the chemical equation for a reversible reaction is written. *Thus, we must write the balanced chemical equation when citing a value for K_c.*

Example 14.2

The equilibrium constant for the reaction

$$\tfrac{1}{2}H_2(g) + \tfrac{1}{2}I_2(g) \rightleftharpoons HI(g)$$

at 718 K is 7.07. **(a)** What is the value of K_c at 718 K for the reaction $HI(g) \rightleftharpoons \tfrac{1}{2}H_2(g) + \tfrac{1}{2}I_2(g)$? **(b)** What is the value of K_c at 718 K for the reaction $H_2(g) + I_2(g) \rightleftharpoons 2HI(g)$?

SOLUTION

(a) Because the reaction in question is the reverse of the one for which the equilibrium constant is 7.07, the equilibrium constant we seek is the inverse of 7.07:

$$K_c = \frac{1}{7.07} = 0.141$$

(b) In this chemical equation, the coefficients of the original equation have been doubled. Thus, we need to raise the value of the original equilibrium constant to the power of 2—in other words, we must square the original value:

$$K_c = (7.07)^2 = 50.0$$

EXERCISE 14.2A

The equilibrium constant for the reaction $SO_2(g) + \frac{1}{2} O_2(g) \rightleftharpoons SO_3(g)$ is 20.0 at 973 K. Calculate K_c at 973 K for the reaction

$$2 SO_3(g) \rightleftharpoons 2 SO_2(g) + O_2(g)$$

EXERCISE 14.2B

The equilibrium constant for the reaction $\frac{1}{2} N_2(g) + \frac{3}{2} H_2O(g) \rightleftharpoons NH_3(g) + \frac{3}{4} O_2(g)$ at 900 K is 1.97×10^{-20}. Calculate K_c at 900 K for the reaction

$$4 NH_3(g) + 3 O_2(g) \rightleftharpoons 2 N_2(g) + 6 H_2O(g)$$

The Equilibrium Constant for an Overall Reaction

In Section 6.6, we combined the equations for individual reactions to obtain an overall equation. At the same time, we used Hess's law to combine the enthalpy changes for the individual reactions to obtain the enthalpy change for the overall reaction. We use a similar approach to obtain the equilibrium constant for an overall reaction.

Suppose we want to know the equilibrium constant at 298 K for this reaction:

(a) $$N_2O(g) + \tfrac{3}{2} O_2(g) \rightleftharpoons 2 NO_2(g)$$

If we know values at 298 K for two other reactions involving these substances, we can add these equations to get the equation for reaction (a) as the overall equation:

(b) $N_2O(g) + \tfrac{1}{2} O_2(g) \rightleftharpoons 2 \cancel{NO(g)}$ $K_c(b) = 1.7 \times 10^{-13}$

(c) $2 \cancel{NO(g)} + O_2(g) \rightleftharpoons 2 NO_2(g)$ $K_c(c) = 4.67 \times 10^{13}$

Overall: $N_2O(g) + \tfrac{3}{2} O_2(g) \rightleftharpoons 2 NO_2(g)$ $K_c(a) = ?$

Now we can find the relationship between the unknown $K_c(a)$ and the known $K_c(b)$ and $K_c(c)$:

$$\frac{[\cancel{NO}]^2}{[N_2O][O_2]^{1/2}} \times \frac{[NO_2]^2}{[\cancel{NO}]^2[O_2]} = \frac{[NO_2]^2}{[N_2O][O_2]^{3/2}}$$

$$K_c(b) \qquad\quad \times K_c(c) \qquad = K_c(a)$$

$$1.7 \times 10^{-13} \quad \times 4.67 \times 10^{13} = 7.9$$

The preceding illustrates yet another general rule:

When we __add__ the equations for individual reactions to obtain an overall equation, we __multiply__ their equilibrium constants to obtain the equilibrium constant for the overall reaction.

Equilibria Involving Gases

In reactions involving gases, it is often convenient to measure partial pressures rather than molarities. Consider the general gas-phase reaction

$$a\,A(g) + b\,B(g) + \cdots \rightleftharpoons g\,G(g) + h\,H(g) + \cdots$$

We can define a **partial pressure equilibrium constant, K_p,** as

$$K_p = \frac{(P_G)^g (P_H)^h \cdots}{(P_A)^a (P_B)^b \cdots} \tag{14.2}$$

where each P is a partial pressure.

At times, we have a value of K_c for a reaction and need to know K_p, or vice versa. Let us use the following reaction at 298 K to derive a relationship between K_c and K_p:

$$2 NO(g) + O_2(g) \rightleftharpoons 2 NO_2(g) \qquad K_c = 4.67 \times 10^{13}$$

Suppose we apply the ideal gas law ($PV = nRT$) to NO_2 and solve for P_{NO_2}:

$$P_{NO_2} = \frac{n_{NO_2}}{V} RT$$

Then, with the gas volume measured in liters, we can replace n_{NO_2}/V by its equivalent, the molarity of NO_2:

$$P_{NO_2} = \frac{n_{NO_2}}{V} \times RT = [NO_2]RT$$

We can do the same for NO and O_2 and write the K_p expression for the reaction, which is

$$K_p = \frac{(P_{NO_2})^2}{(P_{NO})^2(P_{O_2})} = \frac{([NO_2]RT)^2}{([NO]RT)^2[O_2]RT} = \frac{[NO_2]^2 \cancel{(RT)^2}}{[NO]^2 \cancel{(RT)^2}[O_2](RT)}$$

$$= \frac{[NO_2]^2}{[NO]^2[O_2]} \times \frac{1}{RT}$$

The expression shown in red is simply K_c for the reaction. The relationship between K_p and K_c is therefore

$$K_p = \frac{K_c}{RT} = K_c(RT)^{-1}$$

Now, let's consider again our general gas-phase reaction,

$$a\,A(g) + b\,B(g) + \cdots \rightleftharpoons g\,G(g) + h\,H(g) + \cdots$$

Following the same procedure we just used for the $NO(g) + O_2(g)$ reaction, we obtain an equation for the general reaction:

$$K_p = K_c(RT)^{\Delta n_{gas}} \tag{14.3}$$

The exponent Δn_{gas} is the change in the number of moles of *gas* as the reaction occurs in the forward direction. That is, $\Delta n_{gas} = (g + h + \cdots) - (a + b + \cdots)$.

Returning to the reaction

$$2\,NO(g) + O_2(g) \rightleftharpoons 2\,NO_2(g)$$

we see that $\Delta n_{gas} = 2 - (2 + 1) = 2 - 3 = -1$, and $K_p = K_c(RT)^{-1}$. To evaluate K_p from the known value of K_c (4.67×10^{13}), we use $R = 0.08206$ and $T = 298$:

$$K_p = K_c(RT)^{-1} = 4.67 \times 10^{13} \times (0.08206 \times 298)^{-1} = 1.91 \times 10^{12}$$

Although units are not used for equilibrium constants, it is necessary that we use the dimensionless value of the ideal gas constant that corresponds to concentrations and pressures ($R = 0.08206$ L atm mol^{-1} K^{-1}) and use temperatures expressed in units of Kelvin. Unless otherwise indicated, K_p values are based on pressures expressed in atmospheres. In most respects, we can deal with K_p expressions as we did with K_c expressions, as illustrated in Example 14.3.

Example 14.3

Consider the equilibrium between dinitrogen tetroxide and nitrogen dioxide:

$$N_2O_4(g) \rightleftharpoons 2\,NO_2(g) \qquad K_p = 0.660 \text{ at } 319 \text{ K}$$

(a) What is the value of K_c for this reaction? **(b)** What is the value of K_p for the reaction $2\,NO_2(g) \rightleftharpoons N_2O_4(g)$? **(c)** If the equilibrium partial pressure of $NO_2(g)$ is 0.332 atm, what is the equilibrium partial pressure of $N_2O_4(g)$?

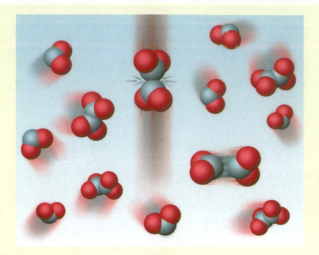

◀ Dinitrogen tetroxide gas and nitrogen dioxide gas exist together in equilibrium. At a given temperature, the amounts of NO_2 and N_2O_4 do not change, but some pairs of NO_2 molecules are combining to form N_2O_4, and at the same time, some N_2O_4 molecules are splitting into two NO_2 molecules. The rates of the two processes are equal, indicating that dinitrogen tetroxide and nitrogen dioxide exist in a dynamic equilibrium.

SOLUTION

(a) To find K_c, we will use Equation (14.3), which relates K_c and K_p.

$$K_p = K_c(RT)^{\Delta n_{gas}}$$

For this reaction, Δn_{gas} is equal to the difference between the number of moles of NO_2 and of N_2O_4 in the balanced equation.

$$\Delta n_{gas} = 2 - 1 = 1 \quad \text{and}$$
$$K_p = K_c(RT)^1$$

We can now solve for K_c by using the known values of K_p, T, and R.

$$K_c = \frac{K_p}{(RT)^1} = \frac{0.660}{0.08206 \times 319} = 0.0252$$

(b) The constant K_p for the reaction specified in this part is the inverse of the K_p value for the reaction initially given. We therefore must invert the known value (0.660).

Given: $\quad N_2O_4(g) \rightleftharpoons 2\,NO_2(g) \quad K_p = 0.660$ at 319 K

Specified: $\quad 2\,NO_2(g) \rightleftharpoons N_2O_4(g)$

$$K_p = \frac{1}{0.660} = 1.52$$

(c) To find the partial pressure of $N_2O_4(g)$, we can use either the original K_p expression or the one derived in part (b). Here we use the original one and substitute in the equilibrium partial pressure of NO_2.

$$N_2O_4(g) \rightleftharpoons 2\,NO_2(g)$$

$$K_p = \frac{(P_{NO_2})^2}{(P_{N_2O_4})} = 0.660$$

$$= \frac{(0.332)^2}{(P_{N_2O_4})} = 0.660$$

$$P_{N_2O_4} = \frac{(0.332)^2}{0.660} = 0.167 \text{ atm}$$

ASSESSMENT

Note that in parts (a) and (b) we did not substitute units into the equilibrium constant expressions; we were seeking the values of K_c and K_p, which are dimensionless quantities. In part (c), we started with a dimensionless quantity (K_p), but we did show a unit in our final result (0.167 atm) because the value $K_p = 0.660$ is valid only if the pressure unit used for the gases is atmospheres.

EXERCISE 14.3A

Given $K_c = 1.8 \times 10^{-6}$ for the reaction $2\,NO(g) + O_2(g) \rightleftharpoons 2\,NO_2(g)$ at 457 K, determine the value of K_p at 457 K for the reaction $NO_2(g) \rightleftharpoons NO(g) + \frac{1}{2}\,O_2(g)$.

EXERCISE 14.3B

For the reaction $NO(g) + \frac{3}{2}\,H_2O(g) \rightleftharpoons NH_3(g) + \frac{5}{4}\,O_2(g)$, $K_p = 2.6 \times 10^{-16}$ at 900 K. What is K_c at 900 K for the reaction $4\,NH_3(g) + 5\,O_2(g) \rightleftharpoons 4\,NO(g) + 6\,H_2O(g)$?

Equilibria Involving Pure Solids and Liquids

So far in this chapter we have considered only *homogeneous* reactions in which all reactants and products are gases. The reaction medium is a single gaseous phase. In *heterogeneous* reactions, the reactants and products are not in the same phase, and we need to make some accommodations in equilibrium constant expressions.

A general feature of equilibrium constant expressions is that they conform to the following idea:

> *Equilibrium constant expressions do not include terms for any reactants or products present as pure solids or pure liquids because their concentrations do not change in a reaction.*

Although the *amounts* of pure solid and liquid reactants and products change during a reaction, their *concentrations* do not change.

Consider the reversible decomposition of calcium carbonate (the chief constituent of limestone):

$$CaCO_3(s) \rightleftharpoons CaO(s) + CO_2(g)$$

If $CaCO_3(s)$ is heated in a closed container, it decomposes to produce a second solid phase, $CaO(s)$, and a gaseous phase, $CO_2(g)$. In the reverse reaction, some of the newly formed $CaO(s)$ and $CO_2(g)$ recombine to form $CaCO_3(s)$. When equilibrium is reached, there is a smaller amount of $CaCO_3(s)$ than was present initially, together with the two new phases, $CaO(s)$ and $CO_2(g)$. The two solids coexist as a heterogeneous mixture of completely separate phases—one phase consists of $CaCO_3$ only, and the other of CaO only. In other words, each of the solid phases is pure. Because the compositions of these solids remain fixed throughout the reaction, we do not include them in the equilibrium constant expression.

The $CO_2(g)$ that forms is also pure. However, as the amount of $CO_2(g)$ increases, so do the concentration and pressure of CO_2 in the closed container. Thus, CO_2 does appear in the equilibrium constant expression. In fact, it is the only substance that does:

$$K_c = [CO_2(g)] \quad \text{and} \quad K_p = P_{CO_2}$$

As noted on page 580, the thermodynamic equilibrium constant is written in terms of activities. Because the activities of pure solids and pure liquids are defined as equal to unity, there is no need to include these terms in the K_c and K_p expressions.

We can write similar expressions for the equilibrium between pure liquid water and water vapor:

$$H_2O(l) \rightleftharpoons H_2O(g)$$

$$K_c = [H_2O(g)] \quad \text{and} \quad K_p = P_{H_2O}$$

Note that K_p for a liquid–vapor equilibrium is simply the vapor pressure of the liquid.

Example 14.4

The reaction of steam and coke (a form of carbon) produces a mixture of carbon monoxide and hydrogen, called water-gas. This reaction has long been used to make combustible gases from coal:

$$C(s) + H_2O(g) \rightleftharpoons CO(g) + H_2(g)$$

Write the equilibrium constant expression for K_c for this reaction.

SOLUTION

The products CO and H_2 and the reactant H_2O are all gases and are represented in the equilibrium constant expression, but C(s), a solid, is not. The equilibrium constant expression is therefore

$$K_c = \frac{[CO][H_2]}{[H_2O]}$$

Review of Equilibrium Constant
Expressions activity

EXERCISE 14.4A

Write the partial pressure equilibrium constant expression for the reaction in Example 14.4.

EXERCISE 14.4B

The reaction of steam with iron is an old method of producing hydrogen gas:

$$3 \, Fe(s) + 4 \, H_2O(g) \rightleftharpoons Fe_3O_4(s) + 4 \, H_2(g)$$

Write the equilibrium constant expressions, K_c and K_p, for this reaction.

Equilibrium Constants: When Do We Need Them and When Do We Not?

In principle, every reaction is reversible, at least to some extent, and can be described through an equilibrium constant expression. In many cases, however, we do not need to use equilibrium constants in calculations. How can this be? Let's answer this question by considering three specific cases.

First, consider the reaction of hydrogen and oxygen gases at 298 K:

$$2 \, H_2(g) + O_2(g) \rightleftharpoons 2 \, H_2O(l)$$

$$K_p = \frac{1}{(P_{H_2})^2(P_{O_2})} = 1.4 \times 10^{83}$$

Starting with a 2 : 1 mole ratio of hydrogen to oxygen, the equilibrium partial pressures of $H_2(g)$ and $O_2(g)$ must become extremely small—approaching zero—in order for the K_p value to be so large. We conclude that, for all practical purposes, the hydrogen and oxygen are totally consumed in the reaction. We say that a reaction *goes to completion* if one or more reactants is totally consumed, and we can do calculations by using just the principles of stoichiometry (Chapter 3).

> *A very large numerical value of K_c or K_p signifies that a reaction goes to completion, or essentially so. (In effect, the reaction is not reversible.)*

As a second case, let us consider a reaction with a very different outcome, the decomposition of limestone at 298 K:

$$CaCO_3(s) \rightleftharpoons CaO(s) + CO_2(g) \qquad K_p = P_{CO_2} = 1.9 \times 10^{-23}$$

Intuitively, we know that limestone, which is mainly $CaCO_3(s)$, does not decompose to any great extent at normal temperatures. The K_p value tells us that the partial pressure of $CO_2(g)$ in equilibrium with $CaCO_3(s)$ and $CaO(s)$ is exceedingly small—$P_{CO_2} = K_p = 1.9 \times 10^{-23}$ atm.

> *A very small numerical value of K_c or K_p signifies that the forward reaction, as written, occurs only to a very slight extent.*

In fact, in many such cases, we say that the forward reaction does not occur. Therefore, we sometimes describe a reaction that has a very small equilibrium constant as "no reaction":

$$CaCO_3(s) \xrightarrow{298 \text{ K}} \text{"no reaction"}$$

The situation is quite different when we consider the decomposition of limestone at about 1300 K:

$$CaCO_3(s) \rightleftharpoons CaO(s) + CO_2(g) \qquad K_p \approx 1$$

Here the forward and reverse reactions are both significant, and we do indeed need to use the K_p expression in calculations.

Finally, we consider a case in which chemical kinetics strongly influences the progress of a reaction. We must keep in mind that an equilibrium constant expression

It is difficult to give a precise meaning to "very large," but K values with double-digit powers of 10 generally meet the requirement.

Double-digit *negative* powers of 10 generally meet the requirement for "very slight."

Equilibrium Constant activity

applies only to a reversible reaction *at equilibrium*. Reaction rates determine how long a system takes to reach equilibrium and thus, indirectly, when the equilibrium constant expression can be used. Although K_p for the reaction of $H_2(g)$ and $O_2(g)$ at 298 K described on page 586 is very large, suggesting that the reaction should go to completion, the reaction proceeds at an immeasurably slow rate because of its high activation energy. The reaction never reaches equilibrium at 298 K. It is only when the mixture is catalyzed, strongly heated, or ignited by a spark that the reaction occurs at an explosive speed. Thus, in this and similar cases, chemists say that the reaction is *thermodynamically favorable* (meaning that the equilibrium constant is large) but is *kinetically controlled* (meaning that the exceedingly slow reaction rate prevents any significant reaction from occurring).

> An equilibrium constant expression does not fully describe the actual extent of a chemical reaction under all conditions at a given time. In some cases, the reaction is limited by chemical kinetics.

Example 14.5

Is the reaction $CaO(s) + CO_2(g) \rightleftharpoons CaCO_3(s)$ likely to occur to any appreciable extent at 298 K?

SOLUTION

This reaction is the reverse of that describing the decomposition of limestone at 298 K. Its K_p value is therefore the reciprocal of that for the 298 K decomposition of limestone: $K_p = 1/(1.9 \times 10^{-23}) = 5.3 \times 10^{22}$. The large value of K_p leads us to expect the forward reaction to occur to a very significant extent. In fact, over time, the reaction should go essentially to completion.

EXERCISE 14.5A

Refer to Example 14.1, and determine whether we can assume that the reaction $CO(g) + Cl_2(g) \rightleftharpoons COCl_2(g)$ goes essentially to completion at 395 °C. Explain your reasoning.

EXERCISE 14.5B

The water-gas reaction, $C(s) + H_2O(g) \rightleftharpoons CO(g) + H_2(g)$, occurs hardly at all in the forward direction at room temperature, but at 1100 K, significant amounts of all reactants and products are found at equilibrium. Which seems the most plausible value of K_p at 1100 K: 1×10^{-30}; 1×10^{-21}; 10.0; 1×10^{15}; or 1×10^{35}? Explain.

▲ That we are today still able to see every exquisite detail of Michelangelo's marble sculpture *David*, finished in 1504, is testimony to the very, very limited decomposition of $CaCO_3(s)$ that occurs at 298 K. Many types of marble are nearly pure $CaCO_3(s)$. *Source:* Michelangelo (1475–1564), David p. (testa di profilo). Accademia Firenze. Scala/Art Resource.

The Reaction Quotient, *Q*: Predicting the Direction of Net Change

As we have noted, *at equilibrium*, concentrations and partial pressures of reactants and products have a fixed relationship for a given reaction. That is, concentrations must be in accord with the concentration equilibrium constant K_c, and partial pressures must be in agreement with the partial pressure equilibrium constant K_p. Initially, however, we can bring together reactants and/or products in just about any concentrations or partial pressures. For these *nonequilibrium* conditions, the expression having the same form as the concentrations ratio that gives us K_c or the partial pressures ratio that gives us K_p is called the **reaction quotient**, Q_c or Q_p. The reaction quotient is *not* constant for a reaction, but it is quite useful because it allows us to predict the direction in which a net change must occur in order for equilibrium to be established. To illustrate, let us turn again to the decomposition of $HI(g)$,

Direction of Net Change activity

$$2\,HI(g) \rightleftharpoons H_2(g) + I_2(g)$$

and the 698 K data listed in Table 14.1.

The Value of Q_c for the Initial Conditions in Experiment 1 In this experiment, we start with only the reactant. Because there are no products initially, a net change must occur in the forward direction (to the right in our written equation). When we substitute the initial concentrations $[HI] = 1.000$ M, $[H_2] = [I_2] = 0.000$ M into the reaction quotient expression, we find that

$$Q_c = \frac{[H_2][I_2]}{[HI]^2} = \frac{(0) \times (0)}{(1.000)^2} = 0$$

The initial value of Q_c is 0, but as the reaction proceeds in the forward direction, the numerator of this ratio—$[H_2][I_2]$—increases in value, and the denominator—$[HI]^2$—gets smaller. Both of these changes cause the value of Q_c to increase. Equilibrium is reached when $Q_c = K_c$. This analysis suggests that

If $Q_c < K_c$, a net change occurs in the <u>forward</u> direction, that is, from left to right in the written equation for the reaction. (The rate of the forward reaction exceeds that of the reverse reaction until equilibrium is reached.)

The Value of Q_c for the Initial Conditions in Experiment 2 Here we start with products only, which means a net change must occur in the reverse direction (to the left in our equation) if any HI is to form. We can calculate the value of Q_c for the concentrations $[HI] = 0.000$ M, and $[H_2] = [I_2] = 1.000$ M.

$$Q_c = \frac{[H_2][I_2]}{[HI]^2} = \frac{(1.000) \times (1.000)}{(0.000)^2} \longrightarrow \infty$$

As the reaction proceeds in the reverse direction, the numerator of this ratio decreases and the denominator gets larger, both of which cause the value of Q_c to decrease. Again, equilibrium is reached when $Q_c = K_c$. This analysis suggests another criterion:

If $Q_c > K_c$, a net change proceeds in the <u>reverse</u> direction, that is, from right to left in the written equation for the reaction. (The rate of the reverse reaction exceeds that of the forward reaction until equilibrium is reached.)

For both of the preceding cases, we could have predicted the direction of a net change without evaluating Q_c. A net reaction occurred to produce some of the reactant missing in the initial reaction mixture. Sometimes, however, we need to compare the reaction quotient and the equilibrium constant in order to predict the direction of a net change. Experiment 3 of Table 14.1 provides such an instance, as illustrated in Example 14.6.

Figure 14.4 summarizes the relationship between reaction quotients and equilibrium constants.

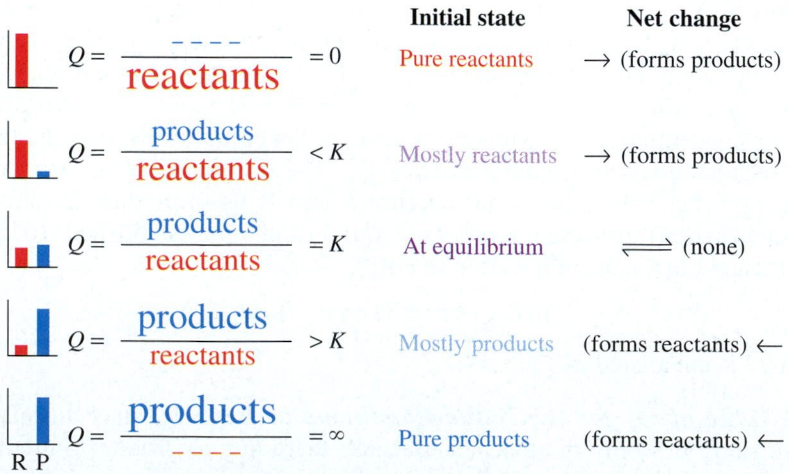

▶ **FIGURE 14.4 Relating Q and K and predicting the direction of net reaction**
When reactants predominate in the reaction mixture (top), Q is less than K and the reaction proceeds in the direction that forms products. When products predominate (bottom), Q is greater than K and the reaction proceeds in the direction that forms reactants. At equilibrium (center), Q equals K. Note that the heights of the bars shown on the left only illustrate the trend; their absolute heights would be determined by the value of K.

	Initial state	Net change
$Q = \dfrac{----}{\text{reactants}} = 0$	Pure reactants	$\rightarrow$ (forms products)
$Q = \dfrac{\text{products}}{\text{reactants}} < K$	Mostly reactants	$\rightarrow$ (forms products)
$Q = \dfrac{\text{products}}{\text{reactants}} = K$	At equilibrium	$\rightleftharpoons$ (none)
$Q = \dfrac{\text{products}}{\text{reactants}} > K$	Mostly products	(forms reactants) $\leftarrow$
$Q = \dfrac{\text{products}}{----} = \infty$	Pure products	(forms reactants) $\leftarrow$

R P

Example 14.6

Predict the direction of net change for Experiment 3 in Table 14.1.

SOLUTION

First, we substitute the initial concentrations for Experiment 3, $[HI] = [H_2] = [I_2] = 1.000$ M, into the expression for the reaction quotient, Q_c. Then we can compare the value of Q_c with the value $K_c = 1.84 \times 10^{-2}$ from page 578.

$$Q_c = \frac{[H_2][I_2]}{[HI]^2} = \frac{(1.000) \times (1.000)}{(1.000)^2} = 1.000$$

Because $Q_c > K_c$ ($1.000 > 1.84 \times 10^{-2}$), we conclude that a net change occurs in the *reverse* direction, from *right to left*.

ASSESSMENT

Notice how our prediction is supported by the equilibrium data for Experiment 3 in Table 14.1: $[HI] = 2.360$ M and $[H_2] = [I_2] = 0.320$ M. The equilibrium concentration of HI is greater than its initial concentration, and the equilibrium concentrations of H_2 and I_2 are less than their initial concentrations. These changes in concentration correspond to a net change from right to left.

EXERCISE 14.6A

In which direction would a net change occur if the initial conditions in the reaction of Example 14.6 were $[HI] = 1.00$ M and $[H_2] = [I_2] = 0.100$ M?

EXERCISE 14.6B

For the reaction $H_2S(g) + I_2(s) \rightleftharpoons 2 HI(g) + S(s)$, $K_p = 1.34 \times 10^{-5}$ at 60 °C. Initially, we bring together $H_2S(g)$ at a partial pressure of 0.010 atm, HI(g) at 0.0010 atm, $I_2(s)$, and S(s). When equilibrium is established, which gas will show an increase in its partial pressure? Which gas will show a decrease in its partial pressure? Which solid will increase in amount, and which will decrease?

14.4 Qualitative Treatment of Equilibrium: Le Châtelier's Principle

In working with the condition of equilibrium, we do not always need precise numerical results. Sometimes either nonnumerical answers or simple ballpark estimates are enough. A useful qualitative guide to equilibrium, called **Le Châtelier's principle,** was framed by Henri Le Châtelier in 1888. Le Châtelier stated his principle in a rather lengthy manner, but this rough paraphrase will serve our purposes:

> *When any __change__ in concentration, temperature, pressure, or volume is imposed on a system __at equilibrium__, the system responds by attaining a new equilibrium condition that __minimizes__ the impact of the imposed change.*

Changing the Amounts of Reacting Species

To begin our work with Le Châtelier's principle, we will consider the formation of the ester octyl acetate from 1-octanol and acetic acid, catalyzed by hydrogen ion. Under most reaction conditions, the three organic liquids, together with the small quantity of water that also forms, coexist in a single liquid phase. We suggest this by using the state designation "(soln)" in the equation below:

▲ The distinctive aroma and flavor of oranges are due to the ester octyl acetate, $CH_3(CH_2)_6CH_2OCOCH_3$. This compound can easily be synthesized to produce artificial orange flavorings.

$$\underset{\text{1-Octanol}}{CH_3(CH_2)_6CH_2OH(soln)} + \underset{\text{Acetic acid}}{CH_3COOH(soln)} \underset{}{\overset{H^+}{\rightleftharpoons}} \underset{\text{Octyl acetate}}{CH_3(CH_2)_6CH_2OCOCH_3(soln)} + H_2O(soln)$$

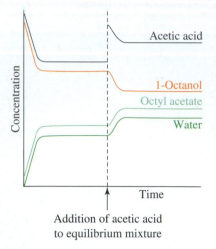

Addition of acetic acid
to equilibrium mixture

▲ A system at equilibrium responds to an increase in the concentration of one component through additional reactions that lead to new equilibrium concentrations.

Le Châtelier's Principle movie

Let us start with a mixture that is initially at equilibrium, at which point the reaction quotient is equal to the equilibrium constant:

$$Q_c = \frac{[\text{octyl acetate}][\text{H}_2\text{O}]}{[\text{1-octanol}][\text{CH}_3\text{COOH}]} = K_c$$

Notice that we have included the term $[\text{H}_2\text{O}]$ in the Q_c and K_c expressions. This is because the H_2O is present as one component in the solution, not as a pure liquid phase. Thus, the term $[\text{H}_2\text{O}]$ varies from one reaction mixture to another, just as do the concentrations of the organic components.

Now let us disturb the equilibrium by adding more acetic acid to the reaction flask. We indicate the resulting increase in concentration by using red type for acetic acid in the following expression:

$$Q_c = \frac{[\text{octyl acetate}][\text{H}_2\text{O}]}{[\text{1-octanol}][\text{CH}_3\text{COOH}]} < K_c$$

Because we have increased the denominator, the ratio of concentrations, Q_c, is now less than K_c, but this condition exists only temporarily. The concentrations must change in such a way as to make Q_c once again equal to K_c. This requires the numerator to become larger, which happens if some (but not all) of the added CH_3COOH is consumed in a net forward reaction. Additional octyl acetate and water are produced. At the same time, however, 1-octanol is consumed. When equilibrium is reestablished, the concentrations of acetic acid, octyl acetate, and water will all be greater than in the original equilibrium (red) and that of 1-octanol will be less than in the original equilibrium (blue).

$$Q_c = \frac{[\text{octyl acetate}][\text{H}_2\text{O}]}{[\text{1-octanol}][\text{CH}_3\text{COOH}]} = K_c$$

We can use Le Châtelier's principle to arrive at the same conclusion without having to work through the reaction quotient. To counter the effect of an added reactant,

The Significance of Chemical Equilibrium, in the Words of Henri Le Châtelier

Henri Le Châtelier was one of the first to appreciate the power of thermodynamics in dealing with chemical problems. For example, he fully understood the difference between a reaction that goes to completion and one that can only reach a state of equilibrium. He stated this distinction rather nicely in a journal article in 1888, from which we have taken the following quotation. In reading Le Châtelier's account, think of a *limited reaction* as a reversible reaction at equilibrium.

"It is known that in the blast furnace the reduction of iron oxide is produced by carbon monoxide, according to the reaction $\text{Fe}_2\text{O}_3(s) + 3\,\text{CO}(g) \rightleftharpoons 2\,\text{Fe}(s) + 3\,\text{CO}_2(g)$, but the gas leaving the chimney contains a considerable proportion of carbon monoxide.... Because this incomplete reaction was thought to be due to an insufficiently prolonged contact between carbon monoxide and the iron ore, the dimensions of the furnaces have been increased. In England they have been made as high as thirty meters. But the proportion of carbon monoxide escaping has not diminished, thus demonstrating, by an experiment costing several hundred thousand francs, that the reduction of iron oxide by carbon monoxide is a limited reaction. Acquaintance with the laws of chemical equilibrium would have

permitted the same conclusion to be reached more rapidly and far more economically."

◄ Henri Le Châtelier (1850–1936), a French chemist, formulated a principle that serves as a useful qualitative guide to equilibrium.

the reaction that can consume some of that reactant is stimulated. In this case, acetic acid is consumed in the forward reaction. To establish the new equilibrium, the reaction must go farther in the forward direction, a net reaction to the right. Le Châtelier's principle predicts the following results for each species:

- *Acetic acid, CH_3COOH.* Some, but not all, of the added acetic acid is consumed in the forward reaction. There will still be more acetic acid in the new equilibrium mixture than in the original equilibrium state.

- *1-Octanol.* There will be less 1-octanol than in the original equilibrium state because some of the 1-octanol present in the original equilibrium mixture reacts with some of the added acetic acid.

- *Octyl acetate and water.* There will be more of each of these products in the new equilibrium state because both are formed as the forward reaction is stimulated.

 **Le Châtelier's Principle activity**

In industrial settings, chemists often use an excess of acetic acid, as much as 10 mol acetic acid to 1 mol of the more expensive 1-octanol. This drives the equilibrium toward the product side, giving a good yield of octyl acetate. Another method of improving the equilibrium yield of octyl acetate is to remove water, as illustrated in Example 14.7.

Example 14.7

Water can be removed from an equilibrium mixture in the reaction of 1-octanol and acetic acid, for example, by using a solid drying agent that is insoluble in the reaction mixture. Describe how the removal of a small quantity of water affects the equilibrium.

$$CH_3(CH_2)_6CH_2OH(soln) + CH_3COOH(soln) \xrightleftharpoons{H^+}$$
$$CH_3(CH_2)_6CH_2OCOCH_3(soln) + H_2O(soln)$$

SOLUTION

As water is removed, the reverse reaction is slowed and the forward, water-forming reaction is stimulated. Not all of the water that is removed can be replaced, however, because a new equilibrium could not tolerate a constant concentration of H_2O at the same time that the other three concentrations change. Thus, in the new equilibrium the amount of water is somewhat less than in the original equilibrium, the amount of octyl acetate is greater, and the amounts of both 1-octanol and acetic acid are less.

EXERCISE 14.7A

What should be the effect of (a) adding $H_2(g)$, (b) removing $N_2(g)$, and (c) removing $NH_3(g)$ on a constant-volume equilibrium mixture of N_2, H_2, and NH_3? The reaction is

$$N_2(g) + 3 H_2(g) \rightleftharpoons 2 NH_3(g)$$

EXERCISE 14.7B

Describe the possible changes that might occur if an aqueous solution of acetic acid is added to an equilibrium mixture for the reaction in Example 14.7.

When we add or remove a reacting species from a *homogeneous* equilibrium mixture, we change the concentration of the species. If the concentration of one reactant changes, so too must the concentrations of all the others change, in order to reestablish the constant value of K_c. If the component added or removed is a pure solid or liquid in a *heterogeneous* equilibrium mixture, there is *no change* in the equilibrium condition. As we have seen, liquids and solids do not appear in the equilibrium constant expression. Thus, in the reaction

$$CaCO_3(s) \rightleftharpoons CaO(s) + CO_2(g) \qquad K_p = P_{CO_2}$$

the pressure of the $CO_2(g)$ in equilibrium with $CaCO(s)$ and $CaCO_3(s)$ is unaffected by the amounts of the two solids present (Figure 14.5).

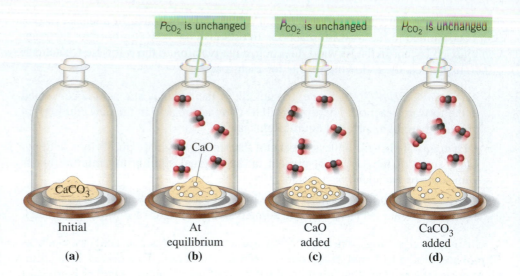

▶ **FIGURE 14.5   Equilibrium in the reaction**
$CaCO_3(s) \rightleftharpoons CaO(s) + CO_2(g)$
(a) Initially, only $CaCO_3(s)$ is present.
(b) Decomposition of $CaCO_3(s)$ yields $CaO(s)$ (white circles) and $CO_2(g)$. The CO_2 exerts its equilibrium partial pressure. Neither (c) additional $CaO(s)$ nor (d) additional $CaCO_3(s)$ has any effect on the partial pressure of the CO_2.

QUESTION: What takes place if $CaCO_3(s)$ is heated in an open system? How does the situation differ from that of part (b)?

Similarly, in the following equilibrium, the addition or removal of liquid water does not affect the vapor pressure of water:

$$H_2O(l) \rightleftharpoons H_2O(g) \qquad K_p = P_{H_2O}$$

Changing External Pressure or Volume in Gaseous Equilibria

We can increase the partial pressure of a gaseous component in a constant-volume equilibrium mixture by adding more of the gas to the mixture, or we can decrease the partial pressure by removing some of the gas from the mixture. A net reaction occurs to the left or to the right in the manner that we have already described (see Exercise 14.7A).

We can increase the partial pressures of *all* the gases in an equilibrium mixture by increasing the external pressure and thereby reducing the reaction volume. Likewise, we can reduce the partial pressures of all the gases by reducing the external pressure and thereby increasing the reaction volume. (We can also reduce partial pressures by transferring the reaction mixture into an evacuated container of larger volume.)

Consider the decomposition of $N_2O_4(g)$ to $NO_2(g)$ at 298 K:

$$N_2O_4(g) \rightleftharpoons 2\,NO_2(g) \qquad K_p = 0.145$$

Figure 14.6a depicts an equilibrium mixture under an external pressure of 1 atm. Now suppose we quickly increase the external pressure to 2 atm. First, let us predict what should happen by comparing Q_p and K_p.

Initial Equilibrium Mixture Suppose $P_{N_2O_4}$ and P_{NO_2} are the equilibrium partial pressures when the total pressure is 1 atm, as in Figure 14.6a. This initial equilibrium is described by the expression

$$Q_p = K_p = \frac{(P_{NO_2})^2}{P_{N_2O_4}}$$

Disturbed Equilibrium Mixture Now imagine that the total pressure is increased to 2 atm, as in Figure 14.6b. Because the volume has been reduced to one-half its original value, each partial pressure doubles, as does the value of Q_p:

$$Q_p = \frac{2\,P_{NO_2} \times 2\,P_{NO_2}}{2\,P_{N_2O_4}} = 2 \times \frac{(P_{NO_2})^2}{P_{N_2O_4}} = 2 \times K_p$$

New Equilibrium Mixture Because Q_p is now greater than K_p, a net reaction will occur in the reverse direction, consuming some of the NO_2 and producing more N_2O_4. This change reduces the numerator in the reaction quotient, increases the denominator, and therefore reduces Q_p so that it is once again equal to K_p. We say that equilibrium shifts to the left, as represented in Figure 14.6c.

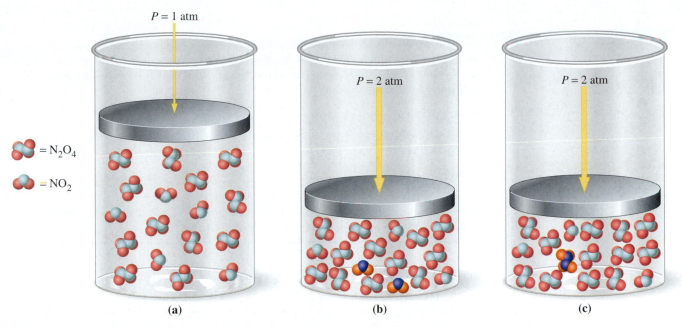

$P = 1$ atm

= N_2O_4

= NO_2

(a)

$P = 2$ atm

(b)

$P = 2$ atm

(c)

▲ **FIGURE 14.6   Illustrating Le Châtelier's principle in the reaction $N_2O_4(g) \rightleftharpoons 2\,NO_2(g)$ at 298 K**
(a) Equilibrium is established at a total pressure of 1 atm. For every 17 molecules, 5 are NO_2 and 12 are N_2O_4. (b) The total pressure is increased to 2 atm. Momentarily, the same 17 molecules are present. (c) The system accommodates to the reduced volume through the reaction of two NO_2 molecules to form one N_2O_4 molecule (reaction highlighted by dark blue N atoms). The new equilibrium has 16 molecules in place of the original 17. Of these, 3 are NO_2 and 13 are N_2O_4. Notice that in every case, the same total number of atoms is present: 29 N atoms and 58 O atoms.

QUESTION: How would the mixture in (a) be influenced by decreasing the pressure to 0.5 atm?

Now let us consider this issue from the standpoint of Le Châtelier's principle. When we decrease the volume of an equilibrium mixture by increasing the external pressure, we crowd the molecules more closely together. To minimize the effect of this overcrowding, a net reaction occurs, producing a smaller number of molecules. Because two moles of $NO_2(g)$ yield just one mole of $N_2O_4(g)$, the reverse reaction is stimulated and equilibrium shifts to the left, just as we concluded in the assessment based on Q_p and K_p.

NO_2–N_2O_4 Equilibrium animation

The following statements summarize the effect of changes in external pressure (or system volume) on an equilibrium involving gases.

- When the external pressure is increased (or the system volume is reduced), an equilibrium shifts in the direction producing the smaller number of moles of gas.

- When the external pressure is decreased (or the system volume is increased), an equilibrium shifts in the direction producing the larger number of moles of gas.

- If there is no change in the number of moles of gas in a reaction, changes in external pressure (or system volume) have no effect on an equilibrium.

If changes in gas pressures or volumes are produced by adding an inert gas to an equilibrium mixture, the effects are somewhat different. If the inert gas is added at a *constant external pressure,* the volume expands to accommodate the added gas. This has the same effect as transferring the mixture to a container of larger volume. If the inert gas is added to a *constant-volume* mixture, the concentrations and partial pressures of reactants and products do not change, and the inert gas does not affect the equilibrium.

Example 14.8

An equilibrium mixture of $SO_2(g)$, $O_2(g)$, and $SO_3(g)$ is transferred from a 1.00-L flask to a 2.00-L flask. In which direction does a net reaction proceed to restore equilibrium? The balanced equation for the reaction is

$$2\,SO_3(g) \rightleftharpoons 2\,SO_2(g) + O_2(g)$$

SOLUTION

Transferring the mixture to a larger flask increases the volume to be filled with molecules. We expect the equilibrium to shift in the direction that produces the larger number of moles of gas—in this case, the forward reaction. Some of the $SO_3(g)$ is converted to $SO_2(g)$ and $O_2(g)$, and equilibrium shifts to the right.

EXERCISE 14.8A

Consider the reaction $H_2(g) + I_2(g) \rightleftharpoons 2 HI(g)$. How is the equilibrium amount of $HI(g)$ changed by compressing an equilibrium mixture into a smaller volume? Explain.

EXERCISE 14.8B

For the reaction $2 NO(g) + O_2(g) \rightleftharpoons 2 NO_2(g)$, in which direction will a net reaction occur if additional $NO_2(g)$ is added to an equilibrium mixture at the same time that the mixture is transferred from a 1.00-L to a 1.50-L flask? Explain.

Changing the Equilibrium Temperature

The changes we have described thus far do not change the value of the equilibrium constant, but changing the temperature of an equilibrium mixture does change the value of K_p or K_c. If the equilibrium constant becomes larger with a change in temperature, the forward reaction is favored and equilibrium shifts to the right. If it becomes smaller, the reverse reaction is favored and equilibrium shifts to the left. Le Châtelier's principle enables us to assess, qualitatively, how temperature affects equilibrium (Figure 14.7).

To change the temperature of a reaction mixture, we must either add heat to raise the temperature or remove heat to lower the temperature. Adding heat to an equilibrium mixture will stimulate the reaction that can absorb some of the heat—the endothermic reaction. Removing heat will stimulate the exothermic reaction.

> *Raising* the temperature of an equilibrium mixture shifts equilibrium in the direction of the *endothermic* reaction; *lowering* the temperature shifts equilibrium in the direction of the *exothermic* reaction.

Recall from Section 12.4 that the majority of solid solutes (95% or more) have aqueous solubilities that increase with temperature. In these cases, solution formation is an endothermic process, and an endothermic process is favored at higher temperatures. This means that at higher temperatures, more of the solute dissolves before the solution becomes saturated, that is, before equilibrium is reached.

In Example 14.9, we will show that we can also consider heat as if it were a reactant or a product of a reaction. Then we can reason as we would for changing the amount of a reacting species.

▶ **FIGURE 14.7 Effect of increasing temperature on equilibrium**

(a) Heating an equilibrium mixture in which the forward reaction is endothermic shifts the equilibrium toward more products. (b) If the forward reaction is exothermic, equilibrium in the reaction mixture shifts toward more reactants.

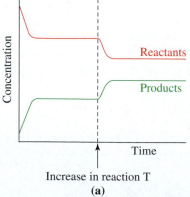

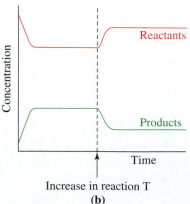

Chemical Equilibrium and the Synthesis of Ammonia

A mmonia is an important industrial chemical, ranking about sixth in quantity of the chemicals produced in the United States. It is used directly as a fertilizer, to make other nitrogen-containing fertilizers, and in the production of explosives and plastics. Following are typical conditions used in its manufacture.

$$N_2(g) + 3 H_2(g) \rightleftharpoons 2 NH_3(g) \qquad \Delta H° = -92.22 \text{ kJ}$$

Reactants: 3:1 mol ratio of H_2 to N_2

Temperature: 400–600 °C

Pressure: 140–340 atm

Catalyst: Fe_3O_4 with small amounts of Al_2O_3, MgO, CaO, and K_2O. The Fe_3O_4 is reduced to metallic iron before use.

Based on the chemical equation and Le Châtelier's principle, we would conclude that high yields of $NH_3(g)$ are favored by the following conditions.

1. *Low temperatures* because the forward reaction is exothermic.

2. *High pressures* because the forward reaction results in a decrease in the number of moles of gas.

3. *Continuous removal of NH_3* because the removal of a product stimulates the forward reaction to form additional product.

The synthesis reaction is indeed carried out at high pressures. The temperatures used, however, are moderately high, not low. The theoretical percent yield for converting a 3 mol H_2:1 mol N_2 mixture to NH_3 is more than 90% at high pressures and room temperature, but with these conditions it would take far too long to reach equilibrium. Instead, the reaction is kinetically controlled (page 587). Even though the NH_3 yield is only 20% in an equilibrium mixture in the presence of a catalyst at 500 °C and 200 atm, that equilibrium is reached in less than a minute.

The equilibrium mixture of gases is cooled to the point at which the ammonia liquefies. The $NH_3(l)$ is removed, and the unreacted $H_2(g)$ and $N_2(g)$, still in a 3:1 mole ratio, are recycled through the process.

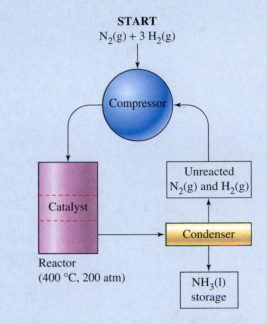

START
$N_2(g) + 3 H_2(g)$

Compressor

Unreacted $N_2(g)$ and $H_2(g)$

Catalyst

Condenser

Reactor
(400 °C, 200 atm)

$NH_3(l)$
storage

▲ **FIGURE 14.8 Ammonia synthesis by the Haber process**
A mixture of N_2 and H_2 is sent through a compressor and into a reactor that contains a catalyst and maintains the mixture at a high temperature and pressure. The equilibrium mixture is then removed from the reactor and cooled in a condenser. Liquid NH_3 is removed, and the unreacted N_2 and H_2 are returned to the cycle.

Example 14.9

Is the amount of NO(g) formed from given amounts of $N_2(g)$ and $O_2(g)$,

$$N_2(g) + O_2(g) \rightleftharpoons 2 NO(g) \qquad \Delta H° = +180.5 \text{ kJ}$$

greater at high or low temperatures?

SOLUTION

As written, the $\Delta H°$ value is for the forward reaction. Because $\Delta H°$ is positive, the forward reaction is endothermic. Thus, the equilibrium shifts to the right as the temperature is raised. The conversion of $N_2(g)$ and $O_2(g)$ to NO(g) is favored at high temperatures.

Alternatively, we can think of the reaction in this way:

$$N_2(g) + O_2(g) + \text{heat} \rightleftharpoons 2 NO(g)$$

We can then reason as follows: Raising the temperature, that is, adding heat (a "reactant") stimulates the forward reaction. Thus, equilibrium is shifted to the right.

Application Note

Because conversion of $N_2(g)$ and $O_2(g)$ to NO(g) is favored at high temperatures, NO(g) is found in the exhaust of high-compression automobile engines—they operate at high temperatures.

EXERCISE 14.9A

Is the conversion of $SO_2(g)$ to $SO_3(g)$,

$$2 SO_2(g) + O_2(g) \rightleftharpoons 2 SO_3(g) \qquad \Delta H° = -198 \text{ kJ}$$

more nearly complete at high or low temperatures?

EXERCISE 14.9B

Gaseous nitrogen monoxide and gaseous nitrogen dioxide react reversibly to form gaseous dinitrogen trioxide, which has a standard enthalpy of formation at 298.15 K of 83.72 kJ/mol. Will the equilibrium partial pressure of dinitrogen trioxide be greater at the freezing point or boiling point of water? Explain.

Adding a Catalyst

The reaction of $SO_2(g)$ with $O_2(g)$ to produce $SO_3(g)$,

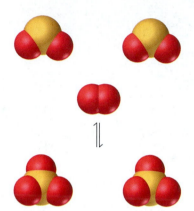

$$2 SO_2(g) + O_2(g) \xrightarrow{\text{Pt}} 2 SO_3(g) \qquad K_c = 2.8 \times 10^2 \text{ at 1000 K}$$

is greatly accelerated by a catalyst (such as platinum metal). However, the reverse reaction—the decomposition of $SO_3(g)$ to $SO_2(g)$—is also greatly speeded by the catalyst. Because the rates of the forward and reverse reactions are increased to the same extent, the proportion of $SO_3(g)$ in the equilibrium mixture is the same as if no catalyst were present. That is, a catalyst does not shift an equilibrium to the right or left, nor does it affect the value of the equilibrium constant. The catalyst merely causes the system to reach equilibrium more quickly.

The role of a catalyst is to change the mechanism of a reaction to one of lower activation energy. Because a catalyst does not affect an equilibrium state, we can conclude that equilibrium is a function only of the states of the reactants and products and not the reaction path. We will explore this concept further in Chapter 17.

Example 14.10 A Conceptual Example

Flask A, pictured below, initially contains an equilibrium mixture of the reactants and products of the reaction

$$CO(g) + H_2O(g) \rightleftharpoons CO_2(g) + H_2(g) \qquad \Delta H = -41 \text{ kJ}; K_c = 9.03 \text{ at 698 K}$$

It is isolated from flask B by a closed valve. When the valve is opened, a new equilibrium is established as the contents of the two flasks mix. Describe, qualitatively, how the amounts of CO, H_2O, CO_2, and H_2 in the new equilibrium compare with the amounts in the initial equilibrium if **(a)** flask B initially contains Ar(g) at 1 atm pressure; **(b)** flask B initially contains 1.0 mol CO_2; **(c)** flask B initially contains 1.0 mol CO and the temperature of the A–B mixture is raised by 100 °C. If you are uncertain of the result in any of the three cases, explain why.

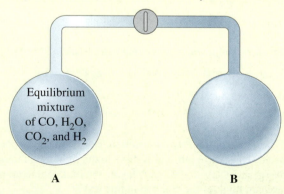

Equilibrium mixture of CO, H_2O, CO_2, and H_2

A   B

ANALYSIS AND CONCLUSIONS

(a) Ar(g) is an inert gas and has no effect on the reaction. Because the reaction involves the same number of reactant and product molecules, the equilibrium is unaffected by the change in volume. The amounts of CO, H_2O, CO_2, and H_2 are all unchanged.

(b) As in part (a), increasing the volume has no effect on the equilibrium, but having more $CO_2(g)$ present stimulates the reverse reaction. Some of the CO_2 present in the A–B mixture is converted to CO and H_2. In the new equilibrium, therefore, the amounts of CO, H_2O, and CO_2 will all be greater than in the initial equilibrium in flask A. The amount of H_2 will be less than in the initial equilibrium because it is consumed in the reverse reaction.

(c) Adding more CO(g) favors the forward reaction, but raising the temperature favors the endothermic reverse reaction. Because these factors work in opposition, we cannot make a qualitative prediction.

EXERCISE 14.10A

Repeat Example 14.10 for the case where (a) flask B initially contains 1.0 mol $H_2(g)$ at 1 atm pressure; (b) flask B initially contains 1.0 mol $H_2(g)$ and 1.0 mol $H_2O(g)$; (c) flask B initially contains 1.0 mol H_2O and the temperature of the A–B mixture is lowered by 100 °C.

EXERCISE 14.10B

The principal source of hydrogen for use in the manufacture of ammonia is a process called "methane gas reforming"—a reaction in which the gases methane and water react to produce the gases carbon dioxide and hydrogen. Will the formation of products be favored if equilibrium in this reaction is established at (a) high or low temperatures, (b) high or low pressures?

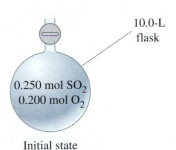

10.0-L flask

0.250 mol SO_2
0.200 mol O_2

Initial state

14.5 Some Illustrative Equilibrium Calculations

We will conclude this chapter by showing how to use equilibrium constants to solve problems. The knowledge gained here will be quite useful in solving many more such problems in the next two chapters. For convenience, we will divide the problems into two basic types: those in which we use experimental data to determine equilibrium constants and those in which we use equilibrium constants to calculate equilibrium concentrations or partial pressures.

Determining Equilibrium Constants from Experimental Data

In Example 14.11, we seek a K_c value. As suggested by Figure 14.9, we are given initial amounts of reactants and the equilibrium amount of product. From these data, we can establish the amounts of all species present in the equilibrium state and then the equilibrium concentrations. Finally, we can calculate K_c from those concentrations.

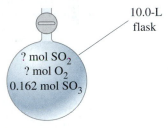

10.0-L flask

? mol SO_2
? mol O_2
0.162 mol SO_3

Equilibrium state

▲ **FIGURE 14.9   Example 14.11 illustrated**

The key in Example 14.11 is to determine the amounts of SO_2 and O_2 consumed to reach equilibrium.

Example 14.11

In a 10.0-L vessel at 1000 K, 0.250 mol SO_2 and 0.200 mol O_2 react to form 0.162 mol SO_3 at equilibrium. What is K_c, at 1000 K, for the reaction that is shown here?

$$2 SO_2(g) + O_2(g) \rightleftharpoons 2 SO_3(g)$$

STRATEGY

To implement the general strategy alluded to in the paragraph preceding this example, let us tabulate (a) the concentrations of substances present initially, (b) the changes in these concentrations that occur in reaching equilibrium, and (c) the equilibrium concentrations. Often the key step is (b), in which we identify the changes that occur and determine their relationship to one another. Think of this three-step format as the "ICE" approach—initial/change/equilibrium.

SOLUTION

Let us begin by listing the initial concentrations of the three gases.

$$[SO_2] = \frac{0.250 \text{ mol}}{10.0 \text{ L}} = 0.0250 \text{ M} \quad [O_2] = \frac{0.200 \text{ mol}}{10.0 \text{ L}} = 0.0200 \text{ M} \quad [SO_3] = 0$$

From the given information, we can also calculate the equilibrium concentration of SO_3.

$$[SO_3] = \frac{0.162 \text{ mol}}{10.0 \text{ L}} = 0.0162 \text{ M}$$

Next, we arrange these quantities in the ICE format.

The reaction:	$2\,SO_2(g)$	$+\;O_2(g)$	$\rightleftharpoons$	$2\,SO_3$
Initial concentrations, M:	0.0250	0.0200		0
Changes, M:	?	?		?
Equilibrium concentrations, M:	?	?		0.0162

Now, we must replace the question marks with numerical values. We started with no SO_3 and produced an equilibrium concentration of 0.0162 M. The change in $[SO_3]$ must therefore be equal to the difference between final and initial conditions. We use the plus sign to signify that something is formed.

$$\Delta[SO_3] = +0.0162 \text{ M}$$

From the chemical equation, we see that the number of moles per liter of SO_2 consumed must be the same as the number of moles per liter of SO_3 produced. Here we use the minus sign to signify that something is consumed.

$$\Delta[SO_2] = -0.0162 \text{ M}$$

Because only one mole per liter of O_2 is required for every two moles per liter of SO_3 produced, the change in the concentration of O_2 is one-half the change in concentration of SO_3.

$$\Delta[O_2] = -\tfrac{1}{2} \times 0.0162 \text{ M} = -0.00810 \text{ M}.$$

Now we can complete the table by algebraically adding these changes and the initial concentrations to get equilibrium concentrations.

The reaction:	$2\,SO_2(g)$	$+\;O_2(g)$	$\rightleftharpoons$	$2\,SO_3(g)$
Initial concentrations, M:	0.0250	0.0200		0
Changes, M:	−0.0162	−0.00810		+0.0162
Equilibrium concentrations, M:	0.0088	0.0119		0.0162

Finally, we substitute the equilibrium concentrations into the equilibrium constant expression.

$$K_c = \frac{[SO_3]^2}{[SO_2]^2[O_2]} = \frac{(0.0162)^2}{(0.0088)^2(0.0119)} = 2.8 \times 10^2$$

ASSESSMENT

Keep in mind that the concentration changes that occur in establishing equilibrium are governed by the stoichiometry of the reaction (that is, by the coefficients in the balanced equation). Also, although some concentration changes are negative and some are positive, the equilibrium concentrations themselves must always be positive quantities.

EXERCISE 14.11A

A reaction starts with 1.00 mol each of PCl_3 and Cl_2 in a 1.00-L flask. When equilibrium is established at 250 °C in the reaction $PCl_3(g) + Cl_2(g) \rightleftharpoons PCl_5(g)$, the amount of PCl_5 present is 0.82 mol. What is K_c for this reaction?

EXERCISE 14.11B

A 1.00-kg sample of $Sb_2S_3(s)$ and a 10.0-g sample of $H_2(g)$ are allowed to react in a 25.0-L container at 713 K. At equilibrium, 72.6 g $H_2S(g)$ is present. What is the value of K_p at 713 K for the following reaction?

$$Sb_2S_3(s) + 3\,H_2(g) \rightleftharpoons 2\,Sb(s) + 3\,H_2S(g)$$

Calculating Equilibrium Quantities from K_c and K_p Values

One of the most common types of equilibrium calculations is illustrated in Example 14.12. We start with initial reactants and no products and with the known value of the equilibrium constant. Then we use those data to calculate the amounts of substances present at equilibrium. Typically we use the symbol x to identify one of the changes in concentration that occurs in establishing equilibrium. Then we relate all the other concentration changes to x, substitute appropriate terms into the equilibrium constant expression, and solve for x.

Example 14.12

Consider the reaction

$$H_2(g) + I_2(g) \rightleftharpoons 2\,HI(g) \qquad K_c = 54.3 \text{ at } 698 \text{ K}$$

If we start with 0.500 mol $H_2(g)$ and 0.500 mol $I_2(g)$ in a 5.25-L vessel at 698 K, how many moles of each gas will be present at equilibrium?

STRATEGY

As noted in the paragraph preceding this example, we can let x represent a change in one concentration and relate the other changes to the chosen x. In particular, let us say that $-x$ is the change in $[H_2]$ and in $[I_2]$. Then, through the reaction stoichiometry, the change in $[HI]$ is $+2x$. We enter these values and other data into the ICE format to obtain equilibrium concentrations in terms of initial concentrations and the variable x. Then we enter these concentrations into the equilibrium constant expression and solve for x. Once we have a value for x, we can calculate the equilibrium concentrations and, from them, the actual equilibrium amounts in moles.

SOLUTION

First, let us calculate the initial concentrations in the 5.25-L flask.

$$[H_2] = [I_2] = \frac{0.500 \text{ mol}}{5.25 \text{ L}} = 0.0952 \text{ M} \quad [HI] = 0$$

We enter these initial concentrations, the changes in concentrations in terms of x, and the equilibrium concentrations into the ICE format.

The reaction:	$H_2(g)$	$+$	$I_2(g)$	$\rightleftharpoons$	$2\,HI(g)$
Initial concentrations, M:	0.0952		0.0952		0
Changes, M:	$-x$		$-x$		$+2x$
Equilibrium concentrations, M:	$(0.0952 - x)$		$(0.0952 - x)$		$2x$

Next, we can enter the equilibrium concentrations into the K_c expression.

$$K_c = \frac{[HI]^2}{[H_2][I_2]} = \frac{(2x)^2}{(0.0952 - x)(0.0952 - x)} = \frac{(2x)^2}{(0.0952 - x)^2} = 54.3$$

From this point, the most direct approach is to take the square root of each side of the equation and solve for x. (We can easily do this because both numerator and denominator in the term on the left are squared quantities.)

$$\left[\frac{(2x)^2}{(0.0952 - x)^2} \right]^{1/2} = \frac{2x}{(0.0952 - x)} = (54.3)^{1/2}$$

$$2x = (54.3)^{1/2} \times (0.0952 - x)$$

$$2x = 7.37 \times (0.0952 - x) = 0.702 - 7.37x$$

$$9.37x = 0.702$$

$$x = 0.0749$$

We can now calculate the equilibrium concentrations.

$$[H_2] = [I_2] = 0.0952 - x = 0.0952 - 0.0749 = 0.0203 \text{ M}$$

$$[HI] = 2x = 2 \times 0.0749 = 0.150 \text{ M}$$

To determine the equilibrium amounts, we multiply the equilibrium concentrations by the volume.

$$\text{mol } H_2 = \text{mol } I_2 = 5.25 \text{ L} \times 0.0203 \text{ mol/L} = 0.107 \text{ mol}$$

$$\text{mol } HI = 5.25 \text{ L} \times 0.150 \text{ mol/L} = 0.788 \text{ mol}$$

ASSESSMENT

A useful check on this result is to substitute the calculated equilibrium concentrations into the K_c expression to see what value of K_c they yield:

$$K_c = \frac{[HI]^2}{[H_2][I_2]} = \frac{(0.150)^2}{(0.0203)(0.0203)} = 54.6$$

This is close enough to the given value of $K_c = 54.3$ to confirm that our calculation is correct.

EXERCISE 14.12A

Starting with 0.100 mol each of CO and H_2O in a 5.00-L flask, equilibrium is established in the following reaction at 600 K:

$$CO(g) + H_2O(g) \rightleftharpoons CO_2(g) + H_2(g) \qquad K_c = 23.2 \text{ at } 600 \text{ K}$$

When equilibrium is established, what will be **(a)** the number of moles of $H_2(g)$ and **(b)** the partial pressure of $H_2(g)$?

EXERCISE 14.12B

Show that for the reaction in Example 14.12, $H_2(g) + I_2(g) \rightleftharpoons 2\,HI(g)$, the equilibrium amounts of reactants and products are independent of the volume of the reaction flask. Would you expect this to be the case for all reversible reactions at equilibrium? Explain.

Example 14.13

Suppose that in the reaction of Example 14.12, the initial amounts are 0.800 mol H_2 and 0.500 mol I_2. What will be the amounts of reactants and products when equilibrium is attained?

STRATEGY

The overall strategy here is just the same as in Example 14.12, but because the initial amounts of reactant are unequal, the algebraic solution is somewhat more difficult. It uses the quadratic formula.

SOLUTION

As in Example 14.12, let us first determine the initial concentrations.

$$[H_2] = \frac{0.800 \text{ mol}}{5.25 \text{ L}} = 0.152 \text{ M} \quad [I_2] = \frac{0.500 \text{ mol}}{5.25 \text{ L}} = 0.0952 \text{ M} \quad [HI] = 0$$

Then we can set up an ICE tabular format as usual.

The reaction:	$H_2(g)$	+	$I_2(g)$	$\rightleftharpoons$	2 HI(g)
Initial concentrations, M:	0.152		0.0952		0
Changes, M:	$-x$		$-x$		$+2x$
Equilibrium concentrations, M:	$(0.152 - x)$		$(0.0952 - x)$		$2x$

Next, we can enter the equilibrium concentrations into the K_c expression.

$$K_c = \frac{[HI]^2}{[H_2][I_2]} = 54.3$$

We cannot extract the square root of each side of this equation, because the two terms in the denominator of the fraction are not identical. Instead, we have to multiply both sides of the equation by the factor $(0.152 - x)(0.0952 - x)$.

$$\frac{(2x)^2}{(0.152 - x)(0.0952 - x)} = 54.3$$

$$4x^2 = 54.3 \times (0.152 - x)(0.0952 - x)$$

$$4x^2 = 54.3 \times (0.0145 - 0.247x + x^2)$$

$$4x^2 = 54.3x^2 - 13.4x + 0.787$$

Now, gathering the terms into the form $ax^2 + bx + c = 0$, we note that $a = 50.3$, $b = -13.4$, and $c = 0.787$.

$$50.3x^2 - 13.4x + 0.787 = 0$$

The solutions of this equation are given by the quadratic formula.

$$x = \frac{-b \pm \sqrt{b^2 - 4ac}}{2a}$$

We then substitute for a, b, and c.

$$x = \frac{-(-13.4) \pm \sqrt{(-13.4)^2 - (4 \times 50.3 \times 0.787)}}{2 \times 50.3}$$

$$= \frac{13.4 \pm \sqrt{21.22}}{100.6} = \frac{13.4 \pm 4.61}{100.6}$$

Now we are faced with a choice; the solution gives two values for x. The correct answer must be $x = 0.0874$ and not 0.179, because the decrease in $[I_2]$ cannot be greater than 0.0952 M, the initial concentration.

$$x = \frac{13.4 + 4.61}{100.6} = 0.179$$

$$x = \frac{13.4 - 4.61}{100.6} = 0.0874$$

We can now calculate the equilibrium concentrations.

$$[H_2] = 0.152 - x = 0.152 - 0.0874 = 0.065 \text{ M}$$

$$[I_2] = 0.0952 - x = 0.0952 - 0.0874 = 0.0078 \text{ M}$$

$$[HI] = 2x = 2 \times 0.0874 = 0.175 \text{ M}$$

Then we determine the equilibrium amounts by multiplying the equilibrium concentrations by the volume.

? mol H_2 = 5.25 L × 0.065 mol H_2/L = 0.34 mol H_2

? mol I_2 = 5.25 L × 0.0078 mol I_2/L = 0.041 mol I_2

? mol HI = 5.25 L × 0.175 mol HI/L = 0.919 mol HI

ASSESSMENT

To check the validity of the calculated equilibrium concentrations, we can proceed as in Example 14.12—substitute the calculated concentrations into the K_c expression and see if they yield a value close to that given for K_c. In this example, that would be a calculated value of 60 compared to the given $K_c = 54.3$, a good match.

EXERCISE 14.13A

Starting with 0.100 mol CO and 0.200 mol Cl_2 in a 25.0-L flask, how many moles of $COCl_2$ will be present at equilibrium in the following reaction?

$$CO(g) + Cl_2(g) \rightleftharpoons COCl_2(g) \qquad K_c = 1.2 \times 10^3 \text{ at 668 K}$$

EXERCISE 14.13B

Starting with 0.78 mol N_2 and 0.21 mol O_2 (their proportions in one mole of air), what will be the mole fraction of NO(g) at equilibrium in this reaction?

$$N_2(g) + O_2(g) \rightleftharpoons 2\,NO(g) \qquad K_c = 2.1 \times 10^{-3} \text{ at 2500 K}$$

(*Hint*: Do you need to know the volume of the reaction mixture? Recall Exercise 14.12B.)

If we start with a mixture in which a reactant or product is absent, we know that a net reaction must occur in the direction in which some of that species is produced. This was the case in Examples 14.12 and 14.13. If the initial mixture contains all of the reactants and products, however, we do not immediately know in which direction a net reaction will occur, but we can find out easily enough. We do this by evaluating the reaction quotient Q_c for the initial conditions and comparing its value to that of the equilibrium constant (page 588).

Example 14.14

Carbon monoxide and chlorine react to form phosgene, $COCl_2$, which is used in the manufacture of pesticides, herbicides, and plastics:

$$CO(g) + Cl_2(g) \rightleftharpoons COCl_2(g) \qquad K_c = 1.2 \times 10^3 \text{ at 668 K}$$

How much of each substance, in moles, will there be at equilibrium in a reaction mixture that initially has 0.0100 mol CO, 0.0100 mol Cl_2, and 0.100 mol $COCl_2$ in a 10.0-L flask?

STRATEGY

Because all reactant and products are present initially, it is not immediately obvious in which direction a net change will occur to establish equilibrium. We can easily find out, though, by evaluating the reaction quotient Q_c for the initial concentrations and comparing it with K_c. Knowing the direction of net change will help us to choose an appropriate quantity to represent as x. Once we have done this, the calculation is very much as in previous examples.

SOLUTION

Let us first determine the initial concentrations and evaluate Q_c.

$$[CO]_{initial} = [Cl_2]_{initial} = \frac{0.0100 \text{ mol}}{10.0 \text{ L}} = 0.00100 \text{ M}$$

$$[COCl_2]_{initial} = \frac{0.100 \text{ mol}}{10.0 \text{ L}} = 0.0100 \text{ M}$$

$$Q_c = \frac{[COCl_2]_{initial}}{[CO]_{initial}[Cl_2]_{initial}} = \frac{0.0100}{(0.00100)(0.00100)} = 1.00 \times 10^4$$

Because $Q_c > K_c$ (that is, $1.00 \times 10^4 > 1.2 \times 10^3$), a net reaction must occur in the reverse direction, which means that at equilibrium the concentrations of CO and Cl_2 will be greater than initially and the concentration of $COCl_2$ will be less. Now we are ready to calculate the equilibrium concentrations.

Direction of net change: $\longleftarrow$

The reaction:	CO(g)	+	$Cl_2(g)$	$\rightleftharpoons$	$COCl_2(g)$
Initial concentrations, M:	0.00100		0.00100		0.0100
Changes, M:	$+x$		$+x$		$-x$
Equilibrium concentrations, M:	$(0.00100 + x)$		$(0.00100 + x)$		$(0.0100 - x)$

At this point, we substitute the equilibrium concentrations into the equilibrium constant expression and simplify.

$$K_c = \frac{[COCl_2]}{[CO] \times [Cl_2]} = \frac{(0.0100 - x)}{(0.00100 + x)(0.00100 + x)} = 1.2 \times 10^3$$

$$0.0100 - x = 1.2 \times 10^3 \times (0.00100 + x)(0.00100 + x)$$

$$0.0100 - x = 1.2 \times 10^3 \times (1.00 \times 10^{-6} + 0.00200x + x^2)$$

$$0.0100 - x = 1.2 \times 10^{-3} + 2.4x + (1.2 \times 10^3)x^2$$

Now let us put the equation into the quadratic form: $ax^2 + bx + c = 0$.

$$(1.2 \times 10^3)x^2 + 3.4x - 0.0088 = 0$$

From this form, we use the quadratic formula to solve for x.

$$x = \frac{-3.4 \pm \sqrt{(3.4)^2 - [4 \times 1.2 \times 10^3 \times (-0.0088)]}}{2(1.2 \times 10^3)}$$

Only one of the two possible roots to this quadratic equation is acceptable—the positive root.

$$x = \frac{-3.4 \pm \sqrt{53.8}}{2.4 \times 10^3} = \frac{-3.4 + 7.33}{2.4 \times 10^3} = 0.0016$$

We can now calculate the equilibrium concentrations.

$$[CO] = [Cl_2] = 0.00100 + x = 0.00100 + 0.0016 = 0.0026 \text{ M}$$

$$[COCl_2] = 0.0100 - x = 0.0100 - 0.0016 = 0.0084 \text{ M}$$

To determine the equilibrium amounts, we multiply the equilibrium concentration of each species by the volume of the equilibrium mixture.

$$\text{mol CO} = \text{mol Cl}_2 = 10.0 \text{ L} \times 0.0026 \text{ mol/L} = 0.026 \text{ mol}$$

$$\text{mol COCl}_2 = 10.0 \text{ L} \times 0.0084 \text{ mol/L} = 0.084 \text{ mol}$$

ASSESSMENT

In evaluating $\pm\sqrt{53.8}$ in the quadratic formula, we used only the positive root. If we had used the negative root, we would have obtained a negative value of x and therefore negative concentrations of CO and Cl_2, producing invalid solutions. In checking for the correctness of our answer, as was done in previous examples, we find excellent agreement between the value of K_c calculated from the equilibrium concentrations and the given value:

$$K_c = [COCl_2]/[CO][Cl_2] = 0.0084/(0.0026)^2 = 1.2 \times 10^3$$

EXERCISE 14.14A

Determine the number of moles of each reactant and product present when equilibrium is established in a gaseous mixture that initially contains 0.0100 mol H_2 and 0.100 mol HI in a 5.25-L volume at 698 K in the reaction

$$H_2(g) + I_2(g) \rightleftharpoons 2 HI(g) \qquad K_c = 54.3 \text{ at } 698 \text{ K}$$

EXERCISE 14.14B

How many moles of each reactant and product will be present when equilibrium is established in the reaction of Exercise 14.14A if the mixture initially contains 0.0100 mol H_2, 0.0100 mol I_2, and 0.100 mol HI?

As shown in Figure 14.10 and Example 14.15, in a calculation based on K_p, we emphasize the partial pressures of the various gases and the total gas pressure. The pressure of a gas in a mixture at constant temperature and constant volume is propor-

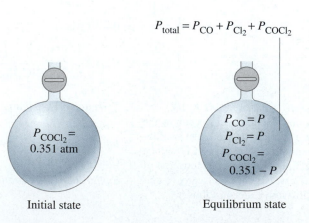

$$P_{total} = P_{CO} + P_{Cl_2} + P_{COCl_2}$$

$$P_{COCl_2} = 0.351 \text{ atm}$$

Initial state

$$P_{CO} = P$$
$$P_{Cl_2} = P$$
$$P_{COCl_2} = 0.351 - P$$

Equilibrium state

▶ **FIGURE 14.10   Example 14.15 illustrated**

tional to the concentration of the gas. Thus, we can use changes in partial pressures (represented by P in Example 14.15) instead of the changes in concentrations x we used in previous examples.

Example 14.15

A sample of phosgene, $COCl_2(g)$, is introduced into a constant-volume vessel at 395 °C and observed to exert an initial pressure of 0.351 atm. When equilibrium is established for the

$$\text{Reaction } CO(g) + Cl_2(g) \rightleftharpoons COCl_2(g) \qquad K_p = 22.5$$

what will be the partial pressure of each gas and the total gas pressure?

STRATEGY

As indicated in the paragraph preceding this example, when working with K_p, we use partial pressures in place of molar concentrations and base the equilibrium constant expression on an appropriately chosen partial pressure P. Although we can work with the equilibrium reaction as written and base our calculation on a net change occurring to the left, it may be easier to reverse the equation and think about a net change occurring to the right. When we reverse the equation, of course, we must take the reciprocal of the given K_p value as the new K_p, that is, 1/22.5, or 0.0444. We use the ICE format and conclude by solving a quadratic equation.

SOLUTION

First, we use the ICE tabular format for the reverse of the given equilibrium equation, with $-P$ representing the change in partial pressure of $COCl_2(g)$.

The reaction:	$COCl_2(g)$	$\rightleftharpoons$	$CO(g)$	+	$Cl_2(g)$ $K_p = 0.0444$
Initial pressures, atm:	0.351		0		0
Changes, atm:	$-P$		$+P$		$+P$
Equilibrium pressures, atm:	$(0.351 - P)$		P		P

Now we can substitute into the K_p expression and obtain a quadratic equation in P.

$$K_p = \frac{(P_{CO})(P_{Cl_2})}{P_{COCl_2}} = 0.0444$$

$$= \frac{P \times P}{0.351 - P} = 0.0444$$

$$P^2 = 0.0156 - 0.0444P$$

$$P^2 + 0.0444P - 0.0156 = 0$$

Next, we use the quadratic formula to obtain a value of P.

$$P = \frac{-0.0444 \pm \sqrt{(0.0444)^2 - [4 \times 1 \times (-0.0156)]}}{2}$$

The correct root to use in doing this is the positive root.

$$= \frac{-0.0444 \pm 0.254}{2} = 0.105$$

We can calculate the equilibrium partial pressures from the relationships given in the ICE table.

$$P_{CO} = P_{Cl_2} = 0.105 \text{ atm}$$

$$P_{COCl_2} = 0.351 - 0.105 = 0.246 \text{ atm}$$

We then find the total pressure by summing the individual partial pressures.

$$P_{total} = P_{CO} + P_{Cl_2} + P_{COCl_2}$$

$$= 0.105 + 0.105 + 0.246 = 0.456 \text{ atm}$$

ASSESSMENT

Note that in solving the quadratic equation, we rejected the negative root because negative partial pressures have no physical significance. Because two moles of gaseous products are produced for every mole of reactant in this reaction, we expect a higher total pressure at equilibrium than initially, and we found just that. Also, because $COCl_2$ must be consumed to produce CO and Cl_2, we should expect P_{COCl_2} at equilibrium to be less than the initial P_{COCl_2}, as it is. Finally, we can check the correctness of the calculated partial pressures in the usual way. The agreement with the given value of $K_p = 0.0444$ is excellent:

$$K_p = (P_{CO})(P_{Cl_2})/P_{COCl_2} = (0.105)^2/0.246 = 0.0448$$

EXERCISE 14.15A

What are the equilibrium partial pressures for each species, and the total pressure for the reaction in Example 14.15 if initially $P_{CO} = 1.00$ atm and $P_{Cl_2} = 1.00$ atm?

EXERCISE 14.15B

Ammonium hydrogen sulfide dissociates readily, even at room temperature, in the reaction

$$NH_4HS(s) \rightleftharpoons NH_3(g) + H_2S(g) \qquad K_p = 0.108 \text{ at } 25°C$$

What is the total pressure of the gases in equilibrium with $NH_4HS(s)$ at 25 °C?

We chose our examples and exercises carefully so that we could solve them with the quadratic formula, and we will continue to do so in the next two chapters. However, you may occasionally encounter situations that require solving more complex equations. Scientists and engineers routinely use graphing calculators or computers to solve such equations, but the equations can also be solved by an approximation method that is illustrated in Appendix A. Even with the availability of computers, scientists still look for opportunities to make valid assumptions that will simplify problems.

Cumulative Example

A mixture of $H_2S(g)$ and $CH_4(g)$ in the mole ratio 2:1 was brought to equilibrium at 700 °C and a total pressure of 1.00 atm. The equilibrium mixture was analyzed and found to contain 9.54×10^{-3} mol H_2S. The CS_2 present at equilibrium was converted, first to H_2SO_4 and then to $BaSO_4$, with 1.42×10^{-3} mol $BaSO_4$ being obtained. Use these data to determine K_p at 700 °C for the reaction

$$2 H_2S(g) + CH_4(g) \rightleftharpoons CS_2(g) + 4 H_2(g)$$

STRATEGY

First, we must identify the equilibrium constant expression for which we seek the K_p value. It is

$$K_p = \frac{P_{CS_2}(P_{H_2})^4}{P_{CH_4}(P_{H_2S})^2}$$

This expression indicates that we need to extract the partial pressures of all four gases from the given data. A logical starting point is with the stated equilibrium amount of H_2S. The equilibrium amount of CS_2 is also easily obtainable from the amount of $BaSO_4$ to which the CS_2 is converted. We can find the equilibrium amounts of the other two gases with simple stoichiometric relationships. Then we can convert the amounts of the four gases, first to mole fractions and then to partial pressures. At that point we can calculate the numerical value of K_p.

SOLUTION

The equilibrium amount of H_2S is given.

9.54×10^{-3} mol H_2S

Next, we use two simple stoichiometric factors to find the equilibrium amount of CS_2.

$$? \text{ mol } CS_2 = 1.42 \times 10^{-3} \text{ mol } BaSO_4 \times \frac{1 \text{ mol } S}{1 \text{ mol } BaSO_4} \times \frac{1 \text{ mol } CS_2}{2 \text{ mol } S}$$

$$= 7.10 \times 10^{-4} \text{ mol } CS_2$$

Because the H_2S and CH_4 are present initially in a 2:1 mole ratio and react in that same ratio, the amount of CH_4 at equilibrium is one-half that of H_2S.

$$? \text{ mol } CH_4 = 9.54 \times 10^{-3} \text{ mol } H_2S \times \frac{1 \text{ mol } CH_4}{2 \text{ mol } H_2S}$$

$$= 4.77 \times 10^{-3} \text{ mol } CH_4$$

The amount of H_2 at equilibrium is four times that of CS_2.

$$? \text{ mol } H_2 = 7.10 \times 10^{-4} \text{ mol } CS_2 \times \frac{4 \text{ mol } H_2}{1 \text{ mol } CS_2}$$

$$= 2.84 \times 10^{-3} \text{ mol } H_2$$

The total number of moles of gas in the equilibrium mixture is the sum of the number of moles of H_2S, CS_2, CH_4, and H_2.

9.54×10^{-3} mol H_2S + 7.10×10^{-4} mol CS_2 + 4.77×10^{-3} mol CH_4

$+ 2.84 \times 10^{-3}$ mol H_2 = 1.786×10^{-2} mol gas

We can determine the mole fraction of each gas by using Equation (5.14), $x_i = n_i/n_{total}$, and we can determine the partial pressures by using Equation (5.16), $P_i = x_i P_{total}$.

$$P_{H_2S} = x_{H_2S}P_{total} = \frac{9.54 \times 10^{-3} \text{ mol } H_2S}{1.786 \times 10^{-2} \text{ mol total}} \times 1.00 \text{ atm} = 0.534 \text{ atm}$$

$$P_{CH_4} = x_{CH_4}P_{total} = \frac{4.77 \times 10^{-3} \text{ mol } CH_4}{1.786 \times 10^{-2} \text{ mol total}} \times 1.00 \text{ atm} = 0.267 \text{ atm}$$

$$P_{CS_2} = x_{CS_2}P_{total} = \frac{7.10 \times 10^{-4} \text{ mol } CS_2}{1.786 \times 10^{-2} \text{ mol total}} \times 1.00 \text{ atm} = 0.0398 \text{ atm}$$

$$P_{H_2} = x_{H_2}P_{total} = \frac{2.84 \times 10^{-3} \text{ mol } H_2}{1.786 \times 10^{-2} \text{ mol total}} \times 1.00 \text{ atm} = 0.159 \text{ atm}$$

We get the value of K_p by substituting these partial pressures into the equilibrium constant expression.

$$K_p = \frac{P_{CS_2}(P_{H_2})^4}{P_{CH_4}(P_{H_2S})^2} = \frac{0.0398 \times (0.159)^4}{0.267 \times (0.534)^2} = 3.34 \times 10^{-4}$$

ASSESSMENT

Here are some key checkpoints in this problem. Ensure that (1) the K_p expression matches the balanced chemical equation, (2) the number of moles of each gas and the total number of moles of gas are correctly determined, and (3) the sum of the partial pressures is equal to the total gas pressure (1 atm). Note that the value of K_p is quite small, which is consistent with the predominance of reactants (0.267 atm and 0.534 atm) over products (0.0398 atm and 0.159 atm).

Concept Review with Key Terms

14.1 The Dynamic Nature of Equilibrium—In a reversible reaction at **equilibrium**, the concentrations of all reactants and products remain constant with time as a result of the forward and reverse reactions occurring at equal rates.

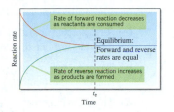

14.2 The Equilibrium Constant Expression—For the general reaction represented by the equation

$$a\,A + b\,B + \cdots \rightleftharpoons g\,G + h\,H + \cdots$$

the concentrations of reactants and products at equilibrium must conform to the **equilibrium constant expression** for the **concentration equilibrium constant (K_c)**.

$$K_c = \frac{[G]^g[H]^h \cdots}{[A]^a[B]^b \cdots}$$

The exponents in this expression correspond to the coefficients in the chemical equation.

14.3 Modifying Equilibrium Constant Expressions—For equilibria involving gases, a **partial pressure equilibrium constant expression (K_p)** can be written that uses partial pressures in place of concentrations. For the general reaction above, where reactants and products exist in the gas phase,

$$K_p = \frac{(P_G)^g(P_H)^h \cdots}{(P_A)^a(P_B)^b \cdots}$$

If the chemical equation for a reversible reaction is modified, the equilibrium constant expression must also be modified. If the equation is reversed, the K_c or K_p expression is inverted. If the coefficients of the equation are multiplied by a common factor, the K_c or K_p expression is raised to the corresponding power. K_c and K_p are expressed as dimensionless numbers, and pure solid and liquid phases are not represented in equilibrium constant expressions.

In general, if K_c or K_p is very large, the forward reaction goes to completion; if K_c or K_p is very small, the forward reaction occurs to a very limited extent. Usually, calculations based on the equilibrium constant expression are necessary only when K_c or K_p values lie between these extremes.

The **reaction quotient** is a ratio of concentrations (Q_c) or partial pressures (Q_p) having the same form as an equilibrium constant expression but using *nonequilibrium* concentrations or partial pressures. By comparing the values of Q and K, we can determine whether a net reaction will occur in the forward direction ($Q < K$) or in the reverse direction ($Q > K$) to establish equilibrium.

14.4 Qualitative Treatment of Equilibrium: Le Châtelier's Principle—Qualitative predictions about the effect of various changes—amounts of reactants or products, volume, external pressure, or temperature—can be based on **Le Châtelier's principle**. The principle states that when a change is imposed on a system at equilibrium, the system responds by attaining a new equilibrium in which the impact of the change is minimized. Catalysts speed the attainment of equilibrium but have no effect on the equilibrium state, that is, on reactant and product concentrations or partial pressures.

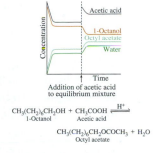

14.5 Some Illustrative Equilibrium Calculations—The most common types of equilibrium calculations are (a) determining the value of an equilibrium constant from initial and equilibrium conditions, and (b) using initial conditions and the equilibrium constant to determine equilibrium conditions. We need to solve algebraic equations to carry out some of these calculations. Several techniques for doing so are introduced in Section 14.5.

Assessment Goals

When you have mastered the material in this chapter, you will be able to:

- Describe the nature of dynamic equilibrium in terms of forward and reverse processes.
- Write the equilibrium constant expression for reversible chemical reactions.
- Express equilibrium constants in terms of rate constants for forward and reverse reactions.
- Derive equilibrium constants for other reactions related to a given reaction with a known equilibrium constant (that is, reactions reversed or with coefficients in the chemical equation multiplied by a common factor).
- Derive the equilibrium constant for an overall reaction obtained by combining two or more reactions and their equilibrium constants.

- Distinguish between K_c and K_p, and obtain a value for one of them from a known value of the other.
- Describe nonequilibrium conditions through a reaction quotient, and use this value to predict the direction in which a net reaction must occur to achieve equilibrium.
- Use LeChâtelier's principle to predict shifts in equilibria following the addition or removal of a species to an equilibrium mixture.
- Use LeChâtelier's principle to predict the effect of changes in pressure and temperature, or the presence of a catalyst, on a reaction mixture at equilibrium.
- Determine the values of equilibrium constants from experimental concentration or partial-pressure data.
- Given initial conditions, calculate equilibrium concentrations and partial pressures from K_c and K_p values.

Self-Assessment Questions

1. Why is equilibrium in a reversible reaction described as *dynamic*? Cite some experimental evidence to support this idea.

2. Consider the following reaction at a particular temperature:

$$NO_2(g) + N_2O(g) \rightleftharpoons 3\ NO(g)$$

Given that the equilibrium concentrations are $[NO_2] = 1.25$ M, $[N_2O] = 1.80$ M, and $[NO] = 0.0015$ M, what is the value of the equilibrium constant K_c?

(a) 6.7×10^8 (c) 6.7×10^{-4}

(b) 2.0×10^{-3} (d) 1.5×10^{-9}

3. For the following reaction, to what exponent must (RT) be raised in converting from K_c to K_p?

$$2\ SO_2(g) + O_2(g) \rightleftharpoons 2\ SO_3(g)$$

(a) 1   (b) −1   (c) 2   (d) 0

4. Equilibrium is established in the following reaction.

$$A + B \rightleftharpoons C + D \qquad K_c = 10.0$$

At this point, which of the following must be true?

(a) $[C][D] = [A][B]$

(b) $[C] = [A]$ and $[D] = [B]$

(c) $[A][B] = 0.10 \times [C][D]$

(d) $[A] = [B] = [C] = [D] = 10.0$

5. At some point during a reaction, the following concentrations are measured.

$$2\ ICl\ (g) \rightleftharpoons I_2(g) + Cl_2(g) \qquad K_c = 8.33 \times 10^{-4}$$
$$0.15\ M \qquad 0.00125\ M \quad 0.0075\ M$$

Which of the following represents a true statement?

(a) The reaction is at equilibrium.

(b) The reaction will proceed until all ICl is consumed.

(c) I_2 will be consumed in reaching equilibrium.

(d) The concentration of Cl_2 will increase in reaching equilibrium.

6. Describe how the equilibrium constant for an overall reaction is related to the equilibrium constants for the individual reactions that yield the overall reaction.

7. Which of the following are true of a reaction system at equilibrium?

(a) All concentrations of reactants and products are the same.

(b) The product of the concentrations of the reactants is equal to the product of the concentrations of the products.

(c) The sum of the concentrations of the reactants is equal to the sum of the concentrations of the products.

(d) Reactants are reacting to form products at the same rate as products are reacting to form reactants.

(e) There is one unique set of equilibrium concentrations that satisfies a particular K_c value.

(f) There are many sets of equilibrium concentrations that satisfy the K_c value.

8. What role does a catalyst play in a reversible chemical reaction? How does a catalyst affect the value of the equilibrium constant for a reaction?

9. In the reaction $2\ SO_2(g) + O_2(g) \rightleftharpoons 2\ SO_3(g)$, equilibrium is established at a temperature at which $K_c = 100$. Which of the following is(are) true if the number of moles of SO_3 in the equilibrium mixture is equal to the number of moles of SO_2?

(a) The number of moles of O_2 is also equal to the number of moles of SO_2.

(b) The number of moles of O_2 is half the number of moles of SO_2.

(c) $[O_2] = 0.01$ M

(d) $[O_2]$ may have any of several different values.

10. Write equilibrium constant expressions, K_p, for the following reactions.

(a) $CO(g) + H_2O(g) \rightleftharpoons CO_2(g) + H_2(g)$

(b) $N_2(g) + 3\ H_2(g) \rightleftharpoons 2\ NH_3(g)$

(c) $NH_4HS(s) \rightleftharpoons NH_3(g) + H_2S(g)$

11. Write an equilibrium constant expression, K_p, based on the formation of one mole of each of the following gaseous compounds from its elements at 25 °C.

(a) $NO(g)$ (b) $NH_3(g)$ (c) $NOCl(g)$

12. Write an equilibrium constant expression, K_c, for each of the following reversible reactions.

(a) Carbon monoxide gas reduces nitrogen monoxide gas to gaseous nitrogen; carbon dioxide gas is the other product.

(b) Oxygen gas oxidizes gaseous ammonia to gaseous nitrogen monoxide; water vapor is the other product.

(c) Solid sodium hydrogen carbonate decomposes to form solid sodium carbonate, water vapor, and carbon dioxide gas.

13. Equilibrium is established in the reversible reaction.

$$4 \, HCl(g) + O_2(g) \rightleftharpoons 2 \, H_2O(g) + 2 \, Cl_2(g)$$
$$\Delta H^\circ = -114.4 \, kJ$$

Describe four changes that can be made to this mixture to increase the amount of $Cl_2(g)$ at equilibrium.

14. In which of the following reactions will the amount of product at equilibrium increase if the total gas pressure is raised from 1 to 10 atm? Explain.

(a) $SO_2(g) + Cl_2(g) \rightleftharpoons SO_2Cl_2(g)$

(b) $N_2(g) + O_2(g) \rightleftharpoons 2 \, NO(g)$

(c) $SO_2(g) + \frac{1}{2} O_2(g) \rightleftharpoons SO_3(g)$

15. The extent of dissociation that occurs in one of the following reactions depends on the volume of the reaction vessel, and in the other it does not. Identify the situation for each reaction, and explain why they are not the same.

(a) $2 \, NO(g) \rightleftharpoons N_2(g) + O_2(g)$

(b) $2 \, NOCl(g) \rightleftharpoons 2 \, NO(g) + Cl_2(g)$

16. In one of the following reactions, the extent of dissociation increases with an increase in temperature, while in the other it de-creases with an increase in temperature. Identify the situation for each reaction, and explain why they are not the same.

(a) $2 \, NO(g) \rightleftharpoons N_2(g) + O_2(g)$ $\Delta H^\circ = -180.5 \, kJ$

(b) $2 \, SO_3(g) \rightleftharpoons 2 \, SO_2(g) + O_2(g)$ $\Delta H^\circ = 197.8 \, kJ$

17. A mixture of 1.00 mole *each* of $CO(g)$, $H_2O(g)$, and $CO_2(g)$ is introduced into a 10.0-L flask at a temperature at which $K_c = 10.0$ for the reaction

$$CO(g) + H_2O(g) \rightleftharpoons CO_2(g) + H_2(g)$$

Which of the following statements is true when the reaction reaches a state of equilibrium?

(a) The amount of H_2 will be 1.00 mole.

(b) The amount of $CO_2(g)$ will be greater than 1.00 mole, and the amounts of $CO(g)$, $H_2O(g)$, and $H_2(g)$ will each be less than 1.00 mole.

(c) The amounts of all reactants and products will be greater than 1.00 mole.

(d) The amounts of $CO_2(g)$ and $H_2O(g)$ will each be less than 1.00 mole.

18. A mixture of 3.00 mol of chlorine and 2.10 mol of iodine is *initially* placed into a rigid 1.000-L container at 350 °C. When the mixture has come to equilibrium, the concentration of iodine monochloride is 2.32 M. What is the equilibrium constant for this reaction at 350 °C?

$$Cl_2(g) + I_2(g) \rightleftharpoons 2 \, ICl(g)$$

(a) 7.37   (b) 5.85   (c) 8.54   (d) 3.11

Problems

Equilibrium Constant Relationships

19. Determine the values of K_c that correspond to the following values of K_p.

(a) $SO_2Cl_2(g) \rightleftharpoons SO_2(g) + Cl_2(g)$
$$K_p = 2.9 \times 10^{-2} \text{ at } 303 \, K$$

(b) $2 \, NO_2(g) \rightleftharpoons 2 \, NO(g) + O_2(g)$ $K_p = 0.275 \text{ at } 700 \, K$

(c) $CO(g) + Cl_2(g) \rightleftharpoons COCl_2(g)$ $K_p = 22.5 \text{ at } 395 \, ^\circ C$

20. Determine the values of K_p that correspond to the following values of K_c.

(a) $CO(g) + Cl_2(g) \rightleftharpoons COCl_2(g)$
$$K_c = 1.2 \times 10^3 \text{ at } 668 \, K$$

(b) $2 \, NO(g) + Br_2(g) \rightleftharpoons 2 \, NOBr(g)$
$$K_c = 1.32 \times 10^{-2} \text{ at } 1000 \, K$$

(c) $2 \, COF_2(g) \rightleftharpoons CO_2(g) + CF_4(g)$ $K_c = 2.00 \text{ at } 1000 \, ^\circ C$

21. For the reaction $SO_3(g) \rightleftharpoons SO_2(g) + \frac{1}{2} O_2(g)$, $K_c = 16.7$ at 1000 K. What is the value of K_c at 1000 K for the reaction $2 \, SO_2(g) + O_2(g) \rightleftharpoons 2 \, SO_3(g)$?

22. For the reaction $N_2(g) + O_2(g) \rightleftharpoons 2 \, NO(g)$, $K_c = 4.08 \times 10^{-4}$ at 2000 K. What is the value of K_c at 2000 K for the reaction $NO(g) \rightleftharpoons \frac{1}{2} N_2(g) + \frac{1}{2} O_2(g)$?

23. Given that

$$2 \, Cl_2(g) + 2 \, H_2O(g) \rightleftharpoons 4 \, HCl(g) + O_2(g)$$
$$K_p = 0.0366 \text{ at } 450 \, ^\circ C$$

determine K_c at 450°C for the reaction

$$HCl(g) + \frac{1}{4} O_2(g) \rightleftharpoons \frac{1}{2} Cl_2(g) + \frac{1}{2} H_2O(g)$$

24. Given that

$$N_2(g) + 3 \, H_2(g) \rightleftharpoons 2 \, NH_3(g)   K_c = 155 \text{ at } 300 \, ^\circ C$$

determine K_p at 300°C for the reaction

$$\tfrac{2}{3} NH_3(g) \rightleftharpoons \tfrac{1}{3} N_2(g) + H_2(g)$$

25. What is the value of K_p at 298 K for the reaction $\frac{1}{2} CH_4(g) + H_2O(g) \rightleftharpoons \frac{1}{2} CO_2(g) + 2 \, H_2(g)$, given the following data at 298 K?

$$CO_2(g) + H_2(g) \rightleftharpoons CO(g) + H_2O(g)$$
$$K_p = 9.80 \times 10^{-6}$$

$$CO(g) + 3 \, H_2(g) \rightleftharpoons CH_4(g) + H_2O(g)$$
$$K_p = 8.00 \times 10^{24}$$

26. What is the value of K_p at 298 K for the reaction $N_2(g) + 2 \, O_2(g) \rightleftharpoons N_2O_4(g)$, given the following data at 298 K?

$$\tfrac{1}{2} N_2(g) + \tfrac{1}{2} O_2(g) \rightleftharpoons NO(g)   K_p = 6.9 \times 10^{-16}$$

$$NO_2(g) \rightleftharpoons NO(g) + \tfrac{1}{2} O_2(g)   K_p = 6.7 \times 10^{-7}$$

$$2 \, NO_2(g) \rightleftharpoons N_2O_4(g)   K_p = 6.7$$

27. Determine K_c at 298 K for the reaction $2\,CH_4(g) \rightleftharpoons C_2H_2(g) + 3\,H_2(g)$, given the following data at 298 K.

$$CH_4(g) + H_2O(g) \rightleftharpoons CO(g) + 3\,H_2(g)$$
$$K_p = 1.2 \times 10^{-25}$$

$$2\,C_2H_2(g) + 3\,O_2(g) \rightleftharpoons 4\,CO(g) + 2\,H_2O(g)$$
$$K_p = 1.1 \times 10^2$$

$$H_2(g) + \tfrac{1}{2}O_2(g) \rightleftharpoons H_2O(g)$$
$$K_p = 1.1 \times 10^{40}$$

(*Hint:* How are K_c and K_p for the reaction related?)

28. Determine K_c at 298 K for the reaction $\tfrac{1}{2}N_2(g) + \tfrac{1}{2}O_2(g) + \tfrac{1}{2}Cl_2(g) \rightleftharpoons NOCl(g)$, given the following data at 298 K.

$$\tfrac{1}{2}N_2(g) + O_2(g) \rightleftharpoons NO_2(g) \qquad K_p = 1.0 \times 10^{-9}$$

$$NOCl(g) + \tfrac{1}{2}O_2(g) \rightleftharpoons NO_2Cl(g) \qquad K_p = 1.1 \times 10^2$$

$$NO_2(g) + \tfrac{1}{2}Cl_2(g) \rightleftharpoons NO_2Cl(g) \qquad K_p = 0.3$$

(*Hint:* How are K_c and K_p for the reaction related?)

29. Is K_c for the reaction $2\,ICl(g) \rightleftharpoons I_2(g) + Cl_2(g)$ necessarily greater than K'_c for the reaction $ICl(g) \rightleftharpoons \tfrac{1}{2}I_2(g) + \tfrac{1}{2}Cl_2(g)$? Explain.

30. The reversible reaction $N_2O_4(g) \rightleftharpoons 2\,NO_2(g)$ has a value of $K_p = 0.145$ at 25°C. *Without doing a calculation,* state whether the numerical value of K_p at 25°C for the reaction $\tfrac{1}{2}N_2O_4(g) \rightleftharpoons NO_2(g)$ is greater than, equal to, or less than 0.145.

31. If the equilibrium concentrations in the reaction $2\,A(g) + B(g) \rightleftharpoons C(g)$ are $[A] = 2.4 \times 10^{-2}$ M, $[B] = 4.6 \times 10^{-3}$ M, and $[C] = 6.2 \times 10^{-3}$ M, calculate the value of K_c.

32. If the equilibrium concentrations in the reaction $A(g) + 2\,B(g) \rightleftharpoons 2\,C(g)$ are $[A] = 0.025$ M, $[B] = 0.15$ M, and $[C] = 0.55$ M, calculate the value of K_c.

33. In the reaction $2\,H_2(g) + S_2(g) \rightleftharpoons 2\,H_2S(g)$, $K_c = 6.28 \times 10^3$ at 900 K. What is the equilibrium value of $[H_2]$ if at equilibrium, $[H_2S] = [S_2]^{1/2}$?

34. In the reaction $CO(g) + Cl_2(g) \rightleftharpoons COCl_2(g)$, $K_c = 1.2 \times 10^3$ at 395°C. What is the equilibrium value of $[COCl_2]$ if at equilibrium, $[CO] = 2\,[Cl_2] = \tfrac{1}{2}[COCl_2]$?

35. For the reaction $H_2S(g) + I_2(s) \rightleftharpoons S(s) + 2\,HI(g)$, $K_p = 1.33 \times 10^{-5}$ at 333 K. What will be the *total* pressure of the gases above an equilibrium mixture if $P_{HI} = 0.010 \times P_{H_2S}$?

36. For the reaction $C(s) + CO_2(g) \rightleftharpoons 2\,CO(g)$, $K_p = 63$. What will be the *total* pressure of the gases above an equilibrium mixture if $P_{CO} = 10\,P_{CO_2}$?

37. In the reaction $N_2O_4(g) \rightleftharpoons 2\,NO_2(g)$, equilibrium is reached at a temperature at which $P_{NO_2} = 3\,(P_{N_2O_4})^{1/2}$. What must be the value of K_p at this temperature?

38. For the reaction $CO(g) + H_2O(g) \rightleftharpoons CO_2(g) + H_2(g)$, $K_p = 23.2$ at 600 K. Is it possible to have an equilibrium mixture at 600 K in which **(a)** $P_{CO} = P_{H_2O} = P_{CO_2} = P_{H_2}$? **(b)** $P_{H_2}/P_{H_2O} = P_{CO_2}/P_{CO}$? **(c)** $(P_{CO_2})(P_{H_2}) = (P_{CO})(P_{H_2O})$? **(d)** $P_{CO_2}/P_{H_2O} = P_{H_2}/P_{CO}$? Explain.

39. *Without doing detailed calculations,* determine which of the following statements is(are) correct in describing the equilibrium

established when equal numbers of moles of H_2, N_2, and HCN and an excess of graphite are allowed to come to equilibrium in a constant-volume reaction vessel at 2025 K in the following reaction.

$$H_2(g) + N_2(g) + 2\,C(graphite) \rightleftharpoons 2\,HCN(g)$$
$$K_c = 3.43 \times 10^{-3} \text{ at } 2025K$$

(a) The equilibrium amount of HCN can be increased by reducing the reaction volume; **(b)** the amounts of H_2 and N_2 increase from their initial amounts but remain in equimolar proportions; **(c)** the amount of graphite at equilibrium is the same as initially; **(d)** K_p for the reaction is less than K_c; **(e)** the change in partial pressure of HCN from its initial value is twice as great as the changes in partial pressures of H_2 and N_2.

40. *Without doing detailed calculations,* determine which of the following statements is(are) correct in describing the equilibrium established when an initial mixture containing an equal number of moles of SO_2 and O_2 is brought to equilibrium in a constant-volume vessel at a temperature of 1030 °C in the following reaction.

$$2\,SO_2(g) + O_2(g) \rightleftharpoons 2\,SO_3(g) \qquad K_c = 1.98 \text{ at } 1030\,°C$$

(a) The pressure of the mixture increases; **(b)** the number of moles of SO_3 at equilibrium is twice the number of moles of O_2 present initially; **(c)** the number of moles of SO_3 at equilibrium depends on the volume of the reaction vessel; **(d)** the equilibrium mixture no longer has equal numbers of moles of SO_2 and O_2; **(e)** K_p for the reaction is greater than K_c.

41. The sketch (shown below) represents an initial nonequilibrium mixture of NO(g), $Br_2(g)$, and NOBr(g) at a temperature at which $K_c = 3.0$ in the reversible reaction

$$2\,NO(g) + Br_2(g) \rightleftharpoons 2\,NOBr(g)$$

Assume that the concentrations of the species in the reaction are in the same proportions as the numbers of molecules represented, and determine which of the three lettered sketches best represents an equilibrium mixture of this reaction. Explain your reasoning.

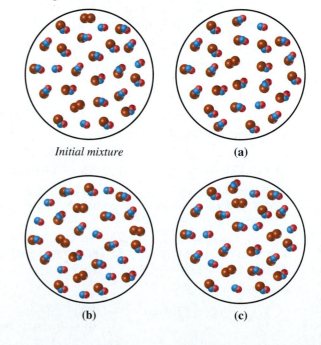

Initial mixture   (a)

(b)   (c)

42. The sketch (shown on the right) represents an initial nonequilibrium mixture of $SO_2(g)$, $Cl_2(g)$, and $SO_2Cl_2(g)$ at a temperature at which $K_c = 4.0$ in the reversible reaction

$$SO_2(g) + Cl_2(g) \rightleftharpoons SO_2Cl_2(g)$$

Assume that the concentrations of the species in the reaction are in the same proportions as the numbers of molecules represented, and determine which of the three lettered sketches best represents an equilibrium mixture of this reaction. Explain your reasoning.

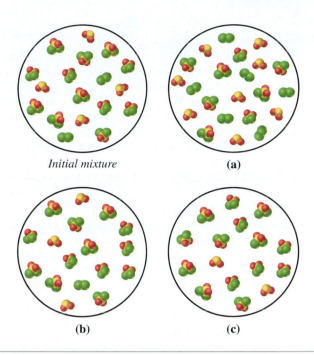

Initial mixture   **(a)**

(b)   **(c)**

Le Châtelier's Principle

43. The water-gas reaction is used to produce combustible gases from carbon (coal) and steam.

$$C(s) + H_2O(g) \rightleftharpoons CO(g) + H_2(g)$$

$$K_p = 9.7 \times 10^{-17} \text{ at 298 K} \qquad \Delta H^\circ = 131 \text{ kJ}$$

What will be the effect on the final equilibrium amount of $H_2(g)$ if a gaseous mixture originally at equilibrium with a large excess of $C(s)$ at 298 K is subjected to the following changes: **(a)** more $H_2O(g)$ is added; **(b)** a catalyst is added; **(c)** the mixture is transferred to a reaction vessel of greater volume; **(d)** more $CO(g)$ is added; **(e)** an inert gas is added to the reaction vessel to increase the total pressure; **(f)** the temperature is raised to 1000 K; **(g)** a small amount of $C(s)$ is removed?

44. Of the N_2O_4 molecules present in a sample of the gas, 12.5% are dissociated into NO_2 when equilibrium is established at 25 °C.

$$N_2O_4(g) \rightleftharpoons 2 NO_2(g) \qquad \Delta H = 57.2 \text{ kJ}$$

Will the percent dissociation be greater than or less than 12.5% if **(a)** the reaction mixture is transferred to a vessel of twice the volume; **(b)** the temperature is raised to 50 °C; **(c)** a catalyst is added to the reaction vessel; **(d)** neon gas is added to the reaction mixture in a constant-volume flask?

45. Write equations for the dissociation of **(a)** $Cl_2(g)$, **(b)** $S_2(g)$, and **(c)** $H_2(g)$ into gaseous atoms. Must the extent of dissociation of these diatomic molecules into atoms increase with an increase in temperature? Explain.

46. Write equations for the dissociation of **(a)** $H_2O(g)$ into $H_2(g)$ and $O_2(g)$; and **(b)** $NO(g)$ into $N_2(g)$ and $O_2(g)$. Must the extent of dissociation of molecules of *compounds* into their constituent elements increase with an increase in temperature? Explain. (*Hint:* Use data from Appendix C if necessary.)

47. In the formation of the following gaseous compounds from their gaseous elements, identify the reactions that occur to a greater extent at high pressures and the reactions that are unaffected by the total pressure: **(a)** $NO(g)$ from $N_2(g)$ and $O_2(g)$, **(b)** $NH_3(g)$ from $N_2(g)$ and $H_2(g)$, **(c)** $HI(g)$ from $H_2(g)$ and $I_2(g)$, **(d)** $H_2S(g)$ from $H_2(g)$ and $S_2(g)$?

48. Hemoglobin (Hb) in the blood absorbs oxygen to form oxyhemoglobin $[Hb(O_2)_2]$, which transports oxygen to the cells. The equilibrium can be represented as: $Hb + 2 O_2(g) \rightleftharpoons Hb(O_2)_2$. Consider the following data as necessary. Atmospheric pressure at sea level is 101 kPa; on top of Mt. Everest it is 33.7 kPa; in a hyperbaric chamber it may be about 400 kPa. What is the effect on the oxygen supply to the cells of **(a)** going to the top of Mt. Everest and **(b)** going into a hyperbaric chamber?

The Reaction Quotient and the Direction of Net Change

49. If a gaseous mixture at 588 K contains 20.0 g each of CO and H_2O and 25.0 g each of CO_2 and H_2, in what direction will a net reaction occur to reach equilibrium? Explain.

$$CO(g) + H_2O(g) \rightleftharpoons CO_2(g) + H_2(g)$$
$$K_c = 31.4 \text{ at 588 K}$$

50. If a 2.50-L reaction vessel at 1000 °C contains 0.525 mol CO_2, 1.25 mol CF_4, and 0.75 mol COF_2, in what direction will a net reaction occur to reach equilibrium? Explain.

$$CO_2(g) + CF_4(g) \rightleftharpoons 2 COF_2(g) \qquad K_c = 0.50 \text{ at 1000 °C}$$

51. The following amounts of substances are added to a 7.25-L reaction vessel at 773 K: 0.103 mol CO, 0.205 mol H_2, 2.10 mol

CH_4, 3.15 mol H_2O. In what direction will a net reaction occur to reach equilibrium? Explain.

$$CO(g) + 3 H_2(g) \rightleftharpoons CH_4(g) + H_2O(g)$$
$$K_p = 102 \text{ at 773 K}$$

52. The following quantities of substances are added to a 15.5-L reaction vessel at 773 K: 25.2 g CO, 15.1 g H_2, 130.2 g CH_4, 125.0 g H_2O. In what direction will a net reaction occur to reach equilibrium? Explain.

$$CO(g) + 3 H_2(g) \rightleftharpoons CH_4(g) + H_2O(g)$$
$$K_p = 102 \text{ at 773 K}$$

Experimental Determination of Equilibrium Constants

53. In a sealed 10.5-L vessel at 184 °C, equilibrium is established between $NO_2(g)$ and its dissociation products, $NO(g)$ and $O_2(g)$. The quantities found at equilibrium are 1.353 g NO_2, 0.0960 g NO, and 0.0512 g O_2. What is the value of K_c for the reaction $2 NO_2(g) \rightleftharpoons 2 NO(g) + O_2(g)$? What is the value of K_p?

54. In a sealed 1.75-L vessel at 250 °C, equilibrium is established between $PCl_5(g)$ and its dissociation products, $PCl_3(g)$ and $Cl_2(g)$. The quantities found at equilibrium are 0.562 g PCl_5, 1.950 g PCl_3, and 1.007 g Cl_2. What is the value of K_c for the reaction $PCl_5(g) \rightleftharpoons PCl_3(g) + Cl_2(g)$? What is the value of K_p?

55. Ammonium carbamate dissociates into ammonia gas and carbon dioxide gas. If we start with a sample of pure $NH_2COONH_4(s)$ at 30 °C, the *total* pressure of the gases is 0.164 atm when equilibrium is established. Write an equation for the dissociation reaction, and determine the value of K_p.

56. Ammonium hydrogen sulfide dissociates into ammonia gas and hydrogen sulfide gas. If we start with a sample of pure $NH_4HS(s)$ at 25 °C, the *total* pressure of the gases is 0.658 atm when equilibrium is established. Write an equation for the dissociation reaction and determine the value of K_p.

57. A 0.682-g sample of $ICl(g)$ is placed in a 625-mL reaction vessel at 682 K. When equilibrium is reached in the reaction $2 ICl(g) \rightleftharpoons I_2(g) + Cl_2(g)$, 0.0383 g I_2 is found in the mixture. What is K_c for this reaction?

58. A mixture of 5.25 g I_2 and 2.15 g Br_2 is brought to equilibrium in a 3.15-L reaction vessel at 115 °C. At equilibrium, 1.98 g of I_2 is present. What is K_c at 115 °C for the reaction $I_2(g) + Br_2(g) \rightleftharpoons 2 IBr(g)$?

Calculations Based on K_c

59. Assume that at 25 °C, with $AlCl_3$ as a catalyst, the following equilibrium can be established between the liquids cyclohexane and methylcyclopentane.

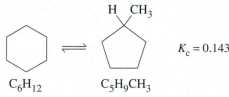

$K_c = 0.143$

C_6H_{12} $C_5H_9CH_3$

Cyclohexane   Methylcyclopentane

If initially 1.00×10^2 g cyclohexane is present, what mass of methylcyclopentane will be present in the equilibrium mixture? (*Hint:* Does the volume of solution matter?)

60. The equilibrium constant for the isomerization of butane at 25 °C is $K_c = 7.94$.

$$CH_3CH_2CH_2CH_3 \rightleftharpoons CH_3\overset{\overset{\displaystyle CH_3}{|}}{C}HCH_3$$

Butane   Isobutane

If 5.00 g butane is introduced into a 12.5-L flask at 25 °C, what mass of isobutane will be present when equilibrium is reached?

61. For the water-gas reaction (Problem 43), $K_c = 0.111$ at about 1100 K. If 0.100 mol $H_2O(g)$ and 0.100 mol $H_2(g)$ are mixed with excess $C(s)$ at this temperature and equilibrium is established, how many moles of $CO(g)$ will be present? No $CO(g)$ is present initially.

62. For the reaction $CO(g) + H_2O(g) \rightleftharpoons CO_2(g) + H_2(g)$, $K_c = 23.2$ at 600 K. If 0.250 mol each of CO and H_2O are introduced into a reaction vessel and equilibrium is established, how many moles each of CO_2 and H_2 will be present? (*Hint:* Does the volume of the reaction mixture matter?)

63. For the synthesis of phosgene at 395 °C, $CO(g) + Cl_2(g) \rightleftharpoons COCl_2(g)$, $K_c = 1.2 \times 10^3$. If 20.0 g CO and 35.5 g Cl_2 are placed in an 8.05-L reaction vessel at 395 °C and equilibrium is established, how many grams of $COCl_2$ will be present?

64. For the decomposition of carbonyl fluoride, $2 COF_2(g) \rightleftharpoons CO_2(g) + CF_4(g)$, $K_c = 2.00$ at 1000 °C. If 0.500 mol $COF_2(g)$ is placed in a 3.23-L reaction vessel at 1000 °C, how many moles of $COF_2(g)$ will remain *undissociated* when equilibrium is reached?

65. To establish equilibrium in the following reaction at 250 °C,

$$PCl_3(g) + Cl_2(g) \rightleftharpoons PCl_5(g) \qquad K_c = 26 \text{ at } 250 \text{ °C}$$

0.100 mole each of PCl_3 and Cl_2 and 0.0100 mol PCl_5 are introduced into a 6.40-L reaction flask. How many moles of each of the gases should be present when equilibrium is established?

66. If 0.250 mol SO_2Cl_2, 0.150 mol SO_2, and 0.0500 mol Cl_2 come to equilibrium in a 12.0-L reaction vessel at 102 °C, how many moles of each gas will be present?

$$SO_2Cl_2(g) \rightleftharpoons SO_2(g) + Cl_2(g)$$

$$K_c = 7.77 \times 10^{-2} \text{ at } 102 \text{ °C}$$

67. At 300 °C, for the reaction $2 NH_3(g) \rightleftharpoons N_2(g) + 3 H_2(g)$, $K_c = 6.46 \times 10^{-3}$. If an initial 1.00 mol NH_3 is brought to equilibrium in a 10.0-L reaction vessel at 300 °C, what percent of the NH_3 will remain *undissociated*? (*Hint:* Use the method of successive approximations in Appendix A.)

68. At 1030 °C, for the reaction $2 SO_3(g) \rightleftharpoons 2 SO_2(g) + O_2(g)$, $K_c = 0.504$. If an initial 1.00 mol SO_3 is brought to equilibrium in a 5.00-L reaction vessel at 1030 °C, what percent of the SO_3 will remain *undissociated*? (*Hint:* Use the method of successive approximations in Appendix A.)

Calculations Based on K_p

69. In the reaction $C(s) + S_2(g) \rightleftharpoons CS_2(g)$, $K_p = 5.60$ at 1009 °C. If, at equilibrium, $P_{CS_2} = 0.152$ atm, find **(a)** P_{S_2} and **(b)** the total gas pressure, P_{total}.

70. In the reaction $Sb_2S_3(s) + 3 H_2(g) \rightleftharpoons 2 Sb(s) + 3 H_2S(g)$, $K_p = 0.429$ at 713 K. If, at equilibrium, $P_{H_2S} = 0.200$ atm, find **(a)** P_{H_2} and **(b)** the total gas pressure, P_{total}.

71. For the reaction $C(s) + 2 H_2(g) \rightleftharpoons CH_4(g)$, $K_p = 0.263$ at 1000 °C. Calculate the total pressure when 0.100 mol CH_4 and an excess of $C(s)$ are brought to equilibrium at 1000 °C in a 4.16-L reaction vessel.

72. Solid molybdenum is kept in contact with $CH_4(g)$, initially at a pressure of 1.00 atm, in a reaction at 973 K. Calculate the total pressure when equilibrium is established.

$$2 Mo(s) + CH_4(g) \rightleftharpoons Mo_2C(s) + 2 H_2(g)$$
$$K_p = 3.55 \text{ at } 973 \text{ K}$$

73. For the synthesis of methanol at 483 K, $CO(g) + 2 H_2(g) \rightleftharpoons CH_3OH(g)$, $K_p = 9.23 \times 10^{-3}$. A gaseous mixture containing H_2 and CO in a $2:1$ mole ratio and at an initial total pressure of 12.0 atm is allowed to come to equilibrium at 483 K. What will be the partial pressure of $CH_3OH(g)$ in the equilibrium mixture? (*Hint:* Use the method of successive approximations of Appendix A.)

74. A gaseous mixture is prepared at 900 K in which the initial partial pressures are $P_{SO_3} = 1.50$ atm, $P_{SO_2} = 1.00$ atm, $P_{O_2} = 0.500$ atm. What is the partial pressure of SO_3 at equilibrium?

$$2 SO_3(g) \rightleftharpoons 2 SO_2(g) + O_2(g) \qquad K_p = 0.023 \text{ at } 900 \text{ K}$$

(*Hint:* Use the method of successive approximations of Appendix A.)

Additional Problems

Problems marked with an * may be more challenging than others.

75. A mixture of ammonia, oxygen, water vapor, and nitrogen gas is at equilibrium:

$$4 NH_3(g) + 3 O_2(g) \rightleftharpoons 6 H_2O(g) + 2 N_2(g)$$
$$\Delta H = -1267 \text{ kJ}$$

In what direction will a net change occur in reestablishing equilibrium if (a) oxygen gas is removed; (b) the pressure on the system is increased; (c) the reaction mixture is cooled; (d) ammonia gas is added; (e) the reaction mixture is cooled to 20 °C *and* allowed to reach equilibrium at that temperature, at which water is a liquid product, and the pressure on the system is increased; (f) the reaction mixture is cooled to −120 °C *and* allowed to reach equilibrium at that temperature, at which both the water and ammonia are solids, and the pressure on the system is increased?

76. For the equilibrium $C(s) + CO_2(g) \rightleftharpoons 2 CO(g)$ at 1000 K, $K_p = 1.6 \times 10^2$. At equilibrium, a 10.0-L container has 3.7 g of carbon, and a partial pressure of CO of 0.54 atm. What is the partial pressure of CO_2 at equilibrium?

77. At 500 °C, an equilibrium mixture in the reaction $CO(g) + H_2O(g) \rightleftharpoons CO_2(g) + H_2(g)$ contains 0.021 mol CO, 0.121 mol H_2O, 0.179 mol CO_2, and 0.079 mol H_2. If an additional 0.100 mol H_2 is added to the mixture, (a) in what direction must a net change occur to restore equilibrium, and (b) what will be the amounts of the four substances when equilibrium is reestablished?

78. For the isomerization of butane to isobutane, $K_c = 7.94$ (see Problem 60), sketch a graph of the concentrations of butane and of isobutane as a function of time, in the form of Figure 14.3. Start with any initial concentration of butane (but no isobutane), and show the equilibrium concentrations in their expected relative proportions.

*79. An analysis of the gaseous phase $[S_2(g)$ and $CS_2(g)]$ present at equilibrium at 1009 °C in the reaction $C(s) + S_2(g) \rightleftharpoons CS_2(g)$ shows it to be 13.71% C and 86.29% S, by mass. What is K_c for this reaction?

80. A 0.100-mol sample of N_2O_4 is placed in a 2.50-L flask and brought to equilibrium at 348 K. The mole fraction of NO_2 at equilibrium is 0.746. Determine K_p at 348 K for the reaction $N_2O_4(g) \rightleftharpoons 2 NO_2(g)$. Under what conditions could you obtain the pure 0.100 mol N_2O_4 present initially (that is, not contaminated with NO_2)?

*81. Consider the dissociation of $H_2S(g)$ at 750 °C.

$$2 H_2S(g) \rightleftharpoons 2 H_2(g) + S_2(g) \qquad K_c = 1.1 \times 10^{-6}$$

If 1.00 mol H_2S is placed in a 7.50-L reaction vessel and heated to 750 °C, how many moles of H_2 and S_2 will be present when equilibrium is reached? (*Hint:* Look for an assumption that can greatly simplify the algebraic solution.)

82. The decomposition of calcium sulfate, $2 CaSO_4(s) \rightleftharpoons 2 CaO(s) + 2 SO_2(g) + O_2(g)$, has $K_p = 1.45 \times 10^{-5}$ at 1625 K. A sample of $CaSO_4(s)$ is introduced into a reaction vessel that is filled with air at 1.0000 atm pressure at 1625 K. What will be the partial pressure of $SO_2(g)$ when equilibrium is established? Recall that the mole percent O_2 in air is 20.95%.

83. If, at 25 °C, 12.35 g of anhydrous (dry) $Na_2HPO_4(s)$ is added to a 3.45-L container containing 42 mg $H_2O(g)$, will any of the hydrate $Na_2HPO_4 \cdot 2 H_2O(s)$ form?

$$Na_2HPO_4(s) + 2 H_2O(g) \rightleftharpoons Na_2HPO_4 \cdot 2 H_2O(s)$$
$$K_p = 6.01 \times 10^3$$

If so, how many moles will form?

*84. Benzene can be produced from hexane in the reversible reaction $C_6H_{14}(g) \rightleftharpoons C_6H_6(g) + 4 H_2(g)$. The partial pressure equilibrium constant, K_p, for this reaction varies with Kelvin temperature according to the following equation.

$$\log_{10} K_p = 23.45 - 13{,}941/T$$

Equilibrium is established by starting with pure $C_6H_{14}(g)$. What must be the temperature if the *initial* gas pressure is 1.00 atm and the *equilibrium* partial pressure of C_6H_6 is 0.100 atm?

*85. What is the mole percent $COCl_2$ in the equilibrium gaseous mixture when $COCl_2(g)$ dissociates at 395 °C and the *total* equilibrium pressure is 3.00 atm?

$$COCl_2(g) \rightleftharpoons CO(g) + Cl_2(g)$$
$$K_p = 4.44 \times 10^{-2} \text{ at } 395 °C$$

*86. Suppose that in the reaction $PCl_5(g) \rightleftharpoons PCl_3(g) + Cl_2(g)$, the fraction of the PCl_5 molecules that dissociate to reach equilibrium is α and the *total* gas pressure at equilibrium is P.

(a) Show that

$$K_p = \frac{\alpha^2 P}{1 - \alpha^2}$$

(*Hint:* Start with 1.00 mol PCl_5 in a container of volume V. Use the ideal gas law, as needed.) (b) At 250 °C, $K_p = 1.78$. What is the fraction of PCl_5 that dissociates at this temperature when the total equilibrium pressure is 2.50 atm? (c) Under what total pressure must the gaseous mixture be maintained to limit dissociation of PCl_5 to 10.0% at 250 °C?

* **87.** Concerning the Cumulative Example on page 604, what would be the equilibrium amounts of the reactants and products if 0.100 mole each of $H_2S(g)$, $CH_4(g)$, $CS_2(g)$, and $H_2(g)$ were allowed to come to equilibrium in a 1.00-L vessel at 700 °C?

* **88.** The hydrogen required in the synthesis of ammonia is generally produced at the plant site from natural gas (methane) in the following reaction conducted at a high temperature.

$$CH_4(g) + 2\,H_2O(g) \rightleftharpoons CO_2(g) + 4\,H_2(g)$$

The following data are also given for 1000 K.

(a) $CO(g) + H_2O(g) \rightleftharpoons CO_2(g) + H_2(g)$
$$\Delta H° = -40\ \text{kJ};\ K_c = 1.4$$

(b) $CO(g) + 3\,H_2(g) \rightleftharpoons H_2O(g) + CH_4(g)$
$$\Delta H° = -230\ \text{kJ};\ K_c = 190$$

If 1.00 mole each of CH_4 and H_2O are allowed to react in a 10.0-L vessel at 1000 K, how many moles of H_2 will be present when equilibrium is established? How would the amount of H_2 change if the equilibrium temperature were raised above 1000 K?

89. Show that the equilibrium constant expression for the reaction $H_2(g) + I_2(g) \rightleftharpoons 2\,HI(g)$ is consistent with this proposed two-step mechanism for the reaction.

Fast: $\qquad\qquad I_2(g) \underset{k_{-1}}{\overset{k_1}{\rightleftharpoons}} 2\,I(g)$

Slow: $\qquad 2\,I(g) + H_2(g) \underset{k_{-2}}{\overset{k_2}{\rightleftharpoons}} 2\,HI(g)$

Apply Your Knowledge

* **90.** **[Historical]** Classical studies on the subject of this problem helped form the empirical basis for equilibrium constants in the 1870s. In one experiment, equilibrium was established in the homogeneous reaction of acetic acid and ethanol by starting with 1.51 mol CH_3COOH and 1.66 mol CH_3CH_2OH.

$$CH_3COOH + CH_3CH_2OH \rightleftharpoons CH_3COOCH_2CH_3 + H_2O$$

$\quad$ Acetic acid $\qquad$ Ethanol $\qquad\qquad$ Ethyl acetate

After equilibrium is reached, exactly one-hundredth of the equilibrium mixture is titrated with $Ba(OH)_2(aq)$ to determine the amount of acetic acid present. The volume of 0.1025 M $Ba(OH)_2(aq)$ required for the titration is 22.44 mL. Use these data to calculate K_c for the formation of ethyl acetate.

* **91.** **[Laboratory]** The photograph shows saturated $I_2(aq)$ floating on colorless $CCl_4(l)$. After the mixture is vigorously shaken and equilibrium is established, most of the iodine has passed from the water layer to the CCl_4 layer, which now has a purple color (right). For this equilibrium, $I_2(aq) \rightleftharpoons I_2$ (in CCl_4),

$$K_c = \frac{[I_2]_{CCl_4}}{[I_2]_{aq}} = 85.5$$

The concentration of saturated $I_2(aq)$ is 1.33×10^{-3} M.

(a) If 25.0 mL of saturated $I_2(aq)$ is shaken with 25.0 mL of $CCl_4(l)$, how many milligrams of I_2 remain in the aqueous layer at equilibrium? **(b)** If the 25.0-mL sample of saturated $I_2(aq)$ is shaken with 50.0 mL of $CCl_4(l)$ instead of 25.0 mL, would you expect the mass of I_2 remaining in the aqueous layer to be the same, more, or less than that calculated in (a)? Explain, but without doing a calculation. **(c)** If the aqueous layer at equilibrium in (a) is brought to equilibrium with a second 25.0-mL sample of $CCl_4(l)$, would you expect the mass of I_2 remaining in the aqueous layer in this second equilibrium to be the same, more, or less than that in (b)? Explain, but without doing a calculation.

* **92.** **[Laboratory]** The decomposition of HI(g) is represented by the equation $2\,HI(g) \rightleftharpoons H_2(g) + I_2(g)$ and the data given below. HI(g) is introduced into five identical 400-cm³ glass bulbs, and the five bulbs are maintained at 623 K. Each bulb is opened after a period of time and analyzed for I_2 by titration with 0.0150 M $Na_2S_2O_3(aq)$.

$$I_2(aq) + 2\,Na_2S_2O_3(aq) \longrightarrow Na_2S_4O_6(aq) + 2\,NaI(aq)$$

What is the value of K_p at 623 K?

Bulb no.	Original Mass of HI(g), g	Bulb Opened After, h	Volume of 0.0150 M $Na_2S_2O_3$ Required for Titration, mL
1	0.300	2	20.96
2	0.320	4	27.90
3	0.315	12	32.31
4	0.406	20	41.50
5	0.280	40	28.68

e-Media Problems

The activities described in these problems can be found in the e-Media Activities and Interactive Student Tutorial (IST) modules of the Companion Website, *http://chem.prenhall.com/hillpetrucci*.

93. The physical equilibrium found in saturated salt solutions is illustrated in the **Dynamic Equilibrium** animation (*Section 14-1*). **(a)** What is dynamic about this process? **(b)** Describe the equilibrium state of this system in terms of the rate of dissolution and rate of crystallization.

94. Four generic reactions are represented in the **Equilibrium Constant** activity (*Section 14-3*). For each reaction, set the initial concentrations of all species to an identical nonzero value and "equilibrate," using an equilibrium constant value near 1.0. **(a)** In terms of the concentration of products, describe the influence of increasing the value of the equilibrium constant by a factor of 1000. **(b)** Is the effect the same for each reaction?

95. In the **Direction of Net Change** activity (*Section 14-3*), observe the progress toward equilibrium in terms of the changes in concentration of NO_2 and N_2O_4 for the three experimental runs.

(a) For which of the experiments does the concentration of NO_2 decrease with time? **(b)** Must the concentration of NO_2 always be greater than that of N_2O_4 to have a decrease in the NO_2 concentration over time? **(c)** Describe the changes in concentration for each experimental run in terms of a reaction quotient and the equilibrium constant.

96. View the **Le Châtelier's Principle** movie (*Section 14-4*). **(a)** List the factors seen in the movie that cause the equilibrium of the reaction to shift toward the products. Suggest other factors that will also shift the equilibrium in that direction. **(b)** Sketch a graph of concentration versus time for both reactants and products showing the influence of these factors on equilibrium.

97. For the reaction shown in the **NO_2–N_2O_4 Equilibrium** animation (*Section 14-4*), **(a)** describe the effect on the equilibrium composition when the total system pressure is reduced by a factor of 2 from its initial value. **(b)** Sketch a graph of partial pressure versus time for both NO_2 and N_2O_4, depicting the effect of this pressure change.

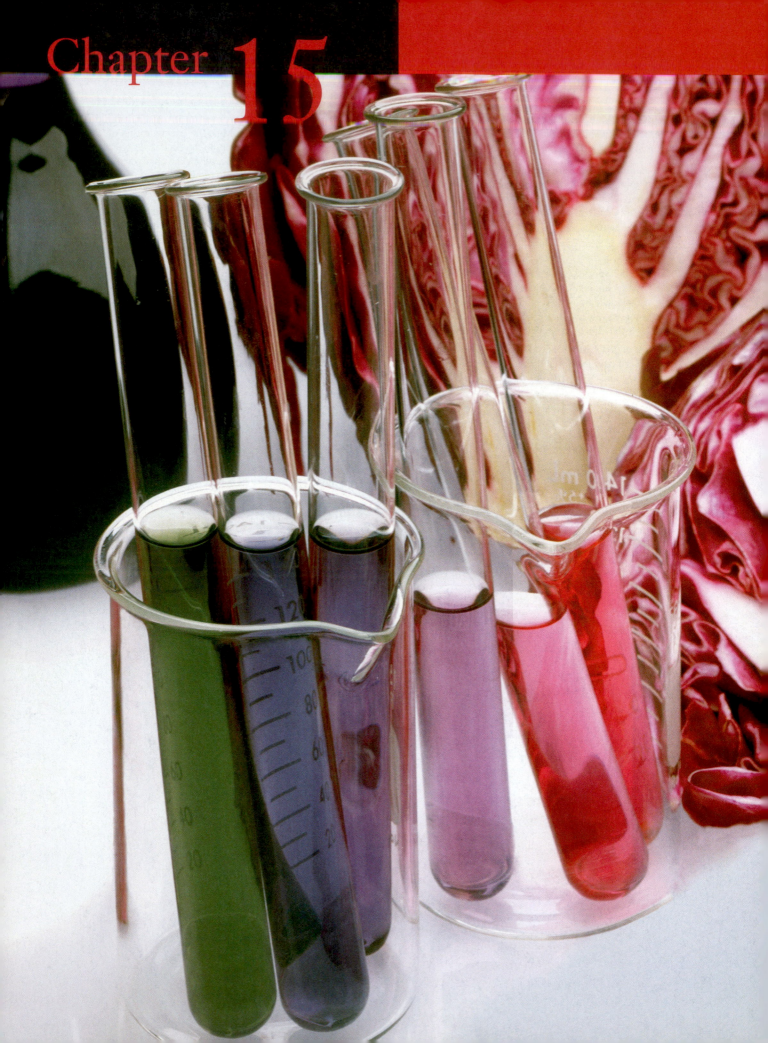

Chapter 15

Acids, Bases, and Acid–Base Equilibria

ACIDS AND BASES are such important compounds that we introduced them in Chapter 2 and have used them by name and formula often in following chapters. In Chapter 4, we classified acids and bases in aqueous solutions as *strong* or *weak* according to their degree of ionization. We also presented acid–base reactions as one of the three main types of reactions that occur in aqueous solutions, and we introduced the stoichiometry of acid–base titrations.

In this chapter, we will examine factors that affect acid and base strength. We will propose answers to such questions as

- What do we mean by the pH of a solution—such as that of an acid rain sample with a pH of 4.2?
- How can we calculate the pH of a solution?
- How can we maintain a constant pH in a solution even though acid or base is added to it?
- How does the body normally maintain a pH of 7.4 in blood even though cellular metabolism adds acids to the bloodstream?
- How do acid–base indicators work?

We will also examine neutralization titrations in greater depth.

CONTENTS

15.1 The Brønsted–Lowry Theory of Acids and Bases

Arrhenius proposed the first successful theory of acids and bases, which we introduced in Section 2.9. According to the Arrhenius theory, an acid in aqueous solution produces hydrogen ions, $H^+(aq)$, and a base produces hydroxide ions, $OH^-(aq)$. Later extensions of the theory also distinguished between *strong* acids and bases and *weak* acids and bases. A strong acid ionizes essentially completely into $H^+(aq)$ and accompanying anions, and a strong base dissociates essentially completely into $OH^-(aq)$ and

◀ The substances that produce the brilliant colors found in plants often have surprising properties. Anthocyanin, the complex organic compound responsible for the color of purple cabbage, is an acid-base indicator that is easily extracted by boiling the cabbage in water. As can be seen here, the extract turns green in household ammonia (base) but is pink in vinegar (acid). Acid-base chemistry discussed in this chapter is an important part of many processes and phenomena, from the hardening of cement to the action of the local anesthetic Novocaine (see page 626).

▲ Solutions of $NH_3(aq)$ are often labeled NH_4OH and called ammonium hydroxide.

accompanying cations. (Table 4.1 lists the common strong acids and strong bases.) Weak acids and bases ionize reversibly, reaching a state of equilibrium in which only a small fraction of the acid or base has ionized.

The Arrhenius theory is limited, however, in that it applies only to aqueous solutions and does not adequately explain why such compounds as ammonia (NH_3) are bases. Arrhenius's theory seems to require that a base *contain* either OH^- or at least an OH group that can become OH^-. There is no such ion or group in NH_3, but as we saw in Chapter 4, ammonia is nevertheless a weak base. In fact, because Arrhenius and his contemporaries thought that a base had to contain OH^-, aqueous ammonia was long known as ammonium hydroxide and its formula written NH_4OH. That name and formula are still sometimes used today, but it is unlikely that discrete molecules of NH_4OH exist in an aqueous solution. It is not possible, for example, to write a Lewis structure for a molecule of NH_4OH.

The Brønsted–Lowry Theory

To a large degree, a theory proposed independently by J. N. Brønsted in Denmark and T. M. Lowry in Great Britain in 1923 overcame the shortcomings of the Arrhenius theory. According to the Brønsted–Lowry theory, an acid is a **proton donor** and a base is a **proton acceptor.** A proton in this sense is simply an ionized hydrogen atom, H^+. However, a substance cannot simply shed a proton, because the proton cannot exist in a solution as a free, independent particle. The positively charged proton attaches itself to a center of negative charge, such as a lone pair of electrons on an atom in some other species. Thus, for every acid, there must be an available base. For example, when HCl ionizes in water, HCl molecules are proton donors and H_2O molecules are proton acceptors. The ionization is an *acid–base* reaction.

The red color in this arrow represents loss of a proton by an acid, and the blue represents gain of the proton by a base.

 Aqueous Acids animation

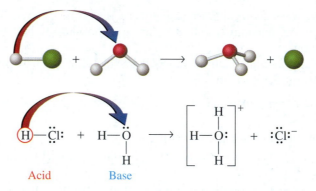

In the Brønsted–Lowry theory, H_3O^+ takes the place of the H^+ of the Arrhenius theory. The H_3O^+ ion, called the *hydronium ion,* forms when an H_2O molecule, acting as a Bronsted–Lowry base, uses a lone pair of electrons on its O atom to form a coordinate covalent bond with a proton, H^+.*

Brønsted–Lowry theory is not limited to reactions in aqueous solutions. For example, HCl is an acid in pure liquid ammonia, $NH_3(l)$, just as it is in water. In this case, the base that accepts a proton from HCl is the ammonia molecule. In fact, the transfer of a proton from HCl to NH_3 can occur even in the gaseous state, as we saw in Figure 5.18:

*Each proton, H^+, is probably associated with several H_2O molecules—for example, four H_2O molecules in $H(H_2O)_4{}^+$. However, in applying the Brønsted–Lowry theory, we will always assume that the singly hydrated proton, H_3O^+, is the species present in aqueous solutions.

The Brønsted–Lowry theory also successfully explains how ammonia acts as a base in water, which the Arrhenius theory fails to do. In the ionization of NH_3, water is the acid:

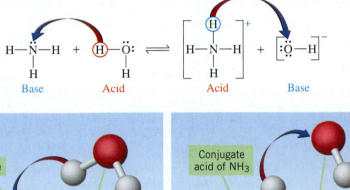

Recall that NH_3 is a weak base; it does not ionize completely. The equation we have just written for the ionization of NH_3 is therefore incomplete because it does not show that the ionization is reversible and reaches an equilibrium state. In the reverse reaction, NH_4^+ is the acid and OH^- is the base. This reversible acid–base reaction is represented below and in Figure 15.1.

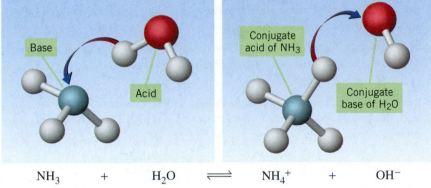

🌐 **Aqueous Bases animation**

$$NH_3 \quad + \quad H_2O \quad \rightleftharpoons \quad NH_4^+ \quad + \quad OH^-$$

◀ FIGURE 15.1   Ionization of NH_3 as a Brønsted–Lowry base
Proton transfer is represented by a red-to-blue arrow. The red indicates loss of the proton by an acid and the blue, gain of the proton by a base. Because ammonium ion is a stronger acid than water and because hydroxide ion is a stronger base than ammonia, the equilibrium condition actually lies quite far to the left.

The labels in the reversible equation above designate the acid and base for the forward reaction *and* for the reverse reaction. Notice that the forward-direction acid (H_2O) differs from the reverse-direction base (OH^-) by only a proton H^+. And the same is true for the forward-reaction base (NH_3) and the reverse-direction acid (NH_4^+). The combinations NH_4^+/NH_3 and H_2O/OH^-, where the two members of each pair differ by only one H^+, are called *conjugate acid–base pairs*. The **conjugate acid** of a base consists of the base *plus* an attached proton, H^+. Thus, NH_4^+ is the conjugate acid of the base NH_3, and the combination NH_4^+/NH_3 is a conjugate acid–base pair. The **conjugate base** of an acid consists of the acid *minus* a proton, H^+. Thus, OH^- is the conjugate base of the acid H_2O, and the combination H_2O/OH^- is a conjugate acid–base pair.

Acetic acid also ionizes reversibly, a reaction we can represent as

🌐 **Conjugate Acids and Bases activity**

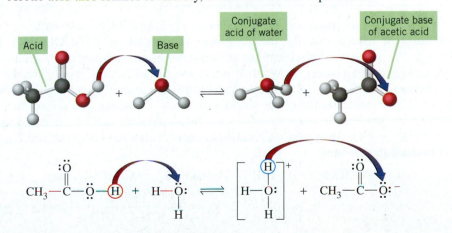

The conjugate acid–base pairs are CH_3COOH/CH_3COO^- and H_3O^+/H_2O. Notice that in this reaction, H_2O is a base, whereas in the ionization of ammonia, H_2O is an acid. This dual role of H_2O is suggested here:

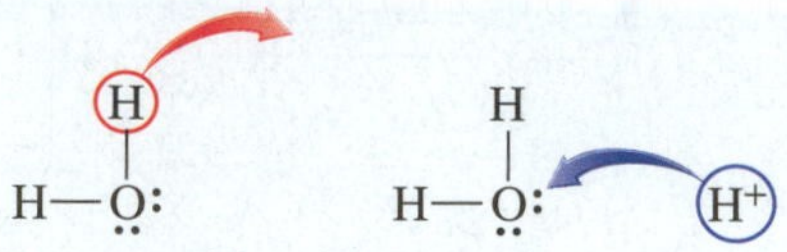

H₂O as an acid   H₂O as a base

A substance that can act either as an acid or a base is **amphiprotic.** We identify Brønsted–Lowry acids and bases and amphiprotic species in Example 15.1 and Exercises 15.1A and B.

Example 15.1

Identify the Brønsted–Lowry acids and bases and their conjugates in

(a) $H_2S + NH_3 \rightleftharpoons NH_4^+ + HS^-$ (b) $OH^- + H_2PO_4^- \rightleftharpoons H_2O + HPO_4^{2-}$

SOLUTION

To identify Brønsted–Lowry acids and bases, we look for the proton donors and proton acceptors in each reaction.

(a) Because H_2S is converted to HS^- by donating a proton, H_2S is an acid; HS^- is its conjugate base. Because NH_3 accepts the proton that H_2S donates, NH_3 is a base; NH_4^+ is its conjugate acid.

$$H_2S + NH_3 \rightleftharpoons NH_4^+ + HS^-$$

Acid   Base   Acid   Base

(b) Because OH^- accepts a proton from $H_2PO_4^-$, OH^- is a base; H_2O is its conjugate acid. Because $H_2PO_4^-$ donates a proton to OH^-, $H_2PO_4^-$ is an acid; HPO_4^{2-} is its conjugate base.

$$OH^- + H_2PO_4^- \rightleftharpoons H_2O + HPO_4^{2-}$$

Base   Acid   Acid   Base

EXERCISE 15.1A

Identify the Brønsted–Lowry acids and bases and their conjugates in

(a) $NH_3 + HCO_3^- \rightleftharpoons NH_4^+ + CO_3^{2-}$

(b) $H_3PO_4 + H_2O \rightleftharpoons H_2PO_4^- + H_3O^+$

EXERCISE 15.1B

Identify the amphiprotic species in each reaction in Exercise 15.1A, and specify whether the species acts as an acid or a base in that reaction. For each amphiprotic species acting as an acid in Exercise 15.1A, write an equation showing it acting as a base in a reaction with HCl. For each amphiprotic species acting as a base in Exercise 15.1A, write an equation showing it acting as an acid in a reaction with NH_3.

As we did for reversible reactions in general in Chapter 14, we can write equilibrium constant expressions for the reversible ionizations of $CH_3COOH(aq)$ and $NH_3(aq)$. We make two modifications in the equilibrium constant expressions for weak acids and weak bases: (1) We treat water, the solvent, as if it were pure $H_2O(l)$ and assume that its activity (effective concentration) is essentially unchanged between the pure liquid state and the dilute aqueous solution. Therefore, we do not include $[H_2O]$ in the equilibrium constant expression. (2) Instead of calling the equilibrium constants K_c, we use the symbols K_a, called the **acid ionization constant,** and K_b, the **base ionization constant:**

$$CH_3COOH(aq) + H_2O(l) \rightleftharpoons H_3O^+(aq) + CH_3COO^-(aq)$$

$$K_a = \frac{[H_3O^+][CH_3COO^-]}{[CH_3COOH]} = 1.8 \times 10^{-5}$$

$$NH_3(aq) + H_2O(l) \rightleftharpoons NH_4^+(aq) + OH^-(aq)$$

$$K_b = \frac{[NH_4^+][OH^-]}{[NH_3]} = 1.8 \times 10^{-5}$$

Values of K_a and K_b can be established by several different methods, including two considered later in the chapter—measurement of pH (page 636) and titration (page 660). The fact that K_a for acetic acid and K_b for ammonia have the same value is of no significance; it is purely coincidental.

Strengths of Conjugate Acid–Base Pairs

Recall from Chapter 4 that electrical conductivity experiments and other evidence indicate that HCl is a *strong* acid; it is essentially completely ionized in aqueous solution. Similar evidence shows that acetic acid is a *weak* acid. Only a fraction of the CH_3COOH molecules ionize in an acetic acid solution.

Let us compare the relative strengths of these two acids: HCl and CH_3COOH:

$$\underset{\text{Acid}}{HCl} + \underset{\text{Base}}{H_2O} \longrightarrow H_3O^+ + Cl^-$$

$$\underset{\text{Acid}}{CH_3COOH} + \underset{\text{Base}}{H_2O} \rightleftharpoons \underset{\text{Acid}}{H_3O^+} + \underset{\text{Base}}{CH_3COO^-}$$

The single arrow in the HCl equation reflects what we know to be true about a strong acid: Its ionization is virtually complete. And the double arrow in the CH_3COOH equation reflects what we know to be true about a weak acid: Its ionization is only partial. To say that HCl is a stronger acid than CH_3COOH means that a HCl molecule donates a proton to a water molecule much more readily than does an CH_3COOH molecule. From the standpoint of the reverse reactions, we conclude that Cl^- is a far weaker base than CH_3COO^-. Equilibrium in the HCl reaction lies very far to the right; chloride ion stays as chloride ion and has almost no tendency to acquire a proton to form a HCl molecule. In the acetic acid reaction, equilibrium lies to the left, demonstrating the tendency of an acetate ion to accept a proton and become an CH_3COOH molecule.

We can summarize these observations in three generalizations about conjugate acid–base pairs:

- *The stronger an acid, the weaker is its conjugate base.*
- *The stronger a base, the weaker is its conjugate acid.*
- *An acid–base reaction is favored in the direction from the stronger member to the weaker member of each conjugate acid–base pair.*

Let us use these ideas to determine the direction of net change in the reaction

$$CH_3COOH + Br^- \rightleftharpoons HBr + CH_3COO^-$$

Like HCl, HBr is a strong acid, whereas CH_3COOH is a weak acid. The reaction from stronger acid to weaker acid, the reverse reaction, occurs extensively. Regarding the forward reaction, acetic acid, a weak acid, is not very successful in transferring protons to bromide ion, the conjugate base of the strong acid HBr and consequently a very weak base. Thus the direction of net change in this reversible reaction is the reverse direction.

To apply this line of reasoning requires that we have a relative ranking of acid and base strengths, such as that shown in Table 15.1. The strongest acids and weakest conjugate bases are at the top of the table. At the bottom of the table, we find the weakest acids and the strongest conjugate bases.

In constructing a ranking such as that of Table 15.1, we can use K_a values to compare the strengths of weak acids. Thus, hydrofluoric acid ($K_a = 6.6 \times 10^{-4}$) is a somewhat stronger proton donor than acetic acid ($K_a = 1.8 \times 10^{-5}$). For strong acids, water has a *leveling effect*. All acids above hydronium ion in Table 15.1 are strong acids. When placed in water, they ionize essentially completely, producing

Table 15.1 Relative Strengths of Some Brønsted–Lowry Acids and Their Conjugate Bases

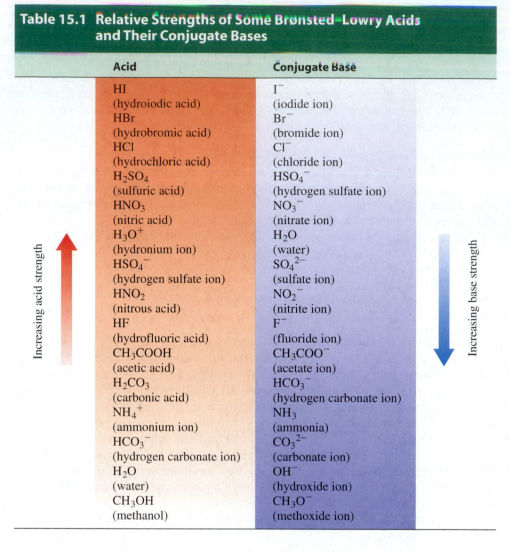

Acid	Conjugate Base
HI (hydroiodic acid)	I⁻ (iodide ion)
HBr (hydrobromic acid)	Br⁻ (bromide ion)
HCl (hydrochloric acid)	Cl⁻ (chloride ion)
H_2SO_4 (sulfuric acid)	HSO_4^- (hydrogen sulfate ion)
HNO_3 (nitric acid)	NO_3^- (nitrate ion)
H_3O^+ (hydronium ion)	H_2O (water)
HSO_4^- (hydrogen sulfate ion)	SO_4^{2-} (sulfate ion)
HNO_2 (nitrous acid)	NO_2^- (nitrite ion)
HF (hydrofluoric acid)	F⁻ (fluoride ion)
CH_3COOH (acetic acid)	CH_3COO^- (acetate ion)
H_2CO_3 (carbonic acid)	HCO_3^- (hydrogen carbonate ion)
NH_4^+ (ammonium ion)	NH_3 (ammonia)
HCO_3^- (hydrogen carbonate ion)	CO_3^{2-} (carbonate ion)
H_2O (water)	OH^- (hydroxide ion)
CH_3OH (methanol)	CH_3O^- (methoxide ion)

Increasing acid strength (left) — *Increasing base strength* (right)

hydronium ion, a weaker acid. In effect, all strong acids in water are *leveled* to the same strength, that of H_3O^+. To compare the strengths of strong acids, we need to study them in a solvent that is a weaker base than water, such as diethyl ether or acetone. Because ionization in these solvents is not complete, we can rank strong acids according to their acid ionization constants in certain nonaqueous solutions.

15.2 Molecular Structure and Strengths of Acids and Bases

As we have noted before, a central challenge in chemistry is to explain observable properties of a substance in terms of its molecular structure. Referring to Table 15.1, we would like to explain, for example, why HF is a weak acid whereas HCl is strong, why HNO_3 is a stronger acid than HNO_2, and so on. Several factors affect acid strength, and sorting out these factors can be difficult. There are, however, two characteristics of chemical bonds that play a notable role: bond strength and bond polarity. Let us see how these characteristics affect the acid strength of binary acids. Molecules of a *binary acid* are composed of atoms of hydrogen and another nonmetal, X. The formula of the binary acid is HX if X is in group 7A, H_2X if X is in group 6A, and H_3X if X is in group 5A. First, we will consider the primary factor affecting the strengths of binary acids within the same group of the periodic table and then within the same period.

Variations in Strength of Same-Group Binary Acids

As an example, let us consider the binary acids in group 7A, the hydrogen halides. First, we can write the general expression (15.1):

$$H-\overset{\cdot\cdot}{\underset{\cdot\cdot}{X}}: \; + \; H-\overset{\cdot\cdot}{O}: \; \rightleftharpoons \; \left[H-\overset{\cdot\cdot}{O}: \right]^+ \; + \; :\overset{\cdot\cdot}{\underset{\cdot\cdot}{X}}:^- \qquad (15.1)$$

Acid   Base   Acid   Base

The base/acid conjugate pair H_2O/H_3O^+ is the same for all four acids in group 7A. Any differences in acid strengths must therefore be related to the molecule HX and the ion X^-.

In Equation (15.1), the greater the tendency for the transfer of a proton from HX to H_2O, the more the forward reaction is favored and the *stronger* the acid HX. The greater the tendency for the transfer of a proton from H_3O^+ to X^-, the more the reverse reaction is favored and the *weaker* the acid HX.

We can consider the proton transfer from HX to H_2O to be somewhat related to the process of breaking a bond in a gaseous molecule to produce gaseous atoms, even though in aqueous solutions ions are formed, not atoms. The bond-dissociation energy (Chapter 9) is a measure of the difficulty of breaking a bond. A lower bond-dissociation energy means the H–X bond is easier to break, which in turn means that HX donates a proton more readily. A weaker H–X bond means a stronger acid HX; a stronger H–X bond means a weaker acid HX:

Bond-dissociation energy
(kJ/mol) decreases

	569	>	431	>	368	>	297
	HF		**HCl**		**HBr**		**HI**

Acid strength increases K_a 6.6×10^{-4} $<$ $\approx 10^6$ $<$ $\approx 10^8$ $<$ $\approx 10^9$

We must also evaluate the tendency for proton transfer in the reverse reaction. Protons (H^+) are attracted to the negatively charged ion X^-. The smaller the X^- ion, the greater its negative charge density, and the more strongly the proton is attracted. As a result, the tendency for the reverse reaction to occur increases, leading to a *weaker* acid HX. Conversely, the larger the X^- ion, the weaker is the H^+/X^- attraction, the less is the tendency for the reverse reaction to occur, and the *stronger* is the acid HX:

Anion radius (pm) increases

	136	<	181	<	195	<	216
	F⁻		**Cl⁻**		**Br⁻**		**I⁻**

Acid strength increases K_a 6.6×10^{-4} $<$ $\approx 10^6$ $<$ $\approx 10^8$ $<$ $\approx 10^9$

	HF		**HCl**		**HBr**		**HI**

Thus, the strengths of binary acids in the same group of the periodic table correlate well with both bond-dissociation energy and anion size, leading to the conclusion:

The strengths of binary acids increase from top to bottom in a group of the periodic table:

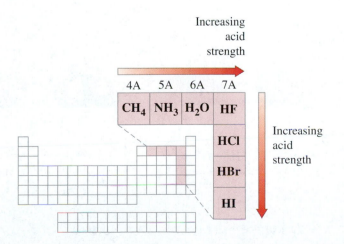

◀ Representative trends in strengths of binary acids.

Variations in Strength of Same-Period Binary Acids

Variations in acid strengths in the same period correlate well with the *electronegativity difference* (ΔEN) between H and the element to which it is bonded. Recall from Chapter 9 that a small ΔEN signifies a bond that is largely covalent in character, whereas a large difference indicates partial charge separation. An acid molecule that has partial ionic charges loses H^+ to a base more readily and is thus a stronger acid than an acid molecule without partial charges. We can use a ranking for the second-period binary compounds of hydrogen to make the following generalization:

> *The strengths of binary acids increase from left to right across a period of the periodic table.*

ΔEN *increases*	0.4	<	0.9	<	1.4	<	1.9
Acid strength increases	**CH_4**	<	**NH_3**	<	**H_2O**	<	**HF**

Nonpolar methane (CH_4) does not ionize as an acid. In aqueous solution, NH_3 ionizes weakly, but as a base, not as an acid. The ability of H_2O to ionize as an acid is quite limited, as we shall see in Section 15.3. In contrast, HF is a weak acid with $K_a = 6.6 \times 10^{-4}$. Bond dissociation energies of the binary molecules, all of which are quite large, do not correlate with these observations. However, the tendency of a proton (H^+) to leave the highly polar H—F bond to bond to a water molecule overcomes to some extent the high bond dissociation energy of HF.

Strengths of Oxoacids

Most acids contain oxygen as well as hydrogen and a third nonmetal (which we denote by the symbol E). In oxoacids, there is at least one H atom bonded to an O atom:

$$H—O—E$$

The broken lines in this structure suggest that other groups may also be bonded to the E atom. The strength of the O—H bond is affected by how strongly the electrons of the bond are attracted to E. If E strongly attracts electrons, then to some extent, electron density is withdrawn from the O—H bond. We would expect this electron-withdrawing effect to increase as the electronegativity of E increases. This weakens the O—H bond, causes a proton to be more easily released, and increases the acid strength. The following data show the effect of the electronegativity of E on acid strength, namely that acid strength of oxoacids increases as we move *upward* in any group of the periodic table:

Electronegativity of E *increases*	2.5		<		2.8		<		3.0
	HOI				**HOBr**				**HOCl**
Acid strength increases	$K_a = 2.3 \times 10^{-11}$	<		2.5 $\times$ 10^{-9}	<		2.9 $\times$ 10^{-8}		

For a given E atom, the number of terminal O atoms in the molecule profoundly influences the strength of an oxoacid. For instance, when E is Cl, as in Figure 15.2, the

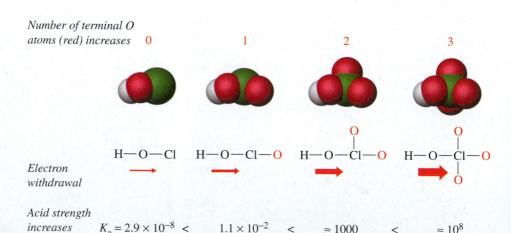

Number of terminal O atoms (red) increases   0   1   2   3

Electron withdrawal

Acid strength increases $K_a = 2.9 \times 10^{-8}$ < 1.1×10^{-2} < ≈ 1000 < $\approx 10^8$

▶ **FIGURE 15.2** **Electron withdrawal and acid strength**
As the number of terminal oxygen atoms on the chlorine increases, the withdrawal (red arrow) of electron density from the H—O bond increases. This weakens the H—O bond, making it easier for the acid to donate its proton.

value of K_a increases by several powers of 10 for each additional terminal O atom. Oxygen is the second most electronegative of all the elements. Thus, terminal O atoms act together with the E atom to withdraw electrons from the O—H bond. This weakens the bond and increases the strength of the acid by making it easier for the acid to donate its proton.

Strengths of Carboxylic Acids

As with other oxoacids, the strength of a carboxylic acid (RCOOH) depends on the ease with which electrons can be withdrawn from an O—H bond, thus facilitating the liberation of a proton. Because all carboxylic acids share the COOH group, we must look to differences in the R groups to explain variations in acid strength:

$$\begin{matrix} & O \\ & \| \\ R-&C-O-H \end{matrix}$$

If the R group is simply a hydrocarbon chain, it has little effect on acid strength, as we can see by comparing the K_a values of a two-carbon carboxylic acid and its five-carbon homolog:

CH_3COOH $K_a = 1.8 \times 10^{-5}$ $CH_3(CH_2)_3COOH$ $K_a = 1.5 \times 10^{-5}$

  Acetic acid (ethanoic acid)   Valeric acid (pentanoic acid)

Any atoms of high electronegativity in the R groups can withdraw electrons from the O–H bond, thereby weakening the bond and increasing the acid strength. In these cases, we need to consider two factors when predicting the strength of the carboxylic acid: (1) the number of highly electronegative atoms in the R group and (2) how close these atoms are to the COOH group. In Figure 15.3, we can see the effects of electronegative substituents and their numbers and distances from the COOH group. The more electronegative the substituent(s) and the closer they are to the COOH group, the stronger the acid.

Strengths of Carboxylic Acids animation

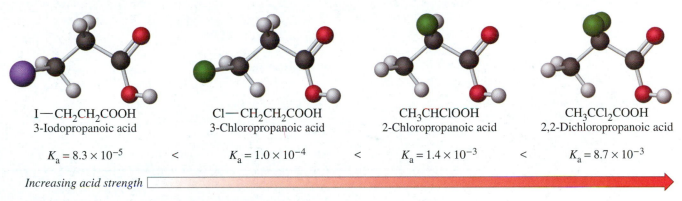

I—CH₂CH₂COOH		Cl—CH₂CH₂COOH		CH₃CHClOOH		CH₃CCl₂COOH
3-Iodopropanoic acid		3-Chloropropanoic acid		2-Chloropropanoic acid		2,2-Dichloropropanoic acid
$K_a = 8.3 \times 10^{-5}$	<	$K_a = 1.0 \times 10^{-4}$	<	$K_a = 1.4 \times 10^{-3}$	<	$K_a = 8.7 \times 10^{-3}$

Increasing acid strength

▲ **FIGURE 15.3   Substituents and carboxylic acid strength**
Increasing electronegativity of the halogen substituents and increasing proximity to the carboxyl group increase the strength of the carboxylic acid.
QUESTION: Give the structure or name of a carboxylic acid even stronger than 2,2-dichloropropanoic acid.

Example 15.2

Select the stronger acid in each pair:

(a) nitrous acid, HNO_2, and nitric acid, HNO_3

(b) $BrCH_2COOH$ and Cl_3CCOOH

SOLUTION

(a) We need to replace the molecular formulas by Lewis structures to assess the numbers of terminal atoms in the two acids. From the structures in the margin, we expect HNO_3 to be the stronger acid because it has two terminal O atoms, whereas HNO_2 has only one. In fact, HNO_3 is a strong acid (ionizes completely), and HNO_2 is a weak acid (ionizes only partially).

$$H-\overset{..}{\underset{..}{O}}-\overset{..}{N}=\overset{..}{\underset{..}{O}}:$$
Nitrous acid

$$H-\overset{..}{\underset{..}{O}}-\overset{\overset{\displaystyle \overset{..}{O}:}{\|}}{N}-\overset{..}{\underset{..}{O}}:$$
Nitric acid

(b) We should expect $BrCH_2COOH$ to be a somewhat stronger acid than CH_3COOH (acetic acid) because of the presence of the electronegative Br atom on the C atom adjacent to the COOH group. However, CCl_3COOH, with *three* of the more electronegative Cl atoms adjacent to the C atom of the COOH group, should be much stronger than either CH_3COOH or $BrCH_2COOH$.

ASSESSMENT

(a) We can verify the correctness of our conclusion with the knowledge from Table 4.1 that HNO_3 is a strong acid and HNO_2 (because it is not listed in that table) is a weak acid.

(b) The correctness of our conclusion is confirmed by the K_a values, which are reported in the chemical literature: 1.8×10^{-5} for CH_3COOH, 1.3×10^{-3} for $BrCH_2COOH$, and 3.0×10^{-1} for CCl_3COOH.

EXERCISE 15.2A

Select the stronger acid in each pair:

(a) H_2S and H_2Te

(b) $CH_3CH_2CH_2CHBrCOOH$ and $ClCH_2CH_2CH_2CH_2COOH$

EXERCISE 15.2B

Arrange the following acids in order of increasing strength (that is, from the acid you expect to be weakest to the acid you expect to be strongest). Explain the basis of your order.

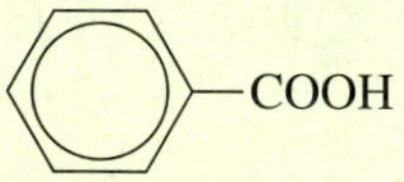

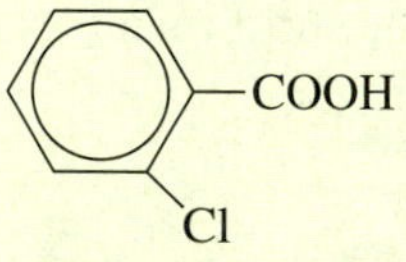

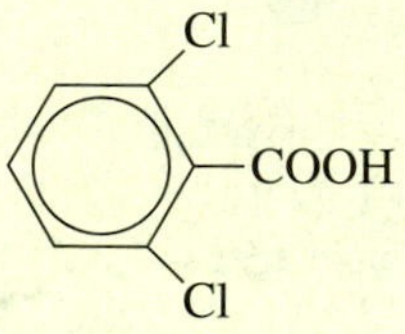

Strengths of Amines as Bases

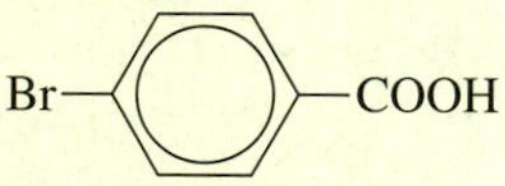

Now let us take a brief look at molecular structure and base strength. We have just seen that any atom or group that withdraws electrons from the bond where a proton (H^+) is to be released makes an acid stronger. As we might expect, any atom or group that withdraws electrons from the atom to which a proton (H^+) bonds makes a base weaker. For example, an electronegative Br atom in $BrNH_2$ attracts the lone-pair electrons of the N atom in NH_3, making them less available for bonding with a proton. Thus, $BrNH_2$ ($K_b = 2.5 \times 10^{-8}$) is a weaker base than NH_3 ($K_b = 1.8 \times 10^{-5}$).

Aromatic amines are much weaker bases than aliphatic (nonaromatic) amines. For aniline, $C_6H_5NH_2$, for example, $K_b = 7.4 \times 10^{-10}$. Recall from Chapter 10 that, via resonance, the π electrons of the benzene ring are *delocalized* (spread out) over the entire ring. In aniline, the lone-pair electrons on the N atom are also delocalized. As a result, these electrons are much less likely to accept a proton. Delocalization of these electrons is seen in resonance structures III-V (structures I and II are the usual Kekulé structures for benzene rings):

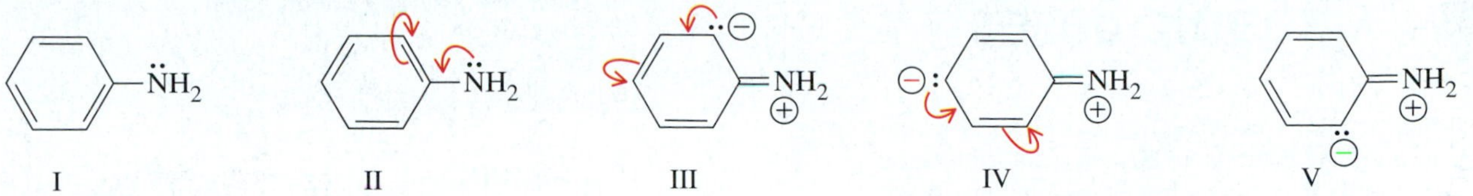

I II III IV V

Electron-withdrawing groups on the ring further diminish the basicity of aromatic amines relative to aniline, as illustrated in Example 15.3.

Example 15.3

Select the weaker base in each pair:

(a) $HOCH_2CH_2NH_2$ and $CH_3CH_2NH_2$

(b) 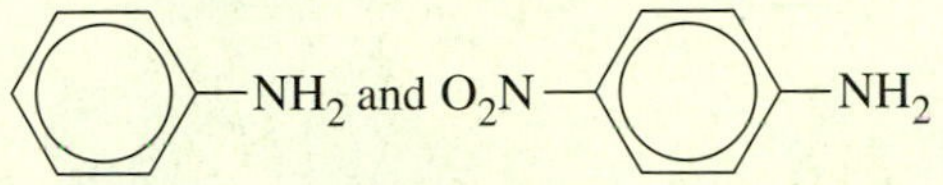

SOLUTION

(a) Note that $HOCH_2CH_2NH_2$ has an electronegative oxygen atom on the second carbon atom from the NH_2 group. Thus, we expect $HOCH_2CH_2NH_2$ to be a somewhat weaker base than $CH_3CH_2NH_2$ because of the presence of the oxygen atom. (The tabulated K_b values are 2.5×10^{-5} for $HOCH_2CH_2NH_2$ and 4.3×10^{-4} for $CH_3CH_2NH_2$.)

(b) The NO_2 group, with three electronegative atoms, withdraws electrons from the NH_2 group, making *p*-$NO_2C_6H_4NH_2$ less likely to attract an H^+ ion and therefore a weaker base than $C_6H_5NH_2$.

EXERCISE 15.3A

Arrange the following bases in order of increasing strengths (that is, from the base you expect to be weakest to the base you expect to be strongest). Explain the basis of your order.

(a) 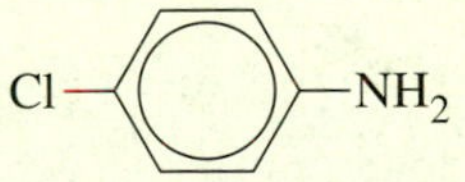

(b) $CH_3(CH_2)_3NH_2$

(c) $CH_3CHCl(CH_2)_2NH_2$

(d) 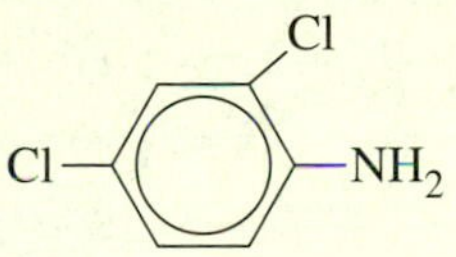

EXERCISE 15.3B

From the five R groups

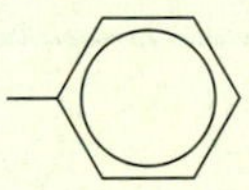 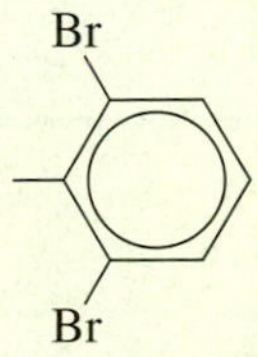 $-CH_2CH_2CH_3$ $-CH_2CHClCH_3$ 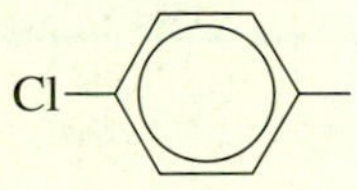

choose the one that makes the amine $R{-}NH_2$ **(a)** most strongly basic and **(b)** most weakly basic, and explain your choices.

Organic Bases

As we first noted in Section 4.2, amines are compounds in which one of the H atoms in NH_3 is replaced by a hydrocarbon group, R. Amines are bases, and water-soluble amines can accept a proton from water. For example, the reaction between trimethylamine, $(CH_3)_3N$, and water is

$$(CH_3)_3N(aq) + H_2O(l) \rightleftharpoons [(CH_3)_3N{-}H]^+(aq) + OH^-(aq)$$

$$K_b = 6.3 \times 10^{-5}$$

Nearly all amines, including those that are not very soluble in water, react with strong acids to form water-soluble salts. For example, only 3.5 g of aniline ($C_6H_5NH_2$) dissolves in 100 mL of water at 25 °C. However, the reaction between aniline and hydrochloric acid forms the salt commonly called aniline hydrochloride, which is soluble to the extent of about 100 g per 100 mL of water:

$$C_6H_5NH_2(l) + HCl(aq) \rightleftharpoons C_6H_5NH_3^+(aq) + Cl^-(aq)$$

   Aniline   Aniline hydrochloride

Many medicines, both synthetic and naturally occurring, contain the amine functional group. *Alkaloids* are basic nitrogen-containing compounds produced by plants. Familiar alkaloids include caffeine, cocaine, morphine, nicotine, and strychnine. Many alkaloids have important physiological effects: They are medicines, poisons, and drugs of abuse.

Medicines that are amines are often converted to salts to enhance their water solubility. For instance, procaine is soluble only to the extent of 0.5 g in 100 g of water. In contrast, its hydrochloride salt is soluble to the remarkable degree of 100 g in 100 g of water. Procaine hydrochloride, often called by the trade name Novocain®, is a well-known local anesthetic.

We also make use of the chemistry of amines when we put lemon juice on fish. The unpleasant fishy odor is due to amines. The citric acid in the lemon juice converts the amines to non-volatile salts, thus reducing the odor.

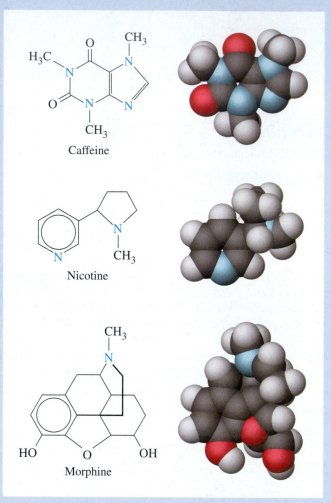

Caffeine

Nicotine

Morphine

▲ Some common alkaloids.

Procaine

Procaine hydrochloride (Novocain®)

▲ An amine and its hydrochloride salt.

Application Note

Electrical conductivity is used to measure purity in some water-treatment systems. Lower conductivity generally means purer water, although the presence of microorganisms cannot be detected through electrical conductivity.

15.3   Self-Ionization of Water—the pH Scale

Even the purest water conducts electricity, though it takes exceptionally sensitive instruments to detect it. Electrical conductivity requires the presence of ions, but where do they come from in pure water? The Brønsted–Lowry theory helps us to see how they form. Recall that water is *amphiprotic*. Therefore, to a slight extent, water molecules can transfer protons among themselves. In the self-ionization of water, for every H_2O molecule that loses a proton, another H_2O molecule gains it. In the reaction,

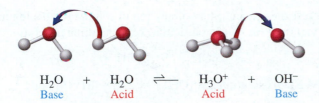

$$H_2O \; + \; H_2O \; \rightleftharpoons \; H_3O^+ \; + \; OH^-$$

Base   Acid   Acid   Base

H_3O^+ is a much stronger acid than H_2O, and OH^- is a much stronger base than H_2O. As a result, the *reverse* reaction is strongly favored, and the equilibrium state lies *far to the left,* which for emphasis we have designated here by the unequal arrow lengths.

We can write the equilibrium constant expression for the self-ionization of water in the usual manner by assuming an activity of 1 for H_2O and replacing the activities of H_3O^+ and OH^- by molarities. We call the equilibrium constant the **ion product of water** and represent it by the symbol K_w:

$$2\,H_2O(l) \rightleftharpoons H_3O^+(aq) + OH^-(aq) \qquad K_w = [H_3O^+][OH^-]$$

At 25 °C, the experimentally determined equilibrium concentrations in pure water are

$$[H_3O^+] = [OH^-] = 1.0 \times 10^{-7}\,M$$

We can then calculate the value of K_w:

$$K_w = [H_3O^+][OH^-] = (1.0 \times 10^{-7})(1.0 \times 10^{-7}) = 1.0 \times 10^{-14} \text{ (at 25 °C)} \quad \textbf{(15.2)}$$

The equilibrium constant K_w is of great significance because it applies not just to pure water but to *all* aqueous solutions—that is, to solutions of acids, bases, salts, and nonelectrolytes. Consider, for example, a solution that is 0.00015 M HCl. Because HCl is a strong acid, its ionization is complete:

$$HCl + H_2O \longrightarrow H_3O^+ + Cl^-$$

Complete ionization produces a H_3O^+ concentration of 0.00015 M. This is more than 1000 times greater than the 1×10^{-7} mol H_3O^+ per liter found in pure water. Thus, we can ignore the self-ionization of water and state that $[H_3O^+]$ in 0.00015 M HCl = 0.00015 M = 1.5×10^{-4} M. With this information and the K_w expression, we can calculate $[OH^-]$ in the solution:

$$[OH^-] = \frac{K_w}{[H_3O^+]} = \frac{1.0 \times 10^{-14}}{1.5 \times 10^{-4}} = 6.7 \times 10^{-11}\,M$$

pH and pOH

In 1909, the Danish biochemist Søren Sørenson proposed a convenient convention that is still used today. He let the term pH refer to "the power of the hydrogen ion." By this, he meant the hydrogen ion concentration expressed as a negative power of 10: $[H^+] = 10^{-pH}$. Using $[H_3O^+]$ rather than $[H^+]$ and a common logarithm (log) in place of a power of 10, we define the **pH** of a solution as the *negative* of the common logarithm of $[H_3O^+]$:

$$pH = -\log[H_3O^+] \qquad \textbf{(15.3)}$$

To determine the pH of 0.00015 M HCl, we first determine $[H_3O^+]$ in the solution. As we saw earlier, $[H_3O^+] = 1.5 \times 10^{-4}$ M.

$$pH = -\log[H_3O^+] = -\log(1.5 \times 10^{-4}) = 3.82$$

We can also calculate the value of $[H_3O^+]$ that corresponds to a given pH. Such problems require an *inverse* calculation. For example, the value of $[H_3O^+]$ in a solution that has a pH of 2.19 is

$$-\log[H_3O^+] = pH = 2.19$$

$$\log[H_3O^+] = -2.19$$

$$[H_3O^+] = \text{antilog}(-2.19) = 10^{-2.19} = 6.5 \times 10^{-3}$$

Self-Ionization of Water animation

Like other equilibrium constants, the value of K_w depends on temperature (see Problem 111).

Problem-Solving Note

Note that pH is defined in terms of the common logarithm, log (the base 10 logarithm), and *not* in terms of the natural logarithm, ln (the base e logarithm). (Also, see Appendix A.) The minus sign in the definition makes a typical pH value a *positive* number.

Problem-Solving Note

The number 1.5×10^{-4} has only two significant figures; the exponent 4 is simply the decimal point locator in the decimal equivalent: 0.00015. Similarly, in log $1.5 \times 10^{-4} = 3.82$, the digit 3 does not enter into the assessment of significant figures; only the two digits following the decimal point are significant.

We can express the concentration of any ion in solution with a logarithmic expression similar to that used for pH. In particular, we can define **pOH** as

$$pOH = -\log[OH^-] \tag{15.4}$$

A 2.5×10^{-3} M NaOH solution has $[OH^-] = 2.5 \times 10^{-3}$ M and

$$pOH = -\log(2.5 \times 10^{-3}) = -(-2.60) = 2.60$$

Logarithmic expressions can also be used to replace exponential numbers in equilibrium constants. Thus, **pK_w** is defined as the negative logarithm of K_w. At 25 °C, $pK_w = 14.00$.

This definition of pK_w allows us to derive a simple relationship between the pH and pOH of a solution:

$$K_w = [H_3O^+][OH^-] = 1.0 \times 10^{-14}$$

$$-\log K_w = -\log([H_3O^+][OH^-]) = -\log(1.0 \times 10^{-14})$$

$$pK_w = -\log[H_3O^+] - \log[OH^-] = 14.00$$

$$pK_w = pH + pOH = 14.00 \tag{15.5}$$

Thus, knowing the value of either pH or pOH in a solution, we can calculate the value of the other. Recall that the pOH corresponding to 2.5×10^{-3} M NaOH is 2.60. Now we can readily calculate the pH of this solution:

$$pH = pK_w - pOH = 14.00 - pOH = 14.00 - 2.60 = 11.40$$

In pure water, the concentrations of H_3O^+ and OH^- are equal: $[H_3O^+] = [OH^-] = 1.0 \times 10^{-7}$ M at 25 °C. Thus, pH and pOH are both 7.00. Pure water and all aqueous solutions with pH = 7.00 at 25 °C are said to be *neutral solutions*. If the pH is less than 7.00, a solution is *acidic*; if the pH is above 7.00, a solution is *basic*, or *alkaline*. As a solution becomes more acidic, $[H_3O^+]$ increases and pH decreases. As a solution becomes more basic, $[H_3O^+]$ decreases and pH increases.

 **Acid-Base Strength activity**

Figure 15.4 gives the pH values of a number of familiar materials. Keep in mind that because pH is on a logarithmic scale, every unit change in pH represents a *tenfold* change in $[H_3O^+]$. Thus, cola drinks (pH ≈ 2.5) are approximately 10 times more acidic than orange juice (pH ≈ 3.5) and about 100 times more acidic than tomato juice (pH ≈ 4.5).

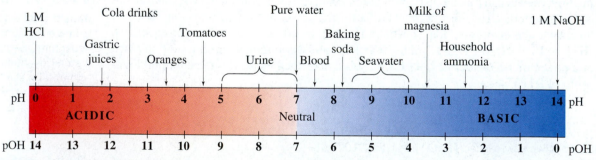

▲ **FIGURE 15.4**   **The pH scale**

The pH values of common substances range from about 0 to 14. Negative numbers are occasionally encountered ($[H_3O^+] = 10$ M corresponds to pH = −1), as are values somewhat greater than 14 ($[OH^-] = 10$ M corresponds to pH = 15). However, pH measurements are less accurate outside the range of pH 2 to 10. Note that the pOH values decrease as you move from left to right: Because the sum of pH and pOH is a constant, the pOH must decrease as the pH increases.

QUESTION: What is the approximate $[H_3O^+]$ in freshly squeezed orange juice?

Example 15.4

By the method suggested in Figure 15.5, a student determines the pH of Milk of magnesia, a suspension of solid magnesium hydroxide in its saturated aqueous solution, and obtains a value of 10.52. What is the molarity of $Mg(OH)_2$ in its saturated aqueous solution? The suspended, undissolved $Mg(OH)_2(s)$ does not affect the measurement.

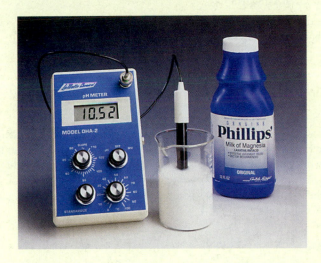

◀ **FIGURE 15.5   Measurement of pH with a pH meter—Example 15.4 illustrated**

No modern laboratory is complete without the electrical measuring instrument known as a pH meter.

STRATEGY

The pH meter measures the pH of the Milk of magnesia. Equation (15.5) relates pOH to pH, and Equation (15.4) relates $[OH^-]$ to pOH. Once we have $[OH^-]$, we can use the conversion factor 1 mol $Mg(OH)_2$/2 mol OH^- to determine the molarity of saturated $Mg(OH)_2(aq)$.

SOLUTION

We will assume that the dissolved magnesium hydroxide is a strong base, with its dissociation into ions represented by the equation to the right.

$$Mg(OH)_2(aq) \longrightarrow Mg^{2+}(aq) + 2\ OH^-(aq)$$

The pOH of the Milk of magnesia is given by Equation (15.5).

$$pOH = 14.00 - pH = 14.00 - 10.52 = 3.48$$

We obtain $[OH^-]$ from the pOH value.

$$\log[OH^-] = -3.48$$
$$[OH^-] = 10^{-3.48} = 3.3 \times 10^{-4}\ M$$

Now we need only to relate the molarity of $Mg(OH)_2$ to the hydroxide ion concentration.

$$Molarity = \frac{3.3 \times 10^{-4}\ mol\ OH^-}{1\ L} \times \frac{1\ mol\ Mg(OH)_2}{2\ mol\ OH^-}$$

$$= \frac{1.7 \times 10^{-4}\ mol\ Mg(OH)_2}{1\ L} = 1.7 \times 10^{-4}\ M\ Mg(OH)_2$$

ASSESSMENT

A straightforward way to verify the answer is to start with the calculated molarity of the $Mg(OH)_2$ and successively calculate $[OH^-]$, pOH, and pH. The pH obtained should be essentially 10.52, and it is.

EXERCISE 15.4A

What is the pH of a solution prepared by dissolving 0.0105 mol HNO_3 in 225 L of water?

EXERCISE 15.4B

What is the pH of a solution containing 2.65 g $Ba(OH)_2$ in 735 mL of aqueous solution? Assume that the $Ba(OH)_2$ is completely dissociated.

Example 15.5   A Conceptual Example

Is the solution 1.0×10^{-8} M HCl acidic, basic, or neutral?

ANALYSIS AND CONCLUSIONS

Because the solute is a strong acid, our first thought is that the solution should be acidic. However, let us check our assumption by calculating the pH:

$$HCl + H_2O \longrightarrow H_3O^+ + Cl^-$$

$$[H_3O^+] = 1.0 \times 10^{-8} \text{ M} \quad \text{and} \quad pH = -\log[H_3O^+] - \log(1.0 \times 10^{-8}) = 8.00$$

Because the pH is greater than 7.00, the solution appears to be basic.

A quandary arises here because the HCl(aq) is so dilute that the self-ionization of water produces more H_3O^+ than does the strong acid. Still, the total $[H_3O^+]$ from the *two* sources must be somewhat greater than 1.0×10^{-7} M. The pH is therefore somewhat less than 7.00; the solution is *acidic*. We conclude that we can generally ignore the slight self-ionization of water when other ionization processes predominate. However, we cannot ignore it when the pH is within one unit or so of pH = 7.

EXERCISE 15.5A

Is a solution that is 1.0×10^{-8} M NaOH acidic, basic, or neutral? Explain.

EXERCISE 15.5B

Is a solution formed by mixing 25.0 mL of 1.0×10^{-8} M HCl and 25.0 mL of 1.0×10^{-8} M NaOH with 25.0 mL of pure water acidic, basic, or neutral? Explain.

Problem-Solving Note

An exact calculation of the pH is the subject of Problem 109.

15.4   Equilibrium in Solutions of Weak Acids and Weak Bases

To calculate the pH of a solution of a weak acid or a weak base, we first must determine $[H_3O^+]$ or $[OH^-]$; and for that, we use an equilibrium calculation based on the acid or base ionization constant, K_a or K_b. Table 15.2 is a brief list of K_a and K_b values, and Appendix C includes a more extensive list. Table 15.2 also includes **pK_a** and **pK_b** values, which are logarithmic expressions of ionization constants:

$$pK_a = -\log K_a \tag{15.6}$$

$$pK_b = -\log K_b \tag{15.7}$$

Just as low values of pH and pOH correspond to high concentrations of H_3O^+ and OH^-, respectively, low values of pK_a and pK_b correspond to large values of K_a and K_b. The entries in Table 15.2 are arranged by increasing value of pK_a or pK_b, and therefore in the order of decreasing acid strength and decreasing base strength. Thus, HNO_2 with $pK_a = 3.14$ is a stronger acid than HOCl with $pK_a = 7.54$.

 **Acid Equilibrium activity**

Some Acid–Base Equilibrium Calculations

The equilibrium calculations we do in this chapter are much like those of the preceding chapter, and much of what we did there applies here as well. We write an equation for the reversible reaction, organize data under the equation, assess the changes that occur in establishing equilibrium, and then calculate the equilibrium concentrations.

Table 15.2 Ionization Constants of Some Weak Acids and Weak Bases in Water at 25 °C

	Ionization Equilibrium	Ionization Constant, K	pK
Inorganic Acids		$K_a =$	
Chlorous acid	$HClO_2 + H_2O \rightleftharpoons H_3O^+ + ClO_2^-$	1.1×10^{-2}	1.96
Nitrous acid	$HNO_2 + H_2O \rightleftharpoons H_3O^+ + NO_2^-$	7.2×10^{-4}	3.14
Hydrofluoric acid	$HF + H_2O \rightleftharpoons H_3O^+ + F^-$	6.6×10^{-4}	3.18
Hypochlorous acid	$HOCl + H_2O \rightleftharpoons H_3O^+ + OCl^-$	2.9×10^{-8}	7.54
Hypobromous acid	$HOBr + H_2O \rightleftharpoons H_3O^+ + OBr^-$	2.5×10^{-9}	8.60
Hydrocyanic acid	$HCN + H_2O \rightleftharpoons H_3O^+ + CN^-$	6.2×10^{-10}	9.21
Carboxylic Acids		$K_a =$	
Chloroacetic acid	$CH_2ClCOOH + H_2O \rightleftharpoons H_3O^+ + CH_2ClCOO^-$	1.4×10^{-3}	2.85
Formic acid	$HCOOH + H_2O \rightleftharpoons H_3O^+ + HCOO^-$	1.8×10^{-4}	3.74
Benzoic acid	$C_6H_5COOH + H_2O \rightleftharpoons H_3O^+ + C_6H_5COO^-$	6.3×10^{-5}	4.20
Acetic acid	$CH_3COOH + H_2O \rightleftharpoons H_3O^+ + CH_3COO^-$	1.8×10^{-5}	4.74
Inorganic Bases		$K_b =$	
Ammonia	$NH_3 + H_2O \rightleftharpoons NH_4^+ + OH^-$	1.8×10^{-5}	4.74
Hydrazine	$H_2NNH_2 + H_2O \rightleftharpoons H_2NNH_3^+ + OH^-$	8.5×10^{-7}	6.07
Hydroxylamine	$HONH_2 + H_2O \rightleftharpoons HONH_3^+ + OH^-$	9.1×10^{-9}	8.04
Amines		$K_b =$	
Dimethylamine	$(CH_3)_2NH + H_2O \rightleftharpoons (CH_3)_2NH_2^+ + OH^-$	5.9×10^{-4}	3.23
Ethylamine	$CH_3CH_2NH_2 + H_2O \rightleftharpoons CH_3CH_2NH_3^+ + OH^-$	4.3×10^{-4}	3.37
Methylamine	$CH_3NH_2 + H_2O \rightleftharpoons CH_3NH_3^+ + OH^-$	4.2×10^{-4}	3.38
Pyridine	$C_5H_5N + H_2O \rightleftharpoons C_5H_5NH^+ + OH^-$	1.5×10^{-9}	8.82
Aniline	$C_6H_5NH_2 + H_2O \rightleftharpoons C_6H_5NH_3^+ + OH^-$	7.4×10^{-10}	9.13

Example 15.6

Ordinary vinegar is approximately 1 M CH_3COOH, and as shown in Figure 15.6, it has a pH of about 2.4. Calculate the expected pH of 1.00 M $CH_3COOH(aq)$, and show that the calculated and measured pH values are in good agreement.

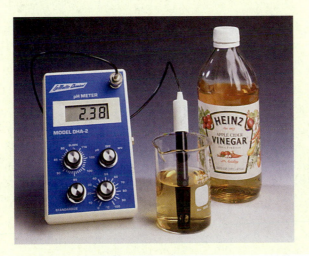

◀ **FIGURE 15.6 Demonstrating that acetic acid is a weak acid: Measuring the pH of vinegar**

The experimentally determined pH of vinegar—an aqueous solution of about 1 M acetic acid—is considerably higher than that of a strong acid of the same molarity (pH ≈ 1.0).

STRATEGY

To calculate the pH of a solution, we must first calculate $[H_3O^+]$. To calculate $[H_3O^+]$, we need to identify the sources of $[H_3O^+]$. In 1.00 M $CH_3COOH(aq)$, H_3O^+ comes from (1) the ionization of CH_3COOH and (2) the self-ionization of H_2O. We can assess the relative importance of these two sources by comparing K_a of acetic acid and K_w of water:

$$CH_3COOH + H_2O \rightleftharpoons H_3O^+ + CH_3COO^- \qquad K_a = 1.8 \times 10^{-5}$$

$$H_2O + H_2O \rightleftharpoons H_3O^+ + OH^- \qquad K_w = 1.0 \times 10^{-14}$$

Because K_w is so much smaller than K_a of acetic acid, the self-ionization of water is a negligible source of H_3O^+, and we can ignore it in the equilibrium calculation. We need to consider only the ionization of acetic acid. In the equilibrium calculation, we can employ the ICE format introduced in Chapter 14, solve for $x = [H_3O^+]$, and then calculate pH from our $[H_3O^+]$ value.

SOLUTION

In our ICE format, the initial concentrations are based on how the solution was prepared and before the ionization of acetic acid is considered. There is no acetate ion added to the solution initially, but there is a trace of H_3O^+ from the water. So we write $[CH_3COO^-] = 0$, but $[H_3O^+] \approx 0$, where the symbol $\approx$ means "approximately equal to."

The reaction:	CH_3COOH	$+ H_2O \rightleftharpoons$	H_3O^+	$+ CH_3COO^-$
Initial concentrations, M:	1.00		≈ 0	0
Changes, M:	$-x$		$+x$	$+x$
Equilibrium concentrations, M:	$(1.00 - x)$		x	x

Next, we substitute data from the ICE table into the K_a expression.

$$K_a = \frac{[H_3O^+][CH_3COO^-]}{[CH_3COOH]}$$

$$= \frac{x \cdot x}{1.00 - x} = \frac{x^2}{1.00 - x} = 1.8 \times 10^{-5}$$

Because the highest power is x^2, this is a quadratic equation. We could rearrange the equation to the familiar form $ax^2 + bx + c = 0$ and solve for x, but there may be a simpler way. If we assume that ionization of the acetic acid is quite limited, x should be much smaller than 1 (that is, $x \ll 1$) and we can replace the term $(1.00 - x)$ by 1.00. Then we can solve the resulting equation for x.

$$\frac{x^2}{1.00 - x} = \frac{x^2}{1.00} = 1.8 \times 10^{-5}$$

$$x^2 = 1.8 \times 10^{-5}$$

$$x = [H_3O^+] = (1.8 \times 10^{-5})^{1/2} = 4.2 \times 10^{-3} \text{ M}$$

Before proceeding, we must prove that this simplifying assumption is valid. We can do so by using the calculated value of x to see if the value of $(1.00 - x)$ is indeed very nearly equal to 1.00.

$$1.00 - x = 1.00 - 4.2 \times 10^{-3} = 1.00 - 0.0042 = 0.9958 \approx 1.00$$

Our assumption is valid for the precision allowed by the calculation—two decimal places—and we can proceed to calculate the pH using Equation (15.3).

$$pH = -\log[H_3O^+] = -\log(4.2 \times 10^{-3}) = 2.38$$

ASSESSMENT

The agreement between the calculated pH and the experimentally determined pH shown in Figure 15.6 is remarkably good—both are pH = 2.38. This excellent agreement is partly fortuitous because the acetic acid concentration in commercial vinegar is somewhat variable. We can also check the calculation in the usual way: Substitute the value of x back into the K_a expression and solve for K_a. The value obtained should be 1.8×10^{-5} or very nearly so.

EXERCISE 15.6A

Determine the pH of 0.250 M CH_3CH_2COOH (propionic acid). Obtain the K_a value from Appendix C.

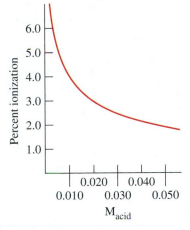
The solution we considered in Example 15.6 is 1.00 M $CH_3COOH(aq)$. Let us call 1.00 M the *nominal* molarity of the solution. It describes how to prepare the solution—dissolve 1.00 mole of pure CH_3COOH (60.05 g) in enough water to make one liter of solution. Once the acetic acid is in solution, some of it ionizes, so that the concentration of CH_3COOH molecules—that is, $[CH_3COOH]$—is no longer 1.00 M. Instead, $[CH_3COOH] = (1.00 - x)$ M, where $x = [H_3O^+] = [CH_3COO^-]$. Recall that we made the assumption that $x \ll 1$ and therefore $(1.00 - x) \approx 1.00$. This assumption implies that so little of the CH_3COOH ionizes that the concentration of nonionized acetic acid $[CH_3COOH]$ is equal to the nominal concentration, 1.00 M.

Figure 15.7 shows the relationship between the nominal molarity of acetic acid and its percent ionization. We can see from the figure that at nominal concentrations greater than 0.010 M, acetic acid is less than 5% ionized. The simplifying assumption of negligible ionization usually yields an acceptable result, say a pH within a few hundredths of a unit of the value calculated by more exact means, if the condition

$$x < 5\% \text{ of } M_{acid}$$

is met. In this statement, $x = [H_3O^+]$, and M_{acid} is the nominal molarity of the acid. We can see this was the case in Example 15.6:

$$\frac{x}{M_{acid}} \times 100\% = \frac{4.2 \times 10^{-3} \text{ M}}{1.00 \text{ M}} \times 100\% = 0.42\%$$

Whether the "5% rule" is met depends on both the nominal molarity of the acid and the value of K_a. We can generally expect the rule to be met if the ratio M_{acid}/K_a is greater than 100.

▲ **FIGURE 15.7   Percent ionization of acetic acid as a function of nominal molarity**

Example 15.7

What is the pH of 0.00200 M $CH_2ClCOOH(aq)$?

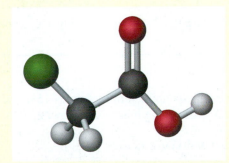

◀ Chloroacetic acid

STRATEGY

This problem can be approached in the same way as Example 15.6. However, at the point where a simplifying assumption might ordinarily be attempted, let us use the criterion just stated above, namely, the ratio M_{acid}/K_a. Then we can decide whether to make the simplifying assumption or proceed with the quadratic formula.

SOLUTION

We begin by tabulating the data as we did in Example 15.6.

We find the acid ionization constant for chloroacetic acid, $CH_3ClCOOH$, in Table 15.2 and use it in solving for the equilibrium concentrations.

The reaction:	$CH_2ClCOOH + H_2O \rightleftharpoons H_3O^+ + CH_2ClCOO^-$		
Initial concentrations, M:	0.00200	≈ 0	0
Changes, M:	$-x$	$+x$	$+x$
Equilibrium concentrations, M:	$(0.00200 - x)$	x	x

$$K_a = \frac{[H_3O^+][CH_2ClCOO^-]}{[CH_2ClCOOH]}$$

$$1.4 \times 10^{-3} = \frac{x \cdot x}{0.00200 - x} = \frac{x^2}{0.00200 - x}$$

Now let us evaluate the ratio M_{acid}/K_a.

$$\frac{M_{acid}}{K_a} = \frac{0.00200}{1.4 \times 10^{-3}} = 1.4$$

Because this ratio is much less than 100, we expect the assumption that $x \ll 0.0200$ will fail. Therefore, we must use the quadratic formula to solve the equation.

$$x^2 = (0.00200 - x) \times 1.4 \times 10^{-3}$$

$$x^2 + 1.4 \times 10^{-3}x - 2.8 \times 10^{-6} = 0$$

$$x = \frac{-1.4 \times 10^{-3} \pm \sqrt{(1.4 \times 10^{-3})^2 - 4 \times 1 \times (-2.8 \times 10^{-6})}}{2}$$

$$x = \frac{-1.4 \times 10^{-3} \pm \sqrt{1.3 \times 10^{-5}}}{2}$$

$$x = [H_3O^+] = \frac{-1.4 \times 10^{-3} \pm 3.6 \times 10^{-3}}{2} = \frac{2.2 \times 10^{-3}}{2} = 1.1 \times 10^{-3}$$

From the result $[H_3O^+] = 1.1 \times 10^{-3}$ M, we can now determine the pH. $pH = -\log[H_3O^+] = -\log(1.1 \times 10^{-3}) = 2.96$

ASSESSMENT

Notice that if we had made the simplifying assumption, we would have obtained

$$x = \sqrt{0.00200 \times 1.4 \times 10^{-3}} = [H_3O^+] = 1.7 \times 10^{-3} \text{ M}$$

which is about 55% larger than the more accurate answer, $[H_3O^+] = 1.1 \times 10^{-3}$ M, and well beyond the 5% rule limitation.

EXERCISE 15.7A

Calculate the pH of 0.0100 M chlorous acid. Obtain K_a from Table 15.2.

EXERCISE 15.7B

Determine the minimum molarity of chloroacetic acid for which the 5% rule would apply in Example 15.7. (Hint: What two relationships must be satisfied?)

We can use the methods applied in Examples 15.6 and 15.7 to calculate pH values of solutions of weak bases if we make the following adjustments:

• Use a K_b expression to calculate $[OH^-]$.
• Calculate pOH from $[OH^-]$.
• Calculate pH by using Equation (15.5): $pH + pOH = 14.00$.
• Use the criterion $M_{base}/K_b > 100$ to decide whether a quadratic equation may be simplified.

Example 15.8

What is the pH of 0.500 M $NH_3(aq)$?

STRATEGY

In Example 15.6, we explained why the self-ionization of water is an insignificant source of H_3O^+ in $CH_3COOH(aq)$. Likewise, the self-ionization of water is an insignificant source of OH^- in $NH_3(aq)$. Thus, we can calculate $[OH^-]$ by applying the ICE format to the ionization of $NH_3(aq)$. From this $[OH^-]$, we can obtain pOH and then pH.

SOLUTION

We begin by tabulating the relevant data.

The reaction:	NH_3	$+ H_2O \rightleftharpoons$	NH_4^+	$+ OH^-$
Initial concentrations, M:	0.500		0	≈ 0
Changes, M:	$-x$		$+x$	$+x$
Equilibrium concentrations, M:	$(0.500 - x)$		x	x

We find the base ionization constant for ammonia in Table 15.2 and use it to solve for the equilibrium concentrations.

$$K_b = \frac{[NH_4^+][OH^-]}{[NH_3]} = 1.8 \times 10^{-5}$$

$$1.8 \times 10^{-5} = \frac{x \cdot x}{0.500 - x} = \frac{x^2}{0.500 - x}$$

To determine whether we can make the assumption that $x \ll 0.500$, we evaluate the ratio of the base concentration to base ionization constant.

$$\frac{M_{base}}{K_b} = \frac{0.500}{1.8 \times 10^{-5}} = 2.8 \times 10^4$$

Because this ratio is much greater than 100, the assumption should be valid. We can solve the simplified equation.

$$\frac{x^2}{0.500 - x} \approx \frac{x^2}{0.500} = 1.8 \times 10^{-5}$$

$$x^2 = 0.500 \times 1.8 \times 10^{-5} = 9.0 \times 10^{-6}$$

$$x = [OH^-] = (9.0 \times 10^{-6})^{1/2} = 3.0 \times 10^{-3} \text{ M}$$

From this $[OH^-]$, we can calculate the pOH.

$$pOH = -\log[OH^-] = -\log(3.0 \times 10^{-3}) = 2.52$$

Finally, we calculate the pH.

$$pH + pOH = 14.00$$

$$pH = 14.00 - pOH = 14.00 - 2.52 = 11.48$$

ASSESSMENT

We have already justified the assumption that $x \ll 0.500$, and we can establish just how valid it is with a simple calculation: $0.500 - x = 0.500 - 0.0030 = 0.497$. The assumption produces an error of only 3 parts in 500, or 0.6%.

EXERCISE 15.8A

What is the pH of a solution that is 0.200 M in methylamine, CH_3NH_2? Obtain a value of K_b from Table 15.2.

EXERCISE 15.8B

What is the pH of the solution obtained by diluting 5.00 mL of 0.0100 M $NH_3(aq)$ to 1.000 L?

In Chapter 14, we showed how to calculate an equilibrium constant from experimental data. The K_a of a weak acid or K_b of a weak base can be established in several ways. Perhaps the simplest way is to measure the pH and calculate $[H_3O^+]$ or $[OH^-]$ from the pH value.

Example 15.9

The pH of a 0.164 M aqueous solution of dimethylamine is 11.98. What are the values of K_b and pK_b? The ionization equation is

$$(CH_3)_2NH + H_2O \rightleftharpoons (CH_3)_2NH_2^+ + OH^- \qquad K_b = ?$$

Dimethylamine   Dimethylammonium ion

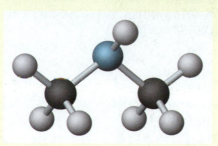

◀ Dimethylamine

STRATEGY

This problem is similar to Example 14.11. There, the starting point was the equilibrium concentration of sulfur trioxide $[SO_3]$, and here it will be the equilibrium concentration of hydroxide ion, $[OH^-]$, which we can obtain from the measured pH of the solution.

SOLUTION

Let us begin with an ICE tabular format.

The reaction	$(CH_3)_2NH$	+ H_2O $\rightleftharpoons$	$(CH_3)_2NH_2^+$	+ OH^-
Initial concentrations, M:	0.164		0	≈0
Changes, M:	?		?	?
Equilibrium concentrations, M:	?		?	?

We can get $[OH^-]$ at equilibrium (?) from the pH of the solution.

$$pOH = 14.00 - pH = 14.00 - 11.98 = 2.02$$

$$-\log[OH^-] = pOH = 2.02$$

$$\log[OH^-] = -2.02$$

$$[OH^-] = 10^{-2.02} = 9.5 \times 10^{-3} \text{ M}$$

Now let us go back and substitute a numerical value for each of the missing items in the tabulated data.

The reaction:	$(CH_3)_2NH$	+ H_2O $\rightleftharpoons$	$(CH_3)_2NH_2^+$	+ OH^-
Initial concentrations, M:	0.164		0	≈0
Changes, M:	-9.5×10^{-3}		$+9.5 \times 10^{-3}$	$+9.5 \times 10^{-3}$
Equilibrium concentrations, M:	$(0.164 - 9.5 \times 10^{-3})$		9.5×10^{-3}	9.5×10^{-3}

Finally, we substitute the equilibrium concentrations into the K_b expression and solve for K_b. Notice that no assumptions are needed here; we know each of the concentration terms in the K_b expression.

$$K_b = \frac{[(CH_3)_2NH_2^+][OH^-]}{[(CH_3)_2NH]} = \frac{(9.5 \times 10^{-3})(9.5 \times 10^{-3})}{(0.164 - 0.0095)} = 5.8 \times 10^{-4}$$

$$pK_b = -\log K_b = -\log(5.8 \times 10^{-4}) = 3.24$$

ASSESSMENT

Unlike the three preceding examples, the question of whether to make a simplifying assumption does not arise in this example. That is, we do not have to consider whether to replace the term $M - x$ by M; we have a value for each: M = 0.164 and x = 0.0095.

EXERCISE 15.9A

Suppose you discovered a new acid, HZ, and found that the pH of a 0.0100 M solution is 3.12. What are K_a and pK_a for HZ?

$$HZ + H_2O \rightleftharpoons H_3O^+ + Z^- \qquad K_a = ?$$

EXERCISE 15.9B

What is the molarity of $NH_3(aq)$ if the ammonia solution has the same pH as 0.200 M $(CH_3)_2NH(aq)$? Use data from Table 15.2.

Example 15.10   A Conceptual Example

Without doing detailed calculations, indicate which solution has the greater $[H_3O^+]$, 0.030 M HCl or 0.050 M CH_3COOH.

ANALYSIS AND CONCLUSIONS

HCl(aq), being a strong acid, is essentially 100% ionized, and in 0.030 M HCl, $[H_3O^+]$ = 0.030 M. Acetic acid, CH_3COOH, is a weak acid. The 0.050 M CH_3COOH would have to be more than 50% ionized to have $[H_3O^+]$ equal to that in 0.030 M HCl. On page 633 and in Figure 15.7, we see that a solution as dilute as 0.005 M CH_3COOH is only about 5% ionized; the percent ionization in 0.050 M CH_3COOH is much smaller still. Also, just by reflecting on the magnitude of K_a for acetic acid (1.8×10^{-5}), we expect the ionization not to go very far before equilibrium is reached, certainly not to the point at which more than half the molecules are ionized. The 0.030 M HCl has the greater $[H_3O^+]$.

EXERCISE 15.10A

Without doing detailed calculations, determine which of these solutions has the higher pH: 0.025 M NH_3(aq) or 0.030 M methylamine, CH_3NH_2(aq). (*Hint:* Refer to Table 15.2.)

EXERCISE 15.10B

Without doing detailed calculations, determine which of these solutions has the higher pH: 0.0010 M HCl(aq) or 0.10 M CH_3COOH(aq).

15.5   Polyprotic Acids

A **monoprotic acid** has one ionizable H atom per molecule. Hydrochloric acid, HCl, is a monoprotic acid. Acetic acid, CH_3COOH, has four H atoms per molecule, but only one of them is ionizable; therefore it is also a monoprotic acid. A **polyprotic acid** has more than one ionizable H atom per molecule. Carbonic acid, H_2CO_3, has two H atoms, both ionizable; it is a *diprotic* acid. Phosphoric acid, H_3PO_4, has three H atoms, all ionizable; it is a *triprotic* acid. In this section, we will consider three polyprotic acids—phosphoric acid, carbonic acid, and sulfuric acid. Additional polyprotic acids are listed in Appendix C.

Phosphoric Acid

A key feature of polyprotic acids is that each ionizable H atom ionizes separately. For example, a molecule of H_3PO_4 does not give up its three ionizable H atoms in a single action. Instead, the complete ionization takes place in three distinct steps. Each step involves a reversible reaction with its own specific K_a value.

(1)
$$H_3PO_4 + H_2O \rightleftharpoons H_3O^+ + H_2PO_4^-$$

$$K_{a_1} = \frac{[H_3O^+][H_2PO_4^-]}{[H_3PO_4]} = 7.1 \times 10^{-3}$$

(2)
$$H_2PO_4^- + H_2O \rightleftharpoons H_3O^+ + HPO_4^{2-}$$

$$K_{a_2} = \frac{[H_3O^+][HPO_4^{2-}]}{[H_2PO_4^-]} = 6.3 \times 10^{-8}$$

(3)
$$HPO_4^{2-} + H_2O \rightleftharpoons H_3O^+ + PO_4^{3-}$$

$$K_{a_3} = \frac{[H_3O^+][PO_4^{3-}]}{[HPO_4^{2-}]} = 4.3 \times 10^{-13}$$

It is easy to see why the second ionization constant, K_{a_2}, is much smaller than the first, K_{a_1}. The first ionization involves separation of a positive H^+ and a negative $H_2PO_4^-$ ion. The second ionization requires separation of H^+ and a negative HPO_4^{2-} ion, which is more negative than the $H_2PO_4^-$ ion. That second separation is more difficult than the first. Likewise, the third ionization requires separating H^+ and PO_4^{3-}, which is even more

Application Note

The principal use of phosphoric acid is in the manufacture of phosphate fertilizers. Phosphoric acid and its salts are also widely used in the food industry.

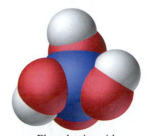

Phosphoric acid

difficult, making K_{a_3} smaller still than K_{a_2}. Thus, ionization constants for a polyprotic acid progressively decrease: $K_{a_1} > K_{a_2} > K_{a_3} \dots$. Often, each K_a is hundreds or even thousands of times smaller than the preceding one, and in these cases we can make two additional generalizations about the ionization constants of polyprotic acids:

- Because K_{a_2}, K_{a_3}, ... are so small, few of the anions produced in the first ionization step ionize further.
- In all but extremely dilute solutions, essentially all the H_3O^+ ions come from the first ionization step alone.

In the first ionization step of phosphoric acid, H_3O^+ and $H_2PO_4^-$ ions are produced in equal number. If few of the $H_2PO_4^-$ ions undergo further ionization, we see that $[H_3O^+] \approx [H_2PO_4^-]$. Note the interesting result when we incorporate this idea into the expression for K_{a_2}:

$$K_{a_2} = \frac{\cancel{[H_3O^+]}[HPO_4^{2-}]}{\cancel{[H_2PO_4^-]}} = 6.3 \times 10^{-8}$$

Because they are essentially equal, the molarities of H_3O^+ and $H_2PO_4^-$ cancel, leaving us with

$$[HPO_4^{2-}] = K_{a_2} = 6.3 \times 10^{-8}$$

This relationship holds regardless of the values of $[H_3O^+]$ and $[H_2PO_4^-]$ and, thus, regardless of the molarity of the original phosphoric acid solution.

Example 15.11

Calculate the following concentrations in an aqueous solution that is 5.0 M H_3PO_4: (a) $[H_3O^+]$, (b) $[H_2PO_4^-]$, (c) $[HPO_4^{2-}]$, (d) $[PO_4^{3-}]$.

STRATEGY

We can obtain these ion concentrations by using the general ideas about polyprotic acids we have developed. The key idea is indicated in each part of the solution.

SOLUTION

(a) Because K_{a_1} exceeds K_{a_2} by more than 10^5, essentially all the H_3O^+ ions come from the first ionization of phosphoric acid. Our ICE format for this ionization is

The reaction:	$H_3PO_4 + H_2O \rightleftharpoons$	H_3O^+	$+$	$H_2PO_4^-$
Initial concentrations, M:	5.0	≈ 0		0
Changes, M:	$-x$	$+x$		$+x$
Equilibrium concentrations, M:	$(5.0 - x)$	x		x

$$K_{a_1} = \frac{[H_3O^+][H_2PO_4^-]}{[H_3PO_4]} = \frac{x \cdot x}{(5.0 - x)} = 7.1 \times 10^{-3}$$

If we assume that $x \ll 5.0$, so that $(5.0 - x) \approx 5.0$, we can write

$$x^2 = 5.0 \times 7.1 \times 10^{-3} = 3.6 \times 10^{-2}$$

$$x = [H_3O^+] = 0.19 \text{ M}$$

(b) Because K_{a_2} is so small, we know that little of the $H_2PO_4^-$ produced in the first ionization undergoes further ionization. For this reason, $[H_2PO_4^-] \approx [H_3O^+]$ and $[H_2PO_4^-] = 0.19$ M.

(c) We established on page 638 that $[HPO_4^{2-}] = K_{a_2}$ regardless of the molarity of a phosphoric acid solution. Thus, $[HPO_4^{2-}] = K_{a_2} = 6.3 \times 10^{-8}$ M.

(d) $[PO_4^{3-}]$ is formed only through the very limited third ionization. We now have all the data needed to calculate $[PO_4^{3-}]$:

$$\frac{[H_3O^+][PO_4^{3-}]}{[HPO_4^{2-}]} = K_{a_3} = 4.3 \times 10^{-13}$$

From previous steps, we substitute $[H_3O^+] = 0.19$ and $[HPO_4^{2-}] = 6.3 \times 10^{-8}$ in this expression:

$$\frac{0.19 \times [PO_4^{3-}]}{6.3 \times 10^{-8}} = 4.3 \times 10^{-13}$$

$$[PO_4^{3-}] = 1.4 \times 10^{-19} \text{ M}$$

ASSESSMENT

We should use the 5% rule to check the validity of the assumption in part (a) that $x \ll 5.0$:

$$\frac{x}{M_{acid}} \times 100\% = \frac{0.19 \text{ M}}{5.0 \text{ M}} \times 100\% = 3.8\%$$

We see that the assumption meets the requirement of the 5% rule, but just barely.

EXERCISE 15.11A

What is the pH of a solution that is 0.125 M in maleic acid, a diprotic acid used as an additive to retard rancidity in fats and oils? The equations for the two ionizations are

$$HOOCCH{=}CHCOOH + H_2O \rightleftharpoons H_3O^+ + HOOCCH{=}CHCOO^-$$

$$K_{a_1} = 1.2 \times 10^{-2}$$

$$HOOCCH{=}CHCOO^- + H_2O \rightleftharpoons H_3O^+ + {}^-OOCCH{=}CHCOO^-$$

$$K_{a_2} = 4.7 \times 10^{-7}$$

EXERCISE 15.11B

Acids are added to cola drinks to lower the pH to about 2.5 to impart tartness. Show that a cola that contains from 0.057 to 0.084% of 75% phosphoric acid (H_3PO_4), by mass, meets this pH requirement.

▲ A typical cola drink contains phosphoric acid (see Exercise 15.11B).

Carbonic Acid

When carbon dioxide dissolves in water, a reaction occurs, producing carbonic acid:

$$CO_2(aq) + H_2O(l) \rightleftharpoons H_2CO_3(aq)$$

The reaction is reversible. Carbonic acid is weak, diprotic, and unstable. It readily reverts to $CO_2(g)$ and H_2O. In an open vessel, $CO_2(g)$ escapes, and as a result, the reaction goes to completion *to the left*, just as we would expect from LeChâtelier's principle.

The two ionizations of H_2CO_3 and their K_a values are

(1) $H_2CO_3 + H_2O \rightleftharpoons H_3O^+ + HCO_3^-$ $\quad K_{a_1} = \dfrac{[H_3O^+][HCO_3^-]}{[H_2CO_3]} = 4.2 \times 10^{-7}$

(2) $HCO_3^- + H_2O \rightleftharpoons H_3O^+ + CO_3^{2-}$ $\quad K_{a_2} = \dfrac{[H_3O^+][CO_3^{2-}]}{[HCO_3^-]} = 4.7 \times 10^{-11}$

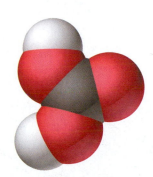

Carbonic acid

Because K_{a_1} is much greater than K_{a_2}, carbonic acid conforms to our generalizations about polyprotic acids (page 638): Few of the HCO_3^- ions produced in the first ionization step ionize further.

Neutralization of H_2CO_3 with one mole of a strong base, such as NaOH, produces salts such as $NaHCO_3$, sodium hydrogen carbonate. Neutralization with two moles of a strong base produces carbonate salts, such as Na_2CO_3, sodium carbonate.

Equilibria based on the ionization of carbonic acid are important in several natural phenomena, including formation of hard water and limestone caves. These equilibria are also essential in maintaining the proper pH of blood.

Sulfuric Acid

Sulfuric acid is an unusual diprotic acid in that its first ionization step goes essentially to completion, but its second step does not:

(1) $\quad H_2SO_4 + H_2O \longrightarrow H_3O^+ + HSO_4^- \quad K_{a_1} \approx 10^3$

(2) $\quad HSO_4^- + H_2O \rightleftharpoons H_3O^+ + SO_4^{2-} \quad K_{a_2} = \dfrac{[H_3O^+][SO_4^{2-}]}{[HSO_4^-]} = 1.1 \times 10^{-2}$

For purposes of calculations, we can consider three categories of sulfuric acid solutions, described here in increasing order of complexity.

1. *Concentrated solutions* (greater than about 0.50 M H_2SO_4). In these solutions, essentially all the H_3O^+ is produced in the first ionization because this step goes to completion, whereas the second ionization is reversible. Thus, in 1.00 M H_2SO_4, we expect $[H_3O^+] \approx 1.00$ M. (See also Example 15.12.)

2. *Very dilute solutions* (less than about 0.0010 M H_2SO_4). Again, the first ionization goes essentially to completion. Because the value of $K_{a_2}(1.1 \times 10^{-2})$ is relatively large for a weak acid, and because the solution is so dilute, the second ionization also goes essentially to completion. We predict that two H_3O^+ ions are produced for every H_2SO_4 molecule present originally, leaving essentially no H_2SO_4 molecules or HSO_4^- ions in solution. Thus, in 0.0010 M H_2SO_4, $[H_3O^+] \approx 0.0020$ M and $[SO_4^{2-}] \approx 0.0010$ M. (See Exercise 15.12A.)

3. *Intermediate concentrations* (between about 0.0010 M H_2SO_4 and 0.50 M H_2SO_4). Both ionization steps must be considered when calculating ion concentrations. (See Exercise 15.12B.)

Application Note

Sulfuric acid is produced in greater quantity than any other manufactured chemical. Its main use is in the production of fertilizers.

Sulfuric acid

Example 15.12

What is the approximate pH of 0.71 M H_2SO_4?

SOLUTION

This solution fits the first of the three categories. Essentially all the H_3O^+ comes from the first ionization step, and that step goes to completion:

$$H_2SO_4 + H_2O \longrightarrow H_3O^+ + HSO_4^-$$

Thus, 0.71 M H_2SO_4 has $[H_3O^+] \approx 0.71$ M.

$$pH = -\log[H_3O^+] \approx -\log 0.71 = 0.15$$

EXERCISE 15.12A

What is the approximate pH of 8.5×10^{-4} M H_2SO_4?

EXERCISE 15.12B

Without doing detailed calculations, indicate which of the following is most likely to be closest to the measured $[H_3O^+]$ in 0.020 M H_2SO_4: (a) 0.020 M, (b) 0.025 M, (c) 0.039 M, (d) 0.045 M. Explain your reasoning.

15.6 Ions as Acids and Bases

A package of washing soda carries a common warning found on products that are either strongly acidic or strongly basic: Avoid contact with eyes and prolonged contact with skin. However, the principal component of this cleaner is listed as sodium carbonate, Na_2CO_3, and therefore our first thought might be that Na_2CO_3 is neither an acid nor a base because we see no H atoms, no OH groups, and no N atoms with lone-pair electrons. Yet, a 1.0 M $Na_2CO_3(aq)$ solution of washing soda is quite basic, having a pH of about 12.

When the ionic compound $Na_2CO_3(s)$ dissolves, it dissociates entirely into Na^+ and CO_3^{2-} ions:

$$Na_2CO_3(s) \xrightarrow{H_2O} 2\,Na^+(aq) + CO_3^{2-}(aq)$$

The Brønsted–Lowry theory explains how the carbonate ions then react to produce OH^- ions:

$$CO_3^{2-} + H_2O \rightleftharpoons HCO_3^- + OH^-$$
$$\text{Base} \qquad \text{Acid} \qquad\quad \text{Acid} \qquad \text{Base}$$

This reaction raises $[OH^-]$ to a value much higher than 1.0×10^{-7} M, and $[H_3O^+]$ decreases accordingly. The pH therefore rises well above 7.0. Because sodium ions do not react with water, they do not affect the pH of the solution:

$$Na^+(aq) + H_2O \longrightarrow \text{no reaction}$$

Although acid–base reactions between ions and water molecules are fundamentally no different from other acid–base reactions, they are often called **hydrolysis** reactions. In $Na_2CO_3(aq)$, we say that CO_3^{2-} hydrolyzes and Na^+ does not. Cations of groups 1A and 2A do not hydrolyze, but many other metal cations do hydrolyze, particularly those that are small and carry a high charge, as we will discover in the next chapter.

Now let us consider some useful generalizations about hydrolysis of the salt solutions. In the numbered items that follow, each salt printed in red corresponds to one of the solutions shown in Figure 15.8.

1. *Salts of strong acids and strong bases form <u>neutral</u> solutions* (pH = 7). Examples of such salts are NaCl, KNO_3, and BaI_2. The anions Cl^-, NO_3^-, and I^- are conjugate bases of strong acids and are therefore all extremely weak bases. They do not hydrolyze, nor do group 1A and group 2A cations.

2. *Salts of weak acids and strong bases form <u>basic</u> solutions* (pH > 7). The anion hydrolyzes as a base. Examples: Na_2CO_3, KNO_2, CH_3COONa. The anions CO_3^{2-},

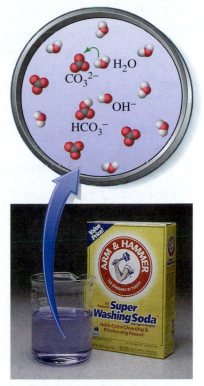

▲ This sodium carbonate solution contains a few drops of thymolphthalein indicator, which is colorless below pH 9.4 and blue above pH 10.6. The blue color tells us that the solution is basic: Sodium carbonate dissociates completely into Na^+ (omitted here for clarity) and CO_3^{2-}, and the partial hydrolysis of CO_3^{2-} forms HCO_3^- and OH^-, as suggested in the microscopic view.

(a) NaCl(aq)	**(b)** $CH_3COONa(aq)$	**(c)** $NH_4Cl(aq)$	**(d)** $CH_3COONH_4(aq)$

▲ **FIGURE 15.8 The pH of aqueous solutions of salts**

Each solution contains a few drops of bromthymol blue indicator, which is yellow in solutions having a pH below 7, green in solutions for which the pH is 7, and blue in solutions having a pH above 7. (a) A solution of sodium chloride is neutral. (b) A solution of sodium acetate is basic. (c) A solution of ammonium chloride is acidic. (d) A solution of ammonium acetate is neutral.

NO_3^-, and CH_3COO^- are the conjugate bases of weak acids and are therefore considerably stronger bases than the anions Cl^-, NO_3^-, and I^- referred to in the preceding case. Again, group 1A and group 2A cations do not hydrolyze.

3. *Salts of strong acids and weak bases form acidic solutions* (pH < 7). The cation hydrolyzes as an acid. Examples: NH_4Cl, NH_4NO_3, and NH_4Br. The cation NH_4^+ is the conjugate acid of the weak base NH_3. As in the strong acid–strong base case, anions such as Cl^-, NO_3^-, and Br^- are the conjugate bases of strong acids and do not hydrolyze.

4. *Salts of weak acids and weak bases form solutions that are acidic in some cases, neutral or basic in others.* The cations hydrolyze as acids and the anions as bases. The solution pH depends on the relative acid and base strengths. Examples: NH_4CN, NH_4NO_2, and CH_3COONH_4.

We can summarize all four of these cases in a single statement:

Only ions that are the conjugates of weak acids or weak bases hydrolyze appreciably.

Example 15.13 A Conceptual Example

(a) Is $NH_4I(aq)$ acidic, basic, or neutral? **(b)** What conclusion can you draw from Figure 15.8d about the equilibrium constants for the hydrolysis reactions in $CH_3COONH_4(aq)$?

ANALYSIS AND CONCLUSIONS

(a) Ammonium iodide is the salt of a strong acid, HI, and a weak base, NH_3, an example of case 3 above. The cation NH_4^+ hydrolyzes, and the solution is acidic:

$$NH_4^+ + H_2O \rightleftharpoons NH_3 + H_3O^+$$

The anion I^-, a very weak base, does not hydrolyze appreciably.

(b) Ammonium acetate is the salt of a weak acid, CH_3COOH, and a weak base, NH_3. It is an example of case 4; both ions hydrolyze:

$$NH_4^+ + H_2O \rightleftharpoons NH_3 + H_3O^+$$

$$CH_3COO^- + H_2O \rightleftharpoons CH_3COOH + OH^-$$

From the observation that the pH of aqueous ammonium acetate is 7, we expect the equilibrium constants for the two hydrolysis reactions to have about the same value.

EXERCISE 15.13A

Indicate whether the solutions **(a)** $NaNO_3(aq)$ and **(b)** $CH_3CH_2CH_2COOK(aq)$ are acidic, basic, or neutral. Explain.

EXERCISE 15.13B

Arrange the following 0.1 M solutions in the expected order of increasing pH, and state your reasoning for doing so: $NaCl(aq)$, $HCl(aq)$, $NaOH(aq)$, $KNO_2(aq)$, $NH_4Br(aq)$.

To make a quantitative prediction of the pH of a solution in which hydrolysis occurs, we need an equilibrium constant for the hydrolysis reaction. Consider, for example, the hydrolysis of acetate ion:

$$CH_3COO^- + H_2O \rightleftharpoons CH_3COOH + OH^- \qquad K_b = \frac{[CH_3COOH][OH^-]}{[CH_3COO^-]} = ?$$

Two of the concentration terms in the K_b expression are the same as in the K_a expression for the ionization of acetic acid, the conjugate acid of CH_3COO^-. It seems, then, that K_b for CH_3COO^- and K_a for CH_3COOH should be related to each other. Suppose we multiply both the numerator and denominator of the K_b expression by $[H_3O^+]$:

$$K_b = \frac{[CH_3COOH][OH^-][H_3O^+]}{[CH_3COO^-][H_3O^+]} = \frac{K_w}{K_a} = \frac{1.0 \times 10^{-14}}{1.8 \times 10^{-5}} = 5.6 \times 10^{-10}$$

Note that the factors in blue print are equivalent to K_w. Those in red give the *inverse* of K_a for acetic acid, that is, $1/K_a$. The result shows that the product of K_a and K_b of a conjugate acid–base pair equals the ion product of water, which at 25 °C is $K_w = 1.0 \times 10^{-14}$:

$$K_a \times K_b = K_w \qquad (15.8)$$

Many tables of ionization constants list only pK_a values, but we can get a K_b value by converting the K values in Equation (15.8) to pK values and then writing

$$pK_a + pK_b = pK_w \qquad (15.9)$$

At 25 °C, $pK_w = 14.00$. Ionization constants obtained in this way can be used to calculate the pH of solutions of salts that hydrolyze, as illustrated in Example 15.14.

Example 15.14

Calculate the pH of a solution that is 0.25 M CH_3COONa(aq).

STRATEGY

Acetate ion hydrolyzes as a base, and we just derived a numerical value of K_b for that hydrolysis. We can use the usual ICE format and $K_b = 5.6 \times 10^{-10}$ to obtain a quadratic equation that we can solve for $[OH^-]$. The pH of the solution follows easily from that point.

SOLUTION

We present the data under the equation for the hydrolysis reaction in the usual way.

The reaction:	$CH_3COO^- + H_2O$	$\rightleftharpoons$	CH_3COOH	$+ OH^-$
Initial concentrations, M:	0.25		0	≈ 0
Changes, M:	$-x$		$+x$	$+x$
Equilibrium concentrations, M:	$(0.25 - x)$		x	x

$$K_b = \frac{[CH_3COOH][OH^-]}{[CH_3COO^-]} = \frac{K_w}{K_a} = \frac{1.0 \times 10^{-14}}{1.8 \times 10^{-5}} = 5.6 \times 10^{-10}$$

$$\frac{x \cdot x}{0.25 - x} = 5.6 \times 10^{-10}$$

Let us assume that $x \ll 0.25$, so that $(0.25 - x) = 0.25$. Then

$$\frac{x^2}{0.25} = 5.6 \times 10^{-10}$$

$$x^2 = 1.4 \times 10^{-10}$$

$$x = [OH^-] = (1.4 \times 10^{-10})^{1/2} = 1.2 \times 10^{-5} \text{ M}$$

$$pOH = -\log[OH^-] = -\log(1.2 \times 10^{-5}) = 4.92$$

$$pH = 14.00 - pOH = 14.00 - 4.92 = 9.08$$

ASSESSMENT

First, we see that the assumption $x \ll 0.25$ is justified because $x = 1.2 \times 10^{-5}$. To see that our answer is at least qualitatively correct, note that pH > 7. Hydrolysis of acetate ion should produce a basic solution. A common error is to confuse pOH (4.92) with pH, but of course the pH of a basic solution cannot be less than 7.

EXERCISE 15.14A

Calculate the pH of a 0.052 M NH_4Cl aqueous solution.

EXERCISE 15.14B

A 50.00-mL sample of 0.120 M CH_3COOH(aq) is neutralized with 18.75 mL of 0.320 M NaOH. What is the pH of the neutralized solution? (*Hint:* What is the neutralization reaction, and what are the concentrations of its products?)

Example 15.15

What is the molarity of an $NH_4NO_3(aq)$ solution that has a pH = 4.80?

STRATEGY

Ammonium nitrate is the salt of a strong acid (HNO_3) and a weak base (NH_3). In $NH_4NO_3(aq)$, NH_4^+ hydrolyzes and NO_3^- does not. The ICE format must be based on the hydrolysis equilibrium for $NH_4^+(aq)$. In that format $[H_3O^+]$, derived from the pH, will be a known quantity, and the initial concentration of NH_4^+ will be the unknown.

SOLUTION

We begin by writing the equation for the hydrolysis equilibrium and the equation for K_a in terms of K_w and K_b.

$$NH_4^+ + H_2O \rightleftharpoons H_3O^+ + NH_3$$

$$K_a = \frac{K_w}{K_b} = \frac{1.0 \times 10^{-14}}{1.8 \times 10^{-5}} = 5.6 \times 10^{-10}$$

As usual, we can calculate $[H_3O^+]$ from the pH of the solution.

$$\log[H_3O^+] = -pH = -4.80$$

$$[H_3O^+] = 10^{-4.80} = 1.6 \times 10^{-5}\,M$$

If we assume that all the hydronium ion comes from the hydrolysis reaction, we can set up an ICE format in which x represents the unknown initial concentration of NH_4^+.

The reaction	NH_4^+	+ H_2O $\rightleftharpoons$	H_3O^+	+	NH_3
Initial concentrations, M:	x		≈ 0		0
Changes, M:	-1.6×10^{-5}		$+1.6 \times 10^{-5}$		$+1.6 \times 10^{-5}$
Equilibrium concentrations, M:	$(x - 1.6 \times 10^{-5})$		1.6×10^{-5}		1.6×10^{-5}

We now substitute equilibrium concentrations into the ionization constant expression for the hydrolysis reaction.

$$K_a = \frac{[H_3O^+][NH_3]}{[NH_4^+]} = \frac{(1.6 \times 10^{-5})(1.6 \times 10^{-5})}{(x - 1.6 \times 10^{-5})} = 5.6 \times 10^{-10}$$

We can assume that the ammonium ion is mostly *nonhydrolyzed* and that the change in $[NH_4^+]$ is much smaller than the initial $[NH_4^+]$, so that $1.6 \times 10^{-5} \ll x$ and we can replace $(x - 1.6 \times 10^{-5})$ by x. Then we can solve for x.

$$\frac{(1.6 \times 10^{-5})^2}{x} = 5.6 \times 10^{-10}$$

$$x = \frac{(1.6 \times 10^{-5})^2}{5.6 \times 10^{-10}} = [NH_4^+] = 0.46\,M$$

The solution is 0.46 M NH_4NO_3.

ASSESSMENT

Let us verify the two assumptions we made in this problem: $[H_3O^+]$ in the solution $(1.6 \times 10^{-5}\,M)$ is 160 times greater than $[H_3O^+]$ produced by the self-ionization of pure water $(1.0 \times 10^{-7}\,M)$. (1) Our decision to ignore the self-ionization of water is justified. (2) We are also justified in replacing the quantity $(x - 1.6 \times 10^{-5})$ by x because the value $x = 0.46$ is very much larger than 1.6×10^{-5}.

EXERCISE 15.15A

What is the molarity of a $CH_3COONa(aq)$ solution that has a pH = 9.10?

EXERCISE 15.15B

Without doing detailed calculations, determine which of the following solutions should have the higher pH: 0.10 M NH_4NO_2 or 0.10 M NH_4CN. Explain your reasoning.

Problem-Solving Note

Because K_a for the hydrolysis of NH_4^+ derived in Example 15.15 and K_b for the hydrolysis of CH_3COO^- derived in Example 15.14 are equal, we can see why $CH_3COONH_4(aq)$ should be pH-neutral, as predicted in part (b) of Example 15.13 and seen in Figure 15.8.

15.7   The Common Ion Effect

The two solutions pictured in Figure 15.9 both contain acetic acid, CH_3COOH, at the same molarity. Both also contain an indicator known as bromphenol blue. The solution at the right, however, also contains sodium acetate, CH_3COONa, as a second solute. As is evident from their different colors, the two solutions have different pH values. The solution having both acetic acid and sodium acetate as solutes has a pH that is higher by about two units, and $[H_3O^+]$ in this solution is therefore only about one-hundredth of that in the solution containing only acetic acid. There is no mystery here, however. This is a classic example of Le Châtelier's principle.

When sodium acetate is added to an acetic acid solution, the sodium acetate dissociates completely to yield $CH_3COO^-(aq)$ and $Na^+(aq)$. Thus, the concentration of $CH_3COO^-(aq)$, one of the products of the ionization of acetic acid, increases. According to Le Châtelier's principle, if we increase the concentration of one of the products of a reversible reaction, in this case CH_3COO^-, equilibrium shifts to the left:

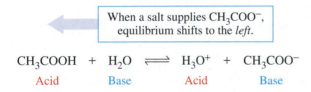

When a salt supplies CH_3COO^-, equilibrium shifts to the *left*.

$$CH_3COOH + H_2O \rightleftharpoons H_3O^+ + CH_3COO^-$$

Acid   Base   Acid   Base

As CH_3COO^-, a base, is consumed in the reverse reaction, so too is H_3O^+, an acid. Therefore $[H_3O^+]$ decreases, and the pH increases accordingly. Because it is found both in aqueous acetic acid and in sodium acetate, the acetate ion in this case is called a *common ion*. The **common ion effect** is the suppression of the ionization of a weak acid or a weak base by the presence of a common ion from a strong electrolyte.

In Example 15.16, we calculate the pH of the acetic acid–sodium acetate solution we have just described. Exercise 15.16A shows the effect of NH_4^+ as a common ion on the ionization of $NH_3(aq)$. In Exercise 15.16B, we explore the common ion effect in a solution containing both a strong acid and a weak acid.

Acid Equilibrium activity

◀ **FIGURE 15.9   The common ion effect**

The solutions contain bromphenol blue indicator. The yellow color in the left beaker indicates a pH < 3.0, and the blue-violet color in the right beaker indicates a pH < 4.6. In the microscopic views, the additional acetate ion on the right (sodium ion has been omitted for clarity) suppresses ionization of the acetic acid, with the result that less H_3O^+ is produced and the pH increases. The microscopic views are merely suggestive. Only a tiny proportion of acetic acid ionizes in either case, and the $[H_3O^+]$ on the right is actually about a hundred times lower than it is on the left.

□ CH_3COO^-
● CH_3COOH
● H_3O^+

Example 15.16

Calculate the pH of an aqueous solution that is both 1.00 M CH_3COOH and 1.00 M CH_3COONa.

STRATEGY

The pH of the solution is independent of how the solution is prepared, as long as it has a nominal concentration of 1.00 M in each solute. That is, we can dissolve sodium acetate in an aqueous solution of acetic acid, or we can dissolve acetic acid in an aqueous solution of sodium acetate. In our calculation, we will begin with 1.00 M $CH_3COOH(aq)$ and assume (hypothetically, of course) that no ionization occurs until we have added enough of the soluble ionic compound CH_3COONa to make the solution also 1.00 M in $CH_3COO^-(aq)$:

$$CH_3COONa(s) \xrightarrow{H_2O} CH_3COO^-(aq) + Na^+(aq)$$

SOLUTION

First, we write an ICE format for the equilibrium established in a solution in which the initial concentration of both CH_3COO^- and CH_3COOH is 1.00 M.

The reaction:	CH_3COOH	+	H_2O	$\rightleftharpoons$	H_3O^+	+	CH_3COO^-
Initial concentrations, M:	1.00				≈ 0		1.00
Changes, M:	$-x$				$+x$		$+x$
Equilibrium concentrations, M:	$(1.00 - x)$				x		$(1.00 + x)$

Next, we write the K_a expression and substitute numerical values from our ICE table and the value of K_a for acetic acid from Table 15.2.

$$K_a = \frac{[H_3O^+][CH_3COO^-]}{[CH_3COOH]} = 1.8 \times 10^{-5}$$

$$\frac{x(1.00 + x)}{1.00 - x} = 1.8 \times 10^{-5}$$

Now let us assume that x is very small and that therefore $(1.00 - x) \approx (1.00 + x) \approx 1.00$.

$$\frac{x(1.00)}{1.00} = 1.8 \times 10^{-5} \qquad x = [H_3O^+] = 1.8 \times 10^{-5} \text{ M}$$

We calculate the pH from $[H_3O^+]$ in the usual way.

$$pH = -\log[H_3O^+] = -\log(1.8 \times 10^{-5}) = 4.74$$

ASSESSMENT

First, note that the assumption that x can be ignored in comparison to 1.00 is valid:

$$(1.00 - 1.8 \times 10^{-5}) = 1.00 \quad \text{and} \quad (1.00 + 1.8 \times 10^{-5}) = 1.00.$$

Also noteworthy is a comparison of the calculated pH (4.74) with that found in Example 15.6, where there was no common ion (pH 2.38). This shows just how effective the presence of acetate ion is in suppressing the ionization of acetic acid.

EXERCISE 15.16A

Calculate the pH of an aqueous solution that is 0.15 M NH_3 and 0.35 M NH_4NO_3.

$$NH_3 + H_2O \rightleftharpoons NH_4^+ + OH^- \qquad K_b = 1.8 \times 10^{-5}$$

EXERCISE 15.16B

What is the common ion in an aqueous solution that is 0.10 M in HCl and 0.10 M in CH_3COOH? What is $[H_3O^+]$ in this solution? What is $[CH_3COO^-]$?

15.8 Buffer Solutions

The acetic acid–sodium acetate solution of Example 15.16 is a **buffer solution,** a solution that changes pH *only slightly* when small amounts of a strong acid or a strong base are added.

Buffer solutions have many important applications in industry, in the laboratory, and in living organisms. Some chemical reactions consume acids, others produce acids, and many are catalyzed by H_3O^+. If we want to study the kinetics of these reactions or simply to control their reaction rates, we often need to control the pH. We can

Water

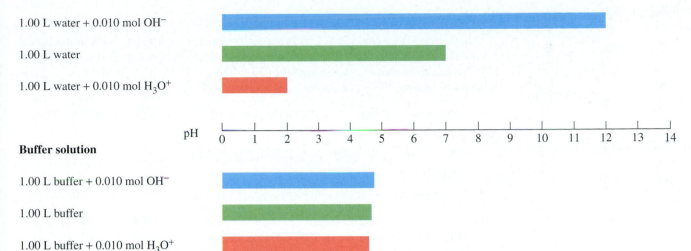

Buffer solution

▲ **FIGURE 15.10** **Depicting buffer action**

The pH of pure water undergoes very large changes in pH when a small amount of either an acid or a base is added. Water has no buffering ability. In contrast, the corresponding pH changes in a buffer solution that is 1.00 M in CH_3COOH and 1.00 M in CH_3COONa are almost imperceptible.

QUESTION: What are the chemical equations that describe the addition of (a) 0.010 mol OH^- and (b) 0.010 mol H_3O^+ to the buffer solution?

minimize pH changes by conducting the reactions in buffered solutions. Enzyme-catalyzed reactions are particularly sensitive to pH changes. Studies that involve proteins are usually performed in buffered media because the magnitude and sign of the electric charge carried by the protein molecules depend on the pH.

Figure 15.10 shows that pure water does not buffer at all. Even tiny amounts of acid or base produce large changes in the pH of water. To be a buffer, a solution usually must contain

- a *weak acid and its salt (conjugate base)* or
- a *weak base and its salt (conjugate acid)*.

Buffer Solutions activity

The acid component of the buffer can neutralize small added amounts of OH^-, and the basic component can neutralize small added amounts of H_3O^+.

How a Buffer Solution Works

Let us look more closely at the buffer solution of Example 15.16, which is 1.00 M in both CH_3COOH and CH_3COONa. For this solution, we can write

$$CH_3COOH + H_2O \rightleftharpoons H_3O^+ + CH_3COO^-$$

$$K_a = \frac{[H_3O^+][CH_3COO^-]}{[CH_3COOH]} = 1.8 \times 10^{-5}.$$

At equilibrium, $[H_3O^+]$ is given by the expression

$$[H_3O^+] = K_a \times \frac{[CH_3COOH]}{[CH_3COO^-]}$$

$$= 1.8 \times 10^{-5} \times \frac{[CH_3COOH]}{[CH_3COO^-]}$$

Now let us substitute the equilibrium concentrations from Example 15.16 into this expression, including the simplifying assumption that $x \ll 1.00$:

$$[H_3O^+] = K_a \times \frac{[CH_3COOH]}{[CH_3COO^-]}$$

$$= 1.8 \times 10^{-5} \times \frac{(1.00 - x)}{(1.00 + x)} = 1.8 \times 10^{-5} \times \frac{1.00}{1.00} = 1.8 \times 10^{-5} \text{ M}$$

From this value of $[H_3O^+]$, we can get the pH of the buffer solution:

$$pH = -\log[H_3O^+] = -\log(1.8 \times 10^{-5}) = 4.74$$

Now, suppose we add to the buffer solution enough of a strong base to neutralize 2% of the acetic acid. This will reduce the concentration of the acid from 1.00 M to 0.98 M and raise that of acetate ion from 1.00 M to 1.02 M. A neutralization reaction in a buffer solution always converts some of one buffer component to the other.

The buffer reaction:	CH_3COOH	$+$	OH^-	$\longrightarrow$ H_2O $+$	CH_3COO^-
Initial buffer:	1.00 M		≈ 0 M		1.00 M
Add:			0.02 M		
Changes:	-0.02 M		-0.02 M		$+0.02$ M
After neutralization:	0.98 M		≈ 0 M		1.02 M

Now let us calculate $[H_3O^+]$ and pH in a solution that is 0.98 M CH_3COOH and 1.02 M CH_3COONa. We can do this in the same way that we determined $[H_3O^+]$ in the solution of Example 15.16, which was 1.00 M in both CH_3COOH and CH_3COONa. However, let us simplify our calculation by anticipating that $[H_3O^+]$ (represented by x) is very much smaller than 0.98 M, the CH_3COOH concentration, and also very much smaller than 1.02 M, the CH_3COO^- concentration. With this simplification, our expression for $[H_3O^+]$ becomes

$$[H_3O^+] = K_a \times \frac{[CH_3COOH]}{[CH_3COO^-]}$$

$$= 1.8 \times 10^{-5} \times \frac{(0.98 - x)}{(1.02 + x)} = 1.8 \times 10^{-5} \times \frac{0.98}{1.02} = 1.7 \times 10^{-5} \text{ M}$$

$$pH = -\log[H_3O^+] = -\log(1.7 \times 10^{-5}) = 4.77$$

Note that the $[H_3O^+]$ and pH of the buffer solution have hardly been affected: 1.8×10^{-5} M and 4.74 before addition of the base and 1.7×10^{-5} M and 4.77 after the base was added.

By comparison, if 0.020 mole of a strong base is added to 1 L of pure water, the pH increases from 7.00 to 12.30.

When we add an acid to a buffer solution, the situation is quite similar. If 2% of the conjugate base (acetate ion) is neutralized by added H_3O^+, we have

The buffer reaction:	CH_3COO^-	$+$	H_3O^+	$\longrightarrow$ H_2O $+$	CH_3COOH
Initial buffer:	1.00 M		≈ 0 M		1.00 M
Add:			0.02 M		
Changes:	-0.02 M		-0.02 M		$+0.02$ M
After neutralization:	0.98 M		≈ 0 M		1.02 M

To calculate $[H_3O^+]$ and pH in a solution that is 1.02 M CH_3COOH and 0.98 M CH_3COONa, we can proceed as before:

$$[H_3O^+] = K_a \times \frac{[CH_3COOH]}{[CH_3COO^-]}$$

$$= 1.8 \times 10^{-5} \times \frac{(1.02 - x)}{(0.98 + x)} = 1.8 \times 10^{-5} \times \frac{1.02}{0.98} = 1.9 \times 10^{-5} \text{ M}$$

$$pH = -\log[H_3O^+] = -\log(1.9 \times 10^{-5}) = 4.72$$

Again, the $[H_3O^+]$ and pH are hardly affected.

We have shown that a buffer solution resists a change in pH when either an acid or a base is added to the solution. However, we should note that when an acid is added, the pH is lowered very slightly, and when a base is added, the pH is raised a tiny bit. We can use this fact as a helpful check on buffer calculations.

An Equation for Buffer Solutions

In certain applications, we need to calculate the pH of a buffer solution many times. We can do this with a single, simple equation, *but we should be aware of its limitations*.

To establish such an equation, we begin with the equation for $[H_3O^+]$ in a buffer solution containing acetic acid and sodium acetate—an equation we have used before:

$$[H_3O^+] = K_a \times \frac{[CH_3COOH]}{[CH_3COO^-]}$$

Now we take the negative logarithm of each side of the equation:

$$-\log[H_3O^+] = -\log K_a - \log \frac{[CH_3COOH]}{[CH_3COO^-]}$$

In the equation above, we substitute pH for $-\log[H_3O^+]$, pK_a for $-\log K_a$ and replace $-\log([CH_3COOH]/[CH_3COO^-])$ by $+\log([CH_3COO^-]/[CH_3COOH])$ to obtain:

$$pH = pK_a + \log \frac{[CH_3COO^-]}{[CH_3COOH]}$$

This equation for the pH of an acetic acid–acetate ion buffer is a special case of a more general equation known as the **Henderson–Hasselbalch equation**:

$$pH = pK_a + \log \frac{[\text{conjugate base}]}{[\text{weak acid}]} \qquad \textbf{(15.10)}$$

The Henderson–Hasselbalch equation has an important limitation. It is useful only if we can substitute the measured or nominal molarities for equilibrium molarities of weak acids and conjugate bases. When we use the equation, we treat weak acids and conjugate bases *as if* their molarities were unchanged by the ionization equilibrium. This is the same assumption that we discussed on page 663. It results in substitutions of this sort: M_{acid} for $(M_{acid} - x)$, M_{base} for $(M_{base} + x)$, and so on.

The net effect of the Henderson–Hasselbalch equation limitation is that the equation works best for buffer solutions that fit the following criteria:

- The ratio [conjugate base]/[weak acid] has a value *between 0.10 and 10.*
- Both [conjugate base] and [weak acid] exceed K_a by a factor of *100 or more.*

Problem-Solving Note

The net effect of the manipulations with the log terms is to invert the original ratio and change the sign of its log from negative to positive.

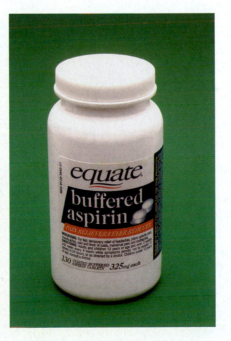

◀ Buffers see widespread use in industry and medicine. Some types of industrial adhesives used for plywood are buffered to control hardening time and aging characteristics. In some individuals, buffered aspirin is less likely to produce stomach upset than is regular aspirin.

Example 15.17

A buffer solution is 0.24 M NH_3 and 0.20 M NH_4Cl. **(a)** What is the pH of this buffer? **(b)** If 0.0050 mol NaOH is added to 0.500 L of this solution, what will be the pH?

STRATEGY

To determine the pH in part (a), we will use the same method that we used in introducing the common ion effect. We could use the Henderson–Hasselbalch equation instead, and that is what we will do in part (b). Also, in part (b) we need to do a preliminary stoichiometry calculation, for which we will find it convenient to use the ICE format.

SOLUTION

(a) We begin with the usual ICE format describing the equilibrium.

The reaction:	NH_3	$+ H_2O \rightleftharpoons$	NH_4^+	$+$	OH^-
Initial concentrations, M:	0.24		0.20		≈ 0
Changes, M:	$-x$		$+x$		$+x$
Equilibrium concentrations, M:	$(0.24 - x)$		$(0.20 + x)$		x

In substituting equilibrium concentrations into the K_b expression, we assume that $x \ll 0.20$. This allows us to substitute 0.20 for $(0.20 + x)$ and 0.24 for $(0.24 - x)$.

$$K_b = \frac{[NH_4^+][OH^-]}{[NH_3]}$$

$$1.8 \times 10^{-5} = \frac{(0.20 + x)x}{(0.24 - x)}$$

$$1.8 \times 10^{-5} = \frac{0.20x}{0.24}$$

Now we solve for x, which is equal to $[OH^-]$.

$$0.20x = (1.8 \times 10^{-5}) \times 0.24$$

$$x = [OH^-] = 1.8 \times 10^{-5} \times \frac{0.24}{0.20} = 2.2 \times 10^{-5} \text{ M}$$

We can use the $[OH^-]$ to calculate the pOH and the pH of the solution.

$$pOH = -\log[OH^-] = -\log(2.2 \times 10^{-5}) = 4.66$$

$$pH = 14.00 - pOH = 14.00 - 4.66 = 9.34$$

(b) First we need to determine how adding the NaOH affects the buffer solution. We begin by calculating the number of moles of NH_3 and NH_4^+ in the original buffer solution.

$$0.24 \text{ M} \times 0.500 \text{ L} = 0.12 \text{ mol of } NH_3$$

$$0.20 \text{ M} \times 0.500 \text{ L} = 0.10 \text{ mol of } NH_4^+$$

The 0.005 mol OH^- added will react with the acidic NH_4^+.

The buffer reaction:	NH_4^+	$+$	OH^-	$\rightleftharpoons H_2O$	$+$	NH_3
Initial buffer:	0.10 mol		≈ 0 mol			0.12 mol
Add:			0.005 mol			
Changes:	-0.005 mol		-0.005 mol			$+0.005$ mol
After neutralization:	0.095 mol		≈ 0 mol			0.125 mol
M (divide by 0.500 L):	0.19 M		≈ 0 M			0.25 M

We now use Equation (15.10) to calculate the pH.

$$pH = pK_a + \log \frac{[\text{conjugate base}]}{[\text{weak acid}]}$$

The weak acid in this buffer is NH_4^+ derived from the salt NH_4Cl. The conjugate base is NH_3. We need the pK_a value for NH_4^+. We can get this from the K_b value of NH_3.

$$pK_a + pK_b = 14.00$$

$$pK_a = 14.00 - pK_b = 14.00 - [-\log(1.8 \times 10^{-5})]$$

$$= 14.00 - 4.74 = 9.26$$

The concentrations are 0.25 M for $[NH_3]$, the conjugate base, and 0.19 M for NH_4^+, the weak acid.

$$pH = 9.26 + \log \frac{0.25}{0.19} = 9.26 + 0.12 = 9.38$$

ASSESSMENT

As expected, our simplifying assumption in part (a) was valid: $x = 2.2 \times 10^{-5}$ M is very much smaller than 0.20 M and 0.24 M. Note that equilibrium calculations can often be done most easily on a molarity basis, as in part (a), but stoichiometric calculations are generally easier to follow on a mole basis, as in part (b). Finally, consider the relatively small increase in pH, from 9.34 in part (a) to 9.38 in part (b), that followed the addition of a rather substantial amount of OH^- to the buffer solution. The 0.24 M NH_3/0.20 M NH_4Cl solution is strongly buffered.

EXERCISE 15.17A

What is the final pH if 0.03 mol HCl is added to 0.500 L of a buffer solution that is 0.24 M NH_3 and 0.20 M NH_4Cl?

EXERCISE 15.17B

In Example 15.17, we saw that the addition of 0.0050 mol NaOH to 0.500 L of a solution that is 0.24 M NH_3 and 0.20 M NH_4Cl raised the pH from 9.34 to 9.38. How many moles of NaOH should have been added to raise the pH to 9.50? (*Hint:* What must be the value of the ratio $[NH_3]/[NH_4^+]$?)

Buffers in Blood

Buffers are of utmost importance in our blood and other body fluids. Cells in living organisms must maintain a proper pH in order to carry out essential life processes, primarily because enzyme function (Section 13.11) is sharply dependent on pH. The normal pH value of blood plasma is 7.4. Sustained variations of a few tenths of a pH unit can cause severe illness or death.

Acidosis, a condition in which the pH of blood decreases, can be brought on by heart failure, kidney failure, diabetes mellitus, persistent diarrhea, a long-term high-protein diet, or other factors. Prolonged, intense exercise can cause temporary acidosis.

Alkalosis, characterized by an increase in the pH of blood, may result from severe vomiting, hyperventilation (excessive breathing, sometimes caused by anxiety or hysteria), or exposure to high altitudes (altitude sickness). Arterial blood samples taken from climbers who reached the summit of Mt. Everest (8848 m) without supplemental oxygen had pH values between 7.7 and 7.8. This alkalosis was caused by hyperventilation. To compensate for the very low partial pressures of O_2 at this altitude (about 43 mmHg versus about 159 mmHg at sea level), the climbers had to breathe extremely rapidly.

A good measure of the buffer capacity of human blood is that the addition of 0.01 mol HCl to 1 L of blood lowers the pH only from 7.4 to 7.2. In contrast, the same amount of HCl added to a saline (NaCl) solution isotonic with blood lowers the pH from 7.0 to 2.0. The saline solution has no buffer capacity.

There are several buffers in the blood. Perhaps the most important is the one made of bicarbonate ions, HCO_3^-, and carbonic acid, H_2CO_3. In the discussion of this buffer system, we treat the H_2CO_3 as originating from the complete conversion of carbon dioxide gas created in the body as a by-product of metabolic reactions. We consider only the first ionization of the diprotic acid H_2CO_3:

(1) $$H_2CO_3 + H_2O \rightleftharpoons H_3O^+ + HCO_3^-$$

The blood buffers must be able to neutralize excess acid, such as the lactic acid produced by exercise. A relatively high concentration of HCO_3^- helps in this regard because this ion reacts with excess acid to reverse the ionization reaction shown in Equation (1).

Excess alkalinity is much less common than excess acidity. Should it occur, however, additional H_2CO_3 can be formed in the lungs by reabsorbing $CO_2(g)$ to build up the H_2CO_3 content of the blood:

$$CO_2(g) + H_2O(l) \rightleftharpoons H_2CO_3(aq)$$

The H_2CO_3 then ionizes as needed to neutralize the excess base.

Other blood buffers include the dihydrogen phosphate ($H_2PO_4^-$)/hydrogen phosphate (HPO_4^{2-}) system:

(2) $$H_2PO_4^- + H_2O \rightleftharpoons H_3O^+ + HPO_4^{2-}$$

The HPO_4^{2-} reacts with excess acid in the reverse of reaction (2). Excess alkali entering the blood can be neutralized in part by reaction with $H_2PO_4^-$.

Some plasma proteins also act as buffers. The COO^- groups on the protein molecules can react with excess acid:

$$Protein\text{---}COO^- + H_3O^+ \longrightarrow Protein\text{---}COOH + H_2O$$

The NH_3^+ groups on the protein molecules can neutralize excess base:

$$Protein\text{---}NH_3^+ + OH^- \longrightarrow Protein\text{---}NH_2 + H_2O$$

The blood buffers do a remarkable job of maintaining the pH of blood at 7.4. They can be overwhelmed, however, if the body's metabolism goes badly amiss.

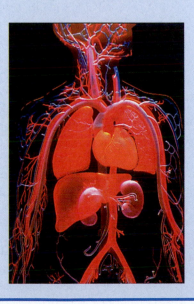

◀ Blood is highly buffered. It is carried in the human circulatory system at a pH of 7.4.

Quite often in the laboratory, we need to prepare a buffer solution of a particular pH. In Example 15.18, we use the Henderson–Hasselbalch equation to do the necessary calculation, although we could use the unmodified equilibrium constant expression instead.

Example 15.18

What concentration of acetate ion in 0.500 M $CH_3COOH(aq)$ produces a buffer solution with pH = 5.00?

SOLUTION

We begin with the Henderson–Hasselbalch equation for an acetic acid–acetate ion buffer solution:

$$pH = pK_a + \log \frac{[CH_3COO^-]}{[CH_3COOH]}$$

We can replace pH by 5.00, pK_a by $-\log K_a$ and $[CH_3COOH]$ by 0.500 and then solve for $[CH_3COO^-]$:

$$5.00 = -\log(1.8 \times 10^{-5}) + \log \frac{[CH_3COO^-]}{0.500}$$

$$5.00 = 4.74 + \log \frac{[CH_3COO^-]}{0.500}$$

$$\log \frac{[CH_3COO^-]}{0.500} = 5.00 - 4.74 = 0.26$$

$$\frac{[CH_3COO^-]}{0.500} = 10^{0.26} = 1.82$$

$$[CH_3COO^-] = 0.500 \times 1.82 = 0.910 \text{ M}$$

ASSESSMENT

A simple check on our answer is to calculate the pH of a solution that is 0.500 M $CH_3COOH(aq)$ and 0.910 M $CH_3COO^-(aq)$. The result we obtain, of course, should be pH = 5.00.

EXERCISE 15.18A

What concentration of acetic acid in 0.250 M $CH_3COONa(aq)$ is needed to produce a buffer solution with pH = 4.50?

EXERCISE 15.18B

What mass of NH_4Cl must be present in 0.250 L of 0.150 M $NH_3(aq)$ to produce a buffer solution with pH = 9.05?

Buffer Capacity and Buffer Range

In Example 15.17, we examined the buffering ability of a solution of 0.24 M NH_3 and 0.20 M NH_4Cl. We saw that when 0.0050 mol OH^- was added to 0.500 L of this solution, the pH increased only from 9.34 to 9.38. If we had considered the addition of 0.050 mol OH^- (10 times as much), we would have found the buffer less effective; the pH would have increased to 9.79.

There is a limit to the capacity of a buffer solution to neutralize added acid or base, and this limit is reached before all of one of the buffer components has been consumed. In the case of the 0.24 M NH_3/0.20 M NH_4Cl buffer, the limits are just less than 0.20 mol OH^-/L (which would neutralize all the NH_4^+) and just less than 0.24 mol H_3O^+/L (which would neutralize all the NH_3).

In general, the more concentrated the buffer components in a solution, the more added acid or base the solution can neutralize. And, as a rule, a buffer is most effective if the concentrations of the buffer acid and its conjugate base are equal to each other. In the Henderson–Hasselbalch equation, the equal concentrations cancel, and because

Notice that the conditions that govern a high buffer capacity are the same as those required in the Henderson–Hasselbalch equation.

$\log 1 = 0$, we then have $pH = pK_a$. For an acetic acid–sodium acetate buffer, for example, we get $pH = pK_a = 4.74$:

$$pH = pK_a + \log \frac{[CH_3COO^-]}{[CH_3COOH]}$$

$$pH = 4.74 + \log 1$$

$$pH = 4.74$$

In general, the pH range over which a buffer solution is most effective is about one pH unit on either side of $pH = pK_a$. This corresponds to the ratios

$$0.10 < \frac{[conjugate\ base]}{[weak\ acid]} < 10.0$$

$$-1.0 < \log \frac{[conjugate\ base]}{[weak\ acid]} < 1.0$$

Thus, for an acetic acid–acetate ion buffer, the effective pH range is 3.74–5.74. For an NH_3/NH_4^+ buffer, where pK_a for NH_4^+ is 9.26, the effective pH range is 8.26–10.26. Note that we use pK_a for NH_4^+ and not pK_b for NH_3.

15.9   Acid–Base Indicators

An **acid–base indicator** is a combination of a weak acid having one color and the conjugate base of the acid having a different color. (One of the colors may be "colorless.") The color of a solution containing the indicator depends on the relative proportions of the weak acid and conjugate base; in turn, these proportions depend on the pH.

In Figure 15.11, when a few drops of phenol red indicator are added to pure water, the color is orange. A few drops of lemon juice—an acid—change the color to yellow, whereas a few drops of household ammonia—a base—change the color to red. To explain these color changes, let us call the yellow form of phenol red the indicator acid, HIn, and the red form the indicator base, In^-. The equilibrium between the two is

$$HIn + H_2O \rightleftharpoons H_3O^+ + In^-$$

Yellow   Red

In an acidic solution, $[H_3O^+]$ is high. In acidic solutions containing the phenol red indicator, the H_3O^+ originating from the acid is common to the indicator ionization equilibrium and therefore suppresses the ionization of the indicator acid. In accordance with Le Châtelier's principle, equilibrium shifts to the left, favoring the yellow form of the indicator. When phenol red indicator is added to a basic solution, such as $NH_3(aq)$, H_3O^+ from the indicator acid is neutralized. Equilibrium shifts to the right, favoring the red indicator anion. In a solution that is close to neutral, the indicator acid

Natural Indicators movie

◀ **FIGURE 15.11   Phenol red—A pH indicator**

In an acidic solution, phenol red has a yellow color (left). In pure water it is orange (center), and in an alkaline solution it is red (right).

QUESTION: If bromthymol blue (page 654) were used as the indicator, what colors would appear in the three beakers?

▶ **FIGURE 15.12** **Litmus: A general-purpose indicator**
We often use litmus paper as a rough measure to determine whether a material is acidic or basic. The soil sample on the left turns red litmus blue; it is basic. The soil sample on the right turns blue litmus red; it is acidic.

and anion are present in about equal concentrations, and the observed color is orange—a mixture of red and yellow.

Phenol red indicator:

pH < 6.6	6.6 < pH < 8.2	pH > 8.2
Yellow	Orange	Red

One of the most commonly used acid–base indicators in the introductory chemistry laboratory is litmus, a natural dye extracted from lichens. It is often used in the form of paper strips that have been impregnated with an aqueous solution of litmus and then allowed to dry. The litmus paper is moistened with the solution being tested. The color change of litmus occurs over a much broader pH range than of other indicators, and it is used only to give a general indication of whether a solution is acidic or basic (Figure 15.12).

Litmus indicator:

pH < 4.5	pH > 8.3
Red	Blue

Figure 15.13 shows the pH range of color change for numerous other common acid–base indicators. To either side of the illustrated pH range, the indicators will exhibit the color shown to that side.

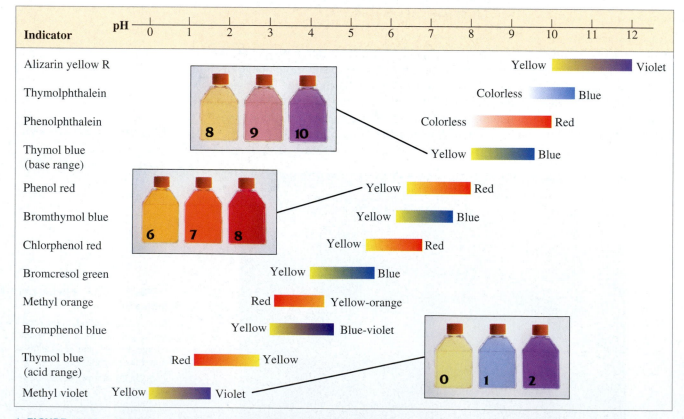

▲ **FIGURE 15.13** **pH ranges and colors of common acid–base indicators**

Acid–base indicators are often used for applications in which a precise pH reading is not necessary. A familiar use is in acid–base titrations, as we will describe in the next section.

Example 15.19   A Conceptual Example

Explain the series of color changes of thymol blue indicator produced by the actions pictured in Figure 15.14:

(a) A few drops of thymol blue are added to HCl(aq).

(b) Some sodium acetate is added to solution (a).

(c) A small quantity of sodium hydroxide is added to solution (b).

(d) An additional, larger quantity of sodium hydroxide is added to solution (c).

(a) (b) (c) (d)

▲ FIGURE 15.14   Example 15.19 illustrated

ANALYSIS AND CONCLUSIONS

(a) The red color of the indicator shows the solution to be strongly acidic: pH < 1.2 (see Figure 15.13). The HCl is completely ionized:

$$HCl + H_2O \longrightarrow H_3O^+ + Cl^-$$

(b) The yellow color shows that the pH has risen to a value greater than 3.0. This corresponds to $[H_3O^+] < 1 \times 10^{-3}$ M. A reaction between excess acetate ion and H_3O^+ (limiting reactant) goes nearly to completion:

From HCl   From CH₃COONa

$$H_3O^+ + CH_3COO^- \longrightarrow CH_3COOH + H_2O$$

The resulting solution is one of a weak acid, CH_3COOH, and its conjugate base, CH_3COO^-. It is a buffer solution with a pH of about pH = pK_a = 4.74.

(c) The small quantity of added NaOH is neutralized by the weak acid:

$$CH_3COOH + OH^- \longrightarrow H_2O + CH_3COO^-$$

This produces only a small change in the ratio $[CH_3COO^-]/[CH_3COOH]$ and correspondingly small changes in $[H_3O^+]$ and pH. The solution color remains yellow.

(d) The quantity of NaOH added is enough to neutralize all the CH_3COOH in the buffer. The buffer capacity has been exceeded, and the buffer action is destroyed. Unreacted OH^- raises the pH to a value above 9.6, at which point thymol blue indicator is blue:

Excess

$$CH_3COOH + OH^- \longrightarrow H_2O + CH_3COO^-$$

EXERCISE 15.19A

An aqueous solution is known to be one of the following four: **(a)** 1.00 M NH_4Cl, **(b)** 1.00 M NH_4Cl/1.00 M NH_3, **(c)** 1.00 M HCl/1.00 M HNO_3, **(d)** 1.00 M CH_3COOH/1.00 M CH_3COONa. A few drops of bromcresol green indicator added to the solution produce a green color (see Figure 15.13). Which solution(s) does this observation eliminate? What additional simple test involving a common laboratory acid and/or base could you perform to determine which of the four it is?

EXERCISE 15.19B

Suppose the solution in Figure 15.14b is 100.0 mL of a buffer in which $[CH_3COOH]$ = $[CH_3COO^-]$ = 0.200 M. Approximately how many moles of NaOH, at a minimum, must be added to get the color to change to the blue of Figure 15.14d? (*Hint:* Refer to Figure 15.13.)

15.10 Neutralization Reactions and Titration Curves

Application Note

Medical doctors sometimes "titrate" a patient's disease symptoms. Patients taking prednisone, a corticosteroid used to treat skeletomuscular pain, may be told to increase the dosage very slowly until the symptoms disappear. Fortunately, endpoints in an acid–base titration are much easier to detect than those in a medical titration.

In Section 4.6, we noted that *neutralization* is the reaction of an acid and a base and that *titration* is a common technique for conducting a neutralization. At the *equivalence point* in a titration, the acid and base have been brought together in exact stoichiometric proportions.

We can use an acid–base indicator to locate the equivalence point. As noted in Section 4.6, the point in the titration at which the indicator changes color is called the *endpoint* of the titration, and the trick is to match the indicator endpoint and the equivalence point of the neutralization. Specifically, we need an indicator whose color change occurs over a pH range that includes the pH at the equivalence point.

We can best match the indicator endpoint and the equivalence point for a neutralization reaction by plotting a **titration curve,** a graph of pH versus volume of titrant added. We can measure the pH with a pH meter, and a recorder connected to the meter can automatically plot the titration curve. In this section, we will calculate the pH expected at some characteristic points on two types of titration curves and use these calculations to review some of the acid–base equilibria presented earlier in this chapter.

🌀 **Acid-Base Titration animation**

In a typical titration, we use less than 50 mL of titrant that is 1 M or less. The amount of H_3O^+ or OH^- delivered from the buret is generally only a few thousandths of a mole—for example, 2.00×10^{-3} mol. Therefore, we can avoid most of the exponential terms in calculations if we work with the unit *millimole* (mmol) rather than mole. Because 1 mmol is 0.001 mole, we can get millimoles by dividing moles by 1000. We also get milliliters by dividing liters by 1000, and as a result, we can redefine molarity in this way:

$$M = \frac{mol}{L} = \frac{mol/1000}{L/1000} = \frac{mmol}{mL}$$

Thus, a solution that is 0.500 M HCl has 0.500 mol HCl per liter of solution or 0.500 mmol HCl per milliliter of solution.

Titration of a Strong Acid with a Strong Base

Suppose we place 20.00 mL of 0.500 M HCl, a strong acid, in a small flask and slowly add to it 0.500 M NaOH, a strong base. To establish data for a titration curve, we can calculate the pH of the accumulated solution at different points in the titration. Then we can plot these pH values versus the volume of NaOH(aq) added. From the titration curve, we can establish the pH at the equivalence point and identify appropriate indicators for the titration.

The photographs in Figure 4.16 illustrate the technique of acid–base titration.

In Example 15.20, we calculate four representative points on the titration curve. In Exercise 15.20A, we focus on the region near the equivalence point to illustrate the very rapid rise in pH as titrant is added.

Example 15.20

Calculate the pH at the following points in the titration of 20.00 mL of 0.500 M HCl with 0.500 M NaOH:

$$H_3O^+ + Cl^- + Na^+ + OH^- \longrightarrow Na^+ + Cl^- + 2\,H_2O$$

(a) before the addition of any NaOH

(b) after the addition of 10.00 mL of 0.500 M NaOH

(c) after the addition of 20.00 mL of 0.500 M NaOH

(d) after the addition of 20.20 mL of 0.500 M NaOH

STRATEGY

(a) Before any NaOH(aq) has been added, the solution is 0.500 M HCl(aq). We calculate its pH in the usual way. This is the *initial* point of the titration.

(b) When 10.00 mL of 0.500 M NaOH has been added, half of the HCl has been neutralized. We use a stoichiometric calculation in a format resembling the ICE format for equilibrium calculations to determine the remaining amount of H_3O^+, and from $[H_3O^+]$ we calculate pH. This is the *half-neutralization* point of the titration.

(c) Complete neutralization occurs here because the HCl and NaOH have been brought together in their stoichiometric proportions. The pH is that of the NaCl(aq) produced. This is the *equivalence point* of the titration.

(d) Beyond the equivalence point, OH^- is in excess and the solution is basic; but the calculations are similar to part (b) except that now we convert, consecutively, from $[OH^-]$ to pOH to pH.

SOLUTION

(a) The pH of 0.500 M HCl, a strong acid solution, is

$$pH = -\log[H_3O^+] = -\log(0.500) = 0.301$$

(b) The total amount of H_3O^+ to be titrated is

$$20.00\ mL \times 0.500\ mmol\ H_3O^+/mL = 10.0\ mmol\ H_3O^+$$

The amount of OH^- in 10.00 mL of 0.500 M NaOH is

$$10.00\ mL \times 0.500\ mmol\ OH^-/mL = 5.00\ mmol\ OH^-$$

We can represent the progress of the neutralization reaction as follows:

The reaction:	H_3O^+	$+ \quad OH^-$	$\longrightarrow 2 H_2O$
Initial amounts, mmol:	10.0	≈ 0	
Add, mmol:		5.00	
Changes, mmol:	-5.00	-5.00	
After reaction, mmol:	5.0	≈ 0	

The total volume is 20.00 mL + 10.00 mL = 30.00 mL, and

$$[H_3O^+] = \frac{5.0 \text{ mmol } H_3O^+}{30.00 \text{ mL}} = 0.17 \text{ M}$$

$$pH = -\log[H_3O^+] = -\log 0.17 = 0.77$$

(c) Because neither Na^+ nor Cl^- hydrolyzes, at the equivalence point the pH of the solution is 7.00.

(d) The amount of OH^- in 20.20 mL of 0.500 M NaOH is

$$20.20 \text{ mL} \times \frac{0.500 \text{ mmol } OH^-}{1 \text{ mL}} = 10.1 \text{ mmol } OH^-$$

Again, we can represent the progress of the neutralization as follows:

The reaction:	H_3O^+	$+ \quad OH^-$	$\longrightarrow 2 H_2O$
Initial amounts, mmol:	10.0	≈ 0	
Add, mmol:		10.1	
Changes, mmol:	-10.0	-10.0	
After reaction, mmol:	≈ 0	0.1	

We can calculate $[OH^-]$, pOH, and pH as follows:

$$[OH^-] = \frac{0.1 \text{ mmol } OH^-}{(20.00 + 20.20) \text{ mL}} = 0.002 \text{ M}$$

$$pOH = -\log[OH^-] = -\log 0.002 = 2.7$$

$$pH = 14.00 - pOH = 14.00 - 2.7 = 11.3$$

EXERCISE 15.20A

For the titration described in Example 15.20, determine the pH after the addition of the following volumes of 0.500 M NaOH: **(a)** 19.90 mL **(b)** 19.99 mL **(c)** 20.01 mL **(d)** 20.10 mL.

EXERCISE 15.20B

In the titration of 25.00 mL of 0.220 M NaOH, what is the pH of the solution after 8.10 mL of 0.252 M HCl has been added?

Figure 15.15 illustrates the following features of the titration curve for a strong acid titrated with a strong base:

- The pH is low at the beginning of the titration.
- The pH changes slowly until just before the equivalence point.
- Just before the equivalence point, the pH rises sharply.
- At the equivalence point, the pH is 7.00.
- Just past the equivalence point, the pH continues its sharp rise.
- Further beyond the equivalence point, the pH continues to increase, but much more slowly.
- Any indicator whose color changes in the pH range from about 4 to 10 can be used when a strong acid is being titrated with a strong base.

pH during Titration activity

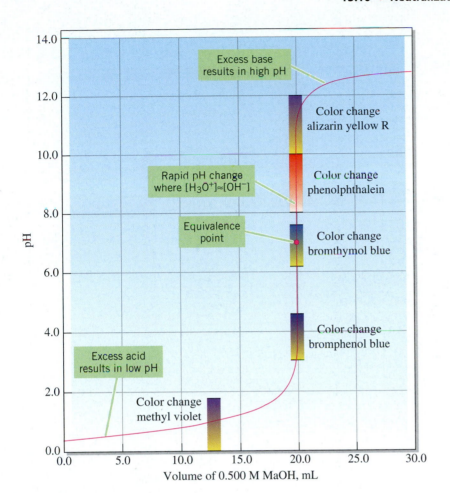

◀ **FIGURE 15.15** **Titration curve for a strong acid titrated with a strong base: 20.00 mL of 0.500 M HCl titrated with 0.500 M NaOH**

Any indicator that changes color along the steep portion of the titration curve is suitable for the titration. Methyl violet changes color too soon, and alizarin yellow R too late.

Titration of a Weak Acid with a Strong Base

If we use the same strong base to titrate two different solutions of equal molarity—one a strong acid and the other a weak acid—we get titration curves with two features in common: (1) The same volume of base is required to reach the equivalence point in both cases, and (2) the portions of the curves after the equivalence points are substantially the same. However, there are some important differences between the two titration curves prior to the equivalence point, as we suggest in Figure 15.16.

In contrast to the curve for the titration of a strong acid with a strong base, in the titration of a weak acid with a strong base, we find these features in the titration curve:

- The initial pH is higher because the weak acid is only partially ionized.
- At the half-neutralization point, pH = pK_a. The solution at this point is a buffer solution in which the concentrations of the weak acid and its conjugate base (the anion) are equal.
- The pH is greater than 7 at the equivalence point because the anion of the weak acid hydrolyzes.
- The steep portion of the curve just prior to and just beyond the equivalence point is confined to a smaller pH range.
- An indicator must be chosen whose color change brackets the pH at the equivalence point, which occurs in a basic solution. Generally, the midpoint of the pH range in which the indicator changes color must be well above pH 7.

The calculations in Example 15.21 should clarify some of these points. As suggested by Figure 15.17, the calculations required for the titration of a weak acid with a strong base include most of the types of calculations considered throughout the chapter.

► **FIGURE 15.16** **Titration curve for a weak acid titrated with a strong base: 20.00 mL of 0.500 M CH$_3$COOH titrated with 0.500 M NaOH**

Phenolphthalein can be used as an indicator for this titration; bromphenol blue cannot. When exactly half the acid is neutralized, $[CH_3COOH] = [CH_3COO^-]$ and $pH = pK_a = 4.74$.

QUESTION: Why does the equivalence point of this titration occur at a pH greater than 7?

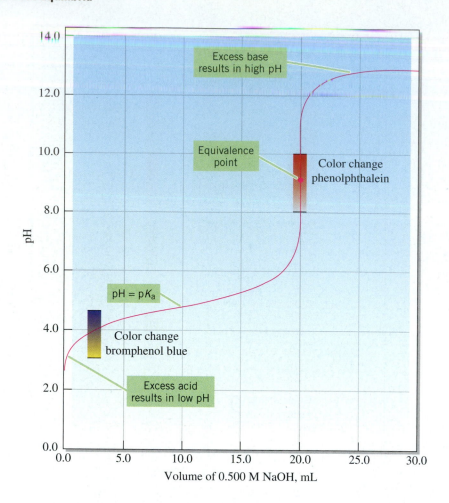

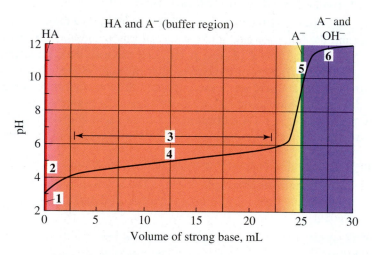

▲ **FIGURE 15.17** **Summarizing the types of equilibrium calculations needed to create a titration curve for a weak acid titrated with a strong base**

The six types of calculations needed to establish the graph are

1. Initial pH: ionization of a weak acid.
2. Early stages of titration: ionization of weak acid suppressed by common ion.
3. In pH range, $pH = pK_a \pm 1$: buffer solutions (Henderson–Hasselbalch equation).
4. At half-neutralization point: $pH = pK_a$.
5. At equivalence point: hydrolysis of an anion (conjugate base of the weak acid).
6. Beyond equivalence point: strong base in aqueous solution.

Example 15.21

Calculate the pH at the following points in the titration of 20.00 mL of 0.500 M CH_3COOH with 0.500 M NaOH:

$$CH_3COOH + Na^+ + OH^- \longrightarrow Na^+ + CH_3COO^- + H_2O$$

(a) before the addition of any NaOH

(b) after the addition of 8.00 mL of 0.500 M NaOH

(c) after the addition of 10.00 mL of 0.500 M NaOH

(d) after the addition of 20.00 mL of 0.100 M NaOH

(e) after the addition of 21.00 mL of 0.100 M NaOH

STRATEGY

(a) This is the pH of 0.500 M CH_3COOH, the initial pH. Because the acid is weak, we must do an equilibrium calculation for $[H_3O^+]$ using the ICE format. The pH follows directly from $[H_3O^+]$.

(b) Substantial concentrations of both acetic acid and acetate ion are found in this buffer region. After a brief stoichiometry calculation, we can use the Henderson–Hasselbalch equation (15.10) to determine the pH.

(c) This is the point in the titration where the weak acid is half-neutralized. We will find that at this point $pH = pK_a$.

(d) The pH at the equivalence point is established by hydrolysis of acetate ion (recall Example 15.14).

(e) Beyond the equivalence point, we have a solution of $CH_3COONa(aq)$ containing excess OH^-. From $[OH^-]$, we can determine pOH and pH.

SOLUTION

(a) This calculation is similar to that of Example 15.6, except that here the acid is 0.500 M.

The reaction:	CH_3COOH	$+ H_2O \rightleftharpoons$	H_3O^+	$+ CH_3COO^-$
Initial concentrations, M:	0.500		≈ 0	0
Changes, M:	$-x$		$+x$	$+x$
Equilibrium concentrations, M:	$(0.500 - x)$		x	x

$$K_a = \frac{[H_3O^+][CH_3COO^-]}{[CH_3COOH]}$$

$$1.8 \times 10^{-5} = \frac{x \cdot x}{0.500 - x}$$

We can make the usual assumption, that is, that $x \ll 0.500$:

$$K_a = \frac{x^2}{0.500} = 1.8 \times 10^{-5}$$

$$x^2 = 9.0 \times 10^{-6}$$

$$x = [H_3O^+] = 3.0 \times 10^{-3} \text{ M}$$

$$pH = -\log[H_3O^+] = -\log(3.0 \times 10^{-3}) = 2.52$$

(b) The addition of 8.00 mL of 0.500 M NaOH represents the addition of

$$8.00 \text{ mL} \times \frac{0.500 \text{ mmol OH}^-}{1 \text{ mL}} = 4.00 \text{ mmol OH}^-$$

At this point in the titration, we have the following data.

The reaction: $\qquad CH_3COOH + OH^- \longrightarrow H_2O + CH_3COO^-$

Initial amounts, mmol: $\qquad 10.0 \qquad\qquad \approx 0 \qquad\qquad\qquad \approx 0$

Add, mmol: $\qquad\qquad\qquad\qquad 4.00$

Changes, mmol: $\qquad -4.00 \qquad\qquad -4.00 \qquad\qquad\qquad +4.00$

After reaction, mmol: $\qquad 6.0 \qquad\qquad \approx 0 \qquad\qquad\qquad 4.00$

We can apply Equation (15.10) with $pK_a = 4.74$, $[CH_3COO^-] = 4.00$ mmol/28.00 mL, and $[CH_3COOH] = 6.0$ mmol/28.00 mL:

$$pH = pK_a + \log \frac{[CH_3COO^-]}{[CH_3COOH]}$$

$$pH = 4.74 + \log \frac{(4.00/28.00)}{(6.0/28.00)}$$

$$pH = 4.74 + \log 0.67 = 4.74 - 0.17 = 4.57$$

Problem-Solving Note

The concentration units in the logarithmic expression are mmol/mL = M. There is no need to include them here, however, because they appear in a ratio and cancel out.

(c) At the half-neutralization point, the titration has progressed as follows.

The reaction: $\qquad CH_3COOH + OH^- \longrightarrow H_2O + CH_3COO^-$

Initial amounts, mmol: $\qquad 10.0 \qquad\qquad \approx 0 \qquad\qquad\qquad \approx 0$

Add, mmol: $\qquad\qquad\qquad\qquad 5.00$

Changes, mmol: $\qquad -5.00 \qquad\qquad -5.00 \qquad\qquad\qquad +5.00$

After reaction, mmol: $\qquad 5.0 \qquad\qquad \approx 0 \qquad\qquad\qquad 5.00$

The CH_3COOH and CH_3COO^- are present in equal amounts—5.0 mmol—and in the same 30.00 mL of solution; therefore, their concentrations are equal. This means that

$$pH = pK_a + \log \frac{\cancel{[CH_3COO^-]}}{\cancel{[CH_3COOH]}} = pK_a + \log(1)$$

$$pH = pK_a = 4.74$$

(d) At the equivalence point, 10.0 mmol each of CH_3COOH and NaOH have reacted to produce 10.0 mmol of CH_3COONa in 40.00 mL of solution. The solution molarity is

$$\frac{10.0 \text{ mmol } CH_3COONa}{40.00 \text{ mL}} = 0.250 \text{ M } CH_3COONa$$

We calculated the pH of 0.250 M CH_3COONa in Example 15.14 and found pH = 9.08.

(e) The amount of OH^- is calculated from the added volume and the NaOH solution concentration:

$$21.00 \text{ mL} \times \frac{0.500 \text{ mmol } OH^-}{1 \text{ mL}} = 10.5 \text{ mmol } OH^-$$

At this point beyond the equivalence point, we have the following data.

The reaction: $\qquad CH_3COOH + OH^- \longrightarrow H_2O + CH_3COO^-$

Initial amounts, mmol: $\qquad 10.0 \qquad\qquad \approx 0 \qquad\qquad\qquad \approx 0$

Add, mmol: $\qquad\qquad\qquad\qquad 10.5$

Changes, mmol: $\qquad -10.0 \qquad\qquad -10.0 \qquad\qquad\qquad +10.0$

After reaction, mmol: $\qquad \approx 0 \qquad\qquad 0.5 \qquad\qquad\qquad 10.0$

Acetate ion is a very weak base compared to OH^-. The hydroxide ion concentration is simply

$$[OH^-] = \frac{0.5 \text{ mmol}}{(20.00 + 21.00)\text{mL}} = 0.01 \text{ M}$$

$$pOH = -\log[OH^-] = -\log(1 \times 10^{-2}) = 2.0$$

$$pH = 14.00 - pOH = 14.00 - 2.0 = 12.0$$

Example 15.22 A Conceptual Example

This titration curve shown in Figure 15.18 involves 1.0 M solutions of an acid and a base. Identify the type of titration it represents.

Tritration Curves activity

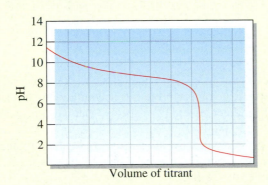

▲ FIGURE 15.18   Example 15.22 illustrated

ANALYSIS AND CONCLUSIONS

We can identify these features of the titration curve and draw certain conclusions from them:

- The pH starts high and decreases during the titration. The solution being titrated is therefore a base, and the titrant is an acid.

- The initial pH is about 11.5, but a 1.0 M solution of a strong base would dissociate completely so that $[OH^-] = 1.0 \, M$, $[H_3O^+] = 1.0 \times 10^{-14}$, and pH = 14.00. We conclude that the base must be a weak base.

- The pH at the equivalence point is less than 7. This is the pH expected for the hydrolysis of the salt of a strong acid and a weak base.

- The pH beyond the equivalence point drops to a low value (pH < 1), again indicating that the titrant is a strong acid.

From these observations, we conclude that the curve represents the titration of a *weak base* with a *strong acid*.

15.11 Lewis Acids and Bases

We have devoted most of this chapter to the Brønsted–Lowry theory of acids and bases, which works well for the reactions in aqueous solutions that we have emphasized here. However, there are reactions in nonaqueous solvents, in the gaseous state, and even in the solid state that can be considered acid–base reactions. Chemists have devised several more general acid–base theories. In this section, we will briefly consider one of the more important alternative theories.

In 1923, G. N. Lewis proposed an acid–base theory that focuses on the role of electron pairs, in contrast to the protons of the Brønsted–Lowry theory. A **Lewis acid** is a species that is an electron-pair *acceptor*. **A Lewis base** is a species that is an electron-pair *donor*. In a Lewis acid–base reaction, a covalent bond is formed between the acid and base. We encountered an example earlier in the text (page 362) when we looked at the formation of a coordinate covalent bond between BF_3 and F^-:

$$\ddot{:}\overset{\displaystyle :\ddot{F}:}{\underset{\displaystyle :\ddot{F}:}{\ddot{F}-B}} \quad + \quad :\ddot{F}:^- \quad \longrightarrow \quad \left[\overset{\displaystyle :\ddot{F}:}{\underset{\displaystyle :\ddot{F}:}{:\ddot{F}-B-\ddot{F}:}}\right]^-$$

Lewis acid   Lewis base

This example suggests that in a Lewis acid–base reaction, we should look for (1) a species that has an available empty orbital to accommodate an electron pair (such as the B atom in the BF_3 molecule) and (2) a species that has lone-pair electrons (such as the F^- ion).

The definition of a Lewis base allows us to consider typical Brønsted–Lowry bases, such as OH^-, NH_3, and H_2O, as also being Lewis bases because they all have electron pairs available for sharing:

$$\left[:\ddot{O}-H\right]^- \qquad H-\overset{\displaystyle |}{\underset{\displaystyle H}{\ddot{N}}}-H \qquad H-\overset{\displaystyle |}{\underset{\displaystyle H}{\ddot{O}:}}$$

The definition of a Lewis acid, however, does not seem to fit a Brønsted–Lowry acid like HCl because a HCl molecule cannot gain an electron pair. However, HCl ionizes to produce H^+, and the H^+ ion *is* a Lewis acid because it accepts an electron pair when it bonds to a lone pair of electrons in a H_2O molecule, forming the hydronium ion H_3O^+. Thus, we can think of HCl as a *source* of a Lewis acid.

Now let us consider a reaction that is an acid–base reaction according to the Lewis theory but not the Brønsted–Lowry theory—no hydrogen atoms are involved:

$$CaO(s) + SO_2(g) \longrightarrow CaSO_3(s)$$

The best way to see this as a Lewis acid–base reaction is to write Lewis structures and to identify an electron pair (shown in red here) that is supplied by O^{2-} (the base) to SO_2 (the acid). A shift of an electron pair (in blue) out of the sulfur-to-oxygen double bond to a lone-pair position is also involved:

$$Ca^{2+} \quad :\ddot{O}:^{2-} \quad + \quad \overset{\displaystyle \ddot{O}:}{\underset{\displaystyle :\ddot{O}:}{S:}} \quad \longrightarrow \quad Ca^{2+} \left[\overset{\displaystyle :\ddot{O}:}{\underset{\displaystyle :\ddot{O}:}{:\ddot{O}-S}}\right]^{2-}$$

In organic chemistry, Lewis acids are often called *electrophiles:* species that seek electrons. Lewis bases are called *nucleophiles:* species that seek a positive site. We consider applications to organic chemistry in Chapter 23.

Lewis Acid-Base Theory animation

Cumulative Example

A 0.0500 M aqueous solution of cyanoacetic acid, $CNCH_2COOH$, has a freezing point of $-0.11\,°C$. Calculate the freezing point of a 0.250 M aqueous solution of cyanoacetic acid.

STRATEGY

To determine the freezing point of the 0.250 M solution, we must determine the total concentration of species in the solution. We can assume that for dilute aqueous solutions, molarity and molality concentrations are essentially the same (see page 507). Then we can use the freezing-point data in Equation (12.7) to determine the total molarity of the molecules and ions in 0.0500 M cyanoacetic acid. With this information, we will be able to evaluate the acid ionization constant, K_a, for cyanoacetic acid. Next, we can apply this K_a value to calculate the total molarity of the molecules and ions in the 0.250 M solution and, again using Equation (12.7), find the freezing point of the solution.

SOLUTION

We can show the ionization equation for cyanoacetic acid, the nominal molarity of the solution C, and the molarities of the ions x in the standard ICE format used throughout this chapter.

The reaction: $CNCH_2COOH(aq) + H_2O(l) \rightleftharpoons H_3O^+(aq) + CNCH_2COO^-(aq)$

Initial molarities: C ≈0 0

Changes: $-x$ $+x$ $+x$

Equilibrium molarities: $C - x$ x x

Now we can represent the total molarity of particles (molecules and ions) in the solution.

Total molarity $= (C - x) + x + x = C + x$

We find the total molality (molarity) of the 0.0500 M solution by rearranging Equation (12.7) and solving for m.

$$m = \frac{\Delta T_f}{K_f} = \frac{-0.11\,°C}{-1.86\,°C\,m^{-1}} = 0.059\,m$$

The value $0.059\,m$ (or 0.059 M) is equal to $C + x$, where $C = 0.0500$ M. Thus, we can solve for x.

$0.059\,M = C + x$ and $x = 0.059\,M - 0.0500\,M = 0.009\,M$

We obtain K_a for cyanoacetic acid by substituting known values into the equilibrium expression.

$$K_a = \frac{[H_3O^+][CNCH_2COO^-]}{[CNCH_2COOH]}$$
$$= \frac{(x)(x)}{C - x} = \frac{(0.009)^2}{0.0500 - 0.009} = 2 \times 10^{-3}$$

To obtain the total molarity of the molecules and ions in 0.250 M cyanoacetic acid, we again use the K_a expression, this time with $C = 0.250$ M and solving for x.

$$K_a = \frac{[H_3O^+][CNCH_2COO^-]}{[CNCH_2COOH]}$$
$$= \frac{(x)(x)}{C - x} = \frac{x^2}{0.250 - x} = 2 \times 10^{-3}$$
$$x^2 + 0.002x - 5 \times 10^{-4} = 0$$
$$x = \frac{-0.002 \pm \sqrt{(0.002)^2 + (4 \times 5 \times 10^{-4})}}{2}$$
$$x = 0.02$$

The total molarity (molality) in 0.250 M cyanoacetic acid is the sum of the nominal molarity (C) and that of the ions (x).

$C + x = (0.250 + 0.02)\,M = 0.27\,M$

Now we use Equation (12.7) to calculate the freezing-point depression in this solution:

$$\Delta T_f = K_f \times m = -1.86\,°C\,m^{-1} \times 0.27\,m = -0.50\,°C$$

Because the freezing point of water is 0 °C, the freezing point of the 0.250 M cyanoacetic acid solution is $-0.50\,°C$.

Concept Review with Key Terms

15.1 The Brønsted–Lowry Theory of Acids and Bases—In the Brønsted–Lowry theory, an acid is a **proton donor** and a base is a **proton acceptor**. Every acid has a **conjugate base**, which is the acid minus a proton, and every base has a **conjugate acid**, which is the base plus a proton. In an acid–base reaction, the forward reaction is between an acid and a base and the reverse reaction is between the conjugate base and conjugate acid.

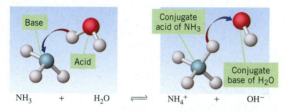

$$NH_3 \quad + \quad H_2O \quad \rightleftharpoons \quad NH_4^+ \quad + \quad OH^-$$

If a Brønsted–Lowry acid is strong, its conjugate base is weak; if a Brønsted–Lowry base is strong, its conjugate acid is weak. An acid–base reaction is favored in the direction from the stronger member to the weaker member of each conjugate acid–base pair. A substance that can act as either an acid or a base is **amphiprotic**. For reversible acid–base reactions, the equilibrium constants are called the **acid ionization constant, K_a,** and the **base ionization constant, K_b,** respectively, and are calculated by the same approach used to calculate K_c.

15.2 Molecular Structure and Strengths of Acids and Bases—Acid ionization requires bond breakage and release of a proton (H^+). The ease of bond breakage is affected by such factors as electronegativity, number of terminal O atoms in oxoacids, and placement of substituent groups on carbon chains in carboxylic acids.

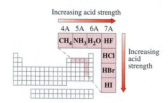

15.3 Self-Ionization of Water—the pH Scale—Water is amphiprotic, acting either as an acid or as a base, depending on conditions in a given solution. It undergoes limited self-ionization, producing H_3O^+ and OH^-. At equilibrium, the concentrations of H_3O^+ and OH^- are in accordance with the **ion product of water, K_w**.

$$K_w = [H_3O^+][OH^-] = 1.0 \times 10^{-14} \text{ at } 25\ °C$$

The concentrations of these ions are commonly expressed in the notation of negative logarithms:

$$\textbf{pH} = -\log[H_3O^+]$$
$$\textbf{pOH} = -\log[OH^-]$$

Also, the ion product of water can be expressed in a similar notation scheme:

$$\textbf{p}K_\text{w} = -\log K_w$$

The pH in both pure water and in neutral solutions is 7. The pH in acidic solutions is less than 7, and the pH in basic solutions is greater than 7. The pH and the pOH of a solution are related through pK_w, and in aqueous solutions at 25 °C,

$$pK_\text{w} = 14.00 = \text{pH} + \text{pOH}$$

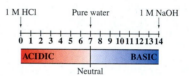

15.4 Equilibrium in Solutions of Weak Acids and Weak Bases—Calculations involving weak acids and bases aim to determine equilibrium concentrations of all species involved and often begin with information of initial conditions and acid or base ionization constants. These constants can be expressed on the logarithmic scale as pK_a and pK_b, where

$$pK_a = -\log K_a$$
$$pK_b = -\log K_b$$

15.5 Polyprotic Acids—Acids possessing one ionizable H per molecule are called **monoprotic acids**; those possessing more than one are called **polyprotic acids**. The ionization of the different H atoms in polyprotic acids occurs stepwise, and these steps can be described as distinct reversible reactions with separate equilibrium constant expressions.

Phosphoric acid

15.6 Ions as Acids and Bases—**Hydrolysis** reactions cause salt solutions to be acidic, neutral, or basic, depending on the ions making up the salts. Ions that hydrolyze include conjugates of weak acids and conjugates of weak bases. In calculating the pH of a salt solution, it is often necessary to establish the ionization constant of an acid or a base from that of its conjugate:

$$pK_a(\text{acid}) + pK_b(\text{conjugate base}) = 14.00$$

15.7 The Common Ion Effect—A strong electrolyte that produces an ion also involved in the ionization equilibrium of a weak acid or a weak base suppresses the ionization of that weak electrolyte in accordance with Le Châtelier's principle. This phenomenon is known as the **common ion effect**.

15.8 Buffer Solutions—A **buffer solution** is a mixture of a weak acid and its conjugate base or a weak base and its conjugate acid. A buffer solution maintains an essentially constant pH when small amounts of a strong acid or strong base are added. The pH of buffer solutions is described by the **Henderson–Hasselbalch equation**.

$$pH = pK_a + \log \frac{[\text{conjugate base}]}{[\text{weak acid}]}$$

15.9 Acid–Base Indicators—**Acid–base indicators** are weak acids for which the acid and its conjugate base have different colors. The concentration of H_3O^+ ions in the solution being tested affects the equilibrium of the indicator acid and its conjugate base and therefore affects the color exhibited by the indicator.

15.10 Neutralization Reactions and Titration Curves—An appropriate indicator for an acid–base titration undergoes a color change at the pH of the equivalence point. A **titration curve**, a graph of pH versus volume of titrant added, can be used to establish the equivalence

point. Many of the concepts introduced throughout the chapter can be used to calculate the pH at different points on a titration curve.

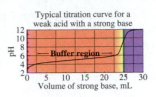
Typical titration curve for a weak acid with a strong base

15.11 Lewis Acids and Bases—A **Lewis acid** is an electron pair acceptor, and a **Lewis base** is an electron-pair donor. In a Lewis acid–base reaction, new covalent bonds are formed.

Assessment Goals

When you have mastered the material in this chapter, you will be able to:

- Describe the Brønsted–Lowry theory, and identify acids, bases, and their conjugate pairs in terms of proton acceptors and proton donors.

- Write equilibrium constant expressions for the reversible ionization of Brønsted–Lowry acids and bases.

- Predict the influence of molecular structure on the strengths of acids and bases.

- Calculate values of the logarithmic terms, pH, pOH, pK_a, and pK_b, from the appropriate ion concentrations and ionization constants.

- Calculate $[H_3O^+]$ and $[OH^-]$ from pH or pOH values.

- Relate equilibrium values of $[H_3O^+]$ and $[OH^-]$ to one another by using the ion product of water.

- Calculate equilibrium concentrations and pH values for weak acid and weak base solutions, for polyprotic acid solutions, and for solutions containing ions that act as acids and bases in solutions.

- Determine values of conjugate acid and base ionization constants through their relationship to K_w.

- Describe the common ion effect, and predict the influence on acid–base equilibria of adding common ions to a solution.

- Calculate equilibrium concentrations and pH values for solutions of weak acids and weak bases containing common ions.

- Describe the action of a buffer solution, and calculate equilibrium concentrations and pH values of such solutions, using K_a or K_b expressions or the Henderson–Hasselbalch equation.

- Calculate the effect on the pH produced by adding a strong acid or strong base to a buffer solution, and describe the pH range and buffer capacity of the solution.

- Describe the function of an acid–base indicator and the relationship between pH and the color of the indicator.

- From tabulated values, select appropriate indicators for a given pH measurement and determine the color of an indicator in a given solution.

- Describe the regions of a titration curve in terms of the equilibria that exist at different points along the curve. Calculate equilibrium concentrations and pH values at various points during titrations involving strong or weak acids and bases.

- Contrast the Lewis description of acids and bases to that provided by the Brønsted–Lowry theory.

Self-Assessment Questions

1. Write equations to represent the ionization of HI as an acid according to both the Arrhenius and Brønsted–Lowry theories.

2. Can a substance be a Brønsted–Lowry acid if it does not contain H atoms? Are there any characteristic atoms that must be present in a Brønsted–Lowry base?

3. Which of the following is incorrectly paired with its Brønsted–Lowry conjugate base or conjugate acid?
   (a) H_2O, H_3O^+
   (b) CH_3COO^-, CH_3COOH
   (c) NH_3, H_3O^+
   (d) F^-, HF

4. Which of the following is(are) *amphiprotic* species?
   (a) H_3PO_4
   (b) $H_2PO_4^-$
   (c) NH_4^+
   (d) H_2O

5. Write equations for the ionizations and K_a expressions for each of the following as Brønsted–Lowry weak acids.
   (a) HOClO
   (b) CH_3CH_2COOH
   (c) HCN
   (d) C_6H_5OH

6. An aqueous solution is 0.10 M in the weak base methylamine, CH_3NH_2. Which of the following is true for this solution?
   (a) $[H_3O^+] = 0.10$ M
   (b) $[OH^-] = 0.10$ M
   (c) pH < 7
   (d) pH > 7

7. Explain how the strengths of
   (a) *binary* acids are affected by bond energies and the ionic radii of the anions they produce
   (b) *oxoacids* are affected by the electronegativity of the central nonmetal atom and the number of terminal O atoms

8. Describe (a) two factors that affect the acidic strength of a carboxylic acid, and (b) two factors that affect the basic strength of amines.

9. A solution has a pH of 5.00. In this solution, $[OH^-]$ must be
   (a) 1.0×10^{-9} M
   (b) 1.0×10^{-7} M
   (c) greater than 1.0×10^{-5} M
   (d) 1.0×10^{-5} M

10. What is meant by the *percent ionization* of an acid? How is the percent ionization of a weak acid affected by the concentration of the acid?

11. What are the characteristic features of a *polyprotic acid*? Is C_6H_6 a polyprotic acid? Explain.

12. For a 0.10 M solution of H_2SO_3, which is a *diprotic* acid, $K_{a_1} = 1.3 \times 10^{-2}$ and $K_{a_2} = 6.3 \times 10^{-8}$. Which of the following is true for this solution?
(a) $[H_3O^+] = 0.10\,M$ **(c)** $[HSO_3^-] = 0.013\,M$
(b) $[H_3O^+] = 0.013\,M$ **(d)** $[SO_3^{2-}] = 6.3 \times 10^{-8}\,M$

13. What is the relationship between K_a for a Brønsted–Lowry acid and K_b for its conjugate base?

14. To increase the ionization of formic acid, HCOOH(aq), which of the following should be added to the solution?
(a) NaCl   **(c)** H_2SO_4
(b) HCOONa   **(d)** $NaHCO_3$

15. What do the terms *capacity* and *range* mean when describing a buffer solution? Does either of these depend on the concentrations of the buffer components? Explain.

16. Describe the difference between the *equivalence point* and the *endpoint* of an acid–base titration carried out with an acid–base indicator.

17. If an indicator is to be used in an acid–base titration having an equivalence point in the pH range 8 to 10, the indicator must
(a) be a weak base
(b) have a $K_a \approx 1.0 \times 10^{-9}$
(c) ionize in two steps, with the product $K_{a_1} \times K_{a_2} \approx 1 \times 10^{-9}$
(d) be added to the solution only after the solution has become alkaline

18. If the titration of a *weak base* is carried out using a *strong acid* as the titrant, at what point on the titration curve will the pH be lowest and at what point will it be highest?

19. Would you expect the equivalence point of each of the following titrations to be below, above, or at pH 7?
(a) $NaHCO_3$(aq) is titrated with NaOH(aq)
(b) HCl(aq) is titrated with NH_3(aq)
(c) KOH(aq) is titrated with HI(aq)

20. Is it possible for a substance to be a Lewis acid without being a Brønsted–Lowry acid? Explain.

Problems

Note: Unless otherwise indicated, assume a temperature of 25 °C.

Brønsted–Lowry Acids and Bases

21. For each of the following pairs, write an equation in which the first species given acts as a Brønsted–Lowry acid.
(a) HIO_4, NH_3 **(b)** H_2O, NH_2OH **(c)** H_3BO_3, NH_2^-

22. For each of the following pairs, write an equation in which the first species given acts as a Brønsted–Lowry acid.
(a) HBr. $CH_3CH_2NH_2$ **(b)** $ClCH_2COOH$, HS^- **(c)** HSO_4^-, F^-

23. For each of the following, identify the conjugate acid–base pairs.
(a) $HOClO_2 + H_2O \rightleftharpoons H_3O^+ + OClO_2^-$
(b) $HSeO_4^- + NH_3 \rightleftharpoons NH_4^+ + SeO_4^{2-}$
(c) $HCO_3^- + OH^- \rightleftharpoons CO_3^{2-} + H_2O$
(d) $C_5H_5NH^+ + H_2O \rightleftharpoons C_5H_5N + H_3O^+$

24. For each of the following, identify the conjugate acid–base pairs. Then use Table 15.1 to rank the four reactions in order of increasing tendency for the reaction to go to completion (to the right).
(a) $HSO_4^- + F^- \rightleftharpoons HF + SO_4^{2-}$
(b) $NH_4^+ + Cl^- \rightleftharpoons NH_3 + HCl$
(c) $HCl + CH_3COO^- \rightleftharpoons CH_3COOH + Cl^-$
(d) $CH_3OH + Br^- \rightleftharpoons HBr + CH_3O^-$

25. Identify the species that is amphiprotic, and write two equations for its reaction with H_2O(l) that illustrate its amphiprotic character: HI, $H_2PO_4^-$, NH_4^+, H_2CO_3, CO_3^{2-}.

26. Identify the species that is amphiprotic, and write one equation for its reaction with OH^-(aq) and another for its reaction with HBr(aq): H_2S, SO_2, HSO_3^-, H_2CO_3.

27. With which of the following acids will the reaction of sulfate ion, SO_4^{2-}, proceed farthest toward completion (to the right): HF, HCl, HCO_3^-, or CH_3COOH? Explain.

28. With which of the following bases will the reaction of acetic acid, CH_3COOH, proceed farthest toward completion (to the right)?
(a) CO_3^{2-} **(b)** F^- **(c)** Cl^- **(d)** NO_3^-

29. Aniline, $C_6H_5NH_2$, is a weak base in water, but it is a strong base when dissolved in pure acetic acid, CH_3COOH(l). Explain this difference in behavior.

30. Pure H_2SO_4, like water, is an amphiprotic solvent. Write an equation for the self-ionization of H_2SO_4. Draw a Lewis structure for the acid product.

Molecular Structure and the Strengths of Acids and Bases

31. Identify the stronger acid in each pair, and explain your choice.
(a) H_2S or H_2Se **(d)** H_2Se or HBr
(b) $HClO_3$ or HIO_3 **(e)** HN_3 or HCN
(c) H_3AsO_4 or $H_2PO_4^-$ **(f)** HNO_3 or HSO_4^-

32. Explain why perchloric acid, $HClO_4$, **(a)** is a strong acid, whereas chlorous acid, $HClO_2$, is a weak acid; and **(b)** is a stronger acid than sulfuric acid, H_2SO_4.

For Problems 33–36, refer to ideas in the text and data presented in Table 15.2.

33. Estimate a value of K_a for **(a)** 2,3-dichloropropanoic acid, $ClCH_2CHClCOOH$, and **(b)** 4-chlorobutanoic acid, $ClCH_2CH_2CH_2COOH$.

34. Estimate the pK_a of dichloroacetic acid. How would you expect pK_a of trichloroacetic acid to compare with that of dichloroacetic acid?

For Problems 35 and 36, you can review the discussion of aromatic compounds (Section 10.9) if you need help in writing the formulas.

35. Phenol, C_6H_5OH, used as a disinfectant and in the manufacture of plastics, dyes, and indicators, ionizes as an acid.

$$C_6H_5OH + H_2O \rightleftharpoons C_6H_5O^- + H_3O^+$$
$$K_a = 1.0 \times 10^{-10}$$

Arrange the following substituted phenols in the order in which you would expect their K_a values to *increase*. Where would you expect phenol itself to fit into this ranking?

(a) 3-chlorophenol   **(c)** 2,4,6-trichlorophenol

(b) 2,4-dichlorophenol   **(d)** 4-chlorophenol

36. Aniline, $C_6H_5NH_2$, a starting material for many dyes and pharmaceuticals, ionizes as a base.

$$C_6H_5NH_2 + H_2O \rightleftharpoons C_6H_5NH_3^+ + OH^-$$
$$K_b = 7.4 \times 10^{-10}$$

Arrange the following substituted anilines in the order in which you would expect their K_b values to *increase*. Where would you expect aniline itself to fit into this ranking?

(a) 2-nitroaniline   **(c)** 2,6-dinitroaniline

(b) 2,4-dinitroaniline   **(d)** 3-nitroaniline

The pH Scale

37. What is the $[H_3O^+]$ in **(a)** lemonade, pH 2.91; **(b)** shampoo, pH 9.26; **(c)** tomato puree, pH 4.35; **(d)** tea, pH 3.94?

38. What is the $[OH^-]$ in **(a)** paint stripper, pH 13.70; **(b)** rhubarb, pH 3.65 **(c)** blood plasma, pH 7.42?

39. What is the pH of each of the following aqueous solutions?

(a) 0.039 M HCl   **(c)** 0.65 M HBr

(b) 0.070 M KOH   **(d)** 2.5×10^{-4} M $Ca(OH)_2$

40. What is the pOH of each solution in Problem 39?

41. What is the pOH of each of the following aqueous solutions?

(a) 0.073 M LiOH   **(c)** 0.045 M $Ba(OH)_2$

(b) 1.75 M NaOH   **(d)** 9.1×10^{-2} M $HClO_4$

42. What is the pH of each solution in Problem 41?

43. Describe how you would prepare 5.00 L of an aqueous solution having a pH = 10.70, if you had a supply of 0.250 M NaOH available.

44. Describe how you would prepare 2.00 L of an aqueous solution having a pH = 3.60, if you had a supply of 0.100 M HCl available.

45. Which has the higher pH, 0.0062 M $Ba(OH)_2(aq)$ or an ammonia cleanser having a pH = 11.65? Explain.

46. Which has the lower pH, 0.00048 M H_2SO_4 or a vinegar solution having pH = 2.42? Explain.

Equilibria in Solutions of Weak Acids and Weak Bases

(Where necessary, obtain K_a and K_b values from Table 15.2 or Appendix C.2.)

47. Calculate the pH of **(a)** a 0.22 M aqueous solution of phenylacetic acid, $HC_8H_7O_2$ and **(b)** an aqueous solution containing 32.9 g formic acid, HCOOH, per liter.

48. Calculate the pH of the following: **(a)** a 0.084 M aqueous solution of quinoline, C_9H_7N, and **(b)** an aqueous solution containing 5.65 g hydroxylamine, NH_2OH, in 226 mL of solution.

49. Phenol, C_6H_5OH ($pK_a = 10.00$), is widely used in synthesizing organic chemicals, but it is more familiarly known as a general disinfectant (carbolic acid). A saturated aqueous solution of phenol has a pH = 4.90. What is the molarity of phenol in this solution?

50. Hydrazoic acid, HN_3 ($pK_a = 4.72$), is perhaps best known through its sodium salt, sodium azide, NaN_3, the gas-forming substance in automobile air-bag systems. What molarity of HN_3 is required to produce an aqueous solution with pH = 3.10?

51. A 1.00-g sample of aspirin (acetylsalicylic acid) is dissolved in 0.300 L of water at 25 °C, and its pH is found to be 2.62. What is the K_a of aspirin?

$$o\text{-}C_6H_4(OCOCH_3)COOH + H_2O \rightleftharpoons$$
$$H_3O^+ + o\text{-}C_6H_4(OCOCH_3)COO^- \quad K_a = ?$$

52. Codeine, $C_{18}H_{21}NO_3$, a commonly prescribed painkiller, is a weak base. A saturated aqueous solution contains 1.00 g codeine in 120 mL of solution and has a pH = 9.8. What is the K_b of codeine?

$$C_{18}H_{21}NO_3 + H_2O \rightleftharpoons [C_{18}H_{21}NHO_3]^+ + OH^- \quad K_b = ?$$

53. Calculate the pH of 0.105 M CCl_3COOH (trichloroacetic acid, $pK_a = 0.52$).

54. Piperidine, $C_5H_{11}N$ ($pK_b = 2.89$), is a colorless liquid having the odor of pepper. Calculate the pH of 0.00250 M piperidine.

55. What is the molarity of a formic acid solution, HCOOH(aq), that has the same pH as 0.150 M CH_3COOH(aq)?

56. What is the molarity of a methylamine solution, CH_3NH_2(aq), that has the same pH as 0.0850 M NH_3(aq)?

Polyprotic Acids

57. *Without doing detailed calculations*, determine which of the following polyprotic solutions will have the *lower pH*: 0.0045 M H_2SO_4 or 0.0045 M H_3PO_4.

58. *Without doing detailed calculations,* determine which of the following is the most likely pH for 0.010 M H_2SO_4: **(a)** 2.00, **(b)** 1.85, **(c)** 1.70, or **(d)** 1.50.

59. Oxalic acid, HOOCCOOH, is a weak dicarboxylic acid found as its potassium or calcium salt in the cell sap of many plants (for example, rhubarb leaves). The solubility of oxalic acid at 25 °C is about 83 g/L. What are the **(a)** pH and **(b)** [$^-$OOCCOO$^-$] in the saturated solution?

$$HOOCCOOH + H_2O \rightleftharpoons H_3O^+ + HOOCCOO^-$$
$$K_{a_1} = 5.4 \times 10^{-2}$$

$$HOOCCOO^- + H_2O \rightleftharpoons H_3O^+ + {}^-OOCCOO^-$$
$$K_{a_2} = 5.3 \times 10^{-5}$$

60. Citric acid ($pK_{a_1} = 3.13$, $pK_{a_2} = 4.76$, $pK_{a_3} = 6.40$) is found in citrus fruit. Calculate the approximate pH of lemon juice, which is about 5% citric acid by mass.

$$
\begin{array}{c}
CH_2COOH \\
| \\
HO-CCOOH \\
| \\
CH_2COOH
\end{array}
$$
Citric acid

61. For a 0.15 M H_3PO_4(aq) solution, determine **(a)** pH, **(b)** [H_3PO_4], **(c)** [$H_2PO_4^-$], **(d)** [HPO_4^{2-}], and **(e)** [PO_4^{3-}].

62. In thiophosphoric acid (H_3PO_3S), a S atom substitutes for one of the O atoms in the H_3PO_4 molecule. The K_a values for this acid are

$$K_{a_1} = 1.6 \times 10^{-2}, K_{a_2} = 3.7 \times 10^{-6}, K_{a_3} = 8.3 \times 10^{-11}$$

For a solution that is 0.50 M H_3PO_3S(aq), determine **(a)** pH, **(b)** [H_3PO_3S], **(c)** [$H_2PO_3S^-$], **(d)** [HPO_3S^{2-}], and **(e)** [PO_3S^{3-}].

Hydrolysis

63. Predict whether each of the following solutions is acidic, basic, or neutral. Write balanced equation(s) for any hydrolysis reaction(s) that occur.

(a) $RbClO_4$(aq) **(b)** $CH_3CH_2NH_3Br$(aq) **(c)** $HCOONH_4$(aq)

64. Predict whether each of the following solutions is acidic, basic, or neutral. Write balanced equation(s) for any hydrolysis reaction(s) that occur.

(a) CH_3CH_2COOK(aq) **(b)** $Mg(NO_3)_2$(aq) **(c)** NH_4CN(aq)

65. Which of the following 0.100 M aqueous solutions has the *lowest* pH: **(a)** $NaNO_3$, **(b)** CH_3COOK, **(c)** NH_4I, or **(d)** Na_3PO_4?

66. Which of the solutions in Problem 65 has the *highest* pH?

67. For a solution that is 0.080 M NaOCl, **(a)** write an equation for the hydrolysis reaction that occurs, and determine **(b)** the equilibrium constant for this hydrolysis and **(c)** the pH.

68. For a solution that is 0.602 M NH_4Cl, **(a)** write an equation for the hydrolysis reaction that occurs, and determine **(b)** the equilibrium constant for this hydrolysis and **(c)** the pH.

69. What is the molarity of a sodium acetate solution that has a pH = 9.05?

70. What is the molarity of an ammonium bromide solution that has a pH = 5.75?

Common Ion Effect

71. Which of the following solutions and substances will *suppress* the ionization of formic acid, HCOOH(aq): **(a)** NaCl, **(b)** KOH(aq), **(c)** $HClO_4$, **(d)** $(HCOO)_2Ca$, **(e)** Na_2CO_3? Explain.

72. Which of the following solutions and substances will *increase* the ionization of formic acid, HCOOH(aq): **(a)** KI, **(b)** NaOH(aq), **(c)** HNO_3, **(d)** HCOONa, **(e)** NaCN? Explain.

73. Calculate [NH_4^+] in a solution that is 0.15 M NH_3 and 0.015 M KOH.

74. Calculate [$C_6H_5COO^-$] in a solution that is 0.015 M C_6H_5COOH and 0.051 M HCl.

75. Calculate the pH of a solution that is 0.350 M CH_3CH_2COOH and 0.0786 M in the salt CH_3CH_2COOK. (For CH_3CH_2COOH, $K_a = 1.3 \times 10^{-5}$.)

76. Calculate the pH of a solution that is 0.132 M in diethylamine, $(CH_3CH_2)_2NH$, and 0.145 M in diethylammonium chloride, $(CH_3CH_2)_2NH_2Cl$. [For $(CH_3CH_2)_2NH$, $K_b = 6.9 \times 10^{-4}$.]

More Acid–Base Equilibria

77. *Without doing detailed calculations*, determine which of the following solutions should have a pH higher than 7.00, and which lower than 7.00: **(a)** 0.0050 M C_6H_5COOH, **(b)** 1.0×10^{-5} M NH_3, **(c)** 0.022 M CH_3NH_3Cl, **(d)** 0.10 M KH_2PO_4, **(e)** saturated $Ba(OH)_2$(aq), **(f)** 0.25 M $NaNO_2$.

78. *Without doing detailed calculations*, arrange the following 0.10 M aqueous solutions in order of *increasing* pH: **(a)** HCl, **(b)** CH_3COONa, **(c)** KCl, **(d)** H_3PO_4, **(e)** NH_3, **(f)** H_2SO_4, **(g)** NH_4NO_3, **(h)** KOH.

79. Write two equations—one to show the ionization of HPO_4^{2-} as an acid and one to show its ionization (hydrolysis) as a base. Would you expect an aqueous solution of Na_2HPO_4 to be acidic or basic? (*Hint:* What are the relevant K_a and K_b values?)

80. Predict whether an aqueous solution can be made up to have the following concentrations or whether a chemical reaction must occur.

  (a) 0.25 M CH_3COONa and 0.15 M HI

  (b) 0.050 M KNO_2 and 0.18 M KNO_3

  (c) 0.0050 M $Ca(OH)_2$ and 0.0010 M HNO_3

  (d) 0.20 M NH_4Cl and 0.35 M NaOH

Buffer Solutions

81. A solution is 0.405 M HCOOH (formic acid) and 0.326 M in the salt HCOONa (sodium formate). What is the pH of this buffer solution?

82. What is the pH of a buffer solution that is 0.245 M CH_3NH_2 (methylamine) and 0.186 M CH_3NH_3Cl (methylammonium chloride)?

83. What mass of $(NH_4)_2SO_4(s)$ must be dissolved in 0.100 L of 0.350 M NH_3(aq) to produce a buffer solution with pH = 10.05?

84. The only significant solute in white vinegar is acetic acid. Suppose you wanted to make a buffer solution from a particular white vinegar ($d = 1.01$ g/mL, 4.53% CH_3COOH by mass). What would be the pH of the buffer obtained by dissolving 10.0 g of sodium acetate in 0.250 L of the vinegar?

85. If 1.00 mL of 0.250 M HCl is added to 50.0 mL of the buffer solution described in Problem 81, what will be the pH of the final solution?

86. If 2.00 mL of 0.0850 M NaOH is added to 75.0 mL of the buffer solution described in Problem 82, what will be the pH of the final solution?

87. A buffer solution contains an acid as one component and a base as the other. Can these two components be HCl and NaOH? Explain.

88. Aqueous solutions of acetic acid contain both an acid, CH_3COOH, and its conjugate base, CH_3COO^-. Can an acetic acid solution alone be considered a buffer? Explain.

Acid–Base Indicators

89. Why are so many more acid–base indicators suitable for the titration of a strong acid with a strong base than are suitable for the titration of a weak acid with a strong base?

90. Why can a single acid–base indicator suffice to determine the equivalence point in a titration, whereas more than one indicator is often needed to determine the pH of a solution?

91. Refer to Figure 15.13, and determine the color you would expect for the following acid–base indicators in the given aqueous solution.

  (a) bromthymol blue in 0.10 M NH_4Cl

  (b) bromphenol blue in 0.10 M CH_3COOK

  (c) phenolphthalein in 0.10 M Na_2CO_3

  (d) methyl violet in 0.10 M CH_3COOH

92. A particular buffer solution is thought to have a pH greater than 9.5. How could you use a pair of acid–base indicators to verify this? Which two indicators from Figure 15.13 would you use? Explain.

93. A 0.100 M HCl(aq) solution contains thymol blue indicator and has a red color. A 0.100 M NaOH(aq) solution with phenolphthalein indicator also has a red color. What should be the color when equal volumes of the two solutions (with their indicators) are mixed?

94. When added to an unknown solution, bromcresol green indicator has a yellow color. When added to another sample of the same solution, thymol blue indicator also turns yellow. To within about one pH unit, what is the pH of the solution?

Neutralization Reactions and Titration Curves

95. Produce a listing comparable to that on page 658 to describe the titration of a strong base (for example, NaOH) with a strong acid (for example, HCl). Can the same indicator be used in the titration of a strong base with a strong acid as in the titration of a strong acid with a strong base? Explain.

96. Produce a listing comparable to that on page 659 to describe the titration of a weak base (for example, NH_3) with a strong acid (for example, HCl). Can the same indicator be used in the titration of a weak base with a strong acid as in the titration of a weak acid with a strong base? Explain.

97. In the titration of 20.00 mL of 0.500 M CH_3COOH by 0.500 M NaOH, calculate the pH of the solution at the point in the titration where 7.45 mL of 0.500 M NaOH has been added.

98. In the titration of 20.00 mL of 0.500 M HCl by 0.500 M NaOH, calculate the volume of 0.500 M NaOH required to reach a pH of 2.0. (*Hint:* Set up an equation in the unknown volume, V, and solve for V.)

99. Construct, by calculation, a titration curve for the titration of 40.00 mL of 0.200 M NH_3 with 0.500 M HCl. Specifically, what is **(a)** the volume of titrant required to reach the equivalence point; and what is the pH **(b)** initially, **(c)** after the addition of 5.00 mL of 0.500 M HCl, **(d)** at the half-neutralization point, **(e)** after the addition of 10.00 mL of 0.500 M HCl, **(f)** at the equivalence point, and **(g)** after the addition of 20.00 mL of 0.500 M HCl?

100. Construct, by calculation, a titration curve for the titration of 25.00 mL of 0.250 M CH_3CH_2COOH ($pK_a = 4.89$) with 0.330 M NaOH. Specifically, what is **(a)** the volume of titrant required to reach the equivalence point; and what is the pH **(b)** initially, **(c)** after the addition of 2.00 mL of 0.330 M NaOH, **(d)** at the half-neutralization point, **(e)** after the addition of 12.00 mL of 0.330 M NaOH, **(f)** at the equivalence point, and **(g)** after the addition of 20.00 mL of 0.330 M NaOH?

Lewis Acids and Bases

101. Identify the Lewis acid and Lewis base in each of the following reactions.

(a) $OH^-(aq) + Al(OH)_3(s) \longrightarrow Al(OH)_4^-(aq)$

(b) $Cu^{2+}(aq) + 4\,NH_3(aq) \longrightarrow [Cu(NH_3)_4]^{2+}(aq)$

(c) $CO_2(g) + OH^-(aq) \longrightarrow HCO_3^-(aq)$

102. Identify the Lewis acid and Lewis base in each of the following reactions.

(a) $CaO(s) + CO_2(g) \longrightarrow CaCO_3(s)$

(b) $Ag^+(aq) + 2\,CN^-(aq) \longrightarrow Ag(CN)_2^-(aq)$

(c) $B(OH)_3(s) + OH^-(aq) \longrightarrow B(OH)_4^-(aq)$

Additional Problems

Problems marked with an * may be more challenging than others.

103. How many grams of (a) lactic acid, $CH_3CHOHCOOH$, together with 2.00 g NaOH, must be used to make 225 mL of lactic acid–lactate ion buffer at pH 4.00; and (b) how many grams of acetic acid and sodium hydroxide are required to prepare 1.00 L of 0.0500 M acetic acid–acetate ion buffer at pH 3.80?

104. Select an appropriate conjugate acid–base pair for preparing a buffer of (a) pH 3.00, (b) pH 7.00, (c) pH 12.00.

105. If you have 1.00 M stock solutions each of H_3PO_4, NaH_2PO_4, Na_2HPO_4, and Na_3PO_4, describe how you would make (a) a pH 7.20 buffer that is 0.10 M in one species, and (b) a pH 6.80 buffer that is 0.25 M in one species.

106. When 10.0 mL of 0.0900 M HZ acid is mixed with 20.0 mL of 0.150 M KZ salt, a solution of pH 5.90 is obtained. What is the pK_a of HZ?

107. What volume of concentrated nitric acid (70.4% HNO_3 by mass; $d = 1.42$ g/mL) must be dissolved in water to prepare 0.500 L of a solution having pH = 2.20?

* **108.** Can a solution have $[H_3O^+] = 2 \times [OH^-]$? Can a solution have pH = 2 × pOH? If so, will the two solutions be the same?

* **109.** Calculate the pH of a solution that is 1.0×10^{-8} M HCl.

* **110.** Determine $[H_3O^+]$ in 0.020 M $H_2SO_4(aq)$, and compare your result with that obtained in Exercise 15.12B.

111. Use data from Appendix C for the neutralization reaction $H^+(aq) + OH^-(aq) \rightarrow H_2O(l)$ to determine whether the ion product for water, K_w, increases or decreases with increasing temperature.

* **112.** A handbook lists the following formula for the percent ionization of a weak acid.

$$\% \text{ Ionized} = \frac{100}{1 + 10^{(pK - pH)}}$$

(a) Derive this equation. Are any assumptions required in this derivation? Explain.

(b) Use the equation to determine the percent ionization of a formic acid solution, $HCOOH(aq)$, with a pH of 2.75.

(c) A 0.0100 M solution of p-nitrophenol has a pH of 4.58. What is K_a of p-nitrophenol?

113. Sodium hydrogen sulfate is an acidic salt with a number of uses, such as metal pickling (removal of surface deposits). It is made by the reaction of sodium chloride and sulfuric acid. To deter-

mine the percent sodium chloride impurity in sodium hydrogen sulfate, a 1.016-g sample is titrated with 36.56 mL of a 0.225 molar aqueous solution of sodium hydroxide. What is the percent sodium chloride in the sample titrated? What is a suitable indicator for this titration?

* **114.** Explain why the *difference* in values between pK_{a_1} and pK_{a_2} is considerably greater for maleic acid than for fumaric acid.

Maleic acid:
$pK_{a_1} = 1.91$; $pK_{a_2} = 6.33$

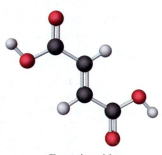

Fumaric acid:
$pK_{a_1} = 3.10$; $pK_{a_2} = 4.60$

* **115.** A handbook lists for hydrazine, N_2H_4, that $pK_{b_1} = 6.07$ and $pK_{b_2} = 15.05$. Draw a structural formula for hydrazine, and write equations to show how hydrazine can ionize as a base in two distinctive steps. Explain why pK_{b_2} is so much larger than pK_{b_1}. Calculate the pH of 0.150 M N_2H_4.

* **116.** A 275-mL sample of vapor in equilibrium with 1-propylamine at 25 °C is removed and dissolved in 0.500 L of H_2O. For 1-propylamine, $pK_b = 3.43$ and the vapor pressure at 25.0 °C is 316 Torr. (a) What should be the pH of the aqueous solution? (b) How many milligrams of NaOH dissolved in 0.500 L of water give the same pH?

117. The titration curve in the figure shown below involves 1.00 M solutions of an acid and a base. In the manner of Example 15.22 and Exercise 15.22A, **(a)** identify the type of acid–base titration represented by this curve, **(b)** estimate the pK value of the *titrant*, and **(c)** obtain a value of the pH at the equivalence point, by calculation and by estimation from the titration curve.

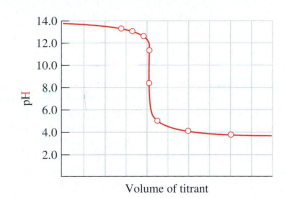

Volume of titrant

118. Assume that the color of an acid–base indicator is the acid color if at least 90% of the indicator is in the form of HIn, and the base color if at least 90% of the indicator is in the form of In⁻. Show that the pH range for the indicator to change color is roughly $pH = pK_{HIn} \pm 1$

* **119.** We have shown how to determine the pH of an aqueous solution in which hydrolysis of an ion occurs. Calculating the pH is more difficult if a given ion can undergo *both* hydrolysis and further ionization. **(a)** Write one equation to represent the ionization of $H_2PO_4^-$ as an acid, and another to represent its hydrolysis (ionization as a base). **(b)** Show that the pH of an aqueous solution containing $H_2PO_4^-$ is *independent* of the molarity, M, as long as the solution is sufficiently concentrated (greater than about 0.010 M). Show that the pH corresponds to the formula $pH = \frac{1}{2}(pK_{a_1} + pK_{a_2})$. **(c)** What is the pH of $NaHCO_3(aq)$?

120. *o*-Phthalic acid [o-$C_6H_4(COOH)_2$] is used to make phenolphthalein indicator. A saturated aqueous solution of *o*-phthalic acid has 0.6 g/100 mL and a pH = 2.33. A 0.10 M solution of potassium hydrogen *o*-phthalate has a pH = 4.19. Determine pK_{a_1} and pK_{a_2} for *o*-phthalic acid. (*Hint:* Refer to Problem 119.)

121. Show that when [H_3O^+] of a solution is reduced to half its original value, the pH increases by 0.30, *regardless of the initial pH*. Is it also true that when any solution is diluted to half its original concentration, its pH increases by 0.30? Explain.

* **122.** An acetic acid–acetate buffer solution can be prepared from (a) acetic acid and sodium acetate, (b) acetic acid and NaOH, or (c) hydrochloric acid and sodium acetate. Determine the relative molar proportions needed of the two reagents in each case for a buffer that is 1.00 M in acetic acid and 0.500 M in acetate ion. The buffer solution prepared by method (c) will likely have a slightly different pH from the ones prepared by (a) and (b). Explain why.

123. Both pure acetic acid and pure (liquid) ammonia are capable of self-ionization. Write the equation for self-ionization of each. Draw the Lewis structures for the acid and base produced in each case.

* **124.** The container represents a solution formed by mixing 190 mL of 0.100 M NaOH with 120 mL of a solution that is both 0.200 M in CH_3COOH and 0.050 M in HBr. The species that may be present in the mixture include NaOH, CH_3COOH, HBr, Na^+, OH^-, CH_3COO^-, H_3O^+, and Br^-. Identify (by color) the four species shown as present in the solution. Explain why the other four species were *not* shown in the figure.

* **125.** What is the pH of a solution that is 0.250 M CH_3COOH and also 0.150 M HCOOH? (*Hint:* You will need to solve two equations simultaneously.)

* **126.** To determine [H_3O^+] in 0.00250 M $HNO_2(aq)$, the first equation we would obtain in solving for x = [H_3O^+] is

$$K_a = \frac{[H_3O^+][NO_2^-]}{[HNO_2]} = \frac{x^2}{0.00250 - x} = 7.2 \times 10^{-4}$$

From the criteria on page 633, we realize that we must use the quadratic formula. However, we can also use the approximation procedure outlined here. Assume that x is very small compared to 0.00250, to obtain $x = (0.00250 \times 7.2 \times 10^{-4})^{1/2} = 0.0013$. Even though the assumption is not valid, it seems likely that [HNO_2] is closer to $(0.00250 - 0.0013)$ than to 0.00250 M. Substitute this value into original equation to obtain

$$\frac{x^2}{(0.00250 - 0.0013)} = 7.2 \times 10^{-4}$$

and $x = 9.3 \times 10^{-4}$. At this point, we can say this about the exact value of x:

$$9.3 \times 10^{-4} < x < 1.3 \times 10^{-3}$$

Use the new value of x to write that [HNO_2] = $(0.00250 - 0.00093)$ M, and solve the original equation once again for x. Continue this procedure until x attains a constant value.

(a) Show that the constant value of x = [H_3O^+] obtained in this way is the same as that obtained by using the quadratic formula.

(b) Use this method of successive approximations to determine the pH of 0.500 M $HClO_2$.

(c) Use this method to obtain the pH of 0.00200 M $CH_2ClCOOH$. What factor seems to determine how many successive approximations are needed to obtain a final answer?

Apply Your Knowledge

127. **[Laboratory]** Suppose that after measuring the pH of 0.05 M HCl, two drops (0.10 mL) of the solution remain behind on the pH electrode. When this electrode is then dipped into 100 mL of pure distilled water, what will be the pH reading of the water? Why is distilled water a poor choice for setting the pH = 7 reading of a pH meter? Describe a specific solution that would be a much better choice for this pH = 7 calibration.

* **128.** **[Laboratory]** The accompanying titration curve is that of 10.0 mL of 0.100 M H_3PO_4 titrated with 0.100 M NaOH.

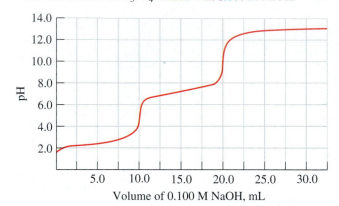

Volume of 0.100 M NaOH, mL

(a) Write a net ionic equation for the neutralization reaction that occurs prior to the first equivalence point. What substance is present in solution at the first equivalence point? **(b)** Write a net ionic equation for the neutralization reaction that occurs between the first and second equivalence points. What substance is present in solution at the second equivalence point? **(c)** Select an appropriate indicator for titration to each of the equivalence points. **(d)** H_3PO_4 is a *tri*protic acid. Why do you suppose the titration curve is lacking a third equivalence point? **(e)** How would the titration curve be altered if the solution being titrated were 0.100 M in *both* H_3PO_4 and HCl? **(f)** Suppose that 10.00 mL of a solution containing HCl and H_3PO_4 is titrated with 0.2080 M NaOH and requires 20.10 mL to reach the first equivalence point and 34.30 mL to reach the second. What are the nominal molarities of HCl and H_3PO_4 in the solution?

129. **[Environmental]** A batch of sewage sludge is dewatered to 28% solids; the remaining water has a pH of 6.0. The sludge is made quite alkaline to kill disease-causing microorganisms and make the sludge acceptable for spreading on farmland. How much quicklime (CaO) is needed to raise the pH of 1.0 ton of this sludge to 12.0?

$$CaO(s) + H_2O(l) \longrightarrow Ca(OH)_2(aq)$$

* **130.** **[Laboratory]** A 7.182-g sample of an unknown monoprotic weak acid is diluted to 250.0 mL in a volumetric flask. A 25.00-mL portion is pipetted into a separate flask and about 50 mL of distilled water is added. This mixture requires 40.83 mL of 0.1012 M NaOH for titration. At the endpoint, a second 25-mL portion of the weak acid is pipetted into the flask and mixed thoroughly. The pH of this solution is found to be 3.84. Use the data as needed to determine K_a of the weak acid.

131. **[Biochemical]** Tris(hydroxymethyl)aminomethane (TRIS) is a solid used to make buffers for use in biochemistry. TRIS reacts with HCl to form a salt called TRIS hydrochloride. Describe

how you would prepare 0.225 L of a pH 7.62 buffer using 0.100 M TRIS and 1.00 M HCl. The pK_b of TRIS is 5.92.

$$HOCH_2-\overset{\overset{\displaystyle CH_2OH}{|}}{\underset{\underset{\displaystyle CH_2OH}{|}}{C}}-NH_2$$

TRIS

* **132.** **[Laboratory]** The preparation of standard solutions (concentration accurately known) often requires that the desired solution first be prepared to an approximate concentration. Then, to determine its precise concentration, the solution is standardized by titration against a *primary standard*. Characteristics of a good primary standard include high purity, easily weighed, complete and rapid reaction with the solution being standardized. Suppose you want to standardize approximately 1 M HCl.

(a) Which of the following: sodium phosphate, TRIS (see Problem 131), concentrated aqueous ammonia (about 15 M), or aniline is the best candidate for a primary standard? Give one reason each for those not selected.

(b) In practice, Na_2CO_3 is often used as a primary standard. In the titration of Na_2CO_3(aq), the solution is brought nearly to the equivalence point with HCl. The titration is stopped and the solution is boiled for a minute or so. Then the solution is cooled to room temperature where the titration is completed. Write a net ionic equation for the titration reaction. What is accomplished by boiling the solution, and why is that necessary? Would it be necessary to boil the solution if sodium phosphate were used as the primary standard? Why or why not? Explain why an error would be introduced into the titration if concentrated NH_3(aq) were used as the primary standard.

(c) Calculate the mass range of TRIS needed if each titration uses 30–45 mL of the HCl(aq). Calculate the mass range of Na_2CO_3 if it were used instead of TRIS.

(d) Select an appropriate indicator from Figure 15.13 when TRIS is used for the titration. Assume that the TRIS is dissolved in 45 mL of water and that about 45 mL of HCl is used for the titration.

133. **[Environmental]** Normal rainfall is made slightly acidic from the dissolving of atmospheric CO_2 in the rainwater and ionization of the resulting carbonic acid, H_2CO_3. At a CO_2(g) partial pressure of 1 atm and a temperature of 25 °C, the solubility of CO_2(g) in water is 0.759 mL CO_2 at STP per mL H_2O. Given that air contains 0.037% CO_2 by volume, estimate the pH of rainwater that is saturated with CO_2(g). (*Hint:* Recall Henry's law [Equation (12.5)], and assume that the dissolved CO_2 is present as carbonic acid.)

134. **[Biochemical]** The two most important inorganic blood buffers are the phosphoric acid and carbonic acid systems. At the pH of blood (7.4), what is the ratio of the following:
(a) $[PO_4^{3-}]/[HPO_4^{2-}]$, **(b)** $[HPO_4^{2-}]/[H_2PO_4^{-}]$,
(c) $[H_2PO_4^{-}]/[H_3PO_4]$, **(d)** $[CO_3^{2-}]/[HCO_3^{-}]$,
and **(e)** $[HCO_3^{-}]/[H_2CO_3]$?
Which conjugate acid–base pair is most responsible for the buffering at normal blood pH?

e-Media Problems

The activities described in these problems can be found in the e-Media Activities and Interactive Student Tutorial (IST) modules of the Companion Website, *http://chem.prenhall.com/hillpetrucci.*

135. In the **Aqueous Bases** animation (*Section 15-1*), **(a)** which molecules can be classified as Arrhenius bases? **(b)** Which molecules can be described as Brønsted–Lowry bases?

136. View the **Strength of Carboxylic Acids animation** (*Section 15-2*). For this series of acids, use the variation in electron density distributions to describe the variation in acid strength.

137. Use the **Acid Equilibrium** activity (*Section 15-4*) to explore the strength of a series of acids. Start with 1.0 M concentrations of each of the inorganic acids listed in Table 15.2. "Equilibrate" the solutions and calculate the pH of each of the equilibrated solutions. What is the range of pH values in 1.0 M solutions of this series of acids?

138. **(a)** From the titration curve in the **Acid–Base Titration** animation (*Section 15-10*), estimate the initial volume of the 0.100 M acid solution. **(b)** On the same plot, sketch the titration curve seen in the animation and the titration curve produced when the same volume of 0.100 M acid is titrated with a 0.200 M solution of a strong base.

139. View the **Lewis Acid–Base Theory** animation (*Section 15-11*). **(a)** How do the Brønsted–Lowry and Lewis descriptions of acid–base behavior differ? **(b)** Explain why we cannot use the Brønsted–Lowry theory to describe the acidic behavior of BF_3 or CO_2.

Chapter 16

More Equilibria in Aqueous Solutions: Slightly Soluble Salts and Complex Ions

IN CHAPTER 14, we introduced general ideas about chemical equilibrium and applied them mostly to reactions involving gases. Much of Chapter 15 dealt with equilibria in aqueous solutions of weak acids and weak bases. In this chapter, we will focus on equilibria between slightly soluble salts and their ions in solution, and on equilibria that involve complex ions. However, as we shall see, these equilibria are often interwoven with equilibria involving acids and bases.

Applications of the concepts in this chapter are as familiar as our teeth and the film in our cameras: Tooth decay is related to the solubility of slightly soluble substances, and the developing of photographic film makes use of complex ion formation. We will consider both of these applications and many others as well.

CONTENTS

16.1 The Solubility Product Constant, K_{sp}

Many important ionic compounds are only slightly soluble in water. In fact, we often use the term "insoluble" to describe such compounds. For example, the solubility of barium sulfate, $BaSO_4$, is only 0.000246 g per 100 g water at 25 °C—a limited solubility that makes possible an unusual application for suspensions of barium sulfate in water (Figure 16.1).

We can use the following equation to represent the equilibrium between $BaSO_4(s)$ and the ions present in a *saturated* aqueous solution:

$$BaSO_4(s) \rightleftharpoons Ba^{2+}(aq) + SO_4^{2-}(aq)$$

As usual, the symbol (aq) indicates that water is the solvent, but H_2O does not participate in the chemical reaction and does not appear in the chemical equation. As we did in several instances in Chapter 15, we use a special name and symbol for the equilibrium constant representing solubility equilibrium. We call it the *solubility product constant* and represent it with the symbol K_{sp}:

$$K_{sp} = [Ba^{2+}][SO_4^{2-}] = 1.1 \times 10^{-10} \text{ (at 25 °C)}$$

Equilibrium in Solution Formation animation

◄ Although we normally consider calcium carbonate, $CaCO_3$, to be insoluble in water, it is in fact very slightly soluble. Only a few milligrams of $CaCO_3$ will dissolve in a liter of water, but that slight solubility is sufficient, over many centuries, to deposit $CaCO_3$ in the form of stalactites, stalagmites, and other beautiful and intricate cave formations. In this chapter we will examine the chemistry associated with such slightly soluble salts.

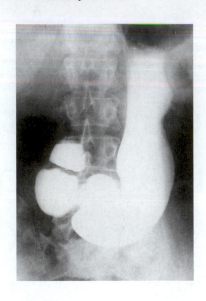

▶ **FIGURE 16.1   Barium sulfate is sparingly soluble in water**
A suspension of $BaSO_4(s)$ in water coats the stomach of a patient in this X-ray photograph. The stomach is rendered visible because $BaSO_4(s)$ strongly absorbs X-rays. The patient is spared from the poisonous action of Ba^{2+} ion because the $BaSO_4(s)$ is only very slightly soluble in water. The $BaSO_4(s)$ is later eliminated from the digestive tract.

Recall that pure liquids and solids do not appear in equilibrium constant expressions, and therefore this expression contains only the terms $[Ba^{2+}]$ and $[SO_4^{2-}]$. Formally, we say that the **solubility product constant, K_{sp},** is the product of the concentrations of the ions involved in a solubility equilibrium, each raised to a power equal to the stoichiometric coefficient of that ion in the chemical equation for the equilibrium. Like other equilibrium constants, K_{sp} depends on temperature. Table 16.1 lists some typical solubility equilibria and their solubility product constants at about room temperature. A more extensive listing of K_{sp} values is given in Appendix C.

Table 16.1   Some Solubility Product Constants at 25 °C

Solute	Solubility Equilibrium	K_{sp}
Aluminum hydroxide	$Al(OH)_3(s) \rightleftharpoons Al^{3+}(aq) + 3\,OH^-(aq)$	1.3×10^{-33}
Barium carbonate	$BaCO_3(s) \rightleftharpoons Ba^{2+}(aq) + CO_3^{2-}(aq)$	5.1×10^{-9}
Barium sulfate	$BaSO_4(s) \rightleftharpoons Ba^{2+}(aq) + SO_4^{2-}(aq)$	1.1×10^{-10}
Calcium carbonate	$CaCO_3(s) \rightleftharpoons Ca^{2+}(aq) + CO_3^{2-}(aq)$	2.8×10^{-9}
Calcium fluoride	$CaF_2(s) \rightleftharpoons Ca^{2+}(aq) + 2\,F^-(aq)$	5.3×10^{-9}
Calcium sulfate	$CaSO_4(s) \rightleftharpoons Ca^{2+}(aq) + SO_4^{2-}(aq)$	9.1×10^{-6}
Calcium oxalate	$CaC_2O_4(s) \rightleftharpoons Ca^{2+}(aq) + C_2O_4^{2-}(aq)$	2.7×10^{-9}
Chromium(III) hydroxide	$Cr(OH)_3(s) \rightleftharpoons Cr^{3+}(aq) + 3\,OH^-(aq)$	6.3×10^{-31}
Copper(II) sulfide	$CuS(s) \rightleftharpoons Cu^{2+}(aq) + S^{2-}(aq)$	8.7×10^{-36}
Iron(III) hydroxide	$Fe(OH)_3(s) \rightleftharpoons Fe^{3+}(aq) + 3\,OH^-(aq)$	4×10^{-38}
Lead(II) chloride	$PbCl_2(s) \rightleftharpoons Pb^{2+}(aq) + 2\,Cl^-(aq)$	1.6×10^{-5}
Lead(II) chromate	$PbCrO_4(s) \rightleftharpoons Pb^{2+}(aq) + CrO_4^{2-}(aq)$	2.8×10^{-13}
Lead(II) iodide	$PbI_2(s) \rightleftharpoons Pb^{2+}(aq) + 2\,I^-(aq)$	7.1×10^{-9}
Magnesium carbonate	$MgCO_3(s) \rightleftharpoons Mg^{2+}(aq) + CO_3^{2-}(aq)$	3.5×10^{-8}
Magnesium fluoride	$MgF_2(s) \rightleftharpoons Mg^{2+}(aq) + 2\,F^-(aq)$	3.7×10^{-8}
Magnesium hydroxide	$Mg(OH)_2(s) \rightleftharpoons Mg^{2+}(aq) + 2\,OH^-(aq)$	1.8×10^{-11}
Magnesium phosphate	$Mg_3(PO_4)_2(s) \rightleftharpoons 3\,Mg^{2+}(aq) + 2\,PO_4^{3-}(aq)$	1×10^{-25}
Mercury(I) chloride	$Hg_2Cl_2(s) \rightleftharpoons Hg_2^{2+}(aq) + 2\,Cl^-(aq)$	1.3×10^{-18}
Mercury(II) sulfide	$HgS(s) \rightleftharpoons Hg^{2+}(aq) + S^{2-}(aq)$	2×10^{-53}
Silver bromide	$AgBr(s) \rightleftharpoons Ag^+(aq) + Br^-(aq)$	5.0×10^{-13}
Silver chloride	$AgCl(s) \rightleftharpoons Ag^+(aq) + Cl^-(aq)$	1.8×10^{-10}
Silver iodide	$AgI(s) \rightleftharpoons Ag^+(aq) + I^-(aq)$	8.5×10^{-17}
Strontium carbonate	$SrCO_3(s) \rightleftharpoons Sr^{2+}(aq) + CO_3^{2-}(aq)$	1.1×10^{-10}
Strontium sulfate	$SrSO_4(s) \rightleftharpoons Sr^{2+}(aq) + SO_4^{2-}(aq)$	3.2×10^{-7}
Zinc sulfide	$ZnS(s) \rightleftharpoons Zn^{2+}(aq) + S^{2-}(aq)$	1.6×10^{-24}

As illustrated in Example 16.1, we must base a K_{sp} expression on a balanced equation, and the equation is always based on one mole of ionic solid. That is, the equation has a single term on the left—the formula of the ionic solid—and the coefficient of that term is 1.

Example 16.1

Write a solubility product constant expression for equilibrium in a saturated aqueous solution of the slightly soluble salts (a) iron(III) phosphate, $FePO_4$ and (b) chromium(III) hydroxide, $Cr(OH)_3$.

STRATEGY

We first write chemical equations for the solubility equilibrium, basing each equation on one mole of solid on the left side. The coefficients on the right side are then the numbers of moles of cations and anions per mole of the compound. As with other equilibrium constant expressions, coefficients in the balanced equation appear as exponents in the K_{sp} expression.

SOLUTION

(a) $FePO_4(s) \rightleftharpoons Fe^{3+}(aq) + PO_4^{3-}(aq)$ $K_{sp} = [Fe^{3+}][PO_4^{3-}]$

(b) $Cr(OH)_3(s) \rightleftharpoons Cr^{3+}(aq) + 3\,OH^-(aq)$ $K_{sp} = [Cr^{3+}][OH^-]^3$

EXERCISE 16.1A

Write a K_{sp} expression for equilibrium in a saturated aqueous solution of (a) MgF_2, (b) Li_2CO_3, and (c) $Cu_3(AsO_4)_2$.

EXERCISE 16.1B

Write a K_{sp} expression for equilibrium in a saturated solution of (a) magnesium hydroxide (commonly known as Milk of magnesia), (b) scandium fluoride, ScF_3 (used in the preparation of scandium metal), and (c) zinc phosphate (used in dental cements).

16.2   The Relationship Between K_{sp} and Molar Solubility

The solubility product constant is related to the solubility of an ionic solute, but K_{sp} and molar solubility—the molarity of a solute in a saturated aqueous solution—are not the same thing. However, we can determine K_{sp} from molar solubility and vice versa.

Although K_{sp} calculations are often easier than other equilibrium calculations, they are much more subject to error. One reason is that some K_{sp} values are not precisely known. Another is that interionic attractions complicate equilibrium relationships unless ion concentrations are quite low. Consequently, we ordinarily use the solubility product concept only for slightly soluble ionic solutes. Even for slightly soluble solutes, some calculations give only ballpark estimates.

Like other equilibrium calculations, those dealing with solubility equilibrium fall into two broad categories: determining a value of K_{sp} from experimental data, and calculating equilibrium concentrations when a value of K_{sp} is known. The first type of calculation is illustrated by Example 16.2.

Example 16.2

At 20 °C, a saturated aqueous solution of silver carbonate contains 32 mg of Ag_2CO_3 per liter of solution. Calculate K_{sp} for Ag_2CO_3 at 20 °C. The balanced equation is

$$Ag_2CO_3(s) \rightleftharpoons 2\,Ag^+(aq) + CO_3^{2-}(aq) \qquad K_{sp} = ?$$

STRATEGY

To evaluate K_{sp}, we need the molarities of the individual ions in the saturated solution, but we are given only the solubility of the solute. Thus, the heart of the calculation is to obtain ion molarities from a concentration in milligrams of solute per liter. Once we have these ion molarities, we can get a value for K_{sp} by substituting them into the solubility constant expression.

SOLUTION

From the equation for the solubility equilibrium, we see that 2 mol Ag^+ and 1 mol CO_3^{2-} appear in solution for every 1 mol Ag_2CO_3 that dissolves. We can represent this fact by conversion factors (shown in blue).

$$[Ag^+] = \frac{32 \text{ mg } Ag_2CO_3}{L} \times \frac{1 \text{ g } Ag_2CO_3}{1000 \text{ mg } Ag_2CO_3} \times \frac{1 \text{ mol } Ag_2CO_3}{275.8 \text{ g } Ag_2CO_3}$$

$$\times \frac{2 \text{ mol } Ag^+}{1 \text{ mol } Ag_2CO_3} = 2.3 \times 10^{-4} \text{ M}$$

$$[CO_3^{2-}] = \frac{32 \text{ mg } Ag_2CO_3}{L} \times \frac{1 \text{ g } Ag_2CO_3}{1000 \text{ mg } Ag_2CO_3} \times \frac{1 \text{ mol } Ag_2CO_3}{275.8 \text{ g } Ag_2CO_3}$$

$$\times \frac{1 \text{ mol } CO_3^{2-}}{1 \text{ mol } Ag_2CO_3} = 1.2 \times 10^{-4} \text{ M}$$

Now we can write the K_{sp} expression for Ag_2CO_3 and substitute in the equilibrium concentrations of the ions.

$$K_{sp} = [Ag^+]^2[CO_3^{2-}] = (2.3 \times 10^{-4})^2(1.2 \times 10^{-4}) = 6.3 \times 10^{-12}$$

EXERCISE 16.2A

In Example 15.4, we determined the molar solubility of magnesium hydroxide from the measured pH of its saturated solution. Use data from that example to determine K_{sp} for $Mg(OH)_2$.

EXERCISE 16.2B

A saturated aqueous solution of silver(I) chromate contains 14 ppm of Ag^+ by mass. Determine K_{sp} for silver(I) chromate. Assume the solution has a density of 1.00 g/mL.

In a second type of calculation, illustrated in Example 16.3, we determine the molar solubility of a solute from a tabulated value of its K_{sp}. This is a practical type of calculation because many reference books list only K_{sp} values and not solubilities for slightly soluble solutes.

Example 16.3

From the K_{sp} value for silver sulfate, calculate its molar solubility at 25 °C.

$$Ag_2SO_4(s) \rightleftharpoons 2 Ag^+(aq) + SO_4^{2-}(aq) \qquad K_{sp} = 1.4 \times 10^{-5} \text{ at } 25 \text{ °C}$$

STRATEGY

The equation shows that when 1 mol $Ag_2SO_4(s)$ dissolves, 1 mol SO_4^{2-} and 2 mol Ag^+ appear in solution. If we let s represent the number of moles of Ag_2SO_4 that dissolve per liter of solution, the two ion concentrations are

$$[SO_4^{2-}] = s \quad \text{and} \quad [Ag^+] = 2s$$

We then substitute these concentrations into the K_{sp} expression and solve for s, which, because 1 mol of Ag_2SO_4 produces 1 mol of SO_4^{2-}, is the molar solubility we are seeking.

SOLUTION

First, we write the K_{sp} expression for Ag_2SO_4 based on the solubility equilibrium equation.

$$K_{sp} = [Ag^+]^2[SO_4^{2-}] = 1.4 \times 10^{-5}$$

Then, we substitute the values $[Ag^+] = 2s$ and $[SO_4^{2-}] = s$ to obtain the equation that must be solved.

$$1.4 \times 10^{-5} = (2s)^2(s) = 4s^3$$

$$s^3 = \frac{1.4 \times 10^{-5}}{4} = 3.5 \times 10^{-6}$$

$$s = (3.5 \times 10^{-6})^{1/3} = 1.5 \times 10^{-2} \text{ M}$$

Now, by recalling that $s = [SO_4^{2-}]$ and that $[SO_4^{2-}]$ is the same as the molar solubility of Ag_2SO_4, we obtain our final result.

$$\frac{1.5 \times 10^{-2} \text{ mol } SO_4^{2-}}{L} \times \frac{1 \text{ mol } Ag_2SO_4}{1 \text{ mol } SO_4^{2-}} = 1.5 \times 10^{-2} \text{ M } Ag_2SO_4$$

ASSESSMENT

In substituting ion concentrations into the K_{sp} expression, it is important to note that (i) the factor 2 appears in 2s because the concentration of Ag^+ is *twice* that of SO_4^{2-}, and (ii) the exponent 2 appears in the term $(2s)^2$ because the concentration of Ag^+ (that is, 2s) must be raised to the *second* power in the K_{sp} expression. Be alert for similar situations that might arise in other problems.

EXERCISE 16.3A

Calculate the molar solubility of silver arsenate, given that

$$Ag_3AsO_4(s) \rightleftharpoons 3\,Ag^+(aq) + AsO_4^{3-}(aq) \qquad K_{sp} = 1.0 \times 10^{-22}$$

EXERCISE 16.3B

Using the K_{sp} value in Table 16.1, determine the quantity of I^-, in parts per million, in a saturated aqueous solution of PbI_2. Assume the solution has a density of 1.00 g/mL.

At times, instead of calculating molar solubilities, we need only to make some qualitative judgments. Often we can do this by comparing K_{sp} values, as illustrated in Example 16.4.

Example 16.4   A Conceptual Example

Without doing detailed calculations, but using data from Table 16.1, establish the order of *increasing* solubility of these silver halides in water: AgCl, AgBr, AgI.

ANALYSIS AND CONCLUSIONS

The key to solving this problem is to recognize that all three solutes are of the same type; that is, the ratio of number of cations to anions in each is 1:1. As suggested by the following equation for the solubility equilibrium of AgX, the ion concentrations are equal to each other and to the molar solubility, s.

$$AgX(s) \rightleftharpoons Ag^+(aq) + X^-(aq)$$
$$s \text{ mol } Ag^+/L \qquad s \text{ mol } X^-/L$$
$$K_{sp} = [Ag^+][X^-] = (s)(s) = s^2$$

The solute solubility $s = \sqrt{K_{sp}}$.

The order of increasing molar solubility is the same as the order of increasing K_{sp} values:

Increasing molar solubilities: $\qquad$ AgI $\quad<\quad$ AgBr $\quad<\quad$ AgCl

Increasing K_{sp}: $\qquad 8.5 \times 10^{-17} \quad<\quad 5.0 \times 10^{-13} \quad<\quad 1.8 \times 10^{-10}$

This method works only if we are comparing solutes that are all of the same general formula, that is, all MX, all MX_2, and so on.

EXERCISE 16.4A

Without doing detailed calculations, but using data from Table 16.1, arrange the following solutes in order of increasing molar solubility: MgF_2, CaF_2, $PbCl_2$, PbI_2.

EXERCISE 16.4B

Without doing detailed calculations, but using data from Table 16.1, indicate which of the following compounds has the greatest molar solubility: $BaSO_4$, CaF_2, PbI_2.

16.3   The Common Ion Effect in Solubility Equilibria

The common ion effect we studied in Chapter 15 also affects solubility equilibria. For example, sodium sulfate dissociates completely in water to form Na^+ and SO_4^{2-} ions:

$$Na_2SO_4(s) \xrightarrow{H_2O} 2\,Na^+(aq) + SO_4^{2-}(aq)$$

If we add some sodium sulfate solution to a saturated solution of silver sulfate, as in Figure 16.2, sulfate ion from the sodium sulfate increases $[SO_4^{2-}]$ and disturbs the solubility equilibrium:

$$Ag_2SO_4(s) \rightleftharpoons 2\,Ag^+(aq) + SO_4^{2-}(aq)$$

Common Ion Effect in Solubility activity

Le Châtelier's Principle movie

Common Ion Effect animation

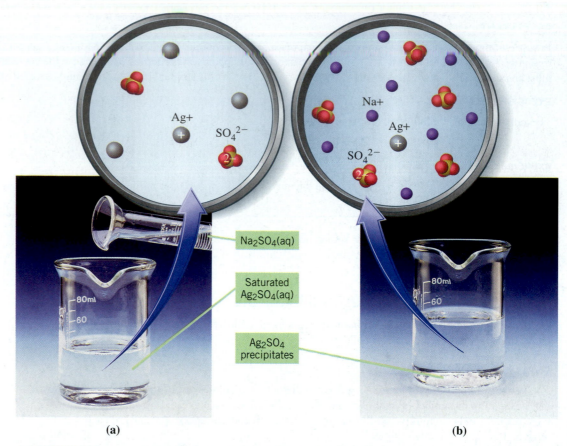

▲ **FIGURE 16.2** **The common ion effect in solubility equilibrium**

(a) A saturated $Ag_2SO_4(aq)$ solution. (b) Following the addition of Na_2SO_4, most of the Ag^+ ions originally present (about seven in eight) have precipitated.

QUESTION: What influence do the Na^+ ions have on the solubility of silver sulfate in solution (b)?

The added common ion creates a stress that, according to Le Châtelier's principle, is relieved as equilibrium shifts to the left. To establish a new equilibrium, several things happen.

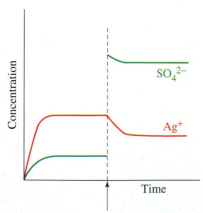

▲ A graphical representation of the response of a saturated solution of slightly soluble silver sulfate to the addition of sulfate ion. The solubility equilibrium is forced in the reverse direction until a new equilibrium is established with $[Ag^+]$ greatly reduced.

- Some $Ag_2SO_4(s)$ precipitates.
- $[Ag^+]$ is *smaller than* its original equilibrium value.
- $[SO_4{}^{2-}]$ is *greater than* its original equilibrium value.

We can achieve a similar effect by adding $AgNO_3(aq)$ to saturated $Ag_2SO_4(aq)$. In this case, we predict these effects:

- Some $Ag_2SO_4(s)$ precipitates.
- $[Ag^+]$ is *greater than* its original equilibrium value.
- $[SO_4{}^{2-}]$ is *smaller than* its original equilibrium value.

We can diagram the overall effect of the common ions Ag^+ and $SO_4{}^{2-}$ as follows:

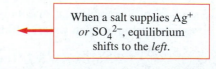

$$Ag_2SO_4(s) \rightleftharpoons 2\,Ag^+(aq) + SO_4{}^{2-}(aq)$$

To summarize the common ion effect on a solubility equilibrium:

The solubility of a slightly soluble ionic compound is lowered when a second solute that furnishes a common ion is added to the solution.

Qualitatively, both common ions, Ag^+ and SO_4^{2-}, have the same effect on the solubility of Ag_2SO_4: They reduce it. Quantitatively, however, Ag^+ has a greater effect than SO_4^{2-}, as can be seen by comparing the result of Exercise 16.5A with that of Example 16.5.

Example 16.5

Calculate the molar solubility of Ag_2SO_4 in 1.00 M $Na_2SO_4(aq)$.

STRATEGY

The molar solubility of Ag_2SO_4 will be influenced by the concentration of SO_4^{2-} in solution coming from the 1.00 M $Na_2SO_4(aq)$. We will assume that the Na_2SO_4 is completely dissociated and that the presence of Ag_2SO_4 has no effect on this dissociation. As in Example 16.3, we will use s to represent the number of moles of Ag_2SO_4 dissolved per liter of saturated solution. We will then represent $[Ag^+]$ and $[SO_4^{2-}]$ in terms of s and solve the K_{sp} expressions for s.

SOLUTION

We can tabulate the relevant data, including the already present 1.00 mol SO_4^{2-}/L, in the ICE format introduced in Chapter 14.

The reaction: $Ag_2SO_4(s) \rightleftharpoons 2\,Ag^+(aq) + SO_4^{2-}(aq)$ $K_{sp} = 1.4 \times 10^{-5}$

Initial concentrations, M:	0	1.00 (from the Na_2SO_4)
From Ag_2SO_4:	$2s$	s
Equilibrium concentrations, M:	$2s$	$(1.00 + s)$

The equilibrium concentrations from the ICE table must satisfy the K_{sp} expression.

$$K_{sp} = [Ag^+]^2[SO_4^{2-}]$$
$$1.4 \times 10^{-5} = (2s)^2(1.00 + s)$$

To simplify the equation, let us assume that s is much smaller than 1.00 M, so that $(1.00 + s) \approx 1.00$.

$$(2s)^2(1.00) = 1.4 \times 10^{-5}$$
$$4s^2 = 1.4 \times 10^{-5}$$

We obtain our final result by solving for s.

$$s^2 = 3.5 \times 10^{-6}$$
$$s = \text{molar solubility} = (3.5 \times 10^{-6})^{1/2}$$
$$= 1.9 \times 10^{-3} \text{ mol } Ag_2SO_4/L$$

ASSESSMENT

When we assumed that $(1.00 + s) \approx 1.00$, we were saying that essentially all the common ion comes from the added source (1.00 M) and essentially none comes from the slightly soluble solute (s). The assumption is valid here: $(1.00 + 1.9 \times 10^{-3}) = 1.00$ (to two decimal places). This type of assumption will often be valid in calculations of this type.

Notice that the solubility of Ag_2SO_4 found here is only about one-eighth of that found for Ag_2SO_4 in pure water in Example 16.3, illustrating, as we had intended, the common ion effect.

EXERCISE 16.5A

Calculate the molar solubility of Ag_2SO_4 in 1.00 M $AgNO_3(aq)$.

EXERCISE 16.5B

How many grams of silver nitrate must be added to 250.0 mL of a saturated solution of $Ag_2SO_4(aq)$ to reduce the solubility of the Ag_2SO_4 to 1.0×10^{-3} M (K_{sp} of $Ag_2SO_4 = 1.4 \times 10^{-5}$)?

Problem-Solving Note

In Exercise 16.5A, $[Ag^+]$ at equilibrium will be $1.00 + 2s$. You must square this entire term in the K_{sp} expression, but you will find that the factor $2s$ is negligible and drops out.

The common ion effect is frequently used in analytical chemistry, as illustrated in the following method of determining the percent calcium in limestone. Limestone is principally calcium carbonate, $CaCO_3$, with small amounts of other constituents.

First, we obtain calcium ion in solution by dissolving the limestone sample in $HCl(aq)$:

$$CaCO_3(s, \text{impure}) + 2\,H_3O^+(aq) \longrightarrow Ca^{2+}(aq) + 3\,H_2O(l) + CO_2(g)$$

▲ In the analytical laboratory, decomposition of calcium oxalate is usually performed in a temperature-controlled furnace. The temperature must be high enough so that calcium oxalate decomposes completely to pure $CaCO_3$ but not so high that the $CaCO_3$ decomposes to CaO.

After neutralizing any unreacted HCl(aq), we add an *excess* of ammonium oxalate solution, $(NH_4)_2C_2O_4(aq)$. By an excess, we mean that we use enough $C_2O_4^{2-}$ so that no further calcium oxalate precipitate forms in the reaction

$$Ca^{2+}(aq) + C_2O_4^{2-}(aq) \longrightarrow CaC_2O_4(s)$$

and a significant excess concentration of $C_2O_4^{2-}$ remains in the saturated solution. The presence of excess oxalate ion in the solubility equilibrium greatly reduces $[Ca^{2+}]$ in the saturated $CaC_2O_4(aq)$. Without this excess common ion, a small but significant percentage of Ca^{2+} would remain in solution. We separate the $CaC_2O_4(s)$ precipitate by filtration and wash it to remove residual impurities. For the washing, we use $(NH_4)_2C_2O_4(aq)$ rather than pure water—again, so the common ion, $C_2O_4^{2-}$, will keep the solubility of the calcium oxalate extremely low.

Finally, we heat the precipitate to 500 °C to decompose the calcium oxalate to *pure* $CaCO_3(s)$:

$$CaC_2O_4(s) \xrightarrow{\Delta} CaCO_3(s, pure) + CO(g)$$

We can then determine the mass of $CaCO_3(s)$ obtained and calculate the percent Ca in the original limestone. (See Problem 104.)

Solubility and Activities

We have just seen how the solubility of a slightly soluble solid is reduced in the presence of an excess of a common ion. What about a case where ions are introduced that are *not* common to the solid? Consider the molar solubility of calcium fluoride. In pure water, its solubility is 2.15×10^{-4} mol/L, whereas in 0.010 M sodium sulfate, its solubility is 3×10^{-4} mol/L, an increase of nearly 50%. We might expect CaF_2 to be *as* soluble in 0.010 M Na_2SO_4 as it is in pure water because there are no common ions to affect its solubility, but why should it be nearly 50% *more* soluble in 0.010 M Na_2SO_4 than in pure water?

Even though the ions from sodium sulfate are not common to calcium fluoride, they do have an effect on the solubility of the latter. Because of interionic attractions, there are more anions than cations present in the vicinity of each calcium ion in solution, and when the solvent is $Na_2SO_4(aq)$ rather than pure water, most of these anions are SO_4^{2-}. Around each fluoride ion, on average, there are more cations than anions, and most of these cations are Na^+. As a result, the Ca^{2+} and F^- ions are hindered somewhat from combining with each other—and more of them remain uncombined in the equilibrium solution than if Na_2SO_4 were not present. Thus, the solubility of CaF_2 is greater in $Na_2SO_4(aq)$ than in water alone, where no foreign ions are present to disrupt the ionic atmosphere of the Ca^{2+} and F^- ions.

Regarding aqueous solutions of ionic compounds, it is almost as if each ion has two "concentrations." One is the stoichiometric concentration, based simply on the total amount of solute in the solution. The other, called an *activity*, takes into account interionic attractions and is an "effective" concentration. The stoichiometric concentration is applicable in stoichiometric calculations, such as those based on titration data. The activity is the appropriate quantity for use in equilibrium calculations. We have been substituting molarities for activities in equilibrium calculations and in most cases have obtained reasonably good results. Sometimes, though, molarities deviate significantly from true activities, and the results are prone to error. We will look further into the use of activities in equilibrium constant expressions in the next chapter.

▲ **FIGURE 16.3** **Precipitation of anions from municipal tap water**

When a few drops of $AgNO_3(aq)$ are added to municipal tap water, a white precipitate forms. The precipitate forms because certain ions present in the tap water (for example, Cl^-, CO_3^{2-}, and SO_4^{2-}) form slightly soluble silver compounds.

16.4   Will Precipitation Occur? Is It Complete?

Sometimes a student mistakenly uses tap water instead of deionized water in preparing a solution. If the solute is $NaNO_3$, this may not cause a problem. However, if the solute is $AgNO_3$, a cloudy solution will probably result (Figure 16.3). The cloudiness is due to a trace of a precipitate, principally AgCl. What concentration of Cl^- can be tolerated in water and not form a precipitate with Ag^+? Let us establish a criterion to answer this and similar questions.

To Determine Whether Precipitation Occurs

We can begin with the expressions that describe the equilibrium between AgCl(s) and its ions:

$$AgCl(s) \rightleftharpoons Ag^+(aq) + Cl^-(aq) \qquad K_{sp} = [Ag^+][Cl^-] = 1.8 \times 10^{-10}$$

Suppose we want to prepare 0.100 M $AgNO_3$(aq) using municipal tap water as the solvent, and suppose the water has $[Cl^-] = 1 \times 10^{-6}$ M. Let us write, for a solution prepared under these conditions, a reaction quotient in the form of an ion product denoted by the symbol Q_{ip}. Then we can compare Q_{ip} to K_{sp}.

$$Q_{ip} = [Ag^+]_{initial} \times [Cl^-]_{initial} = 0.100 \times (1 \times 10^{-6}) = 1 \times 10^{-7}$$

We see that Q_{ip} greatly exceeds $K_{sp}(1 \times 10^{-7} > 1.8 \times 10^{-10})$. Recall from Section 14.3 that when a reaction quotient exceeds the value of the corresponding K, a net reaction should occur in the reverse direction; Ag^+ and Cl^- should combine to form AgCl(s). That is, *precipitation should occur*, and it should continue until Q_{ip} falls to a value equal to K_{sp}. At this point, the solid solute will be in equilibrium with its ions in a saturated solution.

Precipitation Reactions movie

Next, suppose we prepare 0.100 M $AgNO_3$, using deionized water in which the chloride ion concentration has been lowered to 1×10^{-10} M. Again, we can calculate a Q_{ip} value:

$$Q_{ip} = [Ag^+]_{initial} \times [Cl^-]_{initial} = 0.100 \times (1 \times 10^{-10}) = 1 \times 10^{-11}$$

Now Q_{ip} is smaller than $K_{sp}(1 \times 10^{-11} < 1.8 \times 10^{-10})$. When a reaction quotient is less than the value of the corresponding K, a net reaction should occur in the forward direction. However, this is not possible because there is no AgCl(s) present to dissolve. The solution is unsaturated, and certainly no AgCl(s) will precipitate.

Based on examples like these, we can formulate general rules that allow us to predict what should happen when we mix solutions containing ions capable of precipitating as a slightly soluble ("insoluble") ionic solid:

- Precipitation *should occur* if $Q_{ip} > K_{sp}$.
- Precipitation *cannot occur* if $Q_{ip} < K_{sp}$.
- A solution is *just saturated* if $Q_{ip} = K_{sp}$.

As Example 16.6 illustrates, we can apply these criteria in three steps:

Step 1: Determine the initial concentrations of ions.
Step 2: Evaluate the reaction quotient Q_{ip}.
Step 3: Compare Q_{ip} and K_{sp} to determine whether precipitation will occur.

Example 16.6

If 1.00 mg of Na_2CrO_4 is added to 225 mL of 0.00015 M $AgNO_3$(aq), will a precipitate form?

$$Ag_2CrO_4(s) \rightleftharpoons 2 Ag^+(aq) + CrO_4^{2-}(aq) \qquad K_{sp} = 1.1 \times 10^{-12}$$

STRATEGY

We can apply the three-step strategy just outlined.

SOLUTION

Step 1: *Initial concentrations of ions:* The initial concentration of Ag^+ is simply 1.5×10^{-4} M. To establish CrO_4^{2-}, we first need to determine the number of moles of CrO_4^{2-} placed in solution.

$$? \text{ mol } CrO_4^{2-} = 1.00 \times 10^{-3} \text{ g } Na_2CrO_4 \times \frac{1 \text{ mol } Na_2CrO_4}{162.0 \text{ g } Na_2CrO_4} \times \frac{1 \text{ mol } CrO_4^{2-}}{1 \text{ mol } Na_2CrO_4}$$

$$= 6.17 \times 10^{-6} \text{ mol } CrO_4^{2-}$$

We then determine $[CrO_4^{2-}]$ in the 225 mL (0.225 L) of solution that contains 6.17×10^{-6} mol CrO_4^{2-}.

$$[CrO_4^{2-}] = \frac{6.17 \times 10^{-6} \text{ mol } CrO_4^{2-}}{0.225 \text{ L}} = 2.74 \times 10^{-5} \text{ M}$$

Step 2: *Evaluation of Q_{ip}:*

$$Q_{ip} = [Ag^+]_{initial}{}^2[CrO_4{}^{2-}]_{initial} = (1.5 \times 10^{-4})^2(2.74 \times 10^{-5}) = 6.2 \times 10^{-13}$$

Step 3: *Comparison of Q_{ip} and K_{sp}:*

$$Q_{ip} = 6.2 \times 10^{-13} \text{ is less than } K_{sp} = 1.1 \times 10^{-12}$$

Because $Q_{ip} < K_{sp}$, we conclude that no precipitation occurs.

EXERCISE 16.6A

If 1.00 g $Pb(NO_3)_2$ and 1.00 g MgI_2 are both added to 1.50 L of H_2O, should a precipitate form (K_{sp} for $PbI_2 = 7.1 \times 10^{-9}$)?

EXERCISE 16.6B

Will precipitation occur from a solution that has 2.5 ppm $MgCl_2$ and a pH of 10.35? Assume the solution density is 1.00 g/mL [K_{sp} for $Mg(OH)_2 = 1.8 \times 10^{-11}$].

▲ **FIGURE 16.4   Example 16.7 illustrated**

Top: A few drops of concentrated KI(aq) are added to dilute $Pb(NO_3)_2$(aq). Bottom: The mixture after a few moments.

Example 16.7   A Conceptual Example

Figure 16.4 shows the result of adding a few drops of concentrated KI(aq) to a dilute solution of $Pb(NO_3)_2$(aq). What is the yellow solid that appears at first? Explain why it then disappears.

ANALYSIS AND CONCLUSIONS

Where the drops of concentrated KI(aq) first enter the $Pb(NO_3)_2$ solution, there is a localized excess of iodide ion. In this region of the solution, [I^-] is large enough so that

$$Q_{ip} = [Pb^{2+}][I^-]^2 > K_{sp}$$

Because $Q_{ip} > K_{sp}$, yellow PbI_2(s) precipitates. When the PbI_2(s) that initially formed settles below the point of entry of the KI(aq), however, [I^-] is no longer large enough to maintain a saturated solution for the concentration of Pb^{2+} present, and therefore the solid redissolves. Based on a *uniform* distribution of I^- ion throughout the solution, the following holds true:

$$Q_{ip} = [Pb^{2+}][I^-]^2 < K_{sp}$$

EXERCISE 16.7A

Describe how you might use observations of the type suggested by the photographs of Figure 16.4 to estimate the value of K_{sp} for PbI_2.

EXERCISE 16.7B

Consider the following situation similar to Example 16.7: Initially, the 100.0 mL of dilute solution in the beaker is 1×10^{-4} M and the more concentrated solution added dropwise is 0.10 M. **(a)** If the dilute solution is $Pb(NO_3)_2$(aq) and the more concentrated one is KI(aq), will the observations made resemble Figure 16.4? Explain. **(b)** What should be observed if the dilute solution is KI(aq) and the more concentrated one is $Pb(NO_3)_2$? Explain.

Example 16.7 illustrates qualitatively that in applying the precipitation criteria, we must consider the effect of dilution when solutions are mixed. Example 16.8 demonstrates the dilution effect quantitatively.

Example 16.8

If 0.100 L of 0.0015 M $MgCl_2$ and 0.200 L of 0.025 M NaF are mixed, should a precipitate of MgF_2 form?

$$MgF_2(s) \rightleftharpoons Mg^{2+}(aq) + 2\,F^-(aq) \qquad K_{sp} = 3.7 \times 10^{-8}$$

STRATEGY

We can apply the same three-step strategy as in Example 16.6, but now we will incorporate into Step 1 the assumption that the volume of the mixture of solutions is 0.100 L + 0.200 L = 0.300 L.

SOLUTION

Step 1: *Initial concentrations of ions:* First, we need to determine $[Mg^{2+}]$ and $[F^-]$ as they initially exist in this mixed solution.

$$[Mg^{2+}] = \frac{0.100 \text{ L} \times \dfrac{0.0015 \text{ mol MgCl}_2}{\text{L}} \times \dfrac{1 \text{ mol Mg}^{2+}}{1 \text{ mol MgCl}_2}}{0.300 \text{ L}} = 5.0 \times 10^{-4} \text{ M}$$

$$[F^-] = \frac{0.200 \text{ L} \times \dfrac{0.025 \text{ mol NaF}}{\text{L}} \times \dfrac{1 \text{ mol F}^-}{1 \text{ mol NaF}}}{0.300 \text{ L}} = 1.7 \times 10^{-2} \text{ M}$$

Step 2: *Evaluation of Q_{ip}:*

$$Q_{ip} = [Mg^{2+}][F^-]^2 = (5.0 \times 10^{-4})(1.7 \times 10^{-2})^2 = 1.4 \times 10^{-7}$$

Step 3: *Comparison of Q_{ip} and K_{sp}:* Because $Q_{ip} > K_{sp}$, we conclude that precipitation should occur.

$Q_{ip} = 1.4 \times 10^{-7}$ is greater than $K_{sp} = 3.7 \times 10^{-8}$.

EXERCISE 16.8A

Should a precipitate of $MgF_2(s)$ form when equal volumes of 0.0010 M $MgCl_2(aq)$ and 0.020 M $NaF(aq)$ are mixed?

EXERCISE 16.8B

How many grams of $KI(s)$ should be dissolved in 10.5 L of 0.00200 M $Pb(NO_3)_2(aq)$ so that precipitation of $PbI_2(s)$ *will just begin?*

To Determine Whether Precipitation Is Complete

We have criteria for determining whether or not precipitation will occur, but how can we determine whether precipitation goes to completion if it does occur? That is, what percentage of the ion in question is precipitated, and what percentage remains in solution? A slightly soluble solid never totally precipitates from solution, but we generally consider precipitation to be essentially complete if about 99.9% of the target ion is precipitated and only 0.1% or less is left in solution.

Three conditions that generally *favor completeness of precipitation* are

1. a very small value of K_{sp}.
2. a high initial concentration of the target ion.
3. a concentration of common ion that greatly exceeds that of the target ion.

With these conditions, the solubility of the precipitate is suppressed, meaning that the concentration of the target ion remaining in solution is quite low—only a tiny percentage of its initial concentration.

Example 16.9

To a solution with $[Ca^{2+}] = 0.0050$ M, we add sufficient solid ammonium oxalate, $(NH_4)_2C_2O_4(s)$, to make the initial $[C_2O_4{}^{2-}] = 0.0051$ M. Will the precipitation of Ca^{2+} as $CaC_2O_4(s)$ be complete?

$$CaC_2O_4(s) \rightleftharpoons Ca^{2+}(aq) + C_2O_4{}^{2-}(aq) \qquad K_{sp} = 2.7 \times 10^{-9}$$

STRATEGY

We can begin with the three-step approach to determine whether precipitation occurs. If no precipitation occurs, we need not proceed further—precipitation obviously will not be complete. If precipitation does occur, we will need to make further calculations, which we will do in two parts: First, we will treat the situation *as if* complete precipitation occurred. This will enable us to determine the excess $[C_2O_4{}^{2-}]$ present. Then, we will solve for the solubility of a slightly soluble solute in the presence of a common ion in a calculation similar to that of Example 16.5.

▲ Some types of kidney stones are mainly calcium oxalate, $CaC_2O_4(s)$

SOLUTION

First, we use our usual three-step approach to determine if precipitation occurs.

Step 1: *Initial concentrations of ions:* These are given. $[Ca^{2+}] = 0.0050$ M, and $[C_2O_4^{2-}] = 0.0051$ M.

Step 2: *Evaluation of Q_{ip}:* $Q_{ip} = [Ca^{2+}][C_2O_4^{2-}] = (5.0 \times 10^{-3})(5.1 \times 10^{-3}) = 2.6 \times 10^{-5}$

Step 3: *Comparison of Q_{ip} and K_{sp}:* $Q_{ip} = 2.6 \times 10^{-5}$ exceeds $K_{sp} = 2.7 \times 10^{-9}$

Because $Q_{ip} > K_{sp}$, precipitation should occur.

Next, we follow the two-part approach outlined in our basic strategy.

Assume complete precipitation.

We describe the stoichiometry of the precipitation in a format resembling the ICE format for equilibrium calculations (see Example 15.17b).

The reaction:	$Ca^{2+}(aq)$	$+$	$C_2O_4^{2-}(aq)$	$\longrightarrow$	$CaC_2O_4(s)$
Initial concentrations, M:	0.0050		0.0051		
Changes, M:	-0.0050		-0.0050		
After precipitation, M:	0		0.0001		

The solubility equilibrium.

Here we enter the appropriate data into the ICE format for solubility equilibrium in the presence of a common ion.

The reaction:	$CaC_2O_4(s)$	$\rightleftharpoons$	$Ca^{2+}(aq)$	$+$	$C_2O_4^{2-}(aq)$
Initial concentrations, M:			0		0.0001
Changes, M:			$+s$		$+s$
Equilibrium concentrations, M:			s		$(0.0001 + s)$

As usual, we assume that $s \ll 0.0001$ and that $(0.0001 + s) \approx 0.0001$. (We test the validity of this assumption in the Assessment.)

$$K_{sp} = [Ca^{2+}][C_2O_4^{2-}] = s(0.0001 + s)$$
$$\approx s \times 0.0001 = 2.7 \times 10^{-9}$$
$$s = [Ca^{2+}] = 3 \times 10^{-5} \text{ M}$$

We calculate the percentage of Ca^{2+} remaining in solution as the ratio of the remaining Ca^{2+} concentration to the initial Ca^{2+} concentration $(5 \times 10^{-3}$ M).

$$\frac{3 \times 10^{-5} \text{ M}}{5 \times 10^{-3} \text{ M}} \times 100 = 0.6\%$$

Our rule for complete precipitation requires that less than 0.1% of an ion should remain in solution. We therefore conclude that precipitation is incomplete.

ASSESSMENT

In this problem, we made the assumption that the contribution of s to the term $0.0001 + s$ was negligible. Such assumptions are usually valid in common ion situations because the concentration of the common ion is generally relatively large, but 0.0001 M is not a high concentration. If we had not made the simplifying assumption and solved a quadratic equation, we would have obtained $s = 2 \times 10^{-5}$ M. From this result, the percent Ca^{2+} remaining in solution is 0.4%, but our conclusion that precipitation is incomplete is still valid. If our task had been to calculate $[Ca^{2+}]$ remaining in solution, we would not have made the simplifying assumption.

EXERCISE 16.9A

Consider a solution in which $[Ca^{2+}] = 0.0050$ M. If we add sufficient solid ammonium oxalate, $(NH_4)_2C_2O_4(s)$, so that the solution is also $[C_2O_4^{2-}] = 0.0100$ M, will the precipitation of Ca^{2+} as $CaC_2O_4(s)$ be complete?

EXERCISE 16.9B

The two members of each pair of ions are brought together in solution. *Without doing detailed calculations,* arrange the pairs according to how close to completeness the cation precipitation is—farthest from completeness first and closest to completeness last.

(a) $[Ca^{2+}] = 0.110$ M and $[SO_4^{2-}] = 0.090$ M

(b) $[Pb^{2+}] = 0.12$ M and $[Cl^-] = 0.25$ M

(c) $[Mg^{2+}] = 0.12$ M and $[SO_4^{2-}] = 0.15$ M

(d) $[Ag^+] = 0.050$ M and $[Cl^-] = 0.055$ M

Selective Precipitation: Precipitating One Ion and Not Another

In Example 16.4, we compared the solubilities of AgCl, AgBr, and AgI. We found AgI to be the least soluble and AgCl most soluble. Now let us consider the series of questions in Example 16.10 to see how the difference in solubility of AgI and AgCl enables us to separate I^- from Cl^- in aqueous solution.

Example 16.10

An aqueous solution that is 2.00 M in $AgNO_3$ is slowly added from a buret to an aqueous solution that is 0.0100 M in Cl^- and 0.0100 M in I^- (Figure 16.5).

$$AgCl(s) \rightleftharpoons Ag^+(aq) + Cl^-(aq) \qquad K_{sp} = 1.8 \times 10^{-10}$$

$$AgI(s) \rightleftharpoons Ag^+(aq) + I^-(aq) \qquad K_{sp} = 8.5 \times 10^{-17}$$

(a) Which ion, Cl^- or I^-, is the first to precipitate from solution?

(b) When the second ion begins to precipitate, what is the remaining concentration of the first ion?

(c) Is separation of the two ions by selective precipitation feasible?

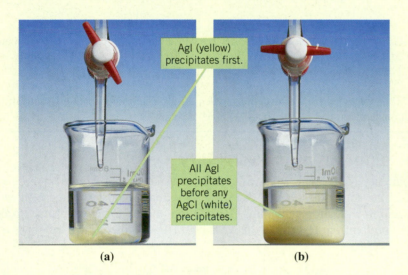

AgI (yellow) precipitates first.

All AgI precipitates before any AgCl (white) precipitates.

(a)　　　　**(b)**

◀ **FIGURE 16.5 Selective precipitation—Example 16.10 illustrated**

(a) The first precipitate to form when $AgNO_3(aq)$ is added to an aqueous solution containing $[Cl^-]$ and $[I^-]$ is yellow AgI(s). (b) Essentially all the I^- has precipitated before the precipitation of white AgCl(s) begins.

STRATEGY

The molarity of the silver nitrate solution is 200 times greater than $[Cl^-]$ or $[I^-]$, which means that only a small volume of $AgNO_3(aq)$ is required in the precipitation. Because of this, we can neglect throughout the calculation the slight dilution that occurs as the $AgNO_3(aq)$ is added to the beaker.

SOLUTION

(a) We must find the $[Ag^+]$ necessary to precipitate each anion. Precipitation will begin when the solution has just been saturated, a point at which $Q_{ip} = K_{sp}$.

To precipitate Cl^-.

When Q_{ip} is equal to K_{sp} for silver chloride, $[Cl^-] = 0.0100$ M. We then solve the K_{sp} expression for $[Ag^+]$.

$$Q_{ip} = [Ag^+][Cl^-] = K_{sp} = 1.8 \times 10^{-10}$$

$$[Ag^+] = \frac{K_{sp}}{[Cl^-]} = \frac{1.8 \times 10^{-10}}{0.0100} = 1.8 \times 10^{-8} \text{ M}$$

To precipitate I^-.

When Q_{ip} is equal to K_{sp} for silver iodide, $[I^-] = 0.0100$ M. We then solve the K_{sp} expression for $[Ag^+]$.

$$Q_{ip} = [Ag^+][I^-] = K_{sp} = 8.5 \times 10^{-17}$$

$$[Ag^+] = \frac{K_{sp}}{[I^-]} = \frac{8.5 \times 10^{-17}}{0.0100} = 8.5 \times 10^{-15} \text{ M}$$

Precipitation of AgI(s) begins as soon as $[Ag^+]$ = 1.8×10^{-15} M, a concentration much too low for AgCl(s) to precipitate. Therefore I^- is the first ion to precipitate.

(b) Next, we determine $[I^-]$ remaining in solution when $[Ag^+]$ = 1.8×10^{-8} M, the $[Ag^+]$ at which AgCl(s) begins to precipitate. To determine this quantity, we rearrange the K_{sp} expression for silver iodide and solve for $[I^-]$.

$$[I^-] = \frac{K_{sp}}{[Ag^+]} = \frac{8.5 \times 10^{-17}}{1.8 \times 10^{-8}} = 4.7 \times 10^{-9} \text{ M}$$

(c) Now, let us see what percent of the I^- is still in solution when AgCl(s) begins to precipitate.

$$? \% = \frac{4.7 \times 10^{-9} \text{ M}}{0.0100 \text{ M}} \times 100\% = 4.7 \times 10^{-5}\%$$

The amount of I^- remaining is far below the 0.1% we require for completeness of precipitation. Thus, complete precipitation of I^- occurs before Cl^- precipitation begins. Separation of the two anions is feasible.

ASSESSMENT

This example illustrates a general rule: The greater the difference in magnitude of the relevant K_{sp} values, the more likely it is that ion separations by selective precipitation can be achieved. The K_{sp} values of AgI and AgCl differ by a factor of about 2×10^6.

EXERCISE 16.10A

Answer the three questions posed in Example 16.10 for a solution in which the concentrations of the ions to be precipitated are $[Cl^-]$ = $[Br^-]$ = 0.0100 M, the titrant is again 2.00 M $AgNO_3$, and K_{sp} for AgBr is 5.0×10^{-13}.

EXERCISE 16.10B

Use the results of Example 16.10 and Exercise 16.10A and, *without doing detailed calculations*, determine whether I^- and Br^- can be separated by selective precipitation from a solution in which $[I^-]$ = $[Br^-]$ = 0.0100 M. Explain.

16.5   Effect of pH on Solubility

The solubility of an ionic solute is greatly affected by pH if an acid–base reaction occurs as the solute dissolves. Let us see why this should be the case. If we add a strong acid to a solution containing fluoride ion, the F^- ion, a relatively strong base, accepts a proton from H_3O^+ and the weak acid HF is formed. This acid–base reaction, which we write as

$$H_3O^+(aq) + F^-(aq) \rightleftharpoons HF(aq) + H_2O(l)$$

$$\text{Acid} \qquad \text{Base} \qquad \text{Acid} \qquad \text{Base}$$

is the reverse of the ionization of HF. Now suppose the solution containing F^- ion is a saturated solution of Ca^{2+} and F^- ions in equilibrium with $CaF_2(s)$. According to Le Châtelier's principle, as $F^-(aq)$ is converted to HF(aq), the equilibrium

$$CaF_2(s) \rightleftharpoons Ca^{2+}(aq) + 2\,F^-(aq)$$

shifts to the right. At the same time, the equilibrium involving the added acid,

$$2\,H_3O^+(aq) + 2\,F^-(aq) \rightleftharpoons 2\,HF(aq) + 2\,H_2O(l)$$

also shifts to the right as more acid is added. An overall reaction occurs in which the dissolution of $CaF_2(s)$ is stimulated:

$$CaF_2(s) + 2\,H_3O^+(aq) \rightleftharpoons Ca^{2+}(aq) + 2\,HF(aq) + 2\,H_2O(l)$$

In contrast to calcium fluoride, the solubility of silver chloride is *independent* of pH. For example, AgCl(s) does not dissolve in dilute $HNO_3(aq)$ because chloride ion, the very weak conjugate base of the strong acid HCl, does not accept a proton from H_3O^+:

$$H_3O^+(aq) + Cl^-(aq) \longrightarrow \text{no reaction}$$

Thus, the solubility of AgCl(s) does not increase as the pH is lowered.

Carbonate ion is a moderately strong base, and the solubilities of carbonates therefore depend strongly on pH. When an acid is added to an insoluble carbonate, such as $CaCO_3$, carbonate ions are rapidly converted to hydrogen carbonate ions, HCO_3^-.

These ions react further to form carbonic acid, H_2CO_3, which then dissociates into $CO_2(g)$ and water.

$$CO_3^{2-}(aq) + H_3O^+(aq) \longrightarrow HCO_3^-(aq) + H_2O(l)$$

$$HCO_3^-(aq) + H_3O^+(aq) \longrightarrow H_2CO_3(aq) + H_2O(l)$$

$$H_2CO_3(aq) \longrightarrow H_2O(l) + CO_2(g)$$

Because $CO_2(g)$ escapes, there is little opportunity for the reverse reactions to occur.

To say that the solubility of an "insoluble" carbonate increases as a solution is made more acidic means that the cation concentration in the solution increases. As we see from the reaction of $CaCO_3(s)$ and $HCl(aq)$,

$$CaCO_3(s) + 2\,HCl(aq) \longrightarrow Ca^{2+}(aq) + 2\,Cl^-(aq) + H_2O(l) + CO_2(g)$$

an overall reaction occurs in which carbonate ions decompose. The primary solute in the acidic solution is not calcium carbonate but rather calcium chloride. Figure 16.6 shows a simple geological field test based on this reaction.

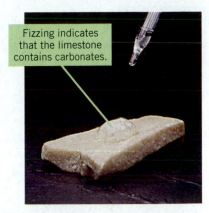

Fizzing indicates that the limestone contains carbonates.

▲ **FIGURE 16.6** **The reaction of an acid and a carbonate**
When a drop of $HCl(aq)$ is placed on a piece of limestone, a characteristic fizzing occurs as $CO_2(g)$ is formed.

Example 16.11

What is the molar solubility of $Mg(OH)_2(s)$ in a buffer solution having $[OH^-] = 1.0 \times 10^{-5}$ M, that is, pH = 9.00?

$$Mg(OH)_2(s) \rightleftharpoons Mg^{2+}(aq) + 2\,OH^-(aq) \qquad K_{sp} = 1.8 \times 10^{-11}$$

STRATEGY

First, note that because the solution is a buffer, any OH^- coming from the dissolution of $Mg(OH)_2(s)$ is neutralized by the acidic component of the buffer. Therefore, the pH will remain at 9.00 and the $[OH^-]$ at 1.0×10^{-5} M. We will use the K_{sp} expression and $[OH^-] = 1.0 \times 10^{-5}$ M to calculate $[Mg^{2+}]$ in the solution.

SOLUTION

$$K_{sp} = [Mg^{2+}][OH^-]^2 = 1.8 \times 10^{-11}$$

$$K_{sp} = [Mg^{2+}] \times (1.0 \times 10^{-5})^2 = 1.8 \times 10^{-11}$$

$$[Mg^{2+}] = \frac{1.8 \times 10^{-11}}{1.0 \times 10^{-10}} = 0.18\text{ M}$$

For every mole of $Mg^{2+}(aq)$ appearing in solution, 1 mol of $Mg(OH)_2(s)$ must have dissolved. Thus the molar solubility of $Mg(OH)_2$ and the equilibrium concentration of Mg^{2+} are the same, meaning that the molar solubility of $Mg(OH)_2$ at pH 9.00 is 0.18 mol $Mg(OH)_2$/L.

EXERCISE 16.11A

What is the molar solubility of $Fe(OH)_2$ in a buffer solution with pH 6.50 [K_{sp} for $Fe(OH)_2 = 8.0 \times 10^{-16}$]?

EXERCISE 16.11B

Can a buffer solution that is 0.520 M CH_3COOH and 0.180 M CH_3COONa also have $[Fe^{3+}] = 1.0 \times 10^{-3}$ M without forming a precipitate of $Fe(OH)_3(s)$? Explain [K_a for $CH_3COOH = 1.8 \times 10^{-5}$, K_{sp} for $Fe(OH)_3 = 4 \times 10^{-38}$].

If we were to use the method of Example 16.11 to calculate a molar solubility of $Mg(OH)_2$ at pH 3.00, we would get a solubility of 1.8×10^{11} M—an impossible result. We can understand why it is impossible with the help of Figure 12.7. There, we see several H_2O molecules for every ion; the H_2O molecules provide the ion–dipole forces that keep the ions in solution. This suggests that there cannot be as many ions in solution as there are H_2O molecules, and so the maximum molar solubility of an ionic solute should be less than the 55.5 mol of H_2O to be found in one liter of pure water. If the pH of 3.00 is supplied by $HCl(aq)$, the true molar solubility is that of the $MgCl_2$ formed in the reaction

Dissolution of $Mg(OH)_2$ by Acid animation

$$Mg(OH)_2(s) + 2\,H_3O^+(aq) + 2\,Cl^-(aq) \longrightarrow Mg^{2+}(aq) + 2\,Cl^-(aq) + 4\,H_2O(l)$$

pH, Solubility, and Tooth Decay

Tooth enamel, the hard outer part of human teeth, is composed principally of the mineral *hydroxyapatite*, $Ca_5(PO_4)_3OH$. Hydroxyapatite is insoluble in water but somewhat soluble in acidic solution, and therein lie the origins of tooth decay.

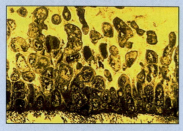

◀ Plaque on tooth enamel as seen in a false color image made with an electron microscope.

A carbohydrate–protein combination called *mucin* causes a film called *plaque* to form on teeth. If not removed by brushing and flossing, the buildup of plaque traps food particles. Bacteria ferment carbohydrates in these particles, producing lactic acid ($CH_3CHOHCOOH$). Saliva cannot penetrate the plaque and thus cannot buffer against the buildup of lactic acid. As a result, the pH on the tooth surface drops to as low as 4.5. The H_3O^+ present in this acidic environment neutralizes the OH^- in

hydroxyapatite and converts some of the PO_4^{3-} to HPO_4^{2-}. Because the calcium salt of HPO_4^{2-} is water-soluble, the result of HPO_4^{2-} formation is that some hydroxyapatite dissolves:

$$Ca_5(PO_4)_3OH(s) + 4\,H_3O^+(aq)$$
$$\longrightarrow 5\,Ca^{2+}(aq) + 3\,HPO_4^{2-}(aq) + 5\,H_2O(l)$$

If unchecked, this dissolution of tooth enamel produces cavities.

Tooth erosion is even more rapid in people with *bulimia*, an eating disorder characterized by binge eating followed by vomiting. Hydrochloric acid vomited from the stomach drops the pH in the mouth to as low as 1.5, thus stimulating the dissolution of hydroxyapatite.

One way to fight tooth decay is to add fluorides to drinking water or to toothpaste. Fluorides convert some of the hydroxyapatite in enamel to *fluorapatite*:

$$Ca_5(PO_4)_3OH(s) + F^-(aq) \rightleftharpoons Ca_5(PO_4)_3F(s) + OH^-(aq)$$

Hydroxyapatite   Fluorapatite

Fluorapatite is less soluble in acidic media than is hydroxyapatite, because F^- is a weaker base than OH^-. As a result, tooth decay is slowed.

Example 16.12   A Conceptual Example

Without doing detailed calculations, determine in which of the following solutions $Mg(OH)_2(s)$ is most soluble: **(a)** 1.00 M NH_3, **(b)** 1.00 M NH_3/1.00 M NH_4^+, **(c)** 1.00 M NH_4Cl.

$$Mg(OH)_2(s) \rightleftharpoons Mg^{2+}(aq) + 2\,OH^-(aq) \qquad K_{sp} = 1.8 \times 10^{-11}$$

ANALYSIS AND CONCLUSIONS

At high pH values, OH^- in the solution will act as a common ion and reduce the solubility of the $Mg(OH)_2$. At lower pH values, OH^- produced by the dissolution of $Mg(OH)_2$ is neutralized, and this causes more $Mg(OH)_2$ to dissolve, thus increasing the solubility. We need to compare the pH values of the three solutions, and the solution having the lowest pH value will dissolve the greatest amount of the $Mg(OH)_2(s)$.

(a) We know that the pH of 1.00 M NH_3, a weak base, is greater than 7.

(b) Because $[NH_3] = [NH_4^+]$, the 1.00 M NH_3/1.00 M NH_4^+ buffer solution has a pH equal to the pK_a of NH_4^+. For NH_3, $K_b = 1.8 \times 10^{-5}$ and $pK_b = -\log(1.8 \times 10^{-5})$. For NH_4^+, $pK_a = 14.00 - pK_b = 14.00 - 4.74 = 9.26$. A buffer solution with pH = 9.26 is moderately basic.

(c) In $NH_4Cl(aq)$, NH_4^+ hydrolyzes and Cl^- does not:

$$NH_4^+(aq) + H_2O(l) \rightleftharpoons H_3O^+(aq) + NH_3(aq) \qquad K_a = K_w/K_b$$

This solution should be acidic with a pH less than 7. Being the solution with the lowest pH, the 1.00 M NH_4Cl solution should dissolve the greatest amount of $Mg(OH)_2$.

16.6 Equilibria Involving Complex Ions

At first sight, the data in Table 16.2 are puzzling. They deal with the solubility of $AgCl(s)$ in solutions containing the common ion Cl^-. The solubilities we might *predict* based on calculations such as in Example 16.5 are much smaller than the solubility in water, and they become progressively smaller as $[Cl^-]$ increases. In contrast, the *measured* solubilities are not nearly as small as predicted, and in solutions with $[Cl^-]$ greater than about 0.03 M, the solubility of $AgCl(s)$ actually increases as $[Cl^-]$ increases. We need an explanation other than the common ion effect for these unexpected data, and we find it in complex ion formation.

Table 16.2	Solubility of Silver Chloride in NaCl(aq)		
$[Cl^-]$, M	**Predicted Solubility (mol AgCl/L) $\times 10^5$**	**Measured Solubility (mol AgCl/L) $\times 10^5$**	
0.000	1.3	1.3	Decreasing solubility
0.0039	0.0046	0.072	
0.036	0.00050	0.19	
0.35	0.000051	1.7	
1.4	0.000013	18	Increasing solubility
2.9	0.0000063	1000	

Complex Ion Formation

A **complex ion** is a polyatomic cation or anion consisting of a central metal atom or ion that has other groups called **ligands** bonded to it. Four common ligands featured in this section are the anions Cl^- and OH^- and the molecules H_2O and NH_3.

The Lewis structures of these ligands show that they all have at least one lone pair of electrons on the atom through which bonding occurs:

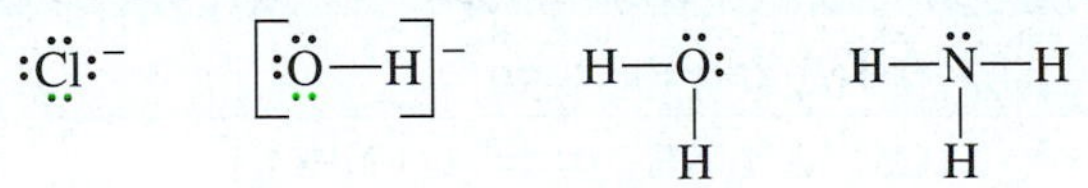

$$:\ddot{C}l:^- \qquad \left[:\ddot{O}-H\right]^- \qquad H-\ddot{O}: \qquad H-\ddot{N}-H$$
$$\qquad\qquad\qquad\qquad\qquad\qquad | \qquad\qquad |$$
$$\qquad\qquad\qquad\qquad\qquad\qquad H \qquad\qquad H$$

One way to describe complex ion formation is to say that the metal center is a *Lewis acid* that accepts electron pairs from the ligands, which are *Lewis bases*. For the ligands shown above, bonding to the metal center occurs through the sharing of one pair of electrons from the ligand, resulting in a coordinate covalent bond between the ligand and metal center.

▲ FIGURE 16.7　The structure of the complex ion [AgCl₂]⁻

This illustration suggests that lone-pair electrons on the Cl⁻ ions are used to form coordinate covalent bonds with the central Ag⁺ ion. The complex ion has a linear structure. Bonding and structures of complex ions are discussed in Chapter 22.

The complex ion responsible for the increased solubility of AgCl(s) at high [Cl⁻] consists of a central Ag⁺ ion plus two Cl⁻ ions as ligands. The formation of this complex ion, [AgCl₂]⁻, is represented by the equation

$$Ag^+(aq) + 2\,Cl^-(aq) \rightleftharpoons [AgCl_2]^-(aq)$$

and the structure of the complex ion is suggested in Figure 16.7. This complex ion, an anion, has a net charge of 1− because it is made of one Ag⁺ cation and two Cl⁻ anions.

Copper(II) ions and ammonia molecules form a complex cation; the central copper ion has a charge of 2+, and the ammonia molecules are neutral:

$$\underset{\text{Blue}}{Cu^{2+}(aq)} + 4\,NH_3(aq) \rightleftharpoons \underset{\text{Deep blue-violet}}{[Cu(NH_3)_4]^{2+}(aq)}$$

The formation of this complex ion is shown in Figure 16.8.

▶ FIGURE 16.8　Complex ion formation

The dilute CuSO₄(aq) solution on the left owes its pale blue color to Cu²⁺(aq). When we add NH₃(aq) (here labeled "conc ammonium hydroxide"), the solution color changes to a deep blue-violet (right) as Cu²⁺(aq) is converted to [Cu(NH₃)₄]²⁺. The molecular view shows the ions [Cu(NH₃)₄]²⁺ and SO₄²⁻.

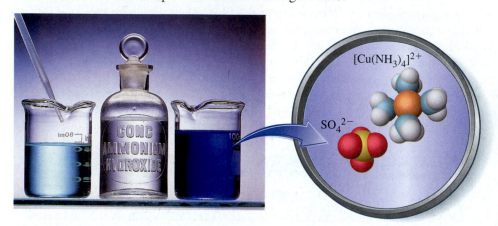

Not all the Ag⁺ in a solution of Cl⁻(aq) exists as [AgCl₂]⁻, and not all the Cu²⁺ in a solution of NH₃(aq) exists as [Cu(NH₃)₄]²⁺. This is true because the formation reaction of a complex ion is reversible, and we describe the equilibrium state in the reaction through an equilibrium constant called a **formation constant,** represented by the symbol K_f. Thus, for the formation of the complex ions [AgCl₂]⁻ and [Cu(NH₃)₄]²⁺, we have

$$Ag^+(aq) + 2\,Cl^-(aq) \rightleftharpoons [AgCl_2]^-(aq)$$

$$K_f = \frac{[[AgCl_2]^-]}{[Ag^+][Cl^-]^2} = 1.2 \times 10^8$$

$$Cu^{2+}(aq) + 4\,NH_3(aq) \rightleftharpoons [Cu(NH_3)_4]^{2+}(aq)$$

$$K_f = \frac{[[Cu(NH_3)_4]^{2+}]}{[Cu^{2+}][NH_3]^4} = 1.1 \times 10^{13}$$

Table 16.3 lists some representative formation constants. A more extensive tabulation is given in Appendix C.

Complex ions form in stepwise fashion. For example, Cu²⁺(aq) + NH₃(aq) ⇌ [CuNH₃]²⁺(aq) is described by K_1; [CuNH₃]²⁺(aq) + NH₃(aq) ⇌ [Cu(NH₃)₂]²⁺ is described by K_2, and so on. For our purposes, we will use an overall value K_f for formation of the final complex.

Application Note

The copper-ammonia complex ion depicted in Figure 16.8 is used in the manufacture of rayon and cellophane. The iron(III) salt of the ferrocyanide ion [Fe(CN)₆]³⁻, listed in Table 16.3, is the artist's pigment known as Prussian Blue.

Table 16.3　Some Formation Constants for Complex Ions

Complex	Equilibrium Reaction	K_f
$[Co(NH_3)_6]^{3+}$	$Co^{3+} + 6\,NH_3 \rightleftharpoons [Co(NH_3)_6]^{3+}$	4.5×10^{33}
$[Cu(NH_3)_4]^{2+}$	$Cu^{2+} + 4\,NH_3 \rightleftharpoons [Cu(NH_3)_4]^{2+}$	1.1×10^{13}
$[Fe(CN)_6]^{4-}$	$Fe^{2+} + 6\,CN^- \rightleftharpoons [Fe(CN)_6]^{4-}$	1×10^{37}
$[Fe(CN)_6]^{3-}$	$Fe^{3+} + 6\,CN^- \rightleftharpoons [Fe(CN)_6]^{3-}$	1×10^{42}
$[PbCl_3]^-$	$Pb^{2+} + 3\,Cl^- \rightleftharpoons [PbCl_3]^-$	2.4×10^1
$[Ag(NH_3)_2]^+$	$Ag^+ + 2\,NH_3 \rightleftharpoons [Ag(NH_3)_2]^+$	1.6×10^7
$[AgCl_2]^-$	$Ag^+ + 2\,Cl^- \rightleftharpoons [AgCl_2]^-$	1.2×10^8
$[Ag(S_2O_3)_2]^{3-}$	$Ag^+ + 2\,S_2O_3^{2-} \rightleftharpoons [Ag(S_2O_3)_2]^{3-}$	1.7×10^{13}
$[Zn(NH_3)_4]^{2+}$	$Zn^{2+} + 4\,NH_3 \rightleftharpoons [Zn(NH_3)_4]^{2+}$	4.1×10^8
$[Zn(CN)_4]^{2-}$	$Zn^{2+} + 4\,CN^- \rightleftharpoons [Zn(CN)_4]^{2-}$	1×10^{18}
$[Zn(OH)_4]^{2-}$	$Zn^{2+} + 4\,OH^- \rightleftharpoons [Zn(OH)_4]^{2-}$	4.6×10^{17}

▲ **FIGURE 16.9**   **Complex ion formation and solute solubility**
A precipitate of AgCl(s) (left beaker) readily dissolves in an aqueous solution of NH_3 because of the formation of the complex ion $[Ag(NH_3)_2]^+$ (right beaker). The molecular view of the final solution shows $[Ag(NH_3)_2]^+$ and Cl^- ions and NH_3 molecules.
QUESTION: What other ions are present but not shown for the solution in the right beaker?

Complex Ion Formation and Solubilities

Figure 16.9 shows that water-insoluble silver chloride does dissolve in aqueous ammonia. One Ag^+ ion from AgCl(s) combines with two NH_3 molecules to form the complex ion $[Ag(NH_3)_2]^+$. The Cl^- from AgCl(s) goes into solution as $Cl^-(aq)$. This dissolution is best described as the overall reaction that results by combining two equilibrium processes:

(1) $\qquad\qquad AgCl(s) \rightleftharpoons Ag^+(aq) + Cl^-(aq) \qquad K_{sp} = 1.8 \times 10^{-10}$

(2) $\qquad Ag^+(aq) + 2\,NH_3(aq) \rightleftharpoons [Ag(NH_3)_2]^+(aq) \qquad K_f = 1.6 \times 10^7$

Overall reaction: $\quad AgCl(s) + 2\,NH_3(aq) \rightleftharpoons [Ag(NH_3)_2]^+(aq) + Cl^-(aq)$

$$K_c = K_{sp} \times K_f = 2.9 \times 10^{-3}$$

The equilibrium constant for the dissolution of AgCl(s) in $NH_3(aq)$ is not large ($K_c = 2.9 \times 10^{-3}$). Nevertheless, in the presence of a high concentration of $NH_3(aq)$, equilibrium lies far enough to the right that a significant amount of AgCl(s) dissolves.

We can write similar equations and equilibrium constants for the action of $NH_3(aq)$ on AgBr(s) and AgI(s). In each case, $K_c = K_{sp} \times K_f$. The equilibrium constants K_c for the dissolution of AgCl(s), AgBr(s), and AgI(s) in $NH_3(aq)$ are 2.9×10^{-3}, 8.0×10^{-6}, and 1.4×10^{-9}, respectively. These values are consistent with the observation that AgCl(s) is moderately soluble, AgBr(s) is slightly soluble, and AgI(s) is only very slightly soluble in aqueous ammonia. The K_c values are directly proportional to values of K_{sp} because K_f is the same for each reaction. Thus, the solubilities of AgCl, AgBr, and AgI in $NH_3(aq)$ parallel their K_{sp} values:

Solubility in $NH_3(aq)$ decreases: $\quad$ AgCl $\quad > \quad$ AgBr $\quad > \quad$ AgI

K_{sp} *decreases:* $\qquad\qquad 1.8 \times 10^{-10} \quad 5.0 \times 10^{-13} \quad 8.5 \times 10^{-17}$

Silver bromide, as an emulsion in gelatin, is used in photographic film. The photographic image is a deposit of metallic silver in the emulsion, a deposit produced when the film is first exposed to light and then treated with a mild reducing agent, such as hydroquinone, $C_6H_4(OH)_2$. After the film has been developed in this way, it is necessary to "fix" it—to remove the unexposed AgBr(s) so that the film will not darken on further exposure to light. For this purpose, photographers can use sodium thiosulfate, $Na_2S_2O_3$ (also called sodium hyposulfite, or "hypo").

The dissolving of AgBr(s) in $Na_2S_2O_3(aq)$, like that of AgBr(s) in $NH_3(aq)$, is the overall result of two simultaneous equilibrium processes:

(1) $\qquad\qquad AgBr(s) \rightleftharpoons Ag^+(aq) + Br^-(aq) \qquad K_{sp} = 5.0 \times 10^{-13}$

(2) $\qquad Ag^+(aq) + 2\,S_2O_3^{2-}(aq) \rightleftharpoons [Ag(S_2O_3)_2]^{3-}(aq) \qquad K_f = 1.7 \times 10^{13}$

Overall reaction: $\quad AgBr(s) + 2\,S_2O_3^{2-}(aq) \rightleftharpoons [Ag(S_2O_3)_2]^{3-}(aq) + Br^-(aq)$

$$K_c = K_{sp} \times K_f = 5.0 \times 10^{-13} \times 1.7 \times 10^{13} = 8.5$$

Now we can see why sodium thiosulfate is used as a photographic fixer and ammonia is not. The equilibrium constant for the dissolution of $AgBr(s)$ in $S_2O_3^{2-}(aq)$ ($K_c = 8.5$) is about a million times larger than for its dissolution in $NH_3(aq)$ ($K_c = 8.0 \times 10^{-6}$). Sodium thiosulfate solutions can dissolve the excess $AgBr(s)$ from photographic film more completely than ammonia solutions can.

The three examples that follow illustrate three common types of calculations involving complex ions:

- In Example 16.13, we illustrate the use of a K_f expression to calculate equilibrium concentrations.
- In Example 16.14, we apply precipitation criteria to solutions containing complex ions.
- In Example 16.15, we calculate a solubility when complex ions are formed.

Example 16.13

Calculate the concentration of free silver ion, $[Ag^+]$, in an aqueous solution prepared as 0.10 M $AgNO_3$ and 3.0 M NH_3.

$$Ag^+(aq) + 2 NH_3(aq) \rightleftharpoons [Ag(NH_3)_2]^+(aq) \qquad K_f = 1.6 \times 10^7$$

STRATEGY

First, because K_f is such a large number, we can assume that the formation of $[Ag(NH_3)_2]^+$ goes to completion. We can then establish the concentrations of $[Ag(NH_3)_2]^+$ and free NH_3 through a straightforward stoichiometric calculation. Next, we can return to the formation reaction for the complex ion and consider the extent to which it proceeds in the reverse direction. For this calculation, we will use the K_f expression and solve for $x = [Ag^+]$.

SOLUTION

We complete the bottom row of our ICE tabulation to establish the concentrations of complex ion and free ammonia, assuming that the formation reaction goes to completion.

The reaction:	$Ag^+(aq)$	+	$2 NH_3(aq)$	$\longrightarrow$	$[Ag(NH_3)_2]^+(aq)$
Initial concentrations, M:	0.10		3.0		0
Changes, M:	−0.10		−0.20		+0.10
Final concentrations, M:	0		2.8		0.10

Now, we restate the question in this way: What is $[Ag^+]$ in a solution that is 0.10 M $[Ag(NH_3)_2]^+$ and 2.8 M NH_3? To answer this question, we need to focus on the reverse direction in the reaction for $[Ag(NH_3)_2]^+$ formation.

The reaction:	Ag^+	+	$2 NH_3$	$\rightleftharpoons$	$[Ag(NH_3)_2]^+$
Initial concentrations, M:	0		2.8		0.10
Changes, M:	+x		+2x		−x
Equilibrium concentrations, M:	x		(2.8 + 2x)		(0.10 − x)

When we substitute the equilibrium concentrations into the K_f expression, we obtain an algebraic equation that we must solve for $x = [Ag^+]$.

$$\frac{[[Ag(NH_3)_2]^+]}{[Ag^+][NH_3]^2} = K_f$$

$$\frac{(0.10 - x)}{x(2.8 + 2x)^2} = 1.6 \times 10^7$$

Because the formation constant is so large, practically all the silver will be present as $[Ag(NH_3)_2]^+$ and very little as Ag^+. We can therefore assume that x is very much smaller than 0.10 and that $2x$ is very much smaller than 2.8.

$$\frac{0.10}{x(2.8)^2} = 1.6 \times 10^7$$

$$0.10 = x(2.8)^2 \times 1.6 \times 10^7$$

$$x = \frac{0.10}{(2.8)^2 \times 1.6 \times 10^7}$$

$$x = [Ag^+] = 8.0 \times 10^{-10} \text{ M}$$

EXERCISE 16.13A

Calculate the concentration of free silver ion, $[Ag^+]$, in an aqueous solution that is 0.10 M $AgNO_3$ and 1.0 M $Na_2S_2O_3$.

$$Ag^+(aq) + 2 S_2O_3^{2-}(aq) \rightleftharpoons [Ag(S_2O_3)_2]^{3-}(aq) \qquad K_f = 1.7 \times 10^{13}$$

EXERCISE 16.13B

What must be the *total* NH_3 concentration in a solution that is initially 0.050 M $AgNO_3$ if $[Ag^+]$ is to be 1.0×10^{-8} M (K_f for $[Ag(NH_3)_2]^+ = 1.6 \times 10^7$)? (*Hint:* Total NH_3 concentration refers to the NH_3 that is free in solution *plus* that present as ligands in $[Ag(NH_3)_2]^+$.)

Example 16.14

If 1.00 g KBr is added to 1.00 L of the solution described in Example 16.13, should any AgBr(s) precipitate from the solution?

$$AgBr(s) \rightleftharpoons Ag^+(aq) + Br^-(aq) \qquad K_{sp} = 5.0 \times 10^{-13}$$

STRATEGY

We need to compare Q_{ip} for the described solution with K_{sp} for silver bromide. In the Q_{ip} expression, $[Ag^+]$ is that calculated in Example 16.13. We can establish $[Br^-]$ from data given in this example.

SOLUTION

From Example 16.13, we know that $[Ag^+] = 8.0 \times 10^{-10}$ M. We determine $[Br^-]$ when 1.00 g KBr is dissolved in 1.00 L of the solution with the equation

$$[Br^-] = \frac{1.00 \text{ g KBr} \times \dfrac{1 \text{ mol KBr}}{119.0 \text{ g KBr}} \times \dfrac{1 \text{ mol Br}^-}{1 \text{ mol KBr}}}{1.00 \text{ L}} = 8.40 \times 10^{-3} \text{ M}$$

$$Q_{ip} = [Ag^+]_{initial} \times [Br^-]_{initial} = (8.0 \times 10^{-10})(8.40 \times 10^{-3}) = 6.7 \times 10^{-12}$$

Because $Q_{ip} > K_{sp}$, we conclude that some AgBr(s) should precipitate from the solution.

EXERCISE 16.14A

If 1.00 g KI is added to 1.00 L of the solution described in Exercise 16.13A, should any AgI(s) precipitate from the solution?

$$AgI(s) \rightleftharpoons Ag^+(aq) + I^-(aq) \qquad K_{sp} = 8.5 \times 10^{-17}$$

EXERCISE 16.14B

What is the maximum mass of KBr that can be added to 250.0 mL of a solution that has $[NH_3] = 1.00$ M and $[Ag(NH_3)_2]^+ = 0.050$ M without forming a precipitate of AgBr(s) (K_f for $[Ag(NH_3)_2]^+ = 1.6 \times 10^7$, K_{sp} for AgBr $= 5.0 \times 10^{-13}$)?

▲ The silver–ammonia complex is sometimes used to coat glass with silver in making mirrors, as illustrated here by the silver-coated flask. The $[Ag(NH_3)_2]^+$ is reduced to silver metal more slowly than is Ag^+, giving a much more uniform silver coat than is given by Ag^+ alone.

Example 16.15

What is the molar solubility of AgBr(s) in 3.0 M NH_3?

$$AgBr(s) + 2 NH_3(aq) \rightleftharpoons [Ag(NH_3)_2]^+(aq) + Br^-(aq) \qquad K_c = 8.0 \times 10^{-6}$$

STRATEGY

The key relationship in the dissolution equilibrium is that for every formula unit of AgBr(s) that is dissolved, one ion each of $[Ag(NH_3)_2]^+$ and Br^- is formed. The molar solubility is the same as the molarity of either of these ions at equilibrium. We calculate these ion molarities in an equilibrium calculation based on the K_c expression.

SOLUTION

We can use the ICE format to establish equilibrium concentrations in terms of the initial concentration of $NH_3(aq)$ and the unknown equilibrium concentration of $[Ag(NH_3)_2]^+(aq) = s$.

The reaction:	AgBr(s) +	2 NH$_3$(aq) $\rightleftharpoons$	[Ag(NH$_3$)$_2$]$^+$ +	Br$^-$(aq)
Initial concentrations, M:		3.0	0	0
Changes, M:		$-2s$	$+s$	$+s$
Equilibrium concentrations, M:		$(3.0 - 2s)$	s	s

Now we write the equilibrium constant expression for the reaction and substitute the above data.

$$K_c = \frac{[[Ag(NH_3)_2]^+][Br^-]}{[NH_3]^2} = \frac{s \cdot s}{(3.0 - 2s)^2} = 8.0 \times 10^{-6}$$

At this point, we could assume that $s \ll 3.0$ and proceed in the usual fashion, but we can simplify our calculation just by taking the square root of each side of the equation.

$$\frac{s^2}{(3.0 - 2s)^2} = \left(\frac{s}{3.0 - 2s}\right)^2 = 8.0 \times 10^{-6}$$

$$\frac{s}{3.0 - 2s} = (8.0 \times 10^{-6})^{1/2} = 2.8 \times 10^{-3}$$

Then, we can complete the calculation.

$$s = (3.0 - 2s)(2.8 \times 10^{-3}) = (8.4 \times 10^{-3}) - 5.6 \times 10^{-3}s$$

$$1.0056s = 8.4 \times 10^{-3}$$

$$s = [Ag(NH_3)_2]^+ = [Br^-] = 8.4 \times 10^{-3} \text{ M}$$

The molar solubility is therefore 8.4×10^{-3} mol AgBr/L.

EXERCISE 16.15A

What is the molar solubility of AgBr(s) in 0.500 M $Na_2S_2O_3$(aq)?

$$AgBr(s) + 2\,S_2O_3^{2-}(aq) \rightleftharpoons [Ag(S_2O_3)_2]^{3-}(aq) + Br^-(aq) \qquad K_c = 8.5$$

EXERCISE 16.15B

Without doing detailed calculations, determine in which of the following 0.100 M solutions AgI(s) should be most soluble: NH_3(aq), $Na_2S_2O_3$(aq), or NaCN(aq). Explain.

Example 16.16 A Conceptual Example

Figure 16.10 shows that a precipitate forms when HNO_3(aq) is added to the solution in the beaker on the right in Figure 16.9. Write the equation(s) to show what happens.

◀ **FIGURE 16.10 Destruction of a complex ion—Example 16.16 illustrated**

The beaker on the left contains $[Ag(NH_3)_2]^+$(aq), NH_3(aq), and a trace of Ag^+(aq). When HNO_3(aq) is added to this solution (right), a precipitate forms.

ANALYSIS AND CONCLUSIONS

The principal solute species in the beaker on the right in Figure 16.9 are $[Ag(NH_3)_2]^+$(aq), Cl^-(aq), and free NH_3(aq), and in addition there is a trace of free Ag^+. The added HNO_3(aq) is a source of H_3O^+. A proton is then transferred from H_3O^+, an acid, to the free NH_3, a base:

(a) H_3O^+(aq) + NH_3(aq) $\longrightarrow$ NH_4^+(aq) + H_2O(l)

According to Le Châtelier's principle, the stress imposed by this removal of NH_3(aq) stimulates the decomposition of $[Ag(NH_3)_2]^+$(aq). This decomposition produces more NH_3(aq) and, simultaneously, more Ag^+(aq):

(b) $[Ag(NH_3)_2]^+$(aq) $\longrightarrow$ Ag^+(aq) + 2 NH_3(aq)

As $[Ag^+]$ increases, the ion product $[Ag^+][Cl^-]$ soon exceeds K_{sp}, and AgCl(s) precipitates:

(c) Ag^+(aq) + Cl^-(aq) $\longrightarrow$ AgCl(s)

The overall equation, then, is 2(a) + (b) + (c):

$$[Ag(NH_3)_2]^+(aq) + Cl^-(aq) + 2\,H_3O^+(aq) \longrightarrow AgCl(s) + 2\,NH_4^+(aq) + 2\,H_2O(l)$$

EXERCISE 16.16A

Would you expect AgCl(s) to precipitate in Figure 16.10 if NH_4NO_3(aq) were used instead of HNO_3(aq)? Explain.

EXERCISE 16.16B

Write net ionic equations to describe what happens when **(a)** a few drops of concentrated ammonia solution are added to 1.00 M $Zn(NO_3)_2$ solution, **(b)** a larger quantity of concentrated ammonia solution is added to the product of (a), and **(c)** a large excess of concentrated acetic acid is added to the product of (b).

Complex Ions in Acid–Base Reactions

In discussing hydrolysis reactions in Chapter 15, we stated that cations of groups 1A and 2A do not hydrolyze. However, we also noted that many other metal cations do hydrolyze, particularly those that are small and carry a high charge. Aqueous solutions of iron(III) compounds, for example, are about as acidic as acetic acid solutions. We can use complex ion formation to explain why this is so.

To explain how H_3O^+ ions might be produced in these solutions, let us begin by noting that most of the Fe(III) is present as the complex ion $[Fe(H_2O)_6]^{3+}$. Water molecules are quite commonly found as ligands in complex ions because H_2O is a Lewis base and there are many, many H_2O molecules in every aqueous solution.

Now, consider that the electron-withdrawing power of the small, highly charged Fe^{3+} ion weakens an O–H bond in a *ligand* water molecule, causing it to give up a proton to a (non-ligand) water molecule in the solution. The result is the formation of an H_3O^+ ion in solution and the conversion of an H_2O ligand to OH^-. This process is suggested by Figure 16.11 and by the equation

$$[Fe(H_2O)_6]^{3+}(aq) + H_2O(l) \rightleftharpoons [FeOH(H_2O)_5]^{2+}(aq) + H_3O^+(aq)$$

$$K_a = 9 \times 10^{-4}$$

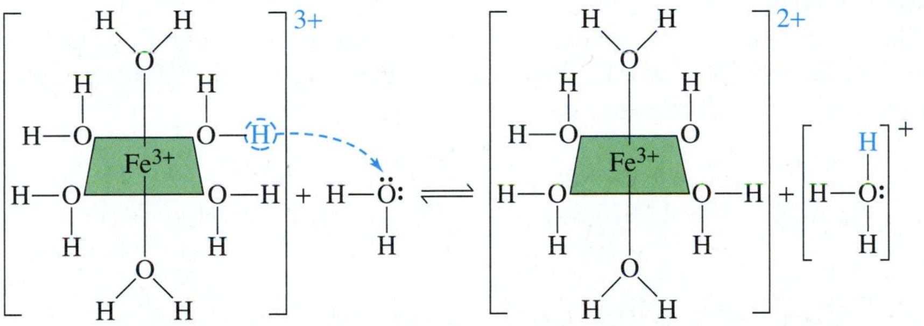

◀ FIGURE 16.11 Ionization of $[Fe(H_2O)_6]^{3+}(aq)$ as an acid
The transfer of protons from ligand water molecules in $[Fe(H_2O)_6]^{3+}$ to solvent water molecules produces an increase in $[H_3O^+]$ in the solution.

In the iron(II) complex ion $[Fe(H_2O)_6]^{2+}$, the electron-withdrawing effect of the larger, less highly charged Fe^{2+} central ion is weaker. As a consequence, $[Fe(H_2O)_6]^{2+}$ does not ionize as extensively as $[Fe(H_2O)_6]^{3+}$:

$$[Fe(H_2O)_6]^{2+}(aq) + H_2O(l) \rightleftharpoons [FeOH(H_2O)_5]^{+}(aq) + H_3O^+(aq)$$

$$K_a = 1 \times 10^{-7}$$

As shown in Figure 16.12, $Al(OH)_3(s)$, which is insoluble in water, reacts with and dissolves in *both* HCl(aq) and NaOH(aq). Complex ion formation helps us to explain this behavior.

When $Al(OH)_3(s)$ reacts with HCl(aq) (Figure 16.12a), H_3O^+ from HCl(aq) reacts with OH^- from $Al(OH)_3(s)$ to form H_2O. The H_2O molecules become ligands in the complex ion $[Al(H_2O)_6]^{3+}$. The overall reaction is

$$Al(OH)_3(s) + 3\,H_3O^+(aq) \longrightarrow [Al(H_2O)_6]^{3+}(aq)$$

When $Al(OH)_3(s)$ reacts with NaOH(aq) (Figure 16.12b), OH^- from the NaOH(aq) and the OH^- ions in $Al(OH)_3(s)$ together become the ligands in the complex ion $[Al(OH)_4]^-$, with the overall reaction being

$$Al(OH)_3(s) + OH^-(aq) \longrightarrow [Al(OH)_4]^-(aq)$$

▲ **FIGURE 16.12** **Amphoteric behavior of Al(OH)₃(s)**

(a) Aluminum hydroxide, Al(OH)₃(s), reacts with HCl(aq). In the resulting solution (right), the aluminum is present as the complex ion, [Al(H₂O)₆]³⁺(aq). (b) Freshly precipitated Al(OH)₃(s) reacts with NaOH(aq) to form a solution (right) of the complex ion [Al(OH)₄]⁻(aq).

Aluminum hydroxide is *amphoteric*, which means, as we saw in Section 8.9, that it can react with either an acid or a base. The hydroxides of zinc and chromium(III) are also amphoteric. The ions produced in acidic solution are $[Zn(H_2O)_4]^{2+}$(aq) and $[Cr(H_2O)_6]^{3+}$(aq), respectively. In basic solution, they are $[Zn(OH)_4]^{2-}$ and $[Cr(OH)_4]^-$. The hydroxide of iron(III), on the other hand, is not amphoteric. It dissolves in acidic solutions to produce the cation $[Fe(H_2O)_6]^{3+}$(aq), but it does not react in basic solutions.

The oxides Al_2O_3, ZnO, and Cr_2O_3 behave toward acids and bases much as their hydroxides do—in other words, the oxides are also amphoteric.* For example, consider the reactions of aluminum oxide:

$$Al_2O_3(s) + 6 H_3O^+(aq) + 3 H_2O(l) \longrightarrow 2[Al(H_2O)_6]^{3+}(aq)$$

$$Al_2O_3(s) + 2 OH^-(aq) + 3 H_2O(l) \longrightarrow 2[Al(OH)_4]^-(aq)$$

16.7   Qualitative Inorganic Analysis

Many concepts that we have studied in this text—particularly in this and the preceding chapter—have important applications in analytical chemistry. For example, acid–base chemistry, precipitation reactions, oxidation–reduction, and complex ion formation all come into sharp focus in an area of analytical chemistry called *classical qualitative inorganic analysis.*

"Qualitative" signifies that we are interested in determining what is present but not how much (that would be *quantitative* analysis). "Inorganic" indicates that we are analyzing for inorganic constituents, usually ions. By "classical," we mean analyses based on chemical reactions, as opposed to more modern *instrumental* analyses. Although classical qualitative analysis is not as widely used today as instrumental methods, it is still a good vehicle for applying all the basic concepts of equilibria in aqueous solutions. Also, certain qualitative tests, such as that to distinguish between deionized water and municipal tap water shown in Figure 16.3, are so simple that we still favor them over more modern, sophisticated tests.

---

*The term *amphiprotic* (page 618) refers to ions or molecules that are able to either donate or accept a proton. It is best to use the term *amphoteric,* however, when describing the ability of substances, particularly oxides and hydroxides, to react with both acids and bases. For example, although Al_2O_3 is amphoteric, it cannot be amphiprotic because it contains no H atoms.

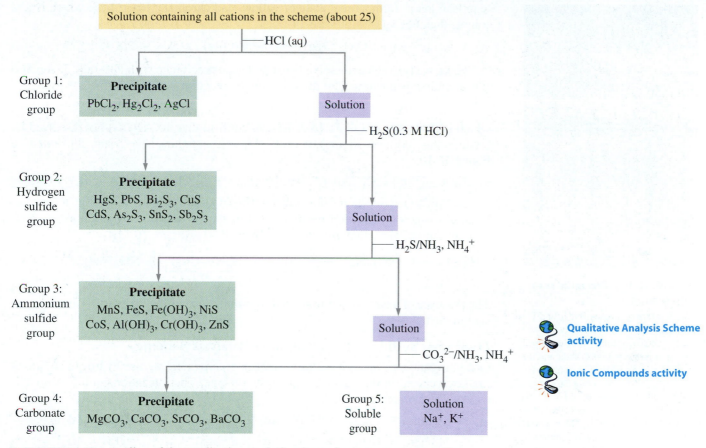

▲ FIGURE 16.13   **Outline of the qualitative analysis scheme for some common cations**
The text describes how we can use differences in solubilities of ionic compounds to separate and, ultimately, to confirm the presence or absence of various cations.

Figure 16.13 outlines a general scheme for analyzing an aqueous solution for the possible presence of any or all of about 25 common cations. We can use differences in solubilities of certain compounds of these cations to separate the cations into five groups. Then we do additional separations in each group. Ultimately we obtain each cation, if present, in a solution from which other cations have been eliminated. We then perform a test for that cation in the solution. A separate scheme is used to test for anions. In this discussion, we will consider the group of cations called group 1 in some detail. We will consider the remainder of the cation analysis scheme more briefly.

The group numbers used in the qualitative analysis scheme are unrelated to the group numbers in the periodic table. For example, group 4 of the qualitative analysis scheme consists of group 2A cations, and group 3 of the scheme contains one group 3A cation (Al^{3+}) and several transition metal cations.

Cation Group 1

To an aqueous solution that might contain up to 25 different cations—this solution being analyzed is called our *unknown*—we add HCl(aq). If a precipitate forms, we know that the unknown contains one or more of these cations, Pb^{2+}, Hg_2^{2+}, or Ag^+, because they are the only ones that form insoluble chlorides. If there is no group 1 precipitate, we know that Pb^{2+}, Hg_2^{2+}, and Ag^+ are absent from the unknown. If there is a precipitate, we filter it off and save it. We save the solution separated from the precipitate and analyze it for the presence of cations from the other groups.

If we do get a group 1 precipitate, it could be $PbCl_2$, Hg_2Cl_2, AgCl, or a mixture of them. Of the three, $PbCl_2(s)$ is the most soluble in water, and some $PbCl_2$ dissolves when we wash the precipitate with *hot* water. To test for the presence of Pb^{2+}, we add K_2CrO_4(aq) to the hot-water washings. If Pb^{2+} is present, chromate ion combines

The K_{sp} values of the precipitates for group 1 cations are $PbCl_2$, 1.6×10^{-5}; Hg_2Cl_2, 1.3×10^{-18}; and AgCl, 1.8×10^{-10}.

▲ **FIGURE 16.14   Precipitates in cation group 1**

(Left): Mixture of the possible group 1 precipitates, all white solids: $PbCl_2$, Hg_2Cl_2, AgCl. (Middle). After any $PbCl_2$ has been removed with hot water and the remaining white solid has been treated with $NH_3(aq)$, formation of a dark gray solid indicates the presence of Hg_2^{2+}; the gray solid is a mixture of Hg (black) and $HgNH_2Cl$ (white). (Right): The hot-water solution obtained when the precipitate in the left tube that was treated to remove $PbCl_2$ is treated with $K_2CrO_4(aq)$; the yellow precipitate is $PbCrO_4$.

with lead ion to form a precipitate of yellow lead chromate, $PbCrO_4(s)$, which is less soluble than $PbCl_2(s)$ (Figure 16.14):

$$Pb^{2+}(aq) + CrO_4^{2-}(aq) \longrightarrow PbCrO_4(s) \qquad K_{sp} = 2.8 \times 10^{-13}$$

Next, we treat the undissolved portion of the precipitate with $NH_3(aq)$. If AgCl(s) is present in this precipitate, it dissolves by the reaction

$$AgCl(s) + 2\ NH_3(aq) \longrightarrow [Ag(NH_3)_2]^+(aq) + Cl^-(aq)$$

which is illustrated in Figure 16.9. We can confirm that the AgCl(s) has dissolved by treating the $[Ag(NH_3)_2]^+(aq)$ with $HNO_3(aq)$, obtaining the result illustrated in Figure 16.10:

$$[Ag(NH_3)_2]^+(aq) + Cl^-(aq) + 2\ H_3O^+(aq) \longrightarrow AgCl(s) + 2\ NH_4^+(aq) + 2\ H_2O(l)$$

As AgCl(s) dissolves in $NH_3(aq)$, any $Hg_2Cl_2(s)$ mixed in with the AgCl(s) undergoes an oxidation–reduction reaction. One of the products of the reaction is a dark gray mixture of elemental mercury and $HgNH_2Cl(s)$ (Figure 16.14):

$$Hg_2Cl_2(s) + 2\ NH_3(aq) \longrightarrow \underbrace{Hg(l) + HgNH_2Cl(s)}_{\text{dark gray}} + NH_4^+(aq) + Cl^-(aq)$$

The presence of a dark gray mixture at this point indicates that mercury(I) ions are present in the unknown.

Hydrogen Sulfide in the Qualitative Analysis Scheme

Once the cation group 1 chlorides have been precipitated, we use hydrogen sulfide as the next reagent in the qualitative analysis scheme. The ionization equilibria for this *weak diprotic acid* are

$$H_2S(aq) + H_2O \rightleftharpoons H_3O^+(aq) + HS^-(aq) \qquad K_{a_1} = 1.0 \times 10^{-7}$$
$$HS^-(aq) + H_2O \rightleftharpoons H_3O^+(aq) + S^{2-}(aq) \qquad K_{a_2} = 1.0 \times 10^{-19}$$

From the extremely small value of K_{a_2}, we expect very little ionization of hydrogen sulfide ion, HS^-. This suggests that HS^- and not S^{2-} is the precipitating agent. The precipitation of a metal sulfide MS can be represented by the overall equation

$$H_2S(aq) + H_2O(l) \rightleftharpoons H_3O^+(aq) + HS^-(aq)$$
$$M^{2+}(aq) + HS^-(aq) + H_2O(l) \rightleftharpoons MS(s) + H_3O^+(aq)$$

Overall reaction: $M^{2+}(aq) + H_2S(aq) + 2\ H_2O(l) \rightleftharpoons MS(s) + 2\ H_3O^+(aq)$

Hydrogen sulfide gas has a familiar rotten egg odor that is especially noticeable in volcanic areas and near sulfur hot springs. At levels of 10 ppm, the gas can produce headaches and nausea, and it can be lethal at about 100 ppm. Because of its toxicity, $H_2S(g)$ is generally produced only in small quantities in the laboratory, and this is done directly in the solution where it is to be used. For example, it is slowly released when thioacetamide is heated in aqueous solution:

$$\underset{\text{Thioacetamide}}{CH_3\overset{\overset{\displaystyle S}{\|}}{C}NH_2(aq)} + H_2O(l) \longrightarrow \underset{\text{Acetamide}}{CH_3\overset{\overset{\displaystyle O}{\|}}{C}NH_2(aq)} + H_2S(aq)$$

Qualitative Analysis simulation

Cation Groups 2, 3, 4, and 5

When we apply Le Châtelier's principle to the overall reaction for the precipitation of metal sulfides, we see that the reverse reaction is favored in acidic solutions, where $[H_3O^+]$ is high, and the forward reaction is favored in basic solutions, where $[H_3O^+]$ is low. Thus, metal sulfides are more soluble in acidic solutions and less soluble in basic solutions:

$$M^{2+}(aq) + H_2S(aq) + 2\ H_2O(l) \underset{\text{acidic solution}}{\overset{\text{basic solution}}{\rightleftharpoons}} MS(s) + 2\ H_3O^+(aq)$$

The concentration of HS^- is so low in a strongly acidic solution of $H_2S(aq)$ that only the most insoluble sulfides precipitate. These include the eight metal sulfides in cation group 2.* They precipitate from $H_2S(aq)$ that is also 0.3 M in HCl.

Of the eight cations in group 3, five form sulfides that are soluble in acidic solution but insoluble in an alkaline NH_3/NH_4Cl buffer solution. The other three group 3 cations form hydroxide precipitates in the alkaline solution.

The cations of groups 4 and 5 form soluble sulfides, even in basic solution. Their hydroxides, except $Mg(OH)_2$, are either moderately or highly soluble. The group 4 cations are precipitated as carbonates from a buffered alkaline solution. The cations of group 5 remain soluble in the presence of all common reagents.

Within each group, we need to use additional reactions to dissolve group precipitates and to separate and selectively precipitate individual cations for identification and confirmation. These reactions include oxidation–reduction, complex ion formation, and amphoteric behavior, as well as precipitation. Flame tests are a prominent feature for several of the ions in groups 4 and 5 (see Figure 8.19).

Consider, for example, how we might detect the presence of Cr^{3+} in an unknown. When the group 3 cations are precipitated, groups 1 and 2 have already been removed, and cations of groups 4 and 5 remain in the filtrate (the solution that passes through the filter paper when the precipitates are filtered out). We dissolve the group 3 precipitate in a mixture of HCl(aq) and HNO_3(aq), obtaining a solution in which the ions possibly present are

$$Fe^{3+}, Mn^{2+}, Co^{2+}, Ni^{2+}, Al^{3+}, Zn^{2+}, \text{ and } Cr^{3+}$$

Now we make the solution basic with NaOH(aq) and also add hydrogen peroxide, H_2O_2(aq). Although all the group 3 hydroxides are insoluble in water, three of them are amphoteric. In addition to being basic, they are acidic enough to dissolve in a strongly alkaline solution:

$$Al(OH)_3(s) + OH^-(aq) \longrightarrow [Al(OH)_4]^-(aq)$$
$$Zn(OH)_2(s) + 2\,OH^-(aq) \longrightarrow [Zn(OH)_4]^{2-}(aq)$$
$$Cr(OH)_3(s) + OH^-(aq) \longrightarrow [Cr(OH)_4]^-(aq)$$

The hydrogen peroxide, an oxidizing agent, brings about several oxidations, including that of $[Cr(OH)_4]^-$ to CrO_4^{2-}:

$$3\,H_2O_2(aq) + 2\,[Cr(OH)_4]^-(aq) + 2\,OH^-(aq) \longrightarrow 2\,CrO_4^{2-}(aq) + 8\,H_2O(l)$$
$$\text{Green} \qquad\qquad\qquad\qquad\qquad\qquad\qquad\qquad \text{Yellow}$$

Group 3 is thus separated into two subgroups: a precipitate of hydroxides and a filtrate that may or may not contain $[Al(OH)_4]^-$, $[Zn(OH)_4]^{2-}$, and CrO_4^{2-}. The appearance of a yellow color in the filtrate is a strong indication that CrO_4^{2-} is present, because the other ions that may be present in the filtrate—$[Al(OH)_4]^-$ and $[Zn(OH)_4]^{2-}$—are colorless. The presence of CrO_4^{2-} can also be confirmed by other tests, such as the precipitation of yellow $BaCrO_4$(s).

Cumulative Example

A solid mixture containing 1.00 g of ammonium chloride and 2.00 g of barium hydroxide is heated to expel ammonia. The liberated NH_3(g) is then dissolved in 0.500 L of water containing 225 ppm Ca^{2+} as calcium chloride. Will a precipitate form in this water?

STRATEGY

First, the precipitate, if there is one, will be $Ca(OH)_2$(s). The Ca^{2+} is present in the water, and after the liberated NH_3(g) dissolves in the water, the OH^- is produced by ionization of the weak base NH_3(aq):

$$NH_3(aq) + H_2O(l) \rightleftharpoons NH_4^+(aq) + OH^-(aq) \qquad K_b = 1.8 \times 10^{-5} \qquad \text{(a)}$$

*Note that Figure 16.13 shows Pb^{2+} both in group 1 and in group 2. This is because the solubility of $PbCl_2$(s) is high enough to allow some Pb^{2+} to remain in the filtrate from the group 1 precipitation. When this filtrate is then treated with H_2S to test for group 2 cations, the Pb^{2+} precipitates as PbS(s), which is much less soluble than $PbCl_2$(s).

If a precipitate is to form, the final ion product in the water, $Q_{ip} = [Ca^{2+}][OH^-]^2$, must exceed K_{sp} for $Ca(OH)_2$:

$$Ca(OH)_2(s) \rightleftharpoons Ca^{2+}(aq) + 2\,OH^-(aq) \quad K_{sp} = 5.5 \times 10^{-6} \tag{b}$$

To determine the amount of $NH_3(g)$ liberated, we must perform a limiting-reactant calculation (recall Example 3.20) for the reaction

$$2\,NH_4Cl(s) + Ba(OH)_2(s) \rightleftharpoons BaCl_2(s) + 2\,H_2O(l) + 2\,NH_3(g) \tag{c}$$

The final steps in the calculation involve obtaining $[OH^-]$ from equation (a), converting ppm Ca^{2+} to $[Ca^{2+}]$, and comparing Q_{ip} and K_{sp} for reaction (b).

SOLUTION

To determine the amount of $NH_3(g)$ available, we complete two stoichiometric calculations based on reaction (c).

Assuming NH₄Cl is the limiting reactant:

$$? \text{ mol } NH_3 = 1.00 \text{ g } NH_4Cl \times \frac{1 \text{ mol } NH_4Cl}{53.49 \text{ g } NH_4Cl} \times \frac{2 \text{ mol } NH_3}{2 \text{ mol } NH_4Cl}$$

$$= 0.0187 \text{ mol } NH_3$$

Assuming Ba(OH)₂ is the limiting reactant:

$$? \text{ mol } NH_3 = 2.00 \text{ g } Ba(OH)_2 \times \frac{1 \text{ mol } Ba(OH)_2}{171.3 \text{ g } Ba(OH)_2} \times \frac{2 \text{ mol } NH_3}{1 \text{ mol } Ba(OH)_2}$$

$$= 0.0234 \text{ mol } NH_3$$

Based on the smaller of the two values just calculated, we determine the molarity of NH_3.

$$\frac{0.0187 \text{ mol } NH_3}{0.500 \text{ L}} = 0.0374 \text{ M } NH_3$$

Now we use the method of Example 15.8 to determine $[OH^-]$, where $[OH^-] = x$ and $[NH_3] = 0.0374 - x$.

$$K_b = \frac{[NH_4^+][OH^-]}{[NH_3]} = \frac{x \cdot x}{0.0374 - x} = 1.8 \times 10^{-5}$$

We assume that $x \ll 0.0374$ and solve for $x = [OH^-]$.

$$x^2 = 0.0374 \times 1.8 \times 10^{-5} = 6.7 \times 10^{-7}$$
$$\text{and} \quad x = \sqrt{6.7 \times 10^{-7}} = 8.2 \times 10^{-4}$$

The quantity of liquid into which we inject the $NH_3(g)$ is 0.500 L. Because this liquid is nearly pure water with $d = 1.00$ g/mL, its mass is 500 g. We can then calculate the mass of Ca^{2+} in this 500 g water, based on the literal meaning of 225 ppm.

$$? \text{ g } Ca^{2+} = 500 \text{ g water} \times \frac{225 \text{ g } Ca^{2+}}{1,000,000 \text{ g water}} = 0.113 \text{ g } Ca^{2+}$$

By converting from mass to moles of Ca^{2+} and dividing by the solution volume, we determine $[Ca^{2+}]$.

$$[Ca^{2+}] = \frac{0.113 \text{ g } Ca^{2+}}{0.500 \text{ L}} \times \frac{1 \text{ mol } Ca^{2+}}{40.08 \text{ g } Ca^{2+}} = 5.64 \times 10^{-3} \text{ M}$$

Now we can evaluate the ion product, Q_{ip}, and compare it with K_{sp} for $Ca(OH)_2$.

$$Q_{ip} = [Ca^{2+}][OH^-]^2 = 5.64 \times 10^{-3} \times (8.2 \times 10^{-4})^2 = 3.8 \times 10^{-9}$$

When we compare Q_{ip} with K_{sp}, we conclude that no precipitate will form.

$$Q_{ip} = 3.8 \times 10^{-9} \text{ is considerably smaller than } K_{sp} = 5.5 \times 10^{-6}.$$

ASSESSMENT

In determining $[OH^-]$, we made the assumption that it would be much smaller than $[NH_3]$, and this assumption is valid: $x = 8.2 \times 10^{-4}$ is only about 2% of 0.0374.

Note that even if a precipitate had formed, its quantity would have been very small. The maximum yield from 0.113 g Ca^{2+} would be 0.209 g $Ca(OH)_2$ [that is, 0.113 g $Ca^{2+} \times (74.1 \text{ g } Ca(OH)_2/40.1 \text{ g } Ca^{2+})$]. With incomplete precipitation, the quantity would have been smaller still. Yet even a small amount of precipitate may be discernible as turbidity in a solution.

Concept Review with Key Terms

16.1 The Solubility Product Constant, K_{sp}—The **solubility product constant, K_{sp}**, represents the equilibrium between a slightly soluble ionic compound and its ions in a saturated aqueous solution. For a saturated solution of barium sulfate, equilibrium is described by the equation

$$BaSO_4(s) \rightleftharpoons Ba^{2+}(aq) + SO_4^{2-}(aq) \quad \text{and} \quad K_{sp} = [Ba^{2+}][SO_4^{2-}]$$

16.2 The Relationship Between K_{sp} and Molar Solubility—The molar solubility of a compound is defined as its molarity in a saturated aqueous solution. The K_{sp} and molar solubility of an ionic solute are related in such a way that either one can be established from a value of the other.

16.3 The Common Ion Effect in Solubility Equilibria—The common ion effect causes the solubility of a slightly soluble ionic compound to decrease, often significantly, in the presence of an excess of one of the ions involved in the solubility equilibrium. This effect is particularly useful in procedures involving precipitates in quantitative analysis. For example,

> When a salt supplies Ag^+ or SO_4^{2-}, equilibrium shifts to the *left*.

$$Ag_2SO_4(s) \rightleftharpoons 2\,Ag^+(aq) + SO_4^{2-}(aq)$$

16.4 Will Precipitation Occur? Is It Complete?—We can determine what will happen when certain ions are brought together in solution by comparing the reaction quotient or ion product, Q_{ip}, with K_{sp}:

- If $Q_{ip} > K_{sp}$, a precipitate should form.
- If $Q_{ip} < K_{sp}$, a precipitate will not form.
- If $Q_{ip} = K_{sp}$, a solution is just saturated.

We assume that precipitation is complete if no more than 0.1% of the target ion remains in solution after precipitation has occurred. A small value of K_{sp}, a high initial ion concentration, and the presence of a common ion favor complete precipitation. A mixture of ions in solution can be separated by the addition of an ion with which similar precipitates are formed, provided that the K_{sp} values of the possible precipitates differ widely.

16.5 Effect of pH on Solubility—The solubilities of some slightly soluble compounds depend strongly on pH. For example, solutes become more soluble in acidic solutions if their anions are sufficiently basic to accept protons from $H_3O^+(aq)$.

16.6 Equilibria Involving Complex Ions—A **complex ion** is a collection of **ligands** joined to a central metal ion through coordinate covalent bonds involving lone-pair electrons on the ligands:

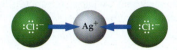

The **formation constant** of a complex ion, K_f, describes the equilibrium between the complex ion, the free cation, and the free ligands. Certain solutes become more soluble in the presence of species that can serve as ligands in complex ions. The extent to which a solute dissolves in the presence of complexing ligands depends on the values of K_{sp} and K_f. The ability of ligand H_2O molecules to donate protons to free H_2O molecules accounts for the acidic character of some complex ions:

The formation of complex ions with OH^- ions as ligands accounts for the amphoterism of some oxides and hydroxides. An amphoteric oxide or hydroxide can react with either an acid or a base.

16.7 Qualitative Inorganic Analysis—Precipitation, acid–base, and oxidation–reduction reactions, together with complex ion formation and reactions of amphoteric species, are all used extensively in the classical scheme for the qualitative analysis of common cations. This procedure, summarized in Figure 16.13 for metal cations, uses differences in the solubilities of ionic compounds in the presence of selected reagents to separate and ultimately confirm the presence or absence of various cations in a solution of unknown composition.

Assessment Goals

When you have mastered the material in this chapter, you will be able to:

- Write expressions for K_{sp} describing solubility equilibria of slightly soluble salts.
- Calculate K_{sp} values of slightly soluble salts from equilibrium concentrations of saturated solutions.
- Use K_{sp} to calculate the molar solubility of a slightly soluble salt and the equilibrium concentrations of its ions in solution.
- Predict and calculate the effect on the solubility of a slightly soluble salt of adding a common ion to the saturated solution.
- Calculate ion products, Q_{ip}, and predict whether precipitation occurs by comparing them with K_{sp} values.

- Determine the conditions required for complete precipitation of a slightly soluble salt.
- Predict and calculate the effect of pH on the solubilities of slightly soluble salts.
- Write equations for the formation of complex ions.
- Using formation constants, K_f, determine the concentrations of ions in solutions containing complex ions.
- Write equations to illustrate the amphoteric nature of some complex ions, metal hydroxides, and metal oxides.
- Interpret the results of a qualitative analysis scheme in order to establish the presence or absence of common cations.

Self-Assessment Questions

1. Write the solubility product constant expression for equilibrium in a saturated solution of (a) iron(III) hydroxide, and (b) gold(III) oxalate.

$$Fe(OH)_3(s) \rightleftharpoons Fe^{3+}(aq) + 3\ OH^-(aq)$$

$$Au_2(C_2O_4)_3(s) \rightleftharpoons 2\ Au^{3+}(aq) + 3\ C_2O_4{}^{2-}(aq)$$

2. Pure water is saturated with slightly soluble PbI_2. The most accurate statement that we can make about $[Pb^{2+}]$ in this solution is which of the following?

   (a) $[Pb^{2+}] = [I^-]$ (c) $[Pb^{2+}] = \sqrt{K_{sp}}$

   (b) $[Pb^{2+}] = K_{sp}$ (d) $[Pb^{2+}] = 0.5\ [I^-]$

3. Why is the solubility product concept limited to ionic compounds that are only slightly soluble in water? Why is it not useful in describing saturated solutions of NaCl or $NaNO_3$, for example?

4. Which compound in each set is more soluble? Why?

   (a) $BaSO_4$ ($K_{sp} = 1.1 \times 10^{-10}$) or $PbSO_4$ ($K_{sp} = 1.6 \times 10^{-8}$)

   (b) PbI_2 ($K_{sp} = 7.1 \times 10^{-9}$) or CaF_2 ($K_{sp} = 5.3 \times 10^{-9}$)

5. Which of the following is most likely to occur when 1.058 g of Na_2SO_4 is added to 375.0 mL of saturated $BaSO_4$(aq)?

   (a) $[Ba^{2+}]$ is reduced.

   (b) $[SO_4{}^{2-}]$ is reduced.

   (c) The solubility of $BaSO_4$ is increased.

   (d) $[Ba^{2+}]$ does not change.

6. Which of the following is most likely to lead to the complete precipitation of a metal ion M^{2+}, as its sulfide, MS(s), from saturated H_2S(aq)?

   (a) Add an acid.

   (b) Increase the $[H_2S]$ in solution.

   (c) Raise the pH.

   (d) Heat the solution.

7. Describe how ions in solution can be separated by *selective precipitation*. What conditions are necessary for a successful separation?

8. When does pH affect the solubility of a slightly soluble solute, and when does it not? Give some examples.

9. Which of the following solutions could be used to increase the aqueous solubility of $Fe(OH)_3$(s)? That is, which would produce a solution with a greater $[Fe^{3+}]$ than found in saturated $Fe(OH)_3$(aq): HCl(aq), NaOH(aq), CH_3COOH(aq), NH_3(aq)? Explain.

10. What is a complex ion? What are the ligands in a complex ion? Describe complex ion formation as a Lewis acid–base reaction.

11. What are the concentration terms that appear in the K_f expressions of each of the following complex ions?

    (a) $[Ag(NH_3)_2]^+$ (b) $[Zn(NH_3)_4]^{2+}$ (c) $[Ag(S_2O_3)_2]^{3-}$

12. Explain why $PbCl_2$(s) is less soluble in 1 M $Pb(NO_3)_2$ than in pure water, but somewhat more soluble in 1 M HCl(aq) than in pure water.

13. $Cu(OH)_2$ is highly soluble in all of the following except one. Which of these is the exception?

    (a) H_2O(l) (c) HCl(aq)

    (b) NH_3(aq) (d) HNO_3(aq)

14. A solution contains both Cu^{2+}(aq) and Ag^+(aq). To precipitate one of these ions and leave the other in solution, which of the following should be added?

    (a) H_2S(aq) (c) HNO_3(aq)

    (b) HCl(aq) (d) NH_4NO_3(aq).

15. Write an equation involving $[Al(H_2O)_6]^{3+}$ to account for the fact that aqueous solutions of Al^{3+} are acidic.

16. What reagent is used to separate group 1 cations from other cations in the qualitative analysis scheme? What reagents are used to separate group 2 cations from those of groups 3, 4, and 5?

17. In qualitative analysis of group 1 cations, what reagent is used to separate $PbCl_2$(s) from the other chlorides? What reagent is used to separate AgCl(s) from Hg_2Cl_2(s)?

18. Which of the following solids is(are) likely to be more soluble in acidic solution, and which in basic solution? Which is(are) likely to have a solubility that is essentially independent of pH?

    (a) $H_2C_2O_4$ (c) CdS (e) $NaNO_3$

    (b) $MgCO_3$ (d) KCl (f) $Ca(OH)_2$

Problems

Use data from Tables 16.1 and 16.3 or Appendix C, as necessary.

The Solubility Product Constant, K_{sp}

19. Write a chemical equation representing solubility equilibrium for (a) $Hg_2(CN)_2$, $K_{sp} = 5 \times 10^{-40}$; and (b) Ag_3AsO_4, $K_{sp} = 1.0 \times 10^{-22}$.

20. Write a chemical equation representing solubility equilibrium for (a) YF_3, $K_{sp} = 6.6 \times 10^{-13}$; and (b) $Fe_4[Fe(CN)_6]_3$, $K_{sp} = 3.3 \times 10^{-41}$.

21. Write the solubility product constant expression for each compound in Problem 19.

22. Write the solubility product constant expression for each compound in Problem 20.

The Relationship Between Solubility and K_{sp}

23. Can the numerical values of the molar solubility and the solubility product constant of a slightly soluble ionic compound ever be the same? Which of the two is usually the larger value? Explain.

24. The K_{sp} values of $CuCO_3$ and $ZnCO_3$ are 1.4×10^{-10} and 1.4×10^{-11}, respectively. Does this mean that $CuCO_3$ is 10 times as soluble as $ZnCO_3$? Explain.

25. Calculate K_{sp} values of (a) CaF_2, solubility 3.32×10^{-4} mol/L; (b) aluminum hydroxide, solubility 2.6×10^{-9} mol/L; and (c) cadmium iodate, solubility 0.097 g $Cd(IO_3)_2/100$ mL H_2O.

26. Calculate K_{sp} values of the following, for which a reference book lists the indicated solubilities: (a) $Ce(IO_3)_4$, solubility 1.8×10^{-4} mol/L; (b) Hg_2SO_4, solubility 8.9×10^{-4} mol/L; and (c) barium chromate, 0.0010 g $BaCrO_4/100$ mL H_2O.

27. For each set, determine which slightly soluble solute has the greater molar solubility: (a) AgCl or Ag_2CrO_4, and (b) $Mg(OH)_2$ or $MgCO_3$.

28. For each set, which salt gives the highest concentration of the indicated metal ion in water at equilibrium: (a) Sr^{2+}: $SrCO_3$, $SrSO_4$, or SrF_2; (b) Mg^{2+}: $Mg_3(AsO_4)_2$, ($K_{sp} = 2.0 \times 10^{-20}$) $Mg(OH)_2$, or MgC_2O_4 ($K_{sp} = 9.3 \times 10^{-3}$)?

29. A 50.0-mL sample of a saturated aqueous solution of barium thiosulfate was treated with HCl(aq), and 6.4 mg of sulfur was obtained in the reaction that follows.

$$S_2O_3{}^{2-}(aq) + 2\, H_3O^+ \longrightarrow 3\, H_2O + SO_2(g) + S(s)$$

Use these data to determine K_{sp} for barium thiosulfate.

30. A 250.0-mL sample of a saturated aqueous solution of lanthanum(III) hydroxide required 2.3 mL of 0.0010 M HCl(aq) for its neutralization.

$La(OH)_3$(saturated aq) + 3 HCl(aq)

$\qquad\qquad \longrightarrow LaCl_3(aq) + 3\, H_2O(l)$

Use this information to determine K_{sp} for lanthanum hydroxide.

31. *Without doing detailed calculations*, indicate which of the following saturated aqueous solutions has the highest concentration of Ca^{2+} ion: $CaCO_3$, CaF_2, $CaSO_4$. Explain.

32. *Without doing detailed calculations*, indicate which of the following saturated aqueous solutions has the highest concentration of $PO_4{}^{3-}$ ion: $Ca_3(PO_4)_2$, $FePO_4$, $Mg_3(PO_4)_2$. Explain.

33. Fluoridated drinking water contains about 1 part per million (ppm) of F^-. Is MgF_2 sufficiently soluble in water to be used as the source of fluoride ion for this fluoridation? Explain. (*Hint:* Recall that 1 ppm signifies 1 g F^- per 10^6 g solution.)

34. Calculate the concentration of Cu^{2+} in parts per billion (ppb) in a saturated solution of copper(II) arsenate, $Cu_3(AsO_4)_2$(aq). (*Hint:* Recall that 1 ppb signifies 1 g Cu^{2+} per 10^9 g solution.)

35. Given the K_{sp} values for $PbCl_2$ of 1.6×10^{-5} at 25 °C and 3.3×10^{-3} at 80 °C, if 1.00 mL of saturated $PbCl_2$(aq) at 80 °C is cooled to 25 °C, will a sufficient amount of $PbCl_2$(s) precipitate to be visible? Assume that you can detect as little as 1 mg of the solid.

36. When a 1.00-L sample of a saturated aqueous solution of calcium tungstate is heated from 15 °C to 100 °C, 5.2 mg $CaWO_4$ deposits from solution. Given that

$$CaWO_4(s) \rightleftharpoons Ca^{2+}(aq) + WO_4{}^{2-}(aq)$$

$$K_{sp} = 4.9 \times 10^{-10} \text{ at } 15 \text{ °C}$$

what is the value of K_{sp} at 100 °C?

The Common Ion Effect in Solubility Equilibria

37. In which of the following is the slightly soluble AgBr likely to be *most soluble:* pure water, 0.050 M KBr, or 1.25 M $AgNO_3$? Explain.

38. In which of the following is the slightly soluble Ag_2CrO_4 likely to be *least soluble:* pure water, 0.10 M K_2CrO_4, or 0.10 M $AgNO_3$? Explain.

39. Calculate the molar solubility in 0.10 M $MgCl_2$ of (a) $Mg(OH)_2$ and (b) $MgNH_4PO_4$ (a slightly soluble salt containing ammonium ion, $K_{sp} = 2.5 \times 10^{-13}$).

40. Calculate the molar solubility of (a) $Mg(OH)_2$ in 0.25 M NaOH(aq) and (b) $BaSO_4$ in 0.010 M Na_2SO_4.

41. A 15.0-g sample of NaI is dissolved in water to make 0.250 L of solution. The solution is then saturated with PbI_2. What will be $[Pb^{2+}]$ and $[I^-]$ in the saturated solution?

42. A 1.05-g sample of LiCl is used to make 0.125 L of an aqueous solution. The solution is then saturated with Li_3PO_4. What will be $[Li^+]$ and $[PO_4{}^{3-}]$ in the saturated solution?

43. A solution is saturated with Ag_2SO_4. (a) Calculate $[Ag^+]$ in this saturated solution. (b) What mass of Na_2SO_4 must be added to 0.500 L of the solution to decrease $[Ag^+]$ to 4.0×10^{-3} M?

44. A solution is saturated with Ag_2CrO_4. (a) Calculate $[CrO_4{}^{2-}]$ in this saturated solution. (b) What mass of $AgNO_3$ must be added to 0.635 L of the solution to reduce $[CrO_4{}^{2-}]$ to 1.0×10^{-8} M?

45. Calculate the molar solubility of barium thiosulfate in 0.0033 M $Na_2S_2O_3$(aq). (*Hint:* Check any simplifying assumptions that you make.)

46. Calculate the molar solubility of strontium chromate in 0.0025 M Na_2CrO_4(aq). (*Hint:* Check any simplifying assumptions that you make.)

Precipitation Criteria

47. What $[CrO_4^{2-}]$ must be present in 0.00105 M $AgNO_3$(aq) to just cause Ag_2CrO_4(s) to precipitate?

48. What must be the pH of a solution that is 0.050 M Fe^{3+} to just cause $Fe(OH)_3$(s) to precipitate?

49. Will a precipitate form, and if so, what is the precipitate when (a) 235 mL of 0.0022 M $MgCl_2$ and 485 mL of 0.0055 M NaF are mixed; (b) 136 mL of 0.00015 M $Pb(NO_3)_2$ and 234 mL of 0.00028 M Na_3AsO_4 are mixed?

50. Will a precipitate form if (a) 0.0010 mol $Hg_2(NO_3)_2$ and 0.0010 mol NaCl are added to 20.00 L of water; (b) 0.48 mg $MgCl_2$ and 12.2 mg Na_2CO_3 are added to 225 mL of water?

51. In hard water, $[Ca^{2+}]$ is about 2.0×10^{-3} M. Water is fluoridated with 1.0 g F^- per 1.0×10^3 L of water. Will CaF_2(s) precipitate from hard water upon fluoridation?

52. A municipal water sample contains 46.1 mg SO_4^{2-}/L and 30.6 mg Cl^-/L. How many drops (1 drop = 0.05 mL) of 0.0010 M $AgNO_3$ must be added to 1.00 L of this water to just produce a precipitate? What will this precipitate be?

Completeness of Precipitation and Selective Precipitation

53. Aqueous calcium chloride is slowly added to a solution that is originally 0.010 M in carbonate ion and 0.010 M in sulfate ion. What calcium salt precipitates first? What must be the concentration of the first anion when the second anion just begins to precipitate?

54. Consider an aqueous solution that is 0.10 M in Zn(II) ion and 0.10 M in Ni(II) ion. Is it possible to quantitatively precipitate one of these with carbonate ion without precipitating the other? Show calculations to justify your answer.

55. The first step in the extraction of magnesium metal from seawater is the precipitation of Mg^{2+} as $Mg(OH)_2$(s). $[Mg^{2+}]$ in seawater is about 0.059 M. If a seawater sample is treated so that its $[OH^-]$ is kept constant at 2.0×10^{-3} M, what will be $[Mg^{2+}]$ remaining in solution after precipitation has occurred? Is precipitation of $Mg(OH)_2$(s) complete under these conditions?

56. Is precipitation of $CaSO_4$ complete if a solution that is 0.0250 M in Ca^{2+} is also made 0.500 M in K_2SO_4(aq)?

57. What pH must be maintained by a buffer solution so that no more than 0.010% of the Mg^{2+} present in 0.360 M $MgCl_2$(aq) remains in solution following the precipitation of $Mg(OH)_2$(s)?

58. Nineteen centuries ago, the Romans added calcium sulfate to wine. It clarifies the wine and removes dissolved lead.

(a) Calculate $[SO_4^{2-}]$ in wine saturated with $CaSO_4$. (b) What $[Pb^{2+}]$ remains in wine that is saturated with $CaSO_4$?

59. Concentrated NaF(aq) is slowly added to a solution that is 0.010 M $CaCl_2$ and 0.010 M $MgCl_2$. (a) What is the first precipitate to form? (b) What $[F^-]$ is needed to start precipitating the second precipitate? (c) Can Ca^{2+} and Mg^{2+} be separated by selective precipitation of their fluorides? Explain.

60. Concentrated $Pb(NO_3)_2$(aq) is slowly added to a solution that is 0.15 M Na_2CrO_4(aq) and 0.15 M Na_2SO_4. (a) What is the first precipitate to form? (b) What is the $[Pb^{2+}]$ needed to start precipitating the second precipitate? (c) Can CrO_4^{2-} and SO_4^{2-} be separated by selective precipitation with Pb^{2+}? Explain.

61. A 10.00-mL portion of a 0.50 M $AgNO_3$(aq) solution is added to 100.0 mL of a solution that is 0.010 M in Cl^- and 0.010 M in SO_4^{2-}. (a) Will AgCl(s) precipitate from this solution? If so, how many moles? (b) Will Ag_2SO_4(s) precipitate? If so, how many moles?

62. A 10.00-mL portion of 0.25 M $AgNO_3$(aq) solution is added to 100.0 mL of a solution that is 0.010 M in Cl^- and 0.010 M in Br^-. (a) Will AgCl(s) precipitate from this solution? If so, how many moles? (b) Will AgBr(s) precipitate from this solution? If so, how many moles?

Effect of pH on Solubility

63. Which of the following would you add to a mixture of $CaCO_3$(s) and its saturated solution to increase the molar solubility of $CaCO_3$: more water, Na_2CO_3, NaOH, $NaHSO_4$? Explain.

64. Which of the following solids are likely to be more soluble in acidic solution, and which in basic solution? Which are likely to have a solubility that is essentially independent of pH? Explain.

(a) C_6H_5COOH (c) CaC_2O_4 (e) $CaCl_2$

(b) $BaCO_3$ (d) $LiNO_3$ (f) $Sr(OH)_2$

65. Calculate the solubility of $Co(OH)_3$(s) in a buffer solution with pH = 4.60.

66. Calculate the solubility of $Mg(OH)_2$ in a buffer solution that is 0.75 M NH_3 and 0.50 M NH_4Cl.

67. Boiler scale (mostly $CaCO_3$) is insoluble in water. Hydrochloric acid is sometimes used to remove the scale from boilers in commercial power plants. Write an equation to show what happens. Boiler scale is also produced in automatic coffeemakers. The recommended way to remove it is to run vinegar (acetic acid) through the machine. Write an equation to show what happens.

68. If a concentrated aqueous solution of sodium acetate is mixed with an aqueous solution of silver nitrate, a precipitate of silver acetate forms. The precipitate dissolves readily when nitric acid is added. Write equations that explain what happens.

Complex Ion Formation

69. A complex ion has Cr^{3+} as its central ion and four NH_3 molecules and two Cl^- ions as ligands. Write the formula of this complex ion.

70. A complex ion has Fe^{2+} as the central ion and six cyanide ions as ligands. Write the formula of this complex ion.

71. Which of the following reagents will increase significantly the concentration of free Zn^{2+} ions when it is added to a solution of the complex ion $[Zn(NH_3)_4]^{2+}$? Explain.

(a) $H_2O(l)$ (c) $NH_3(aq)$

(b) $NaHSO_4(aq)$ (d) $HCl(aq)$

72. Which of the following complex ions would you expect to have the *lowest* $[Ag^+]$ in a solution that is 0.10 M in the complex ion and 1.0 M in the free ligand? Explain.

(a) $[Ag(NH_3)_2]^+$ (b) $[Ag(CN)_2]^-$ (c) $[Ag(S_2O_3)_2]^{3-}$

73. When a few drops of concentrated $Na_2SO_4(aq)$ are added to a dilute solution of $AgNO_3(aq)$, a white precipitate forms. When a small quantity of concentrated $NH_3(aq)$ is added to the mixture, the precipitate redissolves, resulting in a colorless solution. When this solution is made acidic with $HNO_3(aq)$, a white precipitate again appears. Write equations to represent these three observations.

74. Figure 16.9 shows that $AgCl(s)$ can be dissolved in $NH_3(aq)$, and Figure 16.10 shows that $AgCl(s)$ can be reprecipitated by adding $HNO_3(aq)$. $AgCl(s)$ can also be dissolved in concentrated $HCl(aq)$, but it cannot be reprecipitated by the addition of $HNO_3(aq)$. Explain this difference.

75. Write plausible net ionic equations to represent what happens, if anything, in each of the following. If no reaction occurs, write N. R.

(a) $Fe(OH)_3(s) + HCl(concd\ aq) \longrightarrow$

(b) $NaCl(aq) + NaOH(concd\ aq) \longrightarrow$

(c) $CrCl_3(aq) + NaOH(concd\ aq) \longrightarrow$

76. Write plausible net ionic equations to represent what happens, if anything, in each of the following. If no reaction occurs, write N. R.

(a) $ZnCl_2(aq) + NaOH(concd\ aq) \longrightarrow$

(b) $Al_2O_3(s) + HCl(concd\ aq) \longrightarrow$

(c) $Fe(OH)_3(s) + NaOH(concd\ aq) \longrightarrow$

Complex Ion Equilibria

77. What is $[Zn^{2+}]$ in a solution that is 0.25 M in $[Zn(NH_3)_4]^{2+}$ and has $[NH_3]$ equal to 1.50 M?

78. What must be $[NH_3]$ in a solution that is 0.15 M in $[Ag(NH_3)_2]^+$ if $[Ag^+]$ is to be kept at 1.0×10^{-8} M?

79. Should $PbI_2(s)$ precipitate if 2.00 mL of 2.00 M $KI(aq)$ is added to 0.300 L of an aqueous solution that is 0.010 M in $[PbCl_3]^-$ and has $[Cl^-]$ equal to 1.50 M?

80. Should $BaS_2O_3(s)$ precipitate if 2.50 mL of 1.50 M $Ba(NO_3)_2(aq)$ is added to 0.275 L of an aqueous solution that is 0.066 M in $[Ag(S_2O_3)_2]^{3-}$ and has $[Ag^+]$ equal to 1.25 M?

81. How many milligrams of KBr can be added before $AgBr(s)$ precipitates from 1.42 L of an aqueous solution that is 0.220 M in $[Ag(NH_3)_2]^+$ and has $[NH_3]$ equal to 0.805 M?

82. What is the minimum $[CN^-]$ required in 0.750 L of an aqueous solution that is 0.250 M in $[Ag(CN)_2]^-$ so that $AgI(s)$ will not precipitate when 12.5 g KI is later added?

83. Calculate the molar solubility of $AgBr(s)$ in 0.100 M $Na_2S_2O_3(aq)$. (*Hint:* Write an equation for the dissolution reaction, and determine its K_c value as shown in the text.)

84. Calculate the molar solubility of $AgI(s)$ in 0.100 M $NaCN(aq)$. (*Hint:* Write an equation for the dissolution reaction, and determine its K_c value as shown in the text.)

Qualitative Inorganic Analysis

85. Both Pb^{2+} and Ag^+ form insoluble chlorides and sulfides. In the qualitative analysis scheme, Pb^{2+} appears both in cation group 1 and in group 2, whereas Ag^+ appears only in group 1. Give a plausible explanation for this observation.

86. The test for NH_4^+ in the qualitative analysis scheme is suggested in the accompanying photograph. A sample of the original unknown is treated with concentrated $NaOH(aq)$, and a strip of moistened red litmus paper is suspended in the vapor above the solution. In a positive test for NH_4^+, the litmus turns blue. (a) Explain how this test for NH_4^+ works, by writing ionic equations for the reactions involved. (b) Why is it possible to test for NH_4^+ on the *original* sample, whereas all the other cations must be separated into groups before testing?

87. A cation group 1 unknown is treated with $HCl(aq)$ and yields a white precipitate. When $NH_3(aq)$ is added to the precipitate, it turns a gray color. According to these tests, indicate for each of the group 1 cations whether it is definitely present, definitely absent, or whether its presence remains uncertain.

88. A cation group 3 unknown is treated with $NaOH(aq)$ and H_2O_2. No precipitate forms, but the solution becomes bright yellow in color. Indicate for each of the group 3 cations whether it is probably present, most likely absent, or whether its presence remains uncertain.

Additional Problems

Use data from Tables 16.1 and 16.3 or Appendix C, as necessary. Problems marked with an * may be more challenging than others.

89. Write equations to show why $Zn(OH)_2(s)$ is readily soluble in each of the following dilute solutions: $HCl(aq)$, $CH_3COOH(aq)$, $NH_3(aq)$, and $NaOH(aq)$.

90. Show that a precipitate of $Mg(OH)_2(s)$ forms in an aqueous solution that is 0.350 M $MgCl_2$ and 0.750 M NH_3. Explain why the precipitate can be kept from forming by adding NH_4Cl to the solution. What minimum mass of NH_4Cl should be present in 0.500 L of the solution to prevent the precipitation of $Mg(OH)_2(s)$?

91. Will lead(II) oxalate precipitate if 50.0 mg of $Pb(NO_3)_2$ is added to 0.500 L of 0.0100 M $H_2C_2O_4$ solution? (For PbC_2O_4, $K_{sp} = 4.8 \times 10^{-10}$.)

92. A 0.100-L sample of saturated $SrC_2O_4(aq)$ is analyzed for its oxalate ion content by titration with permanganate ion in acidic solution. The volume of 0.00500 M $KMnO_4$ required is 3.22 mL. Use these data to determine K_{sp} for SrC_2O_4. The balanced equation for the titration reaction is

$$2\,MnO_4^-(aq) + 5\,C_2O_4^{2-}(aq) + 16\,H^+(aq)$$
$$\longrightarrow 2\,Mn^{2+}(aq) + 10\,CO_2(g) + 8\,H_2O(l)$$

93. Calculate the number of moles of NH_3 that must be added to prevent the precipitation of $AgCl(s)$ from 0.250 L of 0.100 M $AgNO_3$ when 0.100 mol NaCl is subsequently added.

94. Obtain K_a for hydrazoic acid, HN_3, and K_{sp} for lead azide, $Pb(N_3)_2$, from Appendix C, and use these to determine a value of K_c for the dissolution of $Pb(N_3)_2(s)$ in an acidic solution. Then calculate the molar solubility of $Pb(N_3)_2(s)$ in a buffer solution with pH = 2.85.

* 95. Exercise 16.12 asks for an equation describing the dissolution of $Mg(OH)_2(s)$ in $NH_4Cl(aq)$. Determine a value of K_c for this reaction, and then calculate the molar solubility of $Mg(OH)_2$ in 0.500 M $NH_4Cl(aq)$.

* 96. The text gives the following equation for the precipitation of a metal sulfide in the qualitative analysis scheme.

$$M^{2+}(aq) + H_2S(aq) + 2\,H_2O \rightleftharpoons MS(s) + 2\,H_3O^+(aq)$$

Use the description of the solubility product constant of $MS(s)$ given in a footnote to Table C.2C (Appendix C), together with other relevant data, to show that the equilibrium constant for the precipitation reaction has the form: $K_c = K_w K_{a_1}/K_{sp}$. What is the value of K_c if the metal sulfide is FeS?

* 97. The dissolution of a metal sulfide in an acidic solution can be represented by the reverse of the equation written in Problem 96.

With the aid of such an equation and its equilibrium constant, determine the molar solubility of $MnS(s)$ in a buffer solution that is 1.50 M in CH_3COOH and 0.45 M in CH_3COO^-.

98. How many milliliters of 0.0050 M $NaBr(aq)$ can be added before any $AgBr(s)$ precipitates from 165 mL of an aqueous solution that is 0.62 M in $[Ag(NH_3)_2]^+$ and has $[NH_3]$ equal to 0.50 M?

* 99. A white solid mixture consists of two compounds, and the two compounds contain different cations. When treated with water, part of the mixture dissolves and part does not. The solution obtained is treated with $NH_3(aq)$ and yields a white precipitate. The part of the original solid mixture that is insoluble in water dissolves in $HCl(aq)$ with the evolution of a gas. The resulting solution is treated with $(NH_4)_2SO_4(aq)$ and produces a white precipitate. State which of the following cations *might* have been present and which *could not* have been present in the original solid mixture. Explain your reasoning.

   (a) Mg^{2+} **(c)** NH_4^+ **(e)** Na^+

   (b) Ba^{2+} **(d)** Cu^{2+}

* 100. Calculate the pH of a solution that is saturated with lithium phosphate, Li_3PO_4. (*Hint:* What reaction establishes $[H_3O^+]$ in the solution?)

* 101. A mixture of $PbSO_4(s)$ and $PbS_2O_3(s)$ is shaken with pure water until a saturated solution is formed. Both solids remain in excess. What is $[Pb^{2+}]$ in the saturated solution? (*Hint:* You must use the solubility product expressions for both $PbSO_4$ and PbS_2O_3 and a third relationship as well.)

102. Following up on the margin note on page 694 concerning the stepwise formation of $[Cu(NH_3)_4]^{2+}(aq)$, the value of $K_1 = 1.9 \times 10^4$ and $K_2 = 3.9 \times 10^3$. The value of $K_3 = 1.0 \times 10^3$. What must be the value of K_4?

103. As we will see in Chapter 20, most ordinary soaps are water-soluble sodium or potassium salts of long-chain fatty acids. However, soaps of divalent cations, such as Ca^{2+}, are only slightly water-soluble and are often seen in the common soap scum formed in hard water. A handbook lists the solubility of the typical calcium soap, calcium palmitate, $Ca[CH_3(CH_2)_{14}COO]_2$, as 0.003 g/100 mL at 25 °C.

   (a) If sufficient sodium soap is used to produce a concentration of palmitate ion equal to 0.10 M in a water sample having 25 ppm Ca^{2+}, would you expect any soap scum to form?

   (b) If a soap scum does form under the conditions in part (a), how many grams of calcium palmitate would precipitate in a bowl containing 6.5 L of the water?

Apply Your Knowledge

* 104. **[Laboratory]** A student is provided with a sample of limestone (see page 684) weighing 0.7291 g. The sample is dissolved in HCl, then precipitated by adding excess ammonium oxalate solution. The calcium oxalate is filtered, dried, then heated to 450 °C to convert the calcium oxalate to calcium carbonate. The calcium carbonate is found to weigh 0.6248 g.

   (a) Calculate the percentage of calcium in the limestone sample.

   (b) Suppose that, instead of an excess of ammonium oxalate, a stoichiometric amount of ammonium oxalate was added to precipitate calcium oxalate. If the total solution volume were 325 mL, what mass of calcium carbonate would have been obtained and what percentage of calcium in the limestone would have been calculated?

105. [Biochemical] Ba^{2+}(aq) is poisonous when ingested. The lethal dosage in mice is about 12 mg Ba^{2+} per kg of body mass. Despite this fact, $BaSO_4$ is widely used in medicine to obtain X-ray photographs of the gastrointestinal tract. **(a)** Explain why $BaSO_4$(s) is safe to take internally, even though Ba^{2+}(aq) is poisonous. **(b)** What is the concentration of Ba^{2+}, in milligrams per liter, in saturated $BaSO_4$(aq)? **(c)** $MgSO_4$ can be mixed with $BaSO_4$(s) in this medical procedure. What function does the $MgSO_4$ serve?

106. [Biochemical] An old remedy for constipation (no longer used!) was a dose of calomel, Hg_2Cl_2. Calculate the Hg_2^{2+} concentration in ppb in 1.00 L of stomach acid that is 1.0 M in HCl and is saturated with calomel. Assume the stomach acid has a density of 1.00 g/mL.

* 107. [Laboratory] *Argentiometric* titrations employ silver ion as a titrant (see Figure 4.17) or as the species determined. The concentration of silver ion, $[Ag^+]$, can be followed during the titration in much the same way as $[H_3O^+]$ is followed in an acid–base titration. Moreover, we can define $pAg = -\log[Ag^+]$ as a parallel to $pH = -\log[H_3O^+]$. For the titration of 25.0 mL of 0.100 M KSCN with 0.0500 M $AgNO_3$, calculate and plot the pAg of the solution versus volume of $AgNO_3$ at the following volumes: 1.0, 10.0, 25.0, 49.0, 50.0, 51.0, and 60.0 mL of $AgNO_3$. Comment on the value of pAg before any silver nitrate is added.

* 108. [Laboratory] A common laboratory experiment is that of determining K_{sp} of lead iodide. For this purpose, 0.020 M solutions of KI(aq) and $Pb(NO_3)_2$(aq) are prepared.

(a) To 10.0 mL of the lead(II) solution, the KI solution is added drop by drop from a special fine dropper, shaken vigorously each time, until a permanent yellow cloudiness is seen. In a separate experiment, it was found that 100 drops from the dropper had a volume of 1.12 mL. Calculate the number of drops (to the nearest drop) of KI solution that should be needed to reach the endpoint.

(b) A student is assigned the experiment described in part (a) except that he reverses the procedure, measuring 10.0 mL of 0.020 M KI solution and adding the lead nitrate to this solution. The number of drops of titrant required here would probably differ from the number found in part (a). Give a reason why this might be the case.

109. [Laboratory] In a qualitative analysis procedure known as *carbonate transposition*, anions from a slightly soluble compound, (such as a sulfate or hydroxide) are replaced by carbonate ion. That is, the carbonate ion appears in the solid and the other anion is freed in the solution. Suppose we use 3 M Na_2CO_3(aq) and that an anion concentration of 0.050 M is sufficient for detection of the anion in a qualitative analysis test. Predict whether carbonate transposition will be effective in the following conversions.

(a) $BaSO_4(s) + CO_3^{2-}(aq) \rightleftharpoons BaCO_3(s) + SO_4^{2-}(aq)$

(b) $Mg(OH)_2(s) + CO_3^{2-}(aq) \rightleftharpoons MgCO_3(s) + 2\,OH^-(aq)$

(c) $2\,AgCl(s) + CO_3^{2-}(aq) \rightleftharpoons Ag_2CO_3(s) + 2\,Cl^-(aq)$

110. [Environmental] The chief compound in marble is calcium carbonate. Marble has been widely used for statues and ornamental work on buildings, but marble is readily attacked by acids, as is seen in the accompanying photographs. Assume that the following is the overall reaction that occurs.

$$CaCO_3(s) + H_3O^+(aq) \rightleftharpoons Ca^{2+}(aq) + HCO_3^-(aq) + H_2O(l)$$

Determine the equilibrium constant for this reaction, and then determine the solubility of the marble (that is, $[Ca^{2+}]$ in a saturated solution) in **(a)** normal rainwater with pH = 5.6, and **(b)** acid rainwater with pH = 4.22. (*Hint:* What combination of reactions yields the overall reaction?)

▲ These photographs show the effect of acid rain on a marble statue of George Washington. The photograph on the left was made in 1935, and the one on the right in the mid-1990s. The primary constituent of marble is $CaCO_3$(s).

 e-Media Problems

The activities described in these problems can be found in the e-Media Activities and Interactive Student Tutorial (IST) modules of the Companion Website, *http://chem.prenhall.com/hillpetrucci*.

111. In the **Common Ion Effect** animation (*Section 16-3*), **(a)** what causes the decrease in Ag^+ ion concentration in solution? **(b)** How do the Na^+ ions influence the solubility equilibrium? **(c)** Predict the effect of adding solid $AgNO_3$ to the original solution on the concentrations of Ag^+ and I^-.

112. (a) Write net ionic equations for each of the reactions in the **Precipitation Reactions** movie (*Section 16-4*). **(b)** Using observations made from the movie, describe a scheme to test for Ag^+ ions in solution.

113. View the **Dissolution of Mg(OH)₂ by Acid** animation (*Section 16-5*). **(a)** Describe, in qualitative terms, the pH dependence of the solubility of $Mg(OH)_2$. **(b)** In chemical terms, describe the function of milk of magnesia, a product commonly used to treat acid indigestion and heartburn.

114. Use the tests provided in the **Qualitative Analysis** simulation (*Section 16-7*) to determine the identity of Unknown Compound #2.

115. Use the data presented in the **Ionic Compounds** activity (*Section 16-7*) to write a specific set of solubility rules for the Fe^{3+} cation. Consider only the anions present in the activity.

Chapter 17

Thermodynamics: Spontaneity, Entropy, and Free Energy

THERMODYNAMICS EXAMINES RELATIONSHIPS involving heat and work. In Chapter 6, we focused on thermochemistry, specifically on the enthalpy changes (heats) of chemical reactions; thermochemistry is a branch of thermodynamics. In subsequent chapters, we considered heat effects in relation to other concepts, such as phase changes and solution formation.

In this chapter, we will consider three main ideas:

- *Spontaneity* is the notion of whether a process can take place unassisted.
- *Entropy* is a measure of how energy is spread out among the atoms and molecules of a system.
- *Free energy* is a thermodynamic function that relates enthalpy and entropy to spontaneity.

Perhaps most important of all, we will relate free energy changes to equilibrium constants. Doing so makes it possible to calculate equilibrium constants from tabulated data.

17.1 Why Study Thermodynamics?

Having the right kind of knowledge can bring great rewards. With a knowledge of thermodynamics and by making a few calculations before embarking on a new venture, scientists and engineers can save themselves a great deal of time, money, and frustration. Recall Le Châtelier's statement about the importance of understanding chemical equilibrium, a key thermodynamics concept (page 590).

◄ Spontaneity—the ability of a reaction to take place unassisted—may have desirable consequences. For example, the spontaneous reaction that generates ammonia from nitrogen and hydrogen gases produces much of the world's artificial fertilizers. In other cases, a spontaneous reaction may have disastrous consequences. About 4000 people had to be evacuated from the town of Glenwood Springs, Colorado, in June 2002, due to the hillside fire seen here. This fire was ignited by another example of a spontaneous reaction—a seam of underground coal that had been burning for years.

The practical value of thermodynamics was well recognized early in the twentieth century by two American chemists, G. N. Lewis and Merle Randall, who wrote in the introduction to their landmark textbook,*

"To the manufacturing chemist thermodynamics gives information concerning the stability of his substances, the yield which he may hope to attain, the methods of avoiding undesirable substances, the optimum range of temperature and pressure, the proper choice of solvent. . . . "

With considerable foresight, they further offered the opinion that "the greatest development of applied thermodynamics is still to come." Even Lewis and Randall may have been surprised at how widely thermodynamics is now applied. In our study, we will find that we can apply thermodynamics to systems as complex as living organisms, not just to reactions in beakers, batteries, and blast furnaces.

17.2   Spontaneous Change

We know from experience that some processes occur on their own, without requiring us to do anything. The ice melts as a glass of iced tea stands at room temperature. An iron nail rusts when exposed to the oxygen and water vapor of moist air. A violent reaction occurs when sodium metal and chlorine gas come in contact (Figure 9.3):

$$2\,Na(s) + Cl_2(g) \longrightarrow 2\,NaCl(s)$$

These are all examples of spontaneous processes. A **spontaneous process** is one that can occur in a system left to itself; no action from outside the system is necessary in order for the process to occur.

We know equally well that certain other processes are *nonspontaneous;* they do not occur by themselves. Water does not freeze to form ice at room temperature. The iron oxide of a rusty nail does not revert to iron metal and oxygen gas. Solid sodium chloride does not decompose into sodium metal and chlorine gas. A **nonspontaneous process** is one that cannot take place in a system left to itself. We can summarize these observations by the statement

If a process is spontaneous, the reverse process is nonspontaneous and vice versa.

Some nonspontaneous processes are impossible. For example, we cannot freeze liquid water at atmospheric pressure and a constant temperature of 25 °C. Other nonspontaneous processes are possible but require some action from outside the system. For example, we can convert sodium chloride to sodium metal and chlorine gas by melting solid NaCl and passing an electric current through it, a process called electrolysis. Only continuous application of electric current causes the nonspontaneous reaction to occur.

The term "spontaneous" signifies nothing about how fast a process occurs. The spontaneous reaction between sodium and chlorine is extremely fast; the spontaneous rusting of iron is much slower. If we mix H_2 and O_2 gases at room temperature, we see no evidence of a chemical reaction. Yet, thermodynamic criteria indicate that this reaction is indeed spontaneous. Thus, thermodynamics can tell us if a process is *possible*, but only chemical kinetics (Chapter 13) can tell us *how fast* the process will occur. Chemical kinetics dictates that the reaction between H_2 and O_2 will be immeasurably slow at room temperature because of its high activation energy. At high temperatures, thermodynamics again predicts that the reaction will be spontaneous, and chemical kinetics forecasts that the reaction will be fast. In line with these predictions, hydrogen and oxygen combine with explosive speed at high temperatures.

- Thermodynamics determines the equilibrium *state* of a system. We use thermodynamics to predict the proportions of products and reactants present at equilibrium. Will the forward reaction go essentially to completion? Will it take place only to a slight extent? Will there be some intermediate balance between products and reactants?

▲ **Two meanings of the word "spontaneous"**

In "spontaneous combustion," a combustible material suddenly bursts into flame, as in the sudden ignition of this shipment of latex gloves that caused a fire at the Brooklyn Naval Yard in 1995. In common usage, we say that the combustion is spontaneous because it happens suddenly and without forewarning. In scientific usage, a spontaneous reaction is one that has a natural tendency to occur. However, to say that a reaction is spontaneous says nothing about whether the reaction will occur rapidly or be immeasurably slow.

 Nitrogen Triiodide animation

*G. N. Lewis and M. Randall, *Thermodynamics and the Free Energy of Chemical Substances*, McGraw-Hill, 1923.

• Kinetics determines the *pathway* by which equilibrium is reached. A high activation energy can effectively block a reaction that is thermodynamically favored. Combustion reactions generally are thermodynamically favored, but fortunately for life on Earth, most also have a high activation energy. Were this not so, most of Earth's organic matter would rapidly burn upon exposure to the oxygen in the atmosphere. Thus, both thermodynamics and kinetics are important in answering the question: Will a reaction occur?

In Example 17.1, we show, even this early in our discussion, how we can identify some processes as spontaneous or nonspontaneous. However, there are many other cases where we cannot be sure.

Example 17.1

Indicate whether each of the following processes is spontaneous or nonspontaneous. Comment on cases where a clear determination cannot be made.

(a) The action of toilet bowl cleaner, HCl(aq), on "lime" deposits, $CaCO_3$(s).

(b) The boiling of water at normal atmospheric pressure and 65 °C.

(c) The reaction of N_2(g) and O_2(g) to form NO(g) at room temperature.

(d) The melting of an ice cube.

SOLUTION

(a) When we add the acid, the fizzing that occurs—the escape of a gas—indicates that the reaction occurs without any further action on our part. The net ionic equation is

$$CaCO_3(s) + 2 H_3O^+(aq) \longrightarrow Ca^{2+}(aq) + 3 H_2O(l) + CO_2(g)$$

The reaction is *spontaneous*.

(b) We know that the normal boiling point of a liquid is the temperature at which the vapor pressure is equal to 1 atm. For water, this temperature is 100 °C. Thus, the boiling of water at 65 °C and 1 atm pressure is *nonspontaneous* and not possible.

(c) Nitrogen and oxygen gases occur mixed in air, and there is no evidence of their reaction to form toxic NO at room temperature. This makes their reaction to form NO(g) appear to be *nonspontaneous*. However, this could be an example of a spontaneous reaction that occurs extremely slowly, like that of H_2 and O_2 to form H_2O. We need additional criteria before we can answer this question.

(d) We know that at 1 atm pressure ice melts spontaneously at temperatures above its normal melting point of 0 °C. Below this temperature, it does not. To answer this question, we would have to know the temperature.

EXERCISE 17.1A

Indicate whether each of the following processes is spontaneous or nonspontaneous. Comment on cases where a clear determination cannot be made.

(a) The decay of a piece of wood buried in soil.

(b) The formation of sodium, Na(s), and chlorine, Cl_2(g), in a vigorously stirred aqueous solution of sodium chloride, NaCl(aq).

(c) The ionization of hydrogen chloride when HCl(g) dissolves in liquid water.

EXERCISE 17.1B

Indicate whether each of the following processes is spontaneous or nonspontaneous. Comment on cases where a clear determination cannot be made.

(a) The dissolution of 1.00 mL of liquid ethanol in 100 mL of liquid water.

(b) The formation of lime, CaO(s), and carbon dioxide gas at 1 atm pressure from limestone, $CaCO_3$(s), at 600 °C.

(c) The condensation of CO_2(g) at 0.50 atm to produce CO_2(s).

(d) The displacement of H_2(g) at 1 atm from 1 M HCl(aq) by Cu(s).

▶ **FIGURE 17.1** **Direction of decreasing energy: Is it a criterion for spontaneous change?**

Water in the waterfall flows spontaneously to a lower elevation, where the potential energy is lower. The enthalpy diagram for the conversion of H_2 and O_2 to H_2O shows the lowering of enthalpy when this spontaneous conversion takes place at 25 °C and 1 atm. Thus, this situation is analogous to that of the waterfall. The enthalpy diagram for the vaporization of water to produce $H_2O(g)$ at its equilibrium vapor pressure at 25 °C shows that for this process, *which is also spontaneous*, the enthalpy *increases*. Thus, the waterfall analogy does not work in this case.

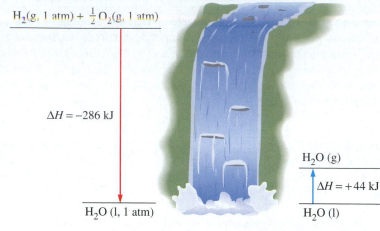

The formation of water at 25 °C and 1 atm: a spontaneous process that is exothermic

The vaporization of water at 25 °C and pressures up to 0.0313 atm: a spontaneous process that is endothermic

The unresolved cases in Example 17.1 indicate that it would be helpful to have criteria that tell us whether a process proceeds spontaneously. Common knowledge can help. You know that if you leave your car in neutral gear and forget to set the parking brake, the car will roll downhill. You also know that water in a stream always flows only one way—downhill. And you probably know why these things happen: If no counteracting forces are present, the force of gravity attracts objects toward the center of Earth, that is, in the direction of *decreasing* potential energy.

By similar reasoning, early chemists proposed that spontaneous chemical reactions should occur in the direction of decreasing energy. In Chapter 6, we defined internal energy, U, as the total energy content of a system and ΔU as the change in internal energy accompanying a process such as a chemical reaction. Because most chemical reactions are carried out in vessels open to the atmosphere, we found it convenient to use the closely related energy quantity *enthalpy, H*. The enthalpy change, ΔH, during a chemical reaction is equal to the heat of a reaction, q_p, which is the quantity of heat that enters or leaves a system in which a reaction occurs at constant pressure.

Figure 17.1 compares a spontaneous mechanical process—the fall of water—with two chemical processes. The direction of spontaneous flow in the waterfall is to a lower elevation, to a lower potential energy. Internal energy in a chemical system is analogous to potential energy in a mechanical system. However, in place of internal energy, let us use the familiar and closely related enthalpy, H. By the waterfall analogy, we might expect reactions in which enthalpy decreases to be spontaneous. Recall from Chapter 6 that reactions with $\Delta H < 0$ are *exothermic*. We might therefore expect reactions in which enthalpy increases to be nonspontaneous. Recall also that reactions with $\Delta H > 0$ are *endothermic*. Figure 17.1 indicates that the reaction of $H_2(g)$ and $O_2(g)$ to form $H_2O(l)$ is exothermic and spontaneous. As we also know, at 25 °C the vaporization of water is spontaneous until the vapor pressure reaches 23.8 mmHg (0.0313 atm). But this process is *endothermic*.

The idea that exothermic reactions are spontaneous and that endothermic reactions are not spontaneous works in many cases. However, as in the vaporization of water just cited, enthalpy change is not a sufficient criterion for predicting spontaneous change. We need to look at some additional factors.

17.3 The Concept of Entropy

Consider a tank of propane connected to a barbecue grill. The valve is opened, propane gas exits the tank and flows out of the burner, where it mixes with the oxygen in the air and can be ignited. It is readily apparent that there is a driving force behind the process. The propane pressure is considerably higher than atmospheric pressure, and therefore propane is emitted from the tank. Now, let us consider a thought experi-

The Origins of Thermodynamics

English inventor Thomas Newcomen (1663–1729) developed the first steam engine about 1711. The Newcomen engine expelled used steam to the atmosphere and was quite inefficient. However, it was successfully used to pump water out of coal mines. In 1763, James Watt (1736–1819) was sent a Newcomen engine to repair. While doing so, Watt discovered that he could make the engine more efficient by condensing the used steam in a condenser outside the main cylinder and returning the condensed steam to the steam generator driving the engine. Watt's steam engine was quite popular because it was four times more powerful than the Newcomen engine, but the efficiency of the engine was still quite low.

Nicolas Léonard Sadi Carnot (1796–1832), a French physicist, military engineer, and statesman, tried to deduce on a theoretical basis how Watt's engine might be improved. In 1824, Carnot described his conception of the perfect engine, now called a *Carnot engine*, in which heat energy is used to the maximum extent possible. He showed that heat cannot pass from a colder body to a warmer one and that the efficiency of an engine depends on the temperature of the steam entering the engine and on the temperature of the condenser. Carnot's discoveries later became the basis for the second law of thermodynamics.

▲ Watt's first steam engine, now on display in the Science Museum, London, England.

ment suggested by Figure 17.2a, with propane molecules at 1 atm in the bulb on the left and oxygen molecules at 1 atm in the bulb on the right. Suppose that the stopcock between the two gases is opened. What will happen? Even though there is no pressure difference, the two gases will mix spontaneously, and shortly the situation shown in Figure 17.2b will be established. If this is not intuitively apparent, suppose instead that we begin with a mixture as in Figure 17.2b. We would not expect the two gases to separate spontaneously to produce the arrangement in Figure 17.2a. That reverse process is nonspontaneous, which means the mixing process must be spontaneous.

A moment's thought will reveal that the mixing process is neither exothermic nor endothermic to any significant extent. For the conditions described here, propane, oxygen, and propane–oxygen mixtures are essentially ideal gases, and intermolecular forces are negligible. Therefore, there should be no significant enthalpy change in the process. Clearly, there is another factor involved in this process.

The other factor is a thermodynamic quantity called *entropy*, a mathematical concept that is difficult to portray visually, but a bit of historical background might help. The entropy concept was developed by Sadi Carnot in the 1820s and expressed through a mathematical equation for calculating the maximum efficiency of a heat engine (for example, a steam engine). No notion of the structure of matter at the microscopic level enters into this historical view of entropy. The entropy concept was given a molecular interpretation by Ludwig Boltzmann several decades later.

According to Boltzmann's ideas, the *total energy* of the system remains unchanged in the mixing of ideal gases. As the mixing takes place, however, there is an increase in the number of possibilities for the *distribution* of that energy within the system. Energy spread out over a greater number of available energy levels means an

 **Mixing of Gases animation**

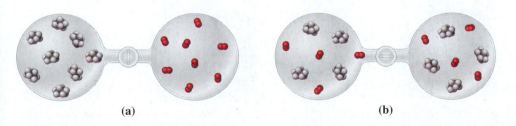

(a)   (b)

◄ **FIGURE 17.2 The spontaneous mixing of gases**

(a) Propane and oxygen gases in separate identical flasks connected by a stopcock. (b) The same two gases when the stopcock is opened. The gas molecules mix spontaneously.

increase in entropy. For example, in the mixing of gases shown in Figure 17.2 the increased volume available to the molecules makes available a greater number of energy levels among which a system's energy can be dispersed, producing an increase in entropy. Unfortunately, it is not easy to represent this dispersal, or spreading, of energy at the molecular level with physical pictures. On the other hand, there are certain indications of a change in entropy for which we can look, and we will present them shortly (page 719).

Since the time of Boltzmann, scientists have observed that in many cases the spreading of the energy and increase of entropy correlate with a greater physical disorder, or randomness, at the microscopic level. For example, the availability of energy levels for dispersal of a system's energy and the degree of disorder both increase as molecules become more widely scattered, undergo a greater variety of internal motions, perhaps undergo dissociation into smaller molecules or individual atoms, and so on. Physical pictures based on a degree of disorder are thus a crude but sometimes successful means of viewing entropy change. Remember, however, that although an increase in disorder may parallel an increase in entropy, the true meaning of entropy still relates to the availability and occupation of energy levels.

At this point, we can conclude that there are two natural tendencies behind spontaneous processes: the tendency for a system to achieve a lower energy state and the tendency for the system's energy to become dispersed among a larger number of energy levels. Thus, we need to consider *two* questions when judging whether a change will be spontaneous:

- Does the enthalpy of the system increase or decrease, and by how much?
- Does the entropy of the system increase or decrease, and by how much?

When a process involves both a decrease in energy and an increase in entropy, we can easily predict the direction of spontaneous change. Consider the conversion of ozone (O_3) to ordinary oxygen (O_2):

$$2\,O_3 \longrightarrow 3\,O_2 \qquad \Delta H = -285 \text{ kJ}$$

The reaction is exothermic, which favors reaction to the right. There is also an increase in entropy favoring reaction to the right. The system of six O atoms arranged in three O_2 molecules has more available energy levels among which the system's energy can be distributed than does the system of six O atoms arranged in two O_3 molecules. Thus, we predict that the direction of spontaneous change in the reaction is to the right.

In some cases, either enthalpy change or entropy change is negligible and the other factor takes precedence. An example is the formation of an ideal solution of benzene and toluene, shown in Figure 17.3. The enthalpy change $\Delta H \approx 0$ because no

Thermodynamics of Dissolution activity

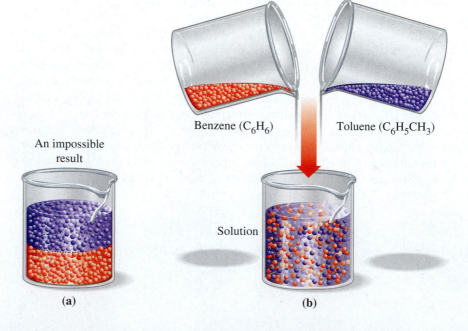

Benzene (C_6H_6)   Toluene ($C_6H_5CH_3$)

An impossible result

Solution

▶ **FIGURE 17.3   Formation of an ideal solution**

(a) A mixture of benzene and toluene in which the molecules remain segregated is impossible. (b) Formation of the solution is driven by an increase in entropy.

QUESTION: Name two liquids that could give rise to the result described as impossible for benzene and toluene.

(a)   (b)

chemical reaction occurs and also because the intermolecular forces in the solution are essentially the same as in the pure liquids. Because enthalpy change is not important, the change in entropy determines the direction of spontaneous change. The condition in which the two types of molecules remain segregated from one another, as in Figure 17.3a, provides fewer energy levels for dispersal of the system's energy than in the solution shown in Figure 17.3b. Mixing to form a solution is spontaneous, a situation quite similar to the spontaneous mixing of ideal gases that we considered earlier.

In many cases, however, the two factors work in opposition. That is, enthalpy and entropy may both decrease, or they may both increase. In these cases, we need to determine which factor predominates.

Entropy

We have already considered some general ideas about entropy, but now let us look at the concept in more specific ways. First, to recapitulate, **entropy (S)** is a thermodynamic property related to the way in which the energy of a system is distributed among the available energy levels:

> *The greater the number of configurations of the microscopic particles (atoms, ions, molecules) among the energy levels in a particular state of a system, the greater is the entropy of the system.*

Recall that both enthalpy (H) and enthalpy change (ΔH) are *state functions* (Chapter 6). We will show shortly that entropy is also a state function. That is, the entropy of a system has a unique value when the composition, temperature, and pressure of the system are specified. Also, the difference in entropy between two states, the **entropy change (ΔS),** has a unique value.

We have noted that entropy increases as the ideal solution of Figure 17.3 forms. We represent the entropy increase ΔS that accompanies formation of an ideal solution by the expression

$$\Delta S = S_{\text{mixt}} - [S_{\text{A(l)}} + S_{\text{B(l)}}] > 0$$

where S_{mixt} is the entropy of the solution, S_{A} is the entropy of the solvent A, and S_{B} is the entropy of the solute B. Another example of an increase in entropy is that accompanying the vaporization of water, as suggested by Figure 17.4.

In assessing entropy changes, look for the following indicators. Entropy generally *increases* when

- Solids melt to form liquids.
- Solids or liquids vaporize to form gases.
- Solids or liquids dissolve in a solvent to form nonelectrolyte solutions.
- A chemical reaction produces an increase in the number of molecules of *gases*.
- A substance is heated.

 **Changes of State simulation**

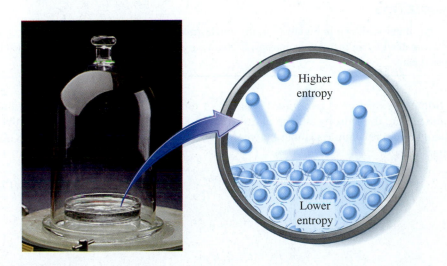

◀ **FIGURE 17.4   Increase of entropy in the vaporization of water**

The photograph depicts the vaporization of water. At the macroscopic level, nothing appears to be taking place. However, in the molecular view, we see that all the molecules are in motion and that the molecules are much more widely spaced in the vapor state (higher entropy) than in the liquid state (lower entropy).

QUESTION: Which is expected to produce a greater change in entropy, the vaporization of one mole of water or the sublimation of one mole of ice?

Note that the last of the listed items—that the entropy of a substance increases with temperature—does not correlate well with the idea of increased disorder, or randomness. When a fixed quantity of gas is heated in a closed volume, the spatial distribution of the gas molecules is no more random than at a lower temperature. However, along with the increased energy in the heated gas comes a very large increase in the number of energy levels among which the energy is dispersed—an increase in entropy.

Example 17.2

Predict whether each of the following leads to an increase or decrease in the entropy of a system. If in doubt, explain why.

(a) The synthesis of ammonia:

$$N_2(g) + 3 H_2(g) \longrightarrow 2 NH_3(g)$$

(b) Preparation of a sucrose solution:

$$C_{12}H_{22}O_{11}(s) \xrightarrow{H_2O(l)} C_{12}H_{22}O_{11}(aq)$$

(c) Evaporation to dryness of a solution of urea, $CO(NH_2)_2$, in water:

$$CO(NH_2)_2(aq) \longrightarrow CO(NH_2)_2(s)$$

SOLUTION

(a) Four moles of gaseous reactants produce two moles of gaseous product. Because the reaction leads to a decrease in the number of gas molecules, we predict a decrease in entropy.

(b) The sucrose molecules are confined in highly ordered crystals in the solid state, whereas they become widely distributed in the aqueous solution. The number of available energy levels for dispersal of the system's energy increases, and we therefore predict an increase in entropy.

(c) Crystallization of the urea from an aqueous solution is the reverse of the process described in part (b), thus indicating a decrease in entropy. At the same time, however, in the evaporation of water a liquid is converted to a gas, suggesting an increase in entropy (recall Figure 17.4). Without further information, we cannot say whether the entropy of the system increases or decreases.

EXERCISE 17.2A

Predict whether each of the following leads to an increase or decrease in the entropy of a system. If in doubt, explain why.

(a) $NH_3(g) + HCl(g) \longrightarrow NH_4Cl(s)$
(b) $2 KClO_3(s) \longrightarrow 2 KCl(s) + 3 O_2(g)$
(c) $CO(g) + H_2O(g) \longrightarrow CO_2(g) + H_2(g)$

EXERCISE 17.2B

In the situation shown in Figure 17.4, the transformation of liquid water to water vapor is accompanied by an increase in entropy. Explain why the liquid water does not completely disappear.

Sometimes we need *quantitative* values of entropy changes. To get these values, we must relate the entropy change of a system, ΔS, to the quantity of heat, q_{rev}, that is exchanged in a reversible manner between a system and its surroundings at the Kelvin temperature, T:

$$\Delta S = \frac{q_{rev}}{T} \tag{17.1}$$

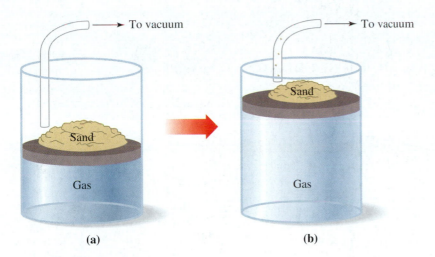

(a)   (b)

▲ **FIGURE 17.5   A nearly reversible process**
A reversible process can be made to reverse direction with just an infinitesimal change in a system variable. (a) A system in equilibrium, consisting of a gas in a chamber sealed by a movable piston weighted down with a pile of sand. (b) Grains of sand are removed one at a time from the pile on top of the piston, so that the piston rises very slowly as the gas expands. If at any time a grain of sand were *added* to the pile, the piston would reverse direction. In reality, the process is not quite reversible because grains of sand have more than an infinitesimal mass.

Just what do we mean by q_{rev}? As we noted on page 221, ΔU and ΔH are state functions but q and w are not. To make the entropy change ΔS a state function, we have to specify a particular path in order to give q a unique value. That path must be a *reversible* process. A reversible process is never more than an infinitesimal step from equilibrium. That is, it is a process that can be made to reverse its direction by just a minute change in some variable. For the reversible flow of heat, there can be only an infinitesimal difference in temperature between the system and its surroundings. Perhaps it is easier to imagine a process in which work is performed reversibly, as suggested in Figure 17.5.

We can rationalize the direct relationship between ΔS and q_{rev} in the equation $\Delta S = q_{rev}/T$ by noting that heating a substance increases both atomic and molecular motion. This increased motion increases the number of available energy levels among which the system's energy can be dispersed, and therefore entropy increases. This is true regardless of whether the substance is a solid, liquid, or gas. The increased number of available energy levels depends on the quantity of heat, and hence ΔS is directly proportional to q_{rev}.

The inverse relationship between ΔS and the Kelvin temperature, T, reflects the fact that the proportional increase with temperature in the number of energy levels in a system is greater at low temperatures than at high temperatures. Thus, a given quantity of heat absorbed by a solid initially at a low temperature increases the number of available energy levels proportionally more than does the same quantity of heat when absorbed by a gas initially at a high temperature. Figure 17.6 describes an analogy to which many people can relate.

The enthalpy change for the transition between two phases of matter at their equilibrium temperature has a unique value. This enthalpy change is the q_{rev} we need in calculating the entropy change for the transition. For example, we can represent the fusion of ice at 273 K by the equation

$$H_2O(s) \rightleftharpoons H_2O(l) \qquad \Delta H_{fusion} = 6.01 \text{ kJ} \qquad \Delta S_{fusion} = ?$$

▲ **FIGURE 17.6   An analogy to the relationship between entropy change and temperature**
Low-temperature analogy: At low temperatures, the number of accessible energy states for a system of particles is limited, and most of these states are occupied. Raising the temperature somewhat produces a large increase in the number of occupied energy states. This condition is much like well-ordered shelves in a supermarket. An earthquake of 5.0 magnitude on the Richter scale would produce considerable disorder in a supermarket. It is as if the earthquake makes many more energy states available, spreading the energy of the system over a larger number of states (entropy increase). *High-temperature analogy:* At high temperatures, the number of accessible energy states for a system of particles is high, and raising the temperature somewhat produces a relatively smaller increase in the number of occupied energy states than the same temperature increase at a lower temperature. This condition is analogous to a 5.0 aftershock hitting the same supermarket following an earlier, larger earthquake; little additional disorder would be produced.

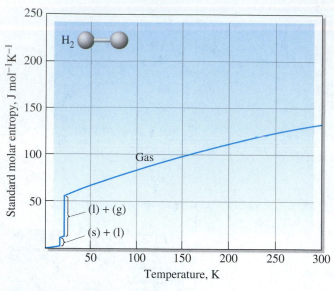

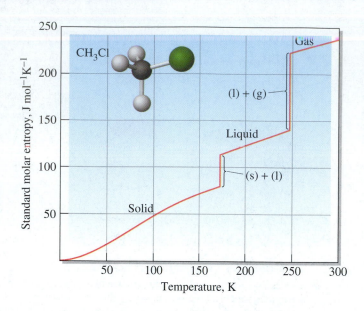

▲ **FIGURE 17.7** **Entropy as a function of temperature**

The entropies of hydrogen, H_2, and methyl chloride, CH_3Cl, are plotted as a function of temperature from 0 K to 300 K. Data are for molar quantities. The phases present at various temperatures are noted. The vertical line segment labeled (s) + (l) on each graph represents fusion, and the line segment labeled (l) + (g) represents vaporization.

QUESTION: From about 175 K to 250 K, why is the molar entropy of liquid methyl chloride greater than the molar entropy of gaseous hydrogen?

Temperature Dependence of Entropy simulation

We can then calculate the entropy change, ΔS_{fusion}, as follows:

$$\Delta S_{fusion} = \frac{q_{rev}}{T} = \frac{\Delta H_{fusion}}{T} = \frac{6.01 \text{ kJ}}{273 \text{ K}} = 0.0220 \text{ kJ K}^{-1} = 22.0 \text{ J K}^{-1}$$

In this calculation, the units of entropy change are those of energy divided by temperature: $kJ \ K^{-1}$ or $J \ K^{-1}$.

We will shortly use entropy to formulate the *second law of thermodynamics*, and that will provide us with a criterion for spontaneous change. First, however, let us look more closely at evaluating entropy and entropy changes.

Standard Molar Entropies

Lattice Vibrations animation

By relating entropy to the number of energy levels available for the dispersal of a system's energy, we might conceive of a condition in which the energy of a system is confined to a single level, with no opportunity for dispersal. We can consider this a condition of zero entropy. An example is that found in a crystalline substance at 0 K, where atomic and molecular motion ceases. We can state the **third law of thermodynamics** as follows:

The entropy of a pure, perfect crystal can be taken to be zero at 0 K.

Recall that with internal energy and enthalpy, we are not able to work with absolute values. We can work only with *differences*, that is, ΔU and ΔH.

We can use the third law of thermodynamics to obtain the entropy of a substance from experimental data. Beginning at temperatures very near 0 K, heat is slowly added to a substance and the temperature is measured. Entropy changes are calculated in increments, using the expression $\Delta S = q_{rev}/T$. Because the entropy at 0 K is zero, when these incremental entropy changes are added together, the result is an *absolute* entropy.

Figure 17.7 plots entropy as a function of Kelvin temperature for two substances, hydrogen and methyl chloride. Each graph has some interesting features. There is a sharp increase in entropy at the melting point [labeled (s) + (l) in both graphs], and at the boiling point [labeled (l) + (g)], there is an even sharper increase. There is a large increase in the number of available energy levels whenever a solid is converted to a liquid, but an even larger increase when a liquid is converted to a gas.

In comparing hydrogen and methyl chloride at 298 K, we see that the entropy of $CH_3Cl(g)$ is significantly greater than that of $H_2(g)$. One important factor responsible

Entropy and Probability

Although the concept of entropy applies only to the microscopic world of atoms, molecules, and ions, as we have noted, macroscopic analogies can help us understand the concept. At the microscopic level, there are more possibilities, and hence a greater probability, for certain states than for others. Probability plays a major role in determining the likelihood of events in the molecular world.

To illustrate, let us consider the spilling of four playing cards. There are two possible ways a given card may land—faceup or facedown—and the number of possible arrangements of the four cards is $2^4 = 16$. These 16 possibilities are pictured in the figure, where they are divided into five categories:

(a) all four cards faceup

(b) three cards faceup and one facedown

(c) two cards faceup and two facedown

(d) one card faceup and three facedown

(e) all four cards facedown

From the figure, we see that there is a probability of 1/16 that the cards will be in arrangement (a) and 1/16 in arrangement (e). These are the most ordered arrangements. The probability of arrangement (b) is 4/16, as is that of arrangement (d). The probability of arrangement (c) is the greatest, 6/16. This is the arrangement that we would expect to see most often if we observed the spilling of four cards over a large number of observations.

Early workers in thermodynamics developed the concept of entropy without regard to the ultimate nature of matter—some did not even believe in the existence of atoms. Ludwig Boltzmann (1844–1906) was the first to promote the idea that macroscopic observations in thermodynamics are governed by mechanical laws and laws of chance applied to atoms and molecules. Toward this end, he derived an equation relating entropy and probability:

$$S = k \ln W$$

In this equation, S is the entropy of a system in a particular state and W is the number of different configurations of equal energy in which the microscopic particles can be found in that state. The proportionality constant, k, is called the Boltzmann constant and is equal to R/N_A, the gas constant divided by Avogadro's number.

If we apply Boltzmann's equation to the card analogy, we get the following results. Because there is only one possibility for arrangements (a) and (e) in the figure, $W = 1$, $\ln 1 = 0$, and $S = 0$. Arrangements (a) and (e) represent perfect order (zero entropy). For arrangements (b) and (d), there are four possibilities each, and the entropy is proportional to $\ln 4 (1.39)$. For arrangement (c), with six possibilities, the entropy is proportional to $\ln 6 (1.79)$. Thus, the more possible arrangements, the greater is the entropy.

Notice that Boltzmann's equation also predicts that the entropy of a pure, perfect crystal at 0 K should be zero; in this state $W = 1$. Thus, Boltzmann's equation is consistent with the third law of thermodynamics.

▲ **The 16 possible arrangements of 4 spilled playing cards**
Four playing cards—a heart, a diamond, a spade, and a club—are allowed to fall to the floor. The cards may land faceup or facedown. The number of possibilities for each arrangement is shown in parentheses on the right.

for this difference in entropies is the difference in molecular structure: CH_3Cl molecules have five atoms, whereas H_2 molecules have two. The five atoms in CH_3Cl have more ways of moving with respect to one another—more vibrational modes—than do the two atoms in H_2. The more ways that molecules can absorb energy, the more energy levels available for dispersal of this energy, and the greater the entropy. This observation suggests that we add this generalization about entropy to the list on page 719:

In general, the more atoms in its molecules, the greater is the entropy of a substance.

Another point implied by Figure 17.7 is that entropy, like internal energy and enthalpy, is an *extensive* property—its magnitude depends on the amount of substance. The data in Figure 17.7 are standard molar entropies. The **standard molar entropy, $S°$,** of a substance is the entropy of one mole of the substance in its standard state. Usually, standard molar entropies are tabulated for a temperature of 25.00 °C (298.15 K). A listing of typical data is given in Appendix C.

We will use standard molar entropies mainly to determine standard entropy changes for chemical reactions. We can relate the entropy change in any reaction to the entropies of the reactants and products in the same way that we relate enthalpy change to enthalpies of formation [Equation (6.21)]. That is, we can write the expression:

$$\Delta S = \sum \nu_p S°(\text{products}) - \sum \nu_r S°(\text{reactants}) \tag{17.2}$$

In this equation, Σ refers to a sum and ν_p and ν_r are the *stoichiometric numbers*, that is, the coefficients of the products and reactants in the equation for the reaction. We apply this expression and data from Appendix C in Example 17.3.

Example 17.3

Use data from Appendix C to calculate the standard entropy change at 25 °C for the Deacon process, a high-temperature, catalyzed reaction used to convert hydrogen chloride (a by-product from organic chlorination reactions) into chlorine:

$$4 \, HCl(g) + O_2(g) \longrightarrow 2 \, Cl_2(g) + 2 \, H_2O(g)$$

STRATEGY

This problem requires a straightforward application of Equation (17.2), keeping in mind that we should expect a decrease in entropy ($\Delta S° < 0$) because five moles of gaseous reactants yield four moles of gaseous products.

SOLUTION

We relate $\Delta S°$ for the reaction to the standard molar entropies, $S°$, of reactants and products as follows:

$$\Delta S° = \sum \nu_p S°(\text{products}) - \sum \nu_r S°(\text{reactants})$$

$$= [2 \, \text{mol} \times S°_{Cl_2(g)} + 2 \, \text{mol} \times S°_{H_2O(g)}]$$

$$\quad -[4 \, \text{mol} \times S°_{HCl(g)} + 1 \, \text{mol} \times S°_{O_2(g)}]$$

$$= (2 \, \text{mol} \times 223.0 \, \text{J mol}^{-1} \, \text{K}^{-1}) + (2 \, \text{mol} \times 188.7 \, \text{J mol}^{-1} \, \text{K}^{-1})$$

$$\quad -(4 \, \text{mol} \times 186.8 \, \text{J mol}^{-1} \, \text{K}^{-1}) - (1 \, \text{mol} \times 205.0 \, \text{J mol}^{-1} \, \text{K}^{-1})$$

$$= -128.8 \, \text{J K}^{-1}$$

ASSESSMENT

Just as we expected, $\Delta S°$ is indeed negative; entropy decreases.

EXERCISE 17.3A

Use data from Appendix C to calculate the standard molar entropy change at 25 °C for the reaction

$$CO(g) + H_2O(g) \longrightarrow CO_2(g) + H_2(g)$$

Note that this is reaction (c) in Exercise 17.2A, for which your answer should have been that you could not predict whether entropy increases or decreases.

EXERCISE 17.3B

The following reaction may occur in the high-temperature, catalyzed oxidation of $NH_3(g)$. Use data from Appendix C to calculate the standard molar entropy change for this reaction at 25 °C.

$$NH_3(g) + O_2(g) \longrightarrow N_2(g) + H_2O(g) \qquad \text{(not balanced)}$$

The Second Law of Thermodynamics

We have seen that we cannot use the enthalpy change in a reaction (ΔH) as the sole criterion for spontaneous change. Can we use the tendency toward increasing entropy ($\Delta S > 0$) as a sole criterion?

In fact, we can, but only if we consider the entropy change of a system *and of its surroundings*. We call this total entropy change the *entropy change of the universe:*

$$\Delta S_{total} = \Delta S_{universe} = \Delta S_{system} + \Delta S_{surroundings} \qquad (17.3)$$

The **second law of thermodynamics** establishes that *all spontaneous processes increase the entropy of the universe*. In other words, for any spontaneous process,

$$\Delta S_{univ} > 0$$

If during any process, entropy increases in both the system and its surroundings, the process is surely spontaneous. Just as surely, if entropy decreases in both the system and its surroundings, the process is nonspontaneous and cannot occur on its own. What will ΔS_{univ} be if one term is positive and the other is negative? To find out, let us consider the freezing of liquid water at $-15\,°C$:

$$H_2O(l) \longrightarrow H_2O(s)$$

We can see that $\Delta S_{sys} < 0$ because ice has a lower entropy than liquid water. Let us not forget, however, that when water freezes, it gives off heat—the heat of fusion. Because this heat is absorbed by the surroundings, $\Delta S_{surr} > 0$. By calculation, it can be shown that the magnitude of ΔS_{surr} exceeds that of ΔS_{sys}.* The total entropy change, ΔS_{univ}, is positive, and the overall process is therefore spontaneous, just as we intuitively know it should be.

Although ΔS_{univ} gives a single criterion for spontaneous change, we find it difficult to apply because the interactions between a system and its surroundings are often complex. In Section 17.4, we will see that there is an easier approach.

17.4   Free Energy and Free Energy Change

In this section, we will introduce a new thermodynamic function and a new criterion for spontaneous change based *just on the system*, without regard to the surroundings.

To do so, imagine a process conducted at constant temperature and pressure and with work limited to pressure–volume work (page 226). For this process, $\Delta H_{sys} = q_p$. Because the process is carried out at constant temperature, this heat must be exchanged with the surroundings; that is, $q_{surr} = -q_p = -\Delta H_{sys}$. We can use this information to evaluate ΔS_{surr}:[†]

$$\Delta S_{surr} = \frac{q_{surr}}{T} = \frac{-\Delta H_{sys}}{T} \qquad (17.4)$$

Now let us turn to the expression

$$\Delta S_{univ} = \Delta S_{sys} + \Delta S_{surr}$$

and substitute for ΔS_{surr} from Equation (17.4):

$$\Delta S_{univ} = \Delta S_{sys} - \frac{\Delta H_{sys}}{T}$$

---

*These entropy changes must be calculated for a hypothetical reversible process having the same initial and final states as the process in which supercooled water freezes at $-15\,°C$. However, such calculations are beyond the scope of the text.

[†]For this equation to be valid, the surroundings must gain or lose heat by a *reversible* path, that is, $q_{surr} = q_{rev}$.

▲ J. Willard Gibbs (1839–1903), a professor of mathematical physics at Yale University, was not well known to his contemporaries, mostly because his ideas were quite abstract and published in obscure journals. Eventually, though, his ideas were understood, and he was recognized as a giant in the emerging field of chemical thermodynamics.

Next we can multiply both sides by T:

$$T\Delta S_{univ} = T\Delta S_{sys} - \Delta H_{sys} = -(\Delta H_{sys} - T\Delta S_{sys})$$

Then we can multiply by -1, changing signs:

$$-T\Delta S_{univ} = \Delta H_{sys} - T\Delta S_{sys} \qquad (17.5)$$

The significance of this equation is that it relates the elusive ΔS_{univ} to two quantities, ΔH_{sys} and $-T\Delta S_{sys}$, that are based only on the system itself. We do not have to consider the surroundings at all.

At this point, let us introduce a new thermodynamic function called the **Gibbs free energy, G,** and defined by the equation

$$G = H - TS$$

If we set the term on the left in Equation (17.5) equal to ΔG, which is called the **free energy change,** we obtain the result

$$\Delta G_{sys} = -T\Delta S_{univ} \qquad (17.6)$$

Now, by combining Equations (17.5) and (17.6), we see that the free energy change for a process at constant temperature and pressure is given by

$$\Delta G_{sys} = \Delta H_{sys} - T\Delta S_{sys}$$

Because each term is now based on the system, we can drop the subscripts. This relationship is commonly called the *Gibbs equation:*

$$\Delta G = \Delta H - T\Delta S \qquad (17.7)$$

Because the criterion for spontaneous change is that $\Delta S_{univ} > 0$ and because $\Delta G = -T\Delta S_{univ}$, a second criterion for spontaneous change is that $\Delta G < 0$. That is, the change in free energy of the system must be negative. In summary, these are the criteria that apply to a process at constant temperature and pressure:

- If $\Delta G < 0$, a process is spontaneous.
- If $\Delta G > 0$, a process is nonspontaneous.
- If $\Delta G = 0$, neither the forward nor the reverse process is favored; there is no net change, and the process is at equilibrium.

Using ΔG as a Criterion for Spontaneous Change

We will first consider some qualitative applications of the Gibbs equation:

$$\Delta G = \Delta H - T\Delta S$$

Gibbs Free Energy simulation

In the equation, we see that if ΔH is negative and ΔS is positive, then ΔG must be negative and a reaction will be spontaneous. We can just as easily see that if ΔH is positive and ΔS is negative, then ΔG must be positive and a reaction will be nonspontaneous. These two simple cases are represented as cases 1 and 4 in Table 17.1.

Table 17.1 Criterion for Spontaneous Change: $\Delta G = \Delta H - T\Delta S$

Case	ΔH	ΔS	ΔG	Result	Example
1	−	+	−	Spontaneous at all T	$2\,O_3(g) \longrightarrow 3\,O_2(g)$
2	− −	− −	− +	Spontaneous at low T Nonspontaneous at high T	$N_2(g) + 3\,H_2(g) \longrightarrow 2\,NH_3(g)$
3	+ +	+ +	+ −	Nonspontaneous at low T Spontaneous at high T	$2\,H_2O(g) \longrightarrow 2\,H_2(g) + O_2(g)$
4	+	−	+	Nonspontaneous at all T	$2\,C(graphite) + 2\,H_2(g) \longrightarrow C_2H_4(g)$

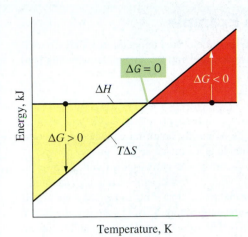

◀ **FIGURE 17.8** ΔG **as a criterion for spontaneous change**

ΔG, ΔH, and $T\Delta S$ all have the units of energy. At all temperatures, the value of ΔG is the value on the ΔH line minus the value on the $T\Delta S$ line. That is, ΔG is the distance between the two lines. At the temperature at which the lines intersect, this distance is zero ($\Delta G = 0$) and the system is at equilibrium. Below this temperature, $\Delta G > 0$ and the reaction is nonspontaneous. Above this temperature, $\Delta G < 0$ and the reaction is spontaneous. The situation described here is that of case 3 in Table 17.1.

QUESTION: How would the plot of Gibbs free energy versus temperature differ for case 2 in Table 17.1?

Situations in which ΔH and ΔS are both negative—case 2 in Table 17.1—or both positive—case 3 in Table 17.1—require more thought. Specifically, whether a reaction is spontaneous or not—that is, whether ΔG is negative or positive—depends on the temperature. In these cases, ΔG will have the same sign as ΔH at lower temperatures and the same sign as $-T\Delta S$ at higher temperatures, as suggested by Figure 17.8. Example 17.4 shows how we determine the sign of ΔG in two situations.

Example 17.4

Predict which of the four cases in Table 17.1 you expect to apply to the following reactions.

(a) $C_6H_{12}O_6(s) + 6\,O_2(g) \longrightarrow 6\,CO_2(g) + 6\,H_2O(g)$ $\Delta H = -2540\ \text{kJ}$

(b) $Cl_2(g) \longrightarrow 2\,Cl(g)$

SOLUTION

(a) The reaction is exothermic ($\Delta H < 0$). Because 12 moles of gaseous products replace 6 moles of gaseous reactant, we expect $\Delta S > 0$. Because $\Delta H < 0$ and $\Delta S > 0$, we expect this reaction to be spontaneous at all temperatures (case 1 of Table 17.1).

(b) One mole of gaseous reactant produces two moles of gaseous product, and therefore we expect $\Delta S > 0$. No value is given for ΔH. However, we can determine the sign of ΔH by examining what occurs during the reaction. Bonds are broken, but no new bonds are formed. Because molecules must absorb energy for bonds to break, the reaction must be endothermic. Thus, ΔH must be positive. Because ΔH and ΔS are both positive, this reaction is an example of case 3 in Table 17.1. We expect the reaction to be nonspontaneous at low temperatures and spontaneous at high temperatures.

EXERCISE 17.4A

Predict which of the four cases in Table 17.1 is likely to apply to each of the following reactions.

(a) $N_2(g) + 2\,F_2(g) \longrightarrow N_2F_4(g)$ $\Delta H = -7.1\ \text{kJ}$
(b) $COCl_2(g) \longrightarrow CO(g) + Cl_2(g)$ $\Delta H = +110.4\ \text{kJ}$

EXERCISE 17.4B

Predict which of the four cases in Table 17.1 is likely to apply to the following reaction. Use the value $-93.85\ \text{kJ/mol}$ for ΔH_f° for BrF(g) and additional data from Appendix C as necessary.

$$Br_2(g) + BrF_3(g) \longrightarrow 3\,BrF(g)$$

Example 17.5   A Conceptual Example

Molecules exist from 0 K to a few thousand kelvins. At elevated temperatures, they dissociate into atoms. Use the relationship between enthalpy and entropy to explain why this is to be expected.

ANALYSIS AND CONCLUSIONS

To cause a molecule to dissociate into its atoms, we must supply the molecule with enough energy to induce such vigorous vibrations that its atoms fly apart, an endothermic process ($\Delta H > 0$). To assess the entropy change, we first note that a molecule has a greater number of available energy levels than does any one of its constituent atoms taken alone. However, because two or more atoms are produced for every molecule dissociated, we find a greater number of available energy levels in a system of individual atoms than if the same atoms are united into molecules ($\Delta S > 0$). The key factor, then, is the temperature, T. At low temperatures, ΔH is the determining factor, dissociation into atoms is a nonspontaneous process, and therefore molecules are generally stable with respect to uncombined atoms. However, no matter how large the value of ΔH, eventually a temperature is reached at which the magnitude of $T\Delta S$ exceeds that of ΔH. Then ΔG is negative, and the dissociation becomes a spontaneous process. For all known molecules, this high temperature limit is no more than a few thousand kelvins.

EXERCISE 17.5A

If the process described in Figure 17.8 is $CO_2(s, 1 \text{ atm}) \longrightarrow CO_2(g, 1 \text{ atm})$, at what temperature will the two lines intersect? Explain. (*Hint:* What information do you need from an earlier chapter?)

EXERCISE 17.5B

With the aid of data from Appendix C, explain why you would expect the dissociation of nitrogen dioxide to nitrogen and oxygen (with all gases at 1 atm) to be spontaneous at all temperatures and the dissociation of ammonia into hydrogen and nitrogen (with all gases at 1 atm) to be spontaneous only at certain temperatures.

17.5   Standard Free Energy Change, $\Delta G°$

The **standard free energy change, $\Delta G°$,** of a reaction is the free energy change when reactants and products are in their standard states. The convention for standard states is the same as that used for $\Delta H°$ on page 240: The standard state of a solid or liquid is the pure substance at 1 atm pressure* and at the temperature of interest. For a gas, the standard state is the pure gas behaving as an ideal gas at 1 atm pressure and the temperature of interest.

One way to evaluate the standard free energy change of a reaction is to use standard enthalpy and standard entropy changes in the Gibbs equation:

$$\Delta G° = \Delta H° - T\Delta S° \tag{17.8}$$

In using Equation (17.8), we must be careful to use the same energy unit for $\Delta H°$ and $\Delta S°$. For example, tabulated data generally use *kilojoules* for enthalpy changes and *joules* per kelvin for entropies.

Another way to evaluate $\Delta G°$ is to use tabulated standard free energies of formation. The **standard free energy of formation, $\Delta G_f°$,** is the free energy change that occurs in the formation of one mole of a substance in its standard state from the reference forms of its elements in their standard states. The reference forms of the elements generally are their most stable forms. Like standard enthalpies of formation, the standard free energies of formation of the *elements* in their reference forms have a value of zero. Appendix C contains a listing of the standard free energies of formation of many substances at 25.00 °C (298.15 K), the temperature usually used for tabulations.

To obtain a standard free energy of reaction from standard free energies of formation, we use the same type of relationship that we used for enthalpies (see Equation 6.21):

$$\Delta G° = \sum \nu_p \, \Delta G_f°(\text{products}) - \sum \nu_r \, \Delta G_f°(\text{reactants}) \tag{17.9}$$

---

*The IUPAC standard-state pressure is 1 bar. Because the difference between 1 bar (exactly 100,000 Pa) and 1 atm (101,325 Pa) is slight, we will continue to use 1 atm.

Both methods of calculating a standard free energy change are illustrated in Example 17.6.

Example 17.6

Calculate ΔG° at 298 K for the reaction

$$4\,HCl(g) + O_2(g) \longrightarrow 2\,Cl_2(g) + 2\,H_2O(g) \qquad \Delta H^\circ = -114.4\,kJ$$

(a) using the Gibbs equation (17.8) and (b) from standard free energies of formation.

STRATEGY

(a) To use the standard-state form of the Gibbs equation, we need values of both ΔH° and ΔS°. The value of ΔH° is given, and we can obtain ΔS° from standard molar entropies, which we did in Example 17.3.

(b) We must look up standard free energies of formation in Appendix C and use them in Equation (17.9).

SOLUTION

(a) The value of ΔS° we determined in Example 17.3 was $-128.8\,J\,K^{-1}$. To express ΔH° and ΔS° in the same energy unit, we can convert $-128.8\,J\,K^{-1}$ to $-0.1288\,kJ\,K^{-1}$. Then

$$\Delta G^\circ = \Delta H^\circ - T\Delta S^\circ = -114.4\,kJ - \left[298\,K \times (-0.1288\,kJ\,K^{-1})\right]$$

$$= -114.4\,kJ + 38.4\,kJ = -76.0\,kJ$$

(b) Equation (17.9) takes the form

$$\Delta G^\circ = 2\,\Delta G_f^\circ[Cl_2(g)] + 2\,\Delta G_f^\circ[H_2O(g)] - 4\,\Delta G_f^\circ[HCl(g)] - \Delta G_f^\circ[O_2(g)]$$

$$= 2\,mol \times (0.00\,kJ/mol) + 2\,mol \times (-228.6\,kJ/mol)$$

$$\quad - 4\,mol \times (-95.30\,kJ/mol) - 1\,mol \times (0.00\,kJ/mol)$$

$$= (-457.2 + 381.2)\,kJ = -76.0\,kJ$$

ASSESSMENT

The principal check on part (a) is that the $T\Delta S^\circ$ product should be about 40 kJ (that is, 300×0.13), which, when added to the value of ΔH°, should yield a value of $\Delta G^\circ \approx -75\,kJ$. The principal check on part (b) is that it should give the same answer as part (a), and it does.

EXERCISE 17.6A

Use the standard-state form of the Gibbs equation to determine the standard free energy change at 25 °C for these reactions:

(a) $2\,NO(g) + O_2(g) \longrightarrow 2\,NO_2(g) \qquad \Delta H^\circ = -114.1\,kJ \qquad \Delta S^\circ = -146.2\,J\,K^{-1}$

(b) $2\,CO(g) + 2\,NH_3(g) \longrightarrow C_2H_6(g) + 2\,NO(g) \qquad \Delta H^\circ = +409.0\,kJ$
$$\Delta S^\circ = -129.1\,J\,K^{-1}$$

EXERCISE 17.6B

Use standard free energies of formation from Appendix C to determine the standard free energy change at 25 °C for these reactions:

(a) $CS_2(l) + 2\,S_2Cl_2(g) \longrightarrow CCl_4(l) + 6\,S(s)$

(b) $NH_3(g) + O_2(g) \longrightarrow N_2(g) + H_2O(g)$ *(not balanced)*

17.6 Free Energy Change and Equilibrium

So far, we have considered conditions in which $\Delta G < 0$ (spontaneous process) and in which $\Delta G > 0$ (nonspontaneous process). Now let us consider a process that is neither spontaneous nor nonspontaneous. In this case, there is no net change because the forward and reverse processes occur at the same rate. This is the equilibrium condition, at which $\Delta G = 0$:

$$\Delta G = \Delta H - T\Delta S = 0 \quad \text{(at equilibrium)}$$

and

$$\Delta H = T\Delta S \quad (at\ equilibrium)$$

At the equilibrium temperature, we can use this latter expression to determine either ΔH from a value of ΔS or ΔS from a value of ΔH.

When we relate the enthalpy and entropy of vaporization at the normal boiling point of a liquid, the liquid and vapor phases are essentially in their standard states (1 atm $\approx$ 1 bar). Thus, we can work with standard enthalpy and entropy changes. Consider the boiling of benzene at 80.10 °C:

$$C_6H_6(l,\ 1\ atm) \rightleftharpoons C_6H_6(g,\ 1\ atm) \qquad \Delta H° = 30.76\ kJ/mol$$

$$\Delta S°_{vapn} = \frac{\Delta H°_{vapn}}{T_{bp}} = \frac{30.76\ kJ\ mol^{-1}}{(80.10 + 273.15)K} = 0.08708\ kJ\ mol^{-1}\ K^{-1}$$

We can also express the standard molar entropy of vaporization of benzene at 80.10 °C in the units $J\ mol^{-1}\ K^{-1}$.

$$\Delta S°_{vapn} = 0.08708\ kJ\ mol^{-1}\ K^{-1} \times \frac{1000\ J}{1\ kJ} = 87.08\ J\ mol^{-1}\ K^{-1}$$

Many liquids have about the same standard molar entropy of vaporization at their normal boiling points—approximately 87 $J\ mol^{-1}\ K^{-1}$. These observations are summarized by **Trouton's rule:**

$$\Delta S°_{vapn} = \frac{\Delta H°_{vapn}}{T_{bp}} \approx 87\ J\ mol^{-1}\ K^{-1} \qquad (17.10)$$

Trouton's rule implies that the evaporation of one mole of liquid at the normal boiling point is accompanied by about the same increase in the number of accessible energy levels from one liquid to another. The rule works best for nonpolar substances, such as those in Figure 17.9. It generally fails for liquids with extensive hydrogen bonding. Hydrogen bonding among molecules leads to a lower entropy than would otherwise be expected in the liquid state. In turn, this leads to a greater-than-expected entropy of vaporization, that is, to $\Delta S°_{vapn} > 87\ J\ mol^{-1}\ K^{-1}$. For example, for the vaporization of H_2O at 100 °C, $\Delta S°_{vapn} = 109\ J\ mol^{-1}\ K^{-1}$.

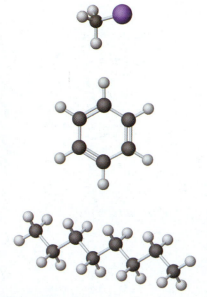

▲ (Top to bottom): Molecular models of iodomethane, benzene, and octane.

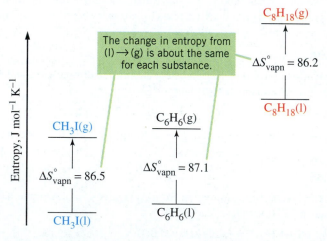

▲ **FIGURE 17.9   Illustrating Trouton's rule**

Entropies of vaporization are given in $J\ K^{-1}$ per mole of liquid vaporized. The three liquids all have a standard molar entropy of vaporization of about 87 $J\ mol^{-1}\ K^{-1}$. However, the following tabulation suggests that the liquids have little else in common.

	Molecular Mass, u	Molar Entropy, $S°_{298}$, $J\ mol^{-1}\ K^{-1}$	At the Normal Boiling Point		
			bp, K	$\Delta H°_{vapn}$ kJ/mol	$\Delta S°_{vapn}$ $J\ mol^{-1}\ K^{-1}$
CH_3I, iodomethane	142	163	315.6	27.3	86.5
C_6H_6, benzene	78	173	353.3	30.8	87.1
C_8H_{18}, octane	114	358	398.9	34.4	86.2

Example 17.7

At its normal boiling point, the enthalpy of vaporization of pentadecane, $CH_3(CH_2)_{13}CH_3$, is 49.45 kJ/mol. What should its approximate normal boiling point temperature be?

STRATEGY

Trouton's rule provides an approximate value $(87 \text{ J mol}^{-1} \text{ K}^{-1})$ for the molar entropy of vaporization of a *nonpolar* liquid at its normal boiling point. Pentadecane, an alkane, is a nonpolar liquid. When we apply Equation (17.10), we will solve for the boiling point, T_{bp}.

SOLUTION

We solve Equation (17.10) for T_{bp} and then substitute the known value of ΔH°_{vapn}, together with $\Delta S^{\circ}_{vapn} \approx 87 \text{ J mol}^{-1} \text{ K}^{-1}$:

$$\Delta S^{\circ}_{vapn} = \frac{\Delta H^{\circ}_{vapn}}{T_{bp}} \approx 87 \text{ J mol}^{-1} \text{ K}^{-1}$$

$$T_{bp} \approx \frac{\Delta H^{\circ}_{vapn}}{87 \text{ J mol}^{-1} \text{ K}^{-1}} \approx \frac{49.45 \times 10^3 \text{ J mol}^{-1}}{87 \text{ J mol}^{-1} \text{ K}^{-1}} \approx 570 \text{ K}$$

ASSESSMENT

To test the validity of our estimate, we could consult a chemistry handbook. We would find the experimentally determined boiling point of pentadecane to be 543.8 K. Our estimate is within about 5% of the true value.

EXERCISE 17.7A

The normal boiling point of styrene, $C_6H_5CH{=}CH_2$, is 145.1 °C. Estimate the molar enthalpy of vaporization of styrene at its normal boiling point.

EXERCISE 17.7B

Would you expect ΔS°_{vapn} for methanol, CH_3OH, to be greater, less than, or about equal to $87 \text{ J mol}^{-1} \text{ K}^{-1}$? Explain. Use the normal boiling point of 64.7 °C and data from Appendix C to estimate a value of ΔS°_{vapn}.

Raoult's Law Revisited

Recall Raoult's law from Section 12.6:

$$P_{solv} = x_{solv} \cdot P^{\circ}_{solv}$$

Because the mole fraction of solvent in a solution (x_{solv}) is less than 1, the vapor pressure of the solvent (P_{solv}) in an ideal solution is lower than that of the pure solvent (P°_{solv}). The concept of entropy gives us a way to explain Raoult's law, as illustrated in Figure 17.10.

In describing the formation of an ideal solution (Figure 17.3), we saw that a solution has a higher entropy than the pure solvent. Because the intermolecular forces in the solution and in the pure solvent are comparable, we expect the enthalpy of vaporization (ΔH_{vapn}) to be essentially the same for evaporation from the solution and evaporation from the pure solvent. The entropy of vaporization should also be the same because $\Delta S_{vapn} = \Delta H_{vapn}/T$. However, because the ideal solution has a higher entropy than the pure solvent, the vapor from the solution must have a higher entropy than the vapor from the pure solvent. The entropy of a vapor increases as its molecules expand into a larger volume, that is, if they are at a *lower* pressure. Thus, solutes in a solution reduce the vapor pressure of the solvent.

Relationship of ΔG° to the Equilibrium Constant, K_{eq}

Liquid water is in equilibrium with water vapor at 1 atm pressure and 100 °C. As we have seen, ΔG for this (or any other) equilibrium process is zero. Moreover, because the liquid and vapor are both in their standard states, we can write

$$H_2O(l, 1 \text{ atm}) \rightleftharpoons H_2O(g, 1 \text{ atm}) \qquad \Delta G^{\circ}_{373} = 0$$

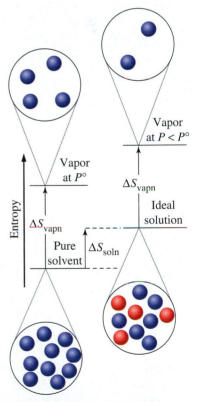

▲ **FIGURE 17.10   An entropy-based explanation of Raoult's law**

The features brought out by the diagram are that

- the entropy of the solution is greater than that of the pure solvent.

- the entropy change for vaporization of a quantity of the solvent, ΔS_{vapn}, is the same in the pure solvent and in the ideal solution.

- the vapor phase above the solution is less dense, and therefore at a lower pressure, than above the pure solvent.

We can write the same equation for the vaporization of water at 25 °C and determine ΔG°_{298} from tabulated standard free energies of formation.

$$H_2O(l, 1 \text{ atm}) \rightleftharpoons H_2O(g, 1 \text{ atm}) \qquad \Delta G^\circ_{298} = +8.590 \text{ kJ}$$

The positive value of ΔG°_{298} shows that the process is nonspontaneous. This does not mean that water will not vaporize at 25 °C; instead, it just means that the evaporating water will not produce a vapor that is *at 1 atm pressure*. Rather, equilibrium is displaced *to the left*—toward liquid water—resulting in a vapor pressure less than 1 atm. From Table 11.2 we see that the equilibrium vapor pressure of water at 25 °C is 23.8 mmHg (0.0313 atm), a fact we can represent by the equation

$$H_2O(l) \rightleftharpoons H_2O(g, 0.0313 \text{ atm}) \qquad \Delta G_{298} = 0$$

To summarize, $\Delta G^\circ = 0$ is a criterion for equilibrium at a single temperature, the one temperature at which the equilibrium state has all reactants and products in their standard states. At equilibrium at every other temperature, some or all of the reactants and products must be in a *nonstandard* state. For these nonstandard conditions, the criterion for equilibrium is that $\Delta G = 0$ (*not* $\Delta G^\circ = 0$). This seems to make ΔG° of limited value, but we will now see that it really is quite useful.

Although it is beyond the scope of this text to demonstrate this fact, the quantities ΔG and ΔG° are related to each other through the *reaction quotient, Q*, by the equation

$$\Delta G = \Delta G^\circ + RT \ln Q \qquad \textbf{(17.11)}$$

Now consider a reaction at equilibrium, in which $\Delta G = 0$ and $Q = K_{eq}$. This leads to the expression

$$0 = \Delta G^\circ + RT \ln K_{eq} \quad \textit{(at equilibrium)}$$

and

$$\Delta G^\circ = -RT \ln K_{eq} \quad \textit{(at equilibrium)} \qquad \textbf{(17.12)}$$

This is one of the most important equations in all of chemical thermodynamics. It says that if we have ΔG° for a reaction at a given temperature, we can calculate the equilibrium constant at that temperature. With the equilibrium constant, we can then calculate equilibrium concentrations or partial pressures, just as we did in earlier chapters. In applying this equation, we should note the following:

- R, the gas constant, is expressed as 8.3145 J mol^{-1} K^{-1}.
- T, the temperature, is expressed in kelvins, K.
- K_{eq} must have the format described next.

The Equilibrium Constant, K_{eq}

Recall that in Chapter 14 we wrote a K_c expression when we used concentrations to describe reactants and products and a K_p expression when we used partial pressures to describe reactants and products. In the expression $\Delta G^\circ = -RT \ln K_{eq}$, we write the **equilibrium constant K_{eq}** rather than K_c or K_p, and K_{eq} may differ from both of them. Because the equation relating the standard free energy change and the equilibrium constant includes the term $\ln K_{eq}$, there is a special requirement of K_{eq}: It must be a *dimensionless* number because we cannot take the logarithm of a unit.

Activities are the dimensionless quantities needed in the equilibrium constant, K_{eq}. Recall from previous discussions (page 684) that activities are effective concentrations. However, we can continue to replace activities by the magnitudes (that is, numerical values without units) of molarities and partial pressures if we use the following conventions:

- *For pure solid and liquid phases,* we define the activity as being $a = 1$.
- *For a gas,* we assume ideal gas behavior, and then we drop the unit from the value of the gas partial pressure expressed in atmospheres and use that unitless number for the gas activity.
- *For a solute in aqueous solution,* we assume that intermolecular or interionic attractions are negligible (that is, that the solution is dilute), and then we drop the units from the solute molarity and use that unitless number for the solute activity.

Example 17.8

Write the equilibrium constant expression, K_{eq}, for the oxidation of chloride ion by manganese dioxide in an acidic solution:

$$MnO_2(s) + 4 H^+(aq) + 2 Cl^-(aq) \rightleftharpoons Mn^{2+}(aq) + Cl_2(g) + 2 H_2O(l)$$

STRATEGY

First, we can write the K_{eq} expression in terms of activities (a), and then we can substitute the numerical values of the appropriate quantities for these activities in accordance with the conventions just introduced.

SOLUTION

We start with the equation

$$K_{eq} = \frac{a_{Mn^{2+}} \times a_{Cl_2} \times (a_{H_2O})^2}{a_{MnO_2} \times (a_{H^+})^4 \times (a_{Cl^-})^2}$$

In substituting for these activities according to our conventions,

1. No term will appear for $MnO_2(s)$ because it is a pure solid phase with $a = 1$.
2. For Mn^{2+}, H^+, and Cl^-, ionic species in dilute aqueous solution, activities are replaced by the numerical values of the molarities.
3. For $Cl_2(g)$, the activity is replaced by the numerical value of the partial pressure.
4. Because H_2O is the preponderant species in the dilute aqueous solution, the activity of H_2O is essentially the same as in pure liquid water, which means that $a = 1$.

Making these substitutions, we get

$$K_{eq} = \frac{[Mn^{2+}](P_{Cl_2})}{[H^+]^4[Cl^-]^2}$$

ASSESSMENT

Notice that this K_{eq} expression contains both molarities and a partial pressure. It is neither a K_c expression nor a K_p expression.

EXERCISE 17.8A

Write the K_{eq} expression for the reaction

$$2 Al(s) + 6 H^+(aq) \rightleftharpoons 2 Al^{3+}(aq) + 3 H_2(g)$$

EXERCISE 17.8B

Write an equation for the dissolution of magnesium hydroxide in an acidic solution, and then write the K_{eq} expression for this reaction.

Calculating Equilibrium Constants, K_{eq}

Now we are ready to combine several ideas to calculate an equilibrium constant, K_{eq}, and to consider its significance. Let us calculate K_{eq} for the vaporization of water at 25 °C, for which we have previously written

$$H_2O(l) \rightleftharpoons H_2O(g) \qquad \Delta G^\circ_{298} = +8.590 \text{ kJ}$$

We can begin by rearranging the equation $\Delta G^\circ = -RT \ln K_{eq}$ to

$$\ln K_{eq} = \frac{-\Delta G^\circ}{RT}$$

Then we substitute values of ΔG°, R, and T. In making these substitutions, we change the units of ΔG° from kilojoules to kilojoules per mole, 8.590 kJ/mol. The mol^{-1} portion of the unit kJ/mol signifies that the value of ΔG° corresponds to the specific stoichiometric amounts implied in the balanced equation (where coefficients describe numbers of moles).* Finally, we convert this value to 8.590×10^3 J/mol to provide the proper cancellation of units in our calculation:

$$\ln K_{eq} = \frac{-8.590 \times 10^3 \text{ J mol}^{-1}}{8.3145 \text{ J mol}^{-1} \text{ K}^{-1} \times 298.15 \text{ K}} = -3.465$$

$$K_{eq} = e^{-3.465} = 0.0313$$

Now let us write the K_{eq} expression for the liquid–vapor equilibrium for water. According to the conventions regarding activities, the activity of $H_2O(l)$ is $a = 1$, and as a result, the appropriate expression for K_{eq} is

$$K_{eq} = P_{H_2O(g)} = 0.0313$$

which tells us that the equilibrium vapor pressure of water at 25 °C is 0.0313 atm (23.8 mmHg). This calculated vapor pressure is the same as the experimentally measured value listed in Table 11.2.

Example 17.9

Determine the value of K_{eq} at 25 °C for the reaction $2 NO_2(g) \rightleftharpoons N_2O_4(g)$.

STRATEGY

Our first task is to obtain a value of $\Delta G°$, which we can do from tabulated standard free energies of formation in Appendix C. Then we can calculate K_{eq} using Equation (17.12).

SOLUTION

The standard free energies of formation of $N_2O_4(g)$ and $NO_2(g)$ from Appendix C are substituted into Equation (17.9).

$$\Delta G° = \Delta G_f°[N_2O_4(g)] - 2 \times \Delta G_f°[NO_2(g)]$$

$$\Delta G° = \left(1 \text{ mol N}_2O_4 \times \frac{97.82 \text{ kJ}}{1 \text{ mol N}_2O_4} \right) - \left(2 \text{ mol NO}_2 \times \frac{51.30 \text{ kJ}}{1 \text{ mol NO}_2} \right)$$

$$= -4.78 \text{ kJ}$$

Next we list the data required for Equation (17.12), first changing the unit of $\Delta G°$ from kJ to J/mol. Then we can rearrange Equation (17.12) to solve for K_{eq}.

$$\Delta G° = -4780 \text{ J mol}^{-1}, \quad R = 8.3145 \text{ J mol}^{-1} \text{ K}^{-1}, \quad \text{and} \quad T = 298.15 \text{ K}$$

$$\Delta G° = -RT \ln K_{eq}$$

$$\ln K_{eq} = \frac{-\Delta G°}{RT} = \frac{-(-4780 \text{ J mol}^{-1})}{8.3145 \text{ J mol}^{-1} \text{ K}^{-1} \times 298.15 \text{ K}} = 1.93$$

$$K_{eq} = e^{1.93} = 6.9$$

ASSESSMENT

In this problem, the value of $\Delta G°$ is negative, and so $\ln K_{eq}$ must be greater than 1. If we had made a rather common error and forgotten about the minus sign on the right in Equation (17.12), the value of K_{eq} would have been less than 1 (that is, $e^{-1.93} = 0.145$)—a tip-off that an error had been made.

EXERCISE 17.9A

Use data from Appendix C to determine K_{eq} at 25 °C for the reaction $2 HgO(s) \rightleftharpoons 2 Hg(l) + O_2(g)$.

EXERCISE 17.9B

Use data from Appendix C to determine K_{eq} at 25 °C for the reaction $2 NO(g) + Br_2(l) \rightleftharpoons 2 NOBr(g)$. If $NO(g)$ at 1 atm is allowed to react with an excess of $Br_2(l)$, what will be the partial pressures of $NO(g)$ and $NOBr(g)$ at equilibrium?

---

*In the expression $2 O_3(g) \rightleftharpoons 3 O_2(g) \, \Delta G° = -326.4$ kJ/mol, the value of $\Delta G°$ corresponds specifically to the reaction of 2 mol of ozone to produce 3 mol of oxygen. If instead we consider the reaction $O_3(g) \rightleftharpoons \frac{3}{2} O_2(g)$, then the value of $\Delta G° = -\frac{1}{2} \times 326.4 = -163.2$ kJ/mol. Note that the value of K_{eq} obtained for the second formulation is simply the square root of K_{eq} for the first formulation.

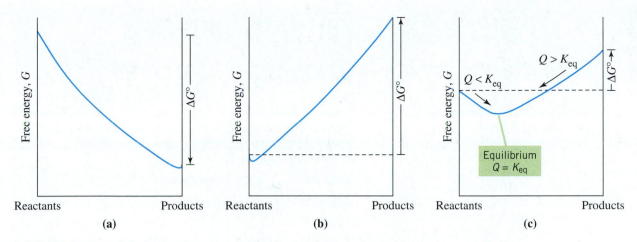

Reactants   Products   Reactants   Products   Reactants   Products

(a)   (b)   (c)

▲ **FIGURE 17.11** $\Delta G°$ **and the direction and extent of spontaneous change**
(a) $\Delta G°$ is large and negative: Equilibrium lies far to the right, and the reaction goes essentially to completion. (b) $\Delta G°$ is large and positive: Equilibrium lies far to the left, and the reaction hardly occurs at all. (c) $\Delta G°$ is neither very large nor very small and either positive or negative: The equilibrium point lies well within the profile. If $Q < K_{eq}$, a net reaction proceeds in the forward direction; if $Q > K_{eq}$, a net reaction proceeds in the reverse direction.

The Significance of the Sign and Magnitude of $\Delta G°$

In Chapter 14 (page 586), we found that we could use the magnitude of an equilibrium constant to decide, roughly, whether a reaction would (a) go essentially to completion, (b) occur hardly at all in the forward direction, or (c) reach an equilibrium condition that must be described through a value of K. Now that we have Equation (17.12) linking K_{eq} and $\Delta G°$,

$$\Delta G° = -RT \ln K_{eq}$$

we can use $\Delta G°$ alone to decide among these three possibilities. In Figure 17.11, we have plotted three reaction profiles suggesting how free energy can change during a reaction, starting with reactants in their standard states on the left and proceeding to the products in their standard states on the right. The minimum point in a reaction profile represents chemical equilibrium. This minimum point can be approached from either direction, depending on the value of the reaction quotient Q. Once this minimum is reached, however, the reaction does not proceed any further along the profile.

In case (a) of Figure 17.11, the free energy of the products is much lower than that of the reactants. The difference in free energy between them—the standard free energy change $\Delta G°$—is a *large, negative* quantity, and equilibrium is so far to the right that we say the reaction goes to completion. In case (b), the situation is reversed: The free energy of the products is much higher than that of the reactants, the standard free energy change is a *large positive* quantity, and equilibrium is so far to the left that we say no reaction occurs. In case (c), the difference in free energies of the reactants and products is small, and equilibrium lies more toward the interior of the reaction profile.

We cannot be highly specific as to what constitutes a "large" value of $\Delta G°$ and what constitutes a "small" value. As a general rule, however, if $\Delta G°$ is several hundred kilojoules per mole and negative, it is quite likely that a reaction will go essentially to completion. If $\Delta G°$ is several hundred kilojoules per mole and positive, it is just as likely that no significant reaction will occur. When the negative or positive value of $\Delta G°$ is much closer to zero, we usually have to do an equilibrium calculation.

🌐 **Energy Diagrams and Equilibrium Constants activity**

Coupled Reactions

Consider the decomposition of mercury(II) oxide:

$$HgO(s) \longrightarrow Hg(l) + \tfrac{1}{2} O_2(g) \qquad \Delta G°_{298} = +58.56 \text{ kJ} \qquad \Delta H°_{298} = +90.83 \text{ kJ}$$

Because the reaction produces a gas from a solid, entropy increases: $\Delta S° > 0$. When $\Delta H° > 0$ and $\Delta S° > 0$, the forward reaction is favored at higher temperatures (case 3 of Table 17.1). Because $\Delta G°_{298}$ is not particularly large, the reaction to produce

Thermodynamics and Living Organisms

In living organisms, the energy stored in food—potential energy—is changed into the kinetic energy associated with breathing, jogging, and twiddling thumbs. How do these energy changes come about? Do they obey the laws of thermodynamics?

A reaction with a decrease in free energy is *exergonic* ($\Delta G < 0$), and one with a free energy increase is *endergonic* ($\Delta G > 0$). Living organisms carry out exergonic reactions to maintain life processes, but there must also be endergonic processes that store free energy to provide for the free energy needed in the exergonic reactions.

In living organisms, complex molecules such as proteins and DNA are produced from simpler ones. These processes usually involve a decrease in entropy ($\Delta S < 0$), an increase in enthalpy ($\Delta H > 0$), and subsequently an increase in free energy ($\Delta G > 0$):

$$\Delta G = \Delta H - T\Delta S$$

We expect these processes to be nonspontaneous, but how can nonspontaneous reactions essential to life proceed in directions that are not natural?

Consider this: If you drop a book, it falls to the floor—a spontaneous process. The book will not leap back into your hands—a nonspontaneous process. Suppose, however, the book on the floor is connected by a rope over a pulley to another, larger book. Now, when you release the rope and let the larger book fall, the smaller book rises (see Figure 17.12). The spontaneous process (fall of the larger book) drives the nonspontaneous process (rise of the smaller book). This is a *coupled* process, analogous to coupled chemical reactions.

The reaction by which plants combine the simple sugars glucose and fructose into sucrose has a positive value of ΔG, meaning the reaction is endergonic. The reaction must be coupled to one that is exergonic, such as the hydrolysis of ATP (adenosine triphosphate) to ADP (adenosine diphosphate). The sum of the two reactions has $\Delta G < 0$:

Glucose + fructose $\longrightarrow$ sucrose + H_2O $\Delta G = +29.3$ kJ

ATP + H_2O $\longrightarrow$ ADP + HPO_4^{2-} $\Delta G = -30.5$ kJ
___
Glucose + fructose + ATP $\longrightarrow$ sucrose + ADP + HPO_4^{2-}

$$\Delta G = -1.2 \text{ kJ}$$

The reactions are coupled through the *intermediate* glucose-1-phosphate. First, ATP reacts with glucose to form this intermediate, and some of the energy stored in the bonds of ATP is transferred to bonds in the intermediate:

(a) Glucose + ATP $\longrightarrow$ glucose-1-phosphate + ADP

When the intermediate reacts with fructose, the stored energy is transferred to the sucrose molecule:

(b) Glucose-1-phosphate + fructose $\longrightarrow$ sucrose + HPO_4^{2-}

Glucose-1-phosphate, an intermediate, drops out of the equation for the overall reaction [(a) + (b)]:

Glucose + fructose + ATP

$$\longrightarrow \text{sucrose} + \text{ADP} + HPO_4^{2-}$$
$$\Delta G = -1.2 \text{ kJ}$$

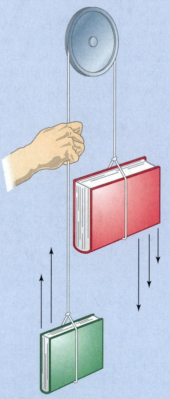

◀ **FIGURE 17.12 An analogy to coupled reactions**

If it were not tied to the larger book, the small book would not rise on its own, because such motion is a nonspontaneous process. When coupled by the rope to a larger book that is allowed to fall (a spontaneous process), the small book rises. The coupled process is spontaneous.

$O_2(g)$ at 0.25 atm becomes spontaneous at a moderate temperature of about 500 °C. (The 0.25-atm pressure is slightly higher than the partial pressure of $O_2(g)$ in the atmosphere.) The rather favorable thermodynamics of this reaction made it possible for Joseph Priestley (1733–1804) to isolate oxygen in 1774 just by heating mercury(II) oxide.

Now consider the reaction in which copper(I) oxide decomposes to produce metallic copper and oxygen gas:

$$Cu_2O(s) \longrightarrow 2\ Cu(s) + \tfrac{1}{2} O_2(g)$$

$$\Delta G^\circ_{298} = +149.9 \text{ kJ} \qquad \Delta H^\circ_{298} = +170.7 \text{ kJ}$$

Here, $\Delta G°$ is considerably more positive than in the reaction of mercury(II) oxide. Consequently, the $Cu_2O(s)$ must be heated to a much higher temperature than $HgO(s)$—to more than 2500 °C—before its decomposition under standard-state conditions becomes spontaneous. It is not feasible to produce copper metal simply by heating the oxide.

Suppose, however, that we heat $Cu_2O(s)$ in the presence of carbon. The carbon combines with oxygen that comes from the $Cu_2O(s)$. Here, we can think in terms of two individual reactions plus an overall reaction. We obtain the equation for the overall reaction by combining the equations for the two individual reactions, and we obtain ΔG for the overall reaction as the sum of the two individual ΔG values. This is the same way we treat ΔH values when applying Hess's law (Section 6.6):

$$Cu_2O(s) \longrightarrow 2\ Cu(s) + \tfrac{1}{2} O_2(g) \qquad \Delta G°_{298} = +149.9\ \text{kJ}$$

$$\underline{C(\text{graphite}) + \tfrac{1}{2} O_2(g) \longrightarrow CO(g) \qquad \Delta G°_{298} = -137.2\ \text{kJ}}$$

$$Cu_2O(s) + C(\text{graphite}) \longrightarrow 2\ Cu(s) + CO(g) \qquad \Delta G°_{298} = +12.7\ \text{kJ}$$

The overall reaction with reactants and products in their standard states is not spontaneous at 298 K, but it is much closer to being so than is the unaided decomposition of $Cu_2O(s)$. Although the commercial production of copper by this reaction is carried out at much higher temperatures, the overall reaction actually becomes spontaneous for standard-state conditions at about 100 °C.

Any combination of two or more simpler reactions is a **coupled reaction.** However, the term is usually used to refer to a combination of one spontaneous reaction and one nonspontaneous reaction to produce an overall reaction that is spontaneous.

17.7 The Dependence of $\Delta G°$ and K_{eq} on Temperature

We have now established criteria for spontaneous change, and we have examined a number of qualitative and quantitative assessments that we can make with just two equations:

$$\Delta G = \Delta H - T\Delta S$$

$$\Delta G° = -RT \ln K_{eq}$$

There appears to be one serious limitation, however. All our quantitative calculations have been at 25 °C because tabulated thermodynamic data are given mostly for that temperature. Yet we know that for practical purposes, we must carry out reactions at a variety of temperatures.

To obtain equilibrium constants at different temperatures, we will assume that ΔH and ΔS do not change much with temperature. (Notice that we made this assumption in Figure 17.8.) In particular, we will assume that values of $\Delta H°$ and $\Delta S°$ at 25 °C will apply at other temperatures as well. This assumption works reasonably well because standard enthalpies of formation and standard molar entropies of the products and reactants all change in roughly the same way with changes in temperature.

We now show a procedure that allows us to derive an expression relating K_{eq} to temperature.

To obtain a value of $\Delta G°$, we can substitute the 25 °C values of $\Delta H°$ and $\Delta S°$ and the desired temperature into Equation (17.8).

$$\Delta G° = \Delta H° - T\Delta S°$$

Then, to obtain K_{eq} at the desired temperature, we can use the relationship between standard free energy and the equilibrium constant, Equation (17.12).

$$\Delta G° = -RT \ln K_{eq}$$

Because we most often wish to calculate K_{eq}, we can combine Equations (17.8) and (17.12) into a single equation relating K_{eq} and temperature.

$$\Delta H° - T\Delta S° = \Delta G° = -RT \ln K_{eq}$$

Rearranging, we obtain a single equation allowing us to use enthalpy and entropy values at 25 °C to calculate K_{eq} at the desired temperature, T:

$$\ln K_{eq} = \frac{-\Delta H°}{RT} + \frac{\Delta S°}{R}$$

If we assume that $\Delta H°$ and $\Delta S°$ remain essentially constant over a range of temperatures, we can replace the term $\Delta S°/R$ by a constant and we can write the equation in a familiar form $(y = mx + b)$.

$$\ln K_{eq} = \frac{-\Delta H°}{RT} + \text{constant} \qquad (17.13)$$

If we plot $\ln K_{eq}$ against $1/T$, we get a straight line with a slope of $-\Delta H°/R$ and an intercept of $\Delta S°/R$. The best test of the assumption that $\Delta H°$ and $\Delta S°$ are independent of temperature is to plot some equilibrium data and see if they yield a straight line. This is illustrated in Figure 17.13.

By the method shown in Appendix A, we can replace Equation (17.13) with

$$\ln \frac{K_2}{K_1} = \frac{\Delta H°}{R}\left(\frac{1}{T_1} - \frac{1}{T_2}\right) \qquad (17.14)$$

Note the similarity between the van't Hoff equation and the Arrhenius equation we encountered in Chapter 13.

This equation, called the *van't Hoff equation*, relates the equilibrium constant, K_{eq}, at two temperatures. The value at temperature T_1 is K_1, and the value at temperature T_2 is K_2. In this expression, $\Delta H°$ is the standard enthalpy change in the reaction and R is the gas constant.

If the equilibrium is between a liquid or solid and its vapor, the equilibrium constant is simply the pressure of the vapor. Consequently, we can substitute P (vapor pressure) for K (equilibrium constant) and $\Delta H°_{vapn}$ or $\Delta H°_{subl}$ for $\Delta H°$. The equation is then called the *Clausius–Clapeyron equation*:

Problem-Solving Note

For solving problems, some prefer one of the following forms of the van't Hoff equation:

$$\ln \frac{K_2}{K_1} = \frac{-\Delta H°}{R}\left(\frac{1}{T_2} - \frac{1}{T_1}\right)$$

$$\ln \frac{K_2}{K_1} = \frac{\Delta H°}{R}\left(\frac{T_2 - T_1}{T_1 \cdot T_2}\right)$$

Vaporization

$$\ln \frac{P_2}{P_1} = \frac{\Delta H°_{vapn}}{R}\left(\frac{1}{T_1} - \frac{1}{T_2}\right) \qquad (17.15)$$

Sublimation

$$\ln \frac{P_2}{P_1} = \frac{\Delta H°_{subl}}{R}\left(\frac{1}{T_1} - \frac{1}{T_2}\right) \qquad (17.16)$$

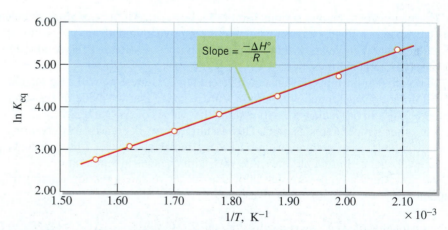

▲ **FIGURE 17.13** **Temperature dependence of K_{eq} for the reaction**
$$CO(g) + H_2O(g) \longrightarrow CO_2(g) + H_2(g)$$
The value of $\Delta H°$ obtained from the slope of this graph is about -40 kJ, a value reasonably close to that of -41.2 kJ for $\Delta H°_{298}$, obtained from data in Appendix C. Some of the data used in the graph are shown here.

T, K	1/T	K_{eq}	$\ln K_{eq}$
478	2.09×10^{-3}	210	5.35
533	1.88×10^{-3}	73	4.29
588	1.70×10^{-3}	31	3.43
643	1.56×10^{-3}	16	2.77

Example 17.10

Consider this reaction at 298 K:

$$CO(g) + H_2O(g) \rightleftharpoons CO_2(g) + H_2(g) \qquad \Delta H_{298}° = -41.2 \text{ kJ}$$

Determine K_{eq} for the reaction at 725 K.

STRATEGY

This is the reaction for which K_{eq} is plotted as a function of $1/T$ in Figure 17.13. We cannot simply use the graph to obtain ln K_{eq} when $1/T = 1/725$ K because this point falls outside the range of data plotted. However, we can select a data point listed in the caption of Figure 17.13 as our K_1 and T_1 for use in Equation (17.14). The other required data are given in the statement of the problem.

SOLUTION

Suppose these are the data we choose for the van't Hoff equation:

$$K_2 = ? \quad T_2 = 725 \text{ K} \qquad \Delta H° = \Delta H_{298}° = -41.2 \text{ kJ/mol}$$

$$K_1 = 16 \quad T_1 = 643 \text{ K}$$

The solution of Equation (17.14) for K_2 then follows:

$$\ln\frac{K_2}{K_1} = \frac{\Delta H°}{R}\left(\frac{1}{T_1} - \frac{1}{T_2}\right) = \frac{(-41.2 \times 1000) \text{ J mol}^{-1}}{8.3145 \text{ J mol}^{-1} \text{ K}^{-1}}\left(\frac{1}{643 \text{ K}} - \frac{1}{725 \text{ K}}\right)$$

$$\ln\frac{K_2}{K_1} = -4.96 \times 10^3 \text{ K} \times (1.56 \times 10^{-3} - 1.38 \times 10^{-3})\text{K}^{-1}$$

$$\ln\frac{K_2}{K_1} = -0.89$$

$$\frac{K_2}{K_1} = e^{-0.89} = 0.41$$

$$K_2 = 0.41 \times K_1 = 0.41 \times 16 = 6.6$$

ASSESSMENT

Notice that in the data table accompanying Figure 17.13, the value of K_{eq} becomes progressively smaller as the temperature increases. Because T_2 is greater than the highest temperature in the table, K_2 must be somewhat less than 16. The calculated value of 6.6 seems reasonable. Another point to notice in this problem is that in order to avoid an inadvertent loss of significant figures, it is best to store the value of $(1/T_1 - 1/T_2)$ in your calculator rather than to write intermediate results, such as the 1.56×10^{-3} and 1.38×10^{-3} shown in the solution.

EXERCISE 17.10A

In Example 17.9, we determined the value of K_{eq} at 25 °C for the reaction $2 \text{ NO}_2(g) \rightleftharpoons \text{N}_2\text{O}_4(g)$. What is the value of K_{eq} for this reaction at 65 °C? (*Hint:* You will have to use tabulated data to establish the value of $\Delta H°$.)

EXERCISE 17.10B

At 25 °C, the vapor pressure of water is 23.8 mmHg and the enthalpy of vaporization is 44.0 kJ/mol. Use the Clausius–Clapeyron equation to calculate the vapor pressure of water at 40.0 °C. Compare your result with data from Table 11.2.

Cumulative Example

Waste silver from photographic solutions or laboratory operations can be recovered using an appropriate redox reaction. A 100.0-mL sample of silver waste is 0.200 M in Ag^+(aq) and 0.0200 M in Fe^{3+}(aq). Enough iron(II) sulfate is added to make the solution 0.200 M in Fe^{2+}(aq). When equilibrium is established at 25 °C, how many moles of solid silver will be present?

$$\text{Ag}^+(aq) + \text{Fe}^{2+}(aq) \rightleftharpoons \text{Ag}(s) + \text{Fe}^{3+}(aq)$$

STRATEGY

The Ag(s) must come at the expense of some of the $Ag^+(aq)$ originally present. We can determine $[Ag^+]$ at equilibrium, and the amount of Ag(s) that forms will be related to the difference $[Ag^+]_{init} - [Ag^+]_{equil}$. The core of the equilibrium calculation will be the ICE tabular format introduced in Chapter 14, but first we need an equilibrium constant for the reaction. Because this information is not given, we must turn to Equation (17.12) relating $\ln K_{eq}$ to $\Delta G°$ for the reaction. We can establish $\Delta G°$ from data in Appendix C. In writing out the solution to this problem, we will reverse the order of the steps described here.

SOLUTION

To determine $\Delta G°$, we write the appropriate data from Appendix C under the chemical equation and then apply Equation (17.9).

$$Ag^+(aq) \;+\; Fe^{2+}(aq) \;\rightleftharpoons\; Ag(s) \;+\; Fe^{3+}(aq)$$

$\Delta G_f°$, kJ/mol   77.11   −78.90   −4.7

$$\Delta G° = 1 \text{ mol Fe}^{3+} \times (-4.7 \text{ kJ/mol Fe}^{3+})$$
$$- [(1 \text{ mol Ag}^+ \times 77.11 \text{ kJ/mol Ag}^+) + 1 \text{ mol Fe}^{2+} \times (-78.90 \text{ kJ/mol Fe}^{2+})]$$
$$= -2.9 \text{ kJ}$$

Then we can use this value of $\Delta G°$ in Equation (17.12) to solve for $\ln K_{eq}$.

$$\Delta G° = -RT \ln K_{eq} \quad \text{and} \quad \ln K_{eq} = \frac{-\Delta G°}{RT}$$

$$\ln K_{eq} = \frac{2900 \text{ J mol}^{-1}}{8.3145 \text{ J mol}^{-1} \text{ K}^{-1} \times 298 \text{ K}} = 1.2$$

We solve for K_{eq} by taking the exponential of both sides of the previous expression.

$$K_{eq} = e^{\ln K_{eq}} = e^{1.2} = 3.3$$

We could establish the direction of net change by calculating Q_c, but that is not necessary. Because there is no silver metal initially, the reaction *must* proceed to the right to form silver metal. Now, we enter the solution concentration data into the ICE format.

The reaction:	$Ag^+(aq)$	$+$	$Fe^{2+}(aq)$	$\rightleftharpoons$	$Ag(s)$	$+$	$Fe^{3+}(aq)$
Initial concentrations, M:	0.200		0.200		—		0.0200
Changes, M:	$-x$		$-x$				$+x$
Equilibrium concentrations, M:	$0.200 - x$		$0.200 - x$				$0.0200 + x$

Next we substitute the equilibrium concentrations from the ICE table into the K_{eq} expression and express the equation in the customary quadratic format.

$$K_{eq} = \frac{[Fe^{3+}]}{[Ag^+][Fe^{2+}]} = \frac{0.0200 + x}{(0.200 - x)^2} = \frac{0.0200 + x}{0.0400 - 0.400x + x^2} = 3.3$$

$$0.0200 + x = 0.13 - 1.3x + 3.3x^2$$

$$3.3x^2 - 2.3x + 0.11 = 0$$

We now solve the quadratic equation for x, keeping in mind that x must be less than 0.200.

$$x = \frac{2.3 \pm \sqrt{(2.3)^2 - 4 \times 0.11 \times 3.3}}{2 \times 3.3} = \frac{2.3 \pm \sqrt{5.3 - 1.5}}{6.6} = \frac{2.3 \pm 1.9}{6.6} = 0.06$$

Then we determine the equilibrium concentrations of the ions.

$$[Ag^+] = [Fe^{2+}] = 0.200 - x = 0.200 - 0.06 = 0.14 \text{ M}$$

$$[Fe^{3+}] = 0.0200 + x = 0.0200 + 0.06 = 0.08 \text{ M}$$

Finally, we can determine the number of moles of Ag(s) displaced from solution and appearing as Ag(s) by taking the difference between initial and equilibrium concentrations and multiplying by the volume.

$$? \text{ mol Ag(s)} = ([Ag^+]_{init} - [Ag^+]_{equil}) \times 0.1000 \text{ L}$$
$$= (0.200 - 0.14) \text{ mol/L} \times 0.1000 \text{ L}$$
$$= 0.006 \text{ mol Ag(s)}$$

ASSESSMENT

In choosing the correct root of the quadratic equation, we were guided by the fact that $x < 0.200$. That is, $[Ag^+]$ at equilibrium is $(0.200 - x)$ M, and this value must be a positive quantity. The rejected root had $x = 0.6$. We can verify that the equilibrium concentrations are correct by showing that Q_c based on equilibrium concentrations is equal to the value of K_{eq} used in the calculation. Here, we obtain $Q_c = 0.08/(0.14)^2 = 4$. This matches $K_{eq} = 3.3$ quite well, given the rather limited precision of the calculation. The limitation in precision was introduced in calculating $\Delta G°$, where we had to work with the difference between two numbers of nearly equal magnitude.

Concept Review with Key Terms

17.1 Why Study Thermodynamics?—The study of **thermodynamics** offers the opportunity to predict the stability of chemical substances and the extent and direction of chemical reactions.

17.2 Spontaneous Change—A **spontaneous process** is one that occurs by itself without outside intervention. A **nonspontaneous process** is one that cannot occur on its own.

17.3 The Concept of Entropy—**Entropy**, S, is a measure of the energy distribution within a system. **Entropy changes**, ΔS, are associated with changes in the way in which a system's energy is distributed among available energy levels. The **third law of thermodynamics** states that the entropy of a pure, perfect crystal at 0 K can be taken to be zero. This serves as a starting point for the experimental determination of **standard molar entropies**, which can be used to calculate entropy changes in chemical reactions.

The direction of spontaneous change is the direction in which *total* entropy increases. Total entropy change, also called the entropy change of the universe, is the sum of the entropy change of a system *and* of its surroundings:

$$\Delta S_{univ} = \Delta S_{sys} + \Delta S_{surr}$$

According to the **second law of thermodynamics**, the entropy of the universe, S_{univ}, must always increase for a spontaneous process, that is, $\Delta S_{univ} > 0$.

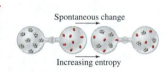

Spontaneous change

Increasing entropy

17.4 Free Energy and Free Energy Change—The **Gibbs free energy**, G, is used to describe the spontaneity of a process.

$$G = H - T\Delta S$$

The **free energy change**, ΔG, is equal to $-T\Delta S_{univ}$, and it applies *just to a system itself*, without regard for the surroundings. It is defined by the Gibbs equation:

$$\Delta G = \Delta H - T\Delta S$$

For $\Delta H > 0$ and $\Delta S > 0$

For a spontaneous process at constant temperature and pressure, ΔG must be negative. In many cases, we can predict the sign of ΔG from the signs of ΔH and ΔS.

17.5 Standard Free Energy Change, $\Delta G°$—The **standard free energy change**, $\Delta G°$, can be calculated (1) by substituting standard enthalpies and entropies of reaction and a Kelvin temperature into the

Gibbs equation or (2) by combining **standard free energies of formation** through the expression

$$\Delta G° = \left[\sum \nu_p \, \Delta G_f°(products)\right] - \left[\sum \nu_r \, \Delta G_f°(reactants)\right]$$

17.6 Free Energy Change and Equilibrium—The condition of equilibrium is one for which $\Delta G = 0$. Based on this condition, **Trouton's rule** states that the entropy change for the equilibrium vaporization of nonpolar liquids at their normal boiling points is approximately constant:

$$\Delta S_{vapn}° = \frac{\Delta H_{vapn}°}{T_{bp}} \approx 87 \text{ J mol}^{-1} \text{ K}^{-1}$$

The standard free energy change is a particularly useful property to use in describing equilibrium because of its relationship to the **equilibrium constant, K_{eq}:**

$$\Delta G° = -RT \ln K_{eq}$$

One of the requirements of this equation is that the K_{eq} expression be based on *dimensionless* quantities called activities. However, in place of activities, we can generally substitute the numerical values of molarities for solutes, the numerical values of partial pressures expressed in atmospheres for gases, and a value of 1 for pure solids and liquids.

The value of $\Delta G°$ is in itself often sufficient to determine whether a reaction is likely to (a) go to completion ($\Delta G°$ very large and negative), (b) occur hardly at all ($\Delta G°$ very large and positive), or (c) reach an equilibrium condition requiring an equilibrium calculation ($\Delta G°$ not very large and either positive or negative).

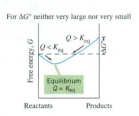

For $\Delta G°$ neither very large nor very small

Sometimes a nonspontaneous reaction can be made to occur by combining it with a spontaneous reaction. Together, these two **coupled reactions** produce a spontaneous overall reaction.

17.7 The Dependence of $\Delta G°$ and K_{eq} on Temperature—Values of $\Delta G_f°$, $\Delta H_f°$, and $S°$ are usually tabulated for 25 °C. To obtain values of K_{eq} at other temperatures, we generally assume that $\Delta H°$ and $\Delta S°$ are independent of temperature. With this assumption, we can derive the van't Hoff equation to relate the equilibrium constant and temperature.

$$\ln \frac{K_2}{K_1} = \frac{\Delta H°}{R}\left(\frac{1}{T_1} - \frac{1}{T_2}\right)$$

Assessment Goals

When you have mastered the material in this chapter, you will be able to:

- Describe the difference between spontaneous and nonspontaneous processes.

- Categorize everyday processes as spontaneous or nonspontaneous.

- Define entropy in terms of the dispersal of a system's energy among the available microscopic energy states of a system.

- Recognize processes associated with increases and decreases in entropy.

- Relate absolute entropies to reversible heat exchanges between a system and its surroundings.

- State the second and third laws of thermodynamics.

- Describe the influence of temperature and phase changes on the entropy of a system.

- Use the equation for Gibbs free energy change to predict the spontaneity of a chemical reaction.

- Describe the conditions distinguishing ΔG from $\Delta G°$.

- Calculate the standard free energy change of a reaction from the standard free energies of formation of the reactants and products.

- Describe the uses and limitations of Trouton's rule for the entropy of vaporization of liquids.

- Calculate $\Delta G°$ from K_{eq} and vice versa.

- Calculate ΔG under nonequilibrium conditions from the standard free energy change and the reaction quotient.

- Describe the relationship between the magnitude and sign of $\Delta G°$ and the position of equilibrium of a chemical reaction.

- Calculate changes in equilibrium constants with temperature using the van't Hoff equation, and the variation of vapor pressures with temperature using the Clausius–Clapeyron equation.

Self-Assessment Questions

1. From common knowledge, predict which of the following are spontaneous changes: **(a)** the souring of milk, **(b)** obtaining copper metal from copper ore, **(c)** the rusting of a steel can in moist air. Explain your predictions.

2. For a reaction to occur spontaneously as written, which of the following must be true?
   (a) The entropy of the system must increase.
   (b) The enthalpy of the surroundings must increase.
   (c) The free energy change must be negative.
   (d) The value of the reaction quotient, Q, must exceed that of the equilibrium constant, K_{eq}.

3. Which of the following changes represents an increase and which a decrease in the entropy of the system: **(a)** the freezing of acetic acid **(b)** the sublimation of the moth repellent, *para*-dichlorobenzene ($C_6H_4Cl_2$) **(c)** the burning of gasoline?

4. Why is the notion of a reversible process so important to the concept of entropy?

5. Why is the entropy change in a system not always a reliable predictor of whether the process producing the change is spontaneous?

6. What does the second law of thermodynamics tell us about the concept of entropy?

7. What is the basic idea underlying the third law of thermodynamics?

8. Why do we tabulate standard entropies of substances, $S°$, rather than their entropies of formation, $\Delta S_f°$, as we do for values of $\Delta H_f°$ and $\Delta G_f°$?

9. Which would you expect to have the higher absolute molar entropy at 25 °C, $NOF_3(g)$ or $NO_2F(g)$? Explain.

10. The free energy change of a reaction is a measure of which of the following?
    (a) the quantity of heat given off to the surroundings
    (b) the direction in which a net reaction occurs
    (c) the increased molecular disorder produced in the reaction
    (d) how rapidly the reaction occurs

11. Under what conditions of temperature—low or high—would you expect a reaction to proceed furthest in the forward direction if the reaction has $\Delta H < 0$ and $\Delta S < 0$?

12. What is the special significance of a reaction for which **(a)** $\Delta G = 0$, **(b)** $\Delta G° = 0$?

13. For the reaction. $Br_2(g) \rightarrow 2\,Br(g)$, we should expect what to be true at all temperatures?
    (a) $\Delta H < 0$ **(c)** $\Delta G < 0$
    (b) $\Delta S < 0$ **(d)** $\Delta S > 0$

14. Which of the following is not accurately described by Trouton's rule?
    (a) benzene, C_6H_6 **(c)** ammonia, NH_3
    (b) toluene, $C_6H_5CH_3$ **(d)** carbon disulfide, CS_2

15. What do we mean by the term *coupled reactions*? How are coupled reactions used?

16. How does the equilibrium constant K_{eq} resemble, and how does it differ from, K_c and K_p?

17. How does thermodynamics make it possible to *calculate* the value of an equilibrium constant, K_{eq}? What kind of data are needed for this calculation?

18. If $\Delta G° = 0$ for a reaction, then which of the following is true?
    (a) $\Delta H° = 0$ **(c)** $K = 0$
    (b) $\Delta S° = 0$ **(d)** $K = 1$

Problems

Disorder, Entropy, and Spontaneous Change

19. For each of the following reactions, indicate whether you would expect the entropy of the system to increase or decrease. If you cannot tell just by inspecting the equation, explain why.
    (a) $CH_3OH(l) \longrightarrow CH_3OH(g)$
    (b) $N_2O_4(g) \longrightarrow 2\,NO_2(g)$
    (c) $CO(g) + H_2O(g) \longrightarrow CO_2(g) + H_2(g)$
    (d) $2\,KClO_3(s) \longrightarrow 2\,KCl(s) + 3\,O_2(g)$
    (e) $CH_3COOH(l) \longrightarrow CH_3COOH(s)$
    (f) $N_2(g) + O_2(g) \longrightarrow 2\,NO(g)$
    (g) $N_2H_4(l) \longrightarrow N_2(g) + 2\,H_2(g)$
    (h) $2\,NH_3(g) + H_2SO_4(aq) \longrightarrow (NH_4)_2SO_4(aq)$

20. Which of the following processes should have the *largest* absolute value of ΔS? Which should have the smallest value? Explain your choices.
    (a) $H_2O(s, 1\text{ atm}) \longrightarrow H_2O(l, 1\text{ atm})$
    (b) $H_2O(s, 1\text{ atm}) \longrightarrow H_2O(g, 4.58\text{ mmHg})$
    (c) $C(\text{diamond}, 1\text{ atm}) \longrightarrow C(\text{graphite}, 1\text{ atm})$
    (d) $CO_2(g, 25\text{ atm}) \longrightarrow CO_2(l, 25\text{ atm})$

21. In the manner used to describe Figure 17.3, describe the situation if the two liquids are water and octane (C_8H_{18}, a component of gasoline). That is, what would you expect to find for the final condition of the mixture? Explain.

22. In the manner suggested by Figure 17.4, make sketches to represent entropy changes occurring in **(a)** the melting of a solid, and **(b)** the condensation of a vapor.

23. Three possible completions of a statement are given. Explain what is wrong with each one, and then complete the statement correctly. *For a process to occur spontaneously,*
    (a) the entropy of the system must increase.
    (b) the entropy of the surroundings must increase.
    (c) both the entropy of the system and of the surroundings must increase.

24. Explain why:
    (a) some exothermic reactions do not occur spontaneously.
    (b) some reactions in which the entropy of the system increases do not occur spontaneously.

25. At one atmosphere, carbon dioxide undergoes a transition from solid to gas at 194.5 K. Discuss the appearance of a graph similar to Figure 17.7 for carbon dioxide as compared to those of hydrogen and of CH_3Cl.

26. Consider the following reaction.

$$2 S(s, rhombic) + Cl_2(g) \longrightarrow S_2Cl_2(g)$$

Comment on the difficulty in deciding whether $\Delta S°$ is positive or negative when using the five generalizations on page 719. Describe how the additional generalization on page 723 can help you decide. Calculate the actual value of $\Delta S°$ using data from Appendix C.

27. An environmental problem posed by the use of aluminum cans is that their disintegration under environmental conditions appears to take an almost endless period of time. Can we say that the environmental disintegration of aluminum is a nonspontaneous process? Explain.

28. Based on a consideration of entropy changes, why is it so difficult to eliminate environmental pollution such as the contamination of groundwater with water-soluble methyl *tert*-butyl ether?

Free Energy and Spontaneous Change

29. Explain why free energy change is easier to use as a criterion for spontaneous change in a system than is entropy change alone.

30. Explain why in a reversible reaction, both the forward and the reverse reactions can be spontaneous, but not under the same conditions.

31. Make a sketch similar to Figure 17.8 that corresponds to case 1 of Table 17.1. Explain the significance of the plot in relation to the prediction in Table 17.1.

32. Make a sketch similar to Figure 17.8 that corresponds to case 2 of Table 17.1. Explain the significance of the plot in relation to the prediction in Table 17.1.

33. Three possible completions of a statement are given. Explain what is wrong with each one, and then complete the statement correctly. *The free energy change of a reaction indicates*

  (a) whether the reaction is endothermic or exothermic.

  (b) whether the reaction is accompanied by an increase or decrease in molecular disorder.

  (c) whether equilibrium in the reaction is favored at high or low temperatures.

34. Most if not all combustion reactions fall into which case in Table 17.1? Explain.

35. Give an example of a phase change that is **(a)** nonspontaneous at low temperatures and spontaneous at higher temperatures, and **(b)** spontaneous at low temperatures and nonspontaneous at higher temperatures. What is the equilibrium temperature in each case?

36. Would you expect each of the following reactions to be spontaneous at low temperatures, high temperatures, all temperatures, or not at all? Explain.

  (a) $PCl_3(g) + Cl_2(g) \longrightarrow PCl_5(g) \qquad \Delta H° = -87.9 \text{ kJ}$

  (b) $2 NH_3(g) \longrightarrow N_2(g) + 3 H_2(g) \qquad \Delta H° = +92.2 \text{ kJ}$

  (c) $2 N_2O(g) \longrightarrow 2 N_2(g) + O_2(g) \qquad \Delta H° = -164.1 \text{ kJ}$

  (d) $H_2O(g) + \frac{1}{2} O_2(g) \longrightarrow H_2O_2(g) \qquad \Delta H° = +105.5 \text{ kJ}$

  (e) $CH_4(g) + 2 O_2(g) \longrightarrow CO_2(g) + 2 H_2O(g)$

$$\Delta H° = -802.3 \text{ kJ}$$

  (f) $2 CO(g) + O_2(g) \longrightarrow 2 CO_2(g) \qquad \Delta H° = -566.0 \text{ kJ}$

37. The following reaction is spontaneous under standard-state conditions at room temperature.

$$CO(g) + Cl_2(g) \longrightarrow COCl_2(g)$$

To make it a nonspontaneous reaction, would you raise or lower the temperature? Explain.

38. Would you expect the following reaction to be spontaneous under standard-state conditions at room temperature?

$$3 O_2(g) \longrightarrow 2 O_3(g) \qquad \Delta H° = +285.4 \text{ kJ}$$

What effect will changing the temperature have on the spontaneity of this reaction? Explain.

Standard Free Energy Change

39. Use data from Appendix C to determine $\Delta G°$ values for the following reactions at 25 °C.

  (a) $FeO(s) + H_2(g) \longrightarrow Fe(s) + H_2O(g)$

  (b) $CdO(s) + 2 HCl(g) \longrightarrow CdCl_2(s) + H_2O(l)$

40. Use data from Appendix C to determine $\Delta G°$ values for the following reactions at 25 °C.

  (a) $C_2H_4(g) + H_2(g) \longrightarrow C_2H_6(g)$

  (b) $SO_3(g) + CaO(s) \longrightarrow CaSO_4(s)$

41. Use data from Appendix C to determine $\Delta H°$ and $\Delta S°$, at 298 K, for the following reaction. Then determine $\Delta G°$ in two ways and compare the results.

$$CS_2(l) + 3 O_2(g) \longrightarrow CO_2(g) + 2 SO_2(g)$$

42. Use data from Appendix C to determine $\Delta H°$ and $\Delta S°$, at 298 K, for the following reaction. Then determine $\Delta G°$ in two ways and compare the results.

$$C(graphite) + H_2O(g) \longrightarrow CO(g) + H_2(g)$$

43. Why is the standard free energy change, $\Delta G°$, so important in dealing with the question of spontaneous change, even though the conditions in a chemical reaction are usually *nonstandard*?

44. Under what conditions will ΔG for a reaction be negative even though $\Delta G°$ is positive?

Free Energy Change and Equilibrium

45. Estimate the normal boiling point of heptane, C_7H_{16}, given that at this temperature $\Delta H°_{vapn} = 31.69$ kJ/mol.

46. A reference book lists the following values for chloroform (trichloromethane) at 298 K: $\Delta H°_f[CHCl_3(l)] = -132.3$ kJ/mol, $\Delta H°_f[CHCl_3(g)] = -102.9$ kJ/mol. Estimate the normal boiling point of chloroform.

47. The normal boiling point of sulfuryl chloride, $SO_2Cl_2(l)$, is 69.3 °C. Estimate $\Delta H°_{vapn}$ of sulfuryl chloride. Compare your result with a value based on data from Appendix C.

48. The normal boiling point of $Br_2(l)$ is 59.47 °C. Estimate $\Delta H°_{vapn}$ of bromine. Compare your result with a value based on data from Appendix C.

49. At 850 K, for the reaction, $2 SO_2(g) + O_2(g) \rightleftharpoons 2 SO_3(g)$, $\Delta G° = -36.3$ kJ. What is **(a)** ΔG for a mixture at equilibrium at 850 K, and **(b)** K_p for this reaction at 850 K?

50. At 765 K, $K_p = 46.0$ for the reaction $H_2(g) + I_2(g) \rightleftharpoons 2 HI(g)$. What is **(a)** ΔG for a mixture at equilibrium at 765 K, and **(b)** $\Delta G°$ for this reaction at 765 K?

51. Write K_{eq} expressions for the following reactions. Which, if any, of these expressions correspond to equilibrium constants that we have previously denoted as K_c, K_p, K_a, and so on?

(a) $2 NaHSO_3(s) \rightleftharpoons Na_2SO_3(s) + H_2O(g) + SO_2(g)$

(b) $Mg(OH)_2(s) \rightleftharpoons Mg^{2+}(aq) + 2 OH^-(aq)$

(c) $CH_3COO^-(aq) + H_2O(l) \rightleftharpoons CH_3COOH(aq) + OH^-(aq)$

52. Write K_{eq} expressions for the following reactions. Which, if any, of these expressions correspond to equilibrium constants that we have previously denoted as K_c, K_p, K_a, and so on?

(a) $2 NO(g) + O_2(g) \rightleftharpoons 2 NO_2(g)$

(b) $MgSO_3(s) \rightleftharpoons MgO(s) + SO_2(g)$

(c) $HCN(aq) + H_2O(l) \rightleftharpoons H_3O^+(aq) + CN^-(aq)$

53. Use data from Appendix C to determine K_p at 298 K for these reactions.

(a) $2 SO_2(g) + O_2(g) \rightleftharpoons 2 SO_3(g)$

(b) $CH_4(g) + 2 H_2O(g) \rightleftharpoons CO_2(g) + 4 H_2(g)$

54. Use data from Appendix C to determine K_p at 298 K for these reactions.

(a) $2 N_2O(g) + O_2(g) \rightleftharpoons 4 NO(g)$

(b) $2 NH_3(g) + 2 O_2(g) \rightleftharpoons N_2O(g) + 3 H_2O(g)$

55. The following data are given for the sublimation of naphthalene at 298 K: $\Delta S° = 168.7$ J mol^{-1} K^{-1}, $\Delta H° = 73.6$ kJ mol^{-1}. Calculate the pressure of naphthalene vapor in equilibrium with solid naphthalene at 298 K.

$$C_{10}H_8(s) \rightleftharpoons C_{10}H_8(g)$$

56. The following are data for the vaporization of toluene, $C_6H_5CH_3(l)$, at 298 K: $\Delta S° = 99.7$ J mol^{-1} K^{-1}, $\Delta H° = 38.0$ kJ mol^{-1}. Calculate the equilibrium vapor pressure of toluene at 298 K.

57. In an *equilibrium* mixture in the following reaction at 345 °C,

$$CO(g) + H_2O(g) \rightleftharpoons CO_2(g) + H_2(g)$$

the mole fractions of the gases were found to be $x_{CO_2} = x_{H_2} = 0.320$, $x_{CO} = 0.0133$, and $x_{H_2O} = 0.347$. **(a)** What is $\Delta G°$ for this reaction at 345 °C? **(b)** In what direction will a net reaction occur if one brings together 0.085 mol CO, 0.112 mol H_2O, 0.145 mol CO_2 and 0.226 mol H_2 and allows them to come to equilibrium? **(c)** What is the composition of the equilibrium mixture obtained by the reaction in **(b)**?

58. At 440 °C, for the reaction $CO_2(g) + 4 H_2(g) \rightleftharpoons CH_4(g) + 2 H_2O(g)$, $\Delta G° = -32$ kJ. **(a)** What is K_p for this reaction at 440 °C? **(b)** In what direction will a net reaction occur to establish equilibrium in a mixture in which initially $P_{CO_2} = 0.25$ atm, $P_{H_2} = 0.65$ atm, $P_{CH_4} = 1.12$ atm, and $P_{H_2O} = 2.52$ atm? **(c)** What are the partial pressures of the gases at equilibrium in **(b)**?

59. Use data from Appendix C to determine which of the following can be coupled with the reaction

$$CoO(s) \longrightarrow Co(s) + \tfrac{1}{2}O_2(g) \qquad \Delta G° = 237.9 \text{ kJ}$$

to make the reduction of CoO(s) to Co(s) spontaneous for standard-state conditions at 298 K: **(a)** oxidation of C(graphite) to CO(g), **(b)** oxidation of $H_2(g)$ to $H_2O(g)$, **(c)** oxidation of CO(g) to $CO_2(g)$? Explain.

60. Use data from Appendix C to determine which of the reactions, **(a)** $FeO(s) \longrightarrow Fe(s) + \tfrac{1}{2}O_2(g)$, **(b)** $CaO(s) \longrightarrow Ca(s) + \tfrac{1}{2}O_2(g)$, or **(c)** $MnO_2(s) \longrightarrow Mn(s) + O_2(g)$, can be coupled with the oxidation of Al(s) to $Al_2O_3(s)$ to give an overall reaction that is spontaneous for standard-state conditions at 298 K. Explain.

$\Delta G°$ and K_{eq} as Functions of Temperature

61. The following thermodynamic data are for 298 K. Use these data and data from Appendix C to determine K_{eq} at 45 °C for the reaction

$$CO_2(g) + SF_4(g) \rightleftharpoons CF_4(g) + SO_2(g)$$

	$\Delta H_f°$, kJ/mol	$S°$, J mol^{-1} K^{-1}
$SF_4(g)$	−763	299.6
$CF_4(g)$	−925	261.6

62. The following reaction is carried out on an industrial scale for the production of thionyl chloride, a chemical used in the manufacture of pesticides.

$$SO_3(g) + SCl_2(l) \rightleftharpoons OSCl_2(l) + SO_2(g)$$

The following thermodynamic data are for 298 K. Use these data and data from Appendix C to determine the temperature at which $K_{eq} = 1.0 \times 10^{15}$ for the reaction.

	$\Delta H_f°$, kJ/mol	$S°$, J mol^{-1} K^{-1}
$SCl_2(l)$	−50.0	184
$OSCl_2(l)$	−245.6	121

63. The normal boiling point of 1-butanol is 117.8 °C.

$$CH_3(CH_2)_2CH_2OH(l, 1 \text{ atm}) \rightleftharpoons CH_3(CH_2)_2CH_2OH(g, 1 \text{ atm})$$
$$\Delta H° = 43.82 \text{ kJ}$$

Calculate the boiling point of 1-butanol when the barometric pressure is 747 mmHg.

64. The normal boiling point of diethyl ether is 34.66 °C.

$$CH_3CH_2OCH_2CH_3(l, 1 \text{ atm}) \rightleftharpoons CH_3CH_2OCH_2CH_3(g, 1 \text{ atm})$$
$$\Delta H° = 26.70 \text{ kJ}$$

Calculate the vapor pressure of diethyl ether at 32.50 °C.

65. Use data from Appendix C to calculate K_p at 155 °C for the following reaction.

$$2 NO(g) + O_2(g) \rightleftharpoons 2 NO_2(g)$$

66. Use data from Appendix C to calculate K_{eq} at 375 °C for the reversible reaction in which hydrogen sulfide and carbon dioxide gases produce gaseous water and carbonyl sulfide, COS(g).

67. With data from Appendix C, estimate the normal boiling point of carbon tetrachloride.

68. With data from Appendix C, estimate the temperature at which the sublimation pressure of $P_4(g)$ above solid red phosphorus is 1 atm.

69. With data from Appendix C, determine the temperature to which $Ag_2O(s)$ must be heated to produce $O_2(g)$ at an equilibrium partial pressure of 0.21 atm, the same as found in the atmosphere.

$$2\,Ag_2O(s) \rightleftharpoons 4\,Ag(s) + O_2(g)$$

70. Equilibrium is established in the following reaction at 60.0 °C.

$$2\,BrCl(g) \rightleftharpoons Br_2(g) + Cl_2(g)$$

Starting with an initial pressure of $BrCl(g)$ of 1.00 atm, what is the partial pressure of $BrCl(g)$ when equilibrium is reached?

Additional Problems

Problems marked with an * may be more challenging than others.

71. The reversible expansion of a gas is suggested in Figure 17.5. Is the expansion of a gas pictured in Figure 6.8 reversible? Explain.

*72. On several occasions, we have made the assumption that $\Delta H°$ and $\Delta S°$ undergo little change with temperature. Why can we not make the same assumption about $\Delta G°$? If we assume that $\Delta H_{298}°$ and $\Delta S_{298}°$ do not change with temperature, is it possible for a reaction that is nonspontaneous under standard-state conditions to become spontaneous both at some higher temperature and at some lower temperature? Explain.

73. Arrange the following substances in order from greatest to least deviation from Trouton's rule: **(a)** $CH_3(CH_2)_6CH_2OH$, **(b)** CH_3CHO, **(c)** CH_3CH_2OH, **(d)** $CH_3(CH_2)_6CH_3$. Explain.

74. What is the relationship between the effect of temperature on equilibrium that we predicted qualitatively in Chapter 14 and the effect of temperature on K_{eq} discussed in Section 17.7?

*75. For the decomposition of $NaHCO_3(s)$, estimate the temperature at which the total pressure of the gases above the solids is 1 atm.

$$2\,NaHCO_3(s) \rightleftharpoons Na_2CO_3(s) + H_2O(g) + CO_2(g)$$

76. The vapor pressure of hydrazine, N_2H_4, at 35.0 °C is 25.7 mmHg; at 50.0 °C, it is 57.0 mmHg. Estimate a value of the standard free energy change at 25 °C for the vaporization, and determine a more exact value with data from Appendix C. Compare the results.

$$N_2H_4(l) \longrightarrow N_2H_4(g)$$

*77. An equilibrium mixture at 727 °C contains 0.276 mol H_2, 0.276 mol CO_2, 0.224 mol CO, and 0.224 mol H_2O.

$$CO_2(g) + H_2(g) \rightleftharpoons CO(g) + H_2O(g)$$

(a) What is the value of $\Delta G°$ at 727 °C? **(b)** In what direction will a net chemical change occur if one brings together 0.0500 mol CO_2, 0.0700 mol H_2, 0.0400 mol CO, and 0.0850 mol H_2O? **(c)** What will be the composition of the gaseous mixture in (b) when equilibrium is reestablished in the reaction?

*78. When ethane is passed over a catalyst at 900 K and 1 atm, it is partly decomposed to ethylene and hydrogen.

$$CH_3CH_3(g) \rightleftharpoons CH_2{=}CH_2(g) + H_2(g)$$

Calculate the mole percent $H_2(g)$ at equilibrium.

79. Following are values of K_p at different temperatures for the reaction, $2\,SO_2(g) + O_2(g) \rightleftharpoons 2\,SO_3(g)$. At 800 K, $K_p = 9.1 \times 10^2$; at 900 K, $K_p = 4.2 \times 10^1$; at 1000 K, $K_p = 3.2$; at 1100 K, $K_p = 0.39$; and at 1170 K, $K_p = 0.12$. Construct a plot to determine $\Delta H°$ for this reaction.

*80. An ultrahigh vacuum system is capable of reducing the pressure in a system to less than 10^{-9} mmHg. Do you think it is possible to achieve such a high vacuum at 25 °C if a sample of $NH_4Cl(s)$ is present in the system (see illustration)? If not, can the ultralow pressure be achieved by changing the temperature? Explain.

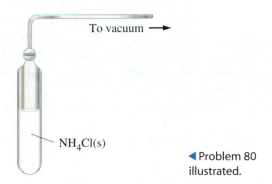

◄ Problem 80 illustrated.

81. Use data from Appendix C, together with the molar enthalpy of fusion of mercury of 2.30 kJ/mol at the melting point of −38.86 °C, to estimate the vapor pressure of solid mercury at the sublimation temperature of dry ice [$CO_2(s)$], −78.5 °C.

*82. Hydrogen cyanide is produced in large quantities for use in the manufacture of plastics. One process for its manufacture involves the reaction of methane (natural gas) and ammonia at about 1200 °C.

$$CH_4(g) + NH_3(g) \rightleftharpoons HCN(g) + 3\,H_2(g)$$

A reaction vessel contains $CH_4(g)$ and $NH_3(g)$, each initially at a partial pressure of 1.00 atm, at 1200 °C. What will be the total gas pressure when the system is brought to equilibrium?

*83. The decomposition of phosgene gas is represented by the following equation.

$$COCl_2(g) \rightleftharpoons CO(g) + Cl_2(g)$$

At what temperature will a sample of $COCl_2(g)$ be 10.0% dissociated at a total pressure of 1.00 atm? (*Hint:* Refer to Problem 86 in Chapter 14)

*84. Use thermodynamic data from Appendix C to obtain K_{eq} for the following reaction.

$$Mg(OH)_2(s) + 2\,NH_4{}^+(aq)$$
$$\rightleftharpoons Mg^{2+}(aq) + 2\,NH_3(aq) + 2\,H_2O(l)$$

Then obtain K_{eq} from other tabulated equilibrium constants in Appendix C, and compare the results.

*85. Use thermodynamic data from Appendix C to obtain a value of K_{sp} for Ag_2SO_4, and compare your result with the one found in the table of solubility product constants in Appendix C.

86. Sketch a plot of ΔG as a function of absolute temperature for (a) an exothermic spontaneous reaction, (b) an exothermic nonspontaneous reaction, (c) an endothermic spontaneous reaction, (d) an endothermic nonspontaneous reaction. What information can be obtained from the slope and intercept of each plot?

* 87. Consider the vaporization of water: $H_2O(l) \longrightarrow H_2O(g)$ at 100 °C, with $H_2O(l)$ in its standard state, but with the partial pressure of $H_2O(g)$ at 1.50 atm. Which of the following statements are true concerning this vaporization at 100 °C: (a) $\Delta G° = 0$, (b) $\Delta G = 0$, (c) $\Delta G° > 0$, (d) $\Delta G > 0$? Explain.

* 88. Titanium is obtained by the reduction of $TiCl_4(l)$, which in turn is produced from the mineral rutile (TiO_2).

   (a) With data from Appendix C, determine $\Delta G°$ at 298 K for the reaction

   $$TiO_2(s) + 2\,Cl_2(g) \longrightarrow TiCl_4(l) + O_2(g)$$

   (b) Show that the conversion of $TiO_2(s)$ to $TiCl_4(l)$, with reactants and products in their standard states, is spontaneous at 298 K if the reaction in (a) is coupled with the reaction

   $$2\,CO(g) + O_2(g) \longrightarrow 2\,CO_2(g)$$

* 89. Introduced into a 1.50-L flask is 0.100 mol $PCl_5(g)$. The flask is held at 227 °C until equilibrium is established. What are the partial pressures and the total pressure of the gases in the flask at equilibrium?

   $$PCl_5(g) \rightleftharpoons PCl_3(g) + Cl_2(g)$$

   (*Hint*: Use data from Appendix C and appropriate relationships from this chapter.)

90. Analogies are commonly used to relate thermodynamic concepts to ordinary experiences. In Chapter 6, we used a mountain-climbing analogy to illustrate the concept of a state function (Figure 6.9), and we used a water cycle as an analogy to illustrate the first law of thermodynamics (Figure 6.10). Formulate an analogy to spontaneous change based on the accompanying photograph. In your analogy, touch upon both thermodynamic and kinetic factors.

▲ Problem 90 illustrated

Apply Your Knowledge

91. **[Collaborative]** A round-trip spaceship voyage to Mars carrying astronauts would require a great deal of fuel. One proposal to reduce the quantity of fuel is to take along a small quantity of hydrogen. On Mars, this hydrogen would be allowed to react with the abundant carbon dioxide in the Martian atmosphere to produce methane and water. The water would then be decomposed to hydrogen and oxygen by electrolysis. The hydrogen would be recycled into the first reaction, and the oxygen would be stored. On the return trip, the spacecraft would be powered by the reaction of the methane and oxygen produced on Mars. Discuss the feasibility of the proposed plan in thermodynamic terms.

92. **[Historical]** The reaction used by Joseph Priestley in his discovery of oxygen in 1774 was described on page 735 and is pictured below.

   (a) Verify the statement in the text that the reaction should yield an oxygen partial pressure of 0.25 atm at about 500 °C.

   (b) In principle, which of the following substances could be used in a coupled reaction at 1 atm pressure and 298.15 K to bring about the decomposition of mercury(II) oxide to produce liquid mercury: hydrogen, graphite, carbon monoxide, carbon dioxide, copper(I) oxide, copper(II) oxide, nitrogen,

◄ Problem 91 illustrated

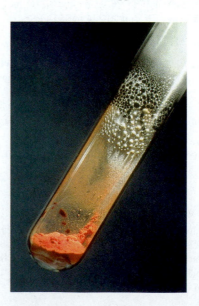

◄ Problem 92 illustrated

dinitrogen monoxide, nitrogen monoxide, sulfur dioxide, sulfur trioxide, chlorine, aluminum metal, silver metal, magnesium oxide?

93. **[Biochemical]** The free energy of solution formation is related to concentration by the expression: $\Delta G = \Delta G_f^\circ + RT \ln c$, where c is the solute molarity. The concentration of creatine, an amino acid derivative formed during muscle contraction, is 2.0 mg/100 mL in blood and 75 mg/100 mL in urine. Calculate the free energy change for the transfer of creatine from blood to urine at 37 °C. (*Hint:* Can you see that it is not necessary to convert concentrations from mg/100 mL to molarity? Why not?)

94. **[Biochemical]** The ΔG° of ATP hydrolysis is -30.5 kJ/mol (page 736). The ΔG° of glucose-6-phosphate hydrolysis is -13.9 kJ/mol. What is the ΔG° of the following reaction?

$$\text{ATP} + \text{glucose} \longrightarrow \text{glucose-6-phosphate} + \text{ADP}$$

What concentration of glucose-6-phosphate relative to glucose would cause the reverse reaction to occur spontaneously?

95. **[Historical]** Explain the meaning of the celebrated remark attributed to Rudolf Clausius in 1865. "*Die Energie der Welt ist konstant; die Entropie der Welt strebt einem Maximum zu.*" ("The energy of the world is constant; the entropy of the world increases toward a maximum.")

96. **[Biochemical]** Given the following data, calculate ΔG° and K_{eq} for the isomerization of glucose-1-phosphate to fructose-6-phosphate.

$$\text{Glucose-1-phosphate} \longrightarrow \text{glucose-6-phosphate}$$
$$\Delta G^\circ = -7.28 \text{ kJ/mol}$$

$$\text{Fructose-6-phosphate} \longrightarrow \text{glucose-6-phosphate}$$
$$\Delta G^\circ = -1.67 \text{ kJ/mol}$$

97. **[Biochemical]** Calculate the actual ΔG of ATP hydrolysis in muscle if the concentrations of reactants and products are ATP,

3.0 mM; ADP, 6.0 mM; and $HPO_4{}^{2-}$, 4.0 mM. Remember to use body temperature (37 °C).

$$\text{ATP} + H_2O \longrightarrow \text{ADP} + HPO_4{}^{2-} \qquad K_{eq} = 2.22 \times 10^5$$

98. **[Historical]** The efficiency of a steam engine is given by the ratio w/q_h, where w is the work done by the engine and q_h is the amount of heat transferred to the high-temperature steam through the combustion of a fuel. If all the heat of the high-temperature steam could be converted to work, the ratio w/q_h would equal 1 and the engine would be 100% efficient. However as Carnot deduced (page 717), this is not possible because the cooling water in the condenser must always carry away some of the heat conveyed to the high-temperature steam in the boiler of the engine. Carnot's theorem, later restated as the second law of thermodynamics, expresses this maximum efficiency, shown below on a percentage basis.

$$\text{Efficiency} = \frac{w}{q_h} = \frac{T_h - T_1}{T_h} \times 100\%$$

In this equation, T_h is the temperature of the high-temperature steam and T_1 is that of the condenser. If the condenser temperature of a steam engine with 36% efficiency is 41 °C,

(a) What is the minimum temperature of the high-temperature steam, and why is the actual temperature probably higher?

(b) If the high-temperature steam in (a) is in equilibrium with liquid water, estimate the pressure of the steam. Assume that $\Delta H_{vap} \approx 40$ kJ/mol.

(c) Describe how the operation of a heat engine relates to the common statement in gamblers' parlance that "You can't win and you can't break even."

* 99. **[Laboratory]** Two drops each of 2.0 M $Pb(NO_3)_2$ and K_2CrO_4 are mixed, and a yellow precipitate forms. Based on this information and on data from Appendix C.1, estimate a value for ΔG_f° for lead(II) chromate.

e-Media Problems

The activities described in these problems can be found in the e-Media Activities and Interactive Student Tutorial (IST) modules of the Companion Website, *http://chem.prenhall.com/hillpetrucci*.

100. Observe the motion of atoms in the **Mixing of Gases** animation (*Section 17-3*). **(a)** Describe the process by which the change in entropy occurs for this system. **(b)** When is the system in a state of minimum entropy, and when is it in a state of maximum entropy?

101. In the **Changes of State** simulation (*Section 17-3*), **(a)** in which portion [(a), (b), (c), or (d)] of the temperature-versus-time curve is the greatest change in entropy occurring? **(b)** How is the temperature changing in this portion of the curve?

102. In the **Lattice Vibrations** animation (*Section 17-3*), **(a)** describe, on an atomic scale, how temperature reflects a change you see. **(b)** What property of the crystal is changing with the degree of order?

103. In the **Gibbs Free Energy** simulation (*Section 17-4*), **(a)** for which of the reactions does ΔG become more positive with increasing temperature? **(b)** Define the slope and intercept of the plot of Gibbs free energy versus temperature. **(c)** Which thermodynamic quantity more strongly influences free energy at the low and high temperature limits of the plot?

Chapter 18

Electrochemistry

NO ONE NEEDS to be reminded of how useful electricity is. We use it to cook foods, heat and cool homes, run the motors in vacuum cleaners and other household devices, and power television sets and computers. In the most common method of generating electricity, the heat of combustion of a fuel is converted to mechanical work and then to electricity. However, the second law of thermodynamics imposes a restriction on how much of a given quantity of heat energy can be converted to work: Such conversions must always be accompanied by waste heat. In a typical power plant, for example, less than half the heat of combustion is converted to electricity, and in an automobile's internal combustion engine, the conversion is even less efficient.

A much more efficient way to generate electricity is to avoid heat-to-work conversions and go directly from chemical energy to electric energy. **Electrochemistry** deals with the links between chemical reactions and electricity. Because electricity involves the flow of electrons, the types of chemical reactions associated with electricity are those in which electrons are transferred from one atom to another, which, as we know from Chapter 4, means *oxidation–reduction* reactions.

First we will consider a useful method for balancing oxidation–reduction equations. We will then see how spontaneous chemical reactions can produce electricity and how electricity can cause nonspontaneous reactions to occur. We will also examine practical applications of redox reactions, from batteries and fuel cells as electric power sources, to methods of controlling corrosion, to the manufacture of key chemicals and metals.

Oxidation–Reduction: The Transfer of Electrons

An aqueous solution containing copper(II) ions has a blue color (recall Figure 16.6), and solid silver metal has a silvery color. We can use these facts to describe the result of immersing a coil of copper wire in an aqueous solution containing Ag^+ ions

◄ A futuristic General Motors concept car called Hywire is displayed at the Great Wall, north of Beijing, in November 2003. Hywire is an electric car with a difference. Instead of using conventional batteries, electricity is generated by a chemical reaction in a fuel cell. Chemical reactions that generate electricity, and chemical reactions that are caused by electricity, are explored in this chapter.

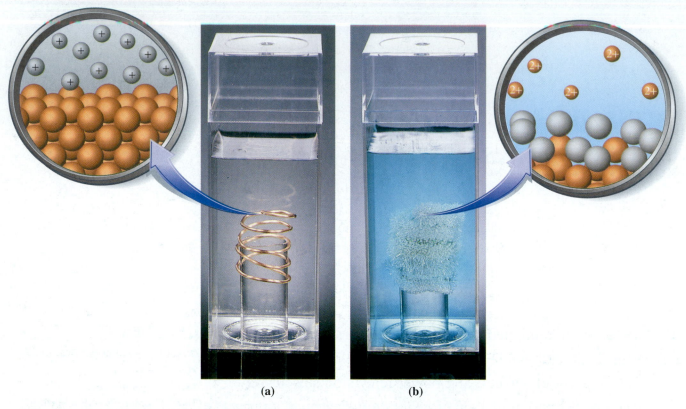

▲ **FIGURE 18.1** **The displacement of $Ag^+(aq)$ by $Cu(s)$: An oxidation–reduction reaction**
(a) A coil of copper wire is suspended in colorless $AgNO_3(aq)$. (b) After a period of time, a blue color has developed in the solution and a silvery deposit has covered the copper coil. In the microscopic view of (a), well-ordered layers of copper atoms in the coil are in contact with a colorless solution containing silver ions (gray). In the microscopic view of (b), copper(II) ions have entered the solution and silver atoms have deposited from solution.

**Oxidation-Reduction Reactions
Part II animation**

(Figure 18.1). Our casual description might be, "Copper goes into solution, and silver comes out of solution." Let us try to explain what happens more scientifically.

18.1 Half-Reactions

In Figure 18.1, we see that Cu *atoms* give up electrons to become Cu^{2+} *ions*. The electrons remain on the copper wire, and the Cu^{2+} ions go into solution. Because the oxidation state of the copper increases from 0 to +2, this process is an oxidation:

$$\textit{Oxidation:} \quad Cu(s) \longrightarrow Cu^{2+}(aq) + 2\,e^-$$

At the same time, Ag^+ ions in the solution come into contact with the copper wire, gain electrons left as Cu atoms oxidized to Cu^{2+} ions, and deposit as Ag atoms. Because the oxidation state of the silver decreases from +1 to 0, this process is a reduction:

$$\textit{Reduction:} \quad Ag^+(aq) + e^- \longrightarrow Ag(s)$$

We call these two processes half-reactions. A **half-reaction** is either an oxidation or a reduction process, which we can represent by a *half-equation*. Taken together, two half-reactions—one an oxidation, one a reduction—constitute an oxidation–reduction reaction.

18.2 The Half-Reaction Method of Balancing Redox Equations

We can combine two half-equations into an overall oxidation–reduction equation, but in doing so we must make sure that the same number of electrons is involved in the oxidation half-reaction as in the reduction half-reaction. In the reaction between Cu(s) and $Ag^+(aq)$, for instance, each Cu atom loses two electrons but each Ag^+ gains only one. We must therefore multiply the reduction half-equation by the factor 2 before combining the half-equations into the overall redox equation:

$$Oxidation: \qquad Cu(s) \longrightarrow Cu^{2+}(aq) + \cancel{2e^-}$$
$$Reduction: \qquad 2\,Ag^+(aq) + \cancel{2e^-} \longrightarrow 2\,Ag(s)$$
$$\overline{Overall: \qquad Cu(s) + 2\,Ag^+(aq) \longrightarrow Cu^{2+}(aq) + 2\,Ag(s)}$$

We have just demonstrated a fundamental principle underlying the half-reaction method of balancing an oxidation–reduction equation:

Balancing Redox Equations activity

All the electrons "lost" in an oxidation half-reaction must be "gained" in a reduction half-reaction. That is, although electrons can move around in a chemical reaction, they cannot be created or destroyed.

The approach we used to balance the equation for the reaction between Cu(s) and $Ag^+(aq)$ follows this general multistep strategy:

Step 1: Separate a redox equation into two half-equations, one for oxidation and one for reduction.

Step 2: Balance the number of atoms of each element in each half-equation. In general, this entails first balancing all atoms except O and H. After this, balance the O and H atoms.

Step 3: Balance each half-equation for electric charge by adding the number of electrons needed to establish the same net charge on each side of the half-equation. Electrons are gained and appear on the left in the reduction half-equation; electrons are lost and appear on the right in the oxidation half-equation.

Step 4: Adjust the coefficients in the half-equations so that the same number of electrons appears in each half-equation. This may require multiplying one or both half-equations by an appropriate factor.

Step 5: Add the two adjusted half-equations to obtain an overall redox equation.

Step 6: Simplify the overall redox equation as necessary. In some cases, to simplify means to divide all coefficients by a common divisor; in other cases it means to adjust certain coefficients so that each participant in the reaction appears only on one side of the overall equation.

Redox Reactions in Acidic Solution

We can implement the multistep strategy for specific reactions occurring under different conditions. First, we will consider redox reactions in acidic solution. In these reactions, water molecules and hydrogen ions can participate as reactants or products, and we should expect H^+ and/or H_2O to appear in one or both half-equations and in the overall equation. In acidic solutions, we will balance O atoms by adding as many molecules of H_2O as needed and balance H atoms by adding as many H^+ ions as needed.

Problem-Solving Note

To simplify the redox equations we write in this chapter, we will represent hydronium ion as H^+ rather than H_3O^+. Also, when balancing redox equations, we will omit state designations such as (l) and (aq), except in the final balanced equation.

▲ When a cellular phone is used, a redox reaction occurs in its battery. When the phone is recharged, the reverse reaction occurs in the battery.

Example 18.1

Permanganate ion, MnO_4^-, is used in the laboratory as an oxidizing agent; thiosulfate ion, $S_2O_3^{2-}$, is used as a reducing agent. Write a balanced equation for the reaction of these ions in an acidic aqueous solution to produce manganese(II) ion and sulfate ion.

STRATEGY

We must first write a skeleton equation, which includes just the principal reactants and products in the redox reaction. The reactants are MnO_4^- and $S_2O_3^{2-}$; the products are Mn^{2+} and SO_4^{2-}, and we can assign an oxidation number to each atom:

$$\overset{+7\ -2}{MnO_4^-} + \overset{+2\ -2}{S_2O_3^{2-}} \longrightarrow \overset{+2}{Mn^{2+}} + \overset{+6\ -2}{SO_4^{2-}} \quad (not\ balanced)$$

To balance this redox equation, we can follow the multistep approach outlined above.

SOLUTION

Step 1: *Separate the redox equation into two half-equations.* MnO_4^- *is reduced to* Mn^{2+} (the oxidation number of Mn decreases).

$$MnO_4^- \longrightarrow Mn^{2+} \quad \text{(skeleton half-equation for reduction)}$$

$S_2O_3^{2-}$ *is oxidized to* SO_4^{2-} (the oxidation number of S increases).

$$S_2O_3^{2-} \longrightarrow SO_4^{2-} \quad \text{(skeleton half-equation for oxidation)}$$

Step 2: *Balance the numbers of atoms in each half-equation, first other atoms and then O and H.* The Mn atoms are balanced.

$$MnO_4^- \longrightarrow Mn^{2+} \quad \text{(1 Mn atom on each side)}$$

To balance the S atoms, we place the coefficient 2 on the right side.

$$S_2O_3^{2-} \longrightarrow 2\,SO_4^{2-} \quad \text{(2 S atoms on each side)}$$

To balance O atoms, we add 4 H_2O on the right side of the reduction half-equation and 5 H_2O on the left side of the oxidation half-equation.

$$MnO_4^- \longrightarrow Mn^{2+} + 4\,H_2O \quad \text{(4 O atoms on each side)}$$
$$S_2O_3^{2-} + 5\,H_2O \longrightarrow 2\,SO_4^{2-} \quad \text{(8 O atoms on each side)}$$

To balance H atoms, we add 8 H^+ to the left side of the reduction half-equation and 10 H^+ to the right side of the oxidation half-equation.

$$MnO_4^- + 8\,H^+ \longrightarrow Mn^{2+} + 4\,H_2O \quad \text{(8 H atoms on each side)}$$
$$S_2O_3^{2-} + 5\,H_2O \longrightarrow 2\,SO_4^{2-} + 10\,H^+ \quad \text{(10 H atoms on each side)}$$

Step 3: *Balance each half-equation for electric charge by adding electrons.* Electrons are gained in a reduction and therefore appear on the left side of the reduction half-equation. We add 5 electrons to the left to reduce the net charge from 7+ to 2+, the same as the 2+ net charge on the right.

Charges: $\quad\quad\quad\quad 1- \quad\quad 8+ \quad\quad 5- \quad\quad\quad\quad 2+$
Reduction: $\quad MnO_4^- + 8\,H^+ + 5\,e^- \longrightarrow Mn^{2+} + 4\,H_2O$
Net charges: $\quad\quad left: -1 + 8 - 5 = 2+ \quad\quad\quad right: 2+$

Electrons are lost in an oxidation and therefore appear on the right side of the oxidation half-equation. We add 8 electrons to the right to reduce the net charge from 6+ to 2−, the same as the net charge on the left.

Charges: $\quad\quad\quad\quad 2- \quad\quad\quad\quad\quad\quad 4- \quad\quad 10+ \quad\quad 8-$
Oxidation: $\quad S_2O_3^{2-} + 5\,H_2O \longrightarrow 2\,SO_4^{2-} + 10\,H^+ + 8\,e^-$
Net charges: $\quad\quad left: 2- \quad\quad\quad right: -4 + 10 - 8 = 2-$

Step 4: *Adjust the coefficients in the half-equations so that the same number of electrons appears in each half-equation.* The reduction half-equation has 5 e^-, and the oxidation half-equation has 8 e^-. Thus, we multiply the reduction half-equation by 8 and the oxidation half-equation by 5.

Reduction: $\quad 8\,MnO_4^- + 64\,H^+ + 40\,e^- \longrightarrow 8\,Mn^{2+} + 32\,H_2O$
Oxidation: $\quad\quad\quad 5\,S_2O_3^{2-} + 25\,H_2O \longrightarrow 10\,SO_4^{2-} + 50\,H^+ + 40\,e^-$

Step 5: *Add the two adjusted half-equations to obtain an overall redox equation.*

Overall: $\quad 8\,MnO_4^- + 5\,S_2O_3^{2-} + 64\,H^+ + 25\,H_2O$
$\quad\quad\quad\quad\quad\quad \longrightarrow 8\,Mn^{2+} + 10\,SO_4^{2-} + 50\,H^+ + 32\,H_2O$

Step 6: *Simplify the overall equation.* With 64 H^+ on the left and 50 H^+ on the right, we see that we can simplify by subtracting 50 H^+ from each side. Likewise, with 25 H_2O on the left and 32 on the right, we can simplify H_2O by subtracting 25 H_2O from each side.

$$8\,MnO_4^- + 5\,S_2O_3^{2-} + (64 - 50)H^+ + (25 - 25)H_2O$$
$$\longrightarrow 8\,Mn^{2+} + 10\,SO_4^{2-} + (50 - 50)H^+ + (32 - 25)H_2O$$

At this point, the equation is balanced and in its simplest form.

$$8\,MnO_4^- + 5\,S_2O_3^{2-} + 14\,H^+ \longrightarrow 8\,Mn^{2+} + 10\,SO_4^{2-} + 7\,H_2O$$

ASSESSMENT

Usually it is not necessary to assign oxidation numbers as a preliminary step. In this case, obviously one of the half-equations involves the Mn-containing species ($MnO_4^- \longrightarrow Mn^{2+}$) and the other involves the S-containing species ($S_2O_3^{2-} \longrightarrow SO_4^{2-}$). After the balanced half-equations are written, it becomes apparent which is the oxidation and which is the reduction.

To verify that the overall equation is correctly balanced, check to ensure that both atoms and electric charge are balanced. This check is shown here in a tabular format.

	Reactants (left)	Products (right)
Mn	8	8
S	10	10
O	47	47
H	14	14
Charge	$(8 \times 1-) + (5 \times 2-) + (14 \times 1+)$ $= 4-$	$(8 \times 2+) + (10 \times 2-)$ $= 4-$

Because the charges are equal on both sides and the same number of atoms of each element appear on each side, the equation is balanced.

EXERCISE 18.1A

Write a balanced equation for the reaction in which zinc metal is oxidized to zinc(II) ion by a dilute acidic solution containing nitrate ion. Dinitrogen monoxide gas is also formed.

EXERCISE 18.1B

Write a balanced equation for the oxidation of phosphorus by nitric acid, which is described by

$$P_4(s) + H^+(aq) + NO_3^-(aq) \longrightarrow H_2PO_4^-(aq) + NO(g)$$

Redox Reactions in Basic Solution

For a reaction in basic solution, $OH^-(aq)$ should appear instead of $H^+(aq)$ in the balanced equation. A method commonly used to balance such an equation is to balance the equation *as if* the reaction occurs in acidic solution. Then, to *each* side of the overall equation, add as many OH^- as there are H^+ in the equation. As a result, one side of the equation will have H^+ and OH^- ions in equal number, which can be combined into H_2O molecules and replaced in the equation by that number of H_2O molecules. The other side of the equation will have OH^- ions. We apply this method in Example 18.2.

Problem-Solving Note

As we learned on page 145, in a *disproportionation* reaction a substance is both oxidized and reduced in the same reaction. We consider a disproportionation reaction in Example 18.2.

Example 18.2

In basic solution, Br_2 disproportionates to bromide ions and bromate ions. Use the half-reaction method to balance the equation for this reaction:

$$Br_2(l) \longrightarrow Br^-(aq) + BrO_3^-(aq)$$

STRATEGY

As in Example 18.1, we can begin by assigning oxidation numbers:

$$\overset{0}{Br_2} \longrightarrow \overset{-1}{Br^-} + \overset{+5}{BrO_3^-}$$

In this disproportionation reaction, Br_2, the only source of Br atoms, must appear on the left side in both half-equations; Br^- appears on the right side of one half-equation, and BrO_3^- appears on the right side of the other. These facts provide the basis for the first step, and other steps proceed as in Example 18.1. As a final step, we make the adjustment from an acidic to a basic solution.

SOLUTION

Step 1: *Separate the redox equation into two half-equations.*

Reduction: $Br_2 \longrightarrow Br^-$

Oxidation: $Br_2 \longrightarrow BrO_3^-$

Step 2: *Balance the numbers of atoms in each half-equation, first other atoms and then O and H.* First, we balance the Br atoms.

$Br_2 \longrightarrow 2\,Br^-$

$Br_2 \longrightarrow 2\,BrO_3^-$

Then, we balance the O atoms.

$Br_2 \longrightarrow 2\,Br^-$

$Br_2 + 6\,H_2O \longrightarrow 2\,BrO_3^-$

Finally, we balance the H atoms.

$Br_2 \longrightarrow 2\,Br^-$

$Br_2 + 6\,H_2O \longrightarrow 2\,BrO_3^- + 12\,H^+$

Step 3: *Balance each half-equation for electric charge by adding electrons.*

Reduction: $Br_2 + 2\,e^- \longrightarrow 2\,Br^-$

Oxidation: $Br_2 + 6\,H_2O \longrightarrow 2\,BrO_3^- + 12\,H^+ + 10\,e^-$

Step 4: *Adjust the coefficients in the half-equations so that the same number of electrons appears in each half-equation.* Multiply the reduction half-equation by 5; leave the oxidation half-equation as is.

Reduction: $5\,(Br_2 + 2\,e^- \longrightarrow 2\,Br^-)$

$5\,Br_2 + 10\,e^- \longrightarrow 10\,Br^-$

Oxidation: $Br_2 + 6\,H_2O \longrightarrow 2\,BrO_3^- + 12\,H^+ + 10\,e^-$

Step 5: *Add the two adjusted half-equations to obtain an overall redox equation.*

Overall: $6\,Br_2 + 6\,H_2O \longrightarrow 10\,Br^- + 2\,BrO_3^- + 12\,H^+$

Step 6: *Simplify the overall equation.* All the coefficients in the overall equation in step 5 are divisible by 2.

$3\,Br_2 + 3\,H_2O \longrightarrow 5\,Br^- + BrO_3^- + 6\,H^+$

To make the adjustment from acidic to basic solution, because there are $6\,H^+$ in the simplified balanced overall equation, we add $6\,OH^-$ ions to each side.

$3\,Br_2 + 3\,H_2O + 6\,OH^- \longrightarrow 5\,Br^- + BrO_3^- + 6\,H^+ + 6\,OH^-$

On the right side, we combine $6\,H^+$ and $6\,OH^-$ into $6\,H_2O$, giving us $3\,H_2O$ on the left and $6\,H_2O$ on the right.

$3\,Br_2 + 3\,H_2O + 6\,OH^- \longrightarrow 5\,Br^- + BrO_3^- + 6\,H_2O$

Thus, our final step is to simplify by subtracting $3\,H_2O$ molecules from each side to yield the final balanced equation in basic solution.

$3\,Br_2(l) + 6\,OH^-(aq) \longrightarrow 5\,Br^-(aq) + BrO_3^-(aq) + 3\,H_2O(l)$

ASSESSMENT

Although we labeled the half-equations in step 1 as reduction and oxidation based on changes in oxidation number, we did not need to identify them at that point. We could have waited to apply those labels until we balanced the half-equations for numbers of atoms and for electric charge. At that point, our oxidation–reduction identification would be based on whether electrons were gained (reduction) or lost (oxidation).

To verify that the final equation is correctly balanced, we see that it has six atoms each of Br, H, and O on the left and on the right, and that the net charge on each side is 6−.

EXERCISE 18.2A

Cyanate ion in waste solutions from gold-mining operations can be destroyed by treatment with hypochlorite ion in basic solution. Write a balanced oxidation–reduction equation for this reaction.

$$OCN^-(aq) + OCl^-(aq) + OH^-(aq) \longrightarrow CO_3^{2-}(aq) + N_2(g) + Cl^-(aq) + H_2O(l)$$

(*Hint:* Notice that in the oxidation half-reaction, OCN^- yields two products.)

EXERCISE 18.2B

In basic solution, permanganate ion oxidizes ethanol to acetate ion and is itself reduced to solid manganese(IV) oxide. Write a balanced equation for this redox reaction.

Voltaic Cells

The "lemon battery" in Figure 18.2—strips of zinc and copper stuck into a lemon—is an example of a **voltaic cell,** which, as we will see in upcoming sections, is a device that uses a spontaneous oxidation–reduction reaction to produce electricity.

18.3   A Qualitative Description of Voltaic Cells

When we dip a strip of zinc metal into an aqueous solution of zinc sulfate as shown in Figure 18.3, some zinc atoms are oxidized. Each Zn atom that is oxidized leaves behind two electrons and enters the solution as a Zn^{2+} ion:

$$Oxidation: \quad Zn(s) \longrightarrow Zn^{2+}(aq) + 2\,e^-$$

At the same time, some Zn^{2+} ions in solution gain two electrons from the zinc strip and deposit on the strip as Zn atoms. They are reduced:

$$Reduction: \quad Zn^{2+}(aq) + 2\,e^- \longrightarrow Zn(s)$$

The opposing oxidation and reduction processes quickly come to equilibrium.

$$Zn(s) \underset{\text{Reduction}}{\overset{\text{Oxidation}}{\rightleftharpoons}} Zn^{2+}(aq) + 2\,e^-$$

A strip of metal used in an electrochemical experiment is called an **electrode.** The equilibrium established at the surface of the electrode when it is immersed in a solution of its ions is called an *electrode equilibrium.* Another example of an electrode equilibrium is that between a copper electrode and an aqueous solution of Cu^{2+}:

$$Cu(s) \underset{\text{Reduction}}{\overset{\text{Oxidation}}{\rightleftharpoons}} Cu^{2+}(aq) + 2\,e^-$$

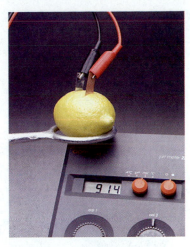

▲ **FIGURE 18.2   A simple voltaic cell**

Two dissimilar metal strips (copper and zinc) and an electrolyte (lemon juice) are the key components of a device that converts chemical energy to electricity—a *voltaic cell.* Electrons flow through the wires and the electric meter. Ions in the lemon juice flow between the metal strips. Together, the electron and ion flows constitute an electric current. The significance of the meter reading is described on page 757.

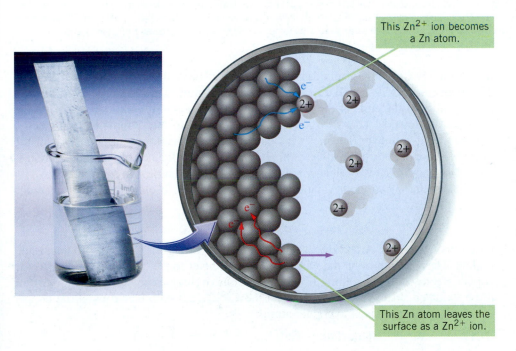

This Zn^{2+} ion becomes a Zn atom.

This Zn atom leaves the surface as a Zn^{2+} ion.

◀ **FIGURE 18.3   Electrode equilibrium**

The zinc metal strip, called an electrode, is partially immersed in a solution containing Zn^{2+} ions. Oxidation of Zn to Zn^{2+} (near the bottom of the microscopic view) and reduction of Zn^{2+} to Zn (near the top) occur at the electrode until a condition of equilibrium is reached. For clarity, the anions that are needed to produce an electrically neutral solution are not shown.

Voltaic Cells (I): The Copper-Zinc Cell animation

In Chapter 4, we used the activity series of the metals (Figure 4.15) to show that zinc is rather easily oxidized and therefore a good reducing agent. In contrast, copper is not so readily oxidized and is a poor reducing agent. We can use the voltaic cell pictured in Figure 18.4 to demonstrate that the forward direction (oxidation) is more strongly favored in the electrode equilibrium at a zinc electrode than at a copper electrode.

A **half-cell** consists of an electrode immersed in a solution of ions. Figure 18.4 represents two half-cells: one containing Zn(s) in $ZnSO_4(aq)$ and the other containing

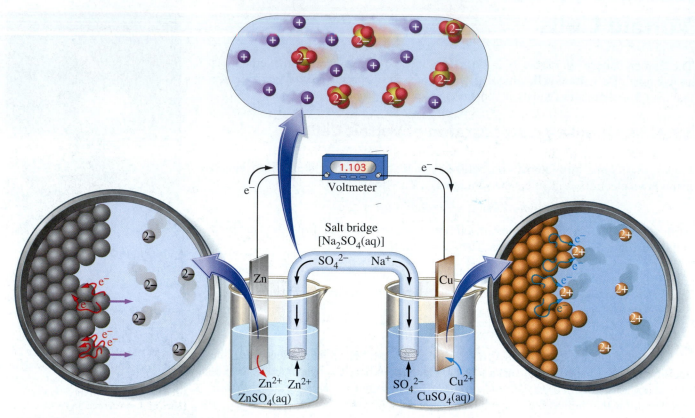

▲ FIGURE 18.4   A zinc–copper voltaic cell

(Left microscopic view) At the zinc electrode, Zn atoms lose electrons and enter the solution as Zn^{2+} ions. The electrons flow through an external circuit from the Zn electrode to the Cu electrode. (Right microscopic view) At the copper electrode, Cu^{2+} ions gain electrons and deposit on the electrode as Cu atoms. (Top microscopic view) The migration of Na^+ and SO_4^{2-} ions. Also migrating into the salt bridge are Zn^{2+} ions from the left beaker and SO_4^{2-} ions from the right beaker. Voltaic cells similar to this one were used to power telegraph lines in the mid-nineteenth century.

QUESTION: In which half-cell does the electrode decrease in mass as the reaction progresses?

Cu(s) in $CuSO_4$(aq). The solutions in the two half-cells are joined by a **salt bridge.** This is an inverted U-shaped tube containing a salt solution, for example, Na_2SO_4(aq). Porous plugs at the ends of the tube prevent the bulk flow of solution but permit the migration of ions. Metal wires connect the electrodes to the terminals of an electric meter called a voltmeter. The meter indicates that electrons flow continuously from the zinc to the copper electrode. (The voltmeter and the significance of its reading are discussed on page 757.) We can account for the observed electric current in this way.

At the zinc electrode, oxidation occurs. Zinc atoms lose electrons and enter the solution as Zn^{2+} ions:

$$Oxidation: \quad Zn(s) \longrightarrow Zn^{2+}(aq) + 2\ e^-$$

Electrons left behind on the zinc electrode by the departing Zn^{2+} ions do not just accumulate. They flow through the wires and the voltmeter to the copper electrode. At the copper electrode, the incoming flood of electrons overcomes any tendency for copper to be oxidized. Instead, Cu^{2+} ions from the $CuSO_4$(aq) gain electrons at the copper electrode and are reduced to copper atoms:

$$Reduction: \quad Cu^{2+}(aq) + 2\ e^- \longrightarrow Cu(s)$$

As a result of the half-reactions at the electrodes, the following oxidation–reduction reaction occurs:

$$Overall\ reaction: \quad Zn(s) + Cu^{2+}(aq) \longrightarrow Zn^{2+}(aq) + Cu(s)$$

Our description to this point tells only part of the story, however, for there can be no electric current unless electric charge moves not only through the connecting wire but also through the solutions in the half-cells. Electrons constitute the electric current through the external circuit, but they cannot move through the solutions because any electrons arriving at the electrode surfaces are exchanged in the oxidation and reduction half-reactions. Instead, *cations* and *anions* carry electric charge through the solutions.

As zinc is oxidized, the number of Zn^{2+} ions in the zinc half-cell increases, tending to build up a positive charge in the solution. Simultaneously, as Cu^{2+} is reduced, the number of Cu^{2+} ions decreases in the copper half-cell, tending to build up a negative charge in the solution. The salt bridge prevents the buildup of charges. Sulfate ions migrate out of the salt bridge into the zinc half-cell to neutralize the excess positive charge associated with the Zn^{2+} ions produced in the oxidation. Also, some SO_4^{2-} ions migrate out of the copper half-cell into the salt bridge. At the same time, Na^+ ions from the salt bridge migrate into the copper half-cell to replace Cu^{2+} ions that have been reduced. Also, some Zn^{2+} ions migrate out of the zinc half-cell into the salt bridge.

Some Important Electrochemical Terms

The general name for an electrochemical device that combines two half-cells, with the appropriate connections between electrodes and solutions, is **electrochemical cell.** The voltaic cell, the type of electrochemical cell we have just been talking about, is, as noted above, one in which electric current is generated from a spontaneous redox reaction. Let us consider a few other important terms used to describe voltaic cells.

The **anode** is the electrode at which oxidation occurs. The **cathode** is the electrode at which reduction occurs. Figure 18.5 employs these terms using the reaction from Figure 18.4. Because the anode is the source of electrons in a voltaic cell, it is also the *negative* electrode. The cathode is the receiver of electrons and is therefore the *positive* electrode. In a real cell, however, the extent of electron buildup at the anode is slight, so that there is not much difference in the net charge present at the two electrodes.

Another way to describe the difference between anode and cathode is to say that the potential energy of the electrons at the anode is greater than at the cathode. Thus, when the two electrodes are joined in a voltaic cell, electrons flow from the anode to the cathode. This is the same as describing the flow of water over a waterfall—from a point where the water has high potential energy to a point where it has lower potential energy. The change in potential energy depends on the height of the waterfall. The property in an electric circuit analogous to the height of a waterfall is the *electric potential,* which is measured in *volts* and is defined as energy per unit of charge that flows. As mentioned in our Chapter 7 discussion of Thomson's cathode-ray experiments, the **coulomb (C)** is the unit of electric charge, and an electric potential of one **volt (V)** is defined as one joule per coulomb:

$$1 \text{ V} = 1 \text{ J/C} \qquad (18.1)$$

A *voltmeter* measures a difference in electric potential between two points in an electric circuit. If the two points are the electrodes in a voltaic cell, the *potential difference* is the driving force that propels electrons from the anode to the cathode; this potential difference is also called the **cell potential** (E_{cell}). Because the measurements are in volts, the cell potential is also called the *cell voltage.* The cell potential of the Zn–Cu voltaic cell of Figure 18.4 is 1.103 V, and that of the zinc/copper/lemon voltaic cell shown in Figure 18.2 is 0.914 V, or 914 *milli*volts (mV). The overall oxidation–reduction reaction that occurs in a voltaic cell is called the *cell reaction.*

Credit for the discovery of current electricity is generally given to Luigi Galvani (1737–1798). However, Alessandro Volta (1745–1827) was the first to construct cells similar to the one in Figure 18.4 for the production of electricity, hence the name *voltaic* cell. However, the same cells are sometimes referred to as *galvanic* cells.

The following mnemonic (memory) device may help you remember the difference between cathode and anode.*

Cathode	**A**node
A	U
Reduction	T
	Oxidation

* Kindly supplied by Professor Richard S. Treptow, Chicago State University.

Summary of Electrochemical Terms activity

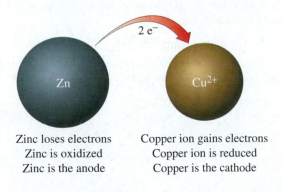

Zinc loses electrons	Copper ion gains electrons
Zinc is oxidized	Copper ion is reduced
Zinc is the anode	Copper is the cathode

◀ **FIGURE 18.5** **The electrode processes in the voltaic cell of Figure 18.4 summarized**

▲ In 1991, Virginia Tech students created the world's longest human salt bridge. Electrolytes in the bodies of 1500 students holding hands allowed the measurement of a cell potential. Conductivity between hands was enhanced with salty potato chip crumbs and soft drinks.

Cell Diagrams activity

Cell Diagrams

It is more convenient to represent an electrochemical cell by a **cell diagram** than by a drawing like Figure 18.4. By international agreement, the following conventions are used for cell diagrams:

- The *anode* is placed on the *left* side of the diagram.
- The *cathode* is placed on the *right* side of the diagram.
- A *single* vertical line (|) is used to represent the boundary between different phases, such as between an electrode and a solution.
- A *double* vertical line (‖) is used to represent a salt bridge or other type of porous barrier separating two half-cells.

Following these conventions, the cell diagram for Figure 18.4 is

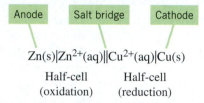

$$Zn(s)|Zn^{2+}(aq)\|Cu^{2+}(aq)|Cu(s)$$

Half-cell (oxidation)   Half-cell (reduction)

In many cases, we use an electrode that does not participate in an oxidation–reduction equilibrium. Rather, the inert electrode simply furnishes the surface on which an electric potential is established. For example, to establish electrode equilibrium between chlorine gas and chloride ions, we can immerse a strip of the inert metal platinum into a solution of $Cl^-(aq)$ and bubble chlorine gas over the metal surface. The redox equilibrium

$$2\,Cl^-(aq) \xrightarrow{\text{on Pt}} Cl_2(g) + 2\,e^-$$

is established on the surface of the platinum and imparts a characteristic electric potential to the metal. The cell diagrams for any half-cells using this Cl_2/Cl^- electrode would be

As an anode *(oxidation):* $Pt(s)|Cl_2(g)|Cl^-(aq)$

As a cathode *(reduction):* $Cl^-(aq)|Cl_2(g)|Pt(s)$

Example 18.3

Describe the half-reactions and the overall reaction that occur in the voltaic cell represented by the cell diagram

$$Pt(s)|Fe^{2+}(aq),\ Fe^{3+}(aq)\|Cl^-(aq)|Cl_2(g)|Pt(s)$$

SOLUTION

The half-reactions occur on the surface of the platinum, but because platinum is inert, other species are involved in the reactions. According to the cell diagram conventions, the electrode on the left is the anode, where oxidation occurs. In this case, Fe^{2+} is oxidized to Fe^{3+}:

Oxidation: $Fe^{2+}(aq) \longrightarrow Fe^{3+}(aq) + e^-$

The electrode on the right is the cathode, where reduction occurs, which in this case means chlorine gas is reduced to chloride ion:

Reduction: $Cl_2(g) + 2\,e^- \longrightarrow 2\,Cl^-(aq)$

To combine these half-equations into a redox equation, we must multiply the oxidation half-equation by the factor 2. Then we can add it to the reduction half-equation, and the electrons will cancel.

Oxidation:	$2\,Fe^{2+}(aq) \longrightarrow 2\,Fe^{3+}(aq) + \cancel{2e^-}$
Reduction:	$Cl_2(g) + \cancel{2e^-} \longrightarrow 2\,Cl^-(aq)$
Cell reaction:	$2\,Fe^{2+}(aq) + Cl_2(g) \longrightarrow 2\,Fe^{3+}(aq) + 2\,Cl^-(aq)$

Remember that the left-anode, right-cathode convention for cell diagrams is just that—a convention—and therefore arbitrary. When we assemble a voltaic cell of unknown half-cells in the laboratory, the one on the left will not necessarily be the anode. We can say it is the anode only if we determine *by experiment* that oxidation occurs in that half-cell. And even if the half-cell on the left is the anode, someone on the opposite side of the lab bench would see the same anode half-cell as being on their right. Left and right in the three-dimensional world are not as unambiguous as on a sheet of paper.

18.4 Standard Electrode Potentials

We have described how to construct a voltaic cell by combining two half-cells and how to measure the cell voltage with a voltmeter. However, it is also possible to calculate cell voltages without making specific reference to measurements. To do so, we establish characteristic potentials for individual half-cells and then determine differences in potential (cell voltages) when the half-cells are combined into voltaic cells. But there is a difficulty. We cannot measure an individual electrode potential; instead, the only thing we can measure is a potential *difference*. The situation is similar to determining the elevation of a point on Earth. We need to give the elevation relative to an arbitrarily assigned zero point of elevation—usually mean sea level.

The zero point of electrode potentials is taken to be that of the half-cell pictured in Figure 18.6 and called the *standard hydrogen electrode*. Other electrode potentials are reported in relation to this arbitrarily assigned zero.

In the **standard hydrogen electrode (SHE),** hydrogen gas at exactly 1 bar pressure is bubbled over an inert platinum electrode and into an aqueous solution in which the concentration has been adjusted so that the activity of H_3O^+ is exactly $a = 1$. For simplicity, we will use the symbol H^+ in place of H_3O^+, replace $a_{H^+} = 1$ by $[H^+] = 1\ M$,* and

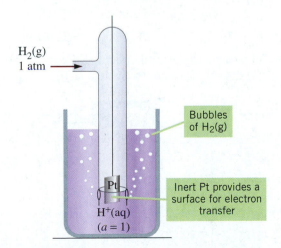

$H_2(g)$
1 atm

Bubbles
of $H_2(g)$

Inert Pt provides a
surface for electron
transfer

Pt

$H^+(aq)$
$(a = 1)$

◀ **FIGURE 18.6   The standard hydrogen electrode (SHE)**

The inert platinum strip acquires a potential that is determined by the equilibrium $2\ H^+(a = 1) + 2\ e^- \rightleftharpoons H_2(g, 1\ bar)$. The condition of $a_{H^+} = 1$ can be approximated by $[H^+] = 1\ M$ and that of 1 bar by 1 atm.

*As we noted in Chapter 17, activities should be used instead of molarities for precise work dealing with solution properties, equilibrium constants, and the like. To keep things simple, however, we will continue to use molarities and partial pressures in place of activities, but we will report results *as if* we had used activities.

replace $P_{H_2} = 1$ bar by $P_{H_2} = 1$ atm. Equilibrium between H^+ ions and H_2 molecules is established on the platinum surface as H^+ is reduced to H_2 and H_2 is oxidized to H^+:

$$2\,H^+(1\,M) + 2\,e^- \xrightleftharpoons{\text{on Pt}} H_2\,(g,\,1\,atm) \qquad E° = 0\,V \;(\textit{exactly})$$

Because standard electrode potentials are based on reduction tendencies, they are also called standard *reduction* potentials.

By international agreement, a **standard electrode potential, $E°$,** for any half-reaction is based on the tendency for *reduction* to occur at an electrode. These standard potentials are measured with all solution species present at unit activity $(a = 1)$, which is about 1 M, and all gases at 1 bar pressure, which is about 1 atm. When no other metal is indicated as the electrode material, the potential is that for equilibrium established on an inert surface, such as platinum metal. The standard electrode potential for the standard hydrogen electrode, as we have seen, is arbitrarily set at *exactly* zero volts.

To establish standard electrode potentials for other half-reactions, we can construct a voltaic cell like that shown in Figure 18.7, where a standard hydrogen electrode is joined with a standard copper electrode. The cell diagram for this setup is

$$\text{Pt}|H_2(g,\,1\,atm)|H^+(1\,M)\|Cu^{2+}(1\,M)|Cu(s)$$
$$\text{(Anode)} \qquad\qquad\qquad \text{(Cathode)}$$

We find that electrons flow from the hydrogen electrode (anode) to the copper electrode (cathode) at a measured voltage of 0.340 V. This cell voltage, called the **standard cell potential $(E°_{cell})$,** is the difference between the standard potential of the cathode and that of the anode:

$$E°_{cell} = E°(\text{cathode}) - E°(\text{anode}) \qquad\qquad \textbf{(18.2)}$$

Or, using the conventional notation of cell diagrams, where the anode is on the left and the cathode is on the right, we can say

$$E°_{cell} = E°(\text{right}) - E°(\text{left})$$

We will represent the standard electrode potential $E°$ of the electrode on the right—the standard Cu^{2+}/Cu electrode—in the form

$$E°_{Cu^{2+}/Cu}$$

The electrode on the left is the standard hydrogen electrode, for which the defined potential is

$$E°_{H^+/H_2} = 0.000\,V$$

To find the standard electrode potential for the reduction of Cu^{2+} to $Cu(s)$, we can write

$$E°_{cell} = E°_{Cu^{2+}/Cu} - E°_{H^+/H_2} = +0.340\,V$$
$$E°_{Cu^{2+}/Cu} - 0.000\,V = +0.340\,V$$
$$E°_{Cu^{2+}/Cu} = +0.340\,V$$

The positive value of the standard electrode potential means that Cu^{2+} ions are more readily reduced to $Cu(s)$ than H^+ ions are reduced to $H_2(g)$.

Standard Reduction Potentials animation

▶ **FIGURE 18.7   Measuring the standard potential of the Cu^{2+}/Cu electrode**
The standard hydrogen electrode is the anode, and the Cu^{2+}/Cu electrode is the cathode. Contact between the solutions in the two half-cells is through a salt bridge, which permits the migration of ions but prevents bulk flow of the solutions. The direction of electron flow and the volt-meter reading are shown.

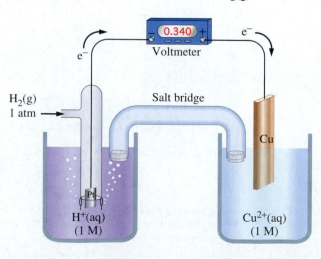

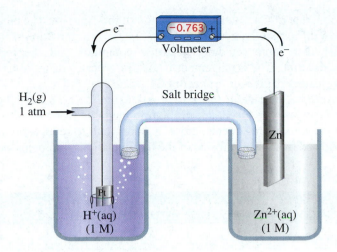

◀ FIGURE 18.8 Measuring the standard potential of the Zn^{2+}/Zn electrode

As in Figure 18.7, the standard hydrogen electrode appears on the left and the metal electrode on the right. However, as signified by the direction of electron flow and the negative voltage reading, the standard hydrogen electrode is now the *cathode*. Therefore the Zn^{2+}/Zn electrode is the *anode*. When voltaic cells are assembled as in Figures 18.7 and 18.8, correct magnitudes and signs of all standard electrode potentials can be established.

Voltaic Cells (II): The Zinc-Hydrogen Cell animation

Now consider what happens in the cell in Figure 18.8, where we replace the Cu^{2+}/Cu half-cell by a Zn^{2+}/Zn half-cell. The voltmeter in the circuit registers -0.763 V, and the cell diagram is

$$Pt|H_2(g, 1 \text{ atm})|H^+(1 \text{ M})\|Zn^{2+}(1 \text{ M})|Zn(s) \qquad E^{\circ}_{cell} = -0.763 \text{ V}$$

This standard cell potential and the standard electrode potentials are related as in the Cu^{2+}/Cu case:

$$E^{\circ}_{cell} = E^{\circ}(\text{right}) - E^{\circ}(\text{left})$$
$$E^{\circ}_{cell} = E^{\circ}_{Zn^{2+}/Zn} - E^{\circ}_{H^+/H_2} = -0.763 \text{ V}$$
$$E^{\circ}_{Zn^{2+}/Zn} - 0.000 \text{ V} = -0.763 \text{ V}$$
$$E^{\circ}_{Zn^{2+}/Zn} = -0.763 \text{ V}$$

The negative value of the standard electrode potential means that $Zn^{2+}(aq)$ is less readily reduced to $Zn(s)$ than $H^+(aq)$ is reduced to $H_2(g)$.

What is the physical significance of the negative value of E°_{cell} in Figure 18.8? It has to do with how the electrodes are connected to the voltmeter. The proper connection for a positive reading is that the anode be connected to the negative terminal of the voltmeter and the cathode to the positive terminal. If the connections are reversed, as they are in Figure 18.8, the voltmeter reading will be negative. In the cell that we have been describing, we assumed that the Zn electrode is the cathode, but it is actually the anode.

Figure 18.9 is a diagram of the three standard electrode potentials we have been discussing, arranged in terms of decreasing value of E°. The electrode potential for the

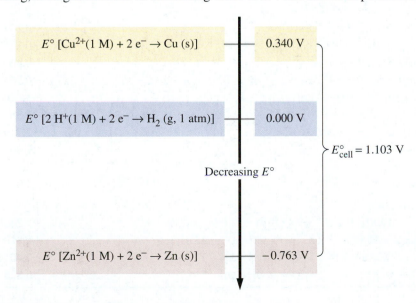

◀ FIGURE 18.9 A representation of standard electrode potentials

The standard potentials for the copper and zinc electrodes are shown in relation to the standard hydrogen electrode potential. The 1.103-V potential difference between the copper and zinc electrodes is the cell voltage for the voltaic cell of Figure 18.4.

easiest reduction, that of Cu^{2+} to Cu, is at the top of the diagram. The electrode potential for the most difficult reduction, that of Zn^{2+} to Zn, is at the bottom.

Table 18.1 lists some standard electrode potentials and the reduction half-reactions they describe. The arrangement is also in order of decreasing $E°$ values. A more extensive listing is given in Appendix C.

Viewed from the left side of the half-equations in Table 18.1, the best oxidizing agents, those most easily reduced, appear high in the table (for example, F_2, O_3, and so on). The poorest oxidizing agents, those most difficult to reduce, are at the bottom of the table. Viewed from the right side of the half-equations—in other words, looking

Tables of standard electrode potentials are sometimes written in the reverse order, with the best reducing agents—those most easily oxidized—at the top of the table.

Table 18.1 Selected Standard Electrode Potentials at 25 °C

Reduction Half-Reaction	$E°$, Volts
Acidic Solution	
$F_2(g) + 2\,e^- \longrightarrow 2\,F^-(aq)$	+2.866
$O_3(g) + 2\,H^+(aq) + 2\,e^- \longrightarrow O_2(g) + H_2O(l)$	+2.075
$S_2O_8{}^{2-}(aq) + 2\,e^- \longrightarrow 2\,SO_4{}^{2-}(aq)$	+2.01
$H_2O_2(aq) + 2\,H^+(aq) + 2\,e^- \longrightarrow 2\,H_2O(l)$	+1.763
$MnO_4{}^-(aq) + 8\,H^+(aq) + 5\,e^- \longrightarrow Mn^{2+}(aq) + 4\,H_2O(l)$	+1.51
$PbO_2(s) + 4\,H^+(aq) + 2\,e^- \longrightarrow Pb^{2+}(aq) + 2\,H_2O(l)$	+1.455
$Cl_2(g) + 2\,e^- \longrightarrow 2\,Cl^-(aq)$	+1.358
$Cr_2O_7{}^{2-}(aq) + 14\,H^+(aq) + 6\,e^- \longrightarrow 2\,Cr^{3+}(aq) + 7\,H_2O(l)$	+1.33
$MnO_2(s) + 4\,H^+(aq) + 2\,e^- \longrightarrow Mn^{2+}(aq) + 2\,H_2O(l)$	+1.23
$O_2(g) + 4\,H^+(aq) + 4\,e^- \longrightarrow 2\,H_2O(l)$	+1.229
$2\,IO_3{}^-(aq) + 12\,H^+(aq) + 10\,e^- \longrightarrow I_2(s) + 6\,H_2O(l)$	+1.20
$Br_2(l) + 2\,e^- \longrightarrow 2\,Br^-(aq)$	+1.065
$NO_3{}^-(aq) + 4\,H^+(aq) + 3\,e^- \longrightarrow NO(g) + 2\,H_2O(l)$	+0.956
$Ag^+(aq) + e^- \longrightarrow Ag(s)$	+0.800
$Fe^{3+}(aq) + e^- \longrightarrow Fe^{2+}(aq)$	+0.771
$O_2(g) + 2\,H^+(aq) + 2\,e^- \longrightarrow H_2O_2(aq)$	+0.695
$I_2(s) + 2\,e^- \longrightarrow 2\,I^-(aq)$	+0.535
$Cu^{2+}(aq) + 2\,e^- \longrightarrow Cu(s)$	+0.340
$SO_4{}^{2-}(aq) + 4\,H^+(aq) + 2\,e^- \longrightarrow 2\,H_2O(l) + SO_2(g)$	+0.17
$Sn^{4+}(aq) + 2\,e^- \longrightarrow Sn^{2+}(aq)$	+0.154
$S(s) + 2\,H^+(aq) + 2\,e^- \longrightarrow H_2S(g)$	+0.14
$2\,H^+(aq) + 2\,e^- \longrightarrow H_2(g)$	0
$Pb^{2+}(aq) + 2\,e^- \longrightarrow Pb(s)$	−0.125
$Sn^{2+}(aq) + 2\,e^- \longrightarrow Sn(s)$	−0.137
$Co^{2+}(aq) + 2\,e^- \longrightarrow Co(s)$	−0.277
$Fe^{2+}(aq) + 2\,e^- \longrightarrow Fe(s)$	−0.440
$Zn^{2+}(aq) + 2\,e^- \longrightarrow Zn(s)$	−0.763
$Al^{3+}(aq) + 3\,e^- \longrightarrow Al(s)$	−1.676
$Mg^{2+}(aq) + 2\,e^- \longrightarrow Mg(s)$	−2.356
$Na^+(aq) + e^- \longrightarrow Na(s)$	−2.713
$Ca^{2+}(aq) + 2\,e^- \longrightarrow Ca(s)$	−2.84
$K^+(aq) + e^- \longrightarrow K(s)$	−2.924
$Li^+(aq) + e^- \longrightarrow Li(s)$	−3.040
Basic Solution	
$O_3(g) + H_2O(l) + 2\,e^- \longrightarrow O_2(g) + 2\,OH^-(aq)$	+1.246
$OCl^-(aq) + H_2O(l) + 2\,e^- \longrightarrow Cl^-(aq) + 2\,OH^-(aq)$	+0.890
$O_2(g) + 2\,H_2O(l) + 4\,e^- \longrightarrow 4\,OH^-(aq)$	+0.401
$2\,H_2O(l) + 2\,e^- \longrightarrow H_2(g) + 2\,OH^-(aq)$	−0.828

at the reverse reactions—the best reducing agents are at the bottom of the table (Li, K, and so forth), and the poorest reducing agents are at the top.

We will find Table 18.1 very useful. First we will apply it to the rather simple task of determining a standard cell potential, E°_{cell}, from standard electrode potentials, E°. Consider, for example, how we can calculate E°_{cell} for the voltaic cell in Figure 18.4:

$$Zn(s)|Zn^{2+}(aq)\|Cu^{2+}(aq)|Cu(s)$$

All we need to do is substitute standard electrode potentials from Table 18.1 into Equation (18.2):

$$E^\circ_{cell} = E^\circ(\text{cathode}) - E^\circ(\text{anode})$$
$$= E^\circ_{Cu^{2+}/Cu} - E^\circ_{Zn^{2+}/Zn}$$
$$= 0.340\ V - (-0.763\ V) = 1.103\ V$$

In this type of calculation, three quantities will always be involved: E°_{cell}, E° for the cathode, and E° for the anode. If we have values for any two, we can calculate the third. In Example 18.4, we use this idea to determine the value of an unknown standard electrode potential.

Example 18.4

Determine E° for the reduction half-reaction $Ce^{4+}(aq) + e^- \longrightarrow Ce^{3+}(aq)$, given that the cell voltage for the voltaic cell

$$Co(s)|Co^{2+}(1\ M)\|Ce^{4+}(1\ M), Ce^{3+}(1\ M)|Pt(s)$$

is $E^\circ_{cell} = 1.887\ V$.

STRATEGY

The reduction half-reaction occurs in the cathode half-cell, shown on the right in the cell diagram. We are seeking E° for the cathode—that is, $E^\circ_{Ce^{4+}/Ce^{3+}}$. The oxidation half-reaction, $Co(s) \longrightarrow Co^{2+} + 2\ e^-$, occurs in the anode half-cell on the left. From Table 18.1, we find that $E^\circ_{Co^{2+}/Co} = -0.277\ V$.

SOLUTION

In the same manner as illustrated for the Zn–Cu voltaic cell, we can base our calculation on Equation (18.2):

$$E^\circ_{cell} = E^\circ(\text{cathode}) - E^\circ(\text{anode})$$
$$= E^\circ_{Ce^{4+}/Ce^{3+}} - E^\circ_{Co^{2+}/Co}$$
$$1.887\ V = E^\circ_{Ce^{4+}/Ce^{3+}} - (-0.277\ V)$$
$$1.887\ V - 0.277\ V = E^\circ_{Ce^{4+}/Ce^{3+}} = 1.610\ V$$

ASSESSMENT

Perhaps the main source of error in using Equation (18.2) is the incorrect use of signs, and so that is one matter we should always check. The difference between our calculated Ce^{4+}/Ce^{3+} potential and the given Co^{2+}/Co potential is the measured cell voltage of 1.887 V. Because the E° value for the Co^{2+}/Co half-cell ($-0.277\ V$) is below that of the SHE (0.000 V), the Ce^{4+}/Ce^{3+} potential must be a positive value, but it must also be less than 1.887 V, just as we found it to be.

EXERCISE 18.4A

Use the following information and data from Table 18.1 to determine $E^\circ_{Sm^{2+}/Sm}$.

$$Sm(s)|Sm^{2+}(1\ M)\|I^-(1\ M)|I_2(s)|Pt(s) \qquad E^\circ_{cell} = 3.21\ V$$

EXERCISE 18.4B

Use the following information and data from Table 18.1 to determine $E^\circ_{ClO_4^-/Cl_2}$.

$$Pt(s)|Cl_2(g)|ClO_4^-(1\ M)\|Cl^-(1\ M)|Cl_2(g)|Pt(s) \qquad E^\circ_{cell} = -0.034\ V$$

▲ **FIGURE 18.10 Cell potential E_{cell}° is an intensive property**

That cell potential E_{cell}° is an intensive property can be seen in the constant voltage (1.5 V) of dry cell batteries, whether they are large D batteries or small AA batteries. The voltage does not depend on the amounts of substances involved in the cell reaction.

QUESTION: What is one way to produce a dry cell battery having a voltage other than 1.5 V?

Two additional important points about electrode and cell potentials are

- Electrode potentials and cell potentials are intensive properties, which means their magnitudes are fixed once the particular species and concentrations are specified. Thus, the magnitude of electrode potential does *not* depend on the size of a half-cell or voltaic cell. Figure 18.10 illustrates this fact with a familiar example.
- Cell potentials can be calculated for oxidation–reduction reactions even when no voltaic cells are involved. Specifically, we can calculate E_{cell}° from the equation for a cell reaction without writing a cell diagram. This idea is illustrated in Example 18.5.

Example 18.5

Balance the following oxidation–reduction equation, and determine E_{cell}° for the reaction.

$$O_2(g) + H^+(aq) + I^-(aq) \longrightarrow H_2O(l) + I_2(s)$$

STRATEGY

First, we need to separate the redox equation into half-equations for oxidation and reduction. Then, we obtain a balanced equation by the half-reaction method of Section 18.1. We complete the answer by obtaining E° values for the two half-reactions from Table 18.1 and using them in Equation (18.2).

SOLUTION

The balanced half-equations and their combination into the balanced redox equation are

Reduction:	$O_2 + 4 H^+ + 4 e^- \longrightarrow 2 H_2O$	
Oxidation:	$2 \{2 I^- \longrightarrow I_2 + 2 e^-\}$	
Overall:	$O_2(g) + 4 H^+(aq) + 4 I^-(aq) \longrightarrow 2 H_2O(l) + 2 I_2(s)$	

If the reaction were carried out in a voltaic cell, the half-reaction at the cathode would be a reduction and the half-reaction at the anode would be an *oxidation*. With this in mind, we can write

$$E_{cell}^\circ = E^\circ(\text{cathode}) - E^\circ(\text{anode})$$

$$= E^\circ(\text{reduction}) - E^\circ(\text{oxidation})$$

Then, we can substitute the E° values for the reduction and oxidation half-equations from Table 18.1:

$$E_{cell}^\circ = E_{O_2/H_2O}^\circ - E_{I_2/I^-}^\circ = +1.229 \text{ V} - 0.535 \text{ V} = +0.694 \text{ V}$$

ASSESSMENT

First, notice that although we doubled the oxidation half-equation before combining it with the reduction half-equation, we did not alter the E_{I_2/I^-}° value in any way, signifying that E° is an intensive property. To assure ourselves that we used the correct signs in calculating E_{cell}°, we notice that both E° values are positive quantities and that the difference between the larger one (for the reduction process) and the smaller one (for the oxidation process) is also a positive quantity.

EXERCISE 18.5A

Determine the E_{cell}° value for each of the following redox reactions.

(a) $2 \text{ Al}(s) + 3 \text{ Cu}^{2+}(aq) \longrightarrow 2 \text{ Al}^{3+}(aq) + 3 \text{ Cu}(s)$

(b) $2 \text{ NO}_3^-(aq) + 3 \text{ Pb}^{2+}(aq) + 2 \text{ H}_2O(l) \longrightarrow 2 \text{ NO}(g) + 3 \text{ PbO}_2(s) + 4 \text{ H}^+(aq)$

EXERCISE 18.5B

Write a balanced equation for the oxidization in acidic solution of $\text{Mn}^{2+}(aq)$ to $\text{MnO}_4^-(aq)$ by $\text{S}_2\text{O}_8^{2-}(aq)$, which is reduced to $\text{SO}_4^{2-}(aq)$, and determine E_{cell}° for the reaction.

18.5   Electrode Potentials, Spontaneous Change, and Equilibrium

When a reaction takes place in a voltaic cell, work is done. We can think of this electrical work as the work done by electric charges in motion. The total work done is the product of (1) the cell voltage E_{cell}; (2) the number of moles of electrons, n, transferred between electrodes; and (3) the electric charge per mole of electrons—a quantity called the **Faraday constant, F,** and equal to 96,485 coulombs per mole:

$$w_{elec} = nFE_{cell} \qquad\qquad (18.3)$$

The product nFE_{cell} has the units volt × coulomb. From the definition of the volt, $1\ V = 1\ J/C$ (page 757), we determine that the unit of electrical work is the joule: $1\ J = 1\ V\ C$.

Electrical work is related to free energy change in the following way: The maximum amount of useful work that a system can do is $-\Delta G$, and this maximum amount of work can be realized as electrical work. As a consequence, the electrical work done by a voltaic cell is

$$-\Delta G = w_{elec} = nFE_{cell}$$

Thus, for an oxidation–reduction reaction we can write

$$\Delta G = -nFE_{cell} \qquad\qquad (18.4)$$

Notice that in this equation we have not written the superscript "°". The notation E_{cell} with no superscript signifies that the conditions at the electrodes are *not* standard conditions—solute concentrations may *not* be 1 M, and gas pressures may *not* be 1 atm. Similarly, the notation ΔG instead of $\Delta G°$ also implies nonstandard conditions. Equation (18.4) is perfectly general, however, and therefore we can apply it to a cell in which all substances *are* in the standard state. In that case, we should use the superscript "°" and write

$$\Delta G° = -nFE°_{cell} \qquad\qquad (18.5)$$

Criteria for Spontaneous Change in Redox Reactions

We know from Chapter 17 that ΔG must be negative in order for a reaction to proceed spontaneously. Equation (18.4) tells us that if ΔG is negative, then E_{cell} must be positive. Putting these two facts together leads to some important new ideas concerning spontaneous change:

- If E_{cell} is positive, then ΔG is negative and the reaction in the forward direction (from left to right) is spontaneous.

- If E_{cell} is negative, then ΔG is positive and the reaction in the forward direction is nonspontaneous.

- If $E_{cell} = 0$, the system is at equilibrium.

- When a cell reaction is reversed, E_{cell} and ΔG change signs.

When we obtain an E_{cell} value by combining standard electrode potentials from Table 18.1, we get a *standard* cell potential, $E°_{cell}$. Any predictions we make will therefore be for reactions having reactants and products in their standard states. Usually, however, *qualitative* predictions based on standard-state conditions apply over a wide range of nonstandard conditions as well.

In Example 18.6, we apply our new E_{cell} criteria for spontaneous change.

Example 18.6

Will copper metal displace silver ion from aqueous solution? That is, does the reaction

$$Cu(s) + 2\,Ag^+(1\ M) \longrightarrow Cu^{2+}(1\ M) + 2\,Ag(s)$$

occur spontaneously from left to right?

STRATEGY

We must separate the overall equation into its two half-equations, assign the appropriate $E°$ values, and then recombine the half-equations to obtain $E°_{cell}$. If the resulting $E°_{cell}$ is positive, then the reaction is spontaneous in the forward direction.

SOLUTION

Reduction:	$2\,\{Ag^+(1\ M) + e^- \longrightarrow Ag(s)\}$
Oxidation:	$Cu(s) \longrightarrow Cu^{2+}(1\ M) + 2\,e^-$

Overall: $Cu(s) + 2\,Ag^+(1\ M) \longrightarrow Cu^{2+}(1\ M) + 2\,Ag(s)$

$$\begin{aligned} E°_{cell} &= E°(\text{cathode}) - E°(\text{anode}) \\ &= E°(\text{reduction}) - E°(\text{oxidation}) \\ &= E°_{Ag^+/Ag} - E°_{Cu^{2+}/Cu} \\ &= 0.800\ V - 0.340\ V = 0.460\ V \end{aligned}$$

Because $E°_{cell}$ is positive, the forward direction should be the direction of spontaneous change. Copper metal should displace silver ions from solution.

EXERCISE 18.6A

Should the reaction

$$Cu^{2+}(aq) + 2\,Fe^{2+}(aq) \longrightarrow 2\,Fe^{3+}(aq) + Cu(s)$$

occur spontaneously as written for standard-state conditions?

EXERCISE 18.6B

In the reaction

$$2\,Mn^{2+}(aq) + 2\,IO_3^-(aq) + 2\,H_2O(l) \rightleftharpoons 2\,MnO_4^-(aq) + I_2(s) + 4\,H^+(aq)$$

all the reactants are in their standard states. Which is the direction of spontaneous change? Explain.

Redox Chemistry of Iron and Copper movie

Note that the spontaneous reaction predicted in Example 18.6 is the one illustrated in Figure 18.1. There, we saw that the reaction occurs when a coil of copper wire is immersed in an aqueous solution of silver nitrate.

It is important to understand that although we use cell terminology to make predictions about the direction of spontaneous change in redox reactions, these predictions apply *whether the reactions are carried out in voltaic cells or simply by mixing the reactants*.

The Activity Series of the Metals Revisited

We can now give a theoretical explanation of the activity series of the metals introduced in Section 4.4. According to the relationship between E_{cell} and the direction of spontaneous change, a metal A in contact with a solution of ions of metal B will displace those ions from solution if metal B lies above metal A in a listing of electrode potentials arranged by decreasing value, such as in Table 18.1. And treating the listing for hydrogen as just another entry in the table, we can say that a metal appearing below the standard hydrogen electrode (for example, Fe, Zn, and Al) will react with a mineral acid (a solution having H^+ as the only oxidizing agent) to produce $H_2(g)$. A metal appearing above the standard hydrogen electrode (for example, Cu and Ag) will not displace $H_2(g)$ from a solution of $H^+(aq)$. This assessment yields the same conclusions we previously stated for the activity series of metals (Figure 4.13).

Example 18.7   A Conceptual Example

The photograph in Figure 18.11 shows strips of copper and zinc joined together and then dipped in HCl(aq). Explain what happens. That is, what are the gas bubbles on the zinc and on the copper, and how did they get there?

ANALYSIS AND CONCLUSIONS

A metal should react with HCl(aq), displacing from solution the $H_2(g)$ formed in the reduction of H^+, if the metal lies above $H_2(g)$ in the activity series of the metals (Figure 4.13) or below $H_2(g)$ in the listing of standard electrode potentials in Table 18.1. Zinc fits this requirement, as we can establish by using $E°$ data:

▲ **FIGURE 18.11**   **Copper–zinc assembly in HCl(aq)**

Reduction: $2 H^+(1 M) + 2 e^- \longrightarrow H_2(g, 1\ atm)$

Oxidation: $Zn(s) \longrightarrow Zn^{2+}(1 M) + 2 e^-$

Overall: $Zn(s) + 2 H^+(1 M) \longrightarrow Zn^{2+}(1 M) + H_2(g, 1\ atm)$

$$E°_{cell} = E°(\text{reduction}) - E°(\text{oxidation})$$
$$= E°_{H^+/H_2} - E°_{Zn^{2+}/Zn}$$
$$= 0.000\ V - (-0.763\ V) = 0.763\ V$$

Thus, the bubbles on the zinc are bubbles of H_2 gas formed in the reaction of Zn with HCl(aq). But how does a gas form on the copper, when Cu(s) lies below $H_2(g)$ in the activity series of the metals and above $H_2(g)$ in the listing of standard electrode potentials in Table 18.1? Because of these relative positions, Cu should not displace $H_2(g)$ from HCl(aq).

Reduction: $2 H^+(1 M) + 2 e^- \longrightarrow H_2(g, 1\ atm)$

Oxidation: $Cu(s) \longrightarrow Cu^{2+}(1 M) + 2 e^-$

Overall: $Cu(s) + 2 H^+(1 M) \longrightarrow Cu^{2+}(1 M) + H_2(g, 1\ atm)$

$$E°_{cell} = E°(\text{reduction}) - E°(\text{oxidation})$$
$$= E°_{H^+/H_2} - E°_{Cu^{2+}/Cu}$$
$$= 0.000\ V - (0.340\ V) = -0.340\ V$$

The negative value for $E°_{cell}$ tells us that the reaction of Cu(s) with HCl(aq) is nonspontaneous.

Here is a plausible explanation for the appearance of gas bubbles on the copper. The gas *is* H_2 created in the reduction of H^+, but it is *not* formed by a reaction involving Cu(s). The electrons required for the reduction of H^+ to H_2 on the copper come from the oxidation of the zinc. Copper, an excellent electrical conductor, merely acts as an inert extension of the zinc electrode. Thus, although H_2 will form on the zinc even if the copper is not present, H_2 will *not* form on the copper *unless* the copper is connected to a metal more active than hydrogen, such as zinc.

The device in Figure 18.11 is actually a voltaic cell. The portion of the reduction half-reaction that occurs on the copper is physically separated from the oxidation half-reaction that occurs only on the zinc. Electrons flow from the zinc to the copper.

EXERCISE 18.7A

Offer a plausible explanation of what occurs in the lemon battery in Figure 18.2.

EXERCISE 18.7B

How would you expect the observations in Figure 18.11 to change if **(a)** the HCl(aq) were replaced by $CH_3COOH(aq)$; **(b)** the HCl(aq) were replaced by $HNO_3(aq)$; **(c)** the copper strip were replaced by a pool of Hg(l) at the bottom of the beaker of HCl(aq), with the zinc strip touching the Hg(l)?

Equilibrium Constants for Redox Reactions

Equation (17.12) gives us an important relationship between $\Delta G°$ and K_{eq} for any reaction, $\Delta G° = -RT \ln K_{eq}$, and because we know from Equation (18.5) that $\Delta G°$ is also equal to $-nFE°_{cell}$, we have a relationship between K_{eq} and $E°_{cell}$:

$$\Delta G° = -RT \ln K_{eq} = -nFE°_{cell}$$

This allows us to write

$$E°_{cell} = \frac{RT \ln K_{eq}}{nF} \tag{18.6}$$

In the above equation, $E°_{cell}$ is the standard cell potential, R is the ideal gas constant $(8.3145 \text{ J mol}^{-1} \text{ K}^{-1})$, T is the Kelvin temperature, n is the number of moles of electrons involved in the reaction, and F is the Faraday constant. Generally, $E°_{cell}$ values are obtained from tables of $E°$ at 25 °C, and the equation is almost always used at 25 °C. As a result, the quantity RT/F has the same value from one calculation to another and is therefore easily replaced by its numerical equivalent, 0.025693 V. (Note that the symbol V [volt] replaces the ratio J/C [joule/coulomb].)

The equivalent expression using common logarithms is
$E°_{cell} = (0.059160 \ V/n) \log K_{eq}$.
Note that the constant is the same as the one in the Nernst equation (page 770).

$$E°_{cell} = \frac{RT}{nF} \ln K_{eq} = \frac{8.3145 \text{ J mol}^{-1} \text{ K}^{-1} \times 298.15 \text{ K}}{n \times 96485 \text{ C mol}^{-1}} \ln K_{eq}$$

$$= \frac{0.025693 \text{ V}}{n} \ln K_{eq} \tag{18.7}$$

We can establish n from the half-cell reactions: It is the number of electrons in either the reduction or oxidation half-equation *after* the coefficients have been adjusted to balance the equation. That is, n is the number of electrons that cancel out on each side when the half-equations are combined to give the balanced redox equation.

Example 18.8

Calculate the values of $\Delta G°$ and K_{eq} at 25 °C for the reaction

$$Cu(s) + 2 Ag^+(1 M) \rightleftharpoons Cu^{2+}(1 M) + 2 Ag(s)$$

STRATEGY

Looking at the half-equations and overall equation we wrote for this reaction in Example 18.6, we see that we had two electrons in each half-equation before we combined them; therefore, $n = 2$. We found $E°_{cell}$ for the reaction to be +0.460 V. We can get $\Delta G°$ by substituting $n = 2$ and known values of F and $E°_{cell}$ into Equation (18.5). To get K_{eq}, we can use Equation (18.7).

SOLUTION

To obtain $\Delta G°$, we substitute the appropriate data into Equation (18.5):

$$\Delta G° = -nFE°_{cell}$$

$$\Delta G° = -2 \text{ mol e}^- \times 96{,}485 \text{ C/mol e}^- \times 0.460 \text{ V} = -8.88 \times 10^4 \text{ J (or } -88.8 \text{ kJ)}$$

To obtain K_{eq}, we substitute $E°_{cell}$ into Equation (18.7) and solve first for $\ln K_{eq}$ and then for K_{eq}:

$$E°_{cell} = \frac{0.025693 \text{ V}}{n} \ln K_{eq}$$

$$0.460 \text{ V} = \frac{0.025693 \text{ V}}{2} \ln K_{eq}$$

$$\ln K_{eq} = \frac{2 \times 0.460 \text{ V}}{0.025693 \text{ V}} = 35.8$$

$$K_{eq} = e^{35.8} = 4 \times 10^{15}$$

ASSESSMENT

In a problem of this sort, we must ensure that the values of $\Delta G°$ and K_{eq} are compatible. If we had mistakenly omitted the negative sign in Equation (18.5), we would have obtained a large positive value for $\Delta G°$, implying almost no tendency for the forward reaction to occur. The large positive power of 10 for the value of K_{eq}, on the other hand, suggests a reaction that goes essentially to completion. If such were the case, clearly one of the results would be wrong.

EXERCISE 18.8A

Calculate the values of $\Delta G°$ and K_{eq} at 25 °C for the reaction

$$3 \text{ Mg(s)} + 2 \text{ Al}^{3+}(1 \text{ M}) \rightleftharpoons 3 \text{ Mg}^{2+}(1 \text{ M}) + 2 \text{ Al(s)}$$

EXERCISE 18.8B

Determine K_{eq} for the reaction of silver metal with nitric acid. The silver is oxidized to $\text{Ag}^+(\text{aq})$, and nitrate ion is reduced to NO(g).

Thermodynamics, Equilibrium, and Electrochemistry: A Summary

Let us pause for a moment and use Figure 18.12 as an aid to reviewing some important relationships. At the center of the figure, we have grouped the three related properties $\Delta G°$, K_{eq}, and $E°_{cell}$. We have joined them by double-headed arrows to show that the quantitative relationships can be applied in either direction. Noted around the edges of the figure are the principal experimental methods used to establish values of the three properties.

Calorimetric measurements can be quite precise, and they are important in establishing $\Delta G°$ values. Electrical measurements are among the most precise that can be made in the chemical laboratory, and they are an important source of $\Delta G°$ and K_{eq} values as well as $E°_{cell}$ values. In contrast, equilibrium composition measurements are used to establish some K_{eq} values but are less often used to generate $\Delta G°$ and $E°_{cell}$ values. Table 18.2 gives additional comparative information for $\Delta G°$, K_{eq}, and $E°_{cell}$.

Thermodynamics, Equilibrium and Electrochemistry activity

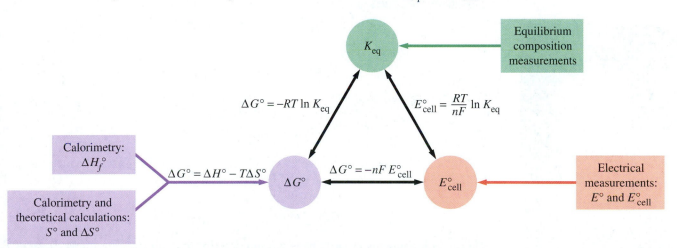

▲ **FIGURE 18.12**   **Summarizing important relationships from thermodynamics, equilibrium, and electrochemistry**

Table 18.2 Comparing $\Delta G°$, K_{eq}, and $E°_{cell}$

	$\Delta G°$	K_{eq}	$E°_{cell}$
Equilibrium under standard-state conditions	0	1	0
Spontaneous reaction under standard-state conditions	<0	>1	>0
Approximate range of values encountered	±several hundred kJ	$10^{\pm 100}$	±a few volts

18.6 Effect of Concentrations on Cell Voltage

We can do quite a bit with standard cell voltages, E°_{cell}, but most cell measurements—and many calculations—involve nonstandard conditions. First, let us look qualitatively at the relationship between E°_{cell} and E_{cell}. To do this, we turn again to the voltaic cell of Figure 18.4.

According to Le Châtelier's principle, the forward reaction will be favored by an increased $[Cu^{2+}]$ and/or decreased $[Zn^{2+}]$. When the forward reaction is favored, $-\Delta G$ and E_{cell} increase. The reverse reaction will be favored by a decreased $[Cu^{2+}]$ and/or by an increased $[Zn^{2+}]$. When the reverse reaction is favored, $-\Delta G$ and E_{cell} decrease. Cell reactions and cell voltages for standard conditions and two nonstandard conditions are summarized as follows.

Increase in $[Cu^{2+}]$ or decrease in $[Zn^{2+}]$ favors forward reaction
→

$$Zn(s) + Cu^{2+}(1.5\ M) \rightleftharpoons Zn^{2+}(0.075\ M) + Cu(s) \qquad E_{cell} = 1.142\ V$$
$$Zn(s) + Cu^{2+}(1.0\ M) \rightleftharpoons Zn^{2+}(1.0\ M) + Cu(s) \qquad E^\circ_{cell} = 1.103\ V$$
$$Zn(s) + Cu^{2+}(0.075\ M) \rightleftharpoons Zn^{2+}(1.5\ M) + Cu(s) \qquad E_{cell} = 1.064\ V$$

←
Decrease in $[Cu^{2+}]$ or increase in $[Zn^{2+}]$ favors reverse reaction

Now, let us look at the matter quantitatively by combining several familiar expressions. We start with Equation (17.11), which relates ΔG, ΔG°, and the reaction quotient Q:

$$\Delta G = \Delta G^\circ + RT \ln Q$$

Into this equation, we can substitute the expressions

$$\Delta G = -nFE_{cell} \quad \text{and} \quad \Delta G^\circ = -nFE^\circ_{cell}$$

This gives the equation

$$-nFE_{cell} = -nFE^\circ_{cell} + RT \ln Q$$

Next, we change signs,

$$nFE_{cell} = nFE^\circ_{cell} - RT \ln Q$$

and then solve for E_{cell}:

$$E_{cell} = E^\circ_{cell} - \frac{RT}{nF} \ln Q$$

If we limit our calculations to 25.00 °C (298.15 K), we can replace the factor RT/F by 0.025693 V, as we did when relating E°_{cell} and K_{eq}:

$$E_{cell} = E^\circ_{cell} - \frac{0.025693\ V}{n} \ln Q$$

This equation has customarily been written in terms of common logarithms. To switch from natural to common logarithms, we have to multiply 0.025693 V by ln 10, which is equal to 2.3026. Doing so gives us 0.025693 V × 2.3026 = 0.059161 V, usually rounded off to 0.0592 V. As a result, the form of the equation we will use is

$$E_{cell} = E^\circ_{cell} - \frac{0.0592\ V}{n} \log Q \qquad (18.8)$$

This equation, called the **Nernst equation,** relates a cell voltage for *nonstandard* conditions, E_{cell}, to a standard cell voltage, E°_{cell}, and to the concentrations of reactants and products expressed through the reaction quotient, Q. In the equation, n is the number of moles of electrons involved in the cell reaction.

The Nernst equation is especially useful for determining the concentration of a species in a voltaic cell through a measurement of E_{cell}. Also, the equation helps us to

▲ Walther Nernst (1864–1941) is well known for several important achievements in chemistry. He formulated the Nernst equation in 1889, when he was only 25 years old. In the same year, he suggested the solubility product concept. In 1906, he proposed a hypothesis that we now know as the third law of thermodynamics.

understand why E_{cell} does not remain constant as electric current is drawn from a cell. As the cell reaction proceeds, the concentrations of products increase and the concentrations of reactants decrease; consequently, Q increases and E_{cell} falls, reaching zero at equilibrium.

Example 18.9

Calculate the expected voltmeter reading for the voltaic cell pictured in Figure 18.13.

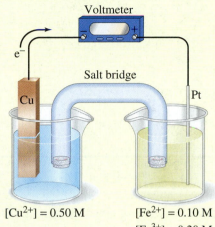

$[Cu^{2+}] = 0.50$ M $[Fe^{2+}] = 0.10$ M   ◀ **FIGURE 18.13**
$[Fe^{3+}] = 0.20$ M   **Example 18.9 illustrated**

STRATEGY

First, we will use data from Table 18.1 to determine E_{cell}° for the cell reaction. Then we will formulate the reaction quotient Q and determine E_{cell}, which is the reading on the voltmeter.

SOLUTION

The direction of electron flow shown in Figure 18.13 enables us to identify the anode and cathode for the *standard* half-reactions:

Cathode (reduction): $2 \{Fe^{3+}(1 \text{ M}) + e^- \longrightarrow Fe^{2+}(1 \text{ M})\}$

Anode (oxidation): $Cu(s) \longrightarrow Cu^{2+}(1 \text{ M}) + 2 e^-$

Overall: $Cu(s) + 2 Fe^{3+}(1 \text{ M}) \longrightarrow Cu^{2+}(1 \text{ M}) + 2 Fe^{2+}(1 \text{ M})$

$$E_{cell}^{\circ} = E^{\circ}(\text{cathode}) - E^{\circ}(\text{anode})$$
$$= E_{Fe^{3+}/Fe^{2+}}^{\circ} - E_{Cu^{2+}/Cu}^{\circ}$$
$$= 0.771 \text{ V} - 0.340 \text{ V} = 0.431 \text{ V}$$

Thus, we seek E_{cell} for a voltaic cell in which $E_{cell}^{\circ} = +0.431$ V, $n = 2$, and the cell reaction is

$$Cu(s) + 2 Fe^{3+}(0.20 \text{ M}) \longrightarrow Cu^{2+}(0.50 \text{ M}) + 2 Fe^{2+}(0.10 \text{ M})$$

We begin with the Nernst equation (18.8):

$$E_{cell} = E_{cell}^{\circ} - \frac{0.0592 \text{ V}}{n} \log Q$$

$$= 0.431 \text{ V} - \frac{0.0592 \text{ V}}{2} \log\left(\frac{[Cu^{2+}][Fe^{2+}]^2}{[Fe^{3+}]^2}\right)$$

$$= 0.431 \text{ V} - \frac{0.0592 \text{ V}}{2} \log\left(\frac{0.50(0.10)^2}{(0.20)^2}\right)$$

$$= 0.431 \text{ V} - (0.0296 \text{ V} \times \log 0.125) = 0.431 \text{ V} - (-0.027) \text{ V} = +0.458 \text{ V}$$

Concentration Cells

The voltaic cell in Figure 18.14 differs from others that we have considered in that the two electrodes are identical to each other. However, the solutions in the half-cells have different concentrations, and this concentration difference creates a potential difference. If the potential of a voltaic cell is determined solely by a difference in the concentration of solutes in equilibrium with identical electrodes, that cell is called a **concentration cell.** Let us determine the source of this voltage and the nature of the cell reaction.

Under standard-state conditions, the concentration of solute would be the same in both electrode compartments. There would be no current flow, meaning that the cell potential $E^{\circ}_{cell} = 0.0$ V. With $E^{\circ}_{cell} = 0$, the Nernst equation (18.8) for a concentration cell is reduced to the form

$$E_{cell} = -\frac{0.0592 \text{ V}}{n} \log Q \qquad (18.9)$$

Let us use Equation (18.9) to calculate the cell voltage of the following concentration cell:

$$Cu^{2+}(1.50 \text{ M}) \longrightarrow Cu^{2+}(0.025 \text{ M}) \qquad E_{cell} = ?$$

▶ **FIGURE 18.14   A concentration cell**

The electrodes are identical in the two half-cells, but the solution concentrations differ. The driving force for the cell reaction is the tendency for the solution concentrations to become equalized: $Cu^{2+}(1.50 \text{ M}) \longrightarrow Cu^{2+}(0.025 \text{ M})$.

QUESTION: Would electrons flow in the direction shown in the drawing if the Cu^{2+} concentrations in the two half-cells were switched?

Voltmeter

0.025 M $CuSO_4$(aq)   1.50 M $CuSO_4$(aq)

The Electrochemistry of a Heartbeat

The heart is a pump made mostly of muscle, and like all other muscles, it contracts in response to an electric signal. In muscle, electric current consists of the movement of ions across cell membranes. Sodium ion concentrations are high outside the cell and low inside; potassium ion concentrations are low outside the cell and high inside. These concentration gradients lead to a voltage difference called the *membrane potential.*

Unlike other muscles, the heart has its own stimulator and built-in circuitry—it keeps on beating even if outside nerve connections are cut. Cut the nerve connections to the pectoral muscles, in contrast, and your arms are paralyzed for good.

The pacemaker cells controlling heart rate have a natural voltage that depends mainly on the concentrations of sodium and potassium ions. All cell membranes have a mechanism that moves these ions across the membranes, keeping K^+ inside the cell and expelling Na^+. Some K^+ leaks out, however, and this gives the interior of the cell a negative potential with respect to the outside. When the potential difference reaches a critical value, channels open up in the pacemaker cell membranes and Na^+ ions rush in. This results in an electric discharge, a current that spreads almost instantaneously to other heart muscle cells. The cells act in concert, and the heart beats.

Calcium ions also play a vital role in the heartbeat. Following the electric discharge just described, Ca^{2+} ions flow into the cells for a fraction of a second. When the flow of Ca^{2+} ions into the cells ceases, K^+ ions begin to leak out again, reestablishing a potential difference across the membrane.

The proper concentration of K^+ is crucial. Normal blood concentrations are 3.5 to 5.0 mmol K^+/L blood. The normal ratio of $[K^+]$ between the inside and outside of the pacemaker cells is about 30 : 1. If $[K^+]$ in the blood is too low, the ionic imbalance goes beyond the critical stage, and the pacemaker cells are overcharged; they remain cocked like broken pistols but do not fire. In other words, the heart does not beat. If $[K^+]$ in the blood is too high, the potential difference never gets high enough to fire off a proper signal, and consequently the heartbeat diminishes to nothingness. Wasting of tissue from disease or starvation, loss of fluids through prolonged diarrhea or excessive vomiting, and long-term use of diuretic drugs all lead to dangerously low K^+ levels. Excessive dietary supplements and kidney diseases in which the body does not excrete enough potassium lead to K^+ levels that are too high. Both extremes—$[K^+]$ too low or too high—lead to cardiac arrest.

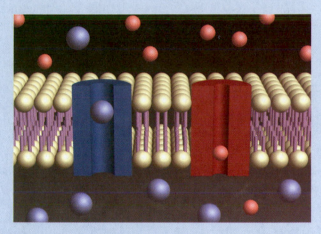

▲ This computer-generated model of a portion of a cell membrane shows a K^+ channel (blue) and an Na^+ channel (red). The area outside the cell (top) is rich in Na^+ and low in K^+. Inside the cell, the fluids are relatively rich in K^+ and low in Na^+.

In this case, $n = 2$ and $Q = 0.025/1.50$. Therefore

$$E_{cell} = -\frac{0.0592 \text{ V}}{2} \log \frac{0.025}{1.50} = -0.0296 \text{ V} \times \log 0.017$$

$$= -0.0296 \text{ V} \times (-1.77) = 0.0524 \text{ V}$$

The cell reaction in a concentration cell indicates that the only change is that the more concentrated solution is diluted and the more dilute one becomes more concentrated, just as would happen if we simply brought two solutions of different concentrations into contact with each other. Entropy increases, and the process is spontaneous.

A concentration cell is unusual in that it uses the natural tendency for solutions to mix as a basis of generating electricity.

pH Measurement

Suppose we construct a concentration cell consisting of two hydrogen electrodes. One is a standard hydrogen electrode, and the other is in a solution having an unknown pH; that is, with $[H^+] = x$. If $x < 1.0$ M—and this is generally the case—oxidation occurs at the nonstandard hydrogen electrode (the anode) and reduction occurs at the standard hydrogen electrode (the cathode):

Reduction:	$2\,H^+(1\,M) + 2\,e^-$	$\longrightarrow$ ~~$H_2(g, 1\,atm)$~~
Oxidation:	~~$H_2(g, 1\,atm)$~~	$\longrightarrow 2\,H^+(x\,M) + 2\,e^-$
Overall:	$2\,H^+(1\,M)$	$\longrightarrow 2\,H^+(x\,M)$

The voltage of this concentration cell is given by the modified Nernst equation (18.9):

$$E_{cell} = -\frac{0.0592\ \text{V}}{2}\log\frac{x^2}{1^2}$$

$$= -0.0296\ \text{V}\log x^2$$

$$= -0.0296\ \text{V}\,(2\log x) = -(0.0592\log x)\ \text{V}$$

Because we often represent the acidity of a solution by the expression $pH = -\log[H^+] = -\log x$, we can write for the concentration cell

$$E_{cell} = (0.0592\ pH)\ \text{V}$$

Thus, if we measure an E_{cell} value of, say, 0.225 V in such a concentration cell at 25 °C, the unknown solution has a pH of 0.225 V/0.0592 V = 3.80.

The pH Meter

The preceding discussion shows that it is possible to use a concentration cell consisting of a standard hydrogen electrode and a nonstandard hydrogen electrode to measure pH values. However, a standard hydrogen electrode is difficult to construct and maintain, and so in practice we generally employ some other reference electrode having a precisely known $E°$. We also replace the nonstandard hydrogen electrode by a special electrode known as a *glass electrode*. This electrode consists of a silver wire coated with AgCl and immersed in HCl(aq), with the wire and the solution enclosed in a thin glass membrane.

When the electrode is dipped into a solution of unknown pH, H^+ ions are exchanged at the glass surface. The potential developed on the silver wire depends on $[H^+]$ in the unknown solution. The potential difference between the glass electrode and the reference electrode similarly depends on $[H^+]$ in the unknown solution. Figure 18.15 shows a typical pH meter.

▲ FIGURE 18.15   The glass electrode and the pH meter

A reference electrode and a glass electrode are both enclosed in the combination electrode shown. The potential difference between the two is converted to the pH value of the solution (3.12) displayed by the meter.

🌐 **Cross Section of a Dry Cell activity**

18.7   Batteries: Using Chemical Reactions to Make Electricity

In everyday life, we call any device that stores chemical energy for later release of electricity a *battery*. In more accurate terms, however, a *battery* is defined as an assembly of two or more voltaic cells connected together. Therefore a flashlight "battery," which is a single voltaic cell with two electrodes in contact with one or more electrolytes, is not technically a battery and is more accurately referred to as a *cell*. An automobile battery, on the other hand, is a true battery because it is a collection of (usually) six voltaic cells. In this section, we consider three different types of cells or batteries.

The Dry Cell

In most flashlights and portable electronic devices, we use *primary cells*. The reactions in primary cells are irreversible, which means the cells cannot be recharged. During use, reactants are converted to products, and when the reactants are used up, the cell is "dead." A typical example, the *Leclanché cell*—better known as a dry cell—is diagrammed in Figure 18.16.

A zinc case is the anode, and an inert carbon (graphite) rod in contact with $MnO_2(s)$ is the cathode. The electrolyte is a moist paste of NH_4Cl and $ZnCl_2$. The cell is called a dry cell because there is no free-flowing liquid. The concentrated electrolyte solution is thickened into a paste by an agent such as starch. The cell diagram for a Leclanché cell is

$$Zn(s)|ZnCl_2(aq), NH_4Cl(aq)|MnO(OH)(s)|MnO_2(s)|C(graphite)$$

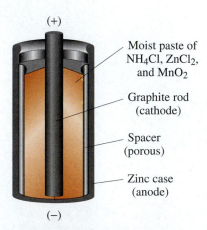

(+)

Moist paste of NH_4Cl, $ZnCl_2$, and MnO_2

Graphite rod (cathode)

Spacer (porous)

Zinc case (anode)

(−)

▲ FIGURE 18.16   Cross section of a Leclanché (dry) cell

The moist paste consists of $NH_4Cl(aq)$ and $ZnCl_2(aq)$; carbon black (a finely divided form of carbon) and $MnO_2(s)$ are also present in the paste.

The reactions that occur when electric current is drawn from a dry cell are quite complex and not completely understood. We will give only a simplified description. As suggested by the cell diagram, zinc metal is oxidized to Zn^{2+} at the anode:

$$Anode: \quad Zn(s) \longrightarrow Zn^{2+}(aq) + 2\,e^-$$

The reduction reaction at the cathode appears to be

$$Cathode: \quad MnO_2(s) + H^+(aq) + e^- \longrightarrow MnO(OH)(s)$$

followed by

$$2\,MnO(OH)(s) \longrightarrow Mn_2O_3(s) + H_2O(l)$$

The Leclanché cell is cheap to make and has a maximum cell voltage of 1.55 V, but it has two disadvantages: (1) The cell voltage drops rapidly during use because of the buildup of Zn^{2+} near the anode and the change of pH at the cathode, and (2) the zinc metal slowly reacts with the electrolyte, even when the cell is disconnected.

Alkaline cells cost more than ordinary dry cells but last longer. They have a longer shelf life and can be kept in service longer. In an alkaline cell, NaOH or KOH replaces NH_4Cl as the electrolyte. The reduction half-reaction is the same as in the ordinary dry cell, but in an alkaline medium. Zinc ion produced in the oxidation half-reaction combines with $OH^-(aq)$ to form $Zn(OH)_2(s)$:

Cathode:	$2\,MnO_2(s) + H_2O(l) + 2\,e^- \longrightarrow Mn_2O_3(s) + 2\,OH^-(aq)$
Anode:	$Zn(s) + 2\,OH^-(aq) \longrightarrow Zn(OH)_2(s) + 2\,e^-$

Cell reaction:	$Zn(s) + 2\,MnO_2(s) + H_2O(l) \longrightarrow Zn(OH)_2(s) + Mn_2O_3(s)$

$$\begin{aligned}
E^\circ_{cell} &= E^\circ(\text{cathode}) - E^\circ(\text{anode}) \\
&= E^\circ_{MnO_2/Mn_2O_3} - E^\circ_{Zn(OH)_2/Zn} \\
&= 0.118\ V - (-1.246\ V) = 1.364\ V
\end{aligned}$$

In practice, because the conditions in the cell are nonstandard, the cell potential is not equal to 1.364 V—it is actually somewhat higher, about 1.5 V.

Another important advantage of the alkaline cell over the ordinary dry cell is that the voltage in the alkaline cell does not drop as rapidly when current is drawn. This is because the concentration of zinc ion does not build up at the anode and also because the pH remains more nearly constant at the cathode.

The Lead–Acid Storage Battery

The lead–acid storage battery used in automobiles employs *secondary* cells, which are rechargeable. The cell reaction can be reversed and the battery restored to its original condition. The battery can be used through repeated cycles of discharging and recharging. Figure 18.17 shows a portion of one cell in the lead–acid battery. Several anodes

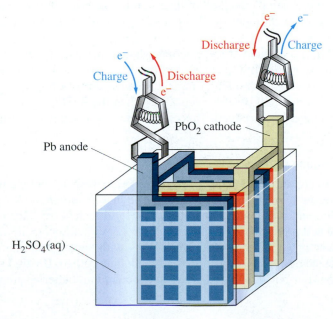

Discharge
Charge
e⁻   e⁻
Charge
Discharge
e⁻
PbO₂ cathode
Pb anode
H₂SO₄(aq)

◀ **FIGURE 18.17   A lead–acid (storage) cell**

The composition of the electrodes, the cell reaction, and the cell voltage are described in the text. Shown here are two anode plates (blue) and two cathode plates (red) connected in parallel. The parallel connection increases the surface area of the electrodes and the capacity of the cell to deliver current.

and several cathodes are connected together in each cell to increase its current-delivering capacity. Each cell has a voltage slightly over 2 V, and six cells are connected together in series, the positive terminal of one cell connected to the negative terminal of a neighboring cell, to form a 12-V battery.

The anodes in the lead–acid storage cell are made of a lead alloy, and the cathodes are made of a lead alloy impregnated with red lead dioxide. The electrolyte is dilute sulfuric acid. The half-reactions and the cell reaction are

Cathode: $PbO_2(s) + 3 H^+(aq) + HSO_4^-(aq) + 2 e^- \longrightarrow PbSO_4(s) + 2 H_2O(l)$

Anode: $Pb(s) + HSO_4^-(aq) \longrightarrow PbSO_4(s) + H^+(aq) + 2 e^-$

Cell reaction: $Pb(s) + PbO_2(s) + 2 H^+(aq) + 2 HSO_4^-(aq) \longrightarrow 2 PbSO_4(s) + 2 H_2O(l)$

$$E^\circ_{cell} = E^\circ(\text{cathode}) - E^\circ(\text{anode})$$
$$= E^\circ_{PbO_2/PbSO_4} - E^\circ_{PbSO_4/Pb}$$
$$= 1.690\text{ V} - (-0.356\text{ V}) = 2.046\text{ V}$$

Even though the conditions in a lead–acid cell are nonstandard, the voltage delivered by the cell is just about the same as the E°_{cell} value.

As the cell reaction proceeds, $PbSO_4(s)$ precipitates and partially coats both electrodes, and the water that is formed dilutes the $H_2SO_4(aq)$. In this condition, the cell is *discharged*. By connecting the cell to an external electric energy source, we can force electrons to flow in the opposite direction. The half-cell reactions and the cell reaction are reversed, and the battery is recharged.

Recharging reaction:

$$2 PbSO_4(s) + 2 H_2O(l) \longrightarrow Pb(s) + PbO_2(s) + 2 H^+(aq) + 2 HSO_4^-(aq)$$
$$E^\circ_{cell} = -2.046\text{ V}$$

The negative value of E°_{cell} signifies that the recharging reaction is nonspontaneous. The external source used to recharge the battery must provide a voltage in excess of 2.046 V for each cell in the battery, thus in excess of 6 × 2.046 V for a typical 12-V battery.

In an automobile, the battery is discharged when the engine is started. While running, the engine powers the alternator, which produces the electric energy required to recharge the battery. Thus, the battery is constantly recharged as the automobile is driven. When a dead battery is unable to start the engine, the engine can often be jump-started by connecting the weak battery to a fully charged battery. The engine gets its start from the good battery, and immediately the weak battery begins recharging.

As a battery is discharged, the $H_2SO_4(aq)$ is diluted, but the acid is reconcentrated during recharging. A simple method of determining the state of charge in a lead–acid battery is to measure the density of the acid. The more concentrated the acid, the greater its density.

In principle, a lead–acid storage battery should last indefinitely, but it does not because fine deposits of $PbSO_4(s)$ form on the electrodes as the cells discharge. If a discharged battery stands for a long time, the small grains of $PbSO_4(s)$ may grow into large crystals that then fall off the electrodes. When the $PbSO_4(s)$ is in this form, the battery can no longer be recharged. This condition of "sulfating" is an important cause of battery failure.

Other Secondary Cells

There are too many different kinds of cells to describe them all, but we will mention a few that are seeing a great deal of use today. The *nickel–cadmium (NiCd) cell* uses a cadmium anode and a cathode containing $Ni(OH)_2$; it can be recharged hundreds of times. A NiCd cell produces 1.2 V rather than the 1.5 V of a Leclanché cell or alkaline cell. This lower potential can create difficulties with some electronic devices. For example, a device designed to operate on four alkaline cells (6.0 volts) may not operate properly on four NiCd cells (4.8 V). Nickel–cadmium batteries are used in many cordless tools.

Air Batteries

An air battery uses $O_2(g)$ from air as an oxidizing agent. The reducing agent is typically a metal, such as zinc or aluminum. In an aluminum–air battery, oxidation occurs at an aluminum anode and reduction at a carbon cathode. The electrolyte circulated through the battery is $NaOH(aq)$. Aluminum is oxidized to Al^{3+}, but because of the high concentration of OH^-, the complex anion $[Al(OH)_4]^-$ is formed (page 700):

Cathode: $\quad\quad\quad\quad 3\,\{O_2(g) + 2\,H_2O(l) + 4\,e^- \longrightarrow 4\,OH^-(aq)\}$

Anode: $\quad\quad\quad\quad 4\,\{Al(s) + 4\,OH^-(aq) \longrightarrow [Al(OH)_4]^-(aq) + 3\,e^-\}$

Overall: $\quad 4\,Al(s) + 3\,O_2(g) + 6\,H_2O(l) + 4\,OH^-(aq) \longrightarrow 4[Al(OH)_4]^-(aq)$

$$E^\circ_{cell} = E^\circ(\text{cathode}) - E^\circ(\text{anode})$$
$$= E^\circ_{O_2/OH^-} - E^\circ_{Al(OH)_4^-/Al}$$
$$= 0.401\ V - (-2.310\ V) = 2.711\ V$$

When the battery is being used, it is fed chunks of aluminum metal and water. The electrolyte is circulated outside the battery, where the $[Al(OH)_4]^-(aq)$ is precipitated as $Al(OH)_3(s)$. Aluminum is later recovered from the $Al(OH)_3(s)$ in an aluminum manufacturing facility. An aluminum–air battery can power an automobile for several hundred miles before requiring refueling and removal of $Al(OH)_3(s)$.

▲ Aluminum–air batteries use oxygen gas from the air as an oxidizing agent and aluminum metal as the reducing agent. The battery consumes aluminum metal and water and produces aluminum hydroxide.

Nickel–metal hydride cells (NiMH) use a metal alloy anode that contains hydrogen in the form of a nonstoichiometric metallic hydride such as $LaNi_5H_x$. During use, the anode releases the hydrogen and water is formed. Because hydrogen has a very low molecular mass, a NiMH cell is lighter than a NiCd cell of the same size. Because it does not contain cadmium, the NiMH cell is less of an environmental problem when it is time for disposal. Like the NiCd cell, a NiMH cell produces 1.2 V.

Lithium-ion cells take advantage of lithium's low density and high negative reduction potential. A typical lithium-ion cell produces 3.0–4.0 V, depending on the construction, and weighs significantly less than an alkaline cell of similar size. Today's lithium-ion cells normally do not contain metallic lithium, but instead use a lithium–cobalt oxide or lithium–manganese oxide as the anode. The electrolyte is an organic solvent containing a dissolved lithium salt. Many modern laptop computers and cellular phones use lithium-ion cells because they are light and produce more energy for their mass than do NiCd cells. Nonrechargeable lithium cells are now available that use iron as an additive. The iron reduces the cell potential to 1.5 V, so that the cell can be used as a direct replacement for alkaline cells or Leclanché cells.

Fuel Cells

Modern civilization runs mainly on fossil fuels, but both fossil-fuel combustion and conversion of the evolved heat to electricity are inefficient processes, the overall efficiency being approximately 30%. A voltaic cell, in contrast, is able to convert chemical energy to electricity with an efficiency of 90% or more. Because combustion reactions and voltaic cell reactions both involve oxidation and reduction, why not design a voltaic cell in which the cell reaction is equivalent to a combustion reaction? This has already been done, and scientists have been studying such devices for more than 50 years. They are called *fuel cells*, and the reaction in them is

$$\text{Fuel} + \text{oxygen} \longrightarrow \text{oxidation products}$$

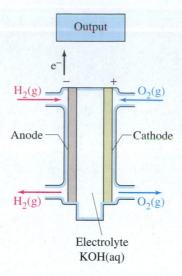

◄ FIGURE 18.18 A hydrogen–oxygen fuel cell
The electrodes are porous to allow the gaseous reactants easy access to the electrolyte. The electrode material also catalyzes the electrode reactions.

QUESTION: Why does a fuel cell not "go dead," as so often happens with a dry cell?

A fuel cell is like a battery with the contents outside the battery case rather than inside. It is a flow battery. In principle, a fuel cell will continue to operate as long as reactants are fed into it.

In a hydrogen–oxygen fuel cell, hydrogen and oxygen gases flow over separate inert electrodes in contact with an electrolyte such as $KOH(aq)$ (Figure 18.18). The cell reactions are

Cathode:	$O_2(g) + 2\,H_2O + 4\,e^- \longrightarrow 4\,OH^-$
Anode:	$2\,\{H_2(g) + 2\,OH^- \longrightarrow 2\,H_2O + 2\,e^-\}$
Overall:	$2\,H_2(g) + O_2(g) \longrightarrow 2\,H_2O(l)$

$$E^\circ_{cell} = E^\circ(\text{cathode}) - E^\circ(\text{anode})$$
$$= E^\circ_{O_2/OH^-} - E^\circ_{H_2O/H_2}$$
$$= 0.401\ \text{V} - (-0.828\ \text{V}) = 1.229\ \text{V}$$

This E°_{cell} value is for 25 °C. Hydrogen–oxygen fuel cells are generally operated at nonstandard conditions and at temperatures often considerably higher than 25 °C, making their operating voltages about 1.0 to 1.1 V. These fuel cells have been widely used in space vehicles, where, in addition to the electricity, the water formed is a valuable by-product.

The $H_2(g)$ used in a hydrogen–oxygen fuel cell comes from a hydrocarbon fuel, usually methane. In principle, however, a fuel cell can run directly on the hydrocarbon fuel, in which case the reactions are

Cathode:	$2\,\{O_2(g) + 4\,H^+ + 4\,e^- \longrightarrow 2\,H_2O\}$
Anode:	$CH_4(g) + 2\,H_2O \longrightarrow CO_2(g) + 8\,H^+ + 8\,e^-$
Overall:	$CH_4(g) + 2\,O_2(g) \longrightarrow CO_2(g) + 2\,H_2O(l)$

An indirect method that accomplishes the same objective converts methane to hydrogen in the reaction:

$$CH_4(g) + H_2O(g) \longrightarrow CO(g) + 3\,H_2(g)$$

▲ The Daimler-Benz NECAR 3 prototype runs on hydrogen–oxygen fuel cells. Hydrogen is produced on board from methanol.

The $H_2(g)$ produced is then used in a hydrogen–oxygen fuel cell.

Hydrogen–oxygen fuel cells show great promise for powering automobiles because they produce little in the way of undesirable products (beyond humid air). Government agencies classify such vehicles as zero-emission.

18.8 Corrosion: Metal Loss Through Voltaic Cells

When carried out in voltaic cells, redox reactions are important sources of electricity. At the same time, some redox reactions produce electricity that underlies the destruction caused by corrosion, a matter of great economic importance. It is estimated that in the United States alone, corrosion costs $70 billion a year. Perhaps 20% of all the iron and steel produced in the United States each year goes to replace corroded material.

Let us look first at the corrosion of iron. In moist air, iron metal can be oxidized to Fe^{2+}, particularly at scratches, nicks, or dents. The energy state of these marred regions of any iron surface, called *anodic* areas, is slightly higher than the energy state of unmarred regions, and oxidation occurs a bit more readily at the high-energy regions than elsewhere on the surface. The anodic areas become pitted. Electrons lost in the oxidation are conducted along the iron to other regions, called *cathodic* areas, where atmospheric $O_2(g)$ is reduced to OH^-:

Cathode: $\quad O_2(g) + 2 H_2O(l) + 4 e^- \longrightarrow 4 OH^-(aq)$

Anode: $\qquad\qquad\qquad 2 \{Fe(s) \longrightarrow Fe^{2+}(aq) + 2 e^-\}$

Overall: $\quad 2 Fe(s) + O_2(g) + 2 H_2O(l) \longrightarrow 2 Fe^{2+}(aq) + 4 OH^-(aq)$

$$E^\circ_{cell} = E^\circ(\text{cathode}) - E^\circ(\text{anode})$$
$$= E^\circ_{O_2/OH^-} - E^\circ_{Fe^{2+}/Fe}$$
$$= 0.401 \text{ V} - (-0.440 \text{ V}) = 0.841 \text{ V}$$

▲ Corrosion causes great economic loss, as shown by this abandoned ship.

The corrosion of iron is a spontaneous redox reaction like that occurring in a voltaic cell. Oxidation and reduction occur at separate points on the metal. Electrons are conducted through the metal, and the circuit is completed by an electrolyte in aqueous solution. In the northern states, this solution is often the slush from road salt and melting snow; in coastal areas, it can be ocean spray.

Iron(II) ions migrate from anodic areas to cathodic areas, where they combine with OH^- ions to form insoluble iron(II) hydroxide:

$$Fe^{2+}(aq) + 2 OH^-(aq) \longrightarrow Fe(OH)_2(s)$$

Usually, iron(II) hydroxide is further oxidized to iron(III) hydroxide by atmospheric oxygen:

$$4 Fe(OH)_2(s) + O_2(g) + 2 H_2O(l) \longrightarrow 4 Fe(OH)_3(s)$$

Iron(III) hydroxide is generally represented as a hydrated iron(III) oxide of variable composition: $Fe_2O_3 \cdot x H_2O$. This material is the familiar *iron rust*. The overall process described here is illustrated in Figure 18.19. In order to emphasize the resemblance to a voltaic cell, we have focused in Figure 18.19 on a situation in which the anodic and cathodic regions are some distance apart and intermediate stages of the corrosion are identifiable. Often, though, all the reactions occur simultaneously on the corroding portion of the iron. Then, all we really notice over time is the accumulation of rust.

Steps to Corrosion activity

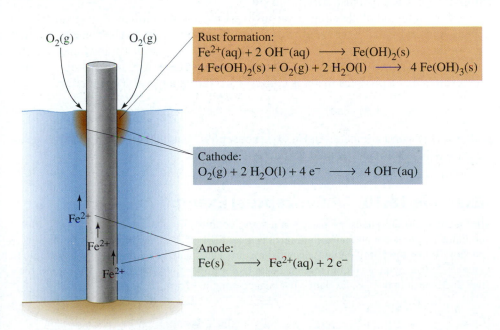

Rust formation:
$Fe^{2+}(aq) + 2 OH^-(aq) \longrightarrow Fe(OH)_2(s)$
$4 Fe(OH)_2(s) + O_2(g) + 2 H_2O(l) \longrightarrow 4 Fe(OH)_3(s)$

Cathode:
$O_2(g) + 2 H_2O(l) + 4 e^- \longrightarrow 4 OH^-(aq)$

Anode:
$Fe(s) \longrightarrow Fe^{2+}(aq) + 2 e^-$

◀ **FIGURE 18.19 Corrosion of an iron piling: An electrochemical process**

This schematic drawing illustrates the anodic and cathodic regions, their half-reactions, and the final formation of rust. The cathodic region is near the air–water interface, where the availability of $O_2(g)$ is greatest. The anodic region is at greater depths below the water's surface. The $Fe^{2+}(aq)$ formed at the anodic region migrates to the cathodic region, where rust formation occurs.

QUESTION: Why would the placement of a tight plastic sleeve at the air–water interface have the effect of reducing rust?

Aluminum's resistance to corrosion can be enhanced by building up the oxide coating through a process called *anodizing*, discussed in Section 21.3.

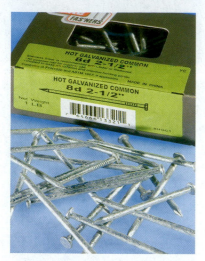

▲ Galvanized nails are iron nails that have been dipped in molten zinc for longer life.

Sacrificial Mg anodes

▲ Sacrificial magnesium anodes attached to the steel structure of a ship provide cathodic protection against corrosion.

Aluminum is more easily oxidized than iron. Knowing this, we might expect aluminum to corrode more rapidly than iron, but the longevity of beer and soda pop cans thrown along roadsides attests to the fact that it does not. How can this be? The answer is that when freshly prepared aluminum is exposed to air, an oxide coating quickly forms on its surface. This thin, hard film of aluminum oxide is impervious to air and protects the underlying metal from further oxidation. In contrast, iron oxide (rust) is porous; it readily flakes off, exposing the fresh iron surface to further corrosion. The difference in behavior of the corrosion products of iron and aluminum explains why steel cans eventually disintegrate in the environment whereas aluminum cans seem to last forever.

The simplest line of defense against iron corrosion is to paint the iron to exclude oxygen from the surface. Another approach is to coat the iron with a thin layer of another, less active metal, as in coating a steel can with tin. Either way, however, the surface is protected only as long as the coating does not crack, chip, or peel. A car body usually rusts first where the paint is scratched or chipped. A tin can rusts rapidly if it is dented and the tin coating cracked. Corrosion is promoted by the less active tin, which now becomes a large cathodic area supporting reduction half-reactions fed by electrons from the underlying iron, the anodic area.

An entirely different approach is to protect iron with a *more* active metal, as in the zinc-clad iron known as *galvanized iron.* The oxide coating on the zinc surface covering the iron is like that on aluminum—thin, hard, and impervious to air. And it does not matter if a break occurs in the zinc plating. The zinc continues to corrode rather than the iron because zinc is more easily oxidized than iron.

Another method that works on the same general principle as galvanized iron is known as *cathodic protection.* The iron or steel object to be protected—a ship, a pipeline, a storage tank, a water heater, or the plumbing system in a swimming pool—is connected to a chunk of an active metal such as magnesium, aluminum, or zinc, either directly or with a wire. Oxidation occurs at the active metal, and that metal slowly dissolves. The iron surface acquires electrons from the active metal and supports the cathodic half-reaction (a reduction). As long as some of the active metal remains, the iron is protected. The active metal is called a *sacrificial anode.* One of the more important uses of magnesium metal is in sacrificial anodes (about 12 million tons annually in the United States).

Another example of the electrochemical control of corrosion is seen in a method for the removal of tarnish from silver. The tarnish is generally black insoluble silver sulfide formed when silver comes in contact with sulfur compounds, either in the atmosphere or in foods such as eggs. We can use a silver polish to remove the tarnish, but in doing so, we also lose part of the silver.

Another method of cleaning silver involves placing the tarnished silver in contact with aluminum foil that is covered with $NaHCO_3(aq)$ and heated. The concentration of Ag^+ in the aqueous solution in contact with $Ag_2S(s)$ is extremely low, and yet aluminum is a sufficiently strong reducing agent to convert this Ag^+ back to silver metal. The result is that a precious metal is conserved at the expense of a cheaper one:

$$3\,Ag^+(aq) + Al(s) \longrightarrow 3\,Ag(s) + Al^{3+}(aq)$$

Sodium hydrogen carbonate ($NaHCO_3$) (household "baking soda") works well as an electrolyte to sustain this desired voltaic cell reaction.

Example 18.10   A Conceptual Example

In Figure 18.20, iron nails are placed in a warm colloidal dispersion of agar in water. Phenolphthalein indicator and potassium ferricyanide, $K_3[Fe(CN)_6]$, are also present in the dispersion. When the faintly yellow dispersion cools and stands for a few hours, it solidifies into a gel, and soon the colored regions begin to develop—pink along the middle part of the nail and blue at both ends. Explain what is happening in Figure 18.20a.

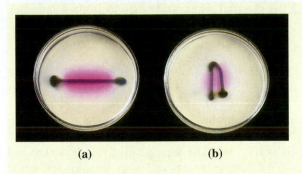

▲ **FIGURE 18.20**
Demonstrating the corrosion of iron

ANALYSIS AND CONCLUSIONS

We learned in Chapter 15 that phenolphthalein, a pH indicator, is colorless in acidic and neutral solutions but pink above pH 10. The source of the OH^- ions responsible for the pink color in Figure 18.19a is the cathodic half-reaction in the corrosion of iron:

$$Cathode: \quad O_2(g) + 2 H_2O(l) + 4 e^- \longrightarrow 4 OH^-(aq)$$

We cannot have a cathodic half-reaction without an anodic half-reaction, and here that half-reaction is

$$Anode: \quad Fe(s) \longrightarrow Fe^{2+}(aq) + 2 e^-$$

The $Fe^{2+}(aq)$ reacts with the $[Fe(CN)_6]^{3-}(aq)$ to form a complex precipitate having a blue color (called Turnbull's blue). The blue color at the head and tip of the nail tells us these are anodic areas; they are preferentially oxidized because the metal is more strained (and hence more energetic) in these regions. The pink along the nail body indicates that this is a *cathodic* region.

EXERCISE 18.10A

Explain what is happening in Figure 18.20b and why this situation differs in one detail from what is seen in Figure 18.20a.

EXERCISE 18.10B

Suppose that in the preparation of the samples shown in Figure 18.20, the phenolphthalein and potassium ferricyanide were omitted. What observations would you expect to make over time?

Electrolytic Cells

In Section 18.3, we noted that a combination of two half-cells with the appropriate connections between electrodes and solutions is called an electrochemical cell. If the cell produces electric current through a spontaneous reaction, it is a voltaic cell. Now let us look at the opposite case—*nonspontaneous* reactions in electrochemical cells.

When electric current from an external source is passed through an electrochemical cell, a nonspontaneous reaction called **electrolysis** occurs within the cell, and the cell is called an **electrolytic cell.** In the remainder of this chapter, we consider some electrolysis reactions and the electrolytic cells in which they are carried out.

If we use a source of direct electric current—a battery or an electric generator—to pass electricity through molten sodium chloride, we observe that yellow-green chlorine gas forms at one electrode and silvery liquid metallic sodium forms at the other electrode and floats on the molten salt (Figure 18.21). The external source of electricity acts like an "electron pump." It pulls electrons away from the electrode where the oxidation of Cl^- occurs:

$$Oxidation: \quad 2 Cl^- \longrightarrow Cl_2(g) + 2 e^-$$

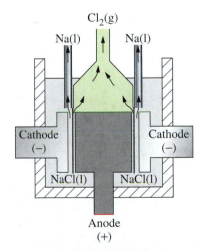

▲ **FIGURE 18.21   The electrolysis of molten sodium chloride**

The electrolysis cell pictured here is called a *Downs cell.* The electrolyte is molten NaCl(l) with a small amount of $CaCl_2$ added to lower the melting point of the NaCl(s). Liquid sodium metal forms at the steel cathode, and gaseous chlorine forms at the graphite anode. A steel gauze diaphragm keeps the sodium and chlorine from recombining to form sodium chloride.

and pushes those electrons through the external circuit onto the electrode where they are used continuously in the reduction of sodium ion:

$$\text{Reduction:} \quad Na^+ + e^- \longrightarrow Na(l)$$

The overall reaction is

$$2\,NaCl(l) \xrightarrow{\text{electrolysis}} 2\,Na(l) + Cl_2(g)$$

Electrolysis of Water animation

In describing an electrolytic cell, we retain these two terms that we introduced in our discussion of voltaic cells: *anode* for the electrode where oxidation occurs and *cathode* for the electrode where reduction occurs. The fact that electrons leave the cell through the anode and enter the cell through the cathode also remains the same between the two types of cells. However, the polarities of the electrodes in an electrolytic cell are reversed from those in the voltaic cell. This happens because in the electrolytic cell, the external source of electricity controls the flow of electrons. Because the external electric source withdraws electrons from the electrode where oxidation occurs, the anode of an electrolytic cell has a deficiency of electrons and acquires a *positive* charge relative to the cathode. Because of the excess of electrons forced onto it by the external electric source, the cathode of the electrolytic cell, where reduction occurs, has a *negative* charge relative to the anode.

18.9   Predicting Electrolysis Reactions

As long as the potential difference applied to the electrodes in Figure 18.21 is sufficiently great (in excess of about 4 V), the electrolysis of NaCl(l) occurs and produces Na(l) and $Cl_2(g)$ as the only products. If we apply the same voltage in an electrolytic cell containing an *aqueous* solution of NaCl, however, we do not get the same products, as we shall see shortly.

We can often predict the products of an electrolysis reaction by considering various oxidation and reduction half-reactions that might occur and then selecting the ones that can be accomplished at the lowest voltage.

The Electrolysis of NaCl(aq)

Imagine an electrolytic cell in which two platinum electrodes are immersed in the same aqueous solution of NaCl and then connected to a battery. The battery will make one of the electrodes an anode and the other a cathode. Two possible half-reactions are essentially the ones observed in the cell reaction occurring in Figure 18.21. However, two other half-reactions are possible in an aqueous solution: the oxidation and reduction of water molecules. Thus we have, in order of decreasing standard reduction potential, $E°$,

Anode (oxidation):

$$2\,Cl^-(aq) \longrightarrow Cl_2(g) + 2\,e^- \qquad E°_{Cl_2/Cl^-} = 1.358\ V$$

$$2\,H_2O(l) \longrightarrow 4\,H^+(aq) + O_2(g) + 4\,e^- \qquad E°_{O_2/H_2O} = 1.229\ V$$

Cathode (reduction):

$$2\,H_2O(l) + 2\,e^- \longrightarrow 2\,OH^-(aq) + H_2(g) \qquad E°_{H_2O/H_2} = -0.828\ V$$

$$Na^+(aq) + e^- \longrightarrow Na(s) \qquad E°_{Na^+/Na} = -2.713\ V$$

Suppose we determine $E°_{cell}$ in the customary fashion, using the equation

$$E°_{cell} = E°(\text{cathode}) - E°(\text{anode})$$

All combinations of cathode and anode half-reactions will give negative values of $E°_{cell}$, but this simply means that an electrolysis reaction does not occur spontaneously. Still, it is reasonable to expect that the reaction most likely to occur is the one with the least negative value of $E°_{cell}$, because that reaction would require the lowest applied voltage from the external electricity source. The favored electrolysis reaction at the cathode should therefore involve the reduction of H_2O ($E° = -0.828\ V$) rather than the reduction of Na^+ ($E° = -2.713\ V$). It would also seem to involve the oxidation of H_2O ($E° = 1.229\ V$) rather than Cl^- ($E° = 1.358\ V$) at the anode.

There are other factors involved, however. One is the fact that the reactants and products may not be in their standard states. For example, in the industrial electrolysis of NaCl(aq) the pH is adjusted to about 4. This is not the pH of 0 associated with $[H^+] \approx 1$ M. An adjustment for this difference in pH gives $E = 0.99$ V as the *nonstandard* reduction potential of the half-reaction $2\,H_2O(l) \longrightarrow 4\,H^+(aq) + O_2(g) + 4\,e^-$. Lowering this reduction potential from 1.229 V to 0.99 V makes it even more likely that the oxidation of H_2O to O_2 is favored over the oxidation of Cl^- to Cl_2.

A second factor works in the opposite direction. In many half-reactions, particularly those involving gases, various interactions at electrode surfaces make the required voltage for electrolysis higher than the voltage calculated from $E°$ data. **Overvoltage** is the excess voltage above the voltage calculated from $E°$ values that is required in electrolysis. The overvoltage for the formation of $H_2(g)$ at the cathode is almost zero, but the overvoltage for the formation of $O_2(g)$ at the anode is high, considerably higher than for the formation of $Cl_2(g)$.

The net effect of these two confounding factors is that $O_2(g)$ is formed in the electrolysis of very dilute NaCl(aq), but in more concentrated NaCl(aq), the product at the anode is almost exclusively $Cl_2(g)$. In the commercial electrolysis of NaCl(aq), which we further describe in Section 18.11, the overall electrolysis reaction is

$$2\,Cl^-(aq) + 2\,H_2O(l) \xrightarrow{\text{electrolysis}} 2\,OH^-(aq) + H_2(g) + Cl_2(g)$$

Inert Electrodes and Active Electrodes

The electrodes used in the electrolysis of NaCl(l) and NaCl(aq) we just described were inert, which means their only function was to provide a surface on which the exchange of electrons, and hence the oxidation and reduction half-reactions, could occur. Otherwise, they did not participate in the electrolysis reaction. In contrast, some electrodes are *active* electrodes; they directly participate in a half-reaction. Thus, in predicting electrolysis reactions, we must always consider the nature of the electrodes, as illustrated in Example 18.11.

Example 18.11

Predict the electrolysis reaction when $AgNO_3(aq)$ is electrolyzed **(a)** using platinum electrodes and **(b)** using a silver anode and a platinum cathode.

STRATEGY

Because platinum electrodes are inert, in part (a) we need to consider only the species present in the solution, Ag^+ ions, NO_3^- ions, and H_2O molecules. In part (b) we must consider the same solution species as in (a) and also consider that the silver anode might be an active electrode. Whenever predicting electrolysis reactions in aqueous solutions, keep in mind that if no other half-reactions are possible, oxidation of H_2O may occur at the anode and reduction of H_2O at the cathode, resulting in the electrolysis of water.

$$2\,H_2O(l) \xrightarrow{\text{electrolysis}} 2\,H_2(g) + O_2(g)$$

SOLUTION

(a) Silver ion is easily reduced, as we can see from its half-reaction ranking in Table 18.1:

$$Ag^+(aq) + e^- \longrightarrow Ag(s) \qquad E°_{Ag^+/Ag} = 0.800 \text{ V}$$

This reduction should occur exclusively over that of water molecules, for which

$$2\,H_2O(l) + 2\,e^- \longrightarrow H_2(g) + 2\,OH^-(aq) \qquad E°_{H_2O/H_2} = -0.828 \text{ V}$$

On the oxidation side, because NO_3^- has the N atom in its highest possible oxidation state $(+5)$, it cannot be oxidized. Therefore, the only likely oxidation is that of H_2O molecules:

$$2\,H_2O(l) \longrightarrow O_2(g) + 4\,H^+(aq) + 4\,e^- \qquad E°_{O_2/H_2O} = 1.229 \text{ V}$$

The calculated voltage for the electrolytic cell reaction is

$$E°_{cell} = E°(\text{cathode}) - E°(\text{anode})$$

$$= 0.800 \text{ V} - 1.229 \text{ V} = -0.429 \text{ V}$$

and the electrolysis reaction is

$$4 \text{ Ag}^+(\text{aq}) + 2 \text{ H}_2\text{O}(\text{l}) \xrightarrow{\text{electrolysis}} 4 \text{ Ag}(\text{s}) + 4 \text{ H}^+(\text{g}) + \text{O}_2(\text{g})$$

(b) The reduction half-reaction at the cathode is the same as in part (a):

$$\text{Ag}^+(\text{aq}) + \text{e}^- \longrightarrow \text{Ag}(\text{s}) \qquad E°_{\text{Ag}^+/\text{Ag}} = 0.800 \text{ V}$$

In contrast to the platinum electrode, however, the silver anode is active, and the silver undergoes oxidation:

$$\text{Ag}(\text{s}) \longrightarrow \text{Ag}^+(\text{aq}) + \text{e}^- \qquad E°_{\text{Ag}^+/\text{Ag}} = 0.800 \text{ V}$$

The calculated voltage of the electrolytic cell is therefore

$$E°_{cell} = E°(\text{cathode}) - E°(\text{anode})$$

$$= 0.800 \text{ V} - 0.800 \text{ V} = 0 \text{ V}$$

ASSESSMENT

The $E°_{cell} = -0.429$ V calculated in part (a) signifies that the electrolysis reaction is nonspontaneous, just as we expect it to be. In order to accomplish the electrolysis of the $\text{AgNO}_3(\text{aq})$, the applied voltage from the external source would have to exceed 0.429 V. In part (b), the $E°_{cell} = 0$ V does not mean that electrolysis can be accomplished just by placing the electrodes in the solution without using a battery or other source of electricity. A small voltage is required to drive electrons through the external circuit and ions through the solution. The overall electrolysis simply involves the transport of silver, as Ag^+, through the solution from the anode to the cathode:

$$\text{Ag}(\text{s})_{\text{anode}} \xrightarrow{\text{electrolysis}} \text{Ag}(\text{s})_{\text{cathode}}$$

EXERCISE 18.11A

Predict the electrolysis reaction when $\text{KBr}(\text{aq})$ is electrolyzed using platinum electrodes.

EXERCISE 18.11B

Predict the electrolysis reaction when $\text{AgNO}_3(\text{aq})$ is electrolyzed using a silver anode and a copper cathode. What is the approximate external voltage required for the electrolysis?

Example 18.12   A Conceptual Example

Two electrochemical cells are connected as shown. Specifically, the zinc electrodes of the two cells are joined, as are the copper electrodes. Will there be **(a)** no current, **(b)** a flow of electrons in the direction of the red arrows, or **(c)** a flow of electrons in the direction of the blue arrows?

$$\text{Zn}(\text{s}) \mid \text{Zn}^{2+}(1 \text{ M}) \parallel \text{Cu}^{2+}(0.10 \text{ M}) \mid \text{Cu}(\text{s}) \qquad \text{Zn}(\text{s}) \mid \text{Zn}^{2+}(0.10 \text{ M}) \parallel \text{Cu}^{2+}(1.0 \text{ M}) \mid \text{Cu}(\text{s})$$

Cell A   Cell B

ANALYSIS AND CONCLUSIONS

Notice that these are two voltaic cells like those in Figure 18.4, but set up in opposition to each other. That is, a stream of electrons coming out of the zinc electrode of one cell encounters a stream of electrons coming from the zinc electrode of the other cell. The net flow of electrons will be in the direction that has the greater push (voltage) behind it.

If the two cells were identical, their E_{cell} values would be the same and there would be no net electron flow. Because the cells are not identical, however, we can eliminate possibility (a).

When the cells operate as voltaic cells, the cell reaction, as we have seen before, is

$$Zn(s) + Cu^{2+}(aq) \longrightarrow Zn^{2+}(aq) + Cu(s)$$

Cell A has $[Zn^{2+}] = 1.0$ M (standard state) and $[Cu^{2+}] = 0.10$ M (more dilute than the standard state). From Le Châtelier's principle, we would expect the reaction to be less favorable than under standard-state conditions, and so for this cell we can say $E_{cell} < E_{cell}^{\circ}$.

Cell B has $[Zn^{2+}] = 0.10$ M (more dilute than the standard state) and $[Cu^{2+}] = 1.0$ M (standard state). From Le Châtelier's principle, we would expect the forward reaction to be more favorable than under standard-state conditions and therefore $E_{cell} > E_{cell}^{\circ}$.

We conclude that there is a net flow of electrons from the Zn electrode of cell B to the Zn electrode of cell A. That is, cell B is a voltaic cell, forcing electrons into the Zn cathode of cell A, which is an electrolytic cell. Also, electrons are withdrawn from the Cu anode of cell A. The direction of electron flow is that of the red arrows.

EXERCISE 18.12A

Write net ionic equations for the cell reactions in cell A and cell B when they are connected as shown in Example 18.12.

EXERCISE 18.12B

Explain why electric current cannot flow indefinitely in the cell A/cell B combination of Example 18.12.

18.10   Quantitative Electrolysis

The quantitative basis of electrolysis was largely established by Michael Faraday. The quantity of reactant consumed or product formed during electrolysis is related to (1) the molar mass of the substance, (2) the quantity of electric charge used, and (3) the number of electrons transferred in the electrode reaction.

The charge carried by one electron is -1.6022×10^{-19} C. Electric *current*, expressed in **amperes (A),** is the rate of flow of electric charge. One ampere of current is a flow of one coulomb per second:

$$1 \text{ A} = 1 \text{ C/s} \qquad \textbf{(18.10)}$$

The total electric charge involved in an electrolysis reaction, then, is the product of the current and the length of time that the current flows:

$$\text{Electric charge (C)} = \text{current (C/s)} \times \text{time (s)} \qquad \textbf{(18.11)}$$

We need this expression and the Faraday constant from Section 18.5 (96,485 C/mol e⁻) to calculate the quantitative outcomes of electrolysis reactions. Our usual strategy will be

Step 1: *Determine the amount of charge—the product of current and time.*

Step 2: *Convert the amount of charge to moles of electrons:* 1 mol e⁻ = 96,485 C.

Step 3: *Use a half-equation to relate moles of electrons to moles of a reactant or product.* For example, the half-equation $Cu^{2+}(aq) + 2 e^- \longrightarrow Cu(s)$ yields the factor

$$\frac{1 \text{ mol Cu}}{2 \text{ mol e}^-}$$

Step 4: *Convert from moles of reactant or product to the final quantity desired—* mass, volume of a gas, and so on.

For some problems, we need to use variations of this four-step strategy. For example, in Exercise 18.13A, each step of the strategy is inverted, and the steps are performed in the reverse order: 4, 3, 2, 1.

▲ Michael Faraday (1791–1867) is known for his discoveries in electricity and magnetism and, additionally, in electrochemistry. Faraday was unschooled in mathematics but had an unrivaled intuitive ability to visualize physical phenomena.

 **Electrolysis activity**

Example 18.13

We can use electrolysis to determine the gold content of a sample. The sample is dissolved, and all the gold is converted to $Au^{3+}(aq)$, which is then reduced back to $Au(s)$ on an electrode of known mass. The reduction half-reaction is $Au^{3+}(aq) + 3\ e^- \longrightarrow Au(s)$. What mass of gold will be deposited at the cathode in 1.00 hour by a current of 1.50 A?

STRATEGY

We will apply the four-step procedure just outlined.

SOLUTION

Step 1: *Determine the amount of charge.* We calculate the total electric charge involved from the known values of time and current.

$$\frac{1.50\ \text{C}}{\text{s}} \times 1\ \text{h} \times \frac{60\ \text{min}}{1\ \text{h}} \times \frac{60\ \text{s}}{1\ \text{min}} = 5.40 \times 10^3\ \text{C}$$

Step 2: *Convert the amount of charge to moles of electrons.* Next, we use the Faraday constant to determine the number of moles of electrons.

$$5.40 \times 10^3\ \text{C} \times \frac{1\ \text{mol}\ e^-}{96{,}485\ \text{C}} = 0.0560\ \text{mol}\ e^-$$

Step 3: *Use a half-equation to relate moles of electrons to moles of the product.* According to the reduction half-equation, 3 mol e^- are required for every mole of Au deposited.

$$0.0560\ \text{mol}\ e^- \times \frac{1\ \text{mol Au}}{3\ \text{mol}\ e^-} = 0.0187\ \text{mol Au}$$

Step 4: *Convert from moles to grams of product.* The mass of Au, in grams, is simply the product of the amount in moles and the molar mass.

$$0.0187\ \text{mol Au} \times \frac{197.0\ \text{g Au}}{1\ \text{mol Au}} = 3.68\ \text{g Au}$$

ASSESSMENT

Notice that this calculation could easily have been done in a single setup combining all four steps. This single-setup approach has the advantage of eliminating all intermediate rounding of results, even though in this case the final result is the same as in the stepwise approach:

$$?\ \text{g Ag} = 1.50\frac{\text{C}}{\text{s}} \times 1\ \text{h} \times \frac{60\ \text{min}}{1\ \text{h}} \times \frac{60\ \text{s}}{1\ \text{min}} \times \frac{1\ \text{mol}\ e^-}{96{,}485\ \text{C}} \times \frac{1\ \text{mol Au}}{3\ \text{mol}\ e^-} \times \frac{197.0\ \text{g Au}}{1\ \text{mol Au}} = 3.68\ \text{g Au}$$

EXERCISE 18.13A

For how many minutes must the electrolysis of a solution of $CuSO_4(aq)$ be carried out, at a current of 2.25 A, to deposit 1.00 g of $Cu(s)$ at the cathode?

EXERCISE 18.13B

A coulometer is a device for measuring quantities of electric charge. One type of coulometer is an electrolytic cell with platinum electrodes in $AgNO_3(aq)$. If 2.175 g $Ag(s)$ is deposited on the cathode in an electrolysis lasting 21 min and 12 s, **(a)** what is the amount of charge, in coulombs, passing through the cell, and **(b)** what is the magnitude of the current in amperes?

Example 18.14   An Estimation Example

Without doing detailed calculations, determine which of the following solutions will yield the greatest mass of metal at a platinum cathode during electrolysis by a 1.50-A electric current for 30.2 min: $CuSO_4(aq)$, $AgNO_3(aq)$, or $AuCl_3(aq)$.

ANALYSIS AND CONCLUSIONS

We can begin by comparing the cathode half-reactions:

$$Cu^{2+} + 2\ e^- \longrightarrow Cu(s) \qquad Ag^+ + e^- \longrightarrow Ag(s) \qquad Au^{3+} + 3\ e^- \longrightarrow Au(s)$$

From these half-equations, we see that for every 1 mol e^-, it is possible to obtain

$$\tfrac{1}{2}\ \text{mol Cu(s)} \quad \text{or} \quad 1\ \text{mol Ag(s)} \quad \text{or} \quad \tfrac{1}{3}\ \text{mol Au(s)}$$

With just a glance at a table of atomic masses (Cu = 63.5 u; Ag = 108 u; Au = 197 u), we can conclude that the

$$\text{Mass of } \tfrac{1}{2}\ \text{mol Cu} \ < \ \text{mass of } \tfrac{1}{3}\ \text{mol Au} \ < \ \text{mass of 1 mol Ag}$$

Electrodeposition of Metals activity

The $AgNO_3(aq)$ yields the greatest mass of metal. Notice that we did not need to use the current and time data because we did not have to calculate actual masses of the three metals.

EXERCISE 18.14A

Without doing detailed calculations, determine which of the following solutions will yield the greatest volume of gas (at STP) at a platinum anode when the same amount of electric charge is passed through each: $Mg(NO_3)_2(aq)$, $NaCl(aq)$, or $KI(aq)$.

EXERCISE 18.14B

Determine which of the following aqueous solutions will yield 1.00 g of solid metal at a platinum cathode in the shortest time in an electrolysis using a 2.50-A current: (a) 100.0 mL of 1.00 M $NaCl(aq)$, (b) 50.0 mL of 0.50 M $CuSO_4(aq)$, (c) 150.0 mL of 0.25 M $AgNO_3(aq)$, (d) 125 mL of 1.00 M $NiCl_2$.

18.11   Applications of Electrolysis

There are many practical applications of electrolysis. Let us conclude this chapter by considering a few of them.

Producing Chemicals by Electrolysis

Electrolysis plays an important role in the manufacture and purification of a host of substances: sodium (Figure 18.21), calcium, magnesium, aluminum (Section 21.3), copper, zinc, silver, hydrogen, chlorine, fluorine, hydrogen peroxide, sodium hydroxide, potassium dichromate, and potassium permanganate, among others. One of the most commercially important electrolyses is that of $NaCl(aq)$. We wrote a net ionic equation for this electrolysis on page 783, but we can also include the spectator ion, Na^+:

$$2\,Na^+(aq) + 2\,Cl^- + 2\,H_2O(l) \longrightarrow 2\,Na^+(aq) + 2\,OH^-(aq) + H_2(g) + Cl_2(g)$$

Sodium hydroxide   Hydrogen   Chlorine

The products most sought are the chlorine gas and sodium hydroxide solution. Hence, the electrolysis of $NaCl(aq)$ is called the *chlor-alkali* process.

One common type of chlor-alkali electrolytic cell, called a diaphragm cell, is pictured in Figure 18.22. The major challenge in a chlor-alkali process is to keep $Cl_2(g)$ from coming into contact with the $NaOH(aq)$, because in an alkaline solution Cl_2 disproportionates to ClO^- and Cl^-. The diaphragm prevents this contact. A chlor-alkali cell must also prevent the mixing of Cl_2 and H_2 because mixtures of these gases can be explosive.

Electroplating movie

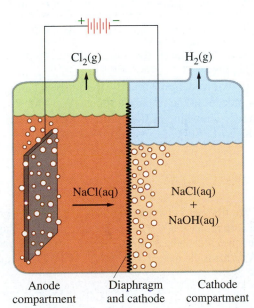

◀ **FIGURE 18.22   A diaphragm chlor-alkali cell**

The anode is a specially treated titanium metal. The diaphragm and cathode are a composite unit consisting of an asbestos–polymer mixture deposited on a steel wire mesh. A difference in the solution levels is maintained so that $NaCl(aq)$ moves through the diaphragm during the electrolysis.

Application Note

The deposition of silver is more uniform from [Ag(CN)$_2$]$^-$(aq) than from Ag$^+$(aq). That is why cyanide solutions are often used in electroplating silver.

▲ Electroplating is used to enhance the appearance and durability of a wide variety of objects, from baby shoes to automobile bodies.

Electroplating

Electrolysis can be used to coat one metal onto another in a process known as *electroplating*. Usually, the object to be electroplated, such as a spoon, is cast from an inexpensive metal. It is then coated with a thin layer of a more attractive, corrosion-resistant, expensive metal, such as silver or gold. The finished product costs far less than if the object were made entirely of the more expensive metal.

A cell for electroplating silver is shown in Figure 18.23. A piece of pure silver is the anode, the spoon is the cathode, and a solution of silver nitrate is the electrolyte. Think of the spoon as replacing the inert platinum cathode, and you will see that the electrolysis is the same as that described in Example 18.11b:

$$Ag(s, \text{anode}) \longrightarrow Ag(s, \text{cathode})$$

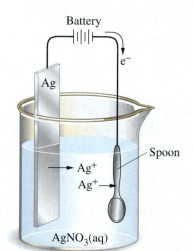

◀ **FIGURE 18.23**

An electrochemical cell for silver plating

The anode is a silver bar, and the cathode is an iron spoon.

A similar but larger-scale industrial use of electroplating is in the refining of copper. High-purity copper is required for applications that depend on copper's high electrical conductivity. A massive anode of impure copper and a thin cathode of pure copper are suspended in an electrolyte such as CuSO$_4$(aq). Just as in the case of silver plating, the overall process is the transfer of copper atoms from the anode to the cathode:

Anode:	$Cu(s, \text{impure}) \longrightarrow Cu^{2+}(aq) + 2\ e^-$
Cathode:	$Cu^{2+}(aq) + 2\ e^- \longrightarrow Cu(s, \text{pure})$
Overall:	$Cu(s, \text{impure}) \longrightarrow Cu(s, \text{pure})$

Electrolysis is continued until most of the anode has been consumed and the cathode has built up into a thick sheet.

Because copper is oxidized more readily than silver and gold, these latter two metals, which may be present in the impure copper, drop to the bottom of the electrolysis cell below the anode. The value of the precious metals in this *anode mud* is often sufficient to pay for the cost of the electrorefining.

Cumulative Example

An important source of silver is the process in which lead is purified from lead-containing ores. The percent Ag in a 1.050-g lead sample is determined as follows: The sample is dissolved in nitric acid to produce 500.0 mL of a solution of Pb^{2+}(aq) containing a small quantity of Ag$^+$(aq). A strip of pure Ag(s) is immersed in the solution, and the potential difference between this half-cell as the cathode and a standard silver–silver chloride anode is found to be 0.281 V. What is the mass percent silver in the lead sample?

STRATEGY

Our first task is to identify (i) the two half-reactions that occur in the cell reaction, (ii) the standard electrode potentials of these half-reactions, and (iii) the value of $E°_{cell}$. At that point, we can write an equation for the cell reaction and use the Nernst equation (18.8) to relate $E°_{cell}$ and the measured E_{cell} to the unknown silver ion concentration, $[Ag^+]$. Once we have $[Ag^+]$, we can determine the number of moles, and then grams, of Ag in the sample. Then we can calculate percent Ag in the usual way.

SOLUTION

We obtain the two half-reactions and their standard electrode potentials from Appendix C.

$$Ag^+(aq) + e^- \longrightarrow Ag(s) \qquad E° = 0.800 \text{ V}$$
$$AgCl(s) + e^- \longrightarrow Ag(s) + Cl^-(aq) \qquad E° = 0.2223 \text{ V}$$

In the cell reaction, reduction occurs at the silver–silver ion cathode, and oxidation occurs at the silver–silver chloride anode.

Reduction: $\qquad Ag^+(aq) + e^- \longrightarrow Ag(s)$
Oxidation: $\qquad Ag(s) + Cl^-(aq) \longrightarrow AgCl(s) + e^-$
_____
Overall: $\qquad Ag^+(aq) + Cl^-(aq) \longrightarrow AgCl(s)$

Then we determine the value of $E°_{cell}$ for the overall cell reaction.

$$E°_{cell} = E°(\text{cathode}) - E°(\text{anode})$$
$$= E°(\text{reduction}) - E°(\text{oxidation})$$
$$= 0.800 \text{ V} - 0.2223 \text{ V} = 0.578 \text{ V}$$

Now we write the cell reaction based on the measured value of E_{cell}, with the appropriate ion concentrations, including $[Ag^+] = x$. Note that $[Cl^-] = 1$ M because the silver–silver chloride electrode is a standard electrode.

$$Ag^+(x \text{ M}) + Cl^-(1 \text{ M}) \longrightarrow AgCl(s) \qquad E_{cell} = 0.281 \text{ V}$$

Next we write the Nernst equation based on this cell reaction.

$$E_{cell} = E°_{cell} - \frac{0.0592 \text{ V}}{n} \log Q$$

$$0.281 \text{ V} = 0.578 \text{ V} - \frac{0.0592 \text{ V}}{1} \log \frac{1}{[Ag^+][Cl^-]}$$

$$0.281 = 0.578 - 0.0592 \log \frac{1}{x \cdot 1}$$

We solve this equation for the unknown $x = [Ag^+]$.

$$\log \frac{1}{x} = \frac{-0.297}{-0.0592} = 5.02$$

$$\frac{1}{x} = 10^{5.02} = 1.0 \times 10^5$$

$$x = 1.0 \times 10^{-5} \text{ M}$$

We use this value of x as $[Ag^+]$ and determine the mass of silver ion in the 500.0 mL of solution.

$$? \text{ g Ag} = 0.500 \text{ L} \times \frac{1.0 \times 10^{-5} \text{ mol Ag}^+}{1 \text{ L}} \times \frac{107.87 \text{ g Ag}^+}{1 \text{ mol Ag}^+} = 5.4 \times 10^{-4} \text{ g Ag}^+$$

Finally, in calculating the mass percent silver in the lead sample, we note that all the Ag in the sample appeared as Ag^+ in solution.

$$\% \text{ Ag} = \frac{5.4 \times 10^{-4} \text{ g Ag}^+ \times \frac{1 \text{ g Ag}}{1 \text{ g Ag}^+}}{1.050\text{-g sample}} \times 100\% = 0.051\% \text{ Ag}$$

ASSESSMENT

Because the silver is present as an impurity in the lead, we should expect the percent silver to be quite small. Our answer seems reasonable in this regard. The most likely errors in a calculation of this type are in the signs of various quantities—standard electrode potentials, E_{cell} and $E°_{cell}$ values, and logarithmic terms. Each step must be carefully checked for correct signs in the course of the calculation.

Concept Review with Key Terms

18.1 Half-Reactions—**Electrochemistry** involves chemical reactions in which the flow of electrons leads to oxidation–reduction reactions. A redox reaction can be separated into two **half-reactions**, one for oxidation and one for reduction. For example:

Oxidation: $\qquad\qquad$ $Cu(s) \longrightarrow Cu^{2+}(aq) + 2\,e^-$

Reduction: $\quad$ $2\,Ag^+(aq) + 2\,e^- \longrightarrow 2\,Ag(s)$

Overall: $\quad$ $Cu(s) + 2\,Ag^+(aq) \longrightarrow Cu^{2+}(aq) + 2\,Ag(s)$

18.2 The Half-Reaction Method of Balancing Redox Equations—Half-equations for half-reactions can be balanced separately. When the half-equations are combined to yield a balanced redox equation, the electrons lost in the oxidation half-reaction must equal the electrons gained in the reduction half-reaction.

18.3 A Qualitative Description of Voltaic Cells—In a **voltaic cell**, a spontaneous redox reaction is used to produce electricity. Half-reactions can be conducted in **half-cells**. The **electrodes** in two half-cells are joined by a wire, and electrolyte solutions may be joined by a **salt bridge**.

The appropriate combination of two half-cells produces an **electrochemical cell**. Oxidation occurs in the **anode** half-cell, and reduction occurs in the **cathode** half-cell. The force driving the chemical reaction in an electrochemical cell is known as the **cell potential, E_{cell}**. Cell potentials are recorded in **volts (V)**, a measure of energy (J) per unit charge, where the unit of charge is the **coulomb (C)**. A **cell diagram** for a voltaic cell is written with the anode on the left and the cathode on the right. Phase boundaries are represented by a vertical line (|), and a salt bridge or porous barrier is represented by a double vertical line (‖). Solution concentrations may also be noted, as in this example of notation for a voltaic cell:

$$Zn(s)|Zn^{2+}(0.10\ M)\|Cu^{2+}(0.50\ M)|Cu(s)$$

18.4 Standard Electrode Potentials—A **standard hydrogen electrode** has H^+ ion at unit activity ($\approx 1\ M$) in equilibrium with H_2 (g, 1 bar $\approx$ 1 atm) on an inert platinum electrode. **Standard electrode potentials, $E°$**, for reduction half-reactions are obtained by comparison to the standard hydrogen electrode, which is assigned a standard electrode potential of zero.

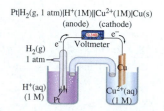

The **standard cell potential, $E_{cell}°$**, for a reaction is the difference between the standard potentials of the cathode and anode:

$$E_{cell}° = E°(\text{cathode}) - E°(\text{anode})$$

18.5 Electrode Potentials, Spontaneous Change, and Equilibrium—A redox reaction for which $E_{cell} > 0$ occurs spontaneously. If $E_{cell} < 0$, the reaction is nonspontaneous. Often these criteria are applied for standard conditions, and $E_{cell}°$ values are obtained from tables of standard electrode potentials. Another important use of $E_{cell}°$ is in determining values of $\Delta G°$ and K_{eq}:

$$\Delta G° = -nFE_{cell}° = -RT \ln K_{eq}$$

Here the standard free energy change ($\Delta G°$) is related to the number of moles of electrons (n) transferred in the overall reaction, the **Faraday constant (F)**, and the standard cell potential ($E_{cell}°$).

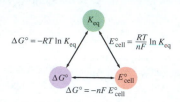

18.6 Effect of Concentrations on Cell Voltage—The **Nernst equation** relates a cell voltage under nonstandard conditions to $E_{cell}°$, the number of electrons transferred (n), and the reaction quotient (Q). At 25 °C, in volts,

$$E_{cell} = E_{cell}° - \frac{0.0592}{n} \log Q$$

In a **concentration cell**, the half-cells have identical electrodes and solutions of the same electrolyte but at different concentrations. The cell voltage depends only on these concentrations. That is, in volts, the Nernst equation reduces to

$$E_{cell} = \frac{-0.0592}{n} \log Q \ (\text{at } 25\ °C)$$

18.7 Batteries: Using Chemical Reactions to Make Electricity—Commercially, voltaic cells are used singly or joined together to yield higher voltages or higher currents. In primary cells, the electrode reactions are irreversible; in secondary cells, they are reversible. Thus primary cells cannot be recharged, but secondary cells can be. In a fuel cell, a fuel and oxygen (from air) combine to produce oxidation products, and chemical energy is released as electricity.

18.8 Corrosion: Metal Loss Through Voltaic Cells—A corroding metal consists of anodic areas, at which dissolution of the metal occurs, and cathodic areas, where atmospheric oxygen is reduced to hydroxide ion. A metal can be protected against corrosion by plating it with a second metal that corrodes less readily. In another method, called cathodic protection, an active metal is sacrificed to protect a less active metal to which it is joined.

18.9 Predicting Electrolysis Reactions—In an **electrolytic cell**, direct electric current from an external source drives nonspontaneous chemical reactions. Electrode potential data can be used to predict the probable products of an **electrolysis** reaction, although other factors may also have to be considered. In some instances, a voltage greater than that calculated from $E°$ data, defined as the **overvoltage**, is required for electrolysis to occur.

18.10 Quantitative Electrolysis—The amount of chemical change produced in electrolysis depends on the amount of charge that is transferred at the electrodes. The amount of charge, in turn, depends on the magnitude of the electric current used, measured in **amperes (A)**, and the time required for the electrolysis. The ampere is defined as

$$1\ A = 1\ C/s$$

18.11 Applications of Electrolysis—Electrolysis is used to manufacture chemicals, refine metals, and deposit metal coatings—a process known as electroplating.

Assessment Goals

When you have mastered the material in this chapter, you will be able to:

- Describe the basis of electrochemical cells in terms of oxidation–reduction reactions.
- Write cell diagrams for, and identify the components of, electrochemical cells.
- Use standard electrode potentials to determine standard cell potentials.
- Use standard electrode potentials to predict the spontaneity of cell reactions.
- Calculate values of $\Delta G°$ and K_{eq} from standard cell potentials.
- Determine cell potentials under nonstandard conditions, using the Nernst equation.

- Use the Nernst equation to calculate cell potentials corresponding to given half-cell concentrations or to calculate half-cell concentrations producing a desired cell potential.
- Describe the characteristics of primary and secondary batteries and fuel cells.
- Describe corrosion in terms of coupled oxidation and reduction half-reactions.
- Predict the products of electrolysis reactions from standard cell potentials.
- Calculate the amounts of substances consumed or produced in electrolysis reactions from time and current data.

Self-Assessment Questions

1. What is the difference between an oxidation–reduction reaction and a half-reaction?

2. What is meant by each of the following terms?

   (a) electrode
   (d) cathode

   (b) half-cell
   (e) anode

   (c) voltaic cell

3. What is the function of a salt bridge in an electrochemical cell?

4. What are the similarities and differences between voltaic cells and electrolytic cells?

5. Which of these is true of the conversion of NpO_2^+ to Np^{4+}?

   (a) It is an oxidation process.

   (b) It can occur only in an electrochemical cell.

   (c) It is a reduction process.

   (d) It is an oxidation–reduction process.

6. In an electrochemical cell that functions as a *voltaic* cell, which of the following is true?

   (a) Electrons move from the cathode to the anode.

   (b) Electrons move through a salt bridge.

   (c) Electrons can move either from the cathode to the anode or from the anode to the cathode.

   (d) Reduction occurs at the cathode.

7. Which of the following pairs of terms have the same meaning, and which have a different meaning? Explain.

   (a) cell voltage and cell potential

   (b) electrode potential and standard electrode potential

   (c) half-cell and half-reaction

8. Some standard electrode potentials have positive values and some have negative values. Can electrode potentials be defined so that all have the same sign? Explain.

9. The value of $E°_{cell}$ for the oxidation–reduction reaction $Zn(s) + Pb^{2+}(1.0\ M) \longrightarrow Zn^{2+}(1.0M) + Pb(s)$ is +0.66 V. What is E_{cell} for the reaction $Zn(s) + Pb^{2+}(0.01\ M) \longrightarrow Zn^{2+}(0.1\ M) + Pb(s)$?

   (a) +0.72 V
   (c) +0.66 V

   (b) +0.69 V
   (d) +0.63 V

10. What is the criterion for spontaneous chemical change based on cell potentials? Explain.

11. Which two of the following metals will *not* dissolve in a 1.0 M H^+ solution: Mg, Ag, Zn, Fe, Au?

12. Explain why electrode potentials can be used to make predictions of spontaneous change even for reactions that are not carried out in electrochemical cells. Are there some reactions for which they cannot be used? Explain.

13. If $E°_{cell} < 0$, is a spontaneous cell reaction impossible regardless of the conditions? Explain.

14. What is a concentration cell? What is the value of $E°_{cell}$ for a concentration cell? What is the overall change that occurs as electric current is drawn from the cell?

15. A certain concentration cell constructed from a standard hydrogen electrode and a hydrogen electrode of unknown concentration registers an $E_{cell} = +0.256$ V. What is the pH of the unknown solution?

16. What is the difference between a primary and a secondary cell? Give an example of each.

17. What is a fuel cell? What is the basic principle involved in its operation?

18. Describe the electrochemical nature of corrosion. Use the corrosion of iron as an example.

19. Explain why iron corrodes much more readily than does aluminum, even though iron is the less active of the two metals.

20. What is cathodic protection? How does it provide protection against corrosion?

21. How are standard electrode potentials involved in determining the voltage required to carry out an electrolysis?

22. The reaction $Cu^{2+}(aq) + 2\ Cl^-(aq) \longrightarrow Cu(s) + Cl_2(g)$ has $E°_{cell} = -1.02$ V. Which of these statements is true?

    (a) The reaction can be made to occur in an electrolysis cell.

    (b) The reaction can be made to produce electricity in a voltaic cell.

    (c) The reaction occurs whenever Cu^{2+} and Cl^- are brought together in aqueous solution.

    (d) The reaction can occur in acidic solution but not in basic solution.

23. Should an object to be electroplated with a metal be made the anode or the cathode in an electrolytic cell? Why does this matter?

24. A quantity of electrical charge that brings about the deposition of 4.5 g Al from $Al^{3+}(aq)$ at a cathode will also produce what volume (STP) of $H_2(g)$ from $H^+(aq)$ at a cathode?

    (a) 44.8 L
    (c) 11.2 L

    (b) 22.4 L
    (d) 5.6 L

Problems

Oxidation–Reduction Reactions

25. Complete and balance the following half-equations, and indicate whether oxidation or reduction is involved.

   (a) $HNO_2(aq) \longrightarrow NO_2(g)$ (acidic solution)

   (b) $PbO_2(s) \longrightarrow PbO(s)$ (acidic solution)

   (c) $CH_3CH_2OH(aq) \longrightarrow CO_2(g)$ (basic solution)

26. Complete and balance the following half-equations, and indicate whether oxidation or reduction is involved.

   (a) $ClO_2(g) \longrightarrow ClO_3^-(aq)$ (acidic solution)

   (b) $MnO_4^-(aq) \longrightarrow MnO_2(s)$ (basic solution)

   (c) $SbH_3(g) \longrightarrow Sb(s)$ (basic solution)

27. Use the half-reaction method to balance the following redox equations in acidic solution.

   (a) $Fe^{2+}(aq) + Cr_2O_7^{2-}(aq) \longrightarrow Fe^{3+}(aq) + Cr^{3+}(aq)$

   (b) $S_8(s) + O_2(g) \longrightarrow SO_4^{2-}(aq)$

   (c) $Fe^{3+}(aq) + NH_2OH_2^+(aq) \longrightarrow Fe^{2+}(aq) + N_2O(g)$

28. Use the half-reaction method to balance the following equations in acidic solution.

   (a) $Ag(s) + NO_3^-(aq) \longrightarrow Ag^+(aq) + NO(g)$

   (b) $H_2O_2(aq) + MnO_4^-(aq) \longrightarrow Mn^{2+}(aq) + O_2(g)$

   (c) $Cl_2(g) + I^-(aq) \longrightarrow Cl^-(aq) + IO_3^-(aq)$

29. Use the half-reaction method to balance the following equations in basic solution.

   (a) $Fe(OH)_2(s) + O_2(g) \longrightarrow Fe(OH)_3(s)$

   (b) $S_8(s) \longrightarrow S_2O_3^{2-}(aq) + S^{2-}(aq)$

   (c) $CrI_3(s) + H_2O_2(aq) \longrightarrow CrO_4^{2-}(aq) + IO_4^-(aq)$

30. Use the half-reaction method to balance the following equations in basic solution.

   (a) $CrO_4^{2-}(aq) + AsH_3(g) \longrightarrow Cr(OH)_3(s) + As(s)$

   (b) $CH_3OH(aq) + MnO_4^-(aq) \longrightarrow HCOO^-(aq) + MnO_2(s)$

   (c) $[Fe(CN)_6]^{3-}(aq) + N_2H_4(aq)$
   $$\longrightarrow [Fe(CN)_6]^{4-}(aq) + N_2(g)$$

31. Write balanced equations for (a) the reaction of oxalic acid (HOOCCOOH) and permanganate ion in acidic solution to produce manganese(II) ion and carbon dioxide gas, (b) the reaction of $Cr_2O_7^{2-}$ and UO^{2+} to produce UO_2^{2+} and Cr^{3+} in an acidic aqueous solution, and (c) the reaction in basic solution of nitrate ion and zinc to produce zinc(II) ion and gaseous ammonia.

32. Write balanced equations for (a) the reaction of thiosulfate and permanganate ions in acidic solution to form sulfate and manganese(II) ions, (b) the action of permanganate ion on acetaldehyde (CH_3CHO) in basic solution to produce manganese(IV) oxide and acetate ion, and (c) the disproportionation of manganate ion (MnO_4^{2-}) to permanganate ion and solid manganese(IV) oxide in a basic solution.

Electrode Potentials and Voltaic Cells

Use data from Table 18.1 and Appendix C, as necessary.

33. For the voltaic cell pictured, write an equation for the cell reaction that occurs, and determine the voltmeter reading if the metal, M, is (a) Sn, (b) Zn, and (c) Cu. Note that the voltmeter can display both positive and negative voltages.

34. From the indicated voltages for the voltaic cell pictured, determine the standard electrode potential, $E^\circ_{M^{3+}/M}$, if the metal, M, is (a) In, $E^\circ_{cell} = 0.086$ V; (b) Cr, $E^\circ_{cell} = -0.32$ V; (c) Tl, $E^\circ_{cell} = 1.14$ V.

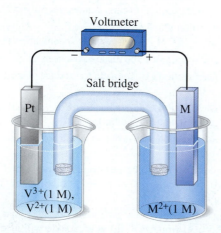

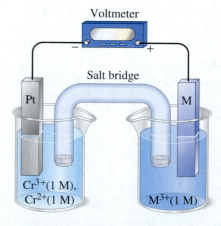

35. For the reaction

$$3\,V(s) + 2\,SbO^+(aq) + 4\,H^+(aq)$$
$$\longrightarrow 3\,V^{2+}(aq) + 2\,Sb(s) + 2\,H_2O(l) \qquad E^\circ_{cell} = 1.334\,V$$

given that $E^\circ_{V^{2+}/V} = -1.13\,V$, determine the value of E° for the half-reaction

$$SbO^+(aq) + 2\,H^+(aq) + 3\,e^- \longrightarrow Sb(s) + H_2O(l)$$

36. For the reaction

$$2\,CuI(s) + Cd(s) \longrightarrow Cd^{2+}(aq) + 2\,I^-(aq) + 2\,Cu(s)$$
$$E^\circ_{cell} = +0.23\,V$$

given that $E^\circ_{Cd^{2+}/Cd} = -0.403\,V$, determine the value of E° for the half-reaction

$$2\,CuI(s) + 2\,e^- \longrightarrow 2\,Cu(s) + 2\,I^-(aq)$$

37. $E^\circ_{cell} = 1.47\,V$ for the voltaic cell

$$V(s)|V^{2+}(1\,M)\|Cu^{2+}(1\,M)|Cu(s)$$

Determine the value of $E^\circ_{V^{2+}/V}$.

38. $E^\circ_{cell} = 3.73\,V$ for the voltaic cell

$$Y(s)|Y^{3+}(1\,M)\|Cl^-(1\,M)|Cl_2(g, 1\,atm)|Pt$$

Determine the value of $E^\circ_{Y^{3+}/Y}$.

39. Write equations for the half-reactions and the overall cell reaction, and calculate E°_{cell} for each of the voltaic cells diagrammed below.

(a) $Pt|Fe^{2+}(aq), Fe^{3+}(aq)\|Cr_2O_7{}^{2-}(aq), Cr^{3+}(aq)|Pt$
(b) $Pt|NO(g)|NO_3{}^-(aq), H^+(aq)\|H^+(aq), H_2O_2(aq)|Pt$

40. Write equations for the half-reactions and the overall cell reaction, and calculate E°_{cell} for each of the voltaic cells diagrammed below.

(a) $Pt|I_2(s)|I^-(aq)\|Cl^-(aq)|Cl_2(g)|Pt$
(b) $Pt|PbO_2(s)|Pb^{2+}(aq), H^+(aq)\|S_2O_8{}^{2-}(aq), SO_4{}^{2-}(aq)|Pt$

41. Each of the following reactions take places in a voltaic cell. Write equations for the half-reactions and the overall cell reaction. Write a cell diagram for the voltaic cell, and calculate the value of E°_{cell}.

(a) $Fe^{3+}(aq) + Sn^{2+}(aq) \longrightarrow Fe^{2+}(aq) + Sn^{4+}(aq)$
(b) $Cu(s) + H^+(aq) + NO_3{}^-(aq)$
$$\longrightarrow Cu^{2+}(aq) + H_2O(l) + NO(g)$$

42. Each of the following reactions takes place in a voltaic cell. Write equations for the half-reactions and the overall cell reaction. Write a cell diagram for the voltaic cell, and calculate the value of E°_{cell}.

(a) $Zn(s) + Ag^+(aq) \longrightarrow Ag(s) + Zn^{2+}(aq)$
(b) $Fe^{2+}(aq) + O_2(g) + H^+(aq) \longrightarrow Fe^{3+}(aq) + H_2O(l)$

E°_{cell} and the Spontaneity of Redox Reactions

Use data from Table 18.1 and Appendix C, as necessary.

43. Predict whether a spontaneous reaction will occur in the forward direction in each of the following. Assume that all reactants and products are in their standard states.

(a) $Sn^{4+}(aq) + 2\,I^-(aq) \longrightarrow Sn^{2+}(aq) + I_2(s)$
(b) $2\,MnO_2(s) + 3\,ClO^-(aq) + 2\,OH^-(aq)$
$$\longrightarrow 2\,MnO_4{}^-(aq) + 3\,Cl^-(aq) + H_2O(l)$$

44. Predict whether a spontaneous reaction will occur in the forward direction in each of the following. Assume that all reactants and products are in their standard states.

(a) $Sn(s) + Co^{2+}(aq) \longrightarrow Sn^{2+}(aq) + Co(s)$
(b) $6\,Br^-(aq) + Cr_2O_7{}^{2-}(aq) + 14\,H^+(aq)$
$$\longrightarrow 2\,Cr^{3+}(aq) + 7\,H_2O(l) + 3\,Br_2(l)$$

45. Predict whether each of the following processes will proceed in the forward direction to any appreciable extent.

(a) the reduction of $Sn^{4+}(aq)$ to $Sn^{2+}(aq)$ by $Cu(s)$
(b) the oxidation of $I_2(s)$ to $IO_3{}^-(aq)$ by $O_3(g)$ in acidic solution
(c) the oxidation of $Cr(OH)_3(s)$ to $CrO_4{}^{2-}(aq)$ by H_2O_2 in basic solution

46. Predict whether each of the following processes will proceed in the forward direction to any appreciable extent.

(a) the displacement of $Cd^{2+}(aq)$ by $Al(s)$
(b) the oxidation of $Cl^-(aq)$ to $Cl_2(g)$ by $Br_2(l)$
(c) the oxidation of $Cl^-(aq)$ to $ClO_3{}^-(aq)$ by H_2O_2 in basic solution

47. Silver does not react with $HCl(aq)$, but it does react with $HNO_3(aq)$. (a) Explain the difference in the behavior of silver toward these two acids. (b) Write a plausible net ionic equation for the reaction of silver with $HNO_3(aq)$.

48. Can we use sodium metal to displace Mg^{2+} from aqueous solution? If the reaction does occur, write the half-equations and the overall equation. If the displacement reaction does *not* occur, write the equation for the reaction that does occur.

49. Palladium is a rare metal used as a catalyst. Copper and silver will both displace Pd^{2+} from aqueous solution. The metal itself will react with $HNO_3(aq)$ producing $Pd^{2+}(aq)$ and $NO(g)$. Estimate a value of $E^\circ_{Pd^{2+}/Pd}$.

50. Rhodium is a rare metal used as a catalyst. The metal does not react with $HCl(aq)$, but it does react with $HNO_3(aq)$, producing $Rh^{3+}(aq)$ and $NO(g)$. Copper will displace Rh^{3+} from aqueous solution, but silver will not. Estimate a value of $E^\circ_{Rh^{3+}/Rh}$.

$E^\circ_{cell}, \Delta G^\circ, K_{eq}$

51. Determine the values of E°_{cell} and ΔG° for the following reactions.

(a) $O_2(g) + 4\,I^-(aq) + 4\,H^+(aq) \longrightarrow 2\,H_2O(l) + 2\,I_2(s)$
(b) $Cr_2O_7{}^{2-}(aq) + 3\,Cu(s) + 14\,H^+(aq)$
$$\longrightarrow 2\,Cr^{3+}(aq) + 3\,Cu^{2+}(aq) + 7\,H_2O(l)$$

52. Determine the values of E°_{cell} and ΔG° for the following reactions.

(a) $Al(s) + 3\,Ag^+(aq) \longrightarrow Al^{3+}(aq) + 3\,Ag(s)$
(b) $4\,IO_3{}^-(aq) + 4\,H^+(aq)$
$$\longrightarrow 2\,I_2(s) + 2\,H_2O(l) + 5\,O_2(g)$$

53. Write the equilibrium constant expression for each of the following reactions, and determine the numerical value of K_{eq} at 25 °C.

(a) $PbO_2(s) + 4 H^+(aq) + 2 Cl^-(aq)$
$$\rightleftharpoons Pb^{2+}(aq) + 2 H_2O(l) + Cl_2(g)$$

(b) $3 O_2(g) + 2 Br^-(aq) \rightleftharpoons 2 BrO_3^-(aq)$ (basic solution)

54. Write the equilibrium constant expression for each of the following reactions, and determine the numerical value of K_{eq} at 25 °C.

(a) $Ag^+(aq) + Fe^{2+}(aq) \rightleftharpoons Fe^{3+}(aq) + Ag(s)$

(b) $MnO_2(s) + 4 H^+(aq) + 2 Cl^-(aq)$
$$\rightleftharpoons Mn^{2+}(aq) + 2 H_2O(l) + Cl_2(g)$$

(c) $2 OCl^-(aq) \rightleftharpoons 2 Cl^-(aq) + O_2(g)$ (basic solution)

55. A strip of copper metal is immersed in 1.00 M $Ag^+(aq)$ at 25 °C. What reaction occurs? What will the cation concentrations in solution be at equilibrium?

56. A strip of tin metal is immersed in 1.00 M $Pb^{2+}(aq)$ at 25 °C. What reaction occurs? What will the cation concentrations in solution be at equilibrium?

Effect of Concentration on E_{cell}

Use data from Table 18.1 and Appendix C, as necessary.

57. What is the value of E_{cell} of each of the following reactions when carried out in a voltaic cell?

(a) $Pb(s) + 2 H^+(0.0025 M)$
$$\longrightarrow Pb^{2+}(0.85 M) + H_2(g, 0.95 atm)$$

(b) $ClO_3^-(0.65 M) + 3 Mn^{2+}(0.25 M) + 3 H_2O(l)$
$$\longrightarrow Cl^-(1.50 M) + 3 MnO_2(s) + 6 H^+(1.25 M)$$

58. What is the value of E_{cell} of each of the following reactions when carried out in a voltaic cell?

(a) $Fe(s) + 2 Ag^+(0.0015 M) \longrightarrow Fe^{2+}(1.33 M) + 2 Ag(s)$

(b) $4 VO^{2+}(0.050 M) + O_2(g, 0.25 atm) + 2 H_2O(l)$
$$\longrightarrow 4 VO_2^+(0.75 M) + 4 H^+(0.30 M)$$

59. What is E_{cell} for the voltaic cell diagrammed below?

$$Pt|H_2(g, 1 atm)|0.0025 M HCl\|H^+(1 M)|H_2(g, 1 atm)|Pt$$

60. What is E_{cell} for the voltaic cell diagrammed below?

$$Pt|H_2(g, 1 atm)|0.0675 M HCl\|0.0250 M KOH|H_2(g, 1 atm)|Pt$$

61. The voltaic cell diagrammed below registers $E_{cell} = 0.108$ V.

$$Pt, H_2(g, 1 atm)|H^+(x M)\|H^+(1 M)|H_2(g, 1 atm),Pt$$

What is the pH of the unknown solution?

62. A voltaic cell represented by the following cell diagram has $E_{cell} = -0.015$ V. Calculate $[Ag^+]$ in the cell.

$$Pt|Fe^{2+}(0.125 M),Fe^{3+}(0.068 M)\|Ag^+(x M)|Ag(s)$$

63. *Without doing detailed calculations,* determine which of the following voltaic cells should have the higher cell voltage. Explain your reasoning.

(a) $Cu(s)|Cu^{2+}(1.25 M)\|Ag^+(0.55 M)|Ag(s)$
(b) $Cu(s)|Cu^{2+}(0.12 M)\|Ag^+(0.60 M)|Ag(s)$

64. *Without doing detailed calculations,* determine which of the following voltaic cells has the highest concentration of Cu^{2+}. Explain your reasoning.

(a) $Zn(s)|Zn^{2+}(1.00 M)\|Cu^{2+}(? M)|Cu(s)$ $E_{cell} = 1.06$ V
(b) $Zn(s)|Zn^{2+}(0.10 M)\|Cu^{2+}(? M)|Cu(s)$ $E_{cell} = 1.15$ V
(c) $Zn(s)|Zn^{2+}(1.0 \times 10^{-3} M)\|Cu^{2+}(? M)|Cu(s)$
$$E_{cell} = 1.16 V$$

Batteries

65. Describe plausible electrode reactions and the cell reaction for a cell having a Mg anode and an O_2 cathode. Determine its E_{cell}°.

66. Describe plausible electrode reactions and the cell reaction for a cell having a Zn anode and a Cl_2 cathode. Determine its E_{cell}°.

67. Silver–zinc cells, called button batteries, are tiny cells used in watches, electronic calculators, hearing aids, and cameras. The battery has the following cell diagram.

$$Zn(s),ZnO(s)|KOH(sat'd aq)|Ag_2O(s),Ag(s)$$

Its storage capacity is about six times that of a lead–acid cell of the same size. Write equations for the half-reactions and the overall reaction that occur when the cell is discharged.

68. A mercury battery, once widely used, is being phased out because of the problem of disposal of the toxic mercury. A simplified cell diagram for the battery is

$$Zn(s),ZnO(s)|KOH(sat'd aq)|HgO(s),Hg(l)$$

Write equations for the half-reactions and the overall reaction that occur when the cell is discharged.

Corrosion

69. Why are water, an electrolyte, and oxygen all required for the corrosion of iron?

70. Iron can be protected from corrosion by copper plating or by zinc plating. Both methods are effective as long as the plating remains intact. If a break occurs in the plating, however, zinc plating proves to be far superior to copper plating in providing protection to the underlying iron. Explain.

71. Explain why the term "sacrificial anode" is appropriate in describing one method of protecting iron from corrosion.

72. Describe how silver tarnish can be removed without the loss of silver.

73. Example 18.10 and Exercises 18.10A and 18.10B dealt with interpreting parts **(a)** and **(b)** of Figure 18.20. The metal shown with the iron nail in part **(c)** of photograph at the right is zinc.

Explain how and why part **(c)** of this photograph differs from Figure 18.20a.

74. Explain how and why part **(d)** of the photograph below differs from Figure 18.20a. The metal shown with the iron nail is copper.

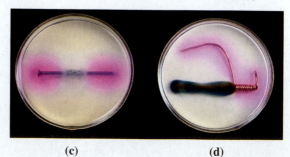

(c)   **(d)**

Electrolysis

75. Write a net ionic equation for the expected reaction when the electrolysis of $Cu(NO_3)_2(aq)$ is conducted using **(a)** a copper anode and a copper cathode, **(b)** an inert platinum anode and an iron cathode, **(c)** inert platinum for both electrodes.

76. Write a net ionic equation for the expected reaction when the electrolysis of $NiSO_4(aq)$ is conducted using **(a)** a nickel anode and an iron cathode, **(b)** a nickel anode and an inert platinum cathode, **(c)** an inert platinum anode and a nickel cathode.

77. Use electrode potential data to predict the probable products and the minimum voltage required in the electrolysis with inert platinum electrodes of each of the following.

  (a) $BaCl_2(l)$ **(b)** $HBr(aq)$ **(c)** $NaNO_3(aq)$

78. Use electrode potential data to predict the probable products and the minimum voltage required in the electrolysis with inert platinum electrodes of each of the following.

  (a) $ZnSO_4(aq)$ **(b)** $MgBr_2(l)$ **(c)** $NiCl_2(aq)$

79. How many grams of silver are deposited at a platinum cathode in the electrolysis of $AgNO_3(aq)$ by 1.73 A of electric current in 2.05 h?

80. How many mL of $H_2(g)$, measured at 23.5 °C and 749 mmHg, are produced at a platinum cathode in the electrolysis of $H_2SO_4(aq)$ by 2.45 A of electric current in 5.00 min?

81. How many coulombs of electric charge are required to deposit 25.0 g Cu(s) at the cathode in the electrolysis of $CuSO_4(aq)$?

82. What is the electric current, in amperes, if 212 mg Ag(s) is deposited at the cathode in 1435 s in the electrolysis of $AgNO_3(aq)$?

83. *Without doing detailed calculations*, determine which of the following will yield the greatest mass of metal deposited on a platinum cathode when the solution is electrolyzed for exactly 1 h with a current of 1.00 A. Explain your choice.

  (a) $Cu(NO_3)_2(aq)$ **(c)** $Zn(NO_3)_2(aq)$

  (b) $AgNO_3(aq)$ **(d)** $NaNO_3(aq)$

84. *Without doing detailed calculations,* determine which of the following solutions will take the longest time to electrolyze under these conditions: Equal volumes of the solutions are used, and electrolysis is carried out with inert platinum electrodes and a current of 1.00 A. Electrolysis is stopped when the solution concentration falls to half its initial value. Explain your choice.

  (a) 0.50 M $Cu(NO_3)_2$ **(c)** 0.80 M $AgNO_3$

  (b) 0.75 M HCl   **(d)** 0.30 M $Zn(NO_3)_2$

Additional Problems

Problems marked with an * may be more challenging than others.

85. When $P_4(s)$ is heated with water, it disproportionates to phosphine, $PH_3(g)$, and phosphoric acid. Write a balanced equation for this reaction.

86. Cyanide wastes can be detoxified by adding chlorine gas to a basic solution of the wastes. The cyanide ion is converted to cyanate ion (OCN^-), and the chlorine is reduced to chloride ion. Following the addition of some acid so that the solution is not quite so basic, further reaction with chlorine gas converts cyanate ion to hydrogen carbonate (bicarbonate) ion and nitrogen gas. Write balanced equations for the two reactions just described.

*87. Balance the following redox equations by the half-reaction method.

  (a) $B_2Cl_4 + OH^- \longrightarrow BO_2^- + Cl^- + H_2O + H_2(g)$

  (b) $CH_3CH_2ONO_2 + Sn + H^+$
$$\longrightarrow CH_3CH_2OH + NH_2OH + Sn^{2+} + H_2O$$

  (c) $F_5SeOF + OH^- \longrightarrow SeO_4^{2-} + F^- + H_2O + O_2(g)$

  (d) $As_2S_3 + OH^- + H_2O_2 \longrightarrow AsO_4^{3-} + SO_4^{2-} + H_2O$

  (e) $XeF_6 + OH^-$
$$\longrightarrow XeO_6^{4-} + F^- + H_2O + Xe(g) + O_2(g)$$

88. Although we have used the half-reaction method only to balance equations for redox reactions occurring in aqueous solution, the method can actually be used more widely. Balance the following equation by the half-reaction method, and explain why the method works.

$$NO(g) + H_2(g) \longrightarrow NH_3(g) + H_2O(g)$$

89. Figure 18.3 describes what happens when a zinc electrode is partially immersed in $ZnSO_4(aq)$. Will a similar description apply to a sodium electrode partially immersed in NaCl(aq)? Explain.

90. Consider the silver plating of an iron spoon in Figure 18.23. Show that, in principle, Ag^+ should be reduced to Ag(s) simply by immersing the spoon in $AgNO_3(aq)$. Why do you suppose that electroplating is used rather than displacement of Ag^+ from solution?

*91. How many milliliters of $O_2(g)$, collected over water and measured at 20.0 °C and a barometric pressure of 761.5 mmHg, should be liberated at a platinum anode at the same time that 1.02 g Ag(s) is deposited at a platinum cathode in the electrolysis of $AgNO_3(aq)$?

92. The electrolysis of 0.100 L of 0.785 M $AgNO_3(aq)$ using platinum electrodes is carried out with a current of 1.75 A. What is the molarity of the $AgNO_3(aq)$ 25.0 min after electrolysis is begun?

93. Use standard electrode potential data to show that $MnO_2(s)$ should not react with HCl(aq) to liberate $Cl_2(g)$. Yet when $MnO_2(s)$ and concentrated HCl(aq) are heated together, $Cl_2(g)$ does form. In fact, this is a common laboratory method of generating small quantities of $Cl_2(g)$. Explain why the reaction occurs.

94. What minimum voltage is required to recharge a lead–acid storage battery? If a voltage much greater than this minimum is used, a potentially explosive mixture of gases could accumulate in the battery. Explain why this is so.

95. Calculate E_{cell} for the following voltaic cell.

$$Pt, H_2(g, 1 \text{ atm})|CH_3COOH(1 \times 10^{-4} \text{ M})\|$$
$$H^+(0.010 \text{ M})|H_2(g, 1 \text{ atm}), Pt$$

96. Calculate E_{cell} for the following voltaic cell.

$$Pt|H_2(g, 1 \text{ atm})|H^+(0.010 \text{ M})\|$$
$$NH_3(0.45 \text{ M}), NH_4^+(0.15 \text{ M})|H_2(g, 1 \text{ atm})|Pt$$

*97. What is E_{cell} of the following voltaic cell?

$$Cu(s)|Cu^{2+}(0.10 \text{ M})\|Ag_2CrO_4(\text{sat'd aq})|Ag(s)$$

*98. What $[Cl^-]$ should be maintained in the anode half-cell if the following voltaic cell is to have $E_{cell} = 0.100$ V?

$$Ag(s)|AgCl(s)|Cl^-(x \text{ M})||Cu^{2+}(0.25 \text{ M})|Cu(s)$$

*99. What happens to the voltage of the concentration cell in Figure 18.14 as it operates over a period of time? That is, does the voltage increase, decrease, or remain constant? If the cell operates continuously, does it stop producing electricity at some point? If so, what is the condition in each half-cell compartment at this point?

*100. The *efficiency value*, ε, of a fuel cell reaction is $\varepsilon = \Delta G°/\Delta H°$. What is the efficiency value of the methane–oxygen fuel cell described on page 778? What is the theoretical voltage, $E°_{cell}$, of this fuel cell?

101. In the original construction of the Statue of Liberty, a framework of iron ribs was covered with thin sheets of copper less than 2.5 mm thick. The copper skin and iron framework were separated by a layer of asbestos. The asbestos wore away with time, and the iron ribs corroded to the extent that some of them lost more than half their mass in the 100 years before the statue was restored. The copper skin lost only about 4% of its thickness in those 100 years. Use electrochemical principles to explain these observations.

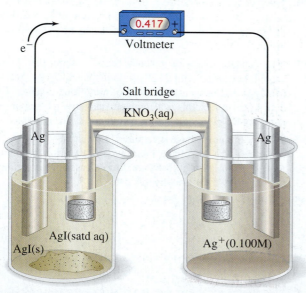

*102. The graphic summary given in Figure 18.12 indicates that K_{eq} values can sometimes be obtained from electrical measurements. For the voltaic cell pictured below, (a) write a cell diagram, and (b) with data from the figure and elsewhere in the text, calculate a value of K_{sp} for AgI.

103. Refer to the discussion of the aluminum–air battery on page 777. (a) How many grams of aluminum are consumed if a 10.0-A electric current is drawn from the battery for 4.00 h?

(b) Use data from page 777 and Appendix C to obtain a value of $\Delta G°_f\{[Al(OH)_4]^-\}$.

*104. A 250.0-mL sample of 0.1000 M $CuSO_4(aq)$ is electrolyzed with a current of 3.512 A for 1368 s. Sufficient NH_3 is added to complex any remaining Cu^{2+} and to maintain a free $[NH_3] = 0.10$ M. If the blue color of $[Cu(NH_3)_4]^{2+}$ is detectable at concentrations as low as 1×10^{-5} M, will the blue color be seen in this case?

*105. Refer to Example 18.12 and the electrochemical cells described there through cell diagrams. What will be the concentrations of the ions in each half-cell of each electrochemical cell when electric current no longer flows between them?

*106. Consider the following reversible reaction with the indicated initial concentrations. What will the ion concentrations be when equilibrium is reached?

$$Hg^{2+}(0.250 \text{ M}) + 2 Fe^{2+}(0.180 \text{ M})$$
$$\rightleftharpoons 2 Fe^{3+}(0.210 \text{ M}) + Hg(l)$$

*107. When AgCl(s) is added to a solution of $Br^-(aq)$, this reversible reaction occurs.

$$AgCl(s) + Br^-(aq) \rightleftharpoons AgBr(s) + Cl^-(aq)$$

If the initial $[Br^-] = 0.4000$ M, what will be the concentration of Br^- at equilibrium?

*108. The potential of a copper metal electrode in Cu(II) solution is measured (against a standard hydrogen electrode) at several different concentrations of copper(II) ion. What variables should be plotted to obtain a straight line? What will be the slope of that line? Would the answer be different for a silver electrode in solutions of different concentrations of silver ion?

109. Using electrochemical techniques, chemists can detect tiny amounts of certain metal ions. One type of analysis, involving electrolysis of Cd^{2+}, can detect a signal from 1 picoampere (pA) of current that lasts only 1 ms. How much Cd^{2+}, in moles, is detected by this technique? How many Cd^{2+} ions is this?

110. Calculate the mass of lead metal that dissolves in an automobile battery when the starter is cranked for 30 s. A typical starter draws 80-A current. Use the discussion of the lead–acid battery to determine the appropriate half-reaction.

111. The electrolysis of $Na_2SO_4(aq)$ by a 6-V battery is conducted in two separate half-cells joined by a salt bridge also containing $Na_2SO_4(aq)$. The cell diagram for the electrolysis is

$$Pt|Na_2SO_4(aq)||Na_2SO_4(aq)|Pt$$

Phenolphthalein indicator is added to each half-cell.

(a) What are the likely half-reactions occurring at the anode and at the cathode of the electrolysis cell? Describe any color changes occurring in the half-cell compartments.

(b) Could the electrolysis also be carried out with a 1.5-V dry cell battery? Explain.

(c) After electrolysis is stopped, the solutions of the two half-cells are mixed. Describe and explain any color changes that occur.

(d) Explain why your answer in (c) does not depend on the concentration of the $Na_2SO_4(aq)$, the volumes of the solutions in the half-cells, the amount of current used in the electrolysis, or on how long the electrolysis is carried out.

(e) In one experiment, a 10.00-mL sample of HCl(aq) of unknown concentration is added to the cathode compartment along with phenolphthalein. Electrolysis is carried out with a 23.2-mA (milliampere) current, and the solution color becomes pink after 8 min, 22 s. What was the molarity of the HCl(aq) sample?

112. Hydrogen–oxygen fuel cells are used on spacecraft to produce electricity.

   (a) Write equations for the two half-reactions, label them as occurring at the anode and cathode, and give the equation for the overall reaction.

   (b) Calculate E_{cell} assuming $P_{O_2} = 2.0$ atm and $P_{H_2} = 2.0$ atm.

   (c) If the spacecraft carries two 225-L cylinders of H_2 and one 225-L cylinder of O_2, each initially at a pressure of 205 atm and maintained at a temperature of 298 K, how much ener-

gy, in kilojoules, can a fuel cell produce if it achieves 82% efficiency?

   (d) If the fuel cell is to supply energy for 10 days, what is its average power output in watts?

   (e) What assumptions limit the accuracy of this calculation?

* **113.** At 25 °C, the following cell was reported to have a voltage of 1.172 V.

$$Pt|H_2(1 \text{ atm})|NaOH(1 \text{ M})|Ag_2O(s)|Ag(s)$$

Use this information and any other necessary data from Appendix C to calculate the partial pressure of oxygen in equilibrium with silver oxide and silver at 25 °C. Does pure silver spontaneously oxidize in air? Explain.

Apply Your Knowledge

114. **[Laboratory]** In practical use, the saturated calomel electrode and the silver–silver chloride electrode are more convenient reference electrodes than the standard hydrogen electrode. The half-reactions are $Hg_2Cl_2(s) + 2 e^- \longrightarrow 2 Hg(l) + 2 Cl^-(aq)$ and $AgCl(s) + e^- \longrightarrow Ag(s) + Cl^-(aq)$. Both electrodes employ a saturated solution of KCl. Explain why their potentials remain constant. What should happen to the potential—more positive or more negative—if the temperature increases?

115. **[Laboratory]** In a *coulometric titration*, a reagent is generated electrochemically; the reagent then reacts with the substance being determined. Since the reagent reacts as it is generated, unstable reagents can be used. In the determination of trace amounts of cyclohexene in the effluent from a factory, a coulometer used a solution of bromide ion to generate elemental bromine, which then reacted with cyclohexene:

$$C_6H_{10} + Br_2 \longrightarrow C_6H_{10}Br_2$$

The coulometer produced 9.65 mA, and it took 235 s to generate enough bromine to react stoichiometrically with the cyclohexene. Calculate the mass of cyclohexene in the solution.

116. **[Biochemical]** Nitroglycerin (1,2,3-glyceryl trinitrate) is a high explosive, but it is also used in tiny amounts to treat chest pains

(angina) caused by clogged heart arteries. The compound is reduced in the body to 1,2-glyceryl dinitrate and nitrite ions. The nitrite ions are reduced in mitochondria to nitrogen monoxide, a signaling molecule that dilates blood vessels, increasing the flow of blood to heart muscle. (a) Convert the descriptions to half-equations (acid solution). (b) Assuming that the other half reaction in each case is

$$NADH \longrightarrow NAD^+ + H^+ + 2 e^-$$

write and balance the two redox equations. (*Hint:* 1,2-glyceryl trinitrate and 1,2,3-glyceryl trinitrate have the same structure as glycerol, but with two and three $-ONO_2$ groups, respectively, in place of the hydroxyl groups.)

* **117.** **[Laboratory]** An acid–base titration is carried out in an electrochemical cell in which 100.00 mL of 0.0350 M NaOH(aq) and a hydrogen electrode are the anode half-cell, and a standard hydrogen electrode is the cathode half-cell. Determine the pH in the anode half-cell and E_{cell} after the following volumes of 0.150 M HCl(aq) are added: (a) 0.00 mL, (b) 5.00 mL, (c) 10.00 mL, (d) 15.00 mL, (e) 22.00 mL, (f) 23.00 mL, (g) 24.00 mL, (h) 25.00 mL. Sketch two titration curves, one based on pH and the other on E_{cell}. How do they compare?

 # e-Media Problems

The activities described in these problems can be found in the e-Media Activities and Interactive Student Tutorial (IST) modules of the Companion Website, *http://chem.prenhall.com/hillpetrucci*.

118. In the **Oxidation–Reduction Reactions (Part II)** animation (*Section 18-1*), (a) for each of the reactions, would the mass of the zinc rod increase or decrease with the progress of the reaction? (b) For a fixed number of electrons transferred, will the change in mass of zinc be greater for the first or the second reaction?

119. In the **Standard Reduction Potentials** animation (*Section 18-4*), (a) what spontaneous reaction would occur if the two metal half-cells were joined? (b) Write the balanced equation for this reaction. (c) Determine the standard cell potential for this spontaneous reaction.

120. For the first electrochemical reaction shown in the **Voltaic Cells (II): The Zinc–Hydrogen Cell** animation (*Section 18-4*), predict the qualitative change in E_{cell} if (a) the initial $Zn^{2+}(aq)$ concentration was increased and (b) the initial $H_3O^+(aq)$ concentration was decreased. For the second reaction, what would be the effect on E_{cell} of eliminating the presence of hydrogen gas above the solution in the hydrogen cell?

121. In the **Redox Chemistry of Iron and Copper** movie (*Section 18-5*), (a) write the half-equations for the process shown. (b) Draw an atomic-scale picture of the solid–liquid interface that illustrates the changes occurring at the interface as a result of the two half-reactions.

122. For the reaction seen in the **Electroplating** movie (*Section 18-11*), (a) calculate the current required to deposit 0.15 g of metallic Cr in 10.0 min, and (b) indicate on which electrode (cathode or anode) the chromium is deposited.

Chapter 19

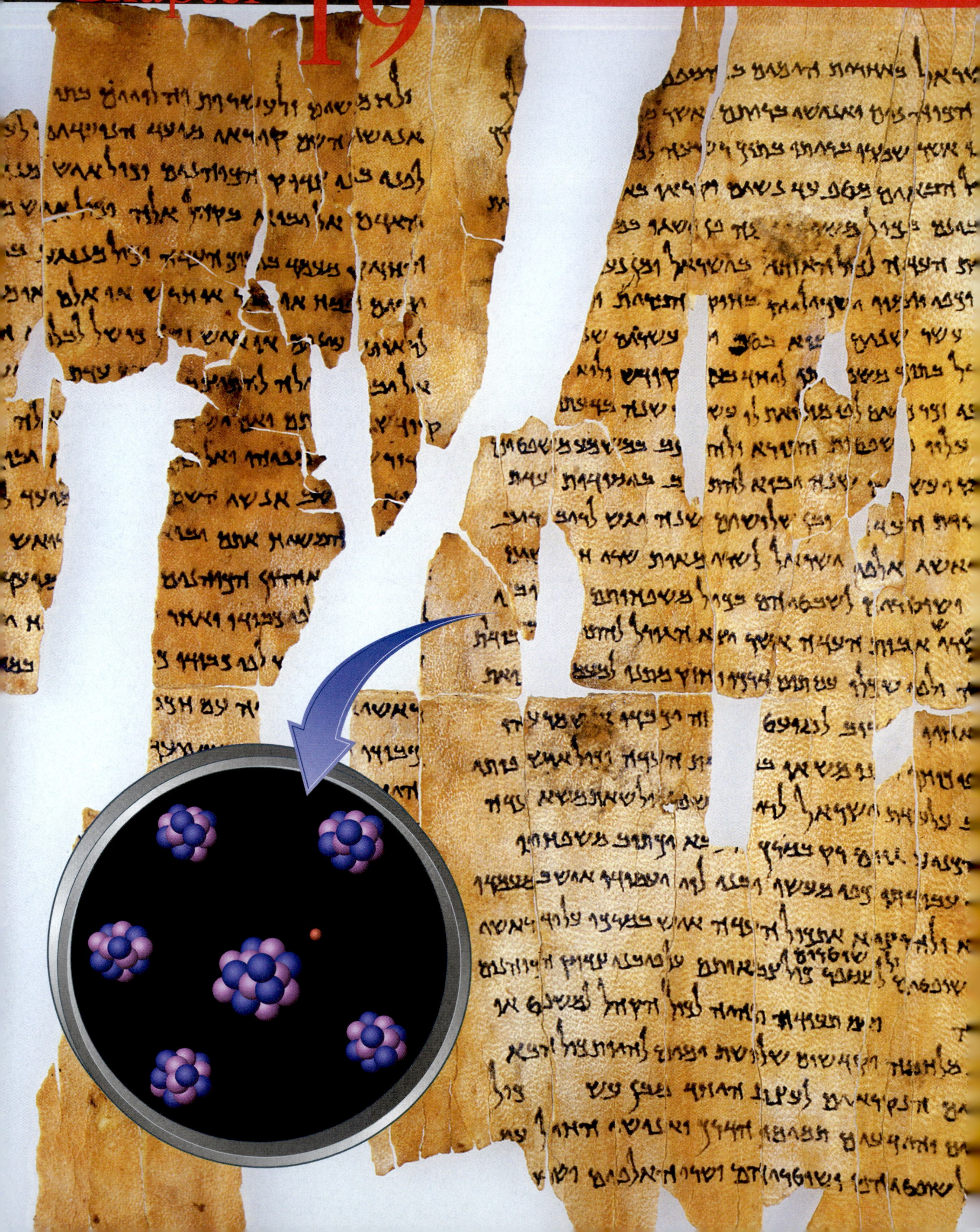

Nuclear Chemistry

RADIOACTIVITY, OR RADIOACTIVE DECAY, is the spontaneous change of the nuclei of certain atoms, accompanied by the emission of subatomic particles and/or high-frequency electromagnetic radiation. Our main concern in this chapter will be radioactivity and its applications to chemistry, the life sciences, and medicine. We will also consider such topics as nuclear energy and the effect of ionizing radiation on matter.

19.1 Radioactivity and Nuclear Equations

There are five principal ways in which atomic nuclei may display radioactivity. That is, there are five principal types of radioactive decay, and we can describe them in terms of the radiation emitted. In our discussion, we will write *nuclear equations* to represent nuclear changes. A nucleus with a specified number of protons and neutrons and a specified energy is called a **nuclide.*** In a nuclear equation, nuclides and emitted particles are represented in the form $^A_Z E$, where E is the chemical symbol for the element undergoing radioactive decay, Z is the atomic number of the element, and A is the mass number of the isotope. Protons and neutrons are collectively called **nucleons,** and the basic principle in writing a nuclear equation is that nucleons are conserved in a nuclear reaction (just as atoms are conserved in a chemical reaction). That is,

> *The two sides of a nuclear equation must have the same totals of atomic numbers and mass numbers.*

---

*We have previously noted that two or more nuclides of the same element are called isotopes, for example, $^{12}_6 C$ and $^{13}_6 C$.

◄ The Dead Sea scrolls were discovered in Palestine in 1947. How old are they? Nature left an imprint on the linen wrappings of the scrolls, an imprint that is almost as definite as a date stamp, but we must learn how to read it. Reading this particular date stamp requires that we determine the quantity of carbon-14 present in the linen. No practical *chemical* test can distinguish the very rare carbon-14 from the more abundant carbon-12 and carbon-13, for the three isotopes are virtually identical from a chemical perspective. However, there is an important difference in the *nuclear* properties of the isotopes: carbon-14 is *radioactive*. The inset shows stable carbon-12 and carbon-13 nuclei, along with the decay of a nucleus of carbon-14 by emission of a *beta particle*. Using the proper instruments, the rate of beta particle emission can be detected to determine the age of the linen. The phenomena described here—radioactivity, beta emission, detection of radiation—are some of the topics discussed in this chapter.

Types of Radioactive Decay
activity

Table 19.1 Types of Radioactive Decay: A Summary

Mode of decay	Radiation emitted	Changes in nucleus	
		Atomic number	**Mass number**
Alpha emission (α)	$^{4}_{2}\text{He}$	-2	-4
Beta emission (β^-)	$^{0}_{-1}\text{e}$	$+1$	0
Gamma emission (γ)	$^{0}_{0}\gamma$	0	0
Positron emission (β^+)	$^{0}_{1}\text{e}$	-1	0
Electron capture (EC)	X rays	-1	0

The common products of radioactive decay are summarized in Table 19.1 and are discussed more fully in the following paragraphs.

Alpha-particle emission: An **alpha (α) particle** has the same composition as a helium nucleus: two protons and two neutrons. Thus an α particle has a mass of 4 u and a charge of 2+. Because they carry a positive charge, α particles are deflected in electric and magnetic fields. Their deflection in an electric field is illustrated in Figure 19.1. The penetrating power of α particles through matter is so low that the particles can generally be stopped by a sheet of paper. The symbol for an α particle is $^{4}_{2}\text{He}$. We can represent α-particle emission with a nuclear equation, as in the radioactive decay of uranium-238 to thorium-234:

$$\textit{Mass numbers:} \quad 238 = 234 + 4$$
$$^{238}_{92}\text{U} \longrightarrow {}^{234}_{90}\text{Th} + {}^{4}_{2}\text{He}$$
$$\textit{Atomic numbers:} \quad 92 = 90 + 2$$

When a nuclide emits an α particle, *the atomic number of the nuclide decreases by 2 and its mass number decreases by 4.* Because the number of protons in the nuclide changes as a result of the decay, the newly formed nuclide is of a different element (here, thorium).

Beta-particle emission: **Beta (β^-) particles** are electrons. Like all other electrons, β particles have very little mass and carry a charge of 1−. They are deflected in electric and magnetic fields, but in the direction opposite the direction of α-particle deflection (Figure 19.1). Beta particles are more penetrating than α particles and can pass through an aluminum sheet 2–3 mm thick. The symbol for a β particle is $^{0}_{-1}\text{e}$.

Although an atomic nucleus contains the protons and neutrons that make up an α particle, a nucleus does not contain electrons. Instead, the electrons emitted as β particles are created as a neutron is converted to a proton plus an electron:

$$^{1}_{0}\text{n} \longrightarrow {}^{1}_{1}\text{p} + {}^{0}_{-1}\text{e}$$

Separation of Alpha, Beta, and
Gamma Rays animation

Because the atomic number is the number of protons, it also represents the positive charge on a nucleus. We therefore assign the neutron an atomic number of 0 (no charge). The electron has the equivalent of an atomic number of −1; it carries the

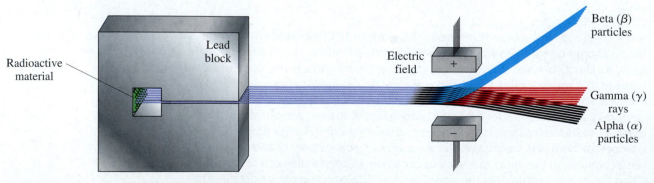

▲ **FIGURE 19.1   Three types of radiation from radioactive materials**

QUESTION: How would the products of positron emission and electron capture be affected by the electric field shown in this drawing?

same charge as a proton, but negative in sign. An example of a radioactive decay that produces β particles is the decay of thorium-234 to protactinium-234:

$$
\begin{array}{lcccc}
\textit{Mass numbers:} & 234 & = & 234 & + & 0 \\
& {}^{234}_{90}\text{Th} & \longrightarrow & {}^{234}_{91}\text{Pa} & + & {}^{0}_{-1}\text{e} \\
\textit{Atomic numbers:} & 90 & = & 91 & + & -1
\end{array}
$$

When a nuclide emits a β particle, *the atomic number of the nuclide increases by 1 and its mass number is unchanged.* As with α decay, the decaying nuclide in β decay is of one element (here, thorium), and the new nuclide formed is of a different element (here, protactinium).

Gamma-ray emission: **Gamma (γ) rays** are a highly penetrating form of electromagnetic radiation. They consist of photons, and thus they are not particles of matter. They are emitted by energetic nuclei as a means of reaching a lower energy state. In a nuclear equation for γ-ray emission, we represent the energetic nucleus by affixing a symbol such as * to its chemical symbol. For example, in the radioactive decay of uranium-238 by first an α-particle emission, then two β-particle emissions, and finally another α-particle emission, 23% of the thorium-230 nuclei formed are in an excited state: ${}^{230}_{90}\text{Th}^*$. These nuclei then emit energy as γ rays:

$$
\begin{array}{lcccc}
\textit{Mass numbers:} & 230 & = & 230 & + & 0 \\
& {}^{230}_{90}\text{Th}^* & \longrightarrow & {}^{230}_{90}\text{Th} & + & \gamma \\
\textit{Atomic numbers:} & 90 & = & 90 & + & 0
\end{array}
$$

Gamma-ray emission may also accompany other forms of radioactive decay, such as α, β, or positron emission.

When a nuclide emits a γ ray, *both the atomic number of the nuclide and its mass number remain unchanged.* The new and old nuclides are of the same element. As we would expect for a form of electromagnetic radiation, γ rays are unaffected by electric and magnetic fields (Figure 19.1).

Positron emission: **Positrons** are particles having the same mass as electrons but carrying a charge of $1+$. They are sometimes called *positive electrons* and referred to as β^+ particles. Their penetrating power through matter is very limited because when a positron comes into contact with an electron, the two particles annihilate each other and are converted to two γ rays. Positrons are formed in the nucleus through the conversion of a proton to a neutron and a positron:

$$
{}^{1}_{1}\text{p} \longrightarrow {}^{1}_{0}\text{n} + {}^{0}_{1}\text{e}
$$

Positrons are most commonly emitted in the radioactive decay of certain nuclides of the lighter elements. About 82% of the radioactive decay of aluminum-26 to magnesium-26 is by positron emission:

$$
\begin{array}{lcccc}
\textit{Mass numbers:} & 26 & = & 26 & + & 0 \\
& {}^{26}_{13}\text{Al} & \longrightarrow & {}^{26}_{12}\text{Mg} & + & {}^{0}_{1}\text{e} \\
\textit{Atomic numbers:} & 13 & = & 12 & + & 1
\end{array}
$$

When a decaying nuclide emits a positron, *the atomic number of the nuclide decreases by 1 and its mass number is unchanged.* The new nuclide is that of an element different from the decaying nuclide.

Electron capture: **Electron capture (EC)** is a process in which the nucleus absorbs an electron from an inner electron shell, usually the first or second. An X ray is then released when an electron drops from a higher quantum level to fill the level vacated by the captured electron. Once inside the nucleus, the captured electron combines with a proton to form a neutron:

$$
{}^{0}_{-1}\text{e} + {}^{1}_{1}\text{p} \longrightarrow {}^{1}_{0}\text{n}
$$

Nuclear equations for electron capture usually show the captured electron as a reactant. Iodine-125, used in medicine to diagnose pancreatic function and intestinal fat absorption, decays to tellurium-125 by electron capture:

$$
\begin{array}{lccccc}
\textit{Mass numbers:} & 125 & + & 0 & = & 125 \\
& {}^{125}_{53}\text{I} & + & {}^{0}_{-1}\text{e} & \longrightarrow & {}^{125}_{52}\text{Te} \\
\textit{Atomic numbers:} & 53 & + & -1 & = & 52
\end{array}
$$

The result of electron capture is the same as that of positron emission. *The atomic number of the decaying nuclide decreases by 1, and its mass number is unchanged.* The new nuclide is that of an element different from the decaying nuclide.

Example 19.1

Radon-222 is enclosed in capsules as a radiation source for treatment of some types of cancer; phosphorus-32 is used to label red blood cells for blood volume determinations; aluminum-28 is produced in the bombardment of aluminum-27 by neutrons. Write a nuclear equation for (a) α-particle emission by radon-222, (b) β^- decay of phosphorus-32, (c) γ decay of aluminum-28.

STRATEGY

We need to recognize that nucleons are conserved in a nuclear reaction and that the changes nuclei undergo in radioactive decay are those summarized in Table 19.1. In this way, if we have data on all but one species in the nuclear equation—reactant or product—we can deduce data for the unknown.

SOLUTION

(a) We identify two of the species from the information given—radon-222 and $_{2}^{4}\text{He}$.

$$_{86}^{222}\text{Rn} \longrightarrow ? + {}_{2}^{4}\text{He}$$

The missing nuclide must have atomic number $Z = 86 - 2 = 84$ and mass number $A = 222 - 4 = 218$. The element with $Z = 84$ is polonium:

$$_{86}^{222}\text{Rn} \longrightarrow {}_{84}^{218}\text{Po} + {}_{2}^{4}\text{He}$$

(b) The atomic number increases by one unit, and the mass number remains constant in β^- emission. The two species described in the statement are phosphorus-32 and $_{-1}^{0}\text{e}$.

$$_{15}^{32}\text{P} \longrightarrow ? + {}_{-1}^{0}\text{e}$$

The missing nuclide must have $Z = 15 + 1 = 16$ and $A = 32$. The element with $Z = 16$ is sulfur:

$$_{15}^{32}\text{P} \longrightarrow {}_{16}^{32}\text{S} + {}_{-1}^{0}\text{e}$$

(c) Both the mass number and the atomic number remain constant in γ decay. On the right side of the equation, we simply remove the symbol * from Al and add a γ ray:

$$_{13}^{28}\text{Al*} \longrightarrow {}_{13}^{28}\text{Al} + \gamma$$

EXERCISE 19.1A

Write a nuclear equation for the decay of (a) radon-212 by α-particle emission, (b) argon-37 by electron capture, and (c) nickel-60 by γ-ray emission.

EXERCISE 19.1B

Write a nuclear equation for (a) the α-particle decay of an isotope to produce lead-214 and (b) the production of sulfur-36 through positron emission.

19.2 Naturally Occurring Radioactivity

Let us first consider a few of the naturally occurring radioactive nuclides among the lighter elements. Hydrogen-3 (tritium) and carbon-14 are formed by cosmic radiation entering Earth's upper atmosphere. They are found in trace amounts in the atmosphere and in compounds that incorporate H and C atoms—including living matter. Both hydrogen-3 and carbon-14 are β^- emitters:

$$_{1}^{3}\text{H} \longrightarrow {}_{2}^{3}\text{He} + {}_{-1}^{0}\text{e}$$
$$_{6}^{14}\text{C} \longrightarrow {}_{7}^{14}\text{N} + {}_{-1}^{0}\text{e}$$

Another naturally occurring radioactive isotope is potassium-40, which makes up 0.0118% of naturally occurring potassium atoms. There are three modes of radioactive decay of this nuclide—β^- emission, positron emission, and electron capture:

$$_{19}^{40}\text{K} \longrightarrow {}_{20}^{40}\text{Ca} + {}_{-1}^{0}\text{e} \qquad {}_{19}^{40}\text{K} \longrightarrow {}_{18}^{40}\text{Ar} + {}_{+1}^{0}\text{e} \qquad {}_{19}^{40}\text{K} + {}_{-1}^{0}\text{e} \longrightarrow {}_{18}^{40}\text{Ar}$$

When Earth was young, potassium-40 was much more abundant than it is today. Practically all the argon in Earth's atmosphere is argon-40, and scientists think that it has come from the radioactive decay of potassium-40.

Radioactive Decay Series

Most of the naturally occurring nuclides of the lighter elements have stable nuclei; these nuclides are not radioactive. In contrast, *all* nuclides of the heaviest elements are radioactive. A dividing line is found in the element bismuth ($Z = 83$). Thus, all nuclides with atomic number greater than 83 are radioactive. We will have more to say about this difference in nuclear properties between the light and heavy elements when we discuss nuclear stability in Section 19.6.

Even though they are radioactive, many nuclides of high atomic number are found in natural sources. And a few of them decay very slowly, as reflected in their half-lives. The half-life for radioactive decay is similar to the half-life of a chemical reaction. The **half-life ($t_{1/2}$)** of a radioactive nuclide is the time required for one-half of a statistically large number of nuclei in a sample of the nuclide to decay. Half-lives range from microseconds—radium-218 has a half-life of 14 μs—to billions of years—the value for uranium-238 is 4.51×10^9 years.

A series of radioactive decays beginning with a long-lived radioactive nuclide and ending with a nonradioactive one is called a **radioactive decay series.** As an example, nuclear equations for the first few steps in the uranium-238 series are written in the margin here, and the entire series is summarized in Figure 19.2.

This natural decay series helps us to understand a number of interesting observations. Even though some of the nuclides in the decay series have very short half-lives, all the nuclides in the series exist because they are constantly being formed. The 1600-year half-life of radium-226 gives an expectation of less than 1 gram of radium in the several tons of uranium ore processed by Marie Curie in 1898. Thus we can truly marvel at her feat of extracting 120 milligrams of $RaCl_2$. For the nuclide polonium-210, which has a half-life of only 138 days, the expectation is that there was only a fraction of a milligram of polonium present, and it is understandable that she could not isolate it.

$$^{238}_{92}U \longrightarrow {}^{234}_{90}Th + {}^{4}_{2}He$$
$$^{234}_{90}Th \longrightarrow {}^{234}_{91}Pa + {}^{0}_{-1}e$$
$$^{234}_{91}Pa \longrightarrow {}^{234}_{92}U + {}^{0}_{-1}e$$
$$^{234}_{92}U \longrightarrow {}^{230}_{90}Th + {}^{4}_{2}He$$
$$\vdots \qquad \vdots \qquad \vdots$$

Radioactive Decay Series activity

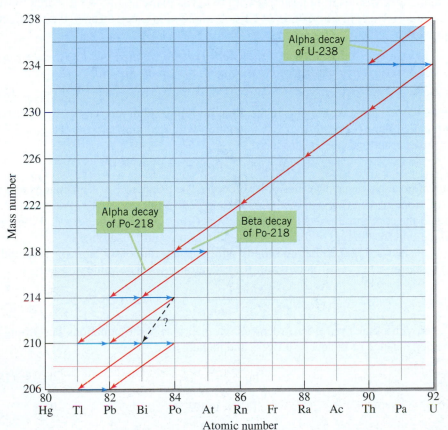

◀ **FIGURE 19.2 The natural radioactive decay series for $^{238}_{92}U$**

The long red arrows pointing down and to the left correspond to α-particle emissions. The short blue horizontal arrows represent β^--particle emissions.

QUESTION: Can radioactive decay follow the path of the line marked with a question mark?

Kinetics of Radioactive Decay activity

The Age of Earth

Radioactive decay schemes give us a way to approach an intriguing question: How old is Earth? The starting nuclides of four decay series that occur naturally on Earth are

$$^{232}_{90}\text{Th} \qquad ^{238}_{92}\text{U} \qquad ^{235}_{92}\text{U} \qquad ^{232}_{93}\text{Np}$$

$$t_{1/2} = 1.4 \times 10^{10} \text{ y} \qquad 4.5 \times 10^9 \text{ y} \qquad 7.1 \times 10^8 \text{ y} \qquad 2.2 \times 10^6 \text{ y}$$

Let us assume that the lowest detectable level of radioactivity occurs after about 30 half-lives, which comes to about one radioactive atom remaining for every 1 billion radioactive atoms present initially, that is $(1/2)^{30} = 9.3 \times 10^{-10} \approx 1 \times 10^{-9}$. For neptunium-237, 30 half-lives is about $30 \times 2.2 \times 10^6 \text{ y} = 7 \times 10^7 \text{ y}$. Because there is essentially no neptunium-237 to be found on Earth, this planet must be older than 7×10^7 y. However, because uranium-235 can still be found, Earth is probably not as old as $30 \times 7.1 \times 10^8 \text{ y} = 2 \times 10^{10}$ y.

Our best estimate of the age of Earth comes from comparing the percent natural abundance of uranium-238 (99.28%) with that of uranium-235 (0.72%). It would have taken about 6 billion years for these percent relative abundances to develop if the two isotopes had initially been present in equal abundance. There are reasons to believe that the initial proportion of uranium-235 might not have been as great as that of uranium-238, and so 6 billion years is an upper limit. Other dating methods suggest a probable age of about 5 billion years for Earth.

19.3 Radioactive Decay Rates

In chemical kinetics, we can relate the half-life of a reaction to the rate constant and rate of reaction if we know the order of the reaction (Chapter 13.) We can do something similar for radioactive decay.

In a radioactive sample, there is no way to predict when any *particular* atom will decay. It could happen within the next second or not for a million years. The decay process is random. However, if we have a sample with a large number of atoms, we can measure how many of the atoms decay in a unit of time, and that number is quite reproducible. Moreover, in a sample with twice as many atoms, we find that twice as many atoms decay in a unit of time. This observation is stated through the **radioactive decay law:**

> *The rate of disintegration of a radioactive nuclide, called the decay rate or activity, A, is directly proportional to the number of atoms present:*

$$\text{Decay rate} = A = \lambda N \qquad \textbf{(19.1)}$$

where N is the number of atoms in the decaying sample and λ is a proportionality constant called the *decay constant*. This equation is quite similar to the rate law for the first-order reaction A $\longrightarrow$ products:

$$\text{Rate of reaction} = k[\text{A}]$$

Thus radioactive decay is a first-order process. The decay rate A is analogous to a rate of reaction; the decay constant λ, is analogous to k; and the number of atoms is analogous to the concentration [A]. The unit of λ is $(\text{time})^{-1}$, for example, s^{-1}, d^{-1}, or y^{-1}; and the unit for the rate of decay is atoms/time, for example, atoms/s.

Consider a 5,500,000-atom sample with a decay rate of 85 atoms per second. In this case, $N = 5.5 \times 10^6$ atoms and $A = 85$ atoms s^{-1}. The value of λ is

$$\lambda = \frac{A}{N} = \frac{85 \text{ atoms s}^{-1}}{5.5 \times 10^6 \text{ atoms}} = 1.5 \times 10^{-5} \text{ s}^{-1}$$

As with other first-order processes, we can also write equations describing the time dependence of the decay process and its half-life:

$$\ln \frac{N_t}{N_0} = -\lambda t \qquad \textbf{(19.2)}$$

$$t_{1/2} = \frac{0.693}{\lambda} \qquad \textbf{(19.3)}$$

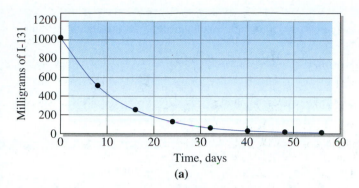
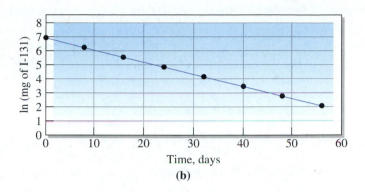

▲ **FIGURE 19.3   Graphic representation of decay of I-131**
(a) Each plotted point represents the amount of radioactive iodine-131 that remains after a time interval equal to one half-life. (b) Plotting the natural logarithm of the amount remaining gives a straight line.

QUESTION: What quantity can be obtained from the slope of this straight line?

In these equations, N_0 represents the initial number of atoms at $t = 0$, that is, at the time we begin our measurement; N_t is the number of atoms at a later time, t; λ is the decay constant; and $t_{1/2}$ is the half-life of the decay process. Note that the shorter the half-life, the larger the value of λ and the faster the decay proceeds.

Iodine-131, a radioactive nuclide used in studies of the thyroid gland, has a half-life of eight days. Thus, half the atoms in a sample undergo decay in an eight-day period during which time the decay rate A falls to half its initial value. The decay rate is down to one-fourth of its initial value in 16 days, to one-eighth in 24 days, and so on. Figure 19.3 shows, for iodine-131, the relationship between amount of iodine-131 and time (Figure 19.3a) and the relationship between the logarithm of amount of iodine-131 and time (Figure 19.3b).

Table 19.2 lists a number of radioactive nuclides, their half-lives, and some uses.

First Order Process animation

Table 19.2 Half-lives of Representative Radioactive Nuclides

Nuclide	Half-life[a]	Typical Use
Hydrogen-3	12.26 y	Biochemical tracer
Carbon-11	20.39 min	PET scans
Carbon-14	5730 y	Dating of artifacts
Sodium-24	14.659 h	Tracer, cardiovascular system
Phosphorus-32	14.3 d	Biochemical tracer
Potassium-40	1.25×10^9 y	Dating of rocks
Iron-59	44.496 d	Tracer, red blood cell lifetime
Cobalt-60	5.271 y	Radiation treatment of cancer
Strontium-90	28.5 y	No present use (component of radioactive fallout)
Iodine-131	8.040 d	Tracer, thyroid studies
Radium-226	1.60×10^3 y	Radiation therapy for cancer
Uranium-238	4.51×10^9 y	Dating of rocks and Earth's crust

[a] min = minute, h = hour, d = day, y = year.

Example 19.2

The nuclide sodium-24 (Table 19.2) is used to detect constrictions and obstructions in the human circulatory system. It emits β^- particles. **(a)** What is the decay constant, in s^{-1}, for sodium-24? **(b)** What is the activity of a freshly synthesized 1.00-mg sample of sodium-24? **(c)** What will be the rate of decay of the 1.00-mg sample after one week?

STRATEGY

(a) We find the half-life of sodium-24 in Table 19.2 and convert it from hours to seconds. Then we use Equation (19.3) to relate the half-life to the decay constant.

(b) Equation (19.1) relates the activity (A) of the sample A to the decay constant (λ), which we obtained in part (a), and to the number of atoms (N), which we can establish from the mass of sodium-24.

(c) Here we need Equation (19.2). The required data comes from parts (a) and (b). Also, when using Equation (19.2), we will find it convenient to use the expression $N = A/\lambda$.

SOLUTION

(a) The value of $t_{1/2}$ for sodium-24 in Table 19.2 is 14.659 h. In order that the unit of λ be s^{-1}, we express the half-life in seconds.

$$t_{1/2} = 14.659 \text{ h} \times \frac{60 \text{ min}}{1 \text{ h}} \times \frac{60 \text{ s}}{1 \text{ min}} = 5.2772 \times 10^4 \text{ s}$$

Next we use Equation (19.3) to relate the decay constant to $t_{1/2}$.

$$\lambda = \frac{0.693}{t_{1/2}} = \frac{0.693}{5.2772 \times 10^4 \text{ s}} = 1.31 \times 10^{-5} \text{ s}^{-1}$$

(b) We use molar mass and Avogadro's number as conversion factors to obtain the number of atoms in 1.00 mg of sodium-24.

$$N = 1.00 \times 10^{-3} \text{ g } ^{24}\text{Na} \times \frac{1 \text{ mol } ^{24}\text{Na}}{24.0 \text{ g } ^{24}\text{Na}} \times \frac{6.022 \times 10^{23} \text{ atoms } ^{24}\text{Na}}{1 \text{ mol } ^{24}\text{Na}}$$

$$= 2.51 \times 10^{19} \text{ atoms } ^{24}\text{Na}$$

The activity of the 1.00-mg sample of sodium-24 is the product of the decay constant and number of atoms in the sample.

$$A = \lambda N = 1.31 \times 10^{-5} \text{ s}^{-1} \times 2.51 \times 10^{19} \text{ atoms}$$

$$= 3.29 \times 10^{14} \text{ atoms/s}$$

(c) Because the unit of λ is s^{-1}, we need to express t in seconds.

$$t = 1 \text{ week} \times \frac{7 \text{ d}}{1 \text{ week}} \times \frac{24 \text{ h}}{1 \text{ d}} \times \frac{60 \text{ min}}{1 \text{ h}} \times \frac{60 \text{ s}}{1 \text{ min}} = 6.05 \times 10^5 \text{ s}$$

Next, we substitute activities for numbers of atoms, that is, $N = A/\lambda$, in Equation (19.2).

$$\ln \frac{N_t}{N_0} = -\lambda t$$

$$\ln \frac{A_t/\lambda}{A_0/\lambda} = -\lambda t$$

$$\ln \frac{A_t}{A_0} = -\lambda t$$

At this point, we substitute known values for λ and t and solve for $\ln(A_t/A_0)$.

$$\ln \frac{A_t}{A_0} = -1.31 \times 10^{-5} \text{ s}^{-1} \times 6.05 \times 10^5 \text{ s} = -7.63$$

$$A_t/A_0 = e^{-7.93} = 3.6 \times 10^{-4}$$

Finally, we substitute the known value of A_0 and solve for A_t.

$$A_t = 3.6 \times 10^{-4} \times A_0 = 3.6 \times 10^{-4} \times 3.29 \times 10^{14} \text{ atoms/s}$$

$$= 1.2 \times 10^{11} \text{ atoms/s}$$

EXERCISE 19.2A

If the current rate of decay of a sample of phosphorus-32 is 2.50×10^{10} atoms/s, what will be the decay rate 1.00 year from now? (See Table 19.2.)

EXERCISE 19.2B

The half-life of plutonium-239 is 2.411×10^4 y. How long would it take for a sample of plutonium-239 to decay to 1.00% of its present activity?

Example 19.3   An Estimation Example

The half-life of iodine-131 is 8.040 days. Following the release of large quantities of iodine-131 during the Chernobyl nuclear disaster of 1986, approximately how long did it take for the activity of this isotope to fall to 1% of its initial value?

ANALYSIS AND CONCLUSIONS

We could use the methods of Example 19.2 to do an exact calculation, but note that 1% represents 1/100 of the initial activity. During one half-life, the activity fell to half its initial value; during the next half-life period, to one-quarter; and so on.

Time:	$t_{1/2}$	$2(t_{1/2})$	$3(t_{1/2})$	$4(t_{1/2})$	$5(t_{1/2})$	$6(t_{1/2})$	$7(t_{1/2})$
Fraction remaining:	1/2	1/4	1/8	1/16	1/32	1/64	1/128

Thus, it took between 6 and 7 half-lives for the activity to fall to 1/100 of its initial value. If we estimate 6.5 half-lives, the time required was $6.5 \times 8\ \text{d} \approx 52\ \text{d}$.

EXERCISE 19.3A

As a rough rule, a radioactive isotope is deemed no longer radioactive after 30 half-lives. What fraction of the iodine-131 atoms initially present in Example 19.3 will remain after 30 half-lives? How long would this decay take?

EXERCISE 19.3B

Without doing detailed calculations, determine which of these samples has the greatest rate of decay: **(a)** 1 μmol sodium-24, **(b)** 1 μg carbon-11, **(c)** 1 g uranium-238. (See Table 19.2.) Explain your reasoning.

▲ Results of Example 19.3 illustrated. The bars from left to right represent the proportion of iodine remaining after each successive half-life.

Radiocarbon Dating

In Earth's upper atmosphere, carbon-14 is formed at a nearly constant rate as neutrons from cosmic radiation bombard nitrogen-14 atoms. When a neutron is absorbed by the nucleus of a nitrogen-14 atom, a proton is ejected. The result is that the nitrogen-14 atom is converted to a carbon-14 atom. We can show this through a nuclear equation:

$$^{14}_{7}\text{N} + ^{1}_{0}\text{n} \longrightarrow ^{14}_{6}\text{C} + ^{1}_{1}\text{H}$$

The carbon-14 is eventually incorporated into atmospheric carbon dioxide. While carbon-14 is being formed in the upper atmosphere, carbon-14 everywhere is undergoing decay, resulting in an essentially constant concentration in the environment of about one atom of carbon-14 for every 10^{12} atoms of carbon-12. A living plant consumes carbon dioxide, and animals consume plants. Plants and animals incorporate carbon-14 into their tissues as readily as carbon-12, and therefore in the same proportions as found in the environment. Thus, as carbon-14 atoms in living organisms undergo decay, they are constantly replaced with "fresh" carbon-14 atoms.

Carbon-14 in living matter decays by β^- emission at a rate of about 15 disintegrations per minute per gram of carbon. When a tree is cut down, however, it no longer takes in carbon dioxide. The carbon-14 that decays is no longer replaced, and as the concentration of carbon-14 falls, the decay rate falls as well. We can use the reduced decay rate at some later time to estimate the age of an object made of wood from the tree.

Because carbon-14 has a half-life of 5730 years, radiocarbon dating does not work well for objects less than a few hundred years old, because for younger objects, the carbon-14 decay rate in the object will be too close to what it was initially. Nor does the method work well for objects more than about 50,000 years old, because in this case, the level of radioactivity will have fallen to the point where it is not much greater than the background radiation level. Between these limits, however, radiocarbon dating has had some remarkable successes.

► The Shroud of Turin, a linen cloth more than 4 meters long, shows the faint image of a man's face. The image is best visualized in a photographic negative, as shown here. The shroud was thought by some to be the burial shroud of Jesus Christ, in which case it would be about 2000 years old. The cloth was shown by carbon-14 dating in three separate laboratories to be, at most, about 700 years old. This age corresponds well to records that indicate that the shroud first appeared in the historical record in the 1300s. Some people, however, question the validity of the dating results, claiming the sample was obtained from a newer patch, not from the original shroud material.

Example 19.4

A wooden object from an Egyptian tomb is subjected to radiocarbon dating. The decay rate observed for its carbon-14 content is 7.2 dis min^{-1} per g carbon. What is the age of the wood in the object (and, presumably, of the object itself)? The half-life of carbon-14 is 5730 years, and the decay rate for carbon-14 in living organisms is 15 dis min^{-1} per g carbon.

STRATEGY

The age of the wood can be calculated from the integrated rate law for a first-order process, in which we will use a ratio of decay rates in place of a ratio of concentrations according to Equation (19.2). We will also need the decay constant of the process that we will calculate from the half-life using Equation (19.3).

SOLUTION

First, we determine the decay constant, λ:

$$\lambda = \frac{0.693}{t_{1/2}} = \frac{0.693}{5730 \text{ y}} = 1.21 \times 10^{-4} \text{ y}^{-1}$$

As we did in Example 19.2c, let us work with activities (A) instead of numbers of atoms (N). To do so, we can rewrite the radioactive decay law as

$$N_0 = A_0/\lambda \quad \text{and} \quad N_t = A_t/\lambda$$

We need the rate law in the form

$$\ln \frac{N_t}{N_0} = \ln \frac{A_t/\lambda}{A_0/\lambda} = \ln \frac{A_t}{A_0} = -\lambda t$$

Finally, we can substitute the known data and solve for the unknown time, t:

$$\ln \frac{A_t}{A_0} = \ln \frac{7.2}{15} = -\lambda t = -(1.21 \times 10^{-4} \text{ y}^{-1}) \times t$$

$$t = \frac{-\ln(7.2/15)}{1.21 \times 10^{-4} \text{ y}^{-1}} = 6.1 \times 10^3 \text{ y}$$

ASSESSMENT

Note that if we had reversed the values of A_t and A_0, we would have obtained a negative value for the age—a physically meaningless result. The value obtained above for the age of the wood is both positive (physically meaningful) and consistent with the long half-life and cited differences in decay rates.

EXERCISE 19.4A

What will be the decay rate of the carbon-14 in the object described in Example 19.4 at a time 1500 years from now?

EXERCISE 19.4B

Tritium (^{3}H), a β^- emitting hydrogen nuclide, can be used to determine the age of items up to about 100 years. A sample of brandy, stated to be 25 years old and offered for sale at a premium price, has tritium with half the activity of that found in new brandy. Is the claimed age of the beverage authentic? Use data from Table 19.2, and assume that the natural abundance of tritium is a fixed quantity.

19.4   Synthetic Nuclides

For centuries, alchemists tried—without success—to change one element into another, a process called *transmutation*. Modern scientists have learned how to do so. Ernest Rutherford brought about the first transmutation in a laboratory experiment in 1919. By bombarding nitrogen-14 nuclei with α particles, he obtained oxygen-17 and protons as products. We can represent this process with the nuclear equation

$$^{14}_{7}\text{N} + ^{4}_{2}\text{He} \longrightarrow ^{17}_{8}\text{O} + ^{1}_{1}\text{H}$$

Oxygen-17 is a stable, naturally occurring oxygen nuclide with a natural abundance of 0.037%.

Rutherford's experiment was especially important because it established the existence of protons *outside* the nuclei of atoms. Fifteen years later, in an experiment similar to Rutherford's, Irène Curie—daughter of Marie and Pierre—and Frédéric Joliot got a surprisingly different result. They bombarded aluminum-27 with α particles and observed the emission of *two* types of particles: neutrons and positrons. When they stopped the bombardment, neutron emission stopped but positron emission continued. They hypothesized that the nuclear bombardment produces phosphorus-30, which then decays by the emission of positrons:

$$^{27}_{13}\text{Al} + {}^{4}_{2}\text{He} \longrightarrow {}^{30}_{15}\text{P} + {}^{1}_{0}\text{n}$$

$$^{30}_{15}\text{P} \longrightarrow {}^{30}_{14}\text{Si} + {}^{0}_{+1}\text{e}$$

Phosphorus-30 was the first *synthetic* radioactive nuclide. Since its discovery, scientists have synthesized more than a thousand others, with the result that the number of known radioactive nuclides now greatly exceeds the number of nonradioactive ones. We will examine some of the uses of synthetic radioactive nuclides in later sections of this chapter.

▲ Irène Curie met and married Frédéric Joliot when both were assistants in Marie Curie's laboratory. Joliot and the younger Curie just missed out on two great discoveries— the neutron and the positron—but were awarded the Nobel Prize in Chemistry in 1935 for their discovery of artificially induced radioactivity.

Example 19.5

Bombardment of a magnesium-24 nucleus with a deuteron results in the formation of the sodium-22 nucleus. What other particle is produced?

STRATEGY

We can determine the identity of the particle by applying the principles of nuclear equation balancing. (The deuteron is the nucleus of a deuterium atom, $^{2}_{1}\text{H}$.)

SOLUTION

$$^{24}_{12}\text{Mg} + {}^{2}_{1}\text{H} \longrightarrow {}^{22}_{11}\text{Na} + \text{?}$$

The unknown particle must have $A = 26 - 22 = 4$ and $Z = 13 - 11 = 2$. It is therefore an α particle, and the complete nuclear equation is

$$^{24}_{12}\text{Mg} + {}^{2}_{1}\text{H} \longrightarrow {}^{22}_{11}\text{Na} + {}^{4}_{2}\text{He}$$

EXERCISE 19.5A

Write a nuclear equation to represent the bombardment of a chlorine-35 nucleus with a neutron to produce a sulfur-35 nucleus.

EXERCISE 19.5B

Write a nuclear equation to represent the bombardment of a californium-249 nucleus with a nitrogen-15 nucleus to produce a dubnium-260 nucleus and neutrons.

19.5 Transuranium Elements

In 1940, the first of the **transuranium elements**—elements with $Z > 92$—were synthesized by bombarding uranium-238 nuclei with neutrons. First, unstable uranium-239 is formed, and then this nuclide decays by β emission, producing neptunium ($Z = 93$):

$$^{238}_{92}\text{U} + {}^{1}_{0}\text{n} \longrightarrow {}^{239}_{92}\text{U}$$

$$^{239}_{92}\text{U} \longrightarrow {}^{239}_{93}\text{Np} + {}^{0}_{-1}\text{e}$$

Neptunium-239 decays to plutonium ($Z = 94$):

$$^{239}_{93}\text{Np} \longrightarrow {}^{239}_{94}\text{Pu} + {}^{0}_{-1}\text{e}$$

Neutrons are especially effective projectiles for nuclear bombardment because they have no charge and thus are not repelled as they approach a nucleus. However, neutron bombardment produces only small changes in atomic number. To produce large changes in atomic number, relatively massive positive ions are needed. For example, the discovery of element 114 in 1999 was accomplished by bombarding plutonium-242 with calcium-48 nuclei.

Considerable energy must be imparted to a positive ion in order for it to overcome repulsion by a positively charged nucleus. Only in this way can the ion collide with the nucleus and induce a nuclear reaction. Energetic ions are usually produced by accelerating them to extremely high speeds. One type of *charged-particle accelerator*, called a cyclotron, is described in Figure 19.4.

19.6   Nuclear Stability

At some point you may have wondered, "If positive charges repel one another, how is it possible for protons to be packed so closely in the nuclei of atoms?" The answer is that there are attractive nuclear forces that are much stronger than electrostatic forces of repulsion, but the attractive forces are important only at extremely short distances. The strengths of these forces are closely related to the numbers of protons and neutrons in a nucleus. We begin with some observations about the naturally occurring *stable* nuclides, that is, those that do not undergo radioactive decay.

- About 160 stable nuclides have an even number of protons and an even number of neutrons, for example, $^{12}_{6}C$.
- About 50 stable nuclides have an even number of protons and an odd number of neutrons, for example, $^{25}_{12}Mg$.
- About 50 stable nuclides have an odd number of protons and an even number of neutrons, for example, $^{19}_{9}F$.
- Only four stable nuclides have an odd number of protons and an odd number of neutrons: $^{2}_{1}H$, $^{6}_{3}Li$, $^{10}_{5}B$, and $^{14}_{7}N$.

One explanation of nuclear stability is the *nuclear shell theory*. In simplest terms, this theory proposes that the protons and the neutrons exist in shells within the nucleus, much the way electrons are arranged in shells outside the nucleus. The similarity extends to the special stability associated with the closing of shells, similar to what is seen with the noble gases in electron configurations. In the nuclear shell theory, a special stability is associated with nuclei that have any of the following numbers of protons or neutrons:

$$2, 8, 20, 28, 50, 82, 126$$

These numbers are sometimes called *magic numbers* because scientists recognized their significance to nuclear stability before theories were developed to explain them.

One observation consistent with these magic numbers is that α particles are especially stable; they have two protons and two neutrons and are therefore called *doubly magic*. Another observation is that tin ($Z = 50$) has 10 naturally occurring stable nuclides, more than any other element. The magic number of protons (50) seems to

▲ Maria Goeppert-Mayer (1906–1972) followed her chemist husband to several academic locations, working mostly in temporary positions. While at the University of Chicago, stimulated by questions posed by Enrico Fermi, she worked out the shell model of nuclear structure (supposedly in 10 minutes). Her work earned her a share of the 1963 Nobel Prize in Physics.

▶ **FIGURE 19.4   The cyclotron—a charged-particle accelerator**

The two hollow, flat boxes, called dees, are held in a magnetic field and charged electrically. Positive ions are produced between the dees, attracted into the negatively charged dee, and forced into a circular path by the magnetic field. As the ions leave the first dee, the electric charge is reversed and the ions are attracted into the second dee and are again accelerated. The process is repeated until the ions have achieved the desired energy. The ions are then brought out of the accelerator. A playground swing provides an analogy: The rider is given a push at the end of each swing, and the swing travels through an ever-increasing arc.

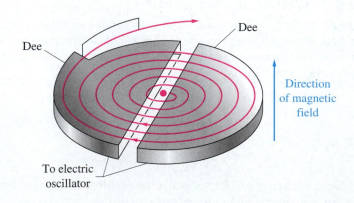

allow for a greater variation in the number of neutrons in the tin nucleus than in others. Also, the uranium-238 radioactive decay series terminates in the nuclide $^{206}_{82}$Pb; the uranium-235 series, in $^{207}_{82}$Pb; and the thorium-232 series, in $^{208}_{82}$Pb. All these terminating stable nuclides have the magic number 82 for the protons, and $^{208}_{82}$Pb is doubly magic, with 82 protons and 126 neutrons.

A crucial factor in the stability of a nucleus is the ratio of neutron number (N) to proton number (Z). Some nuclides of the lightest elements have an N/Z ratio of 1, and as a group these nuclides have an average ratio slightly greater than 1. Examples of nuclides in this group are $^{4}_{2}$He, $^{16}_{8}$O, $^{27}_{13}$Al, $^{39}_{19}$K, and $^{40}_{20}$Ca. Nuclides with $Z > 20$ require a larger number of neutrons than protons to moderate the effect of increasing proton repulsions. For example, the N/Z ratio in $^{56}_{26}$Fe is 30/26 = 1.15, in $^{133}_{55}$Cs, it is 78/55 = 1.42, and in $^{209}_{83}$Bi, 126/83 = 1.52. For nuclides with $Z > 83$, the proton repulsions are too large to be overcome by proton–neutron interactions, and the nuclides are all radioactive.

The general pattern for nuclear stability in terms of neutron and proton numbers is shown in Figure 19.5, a graph of neutron number (N) versus proton number (Z). All the naturally occurring stable (nonradioactive) nuclides are indicated by dots in a region known as the *belt of stability*. The radioactive nuclides in this belt are not shown. (Many radioactive nuclides lie within the belt of stability, but generally these nuclides have much longer half-lives than do radioactive nuclides of the same element that lie outside the belt.)

All nuclides falling outside the belt of stability are radioactive, and their mode of decay is the one that brings the nuclides formed in the decay process into the belt. Nuclides to the left of the belt tend to decay by β emission, and those to the right of the belt decay by positron emission and electron capture. In addition, many of the nuclides above the upper tip of the belt decay by α emission. The synthesis in 1999 of elements 114 and 116 indicates that there is a small island of stability beyond the main belt of stability, as has long been predicted by nuclear theory.

As illustrated in Examples 19.6 and 19.7, we can use data from Figure 19.5 and the list of items on page 810 to answer questions about nuclear stability and radioactive decay.

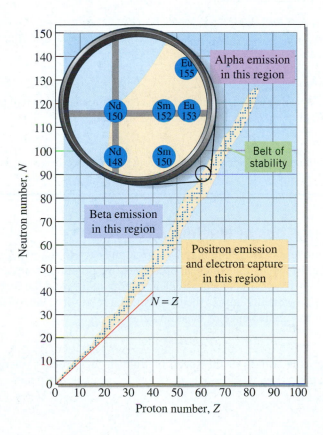

Nuclear Stability activity

◀ **FIGURE 19.5   Neutron–proton ratio and the stability of nuclides**
All the stable nuclides lie within the belt of stability (as do some radioactive ones). At low atomic numbers, the neutron–proton ratios are either 1 : 1 or slightly greater, and as a result, the belt lies close to the line $N = Z$. At higher atomic numbers, the neutron–proton ratios rise to about 1.5 : 1 and the belt moves away from the $N = Z$ line. The belt ends at $Z = 83$. Nuclides outside the belt tend to undergo radioactive decay by the modes indicated.

QUESTION: What would be the expected type(s) of radioactive decay for the hypothetical isotope $^{125}_{60}$E?

Example 19.6

Which of the following would you expect to be radioactive: $^{118}_{50}Sn$, $^{234}_{91}Pa$, $^{54}_{25}Mn$?

SOLUTION

The nuclide $^{118}_{50}Sn$ has 50 protons and $(118 - 50) = 68$ neutrons. This is an even–even combination, the most common for stable nuclides. Also, the neutron–proton ratio of $68:50$ lies within the belt of stability in Figure 19.5. Therefore, $^{118}_{50}Sn$ is not radioactive.

The atomic number 91 in $^{234}_{91}Pa$ exceeds the limit for the naturally occurring stable nuclides $(Z > 83)$. Thus, $^{234}_{91}Pa$ is radioactive.

The nuclide $^{54}_{25}Mn$ has 25 protons and $(54 - 25) = 29$ neutrons. This is an odd–odd combination, found only in four stable nuclides of low atomic numbers. We should therefore expect that this nuclide is radioactive.

EXERCISE 19.6A

Would you expect the isotope $^{74}_{30}Zn$ to be radioactive? Explain.

EXERCISE 19.6B

Using reasoning like that in Example 19.6, cite two isotopes of calcium that you are quite certain are nonradioactive and one isotope that is radioactive. Explain your reasoning.

Example 19.7   A Conceptual Example

What kind of radioactive decay would you expect the nuclide $^{84}_{40}Zr$ to undergo?

ANALYSIS AND CONCLUSIONS

When we check Figure 19.5 at $Z = 40$, we see that a nuclide with $N = 44$ lies below the belt of stability. This confirms that the nuclide is radioactive. We would expect a decay that moves the neutron–proton ratio closer to the belt. This means converting a proton to a neutron. The atomic number goes down by one, and the mass number remains the same. These changes are achieved by either positron emission or electron capture:

$$\text{Positron emission:} \qquad ^{84}_{40}Zr \longrightarrow {}^{84}_{39}Y + {}^{0}_{1}e$$

$$\text{Electron capture:} \qquad ^{84}_{40}Zr + {}^{0}_{-1}e \longrightarrow {}^{84}_{39}Y$$

Notice that in each case, the neutron–proton ratio increases from 44/40 in $^{84}_{40}Zr$ to 45/39 in $^{84}_{39}Y$.

EXERCISE 19.7A

The nuclide $^{84}_{39}Y$ is radioactive. What kind of decay would you expect it to undergo? Is the product of this decay likely to be nonradioactive? Explain.

EXERCISE 19.7B

Write plausible nuclear equations to represent the radioactive decay of the fluorine isotopes ^{17}F and ^{22}F.

19.7   Energetics of Nuclear Reactions

The details of nuclear phenomena may be a great mystery to many people, but almost everyone is keenly aware that nuclear processes are potential sources of enormous quantities of energy. We will consider nuclear energy in this and the following section.

In 1905, while working out his theory of special relativity, Albert Einstein derived the equation for the equivalence of mass and energy:

$$E = mc^2 \qquad\qquad (19.4)$$

The constant c^2, the square of the speed of light in m/s, relates energy in joules to mass in kilograms.

In a typical spontaneous nuclear reaction, a small quantity of matter is transformed into a corresponding quantity of energy. Presumably, this is also true of ordinary chemical reactions, but there the energy changes are so small that the corresponding mass changes are undetectable. This is why we can use the principle of conservation of mass (total mass is unchanged in a chemical reaction) as the basis of stoichiometric calculations.

Nuclear energies are generally expressed in *electronvolts*, where one electronvolt is the energy an electron acquires as it moves through a potential difference of 1 volt. Given the charge on an electron, 1.6022×10^{-19} C, the product of this charge and a 1 V potential difference is

$$1 \text{ eV} = 1.6022 \times 10^{-19} \text{ C} \times 1\text{V} = 1.6022 \times 10^{-19} \text{ VC} = 1.6022 \times 10^{-19} \text{ J}$$

The electronvolt is such an extremely small energy unit that for nuclear energies, we usually use the unit MeV (mega electronvolts) instead, where

$$1 \text{ MeV} = 1 \times 10^6 \text{ eV} = 1 \times 10^6 \times 1.6022 \times 10^{-19} \text{ J} = 1.6022 \times 10^{-13} \text{ J}$$

Another useful relationship is the energy equivalent to 1 atomic mass unit (u). To obtain this, we first find the SI mass equivalent of 1 u, and then we find the energy equivalent of this mass.

Because the mass of a carbon-12 atom is exactly 12 u, it is convenient to express the atomic mass unit in terms of the mass of a single carbon-12 atom.

$$1 \text{ u} = \frac{1}{12} \times (\text{mass of one } {}^{12}\text{C atom})$$

Given that the molar mass of carbon-12 is *exactly* 12 g, the mass of one ^{12}C atom is calculated by dividing by Avogadro's number.

$$\frac{12 \text{ g}}{6.0221 \times 10^{23} \, {}^{12}\text{C atoms}} = 1.9927 \times 10^{-23} \text{g/}{}^{12}\text{C atom}$$

The mass of 1 u is then calculated using the relationship between the atomic mass unit and the mass of a carbon-12 atom.

$$1 \text{ u} = \frac{1}{12} \times 1.9927 \times 10^{-23} \text{ g}$$
$$= 1.6606 \times 10^{-24} \text{ g}$$
$$= 1.6606 \times 10^{-27} \text{ kg}$$

Now, we can apply Einstein's equation, $E = mc^2$, to determine the energy equivalent to 1 u.

$$E = mc^2 = 1.6606 \times 10^{-27} \text{ kg} \times (2.9979 \times 10^8)^2 \text{ m}^2 \text{ s}^{-2}$$
$$= 1.4924 \times 10^{-10} \text{ kg m}^2 \text{ s}^{-2} = 1.4924 \times 10^{-10} \text{ J}$$

Finally, we express the energy in MeV.

$$E = 1.4924 \times 10^{-10} \text{ J} \times \frac{1 \text{ MeV}}{1.6022 \times 10^{-13} \text{ J}} = 931.5 \text{ MeV}$$

In Example 19.8, where we determine the energy change in a nuclear reaction, we use *nuclear* masses. These are easily obtained from atomic masses:

$$\text{Nuclear mass} = \text{atomic mass} - \text{mass of extranuclear electrons}$$

In dealing with energy changes in nuclear reactions, we use the principle that the total mass–energy of the products is equal to the total mass–energy of the reactants. This suggests that

- Mass lost in a nuclear reaction must be replaced by kinetic energy in the products.
- Mass gained in a nuclear reaction must come from the kinetic energy of the reactants.

Problem-Solving Note

To see why 931.5 MeV is expressed only to four significant figures rather than five, review the note on page 227.

Problem-Solving Note

In determining the change in mass in a nuclear reaction, we must use precise values of the nuclear masses because the mass changes are quite small.

Example 19.8

Given the nuclear masses

$$^{241}_{95}\text{Am} = 241.0046 \text{ u} \qquad ^{237}_{93}\text{Np} = 236.9970 \text{ u} \qquad ^{4}_{2}\text{He} = 4.0015 \text{ u}$$

calculate the energy associated with the α decay of americium-241, in MeV.

STRATEGY

We can determine the energy associated with this process by calculating the mass change involved and then converting this mass change to energy units according to the relationship derived on page 813 for 1 u.

SOLUTION

First, we write a nuclear equation and note the masses of the species involved:

$$^{241}_{95}\text{Am} \longrightarrow {}^{237}_{93}\text{Np} + {}^{4}_{2}\text{He}$$

$$241.0046 \text{ u} \qquad 236.9970 \text{ u} \qquad 4.0015 \text{ u}$$

Next, we determine the mass change in the decay of the $^{241}_{95}\text{Am}$ by subtracting the initial mass from the final mass:

$$236.9970 \text{ u} + 4.0015 \text{ u} - 241.0046 \text{ u} = -0.0061 \text{ u}$$

In megaelectronvolts, the energy corresponding to this mass change is

$$-0.0061 \text{ u} \times \frac{931.5 \text{ MeV}}{1 \text{ u}} = -5.7 \text{ MeV}$$

ASSESSMENT

The negative sign signifies that energy is given off.

EXERCISE 19.8A

Neptunium-237 decays by α-particle emission. Determine the energy in MeV associated with this decay. The relevant nuclear masses are $^{237}_{93}\text{Np}$, 236.9970 u; $^{233}_{91}\text{Pa}$, 232.9901 u; and $^{4}_{2}\text{He}$, 4.0015 u.

EXERCISE 19.8B

Determine the energy requirement in MeV for the nuclear reaction

$$^{27}_{13}\text{Al} + {}^{4}_{2}\text{He} \longrightarrow {}^{30}_{15}\text{P} + {}^{1}_{0}\text{n}$$

The atomic masses are $^{27}_{13}\text{Al}$, 26.9815 u; $^{4}_{2}\text{He}$, 4.0026 u; and $^{30}_{15}\text{P}$, 29.9783 u. Use 1.0087 u for the mass of a neutron. Determine this energy in two ways: first using nuclear masses derived from the given atomic masses and then using the atomic masses directly.

Problem-Solving Note

In Exercise 19.8B, we see that the energy calculated for a nuclear reaction is the same whether we use nuclear masses or atomic masses.

Nuclear Binding Energy

We say the mass number of a helium-4 nucleus is 4 because the nucleus is made up of two protons and two neutrons. However, the mass of the nucleus is slightly less than the sum of the nucleon masses. Figure 19.6 depicts the combination of two protons and two neutrons into a nucleus. As the four particles come together and form the nucleus, there is a mass loss of 0.0305 u, a quantity called the *mass defect* of the nucleus. This lost mass is liberated as energy:

$$0.0305 \text{ u} \times \frac{931.5 \text{ MeV}}{1 \text{ u}} = 28.4 \text{ MeV}$$

The energy released in forming a nucleus from its protons and neutrons is called the **nuclear binding energy** and is expressed as a positive quantity. Alternatively, nuclear binding energy is the quantity of energy necessary to separate a nucleus into individual protons and neutrons. If we apportion the nuclear binding energy of $^{4}_{2}\text{He}$

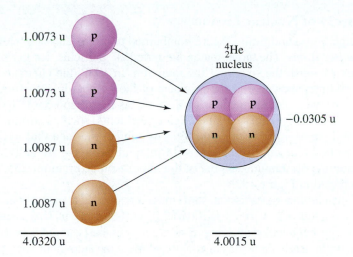

◀ **FIGURE 19.6 Nuclear binding energy in $_2^4$He**

The mass of a helium nucleus, 4.0015 u, is less than that of two protons and two neutrons, 4.0320 u. The energy equivalent of the 0.0305 u mass defect is the nuclear binding energy released when the nucleons bind together.

among the four nucleons, the binding energy per nucleon is $28.4\ \mathrm{MeV}/4 = 7.10\ \mathrm{MeV}$. Figure 19.7 is a graph of binding energy per nucleon as a function of mass number. This graph is useful in explaining nuclear fission and nuclear fusion, discussed in the next section.

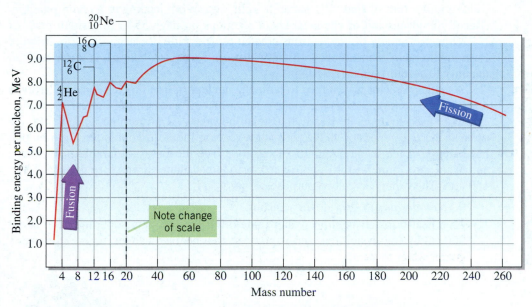

▲ **FIGURE 19.7   Average binding energy per nucleon as a function of mass number**

19.8   Nuclear Fission and Nuclear Fusion

Figure 19.7 shows that the nuclide $_{26}^{56}$Fe has about the maximum in binding energy per nucleon. We can draw two conclusions from the existence of this maximum:

- When light nuclei combine to form a heavier nucleus, mass is converted to energy, and the binding energy per nucleon in the product nucleus increases as mass number A increases, reaching a maximum at about $A = 56$. The process of combining light nuclei into a heavier one is called **nuclear fusion.**

- When heavy nuclei split into lighter ones, mass is converted to energy and the binding energies per nucleon in the product nuclei increase as A decreases, again reaching a maximum at about $A = 56$. The breakup of a heavy nucleus into two lighter fragments is called **nuclear fission.**

Nuclear fission and nuclear fusion are so important in modern life that we will now look at each one in some detail.

Thermal neutrons are so named because their speed is about 2000 m/s, corresponding to a temperature of about 298 K. They induce fission more easily than do "fast" neutrons.

▲ The concrete containment dome of a nuclear reactor is designed to minimize escape of radioactive material in the event of an accident.

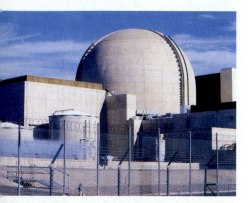

▲ Control rods used in a nuclear reactor to regulate the rate of the chain reaction by absorbing some of the neutrons released during fission.

 Fission Process activity

The Discovery of Nuclear Fission

In 1934, Enrico Fermi and Emilio Segrè bombarded uranium-238 with slow neutrons, called *thermal neutrons*. The target material emitted β^- particles, but Fermi and Segrè did not firmly establish the source of the radiation. In 1938, the chemists Otto Hahn and Fritz Strassman showed that the products of the bombardment of uranium with thermal neutrons were not elements with $Z > 92$, as Fermi and Segrè had expected. They were radioactive nuclides of much lighter elements, such as strontium and barium. In 1939, Lise Meitner became the first to publish the idea that nuclear fission occurs in the bombardment of uranium with thermal neutrons. Actually, the nuclear fission was not that of uranium-238 but of the minor isotope uranium-235. The fission process is depicted in Figure 19.8.

When a nucleus undergoes fission, some mass is converted to energy. On average, the energy release is about 3.2×10^{-11} J (200 MeV) per fission event. This seems like a tiny quantity of energy, but it looks much bigger on a per-gram basis: about 8×10^7 kJ/g. We would have to burn nearly three tons of coal to get the same amount of energy.

Nuclear Reactors

The fission of ^{235}U yields an average of 2.5 neutrons per fission event. These neutrons are important fission products indeed. Those from the first fission event, on average, cause two more ^{235}U nuclei to split. The neutrons from the second round of fission can cause another four or five ^{235}U nuclei to split, and so on, leading to a *chain reaction*. (Recall the discussion of chemical chain reactions on page 554.) In a small mass of ^{235}U, however, many of the neutrons created escape without causing fission. In order for a nuclear explosion to occur, a **critical mass** of ^{235}U is needed—a large enough quantity to sustain a chain reaction.

A nuclear reactor is designed to tame the nuclear fission process so that energy is released in a controlled manner. In the nuclear reactor pictured in Figure 19.9, fission energy is used to generate steam, the steam powers a steam turbine, and the turbine turns an electric generator. In the core of the reactor, fuel rods containing uranium that has been *enriched* in the uranium-235 isotope (see the Essay—"Uranium for Peace and War") are immersed in liquid water maintained under a pressure of from 70 to 150 atm. The water serves two purposes:

- As a *moderator* to slow down neutrons.
- As a heat-transfer medium. Water under high pressure and superheated to about 300 °C by the fission reaction is pumped to a heat exchanger, where it converts colder water to steam.

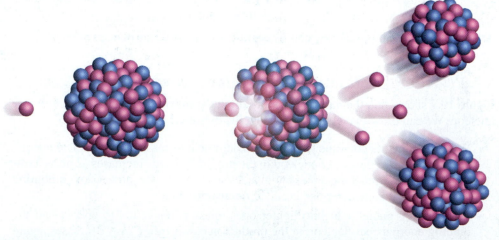

▲ **FIGURE 19.8** **Nuclear fission of a uranium-235 nucleus with thermal neutrons**
A uranium-235 nucleus is struck by a slow neutron. An unstable uranium-236 nucleus is formed, but it breaks into two fragments plus several neutrons. The neutrons can induce the fission of other uranium-235 nuclei.
QUESTION: If the neutrons produced from the fission reaction pictured here each reacted with other uranium-235 nuclei in the manner illustrated, what would be the total number of neutrons produced by those reactions?

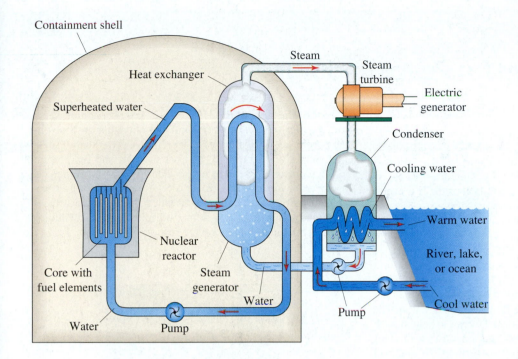

Nuclear Reactors activity

◀ **FIGURE 19.9   Schematic diagram of a nuclear power plant used to generate electricity**

QUESTION: Why must three separate loops of water flow be used in this power generation scheme?

The nuclear reaction is regulated by a set of *control rods,* which readily absorb neutrons; the rods are usually made of cadmium. When the rods are lowered into the reactor, they absorb enough neutrons to slow the fission process. When the rods are raised, more neutrons enter the uranium fuel and the rate of fission increases.

A nuclear reactor used to generate electric power "burns" approximately 1 to 3 kg ^{235}U per day, and a variety of radioactive waste products accumulate. The long-term disposal of radioactive wastes is a controversial issue. At present, most nuclear wastes are kept at the reactor sites, but they cannot remain there permanently. Scientists are studying a proposed burial site in Nevada for the ultimate disposal of radioactive wastes, but the environmental isolation of the site is still uncertain, and the plan has many opponents.

Nuclear Fusion

What would you think of constructing, out in space, one gigantic nuclear reactor that would transmit abundant energy to Earth almost forever? Well, that is exactly what we earthlings have in our Sun, 92 million miles away. The Sun is powered by the *fusion* of atomic nuclei, and its fuel supply—mostly $^1_1$H— will last for billions of years.

On Earth, scientists have unleashed the extraordinary energy of uncontrolled fusion reactions in hydrogen bombs. In a hydrogen bomb, nuclear fusion is initiated by the fission reaction in a fission bomb. However, such a totally uncontrolled fusion reaction cannot be used for practical purposes, and control of fusion reactions as energy sources is probably still decades away.

Scientists face a daunting challenge in developing a fusion energy source. The most promising nuclear reaction is the deuterium–tritium reaction:

$$^2_1\text{H} + {}^3_1\text{H} \longrightarrow {}^4_2\text{He} + {}^1_0\text{n}$$

Before they will fuse, however, the nuclei of deuterium and tritium must be forced extremely close together. Because the positively charged nuclei repel each other so strongly, close approach requires that the nuclei have enormously high thermal energies. At the required temperatures, gases are completely ionized into a mixture of atomic nuclei and electrons known as *plasma.* A temperature higher than 40,000,000 K is necessary to initiate *self-sustaining* fusion—a nuclear reaction that releases more energy than it takes to get it started. Another requirement is that the plasma be confined at an enormously high density long enough for the fusion to occur.

Uranium for Peace and War

Natural uranium is less than 1% ^{235}U, which is the only naturally occurring isotope that is readily fissionable. Nuclear power plants require fuel that is about 3% ^{235}U, and nuclear weapons require about 90% ^{235}U. The process whereby the percentage of ^{235}U in a uranium sample is increased is called *enrichment*, and separation of the less useful ^{238}U during enrichment is difficult and expensive because the two uranium isotopes are essentially identical chemically.

The Manhattan Project, which produced the atomic bombs used by the United States at the end of World War II, used *gaseous diffusion* to separate the isotopes. In this process, uranium is converted to UF_6, which is then allowed to diffuse through porous membranes. Because $^{235}UF_6$ has a molecular mass of 349 u, it diffuses slightly faster than $^{238}UF_6$(352 u). The concentration of $^{235}UF_6$ in the effluent increases very slowly, about 1.004 times per diffusion step. Many thousands of steps are needed to prepare uranium that is sufficiently enriched for use.

In the *centrifugal* process, UF_6 is fed into a spinning cylinder that creates forces a million times greater than gravity (by comparison, ordinary laboratory centrifuges generate a few thousand times gravity's force). Under this force, the $^{238}UF_6$ in the mixture migrates to the wall of the spinning cylinder while the $^{235}UF_6$ tends to remain near the center and can be drawn off. The centrifugal process requires only a few dozen steps to enrich the uranium for power plant use. However, the technical challenge in design and construction of such high-speed centrifuges is daunting.

A process that has received a great deal of attention is *laser separation*. There is a slight difference in the electronic energies of ^{238}U and ^{235}U. Properly tuned, a laser will produce a wavelength that will ionize ^{235}U and leave ^{238}U largely unaffected. The ionized $^{235}U^+$ can then be collected on negatively charged electrodes. The energy needed to run a laser-separation setup should be much less than that required to run either gaseous diffusion or the centrifugal process. In addition, the extent of separation should be greatest in the laser process, and the process could use metallic uranium rather than UF_6 as the feedstock. Nonetheless, work on the U.S. AVLIS (Atomic Vapor Laser Isotope Separation) project was suspended in 1999 due to economic and other factors. Some other countries, including Australia, continue to work on the process.

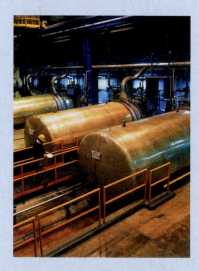

▲ To increase the percentage of ^{235}U in a uranium sample, UF_6 gas containing both ^{235}U and ^{238}U is allowed to diffuse through this series of cylinders. After each trip through a cylinder, the UF_6 is a bit richer in the ^{235}U isotope.

Moreover, this confinement must be done without the plasma coming into contact with the walls of the reactor, where the plasma would immediately cool down and be unable to fuse. One possible method is to confine the plasma in a magnetic field (Figure 19.10).

▶ **FIGURE 19.10   A nuclear fusion device**
Nuclear fusion requires that plasma be confined at exceedingly high temperatures and pressures. A gigantic doughnut-shaped electromagnet called a tokamak is one device designed to contain the plasma.

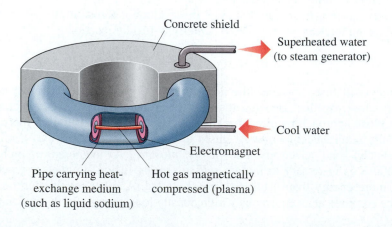

Concrete shield

Superheated water (to steam generator)

Cool water

Electromagnet

Hot gas magnetically compressed (plasma)

Pipe carrying heat-exchange medium (such as liquid sodium)

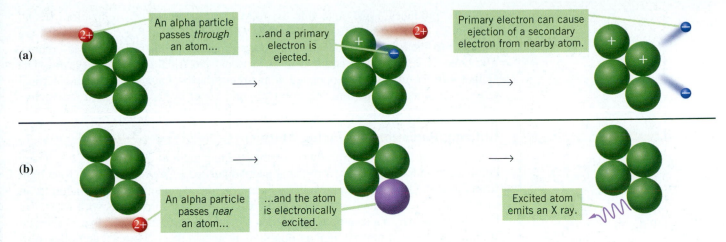

An alpha particle passes *through* an atom...

...and a primary electron is ejected.

Primary electron can cause ejection of a secondary electron from nearby atom.

(a)

An alpha particle passes *near* an atom...

...and the atom is electronically excited.

Excited atom emits an X ray.

(b)

▲ **FIGURE 19.11   Some effects of ionizing radiation**
(a) An α particle passes through an atom and dislodges a primary electron from the atom. The primary electron in turn dislodges a secondary electron from another atom. The atoms that lose electrons become positive ions. (b) An α particle passing near an atom but not colliding with it may excite electrons within the atom to higher energy levels, an event followed by the emission of X rays.

19.9   Effect of Radiation on Matter

With only a few exceptions, α particles, β particles, and γ-rays are not energetic enough to cause other matter to become radioactive. Rather, they dislodge electrons from atoms and molecules to produce ions (Figure 19.11a) and are therefore called *ionizing radiation*. The relatively massive α particle has a high ionizing power and thus produces a large number of ions in passing through matter. The much less massive β^- particles produce far fewer ions, and γ-rays have the lowest ionizing power of all. Electrons freed by ionizing radiation are called *primary* electrons. Some primary electrons are energetic enough to ionize atoms and molecules, producing *secondary* electrons.

Ionizing radiation can also excite electrons to higher energy levels, as shown in Figure 19.11b. The atoms then give off electromagnetic radiation (X rays, ultraviolet light, or visible light) as the electrons return to their ground states.

Radiation Detectors

One of the simplest and oldest ways to detect ionizing radiation is to observe the clouding it produces on photographic film. Antoine Henri Becquerel discovered radioactivity in 1896 as a result of the clouding of film, and clouding is the principle behind the film badges used by those who work with X rays or radioactive materials.

Perhaps the most familiar radiation detector is the *Geiger–Müller counter*, pictured in Figure 19.12. It consists of a cylindrical cathode with a wire anode running along its axis. The electrodes are sealed in a gas-filled tube. Ionizing radiation entering

▲ The photographic film in a film badge worn by people who work around radioactive materials is clouded by radiation. The extent of clouding provides a measure of how much radiation the worker has been exposed to over a given period of time.

🌐 **Effects of Ionizing Radiation activity**

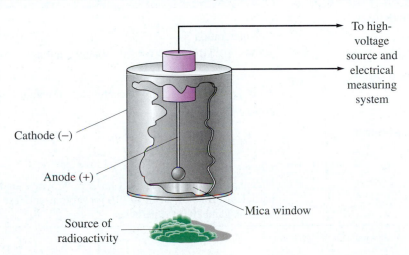

To high-voltage source and electrical measuring system

Cathode (−)

Anode (+)

Mica window

Source of radioactivity

◀ **FIGURE 19.12   A Geiger–Müller counter**
Radiation enters the Geiger–Müller tube through the mica window. Ions produced by the radiation cause an electrical discharge through the gas (usually argon) in the tube. Each pulse of electric current is counted as it passes through the circuit.

the tube produces ions, and this is followed by secondary ionization (recall Figure 19.11a). Positive ions are attracted to the negative cathode, and electrons are attracted to the positive anode. In this way, a pulse of electric current is created. The pulse is detected and counted in an external electrical measuring system. The Geiger–Müller tube must be operated at an appropriate voltage so that each electrical pulse corresponds to one ionizing event brought about by the decay of a radioactive atom in the sample.

Ionizing Radiation and Living Matter

We live in a world where naturally occurring ionizing radiation is all around us. We are bombarded with cosmic rays from the Sun and outer space. We receive ultraviolet light from the Sun. Other radiation reaches us from radioactive nuclides found naturally in rocks, soil, water, air, and even our bodies. The dose we absorb varies with where we live—it is greater, for example, at higher elevations—and how we live—cigarette smokers receive somewhat higher doses than nonsmokers. Human activities also expose us to radiation. About two-thirds of the average radiation exposure for a person comes from natural background radiation, with most of the remaining third coming mainly from medical procedures. Other sources, such as releases by the nuclear industry, occupational exposure, and radioactive fallout, make up only a tiny fraction of the total.

As with other forms of matter, radiation can ionize, excite, or even break apart molecules in living matter. Unusually large dosages are lethal to all living organisms, and even slight exposure can damage DNA and cause birth defects, leukemia, bone cancer, and other forms of cancer.

Radiation Dosage

Radiation dosage can refer either to the amount of radiation that matter absorbs or to the effect radiation has on matter. Table 19.3 lists some of the units used to measure radiation and radiation dosage.

A unit long used in measuring radiation dosages is the **rad** (radiation *a*bsorbed *d*ose), where 1 rad corresponds to the absorption of 1×10^{-2} J of energy per kilogram of matter. A better unit for measuring dosages in biological matter is the **rem** (*r*oentgen *e*quivalent for *m*an), which is the rad multiplied by the *relative biological effectiveness* (Q), a factor that takes into account the different effects of different types of radiation of the same energy:

$$\text{rem} = \text{rad} \times Q$$

Values of Q range from about 1 to 20, as indicated in Table 19.3.

Table 19.3 Radiation Units[a]	
Unit	**Definition**
Radioactivity, A	
becquerel, Bq	1 disintegration/s
curie, Ci	3.70×10^{10} Bq = 3.70×10^{10} disintegrations/s^{-1}
Absorbed Dosage	
gray, Gy	A dosage that deposits 1 J of energy/kg of matter, J kg^{-1}
rad	1 rad = 0.01 Gy
Dose Equivalent	
sievert, Sv	1 Sv = 100 rem
rem	1 rem = 1 rad × Q
	Note that Q is approximately 1 for X rays, γ rays, and β particles; 3 for slow neutrons; 10 for protons and fast neutrons; and 20 for α particles.

[a] SI units are shown in red.

Some representative radiation dosages are

- *1000 rem:* Almost certain to cause death.
- *450 rem:* A short-term dose would kill 50% of a population within 30 days.
- *1 rem:* A short-term dose would likely cause about 100 cases of cancer within 20 to 30 years for every 1 million people exposed.
- *130 mrem/y (130 millirem/year):* The normal average background radiation dosage.
- *20 mrem:* The typical dose in a chest X ray examination.
- *5 mrem/y:* Estimated maximum dose of the general population as a result of nuclear power production.

The potential harm from exposure to radioactive materials depends on several factors. Because the penetrating power of α particles is so low, sources of α radiation are relatively harmless as long as they are outside the body. However, inside the body, as in the lungs or stomach, these sources are extremely hazardous. Sources of X rays and γ rays are hazardous even outside the body because the rays are highly penetrating, as Table 19.4 shows for γ rays. Another factor determining potential harm is the time span over which the dose is absorbed. A 100 rem dose over a few days will usually induce the symptoms of radiation sickness—nausea, vomiting, reduced blood count. The same dose over 20 years is much less likely to induce such symptoms. The United States National Council on Radiation Protection and Measurements recommends that the annual dosage for the general population not exceed 170 mrem above the background level.

Table 19.4 Approximate Stopping Distance for α, β^-, and γ Radiation in Various Materials

Radiation	Energy, MeV	Stopping Distance in						
		Air	Paper	Skin	Water[a]	Aluminum	Concrete	Lead
α	3–9	≈7 cm	2–3 sheets	Outer layer[b]	0.02–0.04 mm			
β^-	0–3		2–3 cm (a book)	Inner layers[c]	0–4 mm	5 mm		1 mm
γ	0.01–10				1–20 cm		2 m	10 cm

[a] Distance at which one-half the radiation is stopped.

[b] Radiation reaches dead cells only.

[c] Radiation reaches and damages living cells; causes burns like a sunburn.

19.10 Applications of Radioactive Nuclides

Since Becquerel's discovery of radioactivity in 1896, its applications have become almost too numerous to count. Some uses are based on the ionizing power of radiation; others are based on the fact that radioactive atoms are "tagged" atoms.

Medical Diagnosis and Therapy

Radioisotopes are widely used to diagnose various disorders. The isotope may be used as is, or it may be incorporated into a biologically active substance that targets a particular organ. An isotope of technetium with a half-life of about six hours is particularly useful for this procedure. The short half-life means that it decays quickly. Also, it is versatile and can be used in a number of diagnoses, including scans of the brain, liver, kidney, and gallbladder. Some other medically useful isotopes were given in Table 19.2.

Although it can induce cancer, ionizing radiation can also be used to treat cancer. The governing principle is that cancerous cells are somewhat more easily destroyed by ionizing radiation than are normal cells. Cobalt-60 is widely used for this purpose; its

5.3-year-half-life decay is accompanied by γ-ray emission. For maximum success with radiation therapy, the γ-ray beam must be carefully focused on the tissue targeted for destruction. Proton beams and neutron beams are also used in cancer therapy. For example, many in the medical community now regard proton-beam therapy as an especially effective treatment of prostate cancer.

In some instances, radioactive chemicals (called radiopharmaceuticals) can be ingested and allowed to find their own way to the targeted tissue. For example, iodine, as iodide ion, concentrates in the thyroid gland. When iodine-131 is used, β^- and γ emissions from this radioactive nuclide destroy cancerous tissue in the thyroid without producing harmful effects elsewhere in the body. As with technetium, the short (eight-day) half-life of the isotope ensures that it does not remain active in the body for very long.

The fact that iodine concentrates in the thyroid gland is also the basis of a novel preventive measure against thyroid cancer. If people—especially children—likely to be exposed to atmospheric fallout from a nearby nuclear mishap are given iodide tablets in time, the nonradioactive iodine saturates the thyroid gland and blocks the uptake of ^{131}I.

Modifying Materials with Radiation

Radiation processing is the treatment of materials with ionizing radiation to alter them in some way. The radiation sources commonly used are γ rays from cobalt-60 or cesium-137 and electron beams. The most prominent application is to kill microorganisms. Irradiation can be used, for example, to sterilize medical supplies, such as sutures, syringes, and hospital clothing. One of its advantages over steam sterilization is that it can be done at room temperature.

Ionizing radiation is also used to sterilize insects. Males rendered sterile by irradiation cause females to lay infertile eggs. If the sterilized males are released in sufficient numbers, the insect population can be kept in check.

Many foods can be sterilized and their shelf life extended through irradiation. Irradiation has been used on various fruits, vegetables, grains, and spices. It is increasingly used in the sterilization of meats to kill disease-causing bacteria, such as the virulent strain of *Escherichia coli* that causes food poisoning. According to the Centers for Disease Control and Prevention, food poisoning by harmful microorganisms kills 5000 people, causes 325,000 hospitalizations, and affects 76 million people each year in the United States. Irradiation does not induce radioactivity, but it does produce some chemical changes, such as breaking down complex carbohydrates into sugars and destroying some of the vitamin B_1 in pork. At this point, however, widespread consumer acceptance of irradiated meat is still uncertain.

Neutron activation is used in chemical analysis. A sample is bombarded with neutrons, some of which are absorbed by nuclei in the sample. In this way, stable nuclei are converted to radioactive ones, which emit γ rays as they decay. Because the energy of the emitted γ rays differs from one element to another, measuring the γ-ray energy tells the analyst which element is contained in the sample. In addition, the *number* of γ rays emitted gives a measure of the concentration of that element. Because the process involves the atomic nuclei and not the extranuclear electrons, an element can be detected regardless of its chemical form. Iron metal and iron oxide, for instance, produce γ rays of the same energy.

Radioactive Tracers

All nuclides of an element have essentially identical physical and chemical properties, but radioactive nuclides are easily identified by the radiation they emit. Radioactive nuclides can be used as **radioactive tracers,** and their atoms can be attached to other substances, which are then said to be "tagged." We can trace the fate of tagged substances much as park rangers follow the movements of a wayward bear by fitting it with a radio transmitter. For example, we can use radioactive tracers to

- Detect leaks in underground piping systems: A compound containing a short-lived nuclide (for example, $^{131}_{53}$I, $t_{1/2} = 8.04$ days) is injected into the system, and leaks are found through radioactivity appearing in the soil.

- Determine frictional wear in piston rings: The ring is subjected to neutron bombardment, which converts some of the carbon in the steel to carbon-14. Wear in the piston ring is assessed by the rate at which the radioactivity of the carbon-14 appears in the engine oil.
- Determine the uptake of phosphorus and its distribution in plants. This can be done by incorporating phosphorus-32, a β^- emitter with a 14.3-day half-life, into phosphate fertilizers fed to the plants.

Reaction Mechanisms

To research chemists, one of the most important uses of radioactive nuclides is in elucidating molecular structures and the mechanisms of chemical reactions. Consider the reaction

$$Br^- + CH_3CH_2Br \longrightarrow CH_3CH_2Br + Br^-$$

which at first sight does not look like a reaction at all because the products are the same as the reactants. However, if we use Br^- containing a small amount of $^{82}_{35}Br^-$, a β^- emitter with a 35-hour half-life, we find some of the radioactivity in the organic product. That is, to some extent, this reaction occurs:

$$^{82}_{35}Br^- + CH_3CH_2Br \longrightarrow CH_3CH_2\,^{82}_{35}Br + Br^-$$

We can measure the reaction rate by monitoring the appearance of radioactivity in the organic product. We find the reaction to be second order, evidence that it proceeds by a mechanism such as

$$Br^- + CH_3CH_2Br \longrightarrow \left[\begin{array}{c} H \quad H \\ \backslash \quad / \\ Br\cdots C\cdots Br \\ | \\ CH_3 \end{array} \right]^- \longrightarrow BrCH_2CH_3 + Br^-$$

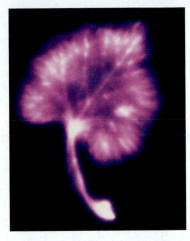

▲ Radiation from phosphorus-32 that has been taken up by a green plant is used to expose photographic film, forming an image showing the distribution of phosphorus in a leaf.

Cumulative Example

The atomic mass of tritium, ^3H, is 3.016 u, and its half-life is 12.29 y. Suppose that hydrogen gas is "spiked" with tritium to the extent of 1.05% by mass. What is the activity, in millicuries, of a 50.08-mL sample of the gas measured at 23.8 °C and 757 mmHg?

STRATEGY

We can break down the solution into three broad parts. (1) Use the ideal gas law (Equation 5.9) to determine first the number of moles of hydrogen in the sample of gas and then the total number of hydrogen atoms. (2) From the mass percent composition of the gas, establish first the mole fraction of tritium atoms in the sample and then the total number of tritium atoms in the sample. (3) Determine the activity of the sample in seconds^{-1}, using Equation (19.1), and then convert this activity to millicuries, using a conversion factor based on data from Table 19.3.

SOLUTION

To calculate the number of moles of hydrogen, we rearrange the ideal gas equation to the form $n = PV/RT$.

$$n = \frac{\left(757\ \text{mmHg} \times \dfrac{1\ \text{atm}}{760\ \text{mmHg}}\right) \times 50.08\ \text{mL} \times \dfrac{1\ \text{L}}{1000\ \text{mL}}}{0.08206\ \text{L atm mol}^{-1}\,\text{K}^{-1} \times (273.15 + 23.8)\ \text{K}}$$

$$= 2.05 \times 10^{-3}\ \text{mol H}_2$$

Then we use the number of atoms per hydrogen molecule and Avogadro's number to get the number of hydrogen atoms in the sample.

$$\text{Number H atoms} = 2.05 \times 10^{-3}\ \text{mol H}_2 \times \frac{2\ \text{mol H}}{1\ \text{mol H}_2} \times \frac{6.022 \times 10^{23}\ \text{H atoms}}{1\ \text{mol H}}$$

$$= 2.47 \times 10^{21}\ \text{H atoms}$$

We can express the mass composition on the basis of 100.00 g of the hydrogen gas.

$$? \text{ g tritium} = 1.05 \text{ g } {}^{3}\text{H}$$

$$? \text{ g naturally occurring } {}^{1}\text{H and } {}^{2}\text{H} = 100.00 \text{ g} - 1.05 \text{ g} = 98.95 \text{ g}$$

Next, we convert these masses to numbers of moles. For the tritium, ${}^{3}\text{H}$, we use 3.016 g/mol, and for the ${}^{1}\text{H}$ and ${}^{2}\text{H}$ together (noted simply as "H"), we use 1.0079 g/mol.

$$? \text{ mol } {}^{3}\text{H} = 1.05 \text{ g } {}^{3}\text{H} \times \frac{1 \text{ mol } {}^{3}\text{H}}{3.016 \text{ g } {}^{3}\text{H}} = 0.348 \text{ mol } {}^{3}\text{H}$$

$$? \text{ mol "H"} = 98.95 \text{ g "H"} \times \frac{1 \text{ mol "H"}}{1.0079 \text{ g "H"}} = 98.17 \text{ mol "H"}$$

We base the calculation of the mole fraction of ${}^{3}\text{H}$ on Equation (12.3).

$$\text{Mole fraction } {}^{3}\text{H} = \frac{0.348 \text{ mol } {}^{3}\text{H}}{0.348 \text{ mol } {}^{3}\text{H} + 98.17 \text{ mol "H"}} = 3.53 \times 10^{-3}$$

Then, we calculate the total number of ${}^{3}\text{H}$ atoms, which we can denote as N.

$$N = 2.47 \times 10^{21} \text{atoms total} \times 3.53 \times 10^{-3} = 8.72 \times 10^{18} \text{ }{}^{3}\text{H atoms}$$

Before determining the activity of these radioactive tritium atoms, we will find it convenient to convert the unit of the half-life from years to seconds.

$$t_{1/2} = 12.29 \text{ y} \times \frac{365 \text{ d}}{1 \text{ y}} \times \frac{24 \text{ h}}{1 \text{ d}} \times \frac{3600 \text{ s}}{1 \text{ h}} = 3.88 \times 10^{8} \text{ s}$$

Now we have the necessary data to obtain the activity, A, from the number of radioactive atoms, N, and the decay constant, λ (which is equal to $0.693/t_{1/2}$).

$$A = \lambda \times N = \frac{0.693}{t_{1/2}} \times N$$

$$= \frac{0.693}{3.88 \times 10^{8} \text{ s}} \times 8.72 \times 10^{18}$$

$$= 1.56 \times 10^{10} \text{ s}^{-1}$$

We conclude the calculation by using the definitions of curie and millicurie (Table 19.3).

$$? \text{ mCi} = 1.56 \times 10^{10} \text{ s}^{-1} \times \frac{1 \text{ Ci}}{3.70 \times 10^{10} \text{ s}^{-1}} \times \frac{1000 \text{ mCi}}{1 \text{ Ci}}$$

$$= 422 \text{ mCi}$$

ASSESSMENT

Note that in this problem we had to deal with all three isotopes of hydrogen. In the calculation of the total number of H atoms in the gaseous sample, we treated them equally because each molecule of gas is accounted for in the same way in the ideal gas law, regardless of whether it is ${}^{1}\text{H}-{}^{1}\text{H}$, ${}^{1}\text{H}-{}^{2}\text{H}$, ${}^{1}\text{H}-{}^{3}\text{H}$, and so on. In the mole fraction calculation, on the other hand, we had to sort the three isotopes of hydrogen into two categories. One category consists of the naturally occurring isotopes ${}^{1}\text{H}$ and ${}^{2}\text{H}$, which together account for hydrogen's atomic mass of 1.00794 u. The other is the synthetic isotope ${}^{3}\text{H}$, whose atomic mass has to be dealt with separately.

The best way to assess whether our final answer is a reasonable one is to estimate the answer in each step to see if it is of the proper magnitude. For example, in the first step the number of moles of H_2 in 50.08 mL of gas should not be greatly different from what it would be at STP $[(0.05008 \text{ L}/22.4 \text{ L}) \times 1 \text{ mol} = 2.24 \times 10^{-3} \text{ mol}]$, and it is not. The total number of H atoms should be of the order of 10^{21}, and it is. The number of ${}^{3}\text{H}$ atoms should be a few powers of 10 less than the total number of atoms, and it is (8.72×10^{18}). A half-life of 12.29 y should be several million times larger when expressed in seconds, and it is (3.88×10^{8} s). The activity, A, should be somewhat less than the number of atoms divided by the half-life, and it is (1.56×10^{10} s^{-1}). In turn, this activity should be somewhat less than 1 curie, and that is what the final answer, 422 mCi, proves to be.

Concept Review with Key Terms

19.1 Radioactivity and Nuclear Equations—**Radioactivity**, or **radioactive decay**, is the process that occurs when radioactive **nuclides** emit **alpha (α) particles**, ${}_{2}^{4}\text{He}$; **beta (β^{-}) particles**, ${}_{-1}^{0}\text{e}$; **gamma (γ) rays** (high-frequency electromagnetic radiation); or **positrons**, ${}_{1}^{0}\text{e}$. In a fifth type of radioactivity, **electron capture (EC)**, a nucleus absorbs one of the atom's inner-shell electrons and then emits an X ray. Nuclear equations representing these modes of decay must have the same total of atomic numbers and of mass numbers on both sides. The mass number is equal to the total number of **nucleons** (protons plus neutrons) in the nucleus. For example:

Mass numbers: 238   =   234   +   4

$${}_{92}^{238}\text{U} \longrightarrow {}_{90}^{234}\text{Th} + {}_{2}^{4}\text{He}$$

Atomic numbers: 92   =   90   +   2

19.2 Naturally Occurring Radioactivity—All known nuclides with $Z > 83$ are radioactive, and many of them occur naturally as members of four radioactive decay series. The relative rates of disappearance of radioactive nuclides are often expressed by **half-life ($t_{1/2}$)**, which is the time required for one-half of a sample of radioactive nuclide to decay. **Radioactive decay series** are used to describe the elemental pathways by which radioactive nuclides are transformed into nonradioactive nuclides.

19.3 Radioactive Decay Rates—The **radioactive decay law** states that the rate of disintegration of radioactive nuclides in a sample, called the activity (A) of the sample, is equal to the number of atoms present (N) times the decay constant (λ):

$$\text{Rate of radioactive decay} = A = \lambda N$$

The time-dependence and half-life of the radioactive decay are described by equations similar to those used in previous discussions of first-order kinetics:

$$\ln \frac{N_t}{N_0} = -\lambda t \qquad t_{1/2} = \frac{0.693}{\lambda}$$

19.4 Synthetic Nuclides—There are a few naturally occurring radioactive nuclides of lower atomic numbers, such as ^3_1H, $^{14}_6\text{C}$, and $^{40}_{19}\text{K}$. For the most part, however, nuclear reactions are required to synthesize radioactive nuclides of elements with $Z < 83$.

19.5 Transuranium Elements—**Transuranium elements**—those for which $Z > 92$—can also be produced synthetically. Massive, highly charged ions have been used as projectiles in charged particle accelerators to synthesize atoms of the heaviest elements.

19.6 Nuclear Stability—Among the factors that establish whether a nuclide is stable or radioactive are the ratio of the neutron number (N) to the proton number (Z), whether the proton and neutron numbers are even or odd, and whether either of them corresponds to a filled nuclear shell.

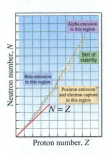

19.7 Energetics of Nuclear Reactions—In the formation of an atomic nucleus from its protons and neutrons, a quantity of mass (*mass defect*) is converted to an equivalent quantity of energy, called the **nuclear binding energy**.

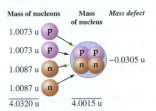

19.8 Nuclear Fission and Nuclear Fusion—The fusing of lighter nuclei to produce heavier ones is the basis of **nuclear fusion** energy, and the splitting of heavier nuclei to produce lighter fragments is the basis of **nuclear fission** energy. The fission of heavier nuclei is used in a controlled manner to generate energy in nuclear reactors. When fission takes place in an uncontrolled manner in the presence of a **critical mass** of a heavy nuclide, a nuclear explosion occurs.

19.9 Effect of Radiation on Matter—Radiation from radioactive materials interacts with matter, principally by forming ions and breaking chemical bonds. Less energetic radiation excites electrons to higher energy states, leading to the emission of electromagnetic radiation. Radiation can be detected in several ways, including the use of photographic film and Geiger—Müller counters. The interaction of radiation with living matter can cause birth defects and various cancers. These effects are linked to the received dosage of radiation, measured in **rad**. The unit **rem** reflects both the dosage (rad) and the biological effect of different types of radiation.

19.10 Applications of Radioactive Nuclides—Radiation and radioisotopes have a number of medical, scientific, and industrial uses. For example, **radioactive tracers** are used to either locate or follow the flow of matter in applications such as leak detection, medical diagnostic exams, and environmental studies.

Assessment Goals

When you have mastered the material in this chapter, you will be able to:

- Describe the basic nature of α, β^-, and γ radiation.

- Write and balance nuclear equations involving the five different decay modes of radioactive nuclides.

- Predict the products of radioactive decays series, given the modes of the decay process.

- Calculate decay rates and half-lives of radioactive isotopes.

- Describe the use of radioactive carbon dating to estimate the age of objects.

- Recognize nuclear reactions leading to the formation of synthetic isotopes, including the transuranium nuclides.

- Predict the nuclear stability of specific isotopes, making use of magic numbers and the belt of stability illustrated in Figure 19.5.

- Relate mass changes to energy changes in nuclear equations, using the relationship $E = mc^2$.

- Describe the use of nuclear fission and nuclear fusion in energy production.

- Describe the effect of radiation on matter and the manner in which radiation exposure is measured.

- Describe the use of radiation and radioisotopes in cancer treatments, sterilization, tracers, and determining reaction mechanisms.

Self-Assessment Questions

1. Of α, β^-, and γ radiation, which has the greatest ability to penetrate through matter? Which has the greatest ability to ionize matter? How does a γ ray differ from an X ray?

2. What type of radioactive decay causes the atomic number of a nucleus to increase by one unit?
   (a) electron capture   (c) α emission
   (b) β^- emission   (d) γ ray emission

3. Which of the following types of radiation is deflected by a magnetic field?
   (a) X ray   (b) γ ray   (c) β^- ray   (d) neutron.

4. Why does radioactive decay by electron capture produce X rays?

5. The most radioactive of the isotopes of an element will be the one with the largest value of
   (a) half-life   (c) mass number
   (b) neutron number   (d) decay constant

6. How is the activity of a radioactive sample related to the amount of sample? How is the activity of the sample related to time?

7. ^{223}Ra has a half-life of 11.4 days. Approximately how long would it take for the radioactivity associated with a sample of ^{223}Ra to decrease to 1% of its initial value?

8. How does radiocarbon dating work? What are its limitations in establishing the ages of objects?

9. What is meant by the term *magic numbers?* What does it mean for a nucleus to be *doubly magic?*

10. Which of the following nuclides is most likely to be radioactive?
    (a) ^{31}P   (b) ^{66}Zn   (c) ^{37}Cl   (d) ^{108}Ag

11. Which of the following nuclides is most likely to decay by positron β^+ emission?
    (a) ^{59}Cu   (b) ^{63}Cu   (c) ^{67}Cu   (d) ^{68}Cu

12. What is a nuclear binding energy? What is a nuclear binding energy per nucleon?

13. Which of the following nuclides has the highest nuclear binding energy per nucleon?
    (a) $^{3}_{1}$H   (b) $^{16}_{8}$O   (c) $^{56}_{26}$Fe   (d) $^{235}_{92}$U

14. Explain why nuclear reactions liberate so much more energy than do chemical reactions.

15. What is the meaning of the term critical mass? To what type of nuclear process is it relevant?

16. What is the basic operating principle of a nuclear fusion reactor, and what is the function of control rods and moderators in such reactors?

17. What is the difference between primary and secondary electrons produced by ionizing radiation?

18. Why is the rem a more satisfactory unit for measuring human radiation dosage than the rad?

19. How is the presence of radiation detected in a Geiger–Müller counter?

20. What basic principles underlie (a) the use of radioactive tracers? (b) Radiation processing?

Problems

Radioactive Decay Processes

21. Each of the following equations represents a radioactive decay process. Supply the missing items.
    (a) $^{219}_{?}\text{Rn} \longrightarrow \text{?} + ^{4}_{2}\text{He}$
    (b) $^{167}_{69}\text{?} + ^{0}_{-1}\text{e} \longrightarrow \text{?}$
    (c) $^{90}_{38}\text{Sr} \longrightarrow ^{0}_{-1}\text{e} + \text{?}$
    (d) $^{79}_{36}\text{Kr} \longrightarrow ^{0}_{1}\text{e} + \text{?}$

22. Each of the following equations represents a radioactive decay process. Supply the missing items.
    (a) $^{39}_{?}\text{Ca} \longrightarrow \text{?} + ^{0}_{1}\text{e}$
    (b) $^{210}_{?}\text{Bi} \longrightarrow \text{?} + ^{0}_{-1}\text{e}$
    (c) $^{63}_{30}\text{Zn} + ^{0}_{-1}\text{e} \longrightarrow \text{?}$
    (d) $^{8}_{4}\text{Be} \longrightarrow ^{4}_{2}\text{He} + \text{?}$

23. What is the final nuclide obtained in each process?
    (a) Indium-123 decays by electron capture.
    (b) A succession of three β^- emissions begins with molybdenum-103.
    (c) A succession of two α emissions, followed by two β^- emissions, begins with astatine-217.

24. What is the final nuclide obtained in each process?
    (a) Polonium-210 decays by α emission.
    (b) Copper-59 decays by β^+ emission.
    (c) Lead-212 decays by β^- emission, followed by α emission and then by β^- emission.

25. Determine the nuclide that (a) undergoes electron capture to produce ^{7}Li, (b) undergoes α decay to form ^{238}U.

26. Determine the nuclide that decays to produce (a) a β^- and ^{191}Ir, (b) a β^+, a γ ray, and ^{58}Ni.

27. The natural radioactive decay series that begins with thorium-232 is represented diagrammatically below. Construct a graph of this series similar to Figure 19.2.

$$^{232}\text{Th}-\alpha-\beta^=-\beta^=-\alpha-\alpha-\alpha-\alpha-\beta^- \begin{smallmatrix} \beta^=-\alpha \\ {}^{208}\text{Pb} \\ \alpha-\beta^- \end{smallmatrix}$$

28. Convert the natural decay series starting with uranium-238 (Figure 19.2) to the format used in Problem 27.

Rate of Radioactive Decay

29. Chemical reactions can be of different orders—zero, first, second, and so forth—whereas radioactive decay is always a first-order process. Why is this so?

30. Many chemical reactions can be catalyzed. Do you think it is equally possible to catalyze a radioactive decay process? Explain.

31. *Without doing detailed calculations*, determine which of the radioactive nuclides in the table **(a)** has the greatest activity, A, for a 1-million-atom sample; **(b)** would undergo a 75% reduction in activity in about 2 days.

Nuclide:	^{15}O	^{18}F	^{28}Mg	^{44}Sc	^{33}P
$t_{1/2}$:	122 s	109.8 min	21 h	3.93 h	25.3 d

32. *Without doing detailed calculations*, determine which of the radioactive nuclides in the table (Problem 31) **(a)** has a decay constant, λ, that is in the middle position when the decay constants are arranged in increasing order; **(b)** would have their activity reduced to less than 1% of their initial value in 1 day.

33. Iodine-131 has a half-life of 8.04 days, and that of phosphorus-32 is 14.3 days, yet the rate of decay of a sample of phosphorus-32 is greater than that of a sample of iodine-131 of the same mass. Explain why.

34. A 1.0-μ mol sample of iodine-131 ($t_{1/2} = 8.04$ d) is incorporated into a 1.0-mol sample of KI(s), and a second 1.0-μ mol sample of iodine-131 is incorporated into a 1.0-mol sample of $PbI_2(s)$. Then 1.0-mmol samples of the two solids are tested for their activity. Will the activities of the two samples be the same? Explain.

35. What is the decay rate in atoms/h of **(a)** a 1-million-atom sample of Pt-200 ($t_{1/2} = 11.5$ h), **(b)** a 1-million-atom sample of C-11?

36. The sample of $RaCl_2$ isolated by Marie Curie contained 91 mg of radium-226, with a half-life of 1600 years. What was the decay rate in s^{-1} of her sample at the time it was isolated?

37. We follow the activity of a sample containing the radioactive isotope sulfur-35. At the start of the experiment, we detect 138 dis per min, and after 20.0 d, the decay rate is 118 dis min^{-1}. What is the half-life of sulfur-35?

38. The rate of decay of selenium-81 by β^- emission is measured in a Geiger–Müller counter. The decay rate, expressed in counts per minute (cpm), is 1242 cpm at $t = 0$ and 277 cpm at $t = 40.0$ min. What is the half-life of selenium-81?

39. The half-life of iron-52 is 8.28 hours. Suppose we measure the activity of a sample containing iron-52 and find it to be 884 dis per min. If we measure its activity again 24.0 hours later, **(a)** *estimate* its activity at the later time, and **(b)** calculate a more exact value of the activity that should be found.

40. The half-life of iodine-131 is 8.04 days. Suppose we follow the activity of a sample of iodine-131 while it falls to 10% of its initial value. **(a)** *Estimate* how long this will take, and **(b)** calculate a more exact value of the time required.

41. What should be the current rate of disintegration of carbon-14, expressed in dis per min per g carbon, for a wooden object that is believed to have been made in 800 BC?

42. What would be the rate of decay of carbon-14, expressed in dis per min per g carbon, for a carbon-containing object 50,000 years old? How difficult do you think it would be to detect this rate of decay? Explain.

Nuclear Reactions and Nuclear Equations

43. Complete the following nuclear equations.

(a) $^{9}_{4}Be + ^{4}_{2}He \longrightarrow ? + ^{12}_{6}C$

(b) $^{130}_{52}Te + ? \longrightarrow ^{130}_{53}I + 2\,^{1}_{0}n$

(c) $^{10}_{5}B + ^{1}_{0}n \longrightarrow ^{7}_{3}Li + ?$

44. Complete the following nuclear equations.

(a) $^{154}_{62}Sm + ^{1}_{0}n \longrightarrow ? + 2\,^{1}_{0}n$

(b) $? + ^{4}_{2}He \longrightarrow ^{133}_{57}La + 4\,^{1}_{0}n$

(c) $^{246}_{96}Cm + ^{13}_{6}C \longrightarrow ^{254}_{102}No + ?$

45. Write nuclear equations to represent **(a)** the reaction of two *deuterons* ($^{2}_{1}H$) to produce helium-3, **(b)** the production of berkelium-243 by the α-particle bombardment of americium-241 nuclei, and **(c)** the bombardment of uranium-238 nuclei by α particles to form plutonium-239.

46. Write nuclear equations to represent the bombardment of **(a)** nitrogen-14 nuclei by α particles to form oxygen-17, **(b)** uranium-238 nuclei by nitrogen-14 nuclei to form einsteinium-246, and **(c)** antimony-121 nuclei by α particles to produce iodine-124, followed by its radioactive decay by positron emission.

47. Technetium-99, a major by-product of nuclear weapons production, has a half-life of 210,000 years. Neutron bombardment converts it to technetium-100, which decays with a 16-second half-life to ruthenium-100. Write nuclear equations for these reactions.

48. When bombarded with nickel-62 nuclei, a lead-208 nucleus yielded a nucleus with $Z = 110$ and $A = 269$ plus one neutron. The nucleus was identified by its decay pattern of three successive α decays ending with nobelium-257. Write nuclear equations for the synthesis and subsequent decay of this nucleus.

Energetics of Nuclear Reactions

49. Calculate the mass defect in the creation of a nucleus of iron-56 from the nucleons it contains. The nuclear mass of iron-56 is 55.92068 u.

50. The mass defect in the formation of a nucleus of neon-20 is 0.173 u. What is the *nuclear* mass of this nuclide, and what is its *atomic* mass?

51. Sulfur-32 has a nuclear mass of 31.96329 u. Determine the binding energy per nucleon in sulfur-32.

52. Determine the binding energy per nucleon for $^{40}_{20}Ca$, which has a nuclear mass of 39.95162 u.

53. Phosphorus-32 decays by β^- emission. Determine the energy associated with this decay, given the nuclear masses: ^{32}P, 31.9657 u and ^{32}S, 31.9633 u. The mass of an electron is 5.486×10^{-4} u.

54. Radium-224 decays by α-particle emission. Determine the energy associated with this decay, given these nuclear masses: ^{224}Ra, 223.9719 u; ^{220}Rn, 219.9642 u; ^4He, 4.00150 u.

55. Thorium-228 decays by α-particle emission. In this decay, 5.50 MeV of energy is given off. What are (a) the *nuclear* mass and (b) the *atomic* mass of thorium-228? The nuclear masses of radium-224 and helium-4 are 223.9719 u and 4.00150 u, respectively.

56. Radium-229 decays by β^- emission. In this decay, 1.14 MeV of energy is given off. What are (a) the *nuclear* mass and (b) the *atomic* mass of actinium-229? The nuclear mass of radium-229 is 228.9866 u, and the mass of an electron is 5.486×10^{-4} u.

57. What is the energy, in megaelectronvolts (MeV), required to bring about the following nuclear reaction?

$$^{232}_{90}Th + {}^4_2He \longrightarrow {}^{235}_{92}U + {}^1_0n$$

The *atomic* masses are ^{232}Th, 232.0382 u; ^4He, 4.0026 u, ^{235}U; 235.0439 u. The mass of a neutron is 1.0087 u.

58. Given that the atomic masses of ^{235}U and ^{236}U are 235.0439 u and 236.0456 u, respectively, explain the point made on page 816 that fission can be initiated with slow neutrons, that is, neutrons with only the kinetic energy associated with ordinary temperatures.

Nuclear Stability

59. Which of the following nuclides do you think are radioactive, and which do you think are stable? Explain.

   (a) $^{10}_6C$ (b) $^{15}_7N$ (c) $^{30}_{14}Si$ (d) $^{36}_{15}P$

60. Suggest a mode of decay for (a) ^{160}Ho, (b) ^{166}Ho, (c) ^{170}Ho. Explain your choices.

61. In some cases, we can identify the most abundant nuclide of an element by rounding off the weighted average atomic mass listed on the inside front cover. Thus, the principal nuclide of sodium (weighted average atomic mass, 22.9898 u) is sodium-23, and that of potassium (weighted average atomic mass, 39.0983 u) is potassium-39. In some cases, though, this method will not work. Cite some examples, and explain why the method fails.

62. In general, the elements with *even* atomic numbers have a greater number of naturally occurring isotopes than those with *odd* atomic numbers. Explain why you would expect this to be the case.

63. There are six naturally occurring nuclides of calcium, with mass numbers 40, 42, 43, 44, 46, and 48. Why do you suppose that calcium-40 is by far the most abundant of these nuclides?

64. For what neutron number, N, would you expect to find the greatest number of stable nuclides? Explain.

Fission and Fusion

65. Suppose that one of the reactions in the fission of uranium-235 is

$$^{235}_{92}U + {}^1_0n \longrightarrow {}^{139}_{54}Xe + {}^{95}_{38}Sr + 2\,{}^1_0n$$

Calculate the energy in MeV released in this fission. The nuclear masses are ^{235}U, 234.9934 u; ^{139}Xe, 138.8891; ^{95}Sr, 94.8985 u; $^1n = 1.0087$ u.

66. Calculate the energy in megaelectronvolts (MeV) released in the nuclear fusion reaction:

$$^2_1H + {}^3_1H \longrightarrow {}^4_2He + {}^1_0n$$

The nuclear masses are ^2H, 2.0135 u; ^3H, 3.0156; ^4He, 4.0015 u; ^1n, 1.0087 u. How does the energy per gram of ^4He produced in this reaction compare with the energy per gram of ^{235}U that undergoes fission, described on page 816?

Radiation Dosage and the Effect of Radiation on Matter

67. Explain why radioactive nuclides with half-lives of intermediate value are generally more hazardous than those with extremely short or extremely long half-lives.

68. Explain why some radioactive substances are hazardous from a distance, whereas others must be taken internally to constitute a hazard.

69. Estimates of the total release of iodine-131 from the Chernobyl nuclear accident of 1986 range from 40 to 50 million curies (Ci). Given the definition of a curie in Table 19.3 and that $t_{1/2} = 8.040$ d for iodine-131, approximately how many grams of iodine-131 were released in the accident?

70. An evaluation of the iodine-131 contamination caused by the Chernobyl nuclear accident placed the activity close to the site at greater than $37{,}000$ kBq/m^2. Given the definition of the becquerel (Bq) in Table 19.3 and that $t_{1/2} = 8.040$ d for iodine-131, what was the minimum level of contamination with iodine-131 near the plant site, expressed in milligrams of iodine-131 per square kilometer?

71. If the positron were included in Table 19.4, what entries would be appropriate for the approximate stopping distances? Explain.

72. Radon-222 is an α-particle emitter with a half-life of 3.82 days. Under what conditions would radon-222 be of little danger, and under what conditions, would it be most hazardous?

Radioactive Tracers

73. Describe how you might use a radioactive tracer to find a leak in the $H_2(g)$ supply line in an ammonia synthesis plant.

74. A small quantity of NaCl(s) containing radioactive $^{24}_{11}Na^+$ is added to $NaNO_3(aq)$. The solution is cooled, and $NaNO_3(s)$ is crystallized from the solution. Would you expect the $NaNO_3(s)$ to be radioactive? Explain.

Additional Problems

Problems marked with an * may be more challenging than others.

75. Refer to Figure 19.1 and indicate the type of deflection expected for a positron. Why is the deflection of the α particle not as great as that of β^-? Draw a sketch of the expected deflections of α, β^-, β^+, and γ rays in a magnetic field rather than an electric field.

76. Francium is such a rare element that it has been estimated that less than about 30 grams are present in Earth's crust at any one time. How do you think such an estimate of the natural abundance of francium is made? Would you expect to find francium in the same natural sources as the other group 1A (alkali) metals? Explain.

77. An aqueous solution of HCl is prepared, incorporating some ^3H as a tracer in HCl. Then the labeled HCl(aq) is used in the following reactions, which are carried out in succession:

$$NH_3(aq) + HCl(aq) \longrightarrow NH_4Cl(aq)$$

$$NH_4Cl(aq) + NaOH(aq) \longrightarrow NaCl(aq) + H_2O(l) + NH_3(g)$$

Would you expect radioactivity to appear **(a)** in the $NH_3(g)$, **(b)** in the NaCl(aq), **(c)** in NaCl(s) crystallized from the solution? Explain.

78. In the decay of uranium-238, some mass is converted to energy carried away by the α particle. What is the energy of the α particle, in MeV? The atomic masses are $^{238}U = 238.0508$ u, $^{234}Th = 234.0436$ u, and $^4He = 4.0026$ u.

79. In the γ-ray decay of $^{230m}_{90}Th$, the energy of the γ rays is 0.050 MeV. What is the wavelength of these γ rays?

80. The uranium-238 decay series (Figure 19.2) is known as the $4n + 2$ series because the mass number of each nuclide in the series conforms to the equation $A = 4n + 2$. The series originating with thorium-232 is the $4n$ series. Other series are the $4n + 1$ and the $4n + 3$ series. To which series does each of the following radioactive nuclides belong?

(a) bismuth-214 **(c)** astatine-215

(b) polonium-216 **(d)** uranium-235

81. Calculate the minimum kinetic energy, in megaelectronvolts (MeV), and the velocity in meters per second (m/s), that deuterons must have in order to bring about this nuclear reaction:

$$^{75}_{33}As + ^2_1H \longrightarrow ^{75}_{34}Se + 2\,^1_0n$$

The *atomic* masses are ^{75}As, 74.9216 u; ^2H, 2.0140 u; ^{75}Se, 74.9225 u. The mass of a neutron is 1.0087 u.

82. Determine the mass, in grams, equivalent to the energy released in the combustion of 1.00 kg of carbon to produce carbon dioxide.

83. The curie, described in Table 19.3, is the rate of radioactive decay of 1.00 g of radium-226. The atomic mass of radium-226 is 226.0254 u. Use this information and the radioactive decay law to obtain a value of Avogadro's number. Would you expect this method to be as accurate as the one based on X-ray crystallography described in Chapter 11 (Problems 67 and 68)? Explain.

* **84.** Tritium, ^{3}H, is a β^- emitter with a 12.26-year half-life. Some HCl with ^{3}H as a tracer is added to HCl(g), which is then used to prepare 50.0 mL of 1.00 M HCl(aq). The activity, A, of the solution is found to be 2.5×10^{10} dis/s. Then a 0.455-g sample of zinc is allowed to react with the 50.0 mL of 1.00 M HCl. What will be the activity of the H_2(g) produced?

* **85.** The percent natural abundance of radioactive potassium-40 ($t_{1/2} = 1.25 \times 10^9$ y) is 0.0117%. The decay of potassium-40 takes place principally by β^- emission. Approximately how many β^- particles are produced per minute by a 50.0-mg sample of the mineral *carnallite*, $KMgCl_3 \cdot 6\ H_2O$?

86. In the first part of the twentieth century, watch dials were coated with radium so that they were visible in the dark. If a watch originally had 40 ng of the radioactive material, which had a decay rate of 2.40×10^2 dis s^{-1}, what will the activity be after **(a)** 50 years, and **(b)** 100 years?

87. Iodine-131 was one of the principal radioisotopes released in the Chernobyl accident. Another was cesium-137. If a 1.00-mg sample of cesium-137 is equivalent to 89.8 millicuries of radiation, what must be the half-life (in years) of cesium-137?

* **88.** For medical uses, gaseous radon-222, produced in the α emission of radium-226, is allowed to accumulate for a period of time and is then withdrawn and sealed into a glass vial. In nuclear chemistry terminology, radium-226 is called the *parent* nuclide and radon-222 is the *daughter*. The half-lives of this parent–daughter pair are 1690 years and 3.82 days, respectively.

(a) Since the half-life of radon-222 is so much shorter than that of radium-226, why does the radon not disappear as fast as it is formed, without ever building up to a significant concentration?

(b) In the following expression, P_0 represents the number of radium-226 atoms present initially and D represents the number of radon-222 atoms present at the time t. The terms λ_P and λ_D are the decay constants for the two nuclides.

$$D = \frac{P_0 \times \lambda_P \times (e^{-\lambda_P \times t} - e^{-\lambda_D \times t})}{\lambda_D - \lambda_P}$$

Show that according to this expression, the number of daughter atoms should begin small, build up to a maximum over time, and eventually decline. Starting with 1.00 g radium-226, estimate whether the time required for the amount of radon-222 to reach a maximum is more nearly 1 day, 10 days, 100 days, 1000 days, or 10 years.

* **89.** An electric power plant burns natural gas that is 92.0% CH_4(g) and 8.0% C_2H_6(g) by volume. Assume complete combustion of the gas, yielding CO_2(g) and H_2O(l) as products. How many liters of this natural gas, measured at 22 °C and 744 mmHg, would have to be burned to liberate as much energy as that produced in the fission of 1.00 mg ^{235}U? Assume the energy release in the fission is that cited on page 816.

* **90.** Procedures used to establish the age of rocks in Earth's crust generally involve determining the ratio of the amounts of two different nuclides in the rock.

(a) Suppose that the molten material that froze to become a rock originally contained 1.00 g ^{238}U. Assume that no lead was present in the rock initially and that none entered or left the rock after it was formed. What mass of ^{206}Pb would you expect to find in the rock after one half-life period of the ^{238}U?

(b) Potassium-40 undergoes radioactive decay by electron capture to argon-40 and by β^- emission to calcium-40. The fraction of the decay that occurs by electron capture is 0.110. The half-life of potassium-40 is 1.25×10^9 y. Assuming that a rock in which potassium-40 has undergone decay retains all of the argon-40 produced, what would be the ^{40}Ar/^{40}K mass ratio in a rock that is 1.5×10^9 years old?

(c) Use the radioactive decay law to verify the statement on page 804 that the percent natural abundances of ^{238}U and ^{235}U correspond to an age of Earth of about 6 billion years, assuming that the two nuclides were originally present in equal abundances. (*Hint:* You will need to solve two equations in two unknowns.)

* **91.** The power of nuclear explosions is often described in kilotons or megatons, meaning equivalent to the explosion of that mass of 2,4,6-trinitrotoluene (TNT). The first atomic bomb tested at Alamogordo, NM, was rated at 20 kilotons. Determine the mass of ^{235}U that underwent fission in the weapon. Use information from Problem 65 as needed. The standard enthalpy of formation of solid 2,4,6-trinitrotoluene is -65.5 kJ/mol, and the explosion reaction can be represented approximately as

$$\text{TNT(s)} \longrightarrow \text{C(graphite)} + $$
$$CO(g) + CO_2(g) + N_2(g) + H_2O(g) \quad \textit{(not balanced)}$$

(*Hint:* Assume that TNT is the sole reactant and that the maximum amount of CO_2 consistent with a balanced equation is formed.)

92. Electron capture is sometimes called K-capture because the electron captured comes from the $n = 1$, or K, shell of the atom. Write the electron configuration for a cobalt-57 atom after it has undergone K-capture. This excited-state ion will give off an X ray when an electron from the L shell ($n = 2$) drops to the K shell. Write the electron configuration for the ion after this L-to-K transition. If the wavelength of the emitted X ray is 0.179 nm, calculate the energy difference in kilojoules per mole between the L and K shells in cobalt.

93. Justify the observation that Cu-64 can undergo β emission, positron emission, or electron capture.

Apply Your Knowledge

94. [Historical] In the early investigations of radioactivity, the members of the uranium decay series were given nonsystematic *radioelement* names, since the amount of element actually formed in decay was often too small to collect and examine. For example, one particular isotope was given the radioelement name *Uranium I*. Uranium I undergoes α decay to uranium X_1, which undergoes β decay to uranium X_2, which undergoes β decay to uranium II. The next step is α decay to ionium, then α decay to radium, then α decay to Ra-emanation. If Ra-emanation is actually radon-222, identify the other radioelements given here by element symbol and mass number.

*95. [Biochemical] Use Table 19.3, along with first ionization energies of carbon (1086 kJ/mol), oxygen (1314 kJ/mol), and hydrogen (1312 kJ/mol) to estimate the fraction of matter that is ionized in a human body that receives a lethal dose (1000 rem) of β radiation. For purposes of calculation, assume the body composition is 43 kg oxygen, 16 kg carbon, and 7 kg hydrogen.

*96. [Collaborative] The *packing fraction* of a nuclide is related to the fraction of the mass of the nucleons that is converted to nuclear binding energy in the formation of the nuclide. It is defined as the fraction $(M - A)/A$, where M is the actual atomic mass of the nuclide and A is the mass number. Use data from a handbook (such as *The Handbook of Chemistry and Physics*, published by the CRC Press) to determine the packing fractions of some representative nuclides. Plot a graph of packing fraction versus mass number, and compare it with Figure 19.5. Explain the relationship between the two.

97. [Environmental] Americium-241 is used in household smoke detectors as a source of α particles. The device actually responds to the presence of ions from a fire, rather than responding to smoke. A particular detector contains 2.0 μCi of ^{241}Am, with a half-life of 458 years. The isotope is encased in a thin aluminum container. Calculate the mass of ^{241}Am in the detector. Suggest a reason why fears of radiation exposure from normal use of such detectors are largely unfounded.

*98. [Environmental] Assume that the explosion at the nuclear power plant at Chernobyl, Ukraine, on April 26, 1986, released 250 g of iodine-131. Determine the numbers of curies of radiation associated with this release 30.0 days after the accident.

99. [Laboratory] A silver coin was irradiated with neutrons to convert some of the silver-107 to silver-108. The decay of this isotope was followed with a Geiger counter, producing the counts per minute (cpm) as shown:

Time, min	Recorded cpm
0	1784
1	1232
2.33	880
3.5	656
4.5	554
5.5	342
6.5	266

The coin was removed from the counter, and background radiation of 27 cpm was measured. Plot the data to determine the half-life of silver-108.

e-Media Problems

The activities described in these problems can be found in the e-Media Activities and Interactive Student Tutorial (IST) modules of the Companion Website, *http://chem.prenhall.com/hillpetrucci*.

100. Observe the illumination pattern on the fluorescent screen in the **Separation of Alpha, Beta, and Gamma Rays** animation (*Section 19-1*). **(a)** Why are three different spots generated on the screen when the source of radioactivity is passed through an electric field? Account for the relative position of each spot in terms of the nature of the particle giving rise to it. **(b)** How would this pattern change if the electric field strength were increased? **(c)** Positron emission is described in the text. Where would this particle impinge on the fluorescent screen?

101. The natural radioactive decay of ^{238}U is illustrated in the **Radioactive Decay Series** activity (*Section 19-2*). Several possible pathways are indicated. Choose one and count the total number and type of particle emissions produced in the series of reactions. What is the final product of the reaction?

102. (a) What aspect of the reactions illustrated in the **First-Order Process** animation (*Section 19-3*) allows the measure of radioactivity levels to be used to determine the age of a sample? **(b)** In such a procedure, what is measured and how is the age of the sample deduced? **(c)** What assumption(s) about the sample must be made in such a determination?

103. Observe the simulated generation of electricity in the **Nuclear Reactors** activity (*Section 19-8*). **(a)** What role do nuclear reactions play in the process of generating electricity? **(b)** In this application, what measures are taken to control the nuclear reactions? **(c)** Why is more than one source of water required for such a reactor? What is the function of each source of water?

Chapter 20

The s-Block Elements

EARLY CHEMICAL KNOWLEDGE was mostly of a practical nature, a how-to of chemical processes. Chemists mainly used trial and error to transform the materials around them, accumulating a wealth of factual information while doing so. In contrast, modern chemical knowledge is based largely on principles that explain facts and answer the "why" as well as the "how-to" of chemical change. So far in this text, we have focused on principles, although we have often referred to their applications. In the remaining chapters, we will emphasize facts and applications, but we will refer to underlying principles repeatedly.

The periodic table provides us with an organizational scheme for a systematic study of the elements and their compounds. In this study, generally called *descriptive chemistry*, we describe the behavior of the elements and their compounds. Where possible, we attempt to rationalize this behavior by using fundamental ideas about atomic and molecular structure, intermolecular forces, thermodynamics, kinetics, oxidation–reduction, acids and bases, and so on. This chapter deals with the s-block elements, and the next two deal with the p-block and d-block elements, respectively. (See Figure 8.6 to review the various block designations on the periodic table.)

Four of the 14 elements that make up the s-block are somewhat unusual cases, although not all for the same reason. We discussed the difficulty in placing hydrogen in the periodic table in Chapter 8, noting there that most periodic tables place hydrogen in group 1A based on its electron configuration of $1s^1$. On the other hand, because hydrogen is a nonmetal with little resemblance to the active metals of group 1A, some periodic tables place it in a special location at the top and near the center of the table.

Because the helium atom has electrons only in its $1s$ orbital $(1s^2)$, it is an s-block element. In its properties, however, helium closely resembles only the other noble gases, which have the electron configuration ns^2np^6 in the outermost occupied shell. They are p-block elements, and we will consider helium with the p-block elements despite its placement in the s block.

◄ The Cone Nebula is one of the best-known objects in the sky. A very dense cloud of gas and dust projects into a glowing pink background. The background is the result of excitation and emission from the most common element in the universe—hydrogen. In this chapter we will examine properties and uses of hydrogen and other group 1A elements, along with the group 2A elements.

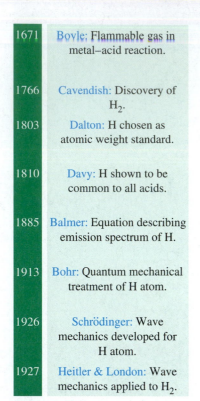

1671	Boyle: Flammable gas in metal–acid reaction.
1766	Cavendish: Discovery of H_2.
1803	Dalton: H chosen as atomic weight standard.
1810	Davy: H shown to be common to all acids.
1885	Balmer: Equation describing emission spectrum of H.
1913	Bohr: Quantum mechanical treatment of H atom.
1926	Schrödinger: Wave mechanics developed for H atom.
1927	Heitler & London: Wave mechanics applied to H_2.

▲ **FIGURE 20.1 Important developments linked to the element hydrogen**

▲ Optical image of the emission nebula NGC 6357, located about 5500 light-years from Earth. The nebula is a cloud of hydrogen 60 light-years (3.5×10^{14} mi) across in which the atoms have been ionized and excited by young, hot blue stars embedded in it.

In their physical and chemical properties, francium and radium do resemble the other members of groups 1A and 2A, respectively. However, because all their isotopes are radioactive, our main interest is in their nuclear properties, and therefore we will not discuss them in this chapter.

Hydrogen

Because hydrogen atoms are the simplest of all the atoms, hydrogen has always been considered rather special. For example, hydrogen was Dalton's choice as an atomic weight standard, it is the element found to be common to all acids, and it was the focus of early quantum mechanical studies (Figure 20.1).

A molecule of H_2 is lower in energy than two H atoms by 436 kJ/mol, and hydrogen therefore exists on Earth mainly as H_2 molecules and not as H atoms. The bond between the two H atoms in H_2 is a single covalent bond. Intermolecular forces in a sample of hydrogen are weak, and therefore hydrogen is a gas that conforms reasonably well to the ideal gas law over a range of temperatures and pressures. Hydrogen gas can be liquefied at 1 atm pressure at 20.39 K, and the liquid freezes at 13.98 K.

20.1   Occurrence and Preparation of Hydrogen

Hydrogen makes up 0.9% of the mass and 15.1% of the atoms in Earth's crust. (Earth's crust is the layer of rock covering the planet to an average depth of 17 km.) If we look beyond our home planet, H atoms are thought to make up 89% of the atoms in the Sun and 85–95% of the atoms in the atmospheres of the outer planets. In the universe as a whole, about 90% of all atoms are H (the rest are mainly He).

Because only trace amounts of free hydrogen (H_2) are found on Earth, this gas must be obtained from its compounds. Although hydrogen occurs in more compounds than does any other element, only a few compounds are economically viable sources of the element. Water, the most abundant hydrogen-containing compound, is usually the first choice. We can obtain H_2 in reactions of steam with carbon (as coal or coke), with carbon monoxide, or with a hydrocarbon. These reactions are carried out at high temperatures, and some require catalysts:

Water–gas reaction: $\quad\quad C(s) + H_2O(g) \xrightarrow{1000\,°C} CO(g) + H_2(g)$

Water–gas shift reaction: $\quad CO(g) + H_2O(g) \xrightarrow[1000\,°C]{catalyst} CO_2(g) + H_2(g)$

Reforming of methane: $\quad CH_4(g) + H_2O(g) \xrightarrow[1000\,°C]{catalyst} CO(g) + 3\,H_2(g)$

The reaction with carbon produces *water gas*, a combustible mixture that consists mainly of CO and H_2. The carbon monoxide is then used in the water–gas shift reaction to increase the yield of H_2. The hydrocarbon reaction—here our example is the reforming of methane—is the principal commercial source of hydrogen. (The term *reforming* is used to signify the restructuring of a hydrocarbon into some other material.)

Hydrogen is an important by-product of petroleum refining operations. For example, in the process called *catalytic reforming*, hydrogen is produced as an alkane hydrocarbon of low octane rating (page 246), such as hexane, and is converted to a hydrocarbon of higher octane rating, such as benzene:

$$C_6H_{14} \xrightarrow{catalyst} C_6H_6 + 4\,H_2$$
$$\text{Hexane} \quad\quad\quad \text{Benzene}$$

The most direct method of producing hydrogen (and oxygen as well) is the decomposition of water. Some compounds can be decomposed into their elements simply by heating strongly enough to break chemical bonds. Even at 2000 °C, though, water is only about 1% decomposed into H_2 and O_2. When decomposition by heating

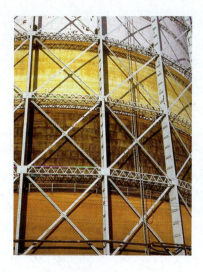

◀ Up until the 1930s, before the advent of inexpensive natural gas, there were more than 11,000 coal gasifiers operating in the United States. They supplied homes and industry with water–gas and related gaseous products. These gases were variously called producer gas, town gas, or city gas and were stored in large tanks.

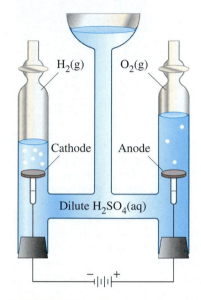

▲ **FIGURE 20.2   The electrolysis of water**

Hydrogen gas forms at the negative electrode (cathode), and oxygen gas forms at the positive electrode (anode). The volume of hydrogen produced is exactly twice the volume of oxygen.

QUESTION: What role does the $H_2SO_4(aq)$ play in the electrolysis of water?

 Electrolysis of Water animation

is not feasible, chemists often use *electrolysis*. Recall that water is a nonelectrolyte and can conduct electric current only when ions of a solute are present (Section 4.1). We learned how to predict the probable products of an electrolysis in Section 18.9. Specifically, we saw that if no other substances are present that are more readily oxidized and reduced than H_2O, the products of the electrolysis are H_2 and O_2.

In dilute sulfuric acid, H^+ ions are reduced to $H_2(g)$ at the cathode, and H_2O molecules are oxidized to $O_2(g)$ at the anode (Figure 20.2):

Reduction (cathode): $2\ \{2\ H^+(aq) + 2\ e^- \longrightarrow H_2(g)\}$

Oxidation (anode): $2\ H_2O(l) \longrightarrow 4\ H^+(aq) + O_2(g) + 4\ e^-$

Overall: $2\ H_2O(l) + 4\ H^+(aq) + 4\ e^- \longrightarrow 2\ H_2(g) + 4\ H^+(aq) + O_2(g) + 4\ e^-$

$$E^\circ_{cell} = E^\circ(\text{cathode}) - E^\circ(\text{anode}) = 0.000\ \text{V} - 1.229\ \text{V} = -1.229\ \text{V}$$

In summary,

$$2\ H_2O(l) \xrightarrow{\text{electrolysis}} 2\ H_2(g) + O_2(g) \qquad E^\circ_{cell} = -1.229\ \text{V}$$

The value of E°_{cell} tells us two things about this reaction: (1) It is nonspontaneous as written, which is why we need an external source of energy, and (2) the required voltage for the electrolysis must exceed 1.229 V. Thus, we should be able to electrolyze water with an ordinary flashlight battery (1.5 V).

The electrolysis of water is an expensive way to make hydrogen. About 0.1 kilowatt hour of electrical energy is required to produce 1 mol $H_2(g)$. It takes as much energy to obtain just two grams of hydrogen by the electrolysis of water as to operate a 100-watt electric lightbulb for one hour. Electrolysis of water is economically feasible only where hydroelectric power is cheap (as in Canada and Norway).

In the laboratory, we often can prepare chemicals by methods that are not economically feasible for commercial production. A common laboratory method for preparing small quantities of hydrogen involves the displacement of $H^+(aq)$ by metals such as zinc that lie below hydrogen in a table of standard electrode potentials (Table 18.1). For example,

$$Zn(s) + 2\ H^+(aq) \longrightarrow Zn^{2+}(aq) + H_2(g)$$

The most active metals, those of group 1A and the heavier members of group 2A, displace $H_2(g)$ even from pure water, where the concentration of H^+ is very low. Here the reactant is H_2O rather than H^+, and ionic hydroxides are formed together with $H_2(g)$:

$2\ M(s) + 2\ H_2O(l) \longrightarrow 2\ MOH(aq) + H_2(g)$ *(where M is any group 1A metal)*

$M(s) + 2\ H_2O(l) \longrightarrow M(OH)_2(aq) + H_2(g)$ *(where M is Ca, Sr, Ba, or Ra)*

▲ This generator, developed by General Motors in 2001, reformulates natural gas, propane, or gasoline to produce hydrogen, which powers a fuel cell to produce electricity.

▲ The effervescence in this beaker is caused by escaping hydrogen gas produced in the reaction of calcium hydride and water. The deep pink color is that of phenolphthalein indicator in the presence of the $OH^-(aq)$ produced in the reaction. The fog above the beaker is condensed water vapor, indicating that the contents of the beaker are much warmer than the surroundings. The reaction is exothermic.

20.2 Binary Compounds of Hydrogen

Hydrogen reacts with other nonmetals to form molecular compounds. We can call these compounds *molecular hydrides*. Two familiar ones are hydrogen chloride and ammonia:

$$H_2(g) + Cl_2(g) \longrightarrow 2\ HCl(g) \quad \Delta H° = -184.62\ kJ;\ K_p = 2.5 \times 10^{33}\ \text{at 298 K}$$

$$3\ H_2(g) + N_2(g) \rightleftharpoons 2\ NH_3(g) \quad \Delta H° = -92.22\ kJ;\ K_p = 6.2 \times 10^5\ \text{at 298 K}$$

The value of K_p for the first reaction is so large that the reaction goes essentially to completion. In fact, the reaction occurs with explosive violence when a mixture of hydrogen and chlorine is exposed to light. The second reaction is reversible. According to Le Châtelier's principle, we should expect the production of ammonia to be favored at high $H_2(g)$ and $N_2(g)$ pressures and at low temperatures. However, because of the high activation energy, equilibrium is reached very slowly at low temperatures. For this reason, ammonia synthesis is carried out at elevated temperatures (page 595).

Among the most important molecular hydrides are those of carbon. *Hydrocarbons* are the customary entry point to the field of organic chemistry (Section 2.9), and they are the fuels derived from petroleum.

Hydrogen reacts with the most active metals to form *ionic hydrides*, in which hydrogen exists as the *hydride ion*, H^-. The formation of many ionic hydrides is energetically favorable, as, for instance,

$$Na(s) + \tfrac{1}{2}\ H_2(g) \longrightarrow NaH(s) \qquad \Delta H_f° = -56.3\ kJ$$

$$Ca(s) + H_2(g) \longrightarrow CaH_2(s) \qquad \Delta H_f° = -186\ kJ$$

Ionic hydrides react with water to liberate $H_2(g)$ and are sometimes used as a source of hydrogen for weather-observation balloons:

$$CaH_2(s) + 2\ H_2O(l) \longrightarrow Ca^{2+}(aq) + 2\ OH^-(aq) + 2\ H_2(g)$$

When hydrides are formed from transition elements, the products, called *metallic hydrides*, retain some metallic properties, such as electrical conductivity. Many of the metallic hydrides are *nonstoichiometric*, meaning that the ratio of H atoms to metal atoms is variable. For example, the formula of titanium hydride can vary from $TiH_{1.8}$ to TiH_2. We can picture H atoms in metallic hydrides as filling the voids between metal atoms in the solid—much like fitting cherries into the holes between oranges in a box (Figure 20.3).* Because some holes fill but others do not, the ratio of H to metal atoms is variable.

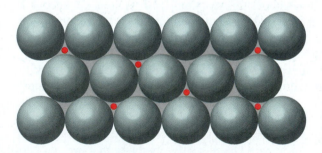

◀ **FIGURE 20.3 Representing a metal hydride**
The small red spheres represent H atoms interspersed in the voids of a close-packed collection of larger metal atoms.

▲ New electrochromic materials based on metal hydrides are being developed. These materials exhibit optical changes when electricity is applied or when exposed to a suitable reactant. The lower half of the "smart window" seen above is reflective because of a thin metallic layer. The upper half has been exposed to hydrogen, forming a transparent metal hydride. When exposed to air the metallic film re-forms.

20.3 Uses of Hydrogen

Nearly half the hydrogen gas produced is used in the manufacture of ammonia (NH_3). Ammonia, in turn, is used in the manufacture of fertilizers, plastics, and explosives. The second most important use of hydrogen is in the petrochemical industry. For example, hydrogen is used to convert benzene to cyclohexane, a cyclic hydrocarbon used as an intermediate in the production of nylon.

Another chemical manufacturing process that uses hydrogen is the synthesis of methanol, an industrial solvent and a raw material for making other organic compounds:

$$CO(g) + 2\ H_2(g) \xrightarrow{\text{catalyst}} CH_3OH(g)$$

*The actual situation may be somewhat more complicated because the arrangement of metal atoms in metallic hydrides is not always the same as in the pure metal.

Flame Tests for Metals movie

In metallurgy, the extraction of pure metals from their mineral sources generally involves an oxidation–reduction reaction. The reducing agent most commonly used is some form of carbon (charcoal, coke, or coal). In some cases, however, carbon is not a sufficiently strong reducing agent, and in others it may react with the metal to form unwanted metal carbides. In such instances, hydrogen is often used, as in the reduction of WO_3 to tungsten metal:

$$WO_3(s) + 3 H_2(g) \xrightarrow{850\,°C} W(s) + 3 H_2O(g)$$

Liquid hydrogen is used as a rocket fuel. The space shuttle rocket engines, for example, use the reaction of hydrogen and oxygen:

$$H_2(g) + \tfrac{1}{2} O_2(g) \longrightarrow H_2O(l) \qquad \Delta H° = -285.8 \text{ kJ}$$

Both reactants are stored as liquids. The propellant tanks holds 1.5×10^6 L of liquid hydrogen and 5.4×10^5 L of liquid oxygen. During liftoff, these propellants power the shuttle's main engines for about 8.5 minutes, with the liquid hydrogen being consumed at a rate of nearly 3000 L/s.

When hydrogen is passed through an electric arc, H_2 molecules absorb energy and dissociate into H atoms. The H atoms then recombine into H_2 molecules in a highly exothermic process. This recombination, coupled with the combustion of H_2, creates an extremely hot flame:

$$2 H(g) \longrightarrow H_2(g) \qquad \Delta H° = -436 \text{ kJ}$$
$$H_2(g) + \tfrac{1}{2} O_2(g) \longrightarrow H_2O(g) \qquad \Delta H° = -241.8 \text{ kJ}$$
$$\overline{2 H(g) + \tfrac{1}{2} O_2(g) \longrightarrow H_2O(g) \qquad \Delta H° = -678 \text{ kJ}}$$

One application of this reaction sequence is the oxyhydrogen welding torch. This tool readily cuts through steel and can be used to melt tungsten, which has a melting point of about 3400 °C.

Group 1A: The Alkali Metals

This group gets its name from the fact that all its metal hydroxides are strongly basic. Table 20.1 lists some properties of the alkali metals. We introduced most of these properties in earlier chapters—for example, electron configurations in Section 8.4, atomic and ionic radii and ionization energies in Section 8.7, flame colors in Section 8.9, electronegativity in Section 9.7, and electrode potentials in Section 18.4.

Table 20.1 Some Properties of the Group 1A Metals

	Li	Na	K	Rb	Cs
Atomic number, Z	3	11	19	37	55
Valence-shell electron configuration	$2s^1$	$3s^1$	$4s^1$	$5s^1$	$6s^1$
Atomic (metallic) radius, pm	152	186	227	248	265
Ionic (M^+) radius, pm	59	99	138	148	169
First ionization energy (I_1), kJ/mol	520	496	419	403	376
Electronegativity	1.0	0.9	0.8	0.8	0.7
Flame color	Carmine	Yellow	Violet	Bluish red	Blue
Melting point, °C	180.54	97.81	63.65	39.1	28.40
Density, g/cm³ at 20 °C	0.534	0.971	0.862	1.532	1.873
Electrode potential, E°, V $M^+(aq) + e^- \longrightarrow M(s)$	-3.040	-2.713	-2.924	-2.924	-2.923
Electrical conductivity[a]	18.6	37.9	25.9	12.7	8.0
Hardness[b]	0.6	0.4	0.5	0.3	0.2

[a] These values are on a scale relative to silver, which is assigned a conductivity of 100; for comparison, the conductivity for copper on this scale is 95.0 and that of gold is 67.7.

[b] On a scale called the Mohs scale that ranks hardness from 1 to 10, with 10 being hardest. A material with a higher Mohs number is able to scratch a material with a lower Mohs number. Examples: talc (1), fingernail (2.5), copper (2.5–3), iron (4–5), glass (5–6), steel (5–8.5), diamond (10). Notice that the alkali metals are softer than talc.

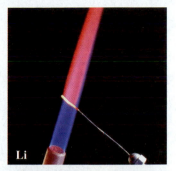
Li

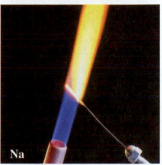
Na

K

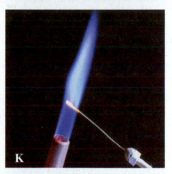
K

Rb

Cs

▲ Flame tests for (top to bottom) Li, Na, K, Rb, and Cs.

A Possible Future: The Hydrogen Economy

The long-term prospect of dwindling supplies and higher prices for coal, natural gas, and petroleum and the threat of global warming mean that we may need a substitute for fossil fuels in future centuries. Hydrogen is often mentioned as a candidate.

Hydrogen is an attractive fuel for several reasons. For example, an automobile engine burning hydrogen is 25–50% more energy-efficient than an engine burning gasoline. Because the only significant combustion product is H_2O, the exhaust from a hydrogen engine is far lower in pollutants than the exhaust from a gasoline engine. The heat of combustion per gram of liquid hydrogen is more than twice that of jet fuel. Aircraft using liquid hydrogen could fly much farther than those using conventional jet fuel.

In addition to taking the place of gasoline and jet fuels for transportation, hydrogen could also replace natural gas for heating homes and other buildings. Because it is an excellent reducing agent, hydrogen could largely replace carbon in metallurgy. Finally, with hydrogen readily available, the cost of producing ammonia and products derived from it would remain low. If we adopt the widespread use of hydrogen, as outlined here, major changes in our way of life would follow, producing what is called a *hydrogen economy*.

Hydrogen looks good for a future fuel, but tough problems must be solved before we move to a hydrogen economy. There are no hydrogen mines, nor can we pump it from the ground by drilling wells. A feasible hydrogen economy will require a cheap way of making hydrogen, storing it, and transporting it. Furthermore, environmental concerns must be addressed. For example, potential leakage of hydrogen into the atmosphere during production and handling could significantly increase its occurrence in the atmosphere (presently about 0.5 ppm by volume) and thereby affect some atmospheric reactions.

The most likely future source of hydrogen will be water. The decomposition of water might be accomplished as the net reaction of a series of thermochemical reactions, or by electrolysis if this can be done cheaply enough. Of course, as much energy is required to break down water into hydrogen and oxygen as is obtained in burning hydrogen as a fuel. What is actually accomplished is a transfer of energy from one form—nuclear, hydroelectric, solar—to another, highly portable and versatile form—hydrogen.

Although hydrogen can be stored as a gas, it occupies a large volume. Liquid hydrogen occupies a much smaller volume, but it must be maintained at extremely low temperatures (below −240 °C). And, in either case, hydrogen must be kept from contact with oxygen (air), into which it diffuses quickly because of its small molecular mass. Mixtures from 4–95% hydrogen in air are explosive. One promising possibility is the use of metal hydrides (Figure 20.3). The gas can be released from the metal by mild heating. In a hydrogen-burning automobile engine, for example, the required heat might come from the engine exhaust.

Planning for a possible hydrogen economy is a source of fascinating challenges for chemists and other natural and social scientists as well.

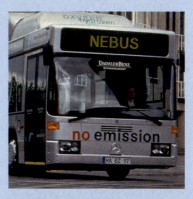

▲ The Daimler-Benz fuel-cell bus NEBUS, which is classified as a zero-emissions vehicle because it produces none of the usual pollutants of gasoline-powered automobiles. Only pure water vapor emerges from the exhaust pipe. The bus is propelled by quiet electric motors, with the necessary power (250 kW) provided by fuel cells in the rear.

Table 20.1 also includes data on two new properties: electrical conductivity and hardness. The numerical values listed for these properties are based on comparisons between substances rather than on particular sets of units.

20.4 Properties and Trends in Group 1A

Sodium and Potassium in Water movie

The group 1A metals exhibit regular trends for a number of properties. As we would expect, the radii of the alkali metal atoms and ions increase regularly from Li to Cs because the number of electron shells increases in proceeding down the group. The regular decrease in first ionization energies occurs because the larger atoms can be stripped of their ns^1 electron more easily than can smaller atoms. The decrease in electronegativities from top to bottom in group 1A signifies an increase in metallic character, which we can also associate with increasing atomic radii and decreasing ionization energies. A continuous increase or decrease of a property—a regular trend—suggests that a single factor is primarily responsible for that property.

Irregular trends suggest that factors are working against each other in determining a property. Consider, for example, the fact that the density d of K is lower than that of the element preceding it (Na) and also lower than that of the one following it (Rb). The factors we need to examine are mass and volume, so let us work with the molar mass and molar volume:

$$d \ (\text{g/cm}^3) = \frac{\text{molar mass (g/mol)}}{\text{molar volume (cm}^3\text{/mol)}}$$

Molar mass increases as atoms become more massive, and molar volume generally increases as atoms become larger.* Seven elements separate Na ($Z = 11$) and K ($Z = 19$). The molar mass added in going across the third period to get from Na to K is less significant than the jump in atomic size in going from three to four electron shells. The molar-volume denominator in our equation increases by a greater factor than the molar-mass numerator, and therefore the density *decreases*. Seventeen elements separate K ($Z = 19$) and Rb ($Z = 37$). The molar mass added in going across the fourth period to get from K to Rb is more significant than the jump in atomic size in going from four to five electron shells. Now the molar-mass numerator increases by a greater factor than the molar-volume denominator, and so the density *increases*.

The alkali metals have two notable physical properties: softness and low melting points. All are soft enough to be scratched with a toothpick and cut with a knife (even a plastic knife). The melting points of cesium and rubidium are low enough to make both of them liquid on a very hot day.

When freshly cut, the alkali metals are bright and shiny—typical metallic properties. The metals quickly tarnish, however, as they react with oxygen in the atmosphere. Another characteristic property is their ability to conduct electric current. In a ranking of electrical conductivity, silver, copper, and gold are numbers 1, 2, and 3, respectively, but sodium is not far behind at number 7. In Chapter 24 we will discuss a theory of bonding in metals that explains electrical conductivity and other metallic properties.

Diagonal Relationships

The first member of a group of the periodic table often differs in significant ways from the other members of its group. The eccentric member of group 1A is lithium (which becomes the first member of the group when we ignore H, a nonmetal). Following are some of the ways that lithium and its compounds differ from the other group 1A metals.

- Lithium carbonate, fluoride, hydroxide, and phosphate are much less soluble in water than are corresponding salts of the other alkali metals.
- Lithium forms a nitride (Li_3N) and is the only alkali metal to do so.
- When it burns in air, lithium forms a normal oxide (Li_2O) rather than a peroxide (M_2O_2) or a superoxide (MO_2).
- Lithium carbonate and lithium hydroxide decompose to form the oxide on heating, whereas the carbonates and hydroxides of the other group 1A metals are thermally stable.

Ions have a property called *charge density,* defined as the ratio of ionic charge to ionic volume, and in large part, lithium's atypical behavior can be attributed to the fact that the charge density of Li^+ is much higher than the charge densities of the other group 1A cations.

In some of its properties, lithium and its compounds resemble magnesium and its compounds. This is an example of the **diagonal relationship,** depicted in Figure 20.4. The similarity between Li and Mg probably results from the roughly equal sizes of the Li and Mg atoms and of the Li^+ and Mg^{2+} ions. The diagonal relationship between beryllium and aluminum is described in Section 20.8.

---

*The way in which atoms pack is also involved in relating molar volume to atomic size. However, the group 1A metals all have the same crystal structure—body-centered cubic—and therefore we can ignore this factor in the present discussion.

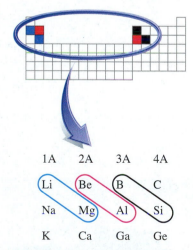

▲ Sodium, a soft metal, can be cut with a knife.

Interactive Periodic Table activity

▲ **FIGURE 20.4 The diagonal relationship in the periodic table**
The elements in each encircled pair have several similar properties.

20.5   Occurrence, Preparation, Uses, and Reactions of Group 1A Metals

The natural abundances of sodium and potassium in Earth's solid crust are 2.27% and 1.84% by mass, respectively; the other alkali metals are much scarcer, ranging from 78 to 18 to 2.6 parts per million of Rb, Li, and Cs, respectively. Francium is exceptionally rare. It is formed by the radioactive decay of heavier elements, and there is probably no more than 15 g of francium in the top 1 km of Earth's crust. Because francium is both rare and highly radioactive, few of its properties have been determined.

Preparation of the Alkali Metals

Although sodium and potassium are found in many minerals in Earth's crust, the principal compounds from which these metals are extracted are NaCl and KCl. These salts can be mined as solids or isolated from natural *brines*, which are aqueous solutions of ionic compounds that usually contain NaCl and often KCl, $MgCl_2$, and other salts. Sodium chloride is also the chief solute in seawater. Lithium is extracted mainly from the mineral *spodumene*, $LiAl(SiO_3)_2$.

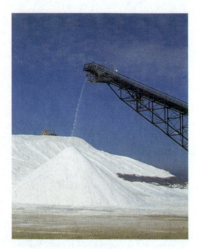

▲ Sodium chloride (table salt) is mined, pumped from brine wells, and, in some seaside locations, harvested from seawater.

To convert an alkali metal ion to an alkali metal atom, the ion must gain an electron, thus undergoing a reduction reaction. These reductions are not easily accomplished spontaneously because the alkali metals are excellent reducing agents. Reduction can be achieved, however, by the electrolysis of molten salts—usually the chlorides.

Potassium metal was the first to be prepared by electrolysis; Humphry Davy electrolyzed molten KOH in 1807. The usual industrial method involves a chemical reaction between sodium metal and potassium chloride:

$$KCl(l) + Na(l) \xrightarrow{850\,°C} NaCl(l) + K(g)$$

In this reaction, Na atoms lose their valence electrons and K^+ ions acquire them. The resulting gaseous potassium escapes from the reaction mixture and is converted, first to a liquid and then to a solid, by cooling. The reverse reaction is actually favored because K atoms have a lower ionization energy and therefore lose electrons more readily than do Na atoms. The method works because, at 850 °C, sodium is a liquid but potassium is a gas. The escape of gaseous potassium stimulates the forward reaction and causes it to go to completion (Le Châtelier's principle).

The alkali metals have a few important uses. Liquid sodium is used as a heat-transfer medium in some types of nuclear reactors and in automobile engine valves. Liquid sodium is especially good for this purpose because it has the following characteristics:

- A higher specific heat than most other liquid metals—it takes a relatively large quantity of heat per gram to change its temperature.
- Good thermal conductivity—heat is readily conveyed through the liquid and delivered to other media.
- A low density and low viscosity—liquid sodium is easy to pump.
- A low vapor pressure—the tendency of liquid sodium to vaporize is limited, even at high temperatures (550 °C).

▲ A self-rescuer used by miners contains potassium superoxide, which produces breathable oxygen for emergency use.

Small quantities of sodium in the form of sodium vapor are used in lamps for outdoor lighting. Perhaps the most important use of sodium metal is as a reducing agent in producing refractory (high melting point) metals, such as titanium, zirconium, and hafnium:

$$MCl_4 + 4\,Na \longrightarrow M + 4\,NaCl \quad (where\ M\ =\ Ti,\ Zr,\ or\ Hf)$$

The use of potassium is limited to a few applications in which sodium, the cheaper metal, cannot be used. A sodium–potassium alloy known as NaK shows promise as a heat-transfer medium. The alloy has a melting point of about −12 °C. Another application of potassium is its conversion to the superoxide, KO_2, which is used in life-support systems, where it acts both to absorb CO_2 and to produce O_2:

$$4\ KO_2(s) + 2\ CO_2(g) \longrightarrow 2\ K_2CO_3(s) + 3\ O_2(g)$$

(Sodium superoxide, NaO_2, can be prepared only under very carefully controlled conditions.)

Lithium is used in lightweight electrical batteries of the type found in clocks and watches, hearing aids, and heart pacemakers. Lithium is a desirable battery electrode because (1) the mass of Li required to release one mole of electrons is so small (6.941 g) and (2) high voltages can be achieved by combining the oxidation of Li(s) ($E^{\circ}_{Li^+/Li} = -3.040$ V) with an appropriate reduction half-reaction. Lithium is also used in alloys with other light metals; when added in small quantities, it imparts high-temperature strength to aluminum and ductility to magnesium. A silver–lithium alloy is used for brazing (welding together) metals. A possible future use of lithium is in the production of tritium (^3H) for use in nuclear fusion reactors.

$$^6_3Li + {}^1_0n \longrightarrow {}^3_1H + {}^4_2He$$

Some Reactions of Lithium, Sodium, and Potassium

The chemistry of the group 1A metals reflects the relative ease of removal of the ns^1 electron to form the metal ions. The alkali metals (M) react directly with elements of group 7A, the halogens (X_2), to form ionic binary halides (MX). As we saw in Section 20.2, they also react with hydrogen to form ionic hydrides (MH). The reaction of $O_2(g)$ with the metals produces various products. Table 20.2 summarizes some of the reaction chemistry of the group 1A metals.

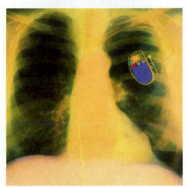

▲ A typical lithium electrochemical cell (top) produces a higher voltage than other electrochemical cells and has a much longer life. These characteristics make lithium cells well-suited for use in heart pacemakers (bottom).

Table 20.2 Some Typical Reactions of the Alkali Metals, M

With halogens (group 7A), X_2

$$2\ M(s) + X_2 \longrightarrow 2\ MX(s)\ (e.g.,\ LiF,\ NaCl,\ KBr,\ CsI)$$

With hydrogen, H_2

$$2\ M(s) + H_2(g) \longrightarrow 2\ MH(s)\ (e.g.,\ LiH,\ NaH)$$

With excess oxygen, $O_2{}^a$

$$4\ Li(s) + O_2(g) \longrightarrow 2\ Li_2O(s)\ (plus\ some\ Li_2O_2)$$
$$2\ Na(s) + O_2(g) \longrightarrow Na_2O_2(s)\ (plus\ some\ Na_2O)$$
$$M(s) + O_2(g) \longrightarrow MO_2(s)\ (where\ M = K,\ Rb,\ or\ Cs)$$

With water, H_2O

$$2\ M(s) + 2\ H_2O(l) \longrightarrow 2\ MOH(aq) + H_2(g)$$

[a] Li_2O is a *normal* oxide; Na_2O_2 is a *peroxide*; MO_2 is a *superoxide*. These oxides are described further in Section 21.9. Under appropriate conditions, all the alkali metals can form M_2O, M_2O_2, and MO_2.

20.6 Important Compounds of Lithium, Sodium, and Potassium

Lithium carbonate is the usual starting material for making other lithium compounds. For example, lithium hydroxide is prepared by the reaction

$$Li_2CO_3(aq) + Ca(OH)_2(aq) \longrightarrow CaCO_3(s) + 2\ LiOH(aq)$$

Solid $CaCO_3$ is removed by filtration, leaving an aqueous solution of lithium hydroxide. The LiOH can be obtained as a solid by evaporating water from the solution.

An interesting use of lithium hydroxide is to remove $CO_2(g)$ from expired air in confined quarters such as submarines and space vehicles, via the reaction

$$2\ LiOH(s) + CO_2(g) \longrightarrow Li_2CO_3(s) + H_2O$$

LiOH is preferred over other ionic hydroxides because, per mole of CO_2 removed from expired air, a smaller mass of LiOH is required. Similarly, in its reaction with

▲ Dubbed the "mailbox", this rectangular lithium hydroxide canister was jury-rigged for emergency service aboard the Apollo 13 spacecraft, following an onboard explosion on April 13, 1970. The mailbox absorbed exhaled carbon dioxide, permitting the crew to survive and land safely a few days later.

Table 20.3 Some Chemicals Produced from NaCl	
Chemical	**U.S. Production, 2002, Thousands of Metric Tons**
Sodium hydroxide, NaOH	8984
Chlorine, Cl_2	11362
Hydrochloric acid, HCl	4033[a]
Sodium sulfate, Na_2SO_4	515[b]

[a] Most HCl is produced as a by-product of the chlorination of hydrocarbons; for example, in the chlorination of methane, $CH_4(g) + Cl_2(g) \longrightarrow CH_3Cl(g) + HCl(g)$.
[b] Includes both Na_2SO_4 and $Na_2SO_4 \cdot 10\,H_2O$.

water, lithium hydride (LiH) produces a greater volume of $H_2(g)$ per unit mass of hydride than do other, cheaper hydrides.

With an annual production in the United States of about 50 million tons, sodium chloride is easily the most important sodium compound. In fact, it is the most used of all minerals in industrial processes that produce other chemicals. Table 20.3 lists several of these NaCl-based chemicals and their estimated annual production in the United States. The substances in Table 20.3 are used to prepare a variety of other materials and consumer products, including plastics, paper, bleach, soap, and laundry detergent.

Figure 20.5 shows how chemists sometimes outline important reactions. A central compound is chosen (NaCl), and the reactant(s) required to convert it to other compounds is(are) noted along arrows. Additional information may be placed along the arrows to indicate required reaction conditions, such as heating (Δ) or catalysts. In Example 20.1, we use a reaction diagram as the basis for writing a series of equations for the reactions that might collectively be used in a laboratory or industrial process.

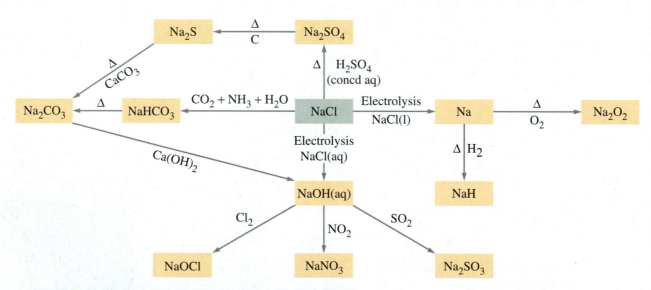

▲ FIGURE 20.5   **Preparation of sodium compounds from NaCl**
The methods of preparation suggested by this diagram are not necessarily the preferred industrial methods.

Example 20.1

The pathway in Figure 20.5 running from NaCl to Na_2SO_4 to Na_2CO_3 is called the Leblanc process. Write equations for the reactions involved.

STRATEGY

Figure 20.5 describes the necessary transformations in an abbreviated form. To translate this information into chemical equations, we need only identify the reactants in each reaction and predict plausible products.

SOLUTION

To convert $NaCl(s)$ to $Na_2SO_4(s)$ requires reaction with H_2SO_4 (concd aq). The other plausible product is $HCl(g)$:

$$2\,NaCl(s) + H_2SO_4(concd\ aq) \longrightarrow Na_2SO_4(s) + 2\,HCl(g)$$

The reaction of $Na_2SO_4(s)$ with carbon yields $Na_2S(s)$, with $CO(g)$ as another plausible product:

$$Na_2SO_4(s) + 4\,C(s) \longrightarrow Na_2S(s) + 4\,CO(g)$$

The final step is the reaction of $Na_2S(s)$ with $CaCO_3(s)$, yielding $Na_2CO_3(s)$ and $CaS(s)$:

$$Na_2S(s) + CaCO_3(s) \longrightarrow Na_2CO_3(s) + CaS(s)$$

Water-soluble Na_2CO_3 is easily extracted from the mixed solid because CaS is only slightly soluble in water.

EXERCISE 20.1A

Write plausible chemical equations for each of the following conversions outlined in Figure 20.5.

(a) $NaCl$ to NaH **(b)** $NaCl$ to $NaOCl$

EXERCISE 20.1B

Use your general knowledge and Figure 20.5 to sketch a diagram showing the simplest conversion of metallic sodium to Na_2SO_3. Write equations for the reactions involved.

Sodium hydroxide is produced by the electrolysis of concentrated $NaCl(aq)$. In the electrolysis, Na^+ is unchanged, Cl^- is converted to $Cl_2(g)$, and H_2O is converted to $H_2(g)$ and OH^-:

$$2\,Na^+(aq) + 2\,Cl^-(aq) + 2\,H_2O(l)$$
$$\xrightarrow{\text{electrolysis}} 2\,Na^+(aq) + 2\,OH^-(aq) + H_2(g) + Cl_2(g)$$

The gases are collected separately. If solid $NaOH$ is desired (rather than an aqueous solution), water is evaporated from the solution after the electrolysis is completed.

Sodium sulfate is obtained both from natural sources and by a synthetic method introduced by J. R. Glauber more than 300 years ago:

$$H_2SO_4(concd\ aq) + 2\,NaCl(s) \xrightarrow{\Delta} Na_2SO_4(s) + 2\,HCl(g)$$

The principle of this reaction is that a volatile acid (HCl) is produced by heating one of its salts ($NaCl$) with a nonvolatile acid (H_2SO_4). Other volatile acids can be produced by similar reactions. The major use of sodium sulfate (Na_2SO_4) is in the paper industry (about 70% of the annual U.S. consumption). In making kraft paper, undesirable lignin is removed from wood by treating the wood with an alkaline solution of sodium sulfide (Na_2S). The Na_2S is produced by the reaction of Na_2SO_4 with carbon. About 100 lb of Na_2SO_4 is used in the production of one ton of paper.

Potassium compounds have some uses similar to their sodium counterparts (for example, K_2CO_3 in glass and ceramics), but their most important use by far is in fertilizers, which account for 95% of the commercial use of potassium compounds. Potassium (as K^+) is one of the three main nutrients required by plants (nitrogen and phosphorus are the other two). KCl is commonly used as a fertilizer because this is the form in which most potassium is obtained from natural sources.

Figure 20.6 shows another way chemists and chemical engineers diagram the industrial production of a chemical. You can think of Figure 20.6 as an elaboration of the part of Figure 20.5 dealing with the conversion of $NaCl$ to $NaHCO_3$. The diagram identifies the raw materials (limestone, brine, and ammonia) and illustrates how they are used in the main reaction. The diagram also shows how substances are recycled. The only ultimate by-product of the Solvay process is $CaCl_2$. The demand for $CaCl_2$ is limited, however, and in the past, much of the $CaCl_2$ was dumped, generally into lakes or streams. Such dumping is now prohibited by environmental regulations.

▲ Gunpowder is a mixture of potassium nitrate (white), sulfur (yellow), and charcoal (black).

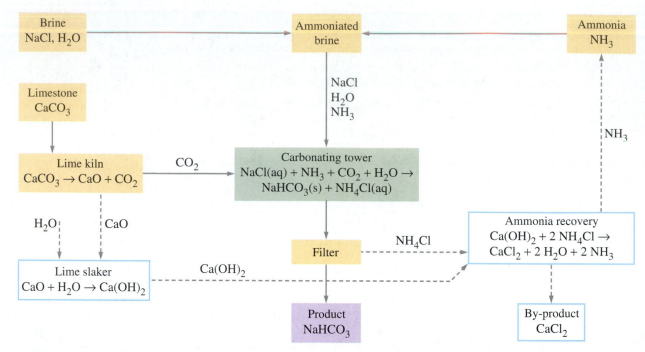

▲ **FIGURE 20.6** **The Solvay process**
The boxes shaded in color and the solid-line arrows outline the steps essential to the main reaction and recovery of the product. The boxes outlined in blue and the dashed-line arrows indicate the recycling steps and formation of the by-product.

Partly for this reason and partly because of the discovery in Wyoming of abundant natural sources of an ore, *trona*, that is rich in sodium carbonate, the Solvay process is now obsolete in North America. It continues to be the main route to Na_2CO_3 elsewhere in the world, however. Both the Solvay process and perhaps the trona source may soon be replaced, as a high-purity sodium bicarbonate ore called *nahcolite* is now being exploited in Colorado. The nahcolite process uses solution mining, an extraction method based on dissolving the bicarbonate salt in hot water rather than the labor-intensive hard-rock mining used to extract trona. With the hot water, a 99.9% pure $NaHCO_3(aq)$ solution is pumped from the ground. It can be converted to sodium carbonate simply by heating.

The changing ways in which industry has obtained sodium carbonate over the years illustrate the importance of economic and environmental factors in establishing a kind of natural progression for industrial chemical processes—advent, modification, decline.

20.7 The Alkali Metals and Living Matter

Hydrogen, oxygen, carbon, and nitrogen are the most abundant elements in the human body, in the order listed. They account for 99% of all the atoms in the body. Sodium and potassium, in the form of Na^+ and K^+, are in a second tier of seven elements that account for about 0.9% of the atoms. Life appears to have evolved using the elements that are most available in the environment, and sodium and potassium have comparatively high abundances in Earth's crust and in fresh water and seawater.

Sodium ions are found primarily in fluids outside cells, for example, in blood plasma. Potassium ions are abundant in fluids within cells. Both Na^+ and K^+ are involved in the transmission of nerve signals and in regulating and controlling body fluids. We commented on their role in the electrochemistry of a heartbeat in Chapter 18 (page 773).

Ordinary table salt (NaCl) is naturally present in most foods and is also often added to foods. It supplies the body with Na^+ and also with chloride ions, which are necessary for the production of stomach acid, HCl(aq). Bananas and orange juice are good sources of K^+, along with potatoes. The recommended daily intake of K^+ is variously given as 2000–5600 mg.

▲ Potatoes are a good dietary source of K^+.

Potassium ion is an essential nutrient for plants. It is generally readily available to plants, except in soil depleted by high-yield agriculture. The usual form of potassium in commercial fertilizers is potassium chloride, KCl.

Because most alkali metal compounds are water-soluble, many acidic drugs are administered in the form of their sodium or potassium salts. For example, "free" penicillin G (benzylpenicillic acid) is only sparingly soluble in water, whereas potassium penicillin G is freely soluble in water.

Lithium carbonate is used in medicine to level out the dangerous manic "highs" that occur in manic-depressive psychoses. Some medical practitioners also recommend lithium carbonate for the depression stage of the cycle. It appears to affect the transport of chemical substances across cell membranes in the brain.

We described earlier the Glauber process for synthesizing Na_2SO_4. One of the hydrates of this salt, sodium sulfate decahydrate, $Na_2SO_4 \cdot 10\ H_2O$, was one of the first synthetic chemicals used in medicine. Still known today as *Glauber's salt*, it acts as a cathartic (a substance that purges the bowels). Glauber's synthesis of this salt helped turn the focus of alchemists from attempts to convert base metals into gold to the synthesis of medicines.

▲ Glauber's salt (sodium sulfate decahydrate) is used in some solar-heated homes as a heat-storage medium (see Problem 90).

Group 2A: The Alkaline Earth Metals

In the early days of chemistry, the term "earth" described substances that are either insoluble or only slightly soluble in water and are not decomposed by heating. The group 2A oxides, MO, and hydroxides, $M(OH)_2$, are not soluble in water to any great extent. In addition, these compounds are basic or *alkaline*. These two properties—water insolubility and alkalinity—account for the group family name: the *alkaline earth* metals. Table 20.4 lists data for the group 2A metals.

Interactive Periodic Table activity

Table 20.4 Some Properties of the Group 2A Metals

	Be	Mg	Ca	Sr	Ba
Atomic number, Z	4	12	20	38	56
Valence-shell electron configuration	$2s^2$	$3s^2$	$4s^2$	$5s^2$	$6s^2$
Atomic (metallic) radius, pm	111	160	197	215	217
Ionic (M^{2+}) radius, pm	31	65	99	113	135
Ionization energy, kJ/mol					
I_1	900	738	590	550	503
I_2	1757	1451	1145	1064	965
Electronegativity	1.5	1.2	1.0	1.0	0.9
Flame color	None	None	Orange-red	Scarlet	Green
Melting point, °C	1278	649	839	769	729
Density, g/cm³ at 20 °C	1.848	1.738	1.550	2.540	3.594
Electrode potential, $E°$, V					
$M^{2+}(aq) + 2\ e^- \longrightarrow M(s)$	−1.70	−2.356	−2.84	−2.89	−2.92
Electrical conductivity[a]	40	36	46	6.9	3.2
Hardness[b]	≈5	2.0	1.5	1.8	≈2

[a] These values are on a scale relative to silver, which is assigned a conductivity of 100; for comparison, the conductivity for copper on this scale is 95.0 and that of gold is 67.7.

[b] On a scale called the Mohs scale that ranks hardness from 1 to 10, with 10 being hardest.

20.8 Properties and Trends in Group 2A

Group 2A shows the same general trends of increasing atomic and ionic sizes and decreasing ionization energies from top to bottom as does group 1A. Table 20.4 lists both the first and second ionization energies because an alkaline earth metal atom loses two electrons when it forms an M^{2+} ion. The ionization energies of the first two members, Be and Mg, are noticeably higher than those of the other three, Ca, Sr, and Ba. A parallel observation is that Be and Mg do not exhibit a flame color, whereas Ca, Sr, and Ba do. The electronegativity values are those we expect for active metals, and the decreasing values down the group are in line with the increasing metallic character of the atoms.

The higher densities of the group 2A metals relative to group 1A are mainly a consequence of the large difference in atomic sizes between the two groups. Molar volumes are significantly smaller in group 2A than in group 1A, and the molar masses are only slightly larger. The standard electrode potentials $\left(E^{\circ}_{M^{2+}/M}\right)$ are large and negative, showing that the alkaline earth metal ions M^{2+} are reduced only with difficulty. As a result, this means the alkaline earth metals are readily oxidized and thus are good reducing agents.

An interesting trend related to data in Table 20.4 is in the molar solubilities of the metal hydroxides in water. At 20 °C we have

$Mg(OH)_2$	$Ca(OH)_2$	$Sr(OH)_2$	$Ba(OH)_2$
0.0002 M	0.021 M	0.066 M	0.23 M

As the cation size increases from left to right in this series, the strength of the interionic attractions that hold the crystalline solid together decreases. As the interionic attractions weaken, the solubility in water increases. Consequently, barium hydroxide is soluble enough to be used as a titrant in acid–base titrations.

The Special Case of Beryllium

Like lithium in group 1A, beryllium—the first member in group 2A—differs in some ways from other members of the group. Some of its distinctive properties are

- BeO does not react with water. [The other group 2A metal oxides do react with water: $MO + H_2O \longrightarrow M(OH)_2$.]
- Be and BeO dissolve in strongly basic solutions to form the ion BeO_2^{2-}. The oxide BeO has acidic properties. The other alkaline earth metal oxides are basic.
- $BeCl_2$ and BeF_2 in the molten state are poor conductors of electricity; they are molecular substances (Figure 20.7). In contrast, the other alkaline earth metal compounds are almost entirely ionic.

In Chapter 21, we will see that in some ways Al, Al_2O_3, and $AlCl_3$ resemble their beryllium counterparts. This is another example of a *diagonal relationship* (Figure 20.4).

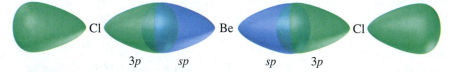

▲ **FIGURE 20.7** **Molecular structure of BeCl₂**

The hybridization scheme for the Be atom is

$$\text{Be:} \quad \underset{1s}{\left[\uparrow\downarrow\right]} \quad \underset{sp}{\left[\uparrow\,\right]\left[\uparrow\,\right]} \quad \underset{2p}{\left[\quad\right]\left[\quad\right]}$$

The *sp* hybrid orbitals are directed outward from the Be atom along a straight line. The $BeCl_2$ molecule is a linear triatomic molecule. The molecular structure in the solid state is more complex, however.

20.9   Occurrence, Preparation, Uses, and Reactions of Group 2A Metals

The natural abundances of calcium and magnesium in Earth's crust are 4.66% and 2.76% by mass, respectively. They rank fifth and sixth among the elements in crustal abundance, just ahead of sodium and potassium. Calcium and magnesium minerals occur in many types of rocks. Limestone rock is mainly $CaCO_3$, and dolomite is a mixed carbonate, $CaCO_3 \cdot MgCO_3$. Strontium and barium are both present in Earth's crust at about 400 ppm, and beryllium is found at 2 ppm. An important mineral source of beryllium is the mineral *beryl*, $Be_3Al_2Si_6O_{18}$. Some familiar gemstones, including aquamarine and emerald, are beryl distinctively colored by impurities.

Preparation of the Alkaline Earth Metals

To obtain beryllium from beryl, the beryl is first converted to BeF_2. Then the BeF_2 is reduced to beryllium, using magnesium as the reducing agent:

$$BeF_2(g) + Mg(l) \xrightarrow{1000\ °C} Be(s) + MgF_2(s)$$

Calcium is generally obtained by electrolysis of molten calcium chloride. Strontium and barium can also be obtained by electrolysis, but more commonly they are obtained by the high-temperature reduction of their oxides, using aluminum as the reducing agent.

The Dow Process for the Production of Magnesium

In the Dow process for magnesium production, calcium carbonate, as either limestone or seashells, is decomposed to $CaO(s)$ and $CO_2(g)$ by heating. Treatment of the $CaO(s)$ with water produces $Ca(OH)_2$, the source of OH^- for the precipitation of $Mg(OH)_2(s)$ from an aqueous solution containing Mg^{2+}:

$$Mg^{2+}(aq) + 2\ OH^-(aq) \longrightarrow Mg(OH)_2(s)$$

The precipitated $Mg(OH)_2(s)$ is washed, filtered, and dissolved in $HCl(aq)$:

$$Mg(OH)_2(s) + 2\ HCl(aq) \longrightarrow MgCl_2(aq) + 2\ H_2O(l)$$

The resulting solution is evaporated to dryness. The $MgCl_2$ is then melted and electrolyzed, yielding pure Mg metal and $Cl_2(g)$:

$$MgCl_2(l) \xrightarrow{electrolysis} Mg(l) + Cl_2(g)$$

The $Cl_2(g)$ is converted to HCl and recycled.

Electrolysis and the heating of materials to high temperatures are among the industrial operations that consume the most energy. The Dow process involves both. The production of 1 kg Mg requires about 300 MJ of energy. In contrast, it takes only about 7 MJ of energy to melt and recast 1 kg of recycled Mg. Recycling of magnesium is especially cost-effective.

Uses of the Alkaline Earth Metals

Alloys of beryllium with other metals have many applications. An alloy of copper with about 2% beryllium is used in springs, clips, and electrical contacts because it is able to withstand metal fatigue (the cracking or rupture of a metal subjected to repeated stress). Because of their low densities, other Be alloys are used in lightweight structural materials. Beryllium is nonsparking, which makes it useful for tools that must be used in flammable atmospheres. The small Be atom has little stopping power for X rays or neutrons; this makes beryllium useful for windows in X-ray tubes and for components in nuclear reactors. Mixtures of beryllium and radium compounds or beryllium and plutonium have long been used as neutron sources. Alpha particles from radium or plutonium induce the nuclear reaction

$$^9_4Be + {^4_2}He \longrightarrow {^{12}_6}C + {^1_0}n$$

Beryllium compounds (and fine beryllium dust) are extremely toxic. Special precautions are necessary when working with them.

Magnesium has a lower density than any other structural metal. Lightweight automobile and aircraft parts are manufactured from magnesium alloyed with aluminum and other metals; these parts are easily fabricated with standard machining and metallurgical methods. Also, magnesium is an important metallurgical reducing agent, for example, in the production of beryllium (above) and titanium:

$$TiCl_4 + 2\ Mg \longrightarrow Ti + 2\ MgCl_2$$

▲ The White Cliffs of Dover, England, are made of chalk, a soft form of limestone, $CaCO_3$.

▲ Emeralds are the mineral *beryl*, a mixed oxide of beryllium, aluminum, and silicon.

There is about 1.3 g Mg^{2+} per kilogram of seawater. Magnesium can be extracted from seawater because Mg^{2+} is the only cation present that forms an insoluble hydroxide.

▲ A prototype of the Lockheed F80 aircraft was made entirely from magnesium, taking advantage of the high strength and low density of the metal.

Magnesium is also used in batteries, in fireworks, and in the synthesis of important reagents for organic chemical reactions.

Calcium is used to reduce the oxides or fluorides of less common metals (Sc, W, Th, U, Pu, and most of the lanthanides) to the free metals. For example,

$$UO_2 + 2\ Ca \longrightarrow U + 2\ CaO$$
$$2\ ScF_3 + 3\ Ca \longrightarrow 2\ Sc + 3\ CaF_2$$

Calcium is also used in the manufacture of batteries and in forming alloys with aluminum, silicon, and lead.

Reactions of the Alkaline Earth Metals

The extent to which the group 2A metals react with water increases as we move down the group. Beryllium fails to react with either cold water or steam. In the case of magnesium, an impervious film of $Mg(OH)_2$ covers the surface and immediately stops the reaction. Magnesium does react with steam, but MgO is formed rather than $Mg(OH)_2$:

$$Mg(s) + H_2O(g) \longrightarrow MgO(s) + H_2(g)$$

Beginning with calcium, we see a slow reaction with cold water, as illustrated in Figure 8.21. Reactivity increases as we move to strontium and barium:

$$Ca(s) + 2\ H_2O(l) \longrightarrow Ca(OH)_2 + H_2(g) \quad \textit{(slow reaction with cold water)}$$
$$Sr(s) + 2\ H_2O(l) \longrightarrow Sr(OH)_2 + H_2(g) \quad \textit{(reaction more rapid than Ca)}$$
$$Ba(s) + 2\ H_2O(l) \longrightarrow Ba(OH)_2 + H_2(g) \quad \textit{(reaction more rapid than Sr)}$$

All the alkaline earth metals react with dilute acids to displace hydrogen:

$$M(s) + 2\ H^+(aq) \longrightarrow M^{2+}(aq) + H_2(g)$$

Some typical reactions of the alkaline earth metals, illustrated below for magnesium, are with the halogens (X_2), with oxygen, and with nitrogen.

$$Mg + X_2 \longrightarrow MgX_2 \quad \textit{(where X = F, Cl, Br, I)}$$
$$2\ Mg + O_2 \longrightarrow 2\ MgO$$
$$3\ Mg + N_2 \longrightarrow Mg_3N_2$$

Example 20.2   A Conceptual Example

The photograph in the margin is a graphic illustration of what happens when we heat a strip of magnesium ribbon in air. If we know that the strip has a mass of 1.000 g, can we calculate with certainty the mass of the product obtained?

ANALYSIS AND CONCLUSIONS

The first requirement in calculating the theoretical yield of a reaction is to know the reactants(s) and product(s) and write an equation for the reaction. Additionally, we will obtain the theoretical yield only if the reaction has a 100% yield. The main difficulty here is that two reactions can occur simultaneously:

$$2\ Mg(s) + O_2(g) \longrightarrow 2\ MgO(s)$$
$$3\ Mg(s) + N_2(g) \longrightarrow Mg_3N_2(s)$$

Given the plentiful supply of $N_2(g)$ and $O_2(g)$ in air and the apparent vigor of the reaction, we can probably assume that magnesium will be completely consumed—it will be the limiting reactant. Because $N_2(g)$ is considerably more abundant in air than is $O_2(g)$, we might expect more of it to react, forming more Mg_3N_2 than MgO. However, this assumption is inappropriate because we know nothing about the relative rates of the two reactions, and rate could be the deciding factor. Thus, we are unable to calculate the mass of the product with certainty.

EXERCISE 20.2A

What is the estimated answer to Example 20.2? That is, what is the maximum and what is the minimum mass of product that might be obtained?

EXERCISE 20.2B

Following up on Exercise 20.2A, what conclusion would you reach if the mass of product obtained were slightly less than the maximum possible?

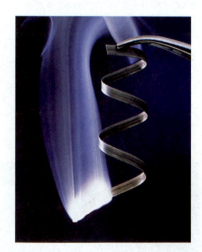

▲ Magnesium burns in air with a bright flame.

20.10 Important Compounds of Magnesium and Calcium

Several magnesium compounds occur naturally, either in mineral form or in brines. These include the carbonate, chloride, hydroxide, and sulfate. Other magnesium compounds can be prepared from these. A few compounds and some of their uses are listed in Table 20.5.

Limestone is a naturally occurring form of calcium carbonate, also containing some clay and other impurities. Because calcium carbonate is the most widely used calcium compound, let us examine the special role of limestone in the chemical industry.

Table 20.5 Important Magnesium Compounds

Compound	Uses
$MgCO_3$	Refractory bricks, glass, inks, rubber reinforcing agents, dentifrices, cosmetics, antacids, laxatives
$MgCl_2$	Magnesium metal, textiles, paper, fireproofing agents, cements, refrigeration brine
MgO	Refractories (furnace linings), ceramics, cements, SO_2 removal from stack gases
$MgSO_4$	Fireproofing, textiles, ceramics, fertilizers, cosmetics, dietary supplements

Limestone: Building Stone and Chemical Raw Material

Some limestone is used as building stone. However, most limestone is used to manufacture other building materials. *Portland cement* is a complex mixture of calcium and aluminum silicates obtained by heating limestone, clay, and sand in a high-temperature rotary kiln. When the cement is mixed with sand, gravel, and water, it solidifies into the familiar material *concrete*. Ordinary *soda–lime glass* (used to make bottles and windows, for example) is a mixture of sodium and calcium silicates formed by heating limestone, sand, and sodium carbonate together.

Limestone is used in the metallurgy of iron and steel to produce an easily liquefied mixture of calcium silicates called *slag*, which carries away impurities from the molten metal. Through chemical reactions producing CaO and $Ca(OH)_2$, limestone is the basis of a large part of the inorganic chemical industry.

In practical applications, the first step is often decomposition of limestone by heating, a process called *calcination*:

$$\text{Calcination:}\quad CaCO_3(s) \xrightarrow{\Delta} \underset{\text{Quicklime}}{CaO(s)} + CO_2(g)$$

Calcination is carried out in a high-temperature kiln (about 1000 °C), with continuous removal of $CO_2(g)$ to promote the forward reaction. The product formed, CaO(s), is called either *lime* or *quicklime*. In the process of *hydration*, the reaction of quicklime with water produces $Ca(OH)_2$, known as *slaked lime*.

$$\text{Hydration:}\quad CaO(s) + H_2O(l) \longrightarrow \underset{\text{Slaked lime}}{Ca(OH)_2(s)}$$

In the process of *carbonation*, $CO_2(g)$ is bubbled through a suspension of $Ca(OH)_2(s)$ in water, and $CaCO_3(s)$ is formed once more:

$$\text{Carbonation:}\quad Ca(OH)_2(s) + CO_2(g) \longrightarrow CaCO_3(s) + H_2O(l)$$

The three steps just described can be combined and used to prepare chemically pure $CaCO_3(s)$ from limestone, an impure material. This pure $CaCO_3(s)$, called *precipitated* calcium carbonate, is used extensively as a filler, providing bulk to materials such as paint, plastics, printing inks, and rubber. It is also used in toothpastes, food, cosmetics, and antacids. Added to paper, calcium carbonate makes the paper bright, opaque, smooth, and capable of absorbing ink well.

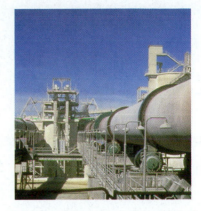

▲ The decomposition (calcination) of limestone is carried out in a long rotary kiln, producing quicklime, CaO, and carbon dioxide. In a cement kiln, the limestone is mixed with clay and sand to produce the complex mixture of calcium silicates and aluminates known as Portland cement.

▲ Glass bottles being manufactured. The principal current use of sodium carbonate is in glassmaking.

Application Note

When heated to a high temperature, CaO (mp 2614 °C) emits light, as seen below. Before the days of electric lighting, lime was heated in a gas flame and the emitted light was focused into a beam for spotlighting, hence the expression "in the limelight." Limelight is a blackbody radiation (page 273).

Quicklime and slaked lime are the cheapest and most widely used bases, and are usually the first choice for neutralizing unwanted acids. Thus, lime is used to neutralize acidic soils in lawns, gardens, and farmland and to treat excess acidity in lakes.

$$CaO(s) + 2\,H^+(aq) \longrightarrow Ca^{2+}(aq) + H_2O(l)$$

Another application of quicklime is in air-pollution control. When coal is burned in an electric power plant, sulfur in the coal is converted to sulfur dioxide gas, a major culprit in acid rain formation. When powdered limestone is mixed with powdered coal before combustion, the limestone decomposes to quicklime, which then reacts with $SO_2(g)$ that would otherwise escape:

$$CaO(s) + SO_2(g) \longrightarrow CaSO_3(s)$$

By reaction with oxygen, the calcium sulfite is converted to calcium sulfate (gypsum), which has a number of uses. Yet another use of quicklime is in treatment of wastewater effluents and sewage.

Slaked lime, $Ca(OH)_2$, is the cheapest commercial base and is used in all applications where high water solubility is not essential. It is used in the manufacture of other alkalis and bleaching powder, in sugar refining, in tanning hides, and in water softening.

A mixture of slaked lime, sand, Portland cement, and water is the familiar mortar used in bricklaying. In the initial setting of the mortar, the bricks absorb excess water from the mortar, and this water is then lost through evaporation. In the final setting, $CO_2(g)$ from air reacts with the $Ca(OH)_2$ in the mortar and converts it back to $CaCO_3$:

$$Ca(OH)_2(s) + CO_2(g) \longrightarrow CaCO_3(s) + H_2O(g)$$

Hydrates

The mineral gypsum has the formula $CaSO_4 \cdot 2H_2O$. Recall that a compound that incorporates water molecules into its crystal structure is called a *hydrate* (Section 2.7). In gypsum, two water molecules are present for every formula unit of $CaSO_4$, and therefore the chemical name is calcium sulfate *dihydrate*.

Another hydrate of calcium sulfate has one water molecule for every two formula units of $CaSO_4$. We could write its formula as $2CaSO_4 \cdot H_2O$, but more commonly we write $CaSO_4 \cdot \frac{1}{2}H_2O$ and call the compound calcium sulfate *hemihydrate*. This hydrate is the well-known *plaster of paris,* so-called because in the Middle ages it was prepared by heating the gypsum of Paris, France:

$$CaSO_4 \cdot 2\,H_2O(s) \xrightarrow{\Delta} CaSO_4 \cdot \tfrac{1}{2}H_2O(s) + \tfrac{3}{2}H_2O(g)$$

Gypsum   Plaster of paris

When mixed with water, plaster of paris reverts to gypsum, expanding slightly as it does so. A mixture of plaster of paris and water is used to make castings where sharp details of an object must be retained, as in dental work and jewelry making.

The most important use of gypsum is in making gypsum board, also known as drywall, which has largely supplanted plaster in the construction industry.

Hydrate formation occurs infrequently in compounds of the alkali metals, but it is common in compounds of the alkaline earth metals. Typical hydrates have the formula $MX_2 \cdot 6\,H_2O$ (where M = Mg, Ca, or Sr, and X = Cl or Br).

▲ These dunes at White Sands National Monument in New Mexico are composed of gypsum.

20.11 The Group 2A Metals and Living Matter

Persons of average size have approximately 25 g of magnesium in their bodies. The magnesium serves a variety of functions in metabolic processes and nerve action. The recommended daily intake of magnesium for adults is 350 mg. In plant life, magnesium plays a critical role in photosynthesis, the process by which plants convert carbon dioxide and water into sugars. The magnesium is present in the green pigment *chlorophyll*, which is necessary for the capture of energy from sunlight. Solar energy is the energy source for the reactions of photosynthesis.

Calcium is essential to all living matter. The human body typically contains from 1 to 1.5 kg of calcium. About 99% of it is in the bones and teeth, principally as the mineral *hydroxyapatite*, $Ca_5(PO_4)_3OH$. The other 1% is involved in several processes: clotting of blood, maintenance of a regular heartbeat, proper functioning of nerves and muscles, and various aspects of metabolism. Children require an adequate daily intake of calcium for the proper development of bones and teeth. Adults—especially older women—require calcium to prevent osteoporosis, a condition in which the bones become porous, brittle, and easily broken. The recommended daily intake of calcium for adults is 1200 mg.

Strontium is not essential to living matter, but it is of interest because of its chemical similarity to calcium. Strontium can follow some of the same chemical pathways in organisms as calcium does. Because of this, the body easily ingests and absorbs the dangerously radioactive isotope strontium-90, a product of fallout from the nuclear fission that occurs in nuclear explosions.

Barium also has no known function in organisms; in fact, the Ba^{2+} ion is toxic. Despite this fact, water suspensions of $BaSO_4(s)$ are sometimes used in X-ray imaging of the gastrointestinal tract: a barium "milkshake" for the upper tract or a barium enema for the lower tract. Barium ions are good absorbers of X rays and make the tract visible in an X-ray photograph. Because it is insoluble ($K_{sp} = 1.1 \times 10^{-10}$), $BaSO_4(s)$ is eliminated by the body with no significant absorption of Ba^{2+} ions.

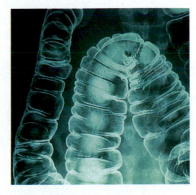

▲ The gastrointestinal tract is rendered visible through the X-ray-absorbing ability of a barium sulfate coating on the tract walls.

Chemistry of Groundwater

Carbonates, especially $CaCO_3$, are involved in a number of natural phenomena. One process begins when rainwater dissolves atmospheric $CO_2(g)$. The $CO_2(g)$ reacts with the water to form carbonic acid, H_2CO_3, which ionizes to form hydrogen carbonate ion, HCO_3^-. We can write an equation for each of these reactions and then combine them into an overall equation that represents rainwater containing dissolved CO_2:

$$CO_2(g) + H_2O(l) \rightleftharpoons H_2CO_3(aq)$$
$$H_2CO_3(aq) + H_2O(l) \rightleftharpoons H_3O^+(aq) + HCO_3^-(aq)$$
$$\overline{CO_2(g) + 2 H_2O(l) \rightleftharpoons H_3O^+(aq) + HCO_3^-(aq)} \quad \text{(rainwater containing } CO_2\text{)}$$

Rainwater containing dissolved CO_2 is acidic. When this water seeps through limestone, water-insoluble $CaCO_3$ is converted to water-soluble $Ca(HCO_3)_2$. In the equations below, the first represents rainwater charged with CO_2 and the second represents an acid–base reaction with H_3O^+ as the acid and CO_3^{2-} (from $CaCO_3$) as the base. The net dissolution equation is the sum of the two.

$$CO_2(g) + H_2O(l) \rightleftharpoons H_3O^+(aq) + HCO_3^-(aq)$$
$$CaCO_3(s) + H_3O^+(aq) \rightleftharpoons Ca^{2+}(aq) + HCO_3^-(aq) + H_2O(l)$$
$$\overline{CaCO_3(s) + H_2O(l) + CO_2(g) \rightleftharpoons Ca^{2+}(aq) + 2 HCO_3^-(aq)}$$

Over time, the dissolving action represented by the forward direction of this equation can produce a large cavity in a limestone bed—a cave. The evaporation of water from $Ca(HCO_3)_2(aq)$ represented by the reverse direction leads to a loss of both H_2O and CO_2. As a result, $Ca(HCO_3)_2(aq)$ is converted back to $CaCO_3(s)$. This process is extremely slow, but over a period of many years, as $Ca(HCO_3)_2(aq)$ drips from the ceiling of a cave, it is converted to icicle-like deposits of $CaCO_3(s)$ called *stalactites*. When some of the dripping $Ca(HCO_3)_2(aq)$ hits the floor of the cave, decomposition occurs there, and limestone ($CaCO_3$) deposits build up from the floor in formations called *stalagmites*. Some of the stalactites and stalagmites grow together into limestone columns (Figure 20.8).

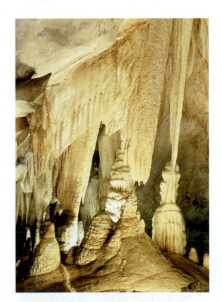

▲ **FIGURE 20.8** Stalactites, stalagmites, and columns in a limestone cavern

The reactions that produce limestone caves also account for the deterioration of limestone buildings and marble statues. The reactions become even more significant under conditions of acid rain (Section 25.7).

20.12   Hard Water and Water Softening

We have just seen how rainwater can become infused with calcium hydrogen carbonate. **Hard water** is groundwater that contains significant concentrations of ions from natural sources, principally Ca^{2+}, Mg^{2+}, and sometimes Fe^{2+}, along with associated anions. If the primary anion is the hydrogen carbonate ion, HCO_3^-, the hardness is said to be **temporary hardness.** If the predominant anions are other than HCO_3^-, for example, Cl^- or SO_4^{2-}, the hardness is called **permanent hardness.**

When water with temporary hardness is heated, hydrogen carbonate ions decompose:

$$2\ HCO_3^-(aq) \xrightarrow{\Delta} CO_3^{2-}(aq) + H_2O(l) + CO_2(g)$$

The CO_3^{2-} ions formed in this way react with $M^{2+}(aq)$ (that is, Ca^{2+}, Mg^{2+}, and Fe^{2+}) to form solid carbonates:

$$M^{2+}(aq) + CO_3^{2-}(aq) \longrightarrow MCO_3(s)$$

The mixed precipitate of $CaCO_3$, $MgCO_3$, and $FeCO_3$, together with any other undissolved solids, is commonly called *boiler scale*. Formation of boiler scale is a serious problem associated with hard water. A boiler clogged with boiler scale heats unevenly, and overheating in parts of the boiler can lead to an explosion. Hard water also interferes with the action of soap (page 854). For many of its uses, then, hard water must be softened.

CaCO₃ deposits on water heater element.

▲ $CaCO_3(s)$ is deposited when water with temporary hardness is heated.

Water Softening

Water softening is the removal of objectionable cations and anions from hard water. Water with temporary hardness can be softened by boiling, but this produces boiler scale. The water can also be softened by addition of a base, which converts HCO_3^- to CO_3^{2-}. The CO_3^{2-} then combines with M^{2+} ions to form the same precipitates as with heating, but in this case they are formed as a gritty solid that can cause problems in washing machines and other devices.

$$HCO_3^-(aq) + OH^-(aq) \longrightarrow CO_3^{2-}(aq) + H_2O(l)$$
$$CO_3^{2-}(aq) + M^{2+}(aq) \longrightarrow MCO_3(s)$$

In this method of water softening, the usual source of $OH^-(aq)$ is slaked lime, $Ca(OH)_2$.

Boiling cannot soften water with permanent hardness. Addition of washing soda (Na_2CO_3) softens the water by precipitating such cations as Ca^{2+}, Mg^{2+}, and Fe^{2+} as carbonates, but excess Na_2CO_3 and salts such as $NaCl$ and Na_2SO_4 remain in solution.

Ion Exchange

An effective way to soften water is through **ion exchange.** In this process, the undesirable ions in hard water are exchanged for ions that are less objectionable. The ion-exchange medium may be either a synthetic resin or a natural or synthetic sodium aluminosilicate called a *zeolite*. In either case, the ion-exchange materials consist of macromolecular (polymer) particles. In contact with water, these materials ionize to produce two types of ions:

1. *fixed ions* that remain attached to the polymer surface
2. *counterions* that have a sign opposite that of the fixed ions and are free to move away from the polymer surface

The total charge of the counterions must equal that of the fixed ions, so that the resin as a whole is electrically neutral. We will describe the process that occurs with an ion-exchange resin, but a similar description also applies to zeolite systems.

In the ion-exchange resin pictured in Figure 20.9, the fixed ions are negatively charged ($—COO^-$) and the counterions are cations. In a fully charged resin, all the

Zeolite 3D model

▲ Zeolites have an open, three-dimensional structure, as suggested by this molecular model, through which ions can migrate freely.

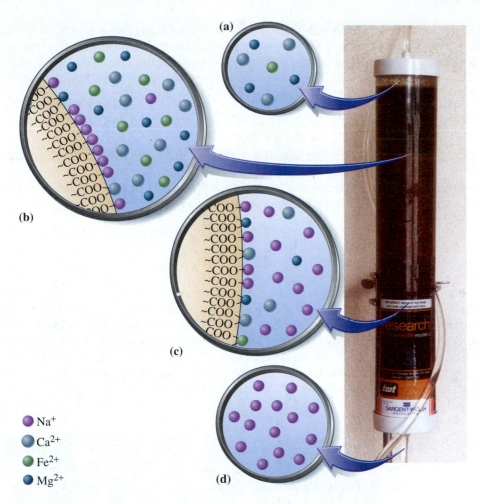

(a)

(b)

(c)

(d)

- 🟣 Na$^+$
- 🔵 Ca^{2+}
- 🟢 Fe^{2+}
- 🔵 Mg^{2+}

Ion Exchange and Water Softening animation

◀ **FIGURE 20.9 Water softening by ion exchange**
(a) Hard water contains Mg^{2+}, Ca^{2+}, and Fe^{2+} ions. (b) A tiny bead of ion-exchange resin has ionic groups (such as COO^-) covalently bonded to it, with sodium ions as counterions. Two sodium ions are replaced by one 2+ ion. (c) As the water passes through the resin bed, more of the 2+ ions in solution replace Na^+ as counterions at the surface of the resin. (d) The effluent water is free of water-hardening ions.

QUESTION: If two moles each of Mg^{2+}, Ca^{2+}, and Fe^{2+} ions are passed through the column and completely adsorbed, how many moles of Na^+ will be liberated?

counterions at the beginning of the process are Na^+. When hard water is passed through, equilibrium is established that can be represented as

$$Na_2R + M^{2+}(aq) \rightleftharpoons MR + 2\,Na^+(aq)$$

where R represents the resin. Because Ca^{2+}, Mg^{2+}, and Fe^{2+} ions in the hard water are preferentially adsorbed to the resin, they displace Na^+ ions as counterions, which enter the water. This water containing Na^+ ions is soft water in that no precipitates form when the water is boiled; nor do the Na^+ ions interfere with the action of soap the way Ca^{2+}, Mg^{2+}, and Fe^{2+} ions do.

To regenerate the resin, concentrated $NaCl(aq)$ is passed through the system, driving the equilibrium in the reverse direction.

Suppose that, instead of NaCl, we use concentrated $HCl(aq)$ to recharge an ion-exchange resin. The counterions will be H^+. Now when a hard-water sample is passed through the resin, the exchange reaction is

$$H_2R + M^{2+}(aq) \longrightarrow MR + 2\,H^+(aq)$$

The water becomes acidic, and we can do the following.

1. Determine the amount of acid in a solution by titration. If we titrate the H^+ eluted (washed) from an ion-exchange column, we can establish the level of hardness in the water (see Problem 84).

2. Pass the effluent from the cation-exchange resin, which contains H^+ as the only cations, through an anion-exchange resin to replace anions by OH^-. The reaction that takes place frees the water of all its ions:

$$H^+(aq) + OH^-(aq) \longrightarrow H_2O(l)$$

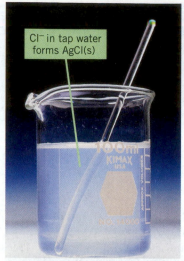

The product is called **deionized water.** We use deionized water rather than tap water in the laboratory because ions present in tap water may interfere with chemical reactions, for example, by forming unwanted precipitates (Figure 20.10). Distilled water for laboratory use is often "polished" by deionization to remove residual ions that may be liberated from plumbing or may splash over from the distilling pot.

20.13   Soaps and Detergents

Recall from earlier discussions that fats are esters of glycerol and long-chain carboxylic acids known as *fatty acids* (pages 63 and 246). **Soaps** are salts of fatty acids formed in the reaction of fats with a base, such as NaOH(aq). Sodium palmitate, a typical soap, is a salt of the 16-carbon palmitic acid. It can be represented as

$$CH_3(CH_2)_{14}COO^- Na^+ \quad or \quad RCOO^- Na^+ \quad [where \; R = CH_3(CH_2)_{14}]$$

Sodium palmitate
(a soap)

A soap acts by dispersing grease and oil films into microscopic droplets. The droplets detach themselves from the surfaces being cleaned, become suspended in water, and are removed by rinsing. As suggested by Figure 20.11, at an oil–water interface, soap molecules orient themselves with their hydrocarbon residues, R, dissolved in the oil and their ionic ends, COO^-, in the water. The soap emulsifies, or solubilizes, the oil.

Soaps that have an alkali metal ion (M^+) as the cation are water-soluble,* but soaps having an M^{2+} cation are not. This means that if an alkali metal soap is used in

▲ **FIGURE 20.10   Deionized water in the chemical laboratory**

(Top) When a solution of $AgNO_3$ is prepared in tap water, a white precipitate forms. Most likely this precipitate is AgCl(s), formed by the reaction of Ag^+(aq) with traces of Cl^-(aq) in the water. (Bottom) No precipitate forms when $AgNO_3$ is dissolved in deionized water.

 Sodium Palmitate 3D model

▶ **FIGURE 20.11   Cleaning action of a soap visualized**

(a) Sodium palmitate, $CH_3(CH_2)_{14}COO^- Na^+$, a typical soap. (b) A microscopic view of soap action. In an oil droplet suspended in water, the hydrocarbon residues (R groups) of soap molecules are immersed in the oil, and the ionic ends extend into the water. Attractive forces between water molecules and the ionic ends cause the oil droplet to be solubilized.

QUESTION: Would you expect sodium octanoate, $CH_3(CH_2)_6COO^- Na^+$, to perform better than or worse than sodium palmitate in solubilizing an oil droplet? Why?

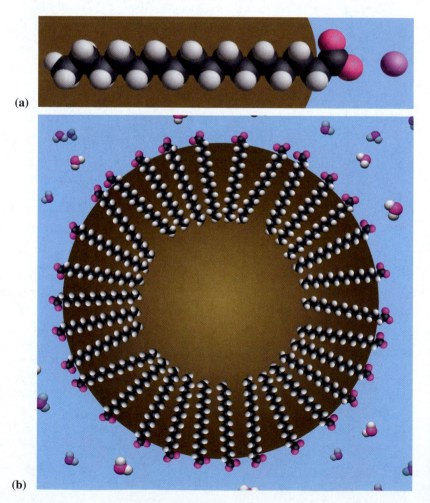

(a)

(b)

*Except when very dilute, alkali metal soaps do not form true solutions in water. Rather, they form colloidal dispersions (Section 12.10).

hard water, calcium, magnesium, and/or iron soaps precipitate. These precipitates are the major components of the familiar soap scum. For example,

$$2\,RCOO^-\,Na^+ + Ca^{2+} \longrightarrow Ca(RCOO)_2(s) + 2\,Na^+(aq)$$

Sodium soap   Calcium soap
(soap scum)

A soap can function well in hard water only after part of it is used up to precipitate all the M^{2+} ions present. In other words, the soap must first soften the water. Doing this, however, leaves an objectionable, dirty, grayish film of soap scum on clothes. One alternative is to soften water by one of the methods previously described before using it with soap. Another alternative is to use a synthetic detergent in place of a soap.

  A synthetic detergent functions in much the same way as a soap, but it does not form precipitates with M^{2+} cations. The detergent can therefore be used in hard water. Early detergents, called *alkylbenzene sulfonates* (*ABS*), had highly branched alkyl groups and were not biodegradable. Later detergents, called *linear alkylbenzene sulfonates* (*LAS*), are unbranched, with a typical formula being

$$CH_3(CH_2)_n\,CH\text{---}\bigcirc\text{---}SO_3^-Na^+ \quad \text{(where } n \text{ is usually 9 to 13)}$$
$$|$$
$$CH_3$$

LAS detergents, like soaps, are biodegradable. Bacteria break them down to carbon dioxide, water, and inorganic ions such as sulfate ion.

 **Surfactant Molecules animation**

Cumulative Example

Assume that all the cations in a container holding 1.00 L of hard water are Ca^{2+}. A 25.00-mL sample of this water is passed through a cation-exchange resin having H_3O^+ as counterions. The water leaving the column is found to have a pH of 2.37. Determine the hardness of the water, expressed as ppm Ca^{2+}.

STRATEGY

There are three key points we should establish at the outset: (1) The hydronium ion in the water leaving the ion-exchange resin is produced by the reaction

$$H_2R + Ca^{2+}(aq) + 2\,H_2O(l) \longrightarrow CaR + 2\,H_3O^+(aq)$$

(2) From the observed pH, we can first calculate $[H_3O^+]$ and then, using a mole ratio from the above equation, calculate $[Ca^{2+}]$ in the sample. (3) Knowing $[Ca^{2+}(aq)]$, we can establish the ppm Ca^{2+}, requiring only the assumption that the density of the hard water is 1.00 g/mL.

SOLUTION

Our first task is to convert pH to $[H_3O^+]$.

$$\log[H_3O^+] = -pH = -2.37$$
$$[H_3O^+] = 10^{-2.37} = 4.27 \times 10^{-3}\,M$$

Now, using a mole ratio from the equation, we obtain the Ca^{2+} ion concentration.

$$[Ca^{2+}] = \frac{4.27 \times 10^{-3}\,mol\,H_3O^+}{1\,L} \times \frac{1\,mol\,Ca^{2+}}{2\,mol\,H_3O^+}$$
$$= 2.1 \times 10^{-3}\,mol\,Ca^{2+}/L$$

The concentration in ppm must be on a mass basis, and so we must convert from moles to grams of Ca^{2+}.

$$?\,g\,Ca^{2+}/L = \frac{2.1 \times 10^{-3}\,mol\,Ca^{2+}}{1\,L} \times \frac{40.08\,g\,Ca^{2+}}{1\,mol\,Ca^{2+}}$$
$$= \frac{0.084\,g\,Ca^{2+}}{1\,L}$$

By assuming 1.00 g/mL for the density of the water, we can replace 1 L water by 1000 g water. Then, by multiplying the numerator and denominator of the above ratio by 1000, we obtain the result in ppm Ca^{2+}.

$$?\,ppm\,Ca^{2+} = \frac{0.084\,g\,Ca^{2+}}{1000\,g\,water} \times \frac{1000}{1000} = \frac{84\,g\,Ca^{2+}}{1 \times 10^6\,g\,water} = 84\,ppm\,Ca^{2+}$$

Here are three noteworthy points in this calculation:

1. The 25.00-mL sample that is passed through the resin and the larger 1.00-L volume do not enter into the calculation at all. The measured pH converts directly to a mole-per-liter basis, *not* to an amount in moles, which would have to be divided by a volume.

2. In converting from pH to $[H_3O^+]$, we carried one more digit in the intermediate result than is significant. We did so for two reasons: If we had combined the steps in which $[H_3O^+]$ and $[Ca^{2+}]$ were determined, which would have been easy enough to do because $[Ca^{2+}] = \frac{1}{2}[H_3O^+]$, we would have obtained the same result as what we show, 2.1×10^{-3} M. If we had rounded 4.27×10^{-3} M to 4.3×10^{-3} M, our value of $[Ca^{2+}]$ would have been 2.2×10^{-3} M, and this would have given a different final answer: 88 ppm Ca^{2+}.

3. To make a rough estimate of the answer, we can begin with the fact that $2 < pH < 3$. This means that 10^{-3} M $< [H_3O^+] < 10^{-2}$ M. The limits on $[Ca^{2+}]$ are one-half those for $[H_3O^+]$, that is, 5×10^{-4} M $< [Ca^{2+}] < 5 \times 10^{-3}$ M. To obtain the limits on a gram-per-liter basis, we multiply the limits on $[Ca^{2+}]$ by the molar mass of Ca (40 g/mol): $0.02 < g\,Ca^{2+}/L < 0.2$. Finally, if we multiply these gram-per-liter limits by 1000 to obtain ppm Ca^{2+}, we see that $20 < ppm\,Ca^{2+} < 200$. Our calculated value of 84 ppm Ca^{2+} falls squarely within this range.

Concept Review with Key Terms

20.1 Occurrence and Preparation of Hydrogen— The main sources of H_2 are the electrolysis of water, the water–gas reactions, and the reforming of hydrocarbons. Small quantities of H_2 gas can be produced by the reaction of very active metals (the group 1A metal plus Ca, Sr, and Ba) with water or by the reaction of less active metals (for example, Zn and Fe) with mineral acids.

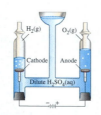

20.2 Binary Compounds of Hydrogen— Hydrogen reacts with nonmetals to form molecular hydrides and with metals to form ionic hydrides or metallic hydrides. Metallic hydrides can be nonstoichiometric.

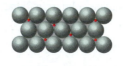

20.3 Uses of Hydrogen—Hydrogen is used (a) in the synthesis of ammonia, (b) in hydrogenation reactions, (c) as a reducing agent in metallurgy, and (d) as a fuel.

20.4 Properties and Trends in Group 1A— Of all the elements, the group 1A metals (the alkali metals) have the largest atomic radii and lowest ionization energies; they also have low densities and low melting points. They form ionic solids with nonmetals and react with water to produce ionic hydroxides and hydrogen gas. In some of its physical and chemical behavior, lithium resembles magnesium, a resemblance called a **diagonal relationship** that is seen with a few other elements as well.

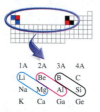

20.5 Occurrence, Preparation, Uses, and Reactions of Group 1A Metals—Sodium is formed in the electrolysis of molten sodium chloride. Sodium hydroxide, chlorine, and hydrogen form in the electrolysis of aqueous sodium chloride. Other chemicals produced from sodium chloride include sodium carbonate, sodium sulfate, and hydrochloric acid. Nearly all group 1A compounds are water-soluble. Sodium and potassium ions are essential to living organisms.

20.6 Important Compounds of Lithium, Sodium, and Potassium— Some common and/or industrially important lithium, sodium, and potassium compounds are lithium carbonate, lithium hydroxide, sodium chloride, sodium hydroxide, sodium sulfate, and potassium chloride.

20.7 The Alkali Metals and Living Matter—Both Na^+ and K^+ serve a number of important roles in the human body. These ions must be supplied in the diet. Many drugs are administered as sodium or potassium salts, which are more soluble than the molecular forms of the drugs.

20.8 Properties and Trends in Group 2A—The group 2A metals (the alkaline earth metals) have smaller atomic radii and greater ionization energies, densities, and melting points than do the group 1A metals. Beryllium is an exception to the general trends in group 2A.

20.9 Occurrence, Preparation, Uses, and Reactions of Group 2A Metals—Calcium and magnesium are abundant elements in Earth's crust. The group 2A metals are usually obtained by electrolysis of the molten chloride. The heavier group 2A metals react with water to liberate hydrogen; magnesium does so only when treated with steam, but beryllium does not react with water, hot or cold. Many alkaline earth compounds are either insoluble or only slightly soluble in water.

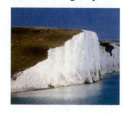

20.10 Important Compounds of Magnesium and Calcium— Among the important group 2A compounds are the carbonates, chlorides, hydroxides, oxides, and sulfates of calcium and magnesium. Calcium and magnesium are found in many minerals, for example, in limestone ($CaCO_3$) and dolomite ($CaCO_3 \cdot MgCO_3$). Also, many occur as hydrates, as in $CaCl_2 \cdot 6\,H_2O$.

20.11 The Group 2A Metals and Living Matter—Magnesium and calcium are essential to all living organisms. Magnesium is involved in a number of metabolic processes, and calcium is a major component of bones and teeth.

20.12 Hard Water and Water Softening—The slight dissolution of minerals (for example, $CaCO_3$ and $CaSO_4$ in acidified rainwater) introduces ions into groundwater. This dissolving action is responsible for natural caverns, limestone formations, and **hard water**. If HCO_3^- is the primary anion in water containing the dissolved minerals, the condi-

tion is called **temporary hardness**; other primary anions, such as Cl^- and SO_4^{2-}, cause **permanent hardness**. Hard water is objectionable because it yields mineral deposits when heated and interferes with the cleaning action of soaps.

Water can be softened by chemical reactions that remove the offending cations, Ca^{2+}, Mg^{2+}, and Fe^{2+}, or by **ion exchange**, a process in which offending ions are replaced by innocuous ones. Ion exchange can also be used to eliminate virtually all ions from water, yielding **deionized water**.

20.13 Soaps and Detergents—**Soaps** are salts of fatty acids that can solubilize oil, but soaps lose their effectiveness in hard water because they precipitate as the Ca^{2+} or Mg^{2+} salts. Detergents do not form such precipitates and can be used in hard water.

Assessment Goals

When you have mastered the material in this chapter, you will be able to:

- Describe the physical properties, occurrences, and uses of hydrogen.

- Recognize the reactions used to form hydrogen as well as the reactions hydrogen undergoes to form binary compounds.

- Describe the physical properties, occurrences, and uses of the alkali and alkaline earth metals.

- Outline the Dow process for preparing magnesium.

- Provide examples of some important compounds of lithium, sodium, magnesium and calcium, as well as some of their uses.

- Define *diagonal relationships*, and provide an example of such a relationship involving alkali metals and alkaline earth metals.

- Identify the ions responsible for the occurrence of hard water.

- Describe how ion exchange can be used for water softening and deionization of water.

- Describe the action of soaps and detergents on a molecular scale.

Self-Assessment Questions

1. Which is the most abundant alkali metal in Earth's crust? Which is the most abundant alkaline earth metal? Name several naturally occurring alkali metal and alkaline earth metal compounds.

2. Which of the group 1A elements are essential to living organisms? Which of the group 2A elements?

3. Why is the percent abundance of hydrogen in Earth's crust greater when based on percent by number of atoms than percent by mass? Would you expect the same relationship to hold for all the elements?

4. In the electrolysis of water pictured in Figure 20.2, does the molarity of the H_2SO_4 change during the electrolysis? If so, in what way?

5. Why is helium, an *s*-block element, not considered a member of group 2A?

6. Which of the group 1A elements react with cold water to produce $H_2(g)$? Which of the group 2A elements?

7. Which of the following is incorrectly matched with an example of the compound type?

    (a) nonstoichiometric hydride, $PdH_{0.6}$

    (b) metallic hydride, TiH_2

    (c) molecular hydride, CaH_2

    (d) ionic hydride, NaH

8. How do metallic hydrides differ from ionic hydrides?

9. What is meant by the term calcination (for example, the calcination of limestone)?

10. What is meant by the diagonal relationship among elements? Which elements display this relationship?

11. In what ways does (a) lithium differ from the other group 1A elements; (b) beryllium differ from the other group 2A elements?

12. What are the principal raw materials required for the manufacture of each of the following products?

    (a) Portland cement   (c) mortar

    (b) soda–lime glass   (d) plaster of paris

13. Which of the following is *not* involved in the Dow process for the production of magnesium?

    (a) the precipitation of $Mg(OH)_2$

    (b) the decomposition of $CaCO_3$

    (c) the electrolysis of H_2O

    (d) the formation $MgCl_2$

14. What functions are served in the human body by each of the following ions?

    (a) sodium   (c) magnesium

    (b) potassium   (d) calcium

15. Which ions are responsible for temporary hardness and permanent hardness in water? What is boiler scale? How is it formed?

16. What are the approaches taken to soften hard water?

17. What is an ion-exchange resin, and how can it be used to produce deionized water?

18. What are the structural features of a soap? How does a calcium soap differ in structure from a sodium soap?

19. What are the structural features of a detergent? What do the symbols ABS and LAS mean when applied to a detergent?

20. How do soaps and detergents function; that is, how do they emulsify oil or grease?

Problems

Group Trends

21. Arrange the following compounds in the expected order of increasing solubility in water, and give the basis for your arrangement: Li_2CO_3, Na_2CO_3, $MgCO_3$.

22. Arrange the following compounds in the expected order of increasing melting point, and give the basis for your arrangement: MgO, $CaCl_2$, $SrBr_2$.

23. Which of the properties in Table 20.1 exhibit an irregular periodic trend? Why do these irregular trends differ from those that are regular?

24. Which of the properties in Table 20.4 exhibit an irregular periodic trend? Why do these irregular trends differ from those that are regular?

Nomenclature

25. Supply a name or formula for each of the following:

(a) K_2O_2
(b) $Ca(HCO_3)_2$
(c) barium phosphate
(d) strontium nitrate tetrahydrate

26. Supply a name or formula for each of the following:

(a) Li_2CO_3
(b) RbO_2
(c) calcium chloride hexahydrate
(d) radium bromide

27. Give formulas and acceptable scientific names of the substances known by these common names: (a) quicklime, (b) Glauber's salt, (c) gypsum, (d) dolomite.

28. Which of the following are pure substances and which are mixtures? For those that are mixtures, what are the main components?

(a) plaster of paris
(b) water gas
(c) soap scum
(d) slaked lime

Industrial Processes

29. Diagram a plausible scheme for the production of magnesium from magnesite ($MgCO_3$).

30. For the Solvay process for the production of sodium hydrogen carbonate, describe the raw materials required, the main product, any by-products, and substances that are recycled. Outline the steps in the process by writing chemical equations.

31. Write an equation for the reaction by which Glauber's salt is produced. What is the underlying principle of this reaction?

32. Write equations to show how methanol (CH_3OH) can be synthesized from coke (carbon) and water.

33. Sodium was first obtained by electrolysis of sodium hydroxide, which has a lower melting point (322 °C) than sodium chloride (801 °C). Why do you suppose that the primary industrial method of producing sodium today is the electrolysis of NaCl rather than of NaOH?

34. According to ionization energies and electrode potentials in Table 20.1, K^+ is more difficult to reduce than Na^+. Does this mean that there is no possible reaction in which sodium metal can reduce a potassium compound to metallic potassium? Explain.

35. In the manner used in Figure 20.5, complete the diagram outlined. Specifically, indicate the reactants and any special conditions needed to produce each substance, using $BaCl_2$ as the starting material.

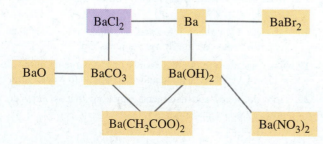

36. In the manner used in Figure 20.5, complete the diagram outlined below. Specifically, indicate the reactants and any special conditions needed to produce each substance, using $MgCO_3$ as the starting material.

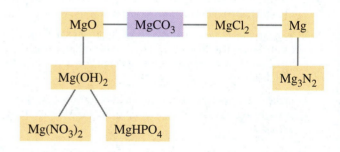

37. Suggest a reason why the magnesium chloride solution obtained midway through the Dow process is not electrolyzed directly, but instead is evaporated to dryness and electrolyzed in a molten state.

38. Construct a diagram for the Dow process similar to that of Figure 20.6.

Chemical Equations

39. Complete the following equations representing the reactions of substances with acids.

(a) $Ca(s) + HCl(aq) \longrightarrow$

(b) $KHCO_3(s) + HCl(aq) \longrightarrow$

(c) $NaF(s) + H_2SO_4(\text{concd aq}) \xrightarrow{\Delta}$

40. Complete the following equations representing the reactions of substances with water.

(a) $Sr(s) + H_2O(l) \longrightarrow$

(b) $LiH(s) + H_2O(l) \longrightarrow$

(c) $BaO(s) + H_2O(l) \longrightarrow$

(d) $C(s) + H_2O(g) \xrightarrow{\Delta}$

41. Write chemical equations to represent each of the following.

(a) the reaction of lithium metal with chlorine gas

(b) the reaction of potassium metal with water

(c) the reaction of cesium metal with liquid bromine

(d) the combustion of potassium metal to form potassium superoxide

42. Write chemical equations to represent each of the following.

(a) the displacement of $H_2(g)$ from HCl(aq) by Al(s)

(b) the reforming of ethane gas (C_2H_6) with steam

(c) the complete hydrogenation of methylacetylene, $CH_3C\equiv CH$

(d) the reduction of $MnO_2(s)$ to Mn(s) with $H_2(g)$

43. Write chemical equations to represent each of the following.

(a) the reaction of BeF_2 with metallic sodium to produce metallic beryllium

(b) the reaction of calcium metal with dilute acetic acid, $CH_3COOH(aq)$

(c) the reaction of plutonium(IV) oxide with calcium to produce metallic plutonium

(d) the calcination of dolomite.

44. Write equations for the reactions you would expect to occur when (a) $BaCO_3(s)$ is heated to a high temperature; (b) $CaCl_2(l)$ is electrolyzed; (c) Ca(s) is added to cold dilute HCl(aq); and (d) an excess of slaked lime is added to aqueous sulfuric acid.

45. Write an equation for the reaction you would expect to occur when (a) $MgCO_3(s)$ is added to HCl(aq); (b) $CO_2(g)$ is bubbled into KOH(aq); and (c) KCl(s) is heated with concentrated $H_2SO_4(aq)$.

46. Write equations to show how each of the following substances can be used in the preparation of $H_2(g)$.

(a) $H_2O(l)$ (c) Mg(s)

(b) HI(aq)   (d) $CH_4(g)$

Use other reactants as necessary—water, acids or bases, metals, and so forth.

47. Write plausible equations to show how you would convert (a) NaCl(s) to $Na_2SO_4(s)$; and (b) $Mg(HCO_3)_2(aq)$ to MgO(s).

48. Write plausible equations to show how you would convert (a) $BaCO_3(s)$ to $Ba(OH)_2(s)$; and (b) NaCl(s) to $Na_2O_2(s)$.

Reaction Stoichiometry

Hint: Write chemical equations, as necessary.

49. How many grams of $CaH_2(s)$ are required to generate sufficient $H_2(g)$ to fill a 679-L weather-observation balloon at 736 mmHg and 11 °C?

50. How many liters of $H_2(g)$, measured at 22 °C and 15.5 atm pressure, are required to convert 175 g propylene ($CH_3CH=CH_2$) to propane (C_3H_8)? What mass of propane is obtained?

51. How many cubic meters of $CO_2(g)$, measured at 748 mmHg and 22 °C, would be produced in the calcination of 1.00×10^3 kg of the mineral dolomite ($CaCO_3 \cdot MgCO_3$)?

52. How many liters of 6.0 M HCl are required to neutralize 55.6 kg $Ca(OH)_2(s)$?

53. Assume that the magnesium content of seawater is 1270 g Mg^{2+}/ton seawater and that the density of seawater is 1.03 g/mL. What minimum volume of seawater, in liters, must be used in the Dow process to obtain 1.00 kg of magnesium? Why is the actual volume required greater than this calculated minimum volume?

54. In the Solvay process, for every 1.00 kg $NaHCO_3(s)$ produced in the main reaction, how many kilograms of $CaCl_2$ are obtained as a by-product of the ammonia recovery step?

55. *Without doing detailed calculations,* determine which of the following hydrates has the greatest mass percent H_2O: (a) $Ba(OH)_2 \cdot 8 H_2O$, (b) $CaCl_2 \cdot 6 H_2O$, (c) $MgSO_4 \cdot 7 H_2O$.

56. *Without doing detailed calculations,* determine which of the following reactions produces the greatest mass of hydrogen per 100 grams of the limiting reactant indicated: (a) Zn(s) with an excess of HCl(aq), (b) Na(s) with an excess of $H_2O(l)$, (c) $CaH_2(s)$ with an excess of $H_2O(l)$.

Hard Water

57. Describe how water with *temporary* hardness can be softened by the addition of $NH_3(aq)$. (*Hint:* What ions are present in the water? What chemical reactions occur?)

58. With reference to Problem 57, do you think water with *permanent* hardness can be softened with $NH_3(aq)$? Explain.

59. A water sample has a hardness expressed as 115 ppm HCO_3^-. Assuming that Ca^{2+} is the only cation present, how many mil-

ligrams of boiler scale would you expect to deposit when 725 mL of the water is evaporated to dryness? (*Hint:* Think of parts per million as $g\ HCO_3^-/10^6\ g$ water $\approx g\ HCO_3^-/10^3$ L water.)

60. A water sample has a hardness of 126 ppm HCO_3^-. How many kilograms of $Ca(OH)_2$ are required to soften 1.50×10^7 L of the water? (*Hint:* Refer to Problem 59 and the equations on page 852.)

61. Describe the chemical reactions that lead to the formation of limestone caves and stalactites and stalagmites in those caves. How are these reactions related to the formation of hard water?

62. To soften water with temporary hardness, Ca^{2+} is precipitated as $CaCO_3$. One way to do this is to add CaO (quicklime) to the water. Why is it that adding a calcium compound does not simply increase $[Ca^{2+}]$ in the water?

63. Describe how a water sample containing the ions Fe^{3+}, Ca^{2+}, Cl^-, and SO_4^{2-} can be *deionized* by the use of ion-exchange resins.

64. Write chemical equations to represent the preparation of deionized water from the following solutions.

   (a) $Ca(HCO_3)_2(aq)$ **(b)** $NaCl(aq)$ **(c)** $NaOH(aq)$

Additional Problems

Problems marked with an * may be more challenging than others.

65. Write chemical equations for the following reactions referred to on page 839.

   (a) the formation of lithium oxide

   (b) the reaction of lithium with nitrogen gas to form lithium nitride

   (c) the calcination of lithium carbonate

66. The use of potassium superoxide in life-support systems is described on page 841. Write a balanced equation to show how sodium peroxide can be used instead.

67. When stored lime is exposed to air for a long time, it loses the properties of CaO, for example, it no longer reacts with water. What reaction(s) must have occurred in the lime?

68. Assuming the availability of water, common reagents (acids, bases, salts), and simple laboratory equipment, give a practical method that could be used to prepare each of the following substances from cesium chloride (CsCl).

   (a) Cs metal   **(c)** $CsNO_3$ **(e)** $CsHCO_3$

   (b) CsOH   **(d)** Cs_2SO_4 **(f)** Cs_2CO_3

69. A 1.00-kg sample of spent uranium and an excess of water and nitric acid react to produce $UO_2(NO_3)_2$ and $H_2(g)$. How many liters of hydrogen gas are formed if the gas is measured at 25 °C and 745 Torr?

* 70. In Example 20.2, we saw that when magnesium is burned in air, the product is a mixture of MgO and Mg_3N_2. When a small quantity of water is added to the mixture, a strong ammonia smell is detected. When the mixture is then strongly heated, pure MgO is obtained. Write equations for the reactions that occur. Do you think this is an economically feasible method of obtaining MgO? If not, why not, and what method would be more feasible?

* 71. Suppose the 1.000-g strip of magnesium described in Example 20.2 yields 1.537 g of product. What is the mass percent of MgO in this product?

72. Use electrode potential data from Appendix C to show that the following metals should all be able to displace H_2 from water (as mentioned on page 835): Li, Na, K, Rb, Cs, Ca, Sr, Ba.

* 73. Determine whether liquid hydrogen or jet fuel has the greater heat of combustion **(a)** on a mass basis and **(b)** on a volume basis. Jet fuel is a mixture of hydrocarbons, but assume $C_{12}H_{26}$ is a representative formula. The densities of $H_2(l)$ and $C_{12}H_{26}(l)$ are 0.0708 g/mL and 0.749 g/mL, respectively. $\Delta H_f^\circ[C_{12}H_{26}(l)] = -350.9$ kJ/mol.

74. A water sample has a hardness of 117 ppm SO_4^{2-}. The cations present are Ca^{2+}. **(a)** Show how this water can be softened with Na_2CO_3 (washing soda). **(b)** How many grams of Na_2CO_3 are required to soften 162 L of this water?

75. How many grams of the soap sodium palmitate are consumed in softening 5.00 L of hard water with 135 ppm HCO_3^-? Assume that the only cation in the water is Ca^{2+}.

76. A sample of water whose hardness is expressed as 185 ppm Ca^{2+} is passed through an ion-exchange column in which the Ca^{2+} is replaced by Na^+. What is the molarity of Na^+ in the treated water?

* 77. Workers who deal with plaster and stucco are aware that considerable heat is given off when slaked lime is produced from quicklime. Determine the maximum temperature reached if 1.00 kg each of quicklime and water are mixed at 25 °C. Use data from Appendix C, and assume specific heats of 0.75 J g^{-1} $°C^{-1}$ for quicklime, 1.18 J g^{-1} $°C^{-1}$ for slaked lime, and 4.18 J g^{-1} $°C^{-1}$ for water.

* 78. Use atomic radii from Table 20.1 and the fact that sodium, potassium, and rubidium all crystallize in a body-centered cubic structure to calculate the densities of these three metals. Show that your results lead to the same trend in densities as found in Table 20.1 and discussed on page 839.

79. In a chemical storeroom, it is not uncommon to find that a chemical that should exist as a crystalline solid at room temperature appears to be a mixture of a solid and liquid. Explain why this might be the case. Should the chemical be discarded, or can it still be used for certain purposes? Explain.

* 80. Paints and other suspensions and emulsions often employ a surface-active agent (surfactant). The surfactant coats the dispersed particles and reduces the viscosity of the mixture, in part by reducing the strength of intermolecular forces between the suspended particles and liquid matrix. One common surfactant is the soap, sodium dodecyl sulfate, $CH_3(CH_2)_{11}OSO_3Na$.

   (a) Consider an oil-based paint having a density of 1.25 g/mL that is 40.0% by mass $TiO_2(s)$ ($d = 4.26$ g/cm^3). Assume that the TiO_2 is present as spherical particles with an average diameter of 1000 nm and that each adsorbed dodecyl sulfate ion covers an area that is 0.4 nm square. What mass of sodium dodecyl sulfate must be added to 1.00 kg of the paint to produce a layer of surfactant that is one dodecyl sulfate ion thick (a monolayer) on the TiO_2 particles?

   (b) Calculate the charge in coulombs produced on a typical TiO_2 particle by the adsorbed dodecyl sulfate ions. How does this charge help to reduce the viscosity of the paint?

   (c) Prolonged, vigorous mixing is needed to coat the pigment in an oil-based paint with this surfactant. In a water-based (latex) paint, the coating process proceeds more quickly. Suggest a reason why.

* 81. Consider the following sequence of observations:

   1. A small chunk of solid CO_2 (dry ice) is added to 0.005 M $Ca(OH)_2(aq)$.

   2. Initially, a white precipitate forms.

   3. After a short time, the precipitate redissolves.

(a) Write chemical equations to explain these observations.

(b) If the 0.005 M $Ca(OH)_2$(aq) is replaced by 0.005 M $CaCl_2$(aq), would you expect a precipitate to form? Explain.

(c) If the 0.005 M $Ca(OH)_2$(aq) is replaced by 0.010 M $Ca(OH)_2$(aq), a precipitate forms, but it does not redissolve. Explain why.

(*Hint:* What equilibrium processes apply, and what are the equilibrium constants?)

Apply Your Knowledge

84. **[Laboratory]** A 100.0-mL sample of hard water is passed through a column of the ion-exchange resin H_2R. The water coming off the column requires 15.17 mL of 0.02650 M NaOH for its titration. What is the hardness of the water, expressed as ppm Ca^{2+}?

* 85. **[Biochemical]** A certain lithium battery has a voltage of 3.0 V. The capacity of the battery is 0.5 ampere · hour (A · h). Assume that to regulate the heartbeat requires 5 microwatts (μW). How many years can the implanted battery operate? What is the minimum mass of lithium that must be present in the battery? (*Hint:* Refer to Appendix B.)

* 86. **[Environmental]** Use data from elsewhere in the text to obtain an equilibrium constant for the reversible reaction involved in the dissolution of $CaCO_3$(s) in CO_2-charged rainwater described on page 851.

87. **[Environmental]** The $[Ca^{2+}]$ and $[SO_4^{2-}]$ in seawater are both found to have a concentration of about 1×10^{-2} M. **(a)** What anion, if any, would precipitate Ca^{2+} from seawater? **(b)** What maximum concentration of Ba^{2+} could coexist with this much sulfate ion?

88. **[Laboratory]** When a 5.00-g sample of impure lime containing some $CaCO_3$ is dissolved in HCl(aq), 143 mL of a gas, measured at 746 mmHg and 22 °C, was liberated. Calculate the percent of $CaCO_3$ in the sample.

* 89. **[Biochemical]** Assume that a person breathes 1.4×10^4 L of air per day at 25 °C and 1.00 atm and that the expired air is 4.0%

by volume CO_2. Calculate the mass of **(a)** lithium hydroxide and **(b)** sodium hydroxide needed to consume exhaled CO_2 for three people on a seven-day space flight. If the cost of sending mass into orbit is about $20,000 per kg, how much more does it cost to use sodium hydroxide than lithium hydroxide for this purpose?

90. **[Environmental]** Solar-heated homes require a heat-storage medium that can release heat at night and on overcast days. A cheap and simple medium is rock, which has a specific heat of about 0.8 J/g °C. An alternative is Glauber's salt (page 843), which undergoes the following transition at 32.384 °C.

$$Na_2SO_4 \cdot 10\,H_2O(s)$$
$$\rightleftharpoons Na_2SO_4(s) + Na_2SO_4(\text{satd aq}) \qquad \Delta H = -112 \text{ kJ}$$

The decahydrate can be converted to the anhydrous salt with solar heat, and the heat is released when the anhydrous salt reverts back to the decahydrate. Suppose that heat is to be stored by warming 3.0×10^3 kg of rock from 70 °F to 120 °F during the day. How many kg of Glauber's salt would be needed to store this same amount of heat?

91. **[Biochemical]** Use the information in the Essay "Fats and Oils: Hydrogenation" on page 372 to determine the number of kg/day of H_2 that must be supplied to a manufacturer of shortening for the hydrogenation of 13600 kg/day of linolenic acid. The product is to contain a 1 : 1 mole ratio of stearic and oleic acids.

82. Comment on the feasibility of using the following reaction as a means of converting Na_2CO_3(aq) to NaOH(aq).

$$Ca(OH)_2(s) + Na_2CO_3(aq) \longrightarrow CaCO_3(s) + 2\,NaOH(aq)$$

For example, assume that the Na_2CO_3(aq) is 2.00 M, and determine the molarity of the NaOH(aq) that can be produced. (*Hint:* Write an ionic equation.)

83. Suggest a method by which astronomers could have determined that the cloud of gas shown in the photograph on page 834 is actually hydrogen and not helium or another element.

🌐 e-Media Problems

The activities described in these problems can be found in the e-Media Activities and Interactive Student Tutorial (IST) modules of the Companion Website, *http://chem.prenhall.com/hillpetrucci*.

92. View the **Electrolysis of Water** animation (*Section 20-1*). **(a)** If a current of 0.75 A was used to carry out the electrolysis reaction, calculate the time required to generate the volume of molecular hydrogen produced in the animation (47.0 mL). Assume a pressure of 1.00 atm. **(b)** Why is this an uneconomical method of producing gaseous hydrogen?

93. The alkali metals are described as the most chemically reactive elements. Examples of their reactions are seen in the **Sodium and Potassium in Water** movie (*Section 20-4*). **(a)** Write a balanced chemical equation for each reaction seen in the movie. **(b)** From your observations, predict the sign of the heats of reaction. **(c)** In these reactions, do the alkali metals act as acids or bases? **(d)** Do they act as oxidizing agents or reducing agents?

(e) Based on the relative reactivities of sodium and potassium in water, predict the reactivities of lithium and cesium in water.

94. In the **Ion Exchange and Water Softening** animation (*Section 20-12*), water molecules are not shown, but they are actively involved in intermolecular interactions. **(a)** Based on the ionic nature of the exchange process, describe the change in entropy of water that occurs in the water-softening process. **(b)** What is the identity of the fixed ions in the animation? Why do they not become solvated during the ion-exchange process?

95. Rotate the **Sodium Palmitate** 3-D model (*Section 20-13*) to observe the molecular structure from different angles by [left- on *Windows*] clicking on the image and moving the mouse. **(a)** What is the general shape of this molecule, as illustrated? **(b)** Sketch two molecules of sodium palmitate, and indicate regions of attractive intermolecular interactions. **(c)** What collective orientation would be expected for sodium palmitate, molecules existing as a colloidal dispersion in water?

Chapter 21

The *p*-Block Elements

IN THIS CHAPTER, we continue the discussion of descriptive chemistry that we began with the *s*-block elements in Chapter 20. Here, we look at the elements of periodic table groups 3A through 8A—the *p*-block elements. The *p*-block includes all the noble gases except helium, all the nonmetals except hydrogen, all the metalloids, and even a few metals, including aluminum, tin, and lead.

Three of the *p*-block elements—O, Si, and Al—are the most abundant elements in Earth's crust. Six *p*-block elements—C, N, O, P, S, and Cl—are among the elements making up the bulk of living matter. Five others—B, F, Si, Se, and I—are required in trace amounts by most plant and animal life. Two of the *p*-block elements—C and S—can occur in the free state and have been known from prehistoric times, and three of them—Sn, Sb, and Pb—have been known for several thousand years. However, most were not discovered until much later.

In our discussion of the *p*-block elements, we will relate the properties of the elements to their positions in the periodic table. We will describe important compounds of the elements, comment on the uses of these compounds, and recount some ways in which we encounter them in daily life. In this discussion, we will use several principles introduced earlier in the text: bonding theory, enthalpy and free energy changes, electrode potentials, and equilibrium constants. In doing so, we will once more see the power of fundamental principles to explain chemical phenomena.

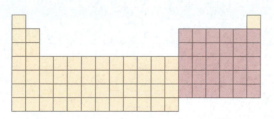

◄ Molten aluminum pours from a furnace into ingot molds for further processing. Aluminum, a *p*-block element, is one of the three most abundant elements in the Earth's crust. It is also one of the most important elements in industry and manufacturing, being used for everything from beverage cans to high-voltage cables. In this chapter we will investigate the sources, properties, trends, and uses of aluminum and other *p*-block elements.

CONTENTS

B	C	N	O	F	Ne
Al	Si	P	S	Cl	Ar
Ga	Ge	As	Se	Br	Kr
In	Sn	Sb	Te	I	Xe
Tl	Pb	Bi	Po	At	Rn

Group 3A

The first two members of group 3A—boron, a nonmetal, and aluminum, a metal—are the most important of the five group members. We will describe only a few aspects of the chemistry of boron, but we will deal more extensively with aluminum. In the following section, we first take a brief look at the group as a whole.

21.1 Properties and Trends in Group 3A

Several atomic and physical properties of the group 3A elements are listed in Table 21.1. Let us consider the valence-shell electron configuration, ns^2np^1, and how it changes when group 3A atoms form compounds. Boron atoms are small, and boron's first ionization energy is higher than the rest of group 3A. Boron therefore tends to use its $2s^22p^1$ valence electrons to form covalent bonds with other nonmetal atoms. Aluminum also forms some covalent bonds, but Al atoms have a tendency to lose the $3s^23p^1$ valence electrons. This happens, for example, in the reaction of aluminum with acids:

$$2\ Al(s) + 6\ H^+(aq) \longrightarrow 2\ Al^{3+}(aq) + 3\ H_2(g)$$

Because Al^{3+} is a small, highly charged ion, it is usually coordinated with other species. Thus, the $Al^{3+}(aq)$ in this equation is mostly $[Al(H_2O)_6]^{3+}$.

Gallium, indium, and thallium also form tripositive ions, but the electron configurations of these ions are not those of noble-gas atoms. The electron configuration of Ga^{3+}, for example, is $[Ar]3d^{10}$. Unlike aluminum, which forms only a 3+ ion, gallium, indium, and thallium also form 1+ ions. Thus, when a gallium atom loses its $4p^1$ electron but retains its $4s^2$ electrons, it becomes the Ga^+ ion and has the electron configuration $[Ar]3d^{10}4s^2$. The heavier elements indium and thallium have an even greater tendency to form 1+ ions. In fact, Tl^+ is more stable in aqueous solution than Tl^{3+}. This is reflected in the large positive potential for the reduction of Tl^{3+} to Tl^+ and in the negative potential for the reduction of Tl^+ to $Tl(s)$:

$$Tl^{3+}(aq) + 2\ e^- \longrightarrow Tl^+(aq) \qquad E° = +1.25\ V$$

$$Tl^+(aq) + e^- \longrightarrow Tl(s) \qquad E° = -0.336\ V$$

Thus, $Tl^{3+}(aq)$ in contact with $Tl(s)$ is readily reduced to $Tl^+(aq)$:

$$Tl^{3+}(aq) + 2\ Tl(s) \longrightarrow 3\ Tl^+(aq) \qquad E°_{cell} = 1.25\ V - (-0.336\ V) = 1.59\ V$$

The valence-shell pair of electrons retained in the Tl^+ ion $(6s^2)$ is called an **inert pair.** This retention of the ns^2 pair in some of the cations of the *p*-block metals of the fourth and higher periods in the periodic table is a common phenomenon.

▲ Gallium is a liquid from 30 °C (just above room temperature) to about 2400 °C. This is one of the largest liquid-state temperature ranges of any substance. Gallium is used in some high-temperature thermometers.

Table 21.1 Some Properties of the Group 3A Elements

	B	Al	Ga	In	Tl
Atomic number	5	13	31	49	81
Valence-shell electron configuration	$2s^22p^1$	$3s^23p^1$	$4s^24p^1$	$5s^25p^1$	$6s^26p^1$
Atomic radius, pm[a]	88	143	122	163	170
Ionic radius, pm[b]	23	50	62	81	95
First ionization energy, kJ/mol	801	577	579	558	589
Electronegativity	2.0	1.6	1.8	1.8	2.0
Density at 20 °C, g/cm³	2.34	2.70	5.91	7.31	11.85
Melting point, °C	2300	660	30	157	304
Electrical conductivity[c]	9×10^{-11}	59.9	5.9	19.0	8.8

[a] For B, single covalent radius; for the other elements, metallic radius.

[b] For the ion M^{3+}.

[c] On a scale in which the conductivity of silver is assigned a value of 100.

One physical property in Table 21.1 that suggests boron's nonmetallic character is its high melting point, which can be attributed to the network covalent bonding in the solid. The other four members of group 3A are good electrical conductors, but boron is a poor conductor. However, because it displays the electrical properties of a semiconductor (Section 24.8), some scientists classify boron as a metalloid.

Boron, the second-period member of group 3A, has a diagonal relationship with silicon, the third-period member of group 4A. Recall our discussion of diagonal relationships in Chapter 20, where we noted some similarities between Li and Mg and between Be and Al. One similarity between B and Si, for example, is that both are semiconductors.

21.2 Boron

Much of the chemistry of boron compounds is based on the lack of an octet of electrons about the central boron atom. These compounds are electron-deficient, and this deficiency causes them to exhibit some unusual bonding features that we explore in our look at boron hydrides.

Boron Hydrides Carbon atoms have four valence electrons, and the simplest hydrocarbon is methane, CH_4. By analogy, because boron atoms have three valence electrons, we might expect the simplest boron–hydrogen compound to be BH_3, *borane*. However, the boron atom in BH_3 lacks a valence-shell octet. It has only *six* electrons in its valence shell:

$$
\begin{array}{c}
H \\
| \\
B - H \\
| \\
H
\end{array}
$$

Borane does not exist as a stable compound.

To complete the octet of the boron atom, the BH_3 group can become part of a more extensive structure by forming a coordinate covalent bond with an atom that has a lone pair of electrons. For example, in the product formed in the reaction between borane and dimethyl ether,

$$
\underset{\text{Dimethyl ether}}{CH_3 - \overset{\cdot\cdot}{\underset{|}{O}} :} + \underset{\text{Borane}}{\overset{H}{\underset{H}{\overset{|}{B} - H}}} \longrightarrow \underset{\text{Dimethyl ether–borane adduct}}{CH_3 - \overset{\cdot\cdot}{\underset{H_3C}{\overset{|}{O}} :} \overset{H}{\underset{H}{\overset{|}{B} - H}}}
$$

this bond is between the B atom of BH_3 and the O atom of the ether. The product, because it results from the "addition" of one structure to another, is called an **adduct.**

The simplest boron hydride that can be isolated is *diborane*, B_2H_6. For many years following its isolation, diborane presented a puzzle because chemists could not figure out what bonds the two BH_3 units together:

$$
\begin{array}{cc}
H & H \\
| & | \\
H - B & ? \ B - H \\
| & | \\
H & H
\end{array}
$$

The problem is that the structure has a total of only 12 valence electrons even though the number required would seem to be 14 (as in ethane, C_2H_6). The difficulty is resolved through a type of bonding we have not seen before: a bond in which a single pair of electrons joins *three* atoms rather than the usual two. As shown in Figure 21.1, diborane has two of these bonds, which are called *three-center bonds*. The H atoms in the three-center bonds are simultaneously bonded to two atoms rather than the usual one atom. We can think of these H atoms as bridging the gap between the two B atoms.

Figure 21.1 shows that the B–H bond lengths are greater in the three-center bonds than in the other B–H bonds. We can explain this fact by noting that two electrons

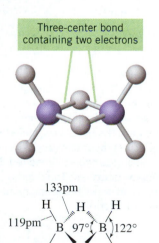

Three-center bond containing two electrons

133pm
119pm
97°
122°
H H H
B B
H H H

▲ **FIGURE 21.1** **Structure of diborane, B_2H_6**
Each of the two middle hydrogen atoms is bonded to two boron atoms.

QUESTION: Do the boron atoms in this structure obey the octet rule?

▲ Diborane and related compounds were once considered for use as rocket fuels but were abandoned because of cost and other practical difficulties. When a boron compound is introduced into a flame, elemental boron imparts a green color to the flame, as seen here.

▲ In the 1880s, borax was hauled from salt flats in Death Valley, California, by a team of 20 mules pulling a pair of wagons—plus a vital 1200-gallon water tank. The borax was hauled 165 miles across mountains and an almost waterless desert to a railroad terminal in Mojave. Borax is still sold today as a laundry aid under the brand name 20 Mule Team Borax®.

cannot bind three atoms as tightly as they bind two atoms. Although sp^3 hybridization of the bonding orbitals of the boron atoms can describe some features of the bonding, the bond angles are not tetrahedral (109.5°). The most satisfactory explanation of bonding in B_2H_6 comes from modern molecular orbital theory, in which orbitals are delocalized among several atoms, as in benzene (Figure 10.30).

In all, about two dozen boron–hydrogen compounds (boranes) are now known, and their structures have been determined by combining molecular orbital theory and experimental measurements. Boranes such as diborane and organic derivatives of borane are frequently used as reagents in the synthesis of organic compounds.

Borax, Boric Oxide, Boric Acid, and Borates Boron compounds are fairly widely distributed on Earth. The average boron concentration is about 9 ppm in Earth's crust and 4.8 ppm in seawater, and it is an essential element for some organisms. However, concentrated mineral deposits of boron compounds are found in only a few locations, such as Turkey and the desert regions of California. *Borax*, $Na_2B_4O_7 \cdot 10\,H_2O$, a hydrated borate, is the primary source of boron.

The first step in the production of boron and boron compounds is the conversion of borax to boric acid. Boric acid is a most unusual acid, as indicated by using the formula $B(OH)_3$, explained later in this section.

$$Na_2B_4O_7 \cdot 10\,H_2O + H_2SO_4 \longrightarrow 4\,B(OH)_3 + Na_2SO_4 + 5\,H_2O$$

Following this, the compound boric oxide, B_2O_3, is prepared by heating $B(OH)_3$ and thereby causing it to dehydrate:

$$2\,B(OH)_3(s) \xrightarrow{\Delta} B_2O_3(s) + 3\,H_2O(g)$$

Elemental boron and boron compounds are prepared from B_2O_3, as illustrated in Example 21.1.

Boric oxide, B_2O_3, reacts with water to form boric acid:

$$B_2O_3(s) + 3\,H_2O(l) \longrightarrow 2\,B(OH)_3(s)$$

We write the formula of boric acid $B(OH)_3$ rather than H_3BO_3 because boric acid is an extremely weak *monoprotic* acid, not triprotic as the formula H_3BO_3 would indicate. Further, the way this compound displays its acidity is different from the way most other weak acids do. Instead of donating a proton, boric acid accepts a hydroxide ion, forming the complex ion $[B(OH)_4]^-$. We can think of a hydroxide ion produced in the self-ionization of water as attaching itself to the B atom of $B(OH)_3$, an electron-deficient structure, through a coordinate covalent bond:

$$OH^- \;+\; \begin{matrix} OH \\ | \\ B-OH \\ | \\ OH \end{matrix} \;\rightleftharpoons\; \left[\, HO-\begin{matrix} OH \\ | \\ B \\ | \\ OH \end{matrix}-OH \,\right]^-$$

The net reaction for the ionization is

$$B(OH)_3(aq) + 2\,H_2O(l) \rightleftharpoons H_3O^+(aq) + [B(OH)_4]^-(aq) \qquad K_a = 5.6 \times 10^{-10}$$

Boric acid is quite toxic if taken internally, but dilute solutions can be used externally as a mild antiseptic—for example, in eyewash solutions. Boric acid is also used as an insecticide against cockroaches and black carpet beetles.

Solutions of the salts of boric acid—borate solutions—are generally quite complex because they contain polymers of the borate anions as well as the simple anions $[B(OH)_4]^-$, BO_3^{3-}, and BO_4^{5-}. Borates hydrolyze in water to give basic solutions, and this is why borax is used in some cleaning agents. Sodium perborate, $NaBO_3 \cdot 4\,H_2O$, crystallizes from an aqueous solution of hydrogen peroxide and borax. It is used in denture cleansers and as a color-safe "oxygen" bleach for clothes that would be harmed by a "chlorine" bleach (sodium hypochlorite). The bleaching action is actually that of H_2O_2 released in the hydrolysis of sodium perborate.

The oxidation number of boron in $NaBO_3 \cdot 4\,H_2O$ is *not* +5, as this empirical formula suggests. A more representative formula is $Na_2[B_2(O_2)_2(OH)_4] \cdot 6\,H_2O$, in which the oxidation number for the B is +3 and "O_2" signifies a peroxide linkage. The structure of the perborate ion is

$$\left[\begin{array}{c} HO \\ HO \end{array} B \begin{array}{c} O-O \\ O-O \end{array} B \begin{array}{c} OH \\ OH \end{array} \right]^{2-}$$

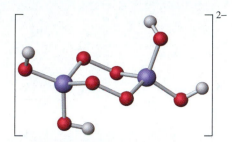

▲ Perborate ion.

Example 21.1

Write a chemical equation to represent **(a)** the high-temperature reduction of B_2O_3 to elemental boron with magnesium as a reducing agent and **(b)** the preparation of boron trichloride by the heating of boric oxide with carbon and chlorine gas. (The carbon is oxidized to carbon monoxide.)

SOLUTION

In order to write an equation for a chemical reaction, we need a complete description of the initial reactants, the final products, and the reaction conditions. Most of this information is provided in the statement of the problem.

(a) $$B_2O_3(s) + 3\,Mg(s) \xrightarrow{\Delta} 2\,B(s) + 3\,MgO(s)$$

(b) $$B_2O_3(s) + 3\,C(s) + 3\,Cl_2(g) \xrightarrow{\Delta} 2\,BCl_3(g) + 3\,CO(g)$$

EXERCISE 21.1A

Write an equation to represent the preparation of pure boron by the reduction of $BCl_3(g)$ with hydrogen gas.

EXERCISE 21.1B

Write an equation for the hydrolysis of perborate ion to form $H_2O_2(aq)$.

21.3 Aluminum

The most important metal of group 3A is aluminum.* Pure aluminum is a malleable, ductile, silvery metal with a density only about one-third that of steel. Aluminum metal is not very strong, but its strength increases when it is alloyed with Cu, Mg, or Si. More than 5 million tons of the metal are produced per year in the United States; most of it is used in lightweight alloys. In this section, we consider the commercial production of the metal, its properties and uses, and some of its important compounds.

Production of Aluminum Earth's crust is 8.3% by mass aluminum. This makes aluminum the third most abundant element and the most abundant metal. Aluminum metal was not isolated until 1825, when Hans Oersted produced it in an impure form. For the next several decades, it remained a semiprecious metal used mostly in jewelry and artwork. It was still rare and expensive in 1884 when an aluminum cap was placed atop the newly completed Washington Monument. Just a couple of years later, however, the situation changed completely. Charles Martin Hall, in the United States, and Paul Heroult, in France, discovered an inexpensive way to make aluminum by electrolysis.

The Hall–Heroult process has two key features. The first is that the process uses *bauxite*, a mixture of hydrates of aluminum oxide, as the source of aluminum and takes advantage of amphoterism to remove the principal impurity, Fe_2O_3, from the bauxite. Aluminum oxide is amphoteric, but Fe_2O_3 is basic rather than amphoteric. Thus, when bauxite is treated with hot, concentrated NaOH(aq), only the Al_2O_3 reacts:

$$Al_2O_3(s) + 2\,OH^-(aq) + 3\,H_2O(l) \longrightarrow 2\,[Al(OH)_4]^-(aq)$$

$$Fe_2O_3(s) + OH^-(aq) \longrightarrow \text{no reaction}$$

*In much of the world, the element of atomic number 13, Al, is spelled alumin*i*um (pronounced al-you-MIN-ee-um). In the United States, though, it is usually spelled aluminum (pronounced a-LOO-min-um).

▲ Charles Martin Hall (1863–1914) was motivated by a professor at Oberlin College who remarked that anyone discovering a cheap method of producing aluminum would become rich and famous. Hall's discovery, in his home laboratory within eight months of his graduation, was the foundation of the aluminum industry in the United States.

▲ Paul Heroult (1863–1914), a student of Le Châtelier's, was, like Hall, 23 years old when he discovered the same method of producing aluminum. Heroult's discovery was the foundation of the aluminum industry in Europe.

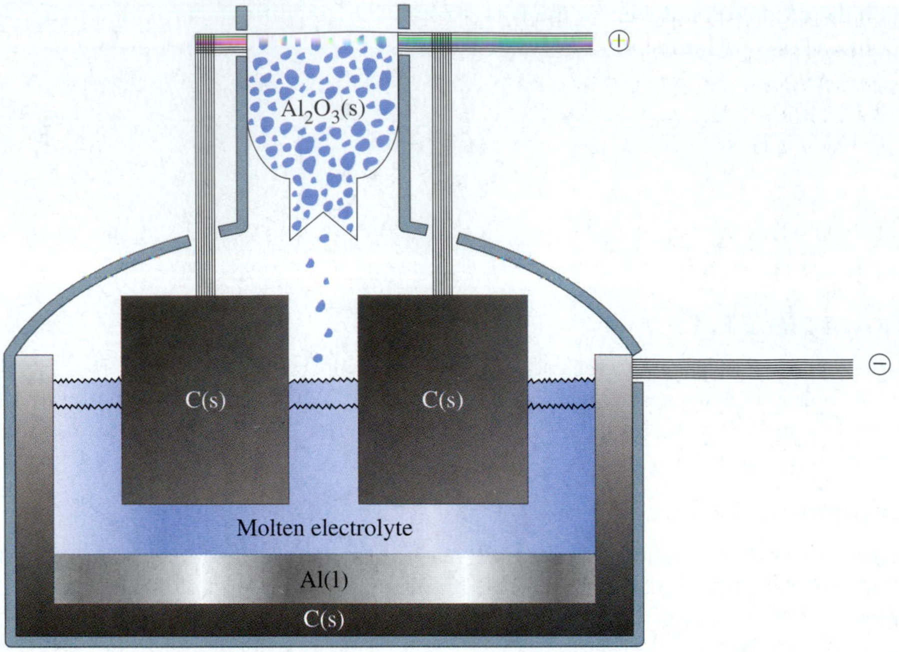

► **FIGURE 21.2** **Electrolysis cell for aluminum production**

The cathode is the carbon lining of a steel tank. The anodes are made of blocks of carbon. Liquid aluminum, which is denser than the molten Na_3AlF_6/Al_2O_3 electrolyte, collects at the bottom of the tank and is removed. Fresh aluminum oxide is continuously added from a hopper above the cell.

Thermite Reaction movie

The solution containing $[Al(OH)_4]^-$ is separated from the solid residue, a so-called red mud (mostly Fe_2O_3), diluted with water, and slightly acidified, with the acidification causing $Al(OH)_3$ to precipitate:

$$[Al(OH)_4]^-(aq) + H_3O^+(aq) \longrightarrow Al(OH)_3(s) + 2\,H_2O(l)$$

The $Al(OH)_3$ is heated to about 1200 °C and decomposes to pure Al_2O_3:

$$2\,Al(OH)_3(s) \longrightarrow Al_2O_3(s) + 3\,H_2O(g)$$

The final step in producing aluminum metal is reduction of Al_2O_3 to Al metal and involves the second key feature of the Hall–Heroult process. The melting point of Al_2O_3 (2045 °C) is much too high to permit reduction by the electrolysis of molten Al_2O_3. Moreover, molten Al_2O_3 is not a particularly good electrical conductor. Hall and Heroult made a crucial innovation by using as the electrolyte in the electrolysis cell a solution of molten *cryolite*, Na_3AlF_6, containing a few percent dissolved Al_2O_3. The mixture is a good electrical conductor and remains liquid at temperatures of about 950 °C, much lower than the melting point of Al_2O_3.

The electrolysis cell pictured in Figure 21.2 uses a carbon-lined steel cell as the cathode and large chunks of carbon as the anodes. The $Al_2O_3(s)$ is continuously added to the cell as liquid aluminum metal is drawn off. The electrode reactions are complex, but the overall electrolysis reaction is

Reduction:	$4\,\{Al^{3+} + 3\,e^- \longrightarrow Al(l)\}$
Oxidation:	$3\,\{C(s) + 2\,O^{2-} \longrightarrow CO_2(g) + 4\,e^-\}$
Overall:	$3\,C(s) + 4\,Al^{3+} + 6\,O^{2-} \longrightarrow 4\,Al(l) + 3\,CO_2(g)$

The production of aluminum requires a great deal of energy, about 15,000 kWh per ton of Al. To produce a ton of steel requires only about one-fifth this much energy. On the other hand, because the density of aluminum (2.70 g/cm^3) is much less than that of iron (7.87 g/cm^3), a smaller mass of aluminum may be required for an application in which aluminum replaces iron. Moreover, because it takes only about one-twentieth as much energy to recycle aluminum as it does to produce it from bauxite, much of the aluminum produced in the United States is recycled.

Properties and Uses of Aluminum The reduction of $Al^{3+}(aq)$ to $Al(s)$ occurs only with difficulty, as the $E°$ value indicates:

$$Al^{3+}(aq) + 3\,e^- \longrightarrow Al(s) \qquad E° = -1.676\ V$$

From this, we deduce that the reverse process, the oxidation of Al(s), occurs readily. Thus, Al is a good reducing agent. One interesting reaction based on the reducing power of aluminum is the highly exothermic *thermite* reaction (Figure 4.10):

$$Fe_2O_3(s) + 2\,Al(s) \longrightarrow 2\,Fe(l) + Al_2O_3(s)$$

The liquid iron produced in the reaction can be used to weld large iron objects.

As we noted in Section 18.8, the oxide that readily forms on aluminum is a thin, impervious film that protects the underlying metal from corrosion. The oxide film can be made thicker by making aluminum the anode in an electrolytic cell with dilute $H_2SO_4(aq)$ as the electrolyte. Equations for the half-reactions and overall reaction are

Anode: $\qquad 2\,Al(s) + 3\,H_2O(l) \longrightarrow Al_2O_3(s) + 6\,H^+ + 6\,e^-$

Cathode: $\quad 3 \times \{2\,H^+(aq) + 2\,e^- \longrightarrow H_2(g)\}$

Overall: $\qquad 2\,Al(s) + 3\,H_2O(l) \longrightarrow Al_2O_3(s) + 3\,H_2(g)$

This *anodized aluminum* can be colored by adding dyes to the electrolyte solution. Bronze and brown anodized aluminum are especially popular in modern buildings and as window frames in homes.

Aluminum is an active metal and readily reacts with acids to produce hydrogen gas:

$$2\,Al(s) + 6\,H^+(aq) \longrightarrow 2\,Al^{3+}(aq) + 3\,H_2(g)$$

However, aluminum also dissolves in basic solutions. We can think of the reaction as a two-step process. First, the Al_2O_3 film on the metal dissolves in $OH^-(aq)$, just as we described for the purification of bauxite:

$$Al_2O_3(s) + 3\,H_2O(l) + 2\,OH^-(aq) \longrightarrow 2[Al(OH)_4]^-(aq)$$

Once the oxide film is removed from the aluminum, the metal is then free to display its true reactivity. It reacts with water in the alkaline solution, displacing hydrogen gas:

$$2\,Al(s) + 6\,H_2O(l) + 2\,OH^-(aq) \longrightarrow 2\,[Al(OH)_4]^-(aq) + 3\,H_2(g)$$

$$E^{\circ}_{cell} = 1.482\ V$$

Some drain cleaners make use of this reaction. They consist of a mixture of solid sodium hydroxide and granules of aluminum metal. When added to water, the heat of solution of the NaOH(s) and the heat given off by the above reaction help to melt fat and grease. Agitation resulting from the formation of the hydrogen gas helps dislodge obstructions and unplug the drain. (Often, though, a plumber's "snake" is more effective than chemical drain cleaners.)

Because its combustion is a highly exothermic reaction, powdered aluminum is used as a component in rocket fuels, explosives, and fireworks.

$$2\,Al(s) + \tfrac{3}{2}\,O_2(g) \longrightarrow Al_2O_3(s) \qquad \Delta H^{\circ} = -1676\ kJ$$

Perhaps most familiar is the use of aluminum in beverage cans, cookware, and wrapping foil. Most aluminum, though, is used in structural materials, usually alloyed with other metals to impart greater strength. Most modern aircraft use aluminum alloys, as do some automobile engines and the exterior trim of modern buildings. Because aluminum has good electrical conductivity (about 63.5% that of an equal volume of copper) and a low density, it is also widely used in the electrical industry. Most high-power transmission lines in the United States are now made of aluminum alloys.

Aluminum Compounds Among the aluminum halides, AlF_3 differs considerably from the others. As an example, let us compare AlF_3 and $AlCl_3$. Although both have a crystal structure based on Al atoms surrounded by six halogen atoms, AlF_3 seems to have the properties normally associated with ionic substances, but $AlCl_3$ does not. AlF_3 has a high melting point (1090 °C) and is therefore nonvolatile. It is very slightly soluble in water and a good electrical conductor in the liquid state. In contrast, solid $AlCl_3$ sublimes readily; is soluble in water and also in some organic solvents, such as ethanol, diethyl ether, and carbon tetrachloride; has a low melting point (193 °C); and is a poor electrical conductor in the liquid state.

Application Note

When the Allies in World War II invaded the Normandy beaches of German-occupied France in June 1944, American Rangers scaled the cliffs overlooking the shore and used thermite grenades to weld the gears of the German heavy guns. The guns, then, could not be aimed down at the Allied troops on the beaches. Their action, taken while the German gunners were at morning formation in a nearby field, probably saved the lives of many Allied soldiers.

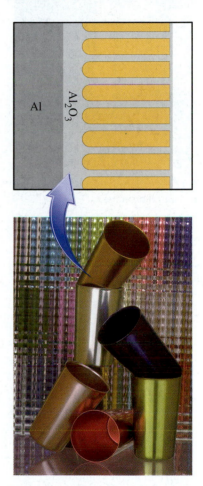

▲ Anodized aluminum. The pores in the electrochemically deposited layer of Al_2O_3 are filled with dye, then sealed. The surface is hard, and the color is permanent.

 Formation of Aluminum Bromide movie

Aluminum Chloride Dimer 3D model

▶ **FIGURE 21.3** **Bonding in Al₂Cl₆**

QUESTION: How does the bonding in this compound differ from that in diborane, a compound with a similar structure?

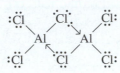

Lewis structure

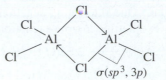

Bonding scheme

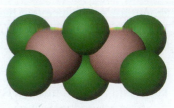

Space-filling model

▲ The reaction between aluminum and liquid bromine.

Application Note

Fatty alcohols such as 1-hexadecanol are used to make specialty detergents for toothpastes and shampoos.

▲ Iron and titanium impurities in aluminum oxide produce blue sapphires.

B	C	N	O	F	Ne
Al	Si	P	S	Cl	Ar
Ga	Ge	As	Se	Br	Kr
In	Sn	Sb	Te	I	Xe
Tl	Pb	Bi	Po	At	Rn

In the liquid and gaseous states, aluminum chloride exists as dimers of $AlCl_3$, that is, as Al_2Cl_6 molecules. In Figure 21.3, the Lewis structure shows that two of the Cl atoms in Al_2Cl_6 are bonded to both Al atoms, forming bridges between the two $AlCl_3$ units. These two bonds between Cl and Al are coordinate covalent bonds in which the Cl atoms supply both electrons. The bonding scheme (middle structure in Figure 21.3) shows that the orbital overlaps in the molecules involve sp^3 hybrid orbitals of Al and $3p$ orbitals of Cl. The chlorine-atom bridges between the two $AlCl_3$ units are also seen in the space-filling model.

Aluminum chloride is an important catalyst in organic chemistry (Chapter 23), where its electron-deficient structure enables it to form complexes with most oxygen-containing organic compounds and other species having lone-pair electrons.

Lithium aluminum hydride, $LiAlH_4$, is used as a reducing agent in organic chemistry. We can think of hydride ion ($H:^-$) as a *pseudohalide* ion. (In fact, some periodic tables place hydrogen at the top of group 7A, as well as at the top of group 1A.) The AlH_4^- ion can be considered an adduct of H^- and the electron-deficient AlH_3 molecule. An example of the use of $LiAlH_4$ in organic chemistry is the reduction of carboxylic acids to alcohols. For palmitic acid, the reduction is*

$$CH_3(CH_2)_{14}COOH \xrightarrow{\text{LiAlH}_4} CH_3(CH_2)_{14}CH_2OH$$

Palmitic acid   1-Hexadecanol
(from palm oil)

Aluminum sulfate is the most important industrial aluminum compound. It is prepared by the action of hot, concentrated $H_2SO_4(aq)$ on $Al_2O_3(s)$. The product that crystallizes from solution is $Al_2(SO_4)_3 \cdot 18\ H_2O$. About half of the 1 million tons of aluminum sulfate produced annually in the United States is used in water treatment. In this application, the pH of the water is raised to the point at which $Al(OH)_3$ precipitates:

$$Al^{3+}(aq) + 3\ OH^-(aq) \longrightarrow Al(OH)_3(s)$$

As it settles, the gelatinous $Al(OH)_3$ traps and removes suspended solids from the water.

We have described the use of aluminum oxide—also known as *alumina*—in the manufacture of aluminum. Another important use is in refractory materials for high-temperature furnaces; the melting point of Al_2O_3 is 2045 °C. Aluminum oxide is also used in the manufacture of ceramic materials (page 874). The mineral *corundum* is a pure form of Al_2O_3, while *emery* is corundum contaminated with iron oxides (Fe_2O_3 and/or Fe_3O_4) and silica (SiO_2). Both of these materials have a Mohs hardness value of 9 (with diamond being 10, the highest value on the scale), and both are used in the manufacture of abrasive materials, such as grinding wheels and sandpaper. Many gemstones are naturally occurring impure forms of aluminum oxide.

Group 4A

The members of group 4A all have the valence-shell electron configuration ns^2np^2. Carbon, the first member, is a nonmetal. Carbon's four valence electrons form four covalent bonds in nearly all carbon compounds. The next two members, silicon and germanium, also form covalent bonds for the most part. Both are metalloids and have

---

*The mechanism of this reaction is quite complex. It proceeds through four steps, and the final mixture must be acidified to obtain the alcohol. In the course of the reaction, $LiAlH_4$ is decomposed to Li^+, Al^{3+}, and $H_2(g)$.

Gemstones: Natural and Artificial

Gemstones are rare in nature. They must contain just the right combination of substances, brought together under the appropriate conditions of temperature, pressure, and concentrations in order to form the proper crystals. One class of gemstones has aluminum oxide (corundum) as the principal constituent and tiny amounts of transition metal oxides as impurities. The particular impurities present determine the colors of the gem (Table 21.2). Corundum gemstones that are pink to dark red are *rubies;* those of all other colors are *sapphires.*

Table 21.2 Common Gemstones Based on Aluminum Oxide

Gem	Impurity
Ruby	Cr
Sapphires	
white	none
orange	Ni, Cr
yellow	Ni
green	Co, V, and/or Ni
blue	Fe, Ti
star	Ti

◀ Manufacture of an artificial ruby.

Scientists have learned how to mimic nature by synthesizing rubies and sapphires in high-temperature furnaces. A mixed powder of Al_2O_3 and the appropriate transition metal oxide(s) is sprayed into the upper region of the furnace. The powder melts in the hottest regions of the furnace, forming a liquid layer that solidifies. The solidified material is gradually withdrawn from the furnace as layer after layer is added to it. Gemstones produced in this way have many industrial applications. The ruby is rare in nature and one of the most valuable of all gemstones. Rubies can now be manufactured for use in jewelry, but good-quality natural rubies are still more valuable than synthetic ones. The natural gems are identified by their slight imperfections; the synthetic ones—as unbelievable as it might seem—are too near to perfection to pass as "natural."

◀ The large rock contains natural rubies, and the small faceted stone is a synthetic ruby.

interesting properties as semiconductors (Section 24.8). Tin and lead are more metallic in their behavior. Both form 2+ and 4+ ions.

21.4 Carbon

The ground-state electron configuration of carbon is $1s^2 2s^2 2p^2$. As we noted in Sections 10.4 and 10.5, there are three ways to hybridize the valence-shell orbitals of a carbon atom: (1) four sp^3 orbitals, as in ethane, C_2H_6, (2) three sp^2 orbitals plus one p orbital, as in ethene (ethylene), C_2H_4, or (3) two sp orbitals plus two p orbitals, as in ethyne (acetylene), C_2H_2. These hybridization possibilities permit the formation of carbon chains and rings having multiple bonds between C atoms, as well as the more common single bonds. In these structures, C atoms also bond to H, O, N, S, halogens, and several other types of atoms. Taken together, these factors lead to a myriad of organic compounds. We have frequently referred to organic compounds since

▲ This jet aircraft is made largely of graphite-fiber composites.

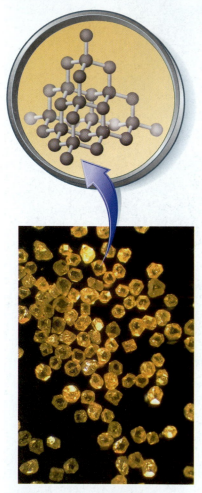

▲ **FIGURE 21.4   Synthetic diamonds**

Synthetic diamonds are made from graphite. Graphite is the more stable form of carbon at room temperature and atmospheric pressure and remains so up to about 3000 °C and 10^4 atm. The conversion of graphite to diamond requires a temperature of 1000 °C to 2000 °C and a pressure of at least 10^5 atm.

Diamond, Graphite, and Buckminsterfullerene 3D models

introducing them in Chapter 2, and we will consider them in more detail in Chapter 23. Our emphasis here will be more on the inorganic chemistry of carbon.

As noted in Section 11.8, elemental carbon exists in nature mainly as the two allotropes diamond and graphite, although two other allotropic forms—fullerenes and nanotubes—are of great interest. Graphite is perhaps most familiar as the writing material in pencils. Because it conducts electric current, graphite is used for electrodes in batteries and industrial electrolysis. It can also withstand high temperatures, leading to its use in foundry molds, furnaces, and other high-temperature devices.

Graphite is also used in the form of fibers. When carbon-based fibers such as rayon are heated to a high temperature, other elements are driven off in gaseous products, leaving behind graphite fibers. These fibers are embedded in plastic materials to make high-strength, lightweight composites, which are used in products as diverse as tennis rackets, canoes, fishing rods, and airplanes.

Only a small proportion of natural diamonds are suitable for gemstones. Most natural diamonds and all synthetic diamonds (Figure 21.4) are used in industry. Because they have a high thermal conductivity (they dissipate heat quickly) and are extremely hard, diamonds are used as abrasives, in drill bits and saws, and in oil-well drilling.

Carbon also exists in amorphous forms, such as coke, charcoal, and carbon black. As previously noted on page 461, *amorphous* implies a noncrystalline solid, one that does not have the long-range order found in crystalline solids. However, amorphous carbon does appear to have some short-range structure. When coal is heated in the absence of air, volatile substances are driven off, leaving a high-carbon residue called *coke*. Coke is the principal metallurgical reducing agent; it is used in the reduction of iron oxide to iron in a blast furnace, for example. A similar *destructive distillation* of wood produces charcoal. As you may have observed in an improperly adjusted Bunsen burner, incomplete combustion of natural gas produces a smoky flame. This smoke can be deposited as a powdery soot called *carbon black*. Carbon black is used as a filler in rubber tires, as a pigment in printing inks, and as the transfer material in carbon paper, copy machines, and laser printers.

Activated carbon is formed by heating carbon black to 800–1000 °C in the presence of steam to expel all volatile matter. This form of carbon is highly porous, like sponges or honeycombs. Because of its high ratio of surface area to volume, activated carbon has a great capacity to adsorb substances from liquids and gases. Activated carbon is used in gas masks to adsorb poisonous gases, in water filters to remove organic contaminants, in sugar processing to remove colored impurities, in air-conditioning systems to control odors, and in industrial plants for the control and recovery of vapors.

Inorganic Carbon Compounds

We have encountered carbon monoxide and carbon dioxide, the two principal oxides of carbon, many times throughout the text. In Table 21.3, we summarize some sources and uses of these gases. In Chapter 25, we will consider some environmental issues concerning CO and CO_2.

Carbon combines with most metals to form compounds called carbides. With active metals, the carbides* are ionic. For example, calcium carbide is formed in the high-temperature reaction of lime and coke:

$$CaO(s) + 3\ C(s) \xrightarrow{2000\ °C} CaC_2(s) + CO(g)$$

Calcium carbide, an ionic compound with a carbon-to-carbon triple bond in the C_2^{2-} anion, reacts with water to produce acetylene, $H-C \equiv C-H$:

$$CaC_2(s) + 2\ H_2O(l) \longrightarrow Ca(OH)_2(s) + C_2H_2(g)$$
$$\text{Acetylene}$$

*The "carbides" of active metals are more properly called acetylides; CaC_2 is calcium acetylide, Na_2C_2 is sodium acetylide, and so on. (See Problems 43 and 44.)

Table 21.3 Sources and Uses of Oxides of Carbon

Source	Uses
Carbon Monoxide, CO	
Incomplete combustion of hydrocarbons: $$2\,CH_4(g) + 3\,O_2(g) \longrightarrow 2\,CO(g) + 4\,H_2O(l)$$	Manufacture of methanol and other organic compounds from synthesis gas (CO/H_2 mixture)
Steam reforming of natural gas: $$CH_4(g) + H_2O(g) \longrightarrow CO(g) + 3\,H_2(g)$$	Metallurgical reducing agent, as in the blast furnace reaction: $$Fe_2O_3(s) + 3\,CO(g) \longrightarrow 2\,Fe(l) + 3\,CO_2(g)$$
Carbon Dioxide, CO_2	
Complete combustion of hydrocarbons: $$CH_4(g) + 2\,O_2(g) \longrightarrow CO_2(g) + 2\,H_2O(l)$$	Refrigerant in freezing, storage, and transport of foods
Decomposition (calcination) of limestone at about 900 °C: $$CaCO_3(s) \longrightarrow CaO(s) + CO_2(g)$$	Production of carbonated beverages
	Petroleum recovery in oil fields
Fermentation by-product in the production of ethanol	Fire extinguisher systems

Calcium carbide is convenient to use. It can be transported as a solid, and the gaseous fuel acetylene can be generated simply by adding water to the solid. Silicon carbide, SiC, is a network covalent solid similar to and nearly as hard as diamond. It is highly resistant to thermal shock and is used as an abrasive and in crucibles for molten metal.

Two other binary compounds of carbon are carbon disulfide, CS_2, and carbon tetrachloride, CCl_4. Carbon disulfide is prepared by the reaction of methane with sulfur vapor in the presence of a catalyst:

$$CH_4(g) + 4\,S(g) \longrightarrow CS_2(l) + 2\,H_2S(g)$$

Carbon disulfide is a flammable, volatile liquid that dissolves sulfur, phosphorus, bromine, iodine, fats, and oils, but its toxicity limits its uses as a solvent. It is an important intermediate in the manufacture of rayon and cellophane (Section 24.9).

Carbon tetrachloride can be prepared by the direct chlorination of methane:

$$CH_4(g) + 4\,Cl_2(g) \longrightarrow CCl_4(l) + 4\,HCl(g)$$

Once extensively used as a solvent, dry-cleaning agent, and fire extinguisher, CCl_4 is declining in importance because it causes liver and kidney damage and is a suspected carcinogen.

Cyanide ion, CN^-, is similar to halide ions, X^-, in several ways: It forms an insoluble silver salt, AgCN, and an acid, HCN, though this acid (hydrocyanic acid) is quite weak. Cyanide ions differ from halide ions in that the former are quite toxic. Despite its toxicity, HCN, a liquid that boils at about room temperature, is widely used in the manufacture of plastics. It is also used—carefully, by well-trained personnel—as a fumigant to kill rodents and insects on ships.

Just as two Cl atoms combine to form the molecule Cl_2, two CN groups can combine to form the *cyanogen* molecule, $(CN)_2$ or C_2N_2. Like chlorine, cyanogen disproportionates in basic solution:

$$(CN)_2(g) + 2\,OH^-(aq) \longrightarrow CN^-(aq) + OCN^-(aq) + H_2O(l)$$

$$Cl_2(g) + 2\,OH^-(aq) \longrightarrow Cl^-(aq) + OCl^-(aq) + H_2O(l)$$

Cyanogen is used as a reagent in organic synthesis and as a fumigant. The cyanogen–oxygen flame is one of the hottest known, producing temperatures of 4500 °C.

▲ This lamp for spelunking (cave exploring) uses calcium carbide to generate acetylene, which burns with a bright white flame for illumination.

High-purity silicon wafers are cut and etched to produce the integrated circuits used in computers and many other electronic devices.

Ceramic insulators of the type commonly seen in electric power substations.

21.5   Silicon

We have seen important differences between the second- and third-period members of groups 1A (Li/Na), 2A (Be/Mg), and 3A (B, Al). In group 4A, the difference is perhaps greatest of all. The most distinctive feature of C atoms—their tendency to bond together into chain and ring structures—is much less significant in Si atoms. The Si–Si bond (bond energy = 226 kJ/mol) and Si–H bond (318 kJ/mol) are weaker than the corresponding C–C (347 kJ/mol) and C–H (414 kJ/mol) bonds. These differences in bond energies are probably not the primary reason Si chains and rings lack the stability of their carbon counterparts. Instead, the activation energies for reactions involving silicon chain and ring compounds are much lower than those of the corresponding carbon compounds; the Si reaction rates are also correspondingly greater. Thus, disilane, Si_2H_6, spontaneously ignites on contact with oxygen (producing SiO_2 and H_2O), whereas the ignition of ethane, C_2H_6, requires a spark or open flame. In any event, strong Si–O bonds (464 kJ/mol) favor silicates as the predominant naturally occurring compounds of silicon. Thus, silicon is the key element of the mineral world just as carbon is the key element of the living world.

A silicon atom, like a carbon atom, forms four bonds in almost all cases. In elemental silicon, each atom uses its four valence electrons ($3s^2 3p^2$) in an sp^3 hybridization scheme. Silicon crystallizes in a cubic network covalent structure similar to that in diamond. There is no allotrope of silicon equivalent to graphite, because the side-by-side overlap of $3p$ orbitals is too limited for π bonding.

Silica, SiO_2

Unlike carbon, silicon forms few multiple covalent bonds. The sidewise overlap of a $3p$ orbital of a Si atom with a $2p$ orbital of an O atom, for example, is too limited to form a strong bond. Considerably less energy is released in forming two Si=O double bonds than in forming four Si–O single bonds. As a result, SiO_2 is not made up of discrete molecules (as is CO_2); instead, it is a network covalent solid (Figure 21.5). Quartz, a form of pure silica, is quite hard (a hardness of 7 on the Mohs scale), has a high melting point (about 1700 °C), and is a nonconductor of electricity. Silica is the basic raw material of the glass, ceramics, and refractory materials industries.

Ceramics

A **ceramic** is an inorganic solid generally produced at high temperatures and characterized by such physical properties as hardness, brittleness, stability at high temperatures, and high melting point. Some ceramics are crystalline materials, and others are amorphous solids. The range of ceramic products is broad and includes structural clay products (bricks and tiles), whiteware (dinnerware and porcelain items), abrasives, refractories (furnace linings), and glass. Silica and silicates are common constituents of ceramics, and glass is the most widely produced ceramic product.

▶ **FIGURE 21.5   The structure of silica, SiO_2**

Each Si atom forms bonds to four O atoms, and each O atom forms bonds to two Si atoms in this three-dimensional network covalent structure.

QUESTION: What are the hybridizations of the Si and O atoms and the resultant bond angles in this structure?

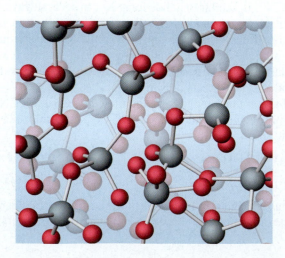

Certain molten materials become viscous when they cool; eventually they cease to flow entirely. Such materials are called *glass*. Unlike a crystalline solid, which has a definite melting point, glass, an amorphous solid, softens and becomes liquid over a range of temperatures.

When a mixture of sodium carbonate, calcium carbonate, and sand (silica) is heated to about 1500 °C, the molten product is a water-insoluble mixture of sodium and calcium silicates called *soda–lime glass*. Because this glass expands and contracts with changes in temperature, it is subject to shattering when its temperature changes rapidly. When the calcium carbonate is replaced by boric oxide, however, a *borosilicate glass* is formed. This glass has a low thermal expansion and is not as subject to thermal shock as is soda–lime glass. Borosilicate glass, perhaps best known by the trade name Pyrex®, is extensively used for laboratory glassware and for ovenware.

Some new ceramic materials have specially designed electrical, magnetic, or optical properties. There are also new methods of preparing ceramic materials. In the *sol–gel* process, the particle size of solid particles is carefully controlled by forming them as a colloidal dispersion (sol). The sol is then converted to a rigid gel, which is the starting point of the fabrication process. Some exceptionally lightweight ceramic materials can be produced by this method. Ceramics with desirable high-temperature mechanical and structural properties are used for gas turbines. Ceramics are being developed for use in automobile engines. In Chapter 24, we will consider superconductors, which are ceramic materials with unusual electrical properties. The promise of ceramic materials has caused some people to speak of our times as being the dawn of a "new stone age."

Silicate Minerals

Silicon is the second most abundant element (after oxygen) in Earth's crust, accounting for 27.2% of its mass. Silicon occurs primarily as silica and silicates. Silicate anions are often quite complex, but most have as a basic structural unit a tetrahedron with a Si atom at the center and O atoms at the four corners (Figure 21.6). The oxidation number of silicon is +4 and that of oxygen is −2. In silicate minerals, SiO_4 tetrahedra are arranged in a variety of ways, leading to a host of mineral forms. Some of the possible arrangements are

- Simple SiO_4^{4-} anions. Typical minerals containing this anion are *thorite* ($ThSiO_4$) and *zircon* ($ZrSiO_4$).

- Combinations of two SiO_4 tetrahedra. In the anion $Si_2O_7^{6-}$, the Si atoms of two tetrahedra share an O atom between them. A typical mineral of this type is the scandium-containing ore *thortveitite* ($Sc_2Si_2O_7$).

- Long chains of SiO_4 tetrahedra in which each Si atom shares two O atoms with Si atoms in adjacent tetrahedra. A typical mineral having this structure is *spodumene*, which has the empirical formula $LiAl(SiO_3)_2$; it is the principal natural source of lithium and lithium compounds.

- Double chains of SiO_4 tetrahedra. Half the tetrahedra share three O atoms, and the other half share two. In *chrysotile asbestos*, the double chains are bound together by cations, chiefly Mg^{2+}. The mineral has a fibrous appearance, and the empirical formula is $Mg_3(Si_2O_5)(OH)_4$.

- Two-dimensional sheets of SiO_4 tetrahedra. Each tetrahedron shares three of its four O atoms with adjacent tetrahedra, leading to two-dimensional sheets, as suggested by Figure 21.7. In *muscovite mica*, the counterions to the silicate anions are mostly K^+ and Al^{3+}; its empirical formula is $KAl_2(AlSi_3O_{10})(OH)_2$. Because bonding within sheets is stronger than between sheets, mica flakes easily. Vermiculite, used as loose-fill insulation, has sheets of mica separated by double water layers.

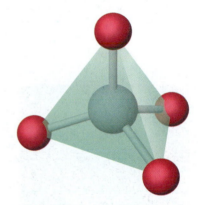

▲ **FIGURE 21.6   The basic SiO_4^{4-} silicate tetrahedron**

The Si atom is at the center of the tetrahedron, and an O atom is at each of the four corners. The green-shaded tetrahedron is the shape typically used to represent SiO_4^{4-} in a mineral structure.

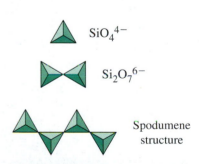

SiO_4^{4-}

$Si_2O_7^{6-}$

Spodumene structure

▲ Muscovite mica separates easily into thin sheets.

▶ **FIGURE 21.7   A two-dimensional sheet in the structure of mica**
Three of the four O atoms in each tetrahedron are shared with another tetrahedron in a two-dimensional array of tetrahedra. The cations found between the planes of silicate anions are mostly K^+ and Al^{3+}.

▲ Petrified wood is largely quartz.

- A three-dimensional array of SiO_4 tetrahedra. In these structures, each tetrahedron shares all four of its O atoms with adjacent tetrahedra. This is the most common arrangement in silicate minerals in Earth's crust. It is the arrangement found in quartz and in such mineral forms as

  - *amethyst:* a purple quartz containing iron as an impurity.
  - *agate:* an impure silica that crystallizes in colored bands.
  - *petrified wood:* very old wood in which colored agate replaces organic matter but the microscopic structure of the wood is retained.

Organosilicon Compounds

So far we have emphasized the inorganic chemistry of silicon, but there is also an organic silicon chemistry that emulates carbon-based organic chemistry. However, organic silicon chemistry is not nearly as rich as that of carbon. Silicon can form Si–Si bonds in chains of up to about a dozen Si atoms:

$$
\begin{array}{ccc}
\underset{\text{Silane}}{\text{H}-\underset{\overset{|}{\text{H}}}{\overset{\overset{\text{H}}{|}}{\text{Si}}}-\text{H}} &
\underset{\text{Disilane}}{\text{H}-\underset{\overset{|}{\text{H}}}{\overset{\overset{\text{H}}{|}}{\text{Si}}}-\underset{\overset{|}{\text{H}}}{\overset{\overset{\text{H}}{|}}{\text{Si}}}-\text{H}} &
\underset{\text{Trisilane}}{\text{H}-\underset{\overset{|}{\text{H}}}{\overset{\overset{\text{H}}{|}}{\text{Si}}}-\underset{\overset{|}{\text{H}}}{\overset{\overset{\text{H}}{|}}{\text{Si}}}-\underset{\overset{|}{\text{H}}}{\overset{\overset{\text{H}}{|}}{\text{Si}}}-\text{H} \cdots}
\end{array}
$$

The *silanes* are thermally unstable. When heated, silanes containing larger numbers of silicon atoms decompose either to silanes containing fewer silicon atoms or to the elements. Like the hydrocarbons, the silanes are combustible; the combustion products are $SiO_2(s)$ and H_2O. As we have noted, however, unlike hydrocarbons, the silanes burst into flame in air. In this regard, silanes are like boranes (Section 21.2), a fact that illustrates the diagonal relationship of the two elements.

Other atoms can be rather easily substituted for H atoms in silanes. For example, if we substitute Cl for H in SiH_4, we obtain successively SiH_3Cl, SiH_2Cl_2, $SiHCl_3$, and $SiCl_4$. The reaction of $(CH_3)_2SiCl_2$ with water produces dimethylsilanol, $(CH_3)_2Si(OH)_2$, a starting material for the production of a class of polymers called *silicones*:

$$(CH_3)_2SiCl_2 + 2\,H_2O \longrightarrow (CH_3)_2Si(OH)_2 + 2\,HCl$$

Dimethylsilanol

Lead Poisoning

The Latin word for lead, *plumbum*, reflects the longtime use of this element in plumbing. Pipes made from this easily forged metal have been used to carry water from the time of the Romans right up to present times. Some older homes still have lead plumbing systems, and many more have copper pipes fitted with lead solder. However, it was not until well into the twentieth century that drinking water was identified as a potential source of lead poisoning. Now the U.S. Environmental Protection Agency requires municipal water utilities to test for the lead content in water, setting a maximum permitted level of 0.015 mg/L (15 ppb).

In addition to contaminated drinking water, we are also exposed to lead through cooking and eating utensils, leaded crystal glass, and pottery glazes. In the past, the chief sources of lead contamination in the environment have been tetraethyllead—an antiknock additive for gasoline—and lead-based paints; in the United States, both of these sources have been phased out in the past 30 or so years.

Mild lead poisoning produces nervousness and mental depression. More severe cases can lead to permanent nerve, brain, and kidney damage, especially in children. Lead also interferes with the biochemical reactions that produce the iron-containing heme group in hemoglobin. Fortunately, average blood lead levels have dropped dramatically since leaded gasoline was phased out. Bismuth alloys are now being used to make lead-free solder for plumbing. However, lead contamination still persists in certain soils and in painted surfaces in some older homes.

▲ The ancient Romans drank wine from lead-containing metal goblets, which could cause lead poisoning. Some historians speculate that lead poisoning may have contributed to the decline and fall of the Roman Empire.

21.6 Tin and Lead

Tin and lead are quite similar to each other. Both are soft and malleable and have a melting point that is considered low for metals. For both, the standard electrode potential for the reduction $M^{2+}(aq) + 2 e^- \longrightarrow M(s)$ is slightly negative. Both are placed just above hydrogen in the activity series of metals (Figure 4.13).

Lead has only one solid form, but tin exists in two allotropic forms. The α (gray), or nonmetallic, form is stable below 13 °C, and the β (white), or metallic, form is stable above 13 °C. When held below 13 °C for a long time, white tin changes to gray tin. The tin expands and crumbles to a powder. This transformation, called tin disease, has led to the disintegration of organ pipes, buttons, medals, and other objects made of tin that are kept below 13 °C.

The main use of tin metal is in the plating of steel for use in cans to store foods. The next most important use is in the manufacture of solders, which are low-melting-point alloys used to join wires or pieces of metal. Other important alloys of tin are bronze, a copper–tin alloy used for casting statues and ornamental work, and pewter, which is mostly tin, with small amounts of other elements.

By far the greatest use of lead worldwide is in lead-acid (storage) batteries. Other uses include the manufacture of solder and other alloys, lead shot, and radiation shields (to protect against X rays and γ rays).

A significant difference between tin and lead is related to the inert-pair effect—the +4 oxidation state is much more stable in tin than in lead. For example, when tin is heated in air, the product is $SnO_2(s)$, whereas the air oxidation of lead produces $PbO(s)$. Metallic tin reacts with $Cl_2(g)$ to form $SnCl_4$ (bp, 115 °C), a reaction that is used to recover tin from scrap tin plate. Lead reacts with $Cl_2(g)$ to form $PbCl_2$ (mp,

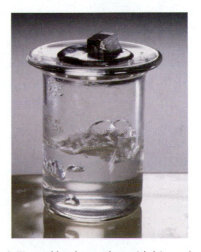

▲ Tin and lead, together with bismuth and cadmium, are common constituents of low-melting-point alloys. The alloy pictured here has a melting point below the boiling point of water.

Table 21.4 Common Compounds of Tin and Lead

Compound	Use(s)
$SnCl_2$	Reducing agent in the laboratory [reduces Fe(III) to Fe(II), Hg(II) to Hg(I), Cu(II) to Cu(I)]; tin plating; catalyst
SnF_2	Cavity preventive in toothpastes (stannous fluoride)
SnO_2	Jewelry abrasive, ceramic glazes, preparation of perfumes and cosmetics, catalyst
SnS_2	Pigment, imitation gilding (tin bronze)
$Pb(NO_3)_2$	Preparation of other lead compounds
PbO	Used in glass, ceramic glazes, and cements
PbO_2	Oxidizing agent, lead-acid battery electrodes, matches, and explosives
$PbCrO_4$	A yellow pigment (chrome yellow) for industrial paints, plastics, and ceramics

501 °C). Tin reacts with sulfur to form SnS, but the reaction can proceed further to produce SnS_2. Lead forms only the sulfide PbS. A few uses of tin and lead compounds are listed in Table 21.4.

Group 5A

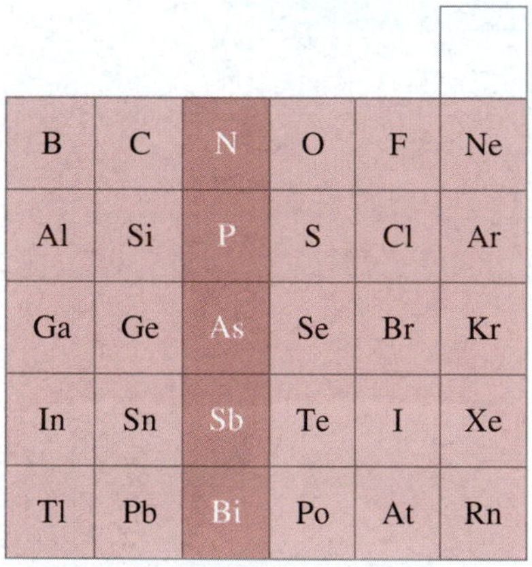

All the members of group 5A have the valence-shell electron configuration ns^2np^3, but their atoms can alter this electron configuration in different ways when forming compounds. In a few cases, a group 5A atom can gain three electrons to produce a 3− ion with the noble-gas electron configuration ns^2np^6. Far more often, though, group 5A atoms acquire this configuration by sharing three electrons. This is especially likely for the smaller atoms, N and P. The larger atoms—As, Sb, and Bi—can give up the p^3 electrons. This leads to an electron configuration of 18 electrons in the shell that is one in from the outermost shell and 2 electrons in the outermost shell. This 18 + 2 configuration is found in several ionic species derived from p-block metals of the fourth and higher periods in the periodic table. In still other cases, all five valence electrons are involved in compound formation, leading to an oxidation number of +5.

Among the group 5A elements, we note the usual decrease in first ionization energy and in electronegativity with increasing atomic number:

Element	N	P	As	Sb	Bi
I_1 (kJ/mol)	1402	1012	947	834	703
Electronegativity	3.0	2.1	2.0	1.9	1.9

These data suggest virtually no metallic character for nitrogen and a progressively increasing metallic character for the heavier members of group 5A. None of the group 5A elements is highly metallic, however. That is, none has a metallic character comparable to that of the group 1A and 2A elements.

21.7   Nitrogen

Nitrogen is found in greater abundance in the atmosphere than anywhere else. The mass percent of nitrogen in Earth's atmosphere is 76.8%, but in Earth's solid crust it is only 0.002%. There are only two important mineral sources of nitrogen: KNO_3 (niter, or saltpeter) and $NaNO_3$ (soda niter, or Chile saltpeter). Because both of these compounds are water-soluble, they are found on Earth's surface mainly in a few desert regions. Nitrogen compounds occur in all living matter, and some organic nitrogen compounds can be extracted from plant or animal sources or from the fossilized remains of ancient plant life, such as coal.

Nitrogen gas is used as an inert blanketing gas in industrial operations, such as in metallurgy and the manufacture of electronics components. Liquid nitrogen is used in

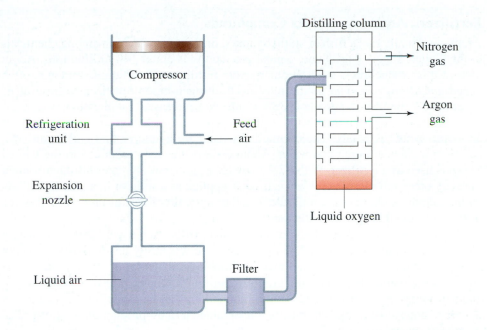

◀ **FIGURE 21.8** **The fractional distillation of liquid air**

Clean air is compressed and then cooled in the refrigeration unit. As it leaves the expansion nozzle, this air expands and so cools even more. Following a series of compressions and expansions, the air cools to a temperature at which it liquefies. The liquid air is filtered to remove $CO_2(s)$ and then distilled. Nitrogen is the most volatile component, with a normal boiling point of 77.4 K; it comes off as a gas. Argon, which boils at 87.5 K, is removed from the middle of the column, and liquid oxygen, the least volatile component with a normal boiling point of 90.2 K, collects at the bottom of the column.

low-temperature applications, such as the fast-freezing of foods. The only important commercial method of producing nitrogen is the fractional distillation of liquid air (Figure 21.8).

Bonding in the N_2 Molecule

When two nitrogen atoms combine to form a molecule of N_2, they acquire valence-shell octets by forming a triple covalent bond. The nitrogen-to-nitrogen triple bond is one of the strongest chemical bonds known. The rupture of one mole of these bonds requires the absorption of 945.4 kJ of energy:

$$N{\equiv}N(g) \longrightarrow 2\,N(g) \qquad \Delta H = +945.4\;\text{kJ}$$

Chemical reactions in which strong bonds are replaced by weaker bonds are endothermic because the energy required to break bonds is not offset by the energy released in forming new bonds. As a result, many nitrogen compounds have positive enthalpies of formation, for example, NO(g):

$$\tfrac{1}{2}\,N_2(g) + \tfrac{1}{2}\,O_2(g) \longrightarrow NO(g) \qquad \Delta H^{\circ}_f = +90.25\;\text{kJ}$$

Generally, highly endothermic reactions do not occur to any appreciable extent, and this is why $N_2(g)$ and $O_2(g)$ can coexist so peacefully in the atmosphere. If the formation of NO from N_2 and O_2 occurred to any appreciable extent at room temperature, which we would expect if the formation reaction were exothermic, the amount of highly noxious NO gas in the atmosphere would be significant. At the same time, the amount of life-sustaining O_2 would be reduced. There would be no life as we know it on Earth.

The Synthesis of Ammonia

In the essay on ammonia synthesis in Section 14.4, we learned that the basic problem in this process is that under most conditions, the reaction

$$N_2(g) + 3\,H_2(g) \rightleftharpoons 2\,NH_3(g)$$

is reversible and does not go to completion. The N_2 for the process is obtained from air. In some cases, pure N_2 obtained by the fractional distillation of air is used, and in other cases, ordinary air (78.1 mol % N_2) is used. Typically, the source of H_2 is the water–gas reaction or the steam reforming of hydrocarbons (Section 20.1). The reaction conditions used to obtain a good yield of NH_3 and an outline of the synthesis reaction were presented on page 595.

▲ Liquid nitrogen can be used for freezing the genetic material of endangered species. This technique aids the New York Zoological Society in its efforts to breed the endangered snow leopard.

▲ Anhydrous liquid ammonia is applied directly into the soil as a fertilizer.

Fertilizers: Ammonia and Its Compounds

Ammonia usually ranks fifth or sixth, by mass, in annual production among chemicals in the United States. Worldwide, annual production is about 140 million tons. About 85% of this production is used to manufacture fertilizers. Pure liquid ammonia, often called *anhydrous ammonia,* is the most used nitrogen-based fertilizer because of its exceptionally high mass percent (82%) of nitrogen and ease of application.

Urea, $CO(NH_2)_2$, is another nitrogen-based fertilizer, often made at the site of ammonia synthesis plants. The required reactants are ammonia and carbon dioxide, and the products are urea and water. About 80% of the urea produced in the United States is used as a fertilizer. With 47% N by mass, urea has a greater nitrogen content than any other solid fertilizer. Urea is usually applied as a solid or in aqueous solution with ammonia and ammonium nitrate. Another agricultural use of urea is as a source of supplemental nitrogen in cattle feed.

The nitrogen-based fertilizer most familiar to the home gardener is probably ammonium sulfate. It has only 21% N by mass, but it is easily handled. Also, it is not highly soluble in water, and so it remains available to plants for a relatively long time and does not readily leach into the groundwater. Furthermore, it also provides sulfur, another necessary plant nutrient. Ammonium sulfate can be made by the reaction of ammonia and sulfuric acid, but mostly it is obtained as a by-product of other chemical manufacturing processes.

The ammonium phosphates are also dual-purpose fertilizers, supplying plants with both nitrogen and phosphorus. They represent one of the fastest-growing segments of the fertilizer industry and now account for about two-thirds of the total production of phosphate fertilizers in the United States.

Ammonium phosphate fertilizers are formed by neutralizing aqueous ammonia with phosphoric acid. The two principal products are ammonium dihydrogen phosphate, $NH_4H_2PO_4$, commonly called monoammonium phosphate (MAP), and ammonium hydrogen phosphate, $(NH_4)_2HPO_4$, commonly called diammonium phosphate (DAP). Together, these two products are the world's leading fertilizers. They are also used in dry-chemical fire extinguishers.

Nitric Acid and Nitrates

Another industrial process that involves nitrogen chemistry, one that is perhaps second in importance only to the ammonia synthesis reaction, is the Ostwald reaction—the catalyzed oxidation of NH_3 to NO:

$$4\,NH_3(g) + 5\,O_2(g) \xrightarrow{\text{Pt/Rh catalyst}} 4\,NO(g) + 6\,H_2O(g)$$

This is the first step in the commercial preparation of nitric acid. The additional steps involve the oxidation of NO to NO_2 and the reaction of NO_2 with water:

$$2\,NO(g) + O_2(g) \longrightarrow 2\,NO_2(g)$$

$$3\,NO_2(g) + H_2O(l) \longrightarrow 2\,HNO_3(aq) + NO(g)$$

The NO formed in the second reaction is recycled in the first reaction.

Some of the industrial uses of nitric acid, such as the manufacture of ammonium nitrate, are based on its acidic properties—namely, its ability to neutralize bases:

$$NH_3(aq) + HNO_3(aq) \longrightarrow NH_4NO_3(aq)$$

Most of the nitric acid produced in the United States goes into the manufacture of ammonium nitrate and other fertilizers.

Nitric acid is also used as an oxidizing agent, with the actual agent being nitrate ion in an acidic solution. When a metal reacts with nitric acid, the metal is oxidized to aqueous metallic ions, but hydrogen gas is rarely obtained as a product. The reduction product has nitrogen atoms with an oxidation number lower than +5: usually NO or

▲ A platinum–rhodium catalyst for the oxidation of ammonia. It is made in the form of a gauze to increase the surface area for the reaction. A mixture of $NH_3(g)$ and air is passed through the gauze very quickly (passage takes about 1 millisecond) to minimize side reactions. Nearly 100% conversion of $NH_3(g)$ to $NO(g)$ is achieved.

NO_2, but even N_2O or NH_4^+ in some instances. The reaction of copper with nitric acid was the subject of Example 4.8:

$$Cu(s) + 4 H^+(aq) + 2 NO_3^-(aq) \longrightarrow Cu^{2+}(aq) + 2 NO_2(g) + 2 H_2O(l)$$

Ammonium nitrate has nitrogen with oxidation numbers -3 (in NH_4^+) and $+5$ (in NO_3^-). This salt can therefore participate in oxidation–reduction reactions all by itself; no other agent is required to react with it. Nitrate ion is an oxidizing agent, and ammonium ion is a reducing agent. When heated gently at about 200 °C, ammonium nitrate forms dinitrogen monoxide and water:

$$NH_4NO_3(l) \xrightarrow{\Delta} N_2O(g) + 2 H_2O(g)$$

Stronger heating or mechanical shock can induce a much more vigorous reaction that is the basis of the use of ammonium nitrate as an explosive:

$$2 NH_4NO_3(l) \xrightarrow{\Delta} 2 N_2(g) + 4 H_2O(g) + O_2(g)$$

Oxides of Nitrogen

The common oxides of nitrogen include nitrogen with every oxidation number from $+1$ to $+5$. Three of these oxides—N_2O, NO, NO_2—were known before Dalton's time. In fact, Dalton cited them in establishing his law of multiple proportions (Section 2.2).

Earlier in this section, we considered methods of producing three of the oxides of nitrogen: N_2O by the mild decomposition of ammonium nitrate, and NO and NO_2 by the reduction of nitrate ion in acidic solution (such as in the reactions of metals with nitric acid). Nitrogen monoxide is produced commercially by the catalytic oxidation of ammonia, as we have already noted. The role of oxides of nitrogen in the production of photochemical smog is well established and is discussed in Chapter 25.

Application Note

The explosive potential of ammonium nitrate in contact with organic materials was dramatically revealed in the ammonium nitrate explosion of a cargo ship in Texas City, Texas, in 1947. Nearly 600 people were killed and several thousand injured. More recently, explosive mixtures of ammonium nitrate fertilizer and fuel oil have been used in terrorist attacks, such as the 1995 Oklahoma City bombing.

Nitrogen Dioxide and Dinitrogen Tetroxide movie

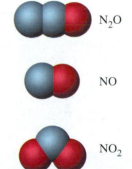

N_2O

NO

NO_2

NO—A Messenger Molecule

In 1992, the simple molecule NO, notorious as an air pollutant, was shown to be a *messenger molecule* that carries signals between cells in the body. Nitric oxide is essential to maintaining blood pressure and establishing long-term memory. It also aids in the immune response to foreign invaders in the body, and it mediates the relaxation phase of intestinal contractions when food is being digested.

The discovery of the physiological role of NO by Louis Ignarro, Robert F. Furchgott, and Ferid Murad was recognized in a 1998 Nobel Prize in Physiology or Medicine. This award has a fortuitous link back to Alfred Nobel, whose invention of dynamite provided the financial basis of the Nobel prizes. Nitroglycerin, the explosive ingredient of dynamite, has long been used in very small doses to relieve the chest pain of heart disease. In his later years, Nobel refused to take nitroglycerin for his heart disease because it causes headaches and he did not think it would relieve his chest pain. Now we know that nitroglycerin acts by releasing NO.

Nitric oxide dilates the blood vessels that allow blood flow into the penis to cause an erection. Research on this role of NO led to the development of anti-impotence drugs for men. Related research has led to drugs for treating shock and a drug for treating high blood pressure in newborn babies.

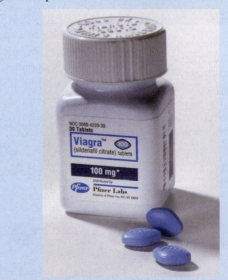

▲ The upfiring thrusters of the U.S. space shuttle, shown here in an in-flight test, use methylhydrazine, CH_3NHNH_2, as a fuel.

Other Nitrogen Compounds

A nitrogen–hydrogen compound related to ammonia is *hydrazine*, NH_2NH_2. Because of its two N atoms, hydrazine can ionize in two steps:

$$NH_2NH_2(aq) + H_2O(l) \rightleftharpoons NH_2NH_3^+(aq) + OH^-(aq) \qquad K_{b_1} = 8.5 \times 10^{-7}$$

$$NH_2NH_3^+(aq) + H_2O(l) \rightleftharpoons [NH_3NH_3]^{2+}(aq) + OH^-(aq) \qquad K_{b_2} = 8.9 \times 10^{-16}$$

Some of hydrazine's most important uses, however, are not as a base but as a reducing agent. For example, it is used to remove dissolved O_2 from boiler water. In the acidic medium used for the redox reaction

$$NH_2NH_3^+(aq) + O_2(g) \longrightarrow 2 H_2O(l) + H^+(aq) + N_2(g) \qquad E^\circ_{cell} = +1.46 \text{ V}$$

hydrazine is found to exist in its cationic form, just as ammonia exists as NH_4^+ in acidic solutions.

The direct reaction of liquid hydrazine with oxygen gas can be used in a fuel cell and also for rocket propulsion:

$$N_2H_4(l) + O_2(g) \longrightarrow N_2(g) + 2 H_2O(l) \qquad \Delta H^\circ = -622.2 \text{ kJ}$$

The oxidation of hydrazine in acidic solution by nitrite ion produces hydrogen azide, HN_3:

$$NH_2NH_3^+(aq) + NO_2^-(aq) \longrightarrow HN_3(aq) + 2 H_2O(l)$$

In aqueous solution, HN_3 is a weak acid ($K_a = 1.9 \times 10^{-5}$) called hydrazoic acid; its salts are called azides. Azides are extremely unstable; lead azide, $Pb(N_3)_2$, for example, is used to make detonators. Sodium azide, NaN_3, also decomposes readily, releasing $N_2(g)$:

$$2 NaN_3(s) \longrightarrow 2 Na(l) + 3 N_2(g)$$

Sodium azide is used in air-bag safety systems in automobiles.

21.8 Phosphorus

Phosphorus is the eleventh most abundant element in Earth's crust. It occurs mainly in phosphate rock, produced in nature from the remains of marine animals deposited millions of years ago. Phosphate rock is a member of the class of minerals called *apatites*. Fluorapatite, $3 Ca_3(PO_4)_2 \cdot CaF_2$, is an example. Elemental phosphorus is prepared by heating calcium phosphate, silica, and coke in an electric furnace.

$$2 Ca_3(PO_4)_2(s) + 10 C(s) + 6 SiO_2(s) \longrightarrow 6 CaSiO_3(l) + 10 CO(g) + P_4(g)$$

The $P_4(g)$ is collected, condensed to a solid, and stored under water.

The solid phosphorus prepared in this way is a waxy, white solid. This *white phosphorus* can be cut with a knife. It melts at 44.1 °C, is a nonconductor of electricity, and ignites spontaneously in air (which is why it is stored under water). White phosphorus is insoluble in water but soluble in nonpolar solvents like carbon disulfide (CS_2). It is highly toxic; a lethal oral dose can be as small as 0.1 g.

As shown in Figure 21.9a, white phosphorus is made up of tetrahedral P_4 molecules, with a P atom at each corner of the tetrahedron. The P–P bonds in P_4 appear to

▲ The glow of white phosphorus. Vapor from the sublimation of white phosphorus reacts slowly with atmospheric oxygen. Energy is evolved as light. The name phosphorus is derived from Greek (*phos*, light; *phoros*, bringing).

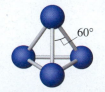

(a) White phosphorus

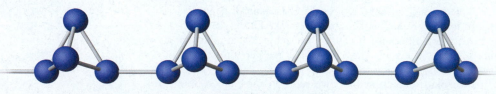

(b) Red phosphorus

▲ **FIGURE 21.9   The molecular structures of white and red phosphorus**

(a) White phosphorus is composed of P_4 molecules. The molecules have a pyramidal shape, with 60° P–P–P bond angles. (b) In red phosphorus, one P–P bond breaks in each P_4 molecule, and the P_4 fragments join together in long chains.

QUESTION: On the basis of bond angles, which P–P bond(s) of red phosphorus would you expect to be most reactive?

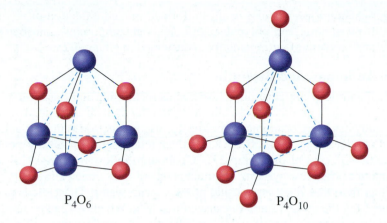

◀ **FIGURE 21.10** Molecular structure of P_4O_6 and P_4O_{10}

In P_4O_6, an O atom is inserted between each pair of P atoms of the basic P_4 structure, leading to a total of six O atoms in the molecule. In P_4O_{10}, an additional O atom is bonded to each P atom, yielding a total of 10 O atoms per molecule.

QUESTION: What is the significance of the blue dashed lines?

involve the overlap of $3p$ orbitals almost exclusively. Normally such overlap produces 90° bond angles, but in P_4 the P–P–P bond angles are 60°. The bonds are said to be *strained,* and species with strained bonds, like P_4, are generally quite reactive.

When white phosphorus is heated to about 300 °C in the absence of air, it is transformed to the allotropic form called *red phosphorus.* At the molecular level, it seems that one P–P bond per P_4 molecule breaks, and the fragments join together into long chains, as shown in Figure 21.9b. The two allotropes differ appreciably in their properties. For example, red phosphorus is less reactive and much less toxic than white phosphorus.

Phosphorus forms two important oxides. In the oxide with the empirical formula P_2O_3, P has the oxidation number +3. In the other, with empirical formula P_2O_5, P has the oxidation number +5. Although these oxides have been commonly called diphosphorus trioxide and diphosphorus pentoxide, respectively, their true molecular formulas are double the empirical formulas, that is, tetraphosphorus hexoxide, P_4O_6, and tetraphosphorus decoxide, P_4O_{10}. The structures of these oxide molecules are shown in Figure 21.10. The hexoxide forms in the reaction of P_4 with a limited quantity of $O_2(g)$, and the decoxide forms with excess $O_2(g)$. An oxide that reacts with water to produce an acid as the sole product is often called an *acid anhydride.* The oxides P_4O_6 and P_4O_{10} are the acid anhydrides of phosphorous acid and phosphoric acid, respectively:

$$P_4O_6(s) + 6\,H_2O(l) \longrightarrow 4\,H_3PO_3(aq) \quad \textit{(phosphorous acid)}$$
$$P_4O_{10}(s) + 6\,H_2O(l) \longrightarrow 4\,H_3PO_4(aq) \quad \textit{(phosphoric acid)}$$

Phosphoric acid is a triprotic acid. In contrast, phosphorous acid is *diprotic* because one of the H atoms in H_3PO_3 is bonded directly to the central P atom and is not ionizable.

Although we represent aqueous phosphoric acid as $H_3PO_4(aq)$, the solution generally contains a complex mixture of species. The simplest phosphoric acid, called *orthophosphoric acid,* can react to form *pyrophosphoric acid:*

$$2\,H_3PO_4 \longrightarrow H_4P_2O_7 + H_2O$$

The structural formulas

H—O—P—O—H   Phosphorous acid

H—O—P—O—H   Phosphoric acid

H—O—P—O—H + H—O—P—O—H ⟶ H—O—P—O—P—O—H + H₂O

Orthophosphoric acid   Orthophosphoric acid   Pyrophosphoric acid

show that the essential reaction in the conversion of orthophosphoric acid to pyrophosphoric acid is the elimination of a molecule of H_2O from two molecules of orthophosphoric acid. A pyrophosphoric acid molecule can react with an additional orthophosphoric acid molecule to form triphosphoric acid:

$$H_4P_2O_7 + H_3PO_4 \longrightarrow H_5P_3O_{10} + H_2O$$

Application Note

In the food industry, phosphoric acid is used to impart tartness to soft drinks. Phosphate salts are used as dietary supplements, in baking powders, in dairy products, and as polishing agents in toothpastes.

Still longer-chain *polyphosphoric acids* are formed in a continuation of this reaction. The salts of phosphoric and polyphosphoric acids are important contributors to the environmental problem of *eutrophication,* considered in Chapter 25.

Arsenic, Antimony, and Bismuth

Arsenic and antimony are generally classified as metalloids, and bismuth as a metal. All are used in the manufacture of alloys. When As and Sb are added to lead, the resulting alloy has more desirable properties for use in electrodes in lead-acid storage batteries than does lead alone. Because of their relatively low melting points, antimony (mp 631 °C) and bismuth (mp 271 °C) are used in low-melting-point alloys for solders and fire-protection systems. Bismuth is one of only three elements (Ga and Ge are the other two) that share an unusual property with water: These elements expand on freezing from the molten state. Because of this, bismuth alloys can be used to make highly detailed castings. The alloys expand slightly on solidification, filling the mold completely and leaving no shrinkage cracks.

An important arsenic compound is the oxide As_4O_6, known as *white arsenic.* It can be made by burning arsenic in air, and it is the acid anhydride of arsenious acid, H_3AsO_3:

$$As_4O_6(s) + 6\,H_2O(l) \longrightarrow 4\,H_3AsO_3(aq)$$

Both inorganic and organic compounds derived from As_4O_6 have found extensive use as insecticides. However, perhaps the most important arsenic compound currently is gallium arsenide, GaAs, because it has some semiconductor and optical properties that make it an important material for supercomputers and communication satellites. (Semiconductors are discussed in Section 24.8.)

Group 6A

The elements of group 6A have the valence-shell electron configuration ns^2np^4. Oxygen and sulfur are clearly nonmetallic, but selenium is less so; tellurium is usually considered a metalloid; polonium is generally classified as a metal.

21.9   Oxygen

Oxygen is one of the most active nonmetals and one of the most important. It forms compounds with all the elements except the light noble gases (He, Ne, and Ar). In general, oxygen forms ionic compounds with metals and covalent compounds with other nonmetals. Oxygen ranks first in abundance in Earth's crust, about 45.5% by mass.

Oxygen and its compounds are ubiquitous, and we have encountered them many times in this text—for example, in combustion reactions and in the reactions of acids and bases in aqueous solutions. Also, oxygen atoms are key components of many functional groups in organic compounds: alcohols, ethers, aldehydes, ketones, carboxylic acids, esters, and so forth. Furthermore, whenever we discuss the chemistry of another element, we discuss its oxygen compounds.

In most of its compounds, oxygen has the oxidation number -2, as it does both in molecular substances such as H_2O and CO_2 and in ionic substances such as Li_2O and MgO. Among the ionic oxides of active metals, there are three possible oxides, with the oxidation number of O changing from -2 in the oxide to -1 in the peroxide and $-\frac{1}{2}$ in superoxides:

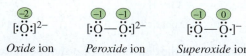

Oxide ion   Peroxide ion   Superoxide ion

In these structures, notice the relationship between the formal charges and the ionic charges, and notice also that the superoxide ion is paramagnetic.

The chief reactions of elemental, atmospheric oxygen are oxidations: combustion (rapid burning), rusting and other forms of corrosion, and respiration. Oxygen reacts

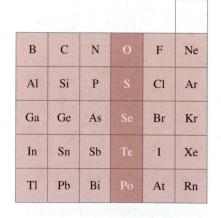

rapidly with active metals. Magnesium, for example, burns with a brilliant white flame when ignited in air (page 848), with the reaction being

$$2 \, Mg(s) + O_2(g) \longrightarrow 2 \, MgO(s) + heat + light$$

This same reaction occurs more slowly on a freshly prepared surface of magnesium at room temperature. Similar reactions occur with aluminum and titanium. The oxides formed in these reactions are thin, transparent coatings that are impervious to air, preventing further oxidation of the metal. Iron, especially in the presence of moisture and electrolytes, typically forms iron(III) oxide (rust; see Section 18.8). This oxide flakes off the surface of the metal, and oxidation (rusting) continues.

Preparation and Uses of Oxygen

Oxygen gas, obtained by the fractional distillation of liquid air (Figure 21.8), is an important commercial chemical. Typically, it ranks third in quantity produced in the United States, following sulfuric acid and nitrogen. Some important uses of oxygen are listed in Table 21.5.

Occasionally, small quantities of oxygen are prepared by the decomposition of compounds containing oxoanions, usually in the presence of a catalyst. For example, the decomposition of potassium chlorate is catalyzed by MnO_2:

$$2 \, KClO_3(s) \xrightarrow{MnO_2(s)} 2 \, KCl(s) + 3 \, O_2(g)$$

The decomposition of certain oxides also yields oxygen gas. Joseph Priestley discovered oxygen in 1774 by heating mercury(II) oxide:

$$2 \, HgO(s) \longrightarrow 2 \, Hg(l) + O_2(g)$$

Lavoisier used this decomposition reaction as the source of oxygen in his early studies of combustion.

Oxygen can also be produced by the decomposition of aqueous solutions of hydrogen peroxide, as we noted in Chapter 13:

$$2 \, H_2O_2(aq) \longrightarrow 2 \, H_2O(l) + O_2(g)$$

The reaction is slow but can be speeded up greatly by using a catalyst, such as $Br^-(aq)$ or $I^-(aq)$ (page 553). Another oxygen-producing reaction uses potassium superoxide (Section 20.5):

$$4 \, KO_2(s) + 2 \, CO_2(g) \longrightarrow 2 \, K_2CO_3(s) + 3 \, O_2(g)$$

Oxygen is also produced, together with hydrogen, during the electrolysis of water (Section 20.1).

Ozone

Ozone, O_3, an allotrope of oxygen, is a familiar substance, known not only for its role in photochemical smog but also through seasonal depletions ("ozone holes") observed in the stratospheric ozone layer over the polar regions. We will consider environmental issues involving ozone in Chapter 25, but ozone is also of chemical interest.

In Chapter 9, we described the ozone molecule as the hybrid of two resonance structures, and with the VSEPR method, we predict for this AX_2E molecule the trigonal planar shape that is observed by experiment (O–O–O bond angle 116.8°):

$$:\ddot{O} \diagdown \overset{\ddot{O}}{\diagup} \ddot{O}: \longleftrightarrow :\ddot{O} \diagup \overset{\ddot{O}}{\diagdown} \ddot{O}:$$

Ozone is a powerful oxidizing agent, especially in acidic solution:

$$O_3(g) + 2 \, H^+(aq) + 2 \, e^- \longrightarrow O_2(g) + H_2O(l) \qquad E° = +2.075 \, V$$

It is used as an oxidizing agent in organic reactions and in the purification of drinking water, treatment of industrial wastes, and bleaching of paper and textiles.

The production of O_3 directly from O_2,

$$3 \, O_2(g) \longrightarrow 2 \, O_3(g) \qquad \Delta H° = +285 \, kJ$$

Table 21.5 Common Uses of Oxygen

Manufacture of iron, steel, and
  other metals
Metal welding and cutting by torch
Manufacture of chemicals
Water treatment
Oxidizer of rocket fuels
Respiration therapy and other
  medical uses

Application Note

The enzyme *peroxidase*, found in blood, catalyzes the decomposition of hydrogen peroxide. Thus, the formation of O_2 is accelerated when hydrogen peroxide is used as an antiseptic on an open wound.

▲ Potassium superoxide was used to replenish the air breathed by cosmonaut V. Kubasov during this 1970 space flight of *Soyuz 6*.

▲ Ozone from this portable generator is used to sanitize foods.

is a highly endothermic reaction and occurs only rarely in the lower atmosphere, chiefly during electrical storms. Ozone, which can be identified by its pungent odor, is also occasionally formed around heavy-duty electrical equipment and xerographic office copiers.

To produce ozone for commercial use, O_2 must be maintained in a high-energy environment by passing either an electric discharge or ultraviolet radiation through it. Because it is unstable and decomposes back to O_2, ozone is usually generated at the point where it is to be used.

21.10   Sulfur

Sulfur and oxygen are similar in several ways, as we expect from the electron configurations of their atoms. Both form ionic compounds with active metals, and they form some similar covalent compounds, for example,

$$H_2O \text{ and } H_2S \qquad CO_2 \text{ and } CS_2 \qquad CH_3CH_2OH \text{ and } CH_3CH_2SH$$

However, oxygen and sulfur compounds also differ in important ways. Let us consider two examples:

- *Hydrogen bonding is often an important intermolecular force in compounds with H–O bonds but not in compounds with H–S bonds.* For example, H_2S is a gas at room temperature; it boils at -61 °C. In contrast, H_2O is a liquid at room temperature, boiling at 100 °C. As another example, ethyl alcohol, CH_3CH_2OH, is a liquid with a boiling point of 78.3 °C, whereas the boiling point of ethyl mercaptan, CH_3CH_2SH, is 35.0 °C.

- *Sulfur can employ an expanded valence shell, but oxygen cannot.* As a consequence, sulfur can form compounds such as SF_4, CH_2SF_4, and SF_6, but there are no comparable compounds of oxygen.

Elemental sulfur exists as several molecular species, some of which are pictured in Figure 21.11. Some familiar forms of sulfur are described next, and a few of these are pictured in Figure 21.12.

- *Rhombic sulfur* (S_α) is a solid made up of S_8 molecules. At 95.5 °C, S_α converts to

- *monoclinic sulfur* (S_β). Monoclinic sulfur, also made up of S_8 molecules, melts at 119 °C, yielding

- *liquid sulfur* (S_λ). This yellow, transparent, mobile liquid is also made up of S_8 molecules. At 160 °C, the S_8 rings open up and join into long spiral-chain molecules, resulting in

- *liquid sulfur* (S_μ). This is a dark, viscous liquid. If liquid S_μ is poured into cold water, the solid first formed is a rubberlike material called *plastic sulfur*. On standing, plastic sulfur slowly reverts to rhombic sulfur. At higher temperatures, the chain molecules in liquid S_μ break up and the liquid flows more freely. At 445 °C liquid S_μ boils, producing

- *sulfur vapor*. The molecules in the vapor range from S_2 to S_{10}, with S_8 predominating at the boiling point and S_2 predominating at higher temperatures.

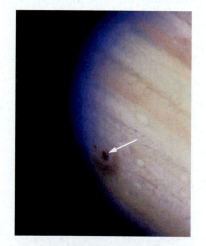

▲ The arrow marks the point at which a fragment of comet Shoemaker–Levy 9 collided with Jupiter in 1994. The impact expelled gases, including $S_2(g)$, from the depths of Jupiter's atmosphere. The molecules were identified by their characteristic spectra.

   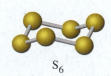   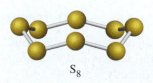   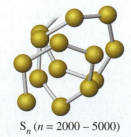

S S_2 S_6 S_8 S_n ($n = 2000 - 5000$)

▲ **FIGURE 21.11   Five molecular forms of sulfur**

(a)   (b)   (c)

▲ **FIGURE 21.12   Four modifications of sulfur**
(a) Rhombic sulfur. (b) Monoclinic sulfur. (c) (Left dish): At its melting point, 119 °C, sulfur is a straw-colored, mobile liquid. (Right dish): Above 160 °C, the liquid becomes dark and viscous (S_μ). (d) Liquid sulfur (S_μ) is poured into water to form plastic sulfur.

(d)

The temperature profile for these forms of sulfur and the changes they undergo is

$$S_\alpha \xrightarrow{95.5\,°C} S_\beta \xrightarrow{119\,°C} S_\lambda \xrightarrow{160\,°C} S_\mu \xrightarrow{445\,°C} S_8(g)$$

$$\longrightarrow S_6(g) \longrightarrow S_4(g) \xrightarrow{1000\,°C} S_2(g) \xrightarrow{2000\,°C} S(g)$$

Sulfur is abundant in Earth's crust. It occurs as elemental sulfur, as mineral sulfides and sulfates, as $H_2S(g)$ in natural gas, and as organic sulfur compounds in oil and coal. In the United States, deposits of elemental sulfur are found in Texas and Louisiana. This sulfur is mined in an unusual way known as the Frasch process (Figure 21.13). A mixture of superheated water and steam (at about 160 °C and 16 atm) is forced down the outermost of three concentric pipes into an underground bed of sulfur-containing rock. The sulfur melts and forms a liquid pool. Compressed air (at 20–25 atm) is pumped down the innermost pipe and forces liquid sulfur up the middle pipe. (The superheated water and steam descending in the outermost pipe help keep the ascending liquid sulfur from solidifying.) The Frasch process is no longer the most important way of obtaining sulfur in the United States. Because of the need to control emissions of sulfur oxides, much of the sulfur now in use is recovered as a by-product of industrial operations.

A small amount of elemental sulfur is used in vulcanizing rubber and as a pesticide (for example, in dusting grapevines). Most, however, is burned to $SO_2(g)$, and the greatest proportion of this is converted by way of $SO_3(g)$ to H_2SO_4.

Sulfur dioxide reacts with water to produce a solution of sulfurous acid, $H_2SO_3(aq)$, but pure H_2SO_3 is unstable; it has never been isolated. Salts of sulfurous acid, called sulfites, are good reducing agents in reactions in which they are oxidized to sulfates, as in the reaction

$$Cl_2(g) + SO_3{}^{2-}(aq) + H_2O(l) \longrightarrow 2\,Cl^-(aq) + SO_4{}^{2-}(aq) + 2\,H^+(aq)$$

However, sulfites can also act as oxidizing agents, as in the reaction

$$2\,H_2S(g) + 2\,H^+(aq) + SO_3{}^{2-}(aq) \longrightarrow 3\,H_2O(l) + 3\,S(s)$$

Sulfur dioxide and sulfites are widely used in the food industry as decolorizing agents and preservatives. The main use of sulfites, however, is in the pulp and paper industry. When wood is digested in an aqueous sulfite solution, chemical reactions occur that separate the cellulosic constituents of wood from unwanted materials, such as lignin. The wood pulp thus formed is used to make paper or rayon.

Sulfuric acid, $H_2SO_4(aq)$, is a strong acid. Dilute aqueous solutions neutralize bases, react with metals to form $H_2(g)$, and react with carbonates to liberate $CO_2(g)$. Concentrated sulfuric acid, on the other hand, has some rather distinctive properties. For one, it is a fairly good oxidizing agent. It will react with copper metal, for example, producing $Cu^{2+}(aq)$ and $SO_2(g)$. Also, it is a very strong dehydrating agent—so

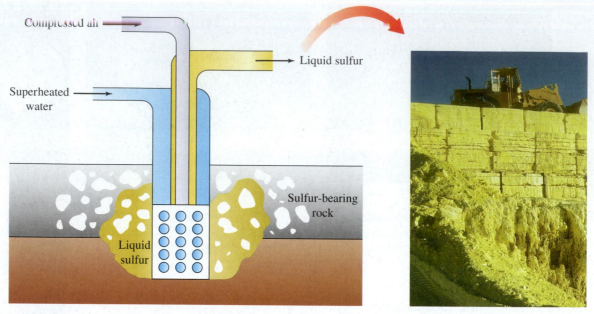

▲ **FIGURE 21.13** **The Frasch process for mining sulfur**
Sulfur is melted by superheated water and steam forced down the outermost of three concentric pipes. Compressed air is blown down the innermost pipe, and the liquid sulfur is forced up the middle pipe. The liquid sulfur is then pumped to collection pools, where it is allowed to freeze into large solid formations (right).

▲ In the left beaker, concentrated sulfuric acid is being added to the sugar sucrose. The carbon produced in the reaction is seen in the beaker on the right.

strong that it will extract H and O atoms in the proportions of H_2O from certain compounds. Concentrated sulfuric acid extracts water from a carbohydrate, leaving a carbon residue:

$$C_{12}H_{22}O_{11}(s) \xrightarrow{\text{H}_2\text{SO}_4\text{(concd)}} 12\ C(s)\ +\ 11\ H_2O(l)$$
Sucrose

Reactions of this type account for the damage to clothing and the severe burns caused by the concentrated acid.

Sulfate salts have many important uses. We discussed the use of *gypsum*, $CaSO_4 \cdot 2\ H_2O$, in the construction industry in Section 20.10, and aluminum sulfate in water treatment earlier in this chapter (page 870). Copper(II) sulfate is used in electroplating, and it is also an effective fungicide and algicide.

Thiosulfates, $S_2O_3^{2-}$, are related to sulfates in that the former have a S atom replacing one of the O atoms in SO_4^{2-} (Figure 21.14). Thiosulfates can be prepared by boiling elemental sulfur in an alkaline solution of sodium sulfite. The sulfur is oxidized to thiosulfate ion, and the sulfite ion is reduced to thiosulfate ion:

$$SO_3^{2-}(aq)\ +\ S(s)\ \longrightarrow\ S_2O_3^{2-}(aq)$$

The thiosulfate ion is a good reducing agent and is used extensively in analytical chemistry. It is also a good complexing agent, that is, it acts as a ligand by attaching to the central ion of a complex ions. This is the property of thiosulfate ion that underlies the use of sodium thiosulfate as a fixer in photography (page 695).

▶ **FIGURE 21.14** **Structures of the sulfate and thiosulfate ions**
(a) Sulfate ion. (b) Thiosulfate ion, formed by replacing one of the terminal O atoms of the sulfate ion with an S atom.

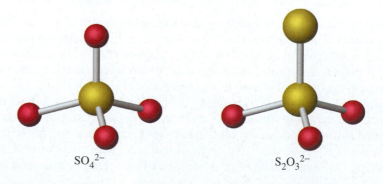

21.11 Selenium, Tellurium, and Polonium

Selenium and tellurium are similar to sulfur but more metallic. For example, sulfur is an electrical insulator, whereas selenium and tellurium are semiconductors. Polonium is an electrical conductor and has other metallic properties, such as the ability to form cations. Because it is highly radioactive, however, polonium has few uses. Polonium-218—an α-particle emitter—has been implicated in the environmental problems associated with radon gas.

Selenium and tellurium occur as selenides and tellurides mixed in sulfide ores (for example, as Cu_2Se and Cu_2Te). The free elements are obtained mainly from the anode mud in the electrolytic refining of copper (page 788). Both Se and Te find some use in alloys, and their compounds are used as additives to control the color of glass. Selenium is an important element in modern technology, primarily because it is a *photoconductor*, which means that its electrical conductivity increases in the presence of light. Photoconductivity is the property that makes selenium useful in photocells for cameras and as the light-sensitive element in photocopying machines.

▲ Selenium-coated light-sensitive element from a photocopier.

Group 7A

Atoms of the group 7A elements—the halogens—have the valence-shell electron configuration ns^2np^5. They have high ionization energies, large negative electron affinities, and high to moderately high electronegativities. This means that they gain electrons readily and lose electrons only with difficulty—just as we expect for a family of nonmetals.

The halogen elements exist as diatomic molecules, X_2 (where X = F, Cl, Br, or I). Much of their reaction chemistry involves redox reactions, and standard electrode potentials are the best guide to their reactivity in aqueous solutions. From the standard electrode potentials listed in Table 21.6, we see clearly that F_2 is the most easily reduced; it is the strongest oxidizing agent of the halogens. In fact, F_2 is the strongest oxidizing agent of all the elements. Because they are so easily reduced, fluorine, chlorine, and bromine occur naturally only as the anions F^-, Cl^-, and Br^-. Iodine also occurs as iodide ion, I^-, but it is found naturally in positive oxidation states as well (for example, as $NaIO_3$).

Astatine, at the bottom of group 7A, is a radioactive element. All its isotopes are short-lived, and only about 0.05 mg has ever been prepared at one time. These samples rapidly undergo radioactive decay, so that few of the properties of astatine have been determined with certainty. We will not consider the element further here.

B	C	N	O	F	Ne
Al	Si	P	S	Cl	Ar
Ga	Ge	As	Se	Br	Kr
In	Sn	Sb	Te	I	Xe
Tl	Pb	Bi	Po	At	Rn

Table 21.6 Standard Electrode Potentials of the Halogens

Reduction half-reaction	$E°$, volts
$F_2(g) + 2\,e^- \longrightarrow 2\,F^-(aq)$	2.866
$Cl_2(g) + 2\,e^- \longrightarrow 2\,Cl^-(aq)$	1.358
$Br_2(l) + 2\,e^- \longrightarrow 2\,Br^-(aq)$	1.065
$I_2(s) + 2\,e^- \longrightarrow 2\,I^-(aq)$	0.535

21.12 Sources and Uses of the Halogens

We will explore major sources and some of the more important uses of fluorine, chlorine, bromine, and iodine in this section. We start with fluorine. As with other families, this first halogen differs in significant ways from the others.

Physical Properties of Halogens movie

Fluorine

The existence of fluorine has been known since early in the nineteenth century, but Henri Moissan was the first to succeed in preparing $F_2(g)$, in 1886. Moissan's method involves the electrolysis of HF dissolved in molten potassium fluoride:

$$2\,H^+ + 2\,F^- \xrightarrow{\text{electrolysis}} H_2(g) + F_2(g)$$

▲ This frosted design was etched on glass by hydrofluoric acid.

▲ Chlorine gas is green-yellow.

This oxidation of F^- to F_2 occurs only with extreme difficulty because F_2 is so easily reduced to F^- in the reverse process. This is yet another example of electrolysis being used to accomplish oxidations or reductions when chemical reactions are not feasible.

Fluorine is used in the production of $UF_6(g)$ for separation of the uranium-235 and uranium-238 isotopes by gaseous diffusion (page 818). Uranium enriched in the uranium-235 isotope is used in nuclear fuels and weapons.

Fluorine is also used to make $SF_6(g)$, which forms when sulfur is burned in fluorine gas. A nonflammable, unreactive gas of low toxicity, SF_6 is used as an insulating gas in high-voltage electrical equipment. Many *interhalogens,* compounds of two (or even three) different halogens, contain fluorine. Examples include ClF_3 and BrF_3, compounds often used as substitutes for fluorine in reactions because they are easier to handle than elemental fluorine.

Fluorine forms compounds with all the elements except the lightest noble gases, He and Ne. Hydrogen fluoride is the starting reagent for preparing most other fluorine compounds. An unusual property of HF is its ability to etch glass, represented here as SiO_2:

$$SiO_2(s) + 4\,HF(aq) \longrightarrow H_2O(l) + SiF_4(g)$$

Etching occurs as the glass in contact with the HF is eroded away. For this reason, HF(aq) must be stored in special containers made of polyethylene or Teflon®.

Chlorine

Chlorine was first produced by Carl Wilhelm Scheele in 1774, using the reaction of MnO_2 and HCl:

$$MnO_2(s) + 4\,HCl(aq) \xrightarrow{\Delta} MnCl_2(aq) + 2\,H_2O(l) + Cl_2(g)$$

This reaction is occasionally used in the laboratory today, but the only significant commercial methods of preparing Cl_2 today involve electrolysis. Some chlorine is made by the electrolysis of molten NaCl or $MgCl_2$, but most comes from the electrolysis of aqueous NaCl (page 787).

Elemental chlorine, with an annual U.S. production of about 13 million tons, generally ranks among the top 10 chemicals. It has three main commercial uses: About 70% is used to produce chlorinated organic compounds; about 20% is used as a bleach in the paper and textile industries and for the treatment of municipal water, sewage, and swimming pools; and about 10% is used to produce chlorine-containing inorganic compounds.

Chlorine forms stable compounds with most of the elements except the noble gases. Especially important are chlorine-containing carbon compounds—that is, organic chlorine compounds. Chlorine reacts with hydrocarbons to produce chlorinated hydrocarbons. For example, in the direct reaction of $Cl_2(g)$ with $CH_4(g)$, chlorine atoms substitute for hydrogen atoms, and products ranging from CH_3Cl to CCl_4 are formed, with HCl as a by-product of the reactions. Similarly, the reaction of ethane, C_2H_6, with chlorine yields products that have one to six Cl atoms substituted for H atoms. Several chlorinated hydrocarbons are used as solvents for nonpolar or slightly polar solutes.

The chlorofluorocarbons (CFCs) have F and Cl atoms bonded to carbon. The most widely used have been $CFCl_3$ and CF_2Cl_2. Because these substances are volatile liquids or easily condensable gases, they work well as refrigerants and as blowing agents to form the pores in plastic foam. We will discuss the role of CFCs in the destruction of stratospheric ozone in Chapter 25.

Bromine and Iodine

Bromine is extracted from subterranean brines. Brines from Arkansas contain 3800–5000 ppm of bromine (as Br^-), and water in the Dead Sea contains about 4500–5000 ppm. The water is treated with $Cl_2(g)$, which oxidizes Br^- to Br_2:

$$Cl_2(g) + 2\,Br^-(aq) \longrightarrow 2\,Cl^-(aq) + Br_2(l) \qquad E^\circ_{cell} = +0.293\ V$$

We first described halogen displacement reactions like this one in Section 8.9.

Iodine was once obtained in small quantities from dried seaweed. (Certain marine plants concentrate I^- selectively in the presence of Cl^- and Br^-.) Today, iodine is also obtained from brines. Wells in Michigan, Oklahoma, and Japan yield brines with 100 ppm I^-. Elemental iodine is obtained from these brines by a process similar to that used for bromine.

Organic bromine compounds are used as pharmaceuticals, dyes, fumigants, and pesticides. Bromine compounds are also used as fire extinguishers and fire retardants. An important inorganic bromine compound is AgBr, the primary light-sensitive agent used in black-and-white photographic film.

Iodine is of less commercial importance than chlorine and bromine, although iodine and its compounds do have applications as catalysts, in medicine, and in photographic emulsions (AgI). Iodine and its compounds also have important uses in analytical chemistry.

Application Note

Iodide ion is used in the thyroid gland to synthesize thyroxin, a substance that helps regulate metabolism. An iodide ion deficiency leads to mental and physical retardation in children. An estimated 10 million children in China have been affected by this deficiency. In adults, iodide ion deficiency causes an enlargement of the thyroid gland, a condition called goiter. Both goiter and retardation can be prevented by the use of iodized salt, a product that is mainly NaCl, with a small quantity of added NaI or KI.

21.13 Hydrogen Halides

In aqueous solution, the hydrogen halides are acids. Except for HF, they are strong acids. Strong hydrogen bonding in HF accounts for the fact that HF has the highest boiling point among the liquid hydrogen halides (Section 11.6), even though it has the lowest molecular mass. Hydrogen bonding may also play a role in making HF(aq) weak rather than strong as an acid.

The hydrogen halides can be prepared by the direct combination of the elements:

$$H_2(g) + X_2(g) \longrightarrow 2\,HX(g)$$

The reaction between H_2 and F_2 occurs with explosive violence. Mixtures of $H_2(g)$ and $Cl_2(g)$ are stable in the dark but react explosively in the presence of light. Both $Br_2(g)$ and $I_2(g)$ react more slowly with $H_2(g)$; a catalyst is required to get a reasonable reaction rate.

Hydrogen halides can also be made by heating a halide salt, such as CaF_2 or NaCl, with a concentrated, *nonvolatile* acid, such as H_2SO_4:

$$CaF_2(s) + H_2SO_4(\text{concd aq}) \longrightarrow CaSO_4(s) + 2\,HF(g)$$

$$2\,NaCl(s) + H_2SO_4(\text{concd aq}) \longrightarrow Na_2SO_4(s) + 2\,HCl(g)$$

However, we cannot use H_2SO_4 to prepare HBr(g) or HI(g) because H_2SO_4 is a sufficiently good oxidizing agent to oxidize Br^- to Br_2 and I^- to I_2. To get around this difficulty, we can use a *nonoxidizing* acid, such as phosphoric acid:

$$NaBr(s) + H_3PO_4(\text{concd aq}) \longrightarrow NaH_2PO_4(s) + HBr(g)$$

Example 21.2 A Conceptual Example

We have mentioned that HBr(g) cannot be prepared by heating a bromide salt with concentrated sulfuric acid because bromine gas is obtained instead.

(a) Show that the reaction

$$2\,NaBr(s) + 2\,H_2SO_4(l) \longrightarrow Na_2SO_4(s) + 2\,H_2O(g) + SO_2(g) + Br_2(g)$$

is not spontaneous at 25 °C and standard-state conditions.

(b) Why does this reaction occur at higher temperatures?

ANALYSIS AND CONCLUSIONS

(a) We need to evaluate the standard free energy change for the reaction. We can do this with standard free energies of formation from Appendix C and the expression $\Delta G^\circ = [\Sigma \nu_p \, \Delta G_f^\circ(\text{products})] - [\Sigma \nu_r \, \Delta G_f^\circ(\text{reactants})]$.

$$2\,NaBr(s) + 2\,H_2SO_4(l) \longrightarrow Na_2SO_4(s) + 2\,H_2O(g) + SO_2(g) + Br_2(g)$$

	$2\,NaBr$	$2\,H_2SO_4$	Na_2SO_4	$2\,H_2O$	SO_2	Br_2
ΔG_f°, kJ/mol	-349.0	-690.1	-1270	-228.6	-300.2	3.14

$$\Delta G^\circ = [[1 \text{ mol} \times -1270 \text{ kJ/mol}] + [2 \text{ mol} \times -228.6 \text{ kJ/mol}]$$
$$+ [1 \text{ mol} \times -300.2 \text{ kJ/mol}]$$
$$+ [1 \text{ mol} \times (3.14 \text{ kJ/mol})]] - \{[2 \text{ mol} \times -349.0 \text{ kJ/mol}]$$
$$+ [2 \text{ mol} \times -690.1 \text{ kJ/mol}]\}$$
$$= -2024 \text{ kJ} - (-2078 \text{ kJ}) = 54 \text{ kJ}$$

The fact that $\Delta G^\circ = +54$ kJ tells us that the reaction will not occur spontaneously with reactants and products in their standard states at 25 °C. This value is so large and positive that we would not expect the reaction to be spontaneous at 25 °C even for most nonstandard conditions.

(b) The simplest approach to showing that the reaction is spontaneous at higher temperatures is to use the Gibbs equation: $\Delta G = \Delta H - T\Delta S$. We do not need to do calculations, however. Just by inspecting the balanced equation, we can see that ΔS is large and positive for this reaction: Two moles each of a solid and a liquid yield one mole of solid and four moles of gas. A reaction with a large positive ΔS is favored at higher temperatures because the high temperature ensures a large negative value of $-T\Delta S$. At some temperature, $-T\Delta S$ is certain to be large enough to overcome a positive value of ΔH. This situation, in which $\Delta H > 0$ and $\Delta S > 0$, corresponds to case 3 in Table 17.1.

EXERCISE 21.2A

Use standard electrode potentials to determine whether the reaction

$$\text{Br}^-(\text{aq}) + 4\,\text{H}^+(\text{aq}) + \text{SO}_4^{2-}(\text{aq}) \longrightarrow 2\,\text{H}_2\text{O}(l) + \text{SO}_2(g) + \text{Br}_2(l)$$

should occur spontaneously as written. Is your result consistent with the results obtained in parts (a) and (b) of Example 21.2? Explain.

EXERCISE 21.2B

Without doing detailed calculations, determine whether the reaction of NaI(s) with $\text{H}_2\text{SO}_4(l)$ should become spontaneous at a lower or higher temperature than the reaction in Example 21.2.

21.14   Oxoacids and Oxoanions of the Halogens

In its compounds, fluorine always has the oxidation number -1. The other halogens, however, can have positive oxidation numbers: $+1$, $+3$, $+5$, and $+7$. These oxidation numbers are found in the oxoacids of Cl, Br, and I listed in Table 21.7. Chlorine forms a complete set of oxoacids, but bromine and iodine do not.

Hypochlorous acid, HOCl, is formed to some extent by the disproportionation of Cl_2 in water:

$$\text{Cl}_2(g) + \text{H}_2\text{O}(l) \rightleftharpoons \text{HOCl}(\text{aq}) + \text{H}^+(\text{aq}) + \text{Cl}^-(\text{aq}) \qquad E^\circ_{\text{cell}} = -0.27 \text{ V}$$

An effective germicide, HOCl(aq) is often used to disinfect water. Hypochlorous acid exists only in solution; it cannot be isolated in the pure state.

Application Note

Alkaline solutions of sodium hypochlorite can be used as a ready source of $\text{Cl}_2(g)$. All that is required is to neutralize the excess alkali; the disproportionation reactions that produce HOCl and OCl^- are then reversed.

Oxidation Number of Halogens activity

Table 21.7 Oxoacids of the Halogens[a]

Halogen Oxidation Number	Chlorine	Bromine	Iodine
$+1$	HOCl	HOBr	HOI
$+3$	HOClO	—	—
$+5$	HOClO_2	HOBrO_2	HOIO_2
$+7$	HOClO_3	HOBrO_3	HOIO_3[b]

[a] In all these acids, the H atom is bonded to an O atom, not to the central halogen atom. Nevertheless, these formulas are often written in the form HClO (for HOCl), HClO_2 (for HOClO), and so on.

[b] $(\text{HO})_5\text{IO}$ is also a component of periodic acid solutions. It is formed in the addition and rearrangement of two H_2O molecules and one of HOIO_3.

If we dissolve Cl_2 in a basic solution, such as NaOH(aq), the equilibrium in the reaction on page 892 is displaced far to the right as HOCl is neutralized. An aqueous solution of a hypochlorite salt is formed:

$$Cl_2(g) + 2\,OH^-(aq) \longrightarrow OCl^-(aq) + Cl^-(aq) + H_2O(l) \qquad E^\circ_{cell} = +0.84\ V$$

Common household bleaches are aqueous alkaline solutions of NaOCl, as are some drain cleaners and a solution commonly used to chlorinate home swimming pools.

Chlorine dioxide, ClO_2, is a bleach for paper, fibers, and flour. When reduced by peroxide ion in aqueous solution, ClO_2 is converted to chlorite ion:

$$2\,ClO_2(g) + O_2^{2-}(aq) \longrightarrow 2\,ClO_2^-(aq) + O_2(g)$$

Sodium chlorite is used to bleach textiles.

When Cl_2 is added to *hot* alkaline solutions, chlorate salts are formed:

$$3\,Cl_2(g) + 6\,OH^-(aq) \longrightarrow 5\,Cl^-(aq) + ClO_3^-(aq) + 3\,H_2O(l)$$

Chlorates are good oxidizing agents. Solid chlorates are used in matches and fireworks. When heated, chlorates decompose to produce oxygen gas, which supports the combustion of the other ingredients.

Electrolysis of chlorate salt solutions yields perchlorates; oxidation of ClO_3^- occurs at a Pt anode in the half-reaction

$$ClO_3^-(aq) + H_2O(l) \longrightarrow ClO_4^-(aq) + 2\,H^+(aq) + 2\,e^-$$

An important laboratory use of perchlorate salts is in solution studies; ClO_4^- has the least tendency of any anion to act as a ligand in complex ion formation. In the presence of readily oxidizable substances, such as most organic compounds, perchlorates often react explosively. Mixtures of ammonium perchlorate, aluminum powder, and a rubberlike binder are used as solid propellants for rockets. Ammonium perchlorate is particularly useful in this application, in part because an oxidizing agent, ClO_4^-, and a reducing agent, NH_4^+, occur in the same compound.

Group 8A

We conclude our survey of *p*-block elements with another brief look at the unusual elements of group 8A, the noble gases. For many years after their discovery, these elements were thought to be inert. The Lewis theory of bonding was based in part on this apparent inertness. We still use the Lewis theory, but we no longer regard the noble gases as inert. The change in thinking came about mainly as a result of some experiments in the 1960s.

In 1962, Neil Bartlett and D. H. Lohman identified a compound of O_2 and PtF_6 in a 1:1 mole ratio. Properties of the compound indicate that it is $(O_2)^+(PtF_6)^-$. It takes 1177 kJ/mol of energy to extract an electron from O_2 to form the ion O_2^+. Noting that this is almost identical to the first ionization energy of Xe (1170 kJ/mol) and that the size of the Xe atom is roughly the same as that of the O_2 molecule, Bartlett was able to substitute Xe for O_2 and obtain a yellow crystalline solid, apparently $XePtF_6$. (We now know the formula to be $Xe(PtF_6)_n$, where n is between 1 and 2.) Before long, chemists around the world had synthesized several additional noble-gas compounds. For example, depending on the reaction conditions, XeF_2, XeF_4, and XeF_6 can be prepared by the direct reaction of Xe and F_2. Some compounds of Kr and Rn exist, as does one compound of Ar (Chapter 8), but no compounds of He or Ne are known.

In Chapter 10, we used the VSEPR method to make qualitative predictions about the molecular geometry of two xenon compounds, XeF_2 and XeF_4. In Table 10.1, we saw that XeF_2 has the VSEPR notation AX_2E_3 and is linear, whereas XeF_4 has the notation AX_4E_2 and is square planar. It is difficult to describe bonding in noble-gas compounds in a way that is consistent with all the experimental facts, and we will not attempt to do so. In fact, the discovery of noble-gas compounds has done much to stimulate the development of modern bonding theories.

▲ Potassium chlorate, an oxidizing agent, and sucrose, a reducing agent, react vigorously, producing a bright flame. The violet color of the flame is caused by K atoms (see Figure 8.19).

Application Note

Some emergency oxygen generators are activated by a reaction between powdered iron and sodium chlorate. The heat of this reaction raises the temperature and causes the rest of the sodium chlorate to decompose into sodium chloride and oxygen. The accidental activation of a cargo of oxygen generators caused the ValuJet crash in Florida in 1996.

B	C	N	O	F	Ne
Al	Si	P	S	Cl	Ar
Ga	Ge	As	Se	Br	Kr
In	Sn	Sb	Te	I	Xe
Tl	Pb	Bi	Po	At	Rn

▲ Crystals of xenon tetrafluoride.

21.15   Occurrence of the Noble Gases

Except for helium and radon, the noble gases are found only in the atmosphere. Helium is found in some natural gas deposits, particularly those underlying the Great Plains of the United States. This helium is formed by the α-particle decay of naturally occurring radioactive isotopes. The α particles ($^4He^{2+}$) acquire two electrons, becoming helium atoms. As we noted in Figure 19.2, radon, a radioactive element, is formed in the radioactive decay series that originates with ^{238}U.

The noble gases were all formed when our planet was formed four billion years ago. Over the eons since that time, most of the original quantities of these gases escaped from Earth's atmosphere. The result is that, except for argon, the present-day atmospheric concentrations of these elements are all extremely low. The fact that argon is much more abundant than the other noble gases can be explained rather easily. It is a product of the radioactive decay of potassium-40, a fairly abundant naturally occurring isotope:

$$^{40}_{19}K \longrightarrow {}^{40}_{20}Ar + {}^{\ 0}_{-1}e$$

Although helium is also constantly being formed through α-particle emissions by radioactive isotopes, it escapes from the atmosphere into outer space at a higher rate than argon because the molar mass of helium is only one-tenth that of argon.

21.16   Properties and Uses of the Noble Gases

Helium is used to fill balloons and dirigibles. Its lifting power is nearly as good as that of hydrogen, but it has the important advantage of being nonflammable. Helium is also used to provide an inert atmosphere for the welding of metals that otherwise might be attacked by oxygen in air.

Liquid helium is used to achieve extremely low temperatures for scientific and commercial purposes. It boils at only 4.2 K, and it exists as a liquid down to temperatures approaching 0 K. All other substances freeze to solids at temperatures well above 0 K; even hydrogen has a freezing point of 14 K. Large quantities of liquid helium are used in *cryogenics,* the study of materials at extremely low temperatures. Some metals become superconductors at liquid-helium temperatures (that is, they essentially lose their resistance to the flow of electric current). Immersing the metal-wire coils of electromagnets in liquid helium can create powerful magnets. These magnets are the key components in nuclear magnetic resonance (NMR) instruments in research laboratories and magnetic resonance imaging (MRI) devices in hospitals.

Another interesting use of helium is in helium–oxygen breathing mixtures for deep-sea divers (page 501). Helium is less soluble in body fluids, and is less likely to cause decompression problems, than is nitrogen. Similar mixtures are also used in the treatment of asthma, emphysema, and other conditions involving respiratory obstruction. The same low molar mass that gives helium its lifting power also permits it to diffuse into partially obstructed areas of the lungs more rapidly than nitrogen does. The strain on the muscles when a helium–oxygen mixture is breathed is less than the strain involved when air is breathed.

Neon is used in lighted advertising signs. A tube with embedded electrodes is shaped into letters or symbols and filled with neon at low pressure. An electric current passed through the gas causes the neon atoms to emit light of characteristic wavelengths. With a spectroscope, we would see individual spectral lines; but with the unaided eye, we see merely the familiar orange-red glow. Other colors can be obtained with mixtures of neon and argon or mercury vapor.

Argon, the most plentiful of the noble gases, can be separated from air rather inexpensively (Figure 21.8). Like helium, argon gas provides an inert atmosphere. In the laboratory, an argon atmosphere is an ideal medium for reactions in which reactants or products are sensitive to air oxidation or to reaction with nitrogen. In industry, argon is used to blanket materials that need to be protected from nitrogen and oxygen, as in certain types of welding and in the preparation of ultrapure semiconductor materials like silicon and germanium. Some incandescent lightbulbs are filled with a mixture of argon and nitrogen to increase their efficiency and life.

▲ The neon lights in the Kodak Pavilion in the Epcot Center are constantly changing in color.

Application Note

A major problem in aluminum recycling is the formation of dross, a solid material that is mainly aluminum oxide formed in the melting furnace when the molten aluminum reacts with oxygen from the air. Argon provides a nonoxidizing atmosphere that reduces dross formation by 60%.

Krypton and xenon are expensive, but they have some commercial applications. They are sometimes used in lasers, photographic flash lamps, and in various kinds of lights, including some bright flashlight bulbs, strobe lights for airport runways, and automobile headlamps. Radon can be collected from the radioactive decay of radium-226:

$$^{226}_{88}\text{Ra} \longrightarrow ^{226}_{86}\text{Rn} + ^{4}_{2}\text{He}$$

The radon can be sealed in small vials and used for radiation therapy of certain malignancies.

Cumulative Example

In the oxidation of $NH_3(g)$ to $NO(g)$ (page 880), temperature, pressure, and contact time with the Pt/Rh catalyst must all be carefully controlled. Depending on the reaction conditions, the reaction can yield $N_2(g)$, $NO(g)$, $N_2O(g)$, or $NO_2(g)$. Which of the four possible reactions is the most thermodynamically favored?

STRATEGY

The most thermodynamically favored reaction is the one with the largest value of K_{eq}, which is the reaction with the smallest (most negative) value of $\Delta G°$. This fact suggests that we can first write balanced equations for the four reactions and then assess $\Delta G°$ for each by using standard free energies of formation from Appendix C.

SOLUTION

The four reactions are all redox reactions, but their equations are balanced easily enough by inspection. Each has $NH_3(g)$ and $O_2(g)$ as reactants and $H_2O(g)$ and a nitrogen-containing substance as products. We use fractional coefficients in the equations so that each equation is based on a single mole of $NH_3(g)$.

(a) $NH_3(g) + \frac{3}{4} O_2(g) \rightleftharpoons \frac{1}{2} N_2(g) + \frac{3}{2} H_2O(g)$

(b) $NH_3(g) + O_2(g) \rightleftharpoons \frac{1}{2} N_2O(g) + \frac{3}{2} H_2O(g)$

(c) $NH_3(g) + \frac{5}{4} O_2(g) \rightleftharpoons NO(g) + \frac{3}{2} H_2O(g)$

(d) $NH_3(g) + \frac{7}{4} O_2(g) \rightleftharpoons NO_2(g) + \frac{3}{2} H_2O(g)$

At this point, we need to apply Equation (17.9) to each reaction, using $\Delta G_f°$ values from Appendix C.

$$\Delta G° = \Sigma \nu_p \, \Delta G_f°(\text{products}) - \Sigma \nu_r \, \Delta G_f°(\text{reactants})$$

We can do four complete calculations, but the fact that each equation has 1 mol of $NH_3(g)$ and 1.5 mol of $H_2O(g)$ leads to a simpler method, based on the generic equation for $\Delta G°$.

$$\Delta G° = \tfrac{3}{2} \times \Delta G_f°[H_2O(g)] + ? \times \Delta G_f°[\text{N-containing species}]$$
$$- \Delta G_f°[NH_3(g)] - ? \times \Delta G_f°[O_2(g)]$$

The terms on the green panels are identical for the four reactions, and their contributions to $\Delta G°$ are identical. Because $\Delta G_f°[O_2(g)] = 0$, the term on the pink panel is always zero. So the determining factor in every case is the value of $? \times \Delta G_f°[\text{N-containing species}]$. We determine these quantities from $\Delta G_f°$ values in Appendix C and the stoichiometry of the reaction.

(a) $\frac{1}{2} \Delta G_f°[N_2(g)] = 0$

(b) $\frac{1}{2} \Delta G_f°[N_2O(g)] = \frac{1}{2}$ mol $\times$ 104.2 kJ/mol = 52.10 kJ

(c) $\Delta G_f°[NO(g)] = 1$ mol $\times$ 86.57 kJ/mol = 86.57 kJ

(d) $\Delta G_f°[NO_2(g)] = 1$ mol $\times$ 51.30 kJ/mol = 51.30 kJ

The smallest of these values is for reaction (a), which means that reaction (a) will have the most negative value of $\Delta G°$ and be the most favored reaction thermodynamically.

ASSESSMENT

An obvious check on the answer is to calculate the actual values of $\Delta G°$ for the four reactions and choose the reaction with the smallest value. If you do this, you should obtain for reactions (a), (b), (c), and (d): $\Delta G° = -326.42$ kJ, -274.32 kJ, -239.85 kJ, and -275.12 kJ, respectively, thereby concluding again that reaction (a) is the most thermodynamically favored. At the same time, you must remember that the actual product obtained depends as much, or perhaps more, on the kinetics of the reaction as on the thermodynamics.

Concept Review with Key Terms

21.1 Properties and Trends in Group 3A—The elements of group 3A exhibit a range of physical and chemical properties. Many of the trends observed in this group can be understood from considerations of their electron configurations and their respective positions in the periodic table. For example, the **inert pair**

of valence-shell electrons retained in the Tl^+ ion is also found in several other cations following a transition series.

21.2 Boron—Because they have only three valence electrons, boron atoms tend to form electron-deficient compounds, leading to some unusual bonding patterns. Compounds resulting from the "addition" of one structure to another, called **adducts**, are common when boron is present. Also, boron atoms form three-center bonds in the boron–hydrogen compounds called boranes. Other common boron compounds include borax, boric oxide, boric acid, and borates.

21.3 Aluminum—Aluminum is the most industrially important element of group 3A. It is an active metal that is protected against corrosion by a film of $Al_2O_3(s)$. Both $Al_2O_3(s)$ and Al react with acids and strong bases. The production of aluminum is based on the amphoterism of $Al_2O_3(s)$ and the electrolysis of Al_2O_3 in molten cryolite. The electron deficiency of aluminum chloride is a useful property in organic syntheses.

21.4 Carbon—Carbon is the key element of organic chemistry, but the free element also has uses. Diamond is prized for hardness and thermal conductivity; while graphite's electrical conductivity and refractory properties have found extensive use. Metal carbides, carbon oxides, and nitrogen-containing carbon compounds are of importance in a wide number of applications.

21.5 Silicon—Silicon is the key element of the mineral world, occurring as silica, SiO_2, and as various minerals based on the silicate anion, SiO_4^{4-}. Silica and silicates are common constituents of **ceramics**— solids that are characterized by hardness and brittleness and are chemically and structurally stable at high temperatures. Some synthetic silicon-containing organic compounds are of commercial importance, including the silicon-containing polymers called silicones.

21.6 Tin and Lead—Tin and lead metals are slightly more active than hydrogen and are widely used in alloys. They can have an oxidation number of either +2 or +4, with Sn(II) being a good reducing agent and Pb(IV) being a good oxidizing agent.

21.7 Nitrogen—Nitrogen is the major constituent of the atmosphere, and essentially all nitrogen compounds—natural and synthetic—are derived from the atmosphere. Some industrially important nitrogen compounds are ammonia, urea, nitric acid, ammonium salts, hydrazine, hydrazoic acids, azides, and oxides of nitrogen.

$$H-N-\underset{\underset{H}{|}}{\overset{\overset{O}{\|}}{C}}-N-H$$
$$\text{Urea}$$

21.8 Phosphorus—The structure of phosphorus is based on the pyramidal molecule, P_4, in both the white and red modifications. The structures of the oxides P_4O_6 and P_4O_{10} are related to that of the P_4 molecule. The principal compounds of phosphorus are the phosphates and polyphosphates.

White phosphorus

21.9 Oxygen—Oxygen forms compounds with all elements except the lighter noble gases. Most oxygen is obtained, together with nitrogen and argon, by the fractional distillation of liquid air. Oxygen forms three types of anions when combined with active metals: oxide (O^{2-}), peroxide (O_2^{2-}), and superoxide (O_2^-). Ozone, O_3, an allotrope of oxygen, is useful as an oxidizing agent, both in the laboratory and in the chemical industry.

21.10 Sulfur—Sulfur differs from oxygen in important ways, such as in its variety of allotropic forms and the changes they undergo. Its important compounds are the oxides, oxoacids, sulfites, sulfates, and thiosulfates, and many of the reactions of these compounds are oxidation–reduction reactions.

21.11 Selenium, Tellurium, and Polonium—Selenium and tellurium are found as minor but important components of technologies ranging from the coloring of glass to the detection of light. Polonium is highly radioactive and has few common uses.

21.12 Sources and Uses of the Halogens—The halogens (group 7A) are nonmetals; fluorine is the most nonmetallic of all elements. Fluorine and chlorine are prepared by electrolysis, and bromine and iodine by displacement reactions. Two halogens can react to form an interhalogen compound. Halogen atoms can substitute for H atoms in hydrocarbons and other organic compounds.

21.13 Hydrogen Halides—Hydrogen halides form by the direct combination of the elements or by the reaction of a halide salt with a nonvolatile acid. In aqueous solution, the hydrogen halides act as acids.

21.14 Oxoacids and Oxoanions of the Halogens—Important classes of halogen compounds include the oxoacids and their salts. The chemical reactions of these compounds are mostly oxidation–reduction reactions.

21.15 Occurrence of the Noble Gases—Most of the noble gases are found in Earth's atmosphere. Some, like He and Ar, are produced in quantity through the decay of radioactive isotopes of other elements. Radon is radioactive.

21.16 Properties and Uses of the Noble Gases—Interest in the noble gases centers on their physical properties and inertness. In contrast, the ability of the heavier noble gases to form some chemical compounds provides important insights into bonding theory.

Assessment Goals

When you have mastered the material in this chapter, you will be able to:

- Describe the chemical properties of the *p*-block elements.

- Describe the chemical properties of the oxoanions and acids of the halogens.

- List some of the chemical compounds formed by noble gases and their properties.

- Apply previously learned techniques, such as VSEPR theory, to the structures of compounds of the *p*-block elements.

- Apply thermodynamic and electrochemical principles to reactions of the *p*-block elements.

Self-Assessment Questions

1. What is the primary feature that characterizes a *p*-block element?

2. Which two *p*-block elements have some similar properties governed by the *diagonal relationship?*

3. Three of the *p*-block elements are the three most abundant elements in Earth's crust. What are they?

4. Which is the most active metal and which is the most active nonmetal among the *p*-block elements?

5. What is a three-center or bridge bond? Describe the bridge bonds in B_2H_6 and Al_2Cl_6. How do those in Al_2Cl_6 differ from those in B_2H_6?

6. What is anodized aluminum? What purposes are served in anodizing aluminum?

7. What purpose is served by each of the following ingredients of common drain cleaners: pellets of $NaOH(s)$ and granules of $Al(s)$?

8. A given copper wire is a better electrical conductor than an aluminum wire of the same diameter and length. Why is it, then, that most high-voltage transmission lines in the United States are made of aluminum rather than copper?

9. In 1866, aluminum metal cost $250 to $750 per kilogram. In 1924, it cost about $0.50/kg. Account for this dramatic difference in price.

10. What is carbon black? What is activated carbon? How is each prepared, and what are some of their uses?

11. What is the basic composition of soda–lime glass, and how is this composition modified in Pyrex® glass?

12. What is the basic structural unit of the silicate anion, and what are some of the ways in which these units are combined in silicate minerals?

13. Cite one similarity and two differences between the peroxide and superoxide ions.

14. Cite one example of allotropy and one example of polymorphism in the element sulfur.

15. Explain why H_2S is a gas at room temperature, whereas H_2O, with only about half the molar mass of H_2S, is a liquid.

16. Explain why copper does not react with $HCl(aq)$, but does react with H_2SO_4 (concd aq).

17. What is an adduct? Name two elements closely associated with adduct formation.

18. What is the inert pair effect? Name some elements in which this effect is seen.

19. What type of compound is suggested by the term interhalogen compound? Give an example.

20. What are chlorinated hydrocarbons? What are chlorofluorocarbons? Give two examples of each.

21. Although we have not specifically considered these compounds in the text, name them by using information from this chapter or principles discussed elsewhere in the text.

(a) AgN_3 (c) At_2O

(b) $KSCN$ (d) H_2TeO_4

22. Although we have not specifically considered these compounds in the text, write plausible formulas for them by using information from this chapter or principles discussed elsewhere in the text.

(a) selenic acid   (c) lead(II) azide

(b) hydrotelluric acid   (d) silver astatide

23. What are the formulas of the following compounds mentioned in this chapter: bauxite, boric oxide, corundum, cyanogen, hydrazine, silica?

24. Use structural formulas to write the equation for the reaction of pyrophosphoric acid and orthophosphoric acid to form triphosphoric acid (page 883).

Problems

Boron

25. Elemental boron can be prepared by heating magnesium metal and boric oxide. Write an equation for this reaction.

26. Boron trifluoride reacts with lithium fluoride to form an ionic compound called lithium tetrafluoborate. Write a plausible equation for this reaction. (*Hint:* The tetrafluoborate ion is related to the borate ion but contains no O atoms.)

27. Although they have similar formulas, phosphoric acid, H_3PO_4, is a weak *triprotic* acid, whereas boric acid, H_3BO_3, is a weak *monoprotic* acid. Explain this difference.

28. Although they have similar formulas, the compound BF_3 exists and the compound BH_3 does not. Offer a plausible explanation of this fact.

29. Use the diagram shown to write equations for the conversion of borax to pure boron.

30. Use the diagram in Problem 29 to write equations for the conversion of borax to BF_3.

31. Diborane can be synthesized in the reaction of boron trichloride with lithium aluminum hydride. Write a plausible equation for this reaction, [*Hint:* Refer to the discussion of lithium aluminum hydride on page 870.]

32. Titanium diboride (mp = 2900 °C) can be prepared from its constituent elements. Write an equation for the reaction. Classify titanium diboride according to Table 11.5, and justify your answer.

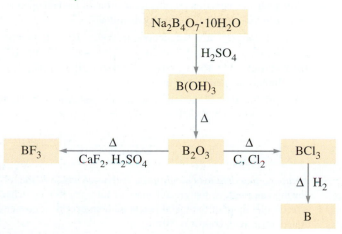

Aluminum

33. A typical baking powder contains baking soda ($NaHCO_3$) and alum [$KAl(SO_4)_2$] as its active ingredients. During baking, the baking powder undergoes a reaction that yields $CO_2(g)$ and $Al(OH)_3(s)$ as two of its products. Write a plausible equation for this reaction.

34. Write equations for a series of reactions that could be used to make the following conversions of aluminum compounds:

Aluminum oxide $\longrightarrow$ aluminum sulfate

$\longrightarrow$ aluminum hydroxide $\longrightarrow$ aluminum chloride

35. Critique the following statement made by a layperson: "Iron rusts, aluminum does not, and that is why aluminum cans last so long outside."

36. What are the two key features that make possible the low-cost production of aluminum from bauxite?

37. Use data from Appendix C to estimate the enthalpy change for the thermite reaction given on page 869. Why is your value only an estimate?

38. Verify the value of $E°_{cell}$ at 298 K for the reaction of aluminum with water in basic solution given on page 869.

39. In the use of aluminum sulfate in water treatment (see page 870), the water to be treated is usually kept between pH 5 and pH 8. Why do you suppose this is the case?

40. Bare aluminum cookware is not recommended for processing tomatoes in home canning, nor is it to be used for soap making (Section 20.13). Explain.

Carbon

41. Write plausible Lewis structures for COS, CN^-, and OCN^-.

42. Write plausible Lewis structures for CS_2, CCl_4, and C_2N_2.

43. Unlike calcium carbide, which yields acetylene, $Al_4C_3(s)$ yields methane when it reacts with water. Write an equation for this reaction.

44. Sodium forms a carbide similar to calcium carbide—that is, containing the C_2^{2-} ion. Write the formula of sodium carbide and an equation for its reaction with water.

Silicon

45. Sketch the geometrical structure and write a Lewis structure for the anion $Si_2O_7^{6-}$.

46. Write a formula and Lewis structure, and sketch the geometric structure for an anion that consists of three SiO_4 units joined through O atoms.

47. Show that the empirical formula given on page 875 is consistent with the expected oxidation numbers of the elements in muscovite mica.

48. Show that the empirical formula given on page 875 is consistent with the expected oxidation numbers of the elements in chrysotile asbestos.

49. In many respects, the naming of organosilicon compounds follows the same rules as organic carbon compounds. Write formulas for **(a)** tetramethylsilane, **(b)** dichlorodimethylsilane, and **(c)** triethylsilane.

50. What is the formula of tetrasilane, a silicon analogue of butane? Write an equation for its combustion in air.

Tin and Lead

51. Use data from Appendix C to determine if $Sn^{2+}(aq)$ is a sufficiently strong reducing agent to carry out the following reductions.

(a) $Cu^{2+}(aq)$ to $Cu^+(aq)$ **(c)** $Ag^+(aq)$ to $Ag(s)$

(b) $V^{3+}(aq)$ to $V^{2+}(aq)$ **(d)** $Cr^{3+}(aq)$ to $Cr^{2+}(aq)$

52. Use data from Appendix C to determine if $PbO_2(s)$ is a sufficiently strong oxidizing agent to carry out the following oxidations in acidic solution.

(a) ClO_3^- to ClO_4^- **(c)** $Ag^+(aq)$ to $Ag^{2+}(aq)$

(b) $H_2O(l)$ to $H_2O_2(aq)$ **(d)** $Sn^{2+}(aq)$ to $Sn^{4+}(aq)$

53. Write plausible equations for the production of tin compounds described in a chemical dictionary as follows.

(a) Stannous chloride [tin(II) chloride]: "By dissolving tin in hydrochloric acid."

(b) Stannic chloride [tin(IV) chloride]: "Treatment of stannous chloride with chlorine."

(c) Stannic oxide [tin(IV) oxide]: "Precipitated from stannic chloride solution by ammonium hydroxide [$NH_3(aq)$]."

54. Write plausible equations for the production of lead compounds described in a chemical dictionary as follows.

(a) Lead acetate: "By the action of acetic acid on litharge (PbO)."

(b) Red lead oxide, Pb_3O_4: "By gently heating litharge (PbO) in a furnace in a current of air."

(c) Lead dioxide: "By action of an alkaline solution of calcium hypochlorite on lead(II) hydroxide."

Nitrogen

55. *Without doing detailed calculations,* place the following in order of increasing mass percent nitrogen: **(a)** dinitrogen monoxide, **(b)** ammonia, **(c)** nitrogen monoxide, **(d)** ammonium chloride, **(e)** hydrazine, **(f)** ammonium sulfate.

56. *Without doing detailed calculations,* determine which of the following can produce the greater mass of nitrogen dioxide when exactly one mole of the metal reacts with an excess of concentrated nitric acid: copper or silver.

57. Complete and balance the following equations based on plausible reaction products.

(a) $NO_2(g) + H_2O(l) \longrightarrow$

(b) $NH_3(g) + O_2(g) \longrightarrow$

(c) $NH_4NO_3(l) \xrightarrow{200\ °C}$

58. Complete and balance the following equations based on plausible reaction products.

(a) $NH_2NH_2(aq) + HCl(aq) \longrightarrow$

(b) $Cu(s) + H^+(aq) + NO_3^-(aq) \longrightarrow$

(c) $NO(g) + O_2(g) \longrightarrow$

59. In acidic solution, hydrazine reduces Fe^{3+} to Fe^{2+}. The hydrazine, which in acidic solution is present as $NH_2NH_3^+$, is oxidized to $N_2(g)$. Write half-equations and a redox equation for this reaction. If E_{cell}° for this reaction is found to be 1.00 V, what is E° for the reduction of $N_2(g)$ to $NH_2NH_3^+(aq)$?

60. The text mentions the combustion of liquid hydrazine as a reaction that can be used for rocket propulsion or as the reaction occurring in a fuel cell (page 882). Use data from Appendix C and Chapter 18 to determine the value of E_{cell}° for this fuel cell.

Phosphorus

61. What are the two principal allotropic forms of phosphorus? Which is the more reactive? How do they differ in molecular structure?

62. What are the formulas of the oxides in which phosphorus has oxidation numbers of +3 and +5, respectively? What are their molecular structures?

63. The industrial preparation of phosphine, $PH_3(g)$, involves the reaction of white phosphorus with KOH(aq). The other reaction product is $KH_2PO_2(aq)$. Write a balanced equation for this redox reaction.

64. Phosphorus trichloride is used to produce many other phosphorus compounds. Write a balanced equation to show how phosphoryl chloride, $POCl_3$, can be made by the reaction of phosphorus trichloride, chlorine, and tetraphosphorus decoxide.

Oxygen

65. Write an equation to represent (a) the action of oxygen on a freshly prepared aluminum surface, (b) the formation of oxygen by the decomposition of potassium chlorate, (c) the formation of oxygen by the action of water on sodium peroxide (sodium hydroxide is the other product), (d) the oxidation of $Pb^{2+}(aq)$ to $PbO_2(s)$ by ozone in an acidic solution (oxygen gas is the other product).

66. Write an equation to represent (a) the formation of ozone by the passage of an electric discharge through oxygen, (b) the formation of oxygen by the thermal decomposition of mercury(II) oxide, (c) the formation of oxygen by the action of water on potassium superoxide (potassium hydroxide is the other product), (d) the oxidation of $Cl^-(aq)$ to $ClO_3^-(aq)$ by ozone in an acidic solution (oxygen gas is the other product).

Sulfur

67. Describe the physical properties and physical principles that make possible the Frasch process for mining sulfur.

68. Outline the phase changes and changes in molecular structure that occur when a sample of rhombic sulfur is heated from room temperature to about 450 °C.

69. The equation given on page 888 for the formation of thiosulfate ion from sulfite ion and sulfur in an alkaline solution is for an oxidation–reduction reaction. Write equations for the half-reactions that occur. Derive the overall equation from these half-equations.

70. When an aqueous solution of sodium thiosulfate is acidified, unstable thiosulfuric acid is formed. The acid decomposes immediately to sulfurous acid and sulfur. Write equations for the two reactions. (*Hint:* What is the formula for thiosulfuric acid?)

The Halogens

71. Which of the following would you add to an aqueous solution containing Br^- to oxidize $Br^-(aq)$ to $Br_2(l)$: I_2, I^-, Cl_2, Cl^-, F^-? Explain your choice(s).

72. Can you think of a reagent that can be used to displace $F^-(aq)$ and produce $F_2(g)$? Explain.

73. If Br^- and I^- occur together in an aqueous solution, I^- can be oxidized to IO_3^- with $Cl_2(aq)$. Simultaneously, Br^- is oxidized to Br_2, which is extracted with $CS_2(l)$. Write chemical equations for the reactions that occur.

74. Write equations to illustrate the oxidation of $Fe^{2+}(aq)$ to $Fe^{3+}(aq)$ by the halogen, X_2. (*Hint:* How many different equations can you write?)

75. One of the following is a *halide*, one is a *halate*, and one fits neither of these categories: BrF_3, $MgBr_2$, $NaClO_3$. Identify each.

76. One of the following is an *interhalogen*, one is a *halide*, one is a *halate*, and one fits none of these categories: $(CN)_2$, ICl, $KBrO_3$, $NaCl$. Identify each.

77. Write equations to represent the reaction of KI with (a) H_2SO_4 (concd aq), and (b) H_3PO_4 (concd aq).

78. Write equations to represent the reaction of Cl_2 with (a) $H_2O(l)$, (b) cold NaOH(aq), and (c) hot NaOH(aq).

Noble Gases

79. Use VSEPR theory to predict the shapes of **(a)** XeF_4 and **(b)** $XeOF_4$.

80. Use VSEPR theory to predict the shapes of **(a)** XeO_3 and **(b)** XeO_4.

81. Why is helium formed in many natural radioactive decay processes, whereas argon is formed in only one radioactive decay?

82. Argon was discovered in 1894 through the observation that the density of nitrogen obtained from air was different from the density of nitrogen obtained from nitrogen compounds. Why did the nitrogen from these two sources appear to have different densities? Which of the two sources do you think yielded the gas with the greater density? Explain.

Additional Problems

Problems marked with an * may be more challenging than others.

83. Predict the likely geometric structures of the interhalogen molecules **(a)** BrF_3 and **(b)** IF_5.

84. Predict the likely geometric structures of the polyatomic ions **(a)** ICl_2^- and **(b)** I_3^-.

85. The reaction of borax, calcium fluoride, and concentrated sulfuric acid yields sodium hydrogen sulfate, calcium sulfate, water, and boron trifluoride as products. Write a balanced equation for this reaction.

86. In one common method used to prepare potassium aluminum alum, $KAl(SO_4)_2 \cdot 12 H_2O$, aluminum foil is dissolved in $KOH(aq)$. The solution thus obtained is treated with $H_2SO_4(aq)$, and the alum is crystallized from the resulting solution. Write plausible equations for these reactions.

87. In 1986, fluorine was prepared by a chemical method (that is, not involving electrolysis). The reactions used were that of hexafluoromanganate(IV) ion, MnF_6^{2-}, with antimony pentafluoride to produce manganese(IV) fluoride and SbF_6^{2-}, followed by the dissociation of manganese(IV) fluoride to manganese(II) fluoride and fluorine gas. Write equations for these two reactions.

88. For the reduction half-reaction,

$$XeF_2(aq) + 2 H^+(aq) + 2 e^- \longrightarrow Xe(g) + 2 HF(aq)$$
$$E° = +2.32 \text{ V}$$

Show that XeF_2 decomposes in aqueous solution, producing $O_2(g)$. Write an equation for this reaction, and calculate $E°_{cell}$. Would you expect this decomposition to be favored in acidic or basic solution? Explain.

*89. The thermite reaction is just one of an entire series of reactions described by Hans Goldschmidt in the late nineteenth century and collectively called *Goldschmidt reactions*. Various combinations include Fe_2O_3, MnO_2, and Cr_2O_3 as oxides and Al and Si as reducing agents. Estimate which of these possible combinations would produce the highest temperature.

*90. The gemstone mineral beryl consists of SiO_4^{4-} tetrahedra arranged in six-membered rings with Be^{2+} and Al^{3+} in a $3:2$ ratio as counterions. Make a sketch of this structure, and deduce the empirical formula of beryl.

*91. In a reaction described on page 887, sulfite ion in acidic solution oxidizes $H_2S(g)$ to free sulfur. What volume of air, at 25 °C and 755 mmHg and containing 1.5 mol % H_2S, could be purged of H_2S by 2.50 L of 1.75 M Na_2SO_3?

92. What masses of $C(s)$ and $Al_2O_3(s)$, in kilograms, are consumed to produce 1.00×10^3 kg Al by the electrolysis reaction of page 868? How many coulombs of electric charge are required for the electrode reactions?

93. Polonium is the only element known to crystallize in the simple cubic structure. In this structure, the interatomic distance is 335 pm. Use this description of the crystal structure to estimate the density of polonium. (*Hint:* How many atoms are in a unit cell? What are the mass and volume of this unit cell?)

94. In aqueous solution, $O_3(g)$ is rapidly decomposed to $O_2(g)$. Write equations for two half-reactions that combine to give the decomposition of $O_3(g)$ as the overall redox reaction. What is the value of $E°_{cell}$ for this reaction?

*95. Magnesium can reduce NO_3^- to $NH_3(g)$ in a basic solution.

$$Mg(s) + NO_3^-(aq) + H_2O(l)$$
$$\longrightarrow Mg(OH)_2(s) + OH^-(aq) + NH_3(g) \quad (not \ balanced)$$

The gaseous NH_3 can be neutralized with excess $HCl(aq)$, and the unreacted $HCl(aq)$ can then be titrated with $NaOH(aq)$. A 25.00-mL sample of a solution with unknown molarity of NO_3^- was treated in this way, and the $NH_3(g)$ was passed into 50.00 mL of 0.1500 M HCl. The excess HCl required 32.10 mL of 0.1000 M NaOH for its titration. What was $[NO_3^-]$ in the sample?

96. Methane and sulfur vapor react to form carbon disulfide and hydrogen sulfide. Carbon disulfide reacts with chlorine gas to form carbon tetrachloride and disulfur dichloride. Further reaction of carbon disulfide and disulfur dichloride produces additional carbon tetrachloride and solid sulfur. Write equations for these three reactions.

*97. The composition of a phosphate mineral can be expressed as % P, % P_2O_5, or % BPL [bone phosphate of lime, $Ca_3(PO_4)_2$].
   (a) Show that % P = $0.436 \times$ (% P_2O_5), and % BPL = $2.185 \times$ (% P_2O_5).
   (b) What is the significance of a % BPL greater than 100%?
   (c) What is the % BPL of a typical phosphate rock?

98. You are given these bond-dissociation energies in kJ/mol at 298 K: O_2, 498; N_2, 946; F_2, 159; Cl_2, 243; ClF, 251; OF (in OF_2), 187; OCl (in Cl_2O), 205; and NF (in NF_3), 280. Calculate $\Delta H°_f$ at 298 K for 1 mol of **(a)** $ClF(g)$; **(b)** $OF_2(g)$, **(c)** $Cl_2O(g)$, **(d)** $NF_3(g)$.

99. You have available elemental sulfur, chlorine gas, metallic sodium, and water. Show how you would use these substances to produce aqueous solutions of **(a)** Na_2SO_3, **(b)** Na_2SO_4, and **(c)** $Na_2S_2O_3$.

* 100. Write plausible half-equations and a balanced redox equation for the reaction of XeF_4 and water to form Xe, XeO_3, and HF. Xe and XeO_3 are produced in a $2:1$ mole ratio, and $O_2(g)$ is also formed.

101. Nitramide and hyponitrous acid both have the formula $H_2N_2O_2$. Hyponitrous acid is a weak diprotic acid; nitramide contains the NH_2 group. Based on this information, write plausible Lewis structures for these two substances.

102. On page 884, we discussed peroxides and superoxides. An old science-fiction story has lunar astronauts using (fictional) "potassium tetroxide," KO_4, to generate the oxygen they breathe. Another (real) potassium–oxygen compound is potassium ozonide, KO_3.

    (a) Write an equation for the reaction of each of these two compounds with water to form oxygen gas and other expected product(s).

    (b) Place KO_4, KO_3, KO_2, Na_2O_2, and LiO_2 in order from smallest to largest mass required to produce 10.0 L of oxygen gas at 25 °C and 0.35 atm.

* 103. On pages 880–881, we mentioned that the reaction of nitric acid with metals can produce several products having nitrogen with a lower oxidation number than in nitric acid. Two of these are $NO(g)$ and $NO_2(g)$. The listing of standard electrode potentials in Appendix C has a value for the reduction of NO_3^- to $NO(g)$, but it does not have one for its reduction to $NO_2(g)$. However, there are sufficient data in the table from which this missing value can be calculated.

    (a) Write a half-equation for the reduction of $NO_3^-(aq)$ to $NO_2(g)$ in acidic solution.

    (b) Show that the half-equation in (a) can also be obtained by a combination of the appropriate half-equations from Appendix C.

(c) To obtain the standard electrode potential for the half-reaction found in (b), we cannot simply take a difference in two electrode potentials as we did so often in Chapter 18. That method works only for combining half-equations in a way that produces an overall redox equation, and the value obtained is an E°_{cell}, not the E° for a reduction half-reaction. What needs to be done instead is to combine ΔG° values and then obtain E° from ΔG° for the half-reaction of interest. Follow this procedure to obtain the standard electrode potential for the reduction of nitric acid to $NO_2(g)$ in acidic solution. (*Hint:* What equation from Chapter 18 is required?)

(d) Show that an alternate method to obtain the desired E° value is to work directly with free energy data from Appendix C, rather than indirectly through E° data as was done in part (c).

* 104. The solubility of $Cl_2(g)$ in water is 6.4 g/L at 25 °C. Some of this chlorine is present as Cl_2, and some is found as HOCl or Cl^-, produced by the following hydrolysis reaction for which $K_c = 4.4 \times 10^{-4}$.

$$Cl_2(aq) + 2\,H_2O(l) \rightleftharpoons HOCl(aq) + H_3O^+(aq) + Cl^-(aq)$$

For a saturated solution of Cl_2 in water, calculate $[Cl_2]$, $[HOCl]$, and $[Cl^-]$.

Apply Your Knowledge

* 105. **[Environmental]** It has been estimated that if all the O_3 in the atmosphere were brought to sea level at STP, the gas would form a layer 0.3 cm thick. Estimate the number of O_3 molecules in Earth's atmosphere. (Assume that Earth's radius is 4000 mi.)

106. **[Laboratory]** Stannous chloride solution is used as a reducing agent in analytical chemistry. (a) After the stannous chloride solution has been prepared, a few pieces of tin metal are dropped into the bottle, to prevent oxidation by air. Use standard electrode potentials to explain the role of metallic tin. (b) In the determination of iron in an ore, the ore (mostly Fe_2O_3) is first dissolved in HCl(aq). A slight excess of Sn^{2+} is added to reduce Fe^{3+} to Fe^{2+} The excess Sn^{2+} is removed by adding $HgCl_2(aq)$, forming $Hg_2Cl_2(s)$, after which the solution is titrated (Figure 4.18). Write net ionic equations for the two redox reactions just described.

* 107. **[Biochemical]** The ability of the iron-containing enzyme peroxidase to catalyze the decomposition of hydrogen peroxide was described in an Application Note on page 885. This catalysis can also be achieved with the simple ion $Fe^{3+}(aq)$. A proposed mechanism for this catalysis involves two reactions. In the first reaction, Fe^{3+} is reduced to Fe^{2+} by H_2O_2. In the second reaction, the Fe^{2+} is oxidized back to Fe^{3+} while the H_2O_2 is reduced. Write an equation for the overall reaction, and show that it is spontaneous. According to this mechanism, what are the minimum and maximum possible E° values for the catalyst? Which of the following might also be suitable catalysts: (a) Cu^{2+}, (b) Br_2, (c) Al^{3+}, (d) Au^{3+}? Explain.

e-Media Problems

The activities described in these problems can be found in the e-Media Activities and Interactive Student Tutorial (IST) modules of the Companion Website, *http://chem.prenhall.com/hillpetrucci*.

108. From the **Formation of Aluminum Bromide** movie (*Section 21-3*) and the molecular structure shown in the **Aluminum Chloride dimer** 3-D model (*Section 21-3*), predict differences in the physical properties of aluminum bromide and aluminum chloride. Compare these properties to those of the aluminum-containing compounds, Al_2O_3 and AlF_3.

109. View the **Diamond, Graphite, and Buckminsterfullerene** 3D models (*Section 21-4*). (a) Which of structures include similar bonding geometries? (b) Account for the physical properties of the different allotropes of carbon in terms of the depicted structures.

110. Two common oxides of nitrogen are seen in the **Nitrogen Dioxide and Dinitrogen Tetroxide** movie (*Section 21-7*). These ox-

ides exist in equilibrium. (a) From the procedure carried out in the movie, predict whether the conversion from NO_2 to N_2O_4 is an endothermic or exothermic reaction. (b) How does your prediction compare to the result obtained from thermodynamic data (Appendix C)?

111. In the **Reactions with Oxygen** movie (*Section 21-9*), several oxidation reactions are shown. (a) Describe the change in oxidation number for each of the elements undergoing reaction with oxygen. (b) Which elements form more than one oxide? (c) Identify the oxidation states of the elements in the other possible compounds.

112. The **Physical Properties of Halogens** movie (*Section 21-12*) illustrates some properties of three halogens. (a) What evidence of a periodic trend do you find in this illustration? What is the underlying basis of this trend? (b) Are similar trends observed in neighboring groups?

Chapter 22

The *d*-Block Elements and Coordination Chemistry

OUR PRIMARY FOCUS in this chapter will be on the *d*-block elements. As in the preceding two chapters, we will relate the physical and chemical properties and uses of this large and varied group of elements to principles of bonding, structure, and reactivity.

Some of the *d*-block elements have many uses and are among the most important of all the elements. Others, because they are rare, have fewer uses. However, even some of the rare ones meet needs that only they can serve. Nine of the *d*-block elements have biological functions essential to human life.

One special characteristic of the *d*-block elements is their ability to form a variety of complex ions. We devote the latter half of this chapter to complex ions and the compounds that contain them, called coordination compounds. We will examine the nature of bonding in complex ions, the structures of complex ions, and properties related to their structures.

The *d*-Block Elements

All the *d*-block elements are metals. Because they provide a transition from the highly metallic *s*-block metals to the less metallic *p*-block metals, the *d*-block (and *f*-block) metals are known as *transition metals*. A more precise definition, however, states that a **transition metal** is an element following groups 1A and 2A in the periodic table and having *d*-orbital vacancies in the next-to-outermost electron shell of either its atoms or its cations. In this sense, zinc, cadmium, and mercury are not transition metals, because they do not have the required *d*-orbital vacancies in their atoms or cations. Our treatment in this chapter, then, is of the transition metals plus Zn, Cd, and Hg.

◀ Compounds containing *d*-block metals, and solutions of those compounds, are well known for their colorful appearance. Shown here are pale green nickel(II) chloride hexahydrate, orange sodium dichromate, purple potassium permanganate, blue-green nickel(II) sulfate hexahydrate, and crimson cobalt(II) chloride hexahydrate. In this chapter we will discuss the reasons for the colors seen in many transition-metal compounds.

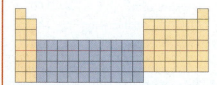

22.1 General Properties of the d-Block Elements and Their Trends

We begin our survey of the d-block elements with a look at some general properties and trends in these properties. We will compare the properties of the d-block elements with those of the s-block and p-block elements.

Some Properties of the Fourth-Period d-Block Elements

Let us look first at some trends in the properties of fourth-period d-block elements. In general, we can also observe these trends in the d-block elements of the fifth, sixth, and seventh periods.

The fourth-period elements with atomic numbers 21–30 are the first series of d-block elements. Table 22.1 lists some of their properties. These elements have their valence electrons in the $n = 4$ shell and a $3d$ subshell that is partially filled, except for Cu and Zn atoms, which have a filled $3d$ subshell.

The fourth-period d-block elements have electronegativity values that range from 1.3 to 1.9, as is characteristic of metals. Note also the gradual increase in electronegativity in moving from left to right across Table 22.1, a general trend described in Section 9.7. Finally, from the relationship between the percent ionic character and the electronegativities of the bonded atoms, we expect bonds between metal and nonmetal atoms to have somewhat less ionic character for d-block metals in general than for s-block metals.

Most of the cations of the fourth-period d-block atoms are produced by the loss of valence-shell electrons and one or more $3d$ electrons. Only zinc atoms fail to lose $3d$ electrons. The variability of oxidation number shown in the fifth row of Table 22.1 is typical for d-block elements. For the fourth-period elements up to and including manganese (group 7B), the highest oxidation number corresponds to the periodic table group number: +3 for Sc (group 3B), +4 for Ti (group 4B), and so on. With the later members of the series (beyond iron), the energy of the $3d$ subshell has fallen so far that the subshell effectively becomes part of the electron core (Figure 22.1). Beyond iron, electrons in the $3d$ subshell are less likely to participate in chemical bonding, and the range of oxidation numbers for elements is more limited.

Periodic Trends: First Row Transition Metals activity

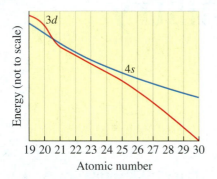

▲ **FIGURE 22.1** **Relative orbital energies in the fourth-period transition series**

With increasing atomic number in the fourth period of the periodic table, the energy level of the $3d$ subshell crosses and then drops below the energy level of the $4s$ subshell.

QUESTION: According to this figure, which element is the first one to obtain a 3d electron?

Table 22.1 Selected Properties of the d-Block Elements of the Fourth Period

	Sc	Ti	V	Cr	Mn	Fe	Co	Ni	Cu	Zn
Atomic number	21	22	23	24	25	26	27	28	29	30
Electron configuration[a]	$3d^14s^2$	$3d^24s^2$	$3d^34s^2$	$3d^54s^1$	$3d^54s^2$	$3d^64s^2$	$3d^74s^2$	$3d^84s^2$	$3d^{10}4s^1$	$3d^{10}4s^2$
Electronegativity	1.4	1.5	1.6	1.7	1.6	1.8	1.9	1.9	1.9	1.7
Common cations	3+	2+, 3+	2+, 3+	2+, 3+	2+, 3+	2+, 3+	2+, 3+	2+	1+, 2+	2+
Common positive oxidation numbers[b]	3	2, 3, 4	2, 3, 4 5	2, 3, 6	2, 3, 4 6, 7	2, 3, 6	2, 3	2, 3	1, 2	2
Atomic radius, pm	161	145	132	125	124	124	125	125	128	133
$E°$, V[c]	−2.03	−1.63	−1.13	−0.90	−1.18	−0.440	−0.277	−0.257	+0.340	−0.763
Melting point, °C	1397	1672	1710	1900	1244	1530	1495	1455	1083	420
Density, g/cm³	3.00	4.50	6.11	7.14	7.43	7.87	8.90	8.91	8.95	7.14
Electrical conductivity[d]	3	4	6	12	1	16	25	23	95	27
Thermal conductivity[d]	4	5	7	22	2	19	23	21	93	27

[a] Each atom has an argon core configuration.

[b] The most important oxidation numbers are printed in red.

[c] For the reduction $M^{2+}(aq) + 2\,e^- \longrightarrow M(s)$ [except for Sc, where the ion is $Sc^{3+}(aq)$].

[d] Electrical and thermal conductivities are on an arbitrary scale relative to 100 for silver, the best metallic conductor.

The negative standard electrode potentials in Table 22.1 show that the $M^{2+}(aq)$ cations are more difficult to reduce than is $H^+(aq)$, with $Cu^{2+}(aq)$ as the only exception:

$$M^{2+}(aq) + 2\,e^- \longrightarrow M(s) \qquad E° < 0\ V$$

$$2\,H^+(aq) + 2\,e^- \longrightarrow H_2(g) \qquad E° = 0\ V$$

$$Cu^{2+}(aq) + 2\,e^- \longrightarrow Cu(s) \qquad E° = +0.340\ V$$

Most of the *s*-block metals react with water to displace $H_2(g)$, but only scandium of the fourth-period *d*-block is active enough to do so.

The physical properties of the fourth-period *d*-block elements are also metallic: They have moderate to high melting points and moderately high densities. For example, chromium (group 6B) melts at 1900 °C and has a density at 25 °C of 7.14 g/cm^3, whereas sulfur (group 6A), a typical solid nonmetal, has a melting point of 119 °C and a density at 25 °C of 2.07 g/cm^3.

Perhaps the most distinctive physical property of the *d*-block metals is their ability to conduct electricity and heat. Although only copper comes close to matching silver in these abilities, the values for the other metals in Table 22.1 are still high compared with the values for nonmetals. For example, the electrical conductivity of silver is 6×10^{18} times that of the nonmetal white phosphorus, and silver's thermal conductivity is almost 2000 times greater.

Other Trends Among the *d*-Block Elements

Now let us look at trends within a group of *d*-block elements. Figure 8.9 shows that atoms generally become larger from the top to the bottom of a group in the periodic table. In comparing atomic radii in groups of *d*-block metals, we see exceptions to this rule (Figure 22.2). Moving from the fourth period to the fifth, we see that atomic radii do increase as expected, but going from the fifth period to the sixth, we see the radii remaining nearly constant or even decreasing slightly. (For example, note the decrease from Zr to its sixth-period counterpart, Hf.)

Why do we see these exceptions? The sixth period includes the 4*f* subshell, which fills before the 5*d* subshell. The shape of *f* orbitals is such that electrons in these orbitals are not very good at screening valence electrons from the nucleus. As a result, the increase in effective nuclear charge with increasing atomic number is greater than we anticipate, and consequently so is the strength of attraction of valence electrons to the nucleus. As a result, atomic size does not increase between the fifth- and sixth-period members of a group of transition elements, a phenomenon known as the **lanthanide contraction.**

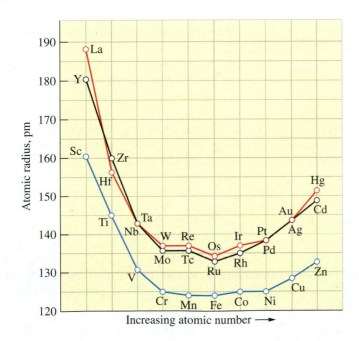

◀ **FIGURE 22.2   Atomic radii of the *d*-block elements of the fourth, fifth, and sixth periods**

The general trend for each period of the *d*-block is similar. The radii of the fourth-period atoms (blue) are smaller than those in the succeeding periods. However, because of the lanthanide contraction, the radii of the fifth-period atoms (black) and the corresponding sixth-period atoms (red) are nearly the same.

Table 22.2 Oxidation Numbers of Some Transition Metals

Period	Common oxidation numbers[a]	
	Group 6B	Group 8B
4	Cr: +2, +3, +6	Fe: +2, +3, +6
5	Mo: +2, +3, +4, +5, +6	Ru: +2, +3, +4, +6, +8
6	W: +2, +3, +4, +5, +6	Os: +2, +3, +4, +6, +8

[a] The most important oxidation numbers are printed in red.

A comparison within the group 6B and 8B transition elements (Table 22.2) shows that the heavier group members display a greater range of oxidation numbers in their compounds than the lighter members. In its oxides, iron has only the oxidation numbers +2 (in FeO) and +3 (in Fe_2O_3). In contrast, ruthenium and osmium, in addition to +2 and +3, also have oxidation numbers +4 (in RuO_2 and OsO_2) and +8 (in RuO_4 and OsO_4). Moreover, compounds in higher oxidation states exhibit more covalent character in their bonding, and therefore relatively low melting points. In line with this tendency, the melting points of FeO and Fe_2O_3 are 1337 and 1462 °C, respectively, whereas those of RuO_4 and OsO_4 are only 25 and 41 °C. Both RuO_4 and OsO_4 are molecular compounds; their melting points depend on the strengths of intermolecular forces rather than on interionic attractions.

Comparing Main-Group and *d*-Block Elements

Among the main-group elements, particularly those of the first and second periods, the *s* and *p* electrons of the valence shell establish the nature of the chemical bonds. In addition, *d* orbitals may also be involved in forming chemical bonds in *p*-block elements but presumably only in compounds in which the central atom has an expanded valence shell, such as SF_4, SF_6, and PF_5. With most of the *d*-block elements, in contrast, *d* orbitals are the ones that are important in forming chemical bonds. Also, differences in behavior between main-group and *d*-block elements regarding complex ion formation, color, magnetic properties, and catalytic activity all reflect the different roles played by *s*, *p*, and *d* orbitals.

22.2   The Elements Scandium Through Manganese

Now we shift our focus from a general overview to a closer look at specific elements.

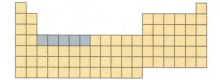

Scandium

Scandium is rather widely distributed in Earth's crust, but there is only one mineral from which it is usually extracted: *thortveitite*, $Sc_2Si_2O_7$. This scarcity of scandium minerals accounts in part for scandium's relatively recent history. Scandium oxide was not isolated until 1879, and the metal itself was not prepared until 1937. Scandium metal can be obtained by the electrolysis of a mixture of molten chlorides that includes $ScCl_3$.

Despite its relative rarity, the physical and chemical properties of scandium have been well characterized. Its chemistry is based mostly on the Sc^{3+} ion. In many ways, scandium resembles not the other transition metals but the main-group metals, particularly aluminum. For example, scandium reacts with either acidic or basic solutions, and even with water, to liberate $H_2(g)$. Another similarity to aluminum is that scandium hydroxide is amphoteric; $Sc(OH)_3(s)$ reacts with acidic solutions to produce Sc^{3+} and with alkaline solutions to form $[Sc(OH)_6]^{3-}$.

Titanium

Titanium is the ninth most abundant element in Earth's crust and the second most abundant transition metal (after iron). Its chief mineral sources are *rutile*, TiO_2, and *ilmenite*, $FeTiO_3$.

Most titanium is produced in a two-step process. First, $TiO_2(s)$ is heated with carbon and chlorine at about 800 °C to form gaseous $TiCl_4$. Then the $TiCl_4(g)$ is reduced with magnesium in an argon atmosphere at about 1000 °C to produce pure $Ti(s)$.

An electrochemical process developed in 2000 at the University of Cambridge in England promises to do for titanium what the Hall–Heroult process did for aluminum. In this process, represented in Figure 22.3, TiO_2 pellets are electrolyzed at the cathode of a cell containing molten calcium chloride. However, unlike the Hall–Heroult process, both the reactant TiO_2 and the product Ti remain in the solid state during the process.

Three desirable properties make titanium a highly useful metal: (1) low density (4.5 g/cm^3); (2) high structural strength, even at high temperatures; and (3) corrosion resistance. The first two properties account for its use in aircraft, racing bicycles, and gas-turbine engines. The third property accounts for its use in the chemical industry, especially for reaction vessels designed to hold mineral acids, alkaline solutions, chlorine, and a host of organic reagents.

The most important compound of titanium is the oxide, TiO_2, and its most important use is as a white pigment in paints, papers, and plastics. Naturally occurring rutile is impure TiO_2. It is usually colored because of impurities, and it requires extensive processing to yield the pure white powder required as a pigment.

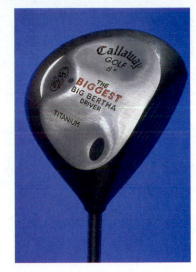

▲ Golf clubs are often made of titanium metal.

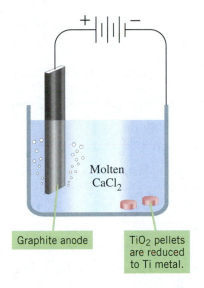

Molten $CaCl_2$

Graphite anode

TiO_2 pellets are reduced to Ti metal.

◀ **FIGURE 22.3   Schematic of a new method of producing titanium**

Pellets of $TiO_2(s)$ dissolve in molten $CaCl_2$, and the $Ti(IV)$ is reduced to a non-adherent, spongelike deposit of titanium metal at the container wall, which serves as the cathode.

QUESTION: The small circles surrounding the anode represent a gas being given off. What is this gas?

Vanadium

Vanadium is reasonably abundant in Earth's crust, ranking nineteenth among the elements. However, it is mostly obtained as a by-product of the production of uranium from the mineral *carnotite*, $K_2(UO_2)_2(VO_4)_2 \cdot 3 H_2O$. Vanadium metal can be prepared in a highly pure form, but mostly it is produced as an iron–vanadium alloy, *ferrovanadium*.

Vanadium is of interest because of (1) its use as an alloying element in steel; (2) the catalytic activity of some of its compounds, principally V_2O_5; and (3) the range of oxidation numbers in its ions. In steel, vanadium imparts strength and toughness, making the steel ideal for use in springs and high-speed machine tools. The catalytic activity of vanadium pentoxide probably involves the reversible loss of O_2 that occurs when V_2O_5 is heated. In the manufacture of sulfuric acid, V_2O_5 is the principal catalyst for the conversion of $SO_2(g)$ to $SO_3(g)$.

In solid compounds and in aqueous solutions, vanadium exhibits the oxidation numbers +2, +3, +4, and +5. The lower oxidation numbers are found in the monatomic cations V^{2+} and V^{3+}. When it is in the higher oxidation numbers, the vanadium is part of a polyatomic cation (VO^{2+} or VO_2^+) or a polyatomic anion (VO_4^{3-}).

▲ The yellow solution contains VO_2^+, in which the oxidation number of vanadium is +5. The blue solution contains VO^{2+}, with an oxidation number of +4 for vanadium. The green solution contains V^{3+}, and the violet solution contains V^{2+}.

▲ Chromium plating for decorative purposes is generally applied in an extremely thin layer—about 10 nm thick. The metal is first plated with a copper or nickel layer about 100 times thicker than the chromium layer. The function of the chromium is to provide an exceptionally bright surface.

Chromium

The most important source of chromium is the mineral called *chromite*, $FeCr_2O_4$. An iron–chromium alloy, *ferrochrome*, is made by reducing chromite ore with carbon:

$$FeCr_2O_4 + 4\,C \xrightarrow{\Delta} \underbrace{Fe + 2\,Cr}_{Ferrochrome} + 4\,CO(g)$$

Ferrochrome and other alloying elements can be added directly to iron to produce stainless steel. When needed, pure chromium can be prepared by reducing Cr_2O_3 in the thermite reaction:

$$Cr_2O_3(s) + 2\,Al(s) \longrightarrow Al_2O_3(s) + 2\,Cr(l)$$

In addition to its use in alloys, chromium can be plated onto other metals, generally by electrolysis from a solution containing CrO_3 in $H_2SO_4(aq)$. Because this electrolysis involves reducing Cr(VI) to Cr(0), it takes six moles of electrons to produce one mole of chromium plate. This means that chromium plating consumes more electric energy and requires a higher current density (amperes per unit area) than do other types of plating. (Silver plating, for instance, requires one mole of electrons per mole of Ag plated.) Despite the greater expense, chromium plating is desirable because the metal is hard and bright. Moreover, a transparent oxide coating covers the chromium, giving it some protection from corrosion.

When chromite ore is heated in molten sodium hydroxide in the presence of air, atmospheric oxygen oxidizes the $FeCr_2O_4$ to sodium chromate, Na_2CrO_4. Most other chromium compounds are prepared from sodium chromate.

Chromate ion in basic or neutral aqueous solutions is bright yellow (Figure 22.4a). When the solution is made acidic, the color changes to a bright orange because of the formation of dichromate ion, $Cr_2O_7^{2-}$ (Figure 22.4b). Chromate and dichromate ions participate in the reversible reaction

$$2\,CrO_4^{2-}(aq) + 2\,H^+(aq) \rightleftharpoons Cr_2O_7^{2-}(aq) + H_2O(l)$$

Dichromate ion is a common oxidizing agent in organic chemistry. In an acidic solution, it oxidizes alcohols to aldehydes, ketones, and carboxylic acids (page 150). It is also used extensively in the analytical chemistry laboratory; its standard electrode potential is large and positive,

$$Cr_2O_7^{2-}(aq) + 14\,H^+(aq) + 6\,e^- \longrightarrow 2\,Cr^{3+}(aq) + 7\,H_2O(l) \qquad E^\circ = 1.33\ V$$

and its aqueous solutions are remarkably stable, remaining unchanged in concentration for months or even years.

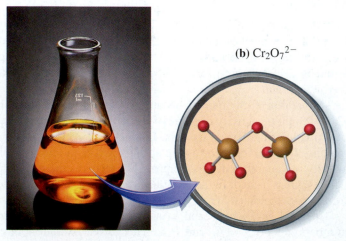

(a) CrO_4^{2-}

(b) $Cr_2O_7^{2-}$

▲ **FIGURE 22.4   Chromate and dichromate ions**
Chromium(VI) exists (a) as CrO_4^{2-} in basic solution and (b) as $Cr_2O_7^{2-}$ in acidic solution.
QUESTION: What is the oxidation number of Cr in each of these ions?

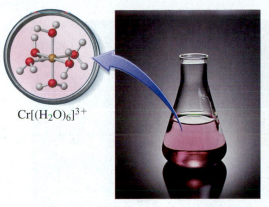

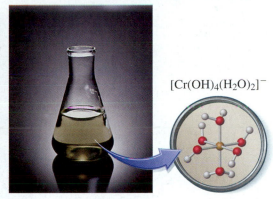

$Cr[(H_2O)_6]^{3+}$

$[Cr(OH)_4(H_2O)_2]^-$

▲ **FIGURE 22.5 Amphoterism of Cr(OH)$_3$(s)**

Freshly precipitated $Cr(OH)_3(s)$ (center) is amphoteric. It reacts in acidic solution, here $HNO_3(aq)$, to produce violet $[Cr(H_2O)_6]^{3+}(aq)$ (left). $Cr(OH)_3(s)$ also reacts in $NaOH(aq)$ to form gray-green $[Cr(OH)_4]^-(aq)$ (right).

QUESTION: Does oxidation–reduction also occur in these acid–base reactions?

In contrast, chromate ion in basic solution is not a good oxidizing agent; its standard electrode potential is slightly negative:

$$CrO_4^{2-}(aq) + 4\,H_2O(l) + 3\,e^- \longrightarrow Cr(OH)_4^-(aq) + 4\,OH^-(aq) \quad E^\circ = -0.13\ V$$

Instead, $CrO_4^{2-}(aq)$ is a precipitating agent. For example, lead chromate and barium chromate precipitates are encountered in the qualitative analysis scheme (Section 16.7). Silver chromate is the indicator for some precipitation titrations (Figure 4.17). Insoluble chromates, particularly $PbCrO_4$ and $ZnCrO_4$, find some use as paint pigments. These inorganic pigments are less affected by the action of sunlight and chemical agents than are organic dyes.

The main oxides of chromium are Cr_2O_3 and CrO_3. Both Cr_2O_3 and the corresponding hydroxide, $Cr(OH)_3$, are amphoteric. The amphoterism of $Cr(OH)_3$ is illustrated in Figure 22.5. The ability of $Cr(OH)_3$ to dissolve in an alkaline solution and the ease with which $[Cr(OH)_4]^-$ can be oxidized to CrO_4^{2-} both play a role in the qualitative analysis scheme.

The oxide CrO_3, in contrast, has only acidic properties. It dissolves in water to produce a strongly acidic solution:

$$2\,CrO_3(s) + 3\,H_2O(l) \longrightarrow 2\,H_3O^+(aq) + Cr_2O_7^{2-}(aq)$$

The name chromium is based on the Greek word *chroma,* which means color. The element is so named because its compounds exhibit many colors (Figures 22.4 and 22.5). The simple ions $Cr^{2+}(aq)$ and $Cr^{3+}(aq)$ form many complex ions that extend the variety of colors considerably. We will explore this matter later in the chapter.

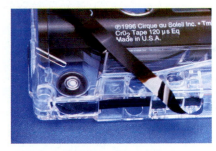

▲ Chromium(IV) oxide, CrO_2, has electrical and magnetic properties that make it useful in magnetic recording tape.

Example 22.1 A Conceptual Example

Write a plausible equation to explain the reaction shown in Figure 22.6, in which pure ammonium dichromate ignited with a match produces *pure* chromium(III) oxide.

ANALYSIS AND CONCLUSIONS

The oxidation number of Cr in ammonium dichromate, $(NH_4)_2Cr_2O_7(s)$, is $+6$. In chromium(III) oxide, the oxidation number of Cr is $+3$. Thus the dichromate ion is reduced. If a reduction occurs, an oxidation must also occur.

In the ammonium ion, NH_4^+, the oxidation number of N is -3, the lowest possible for N. Ammonium ion can be oxidized to any of several higher oxidation states. Because $Cr_2O_3(s)$ is the only product found in the reaction container, any other product(s) must be gaseous. One of the gases is probably $H_2O(g)$, formed from the H and O atoms of the reactant. By assuming that four H_2O's are produced, we can account for all the Cr, H, and O atoms in the reactant:

$$(NH_4)_2Cr_2O_7(s) \longrightarrow Cr_2O_3(s) + 4\,H_2O(g) \quad \text{(equation incomplete)}$$

▲ **FIGURE 22.6**   **Decomposition of ammonium dichromate**

That there is no N in either of these products tells us that there must be a third product, a gas, that contains N. We might postulate it to be N_2, N_2O, NO, or NO_2, but because we have already accounted for all the O initially present, the additional product is probably $N_2(g)$. (If any oxides were to form, the required oxygen would have to come from the air.) A plausible complete equation is therefore

$$(NH_4)_2Cr_2O_7(s) \longrightarrow Cr_2O_3(s) + 4\,H_2O(g) + N_2(g)$$

EXERCISE 22.1A

Describe what you would expect to observe if the dichromate salt used in Figure 22.6 were potassium dichromate.

EXERCISE 22.1B

As noted on page 908, most chromium compounds are made from sodium chromate. The sodium chromate is produced by heating chromite ore in molten NaOH in the presence of air. Write a plausible equation for this reaction. [*Hint:* Both Cr(III) and Fe(II) are oxidized.]

Manganese

Manganese is obtained mainly from the mineral *pyrolusite,* MnO_2. Relatively pure manganese can be made by reducing MnO_2. However, because most manganese is used in steel alloys, common practice is to reduce a mixture of MnO_2 and Fe_2O_3 to obtain an iron–manganese alloy called *ferromanganese:*

$$MnO_2 + Fe_2O_3 + 5\,C \xrightarrow{\Delta} \underbrace{Mn + 2\,Fe}_{\text{Ferromanganese}} + 5\,CO(g)$$

Ferromanganese alloys are wear-resistant and shock-resistant and are used for railroad tracks, bulldozers, and road scrapers. The characteristic color of the Sacagawea "golden dollar" is that of *manganese brass*, a copper–zinc alloy containing a small percentage of manganese.

We can explain the chemical behavior of manganese and its compounds with the help of its electron configuration, $[Ar]3d^54s^2$:

First, by giving up its two valence electrons, a manganese atom acquires the oxidation number +2. Then, by making use of its unpaired 3d electrons, the manganese atom can attain all additional oxidation numbers from +3 to +7.

Manganese dioxide is the starting point for making most other manganese compounds. To obtain the higher oxidation numbers of manganese, MnO_2 is first oxidized to a manganate salt, such as K_2MnO_4, in which the oxidation number of Mn is +6. A strong oxidizing agent ($KClO_3$) in a basic medium (KOH) is required:

$$3\ MnO_2(s) + ClO_3^-(aq) + 6\ OH^-(aq) \longrightarrow 3\ MnO_4^{2-}(aq) + Cl^-(aq) + 3\ H_2O(l)$$

To obtain Mn(VII), K_2MnO_4 is oxidized one step further to $KMnO_4$, with, for instance, Cl_2 as the oxidizing agent.

Potassium permanganate is an important oxidizing agent that is used in both analytical and organic chemistry laboratories. Its oxidizing power also makes it useful as a medical disinfectant and as a substitute for Cl_2 in water purification. In oxidation–reduction reactions in acidic solutions, MnO_4^- is usually reduced to Mn^{2+}; in basic solutions, it is reduced to $MnO_2(s)$. Titrations with permanganate ion are self-indicating because of its deep purple color (Figure 4.18).

The lower oxidation states of manganese can be obtained by reducing $MnO_2(s)$. For example, in the cell reaction in a dry cell (Section 18.7), Zn is oxidized to Zn^{2+} and MnO_2 is reduced to Mn_2O_3. The starting point for the preparation of Mn(II) compounds is the reaction

$$MnO_2(s) + 4\ H^+(aq) + 2\ Cl^-(aq) \longrightarrow Mn^{2+}(aq) + 2\ H_2O(l) + Cl_2(g)$$

▲ The oxidizing power of potassium permanganate is being used here to tarnish a decorative silver shape on a bowl.

Example 22.2

Use data from Appendix C to demonstrate that permanganate ion in acidic solution is a more powerful oxidizing agent than either $O_2(g)$ or $Cr_2O_7^{2-}(aq)$.

STRATEGY

In a redox reaction, an oxidizing agent undergoes reduction. To compare the strengths of oxidizing agents, we can compare the standard electrode potentials for the half-reactions in which the oxidizing agents are reduced.

SOLUTION

We find these values in Appendix C:

$$MnO_4^-(aq) + 4\ H^+(aq) + 3\ e^- \longrightarrow MnO_2(s) + 2\ H_2O(l) \qquad E° = 1.70\ V$$

$$MnO_4^-(aq) + 8\ H^+(aq) + 5\ e^- \longrightarrow Mn^{2+}(aq) + 4\ H_2O(l) \qquad E° = 1.51\ V$$

$$O_2(g) + 4\ H^+(aq) + 4\ e^- \longrightarrow 2\ H_2O(l) \qquad E° = 1.229\ V$$

$$Cr_2O_7^{2-}(aq) + 14\ H^+(aq) + 6\ e^- \longrightarrow 2\ Cr^{3+}(aq) + 7\ H_2O(l) \qquad E° = 1.33\ V$$

The larger $E°$ values for the reduction half-reactions involving $MnO_4^-(aq)$ suggest that in acidic solution, $MnO_4^-(aq)$ is a stronger oxidizing agent than $O_2(g)$ or $Cr_2O_7^{2-}(aq)$.

EXERCISE 22.2A

Show that $MnO_4^-(aq)$ is a good oxidizing agent in basic solution. (*Hint:* What are some oxidations it is able to bring about?)

EXERCISE 22.2B

The oxidation of Mn^{2+} to MnO_4^- in an acidic solution is a confirming test for the presence of manganese in qualitative analysis cation group 3 (Section 16.7). Bismuthate ion from $NaBiO_3$ is used as the oxidizing agent. The bismuthate ion is reduced to BiO^+. **(a)** Write half-equations and an overall equation for this reaction. **(b)** Estimate the standard electrode potential for the reduction half-reaction.

Application Note

In developing countries, solutions of potassium permanganate are used as a disinfectant to wash vegetables, which are often grown with human waste as a fertilizer. In India, these solutions are called *pinky water*.

A brown ring is often found on the inside surface of bottles in which permanganate solutions are stored. The ring is $MnO_2(s)$ formed by the reduction of $MnO_4^-(aq)$. Some $H_2O(l)$ is simultaneously oxidized to $O_2(g)$.

22.3 The Iron Triad: Iron, Cobalt, and Nickel

Iron is the fourth most abundant element in Earth's crust. Cobalt and nickel are not nearly as common, but they are still sufficiently abundant that their annual production is thousands of tons. Cobalt is used primarily in alloys with other metals. About 80% of the U.S. production of nickel goes into alloys, and about 15% is used for electroplating. Smaller quantities of nickel are used as electrodes in batteries and

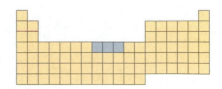

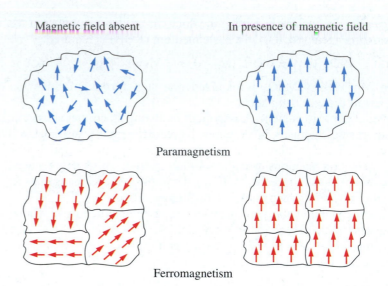

Paramagnetism

Ferromagnetism

Ferromagnetism activity

▲ Color-enhanced image of magnetic domains in a ferromagnetic garnet film.

fuel cells and as a catalyst. Another familiar use is in coinage. The U.S. five-cent coin is 25% nickel.

Ferromagnetism

Most of the transition elements—including iron, cobalt, and nickel—have some unpaired *d* electrons. These unpaired electrons account for *paramagnetism* (Figure 8.7), but Fe, Co, and Ni exhibit a much stronger magnetic effect, known as **ferromagnetism.**

In a paramagnetic solid, a *magnetic moment* is associated with each atom or ion that has unpaired electrons. (Think of a magnetic moment as being like a microscopic bar magnet, having a north pole at one end and a south pole at the other.) As shown in Figure 22.7, the magnetic moments are randomly oriented in the absence of a magnetic field. When the paramagnetic solid is placed in a magnetic field, most of the magnetic moments align in the direction of the field. When the magnetic field is removed, however, the directions of the magnetic moments again become random.

A ferromagnetic solid consists of regions called *domains*. Each domain contains a large number of atoms that have their individual magnetic moments aligned—that is, pointed in the same direction. However, this direction varies from one domain to another, as shown in Figure 22.7. When the ferromagnetic solid is placed in a magnetic field, all the magnetic moments in the domains line up in the direction of the field. The solid becomes magnetized. Moreover, when the field is removed, the orientation of the magnetic moments persists and the solid remains magnetized.

Paramagnetism (a temporary condition) becomes ferromagnetism (a permanent condition) only when the interatomic distances are just right, allowing atoms to be arranged into domains. Iron, cobalt, and nickel meet the requirements, as do certain alloys of other metals, such as manganese in the combinations Al–Cu–Mn, Ag–Al–Mn, and Bi–Mn.

Oxidation States

All three iron-triad elements form 2+ ions:

$$Fe^{2+}[Ar]3d^6 \qquad Co^{2+}[Ar]3d^7 \qquad Ni^{2+}[Ar]3d^8$$

For cobalt and nickel, the oxidation state +2 is the most common, but for iron the most common oxidation state is +3. When an iron atom loses its third electron, forming Fe^{3+}, the electron configuration becomes $[Ar]3d^5$. A half-filled 3*d* subshell with five unpaired electrons is a very stable electron configuration. As a result, iron(II) is oxidized to iron(III) without difficulty:

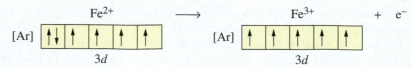

$$Fe^{2+} \longrightarrow Fe^{3+} + e^-$$

[Ar] ⇅ ↑ ↑ ↑ ↑   3d   [Ar] ↑ ↑ ↑ ↑ ↑   3d

Iron(III) can be oxidized further to FeO_4^{2-}, but only with great difficulty. The conversions of Co(II) and Ni(II) to Co(III) and Ni(III) are also difficult because their +3 oxidation states do not correspond to half-filled $3d$ subshells. The +3 oxidation state of cobalt can be stabilized in complex ions, such as $[Co(NH_3)_6]^{3+}$, but the +3 oxidation state is rare in nickel compounds.

22.4 Group 1B: Copper, Silver, and Gold

From earliest times, copper, silver, and gold have been used to make coins. Their use in coins is based on the comparative lack of reactivity of the metals, in keeping with their positive electrode potentials (Table 22.3). Thus, the metal ions are easily reduced to the free metals, and, conversely, the metals are difficult to oxidize and often are found free in nature.

Another insight gained from Table 22.3 concerns atomic radii and densities. That silver and gold have the same atomic radius is another example of the lanthanide contraction. Although the two atoms are about the same size, gold atoms have nearly twice the mass of silver atoms. This largely accounts for the fact that the density of gold is much higher than that of silver.

The group 1B elements are exemplary metals, in terms of certain physical properties. They are exceptionally malleable and ductile, and they have the highest electrical and thermal conductivities of all the metals. Like transition elements in general, the group 1B metals exist in various oxidation states, but the number of states is limited, just as it is in the iron triad. Additional similarities to other transition elements include paramagnetism and color in some of their compounds, along with the ability to form complex ions.

Because of their durability, metallic luster, malleability, and ductility, the group 1B metals are valued in the decorative arts and in jewelry making. Electrical wiring for residential use is almost always made of copper. Silver and gold are used in electronic components. The resistance of copper to corrosion accounts for its use in plumbing, whereas its good thermal conductivity makes it useful in cookware. The primary use of gold, of course, is as a monetary reserve for individuals and nations.

We have noted that the group 1B metals do not displace $H^+(aq)$ from solutions. None of them reacts with HCl(aq). Copper and silver do react with concentrated $H_2SO_4(aq)$ to produce Cu^{2+} or Ag^+ and SO_2, and with concentrated $HNO_3(aq)$ they also react to produce Cu^{2+} or Ag^+ and oxides of nitrogen. However, no H_2 is produced in any of these reactions. Gold does not react with any single acid, but it will react with *aqua regia*, a

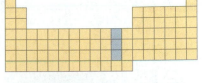

▲ **Several ancient coins**
The group 1B metals—copper, silver, and gold—are called coinage metals because they have long been used in coins.

Table 22.3 Some Properties of Copper, Silver, and Gold

	Cu	Ag	Au
Atomic number	29	47	79
Electron configuration	$[Ar]3d^{10}4s^1$	$[Kr]4d^{10}5s^1$	$[Xe]4f^{14}5d^{10}6s^1$
Atomic radius, pm	128	144	144
Electronegativity	1.9	1.9	2.4
Oxidation numbers[a]	+1, +2	+1, +2, +3	+1, +3
Electrode potential, V			
$\quad M^+(aq) + e^- \longrightarrow M(s)$	+0.522	+0.800	+1.83
$\quad M^{2+}(aq) + 2\,e^- \longrightarrow M(s)$	+0.340	+1.39	—
$\quad M^{3+}(aq) + 3\,e^- \longrightarrow M(s)$	—	—	+1.52
Melting point, °C	1083	962	1064
Density, g/cm^3	8.96	10.5	19.3
Electrical conductivity[b]	95	100	68
Thermal conductivity[b]	93	100	74

[a] The most common oxidation number is printed in red.

[b] Electrical and thermal conductivities are on an arbitrary scale relative to 100 for silver, the best metallic conductor.

d-Block Elements in Living Matter

As indicated in Figure 22.8, relatively few elements are essential to living organisms, and those with low atomic numbers ($Z < 21$) make up the bulk of living matter. Most plant and animal life requires only trace amounts of some elements, and most of these trace elements are members of the fourth-period *d*-block. A 70-kg adult human, for example, contains only about 4 g of iron, but that amount is essential for good health.

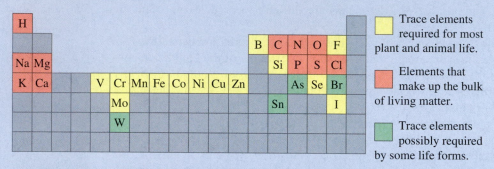

□ Trace elements required for most plant and animal life.

□ Elements that make up the bulk of living matter.

□ Trace elements possibly required by some life forms.

▲ **FIGURE 22.8** **The elements in living matter**

Most of the elements essential to life are relatively abundant in the natural environment. Nine of the 11 most essential elements are also the most abundant in seawater: H, O, Cl, Na, Mg, S, K, Ca, and C. The other two, N and P, are also among the 20 most abundant elements in seawater. It is believed that the life forms on Earth developed from the elements readily available to them.

In biological matter, the transition metals are often incorporated into large organic molecules. The protein hemoglobin, found in red blood cells and responsible for the color of blood, consists of iron-containing heme units attached to polypeptide molecules called globins (Figure 22.9). Hemoglobin transports O_2 molecules from the lungs through the arteries to all parts of the body for use in the metabolism of glucose. It then carries CO_2 produced in this metabolism through the veins back to the lungs.

To perform its essential function, hemoglobin must have iron present as Fe^{2+}. If the iron is oxidized to Fe^{3+}, the resulting substance, called methemoglobin, cannot transport oxygen. This oxidation occurs in the disease called *methemoglobinemia*. High concentrations of nitrate ion in groundwater used for domestic purposes can cause this condition.

Another important biochemical process involving both iron and molybdenum is the fixation of atmospheric nitrogen by certain bacteria through the action of the enzyme *nitrogenase*. Nitrogen molecules are apparently held to Mo atoms in the enzyme, as seen at the right.

The enzyme acts by facilitating the transfer of six electrons to the N≡N molecule, reducing the oxidation numbers of the N atoms from 0 to −3. The N atoms are then released as NH_3 molecules. The electron transfers appear to occur through changes in the oxidation numbers of the Fe and Mo atoms. Thus, the variability of oxidation numbers of *d*-block elements plays an important role in the biological activity of the transition elements.

◀ **FIGURE 22.9**
Heme and hemoglobin

(a) The structure of heme. (b) Four heme units and four coiled polypeptide chains are bonded together in a molecule of hemoglobin.

mixture of 1 part HNO_3 and 3 parts HCl. In this reaction, HNO_3 oxidizes gold to Au^{3+}, and HCl furnishes the Cl^- necessary to form the complex ion $[AuCl_4]^-$. Formation of this stable complex ion drives the reaction in the forward direction by the removal of Au^{3+} ions:

$$Au(s) + 4 H^+(aq) + NO_3^-(aq) + 4 Cl^-(aq)$$
$$\longrightarrow [AuCl_4]^-(aq) + 2 H_2O(l) + NO(g)$$

Gold is resistant to air oxidation, as is silver, but silver tarnishes by reacting with sulfur compounds to produce Ag_2S (page 780). Copper resists corrosion in dry air, but in moist air, it forms green basic copper carbonate, $Cu_2(OH)_2CO_3$.

The principal manufactured compound of copper is copper(II) sulfate pentahydrate, $CuSO_4 \cdot 5H_2O$. It is used in electroplating, in batteries, and to prepare other copper salts. Although essential in trace amounts to living organisms, Cu^{2+} ion is toxic at higher concentrations. As a result, copper sulfate is an important pesticide used against bacteria, algae, and fungi.

Silver nitrate is the chief manufactured compound of silver. Among its varied uses are in silver plating and silvering mirrors, as an antiseptic, and as a reagent in analytical chemistry laboratories. Its most important use, however, is in making silver halides (AgCl, AgBr, AgI) for photography.

Gold compounds are used in electroplating, photography, medicinal chemistry, and in ruby glass and ceramics.

▲ Basic copper carbonate, $Cu_2(OH)_2CO_3$, imparts a characteristic green color to copper roofs and bronze statues.

22.5   Group 2B: Zinc, Cadmium, and Mercury

Zinc, cadmium, and mercury are *d*-block metals, but they resemble transition metals only in certain ways—for example, in forming many complex ions. In other ways, they resemble the alkaline earth metals of group 2A. (Mendeleev combined groups 2A and 2B into group II of his periodic table.) For example, zinc and cadmium form only the 2+ cation, and the cations are diamagnetic and colorless in solution.

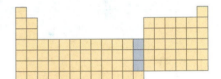

Zinc

Zinc has many uses. Large quantities are used in alloys. For example, *brass* is a copper alloy containing from 20 to 45% zinc plus small quantities of tin, lead, and iron. Brass has an attractive yellow metallic sheen, is corrosion-resistant, and is a good electrical conductor. Zinc oxidizes in air to form a thin, adherent oxide coating that protects the underlying metal from further corrosion. This property explains the use of zinc in making galvanized (zinc-coated) iron. Also, the fact that zinc is more easily oxidized than iron underlies its use as sacrificial anodes that offer cathodic protection to iron (Section 18.8).

Cadmium

Cadmium can substitute for zinc in coating metals for certain applications. Its primary uses are in alloys and as electrodes in batteries. Because of its capacity to absorb neutrons, cadmium is used in control rods in nuclear reactors. The uses of cadmium are severely limited, however. Whereas zinc in trace amounts is an essential element for humans, cadmium is quite toxic. Its effect may be to substitute for zinc in certain enzymes.

Mercury

Mercury differs from zinc and cadmium in at least six significant ways. Mercury is a liquid at room temperature. It does not displace H_2 from acidic solutions. It exists in the +1 oxidation state in the diatomic ion Hg_2^{2+}, which features an Hg–Hg covalent bond. Mercury forms many molecular compounds, such as $HgCl_2$, which are only slightly ionized in aqueous solution. Mercury forms few water-soluble compounds, and most of its compounds are not hydrated. Mercury shows little tendency to oxidize, and its oxide, HgO, readily decomposes to the elements upon heating.

▲ Brass is used extensively on ships because it resists the corrosive action of seawater.

In humans, mercury affects the nervous system and causes brain damage. Mercury compounds were once used to convert fur to felt for felt hats, and the condition known as hatter's disease (which probably afflicted the Mad Hatter in *Alice's Adventures in Wonderland*) is a form of chronic mercury poisoning.

The fact that mercury has a high density and is a metal and a liquid determines many of its uses. It is used in thermometers, barometers, manometers, electric relays and switches, and as electrodes for certain batteries and electrolysis cells. Some uses of mercury are related to its ability to form alloys called *amalgams* with most metals. Dental amalgam used to fill tooth cavities consists primarily of silver and tin alloyed with mercury. The rates at which the amalgam expands and contracts in response to changes in temperature closely match the rates at which teeth expand and contract. Mercury vapor is widely used in fluorescent tubes and street lamps. The quantity in one lamp is so small, however, that the total amount of mercury used for this purpose is quite small.

Liquid mercury was once treated as a laboratory plaything, but we now know that long-term exposure can present a serious health hazard. Mercury vapor is toxic, and levels exceeding 0.05 mg Hg/m^3 are unsafe. Liquid mercury has a low vapor pressure, but even so, the concentration of mercury in the saturated vapor at room temperature is well above the safe limit. Unsafe levels of mercury vapor have been found in various locales in which the element is used—chlor-alkali plants, thermometer factories, smelters, and dental laboratories. In chlor-alkali plants that use mercury, the loss to the environment has been reduced to about 200 mg Hg per ton of Cl_2 produced. This is about 1000 times less than decades ago. In dentistry, various metal powders, porcelain, and plastics are supplanting dental amalgam in many applications.

Mercury poisons the body's systems, in part by interfering with sulfur-containing enzymes. The Hg^{2+} ion reacts with sulfhydryl (SH) groups in an enzyme to change the shape of the enzyme and render it inactive (Figure 13.24). Some microorganisms can convert inorganic mercury to methylmercury (CH_3Hg^+) compounds, which are readily absorbed by organisms. These compounds concentrate in the marine food chain and can lead to unsafe mercury levels in fish.

22.6   The Lanthanide Series (Rare Earths)

The elements cerium ($Z = 58$) through lutetium ($Z = 71$) make up the first series of the f-block. We can think of this first f-block series as an insertion in the sixth-period d-block. The series is preceded by a d-block element, lanthanum, and followed by a d-block element, hafnium. Because the 14 elements in the series immediately follow lanthanum, they are often called the *lanthanide elements*. They are also called *inner-transition elements* because the subshell that is being filled is two principal quantum levels below the valence shell; in the sixth period, the $4f$ subshell fills before the $5d$. The elements of the lanthanide series are often referred to by the general symbol Ln. Their most striking feature is their similarity to one another and to La and the group 3B metals.

These elements are often called the *rare earths*, a name of historical origin but a clear misnomer because several are not rare at all. Cerium (Ce) is the twenty-fifth most abundant element in Earth's crust, more abundant than cobalt, tin, or arsenic. Cerium, neodymium (Nd), and yttrium (Y) are all more abundant than lead, and all the lanthanides except promethium (Pm) are more abundant than such well-known elements as Cd, Ag, Hg, and Au. All promethium isotopes are radioactive.

In spite of their relative abundance, the term "rare earths" seems appropriate for two reasons: (1) For many years after their discovery, there was only one known source for all these elements—*monazite*, a mixed phosphate of La, Th, and the lanthanides; (2) the individual lanthanides are difficult to separate and isolate. (In the nineteenth century, it was extremely difficult to separate them.)

When ions differ sufficiently in a property, we can often separate them in a one-step process. For example, we can readily separate Ag^+ and Cu^{2+} in an aqueous solution by adding Cl^- to the solution; Ag^+ completely precipitates as AgCl(s), and Cu^{2+} remains in solution. If we try to use differences in solubility to separate individual lanthanide cations (Ln^{3+}), however, we get only a slight separation. We have to go through many, many cycles of precipitation and redissolving to achieve total separation. Researchers took nearly seven decades, from 1839 to 1907, to isolate all the lanthanides except promethium (which, because it is found only as a product of nuclear fission, was not isolated until 1945). Recall that in 1913 H. G. J. Moseley related the frequency of X rays to the atomic number of the target element in an X-ray tube (page 309). In just a few days, he was able to identify by atomic number all the lanthanides in a sample of monazite. Today, lanthanides are generally separated by complex ion formation and ion-exchange techniques. Newer techniques, in which the oxidation states are adjusted and the lanthanides are converted to di- and trihalides that differ in volatility, promise to make separation easier and thus less expensive.

For many uses, the lanthanides do not have to be separated from one another. A mixture of the lanthanide metals with about 25% La is used in certain steel and magnesium-based alloys. Lanthanide–cobalt alloys are used to make extremely stable permanent magnets, which are used in such electronic devices as computer hard disk drives. Several lanthanide compounds also have commercial uses. Some of the oxides are used to color glass and ceramic glazes. Neodymium makes glass blue; praseodymium, green; erbium, pink; and holmium, blue. Lanthanide oxides are also used as phosphors in the screens of color television sets and as catalysts. Cerium(IV) compounds are important oxidizing agents in analytical chemistry. The $E°$ value for the reduction of $Ce^{4+}(aq)$ to $Ce^{3+}(aq)$ is greater than the $E°$ value for any reductions employing $Cr_2O_7^{2-}(aq)$ or $MnO_4^-(aq)$.

▲ Rare-earth magnets contain neodymium and are extremely strong for their size.

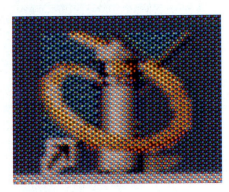

▲ The individual elements of a CRT computer screen are phosphors that contain elements of the lanthanide series. The phosphors glow red, blue, or green, depending on their composition. (™*Netscape*®)

Coordination Chemistry

The two substances in Figure 22.10 have similar formulas, but they are strikingly different in color. They are both *coordination compounds*, which means that each of them is derived from two simpler compounds that are somehow coordinated, or joined together. The formulas given in Figure 22.10 suggest that the two simpler compounds are $CoCl_3$ and NH_3.

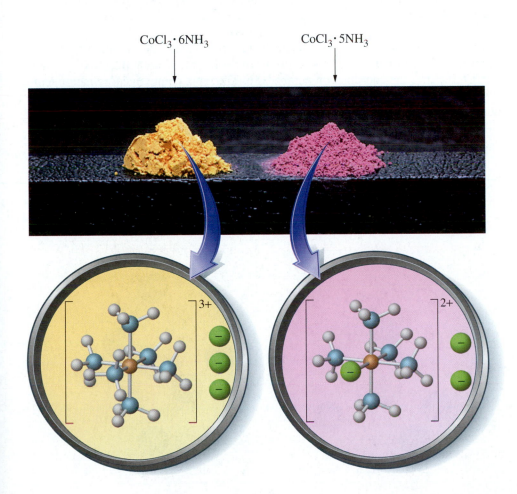

$CoCl_3 \cdot 6NH_3$ $CoCl_3 \cdot 5NH_3$

 Cobalt Coordination Compounds 3D models

◀ **FIGURE 22.10   Two coordination compounds of cobalt(III)**

The two coordination compounds differ in more than color. Each has three moles of Cl⁻ per mole of compound. When one mole of the golden orange compound is dissolved in water and treated with $AgNO_3(aq)$, three moles of $AgCl(s)$ are obtained, just as we might expect. When we treat the purple compound the same way, however, only two moles of $AgCl(s)$ are obtained from one mole of compound. In unraveling the mystery of these two compounds and others of a similar nature, Swiss chemist Alfred Werner established the modern basis of coordination chemistry.

22.7   Werner's Theory of Coordination Chemistry

In 1893, Alfred Werner explained the difference between the golden orange and purple compounds of Figure 22.10 by representing their dissociation in water as follows:

$$[Co(NH_3)_6]Cl_3(s) \xrightarrow{H_2O} [Co(NH_3)_6]^{3+}(aq) + 3\ Cl^-(aq)$$
Golden orange

$$[CoCl(NH_3)_5]Cl_2(s) \xrightarrow{H_2O} [CoCl(NH_3)_5]^{2+}(aq) + 2\ Cl^-(aq)$$
Purple

In the purple compound, one chloride ion is joined to the cobalt ion by a coordinate covalent bond and remains so in solution. Thus, the purple compound yields only two moles of $AgCl(s)$ when treated with an excess of $AgNO_3(aq)$.

Some Terminology for Complexes

We can use Werner's two compounds, $[Co(NH_3)_6]Cl_3$ and $[CoCl(NH_3)_5]Cl_2$, to illustrate some of the terms used to describe complexes. An entity within brackets—for example, $[Co(NH_3)_6]^{3+}$—is properly called a *coordination entity,* but more commonly it is called a complex. As we noted in Section 16.6, a **complex** consists of a central atom, which is usually a metal atom or ion (Co^{3+} in our present example), surrounded by attached groups called **ligands,** which may be either neutral molecules or anions. The ligands may all be of the same type, as in $[Co(NH_3)_6]^{3+}$, or of different types, as in $[CoCl(NH_3)_5]^{2+}$.

The region surrounding the central atom or ion and containing the ligands is called the *coordination sphere.* The **coordination number** is the total number of points at which ligands are attached to the central atom or ion. In both $[Co(NH_3)_6]^{3+}$ and $[CoCl(NH_3)_5]^{2+}$, the coordination number is 6. The most common coordination numbers seen in complexes are 2, 4, and 6, and the geometric shapes they produce are shown in Figure 22.11. Depending on the charges on the metal and on the ligands, the

▲ Alfred Werner (1866–1919) claimed that the basic idea of coordination theory came to him, at age 25, during his sleep. He arose at 2 A.M. and worked out the essentials by 5 A.M. Werner was awarded the 1913 Nobel Prize in Chemistry for his work in coordination chemistry.

Coordination Compound activity

Coordination Number activity

Diamminesilver(I) ion, Tetrammineplatinum(II) ion, Tetramminezinc(II) ion, Hexamminecobalt(III) ion 3D models

▶ **FIGURE 22.11   Four common shapes of complex ions**

In these complex ions, the NH₃ molecules are attached to the central metal ion through the lone-pair electrons on the N atom.

QUESTION: What is the coordination number of each metal ion in these structures?

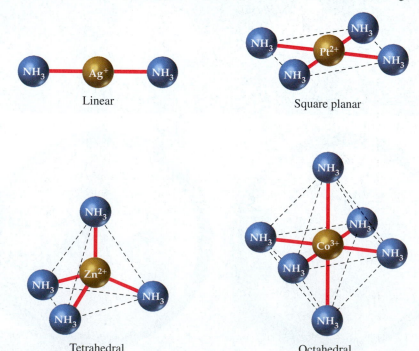

overall charge on a complex may be positive, negative, or zero. If a complex carries a net electric charge, it is called a **complex ion.** For instance, the complex ions $[Co(NH_3)_6]^{3+}$ and $[CoCl(NH_3)_5]^{2+}$ have charges 3+ and 2+, respectively. A compound consisting of one or more complexes is called a **coordination compound.** Golden orange $[Co(NH_3)_6]Cl_3$ and purple $[CoCl(NH_3)_5]Cl_2$ are both coordination compounds.

Example 22.3

What are the coordination number and the oxidation number of the central atom in (a) $[CoCl_4(NH_3)_2]^-$ and (b) $[Ni(CO)_4]$?

STRATEGY

In each complex, the coordination number is equal to the number of points of attachment between ligands and the central atom. The oxidation number of the central atom is determined by considering the overall charge of the complex in light of the individual charges of the ligands.

SOLUTION

(a) The metal atom (cobalt) at the center of this complex ion has six ligands attached—four Cl^- ions and two NH_3 molecules. The coordination number is therefore 6. The complex is an anion with a net charge of 1−. The NH_3 ligands carry no charge, and the four Cl^- ions carry a total charge of 4−. If we let x equal the charge on the cobalt, then $x - 4 = -1$, and $x = +3$. The central ion is therefore Co^{3+}, and its oxidation number is +3.

(b) The central metal in this complex is nickel, and four ligands—all CO molecules—are attached to it, making the coordination number 4. Because both the complex and the CO molecules are electrically neutral, the charge on the Ni must be zero. The oxidation number of Ni is therefore also 0.

EXERCISE 22.3A

What are the coordination number and the oxidation number of the central atom in (a) $[Co(SO_4)(NH_3)_5]^+$ and (b) $[Fe(CN)_6]^{4-}$?

EXERCISE 22.3B

Write the formula of a coordination compound that consists of free chloride anions plus a complex cation containing Cr(III) as the central ion and two chloride ions and four water molecules as ligands.

Ligands

Many metal atoms and ions, particularly those of the transition metals, have empty orbitals that can accommodate electron pairs. The central metal of a complex is a Lewis acid (electron-pair acceptor). Ligands have lone-pair electrons and are Lewis bases (electron-pair donors).

The atom in a ligand that furnishes an electron pair is called a *donor atom.* Ligands with one donor atom have just one point of attachment to the central metal atom or ion. They are **monodentate** (one-toothed) ligands. Chloride ions, hydroxide ions, water molecules, and ammonia molecules are monodentate ligands. When anions are ligands, their names are given an "o" ending, but there is no distinctive ending used in naming neutral molecules as ligands:

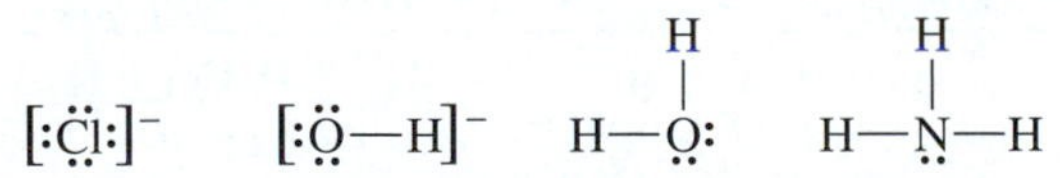

Ligand name: Chloro   Hydroxo   Aqua   Ammine*

---

*The spelling *ammine* with a double m signifies that the ligand is *ammonia*, also spelled with a double m. In contrast, the functional group NH_2 in organic compounds is called the *amine* group and is spelled with a single m.

Ligand 3D models

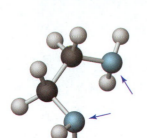

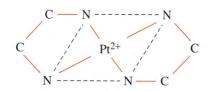

▲ FIGURE 22.12   The chelate [Pt(en)₂]²⁺

The N atoms of the en bond through their lone-pair electrons, pinpointed by the arrows in the ball-and-stick model. They attach at the corners along two edges of the square (black dashed lines), but they *do not* bridge the square by attaching at opposite ends of a diagonal. For simplicity, H atoms on the C and N atoms are not shown.

Another common monodentate ligand is nitrite ion, NO_2^-, but here the donor atom can be either the N atom or one of the O atoms. The ligand is given a different name for each case. The true structure of the nitrite ion is a resonance hybrid of two resonance structures, but we can see the lone-pair electrons on the donor atoms by writing just one of the structures:

$$\left[\ddot{O} \diagdown \underset{N}{} \diagup \ddot{O} \right]^-$$

Ligand name: *Nitrito-N*, if N is the donor atom
               Nitrito-O, if O is the donor atom

Nitrite ion is monodentate because it can bond through only one of its donor atoms at a time. Ethylenediamine also has two donor atoms, but in this molecule the two are so far apart that they can bond independently. The ligand molecule can wrap itself around the central metal atom or ion and attach to it through both donor atoms at the same time. It is a **bidentate** (two-toothed) ligand:

$$H-\overset{H}{\underset{H}{N}}-\overset{H}{\underset{H}{C}}-\overset{H}{\underset{H}{C}}-\overset{H}{\underset{H}{N}}-H$$

Ligand name: Ethylenediamine (en)

Bidentate ligands and ligands having more than two points of attachment are collectively called **polydentate** (many-toothed) ligands. Table 22.4 lists some common monodentate and polydentate ligands.

One of the complex ions whose structure is shown in Figure 22.11 is $[Pt(NH_3)_4]^{2+}$. Suppose we replace each pair of NH_3 molecules along an edge of the square with the bidentate ligand ethylenediamine. The resulting complex ion, $[Pt(en)_2]^{2+}$, has the structure shown in Figure 22.12, where the dashed black lines outline the square-planar structure. Note that this structure has two five-membered rings (pentagons), outlined by the solid red lines (chemical bonds). Each ring consists of one Pt, two N, and two C atoms. When a five- or six-membered ring is produced by the

Table 22.4 Some Common Ligands

MONODENTATE					
Formulaᵃ	**Name as Ligandᵇ**	**Formulaᵃ**	**Name as Ligandᵇ**	**Formulaᵃ**	**Name as Ligandᵇ**
Neutral Molecules					
NH_3	Ammine	NO	Nitrosyl	H_2O	Aqua
CH_3NH_2	Methylamine	CO	Carbonyl	C_5H_5N	Pyridine
Anions					
F^-	Fluoro	OH^-	Hydroxo	NCS^-	Thiocyanato-*N*
Cl^-	Chloro	NO_2^-	Nitrito-*N*	SCN^-	Thiocyanato-*S*
Br^-	Bromo	ONO^-	Nitrito-*O*	OSO_3^{2-}	Sulfato
I^-	Iodo	CN^-	Cyano	SSO_3^{2-}	Thiosulfato

POLYDENTATE		
Name of Ligandᵇ	**Abbreviation**	**Formulaᵃ**
Ethylenediamine	en	$H_2NCH_2CH_2NH_2$
Oxalato	ox	$[OOCCOO]^{2-}$
Ethylenediaminetetraacetato	EDTA	$[(OOCCH_2)_2NCH_2CH_2N(CH_2COO)_2]^{4-}$

ᵃ Donor atoms are shown in red.

ᵇ Most neutral ligands carry the unmodified name, except for aqua, ammine, carbonyl, and nitrosyl. Anion ligand names end in "o," which requires changing the terminal *-e* to *-o* (for example, sulfate ⟶ sulfato). With many common anions, an entire *-ide* ending is changed to *-o* (for example, cyanide ⟶ cyano).

attachment of a polydentate ligand, the complex is a **chelate** (pronounced KEY-late). When a polydentate ligand attaches to a central metal atom or ion, the process is called *chelation* and the ligand is referred to as a *chelating agent*. Chelate comes from the Greek word for *claw*. In the $[Pt(en)_2]^{2+}$ complex ion, it is as though the central Pt^{2+} ion is being pinched by two sets of "molecular claws."

22.8 Nomenclature of Complex Ions and Coordination Compounds

The collection of rules needed to name all possible complexes is extensive and has been revised many times over the years. We will use a simplified list that will help us match an acceptable name with a formula and an appropriate formula with a name.

1. *In naming a complex,* name first the ligands and then the central metal atom or ion, all as a single word. *Example:* $[Cu(NH_3)_4]^{2+}$ is tetraamminecopper(II) ion.

2. (a) *When writing a name from a formula,* name the ligands in alphabetical order, based on the first letters of their names and ignoring Greek numeric prefixes. *Example:* $[CuCl(NH_3)_3]^+$ is triamminechlorocopper(II) ion. That is, the alphabetical order of "triammine" is based on "ammine."
    (b) *When writing a formula from a name,* place anionic ligands *before* neutral ones. The order of the anionic and neutral ligands is alphabetical based on the first letters of their symbols or formulas. *Example:* Diamminediaquadichlorocobalt(III) ion has the formula $[CoCl_2(H_2O)_2(NH_3)_2]^+$.

3. (a) Designate the number of ligands in a complex with a prefix: *mono* = 1 (usually omitted), *di* = 2, *tri* = 3, *tetra* = 4, and so on. *Example:* $[Ag(NH_3)_2]^+$ is *di*ammine silver(I) ion.
    (b) *If the ligand name itself includes a Greek numeric prefix,* such as the prefix "di" in the name ethylene*di*amine, place parentheses around the ligand name and preface it with a prefix indicating the number of ligands in the complex. To distinguish this prefix from the one used in the ligand name, use the prefixes *bis* = 2, *tris* = 3, *tetrakis* = 4. *Example:* $[Pt(en)_2]^{2+}$ is *bis*(ethylene*di*amine) platinum(II) ion.

4. *In naming a complex cation,* use the unmodified name of the central metal. *In naming a complex anion,* add the ending *-ate* to the name of the central metal. For certain common metals in complex anions, use the Latin-based names shown in Table 22.5. In all cases, denote the oxidation number of the central metal by a roman numeral in parentheses. *Examples:* $[Zn(OH)_4]^{2-}$ is tetrahydroxozinc*ate*(II) ion, $[CuCl_4]^{2-}$ is tetrachloro*cuprate*(II) ion, and $[Co(en)_3]^{3+}$ is tris(ethylene*di*amine)cobalt(III) ion.

5. In writing the formula or name of a coordination compound, place the ions in the usual order: cation followed by anion. *Example:* $Na_2[Zn(OH)_4]$ is sodium tetrahydroxozincate(II).

Some coordination compounds have long been known and are still referred to by common names. Two examples are potassium ferr*o*cyanide, $K_4[Fe(CN)_6]$, and potassium ferr*i*cyanide, $K_3[Fe(CN)_6]$. The "o" and "i" indicate that the central ions are

Table 22.5 Names for Some Metals in Complex Anions

Metal	Name in Complex Anion
Copper	*Cuprate*
Gold	*Aurate*
Iron	*Ferrate*
Lead	*Plumbate*
Silver	*Argentate*
Tin	*Stannate*

ferrous (Fe^{2+}) and ferric (Fe^{3+}), respectively. With this information and knowing that the coordination number is 6 in both of these coordination compounds, you can easily relate the common names to the formulas. Nevertheless, their systematic names are even more indicative of the formulas: potassium hexacyanoferrate(II) for $K_4[Fe(CN)_6]$ and potassium hexacyanoferrate(III) for $K_3[Fe(CN)_6]$.

Example 22.4

Name **(a)** $[CrCl_2(NH_3)_4]^+$ and **(b)** $K[PtBrCl_2NH_3]$.

SOLUTION

(a) The ligands are four ammonia molecules (ammine) and two chloride ions (chloro). We refer to them as *tetra*ammine and *di*chloro, and we list them in alphabetical order (ignoring Greek numeric prefixes). The two Cl^- ligands carry a total of two units of negative charge, and the net charge on the complex ion is 1+. The central Cr must therefore be Cr^{3+}. The name is tetraamminedichlorochromium(III) ion.

(b) A formula unit of this coordination compound consists of a K^+ cation and a complex anion, $[PtBrCl_2(NH_3)]^-$. In alphabetical order, the ligands in the complex are one NH_3 molecule (ammine), one Br^- ion (bromo), and two Cl^- ions (dichloro). The central ion must therefore be Pt^{2+}, to account for the net charge of 1− on the complex. The name is potassium amminebromodichloroplatinate(II).

EXERCISE 22.4A

Name **(a)** $[Co(NH_3)_6]^{2+}$ and **(b)** $[CoBr(NH_3)_5]Br_2$.

EXERCISE 22.4B

Name **(a)** $[AuCl_4]^-$ and **(b)** $[CoBr(en)_2H_2O][Ni(CN)_4]$

Example 22.5

Write the formula for **(a)** triamminechlorodinitrito-*O*-platinum(IV) ion and **(b)** sodium hexanitrito-*N*-cobaltate(III).

SOLUTION

(a) This is a complex in which Pt^{4+} is the central ion and the ligands are three (tri) ammonia (ammine) molecules, one chloride (chloro) ion, and two (di) nitrite (nitrito) ions. The net charge on the complex ion is therefore $4 - 1 - 2 = 1$. The nitrito-*O*- signifies that the donor atom of the nitrite ion is the O atom, and Table 22.1 indicates that the ligand formula should be written ONO. The formula of the complex ion is $[PtCl(ONO)_2(NH_3)_3]^+$.

(b) This is a coordination compound made up of monatomic Na^+ ions and complex anions. The cobaltate(III) tells us that the central ion in the complex is Co^{3+}. This central ion and six NO_2^- ions as ligands, with N as the donor atoms, produce a complex anion with a net charge of $+3 - 6 = 3-$ and the formula $[Co(NO_2)_6]^{3-}$. Because there must be three Na^+ cations for each anion, the formula of the coordination compound is $Na_3[Co(NO_2)_6]$.

EXERCISE 22.5A

Write the formula for **(a)** bis(ethylenediamine)oxalatocobalt(III) ion and **(b)** diamminetetrachlorochromate(III) ion.

EXERCISE 22.5B

Write the formula for **(a)** hexaamminechromium(III) hexacyanocobaltate(III) and **(b)** dichlorobis(ethylenediamine)platinum(IV) sulfate.

22.9 Isomerism in Complex Ions and Coordination Compounds

Recall that isomers are compounds that have the same molecular formula but different structures and properties. We find several types of isomers among complex ions and coordination compounds.

Structural Isomers

Structural isomers differ either in the ligands attached to the central atom or in the donor atoms through which the ligands are bonded. Thus, although the complex ions

$$[Co(NO_2)(NH_3)_5]^{2+} \qquad\qquad [Co(ONO)(NH_3)_5]^{2+}$$

Pentaamminenitrito-*N*-cobalt(III) ion   Pentaamminenitrito-*O*-cobalt(III) ion

 **Isomerism animation**

have the same ligands in the same numbers, they are isomers because a different atom acts as donor in the NO_2^- ligand—N in one case and O in the other (the donor atoms are shown in color).

The coordination compounds

$$[Cr(SO_4)(NH_3)_5]Cl \qquad\qquad [CrCl(NH_3)_5]SO_4$$

Pentaamminesulfatochromium(III)   Pentaamminechlorochromium(III)
chloride   sulfate

have the same percent composition by mass and the same molar mass, but they are isomers because they differ in the placement of the anions. The compound on the left has SO_4^{2-} as a ligand attached to the Cr^{3+} central ion and Cl^- as a free anion outside the coordination sphere. The compound on the right has Cl^- as the ligand and SO_4^{2-} as the free anion.

Example 22.6

For each pair, indicate whether the species are isomers.

(a) $[Co(NH_3)_6][Cr(CN)_6]$ and $[Cr(NH_3)_6][Co(CN)_6]$

(b) $[Cr(H_2O)_5(NH_3)]^{3+}$ and $[Cr(NH_3)(H_2O)_5]^{3+}$

(c) $[Co(H_2O)_5(NH_3)]^{3+}$ and $[Co(H_2O)(NH_3)_5]^{3+}$

STRATEGY

The main point to consider in comparing the two members of each pair is whether or not they have identical compositions. If they do not, they represent two different compounds rather than two isomers. If they do, we must still establish that they differ in some structural detail before we call them isomers; they could just be different representations of the same compound.

SOLUTION

(a) These two coordination compounds have the same overall composition and molar mass. They also have the same numbers and types of ligands. They differ in the distribution of ligands between the complex cation and complex anion. The cation in one compound is hexaamminecobalt(III) ion, whereas in the other, it is hexaamminechromium(III) ion. The two compounds are isomers.

(b) Each of these complex ions has one NH_3 and five H_2O molecules as ligands and Cr^{3+} as the central ion. The formulas are written differently, but the structures are identical. These two formulas do *not* represent isomers; they represent the same complex ion.

(c) Each complex ion has Co^{3+} as the central ion and a total of six ligands. However, one ion has five H_2O molecules and one NH_3 molecule, and the other has five NH_3 molecules and one H_2O molecule. These two formulas represent different complex ions. This is not a case of isomerism.

In another type of isomerism, known as **stereoisomerism,** the number and types of ligands and their mode of attachment are the same in two compounds, but the manner in which the ligands occupy the space around the central atom differs. We will consider two important types of stereoisomers.

Geometric Isomerism

Recall the complex ion $[Pt(NH_3)_4]^{2+}$ pictured in Figure 22.11. Suppose we replace two of the NH_3 ligands with Cl^- ions, forming diamminedichloroplatinum(II). Figure 22.13 shows that we can do this in two ways: The two Cl^- ions can be either along the same edge of the square (cis) or on opposite corners (trans). This is similar to the cis–trans isomerism of alkenes (Section 10.5).

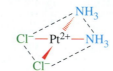

cis-Diamminedichloroplatinum(II)
(cisplatin)

trans-Diamminedichloroplatinum(II)
(transplatin)

▲ **FIGURE 22.13**   **Geometric isomerism in a square-planar complex**

QUESTION: Why is geometric isomerism not possible in $[Pt(NH_3)_3Cl]^+$?

The two geometric isomers of diamminedichloroplatinum(II) offer a striking example of the critical relationship between structure and properties. The cis isomer, known as cisplatin, is a powerful antitumor drug used in chemotherapy. Apparently, in this isomer the two Cl^- ligands are just the right distance apart so that the complex can attach to the DNA in cancerous cells and inhibit further cell growth. In the trans isomer, the Cl^- ligands are too far apart for the complex to be effective.

Cis and trans isomerism can also be found in octahedral complexes. Figure 22.14 shows this to be the case with the octahedral complex, $[CoCl_2(en)_2]^+$. The cis isomer has two Cl^- ions along the same edge of the octahedron. In the trans isomer, the two Cl^- ions are on opposite corners of the octahedron.

Look again at Figure 22.13. If we replace the two NH_3 ligands of cisplatin by Cl^- or the two Cl^- by NH_3, there is no isomerism in either product ($[PtCl_4]^{2-}$ or $[Pt(NH_3)_4]^{2+}$). Now consider a substitution in the $[CoCl_2(en)_2]^+$ of Figure 22.14. If we replace the two Cl^- ligands of the cis isomer by a third ethylenediamine (en), it might appear that there is no isomerism. In fact, however, there are two different tris(ethylenediamine)cobalt(III) ions, as we will see next.

Optical Isomers

To understand optical isomerism, we need to consider the relationship between an object and its mirror image. Think of a plain rubber ball and its image in a mirror. If it were possible to lift the image from the mirror, we would be able to *superimpose* the image over the ball. The real object and its image would be indistinguishable. In contrast, the mirror image of a right hand is a left hand, and one is not superimposable on

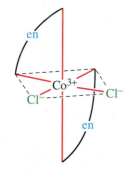

(a) *cis*-Dichlorobis-
(ethylenediamine)cobalt(III) ion

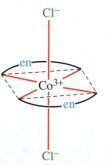

(b) *trans*-Dichlorobis-
(ethylenediamine)cobalt(III) ion

▲ **FIGURE 22.14**   **Geometric isomerism in an octahedral complex**

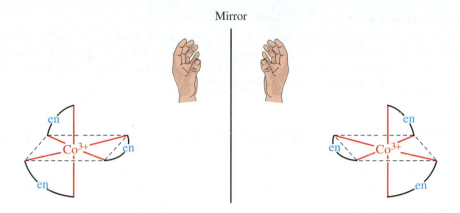

Mirror

◀ **FIGURE 22.15   Optical isomers**
The two structures are nonsuperimposable mirror images of each other. They are like a right hand and a left hand.

the other. As shown in Figure 22.15, two *nonsuperimposable* structures are possible for the tris(ethylenediamine)cobalt(III) ion. Molecules or ions with structures that are nonsuperimposable in the same way that an object and its mirror image are nonsuperimposable are called **enantiomers.**

Enantiomers have identical physical and chemical properties except in a few situations that depend on *handedness* at the molecular level. Think of handedness in this way: Whether persons are right-handed or left-handed, they can all use the same pencils, hammers, and screwdrivers. However, a left-handed baseball player cannot effectively use a baseball glove intended for a right-handed player.

One physical property that is dependent on the handedness of enantiomers is *optical activity.* In Figure 7.10, we pictured an electromagnetic wave as an oscillation of electric and magnetic fields. In normal light, these oscillations occur in all directions perpendicular to the line along which the light is propagated. The light is *unpolarized.* When the oscillations are limited to a single direction, the wave motion occurs in only one plane and the light is *plane-polarized.* Some materials, called polarizers, have the ability to filter out all the oscillations except those in a single plane. They transmit plane-polarized light.

Figure 22.16 shows a light source, the unpolarized light it produces, and the plane-polarized light transmitted by a polarizer. Optically active materials have the ability to *rotate the plane of polarized light.* The optically active substance in the sample tube of Figure 22.16 rotates the plane of polarized light in the clockwise direction as seen by the viewer (that is, to the right). The angle of rotation is measured by rotating an analyzer (a second polarizer) to the extent that the polarized light coming from the sample is completely absorbed.

Optical isomers are isomers that differ from one another in their ability to rotate the plane of polarized light. Enantiomers are optical isomers that rotate the plane of polarized light to the same degree but in opposite directions, that is, one to the right (clockwise as seen by a viewer) and the other to the left (counterclockwise). Measurements of optical activity can sometimes help to establish a correct structure from several hypothetical structures, as illustrated in Example 22.7.

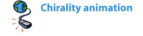
Chirality animation

Optical isomers are common among organic compounds in nature. The compound *carvone* exists as two optical isomers: One gives spearmint its characteristic odor, and the other is oil of caraway seed.

Optical Activity animation

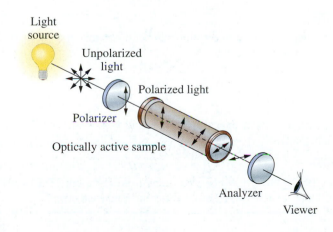

Light source

Unpolarized light

Polarized light

Polarizer

Optically active sample

Analyzer

Viewer

◀ **FIGURE 22.16   Optical activity**
Ordinary light consists of electromagnetic waves that vibrate in all directions perpendicular to the line along which the light is propagated; it is unpolarized. Light having all vibrations in the same plane is *polarized.* Optically active substances are able to rotate the plane of polarized light.

Example 22.7   A Conceptual Example

At one time, a triangular prism was considered as a possible geometric structure for coordination number 6. Assuming that ethylenediamine (en) molecules can link only to adjacent coordination sites, determine the number of possible geometric and/or optical isomers of $[Co(en)_3]^{3+}$, based on the triangular prism structure.

ANALYSIS AND CONCLUSIONS

Four possibilities are

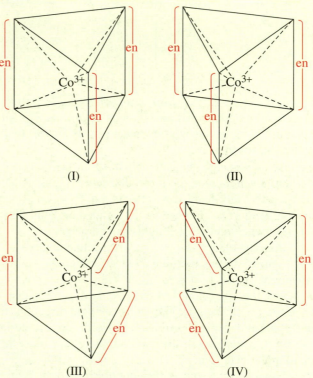

Structures I and II both have (en) ligands along the three vertical edges of the prism; the two structures are identical. Structures III and IV both have one (en) ligand along a vertical edge and the remaining two (en) ligands in the upper and lower triangular faces; these two structures are mirror images of each other. Structure I can of course be superimposed on structure II, and following the appropriate rotation, structure III can be superimposed on its mirror image, structure IV. As a result, there is no optical isomerism.

In contrast, structure I is different from structure III, and these two cannot be superimposed on each other. Adjacent coordination sites along a vertical edge of the prism are different from adjacent sites along an edge of a triangular face. Structures I and III are geometric isomers.

In summary, if the structure of $[Co(en)_3]^{3+}$ were a triangular prism, there would be two geometric isomers and no optical isomers. Because the actual structure of this complex is octahedral, as we saw in Figure 22.15, there are in fact two optical isomers and no geometric isomers.

EXERCISE 22.7A

Another early suggestion for coordination number 6 was that of a planar hexagon with the central atom or ion at its center:

How many geometric and/or optical isomers would you predict for $[Co(en)_3]^{3+}$ based on the hexagonal structure?

EXERCISE 22.7B

The complex $[PtCl_2(NH_3)_2]$ displays cis–trans isomerism, but $[ZnCl_2(NH_3)_2]$ does not. Why do you suppose these two cases are different from each other?

When an optically active complex is synthesized, the two enantiomers are obtained in equal amounts, and the rotation of the plane of polarized light produced by one isomer is just offset by the opposite rotation of the other. As a result, the mixture, called a *racemic mixture,* does not rotate the plane of polarized light at all. Separating the enantiomers in a racemic mixture requires some reaction that distinguishes between a structure and its mirror image. This is the molecular equivalent of separating a huge bin filled with gloves into separate piles for right- and left-hand gloves.

We will encounter optical activity again in Chapter 23.

22.10 Bonding in Complexes: Crystal Field Theory

The Lewis acid–base theory helps explain how a metal ion and ligands join to form a complex ion. However, we need more than the Lewis theory to explain certain properties of complex ions, especially their characteristic colors and magnetic properties. The *crystal field theory* does this quite well.

According to the **crystal field theory**, the attractions between a central atom or ion and its ligands are largely electrostatic. Lone-pair electrons on the ligands are attracted to the positively charged nucleus of the central metal. In addition, however, there are repulsions between the ligand electrons and *d* electrons of the central atom or ion. Crystal field theory focuses on the effect of these repulsions.

We first pictured the five *d* orbitals in Figure 7.28, showing how each is oriented with respect to the *x*-, *y*-, and *z*-axes. We show them again in Figure 22.17, but this time as a group in yellow and red along the same set of axes. Six ligands are shown approaching the set of *d* orbitals along *x*-, *y*-, and *z*-axes. These directions of approach produce a complex that has an octahedral structure. From this figure, we see that the ligands approaching along the vertical *z*-axis encounter most directly the lobes of the d_{z^2} orbitals (yellow) and those approaching along the *x*- and *y*-axes encounter most directly the lobes of the $d_{x^2-y^2}$ orbitals (also yellow). As a result, the strongest repulsions occur between these two orbitals and the approaching ligands. This repulsion raises the energies of these *d* orbitals above what the energies would be in the free ion with no ligands present. The energies of the other *d* orbitals (d_{xy}, d_{xz}, and d_{yz}) are not raised as much. As a result, the *d* energy level of the central atom or ion is split into two groups. We will use the symbol Δ to designate the energy difference between the two groups.

In a similar manner, we could describe how ligands approach a central metal atom or ion to produce tetrahedral or square-planar complexes. The observations we would make, together with that for the octahedral complex, are summarized in Figure 22.18.

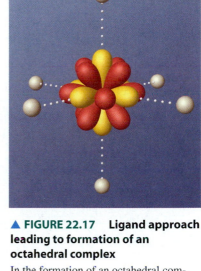

▲ **FIGURE 22.17   Ligand approach leading to formation of an octahedral complex**

In the formation of an octahedral complex, ligands approach the central atom or ion along *x*-, *y*-, and *z*-axes. Maximum interference occurs with the d_{z^2} and $d_{x^2-y^2}$ orbitals (yellow). The energies of these orbitals are raised with respect to the energy levels of the d_{xy}, d_{xz}, and d_{yz} orbitals (red).

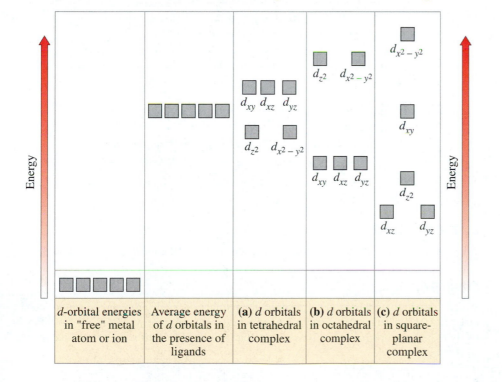

d-orbital energies in "free" metal atom or ion	Average energy of *d* orbitals in the presence of ligands	(a) *d* orbitals in tetrahedral complex	(b) *d* orbitals in octahedral complex	(c) *d* orbitals in square-planar complex

🌐 **Crystal Field Theory activity**

◀ **FIGURE 22.18   Schematic representation of *d*-level splitting**

From left to right in this schematic representation, we see (1) the energy levels of *d* orbitals in a hypothetical central atom or ion free of ligands, (2) the *average* energy level to which the *d* orbitals are raised by the ligands, and (3) the splitting of the energy levels that occurs in the case of (a) tetrahedral, (b) octahedral, and (c) square-planar complexes.

Of what importance is all this? To see, let us consider how the five 3*d* electrons in Fe^{3+} are distributed among the 3*d* orbitals in two of its complex ions. (Remember that iron has $Z = 26$, and the Fe^{3+} ion therefore has 23 electrons in the configuration $[Ar]3d^5$.)

The energy separation (Δ) between the two groups of *d* orbitals is small in $[Fe(H_2O)_6]^{3+}$, and therefore the five 3*d* electrons of Fe^{3+} distribute themselves in the usual way as five unpaired electrons in the 3*d* orbitals, as shown in Figure 22.19a. Because these unpaired electrons have parallel spins, we call $[Fe(H_2O)_6]^{3+}$ a *high-spin complex*. In contrast, in $[Fe(CN)_6]^{3-}$ the energy separation (Δ) between the two groups of *d* orbitals is larger than the energy benefit derived from unpaired electrons. Consequently, the five electrons are all found in the lower-energy *d* orbitals. Four are paired, and only one is unpaired (Figure 22.19b). We say that $[Fe(CN)_6]^{3-}$ is a *low-spin complex*.

To predict the magnetic properties of a complex ion, we need to know its structure (tetrahedral, octahedral, or square planar) and the ability of its ligands to split the *d*-orbital energy levels. The ranking

Field strength   Strong   Weak

$$CN^- > NO_2^- > en > NH_3 > H_2O > OH^- > F^- > Cl^- > Br^- > I^-$$

d-Level splitting, Δ Large   Small

where the red atoms are donor atoms is called the **spectrochemical series**. It shows the relative abilities of some common ligands to split the *d*-orbital energy levels. Ligands that produce large energy separations between groups of *d* orbitals are called *strong-field ligands*, and those that produce small energy separations are called *weak-field ligands*. The energy-level diagrams we drew in Figure 22.19 for the aqua and cyano complex ions of Fe^{3+} are consistent with the order suggested by the spectrochemical series.

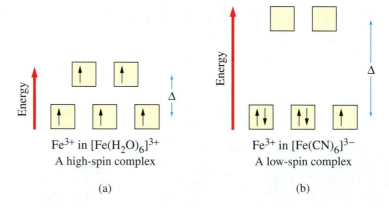

▶ FIGURE 22.19   Electron distribution in two Fe^{3+} complexes
(a) The aqua complex. (b) The cyano complex.

Fe^{3+} in $[Fe(H_2O)_6]^{3+}$
A high-spin complex

(a)

Fe^{3+} in $[Fe(CN)_6]^{3-}$
A low-spin complex

(b)

Example 22.8

How many unpaired electrons would you expect for the octahedral complex ion $[CoF_6]^{3-}$?

STRATEGY

We need to do three things: (1) Determine the number of 3*d* electrons in the central cobalt ion, (2) assess whether the energy difference (Δ) between the two groups of *d* orbitals is likely to be large or small, and (3) use the aufbau principle to distribute electrons among the *d* orbitals.

SOLUTION

1. The complex ion has six F^- ions as ligands and a net charge of $3-$. The central ion must carry a charge of $3+$; it is Co^{3+}. Cobalt has $Z = 27$, and the Co^{3+} ion has 24 electrons in the configuration $[Ar]3d^6$. There are six $3d$ electrons.
2. From the spectrochemical series, we see that F^- is a weak-field ligand. We therefore expect the energy separation (Δ) between the two sets of d orbitals to be small.
3. Because of this small energy separation, we should assign electrons to all five of the d orbitals singly before forming any pairs. Thus, there should be four unpaired electrons:

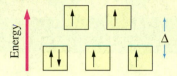

EXERCISE 22.8A

How many unpaired electrons would you expect to find in the octahedral complex ion $[Co(CN)_6]^{3-}$?

EXERCISE 22.8B

How many unpaired electrons would you expect to find in the tetrahedral complex ion $[NiCl_4]^{2-}$?

22.11 Color in Complex Ions and Coordination Compounds

A substance absorbs photons of light if the energies of the photons match the energies required to excite electrons in the atoms of the substance. If the absorbed photons are of wavelengths in the visible-light portion of the electromagnetic spectrum, the light *transmitted* by the substance is colored. That is, the substance absorbs one or more colors of light and transmits the rest. The color of the transmitted light is the *complementary* color of the absorbed light. This idea is illustrated in Figure 22.20. For five solutions we encountered elsewhere in the text, Figure 22.21 shows the color of light absorbed and transmitted.

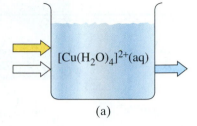

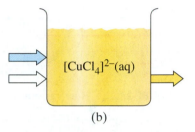

▲ **FIGURE 22.20** **Light absorption and the origin of color**
(a) In $[Cu(H_2O)_4]^{2+}(aq)$, the yellow component of white light is strongly absorbed. The light transmitted through the solution is observed with the *complementary* color of yellow—blue.
(b) In $[CuCl_4]^{2-}(aq)$, the situation is reversed: Blue light is strongly absorbed, and its complementary color—yellow—is observed in the transmitted light.

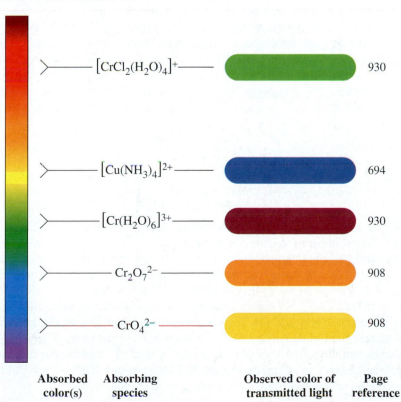

Absorbed color(s)	Absorbing species	Observed color of transmitted light	Page reference
	$[CrCl_2(H_2O)_4]^+$		930
	$[Cu(NH_3)_4]^{2+}$		694
	$[Cr(H_2O)_6]^{3+}$		930
	$Cr_2O_7^{2-}$		908
	CrO_4^{2-}		908

◀ **FIGURE 22.21** **The relationship between an absorbed color and the corresponding observed color of the transmitted light**

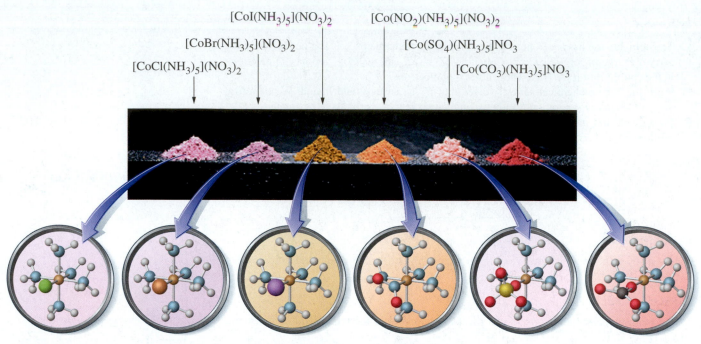

$[CoCl(NH_3)_5](NO_3)_2$
$[CoBr(NH_3)_5](NO_3)_2$
$[CoI(NH_3)_5](NO_3)_2$
$[Co(NO_2)(NH_3)_5](NO_3)_2$
$[Co(SO_4)(NH_3)_5]NO_3$
$[Co(CO_3)(NH_3)_5]NO_3$

▲ **FIGURE 22.22**   **How various ligands may affect colors of coordination compounds**

Ions having the following electron configurations have no electron transitions in the energy range of visible light:

1. a noble-gas electron configuration (for example, Na^+ or Cl^-)
2. an outermost shell with 18 electrons (for example, Zn^{2+})
3. an "18 + 2" configuration (for example, Sn^{2+})

These ions do not absorb visible light; and as a result, their aqueous solutions are colorless.

Many complex ions are colored because the energy differences between *d* orbitals, Δ, match the energies of components of visible light. Substituting one ligand for another can produce subtle changes in the energy levels of the *d* orbitals and striking changes in the colors of complex ions, as seen in Figures 22.22 and 22.23.

We can use crystal field theory to explain the colors of complex ions. For example, to explain why solutions of $[Cr(H_2O)_6]^{3+}$ are violet whereas those of $[Cr(NH_3)_6]^{3+}$ are yellow, let us construct a *d*-orbital energy-level diagram for these octahedral complex ions. We begin by noting that the electron configuration of Cr is $[Ar]3d^5 4s^1$ and that of Cr^{3+} is $[Ar]3d^3$. The distribution of the three 3*d* electrons in the latter is

▲ **FIGURE 22.23   Colors of chromium(III) complex ions**

The vial on the left contains the solution formed by dissolving the green solid $CrCl_3 \cdot 6\ H_2O$ in water. The green color is due to the complex ion $[CrCl_2(H_2O)_4]^+$. The vial on the right shows the same solution after two days. As H_2O molecules displace the Cl^- ions in $[CrCl_2(H_2O)_4]^+$, the solution acquires the violet color of the complex ion $[Cr(H_2O)_6]^{3+}$.

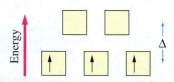

The color of light absorbed in promoting an electron from one orbital to a *d* orbital of higher energy depends on the magnitude of the energy difference, Δ. Because NH_3 is a stronger field ligand than H_2O, Δ is larger for $[Cr(NH_3)_6]^{3+}$ than for $[Cr(H_2O)_6]^{3+}$. Thus, the energy of the light absorbed by the ammine complex ion is higher than the energy of the light absorbed by the aqua complex ion. Higher energy means higher frequency, and higher frequency means shorter wavelength. We should therefore expect $[Cr(NH_3)_6]^{3+}$ to absorb toward the violet end of the spectrum and transmit toward the red end. Thus, the yellow color transmitted by $[Cr(NH_3)_6]^{3+}$ is a reasonable expectation. The light absorbed by $[Cr(H_2O)_6]^{3+}$ is of lesser energy, lower frequency, and longer wavelength. Figure 22.21 shows that $[Cr(H_2O)_6]^{3+}$ absorbs in the green region of the spectrum and transmits in the violet.

22.12   Chelates: Complexes of Special Interest

Recall that *chelation* (page 921) is the formation of five- or six-membered rings of atoms through the attachment of polydentate ligands to central metal atoms or ions. We gave an example of a chelate in Figure 22.12. Now we will focus on a group of common chelating agents—the salts of *ethylenediaminetetraacetic acid* (EDTA). The ligand in these salts is the ethylenediaminetetraacetate ion, $EDTA^{4-}$. The formula of $EDTA^{4-}$ is shown in Table 22.4, and Figure 22.24 shows the structure of a metal–EDTA complex. The five-membered rings (five of them) are outlined in red.

🌐 **$EDTA^{4-}$ ion 3D model**

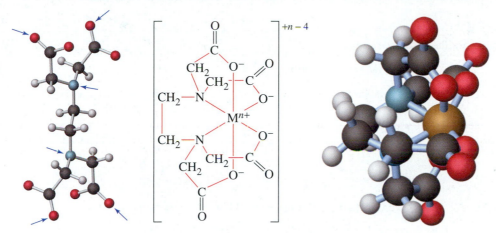

◀ **FIGURE 22.24   Structure of a metal–EDTA complex**
The donor atoms of the $EDTA^{4-}$ ligand are pinpointed by the arrows (left). The central metal ion M^{n+} (orange) can be any of many different cations: Ca^{2+}, Mg^{2+}, Fe^{2+}, Fe^{3+}, and so on (right). The charge on the EDTA anion is $4-$, and the net charge of the complex ion is that of the central ion $(n+)$ plus that of the anion $(4-)$, that is, $+n - 4$.

QUESTION: What would be the charge on a Mg–EDTA complex?

Stability of Chelates

Chelates generally are much more stable than complexes containing monodentate ligands. Consider, for example, the net reaction in which an $EDTA^{4-}$ ion replaces the six NH_3 molecules in the complex ion $[Ni(NH_3)_6]^{2+}$:

$$[Ni(NH_3)_6]^{2+}(aq) + EDTA^{4-}(aq) \rightleftharpoons [NiEDTA]^{2-}(aq) + 6\,NH_3(aq)$$

$$K = 8 \times 10^9$$

In Chapter 17, we learned that (1) an equilibrium constant is related to a standard free energy change: $\Delta G° = -RT \ln K$, and (2) the free energy change depends on the changes in enthalpy and entropy: $\Delta G° = \Delta H° - T\Delta S°$. The large value of K for this EDTA reaction and the corresponding negative value of $\Delta G°$ indicate that the reaction is thermodynamically favorable. Therefore, we say that $[NiEDTA]^{2-}$ is more stable than $[Ni(NH_3)_6]^{2+}$.

The negative value of $\Delta G°$ can be attributed in large part to the increase in entropy in the reaction. For every two particles of reactant—the ions $[Ni(NH_3)_6]^{2+}$ and $EDTA^{4-}$—seven particles are produced: the ion $[NiEDTA]^{2-}$ and six NH_3 molecules. The entropy of the final state is therefore greater than that of the initial state. This increase in entropy during chelation is an important factor in chelate stability.

Sequestering Metal Ions

To sequester a metal ion in solution means to tie it up in a form that effectively removes it from solution. The $EDTA^{4-}$ ligand is one of the best for sequestering metal ions.

To sequester metal ions, EDTA is sometimes added to boiler water. When they are the central ions in chelate structures, Ca^{2+}, Mg^{2+}, and Fe^{3+} are no longer able to form boiler scale (page 852). Another application is to prevent the growth of certain bacteria in liquid soaps, shampoos, and other personal products. Chelation with EDTA removes Ca^{2+} and Mg^{2+} ions, which are important constituents of the cell walls of these bacteria. The cell walls disintegrate, and the bacteria die. Chelation therapy is sometimes used to treat heavy-metal-ion poisoning; the chelated metal ions are more readily excreted via the urinary tract.

Because of its hexadentate character and high negative charge, $EDTA^{4-}$ has the interesting effect of converting a simple cation into the central ion of an *anion*, for

▲ This iron chelate plant food is an iron–EDTA complex.

example, Fe^{3+}(aq) to $[FeEDTA]^-$. For use as a plant food, iron can be effectively transported through soils as $[FeEDTA]^-$(aq). Clay particles in soils have anions on their surfaces, and Fe^{3+}(aq) cannot easily migrate through such a soil; rather, it is immobilized by combining with surface anions. However, complex anions containing iron(III) are not held back by these surface anions, and the iron(III) is thus available to plants. The same phenomenon aids in *phytoremediation* of metal ions (see page 1066).

Cumulative Example

If we try to make CuI_2 by reacting Cu^{2+}(aq) with I^-(aq), we get CuI(s) and I_2(s) instead. Why is this result to be expected?

STRATEGY

Let us begin by writing the net ionic equation for the observed reaction:

$$2\ Cu^{2+}(aq) + 4\ I^-(aq) \longrightarrow 2\ CuI(s) + I_2(s)$$

To show that this reaction is spontaneous, we need to establish either that $\Delta G° < 0$, that $E°_{cell} > 0$, or that K_{eq} is very large. Generally, we do so by substituting $\Delta G°_f$ values from Appendix C into Equation (17.9), but we do not find a $\Delta G°_f$ value for CuI(s) listed in Appendix C. We can turn next to standard electrode potential data in Appendix C, where we find a value for $E°_{I_2/I^-}$ but not a value for $E°_{Cu^{2+}/CuI}$. However, we do find a value for $E°_{Cu^{2+}/Cu^+}$. As we will see, this $E°$ value and the K_{sp} value for CuI are sufficient data for our calculation.

SOLUTION

First, we can express the target reaction, the one of particular interest to us, as the sum of two other reactions for which data are available.

(a) $\qquad 2\ Cu^{2+}(aq) + 2\ I^-(aq) \longrightarrow 2\ \cancel{Cu^+}(aq) + I_2(s)$

(b) $\qquad 2\ \cancel{Cu^+}(aq) + 2\ I^-(aq) \longrightarrow 2\ CuI(s)$

Target reaction: $\quad 2\ Cu^{2+}(aq) + 4\ I^-(aq) \longrightarrow 2\ CuI(s) + I_2(s)$

Reaction (a) is the overall redox reaction resulting from the combination of these two half-reactions.

Reduction: $\qquad 2\ \{Cu^{2+}(aq) + e^- \longrightarrow Cu^+(aq)\}$

Oxidation: $\qquad\qquad 2\ I^-(aq) \longrightarrow I_2(s) + 2\ e^-$

Overall: $\qquad 2\ Cu^{2+}(aq) + 2\ I^-(aq) \longrightarrow 2\ Cu^+(aq) + I_2(s)$

We can establish $E°_{cell}$ for reaction (a) by combining $E°$ values for the two half-reactions.

$$E°_{cell} = E°(\text{reduction}) - E°(\text{oxidation})$$
$$= E°_{Cu^{2+}/Cu^+} - E°_{I_2/I^-}$$
$$= 0.159\ V - 0.535\ V = -0.376\ V$$

Now we can use Equation (18.7) and a value of $n = 2$ to relate $E°_{cell}$ and K_{eq} for reaction (a).

$$E°_{cell} = \frac{0.025693\ V}{n}\ \ln K_{eq}$$

$$-0.376\ V = \frac{0.025693\ V}{2}\ \ln K_{eq}$$

$$\ln K_{eq} = \frac{-2 \times 0.376\ V}{0.025693\ V} = -29.3$$

$$K_{eq} = K_{(a)} = e^{-29.3} = 1.9 \times 10^{-13}$$

Turning to reaction (b), we see that it is the *reverse* of *twice* the solubility equilibrium reaction for copper(I) iodide.

$$2\ CuI(s) \rightleftharpoons 2\ Cu^+(aq) + 2\ I^-(aq) \qquad K_{eq} = (K_{sp})^2$$

We calculate the equilibrium constant for reaction (b) using the value of $K_{sp} = 1.1 \times 10^{-12}$ from Appendix C.

$$2\ Cu^+(aq) + 2\ I^-(aq) \rightleftharpoons 2\ CuI(s)$$

$$K_{(b)} = \frac{1}{(K_{sp(CuI)})^2} = \frac{1}{(1.1 \times 10^{-12})^2} = 8.3 \times 10^{23}$$

Now we are able to obtain a value of K_{eq} for the target reaction.

(a) $\qquad 2\ Cu^{2+}(aq) + 2\ I^-(aq) \longrightarrow 2\ Cu^+(aq) + I_2(s) \qquad K_{(a)} = 1.9 \times 10^{-13}$

(b) $\qquad 2\ Cu^+(aq) + 2\ I^-(aq) \longrightarrow 2\ CuI(s) \qquad K_{(b)} = 8.3 \times 10^{23}$

Target reaction: $\quad 2\ Cu^{2+}(aq) + 4\ I^-(aq) \longrightarrow 2\ CuI(s) + I_2(s)$

$$K_{eq} = 1.9 \times 10^{-13} \times 8.3 \times 10^{23} = 1.6 \times 10^{11}$$

This very large value of K_{eq} is our criterion that the reaction is spontaneous and the one we should expect.

ASSESSMENT

If the necessary data are available, the quickest way to assess the spontaneity of a reaction is to use standard free energies of formation to evaluate $\Delta G°$ for the reaction. The primary goal of this example, however, was to illustrate the interrelated nature of $\Delta G°$, K_{eq}, and $E°_{cell}$, as summarized in Figure 18.12. This enabled us to settle on evaluating K_{eq} as a preferred criterion for spontaneous change in this case.

Concept Review with Key Terms

22.1 General Properties of the *d*-Block Elements and Their Trends—All the elements of the *d* block are **transition metals**. Most exist in several oxidation states and form many complex ions and colored compounds. The leftmost members in a period in the *d* block are active metals, but those to the right in the period are less active. Within the B groups, fifth- and sixth-period members have a greater tendency to exist in their higher oxidation states than do fourth-period members. Also, because of the poor shielding of valence electrons by electrons in the *f* subshell, the atomic radius of the sixth-period element in a given group is about the same as the radius of the fifth-period element of the group. This phenomenon is referred to as the **lanthanide contraction**.

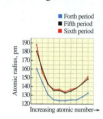

22.2 The Elements Scandium Through Manganese—Scandium resembles aluminum but is not widely used. The uses of titanium depend on its high strength, low density, and corrosion resistance. The oxide TiO_2 is used as a white pigment. Vanadium exists in several oxidation states, and its ions display a variety of colors. Vanadium, chromium, and manganese are all used in the manufacture of steel. Dichromate and permanganate ions ($Cr_2O_7^{2-}$ and MnO_4^-) are widely used oxidizing agents. Dichromate ion converts to chromate ion (CrO_4^{2-}) in a pH-dependent equilibrium. Chromate ion is a precipitating agent.

22.3 The Iron Triad: Iron, Cobalt, and Nickel—These three metals have similar properties. For example, they all exhibit **ferromagnetism** and can be permanently magnetized.

22.4 Group 1B: Copper, Silver, and Gold—These metals are much less active than members of their respective periods that lie farther to the left in the periodic table. They do not displace H_2 from acidic solutions. Their uses are based mostly on their resistance to corrosion and their exceptional abilities to conduct heat and electricity.

22.5 Group 2B: Zinc, Cadmium, and Mercury—These *d*-block metals are not considered transition elements, because their atoms and ions all have filled *d* subshells. Mercury differs from zinc and cadmium in several significant ways, for example, in its inability to displace H_2 from acidic solutions and in its physical properties. Like cadmium, mercury is very toxic.

22.6 The Lanthanide Series (Rare Earths)—The 14 elements in this series, cerium ($Z = 58$) through lutetium ($Z = 71$), follow the *d*-block element lanthanum, giving rise to the common name lanthanide series. These elements make up the first series of *f*-block elements and are sometimes referred to as inner-transition elements. Although these elements were rare and difficult to separate from one another in the years after their discovery, many of the lanthanides are actually more abundant than the more familiar *d*-block elements.

22.7 Werner's Theory of Coordination Chemistry—A **complex** is formed when a metal atom or ion, acting as a Lewis acid, bonds by accepting lone-pair electrons from **ligands**, acting as Lewis bases. A **monodentate** ligand attaches at a single coordination site of the central metal, a **bidentate** ligand at two sites, and a **polydentate** ligand at two or more sites. The **coordination number** represents the total number of points of attachment for a given metal atom or ion. **Chelates** are complexes in which the attachment of polydentate ligands forms ring structures comprising five or six members. Complexes may carry a net charge, depending on their composition, and are then called **complex ions**. A substance consisting of one or more complexes is called a **coordination compound**.

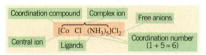

22.8 Nomenclature of Complex Ions and Coordination Compounds—The procedure for naming these substances follows a systematic set of rules that uniquely identify the metal atom, its oxidation state, and the number of and type of ligands.

22.9 Isomerism in Complex Ions and Coordination Compounds—Isomerism among complexes is of two general types. Structural isomers differ in their ligands and/or the donor atoms through which the ligands are bonded to the metal center. **Stereoisomers** differ only in the manner in which ligands occupy the space around the central atom. Geometric isomerism (cis–trans) is one type of stereoisomerism, and optical isomerism is another.

cis-Diamminedichloroplatinum(II) (cisplatin)

trans-Diamminedichloroplatinum(II) (transplatin)

Enantiomers, a type of **optical isomers**, bear the same relationship to one another as an object and its nonsuperimposable mirror image. The only difference in physical properties between a pair of enantiomers is in the direction in which they rotate the plane of polarized light. One isomer rotates it to the right, and the other rotates it to the left.

22.10 Bonding in Complexes: Crystal Field Theory—Interactions between lone-pair electrons on ligands and electrons in the *d* orbitals of the central metal atom or ion produce a splitting of the *d*-orbital energy level. **Crystal field theory** describes the repulsive interactions between these electrons and accurately predicts the characteristic colors and magnetic properties of many complexes.

22.11 Color in Complex Ions and Coordination Compounds—
Electron transitions between *d* orbitals of different energy provide a way for a complex to absorb some wavelength components of visible light and transmit others, giving rise to color. Color is a common property of transition metal complexes, and it is strongly influenced by which ligands are present in the com-

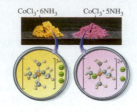

plex. Explanations of the colors and magnetic properties of complexes are facilitated by a listing of common ligands called the **spectrochemical series**.

22.12 Chelates: Complexes of Special Interest—Chelates are generally more stable than complexes involving monodentate and bidentate ligands. The ethylenediaminetetraacetate ion, $EDTA^{4-}$, is a common ligand in chelates. Chelates are useful in sequestering metal ions from solution.

Assessment Goals

When you have mastered the material in this chapter, you will be able to:

- Describe the general chemical and physical properties of the fourth-period *d*-block elements.
- Describe the periodic trend of chemical reactivity within *d*-block elements.
- Apply previously learned concepts, such as oxidation states and electrochemical theory, to the chemistry of the *d*-block elements.
- Locate the position of the lanthanides on the periodic table, and describe their electron configurations and general uses.
- Recognize common ligands, complex ions, and coordination compounds.

- Describe the bonding between ligands and metal atoms or ions in complexes.
- Identify the coordination numbers and oxidation numbers within complexes.
- Name and write formulas for complexes.
- Distinguish between structural isomerism and stereoisomerism in coordination compounds.
- Describe the defining characteristics of geometric isomers and optical isomers.
- Use crystal field theory to explain the colors and magnetic properties of complex ions.
- Explain the stability and sequestering action of chelates.

Self-Assessment Questions

1. Why are *d*-block elements first found in the fourth period, and *f*-block elements in the sixth period?

2. One atom has the electron configuration $[Kr]4d^{10}5s^25p^1$, and another has the electron configuration $[Ar]3d^74s^2$. What are these atoms, and which is a transition element? Explain.

3. Name three properties in which transition metals differ from main-group metals.

4. What is the lanthanide contraction, and among which elements is it observed?

5. Which of the following elements exhibits ferromagnetism?
   (a) Au   (b) Zn   (c) Ni   (d) Ti

6. In its compounds, which element most often has an oxidation state of +3?
   (a) Au   (b) Ag   (c) Ni   (d) Cu

7. Why is the most stable oxidation state of Fe +3, whereas those of Co and Ni are both +2?

8. Give an adequate name for each of the following:
   (a) $ScCl_3$ (b) Fe_2SiO_4 (c) Na_2MnO_4 (d) CrO_3

9. Give the formula for each of the following:
   (a) barium dichromate   (c) mercury(I) bromide
   (b) chromium(III) oxide   (d) vanadium(V) oxide

10. What happens when copper is added to concentrated $HCl(aq)$? To concentrated $HNO_3(aq)$? Why are the results different in the two cases?

11. What are the component parts of a complex ion? What is the relationship between a complex ion and a coordination compound?

12. What is the oxidation number of Ni in the complex ion $[Ni(CN)_4I]^{3-}$?
   (a) −3   (b) −2   (c) +2   (d) +5

13. What is the coordination number of Pt in the complex ion $[Pt(en)_2Cl_2]^{2+}$?
   (a) 3   (b) 4   (c) 5   (d) 6

14. How is the Lewis acid–base theory used to describe complex ion formation? How can ammonia be both a Brønsted–Lowry base and a Lewis base?

15. Explain the meaning and give an example of a monodentate ligand and of a polydentate ligand.

16. What is a chelate, and what conditions are required for its formation?

17. What are the ligands represented by the following names or symbols?
   (a) aqua   (c) ammine   (e) ox
   (b) chloro   (d) en   (f) EDTA

18. What does the ending *-ate* signify when referring to the metal atom in a complex ion, for example, cobaltate(III)?

19. What is the difference between a nitrito-*N* and a nitrito-*O* ligand in a complex ion?

20. Which one of the following complex ions will exhibit isomerism?
   (a) $[Ag(NH_3)_2]^+$ (c) $[Co(NH_3)_5NO_2]^{2+}$
   (b) $[Pt(en)_2Cl_2]^{2+}$ (d) $[Co(NH_3)_5Cl]^{2+}$

21. How does stereoisomerism differ from structural isomerism? How does optical isomerism differ from geometric isomerism?

22. What is meant by the following terms?
   (a) high-spin and low-spin complexes
   (b) strong-field and weak-field ligands
   (c) the spectrochemical series

Problems

Properties of the Transition Elements

23. Both Sc and Ca atoms have two valence-shell electrons ($4s^2$), yet the only oxidation number of Ca in its compounds is +2, whereas Sc has the oxidation number +3. Explain this difference.

24. The +2 oxidation state of silver does exist, but it is much more reactive than the +1 oxidation state. Explain.

25. What oxidation state of chromium might be expected based on its electron configuration but does not commonly occur?

26. In comparing some metal atoms and their ions, we find that Sc is paramagnetic and Sc^{3+} is diamagnetic, whereas both Ti and Ti^{2+} are paramagnetic and both Zn and Zn^{2+} are diamagnetic. Explain these differences.

27. Two adjacent elements in the periodic table, K and Ca, have atomic (metallic) radii of 227 and 197 pm, respectively. Another two adjacent elements, Mn and Fe, both have an atomic radius of 124 pm. Explain why **(a)** the atomic radius of Ca is smaller than that of K; **(b)** the atomic radius of Mn is smaller than that of Ca; and **(c)** the atomic radii of Mn and Fe are the same.

28. In group 5B, Nb and Ta both have an atomic (metallic) radius of 143 pm, whereas that of V, also a member of group 5B, is 131 pm. **(a)** Why are all three radii not the same? **(b)** If there is a reason why all three should not be the same, why are two of the radii still found to be the same?

Scandium Through Manganese

29. Write plausible equations for the following.
  (a) the reaction of Sc(s) with HCl(aq)
  (b) the reaction of $Sc(OH)_3$(s) with HCl(aq)
  (c) the reaction of $Sc(OH)_3$(s) with excess NaOH(aq)

30. Write plausible equations for the following.
  (a) the reaction of Mn(s) with HCl(aq)
  (b) the reaction of $Mn(OH)_2$(s) with HCl(aq)
  (c) the reaction of $Cr(OH)_3$(s) with NaOH(aq)

31. Suggest reactions with common chemicals by which **(a)** $CrCl_3$(s) can be made starting with $K_2Cr_2O_7$(aq), and **(b)** $MnCO_3$(s) can be made starting with $KMnO_4$(aq).

32. Suggest reactions with common chemicals by which **(a)** $BaCrO_4$(s) can be made starting with barium metal and K_2CrO_4(aq), and **(b)** MnO_2(s) can be made starting with $KMnO_4$(aq).

33. The text suggests a two-step process for obtaining $KMnO_4$ from MnO_2(s) (pages 910–911). Write a pair of balanced equations for the process.

34. A qualitative analysis test for Mn^{2+} involves its oxidation to MnO_4^- in acidic solution by sodium bismuthate, $NaBiO_3$. The bismuthate ion is reduced to BiO^+. Write an equation for this redox reaction.

35. When colorless $Pb(NO_3)_2$(aq) and orange $K_2Cr_2O_7$(aq) are mixed, yellow $PbCrO_4$(s) precipitates. Explain.

36. When colorless HCl(aq) is added to orange $K_2Cr_2O_7$(aq), a green solution and a yellow-green gas are formed. Explain.

The Iron Triad

37. Which of the following would you expect to be the best reducing agent: Fe(s), Fe^{2+}(aq), Co^{3+}(aq), or Co(s)? Explain.

38. Which would you expect to be the better oxidizing agent: Fe^{2+}(aq) or Co^{3+}(aq)? Explain.

39. Show that an acidic solution of Co^{3+}(aq) is unstable, and write a plausible balanced redox equation for the reaction that occurs. (For the half-reaction Co^{3+}(aq) $+$ e$^- \longrightarrow Co^{2+}$(aq), $E° = 1.92$ V.)

40. Although iron does not commonly occur with an oxidation number higher than +3, Fe_2O_3(s) can be oxidized to ferrate ion, FeO_4^{2-}, by Cl_2(g) in a strongly basic solution. Write a balanced redox equation for this reaction.

Group 1B

41. Write plausible equations for the following reactions that are described on page 913.
  (a) Ag(s) $+ H_2SO_4$(concd aq) $\longrightarrow$
  (b) Ag(s) $+ HNO_3$(concd aq) $\longrightarrow$

42. Write plausible net ionic equations for the following reactions that are described on page 913.
  (a) Cu(s) $+ H_2SO_4$(concd aq) $\longrightarrow$
  (b) Cu(s) $+ HNO_3$(concd aq) $\longrightarrow$

43. Explain why silver reacts with HNO_3(concd aq) but gold does not.

44. Explain why gold, which reacts with neither HCl(aq) nor HNO_3(aq), does react with a mixture of the two acids. Write a net ionic equation.

Group 2B

45. Describe some important differences between group 2B elements and group 2A elements.

46. Describe some important ways in which mercury differs chemically from zinc and cadmium.

47. Can cadmium plating provide anodic protection to iron in the same manner as does zinc? Explain.

48. Some chemists do not consider zinc or cadmium to be transition metals, even though they appear in the *d*-block. Rationalize this viewpoint.

49. Write plausible net ionic equations to represent (**a**) the reaction of metallic Hg with HNO_3(concd aq) and (**b**) the reaction of ZnO(s) with CH_3COOH(aq).

50. Write plausible net ionic equations to represent (**a**) the reaction of Cd(s) with H_2SO_4(concd aq) and (**b**) the electrolysis of $ZnSO_4$(aq) using Pt electrodes.

Coordination Chemistry

51. What is the coordination number of the central metal atom in (**a**) $[PtCl_3(NH_3)_3]^+$; (**b**) $[Fe(SCN)_2]^+$; (**c**) $[Cu(en)_2]^{2+}$?

52. What is the coordination number of the central metal atom in (**a**) $[Ni(CO)_4]$; (**b**) $[Zn(CN)_6]^{4-}$; (**c**) $[FeEDTA]^{2-}$?

53. How many ligands of the following type would be found in an octahedral complex with Cr^{3+} as the central metal ion?

(**a**) CN^- (**c**) $^-OOCCOO^-$

(**b**) $H_2NCH_2CH_2NH_2$ (**d**) $EDTA^{4-}$

54. How many ligands of the following type would be found in a tetrahedral complex with Cu^{2+} as the central metal ion?

(**a**) CN^- (**b**) en   (**c**) $^-OOCCOO^-$

55. Indicate the oxidation number of the central metal atom in (**a**) $[NiBr_4]^{2-}$; (**b**) $[FeF_5(H_2O)]^{2-}$; (**c**) $[PtCl_2(NH_3)_4]^{2+}$; and (**d**) $[Fe(CO)_5]$.

56. Indicate the oxidation number of the central metal atom in (**a**) $[Pt(en)_2]^{2+}$; (**b**) $[CoCl_2(NH_3)_4]^+$; (**c**) $[Mn(CN)_6]^{3-}$; and (**d**) $[CrCl_4(H_2O)_2]^-$.

Nomenclature

57. Name the following complex ions.

(**a**) $[Zn(NH_3)_6]^{2+}$ (**c**) $[PtCl_2(NH_3)_4]^{2+}$

(**b**) $[FeCl_4]^{2-}$ (**d**) $[Co(en)_3]^{3+}$

58. Name the following complex ions.

(**a**) $[Fe(OH)(H_2O)_5]^+$ (**c**) $[Ag(S_2O_3)_2]^{3-}$

(**b**) $[Co(NO_2)_2(en)_2]^+$ (**d**) $[Fe(ox)_3]^{3-}$

59. Write the formula for each of the following.

(**a**) hexafluorocobaltate(III) ion

(**b**) tetraaquadichlorochromium(III) ion

(**c**) dibromobis(ethylenediamine)cobalt(III) ion

60. Write the formula for each of the following.

(**a**) hexaaquairon(II) ion

(**b**) tetraaminebromochlorochromium(III) ion

(**c**) trioxalatoaluminate(III) ion

61. Name the following coordination compounds.

(**a**) $K_4[Cr(CN)_6]$ (**b**) $K_3[Cr(ox)_3]$

62. Name the following coordination compounds.

(**a**) $K_2[PtCl_6]$ (**b**) $NH_4[Cr(NCS)_4(NH_3)_2]$

63. Write formulas for the following coordination compounds.

(**a**) tetraamminecopper(II) chloride

(**b**) tris(ethylenediamine)chromium(III) hexacyanocobaltate(III)

(**c**) tetraammineplatinum(II) tetrachloroplatinate(II)

64. Write formulas for the following coordination compounds.

(**a**) sodium tetrahydroxozincate(II)

(**b**) tris(ethylenediamine)chromium(III) sulfate

(**c**) dipotassium sodium hexanitrito-*O*-cobaltate(III)

65. Each of the following names is incorrect. Point out the error, and give the correct name.

(**a**) tetraaquacuprate(II) ion

(**b**) pentaamminesulfatocobalt ion.

66. Each of the following names is incorrect. Point out the error, and give the correct name.

(**a**) tetrahydroxozinc(II) ion

(**b**) iron(III)hexafluoride ion

Isomerism

67. Which of the following pairs must be isomers? Explain your conclusion in each case.

(**a**) $[Co(NCS)(NH_3)_5]^+$ and $[Co(SCN)(NH_3)_5]^+$

(**b**) $[PtCl(NH_3)_3]_2[PtCl_4]$ and $[Pt(NH_3)_4][PtCl_3(NH_3)]_2$

(**c**) $K_3[Fe(CN)_6]$ and $K_4[Fe(CN)_6]$

(**d**) $[Co(NO_3)(NH_3)_5]SO_4$ and $[Co(SO_4)(NH_3)_5]NO_3$

68. Which of the following pairs must be isomers? Explain your conclusion in each case.

(**a**) $[Cu(NH_3)_2(H_2O)_2]Cl_2$ and $[Cu(H_2O)_2(NH_3)_2]Cl_2$

(**b**) $[PtCl_2(NH_3)_4]Br_2$ and $[PtBr_2(NH_3)_4]Cl_2$

(**c**) $[Co(NH_3)_6]Cl_3$ and $[Co(NH_3)_6]Cl_2$

(**d**) $[Co(NO_2)(NH_3)_5]^{2+}$ and $[Co(ONO)(NH_3)_5]^{2+}$

69. Draw appropriate sketches to show that the complex ion $[CrCl_4(en)]^-$ *does not* exhibit geometric isomerism.

70. Draw appropriate sketches to show that the complex $[CrCl_3(NH_3)_3]$ *does* exhibit geometric isomerism.

71. Would you expect the octahedral complex ion $[Cr(en)(ox)_2]^-$ to exist as optical isomers? Explain.

72. Would you expect the tetrahedral complex ion $[ZnCl(NH_3)_3]^+$ to exist as optical isomers? Explain.

73. A tetrahedral complex with Br^-, Cl^-, I^-, and CO_3^{2-} ligands can exist as optical isomers but not as geometric isomers. Explain.

74. How many geometric isomers can exist of a tetrahedral complex with Br, Cl, and two SCN^- ligands? Explain.

Magnetism and Color in Complexes

75. Use an orbital diagram to show the expected distribution of electrons in the 3*d* orbitals of the central metal ion in each of the following complex ions. If more than one distribution seems possible, indicate whether you expect the low-spin or the high-spin state to be favored. Also, indicate the number of unpaired electrons in each ion.

(a) octahedral $[Cr(H_2O)_6]^{3+}$ **(c)** octahedral $[Mn(CN)_6]^{3-}$
(b) octahedral $[FeCl_6]^{3-}$ **(d)** tetrahedral $[CoCl_4]^{2-}$

76. Use an orbital diagram to show the expected distribution of electrons in the 3*d* orbitals of the central metal ion in each of the following complex ions. If more than one distribution seems possible, indicate whether you expect the low-spin or the high-spin state to be favored. Also, indicate the number of unpaired electrons in each ion.

(a) octahedral $[PtCl_6]^{2-}$ **(c)** octahedral $[Fe(CN)_6]^{4-}$
(b) tetrahedral $[Zn(H_2O)_4]^{2+}$ **(d)** square planar $[PtCl_4]^{2-}$

77. One of the following compounds has a yellow color, and the other is violet: $[Cr(H_2O)_6]Cl_3$ and $[Cr(NH_3)_6]Cl_3$. Which do you think is the yellow compound, and which is the violet? Explain.

78. One of the following compounds has a green color, and the other has a yellow color: $Fe(NO_3)_2 \cdot 6\,H_2O$ and $K_4[Fe(CN)_6] \cdot 3\,H_2O$. Which do you think is the green compound, and which is the yellow? Explain.

Additional Problems

Problems marked with an * may be more challenging than others.

79. The text states that chromate ion is not a particularly good oxidizing agent. Yet, when $Na_2CrO_4(s)$ is added to hot, concentrated $HCl(aq)$, chlorine gas is formed. Explain why, in fact, this is what you should expect to observe.

80. Write a plausible balanced equation to represent the air oxidation of copper to basic copper carbonate, $Cu_2(OH)_2CO_3$.

*81. Manganate ion, MnO_4^{2-}, is unstable in acidic solution and disproportionates. Write a plausible ionic equation for the disproportionation reaction. (*Hint:* What are the likely products?)

*82. Show that the oxidation of $Fe^{2+}(aq)$ to $Fe^{3+}(aq)$ by atmospheric oxygen in acidic solution is a spontaneous reaction when all reactants and products are in their standard states. Is the reaction also spontaneous if the partial pressure of $O_2(g) = 0.20$ atm, $[Fe^{3+}] = 1.0$ M, $[Fe^{2+}] = 0.10$ M, and pH = 5?

83. You have available as reducing agents $Sn^{2+}(aq)$, $V^{2+}(aq)$, and $Fe^{2+}(aq)$. For each one, determine whether it can carry out the following reductions in acidic solution.

(a) $MnO_2(s)$ to $Mn^{2+}(aq)$ **(c)** $H^+(aq)$ to $H_2(g)$
(b) $I_2(s)$ to $I^-(aq)$

84. You have available as oxidizing agents $Fe^{3+}(aq)$, $NO_3^-(aq)$, and $V^{3+}(aq)$. For each one, determine whether it is capable of carrying out the following oxidations in acidic solution.

(a) $H_2(g)$ to $H^+(aq)$ **(c)** $Hg(l)$ to $Hg^{2+}(aq)$
(b) $Sn^{2+}(aq)$ to $Sn^{4+}(aq)$

85. What mass of chromium can be electrodeposited in 3.25 h with a current of 2.57 A from a chromium-plating bath of the type described on page 908?

*86. The vapor pressure of Hg as a function of temperature is given by the following equation.

$$\log P(\text{mmHg}) = \frac{-a}{T} + b$$

In the equation, $a = 3236$, $b = 8.118$, and $T = $ Kelvin temperature. Show that the concentration of Hg(g) in equilibrium with Hg(l) at 25 °C greatly exceeds the maximum permissible level of 0.05 mg Hg/m^3.

87. Explain why one of the cis–trans isomers of $[CoCl_2(en)_2]$ exhibits optical isomerism and the other does not. Which is which?

88. Does the chelate $[M(EDTA)]^{+n-4}$ pictured in Figure 22.24 display optical isomerism? Explain.

89. The magnetic properties of the octahedral complex ion $[Cr(L)_6]^{3+}$ are independent of the identity of the ligands (L). How do you account for this fact?

90. In some of its complex ions, the paramagnetism of Cr^{2+} corresponds to two unpaired electrons and in others to four. How do you account for this difference?

91. A person in your study group argues that a chloride ion has four unshared pairs of electrons and so should be considered a tetradentate ligand. How should you respond?

*92. The complex ion $[Ni(CN)_4]^{2-}$ is diamagnetic. Use crystal field theory to determine whether the structure of this complex ion is octahedral, square planar, or tetrahedral.

93. Four structures are shown in the accompanying sketch. Indicate whether any of these structures are identical, whether any are geometric isomers, and whether any are optical isomers.

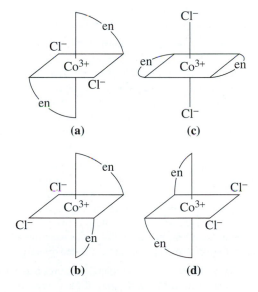

94. The anion form of the amino acid glycine (below) acts as a bidentate ligand. With Cu^{2+}, it forms bis(glycinato)copper(II). Does this coordination compound exhibit geometric isomerism? optical isomerism? Explain.

$$H_2N-CH_2\overset{\overset{\textstyle O}{\|}}{C}-O^-$$

Glycine (anion)

* **95.** Assume that boiler scale is a mixture of the carbonates of Ca^{2+}, Mg^{2+}, and Fe^{3+}, and verify the statement on page 931 that the ligand $EDTA^{4-}$ is able to sequester these cations so that boiler scale does not form. The K_f values are 4×10^{10} for $[CaEDTA]^{2-}$; 4.0×10^{8} for $[MgEDTA]^{2-}$; and 1.7×10^{24} for $[FeEDTA]^{-}$. Assume reasonable values of the total concentration of each cation (say 0.10 M) and of *free* $EDTA^{4-}$ (also 0.10 M).

* **96.** Estimate the total $[Cl^-]$ required in a solution that is initially 0.10 M $CuSO_4$ to produce a visible yellow color.

$$[Cu(H_2O)_4]^{2+} + 4\,Cl^- \rightleftharpoons [CuCl_4]^{2-} + 4\,H_2O$$
$$\text{Blue} \qquad\qquad\qquad \text{Yellow}$$
$$K = 4.2 \times 10^5$$

Assume that 99% conversion of $[Cu(H_2O)_4]^{2+}$ to $[CuCl_4]^{2-}$ is sufficient to produce the yellow color. Also, ignore the presence of any mixed aqua-chloro complex ions.

Apply Your Knowledge

* **97.** [Laboratory] A sample of *trans*-dichlorobis(ethylenediamine) cobalt(III) chloride was prepared by dissolving 5.0 g of $CoCl_2 \cdot 6\,H_2O$ in 12 mL of distilled water, then adding 19 mL of 10.0% by mass ethylenediamine solution ($d = 1.00$ g/mL). A 2.0-mL portion of 30.0% by mass H_2O_2 solution ($d = 1.44$ g/mL) was added to convert cobalt(II) to cobalt(III), after which 12.0 mL of 12 M HCl was added. The solution was heated almost to boiling and allowed to stand for 30 min, during which time the product crystallized from solution (the other product of the reaction is water). The crystals were filtered and dried, and found to weigh 2.77 g. Calculate the percentage yield of the product.

98. [Historical] A compound known before Werner's time was Magnus's green salt, which has the *empirical* formula $PtCl_2 \cdot 2NH_3$. It is actually a coordination compound comprised of a dipositive complex cation and a dinegative complex anion. Propose a formula that is consistent with this information.

* **99.** [Historical] Zeise's salt has the empirical formula $PtCl_2 \cdot KCl \cdot C_2H_4$, but it is actually a coordination compound consisting of a unipositive cation and a complex anion. Propose a formula that is consistent with this information.

* **100.** [Laboratory] A 25.0-mL portion of a nickel(II) solution was treated with 25.0 mL of 0.0200 M EDTA. The excess EDTA required 3.32 mL of 0.0347 M Zn^{2+} for back-titration (page 166). Calculate the concentration of nickel(II) in the solution. Suppose that the same determination was performed on a solution of Fe^{3+} instead of Ni^{2+}. Would the results be different? If so, what would be the concentration of Fe^{3+} in the solution? If the results should remain unchanged, explain why.

* **101.** [Historical] A theory of coordination compounds that predated Werner's was called the chain theory. According to this theory, in a coordination compound like $CoCl_3 \cdot 6\,NH_3$, the Co atom was thought to form three bonds to other groups—either Cl atoms or NH_3 molecules—and NH_3 molecules were believed capable of bonding together into chains. Cl atoms were thought to be ionizable if bonded to NH_3, but not if bonded directly to Co.

(a) Show that this theory predicts structures that account for the behavior toward $AgNO_3(aq)$ of the compounds that Werner designated as $[Co(NH_3)_6]Cl_3$ and $[CoCl(NH_3)_5]Cl_2$.

(b) Does the chain theory also account for the existence of the coordination compounds $[CoCl_3(NH_3)_3]$ and $Na[CoCl_4(NH_3)_2]$? Explain.

* **102.** [Laboratory] A Cu(s) electrode is immersed in a solution that is 1.00 M NH_3 and 1.00 M in $[Cu(NH_3)_4]^{2+}(aq)$. If a standard hydrogen electrode is used as the cathode, E_{cell} is found to be +0.08 V. What is the value obtained by this method for the formation constant, K_f, of $[Cu(NH_3)_4]^{2+}$?

* **103.** [Laboratory] *Absorbance* is a measure of the proportion of monochromatic (single-color) light that is absorbed as the light passes through a solution. An *absorption spectrum* is a graph of absorbance as a function of wavelength. High absorbance corresponds to a large proportion of the light entering a solution being absorbed. Low absorbance signifies that a large proportion of the light is transmitted. The absorption spectrum of $[Ti(H_2O)_6]^{3+}(aq)$ is shown below. The absorption spectrum and the colors in the visible spectrum are shown in (a), and the solution itself in (b).

(a) Describe the color(s) of light that $[Ti(H_2O)_6]^{3+}(aq)$ absorbs most strongly and the color of the solution.

(b) What electron transition in the $[Ti(H_2O)_6]^{3+}$ ion is responsible for the absorption peak, and what is the energy associated with this absorption, expressed in kilojoules per mole?

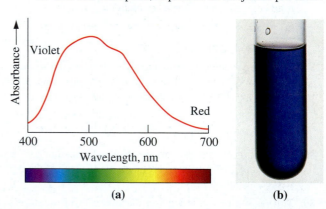

(a)

(b)

e-Media Problems

The activities described in these problems can be found in the e-Media Activities and Interactive Student Tutorial (IST) modules of the Companion Website, *http://chem.prenhall.com/hillpetrucci.*

104. View the **Ligand 3-D** models in (*Section 22-7*). How do the composition and structure of each ligand influence its ability to coordinate to transition metal ions?

105. **(a)** Using examples from the **Isomerism** animation (*Section 22-9*), describe the difference between structural and geometric isomers. **(b)** Which category of isomer would be expected to exhibit the greater chemical difference between isomeric species, and why?

106. After viewing the **Chirality** animation (*Section 22-9*), describe why your two hands are nonsuperimposable mirror images of one another. Repeat this exercise for the coordination compound $[Co(en)_3]^{3+}$.

107. Suggest how the measurement procedures described in the **Optical Activity** animation (*Section 22-9*) could be used to follow the synthesis of enantiomers. How would this approach differ from other methods of characterization based on composition?

108. In the **Crystal Field Theory** activity (*Section 22-10*), **(a)** what is the reason for the splitting of the *d*-orbital energy levels? **(b)** What influence does the splitting have on the properties of transition metal coordination compounds?

Chapter 23

Chemistry and Life: More on Organic, Biological, and Medicinal Chemistry

IN THIS CHAPTER and in much of the two that follow, we extend our study of chemistry into more complex systems. For example, we will explore the ways in which the principles of chemistry apply to reactions in living cells (Sections 23.7–23.10), to the materials that contribute so much to our modern way of life (Chapter 24), and to the environment in which we live (Chapter 25). In doing so, we demonstrate that science is a unified whole and not just a collection of isolated facts.

We begin this chapter by recapitulating some of the organic chemistry discussed in earlier chapters. Then we apply some of the principles we examined in those prior chapters to still more organic compounds. With this background, we move on to the more complex compounds that make up living matter. Then we consider some techniques for determining molecular structures. Finally, we examine some of the chemistry involved in the design and action of common drugs.

Organic compounds do not differ in any fundamental way from inorganic compounds. Indeed, we have used both types to illustrate molecular structures, acid–base chemistry, thermodynamic principles, and other key chemical concepts. However, because there are so many of them, because they can be conveniently organized into families, and for historical reasons, organic compounds are usually studied as a separate branch of chemistry.

Tens of millions of carbon compounds are already known, and new ones are being discovered every day. Why so many carbon compounds? Recall that carbon atoms are unique in their ability to bond to one another to form long chains and that the chains can have branches and can form rings of various sizes. In addition, carbon atoms also bond strongly to atoms of other elements, such as hydrogen, oxygen, and nitrogen, and these atoms can be arranged in many different ways. These facts make it obvious that carbon can form an almost infinite number of molecules of various shapes, sizes, and compositions.

◄ Most of the organic compounds we use today are synthetic, but many are still obtained from nature. The compound halichondrin B (its structural formula is shown in the inset) is obtained from marine sponges such as the one shown here. Halichondrin B is under investigation as an anticancer drug. It is an organic compound both in the original sense of the term (it is derived from a living organism) and in the modern sense (it is a carbon compound).

CONTENTS

We use thousands of carbon compounds every day without even realizing it because they are silently carrying out important chemical reactions within our bodies. Many of these carbon compounds are so vital that we literally could not live without them.

Hydrocarbons

Many of the millions of organic compounds contain only two elements, carbon and hydrogen. These **hydrocarbons** are further classified as follows.

- *Alkanes,* or *saturated hydrocarbons,* first introduced in Section 2.9, have the general formula C_nH_{2n+2}. All the bonds between carbon atoms are single bonds.
- *Alkenes* (Section 9.11) have one or more double bonds between carbon atoms. Those with one double bond have the general formula C_nH_{2n}.
- *Alkynes* (Section 9.11) have one or more triple bonds between carbon atoms. Those with one triple bond have the general formula C_nH_{2n-2}. Alkenes and alkynes collectively are called *unsaturated hydrocarbons.*
- *Aromatic hydrocarbons* have the special kind of bonding noted in Section 10.8.

Some systematic nomenclature of alkanes, alkenes, and alkynes is presented in Appendix D.

23.1   Alkanes

Alkane molecules are essentially nonpolar. For a given molecular mass, alkanes are among the lowest boiling of all the organic compounds (Section 11.5). They are less dense than water and are practically insoluble in it (Section 12.3). Chemically, alkanes are unreactive toward most ionic compounds, mainly because there is no center of either positive or negative charge to attract an ion. Alkanes generally do not react with aqueous acids, bases, or oxidizing agents. Consider butane as an example:

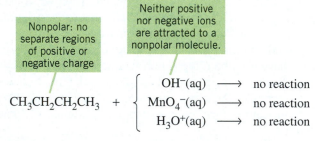

This lack of reactivity toward ionic reagents makes alkanes useful solvents for a variety of organic reactions. Alkanes also are good solvents for other substances of low polarity. For example, hexane is used to extract food oils from soybeans, cottonseed, corn, and other seeds.

Although unreactive toward ionic substances, alkanes react readily at high temperatures with oxygen and with fluorine, chlorine, and bromine. These reactions do not involve ions, but rather proceed by steps that involve free radicals (Section 9.9) as intermediates. The most familiar and important reaction of alkanes is combustion (Section 6.8). We burn alkanes as fuels in automobiles, home heating furnaces, electric power plants, and many other places. An example is the combustion of butane:

$$2\ CH_3CH_2CH_2CH_3(g)\ +\ 13\ O_2(g)\ \longrightarrow\ 8\ CO_2(g)\ +\ 10\ H_2O(l)$$

Like oxygen, halogens are good oxidizing agents and thus react with hydrocarbons. However, the reactions are quite different from combustion. In combustion, the carbon skeleton is completely destroyed. In halogenation, the reaction can be controlled to provide varying degrees of halogenation and leave the carbon skeleton intact. Further, the trend in reactivities of halogens with alkanes is the same as the

Ice floats in water

Solid paraffin sinks in liquid paraffin.

▲ Paraffin wax, a mixture of high-molecular-mass alkanes, is denser as a solid than as a liquid (left). In contrast, ice is less dense than liquid water (right).

trend in the strengths of the halogens as oxidizing agents: $F_2 > Cl_2 > Br_2 > I_2$. Fluorine reacts explosively with hydrocarbons. Thus chemists generally use fluorine compounds rather than $F_2(g)$ to fluorinate hydrocarbons. Chlorination proceeds slowly in the dark at room temperature but quite rapidly at elevated temperatures or in strong light. Bromine is still less reactive, also requiring high temperatures or strong light. Iodine is essentially unreactive toward alkanes.

The reaction of an alkane and a halogen is a **substitution reaction:** A halogen atom substitutes for one or more of the hydrogen atoms of the hydrocarbon. Consider, for example, the reaction of methane and chlorine. With an excess of methane, the main organic product is chloromethane:

$$CH_4(g) + Cl_2(g) \xrightarrow[\text{light}]{\text{Heat or}} CH_3Cl(g) + HCl(g)$$

The mechanism is that of a *chain reaction* initiated by heat or light. The rapid *initiation* reaction occurs when Cl_2 molecules absorb enough energy to dissociate into chlorine atoms ($Cl\cdot$). The reaction is then *propagated* in two steps that are repeated many times: A chlorine atom collides with a CH_4 molecule, extracting an H atom to form $HCl(g)$, and leaving a methyl radical, $H_3C\cdot$. The methyl radical then collides with a Cl_2 molecule, extracting a Cl atom to form $CH_3Cl(g)$* and leaving a Cl atom. The Cl atom attacks another CH_4 molecule, and the propagation steps are repeated. The chain reaction finally ends with three possibilities for a *termination* step, all three of which halt the production of further free radicals. In the equations that follow, only the electrons involved in bond breakage or formation are shown:

Initiation:	$Cl:Cl \xrightarrow[\text{light}]{\text{Heat or}} 2\ Cl\cdot$
Propagation:	$H_3C:H + Cl\cdot \longrightarrow H_3C\cdot + H:Cl$
	$H_3C\cdot + Cl:Cl \longrightarrow H_3C:Cl + Cl\cdot$
Termination:	$2\ Cl\cdot \longrightarrow Cl:Cl$
	or $H_3C\cdot + Cl\cdot \longrightarrow H_3C:Cl$ or $2\ H_3C\cdot \longrightarrow H_3C:CH_3$

Free radicals are often used to initiate polymerization reactions, such as that of ethylene to form polyethylene (Chapter 24). Free-radical reactions are also involved in the destruction of the ozone layer that protects life on Earth from deadly ultraviolet radiation. We will examine this issue further when we study environmental chemistry in Chapter 25.

23.2 Alkenes and Alkynes

Like the alkanes, alkenes and alkynes are essentially nonpolar. All three kinds of hydrocarbons have similar physical properties. They are essentially insoluble in water. Boiling points of alkenes and alkynes are quite close to those of alkanes having the same number of carbon atoms. Chemically, the main similarity among the hydrocarbons is that they all burn. Otherwise, the unsaturated hydrocarbons (alkenes and alkynes) are chemically quite different from the saturated hydrocarbons (alkanes). The difference in reactivity reflects differences in structure. Compare the three two-carbon hydrocarbons:

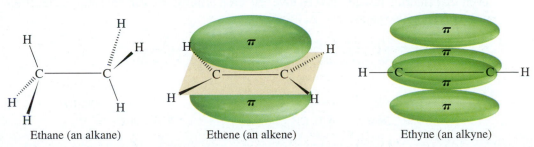

Ethane (an alkane)   Ethene (an alkene)   Ethyne (an alkyne)

▲ To generate electricity, this gas turbine burns natural gas, a mixture of alkanes (mostly methane). Gas turbines are widely used for what in the utility industry is called *peaking*— generating electricity at the times of day when demand is greatest.

Alkane Nomenclature activity

Application Note

Although nearly all organic compounds are combustible, many halogen-containing organic compounds, such as the halons (page 554), are used to put out fires. Carbon tetrachloride, CCl_4, was once widely used in fire extinguishers, but its use has been discontinued because of its toxicity.

---

*The chlorination of methane usually produces a mixture of products. In addition to CH_3Cl, some CH_2Cl_2, $CHCl_3$, and CCl_4 are also formed. Formation of CH_3Cl can be maximized, however, by using a large excess of methane. Traces of ethane, CH_3CH_3, are also formed, just as one of the termination steps of the mechanism predicts.

Application Note

Although alkenes and alkynes burn, they are not generally used as fuels. Rather, alkenes are starting materials for the synthesis of many valuable materials, especially polymers, as we saw in Chapter 9 and will revisit in Chapter 24. Alkynes have fewer uses. Some serve as starting materials for synthesis. Recall that ethyne (acetylene) is used as fuel for the oxyacetylene torch.

In ethane, all the bonds are sigma (σ) bonds and the bonding electrons are localized *between* the atoms that form the bond. In addition to a σ-bond framework, ethene and ethyne have pi (π) bonds. The electrons in π bonds are delocalized and readily accessible to **electrophilic reagents,** that is, to reagents that are deficient in electrons and thus attracted to electrons in other molecules. The π bonds in alkenes and alkynes make them much more chemically reactive than alkanes.

In contrast to the substitution reactions of alkanes, the typical reactions of alkenes are **addition reactions:** Two or more reactant molecules combine to give a single product molecule. These reactions generally involve ions as intermediates; we say they proceed by *ionic* mechanisms. Unlike free-radical reactions, addition reactions usually occur readily even at room temperature and in the dark. An electrophile (electron-seeking molecule or ion) uses the π electrons of the unsaturated hydrocarbon to attach itself to one of the carbon atoms of a multiply bonded pair. A second atom or group then links to the other carbon atom, completing the addition reaction. Consider the addition of HI to ethene:

$$CH_2{=}CH_2 + HI \longrightarrow CH_3CH_2I$$

Ethene   Iodoethane

Because the product molecule no longer has a double bond, it is no longer an alkene; rather, it is a halogenated *alkane*. In this reaction for the formation of iodoethane, the H^+ ion of the HI uses the ethene π electrons to attach itself to one carbon atom, leaving the second carbon atom with a positive charge:

Pair of electrons

Step 1: $H^+ + CH_2{=}CH_2 \longrightarrow CH_3CH_2{}^+$

(The curved arrow indicates the movement of a pair of electrons.) The intermediate $CH_3CH_2{}^+$ is called a **carbocation** because it has a positive charge on a carbon atom. In the second step, the carbocation reacts with I^- to form the product:

Step 2: $CH_3CH_2{}^+ + I^- \longrightarrow CH_3CH_2I$

Addition reactions in which the first step involves an electrophile (such as the H^+ in the iodoethane reaction) are called *electrophilic additions*. Many other reactions take place by a similar mechanism. For example, ethene reacts with chlorine to give 1,2-dichloroethane:

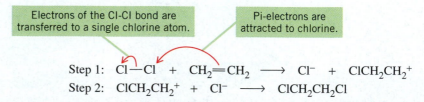

Ethene   1,2-Dichloroethane

In this reaction, it may not be apparent at first that Cl_2 is an electrophile. Recall, however, that chlorine has a high (negative) electron affinity. We can envision the reaction as comprising two steps:

Electrons of the Cl-Cl bond are transferred to a single chlorine atom.   Pi-electrons are attracted to chlorine.

Step 1: $Cl{-}Cl + CH_2{=}CH_2 \longrightarrow Cl^- + ClCH_2CH_2{}^+$
Step 2: $ClCH_2CH_2{}^+ + Cl^- \longrightarrow ClCH_2CH_2Cl$

Testing for Unsaturated Hydrocarbons with Bromine movie

Bromine, Br_2, forms a brownish-red solution in carbon tetrachloride, CCl_4. When such a solution is added to an alkene, the color disappears because the bromine reacts with the alkene to form a colorless dibromoalkane. The decolorization of bromine solutions is frequently used as a simple test for the presence of alkenes:

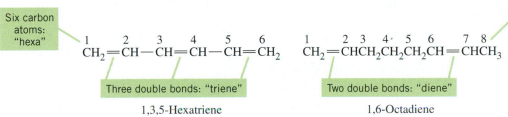

In Chapter 9, we described the reaction of alkenes and alkynes with H_2 in the presence of a catalyst. A common example is the hydrogenation of oils. Unsaturated fats are converted to saturated fats as double bonds are replaced by single bonds.

Another important reaction of alkenes is the addition of a water molecule. This reaction, called *hydration*, requires H^+ as a catalyst. The H^+ is usually furnished by sulfuric acid:

$$CH_2{=}CH_2 + H_2O \xrightarrow{\ H_2SO_4\ } CH_3CH_2OH$$
$$\text{Ethene} \qquad\qquad\qquad \text{Ethanol}$$

Application Note

The large quantities of ethanol used as an industrial solvent are made by a hydration reaction. This ethanol cannot be used in alcoholic beverages, however, because federal law requires that all beverage ethanol be produced by fermentation.

23.3 Conjugated and Aromatic Compounds

Molecules can have more than one multiple bond. Compounds with molecules that have two carbon-to-carbon double bonds are called *dienes;* with three double bonds, *trienes;* and so on:

Six carbon atoms: "hexa"

$$\underset{\text{1}}{CH_2}{=}\underset{\text{2}}{CH}-\underset{\text{3}}{CH}{=}\underset{\text{4}}{CH}-\underset{\text{5}}{CH}{=}\underset{\text{6}}{CH_2}$$

Three double bonds: "triene"

1,3,5-Hexatriene

Eight carbon atoms: "octa"

$$\underset{\text{1}}{CH_2}{=}\underset{\text{2}}{CH}\underset{\text{3}}{CH_2}\underset{\text{4}}{CH_2}\underset{\text{5}}{CH_2}\underset{\text{6}}{CH}{=}\underset{\text{7}}{CH}\underset{\text{8}}{CH_3}$$

Two double bonds: "diene"

1,6-Octadiene

Conjugated Bonding Systems

Note that 1,3,5-hexatriene has alternate double and single bonds between the carbon atoms. A structure that has a series of alternate double and single bonds is said to have a **conjugated bonding system.** Note that 1,6-octadiene is *not* conjugated; the double bonds are separated by more than one single bond. *Aromatic* compounds have conjugated bonding in a ring system, and the most common ones have six π electrons, with the most well-known example being benzene.

Aromatic Compounds

In Section 10.9 we divided organic compounds into two main categories—aliphatics and aromatics—with aromatics being compounds that are related in the broadest sense to the benzene molecule, C_6H_6. Recall from Chapter 10 that the benzene molecule consists of a planar ring of six carbon atoms, each with one hydrogen atom attached, plus six delocalized π electrons located in π molecular orbitals (Figure 10.29). Two common representations of benzene are

Kekulé structures   Molecular orbital representation

▲ The bright red color of these tomatoes is due to lycopene, a molecule with an extended conjugated bonding system (see page 149).

Many common aromatic compounds—toluene ($C_6H_5CH_3$), aniline ($C_6H_5NH_2$), phenol (C_6H_5OH), and benzoic acid (C_6H_5COOH), for example—are considered to be derivatives of benzene.

The other main category of organic compounds, the aliphatics, comprises all open-chain organic compounds plus ring compounds that are not benzene-like. Thus, cycloalkanes have properties much like those of open-chain alkanes and are classified as aliphatic.

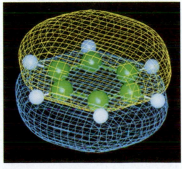

▲ A computer-generated molecular orbital model of benzene.

We defined aromatic compounds as those that resemble benzene, but just which properties of benzene must a compound possess to be aromatic? What properties do all aromatic compounds have in common?

Experimentally, aromatic compounds have molecular formulas that would lead us to expect a high degree of unsaturation. However, aromatic compounds are resistant to the addition reactions generally characteristic of unsaturated compounds. Instead of addition reactions, we often find that aromatic compounds undergo substitution reactions. Presumably, the reason aromatic compounds resist addition reactions is the unusual stability of these molecules. Low heats of hydrogenation and low heats of combustion provide evidence of this stability.

Let us now contemplate some of the theory of aromaticity. Two things we can say at the outset are

- Aromatic molecules are cyclic and planar (flat).
- Aromatic molecules have a conjugated bonding system that produces cyclic clouds of delocalized π electrons above and below the plane of the molecule. The conjugated system ordinarily must extend completely around the ring(s), and the π clouds must contain a total of $(4n + 2)$ electrons, where $n = 0, 1, 2, \ldots$.

Most common aromatic compounds have $n = 1$ so that $4n + 2 = (4 \times 1) + 2 = 6$.

Example 23.1

Which of the following compounds are aromatic and which are aliphatic? Explain.

$$CH_2=CH-CH=CH-CH=CH_2$$

          (a)   (b) 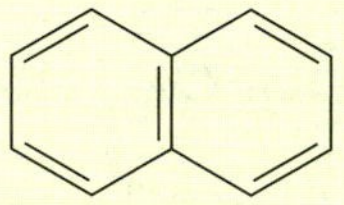   (c) 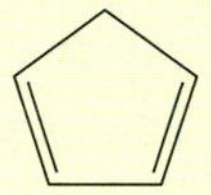   (d)

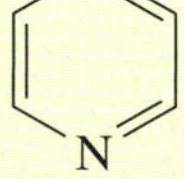

SOLUTION

(a) This compound has six π electrons and a conjugated bonding system, but it is not cyclic. Like all other open-chain compounds, this one is aliphatic.

(b) This compound has six π electrons and a cyclic, conjugated bonding system; it is aromatic.

(c) This compound has 10π electrons ($n = 2$; $[4 \times 2] + 2 = 10$) and a cyclic, conjugated bonding system; it is aromatic.

(d) This compound has a cyclic, conjugated bonding system, but the bonding system does not extend all the way around the ring and has only four π electrons. The compound is aliphatic.

EXERCISE 23.1A

Which of the following compounds are aromatic and which are aliphatic? Explain.

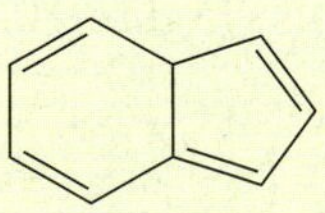

          (a) 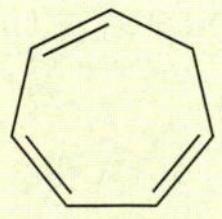   (b) 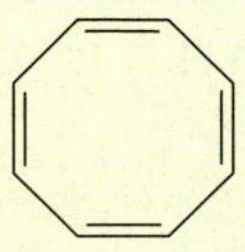   (c) 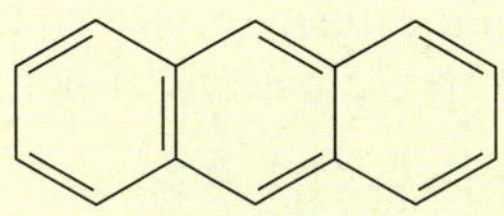   (d)

EXERCISE 23.1B

Despite its similarity to the molecule shown in Example 23.1(d), pyrrole,

Pyrrole   N—H

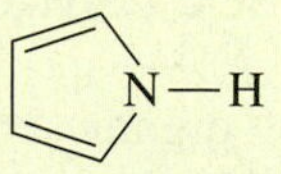

is considered to be aromatic. Explain.

Reactions of Benzene

The molecular formula of benzene suggests that it is an unsaturated hydrocarbon. Six carbon atoms could bond as many as $2n + 2 = 14$ hydrogen atoms, as in hexane, C_6H_{14}. Removing two H atoms to form a ring still leaves 12 hydrogen atoms, as in cyclohexane, C_6H_{12}. Although the benzene molecule is unsaturated, it does not readily undergo the addition reactions typical of alkenes (page 944). Instead, its reactions are mainly substitution reactions. For example, with $FeCl_3$ as a catalyst, benzene reacts with chlorine to form chlorobenzene:

$$C_6H_6 + Cl_2 \xrightarrow{FeCl_3} C_6H_5Cl + HCl$$

To gain a better understanding of this and other aromatic substitution reactions, let us look at the mechanism of the reaction.

The first step is a Lewis acid–base reaction between the catalyst, $FeCl_3$, and Cl_2:

Step 1:

Note that in this case both bonding electrons of Cl_2 go with one of the chlorine atoms. This is quite unlike the free-radical chlorination of methane, in which strong light or high temperatures split Cl_2 into two chlorine atoms, each with an unpaired electron. The electron-deficient chlorine atom* has a positive charge; it is an electrophile. In step 2, the Cl^+ attracts the π electrons of the benzene ring and uses a pair of them to attach itself to a carbon atom of the ring. We use one of the Kekulé structures to represent benzene:

Step 2:

Carbocation
intermediate

The intermediate carbocation gives up a hydrogen ion to $[FeCl_4]^-$ (from step 1), forming the *product* chlorobenzene, the *by-product* HCl, and regenerating the *catalyst* $FeCl_3$.

Step 3:

$$+ [FeCl_4]^- \longrightarrow + HCl + FeCl_3$$

This type of reaction, involving the replacement of one of the hydrogen atoms of an aromatic ring by an electrophile (in this case, Cl^+), is called an **electrophilic aromatic substitution** reaction. Four other examples of this type of reaction are outlined in Figure 23.1.

Aluminum chloride is an especially important Lewis-acid catalyst in organic reactions. Like $FeCl_3$ and other Lewis acids, $AlCl_3$ reacts with many species having lone-pair electrons. For example, it reacts with 2-chloropropane to form a carbocation:

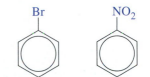

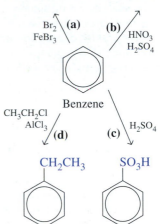

▲ **FIGURE 23.1   Some electrophilic aromatic substitution reactions of benzene**

(a) Halogenation, (b) nitration, (c) sulfonation, and (d) alkylation.

Application Note

Electrophilic substitution reactions are used in the manufacture of detergents. The linear alkylbenzenesulfonate (LAS) detergents described in Section 20.13 are made by alkylation of benzene followed first by sulfonation of the alkylbenzene and then by neutralization of the alkylbenzenesulfonic acid with sodium carbonate.

---

*It is unlikely that an independent Cl^+ ion exists. Rather, as $FeCl_3$ begins to bond to Cl_2, the chlorine atom with a *developing* positive charge ($Cl_3Fe^{\delta-}$—Cl—$Cl^{\delta+}$) begins to form a bond to the benzene ring. Thus, you should not take too literally the idea of the reaction occurring in the distinctly separate steps 1 and 2.

This carbocation is an electrophile that can then attach itself to a benzene ring. The net alkylation reaction is one in which new carbon-to-carbon bonds are formed in a molecule:

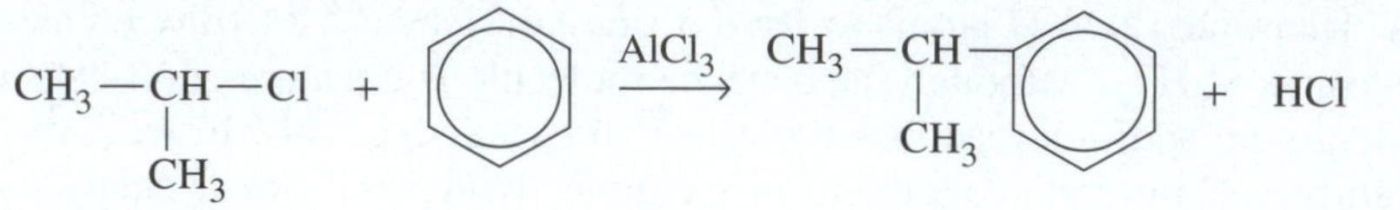

2-Chloropropane   Benzene   Isopropylbenzene
(isopropyl chloride)

Reactions of Other Aromatic Compounds

Some substituted benzenes undergo the same types of reactions as their aliphatic analogues. For example, toluene ($C_6H_5CH_3$) reacts with chlorine at high temperatures, undergoing free-radical substitution in the side chain, CH_3, which reacts like an alkane:

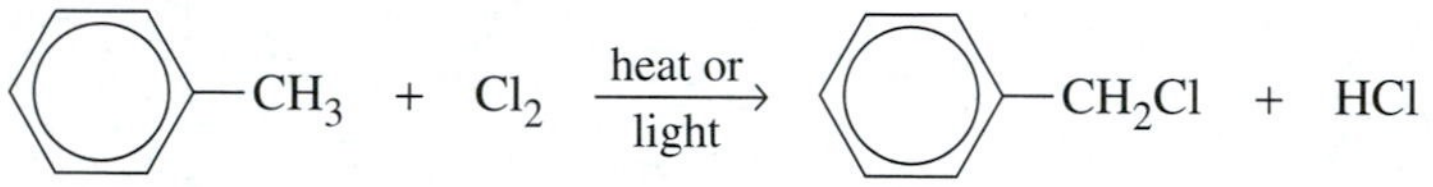

Like benzene, toluene and other substituted benzenes undergo electrophilic aromatic substitution reactions. In the presence of chlorine and $FeCl_3$, for instance, a chlorine atom substitutes for a *ring* hydrogen atom in a substituted benzene. There is a complication, however. Although all six positions on an unsubstituted benzene ring are equivalent, on a substituted benzene ring they are not. Thus, in the reaction of benzene and chlorine, a chlorine atom can substitute for a H atom at any of the six positions, and there is only one possible product, chlorobenzene. In the reaction of toluene and chlorine, however, the substitution of a Cl atom for a H atom can yield three possible substitution products:

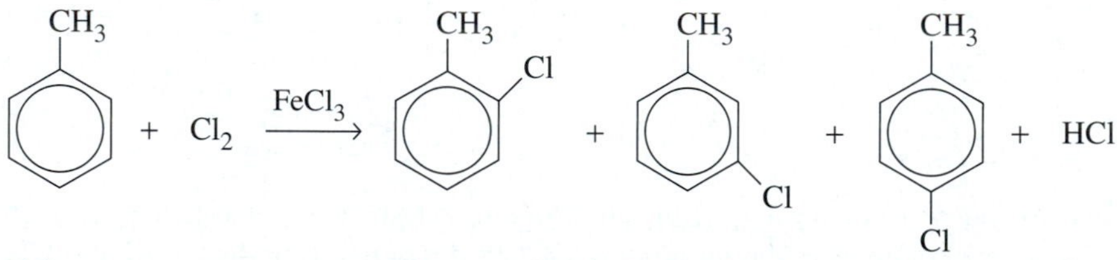

Toluene   *o*-Chlorotoluene   *m*-Chlorotoluene   *p*-Chlorotoluene

In practice, chlorination of toluene gives a mixture composed of mainly *o*- and *p*-chlorotoluene; very little *m*-chlorotoluene is formed. In general, groups present on a substituted benzene ring act like a traffic director, sending an incoming electrophile preferentially either to the meta position or to ortho and para positions. In toluene, the CH_3 group directs the incoming chlorine atoms preferentially to the ortho and para positions. We will not consider the directing influence of substituent effects further here.

Aromatic carboxylic acids, such as benzoic acid (C_6H_5COOH), react much like aliphatic acids, forming salts with bases and esters with alcohols (Section 23.6). Aromatic amines, such as aniline, $C_6H_5NH_2$, undergo typical reactions of amines, such as salt formation in reactions with acids and amide formation with carboxylic acids (Section 23.6). Like other aromatic compounds, benzoic acid and aniline undergo ring substitution reactions.

Recall from page 423 that the two positions adjacent to a substituent on a benzene ring are *ortho* to the substituent, the position directly opposite the substituent is *para*, and the remaining two positions are *meta*.

Application Note

The fact that the methyl group on toluene directs substitution groups to the ortho and para positions on the ring is put to use in making TNT (2,4,6-trinitrotoluene). When toluene is nitrated with excess HNO_3, nitro groups are substituted in both ortho positions and in the para position.

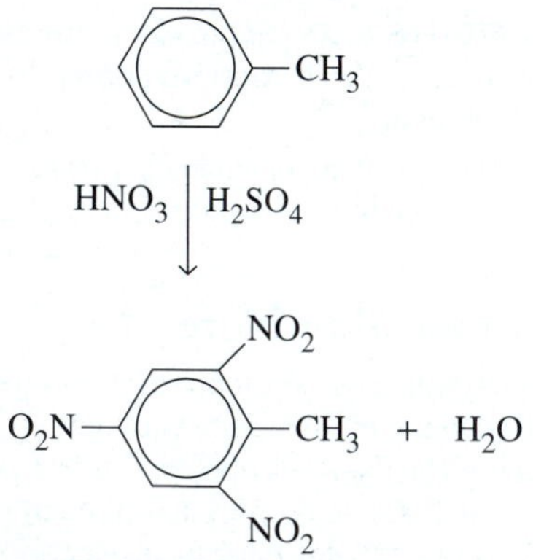

Functional Groups

Recall that many organic molecules have two parts, a hydrocarbon portion that is generally unreactive and a *functional group* where most of the reactions of the molecule take place. The functional group is usually polar, with electron-rich and electron-poor sites where ions are likely to attack. Electrophiles are attracted to an electron-rich site. **Nucleophiles,** ions or molecules attracted to a partially positive carbon atom, are typically negative ions or other Lewis bases. The hydrocarbon portion of the organic molecule is essentially nonpolar, an unlikely location for ionic attack. A molecule of the ketone 2-octanone, which contains a carbonyl functional group (see Appendix D), illustrates these various regions in the typical organic molecule:

Cations (electrophiles) attack here.

Anions (nucleophiles) attack here.

Ions are unlikely to attack the nonpolar hydrocarbon chain.

$$\underset{\delta^+}{CH_3\overset{\overset{\displaystyle O\ \delta^-}{\|}}{C}CH_2CH_2CH_2CH_2CH_2CH_3}$$

23.4 Alcohols and Ethers

In Section 2.10, we noted that alcohols have the general formula ROH, where R is an alkyl group and OH is a hydroxyl group. An alcohol molecule can act as a Brønsted base. One of the lone pairs on the oxygen atom can accept a proton (H^+), just as a water molecule does in forming a hydronium ion:

$$R-\ddot{O}-H \ + \ H-A \ \rightleftharpoons \ \left[R-\underset{\underset{H}{|}}{\ddot{O}}-H\right]^+ \ + \ A^-$$

An alcohol   An acid   A protonated alcohol

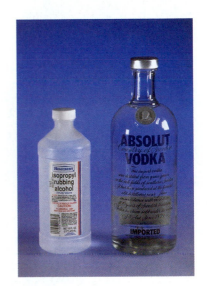

The reaction forms the conjugate acid of the alcohol, and we say that the alcohol has been *protonated*. Many reactions of alcohols occur in acidic solution, and the first step is often one in which the alcohol acts as a Brønsted base.

If the alcohol is 1-propanol and the acid is HI, the overall reaction forms 1-iodopropane and water:

$$CH_3CH_2CH_2OH + HI \longrightarrow CH_3CH_2CH_2I + H_2O$$

1-Propanol   1-Iodopropane

▲ Two familiar alcohols are ethanol (vodka is 40–50% ethanol) and 2-propanol (rubbing alcohol is usually 70–90% 2-propanol in water).

In this reaction, I substitutes for the OH. The substitution is not direct, however, because the OH group is difficult to displace. 1-Propanol does not react directly with I^-:

$$CH_3CH_2CH_2OH + I^- \longrightarrow \text{no reaction}$$

Organic Functionality Nomenclature activity

Only after the alcohol has been protonated does the I^- enter into the reaction, by displacing the OH_2.

To explain these facts, organic chemists propose reaction mechanisms in the manner we described in our study of chemical kinetics (Chapter 13). Consider the following mechanism, which consists of two elementary reactions, the first of which is the protonation:

Acid–base step (fast): $CH_3CH_2CH_2-\ddot{\underset{\cdot\cdot}{O}}:H \ + \ H-I \ \rightleftharpoons \ \left[CH_3CH_2CH_2-\underset{\underset{H}{|}}{\ddot{\underset{\cdot\cdot}{O}}}:H\right]^+ \ + \ I^-$

(a base)   (an acid)

Substitution step (slow): $I^- \ + \ \left[\overset{\overset{\delta^+ \quad \delta^-}{}}{CH_3CH_2CH_2-\underset{\underset{H}{|}}{\ddot{\underset{\cdot\cdot}{O}}}:H}\right]^+ \ \longrightarrow \ CH_3CH_2CH_2I \ + \ H_2O$

The first step is fast—an acid–base reaction between the HI and 1-propanol. Note that the intermediate, $CH_3CH_2CH_2OH_2^+$, has a polar C—O bond. Electrons are displaced toward the more electronegative O atom ($\delta-$) and away from the C atom ($\delta+$). In the slow second step, the negative ion I^- attacks the partially positive C atom from

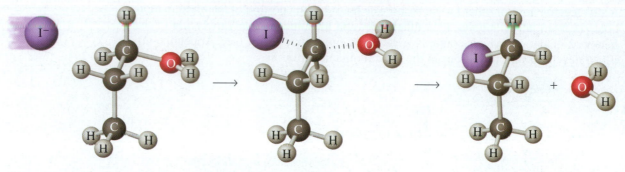

▲ **FIGURE 23.2** **An S$_N$2 Reaction**
The I$^-$ ion displaces OH$_2$ in the second step of the conversion of 1-propanol to 1-iodopropane.

Bimolecular Reaction animation

the back side, that is, from the side opposite the OH$_2$ end of the intermediate. The I$^-$ is a nucleophile, a seeker of positive charge. Because this second step involves two species, the I$^-$ ion and the protonated alcohol, it is a second-order reaction. This reaction mechanism is called a *nucleophilic substitution, second order* and is symbolized as S$_N$2. Figure 23.2 shows the S$_N$2 reaction with molecular models.

Alcohols can also undergo *dehydration* reactions. During dehydration, a OH group and a H atom are removed from the alcohol. In concentrated H$_2$SO$_4$ and at relatively high temperatures, dehydration leads to an alkene. Thus, dehydration of ethanol produces ethene:

$$CH_3CH_2OH \xrightarrow[\Delta]{\text{conc. } H_2SO_4} CH_2{=}CH_2 + H_2O$$

If H$_2$SO$_4$ of a lower concentration is used and the reaction is carried out at a lower temperature, we get a different product—an ether. An **ether** has two hydrocarbon groups joined through an oxygen atom. The general formula is ROR$'$, where R and R$'$ represent hydrocarbon groups that may or may not be identical. At the lower H$_2$SO$_4$ concentration and lower temperature, ethanol forms diethyl ether as the main product:

$$CH_3CH_2OH + HOCH_2CH_3 \xrightarrow{H_2SO_4} CH_3CH_2OCH_2CH_3 + H_2O$$
Diethyl ether

Notice that in alkene formation a H atom and a OH group both are removed from one alcohol molecule. In ether formation, however, a H atom is removed from the OH group of one alcohol molecule, and the OH group is removed from a second alcohol molecule. In both reactions, the first step is protonation of the alcohol:

Acid–base step (fast): CH$_3$CH$_2$OH + H$_2$SO$_4 \rightleftharpoons$ CH$_3$CH$_2$OH$_2^+$ + HSO$_4^-$

What happens next depends on the reaction conditions.

Alkene formation With excess sulfuric acid, nearly all the alcohol is converted to CH$_3$CH$_2$OH$_2^+$. This protonated alcohol then reacts with a base, such as water. If the water molecule attacks at the carbon atom attached to the OH$_2^+$ group, all we get is the reversal of the protonation reaction. However, if the water molecule attacks a hydrogen atom on the adjacent carbon atom, an elimination reaction occurs:

Acid–base step (slow): H$_2\ddot{O}${:}$\rightarrow$H$\diagup$CH$_2\overset{\frown}{=}$CH$_2\diagup\overset{+}{O}$H$_2$

This step involves extracting a proton from a carbon atom and is much slower than the fast acid–base first step. The extraction from the carbon occurs readily only at relatively high temperatures. Because this slow rate-determining step involves both H$_2$O and HCH$_2$CH$_2$OH$_2^+$, the reaction is a second-order elimination reaction, designated E2.

Ether formation If the temperature is relatively low and the ethanol concentration high relative to the H_2SO_4 concentration, an ethanol molecule acts as a nucleophile, displacing water from a protonated ethanol molecule:

Displacement reaction (slow): CH_3CH_2OH + $CH_3CH_2-OH_2^+$ $\longrightarrow$ $\left[\begin{array}{c} CH_3CH_2OCH_2CH_3 \\ | \\ H \end{array} \right]^+$ + H_2O

The product of this step is the conjugate acid of diethyl ether. This protonated ether molecule acts as an acid, readily giving up a proton to a base such as CH_3CH_2OH or H_2O:

Acid-base step (fast): $\left[\begin{array}{c} CH_3CH_2OCH_2CH_3 \\ | \\ H \end{array} \right]^+$ + H_2O $\longrightarrow$ $CH_3CH_2OCH_2CH_3$ + H_3O^+

The slow step is bimolecular, and so the ether-forming reaction is a second-order substitution reaction (S_N2).

Like water (HOH), alcohols (ROH) undergo self-ionization:

$$2\ ROH \rightleftharpoons ROH_2^+ + RO^-$$

In general, alcohols are weaker acids than water, and the pH of a solution of an alcohol in water would be determined mainly by the self-ionization of water. Alcohols are acidic enough, however, to undergo some reactions that resemble those of water. For example, alcohols react with sodium metal to form hydrogen gas and ionic compounds called sodium *alkoxides*. In this oxidation–reduction reaction, only the H atom of the OH group is replaced:

$$2\ ROH(l) + 2\ Na(s) \longrightarrow 2\ RO^-Na^+(\text{in ROH solution}) + H_2(g)$$

The concentration of H^+ produced in the self-ionization of ROH is large enough that a powerful reducing agent such as Na(s) is able to displace the H^+ as $H_2(g)$.

We first described typical oxidation–reduction reactions of alcohols in Section 4.6: Alcohols are oxidized to aldehydes and ketones, and aldehydes are then further oxidized to carboxylic acids:

$$R-CH_2OH \xrightarrow{\text{oxidation}} R-\overset{\overset{\displaystyle O}{\|}}{C}-H \xrightarrow{\text{oxidation}} R-\overset{\overset{\displaystyle O}{\|}}{C}-OH$$
An alcohol   An aldehyde   A carboxylic acid

$$R-\overset{\overset{\displaystyle OH}{|}}{C}H-R' \xrightarrow{\text{oxidation}} R-\overset{\overset{\displaystyle O}{\|}}{C}-R'$$
An alcohol   A ketone

For example, the half-reaction for the oxidation of ethanol is

$$CH_3CH_2OH \longrightarrow CH_3CHO + 2\ H^+ + 2\ e^-$$

In living cells, the oxidizing agent is nicotinamide adenine dinucleotide, which is reduced from NAD^+ to NADH:

$$NAD^+ + H^+ + 2\ e^- \longrightarrow NADH$$

The overall redox reaction is

$$CH_3CH_2OH + NAD^+ \longrightarrow CH_3CHO + NADH + H^+$$

S$_N$2 Displacement Reactions in Organic and Biochemistry

Many organic and biochemical reactions are S$_N$2 displacement reactions. In organic chemistry, S$_N$2 reactions can be used to convert alkyl halides to amines, to sulfur-containing compounds called thiols, and to hundreds of other products:

$$CH_3CH_2Br + 2 NH_3 \longrightarrow CH_3CH_2NH_2 + NH_4Br$$

Bromoethane   Ethylamine

$$CH_3CH_2CH_2CH_2Br + NaSH \longrightarrow CH_3CH_2CH_2CH_2SH + NaBr$$

1-Bromobutane   1-Butanethiol

The thiol 1-butanethiol has a potent skunk odor. A related compound, ethanethiol, CH_3CH_2SH, is used as an odorant in natural gas so that we can detect leaks of this otherwise odorless material.

In biochemistry, many reactions involve an NH$_2$ group on DNA plus some *alkylating agent,* such as an alkyl halide. When such a reaction occurs, the alkylated DNA is *mutated* and, if not repaired, can lead to cancer or a disease that can be passed on to one's descendants. Fortunately, our cells have special enzymes that can ordinarily cut out and replace segments of mutated DNA.

▲ Skunks use thiols similar to 1-butanethiol as a defensive weapon.

23.5   Aldehydes and Ketones

As we have noted, aldehydes and ketones are formed by the oxidation of alcohols. If the OH group of an alcohol is on a terminal carbon atom, an *aldehyde* is formed, and if the OH group is bonded to an interior carbon atom, a *ketone* is produced. Both aldehydes and ketones have a *carbonyl* (C=O) functional group:

An aldehyde   A ketone

An aldehyde has a hydrogen atom attached to the carbonyl carbon atom; a ketone has two hydrocarbon groups attached to the carbonyl carbon atom. These structural formulas are often written on one line as RCHO and RCOR′, in which cases the carbon–oxygen double bond is understood.

The carbonyl group is also part of more complicated functional groups, such as those of carboxylic acids, esters, and amides. It is found in carbohydrates, fats, proteins, nucleic acids, hormones, vitamins, and the host of organic compounds critical to living systems. Right now, we will focus on the carbonyl group in aldehydes and ketones.

Formaldehyde, HCHO, is familiar to many as a biological tissue preservative, as in embalming fluid. It is used industrially in huge quantities, mainly to make phenol–formaldehyde resins (Chapter 24) for bonding plywood. Formaldehyde and other simple aldehydes have a rather foul, irritating odor. Some higher aldehydes, however, are used in flavors and fragrances. The following familiar examples have the carbonyl group highlighted:

▲ Cinnamaldeyde is the main flavor component in cinnamon.

Benzaldehyde
(almond flavor)

Cinnamaldehyde
(cinnamon flavor)

Vanillin
(vanilla flavor)

The simplest ketone, acetone (2-propanone; CH_3COCH_3), is a common solvent used in applications as varied as varnishes and lacquers, rubber cement, and fingernail polish remover. Some other familiar examples of ketones are

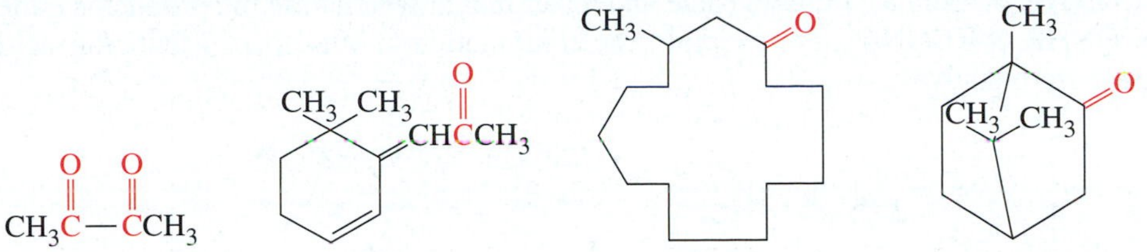

2,3-Butanedione (butter flavoring)

Irone (odor of violets)

Muscone (musk oil, an ingredient in perfumes)

Camphor (an ingredient in some insect repellants)

▲ The delightful aroma of butter is largely due to 2,3-butanedione.

Aldehydes and ketones react at the carbonyl group. Many of these reactions start with a simple addition to the group, often followed by more steps. Consider the addition of HCN. In an acidic solution, the H^+ ion adds to the partially negative O atom of the carbonyl group, and then CN^- adds to the partially positive C of the $C=O$ group:

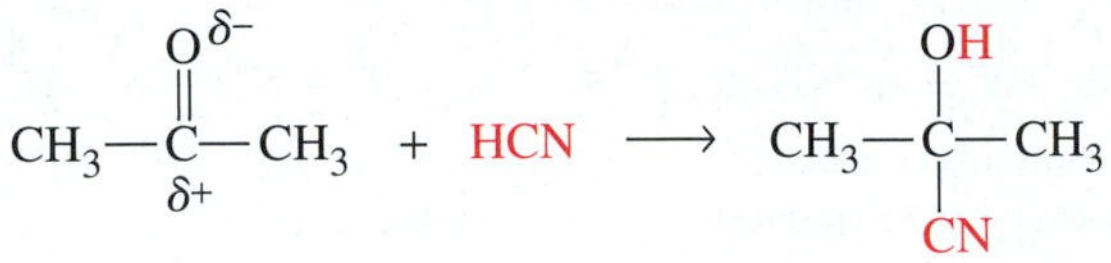

23.6 Carboxylic Acids, Esters, and Amides

We introduced carboxylic acids (RCOOH) in Chapter 2. We then used acetic acid and other carboxylic acids as examples of weak acids in Chapter 4 and in acid–base equilibria calculations in Chapter 15. Carboxylic acids and some compounds derived from them are common in nature and quite important in biochemistry. We will consider two classes of carboxylic acid derivatives here: the *esters* (RCOOR′), which include fats (Section 23.7), and the *amides* (RCONH₂). Proteins (Section 23.9) are perhaps the most impressive example of amides. Here we look at simple carboxylic acids, esters, and amides. Later we turn to the more complex worlds of lipids and proteins.

Carboxylic Acids

The simplest carboxylic acids are widely known by their common names. For example, HCOOH is formic acid and CH_3COOH is acetic acid. Systematic nomenclature is discussed in Appendix D.

By far the most common reaction of the carboxylic acids is as acids. Benzoic acid, the simplest aromatic acid, is a solid that is only slightly soluble in water. However, it reacts with and dissolves in a solution of an aqueous base:

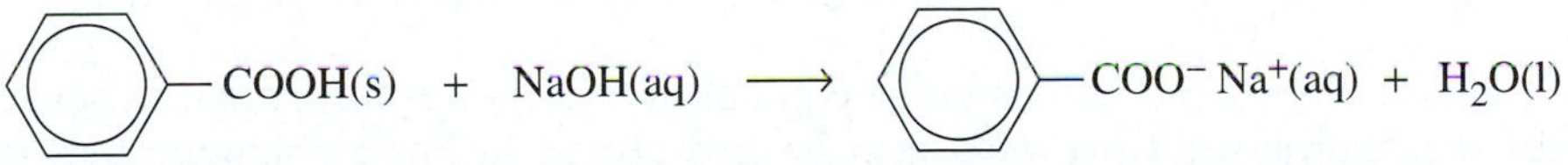

$$\text{COOH(s)} + \text{NaOH(aq)} \longrightarrow \text{COO}^-\text{Na}^+\text{(aq)} + H_2O(l)$$

▲ Carboxylic acids occur in many common household items. Oxalic acid is found in spinach and is used in wood finishing; vinegar contains acetic acid; pantothenic acid is a form of vitamin B_5; oranges and lemons contain citric acid, a triprotic acid; and aspirin is acetylsalicylic acid.

Two other important reactions are the formation of esters and amides. Ester formation closely resembles ether formation (Section 23.4). In ester formation, the OR from an alcohol molecule replaces the OH of a carboxylic acid molecule. A molecule of water is eliminated. As in ether formation, a strong acid such as H_2SO_4 catalyzes the reaction. We can write general and specific reactions as follows:

$$R-\overset{O}{\overset{\|}{C}}-OH + HO-R' \underset{H_2SO_4}{\rightleftharpoons} R-\overset{O}{\overset{\|}{C}}-O-R' + H_2O$$

An acid   An alcohol   An ester

$$CH_3CH_2CH_2COOH + HOCH_2CH_3 \underset{H_2SO_4}{\rightleftharpoons} CH_3CH_2CH_2COOCH_2CH_3 + H_2O$$

Butyric acid   Ethanol   Ethyl butyrate

The reactions are reversible and reach equilibrium.

Just as an ester is derived from a carboxylic acid and an alcohol, an **amide** is derived either from a carboxylic acid and ammonia, in which case the product is $RCONH_2$, or from a carboxylic acid and an amine, in which case the product is either $RCONHR$ or $RCONR_2$. For example, the combination of butyric acid (butanoic acid) and ammonia gives

$$CH_3CH_2CH_2 - \overset{\overset{\displaystyle O}{\|}}{C} - NH_2$$

The amide functional group

In this molecule, the OH group of the carboxylic acid molecule has been replaced by an NH_2 group from an ammonia molecule.

Amides are named by changing the ending of the carboxylic acid name from *-oic acid* to *-amide*. The amide formed from butanoic acid and ammonia is named butyramide (common) or butanamide (IUPAC).

Amides can be made by heating ammonium (or amine) salts of carboxylic acids. For example, acetamide is formed by heating ammonium acetate:

$$CH_3 - \overset{\overset{\displaystyle O}{\|}}{C} - O^-NH_4^+ \xrightarrow{\Delta} CH_3 - \overset{\overset{\displaystyle O}{\|}}{C} - NH_2 + H_2O$$

Ammonium acetate   Acetamide
(ammonium ethanoate)   (ethanamide)

This method is usually carried out in two steps: First, the carboxylic acid is neutralized with ammonia or an amine, and then the resulting salt is heated:

$$CH_3 - \overset{\overset{\displaystyle O}{\|}}{C} - OH + CH_3CH_2NH_2 \longrightarrow CH_3 - \overset{\overset{\displaystyle O}{\|}}{C} - O^-\ H_3N^+ - CH_2CH_3$$

Acetic acid   Ethylamine   Ethylammonium acetate

$$CH_3 - \overset{\overset{\displaystyle O}{\|}}{C} - O^-\ H_3N^+ - CH_2CH_3 \xrightarrow{\Delta} CH_3 - \overset{\overset{\displaystyle O}{\|}}{C} - NHCH_2CH_3 + H_2O$$

N-Ethylacetamide

Reactions of Esters and Amides

The principal reaction of esters is hydrolysis, a process that converts an ester back to the alcohol and carboxylic acid from which it was made:

$$R - \overset{\overset{\displaystyle O}{\|}}{C} - O - R' + H_2O \underset{}{\overset{H_2SO_4}{\rightleftharpoons}} R - \overset{\overset{\displaystyle O}{\|}}{C} - OH + HO - R'$$

An ester   A carboxylic acid   An alcohol

The reaction is reversible and comes to equilibrium. Like ester formation, an acid catalyzes the hydrolysis. Esters can also be hydrolyzed in basic solution, and basic hydrolysis goes to completion:

$$R - \overset{\overset{\displaystyle O}{\|}}{C} - O - R' + NaOH(aq) \xrightarrow{\Delta} R - \overset{\overset{\displaystyle O}{\|}}{C} - O^-Na^+ + HO - R'$$

An ester   A carboxylate salt   An alcohol

The hydrolysis of esters is the basis for making soap (Section 23.7).

The principal reaction of amides is also hydrolysis, a process that converts an amide back to the carboxylic acid and ammonia (or amine) from which it was made. These reactions are catalyzed by either an acid or a base. In basic solution, $NH_3(g)$ is

Application Note

Nylon (Chapter 24), a synthetic polyamide (a polymer with amide functional groups), is made by heating the salt of a dicarboxylic acid and a diamine. Proteins (Section 23.9) are natural polyamides.

▲ Many esters, both natural and synthetic, are used as fragrances and flavors. Ethyl butyrate, made by the reaction of ethyl alcohol with butyric acid, is a pineapple flavoring. Pentyl acetate (also called amyl acetate), made by the reaction of isopentanol with acetic acid, is a banana flavoring.

formed and can be detected either by its odor or by the fact that it turns moist red litmus paper blue. The general reaction is

$$R\overset{\overset{\text{O}}{\|}}{-}C-NH_2 \ + \ NaOH(aq) \ \overset{\Delta}{\longrightarrow} \ R\overset{\overset{\text{O}}{\|}}{-}C-O^-Na^+ \ + \ NH_3(g)$$

An amide   A carboxylate salt

The digestion of proteins (Section 23.9) involves the hydrolysis of amide linkages, a reaction that breaks the protein molecule down into individual amino acids that are then absorbed into the bloodstream.

Biological Molecules

Our bodies are collections of chemicals—exquisitely organized chemicals, to be sure, but chemicals nonetheless. Fats, carbohydrates, proteins, and nucleic acids, together with water, make up most of the mass of the human body. Countless crucial chemical reactions take place within each cell every minute of every day. We depend on the reactions of fats and carbohydrates to furnish energy for our every activity. Proteins provide the building materials that help us to grow and to repair and replace damaged or aging cells and tissues. Perhaps most intriguing of all are the complex molecules called nucleic acids, macromolecules that are involved in the processes of heredity.

To place this discussion in context, we need to describe briefly the *cell*, which is the basic structural unit of living matter in which biomolecules—proteins, nucleic acids, fats, carbohydrates, and so on—are found. Each cell of a plant or animal is enclosed in a cell membrane through which the cell gains nutrients and gets rid of wastes. Inside the cell are a number of structures, called *organelles*, that serve a variety of functions. The largest organelle usually is the cell nucleus, which contains the nucleic acids that control heredity. Protein synthesis, which involves other nucleic acids, takes place at tiny organelles called ribosomes. The energy-producing activities of the cell occur in mitochondria.

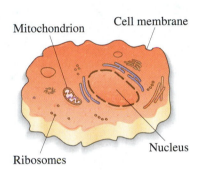

▲ A typical animal cell. Only the cell structures discussed in this text are labeled.

23.7   Lipids

We classify most organic compounds in terms of their chemical composition. Many are characterized by functional groups, and we represent them by general formulas such as ROH for alcohols and RCOOR′ for esters. Lipids, however, are classified on the basis of a physical property, their *solubilities*. Thus, we can give no general structural formula for all lipids. Recall our rule (Chapter 12) that "like dissolves like." A **lipid** is any component of plant or animal tissue that is insoluble in water but soluble in solvents of low polarity, such as diethyl ether, hexane, carbon tetrachloride, and benzene. Lipids include fats and oils (triglycerides), cholesterol, sex hormones, some vitamins (A, D, E, and K), and components of cell membranes called phospholipids.

In Chapter 6, we discussed the heat of combustion for fats, a common source of energy in our diet. Recall that fats yield more energy per gram than other types of biomolecules. Familiar triglycerides include beef fat and olive oil. Here, we will take a closer look at them and briefly examine a few other lipids as well.

Triglycerides: Fats and Oils

Triglycerides* are esters of the trihydroxy alcohol glycerol, $HOCH_2CHOHCH_2OH$, and *fatty acids*. Table 23.1 lists some common fatty acids. If the three fatty acids in a

---

*The systematic name for these compounds is *triacylglycerols,* but the common name *triglycerides* is widely used.

Table 23.1 Some Common Fatty Acids

Common Name	IUPAC Name	Condensed Structural Formula
Saturated acids		
Lauric acid	Dodecanoic acid	$CH_3(CH_2)_{10}COOH$
Myristic acid	Tetradecanoic acid	$CH_3(CH_2)_{12}COOH$
Palmitic acid	Hexadecanoic acid	$CH_3(CH_2)_{14}COOH$
Stearic acid	Octadecanoic acid	$CH_3(CH_2)_{16}COOH$
Monounsaturated acids		
Oleic acid	9-Octadecenoic acid	$CH_3(CH_2)_7CH{=}CH(CH_2)_7COOH$
Polyunsaturated acids		
Linoleic acid	9,12-Octadecadienoic acid	$CH_3(CH_2)_3(CH_2CH{=}CH)_2(CH_2)_7COOH$
Linolenic acid	9,12,15-Octadecatrienoic acid	$CH_3(CH_2CH{=}CH)_3(CH_2)_7COOH$

triglyceride are the same, the molecule is a *simple triglyceride*. Usually, though, the fatty acids are different and the molecule is a *mixed triglyceride:*

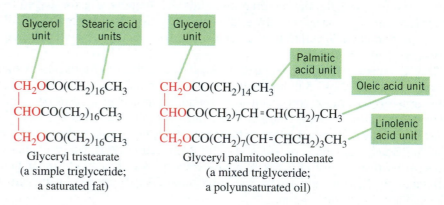

Glyceryl tristearate
(a simple triglyceride;
a saturated fat)

Glyceryl palmitooleolinolenate
(a mixed triglyceride;
a polyunsaturated oil)

As with other esters, a principal reaction of triglycerides is hydrolysis (page 954). Our digestion of triglycerides involves enzyme-catalyzed hydrolysis. In the laboratory, however, hydrolysis is usually done in basic solution and is called **saponification** (Latin, *sapon,* soap; *facere,* to make). For example, saponification of the mixed triglyceride glyceryl palmitooleolinolenate gives glycerol plus salts of the three carboxylic acids from which the triglyceride was made.

CH2OCO(CH2)14CH3
|
CHOCO(CH2)7CH=CH(CH2)7CH3 + 3 NaOH ⟶
|
CH2OCO(CH2)7(CH=CHCH2)3CH3
Glyceryl palmitooleolinolenate

CH2OH
|
CHOH
|
CH2OH
Glycerol

+ CH3(CH2)14COONa
Sodium palmitate

CH3(CH2)7CH=CH(CH2)7COONa
Sodium oleate

CH3(CH2CH=CH)3(CH2)7COONa
Sodium linolenate

The mixture of carboxylate salts is called **soap.** (We discussed the cleansing action of soap in Section 20.13.)

Phospholipids

Phosphorus-containing compounds called *phospholipids* make up another important class of lipids. These compounds are especially important constituents of cell membranes. Many are esters in which the trihydroxy alcohol glycerol has been esterified with two fatty acid molecules and one phosphoric acid unit. Further, an alcohol group is attached to the phosphoric acid unit, forming another ester group. For example, in the fairly simple type of phospholipid called a *phosphatide,* when the phosphoric acid unit is esterified with the aminoalcohol ethanolamine, $HOCH_2CH_2NH_2$, the result is the phospholipid known as phosphatidylethanolamine:

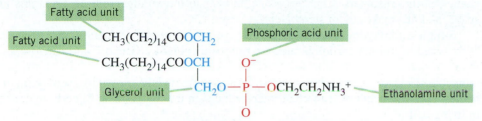

Phosphatidylethanolamine (a cephalin)

Phospholipids like this one, in which the alcohol esterifying the phosphoric unit is ethanolamine, are called *cephalins.* They are important components of brain and nerve cells. Related compounds called lecithins occur in all living cells. They have a choline group ($OCH_2CH_2N(CH_3)_3^+$) in place of the ethanolamine unit of cephalins. Lecithins can be isolated from soybeans and are used as emulsifying agents in salad dressings and other foods that combine oil and water components.

The reason phospholipids are able to make oil and water mix is that phospholipids are *amphipathic,* which means they are attracted both to polar media (usually water) and to nonpolar media (the oil in our example). The phosphoric acid–alcohol portion corresponds to a *hydrophilic* ("water-loving") *head,* and there are two *hydrophobic* ("water-fearing") fatty acid *tails.*

23.8 Carbohydrates

Carbohydrates supply most of the energy needs of our bodies. Carbohydrates were so named because they were once thought to be hydrates of carbon. Indeed, many have formulas of the type $C_x(H_2O)_y$:

Carbohydrate	Molecular Formula	"Hydrate" Formula
Glucose	$C_6H_{12}O_6$	$C_6(H_2O)_6$
Sucrose	$C_{12}H_{22}O_{11}$	$C_{12}(H_2O)_{11}$
Starch	$(C_6H_{10}O_5)_n$	$[C_6(H_2O)_5]_n$

In fact, though, carbohydrates are not merely "carbon hydrates." Their molecular structures are much more complex than those of simple hydrates (Section 2.7), where discrete H_2O molecules are associated with other molecules or ions. **Carbohydrates** are either (a) polyhydroxy aldehydes or polyhydroxy ketones or (b) compounds that are either derived from or can be hydrolyzed to polyhydroxy aldehydes or polyhydroxy ketones. Carbohydrates that have aldehyde groups are called *aldoses,* and those with ketone groups are called *ketoses.* Carbohydrates with five-carbon atom molecules are *pentoses,* and those with six-carbon atom molecules are *hexoses.* These four names can be combined in various ways to give information about a carbohydrate molecule. For example, we can classify the two familiar compounds glucose and fructose as an aldohexose and a ketohexose, respectively:

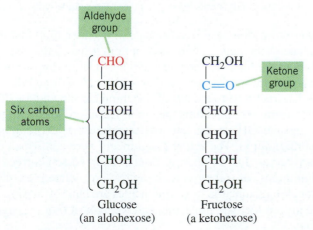

Glucose
(an aldohexose)

Fructose
(a ketohexose)

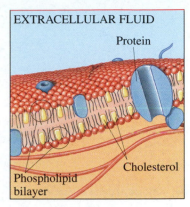

▲ A cell membrane is a double layer (bilayer) of phospholipid molecules with cholesterol and protein molecules embedded in it. Some of the protein molecules provide channels through which polar substances can enter and leave the cell. Nonpolar and slightly polar substances can diffuse directly through the phospholipid bilayer.

Lipid Bilayer 3D model

Carbohydrate Structures activity

Actually, glucose is only one of 16 possible isomers, all of which have the same molecular formula and the same groups in the same order of attachment on the carbon chain. All these isomers differ from one another in the arrangement of the atoms in space, and this makes them *stereoisomers* (page 924).

Simple carbohydrates taste sweet and are called *sugars*. Sugars that cannot be hydrolyzed to simpler carbohydrates are **monosaccharides.** Glucose and fructose are familiar examples. Most monosaccharides have from three to six carbon atoms per molecule. The two three-carbon monosaccharides are widely known by their common names, and they occur as important intermediates in the metabolism of carbohydrates in living cells.

 **Optical Activity animation**

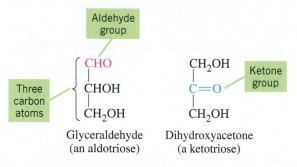

Glyceraldehyde   Dihydroxyacetone
(an aldotriose)   (a ketotriose)

Glyceraldehyde exists as a pair of *enantiomers* (page 925) that can be distinguished experimentally by their action on polarized light. The member of the pair that rotates the plane of polarized light to the right (clockwise from the point of view of the viewer in Figure 22.16) is said to be **dextrorotatory** (designated +). The other member of the pair rotates the plane of polarized light to the left (counterclockwise) and is the **levorotatory** (−) isomer. Most organic compounds that exhibit optical activity have one or more carbon atoms with *four different groups attached.* Such a substituted carbon atom is called a **chiral carbon** atom.

Look again at the structures of the two trioses above. Note that the middle carbon atom of glyceraldehyde has four different groups attached to it—CHO, H, OH, and CH$_2$OH; it is therefore a chiral carbon atom. (There is no chiral carbon atom in dihydroxyacetone; the top and bottom carbon atoms both have two H atoms attached, and the middle one has only three groups attached.)

The arrangement of the groups about a chiral carbon atom is called its **absolute configuration.** Biochemists use a system devised by Emil Fischer more than a century ago to specify the absolute configurations of carbohydrates. The *Fischer projection* of glyceraldehyde (Figure 23.3) has the carbon chain vertical and places the most oxidized group (CHO) at the top, the least oxidized group (CH$_2$OH) at the bottom, and the H and OH along the sides:

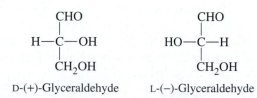

D-(+)-Glyceraldehyde   L-(−)-Glyceraldehyde

In the Fischer projection of glyceraldehyde, it is understood that the chiral carbon atom is in the plane of the page, the top and bottom groups attached to the chiral carbon atom are *behind* the plane of the page, and the side groups on the chiral atom are *out* from the page toward the viewer. Fischer arbitrarily assigned the configuration of glyceraldehyde that has the H to the left of the chiral carbon and the OH to the right to the dextrorotatory form. He called this the D configuration. The configuration with the H on the right and the OH on the left was therefore the levorotatory form, and he called it the L configuration. The actual assignments were confirmed decades later by X-ray diffraction studies. Just by chance, Fischer had guessed correctly.

The configurations of other carbohydrates are related to those of D- and L-glyceraldehyde. We assume the larger molecules are built by adding carbon atoms at the top end of a glyceraldehyde molecule, and we make a D or L assignment based on

▲ Emil Fischer (1852–1919) was awarded the Nobel Prize in Chemistry in 1902 for his research on the structure of sugar molecules. He later figured out how amino acids join together to form proteins.

(a)

(b)

▲ **FIGURE 23.3   Two representations of the structure of D-(+)-glyceraldehyde**
(a) The three-dimensional structure, and (b) a Fischer projection.

the configuration at the bottom chiral carbon atom—the one assumed to have come from D- or L-glyceraldehyde. Nature seems to favor D configurations for carbohydrates, as we see in the four most common hexoses:

Chirality animation

L-alanine and D-alanine 3D models

CHO	CH₂OH	CHO	CHO
H—C—OH	C=O	HO—C—H	H—C—OH
HO—C—H	HO—C—H	HO—C—H	HO—C—H
H—C—OH	H—C—OH	H—C—OH	HO—C—H
H—C—OH	H—C—OH	H—C—OH	H—C—OH
CH₂OH	CH₂OH	CH₂OH	CH₂OH
D-(+)-Glucose	D-(−)-Fructose	D-(+)-Mannose	D-(+)-Galactose

In each of these Fischer projections, the bottom chiral carbon atom is shown in blue.

The name of the second molecule here, D-(−)-fructose, shows that there is no correlation between the absolute configuration (D or L) and the direction in which the molecule rotates polarized light. The absolute configuration is assigned from the Fischer projection; the direction of rotation must be determined by experiment.

Example 23.2

Classify as fully as possible and give the configuration (D or L) and the sign of the rotation (+ or −) for ribose, a sugar found in ribonucleic acid (RNA).

```
        CHO
         |
    H—C—OH
         |
    H—C—OH
         |
    H—C—OH
         |
       CH₂OH
       Ribose
```

SOLUTION

The molecule has an aldehyde group (CHO) and five carbon atoms; it is thus an *aldopentose*. The configuration about the bottom chiral carbon atom has the OH on the right, making this the D configuration. There is no way to tell the sign of the rotation just from the structure.

EXERCISE 23.2A

Classify as fully as possible and give the configuration (D or L) and the sign of the rotation (+ or −) for (**a**) gulose and (**b**) erythrulose.

```
        CHO                  
         |                   
   HO—C—H                    
         |                   
   HO—C—H            CH₂OH
         |             |
    H—C—OH          C=O
         |             |
   HO—C—H         HO—C—H
         |             |
       CH₂OH         CH₂OH
       Gulose        Erythrulose
```

EXERCISE 23.2B

Can D-(+)-fructose exist? If so, draw a Fischer projection. If not, explain why not.

Enantiomers differ from each other in only one physical property: the *direction* in which they rotate polarized light (the two enantiomers rotate the polarized light to the same *extent*). They also differ in only one kind of chemical property: their reaction

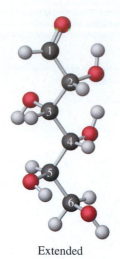

Extended

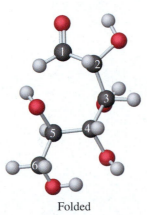

Folded

▲ **FIGURE 23.4** **Two arrangements of the open-chain glucose molecule** The O atom (red) on C-5 in the folded model is much nearer the carbonyl carbon than is the O atom on C-5 in the extended model. The C-5 oxygen and C-1 are the atoms that are joined in the cyclic form.

with other chiral molecules. This may seem like a minor difference, but it is all-important in biochemistry. The enzymes that catalyze nearly all biochemical reactions are chiral; the reactions they catalyze involve a reactant (substrate) that must fit the active site on the enzyme. The substrate must have the proper "handedness" to fit the active site, just as it takes a right-handed glove to fit a right hand. The enzymes that catalyze the metabolism of glucose, for instance, fit only the "right-handed" D-glucose that occurs in nature. Our bodies can derive no energy whatsoever from L-glucose.

Cyclic Structure of Monosaccharides

Although we have represented monosaccharides as straight-chain structures, most exist mainly in a cyclic form. For example, less than 0.02% of D-(+)-glucose exists in solution in the open-chain form. In glucose, the OH group on the fifth carbon (designated C-5 in Figure 23.4) adds to the carbonyl group (C-1), forming a ring made up of five C atoms and one O atom. Ring closure results in a new OH group that can extend either up above the ring or down below it and makes C-1 a new chiral carbon atom. Therefore, there are two cyclic forms of D-glucose. The form in which the OH extends downward is designated α-glucose, and that with the OH pointing upward is β-glucose:

α-Glucose   β-Glucose

Disaccharides

Two monosaccharide molecules can join together to form a **disaccharide.** A OH group on one of the monosaccharides adds to the carbonyl group of the other, and a molecule of water is split off in the process. The three most common disaccharides are *maltose*, *lactose*, and *sucrose* (Figure 23.5). The monosaccharide units from which they are formed are

$$Glucose + glucose \longrightarrow maltose + H_2O$$
$$Glucose + galactose \longrightarrow lactose + H_2O$$
$$Glucose + fructose \longrightarrow sucrose + H_2O$$

In characterizing disaccharides, we must also consider *how* the monosaccharide units are joined together. The two ways of joining are called α linkages and β linkages. In an α *linkage*, the bond to O extends downward (like the OH in α-glucose). In

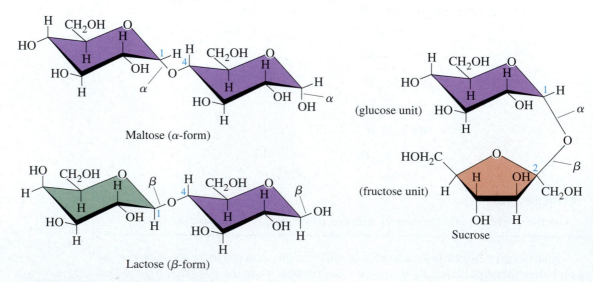

Maltose (α-form)

Lactose (β-form)

Sucrose

▲ **FIGURE 23.5**   **Three disaccharides**

a *β linkage*, the bond to O points upward (like the OH in β-glucose). In maltose (malt sugar), C-1 of one glucose unit is joined to C-4 of the other glucose unit through an α linkage. The configuration at C-1 of the second glucose unit can be either α or β. Lactose (milk sugar) has C-1 of the galactose unit joined to C-4 of the glucose unit by a β linkage. The glucose ring in lactose can have either an α configuration or a β configuration at C-1. Sucrose (cane sugar or beet sugar) has C-1 of the glucose unit joined to C-2 of the fructose unit. Looking at the molecule from the glucose end, the linkage is α; from the fructose unit, the linkage is β.

A disaccharide can be hydrolyzed to monosaccharide units in a reaction catalyzed by acids or by the appropriate enzyme. Thus, sucrose is hydrolyzed to glucose and fructose. Sucrose is dextrorotatory; glucose is also dextrorotatory (accounting for its alternate name dextrose), but fructose is levorotatory (accounting for the alternate name levulose). For equal concentrations of fructose and glucose in a solution, the extent to which the fructose rotates plane-polarized light to the left is much greater than the extent to which the glucose rotates it to the right. Therefore, when sucrose is hydrolyzed in solution, the optical rotation changes from (+) to (−) and the process is known as the *inversion* of sucrose. The product, a mixture of glucose and fructose, is known as *invert sugar.*

Polysaccharides

Complex carbohydrates having more than ten monosaccharide units in a chain are called **polysaccharides** (Figure 23.6). *Starches,* polysaccharides made of glucose units, are the principal energy storage materials of plants, especially cereal grains and potatoes. A simple starch called *amylose* has roughly 60 to 300 glucose units joined by an α linkage between C-1 of one glucose unit and C-4 of the next, just as in the disaccharide maltose. The starch known as *amylopectin* has an amylose-like chain structure but with branching on the chains. Animals store glucose in the form of a starch called *glycogen,* a high-molecular-mass polysaccharide with even more chain branching than amylopectin. Glycogen is stored in animal muscle and liver cells and can be converted back to glucose as needed for energy.

Cellulose, the other polysaccharide shown in Figure 23.6 and also a polymer of glucose, is the main structural material of plants. It is the main component of cotton, flax, wood pulp, and other plant fibers. Cellulose is composed of chains of about 1800–3000 glucose units, with C-1 of one glucose unit joined to C-4 of the next through β linkages. Humans, like most other higher animals, lack the enzymes needed to catalyze cellulose hydrolysis; our enzymes do not fit the β linkages that join the glucose units. Thus, we get no food energy from cellulose. Ruminants (cattle, goats, deer, sheep) have bacteria in their digestive tracts that can hydrolyze cellulose to glucose, and thus these animals can get energy from cellulose, a main component of

▲ The liquid interior of a chocolate-covered cherry is a result of the hydrolysis of sucrose. The cherry is covered with a paste of sucrose containing the enzyme invertase. When the paste hardens, the cherries are dipped in chocolate and stored for a week or two. Invertase catalyzes the hydrolysis of sucrose to a syrupy mixture of glucose and fructose (invert sugar).

▲ Starch forms water-insoluble granules, such as those that make up the bulk of the cells in a potato. The granules appear blue here because they are stained with iodine.

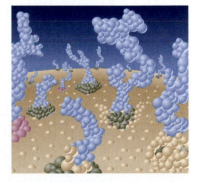

▲ Oligosaccharides—molecules with three to ten monosaccharide units—occur attached to proteins or lipids in cell membranes. These substances (blue in this illustration) arbitrate how cells interact with each other and are involved in the recognition of extracellular materials such as viruses and allergens. They aid or prevent the binding of these foreign bodies. Human blood types differ in the types of oligosaccharides expressed on the cell surface.

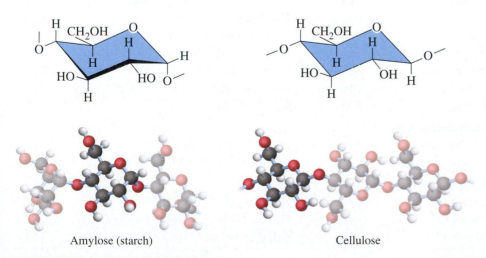

Amylose (starch)   Cellulose

▲ **FIGURE 23.6 Two polysaccharides**
Both amylose and cellulose have glucose as the monomer. In amylose, the glucose units are joined by α linkages; in cellulose they are joined by β linkages.

the grass and other plants the animals eat. Termites also have digestive-tract bacteria that can digest cellulose, allowing the termites to live on a diet of wood.

23.9 Proteins

As we noted in Chapter 9, proteins are polymers of α-amino acids. Recall that an α-amino acid is a carboxylic acid with an amino group on the carbon atom *next to* the carboxyl group:

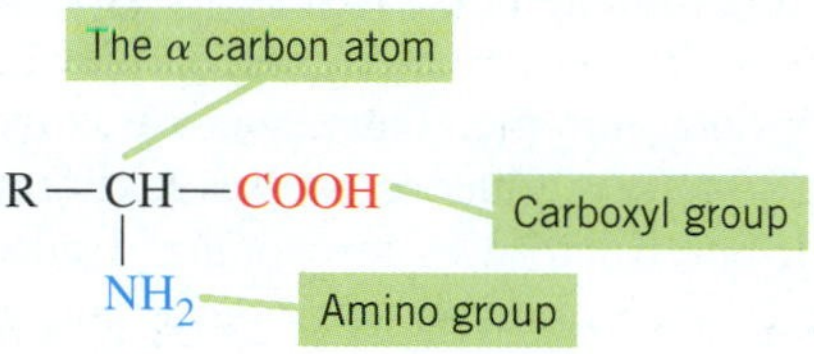

The α carbon atom

$$R-CH-COOH$$

Carboxyl group

$$NH_2$$

Amino group

The 20 amino acids listed in Table 23.2 serve as monomers in protein polymers.

As they occur in nature, 19 of the 20 amino acids in Table 23.2 are optically active and have the L configuration. As with carbohydrates, the configuration is related to that of glyceraldehyde:

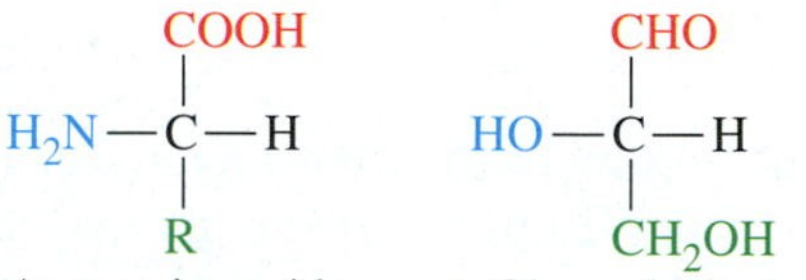

COOH   CHO

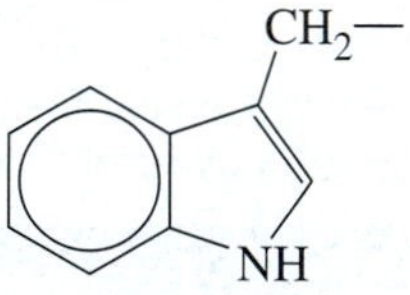

$$H_2N-C-H$$ $$HO-C-H$$

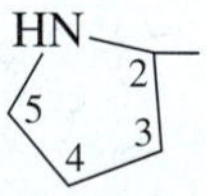

R $$CH_2OH$$

An L amino acid   L-Glyceraldehyde

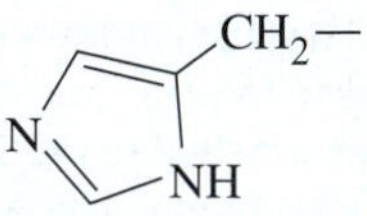

Table 23.2 The 20 Amino Acids Commonly Found in Proteins

Name	Symbol	Condensed Structural Formula	pI[a]
Neutral amino acids			
Glycine	Gly	$HCH(NH_2)COOH$	6.03
Alanine	Ala	$CH_3CH(NH_2)COOH$	6.10
Valine[b]	Val	$(CH_3)_2CHCH(NH_2)COOH$	6.04
Leucine[b]	Leu	$(CH_3)_2CHCH_2CH(NH_2)COOH$	6.04
Isoleucine[b]	Ile	$CH_3CH_2CH(CH_3)CH(NH_2)COOH$	6.04
Phenylalanine[b]	Phe	$C_6H_5CH_2CH(NH_2)COOH$	5.74
Serine	Ser	$HOCH_2CH(NH_2)COOH$	5.70
Threonine[b]	Thr	$CH_3CHOHCH(NH_2)COOH$	5.6
Methionine[b]	Met	$CH_3SCH_2CH_2CH(NH_2)COOH$	5.71
Cysteine[b]	Cys	$HSCH_2CH(NH_2)COOH$	5.05
Tyrosine	Tyr	$p\text{-}HOC_6H_4CH_2CH(NH_2)COOH$	5.70
Tryptophan[b]	Trp	$C_8H_6NCH_2CH(NH_2)COOH$	5.89
Asparagine	Asn	$H_2NCOCH_2CH(NH_2)COOH$	5.4
Glutamine	Gln	$H_2NCOCH_2CH_2CH(NH_2)COOH$	5.7
Proline[c]	Pro	C_4H_8NCOOH	6.21
Acidic amino acids			
Aspartic acid	Asn	$HOOCCH_2CH(NH_2)COOH$	2.96
Glutamic acid	Glu	$HOOCCH_2CH_2CH(NH_2)COOH$	3.22
Basic amino acids			
Lysine[b]	Lys	$H_2N(CH_2)_4CH(NH_2)COOH$	9.74
Arginine	Arg	$H_2NC(=NH)NH(CH_2)_3CH(NH_2)COOH$	10.73
Histidine[b]	His	$C_3H_3N_2CH_2CH(NH_2)COOH$	7.58

[a] pH at the *isoelectric point.*

[b] Essential amino acids.

[c] Proline has a *secondary* amine functional group (NH), making it an *imino* acid.

Glycine is not optically active, because the glycine molecule, H_2NCH_2COOH, has no chiral carbon atom.

Our bodies are unable to synthesize some of the amino acids needed to build functioning protein molecules; these **essential amino acids,** indicated by footnote b in Table 23.2, must be included in our diet. The essential amino acids act as limiting reactants. If there is a shortage of any of these building blocks, the body cannot make enough of the proper protein molecules.

Amino acids are amphoteric, which means they react either with acids or with bases. In highly acidic aqueous solutions, amino acids accept a proton from the aqueous acid and become cations. In highly basic solutions, amino acids donate a proton to the aqueous base and become anions. At some intermediate point, they exist as dipolar ions called *zwitterions:*

$$H_3N^+ \!-\! CH \!-\! COOH \underset{H^+}{\overset{OH^-}{\rightleftharpoons}} H_3N^+ \!-\! CH \!-\! COO^- \underset{H^+}{\overset{OH^-}{\rightleftharpoons}} H_2N \!-\! CH \!-\! COO^-$$

R	R	R
Cation	Zwitterion (dipolar ion)	Anion

The pH at which the amino acid exists mainly as the zwitterion is called the **isoelectric point,** and the pH at that point is called the **pI** of the solution. Recall that in an electric field, cations migrate to the cathode, and anions move toward the anode. At the isoelectric point, the amino acid is overall electrically neutral and so does not migrate in an electric field. At pH values near the pI value, an amino acid acts very much like the salt of a weak acid and a weak base. As such, it acts as a buffer. Added acid reacts with the COO^- group, and added base reacts with the NH_3^+ group.

Amino acids can join together through an amide linkage by eliminating a molecule of water. In biochemistry, the amide linkage is known as a *peptide linkage*, and short chains of amino acid units are called **peptides.** A *dipeptide* has two amino acid units, a *tripeptide* has three, and so on. With a large number of amino acid units, the molecule is called a **polypeptide.**

Peptides are represented by combinations of the three-letter amino-acid abbreviations shown in Table 23.2, starting at the end with the free amino group (the *N-terminal*) and proceeding to the end with a free carboxyl group (the *C-terminal*). Thus, we can represent the pentapeptide

$$H_2N \!-\! CH \!-\! CO \!-\! NH \!-\! CH \!-\! CO \!-\! NH \!-\! CH \!-\! CO \!-\! NH \!-\! CH \!-\! CO \!-\! NH \!-\! CH \!-\! COOH$$

CH_3	CH_2OH	$CH(CH_3)_2$	CH_3	$CH_2CH_2SCH_3$
Alanyl	Seryl	Valyl	Alanyl	Methionine

Proteins and Amino Acids animation

as Ala-Ser-Val-Ala-Met.

The boundary between polypeptides and proteins is quite arbitrary, often drawn at a molecular mass of about 10,000 u. Thus, we usually call molecules with 50 or more amino acid units **proteins** and those with fewer than 50 amino acid units polypeptides.

As we noted in Chapter 9, the **primary structure** of a protein is the exact sequence of amino acid units in the protein molecule. The primary structure of the above pentapeptide is therefore simply Ala-Ser-Val-Ala-Met. We also noted in Chapter 11 that the **secondary structure** of a protein is the shape of an entire protein chain and that the two principal secondary structures are the α-helix and the β-pleated sheet (Figure 11.22).

Many proteins exhibit further structural features. They may have domains of helical structures and domains of pleated sheet structures twisted and tangled together in various arrangements. This level of detail is called the **tertiary structure** of the molecule, defined as the molecule's three-dimensional shape. Figure 23.7 shows the tertiary structure of triose phosphate isomerase, an enzyme involved in sugar metabolism. The folding of a protein chain brings otherwise distant parts of its primary structure close together. Interactions between side chains of the amino acids in these regions stabilize the tertiary structure. Note that the molecule in Figure 23.7 has

The primary structure of β-endorphin (page 397), the opiate-like substance produced in our bodies, is Tyr-Gly-Gly-Phe-Met-Thr-Ser-Glu-Lys-Ser-Gln-Thr-Pro-Leu-Val-Thr-Leu-Phe-Lys-Asn-Ala-Ile-Ile-Lys-Asn-Ala-Tyr-Lys-Lys-Gly-Glu.

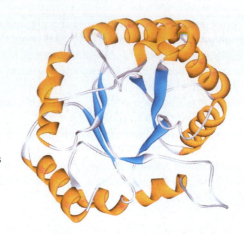

▶ **FIGURE 23.7 Tertiary structure of a protein**

This representation of triose phosphate isomerase, an enzyme involved in sugar metabolism, has a cylindrical arrangement of β strands (blue) surrounded by a garland of α helices (orange). These strands and helices are connected by turns and folds (white and gray) that lack a repeating structure.

QUESTION: Where are the regions of secondary structure in this protein?

regions of β-strands and α-helices connected by turns and folds called *random coils* that do not have repeating structures.

Some protein molecules associate in discrete clusters called a **quaternary structure.** Perhaps the most familiar example of a quaternary structure is hemoglobin. This oxygen-carrying blood protein is an aggregate of four folded chains, as was shown in Figure 22.9.

23.10 Nucleic Acids: Molecules of Heredity

Nucleic acids serve as the information and control centers of the cell. There are two kinds of nucleic acid: **deoxyribonucleic acid (DNA),** found primarily in the cell nucleus, and **ribonucleic acid (RNA),** found in all parts of the cell. Both are polymers of repeating units (monomers) called **nucleotides.** Each nucleotide, in turn, consists of three parts: a sugar, a phosphate unit, and a cyclic amine base. We can represent a nucleotide schematically as shown in Figure 23.8.

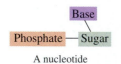

▲ **FIGURE 23.8 Schematic representation of a nuclide**

Every nucleotide is made up of a base, a sugar, and a phosphate.

The Sugars Two sugars are found in nucleotides. Ribose is found in the nucleotides that form ribonucleic acid (RNA), and 2-deoxyribose is found in the nucleotides that form deoxyribonucleic acid (DNA). (The *2-deoxy* indicates that this sugar lacks an oxygen atom on C-2):

$$\text{Ribose} \qquad \text{2-Deoxyribose}$$

The Phosphate Group A phosphate group replaces the OH group on C-5 of the sugar unit in a nucleotide. Think of this replacement as resulting from the reaction of an OH group of a phosphoric acid molecule, $OP(OH)_3$, with the H atom of the OH group on C-5 of the sugar and the elimination of a H_2O molecule:

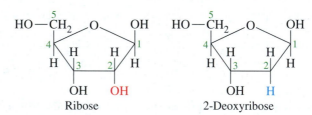

Phosphoric acid   Ribose   A phosphate ester

Note that the bond between the O of the phosphate group and the C-5 of the ribose is an ester linkage.

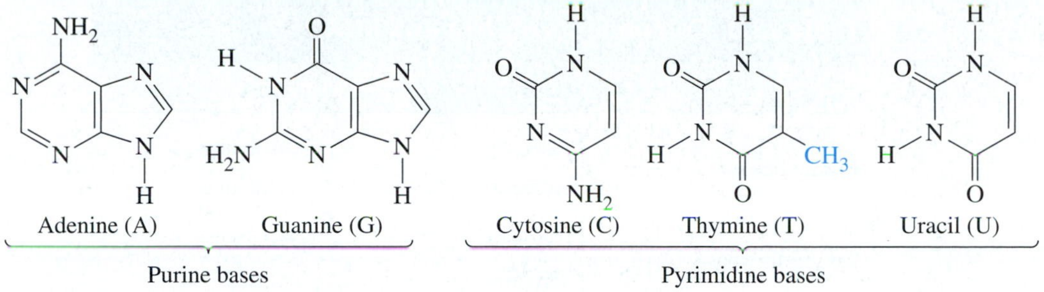

▲ FIGURE 23.9   **The five heterocyclic bases found in nucleotides**

The Bases In a nucleotide, one of five possible heterocyclic amine bases replaces the hydroxyl group on C-1 of the sugar. These bases, shown in Figure 23.9, are the single-ring units cytosine, thymine, and uracil, called *pyrimidines,* and the fused double-ring units adenine and guanine, called *purines.*

Nucleotides are joined to one another through the phosphate groups to form nucleic acid chains. The phosphate unit of one nucleotide is joined to the hydroxyl group on C-3 of the sugar unit of a second nucleotide through an ester linkage. A third nucleotide is attached in the same way, and the process is repeated to build up a long nucleotide chain (Figure 23.10). Schematically, we can represent the nucleic acid chain by extending the diagram shown in Figure 23.8 for a single nucleotide:

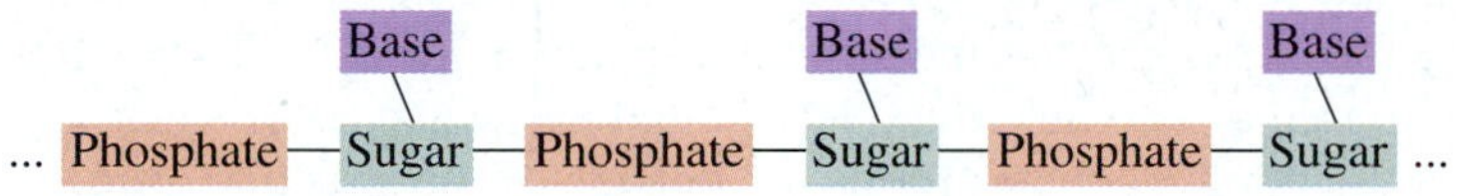

The backbone of the chain consists of alternating sugar and phosphate units. The heterocyclic bases, which branch off this backbone, can be present in various combinations. In the nucleotides found in DNA, the bases are adenine, guanine, cytosine, and thymine. The bases in RNA nucleotides are adenine, guanine, cytosine, and uracil. Table 23.3 summarizes the compositions of DNA and RNA.

Base Sequence in Nucleic Acids

A crucial feature of a nucleic acid molecule is the sequence of the four bases along the strand. Because the molecules are huge, with molecular masses ranging into the billions for mammalian DNA, the four bases may be arranged in an essentially infinite

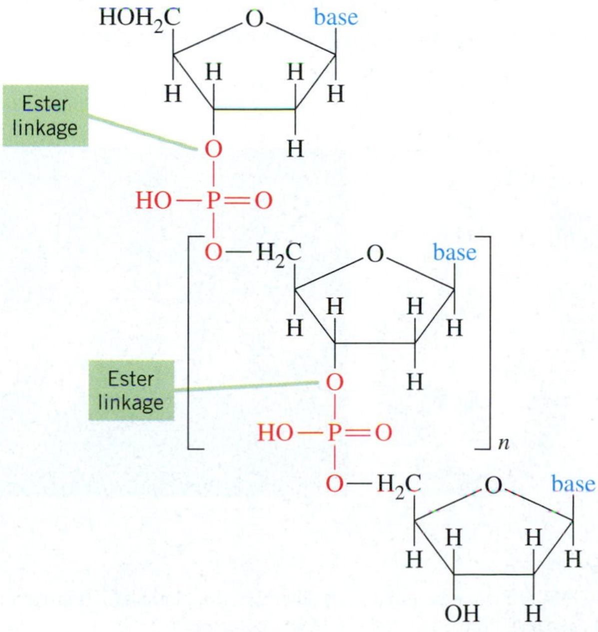

◀ FIGURE 23.10   **The backbone of a deoxyribonucleic acid molecule**

QUESTION: What type of linkage joins each sugar unit to a base unit? Is the linkage the same for all bases?

Table 23.3 Components of DNA and RNA		
	DNA	**RNA**
Purine bases { Adenine	Adenine	Adenine
Guanine	Guanine	Guanine
Pyrimidine bases { Cytosine	Cytosine	Cytosine
Thymine	Thymine	Uracil
Pentose sugar	Deoxyribose	Ribose
Inorganic acid	Phosphoric acid	Phosphoric acid

number of variations. The specific sequence of the bases along the chain is the information storage system needed to build organisms. Before we examine that aspect of nucleic acid chemistry, though, let us consider one more important feature of nucleic acid structure.

Base Pairing in DNA

In DNA, the molar amount of adenine (A) is equal to the molar amount of thymine (T), and the molar amount of guanine (G) is equal to that of cytosine (C). To account for this balance, the bases must be paired, A to T and G to C. But how? James D. Watson and Francis Crick worked out the structure of DNA in 1953. They determined that DNA is composed of two helices coiled around each other. The sugar and phosphate backbone of the polymer chains forms the outside of the structure, rather like a spiral staircase. The heterocyclic amines are paired on the inside, with guanine on one helix always opposite cytosine on the other and adenine on one helix always opposite thymine on the other. In our spiral-staircase analogy, these base pairs are the steps (Figure 23.11).

The two strands of a DNA double helix are not identical. Rather, they are *complementary*. The base A in one chain is always paired to T in the other, and G in one chain to C in the other. This is the only pairing in which hydrogen bonds can effectively join the two strands of nucleotides at the proper internuclear distance for maximum hydrogen bond formation. Figure 23.12 shows the two sets of complementary base pairs. Note that the matching of these complementary pairs, a pyrimidine on one chain and a purine on the other, leads to hydrogen bonding at a fixed interchain distance (1.085 nm). The base pairs fit like pieces of a jigsaw puzzle (but the puzzle is only two pieces wide). Other pairing of bases, such as pyrimidine–pyrimidine or purine–purine, would not produce hydrogen bonds of the correct length.

▲ James D. Watson (b. 1928) (right) and Francis Crick (b. 1916) worked out the double helix model of DNA in 1953 and were awarded the Nobel Prize in Chemistry for this work in 1968.

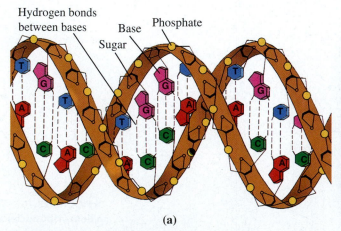

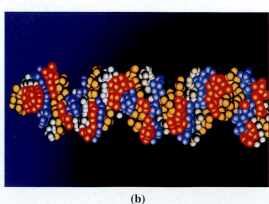

(a)   (b)

▲ **FIGURE 23.11** **The DNA double helix**

(a) A schematic representation. Two strands of DNA are wound around each other in a double helix. Hydrogen bonds between complementary bases hold the strands together. (b) A space-filling model of the DNA molecule.

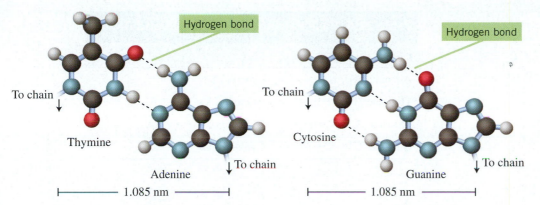

QUESTION: Which do you expect would exhibit a stronger attraction, the thymine–adenine base pair or the cytosine–guanine pair?

DNA segment 3D model

The Watson–Crick structure answers many crucial questions. It explains how cells are able to divide and go on functioning, how genetic data are passed to new generations, and even how proteins are built to required specifications. It all depends on base pairing.

Nucleic Acids, Heredity, and Protein Synthesis

Lions have cubs that grow up to be lions. Eagles lay eggs that produce eaglets that grow up to be eagles. How is it that each species reproduces its own kind? How does a fertilized egg "know" that it should develop into a kangaroo and not a koala?

The physical basis of heredity has been known for a long time. Humans and most other higher organisms reproduce sexually. A sperm cell from the male unites with an egg cell from the female. For human beings, that single fertilized egg cell must carry the information for making legs, liver, lungs, heart, head, hair, and hands—in short, all the instructions needed for growth and maintenance of the individual. Hereditary material is found in the nuclei of all cells, concentrated in elongated, threadlike bodies called *chromosomes*. The number of chromosomes varies with the species. Most human cells have 46 chromosomes, while egg and sperm cells carry half that number. In sexual reproduction, an egg and sperm combine to form the entire complement of chromosomes. A new individual receives half its hereditary material from each parent.

Chromosomes are made of DNA and proteins. The DNA is the primary hereditary material. Arranged along the chromosomes are the basic units of heredity, the genes. Structurally, **genes** are sections of the DNA molecule (although some viral genes contain only RNA).

Replication During cell division, each chromosome produces a duplicate of itself. Transmission of genetic information therefore requires *replication* (copying or duplication) of DNA molecules. The Watson–Crick double helix provides a ready model for replication.

As shown in Figure 23.13, the process of replication begins with the separation of the double helix. This separation is much like the unzipping of a zipper

In 2000, scientists completed a huge scientific venture, called the Human Genome Project. They determined the base sequence of the complete set, or *genome*, of human genes present in a human's chromosomes and mapped the location of most of the 30,000 or so genes on the chromosomes.

(a)   (b)   (c)   (d)

▲ FIGURE 23.13   DNA replication

Table 23.4 Base Pairing Between DNA and RNA	
DNA	**RNA**
Adenine (A) $<\cdots\cdots>$	Uracil (U)
Thymine (T) $<\cdots\cdots>$	Adenine (A)
Cytosine (C) $<\cdots\cdots>$	Guanine (G)
Guanine (G) $<\cdots\cdots>$	Cytosine (C)

(Figure 23.13b). The exposed single strands pick up appropriate nucleotides from the cellular fluid, and enzymes catalyze their attachment to the sugar–phosphate backbone of a new complementary chain (Figure 23.13c). Ultimately, two double helices replace the original one (Figure 23.13d). The information encoded in the original DNA double helix is now contained in each of the two replicates. When a cell divides, each daughter cell gets one of the DNA molecules, and with that one DNA molecule, the daughter cell gets all the information available to the parent cell.

The sequence of bases along the DNA chain encodes the directions for building an organism. Just as in English, where the sequence of letters *a-r-t* means one thing and the same letters in the order *r-a-t* means another, the sequence of bases CGT means one thing and GCT means something else. Although there are only four "letters" in the genetic code of DNA—the four bases—their sequence along the long strands can vary so widely that they can convey essentially unlimited information. Each cell carries in its DNA all the information it needs to determine all the hereditary characteristics of even the most complex organism.

Transcription In the first step of protein synthesis, called **transcription,** the information in a DNA molecule is copied into a molecule of *messenger RNA* (*mRNA*). The base sequence of DNA specifies the base sequence of mRNA. The only difference between transcription and DNA replication is that in transcription, adenine (A) pairs with uracil (U) rather than with thymine (T), as shown in Table 23.4. Each mRNA contains all the information necessary to make one particular protein.

Translation The next step in protein synthesis involves deciphering the code copied by mRNA and **translation** of that code into a specific protein structure. The deciphering occurs when the mRNA travels from the nucleus and attaches itself to a ribosome in the cytoplasm of the cell. Ribosomes are constructed of RNA and proteins.

Transfer RNAs (tRNA) carry amino acids from the cell fluid to the ribosomes. A tRNA molecule has the looped structure shown in Figure 23.14. At the head of the molecule is a set of three base units, called a *base triplet,* that pairs with a set of three complementary bases on mRNA. This triplet determines which amino acid is carried at the tail of the tRNA. To illustrate, the base triplet GUA on a segment of mRNA pairs with the base triplet CAU on a tRNA molecule. All tRNA molecules with the base triplet CAU always carry the amino acid valine. Once the tRNA has paired with the base triplet of mRNA, it releases its amino acid and returns to the cell fluid to pick up another amino acid molecule.

Each of the 61 different tRNA molecules carries a specific amino acid into place on a growing peptide chain. The protein chain gradually built up in this way is released from the tRNA as it is formed. Three base triplets on mRNA are *stop signals* that call for termination of the protein chain.

Figure 23.15 provides an overall summary of protein synthesis.

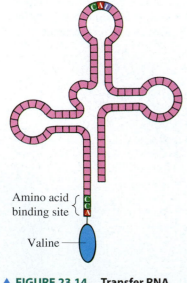

Amino acid binding site

Valine

▲ FIGURE 23.14   Transfer RNA (tRNA)

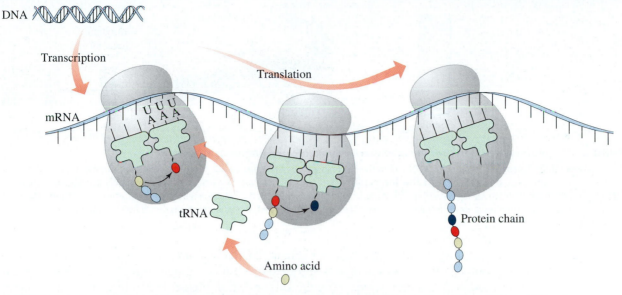

▲ FIGURE 23.15   **Protein synthesis visualized, from DNA through mRNA and tRNA to newly formed protein**

Spectroscopy in Organic Chemistry

We often use the structure of the molecules making up a sample of matter to explain the properties of that matter, and for this reason we have focused much of our study on molecular structure. Molecular structures are determined by **spectroscopy,** a set of methods based on the interaction of electromagnetic radiation with matter. In the sections that follow, we discuss briefly three of the most important methods, especially as they pertain to organic compounds. Each method supplies particular kinds of information, but combining the results of several methods gives us the most complete description of a molecular structure. For example, *infrared (IR) spectroscopy* indicates the functional groups that are present in organic molecules, *ultraviolet–visible (UV–vis) spectroscopy* reveals whether the structure has a series of alternate single and double bonds (conjugated π electron systems), and *nuclear magnetic resonance (NMR) spectroscopy* describes the carbon–hydrogen framework of the molecule.

23.11   The Interaction of Matter with Electromagnetic Radiation

We introduced the electromagnetic spectrum and explored the basis of light emission in Chapter 7. We also noted the basis of emission spectroscopy: The frequencies of emitted radiation depend on energy differences between excited states and ground states of the emitting atoms or molecules. Ground-state atoms or molecules can absorb radiation to become excited. This is the basis of *absorption spectroscopy,* the focus of our discussion here.

Table 23.5 lists several types of electromagnetic radiation and the kinds of changes that their absorption causes in molecules.

Spectroscopy in Organic
Chemistry activity

Table 23.5 **Electromagnetic Radiation and Its Interactions with Molecules**		
Type of Radiation	**Wavelength**	**Type of Interaction**
Vacuum ultraviolet	180–200 nm	Interacts with σ electrons
Ultraviolet	200–400 nm	Interacts with π electrons
Visible	400–800 nm	Interacts with conjugated π electrons
Near infrared	800–2000 nm	Stretches bonds
Infrared	2.0–30.0 μm	Stretches and bends bonds
Radio	1.0–5.0 m	Changes nuclear spin states

▲ **FIGURE 23.16   Schematic diagram of the essential components of a simple spectrometer**
A radiation source illuminates the sample with a portion of the electromagnetic spectrum. After passing through the sample, the radiation is dispersed by a monochromator into a spectrum across the face of a multichannel detector. The detector produces an electric signal proportional in magnitude to the intensity of the radiation. The recorder plots the strength of the electric signal; in an absorption spectrum, peaks occur at wavelengths corresponding to the wavelengths of the absorbed radiation, and the peak heights indicate how much radiation has been absorbed.

In general, a *spectrometer* (Figure 23.16) measures the extent to which radiation of a particular wavelength is absorbed by a sample. Light from the radiation source is limited to a narrow beam (by the slits shown in Figure 23.16) of certain wavelengths (by the monochromator) and is then passed through a sample. If the radiation does not interact with molecules in the sample, the light beam passes through with undiminished intensity. At a wavelength where the molecules do absorb radiation, the transmitted radiation will be either diminished in intensity or completely lacking.

Through the years, chemists have assembled tables of the structural features in molecules that are responsible for absorption at particular wavelengths. With appropriate experimental data and the help of these tables, we can determine the molecular structure of almost any substance.

23.12   Infrared Spectroscopy

Infrared radiation (IR) hitting a sample typically changes the intensity with which the atoms or groups of atoms in the sample vibrate about the bonds that connect them. Like electron transitions from one energy level to another, the vibrational motion of atoms is quantized: When a specific frequency of IR is absorbed by a part of a molecule, the absorbed radiation causes that part of the molecule to vibrate more intensely, rather like the right pitch of a musical note causing a wineglass to vibrate. Figure 23.17 illustrates vibrational changes called *stretching modes* and *bending modes*.

An IR spectrometer records peaks at wavelengths at which absorption occurs. The location of an absorption peak is usually specified by its frequency, but stated in a special way. The **wavenumber** $(\widetilde{\nu})$ is the number of cycles of the wave in a 1-cm length of the IR beam. The unit for wavenumber is reciprocal centimeters, (cm^{-1}):

$$\widetilde{\nu} = \frac{1}{\lambda}(\text{with } \lambda \text{ in centimeters}) \qquad \textbf{(23.1)}$$

Molecular Vibrations animation

▶ **FIGURE 23.17   Vibrational stretching and bending in covalent bonds**

A hypothetical functional group made up of three atoms is attached to a larger molecule (not shown), and the different modes of vibration are illustrated. The symmetric and asymmetric modes of stretching (top left and right, respectively) and the bending mode (bottom) are shown.

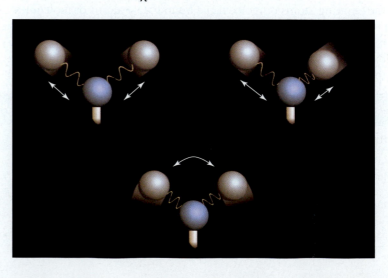

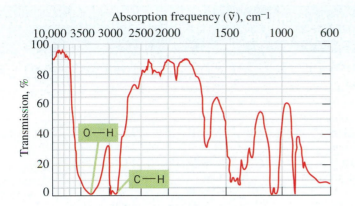

Absorption frequency ($\tilde{v}$), cm^{-1}

◄ FIGURE 23.18 The infrared absorption spectrum of ethanol, CH_3CH_2OH

The absorption peaks due to O–H and C–H stretching frequencies are labeled. Note that in this spectrum, percent transmission is plotted against frequency. Absorption peaks correspond to a low percent transmission, giving rise to deep troughs in the graph.

Consider the oxygen-to-hydrogen bond, O–H. Infrared radiation of frequency 3250–3450 cm^{-1} is absorbed as the bonded atoms vibrate more intensely. Thus, any molecule having a OH group absorbs radiation in the 3250–3450 cm^{-1} range. In which part of this range a particular OH group absorbs varies from one alcohol to another, depending on the overall structure of the molecule. If we find an absorption peak anywhere in this range, however, we know that a compound *could* have a OH group. We cannot be certain, however, because other groups also absorb in this region. For example, the absorption frequency range for the NH group somewhat overlaps that of the OH group. We need other evidence to distinguish between the two possibilities. If an elemental analysis of the compound shows the absence of nitrogen, we can be more confident that an absorption peak in the range 3250–3450 cm^{-1} is due to OH. The infrared absorption spectrum of ethanol in Figure 23.18 shows prominent peaks due to C–H and O–H bonds.

The carbonyl group, C=O, absorbs IR radiation in the range 1630–1780 cm^{-1}. Aldehydes, ketones, carboxylic acids, amides, and esters all have a carbonyl group and therefore absorption peaks in this range:

$$
\underset{\text{Aldehyde}}{R-\overset{\overset{\displaystyle O}{\|}}{C}-H} \qquad
\underset{\text{Ketone}}{R-\overset{\overset{\displaystyle O}{\|}}{C}-R} \qquad
\underset{\text{Carboxylic acid}}{R-\overset{\overset{\displaystyle O}{\|}}{C}-OH} \qquad
\underset{\text{Amide}}{R-\overset{\overset{\displaystyle O}{\|}}{C}-NH_2} \qquad
\underset{\text{Ester}}{R-\overset{\overset{\displaystyle O}{\|}}{C}-OR}
$$

Note that the carboxylic acid has both a C=O group and a OH group. However, the OH absorption peak is somewhat different from that in alcohols.

Figure 23.19 shows the infrared absorption spectrum of 2-hexanone, a ketone.

Infrared absorption peaks of typical functional groups are given in Table 23.6. To illustrate how these data can be used to identify a compound from its infrared absorption

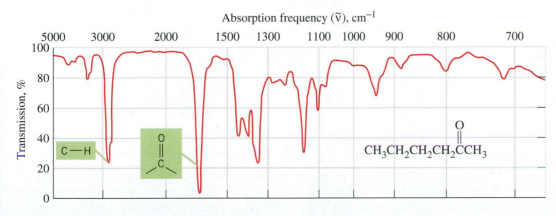

▲ FIGURE 23.19   The infrared absorption spectrum of 2-hexanone
The absorption peaks due to C–H and C=O stretching frequencies are labeled.
QUESTION: What additional peak would you expect to find in the spectrum of hexanoic acid?

Table 23.6 Infrared Absorption by Bonds in Characteristic Functional Groups

Functional Group	Bond	Frequency Range, cm^{-1}
Alkyl	C—H	2850–2960
Aromatic	C—H	~3030
Alcohol or phenol	O—H	3250–3450
Amine	N—H	3300–3500
Ketone	C=O	1680–1750
Ester	C=O	1735–1750
Carboxylic acid	C=O	1710–1780
	O—H	2500–3000
Amide	C=O	1630–1690
	N—H	3140–3500

spectrum, consider the spectrum of the local anesthetic lidocaine, shown in Figure 23.20. The functional groups we expect to have distinctive absorption peaks are the alkyl and aromatic C–H bonds, the N–H bond, and the C=O bond. In Figure 23.20, we see the expected peaks

- at about 3200 cm^{-1}, caused by the amide N–H bond.
- at about 2800 cm^{-1} and 2650 cm^{-1}, caused by the alkyl C–H bonds of the methyl and ethyl groups. (The aromatic C–H absorption, normally seen at 3030 cm^{-1}, is not readily apparent in this spectrum. Aromatic C–H absorption is often weak and can be obscured by other spectral features.)
- at 1680 cm^{-1}, caused by the amide C=O bond.
- at about 1400 cm^{-1}, caused by the C–N bonds.

Infrared spectra are often called the fingerprints of organic compounds. We can say with confidence that two pure samples that give identical IR spectra, matched peak for peak, are the same compound. If the spectra do not match, the samples must be different substances.

The region of an infrared spectrum from about 1200 to 700 cm^{-1} is sometimes called the "fingerprint region" because of its unique appearance from one compound to the next.

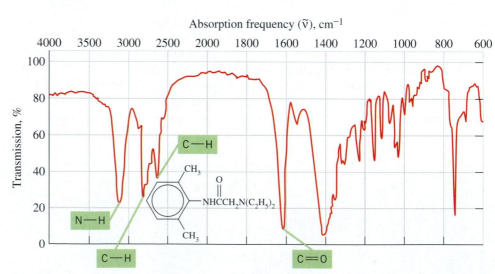

▶ **FIGURE 23.20   The infrared absorption spectrum of lidocaine**
The absorption peaks due to N–H, C–N, C=O, and two kinds of C–H (methyl groups and ethyl groups) bonds are labeled.

23.13   Ultraviolet–Visible Spectroscopy

Ultraviolet (UV) and visible radiation have sufficient energy to boost valence electrons of molecules to higher energy levels. The visible region of the electromagnetic spectrum extends from about 400–760 nm in wavelength. The ultraviolet region ranges from 400 nm down to about 40 nm, but it is the portion from about 200–400 nm that is most commonly used in laboratory studies. Ultraviolet light in this range is especially effective in producing energy-level changes in delocalized π electrons.

As a case in point, let us consider benzene, C_6H_6. Recall that we can represent benzene as the resonance hybrid of two Kekulé structures of alternate double and single bonds.

Molecular orbital theory indicates that the six π electrons reside in three bonding molecular orbitals and that three vacant antibonding molecular orbitals lie at higher energies (Figure 10.29). Ultraviolet radiation can boost electrons from the bonding orbital of highest energy, called the *highest occupied molecular orbital (HOMO),* to the antibonding orbital of lowest energy, called the *lowest unoccupied molecular orbital (LUMO).* Similar absorptions of energy occur in other molecules that have a conjugated bonding arrangement.

We record a UV spectrum by irradiating a sample with UV light that is continuously varied in wavelength. When the wavelength corresponds to the difference in energy between the HOMO and the LUMO of a conjugated system, some of the radiation is absorbed by the sample. The extent of this absorption is measured and plotted on a chart as wavelength versus the percent of radiation absorbed. Ultraviolet spectra often feature a single broad peak, and we label the wavelength at the top of the peak λ_{max}. For benzene, the most prominent peak is at $\lambda_{max} = 203$ nm.

The UV spectrum of a typical conjugated straight-chain molecule, 1,3-butadiene, is shown in Figure 23.21, and that of a typical aromatic molecule, styrene, is shown in Figure 23.22. The λ_{max} value for styrene, 245 nm, comes at a longer wavelength than does that of benzene, 203 nm, because in the styrene molecule, the π electrons of the benzene ring are conjugated with those of the double bond in the side chain:

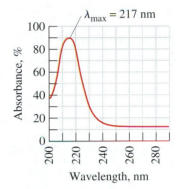

Kekulé structures
of benzene

▲ FIGURE 23.21 The ultraviolet absorption spectrum of 1,3-butadiene

The quantity of energy needed for the π electrons of styrene to be excited from the HOMO to the LUMO is smaller than the quantity of energy needed to excite a similar transition of the π electrons of benzene. This lesser energy can be supplied by radiation of lower frequency. Thus, as the length of conjugation increases in a molecule, λ_{max} of the absorption peak shifts to longer wavelengths. We also see this effect in the three conjugated molecules

$$CH_2=CH-CH=CH_2 \qquad \lambda_{max} = 217 \text{ nm}$$
1,3-Butadiene

$$CH_2=CH-CH=CH-CH=CH_2 \qquad \lambda_{max} = 258 \text{ nm}$$
1,3,5-Hexatriene

$$CH_2=CH-CH=CH-CH=CH-CH=CH_2 \qquad \lambda_{max} = 290 \text{ nm}$$
1,3,5,7-Octatetraene

Substances with no absorption peaks in the visible region are colorless. However, molecules containing an especially long conjugated system absorb visible light, and compounds made up of such molecules are colored. The transmitted light is the complementary color of the color absorbed (Table 23.7). From Table 23.7 we see, for

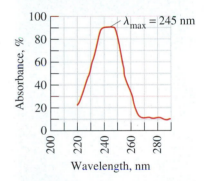

▲ FIGURE 23.22 The ultraviolet absorption spectrum of styrene, a $C_6H_5–CH=CH_2$

Styrene is the monomer from which polystyrene plastics (Chapter 24) are made. The π electrons of the benzene ring are conjugated with those of the double bond in the side chain.

QUESTION: Where would you expect to find λ_{max} for ethylbenzene $C_6H_5–CH_2CH_3$?

Table 23.7 The Wavelengths of Visible Colors[a]

Approximate Wavelength, nm	Color of Light	Complementary Color
425	Violet	Yellow
475	Blue	Orange
500	Blue-green	Red
525	Green	Purple
575	Yellow	Violet
625	Orange	Blue
700	Red	Blue-green

[a] The relationship between the colors of absorbed and transmitted light is also illustrated, in a different way, in Figure 22.21.

example, that a compound that absorbs blue light (475 nm) will appear orange, the complementary color of blue. The pigment β-carotene, prominent in carrots, pumpkins, and some other vegetables, has 11 conjugated double bonds.

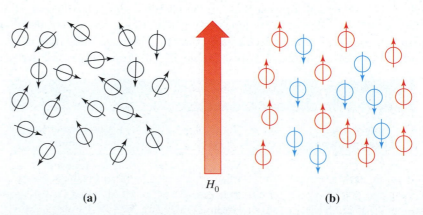

It absorbs at 466 and 497 nm and therefore is orange-red.

Finally, note that no matter how many double bonds are present in a molecule, if they are not part of a *conjugated* system, no absorption will occur in the UV region above 200 nm or in the visible region. Thus, neither 1-hexene nor 1,5-hexadiene has an absorption peak above 200 nm:

$$CH_2\!=\!CHCH_2CH_2CH_2CH_3 \qquad\qquad CH_2\!=\!CHCH_2CH_2CH\!=\!CH_2$$
$$\text{1-Hexene} \qquad\qquad\qquad\qquad \text{1,5-Hexadiene}$$

23.14 Nuclear Magnetic Resonance Spectroscopy

A carbon–hydrogen framework is the backbone of nearly all organic molecules. Nuclear magnetic resonance (NMR) spectroscopy provides more specific information about this framework than either UV–vis or IR spectroscopy.

Nuclear magnetic resonance is based on the absorption of energy by the nuclei of certain atoms in a molecule. If either the atomic number Z or the mass number A of a nucleus is an odd number, the nucleus will be active in NMR spectroscopy. The most important nuclei for NMR studies of organic compounds are hydrogen-1 ($_1^1H$) and carbon-13 ($_6^{13}C$). We will consider only $_1^1H$ spectra, which provide the most structural information.

Atomic nuclei involved in magnetic resonance spin about their axes like tiny tops. Because they are electrically charged, spinning nuclei act like miniature magnets. In the absence of an external magnetic field to line them up, the tiny magnetic fields of the spinning nuclei of the atoms in a sample are randomly oriented. In the presence of a strong external magnetic field of magnitude H_0, however, the nuclei assume preferred orientations, much as a compass needle aligns itself with Earth's magnetic field (Figure 23.23). When a hydrogen-containing compound is placed in a strong magnetic field, the tiny magnetic fields of the nuclei can align either in the same direction

Application Note

Conjugated bonding systems are a factor in producing color in organic compounds such as lycopene, the red pigment of tomatoes, and β-carotene, the orange pigment of carrots. There are other reasons substances have color, though, as we saw in Chapter 22.

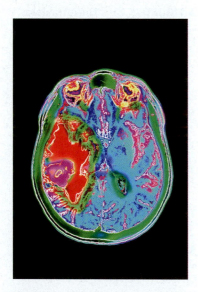

▲ Magnetic resonance imaging (MRI), a medical technique developed from basic research in nuclear magnetic resonance, is now widely used to make images of cross-sections of the human body. The MRI signal originates mainly from the hydrogen atoms of water molecules. Because tissues vary in water content and in the way the water is held in the tissues, signals vary from one type of tissue to another. An MRI scan can therefore precisely locate a tumor or injured tissues without surgery. This MRI of an axial section of a brain shows a tumor (purple). The growing tumor has damaged the surrounding tissue (red), which has filled with fluid.

(a) H_0 **(b)**

▲ **FIGURE 23.23   Nuclear spins illustrated**
(a) In the absence of a strong external magnetic field, nuclear spins are oriented randomly. (b) In the presence of a strong applied magnetic field, there are only two possible orientations. Some nuclear spins are aligned in the direction of the field (red), and some are aligned in the opposite direction (blue). Those with spins in the direction of the applied field are in the lower energy state.

(*parallel*) as the external magnetic field or in the opposite direction (*antiparallel*). The parallel orientation is the ground state, and the antiparallel alignment is an excited state. Because the energy difference between these two states is very small, many nuclei are found in the antiparallel state, though not as many as are found in the parallel state.

If electromagnetic radiation of the proper energy is focused on a sample in a magnetic field, some of the hydrogen nuclei in the ground state absorb the radiation and are boosted to the excited state. We say that these nuclei undergo a "spin flip," changing their orientation in the field from parallel to antiparallel. The required radiation is the low-energy radiation found in the radiofrequency (RF) portion of the electromagnetic spectrum. When the radiation has just the right energy to cause a spin flip, the nuclei are said to be in resonance with the applied frequency; hence the name *nuclear magnetic resonance.*

The frequency of RF energy required* to produce a resonance condition depends on the strength of the applied magnetic field and on the electron environment surrounding the nuclei. Electrons are involved because they, too, are spinning charged particles and have their own tiny magnetic fields. Electron fields in the vicinity of a hydrogen nucleus usually act in opposition to the applied field. The *effective* field that actually acts on a nucleus is therefore a bit weaker than the applied field:

$$H_{actual} = H_0 - H_{local}$$

Just how much weaker the field is depends on the electrons that bond hydrogen to another atom and on electrons in adjacent groups in the molecule.

NMR spectra are displayed on charts with the applied field strength increasing from left to right. The hydrogen nuclei of the compound tetramethylsilane (TMS), $(CH_3)_4Si$, absorb at a higher field strength than the hydrogen nuclei of nearly all other hydrogen-containing compounds. Chemists have therefore adopted TMS as a standard for NMR studies, and its absorption peak appears at the extreme right of an NMR spectrum. The position of an absorption peak for a certain kind of hydrogen nucleus relative to that of the TMS peak is called the **chemical shift** for that kind of hydrogen nucleus. Chemical shifts are stated in units of frequency on an arbitrary *delta scale*, where one delta unit (δ) is equal to one part per million of the operating frequency of the NMR spectrometer. (The chemical shift of TMS, of course, is zero.) Table 23.8 lists typical positions of absorption peaks of several kinds of hydrogen atoms.

In NMR spectroscopy, we detect the absorption of energy and correlate that absorption with the environment of particular hydrogen atoms. If two hydrogen atoms in the same molecule have a different electron environment, the energy of the radiation absorbed by one hydrogen nucleus will be slightly different from the energy of the radiation absorbed by the other hydrogen nucleus, and it is on this difference that NMR spectra are based. For example, in the toluene molecule, the electron environment of the hydrogen atoms in the methyl group (red) is different from the electronic environment of the hydrogen atoms in the benzene ring (blue):

The nuclei of these two "kinds"[†] of hydrogen atoms absorb different frequencies of RF energy in a constant magnetic field. The two kinds of H atoms thus produce two separate peaks in the NMR spectrum of toluene (Figure 23.24).

*Rather than varying the radio frequency, we usually obtain an NMR spectrum by holding the radio frequency constant and varying the magnetic field. It is easier to operate an NMR spectrometer this way, and it yields equivalent results.

[†]In some aromatic compounds, the ortho-, meta-, and para-hydrogen atoms in the ring show up in slightly different positions in the NMR spectra. In toluene, the difference is so slight that we can ignore it here.

Table 23.8 Positions of Absorption of Hydrogen Atoms in NMR Spectroscopy[a]

Type of Hydrogen	Chemical Shift, δ (ppm)
$(CH_3)_4Si$ (TMS)	0
$R-CH_3$	0.8–1.0
$R-CH_2-R$	1.2–1.4
$R-CH-R$ with R below	1.4–1.7
$R-\overset{O}{\underset{\|}{C}}-CH_3$	2.1–2.6
$R-\overset{O}{\underset{\|}{C}}-H$	9.5–9.6
R_2C-CH_2	4.6–5.0
RCH_2OH	3.3–4.0
ArH	6.0–9.5

[a]R stands for an aliphatic (nonaromatic) hydrocarbon group, and Ar represents an aromatic hydrocarbon group.

▶ **FIGURE 23.24** **The NMR spectrum of toluene**

There are two peaks in the toluene spectrum, labeled *a* and *b*. The height of each thin black integral curve superimposed on the red curve in a peak region is a measure of the area under that peak. These areas are in the ratio of 5 (for *b*) to 3 (for *a*), just as we expect from the formula $C_6H_5CH_3$. (The unit of measure, δ ppm, is the chemical shift, and the peak at 0 ppm is that for tetramethylsilane [TMS].)

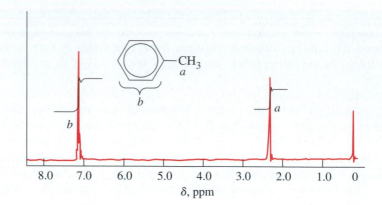

Other features of NMR spectra are helpful in determining molecular structure. One feature is that the area under each peak is proportional to the number of hydrogen nuclei giving rise to that peak. In Figure 23.24, the ratio of the peak areas of aromatic H to methyl H is 5:3. This is just the ratio that we expect from the formula of toluene, $C_6H_5CH_3$.

Another vital feature of NMR spectra (but one that we will not explore in any detail) is the fact that often an absorption peak is split into a collection of several smaller peaks. This splitting occurs whenever the hydrogen nucleus responsible for a peak is bonded to a carbon atom adjacent to another carbon atom having hydrogen atoms bonded to it. For example, in the ethyl group, the peak for the two red H atoms is influenced by the three blue atoms, and vice versa:

| Signal of these two atoms is split into four small peaks. | $\begin{array}{cc} H & H \\ | & | \\ -C-C-H \\ | & | \\ H & H \end{array}$ | Signal of these three atoms is split into three small peaks. |

The splitting of signals into smaller peaks is illustrated in the NMR spectrum of ethylbenzene in Figure 23.25. The splitting arises from a phenomenon called *spin–spin splitting,* caused by the nuclear spin of one H atom interacting with the nuclei spins of nearby H atoms. Spin–spin splitting does not occur when chemically equivalent (the same kind) H atoms are found on the same or adjacent C atoms (as in ethane, CH_3CH_3). Neither does it occur very often when nonequivalent (different kinds) H atoms are two or more C atoms apart. (Note in Figure 23.25 that the aromatic H peak of ethylbenzene is not split by the ethyl H's.)

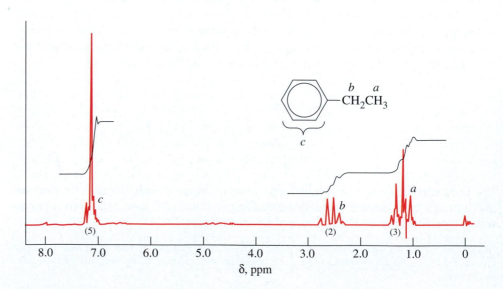

▶ **FIGURE 23.25** **The NMR spectrum of ethylbenzene**

The Chemist's Toolbox

A chemist may be called on to determine such diverse quantities as the concentration of polychlorinated biphenyls (PCB) in soil and the amount of tin compounds in blood. As noted in the chapter, tools available for such work include UV–visible, IR, and NMR spectrometers. However, because organic molecules vary widely in structure and reactivity, there are many other tools the chemist can use to determine the structure or nature of a compound or a mixture.

Often, the first step in analysis of a mixture is *separation*. Chromatography, first developed about a hundred years ago, physically separates a mixture into its components. Dissolved in a liquid or gas (the *mobile phase*), a sample is usually passed through a tube or column containing a powdered material (*stationary phase*). The more strongly a dissolved component is attracted to the stationary phase, the longer it will be retained by the column. In *gas chromatography* the sample is vaporized, and a gas such as helium is the mobile phase. In *liquid chromatography* the sample is dissolved in a liquid solvent that serves as the mobile phase. *Supercritical fluid chromatography* uses a supercritical fluid at high pressure, providing faster and more complete separations of many components.

Once separated, the individual components often must be identified or characterized. *Molecular mass spectroscopy* (MS) is often used. An electron beam fragments the sample and ionizes some of the fragments. The fragment ions are then separated by mass (page 266). The masses and the pattern they make aid in identifying the substance(s). *Electron spin resonance* (ESR) is similar to NMR except that in the former, electrons rather than nuclei are the absorbing particles. *Raman spectrometry* and IR spectrometry are complementary; IR is primarily for polar molecules, and Raman provides information about nonpolar molecules.

A host of other instruments may be used for more specialized analyses and determinations. Ultimately, though, an instrument relies on some sort of a "wet" chemical analysis—an analysis done in solution using ordinary equipment—as the measure of its accuracy, precision, and versatility.

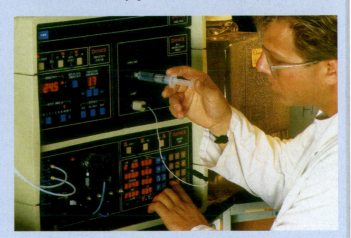

▲ The gas chromatograph is used to separate volatile compounds from one another in a mixture. Separation is often the first step to identification.

Some Organic Chemistry and Biochemistry of Drugs

People have used medicines in attempts to relieve pain and cure illnesses since prehistoric times. However, for most of human history the variety of drugs was limited, and they were discovered mainly through trial and error. Most societies used ethyl alcohol. Some used marijuana, some knew the narcotic effect of the opium poppy, and the native tribes of the Andes Mountains chewed the leaves of the coca plant for the stimulating effect of the cocaine in the leaves.

Only at the beginning of the twentieth century was a scientific rationale for the use of chemical substances to treat diseases developed. In 1904, Paul Ehrlich (1854–1915), a German chemist, realized that certain chemicals were more toxic to disease organisms than to human cells. These chemicals could be used to control or cure infectious diseases. Ehrlich coined the term *chemotherapy*, a shorter version of the term "chemical therapy." He found that certain dyes used to stain bacteria to make them more visible under a microscope also could be used to kill the bacteria. He used dyes against the organism that causes African sleeping sickness. He also synthesized an arsenic compound effective against the organism that causes syphilis. Ehrlich shared the Nobel Prize in Physiology or Medicine in 1908.

23.15 Molecular Shapes and Drug Action

For most of the twentieth century, scientists found useful drugs mainly by chance. After identifying an effective drug and determining its structure, chemists would synthesize dozens of related compounds. Medical research workers would then use laboratory animals to test each compound for toxicity and effectiveness. Those compounds that passed these tests were then tested on humans. This process takes years and is very expensive.

In Chapter 10, we noted that many drugs act by fitting specific receptor sites on cell membranes. In Chapter 13, we described how enzymes act by fitting a substrate molecule into an active site. Drugs and poisons act by blocking or otherwise changing the shapes of certain active sites. It is quite difficult to determine the precise shape of a receptor site and to determine what natural molecule the body produces to activate the receptor. Scientists had to learn the normal biochemistry of cells before they could develop drugs that treat abnormal conditions. Gertrude Elion and George Hitchings of Burroughs Wellcome Research Laboratories in North Carolina and James Black of King's College in London made notable contributions to the basic biochemistry leading to the development of antiviral drugs. Such drugs are used to treat herpes, AIDS, and other conditions caused by viruses. Their work also led to the development of many anticancer drugs. They and other scientists determined the shapes of cell membrane receptors, and they learned much about how normal cells work. With that knowledge, scientists were able to design drugs to block receptors. Today scientists use powerful computers to design molecules to fit receptors. Although drug design often was hit or miss in its early decades, it is now becoming a more precise science. We will consider only a few examples here.

Penicillins: The Enzyme-Inhibition Wars

Penicillin, the first antibiotic, was discovered in 1928 by Alexander Fleming (1881–1955), a Scottish microbiologist then working at the University of London. Antibiotics are water-soluble substances derived from molds or bacteria that inhibit the growth of other microorganisms. Fleming was studying an infectious bacterium, *Staphylococcus aureus,* when one of his cultures became contaminated with blue mold. Contaminated cultures generally are useless, but Fleming noted that the bacterial colonies had been destroyed in the vicinity of the mold. He was able to make crude extracts of the active substance, later called penicillin. It was further purified and improved by Howard Florey (1898–1968), an Australian, and Ernst Boris Chain (1906–1979), a refugee from Nazi Germany. Fleming, Florey, and Chain shared the 1945 Nobel Prize in Physiology or Medicine for their work on penicillin.

Penicillin is not a single compound but a group of compounds with related structures. All have in common a four-membered ring, called a β-lactam; they differ in the nature of the R groups. By varying the R groups, chemists could design penicillin molecules having different properties. The penicillins vary in effectiveness, with the result that bacteria resistant to one penicillin may be killed by another. The structures of several common penicillins are shown in Figure 23.26.

▲ Gertrude Elion shared the 1988 Nobel Prize for Physiology of Medicine.

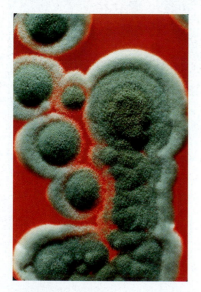

▲ These symmetrical colonies of mold are *Penicillium chrysogenum,* a mutant form that now produces nearly all commercial penicillin.

▶ **FIGURE 23.26** **Some common penicillins**

The β-lactam ring is shown in red. Amoxicillin was the most widely prescribed drug in the United States in 1988, but the development of resistant bacterial strains has caused a dramatic decline in its use.

Penicillin (general formula)

R = ⬡—CH₂—   Penicillin G

= HO—⬡—CH—   Amoxicillin
         NH₂

= ⬡—O—CH₂—   Penicillin V

Penicillins act by inhibiting the enzyme that catalyzes the synthesis of bacterial cell walls. In a reaction analogous to the hydrolysis of an amide, a nucleophilic group on the enzyme attacks the carbonyl carbon atom of the penicillin molecule. The net effect is that the penicillin molecule attaches to the enzyme, changing the active site and inactivating the enzyme:

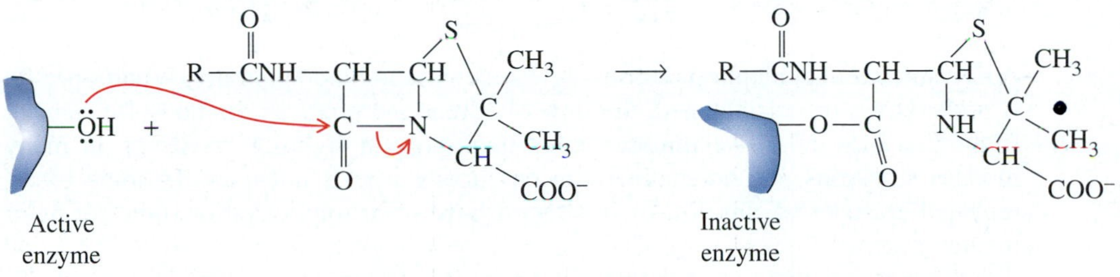

Active enzyme   Inactive enzyme

Human cells have no cell walls and are surrounded only by cell membranes; thus they are not affected by penicillin. This makes it possible for penicillin to destroy bacteria without harming human cells.

Bacteria become resistant to penicillin through mutation. The resistant bacteria produce an enzyme, β-lactamase, that catalyzes the hydrolysis of the amide linkage of the β-lactam:

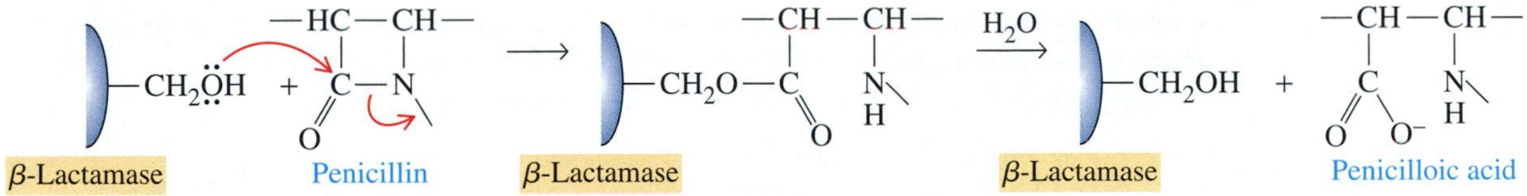

β-Lactamase   Penicillin β-Lactamase β-Lactamase   Penicilloic acid

The hydrolyzed penicillin, called penicilloic acid, is unable to interfere with bacterial cell wall synthesis.

In the continuing fight against infectious bacteria, chemists have developed drugs that inhibit β-lactamase. These drugs can be given along with penicillin, and with β-lactamase inhibited, the penicillin can inhibit the enzyme that catalyzes bacterial cell wall synthesis. Thus, we see how simple organic reactions, such as nucleophilic addition to a carbonyl group and hydrolysis of an amide, become important in the fight against infectious diseases.

Brain Amines: Depression and Mania

Drugs are used to treat diseases of the mind as well as those of the body. The nervous system is made up of about 1.2×10^{10} neurons (nerve cells) with about 10^{13} connections between them. Each neuron is connected to other cells though a long fiber called an *axon* and shorter fibers called *dendrites*. An axon on a given nerve cell may be up to 60 cm long, but there is no continuous pathway to the next cell. Rather, messages must be transmitted across tiny fluid-filled gaps called *synapses*. When an electrical signal from the brain reaches the end of an axon, specific chemicals called *neurotransmitters* are liberated and carry the impulse across the synapse to the next cell. There are perhaps a hundred or so neurotransmitters. Each has a specific function. Messages are carried to other nerve cells, to muscles, and to the endocrine glands (such as the adrenal glands). Each neurotransmitter fits one or more receptor sites on the receptor cell. Many drugs (and some poisons) act by mimicking the action of the neurotransmitter. Others act by blocking the receptor and preventing the neurotransmitter from acting on it. Several of the neurotransmitters are amines, as are some of the drugs that affect the chemistry of our brains.

We all have mood swings. These ups and downs probably result from multiple causes, but it is likely that a variety of chemical compounds formed in the brain are involved. Before we consider the brain chemicals, however, let us take a look at epinephrine, an amine formed in the adrenal glands and sometimes called adrenaline:

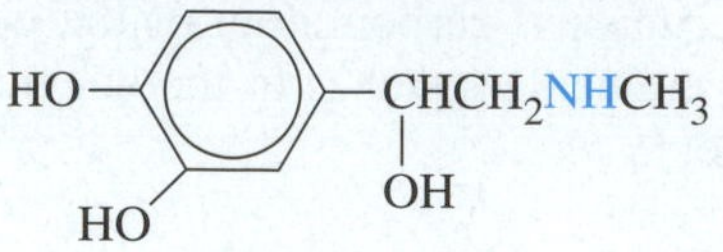

$$HO-\bigcirc-CHCH_2NHCH_3$$
with OH and HO substituents

Epinephrine

A tiny amount of epinephrine causes a great increase in blood pressure. When a person is under stress or is frightened, the flow of adrenaline prepares the body for fight or flight. Because culturally imposed inhibitions prevent fighting or fleeing in many modern situations, the adrenaline-induced supercharge is not used. In some cases, repeated episodes of this sort of frustration have been implicated in some forms of mental illness.

Biochemical theories of mental illness involve brain amines. One is norepinephrine (NE), a relative of epinephrine:

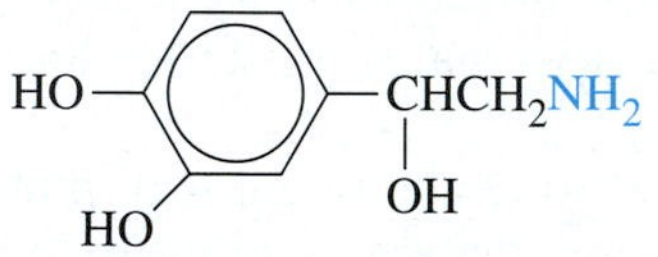

$$HO-\bigcirc-CHCH_2NH_2$$
with OH and HO substituents

Norepinephrine (NE)

Norepinephrine is a neurotransmitter formed in the brain. When formed in excess, NE causes a person to be elated—perhaps even hyperactive. In large excess, NE induces a manic state. A deficiency of NE, in contrast, could cause depression.

Another brain amine is serotonin, also a neurotransmitter:

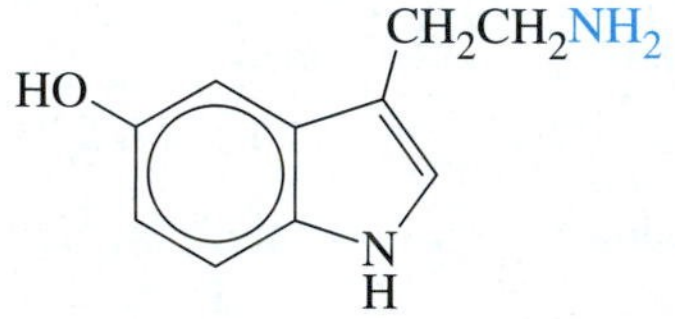

Serotonin structure with CH$_2$CH$_2$NH$_2$ and HO substituents

Serotonin

Serotonin is involved in sleep, sensory perception, and the regulation of body temperature. In the body, serotonin is oxidized to 5-hydroxyindoleacetic acid (5-HIAA):

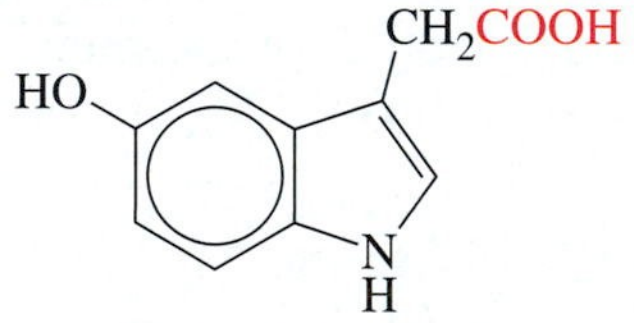

5-HIAA structure with CH$_2$COOH and HO substituents

5-HIAA

Using partial structures, we can represent the oxidation as

$$CH_2CH_2NH_2 \longrightarrow CH_2COOH$$

The compound 5-HIAA is found in unusually low levels in the spinal fluid of violent suicide victims. This indicates that abnormal serotonin metabolism plays a role in depression. Recent research indicates that a low flow of serotonin through the synapses in the frontal lobe of the brain causes depression.

Our human cells have at least six different receptors that are activated by NE and related compounds. Norepinephrine *agonists* (drugs that enhance or mimic its action) are stimulants, and NE *antagonists* (drugs that block the action of NE) slow down various processes. Drugs called beta-blockers are epinephrine and NE antagonists; they reduce the stimulant action of epinephrine and NE on various kinds of cells. Propranolol (Inderal), a well-known beta-blocker, is used to treat cardiac arrhythmias, angina, and hypertension by lessening slightly the force of the heartbeat. Unfortunately, it also causes lethargy and depression. Another beta-blocker, metoprolol (Lopressor), acts selectively on a different receptor found on the cells of the heart. It can be used by hypertensive patients who have asthma, because it does not act on receptors in the bronchi.

Serotonin agonists are used to treat depression, anxiety, and obsessive-compulsive disorder. Serotonin antagonists are used to treat migraine headaches and to relieve the nausea caused by cancer chemotherapy.

Richard Wurtman of the Massachusetts Institute of Technology has found a relationship between diet and serotonin levels in the brain. Serotonin is produced in the body from the amino acid tryptophan (the synthesis involves several steps that are omitted here for simplicity):

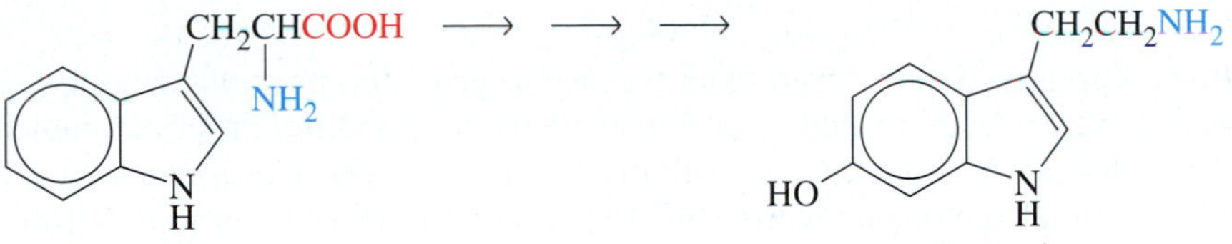

Tryptophan   Serotonin

Wurtman found that diets high in carbohydrates lead to high levels of serotonin. High-protein diets lower the serotonin concentration.

Norepinephrine also is synthesized in the body from an amino acid, tyrosine. Because tyrosine is also a component of our diets, it may well be that our mental state depends somewhat on our diet. The synthesis of NE is complex, but a simplified version of the process is

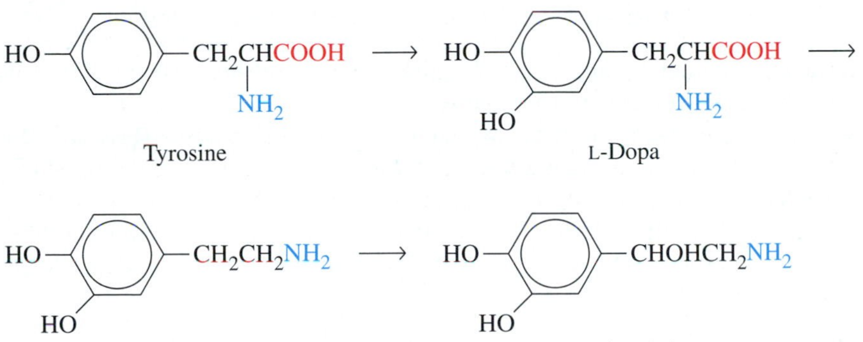

Tyrosine   L-Dopa

Dopamine   Norepinephrine

The intermediate L-dopa is used to treat Parkinson's disease, which is characterized by rigidity and stiffness of muscles. Parkinson's disease results from inadequate dopamine production, but dopamine cannot be used directly to treat the disease because it is not absorbed into the brain. Dopamine has been used to treat hypertension. It also is a neurotransmitter; schizophrenia has been attributed to an overabundance of dopamine in nerve cells.

Stimulant drugs, such as amphetamine and methamphetamine, are related to epinephrine and norepinephrine:

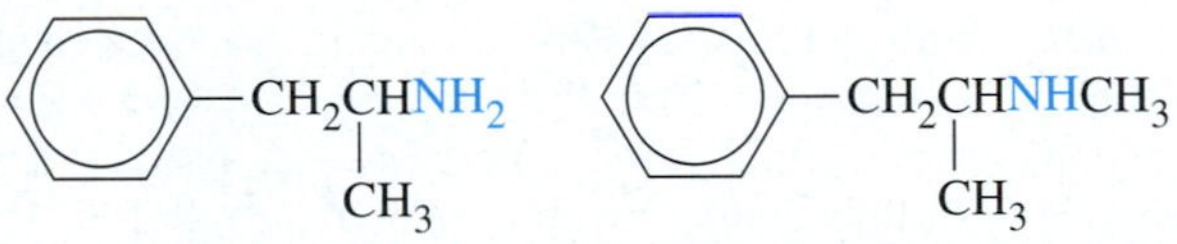

Amphetamine   Methamphetamine

These two synthetic amines are derivatives of β-phenylethylamine. They probably act as stimulants by mimicking the natural brain amines. Methamphetamine has a more pronounced psychological effect than amphetamine.

Some people seek stimulants. Others seek relief from anxiety. A class of drugs called phenothiazines—one member of which is chlorpromazine (Thorazine)—is used to calm psychotic patients. The phenothiazines are dopamine antagonists. Dopamine is important in the control of detailed motion (such as grasping small objects), in memory and emotions, and in exciting the cells of the brain. Some researchers think schizophrenic patients produce too much dopamine; other researchers think that these patients have too many dopamine receptors. In either case, blocking the action of dopamine relieves some of the symptoms of schizophrenia.

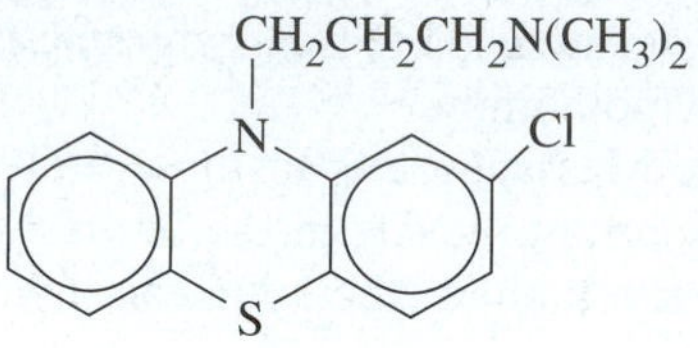

$$CH_2CH_2CH_2N(CH_3)_2$$

Chlorpromazine (Thorazine)

Removing the Cl atom from chlorpromazine gives the phenothiazine known as promazine. Some drugs related to promazine have a very different effect, indicating that slight changes in structure can result in profound changes in properties. Replacing the sulfur atom of promazine with a CH_2CH_2 group produces imipramine (Tofranil), an *antidepressant*:

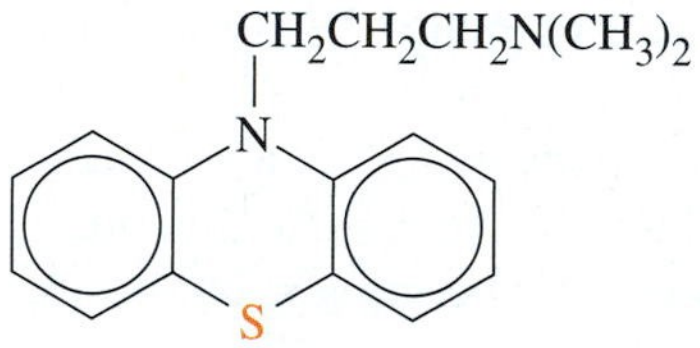

$$CH_2CH_2CH_2N(CH_3)_2$$

Promazine
(a tranquilizer)

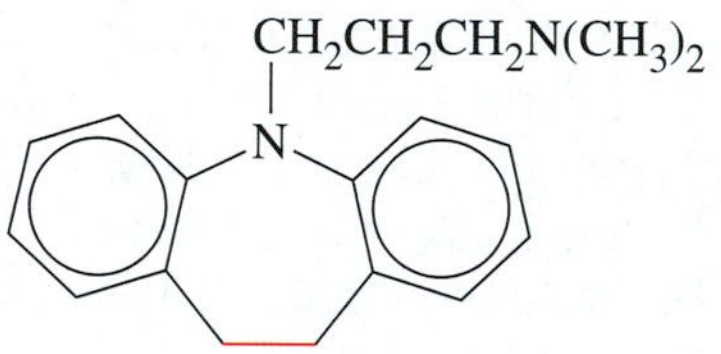

$$CH_2CH_2CH_2N(CH_3)_2$$

Imipramine
(an antidepressant)

These tricyclic antidepressants are only mildly successful because there is a narrow range in which the dose is both safe and effective.

Newer antidepressants are now available, including fluoxetine (Prozac). Doctors prescribe this compound to treat depression, but also to help people cope with gambling problems, obesity, fear of public speaking, premenstrual syndrome (PMS), and many other problems. The drug works by enhancing the effect of serotonin, blocking its reabsorption by the cells. It seems to be safer than the older antidepressants and more easily tolerated.

Love: A Chemical Connection

Can love be chemical in origin? An unsettling idea, perhaps. However, it is possible that the emotions that trigger romantic relationships are governed in part by a chemical called β-phenylethylamine (PEA), which functions as a neurotransmitter in the human brain. It appears to create excited, alert feelings and moods. Increased levels of PEA produce a "high" feeling identical to the feeling people describe as "being in love." Not surprisingly, the chemical structure of PEA resembles that of norepinephrine.

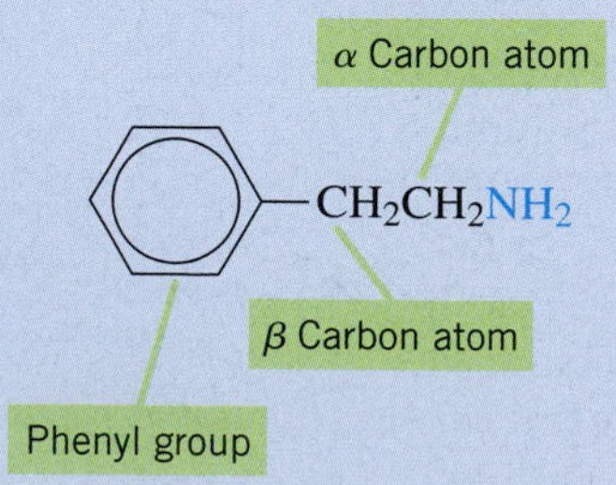

How much PEA does it take to get that high feeling? Levels of PEA in the brain can be estimated by measuring urinary levels of its metabolite, phenylacetic acid ($C_6H_5CH_2COOH$). Low levels of urinary phenylacetic acid have been found to correlate with depression. This discovery has prompted researchers to investigate factors that increase PEA levels in the brain. There are no food sources of PEA, but protein-rich foods contain phenylalanine, an amino acid precursor of PEA. Perhaps a steak dinner is a way to your true love's heart after all.

α Carbon atom

$$CH_2CH_2NH_2$$

β Carbon atom

Phenyl group

The most widely used antianxiety drug, usually self-prescribed, is most likely ethyl alcohol, found in beer, wine, and other alcoholic beverages. Excessive ethanol consumption causes intoxication, and long-term consumption can lead to addiction. Just how ethanol intoxicates is still somewhat of a mystery. Researchers have found that it disrupts a receptor for γ-aminobutyric acid (GABA), a substance the body uses to shut off nerve cell activity. Ethanol thus disrupts the brain's control over muscle activity, causing a drunk person to stagger and fall.

Nearly one out of every ten people in the United States suffers from some degree of mental illness. More than half the patients in hospitals are there because of mental problems. However, many mental illnesses can be alleviated by drugs. When we more fully understand the biochemistry of the brain, such illnesses may be cured by administration of drugs.

23.16   Acid–Base Chemistry and Drug Action

Acid–base chemistry is widely encountered in organic chemistry and biochemistry. As we have seen, acid–base steps are a part of many organic reaction mechanisms. Control of pH is vital in many biochemical reactions because most enzymes work most efficiently at an optimal pH (Chapter 13). The solubility and volatility of chemical compounds, including drugs, are also affected by pH. We will look at a few examples here.

Cocaine

Cocaine was first used in medicine as a local anesthetic, but it also is a powerful stimulant. The drug, an alkaloid, is obtained from the leaves of a shrub that grows on the eastern slopes of the Andes Mountains. Many of the natives living in that area chew coca leaves—mixed with lime and ashes—for their stimulant effect. Cocaine used to arrive in the United States as the salt cocaine hydrochloride. Now much of it comes in the form of broken lumps of the free base, a form called crack cocaine. We can use the molecular formula to write equations for acid–base reactions:

$$C_{17}H_{21}O_4N + HCl \longrightarrow C_{17}H_{21}O_4NH^+Cl^-$$

Cocaine   Cocaine hydrochloride

(free base)

Cocaine hydrochloride can be converted back to the free base by reaction with a strong base:

$$C_{17}H_{21}O_4NH^+Cl^- + OH^- \longrightarrow C_{17}H_{21}O_4N + H_2O + Cl^-$$

Cocaine hydrochloride   Cocaine

(free base)

Cocaine hydrochloride is quite soluble in water (100 g in 40 mL H_2O). It is therefore freely absorbed through the watery mucous membrane of the nose. This is the form used by those who snort cocaine.

Pure cocaine is only somewhat polar and therefore only slightly soluble in water (0.17 g in 100 mL H_2O) but fairly soluble in nonpolar solvents (8.3 g in 100 mL olive oil.). Cocaine hydrochloride has a fairly high melting point (~195 °C), but pure cocaine melts at 98 °C and is quite volatile above 90 °C. Those who smoke cocaine use the free base (crack), which readily vaporizes at the temperature of a burning cigarette. When smoked, cocaine is quickly absorbed through the thin layer of cells that line the lungs, reaching the brain in 15 seconds. Cocaine acts by preventing the neurotransmitter dopamine from being reabsorbed after it is released by nerve cells. High levels of dopamine are therefore available to stimulate the pleasure centers of the brain. After the binge, dopamine is depleted in less than an hour. This leaves the user in a pleasureless state, often craving more cocaine.

Nicotine

Another common stimulant is nicotine. Like cocaine, nicotine is an alkaloid. A high-boiling (247 °C), oily liquid, nicotine is quite soluble in both water and nonpolar solvents. Therefore, it is readily absorbed both through the watery mucous membrane of

▲ Coca leaves.

Nicotine

the mouth from chewing tobacco and through the fatty cells that line the lungs when taken by smoking. Nicotine, which is highly toxic to animals, has been used in agriculture as a contact insecticide. It is especially deadly when injected; the lethal dose for a human is estimated to be about 50 mg.

Because nicotine is toxic, the body must get rid of some of it lest concentrations build up to dangerous levels. The principal route of excretion is in the urine. Excretion is more efficient when the urine is acidic because the nicotine at acidic pH is mainly in the form of its conjugate acid:

$$C_{10}H_{14}N_2 + H^+ \longrightarrow C_{10}H_{14}N_2H^+$$
$$\text{Nicotine} \qquad\qquad \text{Conjugate acid}$$
$$\text{of nicotine}$$

The conjugate acid, with its positive charge, is more soluble in water and much *less* soluble in fatty body tissues than nicotine is. Therefore the lower the pH of the urine, the more rapidly the nicotine is excreted, and the sooner the smoker craves another cigarette. When does the pH of urine drop? After a meal, acids enter the urine from the metabolism of foods. After drinking ethanol, some of the ethanol is oxidized to acetic acid which enters the urine, making it more acidic. When under stress, the body breaks down tissues, producing acids. When does a smoker just have to have another cigarette? After a meal. While drinking alcoholic beverages. When under stress.

Nicotine seems to have a rather transient effect as a stimulant. This initial response is followed by depression, but smokers generally keep a near-constant level of nicotine in their bloodstreams by indulging frequently.

Is nicotine addictive? Consider the 1972 memorandum by a tobacco company scientist who noted that "no one has ever become a cigarette smoker by smoking cigarettes without nicotine." He suggested that the company "think of the cigarette as a dispenser for a dose unit of nicotine."

Cumulative Example

A 75-kg adult smokes a cigarette containing 1.2 mg of nicotine. Assume that all of the nicotine is absorbed and evenly distributed throughout the body, and calculate the mass in μg of nicotine that would be excreted in 375 mL of urine ($d = 1.02$ g/mL) that has a pH of 5.5. What is the concentration of nicotine in the protonated form? The K_b for nicotine is 1.05×10^{-6}.

STRATEGY

We can use the urine density to convert the volume of urine to mass, then use the ratio of nicotine mass to body mass to determine the mass of nicotine in the urine. To determine the mass of protonated nicotine, we can convert the mass of nicotine previously calculated to moles and then to molar concentration in the urine. Then we can use the pH of the urine to find [OH$^-$], substitute that value into the base ionization expression for nicotine, and express the concentrations of the base and acid forms in terms of the molar concentration of nicotine previously calculated. This will provide the molar concentration of the acid form.

SOLUTION

Assuming that the concentration of nicotine in the urine is the same as in the rest of the body, we begin by converting the 375 mL of urine to μg nicotine.

$$? \,\mu\text{g nicotine} = 375 \text{ mL urine} \times \frac{1.02 \text{ g}}{1 \text{ mL}} \times \frac{1 \text{ kg}}{1000 \text{ g}} \times \frac{1.2 \text{ mg nicotine}}{75 \text{ kg}} \times \frac{1000 \,\mu\text{g}}{1 \text{ mg}}$$

$$= 6.1 \,\mu\text{g nicotine}$$

Now we find the molar concentration of nicotine in the urine.

$$? \text{ M nicotine} = \frac{6.1 \,\mu\text{g nicotine}}{375 \text{ mL urine}} \times \frac{1000 \text{ mL}}{1 \text{ L}} \times \frac{1 \text{ g}}{10^6 \,\mu\text{g}} \times \frac{1 \text{ mol } C_{10}H_{14}N_2}{162.2 \text{ g } C_{10}H_{14}N_2}$$

$$= 1.0 \times 10^{-7} \text{ M}$$

This is the *total* concentration of nicotine. We can now assign expressions to the protonated form and the base form, x.

Total concentration of nicotine $= 1.0 \times 10^{-7}$ M

Molar concentration of $C_{10}H_{14}N_2 = x$

Molar concentration of $C_{10}H_{14}N_2H^+ = 1.0 \times 10^{-7} - x$

To obtain the concentration of the protonated form, we use the base ionization expression for nicotine. First we need the [OH$^-$] of the urine.

$$\text{pOH} = 14 - \text{pH} = 14 - 5.5 = 8.5$$

$$[\text{OH}^-] = 10^{-\text{pOH}} = 10^{-8.5} = 3.16 \times 10^{-9} \text{ M}$$

Now we have the expressions and values needed for the base ionization expression for nicotine, and we can solve for x.

$$C_{10}H_{14}N_2 + H_2O \rightleftharpoons C_{10}H_{14}N_2H^+ + OH^-$$

Assuming that x is very small compared to 1.0×10^{-7}, we calculate the value of x from the equilibrium expression.

$$K_b = \frac{[C_{10}H_{14}N_2H^+][OH^-]}{C_{10}H_{14}N_2} = \frac{(1.0 \times 10^{-7} - x)(3.16 \times 10^{-9})}{x} = 1.05 \times 10^{-6}$$

$$x = 3.0 \times 10^{-10}\ M\ C_{10}H_{14}N_2$$

Finally, we calculate the concentration of the protonated form of nicotine, $[C_{10}H_{14}N_2H^+]$, noting that x is small compared to the concentration of the base form.

$$1.0 \times 10^{-7} - x = 1.0 \times 10^{-7} - 3.0 \times 10^{-10} = 1.0 \times 10^{-7}\ M\ C_{10}H_{14}N_2H^+$$

ASSESSMENT

Because nicotine is a base and urine is acidic, we would expect most of the nicotine to be in the protonated form, which indeed is what we found. The value calculated for the concentration of nicotine excreted probably should be treated as an order-of-magnitude value rather than as a truly accurate value. The assumption that the nicotine is distributed evenly throughout the tissues is probably less than perfect. We earlier alluded to the fact that high acidity in urine causes nicotine to be excreted more readily, and so its concentration in urine might be somewhat higher than 1.0×10^{-7} M.

Concept Review with Key Terms

23.1 Alkanes—Hydrocarbons contain only two elements, carbon and hydrogen. Alkanes are hydrocarbons with only single bonds. They are mostly burned as fuels, but they also undergo **substitution reactions** with halogens. Methane and chlorine undergo a chain reaction involving free-radical intermediates.

23.2 Alkenes and Alkynes—Alkenes are hydrocarbons with double bonds, and alkynes have triple bonds. Alkenes typically undergo **addition reactions** in which the π bond is attacked by **electrophilic reagents,** such as halogens and H^+. The reaction mechanisms involve **carbocation** intermediates.

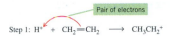

Step 1: $H^+ + CH_2{=}CH_2 \longrightarrow CH_3CH_2^+$

Step 2: $CH_3CH_2^+ + I^- \longrightarrow CH_3CH_2I$

23.3 Conjugated and Aromatic Compounds—**Conjugated bonding systems** contain alternating single and double bonds. Aromatic compounds contain ring structures with conjugated bonding systems. Aromatic compounds react mainly by **electrophilic aromatic substitution**. Lewis acids, such as aluminum chloride, catalyze several of these reactions.

23.4 Alcohols and Ethers—Most of the reactions of organic compounds with alcohol or **ether** functional groups take place at or near the functional group. The hydrocarbon portion of a molecule is generally less reactive than the functional group or groups in the molecule. Functional groups typically react as either electrophiles or as **nucleophiles,** seeking out electron-rich or electron-poor sites, respectively. Alcohols often react by substituting another group (such as a halogen atom) for the OH group. Alcohols can be dehydrated to either alkenes or ethers.

23.5 Aldehydes and Ketones—Aldehydes and ketones contain the carbonyl (C=O) functional group and undergo many reactions that start with an addition to this group.

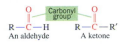

23.6 Carboxylic Acids, Esters, and Amides—Carboxylic acids react as acids, but they also react with alcohols to form esters and with ammonia or amines to form **amides**. Esters and amides can be hydrolyzed in either aqueous acids or aqueous bases.

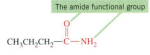

23.7 Lipids—The principal substances in living organisms are lipids, carbohydrates, proteins, and nucleic acids. **Lipids** are components of plant or animal tissue that are insoluble in water. Triglycerides, which are esters of fatty acids and glycerol, are familiar lipids. The esters of saturated fatty acids predominate in fats, and unsaturated fatty acids predominate in oils. **Saponification** is the hydrolysis of fats and oils in basic solutions and produces **soaps**, which are salts of fatty acids. Phospholipids are important components of cell membranes.

23.8 Carbohydrates—**Carbohydrates** include simple sugars, starches, and cellulose. Simple carbohydrates (sugars) that cannot be hydrolyzed to simpler structures are called **monosaccharides**. Many monosaccharides exist as a pair of optically active enantiomers—stereoisomers that rotate the plane of polarized light to either the right [**dextrorotatory** (+)] or to the left [**levorotatory** (−)]. Organic enantiomers typically have a **chiral carbon**, which means a carbon bonded to four different groups. The **absolute configuration** of these groups determines the optical activity.

Two monosaccharide units combine to form a **disaccharide**. **Polysaccharides** are polymers having monosaccharides as monomers. For example, glucose is the monomer in the polysaccharides starch and cellulose.

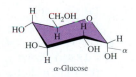

23.9 Proteins—Proteins are polymers of about 20 different α-amino acids. Some of these, the **essential amino acids**, cannot be synthesized by the body and must be included in the diet. Amino acids are amphoteric, and at the **isoelectric point (pI)** exist as dipolar ions (zwitterions).

Amino acids join through **peptide** bonds to form long chains called **polypeptides. Protein** molecules have 50 or more amino acid units. The **primary structure** of a protein refers to the amino acid sequence, the **secondary structure** is the shape of an entire protein chain, and the **tertiary structure** is the protein molecule's three-dimensional shape. Some protein molecules associate in clusters called **quaternary structures**.

23.10 Nucleic Acids: Molecules of Heredity—Both types of nucleic acid, **ribonucleic acid (RNA)** and **deoxyribonucleic acid (DNA)**, have nucleotides as their monomers. **Nucleotides** consist of a sugar, a phosphate group, and a cyclic amine base. The structure of DNA is two long chains of nucleotides joined into a double helix.

Genes are sections of the DNA molecule that contain the basic units of heredity. Protein synthesis takes place at the ribosomes and involves **transcription** of the DNA base sequence by messenger RNA, followed by **translation** of amino acids by transfer RNA to the growing protein structures.

23.11 The Interaction of Matter with Electromagnetic Radiation—**Spectroscopy** involves the measurement of a material's interaction with electromagnetic radiation.

23.12 Infrared Spectroscopy—In infrared spectroscopy, the absorption of infrared radiation by matter causes changes in the vibrational modes of molecules. Adsorption peaks in IR spectra are recorded in units of frequency as **wavenumbers ($\tilde{\nu}$)**, where

$$\tilde{\nu} = \frac{1}{\lambda} \text{(with } \lambda \text{ in centimeters)}$$

23.13 Ultraviolet–Visible Spectroscopy—In ultraviolet–visible spectroscopy, the absorption of radiation by electrons in molecules raises these electrons to higher energy states. In organic compounds, these transitions occur most readily among π electrons in conjugated systems.

23.14 Nuclear Magnetic Resonance Spectroscopy—In nuclear magnetic resonance spectroscopy, spinning nuclei absorb electromagnetic radiation and change their orientation in a magnetic field. When ^{1}H is the target nucleus, an NMR spectrum pinpoints different "kinds" of H atoms for different locations and different electron environments in a molecule. These differences are often described in terms of a **chemical shift**.

23.15 Molecular Shapes and Drug Action—Chemotherapy is the use of chemical substances to control or cure diseases. Many drugs act by fitting a specific site on a cell membrane or by inhibiting a specific enzyme. Penicillin, an antibiotic, acts by inhibiting the synthesis of bacterial cell walls. The brain synthesizes amines that affect mental state. Stimulant drugs, such as the amphetamines, mimic the action of norepinephrine. Other drugs act as either agonists or antagonists to various brain amines.

23.16 Acid–Base Chemistry and Drug Action—Acid–base chemistry is important in organic and biochemical reactions. The powerful stimulant cocaine, as the salt cocaine hydrochloride, is readily absorbed through the watery mucous membranes of the nose. The free-base form of cocaine is absorbed through the cells that line the lungs. Nicotine has a transient stimulant effect. The rate of excretion of nicotine depends on the pH of urine.

Assessment Goals

When you have mastered the material in this chapter, you will be able to:

- Describe the structures and reactions of hydrocarbons, including alkanes, alkenes, and alkynes.
- Identify conjugated and aromatic compounds, including benzene and its derivatives, and describe their general patterns of reactivity.
- Identify different functional groups, including alcohols, ethers, aldehydes, ketones, carboxylic acids, esters, and amides, and describe the different classes of reactions that occur for each.

- Describe the composition of lipids, carbohydrates, proteins, and nucleic acids in terms of representative functional groups and bonding structures.
- Distinguish between the three kinds of spectroscopy commonly used in organic chemistry—infrared (IR), ultraviolet–visible (UV–vis), and nuclear magnetic resonance (NMR)—in terms of the interaction between radiation and matter.
- Describe how molecular shape and acid–base chemistry are related to drug action.

Self-Assessment Questions

1. Briefly identify the important distinctions between (a) an alkane and an alkyl group, (b) aliphatic and aromatic, (c) a hydrocarbon and a carbohydrate, (d) an addition reaction and a substitution reaction, and (e) an electrophile and a nucleophile.

2. Using appropriate examples, distinguish between (a) an aldehyde and a ketone, (b) an ester and an ether, (c) an amide and an amine, and (d) a straight-chain alkane and a branched-chain alkane.

3. Which of the following can be oxidized to produce methyl ethyl ketone?
   (a) 1-butanol
   (b) 2-propanol
   (c) 2-butanol
   (d) *tert*-butyl alcohol

4. Define and illustrate each of the following.
   (a) esterification
   (b) acidic hydrolysis
   (c) saponification

5. Define, describe, illustrate, or explain each of the following terms.
   (a) triglyceride
   (b) aldose
   (c) hexose
   (d) disaccharide
   (e) polysaccharide
   (f) aldopentose
   (g) ketotetrose
   (h) tripeptide
   (i) zwitterion

6. Indicate whether each of the following compounds is aromatic, aliphatic, or both.

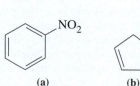

   (a)   (b)   (c)   (d)

7. Distinguish between (a) a fat and an oil, (b) a simple triglyceride and a mixed triglyceride, (c) a polypeptide and a protein, (d) an N-terminal and a C-terminal amino acid, and (e) the primary and secondary structures of a protein.

8. The substance *glyceral trilinoleate* is best described as which of the following?
   (a) a fat   (c) a wax
   (b) an oil   (d) a fatty acid

9. Why are (+)-glucose and (−)-fructose both classified as D sugars?

10. How do amylose and cellulose differ from each other? How are they similar?

11. What is the general structure for an α-amino acid? Do most naturally occurring amino acids have a D or an L configuration? Explain.

12. Name the two kinds of nucleic acids. State two differences between them. Which one is found mainly in the cell nucleus?

13. Which one of the following is not a constituent of a nucleic acid chain?
   (a) purine base   (c) glyceral
   (b) phosphate group   (d) pentose sugar

14. What is meant by base pairing, and what kind of intermolecular force is involved?

15. Which of the following best describes the structure of the DNA molecule?
   (a) a random coil   (c) a pleated sheet
   (b) a double helix   (d) partly coiled

16. We say that DNA controls protein synthesis, yet most DNA resides within the cell nucleus while protein synthesis occurs outside the nucleus. How does DNA exercise its control?

17. What are the similarities and differences between emission and absorption spectroscopy?

18. Describe the meaning of a conjugated system of double bonds, and illustrate with three specific examples.

19. What are the types of changes in molecules produced by the absorption of IR radiation? of UV–Vis radiation?

20. In which of these forms of spectroscopy—UV, Vis, IR, NMR—is the absorbed radiation of greatest energy content? Longest wavelength?

21. Describe the relationship between the frequency of electromagnetic radiation, ν, and its wavenumber, $\tilde{\nu}$.

22. Describe the meaning of the terms HOMO and LUMO and their relationship to the absorption of electromagnetic radiation.

23. What type of change is brought about by the absorption of electromagnetic radiation in NMR spectroscopy?

24. What is the meaning of each of the following terms, and in what spectroscopic method is it encountered?
   (a) chemical shift   (c) parallel and antiparallel
   (b) λ_{max}

25. What is the essential structural feature of a penicillin molecule? How does penicillin kill bacteria without killing human cells?

26. How do bacteria become resistant to penicillin? How can resistant bacteria still be killed by penicillin?

27. How do each of the following affect our mental state?
   (a) norepinephrine   (c) dopamine
   (b) serotonin

28. What are beta-blockers? How do they work?

29. In what way might our mental state be related to our diet?

30. How do drugs counteract anxiety? Do these drugs cure schizophrenia? Explain.

31. How does cocaine act as a stimulant?

32. How does the pH of urine affect a smoker's desire for a cigarette?

Problems

Nomenclature

You should be familiar with the material in Appendix D before attempting this section.

33. Draw the structural formulas for (a) 2-ethyl-1-butene, (b) 4-ethyl-2-methylhexane, (c) 2,4,6,6-tetramethyl-2-heptene, (d) 4-ethyl-3-isopropyloctane, (e) ethylcyclobutane, and (f) 3-ethyl-2-pentene.

34. Draw the structural formulas for (a) 4-methyl-2-hexene, (b) 3-methylpentane, (c) 2,2,5-trimethylhexane, (d) 4-ethyl-3-methyloctane, (e) 3-isopropyl-1-hexyne, and (f) 2,3-dimethyl-2-butene.

35. Name the following compounds by the IUPAC system.
   (a) $CH_3C(CH_3)=CHCH_3$
   (b) $CH_3CHClCH_2C\equiv CCH_2CHCl_2$
   (c) $(CH_3)_3CCH=C(CH_3)CH_2CH_3$
   (d) $CH_2=C(CH_3)CH_2CH_2CH_3$

36. Name the following compounds by the IUPAC system.
   (a) $CH_3CH_2CH(CH_3)CH_2CH_3$
   (b) $CH_3CH(CH_2CH_3)_2$
   (c) $(CH_3)_2CHC\equiv CCH_2CH(CH_3)_2$
   (d) $CH_3CH_2CH=C(CH_3)_2$

37. Name the following compounds by the IUPAC system.

   (a) $CH_3CH_2CHOHC(CH_3)_3$
   (b) $(CH_3)_2CHCH_2OH$

   (c)

   (d)

38. Name the following compounds by the IUPAC system.
   (a) $CH_3CH_2CH_2CH_2CH_2CH_2OH$
   (b) $CH_3CH_2CH_2CH_2CHOHCH_3$

   (c)

   (d)

39. Give structural formulas for (a) 3-hexanol, (b) 3,3-dimethyl-2-butanol, (c) 4,5-dimethyl-3-heptanol, (d) 2-ethyl-1-butanol, (e) 2,3-dibromobenzoic acid, and (f) *m*-isopropylbenzoic acid.

40. Give structural formulas for **(a)** cyclopentanol, **(b)** 4-methyl-2-hexanol, **(c)** heptanoic acid, **(d)** 3-methylbutanoic acid, **(e)** *o*-nitrobenzoic acid, and **(f)** *p*-chlorobenzoic acid.

41. Identify by name each functional group in compounds **(a)**, **(b)**, and **(c)**, whose models are shown here.

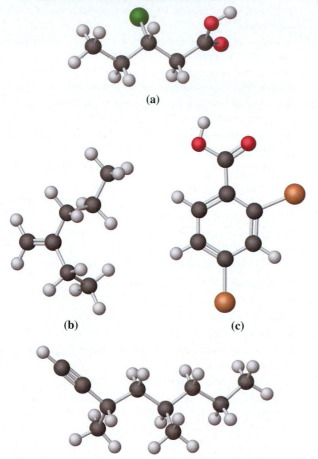

(a)

(b)   **(c)**

(d)

(e)

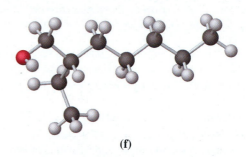

(f)

42. Identify by name each functional group in compounds **(d)**, **(e)**, and **(f)**, whose models are shown here.

43. Write structural formulas for compounds **(a)**, **(b)**, and **(c)**, whose models are shown here. Give IUPAC names to each of the compounds.

44. Write structural formulas for compounds **(d)**, **(e)**, and **(f)**, whose models are shown to the left and above. Give IUPAC names to each of the compounds.

Organic Reactions

45. Classify each of the following conversions as oxidation, dehydration, or hydration. (Only the organic starting material and the product are shown.)

(a) $CH_3CH_2CH_2CHCH_3 \longrightarrow CH_3CH_2CH_2CCH_3$
      | ||
      OH   O

(b)

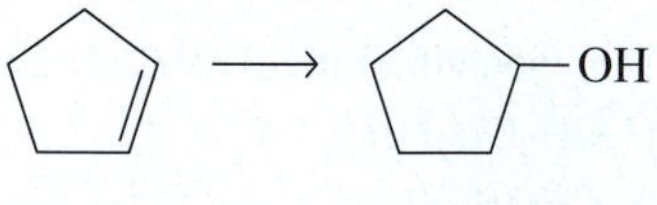

46. Classify each of the following conversions as oxidation, dehydration, or hydration. (Only the organic starting material and the product are shown.)

$CH_3CH_2CHCH_3 \longrightarrow CH_3CH=CHCH_3$
           |
           OH

(a)

$CH_3CH_2CH_2CH_2OH \longrightarrow CH_3CH_2CH_2COH$
                                                ||
                                                O

(b)

47. Complete the following equations by giving the structural formula for the main organic product.

(a) $CH_2=CHCH=CH_2 + 2\,H_2 \xrightarrow{Ni}$

(b) $(CH_3)_2C=C(CH_3)_2 \xrightarrow[H_2SO_4]{H_2O}$

48. Complete the following equations by giving the structural formula for the main organic product.

(a) $(CH_3)_2C=CH_2 + Br_2 \longrightarrow$

(b) $CH_2=C(CH_3)CH_2CH_3 + H_2 \xrightarrow{Ni}$

49. Write an equation for the dehydration of 1-propanol to yield **(a)** an alkene and **(b)** an ether.

50. Write an equation for the dehydration of cyclohexanol to an alkene.

51. Give the structure of the alkene from which each of the following alcohols is made by reaction with water in acidic solution.

CH_3CHCH_3 CH_3CCH_3
      | |
      OH   OH

                            CH_3 (above second structure)

(a)   **(b)**

52. Give the structure of the alkene from which each of the following alcohols is made by reaction with water in acidic solution.

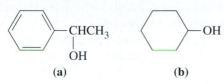

(a)   (b)

53. Write an equation for (a) the reaction of butyric acid with NaHCO₃(aq) and (b) the acid-catalyzed hydrolysis of ethyl acetate.

54. Write equations for (a) the reaction of benzoic acid with NaOH(aq) and (b) the base-catalyzed hydrolysis of butyl acetate.

Lipids

55. Which of these fatty acids are saturated, and which are unsaturated? How many carbon atoms are there in each?

(a) octanoic acid

(b) 6-decenoic acid

(c) 9,11,13-octadecatrienoic acid

56. Which of these fatty acids are saturated, and which are unsaturated? How many carbon atoms are there in each?

(a) decanoic acid

(b) 9-tetradecenoic acid

(c) 9,12-hexadecadienoic acid

57. Give structural formulas for the following.

(a) stearic acid

(b) potassium myristate

(c) glyceryl trioleate

(d) glyceryl tripalmitate

58. Give structural formulas for the following.

(a) sodium oleate

(b) glyceryl oleopalmitostearate

(c) linolenic acid

(d) a highly unsaturated oil

Carbohydrates

59. Identify each of the following as (a) an aldose or a ketose and (b) as a triose, tetrose, pentose, or hexose.

(a) D-glyceraldehyde   (d) D-fructose

(b) D-ribose   (e) L-glucose

(c) L-deoxyribose

60. Identify each of the following as (a) an aldose or a ketose and (b) as a triose, tetrose, pentose, or hexose.

(a) D-glucose   (d) D-galactose

(b) L-fructose   (e) D-deoxyribose

(c) L-mannose   (f) L-glyceraldehyde

61. Specify whether each of the following is a D sugar or an L sugar.

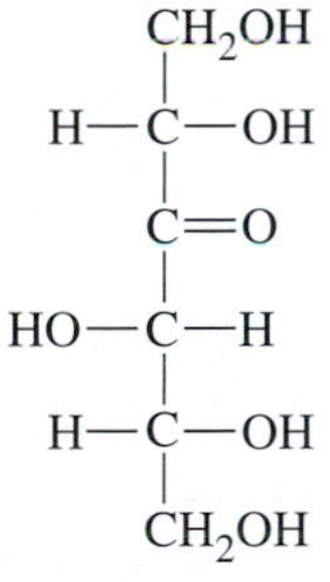 CH₂OH		CHO
H—C—OH	CHO	HO—C—H
C=O	H—C—OH	H—C—OH
HO—C—H	H—C—OH	HO—C—H
H—C—OH	HO—C—H	H—C—OH
CH₂OH	CH₂OH	CH₂OH
(a)	(b)	(c)

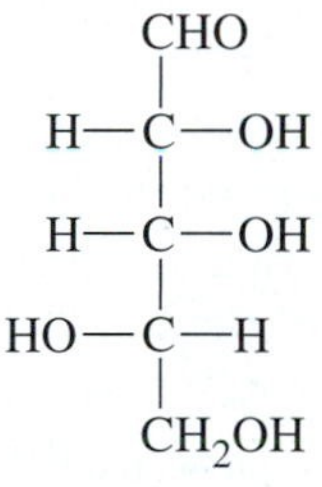

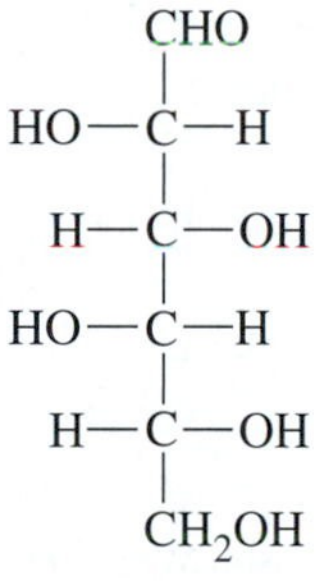

62. Specify whether each of the following is a D sugar or an L sugar.

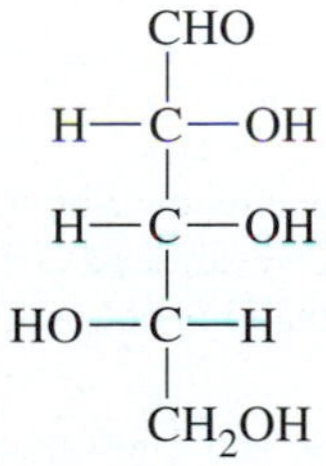 CHO	CH₂OH	
H—C—OH	C=O	CHO
H—C—OH	H—C—OH	H—C—OH
HO—C—H	H—C—OH	HO—C—H
CH₂OH	CH₂OH	CH₂OH
(a)	(b)	(c)

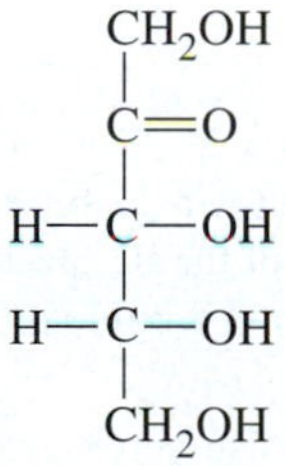

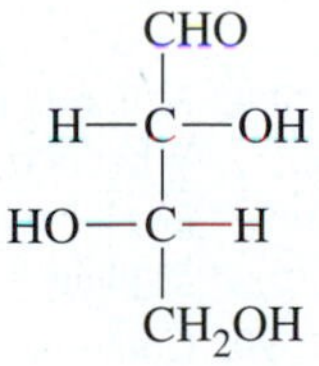

63. Draw cyclic structures for (a) α-D-glucose and (b) β-D-fructose.

64. Draw a cyclic structure for (a) β-D-glucose. (b) Use that structure as a reference to draw the structure for β-D-galactose.

65. For each of these abbreviated sugar formulas, indicate whether the linkage between units is α or β.

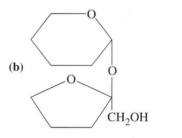

(a)

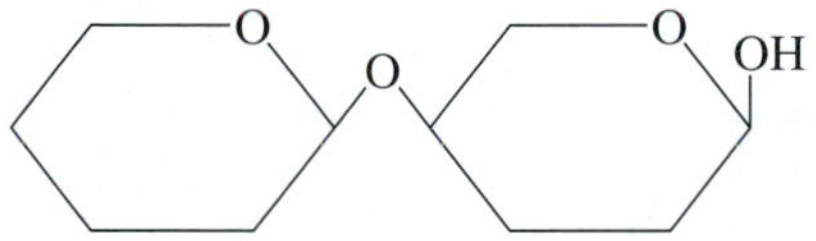

(b)

66. For each of these abbreviated sugar formulas, indicate whether the linkage between units is α or β.

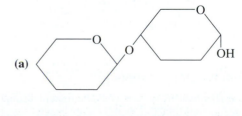

(a)

(b)

Amino Acids and Proteins

67. Draw the side chains of each of the following amino acids.

    (a) aspartic acid   **(d)** serine

    (b) cysteine   **(e)** phenylalanine

    (c) valine

68. Draw the side chains of each of the following amino acids.

    (a) alanine   **(c)** tyrosine

    (b) lysine   **(d)** isoleucine

69. Write a structural formula for an amino acid that has a basic side chain, and give its name.

70. Write a structural formula for an amino acid that has an acidic side chain, and give its name.

71. Write a structural formula for the cation formed when phenylalanine reacts with an acid.

72. Write a structural formula for the anion formed when glycine reacts with a base.

73. Under what conditions does a protein have **(a)** a net positive charge, **(b)** a net negative charge, and **(c)** no net charge?

74. How do the properties of a protein at its isoelectric pH differ from its properties in solutions at other pH values?

Nucleic Acids

75. Identify the sugar and the base in the following nucleotide.

76. Identify the sugar and the base in the following nucleotide.

77. Is the molecule shown a nucleotide? Is the base a purine or a pyrimidine? Would the compound be incorporated in DNA or RNA?

78. Is the molecule shown a nucleotide? Is the base a purine or a pyrimidine? Would the compound be incorporated in DNA or RNA?

79. The base sequence along one strand of DNA is GATTACA. What would be the sequence of the complementary strand of DNA?

80. The base sequence along one strand of DNA is CCTCGGA. What would be the sequence of the complementary strand of DNA?

81. What sequence of bases would appear in the messenger RNA molecule copied from the original DNA strand in Problem 79?

82. What sequence of bases would appear in the messenger RNA molecule copied from the original DNA strand in Problem 80?

83. If the sequence of bases along a messenger RNA strand is UCCGAU, what was the sequence along the DNA template?

84. If the sequence of bases along a messenger RNA strand is AGGUCC, what was the sequence along the DNA template?

85. What are the base triplets on tRNA that pair with the following triplets on mRNA?

    (a) UUU   **(b)** CAU   **(c)** AGC   **(d)** CCG

86. What are the base triplets on mRNA for the following triplets on tRNA molecules?

    (a) UUG   **(b)** GAA

Spectroscopy of Organic Compounds

87. What feature(s) in their IR spectra would be used to distinguish between acetic acid, CH_3COOH, and methyl acetate, CH_3COOCH_3?

88. What feature(s) in their IR spectra would be used to distinguish between phenol, C_6H_5OH, and benzoic acid, C_6H_5COOH? Could visible spectroscopy be used instead? Explain.

89. A compound is known to be either 2-butanol or 2-butanone. What two regions of the IR spectra might be examined to distinguish between them? Would you need to examine both regions? Explain.

90. Could you distinguish between 1-hexamide and 2-hexanone by examining the carbonyl peak of an infrared spectrum? Explain.

91. We have noted that if an organic molecule lacks a conjugated system of double bonds, it will not have an absorption peak in the 200–400 nm range of the spectrum. Will it have an IR spectrum? Explain.

92. Which of the following compounds would you expect to absorb UV radiation between about 215 and 400 nm: **(a)** 2-pentene, **(b)** 1,3-pentadiene, **(c)** 2,4-dimethylpentane, **(d)** 1,5-hexadiene? Explain.

93. The substance *o*-nitrophenol is yellow in color. In what region of the visible spectrum does it absorb light?

94. The aldehyde $CH_3(CH{=}CH)_8CHO$ has $\lambda_{max} = 436$ nm. What color would you expect it to be?

Additional Problems

Problems marked with an * may be more challenging than others.

95. You find an unlabeled jar containing a solid that melts at 48 °C. It ignites readily and burns cleanly. The substance is insoluble in water and floats on the surface of the water. Is the substance likely to be organic or inorganic? Explain.

96. From what alcohol might each of the following acids be prepared via oxidation with acidic dichromate?

(a) CH_3CH_2COOH **(c)** $HCOOH$

(b) $HOOCCOOH$ **(d)** $(CH_3)_2CHCH_2COOH$

97. A lactone is a cyclic ester. What product is formed in each of the following reactions?

(a) 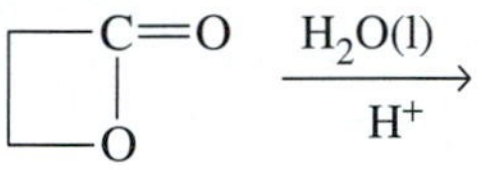 ⌐C=O H₂O(l) / H⁺ →

(b) 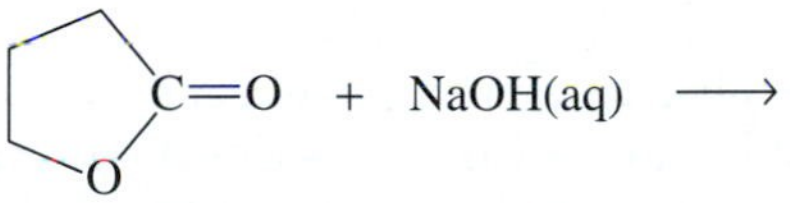 (ring)C=O + NaOH(aq) ⟶

98. Each of the four isomeric butyl alcohols is treated with potassium dichromate in acid. Draw the structural formula of the product (if any) expected from each reaction.

*99. Draw the structural formula of the ether that is formed by the *intra*molecular dehydration of the molecule shown below.

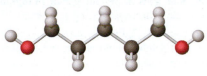

100. If one strand of a DNA molecule has the base sequence AGC, what must be the sequence on the opposite strand? Sketch the structure of this portion of the double helix, showing all hydrogen bonds.

101. What is the energy, in kJ/mol associated with IR radiation having a wavenumber of 3590 cm^{-1}?

102. The structures of two aromatic compounds are shown here.

⬡—N=N—⬡   HO—⬡—OH

One has an orange-red color and the other is white. Indicate which is which, and explain your reasoning.

103. State how infrared absorption spectra would enable you to distinguish between the following pairs of compounds. (*Hint:* Use data from Table 23.6.)

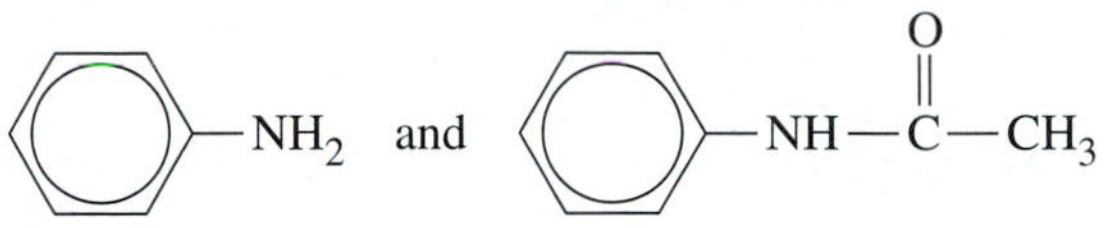

⬡—NH₂   and   ⬡—NH—C(=O)—CH₃

(a)

$CH_3CH_2CHCH_3$ and $CH_3CH_2CCH_3$
with OH on the first, O (double bond) on the second

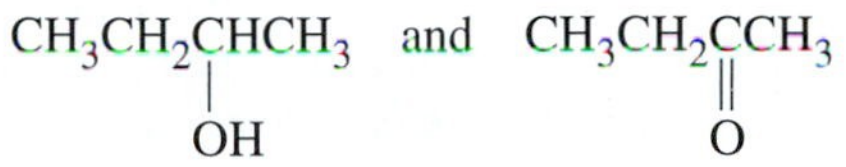

(b)

* 104. How many different "kinds" of H atoms are "seen" in the NMR spectrum of the following molecule? Explain.

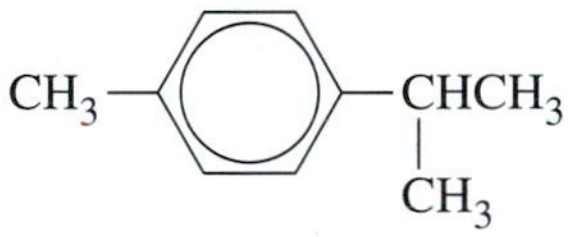

CH_3—⬡—$CHCH_3$ with CH_3 below

* 105. Describe how one of the types of spectroscopy (IR, UV–Vis, or NMR) could be used to distinguish between the compounds in each of the following pairs.

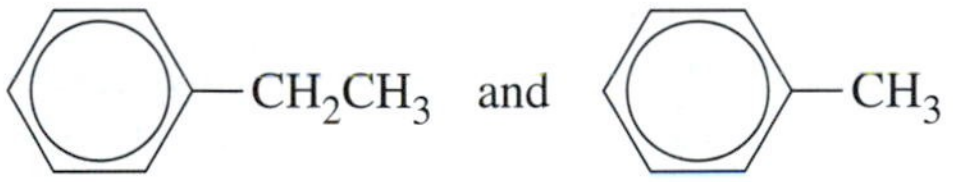

⬡—CH_2CH_3 and   ⬡—CH_3

(a)

CH_3CHCH_3 (with OH)   and CH_3CCH_3 (with O double bond)

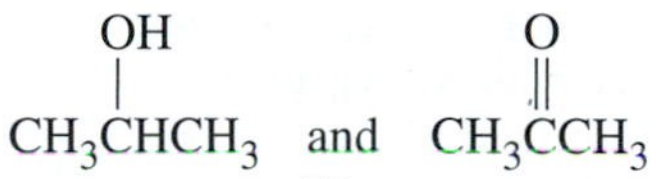

(b)

106. Use information from Table 23.8 to predict the approximate position of NMR absorption peaks (δ values) of the H atoms in boldface type.

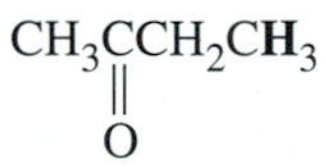

$CH_3CCH_2CH_3$ (with O double bond) $CH_3CHCH_2CH_3$ (with OH)

(a) 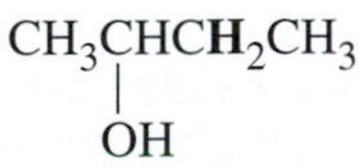   **(b)**

107. What is the meaning of "spin-spin splitting," and in what spectroscopic method is it observed?

108. The $n + 1$ rule in NMR spectroscopy applies to the spin-spin splitting of an NMR absorption into several small peaks: The signal from hydrogen nuclei with n neighboring H atoms is always split into $n + 1$ peaks. Show how this rule applies to NMR absorption of the ethyl group described on page 976.

* 109. Use information from Problem 108 to predict the NMR splitting pattern for each type of H atom in **(a)** 1,1-dibromoethane; **(b)** $CH_3OCH_2CH_2Cl$.

110. Use information from Table 23.8 to predict the approximate position of NMR absorption peaks (δ values) of the H atoms in boldface type, and from Problem 108 to indicate the multiplicity of the signal, that is, whether it will be unsplit (a singlet), split in two (a doublet), in three (a triplet), in four (a quartet), ...

$$CH_3CH_2CH{=}CH_2 \qquad CH_3CH_2C{\equiv}CH$$
(a) **(b)**

(c) $CH_3\overset{O}{\overset{\|}{C}}CH_2CH_3$

(d) (benzene ring)—CHCH₃ with CH₃

111. Aldehydes are named much like the carboxylic acids (Appendix D). Write the structure of **(a)** *cis*-3-hexenal, a compound with an herbal odor, and **(b)** *trans*-2-*cis*-6-nonadienal.

* 112. A particular colorless organic liquid is known to be one of the following compounds: 1-butanol, diethyl ether, methyl propyl ether, butyraldehyde, or propionic acid. Can you identify which it is, based on the following tests? If not, what additional tests would you perform? (1) A 2.50-g sample dissolved in 100.0 g water has a freezing point of $-0.70\ °C$. (2) An aqueous solution of the liquid does not change the color of blue litmus paper. (3) When alkaline $KMnO_4(aq)$ is added to the liquid and the mixture heated, the purple color of the MnO_4^- disappears.

* 113. In the molecule 2-methylbutane, the organic chemist distinguishes the different types of carbon and hydrogen atoms as being primary (1°), secondary (2°), and tertiary (3°). For the monochlorination of hydrocarbons, the following ratio of reactivities is found: $3°/2°/1° = 4.3:3:1$. How many different monochloro derivatives of 2-methylbutane are possible, and what percent of each would you expect to find?

(structure of 2-methylbutane with labeled carbons and hydrogens)

* 114. An organic *acid anhydride* is a carbonyl-containing functional group with the general formula, R—CO—O—CO—R.

(a) Write equations for the reaction of butyric anhydride (butanoic anhydride) with water and with aqueous ammonia.

(b) Which of the two reactions in (a) would you expect to yield the product melting at the higher temperature? Explain.

(c) What is the formula of the new compound formed if the product identified in (b) is collected and heated to about 120 °C?

Apply Your Knowledge

115. **[Laboratory]** For neutralization, 5.10 g of a monocarboxylic acid requires 125 mL of a 0.400 M NaOH solution. Write all possible structural formulas for the acid.

116. **[Biochemical]** Pellagra is a vitamin-deficiency disease. Corn contains the antipellagra factor nicotinamide, which is not readily absorbed in the digestive tract. When corn is treated with slaked lime (calcium hydroxide) to make hominy and masa flour, the vitamin is rendered more available. What reaction takes place? Write the equation.

(structure of Nicotinamide) Nicotinamide

* 117. **[Laboratory]** The most useful solvents for infrared spectrometry are completely nonpolar. Which of the following solvents would be most useful in studying IR spectra: acetone, carbon disulfide, carbon tetrachloride, diethyl ether, chloroform $(CHCl_3)$? Of the five solvents listed, the two that are most useful for infrared spectrometry are also useful solvents in NMR spectrometry, but for a different reason. Explain why. In infrared spectrometry, why are sample containers made of sodium chloride, silver chloride, and potassium bromide more useful than those made of glass, polystyrene, or other plastics?

118. **[Laboratory]** A 10.6-g sample of benzaldehyde was allowed to react with 5.9 g $KMnO_4$ in an excess of $KOH(aq)$. After filtration of the $MnO_2(s)$ and acidification of the solution, 6.1 g of benzoic acid was obtained. What was the percent yield of the reaction?

* 119. **[Biochemical]** Food chemists use the addition of iodine to the double bonds in unsaturated fats to determine the degree of unsaturation. The *iodine number* of a fat is the number of grams of I_2 consumed by 100 g of the fat. Polyunsaturated oils have high iodine numbers, whereas highly saturated fats, such as butter, have low iodine numbers. The addition of bromine to the double bonds in unsaturated fat was illustrated on page 372.

Another way to characterize a fat is through its *saponification value*. This is the number of milligrams of KOH consumed in the hydrolysis of 1.0 g of the fat.

(a) What are the iodine number and saponification value of glyceryl palmitooleolinolenate, whose structural formula is given on page 956?

◄ The fatty acid components of safflower oil, a polyunsaturated cooking oil, are about 6–7% palmitic, 2–3% stearic, 12–14% oleic, 75–80% linoleic, and 0.5–1.5% linolenic acid.

(b) With the information given in the accompanying photograph and in Table 23.1, estimate the range of values expected for the iodine number and the saponification value of safflower oil.

* **120. [Biochemical]** Bradykinin is a nonapeptide that lowers blood pressure and increases the permeability of blood capillaries.

Complete hydrolysis gives the following ratio of amino acids: 3 Pro, 2 Arg, 2 Phe, 1 Gly, and 1 Ser. Both the N-terminal and the C-terminal amino acid units are arginine. Partial hydrolysis gives the following fragments: Gly-Phe-Ser-Pro, Pro-Phe-Arg, Ser-Pro-Phe, Pro-Pro-Gly, Pro-Gly-Phe, Arg-Pro-Pro, and Phe-Arg. What is the amino acid sequence in bradykinin?

e-Media Problems

The activities described in these problems can be found in the e-Media Activities and Interactive Student Tutorial (IST) modules of the Companion Website, *http://chem.prenhall.com/hillpetrucci.*

121. An example of nucleophilic substitution is seen in the **Bimolecular Reaction** animation (*Section 23-4*). Visualize a similar process occurring for the reaction of ethanol and hydroiodic acid described in this section. From the details of this animation, suggest two reasons why the substitution step is the slow step of the second-order reaction.

122. View the **Proteins and Amino Acids** animation (*Section 23-9*) to review the nature of primary, secondary, and tertiary structure. Next view the **DNA Segment** 3-D model (*Section 23-10*). **(a)** Access the Chime submenu [right click on the molecule (Windows) or click-hold on the molecule (Macintosh)], turn off the rotation, display the Wireframe representation, and select the Display Hydrogen Bonds option. Tell what type of structure

(primary, secondary, or tertiary) results from the intermolecular interactions illustrated in this structure. **(b)** Display the ribbon representation and describe the type of secondary structure. How does the ordering of the bases in each chain influence this structure?

123. The stretching and bending motions illustrated in the **Molecular Vibrations** animation (*Section 23-12*) are common to many organic molecules, yet they vary in wavenumber as the identity of the three groups changes. Assuming that the bottom atom of each structure is carbon and the one in the upper left is oxygen, identify regions of the spectrum where you would expect to find absorption peaks if these atoms were found in **(a)** a ketone, **(b)** a carboxylic acid, and **(c)** an amine (see also Table 23.6). **(d)** Identify the third atom or group in the animations for each of the compounds. **(e)** Based on the motions observed in the animation, describe why the peak due to $C=O$ stretching in each of the compounds occurs in a different region of the spectrum.

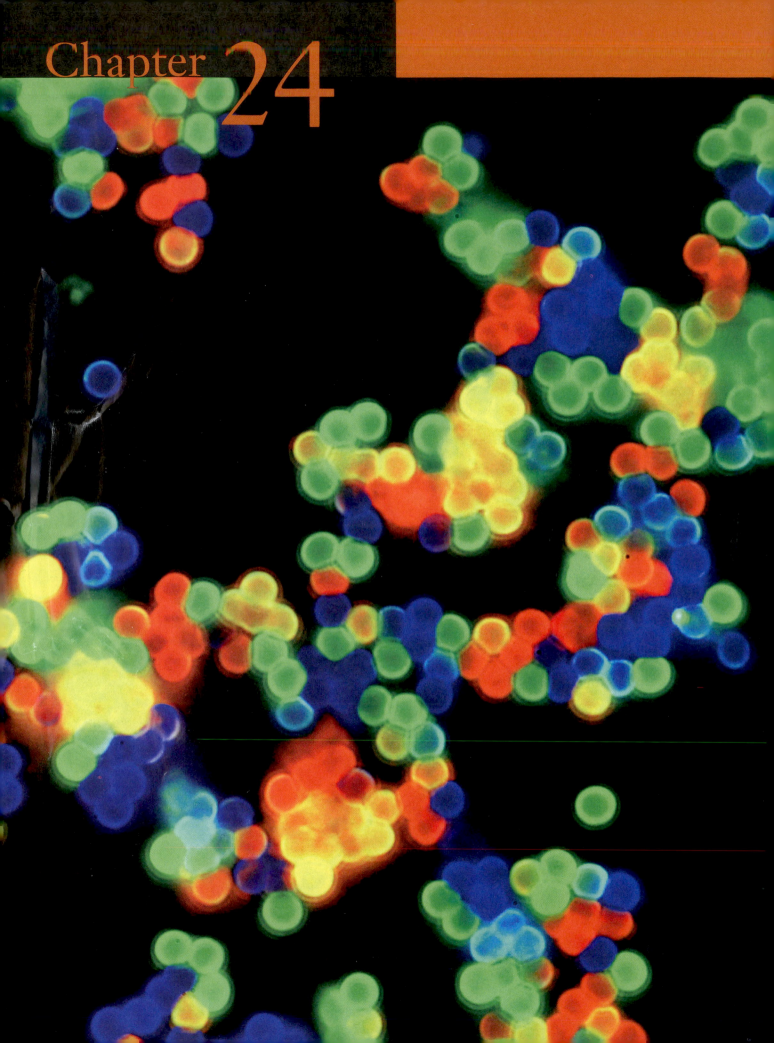

Chapter 24

Chemistry of Materials: Bronze Age to Space Age

PREHISTORIC PEOPLE WERE limited to natural materials that they could use without much alteration. However, the discovery of fire gave these people an effective way of transforming natural materials into new materials. For example, they developed primitive methods of heating minerals to extract tin, lead, and other metals. They fused the tin with naturally occurring copper to make *bronze,* a hard metallic material that they could then fashion into tools. With this discovery, about 7000 years ago, humans passed from the Stone Age to the Bronze Age. The use of iron and the advent of the Iron Age came somewhat later in most areas. Metals, which were so important to ancient people that the metal names are associated with various periods of prehistory, are still materials of great importance today, and we consider them first in this chapter.

In contrast, many materials critical to modern society were all but unknown until well into the twentieth century. We focus on new materials in the remainder of the chapter: semiconductors, polymers, and space-age materials. Semiconductors have revolutionized the electronics industry, which in turn is the driving force behind much of our high-tech world. In many of the materials of our daily lives—ranging from clothing to housing materials to automobiles—polymers have replaced natural fibers, wood, and metals. Half of all industrial chemists in the United States work with polymers in some way. Space-age materials—high-performance alloys, composites, and nanomaterials—constitute an exciting area of research in which new and dramatic discoveries are made almost daily. Because promising discoveries are being made much faster than new applications can be developed, at this time some space-age materials are 'solutions in search of a problem.'

◄ Polymer beads embedded with quantum dots fluoresce in five different colors. Polymers are long, chainlike molecules and are an integral part of our modern society. Quantum dots, made of semiconductor material and only a few nanometers in size, are examples of nanomaterials—substances and materials that take on unique properties when the sample size is reduced to the nanometer scale. Polymers, quantum dots, and semiconductors are all examples of modern materials, which we examine in this chapter.

Metals

We will first consider general approaches for obtaining metals from natural sources. Then we will illustrate these approaches for a few common metals. We will also look at some newer and more specialized methods.

24.1   Metallurgy: From Natural Sources to Pure Metals

Most metals are found in *minerals*, which are crystalline inorganic compounds in Earth's crust. For example, titanium is obtained from the mineral *rutile*, TiO_2, and the metalloid antimony from *stibnite*, Sb_2S_3. An **ore** is a material containing a sufficiently high percentage of some mineral to make extracting a metal from that mineral economically feasible. A titanium ore, for example, has a significant percentage of rutile. Some ores of commercial importance are

- *native ores (free metal):* gold, silver, platinum, and copper
- *oxides:* iron, manganese, aluminum, tin, and titanium
- *sulfides:* copper, nickel, zinc, lead, mercury, antimony, bismuth, and silver
- *carbonates:* sodium, potassium, magnesium, calcium, manganese, iron, and zinc
- *chlorides (often in aqueous solution):* sodium, potassium, magnesium, and calcium

Because silicon is such an abundant element (exceeded only by oxygen in Earth's crust), many metals occur as silicates. For example, aluminum is widely distributed in kaolinite clay, $Al_2Si_2O_5(OH)_4$. However, it is too expensive to extract aluminum from this source. Few silicates are used as ores.

Extractive Metallurgy

Metallurgy is the general study of metals, and *extractive metallurgy* focuses on the activities required to obtain a pure metal from one of its ores. Although there is no single method of extractive metallurgy that applies to all metals, the following processes are all commonly encountered.

- *Mining:* Ores are mined in many ways. Generally, we think of a mine as a hole dug into Earth's crust. Such *deep mines* are the method of choice when the ore is far below the surface. Ores near the surface, however, can be dug up in open-pit mines. Most copper ore in the western United States has been mined in this way. Some mining is done by pumping seawater or brines from salt lakes or subsurface wells to obtain the metals contained therein.

- *Concentration:* Often, the mineral of interest in a mining operation makes up only a small fraction of the material mined. To eliminate as much extraneous material as possible in the metallurgical scheme, ores are generally *concentrated* by physical separation from waste rock. Figure 24.1 illustrates the **flotation** method of

▲ Cinnabar, a mercury ore, principally HgS.

▲ Rutile, a titanium ore, mainly TiO_2, in a matrix of quartz.

Application Note

The iron ore *magnetite*, Fe_3O_4, can be concentrated by a process called magnetic beneficiation. The Fe_3O_4 is attracted to a magnet and thus separated from waste rock, which is composed of nonmagnetic materials.

▲ Gold, as the native ore Au, embedded in quartz.

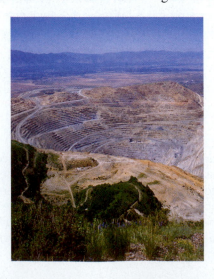

◄ Bingham Canyon Mine, an open-pit copper mine near Salt Lake City, Utah, yields approximately 300,000 tons of copper annually. Measuring about 4 km wide and 800 m deep, it is the largest man-made excavation on Earth.

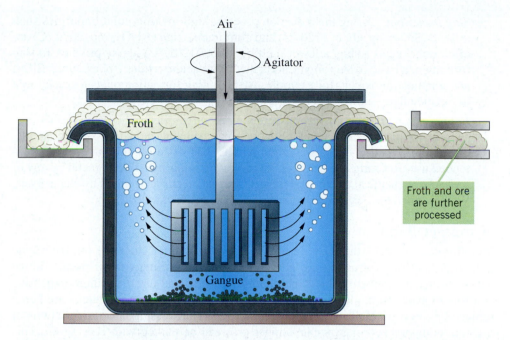

Air

Agitator

Froth

Froth and ore are further processed

Gangue

◀ **FIGURE 24.1** **Concentration of an ore by flotation**
Powdered ore, additives, and water are agitated with air. Air bubbles carry ore particles up into the froth, and the particles of waste rock, called gangue, fall to the bottom of the vat.

Ore Composition activity

concentrating an ore. The ore is ground into a powder, the powder and certain additives are placed in water in a large vat, and the mixture is agitated with air. Particles of ore become attached to air bubbles and rise to the top of the mixture as a froth, which is allowed to overflow. Particles of the undesired waste rock, called *gangue,* fall to the bottom of the vat. The success of the flotation method depends on some critical additives. One additive, called a frother, produces a stable foam. Another additive, called the collector, coats the particles of ore but does not "wet" particles of gangue.

• *Roasting:* Following concentration, ores are often heated to a high temperature in a process called *roasting.* Roasting converts many metal compounds to metal oxides. For example, zinc carbonate yields zinc oxide when heated strongly:

$$ZnCO_3(s) \longrightarrow ZnO(s) + CO_2(g)$$

and lead(II) sulfide produces lead(II) oxide and $SO_2(g)$:

$$2\ PbS(s) + 3\ O_2(g) \longrightarrow 2\ PbO(s) + 2\ SO_2(g)$$

Sulfur dioxide produced in metallurgical processes is generally converted to sulfuric acid. Environmental restrictions preclude venting $SO_2(g)$ into the atmosphere (Section 25.2).

• *Reduction:* The term *reduction* has two somewhat different meanings in extractive metallurgy. In one sense, it refers to the removal of oxygen from a metal oxide. In principle, this can be done by heating the oxide to a sufficiently high temperature. In practice, though, this method of reduction works only for a couple of oxides that decompose at relatively low temperatures. For instance, Ag_2O decomposes rapidly at 250–300°C, and HgO at about 500 °C, with the reaction for the latter being

$$2\ HgO(s) \longrightarrow 2\ Hg(l) + O_2(g)$$

Usually, however, reduction is not so easily accomplished, and a reducing agent is required. The most widely used and inexpensive reducing agent is coke (page 872). In most cases, the partial oxidation of carbon produces CO(g), which serves as the actual reducing agent:

$$SnO_2(s) + 2\ C(s) \longrightarrow Sn(l) + 2\ CO(g)$$
$$SnO_2(s) + 2\ CO(g) \longrightarrow Sn(l) + 2\ CO_2(g)$$

In a few special cases, an especially strong metallic reducing agent, such as aluminum or calcium, is required. This is the case in the thermite reaction (Section 21.3):

$$Cr_2O_3(s) + 2\ Al(s) \longrightarrow Al_2O_3(s) + 2\ Cr(l)$$

Thermite Reaction movie

In *electrolytic* reduction, a metal ion is reduced to the free metal. Examples of this type of metallurgical reduction are the reduction of Na^+ to Na(l) in molten NaCl (Section 18.9), of Mg^{2+} to Mg(l) in molten $MgCl_2$ (Section 20.9), and of Al^{3+} to Al(l) from Al_2O_3 dissolved in molten Na_3AlF_6 (Section 21.3).

Application Note

Slag is sometimes formed into fibers and used as *rock wool* for thermal insulation.

▲ Molten slag being dumped.

▲ Large spherical autoclaves are used to leach ilmenite ore, $FeTiO_3$, with HCl(aq) at 120 °C and 3.5 atm. This is one step in the hydrometallurgical production of pure TiO_2, which is then used either as a pigment or in the metallurgy of titanium.

Hydrometallurgical processes provide another example of green chemistry. The chemical reactions involved are less damaging to the environment than the traditional pyrometallurgical reactions.

Alloy 3D model

- *Slag formation:* During the reduction process, high-melting-point impurities such as sand (SiO_2; mp, about 1720 °C) and aluminum oxide (Al_2O_3, mp, 2072 °C) are often removed as a **slag,** a lower-melting (about 1200 °C) glassy product. In slag formation, a basic oxide (often CaO) reacts with acidic oxides (for example, SiO_2) and amphoteric oxides (for example, Al_2O_3). Slag formation plays a crucial role in the metallurgy of iron.
- *Refining:* A metal obtained by electrolytic reduction often requires no further processing before use. A metal obtained by chemical reduction—for example, through the reaction of a metal oxide with CO(g)—is often too impure for its intended use. **Refining** is the process of removing impurities from a metal by any of a variety of chemical or physical means. Several metals are refined by electrolysis. We described the electrolytic refining of copper in Section 18.11.

Hydrometallurgy

Metallurgical methods that use ore concentration, roasting, chemical or electrolytic reduction, and slag formation are often called **pyrometallurgy** (*pyro* means fire or heat). Pyrometallurgical methods are energy-intensive and usually require expensive controls to avoid polluting the environment. In some cases, these methods are being replaced by **hydrometallurgy,** methods that involve processing aqueous solutions of metallic compounds. For example, copper ores can be treated with H_2SO_4(aq). This reaction converts the copper compounds in the ore to $CuSO_4$(aq), which is then electrolyzed to produce copper. Following are the essential operations of hydrometallurgy.

- *Leaching:* Metal ions are extracted from the ore with a liquid—water, acids, bases, or salt solutions. Sometimes an oxidation–reduction reaction is involved. In some instances, leaching can be done in place, without first digging the ore.

- *Purification and/or concentration:* Impurities are removed from the leach solution, and the solution is concentrated by evaporation to facilitate further processing. Two methods used for the removal of impurities are ion exchange and adsorption on activated carbon (page 872).

- *Precipitation and reduction:* In some processes, the desired metal ions are precipitated as an insoluble ionic solid. In a subsequent step, the metal ions are reduced to the free metal either by displacement from solution by a more active metal or by electrolysis.

Hydrometallurgy has several advantages over pyrometallurgy. For example, hydrometallurgy is a better process for low-grade ores, and it is more energy-efficient. It is also less polluting of the atmosphere. In the pyrometallurgy of zinc, for instance, there are inevitably some emissions of SO_2(g), Hg(g), and mercury compounds, but in the hydrometallurgy of zinc, compounds of sulfur and mercury are retained in solution. Hydrometallurgy presents some difficult problems of its own, however, such as how to contain and dispose of liquid solutions and solid waste. For example, the mercury compounds in the waste solution from zinc hydrometallurgy can be an environmental hazard if the solution is not properly treated before disposal.

Alloys

In many metallurgical processes, the final product desired is not a pure metal but an alloy. An **alloy** is a mixture of two or more metals or of a metal and a nonmetal, formulated to produce properties that make the mixture more desirable than the pure metal. For instance, a small amount of copper added to gold increases the gold's hardness, and the presence of nickel and chromium in iron greatly improves the iron's corrosion resistance.

Some alloys are heterogeneous mixtures, like the familiar lead–tin alloy *solder.* Although the alloy appears homogeneous to the eye, the separate solid phases are visible under a microscope. A few alloys are actual compounds of the constituent metals, such as the amalgam $NaHg_2$. Other alloys are solid solutions, and among solid solutions there are two possibilities. In *substitutional* solid solutions, atoms of one metal

Table 24.1 Several Common Alloys		
Name	**Typical Composition**	**Page of Previous Reference**
Battery plate	94% Pb, 6% Sb	884
Gunmetal bronze	90% Cu, 10% Sn	877
Magnalium	70–90% Al, 10–30% Mg	847
Pewter	85% Sn, 7% Cu, 6% Bi, 2% Sb	877
Plumber's solder	67% Pb, 33% Sn	877
Sterling silver	92.5% Ag, 7.5% Cu	913
Yellow brass	67% Cu, 33% Zn	915

substitute for some of the atoms in the crystal structure of another metal. This type of solid solution generally requires that the atomic radii of the metals in the alloy be closely matched. Thus, silver and gold, both with an atomic radius of 144 pm, form a series of solid solutions whose concentrations can range from pure silver to pure gold.

In *interstitial* solid solutions, small atoms of the minor component(s) occupy voids among the larger atoms of the major component. An example of this type of solid solution is one with about 1% carbon in iron.

In Section 24.2, we will mention several important alloys that have iron as the major component. A few other common alloys, their compositions, and previous references to them are listed in Table 24.1. Later, in our discussion of space-age materials in Section 24.11, we will consider advanced alloys whose compositions have been tailored to provide unique chemical and physical properties.

24.2   Iron and Steel

Among all the metals, iron is most important commercially and second only to aluminum in natural abundance. Iron-containing minerals commonly are present in red soil and red-rock formations.

Pig Iron

For the last 700 years or more, people have used a device called a *blast furnace* to reduce iron ore to iron. A modern blast furnace is pictured in Figure 24.2. Chemical reactions occur as the solid reactants settle from the top of the furnace and gaseous reactants rise from the bottom. The reactions are complex, but we can divide them into four basic categories:

- Formation of reducing agents, principally $CO(g)$ and $H_2(g)$
- Reduction of iron oxide
- Slag formation
- Impurity sources

Representative reactions are noted in Figure 24.2, keyed to the regions of their main occurrence in the blast furnace.

Acidic oxides, such as SiO_2, predominate as impurities in most iron ores, and so iron producers commonly use basic oxides to react with them to form slags. Typically, the basic oxide is CaO formed by the decomposition of limestone ($CaCO_3$). Conversely, if the iron ore contains carbonates as impurities, sand (SiO_2) is used to form the slag.

The iron formed in a blast furnace is called **pig iron.** It is impure iron, generally containing 3–4% C, 0.5–3.5% Si, 0.5–1% Mn, 0.05–2% P, and 0.05–0.15% S. Most pig iron is transferred directly to steelmaking furnaces as a liquid.

The solid metal obtained from liquid pig iron is called **cast iron.** Crude cast iron is brittle when cold, but malleable above 250 °C. Its properties are modified by remelting, reprocessing, and cooling at a controlled rate. Some of the uses of cast iron include automotive engines, boilers, stoves, and cookware.

▲ Iron compounds give soil and rocks a characteristic red color, as in this rock formation in Arches National Park in Utah.

Long ago, steel swords were made by first heating wrought-iron swords in charcoal and then rapidly quenching the hot metal in a cold liquid. The sword cooled so fast that the atoms were trapped in an irregular array that left the steel strong and flexible. Some metalworkers preferred biological fluids—such as urine, a readily available resource—as a quenching medium.

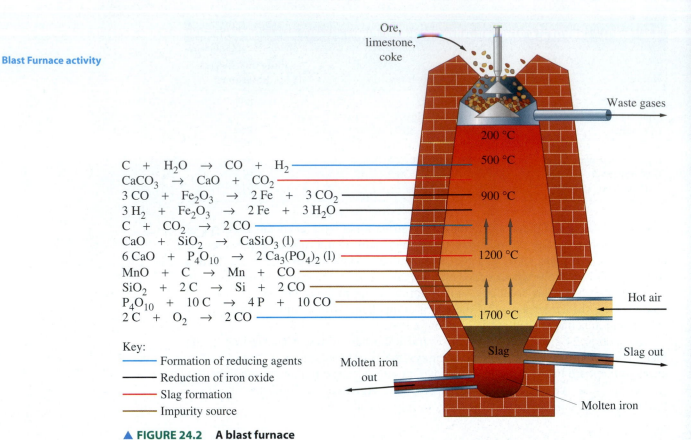

$$C + H_2O \rightarrow CO + H_2$$
$$CaCO_3 \rightarrow CaO + CO_2$$
$$3\,CO + Fe_2O_3 \rightarrow 2\,Fe + 3\,CO_2$$
$$3\,H_2 + Fe_2O_3 \rightarrow 2\,Fe + 3\,H_2O$$
$$C + CO_2 \rightarrow 2\,CO$$
$$CaO + SiO_2 \rightarrow CaSiO_3\,(l)$$
$$6\,CaO + P_4O_{10} \rightarrow 2\,Ca_3(PO_4)_2\,(l)$$
$$MnO + C \rightarrow Mn + CO$$
$$SiO_2 + 2\,C \rightarrow Si + 2\,CO$$
$$P_4O_{10} + 10\,C \rightarrow 4\,P + 10\,CO$$
$$2\,C + O_2 \rightarrow 2\,CO$$

Key:
— Formation of reducing agents
— Reduction of iron oxide
— Slag formation
— Impurity source

▲ **FIGURE 24.2** **A blast furnace**

A modern blast furnace stands up to 90 m high, is computer controlled, and is equipped with environmental control devices.

QUESTION: How could computer control assist in the operation of such a furnace?

In 1985, the wreckage of the *Titanic* was found, and scientists sought to discover why this supposedly unsinkable ship sank so quickly. Metallurgical analysis showed that wrought-iron rivets in the ship's hull contained a high amount of slag. This made the rivets brittle at ice-water temperatures, causing the hull to rip apart rapidly. Heat treatment would have made the rivets less brittle, a fact presumably unknown to the ship's builders.

▲ A particular stainless steel in sinks (Type 304) contains 8–20% Cr, 10.5% Ni, 0.85–1.15% Mn, 0.10% N, 0.04–0.09% P, 0.26–0.35% S, a maximum of 1% Si, and a maximum of 0.15% C.

Cast iron is *wrought* by hammering at 800–900 °C. This process mechanically squeezes out some of the remaining slag and solid impurities and burns out most of the remaining carbon. Wrought iron has many uses, including decorative fences, gates, and grills.

Steel

Most iron is converted to alloys known collectively as **steel.** In general, a steel has more desirable properties (strength, malleability, corrosion resistance) than iron.

Converting pig iron to steel requires two principal steps: reduction of the carbon content to less than 1.5% and removal of major impurities (Si, Mn, P, S) and some minor ones. The resulting steel, because it contains carbon as the principal alloying element, is called *carbon steel* and accounts for the greatest volume of steel production. Low-carbon steel (about 0.25% C) is used for construction beams and girders and for reinforcing rods in concrete. Harder, high-carbon steel (more than 0.7% C) finds use in cutting tools and railroad rails.

The remainder of steel production is in the form of *alloy steel.* In addition to carbon, an alloy steel ordinarily contains Cr, Ni, Mn, V, Mo, Co, and/or W as a major component. For example, stainless steel, an iron alloy with nickel and chromium, is used in flatware and cutlery, chemical plant construction, and ornamental features in architecture. A silicon steel with up to 5% silicon is used in transformers, motors, and generators.

Most older steelmaking methods have been replaced by the basic oxygen process (Figure 24.3). Carbon and sulfur are burned off as gaseous oxides. Silicon, manganese, and phosphorus also are oxidized, but their oxides form slags rather than gases. The SiO_2 and MnO combine to form $MnSiO_3\,(l)$; the P_4O_{10} combines with CaO (from $CaCO_3$ added to the furnace charge) to form $Ca_3(PO_4)_2\,(l)$. The liq-

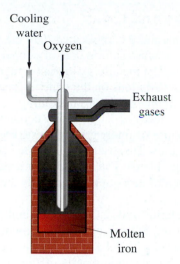

Cooling
water
Oxygen

Exhaust
gases

Molten
iron

◀ **FIGURE 24.3 A basic oxygen
furnace**
Oxygen at about 10 atm and limestone are
added to molten pig iron with 15–30%
scrap iron and steel.

uid slag floats on top of the liquid iron and is poured off. Alloying metals are added
as a final step.

24.3   Tin and Lead

The methods used to extract tin and lead from their ores are good illustrations of the
basic metallurgical processes described in Section 24.1. Their relatively uncompli-
cated extractive metallurgies help account for the discovery of these metals dating
back to ancient times.

Tin

Tin occurs in nature mainly in ores containing *cassiterite,* SnO_2, which can be concentrated by flotation. Roasting of the ore oxidizes metallic impurities and drives off
sulfur and arsenic as volatile oxides. The SnO_2 that remains is then reduced with coke:

$$SnO_2(s) + 2\,C(s) \longrightarrow Sn(l) + 2\,CO(g)$$

The metal is first solidified and then remelted. Tin is fairly low-melting (mp, 232 °C),
and the molten tin is poured off, leaving behind unmelted impurities. Any impurities
that are soluble in liquid tin are oxidized with air, and the oxide film is skimmed off.

Recycling is also an important source of tin. In one recycling method, scrap tin
plate is treated with chlorine gas, which converts the tin to $SnCl_4(l)$ without affecting
the underlying metal (usually steel). The $SnCl_4$ is then converted to SnO_2, and the
SnO_2 is reduced to metallic tin. The tin plate is mainly in the form of "tin" cans—steel
cans with a protective layer of tin (page 780).

▲ These tin-coated steel cans can be
recycled. The tin metal is recovered
through the chemical reactions
described in the text.

Lead

Lead is found chiefly as *galena,* PbS. The ore is concentrated by flotation and then
roasted to the oxide. Reduction of the oxide is carried out with coke. The lead at this
point contains several possible impurities, such as Cu, Ag, Au, Sn, As, and Sb. Lead
producers take advantage of physical properties such as melting points and solubilities
to remove these impurities. When the lead is melted (mp, 327 °C), copper rises to the
top of the liquid as an insoluble solid and is skimmed off. When the temperature is
raised further, Sn, As, and Sb are oxidized, and the oxide film is skimmed off. At this
point, zinc is added to the molten lead. Silver and gold in the molten lead pass into the
molten zinc, in which they are more soluble. When the molten mixture is cooled below
420 °C (the melting point of zinc), the zinc solidifies to a crust containing most of the
Au and Ag; then this crust is skimmed off. Combined with electrolytic refining, this
treatment can produce lead with a purity of 99.99%.

The recycling of used lead is an important alternative to the production of new
lead. Currently, about 70% of manufactured lead is recycled lead.

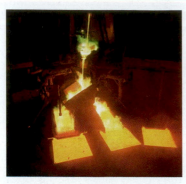

▲ Molten copper being poured into anode casts prior to electrolytic refining.

24.4 Copper, Zinc, Silver, and Gold

The metallurgy of copper is somewhat more complicated than the general procedure described in Section 24.1, primarily because sulfide ores of copper generally contain appreciable quantities of iron sulfides. The metallurgy of zinc is more straightforward, but it also illustrates some interesting variations on the general procedure. The metallurgy of silver and gold are classic examples of hydrometallurgy.

A copper ore typically used to produce copper generally contains only about 0.5% Cu. This means that enormous quantities of finely ground waste rock are generated in concentrating the copper ore by flotation. Disposal of this waste rock in environmentally acceptable ways poses serious problems. The recycling of used copper is now an important alternative to the production of new copper. Currently, nearly half of manufactured copper is recycled copper.

Zinc occurs mainly as ZnS (sphalerite) and $ZnCO_3$ (smithsonite). During roasting, sphalerite is converted to zinc oxide and $SO_2(g)$, and smithsonite is converted to zinc oxide and $CO_2(g)$. Zinc oxide is reduced with coke or powdered coal. The reduction is carried out at about 1100 °C, well above the boiling point of zinc (907 °C). Zinc *vapor* is condensed to a liquid, and the impurities, mostly cadmium and lead, can be removed by fractional distillation of the liquid zinc.

Alternatively, zinc oxide from the roasting step can be dissolved in $H_2SO_4(aq)$. Addition of powdered zinc to the zinc sulfate solution displaces less active metals such as cadmium:

$$Zn(s) + Cd^{2+}(aq) \longrightarrow Zn^{2+}(aq) + Cd(s)$$

Then the $ZnSO_4(aq)$ is electrolytically reduced, and pure zinc metal is deposited at the cathode.

Hydrometallurgy has long been used in extracting silver and gold from their ores. Silver and gold are both found free in nature, but all easily accessible known deposits have been mined. A typical gold ore today contains only about 10 g Au per ton. In one older, environmentally damaging method, low-grade gold ores were treated with mercury. The gold dissolved in the liquid mercury to form an *amalgam* (Section 22.5). The mercury in the amalgam was driven off by heating, leaving behind pure gold.

A method more widely used today is *cyanidation,* a special case of hydrometallurgy in which the gold is leached from the ore by a solution of a cyanide salt. In this method, $O_2(g)$ in air oxidizes the free metal to Au^+, which is then complexed with CN^-:

$$4\,Au(s) + 8\,CN^-(aq) + O_2(g) + 2\,H_2O(l) \longrightarrow 4\,[Au(CN)_2]^-(aq) + 4\,OH^-(aq)$$

(Gold can be oxidized to Au^+ in this reaction only because $[Au(CN)_2]^-$ is such a stable complex ion. Without $CN^-(aq)$ present, the oxidation of Au to Au^+ hardly occurs at all.) The gold is then displaced from $[Au(CN)_2]^-(aq)$ by an active metal such as zinc:

$$2\,[Au(CN)_2]^-(aq) + Zn(s) \longrightarrow 2\,Au(s) + [Zn(CN)_4]^{2-}(aq)$$

▶ Waste solution from the leaching operation at a gold-mining facility in the Mojave Desert of California is stored in a large containment pond.

Cyanidation poses substantial environmental problems because aqueous solutions of cyanides are poisonous and are workplace hazards. Equally important, waste solutions containing cyanide leachates must be held in containment ponds. The ponds must be lined to prevent the solutions from entering groundwater, and migratory birds and other wildlife must be kept away from the ponds. Environmental protection is at least as important a consideration as technical and economic matters when planning any new mining operation.

Example 24.1   A Conceptual Example

Consider the electrolytic method of zinc metallurgy previously described. Why must the ions of metals less active than zinc (for example, Cd^{2+}) be removed before the electrolytic reduction of $ZnSO_4(aq)$ is carried out?

ANALYSIS AND CONCLUSIONS

Recall from Section 18.4 that the value of E_{cell}° for a redox reaction is given by the relationship

$$E_{cell}^{\circ} = E^{\circ}(\text{reduction}) - E^{\circ}(\text{oxidation})$$

Because electrolysis reactions are not spontaneous, the value of E_{cell}° will be negative. The minimum voltage required to bring about the electrolysis is given by

$$E_{electrolysis}^{\circ} = -E_{cell}^{\circ}$$

In the electrolysis we are considering, a metal ion is reduced to the metal at the cathode and an oxidation occurs at the anode. If the metal ion reduced is $Zn^{2+}(aq)$, then

$$E_{cell}^{\circ} = E_{Zn^{2+}/Zn}^{\circ} - E^{\circ}(\text{oxidation})$$
$$= -0.763 \text{ V} - E^{\circ}(\text{oxidation})$$

If the metal ion reduced is $Cd^{2+}(aq)$, then

$$E_{cell}^{\circ} = E_{Cd^{2+}/Cd}^{\circ} - E^{\circ}(\text{oxidation})$$
$$= -0.403 \text{ V} - E^{\circ}(\text{oxidation})$$

Regardless of what oxidation occurs at the anode [actually it is the oxidation of $H_2O(l)$ to $O_2(g)$], the minimum voltage required for the reduction of $Cd^{2+}(aq)$ at the cathode is given by

$$E_{electrolysis}^{\circ} = -E_{cell}^{\circ} = 0.403 \text{ V} + E^{\circ}(\text{oxidation})$$

This voltage is *less than* that required for the reduction of $Zn^{2+}(aq)$, which is

$$E_{electrolysis}^{\circ} = -E_{cell}^{\circ} = 0.763 \text{ V} + E^{\circ}(\text{oxidation})$$

Thus, the voltage we must use to obtain $Zn(s)$ at the cathode is more than adequate to produce $Cd(s)$ as well. Consequently, we must remove $Cd^{2+}(aq)$ and any other metal ions that are more easily reduced than Zn^{2+} *before* we electrolyze the $ZnSO_4(aq)$.

EXERCISE 24.1A

In the displacement of metal ions discussed in Example 24.1, why do you suppose zinc is used rather than some other active metal, such as aluminum?

EXERCISE 24.1B

The hydrometallurgy of silver involves blowing air through an aqueous solution of $CN^{-}(aq)$ in which highly insoluble Ag_2S is suspended. Sulfide ion is oxidized to sulfate ion, and the silver appears in the complex $[Ag(CN)_2]^{-}$. Write a chemical equation for this process.

Bonding in Metals and Semiconductors

In Section 11.10, we used a microscopic view of a metal—its crystal structure—to predict a macroscopic property—the density of the metal. Other macroscopic properties of metals, however, such as electrical conductivity, ductility, and malleability, are determined by the nature of the bonding between metal atoms. In the sections that follow, we will consider two bonding theories for metals. However, we will find that only one (band theory) accounts for the interesting electrical properties of semiconductors, such as silicon and germanium.

24.5   The Free-Electron Model of Metallic Bonding

From the electron configuration $1s^2 2s^1$, we can write a Lewis structure that shows a bond between two Li atoms:

$$Li:Li$$

The molecule Li_2 does exist in the gaseous state, but this fact does not help us to explain how a Li atom is bonded to *eight* nearest neighbors in the solid metal. This example highlights the basic problem in explaining bonding in metals: The atoms do not seem to have enough valence electrons to form all the bonds needed. Nor can the bonding be ionic, because metals seldom form negative ions, and solid ionic compounds do not conduct electricity.

The key to metallic bonding is that certain electrons are *delocalized.* These delocalized electrons are not bound to individual atoms, and they can therefore bind large numbers of metal atoms together.

According to an early theory of metallic bonding, each atom in a metallic crystal loses its valence electrons, releasing them to the crystal as a whole. Because these valence electrons are freed from individual atoms, this theory is known as the **free-electron model** of metallic bonding. We picture the metallic crystal as a lattice of positive ions immersed in a "gas" of electrons. The comparatively massive metal ions undergo lattice vibrations but are otherwise immobile. The valence electrons, in contrast, are highly mobile, zipping about much like atoms or molecules in a gas. The cloud of negative charge associated with the free electrons envelops the positive-ion lattice and is the "glue" that holds the metallic crystal together.

Figure 24.4 suggests how the free-electron model accounts for electrical conductivity. In the absence of an electric field, the motion of the free electrons is completely random (Figure 24.4a). Over a period of time, although electrons may have traveled great distances along zigzag paths, their distribution within a metal remains unchanged. However, if the metal is connected to the terminals of a battery, as in Figure 24.4b, electrons drift toward the positive terminal, even as they continue their zigzag motion. Some electrons leave the metal under the influence of the electric field,

Free-Electron animation

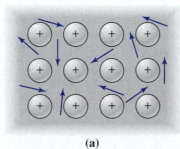

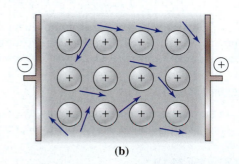

(a)   (b)

▲ **FIGURE 24.4   The free-electron model**

(a) A lattice of cations is immersed in a cloud of negative electric charge made up of the free valence electrons of the metal atoms. In the absence of an electric field, on average, the motion of the free valence electrons in a metal is completely random. (b) In the presence of an electric field, there is a net drift of electrons toward the positive terminal. This electron movement constitutes an electric current.

QUESTION: Why does the free-electron model not apply to ionic or covalent materials?

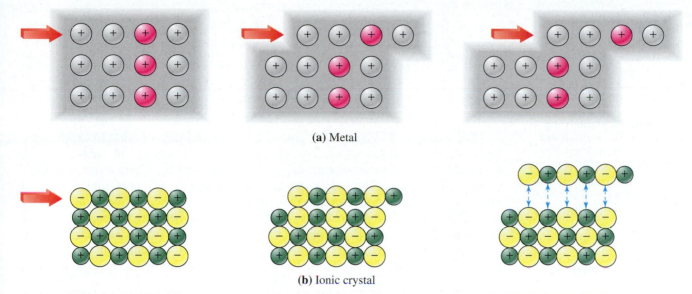

(a) Metal

(b) Ionic crystal

▲ **FIGURE 24.5   Deformation of a metal compared to an ionic solid**
(a) When a metal is deformed, the locations of the positive ions change, but the ions are still surrounded by free electrons; the metal is malleable and ductile. (b) An ionic crystal breaks when distorted, because of the proximity of opposite charges; it is brittle.

and others enter the metal to take their place. This movement of electrons is an electric current in the metal.

Figure 24.5 shows how the free-electron model accounts for the malleability generally found in metals, in contrast to the brittleness of an ionic crystal. When a force is applied to the top layer of ions in the metal shown in Figure 24.5a, the ions shift, but their environment—the surrounding electron gas—is unchanged. The deformation is easily accommodated; the metal is malleable. When a similar force is applied to a layer of ions in the ionic crystal in Figure 24.5b, like-charged ions are brought into proximity of one another, negative to negative and positive to positive. Repulsive forces cause the crystal to rupture; the ionic solid is brittle.

The free-electron model is less successful in explaining the effect of temperature on the electrical conductivities of metals. We expect a gas to flow more readily as its temperature and thus the average speed of its molecules increase. Yet the electrical resistance of a metal increases with temperature, which means that the metal's ability to conduct electric current *decreases*. One possibility, of course, is that the increased vibrations of the cations interfere with the migration of electrons in an electric field, but in any case, the free-electron model cannot account for the details of electrical conductivity.

The fundamental shortcoming of the free-electron theory is that it is a classical theory patterned after the kinetic theory of gases. In principle, classical theory allows for the specification of an electron's position and momentum more precisely than is permitted by Heisenberg's uncertainty principle (Section 7.8). This suggests that a quantum mechanical treatment of bonding in metals should be more satisfactory, as we will see in the next section.

24.6   Band Theory

In Section 10.6, we used molecular orbital theory to describe covalent bonding. We noted that a combination of two atomic orbitals, one from each atom in a bonded pair, produces two molecular orbitals. One of the molecular orbitals, a *bonding* orbital, lies at an energy below that of the atomic orbitals. The other molecular orbital, an *antibonding* orbital, lies at an energy above that of the atomic orbitals. Valence electrons distribute themselves between the molecular orbitals in accordance with the aufbau principle. Figure 24.6 shows the formation of molecular orbitals in Li_2 from the $2s$ orbitals of two Li atoms; the situation in Li_3, in which the $2s$ orbitals of three Li atoms combine; and in Li_4, in which the $2s$ orbitals of four Li atoms combine. In each case, the number of molecular orbitals is equal to the number of atomic orbitals that combine.

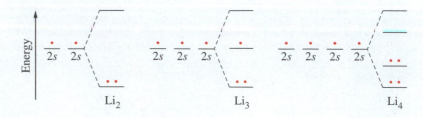

▶ **FIGURE 24.6** **Molecular orbitals for Li₂, Li₃, and Li₄**

Each Li atom contributes a 2s orbital to the formation of molecular orbitals. The molecule Li₂ has one bonding and one antibonding molecular orbital. In Li₃, the additional molecular orbital is at about the same energy as the isolated atomic orbitals; it is essentially a nonbonding molecular orbital. In Li₄, there are two bonding and two antibonding molecular orbitals. Electrons are represented by the red dots.

QUESTION: How would the representations differ if they were constructed for Be₂, Be₃, and Be₄?

 Development of Band Structure activity

▶ **FIGURE 24.7** **The 2s band in lithium metal**

The atomic orbitals of a very large number (n) of Li atoms are combined into a band of molecular orbitals having closely spaced energies. At 0 K, the levels in the bottom half of the band are occupied by electron pairs and those in the top half are empty.

The examples in Figure 24.6 give us a fairly good idea of what to expect for a huge collection of Li atoms, say all the atoms in a crystal of lithium metal. Figure 24.7 depicts molecular orbital energy levels for Li$_n$, where n is a very large number. Here, we see that a huge number, n, of energy levels are contained within an energy range not much greater than the energy difference between the bonding and antibonding molecular orbitals in Li₂. The spacing between the energy levels is so minute that the levels essentially merge into a *band* (much as the individual wavelength components of visible light merge into a continuous spectrum). Because this band is occupied by the valence electrons of the lithium atoms, it is called a **valence band**.

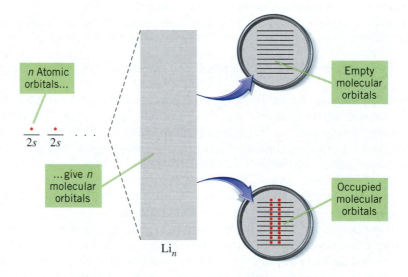

At 0 K, the n valence electrons associated with n Li atoms would occupy the lower half ($n/2$) of the energy levels—two per level. However, the energy difference between levels in the valence band is so small that electrons are easily excited from the highest filled levels to the unfilled levels immediately above them. This excitation can be accomplished by heating the metal or by applying a small voltage. Thus, in the presence of an electric field, electrons are stimulated to move into empty levels within the band, and this motion creates a net drift of electrons in the metal, an electric current.

The band theory description brings out an important requirement for electrical conductivity: the presence of a **conduction band**, a partially filled band of energy levels. Being half-filled, the valence band in lithium meets this requirement for a conduction band, but the requirement seems to present problems when applied to certain other metals. Consider magnesium. In Mg$_n$, the 3s band formed from the 3s orbitals of Mg atoms should be filled because there are two valence electrons ($3s^2$) per atom and thus $2n$ electrons for n molecular orbitals. The band formed by combining 3p atomic orbitals, in contrast, should be empty because magnesium has no 3p electrons. If an energy band is empty, there are no electrons to jump between energy levels, and if a band is filled, there is no place for the electrons to go. It would seem that magnesium should not conduct an electric current at all, but it does, and quite well.

The resolution of this paradox, as illustrated in Figure 24.8, is that the lowest-energy bonding molecular orbitals in the 3p band lie at a *lower* energy than do the

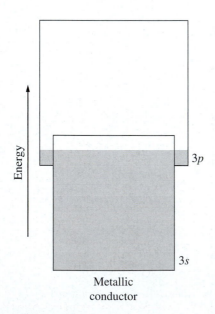

▲ **FIGURE 24.8** **Band overlap in magnesium**

With the overlap in energy of the 3s and 3p bands, a portion of the 3p band is filled because the energy of these levels is lower than the energy of some of the higher levels of the 3s band, which are unfilled. Both the 3s and 3p bands are partially filled and function as conduction bands, making magnesium an electrical conductor.

highest-energy antibonding molecular orbitals in the 3s band—in other words, the bands overlap. As a result, some of the valence electrons in magnesium that we would expect to occupy the top of the 3s band are instead found in the lower levels of the 3p band. This means that both the 3s and 3p bands are only partially filled, and this meets the requirement for electrical conductivity.

Band theory provides a good explanation of metallic luster and metallic colors. Because the energy levels in bands are so closely spaced, there are electronic transitions in a partially filled band that match in energy every component of visible light. Metals absorb the light that falls on them and are therefore opaque. At the same time, electrons that have absorbed energy from incident light are very effective in radiating light of the same frequency. This is why metals are good reflectors of light and have a shiny, often mirrored, appearance.

Superconductors

A toms in a metal lattice vibrate about fixed positions. These vibrations interfere with the flow of electrons and produce an electrical resistance. The more intense the vibrations, the greater the resistance. As a result, the electrons in a metal flow more easily at low temperatures than at high temperatures. Theoretically, the electrical resistance of a metal should be zero at 0 K.

In 1911, H. Kamerlingh Onnes discovered that some metals abruptly lose their electrical resistance at temperatures well above 0 K; when this happens, the metals become superconductors. A **superconductor** is a material that allows electric current to flow indefinitely, with no loss of energy, when the temperature is held below some critical value. The highest critical temperature among the metals is 23 K. This temperature is far below room temperature but is above the boiling point of liquid helium (4.2 K). Superconducting metals in liquid helium are used in powerful electromagnets in NMR spectrometers (Chapter 23), charged-particle accelerators, nuclear fusion research, and magnetic resonance imaging (MRI).

In 1986, researchers discovered a class of materials that become superconductors at much higher temperatures than metals do. For example, the compound $YBa_2Cu_3O_7$ becomes a superconductor at 90 K. Thus, it is a superconductor at the boiling point of liquid nitrogen (77 K), as shown in the photograph.

These new superconductors are not metals; they are ceramic materials (page 874). The structures of ceramic superconductors are complex but are known to involve sheets of copper and oxygen atoms separated by other constituent atoms. An ultimate goal of ceramic superconductor research is to create a superconductor that will function at room temperature and above. Electric transmission lines made of high-temperature superconductors would be able to transmit electricity with no loss of energy. Trains could be propelled at high speed and low-energy cost by using the levitation principle illustrated in the photograph.

The highest critical temperature for a ceramic superconductor is still far below room temperature. Add to this that ceramic superconductors are brittle and not easily made into wires and that their capacity to carry an electric current is limited, and you can see that the development of practical applications involving ceramic superconductors faces a number of challenges.

Superconductivity has also been observed among some fullerene compounds (Figure 11.34). For example, the compound K_3C_{60} at room temperature conducts electricity just as a metal does but becomes superconducting below 19 K. Other fullerenes are superconducting at even higher temperatures, and it seems likely that the upper temperature limit will keep rising as more discoveries are made.

▲ When a magnet is dropped above a superconductor in liquid nitrogen, the falling magnet induces an electric current in the superconductor. In turn, the electric current induces a magnetic field around the superconductor that opposes the field of the falling magnet. The magnet remains suspended at the point where the upward force of the repulsive magnetic field is just matched by the downward force that is due to Earth's gravitational field. The magnet remains suspended as long as the current in the superconductor persists. And this continues as long as the superconductor is maintained at the boiling point of liquid nitrogen (77 K).

▲ FIGURE 24.9 **Metallic luster and color**

Gold and copper have the "metallic" sheen, or luster, associated with reflected light. However, because they absorb and reflect more effectively in the longer-wavelength regions of the visible spectrum than elsewhere, they have characteristic colors: reddish for copper and yellow for gold.

Most metals—magnesium, aluminum, and silver, for example—are equally effective in radiating (or reflecting) light of all wavelengths and, as a result, have a typical metallic (silvery) color. In the case of copper and gold, however, the metal absorbs and reflects light of some wavelengths better than others. Being richer in some wavelength components, the reflected light (and hence the metal) is colored (Figure 24.9).

24.7   Semiconductors

In Figure 24.10a, we see that in an electrical insulator the energy of the conduction band is much higher than the energy of the valence band. Few electrons can acquire enough energy to jump the **energy gap** (E_g) separating the two bands, and thus little or no current flows. Figure 24.10b shows a situation in which the energy gap between the two bands is much smaller. Here, some electrons are able to jump the energy gap, resulting in a limited electrical conductivity. This is the energy band picture for a **semiconductor.**

(For a metal, as we have seen, there is no corresponding energy gap separating a partially filled band and an empty one. Either the valence band is itself a conduction band, as in lithium, or conduction is made possible by the overlap of the valence band and an empty energy band, as in magnesium.)

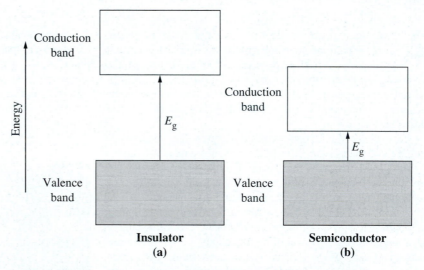

▲ FIGURE 24.10   **Band structure of insulators and semiconductors**
Insulators and semiconductors have an empty conduction band above a filled valence band. (a) In insulators, the energy gap, E_g, is large. Hardly any electrons can jump the gap between the bands. (b) In semiconductors, the energy gap is much smaller, and small numbers of electrons are able to jump the gap, especially as the temperature is increased.

Figure 24.11 suggests two ways of looking at the electrical conductivity of silicon, a typical semiconductor. In the localized picture, an electron (red) is shown leaving a bond between two Si atoms and entering the crystal as a free electron. At the site of the bond rupture, a vacancy (top yellow circle) forms that tends to be filled by an electron (blue) supplied from the rupture of an additional nearby bond. This converts the original one-electron bond back to a two-electron bond. By attracting an electron, the first vacancy acts like a positively charged center; it is called a **positive hole.** Electric current is carried through the semiconductor by both free electrons and positive holes.

In the delocalized picture, every electron that leaves a bond jumps from the valence band to the conduction band. A positive hole is left behind in the valence band. Electric current consists of the movement of electrons between levels in the conduction band plus the movement of positive holes between levels in the valence band.

Raising the temperature of a semiconductor increases its electrical conductivity. With an increase in temperature, electrons acquire more energy and more of them are able to jump the energy gap. This behavior is an important distinction between semi-

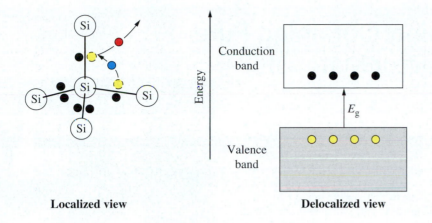

Localized view   **Delocalized view**

◀ **FIGURE 24.11   Two views describing electrical conductivity in semiconductors**

In the localized view, an electron (•) can escape from the bond between two Si atoms and move into the crystal at large. The vacancy it leaves at the bond site (∘) is a positive hole. When an electron (•) from a neighboring bond moves into this positive hole, a new positive hole (∘) forms there, and so on. The net effect is that the positive hole migrates from one bond to the next. In the delocalized view, a positive hole (∘) is produced in the valence band for every electron (•) that jumps from the valence band to the conduction band.

conductors and metals. The electrical conductivity of a metal, as we noted previously, decreases as its temperature is raised.

The conductivity of a semiconductor can also be increased by **doping.** A semiconductor is doped by adding small, carefully controlled amounts of impurities. Two types of semiconductors are produced by doping, as we see next.

n-Type and *p*-Type Semiconductors

Suppose we dope a crystal of silicon with a trace of arsenic. Arsenic atoms have five valence electrons, and silicon has four. To be accommodated into the silicon structure, each As atom must give up one electron. Because they give up electrons, the As atoms in a doped semiconductor are called *donor atoms*. The line labeled "donor level" in Figure 24.12 represents the energy level of the As atoms, which lies quite close to the conduction band. Because the two energy levels are so close to each other, thermal energy alone is enough to cause the "extra" valence electrons of the As atoms to be lost to the conduction band; in the process, the As atoms become immobile positive ions, As^+. Electrical conductivity in this type of semiconductor involves primarily the movement of electrons in the conduction band, with the majority of the electrons coming from the donor atoms. A semiconductor with these characteristics is called an ***n*-type semiconductor;** the *n* stands for the negative charge carried by an electron.

Now consider a silicon semiconductor doped with aluminum. Because it has only three valence electrons, an Al atom can form regular two-electron bonds with only three neighboring Si atoms but only a one-electron bond with the fourth Si atom. The line labeled "acceptor level" in Figure 24.12 shows that the energy level of the aluminum atoms, called *acceptor* atoms, is only slightly above the valence band. Electrons are easily promoted from the valence band into the acceptor level, where they are gained ("accepted") by the Al atoms to form immobile negative ions, Al^-, thereby allowing for a bond to a fourth Si atom. Meanwhile, positive holes have been created in the valence band. Because electrical conductivity in this type of semiconductor involves primarily the migration of positive holes, it is called a ***p*-type semiconductor,** where the *p* stands for the positive charge of a hole.

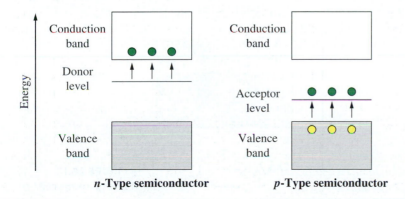

***n*-Type semiconductor**   ***p*-Type semiconductor**

◀ **FIGURE 24.12   Bands in *p*- and *n*-type semiconductors**

In a semiconductor doped with donor atoms, the donor level lies just beneath the conduction band. Electrons (•) are easily promoted into the conduction band, and the semiconductor is of the *n*-type. In a semiconductor doped with acceptor atoms, the acceptor level lies just above the valence band. Electrons (•) are easily promoted to the acceptor level, leaving positive holes (∘) in the valence band. The semiconductor is of the *p*-type.

Zone Refining: Obtaining Pure Semiconductor Materials

The characteristics of a semiconductor are critically dependent on the concentration of dopant atoms present. For example, a *p*-type semiconductor may require only one boron atom for every several million silicon atoms. Furthermore, any impurity atoms in the silicon itself must be kept at levels far below one part per million. A semiconductor manufacturing process may require levels as low as ten impurity atoms per billion. There are not many purification methods that yield such ultrapure materials. Zone refining is one that does.

In **zone refining,** heating coils move along the length of a cylindrical rod of the material to be purified. Sections of the rod are alternately melted and refrozen. Impurities are much more soluble in the molten phase than in the solid, and they concentrate in the liquid zone. Ultimately, almost all the impurity atoms are swept into the liquid zone at the end of the rod, which is cut off and discarded.

In another version of zone refining, a "seed" crystal of silicon is brought into contact with molten silicon. As the seed crystal is slowly withdrawn, surface tension brings some of the molten silicon along with it. Silicon atoms crystallize onto the seed crystal, while the molten material retains the impurities. The result is silicon that is both highly pure and monocrystalline (a single crystal), critical for today's integrated circuits.

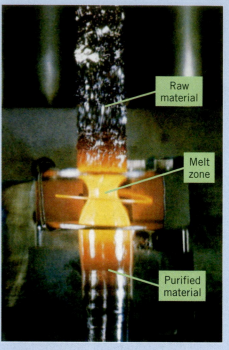

▲ Zone refining.

 Zone Refining animation

A Semiconductor Device: The Photovoltaic Cell

Semiconductor devices have revolutionized the field of electronics. The list of modern technological devices that use semiconductors is almost endless—electronic calculators, computers, radios, television sets, stereo equipment, and cellular telephones, to name but a few.

The semiconductor device pictured in Figure 24.13 converts sunlight (solar energy) to electricity. It is called either a **photovoltaic cell** or a *solar cell*. The cell consists of a thin layer of *p*-type semiconductor, such as Si doped with Al, in contact with an *n*-type semiconductor, such as Si doped with P. When the cell is in the dark, charge does not flow across the *p*–*n* junction. In order for charge to flow, positive holes crossing from the *p* side would have to move away from Al^- ions, and electrons crossing from the *n* side would have to move away from P^+ ions. Separating oppositely charged particles requires energy and is therefore not favored.

▲ Photovoltaic cells provide electricity to a mobile traffic signal.

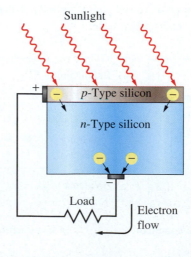

Sunlight

p-Type silicon

n-Type silicon

Load

Electron flow

◄ **FIGURE 24.13**
A photovoltaic cell

Some of the electrons in bonds in the *p*-type semiconductor, however, absorb energy from the sunlight and are promoted to the conduction band. This leaves behind positive holes in the valence band (recall Figure 24.11). Unlike positive holes, conduction electrons can freely cross the *p–n* junction and leave the cell as an electric current. Returning electrons from the electric circuit simply neutralize the positive holes previously formed. Electricity is generated continuously as long as light strikes the cell.

The *p*-type semiconductor in a photovoltaic cell must be very thin—about 1×10^{-4} cm. This thinness is necessary to reduce the tendency for conduction electrons produced by sunlight to be captured by positive holes and immobilized in covalent bonds.

Polymers

As described in Section 9.12, **polymer** molecules, those large molecules made up of repeating sequences of **monomer** units, are the giants of chemistry. The large molecules are often called **macromolecules.** Polymers are made from their monomers much as a brick wall is constructed from individual bricks.

Synthetic polymers are a mainstay of modern life, but nature also makes polymers; they are found in all living matter.

24.8 Natural Polymers

In Chapter 23, we discussed three types of natural polymers: polysaccharides (starch and cellulose), proteins, and nucleic acids. The first attempts to manufacture polymers involved chemical modifications of natural polymers to improve their properties.

Consider cellulose, which has the structure shown in Figure 24.14. It is a polymer with β linkages between glucose monomers (page 960). (The corresponding polymer with α linkages is starch.) The molecular masses of cellulose macromolecules range from about 500,000 to 2,400,000 u. Cellulose can be modified by reactions involving its many hydroxyl (OH) groups.

Recall from page 953 that alcohols and acids react to form esters. Cellulose, with its many alcohol functional groups, readily forms a variety of cellulose esters. With nitric acid, the ester formed is cellulose nitrate:

$$\text{Cell}-\text{OH} + \text{HONO}_2 \longrightarrow \text{Cell}-\text{ONO}_2 + \text{H}_2\text{O}$$

Cellulose   Nitric acid   Cellulose nitrate

Application Note

When dissolved in camphor and ethanol, cellulose nitrate forms *celluloid.* Celluloid was originally developed as a substitute for ivory in billiard balls. However, the balls of celluloid sometimes exploded on impact. Celluloid was also the material used in the film of early movies, but the dangerous flammability of celluloid led to its replacement by safer materials. Celluloid survives today only as Ping-Pong balls.

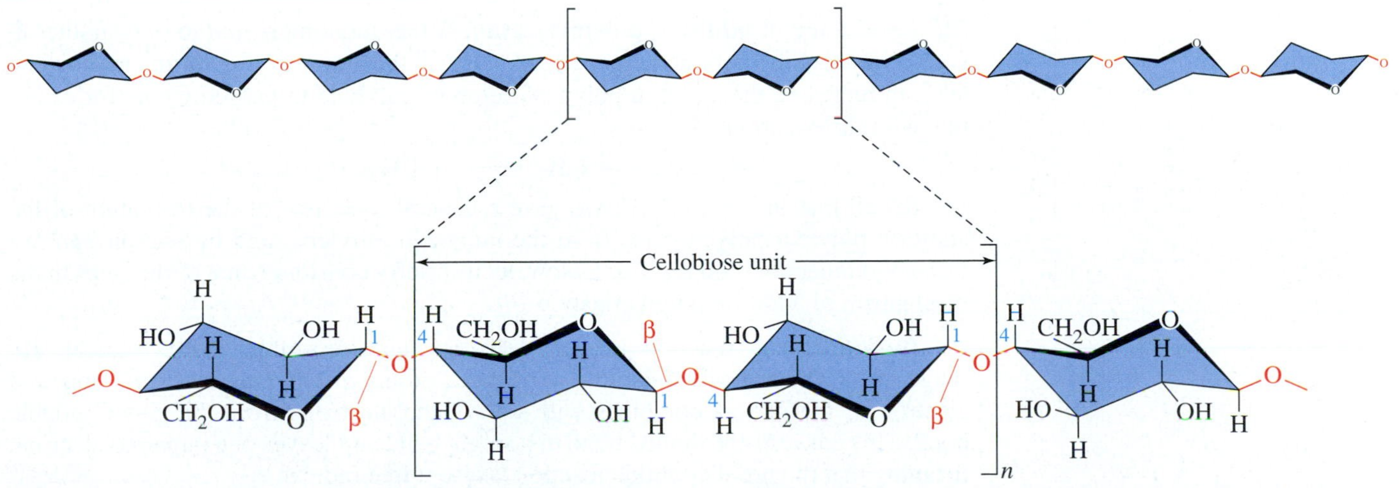

▲ **FIGURE 24.14   The molecular structure of cellulose**
Cellulose is a polymer of the monomer glucose. To highlight the β linkages between monomer units, the disaccharide cellobiose is chosen as the repeating unit in the structure of cellulose shown here.

QUESTION: What natural polymers contain glucose as their monomer?

▲ The wrinkled white garment is made of unmodified cellulose. The blue cellulose acetate garment is sleek and silky.

Isoprene 3D model

In this equation, OH represents one of the many hydroxyl groups in the cellulose molecule. The ONO_2 group is properly called nitrate, but sometimes the first O is ignored, leaving just NO_2 and leading to an alternative name for cellulose nitrate, *nitrocellulose*. When the cellulose nitrate is made from cotton (nearly pure cellulose), it is called *guncotton* and is used to make explosives and smokeless powder for ammunition.

Other modifications of cellulose include *rayon*, in which cellulose is formed into fine cylindrical threads; *cellophane*, in which the cellulose is formed into thin sheets; and the acetate ester *cellulose acetate*, often simply called *acetate*. Cloth made from cellulose acetate feels remarkably like silk.

Natural rubber is a polymer of the simple hydrocarbon monomer *isoprene* (2-methyl-1,3-butadiene):

$$
\begin{array}{ccc}
CH_2 & & CH_2 \\
\| & & \| \\
& C-C & \\
CH_3 & & H
\end{array}
$$

The polymer has the structure

$$
\cdots CH_2 \quad CH_2{-}CH_2 \quad CH_2{-}CH_2 \quad CH_2{-}CH_2 \quad CH_2 \cdots
$$

with the repeating monomer unit shown in red. Notice that the $CH_2{-}CH_2$ units are all cis, that is, on the same side of the $C{=}C$ double bonds. The polymer chains in natural rubber are coiled, twisted, and intertwined with one another. The stretching of rubber involves straightening out the coiled macromolecules.

24.9 Polymerization Processes

In general, whether in a living system or in a laboratory, polymers are made by hooking together many smaller molecules. The result, however, is not quite the same as hooking boxcars together in a train. The polymer is as different from the monomers as long strips of spaghetti are from the particles of flour that make up the spaghetti. For example, polyethylene, the familiar solid, waxy plastic of plastic bags, is a polymer prepared from the gaseous monomer ethylene.

There are two general types of reactions for producing polymers from monomers—*addition polymerization* and *condensation polymerization*.

Addition Polymerization

The key feature of **addition polymerization** is that monomers *add* to one another in such a way that the polymeric product contains all the atoms of the starting monomers. We can represent the addition polymerization of ethylene to polyethylene, for example, with the equation

$$n\, CH_2{=}CH_2 \longrightarrow \text{+}CH_2CH_2\text{+}_n$$

Recall that in Section 9.12 we gave a general overview of the formation of the addition polymer polyethylene from the monomer ethylene, and in Section 23.2 we discussed other addition reactions. Now, let us briefly consider some of the steps in the mechanism of addition polymerization.

Initiation The reaction is started by an *initiator*, often a free radical, **R·** (page 362). The radical bonds to one of the C atoms of ethylene when the unpaired electron of the radical combines with one of the electrons from the $C{=}C$ double bond. This converts the double bond to a single bond and leaves one unpaired electron, meaning that the product of this reaction is also a free radical:

$$
R{\cdot} \;+\;
\begin{array}{c}
H \quad\quad H \\
\diagdown\;\;\diagup \\
C{=}C \\
\diagup\;\;\diagdown \\
H \quad\quad H
\end{array}
\;\longrightarrow\;
\begin{array}{c}
H \;\; H \\
| \;\;\; | \\
R{-}C{-}C{\cdot} \\
| \;\;\; | \\
H \;\; H
\end{array}
$$

Propagation The radical formed in the initiation step joins with another ethylene molecule to form a larger radical:

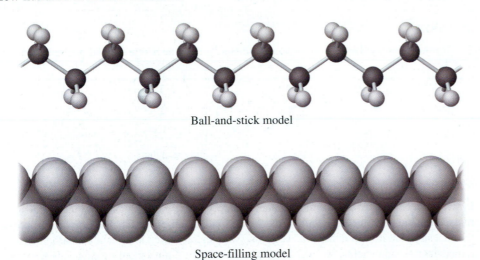

The larger radical forms a still larger one by reacting with another ethylene molecule, and so on, through hundreds of steps.

Termination The propagation finally ends when a molecule is produced that no longer has an unpaired electron. One possible termination step involves the combination of two radicals through an electron-pair bond:

$$R(CH_2)_nCH_2 \cdot \; + \; \cdot CH_2(CH_2)_{\overline{n}}R' \longrightarrow R(CH_2)_nCH_2CH_2(CH_2)_{\overline{n}}R'$$

As a result of the polymerization, a low-molecular-mass gaseous alkene is converted to a high-molecular-mass solid alkane.

Figure 24.15 shows two molecular models of a small portion of a polyethylene molecule. Real polyethylene molecules have huge numbers of carbon atoms—from a few hundred to several thousand.

Ball-and-stick model

Space-filling model

◀ FIGURE 24.15   Molecular models of a segment of a polyethylene molecule

Alkenes commonly serve as the monomers in addition polymerization. By substituting various groups for one or more hydrogen atoms of the simple ethylene molecule, chemists can produce a fantastic array of synthetic polymers. (See again the listing in Table 9.2.)

Condensation Polymerization

In **condensation polymerization,** a small portion of the monomer molecule is *not* incorporated in the final polymer. As an example, we can consider the formation of a type of nylon that has 6-aminohexanoic acid as its monomer. The polymerization involves the reaction of the OH portion of the carboxyl group of one monomer with a H atom of the NH_2 (amine) group of another. The monomers are held together by the newly formed bond—an amide bond—and a molecule of water is also produced:

Synthesis of Nylon 610 movie

$$n\,HO-\overset{O}{\overset{\|}{C}}-CH_2CH_2CH_2CH_2CH_2-\overset{H}{\overset{|}{N}}-H \; + \; n\,HO-\overset{O}{\overset{\|}{C}}-CH_2CH_2CH_2CH_2CH_2-\overset{H}{\overset{|}{N}}-H$$

$$\longrightarrow \left[\overset{O}{\overset{\|}{C}}-CH_2CH_2CH_2CH_2CH_2-\overset{H}{\overset{|}{N}}-\overset{O}{\overset{\|}{C}}-CH_2CH_2CH_2CH_2CH_2-\overset{H}{\overset{|}{N}}\right]_n \; + \; 2n\,H_2O$$

Amide bond

Three key features of condensation polymerization are

- Each monomer molecule contains at least two functional groups.
- The monomers are linked through the functional groups.
- Small molecules are formed as by-products as the monomers are linked.

Several of the commercially important polymers used in fibers, fabrics, and plastics are formed by condensation polymerization, as we will see shortly.

Example 24.2

Write a condensed structural formula for polypropylene, made by the polymerization of propylene (CH_2=$CHCH_3$).

STRATEGY

First, we need to identify the type of polymerization involved. The double bond indicates that the monomer units can join together with no loss of atoms, and we therefore expect polypropylene to be an addition polymer. We also must recognize that linkage occurs through the doubly bonded carbon atoms in the monomers. The methyl groups are attached along the polymer chain as substituent groups.

SOLUTION

The formation of a polymer of n monomer units of propylene is represented by the equation

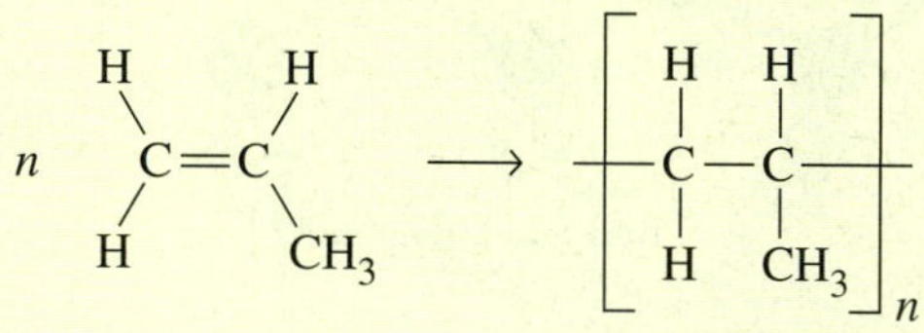

$$n \quad \underset{H}{\overset{H}{}}C=C\underset{CH_3}{\overset{H}{}} \longrightarrow \left[\begin{array}{cc} \overset{H}{\underset{H}{C}} & \overset{H}{\underset{CH_3}{C}} \end{array} \right]_n$$

Polypropylene

ASSESSMENT

In writing a condensed structural formula for a polymer, we must make sure that the usual rules about the bonding capacities of the various atoms are observed. Thus, if we had proposed a structure in which the —CH_3 groups were incorporated into the main carbon chain rather than as side chains, we would have exceeded the limit of only four bonds to each C atom:

$$n\,CH_2=CH-CH_3 \longrightarrow \quad \left[CH_2-CH-CH_3 \right]_n \quad (incorrect)$$

EXERCISE 24.2A

Describe more precisely the source of the error in the condensed structural formula shown in the Assessment section of Example 24.2.

EXERCISE 24.2B

Describe a general structure for the product of the polymerization of propadiene (allene, CH_2=C=CH_2).

Example 24.3

Write the condensed structural formula for the polymer formed from dimethylsilanol, $(CH_3)_2Si(OH)_2$.

STRATEGY

There are no double bonds in dimethylsilanol, and so we do not expect addition polymerization. Rather, we expect a condensation reaction in which a OH group of one molecule and a H atom of another combine to form a molecule of water. Moreover, because there are two OH groups per molecule, each monomer can form bonds with its neighbors on either side.

SOLUTION

We can start by writing the structural formula of the monomer:

$$\text{HO}-\underset{\underset{\text{CH}_3}{|}}{\overset{\overset{\text{CH}_3}{|}}{\text{Si}}}-\text{OH}$$

Then we can represent the polymerization reaction with the equation

$$\text{HO}-\underset{\underset{\text{CH}_3}{|}}{\overset{\overset{\text{CH}_3}{|}}{\text{Si}}}-\text{OH} + \text{HO}-\underset{\underset{\text{CH}_3}{|}}{\overset{\overset{\text{CH}_3}{|}}{\text{Si}}}-\text{OH} + \text{HO}-\underset{\underset{\text{CH}_3}{|}}{\overset{\overset{\text{CH}_3}{|}}{\text{Si}}}-\text{OH} + ... \longrightarrow \left[\underset{\underset{\text{CH}_3}{|}}{\overset{\overset{\text{CH}_3}{|}}{\text{Si}}}-\text{O}\right]_n + n\,\text{H}_2\text{O}$$

ASSESSMENT

Note that, unlike most carbon-based polymers, which contain C–C bonds, there are no Si–Si bonds in this silicon-based polymer, only Si–O–Si linkages.

EXERCISE 24.3A

Write the condensed structural formula for the polymer formed from 3-hydroxypropanoic acid ($\text{HOCH}_2\text{CH}_2\text{COOH}$).

EXERCISE 24.3B

A *homopolymer* is a polymer made from only one type of monomer. Identify four amino acids in Table 23.2 that could conceivably form more than one homopolymer through different polymerization reactions. Explain.

▲ One of the many uses of silicone polymers is in the waterproof caulking compounds used in the installation of window frames and plumbing fixtures.

The polymer in Example 24.3 is a *poly*siloxane but is more commonly known as a *silicone*. The Si–O–Si skeleton of a silicone gives it great high-temperature stability and resistance to oxidation.

24.10 Physical Properties of Polymers

Polymers are classified not only according to their method of synthesis but also according to their physical and mechanical properties. For example, synthetic polymers are often known to the general public as plastics. However, *plastic* has a more restricted meaning to chemists. A **thermoplastic polymer** is really plastic; it can be softened by heating and then formed into desired shapes by applying pressure. In contrast, **thermosetting polymers** become permanently hard at elevated temperatures and pressures. After setting, they cannot be softened and remolded—in other words, they are not at all *plastic*.

We can relate other physical properties of polymers to specific features of their macromolecular structure.

Density

We can illustrate how variations in structure affect polymer properties, by examining two basic kinds of polyethylene, both thermoplastic. Of the two, *high-density polyethylene (HDPE)* has a higher density, greater rigidity, greater strength, and a higher melting point. It is used for such things as threaded bottle caps, radio and television cabinets, toys, and large-diameter pipes. In contrast, *low-density polyethylene (LDPE)* is a waxy, semirigid, translucent material with a low melting point. LDPE is used in insulation for electric wiring, plastic bags, refrigerator dishes, squeeze bottles, and many other common household articles. An essential difference between the two types of polyethylene is shown in Figure 24.16.

What aspects of their structure can account for the differences in properties of HDPE and LDPE? High-density polyethylene consists primarily of linear molecules, that is, of long unbranched chains. The chains can run alongside one another in close contact over relatively great distances. This permits strong intermolecular forces of attraction between adjacent chains; the forces are *dispersion forces* (Section 11.5). The overall effect is to produce an ordered (crystalline) structure that imparts rigidity, strength, and a higher melting temperature to the polymeric material.

▲ **FIGURE 24.16 The effect of heat on polyethylene bottles**

These bottles were heated in the same oven for the same length of time. The one that melted (LDPE) is made of branched-chain polyethylene molecules; the other (HDPE) is composed of straight-chain molecules.

▶ **FIGURE 24.17   Organization of polymer molecules**

(a) Unbranched polymer chains in high-density polyethylene (HDPE).
(b) Branched polymer chains in low-density polyethylene (LDPE).

QUESTION: Which material would experience the greater relative change in mechanical properties if the polymer chains were to be cross-linked? Explain.

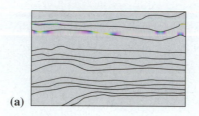

(a)

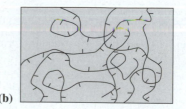

(b)

Low-density polyethylenes have branched chains that prevent the macromolecules from assuming a crystalline structure.* The branches get in the way when two chains try to come into close contact (much as logs with short protruding branch stems are hard to arrange in a neat pile). This decreases the dispersion forces, weakens the attractions between chains, and produces a lower-melting, more flexible material than HDPE.

The two polymer chain arrangements are suggested in Figure 24.17.

Hardness and Rigidity

What if we want a really tough material, something quite rigid and strong, that will not melt or soften up on heating? The answer could well be Bakelite®, the oldest synthetic thermosetting polymer. A condensation polymer prepared by combining the two monomers phenol and formaldehyde, Bakelite is more generally referred to as a phenol–formaldehyde resin. The polymerization involves substituted phenols that are formed as intermediates in the reaction of phenol and formaldehyde:

Phenol   Formaldehyde

The substituted phenols then interact by splitting out water molecules. Water is driven off by heat liberated while the polymer sets. The final product is a huge, complex, three-dimensional network (Figure 24.18). The extensive cross-linking of polymer chains results in rigidity. The final polymer has great strength without having great mass, a useful combination of properties. About half of the United States production of phenol–formaldehyde resins is used as a binder in plywood. Other uses include varnishes for electric coils, molding compounds for the manufacture of electric plugs and switches, and automobile parts such as steering wheels.

Flexibility and Elasticity

Hardness and rigidity are not the only properties we seek in plastics. Frequently we want flexibility and, more particularly, elasticity. *Flexibility* is the ability of a material to yield to forces—to bend or give—without breaking. *Elasticity* is the ability of a material to regain its former shape after a distorting force is removed. **Elastomers** are flexible, elastic materials. The natural polymer *rubber*, described on page 1012, is the prototype for this kind of material.

Natural rubber is soft and tacky when hot. It can be made harder in a reaction with sulfur, called *vulcanization*. In this process, sulfur atoms form cross-links between hydrocarbon chains, as shown in Figure 24.19. As with phenol–formaldehyde resins, this cross-linked three-dimensional structure makes vulcanized rubber a harder, stronger substance than natural rubber. Thus, vulcanized rubber makes excellent automobile tires, whereas natural rubber is totally unsuited for this purpose.

Table 24.2 lists products made of various types of rubber and the degree of cross-linking in each. The smaller the average number of intervening monomer units on each

*One form of branching occurs when the free-radical end of a growing chain coils and attacks a carbon atom several atoms back on the chain. A short branched chain is formed, and new growth continues from the point of attack.

◀ FIGURE 24.18 Bakelite, a phenol–formaldehyde resin

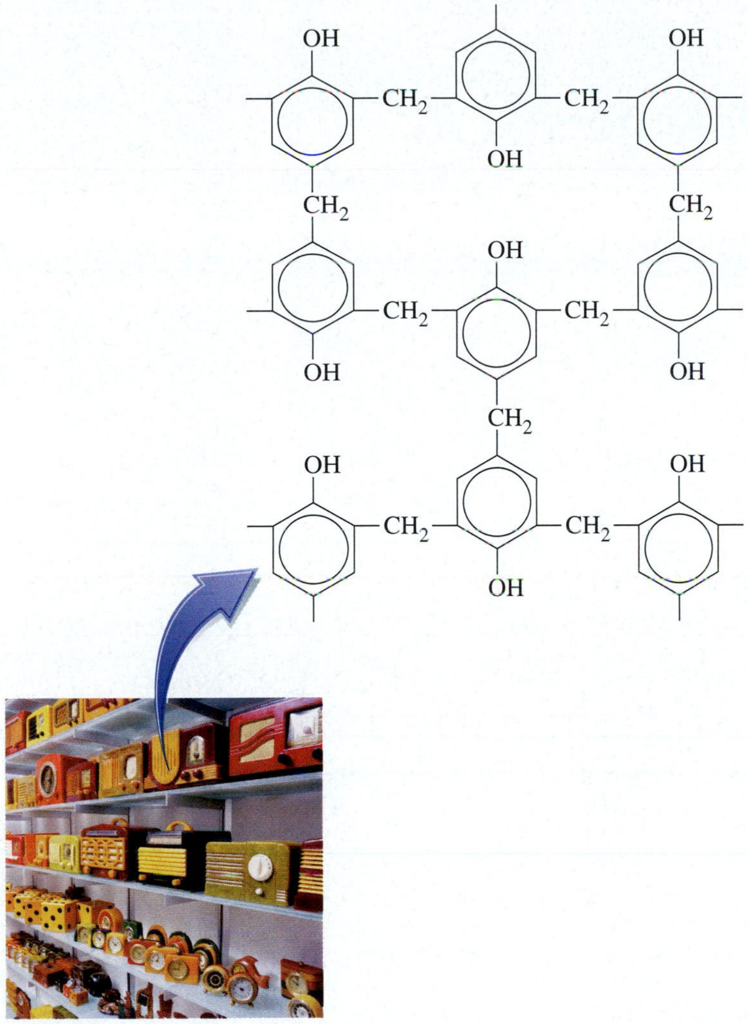

This polymer was used in the 1920s and 1930s to make a variety of products, such as radios, desk clocks, lighters, telephones, and cameras. The highly cross-linked structure gives great rigidity and resistance to heat.

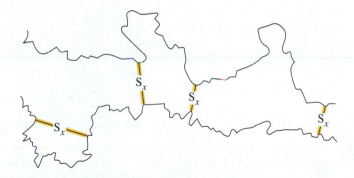

◀ FIGURE 24.19 Vulcanized rubber

The subscript x indicates a variable number of S atoms in the cross-links, usually one to four.

Table 24.2 Extent of Cross-linking in Rubber Products

Product	Monomer Units Between Cross-links	Degree of Cross-linking
Surgical gloves	100–150	Lowest
Kitchen gloves	50–80	↓
Artificial heart membrane	30–40	
Bicycle inner tube	20–30	
Bicycle tire	10–20	Highest

Conducting Polymers

In the polymerization of ethylene, an alkene is converted to an alkane. What kind of product is formed when an *alkyne* is polymerized? As we did for ethylene on page 1012, let us consider a few steps in the polymerization of acetylene:

$$R\cdot \; + \; H-C\equiv C-H \longrightarrow R-\overset{\displaystyle H}{\underset{\displaystyle H}{C}}=C\cdot$$

$$R-\overset{\displaystyle H}{\underset{\displaystyle H}{C}}=C\cdot \; + \; H-C\equiv C-H \longrightarrow R-\overset{\displaystyle H}{\underset{\displaystyle H}{C}}=C-\overset{\displaystyle H}{\underset{\displaystyle H}{C}}=C\cdot$$

$$R-\overset{\displaystyle H}{\underset{\displaystyle H}{C}}=C-\overset{\displaystyle H}{\underset{\displaystyle H}{C}}=C\cdot \; + \; H-C\equiv C-H \longrightarrow R-\overset{\displaystyle H}{\underset{\displaystyle H}{C}}=C-\overset{\displaystyle H}{\underset{\displaystyle H}{C}}=C-\overset{\displaystyle H}{\underset{\displaystyle H}{C}}=C\cdot \;\; \ldots \text{ and so on}$$

We can represent a section of a polyacetylene molecule with the condensed structural formula

$$\cdots-CH=CH-CH=CH-CH=CH-CH=CH-\cdots$$

This formula and the ball-and-stick model presented here both clearly show that the bonding in the backbone of the molecule is a conjugated system (Section 23.13) with alternating single and double carbon-to-carbon bonds. The π electrons in a conjugated system are delocalized, which means that, in effect, an electron can move the entire length of a polyacetylene molecule, making the molecule an electrical conductor.

But how do electrons get from one molecule to another? This is accomplished by doping the polyacetylene with an additive such as iodine. Explanations of conducting polymers are similar to the descriptions given for bonding in metals and semiconductors in Section 24.7. Two important differences, however, are that in conducting polymers (1) electron bands are formed from molecular orbitals of polymer molecules rather than the atomic orbitals of the atoms in the polymer chain and (2) the dopants reside between polymer chains rather than substituting for atoms in the chains.

Conducting polymers were discovered accidentally in the 1970s. They are now beginning to appear in such commercial products as polymer batteries.

▲ Batteries made of foil-like sheets of conducting polymers are light and flexible. They save considerable weight compared to lead-acid batteries and can be crammed into almost any available space.

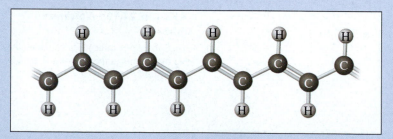

▲ A ball-and-stick model of polyacetylene.

 Polyacetylene 3D model

chain between successive cross-links, the greater the degree of cross-linking. Notice that the harder items in Table 24.2 (those at the bottom of the list) have the most extensive cross-linking.

Just the right degree of cross-linking can improve the elasticity of vulcanized rubber so that it is superior to that of natural rubber, as suggested in Figure 24.20. The individual chains are still relatively free to uncoil and stretch, but when a stretched

● = Carbon

● = Sulfur

▲ **FIGURE 24.20   Elasticity in natural and vulcanized rubber**
(a) In natural (unvulcanized) rubber, the chains slip past one another when the rubber is stretched and tend to remain in their new positions once the stretching force is removed. (b) Vulcanization involves the addition of sulfur cross-links between the chains of natural rubber. (c) When the vulcanized rubber is stretched, the sulfur cross-links allow some deformation, but when the stretching force is removed, the cross-links cause the material to return to its original shape—the property of elasticity.

piece of the rubber is released, the cross-links serve to pull the chains back to their original prestretched arrangement. Rubber bands owe their snap to this sort of molecular structure.

24.11   Modern Synthetic Polymers

Up to this point, we have discussed mostly naturally occurring polymers, such as cellulose and natural rubber, although we have also described some manufactured polymers with histories extending back many decades, such as phenol–formaldehyde resins and silicones. In this section, we will focus specifically on more modern synthetic polymers.

Synthetic Rubber

Several kinds of synthetic rubber were developed during and after World War II. Some of them bear a striking molecular resemblance to nature's own elastomer. For example, polychloroprene (neoprene) is made from a monomer, chloroprene (2-chloro-1,3-butadiene), that is similar to isoprene but has a chlorine atom in place of a methyl group:

Chemists have even learned to make polyisoprene, which has a structure identical to that of natural rubber. The only significant difference between the two is that the "synthetic natural rubber" is not harvested from rubber-tree plantations.

$$\underset{\underset{Cl}{|}}{C}\!=\!\underset{\underset{H}{|}}{C}$$

(with CH$_2$ groups above)

Another kind of synthetic rubber illustrates the principle of *copolymerization*. In this process, a mixture of two different monomers forms a product in which the chain contains both monomers as building blocks. The product is called a *copolymer*. Styrene–butadiene rubber (SBR) is a copolymer of styrene (25%) and butadiene (75%). A segment of an SBR macromolecule might look something like this:

Butadiene unit   Butadiene unit   Styrene unit   Butadiene unit

This synthetic rubber is more resistant to oxidation and abrasion than natural rubber, but it has less satisfactory mechanical properties.

Fibers and Fabrics

A *fiber* is a natural or synthetic material obtained in long, threadlike structures that can be woven into fabric. One factor affecting the quality of a fiber is its *tensile strength*, a measure of the extent to which a material can be subjected to a stretching force without breaking.

Cotton, wool, and silk are natural fibers of great tensile strength that have long been spun and woven into cloth. Now, synthetic fibers have been developed that match the natural fibers in tensile strength and outdo them in their resistance to stretching and shrinking and also to attack by moths. Synthetic polymers have revolutionized the clothing industry.

Polyacrylonitrile (Acrilan®, Creslan®, and the like) is an addition polymer:

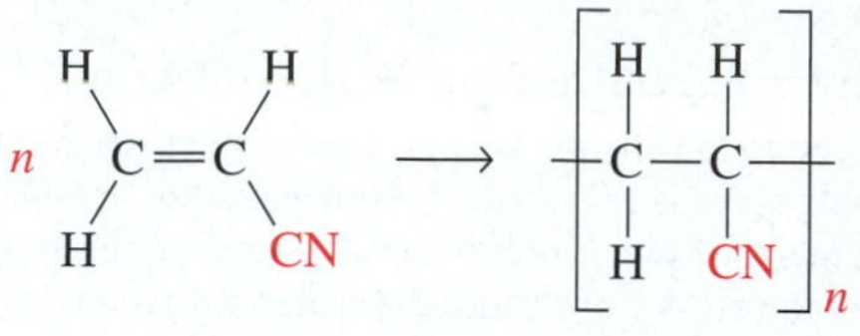

Polyacrylonitrile

Polyesters are condensation polymers. Dacron® polyester, for instance, is made from the condensation of ethylene glycol with terephthalic acid:

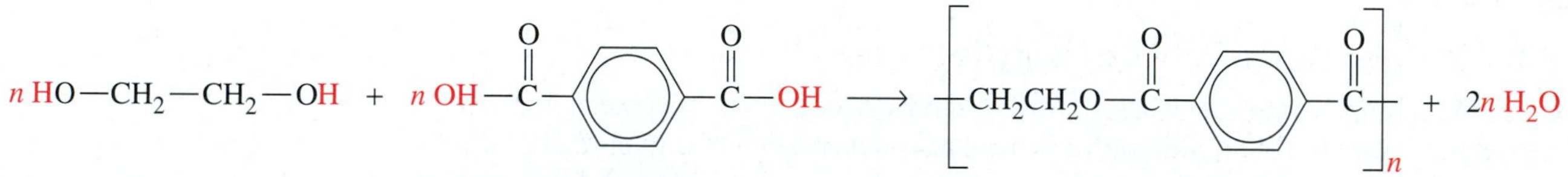

Ethylene glycol   Terephthalic acid   Dacron®

This polymer can be extruded as a film. Coated with a magnetic material, the film is the tape used in tape recorders and videotape machines. Perhaps the most familiar use of this polyester is in plastic bottles for soft drinks.

Polyamides are also condensation polymers. The protein fibers in silk and wool are polyamides. The most common synthetic polyamide is nylon-66. It is made by the condensation of 1,6-hexanediamine and adipic acid:

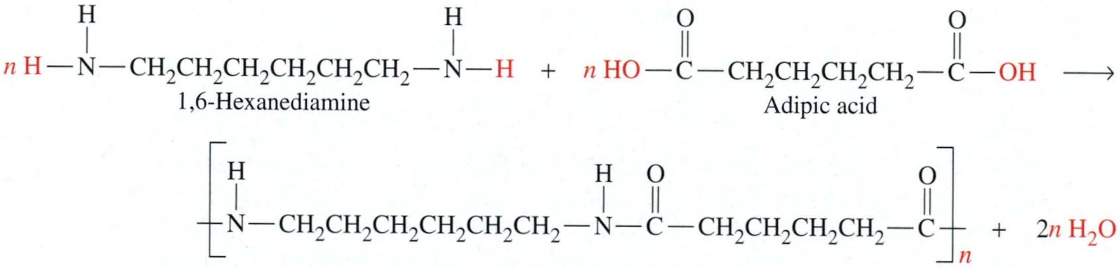

1,6-Hexanediamine   Adipic acid

Nylon-66

Silk, wool, Dacron, and nylon-66, like high-density polyethylene, owe their strength to the ordered, relatively rigid arrangement of their long molecules. However, unlike nonpolar polyethylene, the polyesters and polyamides have polar functional groups. Interactions among these groups give the polymers their unique tensile strengths.

Chemists can now design materials that have seemingly impossible combinations of properties in a single substance. For example, spandex fibers, which are used for stretch fabrics (Lycra®) in ski pants, exercise wear, and swimsuits, combine the elasticity of rubber and the tensile strength of a fiber. How can the coiled, flaccid molecular chains required for elasticity and the fairly rigid, highly ordered chains needed for tensile strength be combined into a single polymer? It can be done by grafting two molecular structures into a polymer chain. Thus, a single macromolecule can have blocks of elastomer components and blocks of fiber components. The fabric made from these macromolecules exhibits flexibility *and* rigidity.

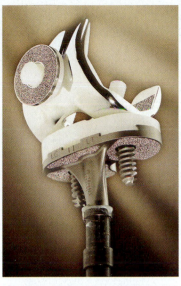

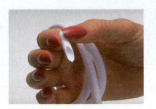

▲ A synthetic artery made of Dacron.

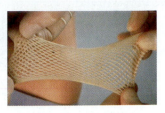

▲ Synthetic skin made from a polymeric material.

▲ Velcro®, the familiar material used in fasteners for shoes, clothes, watchbands, and the like, has hooks on one surface and loops on the other. The hooks and loops are made of the polyamide nylon. This image was made by a scanning electron microscope.

▲ Synthetic polymers are an important part of this hip-joint replacement.

◄ A plastic artificial eye lens made from a synthetic polymeric material.

Biomedical Polymers

One of the most interesting uses of polymers has been in replacements for diseased, worn-out, or missing parts of the human body. About 130,000 artificial ball-and-socket hip joints made of stainless steel (the ball) and plastic (the socket) are installed each year. People crippled by hip injuries or arthritis in the hip are freed from pain and given much more freedom of movement. Patients with heart and circulatory problems can enter a hospital for a "valve job" or for the replacement of other worn-out or damaged parts. Pyrolytic carbon heart valves derived from a polymer are widely used. Knitted Dacron tubes can replace arteries blocked or damaged by atherosclerosis.

Blood begins to clot when it comes in contact with most foreign substances. To prevent clotting, synthetic polymers used inside the body generally must be chemically treated and coated with an anticoagulant before implantation. Another approach that circumvents the problem is to use substances that occur naturally in the body to construct biomedical polymers. For example, polymers of glycolic acid and lactic acid have been used in synthetic films for covering burn wounds.

Synthetic polymers also can trigger a rejection mechanism in the body. To prevent infection and excessive loss of fluids, burns have to be covered while they are healing. When the covering is either human skin from donors or specially treated pigskin, the cover must be changed frequently because the body tends to reject foreign tissues. In contrast, when the cover material is a film made from polymers of glycolic acid and lactic acid, the film is absorbed and metabolized rather than rejected.

The development of biomedical polymers has barely begun. In the future lies the prospect of our being able to replace entire organs of the body. Already, artificial hearts made of synthetic elastomers are used to keep people alive while awaiting a heart transplant.

$HOCH_2COOH$

Glycolic acid

$CH_3CHCOOH$
  $\vert$
  OH

Lactic acid

Space-Age Materials

As we have seen, the physical properties of polymers can be designed through control of their composition and molecular structure. In a similar way, other new materials are being designed and developed to meet the needs of advancing technologies. Examples include new alloys, composite materials, and materials structured on the nanometer scale (nanomaterials). In each of these cases, we will see that the introduction of a specific chemical composition or structure produces materials with unique and novel properties.

Table 24.3 High-Performance Alloys

Name	Approximate Composition	Use	Benefit
Nickel–metal hydride	Ni 87%, La 13%	Batteries	Weight savings, hydrogen storage
Beryllium copper	Cu 98.1%, Be 1.9%, Co trace	Springs, contacts	Elasticity, fracture strength
Waspaloy	Ni 54%, Cr 20%, Co 13%, plus minor elements	Airframe assemblies, missile systems	Temperature resistance
Shape-memory alloy	Ni 50%, Ti 50%, Ni 3–5%, Cu 80–86%, Al 11–14%	Valves, switches, stents, catheters, actuators	Reversible shape change, resistance to plastic deformation
Ti6Al4V	Ti 90%, Al 6%, V 4%	Structural aerospace components Surgical implants	Weight savings/strength Biocompatibility
Platinum–rhodium	Pt 80–90%, Rh 10–20%	Automotive catalytic converters	Redox chemistry
SOFTMAG alloy	Ni 30–78%, Fe 22–70%	Magnetoresistive sensors	Magnetic permeability

24.12   High-Performance Alloys

As discussed in Section 24.1, alloys are mixtures of two or more metals, sometimes containing nonmetals, designed to achieve a physical or chemical property different from that of the parent metal. For example, we discussed in Chapter 20 the use of lithium in formulating lightweight alloy battery electrodes and the use of small percentages of beryllium to enhance the elasticity and fatigue strength of copper. Table 24.3 lists these and other alloys that are being used today in advanced technologies where enhanced performance is required.

Waspaloy™ is a complex alloy containing nickel, chromium, cobalt, aluminum, iron, and a number of other metals and nonmetals. It is heat-resistant, retaining its shape and strength at temperatures well above 600 °C. This property makes Waspaloy useful for high-speed rotating shafts, which generate a great deal of friction heat. Metals that are less resistant to high temperatures might distort and fail quickly in such applications.

Shape-memory alloys are materials that demonstrate the ability to transform to an original shape after being heated or cooled. These alloys can also withstand a large amount of strain without being permanently deformed. Although several alloys have shape memory, alloys composed of nickel and titanium or copper-based alloys, such as CuZnAl and CuAlNi, have been the most extensively used in applications where shape memory is important. The shape-memory effect stems from a temperature-dependent transformation between two different solid phases of the material and can generate a significant force. Alloying of the different metals produces specific structures that will undergo these phase transformations. The unique properties of shape-memory alloys underlie their use as thermomechanical switches and valves in commercial products ranging from coffeepots to automobiles. Alloys of this type have also found biomedical uses in the form of stents, catheters, and orthodontic arch wires.

Titanium alloys are found throughout the aerospace industry and the biomedical community. Structural applications involving titanium alloys take advantage of the strength-to-weight ratio realized when metals such as aluminum and vanadium are alloyed with titanium. Titanium alloys used for dental and prosthetic implants offer the benefit of increased strength-to-weight ratio in addition to biocompatibility. The ability of bone to bond to the alloy surface makes the repaired member stronger, and the biocompatibility also means that the body will not reject the implant as a foreign object. Applications in the marine and power-generation industries have been developed using titanium alloys because of the corrosion resistance of alloys such as Ti6Al4V.

Platinum–rhodium alloys play a crucial role in the transportation industry as the active component of catalytic converters used to reduce the emission of such exhaust pollutants as carbon monoxide, uncombusted hydrocarbons, and nitrogen oxides. The

▲ This ball-and-socket hip joint replacement is constructed entirely of titanium. The ridges in the upper socket allow it to bond strongly to the pelvis to support heavy loads. Titanium is exceptionally compatible with bone and teeth.

complete conversion of these reactants to benign products requires that the catalyst be active as both an oxidation and a reduction catalyst, and the alloying of platinum and rhodium provides the dual nature of the catalyst:

$$2\ CO(g) + O_2(g) \xrightarrow{\text{Pt}} 2\ CO_2(g) \qquad \textit{(oxidation)}$$

$$C_7H_{16}(g) + 11\ O_2(g) \xrightarrow{\text{Pt}} 7\ CO_2(g) + 8\ H_2O(g) \quad \textit{(oxidation)}$$

$$2\ NO(g) + 2\ CO\ (g) \xrightarrow{\text{Rh}} N_2(g) + 2\ CO_2(g) \qquad \textit{(reduction)}$$

Ongoing concern over the emission of environmental pollutants ensures the continued relevance of this technology. However, a significant negative factor is the world's limited supply of rhodium, which increases the cost of producing catalytic converters. This highlights the need to identify alternative materials, likely also to be alloys, that can eventually replace the current alloy of precious metals.

Nickel–iron alloys represent a mixing of two metals that produces a third material that has a property not found in either starting material. This property is enhanced magnetoresistance, which is a measure of how much a material's electrical resistance changes in the presence of a magnetic field. This property has played a central role in the development of the read–write heads found in computer disk drives. The discovery of new alloys with even greater magnetoresistances has stimulated the active and highly competitive research associated with this technology.

24.13   Composites

Composites are made of two or more physically distinct materials that, when combined, exploit the desired structural and mechanical properties of the individual components. Composites can be formed from a wide variety of materials, including metals, ceramics, and polymers. In composite structures, one material often serves as the host, or *matrix*, for a second material that is distributed throughout it. Together, the two materials form a heterogeneous mixture.

One type of composite material widely used today is fiber-reinforced polymer (FRP). The most common FRP consists of either fiberglass cloth or a mat of glass fibers impregnated with either epoxy resin or polyester resin. The glass fibers are very strong but brittle. The polymer resin holds the fibers rigidly in place and keeps them from bending excessively, so that the strength of the fibers can be exploited but their brittleness does not become a problem. The FRP composite, because it is lighter, stronger, and tougher than metal alloys, is now used in such products as automobile bodies, fishing rods, bicycle frames, and bathtubs.

In place of fiberglass, FRP composites may use either graphite fiber (Section 11.8) or polymer fibers such as Kevlar®. The name Kevlar refers to a particular polymer, one composed of crystalline arrays of carbon chains containing aromatic and amide functional groups, and not to the assembled composite. FRP composites containing either graphite fiber or Kevlar in lieu of fiberglass are known for their extreme strength and rigidity. For high-temperature applications, a phenol–formaldehyde resin may replace the epoxy resin. FRP composites in general can be found throughout the aerospace, automotive, and extreme sports industries.

Composite materials can also be composed of mixtures of metals and ceramics. For example, composite materials such as the one shown in Figure 24.21 are being tested for power transmission lines. These novel materials offer the potential for

- increased strength, allowing greater distances to be spanned
- increased rigidity, providing for reduced line sag
- greater conductivity for reduced transmission losses

These future power transmission lines will involve bundled strands of an aluminum–zirconium alloy surrounding a core of ceramic-reinforced composite wires. Within the composite wires, each of which is 2–3 mm in diameter, metallic aluminum acts as the matrix encompassing as many as 25,000 microscopic alumina (Al_2O_3) fibers per strand. The parallel orientation of the fibers along the length of the wire significantly contributes to the strength of the wire, while the metal matrix enhances the conductivity. As with

▲ The fiberglass/epoxy resin of which this bicycle is made is a composite material that takes advantage of the desirable properties of each component. The composite is strong, lightweight, and tough.

▲ On July 7, 2003, Felix Baumgartner "flew" over the English Channel assisted only by an aeronautically designed six-foot-wide carbon-composite wing. Baumgartner was dropped from a plane at 30,000 feet and flew safely over the 21-mile distance in 14 minutes.

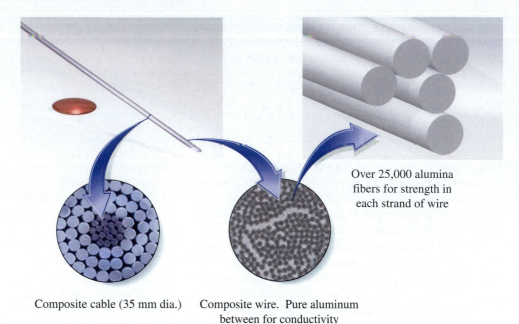

Over 25,000 alumina fibers for strength in each strand of wire

▶ FIGURE 24.21 The 3M™ Composite Conductor:

A metal-ceramic composite. Courtesy 3M.

Composite cable (35 mm dia.)   Composite wire. Pure aluminum between for conductivity

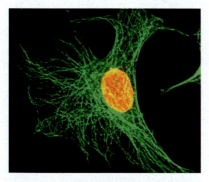

▲ *Quantum dots*—nanometer-sized particles of semiconductor—fluoresce in different colors depending on dot size. Here, red-emitting quantum dots highlight proteins on the surfaces of breast cancer cells, while a conventional blue dye stains the nuclei.

other composites, the individual components combine synergistically to provide a material with unique structural and mechanical properties.

24.14 Nanomaterials

A **nanomaterial** is any material that develops unique physical or chemical properties when the sample size is reduced to a nanometer scale. As we will see, this reduction in scale can occur in a number of forms and can produce an array of results.

Although nanomaterials have been the basis for often overhyped nanotechnologies in recent years, there has been significant debate in the scientific community regarding just what constitutes a nanomaterial. After all, much of our discussion in previous chapters has illustrated that most chemistry, from synthetic reactions to intermolecular interactions, occurs on a nanometer scale. This does not mean that these molecular compounds should be classified as nanomaterials, however. In order for the term *nanomaterial* to apply, there must be a *change* in some property or properties as the scale is reduced to the nanometer level.

One example of a nanomaterial follows closely our description of composites. *Nanocomposites* are composite materials in which one component is present in the form of nanometer-sized fibers or grains. For example, researchers have been able to produce metal oxide coatings that incorporate nanometer-sized grains of one metal oxide in a matrix of a second metal oxide. With the right combination of oxides, the inclusion of nanometer-sized grains has resulted in significantly increased hardness, an important property of protective coatings, by altering the mechanisms of mechanical deformation within the material.

Nanomaterials have also provided the pathway to developing forms of matter having unique electrical properties. For example, in carbon nanotubes composed of graphene sheets rolled into tubes (page 464), the electrical conductivity depends on the way the sheets have been rolled. The electrical conductivity of the nanomaterial differs from that of macroscopic sheets of graphite because of the confinement of electrons within the nanometer-sized structure. Furthermore, carbon nanotubes have been found to exhibit significant changes in electrical conductivity when the tubes are mechanically deformed. Such phenomena may represent the next generation of nanoelectronics, where electronic performance will be determined by properties such as bond angles and molecular orientation.

In a related phenomenon, nanometer-sized particles of semiconducting materials have exhibited optical properties different from expectations based on the band gap of the bulk material. This phenomenon relates to the perturbation of electronic states within a very small sample of a material. These semiconducting nanomaterials offer researchers the ability to tune the color of emitted light by controlling particle size.

Potential applications include spectroscopic methods that are able to probe matter with a spatial resolution of nanometers and chemical sensing in biochemical studies using these optically active nanomaterials as tracers.

Finally, nanometer-sized particles can exhibit unique chemical properties. The rate for reactions involving extremely fine particles of aluminum, just a few nanometers in diameter, is greater than the predicted rate based solely on increased surface area. This "nano" form of aluminum is being investigated for use in high-performance rocket propellants.

Gold is another example of a substance exhibiting a change in chemical properties at the nanometer scale. Recent studies have demonstrated that gold particles between 1 and 3 nanometers in diameter, supported on an oxide substrate, have enhanced catalytic activity in reactions involving the oxidation of carbon monoxide. Although this enhancement has been demonstrated in principle, much work remains before chemists can fabricate uniform distributions of nanometer-sized gold particles and control their size and distribution under industrial conditions. Again, the unique electronic structure of the nanomaterial is thought to produce the observed enhancement in chemical reactivity. Thus, nanomaterials are offering scientists the opportunity to control catalytic reactivity through particle-size selection in addition to the traditional approaches of tailoring the catalyst composition.

Cumulative Example

A *polyurethane* is a polymer involving the reaction of hydroxyl ($-OH$) groups and isocyanate ($-N=C=O$) groups:

$$-OH + -N=C=O \longrightarrow -NH(C=O)O-$$

Suppose a polyurethane is to be prepared using pure 1,6-diisocyanatohexane (hexane diisocyanate, HDI) for the isocyanate functional groups and a 1:2 molar mixture of glycerol (1,2,3-trihydroxypropane) and 1,4-dihydroxybutane for the hydroxyl functional groups. **(a)** How many grams of the hydroxyl mixture must be used for every 100.0 grams of HDI? **(b)** Based on Table 24.2, what general physical properties might be expected for the product?

STRATEGY

One way to approach part (a) is to begin with a fixed mass of the hydroxyl mixture that corresponds to 1 mol of glycerol and 2 mol of 1,4-dihydroxybutane. We can then calculate the mass of HDI that reacts with each part of the mixture. Finally, we can set up a simple conversion factor to determine the mass of hydroxyl mixture that reacts with 100.0 g of HDI. For part (b), we need to determine the general structure of the polymer and deduce the number of carbon atoms between cross-links.

SOLUTION

(a) First, we must determine the structures and formulas of the starting materials. We dealt with the structures of alcohols in Chapter 2, and so can write these structures for glycerol and 1,4-dihydroxybutane.

$$HOCH_2CH(OH)CH_2OH$$
Glycerol
92.10 g/mol

$$HO(CH_2)_4OH$$
1,4-Dihydroxybutane
90.12 g/mol

Based on the structure of the isocyanate group and the compound name, we deduce that HDI is a six-carbon straight chain with an isocyanate group on each end.

$$OCN(CH_2)_6NCO$$
1,6-Diisocyanatohexane
168.2 g/mol

The 1,4-dihydroxybutane has two hydroxyl groups, and HDI has two isocyanate groups, which means the mole ratio between these reactants is 1:1. We can perform an ordinary stoichiometric calculation to find the mass of HDI that reacts with 2 mol 1,4-dihydroxybutane.

$$? \text{ g HDI} = 2 \text{ mol HO(CH}_2)_4\text{OH} \times \frac{1 \text{ mol HDI}}{1 \text{ mol HO(CH}_2)_4\text{OH}} \times \frac{168.2 \text{ g HDI}}{1 \text{ mol HDI}}$$

$$= 336.4 \text{ g HDI}$$

Glycerol has three hydroxyl groups, and we perform a similar calculation to find the mass of HDI that reacts with 1 mol glycerol. In this reaction, however, the HDI:glycerol mole ratio is 1.5:1 or 3:2.

$$? \text{ g HDI} = 1 \text{ mol glycerol} \times \frac{3 \text{ mol HDI}}{2 \text{ mol glycerol}} \times \frac{168.2 \text{ g HDI}}{1 \text{ mol HDI}}$$

$$= 252.3 \text{ g HDI}$$

Now, we can calculate the total mass of the 1:2 molar mixture of hydroxyl compounds and the total mass of HDI that reacted with each component of the mixture.

$$92.10 \text{ g glycerol} + [2 \times 90.12 \text{ g HO(CH}_2)_4\text{OH}] = 272.3 \text{ g hydroxyl mixture}$$

$$336.4 \text{ g HDI} + 252.3 \text{ g HDI} = 588.7 \text{ g HDI}$$

Finally, a simple conversion factor, based on the ratio of total masses involved when using molar quantities, gives the mass of mixture needed to react with 100 g of HDI.

$$? \text{ g hydroxyl mixture} = 100.0 \text{ g HDI} \times \frac{272.3 \text{ g hydroxyl mixture}}{588.7 \text{ g HDI}}$$

$$= 46.25 \text{ g hydroxyl mixture}$$

(b) We know that cross-links occur wherever glycerol occurs in the chain. With our mixture of hydroxyl compounds, an average of two molecules of 1,4-dihydroxybutane and two molecules of HDI will bond between glycerol molecules. This permits us to draw a partial structure. The points at which cross-links may occur are in red. On average, there are 34 atoms between cross-links.

ASSESSMENT

Our product has 34 atoms between cross-links, equivalent to between eight and nine four-carbon monomer chains. Based on Table 24.2, which lists materials in order of increasing hardness, it appears that the degree of cross-linking of our polymer places it at the bottom of the table, probably harder than bicycle tires and likely a very rigid, possibly brittle, product.

Our answer of 46.25 g of mixture per 100.0 g HDI appears sound. HDI is a heavier molecule than either of the two hydroxyl compounds. Furthermore the glycerol has three functional groups instead of two, so that the molar ratio of glycerol to HDI is less than 1, reducing the mass ratio even more.

Concept Review with Key Terms

24.1 Metallurgy: From Natural Sources to Pure Metals—An **ore** is any material containing enough of some mineral to make extraction of the metal economically feasible. Extractive metallurgy involves mining ores, concentrating them through **flotation**, and then roasting; reduction of a metal compound to the free metal, a process that often yields **slag** in addition to the pure metal; and **refining** of the metal. The principal metallurgical approaches include **pyrometallurgy**, based on high-temperature reactions involving solids, and **hydrometallurgy**, which uses lower-temperature reactions involving solutions. An **alloy** is a mixture of two or more metals or of a metal and a nonmetal.

24.2 Iron and Steel—Roasting is not useful in the metallurgy of iron because the principal ores are stable oxides. Coke (carbon) is used to reduce the iron oxides, and the slag is mainly calcium silicate. Impure iron, called **pig iron** in the molten form and **cast iron** in the solid form, is in most cases converted to iron–carbon alloys called **steel**.

24.3 Tin and Lead—These materials are straightforward to mine and extract and have been used since ancient times.

24.4 Copper, Zinc, Silver, and Gold—These metals must be extracted from ores through sometimes complex processes. The cyanidation process for the extraction of gold and silver from their ores is an example of a hydrometallurgical process. Because these metals are valuable and mining them is expensive and disruptive to the environment, a significant effort is made to recycle them.

24.5 The Free-Electron Model of Metallic Bonding—In the **free-electron model** of metals, valence electrons of the atoms in a crystal are combined into an "electron gas" surrounding a network of positive ions. The model accounts for such properties as malleability and ductility, as well as electrical and thermal conductivity.

24.6 Band Theory—Band theory describes bonding both in metals and in semiconductors in terms of bands of closely spaced molecular orbitals. The presence of a band that is only partly filled with electrons is required for electrical conductivity. In some metals, the band obtained by combining the valence orbitals of the metal atoms is only partly occupied. Here, the

valence band is also a **conduction band**. In other metals, an empty conduction band and a filled valence band overlap.

24.7 Semiconductors—Electrical insulators have such a large **energy gap (E_g)** between the valence band and a conduction band that few electrons can make the transition; these materials therefore do not conduct an electric current. In **semiconductors**, the energy gap is much smaller, and a more significant number of electrons can jump the gap, providing a mechanism for electric current to flow through the material. **Superconductors** exhibit zero resistance to the flow of electrons at temperatures below some critical limit.

A semiconductor becomes a much better conductor through the process of **doping**. In an **n-type semiconductor**, the energy level of the donor atoms is close to the energy level of the conduction band, and electric current is carried by electrons in the conduction band. In a **p-type semiconductor**, the energy level of the acceptor atoms is close to the energy level of the valence band, and electric current is carried by **positive holes** in the valence band. Many semiconductors are prepared by a process known as **zone refining**, which reduces impurity levels. Semiconductors have many applications. ranging from microelectronic switches to **photovoltaic cells**.

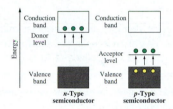

24.8 Natural Polymers—**Polymers** are giant molecules, also called **macromolecules**, that are made from large numbers of smaller molecules called **monomers**:

$$n\, CH_2{=}CH_2 \longrightarrow {+}CH_2CH_2{+}_n$$

Monomer — Polymer

Ball-and-stick model

Space-filling model

24.9 Polymerization Processes—Two basic methods are used to produce polymers from monomers. In **addition polymerization**, monomer units add to one another and the final polymer contains all the atoms of the monomers. In **condensation polymerization**, small molecules such as H_2O are liberated when monomers join through certain functional groups. Each monomer involved in a condensation polymerization must have two or more of the appropriate functional groups.

24.10 Physical Properties of Polymers—Many polymers can be classified in terms of physical properties such as density, hardness, rigidity, elasticity, and flexibility. The physical form of polymers can be divided into three basic groups: fibers, **elastomers**, and plastics. Another classification scheme is based on how changes in temperature affect the polymer properties: **Thermoplastic polymers** can be softened and reshaped; **thermosetting polymers** cannot be softened and reshaped by heat.

24.11 Modern Synthetic Polymers—A wide range of polymer applications are based on synthetic polymeric materials that make use of special properties related to the specific composition and structure of the polymer. Applications include synthetic rubbers, fibers, fabrics, and biocompatible polymers.

24.12 High-Performance Alloys—Specific mixtures of metals in the form of alloys can have unique mechanical, magnetic, and chemical properties.

24.13 Composites—A **composite** is made up of two or more structurally distinct materials in the form of a heterogeneous mixture. The mixture benefits from mechanical properties not found in the individual components. Composites can involve metals, ceramics, and/or polymers.

Courtesy 3M

24.14 Nanomaterials—**Nanomaterials** are materials benefiting from unique properties when the particle dimensions of the sample are reduced to a nanometer scale. This effect has been exploited to develop materials having unique mechanical, electrical, magnetic, optical, and catalytic properties.

Assessment Goals

When you have mastered the material in this chapter, you will be able to:

- List sources of different metals, and describe methods of extraction of metals from ores.
- Outline the process by which steel is produced from iron ore.
- Describe the processes of pyrometallurgy and hydrometallurgy, and give examples of metals produced by each.
- Explain the details of the free-electron theory of metals.
- Describe the electronic energy levels in electrical conductors, semiconductors, and insulators according to band theory.
- Describe the difference between n- and p-type semiconductors in terms of band theory.

- List examples of natural polymers, and describe their composition.
- Write chemical equations describing addition and condensation polymerization processes.
- Identify the formula of a monomer unit from the formula of a polymer and vice versa.
- Relate the physical properties of a polymer to its structure.
- Describe the properties of elastomers, fibers, fabrics and synthetic rubbers.
- List the important characteristics of biomedical polymers.
- Cite examples of space-age materials and relate their unique properties to the composition or structure of the material.

Self-Assessment Questions

1. Which metal is most likely to be found free or uncombined in nature?

   (a) Ca   **(b)** Ba   **(c)** Cu   **(d)** Ce

2. Which form of an ore is least likely to be of commercial importance for the extraction of a metal?

   (a) carbonate   **(c)** oxide

   (b) sulfate   **(d)** chloride

3. What is the flotation process in metallurgy? What is its purpose?

4. In pyrometallurgy, what is the purpose of roasting an ore? Must all ores be roasted? Explain.

5. What is accomplished in the metallurgical operation called reduction? What is the reducing agent most commonly employed? Is it always necessary to use a reducing agent? Explain.

6. Cite two methods that can be used to purify a metal, and give an example of each.

7. Complete the statement: An alloy

   (a) can be a pure metal   **(c)** is a homogeneous mixture

   (b) contains at least one metal   **(d)** cannot be a solution

8. What functions are served by a blast furnace in the metallurgy of iron?

9. What is pig iron? What are its principal impurities? What are carbon steel and alloy steel? In what ways are these two types of steel similar, and how do they differ?

10. Cite several differences between pyrometallurgy and hydrometallurgy.

11. What is meant by the terms amalgamation and cyanidation? Name an element that is obtained by each of these processes.

12. Explain why the ionic and covalent bond descriptions do not work for metals.

13. Describe the meaning of the term *electron gas* when applied to electrons in a metal. Does this term apply to all the electrons or only some? Explain.

14. What is wrong (or incomplete) with the following description of electrical conductivity in the free-electron theory? "The random motion of the electrons in the electron gas in a wire creates an electric current."

15. Describe the band theory of metals. What is the difference between a valence band and a conduction band in a metal? Under what circumstances can they be the same?

16. What is a doped semiconductor? What is a positive hole? What type of semiconductor contains positive holes? How do positive holes move through a semiconductor?

17. Which phrase does *not* correctly complete the statement? An *n*-type semiconductor based on Si

   (a) conducts via the motion of electrons

   (b) may contain P or As

   (c) is more conductive than a *p*-type semiconductor

   (d) has a smaller energy gap than an insulator

18. Describe the meaning of the terms *monomer*, *polymer*, and *copolymer*.

19. In general terms, describe the features that characterize the three broad classes of polymers: plastics, elastomers, and fibers. Can a single polymer be synthesized to fit two of these categories? Explain.

20. Which of the following would *not* be involved in addition or condensation polymerization?

   (a) $CH_2{=}CH_2$ **(c)** $CH_3CH{=}CHCH_3$

   (b) $H_2NCH_2CH_2CH_2NH_2$ **(d)** CH_3COOH

21. How do low-density and high-density polyethylene differ in structure and properties?

22. What is the principal difference between an alloy and a composite?

23. Which of the following could *not* be classified as a composite?

   (a) plywood   **(c)** fiberglass

   (b) polyethylene   **(d)** ceramic reinforced wires

24. What distinguishes a nanomaterial from other materials?

Problems

Metallurgy

25. An ore containing the highest percentage of a metal may not be the best industrial source of the metal. State some reasons why this could be so.

26. Is it correct to say that hydrometallurgy is more environmentally "friendly" than pyrometallurgy? Explain.

27. In the metallurgy of zinc, why is it reasonable to expect to find cadmium as an impurity? Indicate one simple way in which cadmium can be removed from zinc.

28. In the metallurgy of copper, why is it reasonable to expect to find silver and gold as impurities? How can they be removed from the copper?

29. Roasting of a metal sulfide generally converts it to the metal oxide. Explain why this is not the case with cinnabar, HgS.

30. In the metallurgy of silver and gold, an alloy of the two metals is often obtained. The metals can be separated by treating the alloy with nitric acid in a process called *parting*. Explain how this method works.

31. Write the equation for a reaction in which silver metal is displaced from an aqueous solution of $[Ag(CN)_2]^-$.

32. When silver oxide is heated, it decomposes to silver metal and oxygen gas. Write an equation for this reaction.

33. Write chemical equations to describe the following hydrometallurgy of zinc: Zinc oxide is dissolved in $H_2SO_4(aq)$; powdered zinc is added to displace less active metals; and the solution is electrolyzed. What are the electrode half-reactions in the electrolysis? Show how $H_2SO_4(aq)$ is recycled.

34. Write chemical equations showing how pure zinc can be obtained from smithsonite ore, $ZnCO_3$.

35. In one method for recovering tin from scrap tin plate, the metal is treated with $Cl_2(g)$ to form $SnCl_4(l)$. The $SnCl_4(l)$ is hydrolyzed in water to form the hydrated oxide $SnO_2 \cdot xH_2O$. The hydrated tin(IV) oxide is dehydrated by heating, and the product is reduced to metallic tin with carbon. Write plausible equations to represent this method.

36. As an alternative to the method in Problem 35, tin plate is treated with $O_2(g)$ in a basic solution to produce hexahydroxostannate(IV) ion. Then the solution is acidified to precipitate $SnO_2 \cdot xH_2O$. The hydrated oxide is dehydrated by heating, and the product is reduced to metallic tin with carbon. Write plausible equations to represent this method.

37. A blast furnace produces 1.0×10^7 kg of pig iron per day. Assume the pig iron is 95% Fe by mass. How many kilograms of iron ore are consumed in this furnace per day? Assume that the ore is 82% by mass hematite, Fe_2O_3.

38. The blast furnace described in Problem 37 produces 0.5 kg of slag per kg of pig iron. What is the minimum daily requirement of limestone in this furnace? Assume that the limestone is 91% $CaCO_3$, the slag is exclusively $CaSiO_3$, and the limestone is the only source of calcium in the slag.

39. Write plausible chemical equations for the process of isolating lead from galena as outlined in Section 24.3.

40. What current must be supplied for the electrolytic processing of 1.00×10^3 tons of copper per day? Use information from Sections 18.10 and 18.11 as needed.

Band Theory

41. Calcium is a somewhat better electrical conductor than potassium, even though the $4s$ atomic orbitals are filled in Ca atoms and only half filled in K atoms. Explain how this can be.

42. Propose a plausible band structure for aluminum that is consistent with its electrical conductivity.

43. How does band theory account for the lustrous appearance of a clean metal surface? How does it account for the fact that a few metals are colored?

44. According to the band theory, can a bulk metal be transparent to visible light? Explain.

45. How many energy levels are there in the $3p$ conduction band of a 55.5-mg single crystal of magnesium? What is the total number of electrons in the $3s$ and $3p$ bands of this crystal?

46. How many energy levels are there in the valence band of a 35.0-mg single crystal of sodium? How many electrons are in the band?

Semiconductors

47. How does one distinguish between a metallic conductor and a semiconductor in band theory?

48. How does one distinguish between a semiconductor and an insulator in band theory?

49. Classify the following semiconductors as n-type or p-type: **(a)** Ge doped with As; **(b)** Si doped with B.

50. Classify the following semiconductors as n-type or p-type: **(a)** Ge doped with Al; **(b)** Si doped with P.

51. In silicon and germanium, conduction electrons and positive holes are present in equal numbers. Is this also true of n-type and p-type semiconductors based on silicon and germanium? Explain.

52. Would you expect silicon and phosphorus-doped silicon to have about the same electrical conductivity? Explain.

53. Some metals become superconductors at temperatures approaching 0 K. Would you expect the same behavior in semiconductors? Explain.

54. The electrical conductivity of a pure semiconductor material, such as silicon, is strongly dependent on temperature. Would you expect the temperature dependence of the conductivity of a boron-doped silicon to be about the same, greater than, or less than that of silicon? Explain.

55. Certain compounds having the same average number of valence electrons as silicon or germanium are good semiconductors. Which of the following compounds meet this requirement: **(a)** CuS, **(b)** ZnSe, **(c)** PbO, **(d)** GaP?

56. As the percent ionic character of a semiconductor increases, so does the energy gap E_g. Which of the following would you expect to have the greatest energy gap: Ge, CuBr, or GaAs? Explain.

Polymers

57. What is the chemical makeup of each of the following materials, and how is each one made?

    (a) cellophane   **(b)** LDPE

58. What is the chemical makeup of each of the following materials, and how is each one made?

    (a) cellulose nitrate   **(b)** HDPE

59. Why is rubber elastic? How does vulcanization improve the elasticity of rubber?

60. Which of these rubber products require a greater degree of cross-linking, surgical gloves or automobile tires? Explain.

61. Write the structure for *cis*-1,2-dichloroethene, and give the condensed general formula for the polymer made from it.

62. Write the condensed general formula of neoprene, a trans polymer of chloroprene (2-chloro-1,3-butadiene) (page 1019).

63. Represent the structures of addition polymers made from the following monomers, by showing at least four repeating units.

    (a) Styrene, $CH_2{=}CH{-}$⬡

    (b) 1-Hexene, $CH_2{=}CHCH_2CH_2CH_2CH_3$

64. Represent the structures of addition polymers made from the following monomers, by showing at least four repeating units.

(a) Acrylonitrile, $CH_2{=}CH{-}CN$

(b) Vinyl acetate, $CH_2{=}CH{-}O{-}\overset{\overset{\textstyle O}{\|}}{C}{-}CH_3$

65. Represent the structures of the condensation polymers made from the following monomers, by showing at least four repeating units.

(a) Lactic acid, $CH_3\overset{\overset{\textstyle OH}{|}}{CH}{-}\overset{\overset{\textstyle O}{\|}}{C}{-}OH$

(b) Adipic acid, $HO{-}\overset{\overset{\textstyle O}{\|}}{C}(CH_2)_4\overset{\overset{\textstyle O}{\|}}{C}{-}OH$,

and 1,4-diaminobutane, $H_2NCH_2CH_2CH_2CH_2NH_2$

66. Represent the structures of the condensation polymers made from the following monomers, by showing at least four repeating units.

(a) Glycolic acid, $HOCH_2\overset{\overset{\textstyle O}{\|}}{C}{-}OH$

(b) Terephthalic acid, $HO{-}\overset{\overset{\textstyle O}{\|}}{C}$—⟨benzene⟩—$\overset{\overset{\textstyle O}{\|}}{C}{-}OH$,

and 1,4 phenylenediamine, H_2N—⟨benzene⟩—NH_2

67. Based on the partial structural formula of Kevlar® given here, determine (a) the structures of the monomer(s) and (b) whether the polymer is a polyester or a polyamide.

68. Based on the partial structural formula of Qiana® given here, determine (a) the structures of the monomer(s) and (b) whether the polymer is a polyester or a polyamide.

Space-Age Materials

69. Many high-performance alloys are *tempered* to attain maximum strength or hardness. For example, iron can be tempered by heating it to about 1400 °C, where the iron undergoes transition from body-centered cubic to face-centered cubic, then cooling it quickly so that the reverse transition does not occur. Is tempered iron denser or less dense than untempered iron? Explain. Must all tempered metals necessarily follow this pattern of density change? Explain.

70. Titanium(IV) ethoxide, $Ti(OC_2H_5)_4$, is used to make modern ceramic materials that (unlike clays) exhibit very little shrinkage on hardening. What volume of titanium(IV) oxide ceramic ($d = 4.17 \text{ g/cm}^3$) can be made from 1.00 liter of titanium(IV) ethoxide ($d = 1.088 \text{ g/mL}$)?

71. Suppose that a particular process is kinetically first order with respect to surface area of the catalyst, which is iron(III) oxide. Determine the increase in reaction rate that should be obtained by changing from 1.0-μm to 5.0-nm diameter iron(III) oxide catalyst. What is the surface area in m^2/g for the 5.0-nm catalyst? The surface area of a sphere is $4\pi r^2$, the volume of a sphere is $4\pi r^3/3$ and the density of iron(III) oxide is 5.24 g/cm^3.

72. Determine the density of an FRP composite consisting of long cylindrical glass fibers ($d = 2.60 \text{ g/cm}^3$) in a close-packed array similar to Figure 11.41, with epoxy resin ($d = 1.11 \text{ g/cm}^3$) filling the voids between the fibers.

Additional Problems

Problems marked with an * may be more challenging than others.

73. *Vacuum deposition* is a technique used to coat an object such as a piece of glass with a thin layer of metal. The object is placed under high vacuum, and a sample of the metal to be used for coating is electrically heated to vaporize it. Under vacuum, the metal atoms travel in virtually straight lines and coat the object uniformly. Calculate the mass of aluminum needed to produce a coating 1.0-μm thick on the 200-inch-diameter mirror of the Mt. Palomar telescope.

* **74.** A solid-state *diode* is formed when a piece of *p*-type semiconductor and a piece of *n*-type semiconductor are joined. Electricity can be made to flow in one direction through the junction, but not in the other direction. Explain.

* **75.** In practice, it is not possible to make perfectly pure silicon. This means that *p*-type semiconductor always contains a tiny proportion of *n*-type dopant, and vice versa. How might this affect the operation of the diode referred to in Problem 74?

76. The electrical conductivity of a semiconductor increases if either donor or acceptor atoms are incorporated into the structure. However, if both are added in equal number, there is little or no effect on the conductivity. Explain how this can be.

77. The energy gap, E_g, for silicon is 110 kJ/mol. What is the minimum wavelength of light that can promote an electron from the valence band to the conduction band? In what region of the electromagnetic spectrum is this light? Does silicon absorb all visible light, some of it, or none of it? Explain.

78. Isobutylene, $(CH_3)_2C$=CH_2, polymerizes to form polyisobutylene, a sticky polymer used as an adhesive. Copolymerized with 1,3-butadiene, isobutylene forms butyl rubber. Represent a segment of **(a)** the polyisobutylene macromolecule and **(b)** the butyl rubber macromolecule. In each case, show at least four monomer units.

79. The monomers in the polyester Kodel® are terephthalic acid and 1,4-cyclohexanedimethanol. Represent a segment of the Kodel macromolecule containing at least two of each monomer unit. (*Hint:* The methanol group is CH_2OH.)

80. Draw a structure of the likely product(s) if ethanol is substituted for ethylene glycol in the reaction used to make Dacron (page 1020).

81. The monomers 1,2-ethanediol and 1,2-benzenedicarboxylic acid form a soft, tacky polymer. If 1,2,3-propanetriol is substituted for the 1,2-ethanediol, a hard, brittle resin is formed. Explain this difference in behavior.

82. Generally speaking, the molecular masses of polymers formed by addition polymerization are greater than those formed by condensation polymerization. Give a plausible explanation of this observation.

* **83.** The graphs (above right) show the variation of $\Delta G°$ with temperature for the oxidation of C(graphite), Zn, and Mg to CO, ZnO, and MgO, respectively. Graphs of this type are important to metallurgists because they convey information about the feasibility of obtaining metals from their oxides.

(a) Why can magnesium be used to reduce zinc oxide to zinc at all temperatures, whereas zinc cannot be used to reduce magnesium oxide to magnesium at any temperature?

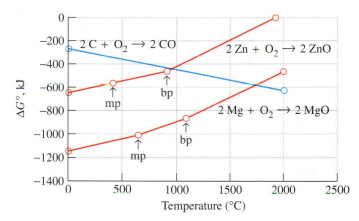

▲ The points on the lines marked by arrows indicate the melting and boiling points of Zn and Mg.

(b) Why can graphite be used to reduce zinc oxide to zinc at some temperatures and not at others?

(c) Is it possible to reduce zinc oxide to zinc by its direct decomposition without requiring a coupled reaction? If so, at what temperatures might this be?

(d) Is it possible to decompose carbon monoxide to graphite and oxygen in a spontaneous reaction? If it is possible, indicate at what temperature(s) this can be accomplished. If it is not possible, explain why not.

(e) Why does the line representing the oxidation of graphite have a negative slope, whereas those for the oxidation of zinc and magnesium have positive slopes?

(f) Describe, roughly, the position of the graphs for the following reactions relative to the graphs presently shown in the figure. Your description should include an indication as to whether each graph has a positive or negative slope.

$$Ti(s) + O_2(g) \longrightarrow TiO_2(s)$$

$$2 Hg(l) + O_2(g) \longrightarrow 2 HgO(s)$$

$$2 Ca(s) + O_2(g) \longrightarrow 2 CaO(s)$$

84. When speaking of the molecular mass of a polymer, chemists have to deal with averages. The *number average molecular mass* of a polymer is analogous to the weighted average atomic mass of an element. What is the number average molecular mass of a sample of polyethylene if 28% of the molecules have a mass of 786 u, 25% have a mass of 702 u, 15% have a mass of 814 u, and 32% have a mass of 758 u?

* **85.** The *functionality* of monomer mixtures tells the average number of reactive functional groups per molecule. For example, 1 mol of glycerol and 2 mol of 1,4-dihydroxybutane (see the Cumulative Example in this chapter) together have a hydroxyl functionality of 2.33, which is moderately high.

(a) Most monomers for condensation polymerization must have a functionality of at least two. Explain why this is so.

(b) A particular monomer has a functionality of exactly two. What kind of polymer might this monomer be expected to produce? Explain.

Chapter 25

Environmental Chemistry

WHEN ASTRONOMERS SPECULATE about life elsewhere in the universe, they look for the possibility of planets within a size and temperature range that allows for liquid water and a gaseous atmosphere. Most of the other worlds in our solar system do not meet this criterion. Instead, they are either barren and airless, like Earth's moon and Mercury, or else have crushing atmospheres—as do Jupiter, Saturn, Uranus, and Neptune—exerting pressures thousands of times greater than the atmospheric pressure on Earth.

Life is intimately dependent on water because only water has the unique properties required to sustain life. The presence of large amounts of liquid water makes our planet unusual in the solar system: It is probably the only one capable of supporting higher forms of life. Scientists who look for life elsewhere in our solar system place their scant hope on the presence of water beneath the dry surface of Mars or the existence of liquid water beneath the frozen seas of Jupiter's moon Europa. If we should someday discover higher forms of life on distant planets of other suns, it will likely be on another watery planet similar to our own.

In this chapter, we look at the composition and properties of Earth's atmosphere and water supplies and at some of the ways in which human activities affect our air and water. Mostly, though, we focus on how knowledge of chemistry can illuminate environmental issues and how chemistry can often be used to alleviate environmental problems. Proper use of scientific knowledge is essential to protecting the only planet that we know to be hospitable to life.

CONTENTS

Earth's Atmosphere

Let us look first at Earth's atmosphere. Is the air we breathe unique? A look at other worlds in our solar system gives us a clue. Astronauts have walked on the dry, dusty, airless surface of the moon. Scientists have sent robotic probes down through clouds of sulfuric acid and a thick blanket of carbon dioxide to land on the hot, inhospitable surface of Venus. Other space probes have descended through the sparse atmosphere

◄ We inhabit the solid surface of planet Earth, but it is the liquid water on its surface and the gaseous atmosphere that makes it hospitable to higher forms of life. In this chapter, we examine Earth's atomsphere and waters and some of the effects of human activities on our environment. In doing so, we apply some of the chemistry we learned in preceding chapters. A knowledge of chemistry is essential to understanding environmental problems and to their solution.

of Mars and used a robot to analyze its rocks. Still other spacecraft have examined the crushing, turbulent, toxic atmospheres of Jupiter, Saturn, Uranus, and Neptune. We do not know much about the atmosphere of distant Pluto because it has not been examined except from vast distances.

Of all the worlds in the solar system, Earth's atmosphere is unique in its ability to support human life. In the sections that follow, we consider first the normal composition of the atmosphere and then some substances introduced into the atmosphere by human activities.

25.1 Atmospheric Composition, Structure, and Natural Cycles

Without food, we can live about a month; without water, a few days; but without air, we would die within minutes. Air is vital because it contains free oxygen (O_2), an element essential to the basic processes of respiration and metabolism. However, life as we know it could not exist in an atmosphere of pure oxygen, because oxidation processes would be greatly accelerated by the increased concentration of O_2. The oxygen in air is diluted with nitrogen, thus lessening the tendency for everything in contact with air to become oxidized. Carbon dioxide and water vapor are but minor components in air, and yet they are the primary raw materials of the plant kingdom; and plants produce the food on which we and all other animal life depend. Even ozone, a gas present only in trace quantities, plays vital roles in shielding Earth's surface from harmful ultraviolet radiation and in maintaining a proper energy balance in the atmosphere.

On a mole percent basis, dry air in the lower atmosphere consists of about 78% N_2, 21% O_2, and 1% Ar. Among the minor constituents of dry air, the most abundant is carbon dioxide. The concentration of CO_2 in air has increased from about 275 ppm in 1880 to 368 ppm in 2001. It continues to rise about 1.5 ppm per year and most likely will continue to rise as more and more fossil fuels (coal, oil, and natural gas) are burned. The composition of dry air is summarized in Table 25.1.

Altogether, the thin blanket of gases that makes up the atmosphere is spread over a surface area of 5.0×10^8 km^2 and has a mass of about 5.2×10^{15} metric tons (1 metric ton = 1000 kg). That is about 10 million metric tons of air over each square kilometer of surface, or 10 metric tons over each square meter, or about 1 metric ton (nearly the mass of one small automobile) over each square foot.

How deep is the atmosphere? It is hard to say because the atmosphere does not end abruptly. Instead, it gradually fades away with increasing distance from Earth's surface. At sea level, the density of air is about 1.3 g/L, but at an altitude of about 50 km, the density drops into the range of micrograms and even nanograms per liter.

Application Note

Even though the atmosphere has an extremely low density at high altitudes, there is still sufficient air present to cause significant drag on large objects. Between October 2000 and February 2001, atmospheric drag caused the International Space Station to drop from an altitude of 390 km to an altitude of 360 km. The drag will increase as more parts are added to the station.

Table 25.1 Composition of Dry Air (near sea level)

Component	Mole Percent[a]
Nitrogen (N_2)	78.084
Oxygen (O_2)	20.946
Argon (Ar)	0.934
Carbon dioxide (CO_2)	0.0368
Neon (Ne)	0.001818
Helium (He)	0.000524
Methane (CH_4)	0.0002
Krypton (Kr)	0.000114
Hydrogen (H_2)	0.00005
Dinitrogen monoxide (N_2O)	0.00005
Xenon (Xe)	0.000009
	Plus traces of: ozone (O_3); sulfur dioxide (SO_2); nitrogen dioxide (NO_2); ammonia (NH_3); carbon monoxide (CO); iodine (I_2)

[a] The compositions of gaseous mixtures are often expressed in percent by volume. Volume percent and mole percent compositions have the same numeric values.

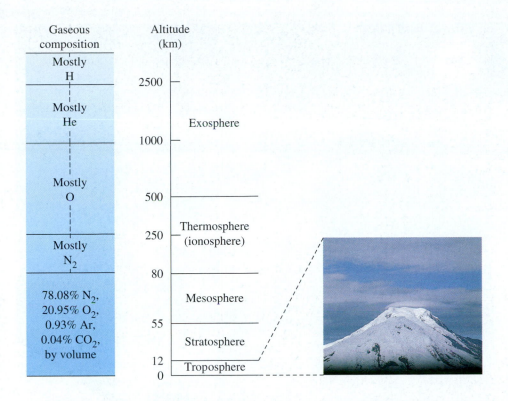

◀ **FIGURE 25.1** **Layers of the atmosphere**

The altitudes of the several layers of the atmosphere are only approximate. For example, the height of the troposphere varies from about 8 km at the poles to 16 km at the equator. The approximate temperatures in these layers are cited in the text.

QUESTION: The troposphere, stratosphere, and mesosphere have the same percent composition by volume. How do these layers differ in absolute composition per unit volume?

Mt. Cayambe in Ecuador is almost on the equator, but its peak is snow-capped, illustrating the decrease in temperature with altitude in the troposphere. Mt. Cayambe's summit, at 5.79 km, is about halfway through the troposphere.

However, 99% of the mass of the atmosphere lies within 30 km of Earth's surface—a thin layer of air indeed, akin to the peel of an apple, only relatively thinner.

Rather arbitrarily, we divide the atmosphere into the layers shown in Figure 25.1. The layer nearest Earth's surface, the *troposphere,* extends about 12 km above the surface and contains about 90% of the mass of the atmosphere. Weather and nearly all human activity occur in the troposphere. Temperatures generally decline as altitude increases in the troposphere, ranging from a maximum of about 320 K at Earth's surface to a minimum of about 220 K in the upper troposphere.

The next layer, the *stratosphere,* extends from about 12 km above Earth's surface to 55 km. The ozone layer that shields living creatures from deadly UV radiation lies in the stratosphere. Supersonic aircraft fly in the lower part of the stratosphere. Temperatures in the stratosphere are fairly constant at about 220 K from 12 to 25 km, then rise with increasing altitude to about 280 K at 50 km.

The layer above the stratosphere, ranging from about 55 to 80 km, is called the *mesosphere.* Here the temperature falls continuously to about 180 K as altitude increases.

The layer above the mesosphere is known as either the *thermosphere* or the *ionosphere.* In this region, molecules can absorb such highly energetic electromagnetic radiation from the Sun that they dissociate into atoms; some of the atoms are further broken down into positive and negative ions and free electrons. Temperatures in the thermosphere rise to about 1500 K as the altitude increases,* but high temperatures in this region do not have the same significance as on Earth's surface. The temperatures are high because the average kinetic energy of the gaseous particles is high. However, because there are relatively few of these particles per unit volume, little energy is transferred through collisions. A cold object brought into this region does not get hot, because there are far too few collisions to bring the object to temperature equilibrium with the gas molecules.

Water Vapor in the Atmosphere

Unless it has been specially dried, air invariably contains water vapor. Atmospheric water vapor plays one of the key roles in the **hydrologic (water) cycle**—the series of natural processes by which water is recycled through the environment (Figure 25.2).

▲ This meteor seen against a starry sky at dusk is an extraterrestrial chunk of matter that has entered Earth's atmosphere. It emits light because it is heated to a high temperature. This heating occurs not from the high temperature in the thermosphere but through frictional resistance the object encounters as it passes through gases in the portion of the atmosphere that lies between about 80 to 100 km.

*The temperature in the thermosphere depends on the amount of solar radiation reaching this region. It varies with sunspot activity and is higher during the day than at night.

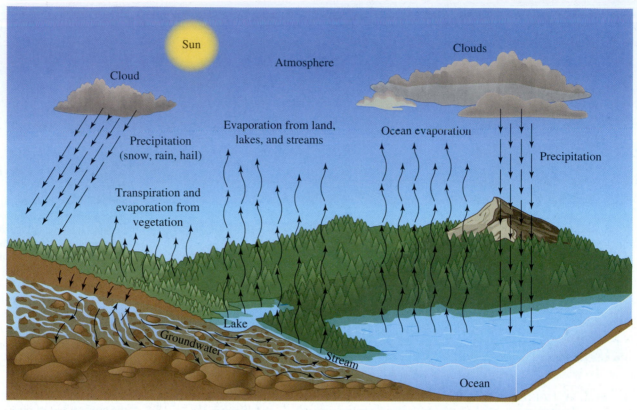

▲ **FIGURE 25.2** **The hydrologic (water) cycle**
The oceans on Earth are vast reservoirs of water. Water evaporates from the oceans, producing moist air masses that move over the land. When the moist air is cooled, the water vapor forms clouds and the clouds produce rain. Rainwater replenishes groundwater and forms lakes and streams, and this water eventually returns to the ocean. Water is also returned to the atmosphere by evaporation throughout the cycle.

 Hydrologic Cycle activity

▲ **FIGURE 25.3** **An old-fashioned indicator of relative humidity**
The strips of filter paper were impregnated with an aqueous solution of cobalt(II) chloride and allowed to dry. Because it is in dry air, the strip on the left is blue, the color of anhydrous $CoCl_2$. In the more humid air above a beaker of water, a portion of the strip on the right is pink, the color of the hexahydrate, $CoCl_2 \cdot 6\,H_2O$. Test strips of this sort have been used in the past to signal large changes in the relative humidity of air.

The proportion of water vapor in air is quite variable, ranging from trace amounts to about 4% by volume. *Humidity* is a general term describing the water vapor content of air. The *absolute humidity* is the quantity of water vapor present in an air sample, usually expressed in g H_2O/m^3 air. The *relative humidity* of air is a measure of water vapor content as a percentage of the maximum possible; it compares the partial pressure of water vapor in an air sample to the maximum partial pressure possible at the given temperature—the vapor pressure of water:

$$\text{Relative humidity} = \frac{\text{partial pressure of water vapor}}{\text{vapor pressure of water}} \times 100\% \qquad \textbf{(25.1)}$$

A number of experimental methods exist for determining the relative humidity of air. A crude but colorful method for estimating relative humidity is illustrated in Figure 25.3.

Example 25.1

The partial pressure of water vapor in a certain air sample at 20.0 °C is 12.8 mmHg. What is the relative humidity of this air?

STRATEGY

To determine the relative humidity using Equation (25.1), we need (1) the partial pressure of water vapor in the sample and (2) the vapor pressure of water at the given temperature. The first quantity is given (12.8 mmHg). For the second, we need the vapor pressure of water at 20.0 °C. We can find it in a tabulation such as Table 11.2.

SOLUTION

$$\text{Relative humidity} = \frac{12.8 \text{ mmHg}}{17.5 \text{ mmHg}} \times 100\% = 73.1\%$$

If we warm the air sample described in Example 25.1, the relative humidity decreases because the vapor pressure of water increases sharply with temperature but the partial pressure of water vapor in the sample remains essentially unchanged. If we cool the air sample, we observe the opposite—the relative humidity increases. By consulting an extensive table of vapor-pressure data, we would find that the vapor pressure of water is 12.8 mmHg at 15.0 °C. When cooled to 15.0 °C, the air sample in Example 25.1 would be *saturated* with water vapor. At temperatures below 15.0 °C, the relative humidity would exceed 100%, and the air would be *supersaturated* with water vapor. This is an unstable situation that is not at equilibrium; it cannot persist. Some of the water vapor would condense as droplets of liquid water that we call *dew*.

The highest temperature at which condensation of water vapor from an air sample can occur is known as the *dew point*. When the dew point is below the freezing point of water (0 °C), the water condenses as *frost* without going through the liquid state.

The water vapor that condenses from air at the dew point is pure water. The liquid in Figure 25.4 also results from the condensation of atmospheric water vapor, but it is not pure water. Rather, it is a solution of calcium chloride, formed in this way: If the partial pressure of water vapor in the air exceeds the vapor pressure of a saturated solution of $CaCl_2 \cdot 6 H_2O$, water vapor condenses on the solid $CaCl_2 \cdot 6 H_2O$ and dissolves some of it, producing a saturated solution. This condensation of water vapor on a solid followed by solution formation is called **deliquescence**; it continues until all the solid has dissolved and the vapor pressure of the solution (now unsaturated) equals the partial pressure of water vapor in the air.

Nitrogen Fixation: The Nitrogen Cycle

Although it exists in large quantities in the atmosphere, nitrogen, an element essential to life, cannot be used directly by higher plants or animals. The N_2 molecules must first be converted to compounds that are more readily usable by living organisms. This conversion of atmospheric nitrogen into nitrogen compounds is called **nitrogen fixation**.

Certain bacteria, such as the cyanobacteria (blue-green algae) found in water and a variety of bacteria that live in root nodules of specific plants, are able to fix atmospheric nitrogen by converting it to ammonia. These nitrogen-fixing bacteria are concentrated in the roots of leguminous plants, such as clover, soybeans, and peas.

Other plants take up nitrogen atoms in the form of nitrate ions or ammonium ions from soil or water. Nitrogen atoms in plants combine with carbon compounds from photosynthesis to form amino acids, the building blocks of proteins. Thus the food chain for animals originates with plant life. The decay of plant and animal life returns nitrogen to the environment as nitrates and ammonia. Eventually the nitrogen in these compounds is returned to the atmosphere as N_2 when *denitrifying* bacteria act on the nitrates and ammonia.

Lightning also fixes some atmospheric nitrogen by creating a high-energy environment in which nitrogen and oxygen can combine. Nitrogen monoxide and nitrogen dioxide are formed in this way:

$$N_2(g) + O_2(g) \xrightarrow{\text{Lightning}} 2 NO(g)$$

$$2 NO(g) + O_2(g) \longrightarrow 2 NO_2(g)$$

The nitrogen dioxide then reacts with water to form nitric acid:

$$3 NO_2(g) + H_2O(l) \longrightarrow 2 HNO_3(aq) + NO(g)$$

▲ Dill covered with morning dew. When the temperature drops to the point where the absolute humidity of the air is greater than the vapor pressure of water, water vapor condenses to the familiar liquid known as dew.

▲ **FIGURE 25.4   Deliquescence of calcium chloride**

Water vapor from the air condenses onto the solid $CaCl_2 \cdot 6 H_2O$ and produces a solution of $CaCl_2(aq)$. Here the solution is saturated, but eventually all the solid will dissolve and the solution will become unsaturated. The deliquescence of $CaCl_2 \cdot 6 H_2O$ occurs only when the relative humidity exceeds 32%. Other water-soluble solids deliquesce under other conditions of relative humidity.

QUESTION: Why will calcium chloride not deliquesce if the relative humidity is less than 32%?

Alfalfa can fix 100 kg/acre or more of atmospheric nitrogen, the equivalent of almost 300 kg of ammonium nitrate fertilizer.

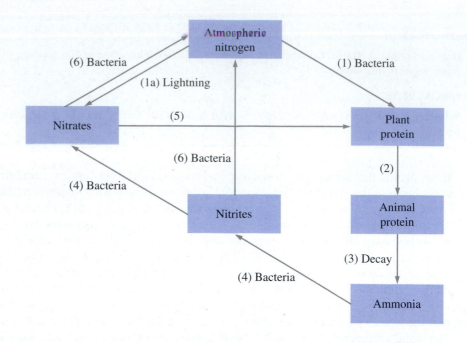

▶ **FIGURE 25.5   The nitrogen cycle**
Bacteria fix atmospheric nitrogen by converting it to plant proteins (1). Nitrates are also converted to proteins by plants (5). Animals feed on plants and other animals (2). The decay of plant and animal proteins produces ammonia (3). Through a series of bacterial actions, ammonia is converted to nitrites and nitrates (4). Denitrifying bacteria decompose nitrites and nitrates, returning N_2O and N_2 to the atmosphere (6). Some atmospheric nitrogen is converted to nitrates during electrical storms (1a). A significant amount of nitrogen fixation and denitrification occurs in oceans. Modern manufacturing and agricultural processes also play important roles in the cycle.

▲ The nitrogen fixation that occurs during electrical storms is an important part of the natural nitrogen cycle.

The nitric acid falls in rainwater, adding to the available nitrates in sea and soil. The net effect of all these natural activities is a constant recycling of nitrogen atoms in the environment in a **nitrogen cycle** (Figure 25.5).

Industrial fixation of nitrogen by the manufacture of inorganic nitrogen fertilizers has modified the nitrogen cycle. These fertilizers have greatly increased the world's food supply because the availability of fixed nitrogen is often the limiting factor in the production of food. Not all the consequences of this interference have been favorable, however. Excessive runoff of dissolved nitrogen fertilizers has led to serious water pollution in some areas, but modern methods of high-yield farming seem to demand synthetic fertilizers.

The Carbon Cycle

Carbon atoms are cycled throughout Earth's solid crust, oceans, and atmosphere by natural processes. In the process of photosynthesis, atmospheric CO_2 is converted to carbohydrates, the chief structural material of plants. For example, the photosynthesis of glucose, one of the simplest carbohydrates, occurs through dozens of sequential steps leading to the net change:

$$6\ CO_2(g) + 6\ H_2O(l) \longrightarrow C_6H_{12}O_6(s) + 6\ O_2(g)$$

Animals acquire carbon compounds by consuming plants or other animals, and they return CO_2 to the atmosphere as a product of respiration. The decay of plant and animal matter also returns CO_2 to the air. Most photosynthesis occurs in the oceans, where algae and related plants convert CO_2 into organic compounds. Some of Earth's carbon is locked up in fossilized forms—in coal, petroleum, and natural gas produced eons ago from decaying organic matter and in limestone from the shells of decayed mollusks from ancient seas. Figure 25.6 shows a simplified **carbon cycle**.

Note in Figure 25.6 that human activities today play a key role in the carbon cycle by releasing carbon atoms as CO and CO_2 when we burn wood and fossil fuels.

25.2   Air Pollution

Human activities affect the atmosphere in many ways. In addition to the release of carbon dioxide into the atmosphere, the combustion of fossil fuels causes regional problems with high levels of carbon monoxide and particles of ash, dust, and soot, and with sulfur dioxide pollution and acid rain. Oxides of nitrogen, formed by high-temperature combustion in air, are involved in the production of photochemical smog. In general, we consider an **air pollutant** to be any substance that is found in air in greater abun-

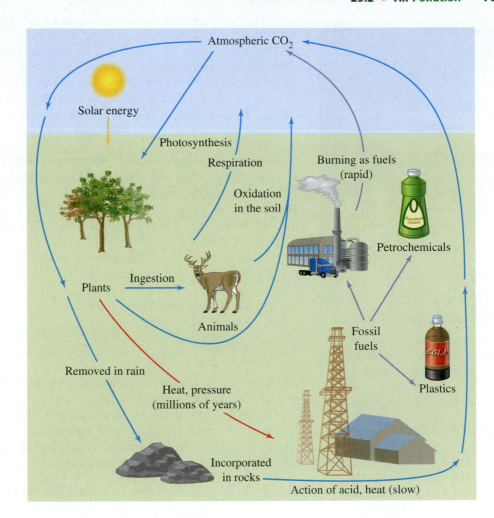

Main cycle
Fossilization tributary
Disruption by human
activities

▲ **FIGURE 25.6   The carbon cycle**

The natural main cycle is indicated by blue arrows. Some carbon atoms are locked up by the formation of fossil fuels and limestone deposits, in what are called fossilization tributaries (red arrow). Disruption of the cycle by human activities is becoming increasingly important (purple arrows).

QUESTION: The carbon released when fossil fuels are burned is isotopically different from the carbon released when wood is burned. Why? (See Section 19.3.)

dance than normally occurs naturally and that has one or more harmful effects on human health or the environment.

Carbon Monoxide

It is estimated that, on a global basis, up to 80% of the carbon monoxide in the atmosphere comes from natural sources. Nature is able to prevent the buildup of CO in the atmosphere, however, because bacteria in soil convert it to CO_2. The other 20% comes from human activity, and this is the part that is the problem. Both carbon monoxide and carbon dioxide are formed in varying quantities when fossil fuels are burned. For instance, the combustion of methane, the major component of natural gas, yields both gases:

$$CH_4(g) + \tfrac{3}{2}O_2(g) \longrightarrow CO(g) + 2\,H_2O(l)$$

$$CH_4(g) + 2\,O_2(g) \longrightarrow CO_2(g) + 2\,H_2O(l)$$

With an excess of O_2, as is the case when air is readily available, the combustion products are almost exclusively $CO_2(g)$ and $H_2O(l)$. If the quantity of air is more limited, as in a dust- or dirt-clogged heater, $CO(g)$ is also formed.

The chief source of $CO(g)$ in polluted air is the incomplete combustion of hydrocarbons in automobile engines. Thus carbon monoxide is a local pollution problem,

▲ Crowded parking garages and tunnels can have dangerously high levels of carbon monoxide gas.

with the greatest threat being in urban areas with heavy vehicular traffic. Millions of tons of this invisible but deadly gas are poured into the atmosphere each year, about 75% of it from automobile exhausts. The U.S. government has set danger levels at 9 ppm CO averaged over an 8-hour period and 35 ppm averaged over a 1-hour period. Even in off-street urban areas, levels often reach 8 ppm or more. On streets and in parking garages, danger levels are exceeded much of the time. Such levels do not cause immediate death, but exposure over a long period can cause physical and mental impairment.

Because CO is an invisible, odorless, tasteless gas, we cannot tell when it is around without using test reagents or instruments. Drowsiness is usually the only symptom of carbon monoxide poisoning, and drowsiness is not always unpleasant. How many auto accidents are caused by drowsiness or sleep induced by CO(g) escaping into the car from a faulty exhaust system? No one knows for sure.

Carbon monoxide exerts its insidious effect by replacing O_2 molecules normally bonded to Fe atoms in hemoglobin in blood. We represented the heme units and polypeptide chains in hemoglobin in Figure 22.9. The form of hemoglobin carrying carbon monoxide is shown in Figure 25.7. In the reversible uptake of O_2 and CO by hemoglobin (Hb) to form HbO_2 and HbCO,

$$(1) \qquad Hb + O_2 \rightleftharpoons HbO_2$$

$$(2) \qquad Hb + CO \rightleftharpoons HbCO$$

The CO is highly effective in displacing O_2 from HbO_2 because reaction (2) proceeds much farther toward product than does reaction (1).

The symptoms of carbon monoxide poisoning are those of oxygen deprivation. All except the most severe cases of the poisoning are reversible, though the process may be very slow. The best antidote is the administration of pure oxygen. At high concentrations, the O_2 can cause the reaction

$$HbCO + O_2 \rightleftharpoons HbO_2 + CO$$

to proceed farther toward products. Artificial respiration may help if pure oxygen is not available.

Chronic exposure to CO, even to low levels, as through cigarette smoking, puts an added strain on the heart and increases the chances of a heart attack. Carbon monoxide impairs the blood's ability to transport oxygen, and consequently the heart has to work harder to supply oxygen to tissues.

Application Note

The Clean Air Act of 1990 instituted maximum limits for emission of CO and of other pollutants from new automobiles. In addition, to reduce local pollution, many large and moderate-sized cities now require periodic testing of emissions from individual automobiles. Some areas also mandate the use of oxygenated fuels to reduce carbon monoxide emissions.

Heme 3D model

▶ **FIGURE 25.7   A molecular view of carbon monoxide poisoning**

The hemoglobin molecule consists of thousands of atoms, but key portions of the molecule are four heme groups, one of which is shown here. Each heme has an iron atom (shown in rust color here) in the center of a square formed by four nitrogen atoms (blue). The heme group is able to bind one small molecule to the iron atom. Ordinarily, it binds an O_2 molecule, but the heme group has a much higher affinity for CO (shown here pointing up from the iron atom) than for O_2. As a result, even when the CO(g) concentration is low, O_2 molecules are easily displaced by CO.

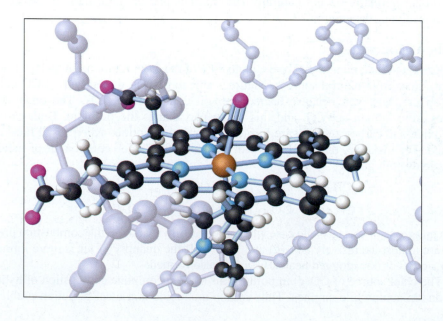

▲ **FIGURE 25.8   Photochemical smog**
(Left) Mexico City on January 9, 1996, when air pollution approached levels classified as "dangerous". (Right) A photograph taken one week earlier, from the same location, on a rare clear day.

Photochemical Smog

We usually think of sunlight as something pleasant. However, when sunlight falls on air containing a mixture of nitrogen oxides, hydrocarbons, and other substances, the light produces a mixture of pollutants called **photochemical smog** (Figure 25.8).

Photochemical smog production starts with the formation of nitrogen monoxide (NO). The reaction between $N_2(g)$ and $O_2(g)$ at ordinary temperatures is obviously extremely slow because these gases coexist in the atmosphere. However, at high temperatures, such as those attained by the combustion of a fuel in air, some NO is formed:

$$N_2(g) + O_2(g) \longrightarrow 2\,NO(g)$$

This reaction takes place in power plants that burn fossil fuels and in incinerators. The greatest source of NO, though, is in the exhaust fumes of automobile engines.

In sufficiently high concentrations, NO can react with hemoglobin in blood and rob it of its oxygen-carrying ability, just as CO does. However, such NO concentrations are rarely reached in polluted air. The main role of NO as an air pollutant is its participation in various reactions that yield several other pollutants.

Nitrogen dioxide, the gas that gives the amber color often seen in polluted air in major urban centers, is formed by the oxidation of NO. Nitrogen dioxide is an irritant to the eyes and respiratory system. Tests with laboratory animals indicate that chronic exposure to levels of NO_2 in the range of 10–25 ppm might lead to emphysema or other degenerative lung diseases. However, as with NO, we are concerned not so much with the direct effects of NO_2 but with its chemical reactions.

In the presence of sunlight, represented here as $h\nu$,* NO_2 decomposes:

$$NO_2(g) + h\nu \longrightarrow NO(g) + O(g)$$

Oxygen *atoms* produced by the photochemical (light-induced) decomposition of NO_2 are highly reactive. They react with many substances that are generally available in polluted air. For example, they react with O_2 molecules to form ozone (O_3):

$$O(g) + O_2(g) \longrightarrow O_3(g)$$

This reaction leads to an ozone concentration in photochemical smog that is considerably higher than normal. Ozone is the main cause of the breathing difficulties some

*The expression $h\nu$ is derived from the equation for the energy of a photon of light, $E = h\nu$ (Chapter 7). Here it represents a photon of sunlight. One photon of the proper energy can split one molecule of $NO_2(g)$ into a molecule of NO(g) and an oxygen atom, represented as O(g).

people experience during smog episodes. Another effect of ozone is that it causes rubber to crack and deteriorate.

In addition to ozone, nitrogen oxides, and hydrocarbons, photochemical smog contains *peroxyacetyl nitrate* (PAN), an organic compound formed by the combination of two free radicals:

$$\underset{\displaystyle \text{CH}_3\overset{\displaystyle \text{O}}{\overset{\|}{\text{C}}}-\text{O}-\text{O}\cdot}{} \; + \; \cdot\text{NO}_2 \longrightarrow \underset{\displaystyle \text{CH}_3\overset{\displaystyle \text{O}}{\overset{\|}{\text{C}}}-\text{O}-\text{ONO}_2}{}$$

(Recall from Chapter 9 that NO_2 is an odd-electron molecule; it has an unpaired electron and is therefore a free radical.) PAN is a powerful *lachrymator;* it makes the eyes form tears.

In addition to their role in smog formation, NO and NO_2 (collectively called NO_x) contribute to the fading and discoloration of fabrics. By forming nitric acid, they contribute to the acidity of rainwater, which accelerates the corrosion of metals and building materials. They also produce crop damage, although specific effects of these gases are difficult to separate from those of other pollutants.

▲ A normal radish plant (left) and one damaged by air pollution (right).

The reactions that form photochemical smog are exceedingly complex and are still not completely understood. We will point out only a few of the more important features of this reaction chemistry.

We have already noted two reactions: the photochemical decomposition of NO_2 and the production of ozone. If ozone formation is to continue, there must be a continuous source of NO_2. In Section 25.1, we noted one reaction for the formation of NO_2:

$$2\,\text{NO(g)} + \text{O}_2(\text{g}) \longrightarrow 2\,\text{NO}_2(\text{g})$$

However, at the low concentrations of NO in a smoggy atmosphere and at normal air temperatures, this reaction is too slow to produce much NO_2. Instead, the NO appears to be converted to NO_2 in another—much faster—reaction that involves hydrocarbons, which come mostly from automotive exhaust. The process begins with a hydrocarbon molecule, RH, reacting with an oxygen atom to produce the free radicals R· and ·OH, followed by the ·OH radical's reaction with another hydrocarbon molecule:

$$\text{RH} + \text{O} \longrightarrow \text{R}\cdot + \cdot\text{OH}$$

$$\text{RH} + \cdot\text{OH} \longrightarrow \text{R}\cdot + \text{H}_2\text{O}$$

The hydrocarbon radicals, R·, can react with O_2 to produce new radicals called *peroxyl* radicals:

$$\text{R}\cdot + \text{O}_2 \longrightarrow \text{RO}_2\cdot$$

Then the peroxyl radicals can react with NO to form NO_2:

$$\text{RO}_2\cdot + \text{NO} \longrightarrow \text{RO}\cdot + \text{NO}_2$$

Automotive exhaust is a major contributor to the production of photochemical smog, but geographic factors are also important. Smog is especially likely to occur in a region like the Los Angeles basin, which is surrounded by mountains. In the absence of strong winds in the basin, the only direction in which mixing and dilution of air pollutants can occur is vertically into the atmosphere. However, at times, the region may also experience a temperature inversion—a mass of warm air overlaying a mass of colder air. The warmer top layer acts like a lid on a reaction vessel, keeping pollutants from moving upward and thereby being diluted, but it does let sunlight through. The sunlight acts on the pollutants to form smog, which is then trapped in the cooler stagnant layer of air. The most severe smog episodes usually occur during strong temperature inversions.

Control of Photochemical Smog

Most measures to reduce levels of photochemical smog focus on automobiles, but potential sources of smog precursors range from power plants to lawn mowers to charcoal lighter fluid. In many parts of the world, automobiles are now equipped with catalytic converters. The first function of these converters is to catalyze the oxidation of carbon monoxide and unburned hydrocarbons to CO_2 and H_2O. The catalyst is usually palladium (Pd) or platinum (Pt) or a mixture of the two. To remove NO from automotive exhaust by reducing it to N_2 requires a reduction catalyst, which is different from the Pt/Pd oxidation catalyst. Some automobiles therefore use a dual catalyst system. Using a fuel–air mixture that is very rich in fuel can also lower the NO content of automotive exhaust because burning such a mixture results in some unburned hydrocarbons and CO(g). The CO can then reduce some of the NO in the exhaust stream to N_2:

$$2\,CO(g) + 2\,NO(g) \longrightarrow 2\,CO_2(g) + N_2(g)$$

Excess unburned hydrocarbons and CO(g) are then oxidized to CO_2 and H_2O in the catalytic converter.

▲ A cutaway view of a dual-bed automobile catalytic converter.

Industrial Smog

Photochemical smog occurs mainly in warm, sunny weather and is characterized by high levels of hydrocarbons, nitrogen oxides, and ozone. The other major kind of smog is usually associated with industrial activities and is therefore called **industrial smog.** It occurs mainly in cool, damp weather and is usually characterized by high levels of sulfur oxides (SO_2 and SO_3, collectively called SO_x) and of particulate matter (dust, smoke, and so on).

A major source of SO_x in some areas is smelters, where sulfide ores are roasted as the first step in the production of copper, lead, and zinc. For example, the smelting of zinc ore to produce ZnO also yields SO_2:

$$2\,ZnS(s) + 3\,O_2(g) \longrightarrow 2\,ZnO(s) + 2\,SO_2(g)$$

Another source of SO_x is coal, especially soft coal from the eastern United States, which has a relatively high sulfur content. When this coal is burned, sulfur compounds in the coal also burn, forming SO_2. This choking, acrid gas is readily absorbed in the respiratory system. It is a powerful irritant and is known to aggravate the symptoms of people who suffer from asthma, bronchitis, emphysema, and other lung diseases.

Some of the sulfur dioxide reacts further with oxygen in air to form sulfur trioxide:

$$2\,SO_2(g) + O_2(g) \longrightarrow 2\,SO_3(g)$$

Sulfur trioxide then reacts with water to form sulfuric acid:

$$SO_3(g) + H_2O(l) \longrightarrow H_2SO_4(aq)$$

Fine droplets of this acid form an aerosol mist that is even more irritating to the respiratory tract than sulfur dioxide.

Particulate matter consists of solid and liquid particles of greater than molecular size (Figure 25.9). The largest particles often are visible in air as dust and smoke. Smaller particles, 1 μm or less in diameter, are called *aerosols* and are often invisible to the naked eye.

Particulate matter consists in part of *soot* (unburned carbon). A larger portion is made up of the mineral matter that occurs in coal but does not burn. In the roaring fire of a huge boiler in a factory or power plant, some of this solid mineral matter is left behind as *bottom ash*. However, a lot of solid matter is carried aloft in the tremendous draft created by the fire. This *fly ash* settles over the surrounding area, covering everything with dust. It is also inhaled, contributing to respiratory problems.

Perhaps the most insidious form of particulate matter is the sulfates. Some of the sulfuric acid in smog reacts with ammonia to form solid ammonium sulfate:

$$2\,NH_3(g) + H_2SO_4(aq) \longrightarrow (NH_4)_2SO_4(s)$$

The solid ammonium sulfate and minute liquid droplets of sulfuric acid are trapped in the lungs, where they can cause considerable damage.

The harmful effects of sulfur dioxide and particulate matter may be magnified by their interaction. A certain level of sulfur dioxide, without the presence of particulate

▲ This copper smelter emitted 900 tons of SO_2 daily before it ceased operation in January 1987.

▲ **FIGURE 25.9 False-color scanning electron micrograph of fly ash from a coal-burning power plant**

Natural Pollutants

Even before there were people, there was pollution. This pollution frequently came from erupting volcanoes, which spewed ash and poisonous gases into the atmosphere. And today, of course, they still do. Kilauea in Hawaii, for example, emits 200–300 tons of SO_2 per day. Acid rain downwind from this volcano has created a barren region called the Kau desert.

The 1991 eruption of Mt. Pinatubo in the Philippines injected such vast quantities of particulates into the stratosphere that the reflection of incoming solar radiation by these particles apparently produced a temporary reversal of the trend toward global warming. Earth's average surface temperature fell from 15.47 °C in 1990 to 15.13 °C in 1992.

Dust storms, especially in arid regions, can add massive amounts of particulate matter to the atmosphere. Smoke and dust from forest fires in Mexico and Central America drift across the United States and into Canada. Dust from the Sahara reaches the Caribbean area and South America. Swamps and marshes emit noxious gases such as hydrogen sulfide, a toxic gas with the odor of rotten eggs.

An important source of particulates in the atmosphere is the sea. Wave action causes seawater droplets to be suspended in the air, and when these droplets evaporate, salt particles are left behind in the air. Tropical thunderstorms carry chloride ions into the stratosphere, where they can be converted to chlorine atoms and perhaps contribute to the destruction of ozone. Thus, the greatest contributor to all the particulate matter in the atmosphere is ordinary salt from a perfectly natural source, reminding us yet again that nature is not always benign.

▲ An eruption at Kilauea volcano on the island of Hawaii.

According to the World Health Organization, Mexico City has the world's worst air pollution. Pollution levels are at least double those considered safe. One five-day episode in 1996 caused 400,000 pollution-related hospital and clinic contacts. The pollution originates from automobiles, industries, and propane used in home heating and cooking. The problem is compounded by location: Mexico City is in a valley surrounded by mountains.

matter, might be reasonably safe. A certain level of particulate matter, without sulfur dioxide around, might also be reasonably safe. Take these same levels of the two together, however, and the effect might be considerably exacerbated, aggravating respiratory problems such as bronchitis or triggering acute asthma attacks. *Synergistic effects* such as this are quite common whenever certain chemicals are brought together: The effect of the two together is much greater than the sum of their individual effects. For example, some forms of asbestos are carcinogenic, and about 35 or 40 of the chemicals in cigarette smoke are carcinogens. Asbestos workers who smoke develop cancer at a much greater rate than do people who are exposed to one of these carcinogens but not the other.

When inhaled deeply into the lungs, the pollutants in industrial smog break down the cells of the tiny air sacs, called alveoli, where oxygen and carbon dioxide exchange ordinarily occurs. The alveoli lose their resilience; as a result, it becomes difficult for them to expel carbon dioxide. Such lung damage contributes to pulmonary emphysema, a condition characterized by an increasing shortness of breath. Emphysema is one of the fastest-growing causes of death in the United States. Although the principal factor in the rise of emphysema is cigarette smoking, air pollution is also a factor.

Not only are the oxides of sulfur and the aerosol mists of sulfuric acid damaging to humans and other animals, they damage plants as well. Leaves become bleached and mottled when exposed to sulfur oxides. The yield and quality of farm crops can be severely affected. In addition, these compounds are also major contributors to acid rain.

Controlling Industrial Smog

Soot and fly ash can be removed from smokestack gases in several ways. One method, the use of electrostatic precipitators, is illustrated in Figure 25.10. The particles in the smokestack gases are given an electric charge, and the charged particles deposit on the oppositely charged collector plate. The energy requirement of this method is high. About 10% of the electricity produced in a power plant is required to operate the electrostatic precipitators. And the ash has to be put somewhere. Ash production in the United States is about 70 million metric tons per year. Some of the ash is used in making cement, and some is melted and formed into fibers called mineral wool, used for insulation. Most of the ash is simply stored on the ground or buried in landfills.

Emissions of SO_x can be reduced by removing sulfur-containing minerals before burning the coal. For example, FeS_2 (*pyrite*) can be concentrated and removed by flotation (Figure 24.1). Another way is to convert the coal to gaseous or liquid hydrocarbons by reaction with $H_2(g)$, leaving minerals such as pyrite behind. Both methods are expensive, however.

An alternative is to remove SO_2 from stack gases after the coal has been burned. In one method, powdered coal and limestone are burned together. The limestone decomposes to CaO and CO_2, and SO_2 reacts with the CaO:

$$CaCO_3(s) \longrightarrow CaO(s) + CO_2(g)$$
$$CaO(s) + SO_2(g) \longrightarrow CaSO_3(s)$$

Sulfur dioxide can also be reacted with hydrogen sulfide, producing elemental sulfur that is easily recovered:

$$2\,H_2S(g) + SO_2(g) \longrightarrow 3\,S(s) + 2\,H_2O(l)$$

Still another possibility is to convert sulfur dioxide to sulfuric acid. However, if all the SO_2 in smokestack gases were converted to sulfuric acid, the quantity produced would greatly exceed current demand. Then the sulfuric acid would present a disposal problem.

25.3 The Ozone Layer

The **ozone layer** is a band of the stratosphere about 20 km thick, centered at an altitude of about 25–30 km. This layer has a much higher ozone concentration than the rest of the atmosphere. Ozone absorbs harmful UV radiation, and the ozone layer thus protects life on Earth.

Ultraviolet radiation, invisible to human eyes, has profound effects on living matter. Proteins, nucleic acids, and other cell components are broken down by any UV radiation of wavelength shorter than about 290 nm. Wavelengths below 230 nm are filtered out by O_2 and other atmospheric components, but ozone alone absorbs in the wavelength range from 230 to 290 nm. In addition, radiation with wavelengths from 290 to 320 nm, called UV-B radiation, produces sunburn and can also cause eye damage and skin cancer; this radiation is only partially absorbed by ozone. Thus, the total amount of UV radiation reaching Earth's surface is critically dependent on the concentration of $O_3(g)$ in the ozone layer. Society therefore faces a challenge to maintain the appropriate concentration of ozone in the ozone layer.

When two opposing processes occur, one producing and the other consuming a substance, the result is a fairly constant concentration. Let us consider the processes that lead to a natural balanced concentration of about 8 ppm of ozone in the ozone layer. Ozone is produced in the upper atmosphere in a sequence of two reactions. First, an O_2 molecule absorbs UV radiation and dissociates into two O atoms:

$$O_2 + h\nu \longrightarrow O + O$$

Then atomic and molecular oxygen react to form ozone. The highly energetic O_3 thus formed would simply decompose back to O_2 and O except for the fact that some excess energy is dissipated when O_3 molecules collide with a third body, M, often a N_2 molecule. This dissipation of excess energy makes an accumulation of O_3 possible:

$$O_2 + O + (M) \longrightarrow O_3 + (M)$$

Fly ash particle

Stack gas flow

50,000 volts

▲ **FIGURE 25.10   An electrostatic precipitator**
Electrons emitted from the negatively charged electrode in the center attach themselves to the particles of fly ash, giving them a negative charge. The negatively charged particles are attracted to the positively charged cylindrical collector plate and are deposited there.

 Statospheric Ozone animation

Catalytic Destruction of Stratospheric Ozone animation

CFCs and Stratospheric Ozone animation

Ozone is decomposed in a sequence of two reactions. First, an ozone molecule decomposes when it absorbs UV radiation:

$$O_3 + h\nu \longrightarrow O_2 + O$$

Then, an atom of oxygen reacts with another ozone molecule, forming two O_2 molecules and releasing heat to the environment:

$$O + O_3 \longrightarrow 2\,O_2 \qquad \Delta H = -391.9\ \text{kJ}$$

The heat released in this reaction accounts for the characteristic temperature increase with altitude in the stratosphere (page 1037).

Other naturally occurring species also contribute to the decomposition of ozone. Atmospheric NO, produced mainly from N_2O released by soil bacteria, decomposes ozone through this sequence of reactions:

(1)	$\cancel{NO} + O_3 \longrightarrow \cancel{NO_2} + O_2$
(2)	$\cancel{NO_2} + O \longrightarrow \cancel{NO} + O_2$

Overall reaction: $\qquad O_3 + O \longrightarrow 2\,O_2$

Note that NO is consumed in the first reaction but regenerated in the second, which means that a little NO goes a long way. Moreover, we should expect that the injection of any additional NO into the stratosphere would increase the destruction of ozone and reduce its steady-state concentration. This additional NO can come, for example, from combustion processes in supersonic jets operating in the stratosphere. Perhaps the best evidence for the depletion of stratospheric ozone is that obtained from studies in Antarctica (Figure 25.11).

Of all human activities that affect the ozone layer, release of chlorofluorocarbons (CFCs) is thought to be the most significant. Because CFCs have a long lifetime in the atmosphere, some of the molecules eventually rise and appear in low concentrations in the stratosphere. There, they absorb UV radiation and decompose to produce atomic and molecular fragments (free radicals):

$$CCl_2F_2 + h\nu \longrightarrow \cdot CClF_2 + Cl\cdot$$

The ozone-destroying cycles set up beyond this point are quite complex, but a simplified representation is

(1)	$\cancel{Cl\cdot} + O_3 \longrightarrow \cancel{ClO\cdot} + O_2$
(2)	$\cancel{ClO\cdot} + O \longrightarrow \cancel{Cl\cdot} + O_2$

Overall reaction: $\qquad O_3 + O \longrightarrow 2\,O_2$

The chlorine atom that reacts in step 1 is regenerated in step 2; one chlorine atom can therefore destroy thousands of ozone molecules. (As in polymerization, the reaction proceeds until ended by the combination of two radicals.)

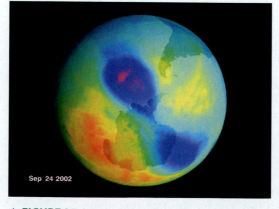
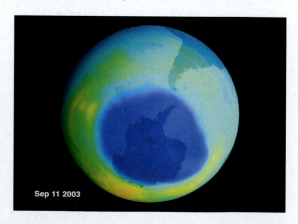

▲ **FIGURE 25.11**

The dark blue region indicates depletion of the ozone layer as measured by satellite on September 24, 2002 (left) and on September 11, 2003 (right). Notice the dramatic difference in area of the ozone hole.

QUESTION: What factors influence the size of the ozone hole over Antartica?

Currently, CFCs and other chlorine- or bromine-containing compounds that might diffuse into the stratosphere and contribute to ozone depletion are being replaced with more benign substances. Hydrofluorocarbons, such as CH_2FCH_3, have no Cl or Br to form radicals and are one kind of replacement. Another kind are hydrochlorofluorocarbons (HCFCs), such as CH_3CCl_2F. These molecules break down more readily in the troposphere, and thus fewer ozone-destroying molecules reach the stratosphere.

25.4 Global Warming: Carbon Dioxide and the Greenhouse Effect

We all exhale carbon dioxide with every breath; it is a normal product of respiration. Consequently, we generally do not think of CO_2 as an air pollutant. Low levels of CO_2 are not toxic, and this gas is a minor component of Earth's atmosphere. However, CO_2 plays a role in determining Earth's climate. Small increases in the concentration of CO_2 could have a profound effect on the environment by producing a significant increase in the average global temperature, an effect called **global warming.**

When electromagnetic radiation from the Sun reaches Earth, some is reflected back into space, some is absorbed by substances in the atmosphere, and some reaches Earth's surface and is absorbed there. The surface then gets rid of some of this absorbed solar energy by emitting infrared radiation toward outer space. Certain atmospheric gases, principally $CO_2(g)$ and $H_2O(g)$, absorb some of this infrared radiation, and as a result, this radiant energy is retained in the atmosphere and warms it. The process, known as the **greenhouse effect** because it resembles the retention of heat in a greenhouse, is summarized in Figure 25.12. The greenhouse effect is natural, and it is crucial to maintaining the proper temperature for life on Earth. Without it, Earth would be an icehouse, permanently covered with snow and ice. Scientists are concerned, however, with the effects of continued increases in the concentrations of greenhouse-enhancing gases.

Increased levels of CO_2 and other greenhouse gases are readily measured. Computer models of the atmosphere indicate that a CO_2 buildup is likely to cause an increase in Earth's average temperature. There are still uncertainties because it is impossible to identify all the factors that should be included in computer models and to know how heavily to weight each factor. For example, warming of the atmosphere could lead to the increased evaporation of water and an accompanying increase in cloud cover. Because clouds reflect some incoming radiation back into space, this

Trees consume water vapor and carbon dioxide gas in the process of photosynthesis. The deforestation of tropical rain forests, as in "slash-and-burn" clearing for agriculture, is a contributing factor in the increase of atmospheric $CO_2(g)$ levels.

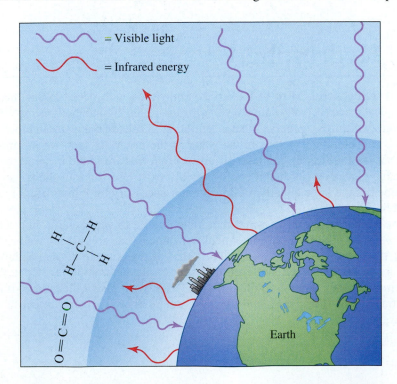

Greenhouse Effect activity

◀ **FIGURE 25.12   The greenhouse effect**

Electromagnetic radiation from the Sun, in the form of visible light, passes through the atmosphere and is absorbed by ground and ocean, warming Earth's surface. The warm surface then emits infrared radiation. Some of this infrared radiation is absorbed by CO_2, H_2O, and other gases and retained in the atmosphere as thermal energy.

QUESTION: By what mechanism do the greenhouse gases absorb infrared radiation? (See Section 23.12.)

▲ The number of huge icebergs that break off the continental ice shelf in Antarctica may increase as a result of global warming.

Motion pictures such as *Deep Impact* and *Armageddon* feature a large comet or asteroid threatening Earth. In reality, if such a comet or asteroid landed in an ocean, it would blast vast quantities of water vapor into the atmosphere. Because water vapor is a powerful greenhouse gas, one result of such an impact could be severe global warming. Even before impact, the frictional heat generated by the object moving through Earth's atmosphere would cause nitrogen and oxygen in the atmosphere to form vast quantities of NO. The NO could lead to substantial destruction of the ozone layer.

could cause global cooling rather than global warming. However, the models are being steadily improved, and almost all predict warming.

In 2001, the United Nations–sponsored Intergovernmental Panel on Climate Change summarized the various forecasts, predicting a global temperature increase of from 1.4 to 5.8 °C by 2100. New evidence suggests that "most of the observed warming" in recent decades has come from gases released as a result of human activities. The panel of scientists predicted that rising temperatures could lead to drastic shifts in weather, with droughts striking farming areas. They expressed fears that melting glaciers could raise sea levels, flooding densely populated coastal areas of China, Egypt, Bangladesh, and other countries. These effects could be experienced as early as the middle of the twenty-first century.

Scientists are not limited to computer modeling to assess the likelihood of global warming. Direct experimental evidence also exists. For example, the ice in the Greenland and Antarctic icecaps is laid down in layers much like the annual growth rings of trees. Analyses of tiny air bubbles trapped in these layers show a strong correlation between atmospheric CO_2 content and estimated global temperatures over the past 160,000 years—lower CO_2 levels correlate with lower temperatures, and higher levels with higher temperatures. Thus it seems reasonable to expect that global temperatures will continue to rise as CO_2 levels increase.

Most atmospheric scientists think that global warming is already under way. Both the atmosphere and the oceans have warmed measurably in the last 50 years, and these changes have been firmly linked to human activities. A few scientists still question the reality of global warming, noting large swings in average temperatures over the years. The uncertainty arises from the fact that the measured increase in average temperature over the past 50 years has been only a few tenths of a degree Celsius, whereas annual variations in certain regions are often as much as 4 or 5 °C.

The main strategy for countering a possible global warming is to curtail the use of fossil fuels, but this may not be enough. For example, several gases—methane, ozone, nitrous oxide (N_2O), and CFCs—are even better absorbers of infrared radiation than is carbon dioxide. (The 1.3 billion cattle in the world alone produce about 20% of the atmospheric methane.) At an international meeting in Kyoto, Japan, in 1998, most of the world's industrial nations agreed to limit emissions of CO_2 over the next several decades. However, the agreement has not yet been implemented. The debate continues about the need for immediate action and how drastic our efforts should be. The debate may continue for years.

The Hydrosphere

Earth is a water world. Most of its surface is covered with oceans, seas, lakes, and streams—waters collectively called the *hydrosphere*. The human body is mostly water, too—about two-thirds by mass. The concentration of salts in the water in blood is similar to the concentration of salts in the water in the ocean. In fact, we are much like walking sacks of seawater. Although primitive life forms are found in some seemingly inhospitable places on Earth and could possibly exist elsewhere in the solar system, Earth alone has the large quantity of liquid water necessary for higher life forms as we know them.

Some of the properties of water that make it able to support life, however, also make it easy to pollute. Many chemical substances are soluble in water. In fact, much of the chemistry we have studied is that of aqueous solutions. Soluble substances are easily dispersed and eventually enter our water supplies. Once they are there, removing them is often quite difficult. A daily supply of clean drinking water is essential to life and good health because contaminated water can cause disease and, in some cases, death.

25.5 Earth's Natural Waters

We have noted the unusual properties of water in previous chapters. The following list provides a quick review of some of the properties and their consequences.

- Water commonly occurs as a liquid, the only prevalent naturally occurring liquid on Earth's surface.
- The solid form of water (ice) is less dense than the liquid form; therefore water expands when it freezes.
- Water has a higher density than most other familiar liquids; hydrocarbons and other organic compounds that are insoluble in water and less dense than water float on its surface.
- Water has a high heat capacity and a high heat of vaporization. Thus, a given quantity of heat produces a much greater temperature increase in a landmass than in a body of water covering the same area. Bodies of water tend to make daily and seasonal temperature variations more moderate along their shores.

Although three-fourths of Earth's surface is covered with water, nearly 98% is salty seawater, unfit for drinking and unsuitable for most industrial purposes. More than 1% of Earth's water is frozen in the polar icecaps, which leaves less than 1% available as fresh water. Fresh water falls on Earth in enormous amounts as rain and snow, but most of it falls into the sea or in areas that are otherwise inaccessible. Thus, available water is not always where the people are. Some areas with adequate rainfall are unsuitable for human habitation because of extremely cold climates or steep mountain slopes. Some areas have adequate *average* rainfall but have periods of drought and flooding. In some places with adequate freshwater supplies, the water is too polluted for many uses.

Natural waters such as rainwater and groundwater are not pure H_2O. Rainwater carries dust particles from the atmosphere and dissolves a little oxygen, nitrogen, and carbon dioxide as it falls through the atmosphere. During electrical storms, traces of nitric acid are found in rainwater as well. Groundwater dissolves minerals from rocks and soil as it moves along on or beneath Earth's surface. Groundwater also dissolves matter from decaying plants and animals. The principal cations in natural groundwater are Na^+, K^+, Ca^{2+}, Mg^{2+}, and sometimes Fe^{2+} or Fe^{3+}. The anions are usually SO_4^{2-}, HCO_3^-, and Cl^-. Table 25.2 provides a summary of substances found in natural waters.

Table 25.2 Some Substances Found in Natural Waters

Substance	Formula	Source
Carbon dioxide	CO_2	Atmosphere
Dust	—	Atmosphere
Nitrogen	N_2	Atmosphere
Oxygen	O_2	Atmosphere
Nitric acid (thunderstorms)	HNO_3	Atmosphere
Sand and soil particles	—	Soil and rock
Sodium ions	Na^+	Soil and rock
Potassium ions	K^+	Soil and rock
Calcium ions	Ca^{2+}	Limestone rock
Magnesium ions	Mg^{2+}	Dolomite rock
Iron(II) ions	Fe^{2+}	Soil and rock
Chloride ions	Cl^-	Soil and rock
Sulfate ions	SO_4^{2-}	Soil and rock
Bicarbonate ions	HCO_3^-	Soil and rock

▲ Modern treatment procedures in the United States produce water that is clean and free of disease-causing organisms.

25.6   Water Pollution

Early people did little to pollute the water and the air, if only because their numbers were few. The Industrial Revolution and a concurrent large increase in population led to serious pollution of the environment. However, the pollution was mostly local and largely biological. Human wastes were dumped on the ground or into the nearest stream. Disease organisms were transmitted through food, water, and direct contact.

Contamination of water supplies by microorganisms from human wastes was a severe problem throughout the world until about 100 years ago. As population increased over the centuries, pollution became more and more of a problem. Then, starting in the 1830s, severe epidemics of cholera swept the Western world, and typhoid fever and dysentery were common decade after decade. In 1900, for example, there were more than 35,000 deaths from typhoid in the United States. It was at about this time that scientists began making the connection between disease and contaminated water. Today, as a result of chemical treatment, municipal water supplies in the more developed nations are generally safe. However, waterborne diseases are still quite common in much of Asia, Africa, and Latin America. Worldwide, an estimated 80% of all the world's sickness is caused by contaminated water. People with waterborne diseases fill half the world's hospital beds and die at a rate of 25,000 per day. Fewer than 10% of the people of the world have access to sufficient clean water, with the consequence that there are still epidemics of cholera, typhoid, and dysentery in many parts of the world.

How much water does one really need? Even though one person needs only about 1.5 L per day for drinking, the average U.S. resident uses about 7 L for drinking and cooking and overall consumes directly almost 400 L each day (Table 25.3). The combined residential, industrial, commercial, and agricultural use of water adds up to an average of 6900 L per person per day, and the rate of use is rapidly increasing. Much of this water is used indirectly in agriculture and industry to produce food and other materials. For example, it takes 800 L of water to produce 1 kg of vegetables and 13,000 L of water to produce a steak from beef cattle fed from irrigated croplands.

We also use water for recreation—for example, swimming, boating, and fishing. For most of these purposes, we need water that is free from bacteria, viruses, and parasitic organisms.

The threat of biological contamination has not been totally eliminated from developed nations. An estimated 30 million people in the United States are at risk because of bacterial contamination of drinking water, for example. Even in the most developed nations, hepatitis A, a viral disease spread through drinking water and contaminated food, at times threatens to reach epidemic proportions. Biological contamination also lessens the recreational value of water, leading to a ban of swimming in many areas.

Table 25.3 Average Daily Per-Person Use of Water in the United States

Use	Amount (L)
Direct use	
Drinking and cooking	7
Flushing toilets	80
Supplying swimming pools and watering lawns	85
Dish washing	14
Bathing	70
Laundry	35
Miscellaneous	90
Total direct use	**381**
Indirect use	
Industrial	3800
Irrigation (agriculture)	2150
Municipal water (nonindustrial)	550
Total indirect use	**6500**
Total overall use	**6881**

Chemical Contamination: From Farm, Factory, and Home

In the past, factories often were built on the banks of streams, and wastes were dumped into the water to be carried away. Currently, fertilizers and pesticides used in agriculture and on lawns and recreational areas such as golf courses have found their way into the water system and have further contaminated it. Transportation of petroleum results in oil spills in oceans, estuaries, and rivers. Acids enter waterways from mines and factories and from acid precipitation. Household chemicals also contribute to water pollution when detergents, solvents, and other chemicals are dumped down drains.

About half the people in the United States drink surface water (from streams and lakes). The other half get their drinking water from groundwater. Toxic chemicals from various sources have been found in both kinds of water supplies. For example, people living close to the Rocky Mountain Arsenal, near Denver, have found their wells contaminated by wastes from the production of pesticides. Wells near Minneapolis are contaminated with creosote, a chemical used as a wood preservative. Water wells in Wisconsin and on Long Island, New York, have been contaminated with aldicarb, a pesticide used on potato crops. Community water supplies in New Jersey have been shut down because of contamination with industrial wastes.

Chemicals buried in dumps—often years ago, before there was much awareness of environmental problems—have now infiltrated groundwater supplies. Often, as at the Love Canal site in Niagara Falls, New York, people built schools and houses on or near old dumpsites. Common contaminants are hydrocarbon solvents, such as benzene and toluene, and chlorinated hydrocarbons, such as carbon tetrachloride (CCl_4), chloroform ($CHCl_3$), and methylene chloride (CH_2Cl_2). Especially common is trichloroethylene (CCl_2CHCl), widely used as a dry-cleaning solvent and as a degreasing compound. These organic compounds dissolve in water only in trace quantities, often in the ppm or ppb range, and it is these tiny amounts that are found in groundwater. However, these chlorinated hydrocarbons are unwanted even in trace amounts, for most of them are suspected carcinogens. Unfortunately, the compounds are notably lacking in reactivity. They therefore decompose so slowly that they are likely to be around for a long time.

Another major source of groundwater contamination is leaking underground storage tanks. Gasoline at service stations has traditionally been stored in buried steel tanks. There are perhaps 2.5 million such tanks in the United States. The tanks last an average of about 15 years before they rust through and begin to leak. As many as 200,000 may now be leaking, many of them at stations that went out of business during the fuel shortages of the 1970s. In many areas of the country, gasoline is now found in water wells near these tanks. Laws now require replacement of old gasoline storage tanks and proper cleanup of any contaminated ground. Groundwater contamination is a serious long-term problem because, once contaminated, an underground aquifer may remain unusable for decades or longer. There is no easy way to remove the contaminants. Pumping out the water and purifying it could take years and cost billions of dollars.

Although groundwater pollution can be a serious problem, it is often overdramatized by the media. The amounts of contaminants are often truly minute. For example, the U.S. Environmental Protection Agency (EPA) has set a safety limit of 10 ppb for aldicarb—that is 10 mg of aldicarb in 1000 L of water. You would have to drink 32,000 L of water to get as much aldicarb as there is aspirin in one tablet. Aldicarb is moderately toxic to mammals, but it breaks down rather quickly in the environment. Contamination of groundwater by toxic substances is serious, but biological contamination is much more widespread and often more deadly than chemical contamination.

Industries in the United States have eliminated a considerable proportion of the water pollution they once produced. Most industries are now in compliance with the Water Pollution Control Act, which requires that they use the best practicable technology. Let us consider a specific case in pollution control.

Waste chromium, in the form of chromate ions (CrO_4^{2-}), and cyanide ions (CN^-) are products of the chromium plating of steel. In the past, these toxic substances were often dumped into waterways. Nowadays, they may be removed to a large extent by

▲ Most of the homes in the Love Canal area near Niagara Falls, New York, have been abandoned because the ground on which they stand has been contaminated by industrial waste.

▲ Rusting gasoline storage tanks can leak gasoline and its additives into the environment. Methyl *tert*-butyl ether (MTBE) is one such additive that is now of environmental concern.

Ironing Out Pollutants

Contaminated groundwater from a leaking landfill often moves in a plume toward a nearby lake or stream. Is there an effective way to remove contaminants as the water moves through the ground? A relatively inexpensive and quite effective way of removing chlorinated compounds is to place a pit containing scrap-iron filings as a barrier in the path of the flow (Figure 25.13).

Recall that iron corrodes through a redox reaction in which dissolved oxygen is the oxidizing agent and iron is the reducing agent:

Oxidation half-reaction:

$$2\{Fe \longrightarrow Fe^{2+} + 2\,e^-\}$$

Reduction half-reaction:

$$O_2 + 4\,H^+ + 4\,e^- \longrightarrow 2\,H_2O$$

Overall reaction:

$$2\,Fe + O_2 + 4\,H^+ \longrightarrow 2\,Fe^{2+} + 2\,H_2O(l)$$

In a reaction with chlorinated compounds, Fe atoms give electrons to the chlorine atoms, converting them to chloride ions. In the reduction half-reaction, a chlorine-containing compound, designated RCl, is converted to a hydrocarbon, RH:

Reduction half-reaction: $RCl + H^+ + 2\,e^- \longrightarrow RH + Cl^-$

The overall reaction can be written

$$Fe + RCl + H^+ \longrightarrow Fe^{2+} + RH + Cl^-$$

The hydrocarbons formed are generally less toxic and more easily degraded by microorganisms than the chlorine-containing compounds.

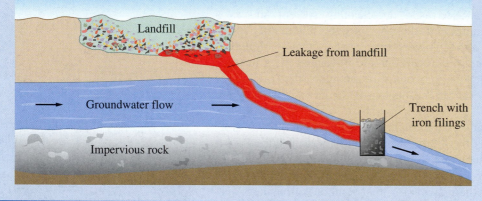

◀ **FIGURE 25.13   Using iron filings to remove chlorinated hydrocarbons from contaminated groundwater**

As contaminated groundwater from a leaking landfill moves toward a nearby lake or stream, it passes through a barrier of iron filings. The iron converts the chlorinated compound to a hydrocarbon plus chloride ions.

chemical treatment. Cyanide ions are removed by a reaction with chlorine in basic solution to form nitrogen gas, bicarbonate ions, and chloride ions:

$$10\,OH^-(aq) + 2\,CN^-(aq) + 5\,Cl_2(g) \longrightarrow N_2(g) + 2\,HCO_3^-(aq) + 10\,Cl^-(aq) + 4\,H_2O(l)$$

Bicarbonate and chloride ions are generally not considered toxic at the levels formed in this reaction.

Chromate ions are removed by reduction with sulfur dioxide, forming Cr^{3+} ion and sulfate ions:

$$2\,CrO_4^{2-}(aq) + 3\,SO_2(g) + 2\,H_2O(l) \longrightarrow 2\,Cr^{3+}(aq) + 3\,SO_4^{2-}(aq) + 4\,OH^-(aq)$$

The Cr^{3+} ions can be precipitated and removed as $Cr(OH)_3(s)$. However, proper control of the pH is essential because $Cr(OH)_3$ is amphoteric; it redissolves in a solution that is either too acidic or too basic. Sulfate ion is usually not a serious pollutant.

Many industries have contributed to water pollution. Wastes from the textile industries include conditioners, dyes, bleaches, oils, dirt, and other organic debris. Most of these can be removed by conventional sewage treatment. Wastes from meat-packing plants include blood and various animal parts. These and other food industry wastes are usually treated by sewage treatment plants.

Some Chemistry and Biology of Sewage

Dumping human sewage into waterways spreads *pathogenic* (disease-causing) microorganisms, but disease is not the only problem. The organic matter in sewage is decomposed by bacteria, and this process depletes dissolved oxygen [O_2(aq)] in the water and overfertilizes the water with plant nutrients. A flowing stream can handle a small amount of waste without difficulty, but large quantities of raw sewage cause undesirable changes.

Most organic material can be degraded (broken down) by microorganisms. This biodegradation can be either *aerobic* or *anaerobic*. **Aerobic oxidation** occurs in the presence of dissolved oxygen. The **biochemical oxygen demand (BOD)** measures the quantity of oxygen, in milligrams, needed for the aerobic oxidation of the organic compounds in 1 L of water. If the BOD is high enough, oxygen is depleted, and higher life forms, such as fish, can no longer survive in the water. However, rapidly flowing streams can regenerate themselves downstream as oxygen from the air enters the moving water and dissolves.

With adequate dissolved oxygen, aerobic bacteria (those that require oxygen) oxidize the organic matter to carbon dioxide, water, and a variety of inorganic ions (Table 25.4). The water is relatively clean, but the ions, particularly the nitrates and phosphates, may serve as nutrients for the growth of algae, which also cause problems. When the algae die, they become organic waste and increase the BOD through a process called **eutrophication.** Algal bloom and die-off are also stimulated by the runoff of fertilizers from farms and lawns and seepage from feedlots. This combination leads to dead and dying streams and lakes that nature cannot purify nearly as quickly as we can pollute them.

When the dissolved oxygen in a body of water is depleted by too much organic matter—whether from sewage, dying algae, or other sources—**anaerobic decay** processes take over. Instead of oxidizing the organic matter, anaerobic bacteria reduce it. Methane (CH_4) is formed. Sulfur is converted to hydrogen sulfide (H_2S) and other foul-smelling organic compounds. Nitrogen is reduced to ammonia and odorous amines. The foul odors indicate that the water is overloaded with organic wastes. No life, other than a few anaerobic organisms, can survive in such water.

Water Treatment: A Drop to Drink

Many cities use water that has been used by other cities upstream. Such water may be polluted with chemicals and pathogenic microorganisms. Making the water safe and palatable involves several steps of physical and chemical treatment (Figure 25.14). The water usually is placed in a settling basin, where it is treated with slaked lime [$Ca(OH)_2$(aq)] and a flocculating agent, such as aluminum sulfate. These materials react to form a gelatinous mass (called *flocs*) of aluminum hydroxide that carries down dirt particles and bacteria:

$$3\ Ca(OH)_2(aq) + Al_2(SO_4)_3(aq) \longrightarrow 2\ Al(OH)_3(s) + 3\ CaSO_4(s)$$

Slaked lime

The water is then filtered through sand and gravel.

▲ An algal bloom leads to oxygen depletion in the water of this pond.

Table 25.4 Some Substances Added to Water by the Breakdown of Organic Matter

Substance	Formula
Aerobic conditions	
Carbon dioxide	CO_2
Nitrate ions	NO_3^-
Phosphate ions	PO_4^{3-}
Sulfate ions	SO_4^{2-}
Bicarbonate ions	HCO_3^-
Anaerobic conditions	
Methane	CH_4
Ammonia	NH_3
Amines	RNH_2
Hydrogen sulfide	H_2S
Methanethiol	CH_3SH

◀ Methane, sometimes called marsh gas, is formed by the anaerobic decay of organic matter. This painting, *Dalton Collecting Marsh Fire Gas,* shows John Dalton, developer of the atomic theory (Chapter 2). *Source:* Ford Madox Brown (1821–1893), "Dalton Collecting Marsh Fire Gas." © Manchester Art Gallery. Photo by John Ingram.

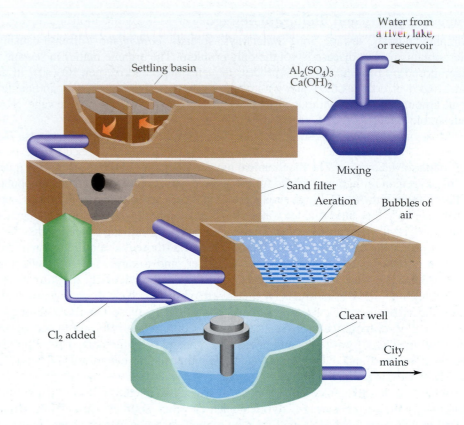

Water Treatment activity

▶ FIGURE 25.14   A diagram of a municipal water purification plant

The next step is usually *aeration:* The water is sprayed into the air to remove odorous compounds and to improve its taste (water without dissolved air tastes flat). Sometimes the water is filtered through charcoal, which adsorbs colored and odorous compounds. In the final step, the water is *chlorinated:* Chlorine is added to kill any remaining bacteria. In some communities that use lake or river water, a lot of chlorine is needed to kill all the bacteria, and the residual chlorine imparts an unpleasant taste to the water.

Some pollutants are difficult to remove from water. For example, nitrates get in groundwater from fertilizers used on farms and lawns, from decomposition of organic wastes in sewage treatment, and from runoff from animal feedlots. Nitrate compounds are highly soluble, and only expensive treatment can remove them from water once they are there.

Many people drink bottled water to avoid real or perceived problems with public water supplies. Bottled water was largely unregulated until 1995, when the Food and Drug Administration set standards to ensure that its minimum quality was equal to that of public water supplies. Much bottled water comes from the same municipal supplies as tap water.

In addition to chlorination to kill bacteria, many communities add fluorides to drinking water to prevent dental caries (tooth decay). Tooth decay was once considered to be the leading chronic disease of childhood, but its incidence has decreased dramatically, thanks in part to fluoridation of drinking water. Only 29% of the nine-year-olds in the United States were cavity-free in 1971, but today more than two-thirds of that age group are cavity-free. Recall that tooth enamel is a complex calcium phosphate called hydroxyapatite (page 692). Fluoride ions replace some of the hydroxide ions, forming a harder mineral called fluorapatite:

$$Ca_5(PO_4)_3OH + F^- \longrightarrow Ca_5(PO_4)_3F + OH^-$$

Enough fluoride (usually as H_2SiF_6 or Na_2SiF_6) is added to the water to give concentrations of 0.7 to 1.1 ppm (by mass) of fluoride. Evidence indicates that such fluoridation results in a reduction in the incidence of dental caries by as much as two-thirds in some areas.

There is some concern about the cumulative effects of consuming fluorides in drinking water, in the diet, in toothpaste, and from other sources. Excessive fluoride consumption during early childhood can cause mottling of the tooth enamel. The enamel becomes brittle in certain areas and gradually discolors. At concentrations higher than those found in drinking water, fluorides may interfere with calcium metabolism, with kidney action, with thyroid function, and with the actions of other glands and organs. In moderate to high concentrations, fluorides are acute poisons. Indeed, sodium fluoride is used as a poison for roaches and rats. However, there is little or no evidence that fluoridation at the levels now used causes any health problems.

Wastewater Treatment Plants

For many years, most communities simply held sewage in settling ponds for a while before discharging it into a stream, lake, or ocean. This constitutes what now is called **primary sewage treatment.** Primary treatment removes some of the solids as *sludge,* but the effluent still has a huge BOD. Often all the dissolved oxygen in the pond is used up, and anaerobic decomposition—with its resulting odors—takes over. Effluent from a primary treatment plant contains considerable dissolved and suspended organic matter. A **secondary sewage treatment** plant passes effluent from a settling tank through sand and gravel filters. There is some aeration in this step, and aerobic bacteria convert most of the organic matter to stable inorganic materials.

A combination of primary and secondary treatment methods, known as the **activated sludge method** (Figure 25.15), is frequently employed. The sewage is placed in tanks and aerated with large blowers. This causes the formation of large, porous flocs, which serve to filter and absorb contaminants. The aerobic bacteria further convert the organic material to sludge. A part of the sludge is recycled to keep the process going, but huge quantities must be removed for disposal. This sludge is stored on land (where it requires large areas), dumped at sea (where it pollutes the ocean), or burned in incinerators (where it requires energy—such as natural gas—and contributes to air pollution). Sometimes the sludge is processed for use as fertilizer.

In many areas, secondary treatment of wastewater is inadequate. **Advanced treatment,** sometimes called *tertiary treatment,* is increasingly required. Several advanced processes are now in use. One process that is becoming more important is charcoal filtration. Charcoal adsorbs organic molecules that are difficult to remove by other methods. There are federal mandates for more communities to treat their water with charcoal, but some cities have not yet complied because of the expense. Even more costly processes, such as reverse osmosis (page 513), may be needed in some cases. Table 25.5 provides a summary of wastewater treatment methods.

The effluent from sewage plants usually is treated with chlorine before being returned to a waterway. Chlorine is added in an attempt to kill any pathogenic microorganisms that might remain. Such treatment has been quite effective in preventing the spread of such waterborne infectious diseases as typhoid fever. Further, some chlorine remains in the water, providing residual protection against pathogenic bacteria. However, chlorination is not effective against viruses, such as those that cause hepatitis.

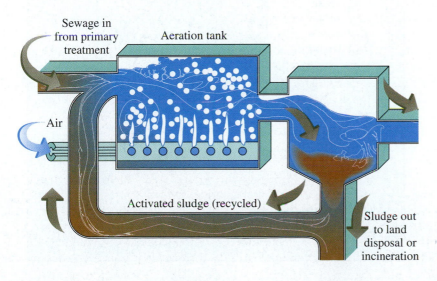

Sewage in from primary treatment

Aeration tank

Air

Activated sludge (recycled)

Sludge out to land disposal or incineration

◄ **FIGURE 25.15**   **A diagram of a secondary sewage treatment plant that uses the activated sludge method**

Table 25.5 Summary of Wastewater Treatment Methods

Method	Cost	Material Removed	Percent Removed
Primary			
Sedimentation	Low	Dissolved organics	25–40
		Suspended solids	40–70
Secondary			
Trickling filters	Moderate	Dissolved organics	80–95
		Suspended solids	70–92
Activated sludge	Moderate	Dissolved organics	85–95
		Suspended solids	85–95
Advanced (tertiary)			
Carbon bed with regeneration	Moderate	Dissolved organics	90–98
Ion exchange	High	Nitrates and phosphates	80–92
Chemical precipitation	Moderate	Phosphates	88–95
Filtration	Low	Suspended solids	50–90
Reverse osmosis	Very high	Dissolved solids	65–95
Electrodialysis	Very high	Dissolved solids	10–40
Distillation	Extremely high	Dissolved solids	90–98

25.7   Acid Rain and Acid Waters

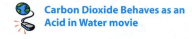

Carbon Dioxide Behaves as an Acid in Water movie

We have seen how sulfur oxides are converted to sulfuric acid, and nitrogen oxides to nitric acid, in the atmosphere. These acids reach Earth as acid rain or acid snow, in the form of deposits from acid fog, or as adsorbents on particulate matter. When rainfall is more acidic than it would be if it contained just dissolved atmospheric $CO_2(g)$, it is called **acid rain.** Some rainfall has been reported that is even more acidic than vinegar or lemon juice.

Acid rain corrodes metals, limestone, and marble, and even ruins the finish on automobiles. It comes mainly from sulfur oxides emitted from power plants and smelters and from nitrogen oxides from automobiles. These acids may be carried for hundreds of kilometers before falling as rain or snow (Figure 25.16). (In addition to

▶ **FIGURE 25.16   Acid rain**

Acids formed from sulfur oxides, mainly from coal-burning power plants, and nitrogen oxides from power plants and automobiles may fall in rain hundreds of kilometers from their sources. These acids cross oceans and international boundaries, causing political as well as environmental problems. This map shows that acid deposition is most severe over the northeastern United States and eastern Canada.

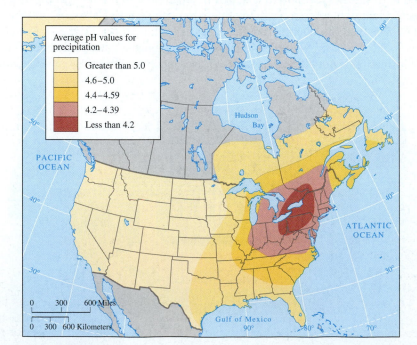

rainfall, another way surface waters are made acidic is from acids flowing into streams from abandoned mines.)

Acid Waters: Dead Lakes

Acid water is detrimental to life in lakes and streams. More than 1000 bodies of water in the eastern United States are acidified, and 11,000 others have only a limited ability to neutralize the acids that enter them. In the Canadian province of Ontario, 48,000 lakes are threatened, and more than 100 lakes in New York's Adirondack Mountains are so acidic that they are devoid of life. The acid rain falling on these areas is thought to originate mainly in the Ohio River Valley and Great Lakes regions.

Acid water has been linked to declining crop and forest yields, but its effects on living organisms are hard to pin down precisely. Probably the greatest effect of acidity is that it causes the release of toxic ions from rocks and soil. For example, aluminum ions, which are tightly bound in clays and other minerals, are released in acidic solution:

▲ The appearance of an acid lake can be quite deceiving. It can be quite beautiful, yet there is not a fish to be found in it. The lake is devoid of life. Acid rain has wiped out entire fish populations and other aquatic life in lakes that lack neutralizing limestone rocks in their watershed.

$$\underset{\text{A clay}}{Al_2Si_2O_5(OH)_4(s)} + 6\,H_3O^+(aq) \longrightarrow 2\,Al^{3+}(aq) + \underset{\text{Sand}}{2\,SiO_2(s)} + 11\,H_2O(l)$$

Aluminum ions have low toxicity to humans, but they seem to be deadly to young fish. Many of the dying lakes have only mature fish; none of the young survive. Ironically, lakes destroyed by excess acidity are often quite beautiful. The water is clear and sparkling—quite a contrast to the water in lakes where fish are killed by oxygen depletion following algal blooms.

Acids are no threat to lakes and streams in areas where the rock is limestone, which can neutralize excess acid. Where rock is principally granite, however, no such neutralization occurs. Acidic waters can be neutralized by adding lime or pulverized limestone. A few such attempts have been carried out, but the process is costly and the results last only a few years. An obvious way to lessen the problem is to either remove the sulfur from coal before burning it or scrub the sulfur oxides from smokestack gases. Considerable progress has been made in this area, but these remedies are expensive, and they add to the cost of electricity.

Toxic Substances in the Biosphere

In concluding this chemistry text, we briefly consider the most important "sphere" on Earth—the **biosphere,** that relatively thin film of air, water, and soil in which almost all life on Earth exists. Nearly all the processes in the biosphere are driven by energy from the Sun. About 23% of the solar energy that reaches Earth drives the water cycle. A tiny portion is absorbed by green plants, which use it to power photosynthesis, directly supplying the energy that sustains life. Let us now turn our attention to some toxic substances that sometimes seem to threaten living things.

One of the main concerns about certain pollutants is their toxicity. Toxic substances, often called poisons, have always been with us. Today our knowledge of poisons is greater than in times past, plus there are more of them than ever before. Industrial accidents, such as that at Bhopal, India, in 1984, which killed 2500 people and injured perhaps 100,000 others, have made the public acutely aware of problems with toxic substances. People are also concerned about long-term exposure to toxic substances in the air, in their drinking water, and in their food. Chemists can detect exceedingly tiny quantities of such substances. It is still quite difficult, however, to determine what effects these trace amounts of toxic materials have on human health. **Toxicology** is the study of the effects of poisons, their identification or detection, and the development and use of antidotes.

▲ Many common plants and plant parts are poisonous, including hydrangeas (top), holly berries (middle), and philodendron (bottom). Plants that produce toxic substances have evolved by the process of natural selection. The toxic substances serve as antifeedants that poison the animals that would otherwise eat the plants.

25.8 Poisons

What is a poison? Perhaps a better question would be, How much is a poison? A substance may be harmless—or even a necessary nutrient—in one amount but injurious or even deadly in another. Even a common substance such as table salt can be poisonous when eaten in abnormally large amounts; too much salt ingested at one time can induce vomiting. There have even been cases of fatal poisoning when salt was accidentally substituted for lactose (milk sugar) in formulas for infants. Some substances are obviously more toxic than others, however. It would take a massive dose of salt to kill the average healthy adult, whereas only a few micrograms of some of the nerve poisons can be fatal. Toxicity depends largely on the chemical nature of the substance.

People also respond differently to the same chemical. To cite an extreme case, 10–20 grams of sugar would cause no acute symptoms in most people but might be dangerous to a diabetic. Excessive amounts of sodium chloride would be especially serious to a person with edema (swelling due to excessive amounts of fluid in the tissues).

Still another complicating factor is that chemicals behave differently when administered in different ways. Nicotine is more than 50 times as toxic when applied intravenously as when taken orally. Good, fresh water is delightful when we drink it, but even water can be deadly when inhaled into the lungs in sufficient quantity. Further complications arise from the fact that even closely related animal species can react differently to a given chemical. Even individuals within a species may react to different degrees.

Poisons Around the House and in the Garden

Many household chemicals are poisonous. Drain cleaners, oven cleaners, and toilet bowl cleaners are highly corrosive to tissues. Some insecticides and rodenticides are quite toxic. Laundry bleach and ammonia are both toxic, and when they are mixed together, the combination can be deadly. Sodium hypochlorite in the bleach reacts with NH_3 to form highly toxic chloramines (NH_2Cl, $NHCl_2$, and NCl_3) and nitrosyl chloride ($NOCl$).

Even seemingly harmless products around the house can be dangerous to young children. A bottle of cough syrup can trigger an emergency visit to the hospital. Toxic substances are also used in home gardens. Herbicides and insecticides are not the only poisons you are likely to find in a garden. Sometimes the plants themselves are toxic. Irises are beautiful, and so are azaleas, hydrangeas, and oleander, but all of these popular perennials are poisonous. Holly berries, wisteria seeds, and the leaves and berries of privet hedges are also among the more poisonous products of the home garden. Some houseplants, such as the philodendron, are also toxic.

Corrosive Poisons

Strong acids and bases and strong oxidizing agents are highly corrosive to human tissue. These chemicals indiscriminately destroy living cells. Both acids and bases, even in dilute solutions, catalyze the hydrolysis of protein molecules in living cells. These reactions involve the breaking of the amide (peptide) linkages in the molecules. In cases of severe exposure, the fragmentation continues until the tissue is completely destroyed.

Acidic air pollutants, such as sulfuric acid aerosols and acids formed in the incineration of plastics and other wastes, are particularly destructive of lung tissue. Other air pollutants also damage living cells. Ozone, peroxyacetyl nitrate (PAN), and the other oxidizing components of photochemical smog probably do their main damage through the deactivation of enzymes. The active sites of enzymes often incorporate —SH groups, and ozone can oxidize these thiol groups to sulfonic acid groups (—SO_3H). This change renders an enzyme inactive and halts vital processes in the cell. No doubt, oxidizing agents can also break bonds in many of the chemical substances in a cell. Such powerful agents as ozone are more likely to make an indiscriminate attack than to react in a highly specific way.

Agents that Block Oxygen Transport and Use

Certain chemical substances block the transport of oxygen in the bloodstream and prevent the oxidation of metabolites by oxygen in the cells. All act on the iron atoms in complex protein molecules. Carbon monoxide binds tightly to the iron atom in hemoglobin, blocking the transport of oxygen (Section 25.2). Nitrates also diminish the ability of hemoglobin to carry oxygen. They are reduced to nitrites by microorganisms in the digestive tract, and in turn the iron atoms in hemoglobin are oxidized from Fe^{2+} to Fe^{3+}. The resulting *methemoglobin* is incapable of carrying oxygen.

Cyanides are among the most notorious poisons in both fact and fiction. Sodium cyanide is used to extract gold and silver from ores and in electroplating baths, and cyanides can enter the environment from these processes. Hydrogen cyanide is used (with great care by specially trained experts) to exterminate insects and rodents in the holds of ships, in warehouses, in railway cars, and on citrus and other fruit trees. Hydrogen cyanide is generated easily enough by treating a cyanide salt with an acid:

$$CN^-(aq) + H_3O^+(aq) \longrightarrow HCN(g) + H_2O(l)$$

NaCN can be accidentally or deliberately ingested, and HCN can be inhaled. In either case, cyanides act almost instantaneously, and it takes only tiny quantities to kill. The average fatal dose for an adult is only 50–200 mg.

Cyanide blocks the oxidation of glucose inside the cell by forming a stable complex with iron- and copper-containing enzymes called *cytochrome oxidases*. The enzymes normally act by providing electrons for the reduction of oxygen in the cell. Cyanide ties up these mobile electrons, rendering them unavailable for the reduction process and bringing an abrupt end to cellular respiration.

Sodium thiosulfate ($Na_2S_2O_3$) is an antidote for cyanide poisoning, but it must be administered quickly. A sulfur atom is transferred from the thiosulfate ion to the cyanide ion, converting cyanide to relatively innocuous thiocyanate ions (SCN^-):

$$S_2O_3^{2-} + CN^- \longrightarrow SCN^- + SO_3^{2-}$$

Unfortunately, few victims of acute cyanide poisoning survive long enough to be treated.

Heavy Metal Poisons

Most metals and their compounds show some toxicity when ingested in large amounts. Even the essential mineral nutrients can be toxic when taken in excessive amounts. In many cases, too much of a metal nutrient (a toxic level) can be as dangerous as too little (deficiency). This concept is illustrated in Figure 25.17. For example, the average adult requires 10–18 mg of iron every day. If less is taken in, the person suffers from anemia. Yet an overdose can cause vomiting, diarrhea, shock, coma, and even death. As few as 10–15 tablets containing 325 mg each of iron (as $FeSO_4$) have been fatal to children.

We are not certain how iron poisoning works. However, the heavy metals—those near the bottom of the periodic table—exert their action primarily by inactivating enzymes (Section 13.11). Both mercury (as Hg^{2+}) and lead (as Pb^{2+}) act in this way. The symptoms of mercury poisoning, which include loss of equilibrium, sight, feeling, and hearing, often do not show up for several weeks. By the time the symptoms become recognizable, extensive damage has already been done to the brain and the rest of the nervous system. This damage is largely irreversible.

Lead poisoning can be treated if detected early enough. As we saw in Chapter 22, it usually is treated by intravenous administration of the calcium salt of EDTA. The Pb^{2+} ions are exchanged for the Ca^{2+} ions, and the lead–EDTA complex is excreted. As with mercury poisoning, the neurological damage done by lead compounds is essentially irreversible. Treatment must be performed early to be effective.

Cadmium (as Cd^{2+}) is also toxic (page 915). Its mode of action is different from that of lead and mercury. Cadmium poisoning leads to loss of calcium ions from the bones, leaving them brittle and easily broken. It also causes severe abdominal pain, vomiting, diarrhea, and a choking sensation.

Nerve Poisons

To understand how nerve poisons work, let us consider the action of acetylcholine as a neurotransmitter, with the help of the following equation:

$$CH_3COOCH_2CH_2N^+(CH_3)_3 + H_2O \underset{\text{Acetylase}}{\overset{\text{Cholinesterase}}{\rightleftharpoons}} CH_3COOH + HOCH_2CH_2N^+(CH_3)_3$$

Acetylcholine   Acetic Acid   Choline

▲ **FIGURE 25.17** **The effect of copper ion on the height of oat seedlings**

From left to right, the concentrations of Cu^{2+} are 0, 3, 6, 10, 20, 100, 500, 2000, and 3000 mg/L. Plants on the left show varying degrees of deficiency; those on the right show copper ion toxicity. The optimum level of Cu^{2+} for oat seedlings is therefore about 100 mg/L.

Acetylcholine carries a signal from one nerve cell to another. After doing so, it hydrolyzes to acetic acid and choline through the action of the enzyme *cholinesterase.* Another enzyme, *acetylase,* converts acetic acid and choline back to acetylcholine, which is then ready to carry the next nerve impulse.

Nerve poisons can disrupt the acetylcholine cycle in three ways:

- Botulin, the deadly toxin given off by the anaerobic bacterium *Clostridium botulinum* found in improperly processed canned food, blocks the synthesis of acetylcholine. No messenger is formed, and no messages are carried between nerve cells. Paralysis sets in and death occurs, usually by respiratory failure.
- Curare, atropine, and some local anesthetics block the acetylcholine receptor sites. In this case, the message is sent but not received. In the case of local anesthetics, this can be good for pain relief in a limited area, but anesthetics, too, can be lethal in sufficient quantity.
- *Anticholinesterase poisons* inhibit the action of cholinesterase. This inhibition keeps the level of acetylcholine high; the message is turned on continuously, overstimulating the receptor cells. Anticholinesterase poisons include the organic phosphorus insecticides and chemical warfare compounds such as tabun, sarin, and soman.

The nerve poisons are among the most toxic synthetic chemicals known. They kill when they are inhaled or absorbed through the skin, resulting in the complete loss of muscular coordination and subsequent death by cessation of breathing. The usual antidote is atropine injection and artificial respiration. Without the antidote, death may occur in 2–10 minutes.

Chlorinated hydrocarbon pesticides, such as DDT, are relatively safe to mammals, but they are also nerve poisons. Acute DDT poisoning causes tremors, convulsions, and cardiac or respiratory failure. Chronic exposure to DDT leads to the degeneration of the central nervous system. Other chlorinated compounds, such as the PCBs, act in a similar manner.

Despite their tremendous potential for death and destruction, the nerve poisons have helped us gain an understanding of the chemistry of the nervous system. That knowledge enables scientists to design antidotes for the nerve poisons. In addition, our increased understanding should contribute to progress along more positive lines—in the control of pain, for example.

Detoxification and Potentiation of Poisons

The human body can handle moderate amounts of some poisons. The liver is able to detoxify some compounds by oxidation, reduction, or coupling with amino acids or other normal body chemicals. Perhaps the most common route is oxidation. Ethanol is detoxified by oxidation to acetaldehyde, which in turn is oxidized to acetic acid, a normal constituent of cells. The acetic acid is then oxidized to carbon dioxide and water.

Highly toxic nicotine from tobacco is detoxified by oxidation to cotinine:

Nicotine   Cotinine

Cotinine is less toxic than nicotine. The added oxygen atom also makes cotinine more water-soluble than nicotine and thus more readily excreted in the urine.

The liver is equipped with a system of enzymes, called P-450, that oxidizes fat-soluble substances. The P-450 system converts these fat-soluble compounds, which are likely to be retained in the body, into water-soluble ones that are readily excreted. It can also join molecules to amino acids. For example, toluene is essentially insoluble in water. The P-450 enzymes oxidize toluene to more soluble benzoic acid. The latter

Application Note

Extremely small amounts of one type of botulism toxin have been found useful in treatment of certain muscle spasms. Injection of the toxin into the muscle blocks synthesis of acetylcholine and can relieve spasms for up to three months. The well-known BOTOX® cosmetic treatment is a form of botulism toxin.

is then coupled with the amino acid glycine to form hippuric acid, which is still more soluble and is readily excreted:

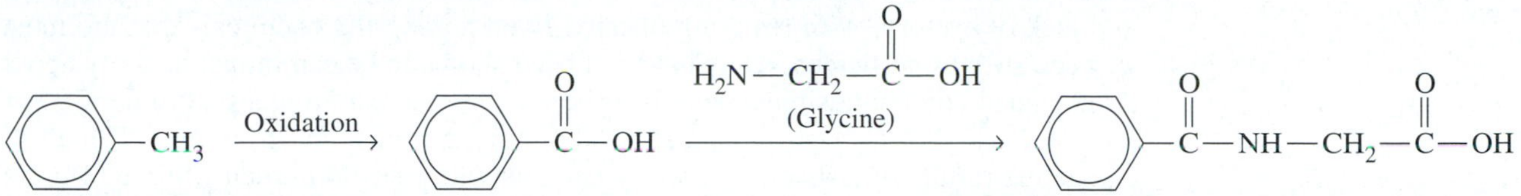

Toluene $\xrightarrow{\text{Oxidation}}$ Benzoic acid $\xrightarrow[\text{(Glycine)}]{\text{H}_2\text{N}-\text{CH}_2-\overset{\text{O}}{\overset{\|}{\text{C}}}-\text{OH}}$ Hippuric acid

The liver enzymes simply oxidize, reduce, or join molecules together. The end product is not necessarily less toxic. For example, methanol is oxidized to the more toxic substance formaldehyde. It is probably the reaction of formaldehyde with the protein in cells that causes the blindness, convulsions, respiratory failure, and death that are characteristic of methanol poisoning. Also, the same enzymes that oxidize alcohols deactivate the male hormone, testosterone. The buildup of these enzymes in a chronic alcoholic leads to a more rapid destruction of testosterone. This appears to be the mechanism for alcoholic impotence, a well-known characteristic of alcoholism.

Benzene, because of its general inertness in the body, is not acted on until it reaches the liver. There, it is slowly oxidized to an epoxide, which is a cyclic ether containing a three-membered ring:

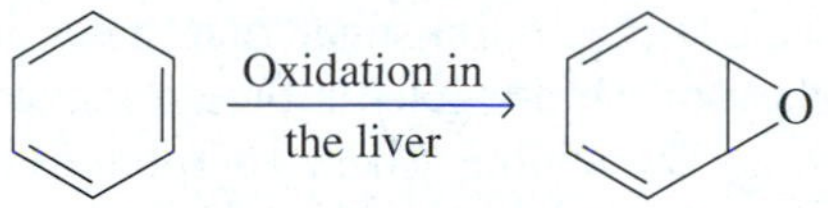

$$\text{(benzene)} \xrightarrow[\text{the liver}]{\text{Oxidation in}} \text{(benzene oxide)}$$

This type of reaction, called *potentiation,* converts a relatively harmless chemical into a much more toxic one. In this benzene potentiation, the product epoxide is a highly reactive molecule that can attack certain key proteins, causing damage that sometimes results in leukemia.

Carbon tetrachloride, CCl_4, is also quite inert in the body. When it reaches the liver, however, it is converted to the reactive trichloromethyl free radical $Cl_3C\cdot$, which attacks unsaturated fatty acids in the body. This action can trigger cancer.

25.9   Carcinogens and Anticarcinogens

Tumors, abnormal growths of new tissue, may be either benign or malignant. Benign tumors are characterized by slow growth; they often regress spontaneously, and they do not invade neighboring tissues. Malignant tumors, often called *cancers,* may grow slowly or rapidly, but their growth is generally irreversible. Malignant growths invade and destroy neighboring tissues. Cancer is not a single disease, but rather a catchall term for more than 200 different afflictions. Many are not even closely related to one another.

What causes cancer? The answers to this seemingly simple question are quite varied and anything but simple. Some chemicals modify DNA, thus scrambling the code for replication and for the synthesis of proteins. For example, aflatoxin B is known to bind to guanine residues in DNA. Just how this initiates cancer, however, is not known for sure.

There is a genetic component to the development of many forms of cancer. Certain genes, called *oncogenes,* seem either to trigger or to sustain the processes that convert normal cells to cancerous ones. Oncogenes arise from ordinary genes that regulate cell growth and cell division and then can be activated by chemical carcinogens, radiation, or perhaps by some viruses. It seems that more than one oncogene must be turned on, perhaps at different stages of the process, before a cancer develops.

We also have suppressor genes that ordinarily prevent the development of cancers. These genes must be inactivated before a cancer develops. Suppressor gene inactivation can occur through mutation, alteration, or loss of the gene. In all, 10 or 15 mutations may be required in a cell before it turns cancerous. Thus, our bodies have some natural protection against cancer.

▲ A surprising number of consumer products warn us about containing suspected carcinogens, such as saccharine.

A **carcinogen** is a material that causes cancer. Many people seem to believe that chemicals are a major cause of cancer, but most cancers are caused by lifestyle factors. Nearly two-thirds of all cancer deaths in the United States are linked to tobacco, diet, and lack of exercise with resulting obesity. Even among the chemicals that are most suspect, such as pesticides, many have not been shown to be carcinogenic. Only about 30 chemical compounds have been identified as human carcinogens. Another 300 or so have been shown to cause cancer in laboratory animals, but it is often difficult to equate the results of tests on laboratory animals with risks to humans. Some of these 300 are widely used and are therefore a subject of some concern.

There are strong correlations between carcinogenicity and certain molecular sizes and shapes, and some of the more notorious carcinogens are polycyclic aromatic hydrocarbons, of which 3,4-benzpyrene is perhaps the best known:

3,4-Benzpyrene

Carcinogenic polycyclic hydrocarbons are formed during the incomplete burning of nearly any organic material. They have been found in charcoal-grilled meats, cigarette smoke, automobile exhausts, coffee, burnt sugar, and many other materials.

Another important class of carcinogens is the aromatic amines. Two prominent ones are β-naphthylamine and benzidine. These compounds once were used widely in the dye industry. They were responsible for a high incidence of bladder cancer among dye-industry workers.

Not all carcinogens are aromatic compounds. Prominent among the aliphatic carcinogens are dimethylnitrosamine [$(CH_3)_2NNO$] and vinyl chloride, the monomer from which the polymer polyvinyl chloride is made. Other aliphatic carcinogens include some epoxides, such as bis(epoxy)butane, and some other three- and four-membered heterocyclic rings containing oxygen or nitrogen.

Few of the known carcinogens are synthetic chemicals. Some, such as safrole in sassafras and the aflatoxins produced by molds on foods, occur naturally. Some researchers estimate that 99.99% of all carcinogens that we ingest are natural ones. Plants produce compounds to protect themselves from fungi, insects, and higher animals, including humans. Some carcinogenic compounds are found in mushrooms, basil, celery, figs, mustard, pepper, fennel, parsnips, and citrus oils—almost every place a curious chemist looks. Carcinogens are also produced during cooking and as products of normal metabolism.

With so many natural carcinogens in food, why do we not all get cancer? Part of the answer to this question is that some substances in food act as **anticarcinogens.** Fiber is believed to protect against colon cancer, for example; the food additive BHT (butylated hydroxytoluene) may protect against stomach cancer, and certain vitamins have anticarcinogenic effects.

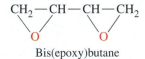

Bis(epoxy)butane

N-Laurylethyleneimine

β-Propiolactone

▲ Three small-ring heterocyclic carcinogens. The top molecule is an epoxide.

▲ Broccoli, cauliflower, and brussels sprouts are among the cruciferous vegetables that have been shown to reduce the incidence of cancer in humans.

BHT

A diet rich in cruciferous vegetables (cabbage, broccoli, brussels sprouts, kale, and cauliflower) has been shown to reduce the incidence of cancer in both animals and humans. We still do not know for sure what components of these foods protect against cancer. Perhaps it is a combination of substances rather than just one. The vitamins that are antioxidants (vitamin C, vitamin E, and β-carotene, a precursor to vitamin A)

How Cigarette Smoking Causes Cancer

The association between cigarette smoking and cancer has been known for decades, but the precise mechanism was not known until 1996. The research scientists who found the link focused on a tumor-suppressor gene called *P53*, a gene that is mutated in about 60% of all lung cancers. They also focused on a metabolite of benzpyrene, a carcinogen found in tobacco smoke.

In the body, benzpyrene is oxidized to an epoxide that is an active carcinogen and binds to specific nucleotides in the gene—sites called hot spots—where mutations frequently occur. It seems quite likely that the benzpyrene metabolite causes many of those mutations in the *P53* gene.

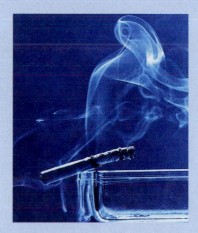

◀ Cigarette smoke contains at least 40 carcinogens, including 3,4-benzpyrene.

seem to exhibit the strongest anticancer properties. Some studies with vitamins A, C, and E, used separately or in combination, have confirmed that each of these vitamins has some ability to lower the incidence of cancer. There probably are many other anticarcinogens in our food that have not yet been identified.

25.10   Hazardous Materials

Stories about toxic substances have caused increasing concern about hazardous materials in the environment. Problems with chemical dumps have made household words out of Love Canal in New York State and Valley of the Drums in Kentucky. Although often overblown in the news media, serious problems do exist.

The U.S. EPA categorizes **hazardous materials** on the basis of their properties:

- **Ignitable materials** are substances that catch fire readily, such as gasoline and other hydrocarbons.

- **Corrosive materials** are substances that corrode storage containers and other equipment, such as strong acids.

- **Reactive materials** are substances that react or decompose readily, possibly producing hazardous by-products. Examples include explosives and materials that react with water to produce toxic fumes, such as powdered bleach (calcium hypochlorite).

- **Toxic materials** are substances that are injurious when inhaled or ingested, such as chlorine, ammonia, formaldehyde, and pesticides.

From this list, we see that hazardous materials can cause fires or explosions, pollute the air, contaminate our food and water, and occasionally poison by direct contact. As long as we want the products our industries produce, however, we will have to deal with the problems of hazardous wastes (Table 25.6).

Many hazardous materials can be rendered less harmful by chemical treatment. For example, acid wastes can be neutralized with inexpensive bases, such as lime. However, the best way to handle hazardous wastes is not to produce them in the first place. Many industries have modified processes to minimize the amount of wastes produced, and some wastes that are produced can be reprocessed to recover energy or materials. Hydrocarbon solvents can be either purified and reused or else burned as fuels. Sometimes, one industry's waste can be a raw material for another industry. For example, waste nitric acid from the metals industry can be converted to fertilizer. Finally, if a hazardous waste cannot be used, incinerated, or treated to render it less

(a)

(b)

▲ (a) A waste dump in 1970 at Malkins Bank, Cheshire, England, with drums leaking chemical wastes. (b) The same site, cleaned up and restored, is now a municipal golf course.

Table 25.6 Industrial Products and Hazardous Waste By-products

Product	Associated Waste
Plastics	Organic chlorine compounds
Pesticides	Organic chlorine compounds, organophosphate compounds
Medicines	Organic solvents and residues, heavy metals (for example, mercury and zinc)
Paints	Heavy metals, pigments, solvents, organic residues
Oil, gasoline	Oil, phenols and other organic compounds, heavy metals, ammonium salts, acids, caustics
Metals	Heavy metals, fluorides, cyanides, acidic and alkaline cleaners, solvents, pigments, abrasives, plating salts, oils, phenols
Leather	Heavy metals, organic solvents
Textiles	Heavy metals, dyes, organic chlorine compounds, solvents

hazardous, it must be stored in a secure landfill. Unfortunately, landfills often leak, contaminating groundwater. When this happens, we clean up one toxic waste dump and move the materials to another, playing a rather macabre shell game.

The best technology at present for treating organic wastes, including chlorinated compounds, appears to be incineration. For example, combustion at 1260 °C eliminates more than 99.9999% of chlorinated compounds, such as PCBs.

Perhaps *biodegradation* of wastes will be the way of the future. Some microorganisms can degrade hydrocarbons in gasoline, for example, and there are bacteria that, when provided with proper nutrients, can degrade chlorinated hydrocarbons.

Weighing Risks and Benefits

Increasingly, we have to decide whether the benefits we gain from hazardous substances are worth the risks we assume by using them. Many issues involving toxic chemicals are emotional, and most of the decisions regarding them are political. Nevertheless, possible solutions to problems posed by toxic chemicals often lie in the field

Phytoremediation

Many of the *d*-block elements and some of the lower *s*-block and *p*-block elements, such as lead and barium, are toxic. Sites of heavy contamination of toxic metals are major problems. A direct method of cleanup is to dig up the contaminated soil and move it elsewhere, but that is very expensive and merely moves a large contamination problem to a new location.

Phytoremediation is a technique that shows promise for ridding contaminated soil of heavy metals as well as petroleum products and other organic compounds. Certain plant species are *hyperaccumulators,* meaning they have a high affinity for various metal ions or other specific chemical species. These plants are cultivated in the problem area. The plants may be harvested and further processed by dehydration and/or ashing. The residue from the plants is far smaller than the mass of the contaminated soil, and disposal is much easier.

The efficacy of phytoremediation has been improved in some cases by watering the plants with a solution of a chelating agent such as EDTA (Section 22.12). Chelated metals are more easily accumulated in the plant. For example, chelation has been

found to increase phytoremediation of lead by a factor of 100 or more in some cases.

▲ The Italian serpentine plant *Alyssum bertolonii* can accumulate so much nickel(II) ion that its sap is green! It is thought that the ability to accumulate heavy metals may be related to the toxic effects of those elements on insects that might otherwise feast on these plants.

of chemistry. We hope the chemistry you have learned in this text will help you make intelligent decisions. Most of all, we hope you will continue to learn more about chemistry, because chemistry affects nearly everything you do. We wish you success and happiness, and may the joy of learning go with you always.

Cumulative Example

A large coal-fired electric plant burns 2500 tons of coal per day. **(a)** Determine the number of homes that this plant can supply with electricity, assuming that coal is nearly pure carbon, that the efficiency of the plant is 41%, and that one home consumes 85 kWh each day. **(b)** The coal that is burned contains 0.65% S by mass. Assume that all the sulfur is converted to SO_2 and that, because of a thermal inversion, the SO_2 remains trapped for one day in a parcel of air that is 45 km × 60 km × 0.40 km. Will the level of SO_2 in this air exceed the primary national air-quality standard of 365 μg SO_2/m^3 air?

STRATEGY

For part (a), we can determine the enthalpy change for the combustion of 2500 tons of solid carbon to carbon dioxide. Next, we must convert that enthalpy change first to kilowatt-hours (using the conversion factor, 1 kWh = 3600 kJ, from the inside back cover) and then to number of homes. In applying the plant efficiency factor, we use the unit kJ(electric)/kJ(thermal) to extract that portion of the energy of combustion of the coal that actually appears as electricity (the rest is waste heat). For part (b), we first calculate the mass of sulfur in the coal and then use simple stoichiometry to determine the mass of sulfur dioxide produced. Next, we convert that mass to micrograms and then divide the mass by the volume of air (in cubic meters).

SOLUTION

(a) First, we write the combustion equation and determine the enthalpy change. Because of our assumption that the coal is pure carbon, the enthalpy change is simply the standard enthalpy of formation of $CO_2(g)$.

$$C(graphite) + O_2(g) \longrightarrow CO_2(g) \qquad \Delta H° = -393.5 \text{ kJ/mol}$$

The next step is to convert 2500 tons of carbon to grams.

$$? \text{ g C} = 2500 \text{ tons C} \times \frac{2000 \text{ lb}}{1 \text{ ton}} \times \frac{453.6 \text{ g}}{1 \text{ lb}} = 2.3 \times 10^9 \text{ g C}$$

Now we can use the molar mass of carbon, the enthalpy change, the plant efficiency factor, and the conversion factors from kJ(thermal) to kWh and from kWh to homes.

$$? \text{ homes} = 2.3 \times 10^9 \text{ g C} \times \frac{1 \text{ mol C}}{12.011 \text{ g C}} \times \frac{393.5 \text{ kJ(thermal)}}{1 \text{ mol C}}$$

$$\times \frac{0.41 \text{ kJ(electric)}}{1 \text{ kJ(thermal)}} \times \frac{1 \text{ kWh}}{3600 \text{ kJ(electric)}} \times \frac{1 \text{ home}}{85 \text{ kWh}}$$

$$= 1.0 \times 10^5 \text{ homes}$$

(b) We begin with a balanced equation for combustion of sulfur.

$$S(s) + O_2(g) \longrightarrow SO_2(g)$$

Next, we use the percent sulfur to find the grams of sulfur associated with the calculated mass of carbon and then convert this mass stoichiometrically to grams of sulfur dioxide.

$$? \text{ g SO}_2 = 2.3 \times 10^9 \text{ g C} \times \frac{0.65 \text{ g S}}{100 \text{ g C}} \times \frac{1 \text{ mol S}}{32.07 \text{ g S}} \times \frac{1 \text{ mol SO}_2}{1 \text{ mol S}} \times \frac{64.06 \text{ g SO}_2}{1 \text{ mol SO}_2}$$

$$= 3.0 \times 10^7 \text{ g SO}_2$$

Finally, we express the concentration as grams of SO_2 per cubic kilometer and convert to the desired units of μg SO_2/m^3.

$$\frac{? \, \mu g \text{ SO}_2}{m^3} = \frac{3.0 \times 10^7 \text{ g SO}_2}{(45 \times 60 \times 0.40) \text{ km}^3} \times \frac{10^6 \, \mu g}{1 \text{ g}} \times \left(\frac{1 \text{ km}}{1000 \text{ m}}\right)^3 = 28 \, \mu g \text{ SO}_2/m^3$$

The $SO_2(g)$ content of the air does not exceed the primary air-quality standard.

ASSESSMENT

It appears that the SO_2 content of the air during the thermal inversion is well below the primary national air-quality standard. Indeed, ten times this concentration of SO_2 still would not exceed the standard. However, we have made a very questionable assumption: that the SO_2 is uniformly dispersed. In practice, it seems almost certain that regions closer to the plant would exhibit much higher levels of SO_2 than more distant regions of this large parcel of air.

Concept Review with Key Terms

25.1 Atmospheric Composition, Structure, and Natural Cycles—Starting at Earth's surface, the primary regions of the atmosphere are called the troposphere, stratosphere, mesosphere, and thermosphere (ionosphere). The chief components of dry air are N_2, O_2, and Ar. Water vapor is an important participant in the **hydrologic (water) cycle**. The humidity of air is a measure of its water vapor content. Dew and frost formation and **deliquescence** are phenomena related to relative humidity. **Nitrogen fixation** describes the conversion of atmospheric nitrogen into nitrogen-containing compounds and is an important step in the **nitrogen cycle**. Chemical fixation of nitrogen to make fertilizers has greatly increased the world's food supply but has also altered the natural nitrogen cycle. Atmospheric CO_2 is the carbon source for carbohydrate synthesis in the **carbon cycle**. Some carbon is locked out of the cycle in fossil fuels (coal, natural gas, and petroleum), but the combustion of these fuels returns CO and CO_2 to the cycle.

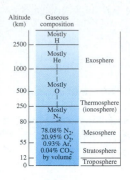

Altitude (km)	Gaseous composition	
	Mostly H	
2500		
	Mostly He	Exosphere
1000		
500	Mostly O	
250		Thermosphere (ionosphere)
80	Mostly N_2	
	78.08% N_2,	Mesosphere
55	20.95% O_2, 0.93% Ar,	
12	0.04% CO_2, by volume	Stratosphere
0		Troposphere

25.2 Air Pollution—Carbon monoxide is an **air pollutant** commonly found in high concentrations in urban areas with heavy vehicular traffic. In the presence of unburned hydrocarbons and sunlight, oxides of nitrogen lead to **photochemical smog**. Temperature inversions also contribute to smog conditions. Smog-control measures focus on catalytic converters for use in automobiles and the control of combustion processes to reduce emissions. **Industrial smog** is associated with industrial processes and practices that produce high levels of sulfur oxides and **particulate matter**.

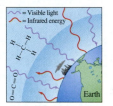

25.3 The Ozone Layer—Ozone, O_3, in the stratosphere protects living organisms by absorbing ultraviolet radiation. The integrity of the **ozone layer** is threatened by human activities, such as the release of chlorofluorocarbons (CFCs) into the atmosphere.

25.4 Global Warming: Carbon Dioxide and the Greenhouse Effect—The continuous buildup of CO_2 in the atmosphere may result in **global warming**. The **greenhouse effect** is a natural process in which infrared radiation emitted by Earth's surface is absorbed by atmospheric gases such as CO_2 and H_2O. Increased levels of these "greenhouse" gases may contribute to increased global warming.

= Visible light
= Infrared energy

Earth

25.5 Earth's Natural Waters—Water covers three-fourths of Earth's surface, but only about 1% of it is available as fresh water. Water has a higher density, specific heat, and heat of vaporization than most other liquids. Unlike most liquids, water expands when it freezes.

25.6 Water Pollution—Fresh water is easily contaminated by various chemicals and microorganisms, some coming from natural processes and others from human activities. Dumping sewage into water increases the amount of organic material in the water. The breakdown of this material by dissolved oxygen is known as **aerobic oxidation**. At high levels of organic material, aerobic oxidation can lead to an increase in **biochemical oxygen demand (BOD)** and eventually harm higher forms of aquatic life. Water pollutants that are nutrients for the growth of algae can lead to **eutrophication** of a lake or stream. When oxygen levels are depleted, **anaerobic decay**—the reduction of organic material to CH_4, H_2S, and NH_3—further harms aquatic life.

Groundwater and surface water each provide drinking water for about half the U.S. population. Municipal water supplies are treated in several physical and chemical steps: settling, filtration, aeration, and chlorination. Fluoridation appears to have greatly reduced the incidence of dental caries.

Wastewater treatment usually includes several processes. **Primary sewage treatment** uses settling ponds for sludge removal. **Secondary sewage treatment** entails sand and gravel filtration. A combination of these methods is known as the **activated sludge method**. **Advanced treatment** methods include charcoal filtration, ion exchange, and reverse osmosis.

25.7 Acid Rain and Acid Waters—Sulfur oxides and nitrogen oxides can react with water in the atmosphere to produce **acid rain**. This form of pollution causes lakes and streams to become so acidic that the acidic water damages fish and other aquatic life.

25.8 Poisons—A number of processes threaten forms of life in the Earth's **biosphere**. **Toxicology** is the study of the responses of living organisms to poisons. Strong acids and bases are corrosive poisons. Substances such as ozone are toxic because they are strong oxidizing agents. Carbon monoxide and nitrites are toxic because they interfere with the blood's ability to transport oxygen. Cyanides are poisons because they shut down cellular respiration. Heavy metal poisons, such as lead and mercury, inactivate enzymes by tying up their SH groups. Nerve poisons, such as organophosphates, interfere with the acetylcholine cycle.

25.9 Carcinogens and Anticarcinogens—**Carcinogens** are slow poisons that trigger the growth of malignant tumors. Some natural substances in foods act to inhibit cancerous growth and are known as **anticarcinogens**.

25.10 Hazardous Materials—**Hazardous materials** are industrial products and by-products that can cause illness or death. The four classes of hazardous materials are **ignitable materials**, **corrosive materials**, **reactive materials**, and **toxic materials**.

Assessment Goals

When you have mastered the material in this chapter, you will be able to:

- List the different layers of the atmosphere and their composition.
- Describe the three natural cycles of our biosphere involving water, nitrogen, and carbon.
- Distinguish between air pollutants, photochemical smog, and industrial smog.

- Describe approaches to controlling the different forms of air pollution.
- Describe the properties of ozone, the importance of the ozone layer, and the effect of human activities on the ozone layer.
- Explain the relationship between carbon dioxide and the greenhouse effect.
- List the sources of natural waters within the hydrosphere.

- Describe different forms of chemical contamination of water and the remediation steps used to address these, including treatment of municipal water, wastewater, and sewage.
- Describe the sources of acid rain and its impact on the environment.
- Describe poisons, their mechanisms of toxicity, their control, and their antidotes.

- Describe natural and synthetic carcinogens, their mechanism of action, and the activity of anticarcinogens.
- Distinguish between the four classes of hazardous materials in terms of their chemical identity and impact.

Self-Assessment Questions

1. Which layer of the atmosphere **(a)** lies nearest the surface of Earth? **(b)** Contains the ozone layer?

2. List the three major components of dry air, and give the approximate (nearest whole number) mole percent of each.

3. What is the nitrogen cycle? How has industrial fixation of nitrogen to make fertilizers affected the nitrogen cycle?

4. Briefly describe each of the following terms dealing with a natural phenomenon.

    (a) deliquescence   **(b)** the greenhouse effect

5. What specific materials are implied by these terms for atmospheric pollutant(s)?

    (a) PAN   **(b)** SO_x **(c)** fly ash

6. By name and/or formula, identify **(a)** two gases able to displace O_2 in blood hemoglobin; **(b)** two "greenhouse" gases, in addition to CO_2 and H_2O; and **(c)** a constituent of acid rain.

7. List two uses of chlorofluorocarbons (CFCs), and describe how CFCs are implicated in the depletion of the ozone layer.

8. What is photochemical smog? What is the role of sunlight in its formation?

9. What is industrial smog? How is it formed?

10. What conditions favor the formation of carbon monoxide during the combustion of gasoline in an automobile engine? How does carbon monoxide exert its poisonous effect?

11. What is synergism? Indicate one specific example of a synergistic effect concerning air pollution.

12. What are the health effects associated with ozone in the stratosphere and at ground level? Why are they not the same?

13. How is each of the following used to reduce air pollution?

    (a) electrostatic precipitator

    (b) catalytic converter

14. Which of the following are important contributors to the formation of photochemical smog, and which are not? Explain.

    (a) NO   **(c)** hydrocarbon vapors

    (b) CO   **(d)** SO_2

15. What is a temperature inversion? How does a temperature inversion contribute to air pollution problems?

16. How are the following terms related to one another regarding air pollution: aerosol, fly ash, and particulate matter?

17. Describe measures that can be used to control the emission of nitrogen oxides in automotive exhaust, and explain why these are not the same measures used to control emissions of hydrocarbons and carbon monoxide.

18. What proportion of Earth's water is seawater? Why are the seas salty?

19. List some waterborne diseases. Why are these diseases no longer common in developed countries?

20. What impurities are present in rainwater?

21. List four cations and three anions present in groundwater.

22. What is BOD? Why is a high BOD undesirable?

23. What are the products of the breakdown in water of organic matter by aerobic bacteria? By anaerobic bacteria?

24. List some ways in which groundwater is contaminated. What are some common industrial contaminants of groundwater?

25. Why do chlorinated hydrocarbons remain in groundwater for such a long time?

26. List two ways by which lakes and streams have become acidic. Why is acidic water especially hazardous to fish?

27. List several ways by which the acidity of rain can be reduced. What kind of rocks tend to neutralize acidic waters? How can we restore (at least temporarily) lakes that are too acidic?

28. List two toxic compounds found in wastes from the chromium plating process. How is each removed?

29. Describe **(a)** a primary sewage treatment plant and **(b)** a secondary sewage treatment plant. What impurities are removed by each? What impurities remain in wastewater after each form of treatment?

30. Describe the activated sludge method of sewage treatment. Why is wastewater chlorinated before it is returned to a waterway?

31. What is meant by advanced treatment of wastewater? What kinds of substances are removed from wastewater by charcoal filtration? Why is it so difficult to remove nitrate ions from water?

32. Why are municipal water supplies **(a)** treated with aluminum sulfate and slaked lime, **(b)** aerated, and **(c)** chlorinated?

33. Give an example that shows how the toxicity of a substance depends on the route of administration.

34. List three corrosive poisons. How do dilute solutions of acids and bases damage living cells? How does ozone damage living cells?

35. How do cyanides exert their toxic effect? How does sodium thiosulfate act as an antidote for cyanide poisoning?

36. Iron (as Fe^{2+}) is a necessary nutrient. What are the effects of too little Fe^{2+}? Of too much?

37. What is acetylcholine? Describe its action.

38. How do **(a)** botulin, **(b)** sarin, and **(c)** organophosphorus compounds affect the acetylcholine cycle?

39. What is the P-450 system? What is its function? Does it always detoxify foreign substances?

40. List two ways that the conversion of nicotine to cotinine in the liver lessens the risk of nicotine poisoning.

41. List two steps in the detoxification of ingested toluene. What is the effect of these steps?

42. What is a tumor? How are benign and malignant tumors different?

43. What are **(a)** oncogenes and **(b)** suppressor genes? How is each involved in the development of cancer?

44. List some conditions under which polycyclic hydrocarbons are formed.

45. What is a hazardous material?

46. Define and give an example of **(a)** a reactive material, **(b)** an ignitable material, and **(c)** a corrosive material.

Problems

Earth's Atmosphere

47. The text states that 99% of the mass of the atmosphere lies within 30 km of the surface of Earth. Which of the following values is a reasonable estimate of air pressure at an altitude of 30 km: **(a)** 0.1 mmHg, **(b)** 1 mmHg, **(c)** 10 mmHg, or **(d)** 100 mmHg? Explain your reasoning. (*Hint:* Recall the basic ideas relating to pressure from Section 5.3.)

48. When present in a very small proportion in air, the concentration of a gas is customarily indicated in parts per million (ppm) rather than in mole percent or volume percent. Use data in Table 25.1 to determine the parts per million in air of the noble gases that are listed there.

49. Suggest a likely reason for the observed change in composition with altitude—from N_2 to O to He to H—shown in Figure 25.1.

50. Part of the design criteria for the International Space Station included a requirement that parts of the station exposed to space be as chemically nonreactive as possible. Suggest a likely reason for this requirement. (*Hint:* Refer to Figure 25.1 and the note on page 1036.)

Water Vapor in the Atmosphere

51. What are the mole percent and ppm of H_2O in an air sample at STP in which the partial pressure of water vapor is 2.00 mmHg?

52. What should be the relative humidity of a sample of air at 25 °C in which the partial pressure of water vapor is 10.5 mmHg? (*Hint:* Use data from Table 11.2.)

53. What is the partial pressure of water vapor in a sample of air having a relative humidity of 75.5% at 20 °C? (*Hint:* Use data from Table 11.2.)

54. A parcel of air has an absolute humidity, expressed as a partial pressure of water vapor, of 18.0 mmHg. At which of the following temperatures does the air have the greatest relative humidity: 25 °C, 30 °C, or 40 °C? Explain.

55. What is the dew point of the parcel of air described in Problem 54? (*Hint:* Use data from Table 11.2.)

56. Why is it that condensed water vapor can be seen above a kettle of boiling water even in a hot kitchen, whereas you can see your breath only on a cold day?

Carbon, CO, and CO$_2$

57. The combustion of a hydrocarbon, especially if the quantity of oxygen is limited, produces a mixture of carbon dioxide and carbon monoxide. The decomposition of a metal carbonate by an acid produces only carbon dioxide, even if the quantity of acid is limited. Explain this difference in behavior.

58. Indicate a natural process or processes by which carbon atoms are **(a)** removed from the atmosphere, **(b)** returned to the atmosphere, **(c)** effectively withdrawn from the carbon cycle.

59. Write an equation that represents the complete combustion of the hydrocarbon hexane, $C_6H_{14}(l)$. Explain why it is not possible to write a unique equation to represent its incomplete combustion.

60. Carbon monoxide is a poisonous gas, even in low concentrations, whereas carbon dioxide is not. Yet, except in some local situations, there is less environmental concern over carbon monoxide than over carbon dioxide. Explain why this is so.

61. The United States leads the world in per capita emissions of $CO_2(g)$, amounting to 19.8 metric tons (t) per person per year (1 t = 1000 kg). What mass, in metric tons, of each of the following fuels would yield this quantity of CO_2? **(a)** CH_4 **(b)** C_8H_{18} **(c)** coal (94.1% C by mass)

62. Tabulations on carbon dioxide emissions often list cement manufacture as one of the sources. Describe two ways in which the manufacture of Portland cement injects carbon dioxide into the atmosphere.

Air Pollution

63. Although alternative automotive fuels can reduce some types of air pollution, they still produce other types of pollution, such as nitrogen oxides. Explain why.

64. Per ton of material consumed, which of the following would you expect to produce the greatest quantity of $SO_2(g)$: **(a)** smelting zinc sulfide, **(b)** smelting lead sulfide, **(c)** burning coal, or **(d)** burning natural gas? Explain.

65. Write equations for the following reactions.
   (a) Sulfur burns in air, forming sulfur dioxide.
   (b) Zinc sulfide, heated in air, yields zinc oxide and sulfur dioxide.
   (c) Sulfur dioxide reacts with oxygen, forming sulfur trioxide.
   (d) Sulfur trioxide reacts with water, forming sulfuric acid.
   (e) Sulfuric acid is completely neutralized by aqueous ammonia.

66. Identify each species, NO, NO_2, Cl, and ClO, as a catalyst or a reactive intermediate in the chemical reactions involved in the ozone layer.

67. Describe how the following particulate matter may be produced.
   (a) sodium chloride from seawater
   (b) sulfate particles in an industrial smog

68. The average person takes 15 breaths per minute, inhaling 0.50 L of air with each breath. What mass of particulates, in milligrams, would the person breathe in a day if the particulate level in air were 75 $\mu g/m^3$?

Water and Water Pollution

69. Give the equation that shows the neutralization of acidic water by limestone.

70. Wastewater disinfected with chlorine must be dechlorinated before it is returned to sensitive bodies of water. The dechlorinating agent often is sulfur dioxide. Write the equation for the reaction. Is the chlorine oxidized or reduced? Identify the oxidizing agent and reducing agent in the reaction.

71. How many kilograms of $Ca(OH)_2$ are needed to neutralize a lake that has been made 2.0×10^{-4} M in sulfuric acid by acid rain and is 300×200 m, with an average depth of 5.0 m?

72. Suggest a reason for the "extremely high" cost of distillation as compared to other methods of tertiary wastewater treatment (Table 25.5).

Additional Problems

Problems marked with an * may be more challenging than others.

73. Assume that typical urban air contains 100 μg of suspended particles per m^3 of air. Assume that the average particle is spherical in shape, with a diameter of 1 μm and a density of 1 g/cm^3. Estimate the number of particles per cm^3 of the air.

*74. The text states that 5.2×10^{15} metric tons of atmospheric gases are spread over a surface area of 5.0×10^8 km^2. Use these facts, together with data from Appendix B, to estimate a value of standard atmospheric pressure.

75. At 20 °C, the vapor pressure observed for a saturated solution of $CaCl_2 \cdot 6\,H_2O$ is 5.67 mmHg. If a quantity of this solution is placed in a large sealed container at 20 °C and the solution is kept saturated by the presence of excess solid, what relative humidity will be maintained in the air in the container? How effective is $CaCl_2 \cdot 6\,H_2O$ in dehumidifying air?

*76. A 12.012-L sample of air becomes saturated with water vapor at 25.0 °C. The air is then cooled to 20.0 °C. What mass of water (dew) will deposit on the walls of the container?

77. There are different ways to assess how much the combustion of various fuels contributes to the buildup of CO_2 in the atmosphere. One relates the mass of CO_2 formed to the mass of fuel burned; another relates the mass of CO_2 to the quantity of heat evolved in the combustion. Which of the three fuels C(graphite), $CH_4(g)$, or $C_4H_{10}(g)$ produces the smallest mass of CO_2 **(a)** per gram of fuel, **(b)** per kJ of heat evolved? (*Hint:* Use data from Appendix C, and assume that all the products of each combustion are gases.)

*78. It has been estimated that if all the ozone in the atmosphere were brought to sea level at STP, the gas would form a layer 0.3 cm thick. **(a)** Estimate the number of O_3 molecules in Earth's atmosphere. **(b)** Comment on the feasibility of a caller's suggestion to a radio science program that depleted ozone in the stratosphere be replaced by transporting unwanted ozone from low altitudes into the stratosphere.

*79. A small gasoline-powered engine with a 1-L storage tank is inadvertently left running overnight in a large warehouse, 95 m $\times$ 38 m $\times$ 16 m. When workers arrive in the morning, are they likely to enter an environment in which the level of CO exceeds the danger level of 35 ppm? (*Hint:* Use C_8H_{18} as a representative formula of gasoline, and make other reasonable assumptions.)

80. To prevent tooth decay, drinking water is usually treated to contain 1.00 ppm of F^- ion. A cylindrical water tank has a diameter of 5.00 m and a depth of 12.0 m. What mass of NaF(s) is needed to establish a F^- ion concentration of 1.00 ppm in a tankful of water? What volume of 0.150 M NaF would be needed to achieve 1.00 ppm of F^-?

81. The workplace standard for $SO_2(g)$ in air is 5 ppm. Approximately what mass of sulfur could be burned in an enclosed workplace, 10.5 m $\times$ 5.4 m $\times$ 3.6 m, before this limit is exceeded?

82. The decomposition of ozone by chlorine atoms can be described by the rate law: rate = $k[Cl][O_3]$.

$$Cl(g) + O_3(g) \longrightarrow ClO(g) + O_2(g)$$
$$k = 7.2 \times 10^9 \text{ M}^{-1}\text{ s}^{-1} \text{ at 298 K}$$

How would the rate of ozone destruction be affected by doubling the concentration of chlorine atoms?

*83. A newspaper article states the following: "Never dump oil down a sewer or water main. One quart of motor oil can make 250,000 gallons of water undrinkable. That is more water than 30 people drink in a lifetime." Estimate the concentration of the oil in the contaminated water, assuming thorough mixing. What other assumptions must you make?

84. The concentration of trichloroethane, CH_3CCl_3, in a sample of groundwater is 34 ppb. What is the concentration in nanomoles per liter?

*85. A leaking tank spills 875 kg of 2-propanol into a lake with a volume of 1.8×10^8 L. How much is the BOD (in mg/L) increased by the spill? Assume that the 2-propanol is oxidized to CO_2.

86. Chlorine, even at very low concentrations, can be quite toxic to aquatic organisms. It also reacts with organic substances in the water to form toxic chlorinated organic compounds such as chloroform and chlorinated phenols. Wastewater that contains chlorine is therefore often dechlorinated before it is discharged. At low pH (below pH 1), chlorine exists principally as $Cl_2(aq)$. At high pH (above pH 8.5), the chlorine largely disproportionates to OCl^- and Cl^-. Two principal reagents are used to dechlorinate wastewater: sulfur dioxide (or salts that release SO_2, such as $NaHSO_3$) and hydrogen peroxide. Write equations for the reaction of OCl^- in wastewater with a pH of 12.3 with **(a)** $NaHSO_3$ and **(b)** H_2O_2.

87. Various regulations often require the reduction of phosphate levels before wastewater can be discharged to waterways. This can be done with any of several different precipitating agents such as **(a)** iron(III) chloride, **(b)** aluminum sulfate, or **(c)** CaO. Write the equation for the precipitation reaction that occurs with each of the reagents. In each case, calculate the cation concentration that must be present if the concentration of PO_4^{3-} is to be reduced to 10 ppm.

88. In investigating an old dump, a student finds a metal container filled with 12 oz of liquid. The label indicates that the can contains a substance that a reference book lists as a carcinogen. The concentration of the substance is given as 8.2 mg/fluid oz. On further investigation, he finds that 20 billion such containers were once filled and distributed in the United States each year. **(a)** How many milligrams of the carcinogen are in each can? **(b)** How many metric tons of the carcinogen were distributed in this manner each year?

* **89.** If the average relative humidity on Earth's surface is 70% at the average Earth temperature of 14 °C, estimate the mass of water vapor in the lowest 1.0 km of the atmosphere. Earth has a land surface area of 1.5×10^6 km^2 and an area of 3.6×10^6 km^2 covered by water. Seawater is about 0.44 M in NaCl and 0.051 M in MgCl$_2$.

* **90.** An air conditioner takes outside air at 1.0 atm, 30 °C, and relative humidity of 85%, and cools it to 10 °C air that is saturated with water vapor. That cool air is exhausted into a room, which maintains the room at 21 °C. The vapor pressure of water at 10 °C is 9.2 Torr. **(a)** If the only source of water vapor is from the cooled air, what is the relative humidity in the room? **(b)** If 100 cubic feet of air is processed each minute, what mass of water vapor is condensed by the air conditioner over 8.0 hours of operation? **(c)** How much energy, in kilojoules, is released by the condensation described in (b)?

Apply Your Knowledge

* **91.** **[Collaborative]** Many industrial wastewater streams are contaminated with metal ions. Many of these ions can be effectively removed by precipitation as insoluble salts. As^{3+}, Cd^{2+}, Cr^{3+}, Cu^{2+}, Fe^{2+}, Mn^{2+}, Ni^{2+}, Pb^{2+}, and Zn^{2+} can be precipitated as hydroxides. Cd^{2+}, Co^{2+}, Cu^{2+}, Fe^{2+}, Hg^{2+}, Mn^{2+}, Ni^{2+}, Ag^{+}, Sn^{2+}, and Zn^{2+} can be precipitated as sulfides. Cd^{2+}, Ni^{2+}, and Pb^{2+} can be precipitated as carbonates. All these reactions are sensitive to the pH of the solution. For each group, write an equation for a representative reaction and state whether the reaction should be carried out at high pH or low pH. Explain.

* **92.** **[Environmental]** For the complete combustion of gasoline, the mass ratio of air to fuel should be about 14.5 to 1. Use ideas from this chapter and elsewhere in the text to show that this is about the ratio that you would predict. (*Hint:* Assume that C$_8$H$_{18}$ is a representative formula of gasoline, and use the composition of air given in Table 25.1.)

93. **[Environmental]** Propose a disposal method for each of the following wastes. Be as specific as possible, and justify your choice.

(a) hydrochloric acid contaminated with iron salts

(b) picric acid (an explosive)

(c) pentane contaminated with residues from penicillin production

* **94.** **[Environmental]** Use the following and other data from the text to show that if all the sulfur in coal used in electric power plants were converted to sulfuric acid, the quantity of acid produced would exceed current demand. (1) Annual U.S. coal consumption by electric power plants: approximately 8.7×10^8 ton. (2) Average SO$_2$ formation in the combustion of coal: 2 mg SO$_2$/kJ heat evolved. (3) Typical annual U.S. production of sulfuric acid: approximately 8.0×10^{10} lb. [*Hint:* You need to estimate the heat of combustion of coal. To do this, assume that coal is 100% C(graphite) and use data from Appendix C.]

* **95.** **[Environmental]** The world's termite population is estimated to be 2.4×10^{17}. Annually, these termites produce an estimated 4.6×10^{16} g CO$_2$. The atmosphere contains 5.2×10^{15} metric tons of air (1 metric ton = 1000 kg). On a number basis, the current CO$_2$ level in the atmosphere is 368 ppm. What fraction of this CO$_2$ level could be attributed to termites if none of the termite-produced CO$_2$ were removed by natural processes?

* **96.** **[Environmental]** An important variable in the combustion of gasoline in an internal combustion engine is the air : fuel ratio. The figure that follows shows how the emission of pollutants is related to the air : fuel ratio. Provide a plausible interpretation of this figure. (*Hint:* Refer to Problem 92. Also, recall that RH represents hydrocarbons.)

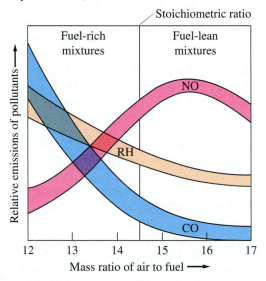

* **97.** **[Environmental]** To grow properly, common species of algae need carbon, nitrogen, and phosphorus atoms in the approximate ratio 106 C : 16 N : 1 P.

(a) Which is the limiting nutrient in a lake that has the following concentrations: C, 505 mg/L; N, 92 mg/L; and P, 14 mg/L?

(b) A scientific team wants to study algal growth in an experimental pond having the following atom ratio of nutrients: 44 C : 7 N : 1 P. They plan to achieve the appropriate nutrient mix by using a particular combination of *three* of the following chemicals: CO(NH$_2$)$_2$, CH$_3$CHOHCOOH, NH$_4$NO$_3$, NH$_4$H$_2$PO$_4$, (NH$_4$)$_2$HPO$_4$. Which three chemicals should they use if they wish to keep the total mass of chemicals *to a minimum?* What is the required *mass* ratio of these three chemicals?

* **98.** **[Environmental]** The herbicide dichlobenil (2,6-dichlorobenzonitrile, C$_7$H$_3$Cl$_2$N) has a solubility of 18 mg/L in water at 20 °C. The herbicide disintegrates with a half-life (first order) of 1.00 year. Suppose that 5.00×10^2 kg of dichlobenil is applied in the fields surrounding a 7.91-acre pond having an average depth of 5.25 ft. Suppose further that half of the dichlobenil permeates into the pond and that this happens within a few days.

(a) After a few days, what will be the concentration of dichlobenil in the pond, in ppm by mass and in molecules of dichlobenil per liter of pond water?

(b) How long would it take for the dichlobenil concentration in the pond to drop to 1.00 ppm following this single application?

(c) Suppose that the fields are treated with the dichlobenil in six-month intervals (fall and spring) each year for a number of years. What will be the concentration of dichlobenil in the pond just before the sixth application of the herbicide is made?

(d) Will the pond water eventually become saturated in dichlobenil? If so, when?

99. [Biochemical] A television mystery story indicated that a corpse contained 12 mg of chloroform. No liquid chloroform was found in the victim's stomach. This quantity supposedly could not have been self-administered as a vapor because the person would have become unconscious before achieving that quantity in her system. The quantity also was said to be less than a lethal dose. The *Merck Index* gives two values for the LD_{50} of chloroform: 2.18 mL/kg of body weight (14 day, orally in rats) and 0.9 mL/kg (organism and duration unspecified). The term LD_{50} refers to the dose that is lethal to half the administered population. Assuming that the corpse weighs 135 lb, how does the quantity of chloroform in her body compare to the lethal dose? What are some of the limitations of these calculations?

100. [Collaborative] The following quotations relating to hydrogen as a fuel are from "Letters to the Editor" in *Chemical and Engineering News*. How is it possible for scientists to have such divergent views on a seemingly straightforward matter?

April 1, 2002, D. C. MacWilliams, Alamo, CA: "The hydrogen fuel cell just shifts the pollution from the vehicle to the manufacturing facility. I believe that methane (or another hydrocarbon) fuel cell should be the focus."

April 1, 2002, George R. Lester, Salem, VA: "'Abundant, pollution-free hydrogen fuel' is a misleading myth unless it has been produced from nonfossil fuels such as solar, nuclear, wind (indirect solar), hydroelectric (very indirect solar), or the like."

April 29, 2002, Jerald A. Cole and Henry Wedaa, California Hydrogen Business Council, Long Beach, CA: "Furthermore, Lester's statement that "'Abundant, pollution-free hydrogen fuel' is a misleading myth" is absolutely untrue. Pollution-free hydrogen is even now being safely manufactured daily in factories around the world by the electrolysis of water—with an impeccable safety record."

June 3, 2002, Jason R. Guth, Chandler, AZ: "Converting fossil fuels to hydrogen can in no way reduce reliance on fossil fuels."

e-Media Problems

The activities described in these problems can be found in the Interactive student Tutorial (IST) module of the Companion Website, *http://chem.prenhall.com/hillpetrucci*.

101. View the **Heme** 3-D model (*Section 25-2*). By using the pop-up menu (right-click for Windows, click and hold for MacIntosh), change the representation of the model from ball-and-stick to space-filling. Why is the heme structure capable of bonding only to small molecules?

102. Consider the chemical reaction $3 O_2(g) \longrightarrow 2 O_3(g)$ described in detail in the **Stratospheric Ozone** animation (*Section 25-3*). (a) What is (are) the intermediate(s) of the reaction converting oxygen to ozone? (b) Construct an energy level diagram of the three elementary reactions leading to the production of ozone in an "energy-stabilized" state.

103. Compare the two sets of reactions seen in the **Catalytic Destruction of Stratospheric Ozone** animation and the **CFCs and Stratospheric Ozone** animation (*Section 25-3*). (a) What feature(s) is (are) common to the reaction mechanisms involving NO and Cl? (b) What extra energetic step is required to initiate the process described in the second animation?

104. View the **Carbon Dioxide Behaves as an Acid in Water** movie (*Section 25-7*). Review the properties of polyprotic acids in Section 15.5. (a) For the process seen in the movie, what species are present at equilibrium following the addition of dry ice to water? Write equations describing the corresponding reactions of (b) $SO_2(g)$ and (c) $SO_3(g)$.

Some Mathematical Operations

A.1 Exponential Notation

A number is in exponential form when it is written as the product of a coefficient—usually with a value between 1 and 10—and a power of ten. Following are two examples of *exponential notation*, a form usually employed by scientists and sometimes called *scientific notation*.

$$4.18 \times 10^3 \quad \text{and} \quad 6.57 \times 10^{-4}$$

Numbers are expressed in exponential form for two reasons: (1) We can write very large or very small numbers in a minimum of printed space and with a reduced chance of typographical error. (2) Numbers in exponential form convey explicit information about the precision of measurements: The number of significant figures in a measured quantity is stated unambiguously.

In the expression 10^n, n is the exponent of ten, and we say that the number ten is raised to the nth power. If n is a *positive* quantity, 10^n has a value *greater than 1*. If n is a *negative* quantity, 10^n has a value *less than 1*. We are particularly interested in cases where n is an integer, as in the following examples.

Positive Powers of 10
$10^0 = 1$
$10^1 = 10$
$10^2 = 10 \times 10 = 100$
$10^3 = 10 \times 10 \times 10 = 1000$
    and so on

The power of ten determines the number of zeros that follow the digit *1*.

Negative Powers of 10
$10^0 = 1$
$10^{-1} = 1/10 = 0.1$
$10^{-2} = 1/(10 \times 10) = 0.01$
$10^{-3} = 1/(10 \times 10 \times 10) = 0.001$
    and so on

The power of ten determines the number of places to the right of the decimal point where the digit *1* appears.

We express (a) 612,000 and (b) 0.000505 in exponential form as follows.

(a) $\quad 612{,}000 = 6.12 \times 100{,}000 = 6.12 \times 10^5$

(b) $\quad 0.000505 = 5.05 \times 0.0001 = 5.05 \times 10^{-4}$

The following approach provides a more direct method for converting numbers to the exponential form.

- Count the number of places a decimal point must be moved to produce a coefficient having a value between 1 and 10.
- The number of places counted then becomes the power of ten.
- The power of ten is *positive* if the decimal point is moved to the *left*.

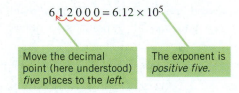

$$6\,1\,2\,0\,0\,0 = 6.12 \times 10^5$$

Move the decimal point (here understood) *five* places to the *left*.

The exponent is *positive five*.

One of the most common questions instructors encounter is "How do I use my calculator to do this?" You may find it a useful exercise to go through the calculations shown in this Appendix, using your calculator, to insure that you understand its use.

- The power of ten is *negative* if the decimal point is moved to the *right*.

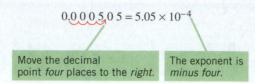

$$0.0\,0\,0\,5\,0\,5 = 5.05 \times 10^{-4}$$

Move the decimal point *four* places to the *right*.   The exponent is *minus four*.

To convert a number from exponential form to the conventional form, move the decimal point in the opposite direction.

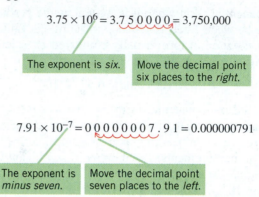

$$3.75 \times 10^6 = 3.7\,5\,0\,0\,0\,0 = 3{,}750{,}000$$

The exponent is *six*.   Move the decimal point six places to the *right*.

$$7.91 \times 10^{-7} = 0\,0\,0\,0\,0\,0\,0\,7\,.\,9\,1 = 0.000000791$$

The exponent is *minus seven*.   Move the decimal point seven places to the *left*.

Your calculator may require different keystrokes than those shown here. Check the specific instructions in the manual supplied with the calculator. Some two-line and graphing calculators may display the entire result, that is, 2.85×10^7 or 1.67×10^{-5}.

It is easy to handle exponential numbers on most calculators. A typical procedure is to enter the number, followed by the key EXP and then the exponent. To enter the number 2.85×10^7, the keystrokes required are [2] [.] [8] [5] [EXP] [7], and the result is displayed as 2.85^{07}. For the number 1.67×10^{-5}, the keystrokes are [1] [.] [6] [7] [EXP] [5] [±], and the result is displayed as 1.67^{-05}.

Many calculators can be set to convert all numbers and calculated results to the exponential form, regardless of the form in which the numbers are entered. Generally, the calculator can also be set to display a fixed number of significant figures in results.

Addition and Subtraction

With modern calculators it is rarely necessary to carry out operations such as addition and subtraction of exponential numbers by hand. However, it is quite important that you understand how such operations are carried out, so that you can correctly evaluate a calculator answer.

To add or subtract numbers in exponential notation with pencil and paper only, we must express each quantity as the same power of ten. This treats the power of ten in the same way as a unit—it is simply "carried along" in the calculation. In the following, each quantity is expressed to the power 10^{-3}.

$$(3.22 \times 10^{-3}) + (7.3 \times 10^{-4}) - (4.8 \times 10^{-4})$$
$$= (3.22 \times 10^{-3}) + (0.73 \times 10^{-3}) - (0.48 \times 10^{-3})$$
$$= (3.22 + 0.73 - 0.48) \times 10^{-3}$$
$$= 3.47 \times 10^{-3}$$

In contrast, most calculators perform these operations automatically, and you generally will not need to convert the numbers to the same power of ten when using a calculator.

Multiplication and Division

To multiply numbers expressed in exponential form, *multiply* all coefficients to obtain the coefficient of the result and *add* all exponents to obtain the power of ten in the result.

$$0.0803 \times 0.0077 \times 455 = (8.03 \times 10^{-2}) \times (7.7 \times 10^{-3}) \times (4.55 \times 10^2)$$
$$= (8.03 \times 7.7 \times 4.55) \times 10^{(-2-3+2)}$$
$$= (2.8 \times 10^2) \times 10^{-3} = 2.8 \times 10^{-1}$$

To divide two numbers in exponential form, *divide* the coefficients to obtain the coefficient of the result and *subtract* the exponent in the denominator from the exponent in the numerator to obtain the power of ten. The following example combines multiplication and division. First, we apply the rule for multiplication separately to the numerator and to the denominator, and then we use the rule for division.

$$\frac{0.015 \times 0.0088 \times 822}{0.092 \times 0.48} = \frac{(1.5 \times 10^{-2})(8.8 \times 10^{-3})(8.22 \times 10^{2})}{(9.2 \times 10^{-2})(4.8 \times 10^{-1})}$$

$$\frac{1.1 \times 10^{-1}}{4.4 \times 10^{-2}} = 0.25 \times 10^{-1-(-2)} = 0.25 \times 10^{1}$$

$$= 2.5 \times 10^{-1} \times 10^{1} = 2.5$$

As with addition and subtraction, most calculators perform multiplication, division, and combinations of the two with no need to record intermediate results.

Raising a Number to a Power and Extracting the Root of an Exponential Number

To raise an exponential number to a given power, raise the coefficient to that power, and multiply the exponent by that power. For example, we can *cube* a number (raise it to the *third* power) in the following manner:

$$(0.0066)^3 = (6.6 \times 10^{-3})^3 = (6.6)^3 \times 10^{(-3) \times 3}$$

| Rewrite in exponential form. | Cube the coefficient. | Multiply the exponent by 3. |

$$= (2.9 \times 10^2) \times 10^{-9} = 2.9 \times 10^{-7}$$

To extract the root of an exponential number, we raise the number to a *fractional* power: one-half power for a square root, one-third power for a cube root, and so on. Most calculators have keys designed for extracting square roots and cube roots. Thus, to extract the square root of 1.57×10^{-5}, enter the number 1.57×10^{-5} into the calculator and use the $(\sqrt{})$ key.

$$\sqrt{1.57 \times 10^{-5}} = 3.96 \times 10^{-3}$$

To extract the cube root of 3.18×10^{10}, enter the number 3.18×10^{10} and use the $(\sqrt[3]{})$ key.

$$\sqrt[3]{3.18 \times 10^{10}} = 3.17 \times 10^{3}$$

Some calculators allow you to extract roots by keying in the root as a fractional exponent.

$$(2.75 \times 10^{-9})^{1/5} = 1.94 \times 10^{-2}$$

We can also extract the roots of numbers by using logarithms.

A.2   Logarithms

The common logarithm (log) of a number (N) is the exponent (x) to which the base 10 must be raised to yield the number.

$$\log N = x$$
$$N = 10^{x}$$
$$N = 10^{\log N}$$

Equivalent expressions

In the following expressions, the numbers N are printed in blue and their logarithms ($\log N$) are printed in red.

$$\log 1 = \log 10^{0} = 0 \qquad\qquad \log 1 = \log 10^{0} = 0$$
$$\log 10 = \log 10^{1} = 1 \qquad\qquad \log 0.1 = \log 10^{-1} = -1$$
$$\log 100 = \log 10^{2} = 2 \qquad\qquad \log 0.01 = \log 10^{-2} = -2$$
$$\log 1000 = \log 10^{3} = 3 \qquad\qquad \log 0.001 = \log 10^{-3} = -3$$

Most of the numbers that we commonly encounter are not integral powers of ten, and their logarithms are not integral numbers. However, the preceding pattern gives us a general idea of what their logarithms might be. Consider, for example, the numbers 655 and 0.0078.

$$100 < \quad 655 \quad < 1000 \qquad\qquad 0.001 < \quad 0.0078 \quad < 0.01$$
$$2 \quad < \log 655 < \quad 3 \qquad\qquad -3 \quad < \log 0.0078 < \quad -2$$

Note that log 655 is between 2 and 3 and that log 0.0078 is between -3 and -2. To get a more exact value, however, we must use a table of logarithms or the [LOG] key on a calculator.

$$\log 655 = 2.816 \qquad\qquad \log 0.0078 = -2.11$$

In working with logarithms, we often need to find the *antilogarithm*, the number that has a certain value for its logarithm. We can think of the antilogarithm in the following terms.

$$\text{If } \log N = 3.576, \text{ then } N = 10^{3.576} = 3.77 \times 10^3.$$

$$\text{If } \log N = -4.57, \text{ then } N = 10^{-4.57} = 2.7 \times 10^{-5}.$$

With a calculator, we simply enter the value of the logarithm (3.576 or -4.57) and then use the $[10^x]$ key.

On some calculators, antilogarithms are found by entering the number, pressing a "second function" ([2nd]) key, then the [LOG] key.

Significant Figures in Logarithms

As they are written, $\log N = 3.576$ appears to have four significant figures and $N = 3.77 \times 10^3$ to have only three, but in reality both values have only *three*. Digits to the *left* of the decimal point in a logarithm simply relate to the power of ten in the exponential form of a number. The only significant digits in a logarithm are those to the *right* of the decimal point. The coefficient of the exponential form of the number should have this same number of digits. Thus, to express the logarithm of 2.5×10^{-12} to two significant figures, we would write log $2.5 \times 10^{-12} = -11.60$.

Some Relationships Involving Logarithms

We can use the definition of logarithms to write $M = 10^{\log M}$ and $N = 10^{\log N}$. For the product $(M \times N)$, we can write either of the following:

$$(M \times N) = 10^{\log M} \times 10^{\log N} = 10^{(\log M + \log N)}$$

$$(M \times N) = 10^{\log(M \times N)}$$

This means that the logarithm of the product of several terms is equal to the sum of the logarithms of the individual terms. Thus,

$$\log(M \times N) = (\log M + \log N) \qquad\qquad \textbf{(A.1)}$$

We can establish two other relationships in a similar manner.

$$\log \frac{M}{N} = (\log M - \log N) \qquad\qquad \textbf{(A.2)}$$

$$\log N^a = a \log N \qquad\qquad \textbf{(A.3)}$$

Equation (A.3) affords a simple method of extracting the roots of numbers. For example, to determine $(2.75 \times 10^{-9})^{1/5}$, write

$$\log(2.75 \times 10^{-9})^{1/5} = 1/5 \times \log(2.75 \times 10^{-9})$$
$$= 1/5 \times (-8.561) = -1.712$$
$$(2.75 \times 10^{-9})^{1/5} = 10^{-1.712} = 0.0194$$

Most scientific calculators have x^y and $\sqrt[y]{x}$ keys to determine powers and roots.

Natural Logarithms

Choosing *10* as the base for common logarithms is arbitrary. Other choices can be made as well. For example, to the base 2, $\log_2 8 = 3$. This simply means that $2^3 = 8$. And $\log_2 10 = 3.322$ means that $2^{3.322} = 10$.

Several of the relationships in this text involve *natural logarithms*. The base for natural logarithms (ln) is the quantity e, which has the value $e = 2.71828\cdots$. We encounter the ln function in circumstances in which the rate of change of a variable is proportional to the value of that variable at the time the rate is measured. Such circumstances are common in physical science, including, for example, the rate of decay of a radioactive material (Section 19.3).

Generally we can work entirely within the natural logarithm system by using the calculator keys [ln] and $[e^x]$ rather than [LOG] and $[10^x]$. However, if we need to convert between natural and common logarithms, we can use the following conversion factor based on the relationship $\log_e 10 = 2.303$:

$$\ln N = 2.303 \log N \qquad \textbf{(A.4)}$$

A.3   Algebraic Operations

To solve an algebraic equation requires that we isolate one quantity—the unknown—on one side of the equation and the known quantities on the other side. This generally requires rearranging terms in the equation, and in these rearrangements, the guiding principle is that *whatever we do to one side of the equation, we must do to the other side as well*. Consider the equation

$$\frac{(5x^2 - 12)}{(x^2 + 4)} = 3$$

1. Multiply both sides of the equation by $(x^2 + 4)$.

$$\cancel{(x^2 + 4)}\frac{(5x^2 - 12)}{\cancel{(x^2 + 4)}} = 3 \times (x^2 + 4)$$

$$5x^2 - 12 = 3x^2 + 12$$

2. Subtract $3x^2$ from each side of the equation.

$$5x^2 - 3x^2 - 12 = \cancel{3x^2} - \cancel{3x^2} + 12$$

$$2x^2 - 12 = 12$$

3. Add 12 to each side of the equation.

$$2x^2 - \cancel{12} + \cancel{12} = 12 + 12 = 24$$

4. Divide each side of the equation by 2.

$$\frac{\cancel{2}x^2}{\cancel{2}} = \frac{24}{2} = 12$$

5. Extract the square root of each side of the equation.

$$\sqrt{x^2} = \pm\sqrt{12} = \pm\sqrt{4} \times \sqrt{3}$$

$$x = \pm 2\sqrt{3}$$

$$x = \pm 3.464$$

Quadratic Equations

A quadratic equation has 2 as the highest power of the unknown x. At times, quadratic equations are of the form

$$(x + n)^2 = m^2$$

To solve for x, extract the square root of each side.

$$(x + n) = \sqrt{m^2} = \pm m$$

and

$$x = m - n \quad \text{or} \quad x = -m - n$$

You will find a quadratic equation of this type in Example 14.12.

More often, however, the quadratic equation will be of the form

$$ax^2 + bx + c = 0$$

where a, b, and c are constants. To solve this equation for x, we can use the *quadratic formula*.

$$x = \frac{-b \pm \sqrt{b^2 - 4ac}}{2a} \qquad \text{(A.5)}$$

Consider the solution of the following quadratic equation.

$$50.3x^2 - 13.4x + 0.787 = 0$$

$$x = \frac{-(-13.4) \pm \sqrt{(-13.4)^2 - (4 \times 50.3 \times 0.787)}}{2 \times 50.3}$$

$$x = \frac{13.4 \times \sqrt{21.22}}{100.6} = \frac{13.4 \pm 4.61}{100.6}$$

$$x = \frac{13.4 + 4.61}{100.6} = 0.179 \quad \text{and} \quad x = \frac{13.4 - 4.61}{100.6} = 0.0874$$

We encounter this particular quadratic equation in Example 14.13, where we find that the only physically significant answer is $x = 0.0874$.

Solving Equations by Approximation

If an equation is of higher degree than quadratic, the most direct solution is often one of successive approximations. We gather the terms involving the unknown on one side of the equation and a constant term on the other side of the equation. Then we apply trial and error with a little reasoning. For example, consider the following equation:

$$\frac{4x^3}{(0.492 - 2x)^2} = 0.023$$

Suppose, based on the physical situation, that x must be a positive quantity smaller than 0.246. Then, let's guess at a possible value of x and see how close the result is to 0.023 when we substitute this value of x into the equation.

Try $x = 0.100$:

$$\frac{4x^3}{(0.492 - 2x)^2} = \frac{4 \times (0.100)^3}{[0.492 - (2 \times 0.100)]^2} = \frac{4 \times 10^{-3}}{[0.492 - 0.200]^2} = 0.0469$$

Because $0.0469 > 0.023$, our guess is not good enough. Now let's make a second approximation.

Try $x = 0.050$:

$$\frac{4x^3}{(0.492 - 2x)^2} = \frac{4 \times (0.050)^3}{[0.492 - (2 \times 0.050)]^2} = \frac{5.0 \times 10^{-4}}{[0.492 - 0.100]^2} = 0.0033$$

Now our result $0.0033 < 0.023$. We seem to be on the other side of the desired answer. We need to try a value $0.050 < x < 0.100$. We need a third approximation.

Try $x = 0.080$:

$$\frac{4x^3}{(0.492 - 2x)^2} = \frac{4 \times (0.080)^3}{[0.492 - (2 \times 0.080)]^2} = \frac{2.0 \times 10^{-3}}{[0.492 - 0.160]^2} = 0.018$$

In this third approximation, we have $0.018 < 0.023$. We are now much closer to an acceptable value of x. In some cases, this might be close enough, but suppose we try one more approximation, with $0.080 < x < 0.10$.

Try $x = 0.085$:

$$\frac{4x^3}{(0.492 - 2x)^2} = \frac{4 \times (0.085)^3}{[0.492 - (2 \times 0.085)]^2} = \frac{2.5 \times 10^{-3}}{[0.492 - 0.170]^2} = 0.024$$

Now we are very close to the correct answer, since $0.024 \approx 0.023$. (A further approximation would show that $0.084 < x < 0.085$.)

This method may seem laborious, but usually it is not. Once the format of the approximations is set up, the calculations can be quickly performed with a calculator.

A.4   Graphs

Suppose we obtain the following data for the quantities x and y as a result of laboratory measurements:

$$x = 0, y = 2 \qquad x = 2, y = 6 \qquad x = 4, y = 10$$
$$x = 1, y = 4 \qquad x = 3, y = 8 \qquad \ldots$$

Just by inspecting these data, you can probably see that they fit the equation

$$y = 2x + 2$$

Sometimes an exact equation cannot be written from the experimental data, or the form of the equation may not be immediately obvious from the data themselves. In these cases, it often proves helpful to graph the data. The data points listed above are plotted in the graph, in which the x values are placed along the horizontal axis (abscissa) and the y values along the vertical axis (ordinate). For each point in the figure, the x and y values are listed in the order (x, y).

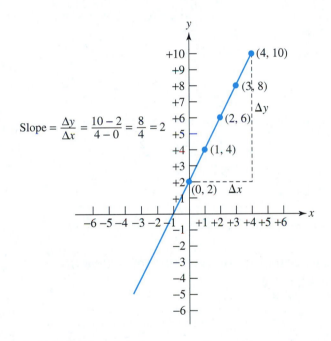

$$\text{Slope} = \frac{\Delta y}{\Delta x} = \frac{10-2}{4-0} = \frac{8}{4} = 2$$

A straight-line graph: $y = mx + b$

We see that the data fall on a straight line. The equation for a straight-line graph is

$$y = mx + b$$

To obtain a value of the *intercept, b*, we set $x = 0$ and obtain the value of y. From the graph, we see that $y = 2$ when $x = 0$. To obtain the slope of the line, we can work with two points on the graph, denoted as points 1 and 2 below.

$$y_2 = mx_2 + b \quad \text{and} \quad y_1 = mx_1 + b$$

The *difference* between these two equations is

$$y_2 - y_1 = m(x_2 - x_1) + \cancel{b} - \cancel{b}$$

Many graphing calculators will solve such equations very easily. Likewise, a computer spreadsheet such as Excel™ or Quattro Pro™ can be used to perform the successive approximations much more readily than can be done with a calculator.

If the points do not fall exactly on the line, we draw the "best" straight line represented by the points, and select two points on it. Alternately, a computer spreadsheet will plot the data and determine the slope and intercept automatically. It is well worth your time to learn to use such tools.

The value of m is

$$m = \frac{y_2 - y_1}{x_2 - x_1} = \frac{\Delta y}{\Delta x}$$

The slope is evaluated in the figure; it is 2. Thus, the equation of the straight line is

$$y = mx + b = 2x + 2$$

A.5   Some Key Equations

On several occasions in the text, we refer to this appendix for details on the derivations of key equations or their manipulation into more useful forms. Abbreviated treatments follow. The first two require some prior knowledge of calculus.

Integrated Rate Equation for First-Order Reaction (Equation (13.7))

For the reaction

$$A \longrightarrow \text{products}$$

having the rate law

$$\text{Rate of reaction} = -(\text{rate of disappearance of A}) = k[A]$$

1. Replace the rate of disappearance of A by the derivative $d[A]/dt$.

$$-\frac{d[A]}{dt} = k[A]$$

2. Rearrange this expression to the form

$$\frac{d[A]}{[A]} = -kdt$$

3. Integrate between the limits A_0 at time $t = 0$ and A_t at time t.

$$\int_{[A]_0}^{[A]_t} \frac{d[A]}{[A]} = -k \int_0^t dt$$

4. The result obtained is

$$\ln \frac{[A]_t}{[A]_0} = -kt$$

Integrated Rate Equation for Second-Order Reaction (Equation (13.11))

For the reaction

$$A \longrightarrow \text{products}$$

having the rate law

$$\text{Rate of reaction} = -(\text{rate of disappearance of A}) = k[A]^2$$

1. Replace the rate of disappearance of A by the derivative $d[A]/dt$.

$$-\frac{d[A]}{dt} = k[A]^2$$

2. Rearrange this expression to the form

$$\frac{d[A]}{[A]^2} = -kdt$$

3. Integrate between the limits A_0 at time $t = 0$ and A_t at time t.

$$\int_{[A]_0}^{[A]_t} \frac{d[A]}{[A]^2} = -k \int_0^t dt$$

4. The result obtained is

$$-\frac{1}{[A]_t} + \frac{1}{[A]_0} = -kt \quad \text{or} \quad \frac{1}{[A]_t} = kt + \frac{1}{[A]_0}$$

The Arrhenius Equation (Equation (13.16))

Our goal is to convert the equation for the straight-line graph of Figure 13.15,

$$\ln k = \frac{-E_a}{RT} + \ln A$$

into an equation that eliminates the constant term, $\ln A$.

1. Write the equation for two different temperatures, T_1 and T_2, at which the rate constants are k_1 and k_2. (E_a and R are constants.)

$$\ln k_2 = \frac{-E_a}{RT_2} + \ln A \qquad \ln k_1 = \frac{-E_a}{RT_1} + \ln A$$

2. Subtract $\ln k_1$ from $\ln k_2$

$$\ln k_2 - \ln k_1 = \frac{-E_a}{RT_2} + \ln A - \left(\frac{-E_a}{RT_1} + \ln A\right)$$

3. Replace $\ln k_2 - \ln k_1$ by $\ln \dfrac{k_2}{k_1}$, and eliminate $\ln A$.

$$\ln \frac{k_2}{k_1} = \frac{E_a}{RT_1} - \frac{E_a}{RT_2} + \cancel{\ln A} - \cancel{\ln A}$$

4. Rearrange the equation to the final form.

$$\ln \frac{k_2}{k_1} = \frac{E_a}{R}\left(\frac{1}{T_1} - \frac{1}{T_2}\right)$$

The van't Hoff Equation (Equation (17.14))

Our goal is to convert the equation for the straight-line graph of Figure 17.13,

$$\ln K_{eq} = \frac{-\Delta H°}{RT} + \text{constant}$$

into an equation that eliminates the term "constant," represented below as A.

1. Write the equation for two different temperatures, T_1 and T_2, at which the equilibrium constants are K_1 and K_2. ($\Delta H°$ and R are constants.)

$$\ln K_2 = \frac{-\Delta H°}{RT_2} + \ln A \qquad \ln K_1 = \frac{-\Delta H°}{RT_1} + \ln A$$

2. Subtract $\ln K_1$ from $\ln K_2$.

$$\ln K_2 - \ln K_1 = \frac{-\Delta H°}{RT_2} + \ln A - \left(\frac{-\Delta H°}{RT_1} + \ln A\right)$$

3. Replace $\ln K_2 - \ln K_1$ by $\ln \dfrac{K_2}{K_1}$, and eliminate $\ln A$.

$$\ln \frac{K_2}{K_1} = \frac{\Delta H°}{RT_1} - \frac{\Delta H°}{RT_2} + \cancel{\ln A} - \cancel{\ln A}$$

4. Rearrange the equation to the final form.

$$\ln \frac{K_2}{K_1} = \frac{\Delta H°}{R}\left(\frac{1}{T_1} - \frac{1}{T_2}\right)$$

Some Basic Physical Concepts

B.1 Velocity and Acceleration

The speed of an object is the distance it travels per unit time. An automobile with a speedometer that reads 105 km/h will, if it continues at this constant speed for exactly one hour, travel a distance of 105 km. For scientific work, *velocity* is a more appropriate term. Velocity has two components: a *magnitude* (speed) and a *direction* (up, down, east, southwest, and so on). The SI units of velocity are distance $\times$ time^{-1} (m s^{-1}).

The velocity of an object changes if its speed or direction of motion changes. The rate of change of velocity is called *acceleration*, which has the units of velocity $\times$ time^{-1} (m s$^{-1} \times$ s$^{-1} =$ m s^{-2}). For an object under a constant acceleration (a), its velocity (v) as a function of time (t) is

$$v = at \tag{B.1}$$

The distance (d) traveled is given by the following equation, which can be established by the methods of calculus.

$$d = \tfrac{1}{2}at^2 \tag{B.2}$$

The *constant acceleration due to gravity* (g) experienced by a freely falling body is 9.0866 m s^{-2}.

B.2 Force and Work

According to Newton's *first law* of motion, an object has a natural tendency—called *inertia*—to remain in motion at a constant velocity if it is moving or to remain at rest if it is not moving. A *force* is required to overcome the inertia of an object—that is, to give motion to an object at rest or to change the velocity of a moving object. Because a change in velocity is an acceleration, we can say that *a force is required to provide acceleration to an object.*

Newton's *second law* of motion describes the force (F) required to produce an acceleration (a) in an object of mass (m).

$$F = ma \tag{B.3}$$

The SI unit of force is the *newton* (N). It is the force required to produce an acceleration of 1 m s^{-2} in a 1-kg mass.

$$1 \text{ N} = 1 \text{ kg} \times 1 \text{ m s}^{-2} = 1 \text{ kg m s}^{-2} \tag{B.4}$$

The weight (W) of an object is the force of gravity on the object. It is the mass of the object multiplied by the acceleration due to gravity.

$$W = F = mg$$

Work is done when a force acts through a distance.

$$\text{Work}(w) = \text{force}(F) \times \text{distance}(d)$$

A joule (J) is the work done when a force of one newton acts through a distance of one meter. When we combine this definition and the SI units of the newton from Equation (B.4), we obtain the SI units of the joule.

$$1 \text{ J} = 1 \text{ N} \times 1 \text{ m} = 1 \text{ N m}$$

$$1 \text{ J} = 1 \text{ kg} \times 1 \text{ m s}^{-2} \times 1 \text{ m} = 1 \text{ kg m}^2 \text{ s}^{-2}$$

B.3   Energy

Energy is the capacity to do work. A moving object has *kinetic* energy as a result of its motion. The work associated with the moving object is given by the previous expressions (page A10).

$$w = F \times d = ma \times d$$

From Equation (B.2), we can substitute the expression for the distance, *d*, that is, $d = \frac{1}{2}at^2$.

$$w = ma \times \tfrac{1}{2}at^2 = \tfrac{1}{2} \times m(at)^2$$

Now, from Equation (B.1), we can substitute velocity (v) for the term at.

$$w = \tfrac{1}{2} \times m \times v^2$$

This is the work required to provide an object of mass *m* with a velocity *v*. This quantity of work appears as the kinetic energy (E_k) of the moving object.

$$E_k = \tfrac{1}{2}mv^2$$

In addition to kinetic energy associated with motion, an object may possess *potential energy*. Potential energy is stored energy that can be released under appropriate circumstances. Think of it as energy that stems from the condition, position, or composition of an object. In principle, equations can be written for the various ways in which potential energy is stored in an object, but we do not specifically use such equations in the text.

B.4   Magnetism

Attractive and repulsive forces associated with magnets are centered in regions of the magnets called *poles*. A magnet has a north pole and a south pole. If two magnets are arranged so that the north pole of one magnet is brought near the south pole of another magnet, there is an attractive force between the magnets. If like poles are brought close together—either both north poles or both south poles—there is a repulsive force. *Unlike poles attract, and like poles repel.*

A magnetic field exists in the region surrounding a magnet in which the influence of the magnet can be felt. For example, a magnetic field can be detected through deflections of a compass needle, or the field can be visualized through the attractive forces that cause a characteristic alignment of iron filings.

▲ **Visualizing a magnetic field of a bar magnet**

The sprinkling of iron filings outlines the magnetic field of a bar magnet.

B.5   Electricity

Electricity is a phenomenon closely related to magnetism. Ultimately, all bulk matter contains electrically charged particles: protons and electrons. However, an object displays a net electric charge—positive or negative—only when the numbers of electrons and protons in the object are unequal. The basic expression dealing with stationary electrically charged particles—static electricity—is Coulomb's law: The magnitude of the force (F) between electrically charged objects is directly proportional to the magnitudes of the charges (Q) and inversely proportional to the *square* of the distance (r) between them.

$$F \propto \frac{Q_1 \times Q_2}{r^2}$$

- *Like charges repel*. Whether both charges are positive or both are negative, their product is a positive quantity. A *positive* force is a *repulsive force*.
- *Unlike charges attract*. If one charge is positive and the other negative, their product is a negative quantity. A *negative* force is an *attractive* force.

An *electric field* exists in the region surrounding an electrically charged object in which the influence of the electric charge is felt. If an uncharged object is brought into the field of a charged object, an electric charge of the opposite sign may be *induced* in the previously uncharged object. This leads to a force of attraction between the two. (See Figure 11.17.)

Electric current is a flow of charged particles—electrons in metallic conductors and positive and negative ions in molten salts and in aqueous salt solutions. The unit of electric charge is the *coulomb* (C). The unit of electric current is the *ampere* (A). A current of one ampere is the flow of one coulomb of electric charge per second.

$$1 \text{ A} = 1 \text{ C}/1 \text{ s} = 1 \text{ C s}^{-1}$$

Electric potential, or voltage, is the energy per unit of charge in an electric current. With coulomb as the unit of charge and joule as the unit of energy, the unit of electrical potential, 1 *volt* (V), is

$$1 \text{ V} = \frac{1 \text{ J}}{1 \text{ C}}$$

Electric *power* is the rate of production (or consumption) of electric energy. The electric power unit, the *watt* (W), signifies the production (or consumption) of one joule of energy per second.

$$1 \text{ W} = 1 \text{ J s}^{-1}$$

Because electric energy in joules is the product (volts × coulombs) and because coulombs per second (C s^{-1}) represents a current in amperes (A), we can also write the following expressions.

$$1 \text{ W} = 1 \text{ V C s}^{-1} = 1 \text{ V} \times 1 \text{ A}$$

As an example, the electric power associated with the passage of 10.0 amp through a 110-volt electric circuit is

$$110 \text{ V} \times 10.0 \text{ A} = 1100 \text{ W}$$

B.6   Electromagnetism

A variety of relationships between electricity and magnetism, collectively called *electromagnetism*, underlie some important practical applications: (1) Magnetic fields are associated with the flow of electrons, as in *electromagnets* (demonstrated in the photograph below). (2) Forces are experienced by current-carrying conductors in a magnetic field, as in *electric motors*. (3) Electric currents are induced when electric conductors are moved through a magnetic field, as in *electric generators*. Several phenomena described in this text are electromagnetic effects.

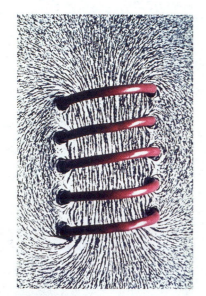

▲ **Visualizing a magnetic field of an electromagnet**

An electric current flowing through a coil of wire produces a magnetic field, outlined here by a sprinkling of iron filings.

▶ **An electromagnet**

Electric current from the battery passes through the coil of wire wrapped around an iron bar. The electric current induces a magnetic field and causes the bar to act as a magnet, attracting small iron objects. When the electric current is cut off, the magnetic field dissipates and the bar loses its magnetism.

Data Tables

C.1 Thermodynamic Properties of Substances at 298.15 K*

Inorganic Substances

	ΔH_f°, kJ mol^{-1}	ΔG_f°, kJ mol^{-1}	S°, J mol^{-1}K^{-1}		ΔH_f°, kJ mol^{-1}	ΔG_f°, kJ mol^{-1}	S°, J mol^{-1}K^{-1}
Aluminum				BrCl(g)	14.6	−0.96	240.0
Al(s)	0	0	28.3	BrF$_3$(g)	−255.6	−229.5	292.4
Al^{3+}(aq)	−531	−485	−321.7	BrF$_3$(l)	−300.8	−240.6	178.2
AlCl$_3$(s)	−705.6	−630.1	109.3	**Cadmium**			
Al$_2$Cl$_6$(g)	−1291	−1221	490	Cd(s)	0	0	51.76
AlF$_3$(s)	−1504	−1425	66.48	Cd^{2+}(aq)	−75.90	−77.61	−73.2
Al$_2$O$_3$(α, s)	−1676	−1582	50.92	CdCl$_2$(s)	−391.5	−344.0	115.3
Al(OH)$_3$(s)	−1276	—	—	CdO(s)	−258	−228	54.8
Al$_2$(SO$_4$)$_3$(s)	−3441	−3100	239	**Calcium**			
Barium				Ca(s)	0	0	41.4
Ba(s)	0	0	62.3	Ca^{2+}(aq)	−542.8	−553.6	−53.1
Ba^{2+}(aq)	−537.6	−560.8	9.6	CaBr$_2$(s)	−682.8	−663.6	130
BaCO$_3$(s)	−1216	−1138	112	CaCO$_3$(s)	−1207	−1128	88.70
BaCl$_2$(s)	−858.1	−810.4	123.7	CaCl$_2$(s)	−795.8	−748.1	105
BaF$_2$(s)	−1209	−1159	96.40	CaF$_2$(s)	−1220	−1167	68.87
BaO(s)	−548.1	−520.4	72.09	CaH$_2$(s)	−186	−147	42
Ba(OH)$_2$(s)	−946.0	−859.4	107	Ca(NO$_3$)$_2$(s)	−938.4	−743.2	193
Ba(OH)$_2\cdot$8 H$_2$O(s)	−3342	−2793	427	CaO(s)	−635.1	−604.0	39.75
BaSO$_4$(s)	−1473	−1362	132	Ca(OH)$_2$(s)	−986.1	−898.6	83.39
Beryllium				Ca$_3$(PO$_4$)$_2$(s)	−4121	−3885	236
Be(s)	0	0	9.54	CaSO$_4$(s)	−1434	−1322	106.7
BeCl$_2$(s)	−496.2	−449.5	75.81	**Carbon** (See also the table of organic substances.)			
BeF$_2$(s)	−1027	−979.5	53.35	C(g)	716.7	671.3	158.0
BeO(s)	−608.4	−579.1	13.77	C (diamond)	1.90	2.90	2.38
Bismuth				C (graphite)	0	0	5.74
Bi(s)	0	0	56.74	CCl$_4$(g)	−102.9	−60.63	309.7
BiCl$_3$(s)	−379	−315	177	CCl$_4$(l)	−135.4	−65.27	216.2
Bi$_2$O$_3$(s)	−573.9	−493.7	151	C$_2$N$_2$(g)	308.9	297.2	242.3
Boron				CO(g)	−110.5	−137.2	197.6
B(s)	0	0	5.86	CO$_2$(g)	−393.5	−394.4	213.6
BCl$_3$(l)	−427.2	−387	206	CO$_3^{2-}$(aq)	−677.1	−527.8	−56.9
BF$_3$(g)	−1137	−1120.3	254.0	C$_3$O$_2$(g)	−93.72	−109.8	276.4
B$_2$H$_6$(g)	36	86.6	232.0	C$_3$O$_2$(l)	−117.3	−105.0	181.1
B$_2$O$_3$(s)	−1273	−1194	53.97	COCl$_2$(g)	−220.9	−206.8	283.8
Bromine				COS(g)	−138.4	−165.6	231.5
Br(g)	111.9	82.43	174.9	CS$_2$(l)	89.70	65.27	151.3
Br$^-$(aq)	−121.6	−104.0	82.4	**Chlorine**			
Br$_2$(g)	30.91	3.14	245.4	Cl(g)	121.7	105.7	165.1
Br$_2$(l)	0	0	152.2	Cl$^-$(aq)	−167.2	−131.2	56.5

* Substances are at 1 atm pressure, and solutes in aqueous solutions are at unit activity ($\approx$1 M). Differences between these data and those based on the IUPAC recommended standard-state pressure of 1 bar, where they exist, are very small.

Inorganic Substances (continued)

	ΔH_f°, kJ mol^{-1}	ΔG_f°, kJ mol^{-1}	S°, J mol^{-1}K^{-1}		ΔH_f°, kJ mol^{-1}	ΔG_f°, kJ mol^{-1}	S°, J mol^{-1}K^{-1}
$Cl_2(g)$	0	0	223.0	$ICl(l)$	−23.89	−13.60	135.1
$ClF_3(g)$	−163.2	−123.0	281.5	**Iron**			
$ClO_2(g)$	102.5	120.5	256.7	$Fe(s)$	0	0	27.28
$Cl_2O(g)$	80.33	97.49	267.9	$Fe^{2+}(aq)$	−89.1	−78.90	−137.7
Chromium				$Fe^{3+}(aq)$	−48.5	−4.7	−315.9
$Cr(s)$	0	0	23.66	$FeCO_3(s)$	−740.6	−666.7	92.88
$Cr_2O_3(s)$	−1135	−1053	81.17	$FeCl_3(s)$	−399.5	−334.1	142.3
$CrO_4^{2-}(aq)$	−881.2	−727.8	50.21	$FeO(s)$	−272	−251.5	60.75
$Cr_2O_7^{2-}(aq)$	−1490	−1301	261.9	$Fe_2O_3(s)$	−824.2	−742.2	87.40
Cobalt				$Fe_3O_4(s)$	−1118	−1015	146
$Co(s)$	0	0	30.0	$Fe(OH)_3(s)$	−823.0	−696.6	107
$CoO(s)$	−237.9	−214.2	52.97	**Lead**			
$Co(OH)_2$ (pink, s)	−539.7	−454.4	79	$Pb(s)$	0	0	64.81
Copper				$Pb^{2+}(aq)$	−1.7	−24.43	10.5
$Cu(s)$	0	0	33.15	$PbI_2(s)$	−175.5	−173.6	174.8
$Cu^{2+}(aq)$	64.77	65.49	−99.6	$PbO_2(s)$	−277	−217.4	68.6
$CuCO_3 \cdot Cu(OH)_2(s)$	−1051	−893.7	186	$PbSO_4(s)$	−919.9	−813.2	148.6
$Cu_2O(s)$	−168.6	−146.0	93.14	**Lithium**			
$CuO(s)$	−157.3	−129.7	42.63	$Li(s)$	0	0	29.12
$Cu(OH)_2(s)$	−450.2	−373	108	$Li^+(aq)$	−278.5	−293.3	13.4
$CuSO_4 \cdot 5\,H_2O(s)$	−2279.6	−1880.1	300.4	$LiCl(s)$	−408.6	−384.4	59.33
Fluorine				$Li_2O(s)$	−597.94	−561.18	37.57
$F(g)$	78.99	61.92	158.7	$LiOH(s)$	−484.9	−439.0	42.80
$F^-(aq)$	−332.6	−278.8	−13.8	$LiNO_3(s)$	−483.1	−381.1	90.0
$F_2(g)$	0	0	202.7	**Magnesium**			
Helium				$Mg(s)$	0	0	32.69
$He(g)$	0	0	126.0	$Mg^{2+}(aq)$	−466.9	−454.8	−138.1
Hydrogen				$MgCl_2(s)$	−641.3	−591.8	89.62
$H(g)$	218.0	203.3	114.6	$MgCO_3(s)$	−1096	−1012	65.7
$H^+(aq)$	0	0	0	$MgF_2(s)$	−1124	−1071	57.24
$H_2(g)$	0	0	130.6	$MgO(s)$	−601.7	−569.4	26.94
$HBr(g)$	−36.40	−53.43	198.6	$Mg(OH)_2(s)$	−924.7	−833.9	63.18
$HCl(g)$	−92.31	−95.30	186.8	$MgSO_4(s)$	−1285	−1171	91.6
$HCl(aq)$	−167.2	−131.3	56.48	**Manganese**			
$HCN(g)$	135	125	201.7	$Mn(s)$	0	0	32.0
$HF(g)$	−271.1	−273.2	173.7	$Mn^{2+}(aq)$	−220.8	−228.1	−73.6
$HI(g)$	26.48	1.72	206.5	$MnO_2(s)$	−520	−465.2	53.05
$HNO_3(l)$	−173.2	−79.91	155.6	$MnO_4^-(aq)$	−541.4	−447.2	191.2
$HNO_3(aq)$	−207.4	−113.3	146.4	**Mercury**			
$H_2O(g)$	−241.8	−228.6	188.7	$Hg(g)$	61.32	31.85	174.9
$H_2O(l)$	−285.8	−237.2	69.91	$Hg(l)$	0	0	76.02
$H_2O_2(g)$	−136.1	−105.5	232.9	$HgO(s)$	−90.83	−58.56	70.29
$H_2O_2(l)$	−187.8	−120.4	110	**Nitrogen**			
$H_2S(g)$	−20.63	−33.56	205.7	$N(g)$	472.7	455.6	153.2
$H_2SO_4(l)$	−814.0	−690.1	156.9	$N_2(g)$	0	0	191.5
$H_2SO_4(aq)$	−909.3	−744.6	20.08	$NF_3(g)$	−124.7	−83.2	260.7
Iodine				$NH_3(g)$	−46.11	−16.48	192.3
$I(g)$	106.8	70.28	180.7	$NH_3(aq)$	−80.29	−26.57	111.3
$I^-(aq)$	−55.19	−51.57	111.3	$NH_4^+(aq)$	−132.5	−79.31	113.4
$I_2(g)$	62.44	19.36	260.6	$NH_4Br(s)$	−270.8	−175	113.0
$I_2(s)$	0	0	116.1	$NH_4Cl(s)$	−314.4	−203.0	94.56
$IBr(g)$	40.84	3.72	258.7	$NH_4F(s)$	−464.0	−348.8	71.96
$ICl(g)$	17.78	−5.44	247.4	$NH_4HCO_3(s)$	−849.4	−666.1	121

Inorganic Substances (continued)

	ΔH_f°, kJ mol^{-1}	ΔG_f°, kJ mol^{-1}	S°, J mol^{-1}K^{-1}		ΔH_f°, kJ mol^{-1}	ΔG_f°, kJ mol^{-1}	S°, J mol^{-1}K^{-1}
$NH_4I(s)$	−201.4	−113	117	**Silver**			
$NH_4NO_3(s)$	−365.6	−184.0	151.1	$Ag(s)$	0	0	42.55
$NH_4NO_3(aq)$	−339.9	−190.7	259.8	$Ag^+(aq)$	105.6	77.11	72.68
$(NH_4)_2SO_4(s)$	−1181	−901.9	220.1	$AgBr(s)$	−100.4	−96.90	107
$N_2H_4(g)$	95.40	159.3	238.4	$AgCl(s)$	−127.1	−109.8	96.2
$N_2H_4(l)$	50.63	149.2	121.2	$AgI(s)$	−61.84	−66.19	115
$NO(g)$	90.25	86.57	210.6	$AgNO_3(s)$	−124.4	−33.5	140.9
$N_2O(g)$	82.05	104.2	219.7	$Ag_2O(s)$	−31.0	−11.2	121
$NO_2(g)$	33.18	51.30	240.0	$Ag_2SO_4(s)$	−715.9	−618.5	200.4
$N_2O_4(g)$	9.16	97.82	304.2	**Sodium**			
$N_2O_4(l)$	−19.6	97.40	209.2	$Na(g)$	107.3	76.78	153.6
$N_2O_5(g)$	11.3	115.1	355.7	$Na(l)$	2.41	0.50	57.86
$NO_3^-(aq)$	−205.0	−108.7	146.4	$Na(s)$	0	0	51.21
$NOBr(g)$	82.17	82.4	273.5	$Na^+(aq)$	−240.1	−261.9	59.0
$NOCl(g)$	51.71	66.07	261.6	$Na_2(g)$	142.0	104.0	230.1
				$NaBr(s)$	−361.1	−349.0	86.82
Oxygen				$Na_2CO_3(s)$	−1131	−1044	135.0
$O(g)$	249.2	231.7	160.9	$NaHCO_3(s)$	−950.8	−851.0	102
$O_2(g)$	0	0	205.0	$NaCl(s)$	−411.1	−384.0	72.13
$O_3(g)$	142.7	163.2	238.8	$NaCl(aq)$	−407.3	−393.1	115.5
$OH^-(aq)$	−230.0	−157.2	−10.75	$NaClO_3(s)$	−365.8	−262.3	123
$OF_2(g)$	24.5	41.8	247.3	$NaClO_4(s)$	−383.3	−254.9	142.3
				$NaF(s)$	−573.7	−543.5	51.46
Phosphorus				$NaH(s)$	−56.27	−33.5	40.02
$P(\alpha, white)$	0	0	41.1	$NaI(s)$	−287.8	−286.1	98.53
$P (red)$	−17.6	−12.1	22.8	$NaNO_3(s)$	−467.9	−367.1	116.5
$P_4(g)$	58.9	24.5	279.9	$NaNO_3(aq)$	−447.4	−373.2	205.4
$PCl_3(g)$	−287.0	−267.8	311.7	$Na_2O_2(s)$	−510.9	−447.7	94.98
$PCl_3(l)$	−319.7	−272.3	217.1	$NaOH(s)$	−425.6	−379.5	64.48
$PCl_5(g)$	−374.9	−305.0	364.5	$NaOH(aq)$	−469.2	−419.2	48.1
$PCl_5(s)$	−443.5	—	—	$NaH_2PO_4(s)$	−1537	−1386	127.5
$PH_3(g)$	5.4	13.4	210.1	$Na_2HPO_4(s)$	−1748	−1608	150.5
$P_4O_{10}(s)$	−2984	−2698	228.9	$Na_3PO_4(s)$	−1917	−1789	173.8
$PO_4^{3-}(aq)$	−1277	−1019	−222	$NaHSO_4(s)$	−1125	−992.9	113
				$Na_2SO_4(s)$	−1387	−1270	149.6
Potassium				$Na_2SO_4(aq)$	−1390	−1268	138.1
$K(g)$	89.24	60.63	160.2	$Na_2SO_4 \cdot 10\,H_2O(s)$	−4327	−3647	592.0
$K(l)$	2.28	0.26	71.46	$Na_2S_2O_3(s)$	−1123	−1028	155
$K(s)$	0	0	64.18	**Sulfur**			
$K^+(aq)$	−252.4	−283.3	102.5	$S (rhombic)$	0	0	31.8
$KBr(s)$	−393.8	−380.7	95.90	$S_8(g)$	102.3	49.16	430.2
$KCN(s)$	−113	−101.9	128.5	$S_2Cl_2(g)$	−18.4	−31.8	331.5
$KCl(s)$	−436.7	−409.2	82.59	$SF_6(g)$	−1209	−1105	291.7
$KClO_3(s)$	−397.7	−296.3	143	$SO_2(g)$	−296.8	−300.2	248.1
$KClO_4(s)$	−432.8	−303.2	151.0	$SO_3(g)$	−395.7	−371.1	256.6
$KF(s)$	−567.3	−537.8	66.57	$SO_4^{2-}(aq)$	−909.3	−744.5	20.1
$KI(s)$	−327.9	−324.9	106.3	$S_2O_3^{2-}(aq)$	−648.5	−522.5	67
$KNO_3(s)$	−494.6	−394.9	133.1	$SO_2Cl_2(g)$	−364.0	−320.0	311.8
$KOH(s)$	−424.8	−379.1	78.87	$SO_2Cl_2(l)$	−394.1	−314	207
$KOH(aq)$	−482.4	−440.5	91.63	**Tin**			
$K_2SO_4(s)$	−1438	−1321	175.6	$Sn (white)$	0	0	51.55
Silicon				$Sn (gray)$	−2.1	0.1	44.14
$Si(s)$	0	0	18.8	$SnCl_4(l)$	−511.3	−440.2	259
$SiH_4(g)$	34	56.9	204.5	$SnO(s)$	−286	−257	56.5
$Si_2H_6(g)$	80.3	127	272.5	$SnO_2(s)$	−580.7	−519.7	52.3
$SiO_2(quartz)$	−910.9	−856.7	41.84				

Inorganic Substances (continued)

	ΔH_f°, kJ mol^{-1}	ΔG_f°, kJ mol^{-1}	S°, J mol^{-1}K^{-1}		ΔH_f°, kJ mol^{-1}	ΔG_f°, kJ mol^{-1}	S°, J mol^{-1}K^{-1}
Titanium				$UF_6(s)$	−2197	−2069	228
$Ti(s)$	0	0	30.6	$UO_2(s)$	−1085	−1032	77.03
$TiCl_4(g)$	−763.2	−726.8	355	**Zinc**			
$TiCl_4(l)$	−804.2	−737.2	252.3	$Zn(s)$	0	0	41.6
$TiO_2(s)$	−944.7	−889.5	50.33	$Zn^{2+}(aq)$	−153.9	−147.1	−112.1
Uranium				$ZnCl_2(s)$	−415.1	−369.4	111.5
$U(s)$	0	0	50.21	$ZnO(s)$	−348.3	−318.3	43.64
$UF_6(g)$	−2147	−2064	378				

Organic Substances

Formula	Name	ΔH_f°, kJ mol^{-1}	ΔG_f°, kJ mol^{-1}	S°, J mol^{-1}K^{-1}
$CH_4(g)$	Methane(g)	−74.81	−50.75	186.2
$C_2H_2(g)$	Acetylene(g)	226.7	209.2	200.8
$C_2H_4(g)$	Ethylene(g)	52.26	68.12	219.4
$C_2H_6(g)$	Ethane(g)	−84.68	−32.89	229.5
$C_3H_8(g)$	Propane(g)	−103.8	−23.56	270.2
$C_4H_{10}(g)$	Butane(g)	−125.7	−17.15	310.1
$C_6H_6(g)$	Benzene(g)	82.93	129.7	269.2
$C_6H_6(l)$	Benzene(l)	48.99	124.4	173.3
$C_6H_{12}(g)$	Cyclohexane(g)	−123.1	31.8	298.2
$C_6H_{12}(l)$	Cyclohexane(l)	−156.2	26.7	204.3
$C_{10}H_8(g)$	Naphthalene(g)	149	223.6	335.6
$C_{10}H_8(s)$	Naphthalene(s)	75.3	201.0	166.9
$CH_2O(g)$	Formaldehyde(g)	−117.0	−110.0	218.7
$CH_3OH(g)$	Methanol(g)	−200.7	−162.0	239.7
$CH_3OH(l)$	Methanol(l)	−238.7	−166.4	126.8
$CH_3CHO(g)$	Acetaldehyde(g)	−166.1	−133.4	246.4
$CH_3CHO(l)$	Acetaldehyde(l)	−191.8	−128.3	160.4
$CH_3CH_2OH(g)$	Ethanol(g)	−234.4	−167.9	282.6
$CH_3CH_2OH(l)$	Ethanol(l)	−277.7	−174.9	160.7
$C_6H_5OH(s)$	Phenol(s)	−165.0	−50.42	144.0
$(CH_3)_2CO(g)$	Acetone(g)	−216.6	−153.1	294.9
$(CH_3)_2CO(l)$	Acetone(l)	−247.6	−155.7	200.4
$CH_3COOH(g)$	Acetic acid(g)	−432.3	−374.0	282.5
$CH_3COOH(l)$	Acetic acid(l)	−484.1	−389.9	159.8
$CH_3COOH(aq)$	Acetic acid(aq)	−488.3	−396.6	178.7
$C_6H_5COOH(s)$	Benzoic acid(s)	−385.1	−245.3	167.6
$CH_3NH_2(g)$	Methylamine(g)	−23.0	32.3	242.6
$C_6H_5NH_2(g)$	Aniline(g)	86.86	166.7	319.2
$C_6H_5NH_2(l)$	Aniline(l)	31.6	149.1	191.3
$C_6H_{12}O_6(s)$	Glucose(s)	−1273.3	−910.4	212.1

C.2 Equilibrium Constants

A. Ionization Constants of Weak Acids at 25 °C

Name of Acid	Formula	K_a	Name of Acid	Formula	K_a
Acetic	$HC_2H_3O_2$	1.8×10^{-5}	Hyponitrous	HONNOH	8.9×10^{-8}
Acrylic	$HC_3H_3O_2$	5.5×10^{-5}		$HONNO^-$	4×10^{-12}
Arsenic	H_3AsO_4	6.0×10^{-3}	Iodic	HIO_3	1.6×10^{-1}
	$H_2AsO_4^-$	1.0×10^{-7}	Lactic	$HC_3H_5O_2$	1.4×10^{-4}
	$HAsO_4^{2-}$	3.2×10^{-12}	Malonic	$H_2C_3H_2O_4$	1.5×10^{-3}
Arsenous	H_3AsO_3	6.6×10^{-10}		$HC_3H_2O_4^-$	2.0×10^{-6}
Benzoic	$HC_7H_5O_2$	6.3×10^{-5}	Nitrous	HNO_2	7.2×10^{-4}
Bromoacetic	$HC_2H_2BrO_2$	1.3×10^{-3}	Oxalic	$H_2C_2O_4$	5.4×10^{-2}
Butyric	$HC_4H_7O_2$	1.5×10^{-5}		$HC_2O_4^-$	5.3×10^{-5}
Carbonic	H_2CO_3	4.4×10^{-7}	Phenol	HOC_6H_5	1.0×10^{-10}
	HCO_3^-	4.7×10^{-11}	Phenylacetic	$HC_8H_7O_2$	4.9×10^{-5}
Chloroacetic	$HC_2H_2ClO_2$	1.4×10^{-3}	Phosphoric	H_3PO_4	7.1×10^{-3}
Chlorous	$HClO_2$	1.1×10^{-2}		$H_2PO_4^-$	6.3×10^{-8}
Citric	$H_3C_6H_5O_7$	7.4×10^{-4}		HPO_4^{2-}	4.2×10^{-13}
	$H_2C_6H_5O_7^-$	1.7×10^{-5}	Phosphorus	H_3PO_3	3.7×10^{-2}
	$HC_6H_5O_7^{2-}$	4.0×10^{-7}		$H_2PO_3^-$	2.1×10^{-7}
Cyanic	HOCN	3.5×10^{-4}	Propionic	$HC_3H_5O_2$	1.3×10^{-5}
Dichloroacetic	$HC_2HCl_2O_2$	5.5×10^{-2}	Pyrophosphoric	$H_4P_2O_7$	3.0×10^{-2}
Fluoroacetic	$HC_2H_2FO_2$	2.6×10^{-3}		$H_3P_2O_7^-$	4.4×10^{-3}
Formic	$HCHO_2$	1.8×10^{-4}		$H_2P_2O_7^{2-}$	2.5×10^{-7}
Hydrazoic	HN_3	1.9×10^{-5}		$HP_2O_7^{3-}$	5.6×10^{-10}
Hydrocyanic	HCN	6.2×10^{-10}	Selenic	H_2SeO_4	strong acid
Hydrofluoric	HF	6.6×10^{-4}		$HSeO_4^-$	2.2×10^{-2}
Hydrogen peroxide	H_2O_2	2.2×10^{-12}	Selenous	H_2SeO_3	2.3×10^{-3}
Hydroselenic	H_2Se	1.3×10^{-4}		$HSeO_3^-$	5.4×10^{-9}
	HSe^-	1×10^{-11}	Succinic	$H_2C_4H_4O_4$	6.2×10^{-5}
Hydrosulfuric	H_2S	1.0×10^{-7}		$HC_4H_4O_4^-$	2.3×10^{-6}
	HS^-	1×10^{-19}	Sulfuric	H_2SO_4	strong acid
Hydrotelluric	H_2Te	2.3×10^{-3}		HSO_4^-	1.1×10^{-2}
	HTe^-	1.6×10^{-11}	Sulfurous	H_2SO_3	1.3×10^{-2}
Hypobromous	HOBr	2.5×10^{-9}		HSO_3^-	6.2×10^{-8}
Hypochlorous	HOCl	2.9×10^{-8}	Thiophenol	HSC_6H_5	3.2×10^{-7}
Hypoiodous	HOI	2.3×10^{-11}	Trichloroacetic	$HC_2Cl_3O_2$	3.0×10^{-1}

B. Ionization Constants of Weak Bases at 25 °C

Name of Base	Formula	K_b
Ammonia	NH_3	1.8×10^{-5}
Aniline	$C_6H_5NH_2$	7.4×10^{-10}
Codeine	$C_{18}H_{21}O_3N$	8.9×10^{-7}
Diethylamine	$(C_2H_5)_2NH$	6.9×10^{-4}
Dimethylamine	$(CH_3)_2NH$	5.9×10^{-4}
Ethylamine	$C_2H_5NH_2$	4.3×10^{-4}
Hydrazine	NH_2NH_2	8.5×10^{-7}
	$NH_2NH_3^+$	8.9×10^{-16}
Hydroxylamine	NH_2OH	9.1×10^{-9}
Isoquinoline	C_9H_7N	2.5×10^{-9}
Methylamine	CH_3NH_2	4.2×10^{-4}
Morphine	$C_{17}H_{19}O_3N$	7.4×10^{-7}
Piperidine	$C_5H_{11}N$	1.3×10^{-3}
Pyridine	C_5H_5N	1.5×10^{-9}
Quinoline	C_9H_7N	6.3×10^{-10}
Triethanolamine	$C_6H_{15}O_3N$	5.8×10^{-7}
Triethylamine	$(C_2H_5)_3N$	5.2×10^{-4}
Trimethylamine	$(CH_3)_3N$	6.3×10^{-5}

C. Solubility Product Constants[a]

Name of Solute	Formula	K_{sp}	Name of Solute	Formula	K_{sp}
Aluminum hydroxide	$Al(OH)_3$	1.3×10^{-33}	Copper(II) carbonate	$CuCO_3$	1.4×10^{-10}
Aluminum phosphate	$AlPO_4$	6.3×10^{-19}	Copper(II) chromate	$CuCrO_4$	3.6×10^{-6}
Barium carbonate	$BaCO_3$	5.1×10^{-9}	Copper(II) ferrocyanide	$Cu_2[Fe(CN)_6]$	1.3×10^{-16}
Barium chromate	$BaCrO_4$	1.2×10^{-10}	Copper(II) hydroxide	$Cu(OH)_2$	2.2×10^{-20}
Barium fluoride	BaF_2	1.0×10^{-6}	Copper(II) sulfide[b]	CuS	6×10^{-37}
Barium hydroxide	$Ba(OH)_2$	5×10^{-3}	Iron(II) carbonate	$FeCO_3$	3.2×10^{-11}
Barium sulfate	$BaSO_4$	1.1×10^{-10}	Iron(II) hydroxide	$Fe(OH)_2$	8.0×10^{-16}
Barium sulfite	$BaSO_3$	8×10^{-7}	Iron(II) sulfide[b]	FeS	6×10^{-19}
Barium thiosulfate	BaS_2O_3	1.6×10^{-5}	Iron(III) arsenate	$FeAsO_4$	5.7×10^{-21}
Bismuthyl chloride	$BiOCl$	1.8×10^{-31}	Iron(III) ferrocyanide	$Fe_4[Fe(CN)_6]_3$	3.3×10^{-41}
Bismuthyl hydroxide	$BiOOH$	4×10^{-10}	Iron(III) hydroxide	$Fe(OH)_3$	4×10^{-38}
Cadmium carbonate	$CdCO_3$	5.2×10^{-12}	Iron(III) phosphate	$FePO_4$	1.3×10^{-22}
Cadmium hydroxide	$Cd(OH)_2$	2.5×10^{-14}	Lead(II) arsenate	$Pb_3(AsO_4)_2$	4.0×10^{-36}
Cadmium sulfide[b]	CdS	8×10^{-28}	Lead(II) azide	$Pb(N_3)_2$	2.5×10^{-9}
Calcium carbonate	$CaCO_3$	2.8×10^{-9}	Lead(II) bromide	$PbBr_2$	4.0×10^{-5}
Calcium chromate	$CaCrO_4$	7.1×10^{-4}	Lead(II) carbonate	$PbCO_3$	7.4×10^{-14}
Calcium fluoride	CaF_2	5.3×10^{-9}	Lead(II) chloride	$PbCl_2$	1.6×10^{-5}
Calcium hydrogen phosphate	$CaHPO_4$	1×10^{-7}	Lead(II) chromate	$PbCrO_4$	2.8×10^{-13}
Calcium hydroxide	$Ca(OH)_2$	5.5×10^{-6}	Lead(II) fluoride	PbF_2	2.7×10^{-8}
Calcium oxalate	CaC_2O_4	2.7×10^{-9}	Lead(II) hydroxide	$Pb(OH)_2$	1.2×10^{-15}
Calcium phosphate	$Ca_3(PO_4)_2$	2.0×10^{-29}	Lead(II) iodide	PbI_2	7.1×10^{-9}
Calcium sulfate	$CaSO_4$	9.1×10^{-6}	Lead(II) sulfate	$PbSO_4$	1.6×10^{-8}
Calcium sulfite	$CaSO_3$	6.8×10^{-8}	Lead(II) sulfide[b]	PbS	3×10^{-28}
Chromium(II) hydroxide	$Cr(OH)_2$	2×10^{-16}	Lead(II) thiosulfate	PbS_2O_3	4.0×10^{-7}
Chromium(III) hydroxide	$Cr(OH)_3$	6.3×10^{-31}	Lithium fluoride	LiF	3.8×10^{-3}
Cobalt(II) carbonate	$CoCO_3$	1.4×10^{-13}	Lithium phosphate	Li_3PO_4	3.2×10^{-9}
Cobalt(II) hydroxide	$Co(OH)_2$	1.6×10^{-15}	Magnesium		
Cobalt(III) hydroxide	$Co(OH)_3$	1.6×10^{-44}	ammonium phosphate	$MgNH_4PO_4$	2.5×10^{-13}
Copper(I) chloride	$CuCl$	1.2×10^{-6}	Magnesium carbonate	$MgCO_3$	3.5×10^{-8}
Copper(I) cyanide	$CuCN$	3.2×10^{-20}	Magnesium fluoride	MgF_2	3.7×10^{-8}
Copper(I) iodide	CuI	1.1×10^{-12}	Magnesium hydroxide	$Mg(OH)_2$	1.8×10^{-11}
Copper(II) arsenate	$Cu_3(AsO_4)_2$	7.6×10^{-36}	Magnesium phosphate	$Mg_3(PO_4)_2$	1×10^{-25}

[a] Data are at various temperatures around room temperature, from 18 to 25 °C.

[b] For a solubility equilibrium of the type $MS(s) + H_2O \rightleftharpoons M^{2+}(aq) + HS^-(aq) + OH^-(aq)$.

C. Solubility Product Constants (continued)

Name of Solute	Formula	K_{sp}	Name of Solute	Formula	K_{sp}
Manganese(II) carbonate	$MnCO_3$	1.8×10^{-11}	Silver nitrite	$AgNO_2$	6.0×10^{-4}
Manganese(II) hydroxide	$Mn(OH)_2$	1.9×10^{-13}	Silver sulfate	Ag_2SO_4	1.4×10^{-5}
Manganese(II) sulfide[b]	MnS	3×10^{-14}	Silver sulfide[b]	Ag_2S	6×10^{-51}
Mercury(I) bromide	Hg_2Br_2	5.6×10^{-23}	Silver sulfite	Ag_2SO_3	1.5×10^{-14}
Mercury(I) chloride	Hg_2Cl_2	1.3×10^{-18}	Silver thiocyanate	$AgSCN$	1.0×10^{-12}
Mercury(I) iodide	Hg_2I_2	4.5×10^{-29}	Strontium carbonate	$SrCO_3$	1.1×10^{-10}
Mercury(II) sulfide[b]	HgS	2×10^{-53}	Strontium chromate	$SrCrO_4$	2.2×10^{-5}
Nickel(II) carbonate	$NiCO_3$	6.6×10^{-9}	Strontium fluoride	SrF_2	2.5×10^{-9}
Nickel(II) hydroxide	$Ni(OH)_2$	2.0×10^{-15}	Strontium sulfate	$SrSO_4$	3.2×10^{-7}
Scandium fluoride	ScF_3	4.2×10^{-18}	Thallium(I) bromide	$TlBr$	3.4×10^{-6}
Scandium hydroxide	$Sc(OH)_3$	8.0×10^{-31}	Thallium(I) chloride	$TlCl$	1.7×10^{-4}
Silver arsenate	Ag_3AsO_4	1.0×10^{-22}	Thallium(I) iodide	TlI	6.5×10^{-8}
Silver azide	AgN_3	2.8×10^{-9}	Thallium(III) hydroxide	$Tl(OH)_3$	6.3×10^{-46}
Silver bromide	$AgBr$	5.0×10^{-13}	Tin(II) hydroxide	$Sn(OH)_2$	1.4×10^{-28}
Silver carbonate	Ag_2CO_3	8.5×10^{-12}	Tin(II) sulfide[b]	SnS	1×10^{-26}
Silver chloride	$AgCl$	1.8×10^{-10}	Zinc carbonate	$ZnCO_3$	1.4×10^{-11}
Silver chromate	Ag_2CrO_4	1.1×10^{-12}	Zinc hydroxide	$Zn(OH)_2$	1.2×10^{-17}
Silver cyanide	$AgCN$	1.2×10^{-16}	Zinc oxalate	ZnC_2O_4	2.7×10^{-8}
Silver iodate	$AgIO_3$	3.0×10^{-8}	Zinc phosphate	$Zn_3(PO_4)_2$	9.0×10^{-33}
Silver iodide	AgI	8.5×10^{-17}	Zinc sulfide[b]	ZnS	2×10^{-25}

D. Complex Ion Formation Constants[a]

Formula	K_f	Formula	K_f
$[Ag(CN)_2]^-$	5.6×10^{18}	$[Fe(en)_3]^{2+}$	5.0×10^9
$[Ag(EDTA)]^{3-}$	2.1×10^7	$[Fe(ox)_3]^{4-}$	1.7×10^5
$[Ag(en)_2]^+$	5.0×10^7	$[Fe(CN)_6]^{3-}$	10^{42}
$[Ag(NH_3)_2]^+$	1.6×10^7	$[Fe(EDTA)]^-$	1.7×10^{24}
$[Ag(SCN)_4]^{3-}$	1.2×10^{10}	$[Fe(ox)_3]^{3-}$	2×10^{20}
$[Ag(S_2O_3)_2]^{3-}$	1.7×10^{13}	$[Fe(SCN)]^{2+}$	8.9×10^2
$[Al(EDTA)]^-$	1.3×10^{16}	$[HgCl_4]^{2-}$	1.2×10^{15}
$[Al(OH)_4]^-$	1.1×10^{33}	$[Hg(CN)_4]^{2-}$	3×10^{41}
$[Al(ox)_3]^{3-}$	2×10^{16}	$[Hg(EDTA)]^{2-}$	6.3×10^{21}
$[Cd(CN)_4]^{2-}$	6.0×10^{18}	$[Hg(en)_2]^{2+}$	2×10^{23}
$[Cd(en)_3]^{2+}$	1.2×10^{12}	$[HgI_4]^{2-}$	6.8×10^{29}
$[Cd(NH_3)_4]^{2+}$	1.3×10^7	$[Hg(ox)_2]^{2-}$	9.5×10^6
$[Co(EDTA)]^{2-}$	2.0×10^{16}	$[Ni(CN)_4]^{2-}$	2×10^{31}
$[Co(en)_3]^{2+}$	8.7×10^{13}	$[Ni(EDTA)]^{2-}$	3.6×10^{18}
$[Co(NH_3)_6]^{2+}$	1.3×10^5	$[Ni(en)_3]^{2+}$	2.1×10^{18}
$[Co(ox)_3]^{4-}$	5×10^9	$[Ni(NH_3)_6]^{2+}$	5.5×10^8
$[Co(SCN)_4]^{2-}$	1.0×10^3	$[Ni(ox)_3]^{4-}$	3×10^8
$[Co(EDTA)]^-$	10^{36}	$[PbCl_3]^-$	2.4×10^1
$[Co(en)_3]^{3+}$	4.9×10^{48}	$[Pb(EDTA)]^{2-}$	2×10^{18}
$[Co(NH_3)_6]^{3+}$	4.5×10^{33}	$[PbI_4]^{2-}$	3.0×10^4
$[Co(ox)_3]^{3-}$	10^{20}	$[Pb(OH)_3]^-$	3.8×10^{14}
$[Cr(EDTA)]^-$	10^{23}	$[Pb(ox)_2]^{2-}$	3.5×10^6
$[Cr(OH)_4]^-$	8×10^{29}	$[Pb(S_2O_3)_3]^{4-}$	2.2×10^6
$[CuCl_3]^{2-}$	5×10^5	$[PtCl_4]^{2-}$	1×10^{16}
$[Cu(CN)_4]^{3-}$	2.0×10^{30}	$[Pt(NH_3)_6]^{2+}$	2×10^{35}
$[Cu(EDTA)]^{2-}$	5×10^{18}	$[Zn(CN)_4]^{2-}$	1×10^{18}
$[Cu(en)_2]^{2+}$	1×10^{20}	$[Zn(EDTA)]^{2-}$	3×10^{16}
$[Cu(NH_3)_4]^{2+}$	1.1×10^{13}	$[Zn(en)_3]^{2+}$	1.3×10^{14}
$[Cu(ox)_2]^{2-}$	3×10^8	$[Zn(NH_3)_4]^{2+}$	4.1×10^8
$[Fe(CN)_6]^{4-}$	10^{37}	$[Zn(OH)_4]^{2-}$	4.6×10^{17}
$[Fe(EDTA)]^{2-}$	2.1×10^{14}	$[Zn(ox)_3]^{4-}$	1.4×10^8

[a] The ligands referred to in this table are monodentate: Cl^-, CN^-, I^-, NH_3, OH^-, SCN^-, $S_2O_3^{2-}$; bidentate: ethylenediamine, en; oxalate ion, ox ($C_2O_4^{2-}$); tetradentate: ethylenediaminetetraacetato ion, $EDTA^{4-}$.

C.3 Standard Electrode (Reduction) Potentials at 25 °C

Reduction Half-Reaction	$E°$, V
$F_2(g) + 2 e^- \longrightarrow 2 F^-(aq)$	+2.866
$OF_2(g) + 2 H^+(aq) + 4 e^- \longrightarrow H_2O(l) + 2 F^-(aq)$	+2.1
$O_3(g) + 2 H^+(aq) + 2 e^- \longrightarrow O_2(g) + H_2O(l)$	+2.075
$S_2O_8{}^{2-}(aq) + 2 e^- \longrightarrow 2 SO_4{}^{2-}(aq)$	+2.01
$Ag^{2+}(aq) + e^- \longrightarrow Ag^+(aq)$	+1.98
$H_2O_2(aq) + 2 H^+(aq) + 2 e^- \longrightarrow 2 H_2O(l)$	+1.763
$MnO_4{}^-(aq) + 4 H^+(aq) + 3 e^- \longrightarrow MnO_2(s) + 2 H_2O(l)$	+1.70
$PbO_2(s) + SO_4{}^{2-}(aq) + 4 H^+(aq) + 2 e^- \longrightarrow PbSO_4(s) + 2 H_2O(l)$	+1.69
$Au^{3+}(aq) + 3 e^- \longrightarrow Au(s)$	+1.52
$MnO_4{}^-(aq) + 8 H^+(aq) + 5 e^- \longrightarrow Mn^{2+}(aq) + 4 H_2O(l)$	+1.51
$2 BrO_3{}^-(aq) + 12 H^+(aq) + 10 e^- \longrightarrow Br_2(l) + 6 H_2O(l)$	+1.478
$PbO_2(s) + 4 H^+(aq) + 2 e^- \longrightarrow Pb^{2+}(aq) + 2 H_2O(l)$	+1.455
$ClO_3{}^-(aq) + 6 H^+(aq) + 6 e^- \longrightarrow Cl^-(aq) + 3 H_2O(l)$	+1.450
$Au^{3+}(aq) + 2 e^- \longrightarrow Au^+(aq)$	+1.36
$Cl_2(g) + 2 e^- \longrightarrow 2 Cl^-(aq)$	+1.358
$Cr_2O_7{}^{2-}(aq) + 14 H^+(aq) + 6 e^- \longrightarrow 2 Cr^{3+}(aq) + 7 H_2O(l)$	+1.33
$MnO_2(s) + 4 H^+(aq) + 2 e^- \longrightarrow Mn^{2+}(aq) + 2 H_2O(l)$	+1.23
$O_2(g) + 4 H^+(aq) + 4 e^- \longrightarrow 2 H_2O(l)$	+1.229
$2 IO_3{}^-(aq) + 12 H^+(aq) + 10 e^- \longrightarrow I_2(s) + 6 H_2O(l)$	+1.20
$ClO_4{}^-(aq) + 2 H^+(aq) + 2 e^- \longrightarrow ClO_3{}^-(aq) + H_2O(l)$	+1.19
$ClO_3{}^-(aq) + 2 H^+(aq) + e^- \longrightarrow ClO_2(g) + H_2O(l)$	+1.175
$NO_2(g) + H^+(aq) + e^- \longrightarrow HNO_2(aq)$	+1.07
$Br_2(l) + 2 e^- \longrightarrow 2 Br^-(aq)$	+1.065
$NO_2(g) + 2 H^+(aq) + 2 e^- \longrightarrow NO(g) + H_2O(l)$	+1.03
$[AuCl_4]^-(aq) + 3 e^- \longrightarrow Au(s) + 4 Cl^-(aq)$	+1.002
$VO_2{}^+(aq) + 2 H^+(aq) + e^- \longrightarrow VO^{2+}(aq) + H_2O(l)$	+1.000
$NO_3{}^-(aq) + 4 H^+(aq) + 3 e^- \longrightarrow NO(g) + 2 H_2O(l)$	+0.956
$Hg^{2+}(aq) + 2 e^- \longrightarrow Hg(l)$	+0.854
$Ag^+(aq) + e^- \longrightarrow Ag(s)$	+0.800
$Fe^{3+}(aq) + e^- \longrightarrow Fe^{2+}(aq)$	+0.771
$O_2(g) + 2 H^+(aq) + 2 e^- \longrightarrow H_2O_2(aq)$	+0.695
$2 HgCl_2(aq) + 2 e^- \longrightarrow Hg_2Cl_2(s) + 2 Cl^-(aq)$	+0.63
$MnO_4{}^-(aq) + e^- \longrightarrow MnO_4{}^{2-}(aq)$	+0.56
$I_2(s) + 2 e^- \longrightarrow 2 I^-(aq)$	+0.535
$Cu^+(aq) + e^- \longrightarrow Cu(s)$	+0.520
$H_2SO_3(aq) + 4 H^+(aq) + 4 e^- \longrightarrow S(s) + 3 H_2O(l)$	+0.449
$C_2N_2(g) + 2 H^+(aq) + 2 e^- \longrightarrow 2 HCN(aq)$	+0.37
$[Fe(CN)_6]^{3-}(aq) + e^- \longrightarrow [Fe(CN)_6]^{4-}(aq)$	+0.361
$Cu^{2+}(aq) + 2 e^- \longrightarrow Cu(s)$	+0.340
$VO^{2+}(aq) + 2 H^+(aq) + e^- \longrightarrow V^{3+}(aq) + H_2O(l)$	+0.337
$PbO_2(s) + 2 H^+(aq) + 2 e^- \longrightarrow PbO(s) + H_2O(l)$	+0.28
$Hg_2Cl_2(s) + 2 e^- \longrightarrow 2 Hg(l) + 2 Cl^-(aq)$	+0.2676
$HAsO_2(aq) + 3 H^+(aq) + 3 e^- \longrightarrow As(s) + 2 H_2O(l)$	+0.240
$AgCl(s) + e^- \longrightarrow Ag(s) + Cl^-(aq)$	+0.2223
$SO_4{}^{2-}(aq) + 4 H^+(aq) + 2 e^- \longrightarrow 2 H_2O(l) + SO_2(g)$	+0.17
$Cu^{2+}(aq) + e^- \longrightarrow Cu^+(aq)$	+0.159
$Sn^{4+}(aq) + 2 e^- \longrightarrow Sn^{2+}(aq)$	+0.154
$S(s) + 2 H^+(aq) + 2 e^- \longrightarrow H_2S(g)$	+0.14
$AgBr(s) + e^- \longrightarrow Ag(s) + Br^-(aq)$	+0.071
$2 H^+(aq) + 2 e^- \longrightarrow H_2(g)$	0
$Pb^{2+}(aq) + 2 e^- \longrightarrow Pb(s)$	−0.125
$Sn^{2+}(aq) + 2 e^- \longrightarrow Sn(s)$	−0.137
$AgI(s) + e^- \longrightarrow Ag(s) + I^-(aq)$	−0.152

Reduction Half-Reaction (continued)	$E°$, V
$V^{3+}(aq) + e^- \longrightarrow V^{2+}(aq)$	-0.255
$Ni^{2+}(aq) + 2\,e^- \longrightarrow Ni(s)$	-0.257
$H_3PO_4(aq) + 2\,H^+(aq) + 2\,e^- \longrightarrow H_3PO_3(aq) + H_2O(l)$	-0.276
$Co^{2+}(aq) + 2\,e^- \longrightarrow Co(s)$	-0.277
$PbSO_4(s) + 2\,e^- \longrightarrow Pb(s) + SO_4{}^{2-}(aq)$	-0.356
$Cd^{2+}(aq) + 2\,e^- \longrightarrow Cd(s)$	-0.403
$Cr^{3+}(aq) + e^- \longrightarrow Cr^{2+}(aq)$	-0.424
$Fe^{2+}(aq) + 2\,e^- \longrightarrow Fe(s)$	-0.440
$2\,CO_2(g) + 2\,H^+(aq) + 2\,e^- \longrightarrow H_2C_2O_4(aq)$	-0.49
$Zn^{2+}(aq) + 2\,e^- \longrightarrow Zn(s)$	-0.763
$Cr^{2+}(aq) + 2\,e^- \longrightarrow Cr(s)$	-0.90
$Mn^{2+}(aq) + 2\,e^- \longrightarrow Mn(s)$	-1.18
$Ti^{2+}(aq) + 2\,e^- \longrightarrow Ti(s)$	-1.63
$U^{3+}(aq) + 3\,e^- \longrightarrow U(s)$	-1.66
$Al^{3+}(aq) + 3\,e^- \longrightarrow Al(s)$	-1.676
$Mg^{2+}(aq) + 2\,e^- \longrightarrow Mg(s)$	-2.356
$Na^+(aq) + e^- \longrightarrow Na(s)$	-2.713
$Ca^{2+}(aq) + 2\,e^- \longrightarrow Ca(s)$	-2.84
$Sr^{2+}(aq) + 2\,e^- \longrightarrow Sr(s)$	-2.89
$Ba^{2+}(aq) + 2\,e^- \longrightarrow Ba(s)$	-2.92
$Cs^+(aq) + e^- \longrightarrow Cs(s)$	-2.923
$K^+(aq) + e^- \longrightarrow K(s)$	-2.924
$Rb^+(aq) + e^- \longrightarrow Rb(s)$	-2.924
$Li^+(aq) + e^- \longrightarrow Li(s)$	-3.040

Basic Solution

$O_3(g) + H_2O(l) + 2\,e^- \longrightarrow O_2(g) + 2\,OH^-(aq)$	$+1.246$
$ClO^-(aq) + H_2O(l) + 2\,e^- \longrightarrow Cl^-(g) + 2\,OH^-(aq)$	$+0.890$
$HO_2{}^-(aq) + H_2O(l) + 2\,e^- \longrightarrow 3\,OH^-(aq)$	$+0.88$
$BrO^-(aq) + H_2O(l) + 2\,e^- \longrightarrow Br^-(aq) + 2\,OH^-(aq)$	$+0.766$
$ClO_3{}^-(aq) + 3\,H_2O(l) + 6\,e^- \longrightarrow Cl^-(aq) + 6\,OH^-(aq)$	$+0.622$
$2\,AgO(s) + H_2O(l) + 2\,e^- \longrightarrow Ag_2O(s) + 2\,OH^-(aq)$	$+0.604$
$MnO_4{}^-(aq) + 2\,H_2O(l) + 3\,e^- \longrightarrow MnO_2(s) + 4\,OH^-(aq)$	$+0.60$
$BrO_3{}^-(aq) + 3\,H_2O(l) + 6\,e^- \longrightarrow Br^-(aq) + 6\,OH^-(aq)$	$+0.584$
$Ni(OH)_3(s) + e^- \longrightarrow Ni(OH)_2(s) + OH^-(aq)$	$+0.48$
$2\,BrO^-(aq) + 2\,H_2O(l) + 2\,e^- \longrightarrow Br_2(l) + 4\,OH^-(aq)$	$+0.455$
$2\,IO^-(aq) + 2\,H_2O(l) + 2\,e^- \longrightarrow I_2(s) + 4\,OH^-(aq)$	$+0.42$
$O_2(g) + 2\,H_2O(l) + 4\,e^- \longrightarrow 4\,OH^-(aq)$	$+0.401$
$Ag_2O(s) + H_2O(l) + 2\,e^- \longrightarrow 2\,Ag(s) + 2\,OH^-(aq)$	$+0.342$
$Co(OH)_3(s) + e^- \longrightarrow Co(OH)_2(s) + OH^-(aq)$	$+0.17$
$NO_3{}^-(aq) + H_2O(l) + 2\,e^- \longrightarrow NO_2{}^-(aq) + 2\,OH^-(aq)$	$+0.01$
$CrO_4{}^{2-}(aq) + 4\,H_2O(l) + 3\,e^- \longrightarrow [Cr(OH)_4]^-(aq) + 4\,OH^-(aq)$	-0.13
$HPbO_2{}^-(aq) + H_2O(l) + 2\,e^- \longrightarrow Pb(s) + 3\,OH^-(aq)$	-0.54
$HCHO(aq) + 2\,H_2O(l) + 2\,e^- \longrightarrow CH_3OH(aq) + 2\,OH^-(aq)$	-0.59
$SO_3{}^{2-}(aq) + 3\,H_2O(l) + 4\,e^- \longrightarrow S(s) + 6\,OH^-(aq)$	-0.66
$AsO_4{}^{3-}(aq) + 2\,H_2O(l) + 2\,e^- \longrightarrow AsO_2{}^-(aq) + 4\,OH^-(aq)$	-0.67
$AsO_2{}^-(aq) + 2\,H_2O(l) + 3\,e^- \longrightarrow As(s) + 4\,OH^-(aq)$	-0.68
$2\,H_2O(l) + 2\,e^- \longrightarrow H_2(g) + 2\,OH^-(aq)$	-0.828
$OCN^-(aq) + H_2O(l) + 2\,e^- \longrightarrow CN^-(aq) + 2\,OH^-(aq)$	-0.97
$As(s) + 3\,H_2O(l) + 3\,e^- \longrightarrow AsH_3(g) + 3\,OH^-(aq)$	-1.21
$[Zn(OH)_4]^{2-}(aq) + 2\,e^- \longrightarrow Zn(s) + 4\,OH^-(aq)$	-1.285
$Sb(s) + 3\,H_2O(l) + 3\,e^- \longrightarrow SbH_3(g) + 3\,OH^-(aq)$	-1.338
$Al(OH)_4{}^-(aq) + 3\,e^- \longrightarrow Al(s) + 4\,OH^-(aq)$	-2.310
$Mg(OH)_2(s) + 2\,e^- \longrightarrow Mg(s) + 2\,OH^-(aq)$	-2.687

Organic Nomenclature

In the early days of organic chemistry, compounds were given common names, often based on natural sources. For example, butyric acid ($CH_3CH_2CH_2COOH$) was so named from the Latin *butyrum*, meaning butter. (The *but-* part of the name survives in the systematic names of compounds with four carbon atoms per molecule: $CH_3CH_2CH_2CH_3$, butane; $CH_3CH_2CH_2CH_2OH$, 1-butanol; and so on.) Today, there are millions of organic compounds, and we could no more memorize individual names for all of them than we could memorize all the listings in the New York City telephone directory. To bring some order to the haphazard naming of the rapidly increasing roster of organic compounds, an international assembly of chemists met in 1892 for the first of many meetings on nomenclature—a system for assigning names. This organization exists today as the International Union of Pure and Applied Chemistry (IUPAC). Here, we will examine some simple rules for naming a few families of organic compounds.

D.1   Alkanes

An IUPAC name of an organic compound often consists of three basic parts: one or more prefixes, a stem, and an ending. The following rules will enable us to name most alkanes.

1. Use the ending *-ane* to indicate that the compound is an alkane.

2. The longest continuous chain (LCC) of carbon atoms is the *parent chain;* it provides the *stem* name. For example, the following alkane is named as a derivative of hexane because there are six carbon atoms in the LCC:

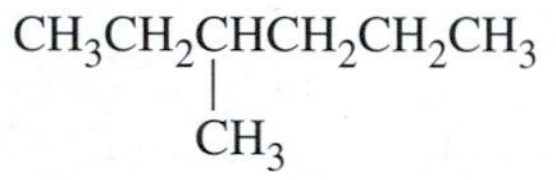

$$CH_3CH_2CHCH_2CH_2CH_3$$
$$|$$
$$CH_3$$

   The stem *hex-* and ending *-ane* combine to give *hexane.* (Recall that the names of the stems for chains with up to 10 carbon atoms were given in Table 2.6.)

3. The prefixes in a name indicate the groups attached to the parent chain. If the group contains only carbon and hydrogen with no double or triple bonds, it is called an *alkyl* group. Alkyl groups are derived from the corresponding alkane by removing one H atom, and the alkyl group is named after the alkane. For example, the *methyl* group $—CH_3$ is derived from *methane*, CH_4. There is only one alkyl group derived from methane and only one from ethane. For the general alkane C_nH_{2n+2} with $n = 3$ (propane) or higher, different alkyl groups are formed depending on which H atom is removed. For example, removal of an end H atom from propane gives a *propyl* group, whereas removal of a H atom from the middle carbon atom of propane yields an *isopropyl* group. The names and formulas of the four simplest alkyl groups are as follows:

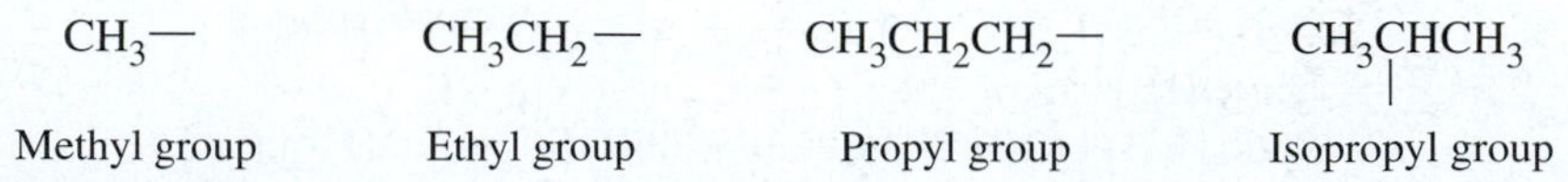

$CH_3—$	$CH_3CH_2—$	$CH_3CH_2CH_2—$	CH_3CHCH_3
Methyl group	Ethyl group	Propyl group	Isopropyl group

4. The smallest possible Arabic numerals are used to indicate the position(s) on the longest carbon chain at which the substituents (alkyl groups, in this case) are

attached. Thus, to complete the name of the alkane in item 2, we again identify the LCC (red):

CH₃CH₂CHCH₂CH₂CH₃
$$CH_3CH_2CHCH_2CH_2CH_3$$
$$|$$
$$CH_3$$

Next, we note that the methyl group —CH_3 is attached to the third carbon atom from the left end. Thus, the compound is *3-methylhexane*. Notice that if we number the carbon atoms from the right end, the methyl group is on the *fourth* carbon atom. The name "4-methylhexane" is incorrect, however, because we must use the *lowest* possible numbers for constituent groups.

5. We use the prefixes *di-* for two, *tri-* for three, and *tetra-* for four to denote two, three, or four identical groups attached to the parent chain. If two identical groups are bonded to the same carbon atom, we must repeat the position number for each group:

$$
\begin{array}{cc}
CH_3 & CH_3 \quad CH_3 \\
| & | \qquad | \\
CH_3CCH_2CH_2CH_2CH_3 & CH_3\!-\!C\!-\!CH_2CHCH_3 \\
| & | \\
CH_3 & CH_3 \\
\text{2,2-Dimethylheptane} & \text{2,2,4-Trimethylpentane}
\end{array}
$$

Commas are used to separate numbers from each other, and hyphens to separate numbers from words.

6. Groups are listed in alphabetical order. Thus, the proper name for the compound

$$
\begin{array}{c}
CH_3\!-\!CHCH_2CH\!-\!CH_2CH_3 \\
| \qquad\quad | \\
CH_3 \quad CH_2CH_3
\end{array}
$$

is 4-ethyl-2-methylhexane, not "2-methyl-4-ethylhexane."

As a final example, let's name the following compound:

$$
\begin{array}{c}
\overset{5}{}\,\overset{4}{}\,\overset{3}{}\quad\overset{2}{}\,\overset{1}{} \\
CH_3CH_2CH\!-\!CHCH_3 \\
| \quad\; | \\
CH_3 \quad CH_3
\end{array}
$$

The LCC has five carbon atoms; the compound is named as a derivative of pentane. There are methyl groups attached to the second and third carbon atoms. The correct name is 2,3-dimethylpentane. Note that we numbered the carbon atoms so that the substituent groups would have the lowest numbers possible.

Example D.1

Give an appropriate IUPAC name for this compound:

$$
\begin{array}{c}
CH_3CH\!-\!CH_2CHCH_3 \\
| \qquad\quad | \\
CH_2 \qquad CH_3 \\
| \\
CH_3
\end{array}
$$

SOLUTION

Let's begin by identifying the LCC. This one is a little tricky. The parent compound is the longest continuous chain, not necessarily the chain drawn straight across the page. The LCC (red) contains *six* carbon atoms.

$$
\begin{array}{c}
\overset{4}{}\qquad\overset{3}{}\;\overset{2}{}\,\overset{1}{} \\
CH_3CH\!-\!CH_2CHCH_3 \\
| \qquad\quad | \\
5\,CH_2 \qquad CH_3 \\
| \\
6\,CH_3
\end{array}
$$

The substituents (black) are two methyl groups, one on the second carbon atom of the LCC and one on the fourth. Again, we use the lowest combination of numbers, in this case requiring that we count from the left end. The correct name is 2,4-dimethylhexane, not "2-ethyl-4-methylpentane."

Example D.2

Give the structural formula for 5-isopropyl-2-methyloctane.

SOLUTION

The compound is derived from octane, so we start with a chain of eight carbon atoms.

$$-C-C-C-C-C-C-C-C-$$

Next, starting on the left, we attach a methyl group to the second C atom.

$$\begin{array}{c} CH_3 \\ | \\ -C-C-C-C-C-C-C-C- \\ 12345678 \end{array}$$

Then, continuing to count from the left, we add an isopropyl group to the fifth C atom. (Remember that an isopropyl group is a three-carbon chain attached by the *middle* carbon atom.)

$$\begin{array}{c} CH_3 \quad\ CH_3CHCH_3 \\ | \qquad\quad | \\ -C-C-C-C-C-C-C-C- \\ 12345678 \end{array}$$

Finally, we add enough H atoms to give each C atom four bonds. The structural formula is

$$\begin{array}{c} CH_3 \quad\ CH_3CHCH_3 \\ | \qquad\quad | \\ CH_3CHCH_2CH_2CHCH_2CH_2CH_3 \end{array}$$

D.2 Alkenes and Alkynes

A few simple alkenes are best known by common names (Section 9.11), but we need systematic names for the many isomers of higher alkenes. Some of the IUPAC rules for alkenes are as follows.

1. All alkenes have names ending in -*ene*.

2. The longest chain of carbon atoms *containing the double bond* is the parent chain. The stem name is the same as that of the alkane with the same number of carbon atoms. The compound $CH_2=CH_2$ has the same number of C atoms as ethane, CH_3CH_3; it is called *ethene*. Similarly, $CH_3CH=CH_2$ has the same number of C atoms as propane, $CH_3CH_2CH_3$; it is called *propene*.

3. For chains of four or more carbon atoms, we must indicate the location of the double bond. To do this, number the carbon atoms from the end of the parent chain that gives the first carbon atom in the double bond the lowest possible number. For example, the compound $CH_3CH=CHCH_2CH_3$ has the double bond between the second and third carbon atoms. Its name is *2-pentene*.

4. Substituent groups are named as in alkanes, and their positions are indicated by a number. Thus, the following alkene is 2-ethyl-5-methyl-1-hexene:

$$\begin{array}{c} \overset{1}{C}H_2\!=\!\overset{2}{C}\overset{3}{C}H_2\overset{4}{C}H_2\overset{5}{C}H\overset{6}{C}H_3 \\ \quad | \qquad\quad | \\ \quad CH_2CH_3 \ \ CH_3 \end{array}$$

Notice that we did not choose the LCC of seven carbon atoms (red) as the parent chain. We chose the *six-carbon chain containing the double bond*.

To name *alkynes*, we use a set of rules almost identical to those for the alkenes. Alkynes differ in that they have a carbon-to-carbon triple bond instead of a double bond, and we use the ending *-yne* rather than the *-ene* ending for alkenes. The IUPAC name of $CH_2 = CH_2$ is *ethyne*, and that of $CH_3C \equiv CH$ is *propyne*. The compounds $CH_3CH_2C \equiv CH$ and $CH_3C \equiv CCH_3$ are *1-butyne* and *2-butyne*, respectively.

Example D.3

Name the following compound:

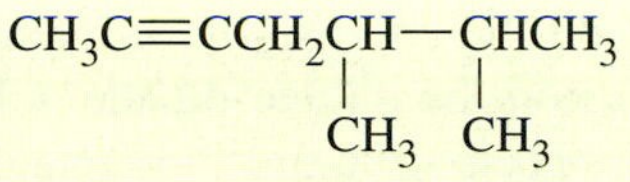

$$CH_3C \equiv CCH_2CH - CHCH_3$$
$$\quad\quad\quad\quad\quad | \quad\quad | $$
$$\quad\quad\quad\quad\quad CH_3 \quad CH_3$$

SOLUTION

The LCC containing the triple bond has seven carbon atoms; the seven-carbon alkyne is hep-tyne. We give the lowest position number to the triple bond, not to the substituent groups. To do this, we number the C atoms as shown here:

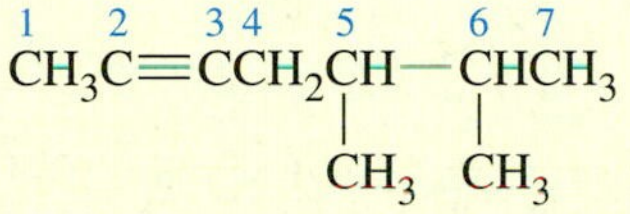

$$\overset{1}{C}H_3\overset{2}{C} \equiv \overset{3}{C}\overset{4}{C}H_2\overset{5}{C}H - \overset{6}{C}H\overset{7}{C}H_3$$
$$\quad\quad\quad\quad\quad | \quad\quad | $$
$$\quad\quad\quad\quad\quad CH_3 \quad CH_3$$

The parent compound is 2-heptyne. There are methyl substituents on the fifth and sixth carbon atoms. The name of the compound is 5,6-dimethyl-2-heptyne.

Example D.4

Give the structural formula for **(a)** 2-methyl-2-pentene and **(b)** 4-methyl-2-hexyne.

SOLUTION

(a) The stem *pent* and ending *-ene* tell us that the LCC containing the double bond has five carbon atoms. The *2* indicates that the double bond is between the second and third carbon atoms.

$$C - C = C - C - C$$

The *2-methyl* locates the $-CH_3$ substituent on the second carbon atom.

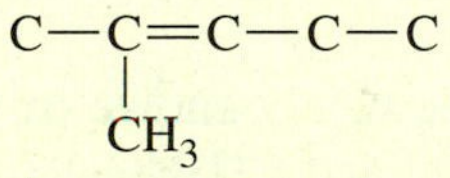

$$C - C = C - C - C$$
$$\quad\quad | $$
$$\quad\quad CH_3$$

Adding enough hydrogen atoms to provide four bonds to each carbon atom gives us the structural formula.

$$CH_3 - C = CH - CH_2 - CH_3$$
$$\quad\quad\quad | $$
$$\quad\quad\quad CH_3$$

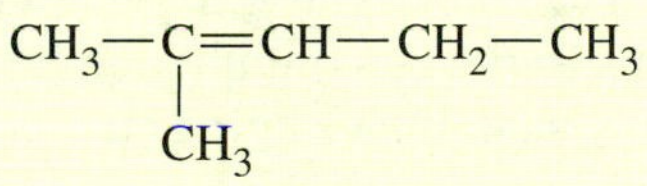

(b) The stem *hex-* and ending *-yne* tell us that the LCC containing a *triple* bond has six carbon atoms. The 2 tells us that the triple bond is between the second and third carbon atoms, and the *4-methyl* locates the $-CH_3$ substituent on the fourth carbon atom. Adding hydrogen atoms gives the structural formula.

$$CH_3C \equiv CCHCH_2CH_3$$
$$\quad\quad\quad\quad | $$
$$\quad\quad\quad\quad CH_3$$

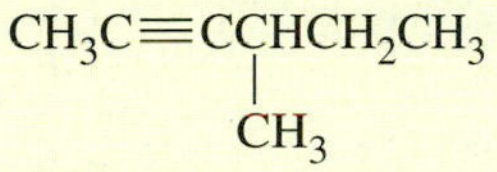

D.3   Alcohols

In Section 2.9, we introduced alcohols as compounds with the general formula ROH, indicating a hydroxyl group (OH) bonded to a carbon atom of an alkyl group (R). The OH group is a substituent on the LCC, but we generally don't name it as a substituent. Rather, we indicate its presence by an ending:

- The presence of the OH functional group is denoted by an *-ol* ending on an alkane stem name (not through the prefix, hydroxyl).
- In numbering the carbon atoms of the LCC, the position of the OH group is given first priority; that is, it is given the lowest number possible.

The two simplest alcohols are known by the common names methyl alcohol and ethyl alcohol. The IUPAC name of CH_3OH is based on the alkane methane, CH_4. We drop the *-e* of methane and add the ending *-ol*; its name is *methanol*. Similarly, the name of CH_3CH_2OH is based on the alkane ethane, CH_3CH_3; its name is *ethanol*. There are two propyl alcohols. Their structures and names are given below.

$$CH_3CH_2CH_2OH$$

1-Propanol

$$CH_3CHCH_3$$
$$|$$
$$OH$$

2-Propanol

Recall that their common names are propyl alcohol and isopropyl alcohol, respectively.

Example D.5

The compound commonly known as *tert*-butyl alcohol, $(CH_3)_3COH$, and its methyl ether, $(CH_3)_3COCH_3$, are used as octane boosters in gasoline. What is the IUPAC name of the alcohol?

SOLUTION

Let's begin by converting the condensed formula to a more complete structural formula. There are three methyl groups and a hydroxyl group, all attached to a single C atom.

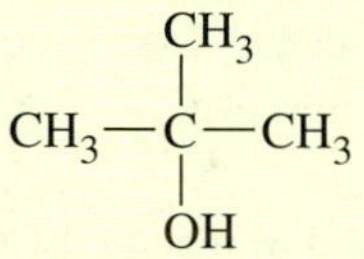

$$\begin{array}{c} CH_3 \\ | \\ CH_3-C-CH_3 \\ | \\ OH \end{array}$$

The LCC is three carbon atoms long, and the hydroxyl group is on the second carbon of this chain, yielding, for the moment, 2-propanol. There is also a methyl group attached to the second carbon of the parent chain, giving the IUPAC name *2-methyl-2-propanol*.

D.4 Carboxylic Acids

The simplest carboxylic acids are widely known by their common names. For example, HCOOH is called formic acid and CH_3COOH is called acetic acid. The IUPAC names are based on alkanes with the same number of carbon atoms. The *-e* ending of the alkane name is replaced by *-oic acid*. Thus, HCOOH is methanoic acid, and CH_3COOH is ethanoic acid. The carboxylic acid with an LCC of eight carbon atoms is octanoic acid. For locating substituents, numbering begins with the carboxylic carbon atom as number 1, as illustrated in Example D.6.

Example D.6

Give the structural formula for 4-ethyl-6-methyloctanoic acid.

SOLUTION

Octanoic tells us that the compound has an LCC of eight carbon atoms, with an end C atom as part of a carboxyl group.

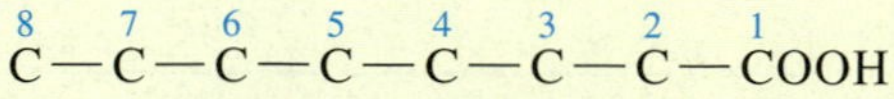

$$\overset{8}{C}-\overset{7}{C}-\overset{6}{C}-\overset{5}{C}-\overset{4}{C}-\overset{3}{C}-\overset{2}{C}-\overset{1}{COOH}$$

The substituents are an ethyl group on the fourth carbon atom (counting *from the carboxyl end*) and a methyl group on the sixth carbon atom. Adding these substituents and enough H atoms to give each C atom four bonds, we get

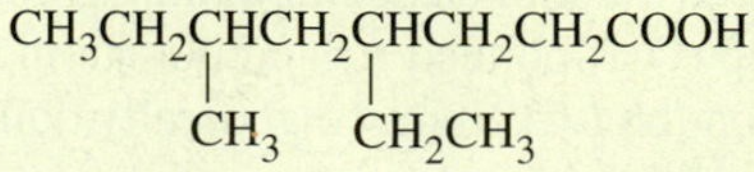

$$\begin{array}{c} CH_3CH_2CHCH_2CHCH_2CH_2COOH \\ | \qquad\quad | \\ CH_3 \quad CH_2CH_3 \end{array}$$

There are many other families of organic compounds. Table D.1 lists several of the most common functional groups and examples of compounds that contain them. With what we have presented here and in Chapters 2 and 23, you should be able to relate the names and formulas of many types of organic compounds.

Table D.1   Some Classes of Organic Compounds and Their Functional Groups

Class	General Structural Formula[a]	Example	Name of Example	Cross Reference
Alkane	$R-H$	$CH_3CH_2CH_2CH_2CH_2CH_3$	hexane	Sections 2.9, 6.8, Chap. 23
Alkene	$\backslash C=C \diagup$	$CH_2=CHCH_2CH_2CH_3$	1-pentene	Section 9.11, Chap. 23
Alkyne	$-C\equiv C-$	$CH_3C\equiv CCH_2CH_2CH_2CH_2CH_3$	2-octyne	Section 9.11, Chap. 23
Alcohol	$R-OH$	$CH_3CH_2CH_2CH_2OH$	1-butanol	Section 2.9, Chap. 23
Alkyl halide	$R-X^{b}$	$CH_3CH_2CH_2CH_2CH_2CH_2Br$	1-bromohexane	Chap. 23
Ether	$R-O-R$	$CH_3-O-CH_2CH_2CH_3$	1-methoxypropane (methyl propyl ether)[c]	Chap. 23
Amine	$R-NH_2$	$CH_3CH_2CH_2-NH_2$	1-aminopropane (propylamine)[c]	Section 4.2, Chap. 15
Aldehyde	$R-\overset{\overset{\displaystyle O}{\|}}{C}-H$	$CH_3CH_2CH_2\overset{\overset{\displaystyle O}{\|}}{C}-H$	butanal (butyraldehyde)[c]	Section 4.5, Chap. 23
Ketone	$R-\overset{\overset{\displaystyle O}{\|}}{C}-R$	$CH_3CH_2\overset{\overset{\displaystyle O}{\|}}{C}CH_2CH_2CH_3$	3-hexanone (ethyl propyl ketone)[c]	Section 4.5, Chap. 23
Carboxylic acid	$R-\overset{\overset{\displaystyle O}{\|}}{C}-OH$	$CH_3CH_2CH_2\overset{\overset{\displaystyle O}{\|}}{C}-OH$	butanoic acid (butyric acid)[c]	Sections 2.9, 4.2, Chap. 15, 23
Ester	$R-\overset{\overset{\displaystyle O}{\|}}{C}-OR$	$CH_3CH_2CH_2\overset{\overset{\displaystyle O}{\|}}{C}-OCH_3$	methyl butanoate (methyl butyrate)[c]	Section 6.8, Chap. 23, Chap. 24 (polymers)
Amide	$R-\overset{\overset{\displaystyle O}{\|}}{C}-NH_2$	$CH_3CH_2CH_2\overset{\overset{\displaystyle O}{\|}}{C}-NH_2$	butanamide (butyramide)[c]	Chap. 23, Chap. 24 (polymers)
Arene	$Ar-H^{d}$	⬡—CH_2CH_3	ethylbenzene	Section 10.8, Chap. 23
Aryl halide	$Ar-X^{b}$	⬡—Br	bromobenzene	Chap. 23
Phenol	$Ar-OH$	Cl—⬡—OH	4-chlorophenol (p-chlorophenol)[c]	Section 10.9, Chap. 23

[a] The functional group is shown in red. R stands for an alkyl group.
[b] X stands for a halogen atom—F, Cl, Br, or I.
[c] Common name.
[d] Ar— stands for an aromatic (*aryl*) group such as the benzene ring.

Glossary

Absolute configuration is the three-dimensional arrangement of groups about a chiral center in a molecule.

Absolute zero is the lowest possible temperature: $-273.15\,°C = 0\,K$.

The **accuracy** of a set of measurements refers to the closeness of the average of the set to the most probable value.

An **acid** is (1) a hydrogen-containing compound that can produce hydrogen ions, H^+, in aqueous solution (Arrhenius theory); (2) a proton donor (Brønsted–Lowry theory); (3) an atom, ion, or molecule that can accept a pair of electrons to form a covalent bond (Lewis theory).

An **acid–base indicator** is a substance added to the reaction mixture in a titration that changes color at or near the equivalence point.

An **acidic oxide** is a nonmetal oxide whose reaction with water produces a ternary acid as its sole product.

The **acid ionization constant** (K_a) is the equilibrium constant for the reversible ionization of a weak acid.

Acid rain is rainfall that is more acidic than is water in equilibrium with atmospheric carbon dioxide.

The **actinide** elements constitute the portion of the f-block of the periodic table in which the $5f$ subshell fills in the aufbau process.

An **activated complex** is an aggregate of atoms in the transition state of a reaction formed by a favorable collision. (*See also* **transition state**.)

The **activated sludge method** of sewage treatment is a process in which sludge from the secondary stage is put into aeration tanks to facilitate decomposition by aerobic microorganisms.

The **activation energy** (E_a) of a reaction refers to the minimum total kinetic energy that molecules must bring into their collisions so that a chemical reaction may occur.

The **active site** on an enzyme is the region of the enzyme molecule where the substrate attaches and a chemical reaction occurs. (*See also* **enzyme** and **substrate**.)

Activity is the effective concentration of a species, and in dilute solutions is often approximated by molar concentration.

The **activity series of the metals** is a listing of the metals in order of their ability to displace one another from solutions of their ions or to displace H^+ as $H_2(g)$ from acidic solutions. (See also Figure 4.13.)

The **actual yield** is the measured quantity of a desired product obtained in a chemical reaction. (*See also* **theoretical yield** and **percent yield**.)

Addition polymerization is a type of polymerization reaction in which monomers add to one another to produce a polymeric product that contains all the atoms of the starting monomers.

In an **addition reaction**, substituent groups join to hydrocarbon molecules at points of unsaturation—double or triple bonds.

An **adduct** is a compound that results from the addition, through a coordinate covalent bond, of one structure to another.

Adhesive forces are intermolecular forces between unlike molecules, for example, between those in a liquid and those in a surface over which the liquid is spread.

Advanced treatment (*tertiary treatment*) is a third stage of sewage treatment in which nitrates, phosphates, and (sometimes) dissolved organic substances are removed from the effluent of a secondary treatment process.

Aerobic oxidation is an oxidation process that occurs in the presence of oxygen.

An **air pollutant** is a substance that is found in air in greater abundance than normally occurs naturally and that has some harmful effect(s) on the environment.

An **alcohol (ROH)** is an organic substance whose molecules contain the hydroxyl group, OH, attached to an alkyl group.

An **aldehyde (RCHO)** is an organic compound whose molecules have a carbonyl functional group with a hydrogen atom attached to the carbonyl carbon.

Aliphatic compounds are those organic compounds with open chains of carbon atoms, or rings of carbon atoms that are similar in structure and properties to the open-chain compounds. (*See also* **aromatic compounds**.)

An **alkane** is a saturated hydrocarbon having the general formula C_nH_{2n+2}. (*See also* **saturated hydrocarbon**.)

An **alkene** is a hydrocarbon whose molecules contain at least one carbon-to-carbon double bond.

An **alkyl group (R)** is a substituent group in organic molecules derived from an alkane molecule by removal of one hydrogen atom.

An **alkyne** is a hydrocarbon whose molecules contain at least one carbon-to-carbon triple bond.

An **allotrope** is one of two or more forms of an element that differ in their basic molecular structure.

An **alloy** is a metallic material consisting of two or more elements.

An **alpha (α) particle** consists of two protons and two neutrons. It is identical to a doubly ionized helium ion, He^{2+}.

An **amide ($RCONH_2$)** is an organic compound in which a nitrogen atom is bonded to the carbon atom of a carbonyl group.

An **amine** is an organic substance in which one or more H atoms of an ammonia molecule are replaced by a hydrocarbon residue.

The **ampere (A)** is the basic unit of electric current. One ampere is a current of 1 coulomb per second: $1\,A = 1\,C/s$.

An **amphiprotic** substance can ionize either as a Brønsted–Lowry acid or base, depending on the acid–base properties of other species in the solution.

An **amphoteric** substance (usually an oxide or hydroxide) can react either with an acid or a base. The central element of the oxide or hydroxide appears in a cation in acidic solutions and in an anion in basic solutions.

Anaerobic decay is decomposition in the absence of oxygen.

The **analyte** is the sought-for substance in an analysis such as a titration.

Analytical chemistry is the branch of chemistry that deals with determination of the composition and properties of substances and mixtures.

An **anion** is a negatively charged ion.

An **anode** is an electrode at which an oxidation half-reaction occurs. It is the negative electrode in a voltaic cell and the positive electrode in an electrolytic cell.

An **antibonding molecular orbital** places a high electron charge density (electron probability) away from the region between bonded atoms. (*See also* **bonding molecular orbital** and **molecular orbital**.)

An **anticarcinogen** is a substance that opposes the action of a carcinogen; it prevents or retards the development of cancer.

An **aqueous solution** is a solution in which water is the solvent.

Aromatic compounds are organic compounds with benzene-like structures. To describe their electronic structures (Lewis structures), resonance theory is used.

The **atmosphere (atm)** is a unit used to measure gas pressure. It is equal to the pressure of a column of mercury having a height of exactly 760 mm. That is, 1 atm = 760 mmHg.

The **atomic mass** of an element is the weighted average of the masses of the atoms of the naturally occurring isotopes of the element.

An **atomic mass unit (u)** is exactly one-twelfth the mass of an atom of carbon-12.

The **atomic number (Z)** of an atom is the number of protons in the atomic nucleus.

An **atomic orbital** is a wave function for an electron corresponding to the assignment of specific values to the n, l, and m_l quantum numbers in a wave equation.

The **atomic radius** is a measure of the size of an atom based on the measurement of internuclear distances. (*See also* **covalent radius**, **ionic radius**, and **metallic radius**.)

Atoms are the smallest distinctive units of a sample of matter. Atoms of one element differ from atoms of all other elements. (*See also* **element**.)

The **aufbau principle** is a hypothetical process for building up an atom from the atom of the preceding atomic number by adding a proton and the requisite number of neutrons to the nucleus and one electron to the appropriate atomic orbital.

Average bond energy is the average of the bond-dissociation energies for a number of different molecular species containing the particular bond. (*See also* **bond-dissociation energy**.)

Avogadro's hypothesis states that equal numbers of molecules of different gases, when compared at the same temperature and pressure, occupy equal volumes.

Avogadro's law states that at a fixed temperature and pressure, the volume of a gas is directly proportional to the amount of gas.

Avogadro's number (N_A) is the number of elementary units in a mole—$6.02214199 \times 10^{23}$ mol^{-1}.

A **band**, in describing bonding in metals and semiconductors, is a collection of a large number of closely spaced molecular orbitals, obtained by combining atomic orbitals of many atoms.

A **barometer** is a device used to measure the pressure exerted by the atmosphere.

A **base** is (1) a compound that produces hydroxide ions, OH$^-$, in aqueous solution (Arrhenius theory); (2) a proton acceptor (Brønsted–Lowry theory); (3) an atom, ion, or molecule that donates a pair of electrons to form a covalent bond (Lewis theory).

The **base ionization constant (K_b)** is the equilibrium constant for the reversible ionization of a weak base.

A **basic oxide** is a metal oxide whose reaction with water produces a base.

A **beta (β^-) particle** is identical to an electron and is emitted by the nuclei of certain radioactive atoms as they undergo decay. In the decaying nucleus, the atomic number increases by one unit and the mass number remains unchanged.

A **bidentate** ligand has two points of attachment to a metal center in a complex.

A **bimolecular reaction** is an elementary reaction in a reaction mechanism that involves the collision of two molecules.

Biochemical oxygen demand (BOD) measures the amount of oxygen needed by aerobic microorganisms to metabolize the organic wastes in water.

The **biosphere** is the part of Earth occupied by living organisms.

A **body-centered cubic (bcc)** crystal structure has as its unit cell a cube with a structural unit at each corner and one in the center of the cell. (*See also* **unit cell**.)

The **boiling point** of a liquid is the temperature at which the liquid boils—the temperature at which the vapor pressure of the liquid is equal to the prevailing atmospheric pressure.

The **bond-dissociation energy (D)** of a particular covalent bond between two atoms is the quantity of energy required to break one mole of bonds of that type in a gaseous species.

A **bonding molecular orbital** places a high electron charge density (electron probability) in the region between two bonded atoms. (*See also* **antibonding molecular orbital** and **molecular orbital**.)

A **bonding pair** is a pair of electrons shared between two atoms in a molecule.

The **bond length** of a particular covalent bond is the distance between the nuclei of two atoms joined by that type of bond.

A **bond moment** describes the extent to which a separation of positive ($\delta+$) and negative ($\delta-$) charges exists in a covalent bond between two atoms.

Bond order refers to the number of electron pairs in a covalent bond—that is, whether a single pair (bond order = 1), two pairs (bond order = 2), or three pairs (bond order = 3). In molecular orbital theory, it is one-half the difference between the number of electrons in bonding molecular orbitals and in antibonding molecular orbitals.

Boyle's law states that for a given amount of gas at a constant temperature, the volume of a gas varies inversely with its pressure. That is, $V \propto 1/P$ or PV = constant.

A **buffer solution** is a solution containing a weak acid and its conjugate base or a weak base and its conjugate acid. Small quantities of added acid are neutralized by one buffer component and small quantities of added base by the other. As a result, the solution pH is maintained nearly constant.

A **buret** is a long graduated tube constructed to deliver precise volumes of a liquid solution through a stopcock valve.

A **calorie (cal)** is the energy needed to raise the temperature of 1 g of water by 1 °C (more precisely, from 14.5 to 15.5 °C). 1 cal = 4.184 J.

A **calorimeter** is a device in which quantities of heat are measured.

A **carbocation** is an intermediate species in certain reactions of organic compounds in which a positive charge is centered on a carbon atom in the species.

A **carbohydrate** is a compound consisting of carbon, hydrogen, and oxygen, generally with twice as many hydrogen atoms as oxygen atoms; a starch, sugar, or cellulose.

The **carbon cycle** refers to the totality of activities in which carbon atoms are cycled through the environment.

A **carboxylic acid (RCOOH)** is an organic substance whose molecules contain the carboxyl group, COOH.

A **carcinogen** is an agent that causes cancer.

Cast iron is a carbon–iron alloy that is cast into shape and contains 1.8–4.5% C.

A **catalyst** is a substance that increases the rate of a reaction without itself being consumed in the reaction. A catalyst changes a reaction mechanism to one with a lower activation energy.

A **cathode** is an electrode at which a reduction half-reaction occurs. It is the positive electrode in a voltaic cell and the negative electrode in an electrolytic cell.

A **cathode ray** is a beam of electrons that travels from the cathode to the anode when an electric discharge is passed through an evacuated tube.

A **cation** is a positively charged ion.

A **cell diagram** is a schematic representation of an electrochemical cell.

Cell potential (E_{cell}) (cell voltage) refers to the potential difference between the electrodes in an electrochemical cell.

A **central atom** in a molecule or polyatomic ion is an atom bonded to two or more other atoms.

A **ceramic** is an inorganic solid generally produced at high temperature, and often containing silica or silicates. Most ceramics are hard, brittle, and stable at high temperatures.

Charles's law states that the volume of a fixed amount of a gas at a constant pressure is directly proportional to its Kelvin temperature. That is, $V \propto T$, or $V = \text{constant} \times T$.

A **chelate** is a five- or six-membered ring structure produced in a complex through the attachment of one or more polydentate ligands to a metal center.

A **chemical bond** is a force that holds atoms together in molecules or ions together in crystals.

Chemical change. *See* **chemical reaction**.

A **chemical equation** is a description of a chemical reaction that uses symbols and formulas to represent the elements and compounds involved in the reaction. Numerical coefficients preceding each symbol or formula and indicating molar proportions may be needed to balance a chemical equation.

A **chemical formula** indicates the composition of a compound through symbols of the elements present and subscripts to indicate the relative numbers of atoms of each element.

Chemical kinetics is the study of the rates of chemical reactions, the factors that affect rates, and reaction mechanisms.

Chemical nomenclature is a systematic way of relating the names and formulas of chemical compounds.

A **chemical property** is a characteristic that matter displays as it undergoes a change in composition.

A **chemical reaction** is a process in which a sample of matter undergoes a change in composition and/or structure of its molecules. One or more original substances (reactants) are changed into one or more new substances (products).

Chemical shift is a term used in nuclear magnetic resonance (NMR) spectroscopy to indicate the location of an absorption peak relative to a standard. The magnitudes of the chemical shifts can be used to determine structural features of a molecule.

A **chemical symbol** is a representation of an element made up of one or two letters derived from the English name of the element (or sometimes from the Latin name of the element or one of its compounds).

Chemistry is a study of the composition, structure, and properties of matter and of the changes that occur in matter.

A **chiral carbon** is a carbon atom that is attached to four different groups.

Cis isomers have two substituent groups attached to the two atoms on the same side of a double bond in an organic molecule, or along the same edge of a square planar or octahedral complex ion. (*See also* **geometric isomers**.)

Cohesive forces are intermolecular forces between like molecules.

A **colligative property** is a physical property—such as vapor pressure lowering, freezing point depression, boiling point elevation, and osmotic pressure—that depends on the concentration of solute in the solution but not on the identity of the solute.

A **colloid** is a dispersion in which the dispersed matter has one or more dimensions (length, width, or thickness) in the range from about 1 nm to 1000 nm.

The **combined gas law** combines Boyle's, Charles's, and Avogadro's laws into a single relationship, written as $\dfrac{P_1 V_1}{n_1 T_1} = \dfrac{P_2 V_2}{n_2 T_2}$.

The **common ion effect** refers to the ability of an ion X to (a) suppress the ionization of a weak acid or weak base that produces X, or (b) reduce the solubility of a slightly soluble ionic compound that produces X.

A **complex** consists of a central atom, which is usually a metal ion, and coordinately covalently bonded groups called ligands.

A **complex ion** is a complex that carries a net electric charge, either positive (a complex cation) or negative (a complex anion).

A **composite** is composed of two or more physically distinct materials that, when combined, exploit the desired structural and mechanical properties of the individual components.

Composition refers to the types of atoms and their relative proportions in a sample of matter.

A **compound** is a substance made up of atoms of two or more elements, with the different atoms joined in fixed proportions.

A **concentration cell** is a voltaic cell having identical electrodes in contact with solutions of different concentrations.

A **concentration equilibrium constant** (K_c) is the numerical value of an equilibrium constant expression in which molarities of products and reactants are used.

Condensation is the conversion of a gas (vapor) to a liquid.

In **condensation polymerization**, monomers with at least two functional groups link together by eliminating small-molecule by-products.

A **conduction band** is a partially filled band of very closely spaced energy levels.

A **conjugate acid** is formed when a Brønsted–Lowry base accepts a proton. Every base has a conjugate acid.

A **conjugate base** is formed when a Brønsted–Lowry acid donates a proton. Every acid has a conjugate base.

A **conjugated bonding system** refers to a molecular structure having a series of alternate single and double bonds. Substances having this feature can absorb UV and/or visible light.

A **continuous spectrum** exhibits a broad range of wavelengths emitted or absorbed without a significant break or gap. (*See also* **emission spectrum** and **line spectrum**.)

Contributing structure. *See* **resonance structure**.

A **conversion factor** is a ratio of terms—equivalent to the number 1—that is used to change the unit(s) in which a quantity is expressed.

A **cooling curve** is a graph of temperature as a function of time, obtained as a substance is cooled. Constant-temperature segments of the curve correspond to phase changes, for example, condensation and freezing.

A **coordinate covalent bond** is a linkage between two atoms in which one atom provides both of the electrons of the shared pair.

A **coordination compound** is a substance made up of one or more complexes.

The **coordination number** of the metal center in a complex is the total number of points around this central atom at which bonding to ligands can occur.

Core electrons are electrons found in the inner electronic shells of atoms. (*See also* **valence electrons**.)

A **corrosive material** is one that degrades a metal or alloy by a chemical reaction. A corrosive material requires a special container because it corrodes conventional container materials.

The **coulomb (C)** is the SI unit of electric charge. The electric charge on an electron, for example, is -1.602×10^{-19} C.

A **coupled reaction** is one that involves two separate processes that can be combined to give a single reaction. In most cases, a thermodynamically unfavorable reaction is combined with another reaction to give an overall reaction that is thermodynamically favorable.

A **covalent bond** is a bond formed by a pair of electrons shared between atoms.

The **covalent radius** of an atom is one-half the distance between the nuclei of two like atoms joined into a molecule.

Critical mass is the minimum mass of a fissionable element that must be present to sustain a chain reaction. This is the mass required to produce an explosion of a nuclear bomb.

The **critical point** refers to the condition at which the liquid and gaseous (vapor) states of a substance become identical. It is the highest temperature point on a vapor pressure curve.

The **critical pressure** of a substance is the pressure at its critical point.

The **critical temperature** of a substance is the temperature at its critical point.

A **crystal** is a structure having plane surfaces, sharp edges, and a regular geometric shape. The fundamental units—atoms, ions, or molecules—are assembled in a regular, repeating manner extending in three dimensions through the crystal.

Crystal field theory is a theory of bonding in complexes that focuses on the abilities of ligands to produce a splitting of a *d*-subshell energy level of the metal center in a complex.

A **cubic close-packed (ccp)** structure has units (atoms, ions, or molecules) arranged in one of the two ways that minimize the voids between the units. The layers are stacked in the arrangement ABCABC. (*See also* **hexagonal close-packed.**)

Dalton's law of partial pressures states that in a mixture of gases, each gas expands to fill the container and exerts its own pressure, called a partial pressure. The total pressure of the mixture is the sum of the partial pressures exerted by the separate gases.

Data are the facts collected by careful observations and measurements made during experiments.

The **d-block** is the portion of the periodic table in which the $(n - 1)d$ subshell (the *d* subshell of the next-to-outermost shell) fills in the aufbau process. The *d*-block comprises the B-group elements in the periodic table.

The **debye (D)** is the unit used to express the dipole moments of polar molecules. One debye is equal to 3.34×10^{-30} C m.

Degenerate orbitals are two or more orbitals that are at the same energy level.

Deionized water is water that has been freed of ions through ion-exchange processes.

Deliquescence is the condensation of water vapor on a solid followed by solution formation.

Delocalized electrons are bonding electrons that are spread out over several atoms, rather than being in a fixed location between two atoms.

The **density (d)** of a sample of matter is its mass per unit volume, that is, the mass of the sample divided by its volume: $d = m/V$.

Deoxyribonucleic acid (DNA) is a polymer of nucleotides. The nucleotides consist of the sugar deoxyribose, a phosphate ester, and a cyclic amine base (adenine, guanine, thymine, or cytosine).

A **dextrorotatory** (+) substance rotates the plane of polarized light to the right.

A **diagonal relationship** refers to the similarity of certain second-period elements in one group of the periodic table with third-period elements of the next group to the right.

Diamagnetism is the weak repulsion by a magnetic field of a substance in which all electrons are paired.

Diffusion is the process by which one substance mixes with one or more other substances as a result of the random motion of molecules.

Dilution is a process of producing a solution of lower concentration from a more concentrated one by the addition of an appropriate quantity of solvent.

The **dipole moment (μ)** of a polar molecule is the product of the magnitude of the charges (δ) and the distance that separates them.

A **disaccharide** is a carbohydrate with molecules that can be hydrolyzed to two monosaccharide units.

A **dispersion force** is an attractive force between an instantaneous dipole and an induced dipole.

A **disproportionation reaction** is an oxidation–reduction reaction in which the same substance is both oxidized and reduced.

Doping refers to the addition of trace amounts of certain elements to a semiconductor to change the semiconducting properties. (*See also* ***n*-type semiconductor** and ***p*-type semiconductor.**)

A **double bond** is a covalent linkage in which two atoms share two pairs of electrons between them.

Dynamic equilibrium occurs when two opposing processes occur at exactly the same rate, with the result that no net change occurs.

The **effective nuclear charge (Z_{eff})** acting on an electron in an atom is the actual nuclear charge less the screening effect of other electrons in the atom.

Effusion is a process in which a gas escapes from its container through a tiny hole (an orifice). (*See also* **Graham's law of effusion.**)

An **elastomer** is a polymeric material that can be stretched significantly by a relatively low stress and upon release of the stress will return substantially to its original length.

An **electrochemical cell** is a combination of two half-cells in which metal electrodes are joined by a wire and the solutions are brought into contact through a salt bridge or by other means. (*See also* **electrolytic cell**, **half-cell**, and **voltaic cell.**)

Electrochemistry is a study of the relationships between electrical energy and chemical reactions.

An **electrode** is a conductive solid dipped into a solution or molten electrolyte to carry electricity to or from the liquid. (*See also* **anode** and **cathode.**)

Electrode potential is a property related to the tendency of a species to be reduced at an electrode.

Electrolysis is the decomposition of compounds by passing electricity through an ionic solution or a molten salt. A nonspontaneous chemical change occurs.

An **electrolyte** is a compound that conducts electricity when molten or in a liquid solution.

An **electrolytic cell** is an electrochemical cell in which electrolysis occurs.

The **electromagnetic spectrum** is the range of wavelengths and frequencies found for electromagnetic waves, extending from very long wavelength radio waves to the shortest gamma rays.

An **electromagnetic wave** originates in the vibrations of electrically charged objects and is propagated through oscillations of electric and magnetic fields.

An **electron** is a particle carrying the fundamental unit of negative electric charge. Electrons have a mass of 0.0005486 u and are found outside the nuclei of atoms.

Electron affinity is the energy change that occurs when an electron is added to an atom in the gaseous state.

Electron capture (EC) is a type of radioactive decay in which a nucleus absorbs an electron from the first or second electronic shell.

The **electron configuration** of an atom describes the distribution of electrons among atomic orbitals in the atom.

The **electronegativity (EN)** of an element is a measure of the tendency of its atoms in molecules to attract bonding electrons to themselves.

An **electron group** is a collection of valence electrons localized in a region around a central atom that exerts repulsions on other groups of valence electrons. It may be a bonding pair of electrons in a single bond, two pairs of electrons in a double bond, three pairs of electrons in a triple bond, a lone pair of electrons, or even an unpaired electron.

The **electron-group geometry** of a molecule or ion is the arrangement of all the electron groups—both bonding and nonbonding—about a central atom.

The **electron spin quantum number** (m_s) is a fourth quantum number (in addition to the three required by the Schrödinger wave equation) needed to complete the orbital designation of an electron. The two possible values of the spin quantum number are $+\frac{1}{2}$ and $-\frac{1}{2}$.

The **electron-Volt** is a unit of energy equal to that acquired by an electron as it passes through a potential difference of 1 V in a vacuum.

Electrophilic aromatic substitution is a reaction type in which an electrophile attaches to an aromatic ring such as benzene, replacing a hydrogen atom on the ring.

An **electrophilic reagent** is an electron-deficient molecule or ion that accepts an electron pair from a molecule, forming a covalent bond.

An **element** is a substance that cannot be broken down into simpler substances by chemical reactions. All atoms of a given element have the same atomic number.

An **elementary reaction** represents, at the molecular level, a single stage in the overall mechanism by which a chemical reaction occurs. (*See also* **bimolecular, termolecular, and unimolecular reaction**.)

An **emission (line) spectrum** is a dispersion of electromagnetic radiation into a discrete set of wavelength components. These components can be rendered as images of a slit (lines) in light from a spectroscope.

An **empirical formula** is the *simplest* formula describing the elements in a compound and the smallest integral (whole number) ratio in which their atoms are combined.

Enantiomers are pairs of mirror-image isomers that differ only in the direction in which they rotate the plane of polarized light. One isomer rotates the plane to the right, and the other rotates the plane to the same degree, but to the left.

An **endothermic reaction** is a reaction in which thermal energy is converted to chemical energy. In an endothermic process, a temperature decrease occurs in an isolated system, or in a nonisolated system, heat is absorbed from the surroundings.

The **endpoint** is the point in a titration at which an added indicator changes color. An indicator is chosen so that its endpoint matches the equivalence point of the reaction. (*See also* **equivalence point**.)

An **energy gap** (E_g) is the energy separation between a valence band and a conduction band that lies above it. In a semiconductor, the gap is relatively small, and in an insulator it is very large. (*See also* **conduction band**) and **valence band**.

An **energy level (shell)** is the state of an atom determined by the location of its electrons

among the various principal shells and subshells.

Enthalpy (H) is a thermodynamic function defined as the sum of the internal energy and the pressure–volume product: $H = E + PV$.

The **enthalpy change** (ΔH) in a chemical reaction is equal to the heat of reaction at constant temperature and pressure, q_P.

An **enthalpy diagram** is a graphical representation of the change in enthalpy that occurs in a chemical reaction.

The **enthalpy (heat) of fusion** (ΔH_{fusion}) is the quantity of heat required to melt a given quantity of a solid.

The **enthalpy (heat) of sublimation** (ΔH_{subl}) is the quantity of heat required to vaporize a given quantity of solid at a constant temperature. It is equal to the sum of the enthalpies of fusion and vaporization.

The **enthalpy (heat) of vaporization** (ΔH_{vapn}) is the quantity of heat required to vaporize a given quantity of liquid at a constant temperature.

Entropy (S) is a property related to the distribution of the energy of a system among the available energy levels.

Entropy change (ΔS) is the difference in entropy between two states of a system, as between the products and reactants of a chemical reaction.

An **enzyme** is a protein that catalyzes reactions occurring in living organisms.

Equilibrium is a condition that is reached when two opposing processes occur at equal rates. As a result, the concentrations (or partial pressures) of the reacting species remain constant with time.

An **equilibrium constant expression** is a particular ratio of concentrations (or partial pressures) of products to reactants in a chemical reaction at equilibrium. The expression has a constant value that is independent of the manner in which equilibrium is reached. (*See also* K_c and K_p.)

The **equilibrium constant** (K_{eq}) is the form of the equilibrium constant based on activities and used in thermodynamic relationships. In the K_{eq} expression, species in solution are usually represented by their molarities and in gases by their partial pressures in atm.

The **equivalence point** of a titration is the point at which two reactants have been introduced into a reaction mixture in their stoichiometric proportions.

An **essential amino acid** is an amino acid that cannot be synthesized in the body and must therefore be included in the diet.

An **ester** ($R'COOR$) is a compound derived from a carboxylic acid and an alcohol. The OH of the acid is replaced by an OR group.

An **ether** ($R'OR$) is a compound having two hydrocarbon groups joined through an oxygen atom.

Eutrophication describes a process in which an overabundance of nutrients leads to an overgrowth of algae in a body of water. The algae then die, and their decay depletes the dissolved oxygen in the water.

An **excited state** of an atom is one in which one or more electrons has been promoted to a higher energy level than in the ground state. (*See also* **ground state**.)

An **exothermic reaction** is a reaction in which chemical energy is converted to thermal energy. In an exothermic process, a temperature increase occurs in an isolated system, or in a nonisolated system, heat is given off to the surroundings.

An **expanded valence shell** of a central atom in a Lewis structure is one that can accommodate more than the usual octet (8) of electrons.

An **experiment** is a carefully controlled procedure devised to test a hypothesis.

An **extensive property** is a physical property, such as mass or volume, that depends on the size or quantity of the sample of matter being considered.

A **face-centered cubic (fcc)** crystal structure has as its unit cell a cube with a structural unit at each of the corners and in the center of each face of the cube. (*See also* **unit cell**.)

The **Faraday constant** (F) is the electric charge, in coulombs, per mole of electrons— 96,485 C/mol.

The **f-block** is the portion of the periodic table in which the $(n - 2)f$ subshell (the f subshell of the second-from-outermost shell) fills in the aufbau process. The f-block consists of the lanthanides and actinides.

Ferromagnetism is a magnetic effect much stronger than paramagnetism and associated with iron, cobalt, nickel, and certain alloys. It requires that atoms be both paramagnetic and of the right size to be able to form magnetic domains.

The **first law of thermodynamics** states that the internal energy of an isolated system is constant, or if a system interacts with its surroundings by exchanging heat and/or work, the exchange must occur in such a way that no energy is created or destroyed. (*See also* **law of conservation of energy**.)

A **first-order reaction** has a rate equation in which the sum of the exponents, $m + n + \cdots = 1$.

Flotation is a metallurgical method by which an ore is separated from waste rock, based on selective wetting of the ore by a surface-active agent.

Formal charge, a concept used in writing Lewis structures, is the number of valence electrons in an isolated atom minus the number of electrons assigned to that atom in a Lewis structure.

The **formation constant** (K_f) describes equilibrium between a complex ion and the cation and ligands from which it is formed.

Formula mass is the mass of a formula unit relative to that of a carbon-12 atom; it is the sum of the masses of the atoms or ions represented by the formula.

A **formula unit** is the simplest combination of atoms or ions consistent with the formula of a compound. In an ionic compound, it is the smallest possible electrically neutral collection of ions.

Fractional crystallization is a method of purifying a solid by dissolving it in a suitable solvent and changing the solution temperature to a value where the solute solubility is lower (usually a lower temperature). Excess solute crystallizes as pure solid, and soluble impurities remain in solution.

Fractional distillation is a method of separating the volatile components of a solution having different vapor pressure and boiling points. It involves repeated vaporizations and condensations occurring continuously in a distillation column.

The **free electron model** of metals considers the metal to be composed of positive ions surrounded by a "sea" of mobile electrons.

The **free energy change** (ΔG) is the difference in free energy between two states of a system, as between the free energies of the products and reactants of a chemical reaction. It is given by the equation $\Delta G = \Delta H - T\Delta S$.

A **free radical** is a highly reactive atom or molecular fragment characterized by having one or more unpaired electrons. Free radicals are encountered as intermediates in some chemical reactions.

Freezing point is the temperature at which a liquid freezes—that is, the liquid comes into equilibrium with solid. For a pure substance, the freezing point and melting point are the same.

The **frequency** (ν) of a wave is the number of cycles of the wave (the number of wavelengths) that pass through a point in a unit of time.

A **functional group** is an atom or grouping of atoms attached to or within a hydrocarbon chain or ring that confers characteristic properties to the molecule as a whole.

Fusion (melting) is the process of changing a solid to a liquid.

Galvanic cell *See* **voltaic cell**.

A **gamma** (γ) **ray** is a highly penetrating form of electromagnetic radiation emitted by the nuclei of certain radioactive atoms as they undergo decay.

The **gas constant** (R) is a proportionality constant used in the ideal gas law and in other relationships, and has a value of 0.082057 L atm/(mol K) or 8.314 J/(mol K).

A **gene** is a section of a DNA molecule found in the chromosomes of cells; genes are the basic units of heredity.

Geometric isomers in organic compounds are isomers (cis, trans) that differ in the positions of attachment of substituent groups at a double bond. In complexes, the isomers differ in the positions of attachment of ligands to the central metal ion.

Gibbs free energy (G), a thermodynamic function used in establishing criteria for equilibrium and for spontaneous change, is defined as $G = H - TS$, where H is the enthalpy, T is the Kelvin temperature, and S is the entropy of a system.

Global warming refers to the anticipated increase in Earth's average temperature resulting from the accumulation of CO_2 and other infrared-absorbing gases in the atmosphere.

Graham's law of effusion states that the rates of effusion of gas molecules are inversely proportional to the square roots of their molar masses.

The **greenhouse effect** refers to the ability of $CO_2(g)$ and certain other gases to absorb and trap energy radiated by Earth's surface as infrared radiation.

The **ground state** of an atom is the atom at its lowest energy level. (*See also* **excited state**.)

Groups of the periodic table are the vertical columns of elements having similar properties.

A **half-cell** is an electrode in a solution of ions; the reaction in the half-cell is either an oxidation or a reduction. (*See also* **electrochemical cell**.)

The **half-life** ($t_{1/2}$) of a chemical reaction is the time required to consume one-half of the initial quantity of a reactant. For radioactive decay, it is the time in which one-half of the atoms of a radioactive nuclide disintegrate.

A **half-reaction** is that portion of an oxidation–reduction reaction that represents either the oxidation process or the reduction process.

Hard water is groundwater containing significant concentrations of doubly charged cations derived from natural sources, such as Ca^{2+}, Mg^{2+}, Fe^{2+}, and associated anions.

A **hazardous material** is one that, when improperly managed, can cause or contribute to death or illness or threaten human health or the environment.

Heat (q) is an energy transfer into or out of a system caused by a difference in temperature between a system and its surroundings.

Heat capacity (C) of a system is the quantity of heat needed to raise the temperature of a system by 1 °C or 1 K.

A **heating curve** is a graph of temperature as a function of time obtained by gradually heating a substance. Constant-temperature segments of the curve correspond to phase changes. (*See also* **cooling curve**.)

The **heat of reaction** (q_{rxn}) is the quantity of heat exchanged between a system and its surroundings when a chemical reaction occurs at a constant temperature and pressure.

The **Henderson–Hasselbalch equation** is used to relate the pH of a solution of a weak acid and its conjugate base to the pK_a of the weak acid and to the ratio of the stoichiometric concentration of the conjugate base to that of the weak acid: pH = pK_a + log ([conjugate base]/[weak acid]).

Henry's law states that the solubility of a gas is directly proportional to the partial pressure of the gas in equilibrium with the solution.

Hess's law states that the enthalpy change of a reaction is constant, whether the reaction is carried out directly in one step or indirectly through a number of steps.

A **heterogeneous mixture** is a mixture in which the composition and/or properties vary from one region to another within the mixture.

A **hexagonal close-packed (hcp)** structure is one of the two crystal arrangements in which the structural units are close-packed. The layers are stacked in the arrangement ABABAB. (*See also* **cubic close-packed**.)

A **homogeneous mixture** is a mixture having the same composition and properties throughout the given mixture.

A **homologous series** is a series of organic compounds whose formulas and structures vary in a regular manner and whose properties are predictable based on this regularity.

Humidity is a measure of the water vapor content of air. The *absolute humidity* is the actual quantity of water vapor present in an air sample, and the *relative humidity* of air is a measure of water vapor content as a percentage of the maximum possible quantity.

Hund's rule states that electrons occupy atomic orbitals of identical energy singly before any pairing of electrons occurs. Furthermore, the electrons in the singly occupied orbitals have parallel spins.

Hybridization is a hypothetical process in which pure atomic orbitals are combined to produce a set of new orbitals called hybrid orbitals to describe covalent bonding by the valence bond method. (*See also* sp, sp^2, sp^3, sp^3d, sp^3d^2 hybridization.)

Hybrid orbitals are formed by a combination of atomic orbitals to produce a set of new orbitals.

A **hydrate** is a compound that incorporates water molecules into its basic solid structure. The formula unit of a hydrate includes a fixed number of water molecules.

A **hydrocarbon** is a compound containing only hydrogen and carbon atoms.

In a **hydrogenation reaction**, $H_2(g)$ is a reactant and H atoms are added to C atoms at a carbon-to-carbon double or triple bond.

A **hydrogen bond** is a type of intermolecular force in which a hydrogen atom covalently bonded in one molecule is simultaneously attracted to a nonmetal atom in a neighboring molecule. In most cases, both the atom to which the hydrogen atom is bonded and the one to which it is attracted must be small atoms of high electronegativity, usually N, O, or F.

The **hydrologic (water) cycle** is the series of natural processes by which water is recycled through the environment—Earth's solid crust, oceans and freshwater bodies, and the atmosphere.

In a general sense, **hydrolysis** is the reaction of a substance with water in which both the substance and the water molecules split apart. In a more limited sense, it is an acid–base reaction between an ion and water.

Hydrometallurgy is the extraction of a metal from its ores by processes that involve water and aqueous solutions.

A **hypertonic** solution is a solution having an osmotic pressure greater than that of body fluids (blood, tears). A hypertonic solution has a greater osmotic pressure than does an isotonic solution.

A **hypothesis** is a tentative explanation or prediction concerning some phenomenon.

A **hypotonic** solution is a solution having an osmotic pressure less than that of body fluids (blood, tears). A hypotonic solution has a lower osmotic pressure than does an isotonic solution.

An **ideal gas** is a hypothetical gas that strictly obeys the simple gas laws and the ideal gas law.

The **ideal gas law** or **ideal gas equation** states that the volume of a gas is directly proportional to the amount of a gas and its Kelvin temperature and is inversely proportional to its pressure: $PV = nRT$.

An **ideal solution** is one for which the heat of solution is zero and the volume of solution is the total of the volumes of the solution components. In general, the physical properties of an ideal solution can be predicted from the properties of its components.

An **ignitable material** is one that burns readily on ignition, presenting a fire hazard.

An **indicator** is a substance added to the reaction mixture in a titration that changes color at or near the equivalence point.

Industrial smog is polluted air associated with industrial activities. The principal pollutants are oxides of sulfur and particulate matter.

An **inert pair** refers to the ns^2 electrons in the valence shell of the posttransition elements of groups 3A, 4A, and 5A. These electrons may remain in the valence shell following the loss of the np electrons, as in Tl^+, Sn^{2+}, Pb^{2+}, and Bi^{3+}.

The **initial rate of reaction** is the rate of a reaction immediately after the reactants are brought together. The rate is generally expressed in terms of the rate of change with time of the concentration of one of the reactants or one of the products.

An **instantaneous rate of reaction** is the rate of a reaction at some particular time in the course of a reaction. It is established through a tangent line to a concentration versus time graph at the time in question.

An **integrated rate law** is an equation derived from the rate law for a reaction that expresses the concentration of a reactant as a function of time. (*See also* **rate law**.)

An **intensive property** is a property of a sample of matter, such as temperature or density, that is independent of the quantity of matter being considered.

An **interhalogen compound** is a compound of two (sometimes three) halogen elements.

An **intermediate** is a substance that is produced in one elementary step in a reaction mechanism and consumed in another. The intermediate does not appear in the chemical equation for the overall reaction.

An **intermolecular force** is a force *between* molecules.

The **internal energy** (U) is the total amount of energy contained in a thermodynamic system. The components of internal energy are energy associated with random molecular motion (thermal energy) and that associated with chemical bonds and intermolecular forces (chemical energy).

An **ion** is an electrically charged particle comprised of one or more atoms.

Ion exchange is the replacement in solution of ions of one type for ions of another type, using an ion-exchange resin or zeolite.

Ionic bonds are attractive forces between positive and negative ions, holding them together in solid crystals.

An **ionic compound** is a compound that consists of oppositely charged ions held together by electrostatic attractions.

Ionic radius is a measure of the size of a cation or anion based on the distance between the centers of ions in an ionic compound.

Ionization energy is the energy required to remove the least tightly bound electron from a ground-state atom (or ion) in the gaseous state.

The **ion product of water** (K_w) is the equilibrium constant for autoionization of water into H_3O^+ and OH^-. At 25 °C, its value is 1.0×10^{-14}.

The **isoelectric point (pI)** is the pH value at which an amino acid exists as a zwitterion.

Isoelectronic species (atoms, ions, molecules) have the same number of electrons.

Isomers are compounds having the same molecular formula but different structural formulas.

An **isotonic** solution is one that has the same osmotic pressure as body fluids (blood, tears).

Isotopes are atoms that have the same number of protons in their nuclei—the same atomic number—but different numbers of neutrons and, therefore, different mass numbers.

The **joule (J)** is the basic unit of energy in SI. It is the work done by a force of 1 newton (N) acting over a distance of 1 meter. That is, $1 \text{ J} = 1 \text{ N m} = 1 \text{ kg m}^2 \text{ s}^{-2}$.

K_c is the numerical value of an equilibrium constant expression in which molarities of products and reactants are used.

K_p is the numerical value of an equilibrium constant expression in which the partial pressures (usually in atm) of gaseous products and reactants are used.

A **kelvin (K)** is the SI base unit of temperature. An interval of 1 kelvin on the Kelvin scale is the same as one degree on the Celsius scale.

The **Kelvin scale** is an absolute temperature scale with its zero at −273.15 °C; its relationship to the Celsius scale is $T(\text{K}) = T(°\text{C}) + 273.15$.

A **ketone** is an organic substance whose molecules have a carbonyl group between two other C atoms.

The **kilogram (kg)** is the SI base unit of mass.

A **kilopascal (kPa)** is 1000 pascals (Pa). (*See also* **pascal**.)

Kinetic energy (E_k) is energy of motion, given by the expression $E_k = \frac{1}{2}mv^2$.

The **kinetic-molecular theory of gases** is a theory based on a small number of postulates concerning gas molecules from which simple gas laws, the ideal gas law, and equations dealing with temperature and molecular speeds can be derived.

The **lanthanide** elements constitute the portion of the f-block of the periodic table in which the $4f$ subshell fills in the aufbau process.

The **lanthanide contraction** describes the general downward trend in the radii of lanthanide atoms and ions with increasing atomic number.

Lattice energy is the enthalpy change that accompanies the formation of one mole of an ionic solid from its gaseous ions.

The **law of combining volumes** states that when gases measured at the same temperature and pressure are allowed to react, the volumes of gaseous reactants and products are in small whole-number ratios.

The **law of conservation of energy** states that in a physical or chemical change, energy can be neither created nor destroyed.

The **law of conservation of mass** states that the total mass remains constant during a reaction. That is, the mass of the products of a reaction is always equal to the total mass of the reactants consumed.

The **law of constant composition** or **law of definite proportions** states that all samples of a particular compound have the same composition. That is, all samples have the same proportions by mass of the elements present.

The **law of multiple proportions** states that when two or more different compounds of the same two elements are compared, the masses of one element that combine with a fixed mass of the second element are in the ratio of small whole numbers.

Le Châtelier's principle is a statement that permits qualitative predictions about the effects produced by changes (amounts of reactants or products, reaction volume, temperature, etc.) imposed on a system at equilibrium. (*See* page 589 for a statement of the principle.)

A **levorotatory** (−) substance rotates the plane of polarized light to the left.

Lewis acid. *See* **acid**.

Lewis base. *See* **base**.

A **Lewis structure** is a representation of covalent bonding through Lewis symbols, shared electron pairs, and lone-pair electrons.

A **Lewis symbol** is a representation of an element in which the chemical symbol stands for the core of the atom and dots placed around the symbol represent its valence electrons.

A **ligand** is a species (atom, molecule, anion, or, rarely, cation) that is bonded to a metal center in a complex by a coordinate covalent bond.

The **limiting reactant** (reagent) is the reactant that is completely consumed in a chemical reaction, thereby limiting the amounts of products formed.

The **line spectrum** of an element reflects the discrete wavelengths of light emitted by the element. (*See also* **emission spectrum**.)

A **liquid crystal** is a physical form of a substance that has the fluid properties of a liquid and the optical properties of a crystalline solid.

A **lipid** is a cellular component that is insoluble in water but soluble in solvents of low polarity such as hexane, diethyl ether, and benzene.

A **liter** (**L**) is a metric unit of volume equal to 1 cubic decimeter or 1000 cubic centimeters: $1 \text{ L} = 1 \text{ dm}^3 = 1000 \text{ cm}^3$.

Lone pairs are electron pairs assigned exclusively to one of the atoms in a Lewis structure. They are not shared and hence are not involved in the chemical bonding.

Macromolecules are giant molecules (polymers) having small molecules (monomers) as their building blocks.

Magic numbers are numbers of protons and neutrons in stable nucleon shells that make up especially stable atomic nuclei.

The **magnetic quantum number** (m_l) is the last of three parameters that must be assigned a specific value to achieve a solution of Schrödinger's wave equation for the hydrogen atom: m_l is an integer between $-l$ and $+l$ (including 0). (*See also* **orbital angular momentum quantum number** and **principal quantum number**.)

A **main-group element** is an element in which the subshell being filled in the aufbau process is either an s or a p subshell of the principal shell of highest principal quantum number (the outermost shell). Main-group elements are located in the s- and p-blocks of the periodic table.

A **manometer** is a device used to measure pressure of a confined gas.

Mass is the quantity of matter in an object. It is related to the force required to move the object or to change its velocity if the object is already in motion.

The **mass percent composition** of a substance is the proportion, by mass, of each element in the substance expressed as a percentage.

The **mass number** (A) is the sum of the number of protons and neutrons in the nucleus of an atom. It is also called the *nucleon number*.

A **mass spectrometer** is a device that separates ions according to their mass-to-charge ratios.

Matter is anything that occupies space and has mass.

The **melting point** of a solid is the temperature at which it melts, that is, the temperature at which it comes into equilibrium with the liquid phase.

A **meniscus** is the interface between a liquid and the air above it.

A **meta director** is a substituent already on a benzene ring, such as —COOH or —NO$_2$,

that causes an incoming electrophile to substitute mainly in the meta position.

A **metal** is an element having a distinctive set of properties: luster, good heat and electrical conductivity, malleability, and ductility. Metal atoms generally have small numbers of valence electrons and a tendency to form cations. Metals are found to the left of the stepped diagonal line in the periodic table.

The **metallic radius** is one-half the distance between the nuclei of adjacent atoms in a solid metal.

A **metalloid** is an element that has the physical appearance of a metal but some nonmetallic properties as well. Metalloids are located along the stepped diagonal line in the periodic table.

The **meter** (**m**) is the SI base unit of length.

The **method of initial rates** is an experimental method of establishing the rate law of a reaction. To establish the order of the reaction with respect to one of the reactants, the initial rates are compared for two different concentrations of that reactant, with the concentrations of all other reactants held constant. (*See also* **initial rate of reaction**, **rate law**, and **order of a reaction**.)

A **millimeter of mercury** (**mmHg**) is a unit used to express gas pressure: $1 \text{ mmHg} = 1/760$ atm (exactly). (*See also* **atmosphere**.)

A **mixture** is a type of matter with composition and properties that may vary from one sample to another. (*See also* **heterogeneous mixture** and **homogeneous mixture**.)

The **molality** (**m**) of a solution is the amount of solute, in moles, per kilogram of solvent (not of solution).

Molar concentration. *See* **molarity**.

Molar heat capacity is the quantity of heat required to change the temperature of one mole of a substance by 1 °C (or 1 K); it is the heat capacity of one mole of substance.

The **molarity** (**M**) of a solution is the amount of solute, in moles, per liter of solution.

The **molar mass** of a substance is the mass of one mole of that substance. It is numerically equal to the atomic mass, molecular mass, or formula mass, and expressed as g/mol.

The **molar volume of a gas** refers to the volume occupied by one mole of gas at a fixed temperature and pressure; it is essentially independent of the identity of the gas. At standard temperature and pressure, the molar volume of an ideal gas is 22.4141 L.

A **mole** (**mol**) is an amount of substance that contains as many elementary units (atoms, molecules, formula units) as there are atoms in exactly 12 g of the isotope carbon-12.

A **molecular compound** has molecules as its smallest characteristic entities, and these molecules determine the properties of the compound.

A **molecular formula** gives the symbol and *exact* number of each kind of atom found in a molecule.

Molecular geometry describes the geometric figure formed when appropriate atomic nuclei in a molecule or polyatomic ion are joined by straight lines. Molecular geometry refers to the geometric shape of a molecule or polyatomic ion.

The **molecularity** of a reaction is the number of molecules (or ions) that come together to form the activated complex.

Molecular mass is the average mass of a molecule of a substance relative to that of a carbon-12 atom; it is the sum of the masses of the atoms represented in the molecular formula.

A **molecular orbital** is a region in a molecule where there is a high electron charge density or a high probability of finding an electron. (*See also* **antibonding molecular orbital** and **bonding molecular orbital**.)

A **molecule** is a group of two or more atoms held together in a definite arrangement by forces called covalent bonds.

The **mole fraction** (x) of a component in a homogeneous mixture (a solution) is the fraction of all the molecules in the mixture contributed by that component.

The **mole percent** of a component in a homogeneous mixture (a solution) is the percentage of all the molecules in the mixture contributed by that component.

Mole ratio. *See* **stoichiometric factor**.

A **monodentate ligand** attaches to the metal center in a complex through one pair of electrons on a donor atom.

Monomers are small molecules that are capable of independent existence, but which under appropriate conditions can join together to form a giant molecule called a polymer. (*See also* **polymerization**.)

A **monoprotic acid** has one ionizable hydrogen atom per molecule.

A **monosaccharide** is a carbohydrate that cannot be hydrolyzed into simpler sugars.

A **multiple bond** is a covalent linkage in which two atoms share either two pairs (double bond) or three pairs (triple bond) of electrons between them.

Nanomaterials are materials possessing unique and desirable properties when the length scale of the sample is reduced to a nanometer scale.

The **Nernst equation** relates a cell voltage under nonstandard conditions, E_{cell}, to the standard cell potential, E°_{cell}, and the concentrations of reactants and products of a redox reaction. Its form at 25 °C is $E_{cell} = E^{\circ}_{cell} - (0.0592/n) \log Q$, where n is the number of moles of electrons transferred in the oxidation and reduction half-reactions of a redox reaction and Q is the reaction quotient.

A **net ionic equation** is an equation that represents the actual molecules or ions that participate in a chemical reaction, eliminating all nonparticipating species (so-called spectator ions).

In a **network covalent solid**, covalent bonds extend throughout the crystalline solid.

A **neutralization** reaction is one in which an acid and a base react in such a manner that there is neither excess acid nor base in the final solution. The products of the reaction are water and a salt.

The **neutron** is a fundamental particle of matter found in the nuclei of atoms. Neutrons have a mass of 1.0087 u and no electric charge.

A **newton (N)** is the basic unit of force in SI. It is the force required to give a 1-kg mass an acceleration of 1 m/s^2. That is, 1 N = 1 kg m s^{-2}.

The **nitrogen cycle** refers to the totality of activities in which nitrogen atoms are cycled through the environment.

Nitrogen fixation refers to the conversion of atmospheric nitrogen (N_2) into nitrogen compounds. This occurs naturally in the nitrogen cycle and artificially in the synthesis of ammonia. (*See also* **nitrogen cycle**.)

A **noble gas** is an element in group 8A of the periodic table. Noble gases have the valence-shell electron configuration ns^2np^6 (except helium, $1s^2$).

A **nonelectrolyte** is a substance that exists exclusively or almost exclusively in molecular form, whether in the pure state or in solution.

A **nonmetal** is an element that lacks metallic properties. Nonmetals are generally poor conductors of heat and electricity and brittle when in the solid state. Nonmetal atoms generally have larger numbers of valence electrons than do metals and they tend to form anions. Nonmetal atoms are confined to the p-block of the periodic table (plus hydrogen and helium).

In a **nonpolar covalent bond**, there is an equal sharing of the electrons between the bonded atoms. The electrons are no closer to one atom than to the other, so there is no charge separation.

A **nonspontaneous process** will not occur in a system left to itself. It can be made to occur only through intervention from outside the thermodynamic system.

The **normal boiling point** of a liquid is the temperature at which the liquid boils at 1 atm pressure.

The **normal melting point** of a solid is the temperature at which the solid melts at 1 atm pressure.

An **n-type semiconductor** is a semiconductor doped with donor atoms that can lose electrons to the conduction band. Electric current in this type of semiconductor is carried primarily by these donor electrons. (*See also* **doping**.)

Nuclear binding energy is the energy released when the nucleons are bound together into the nucleus of an atom. It is the energy equivalent of the mass lost in creating a nucleus from its individual protons and neutrons.

Nuclear fission is the splitting of a large unstable nucleus into two lighter fragments and two or more neutrons. In this process, mass is converted to an equivalent quantity of energy, which is released.

Nuclear fusion is the joining together, or fusing, of lighter nuclei into a heavier one. In the process, some matter is converted to energy, which is released.

Nucleon is the general term for the nuclear particles protons and neutrons.

A **nucleophile** is a molecule or ion that donates a lone pair of electrons to another molecule, forming a covalent bond.

Nucleotides are the structural units, consisting of a sugar, a phosphate ester group, and a cyclic amine base, that make up deoxyribonucleic acid (DNA) and ribonucleic acid (RNA).

The **nucleus** of an atom is the densely packed, positively charged core of an atom, containing the protons, the neutrons, and most of the atom's mass.

Nuclide is a term used to signify an atomic species having a particular atomic number and mass number, such as $^{12}_{6}C$. (*See also* **isotopes**.)

The **octet rule** states that most covalently bonded atoms represented in a Lewis structure have eight electrons in their outermost (valence) shells. In the formation of ionic compounds, the ions of the main-group elements also tend to follow the octet rule.

Optical isomers are molecules or species that are nonsuperimposable mirror images; they differ only in the way they rotate the plane of polarized light. (*See also* **enantiomers**.)

The **orbital angular momentum quantum number** (l) is the second of three parameters that must be assigned a specific value to achieve a solution of Schrödinger's wave equation for the hydrogen atom: $l = 0, 1, 2, 3, \ldots, n - 1$. The value of l establishes a particular sublevel or subshell within a principal energy level.

An **orbital diagram** is a method of denoting an electron configuration in which parentheses or boxes are used to represent orbitals within subshells and arrows are used to represent electrons in the orbitals.

The **order of a reaction** is determined by the exponents of the concentration terms in the rate law for the reaction: Rate of reaction $= k[A]^m[B]^n \dots$. The order of the reaction with respect to A is m; with respect to B, it is n; and so on. The overall order of the reaction is $m + n + \dots$.

An **ore** is a naturally occurring mineral containing a metal in a form and concentration that makes extraction of the metal feasible.

Osmosis is the net flow of a solvent through a semipermeable membrane, from pure solvent into a solution or from a solution of a lower concentration into one of a higher concentration.

The **osmotic pressure** of a solution is the pressure that must be applied to a solution to prevent the flow of solvent molecules into the solution when the solution and pure solvent are separated by a semipermeable membrane.

The **overvoltage** of an electrode reaction is the excess voltage above that calculated from E° values required to bring about the reaction.

Oxidation is a process in which the oxidation number of an element increases. It is the half-reaction of an oxidation–reduction reaction in which electrons are given up.

The **oxidation number** of an element in a compound is a means of designating the number of electrons that its atoms have lost, gained, or shared in forming that compound.

An **oxidizing agent** (oxidant) is a substance that makes possible the oxidation that occurs in an oxidation–reduction reaction. The oxidizing agent itself is reduced.

The **ozone layer** is a band of the stratosphere, about 20 km thick and centered at an altitude of about 25 to 30 km, that has a much higher concentration of ozone than the rest of the atmosphere.

Paramagnetism is the attraction into an external magnetic field of substances that have unpaired electrons.

Partial pressure. *See* **Dalton's law of partial pressure**.

A **partial pressure equilibrium constant** (K_p) is the numerical value of an equilibrium constant expression in which the partial pressures (usually in atm) of gaseous products and reactants are used.

Particulate matter is an air pollutant consisting of solid and liquid particles of greater than molecular size but small enough to remain suspended in air.

Parts per billion (ppb) expresses the composition of a mixture as the number of parts of one component per billion parts of the mixture as a whole, usually on a mass basis for liquid

solutions and a mole basis for gaseous mixtures.

Parts per million (ppm) expresses the composition of a mixture as the number of parts of one component per million parts of the mixture as a whole, usually on a mass basis for liquid solutions and a mole basis for gaseous mixtures.

Parts per trillion (ppt) expresses the composition of a mixture as the number of parts of one component per trillion parts of the mixture as a whole, usually on a mass basis for liquid solutions and a mole basis for gaseous mixtures.

A **pascal (Pa)** is the basic unit of pressure in SI. It is a pressure of 1 newton per square meter, 1 N m^{-2}.

The **Pauli exclusion principle** states that no two electrons in an atom may have all four quantum numbers alike. Consequences of this principle are that there may be no more than two electrons in an orbital and that the two electrons must have opposing spins.

The **p-block** is the portion of the periodic table in which the np subshell (the p subshell of the outer shell) fills in the aufbau process. The p-block elements are all main-group elements.

A **peptide** is composed of a chain of two or more amino acids joined through peptide (amide) linkages in chains of peptides, polypeptides, and proteins.

The **percent yield** is the ratio of the actual yield to the theoretical yield of a chemical reaction, expressed as a percentage.

The **periodic law** states that certain sets of physical and chemical properties recur at regular intervals (periodically) when the elements are arranged according to increasing atomic number.

A **periodic table** is a tabular arrangement of the elements according to increasing atomic number that places elements having similar properties into the same vertical columns. (Mendeleev's original periodic table was arranged according to atomic weights, not atomic numbers.)

Permanent hardness in water is the condition in which the predominant anions are other than HCO_3^-. (*See also* **hard water**.)

The **pH** is the negative of the logarithm of the hydronium ion concentration in a solution: $pH = -\log[H_3O^+]$.

A **phase change** is a change from one phase to another, as in solid to liquid or liquid to gas.

A **phase diagram** is a pressure–temperature plot indicating the conditions under which a substance exists as a solid phase, a liquid, a gas, or some combination of these in equilibrium.

Photochemical smog is air that is polluted with oxides of nitrogen and unburned hydrocarbons, together with ozone and several other components produced by the action of sunlight.

The **photoelectric effect** refers to the emission of electrons from the surface of certain materials when they absorb light of the appropriate frequency.

A **photon** is a quantum of energy in the form of light. The energy of the photon is given by the expression $E = h\nu$.

A **photovoltaic cell** is a device that uses semiconductors to convert solar energy (light) into electricity.

In a **physical change**, a sample of matter undergoes a change in phase or state or other property that is observable but does not involve a change in composition.

A **physical property** is a characteristic that a sample of matter displays without undergoing a change in composition.

A **pi (π) bond** forms by the overlap in a parallel or side-by-side fashion of p orbitals of the bonded atoms. A double bond consists of one σ and one π bond; a triple bond, of one σ and two π bonds.

Pig iron is crude, high-carbon iron produced by reduction of iron ore in a blast furnace.

pK_a is the negative of the logarithm of the ionization constant of an acid: $pK_a = -\log K_a$.

pK_b is the negative of the logarithm of the ionization constant of a base: $pK_b = -\log K_b$.

pK_w is the negative of the logarithm of the ion product of water: $pK_w = -\log K_w = -\log(1.0 \times 10^{-14}) = 14.00$ (at 25 °C). (*See also* **ion product of water**.)

Planck's constant (h) is the numerical constant relating the energy of a photon of light and its frequency: $E = h\nu$. Its value is $6.62606876 \times 10^{-34} \text{ J s}$.

pOH is the negative of the logarithm of the hydroxide concentration in an aqueous solution: $pOH = -\log[OH^-]$.

In a **polar** bond between two atoms, electrons are drawn closer to the more electronegative atom, creating a separation of charge. One end of the bond has a small negative charge, $\delta-$, and the other end, a small positive charge, $\delta+$.

Polarizability is a measure of the ease with which electron charge density in an atom or molecule is distorted by an external electric field. It measures the ease with which a dipole can be induced in an atom or molecule.

In a **polar molecule**, a small separation of positive ($\delta+$) and negative ($\delta-$) charge exists, caused by electronegativity differences and molecular geometry.

A **polyatomic ion** is a charged particle containing two or more covalently bonded atoms.

A **polydentate** ligand attaches to a metal center in a complex at more than one point.

A **polymer** is a giant molecule formed by the combination of smaller molecules (monomers) in a repeating manner.

Polymerization is a type of reaction in which small repeating units (monomers) combine to form giant molecules (polymers).

Polymorphism is the property of a substance crystallizing in two or more forms, such as sulfur in its rhombic and monoclinic forms.

A **polypeptide** is a polymer of amino acids; it is usually of lower molecular mass than a protein.

A **polyprotic acid** has more than one ionizable H atom per molecule. The ionization of a polyprotic acid occurs in discrete steps.

A **polysaccharide** is a carbohydrate, each molecule of which can be hydrolyzed into many monosaccharide units.

A **positive hole** is a "missing electron" in a semiconductor; the vacancy acts like a positive ion.

A **positron** ($\boldsymbol{\beta^+}$) is a positively charged particle having the same mass as a β^- particle.

The **potential difference**, measured in volts, is the difference in electric potential between two points in an electric circuit, for example, between the electrodes in an electrochemical cell.

Potential energy is energy that is due to position or arrangement. It is the energy associated with forces of attraction and repulsion between objects.

A **precipitate** is an insoluble compound formed by a reaction in solution.

A **precipitation reaction** is a chemical reaction between ions in solution that produces an insoluble solid—a precipitate.

The **precision** of a set of measurements refers to how closely members of a set of measurements agree with one another. It reflects the degree of reproducibility of the measurements.

Pressure (**P**) is a force per unit area—that is, $P = F/A$.

Primary sewage treatment involves treatment of sewage in a holding pond intended to remove some of the sewage solids as sludge by simple sedimentation (settling).

The **primary structure** is the amino acid sequence in a protein or of nucleotides in a nucleic acid.

The **principal quantum number** (**n**) is the first of three quantum numbers that must be assigned a specific numerical value to achieve a solution to Schrödinger's wave equation for the hydrogen atom: $n = 1, 2, 3, \ldots$. Its value designates the principal energy level of an electron in an atom.

A **principal shell** (level) refers to the collection of orbitals having the same principal quantum number.

A **product** is a substance that is produced in a chemical reaction. The formulas of products appear on the right side of a chemical equation.

A **protein** is a high-molecular-mass polymer of amino acids.

A **proton** is a nuclear particle carrying the fundamental unit of positive charge and having a mass of 1.0073 u.

A **proton acceptor** is a Brønsted–Lowry base. (*See also* **base**.)

A **proton donor** is a Brønsted–Lowry acid. (*See also* **acid**.)

Pseudohalogens are certain groupings of atoms, such as CN and OCN, that mimic the characteristics of a halogen atom.

A **p-type semiconductor** is a semiconductor that has been doped with acceptor atoms that extract electrons from chemical bonds in the semiconductor, producing positive holes in the valence band. Electric current in these semiconductors is carried primarily by positive holes. (*See also* **doping**.)

Pyrometallurgy includes metallurgical methods based on high-temperature reactions involving solids.

A **quantum** is the smallest quantity of energy that can be emitted or absorbed in a process, as given by the expression $E = h\nu$.

Quantum (wave) **mechanics** is the mathematical description of atomic structure based on the wave properties of subatomic properties.

Quantum numbers are certain integral values assigned to three parameters in a wave equation to obtain acceptable solutions to the equation.

The **quaternary structure** of a protein is the arrangement of protein subunits in geometric shapes.

A **rad** (radiation absorbed dose) corresponds to the absorption of 1×10^{-2} J of energy per kilogram of matter.

The **radioactive decay law** states that the rate of disintegration of a radioactive isotope, called the decay rate or activity, is directly proportional to the number of atoms present.

A **radioactive decay series** is a sequence of nuclear processes involving α and β emissions by which an initial long-lived radioactive nucleus is eventually converted to a stable nonradioactive nucleus.

A **radioactive tracer** is a radionuclide that can be used to follow a physical or chemical process through the ionizing radiation that it emits.

Radioactivity (radioactive decay) is the spontaneous emission of ionizing radiation by the atomic nuclei of certain isotopes.

Raoult's law states that the addition of a solute lowers the vapor pressure of the solvent and that the fractional lowering of the vapor pressure is equal to the mole fraction of the solute.

The **rate constant** (**k**) of a reaction is a numerical constant that relates the rate of the reaction to the concentrations of the reactants. Rate constants are functions of temperature. (*See also* **rate law**.)

The **rate-determining step** in a reaction mechanism is the step (usually the slowest) that is crucial in establishing the rate of an overall reaction.

The **rate law** (rate equation) of a chemical reaction is an expression relating the rate of the reaction to the concentrations of the reactants.

Rate of a reaction is the increase in concentration of a product per unit of time or the decrease in concentration of a reactant per unit of time, usually expressed as M s^{-1}.

A **reactant** is a starting material or substance consumed in a chemical reaction. The formulas of reactants appear on the left side of a chemical equation.

A **reaction mechanism** is a detailed representation of a chemical reaction consisting of a series of elementary reactions. A plausible mechanism must be consistent with the stoichiometry and the rate law of the net reaction.

A **reaction profile** is a schematic representation of changes in energy during the course of a reaction. The profile shows activation energies and enthalpies of reaction and identifies the energies of reactants, transition state(s), and products.

A **reaction quotient** (Q_c) or (Q_p) has the same format as an equilibrium constant (K) but uses initial concentrations rather than equilibrium concentrations.

A **reactive material** is one that tends to react spontaneously or to react vigorously with air or water.

A **reducing agent** (reductant) is a substance that makes possible the reduction that occurs in an oxidation–reduction reaction. The reducing agent itself is oxidized.

Reduction is a process in which the oxidation number of an element decreases. It is the half-reaction of an oxidation–reduction reaction in which electrons are "gained."

Refining is the process of removing impurities from a metal by any of a variety of chemical or physical means.

A **rem** (roentgen equivalent for man) is a *rad* multiplied by a factor that takes into account the fact that different types of radiation of the same energy have different effects on people.

Resonance is a term used to describe a situation in which several plausible Lewis structures can be written to represent a species but in which the true structure cannot be written. The plausible structures are called contributing structures or resonance structures; the true structure, which is a composite of the contributing structures, is called the resonance hybrid.

A **resonance hybrid** is a composite of two or more plausible contributing Lewis structures. The resonance hybrid represents the true structure of a molecule or ion.

A **resonance structure** is one of two or more plausible Lewis structures that can be written to represent a molecule or ion.

The **resultant dipole moment** is the dipole moment of a molecule as a whole based on an assessment of bond moments and the molecular geometry. (*See also* **bond moment** and **dipole moment**.)

Reverse osmosis refers to the net flow of solvent through a semipermeable membrane in the opposite direction from that expected for osmosis. It is produced by applying pressure to a solution in excess of its osmotic pressure. (*See also* **osmosis**.)

Ribonucleic acid (RNA) is a polymer of nucleotides. The nucleotide consists of the sugar ribose, a phosphate ester, and a cyclic amine base (adenine, guanine, uracil, or cytosine).

The **root-mean-square speed** (u_{rms}) of the molecules of a gas is the square root of the average of the squares of the molecular speeds.

A **salt** is an ionic compound in which hydrogen atoms of an acid are replaced by metal ions. Salts are produced in the reaction of an acid and a base.

A **salt bridge** is a salt solution used to connect the two solutions in a voltaic cell. It permits ions to migrate without mixing of the solutions.

Saponification is the alkaline hydrolysis of a fat or other ester.

A **saturated hydrocarbon** has molecules that contain the maximum number of hydrogen atoms for the carbon atoms present. All bonds in the molecules are single covalent bonds.

A **saturated solution** is one in which dynamic equilibrium exists between undissolved solute and the solution. The solution contains the maximum amount of solute that can be dissolved in a particular quantity of solvent at the given temperature.

The ***s*-block** is the portion of the periodic table in which the ns subshell (the s subshell of the outer shell) fills in the aufbau process.

A **scientific law** is a brief statement, sometimes in mathematical terms, used to summarize and describe patterns in large collections of scientific data.

The **second (s)** is the SI base unit of time.

Secondary sewage treatment consists of passing the effluent from a primary treatment

plant through gravel and sand filters to aerate the water and remove finer suspended solids.

The **secondary structure** of a protein is the arrangement of the protein chains with respect to the nearest neighbor amino acid units, for example, a helix or a pleated sheet.

One statement of the **second law of thermodynamics** is that all natural or spontaneous processes are accompanied by an increase in entropy of the universe. That is, $\Delta S_{univ} = \Delta S_{system} + \Delta S_{surroundings} > 0$.

A **second-order reaction** has a rate equation in which the sum of the exponents $m + n + \cdots = 2$.

A **semiconductor** is a substance in which there is only a small energy gap between the valence and conduction band. The electrical conductivity of a semiconductor is not nearly as good as that of a metal, but still much better than that of an insulator.

A **semipermeable membrane** is a material that permits the flow of solvent molecules but severely restricts the flow of solute molecules of a solution.

A **sigma (σ) bond** results from the end-to-end overlap of pure or hybridized atomic orbitals between the bonded atoms. A σ bond exists along a line joining the nuclei of the bonded atoms.

The **significant figures** in a measured quantity are all the digits known with certainty plus the first uncertain digit.

A **simple cubic cell** has an atom, molecule, or ion at each corner of a cube.

A **single bond** is a covalent linkage in which two atoms share one pair of electrons between them.

The **skeletal structure** of a polyatomic species (molecule, ion) indicates the order in which atoms are attached to one another.

Slag is a metallurgical term for a relatively low-melting-point product of the reaction of an acidic oxide and a basic oxide.

Soaps are salts of long-chain carboxylic acids called fatty acids (because they are derived from fats).

The **solubility** of a solute in a particular solvent refers to the concentration of the solute in a saturated solution.

The **solubility product constant** (K_{sp}) describes the equilibrium that exists between a slightly soluble ionic solute and its ions in a saturated aqueous solution.

Solubility rules are a set of generalizations used to classify substances as soluble or insoluble in water.

A **solute** is a solution component that is dissolved in a solvent. A solution may have

several solutes, which are generally present in lesser amounts than is the solvent.

A **solution** is a homogeneous mixture of two or more substances. The composition and properties are uniform throughout a solution.

A **solvent** is the solution component (usually present in greatest amount) in which one or more solutes are dissolved to form the solution.

sp **hybridization** describes a scheme in the valence bond method in which one s and one p orbital are combined into two sp hybrid orbitals oriented in a linear fashion.

spdf **notation** uses numbers to designate a principal shell and the letters s, p, d, and f to identify a subshell.

sp^2 **hybridization** describes a scheme in the valence bond method in which one s and two p orbitals are combined into three sp^2 hybrid orbitals oriented in a trigonal planar fashion.

sp^3 **hybridization** describes a scheme in the valence bond method in which one s and three p orbitals are combined into four sp^3 hybrid orbitals oriented in a tetrahedral fashion.

sp^3d **hybridization** describes a scheme in the valence bond method in which one s, three p, and one d orbital are combined into five sp^3d hybrid orbitals oriented in a trigonal bipyramidal fashion.

sp^3d^2 **hybridization** describes a scheme in the valence bond method in which one s, three p, and two d orbitals are combined into six sp^3d^2 hybrid orbitals oriented in an octahedral fashion.

The **specific heat** of a substance is the quantity of heat required to raise the temperature of 1 gram of substance by 1 °C (or 1 K).

Spectator ions do not participate in an acid–base, precipitation, or redox reaction, but serve only to maintain electrical neutrality. (*See also* **net ionic equation**.)

The **spectrochemical series** is a listing of ligands in order of their abilities to produce a splitting of a d-subshell energy level in a complex. Ligands in the series are referred to as strong field, intermediate field, or weak field, depending on the degree of splitting they produce. (*See also* **crystal field theory**.)

Spectroscopy is a study, using various instrumental methods, of atomic and molecular properties through the absorption or emission of electromagnetic radiation by a substance.

A **spontaneous process** is one that occurs in a system left to itself. Once started, no action from outside the system is required to keep the process going.

A **standard cell potential** (E°_{cell}) (standard cell voltage) is the potential difference (in volts) between the electrodes in a voltaic cell when all species are present in their standard states.

The **standard electrode potential** ($E°$) measures the tendency of a reduction process to occur when all species in a half-cell are present in their standard states. This tendency is measured in volts, relative to an assigned value of zero for the standard hydrogen electrode.

The **standard enthalpy of formation** (ΔH_f°) of a substance is the enthalpy change that occurs in the formation of 1 mol of the substance in its standard state from the reference forms of its elements in their standard states. The reference forms of the elements are generally their most stable forms at the given temperature and 1 atm pressure.

The **standard enthalpy of reaction** ($\Delta H°$) is the enthalpy change for a reaction in which all reactants and products are in their standard states.

The **standard free energy change** ($\Delta G°$) is the free energy change of a process in which the reactants and products are all in their standard states.

The **standard free energy of formation** (ΔG_f°) of a substance is the free energy change that occurs in the formation of one mole of the substance in its standard state from the reference forms of its elements in their standard states. The reference forms of the elements are generally their most stable forms at the given temperature and 1 atm pressure.

A **standard hydrogen electrode (SHE)** has hydrogen gas at 1 atm pressure and hydronium ion at unit activity (about 1 M) in oxidation–reduction equilibrium on an inert platinum electrode. The potential arbitrarily assigned to this electrode is exactly 0 V.

The **standard molar entropy** ($S°$) of a substance is its entropy at standard pressure and a specified temperature.

The **standard state** of a solid or liquid substance is the pure element or compound at 1 atm pressure and at the temperature of interest. For a gaseous substance, the standard state is the (hypothetical) pure gas behaving as an ideal gas at 1 atm pressure and the temperature of interest.

Standard temperature and pressure (STP) for a gas are 273.15 K (0 °C) and 1 atm (760 mmHg).

A **state** of a system is its exact condition, determined by the kinds and amounts of matter present; the structure of this matter at the molecular level; and the prevailing temperature and pressure.

A **state function** is a property that has a unique value that depends only on the present state of a system and not on how that state was reached.

The **states of matter** are the three fundamental conditions in which samples of matter may be obtained: solid, liquid, and gas.

Steel is an alloy of iron containing small amounts of carbon and usually containing other metals such as manganese, nickel, and chromium.

Stereoisomers are isomers that have the same molecular formulas but differ in the arrangement of atoms in three-dimensional space.

A **stoichiometric coefficient** is a coefficient placed in front of a formula in a chemical equation to balance the equation.

A **stoichiometric factor** or mole ratio is a conversion factor relating molar amounts of two species involved in a chemical reaction (that is, a reactant to a product, one reactant to another, and so on). The numerical values used in formulating the factor are the stoichiometric coefficients.

Stoichiometric proportions refer to relative amounts of reactants that are in the mole ratios corresponding to the coefficients in a balanced equation; no reactants are in excess.

Stoichiometry refers to quantitative measurements and relationships involving substances and mixtures of chemical interest.

A **strong acid** is an acid that is essentially completely ionized in solution, that is, an acid that is a strong electrolyte. (*See also* **acid.**)

A **strong base** is a base that is essentially completely ionized in solution, that is, a base that is a strong electrolyte. (*See also* **base.**)

A **strong electrolyte** is a substance that exists almost exclusively in ionic form in solution.

A **structural formula** is a chemical formula that shows how the atoms in a molecule are attached to one another.

Sublimation is the direct passage of molecules from the solid state to the vapor state.

A **sublimation curve** is a graph of the vapor pressure (sublimation pressure) of a solid as a function of temperature. It is analogous to the vapor pressure curve of a liquid.

A **subshell** (sublevel) is the collection of orbitals of a given type (specified by n and l) present in a principal shell. For example, the three $2p$ orbitals constitute the $2p$ subshell.

Subshell (sublevel) notation is a method of denoting an electron configuration that uses numbers to represent the principal shells and the letters s, p, d, and f for subshells. A superscript number following the letter indicates the number of electrons in the subshell.

A **substance** is a type of matter having a definite, or fixed, composition and fixed properties that do not vary from one sample to another. All substances are either elements or compounds.

In a **substitution reaction**, a substituent group replaces a hydrogen atom in a hydrocarbon molecule. This type of reaction is characteristic of alkanes and aromatic hydrocarbons.

The **substrate** in an enzyme-catalyzed reaction is the reactant species that attaches to the active site on an enzyme molecule and undergoes chemical reaction.

A **superconductor** is a metal, alloy, or ceramic material whose resistance to the flow of electricity vanishes at a sufficiently low temperature.

Supercooling is a condition in which a liquid is cooled below its freezing point without the appearance of any solid.

A **supercritical fluid** is a fluid at a temperature above its critical temperature and at a pressure above its critical pressure.

A **supersaturated** solution contains more solute than is present in a saturated solution in equilibrium with undissolved solute.

Surface tension is the amount of work required to extend a liquid surface, usually expressed in joules per square meter, $J\,m^{-2}$.

The **surroundings** refer to that part of the universe with which a system interacts by exchanging heat and/or work and/or matter.

A **system** is that part of the universe chosen for a thermochemical or thermodynamic study. (*See also* **surroundings.**)

Temperature is a physical property, related to the mean kinetic energies of the atoms or molecules in a substance, that indicates the direction of heat flow. Kinetic energy, as heat, is transferred from more energetic (higher temperature) to less energetic (lower temperature) atoms or molecules.

Temporary hardness is present in hard water that has HCO_3^- as its primary anion. (*See also* **hard water.**)

A **terminal atom** in a polyatomic species (molecule, ion) is bonded to just one other atom.

A **termolecular reaction** in a reaction mechanism involves the simultaneous collision of three molecules.

The **tertiary structure** refers to the folds, bends, and twists in a protein or nucleic acid structure.

The **theoretical yield** is the calculated quantity of a product expected in a chemical reaction.

A **theory** provides explanations of observed phenomena and predictions that can be tested by experimentation. It is the intellectual framework for explaining scientific data and scientific laws.

Thermochemistry is the study of energy changes associated with chemical reactions or physical processes, especially energy changes that appear as heat.

Thermodynamics is the science dealing with the relationship between heat and motion (work) and with transformations of energy from one form to another. Thermochemistry is a subfield within thermodynamics.

A **thermoplastic polymer** is one that can be softened by heating and formed into desired shapes by applying pressure.

A **thermosetting polymer** becomes permanently hard at elevated temperatures and pressures.

The **third law of thermodynamics** states that the entropy of a pure, perfect crystal at 0 K is zero. This is the starting point for the experimental determination of absolute molar entropies.

The **titrant** in a titration is the solution, usually of accurately known concentration, that is added to the analyte through a buret.

Titration is a laboratory procedure in which the amount or concentration of one reactant is found by adding a known solution of a second reactant in stoichiometric proportions.

A **titration curve** in a neutralization reaction is a graph of pH versus volume of titrant added from a buret.

A **torr** is a unit used to express gas pressure: 1 Torr = 1 mmHg. (*See also* **millimeter of mercury**.)

A **toxic material** is one that contains or releases poisonous substances in amounts large enough to threaten human health or the environment.

Toxicology is a study of the effects of poisons on the body, their identification and detection, and remedies against them.

Trans isomers have two groups attached to opposite sides of a double bond in an organic molecule, at opposite corners of a square in a square planar complex, or above and below the central plane in an octahedral complex.

Transcription is the process by which DNA directs the synthesis of an mRNA molecule during protein synthesis.

A **transition element** is one in which the subshell being filled in the aufbau process is in a principal shell of less than the highest quantum number (an inner shell). Transition elements are located in the *d*- and *f*-blocks of the periodic table.

A **transition state** is a state that lies between the reactants and products of a chemical reaction. It is produced as a result of collisions between especially energetic molecules.

Translation is the process by which the information contained in a base triplet of an mRNA molecule is converted to a protein structure.

The **transuranium elements** are those with atomic number (*Z*) greater than 92.

A **triglyceride** is an ester formed by the chemical combination of glycerol with three fatty acids. It is also called a *triacylglycerol*.

A **triple bond** is a covalent linkage in which two atoms share three pairs of electrons between them.

A **triple point** is a particular temperature and pressure at which three phases of a pure substance are at equilibrium—solid, liquid, and vapor; or two solid phases and the liquid; or two solid phases and the vapor.

Trouton's rule states that the entropy of vaporization of a nonpolar liquid at its normal boiling point is approximately $87 \text{ J K}^{-1} \text{ mol}^{-1}$.

The **Tyndall effect** is the scattering of light by colloidal particles, which makes a colloidal dispersion distinguishable from a true solution.

Heisenberg's **uncertainty principle** states that the product of the uncertainty in the position of an object and the uncertainty in its momentum (mass, *m*, times speed, *u*) cannot be less than $h/4\pi$. Thus, it is not possible to know with certainty both the position of a subatomic particle and details of its motion.

A **unimolecular reaction** in a reaction mechanism is one in which a single molecule undergoes rearrangement or decomposition.

The **unit cell** of a crystal structure is the simplest parallelepiped that can be used to generate the entire crystalline lattice through straight-line displacements in all three dimensions.

In the **unit-conversion method**, a given quantity is multiplied by a conversion factor to change the unit of the quantity to a different, desired unit.

The **universal gas constant** (*R*) is the numerical constant required to relate pressure, volume, amount, and temperature of a gas in the ideal gas equation, $PV = nRT$. Its numerical value is $0.082057 \text{ L atm mol}^{-1} \text{ K}^{-1}$ or $8.3145 \text{ J mol}^{-1} \text{ K}^{-1}$.

An **unsaturated hydrocarbon** is a carbon–hydrogen compound having one or more multiple bonds (double, triple) between carbon atoms.

An **unsaturated** solution contains less of a solute in a given quantity of solution than is present in a saturated solution. It is a solution having a concentration less than the solubility limit.

A **valence band** is formed by combining atomic orbitals of the valence electrons of a large number of atoms into a set of molecular orbitals very closely spaced in energy. If the band is only partially filled with electrons, it is also a conduction band. (*See also* **band** and **conduction band**.)

The **valence bond theory** describes a covalent bond as a region of high electron charge density that results from the overlap of atomic orbitals between the bonded atoms.

Valence electrons are electrons with the highest principal quantum number. They are found in the outermost electronic shells of atoms. (*See also* **core electrons**.)

The **valence shell** is the outer shell of electrons of an atom, the electrons with the highest principal quantum number.

The **valence-shell electron-pair repulsion (VSEPR) method** is an approach to describing the geometric shapes of molecules and polyatomic ions in terms of the geometrical distribution of electron groups in the valence shell(s) of central atom(s).

van der Waals forces are short-range attractive forces between molecules that include dispersion forces, dipole-dipole forces, and dipole-induced dipole forces.

The **van't Hoff factor** (*i*) is a correction factor that must be incorporated into equations for colligative properties so that the equations may be applied to solutions of strong or weak electrolytes.

A **vapor** is a gas at a temperature below its critical temperature. A vapor can be liquefied by application of pressure without lowering the temperature.

Vaporization, or evaporation, is the process of conversion of a liquid to a gas (vapor).

The **vapor pressure** of a liquid is the pressure exerted by the vapor in dynamic equilibrium with the liquid at a constant temperature.

A **vapor pressure curve** is a graph of the vapor pressure of a liquid as a function of temperature. The curve is the boundary between the liquid and vapor areas in a phase diagram.

The **viscosity** of a liquid substance is a measure of its resistance to flow.

Volt (**V**) is the unit used to measure electrode potentials and electrical potential differences.

A **voltaic cell** is an electrochemical cell that produces electricity through an oxidation–reduction reaction.

A **wave** is a progressive, repeating disturbance propagated from a point of origin to a more distant point.

The **wavelength** is the distance between any two identical points in consecutive cycles of a wave, for example, the distance between the peaks or crests of the wave.

The **wavenumber** ($\tilde{\nu}$) of radiation expresses the frequency of radiation as the number of cycles per centimeter of the wave.

A **weak acid** is an acid that exists partly in ionic form and partly in molecular form in solution, that is, an acid that is a weak electrolyte. (*See also* **acid**.)

A **weak base** is a base that exists partly in ionic form and partly in molecular form in solution, that is, a base that is a weak electrolyte. (*See also* **base**.)

A **weak electrolyte** is a substance that is present partly in molecular form and partly in ionic form in its solutions.

Work is (1) the result of a force acting through a distance, for example, $1 \text{ J} = 1 \text{ N} \times 1 \text{ m}$, or (2) an energy transfer into or out of a thermodynamic system that can be expressed as the product of a force and a distance.

An **X ray** is high-energy electromagnetic radiation produced by the impact of cathode rays (electrons) on a solid, such as on a dense metal anode (a target) in a cathode-ray tube.

A **zero-order reaction** has a rate that is independent of the concentration of reactant(s). The sum of the exponents in its rate equation, $m + n + \cdots = 0$.

Zone refining is a process of purification in which a rod of material is subjected to repeated cycles of melting and freezing. This sweeps impurities in a molten zone to the end of the rod, which is then cut off.

Answers to Selected Problems

These answers are for in-text Exercises and selected end-of-chapter problems. *Note:* Some of your answers may differ slightly from those given here, depending on the number of digits used in expressing atomic masses and other precisely known quantities and whether intermediate results were rounded off.

Chapter 1

Exercises: 1.1A (a) 2.05 μm; **(b)** 4.03 kg; **(c)** 7.06 ns; **(d)** 5.15 cm. **1.1B (a)** 6.217×10^3 g; **(b)** 1.6×10^{-3} s; **(c)** 7.17×10^{-2} g; **(d)** 3.87×10^2 m. **1.2A (a)** 3.55×10^{-4} s; **(b)** 1.885×10^6 m; **(c)** 1.350×10^1 m; **(d)** 4.25×10^{-7} m. **1.2B (a)** 2.28×10^2 kg; **(b)** 8.3×10^{-4} m; **(c)** 4.05×10^{-4} m; **(d)** 1.215×10^3 s. **1.3A (a)** 185 °F; **(b)** 10.0 °F; **(c)** 168 °C; **(d)** −29.3 °C. **1.3B** −459.67 °F. **1.4A** 8.9×10^{-3} m^3. **1.4B** 8.9×10^3 cm^3. **1.5A** 925 g zinc. **1.5B** 5.1 cm^2. **1.6A (a)** 100.5 m; **(b)** 1.50×10^2 g; **(c)** 415 g; **(d)** 6.3 L. **1.6B (a)** 1.80×10^3 m^2; **(b)** 2.33 g/mL; **(c)** 72 kg/m^3; **(d)** 0.63 g/cm^3. No differences between rounding of intermediate results and a final rounding only. **1.7A (a)** 7.63×10^{-2} m; **(b)** 8.56×10^4 mg; **(c)** 1.82×10^3 ft. **1.7B (a)** 5.53×10^4 μg; **(b)** 448 fl oz; **(c)** 23.6 km. **1.8A (a)** 73.8 in.^{2}; **(b)** 0.256 m^3. **1.8B** 1.00×10^3 g/L. **1.9A** 1.034×10^4 kg/m^2. **1.9B** 177 lb. **1.10A** 7.80 g/cm^3. **1.10B** 7.40 g/cm^3. **1.11A** 3.34 gal. **1.11B** 17.2 kg. **1.12A** 20 kg. **1.12B (a)**. **1.13A** Although the mass is the same at 21 °C and 25 °C, volume varies with temperature, and so does density. We are given the density at 25 °C and not 21 °C. **1.13B** 0.69 g/cm^3. **1.14A** Remove the ebony wood and replace it by the block of plastic floating about $\frac{3}{4}$ submerged in the chloroform. **1.14B** The observations: **(a)** 75.0 g; **(b)** plastic rests at the bottom of the beaker filled with ethyl alcohol; **(c)** plastic is about 80% submerged in chloroform; **(d)** plastic weighs 13.5 g when submerged in water. **Self-Assessment Questions: 1.** (d), (f). **8.** (c). **9.** (c). **10.** (a), (c), (e) are elements; the rest are compounds, each made of two elements. **11.** (a). **16.** (b). **20.** (d). **22.** (b). **Problems: 23.** (a), (b), and (d). **25.** (a), (b), (d) are physical, (c) is chemical. **27.** (a) is a substance; (b), (c), (d) are mixtures. **29.** Both mass and weight have changed; body fat has been removed, which has mass (and weight). **31.** Yes, if all the results have the same amount and direction of error. Yes, if the positive and negative errors cancel. **33.** Mass is probably more consistent. Volume varies with particle size, sifting, settling, type of flour, etc. **35. (a)** 8.01 μg; **(b)** 7.9 mL; **(c)** 1.05 km. **37. (a)** 209 °F; **(b)** 576.1 °C; **(c)** −62 °F. **39.** Yes. **41. (a)** 0.0374 L; **(b)** 1.55×10^5 m; **(c)** 0.198 g; **(d)** 1.19×10^4 cm^2; **(e)** 0.078 ms; **(f)** 24.4 m/s. **43. (a)** 6.68 kg; **(b)** $3.8 \times 8.9 \times 235$ cm; **(c)** 0.26 m^3. **45.** (4) < (3) < (1) < (2). **47.** 7.92 in. **49. (a)** 3; **(b)** 3; **(c)** 4; **(d)** 4; **(e)** 3; **(f)** 3. **51. (a)** 2.804×10^3 m; **(b)** 9.01×10^2 s; **(c)** 9.0×10^{-4} cm; **(d)** 2.210×10^2 s. **53. (a)** 505.5 m; **(b)** 2120 s, 3 SF; **(c)** 0.00610, 3 SF; **(d)** 40,000 L, 3 SF (last two zeroes are not significant). **55. (a)** 45.8 cm; **(b)** 167 cm; **(c)** 44.5 g; **(d)** 10.1 L. **57. (a)** 2.32×10^3 mm^3; **(b)** 4.80×10^3 cm^2/g; **(c)** 4.6×10^4 mm^2/mg; **(d)** 1.92×10^{-4} g/mL. **59.** 1.14 g/mL.

61. 5.23 g/cm^3. **63.** 5.4×10^2 g. **65.** 6.73×10^{-6} cm. **67.** 0.0042 g/cm^3. **69.** (b). **71.** 7.1 g/cm^3. **Additional Problems: 74.** Yes, both are linear relationships, and the lines cross at −40 °C **76.** 0.02 m^2. **79.** 116.18 mph. **83.** 2.36×10^5 nails/roll. **86.** 4.9 mg/m^2. **88.** The second (cork). The cork floats and will displace its mass, 8.8 g or 8.8 mL, of water. The brass sinks and will displace its volume, 8.0 mL. **92. (a)** 4.8×10^{-5} cm; **(b)** 0.60 g/cm^3. **93. (a)** Volume is about $30 \times 13 \times 11$ or 4300 cm^3, so $d > 1$ g/cm^3 and the object will not float; **(b)** new $d = 1.16$ g/cm^3; **(c)** 3 more holes. **96.** 66.8 mL chloroform and 33.2 mL bromoform **98.** Floating ice displaces its mass of water, not its volume. No effect on the level. **Apply Your Knowledge: 101.** 10 mg/10 L = 10^2 μg/dL, not 10 μg/dL. **102. (a)** 0.8790 g/mL; **(b)** 7.135 g/cm^3.

Chapter 2

Exercises: 2.1A 45.07 g. The mass of magnesium oxide (the product) is equal to the sum of the masses magnesium and oxygen consumed. The masses of material not involved in the reaction are unchanged. **2.1B** The principal products of the combustion—carbon dioxide gas and water vapor—escape into the air. The mass of the match residue will be less than the original mass. **2.2A** 3.779 g magnesium oxide. **2.2B** 2.361 g magnesium burned; 3.915 g magnesium oxide formed. **2.3A** $^{116}_{50}$Sn. **2.3B** $^{116}_{48}$Cd; 48 p, 48 e, 68 n. **2.4A** 20.18 u. **2.4B** 69.16% copper-63; 30.84% copper-65. **2.5A** Magnesium-24 is most abundant; its mass is closest to 24.305 u. To determine the second-most abundant, at least one of the actual percent abundances would have to be known. **2.5B (a)** No. **(b)** Assuming only a trace of magnesium-26, the weighted-average atomic mass of ^{24}Mg and ^{25}Mg would be 24.305 u if the percent ^{24}Mg were 68.03%. If the amount of ^{26}Mg is more than a trace, the percent ^{24}Mg would have to be larger. **2.6A** N$_2$F$_4$; dinitrogen tetrafluoride. **2.6B** S$_8$O; octasulfur monoxide. **2.7A (a)** P$_4$O$_{10}$; **(b)** heptasulfur dioxide. **2.7B** SO$_2$F$_2$ is a plausible formula based on Figure 2.9 and the nomenclature rules for binary molecular compounds, but this is a ternary molecular compound. **2.8A (a)** K$_2$S; **(b)** Li$_2$O; **(c)** AlF$_3$. **2.8B (a)** Cr$_2$O$_3$; **(b)** FeS; **(c)** Li$_3$N. **2.9A (a)** calcium bromide; **(b)** lithium sulfide; **(c)** iron(II) bromide; **(d)** copper(I) iodide. **2.9B (a)** Cu$_2$S; **(b)** Co$_2$O$_3$; **(c)** Mg$_3$N$_2$. **2.10A (a)** (NH$_4$)$_2$CO$_3$; **(b)** Ca(ClO)$_2$; **(c)** Cr$_2$(SO$_4$)$_3$. **2.10B (a)** KAl(SO$_4$)$_2$; **(b)** MgNH$_4$PO$_4$. **2.11A (a)** potassium hydrogen carbonate; **(b)** iron(III) phosphate; **(c)** magnesium dihydrogen phosphate. **2.11B (a)** sodium selenate; **(b)** iron(III) arsenide; **(c)** disodium hydrogen phosphite. **Self-Assessment Questions: 1.** (c). **3.** No, Cl atoms must form something; they are not in water or carbon dioxide. **7.** (b) are isotopes;

(a) are isobars. **8.** (a) ^8_5B; (b) $^{14}_6C$; (c) $^{235}_{92}U$; (d) $^{60}_{27}Co$. **12.** (a), (d), (f). **19.** molecular formula (a), (b), (d); structural formula (c) and (e). **Problems: 21.** Yes; mass of product & unreacted reactant in excess equals sum of reactant masses. **23.** (a) 104.3 g; (b) 40.3 g MgO; (c) conservation of mass and definite proportions; (d) 80.6 g MgO. **25.** Yes. **27.** 2.61 g. **29.** (a) and (c). **31.** (a) 30p, 34n, 30e; (b) 47p, 62n, 47e; (c) 6p, 8n, 6e; (d) 95p, 148n, 95e. **33.** #2, $^{40}_{20}Ca$; #6, $^{48}_{22}Ti$; #7, $^{48}_{20}Ca$; #2 and #7 are isotopes. **35.** Bromine must consist of isotopes of significant abundance both above and below mass number 80 (in nature, almost equal abundances of isotopes 79 and 81). **37.** 69.7 u. **39.** 28.09 u. **41.** 72.163% Rb-85, 27.387% Rb-87. **43.** group, period, class are: (a) 3A, 4, metal; (b) 5A, 3, nonmetal; (c) 7A, 5, nonmetal; (d) 2A, 7, metal; (e) 1A, 2, metal; (f) 3B, 6, metal; (g) 8A, 5, nonmetal. **45.** (a) N_2; (b) H_2; (c) I_2; (d) P_4. **47.** (a), (c), and (d); (b) is ionic, (e) contains three elements. **49.** (a) $SiCl_4$; (b) SF_6; (c) CS_2; (d) Cl_2O_3; (e) dinitrogen monoxide; (f) diiodine pentoxide; (g) phosphorus pentachloride; (h) diphosphorus trioxide. **51.** (a) Mg^{2+}; (b) Br^-; (c) Ti^{4+}; (d) iodide ion; (e) sulfide ion; (f) chromium(II) ion. **53.** (a) ammonium ion; (b) hypochlorite ion; (c) phosphate ion; (d) hydrogen sulfate ion; (e) OH^-; (f) SO_3^{2-}; (g) CH_3COO^-; (h) CO_3^{2-}. **55.** (a) potassium iodide; (b) magnesium fluoride; (c) nickel(II) sulfate; (d) titanium(IV) bromide; (e) ammonium hydrogen sulfate; (f) aluminum oxide; (g) barium chloride dihydrate; (h) lithium oxalate; (i) sodium sulfate decahydrate. **57.** (a) KCl; (b) $CaCO_3$; (c) Cr_2O_3 (d) $KClO_4$; (e) $NaClO_3$; (f) $FeSO_4 \cdot 7\,H_2O$. **59.** (a) Na_2O; (b) CuOH; (c) BaI_2; (d) $Al_2(SO_4)_3$; (e) sodium bromate; (f) lithium hydrogen sulfate; (g) ammonium dichromate; (h) nickel(II) oxalate. **61.** (a) Cl is not chlorate, NH_4ClO_3; (b) nitrate has three O atoms, KNO_3; (c) SO_4^{2-} needs two sodium ions, Na_2SO_4; (d) Ba^{2+} needs two OH^- ions; $Ba(OH)_2$; (e) O is oxide, ZnC_2O_4; (f) IV signifies 4+ charge, MnO_2; (g) Sr and CrO_4 are both divalent, $SrCrO_4$; (h) II signifies 2+ charge, $Cu_3(PO_4)_2$. **63.** (a) hydrobromic acid; (b) nitrous acid; (c) phosphorous acid; (d) $HClO_2$; (e) KOH; (f) H_2SO_3; (g) barium hydroxide; (h) HI. **65.** IO_3^-, HIO_3, iodic acid.

67. (a) H—C—C—C—C—C—H, $CH_3(CH_2)_3CH_3$;

(b) H—C—C—C—C—O—H, $CH_3(CH_2)_2COOH$;

(c) H—O—C—C—C—H
$HOCH_2CH_2CH_3$

(d) H—C—C—H, CH_2—CH_2
H—C—C—H CH_2—CH_2;

(e) H—C—C—C—H, $CH(CH_3)_3$;

(f) H—C—C—C—O—H, CH_3CH_2COOH.

69. (a) H—C—C—C—C—O—H   H—C—C—C—O—H

(b) $CH_3(CH_2)_2COOH$; $(CH_3)_2CHCOOH$

(c)

(d) isomers **71.** (a) straight chain alkane; (b) alcohol; (c) cyclic alkane; (d) hydrocarbon; (e) carboxylic acid; (f) inorganic; (g) carboxylic acid; (h) inorganic. **73.** (b). **75.** (a) dimethyl ether, methanol; (b) ethyl methyl ether, ethanol and methanol; (c) diethyl ether, ethanol. **77.** $CH_3CH_2OCH_2CH_3$, $CH_3CH_2CH_2OCH_3$, $(CH_3)_2CHOCH_3$.

79. (a)   (b)   (c)

Additional Problems: 82. (c). **86.** H_2, HD, D_2 (least abundant); 10. **89.** No, although the %C and %H are constant, we do not know about %N and %O, only that their sum is constant. **92.** The composition of N_2O_5 is different from those of the other three oxides, while N_2O_4 has the same composition as NO_2. **94.** $^{25}Mg = 10.00\%$; $^{26}Mg = 11.01\%$. **98.** (a) $C_nH_{2n+1}OH$; (b) $C_nH_{2n}O_2$; (c) $C_nH_{2n}O_2$. **103.** empirical formula C_2H_6N; molecular formula $C_4H_{12}N_2$. **105.** (a) $CH_3(CH_2)_{10}COONH_4$; (b) $Ca[CH_3(CH_2)_{16}COO]_2$; (c) $CH_3(CH_2)_7CH{=}CH(CH_2)_7COOK$.
107. (a) methyl acetate; methyl alcohol and acetic acid; (b) propyl acetate; propyl alcohol and acetic acid; (c) isopropyl formate; isopropyl alcohol and formic acid.
110. $CH_3(CH_2)_4CH_3$, hexane; $(CH_3)_2CH(CH_2)_2CH_3$, 2-methylpentane; $CH_3CH_2CH(CH_3)CH_2CH_3$, 3-methylpentane; $(CH_3)_3CCH_2CH_3$, 2,2-dimethylbutane; $(CH_3)_2CHCH(CH_3)_2$, 2,3-dimethylbutane. **111.** (a) $^{40}_{20}Ca$; (b) $^{234}_{90}Th$; (c) $^{120}_{48}Cd$; (d) $^{64}_{30}Zn^{2+}$. **114.** (a) propyl bromide; (b) isobutyl chloride; (c) ethyl iodide; (d) t-butyl fluoride **Apply Your Knowledge: 116.** (a) Physicist's scale, the mixture with heavier isotopes is heavier than O-16 alone, so physicist ratios are slightly larger; (b) Atomic mass of C-12 is divisible by 4, and the new atomic mass of O is 15.9994 which is a change of 0.0006/16 = 4/100,000. **120.** Atomic masses are (mostly) averages of several isotopes rather than single isotopes. Fractional masses can arise

from integral masses and different abundances; i.e., Cl = 35.5 can arise from three Cl-37 and one Cl-35.

Chapter 3

Exercises: 3.1A (a) 124 u, (b) 92.0 u, (c) 98.1 u, (d) 74.1 u. **3.1B** (a) 208.24 u, (b) 108.01 u, (c) 88.106 u, (d) 100.161 u. **3.2A** (a) 29.9 u; (b) 148 u; (c) 234 u; (d) 295 u. **3.2B** 104.06 u; (b) 117.49 u; (c) 392.18 u; (d) 249.69 u. **3.3A** (a) 14.5 mg Ag; (b) 2.47×10^{25} O atoms. **3.3B** (a) 17 mol Al; (b) 173 mL liquid bromine. **3.4A** (a) 3.11×10^{24} Cl atoms; (b) 90.5 g sucrose. **3.4B** 555 mL solution. **3.5A** (d); the sample is about 1/6 mol Mg (about 4 g). **3.5B** (b). C accounts for less than 1/3 the mass of CO_2 (compound a). In propane (compound b), C accounts for the bulk of the mass. Acetic acid (compound c) has the same number of C atoms per molecule as propane, but also two more massive O atoms. Diethyl ether (compound d) has 4 C atoms per molecule, but this is offset by an additional O and 4 H atoms. **3.6A** (a) 6.102% H, 21.20% N, 48.43% O, 24.27% S; (b) 20.00% C, 6.713% H, 46.65% N, 26.64% O; urea has the greatest percent nitrogen. **3.6B** (a) 9.388% N, (b) 10.77% O, (c) 36.08% H_2O **3.7A** 1.37×10^3 mg Na^+. **3.7B** 1.40×10^2 g N. **3.8A** LiH_2PO_4 has the smallest mass of cation (7 g Li^+) per mol P when compared with $Ca(H_2PO_4)_2$ (20 g Ca^{2+}) and $(NH_4)_2HPO_4$ (36 g NH_4^+). Thus, it has the highest %P. **3.8B** Hydrocarbons should have higher percentages of C than the oxygen-containing methanol and acetic acid. The hydrocarbon octane, C_8H_{18}, has a higher proportion of C to H than does butane, C_4H_{10} (equivalent to C_8H_{20}). C_8H_{18} has the highest %C. **3.9A** $C_6H_{12}O$. **3.9B** $C_5H_{10}NO_2$. **3.10A** C_7H_5. **3.10B** (a) Fe_3O_4; (b) CCl_2F_2; (c) $C_7H_5N_3O_6$. **3.11A** ethylene, C_2H_4; cyclohexane, C_6H_{12}; 1-pentene, C_5H_{10}. **3.11B** (a) P_2O_3; (b) C_2H_3; (c) $C_3H_4O_3$; (d) $C_2H_4O_3$; (e) CH_3O; (f) $CuCO_3$. **3.12A** (a) 60.0% C, 13.4% H, 26.6% O; (b) C_3H_8O or C_3H_7OH. **3.12B** (a) 80.24% C; 9.62% H, 10.14% O; (b) $C_{21}H_{30}O_2$. **3.13A** (a) $SiCl_4 + 2 H_2O \longrightarrow SiO_2 + 4 HCl$; (b) $PCl_5 + 4 H_2O \longrightarrow H_3PO_4 + 5 HCl$; (c) $6 CaO + P_4O_{10} \longrightarrow 2 Ca_3(PO_4)_2$. **3.13B** (a) $Pb(NO_3)_2(aq) + 2 KI(aq) \longrightarrow 2 KNO_3(aq) + PbI_2(s)$; (b) $4 HCl(g) + O_2(g) \longrightarrow 2 H_2O(g) + 2 Cl_2(g)$. **3.14A** (a) $2 C_4H_{10} + 13 O_2 \longrightarrow 8 CO_2 + 10 H_2O$; (b) $CH_3CH_2CH_2CHOHCH_2OH + 7 O_2 \longrightarrow 5 CO_2 + 6 H_2O$. **3.14B** $2 C_5H_{12} + 15 O_2 \longrightarrow 10 CO_2 + 12 H_2O$. **3.15A** (a) $FeCl_3 + 3 NaOH \longrightarrow Fe(OH)_3 + 3 NaCl$; (b) $3 Ba(NO_3)_2 + Al_2(SO_4)_3 \longrightarrow 3 BaSO_4 + 2 Al(NO_3)_3$. **3.15B** $3 Ca(OH)_2(s) + 2 H_3PO_4(aq) \longrightarrow Ca_3(PO_4)_2(s) + 6 H_2O(l)$. **3.16A** $2 C_6H_{14}O_4 + 15 O_2 \longrightarrow 12 CO_2 + 14 H_2O$. **3.16B** $2 Pb(NO_3)_2(s) \longrightarrow 2 PbO(s) + 4 NO_2(g) + O_2(g)$. **3.17A** (a) 1.59 mol CO_2; (b) 305 mol H_2O; (c) 0.6060 mol CO_2. **3.17B** (a) 3.83 mol CO_2; (b) 6.22 mol O_2. **3.18A** 21.4 g Mg. **3.18B** 26.4 g N_2 and 15.1 g O_2. **3.19A** 774 ml H_2O. **3.19B** 28.9 mL H_2O. **3.20A** 4.77 g H_2S; 0.9 g FeS in excess. **3.20B** 1.40 g H_2. **3.21A** 26.8 g. **3.21B** 374 g. **3.22A** 1.35 kg $(NH_4)_2HPO_4$. **3.22B** aluminum. **3.23A** (a) 1.26 M; (b) 0.0242 M; 0.0349 M. **3.23B** (a) 6.99×10^{-3} M; (b) 7.19 M; (c) 0.34 M. **3.24A** (a) 673 g KOH; (b) 0.0561 g KOH; (c) 4.91 g KOH. **3.24B** 23.2 mL. **3.25A** 23.5 M HCOOH. **3.25B** 70.4% $HClO_4$. **3.26A** 466 mL. **3.26B** 17.0 mL. **3.27A** 375 mL. **3.27B** (a) 11.9 g CO_2; (b) 0.662 M NaCl. **Self-Assessment Questions: 2.** 6.02×10^{23} molecules O_2, 1.20×10^{24} atoms O. **3.** (c). **4.** (a) 6.02×10^{23} Ca^{2+} ions, 1.20×10^{24} NO_3^- ions; (b) 1.20×10^{24} N atoms, 3.61×10^{24} O atoms. **5.** (a) HO;

(b) CH_2; (c) C_5H_4; (d) C_3H_6O. **8.** (b). **9.** (a) $Hg(NO_3)_2(s) \xrightarrow{\Delta} Hg(l) + 2 NO_2(g) + O_2(g)$ (b) $Na_2CO_3(aq) + 2 HCl(aq) \longrightarrow H_2O(l) + CO_2(g) + 2 NaCl(aq)$ (c) $C_3H_4O_4(s) + 2 O_2(g) \longrightarrow 3 CO_2(g) + 2 H_2O(l)$ **10.** (c). **11.** (a). **14.** They are equal. To change g/L to mg/mL, numerator and denominator are both divided by 1000. **Problems: 17.** (a) 122.99 u (molecular); (b) 146.35 u; (c) 325.33 u; (d) 404.02 u; (e) 383.97 u; (f) 135.04 u (molecular); (g) 102.18 u (molecular); (h) 114.22 u (molecular). **19.** (a) 324.36 u; (b) 152.15 u. **21.** (a) 43.4 g; (b) 42.5 g; (c) 44.5 g; (d) 0.0345 g. **23.** (a) 3.78 mol; (b) 0.204 mol; (c) 0.139 mol; (d) 3.51×10^{-3} mol; (e) 3.78×10^5 mol. **25.** 1.81×10^{24} oxide ions, 1.20×10^{24} iron(III) ions. **27.** (a) 2.82×10^{24} molecules; (b) 2.22×10^{23} ions; (c) 2.119×10^{-22} g; (d) 1.651×10^{-22} g. **29.** mol of atoms: Al (<1 mol) < U (just over 1 mol) < HCl (~1.5 mol) < O_2 (~2 mol). **31.** (a) 45.94% K; 16.46% N; 37.60% O; (b) 64.81% C; 13.60% H; 21.59% O; (c) 10.22% Al; 35.21% P; 54.57% O; (d) 19.05% C; 4.80% H; 76.15% O. **33.** (a) 48.44% O; (b) 13.85% N; (c) 5.03% Be. **35.** (a) N_2O_5; (b) C_5H_{11}. **37.** (a) $C_6H_4Cl_2$; (b) $C_6H_{12}O_6$. **39.** (a) $C_{18}H_{21}NO_3$; (b) $Mg_2P_2O_7$. **41.** $C_6H_6O_2$. **43.** $LiClO_4 \cdot 3 H_2O$. **45.** NH_4NO_2. **47.** CH_3S, $C_2H_6S_2$. **49.** C_4H_4S. **51.** (a) MTBE < ethanol < methanol; (b) yes; (c) 15% MTBE. **53.** (a) $Cl_2O_5 + H_2O \longrightarrow 2 HClO_3$; (b) $V_2O_5 + 2 H_2 \longrightarrow V_2O_3 + 2 H_2O$; (c) $4 Al + 3 O_2 \longrightarrow 2 Al_2O_3$; (d) $TiCl_4 + 2 H_2O \longrightarrow TiO_2 + 4 HCl$; (e) $Sn + 2 NaOH \longrightarrow Na_2SnO_2 + H_2$; (f) $PCl_5 + 4 H_2O \longrightarrow 5 HCl + H_3PO_4$; (g) $CH_3SH + 3 O_2 \longrightarrow CO_2 + SO_2 + 2 H_2O$; (h) $3 Zn(OH)_2 + 2 H_3PO_4 \longrightarrow Zn_3(PO_4)_2 + 6 H_2O$; (i) $3 CH_3CH_2OH + PCl_3 \longrightarrow 3 CH_3CH_2Cl + H_3PO_3$. **55.** $2 Mg(s) + O_2(g) \longrightarrow 2 MgO(s)$; 1 mol $O_2 \backsimeq 2$ mol Mg; 1 mol $O_2 \backsimeq 2$ mol MgO. **57.** (a) $2 HCl(aq) + Zn(s) \longrightarrow H_2(g) + ZnCl_2(aq)$; (b) $C_2H_6(g) + 2 H_2O(g) \longrightarrow 2 CO(g) + 5 H_2(g)$; (c) $P_4O_{10}(s) + 6 H_2O(l) \longrightarrow 4 H_3PO_4(l)$; (d) $Pb(s) + PbO_2(s) + 2 H_2SO_4(aq) \longrightarrow 2 PbSO_4(s) + 2 H_2O(l)$. **59.** $3 Fe_2O_3(s) + H_2(g) \longrightarrow H_2O(g) + 2 Fe_3O_4(s)$. **61.** $4 NH_3 + 5 O_2 \longrightarrow 4 NO + 6 H_2O$. **63.** (a) 3.61×10^3 mol CO_2; (b) 7.31×10^3 mol O_2; (c) 1.69×10^3 mol H_2O; (d) 1.8×10^3 mol C_8H_{18}. **65.** $CaCN_2$. **67.** 1.78×10^4 g CO_2. **69.** (a) 3.77 g CO_2; (b) 68.0% $CaCO_3$. **71.** The reactant that is consumed completely and determines the amount of product formed; when the two reactants are in stoichiometric proportions. **73.** LiOH is limiting; 2.20 mol Li_2CO_3. **75.** (d); 4.00 g is about 1/8 or 0.125 mol O_2, Hg is limiting and produces 0.200 mol HgO, leaving a small excess of O_2. **77.** (a) 516 g KI; (b) $KHCO_3$, 6.7 g. **79.** Yes. **81.** 0.727 g ZnS, 83.4% **83.** 51.3 g NH_4HCO_3. **85.** (a) 2.26 M; (b) 0.0700 M; (c) 0.7802 M; (d) 0.963 M; (e) 0.9462 M; (f) 1.83 M. **87.** (a) 0.0294 mol; (b) 7.66 g; (c) 206 g; (d) 87.9 mL. **89.** (a) 403 mL; (b) 2.49×10^3 mL; (c) 104 mL. **91.** 14.6 M. **93.** 0.0291 M. **95.** (a) 165.2 mL; (b) 16.20 mL. **97.** The final molarity cannot be the simple average (0.15 M) of the two because a greater volume is used of the 0.20 M solution. The answer is (c), 0.17 M. **99.** Pipet 3 × 25 mL and 2 × 10 mL into a 100 mL volumetric flask. Add 1.05 g $AgNO_3$, mix to dissolve, dilute to the mark. Other approaches possible. **101.** 46.5 g $BaSO_4$. **103.** 2.19 M. **105.** 1.91 g CO_2. **107.** (a) 0.2 cm^2; (b) 13 drops. **Additional Problems: 110.** 5 tablets. **113.** $BrCl_3$. **115.** 55.85 u, Fe. **118.** 70.1%. **120.** Empirical CHF; molecular $C_3H_3F_3$; $4 C_3H_3F_3 + 12 O_2 \longrightarrow 3 CF_4 + 6 H_2O + 9 CO_2$. **123.** (a) 48.4 g; (b) 45.9 g; (c) 9.4 g MgO, 40.5 g Mg_3N_2. **124.** $3 C_2H_4O_2 + PCl_3 \longrightarrow H_3PO_3 + 3 C_2H_3OCl$. **Apply Your Knowledge:**

128. (a) 4% P, 4% K; **(b)** 35-0-0; **(c)** KH_2PO_4 and $(NH_4)_2HPO_4$; **(d)** 20 g K_2O requires 42.9 g KNO_3 and contains 5.9 g N, 20 g P_2O_5 requires 33.8 g NaH_2PO_4. Additional 14.1 g N requires 40.2 g NH_4NO_3, total = 116.9 g > 100 g, not possible. **130.** 70.58%. **132.** 1.81 g Na. **133.** 0.4065 g $MgCl_2$, 0.2053 g NaCl.

Chapter 4

Exercises: 4.1A $[Na^+] = 0.438$ M; $[Mg^{2+}] = 0.0512$ M; $[Cl^-] = 0.540$ M. **4.1B** $[Na^+] = 0.093$ M; $[K^+] = 0.020$ M; $[Cl^-] = 0.080$ M; $[C_6H_5O_7^{3-}] = 0.011$ M; $[C_6H_{12}O_6] = 0.111$ M. **4.2A (a)** $Ca(OH)_2(aq) + 2 HCl(aq) \longrightarrow CaCl_2(aq) + 2 H_2O(l)$; **(b)** $Ca^{2+}(aq) + 2 OH^-(aq) + 2 H^+(aq) + 2 Cl^-(aq) \longrightarrow Ca^{2+}(aq) + 2 Cl^-(aq) + 2 H_2O(l)$; **(c)** $OH^-(aq) + H^+(aq) \longrightarrow H_2O(l)$. **4.2B (a)** $2 KHSO_4(aq) + 2 NaOH(aq) \longrightarrow K_2SO_4(aq) + Na_2SO_4(aq) + 2 H_2O(l)$; **(b)** $K^+(aq) + HSO_4^-(aq) + Na^+(aq) + OH^-(aq) \longrightarrow K^+(aq) + Na^+(aq) + SO_4^{2-}(aq) + H_2O(l)$; **(c)** $HSO_4^-(aq) + OH^-(aq) \longrightarrow SO_4^{2-}(aq) + H_2O(l)$. **4.3A** The bulb will be dimly lit with CH_3NH_2 (weak base) and brightly lit with HNO_3 (strong acid). When the solutions are mixed, the $CH_3NH_3^+NO_3^-$ (salt) formed maintains the brightly lit bulb. **4.3B** $Mg(OH)_2(s)$ is essentially insoluble, and the very low ion concentration in $Mg(OH)_2(aq)$ produces only a dim glow in the bulb. The bulb grows brighter as $Mg(CH_3COO)_2(aq)$ forms, and brightest when the $Mg(OH)_2(aq)$ is completely neutralized. Beyond this point, additional $CH_3COOH(aq)$ produces little change because $CH_3COOH(aq)$ is a weak acid. **4.4A (a)** $Mg^{2+}(aq) + 2 OH^-(aq) \longrightarrow Mg(OH)_2(s)$; **(b)** $2 Fe^{3+}(aq) + 3 S^{2-}(aq) \longrightarrow Fe_2S_3(s)$; **(c)** $Sr^{2+}(aq) + SO_4^{2-}(aq) \longrightarrow SrSO_4(s)$. **4.4B (a)** $Zn^{2+}(aq) + SO_4^{2-}(aq) + Ba^{2+}(aq) + S^{2-}(aq) \longrightarrow ZnS(s) + BaSO_4(s)$; **(b)** no reaction; **(c)** $2 HCO_3^-(aq) + Ca^{2+}(aq) + 2 OH^-(aq) \longrightarrow CaCO_3(s) + CO_3^{2-}(aq) + 2 H_2O(l)$. **4.5A** H^+ from the acid combines with OH^- from $Fe(OH)_3$ to form $H_2O(l)$: $Fe(OH)_3(s) + 3 H^+(aq) \longrightarrow Fe^{3+}(aq) + 3 H_2O(l)$. **4.5B** The likely precipitate is calcium palmitate, $Ca[CH_3(CH_2)_{14}COO]_2(s)$. Ionic equation: $2 K^+(aq) + 2 CH_3(CH_2)_{14}COO^-(aq) + Ca^{2+} + 2 Cl^-(aq) \longrightarrow 2 K^+(aq) + 2 Cl^-(aq) + Ca[CH_3(CH_2)_{14}COO]_2(s)$; net ionic equation: $Ca^{2+}(aq) + 2 CH_3(CH_2)_{14}COO^-(aq) \longrightarrow Ca[CH_3(CH_2)_{14}COO]_2(s)$. **4.6A** 42.20% NaCl. **4.6B** 17.4 g AgCl(s); 14.9 g $Mg(OH)_2(s)$. **4.7A (a)** Al, +3, O, −2; **(b)** P, 0; **(c)** C, −2, H, +1, F, −1; **(d)** H, +1, As, +5, O, −2; **(e)** Na, +1, Mn, +7, O, −2; **(f)** Cl, +3, O, −2; **(g)** Cs, +1, O, −1/2. **4.7B (a)** +5; **(b)** +2; **(c)** +5; **(d)** +5/2; **(e)** +4/3; **(f)** +5; **(g)** +3. **4.8A** $Cr_2O_7^{2-}(aq)$ and $HNO_3(aq)$ are oxidizing agents and do not react with each other. HCl(aq) is a reducing agent; $Cl^-(aq)$ is oxidized, probably to $Cl_2(g)$. The $Cr_2O_7^{2-}$ is reduced to Cr^{3+}, accounting for the green color. **4.8B** The solution would be $ZnCl_2(aq)$ in HCl(aq) instead of unreacted HCl(aq), as in (a), or $Cu^{2+}(aq)$ and $Zn^{2+}(aq)$ in $HNO_3(aq)$, as in (b). **4.9A** 74.53 mL. **4.9B** 1.10×10^2 mL. **4.10A** 12.6 mL. **4.10B** 31.7% H_2SO_4. **4.11A** 0.009815 M. **4.11B** 23.42 mL **4.12A** 22.04 mL. **4.12B** 14.23 mL. **Self-Assessment Questions: 1.** (a). **2. (a)** weak acid; **(b)** strong acid; **(c)** strong base; **(d)** weak base; **(e)** salt; **(f)** weak acid. **3.** (c) highest, (b) lowest. **4.** (c). **5.** (a), it is a strong electrolyte and is completely ionized; while (b) and (c) have a higher concentration and are only slightly ionized. **6.** (a). **9.** (c). **11.** (d), most sulfates are soluble but lead is an exception. **12.** (c), $MgCl_2(aq) + Na_2CO_3 \longrightarrow 2 NaCl(aq) + MgCO_3(s)$. **17.** (d). **18. (a)** 0; **(b)** +3; **(c)** −2; **(d)** +6; **(e)** −3; **(f)** +4; **(g)** +4; **(h)** +5;

(i) +5/2; **(j)** −1. **19. (a)** −3; **(b)** +2; **(c)** +1; **(d)** −2; **(e)** +3. **24. (b) Problems: 25.** (a), (b), (c), (e), (f) are strong electrolytes; **(d)** is weak. **27.** All; they are all soluble and are strong electrolytes. **29. (a)** $[Li^+] = [NO_3^-] = 0.647$ M; **(b)** $[Ca^{2+}] = 0.035$ M, $[I^-] = 0.070$ M; **(c)** $[Al^{3+}] = 2.14$ M, $[SO_4^{2-}] = 3.21$ M. **31.** $[Na^+] = 0.0844$ M, $[Cl^-] = 0.0554$ M, $[SO_4^{2-}] = 0.0145$ M. **33.** 3.384 L. **35.** 0.040 M $Al(NO_3)_3$ > 0.1 M KNO_3 > 0.047 M $Ca(NO_3)_2$. **37.** 0.261 M. **39.** 1.92×10^3 mg Cl/L. **41.** 67.5 mL. **43.** (a) < (d) < (c) < (b). **45. (a)** $HBr(aq) \longrightarrow H^+(aq) + Br^-(aq)$; **(b)** $LiOH(aq) \longrightarrow Li^+(aq) + OH^-(aq)$; **(c)** $HF(aq) \rightleftharpoons H^+(aq) + F^-(aq)$. **(d)** $HIO_3(aq) \rightleftharpoons H^+(aq) + IO_3^-(aq)$; **(e)** $(CH_3)_2NH(aq) + H_2O(l) \rightleftharpoons (CH_3)_2NH_2^+(aq) + OH^-(aq)$; **(f)** $HCOOH(aq) \rightleftharpoons H^+(aq) + HCOO^-(aq)$. **47.** (d) < (c) < (a) < (b). **49.** (b); for (a) no reaction; **(b)** $H^+(aq) + OH^-(aq) \longrightarrow H_2O(l)$; **(c)** $NH_3(aq) + H^+(aq) \longrightarrow NH_4^+(aq)$. **51.** $CaCO_3(s) + 2 H^+(aq) \longrightarrow Ca^{2+}(aq) + H_2O(l) + CO_2(g)$. **53. (a)** $2 I^-(aq) + Pb^{2+}(aq) \longrightarrow PbI_2(s)$; **(b)** no reaction; **(c)** $Cr^{3+}(aq) + 3 OH^-(aq) \longrightarrow Cr(OH)_3(s)$; **(d)** no reaction; **(e)** $OH^-(aq) + H^+(aq) \longrightarrow H_2O(l)$; **(f)** $HSO_4^-(aq) + OH^-(aq) \longrightarrow H_2O(l) + SO_4^{2-}(aq)$. **55. (a)** $Mg(OH)_2(s) + 2 H^+(aq) \longrightarrow Mg^{2+}(aq) + 2 H_2O(l)$; **(b)** $HCOOH(aq) + NH_3(aq) \longrightarrow NH_4^+(aq) + HCOO^-(aq)$; **(c)** no reaction; **(d)** $Cu^{2+}(aq) + CO_3^{2-}(aq) \longrightarrow CuCO_3(s)$; **(e)** no reaction. **57.** (a), (b), (c) soluble; (d) insoluble. **59.** No; $MgSO_4$ is soluble in H_2O and $Mg(OH)_2$ reacts with acid to dissolve. Add water to the powder; if it dissolves it is $MgSO_4$. **61.** $Cu^{2+}(aq) + CO_3^{2-}(aq) \longrightarrow CuCO_3(s)$. **63.** Add $Ba^{2+}(aq)$ to each, Na_2SO_4 precipitates $BaSO_4$; add H_2SO_4 to each, $Ba(NO_3)_2$ precipitates $BaSO_4$; the nonreacting solution is NH_3. **65. (a)** oxidation, Cr^{2+} to Cr^{3+}; **(b)** neither, no atom changes oxidation number; **(c)** neither, no atom changes oxidation number. **67. (a)** $4 HCl + O_2 \longrightarrow 2 H_2O + 2 Cl_2$; **(b)** $2 NO + 5 H_2 \longrightarrow 2 H_2O + 2 NH_3$; **(c)** $CH_4 + 4 NO \longrightarrow 2 N_2 + 2 H_2O + CO_2$; **(d)** $3 Ag + 4 H^+ + NO_3^- \longrightarrow NO + 2 H_2O + 3 Ag^+$; **(e)** $IO_4^- + 7 I^- + 8 H^+ \longrightarrow 4 I_2 + 4 H_2O$. **69.** Oxidizing agent and reducing agent, respectively: **(a)** O_2, HCl; **(b)** NO, H_2; **(c)** NO, CH_4; **(d)** NO_3^-, Ag; **(e)** IO_4^-, I^-. **71. (a)** both Pb^{2+} and V^{3+} are being oxidized, but nothing is reduced; **(b)** only S^{2-} is being reduced, nothing is oxidized. **73. (a)** $Zn + 2 H^+ \longrightarrow Zn^{2+} + H_2$; **(b)** no reaction; **(c)** $Fe + 2 Ag^+ \longrightarrow Fe^{2+} + 2 Ag$; **(d)** no reaction. **75. (a)** 46.8 ml; **(b)** 11.9 mL; **(c)** 23.1 mL. **77.** 0.8051 M. **79. (a)** 487 mg $CaCO_3$; **(b)** 195 mg Ca^{2+}. **81.** Tums. **83.** (d); 22.00 mL is about 10% excess OH^-, and the CH_3COOH has been consumed. **85. (a)** 147.7 mL; **(b)** 111.6 mL; **(c)** 29.1 mL. **87.** $Ba^{2+}(aq) + SO_4^{2-}(aq) \longrightarrow BaSO_4(s)$; 0.04329 M. **89. (a)** 12.39 mL; **(b)** 47.67 mL. **91.** 0.1226 M. **93.** 0.235 M Cl^-. **Additional Problems: 96.** 6.4 mL. **99.** 1.5×10^3 kg. **101.** 1.2 M OH^-. **104.** 9.26 g. **107.** +5. **110.** 86.2 mL. **113.** KCl; 3.43 g. **115.** 0.0345% S. **Apply Your Knowledge: 116.** 49.0% protein. **118.** $Ag^+ + SCN^- \longrightarrow AgSCN(s)$; 93.71% Ag **122.** No.

Chapter 5

Exercises: 5.1A (a) 720 mmHg; **(b)** 737 Torr; **(c)** 760.7 Torr; **(d)** 1.01 atm. **5.1B** 2.00×10^2 kPa; 1.97 atm. **5.2A** 6.50 m. **5.2B** 2.90 atm due to the water; 3.90 atm total. **5.3A** 735 mmHg < 745 mmHg < 750 mmHg < 101 kPa < "above 762 mmHg"; but because 103 kPa = 773 mmHg, we are uncertain how to compares it with "above 762 mmHg." That is, we don't know which of the two comes last in the series. **5.3B (a)** When air is sucked through the straw, the pressure of the remaining air drops. Air pressure outside the straw exceeds that within the straw and pushes liquid up the straw. **(b)** The hand-operated water pump acts in the same way as the soda straw. The height limitation arises because of the maximum height of a water column that can be sustained by atmospheric pressure (see Example 5.2). **5.4A** 503 Torr. **5.4B** 17.8 mL. **5.5A** 400 mmHg. **5.5B** (d). The PV product is constant in Exercise 5.5A. Both (a) and (b) use the wrong volume in the PV product, and (c) is not a PV product. Response (d) is a PV product equivalent to 1.2×10^4 Torr L, when properly converted. **5.6A** 234 L. **5.6B** 338 K (65 °C). **5.7A** 7.5 L. **5.7B** (b). The final pressure must be larger than the initial pressure, but not three times as great. Responses (a), (c) and (d) must all be incorrect. Note that the correct answer is roughly 4/3 of the initial pressure (the temperature ratio is 423 K/323 K ~ 4/3). **5.8A** 98.4 g C_3H_8. **5.8B** 3.75×10^3 L CO_2. **5.9A** Greater than 6.7×10^2 °C. **5.9B** $P_2 = (n_2/n_1) \times P_1$. Pressure is directly proportional to amount of gas. With more molecules in the same volume, the frequency of molecular collisions, and hence the pressure, increases. **5.10A** 1.58 mol N_2. **5.10B** 0.93 mol N_2. **5.11A** -139 °C. **5.11B** 21.6 g N_2. **5.12A** 74.3 g/mol. **5.12B** 2.0 atm. **5.13A** 126 u. **5.13B** $C_4H_6O_2$. **5.14A** 1.25 g/L. **5.14B** C_4H_{10}. **5.15A** 94 °C. **5.15B** 92 °C. **5.16A** 2.78 L $O_2(g)$. **5.16B** 207 L $O_2(g)$. **5.17A** 4.16×10^4 L $CO_2(g)$. **5.17B** 746 L $C_5H_{10}(l)$. **5.18A** Partial pressures: O_2, 0.209 atm; Ar, 0.00932 atm; CO_2, 0.0005 atm; total pressure: 0.999 atm. **5.18B** 12.9 atm. **5.19A** Partial pressures: N_2, 0.741 atm; O_2, 0.150 atm; H_2O, 0.060 atm; Ar, 0.009 atm; CO_2, 0.040 atm. **5.19B** Mole fractions: CH_4, 0.636; C_2H_6, 0.253; C_3H_8, 0.054; C_4H_{10}, 0.0562. **5.20A** If only H_2 were added, its partial pressure would be 2.50 atm instead of 2.00 atm. A gas other than H_2 is needed at a partial pressure of 0.50 atm; it can be additional He, but does not have to be. **5.20B** The pressure increases by the factor 673 K/273 K $\approx$ 2.5, even if no N_2 is added. Responses (a) and (b) are incorrect. N_2 must be removed. However, since the amount of N_2 initially is slightly less than 0.10 mol, response (c) is incorrect. The correct response is (d). **5.21A** Total mass of wet gas: 0.0235 g; mass of H_2: 0.0197 g. **5.21B** 0.502 g $KClO_3$. **5.22A** Estimate the value of $\sqrt{T/M}$ for each of the four responses. Choose the one with the largest value. It is (d). **5.22B** The temperature must exceed 273 K multiplied by the ratio of the molar mass of O_2 to H_2. Only response (d), 5000 K, does. **5.23A** N_2 effuses faster by the factor $(39.95/28.01)^{1/2} = 1.194$. **5.23B** NO effuses faster by the factor $(46.07/30.01)^{1/2} = 1.239$, or 23.9% faster. **5.24A** 59 g/mol. **5.24B** 166 s. **Self-Assessment Questions: 3.** (b). **5.** (a) 0.7326 atm; **(b)** 0.762 atm; **(c)** 0.691 atm; **(d)** 1.17 atm. **6.** (a) V decreases; **(b)** V decreases; **(c)** indeterminate; **(d)** V increases. **7.** (a) P increases; **(b)** P increases; **(c)** P increases; **(d)** indeterminate. **9.** (b), (c), (d). **10.** (d). **11.** (a). **12.** (c). **13.** (a) A more dense; **(b)** same density; **(c)** B more dense. **14.** (a). **15.** (b). **16.** (d). **17.** (c). **18.** (a). **19.** (b). **20.** (b). **Problems: 21.** **(a)** 7.10×10^2 Torr; **(b)** 1.01 atm; **(c)** 93.1 kPa; **(d)** 1.73×10^3 mmHg **23.** 121 atm **25.** 0.958 m **27.** 756 mmHg **29.** The height in mm must be greater than the pressure in mmHg, since 1 mm of height = 1 mmHg pressure. **31.** (a) 922 mL; **(b)** 632 mL; **(c)** 707 mL. **33.** 46.4 m³. **35.** (a) 6.54×10^3 L; **(b)** 817 min. **37.** 478 mL. **39.** 150 K. **41.** 3.26 kg CH_4. **43.** 575 mL. **45.** (d). **47.** 2.49×10^3 Torr. **49.** 24.5 L. **51.** 0.489 m³. **53.** (a) 46.9 L; **(b)** 26.9 atm; **(c)** 8.64 mg H_2; **(d)** 248 kPa. **55.** 7.48 L. **57.** 71.5 L. **59.** 178 u. **61.** C_7H_8. **63.** (a) 0.901 g/L; **(b)** 0.968 g/L. **65.** 0.803 atm. **67.** S_8. **69.** (c). **71.** 18.8 L. **73.** 4.50 L. **75.** 28.1 mg Mg. **77.** 122 L. **79.** 123 mL. **81.** 585 mmHg

N_2, 153 mmHg O_2, 24 mmHg CO_2. **83.** (a) $x_{He} = 0.205$, $x_{O_2} = 0.795$; **(b)** $P_{He} = 2.40$ atm; $P_{O_2} = 9.30$ atm; **(c)** $P_{total} = 11.70$ atm. **85.** $P_{O_2} = 736$ mmHg, $x_{O_2} = 0.974$. **87.** 0.154 g O_2, 0.145 g glucose. **89.** (b). **91.** A mixture expands to fill the entire container and so all the gases have the same volume. If the gases of the mixture were at different temperatures, they would quickly equilibrate to the same temperature. Since P_{gas} is directly related to the number of moles of the gas, pressures may be different if the number of moles differ. **93.** (a) 81 u; **(b)** 26 u. **Additional Problems: 95.** 80 g air, 36 lb/in² **97.** 0.181 g He. **100.** 661 mL. **102.** 54.09 u. **105.** No. **107.** 5.75×10^{15} tons. **110.** 44.5% Li, 55.5% Mg. **113.** 1.26×10^4 L air; 1.47×10^3 L CO_2. **117.** (a) 9.78 atm; **(b)** 9.38 atm; **(c)** The a-term, intermolecular attraction, is more important than the b-term, molecular volume, so the real pressure is lower than the ideal pressure. **119.** 2-methylpentane and 3-methylpentane, $(CH_3)_2CHCH_2CH_2CH_3$ and $CH_3CH_2CH(CH_3)CH_2CH_3$. **Apply Your Knowledge: 121.** 144 u. **123.** Slightly over 30 km. **124.** Atomic mass of X = 16.0 u; number of X atoms = 2, 1, 1, 2, respectively; X = oxygen.

Chapter 6

Exercises: 6.1A $\Delta U = -478$ J. **6.1B** The system absorbs heat; $q = +122.6$ J. **6.2A** (a) gas does work ($w < 0$); **(b)** internal energy decreases ($\Delta U < 0$); **(c)** temperature decreases ($\Delta T < 0$). **6.2B** The process can occur. The values of w are different for the expansion and the compression because w is path-dependent. However, because the gas is returned to the identical starting condition, $\Delta U = 0$. **6.3A** $\Delta H = +142.7$ kJ. **6.3B** $\Delta H = -148.9$ kJ. **6.4A** $H_2O(s) \longrightarrow H_2O(l)$ $\Delta H = +6.02$ kJ. **6.4B** $N_2O_3(g) \longrightarrow NO(g) + NO_2(g)$ $\Delta H = +40.5$ kJ, **6.5A** $\Delta H = -1.17 \times 10^3$ kJ. **6.5B** 2.80×10^4 L $CH_4(g)$. **6.6A** 11 J/°C. **6.6B** $q = -153$ kJ. **6.7A** $q = 6.67 \times 10^4$ cal = 66.7 kcal. **6.7B** 4.39×10^5 kg H_2O. **6.8A** 95.1 °C. **6.8B** 1.57×10^3 g Cu. **6.9A** 0.13 J g^{-1} °C^{-1}. **6.9B** 1.64 J g^{-1} °C^{-1}. **6.10A** 60 °C. **6.10B** 27.8 °C. **6.11A** $q_{rxn} = 2.81$ kJ. **6.11B** The result should be substantially the same. The net ionic reactions are the same in the two neutralizations, $H^+(aq)$ + $OH^-(aq) \longrightarrow H_2O(l)$. **6.12A** $\Delta H = -56.2$ kJ **6.12B** 29.9 °C. **6.13A** $\Delta H = -395$ kJ. **6.13B** 5.99 kJ/°C. **6.14A** $\Delta H = -1215$ kJ. **6.14B** $\Delta H = -137.0$ kJ. **6.15A** $\Delta H° = -44.2$ kJ. **6.15B** $2 C_4H_{10}(g) + 13 O_2(g) \longrightarrow 8 CO_2(g) + 10 H_2O(l)$, $\Delta H° = -5755$ kJ. **6.16A** $\Delta H_f° = -52$ kJ/mol $C_2Cl_4(l)$. **6.16B** $\Delta H_f° = +81$ kJ/mol $C_4H_4S(l)$. **6.17A** $\Delta H_f°$ of CH_3CH_2OH and CH_3OH differ only by 39 kJ/mol. In contrast, the enthalpies of the combustion products are much lower for CH_3CH_2OH than for CH_3OH [by 1 mol each of $CO_2(g)$ and $H_2O(l)$]. Per mole, combustion of CH_3CH_2OH gives off the greater amount of heat. **6.17B** The complete combustion of 1 mol $CH_3COOH(l)$ yields 2 mol each of $CO_2(g)$ and $H_2O(l)$. This final state is at a higher enthalpy than the 2 mol $CO_2(g) + 3$ mol $H_2O(l)$ formed in the combustion of 1 mol of $CH_3CH_2OH(l)$. At the same time, the initial state of 1 mol $CH_3COOH(l)$ is at a lower enthalpy than the initial states of 1 mol $C_2H_6(g)$ and 1 mol $CH_3CH_2OH(l)$. The magnitudes of the enthalpies of combustion are: $CH_3COOH(l) <$ $CH_3CH_2OH < C_2H_6$. **6.18A** $\Delta H_f°$ [$BaSO_4$] $= -1473$ kJ/mol. **6.18B** $\Delta H° = +2.4$ kJ [per mole of $Mg(OH)_2(s)$ formed]. **Self-Assessment Questions: 3.** (c). **6.** (a). **8.** (d). **9.** (a). **10.** (b). **12.** (d). **13.** (a). **14.** $2 Fe(s) + 3/2 O_2(g) \longrightarrow Fe_2O_3(s)$. **18.** 53 kcal, 17%. **Problems: 19.** $+442$ J. **21.** 515 J absorbed. **23.** Yes, yes.

25. Endothermic, $+241$ kJ, ΔH because the system is at constant pressure. **27.** $CaO(s) + H_2O(l) \longrightarrow Ca(OH)_2(s)$, $\Delta H = -65.2$ kJ. **29.** (a) -31 kJ; (b) $+138$ kJ. **31.** $1/4\,P_4(s) + 3/2\,Cl_2(g) \longrightarrow PCl_3(g)$, $\Delta H = -287$ kJ. **33.** 1591 kJ evolved. **35.** 416 kJ evolved. **37.** 447 L. **39.** Yes. **41.** 757 g CaO. **43.** 13 J/°C. **45.** (a) 108 kJ; (b) 14.8 kJ. **47.** 52.0 °C. **49.** 0.485 J/(g°C). **51.** No. **53.** Cu. **55.** -53.7 kJ/mol. **57.** $+27$ kJ/mol. **59.** 28.6 kJ evolved per g coal. **61.** 9.32 kJ/°C. **63.** -5.64×10^3 kJ/mol. **65.** -221.0 kJ. **67.** -818.2 kJ. **69.** -1155.4 kJ. **71.** (a) -19 kJ; (b) -1095.9 kJ; (c) $+168.0$ kJ; (d) -1084.3 kJ. **73.** -206 kJ/mol. **75.** -628 kJ/mol.

77. $+30.6$ kJ. **79.**

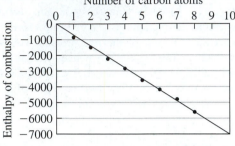

Number of carbon atoms

Slope $= -650.7$ kJ/C atom, intercept $= -260.4$ kJ, so for $C_{10}H_{22}$, $\Delta H_{comb} = -6767$ kJ/mol; for $C_{12}H_{24}$, $\Delta H_{comb} = -8069$ kJ/mol. **81.** 14%. **Additional Problems: 83.** (c). **85.** 27.8 °C. **88.** The aluminum foil; we need only calculate the product (mass × specific heat). **90.** 1.28×10^3 kJ. **93.** About 69 °C. **95.** 33.54 °C. **98.** (a) ΔH is a state function and depends on the concentration of solute; (b) For dilution, Equation (6.21) becomes $\Delta H_f^\circ(\text{dilute}) - \Delta H_f^\circ(\text{concd})$; since ΔH_f° becomes more negative with dilution, the dilution is exothermic. **99.** $C_2H_6O(g) + 3\,O_2(g) \longrightarrow 2\,CO_2(g) + 3\,H_2O(l)$; $\Delta H^\circ = -1.45 \times 10^3$ kJ. **102.** $\Delta H_f^\circ = -156$ kJ/mol. **Apply Your Knowledge: 104.** (a) $2\,C_6H_{14}(l) + 19\,O_2(g) \longrightarrow 12\,CO_2(g) + 14\,H_2O(l)$; $\Delta H^\circ = -8314.0$ kJ; (b) $(CH_3)_2CHCH_2CH_2CH_3$ or $CH_3CH_2CH(CH_3)CH_2CH_3$; (c) 12.1 gal. **107.** (a) 25.2; (b) Plot specific heat vs. $1/(\text{molar mass})$; (c) 59.4 g/mol; (d) about 83 mL. **109.** -74.9 kJ (lab), -82 kJ (theor); Reasons for error: assumption of specific heat of solutions; heat absorbed by calorimeter; solutions not at 1 M concentration; and so on.

Chapter 7

Exercises: 7.1A 2.80×10^{11} Hz. **7.1B** 3.07×10^3 nm. **7.2A** Microwave radiation has longer wavelengths than visible light. **7.2B** (b); comparisons can be made with the information in Figure 7.11. **7.3A** 1.91×10^{-23} J/photon. **7.3B** 8.46×10^{-19} J/photon. **7.4A** 299 kJ/mol photons. **7.4B** infrared ($\approx$1200 nm). **7.5A** $E_6 = -6.053 \times 10^{-20}$ J. **7.5B** No. The allowable energy levels are $n = (-2.179 \times 10^{-18}/E_n)^{1/2}$, where n must be an integer; $(1 \times 10^1)^{1/2} = 3.16$—not an integer. **7.6A** $\Delta E_{level} = 4.086 \times 10^{-19}$ J. **7.6B** Yes. Find a pair of integers such that $(4.269 \times 10^{-20}\text{ J}/2.179 \times 10^{-18}\text{ J}) = (1/n_1^2 - 1/n_2^2)$. The pair must be larger than $n_1 = 5$ and $n_2 = 3$, because the energy difference, 4.269×10^{-20} J, is smaller than the 1.550×10^{-19} J found in Example 7.6. Try combinations such as 5,4; 6,5; 6,4; 7,6; 7,5; and 7,4. The combination 7,5 works. **7.7A** $\nu = 3.083 \times 10^{15}$ s^{-1}. **7.7B** $\lambda = 434.1$ nm (visible region). **7.8A** (a) ΔE_{level} between $n = 1$ and $n = 2$ is $3/4\,B$. The only transitions that would require more energy are from $n = 1$ to $n \geq 3$. **7.8B** No. The energy cited is less than the minimum energy for an emission from the level

$n = 4$, that is, less than $E = 2.179 \times 10^{-18}$ J $\times (1/3^2 - 1/4^2) = 1.2 \times 10^{-19}$ J. **7.9A** $\lambda = 0.105$ nm. **7.9B** $\nu = 1.50 \times 10^3$ m/s. **7.10A** (b), (c) are possible; (a), (d) are not possible because m_l cannot be less than $-l$ nor greater than $+l$. **7.10B** (a) $m_l = -1, 0, 1$; (b) $l = 2$ or 3; (c) $n \geq 4$, $m_l = -3, -2, -1, 0, 1, 2, 3$. **7.11A** (a) three; (b) the f subshells of $n \geq 4$. **7.11B** (a) Sixteen orbitals in $n = 4$ (one $4s$, three $4p$, five $4d$, seven $4f$); (b) total number of orbitals $= n^2$. **Self-Assessment Questions: 1.** (b). **5.** (d). **6.** (b). **7.** (a). **9.** (b). **10.** At the microscopic level, energy changes are very small, and the discontinuities due to quantization are very important. At the macroscopic level, energy changes are much larger and energy transfer appears to be continuous. **12.** The negative sign indicates attractive force. The smaller the value of n, the closer the electron is to the nucleus, and the stronger should be that attractive force. The incorrect expression $E_n = -B \times n^2$ predicts an increase in E as n increases. **13.** (a). **17.** (a) $3d$; (b) $2s$; (c) $4p$; (d) $4f$. **18.** (d). **19.** (a) 3; (b) 2; (c) 4. **21.** (d). **Problems: 23.** Charge 1.602×10^{-19} C, mass 1.672×10^{-27} kg. **25.** 1.6×10^{-19} C, the common factor. **27.** (a) 8.3×10^{-7} kg/C; (b) 9.3×10^{-8} kg/C; (c) 4.1×10^{-7} kg/C; without knowledge of actual isotope masses, we must use mass numbers. **29.** 72.5 u. **31.** (a) 414 m, AM radio; (b) 0.541 m, radio; (c) 2.86×10^{-11} m, X ray. **33.** $\lambda = 700$ nm, $\nu = 4.29 \times 10^{14}$ s^{-1}. **35.** 1.5×10^4 s. **37.** 4.92×10^{-19} J; greater, because 655 nm falls in the red region of the spectrum, and violet light has a shorter wavelength and higher energy. **39.** 2.55×10^{-19} J/photon; 154 kJ/mol photons. **41.** 427 nm. **43.** 246 nm, UV region. **45.** (a) 2.118×10^{-18} J; (b) 7.566×10^{-20} J. **47.** (a) 6.906×10^{14} s^{-1}; (b) 2.923×10^{15} s^{-1}. **49.** No; $n^2 = -B/E = 0.2179$, and there is no integer, squared, that gives 0.2179. **51.** 1312 kJ/mol. **53.** 0.00204 nm. **55.** 1.43×10^8 m/s. **57.** (a) $n = 2$; (b) $n = 4$. **59.** For $n = 4$, $l = 0, 1, 2,$ or 3; for $l = 1$, $m_l = 1, 0,$ or -1. **61.** (a) Magnitude of m_l cannot exceed l; (b) allowed; (c) allowed; (d) n must be 1 or greater. **63.** (a) $n = 4$, $l = 1$, $m_l = 1, 0,$ or -1; (b) $n = 4$, $l = 3$, $m_l = 3, 2, 1, 0, -1, -2,$ or -3; (c) $3d$. **65.** (a) $n = 2$ or greater, $m_s = +\frac{1}{2}$ or $-\frac{1}{2}$; (b) $l = 1$ or 2; (c) $l = 2, 3,$ or 4, $m_s = +\frac{1}{2}$ or $-\frac{1}{2}$; (d) $n = 1$ or greater, $m_l = 0$. **67.** $n = 3$, $l = 2$, $m_l = 0$, $m_s = -\frac{1}{2}$; any set for which $n = 3$, $l = 2$, $m_l = 2, 1, -1,$ or -2, and $m_s = +\frac{1}{2}$ or $-\frac{1}{2}$. **Additional Problems: 69.** 200.6 u, abundances are precisely known but masses are not (mass numbers). **71.** 0.017 s; period $= 1/\nu$. **74.** 2.37×10^{15} Hz; yes, the transition in H from $n = 2$ to $n = 1$ has slightly more energy than is needed, and the additional energy would be converted to kinetic energy of the two N atoms. **77.** maximum $= 121.6$ nm, minimum 91.2 nm. **79.** $n = 4$ to $n = 2$. **82.** Pfund transitions are to $n = 5$, 7400 nm is $n = 6$ to $n = 5$. **84.** 2.41×10^{-34} m. **87.** 7.65×10^3 m/s. **89.** About 700 photons/s. **93.** 0.00251 nm, ten times smaller than observed. **97.** (a) 5.16×10^{14} s^{-1}; (b) no, 1000 nm $= 3 \times 10^{14}$ s^{-1}; (c) 5.25×10^5 m/s. **Apply Your Knowledge: 101.** (a) 7.3×10^5 m/s; (b) 2.4×10^{14} rev/s. **103.** Note that sig. fig. convention is not followed, for clarity. (a) 6.6×10^{-34}%, 100%; (b) 0.02%, 99.98%; (c) 0.003%, 99.997%; (d) Decreasing flame temperature by 500 K decreases the number of emitting atoms, and the emission intensity, by a factor of 12; (e) number of ground-state atoms is virtually the same in (b) and (c).

Chapter 8

Exercises: 8.1A (a) $1s^2 2s^2 2p^6 3s^2 3p^5$; (b) $[Ne]3s^2 3p^5$;

(c)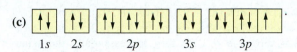

8.1B Only (b) and (c) are valid electron configurations (for oxygen). Response (a) violates Hund's rule; (d) has unpaired electrons with opposing, rather than parallel, spins. **8.2A (a)** Mo: $1s^2 2s^2 2p^6 3s^2 3p^6 3d^{10} 4s^2 4p^6 4d^5 5s^1$; [Kr]$4d^5 5s^1$; **(b)** Bi: $1s^2 2s^2 2p^6 3s^2 3p^6 3d^{10} 4s^2 4p^6 4d^{10} 4f^{14} 5s^2 5p^6 5d^{10} 6s^2 6p^3$; [Xe]$4f^{14} 5d^{10} 6s^2 6p^3$. **8.2B (a)** Sn: [Kr]$4d^{10} 5s^2 5p^2$;

(b) Zr: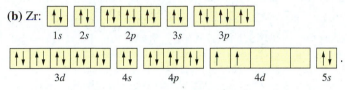

8.3A Se^{2-}, [Ar]$3d^{10} 4s^2 4p^6$; Pb^{2+}: [Xe]$4f^{14} 5d^{10} 6s^2$.

8.3B I^-: [Kr]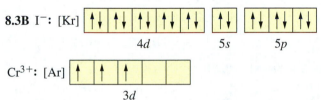

Cr^{3+}: [Ar]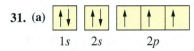

8.4A (a) 2; **(b)** 4; **(c)** 0; **(d)** 0; **(e)** 0; **(f)** 4. **8.4B** The paramagnetic species and odd number of electrons are (a), 1; (d), 3; (f), 2; (g), 1. **8.5A (a)** F < N < Be; **(b)** Be < Ca < Ba; **(c)** F < Cl < S; **(d)** Mg < Ca < K. **8.5B** Based on trends noted in the text: P < Ge < Sn < Pb < Cs. The case of Ca is uncertain. Ca, in the fourth period, is a larger atom than P and Ge, in the third and fourth periods. The Ca atom, far to the left in its period, is probably larger than Sn and Pb, rather far to the right. The Cs atom in the lower left of the periodic table is largest. Probable order: P < Ge < Sn < Pb < Ca < Cs. **8.6A** Y^{3+} < Sr^{2+} < Rb^+ < Br^- < Se^{2-}. **8.6B** Cr^{3+} < Cr^{2+} < Ca^{2+} < K^+ < Cs^+. **8.7A (a)** Be < N < F; **(b)** Ba < Ca < Be; **(c)** S < P < F; **(d)** K < Ca < Mg. **8.7B** I_1 for Al < I_1 for S < I_1 for H < I_2 for Mg < I_1 for Ar < I_2 for K. [I_1 for H (1312) can be calculated from the value of B in the Bohr H atom; other values can be found in the text.]. **8.8A** +400 kJ/mol; Se^{2-} is larger than S^{2-}, for which $EA_2 \approx 450$ kJ/mol. **8.8B (a)** Adding one electron to Si would produce a half-filled p subshell. Adding one electron to Al results in only two electrons in the subshell. **(b)** Adding one electron to P would be adding to a half-filled subshell but adding one electron to Si would produce a half-filled subshell. **8.9A** More nonmetallic: **(a)** O; **(b)** S; **(c)** F. **8.9B** Na (alkali metal) ≈ Ba (alkaline earth metal) > Fe ≈ Co(transition metals) > Sb(metalloid) > Se (nonmetal) > S(nonmetal). In general, group 1A elements are somewhat more metallic than group 2A. However, Na (third period) and Ba (sixth period) are comparable in metallic character. Fe and Co are very similar in atomic radius and ionization energy. The placement of Sb, Se, and S is based on their relative positions in the periodic table. **8.10A (a)** $Z = 52$ (Te); **(b)** $Z = 21$ (Sc); **(c)** lowest I_1, $Z = 39$ (Y); highest I_1, $Z = 48$ (Cd); **(d)** N. **8.10B (a)** Lowest I_1 value: In ($Z = 49$), highest I_1 value: Xe ($Z = 54$); **(b)** K ($Z = 19$), Cr ($Z = 24$), Cu ($Z = 29$); **(c)** Mg ($Z = 12$), Ar ($Z = 18$). **Self-Assessment Questions: 1.** (b), (c). **2.** (c). **3.** The second response. **4. (a)** F; **(b)** Ca; **(c)** N; **(d)** Na; **(e)** Fe. **5. (a)** ns; **(b)** np; **(c)** (n − 1)d; **(d)** (n − 2)f. **7. (a)** 4 valence, 2 core; **(b)** 8, 2; **(c)** 7, 2; **(d)** 3, 10; **(e)** 2, 10. **11.** (c), (e).

12. (d). Yes, odd number of electrons means at least one unpaired electron; No, electrons may be unpaired in degenerate orbitals in a subshell. **15.** (d). **16.** (b). **17.** (a). **19.** (c). **Problems: 21. (a)** Parallel spins in one 2p orbital; **(b)** Three electrons in one 2p orbital; **(c)** Non-parallel spins in three degenerate orbitals. **23. (a)** 3s subshell must fill before 3p; **(b)** 4s subshell not occupied, 3d is occupied; **(c)** 2d subshell does not exist. **25. (a)** Pauli exclusion principle; **(b)** Pauli exclusion and aufbau; **(c)** Rules for quantum numbers n and l. **27. (a)** $1s^2 2s^2 2p^6 3s^2$; **(b)** $1s^2 2s^2 2p^5$; **(c)** $1s^2 2s^2 2p^4$; **(d)** $1s^2 2s^2$; **(e)** $1s^2 2s^2 2p^6 3s^2 3p^3$; **(f)** $1s^2 2s^2 2p^6 3s^2 3p^2$; **(g)** $1s^2 2s^2 2p^2$; **(h)** $1s^2 2s^2 2p^6 3s^2 3p^6 4s^1$; **(i)** $1s^2 2s^2 2p^6 3s^2 3p^6 3d^1 4s^2$. **29. (a)** [Xe] $6s^1$; **(b)** [Kr] $5s^2$; **(c)** [Ar] $3d^2 4s^2$; **(d)** [Kr] $4d^{10} 5s^2 5p^3$; **(e)** [Ar] $3d^{10} 4s^2 4p^5$; **(f)** [Xe] $5d^{10} 6s^2 6p^2$.

31. (a)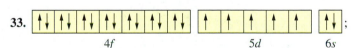

(b)

(c) [Ne]

(d) [Ar]

(e) [Ne]

(f) [Ar]

33.

as the fifth element in the d-block of the sixth period, Re has filled 4f and 6s subshells and a half-filled 5d subshell.

35. (a) [Ar];

(b) [Ar]

(c) [Xe]

(d) [Ne].

37. None; the configuration has 3d but no 4s electrons, and must therefore be a cation, but Co^{3+} would have a $3d^6$ configuration, Ni^{2+} would have $3d^8$, Cu^{2+} would have $3d^9$. **39.** 3+ (loss of 4p electrons) and 5+ (further loss of 4s electrons). **41. (a)** 2, 6A; **(b)** 3, 4A; **(c)** 3, 1A; **(d)** 2, 2A; **(e)** 2, 5A; **(f)** 3, 3A. **43. (a)** 5; **(b)** 32; **(c)** 5; **(d)** 2; **(e)** 10. **45. (a)** diamagnetic, valence 4s electrons paired; **(b)** paramagnetic, unpaired 2p electron;

(c) diamagnetic, [Ar] configuration; **(d)** paramagnetic, unpaired $3p$ electron; **(e)** diamagnetic, pseudo-noble gas configuration. **47.** Ca, Zn, Kr. **49. (a)** Al, lower Z_{eff}; **(b)** Cl^-, isoelectronic but with fewer protons; **(c)** Ba, higher period number (n); **(d)** K (higher n) is larger than Na, which is larger than Na^+ (cation from the atom). **51. (a)** B < Al < Mg < K; B has lowest n, Al has Z_{eff} greater than Mg, K has higher n and lower Z_{eff} than Mg; **(b)** Cl < P < Br < Br^-; Cl slightly higher Z_{eff} than P, Br has higher n than P, Br^- anion larger than Br atom. **53.** At the lower left; highest value of n, lowest Z_{eff}. **55. (a)** Yes, using general trends in size; **(b)** Yes, the species are isoelectronic and F^- has fewer protons, and so is larger. **57. (a)** Ba < Ca < Mg, I decreases with n; **(b)** Al < P < Cl, I increases with Z_{eff}; **(c)** Na < Fe < Cl < F < Ne; both F < Ne and Na < Fe due to Z_{eff}, and F > Cl due to size. **59.** $I_1 < I_2 < I_3 < I_4$ because each successive electron comes from an increasingly positive ion; at I_4 a core electron is removed. **61.** The halogens (7A), because each needs only one electron for a filled shell. **63.** Adding an electron to P produces a pair in one $3p$ orbital, and these electrons repel one another.

65.

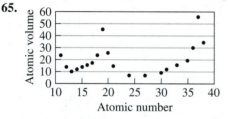

67. Atomic radius (metallic character increases with increase) and ionization energy (metallic character decreases with increase); P < Ge < Bi < Al < Ca < K < Rb. **69.** Tl < Ge < Se < S < Ne. **71. (a)** $Cl_2(g) + 2 Br^-(aq) \longrightarrow 2 Cl^-(aq) + Br_2(aq)$; **(b)** no reaction; **(c)** $Br_2(l) + 2 I^-(aq) \longrightarrow 2 Br^-(aq) + I_2(aq)$ **73. (a)** $N_2O_5(s) + H_2O(l) \longrightarrow 2 HNO_3(aq)$; **(b)** $MgO(s) + 2 CH_3COOH(aq) \longrightarrow Mg(CH_3COO)_2(aq) + H_2O(l)$; **(c)** $Li_2O(s) + H_2O(l) \longrightarrow 2 LiOH (aq)$. **Additional Problems:**

75.

Other answers possible

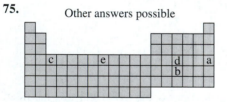

78. $I_1(Cs) < I_1(B) < I_2(Sr) < I_2(In) < I_2(Xe) < I_3(Ca)$; Cs is larger than B; for Sr, In, Xe, I_2 increases from left to right; for Ca, I_3 involves removal of a core electron. **81. (a)** $1s^32s^32p^9 3s^33p^93d^74s^3$; **(b)** $1s^21p^62s^22p^62d^{10}3s^23p^63d^{14}4s^2$; No, in (a) Rb would be a transition metal and Na = $1s^32s^32p^5$, a late p-block element; in (b) Rb would be a transition metal and Na = $1s^21p^62s^22p^1$, an early p-block element. **82.** More protons in He attract the electrons more strongly; -5.23×10^3 kJ/mol for the (one) electron in the He^+ ion, which will have a different energy than that of the (two) electrons in a He atom. **85.** 1312 kJ/mol for (a) and (b), because the Balmer line involves the energy difference between E_∞ and E_2, and the Lyman line involves the energy difference between E_2 and E_1. Ionization energy involves E_∞ and E_1, the sum. **87.** $2 Cl(g) + 2 e^- \longrightarrow 2 Cl^-(g)$, $\Delta H = 2 \times$ electron affinity = -698 kJ; $\Delta H_{overall} = -455$ kJ/mol Cl_2; exothermic. **Apply Your Knowledge: 92. (a)** where Sr is; In_2O_3; where it is now; **(b)** Dulong & Petit atomic mass is 232 u; which gives a formula of $UCl_{3.90}$ which

is about UCl_4; assuming U atomic mass of 240 u gives $UCl_{4.04}$ which is about UCl_4. **94. (a)** A = 2.5×10^{15}, b = 1.18, **(b)** A = 2.5×10^{15}, b = 1.16; they are essentially the same; **(c)** 6.85×10^{-9} cm.

Chapter 9
Exercises: 9.1A (a) :A̤r: **(b)** :B̤r: **(c)** K·

9.1B (a) :A̤s· **(b)** Rb· **(c)** :T̤e·

9.2A $Ba^{2+} + 2\left[:\ddot{I}:\right]^-$ **9.2B** $2 Al^{3+} + 3\left[:\ddot{O}:\right]^{2-}$

9.3A $\Delta H_f^\circ = -617$ kJ/mol LiF(s). **9.3B** Lattice energy = -861 kJ/mol LiCl(s). **9.4A (a)** Ba < Ca < Be; **(b)** Ga < Ge < Se; **(c)** Te < S < Cl. **9.4B** Cl > S > B > Fe > Sc > Na > Rb. **9.5A** C—S < C—H < C—Cl < C—O < C—Mg. **9.5B (a)** Si—F; **(b)** S—Se.

9.6A H—N—N—H with H H on top **9.6B** H—C—C—C̈l: with H H

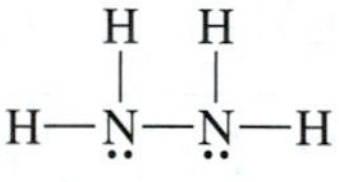

(d)

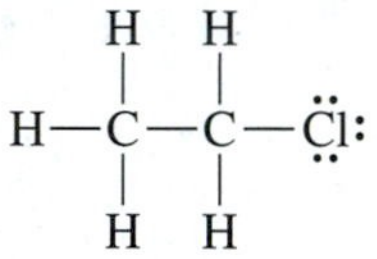

9.7A :S̈=C=Ö: **9.7B** :Ö=N̈—C̈l:.

9.8A $\begin{bmatrix} :\ddot{O}: \\ | \\ :\ddot{O}—Cl—\ddot{O}: \\ | \\ :\ddot{O}: \end{bmatrix}^-$ **9.8B** $\left[:\ddot{O}=N=\ddot{O}:\right]^+$

9.9A H—N—Ö—H ; the second structure has formal charges of +1 (O) and −1 (N), so the first structure with all formal charges of zero is better. **9.9B** H—C—C—Ö—C—H .

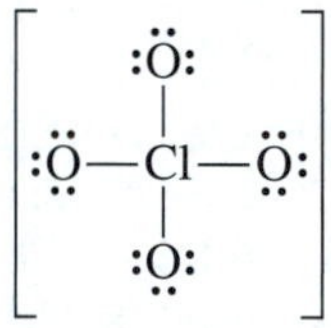

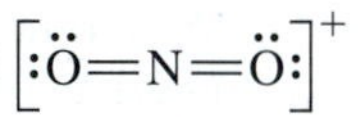

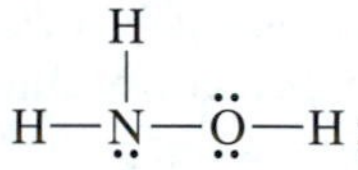

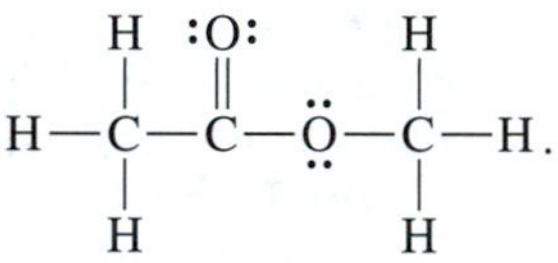

9.10A $\begin{bmatrix} :\ddot{O}: \\ | \\ :\ddot{O}—N=\ddot{O}: \end{bmatrix}^- \longleftrightarrow \begin{bmatrix} :O: \\ || \\ :\ddot{O}—N—\ddot{O}: \end{bmatrix}^- \longleftrightarrow \begin{bmatrix} :\ddot{O}: \\ | \\ :O=N—\ddot{O}: \end{bmatrix}^-$; the resonance hybrid involves equal contributions from the three structures. **9.10B (a)** Two resonance structures for HCO_3^-; **(b)** a single Lewis structure for HOClO; **(c)** a single Lewis structure for CN^-; **(d)** a single Lewis structure for C_2^{2-}; and **(e)** two resonance structures for $HCOO^-$.

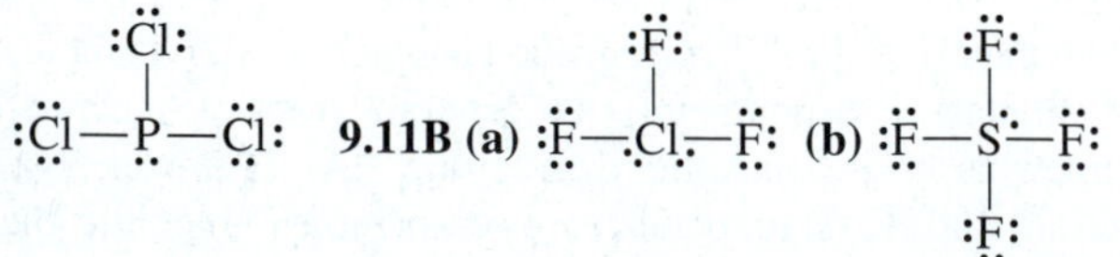

9.11A :C̈l—P—C̈l: **9.11B (a)** 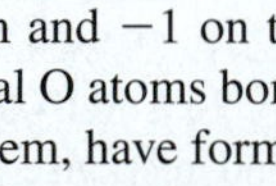 :F̈—C̈l—F̈: **(b)** 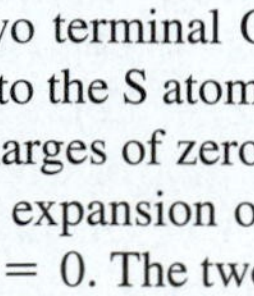 :F̈—S̈—F̈:

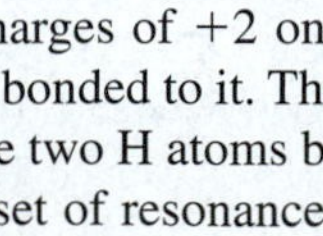

9.12A (a) Incorrect. ClO_2 has 19 valence electrons, but the Lewis structure has 20. **(b)** Correct. **(c)** Incorrect. The structure has 26 electrons rather than the 24 available. **9.12B (1)** One structure limits the valence shell of the central S atom to eight, producing formal charges of +2 on the S atom and −1 on the two terminal O atoms bonded to it. The two central O atoms bonded to the S atom, and the two H atoms bonded to them, have formal charges of zero. **(2)** A set of resonance structures has a valence shell expansion of central S atom to 12 electrons and all formal charges = 0. The two terminal O atoms are doubly bonded to the S atom and the two central O atoms are joined to the S and two H atoms by single bonds. (3) Another set of resonance structures has a valence shell expansion to 10 electrons, one sulfur-to-oxygen double bond, and formal charges of $+1$ and -1. **9.13A** 144 pm. **9.13B** 145 pm. **9.14A** $\Delta H = -113$ kJ. **9.14B** Bond energy, N—F = 278 kJ/mol.

Self-Assessment Questions: 3. (a) Na· **(b)** :Ö:

(c) ·Si· **(d)** :Br· **(e)** ·Ca· **(f)** ·As·

5. (a) Ca· + 2 :Br: $\longrightarrow$ Ca^{2+} + 2 $\left[:\!\overset{..}{\underset{..}{Br}}\!:\right]^{-}$

(b) Ba· + ·Ö: $\longrightarrow$ Ba^{2+} + $\left[:\!\overset{..}{\underset{..}{O}}\!:\right]^{2-}$

(c) 2 ·Al· + 3 ·S: $\longrightarrow$ 2 Al^{3+} + 3 $\left[:\!\overset{..}{\underset{..}{S}}\!:\right]^{2-}$

6. :Ï—Ï:; one bonding pair (dash), six unshared pairs (double dots). **7.** H—$\overset{\delta+}{}$$\overset{\delta-}{F}$:. **9. (a)** N; **(b)** Cl; **(c)** F; **(d)** O. **10. (a)** F; **(b)** Br; **(c)** Cl; **(d)** N. **11. (a)** ionic; **(b)** polar covalent; **(c)** ionic; **(d)** polar covalent; **(e)** ionic; **(f)** ionic; **(g)** nonpolar covalent; **(h)** nonpolar covalent; **(i)** polar covalent. **12. (a)** 1; **(b)** 4; **(c)** 2; **(d)** 1; **(e)** 3; **(f)** 1. **13.** (a); compact structure, low EN atom in center. **14.** (c).

15. (a) :F—P with F atoms; **(c)** :N=Ö: **(e)** :Ö—Br—Ö:

16. (a), (b), (d). **17.** (c). **18.** (b) < (a) < (d) < (c). **19.** (b). **20.** 759 kJ/mol. **21. (a)** alkyne; **(b)** alkene; **(c)** alkyne; **(d)** saturated hydrocarbon. **Problems: 23.** (a), (b), (c), (e).

25. (a) Li^{+}$\left[:\!\overset{..}{\underset{..}{Br}}\!:\right]^{-}$; **(b)** 2 Cs^{+} + $\left[:\!\overset{..}{\underset{..}{S}}\!:\right]^{2-}$;

(c) 3 Na^{+} + $\left[:\!\overset{..}{\underset{..}{N}}\!:\right]^{3-}$; **(d)** 2 Y^{3+} + 3$\left[:\!\overset{..}{\underset{..}{S}}\!:\right]^{2-}$.

27. -420 kJ/mol; Appendix C = -436.7 kJ/mol.
29. -703 kJ/mol. **31.** -1965 kJ/mol.

33. (a) H—Si—H (with H above and below) **(b)** :Cl—N—Cl: (with :Cl: below)

37. Structure (a) should display an ionic bond not a covalent bond, (b) should have covalent bonds not ionic bonds, (c) shows 17 electrons but only has 16 available, (d) has 16 available but only shows 14. **39. (a)** K < As < Br; **(b)** Cs < Ca < Be; **(c)** Pb < Sb < Cl.

41. (a) $\overset{\delta+\ \ \delta-}{H—H}$ < $\overset{\delta+\ \ \delta-}{H—C}$ < $\overset{\delta+\ \ \delta-}{H—N}$ < $\overset{\delta+\ \ \delta-}{H—O}$ < $\overset{\delta+\ \ \delta-}{H—F}$

(b) C—C $\approx$ C—I < $\overset{\delta+\ \ \delta-}{C—Br}$ < $\overset{\delta+\ \ \delta-}{C—Cl}$ < $\overset{\delta+\ \ \delta-}{C—F}$

43. (a) periodic table; **(b)** periodic table; **(c)** values of EN needed; **(d)** periodic table. **45.** From left to right: **(a)** -1, $+1$; **(b)** 0, 0, 0; **(c)** -1, $+1$, -1; **(d)** 0, 0, -1. **47.** From left to right: **(a)** -1, $+1$, 0; **(b)** 0, $+1$, -1; (b) is preferred.

49. $\left[H—\overset{..}{O}—\overset{:O:}{\underset{\parallel}{C}}—\overset{..}{\underset{..}{O}}:\right]^{-}$ $\longleftrightarrow$ $\left[H—\overset{..}{O}—\overset{}{\underset{:O:}{C}}=\overset{..}{\underset{..}{O}}:\right]^{-}$; the resonance

structure is a hybrid of the two structures–the bonds to the terminal oxygen atoms are intermediate between single and double bonds.

51. (a) H—Ö—N=Ö: **(b)** Resonance structures are important only in nitric acid, which has two terminal oxygen atoms rather than the single one seen in nitrous acid.

53. (a) ·N=Ö: :F—Cl—F: :Cl—B—Cl: :F—Se—F: ;
with :F: below, :Cl: below, and :F: / :F: around Se

(b) **(c)** **(d)**

NO is a free radical, the B in BCl$_3$ has an incomplete octet; ClF$_3$ and SeF$_4$ both have expanded valence shells.

55. (a) :S=S—F: (with :F: below) **(b)** [:Ï—Ï—Ï:]$^{-}$

(c) H—Ö—C—Ö—H (with =O below C) **(d)** [:C≡N:]$^{-}$

(e) $\left[\overset{:F:\ \ :F:}{\underset{:F:}{:F—S—F:}}\right]^{-}$ **(f)** $\left[\overset{:O:}{:O—Br—O:}\right]^{-}$

57. (a) H—C—C—C—H (with H atoms on top, :O: and H in middle, H at bottom)

(b) H—C—Ö—H (with =O above C)

(c) H—C—Ö—C—H (with H atoms)

59. In (a) the N is electron deficient, in (b) N cannot have an expanded shell, in (c) there is a negative formal charge on N and positive on O. Better structures are

(a) $\left[\overset{:O:}{\underset{}{Ö=N—Ö}}\right]^{-}$ **(b)** H—N—C≡N: (with H above N)

(c) H—N=C=Ö:

61. The empirical formula C_2H_5O has an odd number of electrons; several unsatisfactory structures can be drawn (below). Doubling the empirical formula to give $C_4H_{10}O_2$ gives many possible structures

·C—C—Ö—H (with H atoms)   H—Ö—C—C—C—C—Ö—H (with H atoms)

Unsatisfactory   Satisfactory

63. (a) 233 pm; **(b)** 149 pm—likely to be high; very polar bond. **65.** HI, with the lowest ΔEN. **67.** $:\!\ddot{F}\!-\!\ddot{N}\!-\!\ddot{N}\!-\!\ddot{F}:$. **69.** -535 kJ. **71.** exothermic, breaking a CH bond and Cl_2 bond, forming a CCl bond and a HCl bond. **73.** 302 kJ/mol bonds. **75.** 416 kJ/mol (experimental) vs. 414 kJ/mol (table). **77. (a)** $H_2C\!=\!CHCH_3$; **(b)** $HC\!\equiv\!CCH_2CH_3$; **(c)** $H_2C\!=\!CHCH_2CH_2CH_3$; **(d)** $CH_3CH_2C\!\equiv\!CCH_2CH_3$ **79. (a)** $H_2C\!=\!CH_2 + H_2 \longrightarrow$ H_3CCH_3; **(b)** $HC\!\equiv\!CH + 2\,H_2 \longrightarrow H_3CCH_3$. **81. (a)** -128 kJ; **(b)** -136.94 kJ. **83. (a)** $\dashv CH_2CH_2CH_2$ $CH_2CH_2CH_2CH_2CH_2\dashv$ **(b)** $\dashv CHClCH_2\!-\!CHClCH_2\!-$ $CHClCH_2\!-\!CHClCH_2\dashv$ **(c)** $\dashv CH(CH_3)CH_2\!-$ $CH(CH_3)CH_2\!-\!CH(CH_3)CH_2\!-\!CH(CH_3)CH_2\dashv$ **Additional Problems: 85.** In an alkane all of the carbon atoms have as many bonds as possible (4), so there are no lone pairs, and the skeletal structure completes the Lewis structure. Organic compounds containing O, N, S, Cl, etc., have lone-pair electrons that do not appear in the structural formula.

87.

91. $H\!-\!\ddot{N}\!-\!N\!\equiv\!N: \longleftrightarrow H\!-\!\ddot{N}\!=\!N\!=\!\ddot{N}:$

93. $+111$ kJ/mol. **95.** -616 kJ/mol; Mg^+ does not have an octet of electrons. **96.** -65 kJ/mol. **98.** Both NO_2 and NO_2^- have resonance structures (two each) with a single and a double NO bond. **102.** 632 kJ/mol (calculated) vs. 590 kJ/mol (table). **Apply Your Knowledge: 104. (a)** 3500 cm; **(b)** 290 pages. **105.** $+97$ kJ/mol. **106. (a)** 106 kJ; **(b)** 0.91; **(c)** 2.96.

Chapter 10

Exercises: 10.1A (a) $SiCl_4$: VSEPR notation, AX_4: electron-group geometry and molecular geometry both tetrahedral. **(b)** $SbCl_5$: VSEPR notation, AX_5: electron-group geometry and molecular geometry both trigonal bipyramidal. **10.1B (a)** BF_4^-: VSEPR notation, AX_4: electron-group geometry and molecular geometry both tetrahedral (Lewis structure on page 362). **(b)** N_3^-: Lewis structure, $\left[\ddot{N}\!=\!N\!=\!\ddot{N}\right]^-$, VSEPR notation, AX_2; electron-group geometry and molecular geometry both linear. **10.2A** SF_4: VSEPR notation, AX_4E. Electron-group geometry is trigonal bipyramidal. In the seesaw structure (Table 10.1), two LP-BP repulsions at 90°; and two at 120°; in the pyramidal structure given, three LP-BP at 90° and one at 180°. Because 90° LP-BP interactions are especially unfavorable, the seesaw structure is adopted. **10.2B** ClF_3: VSEPR notation AX_3E_2: Electron-group geometry is trigonal bipyramidal. In the T-shaped structure (Table 10.1), four LP-BP repulsions at 90°; in the trigonal planar structure, six LP-BP repulsions at 90°. The T-shaped structure is observed. **10.3A** For Lewis structure of $(CH_3)_2O$, see answer to Chapter 9, Problem 57(c). Each C atom bonded to four atoms in tetrahedral fashion (AX_4). The O atom (AX_2E_2) has tetrahedral electron-group geometry, and the C—O—C linkage is angular or bent (bond angle ≈ 109°). **10.3B** The three H atoms in the CH_3 group and the two H atoms in the CH_2 group are in a tetrahedral arrangement about the C atom. The C—C—C bond angle is tetrahedral. The —CHO end of the molecule is trigonal planar, with H—C—O, H—C—C, and C—C—O bond angles of 120°. **10.4A** Polar: SO_2, angular or bent (AX_2E); BrCl, linear but with a small ΔEN between atoms. Nonpolar: BF_3, symmetrical,

trigonal planar (AX_3); N_2, linear, no polarity in bond. **10.4B** SO_3, symmetrical, trigonal planar (AX_3)—nonpolar; SO_2Cl_2, symmetrical, tetrahedral (AX_4), but with different EN for terminal O and Cl atoms—polar; ClF_3 nonsymmetric, T-shaped (AX_3E_2)—polar; BrF_5, nonsymmetric, square pyramidal shape (AX_5E)—polar. **10.5A** NOF, 110°; NO_2F, 118°. LP-BP repulsions from lone-pair electrons on the N atom force O and F atoms closer together in NOF than in NO_2F, with no lone-pair electrons on N atom. **10.5B** COF_2. CS_2 is a symmetrical, linear molecule and SO_3, symmetrical trigonal planar. Neither has a resultant dipole moment. NO should have only a small dipole moment because of the EN difference between N and O is small. NO_2F has $\mu = 0.47$ D, as indicated in Example 10.5. COF_2 should have $\mu > 0.47$ D because its geometrical shape is like that of NOF_2, but the EN of C is less than that of N. **10.6A** $SiCl_4$ is a tetrahedral molecule (AX_4); hybridization scheme for Si: sp^3. **10.6B** I_3^-, a linear ion with trigonal bipyramidal electron group geometry; sp^3d hybridization for central I atom. **10.7A (a)** Tetrahedral molecular geometry around C atom and angular or bent around O atom. **(b)** Hybridization: sp^3 for both C and O. **(c)** 3 σ $C(sp^3)$—$H(1s)$; σ $C(sp^3)$—$O(sp^3)$; σ $O(sp^3)$—$H(1s)$. **10.7B (a)** $:N\!\equiv\!C\!-\!C\!\equiv\!N:$ Linear molecule, **(b)** Hybridization: sp for both C and N. **(c)** σ $C(sp)$—$C(sp)$; 2 σ $C(sp)$—$N(sp)$ + $4\pi C(2p)$—$N(2p)$. **10.8A** The third isomer is 1,1-dichloroethene, $Cl_2C\!=\!CH_2$; it is a polar molecule. **10.8B (a)** Planar $H_2C\!=\!CH_2$ is nonpolar, but replacing one H with F makes $H_2C\!=\!CHF$ polar. **(b)** Because of the very small ΔEN between C and H and the symmetrical shape of *trans*-2-butene (see page 415), the molecule is *nonpolar*. **(c)** Acetylene, $H\!-\!C\!\equiv\!C\!-\!H$, is a symmetrical, linear *nonpolar* molecule. **(d)** Substitute Cl atoms for the two H atoms at the double bond in *cis*-2-butene, and EN differences make the resulting nonsymmetric molecule *polar*. **10.9A** H_2^- should be somewhat stable; two electrons in bonding MO and one, in antibonding MO; bond order is 1/2. **10.9B** The molecule He_2, has two electrons in a bonding MO and two in an antibonding MO. Its bond order is zero and the molecule is not stable. In the ion, He_2^+, there are two electrons in the bonding MO and one in the antibonding MO. The diatomic ion has a bond order of $\frac{1}{2}$ and should be stable. Presumably, He_2 ions with $2+$ or $3+$ charges would also be stable. **10.10A** No effect for Li_2 and Be_2 because the σ_{2p} and π_{2p} molecular orbitals are empty. No effect for N_2 because the σ_{2p} and π_{2p} molecular orbitals are filled. No effect for O_2, F_2 and Ne_2 because the σ_{2p} and π_{2p} molecular orbitals are filled, and the highest energy orbitals with electrons are the π_{2p}^* orbitals (σ_{2s}^* in Ne_2). In B_2, reversal of the σ_{2p} and π_{2p} orbitals would replace the two unpaired electrons in the π_{2p} orbitals by a pair of electrons in the σ_{2p}; the bond order would be unchanged, but the molecule would be diamagnetic instead of paramagnetic. In C_2, the bond order would be unchanged, but the species would be paramagnetic instead of diamagnetic. **10.10B** The 11 valence electrons are distributed in the MO diagram of either N_2 or O_2. In either case, bond order $= (8 - 3)/2 = 2.5$. **Self-Assessment Questions: 3.** (c). **4.** (a). **5.** sp^2, sp^2, sp^3. **6. (a)** 180°; **(b)** 120°; **(c)** 109.5°. **7.** (b). **10.** (a). **14.** (c). **15.** Five σ, one π. **17.** (b).**Problems: 21. (a)** AX_2E_2; **(b)** AX_4; **(c)** AX_3E. **23. (a)** angular; **(b)** tetrahedral; **(c)** trigonal pyramidal. **25. (a)** linear; **(b)** linear; **(c)** tetrahedral; **(d)** trigonal planar. **27. (a)** angular; **(b)** trigonal pyramidal; **(c)** square planar; **(d)** trigonal pyramidal. **29. (a)** Tetrahedral around CH_3 carbon, C-C-N is linear; **(b)** trigonal planar around both nitrogen atoms; **(c)** tetrahedral around CH_3 carbon, C-S-H is angular (109°). **31.** B has 3 electron groups, all bonded, while Cl has 5 electron groups, 3 bonded; AX_3 vs. AX_3E_2. **33.** In CH_4 all four bonds are identical; in $COCl_2$ there are two C—Cl single bonds and one C=O double bond. The difference in bonds is more likely to cause $COCl_2$ to deviate from the predicted angles. **35.** The angle of an angular molecule depends on the electron-group geometry. Tetrahedral EG geometry produces a bond angle of about 109° (in H_2O), and trigonal planar EG geometry produces a bond angle of about 120° (in SO_2). **37.** (a) polar; (b) polar; (c) polar; (d) nonpolar. **39.** Both have the same angular geometry, based on AX_2E_2, with similar bond angles. However, ΔEN for each O—H bond is greater than that for each O—F bond, so water should have the greater dipole moment.

41.

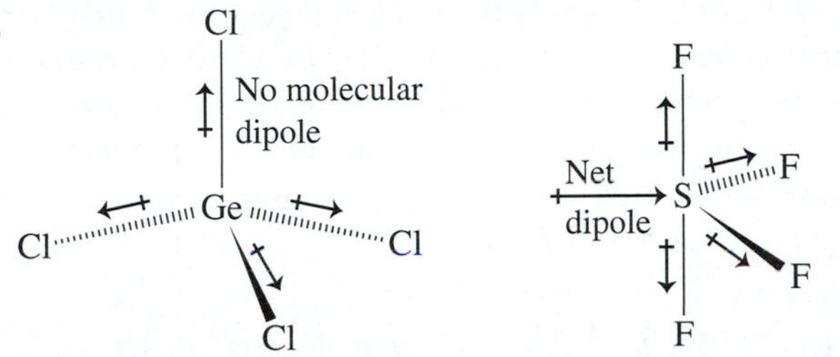

(a)   (b)

43. Li_2 is formed by overlap of two $2s$ orbitals, F_2 is formed by overlap of two $2p$ orbitals. The lobes of $2p$ orbitals can overlap more than the spherical $2s$ orbitals, so the F_2 bond should be stronger. **45.** (a) sp^2; (b) sp; (c) sp^3; (d) sp^3d^2. **47.** (a) N is sp^2, C is sp; (b) CH_3 carbon is sp^3, CN carbon is sp; (c) two CH_3 carbon atoms are sp^3, the two inner carbon atoms are sp; (d) Both C and N are sp^3. **49.** (a) Trigonal planar around N, H—O—N bond is about 109°; (b) Octahedral; (c) Tetrahedral around CH_3 carbon, and H—C≡C—C is linear;

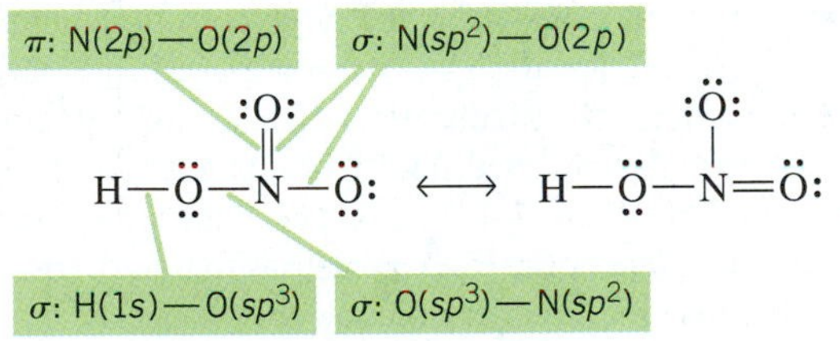

(a)

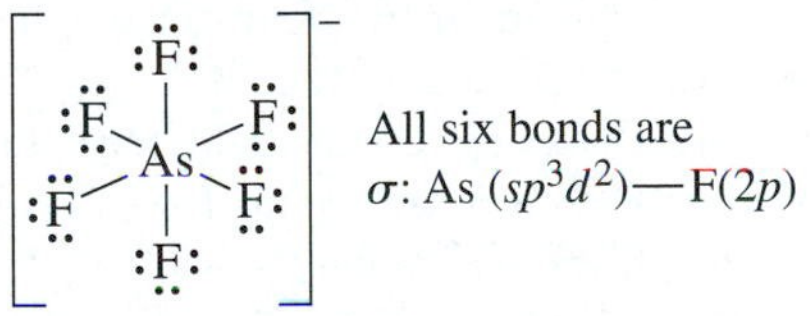

(b)

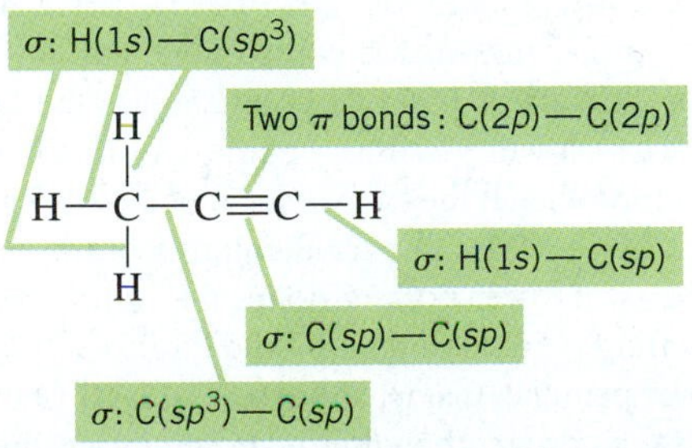

(c)

51. No. There are different numbers of electrons in the two species. Adding two electrons adds an additional lone pair on the central

atom, changing the geometry. **53.** There are four resonance structures, so the two π bonds are delocalized over all four C—O bonds, making the four C—O bonds all the same length and strength. Both carbon atoms are sp^2 hybridized. Selecting one resonance structure for the bonding scheme:

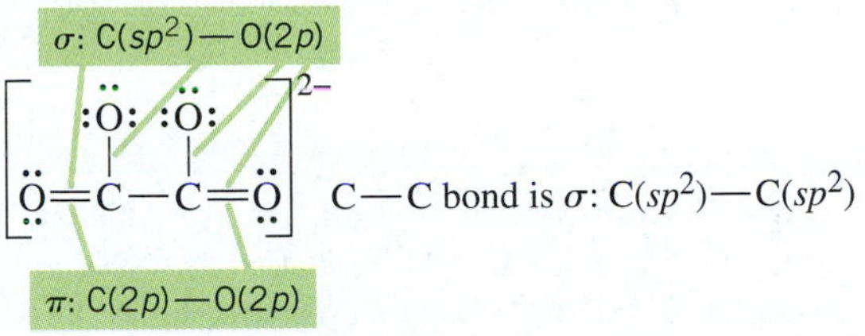

55.

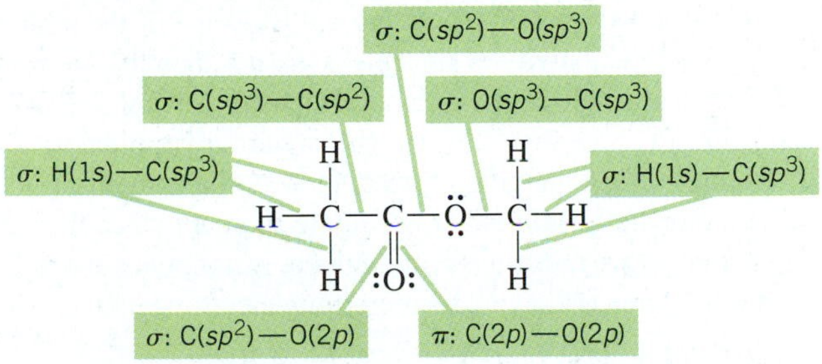

(a)

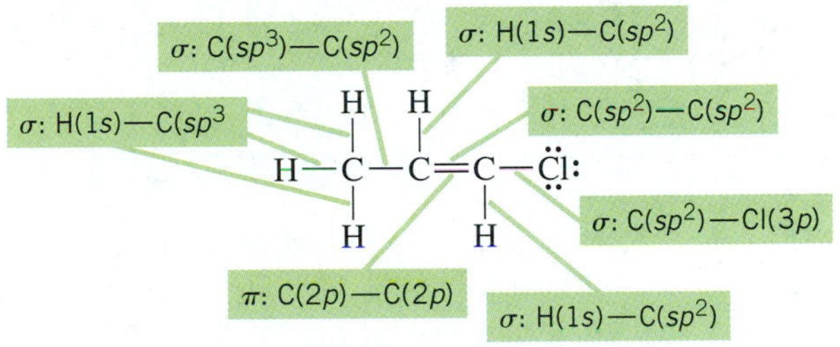

(b)

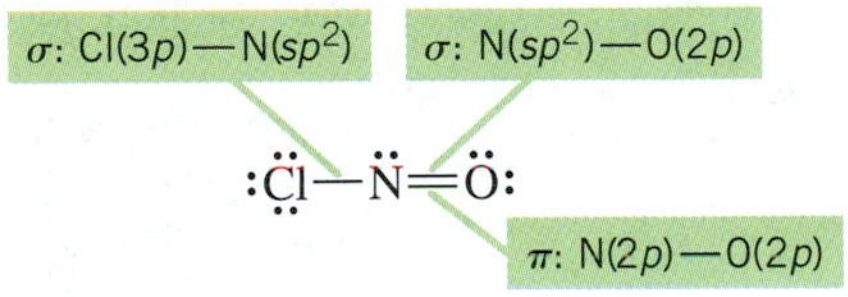

(c)

57. (a), (c). Each side of the double bond must have two different groups or atoms attached. **59.** No, there are two hydrogen atoms on the one side of the double bond. Yes, substituting a chlorine atom as described gives two different groups attached to each side of the double bond. **61.** F_2^+ has greater bond energy because it has fewer antibonding electrons. **63.** Bond order from Lewis structure is 4, and it is 2 from molecular orbital theory. Lewis bonding theory does not account for antibonding electrons. **65.** (a) CN^- (isoelectronic with N_2) has the stronger bond, with a bond order of 3; the bond order of CN^+ (isoelectronic with C_2) is 2; (b) All electrons are paired, so neither is paramagnetic. **67.** O_2^{2-} has a bond order of 1 and is diamagnetic; O_2^- has a bond order of 1.5 and is paramagnetic.

69.

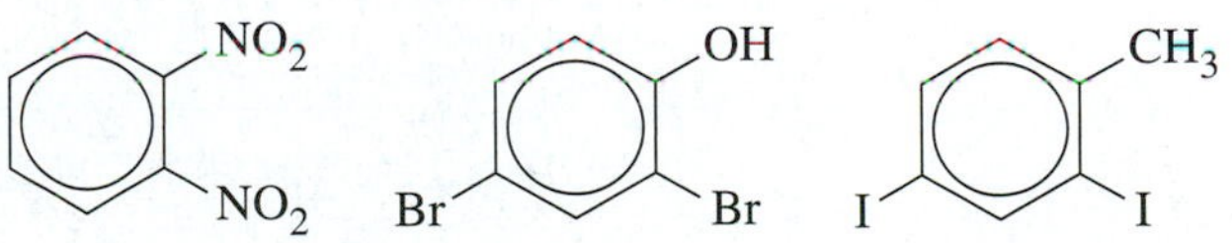

(a)   (b)   (c)

71. **(a)** 2,5-diiodoaniline; **(b)** 1,2,4-trifluorobenzene; **(c)** 3,4,5-tribromotoluene.

73.

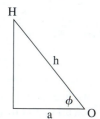

(a)

(b) All C and N atoms are sp^2 hybridized; **(c)** All bond angles (H—C—C, C—C—C, C—C—N, C—N=N) are about 120°, though the C—N—N bond angles are slightly less due to LP-BP repulsion. **Additional Problems: 75.** Not necessarily; bond dipoles may cancel depending on the shape of the molecule. **77.** The electronegativity of fluorine is greater than that of oxygen, resulting in a small shift in electron density toward the fluorine atom. **79.** **(a)** seesaw (AX_4E); **(b)** square pyramidal (AX_5E). **82.** ethane < methylamine < methanol < fluoromethane; the C and H atoms have little effect on dipole moment since their electronegativities are about the same. Ethane is nonpolar, and polarity of the other three is dictated by increasing electronegativity from N to O to F. **84.** Linear molecule;

σ: C(sp)—O($2p$)

σ: C(sp)—C(sp)

$\ddot{O}$=C=C=C=$\ddot{O}$

π: C($2p$)—C($2p$)

π: C($2p$)—O($2p$)

87. The central carbon atom is sp hybridized, and the other two carbon atoms are sp^2 hybridized. **90.** 6.09 D, 17.6% ionic character. **92.** We can draw a right triangle for which the hypotenuse is the magnitude of the O-H dipole, the adjacent side is half the net dipole moment (there are two O—H bonds and both contribute to the net dipole), and the included angle ϕ is half the bond angle of 104.5°, and solve for the adjacent side

(a) $\cos \phi = \dfrac{a}{h}$

$\cos \dfrac{104.5°}{2} = \dfrac{(1.84 \text{ D}/2)}{h}$

$h = 1.50 \text{ D}$

(b) Substitute S for O:

$\cos \phi = \dfrac{a}{h}$

$\cos \phi = \dfrac{(0.93 \text{ D}/2)}{0.67 \text{ D}} = 0.69$

$\phi = 46°$ $2\phi = \text{bond angle} = 92°$

Apply Your Knowledge: 95. **(a)** H—C—C—H   Cyclopentane

(b) H—C—C=C—H   Propene

(d) No. **96.** **(a)** $\text{Li}_2 \longrightarrow \text{Li}_2^+ + e^-$; $\text{Be}_2 \longrightarrow \text{Be}_2^+ + e^-$; $\text{B}_2 \longrightarrow \text{B}_2^+ + e^-$; $\text{C}_2 \longrightarrow \text{C}_2^+ + e^-$; $\text{N}_2 \Longrightarrow \text{N}_2^+ + e^-$; $\text{O}_2 \longrightarrow \text{O}_2^+ + e^-$; $\text{F}_2 \longrightarrow \text{F}_2^+ + e^-$; $\text{Ne}_2 \longrightarrow \text{Ne}_2^+ + e^-$; **(b)** Yes, Ne_2^+ has more bonding electrons than antibonding electrons; BO = 0.5; **(c)** An energy diagram similar to Figure 10.25 for F_2 shows that the electron comes from the π^* antibonding orbital of F_2, which is at a higher energy level than the p orbital in F; **(d)** The electron comes from the π^* antibonding orbital of O_2, which is at a higher energy level than the p orbital in O, so O_2 has a lower first ionization energy than O; **(e)** The electron comes from the σ *bonding* orbital of N_2, which is at a *lower* energy level than the p orbital in N, so N_2 has a *higher* first ionization energy than N; **(f)** The first ionization energy increases as $\text{O}_2 < \text{O} < \text{N} < \text{N}_2$, so N_2 has a much higher ionization energy than does O_2 and is thereby better suited for plasma spectrometry than is O_2. Helium, having an even higher ionization energy because of its smaller size and the fact that it is a noble gas, is also a good candidate.

Chapter 11

Exercises: 11.1A $\Delta H_{vapn} = 34.0$ kJ/mol C_6H_6. **11.1B** $\Delta H_{total} = 92.0$ kJ. **11.2A** CCl_4. A relatively low ΔH_{vapn} and a relatively high molar mass lead to the smallest heat requirement per kilogram. **11.2B** Approximately 97 g. **11.3A** 4.00×10^2 mmHg. **11.3B** 5.33×10^{-3} g H_2O. **11.4A** Room temperature is far above T_c of methane; the methane is present as a gas of relatively low mass, and a pressure gauge is better used to measure its amount. **11.4B** What is observed depends on the initial volume of liquid. To see the disappearance of the meniscus, both liquid and vapor must be present as T_c is reached. Because the densities of the liquid and vapor are equal at T_c, the required mass of substance for the critical phenomena to be observed is the product of the volume of the tube and the critical density. Thus, for example, if the initial mass of liquid is less than this calculated mass, all of the liquid will have vaporized before T_c is reached. **11.5A** If only gas were present, the pressure would be 10 atm—far in excess of the vapor pressure of H_2O at 30.0 °C. Most of the H_2O condenses to liquid. (It cannot all be liquid, because the liquid volume is only about 20 mL and the system volume is 2.61 L.) The final condition is a point on the vapor pressure curve at 30.0 °C. **11.5B** The very cold, solid dry ice (−78.5°C) absorbs heat from the much warmer water. The CO_2(s) sublimes, producing a large volume of cold CO_2(g), which escapes as large bubbles. Water vapor in contact with the cold CO_2(g) condenses to droplets of H_2O(l) ("fog"). This fog of H_2O(l) is carried off by the invisible CO_2(g) and down the walls of the beaker. The CO_2(g) is more dense than air. **11.6A** IBr and BrCl have about the same polarity, but IBr has a greater molar mass. IBr has stronger intermolecular forces and is the solid. BrCl is the gas. **11.6B** **(a)** The molar masses are comparable, but aniline is somewhat polar and toluene is nonpolar. Toluene has the lower boiling point. **(b)** The lower boiling point is that of nonpolar trans-1,2-dichloroethene; the cis isomer is polar. **(c)** The symmetrical placement of Cl atoms in para-dichlorobenzene makes it nonpolar and gives it a lower boiling point; the ortho isomer is polar. **11.7A** The boiling point of butane should be about 36 °C below the boiling point of pentane, that is, about 0 °C. **11.7B** $\text{C}_{11}\text{H}_{24}$ to about $\text{C}_{15}\text{H}_{32}$. **11.8A** Hydrogen bonding is an important intermolecular force in NH_3, $\text{C}_6\text{H}_5\text{OH}$, CH_3COOH, H_2O_2. **11.8B** (d) < (b) < (a) < (c). Isobutane (d) and carbon disulfide (b) are both nonpolar molecules with intermolecular forces limited to dispersion forces.

For these two the order is (d) < (b). Isopropanol (a) and ethylene glycol (c) are both polar and both undergo hydrogen bonding. This suggests that both should have boiling points above that of carbon disulfide, despite their somewhat smaller molar masses. Ethylene glycol has two —OH groups whereas isopropanol has only one —OH group, so we should expect the ethylene glycol to have the highest boiling point. The actual boiling points are (d) isobutane, $-11.7°C$, (b) carbon disulfide, $46.6 °C$, (a) 2-propanol, $82.4 °C$, (c) ethylene glycol, $197.3 °C$. **11.9A** $CsBr < KI < KCl < MgF_2$. **11.9B** MgF_2 is water insoluble. We expect that the interionic forces of attraction in MgF_2 should be considerably greater than in the other three compounds, because of the smaller size and higher charge of Mg^{2+} compared to K^+ and Cs^+, and because of the smaller size of F^- relative to Cl^-, Br^-, and I^-. **11.10A** 286.6 pm. **11.10B** $V = (286.6)^3$ $pm^3 = 2.354 \times 10^{-23} cm^3$; number Fe atoms/unit cell = 2. **11.11A** $N_A = 6.05 \times 10^{23}$. **11.11B** void fraction = $[(2r)^3 - 4/3 \times \pi r^3]/(2r^3)$ = $[(8 - 4\pi/3)/8] = 0.4764$; % voids = $0.4764 \times 100\%$ = 47.64%. **Self-Assessment Questions: 3.** (a). **5.** (d). **7.** (a). **8.** (d). **10.** (c). **13.** (b). **14.** (c). **16.** (b). **Problems: 17.** 1280 kJ. **19.** 199 kJ **21.** 328 g. **23.** When holding your hand in the oven, you are feeling the slow transfer of heat from air to your hand; the air remains in the gas phase. Above the boiling water, steam condenses to liquid water on the skin, with evolution of the very large heat of condensation onto the skin. **25. (a)** ~40 mmHg; **(b)** ~62 °C. **27.** 1.57 L. **29.** 214 mmHg. **31.** The sample cannot be liquid only; 1.82 g $H_2O(l)$ occupies less than 2 mL volume and the container holds 2.55 L. It cannot be vapor only because the $H_2O(g)$ would exert a pressure (749 mmHg) far in excess of the vapor pressure at 30.0 °C. It is a mixture of liquid and gas. **33.** Heat goes into vaporizing water—the water boils at a constant temperature. Because its ignition temperature is greater than 100 °C, the paper does not burn. **35.** A gas cannot be liquefied above T_c, regardless of the pressure applied. A gas can be either liquefied or solidified by a sufficient lowering of the temperature, regardless of its pressure. A gas can always be liquefied by an appropriate combination of pressure and temperature changes. **37.** 7.8 kJ **39.** 8.4 °C. **41.** (c).

43. (a)

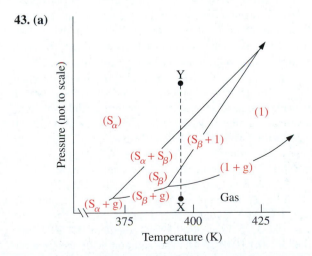

(b) The 2 triple points at the bottom have S_α, S_β and gas present. The triple point at the top has S_α, S_β and liquid present. **(c)** The sulfur gas becomes liquid, then monoclinic sulfur and finally rhombic sulfur as pressure increases.

45.

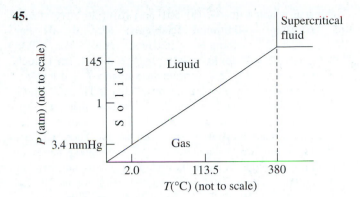

47.

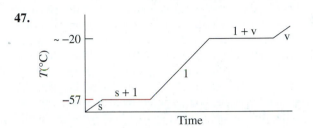

The solid CO_2 becomes a liquid slightly above $-56.7 °C$ (the triple point is 5.1 atm and $-56.7 °C$). The liquid then converts to a gas at a temperature that is probably below room temperature (the critical point is at 72.9 atm and 304.2 K). **49.** Both are nonpolar, CF_4 has fewer electrons and weaker dispersion forces than PBr_5, so CF_4 is lower boiling. **51.** The substances have similar molar mass, but 1-pentanol molecules can form intermolecular hydrogen bonds and 3,3-dimethylpentane molecules cannot. Further, 3,3-dimethylpentane molecules are more compact, have less intermolecular contact and therefore weaker dispersion forces than 1-pentanol. 1-Pentanol has the higher melting point. **53.** BF_3 is the gas. It has small, non-polar molecules; NI_3 is a solid; it has large molecules, large dispersion forces, and is slightly polar; PCl_3 is a liquid with an intermediate molecular mass; CH_3COOH is a liquid–similar mass to BF_3, but it undergoes hydrogen bonding. **55.** $CH_4 < CH_3CH_3 < NH_3 < H_2O$; first two are nonpolar and CH_4 molecules are smaller and thus lower boiling. Both NH_3 and H_2O have hydrogen bonding, but oxygen is more electronegative than nitrogen, so water is higher boiling. **57.** $CH_3OH < C_6H_5OH < NaOH < LiOH$; first two undergo hydrogen bonding, with methanol molecules being smaller (weaker dispersion forces). The other two are ionic with strong interionic forces. The melting point of LiOH is higher than that of NaOH due to the high charge density of the Li^+ ion. **59.** 1-Octanol undergoes hydrogen bonding while octane has only dispersion forces, so 1-octanol has stronger intermolecular forces and a higher surface tension. **61.** Water "wets" glass because the adhesive forces between water molecules and glass exceed the cohesive forces in liquid water (glass surface is polar due to lone pairs on the oxygen atoms in SiO_2). The reverse is true with substances that water does not wet, for example, Teflon® or wax paper. Additives can make water "wetter" if they either reduce cohesive forces or increase adhesive forces (wetting agents usually do the latter). **63. (a)** The larger unit is a unit cell, but not the smaller one. The larger cell can be shifted left, right, up, and down by one unit cell (four symbols) and will coincide with the original pattern. In contrast, when a second small square is stacked on top of the first, the two sides of the interface show different symbols. **(b)** There are 3 hearts, 3 diamonds, and

3 clubs in a unit cell. **65. (a)** 501 pm; **(b)** 1.26×10^{-22} cm^3; **(c)** 3.62 g/cm^3. **Additional Problems: 67. (a)** 404 pm; **(b)** 6.4×10^{23} formula unit/mol. **69.** For Ti^{4+}, $8 \times (1/8) + 1 = 2$; for O^{2-}, $4 \times (1/2) + 2 = 4$. The formula Ti$_2$O$_4$ reduces to TiO$_2$. Ti^{4+} coordination number is 6; that of O^{2-} is 3. No, because there are twice as many O^{2-} as Ti^{4+}, the coordination number of Ti^{4+} should be twice that of O^{2-}. **71.** 1.2×10^{-3}. **73. (a)** 0.213 kJ; **(b)** 28.7 L; **(c)** only gas. **76. (a)** Hydrogen bonding is most important. There are also dipole-dipole and dispersion forces. **(b)** Dipole-dipole forces are most important. There are also dispersion forces. **(c)** Ionic bonding is the most important. There are also dispersion forces. **78.** The density of H$_2$O(l) rises from its value at the melting point to a maximum at 3.98 °C. Above this temperature its density falls with temperature as with most substances. Thus, for every density in the temperature range from 0 °C to 3.98 °C, there is another temperature in a range extending a bit above 3.98 °C at which the same density is observed. **80. (a)** 56.9 mmHg; **(b)** −33.36 °C; **(c)** 108.9 atm. **81.** The Group 4A hydrides show the expected trend: dispersion forces and boiling points increase with increasing molar mass. For the Group 5A, 6A, and 7A hydrides of periods 3, 4, and 5, boiling points also increase with molar mass. Three anomalies are seen for NH$_3$, H$_2$O, HF. The unusually high boiling points for NH$_3$, H$_2$O, and HF result from hydrogen bonding in these liquids. **83.** 4.0×10^2 mmHg. **85.** 27 °C. **90.** 3.516 g/mL. **94.** 1.41×10^3 g. **98. (a)** 599 mmHg; **(b)** 93 °C. **100.** 9.5 mmHg at −101.65 °C. **Apply Your Knowledge: 102.** 0.021 J/m^2. **104.** 22 times.

Chapter 12

Exercises: 12.1A 17.8% glucose, by mass; only the mass ratio of glucose to solution enters into the calculation, not the formula of glucose. **12.1B** 6.98% sucrose, by mass. **12.2A** 34.8% toluene, by volume. **12.2B (a)** 34.4% toluene, by mass; **(b)** $d = 0.874$ g/mL. **12.3A (a)** 0.1 ppb; **(b)** 100 ppt. **12.3B** 69.9 ppm Na$^+$. **12.4A** 0.317 m C$_6$H$_{12}$O$_6$ **12.4B** 3.56 m. **12.5A** 0.418 mL CH$_3$CH$_2$OH. **12.5B** 253 mL H$_2$O. **12.6A (a)** 2.53 m CH$_3$OH; **(b)** 1.86 mol% CO(NH$_2$)$_2$. **12.6B (a)** 8.50% CH$_3$OH, by mass; **(b)** 2.61 M CH$_3$OH; **(c)** 4.97 mol% CH$_3$OH. **12.7A (b).** The solutions are all rather dilute and have densities of approximately 1.0 g/mL. The mole percents of CH$_3$CH$_2$OH are not large, and we expect the largest to be in the solution having the greatest quantity of CH$_3$CH$_2$OH per liter or kilogram of solution. Solution (a) has 0.5 mol CH$_3$CH$_2$OH per liter; (b) has slightly more than 1 mol CH$_3$CH$_2$OH per kilogram of solution; (c) has slightly less than 0.5 mol CH$_3$CH$_2$OH per kilogram of solution; (d) has somewhat less than 1 mol CH$_3$CH$_2$OH per liter. **12.7B (d).** Compare 1.0 kg of each of the four solutions, keeping in mind that 1000 mL H$_2$O $\approx$ 1000 g H$_2$O $\approx$ 55 mol H$_2$O. All the solutions are rather dilute and have H$_2$O as the preponderant component. Solution (a) has about 0.01 mol Na$^+$, whereas solution (b) has about 0.02 mol Na$^+$ (for dilute aqueous solutions molality and molarity are very nearly the same). Solution (c) has 10 g NaNO$_3$ per kg solution, making it about 10/85 m NaNO$_3$ and containing slightly more than 0.10 mol Na$^+$. In solution (d), for every 1000 g H$_2$O present there would be about 55 mol H$_2$O and about 0.55 mol Na$^+$. Solution (d) has the greatest concentration of Na$^+$, whether expressed on a mol or ppm basis. **12.8A** Because of the similarity of its structure to that of benzene, nitrobenzene is more soluble in benzene

than in water. **12.8B** The molecular models; (a) acetic acid, (b) 1-hexanol, (c) hexane, (d) butanoic acid, the order of increasing solubility in water: c < b < d < a. **12.9A** 5.5×10^{-2} mg CO$_2$/100 g H$_2$O. **12.9B** 2.1×10^2 mg. **12.10A** 93.9 mmHg. **12.10B** 8.03 g C$_{10}$H$_8$ **12.11A** Partial and total pressures: $P_{benz} = 51.4$ mmHg; $P_{tol} = 13.0$ mmHg; $P_{total} = 64.4$ mmHg. **12.11B** Mole fractions: 0.230 for benzene and 0.770 for toluene. **12.12A** The solution with equal masses has a greater mole fraction of the more volatile benzene (78 g/mol) than of toluene (92 g/mol); it produces vapor with the greater $x_{benzene}$. **12.12B** $P_{benzene} = 73.2$ mmHg, $P_{toluene} = 6.53$ mmHg. **12.13A** No. Water vapor passes from the more dilute (B) to the more concentrated solution (A) until the two solutions have the same concentration. Then, there is no further net transfer of water between the solutions. **12.13B** There are several possibilities as to what might happen. If the second liquid is miscible with water in all proportions and is also volatile (e.g., methanol, ethanol), vapor escaping from each pure liquid will dissolve into the other liquid; given enough time the two solutions would attain the same mole fraction composition. If the second liquid is miscible with water in all proportions but is essentially nonvolatile (e.g., glycerol), only water vapor will be transferred, and in time all the water would be transferred from beaker A to B. If the second liquid is immiscible with water or nearly so, very little, if any, substance will be transferred through the vapor phase. **12.14A** −2.46 °C. **12.14B** 18 g sucrose. **12.15A** C$_6$H$_6$O$_4$. **12.15B** No. The greatest freezing point depression would occur if the solid were pure glucose (the substance with the smaller molar mass). The molality of the glucose solution would be 0.555 m; the freezing point of the solution would be −1.03 °C. The freezing point of the solution could not be as low as −1.25 °C. **12.16A** 6.86×10^4 g/mol. **12.16B** 9.33 mm H$_2$O. **12.17A** Lowest fp: 0.0080 M HCl; highest fp: 0.010 m C$_6$H$_{12}$O$_6$. **12.17B** They are the results expected for the expression: $\Delta T_f = -i\,K_f\,m$, with $i = 1$ for HCl dissolved in C$_6$H$_6$, and $i = 2$, in H$_2$O. **Self-Assessment Questions: 2.** Molarity involves a volume, which varies with T. Molality does not involve a volume and is therefore independent of T. Mole fraction and mass percent do not involve a volume and are therefore independent of T. Volume percent varies with T. **5.** (c), (e); (Each involves two nonpolar substances.). **6.** (c). **7.** (a). **9.** (b). **10.** Usually not. If only the solvent is volatile, the vapor cannot have the same composition as the solution because the vapor is pure solvent. If all components are volatile, the vapor and solution compositions are likely to differ because of differences in the volatilities of the components. **13.** (a) **17.** No. Osmotic flow is of *solvent* molecules from a *dilute* solution, which has a higher vapor pressure, into a more concentrated one. The concentration of solvent is greater in the dilute solution, so the situation is similar to gases diffusing from higher to lower pressure. **20. (a)** 0.1 M NaHCO$_3$; **(b)** 1 M NaCl; **(c)** 1 M CaCl$_2$; **(d)** 3 M glucose. **Problems: 21.** Mix 112 g NaNO$_3$ and 2.19 kg water. **23. (a)** 17.5% by mass; **(b)** 17.9% by mass; **(c)** 42.7% by mass. **25. a)** 9.28% by volume; **(b)** 11.7% by volume; **(c)** 29.3% by volume. **27.** 12 g. **29. (a)** 5 ppb trichloroethylene; **(b)** 2.5 ppm KI; **(c)** 3.9×10^{-4} M SO$_4$$^{2-}$. **31.** ppt < ppb < ppm < 1 mg/dL < 1%. **33.** 5.38 m. **35.** 0.598 M; 0.624 m. **37.** 10.7 mol. **39. (a)** 0.0434; **(b)** 0.0192. **41.** The solution with the greatest amount of solute per unit mass of solvent—(b) in this case—has the greatest mole fraction of solute. Solution (a) has exactly 1 mol solute per 1000 g H$_2$O. (b) has slightly more than 1 mol solute in 950 g H$_2$O. Solution (c) has about 0.3 mol solute per 900 g H$_2$O.

43. (a) $CHCl_3$, insoluble; **(b)** C_6H_5COOH, slightly soluble; **(c)** $CH_3CHOHCH_2OH$, highly soluble. **45.** Unsaturated. **47. (a)** 21 g water; **(b)** about 44 °C. **49.** Yes. Allow solvent to evaporate. When the solution becomes saturated, crystallization begins to occur. **51. (a)** 1.38×10^{-3} M; **(b)** 7.2 atm. **53. (a)** Partial pressures: pentane, 88.2 mmHg; hexane, 96.8 mmHg; **(b)** Vapor composition: $x_{pent} = 0.477$; $x_{hex} = 0.523$. **55.** 17.4 mmHg. **57. (a)** -0.47 °C; **(b)** 3.70 °C. **59.** 1.28 m. **61.** $K_f = 4.27$ °C m^{-1}. **63.** $C_6H_3O_6N_3$. **65.** The salt solution has the higher osmotic pressure. As it shrinks into a pickle, the cucumber loses water to $NaCl(aq)$. **67.** 9.94 atm. The solution is hypertonic. **69.** To the right (from A to B). **71. (a)** $i = 1$; $T_f = -0.19$ °C; **(b)** $i \approx 3$; $T_f \approx -0.57$ °C; **(c)** i is slightly greater than 1; T_f is slightly less than -0.19 °C; **(d)** $i \approx 2$; $T_f \approx -0.38$ °C. Answer (a) is most precise, since glucose has i of precisely 1. The others are electrolytes and dissociate; i depends on concentration and is less precisely known. **73.** 3; CaO will form $Ca(OH)_2$ in solution, which is partially soluble in water. **75.** Order of decreasing freezing points: (b) > (a) > (e) > (c) > (d). **77.** 2.4×10^2 g NaCl. **79.** The charge on the particles is negative. Al^{3+} is the most highly charged cation of the three, making $AlCl_3(aq)$ the most effective coagulant. **Additional Problems: 81. (a)** 15.1% by volume; **(b)** 12.2% by mass; **(c)** 11.9% mass/volume; **(d)** 7.25 mole %. **84.** 27.0% H_2O by mass. **86.** 486 g sucrose. **88.** 11.7% nitrobenzene by mass; 95% nitrobenzene by mass; They will not have the same boiling points, because the constants are different and the initial boiling points are different. **91. (a)** 0.20 m $(CH_3)_2CO$ has the highest total vapor pressure because x_{water} is as great as in the other solutions and $(CH_3)_2CO$ is volatile; **(b)** NaCl(satd. aq) has the highest concentration of solute particles and the lowest freezing point. **(c)** The vapor pressure of NaCl(satd. aq) remains constant because its concentration does not change as $H_2O(g)$ is lost. The concentrations and vapor pressures of the other solutions do change. **93. (a)** $x_{benzene} = 0.256$; $x_{toluene} = 0.744$; **(b)** $x_{benzene} = 0.455$; $x_{toluene} = 0.545$. **96.** At equilibrium, each solution has a mass fraction of urea of 0.230 and a mole fraction of 0.0822. **100.** 48.6% by mass sucrose, 51.3% by mass glucose. **102.** The liquid is $CaCl_2(aq)$. Water from the air condenses on the solid and dissolves it to form saturated $CaCl_2(aq)$. A solid will not act this way if the vapor pressure of the saturated solution exceeds the partial pressure of water in the atmosphere. **104. (a)** 14 divisions; **(b)** total volume $= 1$ cm^3; total surface area $= 9.82 \times 10^4$ cm^2; The smaller the particle, the larger the surface area to volume ratio. **Apply Your Knowledge: 105.** 9×10^9 Zn atoms **107. (a)** The weight molarity only depends on mass, so it is not dependent on temperature. **(b)** 0.3473 $M_{(wt)}$; 32.40% by mass sulfuric acid **109.** 0.36 L

Chapter 13

Exercises: 13.1A (a) 0.0259 M min^{-1}; **(b)** 8.63×10^{-4} M s^{-1}. **13.1B (a)** 1.05×10^{-5} M s^{-1}; **(b)** 3.15×10^{-5} M s^{-1}. **13.2A (a)** 1.11×10^{-3} M s^{-1}; **(b)** $[H_2O_2]_{310s} = 0.287$ M. **13.2B** about 270 s. **13.3A** 0.0912 M s^{-1}. **13.3B** 3/2 order. **13.4A (a)** $t = 255$ min; **(b)** $[NH_2NO_2] = 0.0139$ M. **13.4B** 3.46×10^{-4} M min^{-1}. **13.5A (a)** $t \approx 480$ s; **(b)** 0.150 g N_2O_5. **13.5B** $P_{tot} = 1850$ mmHg. **13.6A** Slightly above 50 mmHg, perhaps 52 mmHg. **13.6B** 640 mmHg. **13.7A** $k = 0.023$ M^{-1} s^{-1}. **13.7B** The half-life doubles for each succeeding half-life period—55 s for the first $t_{1/2}$, 110 s for the

second $t_{1/2}$, 220 s for the third $t_{1/2}$, and so on. To two significant figures (corresponding to "55 s"), $[A] = 0.20$ M at $t = (55 + 110) = 1.7 \times 10^2$ s, and $[A] = 0.10$ M at $t = (55 + 110 + 220)$ s $= 3.9 \times 10^2$ s. **13.8A** Rate$_2 = 3.2 \times 10^{-3}$ M s^{-1}, rate$_1 = 8.0 \times 10^{-4}$ M s^{-1}, rate$_2$/rate$_1 = 2^n = 4$, and $n = 2$. For straight-line graphs, plot 1/[A] versus t with data from the graphs given. **13.8B** The $[A]_t$ vs. t graph for the first-order reaction starts at $t = 0$ and 2.00 M, above the graph for the second-order reaction having $[A]_0 = 1.00$ M. After 693 s, $[A]_t$ in the first-order reaction has fallen to 1.00 M, and at 2×693 s $= 1386$ s, to $[A]_t = 0.50$ M. For the second-order graph. $[A]_t = 0.50$ M at 1000 s. At 1386 s, the first-order graph still lies above the second-order graph. After $t = 3 \times 693$ s $= 2079$ s, $[A]_t$ in the first-order reaction has fallen to 0.25 M. In the second-order reaction, $[A]_t$ does not reach 0.25 M until $t = 1000$ s $+ 2000$ s $= 3000$ s. The first-order graph falls below the second-order graph some time between 1386 s and 2079 s. The two graphs share one point in common. The time at this intersection of the two graphs can be estimated by trial and error. Substitute different times into the integrated rate equations for the two reaction orders to obtain values of $[A]_t$. Find the time (about 1700 s) at which the $[A]_t$ values of the two graphs are the same. **13.9A** $T = 288$ K (15 °C). **13.9B** $E_a = 163$ kJ/mol. **13.10A** First step: (slow, rate-determining), $NOCl \longrightarrow NO + Cl$; second step (fast), $NOCl + Cl \longrightarrow NO + Cl_2$. Rate of reaction $= k[NOCl]$. **13.10B** The mechanism consists of a fast, reversible first

step: $NO + O_2 \underset{k_{-1}}{\overset{k_1}{\rightleftharpoons}} NO_3$ followed by a slow, rate-determining

step: $NO_3 + NO \overset{k_2}{\longrightarrow} 2\,NO_2$. Overall reaction is $2\,NO + O_2 \longrightarrow 2\,NO_2$. Assume rapid equilibrium in the first step to establish the rate law: rate $= k[NO]^2[O_2]$. **Self-Assessment Questions: 3.** The *average* rate is the rate (change in concentration of reactant or product/change in time) over a finite time period. The *instantaneous* rate is the slope of a tangent line to a plot of concentration of reactant or product vs. time, corresponding to the rate at a specific instant. The *initial* rate is the instantaneous rate at time $= 0$. The average rate approaches the instantaneous rate as the change in time interval decreases. The initial rate is an instantaneous rate, but the reverse is not necessarily true. **6.** (c). **7.** (d). **8.** (a). **9.** (b). **12.** (c). **15.** (c). **Problems: 23.** 0.222 M. **25. (a)** 3.1×10^{-4} M/s; **(b)** 9.3×10^{-4} M/s; **(c)** rate of reaction of A $= 3.1 \times 10^{-4}$ M/s. **27.** It is not zero order. **29. (a)** False, this would be true only for a second-order reaction, which has not been established. **(b)** True, A is consumed twice as fast as B is produced. **31. (a)** First order in $S_2O_8{}^{2-}$, first order in I^-, second order overall; **(b)** 6.1×10^{-3} M^{-1} s^{-1}; **(c)** 5.8×10^{-5} M/s. **33.** Zero order—rate is independent of concentration. **35.** 7.5×10^{-3} M/s. **37. (a)** 0.325 M; **(b)** 84 min. **39. (a)** 2.2×10^{-5} s^{-1}; **(b)** 569 mmHg; **(c)** 22.1 h. **41.** 1.4×10^{14} molecules/L. **43. (a)** 0.096 M; **(b)** 25 min. **45.** For a zero-order reaction, half-life is directly proportional to concentration, so the half-life is longest initially, when the concentration is highest. For a second-order reaction the half-life is inversely proportional to concentration, so the half-life is shortest initially. **47.** Rate $= k[A]^2$; $k = 9.77 \times 10^{-4}$ M^{-1} s^{-1}. **49.** 84 ms. **51.** The calculation involves the frequency of *effective* collisions, which depends on orientation of colliding molecules and is very difficult to determine accurately. **53.** No. Looking at Figure 13.14 for an endothermic process, it is obvious that the activation energy must be greater than the enthalpy change, since the energy barrier must be at least equal to the energy (enthalpy change) required to reach the products. Looking at Figure 13.13 for an exothermic reaction, it is clear that the activation energy "hill" can be virtually zero, but cannot be defined in relation to the enthalpy change. **55.** 91 kJ/mol. **57. (a)** 7.14×10^{-7} min^{-1}; **(b)** 649 K. **59.** A unimolecular step may occur when a molecule acquires enough energy from collisions with other molecules to dissociate. **61. (a)** A + 2 B $\longrightarrow$ C + D; **(b)** Rate = k[A][B]. **63.** 2 NO$_2$ $\longrightarrow$ NO$_3$ + NO (slow); NO$_3$ + CO $\longrightarrow$ NO$_2$ + CO$_2$ (fast). **65.** The rate law derived from this slow equilibrium/fast step mechanism is Rate = k_2 (k_1/k_{-1})([NO]^2[Cl$_2$]/[NOCl]), which does not match the observed rate law. **67.** rate = k_2 [Hg][Tl^{3+}] k_1/k_{-1} = [Hg][Hg^{2+}]/[Hg$_2^{2+}$] or [Hg] = (k_1/k_{-1})[Hg$_2^{2+}$]/[Hg^{2+}] so rate = k_2 (k_1/k_{-1})[Hg$_2^{2+}$][Tl^{3+}]/[Hg^{2+}] **69.** Since I$^-$ is a catalyst, [I$^-$] remains unchanged, and the rate law simplifies to rate = k' [H$_2$O$_2$], where k' depends on the initial [I$^-$]. **71.** An inhibitor may block the active site of the enzyme or it may react with the enzyme to change the shape of the active site. **73.** Both require that reactions occur at active sites. The kinetics of each type of reaction is governed by the availability of these sites. **Additional Problems: 75.** 2.6×10^{-3} M/s. **76.** 0 s, 35.3 mL; 60 s, 27.9 mL; 120 s, 22.6 mL; 180 s, 18.3 mL; 240 s, 14.9 mL; 300 s, 11.9 mL; 360 s, 9.44 mL; 420 s, 7.52 mL; 480 s, 6.08 mL; 540 s, 4.80 mL; 600 s, 3.76 mL. **79. (a)** 1.06 atm; **(b)** 2.12 atm; **(c)** 2.46 atm. **82.** 35.9 min. **85. (a)** 1, 1, −1, 1; **(b)** rate = k[OCl$^-$][I$^-$]/[OH$^-$], k = 60 s^{-1}; **(c)** second step is rate-determining; rate = k_2[I$^-$][HOCl]; the first step is a fast reversible reaction where [HOCl] = (k_1/k_{-1})[OCl$^-$][H$_2$O]/[OH$^-$]; note that [H$_2$O] is constant, so rate = k_{total}[OCl$^-$][I$^-$]/[OH$^-$]; **(d)** No; OH$^-$ is formed in the first step and consumed in the third step, so it is an intermediate. **86.** O$_3$ $\rightleftharpoons$ O + O$_2$ (fast); O + O$_3$ $\longrightarrow$ 2 O$_2$ (slow). **89.** Second order, k = 0.066 g^{-1} h^{-1}. **90.** 173.2 h. **Apply Your Knowledge: 92.** As the solution becomes more acidic, NH$_2$ groups become NH$_3^+$ groups and COOH groups retain their protons. The groups that are necessary for the reaction to occur are NH$_2$ and COO$^-$. **94. (a)** 51.3 kJ/mol; **(b)** 126 chirps/min; **(c)** 71 °F versus 68 °F, so it is close, but not exact. **95. (a)** Rate = $(\Delta[I_3^-]/\Delta t)$ and $\Delta[I_3^-]$ = the constant amount of S$_2$O$_3^{2-}$ initially added, so an increase in Δt means a decrease in rate; **(b)** first, first, second; **(c)** 3.7×10^{-5} M/s; **(d)** 0.0062 M^{-1} s^{-1}; **(e)** S$_2$O$_3^{2-}$ + 3 I$^-$ $\longrightarrow$ 2 SO$_4^{2-}$ + 3 I$_3^-$, rate = k[S$_2$O$_3^{2-}$][I$^-$]; The first step should be slow because the species colliding are negatively charged (repulsive forces). Step 2 (fast) involves unimolecular decomposition of an unstable, highly charged species. Step 3 (fast) involves collision of species that have opposite charges (attractive forces). Steps 2 and 3 should be largely independent of orientation (2 is unimolecular, and 3 involves monatomic ions).

Chapter 14

Exercises: 14.1A No. [COCl]$_2$ = K_c[CO]^2, but there are many possible values for [CO] = [Cl$_2$], and thus for [COCl$_2$]. **14.1B** [SO$_2$] and [SO$_3$] do not have unique values, but the following ratios do: [SO$_3$]^2/[SO$_2$]^2, [SO$_2$]^2/[SO$_3$]^2, [SO$_3$]/[SO$_2$], [SO$_2$]/[SO$_3$]. If [O$_2$] = 1.00 M, [SO$_3$] = 10.0[SO$_2$]. **14.2A** K_c = 2.5×10^{-3}. **14.2B** K_c = 6.64×10^{78}. **14.3A** K_P = 4.6×10^3. **14.3B** K_c = 3.0×10^{60}. **14.4A** K_P = $P_{CO}P_{H_2}/P_{H_2O}$. **14.4B** K_c = [H$_2$]^4/[H$_2$O]^4; K_P = $P_{H_2}^4/P_{H_2O}^4$. **14.5A** Reaction probably does not go to completion.

K_c = 1.2×10^3 is not a particularly large value. **14.5B** K_P = 10.0. For the two smallest values of K_P, the pressure of H$_2$O could be significant, but not those of CO and H$_2$. For the two largest values, the reaction goes essentially to completion. For the value K_P = 10.0, all the gas partial pressures are most likely to be significant. **14.6A** Net reaction goes to the right. **14.6B** Compared to initial values, at equilibrium: P_{H_2S} increases, P_{HI} decreases, amount of I$_2$(s) increases, amount of S(s) decreases. **14.7A** Compared to the original equilibrium, there will be **(a)** more NH$_3$ and H$_2$, less N$_2$ (equilibrium shifts to the right); **(b)** less NH$_3$ and N$_2$, more H$_2$ (to the left); **(c)** less NH$_3$, N$_2$, and H$_2$ (to the right). **14.7B** Addition of CH$_3$COOH(aq) represents adding both a reactant and a product. If the acetic acid is nearly pure, equilibrium should shift to the right—formation of more products. If the acetic acid is very dilute, equilibrium should shift to the left—formation of more reactants. For intermediate concentrations, the result depends on the exact concentration of the CH$_3$COOH(aq). **14.8A** There is no change because no. mol gaseous products = no. mol gaseous reactants. **14.8B** Both changes shift equilibrium to the left. The amounts of NO and O$_2$ increase as well as the amount of NO$_2$. **14.9A** At low temperatures, because the forward reaction is exothermic. **14.9B** Using data from the problem and Appendix C, we find that the formation of N$_2$O$_3$(g) is an exothermic reaction. Thus, its formation is favored at the lower temperature—the freezing point of water. **14.10A (a)** More CO, H$_2$O, and H$_2$ and less CO$_2$ (equilibrium shifts to the left). **(b)** Prediction not possible. Added H$_2$ favors the reverse reaction; added H$_2$O, the forward reaction. **(c)** Both changes shift equilibrium to the right. There will be more CO$_2$ and H$_2$, and less CO. Whether the amount of H$_2$O increases or decreases depends on the original equilibrium condition. **14.10B** The reaction is CH$_4$(g) + 2 H$_2$O(g) $\rightleftharpoons$ CO$_2$(g) + 4 H$_2$(g). Formation of H$_2$(g) is favored at **(a)** higher temperatures (the reaction is endothermic); and **(b)** lower pressures (the number of moles of products is greater than the number of moles of reactants). **14.11A** K_c = 25. **14.11B** K_P = 0.429. **14.12A** 0.0828 mol H$_2$, and P_{H_2} = 0.815 atm. **14.12B** If the equilibrium concentrations are written with the unit mol/L, the unit L will cancel if it appears as many times in the numerator as in the denominator. This occurs only if the same total number of moles of gas appears on each side of the reaction equation. **14.13A** 0.085 mol COCl$_2$. **14.13B** x_{NO} = 0.018. **14.14A** 1.76×10^{-2} mol H$_2$, 7.6×10^{-3} mol I$_2$, 0.085 mol HI. **14.14B** 0.0128 mol H$_2$; 0.0128 mol I$_2$; 0.094 mol HI. **14.15A** P_{CO} = P_{Cl_2} = 0.19 atm; P_{COCl_2} = 0.81 atm; P_{total} = 1.19 atm. **14.15B** P_{total} = 0.658 atm. **Self-Assessment Questions: 2.** (d). **3.** (b). **4.** (c). **5.** (d). **7.** (d), (f). **9.** (c). **10. (a)** K_p = $P_{CO_2}P_{H_2}/P_{CO}P_{H_2O}$; **(b)** K_p = $P_{NH_3}^2/P_{H_2}^3 P_{N_2}$; **(c)** K_p = $P_{NH_3}P_{H_2S}$. **11. (a)** K_p = $P_{NO}/P_{N_2}^{1/2} P_{O_2}^{1/2}$; **(b)** K_p = $P_{NH_3}/P_{H_2}^{3/2} P_{N_2}^{1/2}$; **(c)** K_p = $P_{NOCl}/P_{N_2}^{1/2} P_{O_2}^{1/2} P_{Cl_2}^{1/2}$. **12. (a)** K_c = [CO$_2$]^2[N$_2$]/[CO]^2[NO]^2; **(b)** K_c = [H$_2$O]^6 [NO]^4/[O$_2$]^5[NH$_3$]^4; **(c)** K_c = [H$_2$O][CO$_2$]. **14.** (a), (c). **17.** (b). **18.** (d). **Problems: 19. (a)** 1.2×10^{-3}; **(b)** 4.79×10^{-3}; **(c)** 1.23×10^3. **21.** 3.59×10^{-3}. **23.** 6.35. **25.** 1.13×10^{-10}. **27.** 3.0×10^{66}. **29.** No, it depends on the value of K_c; if K_c < 1, then K_c' > K_c. **31.** 2.3×10^3. **33.** 0.0126 M. **35.** 0.134 atm. **37.** 9. **39.** (b), (e). **41.** (c). **43. (a)** increase; **(b)** none; **(c)** increase; **(d)** decrease; **(e)** none; **(f)** increase; **(g)** none. **45.** All of the reactions are endothermic, because energy must be expended to break stable bonds. Dissociation is favored by increasing the temperature. **47.** (b), (d) greater. **49.** Q_c = 8.88 < K_c, shifts to the right.

51. $Q_P = 97.4 < K_p$, shifts to the right. **53.** $K_p = 1.80 \times 10^{-6}$; $K_p = 6.78 \times 10^{-5}$. **55.** $NH_2COONH_4(s) \rightleftharpoons 2 NH_3(g) + CO_2(g)$; $K_p = 6.7 \times 10^{-4}$. **57.** 1.49×10^{-3}. **59.** 12.5 g. **61.** 0.0436 mol. **63.** 48.2 g. **65.** $PCl_3 = Cl_2 = 0.082$ mol; $PCl_5 = 0.028$ mol. **67.** 46%. **69. (a)** 0.0271 atm; **(b)** 0.179 atm. **71.** 3.65 atm. **73.** 1 atm. **Additional Problems: 75. (a)** shift to the left; **(b)** left; **(c)** right; **(c)** right; **(d)** right; **(e)** right; **(f)** right. **77. (a)** Left; **(b)** 0.164 mol H_2; 0.164 mol CO_2; 0.036 mol CO; 0.136 mol H_2O. **79.** 5.60. **80.** 3.9; high P and low T. **82.** 8.32×10^{-3} atm. **85.** 78.3%. **88.** 0.92 mol; increase. **Apply Your Knowledge: 90.** 3.9. **92.** 0.0149.

Chapter 15

Exercises: 15.1A (a) base: NH_3, conjugate acid, NH_4^+; acid: HCO_3^-, conjugate base, CO_3^{2-}; **(b)** acid: H_3PO_4, conjugate base, $H_2PO_4^-$; base: H_2O, conjugate acid, H_3O^+. **15.1B** Amphiprotic species in 15.1A: HCO_3^- acts as an acid in (a), and as a base in $HCO_3^- + H_3O^+ \rightleftharpoons H_2CO_3 + H_2O$; H_2O acts as a base in (b), and as an acid in $H_2O + NH_3 \rightleftharpoons OH^- + NH_4^+$. As we will discover later in the chapter, $H_2PO_4^-$ is also amphoteric. In reaction (b) it acts as a base in the reverse reaction, but it is an acid in the reaction $H_2PO_4^- + NH_3 \rightleftharpoons NH_4^+ + HPO_4^{2-}$. **15.2A (a)** H_2Te. The Te atom is larger than S, and the H—Te bonds are weaker. **(b)** $CH_3CH_2CH_2CHBrCOOH$. Although Cl is somewhat more electronegative than Br, it is located much farther from the —COOH group, thereby weakening its electron-withdrawing ability. **15.2B** a < d < b < c. The Cl and Br atoms are electron-withdrawing. The effect is weakest with Br opposite the —COOH group, but stronger with Cl adjacent to the group, and stronger still with two Cl atoms adjacent. **15.3A** d < a < c < b. Aromatic amines are much weaker than aliphatic ones. Cl atoms weaken amine bases because they are electron-withdrawing, and more so with more Cl atoms closer to the —NH_2 group. **15.3B (a)** —$CH_2CH_2CH_3$, no electron-withdrawing groups; **(b)** The $C_6H_3Br_2$ group produces an aromatic amine with two electron-withdrawing bromine atoms near the —NH_2 group. **15.4A** pH = 1.331. **15.4B** pH = 12.624. **15.5A** Basic. OH^- from the NaOH raises $[OH^-]$ above the 1.0×10^{-7} M found in water. **15.5B** The pure water has a pH = 7.00. The pH of each solution is also very close to 7.00, one slightly acidic and the other slightly basic. Because equal amounts of excess H_3O^+ and OH^- are present, complete neutralization occurs and the final mixture is pH neutral. **15.6A** pH = 2.74. **15.6B** pH = 4.52. **15.7A** pH = 2.20. **15.7B** The two relationships that must be simultaneously satisfied are $x^2/M = 1.4 \times 10^{-3}$ and $x/M = 0.050$. The minimum molarity is 0.56 M. **15.8A** pH = 11.96. **15.8B** pH = 9.35. **15.9A** $K_a = 6.2 \times 10^{-5}$; $pK_a = 4.21$. **15.9B** $[NH_3] = 6.56$ M. **15.10A** Methylamine. A larger K_b and higher molarity make $[OH^-]$ and pH of CH_3NH_2 greater than those of NH_3(aq). **15.10B** 0.0010 M HCl(aq). In 0.0010 M HCl(aq), a strong acid, $[H_3O^+] = 0.0010$ M, and pH = 3. In 0.10 M CH_3COOH(aq), a weak acid, $[H_3O^+] = \sqrt{K_a \times 0.10} = \sqrt{1.8 \times 10^{-6}} > 1.0 \times 10^{-3}$ M, and pH < 3. **15.11A** pH = 1.48. **15.11B** The molarity of H_3PO_4 in the cola is between about 4.4×10^{-3} M and 6.4×10^{-3} M. The corresponding pH values are 2.52 and 2.39, respectively, and they correlate well the pH $\approx$ 2.5 stated in the problem. **15.12A** Assume complete ionization: pH = 2.77. **15.12B (b)**—complete ionization in the first step and limited in the second. Response (a) has no ionization in the second step and (c), nearly complete ionization. In (d), $[H_3O^+]$ cannot exceed 0.040 M.

15.13A (a) Neutral—salt of strong acid and strong base; **(b)** basic—salt of weak acid and strong base. **15.13B** HCl(aq) < NH_4Br(aq) < NaCl(aq) < KNO_2(aq) < NaOH(aq). **15.14A** pH = 5.27. **15.14B** Neutralization produces 0.0873 M CH_3COONa; pH = 8.84. **15.15A** 0.29 M CH_3COONa(aq). **15.15B** 0.10 M NH_4CN. HCN is a much weaker acid than HNO_2 —the weaker the acid, the stronger the anion as a base. **15.16A** pH = 8.89. **15.16B** H_3O^+ is the common ion; $[H_3O^+] = 0.10$ M, $[CH_3COO^-] = 1.8 \times 10^{-5}$ M. **15.17A** pH = 9.10. **15.17B** 0.020 mol NaOH. **15.18A** $[CH_3COOH] = 0.43$ M. **15.18B** 3.2 g NH_4Cl. **15.19A** The pH $\approx$ 5. This eliminates (b)—a basic buffer—and (c)—strongly acidic. Add a small quantity of acid or base. If there is a color change, the solution is (a); if not, it is the buffer (d). **15.19B** 0.0200 mol NaOH converts all the CH_3COOH to CH_3COO^-. The pH is approximately 9, about the pH at which thymol blue changes color. **15.20A (a)** pH = 2.90; **(b)** pH = 3.90; **(c)** pH = 10.10; **(d)** pH = 11.10. **15.20B** pH = 13.02. **15.21A (a)** pH = 4.96; **(b)** pH = 11.10. **15.21B (a)** pH = 13.22; **(b)** pH = 9.07; **(c)** pH = 5.04; **(d)** pH = 4.74. The titration curve starts at a high pH; has a pH > 7 at the equivalence point; and progresses through a buffer region in a weakly acidic solution beyond the equivalence point. **15.22A** $K_b \approx 1 \times 10^{-5}$; pH at the equiv. point: $\approx$5 (estimated from graph); 4.7 (calculated, assuming $K_b = 1 \times 10^{-5}$). **15.22B** This would not be a satisfactory titration. The change in slope of the titration curve would be too gradual. For instance, the initial pH would be about 11; the pH at the equivalence point would be about 7; and the final pH, about 4. **Self-Assessment Questions: 1.** Arrhenius: HI(aq) $\longrightarrow$ $H^+ + I^-$; Brønsted-Lowry: HI(aq) + H_2O $\longrightarrow$ $H_3O^+ + I^-$. **3.** (c). **4.** (b), (d). **5. (a)** $HClO_2 + H_2O \rightleftharpoons H_3O^+ + ClO_2^-$;

$$K_a = \frac{[H_3O^+][ClO_2^-]}{[HClO_2]}$$

(b) $CH_3CH_2COOH + H_2O \rightleftharpoons H_3O^+ + CH_3CH_2COO^-$;

$$K_a = \frac{[H_3O^+][CH_3CH_2COO^-]}{[CH_3CH_2COOH]};$$

(c) $HCN + H_2O \rightleftharpoons H_3O^+ + CN^-$; $K_a = \dfrac{[H_3O^+][CN^-]}{[HCN]}$;

(d) $C_6H_5OH + H_2O \rightleftharpoons H_3O^+ + C_6H_5O^-$

$$K_a = \frac{[H_3O^+][C_6H_5O^-]}{[C_6H_5OH]}.$$

6. (d). **9.** (a). **12.** (d). **13.** $K_a \times K_b = 10^{-14}$ **16.** Equivalence point: the point at which acid and base are in the exact stoichiometric proportions. Endpoint: the point at which the indicator changes color. The indicator ordinarily is selected so that the endpoint occurs at the equivalence point. **17.** (b). **18.** The pH is highest initially (basic solution), it is lowest when the last of the acid has been added. **19. (a)** above 7; **(b)** below 7; **(c)** at 7. **Problems: 21. (a)** $HIO_4 + NH_3 \rightleftharpoons IO_4^- + NH_4^+$; **(b)** $H_2O + NH_2OH \rightleftharpoons OH^- + NH_2OH_2^+$; **(c)** $H_3BO_3 + NH_2^- \rightleftharpoons H_2BO_3^- + NH_3$.

23. (a) $HOClO_2 + H_2O \rightleftharpoons H_3O^+ + ClO_2^-$;
  acid 1   base 2   acid 2   base 1

(b) $HSeO_4^- + NH_3 \rightleftharpoons NH_4^+ + SeO_4^{2-}$;
  acid 1   base 2   acid 2   base 1

(c) $HCO_3^- + OH^- \rightleftharpoons CO_3^{2-} + H_2O$
  acid 1   base 2   base 1   acid 2.

(d) $C_5H_5NH^+ + H_2O \rightleftharpoons C_5H_5N + H_3O^+$
  acid 1   base 2   base 1   acid 2.

25. $H_2PO_4^- + H_2O \rightleftharpoons H_3PO_4 + OH^-$; $H_2PO_4^- + H_2O$ $\rightleftharpoons HPO_4^{2-} + H_3O^+$. **27.** HCl is the strongest acid, so the reaction with HCl will proceed farthest to the right. **29.** CH_3COOH is a stronger acid than is H_2O, and so aniline accepts a proton more readily from CH_3COOH than from H_2O. **31. (a)** H_2Se is stronger, because Se is larger than S and H_2Se has a lower bond dissociation energy; **(b)** $HClO_3$ is stronger because Cl is more electronegative than I and withdraws more electron density from the H—O bond; **(c)** H_3AsO_4 is stronger because As is larger than P, and because the H^+ is being removed from a neutral molecule, not a 1– ion; **(d)** HBr is stronger because of the greater electronegativity difference; **(e)** HN_3 is stronger because three electronegative nitrogen atoms withdraw electron density more than does one N; **(f)** HNO_3 is stronger because the H^+ is being removed from a neutral molecule and not from a 1– ion. **33.** See p. 623; **(a)** About 4×10^{-3}; less than that of 2,2-dichloropropanoic acid but more than that of 2-chloropropanoic acid; **(b)** About 5×10^{-5}; less than that of 3-chloropropanoic acid but slightly more than that of 1-pentanoic acid or acetic acid; the Cl atom is some distance from the —COOH group and has little effect. **35.** phenol < (d) < (a) < (b) < (c). **37. (a)** 0.0012 M; **(b)** 5.5×10^{-10} M; **(c)** 4.5×10^{-5} M; **(d)** 0.00011 M. **39. (a)** 1.41; **(b)** 12.85; **(c)** 0.19; **(d)** 10.70. **41. (a)** 1.14; **(b)** −0.24; **(c)** 1.05; **(d)** 12.95. **43.** Dilute 10.0 mL of 0.250 M NaOH to 5.00 L. **45.** The $Ba(OH)_2$; its $[OH^-] = 2 \times 0.0062 = 0.0124$ M, pOH = 1.91, pH = 12.09. **47. (a)** 2.48; **(b)** 1.95. **49.** 1.6 M. **51.** 3.6×10^{-4}. **53.** 1.08. **55.** 0.016 M. **57.** 0.0045 M H_2SO_4. **59. (a)** pH = 0.70; **(b)** 5.3×10^{-5} M. **61. (a)** 1.53; **(b)** 0.12 M; **(c)** 0.029 M; **(d)** 6.3×10^{-8} M; **(e)** 9.3×10^{-19} M. **63. (a)** neutral; **(b)** acidic, $CH_3CH_2NH_3^+ + H_2O \rightleftharpoons H_3O^+ + CH_3CH_2NH_2$; **(c)** acidic; both $HCOO^- + H_2O \rightleftharpoons HCOOH + OH^-$ and $NH_4^+ + H_2O \rightleftharpoons H_3O^+ + NH_3$ occur, but $K_a(NH_4^+) > K_b(HCOO^-)$. **65.** (c). **67. (a)** $OCl^- + H_2O \rightleftharpoons HOCl + OH^-$; **(b)** $K_b = 3.4 \times 10^{-7}$; **(c)** pH = 10.22. **69.** 0.23 M. **71.** (c) supplies H_3O^+, and (d) supplies $HCOO^-$. **73.** 1.8×10^{-4} M. **75.** 4.24. **77. (a)** pH < 7; **(b)** pH > 7; **(c)** pH < 7; **(d)** pH < 7; **(e)** pH > 7; **(f)** pH > 7. **79.** $HPO_4^{2-} + H_2O \rightleftharpoons PO_4^{3-} + H_3O^+$, $HPO_4^{2-} + H_2O \rightleftharpoons H_2PO_4^- + OH^-$; basic. **81.** 3.65. **83.** 0.37 g. **85.** 3.63. **87.** No, the components must be a conjugate (weak) acid-base pair. **89.** Suitable endpoints for a strong acid-strong base titration occur over a much wider pH range than for a weak acid-strong base titration (Figures 15.15, 15.16). **91. (a)** yellow; **(b)** blue; **(c)** red; **(d)** violet. **93.** Yellow. **95.** The pH is high at the beginning of the titration; the pH drops slowly until just before the equivalence point; just before the equivalence point the pH drops sharply; at the equivalence point the pH is 7.00; just past equivalence the pH continues its sharp drop; further beyond equivalence the pH continues to drop but much more slowly. The same indicators may be used for both titrations because the steep change in pH occurs over the same span in pH. **97.** 4.52. **99. (a)** 16.0 mL; **(b)** 11.28; **(c)** 9.60; **(d)** 9.26; **(e)** 9.03; **(f)** 5.05; **(g)** 1.48. **101.** Lewis acid and base are, respectively: **(a)** $Al(OH)_3$ and OH^-; **(b)** Cu^{2+} and NH_3; **(c)** CO_2 and OH^-. **Additional Problems: 103. (a)** 7.7 g; **(b)** 3.0 g acetic acid, 0.21 g NaOH. **106.** 5.36. **108.** Yes; yes; no the solutions will not be

the same. **110.** 0.026 M. **113.** 2.8 %NaCl; an indicator that produces an endpoint very near pH = 7. **116. (a)** pH = 11.23; **(b)** 34 mg NaOH. **117. (a)** Strong base titrated with weak acid; **(b)** $pK_a \approx 3.8$; **(c)** pH = 8.75. **120.** $pK_{a_1} = 3.16$; $pK_{a_2} = 5.22$. **122. (a)** 2 mol acetic acid per 1 mol sodium acetate; **(b)** 3 mol acetic acid per 1 mol NaOH; **(c)** 3 mol sodium acetate per 2 mol HCl. Solution (c) forms 2 mol NaCl from the HCl and has a high ionic strength, which will affect activity and pH. **125.** pH = 2.25. **Apply Your Knowledge: 127.** pH − 4.3. Distilled water with even a tiny amount of acidic or basic contaminant will have a pH significantly different from 7.00. One suitable buffer would have $[HPO_4^{2-}]/[H_2PO_4^-] = 0.63$. **129.** 1.8×10^2 g CaO. **130.** $K_a = 1.4 \times 10^{-4}$. **133.** pH = 5.68.

Chapter 16

Exercises: 16.1A (a) $MgF_2(s) \rightleftharpoons Mg^{2+}(aq) + 2 F^-(aq)$, $K_{sp} = [Mg^{2+}][F^-]^2$; **(b)** $Li_2CO_3(s) \rightleftharpoons 2 Li^+(aq) + CO_3^{2-}(aq)$, $K_{sp} = [Li^+]^2[CO_3^{2-}]$; **(c)** $Cu_3(AsO_4)_2(s) \rightleftharpoons 3 Cu^{2+}(aq) + 2 AsO_4^{3-}(aq)$, $K_{sp} = [Cu^{2+}]^3[AsO_4^{3-}]^2$. **16.1B (a)** $MgF_2(s) \rightleftharpoons Mg^{2+}(aq) + 2 F^-(aq)$, $K_{sp} = [Mg^{2+}][F^-]^2$; **(b)** $Li_2CO_3(s) \rightleftharpoons 2 Li^+(aq) + CO_3^{2-}(aq)$, $K_{sp} = [Li^+]^2[CO_3^{2-}]$; **(c)** $Cu_3(AsO_4)_2(s) \rightleftharpoons 3 Cu^{2+}(aq) + 2 AsO_4^{3-}(aq)$, $K_{sp} = [Cu^{2+}]^3[AsO_4^{3-}]^2$. **16.2A** $K_{sp} = 2.0 \times 10^{-11}$. **16.2B** $K_{sp} = 1.1 \times 10^{-12}$ M. **16.3A** 1.4×10^{-6} M. **16.3B** 3.1×10^2 ppm of I^-. **16.4A** The trend in molar solubilities is the same as that of K_{sp} values because all four formulas are of the form MX_2: $CaF_2(K_{sp} = 5.3 \times 10^{-9}) < PbI_2(7.1 \times 10^{-9}) < MgF_2(3.7 \times 10^{-8}) < PbCl_2(1.6 \times 10^{-5})$. **16.4B** For $BaSO_4$, the solubility, $s = \sqrt{K_{sp}} \approx 1 \times 10^{-5}$ M. For CaF_2 and PbI_2, $s = (K_{sp})^{1/3}/4 \approx (K_{sp})^{1/3}$. Both K_{sp} values are to the power 10^{-9}, so $s \approx 10^{-3}$ M. The coefficient in the K_{sp} of PbI_2 is larger than in the K_{sp} of CaF_2, so PbI_2 is most soluble. **16.5A** 1.4×10^{-5} M. **16.5B** 4.9 g $AgNO_3$. **16.6A** Yes, $Q_{ip}(4.6 \times 10^{-8}) > K_{sp}$. **16.6B** No, $Q_{ip}(1.3 \times 10^{-12}) < K_{sp}$. **16.7A** Add KI(aq) dropwise from a buret to a known volume of solution of known $[Pb^{2+}]$. Stir after each drop is added, observing first the appearance and then disappearance of $PbI_2(s)$. Continue until a single drop produces a lasting precipitate. Now, $Q_{ip} = K_{sp}$. Calculate K_{sp} from the $[Pb^{2+}]$ and $[I^-]$. **16.7B (a)** The observations should be similar. At the point of impact of the first droplet K_{sp} of PbI_2 is exceeded and precipitate should first form and then disappear as complete mixing occurs. **(b)** No precipitate should form because the ion product cannot exceed $[Pb^{2+}][I^-]^2 = (0.10)(1 \times 10^{-4})^2 < K_{sp} = 7.1 \times 10^{-9}$. **16.8A** Yes, $Q_{ip}(5 \times 10^{-8}) > K_{sp}$. **16.8B** 3.3 g KI. **16.9A** Yes, only $\approx 0.01\%$ of Ca^{2+} remains in solution. **16.9B** (c) < (a) < (b) < (d). (c) No precipitation of cation occurs; $MgSO_4$ is water-soluble. (a) The mixture becomes $CaSO_4(s)$ with excess $[Ca^{2+}] = 0.020$ M. Slightly less than 0.090/0.110 of the Ca^{2+} precipitates; about 20% of the Ca^{2+} remains in solution. (c) The mixture becomes $PbCl_2(s)$ with excess $[Cl^-] = 0.01$ M. The solubility in this case is slightly less than $S = (K_{sp}/4)^{1/3} \approx 0.016$ M. The fraction of Pb^{2+} precipitating is about $(0.120 - 0.016)/0.120$, and about 13% of the Pb^{2+} remains in solution. (d) The mixture becomes AgCl(s) with excess $[Cl^-] = 0.005$ M. The solubility is slightly less than $S = \sqrt{K_{sp}} \approx 1 \times 10^{-5}$ M. Precipitation of the Ag^+ is essentially complete. **16.10A (a)** Br^- precipitates first; **(b)** $[Br^-] = 2.8 \times 10^{-5}$ M when Cl^- begins to precipitate; **(c)** precipitation of Br^- is not quite complete. **16.10B** I^- and Br^- can be separated. K_{sp} for AgI and AgBr differ by nearly 6000-fold, much greater than 360-fold between AgBr and AgCl. **16.11A** 0.80 M. **16.11B** No, $Q_{ip}(7 \times 10^{-33}) > K_{sp}$. **16.12A** $Mg(OH)_2(s) + 2 NH_4^+(aq) \longrightarrow Mg^{2+}(aq) + 2 H_2O(l) + 2 NH_3(aq)$. **16.12B** Most soluble in 1.00 M $NH_4NO_3(aq)$. Solubility is reduced in 1.00 M $Na_2SO_3(aq)$ due to the common ion effect, and in 1.00 M $NH_3(aq)$ due to the high pH. $NH_4NO_3(aq)$ is slightly acidic and will increase the solubility of $SrSO_3$ over its solubility in pure water, by reacting to form the weak acid HSO_3^-. **16.13A** $[Ag^+] = 9.2 \times 10^{-15}$ M. **16.13B** $[NH_3]_{total} = 0.66$ M. **16.14A** No, $Q_{ip}(5.5 \times 10^{-17}) < K_{sp}$ of AgI. **16.14B** 4.8×10^{-3} g KBr. **16.15A** 0.22 M. **16.15B** NaCN(aq). K_f of $[Ag(CN)_2]^-$ is much larger than that of $[Ag(S_2O_3)_2]^{3-}$ or $[Ag(NH_3)_2]^+$. **16.16A** No; NH_4^+ does not react with NO_3^-. The complex ion is not destroyed and the concentration of free Ag^+ remains too low for AgCl(s) to precipitate. **16.16B** (a) $Zn^{2+}(aq) + 2 NH_3(aq) + 2 H_2O(l) \longrightarrow Zn(OH)_2(s) + 2 NH_4^+(aq)$; (b) $Zn(OH)_2(s) + 4 NH_3(aq) \longrightarrow [Zn(NH_3)_4]^{2+}(aq) + 2 OH^-(aq)$; (c) $[Zn(NH_3)_4]^{2+}(aq) + 2 OH^-(aq) + 6 CH_3COOH(aq) \longrightarrow Zn^{2+}(aq) + 6 CH_3COO^-(aq) + 4 NH_4^+(aq) + 2 H_2O(l)$ **Self-Assessment Questions: 1.** (a) $K_{sp} = [Fe^{3+}][OH^-]^3$; (b) $K_{sp} = [Au^{3+}]^2[C_2O_4^{2-}]^3$ **4.** (a) The compounds are of the same type (MX); $PbSO_4$ has the larger K_{sp} and is the more soluble. (b) The compounds are of the same type (MX_2); PbI_2 has the larger K_{sp} and is the more soluble. **5.** (a). **6.** (c). **9.** HCl and CH_3COOH react with OH^-, increasing $[Fe^{3+}]$. The other solutions reduce $[Fe^{3+}]$ by supplying the common ion OH^-. **11.** (a) $K_f = [[Ag(NH_3)_2]^+]/[Ag^+][NH_3]^2$; (b) $K_f = [[Zn(NH_3)_4]^{2+}]/[Zn^{2+}][NH_3]^4$; (c) $K_f = [[Ag(S_2O_3)_2]^{3-}]/[Ag^+][S_2O_3^{2-}]^2$. **12.** $Pb(NO_3)_2(aq)$ reduces the solubility of $PbCl_2$ through the common-ion effect; HCl(aq) increases it through the formation of the complex ion $[PbCl_3]^-$. **13.** (a). **14.** (b). **15.** $[Al(H_2O)_6]^{3+} + H_2O \rightleftharpoons H_3O^+ + [AlOH(H_2O)_5]^{2+}$. **Problems: 19.** $Hg_2(CN)_2(s) \rightleftharpoons Hg_2^{2+}(aq) + 2 CN^-(aq)$, $K_{sp} = [Hg_2^{2+}][CN^-]^2 = 5 \times 10^{-40}$; **21.** $Ag_3AsO_4(s) \rightleftharpoons 3 Ag^+(aq) + AsO_4^{3-}(aq)$, $K_{sp} = [Ag^+]^3[AsO_4^{3-}] = 1.0 \times 10^{-22}$. **23.** Molar solubility and K_{sp} cannot have the same value. Molar solubility is equal to an ion concentration or some fraction of it; K_{sp} is a *product* of two ion concentrations raised to powers. Molar solubilities are generally larger than K_{sp} because the roots of a number <1 are larger than the number. **25.** (a) $K_{sp} = 1.46 \times 10^{-10}$; (b) $K_{sp} = 1.2 \times 10^{-33}$; (c) $K_{sp} = 3.7 \times 10^{-8}$. **27.** (a) Ag_2CrO_4; (b) $MgCO_3$. **29.** $K_{sp} = 1.6 \times 10^{-5}$. **31.** $CaSO_4$ is more soluble than $CaCO_3$; it has a larger K_{sp}. $[Ca^{2+}]$ is greater in $CaSO_4$(satd aq) than in CaF_2(satd aq), because $(9.1 \times 10^{-6})^{1/2}$ is larger than $[(5.3/4) \times 10^{-9}]^{1/3}$. **33.** Yes. **35.** Yes, 22 mg $PbCl_2$ should precipitate. **37.** Pure water. The other two solutions contain common ions and reduce the solubility. **39.** (a) $s = 6.7 \times 10^{-6}$ M; (b) $s = 1.6 \times 10^{-6}$ M. **41.** $[Pb^{2+}] = 4.4 \times 10^{-8}$ M; $[I^-] = 0.400$ M. **43.** (a) $[Ag^+] = 3.0 \times 10^{-2}$ M; (b) 62 g Na_2SO_4. **45.** 2.7×10^{-3} M. **47.** $[CrO_4^{2-}] = 1.0 \times 10^{-6}$ M. **49.** (a) No; (b) Yes, $Pb_3(AsO_4)_2$ will form. **51.** No, $Q_{ip}(5.5 \times 10^{-12}) < K_{sp}$ of CaF_2. **53.** $CaCO_3$, 3.1×10^{-6} M. **55.** (a) $[Mg^{2+}] = 4.5 \times 10^{-6}$ M; (b) Yes, only 0.0076% of Mg^{2+} remains in solution. **57.** pH $= 10.85$. **59.** (a) $CaF_2(s)$; (b) $[F^-] = 1.9 \times 10^{-3}$ M; (c) No, 15% of the Ca^{2+} remains in solution when $MgF_2(s)$ begins to precipitate. **61.** (a) Essentially complete precipitation of 1.0×10^{-3} mol AgCl(s) occurs. (b) Because $[Ag^+]$ is too low following precipitation of AgCl(s), Ag_2SO_4 does not precipitate. **63.** $NaHSO_4$. The reaction, $HSO_4^- + CO_3^{2-} \longrightarrow HCO_3^- + SO_4^{2-}$, replaces insoluble $CaCO_3$ by the more soluble

$Ca(HCO_3)_2$. **65.** 2.5×10^{-16} M **67.** $CaCO_3(s) + 2 H_3O^+(aq) \longrightarrow Ca^{2+}(aq) + 3 H_2O(l) + CO_2(g)$; $CaCO_3(s) + 2 CH_3COOH(aq) \longrightarrow Ca^{2+}(aq) + 2 CH_3COO^-(aq) + H_2O(l) + CO_2(g)$. **69.** $[CrCl_2(NH_3)_4]^+$. **71.** Both H_3O^+ from HCl (d) and HSO_4^- from $NaHSO_4$ (b) can donate protons to NH_3 in the complex ion, causing the complex ion to dissociate and the concentration of free Zn^{2+} to increase. **73.** $2 Ag^+(aq) + SO_4^{2-}(aq) \longrightarrow Ag_2SO_4(s)$; $Ag_2SO_4(s) + 4 NH_3(aq) \longrightarrow 2[Ag(NH_3)_2]^+(aq) + SO_4^{2-}(aq)$; $[Ag(NH_3)_2]^+(aq) + 2 H_3O^+(aq) \longrightarrow Ag^+(aq) + 2 NH_4^+(aq) + 2 H_2O(l)$. **75.** (a) $Fe(OH)_3(s) + 3 H^+(aq) \longrightarrow Fe^{3+}(aq) + 3 H_2O(l)$; (b) NR; (c) $Cr^{3+}(aq) + 4 OH^-(aq) \longrightarrow [Cr(OH)_4]^-(aq)$. **77.** $[Zn^{2+}] = 1.2 \times 10^{-10}$ M. **79.** A trace of $PbI_2(s)$ should precipitate; $Q_{sp}(2.1 \times 10^{-8})$ is slightly larger than K_{sp}. **81.** 4.0 mg KBr. **83.** 0.043 M. **85.** $K_{sp}(PbS) \ll K_{sp}(PbCl_2)$. Enough Pb^{2+} remains from cation group 1 that K_{sp} of PbS is exceeded. $[Ag^+]$ in equilibrium with AgCl(s) is so low that Ag^+ does not later precipitate as Ag_2S. **87.** Hg_2^{2+} is definitely present; the presence of Ag^+ and Pb^{2+} is uncertain—no tests were performed. **Additional Problems: 91.** Yes. **92.** $K_{sp} = 1.62 \times 10^{-7}$. **94.** $Pb(N_3)_2(s) + 2 H_3O^+(aq) \rightleftharpoons Pb^{2+}(aq) + 2 HN_3(aq) + 2 H_2O(l)$; $K_c = 6.9$. Solubility at pH 2.85 is 0.015 mol $Pb(N_3)_2/L$. **98.** 0.10 mL 0.0050 M NaBr. **101.** 6.5×10^{-4} M. **103.** (a) Yes; (b) 2.3 g calcium palmitate. **Apply Your Knowledge: 104.** (a) 34.32% Ca; (b) 0.6231 g $CaCO_3$, 34.22% Ca. **105.** (a) $[Ba^{2+}]$ in saturated $BaSO_4(aq)$ is too low to be hazardous. (b) 1.4 mg Ba^{2+}/L. (c) SO_4^{2-} from $MgSO_4$ reduces the solubility of $BaSO_4$. **108.** (a) 29-30 drops; (b) In (a) the $[Pb^{2+}]$ is essentially fixed at 0.020 M and the $[I^-]$ is increased until K_{sp} is reached. In (b) the $[I^-]$ is fixed at 0.020 M and the $[Pb^{2+}]$ is gradually increased. Since $[I^-]$ is squared but $[Pb^{2+}]$ is taken to the first power in the expression for K_{sp}, the amount of the second ion needed will differ. **110.** (a) $s = 1.2 \times 10^{-2}$ M; (b) $s = 6.0 \times 10^{-2}$ M.

Chapter 17

Exercises: 17.1A (a) Spontaneous. Cellulose decomposes into simpler molecules, such as CO_2 and H_2O, through the action of microorganisms. (b) Nonspontaneous. A compound cannot be decomposed through a physical change. (c) Spontaneous. HCl(g) dissociates completely simply by dissolving in water. **17.1B** (a) Spontaneous; the liquids are miscible in all proportions at all temperatures at which the liquids exist. (b) Uncertain. All compounds decompose at very high temperatures, but it is uncertain whether this decomposition will produce $CO_2(g)$ at 1 atm at 600 °C. (c) Uncertain. Whether the condensation is spontaneous depends on the temperature; that is, spontaneous at all temperatures below that at which the sublimation pressure of $CO_2(s) = 0.50$ atm. (d) Nonspontaneous. Cu(s) lies below $H_2(g)$ in the activity series (Figure 4.13) and will not displace $H_2(g)$ from acidic solution. **17.2A** (a) Decrease. Two moles of gas are converted to one of solid. (b) Increase. Two moles of solid are converted to two moles of solid and three moles of gas. (c) Uncertain. Two moles of gaseous reactants produce two moles of gaseous products. **17.2B** Vaporization produces an increase in entropy (disorder), but enthalpy also increases. In condensation, entropy and enthalpy both decrease. At equilibrium vaporization and condensation occur at equal rates. **17.3A** $\Delta S° = -42.1$ J/K. **17.3B** $\Delta S° = 131.0$ J/K. **17.4A** (a) $\Delta S < 0$, $\Delta H < 0$, case 2; (b) $\Delta S > 0$, $\Delta H > 0$, case 3. **17.4B** $\Delta S > 0$, $\Delta H° = -56.9$ kJ, case 1. **17.5A** -78.5 °C(See Figure 11.12). **17.5B** Both dissociation reactions have $\Delta S° > 0$, but only the dissociation of $NO_2(g)$ has $\Delta H° < 0$, making it spontaneous at all temperatures. For the dissociation of $NH_3(g)$, $\Delta H° > 0$, and so the dissociation is spontaneous only at higher temperatures. **17.6A (a)** $\Delta G° = -70.5$ kJ; **(b)** $\Delta G° = 447.5$ kJ. **17.6B (a)** $\Delta G° = -66.9$ kJ; **(b)** $\Delta G° = -1305.7$ kJ. **17.7A** $\Delta H°_{vap} = 36$ kJ mol^{-1}. **17.7B** Because of extensive hydrogen bonding in $CH_3OH(l)$, $\Delta S°_{vap} > 87$ J mol^{-1} K^{-1}. $\Delta S° = 112.9$ J mol^{-1} K^{-1} is based on $S°$ values in Appendix C; it is 112.5 J mol^{-1} K^{-1} based on $\Delta H°_f$ values and $\Delta S° = \Delta H°/T_{bp}$. **17.8A** $K_{eq} = [Al^{3+}](P_{H_2})^3/[H^+]^6$. **17.8B** $Mg(OH)_2(s) + 2 H_3O^+(aq) \rightleftharpoons Mg^{2+}(aq) + 4 H_2O(l)$; $K_{eq} = [Mg^{2+}]/[H_3O^+]^2$. **17.9A** $K_{eq} = 3.02 \times 10^{-21}$. **17.9B** $K_{eq} = 27$; $P_{NO} = 0.16$ atm, $P_{NOBr} = 0.84$ atm. **17.10A** $K_{eq} = 0.45$. **17.10B** With the Clausius-Clapeyron equation, $P_{H_2O} = 55.7$ mmHg; from Table 11.2, $P_{H_2O} = 55.3$ mmHg. **Self-Assessment Questions: 1.** (a), (c). **2.** (c). **3. (a)** Decrease (liquid $\longrightarrow$ solid); **(b)** increase (solid $\longrightarrow$ gas); **(c)** increase (liquid + oxygen gas $\longrightarrow$ large volume of gaseous products). **9.** NOF_3 has a greater number of atoms and vibrational modes, and hence a greater entropy, than NO_2F. **10.** (b). **11.** Low temperatures. The ΔH term dominates in the Gibbs equation, offsetting a positive value of $-T\Delta S$ and making $\Delta G < 0$. **13.** (d). **14.** (c). **18.** (d). **Problems: 19. (a)** Increase—system energy dispersed over more levels in gas than in liquid. **(b)** Increase—process produces increased number of molecules of gas. **(c)** Indeterminate—same number of moles of gas on each side of equation. **(d)** Increase—large amount of gas produced from a solid. **(e)** Decrease—system energy is spread among fewer levels in solid than in liquid. **(f)** Indeterminate—The gases are all diatomic and the same number of moles of gas appear on each side of the equation. **(g)** Increase—A liquid decomposes to produce a large amount of gas. **(h)** Entropy decreases when a large volume of gas dissolves in water. **21.** The final state: heterogeneous mixture of octane floating on water. Solution formation would require $\Delta H > 0$ to break hydrogen bonds in the water. The increase in entropy would not be sufficient to overcome the large ΔH. **23.** Correct: For a process to occur spontaneously, the total entropy—S_{univ}—must increase. Errors in other statements: (a) Entropy of the system may increase in some cases and decrease in others; (b) Entropy of the surroundings may also increase or decrease; (c) Entropy of the system and surroundings need not both increase, as long as the increase in one exceeds the decrease in the other. **25.** Rather than two vertical sections, the graph has one vertical section at 194.5 K corresponding to sublimation. **27.** No. The disintegration of an aluminum can is spontaneous, but because a protective coating of Al_2O_3 forms on the surface, complete disintegration takes a very long time. "Spontaneous" does not mean "fast." **29.** A criterion based on ΔS requires assessing both ΔS_{syst} and ΔS_{surr}; the free energy change requires only measurements in the system: $\Delta G = \Delta H - T\Delta S$.

The $T\Delta S$ line does not cross the ΔH line, so ΔG is always negative and the reaction is spontaneous at all temperatures. **33. (a)** The change in *enthalpy* indicates whether the reaction is endothermic or exothermic; **(b)** The *entropy change* indicates whether the reaction involves an increase or decrease in available energy levels; **(c)** The *relative values of* ΔH and $T\Delta S$ indicate whether equilibrium in the reaction is favored at high or low temperatures. **35. (a)** Melting of a solid is nonspontaneous below its melting point and spontaneous above its melting point—0 °C for water. **(b)** The condensation of a vapor at 1 atm pressure to liquid is spontaneous below the normal boiling point and nonspontaneous above the normal boiling point. For water, this temperature is 100 °C. **37.** High temperature, since $\Delta S < 0$ and $\Delta G = \Delta H - T\Delta S$ **39. (a)** $\Delta G° = +22.9$ kJ; **(b)** $\Delta G° = -163$ kJ. **41.** $\Delta H° = -1076.8$ kJ; $\Delta S° = -56.5$ J/K; $\Delta G° = -1060.1$ kJ. **43.** With $\Delta G°$, we can evaluate K_{eq}: $\Delta G° = -RT \ln K_{eq}$. With K_{eq}, we can determine an equilibrium condition. To evaluate ΔG for nonstandard conditions we use $\Delta G = \Delta G° + RT \ln Q$ (where Q is reaction quotient). However, we cannot obtain ΔG without knowing $\Delta G°$. **45.** 364 K (91 °C). **47.** $\Delta H°_{vap} = +29.8$ kJ/mol (Trouton's rule); +30.1 kJ/mol (Appendix C). **49. (a)** $\Delta G = 0$ kJ; **(b)** $K_p = 171$. **51. (a)** $K_{eq} = P_{H_2O}P_{SO_2} = K_p$; **(b)** $K_{eq} = [Mg^{2+}][OH^-]^2 = K_{sp}$; **(c)** $K_{eq} = [CH_3COOH][OH^-]/[CH_3COO^-] = K_b$. **53. (a)** $K_p = 7.2 \times 10^{24}$; **(b)** $K_p = 1.3 \times 10^{-20}$. **55.** $P_{C_{10}H_8} = 0.063$ mmHg. **57. (a)** $\Delta G° = -15.9$ kJ. **(b)** Direction of net reaction: $\longrightarrow$. **(c)** 0.037 mol CO, 0.064 mol H_2O, 0.193 mol CO_2, 0.274 mol H_2. **59.** The reaction to be coupled with the given reaction must have $\Delta G° < -237.9$ kJ. Only (c) will work; it has $\Delta G° = -257.2$ kJ. **61.** $K_{eq} = 3.5 \times 10^{10}$. **63.** 390.4 K. **65.** $K_p = 2.0 \times 10^6$. **67.** Estimated boiling point: 348 K. Use $\Delta H°_{298}$ and $\Delta G°_{298}$ from Appendix C; calculate K_p at 298 K; use the van't Hoff equation to determine T at which $K_p = 1.00$ atm. **69.** 424 K. **Additional Problems: 71.** The expansion in Figure 6.8 is not reversible because the process cannot be reversed by an infinitesimal change. The weights are too large, the change occurs too quickly. **74.** In Chapter 14, we saw that an endothermic reaction would be forced to the right by an increase in temperature. The effect of being forced to the right increases the K_{eq}. From the van't Hoff equation, for positive ΔH (endothermic) an increase in temperature increases the K_{eq}. That is, for $\Delta H > 0$ and $T_2 > T_1$, $K_2 > K_1$. **75.** 393 K. **77. (a)** 3.47 kJ; **(b)** to the left; **(c)** 0.0554 mol CO_2, 0.0754 mol H_2, 0.0346 mol CO, 0.0796 mol H_2O. **81.** Sublimation pressure of Hg(s) at -78.5 °C: $\approx 10^{-9}$ mmHg. **83.** $T = 631$ K. **87. (a)** true from definition of $\Delta G°$, so (c) is false; (d) is true, nonspontaneous process so (b) is false. **89.** $P_{PCl_5} = 1.79$ atm, $P_{PCl_3} = P_{Cl_2} = 0.95$ atm, $P_{total} = 3.69$ atm. **Apply Your Knowledge: 92. (a)** Using van't Hoff's equation, $T = 523$ °C; **(b)** H_2, graphite, CO, Cu_2O, SO_2, and Al. **94.** At body temperature (37 °C), a concentration of glucose-6-phosphate 627 times the glucose concentration. **97.** $\Delta G = -26.3$ kJ. **98. (a)** $T = 491$ K; the temperature is probably higher because some heat is lost to the surroundings; **(b)** 22 atm; **(c)** "You can't win" because you can't get more work out of the engine than the energy put into the engine; "you can't break even" because you can't even get 100% efficiency; the engine always operates "at a loss."

31.

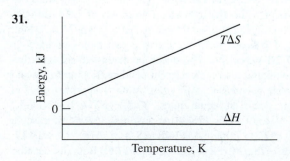

Chapter 18

Exercises: 18.1A $4 Zn(s) + 2 NO_3^-(aq) + 10 H^+(aq) \longrightarrow 4 Zn^{2+}(aq) + N_2O(g) + 5 H_2O(l)$. **18.1B** $3 P_4(s) + 20 NO_3^-(aq)$

$+ 8 H_2O(l) + 8 H^+(aq) \longrightarrow 12 H_2PO_4^-(aq) + 20 NO(g)$. **18.2A** $2 OCN^-(aq) + 3 OCl^-(aq) + 2 OH^-(aq) \longrightarrow$ $2 CO_3^{2-}(aq) + N_2(g) + H_2O(l) + 3 Cl^-(aq)$. **18.2B** $4 MnO_4^-(aq)$ $+ 3 CH_3CH_2OH(aq) \longrightarrow 4 MnO_2(s) + 3 CH_3COO^-(aq) +$ $OH^-(aq) + 4 H_2O(l)$. **18.3A** $2 Al(s) + 6 H^+(aq) \longrightarrow$ $2 Al^{3+}(aq) + 3 H_2(g)$. **18.3B** $Cu(s)|Cu^{2+}(aq)||Ag^+(aq)|Ag(s)$. **18.4A** $E^\circ_{Sm^{2+}/Sm} = -2.67 V$. **18.4B** $E^\circ_{ClO_4^-/Cl_2} = 1.392 V$. **18.5A** **(a)** $E_{cell} = 2.016 V$; **(b)** $E_{cell} = -0.499 V$. **18.5B** $2 Mn^{2+}(aq) + 5 S_2O_8^{2-}(aq) + 8 H_2O(l) \longrightarrow$ $2 MnO_4^-(aq) + 10 SO_4^{2-}(aq) + 16 H^+(aq)$, $E^\circ_{cell} = 0.50 V$. **18.6A** No, $E^\circ_{cell} = -0.431 V$. **18.6B** The reverse reaction is spontaneous; for the forward reaction, $E^\circ_{cell} = -0.31 V$. **18.7A** Similar to Example 18.7, Zn is oxidized, and electrons pass through the voltmeter to the Cu electrode, where H^+ (from lemon juice) is reduced to $H_2(g)$. **18.7B** **(a)** Result should be similar to what is seen in Figure 18.10, but the gas evolution would probably be less vigorous because $[H^+]$ is lower in $CH_3COOH(aq)$, a weak acid, than in $HCl(aq)$, a strong acid. **(b)** $NO_3^-(aq)$ is a stronger oxidizing agent in acidic solutions than is $H^+(aq)$, and the reduction product should be mostly $NO(g)$ or $NO_2(g)$, rather than $H_2(g)$. Moreover, the copper strip would react as well as the zinc, producing a blue solution characteristic of $Cu^{2+}(aq)$ (recall Figure 4.14). **(c)** Result should be similar to Figure 18.10, but with the copious collection of $H_2(g)$ bubbles on the surface of the pool of mercury and a smaller amount on the zinc strip. **18.8A** $\Delta G^\circ = -394 kJ$; $K_{eq} = 1 \times 10^{69}$. **18.8B** For the reaction: $3 Ag(s) + 4 H^+(aq) + NO_3^-(aq) \longrightarrow 3 Ag^+(aq) + NO(g) + 2 H_2O(l)$, $E^\circ_{cell} = 0.156 V$; $K_{eq} = 8.0 \times 10^7$. **18.9A** **(a)** $E_{cell} = 1.056 V$; **(b)** $E_{cell} = 1.150 V$. **18.9B** $E_{cell} = 1.045 V$. **18.10A** The bend, like the head and tip, is strained and more energetic than the body of the nail. It is an anodic area (blue precipitate). **18.10B** The same half-reactions occur even if the early indicators of those half-reactions, the phenolphthalein and potassium ferricyanide, are missing. Over time, in the presence of dissolved $O_2(g)$, the $Fe^{2+}(aq)$ produced in the oxidation will be converted to red-brown $Fe_2O_3(s)$, as outlined in Figure 18.18. The first appearance of this "rust" should be at the same points where the blue precipitate is seen in Figure 18.19. **18.11A** $2 Br^-(aq) + 2 H_2O(l) \longrightarrow Br_2(l)$ $+ 2 OH^-(aq) + H_2(g)$; $E^\circ_{cell} = -1.893 V$. **18.11B** To force the electrolysis, $Ag(s, anode) \longrightarrow Ag(s, cathode)$, the external voltage $> 0.460 V$. Otherwise, this reaction can occur: $Cu(s) + 2 Ag^+(aq) \longrightarrow Cu^{2+}(aq) + 2 Ag(s)$, $E^\circ_{cell} = 0.460 V$. **18.12A** Cell A: $Cu(s) + Zn^{2+}(1.0 M) \longrightarrow Cu^{2+}(0.10 M) + Zn(s)$; cell B: $Zn(s) + Cu^{2+}(1.0 M) \longrightarrow Zn^{2+}(0.10 M) + Cu(s)$. **18.12B** As current flows, $[Zn^{2+}]$ and $[Cu^{2+}]$ change in both cells; when concentrations in the two cells are the same, current stops. **18.13A** 22.5 min. **18.13B** **(a)** 1945 C; **(b)** 1.529 A. **18.14A** $NaCl(aq)$. The oxidation $H_2O \longrightarrow O_2$ produces 0.25 mol O_2 per mol e^-; $Cl^- \longrightarrow Cl_2$ produces 0.5 mol Cl_2 per mol e^-. Oxidation of I^- produces solid I_2. **18.14B** Choice (a) produces $H_2(g)$ rather than metallic sodium. The answer will be the solution requiring the smallest number of moles of electrons transferred in the electrolysis. In solution (b) the number of moles of Cu in 1.00 g is $1.00/63.55 = 0.0157$ mol Cu; the electrolysis requires 2 mol e^- per mol Cu or a total of about 0.03 mol electrons. By contrast, in solution (c) 1.00 g of metal corresponds to about $1.00/107.87 = 0.009$ mol Ag, requiring 0.009 mol e^- for its electrodeposition. The situation with solution (d) is roughly the same as with solution (b). Solution (c) is the answer we seek, provided that there is sufficient Ag^+ in the solution to yield at least 1.00 g

Ag(s). There is, since $0.150 L \times 0.25$ mol $Ag^+/L > 0.009$ mol Ag^+. Note that we did not have to determine whether solutions (b) and (d) have enough solute to yield 1.00 g of the solid metal, nor did we have to use the magnitude of the current since it was the same in all cases. **Self-Assessment Questions: 5.** (c). **6.** (d). **8.** If the reduction most easily achieved is assigned $E^\circ = 0$, all others will have $E^\circ < 0$. If the reduction most difficult to achieve is assigned $E^\circ = 0$, all others will have $E^\circ > 0$. **9.** (d). **11.** Ag, Au. **12.** E°_{cell} values are based on E° values and related to ΔG° values for *redox* reactions. Predictions based on E°_{cell} are the same as if based on ΔG°. They apply regardless of how redox reactions are carried out. **13.** $E^\circ_{cell} < 0$ means that a reaction is nonspontaneous when reactants and products are in their standard states. The reaction may be spontaneous for specific nonstandard conditions. **21.** E° data are used to determine E°_{cell} for the nonspontaneous reaction occurring in electrolysis; the applied voltage must exceed this E°_{cell}. **22.** (a). **23.** An object to be electroplated is the cathode, and the metal to be plated out, in solution as cations. **24.** (d). **Problems: 25.** **(a)** $HNO_2(aq) \longrightarrow NO_2(g) + H^+(aq) + e^-$; **(b)** $PbO_2(s) + 2 H^+(aq) + 2 e^- \longrightarrow PbO(s) + H_2O(l)$; **(c)** $12 OH^-(aq) + CH_3CH_2OH(aq) \longrightarrow 2 CO_2(g) + 9 H_2O(l)$ $+ 12 e^-$. **27.** **(a)** $6 Fe^{2+} + Cr_2O_7^{2-} + 14 H^+ \longrightarrow 6 Fe^{3+} + 2 Cr^{3+} + 7 H_2O$; **(b)** $S_8 + 12 O_2 + 8 H_2O \longrightarrow 8 SO_4^{2-} + 16 H^+$; **(c)** $4 Fe^{3+} + 2 NH_2OH_2^+ \longrightarrow 4 Fe^{2+} + H_2O + N_2O + 6 H^+$. **29.** **(a)** $4 Fe(OH)_2(s) + O_2(g) + 2 H_2O(l) \longrightarrow 4 Fe(OH)_3(s)$; **(b)** $S_8(s) + 12 OH^-(aq) \longrightarrow 2 S_2O_3^{2-}(aq) + 4 S^{2-}(aq) + 6 H_2O(l)$; **(c)** $2 CrI_3(s) + 27 H_2O_2(aq) + 10 OH^-(aq) \longrightarrow 2 CrO_4^{2-}(aq) + 6 IO_4^-(aq) + 32 H_2O(l)$. **31.** **(a)** $5 H_2C_2O_4(aq) + 2 MnO_4^-(aq) + 6 H^+(aq) \longrightarrow 2 Mn^{2+}(aq) + 10 CO_2(g) + 8 H_2O(l)$; **(b)** $Cr_2O_7^{2-}(aq) + 3 UO^{2+}(aq) + 8 H^+(aq) \longrightarrow 2 Cr^{3+}(aq) + 3 UO_2^{2+}(aq) + 4 H_2O(l)$; **(c)** $4 Zn(s) + NO_3^-(aq) + 6 H_2O(l) \longrightarrow 4 Zn^{2+}(aq) + NH_3(g) + 9 OH^-(aq)$. **33.** **(a)** 0.118 V; **(b)** $-0.508 V$; **(c)** 0.595 V. **35.** $E^\circ_{cathode} = 0.20 V$. **37.** $E^\circ_{V^{2+}/V} = -1.13 V$. **39.** **(a)** $Fe^{2+} \longrightarrow Fe^{3+} + e^-$, $Cr_2O_7^{2-} + 14 H^+ + 6 e^- \longrightarrow 2 Cr^{3+} + 7 H_2O$, $6 Fe^{2+} + Cr_2O_7^{2-} + 14 H^+ \longrightarrow 6 Fe^{3+} + 2 Cr^{3+} + 7 H_2O$; $E^\circ_{cell} = 0.56 V$; **(b)** $NO + 2 H_2O \longrightarrow NO_3^- + 4 H^+ + 3 e^-$, $H_2O_2 + 2 H^+ + 2 e^- \longrightarrow 2 H_2O$, $2 NO + 3 H_2O_2 \longrightarrow 2 NO_3^- + 2 H^+ + 2 H_2O$; $E^\circ_{cell} = 2.619 V$. **41.** **(a)** $Fe^{3+} + e^- \longrightarrow Fe^{2+}$, $Sn^{2+} \longrightarrow Sn^{4+} + 2 e^-$, $2 Fe^{3+} + Sn^{2+} \longrightarrow Sn^{4+} + 2 Fe^{2+}$, $Pt|Sn^{2+}, Sn^{4+}||Fe^{2+}, Fe^{3+}|Pt$, $E^\circ_{cell} = 0.617 V$; **(b)** $Cu(s) \longrightarrow Cu^{2+} + 2 e^-$, $4 H^+ + NO_3^- + 3 e^- \longrightarrow 2 H_2O + NO(g)$, $3 Cu(s) + 8 H^+ + 2 NO_3^- \longrightarrow 3 Cu^{2+} + 4 H_2O + 2 NO(g)$, $Cu|Cu^{2+}(aq)||H^+(aq), NO_3^-(aq), NO(g)|Pt$; $E^\circ_{cell} = 0.616 V$. **43.** **(a)** No; **(b)** Yes. **45.** **(a)** No; **(b)** Yes; **(c)** Yes. **47.** **(a)** H^+ is not a good enough oxidizing agent to oxidize $Ag(s)$ to Ag^+; $2 Ag(s) + 2 H^+(aq) \longrightarrow 2 Ag^+(aq) + H_2(g)$, $E^\circ_{cell} = -0.800 V$. **(b)** $NO_3^-(aq)$ oxidizes $Ag(s)$ to Ag^+, $3 Ag(s) + 4 H^+(aq) + NO_3^-(aq) \longrightarrow 3 Ag^+(aq) + NO(g) + 2 H_2O(l)$, $E^\circ_{cell} = 0.156 V$. **49.** $0.800 V < E^\circ_{Pd^{2+}/Pd} < 0.956 V$. **51.** **(a)** $E^\circ_{cell} = 0.694 V$, $\Delta G^\circ = -268 kJ$; **(b)** $E^\circ_{cell} = 0.99 V$, $\Delta G^\circ = -5.7 \times 10^2 kJ$. **53.** **(a)** $K_{eq} = [Pb^{2+}]P_{Cl_2}/([H^+]^4[Cl^-]^2)$ $= 3.6 \times 10^6$; **(b)** $K_{eq} = [BrO_3^-]^2/([Br^-]^2 P_{O_2}^3) = 8.1 \times 10^{-38}$. **55.** $Cu(s) + 2 Ag^+(aq) \longrightarrow Cu^{2+}(aq) + 2 Ag(s)$ at equilibrium $[Ag^+] = 1.2 \times 10^{-8} M$, $[Cu^{2+}] = 0.50 M$. **57.** **(a)** $E_{cell} = -0.026 V$; **(b)** $E_{cell} = 0.19 V$. **59.** $E_{cell} = -0.15 V$. **61.** pH $= 1.82$. **63.** The cell reaction is $Cu(s) + 2 Ag^+(aq) \longrightarrow Cu^{2+}(aq) + 2 Ag(s)$. Cell (b) has a higher E_{cell} because it has a lower ratio of $[Cu^{2+}]/[Ag^+]^2$. **65.** $2 Mg(s) + O_2(g) + 2 H_2O(l) \longrightarrow 2 Mg(OH)_2(s)$, $E^\circ_{cell} = 2.757 V$. **67.** Anode:

$Zn(s) + 2\,OH^-(aq) \longrightarrow ZnO(s) + H_2O(l) + 2\,e^-$; cathode: $Ag_2O(s) + H_2O(l) + 2\,e^- \longrightarrow 2\,Ag(s) + 2\,OH^-(aq)$; cell reaction: $Zn(s) + Ag_2O(s) \longrightarrow ZnO(s) + 2\,Ag(s)$. **69.** Oxygen oxidizes $Fe(s)$ to $Fe^{2+}(aq)$ and $Fe^{3+}(aq)$. Water is involved in the reduction half-reaction and in the conversion of $Fe(OH)_2$ to $Fe(OH)_3$. The electrolyte completes the electric circuit between anodic and cathodic regions. **71.** The sacrificial anode is oxidized. Its sacrifice protects from oxidation the metal to which it is attached. **73.** Zinc is oxidized instead of iron (it is a sacrificial anode). There is a white precipitate of zinc ferricyanide but no Turnbull's blue. **75. (a)** $Cu(s, \text{anode}) \longrightarrow Cu(s, \text{cathode})$; **(b)** $2\,H_2O(l) + 2\,Cu^{2+}(aq) \longrightarrow 4\,H^+(aq) + O_2(g) + 2\,Cu(s)$; **(c)** $2\,H_2O(l) + 2\,Cu^{2+}(aq) \longrightarrow 4\,H^+(aq) + O_2(g) + 2\,Cu(s)$. **77. (a)** $BaCl_2(l) \longrightarrow Ba(l) + Cl_2(g)$, required voltage > 4.28 V; **(b)** $2\,HBr(aq) \longrightarrow H_2(g) + Br_2(l)$, required voltage > 1.065 V; **(c)** $2\,H_2O(l) \longrightarrow 2\,H_2(g) + O_2(g)$, required voltage > 2.057 V. **79.** 14.3 g Ag. **81.** 7.59×10^4 C. **83.** $AgNO_3$. $Ag^+(aq)$ yields 1 mol Ag/mol e^-, and $Cu^{2+}(aq)$ and $Zn^{2+}(aq)$ each yield 0.5 mol/mol e^-. Also, the molar mass of Ag exceeds that of Cu and Zn. **Additional Problems: 86.** $CN^- + 2\,OH^- + Cl_2 \longrightarrow OCN^- + 2\,Cl^- + H_2O$; $2\,OCN^- + 6\,OH^- + 3\,Cl_2 \longrightarrow 2\,HCO_3^- + N_2 + 6\,Cl^- + 2\,H_2O$. **88.** $2\,NO(g) + 5\,H_2(g) \longrightarrow 2\,NH_3(g) + 2\,H_2O(g)$; the method works because the ionic species that are introduced by employing aqueous solutions cancel out. **91.** 58.1 mL $O_2(g)$. **92.** 0.513 M $AgNO_3$. **96.** $E_{cell} = -0.457$ V. **99.** E_{cell} decreases as $[Cu^{2+}]$ increases at the anode and decreases at the cathode. When $[Cu^{2+}] = 0.76$ M in each half-cell, $E_{cell} = 0$ and the current stops. **102.** $Ag(s)|AgI(\text{satd})||Ag^+(0.100\,M)|Ag(s)$, $K_{sp} = 8.2 \times 10^{-17}$. **104.** The blue color will be seen. The Cu^{2+} is converted to $[Cu(NH_3)_4]^{2+}$, $[Cu(NH_3)_4]^{2+} = 4 \times 10^{-4}$ M. **106.** $[Hg^{2+}] = 0.177$ M; $[Fe^{2+}] = 0.034$ M; $[Fe^{3+}] = 0.356$ M. **109.** 3×10^3 ions. **112. (a)** cathode $O_2(g) + 2\,H_2O(l) + 4\,e^- \longrightarrow 4\,OH^-$, anode $H_2(g) + 2\,OH^- \longrightarrow 2\,H_2O(l) + 2\,e^-$, overall $2\,H_2(g) + O_2 \longrightarrow 2\,H_2O$; **(b)** $E_{cell} = 1.256$ V; **(c)** 7.50×10^5 kJ; **(d)** 868 watts; **(e)** Ideal gas behavior assumed; cell is nonstandard. **Apply Your Knowledge: 115.** 9.65×10^{-4} g. **116. (a)** $C_3H_5(NO_3)_3 + H^+ + 2\,e^- \longrightarrow C_3H_5OH(NO_3)_2 + NO_2^-$; $NO_2^- + 2\,H^+ + e^- \longrightarrow NO + H_2O$; **(b)** $C_3H_5(NO_3)_3 + NADH \longrightarrow C_3H_5OH(NO_3)_2 + NO_2^- + NAD^+$; $2\,NO_2^- + 3\,H^+ + NADH \longrightarrow 2\,NO + 2\,H_2O + NAD^+$.

Chapter 19

Exercises: 19.1A (a) $^{212}_{86}Rn \longrightarrow ^{208}_{84}Po + ^4_2He$; **(b)** $^{37}_{18}Ar + ^0_{-1}e \longrightarrow ^{37}_{17}Cl$; **(c)** $^{60}_{27}Co^* \longrightarrow ^{60}_{27}Co + \gamma$. **19.1B (a)** $^{218}_{84}Po \longrightarrow ^{214}_{82}Pb + ^4_2He$; **(b)** $^{36}_{17}Cl \longrightarrow ^{36}_{16}S + ^0_1e$. **19.2A** 5.20×10^2 atoms s^{-1}. **19.2B** $t = 1.60 \times 10^5$ y. **19.3A** Fraction remaining $= 1/2^{30} = 9.3 \times 10^{-10}$; time required, 30×8.040 days $= 241.2$ days. **19.3B (b).** There are about 10 times more ^{24}Na atoms than ^{11}C, but ^{11}C atoms disintegrate about 50 times as fast. There are about $10^4 - 10^5$ times as many ^{238}U atoms as ^{11}C, but $t_{1/2}$ of ^{238}U exceeds that of ^{11}C by a factor of more than 10^9. **19.4A** 6.0 dis min^{-1} per g carbon. **19.4B** The activity falls to one-half its initial value in 12.26 y ($t_{1/2}$ for tritium). The brandy is only about 12 years old, not 25. **19.5A** $^{35}_{17}Cl + ^1_0n \longrightarrow ^{35}_{16}S + ^1_1H$. **19.5B** $^{249}_{98}Cf + ^{15}_7N \longrightarrow ^{260}_{105}Db + 4\,^1_0n$. **19.6A** $^{74}_{30}Zn$ lies above the belt of stability and is radioactive. **19.6B** "Magic numbers" of protons and neutrons in ^{40}Ca and ^{48}Ca make them stable. (Isotopes with $Z = 20$ and $A = 42, 44,$ and 46 are also stable.) ^{39}Ca is below the belt of stability in Figure 19.5 and is radioactive.

19.7A ^{84}Y has 39 protons and 45 neutrons and lies outside the belt of stability (Figure 19.5). It should decay by electron capture or positron emission to yield ^{84}Sr, which is within the belt of stability and could be stable. **19.7B** $^{17}_9F \longrightarrow ^{17}_8O + ^0_1e$ and $^{22}_9F \longrightarrow ^{22}_{10}Ne + ^0_{-1}e$ **19.8A** $\Delta E = -5.0$ MeV. **19.8B** Energy requirement $= \Delta E = 2.7$ MeV. To convert from atomic to nuclear masses subtract the mass of 15 electrons from each side of the equation. The same result is obtained by using either atomic or nuclear masses. **Self-Assessment Questions: 2. (b). 3. (c). 5. (d). 7.** 74 days. **10. (d). 11. (a). 13. (c). 15.** The minimum mass of a fissionable isotope that must be brought into a small volume for a self-sustaining nuclear fission reaction to occur. **Problems: 21. (a)** 86, ^{215}Po; **(b)** Tm, ^{167}Er; **(c)** ^{90}Y; **(d)** ^{79}Br. **23. (a)** Cd-123; **(b)** Rh-103; **(c)** Bi-209. **25. (a)** ^7Be; **(b)** ^{242}Pu.

27.

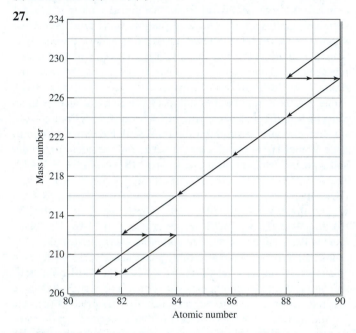

29. Chemical reactions occur as a result of collisions, and their rates depend on the reaction mechanism. Different mechanisms lead to different overall reaction orders. The rate at which the nuclei of a particular nuclide decay depends on how many of them are present. Thus, the process is first order. **31. (a)** ^{15}O has the shortest half-life and the greatest activity. **(b)** A 75% reduction represents two half-life periods. The required $t_{1/2}$ is one day, and the nuclide having about this half-life is ^{28}Mg. **33.** For equal masses, the number of atoms (N) is about four times as great for ^{32}P as for ^{131}I. The decay constant (λ) of ^{32}P is about one-half that of ^{131}I. Because A is proportional to both λ and N, A of ^{32}P is about twice that of $^{131}I(4 \times 1/2 = 2)$. **35. (a)** 6.03×10^4 atoms/h; **(b)** 2.04×10^6 atoms/h. **37.** $t_{1/2} = 88.5$ d. **39. (a)** Not quite 3 half-lives elapse, so the activity should be about 1/8 the original activity or 110 dis/min; **(b)** 119 dis/min. **41.** 11 dis min^{-1} per g carbon. **43. (a)** ^1_0n; **(b)** ^2_1H; **(c)** ^4_2He. **45. (a)** $2\,^2_1H \longrightarrow ^3_2He + ^1_0n$; **(b)** $^{241}_{95}Am + ^4_2He \longrightarrow ^{243}_{97}Bk + 2\,^1_0n$; **(c)** $^{238}_{92}U + ^4_2He \longrightarrow ^{239}_{94}Pu + 3\,^1_0n$. **47.** $^{99}_{43}Tc + ^1_0n \longrightarrow ^{100}_{43}Tc \longrightarrow ^{100}_{44}Ru + ^0_{-1}e$. **49.** 0.5301 u. **51.** 8.520 MeV/nucleon. **53.** -1.8 MeV. **55. (a)** 227.9793 u; **(b)** 228.0287 u. **57.** 11.0 MeV. **59. (a)** is radioactive because there are fewer neutrons than protons; **(d)** should be radioactive with an odd number of neutrons and of protons; **(b)** and **(c)** have a neutron-to-proton ratio near 1 but greater than 1, and should be stable. **61.** ^{36}Cl (17p, 19n) and ^{64}Cu (29p, 35n) do not occur naturally, even though their atomic masses are close to the weighted average atomic mass of the element (35.4527 u for Cl and 63.546 u for Cu). They are "odd-odd" nuclides and are radioactive. **63.** ^{40}Ca is a "doubly magic" nuclide (20p, 20n) with a neutron:proton ratio of $1:1$. These factors contribute to the stability of a nuclide of relatively low atomic number. **65.** $\Delta E = -183.6$ MeV. **67.** A nuclide with a very short half-life decays so quickly that it is not a significant hazard, whereas one with a very long half-life has a very low activity and is less hazardous for that reason. **69.** 320 g $<$ mass of ^{131}I $<$ 400 g. **71.** A few micrometers at most. A positron will annihilate an electron upon encountering one, and there are a lot of electrons in matter, even gaseous matter. **73.** Inject a small amount of ^{3}H into the stream, or add an inert radioactive gaseous element such as ^{41}Ar($t_{1/2}$, 1.83 h) or ^{37}Ar($t_{1/2}$, 35 d) and look for radioactivity around the pipes. **Additional Problems: 78.** -4.3 MeV. **79.** 25 pm. **80. (a)** $4n + 2$; **(b)** $4n$; **(c)** $4n + 3$; **(d)** $4n + 3$. **81.** 4.01 MeV, 1.39×10^7 m/s. **82.** 3.65×10^{-7} g. **86.** 238 dis/s. **89.** 2.1×10^3 L. **92.** cobalt-57: $1s^12s^22p^63s^23p^64s^23d^7$, cobalt-57: $1s^22s^12p^63s^23p^64s^23d^7$, 6.70×10^5 kJ/mol. **93.** The stable isotopes of Cu are ^{63}Cu and ^{65}Cu. Unstable ^{64}Cu lies between the two in Figure 19.4, rather than above or below the stable isotopes. Thus, it attains stability by beta emission to form ^{64}Zn (stable) or by either positron emission or electron capture to form ^{64}Ni. **Apply Your Knowledge: 94.** Uranium I $= ^{238}$U; Uranium X$_1 = ^{234}$Th; Uranium X$_2 = ^{234}$Pa; Uranium II $= ^{234}$U; Ionium $= ^{230}$Th; Radium $= ^{226}$Ra **95.** $\sim 1.1 \times 10^{-5}$%, by calculating each element separately. **99.** Plot ln A versus time in minutes. The slope is -0.305 min^{-1}, the half-life $= -0.693/(-0.305$ min$^{-1}) = 2.27$ min.

Chapter 20

Exercises: 20.1A (a) $2 \text{ NaCl(l)} \xrightarrow{\text{electrolysis}} 2 \text{ Na(l)} + \text{Cl}_2\text{(g)}$, followed by $2 \text{ Na(s)} + \text{H}_2\text{(g)} \longrightarrow 2 \text{ NaH(s)}$; **(b)** $2 \text{ NaCl(aq)} + 2 \text{ H}_2\text{O(l)} \xrightarrow{\text{electrolysis}} 2 \text{ NaOH(aq)} + \text{Cl}_2\text{(g)} + \text{H}_2\text{(g)}$, followed by $2 \text{ NaOH(aq)} + \text{Cl}_2\text{(g)} \longrightarrow \text{NaCl(aq)} + \text{NaOCl(aq)} + \text{H}_2\text{O(l)}$. **20.1B** $2 \text{ Na(s)} + 2 \text{ H}_2\text{O(l)} \longrightarrow 2 \text{ NaOH(aq)} + \text{H}_2\text{(g)}$, $\text{NaOH(aq)} + \text{SO}_2\text{(g)} \longrightarrow \text{Na}_2\text{SO}_3\text{(aq)}$ **20.2A** Maximum mass: 1.658 g (exclusively MgO); minimum mass: 1.384 g (exclusively Mg$_3$N$_2$). **20.2B** Most of the product was MgO. Since N$_2$ is much more abundant in the atmosphere than is O$_2$, it would appear that the rate of reaction of Mg with O$_2$ is much greater than with N$_2$. **Self-Assessment Questions: 1.** Sodium, calcium; some common naturally occurring compounds: NaCl, Na$_2$CO$_3$; CaCl$_2$, CaCO$_3$, CaSO$_4$. **4.** There is no change in amount of H$_2$SO$_4$ because only H$_2$O(l) is electrolyzed. If significant amounts of water are electrolyzed, the concentration of H$_2$SO$_4$ increases. **5.** Because its filled 1s shell gives it properties similar to that of the noble gases. **7.** (c). **13.** (c). **Problems: 21.** Order of increasing solubility: MgCO$_3 <$ Li$_2$CO$_3 <$ Na$_2$CO$_3$. Interionic attractions are strongest between the small, highly charged Mg^{2+} and CO$_3^{2-}$, and are greater for Li$_2$CO$_3$ than Na$_2$CO$_3$ because Li$^+$ is a smaller ion than Na$^+$. **23.** Density, electrode potential, electrical conductivity, and hardness; these properties depend on two or more factors that vary down the periodic table. **25. (a)** potassium peroxide; **(b)** calcium hydrogen carbonate; **(c)** Ba$_3$(PO$_4$)$_2$; **(d)** Sr(NO$_3$)$_2 \cdot$ 4 H$_2$O. **27. (a)** CaO, calcium oxide; **(b)** Na$_2$SO$_4 \cdot$ 10 H$_2$O, sodium sulfate decahydrate; **(c)** CaSO$_4 \cdot$ 2 H$_2$O, calcium sulfate dihydrate; **(d)** CaCO$_3 \cdot$ MgCO$_3$, calcium magnesium carbonate. **29.** $\text{MgCO}_3\text{(s)} + 2 \text{ HCl(aq)} \longrightarrow \text{H}_2\text{O(l)} + \text{CO}_2\text{(g)} + \text{MgCl}_2\text{(aq)}$

$\xrightarrow{\text{evaporation}} \text{MgCl}_2\text{(s)} \xrightarrow{\Delta, \text{electrolysis}} \text{Mg(l)} + \text{Cl}_2\text{(g)}$. **31.** $\text{H}_2\text{SO}_4\text{(conc aq)} + 2 \text{ NaCl(s)} \xrightarrow{\Delta} \text{Na}_2\text{SO}_4\text{(s)} + 2 \text{ HCl(g)}$; $\text{Na}_2\text{SO}_4\text{(s)} + 10 \text{ H}_2\text{O(l)} \longrightarrow \text{Na}_2\text{SO}_4 \cdot 10 \text{ H}_2\text{O(s)}$ (Glauber's salt). Heating the salt of a volatile acid with a nonvolatile acid drives off the volatile acid. **33.** NaOH(s) has to be prepared from NaCl(aq) by electrolysis. To obtain Na from NaOH requires an additional electrolysis. This is a more expensive route than to electrolyze NaCl(l), despite the requirement of a higher temperature.

35.

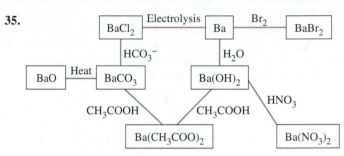

37. Electrolysis of MgCl$_2$(aq) will produce hydrogen gas rather than magnesium metal. **39. (a)** $\text{Ca(s)} + 2 \text{ HCl(aq)} \longrightarrow \text{H}_2\text{(g)} + \text{CaCl}_2\text{(aq)}$; **(b)** $\text{KHCO}_3\text{(s)} + \text{HCl(aq)} \longrightarrow \text{KCl(aq)} + \text{H}_2\text{O(l)} + \text{CO}_2\text{(g)}$; **(c)** $2 \text{ NaF(s)} + \text{H}_2\text{SO}_4\text{(concd aq)} \xrightarrow{\Delta} \text{Na}_2\text{SO}_4\text{(s)} + 2 \text{ HF(g)}$. **41. (a)** $2 \text{ Li(s)} + \text{Cl}_2\text{(g)} \longrightarrow 2 \text{ LiCl(s)}$; **(b)** $2 \text{ K(s)} + 2 \text{ H}_2\text{O(l)} \longrightarrow 2 \text{ KOH(aq)} + \text{H}_2\text{(g)}$; **(c)** $2 \text{ Cs(s)} + \text{Br}_2\text{(l)} \longrightarrow 2 \text{ CsBr(s)}$; **(d)** $\text{K(s)} + \text{O}_2\text{(g)} \longrightarrow \text{KO}_2\text{(s)}$. **43. (a)** $\text{BeF}_2\text{(s)} + 2 \text{ Na(l)} \xrightarrow{\Delta} \text{Be(s)} + 2 \text{ NaF(s)}$; **(b)** $\text{Ca(s)} + 2 \text{ CH}_3\text{COOH(aq)} \longrightarrow \text{Ca(CH}_3\text{COO)}_2\text{(aq)} + \text{H}_2\text{(g)}$; **(c)** $\text{PuO}_2\text{(s)} + 2 \text{ Ca(s)} \xrightarrow{\Delta} \text{Pu(s)} + 2 \text{ CaO(s)}$; **(d)** $\text{CaCO}_3 \cdot \text{MgCO}_3\text{(s)} \xrightarrow{\Delta} \text{CaO(s)} + \text{MgO(s)} + 2 \text{ CO}_2\text{(g)}$. **45. (a)** $\text{MgCO}_3\text{(s)} + 2 \text{ HCl(aq)} \longrightarrow \text{MgCl}_2\text{(aq)} + \text{H}_2\text{O(l)} + \text{CO}_2\text{(g)}$; **(b)** $\text{CO}_2\text{(g)} + 2 \text{ KOH(aq)} \longrightarrow \text{K}_2\text{CO}_3\text{(aq)} + \text{H}_2\text{O(l)}$; **(c)** $2 \text{ KCl(s)} + \text{H}_2\text{SO}_4\text{(conc aq)} \xrightarrow{\Delta} \text{K}_2\text{SO}_4\text{(s)} + 2 \text{ HCl(g)}$. **47. (a)** $2 \text{ NaCl(s)} + \text{H}_2\text{SO}_4\text{(conc aq)} \xrightarrow{\Delta} \text{Na}_2\text{SO}_4\text{(s)} + 2 \text{ HCl(g)}$; **(b)** $\text{Mg(HCO}_3)_2\text{(aq)} \xrightarrow{\Delta} \text{MgCO}_3\text{(s)} + \text{H}_2\text{O(l)} + \text{CO}_2\text{(g)}$, followed by $\text{MgCO}_3\text{(s)} \xrightarrow{\Delta} \text{MgO(s)} + \text{CO}_2\text{(g)}$. **49.** 594 g CaH$_2$. **51.** 267 m^3 CO$_2$. **53.** 694 L seawater. **55.** (c). **57.** NH$_3$ (a base) reacts with HCO$_3^-$ (an acid): $\text{NH}_3\text{(aq)} + \text{HCO}_3^-\text{(aq)} \longrightarrow \text{NH}_4^+\text{(aq)} + \text{CO}_3^{2-}\text{(aq)}$. This is followed by $\text{M}^{2+}\text{(aq)} + \text{CO}_3^{2-}\text{(aq)} \longrightarrow \text{MCO}_3\text{(s)}$, where $\text{M}^{2+} = \text{Mg}^{2+}, \text{Ca}^{2+}$, or Fe^{2+}. **59.** 68.4-mg boiler scale (CaCO$_3$). **61.** A limestone cave and hard water both form through the forward direction of the reaction: $\text{CaCO}_3\text{(s)} + \text{H}_2\text{O(l)} + \text{CO}_2\text{(g)} \rightleftharpoons \text{Ca(HCO}_3)_2\text{(aq)}$. Stalactites and stalagmites form in the reverse of this reaction, as does boiler scale when hard water is boiled. **63.** In a cation exchange, all cations are replaced by H$^+$. Then, in an anion exchange, all anions are replaced by OH$^-$. The H$^+$ and OH$^-$ combine to form H$_2$O, and the water is essentially free of all ions. **Additional Problems: 65. (a)** $4 \text{ Li(s)} + \text{O}_2\text{(g)} \longrightarrow 2 \text{ Li}_2\text{O(s)}$; **(b)** $6 \text{ Li(s)} + \text{N}_2\text{(g)} \longrightarrow 2 \text{ Li}_3\text{N(s)}$; **(c)** $\text{Li}_2\text{CO}_3 \xrightarrow{\Delta} \text{Li}_2\text{O(s)} + \text{CO}_2\text{(g)}$. **66.** $2 \text{ Na}_2\text{O}_2\text{(s)} + 2 \text{ CO}_2\text{(g)} \longrightarrow 2 \text{ Na}_2\text{CO}_3\text{(s)} + \text{O}_2\text{(g)}$. **67.** The CaO(s) is converted to CaCO$_3$(s) by one or both of these routes: $\text{CaO(s)} + \text{CO}_2\text{(g)} \longrightarrow \text{CaCO}_3\text{(s)}$, or $\text{CaO(s)} + \text{H}_2\text{O(l)} \longrightarrow \text{Ca(OH)}_2\text{(s)}$, followed by $\text{Ca(OH)}_2\text{(s)} + \text{CO}_2\text{(g)} \longrightarrow \text{CaCO}_3\text{(s)} + \text{H}_2\text{O(g)}$. **71.** 60.2% MgO(s). **74. (a)** $\text{Ca}^{2+}\text{(aq)} + \text{Na}_2\text{CO}_3\text{(aq)} \longrightarrow \text{CaCO}_3\text{(s)} + 2 \text{ Na}^+\text{(aq)}$; **(b)** 21 g. **80. (a)** 1.7 g sodium dodecyl sulfate; **(b)** 3.1×10^{-12} C; **(c)** The surfactant is a salt and does not dissolve well in oil-based (nonpolar solvent) paint. In latex paint the surfactant dissolves readily and can easily be carried to the particles. **83.** Spectroscopic methods should reveal whether the emitted radiation is that of hydrogen or of other elements. **Apply Your Knowledge: 84.** 80.56 ppm Ca^{2+}. **85.** The battery operates for 34 years and consumes 0.13 g Li(s). **89.** About $300,000. **91.** 246 kg/day.

Chapter 21

Exercises: 21.1A $2 BCl_3(g) + 3 H_2(g) \longrightarrow 2 B(s) + 6 HCl(g)$. **21.1B** $[B_2O_4(OH)_4]^{2-}(aq) + 4 H_2O(l) \longrightarrow 2 H_2O_2(aq) + 2[B(OH)_4]^-(aq)$ **21.2A** The reaction is nonspontaneous for standard-state conditions: $E_{cell}^\circ = -0.90$ V and $\Delta G^\circ = 174$ kJ. The ΔG° differs from Example 21.2 because of different standard states, such as 1 M Br^-(aq) instead of NaBr(s) and 1 M H^+(aq) and 1 M SO_4^{2-}(aq) instead of H_2SO_4(l). **21.2B** The reaction in question differs from that in Example 21.2 only in that 2 NaI(s) is substituted for 2 NaBr(s) on the left and I_2(g) for Br_2(g) on the right. Instead of redoing the complete calculation of Example 21.2, replace 3.14 by 19.36 on the right and $2 \times (-349.0)$ by $2 \times (-286)$ on the left. Thus to the numerical answer for Example 21.2 (54), *add* $(19.36 - 3.14) = 16.22$; and *subtract* $[(-2 \times 286) - (-2 \times 349)] = 126$. The new answer is 54 kJ $+ 16.22$ kJ $- 126$ kJ $= -56$ kJ. The oxidation of sodium iodide to I_2(g) is spontaneous at 25 °C, and so it must become spontaneous at a lower temperature than does the oxidation of sodium bromide to Br_2(g). **Self-Assessment Questions: 3.** Oxygen, silicon, aluminum. **4.** Most active *p*-block metal: Al; nonmetal, F. **8.** Even though a larger cable of Al is required to carry the same amount of current, the mass of the cable is less because of the considerably lower density of Al compared to Cu. **15.** Water has extensive hydrogen bonding between molecules, resulting in a higher boiling point than H_2S, for which hydrogen bonding is unimportant. **21.** (a) silver azide; (b) potassium thiocyanate; (c) astatine oxide, (d) telluric acid. **22.** (a) H_2SeO_4; (b) H_2Te; (c) $Pb(N_3)_2$; (d) AgAt. **23.** bauxite, Al_2O_3; boric oxide, B_2O_3; corundum, Al_2O_3; cyanogen, $(CN)_2$; hydrazine, N_2H_4; silica, SiO_2. **Problems: 25.** $3 Mg(s) + B_2O_3 \xrightarrow{\Delta} 2 B(s) + 3 MgO(s)$. **27.** H_3BO_3 is an acid because it accepts OH^- from water, forming H_3O^+. The $B(OH)_3$ molecule is able to accept only one OH^- through the lone pair of electrons on the boron atom, and is therefore monoprotic. H_3PO_4 donates protons and can lose three H^+ ions from the molecule. **29.** $Na_2B_4O_7 \cdot 10 H_2O + H_2SO_4 \longrightarrow 4 B(OH)_3 + Na_2SO_4 + 5 H_2O; 2 B(OH)_3 \xrightarrow{\Delta} B_2O_3 + 3 H_2O; B_2O_3 + 3 Cl_2 + 3 C \longrightarrow 3 CO + 2 BCl_3; 2 BCl_3 + 3 H_2 \longrightarrow 6 HCl + 2 B$. **31.** $2 BCl_3 + 6 LiAlH_4 \longrightarrow B_2H_6 + 6 AlH_3 + 6 LiCl$. **33.** $12 NaHCO_3 + 4 KAl(SO_4)_2 \xrightarrow{\Delta} 6 Na_2SO_4 + 2 K_2SO_4 + 4 Al(OH)_3 + 12 CO_2$. **35.** Both iron and aluminum oxidize in air, but unlike iron(III) oxide, aluminum's oxide coating is thin and strongly adherent, and protects the underlying metal from further corrosion, so the metal appears not to "rust". **37.** $2 Al(s) + Fe_2O_3(s) \longrightarrow Al_2O_3(s) + 2 Fe(l); \Delta H \approx -852$ kJ. The ΔH value is only an estimate because it is based on data at 298 K, and because it assumes the product Fe(s) rather than Fe(l). **39.** Because the precipitate of $Al(OH)_3$ is amphoteric and will dissolve.

41. $\ddot{S}=C=\ddot{O}:$ $[:C\equiv N:]^-$ $[:\ddot{O}-C\equiv N:]^-$

43. $Al_4C_3(s) + 12 H_2O(l) \longrightarrow 3 CH_4 + 4 Al(OH)_3$. **45.** Geometric structure: two tetrahedra sharing a corner. **47.** $KAl_2(AlSi_3O_{10})(OH)_2$. Oxidation numbers $[1 \times 1](K) + [3 \times 3](Al) + [3 \times 4](Si) + [12 \times (-2)](O) + [2 \times 1](H) = 0$. **49.** (a) $Si(CH_3)_4$; (b) $SiCl_2(CH_3)_2$; (c) $SiH(CH_2CH_3)_3$. **51.** (a) Yes; (b) No; (c) Yes; (d) No. **53.** (a) $Sn(s) + 2 HCl(aq) \longrightarrow SnCl_2(aq) + H_2(g)$; (b) $SnCl_2 + Cl_2(g) \longrightarrow SnCl_4$; (c) $SnCl_4(aq) + 4 NH_3(aq) + 2 H_2O(l) \longrightarrow SnO_2(s) + 4 NH_4^+(aq) + 4 Cl^-(aq)$. **55.** $f < d < c < a < b < e$. **57.** (a) $3 NO_2(g) + H_2O(l) \longrightarrow 2 HNO_3(l) + NO(g)$; (b) $4 NH_3(g) + 5 O_2(g) \longrightarrow 4 NO(g) + 6 H_2O(g)$; (c) $NH_4NO_3(s) \xrightarrow{200 °C} N_2O(g) + 2 H_2O(g)$. **59.** Oxidation: $NH_2NH_3^+(aq) \longrightarrow N_2(g) + 5 H^+(aq) + 4 e^-$; reduction: $Fe^{3+}(aq) + e^- \longrightarrow Fe^{2+}(aq)$; redox reaction: $4 Fe^{3+}(aq) + NH_2NH_3^+(aq) \longrightarrow 4 Fe^{2+}(aq) + N_2(g) + 5 H^+(aq)$; $E_{N_2/NH_2NH_3^+}^\circ = -0.23$ V. **61.** Principal allotropes: white P and red P; structures: white P consists of individual P_4 tetrahedra and red P consists of P_4 units joined together into long chains. **63.** $P_4(s) + 3 KOH(aq) + 3 H_2O(l) \longrightarrow 3 KH_2PO_2(aq) + PH_3(g)$. **65.** (a) $4 Al(s) + 3 O_2(g) \longrightarrow 2 Al_2O_3(s)$; (b) $2 KClO_3(s) \xrightarrow{\Delta} 2 KCl(s) + 3 O_2(g)$; (c) $2 Na_2O_2(s) + 2 H_2O(l) \longrightarrow 4 NaOH(aq) + O_2(g)$; (d) $Pb^{2+}(aq) + H_2O(l) + O_3(g) \longrightarrow PbO_2(s) + 2 H^+(aq) + O_2(g)$. **67.** Because of its low melting point (119 °C) S_8 is melted by superheated water. The water and S_8(l) do not form a solution or react, and the mixture can be brought to the surface with compressed air. The S_8(s) is in a pure state. **69.** Oxidation: $2 S(s) + 6 OH^- \longrightarrow S_2O_3^{2-} + 3 H_2O + 4 e^-$; reduction: $2 SO_3^{2-} + 3 H_2O + 4 e^- \longrightarrow S_2O_3^{2-} + 6 OH^-$; overall: $SO_3^{2-}(aq) + S(s) \longrightarrow S_2O_3^{2-}(aq)$. **71.** Only Cl_2(g) will oxidize Br^-(aq) to Br_2(l). I_2(s) is too poor an oxidizing agent for this, and I^-, Cl^-, and F^- can only act as reducing agents. **73.** $I^-(aq) + 3 Cl_2(g) + 3 H_2O(l) \longrightarrow IO_3^-(aq) + 6 Cl^-(aq) + 6 H^+(aq); 2 Br^-(aq) + Cl_2(g) \longrightarrow 2 Cl^-(aq) + Br_2(l)$; When the mixture, $IO_3^-(aq) + Cl^-(aq) + Br_2(l)$, is extracted with CS_2(l), only the Br_2(l) dissolves in the CS_2(l). **75.** $MgBr_2$ is a halide, $NaClO_3$ is a halate (chlorate) salt, BrF_3 is neither. **77.** (a) $2 KI(s) + 2 H_2SO_4$(concd aq) $\longrightarrow I_2(s) + K_2SO_4(s) + 2 H_2O(l) + SO_2(g)$; (b) $KI(s) + H_3PO_4$(concd aq) $\longrightarrow HI(g) + KH_2PO_4(s)$. **79.** (a) Square planar; (b) square pyramidal. **81.** Argon is produced only by the decay of potassium-40, whereas helium is derived from alpha particles. Alpha particles are emitted by many radioactive isotopes, particularly those of high atomic and mass numbers. **Additional Problems: 83.** (a) BrF_3: 28 valence electrons; VSEPR notation AX_3E_2; electron-group geometry, trigonal bipyramidal; molecular shape, T-shaped; (b) IF_5: 42 valence electrons; VSEPR notation AX_5E; electron-group geometry, octahedral; molecular shape, square pyramidal. **86.** $2 Al(s) + 2 KOH(aq) + 6 H_2O(l) \longrightarrow 2 KAl(OH)_4(aq) + 3 H_2(g); KAl(OH)_4(aq) + 2 H_2SO_4(aq) + 8 H_2O(l) \longrightarrow KAl(SO_4)_2 \cdot 12 H_2O(s)$. **91.** 1.4×10^4 L air. **92.** 334 kg C, 1.89×10^3 kg Al_2O_3, 1.07×10^{10} coulombs. **95.** $[NO_3^-] = 0.1716$ M. **98.** (a) $\Delta H_f^\circ[ClF] \approx -50$ kJ/mol; (b) $\Delta H_f^\circ[OF_2] \approx 34$ kJ/mol; (c) $\Delta H_f^\circ[Cl_2O] \approx 82$ kJ/mol; (d) $\Delta H_f^\circ[NF_3] \approx -129$ kJ/mol. **Apply Your Knowledge: 105.** 5×10^{37} O_3 molecules.

Chapter 22

Exercises: 22.1A Essentially nothing. The match used in an attempt to ignite the dichromate salt might flare a bit initially, with the dichromate, an oxidizing agent, promoting the oxidation of the match. However, a reaction could not be sustained, because there is no reducing agent in the solid. **22.1B** $4 FeCr_2O_4 +$ $16 NaOH + 7 O_2 \xrightarrow{\Delta} 8 Na_2CrO_4 + 4 Fe(OH)_3 + 2 H_2O$. **22.2A** $MnO_4^-(aq) + 2 H_2O(l) + 3 e^- \longrightarrow MnO_2(s) +$ $4 OH^-(aq)$; $E° = 0.60$ V. In basic solution, MnO_4^- will oxidize any species having $E° < 0.60$ V, including $Br^-(aq)$ to $BrO_3^-(aq)$, $Br_2(l)$ to $BrO^-(aq)$, $Ag(s)$ to $Ag_2O(s)$, and $NO_2^-(aq)$ to $NO_3^-(aq)$. **22.2B (a)** *red.* $BiO_3^- + 4 H^+ + 2e^- \longrightarrow BiO^+ + 2 H_2O$; *oxid.* $Mn^{2+} + 4 H_2O \longrightarrow MnO_4^- + 8 H^+ + 5 e^-$; *overall*: $2 Mn^{2+} + 5 BiO_3^- + 4 H^+ \longrightarrow 5 BiO^+ + 2 MnO_4^- + 2 H_2O$; **(b)** The standard electrode potential for the BiO_3^- reduction to BiO^+ must exceed that of the reduction of MnO_4^- to Mn^{2+}, which is 1.51 V. **22.3A** The coordination and oxidation numbers are **(a)** 6 and +3; **(b)** 6 and +2. **22.3B** $[CrCl_2(H_2O)_4]Cl$. **22.4A (a)** hexaamminecobalt(II) ion; **(b)** pentaamminebromocobalt(III) bromide. **22.4B (a)** tetrachloroaurate(III) ion; **(b)** triamminediaquabromocobalt(III) tetracyanonickelate(II). **22.5A (a)** $[Co(en)_3]^{3+}$; **(b)** $[CrCl_4(NH_3)_2]^-$. **22.5B (a)** $[Cr(NH_3)_6]$ $[Co(CN)_6]$; **(b)** $[PtCl_2(en)_2]SO_4$. **22.6A (a)** Isomers. NCS^- is bonded to Fe^{3+} through the S atom in one structure and the N atom in the other. **(b)** Isomers. The ligands, NH_3, are bonded to Zn^{2+} in one structure and Cu^{2+} in the other, similarly with Cl^-. **(c)** Not isomers. One compound has two NH_3 ligands and the other has four. **22.6B (a)** No isomerism. The H_2O molecule can take up any one of four equivalent positions, and the NH_3 molecules, the other three. **(b)** Yes, ligands can be interchanged between the complex cation and complex anion, as in $[Pt(NH_3)_4][CuCl_4]$. **22.7A** Assuming en ligands attach only at adjacent points on the hexagonal ring, all structures are identical to the following:

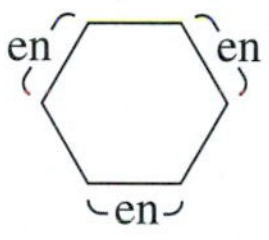

22.7B In $[PtCl_2(NH_3)_2]$, substitution at adjacent corners of the square (*cis*) is different from substitution at the ends of a diagonal (*trans*). In $[ZnCl_2(NH_3)_2]$, all four vertices of a tetrahedron are equivalent and there can be no isomerism. **22.8A** CN^- is a strong field ligand, and the energy separation (Δ) is large. The six *d* electrons of $[Ar]3d^6$ are found as three pairs in the lower-energy set of 3*d* orbitals. **22.8B** In a tetrahedral field, the eight 3*d* electrons of $[Ar]3d^8$ fill the two lower-level and one of the upper-level *d* orbitals. The other two upper level *d* orbitals have single electrons. There are two unpaired electrons. **Self-Assessment Questions: 2.** $[Kr]4d^{10}5s^25p^1$ is the electron configuration of indium, a fifth-period, *p*-block, main-group element; $[Ar]3d^74s^2$ is that of cobalt, a fourth-period, *d*-block, transition element. **5.** (c). **6.** (a). **7.** Co, Ni, and Fe each lose two 4*s* electrons to form 2+ ions, but when iron also loses a 3*d* electron, it obtains a half-filled 3*d* subshell which is especially stable. **8. (a)** scandium chloride; **(b)** iron(II) silicate; **(c)** sodium manganate; **(d)** chromium(VI) oxide. **12.** (c). **13.** (d). **18.** The central atom is part of a complex anion. **20.** (b). **Problems: 23.** To acquire the argon electron configuration, a Ca atom loses two electrons $(4s^2)$, whereas Sc loses three $(3d^14s^2)$. **25.** The +1 oxidation state might be expected because chromium has a $4s^13d^5$ configuration, and should be capable of losing a single valence electron. **27. (a)** Ca < K because $Z_{eff} \approx 2$ for Ca and ≈ 1 for K. The nucleus attracts the valence electrons more strongly in Ca.

(b) Mn < Ca because it has a larger nuclear charge but the same number of valence electrons as Ca. **(c)** The Mn and Fe atoms are the same size because they have the same Z_{eff} and the same number of valence electrons. **29. (a)** $2 Sc(s) + 6 HCl(aq) \longrightarrow$ $2 ScCl_3(aq) + 3 H_2(g)$; **(b)** $Sc(OH)_3(s) + 3 HCl(aq) \longrightarrow$ $ScCl_3(aq) + 3 H_2O(l)$; **(c)** $Sc(OH)_3(s) + 3 Na^+(aq) + 3 OH^-(aq)$ $\longrightarrow 3 Na^+(aq) + [Sc(OH)_6]^{3-}(aq)$. **31. (a)** $K_2Cr_2O_7(aq) +$ $14 HCl(aq) \longrightarrow 3 Cl_2(g) + 2 CrCl_3(aq) + 2 KCl(aq) +$ $7 H_2O(l)$, separate $CrCl_3$ and KCl by fractional crystallization; **(b)** $2 MnO_4^-(aq) + 5 C_2O_4^{2-}(aq) + 16 H^+(aq) \longrightarrow 10 CO_2(g)$ $+ 2 Mn^{2+}(aq) + 8 H_2O(l)$, then $Mn^{2+}(aq) + Na_2CO_3(aq) \longrightarrow$ $MnCO_3(s) + 2 Na^+(aq)$. **33.** $3 MnO_2 + KClO_3 + 6 KOH \longrightarrow$ $3 K_2MnO_4 + KCl + 3 H_2O$, then $2 K_2MnO_4 + Cl_2 \longrightarrow$ $2 KMnO_4 + 2 KCl$ **35.** CrO_4^{2-} in equilibrium with $Cr_2O_7^{2-}$ combines with Pb^{2+} to form $PbCrO_4(s)$; the net reaction goes to completion. **37.** Fe(s); it is most easily oxidized. That is, $Fe^{2+}(aq) + 2e^- \longrightarrow Fe(s)$ has the lowest value of $E°$ $(-0.440$ V). **39.** The reaction $4 Co^{3+}(aq) + 2 H_2O(l) \longrightarrow O_2(g) + 4 H^+(aq) +$ $4 Co^{2+}(aq)$ is spontaneous with $E°_{cell} = 0.69$ V. **41. (a)** $2 Ag(s) +$ $2 H_2SO_4(concd\ aq) \longrightarrow SO_2(g) + 2 H_2O(l) + Ag_2SO_4(s)$; **(b)** $3 Ag(s) + 4 HNO_3(concd\ aq) \longrightarrow NO(g) + 2 H_2O(l) +$ $3 AgNO_3(aq)$. **43.** Compare three reduction half-reactions: (1) $NO_3^- + 4 H^+ + 3 e^- \longrightarrow NO(g) + 2 H_2O$, $E° = 0.956$ V; (2) $Ag^+ + e^- \longrightarrow Ag(s)$, $E° = 0.800$ V; (3) $Au^{3+} + 3 e^-$ $\longrightarrow Au(s)$, $E° = 1.52$ V. The combination of (1) and the reverse of (2) has $E°_{cell} > 0$; (1) and the reverse of (3) has $E°_{cell} < 0$. **45.** Compared to Group 2A, the 2B elements form many complex ions and have smaller atomic radii, higher ionization energies, and $E°$ values that are less negative. **47.** No; although $E°$ for Cd^{2+}/Cd is -0.403 V, that potential is less negative than that of the Fe^{2+}/Fe reduction, and so the iron metal will be oxidized while the cadmium remains unchanged. **49. (a)** $3 Hg(l) +$ $2 NO_3^-(aq) + 8 H^+(aq) \longrightarrow 3 Hg^{2+}(aq) + 2 NO(g) +$ $4 H_2O(l)$; **(b)** $ZnO(s) + 2 CH_3COOH(aq) \longrightarrow Zn^{2+}(aq) +$ $2 CH_3COO^-(aq) + H_2O(l)$. **51. (a)** 6; **(b)** 2; **(c)** 4. **53. (a)** 6; **(b)** 3; **(c)** 3; **(d)** 1. **55. (a)** +2; **(b)** +3; **(c)** +4; **(d)** 0. **57. (a)** hexamminezinc(II); **(b)** tetrachloroferrate(II); **(c)** tetramminedichloroplatinum(IV); **(d)** tris(ethylenediamine)cobalt(III). **59. (a)** $[CoF_6]^{3-}$; **(b)** $[CrCl_2(H_2O)_4]^+$; **(c)** $[CoBr_2(en)_2]^+$. **61. (a)** potassium hexacyanochromate(II); **(b)** potassium tris(oxalato)chromate(III). **63. (a)** $[Cu(NH_3)_4]Cl_2$; **(b)** $[Cr(en)_3]$ $[Co(CN)_6]$; **(c)** $[Pt(NH_3)_4][PtCl_4]$. **65. (a)** It is a cation, so the name ends with "copper(II)": tetraaquacopper(II) ion; **(b)** The oxidation state is not given. The correct name could be a neutral complex, pentaamminesulfatocobalt(II), but is more likely a complex ion, pentaamminesulfatocobalt(III) ion. **67. (a)** Yes, these are structural isomers. The NCS^- ion is bonded through the N atom in one and through the S atom in the other; **(b)** Yes, these are isomers. Both have the formula $Pt_3Cl_6N_6H_{18}$, but the cations and anions differ in composition; **(c)** No, these are different compounds since they have different numbers of K atoms; **(d)** Yes, they are structural isomers, they differ in the ligands on the central metal ion. **69.** The sketch should have Cr^{3+} as the central ion in an octahedron. No matter which pair of adjacent vertices is chosen for linkage of the en ligand, the remaining four vertices for linkage of Cl^- are equivalent. **71.** Yes. Refer to Figure 22.15. Replace any two en ligands by ox, both in the structure on the left and in its mirror image. The resulting structures are still nonsuperimposable mirror images. **73.** Four different groups can be arranged in just two different ways in a tetrahedral complex, and those two ways are mirror images of one another. **75. (a)** $Cr^{3+}[Ar]3d^3$;

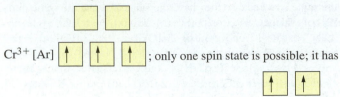

$Cr^{3+}[Ar]$ ⬆ ⬆ ⬆ ; only one spin state is possible; it has

⬆ ⬆

three unpaired e^-. **(b)** $Fe^{3+}[Ar]3d^5$; $Fe^{3+}[Ar]$ ⬆ ⬆ ⬆ because Cl^- is a weak-field ligand, we expect the high-spin state with all five e^- unpaired. **(c)** $Mn^{3+}[Ar]3d^4$;

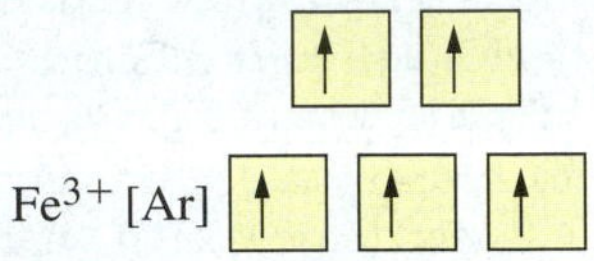

⬜ ⬜

$Mn^{3+}[Ar]$ ⬆⬇ ⬆ ⬆ ; because CN^- is a strong-field ligand, we expect the low-spin state with two unpaired e^-. **(d)** The d-orbital sets in a tetrahedral field are reversed from (a)–(c).

⬆ ⬆ ⬆

$Co^{2+}[Ar]3d^7$; $Co^{2+}[Ar]$ ⬆⬇ ⬆⬇ ; only one spin state is possible, with three unpaired $3d$ e^-. **77.** NH_3 is a stronger field ligand than H_2O. Light of shorter wavelength (higher frequency) is absorbed, and light of longer wavelength is transmitted by the ammine complex than by the aqua complex. **Additional Problems: 80.** $2\,Cu(s) + O_2(g) + H_2O(l) + CO_2(g) \longrightarrow Cu_2(OH)_2CO_3(s)$. **81.** One or both of the following reactions should occur: (1) $5\,MnO_4^{2-} + 8\,H^+ \longrightarrow Mn^{2+} + 4\,MnO_4^- + 4\,H_2O$; (2) $3\,MnO_4^{2-} + 4\,H^+ \longrightarrow MnO_2(s) + 2\,MnO_4^- + 2\,H_2O$. **85.** 2.70 g Cr. **89.** In the complex $[Cr(L)_6]^{3+}$, the three d electrons of $Cr^{3+}([Ar]3d^3)$ remain unpaired regardless of the ligand (see also, answer to Problem 75a). **93.** (a) and (c) are identical, (a) and (b) [or (a) and (d)] are geometric isomers, and (b) and (d) are optical isomers. **Apply Your Knowledge: 97.** 61%. **99.** $K[PtCl_3(C_2H_4)]$. **105.** 0.0154 M. The results would have been the same with iron(III) because EDTA reacts with all metal ions in a one-to-one ratio regardless of charge.

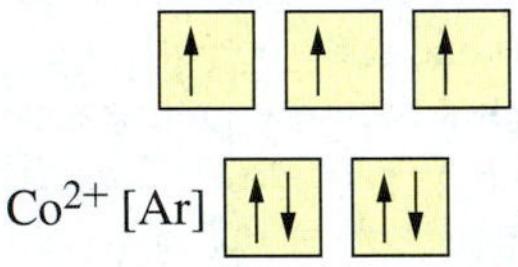

Chapter 23

Exercises: 23.1A (a) Aliphatic. The conjugated system is incomplete, and the π-bonding system does not fit the $4n + 2$ rule. **(b)** Aliphatic. The structure fits the $4n + 2$ rule, but the conjugated system is incomplete. **(c)** Aliphatic. The conjugated system is complete, but the structure does not fit the $4n + 2$ rule. **(d)** Aromatic. The conjugated system extends throughout the structure, and 14 π electrons conform to the $4n + 2$ rule (where $n = 3$). **23.1B** The nitrogen atom is sp^2 hybridized, leaving an unshared pair of electrons in an unhybridized p-orbital to contribute to pi-bonding. Pyrrole therefore has six pi-electrons, which fits the $4n + 2$ rule, and the pi-electrons extend around the entire ring. **23.2A** The sign of the optical rotation cannot be predicted from the formulas alone. **(a)** Gulose is an L-aldohexose; **(b)** erythrulose is an L-ketoterose. **23.2B** No. The name D-(−)-fructose below the Fischer projection on the previous page establishes that the D- configuration of fructose has been found to be levorotatory. The dextrorotatory or (−) configuration must therefore be the L-configuration. **Self-Assessment Questions: 3.** (c). **6. (a)** aromatic; **(b)** aliphatic; **(c)** aliphatic; **(d)** both. **8.** (a). **13.** (c).

15. (b). **18.** A conjugated system has alternating double and single bonds. Examples: 1,3-butadiene, 1,3,5-hexatriene, lycopene. **19.** IR radiation induces various vibrations of bonds; UV and visible radiation causes electronic transitions. **20.** UV; NMR. **21.** $\tilde{\nu} = \nu/c$. **25.** Penicillin molecules have a β lactam, a four-membered ring with one nitrogen atom and three carbon atoms, one an amide. The amide carbon in penicillin attaches to the enzyme that catalyzes the synthesis of bacterial cell walls, deactivating the enzyme. Human cells do not have cell walls and are not harmed. **32.** When the urine has a lower pH (after a meal, after drinking alcohol, or during periods of stress), more nicotine (a base) is excreted because the conjugate acid (ionic) of nicotine is more soluble in urine. The more nicotine excreted, the more quickly the person craves another cigarette. **Problems:**

33. (a) 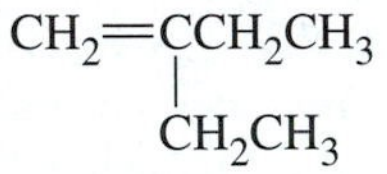 $CH_2{=}CCH_2CH_3$
　　　　　　　$|$
　　　　　　 CH_2CH_3

(b) 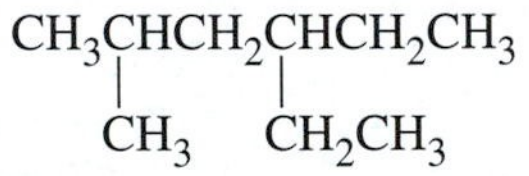 $CH_3CHCH_2CHCH_2CH_3$
　　　　$|$　　　　$|$
　　　 CH_3　　CH_2CH_3

　　　　　　　　　　　　　 CH_3
　　　　　　　　　　　　　 $|$
(c) 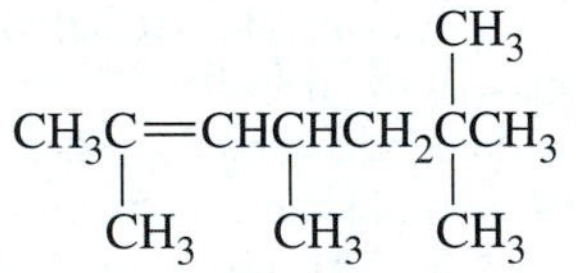 $CH_3C{=}CHCHCH_2CCH_3$
　　　　　　$|$　　$|$　　$|$
　　　　　 CH_3　CH_3　CH_3

(d) 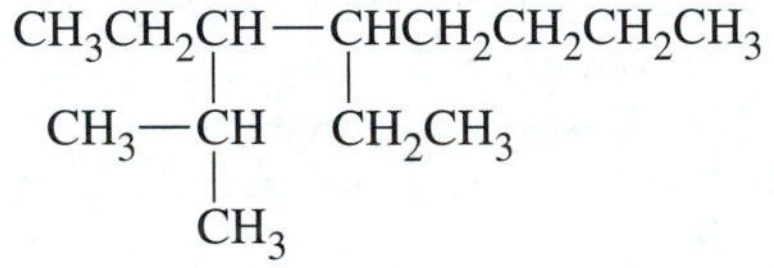 $CH_3CH_2CH{-}CHCH_2CH_2CH_2CH_3$
　　　　　　　$|$　　$|$
　　　　 $CH_3{-}CH$　CH_2CH_3
　　　　　　　$|$
　　　　　　 CH_3

(e) 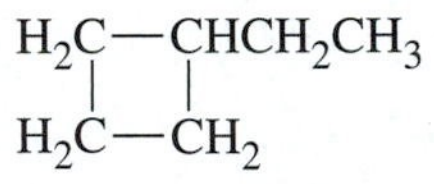 $H_2C{-}CHCH_2CH_3$ 　　**(f)** 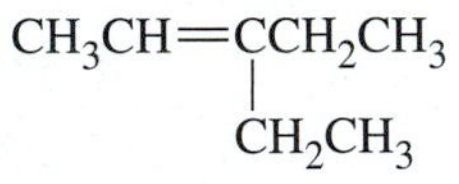 $CH_3CH{=}CCH_2CH_3$
　　　$|$　　　$|$　　　　　　　　　　　　$|$
　　 $H_2C{-}CH_2$　　　　　　　　　 CH_2CH_3

35. (a) 2-methyl-2-butene; **(b)** 1,1,6-trichloro-3-heptyne; **(c)** 2,2,4-trimethyl-3-hexene; **(d)** 2-methyl-1-pentene. **37. (a)** 2,2-dimethyl-3-pentanol; **(b)** 2-methyl-1-propanol; **(c)** 2-chloro-4-iodobenzoic acid; **(d)** 4-chloro-2-nitroaniline. **39. (a)** $CH_3CH_2CH(OH)CH_2CH_2CH_3$; **(b)** $CH_3CH(OH)C(CH_3)_3$; **(c)** $CH_3CH_2CH(OH)CH(CH_3)CH(CH_3)CH_2CH_3$; **(d)** $CH_3CH_2CH(CH_2CH_3)CH_2OH$;

(e)
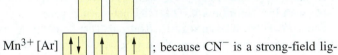

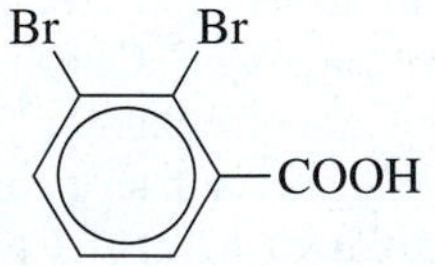

(f) 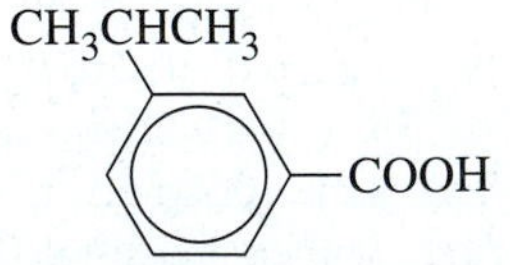 CH_3CHCH_3 attached to a benzene ring with —COOH

41. (a) carboxylic acid, halide; **(b)** alkene; **(c)** carboxylic acid, aromatic ring, halide.

43. (a) $CH_3CH_2CHClCH_2COOH$ 3-chloropentanoic acid

(b) 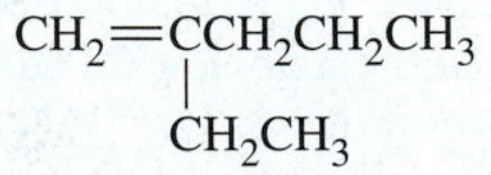 $CH_2{=}CCH_2CH_2CH_3$ 　2-ethyl-1-pentene
　　　　　$|$
　　　 CH_2CH_3

(c) 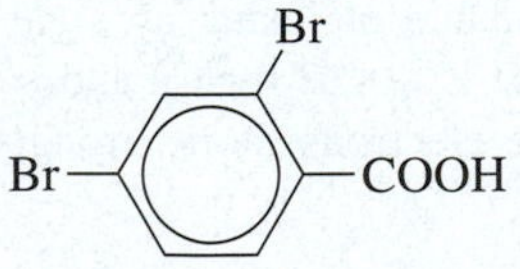 $Br{-}$⬡$-COOH$ with Br 　2,4-dibromobenzoic acid

45. (a) oxidation; **(b)** hydration. **47. (a)** $CH_3CH_2CH_2CH_3$;
(b) $(CH_3)_2C(OH)CH(CH_3)_2$. **49. (a)** $CH_3CH_2CH_2OH$
$\xrightarrow{H_2SO_4, \Delta}$ $CH_3CH{=}CH_2$; **(b)** $2\ CH_3CH_2CH_2OH$
$\xrightarrow{H_2SO_4}$ $(CH_3CH_2CH_2)_2O$. **51. (a)** $CH_3CH{=}CH_2$;
(b) $CH_2{=}C(CH_3)_2$. **53. (a)** $CH_3CH_2CH_2COOH + NaHCO_3(aq)$
$\longrightarrow$ $CH_3CH_2CH_2COONa(aq)$ + $CO_2(g) + H_2O(l)$;
(b) $CH_3COOCH_2CH_3 + H_2O \xrightarrow{H_2SO_4} CH_3COOH +$
CH_3CH_2OH. **55. (a)** Saturated, 8 carbon atoms; **(b)** unsaturated,
10 carbon atoms; **(c)** unsaturated, 18 carbon atoms.
57. $CH_3(CH_2)_{16}COOH$ $CH_2OCO(CH_2)_7CH{=}CH(CH_2)_7CH_3$
         (a) |
$CH_3(CH_2)_{12}COO^-K^+$ $CHOCO(CH_2)_7CH{=}CH(CH_2)_7CH_3$
         (b) |
                                $CH_2OCO(CH_2)_7CH{=}CH(CH_2)_7CH_3$
                                        (c)

$CH_2OCO(CH_2)_{14}CH_3$
|
$CHOCO(CH_2)_{14}CH_3$
|
$CH_2OCO(CH_2)_{14}CH_3$
         (d)

59. (a) aldose, triose; **(b)** aldose, pentose; **(c)** aldose, pentose;
(d) ketose, hexose; **(e)** aldose, hexose. **61. (a)** D-sugar;
(b) L-sugar; **(c)** D-sugar.

63.

(a)   **(b)**

65. (a) β; **(b)** α. **67. (a)** $-CH_2COOH$; **(b)** $-CH_2SH$;
(c) $-CH(CH_3)_2$; **(d)** $-CH_2OH$; **(e)** $-CH_2C_6H_5$ (CH_2 on
phenyl ring). **69.** Lysine, $NH_2(CH_2)_4CH(NH_2)COOH$, also
arginine or histidine (Table 23.2). **71.**

73. (a) pH < pI; **(b)** pH > pI; **(c)** pH = pI. **75.** The sugar is
ribose; the heterocyclic base is the pyrimidine cytosine. **77.** The
molecule is not a nucleotide (no phosphate unit). The base is a
pyrimidine (thymine). The compound is incorporated in DNA.
79. CTAATGT **81.** CUAAUGU **83.** AGGCTA **85. (a)** AAA;
(b) GUA; **(c)** UCG; **(d)** GGC. **87.** CH_3COOH would show a broad
OH peak around 3200–3600 cm^{-1} while CH_3COOCH_3 would
not. **89.** 2-butanol has an $-OH$ band between 3250–3450 cm^{-1},
2-butanone has a $C{=}O$ peak around 1700 cm^{-1}. Either of the two
features would be sufficient to distinguish between the compounds. **91.** Yes. It is likely to have one or more of the structural
features noted in Table 23.6. **93.** Yellow substances absorb blue-
violet light. **Additional Problems: 96. (a)** $CH_3CH_2CH_2OH$;
(b) $HOCH_2CH_2OH$; **(c)** CH_3OH; **(d)** $(CH_3)_2CHCH_2CH_2OH$.

99.

102. The conjugated system of azobenzene [first structure,
$(C_6H_5N)_2$] extends over 14 atoms and should produce a colored
compound, while p-hydroxyphenol (second structure) is colorless;
its conjugated system is no more extensive than that of benzene.
103. (a) The second compound has a C—O stretch and CH_3
stretch; the first does not. **(b)** The first compound has an O—H
stretch, but not the second. The second has a C—O stretch, but not
the first. **113.** There are four isomers: $ClCH_2CH(CH_3)CH_2CH_3$
(31.1%), $CH_3CH(CH_3)CH_2CHCl$ (15.5%), $CH_3CCl(CH_3)$
CH_2CH_3 (22.3%), and $CH_3CH(CH_3)CHClCH_3$ (31.1%).
114. (a) $CH_3CH_2CH_2COOCOCH_2CH_2CH_3 + H_2O \longrightarrow$
$2\ CH_3CH_2CH_2COOH$, $CH_3CH_2CH_2COOCOCH_2CH_2CH_3 +$
$H_2O + 2\ NH_3 \longrightarrow 2\ CH_3CH_2CH_2COO^-NH_4^+$; **(b)** The product of the reaction with $NH_3(aq)$; it is ionic; **(c)** CH_3CH_2
CH_2CONH_2. **Apply Your Knowledge: 115.** The molar
mass is 5.10 g/0.0500 mol = 102 g/mol. Possible formulas
have one COOH group and the molecular formula
$C_5H_{10}O_2$: $CH_3CH_2CH_2CH_2COOH$; $CH_3CH_2CH(CH_3)COOH$;
$(CH_3)_2CHCH_2COOH$; $(CH_3)_3CCOOH$. **117.** Carbon disulfide and
carbon tetrachloride are nonpolar and most useful for infrared
spectrometry. They are also useful for NMR because they have no
hydrogen atoms to produce an interfering signal. Ionic compounds
such as NaCl, AgCl, and KBr have no covalent bonds and no significant vibrational frequencies of absorbance.

Chapter 24
Exercises: 24.1A A metal more active than Zn (for example, Al)
would displace some Zn(s), together with the less active metals.
This would reduce the yield of Zn in the subsequent electrolysis.
Also, a new cation (for example, Al^{3+}) would be introduced,
which might deposit with the Zn during electrolysis.
24.1B $Ag_2S(s) + 4\ CN^-(aq) + 2\ O_2(g) \longrightarrow 2\ Ag(CN)_2^-(aq) +$
$SO_4^{2-}(aq)$ **24.2A** The $-CH_3$ groups are side-chains of the polymer chain; they are not part of the main chain. Also, the repeating
unit shown has one C atom with only three bonds in the $-CH-$
portion and one with five bonds in the $-CH_3-$ portion.
24.2B Since the central carbon atom has no hydrogen atoms, a
branched structure, in which the end of one or two propadiene
molecules attaches to that central carbon, would be expected.
24.3A $-(OCH_2CH_2(C{=}O)-)_n$ **24.3B** Serine, threonine, tyrosine have $-OH$ groups in addition to the amino and carboxyl
groups. These three could form polyesters as well as polyamides.
Tryptophan, asparagine, glutamine, praline, lysine, arginine, and
histidine have at least two basic nitrogen atoms in different locations and could form different polyamides. Aspartic and glutamic
acid have two (different) carboxyl groups each, and either group
could attach to the amino group. **Self-Assessment Questions:**
1. (c). **2.** (b). **5.** Reduction converts a metal oxide to the free metal.
The most common reducing agent is carbon. In a few cases, such
as HgO, the oxide decomposes to the free metal; a reducing agent
is not needed. **7.** (b). **13.** One model of metallic bonding states that
the valence electrons are free to move around the remaining atomic
cores. It is this electron gas (comprised of valence electrons, not all
electrons) in the metal crystal that bonds metal atoms together.
16. A semiconductor that contains a carefully controlled trace of a
specific impurity. A positive hole, characteristic of p-type semiconductor, is a vacancy in the valence band. When a potential is
applied to the p-type semiconductor, an adjacent electron "jumps"
into the hole, leaving another hole. **17.** (c). **20.** (d). **23.** (b).
Problems 25. The metal has to be obtainable in a pure form in an
economically feasible process. The ore richest in the metal may be

scarce, or it may be in a chemical form that is difficult to reduce. **27.** Cd occurs in Zn ores because of the chemical similarity of Zn and Cd. Cd(l) can be separated from Zn(l) by fractional distillation, or the less active Cd(s) can be displaced from a solution of Zn^{2+} and Cd^{2+} by Zn(s). **29.** Any HgO produced by roasting HgS immediately decomposes to Hg(l) and O_2(g) in the overall reaction $HgS(s) + O_2(g) \longrightarrow Hg(l) + SO_2(g)$. **31.** $2[Ag(CN)_2]^-(aq) + Zn(s) \longrightarrow 2\,Ag(s) + [Zn(CN)_4]^{2-}(aq)$. **33.** (1) $ZnO(s) + H_2SO_4(aq) \longrightarrow ZnSO_4(aq) + H_2O(l)$; (2) $Zn(s) + Cd^{2+}(aq) + SO_4{}^{2-}(aq) \longrightarrow Cd(s) + Zn^{2+}(aq) + SO_4{}^{2-}(aq)$; (3) $2\,Zn^{2+}(aq) + 2\,SO_4{}^{2-}(aq) + 2\,H_2O(l) \xrightarrow{\text{electrolysis}} 2\,Zn(s) + 4\,H^+(aq) + 2\,SO_4{}^{2-}(aq) + O_2(g)$. Note that sulfuric acid produced in reaction (3) can be recycled into reaction (1). **35.** (1) $Sn(s) + 2\,Cl_2(g) \longrightarrow SnCl_4(l)$; (2) $SnCl_4(l) + (2 + x)H_2O \longrightarrow SnO_2 \cdot x\,H_2O + 4\,HCl(aq)$; (3) $SnO_2 \cdot x\,H_2O \xrightarrow{\Delta} SnO_2(s) + x\,H_2O(g)$; (4) $SnO_2(s) + 2\,C(s) \xrightarrow{\Delta} Sn(s) + 2\,CO(g)$. **37.** 1.7×10^7 kg ore. **39.** $2\,PbS(s) + 3\,O_2(g) \longrightarrow 2\,PbO(s) + 2\,SO_2(g)$; $PbO(s) + C(s) \longrightarrow Pb(l) + CO(g)$; $PbO(s) + CO(g) \longrightarrow Pb(l) + CO_2(g)$ **41.** The $4p$ band of Ca overlaps with the $4s$ band. The numbers of valence e^- and energy levels are greater in the two partly filled overlapping bands of Ca than in the half-filled $4s$ band in K. **43.** All electron transitions in the visible spectrum are possible between appropriate levels in an electron band. Because most of the incident light is reflected, metals have a luster. If certain wavelength components are reflected more strongly than others, the metal is colored. **45.** 1.37×10^{21} energy levels, 2.74×10^{21} electrons. **47.** In a semiconductor, there is an energy gap between the valence and conduction bands. In a metal, either the valence band is itself a conduction band or it overlaps one; there is no energy gap. **49.** (a) n-type; (b) p-type. **51.** See Figure 24.12. In n-type, electrons in the donor level are readily promoted. Positive holes left in the donor level are immobilized. Electrons predominate as charge carriers. In p-type, electrons are promoted to and immobilized in the low-lying acceptor level. Positive holes predominate as charge carriers. **53.** No, the number of electrons that jump the gap *decreases* as the temperature is lowered. **55.** ZnSe and GaP. **57.** (a) Cellophane is regenerated cellulose, the same material as rayon. It is made by treating cellulose with sodium hydroxide and carbon disulfide, and forcing the viscous solution through a narrow slit to form a film. (b) LDPE, low-density polyethylene, has the formula $\text{+CH}_2\text{CH}_2\text{+}_n$. It is made by the free-radical polymerization of ethylene, $CH_2 = CH_2$. Its chains are branched and pack together poorly. **59.** Rubber is elastic because its coiled polymer chains are straightened out during stretching but return to their coiled state when relaxed. Vulcanization forms crosslinks between chains which pull the chains back into their original shape after stretching.

61.

63.

(a)

(b)

65.

(a)

(b)

67. (b) Terephthalic acid, $HO-\overset{\overset{O}{\|}}{C}-\bigcirc-\overset{\overset{O}{\|}}{C}-OH$,

and 1,4 phenylenediamine, $H_2N-\bigcirc-NH_2$

69. Tempered iron is more dense because fcc has a greater packing efficiency than bcc. All tempered metals need not follow this pattern. For example, if a metal underwent transition from fcc to bcc as the temperature was increased, a decrease in density would be seen. **71.** The rate should increase 200 times; the surface area of the finer catalyst is $2.3 \times 10^2 \text{ m}^2/\text{g}$ **Additional Problems: 73.** 55 g. **75.** In its "nonconducting" mode, the diode will actually conduct a tiny amount of current. The unintended impurity makes the p-type semiconductor behave as n-type to a very limited extent, and vice versa. **77.** Minimum wavelength is $1.09\ \mu\text{m}$, in the infrared region, so silicon should have sufficient energy levels in its conduction band so that it absorbs all visible light. **81.** 1,2-ethanediol has two —OH groups, whereas 1,2,3-propanetriol has three. The polymer of 1,2,3-propanetriol and 1,2-benzenedicarboxylic acid has extensive crosslinking of polymer chains and is much more rigid than the polymer with 1,2-ethanediol. **84.** 760 u. **Apply Your Knowledge: 88.** 12.8% Mn. **91.** 3.60% Ni.

Chapter 25

Exercises: 25.1A 6.74 mmHg. **25.1B** 0.126 g water vapor. **Self-Assessment Questions: 1. (a)** troposphere; **(b)** stratosphere. **6. (a)** carbon monoxide, CO, and nitrogen oxide, NO; **(b)** CH_4 and CCl_2F_2; **(c)** HNO_3 and/or H_2SO_4. **10.** A combustion mixture that is rich in fuel and lean in $O_2(g)$ promotes the formation of CO(g). Carbon monoxide ties up the iron atoms in hemoglobin, preventing the transport of oxygen to the cells of the body. **13. (a)** The electrostatic precipitator applies a charge to particulates, collecting the particulates out of flue gases. **(b)** A catalytic converter dramatically increases the rate of conversion of hydrocarbons and carbon monoxide to water and carbon dioxide; it also speeds the breakdown of nitrogen oxides to nitrogen and oxygen gases. **19.** Cholera, typhoid fever, dysentery. Industrialized nations destroy disease organisms using Cl_2, O_3, or other oxidizing agents. **22.** BOD = biochemical oxygen demand; the amount of oxygen needed to decompose aerobically the organic matter in water. A high BOD means that there is a lot of organic material in the water, which usually means that the oxygen in the water is used up and living organisms will die. **25.** Because chlorinated hydrocarbons are chemically stable and do not react readily with other compounds. **Problems 47.** (c). Air pressure at 30 km is produced by the 1% of air that is at altitudes beyond 30 km. $P_{bar} \approx 0.01 \times 760$ mmHg ≈ 7.6 mmHg ≈ 10 mmHg. **49.** Lower-molecular-mass species predominate at higher altitudes due to their higher root-mean-square velocities. **51.** 0.263 mol% H_2O; 2.36×10^3 ppm H_2O. **53.** 13.2 mmHg. **55.** The dew point is about 20.4 °C, the temperature at which $P_{H_2O(g)} = 18.0$ mmHg. **57.** Combustion is a redox reaction. The oxidation numbers of the C atoms in fuel can be increased to either +4 (CO_2) or +2 (CO). The reaction of an acid and a carbonate is an acid-base reaction.

The oxidation number of C in carbonates is +4. Only CO_2 can form. **59.** $2 C_6H_{14}(l) + 19 O_2(g) \longrightarrow 12 CO_2(g) + 14 H_2O(l)$. If combustion is incomplete, some C atoms combine with two O atoms (CO_2), some with only one (CO), and some with none (C); we cannot write a unique equation. **61.** Masses in metric tons: **(a)** 7.22 t CH_4; **(b)** 6.42 t C_8H_{18}; **(c)** 5.74 t coal. **63.** At high temperatures, some NO (from N_2 and O_2) is formed in a combustion chamber that contains air, regardless of the fuel. **65. (a)** $S_8(s) + 8 O_2(g) \longrightarrow 8 SO_2(g)$; **(b)** $2 ZnS(s) + 3 O_2(g) \longrightarrow 2 ZnO(s) + 2 SO_2(g)$; **(c)** $2 SO_2(g) + O_2(g) \longrightarrow 2 SO_3(g)$; **(d)** $SO_3(g) + H_2O(l) \longrightarrow H_2SO_4(aq)$; **(e)** $H_2SO_4(aq) + 2 NH_3(aq) \longrightarrow (NH_4)_2SO_4(aq)$. **67. (a)** Water evaporates from sea spray and leaves particles of NaCl(s) suspended in the air. **(b)** Sulfur dioxide emissions are converted to $SO_3(g)$, which reacts with water vapor to form $H_2SO_4(aq)$. Droplets of $H_2SO_4(aq)$ are neutralized by a basic substance (e.g., NH_3) to form solid sulfate particles. **69.** $2 H_3O^+(aq) + CaCO_3(s) \longrightarrow Ca^{2+}(aq) + 3 H_2O(l) + CO_2(g)$. **71.** 4.4×10^3 kg. **Additional Problems: 78. (a)** 4×10^{37} molecules O_3; **(b)** It would be difficult to separate and store the ozone; ozone spontaneously breaks down into O_2 gas; the mass of ozone that would need to be transported is extremely large, on the order of millions of tons. **79.** If *all* the C_8H_{18} were converted to CO—an unlikely happening—the CO level would be about 24 ppm CO, well below the danger level of 35 ppm. **85.** 12 mg O_2/L. **90.** 4.5×10^{13} kg. **Apply Your Knowledge 94.** If all the S in coal were converted to H_2SO_4, about 1.8×10^{11} lb H_2SO_4 would be produced, about twice the current annual U.S. production. **97. (a)** C is limiting. **(b)** $CH_3CHOHCOOH$, $CO(NH_2)_2$, and $NH_4H_2PO_4$ in the mass ratio of 10.70 : 1.566 : 1.000. **98. (a)** 4.9 ppm, 1.40×10^{19} molecules/L; **(b)** 2.29 y; **(c)** 9.7 ppm; **(d)** No.

Photo Credits

Index

Hill • Petrucci • McCreary • Perry
Accelerator CD
General Chemistry, 4e
CD License Agreement
© 2005 Pearson Education, Inc.
Pearson Prentice Hall
Pearson Education, Inc.
Upper Saddle River, NJ 07458

YOU SHOULD CAREFULLY READ THE TERMS AND CONDITIONS BEFORE USING THE CD-ROM PACKAGE. USING THIS CD-ROM PACKAGE INDICATES YOUR ACCEPTANCE OF THESE TERMS AND CONDITIONS.

Pearson Education, Inc. provides this program and licenses its use. You assume responsibility for the selection of the program to achieve your intended results, and for the installation, use, and results obtained from the program. This license extends only to use of the program in the United States or countries in which the program is marketed by authorized distributors.

LICENSE GRANT
You hereby accept a nonexclusive, nontransferable, permanent license to install and use the program ON A SINGLE COMPUTER at any given time. You may copy the program solely for backup or archival purposes in support of your use of the program on the single computer. You may not modify, translate, disassemble, decompile, or reverse engineer the program, in whole or in part.

TERM
The License is effective until terminated. Pearson Education, Inc. reserves the right to terminate this License automatically if any provision of the License is violated. You may terminate the License at any time. To terminate this License, you must return the program, including documentation, along with a written warranty stating that all copies in your possession have been returned or destroyed.

LIMITED WARRANTY
THE PROGRAM IS PROVIDED "AS IS" WITHOUT WARRANTY OF ANY KIND, EITHER EXPRESSED OR IMPLIED, INCLUDING, BUT NOT LIMITED TO, THE IMPLIED WARRANTIES OR MERCHANTABILITY AND FITNESS FOR A PARTICULAR PURPOSE. THE ENTIRE RISK AS TO THE QUALITY AND PERFORMANCE OF THE PROGRAM IS WITH YOU. SHOULD THE PROGRAM PROVE DEFECTIVE, YOU (AND NOT PEARSON EDUCATION, INC. OR ANY AUTHORIZED DEALER) ASSUME THE ENTIRE COST OF ALL NECESSARY SERVICING, REPAIR, OR CORRECTION. NO ORAL OR WRITTEN INFORMATION OR ADVICE GIVEN BY PEARSON EDUCATION, INC., ITS DEALERS, DISTRIBUTORS, OR AGENTS SHALL CREATE A WARRANTY OR INCREASE THE SCOPE OF THIS WARRANTY.

SOME STATES DO NOT ALLOW THE EXCLUSION OF IMPLIED WARRANTIES, SO THE ABOVE EXCLUSION MAY NOT APPLY TO YOU. THIS WARRANTY GIVES YOU SPECIFIC LEGAL RIGHTS AND YOU MAY ALSO HAVE OTHER LEGAL RIGHTS THAT VARY FROM STATE TO STATE.

Pearson Education, Inc. does not warrant that the functions contained in the program will meet your requirements or that the operation of the program will be uninterrupted or error-free.

However, Pearson Education, Inc. warrants the CD-ROM(s) on which the program is furnished to be free from defects in material and workmanship under normal use for a period of ninety (90) days from the date of delivery to you as evidenced by a copy of your receipt.
MATTER OF THIS AGREEMENT.

The program should not be relied on as the sole basis to solve a problem whose incorrect solution could result in injury to person or property. If the program is employed in such a manner, it is at the user's own risk and Pearson Education, Inc. explicitly disclaims all liability for such misuse.

LIMITATION OF REMEDIES
Pearson Education, Inc.'s entire liability and your exclusive remedy shall be: 1. the replacement of any CD-ROM not meeting Pearson Education, Inc.'s "LIMITED WARRANTY" and that is returned to Pearson Education, or 2. if Pearson Education is unable to deliver a replacement CD-ROM that is free of defects in materials or workmanship, you may terminate this agreement by returning the program.

IN NO EVENT WILL PEARSON EDUCATION, INC. BE LIABLE TO YOU FOR ANY DAMAGES, INCLUDING ANY LOST PROFITS, LOST SAVINGS, OR OTHER INCIDENTAL OR CONSEQUENTIAL DAMAGES ARISING OUT OF THE USE OR INABILITY TO USE SUCH PROGRAM EVEN IF PEARSON EDUCATION, INC. OR AN AUTHORIZED DISTRIBUTOR HAS BEEN ADVISED OF THE POSSIBILITY OF SUCH DAMAGES, OR FOR ANY CLAIM BY ANY OTHER PARTY.

SOME STATES DO NOT ALLOW FOR THE LIMITATION OR EXCLUSION OF LIABILITY FOR INCIDENTAL OR CONSEQUENTIAL DAMAGES, SO THE ABOVE LIMITATION OR EXCLUSION MAY NOT APPLY TO YOU.

GENERAL
You may not sublicense, assign, or transfer the license of the program. Any attempt to sublicense, assign or transfer any of the rights, duties, or obligations hereunder is void.

This Agreement will be governed by the laws of the State of New York.

Should you have any questions concerning this Agreement, you may contact Pearson Education, Inc. by writing to:
ESM Media Development
Higher Education Division
Pearson Education, Inc.
1 Lake Street
Upper Saddle River, NJ 07458

Should you have any questions concerning technical support, you may write to:
New Media Production
Higher Education Division
Pearson Education, Inc.
1 Lake Street
Upper Saddle River, NJ 07458

YOU ACKNOWLEDGE THAT YOU HAVE READ THIS AGREEMENT, UNDERSTAND IT, AND AGREE TO BE BOUND BY ITS TERMS AND CONDITIONS. YOU FURTHER AGREE THAT IT IS THE COMPLETE AND EXCLUSIVE STATEMENT OF THE AGREEMENT BETWEEN US THAT SUPERSEDES ANY PROPOSAL OR PRIOR AGREEMENT, ORAL OR WRITTEN, AND ANY OTHER COMMUNICATIONS BETWEEN US RELATING TO THE SUBJECT

Selected Physical Constants

Acceleration due to gravity	g	9.80665 m s^{-2}
Speed of light (in vacuum)	c	2.99792458×10^8 m s^{-1}
Gas constant	R	0.0820574 L atm mol^{-1} K^{-1}
		8.314472 J mol^{-1} K^{-1}
Electron charge	e^-	$-1.602176462 \times 10^{-19}$ C
Electron rest mass	m_e	$9.10938188 \times 10^{-31}$ kg
Planck's constant	h	$6.62606876 \times 10^{-34}$ J s
Faraday constant	F	9.64853415×10^4 C mol^{-1}
Avogadro number	N_A	$6.02214199 \times 10^{23}$ mol^{-1}

Some Common Conversion Factors

Length

1 meter (m) = 39.37007874 inches (in.)

1 in. = 2.54 centimeters (cm) (exact)

Mass

1 kilogram (kg) = 2.2046226 pounds (lb)

1 lb. = 453.59237 grams (g)

Volume

1 liter (L) = 1000 mL = 1000 cm^3 (exact)

   1 L = 1.056688 quart (qt)

1 gallon (gal) = 3.785412 L

Force

1 newton (N) = 1 kg m s^{-2}

Energy

1 joule (J) = 1 N m = 1 kg m^2 s^{-2}

1 calorie (cal) = 4.184 J (exact)

1 electronvolt (eV)
   = $1.602176462 \times 10^{-19}$ J

1 eV/atom = 96.485 kJ mol^{-1}

1 kilowatt hour (kWh) = 3600 kJ (exact)

Mass-energy equivalence:

   1 unified atomic mass unit (u)
    = $1.66053873 \times 10^{-27}$ kg
    = 931.4866 MeV

Some Useful Geometric Formulas

Perimeter of a rectangle $= 2l + 2w$

Circumference of a circle $= 2\pi r$

Area of a rectangle $= l \times w$

Area of a triangle $= \frac{1}{2}$ (base $\times$ height)

Area of a circle $= \pi r^2$

Area of a sphere $= 4\pi r^2$

Volume of a parallelepiped $= l \times w \times h$

Volume of a sphere $= \frac{4}{3}\pi r^3$

Volume of a cylinder or prism $=$ (area of base) $\times$ height

$\pi = 3.14159$